THIRD EDITION

ALGEBRA & TRIGONOMETRY

Enhanced with Graphing Utilities

Michael Sullivan
Chicago State University

Michael Sullivan, III
Joliet Junior College

Prentice Hall
Upper Saddle River, New Jersey 07458

Library of Congress Cataloging-in-Publication Data

Sullivan, Michael
   Algebra & Trigonometry: enhanced with graphing utilities / Michael Sullivan, Michael
   Sullivan III.—3rd ed.
      p. cm.
   Includes index.
   ISBN 0-13-065912-6
   1. Algebra. 2. Trigonometry. I. Title: Algebra and trigonometry. II. Sullivan, Michael,
   III. Title.

QA154.3 .S75 2003
512'.13—dc21   2001055350

Editor-in-Chief: Sally Yagan
Acquisitions Editor: Eric Frank
Associate Editor: Dawn Murrin
Vice President/Director of Production and Manufacturing: David W. Riccardi
Executive Managing Editor: Kathleen Schiaparelli
Senior Managing Editor: Linda Mihatov Behrens
Production Editor: Bob Walters
Manufacturing Buyer: Alan Fischer
Manufacturing Manager: Trudy Pisciotti
Executive Marketing Manager: Patrice Lumumba Jones
Marketing Assistant: Rachel Beckman
Assistant Managing Editor, Math Media Production: John Matthews
Editorial Assistant/Supplements Editor: Aja Shevelew
Art Director: Kenny Beck
Interior Designer: Lee Goldstein
Cover Designer: Tom Nery
Creative Director: Carole Anson
Art Editor: Thomas Benfatti
Managing Editor Audio/Video Assets: Grace Hazeldine
Director of Creative Services: Paul Belfanti
Photo Editor: Beth Boyd
Cover Photo: Richard Cummins/CORBIS
Art Studio: Artworks:
            Senior Manager: Patricia Burns
            Production Manager: Ronda Whitson
            Manager, Production Technologies: Matt Haas
            Project Coordinator: Jessica Einsig
            Illustrators: Kathryn Anderson, Mark Landis
            Art Quality Assurance: Timothy Nguyen, Stacy Smith, Pamela Taylor

© 2003, 2000, 1996 by Prentice-Hall, Inc.
Upper Saddle River, New Jersey 07458

Printed in the United States of America

10 9 8 7 6 5 4 3 2 1

ISBN: 0-13-065912-6

Pearson Education Ltd., *London*
Pearson Education Australia Pty. Ltd., *Sydney*
Pearson Education Singapore, Pte. Ltd.
Pearson Education North Asia Ltd., *Hong Kong*
Pearson Education Canada, Inc., *Toronto*
Pearson Educacíon de Mexico, S.A. de C.V.
Pearson Education—Japan, *Tokyo*
Pearson Education Malaysia, Pte. Ltd.

FORMULAS/EQUATIONS

Distance Formula

If $P_1 = (x_1, y_1)$ and $P_2 = (x_2, y_2)$, the distance from P_1 to P_2 is

$$d(P_1, P_2) = \sqrt{(x_2 - x_1)^2 + (y_2 - y_1)^2}$$

Standard Equation of a Circle

The standard equation of a circle of radius r with center at (h, k) is

$$(x - h)^2 + (y - k)^2 = r^2$$

Slope Formula

The slope m of the line containing the points $P_1 = (x_1, y_1)$ and $P_2 = (x_2, y_2)$ is

$$m = \frac{y_2 - y_1}{x_2 - x_1} \qquad \text{if } x_1 \neq x_2$$

$$m \text{ is undefined} \qquad \text{if } x_1 = x_2$$

Point-Slope Equation of a Line

The equation of a line with slope m containing the point (x_1, y_1) is

$$y - y_1 = m(x - x_1)$$

Slope-Intercept Equation of a Line

The equation of a line with slope m and y-intercept b is

$$y = mx + b$$

Quadratic Formula

The solutions of the equation $ax^2 + bx + c = 0, a \neq 0$, are

$$x = \frac{-b \pm \sqrt{b^2 - 4ac}}{2a}$$

If $b^2 - 4ac > 0$, there are two unequal real solutions.
If $b^2 - 4ac = 0$, there is a repeated real solution.
If $b^2 - 4ac < 0$, there are two complex solutions that are not real.

GEOMETRY FORMULAS

Circle

r = Radius, A = Area, C = Circumference
$A = \pi r^2 \qquad C = 2\pi r$

Triangle

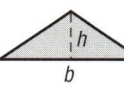

b = Base, h = Altitude (Height), A = area
$A = \frac{1}{2}bh$

Rectangle

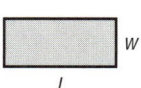

l = Length, w = Width, A = area, P = perimeter
$A = lw \qquad P = 2l + 2w$

Rectangular Box

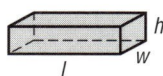

l = Length, w = Width, h = Height, V = Volume
$V = lwh$

Sphere

r = Radius, V = Volume, S = Surface area
$V = \frac{4}{3}\pi r^3 \qquad S = 4\pi r^2$

Right Circular Cylinder

r = Radius, h = Height, V = Volume
$V = \pi r^2 h$

For Yolanda,
Daughter-in-Law and Wife

CONTENTS

CHAPTER 9   **ANALYTIC TRIGONOMETRY**   709

CHAPTER 10   **APPLICATIONS OF TRIGONOMETRIC FUNCTIONS**   779

CHAPTER 11   **POLAR COORDINATES; VECTORS**   829

As professors at both an urban public university and a community college, Michael Sullivan and Michael Sullivan III are aware of the varied needs of Algebra & Trigonometry students, ranging from those who have little mathematical background and a fear of mathematics courses, to those having a strong mathematical education and a high level of motivation. For some of your students, this will be their last course in mathematics, while others will further their mathematical education. This text is written for both groups.

As a teacher, and as an author of precalculus, engineering calculus, finite math, and business calculus texts, Michael understands what students must know if they are to be focused and successful in upper level math courses. However, as a father of four, including the co-author, he also understands the realities of college life. His co-author and son, Michael Sullivan III, believes passionately in the value of technology as a tool for learning that enhances understanding without sacrificing important skills. Together, both authors have taken great pains to ensure that the text contains solid, student-friendly examples and problems, as well as a clear and seamless writing style.

In the Third Edition

The third edition builds upon a strong foundation by integrating new features and techniques that further enhance student interest and involvement. The elements of previous editions that have proved successful remain, while many changes, some obvious, others subtle, have been made. One important benefit of authoring a successful series is the broad-based feedback upon which improvements and additions are ultimately based. Virtually every change to this edition is the result of thoughtful comments and suggestions made by colleagues and students who have used previous editions. This feedback has proved invaluable and has been used to make changes that improve the flow, usability, and accessibility of the text. For example, some topics have been moved to better reflect the way teachers approach the course and problems have been added where more practice was needed. The supplements package has been enhanced through upgrading traditional supplements and adding innovative media components.

Reorganized Content for Algebra and Trigonometry

- Appendix Review
  - Now expanded, this material appears in the beginning of the book as Chapter R
- Chapter 1
  - Scatter diagrams now appear in Section 2.2 Linear Functions and Models
  - Section 1.2 has been split into two sections. In 3/e Section 1.2 contains point plotting, graphing equations on a graphing utility, and intercepts. Section 3.1 covers symmetry and graphing key equations. This adheres to the "just in time approach" by placing symmetry and key equations closer to functions.
  - Section 1.3 from 2/e has been split into two sections—1.3 and 1.5. Section 1.5 has been expanded to include quadratic in form equations.

- Chapter 2
  - Section 2.2 now includes the discussion on scatter diagrams.
  - Quadratic equations now appears earlier as part of Section 1.3.
- Chapter 3
  - Section 3.1 contains the discussion on symmetry and graphing key equations.
  - Section 3.1 from 2/e is now two sections: Section 3.2 covers properties of functions, while Section 3.3 covers the library of functions and piecewise-defined functions.
- Chapter 4
  - This chapter contains Sections 4.1, 4.2, 4.7 and 4.8 from 2/e. The section on rational functions has been split into two sections to accommodate a single lecture for each section.
- Chapter 5
  - This chapter contains the remaining sections of Chapter 4 from 2/e.
- Chapter 6 (Formerly Chapter 5)
  - Section 6.2 now includes a discussion of basic exponential equations.
  - Section 6.3 now includes a discussion of basic logarithmic equations.
- Chapter 7 (Formerly Chapter 6)
- Chapter 8 (Formerly Chapter 7)
  - Graphs of Trigonometric Functions has been split into two sections. Section 8.6 discusses the graphs of the sine and cosine functions, including sinusoidal graphs. Section 8.7 discusses the graphs of the tangent, cotangent, secant and cosecant functions.
- Chapter 9 (Formerly Chapter 8)
  - Section 8.5 has been split into two sections as 9.1 and 9.2 to accommodate a single lecture for each section.
- Chapter 10 (Formerly Chapter 9)
  - Section 10.5 now includes a discussion on combining waves (the method of adding y-coordinates to obtain graphs).
- Chapter 11 (Formerly Chapter 10)
- Chapter 12 (Formerly Chapter 11)
- Chapter 13 (Formerly Chapter 12)
- Chapter 14 (Formerly Chapter 13)

Specific Content Changes

In this edition emphasis is placed on the role of modeling in algebra and trigonometry. To this end, dedicated sections appear on Linear Functions and Models, Quadratic Models, Power Functions and Models, Polynomial Functions and Models, Exponential and Logarithmic Functions and Models, and Trigonometric Models. Many of these applications focus on the areas of business, finance, and economics.

Chapter R review is a robust expansion of the appendix review of the second edition.

New to this edition is a discussion of quadratic in form equations.

A section on combining waves (the method of adding y-coordinates to obtain graphs) has been added.

As a result of these changes, this edition will be an improved teaching device for professors and a better learning tool for students.

Features in the 3rd Edition

- Section *OBJECTIVES* appear in a numbered list to begin each section.
- "Now Work" problems identified by the yellow pencil icon ✎ appear after a concept has been introduced. This directs the student to a problem in the exercises that tests the concept, insuring that the concept has been mastered before moving on.
- References to calculus are identified by a ⚲ calculus icon.
- Historical Perspectives, sometimes with exercises, are presented in context and provide interesting anecdotal information.
- Varied applications and real-world data are abundant in Examples and in Exercises. Many contain sourced data.
- Discussion, Writing, and Research problems appear in each exercise set, identified by an icon 🎺 and red numbers. These problems challenge students to expand the parameters of their understanding by providing the basis for class discussions, writing projects and collaboration.
- An extensive Chapter Review provides a list of important formulas, definitions, theorems, and objectives. Each objective is listed with a page reference and review exercises that test the student's understanding of the objective. The authors' suggestions for practice tests are indicated in blue.
- A cumulative review appears at the end of every chapter, beginning with Chapter 2. These cumulative reviews serve to continually reinforce skills from previous chapters. This makes study for the final examination easier.

Using the 3rd Edition Effectively and Efficiently with Your Syllabus

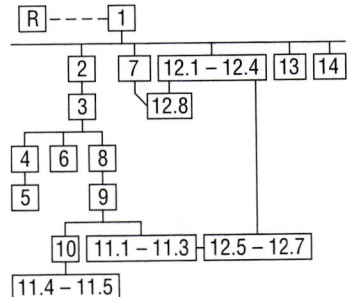

To meet the varied needs of diverse syllabi, this book contains more content than a typical algebra and trigonometry course. The illustration shows the dependencies of chapters on each other. As the chart indicates, this book has been organized with flexibility of use in mind. Even within a given chapter, certain sections can be skipped without fear of future problems.

Chapter R   Review

This chapter is a revision of the old Appendix. It may be used as the first part of the course, or as a "just-in-time" review when the content is required in a later chapter. Specific references to this chapter occur throughout the book to assist in the review process.

Chapter 1   Graphs

This chapter presents an introduction to graphing and the graphing utility. Equations and inequalities are solved algebraically with graphical support. For those who prefer to treat complex numbers and negative discriminants early, Section 5.3 can be covered any time after Section 1.3.

Chapter 2   Linear and Quadratic Functions

This chapter provides an introduction to functions and then discusses two specific types of functions: linear functions and quadratic functions, along with models that utilize these functions.

Chapter 3   Functions and Their Graphs

Perhaps the most important chapter. Section 3.6 can be skipped without adverse effects.

Chapter 4   Polynomial and Rational Functions

Topic selection is dependent on your syllabus.

Chapter 5   The Zeros of a Polynomial Function

Topic selection is dependent on your syllabus. Section 5.1 is not absolutely necessary, but its coverage makes some computation easier.

Chapter 6   Exponential and Logarithmic Functions

Sections 6.1–6.5 follow in sequence; Sections 6.6, 6.7, and 6.8 each require Section 6.3.

Chapter 7   Systems of Equations and Inequalities

Sections 7.1–7.2 follow in sequence; Sections 7.3–7.7 require Sections 7.1 and 7.2, and may be covered in any order.

Chapter 8   Trigonometric Functions

The sections follow in sequence. Section 10.1 on Applications Involving Right Triangles, may be covered immediately after Section 8.3, if so desired.

Chapter 9   Analytic Trigonometry

The sections follow in sequence. Sections 9.2, 9.6, and 9.8 may be skipped in a brief course.

Chapter 10   Applications of Trigonometric Functions

The sections follow in sequence. Sections 10.4 and 10.5 may be skipped in a brief course.

Chapter 11   Polar Coordinates; Vectors

Sections 11.1–11.3 and Sections 11.4–11.5 are independent and may be covered separately.

Chapter 12   Analytic Geometry

Sections 12.1–12.4 follow in sequence. Sections 12.5, 12.6, and 12.7 are independent of each other, but do depend on Sections 12.1–12.4. Section 12.8 is independent of Sections 12.5–12.7, but does depend on Sections 12.1–12.4 as well as Sections 7.1–7.2.

Chapter 13   Sequences; Induction; The Binomial Theorem

There are three independent parts: (1) Sections 13.1–13.3; (2) Section 13.4; (3) Section 13.5

Chapter 14   Counting and Probability

Sections 14.1–14.4 follow in order.

Acknowledgments

Textbooks are written by authors, but evolve from an idea into final form through the efforts of many people. Special thanks to Don Dellen, who first suggested this book and the other books in this series. Don's extensive contributions to publishing and mathematics are well known; we all miss him dearly.

There are many colleagues we would like to thank for their input, encouragement, patience, and support. They have our deepest thanks and appreciation. We apologize for any omissions.

James Africh, *College of DuPage*
Steve Agronsky, *Cal Poly State University*
Grant Alexander, *Joliet Junior College*
Dave Anderson, *South Suburban College*
Joby Milo Anthony, *University of Central Florida*
James E. Arnold, *University of Wisconsin-Milwaukee*

Carolyn Autray, *University of West Georgia*
Agnes Azzolino, *Middlesex County College*
Wilson P Banks, *Illinois State University*
Sudeshna Basu, *Howard University*
Dale R. Bedgood, *East Texas State University*
Beth Beno, *South Suburban College*
Carolyn Bernath, *Tallahassee Community College*

William H. Beyer, *University of Akron*
Annette Blackwelder, *Florida State University*
Richelle Blair, *Lakeland Community College*
Trudy Bratten, *Grossmont College*
Joanne Brunner, *Joliet Junior College*

Warren Burch, *Brevard Community College*
Mary Butler, *Lincoln Public Schools*
William J. Cable, *University of Wisconsin-Stevens Point*
Lois Calamia, *Brookdale Community College*
Jim Campbell, *Lincoln Public Schools*
Roger Carlsen, *Moraine Valley Community College*
Elena Catoiu, *Joliet Junior College*
John Collado, *South Suburban College*
Nelson Collins, *Joliet Junior College*
Jim Cooper, *Joliet Junior College*
Denise Corbett, *East Carolina University*
Theodore C. Coskey, *South Seattle Community College*
John Davenport, *East Texas State University*
Faye Dang, *Joliet Junior College*
Antonio David, *Del Mar College*
Duane E. Deal, *Ball State University*
Timothy Deis, *University of Wisconsin-Platteville*
Vivian Dennis, *Eastfield College*
Guesna Dohrman, *Tallahassee Community College*
Karen R. Dougan, *University of Florida*
Louise Dyson, *Clark College*
Paul D. East, *Lexington Community College*
Don Edmondson, *University of Texas-Austin*
Erica Egizio, *Joliet Junior College*
Christopher Ennis, *University of Minnesota*
Ralph Esparza, Jr., *Richland College*
Garret J. Etgen, *University of Houston*
Pete Falzone, *Pensacola Junior College*
W.A. Ferguson, *University of Illinois-Urbana/Champaign*
Iris B. Fetta, *Clemson University*
Mason Flake, *student at Edison Community College*
Timothy W. Flood, *Pittsburg State University*
Merle Friel, *Humboldt State University*
Richard A. Fritz, *Moraine Valley Community College*
Carolyn Funk, *South Suburban College*
Dewey Furness, *Ricke College*
Dawit Getachew, *Chicago State University*
Wayne Gibson, *Rancho Santiago College*
Robert Gill, *University of Minnesota Duluth*
Sudhir Kumar Goel, *Valdosta State University*
Joan Goliday, *Sante Fe Community College*
Frederic Gooding, *Goucher College*
Sue Graupner, *Lincoln Public Schools*
Jennifer L. Grimsley, *University of Charleston*
Ken Gurganus, *University of North Carolina*
James E. Hall, *University of Wisconsin-Madison*
Judy Hall, *West Virginia University*
Edward R. Hancock, *DeVry Institute of Technology*
Julia Hassett, *DeVry Institute-Dupage*
Michah Heibel, *Lincoln Public Schools*
LaRae Helliwell, *San Jose City College*
Brother Herron, *Brother Rice High School*
Robert Hoburg, *Western Connecticut State University*

Lee Hruby, *Naperville North High School*
Kim Hughes, *California State College-San Bernardino*
Ron Jamison, *Brigham Young University*
Richard A. Jensen, *Manatee Community College*
Sandra G. Johnson, *St. Cloud State University*
Tuesday Johnson, *New Mexico State University*
Moana H. Karsteter, *Tallahassee Community College*
Arthur Kaufman, *College of Staten Island*
Thomas Kearns, *North Kentucky University*
Shelia Kellenbarger, *Lincoln Public Schools*
Keith Kuchar, *Manatee Community College*
Tor Kwembe, *Chicago State University*
Linda J. Kyle, *Tarrant Country Jr. College*
H.E. Lacey, *Texas A & M University*
Harriet Lamm, *Coastal Bend College*
Matt Larson, *Lincoln Public Schools*
Christopher Lattin, *Oakton Community College*
Adele LeGere, *Oakton Community College*
Kevin Leith, *University of Houston*
Jeff Lewis, *Johnson County Community College*
Stanley Lukawecki, *Clemson University*
Janice C. Lyon, *Tallahassee Community College*
Virginia McCarthy, *Iowa State University*
Jean McArthur, *Joliet Junior College*
Tom McCollow, *DeVry Institute of Technology*
Laurence Maher, *North Texas State University*
Jay A. Malmstrom, *Oklahoma City Community College*
Sherry Martina, *Naperville North High School*
Alec Matheson, *Lamar University*
James Maxwell, *Oklahoma State University-Stillwater*
Judy Meckley, *Joliet Junior College*
David Meel, *Bowling Green State University*
Carolyn Meitler, *Concordia University*
Sarnia Metwali, *Erie Community College*
Rich Meyers, *Joliet Junior College*
Eldon Miller, *University of Mississippi*
James Miller, *West Virginia University*
Michael Miller, *Iowa State University*
Kathleen Miranda, *SUNY at Old Westbury*
Thomas Monaghan, *Naperville North High School*
Craig Morse, *Naperville North High School*
Samad Mortabit, *Metropolitan State University*
A. Muhundan, *Manatee Community College*
Jane Murphy, *Middlesex Community College*
Richard Nadel, *Florida International University*
Gabriel Nagy, *Kansas State University*
Bill Naegele, *South Suburban College*
Lawrence E. Newman, *Holyoke Community College*
James Nymann, *University of Texas-El Paso*
Sharon O'Donnell, *Chicago State University*
Seth F. Oppenheimer, *Mississippi State University*

Linda Padilla, *Joliet Junior College*
E. James Peake, *Iowa State University*
Kelly Pearson, *Murray State University*
Thomas Radin, *San Joaquin Delta College*
Ken A. Rager, *Metropolitan State College*
Kenneth D. Reeves, *San Antonio College*
Elsi Reinhardt, *Truckee Meadows Community College*
Jane Ringwald, *Iowa State University*
Stephen Rodi, *Austin Community College*
Bill Rogge, *Lincoln Public Schools*
Howard L. Rolf, *Baylor University*
Phoebe Rouse, *Lousiana State University*
Edward Rozema, *University of Tennessee at Chattanooga*
Dennis C. Runde, *Manatee Community College*
John Sanders, *Chicago State University*
Susan Sandmeyer, *Jamestown Community College*
A.K. Shamma, *University of West Florida*
Martin Sherry, *Lower Columbia College*
Tatrana Shubin, *San Jose State University*
Anita Sikes, *Delgado Community College*
Timothy Sipka, *Alma College*
Lori Smellegar, *Manatee Community College*
John Spellman, *Southwest Texas State University*
Rajalakshmi Sriram, *Okaloosa-Walton Community College*
Becky Stamper, *Western Kentucky University*
Judy Staver, *Florida Community College-South*
Neil Stephens, *Hinsdale South High School*
Christopher Terry, *Augusta State University*
Diane Tesar, *South Suburban College*
Tommy Thompson, *Brookhaven College*
Richard J. Tondra, *Iowa State University*
Marvel Townsend, *University of Florida*
Jim Trudnowski, *Carroll College*
Robert Tuskey, *Joliet Junior College*
Richard G. Vinson, *University of South Alabama*
Mary Voxman, *University of Idaho*
Jennifer Walsh, *Daytona Beach Community College*
Donna Wandke, *Naperville North High School*
Darlene Whitkenack, *Northern Illinois University*
Christine Wilson, *West Virginia University*
Brad Wind, *Florida International University*
Canton Woods, *Auburn University*
Tamara S. Worner, *Wayne State College*
Terri Wright, *New Hampshire Community Technical College, Manchester*
George Zazi, *Chicago State University*

Recognition and thanks are due particularly to the following individuals for their valuable assistance in the preparation of this edition: Sally Yagan, for her continued support and genuine interest; Patrice Jones, for his innovative marketing efforts; Bob Walters, for his organizational skills as production supervisor; Phoebe Rouse, for her specific suggestions for this edition; Teri Lovelace and Cindy Trimble of Laurel Technical Services for their proofreading skill and checking of our answers; and to the entire Prentice-Hall sales staff for their continuing confidence in this book.

Michael Sullivan

Michael Sullivan, III

As you begin your study of Algebra and Trigonometry, you may feel overwhelmed by the numbers of theorems, definitions, procedures, and equations that confront you. You may even wonder whether or not you can learn all of this material in the time allotted. These concerns are normal. Keep in mind that the elements of algebra and trigonometry are all around us as we go through our daily routines. Many of the concepts you will learn to express mathematically, you already know intuitively. For many of you, this may be your last math course, while for others, just the first in a series of many. Either way, this text was written with you in mind. We have spent countless hours teaching Algebra and Trigonometry courses. We know what you're going through. You'll find that we have written a text that doesn't overwhelm, or unnecessarily complicate Algebra and Trigonometry, but at the same time gives you the skills and practice you need to be successful.

This text is designed to help you, the student, master the terminology and basic concepts of Algebra and Trigonometry. These aims have helped to shape every aspect of the book. Many learning aids are built into the format of the text to make your study of the material easier and more rewarding. This book is meant to be a "machine for learning," that can help you focus your efforts, ensuring that you get the most from the time and energy you invest.

Please do not hesitate to contact us through Prentice Hall with any suggestions or comments that would improve this text.

Best Wishes!

Michael Sullivan

Michael Sullivan, III

CHAPTER OPENERS AND CHAPTER PROJECTS

Chapter openers use current articles to set up **Chapter Projects.** Many of the concepts that you encounter in this course relate directly to today's headlines and issues. These chapter projects are designed to give you a chance to use math to better understand the world.

CHAPTER

3

FUNCTIONS AND THEIR GRAPHS

Wednesday February 10, 1999 *The Oregonian* "Ship awaits salvage effort"

COOS BAY—Cleanup crews combed the oil-scarred south coast Tuesday as authorities raced to finish plans to refloat a 639-foot cargo ship mired for six days 150 yards off one of Oregon's most biologically rich beaches.

All day Tuesday, streaks of oil oozed from the cracked hull of the bulk cargo carrier *New Carissa* and spread over six miles of beach. Despite the breached hull, authorities believed they had a better chance of pulling the stricken ship out of beach sands than pumping nearly 400,000 gallons of oil off the ship in the winter surf.

SEE CHAPTER PROJECT 1.

OUTLINE

3.1 Symmetry; Graphing Key Equations
3.2 Properties of Functions
3.3 Library of Functions; Piecewise-Defined Functions
3.4 Graphing Techniques: Transformations

For additional study help, go to
www.prenhall.com/sullivanegu3e

Materials include:
• Graphing Calculator Help
• Chapter Quiz
• Chapter Test
• PowerPoint Downloads
• Chapter Projects
• Student Tips

255

feet?

(d) Graph $A = A(x)$. For what value of x is A smallest?

69. Spheres The volume V of a sphere of radius r is $V = \frac{4}{3}\pi r^3$; the surface area S of this sphere is $S = 4\pi r^2$.

Express the volume V as a function of the surface area S. If the surface area doubles, how does the volume change?

70. Productivity versus Earnings The following data represent the average hourly earnings and productivity (output per hour) of production workers for the years

(96.7, 10.32) on the scatter diagram found in part (a).
(c) Find the average rate of change of hourly earnings for productivity from 94.2 to 96.7.
(d) Interpret the average rate of change found in part (c).
(e) Draw a line through the points (96.7, 10.32) and (100.8, 11.44) on the scatter diagram found in part (a).
(f) Find the average rate of change of hourly earnings for productivity from 96.7 to 100.8.
(g) Interpret the average rate of change found in part (f).
(h) What is happening to the average rate of change of hourly earnings as productivity increases?

Chapter Projects

1. **Oil Spill** An oil tanker strikes a sand bar that rips a hole in the hull of the ship. Oil begins leaking out of the tanker

with the spilled oil forming a circle around the tanker. The radius of the circle is increasing at the rate of 2.2 feet per hour.

(a) Write the area of the circle as a function of the radius r.
(b) Write the radius of the circle as a function of time t.
(c) What is the radius of the circle after 2 hours? What is the radius of the circle after 2.5 hours?
(d) Use the result of part (c) to determine the area of the circle after 2 hours and 2.5 hours.
(e) Determine a function that represents area as a function of time t.
(f) Use the result of part (e) to determine the area of the circle after 2 hours and 2.5 hours.
(g) Compute the average rate of change of the area of the circle from 2 hours to 2.5 hours.
(h) Compute the average rate of change of the area of the circle from 3 hours to 3.5 hours.
(i) Based on the results obtained in parts (g) and (h), what is happening to the average rate of change of the area of the circle as time passes?

Page 320

The Sullivans' **accessible writing style** is apparent throughout, often utilizing various approaches to the same concept. This clear writing style makes potentially difficult concepts intuitive, making class time more productive.

Sometimes it is helpful to think of a function f as a machine that receives as input a number from the domain, manipulates it, and outputs the value. See Figure 6.

Figure 6

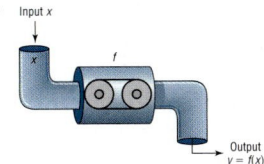

Input x

f

Output
$y = f(x)$

The restrictions on this input/output machine are

1. It only accepts numbers from the domain of the function.
2. For each input, there is exactly one output (which may be repeated for different inputs).

For a function $y = f(x)$, the variable x is called the **independent variable,** because it can be assigned any of the numbers from the domain. The variable y is called the **dependent variable,** because its value depends on x.

Any symbol can be used to represent the independent and dependent variables. For example, if f is the *cube function*, then f can be defined by $f(x) = x^3$ or $f(t) = t^3$ or $f(z) = z^3$. All three functions are the same: Each tells us to cube the independent variable. In practice, the symbols used for the independent and dependent variables are based on common usage, such as using C for cost in business.

Page 200

PREPARING FOR THIS SECTION

Before getting started, review the following:

✓ Pythagorean Theorem (Section R.3, pp. 25–26)   ✓ Functions (Section 2.1, pp. 196–207)

8.2   RIGHT TRIANGLE TRIGONOMETRY

OBJECTIVES
1. Find the Value of Trigonometric Functions of Acute Angles
2. Use the Fundamental Identities
3. Find the Remaining Trigonometric Functions Given the Value of One of Them
4. Use the Complementary Angle Theorem

Figure 20

Hypotenuse
c

b

$90°$

A triangle in which one angle is a right angle ($90°$) is called a **right triangle.** Recall that the side opposite the right angle is called the **hypotenuse,** and the remaining two sides are called the **legs** of the triangle. In Figure 20 we have labeled the hypotenuse as c to indicate that its length is c units, and, in a like manner, we have labeled the legs as a and b. Because the triangle is a right triangle, the Pythagorean Theorem tells us that

$$c^2 = a^2 + b^2$$

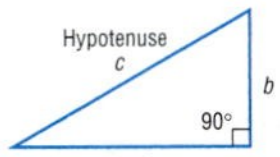

Page 624

Begin each section by reading the **Learning Objectives.** The list of objectives will help you organize your studies and prepare for class. The Learning Objectives also tie the text to MathPro Explorer's Section Objectives.

Step-by-step examples insure that you follow the entire solution process and give you an opportunity to check your understanding of each step.

EXAMPLE 3   **Analyzing the Graph of a Rational Function**

Analyze the graph of the rational function: $R(x) = \dfrac{3x^2 - 3x}{x^2 + x - 12}$

Solution   We factor R to get

$$R(x) = \frac{3x(x - 1)}{(x + 4)(x - 3)}$$

STEP 1:   The domain of R is $\{x \mid x \neq -4, x \neq 3\}$.
STEP 2: R is in lowest terms.
STEP 3:   The graph has two x-intercepts: 0 and 1. The y-intercept is $R(0) = 0$.
STEP 4:   Because

$$R(-x) = \frac{-3x(-x - 1)}{(-x + 4)(-x - 3)} = \frac{3x(x + 1)}{(x - 4)(x + 3)}$$

we conclude that R is neither even nor odd. There is no symmetry with respect to the y-axis or the origin.

STEP 5:   Since R is in lowest terms, the graph of R has two vertical asymptotes: $x = -4$ and $x = 3$.
STEP 6:   Since the degree of the numerator equals the degree of the denominator, the graph has a horizontal asymptote. To find it, we form the quotient of the leading coefficient of the numerator, 3, and the leading coefficient of the denominator, 1. The graph of R has the

Page 358

Real-world data is incorporated into examples and exercise sets to emphasize that mathematics is a tool used to understand the world around us. As you use these problems and examples, you will see the relevance and utility of the skills being covered.

---

39. Federal Income Tax   Two 2001 Tax Rate Schedules are given in the accompanying table. If x equals the amount on Form 1040, line 37, and y equals the tax due, construct a function $y = f(x)$ for each schedule.

2001 TAX RATE SCHEDULES

SCHEDULE X—IF YOUR FILING STATUS IS SINGLE				SCHEDULE Y-1—USE IF YOUR FILING STATUS IS MARRIED FILING JOINTLY OR QUALIFYING WIDOW(ER)			
If the amount on Form 1040, line 37, is: Over—	But not over—	Enter on Form 1040, line 38	of the amount over—	If the amount on Form 1040, line 37, is: Over—	But not over—	Enter on Form 1040, line 38	of the amount over—
$0	$27,050	_____ 15%	$0	$0	$45,200	_____ 15%	$0
27,050	65,550	$4,057.50 + 28%	27,050	45,200	109,250	$6,780.00 + 28%	45,200
65,550	136,750	14,837.50 + 31%	65,550	109,250	166,450	24,714.00 + 31%	109,250
136,750	297,300	36,909.50 + 36%	136,750	166,450	297,300	42,446.00 + 36%	166,450
297,300	_____	94,707.50 + 39.6%	297,300	297,300	_____	89,552.00 + 39.6%	297,300

Page 285

Solution   See Figure 39. We begin by choosing the placement of the coordinate axes so that the x-axis coincides with the road surface and the origin coincides with the center of the bridge. As a result, the twin towers will be vertical (height $746 - 220 = 526$ feet above the road) and located 2100 feet from the center. Also, the cable, which has the shape of a parabola, will extend from the towers, open up, and have its vertex at $(0, 0)$. The choice of placement of the axes enables us to identify the equation of the parabola as $y = ax^2$, $a > 0$. We can also see that the points $(-2100, 526)$ and $(2100, 526)$ are on the graph.

Figure 39

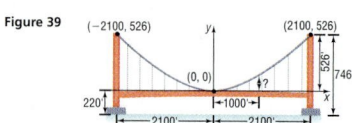

Based on these facts, we can find the value of a in $y = ax^2$.

$$y = ax^2$$
$$526 = a(2100)^2$$
$$a = \frac{526}{(2100)^2}$$

The equation of the parabola is therefore

$$y = \frac{526}{(2100)^2}x^2$$

The height of the cable when $x = 1000$ is

$$y = \frac{526}{(2100)^2}(1000)^2 \approx 119.3 \text{ feet}$$

The cable is 119.3 feet high at a distance of 1000 feet from the center of the bridge.   ∎

━━ NOW WORK PROBLEM **27**.

Fitting a Quadratic Function to Data

③   In Section 2.2, we found the line of best fit for data that appeared to be linearly related. It was noted that data may also follow a nonlinear relation. Figures 40(a) and (b) show scatter diagrams of data that follow a quadratic relation.

Figure 40

$y = ax^2 + bx + c, a > 0$
(a)

$y = ax^2 + bx + c, a < 0$
(b)

Page 242

Many examples end with **"Now Work Problems."** The problems suggested here are similar to the corresponding examples and provide a great way to check your understanding as you work through the chapter. The solutions to all "Now Work" problems can be found in the back of the text as well as the *Student Solutions Manual*.

---

ground, how far has it traveled horizontally?

27. Suspension Bridge   A suspension bridge with weight uniformly distributed along its length has twin towers that extend 75 meters above the road surface and are 400 meters apart. The cables are parabolic in shape and are suspended from the tops of the towers. The cables touch the road surface at the center of the bridge. Find the height of the cables at a point 100 meters from the center. (Assume that the road is level.)

28. Architecture   A parabolic arch has a span of 120 feet and a maximum height of 25 feet. Choose suitable rectangu-

Page 246

PROCEDURES

Procedures, both algebraic and technical, are clearly expressed throughout the text.

Steps for Finding the Real Zeros of a Polynomial Function

STEP 1: Use the degree of the polynomial to determine the maximum number of zeros.

STEP 2: If the polynomial has integer coefficients, use the Rational Zeros Theorem to identify those rational numbers that potentially can be zeros.

STEP 3: Using a graphing utility, graph the polynomial function.

STEP 4: (a) Use eVALUEate, substitution, synthetic division, or long division to test a potential rational zero based on the graph.

(b) Each time that a zero (and thus a factor) is found, repeat Step 4 on the depressed equation. In attempting to find the zeros, remember to use (if possible) the factoring techniques that you already know (special products, factoring by grouping, and so on).

Page 389

3.1 Exercises

In Problems 1–10, plot each point. Then plot the point that is symmetric to it with respect to: (a) the x-axis; (b) the y-axis; (c) the origin.

1. $(3, 4)$ **2.** $(5, 3)$ **3.** $(-2, 1)$ **4.** $(4, -2)$ **5.** $(1, 1)$

6. $(-1, -1)$ **7.** $(-3, -4)$ **8.** $(4, 0)$ **9.** $(0, -3)$ **10.** $(-3, 0)$

In Problems 11–18, the graph of an equation is given. Indicate whether the graph is symmetric with respect to the x-axis, the y-axis, or the origin.

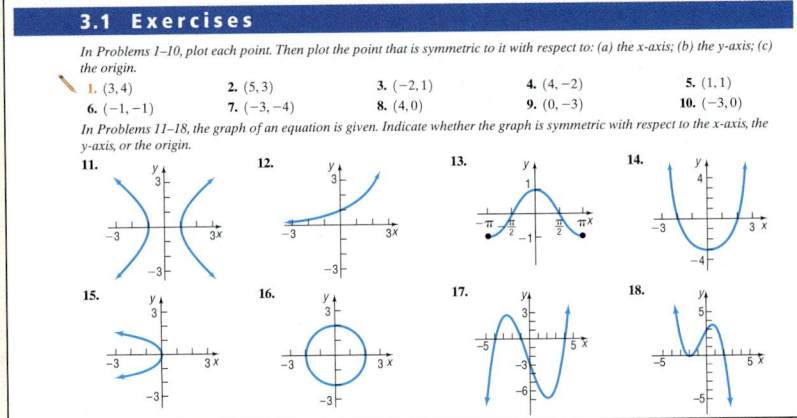

11.   **12.**   **13.**   **14.**

15.   **16.**   **17.**   **18.**

Page 262

END-OF-SECTION EXERCISES

The Sullivans exercises are unparalleled in terms of thorough coverage and accuracy. Each **end-of-section exercise** set begins with visual and concept based problems, starting you out with the basics of the section. Well-thought-out exercises better prepare you for exams.

ALGEBRAIC AND GRAPHING SOLUTIONS

When appropriate, examples are solved using both an **algebraic and graphing approach.** The graphing solution will appear after the algebraic solution to reinforce the concept that graphing utilities often provide approximate, rather than exact, solutions. In this way, the utility is used to support the algebra.

When asked to find the real solutions, if any, of a quadratic equation, always evaluate the discriminant first to see how many real solutions there are.

EXAMPLE 11   **Solving a Quadratic Equation by Using the Quadratic Formula and by Graphing**

Find the real solutions, if any, of the equation $3x^2 - 5x + 1 = 0$.

Algebraic Solution   The equation is in standard form, so we compare it to $ax^2 + bx + c = 0$ to find a, b, and c.

$$3x^2 - 5x + 1 = 0$$
$$ax^2 + bx + c = 0, \quad a = 3, b = -5, c = 1$$

With $a = 3$, $b = -5$, and $c = 1$, we evaluate the discriminant $b^2 - 4ac$.

$$b^2 - 4ac = (-5)^2 - 4(3)(1) = 25 - 12 = 13$$

Since $b^2 - 4ac > 0$, there are two unequal real solutions.

We use the quadratic formula with $a = 3$, $b = -5$, $c = 1$, and $b^2 - 4ac = 13$.

$$x = \frac{-b \pm \sqrt{b^2 - 4ac}}{2a} = \frac{-(-5) \pm \sqrt{13}}{2(3)} = \frac{5 \pm \sqrt{13}}{6}$$

The solution set is $\left\{\dfrac{5 - \sqrt{13}}{6}, \dfrac{5 + \sqrt{13}}{6}\right\}$. These solutions are exact.

Graphing Solution   Figure 34 shows the graph of the equation

$$Y_1 = 3x^2 - 5x + 1$$

Figure 34

As expected, we see that there are two x-intercepts: one between 0 and 1, the other between 1 and 2. The solutions to the equation are 0.23 and 1.43, rounded to two decimal places. These solutions are approximate.

✏ NOW WORK PROBLEMS **81** AND **91.**

Page 123

xxi

25. Per Capita Disposable Income versus Consumption An economist wishes to estimate a linear function that relates per capita consumption expenditures C and disposable income I. Both C and I are measured in dollars. The following data represent the per capita disposable income (income after taxes) and per capita consumption in the United States for 1990 to 1998.

Year	Per Capita Disposable Income (I)	Per Capita Consumption (C)
1990	15,695	14,547
1991	16,700	15,400
1992	17,346	16,035
1993	18,153	16,951
1994	19,711	18,419
1995	20,316	19,061
1996	21,127	19,938
1997	21,871	20,807
1998	22,212	21,385

Source: U.S. Department of Commerce.

Let I represent the independent variable and C the dependent variable.
(a) Use a graphing utility to draw a scatter diagram.
(b) Use a graphing utility to find the line of best fit to the data. Express the solution using function notation.
(c) Interpret the slope. The slope of this line is called the **marginal propensity to consume.**
(d) Predict the consumption of a family whose disposable income is $21,500.

Page 224

MODELING

Many examples and exercises connect real-world situations to mathematical concepts. Learning to work with **models** is a skill that transfers to many disciplines.

GRAPHING UTILITIES AND TECHNIQUES

Increase your understanding, visualize, discover, explore, and solve problems using a **graphing utility.** Sullivan uses the graphing utility to further your understanding of concepts not to circumvent essential math skills.

 2 Graph Functions Using Reflections about the x-Axis or y-Axis
3 Graph Functions Using Compressions and Stretches

At this stage, if you were asked to graph any of the functions defined by
$y = x, y = x^2, y = x^3, y = \sqrt{x}, y = |x|,$ or $y = \frac{1}{x}$, your response should be, "Yes, I recognize these functions and know the general shapes of their graphs." (If this is not your answer, review the previous section and Figures 23 through 28).

Sometimes we are asked to graph a function that is "almost" like one that we already know how to graph. In this section, we look at some of these functions and develop techniques for graphing them. Collectively, these techniques are referred to as **transformations.**

1 Vertical Shifts

EXPLORATION On the same screen, graph each of the following functions:
$$Y_1 = x^2$$
$$Y_2 = x^2 + 2$$
$$Y_3 = x^2 - 2$$
What do you observe?

Figure 33

$Y_2 = x^2 + 2$
$Y_1 = x^2$
$Y_3 = x^2 - 2$

RESULT Figure 33 illustrates the graphs. You should have observed a general pattern. With $Y_1 = x^2$ on the screen, the graph of $Y_2 = x^2 + 2$ is identical to that of $Y_1 = x^2$, except that it is shifted vertically up 2 units. The graph of $Y_3 = x^2 - 2$ is identical to that of $Y_1 = x^2$, except that it is shifted vertically down 2 units.

Page 286

DISCUSSION WRITING AND READING PROBLEMS

These **problems** are designed to get you to "think outside the box," therefore fostering an intuitive understanding of key mathematical concepts. In this example, matching the graph to the functions insures that you understand functions at a fundamental level.

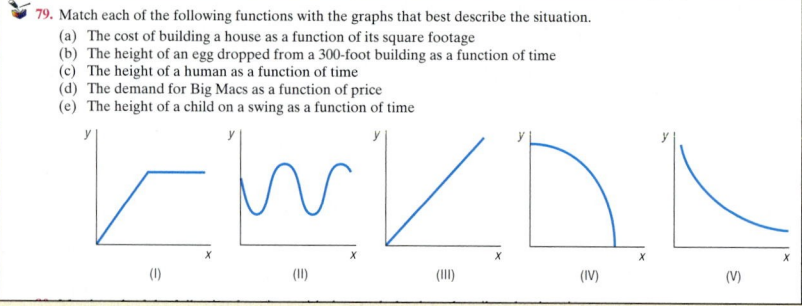

79. Match each of the following functions with the graphs that best describe the situation.
(a) The cost of building a house as a function of its square footage
(b) The height of an egg dropped from a 300-foot building as a function of time
(c) The height of a human as a function of time
(d) The demand for Big Macs as a function of price
(e) The height of a child on a swing as a function of time

(I)   (II)   (III)   (IV)   (V)

Page 213

Chapter Review

Library of Functions

Linear function (p. 276)

$f(x) = mx + b$ Graph is a line with slope m and y-intercept b (see Figure 21)

Constant function (p. 277)

$f(x) = b$ Graph is a horizontal line with y-intercept b (see Figure 22)

Identity function (p. 277)

$f(x) = x$ Graph is a line with slope 1 and y-intercept 0 (see Figure 23)

Square function (p. 277)

$f(x) = x^2$ Graph is a parabola with vertex at $(0, 0)$ (see Figure 24)

Cube function (p. 278)

$f(x) = x^3$ See Figure 25.

Square root function (p. 278)

$f(x) = \sqrt{x}$ See Figure 26.

Reciprocal function (p. 278)

$f(x) = \dfrac{1}{x}$ See Figure 27.

Absolute value function (p. 279)

$f(x) = |x|$ See Figure 28.

Things to Know

Average rate of change of a function (p. 264)	The average rate of change of f from c to x is $$\frac{\Delta y}{\Delta x} = \frac{f(x) - f(c)}{x - c}, \qquad x \neq c$$
Increasing function (p. 266)	A function f is increasing on an open interval I if, for any choice of x_1 and x_2 in I, with $x_1 < x_2$, we have $f(x_1) < f(x_2)$.
Decreasing function (p. 266)	A function f is decreasing on an open interval I if, for any choice of x_1 and x_2 in I, with $x_1 < x_2$, we have $f(x_1) > f(x_2)$.
Constant function (p. 266)	A function f is constant on an open interval I if, for all choices of x in I, the values of $f(x)$ are equal.
Local maximum (p. 267)	A function f has a local maximum at c if there is an open interval I containing c so that, for all $x \neq c$ in I, $f(x) < f(c)$.
Local minimum (p. 268)	A function f has a local minimum at c if there is an open interval I containing c so that, for all $x \neq c$ in I, $f(x) > f(c)$.
Even function f (p. 269)	$f(-x) = f(x)$ for every x in the domain ($-x$ must also be in the domain).
Odd function f (p. 269)	$f(-x) = -f(x)$ for every x in the domain ($-x$ must also be in the domain).
Difference Quotient of f (p. 275)	$$\frac{f(x + h) - f(x)}{h}, \qquad h \neq 0$$

Objectives

Section	You should be able to ...	Review Exercises
3.1	1. Test an equation for symmetry with respect to (a) the x-axis, (b) the y-axis, and (c) the origin (p. 256)	3–10
	2. Know how to graph key equations (p. 258)	11, 12
3.2	1. Find the average rate of change of a function (p. 263)	17–20, 70
	2. Use a graph to determine where a function is increasing, is decreasing, or is constant (p. 266)	1(b), 2(b)
	3. Use a graph to locate local maxima and minima (p. 267)	1(c), 2(c)
	4. Use a graphing utility to approximate local maxima and minima and to determine where a function is increasing or decreasing (p. 268)	41–44
	5. Determine even and odd functions from a graph (p. 269)	1(e), 2(e)
	6. Identify even and odd functions from the equation (p. 270)	21–28
3.3	1. Graph functions in the library (p. 276)	11, 12
	2. Graph piecewise-defined functions (p. 281)	13–16
3.4	1. Graph functions using horizontal and vertical shifts (p. 286)	29, 30, 33–40, 63, 64
	2. Graph functions using reflections about the x-axis or y-axis (p. 289)	35, 36, 40, 63, 64
	3. Graph functions using compressions and stretches (p. 290)	31, 32, 39, 40, 63, 64
3.5	1. Form the sum, difference, product, and quotient of two functions (p. 299)	45–50
	2. Form the composite function and find its domain (p. 301)	51–62
3.6	1. Construct and analyze functions (p. 309)	65–69

Review Exercises

Blue problem numbers indicate the authors' suggestions for use in a Practice Test.

In Problems 1 and 2, use the graph of the function f to find

    (a) The domain and range of f
    (b) The intervals on which f is increasing, decreasing, or constant
    (c) The local minima and local maxima
    (d) Whether the graph is symmetric with respect to the x-axis, the y-axis, or the origin
    (e) Whether the function is even, odd, or neither
    (f) The intercepts, if any

1.

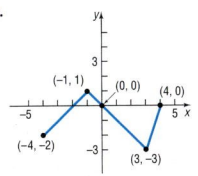

2.

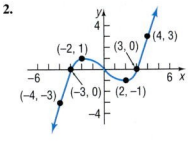

The **Chapter Review** helps check your understanding of the chapter materials in several ways. **"Things to Know"** gives a general overview of review topics. The **"Objectives"** section lists each skill you are expected to have mastered along with page references and review exercises to test your skills. The **"Review Exercises"** then serve as a chance to practice the concepts presented within the chapter. The review materials are designed to make you, the student, confident in knowing the chapter material.

Sullivan M@thP@k
An Integrated Learning Environment

Today's textbooks offer a wide variety of ancillary materials to students, from solutions manuals to tutorial software to text-specific Websites. Making the most of all of these resources can be difficult. Sullivan **M@thP@k** helps students get it together. **M@thP@k** seamlessly integrates the following key products into an **integrated learning environment.**

MathPro 5

MathPro 5 is online, customizable tutorial software integrated with the text at the Learning Objective level for anytime, anywhere learning. MathPro5's "watch" feature integrates lecture videos into the algorithmic tutorial environment. The easy-to-use course management system enables instructors to track and assess student performance on tutorial work, quizzes and tests. A robust reports wizard provides a grade book, individual student reports and class summaries. The customizable syllabus allows instructors to remove and reorganize chapters, sections and objectives. MathPro 5's messaging system enhances communication between students and instructors.

The combination of MathPro5's richly integrated tutorial, testing and robust course management tools provides an unparalleled tutorial experience for students, and new assessment and time-saving tools for instructors.

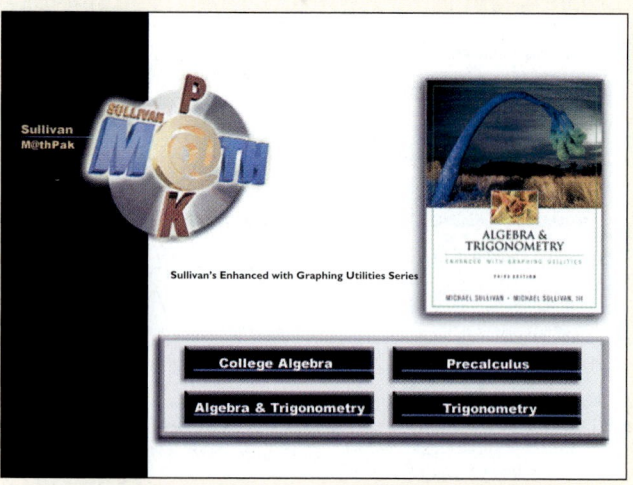

The Sullivan M@thP@k Website

This robust pass-code protected site features quizzes, homework starters, live animated examples, graphing calculator manuals, and much more. It offers the student many, many ways to test and reinforce their understanding of the course material.

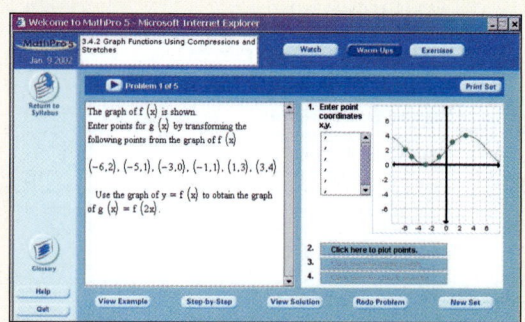

Student Solutions Manual

Written by Mark McCombs, University of North Carolina, Chapel Hill and a long term user of the Sullivan series, the *Student Solutions Manual* offers thorough, accurate solutions that are consistent with the precise mathematics found in the text.

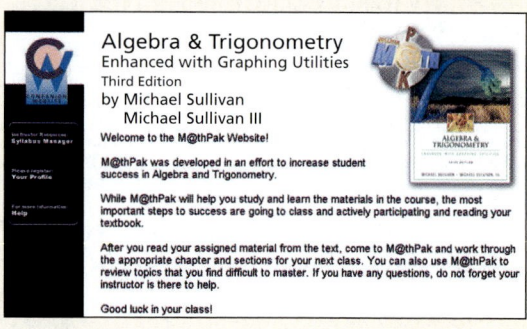

Sullivan M@thP@k.
Helping students *Get it Together.*

Additional Media

Sullivan Companion Website

www.prenhall.com/sullivanegu3e

This text-specific website beautifully complements the text. Here students can find chapter tests, section-specific links, and PowerPoint downloads in addition to other helpful features.

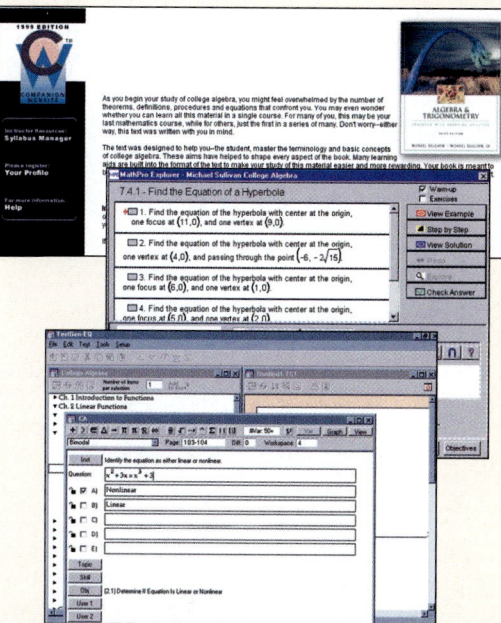

Test Gen-EQ

CD-Rom (Windows/Macintosh)

- Algorithmically driven, text-specific testing program
- Networkable for administering tests and capturing grades
- Edit existing test items or add your own questions to create a nearly unlimited number of tests and drill worksheets

ISBN: 0-13-044705-6

S U P P L E M E N T S

Student Supplements

Student Solutions Manual
Worked solutions to all odd-numbered exercises from the text and complete solutions for chapter review problems and chapter tests. ISBN: 0-13-044717-X

New York Times Themes of the Times
A *free* newspaper from Prentice Hall and *The New York Times.* Interesting and current articles on mathematics which invite discussion and writing about mathematics.

Mathematics on the Internet
Free guide providing a brief history of the Internet, discussing the use of the World Wide Web, and describing how to find your way within the Internet and how to find others on it.

MathPro 5
See description on page xxvi. ISBN: 0-13-044711-0

CD Lecture Series
Lecture videos featuring Michael Sullivan III are available to students and faculty in CD ROM format for added convenience. ISBN: 0-13-044712-9

Prentice Hall Math Tutor Center
The PH Math Tutor provides tutoring for students enrolled in developmental and precalculus mathematics using selected Prentice Hall titles. Registration is required; once registered, students can receive help via toll free phone, fax, and email, only during the hours of operation. (Sunday through Thursday, 5:00 PM – 12:00 AM EST) ISBN: 0-13-064604-0

Instructor Supplements

Instructor's Edition
Offers the instructor a complete set of answers to all problems in the text. Also, for your convenience, the Instructor Resource CD contains the full version of TestGen-EQ, additional Chapter Projects, links to a passcode protected Web site which has the complete Instructor's Solutions Manual and Test Item File. This allows flexibility in preparing for class. ISBN: 0-13-044714-5

Instructor's Resource Manual
Contains complete step-by-step worked-out solutions to all exercises in the textbook. ISBN: 0-13-044715-3

Test Item File
Hard copy of the algorithmic computerized testing materials. ISBN: 0-13-044713-7

MathPro 5
See description on page xxvi. ISBN: 0-13-009307-6

CD Lecture Series
Lecture videos featuring Michael Sullivan III are available to students and faculty in CD ROM format for added convenience. ISBN: 0-13-044712-9

Lecture Videos
These VHS instructional tapes, in a lecture format, feature worked-out examples and exercises, taken from each section of the text. ISBN: 0-13-044716-1

PHOTO AND ILLUSTRATION CREDITS

CHAPTER R   Page 1, Prentice Hall CDA; Page 37, Giraudon/Art Resource, N.Y.

CHAPTER 1   Pages 89 and 192, Prentice Hall CDA

CHAPTER 2   Pages 195 and 252, Najila Feanny/ Stock Boston

CHAPTER 3   Pages 255 and 320, Reuters/Anthony Bolante/Hulton Archive

CHAPTER 4   Pages 323 and 376, Stephen Ferry/ Liaison Agency, Inc.

CHAPTER 5   Pages 379 and 414, Prentice Hall CDA; Page 395, CORBIS

CHAPTER 6   Pages 417 and 508, Amy C. Elra/ PhotoEdit; Page 464, The Granger Collection

CHAPTER 7   Pages 511 and 606, Elena Booraid/ PhotoEdit; Page 574, CORBIS

CHAPTER 8   Pages 609 and 706, Prentice Hall CDA

CHAPTER 9   Pages 709 and 776, Steve Starr/ Stock Boston

CHAPTER 10   Pages 779 and 826, Prentice Hall CDA

CHAPTER 11   Pages 829 and 889, Art Matrix/ Visuals Unlimited; Page 855, CORBIS; Page 864, CORBIS; Page 875, Library of Congress

CHAPTER 12   Pages 891 and 978, CORBIS

CHAPTER 13   Pages 981 and 1029, NASA/GPSC/Tom Stack & Associates, Inc.; Page 1011, The Granger Collection; Page 1024, CORBIS

CHAPTER 14   Pages 1031 and 1071, Myrleen Ferguson/ PhotoEdit; Page 1058, The Granger Collection.

REVIEW

 For additional
study help, go to
www.prenhall.com/sullivanegu3e

Materials include:

- Graphing Calculator Help
- Chapter Quiz
- Chapter Test
- PowerPoint Downloads
- Chapter Projects
- Student Tips

A Look Back, A Look Forward

As the title states, this chapter is a review. The purpose of the chapter is to help you recall facts that you learned in an earlier course. The topics chosen for inclusion here represent information that will be useful to prepare for this course.

Your instructor may decide to cover this chapter or not, depending on the course syllabus. Regardless, as you proceed through the book, references will be made to Chapter R that you can use as a "just in time" review.

PREPARING FOR THIS BOOK

Before getting started, read the Preface to the Student (p. xv).

R.1 REAL NUMBERS

OBJECTIVES
1. Classify Numbers
2. Evaluate Numerical Expressions
3. Work with Properties of Real Numbers

Sets

When we want to treat a collection of similar but distinct objects as a whole, we use the idea of a **set**. For example, the set of *digits* consists of the collection of numbers $0, 1, 2, 3, 4, 5, 6, 7, 8$, and 9. If we use the symbol D to denote the set of digits, then we can write

$$D = \{0, 1, 2, 3, 4, 5, 6, 7, 8, 9\}$$

In this notation, the braces $\{\ \}$ are used to enclose the objects, or **elements**, in the set. This method of denoting a set is called the **roster method**. A second way to denote a set is to use **set-builder notation**, where the set D of digits is written as

$$D = \{\quad x \quad | \quad x \text{ is a digit}\}$$

Read as "D is the set of all x such that x is a digit."

EXAMPLE 1 | **Using Set-builder Notation and the Roster Method**

(a) $E = \{x | x \text{ is an even digit}\} = \{0, 2, 4, 6, 8\}$
(b) $O = \{x | x \text{ is an odd digit}\} = \{1, 3, 5, 7, 9\}$ ■

In listing the elements of a set, we do not list an element more than once because the elements of a set are distinct. Also, the order in which the elements are listed is not relevant. For example, $\{2, 3\}$ and $\{3, 2\}$ both represent the same set.

If every element of a set A is also an element of a set B, then we say that A is a **subset** of B. If two sets A and B have the same elements, then we say that A **equals** B. For example, $\{1, 2, 3\}$ is a subset of $\{1, 2, 3, 4, 5\}$, and $\{1, 2, 3\}$ equals $\{2, 3, 1\}$.

Finally, if a set has no elements, it is called the **empty set**, or the **null set**, and is denoted by the symbol $\emptyset$.

Classification of Numbers

① It is helpful to classify the various kinds of numbers that we deal with as sets. The **counting numbers**, or **natural numbers**, are the numbers in the set $\{1, 2, 3, 4, \dots\}$. (The three dots, called an **ellipsis**, indicate that the pattern continues indefinitely.) As their name implies, these numbers are often used to count things. For example, there are 26 letters in our alphabet; there are 100 cents in a dollar. The **whole numbers** are the numbers in the set $\{0, 1, 2, 3, \dots\}$, that is, the counting numbers together with 0.

> The **integers** are the numbers in the set $\{\dots, -3, -2, -1, 0, 1, 2, 3, \dots\}$.

These numbers are useful in many situations. For example, if your checking account has $10 in it and you write a check for $15, you can represent the current balance as $-\$5$.

Notice that the set of counting numbers is a subset of the set of whole numbers. Each time we expand a number system, such as from the whole numbers to the integers, we do so in order to be able to handle new, and usually more complicated, problems. The integers allow us to solve problems requiring both positive and negative counting numbers, such as profit/loss, height above/below sea level, temperature above/below 0°F, and so on.

But integers alone are not sufficient for *all* problems. For example, they do not answer the question "What part of a dollar is 38 cents?" To answer such a question, we enlarge our number system to include *rational numbers*. For example, $\dfrac{38}{100}$ answers the question "What part of a dollar is 38 cents?"

> A **rational number** is a number that can be expressed as a quotient $\dfrac{a}{b}$ of two integers. The integer a is called the **numerator**, and the integer b, which cannot be 0, is called the **denominator**. The rational numbers are the numbers in the set $\{x \mid x = \dfrac{a}{b}, \text{ where } a, b \neq 0 \text{ are integers }\}$.

Examples of rational numbers are $\dfrac{3}{4}, \dfrac{5}{2}, \dfrac{0}{4}, -\dfrac{2}{3}$, and $\dfrac{100}{3}$. Since $\dfrac{a}{1} = a$ for any integer a, it follows that the set of integers is a subset of the set of rational numbers.

Rational numbers may be represented as **decimals**. For example, the rational numbers $\dfrac{3}{4}, \dfrac{5}{2}, -\dfrac{2}{3}$, and $\dfrac{7}{66}$ may be represented as decimals by merely carrying out the indicated division:

$$\frac{3}{4} = 0.75 \qquad \frac{5}{2} = 2.5 \qquad -\frac{2}{3} = -0.666\dots \qquad \frac{7}{66} = 0.1060606\dots$$

Notice that the decimal representations of $\frac{3}{4}$ and $\frac{5}{2}$ terminate, or end. The decimal representations of $-\frac{2}{3}$ and $\frac{7}{66}$ do not terminate, but they do exhibit a pattern of repetition. For $-\frac{2}{3}$, the 6 repeats indefinitely; for $\frac{7}{66}$, the block 06 repeats indefinitely. It can be shown that every rational number may be represented by a decimal that either terminates or is nonterminating with a repeating block of digits, and vice versa.

On the other hand, there are decimals that do not fit into either of these categories. Such decimals represent **irrational numbers**. Every irrational number may be represented by a decimal that neither repeats nor terminates. In other words, irrational numbers cannot be written in the form $\frac{a}{b}$, where a, $b \neq 0$ are integers.

Irrational numbers occur naturally. For example, consider the isosceles right triangle whose legs are each of length 1. See Figure 1. The length of the hypotenuse is $\sqrt{2}$, an irrational number.

Also, the number that equals the ratio of the circumference C to the diameter d of any circle, denoted by the symbol π (the Greek letter pi), is an irrational number. See Figure 2.

> Together, the rational numbers and irrational numbers form the set of **real numbers**.

Figure 3 shows the relationship of various types of numbers.*

Figure 1

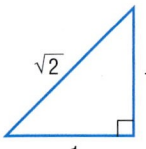

Figure 2

$$\pi = \frac{C}{d}$$

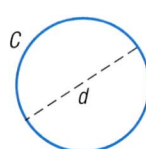

Figure 3

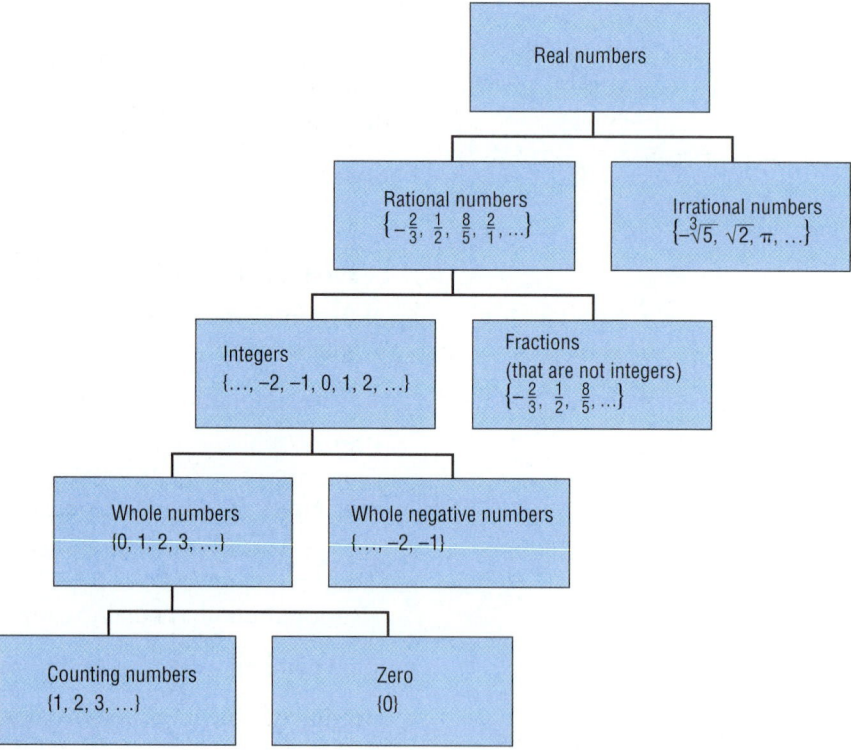

*The set of real numbers is a subset of the set of complex numbers. We study complex numbers in Section 5.3.

| EXAMPLE 2 | **Classifying the Numbers in a Set** |

List the numbers in the set

$$\left\{ -3, \frac{4}{3}, 0.12, \sqrt{2}, \pi, 2.151515\ldots \text{ (where the block 15 repeats)}, 10 \right\}$$

that are

(a) Natural numbers   (b) Integers   (c) Rational numbers
(d) Irrational numbers   (e) Real numbers

Solution
(a) 10 is the only natural number.

(b) -3 and 10 are integers.

(c) $-3, \dfrac{4}{3}, 0.12, 2.151515\ldots$, and 10 are rational numbers.

(d) $\sqrt{2}$ and π are irrational numbers.

(e) All the numbers listed are real numbers. ◼

━ **NOW WORK PROBLEM 3.**

Approximations

Every decimal may be represented by a real number (either rational or irrational), and every real number may be represented by a decimal.

The irrational numbers $\sqrt{2}$ and π have decimal representations that begin as follows:

$$\sqrt{2} = 1.414213\ldots \qquad \pi = 3.14159\ldots$$

In practice, decimals are generally represented by approximations. For example, using the symbol $\approx$ (read as "approximately equal to"), we can write

$$\sqrt{2} \approx 1.4142 \qquad \pi \approx 3.1416$$

In approximating decimals, we either *round* or *truncate* to a given number of decimal places. The number of places establishes the location of the *final digit* in the decimal approximation.

> **Truncation:** Drop all the digits that follow the specified final digit in the decimal.
> **Rounding:** Identify the specified final digit in the decimal. If the next digit is 5 or more, add 1 to the final digit; if the next digit is 4 or less, leave the final digit as it is. Now truncate following the final digit.

| EXAMPLE 3 | **Approximating a Decimal to Two Places** |

Approximate 20.98752 to two decimal places by

(a) Truncating   (b) Rounding

Solution
For 20.98752, the final digit is 8, since it is two decimal places from the decimal point.

(a) To truncate, we remove all digits following the final digit 8. The truncation of 20.98752 to two decimal places is 20.98.

(b) The digit following the final digit 8 is the digit 7. Since 7 is 5 or more, we add 1 to the final digit 8 and truncate. The rounded form of 20.98752 to two decimal places is 20.99. ■

EXAMPLE 4   **Approximating a Decimal to Two and Four Places**

Number	Rounded to Two Decimal Places	Rounded to Four Decimal Places	Truncated to Two Decimal Places	Truncated to Four Decimal Places
(a) 3.14159	3.14	3.1416	3.14	3.1415
(b) 0.056128	0.06	0.0561	0.05	0.0561
(c) 893.46125	893.46	893.4613	893.46	893.4612 ■

 **NOW WORK PROBLEM 7.**

Calculators

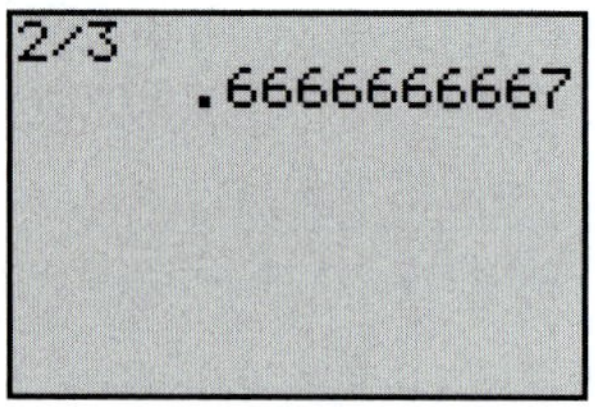

Calculators are finite machines. As a result, they are incapable of displaying decimals that contain a large number of digits. For example, some calculators are capable of displaying only eight digits. When a number requires more than eight digits, the calculator either truncates or rounds. To see how your calculator handles decimals, divide 2 by 3. How many digits do you see? Is the last digit a 6 or a 7? If it is a 6, your calculator truncates; if it is a 7, your calculator rounds.

Figure 4

```
2/3
       .6666666667
```

Figure 4 shows $\dfrac{2}{3}$ on a TI-83 calculator. How many digits are displayed? Does a TI-83 round or truncate? What does your calculator do?

Operations

In algebra, we use letters such as $x, y, a, b,$ and c to represent numbers. The symbols used in algebra for the operations of addition, subtraction, multiplication, and division are $+, -, \cdot,$ and $/$. The words used to describe the results of these operations are **sum**, **difference**, **product**, and **quotient**. Table 1 summarizes these ideas.

TABLE 1		
Operation	**Symbol**	**Words**
Addition	$a + b$	Sum: a plus b
Subtraction	$a - b$	Difference: a less b
Multiplication	$a \cdot b, (a) \cdot b, a \cdot (b), (a) \cdot (b),$ $ab, (a)b, a(b), (a)(b)$	Product: a times b
Division	a/b or $\dfrac{a}{b}$	Quotient: a divided by b

In algebra, we generally avoid using the multiplication sign $\times$ and the division sign $\div$ so familiar in arithmetic. Notice also that when two expressions are placed next to each other without an operation symbol, as in ab, or in parentheses, as in $(a)(b)$, it is understood that the expressions, called **factors**, are to be multiplied.

We also prefer not to use mixed numbers in algebra. When mixed numbers are used, addition is understood; for example, $2\frac{3}{4}$ means $2 + \frac{3}{4}$. In algebra, use of a mixed number may be confusing because the absence of an operation symbol between two terms is generally taken to mean multiplication. The expression $2\frac{3}{4}$ is therefore written instead as 2.75 or as $\frac{11}{4}$.

The symbol $=$, called an **equal sign** and read as "equals" or "is," is used to express the idea that the number or expression on the left of the equal sign is equivalent to the number or expression on the right.

EXAMPLE 5   Writing Statements Using Symbols

(a) The sum of 2 and 7 equals 9. In symbols, this statement is written as $2 + 7 = 9$.

(b) The product of 3 and 5 is 15. In symbols, this statement is written as $3 \cdot 5 = 15$.   ■

━━━ **NOW WORK PROBLEM 19.**

Order of Operations

2   Consider the expression $2 + 3 \cdot 6$. It is not clear whether we should add 2 and 3 to get 5, and then multiply by 6 to get 30; or first multiply 3 and 6 to get 18, and then add 2 to get 20. To avoid this ambiguity, we have the following agreement.

> We agree that whenever the two operations of addition and multiplication separate three numbers the multiplication operation always will be performed first, followed by the addition operation.

For $2 + 3 \cdot 6$, we have

$$2 + 3 \cdot 6 = 2 + 18 = 20$$

EXAMPLE 6   Finding the Value of an Expression

Evaluate each expression.

(a) $3 + 4 \cdot 5$ (b) $8 \cdot 2 + 1$ (c) $2 + 2 \cdot 2$

Solution   (a) $3 + 4 \cdot 5 = 3 + 20 = 23$ (b) $8 \cdot 2 + 1 = 16 + 1 = 17$

                     Multiply first   Multiply first

(c) $2 + 2 \cdot 2 = 2 + 4 = 6$ ■

Figure 5 shows the solutions to Example 6 using a graphing calculator. Notice that the calculator follows the order of operations agreed to.

━━━ **NOW WORK PROBLEM 31.**

Figure 5

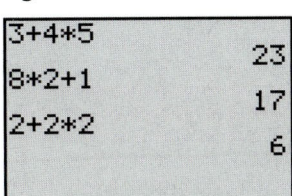

To first add 3 and 4 and then multiply the result by 5, we use parentheses and write $(3 + 4) \cdot 5$. Whenever parentheses appear in an expression, it means "perform the operations within the parentheses first!"

EXAMPLE 7   **Finding the Value of an Expression**

(a) $(5 + 3) \cdot 4 = 8 \cdot 4 = 32$

(b) $(4 + 5) \cdot (8 - 2) = 9 \cdot 6 = 54$ ■

When we divide two expressions, as in

$$\frac{2 + 3}{4 + 8}$$

it is understood that the division bar acts like parentheses; that is,

$$\frac{2 + 3}{4 + 8} = \frac{(2 + 3)}{(4 + 8)}$$

The following list gives the rules for the order of operations.

Rules for the Order of Operations

1. Begin with the innermost parentheses and work outward. Remember that in dividing two expressions the numerator and denominator are treated as if they were enclosed in parentheses.
2. Perform multiplications and divisions, working from left to right.
3. Perform additions and subtractions, working from left to right.

EXAMPLE 8   **Finding the Value of an Expression**

Evaluate each expression.

(a) $8 \cdot 2 + 3$ (b) $5 \cdot (3 + 4) + 2$

(c) $\dfrac{2 + 5}{2 + 4 \cdot 7}$ (d) $2 + [4 + 2 \cdot (10 + 6)]$

Solution   (a) $8 \cdot 2 + 3 = 16 + 3 = 19$

Multiply first

(b) $5 \cdot (3 + 4) + 2 = 5 \cdot 7 + 2 = 35 + 2 = 37$

Parentheses first   Multiply before adding

(c) $\dfrac{2 + 5}{2 + 4 \cdot 7} = \dfrac{2 + 5}{2 + 28} = \dfrac{7}{30}$

(d) $2 + [4 + 2 \cdot (10 + 6)] = 2 + [4 + 2 \cdot (16)]$
$= 2 + [4 + 32] = 2 + [36] = 38$ ■

Figure 6

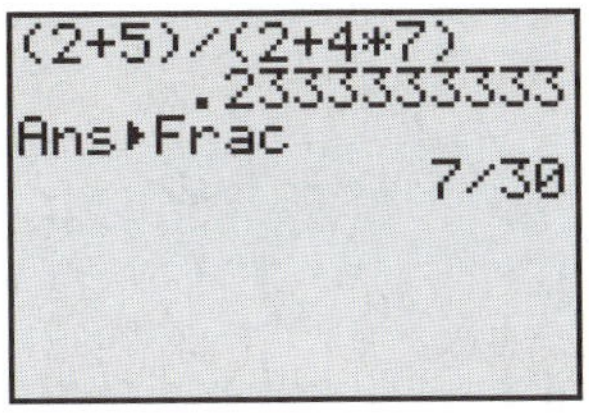

```
(2+5)/(2+4*7)
         .2333333333
Ans▶Frac
              7/30
```

Be careful if you use a calculator. For Example 8(c), you need to use parentheses. See Figure 6.* If you don't, the calculator will compute the expression

$$2 + \frac{5}{2} + 4 \cdot 7 = 2 + 2.5 + 28 = 32.5$$

giving a wrong answer.

➤ NOW WORK PROBLEMS **37** AND **45.**

Properties of Real Numbers

③ We have used the equal sign to mean that one expression is equivalent to another. Four important properties of equality are listed next. In this list, a, b, and c represent real numbers.

> 1. The **reflexive property** states that a number always equals itself; that is, $a = a$.
> 2. The **symmetric property** states that if $a = b$ then $b = a$.
> 3. The **transitive property** states that if $a = b$ and $b = c$ then $a = c$.
> 4. The **principle of substitution** states that if $a = b$ then we may substitute b for a in any expression containing a.

Now, let's consider some other properties of real numbers. We begin with an example.

EXAMPLE 9   **Commutative Properties**

(a) $3 + 5 = 8$ (b) $2 \cdot 3 = 6$
     $5 + 3 = 8$ $3 \cdot 2 = 6$
     $3 + 5 = 5 + 3$ $2 \cdot 3 = 3 \cdot 2$

This example illustrates the **commutative property** of real numbers, which states that the order in which addition or multiplication takes place will not affect the final result.

Commutative Properties

$$a + b = b + a \qquad\qquad (1a)$$
$$a \cdot b = b \cdot a \qquad\qquad (1b)$$

Here, and in the properties listed on pages 10–13, a, b, and c represent real numbers.

---

*Notice that we converted the decimal to its fraction form. Consult your manual to see how your calculator does this.

EXAMPLE 10   **Associative Properties**

(a) $2 + (3 + 4) = 2 + 7 = 9$ (b) $2 \cdot (3 \cdot 4) = 2 \cdot 12 = 24$
    $(2 + 3) + 4 = 5 + 4 = 9$ $(2 \cdot 3) \cdot 4 = 6 \cdot 4 = 24$
    $2 + (3 + 4) = (2 + 3) + 4$ $2 \cdot (3 \cdot 4) = (2 \cdot 3) \cdot 4$ ■

The way we add or multiply three real numbers will not affect the final result. Expressions such as $2 + 3 + 4$ and $3 \cdot 4 \cdot 5$ present no ambiguity, even though addition and multiplication are performed on one pair of numbers at a time. This property is called the **associative property**.

Associative Properties

$$a + (b + c) = (a + b) + c = a + b + c \qquad \text{(2a)}$$
$$a \cdot (b \cdot c) = (a \cdot b) \cdot c = a \cdot b \cdot c \qquad \text{(2b)}$$

The next property is perhaps the most important.

Distributive Property

$$a \cdot (b + c) = a \cdot b + a \cdot c \qquad \text{(3a)}$$
$$(a + b) \cdot c = a \cdot c + b \cdot c \qquad \text{(3b)}$$

The **distributive property** may be used in two different ways.

EXAMPLE 11   **Distributive Property**

(a) $2 \cdot (x + 3) = 2 \cdot x + 2 \cdot 3 = 2x + 6$ *Use to remove parentheses.*
(b) $3x + 5x = (3 + 5)x = 8x$ *Use to combine two expressions.*   ■

✏ **NOW WORK PROBLEM 63.**

The real numbers 0 and 1 have unique properties.

EXAMPLE 12   **Identity Properties**

(a) $4 + 0 = 0 + 4 = 4$ (b) $3 \cdot 1 = 1 \cdot 3 = 3$ ■

The properties of 0 and 1 illustrated in Example 12 are called the **identity properties**.

Identity Properties

$$0 + a = a + 0 = a \qquad \text{(4a)}$$
$$a \cdot 1 = 1 \cdot a = a \qquad \text{(4b)}$$

We call 0 the **additive identity** and 1 the **multiplicative identity**.

For each real number a, there is a real number $-a$, called the **additive inverse** of a, having the following property:

Additive Inverse Property

$$a + (-a) = -a + a = 0 \qquad\qquad (5a)$$

EXAMPLE 13 **Finding an Additive Inverse**

(a) The additive inverse of 6 is -6, because $6 + (-6) = 0$.

(b) The additive inverse of -8 is $-(-8) = 8$, because $-8 + 8 = 0$. ■

The additive inverse of a, that is, $-a$, is often called the *negative* of a or the *opposite* of a. The use of such terms can be dangerous, because they suggest that the additive inverse is a negative number, which it may not be. For example, the additive inverse of -3, or $-(-3)$, equals 3, a positive number.

For each *nonzero* real number a, there is a real number $\dfrac{1}{a}$, called the **multiplicative inverse** of a, having the following property:

Multiplicative Inverse Property

$$a \cdot \frac{1}{a} = \frac{1}{a} \cdot a = 1 \qquad \text{if } a \neq 0 \qquad\qquad (5b)$$

The multiplicative inverse $\dfrac{1}{a}$ of a nonzero real number a is also referred to as the **reciprocal** of a.

EXAMPLE 14 **Finding a Reciprocal**

(a) The reciprocal of 6 is $\dfrac{1}{6}$, because $6 \cdot \dfrac{1}{6} = 1$.

(b) The reciprocal of -3 is $\dfrac{1}{-3}$, because $-3 \cdot \dfrac{1}{-3} = 1$.

(c) The reciprocal of $\dfrac{2}{3}$ is $\dfrac{3}{2}$, because $\dfrac{2}{3} \cdot \dfrac{3}{2} = 1$. ■

With these properties for adding and multiplying real numbers, we can now define the operations of subtraction and division as follows:

The **difference** $a - b$, also read "a less b" or "a minus b," is defined as

$$a - b = a + (-b) \qquad\qquad (6)$$

To subtract b from a, add the opposite of b to a.

If b is a nonzero real number, the **quotient** $\dfrac{a}{b}$, also read as "a divided by b" or "the ratio of a to b," is defined as

$$\frac{a}{b} = a \cdot \frac{1}{b} \qquad \text{if } b \neq 0 \tag{7}$$

EXAMPLE 15   **Working with Differences and Quotients**

(a) $8 - 5 = 8 + (-5) = 3$ (b) $4 - 9 = 4 + (-9) = -5$

(c) $\dfrac{5}{8} = 5 \cdot \dfrac{1}{8}$ ∎

For any number a, the product of a times 0 is always 0; that is,

Multiplication by Zero

$$a \cdot 0 = 0 \tag{8}$$

For a nonzero number a,

Division Properties

$$\frac{0}{a} = 0 \qquad \frac{a}{a} = 1 \qquad \text{if } a \neq 0 \tag{9}$$

Note: Division by 0 is *not defined*. One reason is to avoid the following difficulty: $\dfrac{2}{0} = x$ means to find x such that $0 \cdot x = 2$. But $0 \cdot x$ equals 0 for all x, so there is *no* number x such that $\dfrac{2}{0} = x$.

Rules of Signs

$$a(-b) = -(ab) \qquad (-a)b = -(ab) \qquad (-a)(-b) = ab$$

$$-(-a) = a \qquad \frac{a}{-b} = \frac{-a}{b} = -\frac{a}{b} \qquad \frac{-a}{-b} = \frac{a}{b} \tag{10}$$

EXAMPLE 16   **Applying the Rules of Signs**

(a) $2(-3) = -(2 \cdot 3) = -6$ (b) $(-3)(-5) = 3 \cdot 5 = 15$

(c) $\dfrac{3}{-2} = \dfrac{-3}{2} = -\dfrac{3}{2}$ (d) $\dfrac{-4}{-9} = \dfrac{4}{9}$ (e) $\dfrac{x}{-2} = \dfrac{1}{-2} \cdot x = -\dfrac{1}{2}x$ ∎

If c is a nonzero number, then

Cancellation Properties

$$ac = bc \quad \text{implies} \quad a = b \qquad \text{if } c \neq 0$$

$$\frac{ac}{bc} = \frac{a}{b} \qquad\qquad \text{if } b \neq 0, c \neq 0 \qquad (11)$$

EXAMPLE 17 **Using the Cancellation Properties**

(a) If $2x = 6$, then

$$2x = 6$$
$$2x = 2 \cdot 3 \qquad \text{Factor 6.}$$
$$x = 3 \qquad \text{Cancel the 2's.}$$

(b) $\dfrac{18}{12} = \dfrac{3 \cdot \cancel{6}}{2 \cdot \cancel{6}} = \dfrac{3}{2}$

Cancel the 6's.

Note: We follow the common practice of using slash marks to indicate cancellations.

The next property is used often.

Zero-Product Property

$$\text{If } ab = 0, \text{ then } a = 0 \text{ or } b = 0, \text{ or both.} \qquad (12)$$

EXAMPLE 18 **Using the Zero-Product Property**

If $2x = 0$, then either $2 = 0$ or $x = 0$. Since $2 \neq 0$, it follows that $x = 0$.

Arithmetic of Quotients

$$\frac{a}{b} + \frac{c}{d} = \frac{ad}{bd} + \frac{bc}{bd} = \frac{ad + bc}{bd} \qquad \text{if } b \neq 0, d \neq 0 \qquad (13)$$

$$\frac{a}{b} \cdot \frac{c}{d} = \frac{ac}{bd} \qquad\qquad\qquad \text{if } b \neq 0, d \neq 0 \qquad (14)$$

$$\frac{\dfrac{a}{b}}{\dfrac{c}{d}} = \frac{a}{b} \cdot \frac{d}{c} = \frac{ad}{bc} \qquad\qquad \text{if } b \neq 0, c \neq 0, d \neq 0 \quad (15)$$

EXAMPLE 19 **Adding, Subtracting, Multiplying, and Dividing Quotients**

(a) $\dfrac{2}{3} + \dfrac{5}{2} = \dfrac{2 \cdot 2}{3 \cdot 2} + \dfrac{3 \cdot 5}{3 \cdot 2} = \dfrac{2 \cdot 2 + 3 \cdot 5}{3 \cdot 2} = \dfrac{4 + 15}{6} = \dfrac{19}{6}$

By equation (13)

(b) $\dfrac{3}{5} - \dfrac{2}{3} \underset{\substack{\uparrow \\ \text{By equation (6)}}}{=} \dfrac{3}{5} + \left(-\dfrac{2}{3} \right) \underset{\substack{\uparrow \\ \text{By equation (10)}}}{=} \dfrac{3}{5} + \dfrac{-2}{3}$

$$= \dfrac{3 \cdot 3 + 5 \cdot (-2)}{5 \cdot 3} = \dfrac{9 + (-10)}{15} = \dfrac{-1}{15} = -\dfrac{1}{15}$$

(c) $\dfrac{8}{3} \cdot \dfrac{15}{4} \underset{\substack{\uparrow \\ \text{By equation (14)}}}{=} \dfrac{8 \cdot 15}{3 \cdot 4} = \dfrac{2 \cdot 4 \cdot 3 \cdot 5}{3 \cdot 4 \cdot 1} \underset{\substack{\uparrow \\ \text{By equation (11)}}}{=} \dfrac{2 \cdot 5}{1} = 10$

Note: Slanting the cancellation marks in different directions for different factors, as shown here, is a good practice to follow, since it will help in checking for errors.

(d) $\dfrac{\frac{3}{5}}{\frac{7}{9}} \underset{\substack{\uparrow \\ \text{By equation (15)}}}{=} \dfrac{3}{5} \cdot \dfrac{9}{7} \underset{\substack{\uparrow \\ \text{By equation (14)}}}{=} \dfrac{3 \cdot 9}{5 \cdot 7} = \dfrac{27}{35}$

Note: In writing quotients, we shall follow the usual convention and write the quotient in **lowest terms**; that is, we write it so that any common factors of the numerator and the denominator have been removed using the cancellation properties, equation (11).

$$\dfrac{90}{24} = \dfrac{15 \cdot 6}{4 \cdot 6} = \dfrac{15}{4}$$

$$\dfrac{24x^2}{18x} = \dfrac{4 \cdot 6 \cdot x \cdot x}{3 \cdot 6 \cdot x} = \dfrac{4x}{3} \qquad x \neq 0$$

✏ **NOW WORK PROBLEMS 47, 51, AND 61.**

Sometimes it is easier to add two fractions using *least common multiples* (LCM). The LCM of two numbers is the smallest number that each has as a common multiple.

EXAMPLE 20 | **Finding the Least Common Multiple of Two Numbers**

Find the least common multiple of 15 and 12.

Solution To find the LCM of 15 and 12, we look at multiples of 15 and 12.

15, 30, 45, 60, 75, 90, 105, 120, . . .

12, 24, 36, 48, 60, 72, 84, 96, 108, 120, . . .

The *common* multiples are in blue. The *least* common multiple is 60. ∎

EXAMPLE 21 | **Using the Least Common Multiple to Add Two Fractions**

Find: $\dfrac{8}{15} + \dfrac{5}{12}$

Solution We use the LCM of the denominators of the fractions and rewrite each fraction using the LCM as a common denominator. The LCM of the denominators (12 and 15) is 60. Rewrite each fraction using 60 as the denominator.

$$\dfrac{8}{15} + \dfrac{5}{12} = \dfrac{8}{15} \cdot \dfrac{4}{4} + \dfrac{5}{12} \cdot \dfrac{5}{5} = \dfrac{32}{60} + \dfrac{25}{60} = \dfrac{32 + 25}{60} = \dfrac{57}{60}$$ ∎

✏ **NOW WORK PROBLEM 55.**

HISTORICAL FEATURE

The real number system has a history that stretches back at least to the ancient Babylonians (1800 BC). It is remarkable how much the ancient Babylonian attitudes resemble our own. As we stated in the text, the fundamental difficulty with irrational numbers is that they cannot be written as quotients of integers or, equivalently, as repeating or terminating decimals. The Babylonians wrote their numbers in a system based on 60 in the same way that we write ours based on 10. They would carry as many places for π as the accuracy of the problem demanded, just as we now use

$$\pi \approx 3\frac{1}{7} \quad \text{or} \quad \pi \approx 3.1416 \quad \text{or} \quad \pi \approx 3.14159$$

$$\text{or} \quad \pi \approx 3.14159265358979$$

depending on how accurate we need to be.

Things were very different for the Greeks, whose number system allowed only rational numbers. When it was discovered that $\sqrt{2}$ was not a rational number, this was regarded as a fundamental flaw in the number concept. So serious was the matter that the Pythagorean Brotherhood (an early mathematical society) is said to have drowned one of its members for revealing this terri-

ble secret. Greek mathematicians then turned away from the number concept, expressing facts about whole numbers in terms of line segments.

In astronomy, however, Babylonian methods, including the Babylonian number system, continued to be used. Simon Stevin (1548–1620), probably using the Babylonian system as a model, invented the decimal system, complete with rules of calculation, in 1585. [Others, for example, al-Kashi of Samarkand (d. 1424), had made some progress in the same direction.] The decimal system so effectively conceals the difficulties that the need for more logical precision began to be felt only in the early 1800s. Around 1880, Georg Cantor (1845–1918) and Richard Dedekind (1831–1916) gave precise definitions of real numbers. Cantor's definition, although more abstract and precise, has its roots in the decimal (and hence Babylonian) numerical system.

Sets and set theory were a spin-off of the research that went into clarifying the foundations of the real number system. Set theory has developed into a large discipline of its own, and many mathematicians regard it as the foundation upon which modern mathematics is built. Cantor's discoveries that infinite sets can also be counted and that there are different sizes of infinite sets are among the most astounding results of modern mathematics.

HISTORICAL PROBLEMS

The Babylonian number system was based on 60. Thus 2,30 means $2 + \dfrac{30}{60} = 2.5$, and 4,25,14 means

$$4 + \frac{25}{60} + \frac{14}{60^2} = 4 + \frac{1514}{3600} = 4.42055555\ldots$$

1. What are the following numbers in Babylonian notation?
   (a) $1\dfrac{1}{3}$ (b) $2\dfrac{5}{6}$

2. What are the following Babylonian numbers when written as fractions and as decimals?
   (a) 2,20   (b) 4,52,30   (c) 3,8,29,44

R.1 Concepts and Vocabulary

In Problems 1–3, fill in the blanks.

1. The numbers in the set $\{\,x\,|\,x = \dfrac{a}{b}, \text{where } a, b \neq 0, \text{are integers}\}$ are called _____ numbers.

2. The value of the expression $4 + 5 \cdot 6 - 3$ is _____ .

3. The fact that $2x + 3x = (2 + 3)x$ is a consequence of the _____ Property.

In Problems 4–6, answer True or False to each statement.

4. Rational numbers have decimals that either terminate or are nonterminating with a repeating block of digits.

5. The Zero-Product Property states that the product of any number and zero equals zero.

6. The least common multiple of 12 and 18 is 6.

7. Name three sets of numbers that are subsets of the set of real numbers.

8. Evaluate the expression $(2 + 4) \cdot 4 + 3 \cdot 4$. Explain each step.

9. State the Zero-Product Property.

R.1 Exercises

In Problems 1–6, list the numbers in each set that are (a) Natural numbers, (b) Integers, (c) Rational numbers, (d) Irrational numbers, (e) Real numbers.

1. $A = \left\{ -6, \frac{1}{2}, -1.333\ldots \text{ (the 3's repeat)}, \pi, 2, 5 \right\}$

2. $B = \left\{ -\frac{5}{3}, 2.060606\ldots \text{ (the block 06 repeats)}, 1.25, 0, 1, \sqrt{5} \right\}$

3. $C = \left\{ 0, 1, \frac{1}{2}, \frac{1}{3}, \frac{1}{4} \right\}$

4. $D = \{ -1, -1.1, -1.2, -1.3 \}$

5. $E = \left\{ \sqrt{2}, \pi, \sqrt{2} + 1, \pi + \frac{1}{2} \right\}$

6. $F = \left\{ -\sqrt{2}, \pi + \sqrt{2}, \frac{1}{2} + 10.3 \right\}$

In Problems 7–18, approximate each number (a) rounded and (b) truncated to three decimal places.

7. 18.9526　　**8.** 25.86134　　**9.** 28.65319　　**10.** 99.05249　　**11.** 0.06291　　**12.** 0.05388

13. 9.9985　　**14.** 1.0006　　**15.** $\dfrac{3}{7}$　　**16.** $\dfrac{5}{9}$　　**17.** $\dfrac{521}{15}$　　**18.** $\dfrac{81}{5}$

In Problems 19–28, write each statement using symbols.

19. The sum of 3 and 2 equals 5.

20. The product of 5 and 2 equals 10.

21. The sum of x and 2 is the product of 3 and 4.

22. The sum of 3 and y is the sum of 2 and 2.

23. The product of 3 and y is the sum of 1 and 2.

24. The product of 2 and x is the product of 4 and 6.

25. The difference x less 2 equals 6.

26. The difference 2 less y equals 6.

27. The quotient x divided by 2 is 6.

28. The quotient 2 divided by x is 6.

In Problems 29–62, evaluate each expression.

29. $9 - 4 + 2$　　**30.** $6 - 4 + 3$　　**31.** $-6 + 4 \cdot 3$　　**32.** $8 - 4 \cdot 2$

33. $4 + 5 - 8$　　**34.** $8 - 3 - 4$　　**35.** $4 + \dfrac{1}{3}$　　**36.** $2 - \dfrac{1}{2}$

37. $6 - [3 \cdot 5 + 2 \cdot (3 - 2)]$　　**38.** $2 \cdot [8 - 3(4 + 2)] - 3$　　**39.** $2 \cdot (3 - 5) + 8 \cdot 2 - 1$　　**40.** $1 - (4 \cdot 3 - 2 + 2)$

41. $10 - [6 - 2 \cdot 2 + (8 - 3)] \cdot 2$　　**42.** $2 - 5 \cdot 4 - [6 \cdot (3 - 4)]$

43. $(5 - 3)\dfrac{1}{2}$　　**44.** $(5 + 4)\dfrac{1}{3}$　　**45.** $\dfrac{4 + 8}{5 - 3}$　　**46.** $\dfrac{2 - 4}{5 - 3}$

47. $\dfrac{3}{5} \cdot \dfrac{10}{21}$　　**48.** $\dfrac{5}{9} \cdot \dfrac{3}{10}$　　**49.** $\dfrac{6}{25} \cdot \dfrac{10}{27}$　　**50.** $\dfrac{21}{25} \cdot \dfrac{100}{3}$

51. $\dfrac{3}{4} + \dfrac{2}{5}$　　**52.** $\dfrac{4}{3} + \dfrac{1}{2}$　　**53.** $\dfrac{5}{6} + \dfrac{9}{5}$　　**54.** $\dfrac{8}{9} + \dfrac{15}{2}$

55. $\dfrac{5}{18} + \dfrac{1}{12}$　　**56.** $\dfrac{2}{15} + \dfrac{8}{9}$　　**57.** $\dfrac{1}{30} - \dfrac{7}{18}$　　**58.** $\dfrac{3}{14} - \dfrac{2}{21}$

59. $\dfrac{3}{20} - \dfrac{2}{15}$　　**60.** $\dfrac{6}{35} - \dfrac{3}{14}$　　**61.** $\dfrac{\frac{5}{18}}{\frac{11}{27}}$　　**62.** $\dfrac{\frac{5}{21}}{\frac{2}{35}}$

In Problems 63–74, use the distributive property to remove the parentheses.

63. $6(x + 4)$　　**64.** $4(2x - 1)$　　**65.** $x(x - 4)$　　**66.** $4x(x + 3)$

67. $(x + 2)(x + 4)$　　**68.** $(x + 5)(x + 1)$　　**69.** $(x - 2)(x + 1)$　　**70.** $(x - 4)(x + 1)$

71. $(x - 8)(x - 2)$　　**72.** $(x - 4)(x - 2)$　　**73.** $(x + 2)(x - 2)$　　**74.** $(x - 3)(x + 3)$

75. Explain to a friend how the distributive property is used to justify the fact that $2x + 3x = 5x$.

76. Explain to a friend why $2 + 3 \cdot 4 = 14$, whereas $(2 + 3) \cdot 4 = 20$.

77. Explain why $2(3 \cdot 4)$ is not equal to $(2 \cdot 3) \cdot (2 \cdot 4)$.

78. Explain why $\dfrac{4 + 3}{2 + 5}$ is not equal to $\dfrac{4}{2} + \dfrac{3}{5}$.

79. Is subtraction commutative? Support your conclusion with an example.

80. Is subtraction associative? Support your conclusion with an example.

81. Is division commutative? Support your conclusion with an example.

82. Is division associative? Support your conclusion with an example.

83. If $2 = x$, why does $x = 2$?

84. If $x = 5$, why does $x^2 + x = 30$?

85. Are there any real numbers that are both rational and irrational? Are there any real numbers that are neither? Explain your reasoning.

86. Explain why the sum of a rational number and an irrational number must be irrational.

87. What rational number does the repeating decimal $0.9999\ldots$ equal?

R.2 ALGEBRA REVIEW

OBJECTIVES
1. Graph Inequalities
2. Find Distance on the Real Number Line
3. Evaluate Algebraic Expressions
4. Determine the Domain of a Variable

The Real Number Line

The real numbers can be represented by points on a line called the **real number line**. There is a one-to-one correspondence between real numbers and points on a line. That is, every real number corresponds to a point on the line, and each point on the line has a unique real number associated with it.

Pick a point on the line somewhere in the center, and label it O. This point, called the **origin**, corresponds to the real number 0. See Figure 7. The point 1 unit to the right of O corresponds to the number 1. The distance between 0 and 1 determines the **scale** of the number line. For example, the point associated with the number 2 is twice as far from O as 1 is. Notice that an arrowhead on the right end of the line indicates the direction in which the numbers increase. Figure 7 also shows the points associated with the irrational numbers $\sqrt{2}$ and π. Points to the left of the origin correspond to the real numbers -1, -2, and so on.

Figure 7
Real number line

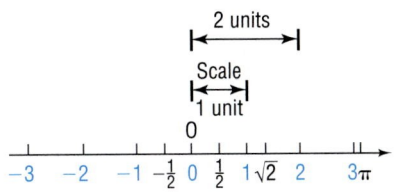

> The real number associated with a point P is called the **coordinate** of P, and the line whose points have been assigned coordinates is called the **real number line**.

✏ **NOW WORK PROBLEM 1.**

The real number line consists of three classes of real numbers, as shown in Figure 8 (on page 18).

Figure 8

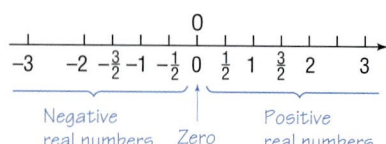

1. The **negative real numbers** are the coordinates of points to the left of the origin O.

2. The real number **zero** is the coordinate of the origin O.

3. The **positive real numbers** are the coordinates of points to the right of the origin O.

Negative and positive numbers have the following multiplication properties:

Multiplication Properties of Positive and Negative Numbers

1. The product of two positive numbers is a positive number.

2. The product of two negative numbers is a positive number.

3. The product of a positive number and a negative number is a negative number.

Inequalities

An important property of the real number line follows from the fact that, given two numbers (points) a and b, either a is to the left of b, a is at the same location as b, or a is to the right of b. See Figure 9.

Figure 9

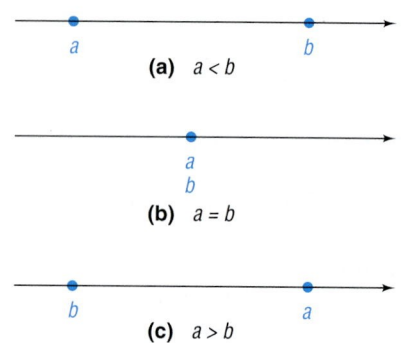

(a) $a < b$

(b) $a = b$

(c) $a > b$

If a is to the left of b, we say that "a is less than b" and write $a < b$. If a is to the right of b, we say that "a is greater than b" and write $a > b$. If a is at the same location as b, then $a = b$. If a is either less than or equal to b, we write $a \le b$. Similarly, $a \ge b$ means that a is either greater than or equal to b. Collectively, the symbols $<$, $>$, $\le$, and $\ge$ are called **inequality symbols**.

Note that $a < b$ and $b > a$ mean the same thing. It does not matter whether we write $2 < 3$ or $3 > 2$.

Furthermore, if $a < b$ or if $b > a$, then the difference $b - a$ is positive. Do you see why?

EXAMPLE 1   **Using Inequality Symbols**

(a) $3 < 7$ (b) $-8 > -16$ (c) $-6 < 0$

(d) $-8 < -4$ (e) $4 > -1$ (f) $8 > 0$ ■

In Example 1(a), we conclude that $3 < 7$ either because 3 is to the left of 7 on the real number line or because the difference $7 - 3 = 4$, is a positive real number.

Similarly, we conclude in Example 1(b) that $-8 > -16$ either because -8 lies to the right of -16 on the real number line or because the difference $-8 - (-16) = -8 + 16 = 8$, is a positive real number.

Look again at Example 1. Note that the inequality symbol always points in the direction of the smaller number.

An **inequality** is a statement in which two expressions are related by an inequality symbol. The expressions are referred to as the **sides** of the inequality. Inequalities of the form $a < b$ or $b > a$ are called **strict inequali-**

ties, whereas inequalities of the form $a \leq b$ or $b \geq a$ are called **nonstrict inequalities**.

Based on the discussion thus far, we conclude that

$$a > 0 \quad \text{is equivalent to} \quad a \text{ is positive}$$
$$a < 0 \quad \text{is equivalent to} \quad a \text{ is negative}$$

We sometimes read $a > 0$ by saying that "a is positive." If $a \geq 0$, then either $a > 0$ or $a = 0$, and we may read this as "a is nonnegative."

✏ **NOW WORK PROBLEMS 5 AND 15.**

① We shall find it useful in later work to graph inequalities on the real number line.

EXAMPLE 2 **Graphing Inequalities**

(a) On the real number line, graph all numbers x for which $x > 4$.

(b) On the real number line, graph all numbers x for which $x \leq 5$.

Figure 10
$x > 4$

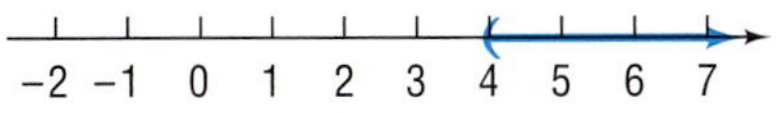

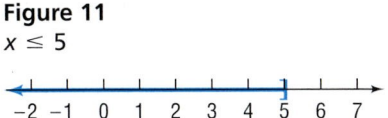

Figure 11
$x \leq 5$

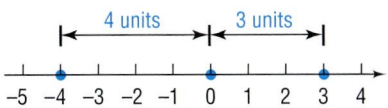

Solution (a) See Figure 10. Notice that we use a left parenthesis to indicate that the number 4 is *not* part of the graph.

(b) See Figure 11. Notice that we use a right bracket to indicate that the number 5 *is* part of the graph. ■

✏ **NOW WORK PROBLEM 21.**

Absolute Value

The *absolute value* of a number a is the distance from 0 to a on the number line. For example, -4 is 4 units from 0; and 3 is 3 units from 0. See Figure 12. Thus, the absolute value of -4 is 4, and the absolute value of 3 is 3.

A more formal definition of absolute value is given next.

Figure 12

The **absolute value** of a real number a, denoted by the symbol $|a|$, is defined by the rules

$$|a| = a \quad \text{if } a \geq 0 \quad \text{and} \quad |a| = -a \quad \text{if } a < 0$$

For example, since $-4 < 0$, the second rule must be used to get $|-4| = -(-4) = 4$.

EXAMPLE 3 **Computing Absolute Value**

(a) $|8| = 8$ (b) $|0| = 0$ (c) $|-15| = -(-15) = 15$ ■

② Look again at Figure 12. The distance from -4 to 3 is 7 units. This distance is the difference $3 - (-4)$, obtained by subtracting the smaller coordinate from the larger. However, since $|3 - (-4)| = |7| = 7$ and $|-4 - 3| = |-7| = 7$, we can use absolute value to calculate the distance between two points without being concerned about which is smaller.

If P and Q are two points on a real number line with coordinates a and b, respectively, the **distance between P and Q**, denoted by $d(P,Q)$, is

$$d(P,Q) = |b - a|$$

Since $|b - a| = |a - b|$, it follows that $d(P,Q) = d(Q,P)$.

EXAMPLE 4   **Finding Distance on a Number Line**

Let P, Q, and R be points on a real number line with coordinates $-5, 7$, and -3, respectively. Find the distance

(a) between P and Q (b) between Q and R

Solution   See Figure 13.

Figure 13

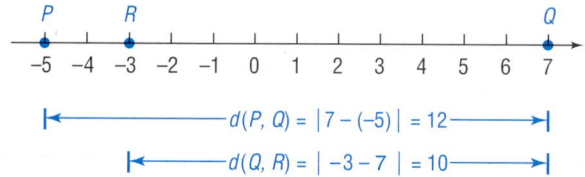

(a) $d(P,Q) = |7 - (-5)| = |12| = 12$
(b) $d(Q,R) = |-3 - 7| = |-10| = 10$ ■

━ **NOW WORK PROBLEM 27.**

Constants and Variables

As we said earlier, in algebra we use letters such as x, y, a, b, and c to represent numbers. If the letter used is to represent *any* number from a given set of numbers, it is called a **variable**. A **constant** is either a fixed number, such as 5 or $\sqrt{3}$, or a letter that represents a fixed (possibly unspecified) number.

Constants and variables are combined using the operations of addition, subtraction, multiplication, and division to form *algebraic expressions*. Examples of algebraic expressions include

$$x + 3 \qquad \frac{3}{1 - t} \qquad 7x - 2y$$

③   To evaluate an algebraic expression, substitute for each variable its numerical value.

EXAMPLE 5   **Evaluating an Algebraic Expression**

Evaluate each expression if $x = 3$ and $y = -1$.

(a) $x + 3y$ (b) $5xy$ (c) $\dfrac{3y}{2 - 2x}$ (d) $|-4x + y|$

Solution   (a) Substitute 3 for x and -1 for y in the expression $x + 3y$.

$$x + 3y = 3 + 3(-1) = 3 + (-3) = 0$$

$$\uparrow$$
$$x = 3, y = -1$$

(b) If $x = 3$ and $y = -1$, then

$$5xy = 5(3)(-1) = -15$$

(c) If $x = 3$ and $y = -1$, then

$$\frac{3y}{2 - 2x} = \frac{3(-1)}{2 - 2(3)} = \frac{-3}{2 - 6} = \frac{-3}{-4} = \frac{3}{4}$$

(d) If $x = 3$ and $y = -1$, then

$$\left|-4x + y\right| = \left|-4(3) + (-1)\right| = \left|-12 + (-1)\right| = \left|-13\right| = 13 \quad \blacksquare$$

A graphing calculator can be used to evaluate an algebraic expression. Figure 14 shows the results of Example 5 using a TI-83.

Figure 14

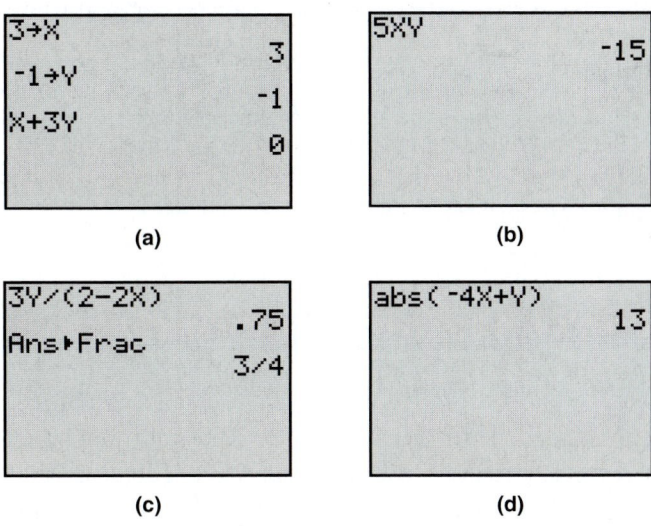

(a) (b)

(c) (d)

NOW WORK PROBLEMS **29** AND **37.**

④ In working with expressions or formulas involving variables, the variables may be allowed to take on values from only a certain set of numbers. For example, in the formula for the area A of a circle of radius r, $A = \pi r^2$, the variable r is necessarily restricted to the positive real numbers. In the expression $\dfrac{1}{x}$, the variable x cannot take on the value 0, since division by 0 is not defined.

> The set of values that a variable may assume is called the **domain of the variable**.

EXAMPLE 6   Finding the Domain of a Variable

The domain of the variable x in the expression

$$\frac{5}{x - 2}$$

is $\{x \mid x \neq 2\}$, since, if $x = 2$, the denominator becomes 0, which is not defined. $\quad \blacksquare$

| EXAMPLE 7 | Circumference of a Circle |

In the formula for the circumference C of a circle of radius r,

$$C = 2\pi r$$

the domain of the variable r, representing the radius of the circle, is the set of positive real numbers. The domain of the variable C, representing the circumference of the circle, is also the set of positive real numbers. ◼

In describing the domain of a variable, we may use either set notation or words, whichever is more convenient.

NOW WORK PROBLEM 47.

HISTORICAL FEATURE

The word *algebra* is derived from the Arabic word *al-jabr*. This word is a part of the title of a ninth century work, "Hisâb al-jabr w'al-muqâbalah," written by Mohammed ibn Mûsâ al-Khowârizmî. The word *al-jabr* means "a restoration," a reference to the fact that, if a number is added to one side of an equation, then it must also be added to the other side in order to "restore" the equality. The title of the work, freely translated, means "The science of reduction and cancellation." Of course, today, algebra has come to mean a great deal more.

R.2 Concepts and Vocabulary

In Problems 1–3, fill in the blanks.

1. A _____ is a letter used in algebra to represent any number from a given set of numbers.

2. On the real number line, the real number zero is the coordinate of the _____.

3. An inequality of the form $a > b$ is called a _____ inequality.

In Problems 4–6, answer True or False to each statement.

4. The product of two negative real numbers is always greater than zero.

5. The distance between two points on the real number line is always greater than zero.

6. The absolute value of a real number is always greater than zero.

7. Give a reason why the statement $5 < 8$ is true.

8. Graph the inequality $x > -3$ on the real number line.

9. What is the domain of the variable in the expression $\dfrac{x}{x-3}$?

R.2 Exercises

1. On the real number line, label the points with coordinates $0, 1, -1, \dfrac{5}{2}, -2.5, \dfrac{3}{4}$, and 0.25.

2. Repeat Problem 1 for the coordinates $0, -2, 2, -1.5, \dfrac{3}{2}, \dfrac{1}{3}$, and $\dfrac{2}{3}$.

In Problems 3–12, replace the question mark by $<$, $>$, or $=$, whichever is correct.

3. $\dfrac{1}{2}$? 0

4. 5 ? 6

5. -1 ? -2

6. -3 ? $-\dfrac{5}{2}$

7. π ? 3.14

8. $\sqrt{2}$? 1.41

9. $\dfrac{1}{2}$? 0.5

10. $\dfrac{1}{3}$? 0.33

11. $\dfrac{2}{3}$? 0.67

12. $\dfrac{1}{4}$? 0.25

In Problems 13–18, write each statement as an inequality.

13. x is positive

14. z is negative

15. x is less than 2

16. y is greater than -5

17. x is less than or equal to 1

18. x is greater than or equal to 2

In Problems 19–22, graph the numbers x on the real number line.

19. $x \geq -2$

20. $x < 4$

21. $x > -1$

22. $x \leq 7$

In Problems 23–28, use the real number line below to compute each distance.

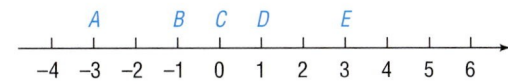

23. $d(C, D)$

24. $d(C, A)$

25. $d(D, E)$

26. $d(C, E)$

27. $d(A, E)$

28. $d(D, B)$

In Problems 29–36, evaluate each expression if $x = -2$ and $y = 3$. Check your answer using a graphing utility.

29. $x + 2y$

30. $3x + y$

31. $5xy + 2$

32. $-2x + xy$

33. $\dfrac{2x}{x - y}$

34. $\dfrac{x + y}{x - y}$

35. $\dfrac{3x + 2y}{2 + y}$

36. $\dfrac{2x - 3}{y}$

In Problems 37–46, find the value of each expression if $x = 3$ and $y = -2$. Check your answer using a graphing utility.

37. $|x + y|$

38. $|x - y|$

39. $|x| + |y|$

40. $|x| - |y|$

41. $\dfrac{|x|}{x}$

42. $\dfrac{|y|}{y}$

43. $|4x - 5y|$

44. $|3x + 2y|$

45. $\big\| 4x| - |5y \big\|$

46. $3|x| + 2|y|$

In Problems 47–54, determine which of the value(s) given below, if any, must be excluded from the domain of the variable in each rational expression:

(a) $x = 3$

(b) $x = 1$

(c) $x = 0$

(d) $x = -1$

47. $\dfrac{x^2 - 1}{x}$

48. $\dfrac{x^2 + 1}{x}$

49. $\dfrac{x}{x^2 - 9}$

50. $\dfrac{x}{x^2 + 9}$

51. $\dfrac{x^2}{x^2 + 1}$

52. $\dfrac{x^3}{x^2 - 1}$

53. $\dfrac{x^2 + 5x - 10}{x^3 - x}$

54. $\dfrac{-9x^2 - x + 1}{x^3 + x}$

In Problems 55–58, determine the domain of the variable x in each expression.

55. $\dfrac{4}{x - 5}$

56. $\dfrac{-6}{x + 4}$

57. $\dfrac{x}{x + 4}$

58. $\dfrac{x - 2}{x - 6}$

In Problems 59–62, use the formula $C = \dfrac{5}{9}(F - 32)$ for converting degrees Fahrenheit into degrees Celsius to find the Celsius measure of each Fahrenheit temperature.

59. $F = 32°$

60. $F = 212°$

61. $F = 77°$

62. $F = -4°$

In Problems 63–72, express each statement as an equation involving the indicated variables.

63. Area of a Rectangle The area A of a rectangle is the product of its length l and its width w.

64. Perimeter of a Rectangle The perimeter P of a rectangle is twice the sum of its length l and its width w.

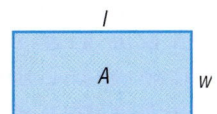

65. Circumference of a Circle The circumference C of a circle is the product of π and its diameter d.

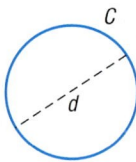

66. Area of a Triangle The area A of a triangle is one-half the product of its base b and its height h.

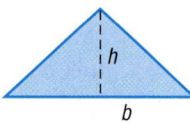

67. Area of an Equilateral Triangle The area A of an equilateral triangle is $\dfrac{\sqrt{3}}{4}$ times the square of the length x of one side.

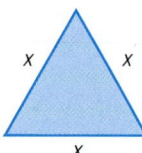

68. Perimeter of an Equilateral Triangle The perimeter P of an equilateral triangle is 3 times the length x of one side.

69. Volume of a Sphere The volume V of a sphere is $\dfrac{4}{3}$ times π times the cube of the radius r.

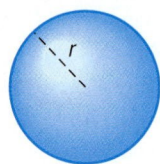

70. Surface Area of a Sphere The surface area S of a sphere is 4 times π times the square of the radius r.

71. Volume of a Cube The volume V of a cube is the cube of the length x of a side.

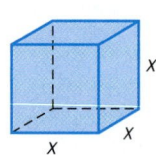

72. Surface Area of a Cube The surface area S of a cube is 6 times the square of the length x of a side.

73. Manufacturing Cost The weekly production cost C of manufacturing x watches is given by the formula $C = 4000 + 2x$, where the variable C is in dollars.
(a) What is the cost of producing 1000 watches?
(b) What is the cost of producing 2000 watches?

74. Balancing a Checkbook At the beginning of the month, Mike had a balance of $210 in his checking account. During the next month, he deposited $80, wrote a check for $120, made another deposit of $25, wrote two checks for $60 and $32, and was assessed a monthly service charge of $5. What was his balance at the end of the month?

75. U.S. Voltage In the United States, normal household voltage is 115 volts. It is acceptable for the actual voltage x to differ from normal by at most 5 volts. A formula that describes this is

$$|x - 115| \le 5$$

(a) Show that a voltage of 113 volts is acceptable.
(b) Show that a voltage of 109 volts is not acceptable.

76. Foreign Voltage In other countries, normal household voltage is 220 volts. It is acceptable for the actual voltage x to differ from normal by at most 8 volts. A formula that describes this is

$$|x - 220| \le 8$$

(a) Show that a voltage of 214 volts is acceptable.
(b) Show that a voltage of 209 volts is not acceptable.

77. Making Precision Ball Bearings The FireBall Company manufactures ball bearings for precision equipment. One of their products is a ball bearing with a stated radius of 3 centimeters (cm). Only ball bearings with a radius within 0.01 cm of this stated radius are acceptable. If x is the radius of a ball bearing, a formula describing this situation is

$$|x - 3| \le 0.01$$

(a) Is a ball bearing of radius $x = 2.999$ acceptable?
(b) Is a ball bearing of radius $x = 2.89$ acceptable?

78. Body Temperature Normal human body temperature is 98.6°F. A temperature x that differs from normal by at least 1.5°F is considered unhealthy. A formula that describes this is

$$|x - 98.6| \ge 1.5$$

(a) Show that a temperature of 97°F is unhealthy.
(b) Show that a temperature of 100°F is not unhealthy.

79. Does $\dfrac{1}{3}$ equal 0.333? If not, which is larger? By how much?

80. Does $\dfrac{2}{3}$ equal 0.666? If not, which is larger? By how much?

81. Is there a positive real number "closest" to 0?

82. I'm thinking of a number! It lies between 1 and 10; its square is rational and lies between 1 and 10. The number is larger than π. Correct to two decimal places, name the number. Now think of your own number, describe it, and challenge a fellow student to name it.

83. Write a brief paragraph that illustrates the similarities and differences between "less than" ($<$) and "less than or equal to" ($\le$).

R.3 GEOMETRY REVIEW

OBJECTIVES   **1** Use the Pythagorean Theorem and Its Converse

2 Know Geometry Formulas

In this section we review some topics studied in geometry that we shall need for our study of algebra.

Pythagorean Theorem

Figure 15

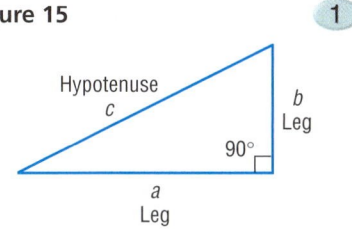

1 The *Pythagorean Theorem* is a statement about *right triangles*. A **right triangle** is one that contains a **right angle**, that is, an angle of 90°. The side of the triangle opposite the 90° angle is called the **hypotenuse**; the remaining two sides are called **legs**. In Figure 15 we have used c to represent the length of the hypotenuse and a and b to represent the lengths of the legs. Notice the use of the symbol ⌐ to show the 90° angle. We now state the Pythagorean Theorem.

Pythagorean Theorem

In a right triangle, the square of the length of the hypotenuse is equal to the sum of the squares of the lengths of the legs. That is, in the right triangle shown in Figure 15,

$$c^2 = a^2 + b^2 \qquad (1)$$

EXAMPLE 1 **Finding the Hypotenuse of a Right Triangle**

In a right triangle, one leg is of length 4 and the other is of length 3. What is the length of the hypotenuse?

Solution Since the triangle is a right triangle, we use the Pythagorean Theorem with $a = 4$ and $b = 3$ to find the length c of the hypotenuse. From equation (1), we have

$$c^2 = a^2 + b^2$$
$$c^2 = 4^2 + 3^2 = 16 + 9 = 25$$
$$c = \sqrt{25} = 5$$

NOW WORK PROBLEM **3**.

The converse of the Pythagorean Theorem is also true.

Converse of the Pythagorean Theorem

In a triangle, if the square of the length of one side equals the sum of the squares of the lengths of the other two sides, then the triangle is a right triangle. The 90° angle is opposite the longest side.

EXAMPLE 2   **Verifying That a Triangle Is a Right Triangle**

Show that a triangle whose sides are of lengths 5, 12, and 13 is a right triangle. Identify the hypotenuse.

Solution   We square the lengths of the sides.

$$5^2 = 25, \quad 12^2 = 144, \quad 13^2 = 169$$

Notice that the sum of the first two squares (25 and 144) equals the third square (169). Hence, the triangle is a right triangle. The longest side, 13, is the hypotenuse. See Figure 16.

Figure 16

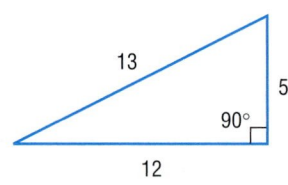

NOW WORK PROBLEM **11.**

EXAMPLE 3   **Applying the Pythagorean Theorem**

The tallest inhabited building in the world is the Sears Tower in Chicago.* If the observation tower is 1450 feet above ground level, how far can a person standing in the observation tower see (with the aid of a telescope)? Use 3960 miles for the radius of Earth. See Figure 17.

Figure 17

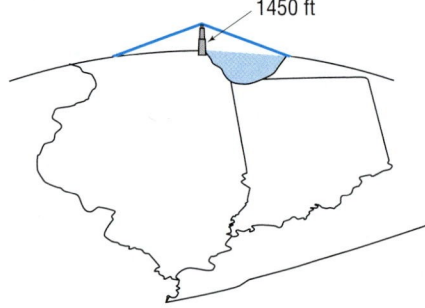

1450 ft

[*Note:*   1 mile = 5280 feet]

Solution   From the center of Earth, draw two radii: one through the Sears Tower and the other to the farthest point a person can see from the tower. See Figure 18. Apply the Pythagorean Theorem to the right triangle.

Since 1450 feet = $\dfrac{1450}{5280}$ miles, we have

Figure 18

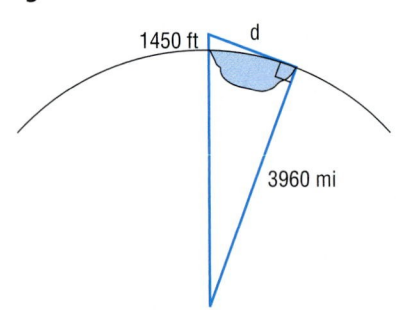

$$d^2 + (3960)^2 = \left(3960 + \frac{1450}{5280}\right)^2$$

$$d^2 = \left(3960 + \frac{1450}{5280}\right)^2 - (3960)^2 \approx 2175.08$$

$$d \approx 46.64$$

A person can see about 47 miles from the observation tower.

NOW WORK PROBLEM **37.**

*Source: Council on Tall Buildings and Urban Habitat (1997): Sears Tower No. 1 for tallest roof (1450 ft) and tallest occupied floor (1431 ft).

Geometry Formulas

② Certain formulas from geometry are useful in solving algebra problems. We list some of these formulas next.

For a rectangle of length l and width w,

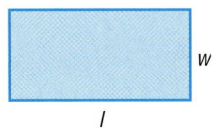

$$\text{Area} = lw \qquad \text{Perimeter} = 2l + 2w$$

For a triangle with base b and altitude h,

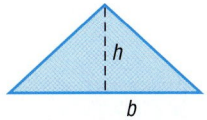

$$\text{Area} = \frac{1}{2}bh$$

For a circle of radius r (diameter $d = 2r$),

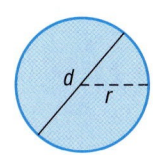

$$\text{Area} = \pi r^2 \qquad \text{Circumference} = 2\pi r = \pi d$$

For a rectangular box of length l, width w, and height h,

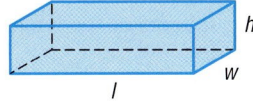

$$\text{Volume} = lwh$$

For a sphere of radius r,

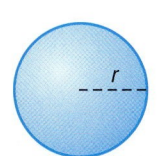

$$\text{Volume} = \frac{4}{3}\pi r^3 \qquad \text{Surface area} = 4\pi r^2$$

For a right circular cylinder of height h and radius r,

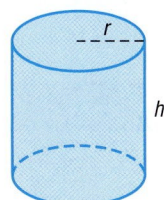

$$\text{Volume} = \pi r^2 h$$

━━━ **NOW WORK PROBLEM 19.**

EXAMPLE 4 **Using Geometry Formulas**

A Christmas tree ornament is in the shape of a semicircle on top of a triangle. How many square centimeters (cm) of copper are required to make the ornament if the height of the triangle is 6 cm and the base is 4 cm?

Figure 19

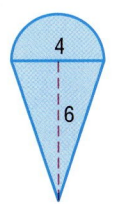

Solution See Figure 19. The amount of copper required equals the shaded area. This area is the sum of the area of the triangle and the semicircle. The triangle has height $h = 6$ and base $b = 4$. The semicircle has diameter $d = 4$, so its radius is $r = 2$.

$$\text{Area} = \text{Area of triangle} + \text{Area of semicircle}$$

$$= \frac{1}{2}bh + \frac{1}{2}\pi r^2 = \frac{1}{2}(4)(6) + \frac{1}{2}\pi 2^2 \qquad b = 4; h = 6; r = 2.$$

$$= 12 + 2\pi \approx 18.28 \text{ cm}^2$$

About 18.28 cm^2 of copper are required.

NOW WORK PROBLEM 33.

R.3 Concepts and Vocabulary

In Problems 1–3, fill in the blanks.

1. A _____ triangle is one that contains an angle of 90 degrees. The longest side is called the _____.
2. For a triangle with base b and altitude h, a formula for the area is _____.
3. The formula for the circumference of a circle of radius r is _____.

In Problems 4–6, answer True or False to each statement.

4. In a right triangle the square of the length of the longest side equals the sum of the squares of the lengths of the other two sides.
5. The triangle with sides of length 6, 8, and 10 is a right triangle.
6. The volume of a sphere of radius r is $\frac{4}{3}\pi r^2$.

7. What is the formula for the area of a circle of radius r?
8. What is the formula for the volume of a sphere of radius r?

R.3 Exercises

In Problems 1–6, the lengths of the legs of a right triangle are given. Find the hypotenuse.

1. $a = 5, b = 12$
2. $a = 6, b = 8$
3. $a = 10, b = 24$
4. $a = 4, b = 3$
5. $a = 7, b = 24$
6. $a = 14, b = 48$

In Problems 7–14, the lengths of the sides of a triangle are given. Determine if the triangle is a right triangle. If it is, identify the hypotenuse.

7. $3, 4, 5$
8. $6, 8, 10$
9. $4, 5, 6$
10. $2, 2, 3$
11. $7, 24, 25$
12. $10, 24, 26$
13. $6, 4, 3$
14. $5, 4, 7$

15. Find the area A of a rectangle with length 4 inches and width 2 inches.
16. Find the area A of a rectangle with length 9 centimeters and width 4 centimeters.
17. Find the area A of a triangle with height 4 inches and base 2 inches.
18. Find the area A of a triangle with height 9 centimeters and base 4 centimeters.
19. Find the area A and circumference C of a circle of radius 5 meters.

20. Find the area A and circumference C of a circle of radius 2 feet.

21. Find the volume V of a rectangular box with length 8 feet, width 4 feet, and height 7 feet.

22. Find the volume V of a rectangular box with length 9 inches, width 4 inches, and height 8 inches.

23. Find the volume V and surface area S of a sphere of radius 4 centimeters.

24. Find the volume V and surface area S of a sphere of radius 3 feet.

25. Find the volume V of a right circular cylinder with radius 9 inches and height 8 inches.

26. Find the volume V of a right circular cylinder with radius 8 inches and height 9 inches.

In Problems 27–30, find the area of the shaded region.

27.
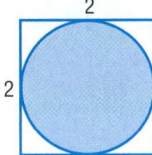
2

2

28.
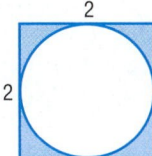
2

2

29.
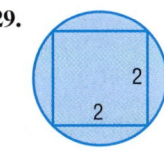
2

2

30.
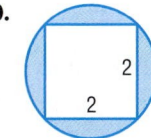
2

2

---

31. How many feet does a wheel with a diameter of 16 inches travel after four revolutions?

32. How many revolutions will a circular disk with a diameter of 4 feet have completed after it has rolled 20 feet?

33. In the figure shown, $ABCD$ is a square, with each side of length 6 feet. The width of the border (shaded portion) between the outer square $EFGH$ and $ABCD$ is 2 feet. Find the area of the border.

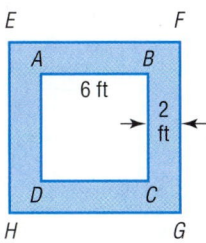

E   *F*

A   *B*

6 ft

2 ft

D   *C*

H   *G*

34. Refer to the figure. Square $ABCD$ has an area of 100 square feet; square $BEFG$ has an area of 16 square feet. What is the area of the triangle CGF?

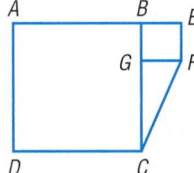

A   *B*   *E*

G   *F*

D   *C*

35. Architecture A Norman window consists of a rectangle surmounted by a semicircle. Find the area of the Norman window shown in the illustration. How much wood frame is needed to enclose the window?

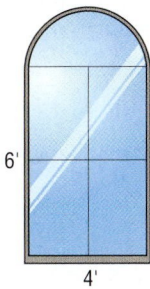

6'

4'

36. Construction A circular swimming pool, 20 feet in diameter, is enclosed by a wooden deck that is 3 feet wide. What is the area of the deck? How much fence is required to enclose the deck?

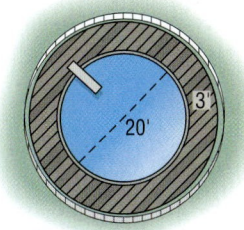

3'

20'

In Problems 37–39, use the facts that the radius of Earth is 3960 miles and 1 mile = 5280 feet.

37. How Far Can You See? The conning tower of the U.S.S. *Silversides*, a World War II submarine now permanently stationed in Muskegon, Michigan, is approximately 20 feet above sea level. How far can you see from the conning tower?

38. How Far Can You See? A person who is 6 feet tall is standing on the beach in Fort Lauderdale, Florida, and looks out onto the Atlantic Ocean. Suddenly, a ship appears on the horizon. How far is the ship from shore?

39. How Far Can You See? The deck of a destroyer is 100 feet above sea level. How far can a person see from the deck? How far can a person see from the bridge, which is 150 feet above sea level?

40. Suppose that m and n are positive integers with $m > n$. If $a = m^2 - n^2$, $b = 2mn$, and $c = m^2 + n^2$, show that a, b, and c are the lengths of the sides of a right triangle. (This formula can be used to find the sides of a right triangle that are integers, such as 3, 4, 5; 5, 12, 13; and so on. Such triplets of integers are called **Pythagorean triples**.)

41. You have 1000 feet of flexible pool siding and wish to construct a swimming pool. Experiment with rectangular-shaped pools with perimeters of 1000 feet. How do their areas vary? What is the shape of the rectangle with the largest area? Now compute the area enclosed by a circular pool with a perimeter (circumference) of 1000 feet. What would be your choice of shape for the pool? If rectangular, what is your preference for dimensions? Justify your choice. If your only consideration is to have a pool that encloses the most area, what shape should you use?

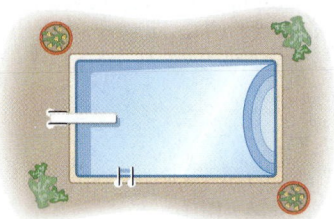 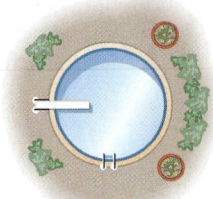

42. The Gibb's Hill Lighthouse, Southampton, Bermuda, in operation since 1846, stands 117 feet high on a hill 245 feet high, so its beam of light is 362 feet above sea level. A brochure states that the light itself can be seen on the horizon about 26 miles distant. Verify the correctness of this information. The brochure further states that ships 40 miles away can see the light and planes flying at 10,000 feet can see it 120 miles away. Verify the accuracy of these statements. What assumption did the brochure make about the height of the ship?

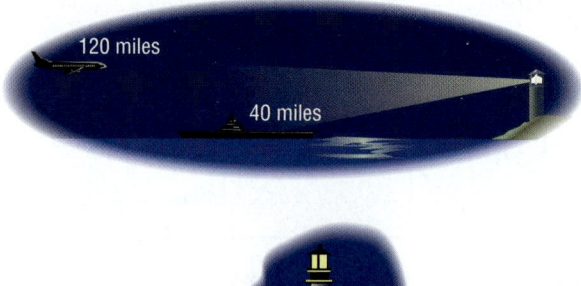

120 miles

40 miles

R.4 INTEGER EXPONENTS

OBJECTIVES
1. Evaluate Expressions Containing Exponents
2. Work with the Laws of Exponents
3. Use a Calculator to Evaluate Exponents
4. Use Scientific Notation

Integer exponents provide a shorthand device for representing repeated multiplications of a real number. For example,

$$3^4 = 3 \cdot 3 \cdot 3 \cdot 3 = 81$$

Additionally, many formulas have exponents. For example,

- The formula for the horsepower rating H of an engine is

$$H = \frac{D^2 N}{2.5}$$

where D is the diameter of a cylinder and N is the number of cylinders.

- A formula for the resistance R of blood flowing in a blood vessel is

$$R = C \frac{L}{r^4}$$

where L is the length of the blood vessel, r is the radius, and C is a positive constant.

If a is a real number and n is a positive integer, then the symbol a^n represents the product of n factors of a. That is,

$$a^n = \underbrace{a \cdot a \cdot \ldots \cdot a}_{n \text{ factors}} \qquad (1)$$

Here it is understood that $a^1 = a$.

In particular, we have

$$a^1 = a$$
$$a^2 = a \cdot a$$
$$a^3 = a \cdot a \cdot a$$

and so on.

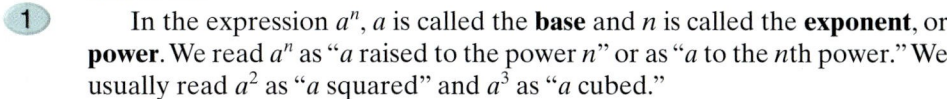

In the expression a^n, a is called the **base** and n is called the **exponent**, or **power**. We read a^n as "a raised to the power n" or as "a to the nth power." We usually read a^2 as "a squared" and a^3 as "a cubed."

EXAMPLE 1 **Evaluating Expressions Containing Exponents**

(a) $2^3 = 2 \cdot 2 \cdot 2 = 8$ (b) $5^2 = 5 \cdot 5 = 25$

(c) $10^1 = 10$ (d) $(-2)^4 = (-2)(-2)(-2)(-2) = 16$

(e) $-2^4 = -(2 \cdot 2 \cdot 2 \cdot 2) = -16$ ■

Notice the difference between Examples 1(d) and 1(e): The exponent applies only to the symbol or parenthetical expression immediately preceding it.

We define a raised to a negative power as follows:

If n is a positive integer and if a is a nonzero real number, then we define

$$a^{-n} = \frac{1}{a^n} \qquad \text{if } a \neq 0 \qquad (2)$$

Whenever you encounter a negative exponent, think "reciprocal."

EXAMPLE 2 **Evaluating Expressions Containing Negative Exponents**

(a) $2^{-3} = \frac{1}{2^3} = \frac{1}{8}$ (b) $x^{-4} = \frac{1}{x^4}$

(c) $\left(\frac{1}{5}\right)^{-2} = \frac{1}{\left(\frac{1}{5}\right)^2} = \frac{1}{\frac{1}{25}} = 25$ ■

NOW WORK PROBLEM **3.**

If a is a nonzero number, we define

$$a^0 = 1 \qquad \text{if } a \neq 0 \qquad (3)$$

Notice that we do not allow the base a to be 0 in a^{-n} or in a^0.

Laws of Exponents

2 Several general rules can be used when dealing with exponents. The first rule is used when multiplying two expressions that have the same base.

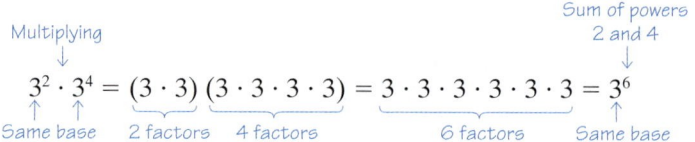

To multiply two expressions having the same base, retain the base and add the exponents.

If a is a real number and m, n are integers, then

$$a^m a^n = a^{m+n} \qquad (4)$$

Equation (4) is true whether the integers m and n are positive, negative, or 0. If m, n, or $m + n$ is 0 or negative, then a cannot be 0, as stated earlier.

EXAMPLE 3 **Using Equation (4)**

(a) $2 \cdot 2^4 = 2^1 \cdot 2^4 = 2^{1+4} = 2^5 = 32$

(b) $(-3)^2(-3)^{-3} = (-3)^{2+(-3)} = (-3)^{-1} = \dfrac{1}{-3} = -\dfrac{1}{3}$

(c) $x^{-3} \cdot x^5 = x^{-3+5} = x^2$

(d) $(-y)^{-1}(-y)^{-3} = (-y)^{-1+(-3)} = (-y)^{-4} = \dfrac{1}{(-y)^4} = \dfrac{1}{y^4}$

NOW WORK PROBLEM **11.**

Another law of exponents applies when an expression containing a power is itself raised to a power.

$$(2^3)^4 = \underbrace{2^3 \cdot 2^3 \cdot 2^3 \cdot 2^3}_{\text{4 factors}} = \underbrace{(2 \cdot 2 \cdot 2)}_{\text{3 factors}}\underbrace{(2 \cdot 2 \cdot 2)}_{\text{3 factors}}\underbrace{(2 \cdot 2 \cdot 2)}_{\text{3 factors}}\underbrace{(2 \cdot 2 \cdot 2)}_{\text{3 factors}} = 2^{12}$$

$$\underbrace{}_{3 \cdot 4 = 12 \text{ factors}}$$

If an expression containing a power is raised to a power, retain the base and multiply the powers.

If a is a real number and m, n are integers, then

$$(a^m)^n = a^{mn} \qquad (5)$$

Equation (5) is true whether the integers m and n are positive, negative, or 0. Again, if m or n is 0 or negative, then a must not be 0.

EXAMPLE 4 **Using Equation (5)**

(a) $\left[(-2)^3\right]^2 = (-2)^{3 \cdot 2} = (-2)^6 = 64$ (b) $(3^{-4})^0 = 3^{(-4)(0)} = 3^0 = 1$

(c) $(x^{-3})^2 = x^{-3 \cdot 2} = x^{-6} = \dfrac{1}{x^6}$ (d) $(y^{-1})^{-2} = y^{(-1)(-2)} = y^2$ ∎

The next law of exponents involves raising a product to a power.

$$(2 \cdot 5)^3 = (2 \cdot 5)(2 \cdot 5)(2 \cdot 5) = (2 \cdot 2 \cdot 2)(5 \cdot 5 \cdot 5) = 2^3 \cdot 5^3$$

If a product is raised to a power, the result equals the product of each factor raised to that power.

If a, b are real numbers and n is an integer, then

$$(a \cdot b)^n = a^n \cdot b^n \qquad (6)$$

Equation (6) is true whether the integer n is positive, negative, or 0. If n is 0 or negative, neither a nor b can be 0.

EXAMPLE 5 **Using Equation (6)**

(a) $(2x)^3 = 2^3 \cdot x^3 = 8x^3$

(b) $(-2x)^0 = (-2)^0 \cdot x^0 = 1 \cdot 1 = 1 \qquad x \neq 0$

(c) $(ax)^{-2} = a^{-2}x^{-2} = \dfrac{1}{a^2} \cdot \dfrac{1}{x^2} = \dfrac{1}{a^2 x^2} \qquad x \neq 0, a \neq 0$

(d) $(ax)^{-2} = \dfrac{1}{(ax)^2} = \dfrac{1}{a^2 x^2} \qquad x \neq 0, a \neq 0$ ∎

✏ **NOW WORK PROBLEM 29.**

Equations (4) and (6) both involve products. Two similar laws involve quotients.

If a and b are real numbers and if m and n are integers, then

$$\frac{a^m}{a^n} = a^{m-n} = \frac{1}{a^{n-m}} \qquad \text{if } a \neq 0 \qquad (7a)$$

$$\left(\frac{a}{b}\right)^n = \frac{a^n}{b^n} \qquad \text{if } b \neq 0 \qquad (7b)$$

Equation (7a) is true whether the integers m and n are positive, negative, or 0. Equation (7b) is true whether the integer n is positive, negative, or 0. In (7b), if n is negative or 0, then a cannot be zero.

EXAMPLE 6   Using Equations (7a) and (7b)

(a) $\dfrac{2^6}{2^4} = 2^{6-4} = 2^2 = 4$ (b) $\left(\dfrac{2}{3}\right)^4 = \dfrac{2^4}{3^4} = \dfrac{16}{81}$

(c) $\dfrac{x^{-2}}{x^{-5}} = x^{-2-(-5)} = x^3$ (d) $\left(\dfrac{5}{2}\right)^{-2} = \dfrac{1}{\left(\dfrac{5}{2}\right)^2} = \dfrac{1}{\dfrac{5^2}{2^2}} = \dfrac{1}{\dfrac{25}{4}} = \dfrac{4}{25}$ ■

You may have observed the following shortcut for problems like Example 6(d):

$$\left(\frac{a}{b}\right)^{-n} = \left(\frac{b}{a}\right)^n \qquad \text{if } a \neq 0, b \neq 0 \qquad (8)$$

EXAMPLE 7   Using Equation (8)

(a) $\left(\dfrac{2}{3}\right)^{-3} = \left(\dfrac{3}{2}\right)^3 = \dfrac{27}{8}$ (b) $\left(\dfrac{3}{4}\right)^{-2} = \left(\dfrac{4}{3}\right)^2 = \dfrac{16}{9}$ ■

— NOW WORK PROBLEM **19.**

EXAMPLE 8   Using the Laws of Exponents

Simplify each expression. Express the answer so that all exponents are positive.

(a) $\dfrac{x^5 y^{-2}}{(x^3 y)^2}, \qquad x \neq 0, y \neq 0$ (b) $\left(\dfrac{x^{-3}}{3y^{-1}}\right)^{-2}, \qquad x \neq 0, y \neq 0$

Solution (a) $\dfrac{x^5 y^{-2}}{(x^3 y)^2} = \dfrac{x^5 y^{-2}}{(x^3)^2 y^2} = \dfrac{x^5 y^{-2}}{x^6 y^2} = \dfrac{x^5}{x^6} \cdot \dfrac{y^{-2}}{y^2}$

$\qquad = x^{5-6} y^{-2-2} = x^{-1} y^{-4} = \dfrac{1}{x} \cdot \dfrac{1}{y^4} = \dfrac{1}{x y^4}$

(b) $\left(\dfrac{x^{-3}}{3y^{-1}}\right)^{-2} = \dfrac{(x^{-3})^{-2}}{(3y^{-1})^{-2}} = \dfrac{x^6}{3^{-2}(y^{-1})^{-2}} = \dfrac{x^6}{\dfrac{1}{9} y^2} = \dfrac{9x^6}{y^2}$ ■

— NOW WORK PROBLEM **47.**

Calculator Use

3 Your graphing calculator has the caret key, $\boxed{\land}$, which is used for computations involving exponents. The next example shows how this key is used.

EXAMPLE 9   Exponents on a Graphing Calculator

Evaluate: $(2.3)^5$

Figure 20

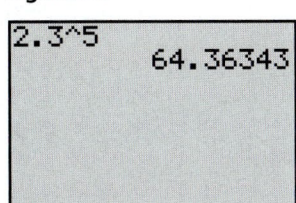

Solution Figure 20 shows the result using a TI-83 graphing calculator. ∎

NOW WORK PROBLEM **63.**

Scientific Notation

4 Measurements of physical quantities can range from very small to very large. For example, the mass of a proton is approximately 0.0000000000000000000000000167 kilogram and the mass of Earth is about 5,980,000,000,000,000,000,000,000 kilograms. These numbers obviously are tedious to write down and difficult to read, so we use exponents to rewrite each.

> When a number has been written as the product of a number x, where $1 \le x < 10$, times a power of 10, it is said to be written in **scientific notation**.

In scientific notation,

$$\text{Mass of a proton} = 1.67 \times 10^{-27} \text{ kilogram}$$
$$\text{Mass of Earth} = 5.98 \times 10^{24} \text{ kilograms}$$

Converting a Decimal to Scientific Notation

To change a positive number into scientific notation:

1. Count the number N of places that the decimal point must be moved in order to arrive at a number x, where $1 \le x < 10$.
2. If the original number is greater than or equal to 1, the scientific notation is $x \times 10^{N}$. If the original number is between 0 and 1, the scientific notation is $x \times 10^{-N}$.

EXAMPLE 10   Using Scientific Notation

Write each number in scientific notation.

(a) 9582   (b) 1.245   (c) 0.285   (d) 0.000561

Solution (a) The decimal point in 9582 follows the 2. We count

$$9\ 5\ 8\ 2\ .$$

stopping after three moves, because 9.582 is a number between 1 and 10. Since 9582 is greater than 1, we write

$$9582 = 9.582 \times 10^3$$

(b) The decimal point in 1.245 is between the 1 and 2. Since the number is already between 1 and 10, the scientific notation for it is $1.245 \times 10^0 = 1.245$.

(c) The decimal point in 0.285 is between the 0 and the 2. We count

$$0 \;.\; 2 \;\; 8 \;\; 5$$

stopping after one move, because 2.85 is a number between 1 and 10. Since 0.285 is between 0 and 1, we write

$$0.285 = 2.85 \times 10^{-1}$$

(d) The decimal point in 0.000561 is moved as follows:

$$0 \;.\; 0 \;\; 0 \;\; 0 \;\; 5 \;\; 6 \;\; 1$$

Thus,

$$0.000561 = 5.61 \times 10^{-4}$$

NOW WORK PROBLEM **69.**

EXAMPLE 11   **Changing from Scientific Notation to Decimals**

Write each number as a decimal.

(a) 2.1×10^4 (b) 3.26×10^{-5} (c) 1×10^{-2}

Solution   (a) $2.1 \times 10^4 = 2 \;.\; 1 \;\; 0 \;\; 0 \;\; 0 \;\; \times 10^4 = 21{,}000$

(b) $3.26 \times 10^{-5} = 0 \;\; 0 \;\; 0 \;\; 0 \;\; 0 \;\; 3 \;.\; 2 \;\; 6 \times 10^{-5} = 0.0000326$

(c) $1 \times 10^{-2} = 0 \;\; 0 \;\; 1 \;.\; \times 10^{-2} = 0.01$

On a graphing calculator, a number such as 3.615×10^{12} is usually displayed as $\boxed{3.615E12}$.

NOW WORK PROBLEM **77.**

EXAMPLE 12   **Using Scientific Notation**

(a) The diameter of the smallest living cell is only about 0.00001 centimeter (cm).* Express this number in scientific notation.
(b) The surface area of Earth is about 1.97×10^8 square miles.† Express the surface area as a whole number.

*Powers of Ten, Philip and Phylis Morrison.
†1998 Information Please Almanac.

Solution (a) $0.00001 \text{ cm} = 1 \times 10^{-5} \text{ cm}$ because the decimal point is moved five places and the number is less than 1.

(b) 1.97×10^{8} square miles $= 197{,}000{,}000$ square miles. ■

NOW WORK PROBLEM **85.**

SUMMARY

We close this section by summarizing the Laws of Exponents. In the list that follows, a and b are real numbers and m and n are integers. Also, we assume that no denominator is 0 and that all expressions are defined.

Laws of Exponents

$$a^{-n} = \frac{1}{a^n} \qquad a^0 = 1, \qquad a \neq 0$$

$$a^m a^n = a^{m+n} \qquad \left(a^m\right)^n = a^{mn} \qquad (a \cdot b)^n = a^n \cdot b^n$$

$$\frac{a^m}{a^n} = a^{m-n} = \frac{1}{a^{n-m}}, \qquad a \neq 0 \qquad \left(\frac{a}{b}\right)^n = \frac{a^n}{b^n}, \qquad b \neq 0$$

HISTORICAL FEATURE

René Descartes
1596–1650

Our method of writing exponents originated with René Descartes (1596–1650), although the concept goes back in various forms to the ancient Babylonians. Even after its introduction by Descartes in 1637, the method took a remarkable amount of time to become completely standardized, and expressions like $aaaaa + 3aaaa + 2aaa - 4aa + 2a + 1$ remained common until 1750. The concept of a rational exponent (see Section R.9) was known by 1400, although inconvenient notation prevented the development of any extensive theory. John Wallis (1616–1703), in 1655, was the first to give a fairly complete explanation of negative and rational exponents, and Sir Isaac Newton's (1642–1727) use of them made exponents standard in their current form.

R.4 Concepts and Vocabulary

In Problems 1–3, fill in the blanks.

1. In the expression 2^4, the number 2 is called the _____ and 4 is called the _____.

2. Using a Law of Exponents, $\left(a^m\right)^n =$ _____.

3. In scientific notation, $1234.5678 =$ _____.

In Problems 4–6, answer True or False to each statement.

4. $(a + b)^2 = a^2 + b^2$.

5. When a number is expressed in scientific notation, it is expressed as the product of a number $x, 0 \leq x < 1$, and a power of 10.

6. To multiply two expressions having the same base, retain the base and multiply the exponents.

7. What does the symbol a^n represent?

8. Write 0.00876 in scientific notation.

9. Use a graphing calculator to evaluate $(1.62)^3$. Round the answer to two decimal places.

R.4 Exercises

In Problems 1–24, simplify each expression.

1. 4^2

2. -4^2

3. 4^{-2}

4. $(-4)^2$

5. -4^{-2}

6. $(-4)^{-2}$

7. $4^0 \cdot 2^{-3}$

8. $(-2)^{-3} \cdot 3^0$

9. $2^{-3} + \left(\dfrac{1}{2}\right)^3$

10. $3^{-2} + \left(\dfrac{1}{3}\right)^2$

11. $3^{-6} \cdot 3^4$

12. $4^{-2} \cdot 4^3$

13. $\dfrac{(3^2)^2}{(2^3)^2}$

14. $\dfrac{(2^3)^3}{(2^2)^3}$

15. $\left(\dfrac{2}{3}\right)^{-3}$

16. $\left(\dfrac{3}{2}\right)^{-2}$

17. $\dfrac{2^3 \cdot 3^2}{2^4 \cdot 3^{-2}}$

18. $\dfrac{3^{-2} \cdot 5^3}{3^2 \cdot 5}$

19. $\left(\dfrac{9}{2}\right)^{-2}$

20. $\left(\dfrac{6}{5}\right)^{-3}$

21. $\dfrac{2^{-2}}{3}$

22. $\dfrac{3^{-2}}{2}$

23. $\dfrac{-3^{-1}}{2^{-1}}$

24. $\dfrac{-2^{-3}}{-1}$

In Problems 25–56, simplify each expression. Express the answer so that all exponents are positive. Whenever an exponent is negative or 0, we assume that the base does not equal 0.

25. $x^0 y^2$

26. $x^{-1} y$

27. xy^{-2}

28. $x^0 y^4$

29. $(8x^3)^{-2}$

30. $(-8x^3)^{-2}$

31. $-4x^{-1}$

32. $(-4x)^{-1}$

33. $3x^0$

34. $(3x)^0$

35. $\dfrac{x^{-2} y^3}{xy^4}$

36. $\dfrac{x^{-2} y}{xy^2}$

37. $x^{-1} y^{-1}$

38. $\dfrac{x^{-2} y^{-3}}{x}$

39. $\dfrac{x^{-1}}{y^{-1}}$

40. $\left(\dfrac{2x}{3}\right)^{-1}$

41. $\left(\dfrac{4y}{5x}\right)^{-2}$

42. $(x^2 y)^{-2}$

43. $x^{-2} y^{-2}$

44. $x^{-1} y^{-1}$

45. $\dfrac{x^{-1} y^{-2} z^3}{x^2 yz^3}$

46. $\dfrac{3x^{-2} yz^2}{x^4 y^{-3} z^2}$

47. $\dfrac{(-2)^3 x^4 (yz)^2}{3^2 xy^3 z^4}$

48. $\dfrac{4x^{-2}(yz)^{-1}}{(-5)^2 x^4 y^2 z^{-2}}$

49. $\dfrac{\left(\dfrac{x}{y}\right)^{-2} \cdot \left(\dfrac{y}{x}\right)^4}{x^2 y^3}$

50. $\dfrac{\left(\dfrac{y}{x}\right)^2}{x^{-2} y}$

51. $\left(\dfrac{3x^{-1}}{4y^{-1}}\right)^{-2}$

52. $\left(\dfrac{5x^{-2}}{6y^{-2}}\right)^{-3}$

53. $\dfrac{(xy^{-1})^{-2}}{xy^3}$

54. $\dfrac{(3xy^{-1})^2}{(2x^{-1}y)^3}$

55. $\left(\dfrac{x}{y^2}\right)^{-2} \cdot (y^2)^{-1}$

56. $\dfrac{(x^2)^{-3} y^3}{(x^3 y)^{-2}}$

57. Find the value of the expression $2x^3 - 3x^2 + 5x - 4$ if $x = 2$. What is the value if $x = 1$?

58. Find the value of the expression $4x^3 + 3x^2 - x + 2$ if $x = 1$. What is the value if $x = 2$?

59. What is the value of $\dfrac{(666)^4}{(222)^4}$?

60. What is the value of $(0.1)^3 (20)^3$?

In Problems 61–68, use a calculator to evaluate each expression. Round your answer to three decimal places.

61. $(8.2)^6$

62. $(3.7)^5$

63. $(6.1)^{-3}$

64. $(2.2)^{-5}$

65. $(-2.8)^6$

66. $-(2.8)^6$

67. $(-8.11)^{-4}$

68. $-(8.11)^{-4}$

In Problems 69–76, write each number in scientific notation.

69. 454.2

70. 32.14

71. 0.013

72. 0.00421

73. 32,155

74. 21,210

75. 0.000423

76. 0.0514

In Problems 77–84, write each number as a decimal.

77. 6.15×10^4

78. 9.7×10^3

79. 1.214×10^{-3}

80. 9.88×10^{-4}

81. 1.1×10^8

82. 4.112×10^2

83. 8.1×10^{-2}

84. 6.453×10^{-1}

85. **Distance from Earth to Its Moon**   The distance from Earth to the Moon is about 4×10^8 meters.* Express this distance as a whole number.

86. **Height of Mt. Everest**   The height of Mt. Everest is 8872 meters.* Express this height in scientific notation.

87. **Wavelength of Visible Light**   The wavelength of visible light is about 5×10^{-7} meter.* Express this wavelength as a decimal.

88. **Diameter of an Atom**   The diameter of an atom is about 1×10^{-10} meter.* Express this diameter as a decimal.

89. **Diameter of Copper Wire**   The smallest commercial copper wire is about 0.0005 inch in diameter.† Express this diameter using scientific notation.

90. **Smallest Motor**   The smallest motor ever made is less than 0.05 centimeter wide.† Express this width using scientific notation.

91. **World Oil Production**   In 1996, world oil production averaged 64,000,000 barrels per day.† How many barrels on average were produced in the month of April 1996? Express your answer in scientific notation.

92. **U.S. Consumption of Oil**   In 1995, the United States consumed 17,640,000 barrels of petroleum products per day.† If there are 42 gallons in one barrel, how many gallons did the United States consume per day? Express your answer in scientific notation.

93. **Astronomy**   One light-year is defined by astronomers to be the distance that a beam of light will travel in 1 year (365 days). If the speed of light is 186,000 miles per second, how many miles are in a light-year? Express your answer in scientific notation.

94. **Astronomy**   How long does it take a beam of light to reach Earth from the Sun, when the Sun is 93,000,000 miles from Earth? Express your answer in seconds, using scientific notation.

95. Look at the summary box where the Laws of Exponents are given. List them in the order of most importance to you. Write a brief position paper defending your ordering.

96. Write a paragraph to justify the definition given in the text that $a^0 = 1, a \neq 0$.

R.5 POLYNOMIALS

OBJECTIVES
1. Recognize Monomials
2. Recognize Polynomials
3. Add and Subtract Polynomials
4. Multiply Polynomials
5. Know Formulas for Special Products
6. Divide Polynomials

We have described algebra as a generalization of arithmetic in which letters are used to represent real numbers. From now on, we shall use the letters at the end of the alphabet, such as x, y, and z, to represent variables and the letters at the beginning of the alphabet, such as a, b, and c, to represent constants. In the expressions $3x + 5$ and $ax + b$, it is understood that x is a variable and that a and b are constants, even though the constants a and b are unspecified. As you will find out, the context usually makes the intended meaning clear.

1   Now we introduce some basic vocabulary.

> A **monomial** in one variable is the product of a constant and a variable raised to a nonnegative integer power. A monomial is of the form
>
> $$ax^k$$
>
> where a is a constant, x is a variable, and $k \geq 0$ is an integer. The constant a is called the **coefficient** of the monomial. If $a \neq 0$, then k is called the **degree** of the monomial.

*Powers of Ten, Philip and Phylis Morrison.
†1998 Information Please Almanac.

EXAMPLE 1 Examples of Monomials

Monomial	Coefficient	Degree	
(a) $6x^2$	6	2	
(b) $-\sqrt{2}x^3$	$-\sqrt{2}$	3	
(c) 3	3	0	Since $3 = 3 \cdot 1 = 3x^0, x \neq 0$
(d) $-5x$	-5	1	Since $-5x = -5x^1$
(e) x^4	1	4	Since $x^4 = 1 \cdot x^4$

Now let's look at some expressions that are not monomials.

EXAMPLE 2 Examples of Nonmonomial Expressions

(a) $3x^{1/2}$ is not a monomial, since the exponent of the variable x is $\dfrac{1}{2}$ and $\dfrac{1}{2}$ is not a nonnegative integer.

(b) $4x^{-3}$ is not a monomial, since the exponent of the variable x is -3 and -3 is not a nonnegative integer.

NOW WORK PROBLEM **1.**

2

Two monomials with the same variable raised to the same power are called **like terms**. For example, $2x^4$ and $-5x^4$ are like terms. In contrast, the monomials $2x^3$ and $2x^5$ are not like terms.

We can add or subtract like terms using the distributive property. For example,

$$2x^2 + 5x^2 = (2 + 5)x^2 = 7x^2 \quad \text{and} \quad 8x^3 - 5x^3 = (8 - 5)x^3 = 3x^3$$

The sum or difference of two monomials that are not like terms is called a **binomial**. The sum or difference of three monomials that are not like terms is called a **trinomial**. For example,

$x^2 - 2$ is a binomial.

$x^3 - 3x + 5$ is a trinomial.

$2x^2 + 5x^2 + 2 = 7x^2 + 2$ is a binomial.

A **polynomial** in one variable is an algebraic expression of the form

$$a_n x^n + a_{n-1} x^{n-1} + \cdots + a_1 x + a_0 \tag{1}$$

where $a_n, a_{n-1}, \ldots, a_1, a_0$ are constants,* called the **coefficients** of the polynomial, $n \geq 0$ is an integer, and x is a variable. If $a_n \neq 0$, it is called the **leading coefficient**, and n is called the **degree** of the polynomial.

---

*The notation a_n is read as "a sub n." The number n is called a **subscript** and should not be confused with an exponent. We use subscripts in order to distinguish one constant from another when a large or undetermined number of constants is required.

The monomials that make up a polynomial are called its **terms**. If all the coefficients are 0, the polynomial is called the **zero polynomial**, which has no degree.

Polynomials are usually written in **standard form**, beginning with the nonzero term of highest degree and continuing with terms in descending order according to degree.

EXAMPLE 3 **Examples of Polynomials**

Polynomial	Coefficients	Degree
$3x^2 - 5 = 3x^2 + 0 \cdot x + (-5)$	$3, 0, -5$	2
$8 - 2x + x^2 = 1 \cdot x^2 + (-2)x + 8$	$1, -2, 8$	2
$5x + \sqrt{2} = 5x^1 + \sqrt{2}$	$5, \sqrt{2}$	1
$3 = 3 \cdot 1 = 3 \cdot x^0$	3	0
0	0	No degree ■

Although we have been using x to represent the variable, letters such as y or z are also commonly used.

$3x^4 - x^2 + 2$ is a polynomial (in x) of degree 4.

$9y^3 - 2y^2 + y - 3$ is a polynomial (in y) of degree 3.

$z^5 + \pi$ is a polynomial (in z) of degree 5.

Algebraic expressions such as

$$\frac{1}{x} \quad \text{and} \quad \frac{x^2 + 1}{x + 5}$$

are not polynomials. The first is not a polynomial because $\dfrac{1}{x} = x^{-1}$ has an exponent that is not a nonnegative integer. Although the second expression is the quotient of two polynomials, the polynomial in the denominator has degree greater than 0, so the expression cannot be a polynomial.

NOW WORK PROBLEM 11.

Adding and Subtracting Polynomials

③ Polynomials are added and subtracted by combining like terms.

EXAMPLE 4 **Adding Polynomials**

Find the sum of the polynomials:

$$8x^3 - 2x^2 + 6x - 2 \quad \text{and} \quad 3x^4 - 2x^3 + x^2 + x$$

Solution We shall find the sum in two ways.

Horizontal Addition: The idea here is to group the like terms and then combine them.

$$(8x^3 - 2x^2 + 6x - 2) + (3x^4 - 2x^3 + x^2 + x)$$
$$= 3x^4 + (8x^3 - 2x^3) + (-2x^2 + x^2) + (6x + x) - 2$$
$$= 3x^4 + 6x^3 - x^2 + 7x - 2$$

Vertical Addition: The idea here is to vertically line up the like terms in each polynomial and then add the coefficients.

$$
\begin{array}{c}
\ \ \overset{x^4}{}\ \ \overset{x^3}{}\ \ \overset{x^2}{}\ \ \overset{x^1}{}\ \ \overset{x^0}{} \\
8x^3 - 2x^2 + 6x - 2 \\
(+)\ \ 3x^4 - 2x^3 + \ x^2 + \ x \\
\hline
3x^4 + 6x^3 - \ x^2 + 7x - 2
\end{array}
$$

We can subtract two polynomials in either of the previous ways.

| EXAMPLE 5 | **Subtracting Polynomials** |

Find the difference: $(3x^4 - 4x^3 + 6x^2 - 1) - (2x^4 - 8x^2 - 6x + 5)$

Solution *Horizontal Subtraction*

$(3x^4 - 4x^3 + 6x^2 - 1) - (2x^4 - 8x^2 - 6x + 5)$

$= 3x^4 - 4x^3 + 6x^2 - 1 + \underline{(-2x^4 + 8x^2 + 6x - 5)}$

Be sure to change the sign of each term in the second polynomial.

$= (3x^4 - 2x^4) + (-4x^3) + (6x^2 + 8x^2) + 6x + (-1 - 5)$

↑
Group like terms.

$= x^4 - 4x^3 + 14x^2 + 6x - 6$

Vertical Subtraction: We line up like terms, change the sign of each coefficient of the second polynomial, and add.

$$
\begin{array}{c}
\ \ \overset{x^4}{}\ \ \overset{x^3}{}\ \ \overset{x^2}{}\ \ \overset{x^1}{}\ \ \overset{x^0}{} \qquad\qquad \overset{x^4}{}\ \ \overset{x^3}{}\ \ \overset{x^2}{}\ \ \overset{x^1}{}\ \ \overset{x^0}{} \\
3x^4 - 4x^3 + 6x^2 \ - 1 \ = \qquad 3x^4 - 4x^3 + \ 6x^2 \ - 1 \\
(-)[2x^4 \ - 8x^2 - 6x + 5] = (+)-2x^4 \ + \ 8x^2 + 6x - 5 \\
\hline
 x^4 - 4x^3 + 14x^2 + 6x - 6
\end{array}
$$

The choice of which of these methods to use for adding and subtracting polynomials is left to you. To save space, we shall most often use the horizontal format.

✏ **NOW WORK PROBLEM 23.**

Multiplying Polynomials

④ Two monomials may be multiplied using the Laws of Exponents and the Commutative and Associative Properties. For example,

$$(2x^3) \cdot (5x^4) = (2 \cdot 5) \cdot (x^3 \cdot x^4) = 10x^{3+4} = 10x^7$$

Products of polynomials are found by repeated use of the Distributive Property and the Laws of Exponents. Again, you have a choice of horizontal or vertical format.

EXAMPLE 6   **Multiplying Polynomials**

Find the product: $(2x + 5)(x^2 - x + 2)$

Solution   *Horizontal Multiplication*

$(2x + 5)(x^2 - x + 2) = 2x(x^2 - x + 2) + 5(x^2 - x + 2)$
↑
Distributive property

$= (2x \cdot x^2 - 2x \cdot x + 2x \cdot 2) + (5 \cdot x^2 - 5 \cdot x + 5 \cdot 2)$
↑
Distributive property

$= (2x^3 - 2x^2 + 4x) + (5x^2 - 5x + 10)$
↑
Law of exponents

$= 2x^3 + 3x^2 - x + 10$
↑
Combine like terms

Vertical Multiplication: The idea here is very much like multiplying a two-digit number by a three-digit number.

$$
\begin{array}{r}
x^2 - x + 2 \\
2x + 5 \\
\hline
2x^3 - 2x^2 + 4x \\
(+)\quad 5x^2 - 5x + 10 \\
\hline
2x^3 + 3x^2 - x + 10
\end{array}
$$

This line is $2x(x^2 - x + 2)$.
This line is $5(x^2 - x + 2)$.
Sum of the above two lines.   ■

✏ **NOW WORK PROBLEM 35.**

Special Products

⑤   Certain products, which we call **special products**, occur frequently in algebra. We can calculate them easily using the **FOIL** (**F**irst, **O**uter, **I**nner, **L**ast) method of multiplying two binomials.

Outer
First

$(ax + b)(cx + d) = ax(cx + d) + b(cx + d)$

Inner
Last

$$
\begin{aligned}
&\quad\quad\quad\ \ \overset{First}{\overbrace{\ \ }}\ \ \overset{Outer}{\overbrace{\ \ }}\ \ \overset{Inner}{\overbrace{\ \ }}\ \ \overset{Last}{\overbrace{\ \ }} \\
&= ax \cdot cx + ax \cdot d + b \cdot cx + b \cdot d \\
&= acx^2 + adx + bcx + bd \\
&= acx^2 + (ad + bc)x + bd
\end{aligned}
$$

EXAMPLE 7   **Using FOIL to Find Products of the Form $(x - a)(x + a)$**

$(x - 3)(x + 3) = x^2 + 3x - 3x - 9 = x^2 - 9$

F O I L   ■

EXAMPLE 8   Using FOIL to Find the Square of a Binomial

(a) $(x + 2)^2 = (x + 2)(x + 2) = x^2 + 2x + 2x + 4 = x^2 + 4x + 4$
(b) $(x - 3)^2 = (x - 3)(x - 3) = x^2 - 3x - 3x + 9 = x^2 - 6x + 9$ ■

EXAMPLE 9   Using FOIL to Find the Product of Two Binomials

(a) $(x + 3)(x + 1) = x^2 + x + 3x + 3 = x^2 + 4x + 3$
(b) $(2x + 1)(3x + 4) = 6x^2 + 8x + 3x + 4 = 6x^2 + 11x + 4$ ■

NOW WORK PROBLEMS **41** AND **49**.

Some products have been given special names because of their form. The following special products are based on Examples 7 and 8.

Difference of Two Squares

$$(x - a)(x + a) = x^2 - a^2 \qquad (2)$$

Squares of Binomials, or Perfect Squares

$$(x + a)^2 = x^2 + 2ax + a^2 \qquad (3a)$$
$$(x - a)^2 = x^2 - 2ax + a^2 \qquad (3b)$$

EXAMPLE 10   Using Special Product Formulas

(a) $(x - 5)(x + 5) = x^2 - 5^2 = x^2 - 25$ Difference of two squares
(b) $(x + 7)^2 = x^2 + 2 \cdot 7 \cdot x + 7^2 = x^2 + 14x + 49$ Square of a binomial ■

EXAMPLE 11   Using Special Product Formulas

(a) $(2x + 1)^2 = (2x)^2 + 2 \cdot 1 \cdot 2x + 1^2 = 4x^2 + 4x + 1$ Notice that we used 2x in place of x in formula (3a).

(b) $(3x - 4)^2 = (3x)^2 - 2 \cdot 4 \cdot 3x + 4^2 = 9x^2 - 24x + 16$ Replace x by 3x in formula (3b). ■

NOW WORK PROBLEM **61**.

Let's look at some more examples that lead to general formulas.

EXAMPLE 12   Cubing a Binomial

(a) $(x + 2)^3 = (x + 2)(x + 2)^2 = (x + 2)(x^2 + 4x + 4)$ Formula (3a)
$$= (x^3 + 4x^2 + 4x) + (2x^2 + 8x + 8)$$
$$= x^3 + 6x^2 + 12x + 8$$

(b) $(x - 1)^3 = (x - 1)(x - 1)^2 = (x - 1)(x^2 - 2x + 1)$ Formula (3b)

$$= (x^3 - 2x^2 + x) - 1(x^2 - 2x + 1)$$

$$= x^3 - 3x^2 + 3x - 1$$ ■

Cubes of Binomials, or Perfect Cubes

$$(x + a)^3 = x^3 + 3ax^2 + 3a^2x + a^3 \qquad \text{(4a)}$$

$$(x - a)^3 = x^3 - 3ax^2 + 3a^2x - a^3 \qquad \text{(4b)}$$

NOW WORK PROBLEM **79**.

EXAMPLE 13 **Forming the Difference of Two Cubes**

$$(x - 1)(x^2 + x + 1) = x(x^2 + x + 1) - 1(x^2 + x + 1)$$

$$= x^3 + x^2 + x - x^2 - x - 1$$

$$= x^3 - 1$$ ■

EXAMPLE 14 **Forming the Sum of Two Cubes**

$$(x + 2)(x^2 - 2x + 4) = x(x^2 - 2x + 4) + 2(x^2 - 2x + 4)$$

$$= x^3 - 2x^2 + 4x + 2x^2 - 4x + 8$$

$$= x^3 + 8$$ ■

Examples 13 and 14 lead to two more special products.

Difference of Two Cubes

$$(x - a)(x^2 + ax + a^2) = x^3 - a^3 \qquad \text{(5)}$$

Sum of Two Cubes

$$(x + a)(x^2 - ax + a^2) = x^3 + a^3 \qquad \text{(6)}$$

Dividing Polynomials

6   The procedure for dividing two polynomials is similar to the procedure for dividing two integers. This procedure should be familiar to you, but we review it briefly next.

EXAMPLE 15   **Dividing Two Integers**

Divide 842 by 15.

Solution

$$
\begin{array}{r}
56 \quad \leftarrow \text{Quotient} \\
\text{Divisor} \rightarrow \quad 15\overline{)842} \quad \leftarrow \text{Dividend} \\
75 \quad \leftarrow 5 \cdot 15 \,(\text{Subtract}) \\
\hline
92 \\
90 \quad \leftarrow 6 \cdot 15 \,(\text{Subtract}) \\
\hline
2 \quad \leftarrow \text{Remainder}
\end{array}
$$

Thus, $\dfrac{842}{15} = 56 + \dfrac{2}{15}$.

In the long division process detailed in Example 15, the number 15 is called the **divisor**, the number 842 is called the **dividend**, the number 56 is called the **quotient**, and the number 2 is called the **remainder**.

To check the answer obtained in a division problem, multiply the quotient by the divisor and add the remainder. The answer should be the dividend.

$$(\text{Quotient})(\text{Divisor}) + \text{Remainder} = \text{Dividend}$$

For example, we can check the results obtained in Example 15 as follows:

$$(56)(15) + 2 = 840 + 2 = 842$$

To divide two polynomials, we first must write each polynomial in standard form. The process then follows a pattern similar to that of Example 15. The next example illustrates the procedure.

EXAMPLE 16   **Dividing Two Polynomials**

Find the quotient and the remainder when

$$3x^3 + 4x^2 + x + 7 \quad \text{is divided by} \quad x^2 + 1$$

Solution   Each polynomial is in standard form. The dividend is $3x^3 + 4x^2 + x + 7$, and the divisor is $x^2 + 1$.

STEP 1:   Divide the leading term of the dividend, $3x^3$, by the leading term of the divisor, x^2. Enter the result, $3x$, over the term $3x^3$, as follows:

$$
\begin{array}{r}
3x \\
x^2 + 1 \overline{)3x^3 + 4x^2 + x + 7}
\end{array}
$$

STEP 2:   Multiply $3x$ by $x^2 + 1$ and enter the result below the dividend.

$$
\begin{array}{r}
3x \\
x^2 + 1 \overline{)3x^3 + 4x^2 + \ x + 7} \\
\underline{3x^3 \qquad\quad + 3x} \qquad \leftarrow 3x \cdot (x^2 + 1) = 3x^3 + 3x
\end{array}
$$

↑
Notice that we align the $3x$ term under the x to make the next step easier.

STEP 3: Subtract and bring down the remaining terms.

$$
\begin{array}{r}
3x \\
x^2 + 1 \overline{)3x^3 + 4x^2 + x + 7} \\
\underline{3x^3 + 3x} \\
4x^2 - 2x + 7
\end{array}
$$

← Subtract (change the signs and add).
← Bring down the $4x^2$ and the 7.

STEP 4: Repeat Steps 1–3 using $4x^2 - 2x + 7$ as the dividend.

$$
\begin{array}{r}
3x + 4 \\
x^2 + 1 \overline{)3x^3 + 4x^2 + x + 7} \\
\underline{3x^3 + 3x} \\
4x^2 - 2x + 7 \\
\underline{4x^2 + 4} \\
-2x + 3
\end{array}
$$

← Divide $4x^2$ by x^2 to get 4.
← Multiply $x^2 + 1$ by 4; subtract.

Since x^2 does not divide $-2x$ evenly (that is, the result is not a monomial), the process ends. The quotient is $3x + 4$, and the remainder is $-2x + 3$.

✔ **CHECK:** (Quotient)(Divisor) + Remainder

$$
\begin{aligned}
&= (3x + 4)(x^2 + 1) + (-2x + 3) \\
&= 3x^3 + 4x^2 + 3x + 4 + (-2x + 3) \\
&= 3x^3 + 4x^2 + x + 7 = \text{Dividend}
\end{aligned}
$$

Thus,

$$
\frac{3x^3 + 4x^2 + x + 7}{x^2 + 1} = 3x + 4 + \frac{-2x + 3}{x^2 + 1}
$$

The next example combines the steps involved in long division.

EXAMPLE 17   **Dividing Two Polynomials**

Find the quotient and the remainder when

$$
x^4 - 3x^3 + 2x - 5 \quad \text{is divided by} \quad x^2 - x + 1
$$

Solution   In setting up this division problem, it is necessary to leave a space for the missing x^2 term in the dividend.

$$
\begin{array}{r}
x^2 - 2x - 3 \quad \text{← Quotient} \\
x^2 - x + 1 \overline{)x^4 - 3x^3 + 2x - 5} \quad \text{← Dividend} \\
\underline{x^4 - x^3 + x^2} \\
-2x^3 - x^2 + 2x - 5 \\
\underline{-2x^3 + 2x^2 - 2x} \\
-3x^2 + 4x - 5 \\
\underline{-3x^2 + 3x - 3} \\
x - 2 \quad \text{← Remainder}
\end{array}
$$

Divisor →
Subtract →
Subtract →
Subtract →

✔ CHECK: (Quotient)(Divisor) + Remainder

$$= (x^2 - 2x - 3)(x^2 - x + 1) + x - 2$$
$$= x^4 - x^3 + x^2 - 2x^3 + 2x^2 - 2x - 3x^2 + 3x - 3 + x - 2$$
$$= x^4 - 3x^3 + 2x - 5 = \text{Dividend}$$

As a result,

$$\frac{x^4 - 3x^3 + 2x - 5}{x^2 - x + 1} = x^2 - 2x - 3 + \frac{x - 2}{x^2 - x + 1}$$

The process of dividing two polynomials leads to the following result:

Theorem

> Let Q be a polynomial of positive degree and let P be a polynomial whose degree is greater than or equal to the degree of Q. The remainder after dividing P by Q is either the zero polynomial or a polynomial of degree less than the degree of the divisor Q.

NOW WORK PROBLEM **89.**

Polynomials in Two Variables

A **monomial in two variables** x and y has the form $ax^n y^m$, where a is a constant, x and y are variables, and n and m are nonnegative integers. The **degree** of a monomial is the sum of the powers of the variables.

For example,

$$2xy^3, \quad x^2y^2, \quad \text{and} \quad x^3y$$

are monomials, each of which has degree 4.

A **polynomial in two variables** x and y is the sum of one or more monomials in two variables. The **degree of a polynomial** in two variables is the highest degree of all the monomials with nonzero coefficients.

EXAMPLE 18 | **Examples of Polynomials in Two Variables**

$$3x^2 + 2x^3y + 5 \qquad \pi x^3 - y^2 \qquad x^4 + 4x^3y - xy^3 + y^4$$

Two variables, degree is 4.   Two variables, degree is 3.   Two variables, degree is 4.

Multiplying polynomials in two variables is handled in the same way as multiplying polynomials in one variable.

EXAMPLE 19 | **Using a Special Product Formula**

To multiply $(2x - y)^2$, use the Squares of Binomials (3b) formula with $2x$ instead of x and y instead of a.

$$(2x - y)^2 = (2x)^2 - 2 \cdot y \cdot 2x + y^2 = 4x^2 - 4xy + y^2$$

NOW WORK PROBLEM **73.**

R.5 Concepts and Vocabulary

In Problems 1–3, fill in the blanks.

1. The polynomial $3x^4 - 2x^3 + 13x^2 - 5$ is of degree _____. The leading coefficient is _____.

2. If $x^3 - 3x^2 + 2x - 1$ is divided by $x + 2$, the remainder is _____ and the quotient is _____.

3. $(x - 2)(x^2 + 2x + 4) =$ _____.

In Problems 4–6 answer True or False to each statement.

4. $4x^{-2}$ is a monomial of degree -2.

5. The degree of the product of two nonzero polynomials equals the sum of their degrees.

6. $(x + a)(x^2 + ax + a) = x^3 + a^3$.

7. Write down a polynomial of degree 4.

8. Is $-3x^2 + 4\sqrt{x}$ a polynomial? Why or why not?

R.5 Exercises

In Problems 1–10, tell whether the expression is a monomial. If it is, name the variable(s) and the coefficient and give the degree of the monomial.

1. $2x^3$
2. $-4x^2$
3. $\dfrac{8}{x}$
4. $-2x^{-3}$
5. $-2xy^2$

6. $5x^2y^3$
7. $\dfrac{8x}{y}$
8. $\dfrac{-2x^2}{y^3}$
9. $x^2 + y^2$
10. $3x^2 + 4$

In Problems 11–20, tell whether the expression is a polynomial. If it is, give its degree.

11. $3x^2 - 5$
12. $1 - 4x$
13. 5
14. $-\pi$
15. $3x^2 - \dfrac{5}{x}$

16. $\dfrac{3}{x} + 2$
17. $2y^3 - \sqrt{2}$
18. $10z^2 + z$
19. $\dfrac{x^2 + 5}{x^3 - 1}$
20. $\dfrac{3x^3 + 2x - 1}{x^2 + x + 1}$

In Problems 21–40, add, subtract, or multiply, as indicated. Express your answer as a single polynomial in standard form.

21. $(x^2 + 4x + 5) + (3x - 3)$
22. $(x^3 + 3x^2 + 2) + (x^2 - 4x + 4)$

23. $(x^3 - 2x^2 + 5x + 10) - (2x^2 - 4x + 3)$
24. $(x^2 - 3x - 4) - (x^3 - 3x^2 + x + 5)$

25. $(6x^5 + x^3 + x) + (5x^4 - x^3 + 3x^2)$
26. $(10x^5 - 8x^2) + (3x^3 - 2x^2 + 6)$

27. $(x^2 - 3x + 1) + 2(3x^2 + x - 4)$
28. $-2(x^2 + x + 1) + (-5x^2 - x + 2)$

29. $6(x^3 + x^2 - 3) - 4(2x^3 - 3x^2)$
30. $8(4x^3 - 3x^2 - 1) - 6(4x^3 + 8x - 2)$

31. $(x^2 - x + 2) + (2x^2 - 3x + 5) - (x^2 + 1)$
32. $(x^2 + 1) - (4x^2 + 5) + (x^2 + x - 2)$

33. $9(y^2 - 3y + 4) - 6(1 - y^2)$
34. $8(1 - y^3) + 4(1 + y + y^2 + y^3)$

35. $x(x^2 + x - 4)$
36. $4x^2(x^3 - x + 2)$

37. $-2x^2(4x^3 + 5)$
38. $5x^3(3x - 4)$

39. $(x + 1)(x^2 + 2x - 4)$
40. $(2x - 3)(x^2 + x + 1)$

In Problems 41–58, multiply the polynomials using the FOIL method. Express your answer as a single polynomial in standard form.

41. $(x + 2)(x + 4)$
42. $(x + 3)(x + 5)$
43. $(2x + 5)(x + 2)$
44. $(3x + 1)(2x + 1)$

45. $(x - 4)(x + 2)$
46. $(x + 4)(x - 2)$
47. $(x - 3)(x - 2)$
48. $(x - 5)(x - 1)$

49. $(2x + 3)(x - 2)$ **50.** $(2x - 4)(3x + 1)$ **51.** $(-2x + 3)(x - 4)$ **52.** $(-3x - 1)(x + 1)$
53. $(-x - 2)(-2x - 4)$ **54.** $(-2x - 3)(3 - x)$ **55.** $(x - 2y)(x + y)$ **56.** $(2x + 3y)(x - y)$
57. $(-2x - 3y)(3x + 2y)$ **58.** $(x - 3y)(-2x + y)$

In Problems 59–82, multiply the polynomials using the special product formulas. Express your answer as a single polynomial in standard form.

59. $(x - 7)(x + 7)$ **60.** $(x - 1)(x + 1)$ **61.** $(2x + 3)(2x - 3)$ **62.** $(3x + 2)(3x - 2)$
63. $(x + 4)^2$ **64.** $(x + 5)^2$ **65.** $(x - 4)^2$ **66.** $(x - 5)^2$
67. $(3x + 4)(3x - 4)$ **68.** $(5x - 3)(5x + 3)$ **69.** $(2x - 3)^2$ **70.** $(3x - 4)^2$
71. $(x + y)(x - y)$ **72.** $(x + 3y)(x - 3y)$ **73.** $(3x + y)(3x - y)$ **74.** $(3x + 4y)(3x - 4y)$
75. $(x + y)^2$ **76.** $(x - y)^2$ **77.** $(x - 2y)^2$ **78.** $(2x + 3y)^2$
79. $(x - 2)^3$ **80.** $(x + 1)^3$ **81.** $(2x + 1)^3$ **82.** $(3x - 2)^3$

In Problems 83–100, find the quotient and the remainder. Check your work by verifying that

$$(Quotient)(Divisor) + Remainder = Dividend$$

83. $4x^3 - 3x^2 + x + 1$ divided by x
84. $3x^3 - x^2 + x - 2$ divided by x
85. $4x^3 - 3x^2 + x + 1$ divided by $x + 2$
86. $3x^3 - x^2 + x - 2$ divided by $x + 2$
87. $4x^3 - 3x^2 + x + 1$ divided by x^2
88. $3x^3 - x^2 + x - 2$ divided by x^2
89. $4x^3 - 3x^2 + x + 1$ divided by $x^2 + 2$
90. $3x^3 - x^2 + x - 2$ divided by $x^2 + 2$
91. $4x^3 - 3x^2 + x + 1$ divided by $2x^3 - 1$
92. $3x^3 - x^2 + x - 2$ divided by $3x^3 - 1$
93. $4x^3 - 3x^2 + x + 1$ divided by $2x^2 + x + 1$
94. $3x^3 - x^2 + x - 2$ divided by $3x^2 + x + 1$
95. $-4x^3 + x^2 - 4$ divided by $x - 1$
96. $-3x^4 - 2x - 1$ divided by $x - 1$
97. $1 - x^2 + x^4$ divided by $x^2 + x + 1$
98. $1 - x^2 + x^4$ divided by $x^2 - x + 1$
99. $x^3 - a^3$ divided by $x - a$
100. $x^5 - a^5$ divided by $x - a$

101. Find the sum of $a, b, c,$ and d if

$$\frac{x^3 - 2x^2 + 3x + 5}{x + 2} = ax^2 + bx + c + \frac{d}{x + 2}$$

102. Explain why the degree of the product of two nonzero polynomials equals the sum of their degrees.

103. Explain why the degree of the sum of two polynomials of different degrees equals the larger of their degrees.

104. Give a careful statement about the degree of the sum of two polynomials of the same degree.

105. Do you prefer adding two polynomials using the horizontal method or the vertical method? Write a brief position paper defending your choice.

106. Do you prefer to memorize the rule for the square of a binomial $(x + a)^2$ or to use FOIL to obtain the product? Write a brief position paper defending your choice.

R.6 FACTORING POLYNOMIALS

OBJECTIVES   **1** Factor the Difference of Two Squares and the Sum and the Difference of Two Cubes
2 Factor Perfect Squares
3 Factor a Second-degree Polynomial: $x^2 + Bx + C$
4 Factor by Grouping
5 Factor a Second-degree Polynomial: $Ax^2 + Bx + C$

Consider the following product:

$$(2x + 3)(x - 4) = 2x^2 - 5x - 12$$

The two polynomials on the left side are called **factors** of the polynomial on the right side. Expressing a given polynomial as a product of other polynomials, that is, finding the factors of a polynomial, is called **factoring**.

We shall restrict our discussion here to factoring polynomials in one variable into products of polynomials in one variable, where all coefficients are integers. We call this **factoring over the integers**.

Any polynomial can be written as the product of 1 times itself or as -1 times its additive inverse. If a polynomial cannot be written as the product of two other polynomials (excluding 1 and -1), then the polynomial is said to be **prime**. When a polynomial has been written as a product consisting only of prime factors, it is said to be **factored completely**. Examples of prime polynomials are

$$2, \quad 3, \quad 5, \quad x, \quad x + 1, \quad x - 1, \quad 3x + 4$$

The first factor to look for in a factoring problem is a common monomial factor present in each term of the polynomial. If one is present, use the distributive property to factor it out.

EXAMPLE 1   Identifying Common Monomial Factors

Polynomial	Common Monomial Factor	Remaining Factor	Factored Form
$2x + 4$	2	$x + 2$	$2x + 4 = 2(x + 2)$
$3x - 6$	3	$x - 2$	$3x - 6 = 3(x - 2)$
$2x^2 - 4x + 8$	2	$x^2 - 2x + 4$	$2x^2 - 4x + 8 = 2(x^2 - 2x + 4)$
$8x - 12$	4	$2x - 3$	$8x - 12 = 4(2x - 3)$
$x^2 + x$	x	$x + 1$	$x^2 + x = x(x + 1)$
$x^3 - 3x^2$	x^2	$x - 3$	$x^3 - 3x^2 = x^2(x - 3)$
$6x^2 + 9x$	$3x$	$2x + 3$	$6x^2 + 9x = 3x(2x + 3)$

Notice that, once all common monomial factors have been removed from a polynomial, the remaining factor is either a prime polynomial of degree 1 or a polynomial of degree 2 or higher. (Do you see why?)

NOW WORK PROBLEM 1.

Special Formulas

① When you factor a polynomial, first check whether you can use one of the special formulas discussed in the previous section.

Difference of Two Squares	$x^2 - a^2 = (x - a)(x + a)$
Perfect Squares	$x^2 + 2ax + a^2 = (x + a)^2$
	$x^2 - 2ax + a^2 = (x - a)^2$
Sum of Two Cubes	$x^3 + a^3 = (x + a)(x^2 - ax + a^2)$
Difference of Two Cubes	$x^3 - a^3 = (x - a)(x^2 + ax + a^2)$

EXAMPLE 2 **Factoring the Difference of Two Squares**

Factor completely: $x^2 - 4$

Solution We notice that $x^2 - 4$ is the difference of two squares, x^2 and 2^2. Thus,

$$x^2 - 4 = (x - 2)(x + 2)$$ ■

EXAMPLE 3 **Factoring the Difference of Two Cubes**

Factor completely: $x^3 - 1$

Solution Because $x^3 - 1$ is the difference of two cubes, x^3 and 1^3, we find that

$$x^3 - 1 = (x - 1)(x^2 + x + 1)$$ ■

EXAMPLE 4 **Factoring the Sum of Two Cubes**

Factor completely: $x^3 + 8$

Solution Because $x^3 + 8$ is the sum of two cubes, x^3 and 2^3, we have

$$x^3 + 8 = (x + 2)(x^2 - 2x + 4)$$ ■

EXAMPLE 5 **Factoring the Difference of Two Squares**

Factor completely: $x^4 - 16$

Solution Because $x^4 - 16$ is the difference of two squares, $x^4 = (x^2)^2$ and $16 = 4^2$, we have

$$x^4 - 16 = (x^2 - 4)(x^2 + 4)$$

But $x^2 - 4$ is also the difference of two squares. Thus,

$$x^4 - 16 = (x^2 - 4)(x^2 + 4) = (x - 2)(x + 2)(x^2 + 4)$$ ■

NOW WORK PROBLEMS **11** AND **29.**

② When the first term and third term of a trinomial are both positive and are perfect squares, such as x^2, $9x^2$, 1, and 4, check to see whether the trinomial is a perfect square.

EXAMPLE 6   Factoring Perfect Squares

Factor completely: $x^2 + 6x + 9$

Solution   The first term, x^2, and the third term, $9 = 3^2$, are perfect squares. Because the middle term $6x$ is twice the product of x and 3, we have a perfect square.

$$x^2 + 6x + 9 = (x + 3)^2$$ ■

EXAMPLE 7   Factoring Perfect Squares

Factor completely: $9x^2 - 6x + 1$

Solution   The first term, $9x^2 = (3x)^2$, and the third term, $1 = 1^2$, are perfect squares. Because the middle term $-6x$ is -2 times the product of $3x$ and 1, we have a perfect square.

$$9x^2 - 6x + 1 = (3x - 1)^2$$ ■

EXAMPLE 8   Factoring Perfect Squares

Factor completely: $25x^2 + 30x + 9$

Solution   The first term, $25x^2 = (5x)^2$, and the third term, $9 = 3^2$, are perfect squares. Because the middle term $30x$ is twice the product of $5x$ and 3, we have a perfect square.

$$25x^2 + 30x + 9 = (5x + 3)^2$$ ■

✏ **NOW WORK PROBLEMS 21 AND 89.**

If a trinomial is not a perfect square, it may be possible to factor it using the technique discussed next.

Factoring a Second-degree Polynomial: $x^2 + Bx + C$

③ The idea behind factoring a second-degree polynomial like $x^2 + Bx + C$ is to see whether it can be made equal to the product of two, possibly equal, first-degree polynomials.

For example, we know that

$$(x + 3)(x + 4) = x^2 + 7x + 12$$

The factors of $x^2 + 7x + 12$ are $x + 3$ and $x + 4$. Notice the following:

$$x^2 + 7x + 12 = (x + 3)(x + 4)$$

the product of 3 and 4 is 12

7 is the sum of 3 and 4

In general, if $x^2 + Bx + C = (x + a)(x + b)$, then $ab = C$ and $a + b = B$.

> To factor a second-degree polynomial $x^2 + Bx + C$, find integers whose product is C and whose sum is B. That is, if there are numbers a, b, where $ab = C$ and $a + b = B$, then
>
> $$x^2 + Bx + C = (x + a)(x + b)$$

EXAMPLE 9 **Factoring Trinomials**

Factor completely: $x^2 + 7x + 10$

Solution First, determine all integers whose product is 10 and then compute their sums.

Integers whose product is 10	1, 10	−1, −10	2, 5	−2, −5
Sum	11	−11	7	−7

The integers 2 and 5 have a product of 10 and add up to 7, the coefficient of the middle term. Thus,

$$x^2 + 7x + 10 = (x + 2)(x + 5)$$ ■

EXAMPLE 10 **Factoring Trinomials**

Factor completely: $x^2 - 6x + 8$

Solution First, determine all integers whose product is 8 and then compute each sum.

Integers whose product is 8	1, 8	−1, −8	2, 4	−2, −4
Sum	9	−9	6	−6

Since −6 is the coefficient of the middle term,

$$x^2 - 6x + 8 = (x - 2)(x - 4)$$ ■

EXAMPLE 11 **Factoring Trinomials**

Factor completely: $x^2 - x - 12$

Solution First, determine all integers whose product is −12 and then compute each sum.

Integers whose product is −12	1, −12	−1, 12	2, −6	−2, 6	3, −4	−3, 4
Sum	−11	11	−4	4	−1	1

Since -1 is the coefficient of the middle term,

$$x^2 - x - 12 = (x + 3)(x - 4)$$ ■

EXAMPLE 12 Factoring Trinomials

Factor completely: $x^2 + 4x - 12$

Solution The integers -2 and 6 have a product of -12 and have the sum 4. Thus,

$$x^2 + 4x - 12 = (x - 2)(x + 6)$$ ■

To avoid errors in factoring, always check your answer by multiplying it out to see if the result equals the original expression.

When none of the possibilities works, the polynomial is prime.

EXAMPLE 13 Identifying Prime Polynomials

Show that $x^2 + 9$ is prime.

Solution First, list the integers whose product is 9 and then compute their sums.

Integers whose product is 9	1, 9	−1, −9	3, 3	−3, −3
Sum	10	−10	6	−6

Since the coefficient of the middle term in $x^2 + 9 = x^2 + 0x + 9$ is 0 and none of the sums equals 0, we conclude that $x^2 + 9$ is prime. ■

Example 13 demonstrates a more general result:

Theorem Any polynomial of the form $x^2 + a^2$, a real, is prime.

■

🖊 **NOW WORK PROBLEMS 35 AND 73.**

Factoring by Grouping

④ Sometimes a common factor does not occur in every term of the polynomial, but in each of several groups of terms that together make up the polynomial. When this happens, the common factor can be factored out of each group by means of the distributive property. This technique is called **factoring by grouping**.

EXAMPLE 14 Factoring by Grouping

Factor completely by grouping: $(x^2 + 2)x + (x^2 + 2) \cdot 3$

Solution Notice the common factor $x^2 + 2$. By applying the distributive property, we have

$$(x^2 + 2)x + (x^2 + 2) \cdot 3 = (x^2 + 2)(x + 3)$$

Since $x^2 + 2$ and $x + 3$ are prime, the factorization is complete. ■

EXAMPLE 15 **Factoring by Grouping**

Factor completely by grouping: $x^3 - 4x^2 + 2x - 8$

Solution To see if factoring by grouping will work, group the first two terms and the last two terms. Then look for a common factor in each group. In this example, we can factor x^2 from $x^3 - 4x^2$ and 2 from $2x - 8$. The remaining factor in each case is the same, $x - 4$. This means that factoring by grouping will work, as follows:

$$
\begin{aligned}
x^3 - 4x^2 + 2x - 8 &= (x^3 - 4x^2) + (2x - 8) \\
&= x^2(x - 4) + 2(x - 4) \\
&= (x^2 + 2)(x - 4)
\end{aligned}
$$

Since $x^2 + 2$ and $x - 4$ are prime, the factorization is complete. ■

EXAMPLE 16 **Factoring by Grouping**

Factor completely by grouping: $3x^3 + 4x^2 - 6x - 8$

Solution Here, $3x + 4$ is a common factor of $3x^3 + 4x^2$ and of $-6x - 8$. Hence,

$$
\begin{aligned}
3x^3 + 4x^2 - 6x - 8 &= (3x^3 + 4x^2) - (6x + 8) \\
&= x^2(3x + 4) - 2(3x + 4) \\
&= (x^2 - 2)(3x + 4)
\end{aligned}
$$

Since $x^2 - 2$ and $3x + 4$ are prime (over the integers), the factorization is complete. ■

NOW WORK PROBLEM **47.**

Factoring a Second-degree Polynomial: $Ax^2 + Bx + C$

⑤ To factor a second-degree polynomial $Ax^2 + Bx + C$ when $A \neq 1$ and A, B, and C have no common factors, follow these steps:

> ***Steps for Factoring $Ax^2 + Bx + C$,***
> ***$A \neq 1, A, B,$ and C Have No Common Factors***
>
> **STEP 1:** Find the value of AC.
> **STEP 2:** Find integers whose product is AC that add up to B. That is, find a and b so that $ab = AC$ and $a + b = B$.
> **STEP 3:** Write $Ax^2 + Bx + C = Ax^2 + ax + bx + C$.
> **STEP 4:** Factor this last expression by grouping.

EXAMPLE 17 **Factoring Trinomials**

Factor completely: $2x^2 + 5x + 3$

Solution Comparing $2x^2 + 5x + 3$ to $Ax^2 + Bx + C$, we find that $A = 2$, $B = 5$, and $C = 3$.

STEP 1: The value of AC is $2 \cdot 3 = 6$.
STEP 2: Determine the integers whose product is $AC = 6$ and compute their sums.

Integers whose product is 6	1, 6	−1, −6	2, 3	−2, −3
Sum	7	−7	5	−5

The integers whose product is 6 that add up to $B = 5$ are 2 and 3.

STEP 3:
$$2x^2 + 5x + 3 = 2x^2 + 2x + 3x + 3$$

STEP 4: Factor by grouping.

$$2x^2 + 2x + 3x + 3 = (2x^2 + 2x) + (3x + 3)$$
$$= 2x(x + 1) + 3(x + 1)$$
$$= (2x + 3)(x + 1)$$

Thus,

$$2x^2 + 5x + 3 = (2x + 3)(x + 1)$$ ■

EXAMPLE 18 Factoring Trinomials

Factor completely: $2x^2 - x - 6$

Solution Comparing $2x^2 - x - 6$ to $Ax^2 + Bx + C$, we find that $A = 2$, $B = -1$, and $C = -6$.

STEP 1: The value of AC is $2 \cdot (-6) = -12$.
STEP 2: Determine the integers whose product is $AC = -12$ and compute their sums.

Integers whose product is −12	1, −12	−1, 12	2, −6	−2, 6	3, −4	−3, 4
Sum	−11	11	−4	4	−1	1

The integers whose product is −12 that add up to $B = -1$ are −4 and 3.

STEP 3:
$$2x^2 - x - 6 = 2x^2 - 4x + 3x - 6$$

STEP 4: Factor by grouping.

$$2x^2 - 4x + 3x - 6 = (2x^2 - 4x) + (3x - 6)$$
$$= 2x(x - 2) + 3(x - 2)$$
$$= (2x + 3)(x - 2)$$

Thus,

$$2x^2 - x - 6 = (2x + 3)(x - 2)$$ ■

NOW WORK PROBLEM **53.**

SUMMARY

We close this section with a capsule summary.

Type of Polynomial	Method	Example
Any polynomial	Look for common monomial factors. (Always do this first!)	$6x^2 + 9x = 3x(2x + 3)$
Binomials of degree 2 or higher	Check for a special product:	
	Difference of two squares, $x^2 - a^2$	$x^2 - 16 = (x - 4)(x + 4)$
	Difference of two cubes, $x^3 - a^3$	$x^3 - 64 = (x - 4)(x^2 + 4x + 16)$
	Sum of two cubes, $x^3 + a^3$	$x^3 + 27 = (x + 3)(x^2 - 3x + 9)$
Trinomials of degree 2	Check for a perfect square, $(x \pm a)^2$.	$x^2 + 8x + 16 = (x + 4)^2$
	Follow the procedures on page 54 or 56.	$6x^2 + x - 1 = (2x + 1)(3x - 1)$
Three or more terms	Grouping	$2x^3 - 3x^2 + 4x - 6 = (2x - 3)(x^2 + 2)$

R.6 Concepts and Vocabulary

In Problems 1 and 2, fill in the blanks.

1. If factored completely, $3x^3 - 12x =$ _____ .

2. If a polynomial cannot be written as the product of two other polynomials (excluding 1 and −1), then the polynomial is said to be _____ .

In Problems 3 and 4, answer True or False to each statement.

3. The polynomial $x^2 + 4$ is prime.

4. $3x^3 - 2x^2 - 6x + 4 = (3x - 2)(x^2 + 2)$.

5. Give an example of a polynomial of degree 2 that is a perfect square.

6. What does "factored completely" mean?+

R.6 Exercises

In Problems 1–10, factor each polynomial by removing the common monomial factor.

1. $3x + 6$
2. $7x - 14$
3. $ax^2 + a$
4. $ax - a$
5. $x^3 + x^2 + x$
6. $x^3 - x^2 + x$
7. $2x^2 - 2x$
8. $3x^2 - 3x$
9. $3x^2y - 6xy^2 + 12xy$
10. $60x^2y - 48xy^2 + 72x^3y$

In Problems 11–18, factor the difference of two squares.

11. $x^2 - 1$
12. $x^2 - 4$
13. $4x^2 - 1$
14. $9x^2 - 1$
15. $x^2 - 16$
16. $x^2 - 25$
17. $25x^2 - 4$
18. $36x^2 - 9$

In Problems 19–28, factor the perfect squares.

19. $x^2 + 2x + 1$
20. $x^2 - 4x + 4$
21. $x^2 + 4x + 4$
22. $x^2 - 2x + 1$
23. $x^2 - 10x + 25$
24. $x^2 + 10x + 25$
25. $4x^2 + 4x + 1$
26. $9x^2 + 6x + 1$
27. $16x^2 + 8x + 1$
28. $25x^2 + 10x + 1$

In Problems 29–34, factor the sum or difference of two cubes.

29. $x^3 - 27$
30. $x^3 + 125$
31. $x^3 + 27$
32. $27 - 8x^3$
33. $8x^3 + 27$
34. $64 - 27x^3$

In Problems 35–46, factor each polynomial.

35. $x^2 + 5x + 6$
36. $x^2 + 6x + 8$
37. $x^2 + 7x + 6$
38. $x^2 + 9x + 8$
39. $x^2 + 7x + 10$
40. $x^2 + 11x + 10$
41. $x^2 - 10x + 16$
42. $x^2 - 17x + 16$
43. $x^2 - 7x - 8$
44. $x^2 - 2x - 8$
45. $x^2 + 7x - 8$
46. $x^2 + 2x - 8$

In Problems 47–52, factor by grouping.

47. $2x^2 + 4x + 3x + 6$ **48.** $3x^2 - 3x + 2x - 2$ **49.** $2x^2 - 4x + x - 2$ **50.** $3x^2 + 6x - x - 2$

51. $6x^2 + 9x + 4x + 6$ **52.** $6x^2 - 9x + 2x - 3$

In Problems 53–64, factor each polynomial.

53. $3x^2 + 4x + 1$ **54.** $2x^2 + 3x + 1$ **55.** $2z^2 + 5z + 3$ **56.** $6z^2 + 5z + 1$

57. $3x^2 + 2x - 8$ **58.** $3x^2 + 10x + 8$ **59.** $3x^2 - 2x - 8$ **60.** $3x^2 - 10x + 8$

61. $3x^2 + 14x + 8$ **62.** $3x^2 - 14x + 8$ **63.** $3x^2 + 10x - 8$ **64.** $3x^2 - 10x - 8$

In Problems 65–112, factor completely each polynomial. If the polynomial cannot be factored, say it is prime.

65. $x^2 - 36$ **66.** $x^2 - 9$ **67.** $2 - 8x^2$ **68.** $3 - 27x^2$

69. $x^2 + 7x + 10$ **70.** $x^2 + 5x + 4$ **71.** $x^2 - 10x + 21$ **72.** $x^2 - 6x + 8$

73. $4x^2 - 8x + 32$ **74.** $3x^2 - 12x + 15$ **75.** $x^2 + 4x + 16$ **76.** $x^2 + 12x + 36$

77. $15 + 2x - x^2$ **78.** $14 + 6x - x^2$ **79.** $3x^2 - 12x - 36$ **80.** $x^3 + 8x^2 - 20x$

81. $y^4 + 11y^3 + 30y^2$ **82.** $3y^3 - 18y^2 - 48y$ **83.** $4x^2 + 12x + 9$ **84.** $9x^2 - 12x + 4$

85. $6x^2 + 8x + 2$ **86.** $8x^2 + 6x - 2$ **87.** $x^4 - 81$ **88.** $x^4 - 1$

89. $x^6 - 2x^3 + 1$ **90.** $x^6 + 2x^3 + 1$ **91.** $x^7 - x^5$ **92.** $x^8 - x^5$

93. $16x^2 + 24x + 9$ **94.** $9x^2 - 24x + 16$ **95.** $5 + 16x - 16x^2$ **96.** $5 + 11x - 16x^2$

97. $4y^2 - 16y + 15$ **98.** $9y^2 + 9y - 4$ **99.** $1 - 8x^2 - 9x^4$ **100.** $4 - 14x^2 - 8x^4$

101. $x(x + 3) - 6(x + 3)$ **102.** $5(3x - 7) + x(3x - 7)$ **103.** $(x + 2)^2 - 5(x + 2)$ **104.** $(x - 1)^2 - 2(x - 1)$

105. $(3x - 2)^3 - 27$ **106.** $(5x + 1)^3 - 1$ **107.** $3(x^2 + 10x + 25) - 4(x + 5)$

108. $7(x^2 - 6x + 9) + 5(x - 3)$ **109.** $x^3 + 2x^2 - x - 2$ **110.** $x^3 - 3x^2 - x + 3$

111. $x^4 - x^3 + x - 1$ **112.** $x^4 + x^3 + x + 1$

113. Show that $x^2 + 4$ is prime.   **114.** Show that $x^2 + x + 1$ is prime.

115. Make up a polynomial that factors into a perfect square.

116. Explain to a fellow student what you look for first when presented with a factoring problem. What do you do next?

R.7 RATIONAL EXPRESSIONS

OBJECTIVES   **1** Reduce a Rational Expression to Lowest Terms
              2 Multiply and Divide Rational Expressions
              3 Add and Subtract Rational Expressions
              4 Use the Least Common Multiple Method
              5 Simplify Mixed Quotients

If we form the quotient of two polynomials, the result is called a **rational expression**. Some examples of rational expressions are

$$\text{(a)}\ \frac{x^3 + 1}{x} \qquad \text{(b)}\ \frac{3x^2 + x - 2}{x^2 + 5} \qquad \text{(c)}\ \frac{x}{x^2 - 1} \qquad \text{(d)}\ \frac{xy^2}{(x - y)^2}$$

Expressions (a), (b), and (c) are rational expressions in one variable, x, whereas (d) is a rational expression in two variables, x and y.

Rational expressions are described in the same manner as rational numbers. In expression (a), the polynomial $x^3 + 1$ is called the **numerator**, and x is called the **denominator**. When the numerator and denominator of a rational expression contain no common factors (except 1 and -1), we say that the rational expression is **reduced to lowest terms**, or **simplified**.

The polynomial in the denominator of a rational expression cannot be equal to 0 because division by 0 is not defined. For example, for the expression $\dfrac{x^3 + 1}{x}$, x cannot take on the value 0. The domain of the variable x is $\{x \mid x \neq 0\}$.

1   A rational expression is reduced to lowest terms by factoring completely the numerator and the denominator and canceling any common factors by using the cancellation property:

$$\frac{a\cancel{c}}{b\cancel{c}} = \frac{a}{b} \qquad \text{if } b \neq 0, c \neq 0 \tag{1}$$

EXAMPLE 1   **Reducing a Rational Expression to Lowest Terms**

Reduce to lowest terms: $\dfrac{x^2 + 4x + 4}{x^2 + 3x + 2}$

Solution   We begin by factoring the numerator and the denominator.

$$x^2 + 4x + 4 = (x + 2)(x + 2)$$

$$x^2 + 3x + 2 = (x + 2)(x + 1)$$

Since a common factor, $x + 2$, appears, the original expression is not in lowest terms. To reduce it to lowest terms, we use the cancellation property:

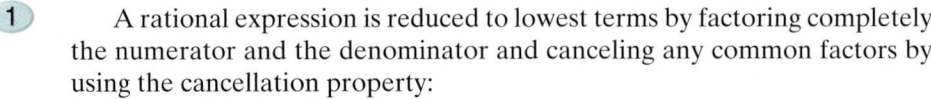

$$\frac{x^2 + 4x + 4}{x^2 + 3x + 2} = \frac{\cancel{(x + 2)}(x + 2)}{\cancel{(x + 2)}(x + 1)} = \frac{x + 2}{x + 1} \qquad x \neq -2, -1 \quad \blacksquare$$

WARNING: Apply the cancellation property only to rational expressions written in factored form. Be sure to cancel only common factors!   ∎

EXAMPLE 2   **Reducing Rational Expressions to Lowest Terms**

Reduce each rational expression to lowest terms.

(a) $\dfrac{x^3 - 8}{x^3 - 2x^2}$
(b) $\dfrac{8 - 2x}{x^2 - x - 12}$

Solution (a) $\dfrac{x^3 - 8}{x^3 - 2x^2} = \dfrac{\cancel{(x-2)}(x^2 + 2x + 4)}{x^2\cancel{(x-2)}} = \dfrac{x^2 + 2x + 4}{x^2}$ $x \neq 0, 2$

(b) $\dfrac{8 - 2x}{x^2 - x - 12} = \dfrac{2(4 - x)}{(x - 4)(x + 3)} = \dfrac{2(-1)\cancel{(x-4)}}{\cancel{(x-4)}(x + 3)} = \dfrac{-2}{x + 3}$ $x \neq -3, 4$

NOW WORK PROBLEM 1.

Multiplying and Dividing Rational Expressions

2   The rules for multiplying and dividing rational expressions are the same as the rules for multiplying and dividing rational numbers. If $\dfrac{a}{b}$ and $\dfrac{c}{d}$, $b \neq 0$, $d \neq 0$, are two rational expressions, then

$$\frac{a}{b} \cdot \frac{c}{d} = \frac{ac}{bd} \qquad \text{if } b \neq 0, d \neq 0 \tag{2}$$

$$\frac{\dfrac{a}{b}}{\dfrac{c}{d}} = \frac{a}{b} \cdot \frac{d}{c} = \frac{ad}{bc} \qquad \text{if } b \neq 0, c \neq 0, d \neq 0 \tag{3}$$

In using equations (2) and (3) with rational expressions, be sure first to factor each polynomial completely so that common factors can be canceled. Leave your answer in factored form.

EXAMPLE 3   **Multiplying and Dividing Rational Expressions**

Perform the indicated operation and simplify the result. Leave your answer in factored form.

(a) $\dfrac{x^2 - 2x + 1}{x^3 + x} \cdot \dfrac{4x^2 + 4}{x^2 + x - 2}$

(b) $\dfrac{\dfrac{x + 3}{x^2 - 4}}{\dfrac{x^2 - x - 12}{x^3 - 8}}$

Solution (a) $\dfrac{x^2 - 2x + 1}{x^3 + x} \cdot \dfrac{4x^2 + 4}{x^2 + x - 2} = \dfrac{(x - 1)^2}{x(x^2 + 1)} \cdot \dfrac{4(x^2 + 1)}{(x + 2)(x - 1)}$

$= \dfrac{(x - 1)^2 (4)\cancel{(x^2 + 1)}}{x\cancel{(x^2 + 1)}(x + 2)\cancel{(x - 1)}}$

$= \dfrac{4(x - 1)}{x(x + 2)}, \qquad x \neq -2, 0, 1$

(b) $\dfrac{\dfrac{x+3}{x^2-4}}{\dfrac{x^2-x-12}{x^3-8}} = \dfrac{x+3}{x^2-4} \cdot \dfrac{x^3-8}{x^2-x-12}$

$$= \dfrac{x+3}{(x-2)(x+2)} \cdot \dfrac{(x-2)(x^2+2x+4)}{(x-4)(x+3)}$$

$$= \dfrac{\cancel{(x+3)}\,\cancel{(x-2)}(x^2+2x+4)}{\cancel{(x-2)}(x+2)(x-4)\cancel{(x+3)}}$$

$$= \dfrac{x^2+2x+4}{(x+2)(x-4)} \qquad x \neq -3, -2, 2, 4$$

NOW WORK PROBLEMS 13 AND 21.

Adding and Subtracting Rational Expressions

③ The rules for adding and subtracting rational expressions are the same as the rules for adding and subtracting rational numbers. If the denominators of two rational expressions to be added (or subtracted) are equal, we add (or subtract) the numerators and keep the common denominator.

If $\dfrac{a}{b}$ and $\dfrac{c}{b}$ are two rational expressions, then

$$\dfrac{a}{b} + \dfrac{c}{b} = \dfrac{a+c}{b} \qquad \dfrac{a}{b} - \dfrac{c}{b} = \dfrac{a-c}{b} \qquad \text{if } b \neq 0 \qquad (4)$$

EXAMPLE 4 **Adding and Subtracting Rational Expressions with Equal Denominators**

Perform the indicated operation and simplify the result. Leave your answer in factored form.

(a) $\dfrac{2x^2-4}{2x+5} + \dfrac{x+3}{2x+5} \qquad x \neq -\dfrac{5}{2}$ (b) $\dfrac{x}{x-3} - \dfrac{3x+2}{x-3} \qquad x \neq 3$

Solution (a) $\dfrac{2x^2-4}{2x+5} + \dfrac{x+3}{2x+5} = \dfrac{(2x^2-4)+(x+3)}{2x+5}$

$$= \dfrac{2x^2+x-1}{2x+5} = \dfrac{(2x-1)(x+1)}{2x+5}$$

(b) $\dfrac{x}{x-3} - \dfrac{3x+2}{x-3} = \dfrac{x-(3x+2)}{x-3} = \dfrac{x-3x-2}{x-3}$

$$= \dfrac{-2x-2}{x-3} = \dfrac{-2(x+1)}{x-3}$$

EXAMPLE 5   **Adding Rational Expressions Whose Denominators Are Additive Inverses of Each Other**

Perform the indicated operation and simplify the result. Leave your answer in factored form.

$$\frac{2x}{x-3} + \frac{5}{3-x} \qquad x \neq 3$$

Solution   Notice that the denominators of the two rational expressions are different. However, the denominator of the second expression is just the additive inverse of the denominator of the first. That is,

$$3 - x = -x + 3 = -1 \cdot (x - 3) = -(x - 3)$$

Then,

$$\frac{2x}{x-3} + \frac{5}{3-x} = \frac{2x}{x-3} + \frac{5}{-(x-3)} = \frac{2x}{x-3} + \frac{-5}{x-3}$$

$$\underset{3-x = -(x-3)}{\uparrow} \qquad \underset{\frac{a}{-b} = \frac{-a}{b}}{\uparrow}$$

$$= \frac{2x + (-5)}{x-3} = \frac{2x-5}{x-3} \qquad \blacksquare$$

✏ **NOW WORK PROBLEMS 33 AND 39.**

If the denominators of two rational expressions to be added or subtracted are not equal, we can use the general formulas for adding and subtracting quotients.

$$\frac{a}{b} + \frac{c}{d} = \frac{a \cdot d}{b \cdot d} + \frac{b \cdot c}{b \cdot d} = \frac{ad + bc}{bd} \qquad \text{if } b \neq 0, d \neq 0 \qquad (5a)$$

$$\frac{a}{b} - \frac{c}{d} = \frac{a \cdot d}{b \cdot d} - \frac{b \cdot c}{b \cdot d} = \frac{ad - bc}{bd} \qquad \text{if } b \neq 0, d \neq 0 \qquad (5b)$$

EXAMPLE 6   **Adding and Subtracting Rational Expressions with Unequal Denominators**

Perform the indicated operation and simplify the result. Leave your answer in factored form.

(a) $\dfrac{x-3}{x+4} + \dfrac{x}{x-2} \qquad x \neq -4, 2$ (b) $\dfrac{x^2}{x^2-4} - \dfrac{1}{x} \qquad x \neq -2, 0, 2$

Solution   (a) $\dfrac{x-3}{x+4} + \dfrac{x}{x-2} = \dfrac{x-3}{x+4} \cdot \dfrac{x-2}{x-2} + \dfrac{x+4}{x+4} \cdot \dfrac{x}{x-2}$

$$\underset{(5a)}{\uparrow}$$

$$= \frac{(x-3)(x-2) + (x+4)(x)}{(x+4)(x-2)}$$

$$= \frac{x^2 - 5x + 6 + x^2 + 4x}{(x+4)(x-2)} = \frac{2x^2 - x + 6}{(x+4)(x-2)}$$

(b) $\dfrac{x^2}{x^2 - 4} - \dfrac{1}{x} = \dfrac{x^2}{x^2 - 4} \cdot \dfrac{x}{x} - \dfrac{x^2 - 4}{x^2 - 4} \cdot \dfrac{1}{x} = \dfrac{x^2(x) - (x^2 - 4)(1)}{(x^2 - 4)(x)}$

(5b)

$$= \dfrac{x^3 - x^2 + 4}{(x - 2)(x + 2)(x)} \qquad \blacksquare$$

✎ NOW WORK PROBLEM 43.

Least Common Multiple (LCM)

4 If the denominators of two rational expressions to be added (or subtracted) have common factors, we usually do not use the general rules given by equations (5a) and (5b). Just as with fractions, we apply the **least common multiple (LCM) method**. The LCM method uses the polynomial of least degree that has each denominator polynomial as a factor.

> ### The LCM Method for Adding or Subtracting Rational Expressions
>
> The Least Common Multiple (LCM) Method requires four steps:
>
> **STEP 1:** Factor completely the polynomial in the denominator of each rational expression.
>
> **STEP 2:** The LCM of the denominators is the product of each of these factors raised to a power equal to the greatest number of times that the factor occurs in occurs in any denominator.
>
> **STEP 3:** Write each rational expression using the LCM as the common denominator.
>
> **STEP 4:** Add or subtract the rational expressions using equation (4).

Let's work an example that requires only Steps 1 and 2.

EXAMPLE 7 **Finding the Least Common Multiple**

Find the least common multiple of the following pair of polynomials:

$$x(x - 1)^2(x + 1) \quad \text{and} \quad 4(x - 1)(x + 1)^3$$

Solution **STEP 1:** The polynomials are already factored completely as

$$x(x - 1)^2(x + 1) \quad \text{and} \quad 4(x - 1)(x + 1)^3$$

STEP 2: Start by writing the factors of the left-hand polynomial. (Or you could start with the one on the right.)

$$x(x - 1)^2(x + 1)$$

Now look at the right-hand polynomial. Its first factor, 4, does not appear in our list, so we insert it.

$$4x(x - 1)^2(x + 1)$$

The next factor, $x - 1$, is already in our list, so no change is necessary. The final factor is $(x + 1)^3$. Since our list has $x + 1$ to the first power only, we replace $x + 1$ in the list by $(x + 1)^3$. The LCM is

$$4x(x - 1)^2(x + 1)^3$$ ■

Notice that the LCM is, in fact, the polynomial of least degree that contains $x(x - 1)^2(x + 1)$ and $4(x - 1)(x + 1)^3$ as factors.

 **NOW WORK PROBLEM 49.**

EXAMPLE 8   **Using the Least Common Multiple to Add Rational Expressions**

Perform the indicated operation and simplify the result. Leave your answer in factored form.

$$\frac{x}{x^2 + 3x + 2} + \frac{2x - 3}{x^2 - 1} \qquad x \neq -2, -1, 1$$

Solution **STEP 1:** Factor completely the polynomials in the denominators.

$$x^2 + 3x + 2 = (x + 2)(x + 1)$$
$$x^2 - 1 = (x - 1)(x + 1)$$

STEP 2: The LCM is $(x + 2)(x + 1)(x - 1)$. Do you see why?

STEP 3: Write each rational expression using the LCM as the denominator.

$$\frac{x}{x^2 + 3x + 2} = \frac{x}{(x + 2)(x + 1)} \uparrow = \frac{x}{(x + 2)(x + 1)} \cdot \frac{x - 1}{x - 1} = \frac{x(x - 1)}{(x + 2)(x + 1)(x - 1)}$$

Multiply numerator and denominator by x − 1 to get the LCM in the denominator.

$$\frac{2x - 3}{x^2 - 1} = \frac{2x - 3}{(x - 1)(x + 1)} \uparrow = \frac{2x - 3}{(x - 1)(x + 1)} \cdot \frac{x + 2}{x + 2} = \frac{(2x - 3)(x + 2)}{(x - 1)(x + 1)(x + 2)}$$

Multiply numerator and denominator by x + 2 to get the LCM in the denominator.

STEP 4: Now we can add by using equation (4).

$$\frac{x}{x^2 + 3x + 2} + \frac{2x - 3}{x^2 - 1} = \frac{x(x - 1)}{(x + 2)(x + 1)(x - 1)} + \frac{(2x - 3)(x + 2)}{(x + 2)(x + 1)(x - 1)}$$

$$= \frac{(x^2 - x) + (2x^2 + x - 6)}{(x + 2)(x + 1)(x - 1)}$$

$$= \frac{3x^2 - 6}{(x + 2)(x + 1)(x - 1)} = \frac{3(x^2 - 2)}{(x + 2)(x + 1)(x - 1)}$$ ■

EXAMPLE 9 **Using the Least Common Multiple to Subtract Rational Expressions**

Perform the indicated operations and simplify the result. Leave your answer in factored form.

$$\frac{3}{x^2 + x} - \frac{x + 4}{x^2 + 2x + 1} \qquad x \neq -1, 0$$

Solution **STEP 1:** Factor completely the polynomials in the denominators.

$$x^2 + x = x(x + 1)$$
$$x^2 + 2x + 1 = (x + 1)^2$$

STEP 2: The LCM is $x(x + 1)^2$.

STEP 3: Write each rational expression using the LCM as the denominator.

$$\frac{3}{x^2 + x} = \frac{3}{x(x + 1)} = \frac{3}{x(x + 1)} \cdot \frac{x + 1}{x + 1} = \frac{3(x + 1)}{x(x + 1)^2}$$

$$\frac{x + 4}{x^2 + 2x + 1} = \frac{x + 4}{(x + 1)^2} = \frac{x + 4}{(x + 1)^2} \cdot \frac{x}{x} = \frac{x(x + 4)}{x(x + 1)^2}$$

STEP 4: Subtract, using equation (4).

$$\frac{3}{x^2 + x} - \frac{x + 4}{x^2 + 2x + 1} = \frac{3(x + 1)}{x(x + 1)^2} - \frac{x(x + 4)}{x(x + 1)^2}$$

$$= \frac{3(x + 1) - x(x + 4)}{x(x + 1)^2}$$

$$= \frac{3x + 3 - x^2 - 4x}{x(x + 1)^2}$$

$$= \frac{-x^2 - x + 3}{x(x + 1)^2} \qquad \blacksquare$$

NOW WORK PROBLEM **59.**

Mixed Quotients

⑤ When sums and/or differences of rational expressions appear as the numerator and/or denominator of a quotient, the quotient is called a **mixed quotient**.* For example,

$$\frac{1 + \dfrac{1}{x}}{1 - \dfrac{1}{x}} \qquad \text{and} \qquad \frac{\dfrac{x^2}{x^2 - 4} - 3}{\dfrac{x - 3}{x + 2} - 1}$$

are mixed quotients. To **simplify** a mixed quotient means to write it as a rational expression reduced to lowest terms. This can be accomplished in either of two ways.

* Some texts use the term **complex fraction**.

> *Simplifying a Mixed Quotient*
>
> **METHOD 1:** Treat the numerator and denominator of the mixed quotient separately, performing whatever operations are indicated and simplifying the results. Follow this by simplifying the resulting rational expression.
>
> **METHOD 2:** Find the LCM of the denominators of all rational expressions that appear in the mixed quotient. Multiply the numerator and denominator of the mixed quotient by the LCM and simplify the result.

We will use both methods in the next example. By carefully studying each method, you can discover situations in which one method may be easier to use than the other.

EXAMPLE 10   **Simplifying a Mixed Quotient**

Simplify: $\dfrac{\dfrac{1}{2}+\dfrac{3}{x}}{\dfrac{x+3}{4}}$ $x \neq -3, 0$

Solution   **METHOD 1:** First, we perform the indicated operation in the numerator, and then we divide.

$$\frac{\dfrac{1}{2}+\dfrac{3}{x}}{\dfrac{x+3}{4}} = \frac{\dfrac{1\cdot x+2\cdot 3}{2\cdot x}}{\dfrac{x+3}{4}} = \frac{\dfrac{x+6}{2x}}{\dfrac{x+3}{4}} = \frac{x+6}{2x}\cdot\frac{4}{x+3}$$

Rule for adding quotients   Rule for dividing quotients

$$= \frac{(x+6)\cdot 4}{2\cdot x\cdot(x+3)} = \frac{\cancel{2}\cdot 2\cdot(x+6)}{\cancel{2}\cdot x\cdot(x+3)} = \frac{2(x+6)}{x(x+3)}$$

Rule for multiplying quotients

METHOD 2: The rational expressions that appear in the mixed quotient are

$$\frac{1}{2},\ \frac{3}{x},\ \frac{x+3}{4}$$

The LCM of their denominators is $4x$. We multiply the numerator and denominator of the mixed quotient by $4x$ and then simplify.

$$\frac{\dfrac{1}{2}+\dfrac{3}{x}}{\dfrac{x+3}{4}} = \frac{4x\cdot\left(\dfrac{1}{2}+\dfrac{3}{x}\right)}{4x\cdot\left(\dfrac{x+3}{4}\right)} = \frac{4x\cdot\dfrac{1}{2}+4x\cdot\dfrac{3}{x}}{\dfrac{4x\cdot(x+3)}{4}}$$

Multiply the numerator and denominator by 4x   Distributive property in numerator

$$= \frac{\cancel{2}\cdot 2x\cdot\dfrac{1}{\cancel{2}}+4\cancel{x}\cdot\dfrac{3}{\cancel{x}}}{\dfrac{\cancel{4}x\cdot(x+3)}{\cancel{4}}} = \frac{2x+12}{x(x+3)} = \frac{2(x+6)}{x(x+3)}$$

Simplify   Factor

EXAMPLE 11   **Simplifying a Mixed Quotient**

Simplify: $\dfrac{\dfrac{x^2}{x-4}+2}{\dfrac{2x-2}{x}-1}$, $x \neq 0, 2, 4$

Solution   We will use Method 1.

$$\frac{\dfrac{x^2}{x-4}+2}{\dfrac{2x-2}{x}-1} = \frac{\dfrac{x^2}{x-4}+\dfrac{2(x-4)}{x-4}}{\dfrac{2x-2}{x}-\dfrac{x}{x}} = \frac{\dfrac{x^2+2x-8}{x-4}}{\dfrac{2x-2-x}{x}}$$

$$= \frac{\dfrac{(x+4)(x-2)}{x-4}}{\dfrac{x-2}{x}} = \frac{(x+4)\cancel{(x-2)}}{x-4} \cdot \frac{x}{\cancel{x-2}}$$

$$= \frac{(x+4)\cdot x}{x-4}$$

NOW WORK PROBLEM **69.**

EXAMPLE 12   **Solving an Application in Electricity**

Figure 21

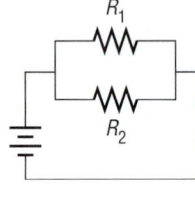

An electrical circuit contains two resistors connected in parallel, as shown in Figure 21. If the resistance of each is R_1 and R_2 ohms, respectively, their combined resistance R is given by the formula

$$R = \frac{1}{\dfrac{1}{R_1}+\dfrac{1}{R_2}}$$

Express R as a rational expression; that is, simplify the right-hand side of this formula. Evaluate the rational expression if $R_1 = 6$ ohms and $R_2 = 10$ ohms.

Solution   We will use Method 2. If we consider 1 as the fraction $\dfrac{1}{1}$, then the rational expressions in the mixed quotient are

$$\frac{1}{1}, \quad \frac{1}{R_1}, \quad \frac{1}{R_2}$$

The LCM of the denominators is R_1R_2. We multiply the numerator and denominator of the mixed quotient by R_1R_2 and simplify.

$$\frac{1}{\dfrac{1}{R_1}+\dfrac{1}{R_2}} = \frac{1 \cdot R_1R_2}{\left(\dfrac{1}{R_1}+\dfrac{1}{R_2}\right)\cdot R_1R_2} = \frac{R_1R_2}{\dfrac{1}{R_1}\cdot R_1R_2 + \dfrac{1}{R_2}\cdot R_1R_2} = \frac{R_1R_2}{R_2+R_1}$$

Thus,

$$R = \frac{R_1R_2}{R_2+R_1}$$

If $R_1 = 6$ and $R_2 = 10$, then

$$R = \frac{6 \cdot 10}{10 + 6} = \frac{60}{16} = \frac{15}{4} \text{ ohms}$$

R.7 Concepts and Vocabulary

In Problems 1 and 2, fill in the blanks.

1. When the numerator and denominator of a rational expression contain no common factors (except 1 and -1), the rational expression is _____ .

2. LCM is an abbreviation for _____ _____ _____ .

In Problems 3 and 4, answer True or False to each statement.

3. The rational expression $\dfrac{2x^3 - 4x}{x - 2}$ is reduced to lowest terms.

4. The LCM of $2x^3 + 6x^2$ and $6x^4 + 4x^3$ is $4x^3(x + 1)$.

5. Define the term *rational expression*.

R.7 Exercises

In Problems 1–12, reduce each rational expression to lowest terms.

1. $\dfrac{3x + 9}{x^2 - 9}$

2. $\dfrac{4x^2 + 8x}{12x + 24}$

3. $\dfrac{x^2 - 2x}{3x - 6}$

4. $\dfrac{15x^2 + 24x}{3x^2}$

5. $\dfrac{24x^2}{12x^2 - 6x}$

6. $\dfrac{x^2 + 4x + 4}{x^2 - 16}$

7. $\dfrac{y^2 - 25}{2y^2 - 8y - 10}$

8. $\dfrac{3y^2 - y - 2}{3y^2 + 5y + 2}$

9. $\dfrac{x^2 + 4x - 5}{x^2 - 2x + 1}$

10. $\dfrac{x - x^2}{x^2 + x - 2}$

11. $\dfrac{x^2 + 5x - 14}{2 - x}$

12. $\dfrac{2x^2 + 5x - 3}{1 - 2x}$

In Problems 13–30, perform the indicated operation and simplify the result. Leave your answer in factored form.

13. $\dfrac{3x + 6}{5x^2} \cdot \dfrac{x}{x^2 - 4}$

14. $\dfrac{3}{2x} \cdot \dfrac{x^2}{6x + 10}$

15. $\dfrac{4x^2}{x^2 - 16} \cdot \dfrac{x - 4}{2x}$

16. $\dfrac{12}{x^2 - x} \cdot \dfrac{x^2 - 1}{4x - 2}$

17. $\dfrac{4x - 8}{-3x} \cdot \dfrac{12}{12 - 6x}$

18. $\dfrac{6x - 27}{5x} \cdot \dfrac{2}{4x - 18}$

19. $\dfrac{x^2 - 3x - 10}{x^2 + 2x - 35} \cdot \dfrac{x^2 + 4x - 21}{x^2 + 9x + 14}$

20. $\dfrac{x^2 + x - 6}{x^2 + 4x - 5} \cdot \dfrac{x^2 - 25}{x^2 + 2x - 15}$

21. $\dfrac{\dfrac{6x}{x^2 - 4}}{\dfrac{3x - 9}{2x + 4}}$

22. $\dfrac{\dfrac{12x}{5x + 20}}{\dfrac{4x^2}{x^2 - 16}}$

23. $\dfrac{\dfrac{8x}{x^2 - 1}}{\dfrac{10x}{x + 1}}$

24. $\dfrac{\dfrac{x - 2}{4x}}{\dfrac{x^2 - 4x + 4}{12x}}$

25. $\dfrac{\dfrac{4 - x}{4 + x}}{\dfrac{4x}{x^2 - 16}}$

26. $\dfrac{\dfrac{3 + x}{3 - x}}{\dfrac{x^2 - 9}{9x^3}}$

27. $\dfrac{\dfrac{x^2 + 7x + 12}{x^2 - 7x + 12}}{\dfrac{x^2 + x - 12}{x^2 - x - 12}}$

28. $\dfrac{\dfrac{x^2 + 7x + 6}{x^2 + x - 6}}{\dfrac{x^2 + 5x - 6}{x^2 + 5x + 6}}$

29. $\dfrac{\dfrac{2x^2 - x - 28}{3x^2 - x - 2}}{\dfrac{4x^2 + 16x + 7}{3x^2 + 11x + 6}}$

30. $\dfrac{\dfrac{9x^2 + 3x - 2}{12x^2 + 5x - 2}}{\dfrac{9x^2 - 6x + 1}{8x^2 - 10x - 3}}$

In Problems 31–48, perform the indicated operations and simplify the result. Leave your answer in factored form.

31. $\dfrac{x}{2} + \dfrac{5}{2}$

32. $\dfrac{3}{x} - \dfrac{6}{x}$

33. $\dfrac{x^2}{2x - 3} - \dfrac{4}{2x - 3}$

34. $\dfrac{3x^2}{2x - 1} - \dfrac{9}{2x - 1}$

35. $\dfrac{x + 1}{x - 3} + \dfrac{2x - 3}{x - 3}$

36. $\dfrac{2x - 5}{3x + 2} + \dfrac{x + 4}{3x + 2}$

37. $\dfrac{3x + 5}{2x - 1} - \dfrac{2x - 4}{2x - 1}$

38. $\dfrac{5x - 4}{3x + 4} - \dfrac{x + 1}{3x + 4}$

39. $\dfrac{4}{x - 2} + \dfrac{x}{2 - x}$

40. $\dfrac{6}{x - 1} - \dfrac{x}{1 - x}$

41. $\dfrac{4}{x - 1} - \dfrac{2}{x + 2}$

42. $\dfrac{2}{x + 5} - \dfrac{5}{x - 5}$

43. $\dfrac{x}{x + 1} + \dfrac{2x - 3}{x - 1}$

44. $\dfrac{3x}{x - 4} + \dfrac{2x}{x + 3}$

45. $\dfrac{x - 3}{x + 2} - \dfrac{x + 4}{x - 2}$

46. $\dfrac{2x - 3}{x - 1} - \dfrac{2x + 1}{x + 1}$

47. $\dfrac{x}{x^2 - 4} + \dfrac{1}{x}$

48. $\dfrac{x - 1}{x^3} + \dfrac{x}{x^2 + 1}$

In Problems 49–56, find the LCM of the given polynomials.

49. $x^2 - 4, \quad x^2 - x - 2$

50. $x^2 - x - 12, \quad x^2 - 8x + 16$

51. $x^3 - x, \quad x^2 - x$

52. $3x^2 - 27, \quad 2x^2 - x - 15$

53. $4x^3 - 4x^2 + x, \quad 2x^3 - x^2, \quad x^3$

54. $x - 3, \quad x^2 + 3x, \quad x^3 - 9x$

55. $x^3 - x, \quad x^3 - 2x^2 + x, \quad x^3 - 1$

56. $x^2 + 4x + 4, \quad x^3 + 2x^2, \quad (x + 2)^3$

In Problems 57–68, perform the indicated operations and simplify the result. Leave your answer in factored form.

57. $\dfrac{x}{x^2 - 7x + 6} - \dfrac{x}{x^2 - 2x - 24}$

58. $\dfrac{x}{x - 3} - \dfrac{x + 1}{x^2 + 5x - 24}$

59. $\dfrac{4x}{x^2 - 4} - \dfrac{2}{x^2 + x - 6}$

60. $\dfrac{3x}{x - 1} - \dfrac{x - 4}{x^2 - 2x + 1}$

61. $\dfrac{3}{(x - 1)^2(x + 1)} + \dfrac{2}{(x - 1)(x + 1)^2}$

62. $\dfrac{2}{(x + 2)^2(x - 1)} - \dfrac{6}{(x + 2)(x - 1)^2}$

63. $\dfrac{x + 4}{x^2 - x - 2} - \dfrac{2x + 3}{x^2 + 2x - 8}$

64. $\dfrac{2x - 3}{x^2 + 8x + 7} - \dfrac{x - 2}{(x + 1)^2}$

65. $\dfrac{1}{x} - \dfrac{2}{x^2 + x} + \dfrac{3}{x^3 - x^2}$

66. $\dfrac{x}{(x - 1)^2} + \dfrac{2}{x} - \dfrac{x + 1}{x^3 - x^2}$

67. $\dfrac{1}{h}\left(\dfrac{1}{x + h} - \dfrac{1}{x}\right)$

68. $\dfrac{1}{h}\left[\dfrac{1}{(x + h)^2} - \dfrac{1}{x^2}\right]$

In Problems 69–78, perform the indicated operations and simplify the result. Leave your answer in factored form.

69. $\dfrac{1 + \dfrac{1}{x}}{1 - \dfrac{1}{x}}$

70. $\dfrac{4 + \dfrac{1}{x^2}}{3 - \dfrac{1}{x^2}}$

71. $\dfrac{x - \dfrac{1}{x}}{x + \dfrac{1}{x}}$

72. $\dfrac{1 - \dfrac{x}{x + 1}}{2 - \dfrac{x - 1}{x}}$

73. $\dfrac{\dfrac{x + 4}{x - 2} - \dfrac{x - 3}{x + 1}}{x + 1}$

74. $\dfrac{\dfrac{x - 2}{x + 1} - \dfrac{x}{x - 2}}{x + 3}$

75. $\dfrac{\dfrac{x - 2}{x + 2} + \dfrac{x - 1}{x + 1}}{\dfrac{x}{x + 1} - \dfrac{2x - 3}{x}}$

76. $\dfrac{\dfrac{2x + 5}{x} - \dfrac{x}{x - 3}}{\dfrac{x^2}{x - 3} - \dfrac{(x + 1)^2}{x + 3}}$

77. $1 - \dfrac{1}{1 - \dfrac{1}{x}}$

78. $1 - \dfrac{1}{1 - \dfrac{1}{1 - x}}$

79. The Lensmaker's Equation The focal length f of a lens with index of refraction n is

$$\frac{1}{f} = (n - 1)\left[\frac{1}{R_1} + \frac{1}{R_2}\right]$$

where R_1 and R_2 are the radii of curvature of the front and back surfaces of the lens. Express f as a rational expression. Evaluate the rational expression for $n = 1.5$, $R_1 = 0.1$ meter, and $R_2 = 0.2$ meter.

80. Electrical Circuits An electrical circuit contains three resistors connected in parallel. If the resistance of each is R_1, R_2, and R_3 ohms, respectively, their combined resistance R is given by the formula

$$\frac{1}{R} = \frac{1}{R_1} + \frac{1}{R_2} + \frac{1}{R_3}$$

Express R as a rational expression. Evaluate R for $R_1 = 5$ ohms, $R_2 = 4$ ohms, and $R_3 = 10$ ohms.

---

  81. The following expressions are called **continued fractions**:

$$1 + \frac{1}{x}, \quad 1 + \frac{1}{1 + \dfrac{1}{x}}, \quad 1 + \frac{1}{1 + \dfrac{1}{1 + \dfrac{1}{x}}}, \quad 1 + \frac{1}{1 + \dfrac{1}{1 + \dfrac{1}{1 + \dfrac{1}{x}}}}, \quad \ldots$$

Each simplifies to an expression of the form

$$\frac{ax + b}{bx + c}$$

Trace the successive values of a, b, and c as you "continue" the fraction. Can you discover the patterns that these values follow? Go to the library and research Fibonacci numbers. Write a report on your findings.

82. Explain to a fellow student when you would use the LCM method to add two rational expressions. Give two examples of adding two rational expressions, one in which you use the LCM and the other in which you do not.

83. Which of the two methods given in the text for simplifying mixed quotients do you prefer? Write a brief paragraph stating the reasons for your choice.

R.8 SQUARE ROOTS; RADICALS

OBJECTIVES

1. Work with Properties of Square Roots
2. Rationalize a Denominator
3. Simplify nth Roots
4. Simplify Radicals

Square Roots

A real number is squared when it is raised to the power 2. The inverse of squaring is finding a **square root**. For example, since $6^2 = 36$ and $(-6)^2 = 36$, the numbers 6 and -6 are square roots of 36.

The symbol $\sqrt{}$, called a **radical sign**, is used to denote the **principal**, or nonnegative, square root. Thus, $\sqrt{36} = 6$.

In general, if a is a nonnegative real number, the nonnegative number b such that $b^2 = a$ is the **principal square root** of a and is denoted by $b = \sqrt{a}$.

The following comments are noteworthy:

1. Negative numbers do not have square roots (in the real number system), because the square of any real number is *nonnegative*. For example, $\sqrt{-4}$ is not a real number, because there is no real number whose square is -4.

2. The principal square root of 0 is 0, since $0^2 = 0$. That is, $\sqrt{0} = 0$.

3. The principal square root of a positive number is positive.

4. If $c \geq 0$, then $(\sqrt{c})^2 = c$. For example, $(\sqrt{2})^2 = 2$ and $(\sqrt{3})^2 = 3$.

EXAMPLE 1 **Evaluating Square Roots**

(a) $\sqrt{64} = 8$ (b) $\sqrt{\dfrac{1}{16}} = \dfrac{1}{4}$ (c) $(\sqrt{1.4})^2 = 1.4$ ∎

Examples 1(a) and (b) are examples of square roots of perfect squares, since $64 = 8^2$; and $\dfrac{1}{16} = \left(\dfrac{1}{4}\right)^2$.

In general, we have the rule

$$\sqrt{a^2} = |a| \qquad (1)$$

Notice the absolute value in equation (1). We need it since the principal square root is nonnegative.

✏ **NOW WORK PROBLEM 1.**

EXAMPLE 2 **Square Roots of Perfect Squares**

(a) $\sqrt{(2.3)^2} = |2.3| = 2.3$ (b) $\sqrt{(-2.3)^2} = |-2.3| = 2.3$
(c) $\sqrt{x^2} = |x|$ ∎

Properties of Square Roots

① We begin with the following observation:

$$\sqrt{4 \cdot 25} = \sqrt{100} = 10 \quad \text{and} \quad \sqrt{4}\,\sqrt{25} = 2 \cdot 5 = 10$$

This suggests the following property of square roots:

Product Property of Square Roots

If a and b are each nonnegative real numbers, then

$$\sqrt{ab} = \sqrt{a}\,\sqrt{b} \qquad (2)$$

When used in connection with square roots, the direction "simplify" means to remove from the square root any perfect squares that occur as factors. We can use equation (2) to simplify a square root that contains a perfect square as a factor, as illustrated by the following examples.

EXAMPLE 3   **Simplifying Square Roots**

(a) $\sqrt{32} = \sqrt{16 \cdot 2} = \sqrt{16}\,\sqrt{2} = 4\sqrt{2}$

$\uparrow$ 16 is a perfect square. $\uparrow$ (2)

(b) $\sqrt{5}\,\sqrt{10} = \sqrt{5 \cdot 10} = \sqrt{50} = \sqrt{25 \cdot 2} = \sqrt{25}\,\sqrt{2} = 5\sqrt{2}$

$\uparrow$ (2)

(c) $-3\sqrt{72} = -3\sqrt{36 \cdot 2} = -3\sqrt{36}\,\sqrt{2} = -3 \cdot 6\sqrt{2} = -18\sqrt{2}$

(d) $\sqrt{75x^2} = \sqrt{(25x^2)(3)} = \sqrt{25x^2}\,\sqrt{3} = \sqrt{25}\,\sqrt{x^2}\,\sqrt{3} = 5|x|\sqrt{3}$

$\uparrow$ (1)   ■

NOW WORK PROBLEMS **13** AND **27**.

Sometimes, to multiply two expressions with square roots, we use the distributive property and the product property of square roots. For example,

$$\sqrt{3}\,(4\sqrt{2} - \sqrt{12}) = \sqrt{3} \cdot 4\sqrt{2} - \sqrt{3} \cdot \sqrt{12} \qquad \text{Distributive Property}$$
$$= 4\sqrt{3 \cdot 2} - \sqrt{3 \cdot 12} \qquad \text{Product Property of Square Roots}$$
$$= 4\sqrt{6} - \sqrt{36} \qquad \text{Simplify.}$$
$$= 4\sqrt{6} - 6$$

EXAMPLE 4   **Multiplying Square Roots**

$$(\sqrt{5} - 5)(\sqrt{5} + 2) = \sqrt{5}(\sqrt{5} + 2) - 5(\sqrt{5} + 2)$$
$$= \sqrt{5} \cdot \sqrt{5} + \sqrt{5} \cdot 2 - 5 \cdot \sqrt{5} - 5 \cdot 2$$
$$= 5 + 2\sqrt{5} - 5\sqrt{5} - 10$$
$$= -5 - 3\sqrt{5} \qquad ■$$

NOW WORK PROBLEM **51**.

Another property of square roots is suggested by the following examples:

$$\sqrt{\frac{36}{9}} = \sqrt{4} = 2 \quad \text{and} \quad \frac{\sqrt{36}}{\sqrt{9}} = \frac{6}{3} = 2$$

Quotient Property of Square Roots

If a is a nonnegative real number and b is a positive real number, then

$$\sqrt{\frac{a}{b}} = \frac{\sqrt{a}}{\sqrt{b}} \qquad (3)$$

EXAMPLE 5   **Simplifying Square Roots**

(a) $\sqrt{\dfrac{81}{25}} = \dfrac{\sqrt{81}}{\sqrt{25}} = \dfrac{9}{5}$

(b) $\dfrac{\sqrt{24}}{\sqrt{3}} = \sqrt{\dfrac{24}{3}} = \sqrt{8} = \sqrt{4\cdot 2} = \sqrt{4}\,\sqrt{2} = 2\sqrt{2}$

NOW WORK PROBLEM **31.**

Rationalizing

② When square roots occur in quotients, it is customary to rewrite the quotient so that the denominator contains no square roots. This process is referred to as **rationalizing the denominator**.

The idea is to multiply by an appropriate expression so that the new denominator contains no square roots. For example:

If Denominator Contains the Factor	Multiply by	To Obtain Denominator Free of Radicals
$\sqrt{3}$	$\sqrt{3}$	$(\sqrt{3})^2 = 3$
$\sqrt{3} + 1$	$\sqrt{3} - 1$	$(\sqrt{3})^2 - 1^2 = 3 - 1 = 2$
$\sqrt{2} - 3$	$\sqrt{2} + 3$	$(\sqrt{2})^2 - 3^2 = 2 - 9 = -7$
$\sqrt{5} - \sqrt{3}$	$\sqrt{5} + \sqrt{3}$	$(\sqrt{5})^2 - (\sqrt{3})^2 = 5 - 3 = 2$

In rationalizing the denominator of a quotient, be sure to multiply both the numerator and the denominator by the expression.

EXAMPLE 6   **Rationalizing Denominators**

Rationalize the denominator: $\dfrac{1}{\sqrt{3}}$

Solution   The denominator contains the factor $\sqrt{3}$, so we multiply the numerator and denominator by $\sqrt{3}$ to obtain

$$\frac{1}{\sqrt{3}} = \frac{1}{\sqrt{3}} \cdot \frac{\sqrt{3}}{\sqrt{3}} = \frac{\sqrt{3}}{(\sqrt{3})^2} = \frac{\sqrt{3}}{3}$$

EXAMPLE 7   **Rationalizing Denominators**

Rationalize the denominator: $\dfrac{5}{4\sqrt{2}}$

Solution   The denominator contains the factor $\sqrt{2}$, so we multiply the numerator and denominator by $\sqrt{2}$ to obtain

$$\frac{5}{4\sqrt{2}} = \frac{5}{4\sqrt{2}} \cdot \frac{\sqrt{2}}{\sqrt{2}} = \frac{5\sqrt{2}}{4(\sqrt{2})^2} = \frac{5\sqrt{2}}{4\cdot 2} = \frac{5\sqrt{2}}{8}$$

EXAMPLE 8 **Rationalizing Denominators**

Rationalize the denominator: $\dfrac{\sqrt{2}}{\sqrt{3} - \sqrt{2}}$

Solution The denominator contains the factor $\sqrt{3} - \sqrt{2}$, so we multiply the numerator and denominator by $\sqrt{3} + \sqrt{2}$ to obtain

$$\frac{\sqrt{2}}{\sqrt{3} - \sqrt{2}} = \frac{\sqrt{2}}{\sqrt{3} - \sqrt{2}} \cdot \frac{\sqrt{3} + \sqrt{2}}{\sqrt{3} + \sqrt{2}} = \frac{\sqrt{2}(\sqrt{3} + \sqrt{2})}{(\sqrt{3})^2 - (\sqrt{2})^2}$$

$$= \frac{\sqrt{2}\,\sqrt{3} + (\sqrt{2})^2}{3 - 2} = \sqrt{6} + 2 \quad\blacksquare$$

NOW WORK PROBLEM 69.

*n*th Roots

The **principal *n*th root of a real number** $a, n \geq 2$ an integer, symbolized by $\sqrt[n]{a}$, is defined as follows:

$$\sqrt[n]{a} = b \quad \text{means} \quad a = b^n$$

where $a \geq 0$ and $b \geq 0$ if $n \geq 2$ is even and a, b are any real numbers if $n \geq 3$ is odd.

Notice that if a is negative and n is even then $\sqrt[n]{a}$ is not defined. When it is defined, the principal *n*th root of a number is unique.

③ The symbol $\sqrt[n]{a}$ for the principal *n*th root of a is sometimes called a **radical**; the integer n is called the **index**, and a is called the **radicand**. If the index of a radical is 2, we call $\sqrt[2]{a}$ the **square root** of a and omit the index 2 by simply writing $\sqrt{a}$. If the index is 3, we call $\sqrt[3]{a}$ the **cube root** of a.

EXAMPLE 9 **Simplifying Principal *n*th Roots**

(a) $\sqrt[3]{8} = \sqrt[3]{2^3} = 2$ (b) $\sqrt[3]{-64} = \sqrt[3]{(-4)^3} = -4$

(c) $\sqrt[4]{\dfrac{1}{16}} = \sqrt[4]{\left(\dfrac{1}{2}\right)^4} = \dfrac{1}{2}$ (d) $\sqrt[6]{(-2)^6} = |-2| = 2$ $\blacksquare$

These are examples of **perfect roots**, since each one simplifies to a rational number. Notice the absolute value in Example 9(d). If n is even, the principal *n*th root must be nonnegative.

In general, if $n \geq 2$ is a positive integer and a is a real number, we have

$$\sqrt[n]{a^n} = a \qquad \text{if } n \geq 3 \text{ is odd} \qquad (4a)$$
$$\sqrt[n]{a^n} = |a| \qquad \text{if } n \geq 2 \text{ is even} \qquad (4b)$$

Radicals provide a way of representing many irrational real numbers. For example, there is no rational number whose square is 2. Using radicals, we can say that $\sqrt{2}$ *is* the positive number whose square is 2.

EXAMPLE 10 **Using a Calculator to Approximate Roots**

Use a calculator to approximate $\sqrt[5]{16}$.

Figure 22

Solution Figure 22 shows the result.

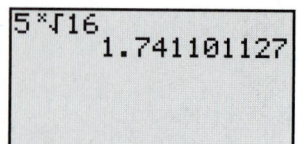

5*√16
           1.741101127

 NOW WORK PROBLEM 81.

Properties of Radicals

Let $n \geq 2$ and $m \geq 2$ denote positive integers, and let a and b represent real numbers. Assuming that all radicals are defined, we have the following properties:

Properties of Radicals

$$\sqrt[n]{ab} = \sqrt[n]{a}\,\sqrt[n]{b} \tag{5a}$$

$$\sqrt[n]{\frac{a}{b}} = \frac{\sqrt[n]{a}}{\sqrt[n]{b}} \tag{5b}$$

$$\sqrt[n]{a^m} = (\sqrt[n]{a})^m \tag{5c}$$

④ When used in reference to radicals, the direction to "simplify" will mean to remove from the radicals any perfect roots that occur as factors. Let's look at some examples of how the preceding rules are applied to simplify radicals.

EXAMPLE 11 **Simplifying Radicals**

(a) $\sqrt[3]{16} = \sqrt[3]{8 \cdot 2} = \sqrt[3]{8} \cdot \sqrt[3]{2} = \sqrt[3]{2^3} \cdot \sqrt[3]{2} = 2\sqrt[3]{2}$
$\quad\quad\quad\quad\uparrow\quad\quad\quad\quad\uparrow$
$\quad\quad\quad$ Factor out $\quad$ (5a)
$\quad\quad\quad$ perfect cube.

(b) $\sqrt[3]{-16x^4} = \sqrt[3]{-8 \cdot 2 \cdot x^3 \cdot x} = \sqrt[3]{(-8x^3)(2x)}$
$\quad\quad\quad\quad\uparrow\quad\quad\quad\quad\quad\quad\quad\uparrow$
$\quad\quad\quad$ Factor perfect $\quad\quad$ Combine perfect
$\quad\quad\quad$ cubes inside radical. $\quad$ cubes.

$\quad\quad = \sqrt[3]{(-2x)^3 \cdot 2x} = \sqrt[3]{(-2x)^3} \cdot \sqrt[3]{2x}$
$\quad\quad\quad\quad\quad\quad\quad\quad\uparrow$
$\quad\quad = -2x\sqrt[3]{2x} \quad\quad$ (5a)

EXAMPLE 12 **Simplifying Radicals**

$$\sqrt[3]{\frac{8x^5}{27}} = \sqrt[3]{\frac{2^3 x^3 x^2}{3^3}} = \sqrt[3]{\left(\frac{2x}{3}\right)^3 \cdot x^2} = \sqrt[3]{\left(\frac{2x}{3}\right)^3} \cdot \sqrt[3]{x^2} = \frac{2x}{3}\sqrt[3]{x^2}$$

 NOW WORK PROBLEMS 19 AND 23.

Two or more radicals can be combined, provided that they have the same index and the same radicand. Such radicals are called **like radicals**.

EXAMPLE 13 **Combining Like Radicals**

$$
\begin{aligned}
-8\sqrt{12} + \sqrt{3} &= -8\sqrt{4 \cdot 3} + \sqrt{3} \\
&= -8 \cdot \sqrt{4}\sqrt{3} + \sqrt{3} \\
&= -16\sqrt{3} + \sqrt{3} = -15\sqrt{3} \quad\blacksquare
\end{aligned}
$$

EXAMPLE 14 **Combining Like Radicals**

$$
\begin{aligned}
\sqrt[3]{8x^4} + \sqrt[3]{-x} + 4\sqrt[3]{27x} &= \sqrt[3]{2^3 x^3 x} + \sqrt[3]{-1 \cdot x} + 4\sqrt[3]{3^3 x} \\
&= \sqrt[3]{(2x)^3} \cdot \sqrt[3]{x} + \sqrt[3]{-1} \cdot \sqrt[3]{x} + 4\sqrt[3]{3^3} \cdot \sqrt[3]{x} \\
&= 2x\sqrt[3]{x} - 1 \cdot \sqrt[3]{x} + 12\sqrt[3]{x} \\
&= (2x + 11)\sqrt[3]{x} \quad\blacksquare
\end{aligned}
$$

✏ **NOW WORK PROBLEM 43.**

HISTORICAL FEATURE

The radical sign, $\sqrt{}$, was first used in print by Coss in 1525. It is thought to be the manuscript form of the letter r (for the Latin word *radix = root*), although this is not quite conclusively proved. It took a long time for $\sqrt{}$ to become the standard symbol for a square root and much longer to standardize $\sqrt[3]{}$, $\sqrt[4]{}$, $\sqrt[5]{}$, and so on. The indexes of the root were placed in every conceivable position, with

$$
\sqrt[3]{8}, \quad \sqrt{③8}, \quad \text{and} \quad \underset{3}{\sqrt{}}8
$$

all being variants for $\sqrt[3]{8}$. The notation $\sqrt{}\sqrt{}16$ was popular for $\sqrt[4]{16}$. By the 1700s, the index had settled where we now put it.

The bar on top of the present radical symbol, as follows,

$$
\sqrt{a^2 + 2ab + b^2}
$$

is the last survivor of the **vinculum**, a bar placed atop an expression to indicate what we would now indicate with parentheses. For example,

$$
a\overline{b + c} = a(b + c)
$$

R.8 Concepts and Vocabulary

In Problems 1 and 2, fill in the blanks.

1. The symbol $\sqrt{}$ is called a _____ _____.

2. In the expression $\sqrt[3]{25}$, the number 3 is called the _____ ; the number 25 is called the _____.

In Problems 3–6, answer True or False to each statement.

3. $\sqrt{a} = a$.

4. The principal square root of a positive real number is always positive.

5. The cube root of a negative real number is always negative.

6. To rationalize $\dfrac{3}{2 - \sqrt{3}}$ you would multiply the numerator and the denominator by $-2 + \sqrt{3}$.

7. Give an example of a perfect cube root.

R.8 Exercises

In Problems 1–12, evaluate each perfect root.

1. $\sqrt{25}$

2. $\sqrt{81}$

3. $\sqrt[3]{27}$

4. $\sqrt[3]{125}$

5. $\sqrt[3]{-64}$

6. $\sqrt[3]{-8}$

7. $\sqrt{\dfrac{1}{9}}$

8. $\sqrt[3]{\dfrac{27}{8}}$

9. $\sqrt{25x^4}$

10. $\sqrt[3]{64x^6}$

11. $\sqrt[3]{8(1+x)^3}$

12. $\sqrt{4(x+4)^2}$

In Problems 13–38, simplify each expression.

13. $\sqrt{8}$

14. $\sqrt{27}$

15. $\sqrt{50}$

16. $\sqrt{72}$

17. $\sqrt[3]{16}$

18. $\sqrt[3]{24}$

19. $\sqrt[3]{-16}$

20. $-\sqrt[3]{16}$

21. $\sqrt{\dfrac{25x^3}{9x}}, \quad x \neq 0$

22. $\sqrt[3]{\dfrac{x}{8x^4}}, \quad x \neq 0$

23. $\sqrt[4]{x^{12}y^8}, \quad x \geq 0, y \geq 0$

24. $\sqrt[5]{x^{10}y^5}$

25. $\sqrt{36x}, \quad x \geq 0$

26. $\sqrt{9x^5}, \quad x \geq 0$

27. $\sqrt{3x^2}\,\sqrt{12x}, \quad x \geq 0$

28. $\sqrt{5x}\,\sqrt{20x^3}, \quad x \geq 0$

29. $\dfrac{\sqrt{3xy^3}\,\sqrt{2x^2y}}{\sqrt{6x^3y^4}}, \quad x > 0, y > 0$

30. $\dfrac{\sqrt[3]{x^2y}\,\sqrt[3]{125x^3}}{\sqrt[3]{8x^3y^4}}, \quad x \neq 0, y \neq 0$

31. $\sqrt{\dfrac{16y^4}{9x^2}}, \quad x > 0, y \geq 0$

32. $\sqrt{\dfrac{9x^4}{16y^6}}, \quad x \geq 0, y > 0$

33. $\left(\sqrt{5}\,\sqrt[3]{9}\right)^2$

34. $\left(\sqrt[3]{3}\,\sqrt{10}\right)^4$

35. $\sqrt{\dfrac{2x-3}{2x^4+3x^3}}\,\sqrt{\dfrac{x}{4x^2-9}}, \quad x > \dfrac{3}{2}$

36. $\sqrt[3]{\dfrac{x-1}{x^2+2x+1}}\,\sqrt[3]{\dfrac{(x-1)^2}{x+1}}, \quad x \neq -1$

37. $\sqrt{\dfrac{x-1}{x+1}}\,\sqrt{\dfrac{x^2+2x+1}{x^2-1}}, \quad x > 1$

38. $\sqrt{\dfrac{x^2+4}{x(x^2-4)}}\,\sqrt{\dfrac{4x^2}{x^4-16}}, \quad x > 2$

In Problems 39–48, simplify each expression.

39. $3\sqrt{2} + 4\sqrt{2}$

40. $6\sqrt{5} - 4\sqrt{5}$

41. $-\sqrt{18} + 2\sqrt{8}$

42. $2\sqrt{12} - 3\sqrt{27}$

43. $5\sqrt[3]{2} - 2\sqrt[3]{54}$

44. $9\sqrt[3]{24} - \sqrt[3]{81}$

45. $\sqrt{8x^3} - 3\sqrt{50x}, \quad x \geq 0$

46. $3x\sqrt{9y} + 4\sqrt{25y}, \quad y \geq 0$

47. $\sqrt[3]{16x^4y} - 3x\sqrt[3]{2xy} + 5\sqrt[3]{-2xy^4}$

48. $8xy - \sqrt{25x^2y^2} + \sqrt[3]{8x^3y^3}, \quad x \geq 0, y \geq 0$

In Problems 49–62, perform the indicated operation and simplify the results.

49. $(3\sqrt{6})(4\sqrt{3})$

50. $(5\sqrt{8})(-3\sqrt{6})$

51. $\sqrt{3}\,(\sqrt{3} - 4)$

52. $\sqrt{5}\,(\sqrt{5} + 6)$

53. $3\sqrt{7}\,(2\sqrt{7} + 3)$

54. $(2\sqrt{6} + 3)(3\sqrt{6})$

55. $(\sqrt{2} - 1)^2$

56. $(\sqrt{3} + \sqrt{5})^2$

57. $(\sqrt[3]{2} - 1)^3$

58. $(\sqrt[3]{4} + 2)^3$

59. $(2\sqrt{x} - 3)(2\sqrt{x} + 5), \quad x \geq 0$

60. $(4\sqrt{x} - 3)(\sqrt{x} + 3), \quad x \geq 0$

61. $\sqrt{1 - x^2} - \dfrac{1}{\sqrt{1 - x^2}}, \quad -1 < x < 1$

62. $\sqrt{1 - x^2} + \dfrac{x^2}{\sqrt{1 - x^2}}, \quad -1 < x < 1$

In Problems 63–78, rationalize the denominator of each expression.

63. $\dfrac{2}{\sqrt{5}}$

64. $\dfrac{\sqrt{3}}{\sqrt{5}}$

65. $\dfrac{8}{\sqrt{6}}$

66. $\dfrac{5}{\sqrt{10}}$

67. $\dfrac{1}{\sqrt{x}}, \quad x > 0$

68. $\dfrac{x}{\sqrt{x^2 + 4}}$

69. $\dfrac{3}{5 + \sqrt{2}}$

70. $\dfrac{2}{\sqrt{7} - 2}$

71. $\dfrac{3}{4 + \sqrt{7}}$

72. $\dfrac{10}{4 - \sqrt{2}}$

73. $\dfrac{\sqrt{5}}{2 + 3\sqrt{5}}$

74. $\dfrac{\sqrt{3}}{2\sqrt{3} + 3}$

75. $\dfrac{\sqrt{3} - \sqrt{2}}{\sqrt{3} + \sqrt{2}}$

76. $\dfrac{\sqrt{5} + \sqrt{3}}{\sqrt{5} - \sqrt{3}}$

77. $\dfrac{1}{\sqrt{x} + 2}, \quad x \geq 0$

78. $\dfrac{1}{\sqrt{x} - 3}, \quad x \geq 0, x \neq 9$

In Problems 79–86, use a calculator to approximate each radical. Round your answer to two decimal places.

79. $\sqrt{2}$

80. $\sqrt{7}$

81. $\sqrt[3]{4}$

82. $\sqrt[3]{-5}$

83. $\dfrac{2 + \sqrt{3}}{3 - \sqrt{5}}$

84. $\dfrac{\sqrt{5} - 2}{\sqrt{2} + 4}$

85. $\dfrac{3\sqrt[3]{5} - \sqrt{2}}{\sqrt{3}}$

86. $\dfrac{2\sqrt{3} - \sqrt[3]{4}}{\sqrt{2}}$

---

87. Calculating the Amount of Gasoline in a Tank An Exxon station stores its gasoline in underground tanks that are right circular cylinders lying on their sides. See the illustration. The volume V of gasoline in the tank (in gallons) is given by the formula

$$V = 40h^2 \sqrt{\frac{96}{h} - 0.608}$$

where h is the height of the gasoline (in inches) as measured on a depth stick.

(a) If $h = 12$ inches, how many gallons of gasoline are in the tank?

(b) If $h = 1$ inch, how many gallons of gasoline are in the tank?

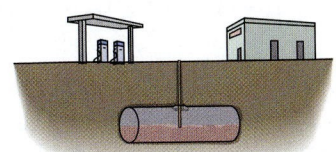

88. Inclined Planes The final velocity v of an object in feet per second (ft/sec) after it slides down a frictionless inclined plane of height h feet is

$$v = \sqrt{64h + v_0^2}$$

where v_0 is the initial velocity (in ft/sec) of the object.

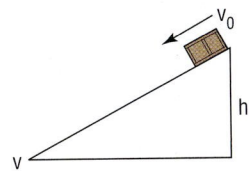

(a) What is the final velocity v of an object that slides down a frictionless inclined plane of height 4 feet? Assume that the initial velocity is 0.

(b) What is the final velocity v of an object that slides down a frictionless inclined plane of height 16 feet? Assume that the initial velocity is 0.

(c) What is the final velocity v of an object that slides down a frictionless inclined plane of height 2 feet with an initial velocity of 4 ft/sec?

In Problems 89–92, use the following information.

Period of a Pendulum The period T, in seconds, of a pendulum of length l, in feet, may be approximated using the formula

$$T = 2\pi \sqrt{\frac{l}{32}}$$

In the following problems, express your answer both as a square root and as a decimal.

89. Find the period T of a pendulum whose length is 64 feet.

90. Find the period T of a pendulum whose length is 16 feet.

91. Find the period T of a pendulum whose length is 8 inches.

92. Find the period T of a pendulum whose length is 4 inches.

93. Give an example to show that $\sqrt{a^2}$ is not equal to a. Use it to explain why $\sqrt{a^2} = |a|$.

R.9 RATIONAL EXPONENTS

OBJECTIVES

1 Evaluate Expressions with Fractional Exponents

2 Simplify Radicals Using Rational Exponents

1 Our purpose in this section is to give a definition for "a raised to the power $\dfrac{m}{n}$," where a is a real number and $\dfrac{m}{n}$ is a rational number. However, we want the definition to obey the laws of exponents stated for integer exponents in Section R.4. For example, if the law of exponents $(a^r)^s = a^{rs}$ is to hold, then it must be true that

$$(3^{1/2})^2 = 3^{\frac{1}{2} \cdot 2} = 3^1 = 3$$

That is, $a^{1/n}$ is a number that, when raised to the power n, is a. But this was the definition we gave of the principal nth root of a in Section R.8. Thus,

$$a^{1/2} = \sqrt{a} \qquad a^{1/3} = \sqrt[3]{a} \qquad a^{1/4} = \sqrt[4]{a}$$

and we can state the following definition:

If a is a real number and $n \geq 2$ is an integer, then

$$a^{1/n} = \sqrt[n]{a} \qquad\qquad (1)$$

provided that $\sqrt[n]{a}$ exists.

Note that if n is even and $a < 0$ then $\sqrt[n]{a}$ and $a^{1/n}$ do not exist.

EXAMPLE 1 **Writing Expressions Containing Fractional Exponents as Radicals**

(a) $4^{1/2} = \sqrt{4} = 2$
(b) $8^{1/2} = \sqrt{8} = 2\sqrt{2}$
(c) $(-27)^{1/3} = \sqrt[3]{-27} = -3$
(d) $16^{1/3} = \sqrt[3]{16} = 2\sqrt[3]{2}$ ■

We now seek a definition for $a^{m/n}$, where m and n are integers containing no common factors (except 1 and -1) and $n \geq 2$. Again, we want the definition to obey the Laws of Exponents stated earlier. For example,

$$a^{m/n} = a^{m(1/n)} = (a^m)^{1/n} \quad \text{and} \quad a^{m/n} = a^{(1/n)m} = (a^{1/n})^m$$

If a is a real number and m and n are integers containing no common factors, with $n \geq 2$, then

$$a^{m/n} = \sqrt[n]{a^m} = (\sqrt[n]{a})^m \qquad\qquad (2)$$

provided that $\sqrt[n]{a}$ exists.

We have two comments about equation (2):

1. The exponent $\dfrac{m}{n}$ must be in lowest terms and $n \geq 2$.
2. In simplifying the rational expression $a^{m/n}$, either $\sqrt[n]{a^m}$ or $(\sqrt[n]{a})^m$ may be used, the choice depending on which is easier to simplify. Generally, taking the root first, as in $(\sqrt[n]{a})^m$, is easier.

EXAMPLE 2 **Using Equation (2)**

(a) $4^{3/2} = (\sqrt{4})^3 = 2^3 = 8$
(b) $(-8)^{4/3} = (\sqrt[3]{-8})^4 = (-2)^4 = 16$
(c) $(32)^{-2/5} = (\sqrt[5]{32})^{-2} = 2^{-2} = \dfrac{1}{4}$
(d) $4^{6/4} = 4^{3/2} = (\sqrt{4})^3 = 2^3 = 8$ ■

✏ ─ **NOW WORK PROBLEM 7.**

Based on the definition of $a^{m/n}$, no meaning is given to $a^{m/n}$ if a is a negative real number and n is an even integer.

The definitions in equations (1) and (2) were stated so that the Laws of Exponents would remain true for rational exponents. For convenience, we list again the Laws of Exponents.

Laws of Exponents

If a and b are real numbers and r and s are rational numbers, then

$$a^r a^s = a^{r+s} \qquad (a^r)^s = a^{rs} \qquad (ab)^r = a^r \cdot b^r$$

$$a^{-r} = \frac{1}{a^r} \qquad \left(\frac{a}{b}\right)^r = \frac{a^r}{b^r} \qquad \frac{a^r}{a^s} = a^{r-s} = \frac{1}{a^{s-r}}$$

where it is assumed that all expressions used are defined.

2 Rational exponents can sometimes be used to simplify radicals.

EXAMPLE 3 **Simplifying Radicals Using Rational Exponents**

Simplify each expression.

(a) $(\sqrt[4]{7})^2$ (b) $\sqrt[9]{x^3}$ (c) $\sqrt[3]{4}\,\sqrt{2}$

Solution (a) $(\sqrt[4]{7})^2 = (7^{1/4})^2 = 7^{\frac{1}{4}\cdot 2} = 7^{\frac{1}{2}} = \sqrt{7}$

(b) $\sqrt[9]{x^3} = x^{3/9} = x^{1/3} = \sqrt[3]{x}$

(c) $\sqrt[3]{4}\,\sqrt{2} = 4^{1/3}\cdot 2^{1/2} = (2^2)^{1/3}\cdot 2^{1/2} = 2^{2/3}\cdot 2^{1/2} = 2^{7/6} = 2\cdot 2^{1/6} = 2\sqrt[6]{2}$

◾ **NOW WORK PROBLEM 17.**

The next example illustrates the use of the Laws of Exponents to simplify.

EXAMPLE 4 **Simplifying Expressions Containing Rational Exponents**

Simplify each expression. Express your answer so that only positive exponents occur. Assume that the variables are positive.

(a) $(x^{2/3}y)(x^{-2}y)^{1/2}$ (b) $\left(\dfrac{2x^{1/3}}{y^{2/3}}\right)^{-3}$ (c) $\left(\dfrac{9x^2 y^{1/3}}{x^{1/3}y}\right)^{1/2}$

Solution (a) $(x^{2/3}y)(x^{-2}y)^{1/2} = (x^{2/3}y)[(x^{-2})^{1/2}y^{1/2}]$

$= x^{2/3}yx^{-1}y^{1/2}$

$= (x^{2/3}\cdot x^{-1})(y\cdot y^{1/2})$

$= x^{-1/3}y^{3/2}$

$= \dfrac{y^{3/2}}{x^{1/3}}$

(b) $\left(\dfrac{2x^{1/3}}{y^{2/3}}\right)^{-3} \underset{\uparrow}{=} \left(\dfrac{y^{2/3}}{2x^{1/3}}\right)^{3} = \dfrac{(y^{2/3})^3}{(2x^{1/3})^3} = \dfrac{y^2}{2^3(x^{1/3})^3} = \dfrac{y^2}{8x}$

Equation (8), page 34

(c) $\left(\dfrac{9x^2 y^{1/3}}{x^{1/3}y}\right)^{1/2} = \left(\dfrac{9x^{2-(1/3)}}{y^{1-(1/3)}}\right)^{1/2} = \left(\dfrac{9x^{5/3}}{y^{2/3}}\right)^{1/2} = \dfrac{9^{1/2}(x^{5/3})^{1/2}}{(y^{2/3})^{1/2}} = \dfrac{3x^{5/6}}{y^{1/3}}$ ■

──── **N O W   W O R K   P R O B L E M 43.**

 The next two examples illustrate some algebra that you will need to know for certain calculus problems.

EXAMPLE 5   **Writing an Expression as a Single Quotient**

Write the following expression as a single quotient in which only positive exponents appear.

$$(x^2 + 1)^{1/2} + x \cdot \dfrac{1}{2}(x^2 + 1)^{-1/2} \cdot 2x$$

Solution $(x^2 + 1)^{1/2} + x \cdot \dfrac{1}{2}(x^2 + 1)^{-1/2} \cdot 2x = (x^2 + 1)^{1/2} + \dfrac{x^2}{(x^2 + 1)^{1/2}}$

$$= \dfrac{(x^2 + 1)^{1/2}(x^2 + 1)^{1/2} + x^2}{(x^2 + 1)^{1/2}}$$

$$= \dfrac{(x^2 + 1) + x^2}{(x^2 + 1)^{1/2}}$$

$$= \dfrac{2x^2 + 1}{(x^2 + 1)^{1/2}}$$ ■

──── **N O W   W O R K   P R O B L E M 51.**

EXAMPLE 6   **Factoring an Expression Containing Rational Exponents**

Factor: $4x^{1/3}(2x + 1) + 2x^{4/3}$

Solution   We begin by looking for factors that are common to the two terms. Notice that 2 and $x^{1/3}$ are common factors.

$$4x^{1/3}(2x + 1) + 2x^{4/3} = 2x^{1/3}[2(2x + 1) + x]$$

$$= 2x^{1/3}(5x + 2)$$ ■

R.9 Concepts and Vocabulary

In Problem 1, fill in the blanks.

1. If a is a real number and $n \geq 2$ is an integer, then $a^{1/n} = $ _____ , provided that _____ exists.

In Problems 2–5, answer True or False to each statement.

2. The Laws of Exponents hold for rational exponents.

3. $\sqrt[3]{16} = (\sqrt[3]{2})^4$.

4. $\sqrt[3]{16} = (\sqrt[3]{4})^2$.

5. $2^{4/3} = 4^{2/3}$.

R.9 Exercises

In Problems 1–34, simplify each expression.

1. $8^{2/3}$

2. $4^{3/2}$

3. $(-27)^{2/3}$

4. $(-64)^{2/3}$

5. $4^{-3/2}$

6. $(-8)^{-5/3}$

7. $9^{-3/2}$

8. $25^{-5/2}$

9. $\left(\dfrac{9}{4}\right)^{3/2}$

10. $\left(\dfrac{27}{8}\right)^{2/3}$

11. $\left(\dfrac{4}{9}\right)^{-3/2}$

12. $\left(\dfrac{8}{27}\right)^{-2/3}$

13. $4^{1.5}$

14. $16^{-1.5}$

15. $\left(\dfrac{1}{4}\right)^{-1.5}$

16. $\left(\dfrac{1}{9}\right)^{1.5}$

17. $(\sqrt{3})^6$

18. $(\sqrt[3]{4})^6$

19. $(\sqrt{5})^{-2}$

20. $(\sqrt[4]{3})^{-8}$

21. $3^{1/2} \cdot 3^{3/2}$

22. $5^{1/3} \cdot 5^{4/3}$

23. $\dfrac{7^{1/3}}{7^{4/3}}$

24. $\dfrac{6^{5/4}}{6^{1/4}}$

25. $2^{1/3} \cdot 4^{1/3}$

26. $9^{1/3} \cdot 3^{1/3}$

27. $\sqrt[3]{3} \cdot \sqrt[3]{27}$

28. $\sqrt[3]{2} \cdot \sqrt[3]{4}$

29. $(\sqrt[4]{2})^{-4}$

30. $(\sqrt[5]{3})^{-5}$

31. $(\sqrt[3]{6})^2$

32. $(\sqrt[4]{5})^3$

33. $\sqrt{2}\,\sqrt[3]{2}$

34. $\sqrt{5}\,\sqrt[3]{5}$

In Problems 35–48, simplify each expression. Express your answer so that only positive exponents occur. Assume that any variables are positive.

35. $\sqrt[8]{x^4}$

36. $\sqrt[6]{x^3}$

37. $\sqrt{x^3}\,\sqrt[4]{x}$

38. $\sqrt[3]{x^2}\,\sqrt{x}$

39. $x^{3/2}x^{-1/2}$

40. $x^{5/4}x^{-1/4}$

41. $(x^3 y^6)^{2/3}$

42. $(x^4 y^8)^{5/4}$

43. $(x^2 y)^{1/3}(x y^2)^{2/3}$

44. $(xy)^{1/4}(x^2 y^2)^{1/2}$

45. $(16x^2 y^{-1/3})^{3/4}$

46. $(4x^{-1} y^{1/3})^{3/2}$

47. $\left(\dfrac{x^{2/5} y^{-1/5}}{x^{-1/3}}\right)^{15}$

48. $\left(\dfrac{x^{1/2}}{y^2}\right)^4 \left(\dfrac{y^{1/3}}{x^{-2/3}}\right)^3$

In Problems 49–62, write each expression as a single quotient in which only positive exponents and/or radicals appear.

49. $\dfrac{x}{(1+x)^{1/2}} + 2(1+x)^{1/2}, \quad x > -1$

50. $\dfrac{1+x}{2x^{1/2}} + x^{1/2}, \quad x > 0$

51. $2x(x^2 + 1)^{1/2} + x^2 \cdot \dfrac{1}{2}(x^2 + 1)^{-1/2} \cdot 2x$

52. $(x+1)^{1/3} + x \cdot \dfrac{1}{3}(x+1)^{-2/3}, \quad x \neq -1$

53. $\sqrt{4x+3} \cdot \dfrac{1}{2\sqrt{x-5}} + \sqrt{x-5} \cdot \dfrac{1}{5\sqrt{4x+3}}, \quad x > 5$

54. $\dfrac{\sqrt[3]{8x+1}}{3\sqrt[3]{(x-2)^2}} + \dfrac{\sqrt[3]{x-2}}{24\sqrt[3]{(8x+1)^2}}, \quad x \neq 2, x \neq -\dfrac{1}{8}$

55. $\dfrac{\sqrt{1+x} - x \cdot \dfrac{1}{2\sqrt{1+x}}}{1+x}, \quad x > -1$

56. $\dfrac{\sqrt{x^2+1} - x \cdot \dfrac{2x}{2\sqrt{x^2+1}}}{x^2+1}$

57. $\dfrac{(x+4)^{1/2} - 2x(x+4)^{-1/2}}{x+4}, \quad x > -4$

58. $\dfrac{(9-x^2)^{1/2} + x^2(9-x^2)^{-1/2}}{9-x^2}, \quad -3 < x < 3$

59. $\dfrac{\dfrac{x^2}{(x^2-1)^{1/2}} - (x^2-1)^{1/2}}{x^2}, \quad x < -1 \quad \text{or} \quad x > 1$

60. $\dfrac{(x^2+4)^{1/2} - x^2(x^2+4)^{-1/2}}{x^2+4}$

61. $\dfrac{\dfrac{1+x^2}{2\sqrt{x}} - 2x\sqrt{x}}{(1+x^2)^2}, \quad x > 0$

62. $\dfrac{2x(1-x^2)^{1/3} + \dfrac{2}{3}x^3(1-x^2)^{-2/3}}{(1-x^2)^{2/3}}, \quad x \neq -1, x \neq 1$

In Problems 63–72, factor each expression. Your answer should contain only positive exponents.

63. $(x+1)^{3/2} + x \cdot \dfrac{3}{2}(x+1)^{1/2}, \quad x \geq -1$

64. $(x^2+4)^{4/3} + x \cdot \dfrac{4}{3}(x^2+4)^{1/3} \cdot 2x$

65. $6x^{1/2}(x^2+x) - 8x^{3/2} - 8x^{1/2}, \quad x \geq 0$

66. $6x^{1/2}(2x+3) + x^{3/2} \cdot 8, \quad x \geq 0$

67. $3(x^2+4)^{4/3} + x \cdot 4(x^2+4)^{1/3} \cdot 2x$

68. $2x(3x+4)^{4/3} + x^2 \cdot 4(3x+4)^{1/3}$

69. $4(3x+5)^{1/3}(2x+3)^{3/2} + 3(3x+5)^{4/3}(2x+3)^{1/2}, \quad x \geq -\dfrac{3}{2}$

70. $6(6x+1)^{1/3}(4x-3)^{3/2} + 6(6x+1)^{4/3}(4x-3)^{1/2}, \quad x \geq \dfrac{3}{4}$

71. $3x^{-1/2} + \dfrac{3}{2}x^{1/2}, \quad x > 0$

72. $8x^{1/3} - 4x^{-2/3}, \quad x \neq 0$

Chapter Review

▪ Things to Know

Classification of numbers (p. 3–5)

Counting numbers	$1, 2, 3, \ldots$
Whole numbers	$0, 1, 2, 3, \ldots$
Integers	$\ldots, -2, -1, 0, 1, 2, \ldots$
Rational numbers	Quotients of two integers (denominator not equal to 0); terminating or repeating decimals
Irrational numbers	Nonrepeating decimals
Real numbers	Rational or irrational numbers

Properties of real numbers (pp. 9–13)

Commutative properties	$a + b = b + a, \quad a \cdot b = b \cdot a$
Associative properties	$a + (b + c) = (a + b) + c,$ $a \cdot (b \cdot c) = (a \cdot b) \cdot c$
Distributive property	$a \cdot (b + c) = a \cdot b + a \cdot c$
Identity properties	$a + 0 = a, \quad a \cdot 1 = a$
Inverse properties	$a + (-a) = 0, \quad a \cdot \dfrac{1}{a} = 1, \quad \text{where } a \neq 0$
Zero-Product property	If $ab = 0$, then $a = 0$ or $b = 0$ or both.

Absolute value (p. 19)

$$|a| = a \text{ if } a \geq 0, \quad |a| = -a \text{ if } a < 0$$

Pythagorean Theorem (p. 25)

In a right triangle, the square of the length of the hypotenuse is equal to the sum of the squares of the lengths of the legs.

Converse of the Pythagorean Theorem (p. 25)

In a triangle, if the square of the length of one side equals the sum of the squares of the lengths of the other two sides, then the triangle is a right triangle.

Exponents (pp. 30–32)

$$a^n = \underbrace{a \cdot a \cdot \ldots \cdot a}_{n\,factors}, n \text{ a positive integer} \qquad a^0 = 1, a \neq 0 \qquad a^{-n} = \frac{1}{a^n}, a \neq 0, n \text{ a positive integer}$$

Laws of exponents (pp. 32–34)

$$a^m \cdot a^n = a^{m+n} \qquad (a^m)^n = a^{mn} \qquad (a \cdot b)^n = a^n \cdot b^n \qquad \frac{a^m}{a^n} = a^{m-n} = \frac{1}{a^{n-m}} \qquad \left(\frac{a}{b}\right)^n = \frac{a^n}{b^n}$$

Polynomial (p. 40)

Algebraic expression of the form $a_n x^n + a_{n-1} x^{n-1} + \cdots + a_1 x + a_0$, n a nonnegative integer

Special products/factoring formulas (pp. 43–45)

$$(x - a)(x + a) = x^2 - a^2$$
$$(x + a)^2 = x^2 + 2ax + a^2, \qquad (x - a)^2 = x^2 - 2ax + a^2$$
$$(x + a)(x + b) = x^2 + (a + b)x + ab$$
$$(ax + b)(cx + d) = acx^2 + (ad + bc)x + bd$$
$$(x + a)^3 = x^3 + 3ax^2 + 3a^2x + a^3, \qquad (x - a)^3 = x^3 - 3ax^2 + 3a^2x - a^3$$
$$(x - a)(x^2 + ax + a^2) = x^3 - a^3, \qquad (x + a)(x^2 - ax + a^2) = x^3 + a^3$$

Radicals; Rational exponents (pp. 75 and 80)

$\sqrt[n]{a} = b$ means $a = b^n$, where $a \geq 0$ and $b \geq 0$ if $n \geq 2$ is even, and a, b are any real numbers if $n \geq 3$ is odd
$$a^{m/n} = \sqrt[n]{a^m} = \left(\sqrt[n]{a}\right)^m$$

Objectives

Section		You should be able to . . .	Review Exercises
R.1	1	Classify numbers (p. 3)	1, 2
	2	Evaluate numerical expressions (p. 7)	3–5
	3	Work with properties of real numbers (p. 9)	7, 8
R.2	1	Graph inequalities (p. 19)	9, 10
	2	Find distance on the real number line (p. 19)	11, 12
	3	Evaluate algebraic expressions (p. 20)	13, 14
	4	Determine the domain of a variable (p. 21)	15, 16
R.3	1	Use the Pythagorean Theorem and its converse (p. 25)	80, 81, 87
	2	Know geometry formulas (p. 27)	84, 85, 86
R.4	1	Evaluate expressions containing exponents (p. 31)	6
	2	Work with the Laws of Exponents (p. 32)	20, 21
	3	Use a calculator to evaluate exponents (p. 35)	22
	4	Use scientific notation (p. 35)	79
R.5	1	Recognize monomials (p. 39)	23
	2	Recognize polynomials (p. 40)	24
	3	Add and subtract polynomials (p. 41)	25, 26
	4	Multiply polynomials (p. 42)	27–32
	5	Know formulas for special products (p. 43)	29, 30
	6	Divide polynomials (p. 45)	33–38
R.6	1	Factor the difference of two squares and the sum and difference of two cubes (p. 51)	45, 46, 49, 50
	2	Factor perfect squares (p. 53)	53, 54
	3	Factor a second-degree polynomial $x^2 + Bx + C$ (p. 53)	39, 40
	4	Factor by grouping (p. 55)	47, 48
	5	Factor a second-degree polynomial $Ax^2 + Bx + C$ (p. 56)	41–44

R.7	①	Reduce a rational expression to lowest terms (p. 60)	55, 56
	②	Multiply and divide rational expressions (p. 61)	57, 58, 63
	③	Add and subtract rational expressions (p. 62)	59–62
	④	Use the least common multiple method (p. 64)	61, 62
	⑤	Simplify mixed quotients (p. 66)	64
R.8	①	Work with properties of square roots (p. 72)	65, 66
	②	Rationalize a denominator (p. 74)	69–74
	③	Simplify nth roots (p. 75)	67, 68
	④	Simplify radicals (p. 76)	19, 65–68
R.9	①	Evaluate expressions with rational exponents (p. 79)	17
	②	Simplify radicals using rational exponents (p. 81)	18

Review Exercises

Blue problem numbers indicate the authors' suggestions for use in a practice test.

In Problems 1 and 2, list the numbers in the set that are (a) Natural numbers, (b) Integers, (c) Rational numbers, (d) Irrational numbers, (e) Real numbers

1. $A = \left\{ -10, 0.65, 1.343434\ldots, \sqrt{7}, \dfrac{1}{9} \right\}$

2. $B = \left\{ 0, -5, \dfrac{1}{3}, 0.59, 1.333\ldots, 2\sqrt{2}, \dfrac{\pi}{2} \right\}$

In Problems 3–6, evaluate each expression

3. $-6 + 4 \cdot (8 - 3)$

4. $\dfrac{5}{18} + \dfrac{1}{12}$

5. $\dfrac{\dfrac{5}{18}}{\dfrac{11}{27}}$

6. $(-3)^5$

In Problems 7 and 8, use the Distributive Property to remove the parentheses.

7. $4(x - 3)$

8. $(x - 2)(3x + 1)$

In Problems 9 and 10, graph the numbers x on the real number line.

9. $x > 3$

10. $x \le 5$

In Problems 11 and 12, let P, Q, and R be points on a real number line with coordinates $-2, 3,$ and 9, respectively.

11. Find the distance between P and Q.

12. Find the distance between Q and R.

In Problems 13 and 14, evaluate each expression if $x = -5$ and $y = 7$.

13. $\dfrac{4x}{x + y}$

14. $|2x - 3y|$

In Problems 15 and 16, determine the domain of the variable.

15. $\dfrac{3}{x - 6}$

16. $\dfrac{x + 1}{x + 5}$

In Problems 17–21, simplify each expression. All exponents must be positive. Assume that $x > 0$ and $y > 0$, when they appear.

17. $16^{1/2}$

18. $(\sqrt[4]{x})^2$

19. $2\sqrt{4x^3} + \sqrt{16x}$

20. $\dfrac{(x^2 y)^{-4}}{(xy)^{-3}}$

21. $(25x^{-4/3} y^{-2/3})^{3/2}$

22. Use a calculator to evaluate $(1.5)^4$.

23. Identify the coefficient and degree of the monomial $-5x^3$.

24. Identify the coefficients and degree of the polynomial $3x^5 + 4x^4 - 2x^3 + 5x - 12$.

In Problems 25–32, perform the indicated operation.

25. $(2x^4 - 8x^3 + 5x - 1) + (6x^3 + x^2 + 4)$

26. $(x^3 + 8x^2 - 3x + 4) - (4x^3 - 7x^2 - 2x + 3)$

27. $(2x + 1)(3x - 5)$

28. $(2x - 5)(3x^2 + 2)$

29. $(4x + 1)(4x - 1)$

30. $(5x + 2)^2$

31. $(x + 1)(x + 2)(x - 3)$

32. $(x + 1)(x + 3)(x - 5)$

In Problems 33–38, find the quotient and remainder. Check your work by verifying that

$$(Quotient)(Divisor) + Remainder = Dividend$$

33. $3x^3 - x^2 + x + 4$ divided by $x - 3$

34. $2x^3 - 3x^2 + x + 1$ divided by $x - 2$

35. $-3x^4 + x^2 + 2$ divided by $x^2 + 1$

36. $-4x^3 + x^2 - 2$ divided by $x^2 - 1$

37. $x^5 + 1$ divided by $x + 1$

38. $x^5 - 1$ divided by $x - 1$

In Problems 39–54, factor each polynomial completely (over the integers). If the polynomial cannot be factored, say it is prime.

39. $x^2 + 5x - 14$

40. $x^2 - 9x + 14$

41. $6x^2 - 5x - 6$

42. $6x^2 + x - 2$

43. $3x^2 - 15x - 42$

44. $2x^3 + 18x^2 + 28x$

45. $8x^3 + 1$

46. $27x^3 - 8$

47. $2x^3 + 3x^2 - 2x - 3$

48. $2x^3 + 3x^2 + 2x + 3$

49. $25x^2 - 4$

50. $16x^2 - 1$

51. $9x^2 + 1$

52. $x^2 - x + 1$

53. $x^2 + 8x + 16$

54. $4x^2 + 12x + 9$

In Problems 55–64, perform the indicated operation and simplify. Leave your answer in factored form.

55. $\dfrac{2x^2 + 11x + 14}{x^2 - 4}$

56. $\dfrac{x^2 - 5x - 14}{4 - x^2}$

57. $\dfrac{9x^2 - 1}{x^2 - 9} \cdot \dfrac{3x - 9}{9x^2 + 6x + 1}$

58. $\dfrac{x^2 - 25}{x^3 - 4x^2 - 5x} \cdot \dfrac{x^2 + x}{1 - x^2}$

59. $\dfrac{x + 1}{x - 1} - \dfrac{x - 1}{x + 1}$

60. $\dfrac{x}{x + 1} - \dfrac{2x}{x + 2}$

61. $\dfrac{3x + 4}{x^2 - 4} - \dfrac{2x - 3}{x^2 + 4x + 4}$

62. $\dfrac{x^2}{2x^2 + 5x - 3} + \dfrac{x^2}{2x^2 - 5x + 2}$

63. $\dfrac{\dfrac{x^2 - 1}{x^2 - 5x + 6}}{\dfrac{x + 1}{x - 2}}$

64. $\dfrac{\dfrac{x + 4}{3}}{\dfrac{1}{4} + \dfrac{2}{x}}$

In Problems 65–68, simplify each expression.

65. $\sqrt{\dfrac{9x^2}{25y^4}}, \quad x \geq 0, y > 0$

66. $\dfrac{\sqrt{2x}\,\sqrt{5xy^3}}{\sqrt{10y}}, \quad x \geq 0, y > 0$

67. $\sqrt[3]{27x^4y^{12}}$

68. $\dfrac{\sqrt[4]{243x^5y}}{\sqrt[4]{3xy^9}}, \quad x > 0, y > 0$

In Problems 69–74, rationalize the denominator of each expression.

69. $\dfrac{4}{\sqrt{5}}$

70. $\dfrac{-2}{\sqrt{3}}$

71. $\dfrac{2}{1 - \sqrt{2}}$

72. $\dfrac{-4}{1 + \sqrt{3}}$

73. $\dfrac{1 + \sqrt{5}}{1 - \sqrt{5}}$

74. $\dfrac{4\sqrt{3} + 2}{2\sqrt{3} + 1}$

In Problems 75–78, write each expression as a single quotient in which only positive exponents and/or radicals appear.

75. $(2 + x^2)^{1/2} + x \cdot \dfrac{1}{2}(2 + x^2)^{-1/2} \cdot 2x$

76. $(x^2 + 4)^{2/3} + x \cdot \dfrac{2}{3}(x^2 + 4)^{-1/3} \cdot 2x$

77. $\dfrac{(x + 4)^{1/2} \cdot 2x - x^2 \cdot \dfrac{1}{2}(x + 4)^{-1/2}}{x + 4}, \quad x > -4$

78. $\dfrac{(x^2 + 4)^{1/2} \cdot 2x - x^2 \cdot \dfrac{1}{2}(x^2 + 4)^{-1/2} \cdot 2x}{x^2 + 4}$

79. U.S. Population   According to the U.S. Census Bureau, the U.S. population in 2000 was 281,421,906. Write the U.S. population in scientific notation.

80. Find the hypotenuse of a right triangle whose legs are of lengths 5 and 8.

81. The lengths of the sides of a triangle are 12, 16, and 20. Is this a right triangle?

82. Manufacturing Cost   The weekly production cost C of manufacturing x hand calculators is given by the formula

$$C = 3000 + 6x - \frac{x^2}{1000}$$

(a) What is the cost of producing 1000 hand calculators?

(b) What is the cost of producing 3000 hand calculators?

83. Quarterly Corporate Earnings   In the first quarter of its fiscal year, a company posted earnings of $1.20 per share. During the second and third quarters, it posted losses of $0.75 per share and $0.30 per share, respectively. In the fourth quarter, it earned a modest $0.20 per share. What were the annual earnings per share of this company?

84. Design   A window consists of a rectangle surmounted by a triangle. Find the area of the window shown in the illustration. How much wood frame is needed to enclose the window?

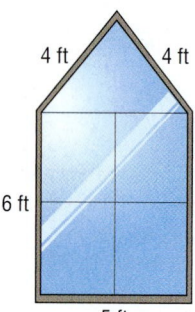

85. Construction   A rectangular swimming pool, 20 feet long and 10 feet wide, is enclosed by a wooden deck that is 3 feet wide. What is the area of the deck? How much fence is required to enclose the deck?

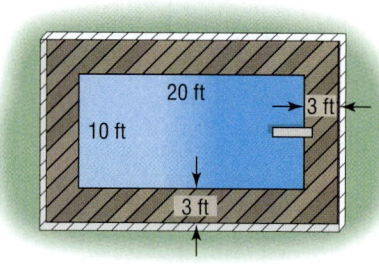

86. Construction   A statue with a circular base of radius 3 feet is enclosed by a circular pond as shown in the illustration. What is the surface area of the pond? How much fence is required to enclose the pond?

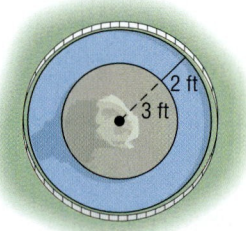

87. How far can a pilot see?   On a recent flight to San Francisco, the pilot announced we were 139 miles from the city, flying at an altitude of 35,000 feet. The pilot claimed he could see the Golden Gate Bridge and beyond. Was he telling the truth? How far could he see?

88. Use the material in this chapter to create a problem that uses each of the following words:
(a) Simplify   (b) Factor   (c) Reduce

89. A rational number is defined as the quotient of two integers. When written as a decimal, the decimal will either repeat or terminate. By looking at the denominator of the rational number, there is a way to tell in advance whether its decimal representation will repeat or terminate. Make a list of rational numbers and their decimals. See if you can discover the pattern. Confirm your conclusion by consulting books on number theory at the library. Write a brief essay on your findings.

90. The current time is 12 noon CST. What time (CST) will it be 12,997 hours from now?

91. Both $\frac{a}{0}$ $(a \neq 0)$ and $\frac{0}{0}$ are undefined, but for different reasons. Write a paragraph or two explaining the different reasons.

GRAPHS

Why Refinance?

There are lots of reasons you might want to refinance, but most people fit into one (or more) of the basic four catagories. Most people want to reduce their monthly payments; some want to consolidate outstanding debt, such as combining a first and second mortgage into a new first mortgage; some want to tap built-up equity in their homes, and some just want to get out of a mortgage product that they don't like, or that's costing too much—going from an ARM to a fixed rate mortgage, for example. Whatever group or groups you fit with, there are certain rules that you must follow.

SEE CHAPTER PROJECT 1.

OUTLINE

 For additional study help, go to

www.prenhall.com/sullivanegu3e

Materials include:

- Graphing Calculator Help
- Chapter Quiz
- Chapter Test
- PowerPoint Downloads
- Chapter Projects
- Student Tips

A Look Back, A Look Forward

In Chapter R, we reviewed skills from Intermediate Algebra. In this chapter we make the connection between algebra and geometry through the rectangular coordinate system. The idea of using a system of rectangular coordinates dates back to ancient times, when such a system was used for surveying and city planning. Apollonius of Perga, in 200 B.C., used a form of rectangular coordinates in his work on conics, although this use does not stand out as clearly as it does in modern treatments. Sporadic use of rectangular coordinates continued until the 1600s. By that time, algebra had developed sufficiently so that René Descartes (1596–1650) and Pierre de Fermat (1601–1665) could take the crucial step, which was the use of rectangular coordinates to translate geometry problems into algebra problems, and vice versa. This step was important for two reasons. First, it allowed both geometers and algebraists to gain critical new insights into their subjects, which previously had been regarded as separate but now were seen to be connected in many important ways. Second, the insights gained made possible the development of calculus, which greatly enlarged the number of areas in which mathematics could be applied and made possible a much deeper understanding of these areas. With the advent of technology, in particular graphing utilities, we are now able not only to visualize the dual roles of algebra and geometry, but we are also able to solve many problems that required advanced methods before this technology.

PREPARING FOR THIS SECTION

Before getting started, review the following:

✓ Real Numbers (Section R.1, pp. 3–5)

✓ Geometry Review (Section R.3, pp. 25–28)

✓ Square Roots (Section R.8, pp. 71–73)

✓ Algebra Review (Section R.2, pp. 17–18)

✓ Integer Exponents (Section R.4, pp. 30–32)

1.1 RECTANGULAR COORDINATES; GRAPHING UTILITIES

OBJECTIVES
1. Use the Distance Formula
2. Use the Midpoint Formula

We locate a point on the real number line by assigning it a single real number, called the *coordinate of the point*. For work in a two-dimensional plane, we locate points by using two numbers.

We begin with two real number lines located in the same plane: one horizontal and the other vertical. We call the horizontal line the **x-axis,** the vertical line the **y-axis,** and the point of intersection the **origin O.** We assign coordinates to every point on these number lines as shown in Figure 1, using a convenient scale. In mathematics, we usually use the same scale on each axis; in applications, a different scale is often used on each axis.

The origin O has a value of 0 on both the x-axis and the y-axis. We follow the usual convention that points on the x-axis to the right of O are associated with positive real numbers, and those to the left of O are associated with negative real numbers. Points on the y-axis above O are associated with positive real numbers, and those below O are associated with negative real numbers. In Figure 1, the x-axis and y-axis are labeled as x and y, respectively, and we have used an arrow at the end of each axis to denote the positive direction.

Figure 1

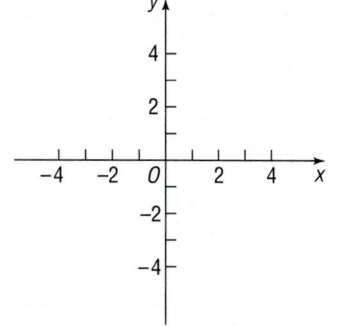

Figure 2

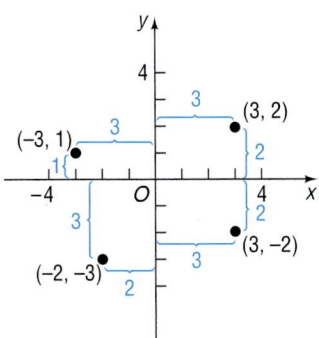

The coordinate system described here is called a **rectangular** or **Cartesian***
coordinate system. The plane formed by the *x*-axis and *y*-axis is sometimes called
the *xy*-**plane,** and the *x*-axis and *y*-axis are referred to as the **coordinate axes.**

Any point *P* in the *xy*-plane can then be located by using an **ordered pair**
(x, y) of real numbers. Let *x* denote the signed distance of *P* from the *y*-axis
(*signed* in the sense that, if *P* is to the right of the *y*-axis, then $x > 0$, and if *P*
is to the left of the *y*-axis, then $x < 0$); and let *y* denote the signed distance
of *P* from the *x*-axis. The ordered pair (x, y), also called the **coordinates** of
P, then gives us enough information to locate the point *P* in the plane.

For example, to locate the point whose coordinates are $(-3, 1)$, go 3
units along the *x*-axis to the left of *O* and then go straight up 1 unit. We **plot**
this point by placing a dot at this location. See Figure 2, in which the points
with coordinates $(-3, 1)$, $(-2, -3)$, $(3, -2)$, and $(3, 2)$ are plotted.

The origin has coordinates $(0, 0)$. Any point on the *x*-axis has coordinates
of the form $(x, 0)$, and any point on the *y*-axis has coordinates of the form $(0, y)$.

If (x, y) are the coordinates of a point *P,* then *x* is called the *x*-**coordinate,**
or **abscissa,** of *P* and *y* is the *y*-**coordinate,** or **ordinate,** of *P.* We identify the
point *P* by its coordinates (x, y) by writing $P = (x, y)$. Usually, we will simply
say "the point (x, y)" rather than "the point whose coordinates are (x, y)."

The coordinate axes divide the *xy*-plane into four sections called
quadrants, as shown in Figure 3. In quadrant I, both the *x*-coordinate and
the *y*-coordinate of all points are positive; in quadrant II, *x* is negative and *y*
is positive; in quadrant III, both *x* and *y* are negative; and in quadrant IV, *x* is
positive and *y* is negative. Points on the coordinate axes belong to no quadrant.

Figure 3

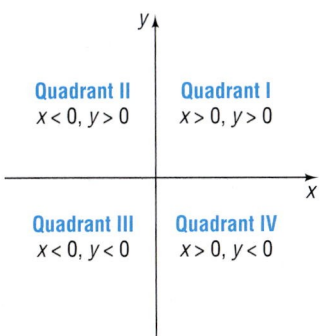

 **NOW WORK PROBLEM 1.**

Graphing Utilities

All graphing utilities, that is, all graphing calculators and all computer software graphing packages, graph equations by plotting points on a screen. The
screen itself actually consists of small rectangles, called **pixels.** The more pixels the screen has, the better the resolution. Most graphing calculators have
48 pixels per square inch; most computer screens have 32 to 108 pixels per
square inch. When a point to be plotted lies inside a pixel, the pixel is turned
on (lights up). The graph of an equation is a collection of lighted pixels. Figure 4 shows how the graph of $y = 2x$ looks on a TI-83 graphing calculator.

The screen of a graphing utility will display the coordinate axes of a rectangular coordinate system. However, you must set the scale on each axis. You
must also include the smallest and largest values of *x* and *y* that you want included in the graph. This is called **setting the viewing rectangle** or **viewing window.** Figure 5 illustrates a typical viewing window.

To select the viewing window, we must give values to the following
expressions:

Figure 4
$y = 2x$

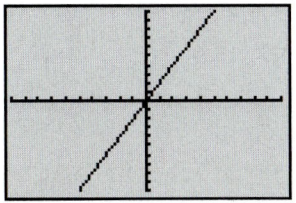

Figure 5

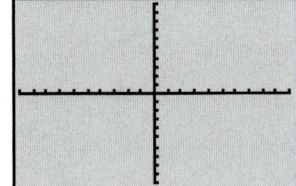

*X*min:	the smallest value of *x*
*X*max:	the largest value of *x*
*X*scl:	the number of units per tick mark on the *x*-axis
*Y*min:	the smallest value of *y*
*Y*max:	the largest value of *y*
*Y*scl:	the number of units per tick mark on the *y*-axis

*Named after René Descartes (1596–1650), a French mathematician, philosopher, and theologian.

Figure 6 illustrates these settings and their relation to the Cartesian coordinate system.

Figure 6

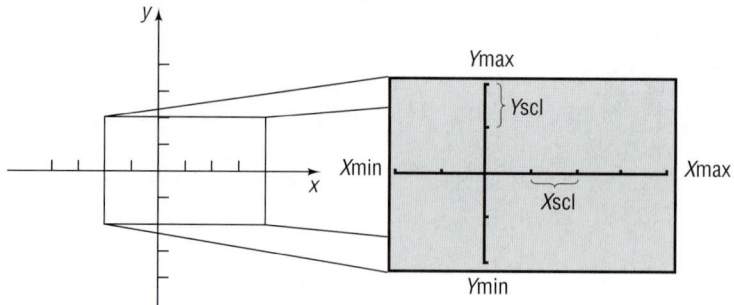

If the scale used on each axis is known, we can determine the minimum and maximum values of x and y shown on the screen by counting the tick marks. Look again at Figure 5. For a scale of 1 on each axis, the minimum and maximum values of x are -10 and 10, respectively; the minimum and maximum values of y are also -10 and 10. If the scale is 2 on each axis, then the minimum and maximum values of x are -20 and 20, respectively; and the minimum and maximum values of y are -20 and 20, respectively.

Conversely, if we know the minimum and maximum values of x and y, we can determine the scales being used by counting the tick marks displayed. We shall follow the practice of showing the minimum and maximum values of x and y in our illustrations so that you will know how the window was set. See Figure 7.

Figure 7

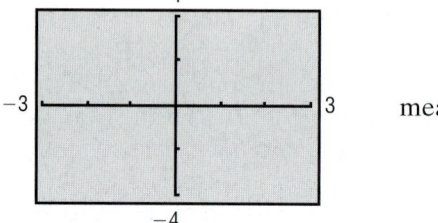

means

$$X\text{min} = -3 \qquad Y\text{min} = -4$$
$$X\text{max} = 3 \qquad Y\text{max} = 4$$
$$X\text{scl} = 1 \qquad Y\text{scl} = 2$$

EXAMPLE 1   **Finding the Coordinates of a Point Shown on a Graphing Utility Screen**

Find the coordinates of the point shown in Figure 8. Assume the coordinates are integers.

Figure 8

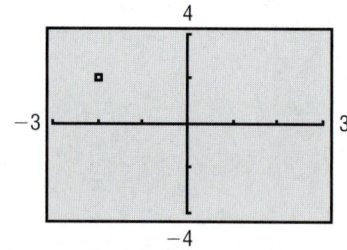

Solution First we note that the viewing window used in Figure 8 is

$$
\begin{array}{ll}
X\text{min} = -3 & Y\text{min} = -4 \\
X\text{max} = 3 & Y\text{max} = 4 \\
X\text{scl} = 1 & Y\text{scl} = 2
\end{array}
$$

The point shown is 2 tick units to the left on the horizontal axis (scale = 1) and 1 tick up on the vertical scale (scale = 2). The coordinates of the point shown are $(-2, 2)$. ■

 NOW WORK PROBLEMS 5 AND 15.

Distance between Points

 If the same units of measurement, such as inches, centimeters, and so on, are used for both the x-axis and the y-axis, then all distances in the xy-plane can be measured using this unit of measurement.

EXAMPLE 2   **Finding the Distance between Two Points**

Find the distance d between the points $(1, 3)$ and $(5, 6)$.

Solution First we plot the points $(1, 3)$ and $(5, 6)$ and connect them with a straight line. See Figure 9(a). We are looking for length d. We begin by drawing a horizontal line from $(1, 3)$ to $(5, 3)$ and a vertical line from $(5, 3)$ to $(5, 6)$, forming a right triangle, as in Figure 9(b). One leg of the triangle is of length 4 (since $|5 - 1| = 4$) and the other is of length 3 (since $|6 - 3| = 3$). By the Pythagorean Theorem, the square of the distance d that we seek is

$$d^2 = 4^2 + 3^2 = 16 + 9 = 25$$
$$d = \sqrt{25} = 5$$

Figure 9

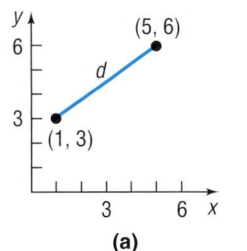   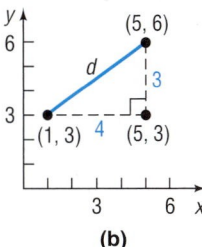

(a)   (b)   ■

The **distance formula** provides a straightforward method for computing the distance between two points.

Figure 10

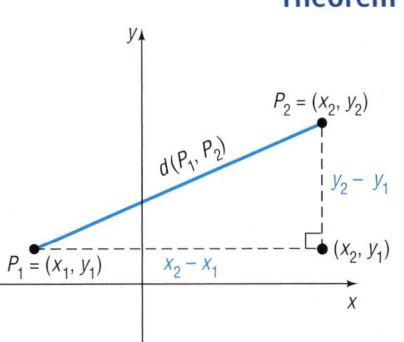

Theorem   **Distance Formula**

The distance between two points $P_1 = (x_1, y_1)$ and $P_2 = (x_2, y_2)$, denoted by $d(P_1, P_2)$, is

$$d(P_1, P_2) = \sqrt{(x_2 - x_1)^2 + (y_2 - y_1)^2} \qquad (1)$$

■

That is, to compute the distance between two points, find the difference of the x-coordinates, square it, and add this to the square of the difference of the y-coordinates. The square root of this sum is the distance. See Figure 10.

Proof of the Distance Formula   Let (x_1, y_1) denote the coordinates of point P_1, and let (x_2, y_2) denote the coordinates of point P_2. Assume that the line joining P_1 and P_2 is neither horizontal nor vertical. Refer to Figure 11(a). The coordinates of P_3 are (x_2, y_1). The horizontal distance from P_1 to P_3 is the absolute value of the difference of the x-coordinates, $|x_2 - x_1|$. The vertical distance from P_3 to P_2 is the absolute value of the difference of the y-coordinates, $|y_2 - y_1|$. See Figure 11(b). The distance $d(P_1, P_2)$ that we seek is the length of the hypotenuse of the right triangle, so, by the Pythagorean Theorem, it follows that

$$[d(P_1, P_2)]^2 = |x_2 - x_1|^2 + |y_2 - y_1|^2$$
$$= (x_2 - x_1)^2 + (y_2 - y_1)^2$$
$$d(P_1, P_2) = \sqrt{(x_2 - x_1)^2 + (y_2 - y_1)^2}$$

Figure 11

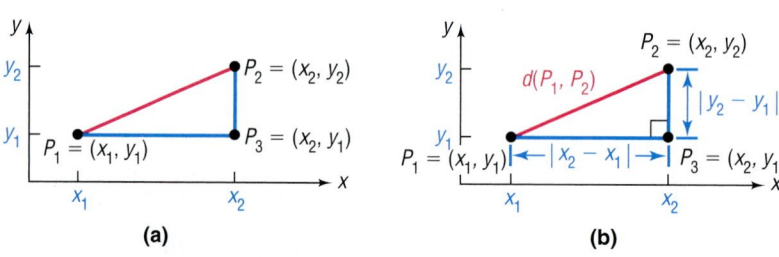

(a)   (b)

Now, if the line joining P_1 and P_2 is horizontal, then the y-coordinate of P_1 equals the y-coordinate of P_2; that is, $y_1 = y_2$. Refer to Figure 12(a). In this case, the distance formula (1) still works, because, for $y_1 = y_2$, it reduces to

$$d(P_1, P_2) = \sqrt{(x_2 - x_1)^2 + 0^2} = \sqrt{(x_2 - x_1)^2} = |x_2 - x_1|$$

Figure 12

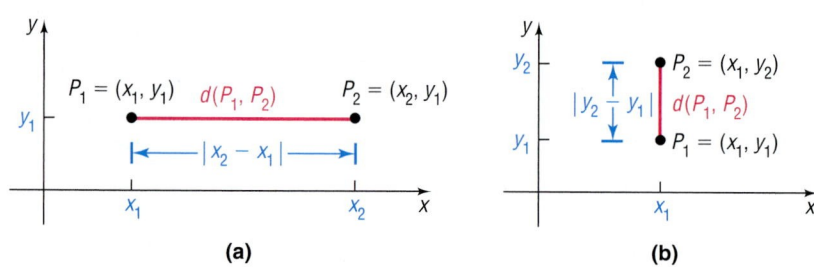

(a)   (b)

A similar argument holds if the line joining P_1 and P_2 is vertical. See Figure 12(b). The distance formula is valid in all cases.   ■

EXAMPLE 3   **Finding the Length of a Line Segment**

Figure 13

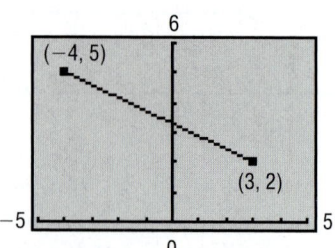

Find the length of the line segment shown in Figure 13.

Solution   The length of the line segment is the distance between the points $(-4, 5)$ and $(3, 2)$. Using the distance formula (1), the length d is

$$d = \sqrt{[3 - (-4)]^2 + (2 - 5)^2} = \sqrt{7^2 + (-3)^2}$$
$$= \sqrt{49 + 9} = \sqrt{58} \approx 7.62$$ ■

NOW WORK PROBLEM **21**.

The distance between two points $P_1 = (x_1, y_1)$ and $P_2 = (x_2, y_2)$ is never a negative number. Furthermore, the distance between two points is 0 only when the points are identical, that is, when $x_1 = x_2$ and $y_1 = y_2$. Also, because $(x_2 - x_1)^2 = (x_1 - x_2)^2$ and $(y_2 - y_1)^2 = (y_1 - y_2)^2$, it makes no difference whether the distance is computed from P_1 to P_2 or from P_2 to P_1; that is, $d(P_1, P_2) = d(P_2, P_1)$.

The introduction to this chapter mentioned that rectangular coordinates enable us to translate geometry problems into algebra problems, and vice versa. The next example shows how algebra (the distance formula) can be used to solve geometry problems.

EXAMPLE 4   **Using Algebra to Solve Geometry Problems**

Consider the three points $A = (-2, 1)$, $B = (2, 3)$ and $C = (3, 1)$.

(a) Plot each point and form the triangle ABC.
(b) Find the length of each side of the triangle.
(c) Verify that the triangle is a right triangle.
(d) Find the area of the triangle.

Solution   (a) Points A, B, C, and triangle ABC are plotted in Figure 14.

Figure 14

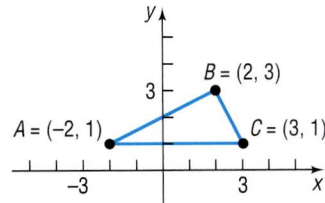

(b) $d(A, B) = \sqrt{[2 - (-2)]^2 + (3 - 1)^2} = \sqrt{16 + 4} = \sqrt{20} = 2\sqrt{5}$

$d(B, C) = \sqrt{(3 - 2)^2 + (1 - 3)^2} = \sqrt{1 + 4} = \sqrt{5}$

$d(A, C) = \sqrt{[3 - (-2)]^2 + (1 - 1)^2} = \sqrt{25 + 0} = 5$

(c) To show that the triangle is a right triangle, we need to show that the sum of the squares of the lengths of two of the sides equals the square of the length of the third side. (Why is this sufficient?) Looking at Figure 14 it seems reasonable to conjecture that the right angle is at vertex B. We shall check to see whether

$$[d(A, B)]^2 + [d(B, C)]^2 = [d(A, C)]^2$$

We find that

$$[d(A, B)]^2 + [d(B, C)]^2 = (2\sqrt{5})^2 + (\sqrt{5})^2$$
$$= 20 + 5 = 25 = [d(A, C)]^2$$

so it follows from the converse of the Pythagorean Theorem that triangle ABC is a right triangle.

(d) Because the right angle is at vertex B, the sides AB and BC form the base and altitude of the triangle. Its area is

$$\text{Area} = \frac{1}{2}(\text{Base})(\text{Altitude}) = \frac{1}{2}(2\sqrt{5})(\sqrt{5}) = 5 \text{ square units} \quad \blacksquare$$

✏️ **NOW WORK PROBLEM 39.**

Midpoint Formula

2   We now derive a formula for the coordinates of the **midpoint of a line segment.** Let $P_1 = (x_1, y_1)$ and $P_2 = (x_2, y_2)$ be the endpoints of a line segment, and let $M = (x, y)$ be the point on the line segment that is the same distance

Figure 15

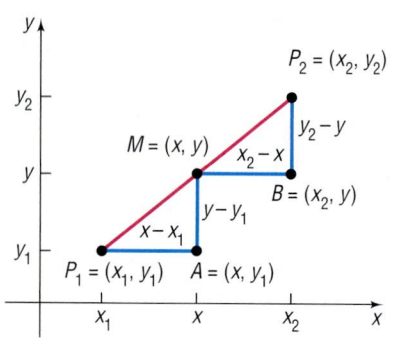

from P_1 as it is from P_2. See Figure 15. The triangles P_1AM and MBP_2 are congruent.* [Do you see why? Angle AP_1M = angle BMP_2,† angle P_1MA = angle MP_2B, and $d(P_1, M) = d(M, P_2)$ is given. So, we have angle–side–angle.] Hence, corresponding sides are equal in length. That is,

$$x - x_1 = x_2 - x \qquad \text{and} \qquad y - y_1 = y_2 - y$$

$$2x = x_1 + x_2 \qquad\qquad\qquad 2y = y_1 + y_2$$

$$x = \frac{x_1 + x_2}{2} \qquad\qquad\qquad y = \frac{y_1 + y_2}{2}$$

Theorem

Midpoint Formula

The midpoint $M = (x, y)$ of the line segment from $P_1 = (x_1, y_1)$ to $P_2 = (x_2, y_2)$ is

$$M = (x, y) = \left(\frac{x_1 + x_2}{2}, \frac{y_1 + y_2}{2} \right) \qquad (2)$$

To find the midpoint of a line segment, we average the x-coordinates and the y-coordinates of the endpoints.

EXAMPLE 5   **Finding the Midpoint of a Line Segment**

Find the midpoint of a line segment from $P_1 = (-5, 5)$ to $P_2 = (3, 1)$. Plot the points P_1 and P_2 and their midpoint. Check your answer.

Solution   We apply the midpoint formula (2) using $x_1 = -5$, $y_1 = 5$, $x_2 = 3$, and $y_2 = 1$. Then the coordinates (x, y) of the midpoint M are

$$x = \frac{x_1 + x_2}{2} = \frac{-5 + 3}{2} = -1 \quad \text{and} \quad y = \frac{y_1 + y_2}{2} = \frac{5 + 1}{2} = 3$$

Figure 16

That is, $M = (-1, 3)$. See Figure 16.

✔ CHECK: Because M is the midpoint, we check the answer by verifying that $d(P_1, M) = d(M, P_2)$:

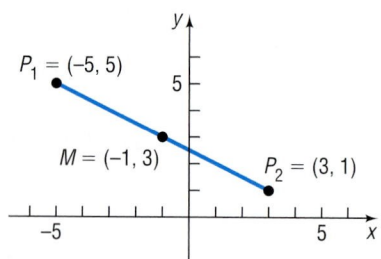

$$d(P_1, M) = \sqrt{[-1 - (-5)]^2 + (3 - 5)^2} = \sqrt{16 + 4} = \sqrt{20}$$

$$d(M, P_2) = \sqrt{[3 - (-1)]^2 + (1 - 3)^2} = \sqrt{16 + 4} = \sqrt{20}$$

NOW WORK PROBLEM **49.**

---

*The following statement is a postulate from geometry. Two triangles are congruent if their sides are the same length (SSS), or if two sides and the included angle are the same (SAS), or if two angles and the included sides are the same (ASA).

†Another postulate from geometry states that the transversal $\overline{P_1P_2}$ forms equal corresponding angles with the parallel line segments $\overline{P_1A}$ and $\overline{MB}$.

1.1 Concepts and Vocabulary

In Problems 1–3, fill in the blanks.

1. If (x, y) are the coordinates of a point P in the xy-plane, then x is called the _____ of P and y is the _____ of P.
2. The coordinate axes divide the xy-plane into four sections called _____.
3. If three distinct points P, Q, and R all lie on a line and if $d(P, Q) = d(Q, R)$ then Q is called the _____ of the line segment from P to R.

In Problems 4–6, answer True or False to each statement.

4. The distance between two points is sometimes a negative number.
5. The point $(-1, 4)$ lies in Quadrant IV of the Cartesian plane.
6. The midpoint of a line segment is found by averaging the x-coordinates and averaging the y-coordinates of the endpoints.

7. Explain what is meant by setting the viewing window.

1.1 Exercises

In Problems 1 and 2, plot each point in the xy-plane. Tell in which quadrant or on what coordinate axis each point lies.

1. (a) $A = (-3, 2)$
   (b) $B = (6, 0)$
   (c) $C = (-2, -2)$
   (d) $D = (6, 5)$
   (e) $E = (0, -3)$
   (f) $F = (6, -3)$

2. (a) $A = (1, 4)$
   (b) $B = (-3, -4)$
   (c) $C = (-3, 4)$
   (d) $D = (4, 1)$
   (e) $E = (0, 1)$
   (f) $F = (-3, 0)$

3. Plot the points $(2, 0)$, $(2, -3)$, $(2, 4)$, $(2, 1)$, and $(2, -1)$. Describe the set of all points of the form $(2, y)$, where y is a real number.
4. Plot the points $(0, 3)$, $(1, 3)$, $(-2, 3)$, $(5, 3)$ and $(-4, 3)$. Describe the set of all points of the form $(x, 3)$, where x is a real number.

In Problems 5–8, determine the coordinates of the points shown. Tell in which quadrant each point lies. Assume the coordinates are integers.

5.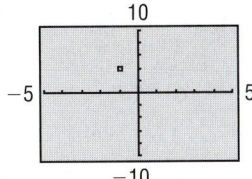

6.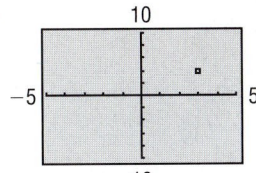

7.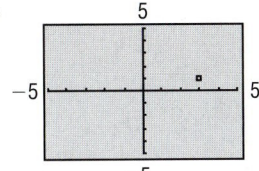

8.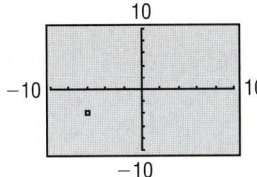

In Problems 9–14, select a setting so that each of the given points will lie within the viewing window.

9. $(-10, 5), (3, -2), (4, -1)$
10. $(5, 0), (6, 8), (-2, -3)$
11. $(40, 20), (-20, -80), (10, 40)$
12. $(-80, 60), (20, -30), (-20, -40)$
13. $(0, 0), (100, 5), (5, 150)$
14. $(0, -1), (100, 50), (-10, 30)$

In Problems 15–20, determine the viewing window used.

15.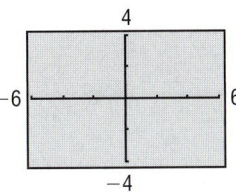

16.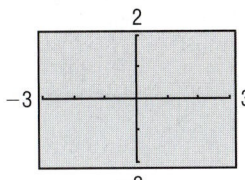

17.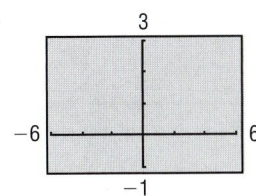

18.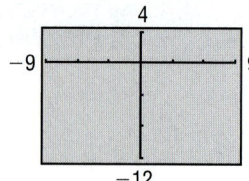

19.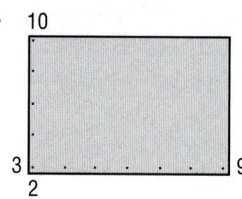

20.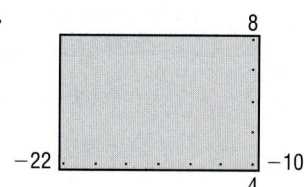

In Problems 21–34, find the distance $d(P_1, P_2)$ between the points P_1 and P_2.

21.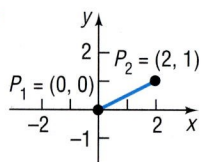

22.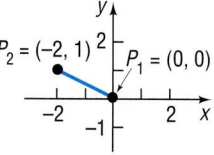

23.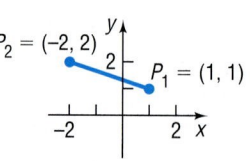

24.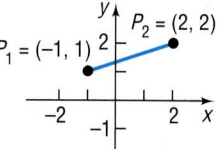

25. $P_1 = (3, 8)$; $P_2 = (5, 4)$

26. $P_1 = (6, 0)$; $P_2 = (2, 4)$

27. $P_1 = (-3, 2)$; $P_2 = (6, 0)$

28. $P_1 = (2, -3)$; $P_2 = (4, 2)$

29. $P_1 = (4, -3)$; $P_2 = (6, 4)$

30. $P_1 = (-4, -3)$; $P_2 = (6, 2)$

31. $P_1 = (-0.2, 0.3)$; $P_2 = (2.3, 1.1)$

32. $P_1 = (1.2, 2.3)$; $P_2 = (-0.3, 1.1)$

33. $P_1 = (a, b)$; $P_2 = (0, 0)$

34. $P_1 = (a, a)$; $P_2 = (0, 0)$

In Problems 35–38, find the length of the line segment. Assume that the end points of each line segment have integer coordinates.

35.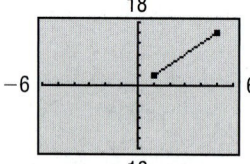

36.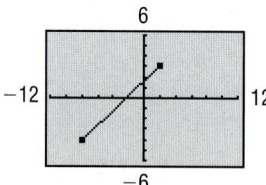

37.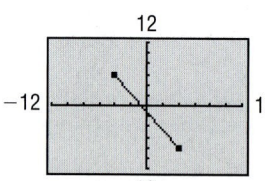

38.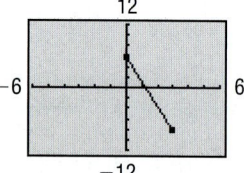

In Problems 39–44, plot each point and form the triangle ABC. Verify that the triangle is a right triangle. Find its area.

39. $A = (-2, 5)$; $B = (1, 3)$; $C = (-1, 0)$

40. $A = (-2, 5)$; $B = (12, 3)$; $C = (10, -11)$

41. $A = (-5, 3)$; $B = (6, 0)$; $C = (5, 5)$

42. $A = (-6, 3)$; $B = (3, -5)$; $C = (-1, 5)$

43. $A = (4, -3)$; $B = (0, -3)$; $C = (4, 2)$

44. $A = (4, -3)$; $B = (4, 1)$; $C = (2, 1)$

45. Find all points having an x-coordinate of 2 whose distance from the point $(-2, -1)$ is 5.

46. Find all points having a y-coordinate of -3 whose distance from the point $(1, 2)$ is 13.

47. Find all points on the x-axis that are 5 units from the point $(4, -3)$.

48. Find all points on the y-axis that are 5 units from the point $(4, 4)$.

In Problems 49–58, find the midpoint of the line segment joining the points P_1 and P_2.

49. $P_1 = (5, 4)$; $P_2 = (3, 2)$

50. $P_1 = (6, 0)$; $P_2 = (2, 4)$

51. $P_1 = (-3, 2)$; $P_2 = (6, 0)$

52. $P_1 = (2, -3)$; $P_2 = (4, 2)$

53. $P_1 = (4, -3)$; $P_2 = (6, 1)$

54. $P_1 = (-4, -3)$; $P_2 = (2, 2)$

55. $P_1 = (-0.2, 0.3)$; $P_2 = (2.3, 1.1)$

56. $P_1 = (1.2, 2.3)$; $P_2 = (-0.3, 1.1)$

57. $P_1 = (a, b)$; $P_2 = (0, 0)$

58. $P_1 = (a, a)$; $P_2 = (0, 0)$

59. The **medians** of a triangle are the line segments from each vertex to the midpoint of the opposite side (see the figure). Find the lengths of the medians of the triangle with vertices at $A = (0,0)$, $B = (6,0)$, and $C = (4,4)$.

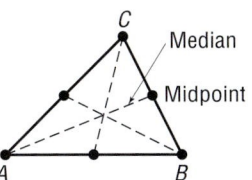

60. An **equilateral triangle** is one in which all three sides are of equal length. If two vertices of an equilateral triangle are $(0,4)$ and $(0,0)$, find the third vertex. How many of these triangles are possible?

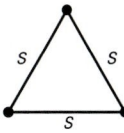

*In Problems 61–64, find the length of each side of the triangle determined by the three points P_1, P_2, and P_3. State whether the triangle is an isosceles triangle, a right triangle, neither of these, or both. (An **isosceles triangle** is one in which at least two of the sides are of equal length.)*

61. $P_1 = (2,1)$; $P_2 = (-4,1)$; $P_3 = (-4,-3)$

62. $P_1 = (-1,4)$; $P_2 = (6,2)$; $P_3 = (4,-5)$

63. $P_1 = (-2,-1)$; $P_2 = (0,7)$; $P_3 = (3,2)$

64. $P_1 = (7,2)$; $P_2 = (-4,0)$; $P_3 = (4,6)$

---

65. Baseball   A major league baseball "diamond" is actually a square, 90 feet on a side (see the figure). What is the distance directly from home plate to second base (the diagonal of the square)?

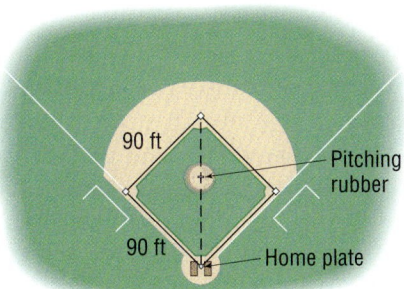

66. Little League Baseball   The layout of a Little League playing field is a square, 60 feet on a side.* How far is it directly from home plate to second base (the diagonal of the square)?

67. Baseball   Refer to Problem 65. Overlay a rectangular coordinate system on a major league baseball diamond so that the origin is at home plate, the positive x-axis lies in the direction from home plate to first base, and the positive y-axis lies in the direction from home plate to third base.

(a) What are the coordinates of first base, second base, and third base? Use feet as the unit of measurement.

(b) If the right fielder is located at $(310, 15)$, how far is it from there to second base?

(c) If the center fielder is located at $(300, 300)$, how far is it from there to third base?

* Source: Little League Baseball, Official Regulations and Playing Rules, 2001.

68. Little League Baseball   Refer to Problem 66. Overlay a rectangular coordinate system on a Little League baseball diamond so that the origin is at home plate, the positive x-axis lies in the direction from home plate to first base, and the positive y-axis lies in the direction from home plate to third base.

(a) What are the coordinates of first base, second base, and third base? Use feet as the unit of measurement.

(b) If the right fielder is located at $(180, 20)$, how far is it from there to second base?

(c) If the center fielder is located at $(220, 220)$, how far is it from there to third base?

69. A Dodge Intrepid and a Mack truck leave an intersection at the same time. The Intrepid heads east at an average speed of 30 miles per hour, while the truck heads south at an average speed of 40 miles per hour. Find an expression for their distance apart d (in miles) at the end of t hours.

70. A hot-air balloon, headed due east at an average speed of 15 miles per hour and at a constant altitude of 100 feet, passes over an intersection (see the figure). Find an expression for its distance d (measured in feet) from the intersection t seconds later.

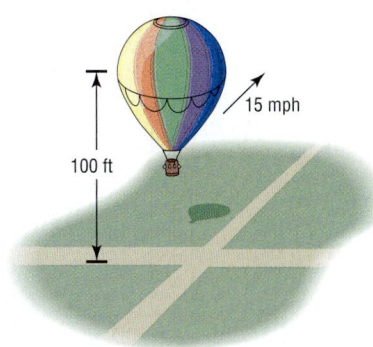

1.2 INTRODUCTION TO GRAPHING EQUATIONS

OBJECTIVES
1. Graph Equations by Hand by Plotting Points
2. Graph Equations Using a Graphing Utility
3. Find Intercepts

An **equation in two variables,** say x and y, is a statement in which two expressions involving x and y are equal. The expressions are called the **sides** of the equation. Since an equation is a statement, it may be true or false, depending on the value of the variables. Any values of x and y that result in a true statement are said to **satisfy** the equation.

For example, the following are all equations in two variables x and y:

$$x^2 + y^2 = 5 \qquad 2x - y = 6 \qquad y = 2x + 5 \qquad x^2 = y$$

The first of these, $x^2 + y^2 = 5$, is satisfied for $x = 1$, $y = 2$, since $1^2 + 2^2 = 1 + 4 = 5$. Other choices of x and y also satisfy this equation. It is not satisfied for $x = 2$ and $y = 3$, since $2^2 + 3^2 = 4 + 9 = 13 \neq 5$.

① The **graph of an equation in two variables** x and y consists of the set of points in the xy-plane whose coordinates (x, y) satisfy the equation.

Graphs play an important role in helping us to visualize the relationships that exist between two variables or quantities. Figure 17 shows the monthly closing price of Microsoft stock from July 31, 1999, through June 30, 2001. For example, the closing price on December 31, 1999, was $116.75.

Figure 17
Monthly closing prices of Microsoft stock, July 31, 1999, to June 30, 2001

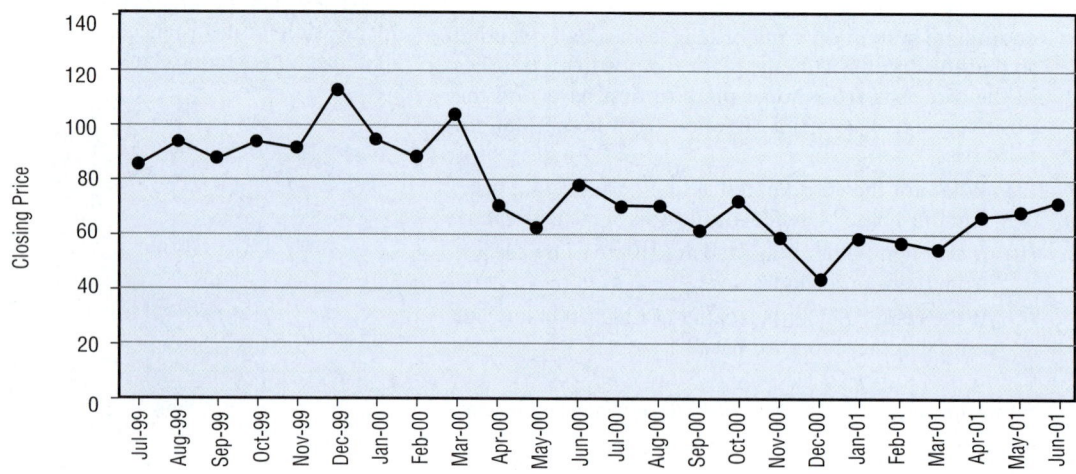

EXAMPLE 1

Determining Whether a Point Is on the Graph of an Equation

Determine if the following points are on the graph of the equation $2x - y = 6$.

(a) $(2, 3)$ (b) $(2, -2)$

Solution   (a) For the point $(2, 3)$, we check to see if $x = 2$, $y = 3$ satisfies the equation $2x - y = 6$.

$$2x - y = 2(2) - 3 = 4 - 3 = 1 \neq 6$$

The equation is not satisfied, so the point $(2, 3)$ is not on the graph.

(b) For the point $(2, -2)$, we have

$$2x - y = 2(2) - (-2) = 4 + 2 = 6$$

The equation is satisfied, so the point $(2, -2)$ is on the graph. ■

 **NOW WORK PROBLEM 1.**

EXAMPLE 2

Graphing an Equation by Hand by Plotting Points

Graph the equation: $y = 2x + 5$

Solution   We want to find all points (x, y) that satisfy the equation. To locate some of these points (and thus get an idea of the pattern of the graph), we assign some numbers to x and find corresponding values for y.

If	Then	Point on Graph
$x = 0$	$y = 2(0) + 5 = 5$	$(0, 5)$
$x = 1$	$y = 2(1) + 5 = 7$	$(1, 7)$
$x = -5$	$y = 2(-5) + 5 = -5$	$(-5, -5)$
$x = 10$	$y = 2(10) + 5 = 25$	$(10, 25)$

By plotting these points and then connecting them, we obtain the graph of the equation (a *line*), as shown in Figure 18.

Figure 18
$y = 2x + 5$

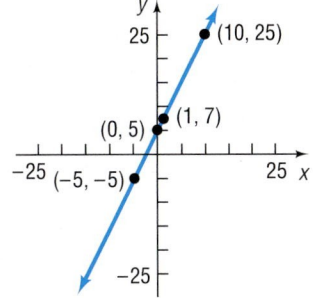

EXAMPLE 3   Graphing an Equation by Hand by Plotting Points

Graph the equation: $y = x^2$

Solution   Table 1 provides several points on the graph. In Figure 19 we plot these points and connect them with a smooth curve to obtain the graph (a *parabola*).

TABLE 1		
x	$y = x^2$	(x, y)
−4	16	(−4, 16)
−3	9	(−3, 9)
−2	4	(−2, 4)
−1	1	(−1, 1)
0	0	(0, 0)
1	1	(1, 1)
2	4	(2, 4)
3	9	(3, 9)
4	16	(4, 16)

Figure 19
$y = x^2$

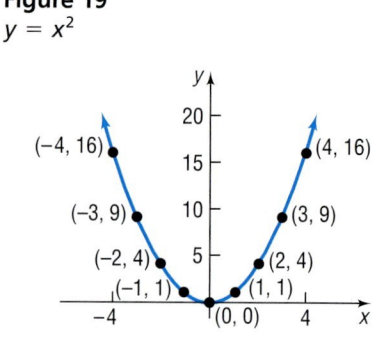

The graphs of the equations shown in Figures 18 and 19 do not show all points. For example, in Figure 18, the point $(20, 45)$ is a part of the graph of $y = 2x + 5$, but it is not shown. Since the graph of $y = 2x + 5$ could be extended out as far as we please, we use arrows to indicate that the pattern shown continues. It is important when illustrating a graph to present enough of the graph so that any viewer of the illustration will "see" the rest of it as an obvious continuation of what is actually there. This is referred to as a **complete graph.**

One way to obtain a complete graph of an equation is to plot a sufficient number of points on the graph until a pattern becomes evident. Then these points are connected with a smooth curve following the suggested pattern. But how many points are sufficient? Sometimes knowledge about the equation tells us. For example, we will learn in Section 1.7 that, if an equation is of the form $y = mx + b$, then its graph is a line. In this case, two points would suffice to obtain the graph.

One purpose of this book is to investigate the properties of equations in order to decide whether a graph is complete. Sometimes we shall graph equations by plotting a sufficient number of points on the graph until a pattern becomes evident; then we connect these points with a smooth curve following the suggested pattern. (Shortly, we shall investigate various techniques that will enable us to graph an equation without plotting so many points.) Other times we shall graph equations using a graphing utility.

Using a Graphing Utility to Graph Equations

From Examples 2 and 3, we see that a graph can be obtained by plotting points in a rectangular coordinate system and connecting them. Graphing utilities perform these same steps when graphing an equation. For example, the TI-83 determines 95 evenly spaced input values,* uses the equation to determine the output values, plots these points on the screen, and finally (if in the connected mode) draws a line between consecutive points.

To graph an equation in two variables x and y using a graphing utility requires that the equation be written in the form $y = \{\text{expression in } x\}$. If the original equation is not in this form, replace it by equivalent equations until the form $y = \{\text{expression in } x\}$ is obtained. In general, there are four ways to obtain equivalent equations.

Procedures That Result in Equivalent Equations

1. Interchange the two sides of the equation:
    Replace $3x + 5 = y$ by $y = 3x + 5$

2. Simplify the sides of the equation by combining like terms, eliminating parentheses, and so on:
    Replace $2y + 2 + 6 = 2x + 5(x + 1)$
    by $2y + 8 = 7x + 5$

3. Add or subtract the same expression on both sides of the equation:
    Replace $y + 3x - 5 = 4$
    by $y + 3x - 5 + 5 = 4 + 5$

4. Multiply or divide both sides of the equation by the same nonzero expression:
    Replace $3y = 6 - 2x$
    by $\dfrac{1}{3} \cdot 3y = \dfrac{1}{3}(6 - 2x)$

EXAMPLE 4   **Expressing an Equation in the Form $y = \{\text{expression in } x\}$**

Solve for y: $2y + 3x - 5 = 4$

Solution   We replace the original equation by a succession of equivalent equations.

$$2y + 3x - 5 = 4$$
$$2y + 3x - 5 + 5 = 4 + 5 \qquad \text{Add 5 to both sides.}$$
$$2y + 3x = 9 \qquad \text{Simplify.}$$
$$2y + 3x - 3x = 9 - 3x \qquad \text{Subtract 3x from both sides.}$$
$$2y = 9 - 3x \qquad \text{Simplify.}$$
$$\frac{2y}{2} = \frac{9 - 3x}{2} \qquad \text{Divide both sides by 2.}$$
$$y = \frac{9 - 3x}{2} \qquad \text{Simplify.}$$

*These input values depend on the values of Xmin and Xmax. For example, if Xmin $= -10$ and Xmax $= 10$, then the first input value will be -10 and the next input value will be $-10 + (10 - (-10))/94 = -9.7872$, and so on.

Now we are ready to graph equations using a graphing utility. Most graphing utilities require the following steps.

Steps for Graphing an Equation Using a Graphing Utility

STEP 1: Solve the equation for y in terms of x.

STEP 2: Get into the graphing mode of your graphing utility. The screen will usually display $Y =$, prompting you to enter the expression involving x that you found in Step 1. (Consult your manual for the correct way to enter the expression; for example, $y = x^2$ might be entered as $x{\wedge}2$ or as $x*x$ or as $x\, x^Y 2$).

STEP 3: Select the viewing window. Without prior knowledge about the behavior of the graph of the equation, it is common to select the **standard viewing window*** initially. The viewing window is then adjusted based on the graph that appears. In this text, the standard viewing window will be

$$X\text{min} = -10 \qquad Y\text{min} = -10$$
$$X\text{max} = 10 \qquad Y\text{max} = 10$$
$$X\text{scl} = 1 \qquad Y\text{scl} = 1$$

STEP 4: Graph.

STEP 5: Adjust the viewing window until a complete graph is obtained.

EXAMPLE 5 **Graphing an Equation on a Graphing Utility**

Graph the equation: $6x^2 + 3y = 36$

STEP 1: We solve for y in terms of x.

$$6x^2 + 3y = 36$$

$$3y = -6x^2 + 36 \qquad \text{Subtract } 6x^2 \text{ from both sides of the equation.}$$

$$y = -2x^2 + 12 \qquad \text{Divide both sides of the equation by 3 and simplify.}$$

STEP 2: From the graphing mode, enter the expression $-2x^2 + 12$ after the prompt $Y =$. Figure 20 shows the expression entered in a TI-83.

STEP 3: Set the viewing window to the standard viewing window.

STEP 4: Graph. The screen should look like Figure 21.

STEP 5: The graph of $y = -2x^2 + 12$ is not complete. The value of Ymax must be increased so that the top portion of the graph is visible.

Figure 20

Figure 21

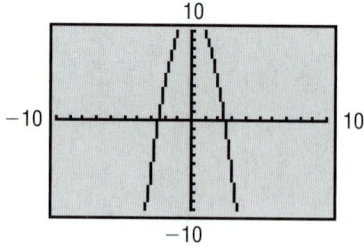

*Some graphing utilities have a ZOOM-STANDARD feature that automatically sets the viewing window to the standard viewing window and graphs the equation.

After increasing the value of Ymax to 12, we obtain the graph in Figure 22.* The graph is now complete.

Figure 22

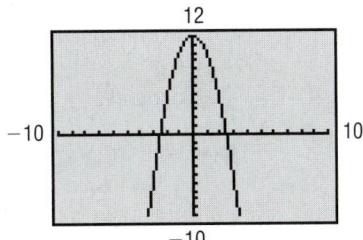

Using a Graphing Utility to Create Tables

In addition to graphing equations, graphing utilities can also be used to create a table of values that satisfy the equation. This feature is especially useful in determining an appropriate viewing window when graphing an equation. Many graphing utilities require the following steps to create a table.

> ### Steps for Creating a Table of Values Using a Graphing Utility
>
> **Step 1:** Solve the equation for y in terms of x.
>
> **Step 2:** Enter the expression in x following the $Y=$ prompt of the graphing utility.
>
> **Step 3:** Set up the table. Graphing utilities typically have two modes for creating tables. In the AUTO mode, the user determines a starting point for the table (TblStart) and ΔTbl (pronounced "delta-table"). The ΔTbl feature determines the increment for x. The ASK mode requires the user to enter values of x and then the utility determines the corresponding value of y.
>
> **Step 4:** Create the table. The user can scroll within the table if the table was created in AUTO mode.

EXAMPLE 6 **Creating a Table Using a Graphing Utility**

Create a table that displays the points on the graph of $6x^2 + 3y = 36$ for $x = -3, -2, -1, 0, 1, 2,$ and 3.

Solution **Step 1:** We solved the equation for y in Example 5 and obtained $y = -2x^2 + 12$.

Step 2: Enter the expression in x following the $Y=$ prompt.

Step 3: We set up the table in the AUTO mode with TblStart $= -3$ and ΔTbl $= 1$.

Step 4: Create the table. The screen should look like Table 2.

In looking at Table 2, we notice that $y = 12$ when $x = 0$. This information could have been used to help to create the initial viewing window by letting us know that Ymax needs to be at least 12 in order to get a complete graph.

TABLE 2

X	Y1
-3	-6
-2	4
-1	10
0	12
1	10
2	4
3	-6

Y1 = -2X²+12

*Some graphing utilities have a ZOOM-FIT feature that determines the appropriate Ymin and Ymax for a given Xmin and Xmax. Consult your owner's manual for the appropriate keystrokes.

Intercepts

3 The points, if any, at which a graph crosses or touches the coordinate axes are called the **intercepts.** See Figure 23. The *x*-coordinate of a point at which the graph crosses or touches the *x*-axis is an ***x*-intercept,** and the *y*-coordinate of a point at which the graph crosses or touches the *y*-axis is a **y-intercept.** In order for a graph to be complete, all of its intercepts must be displayed.

Figure 23

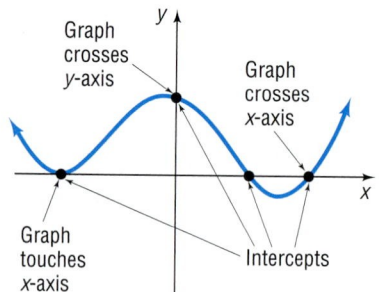

EXAMPLE 7   Finding Intercepts from a Graph

Find the intercepts of the graph in Figure 24. What are its *x*-intercepts? What are its *y*-intercepts?

Solution   The intercepts of the graph are the points

Figure 24

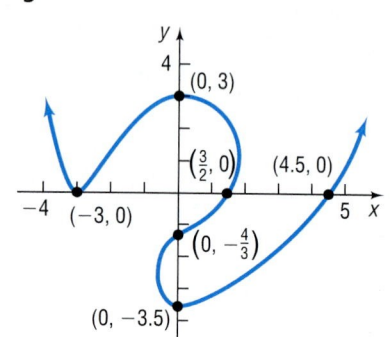

$$(-3, 0), \quad (0, 3), \quad \left(\frac{3}{2}, 0\right), \quad \left(0, -\frac{4}{3}\right), \quad (0, -3.5), \quad (4.5, 0)$$

The *x*-intercepts are -3, $\frac{3}{2}$, and 4.5; the *y*-intercepts are -3.5, $-\frac{4}{3}$, and 3.

🖉 NOW WORK PROBLEM 7.

The intercepts of the graph of an equation can be found by using the fact that points on the *x*-axis have *y*-coordinates equal to 0 and points on the *y*-axis have *x*-coordinates equal to 0.

> **Procedure for Finding Intercepts**
>
> 1. To find the *x*-intercept(s), if any, of the graph of an equation, let $y = 0$ in the equation and solve for *x*.
> 2. To find the *y*-intercept(s), if any, of the graph of an equation, let $x = 0$ in the equation and solve for *y*.

Because the *x*-intercepts of the graph of an equation are those *x*-values for which $y = 0$, they are also called the **zeros** (or **roots**) of the equation.

EXAMPLE 8   Finding Intercepts from an Equation

Find the *x*-intercept(s) and the *y*-intercept(s) of the graph of $y = x^2 - 4$.

Solution To find the *x*-intercept(s), we let $y = 0$ and obtain the equation

$$x^2 - 4 = 0$$

$$(x + 2)(x - 2) = 0 \qquad \text{Factor.}$$

$$x + 2 = 0 \quad \text{or} \quad x - 2 = 0 \qquad \text{Zero-Product Property}$$

$$x = -2 \quad \text{or} \quad x = 2$$

The equation has two solutions, -2 or 2. The *x*-intercepts (or zeros) are -2 and 2.

To find the *y*-intercept(s), we let $x = 0$ in the equation.

$$y = x^2 - 4$$

$$= 0^2 - 4 = -4$$

The *y*-intercept is -4. ■

TABLE 3

X	Y1
-3	5
-2	0
-1	-3
0	-4
1	-3
2	0
3	5

Y₁◾X²-4

Sometimes the TABLE feature of a graphing utility will reveal the intercepts of an equation. Table 3 shows a table for the equation $y = x^2 - 4$. Can you find the intercepts in the table?

Although a table can be used to find the intercepts of the graph of an equation, it is usually easier to graph an equation and use the graph to find the intercepts.

EXAMPLE 9 **Finding Intercepts Using a Graphing Utility**

Use a graphing utility to find the intercepts of the equation $y = x^3 - 16$.

Solution Figure 25(a) shows the graph of $y = x^3 - 16$.

Figure 25

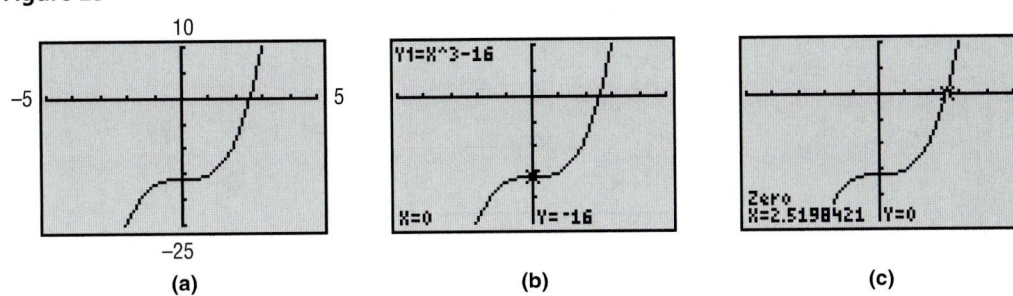

(a) (b) (c)

The eVALUEate feature of a TI-83 graphing calculator* accepts as input a value of *x* and determines the value of *y*. If we let $x = 0$, we find that the *y*-intercept is -16. See Figure 25(b).

The ZERO feature of a TI-83 is used to find the *x*-intercept(s). See Figure 25(c). Rounded to two decimal places, the *x*-intercept is 2.52. ■

━━━━━ **NOW WORK PROBLEMS 27 AND 37.**

*Consult your manual to determine the appropriate keystrokes for these features.

1.2 Concepts and Vocabulary

In Problems 1 and 2, fill in the blanks.

1. The points, if any, at which a graph crosses or touches the coordinate axes are called _____.

2. Because the *x*-intercepts of the graph of an equation are those *x*-values for which $y = 0$, they are also called _____ or _____.

In Problems 3 and 4, answer True or False to each statement.

3. To find the *x*-intercepts of the graph of an equation, let $y = 0$ and solve for *x*.

4. To graph an equation using a graphing utility, we must first solve the equation for *x* in terms of *y*.

5. Explain what is meant by a complete graph.

6. What is the standard viewing window?

7. Draw a graph that contains two *x*-intercepts. At one the graph crosses the *x*-axis, and at the other the graph touches the *x*-axis.

1.2 Exercises

In Problems 1–6 tell whether the given points are on the graph of the equation.

1. Equation: $y = x^4 - \sqrt{x}$
Points: $(0,0); (1,1); (-1,0)$

2. Equation: $y = x^3 - 2\sqrt{x}$
Points: $(0,0); (1,1); (1,-1)$

3. Equation: $y^2 = x^2 + 9$
Points: $(0,3); (3,0); (-3,0)$

4. Equation: $y^3 = x + 1$
Points: $(1,2); (0,1); (-1,0)$

5. Equation: $x^2 + y^2 = 4$
Points: $(0,2); (-2,2); (\sqrt{2}, \sqrt{2})$

6. Equation: $x^2 + 4y^2 = 4$
Points: $(0,1); (2,0); \left(2, \dfrac{1}{2}\right)$

In Problems 7–18, the graph of an equation is given. List the intercepts of the graph.

7.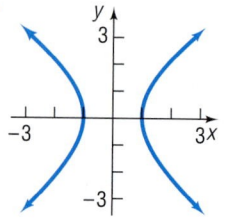

8.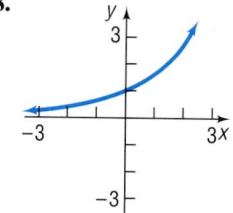

9.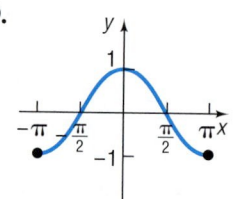

10.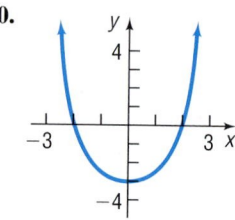

11.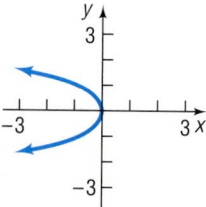

12.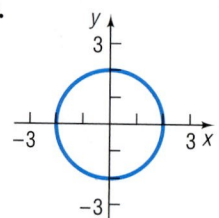

13.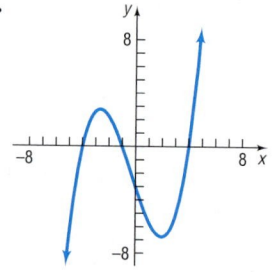

14.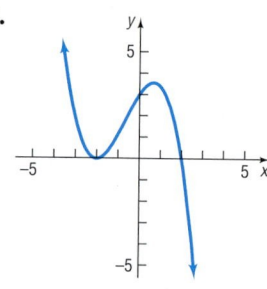

15.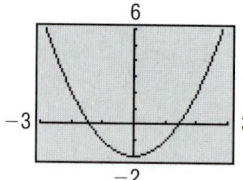

16.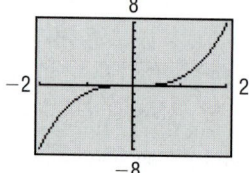

17.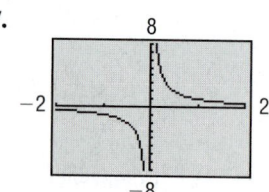

18.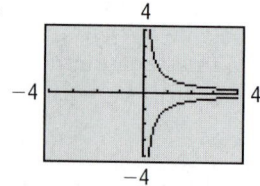

19. If $(a, 2)$ is a point on the graph of $y = 5x + 4$, what is a?

20. If $(2, b)$ is a point on the graph of $y = x^2 + 3x$, what is b?

21. If (a, b) is a point on the graph of $2x + 3y = 6$, write an equation that relates a to b.

22. If $(2, 0)$ and $(0, 5)$ are points on the graph of $y = mx + b$, what are m and b?

In Problems 23–34, determine the intercepts algebraically and graph each equation by hand by plotting points. Be sure to label the intercepts. Verify your results using a graphing utility.

23. $y = x + 2$ **24.** $y = x - 6$ **25.** $y = 2x + 8$ **26.** $y = 3x - 9$

27. $y = x^2 - 1$ **28.** $y = x^2 - 9$ **29.** $y = -x^2 + 4$ **30.** $y = -x^2 + 1$

31. $2x + 3y = 6$ **32.** $5x + 2y = 10$ **33.** $9x^2 + 4y = 36$ **34.** $4x^2 + y = 4$

In Problems 35–42, graph each equation using a graphing utility. Use a graphing utility to approximate the intercepts rounded to two decimal places. Use the TABLE feature to help establish the viewing window.

35. $y = 2x - 13$ **36.** $y = -3x + 14$ **37.** $y = 2x^2 - 15$ **38.** $y = -3x^2 + 19$

39. $3x - 2y = 43$ **40.** $4x + 5y = 82$ **41.** $5x^2 + 3y = 37$ **42.** $2x^2 - 3y = 35$

In Problem 43, you may use a graphing utility, but it is not required.

43. (a) Graph $y = \sqrt{x^2}$, $y = x$, $y = |x|$, and $y = (\sqrt{x})^2$, noting which graphs are the same.

(b) Explain why the graphs of $y = \sqrt{x^2}$ and $y = |x|$ are the same.

(c) Explain why the graphs of $y = x$ and $y = (\sqrt{x})^2$ are not the same.

(d) Explain why the graphs of $y = \sqrt{x^2}$ and $y = x$ are not the same.

44. Make up an equation with the intercepts $(2, 0)$, $(4, 0)$, and $(0, 1)$. Compare your equation with a friend's equation. Comment on any similarities.

45. Draw a graph that contains the points $(-2, -1)$, $(0, 1)$, $(1, 3)$, and $(3, 5)$. Compare your graph with those of other students. Are most of the graphs almost straight lines? How many are "curved"? Discuss the various ways that these points might be connected.

PREPARING FOR THIS SECTION

Before getting started, review the following:

✓ Domain of a variable (Section R.2, p. 21)

✓ Zero-Product Property (Section R.1, p. 13)

✓ Factoring Polynomials (Section R.6, pp 50–58)

✓ Square Roots (Section R.8, pp. 71–74)

1.3   SOLVING EQUATIONS USING A GRAPHING UTILITY; SOLVING LINEAR AND QUADRATIC EQUATIONS

OBJECTIVES

1 Solve Equations in One Variable

2 Solve Equations Using a Graphing Utility

3 Solve Linear Equations

4 Solve Quadratic Equations by Factoring

5 Solve Quadratic Equations by Completing the Square

6 Solve Quadratic Equations Using the Quadratic Formula

1 An **equation in one variable** is a statement in which two expressions, at least one containing the variable, are equal. The expressions are called the **sides** of the equation. Since an equation is a statement, it may be true or false, depending on the value of the variable. Unless otherwise restricted, the

admissible values of the variable are those in the domain of the variable. Those admissible values of the variable, if any, that result in a true statement are called **solutions,** or **roots,** of the equation. To **solve an equation** means to find all the solutions of the equation.

For example, the following are all equations in one variable, x:

$$x + 5 = 9 \qquad x^2 + 5x = 2x - 2 \qquad \frac{x^2 - 4}{x + 1} = 0 \qquad x^2 + 9 = 5$$

The first of these statements, $x + 5 = 9$, is true when $x = 4$ and false for any other choice of x. We say that 4 is a solution of the equation $x + 5 = 9$ or that 4 **satisfies** the equation $x + 5 = 9$, because, when we substitute 4 for x, a true statement results.

Sometimes an equation will have more than one solution. For example, the equation

$$\frac{x^2 - 4}{x + 1} = 0$$

has $x = -2$ and $x = 2$ as solutions.

Usually, we will write the solution of an equation in set notation. This set is called the **solution set** of the equation. For example, the solution set of the equation $x^2 - 9 = 0$ is $\{-3, 3\}$.

Unless indicated otherwise, we will limit ourselves to real solutions, that is, solutions that are real numbers. Some equations have no real solution. For example, $x^2 + 9 = 5$ has no real solution, because there is no real number whose square when added to 9 equals 5.

An equation that is satisfied for every choice of the variable for which both sides are defined is called an **identity.** For example, the equation

$$3x + 5 = x + 3 + 2x + 2$$

is an identity, because this statement is true for any real number x.

In this text, we present two methods for solving equations: algebraic and graphical. We shall see as we proceed through this book that some equations can be solved using algebraic techniques that result in *exact* solutions. For other equations, however, there are no algebraic techniques that lead to an exact solution. For such equations, a graphing utility can often be used to investigate possible solutions. When a graphing utility is used to solve an equation, usually *approximate* solutions are obtained.

One goal of this text is to determine when equations can be solved algebraically. If an algebraic method for solving an equation exists, we shall use it to obtain an exact solution. A graphing utility can then be used to support the algebraic result. However, if an equation must be solved for which no algebraic techniques are available, a graphing utility will be used to obtain approximate solutions.

One method for solving equations algebraically requires that a series of *equivalent equations* be developed from the original equation until an obvious solution results.

> Two or more equations that have precisely the same solutions are called **equivalent equations.**

For example, all the following equations are equivalent, because each has only the solution $x = 5$:

$$2x + 3 = 13$$
$$2x = 10$$
$$x = 5$$

The question, though, is "How do I obtain an equivalent equation?" In general, there are five ways to do so.

Procedures That Result in Equivalent Equations

1. Interchange the two sides of the equation:

    Replace $\quad\quad\quad\quad 3 = x \quad$ by $\quad x = 3$

2. Simplify the sides of the equation by combining like terms, eliminating parentheses, and so on:

    Replace $\quad\quad x + 2 + 6 = 2x + 3(x + 1)$

    by $\quad\quad\quad x + 8 = 5x + 3$

3. Add or subtract the same expression on both sides of the equation:

    Replace $\quad\quad\quad\quad 3x - 5 = 4$

    by $\quad\quad (3x - 5) + 5 = 4 + 5$

4. Multiply or divide both sides of the equation by the same nonzero expression:

    Replace $\quad\quad \dfrac{3x}{x - 1} = \dfrac{6}{x - 1} \quad\quad x \neq 1$

    by $\quad \dfrac{3x}{x - 1} \cdot (x - 1) = \dfrac{6}{x - 1} \cdot (x - 1)$

5. If one side of the equation is 0 and the other side can be factored, then we may use the Zero-Product Property and set each factor equal to 0:

    Replace $\quad\quad\quad\quad x(x - 3) = 0$

    by $\quad\quad x = 0 \quad$ or $\quad x - 3 = 0$

WARNING: Squaring both sides of an equation does not necessarily lead to an equivalent equation. ∎

Whenever it is possible to solve an equation in your head, do so. For example:

The solution of $2x = 8$ is $x = 4$.

The solution of $3x - 15 = 0$ is $x = 5$.

Often, though, some rearrangement is necessary. In the next two examples, we present equations that can be solved algebraically to illustrate the procedures that result in equivalent equations.

EXAMPLE 1   **Solving an Equation Algebraically**

Solve the equation: $3x - 5 = 4$

Solution   We replace the original equation by a succession of equivalent equations.

$$3x - 5 = 4$$
$$(3x - 5) + 5 = 4 + 5 \qquad \text{Add 5 to both sides.}$$
$$3x = 9 \qquad \text{Simplify.}$$
$$\frac{3x}{3} = \frac{9}{3} \qquad \text{Divide both sides by 3.}$$
$$x = 3 \qquad \text{Simplify.}$$

The last equation, $x = 3$, has the single solution 3. All these equations are equivalent, so 3 is the only solution of the original equation, $3x - 5 = 4$.

✔ CHECK: It is a good practice to check the solution by substituting 3 for x in the original equation.

$$3x - 5 = 4$$
$$3(3) - 5 \overset{?}{=} 4$$
$$9 - 5 \overset{?}{=} 4$$
$$4 = 4$$

The solution checks.   ■ ■

✏ **NOW WORK PROBLEM 17.**

In the next examples, we use the Zero-Product Property (procedure 5, listed in the box on page 111).

EXAMPLE 2   **Solving Equations Algebraically by Factoring**

Solve the equations: (a) $x^2 = 4x$ (b) $x^3 - x^2 - 4x + 4 = 0$

Solution   (a) We begin by collecting all terms on one side. This results in 0 on one side and an expression to be factored on the other.

$$x^2 = 4x$$
$$x^2 - 4x = 0$$
$$x(x - 4) = 0 \qquad \text{Factor.}$$
$$x = 0 \quad \text{or} \quad x - 4 = 0 \qquad \text{Apply the Zero-Product property.}$$
$$x = 4$$

The solution set is $\{0, 4\}$.

✔ CHECK: $x = 0$: $0^2 = 4 \cdot 0$ So 0 is a solution.

$x = 4$: $4^2 = 4 \cdot 4$ So 4 is a solution.   ■

(b) Do you recall the method of factoring by grouping? (If not, review pages 55–56.) We group the terms of $x^3 - x^2 - 4x + 4 = 0$ as follows:

$$(x^3 - x^2) - (4x - 4) = 0$$

Factor out x^2 from the first grouping and 4 from the second.

$$x^2(x - 1) - 4(x - 1) = 0$$

This reveals the common factor $(x - 1)$, so we have

$$(x^2 - 4)(x - 1) = 0$$

$$(x - 2)(x + 2)(x - 1) = 0 \qquad \text{Factor again.}$$

$x - 2 = 0 \qquad \text{or} \qquad x + 2 = 0 \qquad \text{or} \qquad x - 1 = 0$ Set each factor equal to 0.

$x = 2 \qquad\qquad\qquad x = -2 \qquad\qquad\qquad x = 1$ Solve.

The solution set is $\{-2, 1, 2\}$.

✔ **CHECK:** $x = -2$: $(-2)^3 - (-2)^2 - 4(-2) + 4 = -8 - 4 + 8 + 4 = 0$ −2 is a solution.

$x = 1$: $1^3 - 1^2 - 4(1) + 4 = 1 - 1 - 4 + 4 = 0$ 1 is a solution.

$x = 2$: $2^3 - 2^2 - 4(2) + 4 = 8 - 4 - 8 + 4 = 0$ 2 is a solution. ■ ■

Steps for Solving Equations Algebraically

STEP 1:   List any restrictions on the domain of the variable.

STEP 2:   Simplify the equation by replacing the original equation by a succession of equivalent equations following the procedures listed on page 111.

STEP 3:   If the result of Step 2 is a product of factors equal to 0, use the Zero-Product Property and set each factor equal to 0 (procedure 5).

STEP 4:   Check your solution(s).

 NOW WORK PROBLEMS 37 AND 57 ALGEBRAICALLY.

Solving Equations Using a Graphing Utility

② When a graphing utility is used to solve an equation, usually *approximate* solutions are obtained. Unless otherwise stated, we shall follow the practice of giving approximate solutions as decimals *rounded to two decimal places.*

The ZERO (or ROOT) feature of a graphing utility can be used to find the solutions of an equation when one side of the equation is 0. In using this feature to solve equations, we make use of the fact that the x-intercepts (or zeros) of the graph of an equation are found by letting $y = 0$ and solving the equation for x. Solving an equation for x when one side of the equation is 0 is equivalent to finding where the graph of the corresponding equation crosses or touches the x-axis.

EXAMPLE 3 **Using ZERO (or ROOT) to Approximate Solutions of an Equation**

Find the solution(s) of the equation $x^3 - x + 1 = 0$. Round answers to two decimal places.

Solution The solutions of the equation $x^3 - x + 1 = 0$ are the same as the x-intercepts of the graph of $Y_1 = x^3 - x + 1$. We begin by graphing the equation. Figure 26 shows the graph.

From the graph there appears to be one x-intercept (solution to the equation) between -2 and -1.

Using the ZERO (or ROOT) feature of our graphing utility, we determine that the x-intercept, and thus the solution to the equation, is $x = -1.32$ rounded to two decimal places. See Figure 27.

Figure 26

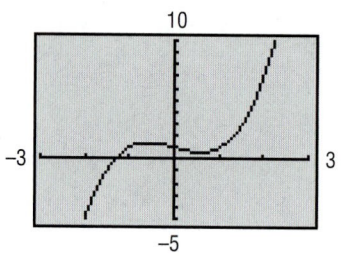

Figure 27

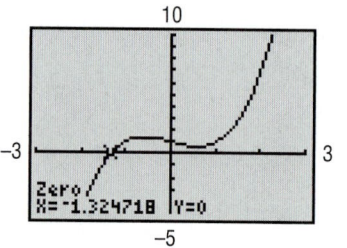

NOW WORK PROBLEM **1.**

A second method for solving equations using a graphing utility involves the INTERSECT feature of the graphing utility. This feature is used most effectively when one side of the equation is not 0.

EXAMPLE 4 **Using INTERSECT to Approximate Solutions of an Equation**

Find the solution(s) to the equation $4x^4 - 3 = 2x + 1$. Round answers to two decimal places.

Solution We begin by graphing each side of the equation as follows: graph $Y_1 = 4x^4 - 3$ and $Y_2 = 2x + 1$. See Figure 28.

At the point of intersection of the graphs, the value of the y-coordinate is the same. Thus, the x-coordinate of the point of intersection represents the solution to the equation. Do you see why? The INTERSECT feature on a graphing utility determines the point of intersection of the graphs. Using this feature, we find that the graphs intersect at $(-0.87, -0.73)$ and $(1.12, 3.23)$ rounded to two decimal places. See Figure 29(a) and (b). The solutions of the equation are $x = -0.87$ and $x = 1.12$, rounded to two decimal places.

Figure 28

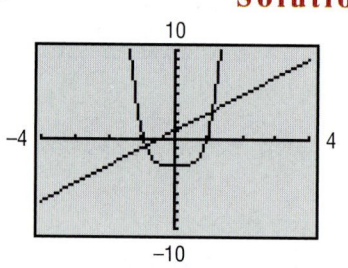

Figure 29

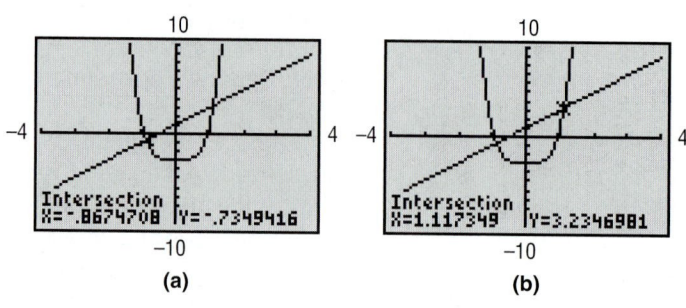

(a)   (b)

NOW WORK PROBLEM **3.**

SUMMARY

The steps to follow for approximating solutions of equations are given next.

Steps for Approximating Solutions of Equations Using ZERO (or ROOT)

STEP 1: Write the equation in the form {expression in x} = 0.

STEP 2: Graph Y_1 = {expression in x}.

STEP 3: Use ZERO (or ROOT) to determine each x-intercept of the graph.

Steps for Approximating Solutions of Equations Using INTERSECT

STEP 1: Graph Y_1 = {expression in x on left side of equation};
Y_2 = {expression in x on the right side of the equation}

STEP 2: Use INTERSECT to determine the x-coordinate of each point of intersection.

We now introduce two specific types of equations that can be solved algebraically to obtain exact solutions: linear equations and quadratic equations. A graphing utility will be used to support the solution.

Linear Equations

③ *Linear equations* are equations such as

$$3x + 12 = 0, \qquad \frac{3}{4}x - \frac{1}{5} = 0, \qquad 0.62x - 0.3 = 0$$

A **linear equation in one variable** is equivalent to an equation of the form

$$ax + b = 0$$

where a and b are real numbers and $a \neq 0$.

Sometimes a linear equation is called a **first-degree equation,** because the left side is a polynomial in x of degree 1.

EXAMPLE 5   **Solving a Linear Equation**

Solve the equation: $3(x - 2) = 5(x - 1)$

Algebraic Solution

$$3(x - 2) = 5(x - 1)$$
$$3x - 6 = 5x - 5 \qquad \text{Use the Distributive Property.}$$
$$3x - 6 - 5x = 5x - 5 - 5x \qquad \text{Subtract 5x from each side.}$$
$$-2x - 6 = -5 \qquad \text{Simplify.}$$

$$-2x - 6 + 6 = -5 + 6 \qquad \text{Add 6 to each side.}$$

$$-2x = 1 \qquad \text{Simplify.}$$

$$\frac{-2x}{-2} = \frac{1}{-2} \qquad \text{Divide each side by } -2.$$

$$x = -\frac{1}{2} \qquad \text{Simplify.}$$

✔ CHECK: Let $x = -\dfrac{1}{2}$ in the expression in x on the left side of the equation and simplify. Let $x = -\dfrac{1}{2}$ in the expression in x on the right side of the equation and simplify. If the two expressions are equal, the solution checks.

$$3(x - 2) = 3\left(-\frac{1}{2} - 2\right) = 3\left(-\frac{5}{2}\right) = -\frac{15}{2}$$

$$5(x - 1) = 5\left(-\frac{1}{2} - 1\right) = 5\left(-\frac{3}{2}\right) = -\frac{15}{2}$$

Since the two expressions are equal, the solution $x = -\dfrac{1}{2}$ checks. ■

Graphing Solution   Graph $Y_1 = 3(x - 2)$ and $Y_2 = 5(x - 1)$. See Figure 30. Using INTER-SECT, we find the point of intersection to be $(-0.5, -7.5)$. The solution of the equation is $x = -0.5$.

Figure 30

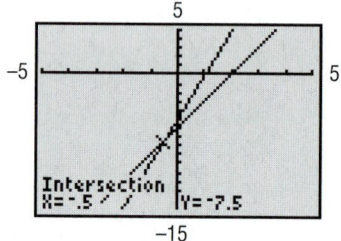

The next example illustrates the solution of an equation that does not appear to be linear, but leads to a linear equation upon simplification.

EXAMPLE 6   **Solving Equations**

Solve the equation: $(2x - 1)(x - 1) = (x - 5)(2x - 5)$

Algebraic Solution

$$(2x - 1)(x - 1) = (x - 5)(2x - 5)$$

$$2x^2 - 3x + 1 = 2x^2 - 15x + 25 \qquad \text{Multiply and combine like terms.}$$

$$2x^2 - 3x + 1 - 2x^2 = 2x^2 - 15x + 25 - 2x^2 \qquad \text{Subtract } 2x^2 \text{ from each side.}$$

$$-3x + 1 = -15x + 25 \qquad \text{Simplify.}$$

$$-3x + 1 - 1 = -15x + 25 - 1 \qquad \text{Subtract 1 from each side.}$$

$$-3x = -15x + 24 \qquad \text{Simplify.}$$

$$-3x + 15x = -15x + 24 + 15x \qquad \text{Add 15x to each side.}$$

$$12x = 24 \qquad \text{Simplify.}$$

$$\frac{12x}{12} = \frac{24}{12} \qquad \text{Divide each side by 12.}$$

$$x = 2 \qquad \text{Simplify.}$$

✔ CHECK: $(2x - 1)(x - 1) = (2 \cdot 2 - 1)(2 - 1) = (3)(1) = 3$

$(x - 5)(2x - 5) = (2 - 5)(2 \cdot 2 - 5) = (-3)(-1) = 3$

Since the two expressions are equal, the solution checks. ■

Graphing Solution   Graph $Y_1 = (2x - 1)(x - 1)$ and $Y_2 = (x - 5)(2x - 5)$. See Figure 31. Using INTERSECT we find the point of intersection to be $(2, 3)$. The solution of the equation is $x = 2$.

Figure 31

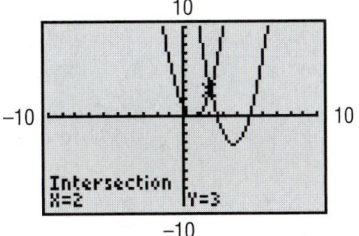

NOW WORK PROBLEM **33**.

Quadratic Equations

4 *Quadratic equations* are equations such as

$$2x^2 + x + 8 = 0, \qquad 3x^2 - 5x = 0, \qquad x^2 - 9 = 0$$

A general definition is given next.

A **quadratic equation** is an equation equivalent to one of the form

$$ax^2 + bx + c = 0 \qquad\qquad (1)$$

where *a, b,* and *c* are real numbers and $a \neq 0$.

A quadratic equation written in the form $ax^2 + bx + c = 0$ is said to be in **standard form.**

Sometimes, a quadratic equation is called a **second-degree equation,** because the left side is a polynomial of degree 2. We shall discuss three algebraic ways of solving quadratic equations: by factoring, by completing the square, and by using the quadratic formula.

When a quadratic equation is written in standard form, $ax^2 + bx + c = 0$, it may be possible to factor the expression on the left side as the product of two first-degree polynomials. Then, by setting each factor equal to 0 and solving the resulting linear equations, we obtain the *exact* solutions of the quadratic equation.

Let's look at an example.

EXAMPLE 7   **Solving a Quadratic Equation by Factoring and by Graphing**

Solve the equation: $x^2 = 12 - x$

Algebraic Solution   We put the equation in standard form by adding $x - 12$ to each side:

$$x^2 = 12 - x$$
$$x^2 + x - 12 = 0$$

The left side of the equation may now be factored as

$$(x + 4)(x - 3) = 0$$

so that

$$x + 4 = 0 \quad \text{or} \quad x - 3 = 0$$
$$x = -4 \qquad\qquad x = 3$$

The solution set is $\{-4, 3\}$.

Graphing Solution   Graph $Y_1 = x^2$ and $Y_2 = 12 - x$. See Figure 32(a). From the graph it appears that there are two points of intersection: one near -4; the other near 3. Using INTERSECT, the points of intersection are $(-4, 16)$ and $(3, 9)$, so the solutions of the equation are $x = -4$ and $x = 3$. See Figure 32(b) and (c).

Figure 32

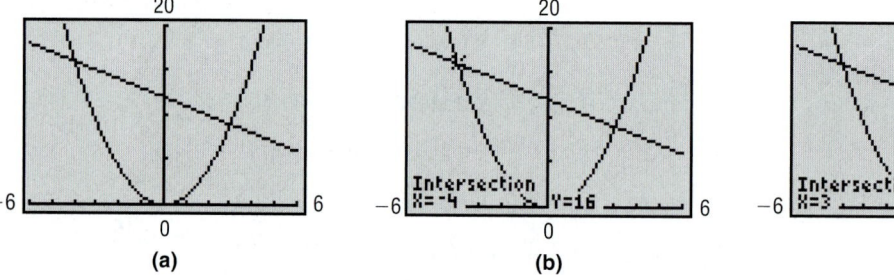

(a)   (b)   (c)

When the left side factors into two linear equations with the same solution, the quadratic equation is said to have a **repeated solution.** We also call this solution a **root of multiplicity 2,** or a **double root.**

EXAMPLE 8   **Solving a Quadratic Equation by Factoring and by Graphing**

Solve the equation: $x^2 - 6x + 9 = 0$

Algebraic Solution   This equation is already in standard form, and the left side can be factored:

$$x^2 - 6x + 9 = 0$$
$$(x - 3)(x - 3) = 0$$

so

$$x = 3 \quad \text{or} \quad x = 3$$

This equation has only the repeated solution 3.

Graphing Solution Figure 33 shows the graph of the equation

$$Y_1 = x^2 - 6x + 9$$

From the graph it appears that there is one x-intercept, 3. Using ZERO (or ROOT), we find that the only x-intercept is $x = 3$. The equation has only the repeated solution 3.

Figure 33

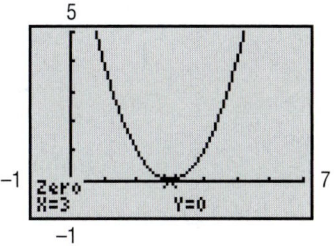

NOW WORK PROBLEM **49.**

The Square Root Method

Suppose that we wish to solve the quadratic equation

$$x^2 = p \qquad\qquad (2)$$

where $p \geq 0$ is a nonnegative number. We proceed as in the earlier examples:

$$x^2 - p = 0 \qquad \text{Put in standard form.}$$
$$(x - \sqrt{p})(x + \sqrt{p}) = 0 \qquad \text{Factor (over the real numbers).}$$
$$x = \sqrt{p} \quad \text{or} \quad x = -\sqrt{p} \qquad \text{Solve.}$$

We have the following result:

$$\text{If } x^2 = p \text{ and } p \geq 0, \text{ then } x = \sqrt{p} \text{ or } x = -\sqrt{p}. \qquad (3)$$

When statement (3) is used, it is called the **Square Root Method.** In statement (3), note that if $p > 0$ the equation $x^2 = p$ has two solutions, $x = \sqrt{p}$ and $x = -\sqrt{p}$. We usually abbreviate these solutions as $x = \pm\sqrt{p}$, read "x equals plus or minus the square root of p." For example, the two solutions of the equation

$$x^2 = 4$$

are

$$x = \pm\sqrt{4}$$

and since $\sqrt{4} = 2$, we have

$$x = \pm 2$$

The solution set is $\{-2, 2\}$.

EXAMPLE 9   **Solving Quadratic Equations by Using the Square Root Method**

Solve each equation.

(a) $x^2 = 5$ (b) $(x - 2)^2 = 16$

Solution (a)
$$x^2 = 5$$
$$x = \pm\sqrt{5} \qquad \text{Use the Square Root Method}$$
$$x = \sqrt{5} \quad \text{or} \quad x = -\sqrt{5}$$

The solution set is $\{-\sqrt{5}, \sqrt{5}\}$.

(b) $(x-2)^2 = 16$
$$x - 2 = \pm\sqrt{16} \qquad \text{Use the Square Root Method}$$
$$x - 2 = \sqrt{16} \quad \text{or} \quad x - 2 = -\sqrt{16}$$
$$x - 2 = 4 \qquad\qquad x - 2 = -4$$
$$x = 6 \qquad\qquad\quad x = -2$$

The solution set is $\{-2, 6\}$.

✔ CHECK: Verify the solutions to Example 9 using a graphing utility. Are the solutions provided by the utility exact?   ■ ■

NOW WORK PROBLEM 63.

Completing the Square

5   We now introduce the method of **completing the square.** The idea behind this method is to "adjust" the left side of a quadratic equation, $ax^2 + bx + c = 0$, so that it becomes a perfect square, that is, the square of a first-degree polynomial. For example, $x^2 + 6x + 9$ and $x^2 - 4x + 4$ are perfect squares because

$$x^2 + 6x + 9 = (x + 3)^2 \quad \text{and} \quad x^2 - 4x + 4 = (x - 2)^2$$

How do we "adjust" the left side? We do it by adding the appropriate number to create a perfect square. For example, to make $x^2 + 6x$ a perfect square, we add 9.

Let's look at several examples of completing the square when the coefficient of x^2 is 1.

Start	Add	Result
$x^2 + 4x$	4	$x^2 + 4x + 4 = (x + 2)^2$
$x^2 + 12x$	36	$x^2 + 12x + 36 = (x + 6)^2$
$x^2 - 6x$	9	$x^2 - 6x + 9 = (x - 3)^2$
$x^2 + x$	$\dfrac{1}{4}$	$x^2 + x + \dfrac{1}{4} = \left(x + \dfrac{1}{2}\right)^2$

Do you see the pattern? Provided that the coefficient of x^2 is 1, we complete the square by adding the square of one-half the coefficient of x.

Start	Add	Result
$x^2 + mx$	$\left(\dfrac{m}{2}\right)^2$	$x^2 + mx + \left(\dfrac{m}{2}\right)^2 = \left(x + \dfrac{m}{2}\right)^2$

NOW WORK PROBLEM 67.

The next example illustrates how the procedure of completing the square can be used to solve a quadratic equation.

EXAMPLE 10 **Solving a Quadratic Equation by Completing the Square**

Solve by completing the square: $x^2 + 5x + 4 = 0$

Solution We always begin this procedure by rearranging the equation so that the constant is on the right side.

$$x^2 + 5x + 4 = 0$$
$$x^2 + 5x = -4$$

Since the coefficient of x^2 is 1, we can complete the square on the left side by adding $\left(\dfrac{1}{2} \cdot 5\right)^2 = \dfrac{25}{4}$. Of course, in an equation, whatever we add to the left side must also be added to the right side. We add $\dfrac{25}{4}$ to *both* sides.

$$x^2 + 5x + \frac{25}{4} = -4 + \frac{25}{4} \qquad \text{\color{teal}Add } \frac{25}{4} \text{ to both sides.}$$

$$\left(x + \frac{5}{2}\right)^2 = \frac{9}{4} \qquad \text{\color{teal}Factor; simplify.}$$

$$x + \frac{5}{2} = \pm\sqrt{\frac{9}{4}} \qquad \text{\color{teal}Use the Square Root Method.}$$

$$x + \frac{5}{2} = \pm\frac{3}{2}$$

$$x = -\frac{5}{2} \pm \frac{3}{2}$$

$$x = -\frac{5}{2} + \frac{3}{2} = -1 \quad \text{or} \quad x = -\frac{5}{2} - \frac{3}{2} = -4$$

The solution set is $\{-4, -1\}$.

✔ CHECK: Verify the solution to Example 10 using a graphing utility.

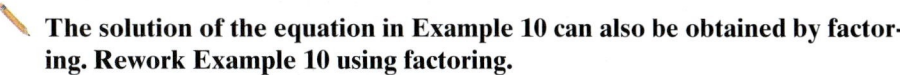 **NOW WORK PROBLEM 71.**

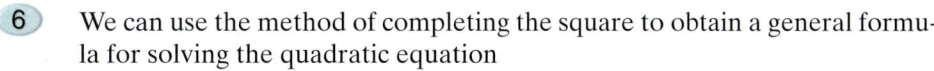 **The solution of the equation in Example 10 can also be obtained by factoring. Rework Example 10 using factoring.**

The Quadratic Formula

⑥ We can use the method of completing the square to obtain a general formula for solving the quadratic equation

$$ax^2 + bx + c = 0, \qquad a \neq 0$$

Note: There is no loss in generality to assume that $a > 0$, since if $a < 0$ we can multiply both sides by -1 to obtain an equivalent equation with a positive leading coefficient.

As in Example 10, we begin by rearranging the terms as

$$ax^2 + bx = -c, \qquad a > 0$$

Since $a > 0$, we can divide both sides by a to get

$$x^2 + \frac{b}{a}x = -\frac{c}{a}$$

Now the coefficient of x^2 is 1. To complete the square on the left side, add the square of $\frac{1}{2}$ the coefficient of x; that is, add

$$\left(\frac{1}{2} \cdot \frac{b}{a}\right)^2 = \frac{b^2}{4a^2}$$

to each side. Then

$$x^2 + \frac{b}{a}x + \frac{b^2}{4a^2} = \frac{b^2}{4a^2} - \frac{c}{a}$$

$$\left(x + \frac{b}{2a}\right)^2 = \frac{b^2 - 4ac}{4a^2} \qquad \frac{b^2}{4a^2} - \frac{c}{a} = \frac{b^2}{4a^2} - \frac{4ac}{4a^2} = \frac{b^2 - 4ac}{4a^2} \qquad (4)$$

Provided that $b^2 - 4ac \geq 0$, we now can use the Square Root Method to get

$$x + \frac{b}{2a} = \pm\sqrt{\frac{b^2 - 4ac}{4a^2}}$$

$$x + \frac{b}{2a} = \frac{\pm\sqrt{b^2 - 4ac}}{2a} \qquad \text{The square root of a quotient equals the quotient of the square roots. Also, } \sqrt{4a^2} = 2a \text{ since } a > 0.$$

$$x = -\frac{b}{2a} \pm \frac{\sqrt{b^2 - 4ac}}{2a} \qquad \text{Add } -\frac{b}{2a} \text{ to both sides.}$$

$$x = \frac{-b \pm \sqrt{b^2 - 4ac}}{2a} \qquad \text{Combine the quotients on the right.}$$

What if $b^2 - 4ac$ is negative? Then equation (4) states that the left expression (a real number squared) equals the right expression (a negative number). Since this occurrence is impossible for real numbers, we conclude that if $b^2 - 4ac < 0$ the quadratic equation has no *real* solution.*

We now state the *quadratic formula*.

Theorem

Quadratic Formula

Consider the quadratic equation

$$ax^2 + bx + c = 0 \qquad a \neq 0$$

If $b^2 - 4ac < 0$, this equation has no real solution.
If $b^2 - 4ac \geq 0$, the real solution(s) of this equation is (are) given by the **quadratic formula:**

$$x = \frac{-b \pm \sqrt{b^2 - 4ac}}{2a}$$

---

*We consider quadratic equations where $b^2 - 4ac$ is negative in Section 5.3, which may be covered anytime after completing this section without any loss of continuity.

The quantity $b^2 - 4ac$ is called the **discriminant** of the quadratic equation, because its value tells us whether the equation has real solutions. In fact, it also tells us how many solutions to expect.

Discriminant of a Quadratic Equation

For a quadratic equation $ax^2 + bx + c = 0$:

1. If $b^2 - 4ac > 0$, there are two unequal real solutions.
2. If $b^2 - 4ac = 0$, there is a repeated real solution, a root of multiplicity 2.
3. If $b^2 - 4ac < 0$, there is no real solution.

When asked to find the real solutions, if any, of a quadratic equation, always evaluate the discriminant first to see how many real solutions there are.

EXAMPLE 11

Solving a Quadratic Equation by Using the Quadratic Formula and by Graphing

Find the real solutions, if any, of the equation $3x^2 - 5x + 1 = 0$.

Algebraic Solution

The equation is in standard form, so we compare it to $ax^2 + bx + c = 0$ to find a, b, and c.

$$3x^2 - 5x + 1 = 0$$
$$ax^2 + bx + c = 0, \qquad a = 3, b = -5, c = 1$$

With $a = 3$, $b = -5$, and $c = 1$, we evaluate the discriminant $b^2 - 4ac$.

$$b^2 - 4ac = (-5)^2 - 4(3)(1) = 25 - 12 = 13$$

Since $b^2 - 4ac > 0$, there are two unequal real solutions.

We use the quadratic formula with $a = 3$, $b = -5$, $c = 1$, and $b^2 - 4ac = 13$.

$$x = \frac{-b \pm \sqrt{b^2 - 4ac}}{2a} = \frac{-(-5) \pm \sqrt{13}}{2(3)} = \frac{5 \pm \sqrt{13}}{6}$$

The solution set is $\left\{ \dfrac{5 - \sqrt{13}}{6}, \dfrac{5 + \sqrt{13}}{6} \right\}$. These solutions are exact.

Graphing Solution

Figure 34 shows the graph of the equation

$$Y_1 = 3x^2 - 5x + 1$$

As expected, we see that there are two x-intercepts: one between 0 and 1, the other between 1 and 2. The solutions to the equation are 0.23 and 1.43, rounded to two decimal places. These solutions are approximate.

Figure 34

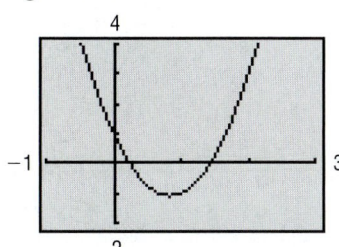

NOW WORK PROBLEMS **81** AND **91.**

EXAMPLE 12   **Solving a Quadratic Equation by Using the Quadratic Formula and by Graphing**

Find the real solutions, if any, of the equation

$$3x^2 + 2 = 4x$$

Algebraic Solution   The equation, as given, is not in standard form.

$$3x^2 + 2 = 4x$$
$$3x^2 - 4x + 2 = 0 \qquad \text{Put in standard form.}$$
$$ax^2 + bx + c = 0 \qquad \text{Compare to standard form.}$$

With $a = 3, b = -4$, and $c = 2$, we find that

$$b^2 - 4ac = (-4)^2 - 4(3)(2) = 16 - 24 = -8$$

Since $b^2 - 4ac < 0$, the equation has no real solution.

Graphing Solution   We use the standard form of the equation and graph

$$Y_1 = 3x^2 - 4x + 2$$

See Figure 35. We see that there are no x-intercepts, so the equation has no real solution, as expected.

Figure 35

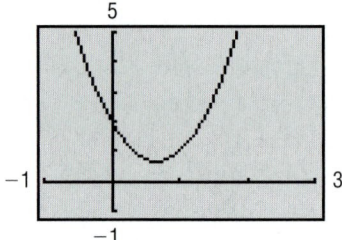

NOW WORK PROBLEM **83.**

SUMMARY   **Procedure for Solving a Quadratic Equation Algebraically**

To solve a quadratic equation, first put it in standard form:

$$ax^2 + bx + c = 0$$

Then:

STEP 1: Identify a, b, and c.
STEP 2: Evaluate the discriminant, $b^2 - 4ac$.
STEP 3: (a) If the discriminant is negative, the equation has no real solution.
(b) If the discriminant is nonnegative, determine whether the left side can be factored. If you can easily spot factors, use the factoring method to solve the equation. Otherwise, use the quadratic formula or the method of completing the square.

1.3 Concepts and Vocabulary

In Problems 1–3, fill in the blanks.

1. When a quadratic equation has a repeated solution, it is called a(n) _____ root or a root of _____ _____.
2. The quantity $b^2 - 4ac$ is called the _____ of a quadratic equation. If it is _____, the equation has no real solution.
3. Linear equations are sometimes called _____ -degree equations. Quadratic equations are sometimes called _____ -degree equations.

In Problems 4–6, answer True or False to each statement.

4. To complete the square in the expression $x^2 + 8x$, add 16.
5. When solving equations using a graphing utility, the solutions are always exact.
6. A quadratic equation that is in the form $ax^2 + bx + c = 0$ is said to be in standard form.

7. Describe four ways to solve a quadratic equation. Which of the methods yield exact solutions?
8. Explain the benefits of evaluating the discriminant of a quadratic equation before attempting to solve it.

1.3 Exercises

In Problems 1–12, use a graphing utility to approximate the real solutions, if any, of each equation rounded to two decimal places. All solutions lie between −10 and 10.

1. $x^3 - 4x + 2 = 0$
2. $x^3 - 8x + 1 = 0$
3. $-2x^4 + 5 = 3x - 2$
4. $-x^4 + 1 = 2x^2 - 3$
5. $x^4 - 2x^3 + 3x - 1 = 0$
6. $3x^4 - x^3 + 4x^2 - 5 = 0$
7. $-x^3 - \frac{5}{3}x^2 + \frac{7}{2}x + 2 = 0$
8. $-x^4 + 3x^3 + \frac{7}{3}x^2 - \frac{15}{2}x + 2 = 0$
9. $-\frac{2}{3}x^4 - 2x^3 + \frac{5}{2}x = -\frac{2}{3}x^2 + \frac{1}{2}$
10. $\frac{1}{4}x^3 - 5x = \frac{1}{5}x^2 - 4$
11. $x^4 - 5x^2 + 2x + 11 = 0$
12. $-3x^4 + 8x^2 - 2x - 9 = 0$

In Problems 13–36, solve each equation algebraically. Verify your solution using a graphing utility.

13. $3x + 2 = 6$
14. $2x + 7 = 5$
15. $2t - 6 = 3 - t$
16. $5y + 6 = -18 - y$
17. $6 - x = 2x + 12$
18. $3 - 2x = 8 - x$
19. $3 + 2n = 5n + 7$
20. $3 - 2m = 3m + 1$
21. $2(3 + 2x) = 3(x - 4)$
22. $3(2 - x) = 2x - 1$
23. $8x - (2x + 1) = 3x - 13$
24. $5 - (2x - 1) = 10 - x$
25. $\frac{2}{3}p = \frac{1}{2}p + \frac{1}{3}$
26. $\frac{1}{2} - \frac{1}{3}p = \frac{4}{3}$
27. $0.9t = 0.4 + 0.1t$
28. $0.9t = 1 + t$
29. $\frac{x + 1}{3} + \frac{x + 2}{7} = 5$
30. $\frac{2x + 1}{3} + 16 = 3x$
31. $\frac{5}{y} + \frac{4}{y} = 3$
32. $\frac{4}{y} - 5 = \frac{18}{2y}$
33. $(x + 7)(x - 1) = (x + 1)^2$
34. $(x + 2)(x - 3) = (x - 3)^2$
35. $x(2x - 3) = (2x + 1)(x - 4)$
36. $x(1 + 2x) = (2x - 1)(x - 2)$

In Problems 37–60, solve each equation by factoring. Verify your solution using a graphing utility.

37. $x^2 - 9x = 0$
38. $x^2 + 4x = 0$
39. $x^2 - 25 = 0$
40. $x^2 - 9 = 0$
41. $z^2 + z - 6 = 0$
42. $v^2 + 7v + 6 = 0$
43. $2x^2 - 5x - 3 = 0$
44. $3x^2 + 5x + 2 = 0$
45. $3t^2 - 48 = 0$
46. $2y^2 - 50 = 0$
47. $x(x + 8) + 12 = 0$
48. $x(x - 4) = 12$
49. $4x^2 + 9 = 12x$
50. $25x^2 + 16 = 40x$
51. $2x^2 - x = 15$

52. $6x^2 + x = 2$

53. $\dfrac{4(x-2)}{x-3} + \dfrac{3}{x} = \dfrac{-3}{x(x-3)}$

54. $\dfrac{5}{x+4} = 4 + \dfrac{3}{x-2}$

55. $x^3 + x^2 - 20x = 0$

56. $x^3 + 6x^2 - 7x = 0$

57. $x^3 + x^2 - x - 1 = 0$

58. $x^3 + 4x^2 - x - 4 = 0$

59. $x^3 - 3x^2 - 4x + 12 = 0$

60. $x^3 - 3x^2 - x + 3 = 0$

In Problems 61–66, solve each equation by the Square Root Method. Verify your solution using a graphing utility.

61. $x^2 = 25$

62. $x^2 = 36$

63. $(x-1)^2 = 4$

64. $(x+2)^2 = 1$

65. $(2x+3)^2 = 9$

66. $(3x-2)^2 = 4$

In Problems 67–70, what number should be added to complete the square of each expression?

67. $x^2 + 8x$

68. $x^2 - 4x$

69. $x^2 - \dfrac{1}{2}x$

70. $x^2 + \dfrac{1}{3}x$

In Problems 71–76, solve each equation by completing the square. Verify your solution using a graphing utility.

71. $x^2 + 4x = 21$

72. $x^2 - 6x = 13$

73. $x^2 - \dfrac{1}{2}x - \dfrac{3}{16} = 0$

74. $x^2 + \dfrac{2}{3}x - \dfrac{1}{3} = 0$

75. $3x^2 + x - \dfrac{1}{2} = 0$

76. $2x^2 - 3x - 1 = 0$

In Problems 77–94, find the real solutions, if any, of each equation. Use the quadratic formula. Verify your solution using a graphing utility.

77. $x^2 - 4x + 2 = 0$

78. $x^2 + 4x + 2 = 0$

79. $x^2 - 4x - 1 = 0$

80. $x^2 + 6x + 1 = 0$

81. $2x^2 - 5x + 3 = 0$

82. $2x^2 + 5x + 3 = 0$

83. $4y^2 - y + 2 = 0$

84. $4t^2 + t + 1 = 0$

85. $4x^2 = 1 - 2x$

86. $2x^2 = 1 - 2x$

87. $4x^2 = 9x + 2$

88. $5x = 4x^2 + 5$

89. $9t^2 - 6t + 1 = 0$

90. $4u^2 - 6u + 9 = 0$

91. $\dfrac{3}{4}x^2 - \dfrac{1}{4}x - \dfrac{1}{2} = 0$

92. $\dfrac{2}{3}x^2 - x - 3 = 0$

93. $4 - \dfrac{1}{x} - \dfrac{2}{x^2} = 0$

94. $4 + \dfrac{1}{x} - \dfrac{1}{x^2} = 0$

In Problems 95–106, find the exact real solutions, if any, of each equation. Use any method. Verify your solution using a graphing utility.

95. $x^2 - 5 = 0$

96. $x^2 - 6 = 0$

97. $16x^2 - 8x + 1 = 0$

98. $4x^2 - 12x + 9 = 0$

99. $10x^2 - 19x - 15 = 0$

100. $6x^2 + 7x - 20 = 0$

101. $2 + z = 6z^2$

102. $2 = y + 6y^2$

103. $x^2 + \sqrt{2}x = \dfrac{1}{2}$

104. $\dfrac{1}{2}x^2 = \sqrt{2}x + 1$

105. $x^2 + x = 4$

106. $x^2 + x = 1$

In Problems 107–112, use the discriminant to determine whether each quadratic equation has two unequal real solutions, a repeated real solution, or no real solution, without solving the equation.

107. $2x^2 - 6x + 7 = 0$

108. $x^2 + 4x + 7 = 0$

109. $9x^2 - 30x + 25 = 0$

110. $25x^2 - 20x + 4 = 0$

111. $3x^2 + 5x - 8 = 0$

112. $2x^2 - 3x - 7 = 0$

In Problems 113–116, solve each equation. The letters a, b, and c are constants.

113. $ax - b = c, \quad a \neq 0$

114. $1 - ax = b, \quad a \neq 0$

115. $\dfrac{x}{a} + \dfrac{x}{b} = c, \quad a \neq 0, b \neq 0, a \neq -b$

116. $\dfrac{a}{x} + \dfrac{b}{x} = c, \quad c \neq 0$

117. Find the number a for which $x = 4$ is a solution of the equation

$$x + 2a = 16 + ax - 6a$$

118. Find the number b for which $x = 2$ is a solution of the equation

$$x + 2b = x - 4 + 2bx$$

Problems 119–122 list some formulas that occur in applications. Solve each formula for the indicated variable.

119. Electricity $\dfrac{1}{R} = \dfrac{1}{R_1} + \dfrac{1}{R_2}$ for R

120. Finance $A = P(1 + rt)$ for r

121. Mechanics $F = \dfrac{mv^2}{R}$ for R

122. Chemistry $PV = nRT$ for T

123. One of the steps in the following list contains an error. Identify it and explain what is wrong.

$$
\begin{aligned}
x &= 2 & (1)\\
3x - 2x &= 2 & (2)\\
3x &= 2x + 2 & (3)\\
x^2 + 3x &= x^2 + 2x + 2 & (4)\\
x^2 + 3x - 10 &= x^2 + 2x - 8 & (5)\\
(x - 2)(x + 5) &= (x - 2)(x + 4) & (6)\\
x + 5 &= x + 4 & (7)\\
1 &= 0 & (8)
\end{aligned}
$$

124. Which of the following pairs of equations are equivalent? Explain.

(a) $x^2 = 9$; $x = 3$
(b) $x = \sqrt{9}$; $x = 3$
(c) $(x - 1)(x - 2) = (x - 1)^2$; $x - 2 = x - 1$

125. The equation

$$\frac{5}{x + 3} + 3 = \frac{8 + x}{x + 3}$$

has no solution, yet when we go through the process of solving it we obtain $x = -3$. Write a brief paragraph to explain what causes this to happen.

126. Make up an equation that has no solution and give it to a fellow student to solve. Ask the fellow student to write a critique of your equation.

127. The word *quadratic* seems to imply four (*quad*), yet a quadratic equation is an equation that involves a polynomial of degree 2. Investigate the origin of the term *quadratic* as it is used in the expression *quadratic equation*. Write a brief essay on your findings.

1.4 SETTING UP EQUATIONS; APPLICATIONS

OBJECTIVES
1. Translate Verbal Descriptions into Mathematical Expressions
2. Set Up Applied Problems
3. Solve Interest Problems
4. Solve Mixture Problems
5. Solve Uniform Motion Problems
6. Solve Constant Rate Job Problems

Applied (word) problems do not come in the form "Solve the equation . . . ." Instead, they supply information using words, a verbal description of the real problem. So, to solve applied problems, we must be able to translate the verbal description into the language of mathematics. We do this by using variables to represent unknown quantities and then finding relationships (such as equations) that involve these variables. The process of doing all this is called **mathematical modeling.**

Any solution to the mathematical problem must be checked against the mathematical problem, the verbal description, and the real problem. See Figure 36 for an illustration of the **modeling process.**

Figure 36

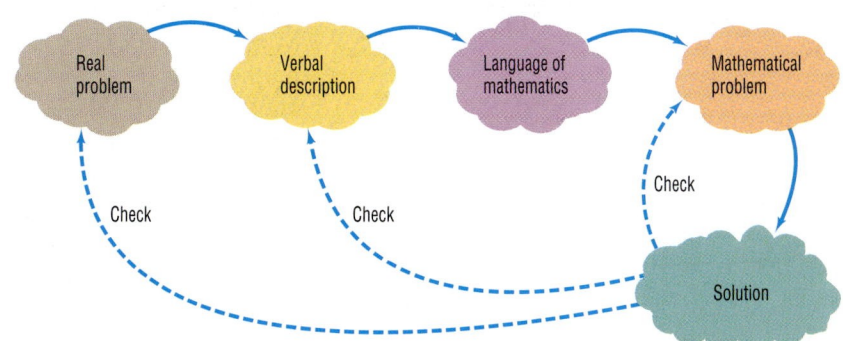

1 Let's look at a few examples that will help you to translate certain words into mathematical symbols.

EXAMPLE 1 **Translating Verbal Descriptions into Mathematical Expressions**

(a) The area of a rectangle is the product of its length and its width.

Translation: If A is used to represent the area, l the length, and w the width, then $A = lw$.

(b) For uniform motion, the velocity of an object equals the distance traveled divided by the time required.

Translation: If v is the velocity, s the distance, and t the time, then $v = s/t$.

(c) A total of $5000 is invested, some in stocks and some in bonds. If the amount invested in stocks is x, express the amount invested in bonds in terms of x.

Translation: If x is the amount invested in stocks, then the amount invested in bonds is $5000 - x$, since their sum is $x + (5000 - x) = 5000$.

(d) Let x denote a number.

The number 5 times as large as x is $5x$.
The number 3 less than x is $x - 3$.
The number that exceeds x by 4 is $x + 4$.
The number that, when added to x, gives 5 is $5 - x$.
■

──── **NOW WORK PROBLEM 1.**

2 Always check the units used to measure the variables of an applied problem. In Example 1(a), if l is measured in feet, then w also must be expressed in feet, and A will be expressed in square feet. In Example 1(b), if v is measured in miles per hour, then the distance s must be expressed in miles and the time t must be expressed in hours. It is a good practice to check units to be sure that they are consistent and make sense.

Although each situation has unique features, we can provide an outline of the steps to follow in setting up applied problems.

> *Steps for Setting Up Applied Problems*
>
> **STEP 1:** Read the problem carefully, perhaps two or three times. Pay particular attention to the question being asked in order to identify what you are looking for. If you can, determine realistic possibilities for the answer.
>
> **STEP 2:** Assign a letter (variable) to represent what you are looking for, and, if necessary, express any remaining unknown quantities in terms of this variable.
>
> **STEP 3:** Make a list of all the known facts, and translate them into mathematical expressions. These may take the form of an equation (or, later, an inequality) involving the variable. If possible, draw an appropriately labeled diagram to assist you. Sometimes a table or chart helps.
>
> **STEP 4:** Solve the equation for the variable, and then answer the question.
>
> **STEP 5:** Check the answer with the facts in the problem. If it agrees, congratulations! If it does not agree, try again.

Let's look at an example.

EXAMPLE 2   Investments

A total of $18,000 is invested, some in stocks and some in bonds. If the amount invested in bonds is half that invested in stocks, how much is invested in each category?

Solution   **STEP 1:**   We are being asked to find the amount of two investments. These amounts must total $18,000. (Do you see why?)

STEP 2:   If we let x equal the amount invested in stocks, then $18,000 - x$ is the amount invested in bonds. [Look back at Example 1(c) to see why.]

STEP 3:   We set up a table:

Amount in Stocks	Amount in Bonds	Reason
x	$18,000 - x$	Total invested is $18,000

We also know that

$$\underbrace{\text{Total amount invested in bonds}}_{18,000 - x} \quad \text{is} \quad \underbrace{\text{one-half that in stocks}}_{\frac{1}{2}(x)}$$

$$\text{=}$$

STEP 4: $18,000 - x = \dfrac{1}{2}x$

$18,000 = x + \dfrac{1}{2}x$ Add x to both sides.

$18,000 = \dfrac{3}{2}x$ Simplify.

$\left(\dfrac{2}{3}\right)18,000 = \left(\dfrac{2}{3}\right)\left(\dfrac{3}{2}x\right)$ Multiply both sides by $\dfrac{2}{3}$.

$12,000 = x$ Simplify.

Thus, $12,000 is invested in stocks and $18,000 - $12,000 = $6000 is invested in bonds.

STEP 5:   The total invested is $12,000 + $6000 = $18,000, and the amount in bonds ($6000) is half that in stocks ($12,000).   ■

 NOW WORK PROBLEM **11.**

EXAMPLE 3   Determining an Hourly Wage

Shannon grossed $435 one week by working 52 hours. Her employer pays time-and-a-half for all hours worked in excess of 40 hours. With this information, can you determine Shannon's regular hourly wage?

Solution   **STEP 1:**   We are looking for an hourly wage. Our answer will be in dollars per hour.

STEP 2:   Let x represent the regular hourly wage; x is measured in dollars per hour.

Step 3:   We set up a table:

	Hours Worked	Hourly Wage	Salary
Regular	40	x	$40x$
Overtime	12	$1.5x$	$12(1.5x) = 18x$

The sum of regular salary plus overtime salary will equal \$435. From the table, $40x + 18x = 435$.

Step 4: $40x + 18x = 435$

$\qquad\quad 58x = 435$

$\qquad\qquad x = 7.50$

Shannon's regular hourly wage is \$7.50 per hour.

Step 5:   Forty hours yields a salary of $40(7.50) = \$300$, and 12 hours of overtime yields a salary of $12(1.5)(7.50) = \$135$, for a total of \$435.

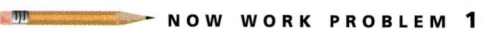 **NOW WORK PROBLEM 15.**

Interest

③   The next example involves **interest.** Interest is money paid for the use of money. The total amount borrowed (whether by an individual from a bank in the form of a loan or by a bank from an individual in the form of a savings account) is called the **principal.** The **rate of interest,** expressed as a percent, is the amount charged for the use of the principal for a given period of time, usually on a yearly (that is, per annum) basis.

> **Simple Interest Formula**
>
> If a principal of P dollars is borrowed for a period of t years at a per annum interest rate r, expressed as a decimal, the interest I charged is
>
> $$I = Prt \qquad\qquad\qquad (1)$$
>
> Interest charged according to formula (1) is called **simple interest.**

EXAMPLE 4   **Finance: Computing Interest on a Loan**

Suppose that Juanita borrows \$500 for 6 months at the simple interest rate of 9% per annum. What is the interest that Juanita will be charged on this loan? How much does Juanita owe after 6 months?

Solution   The rate of interest is given per annum, so the actual time that the money is borrowed must be expressed in years. The interest charged would be the principal ($P = \$500$) times the rate of interest ($r = 9\% = 0.09$) times the time in years $\left(t = \dfrac{6}{12} = \dfrac{1}{2} \right)$:

$$\text{Interest charged} = I = Prt = (500)(0.09)\left(\dfrac{1}{2}\right) = \$22.50$$

Juanita will owe the amount she borrowed, plus interest; that is, she will owe $500 + $22.50 = $522.50. ▪

EXAMPLE 5 **Financial Planning**

Candy has $70,000 to invest and requires an overall rate of return of 9%. She can invest in a safe, government-insured certificate of deposit, but it only pays 8%. To obtain 9%, she agrees to invest some of her money in noninsured corporate bonds paying 12%. How much should be placed in each investment to achieve her goals?

Solution **STEP 1:** The question is asking for two dollar amounts: the principal to invest in the corporate bonds and the principal to invest in the certificate of deposit.

STEP 2: We let x represent the amount (in dollars) to be invested in the bonds. Then $70{,}000 - x$ is the amount that will be invested in the certificate. (Do you see why?)

STEP 3: We set up a table:

	Principal $	Rate %	Time yr	Interest $
Bonds	x	12% = 0.12	1	$0.12x$
Certificate	$70{,}000 - x$	8% = 0.08	1	$0.08(70{,}000 - x)$
Total	70,000	9% = 0.09	1	$0.09(70{,}000) = 6300$

Since the total interest from the investments is equal to $0.09(70{,}000) = 6300$, we have the equation

$$0.12x + 0.08(70{,}000 - x) = 6300$$

(Note that the units are consistent: the unit is dollars on each side.)

STEP 4: $0.12x + 5600 - 0.08x = 6300$

$$0.04x = 700$$
$$x = 17{,}500$$

Candy should place $17,500 in the bonds and $70{,}000 - \$17{,}500 = \$52{,}500$ in the certificate.

STEP 5: The interest on the bonds after 1 year is $0.12(\$17{,}500) = \$2100$; the interest on the certificate after 1 year is $0.08(\$52{,}500) = \$4200$. The total annual interest is $6300, the required amount. ▪

 **NOW WORK PROBLEM 27.**

Mixture Problems

④ Oil refineries sometimes produce gasoline that is a blend of two or more types of fuel; bakeries occasionally blend two or more types of flour for their bread. These problems are referred to as **mixture problems** because they combine two or more quantities to form a mixture.

EXAMPLE 6   **Blending Coffees**

The manager of a Starbucks store decides to experiment with a new blend of coffee. She will mix some B grade Colombian coffee that sells for $5 per pound with some A grade Arabica coffee that sells for $10 per pound to get 100 pounds of the new blend. The selling price of the new blend is to be $7 per pound, and there is to be no difference in revenue from selling the new blend versus selling the other types. How many pounds of the B grade Colombian and A grade Arabica coffees are required?

Solution   **STEP 1:**   The question is asking for the amount of B grade Colombian and A grade Arabica coffee needed to make the new blend of coffee.

STEP 2:   Let x represent the number of pounds of the B grade Colombian coffee. Then $100 - x$ equals the number of pounds of the A grade Arabica coffee. See Figure 37.

Figure 37

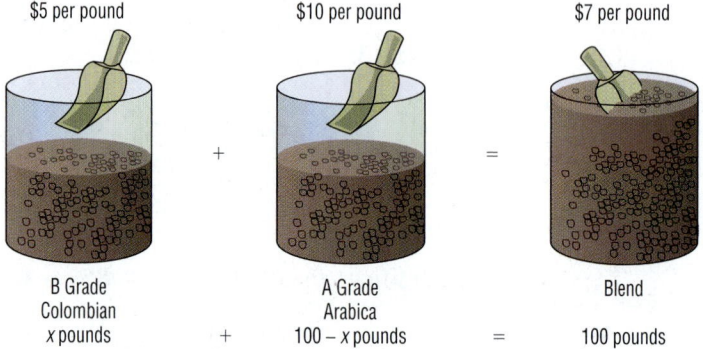

$5 per pound $10 per pound $7 per pound

B Grade   A Grade   Blend
Colombian   Arabica
x pounds   + $100 - x$ pounds   =   100 pounds

STEP 3:   We set up a table.

	Price $/pound	Number of Pounds	Revenue
B grade Colombian	5	x	$5x$
A grade Arabica	10	$100 - x$	$10(100 - x)$
Blend	7	100	700

Since there is to be no difference in revenue between selling the A and B grades separately versus the blend, we have

$$\left(\begin{array}{c}\text{Price per pound}\\\text{of B grade}\end{array}\right)\left(\begin{array}{c}\text{Pounds}\\\text{B grade}\end{array}\right)+\left(\begin{array}{c}\text{Price per pound}\\\text{of A grade}\end{array}\right)\left(\begin{array}{c}\text{Pounds}\\\text{A grade}\end{array}\right)=\left(\begin{array}{c}\text{Price per pound}\\\text{of blend}\end{array}\right)\left(\begin{array}{c}\text{Pound}\\\text{blend}\end{array}\right)$$

$$\$5 \quad \cdot \quad x \quad + \quad \$10 \quad \cdot (100 - x) = \quad (\$7) \quad (100)$$

We have the equation

$$5x + 10(100 - x) = 700$$

STEP 4: $5x + 10(100 - x) = 700$

$5x + 1000 - 10x = 700$

$-5x = -300$

$x = 60$

The manager should blend 60 pounds of B grade Colombian coffee with $100 - 60 = 40$ pounds of A grade coffee to get the desired blend.

STEP 5: The 60 pounds of B grade coffee would sell for $(\$5)(60) = \$300$ and the 40 pounds of A grade coffee would sell for $(\$10)(40) = \$400$; the total revenue, $700, equals the revenue obtained from selling the blend, as desired. ■

▬ **N O W   W O R K   P R O B L E M   31.**

Uniform Motion

⑤ Objects that move at a constant velocity are said to be in **uniform motion.** When the average velocity of an object is known, it can be interpreted as its constant velocity. For example, a bicyclist traveling at an average velocity of 25 miles per hour is in uniform motion.

> **Uniform Motion Formula**
>
> If an object moves at an average velocity v, the distance s covered in time t is given by the formula
>
> $$s = vt \qquad (2)$$

That is, Distance $=$ Velocity $\cdot$ Time.

EXAMPLE 7 **Physics: Uniform Motion**

Tanya, who is a long-distance runner, runs at an average velocity of 8 miles per hour (mph). Two hours after Tanya leaves your house, you leave in your Honda and follow the same route. If your average velocity is 40 mph, how long will it be before you catch up to Tanya? How far will each of you be from your home?

Solution **STEP 1:** The question is asking how long it will be before you catch up to Tanya and how far Tanya and you will be from home at that time.

STEP 2: Refer to Figure 38. We let t represent the time (in hours) that it takes the Honda to catch up with Tanya. When this occurs, the total time elapsed for Tanya is $t + 2$ hours.

Figure 38

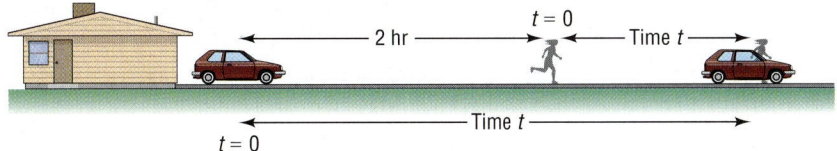

STEP 3: We set up a table.

	Velocity mph	Time hr	Distance mi
Tanya	8	$t + 2$	$8(t + 2)$
Honda	40	t	$40t$

Since the distance traveled is the same, we are led to the following equation:

$$8(t + 2) = 40t$$

STEP 4: Solve this equation for t:

$$8(t + 2) = 40t$$
$$8t + 16 = 40t$$
$$32t = 16$$
$$t = \frac{1}{2} \text{ hour}$$

It will take the Honda $\frac{1}{2}$ hour to catch up to Tanya. Each of you will have gone 20 miles.

STEP 5: In 2.5 hours, Tanya traveled a distance of $(2.5)(8) = 20$ miles. In $\frac{1}{2}$ hour, the Honda travels a distance of $\left(\frac{1}{2}\right)(40) = 20$ miles. ■

EXAMPLE 8   **Physics: Uniform Motion**

A motorboat heads upstream a distance of 24 miles on the Illinois River, whose current is running at 3 miles per hour. The trip up and back takes 6 hours. Assuming that the motorboat maintained a constant speed relative to the water, what was its speed?

Solution   **STEP 1:** The question is asking for the constant speed of the motorboat relative to the water.

STEP 2: See Figure 39. We let v represent the constant speed of the motorboat relative to the water. Then the true speed going upstream is $v - 3$ miles per hour, and the true speed going downstream is $v + 3$ miles per hour. Since Distance = Velocity × Time, then Time = Distance/Velocity.

STEP 3: We set up a table.

Figure 39

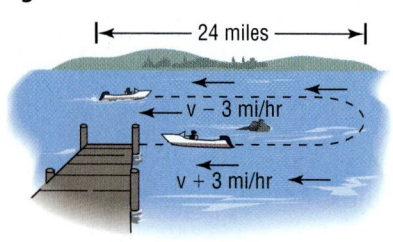

	Velocity mph	Distance mi	Time = Distance/Velocity hr
Upstream	$v - 3$	24	$\dfrac{24}{v - 3}$
Downstream	$v + 3$	24	$\dfrac{24}{v + 3}$

Since the total time up and back is 6 hours, we have

$$\frac{24}{v - 3} + \frac{24}{v + 3} = 6$$

STEP 4: We solve the equation for v:

$$\frac{24}{v-3} + \frac{24}{v+3} = 6$$

$$\frac{24(v+3) + 24(v-3)}{(v-3)(v+3)} = 6 \qquad \text{Add the expressions on the left side.*}$$

$$\frac{48v}{v^2 - 9} = 6 \qquad \text{Simplify}$$

$$48v = 6(v^2 - 9) \qquad \text{Clear fractions.}$$

$$8v = v^2 - 9 \qquad \text{Divide each side by 6.}$$

$$v^2 - 8v - 9 = 0 \qquad \text{Place the quadratic equation in standard form.}$$

$$(v-9)(v+1) = 0 \qquad \text{Factor}$$

$$v = 9 \quad \text{or} \quad v = -1$$

We discard the solution $v = -1$ mile per hour, so the speed of the motorboat relative to the water is 9 miles per hour.

STEP 5: If the speed of the boat is 9 miles per hour, it will take $\dfrac{24}{9-3} = 4$ hours to go upstream and $\dfrac{24}{9+3} = 2$ hours to go downstream, for a total of 6 hours. ∎

 **NOW WORK PROBLEM 37.**

Constant Rate Jobs

⑥ This section involves jobs that are performed at a **constant rate.** Our assumption is that, if a job can be done in t units of time, $\dfrac{1}{t}$ of the job is done in 1 unit of time. Let's look at an example.

EXAMPLE 9 | ## Working Together to Do a Job

At 10 AM Danny is asked by his father to weed the garden. From past experience, Danny knows that this will take him 4 hours, working alone. His older brother, Mike, when it is his turn to do this job, requires 6 hours. Since Mike wants to go golfing with Danny and has a reservation for 1 PM, he agrees to help Danny. Assuming no gain or loss of efficiency, when will they finish if they work together? Can they make the golf date?

Solution **STEP 1:** The question is asking for the time it will take for Mike and Danny to weed the garden working together.

---

*You may wish to review Adding Rational Expressions, pages 62–66, in Section R.7.

STEP 2: Now let t represent the time it takes Mike and Danny to weed the garden working together. Then, in 1 hour, working together, $\dfrac{1}{t}$ of the job is completed.

STEP 3: In 1 hour, working alone, Danny does $\dfrac{1}{4}$ of the job, and in 1 hour, working alone, Mike does $\dfrac{1}{6}$ of the job.

We set up a table.

	Hours to Do Job	Part of Job Done in 1 Hour
Danny	4	$\dfrac{1}{4}$
Mike	6	$\dfrac{1}{6}$
Together	t	$\dfrac{1}{t}$

We reason as follows:

$$\left(\begin{array}{c}\text{Part done by Danny}\\\text{in 1 hour}\end{array}\right) + \left(\begin{array}{c}\text{Part done by Mike}\\\text{in 1 hour}\end{array}\right) = \left(\begin{array}{c}\text{Part done together}\\\text{in 1 hour}\end{array}\right)$$

This leads to the equation: $\dfrac{1}{4} + \dfrac{1}{6} = \dfrac{1}{t}$.

STEP 4: We solve the equation for t.

$$\frac{1}{4} + \frac{1}{6} = \frac{1}{t}$$

$$\frac{3+2}{12} = \frac{1}{t}$$

$$\frac{5}{12} = \frac{1}{t}$$

$$5t = 12$$

$$t = \frac{12}{5}$$

It will take $\dfrac{12}{5} = 2$ hours, 24 minutes to do the job together, so they will finish at 12:24 pm

STEP 5: After 1 hour, working separately, $\dfrac{1}{4} + \dfrac{1}{6} = \dfrac{3}{12} + \dfrac{2}{12} = \dfrac{5}{12}$ of the job will be completed. After 1 hour, working together, $\dfrac{1}{\frac{12}{5}} = \dfrac{5}{12}$ of the job will be completed.

NOW WORK PROBLEM 41.

1.4 Concepts and Vocabulary

In Problems 1–3, fill in the blanks.

1. Using variables to represent unknown quantities and then finding relationships that involve these variables is referred to as _____ _____.

2. The money paid for the use of money is _____.

3. Objects that move at a constant velocity are said to be in _____ _____.

In Problems 4 and 5, answer True or False to each statement.

4. The amount charged for the use of principal for a given period of time is called the rate of interest.

5. If an object moves at an average velocity v, the distance s covered in time t is given by the formula $s = vt$.

6. Suppose that you want to mix two coffees in order to obtain 100 pounds of the blend. If x represents the number of pounds of coffee A, write an algebraic expression that represents the number of pounds of coffee B.

1.4 Exercises

In Problems 1–10, translate each sentence into a mathematical equation. Be sure to identify the meaning of all symbols.

1. **Geometry**   The area of a circle is the product of the number π and the square of the radius.

2. **Geometry**   The circumference of a circle is the product of the number π and twice the radius.

3. **Geometry**   The area of a square is the square of the length of a side.

4. **Geometry**   The perimeter of a square is four times the length of a side.

5. **Physics**   Force equals the product of mass and acceleration.

6. **Physics**   Pressure is force per unit area.

7. **Physics**   Work equals force times distance.

8. **Physics**   Kinetic energy is one-half the product of the mass and the square of the velocity.

9. **Business**   The total variable cost of manufacturing x dishwashers is $150 per dishwasher times the number of dishwashers manufactured.

10. **Business**   The total revenue derived from selling x dishwashers is $250 per dishwasher times the number of dishwashers manufactured.

11. **Finance**   A total of $20,000 is to be invested, some in bonds and some in Certificates of Deposit (CDs). If the amount invested in bonds is to exceed that in CDs by $2000, how much will be invested in each type of instrument?

12. **Finance**   A total of $10,000 is to be divided up between Yani and Diane, with Diane to receive $2000 less than Yani. How much will each receive?

13. **Finance**   An inheritance of $900,000 is to be divided among David, Paige, and Dan in the following manner: Paige is to receive $\frac{3}{4}$ of what David gets, while Dan gets $\frac{1}{2}$ of what David gets. How much does each receive?

14. **Sharing the Cost of a Pizza**   Carole and Canter agree to share the cost of an $18 pizza based on how much each ate. If Carole ate $\frac{2}{3}$ the amount Canter ate, how much should each pay?

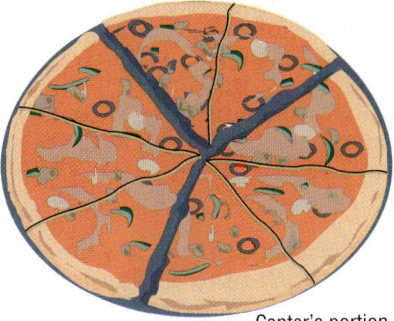

Carole's portion

Canter's portion

15. **Computing Hourly Wages**   Laura, who is paid time-and-a-half for hours worked in excess of 40 hours, had gross weekly wages of $442 for 48 hours worked. What is her regular hourly rate?

16. **Computing Hourly Wages**   Amy is paid time-and-a-half for hours worked in excess of 40 hours and double-time for hours worked on Sunday. If Amy had gross weekly wages of $342 for working 50 hours, 4 of which were on Sunday, what is her regular hourly rate?

17. **Football**   In an NFL football game, the Bears scored a total of 41 points, including one safety (2 points) and two field goals (3 points each). After scoring a touchdown (6 points), a team is given the chance to score 1 or 2 extra points. The Bears, in trying to score 1 extra point after each touchdown, missed 2 extra points. How many touchdowns did the Bears get?

18. **Basketball**   In a basketball game, the Bulls scored a total of 103 points and made three times as many field goals (2 points each) as free throws (1 point each). They also made eleven 3-point baskets. How many field goals did they have?

19. **Geometry**   The perimeter of a rectangle is 60 feet. Find its length and width if the length is 8 feet longer than the width.

20. **Geometry**   The perimeter of a rectangle is 42 meters. Find its length and width if the length is twice the width.

21. **Computing Grades**   Going into the final exam, which will count as two tests, Brooke has test scores of 80, 83, 71, 61, and 95. What score does Brooke need on the final in order to have an average score of 80?

22. **Computing Grades**   Going into the final exam, which will count as two-thirds of the final grade, Mike has test scores of 86, 80, 84, and 90. What score does Mike need on the final in order to earn a B, which requires an average score of 80? What does he need to earn an A, which requires an average of 90?

23. **Business: Discount Pricing**   A builder of tract homes reduced the price of a model by 15%. If the new price is $125,000, what was its original price? How much can be saved by purchasing the model?

24. **Business: Discount Pricing**   A car dealer, at a year-end clearance, reduces the list price of last year's models by 15%. If a certain four-door model has a discounted price of $8000, what was its list price? How much can be saved by purchasing last year's model?

25. **Business: Marking up the Price of Books**   A college book store marks up the price that it pays the publisher for a book by 25%. If the selling price of a book is $56.00, how much did the book store pay for this book?

26. **Personal Finance: Cost of a Car**   The suggested list price of a new car is $12,000. The dealer's cost is 85% of list. How much will you pay if the dealer is willing to accept $100 over cost for the car?

27. **Financial Planning**   Betsy, a recent retiree, requires $6000 per year in extra income. She has $50,000 to invest and can invest in B-rated bonds paying 15% per year or in a Certificate of Deposit (CD) paying 7% per year. How much money should be invested in each to realize exactly $6000 in interest per year?

28. **Financial Planning**   After 2 years, Betsy (see Problem 27) finds that she now will require $7000 per year. Assuming that the remaining information is the same, how should the money be reinvested?

29. **Banking**   A bank loaned out $12,000, part of it at the rate of 8% per year and the rest at the rate of 18% per year. If the interest received totaled $1000, how much was loaned at 8%?

30. **Banking**   Wendy, a loan officer at a bank, has $1,000,000 to lend and is required to obtain an average return of 18% per year. If she can lend at the rate of 19% or at the rate of 16%, how much can she lend at the 16% rate and still meet her requirement?

31. **Blending Teas**   The manager of a store that specializes in selling tea decides to experiment with a new blend. She will mix some Earl Gray tea that sells for $5 per pound with some Orange Pekoe tea that sells for $3 per pound to get 100 pounds of the new blend. The selling price of the new blend is to be $4.50 per pound and there is to be no difference in revenue from selling the new blend versus selling the other types. How many pounds of the Earl Gray tea and Orange Pekoe tea are required?

32. **Business: Blending Coffee**   A coffee manufacturer wants to market a new blend of coffee that will sell for $3.90 per pound by mixing two coffees that sell for $2.75 and $5 per pound, respectively. What amounts of each coffee should be blended to obtain the desired mixture?

   [**Hint:** Assume that the total weight of the desired blend is 100 pounds.]

33. **Business: Mixing Nuts**   A nut store normally sells cashews for $4.00 per pound and peanuts for $1.50 per pound. But at the end of the month, the peanuts had not sold well, so in order to sell 60 pounds of peanuts, the manager decided to mix the 60 pounds of peanuts with some cashews and sell the mixture for $2.50 per pound. How many pounds of cashews should be mixed with the peanuts to ensure no change in the profit?

34. **Business: Mixing Candy**   A candy store sells boxes of candy containing caramels and cremes. Each box sells for $12.50 and holds 30 pieces of candy (all pieces are the same size). If the caramels cost $0.25 to produce and the cremes cost $0.45 to produce, how many of each should be in a box to make a profit of $3?

35. **Chemistry: Mixing Acids**   How many ounces of pure water should be added to 20 ounces of a 40% solution of muriatic acid to obtain a 30% solution of muriatic acid?

36. **Chemistry: Mixing Acids**   How many cubic centimeters of pure hydrochloric acid should be added to 20 cc of a 30% solution of hydrochloric acid to obtain a 50% solution?

37. **Physics: Uniform Motion**   A Metra commuter train leaves Union Station in Chicago at 12 noon. Two hours later, an Amtrak train leaves on the same track, traveling at an average speed that is 50 miles per hour faster than the Metra train. At 3 PM, the Amtrak train is 10 miles behind the commuter train. How fast is each going?

38. **Physics: Uniform Motion**   Two cars enter the Florida Turnpike at Commercial Boulevard at 8:00 AM, each heading for Wildwood. One car's average speed is 10 miles per hour more than the other's. The faster car arrives at Wildwood at 11:00 AM, $\frac{1}{2}$ hour before the other car. What is the average speed of each car? How far did each travel?

39. Physics: Uniform Motion A motorboat can maintain a constant speed of 16 miles per hour relative to the water. The boat makes a trip upstream to a certain point in 20 minutes; the return trip takes 15 minutes. What is the speed of the current? (See the figure.)

40. Physics: Uniform Motion A motorboat heads upstream on a river that has a current of 3 miles per hour. The trip upstream takes 5 hours, while the return trip takes 2.5 hours. What is the speed of the motorboat? (Assume that the motorboat maintains a constant speed relative to the water.)

41. Working Together on a Job Trent can deliver his newspapers in 30 minutes. It takes Lois 20 minutes to do the same route. How long would it take them to deliver the newspapers if they work together?

42. Working Together on a Job Patrick, by himself, can paint four rooms in 10 hours. If he hires April to help, they can do the same job together in 6 hours. If he lets April work alone, how long will it take her to paint four rooms?

43. Emptying Oil Tankers An oil tanker can be emptied by the main pump in 4 hours. An auxiliary pump can empty the tanker in 9 hours. If the main pump is started at 9 AM, when should the auxiliary pump be started so that the tanker is emptied by noon?

44. Using Two Pumps A 5 horsepower (hp) pump can empty a pool in 5 hours. A smaller, 2 hp pump empties the same pool in 8 hours. The pumps are used together to begin emptying this pool. After two hours, the 2 hp pump breaks down. How long will it take the larger pump to empty the pool?

45. Dimensions of a Window The area of the opening of a rectangular window is to be 143 square feet. If the length is to be 2 feet more than the width, what are the dimensions?

46. Dimensions of a Window The area of a rectangular window is to be 306 square centimeters. If the length exceeds the width by 1 centimeter, what are the dimensions?

47. Geometry Find the dimensions of a rectangle whose perimeter is 26 meters and whose area is 40 square meters.

48. Watering a Field An adjustable water sprinkler that sprays water in a circular pattern is placed at the center of a square field whose area is 1250 square feet (see the figure). What is the shortest radius setting that can be used if the field is to be completely enclosed within the circle?

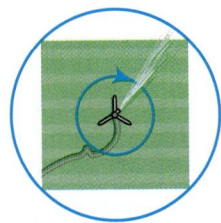

49. Physics: Uniform Motion A motorboat maintained a constant speed of 15 miles per hour relative to the water in going 10 miles upstream and then returning. The total time for the trip was 1.5 hours. Use this information to find the speed of the current.

50. Dimensions of a Patio A contractor orders 8 cubic yards of premixed cement, all of which is to be used to pour a rectangular patio that will be 4 inches thick. If the length of the patio is specified to be twice the width, what will be the patio dimensions? (1 cubic yard = 27 cubic feet)

51. Enclosing a Garden A gardener has 46 feet of fencing to be used to enclose a rectangular garden that has a border 2 feet wide surrounding it (see the figure).
(a) If the length of the garden is to be twice its width, what will be the dimensions of the garden?
(b) What is the area of the garden?
(c) If the length and width of the garden were to be the same, what would be the dimensions of the garden?
(d) What would be the area of the square garden?

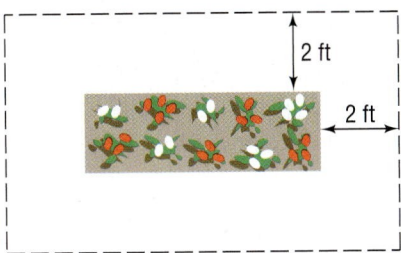

52. Construction A pond is enclosed by a wooden deck that is 3 feet wide. The fence surrounding the deck is 100 feet long.
(a) If the pond is square, what are its dimensions?
(b) If the pond is rectangular and the length of the pond is three times its width, what are the dimensions of the pond?
(c) If the pond is circular, what is the diameter of the pond?
(d) Which pond has the most area?

53. Constructing a Border around a Pool   A pool in the shape of a circle measures 10 feet across. One cubic yard of concrete is to be used to create a circular border of uniform width around the pool. If the border is to have a depth of 3 inches, how wide will the border be? (1 cubic yard = 27 cubic feet)

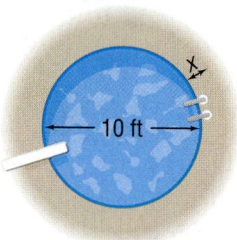

54. Constructing a Border around a Pool   Rework Problem 53 if the depth of the border is 4 inches.

55. Constructing a Border around a Garden   A landscaper, who just completed a rectangular flower garden measuring 6 feet by 10 feet, orders 1 cubic yard of premixed cement, all of which is to be used to create a border of uniform width around the garden. If the border is to have a depth of 3 inches, how wide will the border be? (1 cubic yard = 27 cubic feet)

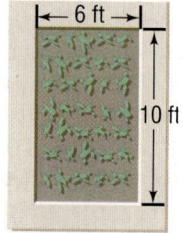

56. Constructing a Coffee Can   A 39 ounce can of Hills Bros.® coffee requires 188.5 square inches of aluminum. If its height is 7 inches, what is its radius? (The surface area A of a right circular cylinder is $A = 2\pi r^2 + 2\pi rh$, where r is the radius and h is the height.)

57. Mixing Water and Antifreeze   How much water should be added to 1 gallon of pure antifreeze to obtain a solution that is 60% antifreeze?

58. Mixing Water and Antifreeze   The cooling system of a certain foreign-made car has a capacity of 15 liters. If the system is filled with a mixture that is 40% antifreeze, how much of this mixture should be drained and replaced by pure antifreeze so that the system is filled with a solution that is 60% antifreeze?

59. Cement Mix   A 20 pound bag of Economy brand cement mix contains 25% cement and 75% sand. How much pure cement must be added to produce a cement mix that is 40% cement?

60. Chemistry: Salt Solutions   How much water must be evaporated from 240 gallons of a 3% salt solution to produce a 5% salt solution?

61. Reducing the Size of a Candy Bar   A jumbo chocolate bar with a rectangular shape measures 12 centimeters in length, 7 centimeters in width, and 3 centimeters in thickness. Due to escalating costs of cocoa, management decides to reduce the volume of the bar by 10%. To accomplish this reduction, management decides that the new bar should have the same 3 centimeter thickness, but the length and width each should be reduced an equal number of centimeters. What should be the dimensions of the new candy bar?

62. Reducing the Size of a Candy Bar   Rework Problem 61 if the reduction is to be 20%.

63. Purity of Gold   The purity of gold is measured in karats, with pure gold being 24 karats. Other purities of gold are expressed as proportional parts of pure gold. Thus, 18 karat gold is $\dfrac{18}{24}$, or 75% pure gold; 12 karat gold is $\dfrac{12}{24}$, or 50% pure gold; and so on. How much 12 karat gold should be mixed with pure gold to obtain 60 grams of 16 karat gold?

64. Chemistry: Sugar Molecules   A sugar molecule has twice as many atoms of hydrogen as it does oxygen and one more atom of carbon than oxygen. If a sugar molecule has a total of 45 atoms, how many are oxygen? How many are hydrogen?

65. Running a Race   Mike can run the mile in 6 minutes, and Dan can run the mile in 9 minutes. If Mike gives Dan a head start of 1 minute, how far from the start will Mike pass Dan? (See the figure.) How long does it take?

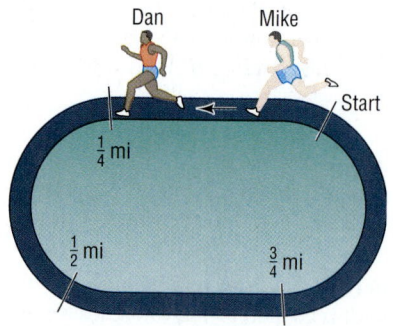

66. Football A tight end can run the 100 yard dash in 12 seconds. A defensive back can do it in 10 seconds. The tight end catches a pass at his own 20 yard line with the defensive back at the 15 yard line. (See the figure.) If no other players were nearby, at what yard line will the defensive back catch up to the tight end?

[**Hint:** At time $t = 0$, the defensive back is 5 yards behind the tight end.]

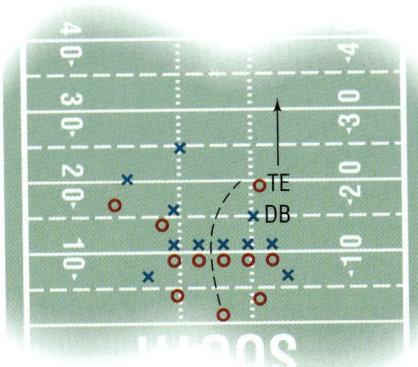

67. Emptying a Tub A bathroom tub will fill in 15 minutes with both faucets open and the stopper in place. With both faucets closed and the stopper removed, the tub will empty in 20 minutes. How long will it take for the tub to fill if both faucets are open and the stopper is removed?

68. Range of an Airplane An air rescue plane averages 300 miles per hour in still air. It carries enough fuel for 5 hours of flying time. If, upon takeoff, it encounters a wind of 30 miles per hour and the direction of the airplane is with the wind in one direction and against it in the other, how far can it fly and return safely? (Assume that the wind remains constant.)

69. Home Equity Loans Suppose that you obtain a home equity loan of $100,000 that requires only a monthly interest payment at 10% per annum, with the principal due after 5 years. You decide to invest part of the loan in a 5 year CD that pays 9% per annum compounded and paid monthly and part in a B+ rated bond due in 5 years that pays 12% per annum compounded and paid monthly. What is the most that you can invest in the CD to ensure that the monthly home equity loan payment is made?

70. Comparing Olympic Heroes In the 1984 Olympics, Carl Lewis of the United States won the gold metal in the 100 meter race with a time of 9.99 seconds. In the 1896 Olympics, Thomas Burke, also of the United States, won the gold medal in the 100 meter race in 12.0 seconds. If they ran in the same race repeating their respective times, by how many meters would Lewis beat Burke?

71. Computing Average Speed In going from Chicago to Atlanta, a car averages 45 miles per hour, and in going from Atlanta to Miami, it averages 55 miles per hour. If Atlanta is halfway between Chicago and Miami, what is the average speed from Chicago to Miami? Discuss an intuitive solution. Write a paragraph defending your intuitive solution. Then solve the problem algebraically. Is your intuitive solution the same as the algebraic one? If not, find the flaw.

72. Speed of a Plane On a recent flight from Phoenix to Kansas City, a distance of 919 nautical miles, the plane arrived 20 minutes early. On leaving the aircraft, I asked the captain, "What was our tail wind?" He replied, "I don't know, but our air speed was 550 knots." How can you determine if enough information is provided to find the tail wind? If possible, find the tail wind. (1 knot = 1 nautical mile per hour)

73. Critical Thinking You are the manager of a clothing store and have just purchased 100 dress shirts for $20.00 each. After 1 month of selling the shirts at the regular price, you plan to have a sale giving 40% off the original selling price. However, you still want to make a profit of $4 on each shirt at the sale price. What should you price the shirts at initially to ensure this? If, instead of 40% off at the sale, you give 50% off, by how much is your profit reduced?

74. Critical Thinking Make up a world problem that requires solving a linear equation as part of its solution. Exchange problems with a friend. Write a critique of your friend's problem.

75. Critical Thinking Without solving, explain what is wrong with the following mixture problem: How many liters of 25% ethanol should be added to 20 liters of 48% ethanol to obtain a solution of 58% ethanol? Now go through an algebraic solution. What happens?

PREPARING FOR THIS SECTION

Before getting started, review the following:

✓ Integer Exponents (Section R.4, pp. 30–33)

✓ Rational Exponents (Section R.9, pp. 79–82)

✓ Square Roots; Radicals (Section R.8, pp. 71–77)

✓ Absolute Value (Section R.2, p. 19)

1.5 RADICAL EQUATIONS; EQUATIONS QUADRATIC IN FORM; ABSOLUTE VALUE EQUATIONS

OBJECTIVES

1. Solve Radical Equations
2. Solve Equations Quadratic in Form
3. Solve Absolute Value Equations

Equations Containing Radicals

When the variable in an equation occurs in a square root, cube root, and so on, that is, when it occurs in a radical, the equation is called a **radical equation.** Sometimes a suitable operation will change a radical equation to one that is linear or quadratic. A commonly used procedure is to isolate the most complicated radical on one side of the equation and then eliminate it by raising each side to a power equal to the index of the radical. Care must be taken, however, because apparent solutions that are not, in fact, solutions of the original equation may result. These are called **extraneous solutions.** In radical equations, extraneous solutions may occur when the index of the radical is even. Therefore, we need to check all answers when working with radical equations.

EXAMPLE 1   **Solving a Radical Equation**

Find the real solutions of the equation $\sqrt[3]{2x - 4} - 2 = 0$.

Algebraic Solution   The equation contains a radical whose index is 3. We isolate it on the left side:

$$\sqrt[3]{2x - 4} - 2 = 0$$

$$\sqrt[3]{2x - 4} = 2$$

Now raise each side to the third power (the index of the radical is 3) and solve.

$$\left(\sqrt[3]{2x - 4}\right)^3 = 2^3 \qquad \text{Raise each side to the power 3.}$$

$$2x - 4 = 8 \qquad \text{Simplify.}$$

$$2x = 12 \qquad \text{Add 4 to each side.}$$

$$x = 6 \qquad \text{Divide both sides by 2.}$$

✔ Check: $\sqrt[3]{2(6) - 4} - 2 = \sqrt[3]{12 - 4} - 2 = \sqrt[3]{8} - 2 = 2 - 2 = 0$ ∎

The solution is $x = 6$.

Graphing Solution   Figure 40 shows the graph of the equation $Y_1 = \sqrt[3]{2x - 4} - 2$. From the graph, we see one x-intercept near 6. Using ZERO (or ROOT), we find that the x-intercept is 6. The only solution is $x = 6$.

Figure 40

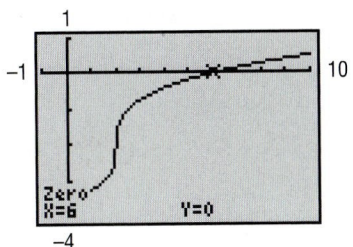

NOW WORK PROBLEM **9.**

EXAMPLE 2   **Solving a Radical Equation**

Find the real solutions of the equation: $\sqrt{x - 1} = x - 7$

Algebraic Solution   We square both sides since the index of a square root is 2.

$$\sqrt{x - 1} = x - 7$$
$$(\sqrt{x - 1})^2 = (x - 7)^2 \qquad \text{Square both sides.}$$
$$x - 1 = x^2 - 14x + 49 \qquad \text{Remove parentheses.}$$
$$x^2 - 15x + 50 = 0 \qquad \text{Put in standard form.}$$
$$(x - 10)(x - 5) = 0 \qquad \text{Factor.}$$
$$x = 10 \quad \text{or} \quad x = 5 \qquad \text{Apply the Zero-Product Property and solve.}$$

✔ CHECK: $x = 10$: $\sqrt{x - 1} = \sqrt{10 - 1} = \sqrt{9} = 3$ and $x - 7 = 10 - 7 = 3$

$x = 5$: $\sqrt{x - 1} = \sqrt{5 - 1} = \sqrt{4} = 2$ and $x - 7 = 5 - 7 = -2$ ∎

The apparent solution $x = 5$ is extraneous; the only solution of the equation is $x = 10$.

Graphing Solution   Graph $Y_1 = \sqrt{x - 1}$ and $Y_2 = x - 7$. See Figure 41. From the graph, there is one point of intersection. Using INTERSECT, the point of intersection is $(10, 3)$, so the solution is $x = 10$.

Figure 41

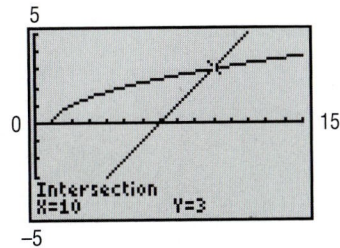

NOW WORK PROBLEM **13.**

Sometimes, we need to raise each side to a power more than once in order to solve a radical equation algebraically.

EXAMPLE 3 | **Solving a Radical Equation**

Find the real solutions of the equation $\sqrt{2x + 3} - \sqrt{x + 2} = 2$.

Algebraic Solution

First, we choose to isolate the more complicated radical expression (in this case, $\sqrt{2x + 3}$) on the left side:

$$\sqrt{2x + 3} = \sqrt{x + 2} + 2$$

Now square both sides (the index of the radical is 2):

$$(\sqrt{2x + 3})^2 = (\sqrt{x + 2} + 2)^2$$

$2x + 3 = (\sqrt{x + 2})^2 + 4\sqrt{x + 2} + 4$ *Remove parentheses.*

$2x + 3 = x + 2 + 4\sqrt{x + 2} + 4$ *Simplify.*

$2x + 3 = x + 6 + 4\sqrt{x + 2}$ *Combine like terms.*

Because the equation still contains a radical, we isolate the remaining radical on the right side and again square both sides.

$x - 3 = 4\sqrt{x + 2}$ *Isolate the radical on the right side.*

$(x - 3)^2 = 16(x + 2)$ *Square both sides.*

$x^2 - 6x + 9 = 16x + 32$ *Remove parentheses.*

$x^2 - 22x - 23 = 0$ *Put in standard form.*

$(x - 23)(x + 1) = 0$ *Factor.*

$x = 23 \quad \text{or} \quad x = -1$

The original equation appears to have the solution set $\{-1, 23\}$. However, we have not yet checked.

✔ CHECK: $x = 23$: $\sqrt{2(23) + 3} - \sqrt{23 + 2} = \sqrt{49} - \sqrt{25} = 7 - 5 = 2$

         $x = -1$: $\sqrt{2(-1) + 3} - \sqrt{-1 + 2} = \sqrt{1} - \sqrt{1} = 1 - 1 = 0$ ■

The equation has only one solution, 23; the apparent solution -1 is extraneous.

Graphing Solution

Graph $Y_1 = \sqrt{2x + 3} - \sqrt{x + 2}$ and $Y_2 = 2$. See Figure 42. From the graph there is one point of intersection. Using INTERSECT, the point of intersection is $(23, 2)$, so the solution is $x = 23$.

Figure 42

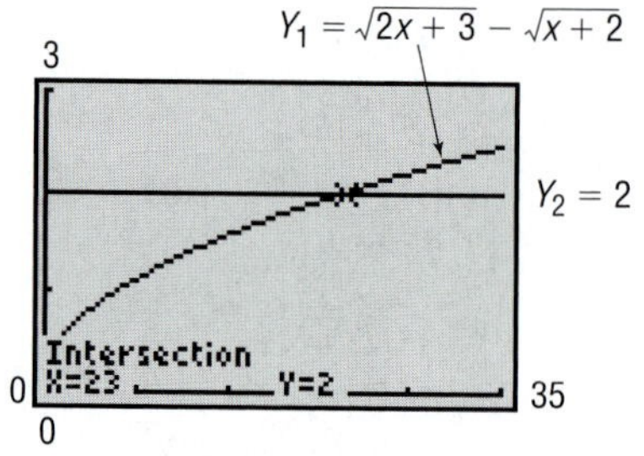

$Y_1 = \sqrt{2x + 3} - \sqrt{x + 2}$

$Y_2 = 2$

Intersection
X=23   Y=2

NOW WORK PROBLEM 23.

Equations Quadratic in Form

The equation $x^4 + x^2 - 12 = 0$ is not quadratic in x, but it is quadratic in x^2. That is, if we let $u = x^2$, we get $u^2 + u - 12 = 0$, a quadratic equation. This equation can be solved for u and, in turn, by using $u = x^2$, we can find the solutions x of the original equation.

In general, if an appropriate substitution u transforms an equation into one of the form

$$au^2 + bu + c = 0, \quad a \neq 0$$

then the original equation is called an **equation of the quadratic type** or an **equation quadratic in form.**

The difficulty of solving such an equation lies in the determination that the equation is, in fact, quadratic in form. After you are told an equation is quadratic in form, it is easy enough to see it, but some practice is needed to enable you to recognize them on your own.

EXAMPLE 4 | **Solving Equations That Are Quadratic in Form**

Find the real solutions of the equation: $(x + 2)^2 + 11(x + 2) - 12 = 0$

Solution   For this equation, let $u = x + 2$. Then $u^2 = (x + 2)^2$, and the original equation,

$$(x + 2)^2 + 11(x + 2) - 12 = 0$$

becomes

$$u^2 + 11u - 12 = 0 \qquad \text{Let } u = x + 2.$$
$$(u + 12)(u - 1) = 0 \qquad \text{Factor.}$$
$$u = -12 \quad \text{or} \quad u = 1 \qquad \text{Solve.}$$

But we want to solve for x. Because $u = x + 2$, we have

$$x + 2 = -12 \quad \text{or} \quad x + 2 = 1$$
$$x = -14 \qquad\qquad x = -1$$

✔ CHECK: $x = -14$: $(-14 + 2)^2 + 11(-14 + 2) - 12$

$$= (-12)^2 + 11(-12) - 12 = 144 - 132 - 12 = 0$$

$x = -1$: $(-1 + 2)^2 + 11(-1 + 2) - 12 = 1 + 11 - 12 = 0$

The original equation has the solution set $\{-14, -1\}$.

✔ CHECK: Verify the solution of Example 4 using a graphing utility. ■

EXAMPLE 5 | **Solving Equations That Are Quadratic in Form**

Find the real solutions of the equation: $x + 2\sqrt{x} - 3 = 0$

Solution   For the equation $x + 2\sqrt{x} - 3 = 0$, let $u = \sqrt{x}$. Then $u^2 = x$, and the original equation,

$$x + 2\sqrt{x} - 3 = 0$$

becomes

$$u^2 + 2u - 3 = 0 \qquad \text{Let } u = \sqrt{x}.$$

$$(u + 3)(u - 1) = 0 \qquad \text{Factor.}$$

$$u = -3 \quad \text{or} \quad u = 1 \qquad \text{Solve.}$$

Since $u = \sqrt{x}$, we have $\sqrt{x} = -3$ or $\sqrt{x} = 1$. The first of these, $\sqrt{x} = -3$, has no real solution, since the square root of a real number is never negative. The second one, $\sqrt{x} = 1$, has the solution $x = 1$.

✔ CHECK: $1 + 2\sqrt{1} - 3 = 1 + 2 - 3 = 0$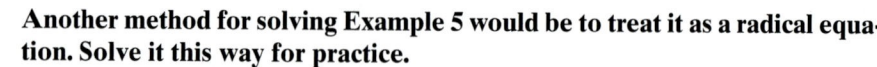

Thus, $x = 1$ is the only solution of the original equation.

✔ CHECK: Verify the solution to Example 5 using a graphing utility.

Another method for solving Example 5 would be to treat it as a radical equation. Solve it this way for practice.

The idea should now be clear. If an equation contains an expression and that same expression squared, make a substitution for the expression. You may get a quadratic equation.

NOW WORK PROBLEM **43**.

Equations Involving Absolute Value

③ Recall that, on the real number line, the absolute value of a equals the distance from the origin to the point whose coordinate is a. For example, there are two points whose distance from the origin is 5 units, -5 and 5. Thus, the equation $|x| = 5$ will have the solution set $\{-5, 5\}$. This leads to the following result:

> **Equations Involving Absolute Value**
>
> If a is a positive real number and if u is any algebraic expression, then
>
> $$|u| = a \quad \text{is equivalent to} \quad u = a \text{ or } u = -a \qquad (1)$$

EXAMPLE 6   **Solving an Equation Involving Absolute Value**

Solve the equation $|x + 4| = 13$.

Algebraic Solution   This follows the form of equation (1), where $u = x + 4$. There are two possibilities:

$$x + 4 = 13 \quad \text{or} \quad x + 4 = -13$$

$$x = 9 \qquad\qquad x = -17$$

The solution set is $\{-17, 9\}$.

Graphing Solution For this equation, graph $Y_1 = |x + 4|$ and $Y_2 = 13$ on the same screen and find their point(s) of intersection, if any. See Figure 43. Using the INTERSECT command (twice), we find the points of intersection to be $(-17, 13)$ and $(9, 13)$. The solution set is $\{-17, 9\}$.

Figure 43

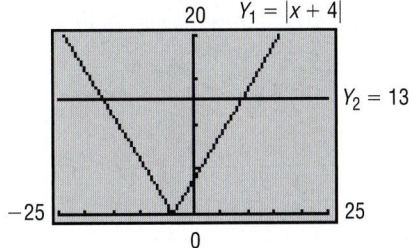

NOW WORK PROBLEM 65.

1.5 Concepts and Vocabulary

In Problems 1–3, fill in the blanks.

1. If $a < 0$, then $|a| = $ _____ .

2. When an apparent solution does not satisfy the original equation, it is called a(n) _____ solution.

3. The equation $|x^2| = 4$ has two real solutions, _____ and _____ .

In Problems 4–6, answer True or False to each statement.

4. The principal square root of any real number is always greater than or equal to zero.

5. The equation $(3x + 1)^4 - 5(3x + 1)^2 + 6$ is quadratic in form.

6. The equation $|x - 4| = 0$ has two real solutions.

7. Which step in the process of solving a radical equation leads to the possibility of extraneous solutions?

8. In your own words, explain what it means for an equation to be quadratic in form.

9. The equation $|x| = -2$ has no solution. Why?

1.5 Exercises

In Problems 1–30, find the real solutions of each equation. Verify your results using a graphing utility.

1. $\sqrt{x} = 3$

2. $\sqrt{z} = 10$

3. $\sqrt{y + 3} = 5$

4. $\sqrt{t - 3} = 7$

5. $\sqrt{2t - 1} = 1$

6. $\sqrt{3t + 4} = 2$

7. $\sqrt{3t + 1} = -6$

8. $\sqrt{5t + 4} = -2$

9. $\sqrt[3]{1 - 2x} - 3 = 0$

10. $\sqrt[3]{1 - 2x} - 1 = 0$

11. $x = 6\sqrt{x}$

12. $x = 4\sqrt{x}$

13. $\sqrt{15 - 2x} = x$

14. $\sqrt{12 - x} = x$

15. $x = 2\sqrt{x - 1}$

16. $x = 2\sqrt{-x - 1}$

17. $\sqrt{x^2 - x - 4} = x + 2$

18. $\sqrt{3 - x + x^2} = x - 2$

19. $3 + \sqrt{3x + 1} = x$

20. $2 + \sqrt{12 - 2x} = x$

21. $\sqrt{2x + 3} - \sqrt{x + 1} = 1$

22. $\sqrt{3x + 7} + \sqrt{x + 2} = 1$

23. $\sqrt{3x + 1} - \sqrt{x - 1} = 2$

24. $\sqrt{3x - 5} - \sqrt{x + 7} = 2$

25. $(3x + 1)^{1/2} = 4$

26. $(3x - 5)^{1/2} = 2$

27. $(5x - 2)^{1/3} = 2$

28. $(2x + 1)^{1/3} = -1$

29. $(x^2 + 9)^{1/2} = 5$

30. $(x^2 - 16)^{1/2} = 9$

In Problems 31–62, find the real solutions of each equation. Verify your results using a graphing utility.

31. $t^4 - 16 = 0$

32. $y^4 - 4 = 0$

33. $x^4 - 5x^2 + 4 = 0$

34. $x^4 - 10x^2 + 25 = 0$

35. $3x^4 - 2x^2 - 1 = 0$

36. $2x^4 - 5x^2 - 12 = 0$

37. $x^6 + 7x^3 - 8 = 0$

38. $x^6 - 7x^3 - 8 = 0$

39. $(x + 2)^2 + 7(x + 2) + 12 = 0$

40. $(2x + 5)^2 - (2x + 5) - 6 = 0$

41. $(3x + 4)^2 - 6(3x + 4) + 9 = 0$

42. $(2 - x)^2 + (2 - x) - 20 = 0$

43. $2(s + 1)^2 - 5(s + 1) = 3$

44. $3(1 - y)^2 + 5(1 - y) + 2 = 0$

45. $x - 4\sqrt{x} = 0$

46. $x + 8\sqrt{x} = 0$

47. $x + \sqrt{x} = 20$

48. $x + \sqrt{x} = 6$

49. $t^{1/2} - 2t^{1/4} + 1 = 0$

50. $z^{1/2} - 4z^{1/4} + 4 = 0$

51. $4x^{1/2} - 9x^{1/4} + 4 = 0$

52. $x^{1/2} - 3x^{1/4} + 2 = 0$

53. $\sqrt[4]{5x^2 - 6} = x$

54. $\sqrt[4]{4 - 5x^2} = x$

55. $\dfrac{1}{(x + 1)^2} = \dfrac{1}{x + 1} + 2$

56. $\dfrac{1}{(x - 1)^2} + \dfrac{1}{x - 1} = 12$

57. $3x^{-2} - 7x^{-1} - 6 = 0$

58. $2x^{-2} - 3x^{-1} - 4 = 0$

59. $2x^{2/3} - 5x^{1/3} - 3 = 0$

60. $3x^{4/3} + 5x^{2/3} - 2 = 0$

61. $\left(\dfrac{v}{v + 1}\right)^2 + \dfrac{2v}{v + 1} = 8$

62. $\left(\dfrac{y}{y - 1}\right)^2 = 6\left(\dfrac{y}{y - 1}\right) + 7$

In Problems 63–84, solve each equation. Verify your results using a graphing utility.

63. $|x| = 6$

64. $|x| = 12$

65. $|2x + 3| = 5$

66. $|3x - 1| = 2$

67. $|1 - 4t| + 8 = 13$

68. $|1 - 2z| + 6 = 9$

69. $|-2x| = 8$

70. $|-x| = 1$

71. $4 - |2x| = 3$

72. $5 - \left|\dfrac{1}{2}x\right| = 3$

73. $\dfrac{2}{3}|x| = 9$

74. $\dfrac{3}{4}|x| = 9$

75. $\left|\dfrac{x}{3} + \dfrac{2}{5}\right| = 2$

76. $\left|\dfrac{x}{2} - \dfrac{1}{3}\right| = 1$

77. $|u - 2| = -\dfrac{1}{2}$

78. $|2 - v| = -1$

79. $|x^2 - 9| = 0$

80. $|x^2 - 16| = 0$

81. $|x^2 - 2x| = 3$

82. $|x^2 + x| = 12$

83. $|x^2 + x - 1| = 1$

84. $|x^2 + 3x - 2| = 2$

In Problems 85–90, use a graphing utility to find the real solutions of each equation. Express any solution rounded to two decimal places.

85. $x - 4x^{1/2} + 2 = 0$

86. $x^{2/3} + 4x^{1/3} + 2 = 0$

87. $x^4 + \sqrt{3}x^2 - 3 = 0$

88. $x^4 + \sqrt{2}x^2 - 2 = 0$

89. $\pi(1 + t)^2 = \pi + 1 + t$

90. $\pi(1 + r)^2 = 2 + \pi(1 + r)$

91. If $k = \dfrac{x + 3}{x - 3}$ and $k^2 - k = 12$, find x.

92. If $k = \dfrac{x + 3}{x - 4}$ and $k^2 - 3k = 28$, find x.

93. Physics: Using Sound to Measure Distance The distance to the surface of the water in a well can sometimes be found by dropping an object into the well and measuring the time elapsed until a sound is heard. If t_1 is the time (measured in seconds) that it takes for the object to strike the water, then t_1 will obey the equation $s = 16t_1^2$, where s is the distance (measured in feet). It follows that $t_1 = \dfrac{\sqrt{s}}{4}$. Suppose that t_2 is the time that it takes for the sound of the impact to reach your ears. Because sound waves are known to travel at a speed of approximately 1100 feet per second, the time t_2 to travel the distance s will be $t_2 = \dfrac{s}{1100}$. See the illustration.

Now $t_1 + t_2$ is the total time that elapses from the moment that the object is dropped to the moment that a sound is heard. Thus, we have the equation

$$\text{Total time elapsed} = \frac{\sqrt{s}}{4} + \frac{s}{1100}$$

Find the distance to the water's surface if the total time elapsed from dropping a rock to hearing it hit water is 4 seconds by graphing

$$Y_1 = \frac{\sqrt{x}}{4} + \frac{x}{1100}, \qquad Y_2 = 4$$

for $0 \le x \le 300$ and $0 \le Y_1 \le 5$ and finding the point of intersection.

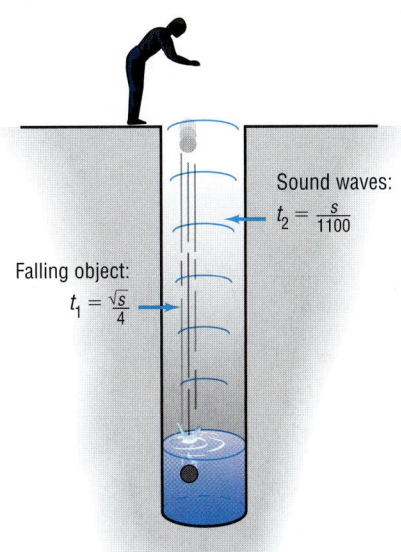

Sound waves: $t_2 = \dfrac{s}{1100}$

Falling object: $t_1 = \dfrac{\sqrt{s}}{4}$

94. Make up a radical equation that has no solution.

95. Make up a radical equation that has an extraneous solution.

PREPARING FOR THIS SECTION

Before getting started, review the following:

✓ Inequalities (Section R.2, pp. 18–19)

✓ Absolute Value (Section R.2, p. 19)

1.6 SOLVING INEQUALITIES

OBJECTIVES

1. Use Interval Notation
2. Use Properties of Inequalities
3. Solve Linear Inequalities Algebraically and Graphically
4. Solve Combined Inequalities Algebraically and Graphically
5. Solve Absolute Value Inequalities Algebraically and Graphically

Suppose that a and b are two real numbers and $a < b$. We shall use the notation $a < x < b$ to mean that x is a number *between* a and b. Thus, the expression $a < x < b$ is equivalent to the two inequalities $a < x$ and $x < b$. Similarly, the expression $a \le x \le b$ is equivalent to the two inequalities $a \le x$ and $x \le b$. The remaining two possibilities, $a \le x < b$ and $a < x \le b$, are defined similarly.

Although it is acceptable to write $3 \ge x \ge 2$, it is preferable to reverse the inequality symbols and write instead $2 \le x \le 3$ so that, as you read from left to right, the values go from smaller to larger.

A statement such as $2 \le x \le 1$ is false because there is no number x for which $2 \le x$ and $x \le 1$. Finally, we never mix inequality symbols, as in $2 \le x \ge 3$.

Intervals

 Let a and b represent two real numbers with $a < b$:

A **closed interval,** denoted by **[a, b],** consists of all real numbers x for which $a \leq x \leq b$.

An **open interval,** denoted by **(a, b),** consists of all real numbers x for which $a < x < b$.

The **half-open,** or **half-closed, intervals** are **(a, b],** consisting of all real numbers x for which $a < x \leq b$, and **[a, b),** consisting of all real numbers x for which $a \leq x < b$.

In each of these definitions, a is called the **left endpoint** and b the **right endpoint** of the interval.

The symbol ∞ (read as "infinity") is not a real number but a notational device used to indicate unboundedness in the positive direction. The symbol $-\infty$ (read as "minus infinity" or "negative infinity") also is not a real number, but a notational device used to indicate unboundedness in the negative direction. Using the symbols ∞ and $-\infty$, we can define five other kinds of intervals:

$[a, \infty)$ consists of all real numbers x for which $x \geq a$ $(a \leq x < \infty)$

(a, ∞) consists of all real numbers x for which $x > a$ $(a < x < \infty)$

$(-\infty, a]$ consists of all real numbers x for which $x \leq a$ $(-\infty < x \leq a)$

$(-\infty, a)$ consists of all real numbers x for which $x < a$ $(-\infty < x < a)$

$(-\infty, \infty)$ consists of all real numbers x $(-\infty < x < \infty)$

Note that ∞ and $-\infty$ are never included as endpoints since they are not real numbers.

Table 4 summarizes interval notation, corresponding inequality notation, and their graphs.

TABLE 4

Interval	Inequality	Graph
The open interval (a, b)	$a < x < b$	
The closed interval $[a, b]$	$a \leq x \leq b$	
The half-open interval $[a, b)$	$a \leq x < b$	
The half-open interval $(a, b]$	$a < x \leq b$	
The interval $[a, \infty)$	$x \geq a$	
The interval (a, ∞)	$x > a$	
The interval $(-\infty, a]$	$x \leq a$	
The interval $(-\infty, a)$	$x < a$	
The interval $(-\infty, \infty)$	All real numbers	

| EXAMPLE 1 | **Writing Inequalities Using Interval Notation** |

Write each inequality using interval notation.

(a) $1 \leq x \leq 3$　　(b) $-4 < x < 0$　　(c) $x > 5$　　(d) $x \leq 1$

Solution　(a) $1 \leq x \leq 3$ describes all numbers x between 1 and 3, inclusive. In interval notation, we write $[1, 3]$.

(b) In interval notation, $-4 < x < 0$ is written $(-4, 0)$.

(c) $x > 5$ consists of all numbers x greater than 5. In interval notation, we write $(5, \infty)$.

(d) In interval notation, $x \leq 1$ is written $(-\infty, 1]$.　　■

| EXAMPLE 2 | **Writing Intervals Using Inequality Notation** |

Write each interval as an inequality involving x.

(a) $[1, 4)$　　(b) $(2, \infty)$　　(c) $[2, 3]$　　(d) $(-\infty, -3]$

Solution　(a) $[1, 4)$ consists of all numbers x for which $1 \leq x < 4$.

(b) $(2, \infty)$ consists of all numbers x for which $x > 2$ $(2 < x < \infty)$.

(c) $[2, 3]$ consists of all numbers x for which $2 \leq x \leq 3$.

(d) $(-\infty, -3]$ consists of all numbers x for which $x \leq -3$ $(-\infty < x \leq -3)$.　　■

> **NOW WORK PROBLEMS 1, 13, AND 21.**

Properties of Inequalities

② The product of two positive real numbers is positive, the product of two negative real numbers is positive, and the product of 0 and 0 is 0. For any real number a, the value of a^2 is 0 or positive; that is, a^2 is nonnegative. This is called the **nonnegative property.**

For any real number a, we have

Nonnegative Property

$$a^2 \geq 0 \tag{1}$$

If we add the same number to both sides of an inequality, we obtain an equivalent inequality. For example, since $3 < 5$, then $3 + 4 < 5 + 4$ or $7 < 9$. This is called the **addition property** of inequalities.

Addition Property of Inequalities

$$\text{If } a < b, \text{ then } a + c < b + c \tag{2a}$$
$$\text{If } a > b, \text{ then } a + c > b + c \tag{2b}$$

The addition property states that the sense, or direction, of an inequality remains unchanged if the same number is added to each side. Figure 44 illustrates the addition property (2a). In Figure 44(a), we see that a lies to the left of b. If c is positive, then $a + c$ and $b + c$ each lie c units to the right of a and b, respectively. Consequently, $a + c$ must lie to the left of $b + c$; that is, $a + c < b + c$. Figure 44(b) illustrates the situation if c is negative.

Figure 44

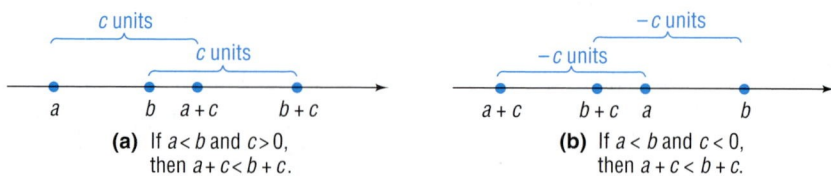

(a) If $a < b$ and $c > 0$,
   then $a + c < b + c$.

(b) If $a < b$ and $c < 0$,
   then $a + c < b + c$.

Draw an illustration similar to Figure 44 that illustrates the addition property (2b).

EXAMPLE 3   **Addition Property of Inequalities**

(a) If $x < -5$, then $x + 5 < -5 + 5$ or $x + 5 < 0$.
(b) If $x > 2$, then $x + (-2) > 2 + (-2)$ or $x - 2 > 0$.   ■

NOW WORK PROBLEM **29.**

We will use two examples to arrive at our next property.

EXAMPLE 4   **Multiplying an Inequality by a Positive Number**

Express as an inequality the result of multiplying each side of the inequality $3 < 7$ by 2.

Solution   We begin with

$$3 < 7$$

Multiplying each side by 2 yields the numbers 6 and 14, so we have

$$6 < 14$$ ■

EXAMPLE 5   **Multiplying an Inequality by a Negative Number**

Express as an inequality the result of multiplying each side of the inequality $9 > 2$ by -4.

Solution   We begin with

$$9 > 2$$

Multiplying each side by -4 yields the numbers -36 and -8, so we have

$$-36 < -8$$ ■

Note that the effect of multiplying both sides of $9 > 2$ by the negative number -4 is that the direction of the inequality symbol is reversed.

Examples 4 and 5 illustrate the following general **multiplication properties** for inequalities:

Multiplication Properties for Inequalities

> If $a < b$ and if $c > 0$, then $ac < bc$.
>
> If $a < b$ and if $c < 0$, then $ac > bc$.   (3a)
>
> If $a > b$ and if $c > 0$, then $ac > bc$.
>
> If $a > b$ and if $c < 0$, then $ac < bc$.   (3b)

The multiplication properties state that the sense, or direction, of an inequality *remains the same* if each side is multiplied by a *positive* real number, while the direction is *reversed* if each side is multiplied by a *negative* real number.

EXAMPLE 6   **Multiplication Property of Inequalities**

(a) If $2x < 6$, then $\dfrac{1}{2}(2x) < \dfrac{1}{2}(6)$ or $x < 3$.

(b) If $\dfrac{x}{-3} > 12$, then $-3\left(\dfrac{x}{-3}\right) < -3(12)$ or $x < -36$.

(c) If $-4x > -8$, then $\dfrac{-4x}{-4} < \dfrac{-8}{-4}$ or $x < 2$.

(d) If $-x < 8$, then $(-1)(-x) > (-1)(8)$ or $x > -8$.   ■

NOW WORK PROBLEM 35.

Solving Inequalities

③ An **inequality in one variable** is a statement involving two expressions, at least one containing the variable, separated by one of the inequality symbols, $<$, $\leq$, $>$, or $\geq$. To **solve an inequality** means to find all values of the variable for which the statement is true. These values are called **solutions** of the inequality.

For example, the following are all inequalities involving one variable, x:

$$x + 5 < 8, \qquad 2x - 3 \geq 4, \qquad x^2 - 1 \leq 3, \qquad \dfrac{x + 1}{x - 2} > 0$$

Two inequalities having exactly the same solution set are called **equivalent inequalities.** As with equations, one method for solving an inequality is to replace it by a series of equivalent inequalities until an inequality with an obvious solution, such as $x < 3$, is obtained. We obtain equivalent inequalities by applying some of the same operations as those used to find equivalent equations. The addition property and the multiplication properties form the basis for the following procedures.

Procedures That Leave the Inequality Symbol Unchanged

1. Simplify both sides of the inequality by combining like terms and eliminating parentheses:

   Replace $\qquad x + 2 + 6 > 2x + 5(x + 1)$

   by $\qquad\qquad x + 8 > 7x + 5$

2. Add or subtract the same expression on both sides of the inequality:

   Replace $\qquad\qquad\quad 3x - 5 < 4$

   by $\qquad\qquad (3x - 5) + 5 < 4 + 5$

3. Multiply or divide both sides of the inequality by the same *positive* expression:

   Replace $\qquad 4x > 16 \quad$ by $\quad \dfrac{4x}{4} > \dfrac{16}{4}$

Procedures That Reverse the Sense or Direction of the Inequality Symbol

1. Interchange the two sides of the inequality:

   Replace $\qquad\qquad 3 < x \quad$ by $\quad x > 3$

2. Multiply or divide both sides of the inequality by the same *negative* expression:

   Replace $\qquad -2x > 6 \quad$ by $\quad \dfrac{-2x}{-2} < \dfrac{6}{-2}$

To solve an inequality using a graphing utility, we follow these steps:

Steps for Solving Inequalities Graphically

STEP 1: Write the inequality in one of the following forms:

$$Y_1 < Y_2, \qquad Y_1 > Y_2, \qquad Y_1 \le Y_2, \qquad Y_1 \ge Y_2$$

STEP 2: Graph Y_1 and Y_2 on the same screen.

STEP 3: If the inequality is of the form $Y_1 < Y_2$, determine on what interval Y_1 is below Y_2.

If the inequality is of the form $Y_1 > Y_2$, determine on what interval Y_1 is above Y_2.

If the inequality is not strict ($\le$ or $\ge$), include the x-coordinates of the points of intersection in the solution.

As the examples that follow illustrate, we solve inequalities using many of the same steps that we would use to solve equations. In writing the solution of an inequality, we may use either set notation or interval notation, whichever is more convenient.

EXAMPLE 7   **Solving an Inequality**

Solve the inequality $4x + 7 \geq 2x - 3$, and graph the solution set.

Algebraic Solution

$$4x + 7 \geq 2x - 3$$

$$4x + 7 - 7 \geq 2x - 3 - 7 \qquad \text{Subtract 7 from both sides.}$$

$$4x \geq 2x - 10 \qquad \text{Simplify.}$$

$$4x - 2x \geq 2x - 10 - 2x \qquad \text{Subtract 2x from both sides.}$$

$$2x \geq -10 \qquad \text{Simplify.}$$

$$\frac{2x}{2} \geq \frac{-10}{2} \qquad \text{Divide both sides by 2. (The direction of the inequality symbol is unchanged.)}$$

$$x \geq -5 \qquad \text{Simplify.}$$

The solution set is $\{x | x \geq -5\}$ or, using interval notation, all numbers in the interval $[-5, \infty)$.

Graphing Solution   We graph $Y_1 = 4x + 7$ and $Y_2 = 2x - 3$ on the same screen. See Figure 45. Using the INTERSECT command, we find that Y_1 and Y_2 intersect at $x = -5$. The graph of Y_1 is above that of Y_2, $Y_1 > Y_2$, to the right of the point of intersection. Since the inequality is not strict, the solution set is $\{x | x \geq -5\}$ or, using interval notation, $[-5, \infty)$.

Figure 45

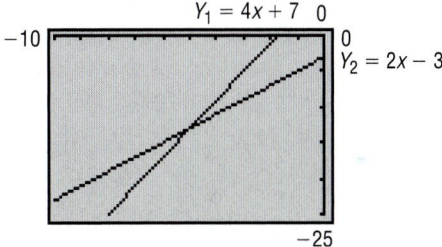

Figure 46
$x \geq -5$ or $[-5, \infty)$

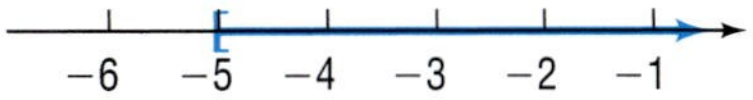

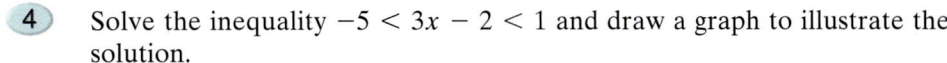

See Figure 46 for the graph of the solution set.

NOW WORK PROBLEM 43.

EXAMPLE 8   **Solving Combined Inequalities**

④ Solve the inequality $-5 < 3x - 2 < 1$ and draw a graph to illustrate the solution.

Algebraic Solution   Recall that the inequality

$$-5 < 3x - 2 < 1$$

is equivalent to the two inequalities

$$-5 < 3x - 2 \quad \text{and} \quad 3x - 2 < 1$$

We will solve each of these inequalities separately.

$$-5 < 3x - 2 \qquad\qquad\qquad\qquad 3x - 2 < 1$$

$$-5 + 2 < 3x - 2 + 2 \quad \text{Add 2 to both sides} \qquad 3x - 2 + 2 < 1 + 2$$

$$-3 < 3x \qquad\qquad \text{Simplify.} \qquad\qquad\qquad 3x < 3$$

$$\frac{-3}{3} < \frac{3x}{3} \qquad\qquad \text{Divide both sides by 3.} \qquad \frac{3x}{3} < \frac{3}{3}$$

$$-1 < x \qquad\qquad \text{Simplify.} \qquad\qquad\qquad x < 1$$

The solution set of the original pair of inequalities consists of all x for which

$$-1 < x \quad \text{and} \quad x < 1$$

This may be written more compactly as $\{x | -1 < x < 1\}$. In interval notation, the solution is $(-1, 1)$.

Graphing Solution   To solve a combined inequality, we graph each part: $Y_1 = -5$, $Y_2 = 3x - 2$, and $Y_3 = 1$. We seek the values of x for which the graph of Y_2 is between the graphs of Y_1 and Y_3. See Figure 47. The point of intersection of Y_1 and Y_2 is $(-1, -5)$, and the point of intersection of Y_2 and Y_3 is $(1, 1)$. The inequality is true for all values of x between these two intersection points. Since the inequality is strict, the solution set is $\{x | -1 < x < 1\}$ or, using interval notation, $(-1, 1)$.

Figure 47

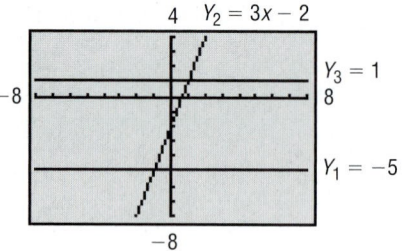

Figure 48
$-1 < x < 1$ or $(-1, 1)$

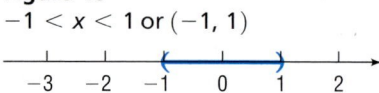

See Figure 48 for the graph of the solution set.

We observe in algebraic solution of Example 8 that the two inequalities that we solved required exactly the same steps. A shortcut to solving the original inequality algebraically is to deal with the two inequalities at the same time, as follows:

$$-5 < \qquad 3x - 2 < 1$$

$$-5 + 2 < \quad 3x - 2 + 2 < 1 + 2 \qquad \text{Add 2 to each part.}$$

$$-3 < \qquad 3x \qquad < 3 \qquad \text{Simplify.}$$

$$\frac{-3}{3} < \qquad \frac{3x}{3} \qquad < \frac{3}{3} \qquad \text{Divide each part by 3.}$$

$$-1 < \qquad x \qquad < 1 \qquad \text{Simplify.}$$

NOW WORK PROBLEM 63.

⑤   Let's look at an inequality involving absolute value.

EXAMPLE 9   **Solving an Inequality Involving Absolute Value**

Solve the inequality: $|x| < 4$

Algebraic Solution   We are looking for all points whose coordinate x is a distance less than 4 units from the origin. See Figure 49 for an illustration. Because any x between -4 and 4 satisfies the condition $|x| < 4$, the solution set consists of all numbers x for which $-4 < x < 4$, that is, all x in the interval $(-4, 4)$.

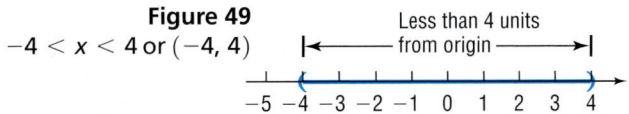

Figure 49
$-4 < x < 4$ or $(-4, 4)$

Less than 4 units from origin

Graphing Solution   We graph $Y_1 = |x|$ and $Y_2 = 4$ on the same screen. See Figure 50. Using the INTERSECT command (twice), we find that Y_1 and Y_2 intersect at $x = -4$ and at $x = 4$. The graph of Y_1 is below that of Y_2, $Y_1 < Y_2$, between the points of intersection. Since the inequality is strict, the solution set is $\{x | -4 < x < 4\}$ or, using interval notation, $(-4, 4)$.

Figure 50

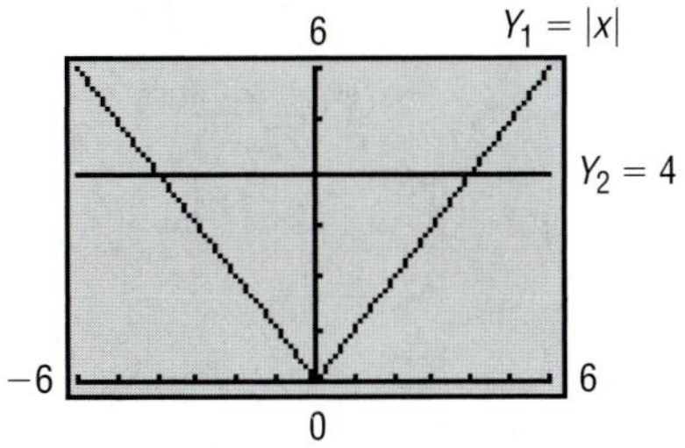

$Y_1 = |x|$

$Y_2 = 4$

We are led to the following results:

Inequalities Involving Absolute Value

If a is any positive number and if u is any algebraic expression, then

$|u| < a$	is equivalent to	$-a < u < a$ (4)
$|u| \leq a$	is equivalent to	$-a \leq u \leq a$ (5)

In other words, $|u| < a$ is equivalent to $-a < u$ and $u < a$.

EXAMPLE 10   **Solving an Inequality Involving Absolute Value**

Solve the inequality $|2x + 4| \leq 3$, and graph the solution set.

Algebraic Solution

$$|2x + 4| \leq 3$$ This follows the form of statement (5); the expression $u = 2x + 4$ is inside the absolute value bars.

$$-3 \leq \quad 2x + 4 \quad \leq 3$$ Apply statement (5).

$$-3 - 4 \leq 2x + 4 - 4 \leq 3 - 4$$ Subtract 4 from each part.

$$-7 \leq \quad 2x \quad \leq -1$$ Simplify.

$$\frac{-7}{2} \leq \quad \frac{2x}{2} \quad \leq \frac{-1}{2}$$ Divide each part by 2.

$$-\frac{7}{2} \leq \quad x \quad \leq -\frac{1}{2}$$ Simplify.

The solution set is $\left\{ x \mid -\dfrac{7}{2} \le x \le -\dfrac{1}{2} \right\}$, that is, all x in the interval $\left[-\dfrac{7}{2}, -\dfrac{1}{2} \right]$.

Graphing Solution   We graph $Y_1 = |2x + 4|$ and $Y_2 = 3$ on the same screen. See Figure 51. Using the INTERSECT command (twice), we find that Y_1 and Y_2 intersect at $x = -3.5$ and at $x = -0.5$. The graph of Y_1 is below that of Y_2, $Y_1 < Y_2$, between the points of intersection. Since the inequality is not strict, the solution set is $\{x \mid -3.5 \le x \le -0.5\}$ or, using interval notation, $[-3.5, -0.5]$.

Figure 51

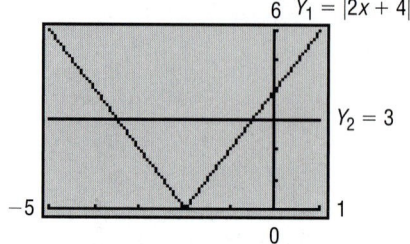

Figure 52

$-\dfrac{7}{2} \le x \le -\dfrac{1}{2}$ or $\left[-\dfrac{7}{2}, -\dfrac{1}{2} \right]$

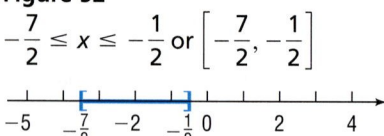

See Figure 52 for a graph of the solution set.

NOW WORK PROBLEM 81.

EXAMPLE 11   **Solving an Inequality Involving Absolute Value**

Solve the inequality $|x| > 3$, and graph the solution set.

Algebraic Solution   We are looking for all points whose coordinate x is a distance greater than 3 units from the origin. Figure 53 illustrates the situation. We conclude that any x less than -3 or greater than 3 satisfies the condition $|x| > 3$. Consequently, the solution set consists of all numbers x for which $x < -3$ or $x > 3$, that is, all x in the intervals $(-\infty, -3)$ or $(3, \infty)$.

Figure 53

$x < -3$ or $x > 3$; $(-\infty, -3)$ or $(3, \infty)$

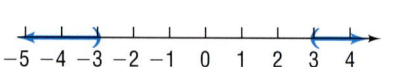

Graphing Solution   We graph $Y_1 = |x|$ and $Y_2 = 3$ on the same screen. See Figure 54. Using the INTERSECT command (twice), we find that Y_1 and Y_2 intersect at $x = -3$ and at $x = 3$. The graph of Y_1 is above that of Y_2, $Y_1 > Y_2$, to the left of $x = -3$ and to the right of $x = 3$. Since the inequality is strict, the solution set is $\{x \mid x < -3 \text{ or } x > 3\}$. Using interval notation, the solution is $(-\infty, -3)$ or $(3, \infty)$.

Figure 54

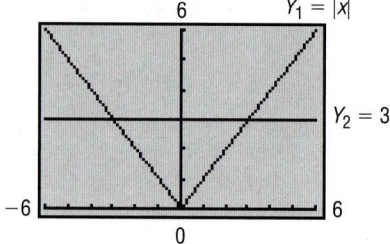

Inequalities Involving Absolute Value

If a is any positive number and u is any algebraic expression, then

$\lvert u \rvert > a$	is equivalent to	$u < -a$ or $u > a$ (6)
$\lvert u \rvert \ge a$	is equivalent to	$u \le -a$ or $u \ge a$ (7)

| EXAMPLE 12 | **Solving an Inequality Involving Absolute Value** |

Solve the inequality $|2x - 5| > 3$, and graph the solution set.

Algebraic Solution

$|2x - 5| > 3$ *This follows the form of statement (6); the expression $u = 2x - 5$ is inside the absolute value bars.*

$$2x - 5 < -3 \quad \text{or} \quad 2x - 5 > 3 \qquad \textit{Apply statement (6).}$$

$$2x - 5 + 5 < -3 + 5 \quad \text{or} \quad 2x - 5 + 5 > 3 + 5 \qquad \textit{Add 5 to each part.}$$

$$2x < 2 \quad \text{or} \quad 2x > 8 \qquad \textit{Simplify.}$$

$$\frac{2x}{2} < \frac{2}{2} \quad \text{or} \quad \frac{2x}{2} > \frac{8}{2} \qquad \textit{Divide each part by 2.}$$

$$x < 1 \quad \text{or} \quad x > 4 \qquad \textit{Simplify.}$$

The solution set is $\{x \mid x < 1 \text{ or } x > 4\}$, that is, all x in the intervals $(-\infty, 1)$ or $(4, \infty)$.

Graphing Solution

We graph $Y_1 = |2x - 5|$ and $Y_2 = 3$ on the same screen. See Figure 55. Using the INTERSECT command (twice), we find that Y_1 and Y_2 intersect at $x = 1$ and at $x = 4$. The graph of Y_1 is above that of Y_2, $Y_1 > Y_2$, to the left of $x = 1$ and to the right of $x = 4$. Since the inequality is strict, the solution set is $\{x \mid x < 1 \text{ or } x > 4\}$. Using interval notation, the solution is $(-\infty, 1)$ or $(4, \infty)$.

Figure 55

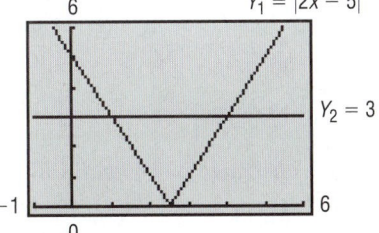
$Y_1 = |2x - 5|$
$Y_2 = 3$

Figure 56
$x < 1$ or $x > 4$; $(-\infty, 1)$ or $(4, \infty)$

$$\xleftarrow{\qquad} \quad -2 \; -1 \;\; 0 \;\; 1 \;\; 2 \;\; 3 \;\; 4 \;\; 5 \;\; 6 \;\; 7 \quad \xrightarrow{\qquad}$$

See Figure 56 for a graph of the solution set. ◼

WARNING: A common error to be avoided is to attempt to write the solution $x < 1$ or $x > 4$ as $1 > x > 4$, which is incorrect, since there are no numbers x for which $1 > x$ and $x > 4$. Another common error is to "mix" the symbols and write $1 < x > 4$, which makes no sense. ◼

NOW WORK PROBLEM **85.**

1.6 Concepts and Vocabulary

In Problems 1–3, fill in the blanks.

1. If each side of an inequality is multiplied by a(n) _____ number, then the sense of the inequality symbol is reversed.

2. A(n) _____ _____, denoted $[a, b]$, consists of all real numbers x for which $a \le x \le b$.

3. The _____ _____ states that the sense, or direction, of an inequality remains the same if each side is multiplied by a positive number, while the direction is reversed if each side is multiplied by a negative number.

In Problems 4–6, answer True or False to each statement. In each statement, assume that $a < b$ and $c < 0$.

4. $a \pm c < b \pm c$

5. $ac > bc$

6. $\dfrac{a}{c} < \dfrac{b}{c}$

7. The inequality $|x| > -0.5$ has all real numbers as a solution. Why?

8. The inequality $x^2 + 1 < -5$ has no solution. Why?

1.6 Exercises

In Problems 1–6, express the graph shown in color using interval notation. Also express each as an inequality involving x.

1.

2.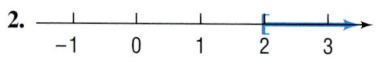

3.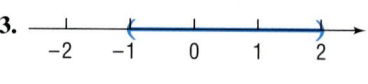

4.

5.

6.

In Problems 7–12, an inequality is given. Write the inequality obtained by:
  (a) *Adding 3 to each side of the given inequality.*
  (b) *Subtracting 5 from each side of the given inequality.*
  (c) *Multiplying each side of the given inequality by 3.*
  (d) *Multiplying each side of the given inequality by −2.*

7. $3 < 5$

8. $2 > 1$

9. $4 > -3$

10. $-3 > -5$

11. $2x + 1 < 2$

12. $1 - 2x > 5$

In Problems 13–20, write each inequality using interval notation, and illustrate each inequality using the real number line.

13. $0 \le x \le 4$

14. $-1 < x < 5$

15. $4 \le x < 6$

16. $-2 < x < 0$

17. $x \ge 4$

18. $x \le 5$

19. $x < -4$

20. $x > 1$

In Problems 21–28, write each interval as an inequality involving x, and illustrate each inequality using the real number line.

21. $[2, 5]$

22. $(1, 2)$

23. $(-3, -2)$

24. $[0, 1)$

25. $[4, \infty)$

26. $(-\infty, 2]$

27. $(-\infty, -3)$

28. $(-8, \infty)$

In Problems 29–42, fill in the blank with the correct inequality symbol.

29. If $x < 5$, then $x - 5$ _____ 0.

30. If $x < -4$, then $x + 4$ _____ 0.

31. If $x > -4$, then $x + 4$ _____ 0.

32. If $x > 6$, then $x - 6$ _____ 0.

33. If $x \ge -4$, then $3x$ _____ -12.

34. If $x \le 3$, then $2x$ _____ 6.

35. If $x > 6$, then $-2x$ _____ -12.

36. If $x > -2$, then $-4x$ _____ 8.

37. If $x \ge 5$, then $-4x$ _____ -20.

38. If $x \le -4$, then $-3x$ _____ 12.

39. If $2x < 6$, then x _____ 3.

40. If $3x \le 12$, then x _____ 4.

41. If $-\dfrac{1}{2}x \le 3$, then x _____ -6.

42. If $-\dfrac{1}{4}x > 1$, then x _____ -4.

In Problems 43–92, solve each inequality algebraically. Express your answer using set notation or interval notation. Graph the solution set. Verify your results using a graphing utility.

43. $x + 1 < 5$

44. $x - 6 < 1$

45. $1 - 2x \le 3$

46. $2 - 3x \le 5$

47. $3x - 7 > 2$

48. $2x + 5 > 1$

49. $3x - 1 \ge 3 + x$

50. $2x - 2 \ge 3 + x$

51. $-2(x + 3) < 8$

52. $-3(1 - x) < 12$

53. $4 - 3(1 - x) \le 3$

54. $8 - 4(2 - x) \le -2x$

55. $\dfrac{1}{2}(x - 4) > x + 8$

56. $3x + 4 > \dfrac{1}{3}(x - 2)$

57. $\dfrac{x}{2} \ge 1 - \dfrac{x}{4}$

58. $\dfrac{x}{3} \ge 2 + \dfrac{x}{6}$

59. $0 \le 2x - 6 \le 4$

60. $4 \le 2x + 2 \le 10$

61. $-5 \le 4 - 3x \le 2$

62. $-3 \le 3 - 2x \le 9$

63. $-3 < \dfrac{2x - 1}{4} < 0$

64. $0 < \dfrac{3x + 2}{2} < 4$

65. $1 < 1 - \dfrac{1}{2}x < 4$

66. $0 < 1 - \dfrac{1}{3}x < 1$

67. $(x + 2)(x - 3) > (x - 1)(x + 1)$

68. $(x - 1)(x + 1) > (x - 3)(x + 4)$

69. $x(4x + 3) \le (2x + 1)^2$

70. $x(9x - 5) \le (3x - 1)^2$

71. $\dfrac{1}{2} \le \dfrac{x + 1}{3} < \dfrac{3}{4}$

72. $\dfrac{1}{3} < \dfrac{x + 1}{2} \le \dfrac{2}{3}$

73. $|x| < 6$

74. $|x| < 9$

75. $|x| > 4$

76. $|x| > 1$

77. $|2x| < 8$

78. $|3x| < 15$

79. $|3x| > 12$

80. $|2x| > 6$

81. $|x - 2| + 2 < 3$

82. $|x + 4| + 3 < 5$

83. $|3t - 2| \le 4$

84. $|2u + 5| \le 7$

85. $|x - 3| \ge 2$

86. $|x + 4| \ge 2$

87. $|1 - 4x| - 7 < -2$

88. $|1 - 2x| - 4 < -1$

89. $|1 - 2x| > |-3|$

90. $|2 - 3x| > |-1|$

91. $|2x + 1| < -1$

92. $|3x - 4| \ge 0$

---

93. Express the fact that x differs from 2 by less than $\dfrac{1}{2}$ as an inequality involving an absolute value. Solve for x.

94. Express the fact that x differs from -1 by less than 1 as an inequality involving an absolute value. Solve for x.

95. Express the fact that x differs from -3 by more than 2 as an inequality involving an absolute value. Solve for x.

96. Express the fact that x differs from 2 by more than 3 as an inequality involving an absolute value. Solve for x.

97. A young adult may be defined as someone older than 21, but less than 30 years of age. Express this statement using inequalities.

98. Middle-aged may be defined as being 40 or more and less than 60. Express the statement using inequalities.

99. Body Temperature   Normal human body temperature is 98.6°F. If a temperature x that differs from normal by at least 1.5° is considered unhealthy, write the condition for an unhealthy temperature x as an inequality involving an absolute value, and solve for x.

100. Household Voltage   In the United States, normal household voltage is 115 volts. However, it is not uncommon for actual voltage to differ from normal voltage by at most 5 volts. Express this situation as an inequality involving an absolute value. Use x as the actual voltage and solve for x.

101. Life Expectancy   Metropolitan Life Insurance Co. reported that an average 25-year-old male in 1996 could expect to live at least 48.4 more years, and an average 25-year-old female in 1996 could expect to live at least 54.7 more years.

(a) To what age can an average 25-year-old male expect to live? Express your answer as an inequality.

(b) To what age can an average 25-year-old female expect to live? Express your answer as an inequality.

(c) Who can expect to live longer, a male or a female? By how many years?

102. General Chemistry   For a certain ideal gas, the volume V (in cubic centimeters) equals 20 times the temperature T (in Kelvin). If the temperature varies from 353K to 393K, inclusive, what is the corresponding range of the volume of the gas?

103. **Real Estate** A real estate agent agrees to sell a large apartment complex according to the following commission schedule: $45,000 plus 25% of the selling price in excess of $900,000. Assuming that the complex will sell at some price between $900,000 and $1,100,000, inclusive, over what range does the agent's commission vary? How does the commission vary as a percent of selling price?

104. **Sales Commission** A used car salesperson is paid a commission of $25 plus 40% of the selling price in excess of owner's cost. The owner claims that used cars typically sell for at least owner's cost plus $70 and at most owner's cost plus $300. For each sale made, over what range can the salesperson expect the commission to vary?

105. **Federal Tax Withholding** The percentage method of withholding for federal income tax (1998)* states that a single person whose weekly wages, after subtracting withholding allowances, are over $517, but not over $1105, shall have $69.90 plus 28% of the excess over $517 withheld. Over what range does the amount withheld vary if the weekly wages vary from $525 to $600, inclusive?

106. **Federal Tax Withholding** Rework Problem 105 if the weekly wages vary from $600 to $700, inclusive.

107. **Electricity Rates** Commonwealth Edison Company's summer charge for electricity is 10.494¢ per kilowatt-hour.[†] In addition, each monthly bill contains a customer charge of $9.36. If last summer's bills ranged from a low of $80.24 to a high of $271.80, over what range did usage vary (in kilowatt-hours)?

108. **Water Bills** The Village of Oak Lawn charges homeowners $21.60 per quarter-year plus $1.70 per 1000 gallons for water usage in excess of 12,000 gallons.[‡] In 1995, one homeowner's quarterly bill ranged from a high of $65.75 to a low of $28.40. Over what range did water usage vary?

109. **Markup of a New Car** The markup over dealer's cost of a new car ranges from 12% to 18%. If the sticker price is $8800, over what range will the dealer's cost vary?

110. **IQ Tests** A standard intelligence test has an average score of 100. According to statistical theory, of the people who take the test, the 2.5% with the highest scores will have scores of more than 1.96σ above the average, where σ (sigma, a number called the *standard deviation*) depends on the nature of the test. If $\sigma = 12$ for this test and there is (in principle) no upper limit to the score possible on the test, write the interval of possible test scores of the people in the top 2.5%.

111. **Computing Grades** In your Economics 101 class, you have scores of 68, 82, 87, and 89 on the first four of five tests. To get a grade of B, the average of the first five test scores must be greater than or equal to 80 and less than

90. Solve an inequality to find the range of the score that you need on the last test to get a B.

What do I need to get a B?

ECONOMICS TEST 5
NEXT PERIOD

68 82
81 89

112. **Computing Grades** Repeat Problem 111 if the fifth test counts double.

113. A car that averages 25 miles per gallon has a tank that holds 20 gallons of gasoline. After a trip that covered at least 300 miles, the car ran out of gasoline. What is the range of the amount of gasoline (in gallons) that was in the tank at the start of the trip?

114. Repeat Problem 113 if the same car runs out of gasoline after a trip of no more than 250 miles.

115. **Arithmetic Mean** If $a < b$, show that $a < \dfrac{a+b}{2} < b$. The number $\dfrac{a+b}{2}$ is called the **arithmetic mean** of a and b.

116. Refer to Problem 115. Show that the arithmetic mean of a and b is equidistant from a and b.

117. **Geometric Mean** If $0 < a < b$, show that $a < \sqrt{ab} < b$. The number $\sqrt{ab}$ is called the **geometric mean** of a and b.

118. Refer to Problems 115 and 117. Show that the geometric mean of a and b is less than the arithmetic mean of a and b.

119. **Harmonic Mean** For $0 < a < b$, let h be defined by

$$\frac{1}{h} = \frac{1}{2}\left(\frac{1}{a} + \frac{1}{b}\right)$$

Show that $a < h < b$. The number h is called the **harmonic mean** of a and b.

120. Refer to Problems 115, 117, and 119. Show that the harmonic mean of a and b equals the geometric mean squared divided by the arithmetic mean.

121. Make up an inequality that has no solution. Make up one that has exactly one solution.

122. How would you explain to a fellow student the underlying reason for the multiplication property for inequalities (page 153); that is, the sense or direction of an inequality remains the same if each side is multiplied by a positive real number, while the direction is reversed if each side is multiplied by a negative real number?

---

*Source: *Employer's Tax Guide.* Department of the Treasury, Internal Revenue Service, 1998.
[†]*Source:* Commonwealth Edison Co., Chicago, Illinois, 1998.
[‡]*Source:* Village of Oak Lawn, Illinois, 1998.

1.7 LINES

OBJECTIVES
1. Calculate and Interpret the Slope of a Line
2. Graph Lines Given a Point and the Slope
3. Find the Equation of Vertical Lines
4. Use the Point-Slope Form of a Line; Identify Horizontal Lines
5. Find the Equation of a Line Given Two Points
6. Write the Equation of a Line in Slope-Intercept Form
7. Identify the Slope and *y*-intercept of a Line from Its Equation
8. Write the Equation of a Line in General Form
9. Define Parallel Lines
10. Find Equations of Parallel Lines
11. Define Perpendicular Lines
12. Find Equations of Perpendicular Lines

In this section we study a certain type of equation that contains two variables, called a *linear equation*, and its graph, a *line*.

Slope of a Line

1. Consider the staircase illustrated in Figure 57. Each step contains exactly the same horizontal **run** and the same vertical **rise.** The ratio of the rise to the run, called the *slope*, is a numerical measure of the steepness of the staircase. For example, if the run is increased and the rise remains the same, the staircase becomes less steep. If the run is kept the same, but the rise is increased, the staircase becomes more steep. This important characteristic of a line is best defined using rectangular coordinates.

Figure 57

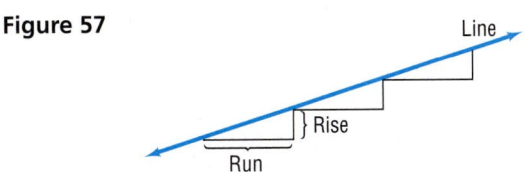

Let $P = (x_1, y_1)$ and $Q = (x_2, y_2)$ be two distinct points. If $x_1 \neq x_2$, the **slope** m of the nonvertical line L containing P and Q is defined by the formula

$$m = \frac{y_2 - y_1}{x_2 - x_1}, \qquad x_1 \neq x_2 \tag{1}$$

If $x_1 = x_2$, L is a **vertical line** and the slope m of L is **undefined** (since this results in division by 0).

Figure 58(a) provides an illustration of the slope of a nonvertical line; Figure 58(b) illustrates a vertical line.

Figure 58

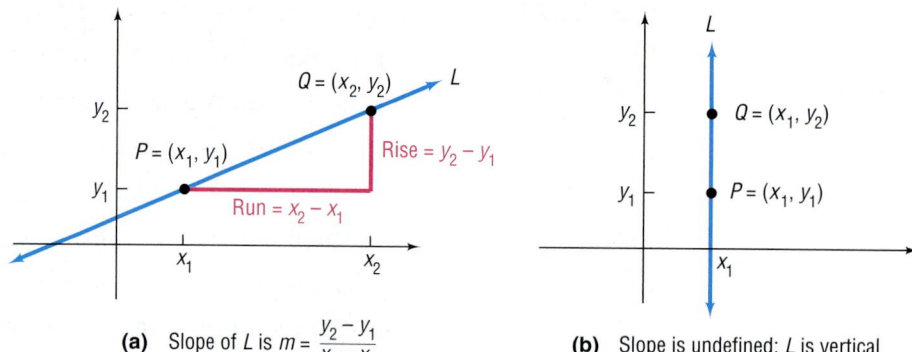

(a) Slope of L is $m = \dfrac{y_2 - y_1}{x_2 - x_1}$

(b) Slope is undefined; L is vertical

As Figure 58(a) illustrates, the slope m of a nonvertical line may be viewed as

$$m = \frac{y_2 - y_1}{x_2 - x_1} = \frac{\text{Rise}}{\text{Run}}$$

We can also express the slope m of a nonvertical line as

$$m = \frac{y_2 - y_1}{x_2 - x_1} = \frac{\text{Change in } y}{\text{Change in } x} = \frac{\Delta y}{\Delta x}$$

That is, the slope m of a nonvertical line L measures the amount that y changes as x changes from x_1 to x_2. This is called the **average rate of change** of y with respect to x.

Two comments about computing the slope of a nonvertical line may prove helpful:

1. Any two distinct points on the line can be used to compute the slope of the line. (See Figure 59 for justification.)

Figure 59

Triangles ABC and PQR are similar (equal angles). Hence, ratios of corresponding sides are proportional so that

Slope using P and $Q = \dfrac{y_2 - y_1}{x_2 - x_1}$

$= $ Slope using A and $B = \dfrac{d(B, C)}{d(A, C)}$

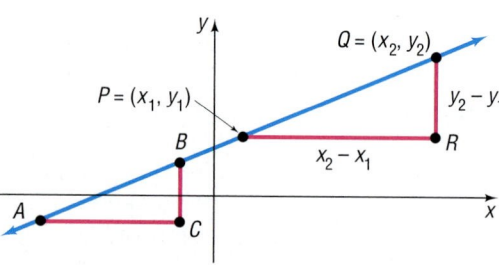

2. The slope of a line may be computed from $P = (x_1, y_1)$ to $Q = (x_2, y_2)$ or from Q to P because

$$\frac{y_2 - y_1}{x_2 - x_1} = \frac{y_1 - y_2}{x_1 - x_2}$$

EXAMPLE 1   **Finding and Interpreting the Slope of a Line Containing Two Points**

The slope m of the line containing the points $(1, 2)$ and $(5, -3)$ may be computed as

$$m = \frac{-3 - 2}{5 - 1} = \frac{-5}{4} = -\frac{5}{4} \quad \text{or as} \quad m = \frac{2 - (-3)}{1 - 5} = \frac{5}{-4} = \frac{-5}{4} = -\frac{5}{4}$$

For every 4 unit change in x, y will change by -5 units. That is, if x increases by 4 units, then y will decrease by 5 units. The average rate of change of y with respect to x is $-\dfrac{5}{4}$.   ■

NOW WORK PROBLEMS **1** AND **7**.

Square Screens

To get an undistorted view of slope on a graphing utility, the same scale must be used on each axis. However, most graphing utilities have a rectangular screen. Because of this, using the same interval for both x and y will result in a distorted view. For example, Figure 60 shows the graph of the line $y = x$ connecting the points $(-4, -4)$ and $(4, 4)$.

We expect the line to bisect the first and third quadrants, but it doesn't. We need to adjust the selections for Xmin, Xmax, Ymin, and Ymax so that a **square screen** results. On most graphing utilities, this is accomplished by setting the ratio of x to y at $3 : 2$.*

Figure 61 shows the graph of the line $y = x$ on a square screen using a TI-83. Notice that the line now bisects the first and third quadrants. Compare this illustration to Figure 60.

To get a better idea of the meaning of the slope m of a line, consider the following:

SEEING THE CONCEPT   On the same square screen graph the following equations:

$Y_1 = 0$	*Slope of line is 0.*
$Y_2 = \dfrac{1}{4}x$	*Slope of line is $\dfrac{1}{4}$.*
$Y_3 = \dfrac{1}{2}x$	*Slope of line is $\dfrac{1}{2}$.*
$Y_4 = x$	*Slope of line is 1.*
$Y_5 = 2x$	*Slope of line is 2.*
$Y_6 = 6x$	*Slope of line is 6.*

See Figure 62.

Figure 60

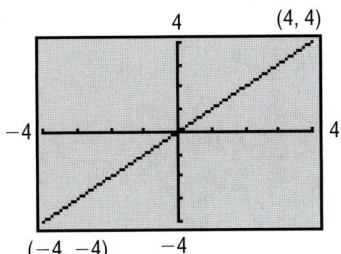

Figure 61

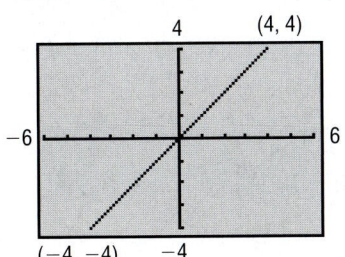

Figure 62

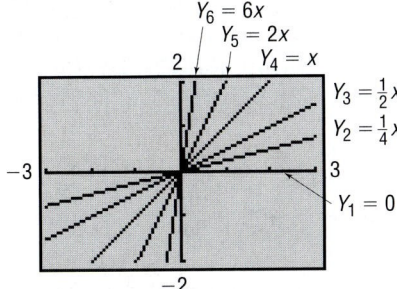

*Most graphing utilities have a feature that automatically squares the viewing window. Consult your owner's manual for the appropriate keystrokes.

SEEING THE CONCEPT   On the same square screen, graph the following equations:

$$Y_1 = 0 \qquad \textcolor{blue}{\text{Slope of line is 0.}}$$

$$Y_2 = -\frac{1}{4}x \qquad \textcolor{blue}{\text{Slope of line is } -\frac{1}{4}.}$$

$$Y_3 = -\frac{1}{2}x \qquad \textcolor{blue}{\text{Slope of line is } -\frac{1}{2}.}$$

$$Y_4 = -x \qquad \textcolor{blue}{\text{Slope of line is } -1.}$$

$$Y_5 = -2x \qquad \textcolor{blue}{\text{Slope of line is } -2.}$$

$$Y_6 = -6x \qquad \textcolor{blue}{\text{Slope of line is } -6.}$$

Figure 63

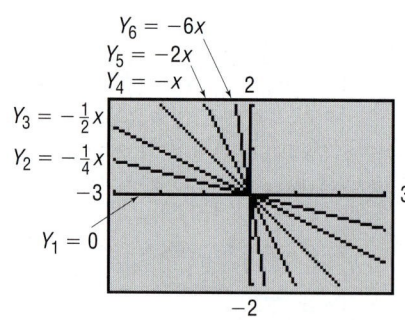

See Figure 63.

Figures 62 and 63 illustrate the following facts:

1. When the slope of a line is positive, the line slants upward from left to right.

2. When the slope of a line is negative, the line slants downward from left to right.

3. When the slope is 0, the line is horizontal.

Figures 62 and 63 also illustrate that the closer the line is to the vertical position, the greater the magnitude of the slope.

2   The next example illustrates how the slope of a line can be used to graph the line.

EXAMPLE 2   **Graphing a Line Given a Point and a Slope**

Draw a graph of the line that contains the point $(3, 2)$ and has a slope of:

(a) $\dfrac{3}{4}$ (b) $-\dfrac{4}{5}$

Solution   (a) Slope $= \dfrac{\text{Rise}}{\text{Run}}$. The fact that the slope is $\dfrac{3}{4}$ means that for every horizontal movement (run) of 4 units to the right there will be a vertical movement (rise) of 3 units. If we start at the given point $(3, 2)$ and move 4 units to the right and 3 units up, we reach the point $(7, 5)$. By drawing the line through this point and the point $(3, 2)$, we have the graph. See Figure 64.

Figure 64

Slope $= \dfrac{3}{4}$

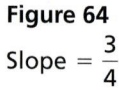
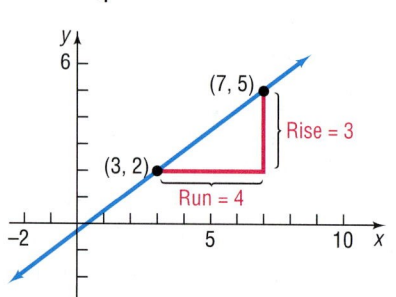

(b) The fact that the slope is

$$-\frac{4}{5} = \frac{-4}{5} = \frac{\text{Rise}}{\text{Run}}$$

means that for every horizontal movement of 5 units to the right there will be a corresponding vertical movement of -4 units (a downward movement). If we start at the given point $(3, 2)$ and move 5 units to the right and then 4 units down, we arrive at the point $(8, -2)$. By drawing the line through these points, we have the graph. See Figure 65.

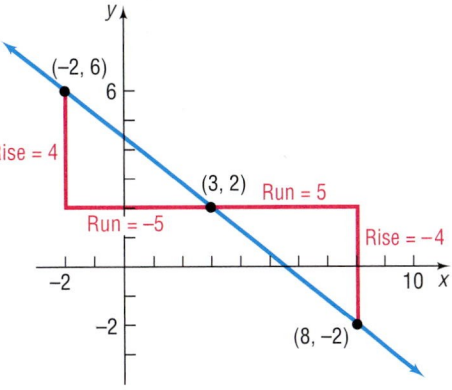

Figure 65

Slope $= -\dfrac{4}{5}$

Alternatively, we can set

$$-\frac{4}{5} = \frac{4}{-5} = \frac{\text{Rise}}{\text{Run}}$$

so that for every horizontal movement of -5 units (a movement to the left) there will be a corresponding vertical movement of 4 units (upward). This approach brings us to the point $(-2, 6)$, which is also on the graph shown in Figure 65.

 **NOW WORK PROBLEM 13.**

Equations of Lines

3   Now that we have discussed the slope of a line, we are ready to derive equations of lines. As we shall see, there are several forms of the equation of a line. Let's start with an example.

EXAMPLE 3   **Graphing a Line**

Graph the equation: $x = 3$

Solution   To graph $x = 3$ by hand, recall that we are looking for all points (x, y) in the plane for which $x = 3$. No matter what y-coordinate is used, the corresponding x-coordinate always equals 3. Consequently, the graph of the equation $x = 3$ is a vertical line with x-intercept 3 and undefined slope. See Figure 66(a).

Figure 66
$x = 3$

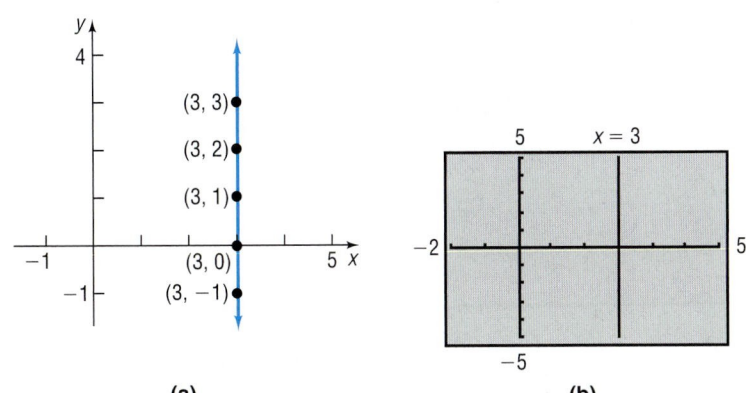

(a)   (b)

To use a graphing utility, we need to express the equation in the form $y = \{$expression in $x\}$. But $x = 3$ cannot be put into this form, so an alternative method must be used. Consult your manual to determine the methodology required to draw vertical lines. Figure 66(b) shows the graph that you should obtain. ■

As suggested by Example 3, we have the following result:

Theorem

Equation of a Vertical Line

A vertical line is given by an equation of the form

$$x = a$$

where a is the x-intercept.

Figure 67

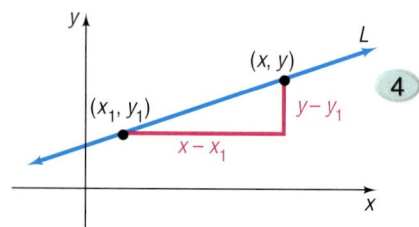

Now let L be a nonvertical line with slope m and containing the point (x_1, y_1). See Figure 67. For any other point (x, y) on L, we have

$$m = \frac{y - y_1}{x - x_1} \quad \text{or} \quad y - y_1 = m(x - x_1)$$

Theorem

Point–Slope Form of an Equation of a Line

An equation of a nonvertical line of slope m that contains the point (x_1, y_1) is

$$y - y_1 = m(x - x_1) \tag{2}$$

EXAMPLE 4   **Using the Point–Slope Form of a Line**

An equation of the line with slope 4 and containing the point $(1, 2)$ can be found by using the point–slope form with $m = 4$, $x_1 = 1$, and $y_1 = 2$.

$$y - y_1 = m(x - x_1)$$
$$y - 2 = 4(x - 1) \qquad m = 4, x_1 = 1, y_1 = 2$$
$$y = 4x - 2$$

See Figure 68.

Figure 68

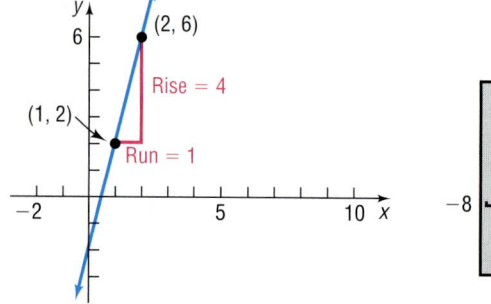

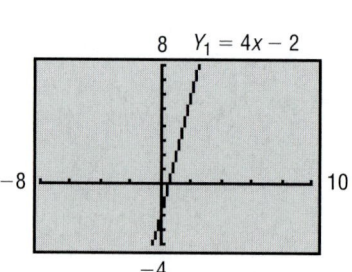

EXAMPLE 5 **Finding the Equation of a Horizontal Line**

Find an equation of the horizontal line containing the point $(3, 2)$.

Solution Because all the y-values are equal, the slope of a horizontal line is 0. To get an equation, we use the point–slope form with $m = 0$, $x_1 = 3$, and $y_1 = 2$.

$$y - y_1 = m(x - x_1)$$
$$y - 2 = 0 \cdot (x - 3) \qquad \text{$m = 0$, $x_1 = 3$, and $y_1 = 2$}$$
$$y - 2 = 0$$
$$y = 2$$

See Figure 69 for the graph.

Figure 69
$y = 2$

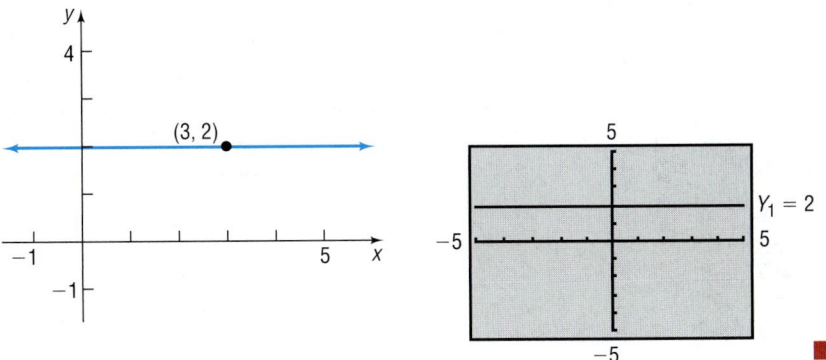

As suggested by Example 5, we have the following result:

Theorem **Equation of a Horizontal Line**

A horizontal line is given by an equation of the form

$$y = b$$

where b is the y-intercept.

EXAMPLE 6 **Finding an Equation of a Line Given Two Points**

⑤ Find an equation of the line L containing the points $(2, 3)$ and $(-4, 5)$. Graph the line L.

Solution We first compute the slope of the line.

$$m = \frac{5 - 3}{-4 - 2} = \frac{2}{-6} = -\frac{1}{3}$$

We use the point $(2, 3)$ and the slope $m = -\dfrac{1}{3}$ to get the point–slope form of the equation of the line.

$$y - 3 = -\frac{1}{3}(x - 2)$$

See Figure 70 for the graph.

Figure 70
$y - 3 = -\dfrac{1}{3}(x - 2)$

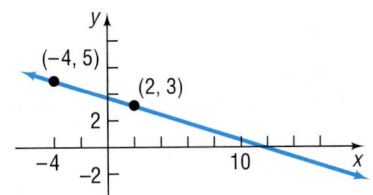

In the solution to Example 6, we could have used the other point, $(-4, 5)$, instead of the point $(2, 3)$. The equation that results, although it looks different, is equivalent to the equation that we obtained in the example. (Try it for yourself.)

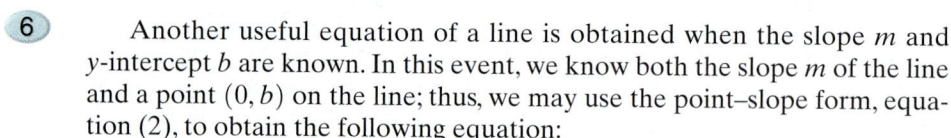 **NOW WORK PROBLEM 27.**

6   Another useful equation of a line is obtained when the slope m and y-intercept b are known. In this event, we know both the slope m of the line and a point $(0, b)$ on the line; thus, we may use the point–slope form, equation (2), to obtain the following equation:

$$y - b = m(x - 0) \quad \text{or} \quad y = mx + b$$

Theorem

Slope–Intercept Form of an Equation of a Line

An equation of a line L with slope m and y-intercept b is

$$y = mx + b \qquad (3)$$

Figure 71
$y = mx + 2$

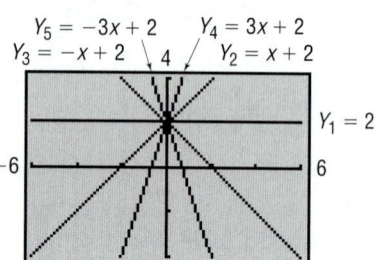

SEEING THE CONCEPT To see the role that the slope m plays, graph the following lines on the same square screen.

$$Y_1 = 2$$
$$Y_2 = x + 2$$
$$Y_3 = -x + 2$$
$$Y_4 = 3x + 2$$
$$Y_5 = -3x + 2$$

See Figure 71. What do you conclude about the lines $y = mx + 2$?

Figure 72
$y = 2x + b$

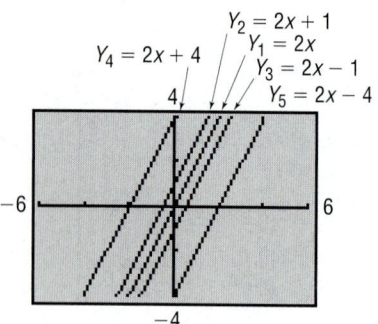

SEEING THE CONCEPT To see the role of the y-intercept b, graph the following lines on the same square screen.

$$Y_1 = 2x$$
$$Y_2 = 2x + 1$$
$$Y_3 = 2x - 1$$
$$Y_4 = 2x + 4$$
$$Y_5 = 2x - 4$$

See Figure 72. What do you conclude about the lines $y = 2x + b$?

7   When the equation of a line is written in slope–intercept form, it is easy to find the slope m and y-intercept b of the line. For example, suppose that the equation of a line is

$$y = -2x + 3$$

Compare it to $y = mx + b$.

$$\begin{array}{cccc} y = & -2x & + & 3 \\ & \uparrow & & \uparrow \\ y = & mx & + & b \end{array}$$

The slope of this line is -2 and its y-intercept is 3.

NOW WORK PROBLEM 59.

EXAMPLE 7 Finding the Slope and y-Intercept

Find the slope m and y-intercept b of the equation $2x + 4y = 8$. Graph the equation.

Solution To obtain the slope and y-intercept, we transform the equation into its slope–intercept form by solving for y.

$$\begin{aligned} 2x + 4y &= 8 \\ 4y &= -2x + 8 \\ y &= -\frac{1}{2}x + 2 \quad y = mx + b \end{aligned}$$

Figure 73
$2x + 4y = 8$

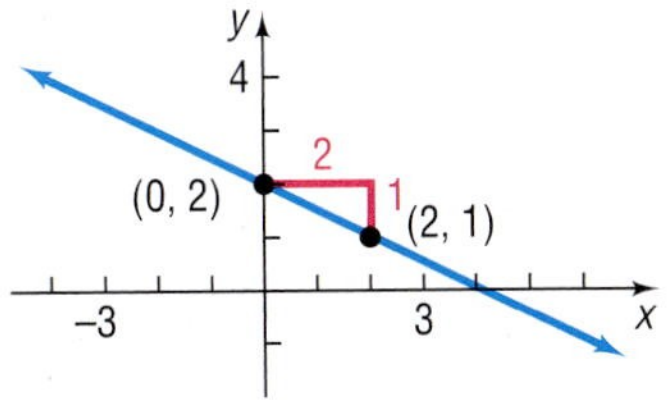

The coefficient of x, $-\frac{1}{2}$, is the slope, and the y-intercept is 2. We can graph the line in two ways:

1. Use the fact that the y-intercept is 2 and the slope is $-\frac{1}{2}$. Then, starting at the point $(0, 2)$, go to the right 2 units and then down 1 unit to the point $(2, 1)$. See Figure 73.

Or:

Figure 74
$2x + 4y = 8$

2. Locate the intercepts. Because the y-intercept is 2, we know that one intercept is $(0, 2)$. To obtain the x-intercept, let $y = 0$ and solve for x. When $y = 0$, we have

$$\begin{aligned} 2x + 4 \cdot 0 &= 8 \\ 2x &= 8 \\ x &= 4 \end{aligned}$$

The intercepts are $(4, 0)$ and $(0, 2)$. See Figure 74. ■

NOW WORK PROBLEM 65.

8 The form of the equation of the line in Example 7, $2x + 4y = 8$, is called the *general form*.

The equation of a line L is in **general form** when it is written as

$$Ax + By = C \qquad (4)$$

where A, B, and C are real numbers, and A and B are not both 0.

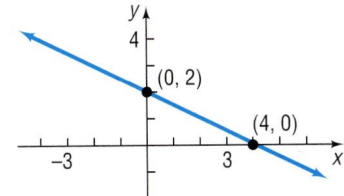

Every line has an equation that is equivalent to an equation written in general form. For example, a vertical line whose equation is

$$x = a$$

can be written in the general form

$$1 \cdot x + 0 \cdot y = a \qquad \text{\textcolor{red}{$A = 1, B = 0, C = a$}}$$

A horizontal line whose equation is

$$y = b$$

can be written in the general form

$$0 \cdot x + 1 \cdot y = b \qquad \text{\textcolor{red}{$A = 0, B = 1, C = b$}}$$

Lines that are neither vertical nor horizontal have general equations of the form

$$Ax + By = C \qquad \text{\textcolor{red}{$A \neq 0$ and $B \neq 0$}}$$

Because the equation of every line can be written in general form, any equation equivalent to (4) is called a **linear equation.**

Parallel and Perpendicular Lines

⑨ When two lines (in the plane) do not intersect (that is, they have no points in common), they are said to be **parallel.** Look at Figure 75. There we have drawn two lines and have constructed two right triangles by drawing sides parallel to the coordinate axes. These lines are parallel if and only if the right triangles are similar. (Do you see why? Two angles are equal.) And the triangles are similar if and only if the ratios of corresponding sides are equal. This suggests the following result:

Figure 75
The distinct lines are parallel if and only if their slopes are equal.

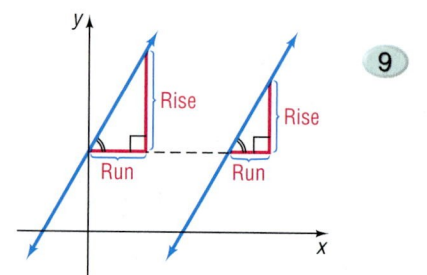

Theorem

> **Criterion for Parallel Lines**
>
> Two nonvertical lines are parallel if and only if their slopes are equal and they have different *y*-intercepts.

The use of the words "if and only if" in the preceding theorem means that actually two statements are being made, one the converse of the other.

If two nonvertical lines are parallel, then their slopes are equal and they have different *y*-intercepts.

If two nonvertical lines have equal slopes and they have different *y*-intercepts, then they are parallel.

EXAMPLE 8   **Showing That Two Lines Are Parallel**

Show that the lines given by the following equations are parallel:

$$L_1: \quad 2x + 3y = 6, \qquad L_2: \quad 4x + 6y = 0$$

Solution   To determine whether these lines have equal slopes and different *y*-intercepts, we write each equation in slope–intercept form:

$$
\begin{array}{ll}
L_1: \quad 2x + 3y = 6 & L_2: \quad 4x + 6y = 0 \\
\qquad\quad 3y = -2x + 6 & \qquad\quad 6y = -4x \\
\qquad\quad y = -\dfrac{2}{3}x + 2 & \qquad\quad y = -\dfrac{2}{3}x
\end{array}
$$

$$\text{Slope} = -\frac{2}{3}; \ y\text{-intercept} = 2 \qquad \text{Slope} = -\frac{2}{3}; \ y\text{-intercept} = 0$$

Because these lines have the same slope, $-\dfrac{2}{3}$, but different y-intercepts, the lines are parallel. See Figure 76.

Figure 76
Parallel lines

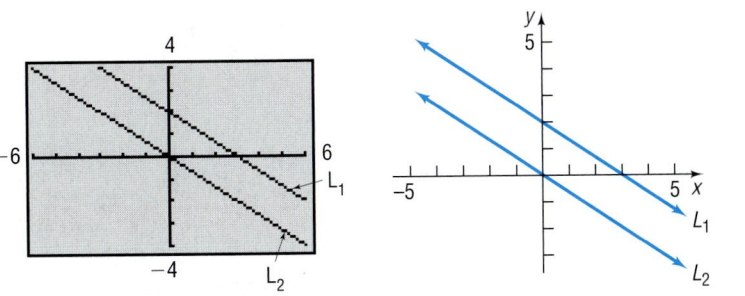

EXAMPLE 9   **Finding a Line That Is Parallel to a Given Line**

⑩ Find an equation for the line that contains the point $(2, -3)$ and is parallel to the line $2x + y = 6$.

Solution   The slope of the line that we seek equals the slope of the line $2x + y = 6$, since the two lines are to be parallel. We begin by writing the equation of the line $2x + y = 6$ in slope–intercept form.

$$2x + y = 6$$
$$y = -2x + 6$$

The slope is -2. Since the line that we seek contains the point $(2, -3)$, we use the point–slope form to obtain

$$y - (-3) = -2(x - 2) \qquad \text{Point–slope form}$$
$$y + 3 = -2x + 4$$
$$y = -2x + 1 \qquad \text{Slope–intercept form}$$
$$2x + y = 1 \qquad \text{General form}$$

This line is parallel to the line $2x + y = 6$ and contains the point $(2, -3)$. See Figure 77.

Figure 77

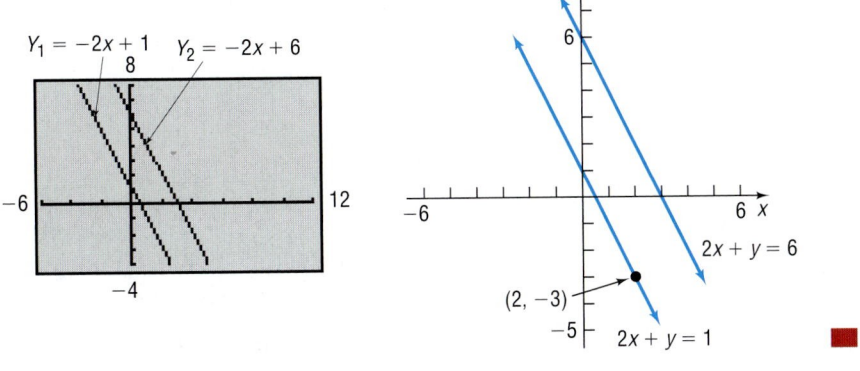

Figure 78
Perpendicular lines

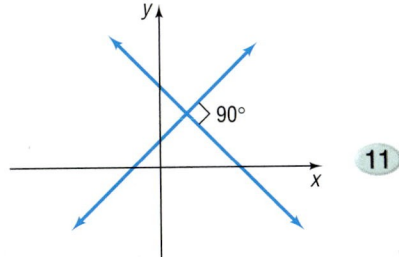

NOW WORK PROBLEM **47.**

⑪   When two lines intersect at a right angle ($90°$), they are said to be **perpendicular.** See Figure 78.
   The following result gives a condition, in terms of their slopes, for two lines to be perpendicular.

Theorem   **Criterion for Perpendicular Lines**

Two nonvertical lines are perpendicular if and only if the product of their slopes is −1.

Here we shall prove the "only if" part of the statement:

If two nonvertical lines are perpendicular, then the product of their slopes is −1.

You are asked to prove the "if" part of the theorem; that is:

If two nonvertical lines have slopes whose product is −1, then the lines are perpendicular.

Proof   Let m_1 and m_2 denote the slopes of the two lines. There is no loss in generality (that is, neither the angle nor the slopes are affected) if we situate the lines so that they meet at the origin. See Figure 79. The point $A = (1, m_2)$ is on the line having slope m_2, and the point $B = (1, m_1)$ is on the line having slope m_1. (Do you see why this must be true?)

Suppose that the lines are perpendicular. Then triangle OAB is a right triangle. As a result of the Pythagorean Theorem, it follows that

Figure 79

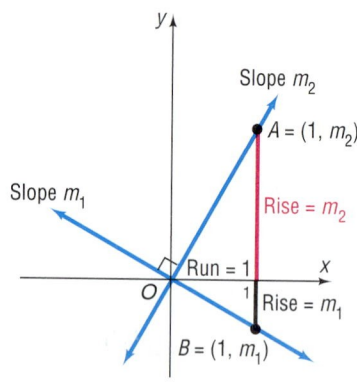

$$[d(O, A)]^2 + [d(O, B)]^2 = [d(A, B)]^2 \qquad (5)$$

By the distance formula, we can write the squares of these distances as

$$[d(O, A)]^2 = (1 - 0)^2 + (m_2 - 0)^2 = 1 + m_2^2$$
$$[d(O, B)]^2 = (1 - 0)^2 + (m_1 - 0)^2 = 1 + m_1^2$$
$$[d(A, B)]^2 = (1 - 1)^2 + (m_2 - m_1)^2 = m_2^2 - 2m_1m_2 + m_1^2$$

Using these facts in equation (5), we get

$$(1 + m_2^2) + (1 + m_1^2) = m_2^2 - 2m_1m_2 + m_1^2$$

which, upon simplification, can be written as

$$m_1m_2 = -1$$

If the lines are perpendicular, the product of their slopes is −1.

You may find it easier to remember the condition for two nonvertical lines to be perpendicular by observing that the equality $m_1m_2 = -1$ means that m_1 and m_2 are negative reciprocals of each other; that is, either

$$m_1 = -\frac{1}{m_2} \text{ or } m_2 = -\frac{1}{m_1}.$$

EXAMPLE 10   **Finding the Slope of a Line Perpendicular to Another Line**

If a line has slope $\dfrac{3}{2}$, any line having slope $-\dfrac{2}{3}$ is perpendicular to it.

EXAMPLE 11   **Finding the Equation of a Line Perpendicular to a Given Line**

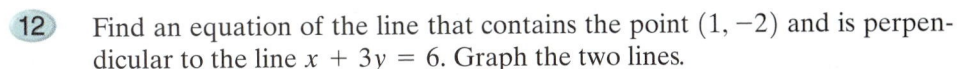 Find an equation of the line that contains the point $(1, -2)$ and is perpendicular to the line $x + 3y = 6$. Graph the two lines.

Solution   We first write the equation of the given line in slope–intercept form to find its slope.

$$x + 3y = 6$$

$$3y = -x + 6 \qquad \text{\color{teal}Proceed to solve for } y.$$

$$y = -\frac{1}{3}x + 2 \qquad \text{\color{teal}Place in the form } y = mx + b.$$

The given line has slope $-\dfrac{1}{3}$. Any line perpendicular to this line will have slope 3. Because we require the point $(1, -2)$ to be on this line with slope 3, we use the point–slope form of the equation of a line.

$$y - (-2) = 3(x - 1) \qquad \text{\color{teal}Point–slope form}$$

To obtain other forms of the equations, we proceed as follows:

$$y + 2 = 3(x - 1)$$

$$y + 2 = 3x - 3$$

$$y = 3x - 5 \qquad \text{\color{teal}Slope–intercept form}$$

$$3x - y = 5 \qquad \text{\color{teal}General form}$$

Figure 80 shows the graphs.

Figure 80

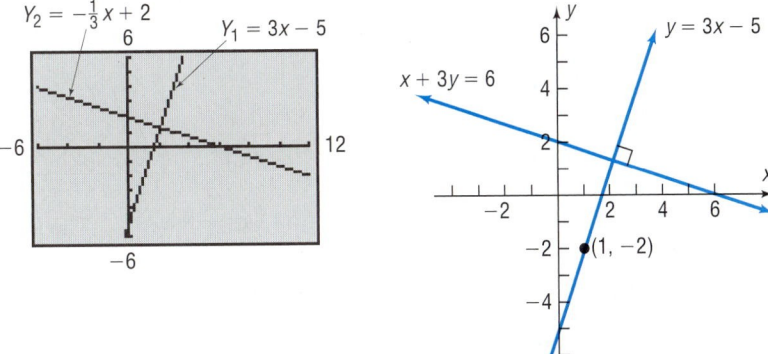

WARNING: Be sure to use a square screen when you graph perpendicular lines. Otherwise, the angle between the two lines will appear distorted.

── **NOW WORK PROBLEM 53.**

1.7   Concepts and Vocabulary

In Problems 1–3, fill in the blanks.

1. The slope of a vertical line is _____; the slope of a horizontal line is _____.

2. Two nonvertical lines have slopes m_1 and m_2, respectively. The lines are parallel if _____ and the _____ are unequal; the lines are perpendicular if _____.

3. A horizontal line is given by an equation of the form _____ where b is the _____.

In Problems 4–6, answer True or False to each statement.

4. Vertical lines have an undefined slope.

5. The slope of the line $2y = 3x + 5$ is 3.

6. Perpendicular lines have slopes that are reciprocals of one another.

7. Can the equation of every line be written in slope–intercept form? Why?

8. Does every line have exactly one x-intercept and one y-intercept? Are there any lines that have no intercepts?

9. What can you say about two lines that have equal slopes and equal y-intercepts?

10. What can you say about two lines with the same x-intercept and the same y-intercept? Assume that the x-intercept is not 0.

11. If two lines have the same slope, but different x-intercepts, can they have the same y-intercept?

12. If two lines have the same y-intercept, but different slopes, can they have the same x-intercept?

1.7   Exercises

In Problems 1–4, (a) find the slope of the line and (b) interpret the slope.

1. 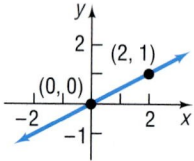   **2.** 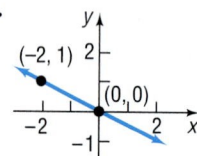   **3.** 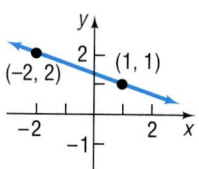   **4.**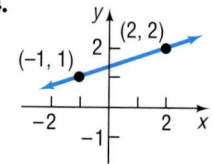

In Problems 5–12, plot each pair of points and determine the slope of the line containing them. Graph the line.

5. $(2, 3); (4, 0)$ **6.** $(4, 2); (3, 4)$ **7.** $(-2, 3); (2, 1)$ **8.** $(-1, 1); (2, 3)$

9. $(-3, -1); (2, -1)$ **10.** $(4, 2); (-5, 2)$ **11.** $(-1, 2); (-1, -2)$ **12.** $(2, 0); (2, 2)$

In Problems 13–20, graph, by hand, the line containing the point P and having slope m.

13. $P = (1, 2); m = 3$ **14.** $P = (2, 1); m = 4$ **15.** $P = (2, 4); m = -\dfrac{3}{4}$

16. $P = (1, 3); m = -\dfrac{2}{5}$ **17.** $P = (-1, 3); m = 0$ **18.** $P = (2, -4); m = 0$

19. $P = (0, 3);$ slope undefined   **20.** $P = (-2, 0);$ slope undefined

In Problems 21–26, the slope and a point on a line are given. Use this information to locate three additional points on the line. Answers may vary. [**Hint:** *It is not necessary to find the equation of the line. See Example 2.*]

21. Slope 4; point $(1, 2)$ **22.** Slope 2; point $(-2, 3)$ **23.** Slope $-\dfrac{3}{2}$; point $(2, -4)$

24. Slope $\dfrac{4}{3}$; point $(-3, 2)$ **25.** Slope -2; point $(-2, -3)$ **26.** Slope -1; point $(4, 1)$

In Problems 27–34, find an equation of the line L.

27.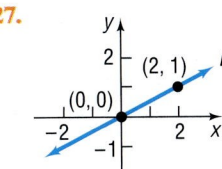

28.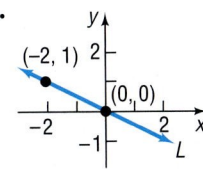

29.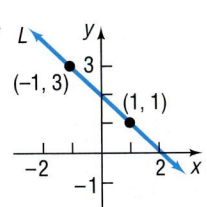

30.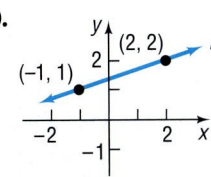

31.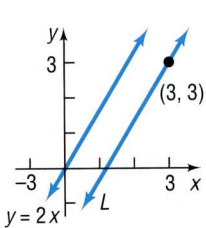

L is parallel to $y = 2x$

32.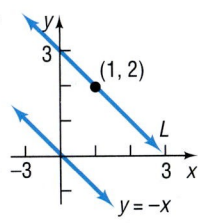

L is parallel to $y = -x$

33.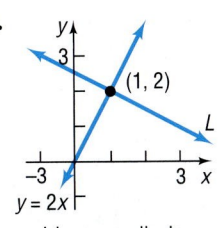

L is perpendicular to $y = 2x$

34.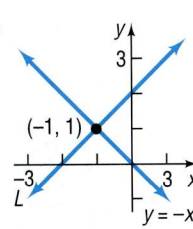

L is perpendicular to $y = -x$

In Problems 35–58, find an equation for the line with the given properties. Express your answer using either the general form or the slope–intercept form of the equation of a line, whichever you prefer.

35. Slope $= 3$; containing the point $(-2, 3)$

36. Slope $= 2$; containing the point $(4, -3)$

37. Slope $= -\dfrac{2}{3}$; containing the point $(1, -1)$

38. Slope $= \dfrac{1}{2}$; containing the point $(3, 1)$

39. Containing the points $(1, 3)$ and $(-1, 2)$

40. Containing the points $(-3, 4)$ and $(2, 5)$

41. Slope $= -3$; y-intercept $= 3$

42. Slope $= -2$; y-intercept $= -2$

43. x-intercept $= 2$; y-intercept $= -1$

44. x-intercept $= -4$; y-intercept $= 4$

45. Slope undefined; containing the point $(2, 4)$

46. Slope undefined; containing the point $(3, 8)$

47. Parallel to the line $y = 2x$; containing the point $(-1, 2)$

48. Parallel to the line $y = -3x$; containing the point $(-1, 2)$

49. Parallel to the line $2x - y = -2$; containing the point $(0, 0)$

50. Parallel to the line $x - 2y = -5$; containing the point $(0, 0)$

51. Parallel to the line $x = 5$; containing the point $(4, 2)$

52. Parallel to the line $y = 5$; containing the point $(4, 2)$

53. Perpendicular to the line $y = \dfrac{1}{2}x + 4$; containing the point $(1, -2)$

54. Perpendicular to the line $y = 2x - 3$; containing the point $(1, -2)$

55. Perpendicular to the line $2x + y = 2$; containing the point $(-3, 0)$

56. Perpendicular to the line $x - 2y = -5$; containing the point $(0, 4)$

57. Perpendicular to the line $x = 8$; containing the point $(3, 4)$

58. Perpendicular to the line $y = 8$; containing the point $(3, 4)$

In Problems 59–78, find the slope and y-intercept of each line. Graph the line by hand. Check your graph using a graphing utility.

59. $y = 2x + 3$

60. $y = -3x + 4$

61. $\dfrac{1}{2}y = x - 1$

62. $\dfrac{1}{3}x + y = 2$

63. $y = \dfrac{1}{2}x + 2$

64. $y = 2x + \dfrac{1}{2}$

65. $x + 2y = 4$

66. $-x + 3y = 6$

67. $2x - 3y = 6$

68. $3x + 2y = 6$

69. $x + y = 1$

70. $x - y = 2$

71. $x = -4$

72. $y = -1$

73. $y = 5$

74. $x = 2$

75. $y - x = 0$

76. $x + y = 0$

77. $2y - 3x = 0$

78. $3x + 2y = 0$

79. Find an equation of the x-axis

80. Find an equation of the y-axis.

In Problems 81–84, match each graph with the correct equation:

(a) $y = x$ (b) $y = 2x$ (c) $y = \dfrac{x}{2}$ (d) $y = 4x$

81. 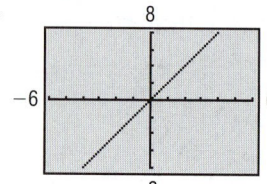   **82.** 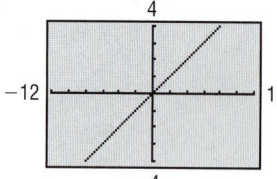   **83.** 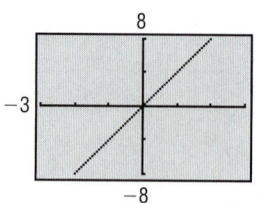   **84.**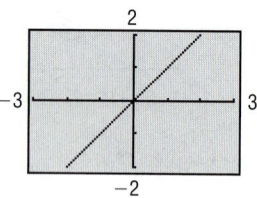

In Problems 85–88, write an equation of each line. Express your answer using either the general form or the slope–intercept form of the equation of a line, whichever you prefer.

85. 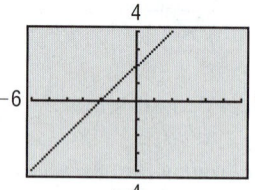   **86.** 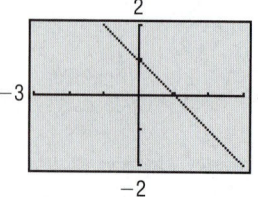   **87.** 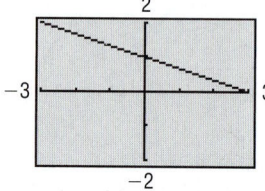   **88.**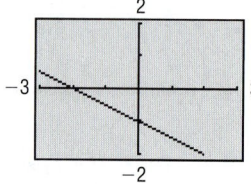

89. Truck Rentals   A truck rental company rents a moving truck for one day by charging $29 plus $0.07 per mile. Write a linear equation that relates the cost C, in dollars, of renting the truck to the number x of miles driven. What is the cost of renting the truck if the truck is driven 110 miles? 230 miles?

90. Cost Equation   The **fixed costs** of operating a business are the costs incurred regardless of the level of production. Fixed costs include rent, fixed salaries, and costs of buying machinery. The **variable costs** of operating a business are the costs that change with the level of output. Variable costs include raw materials, hourly wages, and electricity. Suppose that a manufacturer of jeans has fixed costs of $500 and variable costs of $8 for each pair of jeans manufactured. Write a linear equation that relates the cost C, in dollars, of manufacturing the jeans to the number x pairs of jeans manufactured. What is the cost of manufacturing 400 pairs of jeans? 740 pairs?

91. Measuring Temperature   The relationship between Celsius (°C) and Fahrenheit (°F) degrees of measuring temperature is linear. Find an equation relating °C and °F if 0°C corresponds to 32°F and 100°C corresponds to 212°F. Use the equation to find the Celsius measure of 70°F.

92. Measuring Temperature   The Kelvin (K) scale for measuring temperature is obtained by adding 273 to the Celsius temperature.
(a) Write an equation relating K and °C.
(b) Write an equation relating K and °F (see Problem 91).

93. Business: Computing Profit   Each Sunday, a newspaper agency sells x copies of certain newspaper for $1.00 per copy. The cost to the agency of each newspaper is $0.50. The agency pays a fixed cost for storage, delivery, and so on, of $100 per Sunday.
(a) Write an equation that relates the profit P, in dollars, to the number x of copies sold. Graph this equation.
(b) What is the profit to the agency if 1000 copies are sold?
(c) What is the profit to the agency if 5000 copies are sold?

94. Business: Computing Profit   Repeat Problem 93 if the cost to the agency is $0.45 per copy and the fixed cost is $125 per Sunday.

95. Cost of Electricity   In 1999, Florida Power and Light Company supplied electricity in the summer months to residential customers for a monthly customer charge of $5.65 plus 6.543¢ per kilowatt-hour supplied in the month for the first 750 kilowatt-hours used.* Write an equation that relates the monthly charge, C, in dollars, to the number x of kilowatt-hours used in the month. Graph this equation. What is the monthly charge for using 300 kilowatt-hours? For using 750 kilowatt-hours?

96. Show that the line containing the points (a, b) and (b, a), $a \neq b$, is perpendicular to the line $y = x$. Also show that the midpoint of (a, b) and (b, a) lies on the line $y = x$.

97. The equation $2x - y = C$ defines a **family of lines,** one line for each value of C. On one set of coordinate axes, graph the members of the family when $C = -4$, $C = 0$, and $C = 2$. Can you draw a conclusion from the graph about each member of the family?

98. Rework Problem 97 for the family of lines $Cx + y = -4$.

* Source: Florida Power and Light Co., Miami, Florida, 1999.

99. Which of the following equations might have the graph shown? (More than one answer is possible.)

(a) $2x + 3y = 6$ (e) $x - y = -1$
(b) $-2x + 3y = 6$ (f) $y = 3x - 5$
(c) $3x - 4y = -12$ (g) $y = 2x + 3$
(d) $x - y = 1$ (h) $y = -3x + 3$

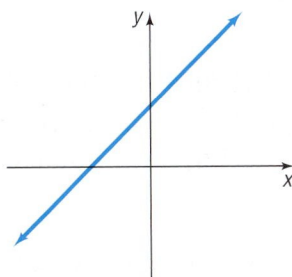

100. Which of the following equations might have the graph shown? (More than one answer is possible.)

(a) $2x + 3y = 6$ (e) $x - y = -1$
(b) $2x - 3y = 6$ (f) $y = -2x - 1$
(c) $3x + 4y = 12$ (g) $y = -\dfrac{1}{2}x + 10$
(d) $x - y = 1$ (h) $y = x + 4$

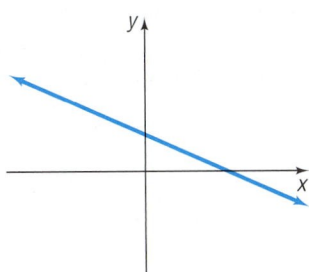

101. The figure below shows the graph of two parallel lines. Which of the following pairs of equations might have such a graph?

(a) $x - 2y = 3$
  $x + 2y = 7$
(b) $x + y = 2$
  $x + y = -1$
(c) $x - y = -2$
  $x - y = 1$
(d) $x - y = -2$
  $2x - 2y = -4$
(e) $x + 2y = 2$
  $x + 2y = -1$

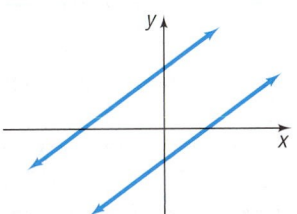

102. The figure below shows the graph of two parallel lines. Which of the following pairs of equations might have such a graph?

(a) $y - 2x = 2$ (c) $2y - x = 2$ (e) $2x + y = -2$
  $y + 2x = -1$ $2y + x = -2$ $2y + x = -2$
(b) $y - 2x = 0$ (d) $y - 2x = 2$
  $2y + x = 0$ $x + 2y = -1$

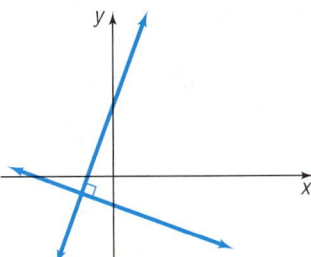

103. The accepted symbol used to denote the slope of a line is the letter m. Investigate the origin of this symbolism. Begin by consulting a French dictionary and looking up the French word *monter*. Write a brief essay on your findings.

104. The term *grade* is used to describe the inclination of a road. How does this term relate to the notion of slope of a line? Is a 4% grade very steep? Investigate the grades of some mountainous roads and determine their slopes. Write a brief essay on your findings.

105. Carpentry Carpenters use the term *pitch* to describe the steepness of staircases and roofs. How does pitch relate to slope? Investigate typical pitches used for stairs and for roofs. Write a brief essay on your findings.

1.8 CIRCLES

OBJECTIVES   ① Write the Standard Form of the Equation of a Circle
② Graph a Circle by Hand and by Using a Graphing Utility
③ Find the Center and Radius of a Circle from an Equation in General Form

 One advantage of a coordinate system is that it enables us to translate a geometric statement into an algebraic statement, and vice versa. Consider, for example, the following geometric statement that defines a circle.

> A **circle** is a set of points in the xy-plane that are a fixed distance r from a fixed point (h, k). The fixed distance r is called the **radius,** and the fixed point (h, k) is called the **center** of the circle.

Figure 81

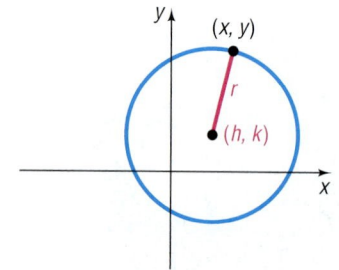

Figure 81 shows the graph of a circle. To find the equation, we let (x, y) represent the coordinates of any point on a circle with radius r and center (h, k). Then the distance between the points (x, y) and (h, k) must always equal r. That is, by the distance formula (Section 1.1),

$$\sqrt{(x - h)^2 + (y - k)^2} = r$$

or, equivalently,

$$(x - h)^2 + (y - k)^2 = r^2$$

> The **standard form of an equation of a circle** with radius r and center (h, k) is
>
> $$(x - h)^2 + (y - k)^2 = r^2 \qquad (1)$$

> The standard form of an equation of a circle of radius r with center at the origin $(0, 0)$ is
>
> $$x^2 + y^2 = r^2$$

> If the radius $r = 1$, the circle whose center is at the origin is called the **unit circle** and has the equation
>
> $$x^2 + y^2 = 1$$

See Figure 82.

Figure 82
Unit circle $x^2 + y^2 = 1$

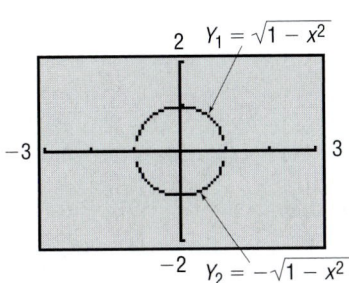

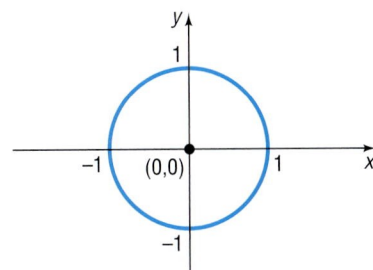

EXAMPLE 1   **Writing the Standard Form of the Equation of a Circle**

Write the standard form of the equation of the circle with radius 5 and center $(-3, 6)$.

Solution   Using the form of equation (1) and substituting the values $r = 5$, $h = -3$, and $k = 6$, we have

$$(x - h)^2 + (y - k)^2 = r^2$$

$$(x + 3)^2 + (y - 6)^2 = 25$$

■

━━━ **NOW WORK PROBLEM 1.**

2   The graph of any equation of the form of equation (1) is that of a circle with radius r and center (h, k).

EXAMPLE 2   **Graphing a Circle by Hand and by Using a Graphing Utility**

Graph the equation: $(x + 3)^2 + (y - 2)^2 = 16$

Solution   The graph of the equation is a circle. To graph the equation by hand, we first compare the given equation to the standard form of the equation of a circle. The comparison yields information about the circle.

$$(x + 3)^2 + (y - 2)^2 = 16$$

$$(x - (-3))^2 + (y - 2)^2 = 4^2$$

$$\underset{\uparrow}{} \qquad \underset{\uparrow}{} \qquad \underset{\uparrow}{}$$

$$(x - h)^2 + (y - k)^2 = r^2$$

Figure 83

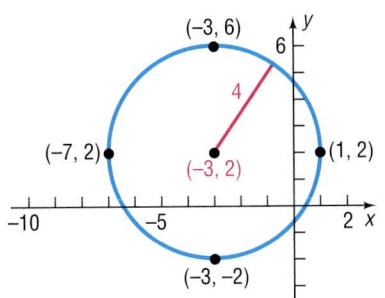

We see that $h = -3$, $k = 2$, and $r = 4$. The circle has center $(-3, 2)$ and a radius of 4 units. To graph this circle, we first plot the center $(-3, 2)$. Since the radius is 4, we can locate four points on the circle by plotting points 4 units to the left, to the right, up, and down from the center. These four points can then be used as guides to obtain the graph. See Figure 83.

To graph a circle on a graphing utility, we must write the equation in the form $y = \{\text{expression involving } x\}$.* We must solve for y in the equation

$$(x + 3)^2 + (y - 2)^2 = 16$$

$$(y - 2)^2 = 16 - (x + 3)^2 \qquad \text{Subtract } (x + 3)^2 \text{ from both sides.}$$

$$y - 2 = \pm\sqrt{16 - (x + 3)^2} \qquad \text{Use the Square Root Method.}$$

$$y = 2 \pm \sqrt{16 - (x + 3)^2} \qquad \text{Add 2 to both sides.}$$

To graph the circle, we graph the top half

$$Y_1 = 2 + \sqrt{16 - (x + 3)^2}$$

and the bottom half

$$Y_2 = 2 - \sqrt{16 - (x + 3)^2}$$

Also, be sure to use a square screen. Otherwise, the circle will appear distorted. Figure 84 shows the graph on a TI-83. ■

Figure 84

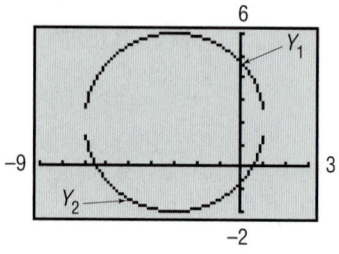

NOW WORK PROBLEM 17.

If we eliminate the parentheses from the standard form of the equation of the circle given in Example 2, we get

$$(x + 3)^2 + (y - 2)^2 = 16$$

$$x^2 + 6x + 9 + y^2 - 4y + 4 = 16$$

which we find, upon simplifying, is equivalent to

$$x^2 + y^2 + 6x - 4y - 3 = 0$$

It can be shown that any equation of the form

$$x^2 + y^2 + ax + by + c = 0$$

has a graph that is a circle, or a point, or has no graph at all. For example, the graph of the equation $x^2 + y^2 = 0$ is the single point $(0, 0)$. The equation $x^2 + y^2 + 5 = 0$, or $x^2 + y^2 = -5$, has no graph, because sums of squares of real numbers are never negative. When its graph is a circle, the equation

$$\boxed{x^2 + y^2 + ax + by + c = 0}$$

is referred to as the **general form of the equation of a circle.**

③ If an equation of a circle is in the general form, we use the method of completing the square to put the equation in standard form so we can identify its center and radius.

EXAMPLE 3 **Graphing a Circle Whose Equation Is in General Form**

Graph the equation $x^2 + y^2 + 4x - 6y + 12 = 0$.

Solution We complete the square in both x and y to put the equation in standard form. Group the expression involving x, group the expression involving y, and put the constant on the right side of the equation. The result is

$$(x^2 + 4x) + (y^2 - 6y) = -12$$

*Some graphing utilities (e.g., TI-82, TI-83, TI-85, and TI-86) have a CIRCLE function that allows the user to enter only the coordinates of the center of the circle and its radius to graph the circle.

Next, complete the square of each expression in parentheses. Remember that any number added on the left side of the equation must be added on the right.

$$(x^2 + 4x + 4) + (y^2 - 6y + 9) = -12 + 4 + 9$$

$$\left(\tfrac{4}{2}\right)^2 = 4 \qquad \left(\tfrac{-6}{2}\right)^2 = 9$$

$$(x + 2)^2 + (y - 3)^2 = 1 \quad \text{Factor}$$

We recognize this equation as the standard form of the equation of a circle with radius 1 and center $(-2, 3)$. To graph the equation by hand, use the center $(-2, 3)$ and the radius 1. See Figure 85(a).

To graph the equation using a graphing utility, we need to solve for y.

$$(y - 3)^2 = 1 - (x + 2)^2$$
$$y - 3 = \pm\sqrt{1 - (x + 2)^2} \quad \text{Use the Square Root Method.}$$
$$y = 3 \pm \sqrt{1 - (x + 2)^2} \quad \text{Add 3 to both sides.}$$

Figure 85(b) illustrates the graph.

Figure 85
$$x^2 + y^2 + 4x - 6y + 12 = 0$$

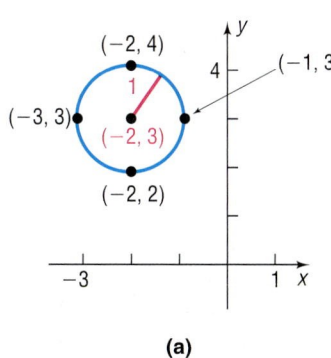

(a)

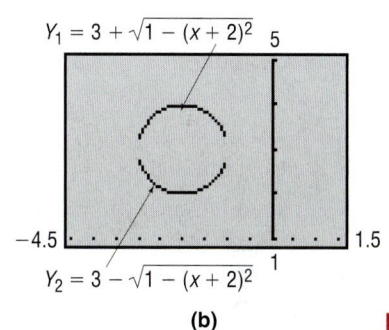

(b)

NOW WORK PROBLEM 19.

EXAMPLE 4   Finding the General Equation of a Circle

Find the general equation of the circle whose center is $(1, -2)$ and whose graph contains the point $(4, -2)$.

Figure 86
$$x^2 + y^2 - 2x + 4y - 4 = 0$$

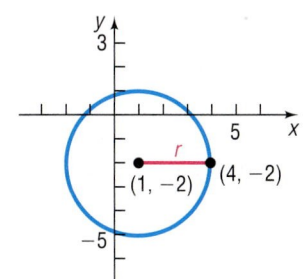

Solution To find the equation of a circle, we need to know its center and its radius. Here, we know that the center is $(1, -2)$. Since the point $(4, -2)$ is on the graph, the radius r will equal the distance from $(4, -2)$ to the center $(1, -2)$. See Figure 86. Thus,

$$r = \sqrt{(4 - 1)^2 + [-2 - (-2)]^2}$$
$$= \sqrt{9} = 3$$

The standard form of the equation of the circle is

$$(x - 1)^2 + (y + 2)^2 = 9$$

Eliminating the parentheses and rearranging terms, we get the general equation

$$x^2 + y^2 - 2x + 4y - 4 = 0$$

Overview

The discussion in Sections 1.7 and 1.8 about lines and circles dealt with two main types of problems that can be generalized as follows:

1. Given an equation, classify it and graph it.

2. Given a graph, or information about a graph, find its equation.

This text deals with both types of problems. We shall study various equations, classify them, and graph them. Although the second type of problem is usually more difficult to solve than the first, in many instances a graphing utility can be used to solve such problems.

1.8   Concepts and Vocabulary

In Problems 1 and 2, fill in the blanks.

1. To complete the square of the expression $x^2 + 5x$, you would _____ the number _____.

2. For a circle, the _____ is the distance from the center to any point on the circle.

In Problems 3 and 4, answer True or False to each statement.

3. The radius of the circle $x^2 + y^2 = 9$ is 3.

4. The center of the circle $(x + 3)^2 + (y - 2)^2 = 13$ is $(3, -2)$.

5. Explain how the center and radius of a circle can be used to graph a circle by hand.

6. Explain how the center and radius of a circle can be used to establish an initial viewing window.

7. If the circumference of a circle is 6π, what is its radius?

1.8   Exercises

In Problems 1–4, find the center and radius of each circle. Write the standard form of the equation.

1.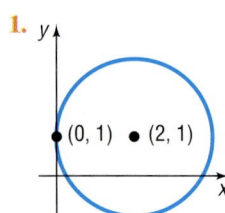

2.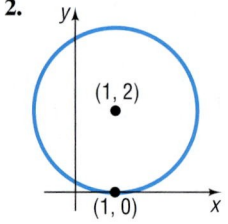

3.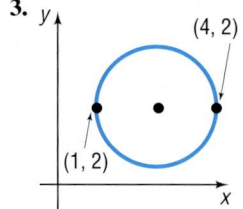

4.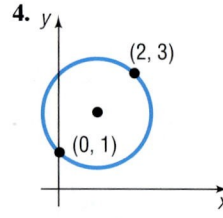

In Problems 5–14, write the standard form of the equation and the general form of the equation of each circle of radius r and center (h, k). By hand, graph each circle.

5. $r = 2$; $(h, k) = (0, 0)$

6. $r = 3$; $(h, k) = (0, 0)$

7. $r = 2$; $(h, k) = (0, 2)$

8. $r = 3$; $(h, k) = (1, 0)$

9. $r = 5;$ $(h, k) = (4, -3)$

10. $r = 4;$ $(h, k) = (2, -3)$

11. $r = 4;$ $(h, k) = (-2, 1)$

12. $r = 7;$ $(h, k) = (-5, -2)$

13. $r = \dfrac{1}{2};$ $(h, k) = \left(\dfrac{1}{2}, 0\right)$

14. $r = \dfrac{1}{2};$ $(h, k) = \left(0, -\dfrac{1}{2}\right)$

In Problems 15–24, find the center (h, k) and radius r of each circle. By hand, graph each circle.

15. $x^2 + y^2 = 25$

16. $x^2 + (y - 1)^2 = 9$

17. $(x - 2)^2 + y^2 = 4$

18. $(x + 3)^2 + (y - 1)^2 = 18$

19. $x^2 + y^2 + 4x - 4y - 1 = 0$

20. $x^2 + y^2 - 6x + 2y + 9 = 0$

21. $x^2 + y^2 - x + 2y + 1 = 0$

22. $x^2 + y^2 + x + y - \dfrac{1}{2} = 0$

23. $2x^2 + 2y^2 - 12x + 8y - 24 = 0$

24. $2x^2 + 2y^2 + 8x + 7 = 0$

In Problems 25–30, find the general form of the equation of each circle.

25. Center at the origin and containing the point $(-2, 3)$

26. Center $(1, 0)$ and containing the point $(-3, 2)$

27. Center $(2, 3)$ and tangent to the x-axis

28. Center $(-3, 1)$ and tangent to the y-axis

29. With end points of a diameter at $(1, 4)$ and $(-3, 2)$

30. With endpoints of a diameter at $(4, 3)$ and $(0, 1)$

In Problems 31–34, match each graph with the correct equation.

(a) $(x - 3)^2 + (y + 3)^2 = 9$ (b) $(x + 1)^2 + (y - 2)^2 = 4$ (c) $(x - 1)^2 + (y + 2)^2 = 4$ (d) $(x + 3)^2 + (y - 3)^2 = 9$

31.
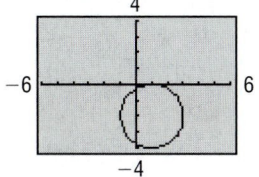

32.
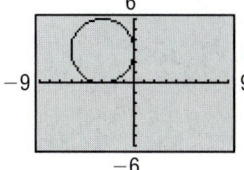

33.
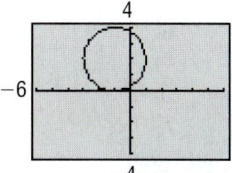

34.
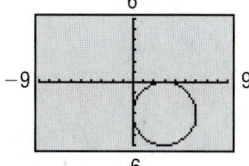

In Problems 35–38, find the standard form of the equation of each circle. Assume that the center has integer coordinates and that the radius is an integer.

35.
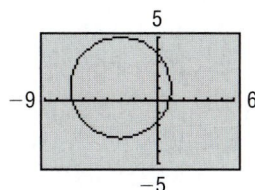

36.
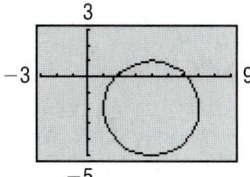

37.
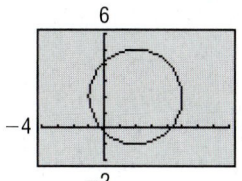

38.
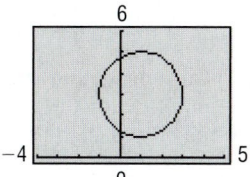

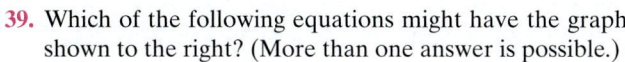

39. Which of the following equations might have the graph shown to the right? (More than one answer is possible.)

(a) $(x - 2)^2 + (y + 3)^2 = 13$

(b) $(x - 2)^2 + (y - 2)^2 = 8$

(c) $(x - 2)^2 + (y - 3)^2 = 13$

(d) $(x + 2)^2 + (y - 2)^2 = 8$

(e) $x^2 + y^2 - 4x - 9y = 0$

(f) $x^2 + y^2 + 4x - 2y = 0$

(g) $x^2 + y^2 - 9x - 4y = 0$

(h) $x^2 + y^2 - 4x - 4y = 4$

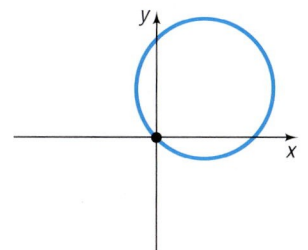

40. Which of the following equations might have the graph shown? (More than one answer is possible.)

(a) $(x - 2)^2 + y^2 = 3$ (e) $x^2 + y^2 + 10x + 16 = 0$

(b) $(x + 2)^2 + y^2 = 3$ (f) $x^2 + y^2 + 10x - 2y = 1$

(c) $x^2 + (y - 2)^2 = 3$ (g) $x^2 + y^2 + 9x + 10 = 0$

(d) $(x + 2)^2 + y^2 = 4$ (h) $x^2 + y^2 - 9x - 10 = 0$

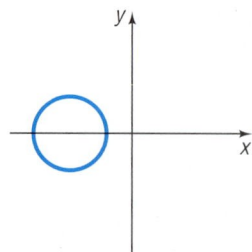

41. Weather Satellites   Earth is represented on a map of a portion of the solar system so that its surface is the circle with equation $x^2 + y^2 + 2x + 4y - 4091 = 0$. A weather satellite circles 0.6 unit above Earth with the center of its circular orbit at the center of Earth. Find the equation for the orbit of the satellite on this map.

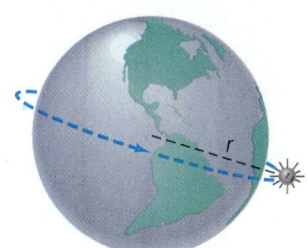

42. The **tangent line** to a circle may be defined as the line that intersects the circle in a single point, called the **point of tangency** (see the figure).

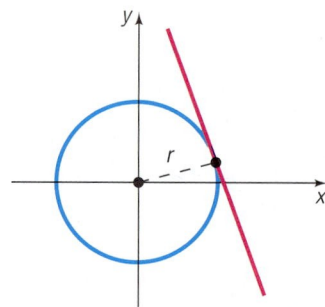

If the equation of the circle is $x^2 + y^2 = r^2$ and the equation of the tangent line is $y = mx + b$, show that:

(a) $r^2(1 + m^2) = b^2$

[**Hint:** The quadratic equation $x^2 + (mx + b)^2 = r^2$ has exactly one solution.]

(b) The point of tangency is $(-r^2m/b, r^2/b)$.

(c) The tangent line is perpendicular to the line containing the center of the circle and the point of tangency.

43. The Greek method for finding the equation of the tangent line to a circle used the fact that at any point on a circle the lines containing the center and the tangent line are perpendicular (see Problem 42). Use this method to find an equation of the tangent line to the circle $x^2 + y^2 = 9$ at the point $(1, 2\sqrt{2})$.

44. Use the Greek method described in Problem 43 to find an equation of the tangent line to the circle $x^2 + y^2 - 4x + 6y + 4 = 0$ at the point $(3, 2\sqrt{2} - 3)$.

45. Refer to Problem 42. The line $x - 2y + 4 = 0$ is tangent to a circle at $(0, 2)$. The line $y = 2x - 7$ is tangent to the same circle at $(3, -1)$. Find the center of the circle.

46. Find an equation of the line containing the centers of the two circles

$$x^2 + y^2 - 4x + 6y + 4 = 0 \qquad \text{and}$$
$$x^2 + y^2 + 6x + 4y + 9 = 0$$

47. If a circle of radius 2 is made to roll along the x-axis, what is an equation for the path of the center of the circle?

Chapter Review

Things to Know

Formulas

Distance formula (p. 93)

$$d = \sqrt{(x_2 - x_1)^2 + (y_2 - y_1)^2}$$

Midpoint formula (p. 96)

$$(x, y) = \left(\frac{x_1 + x_2}{2}, \frac{y_1 + y_2}{2} \right)$$

Slope (p. 163)

$$m = \frac{y_2 - y_1}{x_2 - x_1}, \text{ if } x_1 \neq x_2; \text{ undefined if } x_1 = x_2$$

Parallel lines (p. 172)   Equal slopes ($m_1 = m_2$) with different y-intercepts

Perpendicular lines (p. 174)   Product of slopes is -1 ($m_1 \cdot m_2 = -1$)

Quadratic equation and quadratic formula (p. 122)

If $ax^2 + bx + c = 0$, $a \neq 0$, and if $b^2 - 4ac \geq 0$, then $x = \dfrac{-b \pm \sqrt{b^2 - 4ac}}{2a}$.

Discriminant (p. 123)

If $b^2 - 4ac > 0$, there are two distinct real solutions.

If $b^2 - 4ac = 0$, there is one repeated real solution.

If $b^2 - 4ac < 0$, there are no real solutions.

Equations

Vertical line (p. 168) $x = a$

Horizontal line (p. 169) $y = b$

Point–slope form of the (p. 168) $y - y_1 = m(x - x_1)$; m is the slope of the line, (x_1, y_1) is a
  equation of a line   point on the line

General form of the (p. 171) $Ax + By = C$; A, B, not both 0
  equation of a line

Slope–intercept form of the (p. 170) $y = mx + b$; m is the slope of the line, b is the y-intercept
  equation of a line

Standard form of the (p. 180) $(x - h)^2 + (y - k)^2 = r^2$; r is the radius of the circle, (h, k) is
  equation of a circle   the center of the circle

Equation of the unit circle (p. 180) $x^2 + y^2 = 1$

General form of the
  equation of a circle (p. 182) $x^2 + y^2 + ax + by + c = 0$

Interval notation (p. 150)

$[a, b]$	$\{x \| a \leq x \leq b\}$	$[a, \infty)$	$\{x \| x \geq a\}$
$[a, b)$	$\{x \| a \leq x < b\}$	(a, ∞)	$\{x \| x > a\}$
$(a, b]$	$\{x \| a < x \leq b\}$	$(-\infty, a]$	$\{x \| x \leq a\}$
(a, b)	$\{x \| a < x < b\}$	$(-\infty, a)$	$\{x \| x < a\}$
		$(-\infty, \infty)$	All real numbers

Properties of inequalities

Addition property (p. 151)   If $a < b$, then $a + c < b + c$.

                                           If $a > b$, then $a + c > b + c$.

Multiplication properties (p. 153)   (a) If $a < b$ and if $c > 0$, then $ac < bc$.
                                               If $a < b$ and if $c < 0$, then $ac > bc$.

                                           (b) If $a > b$ and if $c > 0$, then $ac > bc$
                                               If $a > b$ and if $c < 0$, then $ac < bc$.

Absolute value

If $|u| = a$, $a > 0$, then $u = -a$ or $u = a$. (p. 146)

If $|u| \leq a$, $a > 0$, then $-a \leq u \leq a$. (p. 157)

If $|u| \geq a$, $a > 0$, then $u \leq -a$ or $u \geq a$. (p. 158)

Objectives

Section		You should be able to:	Review Exercises
1.1	1	Use the distance formula (p.93)	77, 78, 79, 80, 84
	2	Use the midpoint formula (p.96)	77, 78, 83
1.2	1	Graph equations by hand by plotting points (p.100)	57–62
	2	Graph equations using a graphing utility (p.103)	39, 40, 41, 42
	3	Find intercepts (p.106)	57–62
1.3	1	Solve equations in one variable (p.109)	1–38
	2	Solve equations using a graphing utility (p.113)	39–42
	3	Solve linear equations (p.115)	1–8, 11, 12
	4	Solve quadratic equations by factoring (p.117)	9, 10, 13, 14
	5	Solve quadratic equations by completing the square (p.120)	15, 16
	6	Solve quadratic equations using the quadratic formula (p.121)	15, 16
1.4	1	Translate verbal descriptions into mathematical expressions (p.128)	85, 86
	2	Set up applied problems (p.128)	89–106
	3	Solve interest problems (p.130)	87, 88
	4	Solve mixture problems (p.131)	89–92
	5	Solve uniform motion problems (p.133)	93–96
	6	Solve constant rate job problems (p.135)	97, 98
1.5	1	Solve radical equations (p.142)	17, 18, 23–30, 33, 34
	2	Solve equations quadratic in form (p.145)	29–32
	3	Solve absolute value equations (p.146)	35–38
1.6	1	Use interval notation (p.150)	43–56
	2	Use properties of inequalities (p.151)	43–56
	3	Solve linear inequalities algebraically and graphically (p.153)	43–44
	4	Solve combined inequalities algebraically and graphically (p.155)	45–48
	5	Solve absolute value inequalities algebraically and graphically (p.156)	49–56
1.7	1	Calculate and interpret the slope of a line (p.163)	77, 78, 81
	2	Graph lines (p.166)	67–76
	3	Find the equation of vertical lines (p.167)	69
	4	Use the point–slope form of a line; identify horizontal lines (p.168)	67–76
	5	Find the equation of a line given two points (p.169)	72
	6	Write the equation of a line in slope–intercept form (p.170)	67–76
	7	Identify the slope and y-intercept of a line from its equation (p.170)	67–76
	8	Write the equation of a line in general form (p.171)	67–76
	9	Define parallel lines (p.172)	73, 74
	10	Find equations of parallel lines (p.173)	73, 74
	11	Define perpendicular lines (p.173)	75, 76
	12	Find equations of perpendicular lines (p.175)	75, 76
1.8	1	Write the standard form of the equation of a circle (p.180)	63–66, 82
	2	Graph a circle by hand and by using a graphing utility (p.181)	63–66
	3	Find the center and radius of a circle from an equation in general form and graph it (p.182)	63–66

Review Exercises

Blue problem numbers indicate the authors' suggestions for use in a Practice Test.

In Problems 1–38, find all real solutions, if any, of each equation. Verify your results using a graphing utility.

1. $2 - \dfrac{x}{3} = 8$

2. $\dfrac{x}{4} - 2 = 4$

3. $-2(5 - 3x) + 8 = 4 + 5x$

4. $(6 - 3x) - 2(1 + x) = 6x$

5. $\dfrac{3x}{4} - \dfrac{x}{3} = \dfrac{1}{12}$

6. $\dfrac{4 - 2x}{3} + \dfrac{1}{6} = 2x$

7. $\dfrac{x}{x - 1} = \dfrac{6}{5}, \quad x \neq 1$

8. $\dfrac{4x - 5}{3 - 7x} = 2, \quad x \neq \dfrac{3}{7}$

9. $x(1 - x) = 6$

10. $x(1 + x) = 6$

11. $\dfrac{1}{2}\left(x - \dfrac{1}{3}\right) = \dfrac{3}{4} - \dfrac{x}{6}$

12. $\dfrac{1 - 3x}{4} = \dfrac{x + 6}{3} + \dfrac{1}{2}$

13. $(x - 1)(2x + 3) = 3$

14. $x(2 - x) = 3(x - 4)$

15. $2x + 3 = 4x^2$

16. $1 + 6x = 4x^2$

17. $\sqrt[3]{x^2 - 1} = 2$

18. $\sqrt{1 + x^3} = 3$

19. $x(x + 1) + 2 = 0$

20. $3x^2 - x + 1 = 0$

21. $x^4 - 5x^2 + 4 = 0$

22. $3x^4 + 4x^2 + 1 = 0$

23. $\sqrt{2x - 3} + x = 3$

24. $\sqrt{2x - 1} = x - 2$

25. $x^{3/2} + 5x^{1/2} = 0$

26. $x^{2/3} + x = 0$

27. $\sqrt{x + 1} + \sqrt{x - 1} = \sqrt{2x + 1}$

28. $\sqrt{2x - 1} - \sqrt{x - 5} = 3$

29. $2\sqrt[3]{x^2} - \sqrt[3]{x} = 1$

30. $4\sqrt[3]{x^2} = 1$

31. $x^{-6} - 7x^{-3} - 8 = 0$

32. $6x^{-1} - 5x^{-1/2} + 1 = 0$

33. $\sqrt{x^2 + 3x + 7} - \sqrt{x^2 - 3x + 9} + 2 = 0$

34. $\sqrt{x^2 + 3x + 7} - \sqrt{x^2 + 3x + 9} = 2$

35. $|2x + 3| = 7$

36. $|3x - 1| = 5$

37. $|2 - 3x| + 2 = 9$

38. $|1 - 2x| + 1 = 4$

In Problems 39–42, use a graphing utility to approximate the solutions of each equation rounded to two decimal places. All solutions lie between -10 and 10.

39. $x^3 - 5x + 3 = 0$

40. $-x^3 + 3x + 1 = 0$

41. $x^4 - 3 = 2x + 1$

42. $-x^4 + 7 = x^2 - 2$

In Problems 43–56, solve each inequality. Express your answer using set notation or interval notation. Graph the solution set. Verify your results using a graphing utility.

43. $\dfrac{2x - 3}{5} + 2 \leq \dfrac{x}{2}$

44. $\dfrac{5 - x}{3} \leq 6x - 4$

45. $-9 \leq \dfrac{2x + 3}{-4} \leq 7$

46. $-4 < \dfrac{2x - 2}{3} < 6$

47. $6 > \dfrac{3 - 3x}{12} > 2$

48. $6 > \dfrac{5 - 3x}{2} \geq -3$

49. $|3x + 4| < \dfrac{1}{2}$

50. $|1 - 2x| < \dfrac{1}{3}$

51. $|2x - 5| \geq 9$

52. $|3x + 1| \geq 10$

53. $2 + |2 - 3x| \leq 4$

54. $\dfrac{1}{2} + \left|\dfrac{2x - 1}{3}\right| \leq 1$

55. $1 - |2 - 3x| < -4$

56. $1 - \left|\dfrac{2x - 1}{3}\right| < -2$

In Problems 57–62, graph each equation by hand by plotting points. Be sure to label any intercepts. Verify your results using a graphing utility.

57. $2x - 3y = 6$

58. $2x + 5y = 10$

59. $y = x^2 - 9$

60. $y = x^2 + 4$

61. $x^2 + 2y = 16$

62. $2x^2 - 4y = 24$

In Problems 63–66, find the center and radius of each circle. Graph each circle by hand.

63. $x^2 + y^2 - 2x + 4y - 4 = 0$

64. $x^2 + y^2 + 4x - 4y - 1 = 0$

65. $3x^2 + 3y^2 - 6x + 12y = 0$

66. $2x^2 + 2y^2 - 4x = 0$

In Problems 67–76, find an equation of the line having the given characteristics. Express your answer using either the general form or the slope–intercept form of the equation of a line, whichever you prefer. Graph the line.

67. Slope $= -2$; containing the point $(3, -1)$

68. Slope $= 0$; containing the point $(-5, 4)$

69. Slope undefined; containing the point $(-3, 4)$

70. x-intercept $= 2$; containing the point $(4, -5)$

71. y-intercept $= -2$; containing the point $(5, -3)$

72. Containing the points $(3, -4)$ and $(2, 1)$

73. Parallel to the line $2x - 3y = -4$; containing the point $(-5, 3)$

74. Parallel to the line $x + y = 2$; containing the point $(1, -3)$

75. Perpendicular to the line $x + y = 2$; containing the point $(4, -3)$

76. Perpendicular to the line $3x - y = -4$; containing the point $(-2, 4)$

77. Find the slope of the line containing the points $(7, 4)$ and $(-3, 2)$. What is the distance between these points? What is their midpoint?

78. Find the slope of the line containing the points $(2, 5)$ and $(6, -3)$. What is the distance between these points? What is their midpoint?

79. Show that the points $A = (3, 4)$, $B = (1, 1)$, and $C = (-2, 3)$ are the vertices of an isosceles triangle.

80. Show that the points $A = (-2, 0,)$ $B = (-4, 4)$, and $C = (8, 5)$ are the vertices of a right triangle in two ways:
(a) By using the converse of the Pythagorean Theorem
(b) By using the slopes of the lines joining the vertices

81. Show that the points $A = (2, 5)$, $B = (6, 1)$, and $C = (8, -1)$ lie on a straight line by using slopes.

82. Show that the points $A = (1, 5)$, $B = (2, 4)$, and $C = (-3, 5)$ lie on a circle with center $(-1, 2)$. What is the radius of this circle?

83. The endpoints of the diameter of a circle are $(-3, 2)$ and $(5, -6)$. Find the center and radius of the circle. Write the general equation of this circle.

84. Find two numbers y such that the distance from $(-3, 2)$ to $(5, y)$ is 10.

85. Translate the following statement into a mathematical expression: The perimeter of a rectangle is the sum of two times the length and two times the width.

86. Translate the following statement into a mathematical expression: The total cost of manufacturing x bicycles in one day is $50,000 plus $95 times the number of bicycles manufactured.

87. **Banking** A bank lends out $9000 at 7% simple interest. At the end of 1 year, how much interest is owed on the loan?

88. **Financial Planning** Steve, a recent retiree, requires $5000 per year in extra income. He has $70,000 to invest and can invest in A-rated bonds paying 8% per year or in a certificate of deposit (CD) paying 5% per year. How much money should be invested in each to realize exactly $5000 in interest per year?

89. **Chemistry: Mixing Acids** A laboratory has 60 cubic centimeters (cm^3) of a solution that is 40% HCl acid. How many cubic centimeters of a 15% solution of HCl acid should be mixed with the 60 cm^3 of 40% acid to obtain a solution of 25% HCl? How much of the 25% solution is there?

90. **Business: Blending Coffee** A coffee house has 20 pounds of a coffee that sells for $4 per pound. How many pounds of a coffee that sells for $8 per pound should be mixed with the 20 pounds of $4-per-pound coffee to obtain a blend that will sell for $5 per pound? How much of the $5-per-pound coffee is there to sell?

91. **Chemistry: Salt Solutions** How much water should be added to 64 ounces of a 10% salt solution to make a 2% salt solution?

92. **Chemistry: Salt Solutions** How much water must be evaporated from 64 ounces of a 2% salt solution to make a 10% salt solution?

93. **Extent of Search and Rescue** A search plane has a cruising speed of 250 miles per hour and carries enough fuel for at most 5 hours of flying. If there is a wind that averages 30 miles per hour and the direction of search is with the wind one way and against it the other, how far can the search plane travel?

94. **Extent of Search and Rescue** If the search plane described in Problem 93 is able to add a supplementary fuel tank that allows for an additional 2 hours of flying, how much farther can the plane extend its search?

95. **Rescue at Sea** A life raft, set adrift from a sinking ship 150 miles offshore, travels directly toward a Coast Guard station at the rate of 5 miles per hour. At the time that the raft is set adrift, a rescue helicopter is dispatched from the Coast Guard station. If the helicopter's average speed is 90 miles per hour, how long will it take the helicopter to reach the life raft?

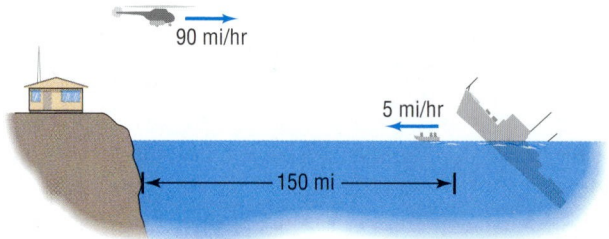

96. **Physics: Uniform Motion** Two bees leave two locations 150 meters apart and fly, without stopping, back and forth between these two locations at average speeds of 3 meters per second and 5 meters per second, respectively. How long is it until the bees meet for the first time? How long is it until they meet for the second time?

97. Working Together to Get a Job Done   Clarissa and Shawna, working together, can paint the exterior of a house in 6 days. Clarissa by herself can complete this job in 5 days less than Shawna. How long will it take Clarissa to complete the job by herself?

98. Emptying a Tank   Two pumps of different sizes, working together, can empty a fuel tank in 5 hours. The larger pump can empty this tank in 4 hours less than the smaller one. If the larger one is out of order, how long will it take the smaller one to do the job alone?

99. Physics: Uniform Motion   A man is walking at an average speed of 4 miles per hour alongside a railroad track. A freight train, going in the same direction at an average speed of 30 miles per hour, requires 5 seconds to pass the man. How long is the freight train? Give your answer in feet.

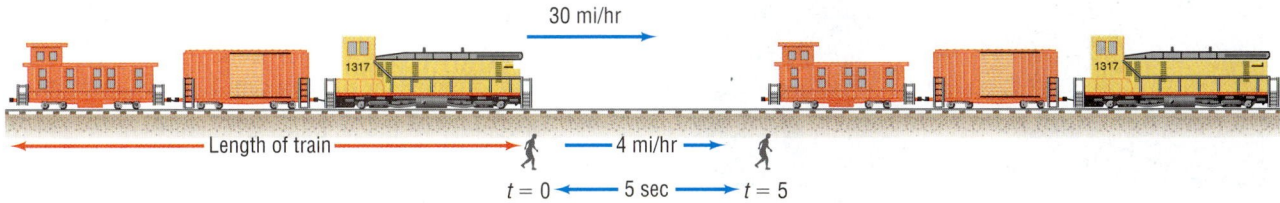

100. Framing a Painting   An artist has 50 inches of oak trim to frame a painting. The frame is to have a border 3 inches wide surrounding the painting.

   (a) If the painting is square, what are its dimensions? What are the dimensions of the frame?
   (b) If the painting is rectangular with a length twice its width, what are the dimensions of the painting? What are the dimensions of the frame?

101. Using Two Pumps   An 8 horsepower (hp) pump can fill a tank in 8 hours. A smaller, 3 hp pump fills the same tank in 12 hours. The pumps are used together to begin filling this tank. After four hours, the 8 hp pump breaks down. How long will it take the smaller pump to fill the tank?

102. Pleasing Proportions   One formula stating the relationship between the length l and width w of a rectangle of "pleasing proportion" is $l^2 = w(l + w)$. How should a 4 foot by 8 foot sheet of plasterboard be cut so that the result is a rectangle of "pleasing proportion" with a width of 4 feet?

103. Business: Determining the Cost of a Charter   A group of 20 senior citizens can charter a bus for a one-day excursion trip for $15 per person. The charter company agrees to reduce the price of each ticket by 10¢ for each additional passenger in excess of 20 who goes on the trip, up to a maximum of 44 passengers (the capacity of the bus). If the final bill from the charter company was $482.40, how many seniors went on the trip, and how much did each pay?

104. Utilizing Copying Machines   A new copying machine can do a certain job in 1 hour less than an older copier. Together they can do this job in 72 minutes. How long would it take the older copier by itself to do the job?

105. Geometry   The hypotenuse of a right triangle measures 13 centimeters. Find the lengths of the legs if their sum is 17 centimeters.

106. Geometry   The diagonal of a rectangle measures 10 inches. If the length is 2 inches more than the width, find the dimensions of the rectangle.

107. Make up four problems that you might be asked to do given the two points $(-3, 4)$ and $(6, 1)$. Each problem should involve a different concept. Be sure that your directions are clearly stated.

108. Describe each of the following graphs. Give justification.
   (a) $x = 0$ (b) $y = 0$
   (c) $x + y = 0$ (d) $xy = 0$
   (e) $x^2 + y^2 = 0$

109. In a 100-meter race, Todd crosses the finish line 5 meters ahead of Scott. To even things up, Todd suggests to Scott that they race again, this time with Todd lining up 5 meters behind the start.
   (a) Assuming that Todd and Scott run at the same pace as before, does the second race end in a tie?
   (b) If not, who wins?
   (c) By how many meters does he win?
   (d) How far back should Todd start so that the race ends in a tie?

After running the race a second time, Scott, to even things up, suggests to Todd that he (Scott) line up 5 meters in front of the start.
   (e) Assuming again that they run at the same pace as in the first race, does the third race result in a tie?
   (f) If not, who wins?
   (g) By how many meters?
   (h) How far up should Scott start so that the race ends in a tie?

110. Explain the difference between the following three problems. Are there any similarities in their solution?
   (a) Write the expression as a single quotient:
$$\frac{x}{x - 2} + \frac{x}{x^2 - 4}$$
   (b) Solve: $\dfrac{x}{x - 2} + \dfrac{x}{x^2 - 4} = 0$
   (c) Solve: $\dfrac{x}{x - 2} + \dfrac{x}{x^2 - 4} < 0$

Chapter Projects

1. **Mortgages**   You Bought a House in July, 1994, for $65,000. You paid $5000 as a down payment and financed the balance at an 8.875% annual percentage rate (APR) over 30 years.

   (a)   The monthly payment is computed using the formula

   $$M = A\left[\frac{1 - (1 + i)^{-n}}{i}\right]^{-1}$$

   where

   M = monthly payment

   A = amount financed

   n = length of the loan, in months

   $i = \dfrac{APR}{12}$, where the APR is expressed as a decimal

   Compute the monthly payment on your mortgage.

   (b)   What is the total value of the payments over the life of the loan?

   (c)   In September, 2001, you decide to refinance your loan because rates have gone down to 6.7% APR for 30 years. However, during the first 10 years of a 30-year mortgage, approximately only 5% of your monthly payment goes toward paying down the original amount financed. Based on this estimate, how much do you need to refinance? (Ignore closing costs and any other costs associated with the refinancing.)

   (d)   If you choose to refinance, what would be the monthly payment based on the current rates available? How much would you end up paying in all for the house?

   If you choose to refinance for 15 years at 6.25% APR instead, what would your monthly payment be? How much would you pay in all for the house? How much would you ultimately save?

2. **Economics**   An **isocost line** represents the various combinations of two inputs that can be used in a production process while maintaining a constant level of expenditures. The equation for an isocost line is

   $$P_L \cdot L + P_K \cdot K = E$$

   where

   P_L represents the price of labor.

   L represents the amount of labor used in the production process.

   P_K represents the price of capital (machinery, etc.).

   K represents the amount of capital used in the production process.

   E represents the total expenditures for the two inputs, labor and capital.

   (a)   What values of L and K make sense as they are defined above?

   (b)   Suppose that P_L = $400 per hour and P_K = $500 per hour. If L = 100 hours, how many hours of capital, K, may be purchased if expenditures are to be $100,000?

   (c)   Suppose that P_L = $400 per hour and P_K = $500 per hour; graph the isocost line if E = $100,000 with the number of hours of labor, L, on the x-axis and the number of hours of capital, K, on the y-axis. In which quadrant is your graph located? Why?

   (d)   Graph the isocost line if E increases to $120,000. What effect does this have on the graph? What do you think would happen if expenditures were to decrease?

   (e)   Determine the slope of the isocost line. Interpret its value.

   (f)   Graph the isocost line with P_L = $500 per hour, P_K = $500 per hour, and E = $100,000. Compare this graph with the graph obtained in (c). What effect does the increase in the price of labor have on the graph?

3. **Cell Phone Service**   In purchasing cell phone service, a person must decide which service will suit them the best. In order to determine the best rates and services that apply to your situation, you have to do some shopping around.

   Suppose you have an old model cell phone and you are looking to upgrade your phone. You have gathered the following information on cell phone services:

   Nokia 5165 $19.99 for the phone

   Ericsson A1228di   Free (after $20 rebate)

   Option 1: $29.99 for 250 "peak time" minutes, unlimited nights and weekends; extra "peak time" minutes cost $0.45 per minute

   Option 2: $39.99 for 400 "peak time" minutes, unlimited nights and weekends; extra "peak time" minutes cost $0.45 per minute

   Option 3: $49.99 for 600 "peak time" minutes, unlimited nights and weekends; extra "peak time" minutes cost $0.35 per minute

To qualify for these phone prices, you must sign up for a two-year contract.

(a) List all of the different combinations of phone and services.

(b) Determine the total cost of each combination described in part (a) for the life of the contract (24 months), assuming that you stay within the allotted "peak time" minutes provided by each contract.

(c) Suppose you expect to use 260 "peak time" minutes per month. Which option provides the best deal? Suppose you expect to use 320 "peak time" minutes per month. Which option provides the best deal?

(d) Suppose you expect to use 410 "peak time" minutes per month. Which option provides the best deal? Suppose you expect to use 450 "peak time" minutes per month. Which option provides the best deal?

(e) Each monthly charge includes a specific number of "peak time" minutes included in the monthly fee. Write a linear equation for each available option where y is the monthly cost and x is the number of "peak time" minutes used if . . . .

   i.   . . . $x \leq 250$
   ii.   . . . $250 < x \leq 400$
   iii. . . . $400 < x \leq 600$
   iv.   . . . $x > 600$

(f) In each case in part (e), graph the equations corresponding to each option in order to determine which option represents the best deal. In each case, remember the restrictions on x, the number of minutes used. Explain your results as though you were describing the deals to a friend.

(g) Suppose you expect to use between 250 and 400 minutes each month. At what point does Option 2 become a better deal than Option 1?

(h) Suppose you expect to use between 400 and 600 minutes each month. At what point does Option 3 become a better deal than Option 2?

This project is based on information given in a "Cingular Wireless" advertisement in the *Denton Record Chronicle* on September 11, 2001.

LINEAR AND QUADRATIC FUNCTIONS

The Use of Statistics in Making Investment Decisions: Using Beta

One of the most common statistics used by investors is beta, which measures a security or portfolio's movement relative to the S&P 500. When determining beta graphically, returns are plotted on a graph, with the horizontal axis representing the S&P returns and the vertical axis representing the security or portfolio returns; the points that are plotted represent the return of the security versus the return of the S&P 500 at the same time period. A regression line is then drawn, which is a straight line that most closely approximates the points in the graph. Beta represents the slope of that line; a steep slope (45 degrees or greater) indicates that when the stock market moved up or down, the security or portfolio moved up or down on average to the same degree or greater, and a flatter slope indicates that when the stock market moved up or down, the security or portfolio moved to a lesser degree. Since beta measures variability, it is used as a measure of risk; the greater the beta, the greater the risk.

SOURCE: Paul E. Hoffman, *Journal of the American Association of Individual Investors*, September, 1991; http://www.aaii.com

SEE CHAPTER PROJECT 1.

OUTLINE

For additional study help, go to

www.prenhall.com/sullivanegu3e

Materials include:

- Graphing Calculator Help
- Chapter Quiz
- Chapter Test
- PowerPoint Downloads
- Chapter Projects
- Student Tips

A Look Back, A Look Forward

Up to now, our discussion has focused on equations. We have solved equations containing one variable and developed techniques for graphing equations containing two variables. In this chapter, we look at a special type of equation involving two variables, called a *function*. This chapter deals with what a function is, how to graph *linear functions* and *quadratic functions,* and how these functions are used in applications. The word function apparently was introduced by René Descartes in 1637. For him, a function simply meant any positive integral power of a variable *x*. Gottfried Wilhelm Leibniz (1646–1716), who always emphasized the geometric side of mathematics, used the word function to denote any quantity associated with a curve, such as the coordinates of a point on the curve. Leonhard Euler (1707–1783) employed the word to mean any equation or formula involving variables and constants. His idea of a function is similar to the one most often seen in courses that precede calculus. Later, the use of functions in investigating heat flow equations led to a very broad definition, due to Lejeune Dirichlet (1805–1859), which describes a function as a rule or correspondence between two sets. It is his definition that we use here.

PREPARING FOR THIS SECTION

Before getting started, review the following:

✓ Intervals (Section 1.6, pp. 150–151)

✓ Evaluating Algebraic Expressions, Domain of a Variable (Section R.2, pp. 20–21)

✓ Linear Equations (Section 1.3, pp. 115–117)

✓ Quadratic Equations (Section 1.3, pp. 117–124)

✓ Solving Inequalities (Section 1.6, pp. 153–156)

✓ Intercepts (Section 1.2, pp. 106–107)

2.1   FUNCTIONS

OBJECTIVES

1. Determine Whether a Relation Represents a Function
2. Find the Value of a Function
3. Find the Domain of a Function
4. Identify the Graph of a Function
5. Obtain Information from or about the Graph of a Function

1   A **relation** is a correspondence between two sets. If x and y are two elements in these sets and if a relation exists between x and y, then we say that x **corresponds** to y or that y **depends on** x, and we write $x \rightarrow y$. We may also write $x \rightarrow y$ as the ordered pair (x, y).

EXAMPLE 1   **An Example of a Relation**

Figure 1 depicts a relation between four individuals and their birthdays. The relation might be named "was born on." Then Katy corresponds to June 20, Dan corresponds to Sept. 4, and so on. Using ordered pairs, this relation would be expressed as

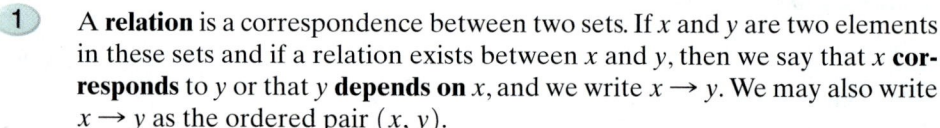

{(Katy, June 20), (Dan, September 4), (Patrick, December 31), (Phoebe, December 31)}

Figure 1

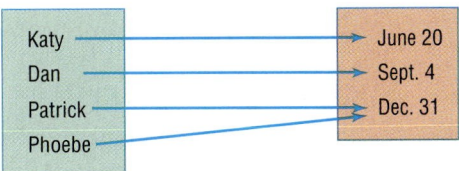

Often, we are interested in specifying the type of relation (such as an equation) that might exist between two variables. For example, the relation between the revenue R resulting from the sale of x items selling for $10 each may be expressed by the equation $R = 10x$. If we know how many items have been sold, then we can calculate the revenue by using the equation $R = 10x$. This equation is an example of a *function*.

As another example, suppose that an icicle falls off a building from a height of 64 feet above the ground. According to a law of physics, the distance s (in feet) of the icicle from the ground after t seconds is given (approximately) by the formula $s = 64 - 16t^2$. When $t = 0$ seconds, the icicle is $s = 64$ feet above the ground. After 1 second, the icicle is $s = 64 - 16(1)^2 = 48$ feet above the ground. After 2 seconds, the icicle strikes the ground. The formula $s = 64 - 16t^2$ provides a way of finding the distance s for any time t $(0 \leq t \leq 2)$. There is a correspondence between each time t in the interval $0 \leq t \leq 2$ and the distance s. We say that the distance s is a function of the time t because:

1. There is a correspondence between the set of times and the set of distances.
2. There is exactly one distance s obtained for any time t in the interval $0 \leq t \leq 2$.

Let's now look at the definition of a function.

Definition of Function

> Let X and Y be two nonempty sets.* A **function** from X into Y is a relation that associates with each element of X exactly one element of Y.

Figure 2

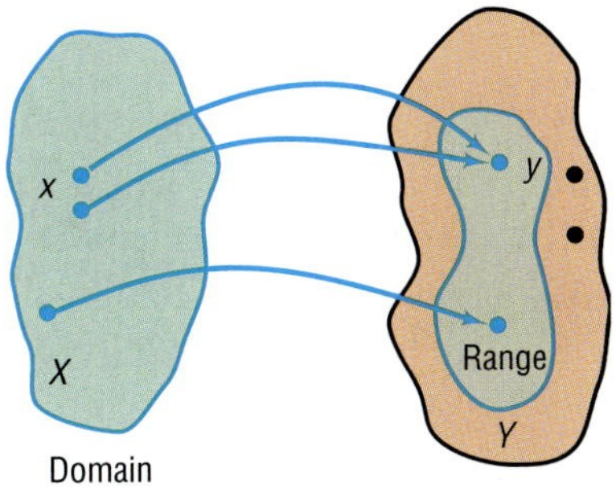

Domain

The set X is called the **domain** of the function. For each element x in X, the corresponding element y in Y is called the **value** of the function at x, or the **image** of x. The set of all images of the elements in the domain is called the **range** of the function. See Figure 2.

Since there may be some elements in Y that are not the image of some x in X, it follows that the range of a function may be a subset of Y, as shown in Figure 2.

Not all relations between two sets are functions. The next example shows how to determine whether a relation is a function or not.

*The sets X and Y will usually be sets of real numbers, in which case a (real) function results. The two sets can also be sets of complex numbers (discussed in Section 5.3), and then we have defined a complex function. In the broad definition (due to Lejeune Dirichlet), X and Y can be any two sets.

EXAMPLE 2 **Determining Whether a Relation Represents a Function**

Determine whether the following relations represent functions.

(a) See Figure 3. For this relation, the domain represents four individuals and the range represents their birthdays.

Figure 3

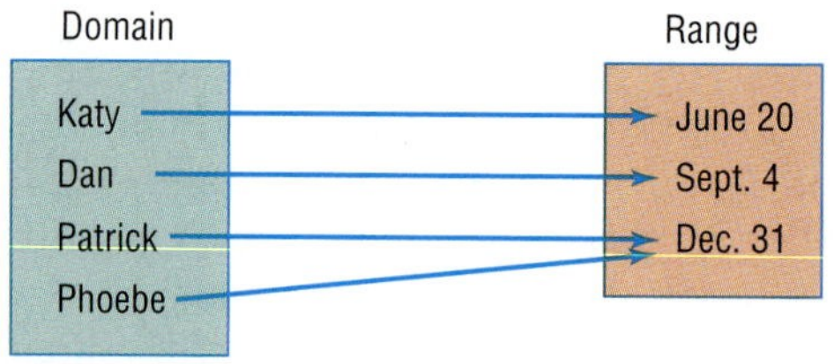

Domain

Katy
Dan
Patrick
Phoebe

Range

June 20
Sept. 4
Dec. 31

(b) See Figure 4. For this relation, the domain represents the employees of Sara's Pre-Owned Car Mart and the range represents their phone number(s).

Figure 4

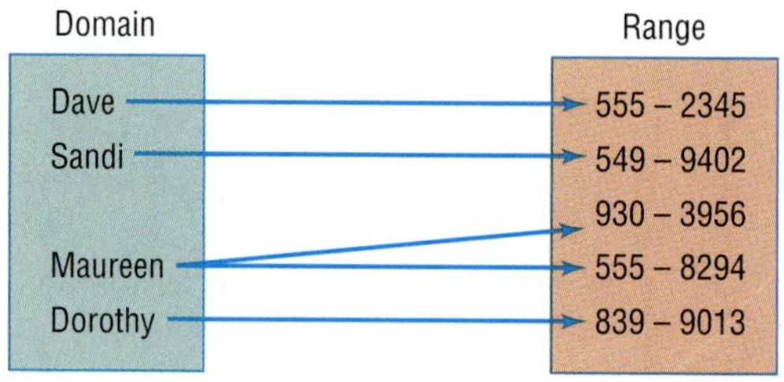

Domain

Dave
Sandi
Maureen
Dorothy

Range

555 – 2345
549 – 9402
930 – 3956
555 – 8294
839 – 9013

Solution (a) The relation is a function because each element in the domain corresponds to exactly one element in the range. Notice that more than one element in the domain can correspond to the same element in the range. (Phoebe and Patrick were born on the same day of the year).

(b) The relation is not a function because each element in the domain does not correspond to exactly one element in the range. Maureen has two telephone numbers; therefore, if Maureen is chosen from the domain, a single telephone number cannot be assigned to her. ■

 **NOW WORK PROBLEM 1.**

We may think of a function as a set of ordered pairs (x, y) in which no two ordered pairs have the same first element, but different second elements. The set of all first elements x is the domain of the function, and the set of all second elements y is its range. Each element x in the domain corresponds to exactly one element y in the range.

EXAMPLE 3 **Determining Whether a Relation Represents a Function**

Determine whether each relation represents a function. If it is a function, state the domain and range.

(a) $\{(1, 4), (2, 5), (3, 6), (4, 7)\}$

(b) $\{(1,4),(2,4),(3,5),(6,10)\}$

(c) $\{(-3,9),(-2,4),(0,0),(1,1),(-3,8)\}$

Solution (a) This relation is a function because there are no ordered pairs with the same first element and different second elements. The domain of this function is $\{1,2,3,4\}$, and its range is $\{4,5,6,7\}$.

(b) This relation is a function because there are no ordered pairs with the same first element and different second elements. The domain of this function is $\{1,2,3,6\}$, and its range is $\{4,5,10\}$.

(c) This relation is not a function because there are two ordered pairs $(-3,9)$ and $(-3,8)$ that have the same first element, but different second elements. ■

In Example 3(b), notice that 1 and 2 in the domain each correspond to 4. This does not violate the definition of a function; two different first elements can have the same second element. A violation of the definition occurs when two ordered pairs have the same first element and different second elements, as in Example 3(c).

 **NOW WORK PROBLEM 5.**

Example 2(a) demonstrates that a function may be defined by a correspondence between two sets. Examples 3(a) and 3(b) demonstrate that a function may be defined by a set of ordered pairs. A function may also be defined by an equation in two variables, usually denoted x and y.

EXAMPLE 4 **Example of a Function**

Consider the equation

$$y = 2x - 5 \qquad 1 \le x \le 6$$

Notice that for each input x there corresponds exactly one output y. For example, if $x = 1$, then $y = 2(1) - 5 = -3$. If $x = 3$, then $y = 2(3) - 5 = 1$. For this reason, the equation is a function. Since we restrict the inputs to the real numbers between 1 and 6, inclusive, the domain of the function is $\{x \mid 1 \le x \le 6\}$. The function specifies that in order to get the image of x we multiply x by 2 and then subtract 5 from this product. ■

Function Notation

Functions are often denoted by letters such as f, F, g, G, and so on. If f is a function, then for each number x in its domain the corresponding image in the range is designated by the symbol $f(x)$, read as "f of x" or as "f at x." We refer to $f(x)$ as the **value of f at the number x**; $f(x)$ is the number that results when x is given and the function f is applied; $f(x)$ does *not* mean "f times x." For example, the function given in Example 4 may be written as $y = f(x) = 2x - 5, 1 \le x \le 6$. Then $f(1) = -3$.

Figure 5 illustrates some other functions. Note that, in every function illustrated, for each x in the domain there is one value in the range.

Figure 5

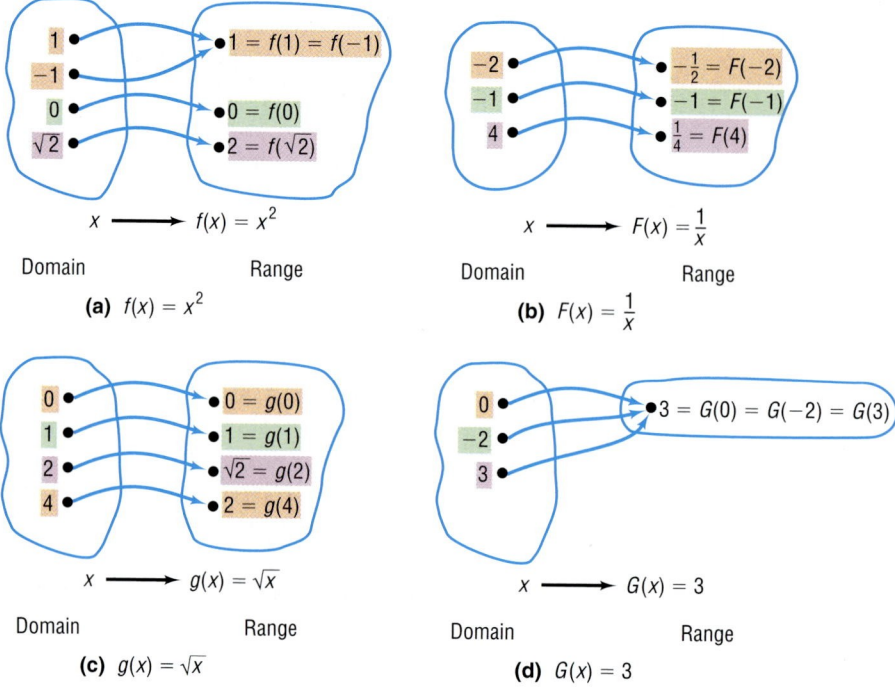

(a) $f(x) = x^2$

(b) $F(x) = \frac{1}{x}$

(c) $g(x) = \sqrt{x}$

(d) $G(x) = 3$

Sometimes it is helpful to think of a function f as a machine that receives as input a number from the domain, manipulates it, and outputs the value. See Figure 6.

Figure 6 Input x

Output
$y = f(x)$

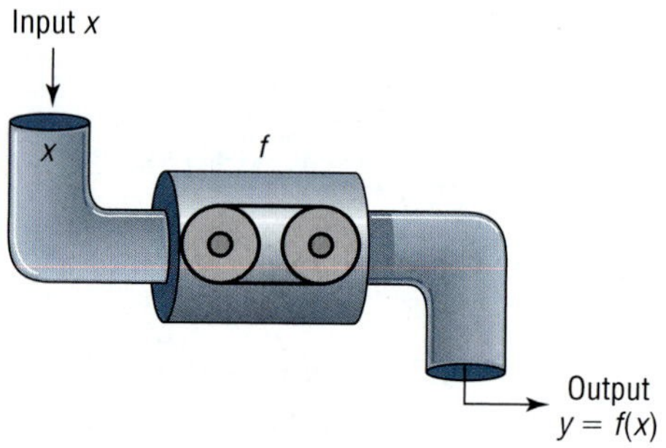

The restrictions on this input/output machine are

1. It only accepts numbers from the domain of the function.

2. For each input, there is exactly one output (which may be repeated for different inputs).

For a function $y = f(x)$, the variable x is called the **independent variable,** because it can be assigned any of the numbers from the domain. The variable y is called the **dependent variable,** because its value depends on x.

Any symbol can be used to represent the independent and dependent variables. For example, if f is the *cube function,* then f can be defined by $f(x) = x^3$ or $f(t) = t^3$ or $f(z) = z^3$. All three functions are the same: Each tells us to cube the independent variable. In practice, the symbols used for the independent and dependent variables are based on common usage, such as using C for cost in business.

2 The independent variable is also called the **argument** of the function. Thinking of the independent variable as an argument can sometimes make it easier to find the value of a function. For example, if f is the function defined by $f(x) = x^3$, then f tells us to cube the argument. Thus, $f(2)$ means to cube 2, $f(a)$ means to cube the number a, and $f(x + h)$ means to cube the quantity $x + h$.

EXAMPLE 5   Finding Values of a Function

For the function f defined by $f(x) = 2x^2 - 3x$, evaluate:

(a) $f(3)$ (b) $f(x) + f(3)$ (c) $f(-x)$

(d) $-f(x)$ (e) $f(x + 3)$ (f) $\dfrac{f(x + h) - f(x)}{h}$

Solution   (a) We substitute 3 for x in the equation for f to get

$$f(3) = 2(3)^2 - 3(3) = 18 - 9 = 9$$

(b) $f(x) + f(3) = (2x^2 - 3x) + (9) = 2x^2 - 3x + 9$

(c) We substitute $-x$ for x in the equation for f:

$$f(-x) = 2(-x)^2 - 3(-x) = 2x^2 + 3x$$

(d) $-f(x) = -(2x^2 - 3x) = -2x^2 + 3x$

(e) $f(x + 3) = 2(x + 3)^2 - 3(x + 3)$ Notice the use of parentheses here.

$$= 2(x^2 + 6x + 9) - 3x - 9$$
$$= 2x^2 + 12x + 18 - 3x - 9$$
$$= 2x^2 + 9x + 9$$

(f) $\dfrac{f(x + h) - f(x)}{h} = \dfrac{[2(x + h)^2 - 3(x + h)] - [2x^2 - 3x]}{h}$

$\uparrow$
$f(x + h) = 2(x + h)^2 - 3(x + h)$

$$= \frac{2(x^2 + 2xh + h^2) - 3x - 3h - 2x^2 + 3x}{h} \quad \text{\small\color{blue}Simplify.}$$

$$= \frac{2x^2 + 4xh + 2h^2 - 3h - 2x^2}{h}$$

$$= \frac{4xh + 2h^2 - 3h}{h}$$

$$= \frac{h(4x + 2h - 3)}{h} \quad \text{\small\color{blue}Factor out } h.$$

$$= 4x + 2h - 3 \quad \text{\small\color{blue}Cancel } h.$$

Notice in this example that $f(x + 3) \neq f(x) + f(3)$ and $f(-x) \neq -f(x)$. Also, the expression in part (f) is called the **difference quotient** of f, an important expression in calculus.

 NOW WORK PROBLEMS 15 AND 73.

Graphing calculators have special keys that enable you to find the value of certain commonly used functions. For example, you should be able to find the square function $f(x) = x^2$, the square root function $f(x) = \sqrt{x}$, the reciprocal function $f(x) = \dfrac{1}{x} = x^{-1}$, and many others that will be discussed later in this book (such as $\ln x$, $\log x$, and so on). Verify the results of Example 6, which follows, on your calculator.

EXAMPLE 6   **Finding Values of a Function on a Calculator**

(a) $f(x) = x^2$; $f(1.23) = 1.5129$

(b) $F(x) = \dfrac{1}{x}$; $F(1.6) = 0.625$

(c) $g(x) = \sqrt{x}$; $g(1.234) \approx 1.110855526$ ■

Graphing calculators can also be used to evaluate any function that you wish. Figure 7 shows the result obtained in Example 5(a) on a TI-83 graphing calculator with the function to be evaluated, $f(x) = 2x^2 - 3x$, in Y_1.*

Figure 7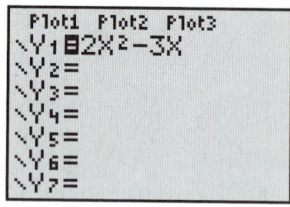

Implicit Form of a Function

In general, when a function f is defined by an equation in x and y, we say that the function f is given **implicitly.** If it is possible to solve the equation for y in terms of x, then we write $y = f(x)$ and say that the function is given **explicitly.** For example,

Implicit Form	Explicit Form
$3x + y = 5$	$y = f(x) = -3x + 5$
$x^2 - y = 6$	$y = f(x) = x^2 - 6$
$xy = 4$	$y = f(x) = \dfrac{4}{x}$

Not all equations in x and y define a function $y = f(x)$. If an equation is solved for y and two or more values of y can be obtained for a given x, then the equation does not define a function.

EXAMPLE 7   **Determining Whether an Equation Is a Function**

Determine if the equation $x^2 + y^2 = 1$ is a function.

*Consult your owner's manual for the required keystrokes.

Solution   To determine whether the equation $x^2 + y^2 = 1$, which defines the unit circle, is a function, we need to solve the equation for y.

$$x^2 + y^2 = 1$$

$$y^2 = 1 - x^2$$

$$y = \pm\sqrt{1 - x^2}$$

For values of x between -1 and 1, two values of y result. For example, if $x = 0$, then $y = \pm\sqrt{1} = \pm1$. This means that the equation $x^2 + y^2 = 1$ is not a function.   ■

 NOW WORK PROBLEM **27.**

COMMENT   The explicit form of a function is the form required by a graphing calculator. Now do you see why it is necessary to graph a circle in two "pieces"? ■

We list next a summary of some important facts to remember about a function f.

SUMMARY   **Important Facts About Functions**

1. For each x in the domain of a function f, there is one and only one image $f(x)$ in the range.

2. f is the symbol that we use to denote the function. It is symbolic of the equation that we use to get from an x in the domain to $f(x)$ in the range.

3. If $y = f(x)$, then x is called the independent variable or argument of f, and y is called the dependent variable or the value of f at x.

Domain of a Function

③   Often the domain of a function f is not specified; instead, only the equation defining the function is given. In such cases, we agree that the domain of f is the largest set of real numbers for which the value $f(x)$ is a real number. The domain of f is the same as the domain of the variable x in the expression $f(x)$.

EXAMPLE 8   **Finding the Domain of a Function**

Find the domain of each of the following functions:

(a) $f(x) = x^2 + 5x$ (b) $g(x) = \dfrac{3x}{x^2 - 4}$ (c) $h(t) = \sqrt{4 - 3t}$

Solution   (a) The function f tells us to square a number and then add five times the number. Since these operations can be performed on any real number, we conclude that the domain of f is all real numbers.

(b) The function g tells us to divide $3x$ by $x^2 - 4$. Since division by 0 is not defined, the denominator $x^2 - 4$ can never be 0, so x can never equal -2 or 2. The domain of the function g is $\{x \mid x \neq -2, x \neq 2\}$.

(c) The function h tells us to take the square root of $4 - 3t$. But only non-negative numbers have real square roots, so the expression under the square root must be greater than or equal to 0. This requires that

$$4 - 3t \geq 0$$
$$-3t \geq -4$$
$$t \leq \frac{4}{3}$$

The domain of h is $\left\{ t \mid t \leq \frac{4}{3} \right\}$ or the interval $\left(-\infty, \frac{4}{3} \right]$. ■

NOW WORK PROBLEM **37**.

If x is in the domain of a function f, we say that f **is defined at** x, or $f(x)$ **exists**. If x is not in the domain of f, we say that f **is not defined at** x, or $f(x)$ **does not exist.** For example, if $f(x) = \dfrac{x}{x^2 - 1}$, then $f(0)$ exists, but $f(1)$ and $f(-1)$ do not exist. (Do you see why?)

We have not said much about finding the range of a function. The reason is that when a function is defined by an equation it is often difficult to find the range.* Therefore, we shall usually be content to find just the domain of a function when only the rule for the function is given. We shall express the domain of a function using inequalities, interval notation, set notation, or words, whichever is most convenient.

The Graph of a Function

In applications, a graph often demonstrates more clearly the relationship between two variables than an equation or table would. For example, Table 1 shows the price per share of Microsoft stock at the end of each month from June 2000 through June 2001.

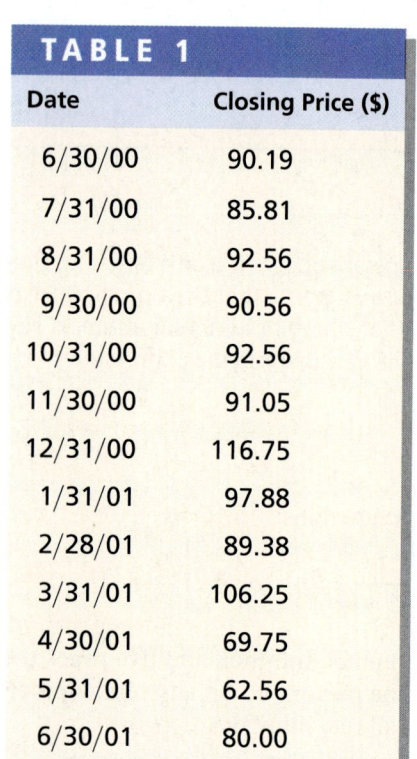

TABLE 1	
Date	**Closing Price ($)**
6/30/00	90.19
7/31/00	85.81
8/31/00	92.56
9/30/00	90.56
10/31/00	92.56
11/30/00	91.05
12/31/00	116.75
1/31/01	97.88
2/28/01	89.38
3/31/01	106.25
4/30/01	69.75
5/31/01	62.56
6/30/01	80.00

Source: Courtesy of A.G. Edwards & Sons, Inc.

Figure 8

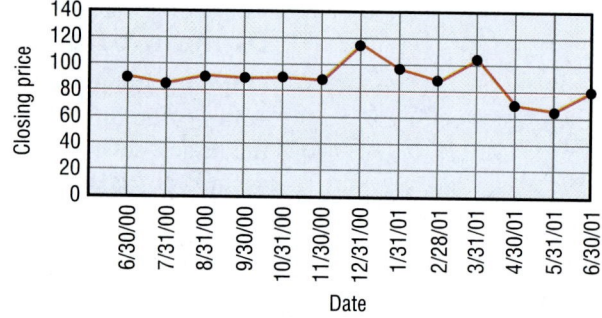

If we plot the data in Table 1, using the date as the x-coordinate and the price as the y-coordinate, and then connect the points, we obtain Figure 8.

We can see from the graph that the price of the stock was falling during January and February and was rising in the month of December. The graph also shows that the lowest price occurred at the end of May, while the highest occurred at the end of December. Equations and tables, on the other hand, usually require some calculations and interpretation before this kind of information can be "seen."

*In Section 6.1, we discuss a way to find the range of a certain class of functions.

Look again at Figure 8. The graph shows that for each date on the horizontal axis there is only one price on the vertical axis. Thus, the graph represents a function, although the exact rule for getting from date to price is not given.

When a function is defined by an equation in x and y, the **graph of the function** is the graph of the equation, that is, the set of points (x, y) in the xy-plane that satisfies the equation.

④ Not every collection of points in the xy-plane represents the graph of a function. Remember, for a function, each number x in the domain has one and only one image y. This means that the graph of a function cannot contain two points with the same x-coordinate and different y-coordinates. Therefore, the graph of a function must satisfy the following **vertical-line test.**

Theorem

Vertical-line Test

A set of points in the xy-plane is the graph of a function if and only if every vertical line intersects the graph in at most one point.

Another way of stating this result is as follows: If any vertical line intersects a graph at more than one point, the graph is not the graph of a function.

EXAMPLE 9   **Identifying the Graph of a Function**

Which of the graphs in Figure 9 are graphs of functions?

Figure 9

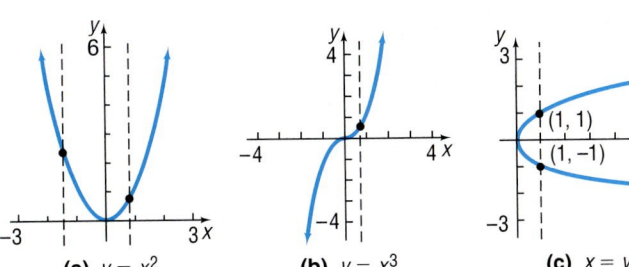

(a) $y = x^2$ **(b)** $y = x^3$ **(c)** $x = y^2$ **(d)** $x^2 + y^2 = 1$

Solution   The graphs in Figures 9(a) and 9(b) are graphs of functions, because every vertical line intersects each graph in at most one point. The graphs in Figures 9(c) and 9(d) are not graphs of functions, because a vertical line intersects each graph in more than one point.

NOW WORK PROBLEM **53.**

⑤ If (x, y) is a point on the graph of a function f, then y is the value of f at x, that is, $y = f(x)$. The next example illustrates how to obtain information about a function if its graph is given.

EXAMPLE 10 Obtaining Information from the Graph of a Function

Figure 10

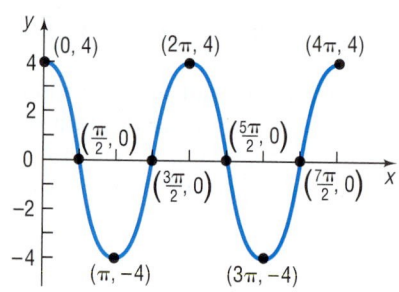

Let f be the function whose graph is given in Figure 10. (The graph of f might represent the distance that the bob of a pendulum is from its at-rest position. Negative values of y mean that the pendulum is to the left of the at-rest position, and positive values of y mean that the pendulum is to the right of the at-rest position.)

(a) What are $f(0), f\left(\dfrac{3\pi}{2}\right)$, and $f(3\pi)$?

(b) What is the domain of f?

(c) What is the range of f?

(d) List the intercepts. (Recall that these are the points, if any, where the graph crosses or touches the coordinate axes.)

(e) How often does the line $y = 2$ intersect the graph?

(f) For what values of x does $f(x) = -4$?

Solution

(a) Since $(0, 4)$ is on the graph of f, the y-coordinate 4 is the value of f at the x-coordinate 0; that is, $f(0) = 4$. In a similar way, we find that when $x = \dfrac{3\pi}{2}$ then $y = 0$, so $f\left(\dfrac{3\pi}{2}\right) = 0$. When $x = 3\pi$, then $y = -4$, so $f(3\pi) = -4$.

(b) To determine the domain of f, we notice that the points on the graph of f will have x-coordinates between 0 and 4π, inclusive; and, for each number x between 0 and 4π there is a point $(x, f(x))$ on the graph. The domain of f is $\{x | 0 \le x \le 4\pi\}$ or the interval $[0, 4\pi]$.

(c) The points on the graph all have y-coordinates between -4 and 4, inclusive; and, for each such number y, there is at least one number x in the domain. The range of f is $\{y | -4 \le y \le 4\}$ or the interval $[-4, 4]$.

(d) The intercepts are $(0, 4), \left(\dfrac{\pi}{2}, 0\right), \left(\dfrac{3\pi}{2}, 0\right), \left(\dfrac{5\pi}{2}, 0\right)$, and $\left(\dfrac{7\pi}{2}, 0\right)$.

(e) Draw the horizontal line $y = 2$ on the graph in Figure 10. Then we find that it intersects the graph four times.

(f) Since $(\pi, -4)$ and $(3\pi, -4)$ are the only points on the graph for which $y = f(x) = -4$, we have $f(x) = -4$ when $x = \pi$ and $x = 3\pi$. ■

When the graph of a function is given, its domain may be viewed as the shadow created by the graph on the x-axis by vertical beams of light. Its range can be viewed as the shadow created by the graph on the y-axis by horizontal beams of light. Try this technique with the graph given in Figure 10.

NOW WORK PROBLEMS 47 AND 51.

EXAMPLE 11   **Obtaining Information about the Graph of a Function**

Consider the function: $f(x) = \dfrac{x}{x + 2}$

(a) Is the point $\left(1, \dfrac{1}{2}\right)$ on the graph of f?

(b) If $x = -1$, what is $f(x)$? What point is on the graph of f?
(c) If $f(x) = 2$, what is x? What point is on the graph of f?

Solution   (a) When $x = 1$, then

$$f(x) = \dfrac{x}{x + 2}$$

$$f(1) = \dfrac{1}{1 + 2} = \dfrac{1}{3}$$

The point $\left(1, \dfrac{1}{3}\right)$ is on the graph of f; the point $\left(1, \dfrac{1}{2}\right)$ is not.

(b) If $x = -1$, then

$$f(x) = \dfrac{x}{x + 2}$$

$$f(-1) = \dfrac{-1}{-1 + 2} = -1$$

The point $(-1, -1)$ is on the graph of f.

(c) If $f(x) = 2$, then

$$f(x) = 2$$

$$\dfrac{x}{x + 2} = 2$$

$$x = 2(x + 2) \qquad \text{Multiply both sides by } x + 2.$$

$$x = 2x + 4 \qquad \text{Remove parentheses.}$$

$$x = -4 \qquad \text{Solve for } x.$$

If $f(x) = 2$, then $x = -4$. The point $(-4, 2)$ is on the graph of f. ■

NOW WORK PROBLEM **63.**

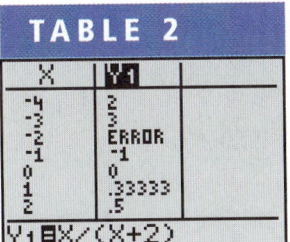

TABLE 2

X	Y1
-4	2
-3	3
-2	ERROR
-1	-1
0	0
1	.33333
2	.5

Y1◻X/(X+2)

We can use the TABLE feature on a graphing calculator to verify the results of Example 11. We enter $Y_1 = \dfrac{x}{x + 2}$ into a graphing utility and create Table 2. For example, when $x = -4$, $Y_1 = 2$, so $f(-4) = 2$. Why does the graphing utility display ERROR when $x = -2$?

Applications

When we use functions in applications, the domain may be restricted by physical or geometric considerations. For example, the domain of the function f defined by $f(x) = x^2$ is the set of all real numbers. However, if f is used to obtain the area of a square when the length x of a side is known, then we must restrict the domain of f to the positive real numbers, since the length of a side can never be 0 or negative.

EXAMPLE 12 **Area of a Circle**

Express the area of a circle as a function of its radius.

Solution

Figure 11

See Figure 11. We know that the formula for the area A of a circle of radius r is $A = \pi r^2$. If we use r to represent the independent variable and A to represent the dependent variable, the function expressing this relationship is

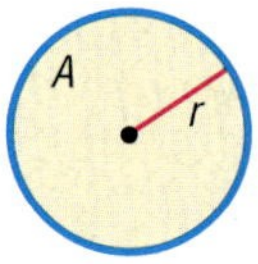

$$A(r) = \pi r^2$$

In this setting, the domain is $\{r|r > 0\}$. (Do you see why?) ∎

NOW WORK PROBLEM **87.**

EXAMPLE 13 **Average Cost Function**

The average cost C of manufacturing x computers per day is given by the function

$$C(x) = 0.56x^2 - 34.39x + 1212.57 + \frac{20,000}{x}$$

Determine the average cost of manufacturing the following:

(a) 30 computers in a day

(b) 40 computers in a day

(c) 50 computers in a day

(d) Graph the function $C = C(x), 0 < x \le 80$.

(e) Create a TABLE with TblStart = 1 and ΔTbl = 1. Which value of x minimizes the average cost?

Solution

(a) The average cost of manufacturing $x = 30$ computers is

$$C(30) = 0.56(30)^2 - 34.39(30) + 1212.57 + \frac{20,000}{30} = \$1351.54$$

(b) The average cost of manufacturing $x = 40$ computers is

$$C(40) = 0.56(40)^2 - 34.39(40) + 1212.57 + \frac{20,000}{40} = \$1232.97$$

(c) The average cost of manufacturing $x = 50$ computers is

$$C(50) = 0.56(50)^2 - 34.39(50) + 1212.57 + \frac{20,000}{50} = \$1293.07$$

(d) See Figure 12 for the graph of $C = C(x)$.

Figure 12

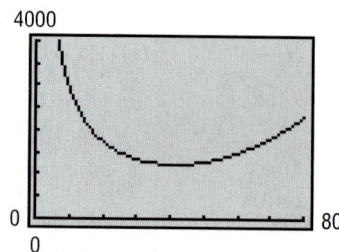

(e) With the function $C = C(x)$ in Y_1, we create Table 3. We scroll down until we find a value of x for which Y_1 is smallest. Table 4 shows that manufacturing $x = 41$ computers minimizes the average cost at $1231.74 per computer.

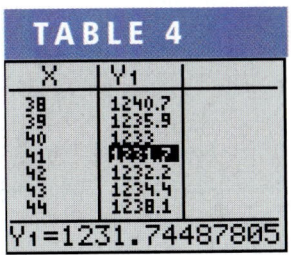

TABLE 3	
X	Y1
1	21179
2	11146
3	7781.1
4	6084
5	5054.6
6	4359.7
7	3856.4
Y1❘.56X²−34.39X...	

TABLE 4	
X	Y1
38	1240.7
39	1235.9
40	1233
41	1231.7
42	1232.2
43	1234.4
44	1238.1
Y1=1231.74487805	

NOW WORK PROBLEM **91.**

SUMMARY

We list here some of the important vocabulary introduced in this section, with a brief description of each term.

Function
A relation between two sets of real numbers so that each number x in the first set, the domain, has corresponding to it exactly one number y in the second set. A set of ordered pairs (x, y) or $(x, f(x))$ in which no two ordered pairs have the same first element, but different second elements.
The range is the set of y values of the function for the x values in the domain.
A function f may be defined implicitly by an equation involving x and y or explicitly by writing $y = f(x)$.

Unspecified domain
If a function f is defined by an equation and no domain is specified, then the domain will be taken to be the largest set of real numbers for which the equation defines a real number.

Function notation
$y = f(x)$
f is a symbol for the function.
x is the independent variable or argument.
y is the dependent variable.
$f(x)$ is the value of the function at x, or the image of x.

Graph of a function
The collection of points (x, y) that satisfies the equation $y = f(x)$.
A collection of points is the graph of a function provided that every vertical line intersects the graph in at most one point (vertical-line test).

2.1 Concepts and Vocabulary

In Problems 1–3, fill in the blanks.

1. If f is a function defined by the equation $y = f(x)$, then x is called the _____ variable and y is the _____ variable.

2. A set of points in the xy-plane is the graph of a function if and only if every _____ line intersects the graph in at most one point.

3. The set of all images of the elements in the domain of a function is called the _____.

In Problems 4–6, answer True or False to each statement.

4. Every relation is a function.

5. The y-intercept of the graph of the function $y = f(x)$, whose domain is all real numbers, is $f(0)$.

6. The independent variable is sometimes referred to as the argument of the function.

7. Describe how you would proceed to find the domain and range of a function if you were given its graph. How would your strategy change if you were given the equation defining the function instead of its graph?

8. How many x-intercepts can the graph of a function have? How many y-intercepts can it have?

9. Is a graph that consists of a single point the graph of a function? Can you write the equation of such a function?

2.1 Exercises

In Problems 1–12, determine whether each relation represents a function. For each function, state the domain and range.

1.

| Domain | Range |

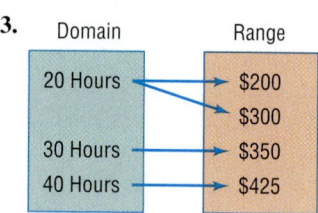

2.

| Domain | Range |

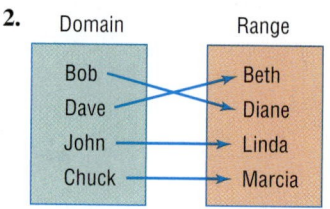

3.

| Domain | Range |

4.

| Domain | Range |

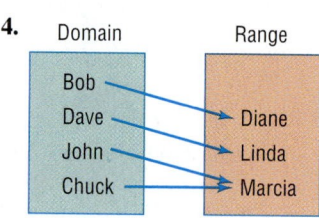

5. $\{(2,6), (-3,6), (4,9), (2,10)\}$

6. $\{(-2,5), (-1,3), (3,7), (4,12)\}$

7. $\{(1,3), (2,3), (3,3), (4,3)\}$

8. $\{(0,-2), (1,3), (2,3), (3,7)\}$

9. $\{(-2,4), (-2,6), (0,3), (3,7)\}$

10. $\{(-4,4), (-3,3), (-2,2), (-1,1), (-4,0)\}$

11. $\{(-2,4), (-1,1), (0,0), (1,1)\}$

12. $\{(-2,16), (-1,4), (0,3), (1,4)\}$

In Problems 13–20, find the following values for each function:

(a) $f(0)$ (b) $f(1)$ (c) $f(-1)$ (d) $f(-x)$ (e) $-f(x)$ (f) $f(x+1)$ (g) $f(2x)$ (h) $f(x+h)$

13. $f(x) = 2x + 5$

14. $f(x) = -3x + 1$

15. $f(x) = 3x^2 + 2x - 4$

16. $f(x) = -2x^2 + x - 1$

17. $f(x) = \dfrac{x}{x^2 + 1}$

18. $f(x) = \dfrac{x^2 - 1}{x + 4}$

19. $f(x) = |x| + 4$

20. $f(x) = \sqrt{x^2 + x}$

In Problems 21–32, determine whether the equation is a function.

21. $y = x^2$

22. $y = x^3$

23. $y = \dfrac{1}{x}$

24. $y = |x|$

25. $y^2 = 4 - x^2$

26. $y = \pm\sqrt{1 - 2x}$

27. $x = y^2$

28. $x + y^2 = 1$

29. $y = 2x^2 - 3x + 4$

30. $y = \dfrac{3x - 1}{x + 2}$

31. $2x^2 + 3y^2 = 1$

32. $x^2 - 4y^2 = 1$

In Problems 33–46, find the domain of each function.

33. $f(x) = -5x + 4$

34. $f(x) = x^2 + 2$

35. $f(x) = \dfrac{x}{x^2 + 1}$

36. $f(x) = \dfrac{x^2}{x^2 + 1}$

37. $g(x) = \dfrac{x}{x^2 - 16}$

38. $h(x) = \dfrac{2x}{x^2 - 4}$

39. $F(x) = \dfrac{x - 2}{x^3 + x}$

40. $G(x) = \dfrac{x + 4}{x^3 - 4x}$

41. $h(x) = \sqrt{3x - 12}$ **42.** $G(x) = \sqrt{1 - x}$ **43.** $f(x) = \dfrac{4}{\sqrt{x - 9}}$ **44.** $f(x) = \dfrac{x}{\sqrt{x - 4}}$

45. $p(x) = \sqrt{\dfrac{2}{x - 1}}$ **46.** $q(x) = \sqrt{-x - 2}$

47. Use the graph of the function f given below to answer parts (a)–(n).

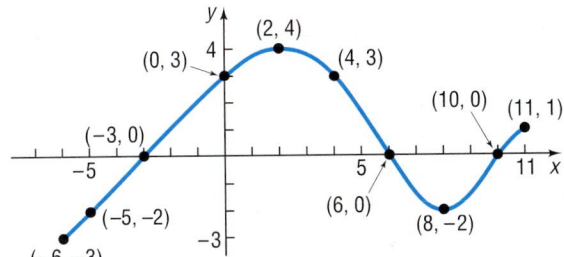

(a) Find $f(0)$ and $f(-6)$.
(b) Find $f(6)$ and $f(11)$.
(c) Is $f(3)$ positive or negative?
(d) Is $f(-4)$ positive or negative?
(e) For what numbers x is $f(x) = 0$?
(f) For what numbers x is $f(x) > 0$?
(g) What is the domain of f?
(h) What is the range of f?
(i) What are the x-intercepts?
(j) What is the y-intercept?
(k) How often does the line $y = \dfrac{1}{2}$ intersect the graph?
(l) How often does the line $x = 5$ intersect the graph?
(m) For what values of x does $f(x) = 3$?
(n) For what values of x does $f(x) = -2$?

48. Use the graph of the function f given below to answer parts (a)–(n).

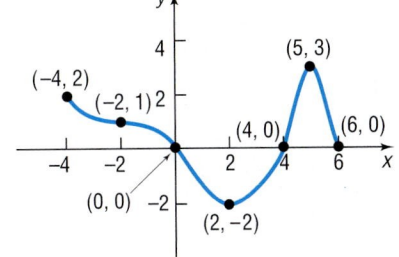

(a) Find $f(0)$ and $f(6)$.
(b) Find $f(2)$ and $f(-2)$.
(c) Is $f(3)$ positive or negative?
(d) Is $f(-1)$ positive or negative?
(e) For what numbers x is $f(x) = 0$?
(f) For what numbers x is $f(x) < 0$?
(g) What is the domain of f?
(h) What is the range of f?
(i) What are the x-intercepts?
(j) What is the y-intercept?
(k) How often does the line $y = -1$ intersect the graph?
(l) How often does the line $x = 1$ intersect the graph?
(m) For what value of x does $f(x) = 3$?
(n) For what value of x does $f(x) = -2$?

In Problems 49–60, determine whether the graph is that of a function by using the vertical-line test. If it is, use the graph to find:
(a) Its domain and range   (b) The intercepts, if any

49.
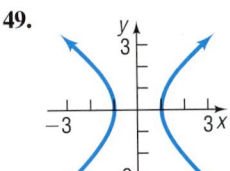

50.
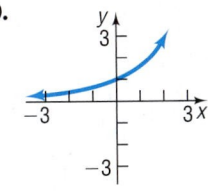

51.
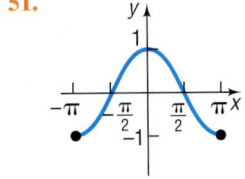

52.
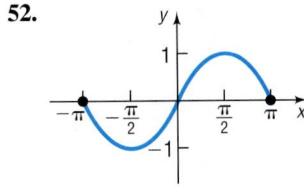

53.
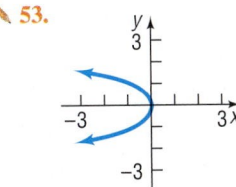

54.
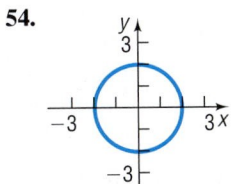

55.
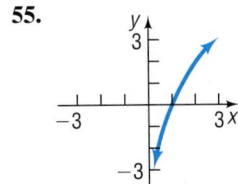

56.
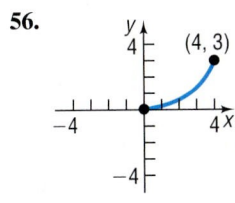

57.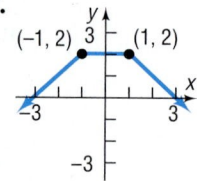

58.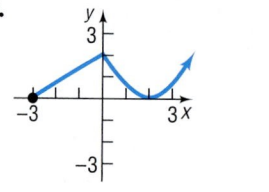

59.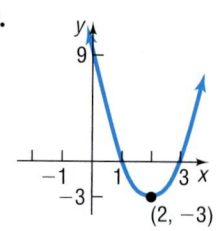

60.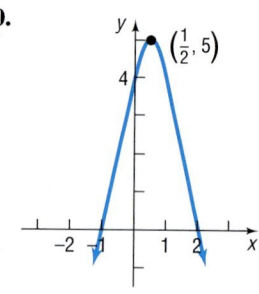

In Problems 61–66, answer the questions about the given function.

61. $f(x) = 2x^2 - x - 1$

(a) Is the point $(-1, 2)$ on the graph of f?

(b) If $x = -2$, what is $f(x)$? What point is on the graph of f?

(c) If $f(x) = -1$, what is x? What point(s) are on the graph of f?

(d) What is the domain of f?

(e) List the x-intercepts, if any, of the graph of f.

(f) List the y-intercept, if there is one, of the graph of f.

62. $f(x) = -3x^2 + 5x$

(a) Is the point $(-1, 2)$ on the graph of f?

(b) If $x = -2$, what is $f(x)$? What point is on the graph of f?

(c) If $f(x) = -2$, what is x? What point(s) are on the graph of f?

(d) What is the domain of f?

(e) List the x-intercepts, if any, of the graph of f.

(f) List the y-intercept, if there is one, of the graph of f.

63. $f(x) = \dfrac{x + 2}{x - 6}$

(a) Is the point $(3, 14)$ on the graph of f?

(b) If $x = 4$, what is $f(x)$? What point is on the graph of f?

(c) If $f(x) = 2$, what is x? What point(s) are on the graph of f?

(d) What is the domain of f?

(e) List the x-intercepts, if any, of the graph of f.

(f) List the y-intercept, if there is one, of the graph of f.

64. $f(x) = \dfrac{x^2 + 2}{x + 4}$

(a) Is the point $\left(1, \dfrac{3}{5}\right)$ on the graph of f?

(b) If $x = 0$, what is $f(x)$? What point is on the graph of f?

(c) If $f(x) = \dfrac{1}{2}$, what is x? What point(s) are on the graph of f?

(d) What is the domain of f?

(e) List the x-intercepts, if any, of the graph of f.

(f) List the y-intercept, if there is one, of the graph of f.

65. $f(x) = \dfrac{2x^2}{x^4 + 1}$

(a) Is the point $(-1, 1)$ on the graph of f?

(b) If $x = 2$, what is $f(x)$? What point is on the graph of f?

(c) If $f(x) = 1$, what is x? What point(s) are on the graph of f?

(d) What is the domain of f?

(e) List the x-intercepts, if any, of the graph of f.

(f) List the y-intercept, if there is one, of the graph of f.

66. $f(x) = \dfrac{2x}{x - 2}$

(a) Is the point $\left(\dfrac{1}{2}, -\dfrac{2}{3}\right)$ on the graph of f?

(b) If $x = 4$, what is $f(x)$? What point is on the graph of f?

(c) If $f(x) = 1$, what is x? What point(s) are on the graph of f?

(d) What is the domain of f?

(e) List the x-intercepts, if any, of the graph of f.

(f) List the y-intercept, if there is one, of the graph of f.

67. If $f(x) = 2x^3 - 4x^2 + 4x + C$ and $f(2) = 5$, what is the value of C?

68. If $f(x) = 3x^2 - 5x + C$ and $f(-1) = 12$, what is the value of C?

69. If $f(x) = \dfrac{3x + 8}{2x - A}$ and $f(0) = 2$, what is the value of A?

70. If $f(x) = \dfrac{2x - B}{3x + 4}$ and $f(2) = \dfrac{1}{2}$, what is the value of B?

71. If $f(x) = \dfrac{2x - A}{x - 3}$ and $f(4) = 0$, what is the value of A? Where is f not defined?

72. If $f(x) = \dfrac{x - B}{x - A}$, $f(2) = 0$, and $f(1)$ is undefined, what are the values of A and B?

In Problems 73–78, find the value of $\dfrac{f(x+h)-f(x)}{h}, h \neq 0,$ for each function. Be sure to simplify.

 73. $f(x) = 4x + 3$

74. $f(x) = -3x + 1$

75. $f(x) = x^2 - x + 4$

76. $f(x) = x^2 + 5x - 1$

77. $f(x) = x^3 - 2$

78. $f(x) = \dfrac{1}{x+3}$

79. Match each of the following functions with the graphs that best describe the situation.

(a) The cost of building a house as a function of its square footage
(b) The height of an egg dropped from a 300-foot building as a function of time
(c) The height of a human as a function of time
(d) The demand for Big Macs as a function of price
(e) The height of a child on a swing as a function of time

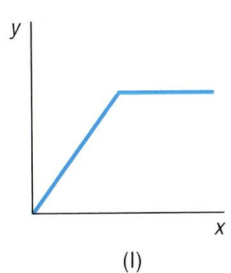 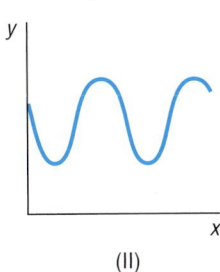 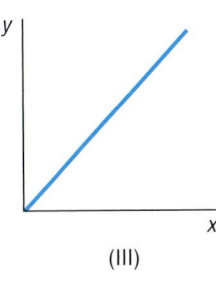 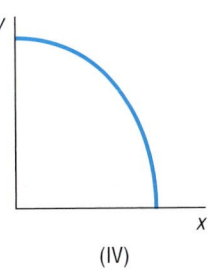 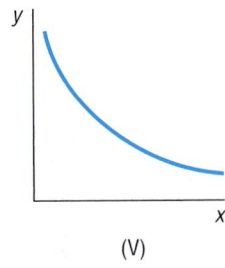

(I)　　　　(II)　　　　(III)　　　　(IV)　　　　(V)

80. Match each of the following functions with the graph that best describes the situation.

(a) The temperature of a bowl of soup as a function of time
(b) The number of hours of daylight per day over a two year period
(c) The population of Florida as a function of time
(d) The distance of a car traveling at a constant velocity as a function of time
(e) The height of a golf ball hit with a 7-iron as a function of time

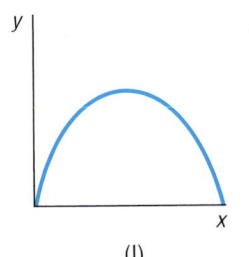 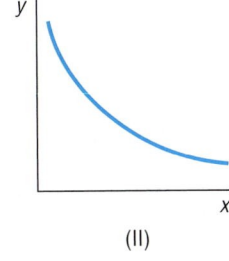 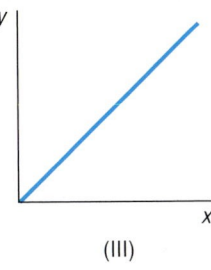 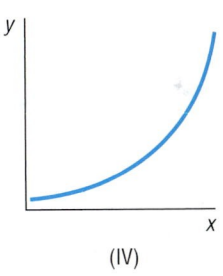 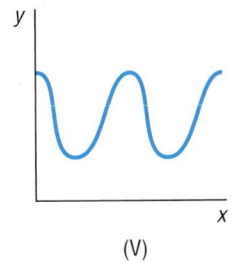

(I)　　　　(II)　　　　(III)　　　　(IV)　　　　(V)

81. Consider the following scenario: Barbara decides to take a walk. She leaves home, walks 2 blocks in 5 minutes at a constant speed, and realizes that she forgot to lock the door. So Barbara runs home in 1 minute. While at her doorstep, it takes her 1 minute to find her keys and lock the door. Barbara walks 5 blocks in 15 minutes and then decides to jog home. It takes her 7 minutes to get home. Draw a graph of Barbara's distance from home (in blocks) as a function of time.

82. Consider the following scenario: Jayne enjoys riding her bicycle through the woods. At the forest preserve, she gets on her bicycle and rides up a 2,000-foot incline in 10 minutes. She then travels down the incline in 3 minutes. The next 5,000 feet is level terrain and she covers the distance in 20 minutes. She rests for 15 minutes. Jayne then travels 10,000 feet in 30 minutes. Draw a graph of Jayne's distance traveled (in feet) as a function of time.

83. The following sketch represents the distance d (in miles) that Kevin is from home as a function of time t (in hours). Answer the questions based on the graph. In parts (a)-(g), how many hours elapsed and how far was Kevin from home during this time?

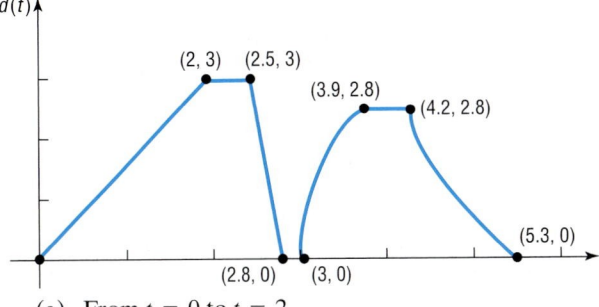

(a) From $t = 0$ to $t = 2$
(b) From $t = 2$ to $t = 2.5$

(c) From $t = 2.5$ to $t = 2.8$
(d) From $t = 2.8$ to $t = 3$
(e) From $t = 3$ to $t = 3.9$
(f) From $t = 3.9$ to $t = 4.2$
(g) From $t = 4.2$ to $t = 5.3$
(h) What is the farthest distance that Kevin is from home?
(i) How many times did Kevin return home?

84. The following sketch represents the speed v (in miles per hour) of Michael's car as a function of time t (in minutes).

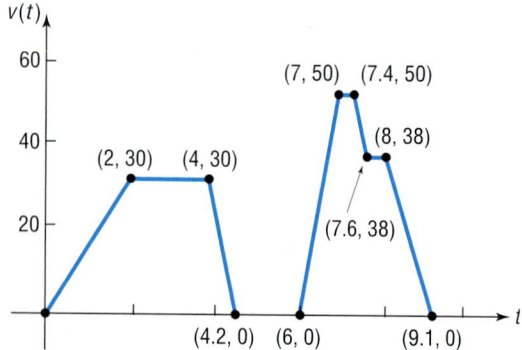

(a) Over what interval of time is Michael traveling fastest?
(b) Over what interval(s) of time is Michael's speed zero?
(c) What is Michael's speed between 2 and 4 minutes?
(d) What is Michael's speed between 4.2 and 6 minutes?
(e) What is Michael's speed between 7 and 7.4 minutes?
(f) When is Michael's speed constant?

85. Effect of Gravity on Earth If a rock falls from a height of 20 meters on Earth, the height H (in meters) after x seconds is approximately

$$H(x) = 20 - 4.9x^2$$

(a) Using a graphing utility, graph $H(x)$.
(b) What is the height of the rock when $x = 1$ second? $x = 1.1$ seconds? $x = 1.2$ seconds? $x = 1.3$ seconds?
(c) When is the height of the rock 15 meters? When is it 10 meters? When is it 5 meters?
(d) When does the rock strike the ground?

86. Effect of Gravity on Jupiter If a rock falls from a height of 20 meters on the planet Jupiter, its height H (in meters) after x seconds is approximately

$$H(x) = 20 - 13x^2$$

(a) Using a graphing utility, graph $H(x)$.
(b) What is the height of the rock when $x = 1$ second? $x = 1.1$ seconds? $x = 1.2$ seconds?
(c) When is the height of the rock 15 meters? When is it 10 meters? When is it 5 meters?
(d) When does the rock strike the ground?

87. Geometry Express the area A of a rectangle as a function of the length x if the length is twice the width of the rectangle.

88. Geometry Express the area A of an isosceles right triangle as a function of the length x of one of the two equal sides.

89. Express the gross salary G of a person who earns $10 per hour as a function of the number x of hours worked.

90. Tiffany, a commissioned salesperson, earns $100 base pay plus $10 per item sold. Express her gross salary G as a function of the number x of items sold.

91. Motion of a Golf Ball A golf ball is hit with an initial velocity of 130 feet per second at an inclination of 45° to the horizontal. In physics, it is established that the height h of the golf ball is given by the function

$$h(x) = \frac{-32x^2}{130^2} + x$$

where x is the horizontal distance that the golf ball has traveled.

(a) Determine the height of the golf ball after it has traveled 100 feet.
(b) 300 feet
(c) 500 feet
(d) Graph the function $h = h(x)$.
(e) Algebraically, determine the distance that the ball has traveled when the height of the ball is 90 feet. Verify your results graphically.
(f) Create a TABLE with TblStart $= 0$ and ΔTbl $= 25$. To the nearest 25 feet, how far does the ball travel before it reaches a maximum height? What is the maximum height?
(g) Adjust the value of ΔTbl until you determine the distance, to within 1 foot, that the ball travels before it reaches a maximum height.
(h) Find the domain of h.

Hint: For what values of x is $h(x) \geq 0$?

92. Cross-sectional Area The cross-sectional area of a beam cut from a log with radius 1 foot is given by the function $A(x) = 4x\sqrt{1 - x^2}$, where x represents the length of half the base of the beam. See the figure.

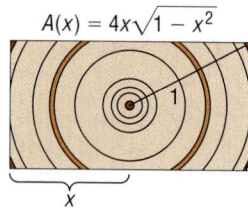

(a) Find the domain of A.
Determine the cross-sectional area of the beam if the length of half the base of the beam is as follows:
(b) One-third foot

(c) One-half of a foot
(d) Two-thirds of a foot
(e) Graph the function $A = A(x)$.
(f) Create a TABLE with TblStart $= 0$ and ΔTbl $= 0.1$ for $0 \le x \le 1$. Which value of x maximizes the cross-sectional area? What should be the length of the base of the beam to maximize the cross-sectional area?

93. **Cost of Trans-Atlantic Travel** A Boeing 747 crosses the Atlantic Ocean (3000 miles) with an airspeed of 500 miles per hour. The cost C (in dollars) per passenger is given by

$$C(x) = 100 + \frac{x}{10} + \frac{36{,}000}{x}$$

where x is the ground speed (airspeed $\pm$ wind).

(a) What is the cost per passenger for quiescent (no wind) conditions?
(b) What is the cost per passenger with a head wind of 50 miles per hour?
(c) What is the cost per passenger with a tail wind of 100 miles per hour?
(d) What is the cost per passenger with a head wind of 100 miles per hour?
(e) Graph the function $C = C(x)$.
(f) Create a TABLE with TblStart $= 0$ and ΔTbl $= 50$. To the nearest 50 miles per hour, what ground speed minimizes the cost per passenger?

94. **Effect of Elevation on Weight** If an object weighs m pounds at sea level, then its weight W (in pounds) at a height of h miles above sea level is given approximately by

$$W(h) = m \left(\frac{4000}{4000 + h} \right)^2$$

(a) If Amy weighs 120 pounds at sea level, how much will she weigh on Pike's Peak, which is 14,110 feet above sea level?
(b) Use a graphing utility to graph the function $W = W(h)$. Use $m = 120$ pounds.
(c) Create a Table with TblStart $= 0$ and ΔTbl $= 0.5$ to see how weight W varies as h changes from 0 to 5 miles.
(d) At what height will Amy weigh 119.95 pounds?
(e) Does your answer to part (d) seem reasonable?

95. Some functions f have the property that $f(a + b) = f(a) + f(b)$ for all real numbers a and b. Which of the following functions have this property?
(a) $h(x) = 2x$
(b) $g(x) = x^2$
(c) $F(x) = 5x - 2$
(d) $G(x) = \dfrac{1}{x}$

96. Draw the graph of a function whose domain is $\{x | -3 \le x \le 8, \quad x \ne 5\}$ and whose range is $\{y | -1 \le y \le 2, y \ne 0\}$. What point(s) in the rectangle $-3 \le x \le 8, -1 \le y \le 2$ cannot be on the graph? Compare your graph with those of other students. What differences do you see?

97. Are the functions $f(x) = x - 1$ and $g(x) = \dfrac{x^2 - 1}{x + 1}$ the same? Explain.

98. Investigate when, historically, the use of function notation $y = f(x)$ first appeared.

PREPARING FOR THIS SECTION

Before getting started, review the following:

✓ Lines (Section 1.7, pp. 163–172)

✓ Plotting Points in the Cartesian Plane (Section 1.1, pp. 90–91)

2.2 LINEAR FUNCTIONS AND MODELS

OBJECTIVES
1. Graph Linear Functions
2. Draw and Interpret Scatter Diagrams
3. Distinguish between Linear and Nonlinear Relations
4. Use a Graphing Utility to Find the Line of Best Fit
5. Construct a Linear Model Using Direct Variation

1 In Section 1.7, we introduced lines. In particular, we discussed the slope–intercept form of the equation of a line $y = mx + b$. When we write the slope–intercept form of a line using function notation, we have a *linear function*.

A **linear function** is a function of the form

$$f(x) = mx + b$$

The graph of a linear function is a line with slope m and y-intercept b.

EXAMPLE 1 | **Graphing a Linear Function**

Graph the linear function: $f(x) = -3x + 7$

Solution   This is a linear function with slope $m = -3$ and y-intercept 7. We find a point on the graph first. When $x = 0$, then $y = 7$, so the point $(0, 7)$ is on the graph. To graph this function, we start at the point $(0, 7)$ and use the slope to find an additional point by moving right 1 unit and down 3 units. See Figure 13.

Figure 13

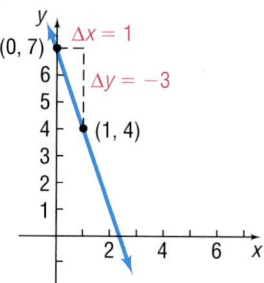

Alternatively, we could have found an additional point by evaluating the function at some $x \neq 0$. For $x = 1$, we find $f(1) = -3(1) + 7 = 4$ and obtain the point $(1, 4)$ on the graph.

 ✏ **NOW WORK PROBLEM 1.**

Many real-world situations may be described using linear functions.

EXAMPLE 2 | **Straight-line Depreciation**

Book value is the value of an asset that a company uses to create its balance sheet. Some companies will depreciate their assets using straight-line depreciation so that the value of the asset declines by a fixed amount each year. The amount of the decline depends on the useful life that the company places on the asset. Suppose that a company just purchased a fleet of new cars for its sales force at a cost of $28,000 per car. The company chooses to depreciate each vehicle using the straight-line method over 7 years. This means that each car will depreciate by $\dfrac{\$28{,}000}{7} = \$4000$ per year.

(a) Write a linear function that expresses the book value of each car as a function of its age.

(b) Graph the linear function.

(c) What is the book value of each car after 3 years?

Solution

(a) If we let $V(x)$ represent the value of each car after x years, then $V(0)$ represents the original value of each car so $V(0) = \$28{,}000$. The y-intercept of the linear function is $\$28{,}000$. Because each car depreciates by $\$4000$ per year, the slope of the linear function is $-\$4000$. The linear function that represents the value of each car after x years is

$$V(x) = -4000x + 28{,}000$$

(b) Figure 14 shows the graph of V.

(c) The book value of the car after 3 years is

$$V(3) = -4000(3) + 28{,}000$$

$$= \$16{,}000$$

Figure 14

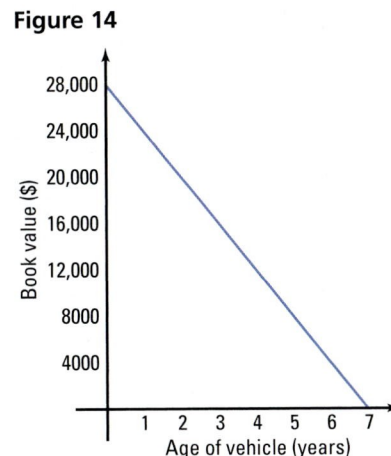

NOW WORK PROBLEM **9**.

Linear functions can also be created by *fitting* a linear function to data. The first step in doing this is to draw a *scatter diagram*.

Scatter Diagrams

② A **relation** is a correspondence between two sets. If x and y are two elements and a relation exists between x and y, then we say that x **corresponds to** y or that y **depends on** x and write $x \rightarrow y$. We may also write $x \rightarrow y$ as the ordered pair (x, y). In this sense, y is referred to as the **dependent** variable and x is called the **independent** variable.

Often we are interested in specifying the type of relation (such as an equation) that might exist between two variables. The first step in finding this relation is to plot the ordered pairs using rectangular coordinates. The resulting graph is called a **scatter diagram.**

EXAMPLE 3   **Drawing a Scatter Diagram**

The data listed in Table 5 represent the apparent temperature versus the relative humidity in a room whose actual temperature is $72°$ Fahrenheit.

TABLE 5

Relative Humidity (%), x	Apparent Temperature, y	(x, y)	Relative Humidity (%), x	Apparent Temperature, y	(x, y)
0	64	(0, 64)	60	72	(60, 72)
10	65	(10, 65)	70	73	(70, 73)
20	67	(20, 67)	80	74	(80, 74)
30	68	(30, 68)	90	75	(90, 75)
40	70	(40, 70)	100	76	(100, 76)
50	71	(50, 71)			

(a) Draw a scatter diagram by hand.

(b) Use a graphing utility to draw a scatter diagram.

(c) Describe what happens to the apparent temperature as the relative humidity increases.

Solution   (a) To draw a scatter diagram by hand, we plot the ordered pairs listed in Table 5, with the relative humidity as the x-coordinate and the apparent temperature as the y-coordinate. See Figure 15(a). Notice that the points in a scatter diagram are not connected.

(b) Figure 15(b) shows a scatter diagram using a graphing utility.

(c) We see from the scatter diagrams that, as the relative humidity increases, the apparent temperature increases.

Figure 15

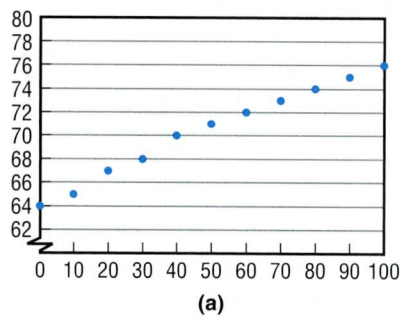

(a)

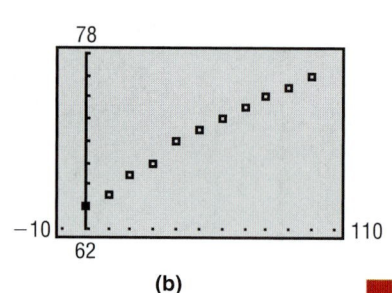

(b)

NOW WORK PROBLEM 19(a).

Curve Fitting

3 Scatter diagrams are used to help us to see the type of relation that exists between two variables. In this text, we will discuss a variety of different relations that may exist between two variables. For now, we concentrate on distinguishing between linear and nonlinear relations. See Figure 16.

Figure 16

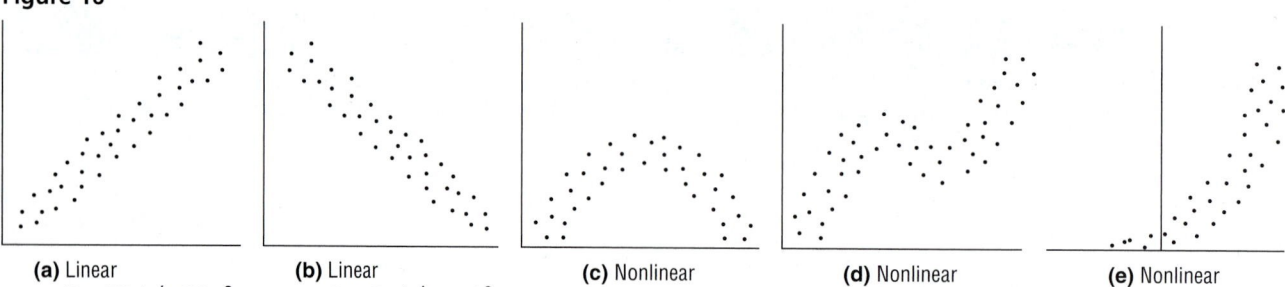

(a) Linear
$y = mx + b, m > 0$

(b) Linear
$y = mx + b, m < 0$

(c) Nonlinear

(d) Nonlinear

(e) Nonlinear

EXAMPLE 4   **Distinguishing between Linear and Nonlinear Relations**

Determine whether the relation between the two variables in Figure 17 is linear or nonlinear.

Figure 17

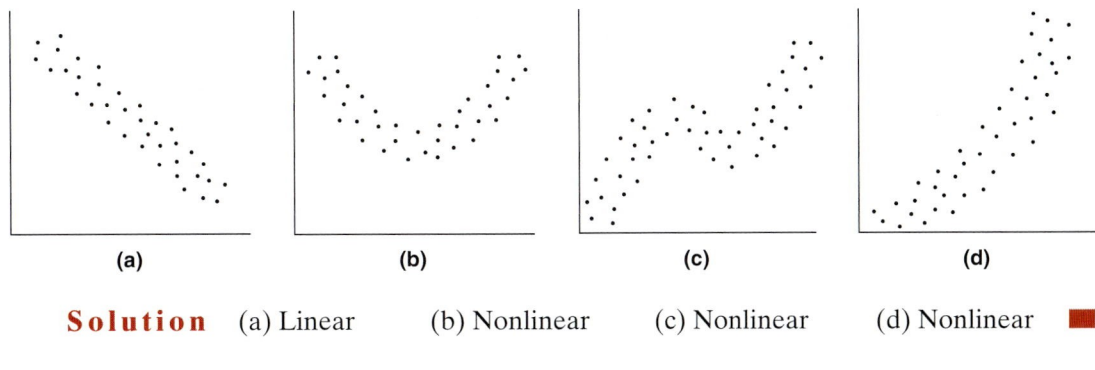

| (a) | (b) | (c) | (d) |

Solution (a) Linear (b) Nonlinear (c) Nonlinear (d) Nonlinear ■

NOW WORK PROBLEM **13.**

In this section we will study data whose scatter diagrams imply that a linear relation exists between the two variables. Nonlinear data will be discussed later.

Suppose that the scatter diagram of a set of data appears to be linearly related as in Figure 16(a) or (b). We might wish to find an equation of a line that relates the two variables. One way to obtain an equation for such data is to draw a line through two points on the scatter diagram and determine the equation of the line.

EXAMPLE 5 **Finding an Equation for Linearly Related Data**

Using the data in Table 5 from Example 3:

(a) Select two points and find an equation of the line containing the points.

(b) Graph the line on the scatter diagram obtained in Example 3(b).

Solution (a) Select two points, say $(10, 65)$ and $(70, 73)$. (You should select your own two points and complete the solution.) The slope of the line joining the points $(10, 65)$ and $(70, 73)$ is

$$m = \frac{73 - 65}{70 - 10} = \frac{8}{60} = \frac{2}{15}$$

The equation of the line with slope $\frac{2}{15}$ and passing through $(10, 65)$ is found using the point–slope form with $m = \frac{2}{15}$, $x_1 = 10$, and $y_1 = 65$.

$$y - y_1 = m(x - x_1)$$

$$y - 65 = \frac{2}{15}(x - 10)$$

$$y = \frac{2}{15}x + \frac{191}{3}$$

(b) Figure 18 shows the scatter diagram with the graph of the line found in part (a). ■

Figure 18

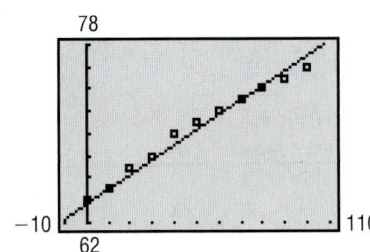

NOW WORK PROBLEMS **19(b)** AND **(c).**

Line of Best Fit

The line obtained in Example 5 depends on the selection of points, which will vary from person to person. So the line that we found might be different from the line you found. Although the line we found in Example 5 appears to fit the data well, there may be a line that "fits it better." Do you think your line fits the data better? Is there a line of *best fit*? As it turns out, there is a method for finding the line that best fits linearly related data (called the *line of best fit*).*

EXAMPLE 6   **Finding the Line of Best Fit**

Using the data in Table 5 from Example 3:

(a) Find the line of best fit using a graphing utility.

(b) Graph the line of best fit on the scatter diagram obtained in Example 3(b).

(c) Interpret the slope.

(d) Use the line of best fit to predict the apparent temperature of a room whose temperature is 72°F and relative humidity is 45%.

Solution   (a) Graphing utilities contain built-in programs that find the line of best fit for a collection of points in a scatter diagram. (Look in your owner's manual for details on how to execute the program.) Upon executing the LINear REGression program, we obtain the results shown in Figure 19. The output that the utility provides shows us the equation $y = ax + b$, where a is the slope of the line and b is the y-intercept. The line of best fit that relates relative humidity to apparent temperature may be expressed as the line $y = 0.121x + 64.409$.

(b) Figure 20 shows the graph of the line of best fit, along with the scatter diagram.

Figure 19

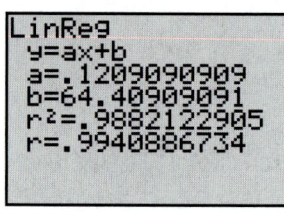

Figure 20

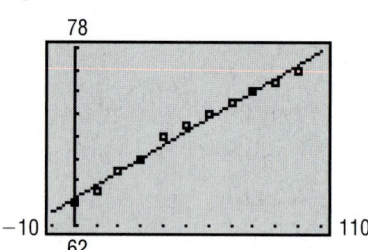

(c) The slope of the line of best fit is 0.121, which means that, for every 1% increase in the relative humidity, apparent room temperature increases 0.121°F.

(d) Letting $x = 45$ in the equation of the line of best fit, we obtain $y = 0.121(45) + 64.409 \approx 70°F$, which is the apparent temperature in the room. ■

 **NOW WORK PROBLEMS 19(d) AND (e).**

---

*We shall not discuss in this book the underlying mathematics of lines of best fit. Most books in statistics and many in linear algebra discuss this topic.

Does the line of best fit appear to be a good fit? In other words, does the line appear to accurately describe the relation between temperature and relative humidity?

And just how "good" is this line of best fit? The answers are given by what is called the *correlation coefficient*. Look again at Figure 19. The last line of output is $r = 0.994$. This number, called the **correlation coefficient, r,** $-1 \leq r \leq 1$, is a measure of the strength of the *linear relation* that exists between two variables. The closer that $|r|$ is to 1, the more perfect the linear relationship is. If r is close to 0, there is little or no *linear* relationship between the variables. A negative value of $r, r < 0$, indicates that as x increases y decreases; a positive value of $r, r > 0$, indicates that as x increases y does also. The data given in Table 5, having a correlation coefficient of 0.994, are indicative of a strong linear relationship with positive slope.

Direct Variation

5   When a mathematical model is developed for a real-world problem, it often involves relationships between quantities that are expressed in terms of proportionality.

> Force is proportional to acceleration.
>
> For an ideal gas held at a constant temperature, pressure and volume are inversely proportional.
>
> The force of attraction between two heavenly bodies is inversely proportional to the square of the distance between them.
>
> Revenue is directly proportional to sales.

Each of these statements illustrates the idea of **variation,** or how one quantity varies in relation to another quantity. Quantities may vary *directly, inversely,* or *jointly.* We discuss direct variation here.

Let x and y denote two quantities. Then y **varies directly** with x, or y is **directly proportional to** x, if there is a nonzero number k such that

$$y = kx$$

The number k is called the **constant of proportionality.**

If y varies directly with x, then y is a linear function of x. The graph in Figure 21 illustrates the relationship between y and x if y varies directly with x and $k > 0, x \geq 0$. Note that the constant of proportionality is, in fact, the slope of the line.

If we know that two quantities vary directly, then knowing the value of each quantity in one instance enables us to write a formula that is true in all cases.

Figure 21
$y = kx, k > 0, x \geq 0$

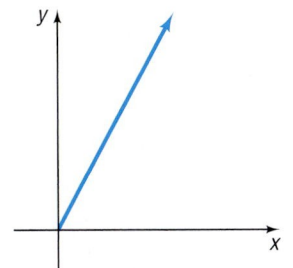

EXAMPLE 7   **Mortgage Payments**

The monthly payment p on a mortgage varies directly with the amount borrowed B. If the monthly payment on a 30-year mortgage is $6.65 for every $1000 borrowed, find a formula that relates the monthly payment p to the amount borrowed B for a mortgage with the same terms. Then find the monthly payment p when the amount borrowed B is $120,000.

Solution   Because p varies directly with B, we know that

$$p = kB$$

for some constant k. Because $p = 6.65$ when $B = 1000$, it follows that

$$6.65 = k(1000)$$
$$k = 0.00665$$

So we have

$$p = 0.00665B$$

We see that p is a linear function of B: $p(B) = 0.00665B$. In particular, when $B = \$120,000$, we find that

$$p(120,000) = 0.00665(\$120,000) = \$798$$

Figure 22 illustrates the relationship between the monthly payment p and the amount borrowed B.

Figure 22

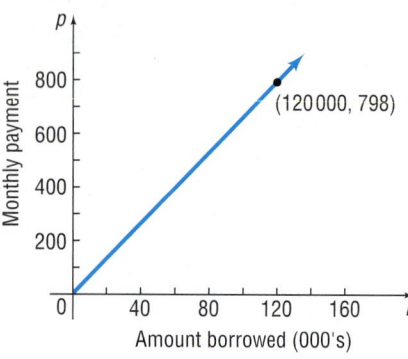

NOW WORK PROBLEM **31.**

2.2   Concepts and Vocabulary

In Problems 1–3, fill in the blanks.

1. For the graph of the linear function $f(x) = mx + b$, m is the _____ and b is the _____.

2. A _____ _____ is used to help us to see the type of relation, if any, that may exist between two variables.

3. If x and y are two quantities, then y is directly proportional to x if there is a nonzero number k such that _____.

In Problems 4–6, answer True or False to each statement.

4. When relations are written as ordered pairs, (x, y), we say that x is related to y. In this sense, x is the dependent variable.

5. The correlation coefficient is a measure of the strength of a linear relation between two variables and must lie between -1 and 1, inclusive.

6. If y varies directly with x, then y is a linear function of x.

7. Find the line of best fit for the ordered pairs $(1, 5)$ and $(3, 8)$. What is the correlation coefficient for these data? Why is this result reasonable?

8. What does a correlation coefficient of 0 imply?

9. What would it mean if a teacher states that a student's average in the class is directly proportional to the amount of time that the student studies?

2.2 Exercises

For Problems 1–8, graph each linear function by hand.

1. $f(x) = 2x + 3$

2. $g(x) = 5x - 4$

3. $h(x) = -3x + 4$

4. $p(x) = -x + 6$

5. $f(x) = \dfrac{1}{4}x - 3$

6. $h(x) = -\dfrac{2}{3}x + 4$

7. $F(x) = 4$

8. $G(x) = -2$

---

9. Straight-line Depreciation   Suppose that a company has just purchased a new computer for $3000. The company chooses to depreciate the computer using the straight-line method over 3 years.
  (a) Write a linear function that expresses the book value of the computer as a function of its age.
  (b) Graph the linear function.
  (c) What is the book value of the computer after 2 years?

10. Straight-line Depreciation   Suppose that a company has just purchased a new machine for its manufacturing facility for $120,000. The company chooses to depreciate the machine using the straight-line method over 10 years.
  (a) Write a linear function that expresses the book value of the machine as a function of its age.
  (b) Graph the linear function.
  (c) What is the book value of the machine after 4 years?

11. Cost Function   The simplest cost function is the linear cost function, $C(x) = mx + b$, where the y-intercept b represents the fixed costs of operating a business and the slope m represents the variable costs. Suppose that a small bicycle manufacturer has daily fixed costs of $1800 and each bicycle costs $90 to manufacture.
  (a) Write a linear function that expresses the cost of manufacturing x bicycles in a day.
  (b) Graph the linear function.
  (c) What is the cost of manufacturing 14 bicycles in a day?

12. Cost Function   Refer to Problem 11. Suppose that the landlord of the building increases the bicycle manufacturer's rent by $100 per month.
  (a) Assuming that the manufacturer is open for business 20 days per month, what are the new daily fixed costs?
  (b) Write a linear function that expresses the cost of manufacturing x bicycles in a day with the higher rent.
  (c) Graph the linear function.
  (d) What is the cost of manufacturing 14 bicycles in a day?

In Problems 13–18, examine the scatter diagram and determine whether the type of relation, if any, that may exist is linear or nonlinear.

13.

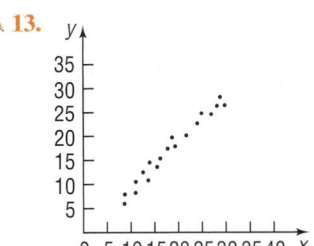

14.

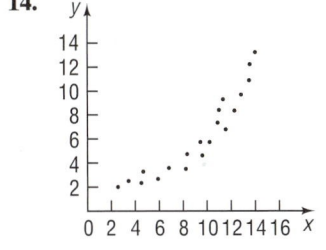

15.

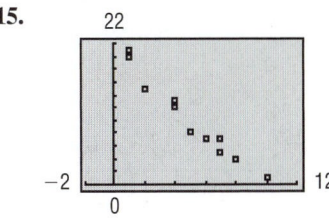

16.

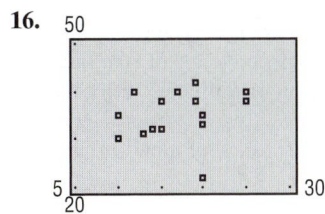

17.

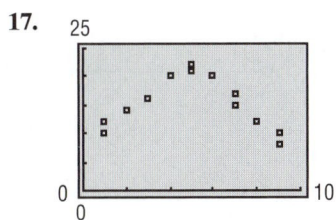

18.

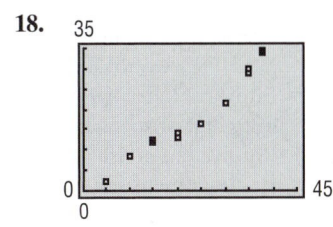

In Problems 19–24:

(a) *Draw a scatter diagram.*
(b) *Select two points from the scatter diagram and find the equation of the line containing the points selected.* *
(c) *Graph the line found in part (b) on the scatter diagram.*
(d) *Use a graphing utility to find the line of best fit.*
(e) *Use a graphing utility to graph the line of best fit on the scatter diagram.*

19.

x	3	4	5	6	7	8	9
y	4	6	7	10	12	14	16

20.

x	3	5	7	9	11	13
y	0	2	3	6	9	11

21.

x	−2	−1	0	1	2
y	−4	0	1	4	5

22.

x	−2	−1	0	1	2
y	7	6	3	2	0

23.

x	−20	−17	−15	−14	−10
y	100	120	118	130	140

24.

x	−30	−27	−25	−20	−14
y	10	12	13	13	18

25. Per Capita Disposable Income versus Consumption An economist wishes to estimate a linear function that relates per capita consumption expenditures C and disposable income I. Both C and I are measured in dollars. The following data represent the per capita disposable income (income after taxes) and per capita consumption in the United States for 1990 to 1998.

Year	Per Capita Disposable Income (I)	Per Capita Consumption (C)
1990	15,695	14,547
1991	16,700	15,400
1992	17,346	16,035
1993	18,153	16,951
1994	19,711	18,419
1995	20,316	19,061
1996	21,127	19,938
1997	21,871	20,807
1998	22,212	21,385

Source: U.S. Department of Commerce.

Let I represent the independent variable and C the dependent variable.
(a) Use a graphing utility to draw a scatter diagram.
(b) Use a graphing utility to find the line of best fit to the data. Express the solution using function notation.
(c) Interpret the slope. The slope of this line is called the **marginal propensity to consume.**
(d) Predict the consumption of a family whose disposable income is $21,500.

26. Height versus Head Circumference A pediatrician wanted to estimate a linear function that relates a child's height, H to their head circumference, C. She randomly selects 9 children from her practice, measures their height and head circumference, and obtains the following data.

Height, H (inches)	Head Circumference, C (inches)
25.25	16.4
25.75	16.9
25	16.9
27.75	17.6
26.5	17.3
27	17.5
26.75	17.3
26.75	17.5
27.5	17.5

Source: Denise Slucki, Student at Joliet Junior College.

Let H represent the independent variable and C the dependent variable.
(a) Use a graphing utility to draw a scatter diagram.
(b) Use a graphing utility to find the line of best fit to the data. Express the solution using function notation.
(c) Interpret the slope.
(d) Predict the head circumference of a child that is 26 inches tall.

* Answers will vary. We will use the first and last data points in the answer section.

27. Gestation Period versus Life Expectancy   A researcher would like to estimate the linear function relating the gestation period of an animal, G, and its life expectancy, L. She collects the following data.

Animal	Gestation (or incubation) Period, G (days)	Life Expectancy, L (years)
Cat	63	11
Chicken	22	7.5
Dog	63	11
Duck	28	10
Goat	151	12
Lion	108	10
Parakeet	18	8
Pig	115	10
Rabbit	31	7
Squirrel	44	9

Source: Time Almanac 2000.

Let G represent the independent variable and L the dependent variable.

(a) Use a graphing utility to draw a scatter diagram.
(b) Use a graphing utility to find the line of best fit to the data. Express the solution using function notation.
(c) Interpret the slope.
(d) Predict the life expectancy of an animal whose gestation period is 89 days.

28. Mortgage Qualification   The amount of money that a lending institution will allow you to borrow mainly depends on the interest rate and your annual income. The following data represent the annual income, I, required by a bank in order to lend L dollars at an interest rate of 7.5% for 30 years.

Annual Income, I ($)	Loan Amount, L ($)
15,000	44,600
20,000	59,500
25,000	74,500
30,000	89,400
35,000	104,300
40,000	119,200
45,000	134,100
50,000	149,000
55,000	163,900
60,000	178,800
65,000	193,700
70,000	208,600

Source: Information Please Almanac, 1999.

Let I represent the independent variable and L the dependent variable.

(a) Use a graphing utility to draw a scatter diagram of the data.
(b) Use a graphing utility to find the line of best fit to the data.
(c) Graph the line of best fit on the scatter diagram drawn in part (a).
(d) Interpret the slope of the line of best fit.
(e) Determine the loan amount that an individual would qualify for if her income is $42,000.

29. Demand for Jeans   The marketing manager at Levi–Strauss wishes to find a function that relates the demand D for men's jeans and p, the price of the jeans. The following data were obtained based on a price history of the jeans.

Price ($/Pair), p	Demand (Pairs of Jeans Sold per Day), D
20	60
22	57
23	56
23	53
27	52
29	49
30	44

(a) Does the relation defined by the set of ordered pairs (p, D) represent a function?
(b) Draw a scatter diagram of the data.
(c) Using a graphing utility, find the line of best fit relating price and quantity demanded.
(d) Interpret the slope.
(e) Express the relationship found in part (c) using function notation.
(f) What is the domain of the function?
(g) How many jeans will be demanded if the price is $28 a pair?

30. Advertising and Sales Revenue   A marketing firm wishes to find a function that relates the sales S of a product and A, the amount spent on advertising the product. The data are obtained from past experience. Advertising and sales are measured in thousands of dollars.

Advertising Expenditures, A	Sales, S
20	335
22	339
22.5	338
24	343
24	341
27	350
28.3	351

(a) Does the relation defined by the set of ordered pairs (A, S) represent a function?

(b) Draw a scatter diagram of the data.

(c) Using a graphing utility, find the line of best fit relating advertising expenditures and sales.

(d) Interpret the slope.

(e) Express the relationship found in part (c) using function notation.

(f) What is the domain of the function?

(g) Predict sales if advertising expenditures are $25,000.

31. Mortgage Payments   The monthly payment p on a mortgage varies directly with the amount borrowed B. If the monthly payment on a 30-year mortgage is $6.49 for every $1000 borrowed, find a linear function that relates the monthly payment p to the amount borrowed B for a mortgage with the same terms. Then find the monthly payment p when the amount borrowed B is $145,000.

32. Mortgage Payments   The monthly payment p on a mortgage varies directly with the amount borrowed B. If the monthly payment on a 15-year mortgage is $8.99 for every $1000 borrowed, find a linear function that relates the monthly payment p to the amount borrowed B for a mortgage with the same terms. Then find the monthly payment when p the amount borrowed B is $175,000.

33. Revenue Function   At the corner Shell station, the revenue R varies directly with the number g of gallons of gasoline sold. If the revenue is $15.84 when the number of gallons sold is 12, find a linear function that relates revenue R to the number of g gallons of gasoline. Then find the revenue R when the number of gallons of gasoline sold is 10.5.

34. Cost Function   The cost C of chocolate-covered almonds varies directly with the number A of pounds of almonds purchased. If the cost is $23.75 when the number of pounds of chocolate-covered almonds purchased is 5, find a linear function that relates the cost C to the number A of pounds of almonds purchased. Then find the cost C when the number of pounds of almonds purchased is 3.5.

35. Physics: Falling Objects   The distance s that an object falls is directly proportional to the square of the time t of the fall. If an object falls 16 feet in 1 second, how far will it fall in 3 seconds? How long will it take an object to fall 64 feet?

36. Physics: Falling Objects   The velocity v of a falling object is directly proportional to the time t of the fall. If, after 2 seconds, the velocity of the object is 64 feet per second, what will its velocity be after 3 seconds?

PREPARING FOR THIS SECTION

Before getting started, review the following:

✓ Intercepts (Section 1.2, pp. 106–107)

✓ Graph of $y = x^2$ (Section 1.2, p. 102)

✓ Solving Quadratic Equations (Section 1.3, pp. 117–124)

2.3   QUADRATIC FUNCTIONS

OBJECTIVES   ① Determine Whether the Graph of a Quadratic Function Opens Up or Down

② Identify the Vertex and Axis of Symmetry of a Quadratic Function

③ Graph Quadratic Functions

A *quadratic function* is a function that is defined by a second-degree polynomial in one variable.

A **quadratic function** is a function of the form

$$f(x) = ax^2 + bx + c \tag{1}$$

where a, b, and c are real numbers and $a \neq 0$. The domain of a quadratic function consists of all real numbers.

Many applications require a knowledge of quadratic functions. For example, suppose that Texas Instruments collects the data shown in Table 6 that relate the number of calculators sold at the price p per calculator. Since the price of a product determines the quantity that will be purchased, we treat price as the independent variable.

TABLE 6	
Price per Calculator, p (Dollars)	Number of Calculators, x
60	11,100
65	10,115
70	9,652
75	8,731
80	8,087
85	7,205
90	6,439

A linear relationship between the number of calculators and the price p per calculator may be given by the equation

$$x = 21,000 - 150p$$

Then the revenue R derived from selling x calculators at the price p per calculator is equal to the unit selling price p of the product times the number x of units actually sold. That is,

$$\begin{aligned} R &= xp \\ &= (21,000 - 150p)p \\ &= -150p^2 + 21,000p \end{aligned}$$

Figure 23

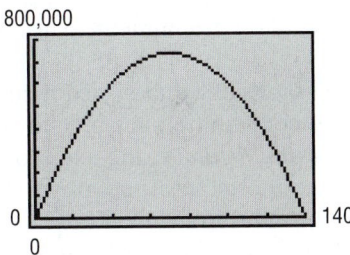

So, the revenue R is a quadratic function of the price p. Figure 23 illustrates the graph of this revenue function, whose domain is $0 \le p \le 140$, since both x and p must be nonnegative.

A second situation in which a quadratic function appears involves the motion of a projectile. Based on Newton's second law of motion (force equals mass times acceleration, $F = ma$), it can be shown that, ignoring air resistance, the path of a projectile propelled upward at an inclination to the horizontal is the graph of a quadratic function. See Figure 24 for an illustration.

Figure 24
Path of a cannonball

Graphing Quadratic Functions

1   We know how to graph the quadratic function $f(x) = x^2$. Figure 25 shows the graph of three functions of the form $f(x) = ax^2$, $a > 0$, for $a = 1$, $a = \dfrac{1}{2}$,

and $a = 3$. Notice that the larger the value of a, the "narrower" the graph is, and the smaller the value of a, the "wider" the graph is.

Figure 25

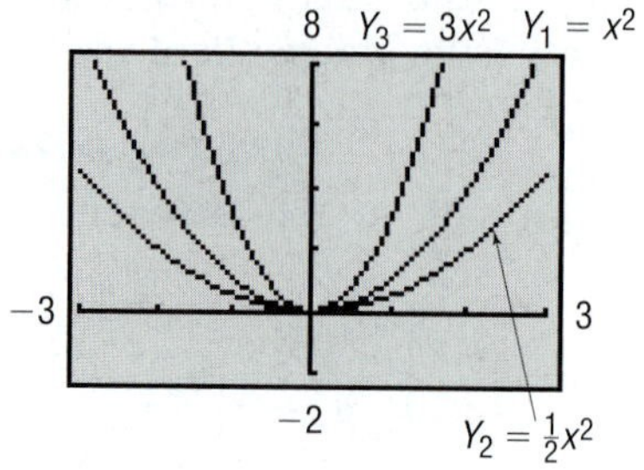

Figure 26

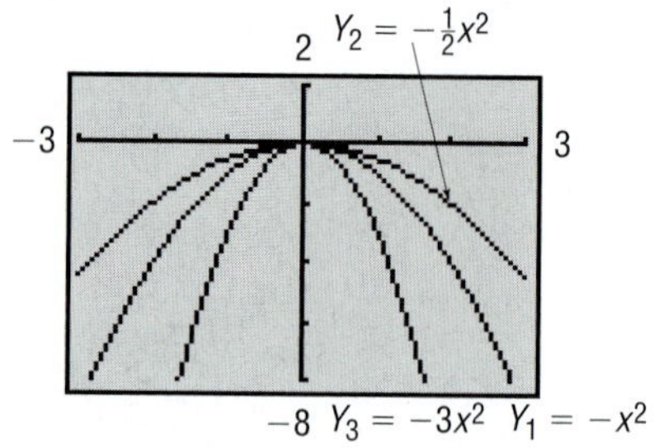

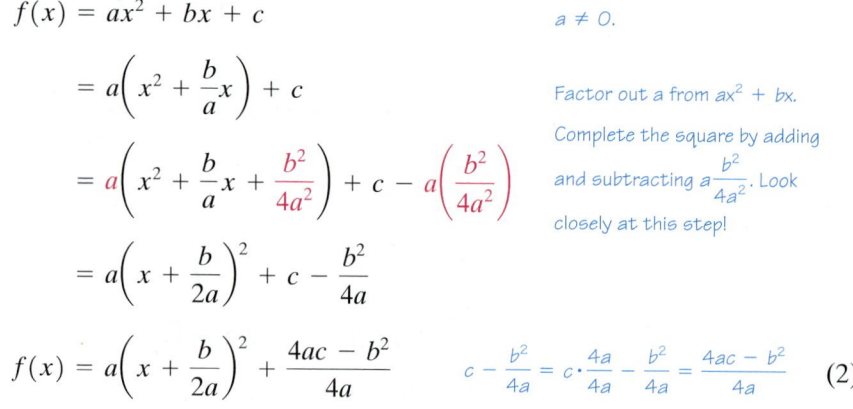

Figure 26 shows the graphs of $f(x) = ax^2$ for $a < 0$. Notice that these graphs are reflections about the x-axis of the graphs in Figure 25. Based on the results of these two figures, we can draw some general conclusions about the graph of $f(x) = ax^2$. First, as $|a|$ increases, the graph becomes "narrower," and as $|a|$ gets closer to zero, the graph gets "wider." Second, if a is positive, then the graph opens "up," and if a is negative, then the graph opens "down."

The graphs in Figures 25 and 26 are typical of the graphs of all quadratic functions, which we call **parabolas**.* Refer to Figure 27, where two parabolas are pictured. The one on the left **opens up** and has a lowest point; the one on the right **opens down** and has a highest point. The lowest or highest point of a parabola is called the **vertex**. The vertical line passing through the vertex in each parabola in Figure 27 is called the **axis of symmetry** (usually abbreviated to **axis**) of the parabola. Because the parabola is symmetric about its axis, the axis of symmetry of a parabola is useful for graphing the parabola by hand.

The parabolas shown in Figure 27 are the graphs of a quadratic function $f(x) = ax^2 + bx + c$, $a \neq 0$. Notice that the coordinate axes are not included in the figure. Depending on the values of a, b, and c, the axes could be placed anywhere. The important fact is that the shape of the graph of a quadratic function will look like one of the parabolas in Figure 27.

A key element in graphing a quadratic function by hand is locating the vertex. To find a formula, we begin with a quadratic function $f(x) = ax^2 + bx + c$, $a \neq 0$, and complete the square in x.

Figure 27

Graphs of a quadratic function,
$f(x) = ax^2 + bx + c, a \neq 0$

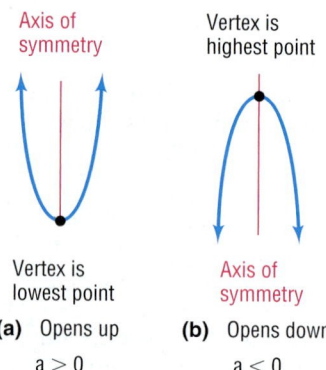

Axis of symmetry

Vertex is highest point

Vertex is lowest point

Axis of symmetry

(a) Opens up
$a > 0$

(b) Opens down
$a < 0$

$$f(x) = ax^2 + bx + c \qquad\qquad a \neq 0.$$

$$= a\left(x^2 + \frac{b}{a}x\right) + c \qquad\qquad \text{Factor out } a \text{ from } ax^2 + bx.$$

$$= a\left(x^2 + \frac{b}{a}x + \frac{b^2}{4a^2}\right) + c - a\left(\frac{b^2}{4a^2}\right) \qquad \begin{array}{l}\text{Complete the square by adding}\\ \text{and subtracting } a\dfrac{b^2}{4a^2}. \text{ Look}\\ \text{closely at this step!}\end{array}$$

$$= a\left(x + \frac{b}{2a}\right)^2 + c - \frac{b^2}{4a}$$

$$f(x) = a\left(x + \frac{b}{2a}\right)^2 + \frac{4ac - b^2}{4a} \qquad c - \frac{b^2}{4a} = c \cdot \frac{4a}{4a} - \frac{b^2}{4a} = \frac{4ac - b^2}{4a} \qquad (2)$$

---

*We shall study parabolas using a geometric definition later in this book.

Suppose that $a > 0$. If $x = -\dfrac{b}{2a}$, then the term $a\left(x + \dfrac{b}{2a}\right)^2 = a\left(-\dfrac{b}{2a} + \dfrac{b}{2a}\right)^2 = a \cdot 0 = 0$. For any other value of x, the term $a\left(x + \dfrac{b}{2a}\right)^2$ will be positive. [Do you see why? $\left(x + \dfrac{b}{2a}\right)^2$ is positive for $x \neq -\dfrac{b}{2a}$ because it is a nonzero quantity squared, a is positive, and the product of two positive quantities is positive]. Because $a\left(x + \dfrac{b}{2a}\right)^2$ is zero if $x = -\dfrac{b}{2a}$, and is positive if $x \neq -\dfrac{b}{2a}$, the value of the function f given in (2) will be smallest when $x = -\dfrac{b}{2a}$. That is, if $a > 0$, the parabola opens up, the vertex is at $\left(-\dfrac{b}{2a}, f\left(-\dfrac{b}{2a}\right)\right)$, and the vertex is a minimum point.

Similarly, if $a < 0$, the term $a\left(x + \dfrac{b}{2a}\right)^2$ is zero for $x = -\dfrac{b}{2a}$ and is negative if $x \neq -\dfrac{b}{2a}$. In this case the largest value of $f(x)$ occurs when $x = -\dfrac{b}{2a}$. That is, if $a < 0$, the parabola opens down, the vertex is at $\left(-\dfrac{b}{2a}, f\left(-\dfrac{b}{2a}\right)\right)$, and the vertex is a maximum point.

We summarize these remarks as follows:

Properties of the Graph of a Quadratic Function
$$f(x) = ax^2 + bx + c$$

$$\text{Vertex} = \left(-\dfrac{b}{2a}, f\left(-\dfrac{b}{2a}\right)\right) \quad \text{Axis of symmetry: the line } x = -\dfrac{b}{2a} \quad (3)$$

Parabola opens up if $a > 0$; the vertex is a minimum point.
Parabola opens down if $a < 0$; the vertex is a maximum point.

EXAMPLE 1 Locating the Vertex without Graphing

Without graphing, locate the vertex and axis of symmetry of the parabola defined by $f(x) = -3x^2 + 6x + 1$. Does it open up or down?

Solution For this quadratic function, $a = -3$, $b = 6$, and $c = 1$. The x-coordinate of the vertex is

$$-\frac{b}{2a} = -\frac{6}{-6} = 1$$

The y-coordinate of the vertex is therefore

$$f\left(-\frac{b}{2a}\right) = f(1) = -3 + 6 + 1 = 4$$

The vertex is located at the point $(1, 4)$. The axis of symmetry is the line $x = 1$. Because $a = -3 < 0$, the parabola opens down. ■

③ The information we gathered in Example 1, together with the location of the intercepts, usually provides enough information to graph $f(x) = ax^2 + bx + c$, $a \neq 0$, by hand.

The y-intercept is the value of f at $x = 0$; that is, $f(0) = c$.

The x-intercepts, if there are any, are found by solving the equation

$$f(x) = ax^2 + bx + c = 0$$

This equation has two, one, or no real solutions, depending on whether the discriminant $b^2 - 4ac$ is positive, 0, or negative. Depending on the value of the discriminant, the graph of f has x-intercepts as follows:

The x-intercepts of a Quadratic Function

1. If the discriminant $b^2 - 4ac > 0$, the graph of $f(x) = ax^2 + bx + c$ has two distinct x-intercepts and so will cross the x-axis in two places.
2. If the discriminant $b^2 - 4ac = 0$, the graph of $f(x) = ax^2 + bx + c$ has one x-intercept and touches the x-axis at its vertex.
3. If the discriminant $b^2 - 4ac < 0$, the graph of $f(x) = ax^2 + bx + c$ has no x-intercept and so will not cross or touch the x-axis.

Figure 28 illustrates these possibilities for parabolas that open up.

Figure 28
$f(x) = ax^2 + bx + c$, $a > 0$

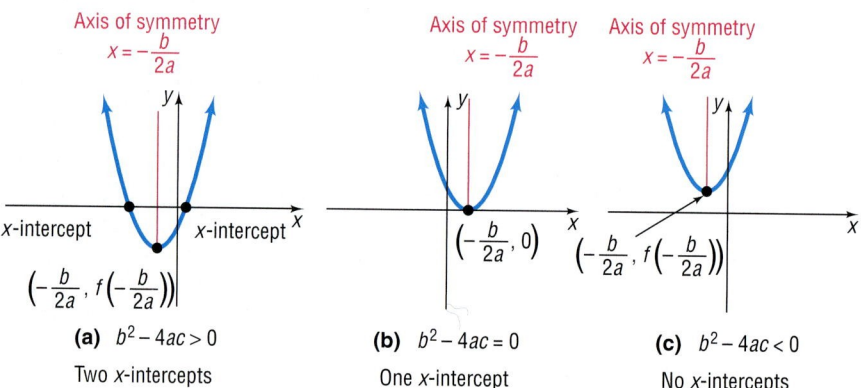

(a) $b^2 - 4ac > 0$
Two x-intercepts

(b) $b^2 - 4ac = 0$
One x-intercept

(c) $b^2 - 4ac < 0$
No x-intercepts

EXAMPLE 2 **Graphing a Quadratic Function by Hand Using Its Vertex, Axis, and Intercepts**

Use the information from Example 1 and the locations of the intercepts to graph $f(x) = -3x^2 + 6x + 1$.

Solution In Example 1, we found the vertex to be at $(1, 4)$ and the axis of symmetry to be $x = 1$. The y-intercept is found by letting $x = 0$. The y-intercept is $f(0) = 1$. The x-intercepts are found by solving the equation $f(x) = 0$. This results in the equation

$$-3x^2 + 6x + 1 = 0 \qquad a = -3, b = 6, c = 1$$

The discriminant $b^2 - 4ac = (6)^2 - 4(-3)(1) = 36 + 12 = 48 > 0$, so the equation has two real solutions and the graph has two x-intercepts. Using the quadratic formula, we find that

$$x = \frac{-b + \sqrt{b^2 - 4ac}}{2a} = \frac{-6 + \sqrt{48}}{-6} = \frac{-6 + 4\sqrt{3}}{-6} \approx -0.15$$

and

$$x = \frac{-b - \sqrt{b^2 - 4ac}}{2a} = \frac{-6 - \sqrt{48}}{-6} = \frac{-6 - 4\sqrt{3}}{-6} \approx 2.15$$

The x-intercepts are approximately -0.15 and 2.15.

The graph is illustrated in Figure 29. Notice how we used the y-intercept and the axis of symmetry, $x = 1$, to obtain the additional point $(2, 1)$ on the graph.

✔ CHECK: Graph $f(x) = -3x^2 + 6x + 1$ using a graphing utility. Use ZERO (or ROOT) to locate the two x-intercepts. ▮ ▮

Figure 29
$f(x) = -3x^2 + 6x + 1$

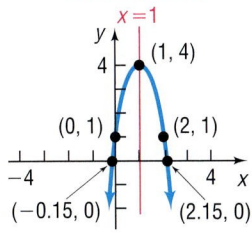

NOW WORK PROBLEM **13.**

If the graph of a quadratic function has only one x-intercept or none, it is usually necessary to plot an additional point to obtain the graph by hand.

EXAMPLE 3 ## Graphing a Quadratic Function by Hand Using Its Vertex, Axis, and Intercepts

Graph $f(x) = x^2 - 6x + 9$ by determining whether the graph opens up or down and by finding its vertex, axis of symmetry, y-intercept, and x-intercept, if any.

Solution For $f(x) = x^2 - 6x + 9$, we have $a = 1$, $b = -6$, and $c = 9$. Since $a = 1 > 0$, the parabola opens up. The x-coordinate of the vertex is

$$-\frac{b}{2a} = -\frac{-6}{2(1)} = 3$$

The y-coordinate of the vertex is

$$f(3) = (3)^2 - 6(3) + 9 = 0$$

So the vertex is at $(3, 0)$. The axis of symmetry is the line $x = 3$. The y-intercept is $f(0) = 9$. Since the vertex $(3, 0)$ lies on the x-axis, the graph touches the x-axis at the x-intercept. By using the axis of symmetry and the y-intercept at $(0, 9)$, we can locate the additional point $(6, 9)$ on the graph. See Figure 30. ▮

Figure 30

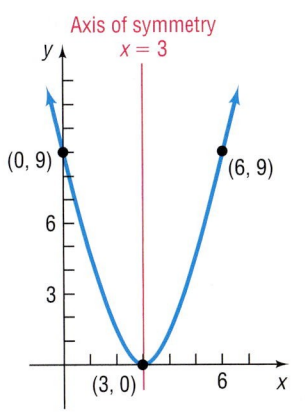

NOW WORK PROBLEM **21.**

EXAMPLE 4 ## Graphing a Quadratic Function by Hand Using Its Vertex, Axis, and Intercepts

Graph $f(x) = 2x^2 + x + 1$ by determining whether the graph opens up or down and by finding its vertex, axis of symmetry, y-intercept, and x-intercepts, if any.

Solution For $f(x) = 2x^2 + x + 1$, we have $a = 2$, $b = 1$, and $c = 1$. Since $a = 2 > 0$, the parabola opens up. The x-coordinate of the vertex is

$$-\frac{b}{2a} = -\frac{1}{4}.$$

The y-coordinate of the vertex is

$$f\left(-\frac{1}{4}\right) = 2\left(\frac{1}{16}\right) + \left(-\frac{1}{4}\right) + 1 = \frac{7}{8}$$

So the vertex is at $\left(-\frac{1}{4}, \frac{7}{8}\right)$. The axis of symmetry is the line $x = -\frac{1}{4}$. The y-intercept is $f(0) = 1$. The x-intercept(s), if any, obey the equation $2x^2 + x + 1 = 0$. Since the discriminant $b^2 - 4ac = (1)^2 - 4(2)(1) = -7 < 0$, this equation has no real solutions, and therefore the graph has no x-intercepts. We use the point $(0, 1)$ and the axis of symmetry $x = -\frac{1}{4}$ to locate the additional point $\left(-\frac{1}{2}, 1\right)$ on the graph. See Figure 31. ■

Figure 31

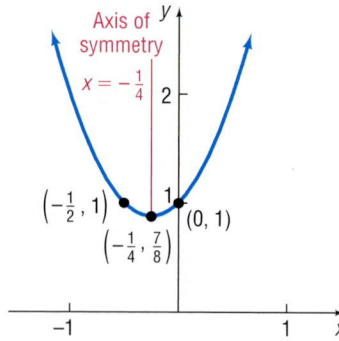

SUMMARY **Steps for Graphing a Quadratic Function $f(x) = ax^2 + bx + c, a \neq 0$ by Hand**

STEP 1: Determine the vertex, $\left(-\frac{b}{2a}, f\left(-\frac{b}{2a}\right)\right)$.

STEP 2: Determine the axis of symmetry, $x = -\frac{b}{2a}$.

STEP 3: Determine the y-intercept, $f(0)$.

STEP 4: (a) If $b^2 - 4ac > 0$, then the graph of the quadratic function has two x-intercepts, which are found by solving $ax^2 + bx + c = 0$.
(b) If $b^2 - 4ac = 0$, the vertex is the x-intercept.
(c) If $b^2 - 4ac < 0$, there are no x-intercepts.

STEP 5: Determine an additional point by using the y-intercept and the axis of symmetry.

STEP 6: Plot the points and draw the graph.

Look back at equation (2) on p. 228. By completing the square in x, we wrote the quadratic function $f(x) = ax^2 + bx + c$ in the form

$$f(x) = a\left(x + \frac{b}{2a}\right)^2 + \frac{4ac - b^2}{4a} \tag{4}$$

If we let h represent the x-coordinate of the vertex, then $h = -\frac{b}{2a}$. If k represents the y-coordinate of the vertex, then $k = f\left(-\frac{b}{2a}\right) = \frac{4ac - b^2}{4a}$. Now (4) becomes

$$f(x) = a(x - h)^2 + k \tag{5}$$

Given the vertex and one additional point, we can use (5) to obtain the quadratic function.

EXAMPLE 5 | **Finding the Quadratic Function Given Its Vertex and One Other Point**

Determine the quadratic function whose vertex is $(1, -5)$ and whose y-intercept is -3. The graph of the parabola is shown in Figure 32.

Solution The vertex is $(1, -5)$, so $h = 1$ and $k = -5$. Substitute these values into equation (5).

$$f(x) = a(x - h)^2 + k$$
$$f(x) = a(x - 1)^2 - 5 \qquad \text{\small{h = 1, k = -5}}$$

To determine the value of a, we use the fact that $f(0) = -3$ (the y-intercept).

$$f(x) = a(x - 1)^2 - 5$$
$$-3 = a(0 - 1)^2 - 5 \qquad \text{\small{x = 0, y = f(0) = -3}}$$
$$-3 = a - 5$$
$$a = 2$$

Figure 32

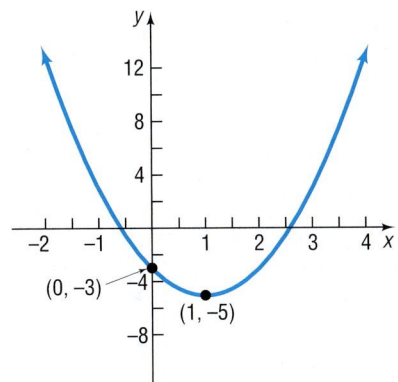

The quadratic function whose graph is shown in Figure 32 is $f(x) = 2(x - 1)^2 - 5 = 2x^2 - 4x - 3$.

✔ CHECK: Figure 33 shows the graph $f(x) = 2x^2 - 4x - 3$ using a graphing utility.

Figure 33

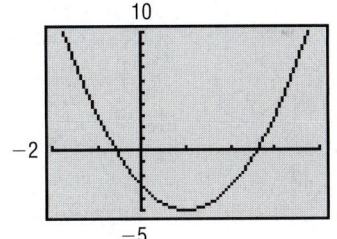

NOW WORK PROBLEM **31.**

Applications

Real-world problems may lead to mathematical problems that involve quadratic functions. Two such problems follow.

EXAMPLE 6 | **Area of a Rectangle with Fixed Perimeter**

A farmer has 50 feet of fence to enclose a rectangular field. Express the area A of the field as a function of the length x of a side.

Figure 34

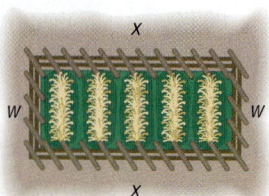

Solution Consult Figure 34. The available fence, 50 feet, represents the perimeter of the rectangle. If the length of the rectangle is x and if w is its width, then the sum of the lengths of the sides is the perimeter, 50.

$$x + w + x + w = 50$$
$$2x + 2w = 50$$
$$x + w = 25$$
$$w = 25 - x$$

The area A is length times width, so

$$A = xw = x(25 - x)$$

The area A as a function of x is

$$A(x) = x(25 - x) = -x^2 + 25x$$

Note that we used the symbol A as the dependent variable and also as the name of the function that relates the length x to the area A. As we mentioned earlier, this double usage is common in applications and should cause no difficulties.

EXAMPLE 7 Economics: Demand Equations

In economics, revenue R is defined as the amount of money derived from the sale of a product and is equal to the unit selling price p of the product times the number x of units actually sold. That is,

$$R = xp$$

In economics, the Law of Demand states that p and x are related: As price p increases, the number x of units sold decreases. Suppose that p and x are related by the following

$$p = -\frac{1}{10}x + 20 \qquad 0 \le x \le 200$$

Express the revenue R as a function of the number x of units sold.

Solution Since $R = xp$ and $p = -\frac{1}{10}x + 20$, it follows that

$$R(x) = xp = x\left(-\frac{1}{10}x + 20\right) = -\frac{1}{10}x^2 + 20x$$

NOW WORK PROBLEM 41.

2.3 Concepts and Vocabulary

In Problems 1 and 2, fill in the blanks.

1. The graph of a quadratic function is called a _____.

2. The vertical line passing through the vertex of a parabola is called the _____.

In Problems 3–5, answer True or False to each statement.

3. The graph of $f(x) = 2x^2 + 3x - 4$ opens up.

4. The y-coordinate of the vertex of $f(x) = -x^2 + 4x + 5$ is $f(2)$.

5. If the discriminant, $b^2 - 4ac$, is equal to zero, the graph of the quadratic function $f(x) = ax^2 + bx + c$ will touch the x-axis at its vertex.

6. State the circumstances that lead the graph of a quadratic function $f(x) = ax^2 + bx + c$ to have no x-intercepts.

7. List the steps necessary to graph a quadratic function by hand.

8. Why does the graph of a quadratic function open up if $a > 0$ and down if $a < 0$?

2.3 Exercises

In Problems 1–8, match each graph to one of the following functions:

1. $f(x) = x^2 - 1$

2. $f(x) = -x^2 - 1$

3. $f(x) = x^2 - 2x + 1$

4. $f(x) = x^2 + 2x + 1$

5. $f(x) = x^2 - 2x + 2$

6. $f(x) = x^2 + 2x$

7. $f(x) = x^2 - 2x$

8. $f(x) = x^2 + 2x + 2$

A.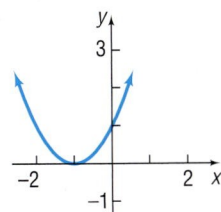

B.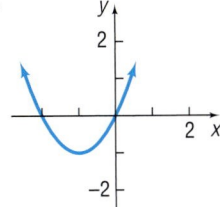

C.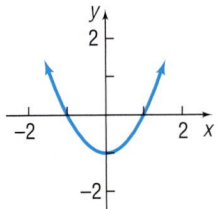

D.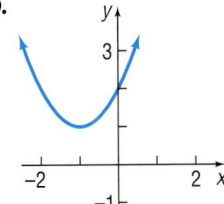

E.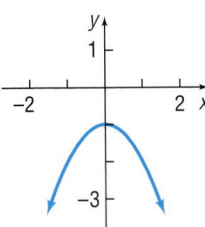

F.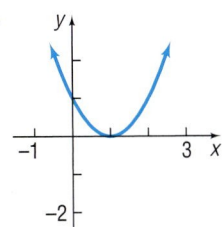

G.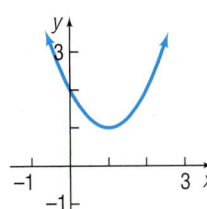

H.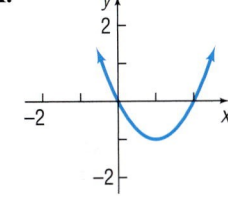

In Problems 9–12, match each graph to one of the following functions.

9. $f(x) = \frac{1}{3}x^2 + 2$

10. $f(x) = 3x^2 + 2$

11. $f(x) = x^2 + 5x + 1$

12. $f(x) = -x^2 + 5x + 1$

A.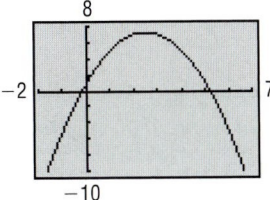

B.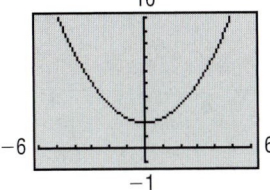

C.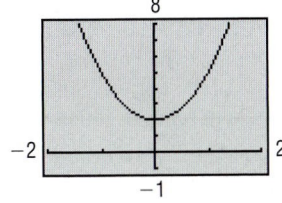

D.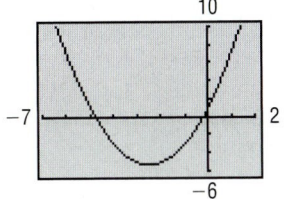

In Problems 13–30, graph each quadratic function by hand by determining whether its graph opens up or down and by finding its vertex, axis of symmetry, y-intercept, and x-intercepts, if any. Verify your results using a graphing utility.

13. $f(x) = x^2 + 2x$ **14.** $f(x) = x^2 - 4x$ **15.** $f(x) = -x^2 - 6x$

16. $f(x) = -x^2 + 4x$ **17.** $f(x) = 2x^2 - 8x$ **18.** $f(x) = 3x^2 + 18x$

19. $f(x) = x^2 + 2x - 8$ **20.** $f(x) = x^2 - 2x - 3$ **21.** $f(x) = x^2 + 2x + 1$

22. $f(x) = x^2 + 6x + 9$ **23.** $f(x) = 2x^2 - x + 2$ **24.** $f(x) = 4x^2 - 2x + 1$

25. $f(x) = -2x^2 + 2x - 3$ **26.** $f(x) = -3x^2 + 3x - 2$ **27.** $f(x) = 3x^2 + 6x + 2$

28. $f(x) = 2x^2 + 5x + 3$ **29.** $f(x) = -4x^2 - 6x + 2$ **30.** $f(x) = 3x^2 - 8x + 2$

In Exercises 31–36, determine the quadratic function whose graph is given.

31.

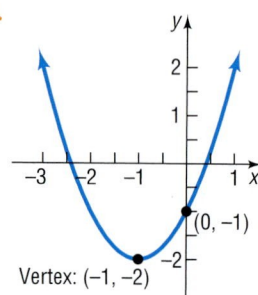

32.

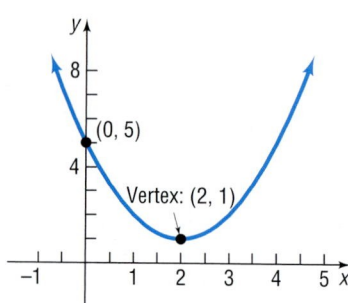

33.

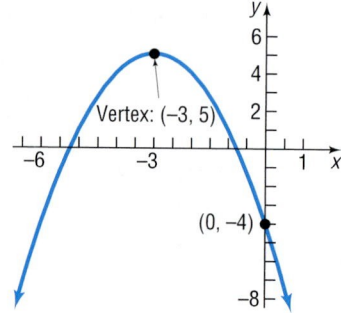

34.

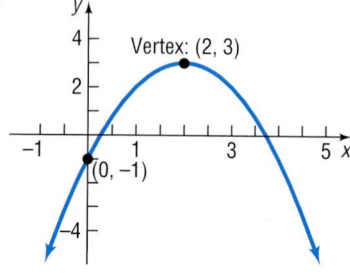

35.

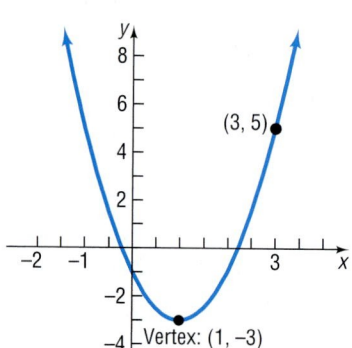

36.

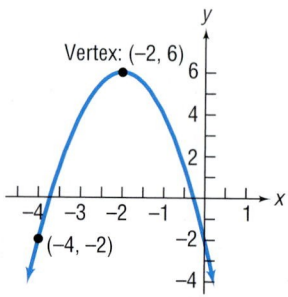

Answer Problems 37 and 38 using the following: A quadratic function of the form $f(x) = ax^2 + bx + c$ with $b^2 - 4ac > 0$ may also be written in the form $f(x) = a(x - r_1)(x - r_2)$, where r_1 and r_2 are the x-intercepts of the graph of the quadratic function.

37. (a) Find a quadratic function whose x-intercepts are -3 and 1 with $a = 1; a = 2; a = -2; a = 5$.
  (b) How does the value of a affect the intercepts?
  (c) How does the value of a affect the axis of symmetry?
  (d) How does the value of a affect the vertex?
  (e) Compare the x-coordinate of the vertex with the midpoint of the x-intercepts. What might you conclude?

38. (a) Find a quadratic function whose x-intercepts are -5 and 3 with $a = 1; a = 2; a = -2; a = 5$.
  (b) How does the value of a affect the intercepts?
  (c) How does the value of a affect the axis of symmetry?
  (d) How does the value of a affect the vertex?
  (e) Compare the x-coordinate of the vertex with the midpoint of the x-intercepts. What might you conclude?

39. The graph of the function $f(x) = ax^2 + bx + c$ has vertex at $(0, 2)$ and passes through the point $(1, 8)$. Find a, b, and c.

40. The graph of the function $f(x) = ax^2 + bx + c$ has vertex at $(1, 4)$ and passes through the point $(-1, -8)$. Find a, b, and c.

41. Demand Equation   The price p and the quantity x sold of a certain product obey the demand equation

$$p = -\frac{1}{6}x + 100 \qquad 0 \le x \le 600$$

Express the revenue R as a function of x. (Remember, $R = xp$.)

42. Demand Equation   The price p and the quantity x sold of a certain product obey the demand equation

$$p = -\frac{1}{3}x + 100 \qquad 0 \le x \le 300$$

Express the revenue R as a function of x.

43. Demand Equation   The price p and the quantity x sold of a certain product obey the demand equation

$$x = -5p + 100 \qquad 0 \le p \le 20$$

Express the revenue R as a function of x.

44. Demand Equation   The price p and the quantity x sold of a certain product obey the demand equation

$$x = -20p + 500 \qquad 0 \le p \le 25$$

Express the revenue R as a function of x.

45. Enclosing a Rectangular Field   David has available 400 yards of fencing and wishes to enclose a rectangular area. Express the area A of the rectangle as a function of the width x of the rectangle.

46. Enclosing a Rectangular Field along a River   Beth has 3000 feet of fencing available to enclose a rectangular field. One side of the field lies along a river, so only three sides require fencing. Express the area A of the rectangle as a function of x, where x is the length of the side parallel to the river.

47. Make up a quadratic function that opens down and has only one x-intercept. Compare yours with others in the class. What are the similarities? What are the differences?

48. On one set of coordinate axes, graph the family of parabolas $f(x) = x^2 + 2x + c$ for $c = -3$, $c = 0$, and $c = 1$. Describe the characteristics of a member of this family.

49. On one set of coordinate axes, graph the family of parabolas $f(x) = x^2 + bx + 1$ for $b = -4$, $b = 0$, and $b = 4$. Describe the general characteristics of this family.

2.4 QUADRATIC MODELS

OBJECTIVES

  1 Find the Maximum or Minimum Value of a Quadratic Function

  2 Use the Maximum or Minimum Value of a Quadratic Function to Solve Applied Problems

  3 Use a Graphing Utility to Find the Quadratic Function of Best Fit

When a mathematical model leads to a quadratic function, the properties of that quadratic function can provide important information about the model. For example, for a quadratic revenue function, we can find the maximum revenue; for a quadratic cost function, we can find the minimum cost.

1

To see why, recall that the graph of a quadratic function $f(x) = ax^2 + bx + c$ is a parabola with vertex at $\left(-\dfrac{b}{2a}, f\left(-\dfrac{b}{2a}\right)\right)$. This vertex is the highest point on the graph if $a < 0$ and the lowest point on the graph if $a > 0$. If the vertex is the highest point $(a < 0)$, then $f\left(-\dfrac{b}{2a}\right)$ is the **maximum value** of f. If the vertex is the lowest point $(a > 0)$, then $f\left(-\dfrac{b}{2a}\right)$ is the **minimum value** of f.

EXAMPLE 1   **Finding the Maximum or Minimum Value of a Quadratic Function**

Determine whether the quadratic function

$$f(x) = x^2 - 4x - 5$$

has a maximum or minimum value. Then find the maximum or minimum value.

Solution   We compare $f(x) = x^2 - 4x - 5$ to $f(x) = ax^2 + bx + c$. We conclude that $a = 1, b = -4$, and $c = -5$. Since $a > 0$, the graph of f opens up, so the vertex is a minimum point. The minimum value occurs at

$$x = -\frac{b}{2a} \uparrow = -\frac{-4}{2(1)} = \frac{4}{2} = 2$$

$$a = 1, b = -4$$

The minimum value is

$$f\left(-\frac{b}{2a}\right) = f(2) = 2^2 - 4(2) - 5 = 4 - 8 - 5 = -9$$

TABLE 7

X	Y1
-1	0
0	-5
1	-8
2	-9
3	-8
4	-5
5	0

Y1■X²-4X-5

✔ CHECK: We can support the algebraic solution using the TABLE feature on a graphing utility. Create Table 7 by letting $Y_1 = x^2 - 4x - 5$. From the table we see that the smallest value of y occurs when $x = 2$, leading us to investigate the number 2 further. The symmetry about $x = 2$ confirms that 2 is the x-coordinate of the vertex of the parabola. We conclude that the minimum value is -9 and occurs at $x = 2$.   ■ ■

NOW WORK PROBLEM 7.

EXAMPLE 2   **Maximizing Revenue**

② The marketing department at Texas Instruments has found that, when certain calculators are sold at a price of p dollars per unit, the revenue R (in dollars) as a function of the price p is

$$R(p) = -150p^2 + 21,000p$$

What unit price should be established in order to maximize revenue? If this price is charged, what is the maximum revenue?

Solution   The revenue R is

$$R(p) = -150p^2 + 21,000p \qquad R(p) = ap^2 + bp + c$$

The function R is a quadratic function with $a = -150, b = 21,000$, and $c = 0$. Because $a < 0$, the vertex is the highest point of the parabola. The revenue R is therefore a maximum when the price p is

$$p = -\frac{b}{2a} \uparrow = -\frac{21,000}{2(-150)} = -\frac{21,000}{-300} = \$70.00$$

$$a = -150, b = 21,000$$

The maximum revenue R is

$$R(70) = -150(70)^2 + 21{,}000(70) = \$735{,}000$$

See Figure 35 for an illustration.

Figure 35

$R(p) = -150p^2 + 21{,}000p$

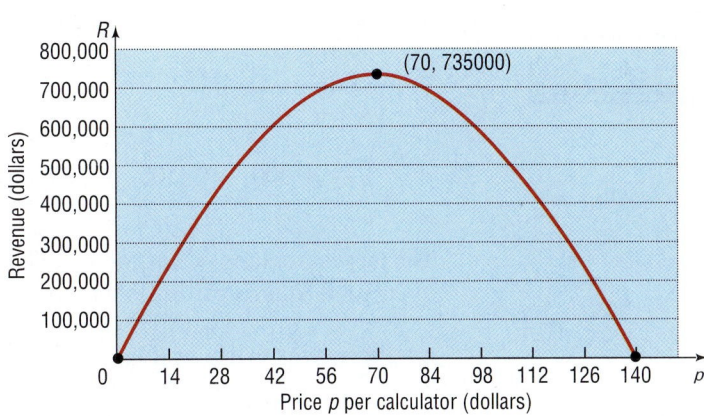

NOW WORK PROBLEM **13.**

EXAMPLE 3 | **Maximizing the Area Enclosed by a Fence**

A farmer has 2000 yards of fence to enclose a rectangular field. What are the dimensions of the rectangle that encloses the most area?

Solution   Figure 36 illustrates the situation. The available fence represents the perimeter of the rectangle. If x is the length and w is the width, then

Figure 36

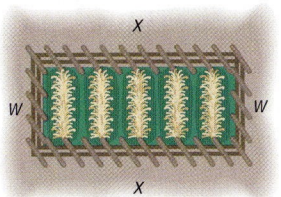

$$2x + 2w = 2000 \qquad (1)$$

The area A of the rectangle is

$$A = xw$$

To express A in terms of a single variable, we solve equation (1) for w and substitute the result in $A = xw$. Then A involves only the variable x. [You could also solve equation (1) for x and express A in terms of w alone. Try it!]

$$2x + 2w = 2000 \qquad \text{Equation (1)}$$
$$2w = 2000 - 2x \qquad \text{Solve for } w.$$
$$w = \frac{2000 - 2x}{2} = 1000 - x$$

Then the area A is

$$A = xw = x(1000 - x) = -x^2 + 1000x$$

Now, A is a quadratic function of x.

$$A(x) = -x^2 + 1000x \qquad a = -1, b = 1000, c = 0$$

Figure 37

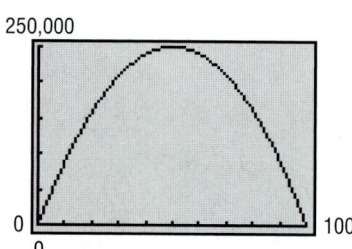

Figure 37 shows a graph of $A(x) = -x^2 + 1000x$ using a graphing utility. Since $a < 0$, the vertex is a maximum point on the graph of A. The maximum value occurs at

$$x = -\frac{b}{2a} = -\frac{1000}{2(-1)} = 500$$

The maximum value of A is

$$A\left(-\frac{b}{2a}\right) = A(500) = -500^2 + 1000(500) = -250,000 + 500,000 = 250,000$$

The largest rectangle that can be enclosed by 2000 yards of fence has an area of 250,000 square yards. Its dimensions are 500 yards by 500 yards. ■

NOW WORK PROBLEM 21.

EXAMPLE 4   **Analyzing the Motion of a Projectile**

A projectile is fired from a cliff 500 feet above the water at an inclination of $45°$ to the horizontal, with a muzzle velocity of 400 feet per second. In physics, it is established that the height h of the projectile above the water is given by

$$h(x) = \frac{-32x^2}{(400)^2} + x + 500$$

where x is the horizontal distance of the projectile from the base of the cliff. See Figure 38.

Figure 38

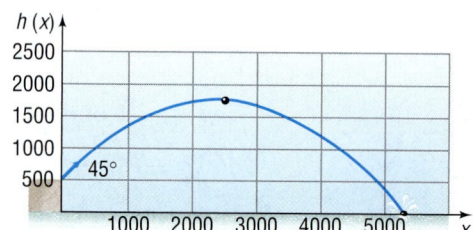

(a) Find the maximum height of the projectile.
(b) How far from the base of the cliff will the projectile strike the water?

Solution   (a) The height of the projectile is given by a quadratic function.

$$h(x) = \frac{-32x^2}{(400)^2} + x + 500 = \frac{-1}{5000}x^2 + x + 500$$

We are looking for the maximum value of h. Since $a < 0$, the maximum value is obtained at the vertex; we compute

$$x = -\frac{b}{2a} = -\frac{1}{2\left(-\dfrac{1}{5000}\right)} = \frac{5000}{2} = 2500$$

The maximum height of the projectile is

$$h(2500) = \frac{-1}{5000}(2500)^2 + 2500 + 500 = -1250 + 2500 + 500 = 1750 \text{ ft}$$

(b) The projectile will strike the water when the height is zero. To find the distance x traveled, we need to solve the equation

$$h(x) = \frac{-1}{5000}x^2 + x + 500 = 0$$

We use the quadratic formula with

$$b^2 - 4ac = 1^2 - 4\left(\frac{-1}{5000}\right)(500) = 1.4$$

$$x = \frac{-b \pm \sqrt{b^2 - 4ac}}{2a} = \frac{-1 \pm \sqrt{1.4}}{2\left(-\dfrac{1}{5000}\right)} \approx \begin{cases} -458 \\ 5458 \end{cases}$$

We discard the negative solution and find that the projectile will strike the water a distance of about 5458 feet from the base of the cliff.

 SEEING THE CONCEPT Graph

$$h(x) = \frac{-1}{5000}x^2 + x + 500 \qquad 0 \le x \le 5500$$

Use MAXIMUM to find the maximum height of the projectile and use ROOT or ZERO to find the distance from the base of the cliff to where it strikes the water. Compare your results with those obtained in Example 4. TRACE the path of the projectile. How far from the base of the cliff is the projectile when its height is 1000 ft? 1500 ft?

 **NOW WORK PROBLEM 25.**

EXAMPLE 5 **The Golden Gate Bridge**

The Golden Gate Bridge, a suspension bridge, spans the entrance to San Francisco Bay. Its 746-foot-tall towers are 4200 feet apart. The bridge is suspended from two huge cables more than 3 feet in diameter; the 90-foot-wide roadway is 220 feet above the water. The cables are parabolic in shape* and touch the road surface at the center of the bridge. Find the height of the cable at a distance of 1000 feet from the center.

---

*A cable suspended from two towers is in the shape of a **catenary**, but when a horizontal roadway is suspended from the cable, the cable takes the shape of a parabola.

Solution See Figure 39. We begin by choosing the placement of the coordinate axes so that the *x*-axis coincides with the road surface and the origin coincides with the center of the bridge. As a result, the twin towers will be vertical (height $746 - 220 = 526$ feet above the road) and located 2100 feet from the center. Also, the cable, which has the shape of a parabola, will extend from the towers, open up, and have its vertex at $(0, 0)$. The choice of placement of the axes enables us to identify the equation of the parabola as $y = ax^2, a > 0$. We can also see that the points $(-2100, 526)$ and $(2100, 526)$ are on the graph.

Figure 39

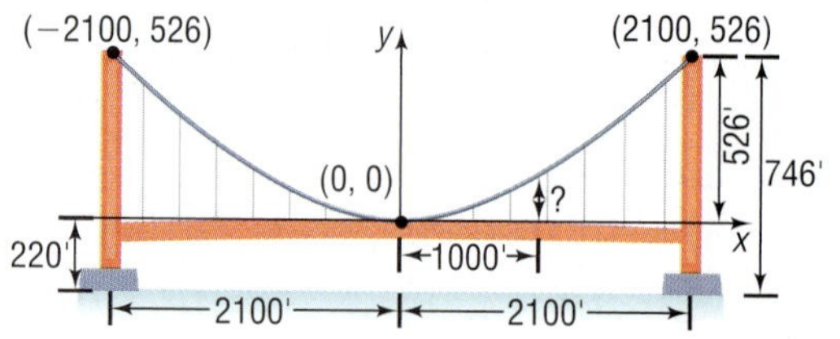

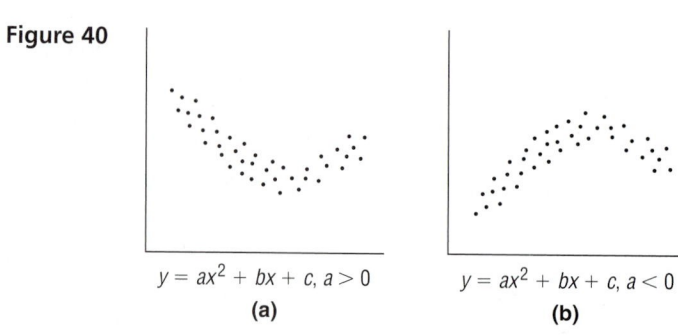

Based on these facts, we can find the value of *a* in $y = ax^2$.

$$y = ax^2$$
$$526 = a(2100)^2$$
$$a = \frac{526}{(2100)^2}$$

The equation of the parabola is therefore

$$y = \frac{526}{(2100)^2}x^2$$

The height of the cable when $x = 1000$ is

$$y = \frac{526}{(2100)^2}(1000)^2 \approx 119.3 \text{ feet}$$

The cable is 119.3 feet high at a distance of 1000 feet from the center of the bridge. ■

NOW WORK PROBLEM 27.

Fitting a Quadratic Function to Data

③ In Section 2.2, we found the line of best fit for data that appeared to be linearly related. It was noted that data may also follow a nonlinear relation. Figures 40(a) and (b) show scatter diagrams of data that follow a quadratic relation.

Figure 40

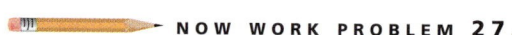

$y = ax^2 + bx + c, a > 0$	$y = ax^2 + bx + c, a < 0$
(a)	**(b)**

EXAMPLE 6   Fitting a Quadratic Function to Data

A farmer collected the data, given in Table 8, which show crop yields Y for various amounts of fertilizer used, x.

(a) With a graphing utility, draw a scatter diagram of the data. Comment on the type of relation that may exist between the two variables.

(b) Use a graphing utility to find the quadratic function of best fit to these data.

(c) Use the function found in part (b) to determine the optimal amount of fertilizer to apply.

(d) Use the function found in part (b) to predict crop yield when the optimal amount of fertilizer is applied.

(e) Draw the quadratic function of best fit on the scatter diagram.

TABLE 8

Plot	Fertilizer, x (Pounds/100 ft^2)	Yield (Bushels)
1	0	4
2	0	6
3	5	10
4	5	7
5	10	12
6	10	10
7	15	15
8	15	17
9	20	18
10	20	21
11	25	20
12	25	21
13	30	21
14	30	22
15	35	21
16	35	20
17	40	19
18	40	19

Solution

(a) Figure 41 shows the scatter diagram, from which it appears that the data follow a quadratic relation, with $a < 0$.

(b) Upon executing the QUADratic REGression program, we obtain the results shown in Figure 42. The output that the utility provides shows us the equation $y = ax^2 + bx + c$. The quadratic function of best fit is $Y(x) = -0.0171x^2 + 1.0765x + 3.8939$, where x represents the amount of fertilizer used and Y represents crop yield.

(c) Based on the quadratic function of best fit, the optimal amount of fertilizer to apply is

$$x = -\frac{b}{2a} = -\frac{1.0765}{2(-0.0171)} \approx 31.5 \text{ pounds of fertilizer per 100 square feet}$$

(d) We evaluate the function $Y(x)$ for $x = 31.5$.

$$Y(31.5) = -0.0171(31.5)^2 + 1.0765(31.5) + 3.8939 \approx 20.8 \text{ bushels}$$

If we apply 31.5 pounds of fertilizer per 100 square feet, the crop yield will be 20.8 bushels according to the quadratic function of best fit.

(e) Figure 43 shows the graph of the quadratic function found in part (b) drawn on the scatter diagram.

Figure 41

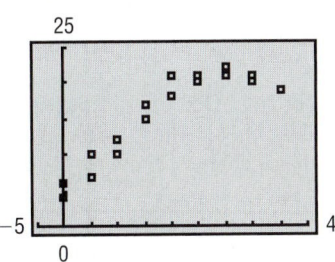

Figure 42

```
QuadReg
y=ax²+bx+c
a=-.0171212121
b=1.076515152
c=3.893939394
```

Figure 43

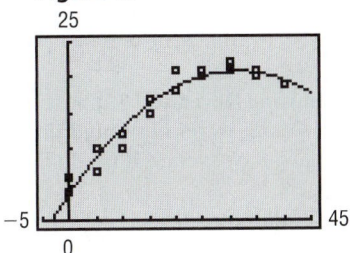

Look again at Figure 42. Notice that the output given by the graphing calculator does not include r, the correlation coefficient. Recall that the correlation coefficient is a measure of the strength of a *linear* relation that exists between two variables. The graphing calculator does not provide an indication of how well the function fits the data in terms of r since a quadratic function cannot be expressed as a linear function.

2.4 Concepts and Vocabulary

In Problems 1 and 2, fill in the blanks.

1. The function $f(x) = ax^2 + bx + c, a \neq 0$, has a minimum value if _____ . The minimum value occurs at $x =$ _____ .

2. The function $f(x) = ax^2 + bx + c, a \neq 0$ has a maximum value if _____ . The maximum value is _____ .

In Problems 3 and 4, answer True or False to each statement.

3. The minimum value of $f(x) = -x^2 + 6x + 2$ is $f(3)$.

4. The scatter diagram shown below indicates that the data follow a quadratic relation.

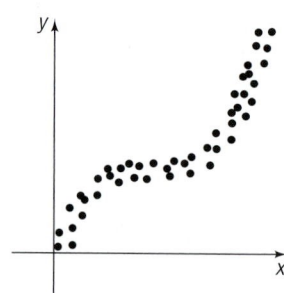

5. Refer to Example 2 on page 238. Notice that if the price charged for the calculators is $0 or $140 the revenue is $0. It is easy to explain why revenue would be $0 if the priced charged is $0, but how can revenue be $0 if the price charged is $140?

2.4 Exercises

In Problems 1–12, determine, without graphing, whether the given quadratic function has a maximum value or a minimum value and then find the value. Use the TABLE feature on a graphing utility to support your answer.

1. $f(x) = x^2 + 2$

2. $f(x) = x^2 - 4$

3. $f(x) = -x^2 - 4x$

4. $f(x) = -x^2 - 6x$

5. $f(x) = 2x^2 + 12x$

6. $f(x) = -2x^2 + 12x$

7. $f(x) = 2x^2 + 12x - 3$

8. $f(x) = 4x^2 - 8x + 3$

9. $f(x) = -x^2 + 10x - 4$

10. $f(x) = -2x^2 + 8x + 3$

11. $f(x) = -3x^2 + 12x + 1$

12. $f(x) = 4x^2 - 4x$

13. Maximizing Revenue   Suppose that the manufacturer of a gas clothes dryer has found that, when the unit price is p dollars, the revenue R (in dollars) is

$$R(p) = -4p^2 + 4000p$$

What unit price should be established for the dryer to maximize revenue? What is the maximum revenue?

14. Maximizing Revenue   The John Deere company has found that the revenue from sales of heavy-duty tractors is a function of the unit price p that it charges. If the revenue R is

$$R(p) = -\frac{1}{2}p^2 + 1900p$$

what unit price p should be charged to maximize revenue? What is the maximum revenue?

15. Minimizing Marginal Cost The marginal cost of a product can be thought of as the cost of producing one additional unit of output. For example, if the marginal cost of producing the 50th product is $6.20, then it cost $6.20 to increase production from 49 to 50 units of output. The marginal cost C (in dollars) to produce x portable 12-inch TVs is given by $C(x) = x^2 - 80x + 2000$. How many TVs should be produced to minimize the marginal cost? What is the minimum marginal cost?

16. Minimizing Marginal Cost (See problem 15) The marginal cost C (in dollars) of manufacturing x cell phones is given by $C(x) = 5x^2 - 200x + 4000$. How many cell phones should be manufactured to minimize the marginal cost? What is the minimum marginal cost?

17. Demand Equation The price p and the quantity x sold of a certain product obey the demand equation

$$p = -\frac{1}{6}x + 100 \qquad 0 \le x \le 600$$

(a) Express the revenue R as a function of x. (Remember, $R = xp$.)
(b) What is the revenue if 200 units are sold?
(c) What quantity x maximizes revenue? What is the maximum revenue?
(d) What price should the company charge to maximize revenue?

18. Demand Equation The price p and the quantity x sold of a certain product obey the demand equation

$$p = -\frac{1}{3}x + 100 \qquad 0 \le x \le 300$$

(a) Express the revenue R as a function of x.
(b) What is the revenue if 100 units are sold?
(c) What quantity x maximizes revenue? What is the maximum revenue?
(d) What price should the company charge to maximize revenue?

19. Demand Equation The price p and the quantity x sold of a certain product obey the demand equation

$$x = -5p + 100 \qquad 0 \le p \le 20$$

(a) Express the revenue R as a function of x.
(b) What is the revenue if 15 units are sold?
(c) What quantity x maximizes revenue? What is the maximum revenue?
(d) What price should the company charge to maximize revenue?

20. Demand Equation The price p and the quantity x sold of a certain product obey the demand equation

$$x = -20p + 500 \qquad 0 \le p \le 25$$

(a) Express the revenue R as a function of x.
(b) What is the revenue if 20 units are sold?
(c) What quantity x maximizes revenue? What is the maximum revenue?
(d) What price should the company charge to maximize revenue?

21. Enclosing a Rectangular Field David has available 400 yards of fencing and wishes to enclose a rectangular area.
(a) Express the area A of the rectangle as a function of the width x of the rectangle.
(b) For what value of x is the area largest?
(c) What is the maximum area?

22. Enclosing a Rectangular Field Beth has 3000 feet of fencing available to enclose a rectangular field.
(a) Express the area A of the rectangle as a function of x, where x is the length of the rectangle.
(b) For what value of x is the area largest?
(c) What is the maximum area?

23. Enclosing the Most Area with a Fence A farmer with 4000 meters of fencing wants to enclose a rectangular plot that borders on a river. If the farmer does not fence the side along the river, what is the largest area that can be enclosed? (See the figure.)

$4000 - 2x$

24. Enclosing the Most Area with a Fence A farmer with 10,000 meters of fencing wants to enclose a rectangular field and then divide it into two plots with a fence parallel to one of the sides (see the figure). What is the largest area that can be enclosed?

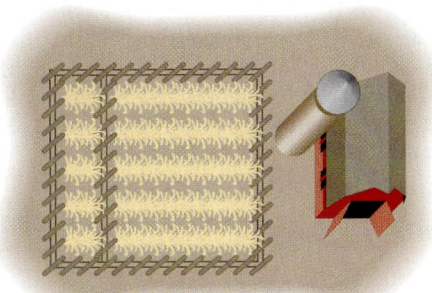

25. Analyzing the Motion of a Projectile A projectile is fired from a cliff 200 feet above the water at an inclination of 45° to the horizontal, with a muzzle velocity of 50 feet per second. The height h of the projectile above the water is given by

$$h(x) = \frac{-32x^2}{(50)^2} + x + 200$$

where x is the horizontal distance of the projectile from the base of the cliff.
(a) How far from the base of the cliff is the height of the projectile a maximum?

(b) Find the maximum height of the projectile.

(c) How far from the base of the cliff will the projectile strike the water?

(d) Using a graphing utility, graph the function h, $0 \le x \le 200$.

(e) Use a graphing utility to verify the solutions found in parts (b) and (c).

(f) When the height of the projectile is 100 feet above the water, how far is it from the cliff?

26. Analyzing the Motion of a Projectile   A projectile is fired at an inclination of 45° to the horizontal, with a muzzle velocity of 100 feet per second. The height h of the projectile is given by

$$h(x) = \frac{-32x^2}{(100)^2} + x$$

where x is the horizontal distance of the projectile from the firing point.

(a) How far from the firing point is the height of the projectile a maximum?

(b) Find the maximum height of the projectile.

(c) How far from the firing point will the projectile strike the ground?

(d) Using a graphing utility, graph the function h, $0 \le x \le 350$.

(e) Use a graphing utility to verify the results obtained in parts (b) and (c).

(f) When the height of the projectile is 50 feet above the ground, how far has it traveled horizontally?

27. Suspension Bridge   A suspension bridge with weight uniformly distributed along its length has twin towers that extend 75 meters above the road surface and are 400 meters apart. The cables are parabolic in shape and are suspended from the tops of the towers. The cables touch the road surface at the center of the bridge. Find the height of the cables at a point 100 meters from the center. (Assume that the road is level.)

28. Architecture   A parabolic arch has a span of 120 feet and a maximum height of 25 feet. Choose suitable rectangular coordinate axes and find the equation of the parabola. Then calculate the height of the arch at points 10 feet, 20 feet, and 40 feet from the center.

29. Constructing Rain Gutters   A rain gutter is to be made of aluminum sheets that are 12 inches wide by turning up the edges 90°. What depth will provide maximum cross-sectional area and hence allow the most water to flow?

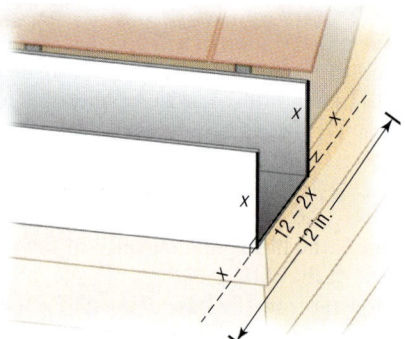

30. Norman Windows   A Norman window has the shape of a rectangle surmounted by a semicircle of diameter equal to the width of the rectangle (see the figure). If the perimeter of the window is 20 feet, what dimensions will admit the most light (maximize the area)?

[**Hint:** Circumference of a circle $= 2\pi r$; area of a circle $= \pi r^2$, where r is the radius of the circle.]

31. Constructing a Stadium   A track and field playing area is in the shape of a rectangle with semicircles at each end (see the figure). The inside perimeter of the track is to be 400 meters. What should the dimensions of the rectangle be so that the area of the rectangle is a maximum?

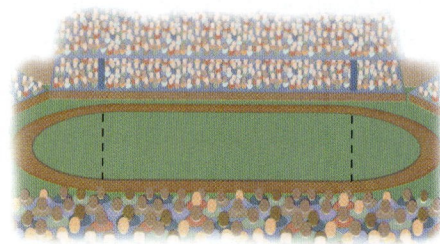

32. Architecture   A special window has the shape of a rectangle surmounted by an equilateral triangle (see the figure). If the perimeter of the window is 16 feet, what dimensions will admit the most light?

[**Hint:** Area of an equilateral triangle $= \dfrac{\sqrt{3}}{2}x^2$, where x is the length of a side of the triangle.]

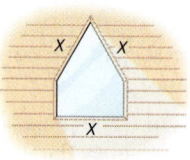

33. Life Cycle Hypothesis   An individual's income varies with his or her age. The table on page 247 shows the median income I of individuals of different age groups within the United States for 1998. For each class, let the class midpoint represent the independent variable x. For the class "65 years and older," we will assume that the class midpoint is 69.5.

Age	Class Midpoint, x	Median Income, I
15–24 years	19.5	$23,564
25–34 years	29.5	$40,069
35–44 years	39.5	$48,451
45–54 years	49.5	$54,148
55–64 years	59.5	$43,167
65 years and older	69.5	$21,729

Source: U.S. Census Bureau

(a) Draw a scatter diagram of the data. Comment on the type of relation that may exist between the two variables.
(b) Use a graphing utility to find the quadratic function of best fit to these data.
(c) Use the function found in part (b) to determine the age at which an individual can expect to earn the most income.
(d) Use the function found in part (b) to predict the peak income earned.
(e) With a graphing utility, graph the quadratic function of best fit on the scatter diagram.

34. **Life Cycle Hypothesis** An individual's income varies with his or her age. The following table shows the median income I of individuals of different age groups within the United States for 1997. For each group, the midpoint represents the independent variable, x. For the age group "65 years and older," we assume that the midpoint is 69.5.

Age	Midpoint, x	Median Income, I
15–24 years	19.5	$22,583
25–34 years	29.5	$38,174
35–44 years	39.5	$46,359
45–54 years	49.5	$51,875
55–64 years	59.5	$41,356
65 years and older	69.5	$20,761

Source: U.S. Census Bureau

(a) Draw a scatter diagram of the data. Comment on the type of relation that may exist between the two variables.
(b) Use a graphing utility to find the quadratic function of best fit to these data.
(c) Use the function found in part (b) to determine the age at which an individual can expect to earn the most income.

(d) Use the function found in part (b) to predict the peak income earned.
(e) With a graphing utility, graph the quadratic function of best fit on the scatter diagram.
(f) Compare the results of parts (c) and (d) in 1997 to those of 1998 in Problem 33.

35. **Miles per Gallon** An engineer collects the following data showing the speed s of a Ford Taurus and its average miles per gallon, M.

Speed, s	Miles per Gallon, M
30	18
35	20
40	23
40	25
45	25
50	28
55	30
60	29
65	26
65	25
70	25

(a) Draw a scatter diagram of the data. Comment on the type of relation that may exist between the two variables.
(b) Use a graphing utility to find the quadratic function of best fit to these data.
(c) Use the function found in part (b) to determine the speed that maximizes miles per gallon.
(d) Use the function found in part (b) to predict miles per gallon for a speed of 63 miles per hour.
(e) With a graphing utility, graph the quadratic function of best fit on the scatter diagram.

36. **Height of a Ball** A physicist throws a ball at an inclination of 45° to the horizontal. The data on page 248 represent the height h of the ball at the instant it has traveled x feet horizontally.
(a) Draw a scatter diagram of the data. Comment on the type of relation that may exist between the two variables.
(b) Use a graphing utility to find the function of best fit to these data.
(c) Use the function found in part (b) to determine how far the ball will travel before it reaches its maximum height.
(d) Use the function found in part (b) to find the maximum height of the ball.
(e) With a graphing utility, graph the quadratic function of best fit on the scatter diagram.

Distance, x	Height, h
20	25
40	40
60	55
80	65
100	71
120	77
140	77
160	75
180	71
200	64

37. Chemical Reactions A self-catalytic chemical reaction results in the formation of a compound that causes the formation ratio to increase. If the reaction rate V is given by

$$V(x) = kx(a - x) \qquad 0 \le x \le a$$

where k is a positive constant, a is the initial amount of the compound, and x is the variable amount of the compound, for what value of x is the reaction rate a maximum?

38. Calculus: Simpson's Rule The figure shows the graph of $y = ax^2 + bx + c$. Suppose that the points $(-h, y_0)$, $(0, y_1)$, and (h, y_2) are on the graph. It can be shown that

the area enclosed by the parabola, the x-axis, and the lines $x = -h$ and $x = h$ is

$$\text{Area} = \frac{h}{3}(2ah^2 + 6c)$$

Show that this area may also be given by

$$\text{Area} = \frac{h}{3}(y_0 + 4y_1 + y_2)$$

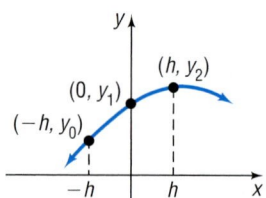

39. Use the result obtained in Problem 38 to find the area enclosed by $f(x) = -5x^2 + 8$, the x-axis, and the lines $x = -1$ and $x = 1$.

40. Use the result obtained in Problem 38 to find the area enclosed by $f(x) = 2x^2 + 8$, the x-axis, and the lines $x = -2$ and $x = 2$.

41. Use the result obtained in Problem 38 to find the area enclosed by $f(x) = x^2 + 3x + 5$, the x-axis, and the lines $x = -4$ and $x = 4$.

42. Use the result obtained in Problem 38 to find the area enclosed by $f(x) = -x^2 + x + 4$, the x-axis, and the lines $x = -1$ and $x = 1$.

Chapter Review

Things to Know

Function (p. 197)

A relation between two sets of real numbers so that each number x in the first set, the domain, has corresponding to it exactly one number y in the second set. The range is the set of y values of the function for the x values in the domain.

x is the independent variable; y is the dependent variable.

A function f may be defined implicitly by an equation involving x and y or explicitly by writing $y = f(x)$.

A function can also be characterized as a set of ordered pairs (x, y) or $(x, f(x))$ in which no two ordered pairs have the same first element, but different second elements.

Function notation (p. 199)

$y = f(x)$

f is a symbol for the function.

x is the argument, or independent variable.

y is the dependent variable.

$f(x)$ is the value of the function at x, or the image of x.

Domain (p. 203)	If unspecified, the domain of a function f is the largest set of real numbers for which $f(x)$ is a real number.
Vertical-line test (p. 205)	A set of points in the plane is the graph of a function if and only if every vertical line intersects the graph in at most one point.
Linear function (p. 216) $f(x) = mx + b$	Graph is a line with slope m and y-intercept b.
Quadratic function (p. 226) $f(x) = ax^2 + bx + c$	Graph is a parabola that opens up if $a > 0$, and opens down if $a < 0$.

Vertex: $\left(-\dfrac{b}{2a}, f\left(-\dfrac{b}{2a} \right) \right)$

Axis of symmetry: $x = -\dfrac{b}{2a}$

y-intercept: $f(0)$

x-intercept(s): If any, found by solving the equation $ax^2 + bx + c = 0$.

Objectives

Section		You should be able to . . .	Review Exercises
2.1	1	Determine whether a relation represents a function (p. 196)	41(a)
	2	Find the value of a function (p. 201)	7–12
	3	Find the domain of a function (p. 203)	13–20
	4	Identify the graph of a function (p. 205)	5
	5	Obtain information from or about the graph of a function (p. 205)	6
2.2	1	Graph linear functions (p. 215)	21–24
	2	Draw and interpret scatter diagrams (p. 217)	41(b), 49(a)
	3	Distinguish between linear and nonlinear relations (p. 218)	41(b), 49(a)
	4	Use a graphing utility to find the line of best fit (p. 220)	41(c)
	5	Construct a linear model using direct variation (p. 221)	42, 43
2.3	1	Determine whether the graph of a quadratic function opens up or down (p. 227)	25–34
	2	Identify the vertex and axis of symmetry of a quadratic function (p. 228)	25–34
	3	Graph quadratic functions (p. 230)	25–34
2.4	1	Find the maximum or minimum value of a quadratic function (p. 237)	35–40
	2	Use the maximum or minimum value of a quadratic function to solve applied problems (p. 238)	44, 46–48, 49(c) and (d)
	3	Use a graphing utility to find the quadratic function of best fit (p. 242)	49(b)

Review Exercises

Blue problem numbers indicate the authors' suggestions for use in a Practice Test.

1. Given that f is a linear function, $f(4) = -5$, and $f(0) = 3$, write the equation that defines f.

2. Given that g is a linear function with slope $= -4$ and $g(-2) = 2$, write the equation that defines g.

3. A function f is defined by

$$f(x) = \frac{Ax + 5}{6x - 2}$$

If $f(1) = 4$, find A.

4. A function g is defined by

$$g(x) = \frac{A}{x} + \frac{8}{x^2}$$

If $g(-1) = 0$, find A.

5. Tell which of the following graphs are graphs of functions.

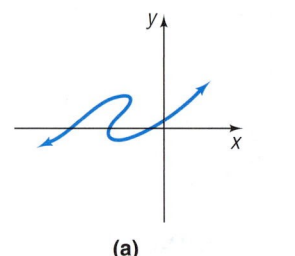

(a)

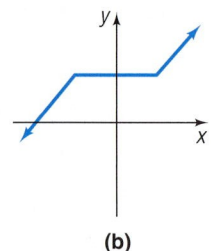

(b)

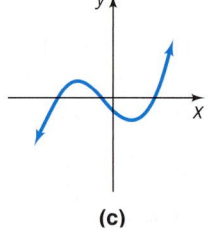

(c)

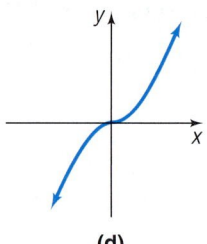

(d)

6. Use the graph of the function f shown to find
   (a) The domain and range of f
   (b) $f(-1)$
   (c) The intercepts of f
   (d) For what value of x does $f(x) = -3$?

In Problems 7–12, find the following for each function:

(a) $f(-x)$ *(b)* $-f(x)$ *(c)* $f(x + 2)$ *(d)* $f(x - 2)$

7. $f(x) = \dfrac{3x}{x^2 - 4}$

8. $f(x) = \dfrac{x^2}{x + 2}$

9. $f(x) = \sqrt{x^2 - 4}$

10. $f(x) = |x^2 - 4|$

11. $f(x) = \dfrac{x^2 - 4}{x^2}$

12. $f(x) = \dfrac{x^3}{x^2 - 4}$

In Problems 13–20, find the domain of each function.

13. $f(x) = \dfrac{x}{x^2 - 9}$

14. $f(x) = \dfrac{3x^2}{x - 2}$

15. $f(x) = \sqrt{2 - x}$

16. $f(x) = \sqrt{x + 2}$

17. $h(x) = \dfrac{\sqrt{x}}{|x|}$

18. $g(x) = \dfrac{|x|}{x}$

19. $f(x) = \dfrac{x}{x^2 + 2x - 3}$

20. $F(x) = \dfrac{1}{x^2 - 3x - 4}$

In Problems 21–24, graph each linear function.

21. $f(x) = 2x - 5$

22. $g(x) = -4x + 7$

23. $h(x) = \dfrac{4}{5}x - 6$

24. $F(x) = -\dfrac{1}{3}x + 1$

In Problems 25–34, graph each quadratic function by hand by determining whether its graph opens up or down and by finding its vertex, axis of symmetry, y-intercept, and x-intercepts, if any. Verify your results using a graphing utility.

25. $f(x) = \dfrac{1}{4}x^2 - 16$

26. $f(x) = -\dfrac{1}{2}x^2 + 2$

27. $f(x) = -4x^2 + 4x$

28. $f(x) = 9x^2 - 6x + 3$

29. $f(x) = \dfrac{9}{2}x^2 + 3x + 1$

30. $f(x) = -x^2 + x + \dfrac{1}{2}$

31. $f(x) = 3x^2 + 4x - 1$

32. $f(x) = -2x^2 - x + 4$

33. $f(x) = x^2 - 4x + 6$

34. $f(x) = x^2 + 2x - 3$

In Problems 35–40, determine whether the given quadratic function has a maximum value or a minimum value, and then find the value.

35. $f(x) = 3x^2 - 6x + 4$

36. $f(x) = 2x^2 + 8x + 5$

37. $f(x) = -x^2 + 8x - 4$

38. $f(x) = -x^2 - 10x - 3$

39. $f(x) = -3x^2 + 12x + 4$

40. $f(x) = -2x^2 + 4$

41. High School versus College GPA   An administrator at Southern Illinois University wants to find a function that relates a student's college grade point average G to the high school grade point average x. She randomly selects eight students and obtains the following data:

High School GPA, x	College GPA, G
2.73	2.43
2.92	2.97
3.45	3.63
3.78	3.81
2.56	2.83
2.98	2.81
3.67	3.45
3.10	2.93

(a) Does the relation defined by the set of ordered pairs (x, G) represent a function?
(b) Draw a scatter diagram of the data.
(c) Using a graphing utility, find the line of best fit relating high school GPA and college GPA.
(d) Interpret the slope.
(e) Express the relationship found in part (c) using function notation.
(f) What is the domain of the function?
(g) Predict a student's college GPA if her high school GPA is 3.23.

42. Mortgage Payments   The monthly payment p on a mortgage varies directly with the amount borrowed B. If the monthly payment on a 30-year mortgage is $854.00 when $130,000 is borrowed, find a linear function that relates the monthly payment p to the amount borrowed B for a mortgage with the same terms. Then find the monthly payment p when the amount borrowed B is $165,000.

43. Revenue Function   At the corner Esso station, the revenue R varies directly with the number g of gallons of gasoline sold. If the revenue is $15.93 when the number of gallons sold is 13.5, find a linear function that relates revenue R to the number g of gallons of gasoline. Then find the revenue R when the number of gallons of gasoline sold is 11.2.

44. Landscaping   A landscape engineer has 200 feet of border to enclose a rectangular pond. What dimensions will result in the largest pond?

45. Geometry   Find the length and width of a rectangle whose perimeter is 20 feet and whose area is 16 square feet.

46. A rectangle has one vertex on the line $y = 10 - x$, $x > 0$, another at the origin, one on the positive x-axis, and one on the positive y-axis. Find the largest area A that can be enclosed by the rectangle.

47. Minimizing Marginal Cost   The marginal cost of a product can be thought of as the cost of producing one additional unit of output. For example, if the marginal cost of

producing the 50th product is $6.20, then it cost $6.20 to increase production from 49 to 50 units of output. Callaway Golf Company has determined that the marginal cost C of manufacturing x Big Bertha golf clubs may be expressed by the quadratic function

$$C(x) = 4.9x^2 - 617.4x + 19{,}600$$

(a) How many clubs should be manufactured to minimize the marginal cost?
(b) At this level of production, what is the marginal cost?

48. Parabolic Arch Bridge   A horizontal bridge is in the shape of a parabolic arch. Given the information shown in the figure, what is the height h of the arch 2 feet from shore?

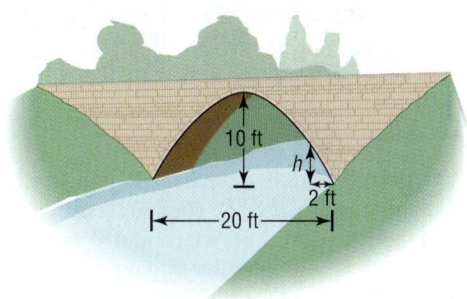

49. Advertising   A small manufacturing firm collected the following data on advertising expenditures A (in thousands of dollars) and total revenue R (in thousands of dollars).

Advertising	Total Revenue
20	$6101
22	$6222
25	$6350
25	$6378
27	$6453
28	$6423
29	$6360
31	$6231

(a) Draw a scatter diagram of the data. Comment on the type of relation that may exist between the two variables.
(b) Use a graphing utility to find the quadratic function of best fit to these data.
(c) Use the function found in part (b) to determine the optimal level of advertising for this firm.
(d) Use the function found in part (b) to find the revenue that the firm can expect if it uses the optimal level of advertising.
(e) With a graphing utility, graph the quadratic function of best fit on the scatter diagram.

Chapter Projects

1. **Analyzing a Stock**   The beta, β, of a stock represents the relative risk of a stock compared with a market basket of stocks, such as Standard and Poor's 500 Index of stocks. Beta is computed by finding the slope of the line of best fit between the rate of return of the stock and rate of return of the S&P 500. The rates of return are computed on a weekly basis.

(a) Find the weekly closing price of your favorite stock and the S&P 500 for 20 weeks. One good source is on the Internet at *http://finance.yahoo.com*.

(b) Compute the rate of return by computing the weekly percentage change in the closing price of your stock and the weekly percentage change in the S&P 500 using the following formula:

$$\text{Weekly \% change} = \frac{P_2 - P_1}{P_1}$$

where P_1 is last week's price and P_2 is this week's price.

(c) Using a graphing utility, find the line of best fit treating the weekly rate of return of the S&P 500 as the independent variable and the weekly rate of return of your stock as dependent variable.

(d) What is the beta of your stock?

(e) Compare your result with that of the Value Line Investment Survey found in your library. What might account for any differences?

2. (a) Consider the linear equation: $y = -2x + 5$. Evaluate the equation when $x = 1, 2, 3, 4, 5$.

(b) For the values in (a), compute $\frac{\Delta y}{\Delta x} = \frac{y_2 - y_1}{x_2 - x_1}$. There will be four values. What do you notice about the values of $\frac{\Delta y}{\Delta x}$? The expression $\frac{\Delta y}{\Delta x}$ is called a first

difference. Recall, $\frac{\Delta y}{\Delta x}$ is also the slope of the line containing the points (x_1, y_1) and (x_2, y_2).

(c) Using the data given in Section 2.2, Problem 28, draw a scatter diagram letting annual income, I, represent the independent variable and loan amount, L, represent the dependent variable.

(d) Compute $\frac{\Delta L}{\Delta I}$ for the data given in the table. What do you notice? Do you think the data is linearly related? Why?

(e) Using a graphing utility, graph the equation $y = 4x^2 - 2x + 3$. Evaluate the equation for $x = -2, -1, 0, 1, 2, 3, 4$. Compute $\frac{\Delta y}{\Delta x}$ (first difference) for each consecutive ordered pair. What happens to $\frac{\Delta y}{\Delta x}$ as x increases? Why is this result reasonable?

(f) Notice that $\Delta x = 1$ each time you compute $\frac{\Delta y}{\Delta x}$. Now compute $\frac{\Delta y}{\Delta x}$ for the $\frac{\Delta y}{\Delta x}$ values you determined in (e). This is called a second difference, denoted $\frac{\Delta^2 y}{\Delta x^2}$, since it is the difference of the first differences. What do you notice about the second differences?

(g) Write a paragraph explaining how first and second differences can be used to identify data that is related through a linear or quadratic function.

3. **CBL Experiment**   Locate the motion detector on a Calculator Based Laboratory (CBL) or a Calculator Based Ranger (CBR) above a bouncing ball.

(a) Plot the data collected in a scatter diagram with time as the independent variable.

(b) Find the quadratic function of best fit for the second bounce.

(c) Find the quadratic function of best fit for the third bounce.

(d) Find the quadratic function of best fit for the fourth bounce.

(e) Compute the maximum height for the second bounce.

(f) Compute the maximum height for the third bounce.

(g) Compute the maximum height for the fourth bounce.

(h) Compute the ratio of the maximum height of the third bounce to the maximum height of the second bounce.

(i) Compute the ratio of the maximum height of the fourth bounce to the maximum height of the third bounce.

(j) Compare the results from parts (h) and (i). What do you conclude?

Cumulative Review

1. Find the distance between the points $P = (-1, 3)$ and $Q = (4, -2)$.

2. Which of the following points are on the graph of $y = x^3 - 3x + 1$?
   (a) $(-2, -1)$ (b) $(2, 3)$ (c) $(3, 1)$

3. Determine the intercepts of $y = 3x^2 + 14x - 5$.

4. Use a graphing utility to find the solution(s) of the equation $x^4 - 3x^3 + 4x - 1 = 0$.

   Note: All the roots are between $x = -10$ and $x = 10$.

5. Solve the inequality $5x + 3 \geq 0$ and graph the solution set.

6. Solve the inequality $|3x - 4| < 5$ and graph the solution set.

7. Find the equation of the line containing the points $(-1, 4)$ and $(2, -2)$. Express your answer in slope–intercept form and graph the line.

8. Find the equation of the line perpendicular to the line $y = 2x + 1$ and containing the point $(3, 5)$. Express your answer in slope–intercept form and graph the line.

9. Graph the equation $x^2 + y^2 - 4x + 8y - 5 = 0$.

10. Determine whether the following relation represents a function: $\{(-3, 8), (1, 3), (2, 5), (3, 8)\}$.

11. For the function f defined by $f(x) = x^2 - 4x + 1$, evaluate:
    (a) $f(2)$
    (b) $f(x) + f(2)$
    (c) $f(-x)$
    (d) $-f(x)$
    (e) $f(x + 2)$
    (f) $\dfrac{f(x + h) - f(x)}{h}, h \neq 0$

12. Find the domain of $h(z) = \dfrac{3z - 1}{z^2 - 6z - 7}$.

13. Determine whether the following graph is the graph of a function.

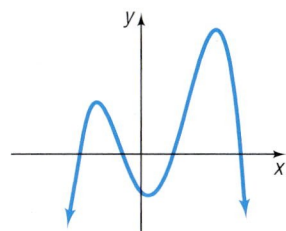

14. Consider the function $f(x) = \dfrac{x}{x + 4}$.

    (a) Is the point $\left(1, \dfrac{1}{4}\right)$ on the graph of f?
    (b) If $x = -2$, what is $f(x)$? What point is on the graph of f?
    (c) If $f(x) = 2$, what is x? What point is on the graph of f?

15. Graph the quadratic function $f(x) = 3x^2 - 5x - 1$ by determining whether its graph opens up or down and by finding its vertex, axis of symmetry, y-intercept, and x-intercepts, if any. Verify your results using a graphing utility.

16. Determine whether the quadratic function $f(x) = 3x^2 - 12x + 1$ has a maximum or minimum value. Then find the maximum or minimum value.

17. **Demand Equation**   The price p and the quantity sold x of a certain product obey the demand equation
    $$x = -4p + 100, \qquad 0 \leq p \leq 25$$
    (a) Express the revenue R as a function of x.
    (b) What is the revenue if 20 units are sold?
    (c) What quantity x maximizes revenue? What is the maximum revenue?
    (d) What price should the company charge to maximize revenue?

FUNCTIONS AND THEIR GRAPHS

Wednesday February 10, 1999 *The Oregonian* "Ship awaits salvage effort"

COOS BAY—Cleanup crews combed the oil-scarred south coast Tuesday as authorities raced to finish plans to refloat a 639-foot cargo ship mired for six days 150 yards off one of Oregon's most biologically rich beaches.

All day Tuesday, streaks of oil oozed from the cracked hull of the bulk cargo carrier *New Carissa* and spread over six miles of beach. Despite the breached hull, authorities believed they had a better chance of pulling the stricken ship out of beach sands than pumping nearly 400,000 gallons of oil off the ship in the winter surf.

SEE CHAPTER PROJECT 1.

OUTLINE

 For additional study help, go to

www.prenhall.com/sullivanegu3e

Materials include:

- Graphing Calculator Help
- Chapter Quiz
- Chapter Test
- PowerPoint Downloads
- Chapter Projects
- Student Tips

A Look Back, A Look Forward

In Chapter 1, we introduced graphing equations by plotting points and finding intercepts. We said that a graph is complete if the illustration presents enough of the graph so that any viewer of the illustration will see the rest of the graph as an obvious continuation of what is actually there. In Chapter 2, we introduced the concept of a function and its graph. In this chapter, we discuss additional techniques for graphing functions, including tests for symmetry and transformations. We will learn that another requirement to obtaining a complete graph is the inclusion of the high and low points on the graph.

PREPARING FOR THIS SECTION

Before getting started, review the following:

✓ Introduction to Graphing Equations (Section 1.2, pp. 100–107)

3.1 SYMMETRY; GRAPHING KEY EQUATIONS

OBJECTIVES   1   Test an Equation for Symmetry with Respect to (a) the *x*-Axis, (b) the *y*-Axis, and (c) the Origin

2   Know How to Graph Key Equations

Symmetry

1   In Chapter 1, we saw the role that intercepts play in obtaining key points on the graph of an equation. Another helpful tool for graphing equations by hand involves *symmetry*, particularly symmetry with respect to the *x*-axis, the *y*-axis, and the origin.

> A graph is said to be **symmetric with respect to the *x*-axis** if, for every point (x, y) on the graph, the point $(x, -y)$ is also on the graph.
>
> A graph is said to be **symmetric with respect to the *y*-axis** if, for every point (x, y) on the graph, the point $(-x, y)$ is also on the graph.
>
> A graph is said to be **symmetric with respect to the origin** if, for every point (x, y) on the graph, the point $(-x, -y)$ is also on the graph.

Figure 1 illustrates the definition. Notice that, when a graph is symmetric with respect to the *x*-axis, the part of the graph above the *x*-axis is a reflection or mirror image of the part below it, and vice versa. And when a graph

Figure 1

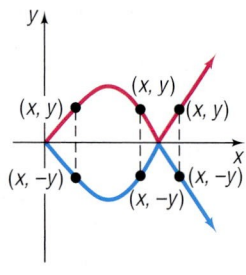

Symmetry with respect to the *x*-axis

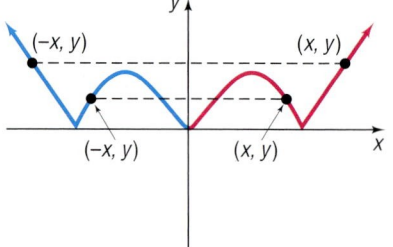

Symmetry with respect to the *y*-axis

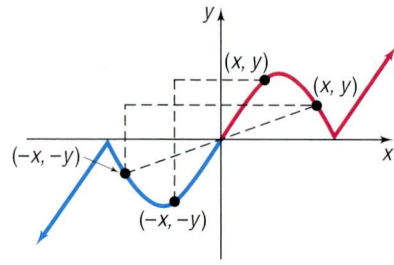

Symmetry with respect to the origin

is symmetric with respect to the y-axis, the part of the graph to the right of the y-axis is a reflection of the part to the left of it, and vice versa. Symmetry with respect to the origin may be viewed in two ways:

1. As a reflection about the y-axis, followed by a reflection about the x-axis

2. As a projection along a line through the origin so that the distances from the origin are equal

EXAMPLE 1 Symmetric Points

(a) If a graph is symmetric with respect to the x-axis and the point $(4, 2)$ is on the graph, then the point $(4, -2)$ is also on the graph.

(b) If a graph is symmetric with respect to the y-axis and the point $(4, 2)$ is on the graph, then the point $(-4, 2)$ is also on the graph.

(c) If a graph is symmetric with respect to the origin and the point $(4, 2)$ is on the graph, then the point $(-4, -2)$ is also on the graph. ■

 **NOW WORK PROBLEM 1.**

When the graph of an equation is symmetric with respect to a coordinate axis or the origin, the number of points that you need to plot in order to see the pattern is reduced. For example, if the graph of an equation is symmetric with respect to the y-axis, then, once points to the right of the y-axis are plotted, an equal number of points on the graph can be obtained by reflecting them about the y-axis. Because of this, before we graph an equation, we first want to determine whether it has any symmetry. The following tests are used for this purpose.

Tests for Symmetry

To test the graph of an equation for symmetry with respect to the

x-Axis　Replace y by $-y$ in the equation. If an equivalent equation results, the graph of the equation is symmetric with respect to the x-axis.

y-Axis　Replace x by $-x$ in the equation. If an equivalent equation results, the graph of the equation is symmetric with respect to the y-axis.

Origin　Replace x by $-x$ and y by $-y$ in the equation. If an equivalent equation results, the graph of the equation is symmetric with respect to the origin.

EXAMPLE 2 Testing an Equation for Symmetry

Test $y = \dfrac{4x^2}{x^2 + 1}$ for symmetry.

Solution　*x-Axis:*　To test for symmetry with respect to the x-axis, replace y by $-y$. Since the result, $-y = \dfrac{4x^2}{x^2 + 1}$, is not equivalent to $y = \dfrac{4x^2}{x^2 + 1}$, we conclude that the graph of the equation is not symmetric with respect to the x-axis.

y-Axis: To test for symmetry with respect to the *y*-axis, replace *x* by −*x*.
Since the result, $y = \dfrac{4(-x)^2}{(-x)^2 + 1} = \dfrac{4x^2}{x^2 + 1}$, is equivalent to $y = \dfrac{4x^2}{x^2 + 1}$, we conclude that the graph of the equation is symmetric with respect to the *y*-axis.

Origin: To test for symmetry with respect to the origin, replace *x* by −*x* and *y* by −*y*.

$$-y = \frac{4(-x)^2}{(-x)^2 + 1} \qquad \text{\small Replace x by −x and y by −y.}$$

$$-y = \frac{4x^2}{x^2 + 1} \qquad \text{\small Simplify.}$$

$$y = -\frac{4x^2}{x^2 + 1} \qquad \text{\small Multiply both sides by −1.}$$

Since the result is not equivalent to the original equation, the graph of the equation $y = \dfrac{4x^2}{x^2 + 1}$ is not symmetric with respect to the origin. ■

Figure 2

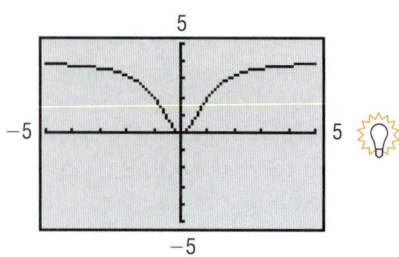

SEEING THE CONCEPT   Figure 2 shows the graph of $y = \dfrac{4x^2}{x^2 + 1}$ using a graphing utility. Do you see the symmetry with respect to the *y*-axis?

▸ **NOW WORK PROBLEM 27.**

Graphs of Key Equations

② The next three examples use intercepts, symmetry, and point-plotting to obtain the graphs of key equations. The first of these is $y = x^3$.

EXAMPLE 3   **Graphing the Equation $y = x^3$ by Finding Intercepts and Checking for Symmetry**

Graph the equation $y = x^3$ by hand by plotting points. Find any intercepts and check for symmetry first.

Solution   First, we seek the intercepts. When $x = 0$, then $y = 0$; and when $y = 0$, then $x = 0$. The origin $(0, 0)$ is the only intercept. Now we test for symmetry.

x-Axis: Replace *y* by −*y*. Since the result, $-y = x^3$, is not equivalent to $y = x^3$, the graph is not symmetric with respect to the *x*-axis.

y-Axis: Replace *x* by −*x*. Since the result, $y = (-x)^3 = -x^3$, is not equivalent to $y = x^3$, the graph is not symmetric with respect to the *y*-axis.

Origin: Replace *x* by −*x* and *y* by −*y*. Since the result, $-y = (-x)^3 = -x^3$, is equivalent to $y = x^3$ (multiply both sides by −1), the graph is symmetric with respect to the origin.

To graph by hand, we use the equation to obtain several points on the graph. Because of the symmetry, we only need to locate points on the graph

Figure 3
$y = x^3$

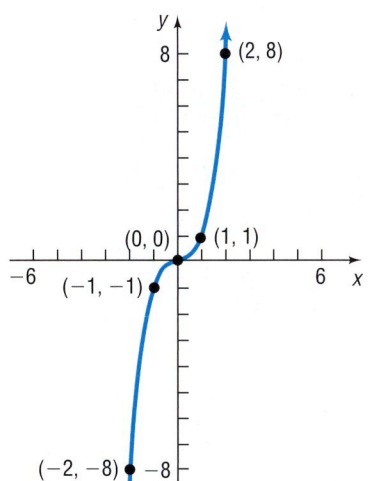

for which $x \geq 0$. See Table 1. Points on the graph could also be obtained using the TABLE feature on a graphing utility. See Table 2. Do you see the symmetry with respect to the origin from the table? Figure 3 shows the graph.

TABLE 1

x	$y = x^3$	(x, y)
0	0	$(0, 0)$
1	1	$(1, 1)$
2	8	$(2, 8)$
3	27	$(3, 27)$

TABLE 2

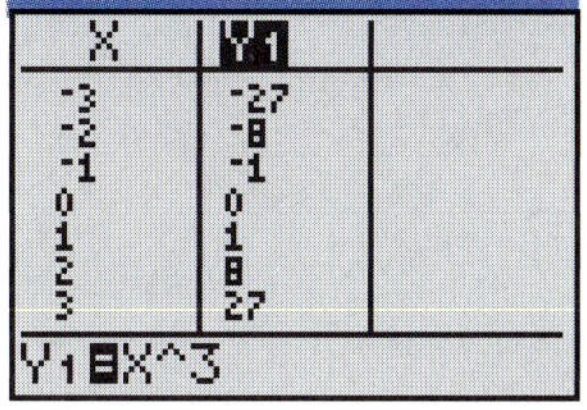

X	Y1
-3	-27
-2	-8
-1	-1
0	0
1	1
2	8
3	27

Y1 ∎X^3

EXAMPLE 4   **Graphing the Equation $x = y^2$**

Graph the equation $x = y^2$. Find any intercepts and check for symmetry first.

Solution   The lone intercept is $(0, 0)$. The graph is symmetric with respect to the x-axis since $x = (-y)^2$ is equivalent to $x = y^2$. The graph is not symmetric with respect to the y-axis or the origin.

To graph $x = y^2$ by hand, we use the equation to obtain several points on the graph. Because the equation is solved for x, it is easier to assign values to y and use the equation to determine the corresponding values of x. See Table 3. Because of the symmetry, we can restrict ourselves to points whose y-coordinates are positive. We then use the symmetry to find additional points on the graph. For example, since $(1, 1)$ is on the graph, so is $(1, -1)$. Since $(4, 2)$ is on the graph, so is $(4, -2)$, and so on. We plot these points and connect them with a smooth curve to obtain Figure 4.

TABLE 3

y	$x = y^2$	(x, y)
0	0	$(0, 0)$
1	1	$(1, 1)$
2	4	$(4, 2)$
3	9	$(9, 3)$

Figure 4
$x = y^2$

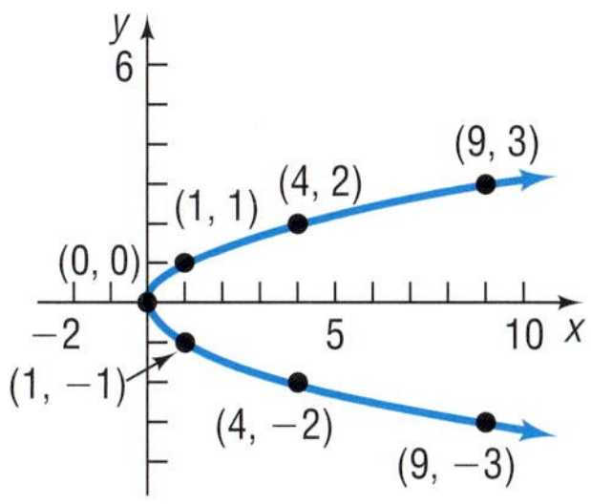

To graph the equation $x = y^2$ using a graphing utility, we must write the equation in the form $y = \{\text{expression in } x\}$. We proceed to solve for y.

$$x = y^2$$
$$y^2 = x$$
$$y = \pm\sqrt{x} \qquad \text{Square Root Method}$$

To graph $x = y^2$, we need to graph both $Y_1 = \sqrt{x}$ and $Y_2 = -\sqrt{x}$ on the same screen. Figure 5 shows the result. Table 4 shows various values of y for a given value of x when $Y_1 = \sqrt{x}$ and $Y_2 = -\sqrt{x}$. Notice that when $x < 0$ we get an error. Can you explain why?

Figure 5

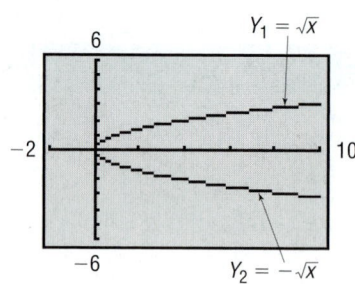

Look again at either Figure 4 or 5. Notice that the graph of the equation fails the vertical-line test, so the equation $x = y^2$ is not a function. However, if we restrict y so that $y \geq 0$, the equation $x = y^2$, $y \geq 0$, may be written as $y = \sqrt{x}$. The portion of the graph of $x = y^2$ in quadrant I is the graph of $y = \sqrt{x}$. See Figure 6. Notice that this graph passes the vertical-line test, so $y = \sqrt{x}$ is a function.

Figure 6
$y = \sqrt{x}$

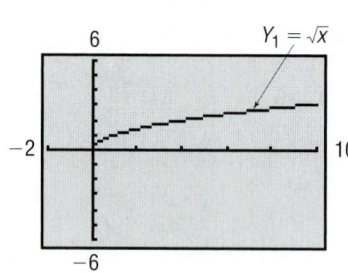

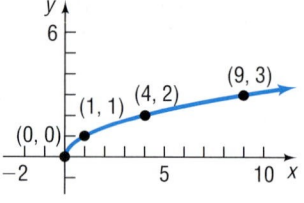

EXAMPLE 5

Graphing the Equation $y = \dfrac{1}{x}$

Consider the equation: $y = \dfrac{1}{x}$

(a) Graph this equation using a graphing utility. Set the viewing window as

$$X\min = -3 \qquad Y\min = -4$$
$$X\max = 3 \qquad Y\max = 4$$
$$X\text{scl} = 1 \qquad Y\text{scl} = 1$$

(b) Use algebra to find any intercepts and test for symmetry.

(c) Draw the graph by hand.

Figure 7

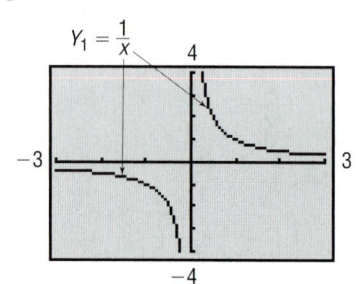

Solution   (a) Figure 7 illustrates the graph. We infer from the graph that there are no intercepts; we may also infer that symmetry with respect to the origin is a possibility. The TRACE feature on a graphing utility can provide further evidence of symmetry with respect to the origin. Using TRACE, we observe that for any ordered pair (x, y) the ordered pair $(-x, -y)$ is also a point on the graph. For example, the points $(0.95744681, 1.0444444)$ and $(-0.95744681, -1.0444444)$ both lie on the graph.

(b) We check for intercepts first. If we let $x = 0$, we obtain a 0 in the denominator, which is not defined. We conclude that there is no y-intercept. If we let $y = 0$, we get the equation $\dfrac{1}{x} = 0$, which has no solution. We conclude that there is no *x*-intercept. The graph of $y = \dfrac{1}{x}$ does not cross the coordinate axes.

Next we check for symmetry.

x-Axis: Replacing *y* by $-y$ yields $-y = \dfrac{1}{x}$, which is not equivalent to $y = \dfrac{1}{x}$.

y-Axis: Replacing *x* by $-x$ yields $y = \dfrac{1}{-x} = -\dfrac{1}{x}$, which is not equivalent to $y = \dfrac{1}{x}$.

Origin: Replacing *x* by $-x$ and *y* by $-y$ yields $-y = -\dfrac{1}{x}$, which is equivalent to $y = \dfrac{1}{x}$.

The graph is symmetric only with respect to the origin.

The inferences drawn in part (a) of the solution are now confirmed.

(c) We can use the equation to form Table 5 and obtain some points on the graph. Because of symmetry, we only find points (x, y) for which *x* is positive. From Table 5 we infer that, if *x* is a large and positive number, then $y = \dfrac{1}{x}$ is a positive number close to 0. We also infer that if *x* is a positive number close to 0 then $y = \dfrac{1}{x}$ is a large and positive number.

Armed with this information, we can graph the equation. Figure 8 illustrates some of these points and the graph of $y = \dfrac{1}{x}$.

TABLE 5

x	$y = \dfrac{1}{x}$	(x, y)
$\dfrac{1}{10}$	10	$\left(\dfrac{1}{10}, 10\right)$
$\dfrac{1}{3}$	3	$\left(\dfrac{1}{3}, 3\right)$
$\dfrac{1}{2}$	2	$\left(\dfrac{1}{2}, 2\right)$
1	1	$(1, 1)$
2	$\dfrac{1}{2}$	$\left(2, \dfrac{1}{2}\right)$
3	$\dfrac{1}{3}$	$\left(3, \dfrac{1}{3}\right)$
10	$\dfrac{1}{10}$	$\left(10, \dfrac{1}{10}\right)$

Figure 8

$y = \dfrac{1}{x}$

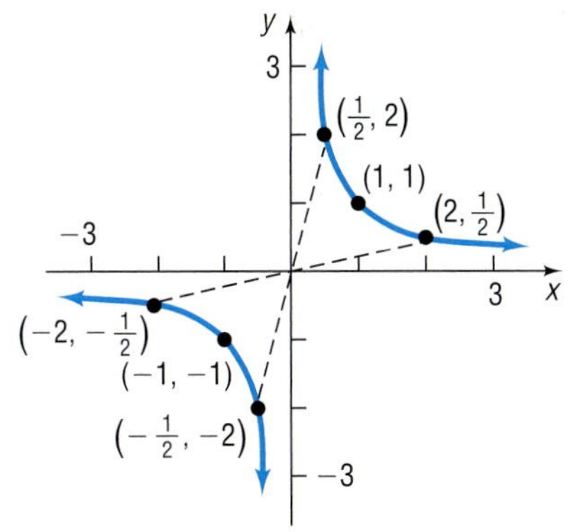

Observe how the absence of intercepts and the existence of symmetry with respect to the origin were utilized.

3.1 Concepts and Vocabulary

In Problems 1 and 2, fill in the blanks.

1. If for every point (x, y) on the graph of an equation, the point $(-x, y)$ is also on the graph, then the graph is symmetric with respect to the _____.

2. The graph of the equation $y = x^3 - 5x$ is symmetric with respect to the _____.

In Problems 3 and 4, answer True or False to each statement.

3. The graph of the equation $y = x^4 + x^2 + 1$ is symmetric with respect to the y-axis.

4. The graph of a circle whose center is at the origin will have symmetry with respect to the x-axis, the y-axis, and the origin.

5. State three types of symmetry that the graph of an equation could have.

6. Given that the point $(1, 2)$ is on the graph of an equation that is symmetric with respect to the origin, what other point is on the graph?

7. If the graph of an equation is symmetric with respect to the y-axis and 6 is an x-intercept of this graph, name another x-intercept.

8. An equation is being tested for symmetry with respect to the x-axis, the y-axis, and the origin. Explain why, if two of these symmetries are present, the remaining one must also be present.

3.1 Exercises

In Problems 1–10, plot each point. Then plot the point that is symmetric to it with respect to: (a) the x-axis; (b) the y-axis; (c) the origin.

1. $(3, 4)$ **2.** $(5, 3)$ **3.** $(-2, 1)$ **4.** $(4, -2)$ **5.** $(1, 1)$

6. $(-1, -1)$ **7.** $(-3, -4)$ **8.** $(4, 0)$ **9.** $(0, -3)$ **10.** $(-3, 0)$

In Problems 11–18, the graph of an equation is given. Indicate whether the graph is symmetric with respect to the x-axis, the y-axis, or the origin.

11.
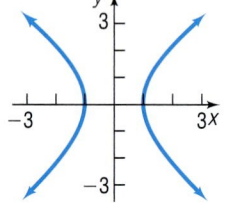

12.
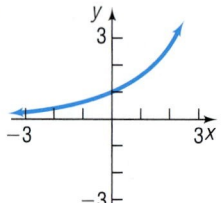

13.
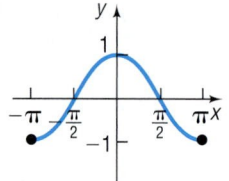

14.
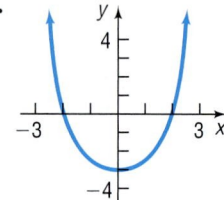

15.
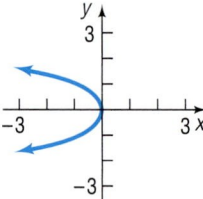

16.
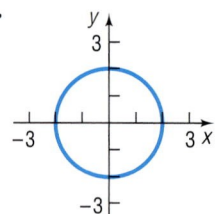

17.
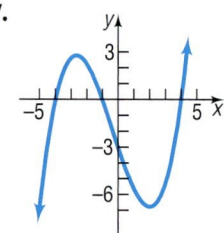

18.
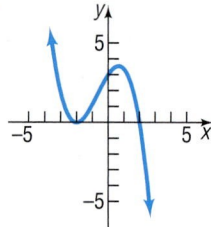

In Problems 19–22, draw a complete graph so that it has the type of symmetry indicated.

19. y-axis
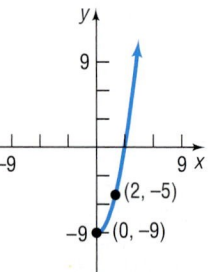

20. x-axis
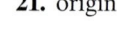

21. origin

22. y-axis
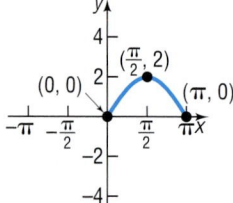

In Problems 23–34, *test each equation for symmetry with respect to the x-axis, the y-axis, and the origin. Graph each equation using a graphing utility to verify your conclusions.*

23. $x^2 = y + 5$

24. $y^2 = x + 3$

25. $y = 3x$

26. $y = -5x$

27. $x^2 + y - 9 = 0$

28. $y^2 - x - 4 = 0$

29. $y = x^3 - 27$

30. $y = x^4 - 1$

31. $y = x^2 - 3x - 4$

32. $y = x^2 + 4$

33. $y = \dfrac{x}{x^2 + 9}$

34. $y = \dfrac{x^2 - 4}{x}$

In Problems 35–38, *draw a quick sketch of the equation by hand. Be sure to label all intercepts.*

35. $y = x^3$

36. $x = y^2$

37. $y = \sqrt{x}$

38. $y = \dfrac{1}{x}$

PREPARING FOR THIS SECTION

Before getting started, review the following:

✓ Intervals (Section 1.6, pp. 150–151)

✓ Slope of a Line (Section 1.7, pp. 163–165)

✓ Point–Slope Form of a Line (Section 1.7, p. 168)

3.2 PROPERTIES OF FUNCTIONS

OBJECTIVES

1 Find the Average Rate of Change of a Function

2 Use a Graph to Determine Where a Function Is Increasing, Is Decreasing, or Is Constant

3 Use a Graph to Locate Local Maxima and Minima

4 Use a Graphing Utility to Approximate Local Maxima and Minima and to Determine Where a Function Is Increasing or Decreasing

5 Determine Even and Odd Functions from a Graph

6 Identify Even and Odd Functions from the Equation

Average Rate of Change

 1

In Section 1.7, we said that the slope of a line could be interpreted as the average rate of change. Often we are interested in the rate at which functions change. To find the average rate of change of a function between any two points on its graph, we calculate the slope of the line containing the two points.

EXAMPLE 1   **Finding the Average Rate of Change of a Function**

A strain of E-coli Beu 397-recA441 is placed into a Petri dish at 30° Celsius and allowed to grow. The data shown in Table 6 are collected. The population is measured in grams and the time in hours.

(a) Draw a scatter diagram of the data, treating time as the independent variable.

(b) Draw a line through the points $(0, 0.09)$ and $(2.5, 0.18)$ on the scatter diagram drawn in part (a).

(c) Find the average rate of change of the population from 0 to 2.5 hours.

(d) Draw a line through the points $(4.5, 0.35)$ and $(6, 0.50)$ on the scatter diagram drawn in part (a).

(e) Find the average rate of change of the population from 4.5 to 6 hours.

(f) What is happening to the average rate of change as time passes?

TABLE 6

Time (hours), x	Population (grams), y
0	0.09
2.5	0.18
3.5	0.26
4.5	0.35
6	0.50

Solution   (a) We plot the ordered pairs $(0, 0.09)$, $(2.5, 0.18)$, and so on, using rectangular coordinates. See Figure 9(a).

(b) Draw the line through $(0, 0.09)$ and $(2.5, 0.18)$. See Figure 9(b).

(c) The average rate of change is found by computing the slope of the line containing the points $(0, 0.09)$ and $(2.5, 0.18)$.

$$\text{Average rate of change} = \frac{0.18 - 0.09}{2.5 - 0} = \frac{0.09 \text{ gram}}{2.5 \text{ hours}} = 0.036 \text{ gram per hour}$$

On average, the population is increasing at the rate of 0.036 gram per hour from 0 to 2.5 hours.

(d) Draw the line through $(4.5, 0.35)$ and $(6, 0.50)$. See Figure 9(c).

Figure 9

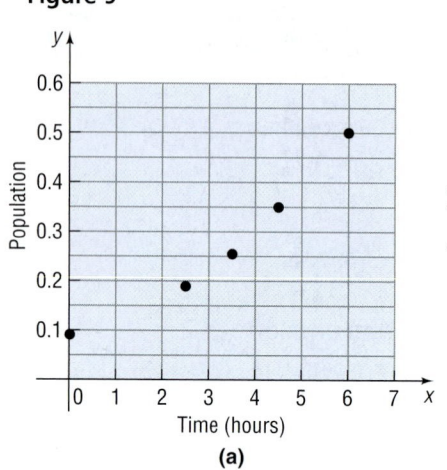
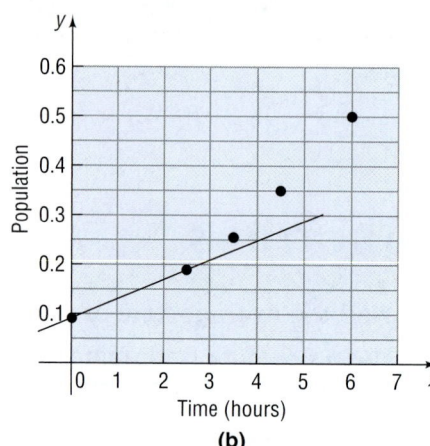
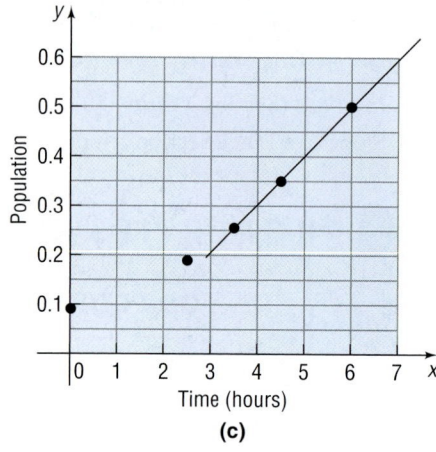

(a)   (b)   (c)

(e) The average rate of change is found by computing the slope of the line containing the points $(4.5, 0.35)$ and $(6, 0.50)$.

$$\text{Average rate of change} = \frac{0.50 - 0.35}{6 - 4.5} = \frac{0.15 \text{ gram}}{1.5 \text{ hours}} = 0.1 \text{ gram per hour}$$

On average, the population is increasing at the rate of 0.1 gram per hour from 4.5 to 6 hours.

 (f) Not only is the size of the population increasing over time, but the rate of increase is also increasing. We say that the population is increasing at an increasing rate.   ■

If we know the function P that relates the time t to the population P, then the average rate of change from 0 hours to 2.5 hours may be expressed as

$$\text{Average rate of change} = \frac{P(2.5) - P(0)}{2.5 - 0} = \frac{0.09}{2.5} = 0.036 \text{ gram per hour}$$

 Expressions like this occur frequently in calculus.

> If c is in the domain of a function $y = f(x)$, the **average rate of change of f** from c to x is defined as
>
> $$\text{Average rate of change} = \frac{\Delta y}{\Delta x} = \frac{f(x) - f(c)}{x - c}, \qquad x \neq c \quad (1)$$

This expression is also called the **difference quotient** of f at c.

The average rate of change of a function has an important geometric interpretation. Look at the graph of $y = f(x)$ in Figure 10. We have labeled two points on the graph: $(c, f(c))$ and $(x, f(x))$. The line containing these two points is called a **secant line**; its slope is

$$m_{\text{sec}} = \frac{f(x) - f(c)}{x - c}$$

Figure 10

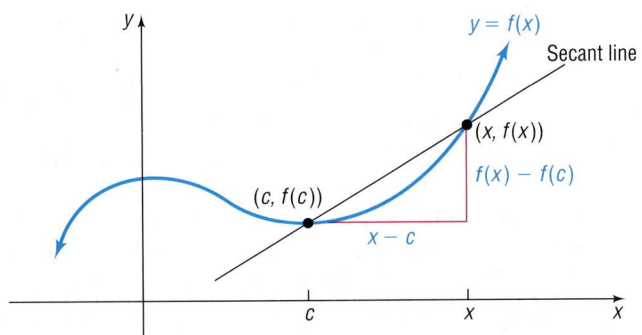

Theorem

Slope of the Secant Line

The average rate of change of a function equals the slope of the secant line containing two points on its graph.

EXAMPLE 2

Finding the Average Rate of Change of a Function

(a) Find the average rate of change of $f(x) = 2x^2 - 3x$ from 1 to x.

(b) Use this result to find the slope of the secant line containing $(1, f(1))$ and $(2, f(2))$.

(c) Find an equation of this secant line.

(d) Graph f and the secant line on the same viewing window.

Solution   (a) The average rate of change of f from 1 to x is

$$\frac{\Delta y}{\Delta x} = \frac{f(x) - f(1)}{x - 1} \qquad x \neq 1.$$

$$= \frac{(2x^2 - 3x) - (-1)}{x - 1} \qquad f(x) = 2x^2 - 3x; \quad f(1) = 2 \cdot 1^2 - 3 \cdot 1 = -1.$$

$$= \frac{2x^2 - 3x + 1}{x - 1} \qquad \text{Simplify.}$$

$$= \frac{(2x - 1)(x - 1)}{x - 1} \qquad \text{Factor the numerator.}$$

$$= 2x - 1 \qquad x \neq 1; \quad \text{Cancel } x - 1.$$

(b) The slope of the secant line containing $(1, f(1))$ and $(2, f(2))$ is the average rate of change of f from 1 to 2. Using $x = 2$ in the solution to part (a), we obtain $m_{\text{sec}} = 2(2) - 1 = 3$.

Figure 11

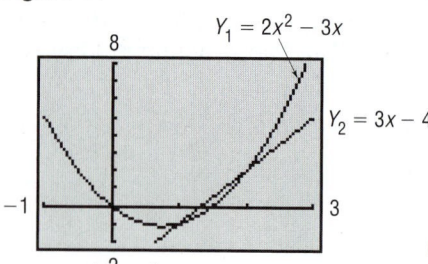

$Y_1 = 2x^2 - 3x$

$Y_2 = 3x - 4$

(c) Use the point–slope formula to find the equation of the secant line.

$$y - y_1 = m_{\text{sec}}(x - x_1) \quad \text{\textcolor{gray}{Point–slope form of the secant line.}}$$
$$y - (-1) = 3(x - 1) \quad \text{\textcolor{gray}{$x_1 = 1, y_1 = f(1) = -1$; \quad $m_{\text{sec}} = 3$.}}$$
$$y + 1 = 3x - 3 \quad \text{\textcolor{gray}{Simplify.}}$$
$$y = 3x - 4 \quad \text{\textcolor{gray}{Slope–intercept form of the secant line.}}$$

(d) See Figure 11 for the graph of f and the secant line.   ■

━━ **NOW WORK PROBLEM 29.**

Figure 12

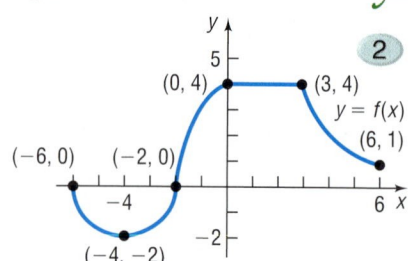

(−6, 0)   (−2, 0)

(0, 4)   (3, 4)

$y = f(x)$

(6, 1)

(−4, −2)

2   Increasing and Decreasing Functions

Consider the graph given in Figure 12. If you look from left to right along the graph of the function, you will notice that parts of the graph are rising, parts are falling, and parts are horizontal. In such cases, the function is described as *increasing*, *decreasing*, or *constant*, respectively.

EXAMPLE 3

Determining Where a Function Is Increasing, Decreasing, or Constant from Its Graph

Where is the function in Figure 12 increasing? Where is it decreasing? Where is it constant?

Solution

To answer the question of where a function is increasing, where it is decreasing, and where it is constant, we use strict inequalities involving the independent variable x, or we use open intervals* of x-coordinates. The graph in Figure 12 is rising (increasing) from the point $(-4, -2)$ to the point $(0, 4)$, so we conclude that it is increasing on the open interval $(-4, 0)$ (or for $-4 < x < 0$). The graph is falling (decreasing) from the point $(-6, 0)$ to the point $(-4, -2)$ and from the point $(3, 4)$ to the point $(6, 1)$. We conclude that the graph is decreasing on the open intervals $(-6, -4)$ and $(3, 6)$ (or for $-6 < x < -4$ and $3 < x < 6$). The graph is constant on the open interval $(0, 3)$ (or $0 < x < 3$).   ■

More precise definitions follow:

> A function f is **increasing** on an open interval I if, for any choice of x_1 and x_2 in I, with $x_1 < x_2$, we have $f(x_1) < f(x_2)$.

> A function f is **decreasing** on an open interval I if, for any choice of x_1 and x_2 in I, with $x_1 < x_2$, we have $f(x_1) > f(x_2)$.

> A function f is **constant** on an open interval I if, for all choices of x in I, the values $f(x)$ are equal.

*The open interval (a, b) consists of all real numbers x for which $a < x < b$.

Figure 13 illustrates the definitions. The graph of an increasing function goes up from left to right, the graph of a decreasing function goes down from left to right, and the graph of a constant function remains at a fixed height.

Figure 13

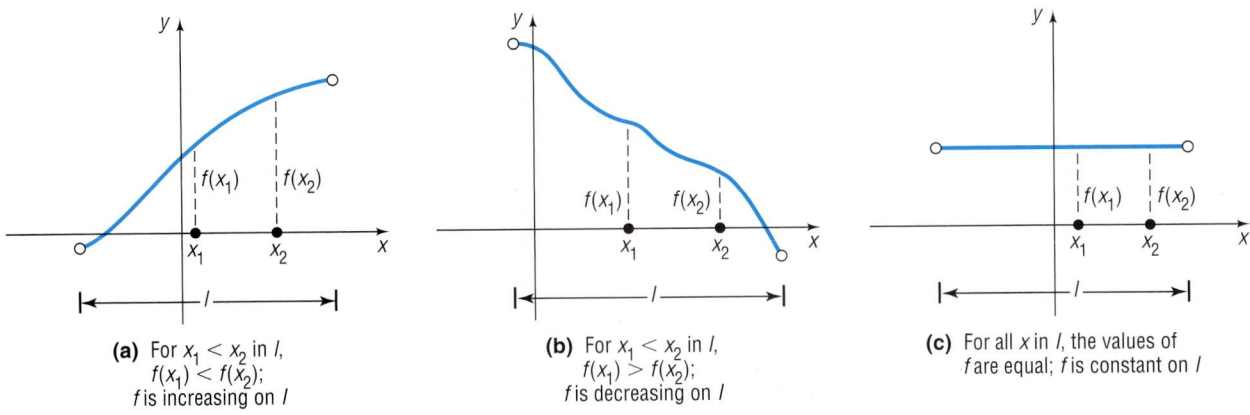

(a) For $x_1 < x_2$ in I,
$f(x_1) < f(x_2)$;
f is increasing on I

(b) For $x_1 < x_2$ in I,
$f(x_1) > f(x_2)$;
f is decreasing on I

(c) For all x in I, the values of f are equal; f is constant on I

 N O W   W O R K   P R O B L E M S   1 ,   3 ,   A N D   5 .

Local Maximum; Local Minimum

3 When the graph of a function is increasing to the left of $x = c$ and decreasing to the right of $x = c$, then at c the value of f is largest. This value is called a *local maximum* of f. See Figure 14(a).

When the graph of a function is decreasing to the left of $x = c$ and is increasing to the right of $x = c$, then at c the value of f is the smallest. This value is called a *local minimum* of f. See Figure 14(b).

Figure 14

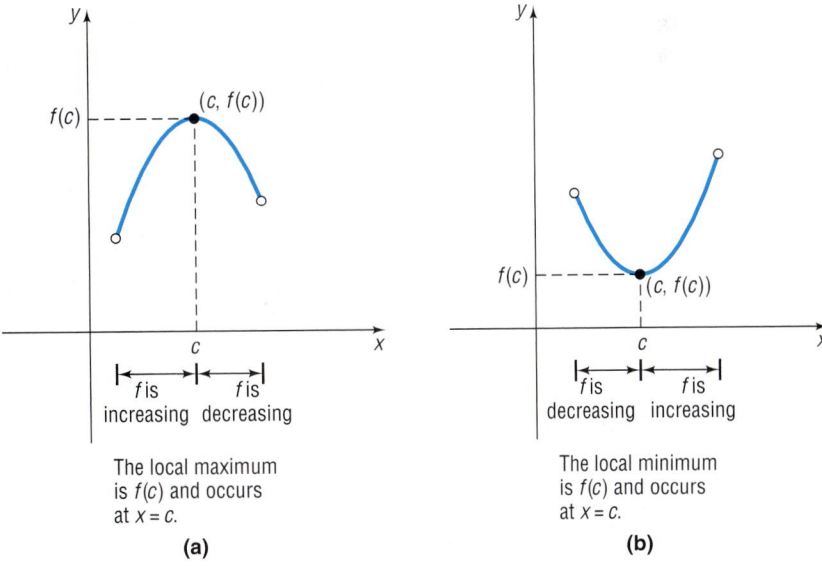

f is increasing f is decreasing

The local maximum is $f(c)$ and occurs at $x = c$.

(a)

f is decreasing f is increasing

The local minimum is $f(c)$ and occurs at $x = c$.

(b)

A function f has a **local maximum at c** if there is an open interval I containing c so that, for all $x \neq c$ in I, $f(x) < f(c)$. We call $f(c)$ a **local maximum of f.**

A function *f* has a **local minimum at *c*** if there is an open interval *I* containing *c* so that, for all $x \neq c$ in *I*, $f(x) > f(c)$. We call $f(c)$ a **local minimum of *f*.**

If *f* has a local maximum at *c* then the value of *f* at *c* is greater than the values of *f* near *c*. If *f* has a local minimum at *c*, then the value of *f* at *c* is less than the values of *f* near *c*. The word *local* is used to suggest that it is only near *c* that the value $f(c)$ is largest or smallest.

EXAMPLE 4   **Finding Local Maxima and Local Minima from the Graph of a Function and Determining Where the Function Is Increasing, Decreasing, or Constant**

Figure 15

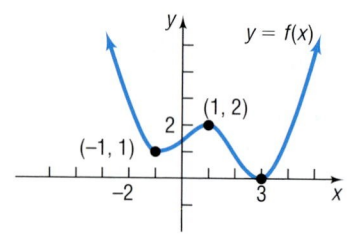

Figure 15 shows the graph of a function *f*.

(a) At what number(s), if any, does *f* have a local maximum?
(b) What are the local maxima?
(c) At what number(s), if any, does *f* have a local minimum?
(d) What are the local minima?
(e) List the intervals on which *f* is increasing. List the intervals on which *f* is decreasing.

Solution   The domain of *f* is the set of real numbers.

(a) *f* has a local maximum at 1, since for all *x* close to 1, $x \neq 1$, we have $f(x) < f(1)$.
(b) The local maximum is $f(1) = 2$.
(c) *f* has a local minimum at -1 and at 3.
(d) The local minima are $f(-1) = 1$ and $f(3) = 0$.
(e) The function whose graph is given in Figure 15 is increasing on the interval $(-1, 1)$. The function is also increasing for all values of *x* greater than 3. That is, the function is increasing on the intervals $(-1, 1)$ and $(3, \infty)$ (or for $-1 < x < 1$ and $x > 3$). The function is decreasing for all values of *x* less than -1. The function is also decreasing on the interval $(1, 3)$. That is, the function is decreasing on the intervals $(-\infty, -1)$ and $(1, 3)$ (or for $x < -1$ and $1 < x < 3$). ■

 **NOW WORK PROBLEMS 7 AND 9.**

 To locate the exact value at which a function *f* has a local maximum or a local minimum usually requires calculus. However, a graphing utility may be used to approximate these values by using the MAXIMUM and MINIMUM features.*

EXAMPLE 5   **Using a Graphing Utility to Approximate Local Maxima and Minima and to Determine Where a Function Is Increasing or Decreasing**

(a) Use a graphing utility to graph $f(x) = 6x^3 - 12x + 5$ for $-2 < x < 2$. Approximate where *f* has a local maximum and where *f* has a local minimum.
(b) Determine where *f* is increasing and where it is decreasing.

*Consult your owner's manual for the appropriate keystrokes.

Solution (a) Graphing utilities have a feature that finds the maximum or minimum point of a graph within a given interval. Graph the function f for $-2 < x < 2$. Using MAXIMUM, we find that the local maximum is 11.53 and it occurs at $x = -0.82$, rounded to two decimal places. See Figure 16(a). Using MINIMUM, we find that the local minimum is -1.53 and it occurs at $x = 0.82$, rounded to two decimal places. See Figure 16(b).

Figure 16

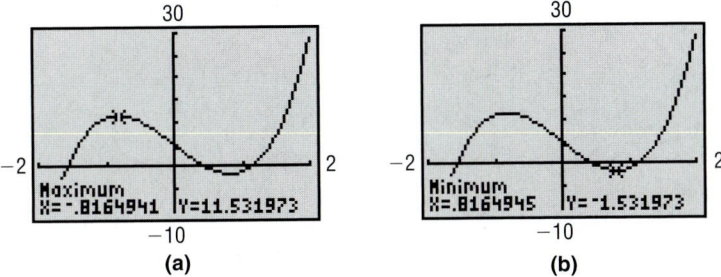

(a)   (b)

(b) Looking at Figures 16(a) and (b), we see that the graph of f is rising (increasing) from $x = -2$ to $x = -0.82$ and from $x = 0.82$ to $x = 2$, so f is increasing on the intervals $(-2, -0.82)$ and $(0.82, 2)$ (or for $-2 < x < -0.82$ and $0.82 < x < 2$). The graph is falling (decreasing) from $x = -0.82$ to $x = 0.82$, so f is decreasing on the interval $(-0.82, 0.82)$ (or for $-0.82 < x < 0.82$). ∎

 **NOW WORK PROBLEM 51.**

Even and Odd Functions

5   A function f is even if and only if whenever the point (x, y) is on the graph of f then the point $(-x, y)$ is also on the graph. Algebraically, we define an even function as follows:

> A function f is **even** if for every number x in its domain the number $-x$ is also in the domain and
>
> $$f(-x) = f(x)$$

A function f is odd if and only if whenever the point (x, y) is on the graph of f then the point $(-x, -y)$ is also on the graph. Algebraically, we define an odd function as follows:

> A function f is **odd** if for every number x in its domain the number $-x$ is also in the domain and
>
> $$f(-x) = -f(x)$$

Refer to Section 3.1, where the tests for symmetry are listed. The following results are then evident.

Theorem
A function is even if and only if its graph is symmetric with respect to the y-axis. A function is odd if and only if its graph is symmetric with respect to the origin.

EXAMPLE 6 **Determining Even and Odd Functions from the Graph**

Determine whether each graph given in Figure 17 is the graph of an even function, an odd function, or a function that is neither even nor odd.

Figure 17

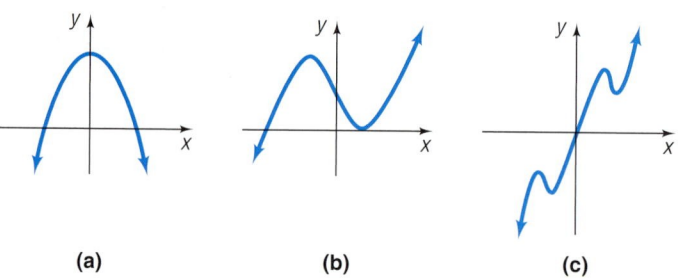

        (a)   (b)   (c)

Solution (a) The graph in Figure 17(a) is that of an even function, because the graph is symmetric with respect to the y-axis.

(b) The function whose graph is given in Figure 17(b) is neither even nor odd, because the graph is neither symmetric with respect to the y-axis nor symmetric with respect to the origin.

(c) The function whose graph is given in Figure 17(c) is odd, because its graph is symmetric with respect to the origin.

 NOW WORK PROBLEM 11.

A graphing utility can be used to conjecture whether a function is even, odd, or neither. As stated, when the graph of an even function contains the point (x, y), it must also contain the point $(-x, y)$. Therefore, if TRACE indicates that both the point (x, y) and the point $(-x, y)$ are on the graph for every x, then we would conjecture that the function is even.*

In addition, the graph of an odd function contains the points $(-x, -y)$ and (x, y). TRACE could be used in the same way to conjecture that the function is odd.*

In the next example, we use a graphing utility to conjecture whether a function is even, odd, or neither. Then we verify our conjecture algebraically.

EXAMPLE 7 **Identifying Even and Odd Functions**

Use a graphing utility to conjecture whether each of the following functions is even, odd, or neither. Verify the conjecture algebraically. Then state whether the graph is symmetric with respect to the y-axis or with respect to the origin.

(a) $f(x) = x^2 - 5$ (b) $g(x) = x^3 - 1$ (c) $h(x) = 5x^3 - x$

*$-X$min and Xmax must be equal for this to work.

Solution (a) Graph the function. Use TRACE to determine different pairs of points (x, y) and $(-x, y)$. For example, the point $(1.957447, -1.168402)$ is on the graph. See Figure 18(a). Now move the cursor to -1.957447 and determine the corresponding y-coordinate. See Figure 18(b). Since $(1.957447, -1.168402)$ and $(-1.957447, -1.168402)$ both lie on the graph, we have evidence that the function is even. Repeating this procedure for additional ordered pairs yields similar results. Therefore, we conjecture that the function is even.

Figure 18

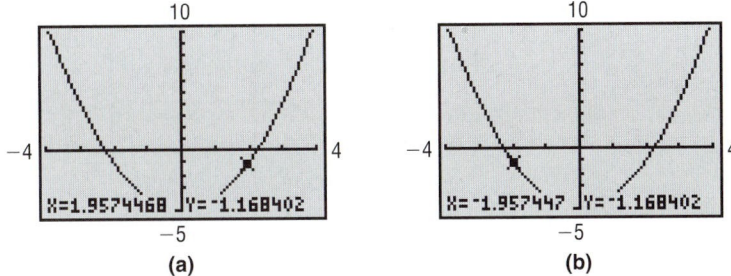

(a)   (b)

To algebraically verify the conjecture, we replace x by $-x$ in $f(x) = x^2 - 5$. Then

$$f(-x) = (-x)^2 - 5 = x^2 - 5 = f(x)$$

Since $f(-x) = f(x)$, we conclude that f is an even function, and the graph is symmetric with respect to the y-axis.

(b) Graph the function. Using TRACE, the point $(1.787234, 4.7087929)$ is on the graph. See Figure 19(a). Now move the cursor to -1.787234 and determine the corresponding y-coordinate, -6.708793. See Figure 19(b). We conjecture that the function is neither even nor odd since (x, y) does not equal $(-x, y)$ (so it is not even) and (x, y) also does not equal $(-x, -y)$ (so it is not odd).

Figure 19

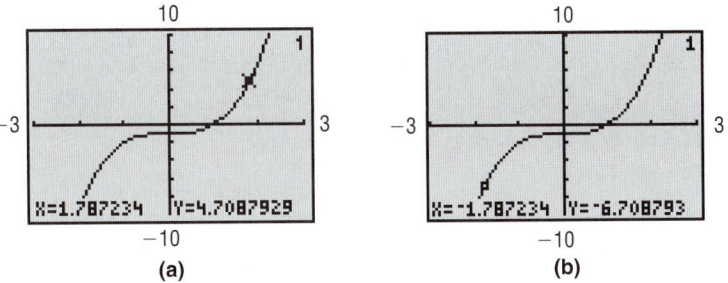

(a)   (b)

To algebraically verify that the function is not even, we find $g(-x)$.

$$g(-x) = (-x)^3 - 1 = -x^3 - 1; \quad g(x) = x^3 - 1$$

Since $g(-x) \neq g(x)$, the function is not even. To algebraically verify that the function is not odd, we find $-g(x)$.

$$-g(x) = -(x^3 - 1) = -x^3 + 1; \quad g(-x) = -x^3 - 1$$

Since $g(-x) \neq -g(x)$, the function is not odd. The graph is not symmetric with respect to the y-axis nor is it symmetric with respect to the origin.

(c) Graph the function. Using TRACE, the point $(1.0212766, 4.3047109)$ is on the graph. See Figure 20(a). We move the cursor to -1.021277 and determine the corresponding y-coordinate, -4.304711. See Figure 20(b). Since (x, y) equals $(-x, -y)$, correct to five decimal places, we have evidence that the function is odd. An examination of other pairs of points leads to the conjecture that the function is odd.

Figure 20

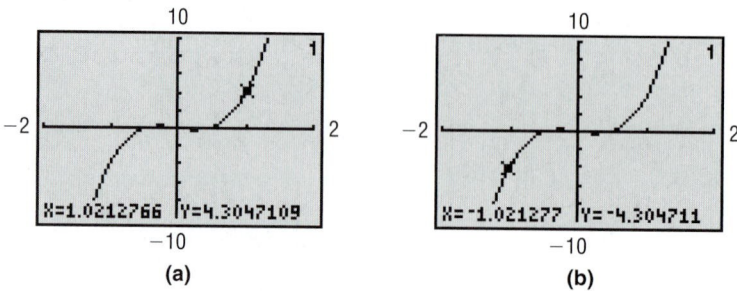

(a)   (b)

To algebraically verify the conjecture, we replace x by $-x$ in $h(x) = 5x^3 - x$. Then

$$h(-x) = 5(-x)^3 - (-x) = -5x^3 + x = -(5x^3 - x) = -h(x)$$

Since $h(-x) = -h(x)$, h is an odd function, and the graph of h is symmetric with respect to the origin. ∎

✏ **NOW WORK PROBLEM 37.**

3.2 Concepts and Vocabulary

In Problems 1–3, fill in the blanks.

1. The average rate of change of a function equals the _____ of the secant line.

2. A function f is _____ on an open interval I if for any choice of x_1 and x_2 in I, with $x_1 < x_2$, we have $f(x_1) < f(x_2)$.

3. An _____ function f is one for which $f(-x) = f(x)$ for every x in the domain of f; an _____ function f is one for which $f(-x) = -f(x)$ for every x in the domain of f.

In Problems 4–6, answer True or False for each statement.

4. A function f is decreasing on an open interval I if, for any choice of x_1 and x_2 in I, with $x_1 < x_2$, we have $f(x_1) > f(x_2)$.

5. A function f has a local maximum at c if there is an open interval I containing c so that, for all $x \neq c$ in I, $f(x) < f(c)$.

6. Even functions have graphs that are symmetric with respect to the origin.

7. Explain how the slope of the secant line and average rate of change of a function are related.

8. Suppose that a friend of yours does not understand the idea of increasing and decreasing functions. Provide an explanation complete with graphs that clarifies the idea.

9. Can a function be both even and odd? Explain.

3.2 Exercises

In Problems 1–10, use the graph of the function f given below.

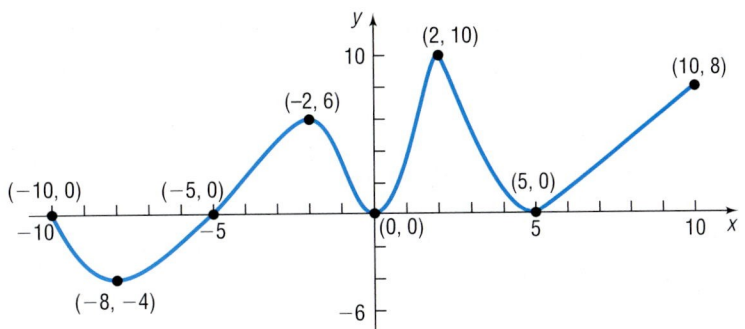

1. Is *f* increasing on the interval $(-8, -2)$?

2. Is *f* decreasing on the interval $(-8, -4)$?

3. Is *f* increasing on the interval $(2, 10)$?

4. Is *f* decreasing on the interval $(2, 5)$?

5. List the interval(s) on which *f* is increasing.

6. List the interval(s) on which *f* is decreasing.

7. Is there a local maximum at 2? If yes, what is it?

8. Is there a local maximum at 5? If yes, what is it?

9. List the numbers at which *f* has a local maximum. What are these local maxima?

10. List the numbers at which *f* has a local minimum. What are these local minima?

In Problems 11–20, the graph of a function is given. Use the graph to find:
  (a) The intercepts, if any
  (b) Its domain and range
  (c) The intervals on which it is increasing, decreasing, or constant
  (d) Whether it is even, odd, or neither

11.
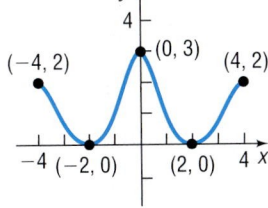

12.
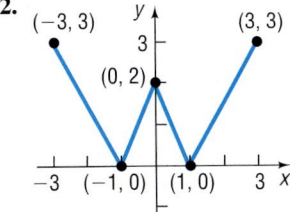

13.
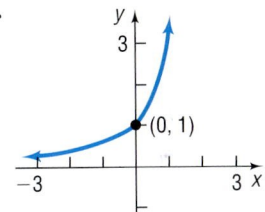

14.
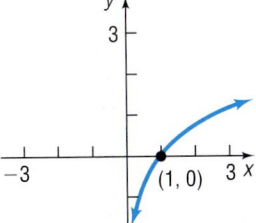

15.
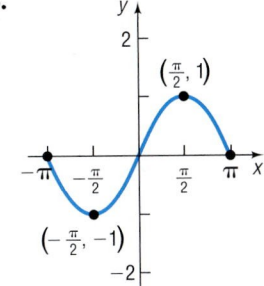

16.
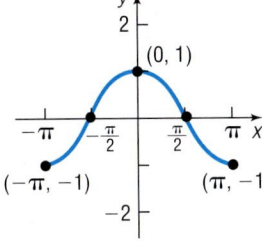

17.
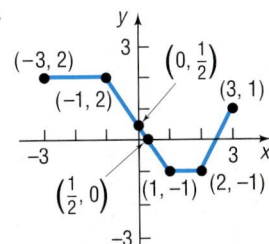

18.
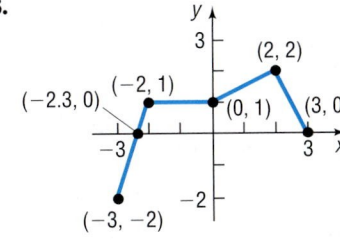

In Problems 19 and 20, assume that the entire graph is shown.

19.

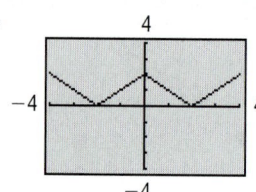

20.

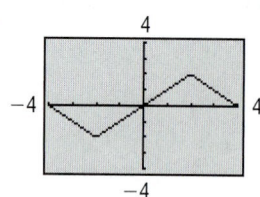

In Problems 21–24, the graph of a function f is given. Use the graph to find

    (a) *The numbers, if any, at which f has a local maximum. What are these local maxima?*
    (b) *The numbers, if any, at which f has a local minimum. What are these local minima?*

21.

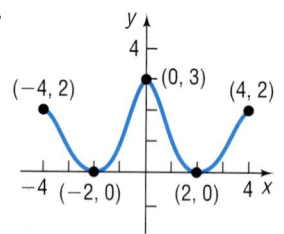

22.

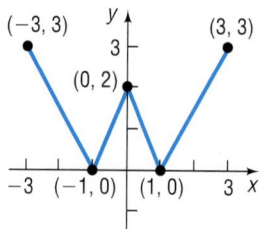

23.

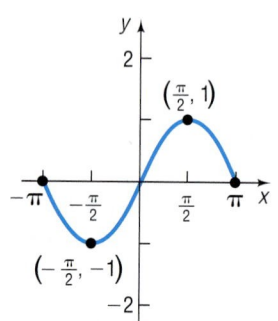

24.

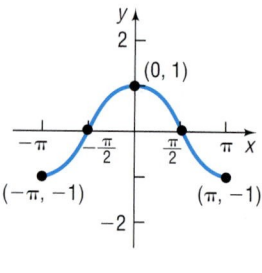

In Problems 25–36, (a) for each function find the average rate of change of f from 1 to x:

$$\frac{f(x) - f(1)}{x - 1}, \qquad x \neq 1$$

    (b) *Use the result from part (a) to compute the average rate of change from $x = 1$ to $x = 2$. Be sure to simplify.*
    (c) *Find an equation of the secant line containing $(1, f(1))$ and $(2, f(2))$.*
    (d) *Graph f and the secant line on the same viewing window.*

25. $f(x) = 5x$　　　　**26.** $f(x) = -4x$　　　　**27.** $f(x) = 1 - 3x$　　　　**28.** $f(x) = x^2 + 1$

29. $f(x) = x^2 - 2x$　　**30.** $f(x) = x - 2x^2$　　**31.** $f(x) = x^3 - x$　　　**32.** $f(x) = x^3 + x$

33. $f(x) = \dfrac{2}{x + 1}$　　**34.** $f(x) = \dfrac{1}{x^2}$　　　**35.** $f(x) = \sqrt{x}$　　　**36.** $f(x) = \sqrt{x + 3}$

In Problems 37–48, use a graphing utility to graph each function; then use TRACE to conjecture whether the function is even, odd, or neither. Finally, algebraically verify the conjecture.

37. $f(x) = 4x^3$　　　　**38.** $f(x) = 2x^4 - x^2$　　**39.** $g(x) = -3x^2 - 5$　　**40.** $h(x) = 3x^3 + 5$

41. $F(x) = \sqrt[3]{x}$　　　**42.** $G(x) = \sqrt{x}$　　　**43.** $f(x) = x + |x|$　　**44.** $f(x) = \sqrt[3]{2x^2 + 1}$

45. $g(x) = \dfrac{1}{x^2}$　　　**46.** $h(x) = \dfrac{x}{x^2 - 1}$　　**47.** $h(x) = \dfrac{-x^3}{3x^2 - 9}$　　**48.** $F(x) = \dfrac{2x}{|x|}$

49. How many x-intercepts can a function defined on an interval have if it is increasing on that interval? Explain.

50. How many y-intercepts can a function have? Explain.

In Problems 51–58, use a graphing utility to graph each function over the indicated interval and approximate any local maxima and local minima. Determine where the function is increasing and where it is decreasing. Round answers to two decimal places.

51. $f(x) = x^3 - 3x + 2$　$(-2, 2)$

52. $f(x) = x^3 - 3x^2 + 5$　$(-1, 3)$

53. $f(x) = x^5 - x^3$　$(-2, 2)$

54. $f(x) = x^4 - x^2$　$(-2, 2)$

55. $f(x) = -0.2x^3 - 0.6x^2 + 4x - 6$　$(-6, 4)$

56. $f(x) = -0.4x^3 + 0.6x^2 + 3x - 2$　$(-4, 5)$

57. $f(x) = 0.25x^4 + 0.3x^3 - 0.9x^2 + 3$　$(-3, 2)$

58. $f(x) = -0.4x^4 - 0.5x^3 + 0.8x^2 - 2$　$(-3, 2)$

⚠ *Problems 59–66 require the following definition of a secant line.*

> *The slope of the secant line containing the two points $(x, f(x))$ and $(x + h, f(x + h))$ on the graph of a function $y = f(x)$ may be given as*
>
> $$m_{sec} = \frac{f(x + h) - f(x)}{(x + h) - x} = \frac{f(x + h) - f(x)}{h}, \qquad h \neq 0$$
>
> *In calculus, this expression is called the **difference quotient of** f.*

(a) *Express the slope of the secant line of each function in terms of x and h. Be sure to simplify your answer.*
(b) *Find m_{sec} for $h = 0.5, 0.1,$ and 0.01 at $x = 1$. What value does m_{sec} approach as h approaches 0?*
(c) *Find the equation for the secant line at $x = 1$ with $h = 0.01$.*
(d) *Graph f and the secant line found in part (c) on the same viewing window.*

59. $f(x) = 2x + 5$ **60.** $f(x) = -3x + 2$ **61.** $f(x) = x^2 + 2x$ **62.** $f(x) = 2x^2 + x$

63. $f(x) = 2x^2 - 3x + 1$ **64.** $f(x) = -x^2 + 3x - 2$ **65.** $f(x) = \dfrac{1}{x}$ **66.** $f(x) = \dfrac{1}{x^2}$

---

67. Maximizing the Volume of a Box An open box with a square base is to be made from a square piece of cardboard 24 inches on a side by cutting out a square from each corner and turning up the sides. (See the illustration.) The volume V of the box as a function of the length x of the side of the square cut from each corner is

$$V(x) = x(24 - 2x)^2$$

Graph V and determine where V is largest.

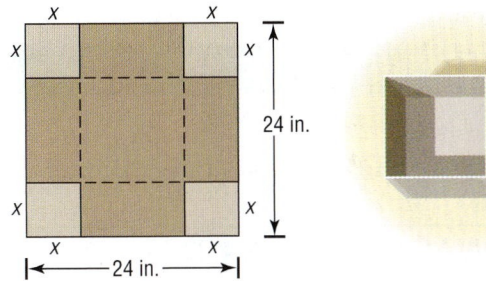

68. Minimizing the Material Needed to Make a Box An open box with a square base is required to have a volume of 10 cubic feet. The amount A of material used to make such a box as a function of the length x of a side of the square base is

$$A(x) = x^2 + \frac{40}{x}$$

Graph A and determine where A is smallest.

69. Maximum Height of a Ball The height s of a ball (in feet) thrown with an initial velocity of 80 feet per second from an initial height of 6 feet is given as a function of the time t (in seconds) by

$$s(t) = -16t^2 + 80t + 6$$

(a) Graph s.
(b) Determine the time at which height is maximum.
(c) What is the maximum height?

70. Minimum Average Cost The average cost of producing x riding lawn mowers per hour is given by

$$C(x) = 0.3x^2 + 21x - 251 + \frac{2500}{x}$$

(a) Graph C.
(b) Determine the number of riding lawn mowers to produce in order to minimize average cost.
(c) What is the minimum average cost?

71. Revenue from Selling Bikes The following data represent the total revenue that would be received from selling x bicycles at Tunney's Bicycle Shop.

Number of Bicycles, x	Total Revenue, R (Dollars)
0	0
25	28,000
60	45,000
102	53,400
150	59,160
190	62,360
223	64,835
249	66,525

(a) Draw a scatter diagram of the data, treating the number of bicycles produced as the independent variable.
(b) Draw a line through the points $(0, 0)$ and $(25, 28000)$ on the scatter diagram found in part (a).
(c) Find the average rate of change of revenue from 0 to 25 bicycles.
(d) Interpret the average rate of change found in part (c).
(e) Draw a line through the points $(190, 62360)$ and $(223, 64835)$ on the scatter diagram found in part (a).
(f) Find the average rate of change of revenue from 190 to 223 bicycles.
(g) Interpret the average rate of change found in part (f).
(h) What is happening to the average rate of change of revenue as the number of bicycles increases?

72. Cost of Manufacturing Bikes   The following data represent the monthly cost of producing bicycles at Tunney's Bicycle Shop.

Number of Bicycles, x	Total Cost of Production, C (Dollars)
0	24,000
25	27,750
60	31,500
102	35,250
150	39,000
190	42,750
223	46,500
249	50,250

(a) Draw a scatter diagram of the data, treating the number of bicycles produced as the independent variable.
(b) Draw a line through the points $(0, 24000)$ and $(25, 27750)$ on the scatter diagram found in part (a).
(c) Find the average rate of change of the cost from 0 to 25 bicycles.
(d) Interpret the average rate of change found in part (c).
(e) Draw a line through the points $(190, 42750)$ and $(223, 46500)$ on the scatter diagram found in part (a).
(f) Find the average rate of change of the cost from 190 to 223 bicycles.

(g) Interpret the average rate of change found in part (f).
(h) What is happening to the average rate of change of cost as the number of bicycles increases?

73. Growth of Bacteria   The following data represent the population of an unknown bacteria.

Time (Days)	Population
0	50
2	234
4	547
6	1280

(a) Draw a scatter diagram of the data, treating time as the independent variable.
(b) Draw a line through the points $(0, 50)$ and $(2, 234)$ on the scatter diagram found in part (a).
(c) Find the average rate of change of the population from 0 to 2 days.
(d) Interpret the average rate of change found in part (c).
(e) Draw a line through the points $(4, 547)$ and $(6, 1280)$ on the scatter diagram found in part (a).
(f) Find the average rate of change of the population from 4 to 6 days.
(g) Interpret the average rate of change found in part (f).
(h) What is happening to the average rate of change of the population as time passes?

PREPARING FOR THIS SECTION

Before getting started, review the following:

✓ Graphs of Certain Equations (Section 1.2: Example 3, p. 102; Section 3.1: Example 3, p. 258; Example 4, p. 259; Example 5, p. 260)

3.3   LIBRARY OF FUNCTIONS; PIECEWISE-DEFINED FUNCTIONS

OBJECTIVES   **1** Graph the Functions Listed in the Library of Functions
2 Graph Piecewise-defined Functions

Library of Functions

1   We now give names to some of the functions that we have encountered. In going through this list, pay special attention to the properties of each function, particularly to the shape of each graph. Knowing these graphs will lay the foundation for later graphing techniques.

Figure 21

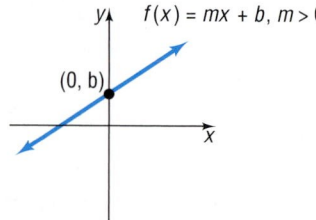

Linear Function

$$f(x) = mx + b \qquad m \text{ and } b \text{ are real numbers}$$

See Figure 21.

The domain of a **linear function** f consists of all real numbers. The graph of this function is a nonvertical line with slope m and y-intercept b. A linear function is increasing if $m > 0$, decreasing if $m < 0$, and constant if $m = 0$.

Constant Function

$$f(x) = b \qquad b \text{ is a real number}$$

Figure 22

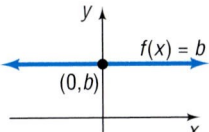

See Figure 22.

A **constant function** is a special linear function ($m = 0$). Its domain is the set of all real numbers; its range is the set consisting of a single number b. Its graph is a horizontal line whose y-intercept is b. The constant function is an even function whose graph is constant over its domain.

Identity Function

$$f(x) = x$$

See Figure 23.

Figure 23

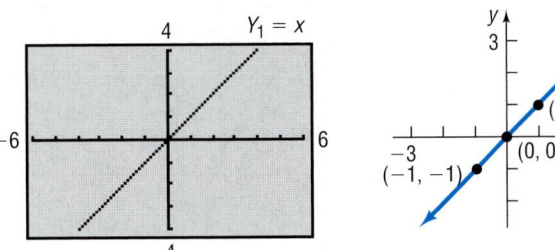

The **identity function** is also a special linear function. Its domain and range are the set of all real numbers. Its graph is a line whose slope is $m = 1$ and whose y-intercept is 0. The line consists of all points for which the x-coordinate equals the y-coordinate. The identity function is an odd function that is increasing over its domain. Note that the graph bisects quadrants I and III.

Square Function

$$f(x) = x^2$$

See Figure 24.

Figure 24

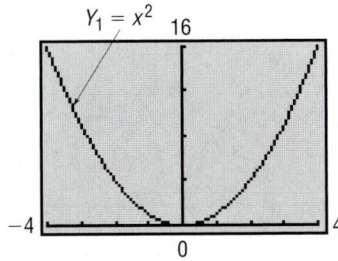
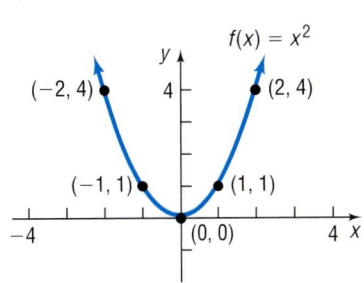

The domain of the **square function** f is the set of all real numbers; its range is the set of nonnegative real numbers. The graph of this function is a parabola whose intercept is at $(0, 0)$. The square function is an even function that is decreasing on the interval $(-\infty, 0)$ and increasing on the interval $(0, \infty)$.

Cube Function

$$f(x) = x^3$$

See Figure 25.

Figure 25

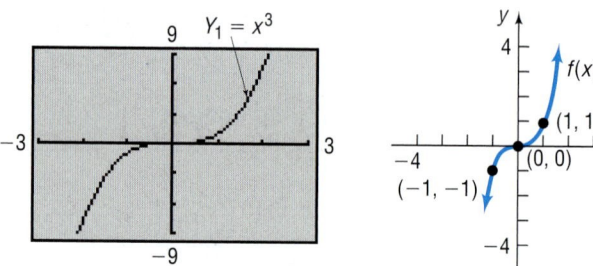

The domain and range of the **cube function** are the set of all real numbers. The intercept of the graph is at $(0, 0)$. The cube function is an odd function that is increasing on the interval $(-\infty, \infty)$.

Square Root Function

$$f(x) = \sqrt{x}$$

See Figure 26.

Figure 26

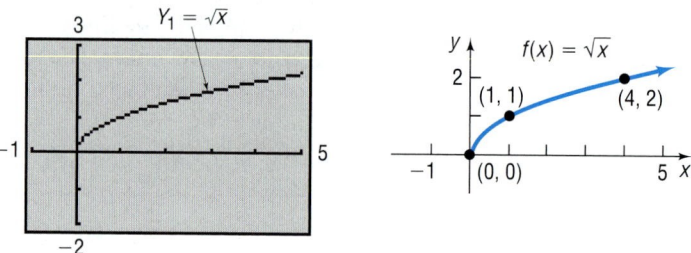

The domain and range of the **square root function** are the set of nonnegative real numbers. The intercept of the graph is at $(0, 0)$. The square root function is neither even nor odd and is increasing on the interval $(0, \infty)$.

Reciprocal Function

$$f(x) = \frac{1}{x}$$

Refer to Example 5, page 260, for a discussion of the equation $y = \dfrac{1}{x}$. See Figure 27.

Figure 27

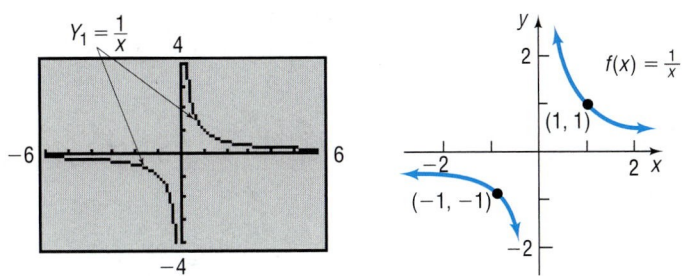

The domain and range of the **reciprocal function** are the set of all nonzero real numbers. The graph has no intercepts. The reciprocal function is an odd function and is decreasing on the intervals $(-\infty, 0)$ and $(0, \infty)$.

Absolute Value Function

$$f(x) = |x|$$

See Figure 28.

Figure 28

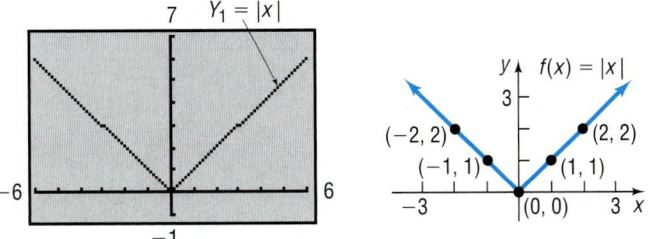

The domain of the **absolute value function** is the set of all real numbers; its range is the set of nonnegative real numbers. The intercept of the graph is at $(0, 0)$. If $x \geq 0$, then $f(x) = x$, and the graph of f is part of the line $y = x$; if $x < 0$, then $f(x) = -x$, and the graph of f is part of the line $y = -x$. The absolute value function is an even function; it is decreasing on the interval $(-\infty, 0)$ and increasing on the interval $(0, \infty)$.

The notation $\text{int}(x)$ stands for the largest integer less than or equal to x. For example,

$$\text{int}(1) = 1 \quad \text{int}(2.5) = 2 \quad \text{int}\left(\frac{1}{2}\right) = 0 \quad \text{int}\left(\frac{-3}{4}\right) = -1 \quad \text{int}(\pi) = 3$$

This type of correspondence occurs frequently enough in mathematics that we give it a name.

Greatest-integer Function

$$f(x) = \text{int}(x) = \text{greatest integer less than or equal to } x$$

Note: Some books use the notation $f(x) = [\![x]\!]$ instead of $\text{int}(x)$.

We obtain the graph of $f(x) = \text{int}(x)$ by plotting several points. See Table 7. For values of x, $-1 \le x < 0$, the value of $f(x) = \text{int}(x)$ is -1; for values of x, $0 \le x < 1$, the value of f is 0. See Figure 29 for the graph.

TABLE 7

X	Y1
-1	-1
-.75	-1
-.5	-1
-.25	-1
0	0
.25	0
.5	0

Y1■int(X)

Figure 29

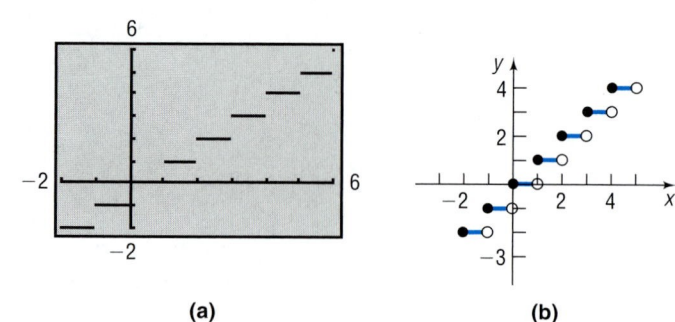

(a)   (b)

COMMENT: When graphing a function using a graphing utility, you can choose either the **connected mode,** in which points plotted on the screen are connected, making the graph appear without any breaks, or the **dot mode,** in which only the points plotted appear. When graphing the greatest-integer function with a graphing utility, it is necessary to be in the **dot mode.** This is to prevent the utility from "connecting the dots" when $f(x)$ changes from one integer value to the next. See Figure 30.

Figure 30

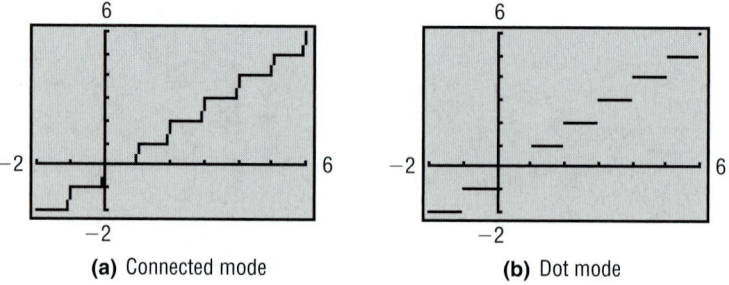

(a) Connected mode   **(b)** Dot mode

The domain of the **greatest-integer function** is the set of all real numbers; its range is the set of integers. The y-intercept of the graph is 0. The x-intercepts lie in the interval $[0, 1)$. The greatest-integer function is neither even nor odd. It is constant on every interval of the form $[k, k + 1)$, for k an integer. In Figure 29(b), we use a solid dot to indicate, for example, that at $x = 1$ the value of f is $f(1) = 1$; we use an open circle to illustrate that the function does not assume the value of 0 at $x = 1$.

From the graph of the greatest-integer function, we can see why it is also called a **step function.** At $x = 0$, $x = \pm 1$, $x = \pm 2$, and so on, this function exhibits what is called a *discontinuity;* that is, at integer values, the graph suddenly "steps" from one value to another without taking on any of the intermediate values. For example, to the immediate left of $x = 3$, the y-coordinates are 2, and to the immediate right of $x = 3$, the y-coordinates are 3.

The functions that we have discussed so far are basic. Whenever you encounter one of them, you should see a mental picture of its graph. For example, if you encounter the function $f(x) = x^2$, you should see in your mind's eye a picture like Figure 24.

NOW WORK PROBLEMS 1–8.

Piecewise-defined Functions

2 Sometimes a function is defined differently on different parts of its domain. For example, the absolute value function $f(x) = |x|$ is actually defined by two equations: $f(x) = x$ if $x \geq 0$ and $f(x) = -x$ if $x < 0$. For convenience, we generally combine these equations into one expression as

$$f(x) = |x| = \begin{cases} x & \text{if } x \geq 0 \\ -x & \text{if } x < 0 \end{cases}$$

When functions are defined by more than one equation, they are called **piecewise-defined** functions.

Let's look at another example of a piecewise-defined function.

EXAMPLE 1   **Analyzing a Piecewise-defined Function**

For the following function f,

$$f(x) = \begin{cases} -x + 1 & \text{if } -1 \leq x < 1 \\ 2 & \text{if } x = 1 \\ x^2 & \text{if } x > 1 \end{cases}$$

(a) Find $f(0)$, $f(1)$, and $f(2)$.   (b) Determine the domain of f.
(c) Graph f.   (d) Use the graph to find the range of f.

Solution   (a) To find $f(0)$, we observe that when $x = 0$ the equation for f is given by $f(x) = -x + 1$. So we have

$$f(0) = -0 + 1 = 1$$

When $x = 1$, the equation for f is $f(x) = 2$. Thus,

$$f(1) = 2$$

When $x = 2$, the equation for f is $f(x) = x^2$. So

$$f(2) = 2^2 = 4$$

(b) To find the domain of f, we look at its definition. We conclude that the domain of f is $\{x \mid x \geq -1\}$, or the interval $[-1, \infty)$.

(c) To graph f by hand, we graph "each piece." First we graph the line $y = -x + 1$ and keep only the part for which $-1 \leq x < 1$. Then we plot the point $(1, 2)$, because when $x = 1$, $f(x) = 2$. Finally, we graph the parabola $y = x^2$ and keep only the part for which $x > 1$. See Figure 31(a).

On a graphing utility, the procedure for graphing a piecewise-defined function varies depending on the particular utility. In general, you need to enter each piece as a function with a restricted domain. See Figure 31(b).

Figure 31

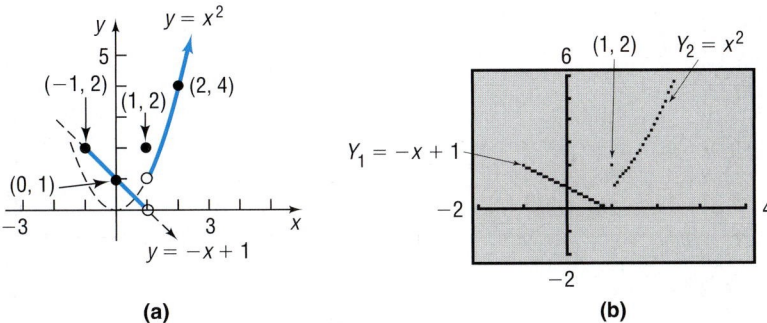

(a)   (b)

In graphing piecewise-defined functions using a graphing utility, it is usually better to be in the dot mode, since such functions may have breaks (discontinuities).

(d) From the graph, we conclude that the range of f is $\{y|y > 0\}$, or $(0, \infty)$.

 **NOW WORK PROBLEM 19.**

| EXAMPLE 2 | **Cost of Electricity** |

In October 2001, Commonwealth Edison Company supplied electricity to residences for a monthly customer charge of $7.57 plus 8.77¢ per kilowatt-hour (kWhr) for the first 400 kWhr supplied in the month and 6.574¢ per kWhr for all usage over 400 kWhr in the month.*

(a) What is the charge for using 300 kWhr in a month?
(b) What is the charge for using 700 kWhr in a month?
(c) If C is the monthly charge for x kWhr, express C as a function of x.
(d) Graph C.

Solution (a) For 300 kWhr, the charge is $7.57 plus 8.77¢ = $0.0877 per kWhr. That is,

$$\text{Charge} = \$7.57 + \$0.0877(300) = \$33.88$$

(b) For 700 kWhr, the charge is $7.57 plus 8.77¢ per kWhr for the first 400 kWhr plus 6.574¢ per kWhr for the 300 kWhr in excess of 400. That is,

$$\text{Charge} = \$7.57 + \$0.0877(400) + \$0.06574(300) = \$62.37$$

(c) If $0 \le x \le 400$, the monthly charge C (in dollars) can be found by multiplying x times $0.0877 and adding the monthly customer charge of $7.57. If $0 \le x \le 400$, then $C(x) = 0.0877x + 7.57$. For $x > 400$, the charge is $0.0877(400) + 7.57 + 0.06574(x - 400)$, since $x - 400$ equals the usage in excess of 400 kWhr, which costs $0.06574 per kWhr. That is, if $x > 400$, then

$$C(x) = 0.0877(400) + 7.57 + 0.06574(x - 400)$$
$$= 42.65 + 0.06574(x - 400)$$
$$= 0.06574x + 16.35$$

The rule for computing C follows two equations:

$$C(x) = \begin{cases} 0.0877x + 7.57 & \text{if } 0 \le x \le 400 \\ 0.06574x + 16.35 & \text{if } x > 400 \end{cases}$$

(d) See Figure 32 for the graph.

Figure 32

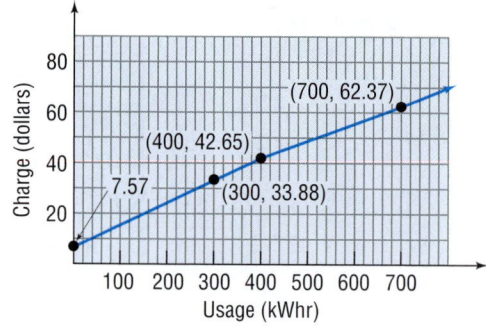

Source: Commonwealth Edison Co., Chicago, Illinois, 2001.

3.3 Concepts and Vocabulary

In Problems 1 and 2, fill in the blanks.

1. The graph of $f(x) = mx + b$ is decreasing if m is _____ than zero.

2. When functions are defined by more than one equation, they are called _____ functions.

In Problems 3 and 4, answer True or False for each statement.

3. The cube function is odd and is increasing on the interval $(-\infty, \infty)$.

4. The domain and the range of the reciprocal function are the set of all real numbers.

5. Describe the square function. Include a graph.

6. Explain why the absolute value function is actually a piecewise-defined function.

3.3 Exercises

In Problems 1–8, match each graph to the function listed whose graph most resembles the one given.

  A. *Constant function*   B. *Linear function*   C. *Square function*
  D. *Cube function*   E. *Square root function*   F. *Reciprocal function*
  G. *Absolute value function*   H. *Greatest-integer function*

1. 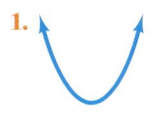   **2.**    **3.** 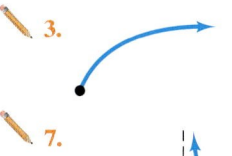   **4.**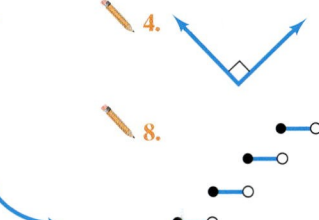

5.    **6.**    **7.**   **8.**

In Problems 9–14, sketch the graph of each function. Label at least three points.

9. $f(x) = x$ **10.** $f(x) = x^2$ **11.** $f(x) = x^3$

12. $f(x) = \sqrt{x}$ **13.** $f(x) = \dfrac{1}{x}$ **14.** $f(x) = |x|$

15. If $f(x) = \begin{cases} x^2 & \text{if } x < 0 \\ 2 & \text{if } x = 0 \\ 2x + 1 & \text{if } x > 0 \end{cases}$
     find: (a) $f(-2)$ (b) $f(0)$ (c) $f(2)$

16. If $f(x) = \begin{cases} x^3 & \text{if } x < 0 \\ 3x + 2 & \text{if } x \geq 0 \end{cases}$
     find: (a) $f(-1)$ (b) $f(0)$ (c) $f(1)$

17. If $f(x) = \text{int}(2x)$, find: (a) $f(1.2)$ (b) $f(1.6)$ (c) $f(-1.8)$

18. If $f(x) = \text{int}\left(\dfrac{x}{2}\right)$, find: (a) $f(1.2)$ (b) $f(1.6)$ (c) $f(-1.8)$

In Problems 19–30:

(a) Find the domain of each function.   *(b) Locate any intercepts.*

(c) Graph each function by hand.   *(d) Based on the graph, find the range.*

(e) Verify your results using a graphing utility.

19. $f(x) = \begin{cases} 2x & \text{if } x \neq 0 \\ 1 & \text{if } x = 0 \end{cases}$

20. $f(x) = \begin{cases} 3x & \text{if } x \neq 0 \\ 4 & \text{if } x = 0 \end{cases}$

21. $f(x) = \begin{cases} -2x + 3 & x < 1 \\ 3x - 2 & x \geq 1 \end{cases}$

22. $f(x) = \begin{cases} x + 3 & x < -2 \\ -2x - 3 & x \geq -2 \end{cases}$

23. $f(x) = \begin{cases} x + 3 & -2 \leq x < 1 \\ 5 & x = 1 \\ -x + 2 & x > 1 \end{cases}$

24. $f(x) = \begin{cases} 2x + 5 & -3 \leq x < 0 \\ -3 & x = 0 \\ -5x & x > 0 \end{cases}$

25. $f(x) = \begin{cases} 1 + x & \text{if } x < 0 \\ x^2 & \text{if } x \geq 0 \end{cases}$

26. $f(x) = \begin{cases} \dfrac{1}{x} & \text{if } x < 0 \\ \sqrt{x} & \text{if } x \geq 0 \end{cases}$

27. $f(x) = \begin{cases} |x| & \text{if } -2 \leq x < 0 \\ 1 & \text{if } x = 0 \\ x^3 & \text{if } x > 0 \end{cases}$

28. $f(x) = \begin{cases} 3 + x & \text{if } -3 \leq x < 0 \\ 3 & \text{if } x = 0 \\ \sqrt{x} & \text{if } x > 0 \end{cases}$

29. $h(x) = 2\operatorname{int}(x)$

30. $f(x) = \operatorname{int}(2x)$

In Problems 31–34, the graph of a piecewise-defined function is given. Write a definition for each function.

31.

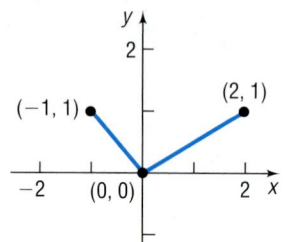

32.

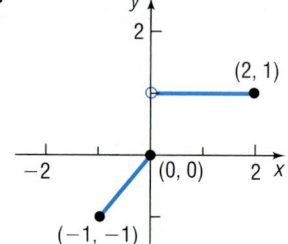

33.

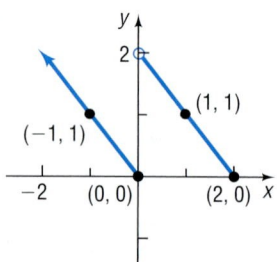

34.

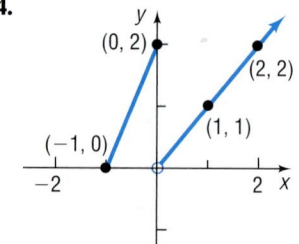

35. Cost of Transporting Goods A trucking company transports goods between Chicago and New York, a distance of 960 miles. The company's policy is to charge, for each pound, $0.50 per mile for the first 100 miles, $0.40 per mile for the next 300 miles, $0.25 per mile for the next 400 miles, and no charge for the remaining 160 miles.

(a) Graph the relationship between the cost of transportation in dollars and mileage over the entire 960-mile route.

(b) Find the cost as a function of mileage for hauls between 100 and 400 miles from Chicago.

(c) Find the cost as a function of mileage for hauls between 400 and 800 miles from Chicago.

36. Car Rental Costs An economy car rented in Florida from National Car Rental® on a weekly basis costs $95 per week. Extra days cost $24 per day until the day rate exceeds the weekly rate, in which case the weekly rate applies. Find the cost C of renting an economy car as a piecewise-defined function of the number x of days used, where $7 \leq x \leq 14$. Graph this function.

Note: Any part of a day counts as a full day.

37. Minimum Payments for Credit Cards Holders of credit cards issued by banks, department stores, oil companies, and so on, receive bills each month that state minimum amounts that must be paid by a certain due date. The minimum due depends on the total amount owed. One such credit card company uses the following rules: For a bill of less than $10, the entire amount is due. For a bill of at least $10 but less than $500, the minimum due is $10. There is a minimum of $30 due on a bill of at least $500 but less than $1000, a minimum of $50 due on a bill of at least $1000 but less than $1500, and a minimum of $70 due on bills of $1500 or more. Find the function f that describes the minimum payment due on a bill of x dollars. Graph f.

38. Interest Payments for Credit Cards Refer to Problem 37. The card holder may pay any amount between the minimum due and the total owed. The organization issuing the card charges the card holder interest of 1.5% per month for the first $1000 owed and 1% per month on any unpaid balance over $1000. Find the function g that gives the amount of interest charged per month on a balance of x dollars. Graph g.

39. **Federal Income Tax**   Two 2001 Tax Rate Schedules are given in the accompanying table. If x equals the amount on Form 1040, line 37, and y equals the tax due, construct a function $y = f(x)$ for each schedule.

2001 TAX RATE SCHEDULES

SCHEDULE X—IF YOUR FILING STATUS IS SINGLE				SCHEDULE Y-I—USE IF YOUR FILING STATUS IS MARRIED FILING JOINTLY OR QUALIFYING WIDOW(ER)			
If the amount on Form 1040, line 37, is: Over—	But not over—	Enter on Form 1040, line 38	of the amount over—	If the amount on Form 1040, line 37, is: Over—	But not over—	Enter on Form 1040, line 38	of the amount over—
$0	$27,050	_____ 15%	$0	$0	$45,200	_____ 15%	$0
27,050	65,550	$4,057.50 + 28%	27,050	45,200	109,250	$6,780.00 + 28%	45,200
65,550	136,750	14,837.50 + 31%	65,550	109,250	166,450	24,714.00 + 31%	109,250
136,750	297,300	36,909.50 + 36%	136,750	166,450	297,300	42,446.00 + 36%	166,450
297,300		94,707.50 + 39.6%	297,300	297,300		89,552.00 + 39.6%	297,300

40. **Cost of Natural Gas**   In April 2000, the Peoples Gas Company had the following rate schedule* for natural gas usage in single-family residences:

Monthly service charge $9.45

Per therm service charge
   1st 50 therms $0.36375/therm
   Over 50 therms $0.11445/therm

Gas charge $0.3128/therm

(a) What is the charge for using 50 therms in a month?
(b) What is the charge for using 500 therms in a month?
(c) Construct a function that gives the monthly charge C for x therms of gas.
(d) Graph this function.

41. **Cost of Natural Gas**   In March 2000, Nicor Gas had the following rate schedule† for natural gas usage in single-family residences:

Monthly customer charge $6.45

Distribution charge
   1st 20 therms $0.2012/therm
   Next 30 therms $0.1117/therm
   Over 50 therms $0.0374/therm

Gas supply charge $0.3209/therm

(a) What is the charge for using 40 therms in a month?
(b) What is the charge for using 202 therms in a month?
(c) Construct a function that gives the monthly charge C for x therms of gas.
(d) Graph this function.

*Source: The peoples Gas Company, Chicago, Illinois.
†Source: Nicor Gas, Aurora, Illinois.

42. **Wind Chill**   The wind chill factor represents the equivalent air temperature at a standard wind speed that would produce the same heat loss as the given temperature and wind speed. One formula for computing the equivalent temperature is

$$W = \begin{cases} t & 0 \le v < 1.79 \\ 33 - \dfrac{(10.45 + 10\sqrt{v} - v)(33 - t)}{22.04} & 1.79 \le v \le 20 \\ 33 - 1.5958(33 - t) & v > 20 \end{cases}$$

where v represents the wind speed (in meters per second) and t represents the air temperature (°C). Compute the wind chill for the following:

(a) An air temperature of 10°C and a wind speed of 1 meter per second (m/sec)
(b) An air temperature of 10°C and a wind speed of 5 m/sec.
(c) An air temperature of 10°C and a wind speed of 15 m/sec.
(d) An air temperature of 10°C and a wind speed of 25 m/sec.
(e) Explain the physical meaning of the equation corresponding to $0 \le v < 1.79$.
(f) Explain the physical meaning of the equation corresponding to $v > 20$.

43. **Exploration**   Graph $y = x^2$. Then on the same screen graph $y = x^2 + 2$, followed by $y = x^2 + 4$, followed by $y = x^2 - 2$. What pattern do you observe? Can you predict the graph of $y = x^2 - 4$? Of $y = x^2 + 5$.

44. Exploration Graph $y = x^2$. Then on the same screen graph $y = (x - 2)^2$, followed by $y = (x - 4)^2$, followed by $y = (x + 2)^2$. What pattern do you observe? Can you predict the graph of $y = (x + 4)^2$? Of $y = (x - 5)^2$?

45. Exploration Graph $y = |x|$. Then on the same screen graph $y = 2|x|$, followed by $y = 4|x|$, followed by $y = \frac{1}{2}|x|$. What pattern do you observe? Can you predict the graph of $y = \frac{1}{4}|x|$? Of $y = 5|x|$?

46. Exploration Graph $y = x^2$. Then on the same screen graph $y = -x^2$. What do you observe? Now try $y = |x|$ and $y = -|x|$. What do you conclude?

47. Exploration Graph $y = \sqrt{x}$. Then on the same screen graph $y = \sqrt{-x}$. What do you observe? Now try $y = 2x + 1$ and $y = 2(-x) + 1$. What do you conclude?

48. Exploration Graph $y = x^3$. Then on the same screen graph $y = (x - 1)^3 + 2$. Could you have predicted the result?

49. Exploration Graph $y = x^2$, $y = x^4$, and $y = x^6$ on the same screen. What do you notice is the same about each graph? What do you notice that is different?

50. Exploration Graph $y = x^3$, $y = x^5$, and $y = x^7$ on the same screen. What do you notice is the same about each graph? What do you notice that is different?

51. Consider the equation

$$y = \begin{cases} 1 & \text{if } x \text{ is rational} \\ 0 & \text{if } x \text{ is irrational} \end{cases}$$

Is this a function? What is its domain? What is its range? What is its y-intercept, if any? What are its x-intercepts, if any? Is it even, odd, or neither? How would you describe its graph?

52. Define some functions that pass through $(0, 0)$ and $(1, 1)$ and are increasing for $x \geq 0$. Begin your list with $y = \sqrt{x}$, $y = x$, and $y = x^2$. Can you propose a general result about such functions?

3.4 GRAPHING TECHNIQUES: TRANSFORMATIONS

OBJECTIVES
1. Graph Functions Using Horizontal and Vertical Shifts
2. Graph Functions Using Reflections about the x-Axis or y-Axis
3. Graph Functions Using Compressions and Stretches

At this stage, if you were asked to graph any of the functions defined by $y = x$, $y = x^2$, $y = x^3$, $y = \sqrt{x}$, $y = |x|$, or $y = \frac{1}{x}$, your response should be, "Yes, I recognize these functions and know the general shapes of their graphs." (If this is not your answer, review the previous section and Figures 23 through 28).

Sometimes we are asked to graph a function that is "almost" like one that we already know how to graph. In this section, we look at some of these functions and develop techniques for graphing them. Collectively, these techniques are referred to as **transformations.**

1 Vertical Shifts

EXPLORATION   On the same screen, graph each of the following functions:

$$Y_1 = x^2$$
$$Y_2 = x^2 + 2$$
$$Y_3 = x^2 - 2$$

What do you observe?

RESULT   Figure 33 illustrates the graphs. You should have observed a general pattern. With $Y_1 = x^2$ on the screen, the graph of $Y_2 = x^2 + 2$ is identical to that of $Y_1 = x^2$, except that it is shifted vertically up 2 units. The graph of $Y_3 = x^2 - 2$ is identical to that of $Y_1 = x^2$, except that it is shifted vertically down 2 units.

Figure 33

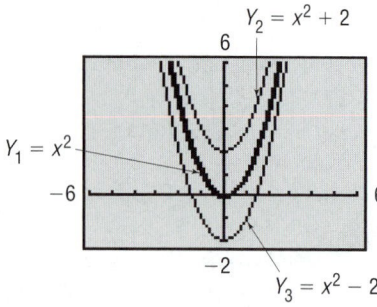

We are led to the following conclusion:

> If a real number k is added to the right side of a function $y = f(x)$, the graph of the new function $y = f(x) + k$ is the graph of f **shifted vertically** up (if $k > 0$) or down (if $k < 0$).

Let's look at an example.

EXAMPLE 1   **Vertical Shift Down**

Use the graph of $f(x) = x^2$ to obtain the graph of $h(x) = x^2 - 4$.

Solution   Table 8 lists some points on the graphs of $f = Y_1$ and $h = Y_2$. Notice that each y-coordinate of h is 4 units less than the corresponding y-coordinate of f.

The graph of h is identical to that of f, except that it is shifted down 4 units. See Figure 34.

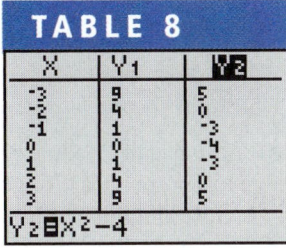

TABLE 8

X	Y$_1$	Y$_2$
-3	9	5
-2	4	0
-1	1	-3
0	0	-4
1	1	-3
2	4	0
3	9	5

Y$_2$■X²-4

Figure 34

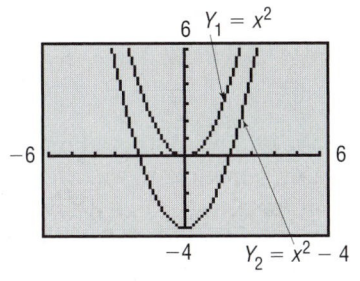

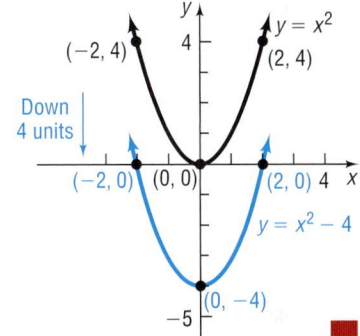

NOW WORK PROBLEM 33.

Horizontal Shifts

EXPLORATION   On the same screen, graph each of the following functions:

$$Y_1 = x^2$$
$$Y_2 = (x - 3)^2$$
$$Y_3 = (x + 2)^2$$

What do you observe?

RESULT   Figure 35 illustrates the graphs.

You should have observed the following pattern. With the graph of $Y_1 = x^2$ on the screen, the graph of $Y_2 = (x - 3)^2$ is identical to that of $Y_1 = x^2$, except it is shifted horizontally to the right 3 units. The graph of $Y_3 = (x + 2)^2$ is identical to that of $Y_1 = x^2$, except it is shifted horizontally to the left 2 units.

Figure 35

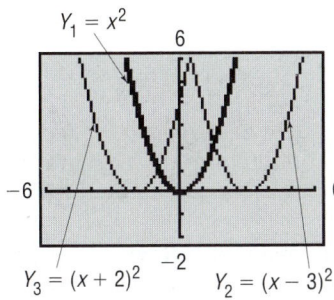

We are led to the following conclusion.

> If the argument x of a function f is replaced by $x - h$, h a real number, the graph of the new function $y = f(x - h)$ is the graph of f **shifted horizontally** left (if $h < 0$) or right (if $h > 0$).

NOW WORK PROBLEM 37.

In Section 2.3 we graphed quadratic functions by identifying points on the graph of the function, plotting the points, and connecting them with a smooth curve. We can also graph quadratic functions using transformations.

EXAMPLE 2 **Graphing a Quadratic Function Using Transformations**

Graph the function $f(x) = x^2 + 6x + 4$.

Solution We begin by completing the square on the right side.

$$f(x) = x^2 + 6x + 4 \qquad \text{Take } \tfrac{1}{2} \text{ the coefficient of } x \text{ and}$$

$$= (x^2 + 6x) + 4 \qquad \text{square the result: } \left(\tfrac{1}{2} \cdot 6\right)^2 = 9.$$

$$= (x^2 + 6x + 9) + (4 - 9) \qquad \text{Add and subtract 9.}$$

$$= (x + 3)^2 - 5 \qquad \text{Factor.}$$

We graph f in stages. First, we note that the rule for f is basically a square function, so we begin with the graph of $y = x^2$ as shown in Figure 36(a). To get the graph of $y = (x + 3)^2$, we shift the graph of $y = x^2$ horizontally 3 units to the left. See Figure 36(b). Finally, to get the graph of $y = (x + 3)^2 - 5$, we shift the graph of $y = (x + 3)^2$ vertically down 5 units. See Figure 36(c). Note the points plotted on each graph. Using key points can be helpful in keeping track of the transformation that has taken place.

Figure 36

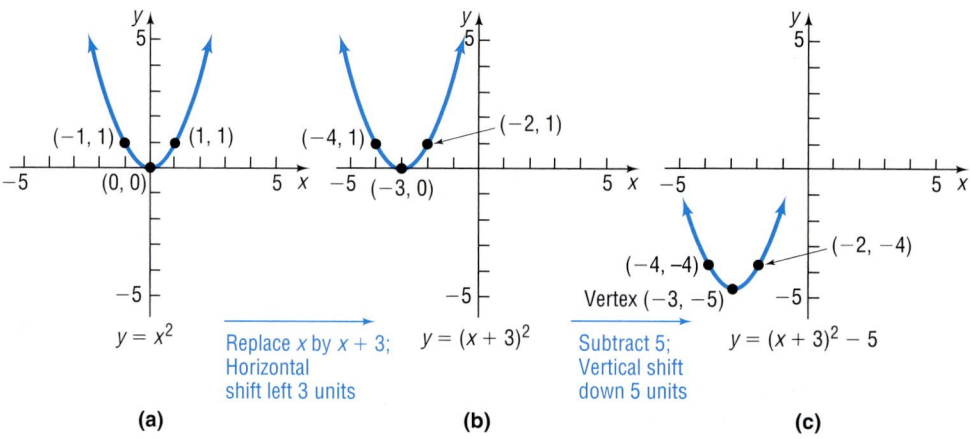

(a) $y = x^2$ — Replace x by $x + 3$; Horizontal shift left 3 units

(b) $y = (x + 3)^2$ — Subtract 5; Vertical shift down 5 units

(c) $y = (x + 3)^2 - 5$

✔ CHECK: Graph $Y_1 = f(x) = x^2 + 6x + 4$ and compare the graph to Figure 36(c).

In Example 2, if the vertical shift had been done first, followed by the horizontal shift, the final graph would have been the same. (Try it for yourself.)

We now have two ways to graph a quadratic function:

1. Complete the square and apply shifting techniques (Example 2).
2. Use the results given in the Summary on page 232 in Section 2.3 to find the vertex and axis of symmetry and to determine whether the graph opens up or down. Then locate additional points (such as the y-intercept and x-intercepts, if any) to complete the graph.

NOW WORK PROBLEMS 51 AND 75.

② Reflections About the x-Axis and the y-Axis

EXPLORATION   Reflection about the x-Axis

(a) Graph $Y_1 = x^2$, followed by $Y_2 = -x^2$.
(b) Graph $Y_1 = |x|$, followed by $Y_2 = -|x|$.
(c) Graph $Y_1 = x^2 - 4$, followed by $Y_2 = -(x^2 - 4) = -x^2 + 4$.

RESULT   See Tables 9(a), (b), and (c) and Figures 37(a), (b), and (c). In each instance, Y_2 is the reflection about the x-axis of Y_1.

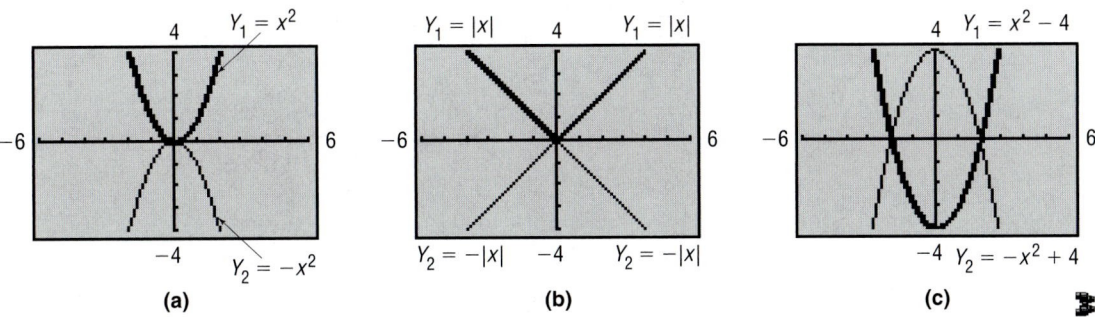

TABLE 9

X	Y₁	Y₂
-3	9	-9
-2	4	-4
-1	1	-1
0	0	0
1	1	-1
2	4	-4
3	9	-9

$Y_2 = -X^2$

(a)

X	Y₁	Y₂
-3	3	-3
-2	2	-2
-1	1	-1
0	0	0
1	1	-1
2	2	-2
3	3	-3

$Y_2 = -abs(X)$

(b)

X	Y₁	Y₂
-3	5	-5
-2	0	0
-1	-3	3
0	-4	4
1	-3	3
2	0	0
3	5	-5

$Y_2 = -X^2 + 4$

(c)

Figure 37

(a) $Y_1 = x^2$, $Y_2 = -x^2$

(b) $Y_1 = |x|$, $Y_2 = -|x|$

(c) $Y_1 = x^2 - 4$, $Y_2 = -x^2 + 4$

> When the right side of the function $y = f(x)$ is multiplied by -1, the graph of the new function $y = -f(x)$ is the **reflection about the x-axis** of the graph of the function $y = f(x)$.

NOW WORK PROBLEM 45.

EXPLORATION   Reflection about the *y*-axis

(a) Graph $Y_1 = \sqrt{x}$, followed by $Y_2 = \sqrt{-x}$.

(b) Graph $Y_1 = x + 1$, followed by $Y_2 = -x + 1$.

(c) Graph $Y_1 = x^4 + x$, followed by $Y_2 = (-x)^4 + (-x) = x^4 - x$.

RESULT   See Tables 10(a), (b), and (c) and Figures 38(a), (b), and (c). In each instance, the second graph is the reflection about the *y*-axis of the first graph.

TABLE 10

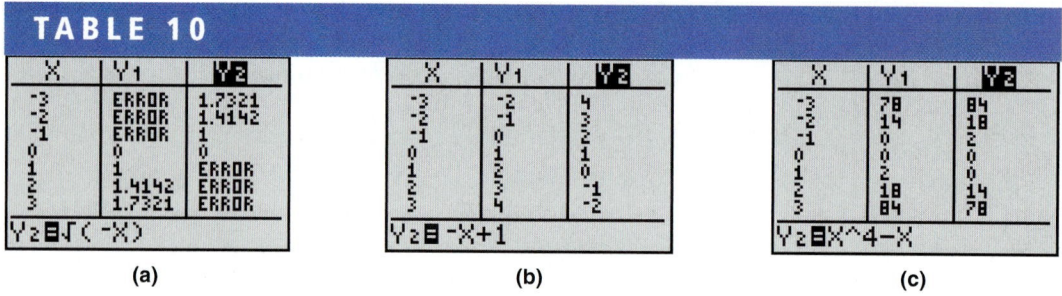

(a)   (b)   (c)

Figure 38

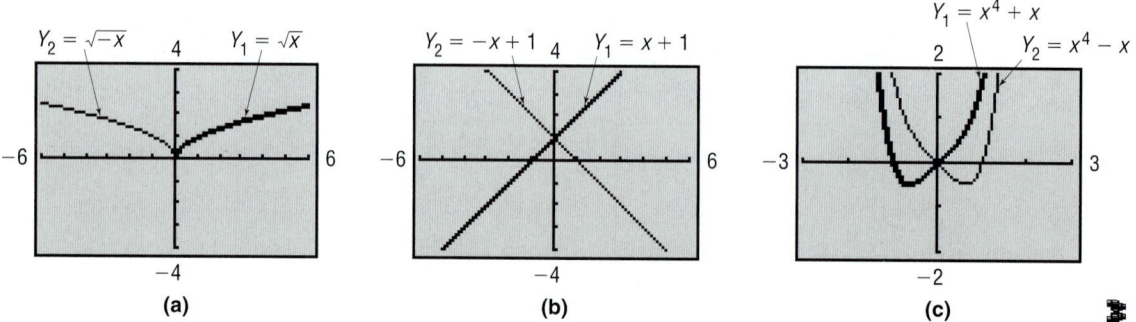

(a)   (b)   (c)

> When the graph of the function $y = f(x)$ is known, the graph of the new function $y = f(-x)$ is the **reflection about the *y*-axis** of the graph of the function $y = f(x)$.

③ Compressions and Stretches

Figure 39

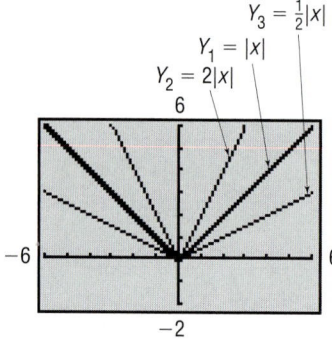

EXPLORATION   On the same screen, graph each of the following functions:

$$Y_1 = |x|$$
$$Y_2 = 2|x|$$
$$Y_3 = \frac{1}{2}|x|$$

RESULT   Figure 39 illustrates the graphs. You should have observed the following pattern. The graph of $Y_2 = 2|x|$ can be obtained from the graph of $Y_1 = |x|$ by multiplying each *y*-coordinate of $Y_1 = |x|$ by 2. This is sometimes referred to as a vertical *stretch* using a factor of 2.

The graph of $Y_3 = \frac{1}{2}|x|$ can be obtained from the graph of $Y_1 = |x|$ by multiplying each y-coordinate by $\frac{1}{2}$. This is sometimes referred to as a vertical *compression* using a factor of $\frac{1}{2}$.

Look at Tables 11 and 12, where $Y_1 = |x|$, $Y_2 = 2|x|$, and $Y_3 = \frac{1}{2}|x|$.

Notice that the values for Y_2 in Table 11 are two times the values of Y_1 for each x-value. Therefore, the graph of Y_2 will be vertically *stretched* by a factor of 2. Likewise, the values of Y_3 in Table 12 are half the values of Y_1 for each x-value.

Therefore, the graph of Y_3 will be vertically *compressed* by a factor of $\frac{1}{2}$.

TABLE 11		
X	Y₁	**Y₂**
-2	2	4
-1	1	2
0	0	0
1	1	2
2	2	4
3	3	6
4	4	8

Y₂ᴇ2abs(X)

TABLE 12		
X	Y₁	**Y₃**
-2	2	1
-1	1	.5
0	0	0
1	1	.5
2	2	1
3	3	1.5
4	4	2

Y₃ᴇ.5abs(X)

> When the right side of a function $y = f(x)$ is multiplied by a positive number a, the graph of the new function $y = af(x)$ is obtained by multiplying each y-coordinate on the graph of $y = f(x)$ by a. The new graph is a **vertically compressed** (if $0 < a < 1$) or a **vertically stretched** (if $a > 1$) version of the graph of $y = f(x)$.

 NOW WORK PROBLEM 41.

What happens if the argument x of a function $y = f(x)$ is multiplied by a positive number a, creating a new function $y = f(ax)$? To find the answer, we look at the following Exploration.

EXPLORATION   On the same screen, graph each of the following functions:
$$Y_1 = f(x) = x^2$$
$$Y_2 = f(4x) = (4x)^2$$
$$Y_3 = f\left(\frac{1}{2}x\right) = \left(\frac{1}{2}x\right)^2$$

RESULT   You should have obtained the graphs shown in Figure 40. Look at Table 13(a).

Figure 40

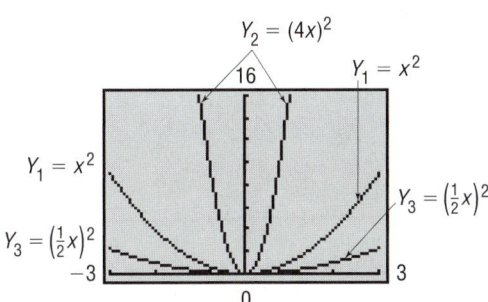

$Y_2 = (4x)^2$
$Y_1 = x^2$
16
$Y_1 = x^2$
$Y_3 = \left(\frac{1}{2}x\right)^2$
$Y_3 = \left(\frac{1}{2}x\right)^2$
−3
3
0

TABLE 13					
X	Y₁	**Y₂**	X	Y₁	**Y₃**
0	0	0	0	0	0
.25	.0625	1	.5	.25	.0625
1	1	16	1	1	.25
4	16	256	2	4	1
16	256	4096	4	16	4
			8	64	16
			16	256	64

Y₂ᴇ(4X)²   Y₃ᴇ(X/2)²

(a)   (b)

Notice that $(1, 1)$, $(4, 16)$, and $(16, 256)$ are points on the graph of $Y_1 = x^2$. Also, $(0.25, 1)$, $(1, 16)$, and $(4, 256)$ are points on the graph of $Y_2 = (4x)^2$. For each y-coordinate, the x-coordinate on the graph of of Y_2 is $\frac{1}{4}$ the x-coordinate on Y_1. The graph of $Y_2 = (4x)^2$ is obtained by multiplying the x-coordinate of each point on the graph of $Y_1 = x^2$ by $\frac{1}{4}$.

The graph of $Y_2 = (4x)^2$ is the graph of $Y_1 = x^2$ compressed horizontally.

Look at Table 13(b). Notice that $(0.5, 0.25)$, $(1, 1)$, $(2, 4)$, and $(4, 16)$ are points on the graph of $Y_1 = x^2$. Also, $(1, 0.25)$, $(2, 1)$, $(4, 4)$, and $(8, 16)$ are points on the graph of $Y_3 = \left(\frac{1}{2}x\right)^2$. For each y-coordinate, the x-coordinate on the graph of Y_3 is 2 times the x-coordinate on Y_1. The graph of $Y_3 = \left(\frac{1}{2}x\right)^2$ is obtained by multiplying the x-coordinate of each point on the graph of $Y_1 = x^2$ by a factor of 2. The graph of $Y_3 = \left(\frac{1}{2}x\right)^2$ is the graph of $Y_1 = x^2$ stretched horizontally. ▰

If the argument of x of a function $y = f(x)$ is multiplied by a positive number a, the graph of the new function $y = f(ax)$ is obtained by multiplying each x-coordinate of $y = f(x)$ by $\frac{1}{a}$. A **horizontal compression** results if $a > 1$, and a **horizontal stretch** occurs if $0 < a < 1$.

Let's look at an example.

Figure 41

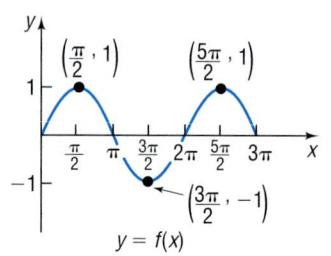

$y = f(x)$

EXAMPLE 3   Graphing Using Stretches and Compressions

The graph of $y = f(x)$ is given in Figure 41. Use this graph to find the graphs of

(a) $y = 2f(x)$ (b) $y = f(3x)$

Solution   (a) The graph of $y = 2f(x)$ is obtained by multiplying each y-coordinate of $y = f(x)$ by a factor of 2. See Figure 42(a).

(b) The graph of $y = f(3x)$ is obtained from the graph of $y = f(x)$ by multiplying each x-coordinate of $y = f(x)$ by a factor of $\frac{1}{3}$. See Figure 42(b).

Figure 42

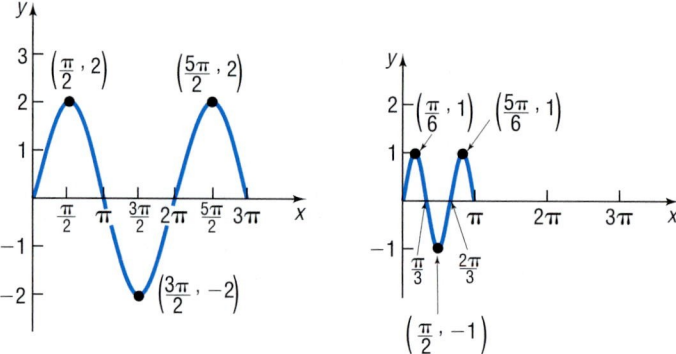

(a) $y = 2f(x)$ (b) $y = f(3x)$

NOW WORK PROBLEMS 63(e) AND (g).

SUMMARY | **Summary of Graphing Techniques**

Table 14 summarizes the graphing procedures that we have just discussed.

TABLE 14

To Graph:	Draw the Graph of f and:	Functional Change to $f(x)$
Vertical shifts		
$y = f(x) + k$, $k > 0$	Add k to each y-coordinate of $y = f(x)$ and raise the graph of f by k units.	Add k to $f(x)$.
$y = f(x) - k$, $k > 0$	Subtract k from each y-coordinate of $y = f(x)$ and lower the graph of f by k units.	Subtract k from $f(x)$.
Horizontal shifts		
$y = f(x + h)$, $h > 0$	Subtract h from each x-coordinate of $y = f(x)$ and shift the graph of f to the left h units.	Replace x by $x + h$.
$y = f(x - h)$, $h > 0$	Add h to each x-coordinate of $y = f(x)$ and shift the graph of f to the right h units.	Replace x by $x - h$.
Compressing or stretching		
$y = af(x)$, $a > 0$	Multiply each y-coordinate of $y = f(x)$ by a. Stretch the graph of f vertically if $a > 1$. Compress the graph of f vertically if $0 < a < 1$.	Multiply $f(x)$ by a.
$y = f(ax)$, $a > 0$	Multiply each x-coordinate of $y = f(x)$ by $\dfrac{1}{a}$. Stretch the graph of f horizontally if $0 < a < 1$. Compress the graph of f horizontally if $a > 1$.	Replace x by ax.
Reflection about the x-axis		
$y = -f(x)$	Multiply each y-coordinate of $y = f(x)$ by -1. Reflect the graph of f about the x-axis.	Multiply $f(x)$ by -1.
Reflection about the y-axis		
$y = f(-x)$	Multiply each x-coordinate of $y = f(x)$ by -1. Reflect the graph of f about the y-axis.	Replace x by $-x$.

The examples that follow combine some of the procedures outlined in this section to get the required graph.

EXAMPLE 4 | **Determining the Function Obtained from a Series of Transformations**

Find the function that is finally graphed after the following three transformations are applied to the graph of $y = |x|$.

1. Shift left 2 units.   2. Shift up 3 units.   3. Reflect about the y-axis.

Solution 1. Shift left 2 units: Replace x by $x + 2$. $y = |x + 2|$
2. Shift up 3 units: Add 3. $y = |x + 2| + 3$
3. Reflect about the y-axis: Replace x by $-x$. $y = |-x + 2| + 3$ ■

 **NOW WORK PROBLEM 25.**

EXAMPLE 5   **Combining Graphing Procedures**

Graph the function: $f(x) = \dfrac{3}{x - 2} + 1$

Solution   We use the following steps to obtain the graph of f:

STEP 1: $y = \dfrac{1}{x}$ Reciprocal function.

STEP 2: $y = \dfrac{3}{x}$ Multiply by 3; vertical stretch of the graph of $y = \dfrac{1}{x}$ by a factor of 3.

STEP 3: $y = \dfrac{3}{x - 2}$ Replace x by x − 2; horizontal shift to the right 2 units.

STEP 4: $y = \dfrac{3}{x - 2} + 1$ Add 1; vertical shift up 1 unit.

See Figure 43.

Figure 43

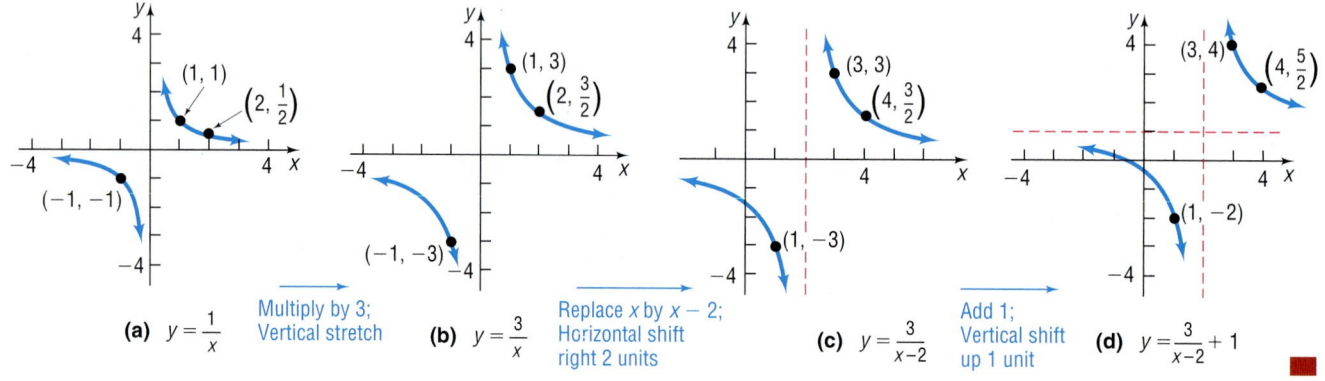

(a) $y = \dfrac{1}{x}$ Multiply by 3; Vertical stretch

(b) $y = \dfrac{3}{x}$ Replace x by x − 2; Horizontal shift right 2 units

(c) $y = \dfrac{3}{x-2}$ Add 1; Vertical shift up 1 unit

(d) $y = \dfrac{3}{x-2} + 1$ ■

Other orderings of the steps shown in Example 5 would also result in the graph of f. For example, try this one:

STEP 1: $y = \dfrac{1}{x}$ Reciprocal function.

STEP 2: $y = \dfrac{1}{x - 2}$ Replace x by x − 2; horizontal shift to the right 2 units.

STEP 3: $y = \dfrac{3}{x - 2}$ Multiply by 3; vertical stretch of the graph of $y = \dfrac{1}{x - 2}$ by a factor of 3.

STEP 4: $y = \dfrac{3}{x - 2} + 1$ Add 1; vertical shift up 1 unit.

| EXAMPLE 6 | **Combining Graphing Procedures** |

Graph the function: $f(x) = \sqrt{1 - x} + 2$

Solution   We use the following steps to obtain the graph of $y = \sqrt{1 - x} + 2$:

STEP 1: $y = \sqrt{x}$ *Square root function.*
STEP 2: $y = \sqrt{x + 1}$ *Replace x by x + 1; horizontal shift left 1 unit.*
STEP 3: $y = \sqrt{-x + 1} = \sqrt{1 - x}$ *Replace x by −x; reflect about y-axis.*
STEP 4: $y = \sqrt{1 - x} + 2$ *Add 2; vertical shift up 2 units.*

See Figure 44.

Figure 44

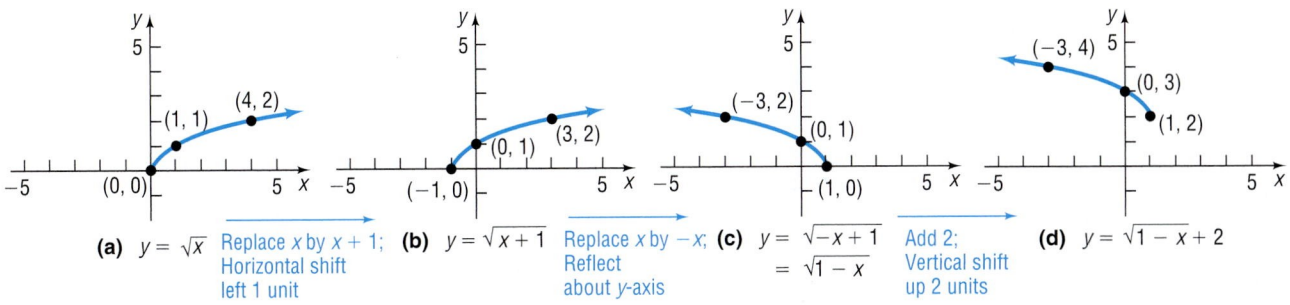

(a) $y = \sqrt{x}$ Replace *x* by *x* + 1;   (b) $y = \sqrt{x + 1}$ Replace *x* by −*x*; (c) $y = \sqrt{-x + 1}$ Add 2;   (d) $y = \sqrt{1 - x} + 2$
Horizontal shift   Reflect $= \sqrt{1 - x}$ Vertical shift
left 1 unit   about *y*-axis   up 2 units

✔ CHECK: Graph $Y_1 = f(x) = \sqrt{1 - x} + 2$ and compare the graph to Figure 44(d).

NOW WORK PROBLEM **55.**

3.4 Concepts and Vocabulary

In Problems 1–3, fill in the blanks.

1. Suppose that the graph of a function f is known. Then the graph of $y = f(x - 2)$ may be obtained by a(n) _____ shift of the graph of f to the _____ a distance of 2 units.

2. Suppose that the graph of a function f is known. Then the graph of $y = f(-x)$ may be obtained by a reflection about the _____-axis of the graph of the function $y = f(x)$.

3. Suppose that the graph of a function f is known. Then the graph of $y = f(3x)$ may be obtained by a horizontal _____ of the graph of f by a factor of _____.

In Problems 4–6, answer True or False to each statement.

4. The graph of $y = -f(x)$ is the reflection about the *x*-axis of the graph of $y = f(x)$.

5. To obtain the graph of $y = f(x + 2) - 3$, shift the graph of $y = f(x)$ horizontally to the right 2 units and vertically down 3 units.

6. To obtain the graph of $y = 4f(x)$, vertically stretch the graph of $y = f(x)$ by a factor of 4. That is, multiply each *y*-coordinate on the graph of $y = f(x)$ by 4.

7. Discuss the two methods for graphing quadratic functions. If asked to graph the quadratic function $f(x) = 3(x - 2)^2 + 1$, which method would you use? Why?

8. Suppose that the graph of a function f is known. Explain how the graph of $y = 4f(x)$ differs from the graph of $y = f(4x)$.

3.4 Exercises

In Problems 1–12, match each graph to one of the following functions without using a graphing utility.

A. $y = x^2 + 2$ B. $y = -x^2 + 2$ C. $y = |x| + 2$ D. $y = -|x| + 2$

E. $y = (x - 2)^2$ F. $y = -(x + 2)^2$ G. $y = |x - 2|$ H. $y = -|x + 2|$

I. $y = 2x^2$ J. $y = -2x^2$ K. $y = 2|x|$ L. $y = -2|x|$

1. 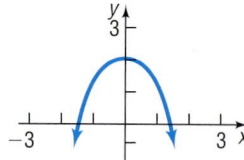   **2.** 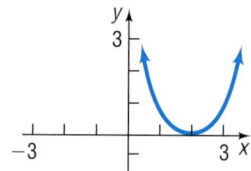   **3.** 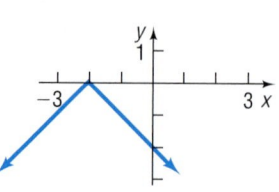   **4.**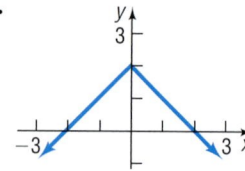

5. 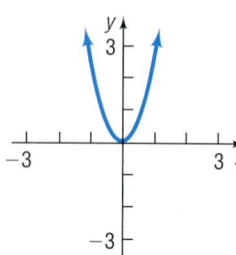   **6.** 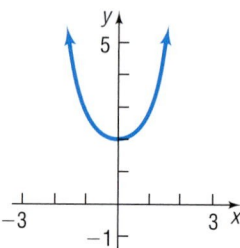   **7.** 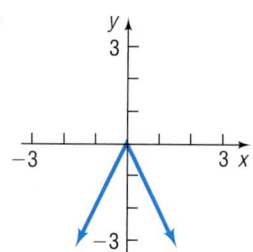   **8.**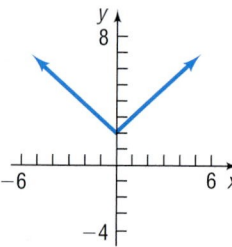

9. 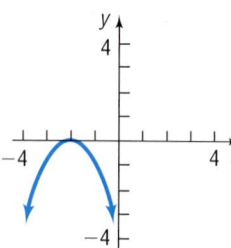   **10.** 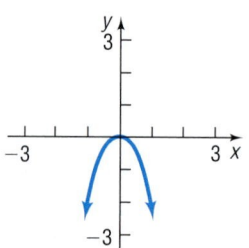   **11.** 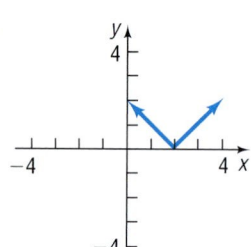   **12.**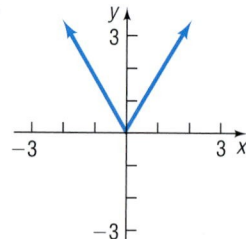

In Problems 13–16, match each graph to one of the following functions without using a graphing utility.

A. $y = x^3$ B. $y = (x + 2)^3$ C. $y = -2x^3$ D. $y = x^3 + 2$

13. 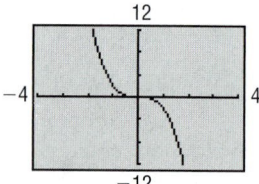   **14.** 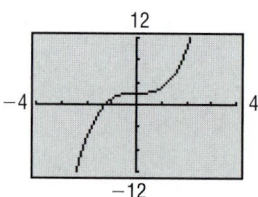   **15.** 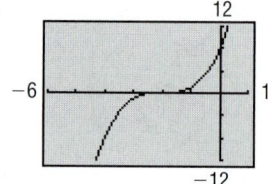   **16.**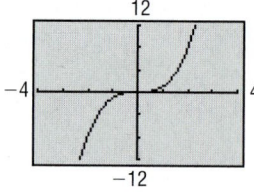

In Problems 17–24, write the function whose graph is the graph of $y = x^3$, but is:

17. Shifted to the right 4 units   **18.** Shifted to the left 4 units

19. Shifted up 4 units   **20.** Shifted down 4 units

21. Reflected about the y-axis   **22.** Reflected about the x-axis

23. Vertically stretched by a factor of 4   **24.** Horizontally stretched by a factor of 4

In Problems 25–28, find the function that is finally graphed after the following transformations are applied to the graph of $y = \sqrt{x}$.

25. (1) Shift up 2 units   **26.** (1) Reflect about the x-axis
       (2) Reflect about the x-axis   (2) Shift right 3 units
       (3) Reflect about the y-axis   (3) Shift down 2 units

27. (1) Reflect about the *x*-axis
(2) Shift up 2 units
(3) Shift left 3 units

28. (1) Shift up 2 units
(2) Reflect about the *y*-axis
(3) Shift left 3 units

29. If $(3, 0)$ is a point on the graph of $y = f(x)$, which of the following must be on the graph of $y = -f(x)$?
  (a) $(0, 3)$ (b) $(0, -3)$
  (c) $(3, 0)$ (d) $(-3, 0)$

30. If $(3, 0)$ is a point on the graph of $y = f(x)$, which of the following must be on the graph of $y = f(-x)$?
  (a) $(0, 3)$ (b) $(0, -3)$
  (c) $(3, 0)$ (d) $(-3, 0)$

31. If $(0, 3)$ is a point on the graph of $y = f(x)$, which of the following must be on the graph of $y = 2f(x)$?
  (a) $(0, 3)$ (b) $(0, 2)$
  (c) $(0, 6)$ (d) $(6, 0)$

32. If $(3, 0)$ is a point on the graph of $y = f(x)$, which of the following must be on the graph of $y = \frac{1}{2}f(x)$?
  (a) $(3, 0)$ (b) $(3/2, 0)$
  (c) $(0, 3/2)$ (d) $(1/2, 0)$

In Problems 33–62 graph each function using transformations (shifting, compressing, stretching, and/or reflecting). Start with the graph of the basic function (for example, $y = x^2$) and show all stages. Verify your results using a graphing utility.

33. $f(x) = x^2 - 1$ **34.** $f(x) = x^2 + 4$ **35.** $g(x) = x^3 + 1$

36. $g(x) = x^3 - 1$ **37.** $h(x) = \sqrt{x - 2}$ **38.** $h(x) = \sqrt{x + 1}$

39. $f(x) = (x - 1)^3 + 2$ **40.** $f(x) = (x + 2)^3 - 3$ **41.** $g(x) = 4\sqrt{x}$

42. $g(x) = \frac{1}{2}\sqrt{x}$ **43.** $h(x) = \frac{1}{2x}$ **44.** $h(x) = \frac{4}{x}$

45. $f(x) = -\frac{1}{x}$ **46.** $f(x) = -\sqrt{x}$ **47.** $g(x) = |-x|$

48. $g(x) = (-x)^3$ **49.** $h(x) = -x^3 + 2$ **50.** $h(x) = \frac{1}{-x} + 2$

51. $f(x) = 2(x + 1)^2 - 3$ **52.** $f(x) = 3(x - 2)^2 + 1$ **53.** $g(x) = \sqrt{x - 2} + 1$

54. $g(x) = |x + 1| - 3$ **55.** $h(x) = \sqrt{-x} - 2$ **56.** $h(x) = \frac{4}{x} + 2$

57. $f(x) = -(x + 1)^3 - 1$ **58.** $f(x) = -4\sqrt{x - 1}$ **59.** $g(x) = 2|1 - x|$

60. $g(x) = 4\sqrt{2 - x}$ **61.** $h(x) = 2\,\text{int}(x - 1)$ **62.** $h(x) = \text{int}(-x)$

In Problems 63–68, the graph of a function f is illustrated. Use the graph of f as the first step toward graphing each of the following functions:

(a) $F(x) = f(x) + 3$ (b) $G(x) = f(x + 2)$ (c) $P(x) = -f(x)$ (d) $H(x) = f(x + 1) - 2$

(e) $Q(x) = \frac{1}{2}f(x)$ (f) $g(x) = f(-x)$ (g) $h(x) = f(2x)$

63.

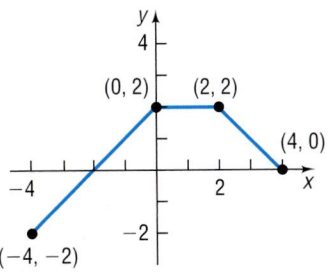

64.

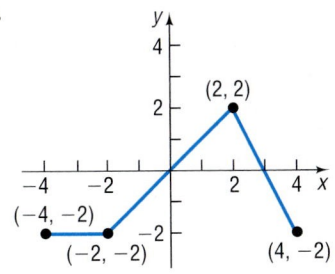

65.

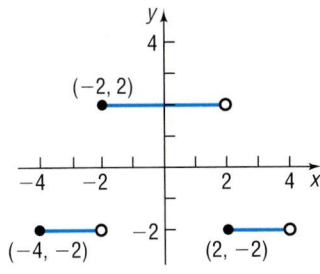

66.

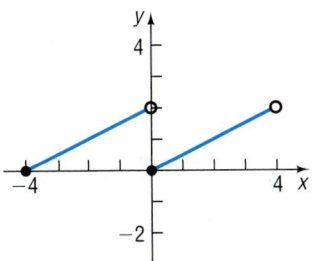

67.

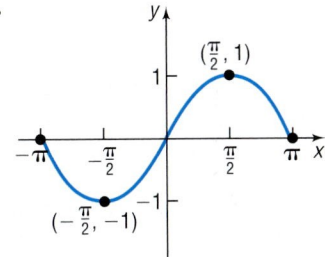

68.

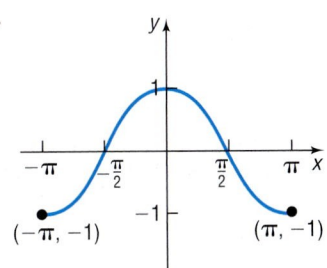

69. Exploration

(a) Use a graphing utility to graph $y = x + 1$ and $y = |x + 1|$.
(b) Graph $y = 4 - x^2$ and $y = |4 - x^2|$.
(c) Graph $y = x^3 + x$ and $y = |x^3 + x|$.
(d) What do you conclude about the relationship between the graphs of $y = f(x)$ and $y = |f(x)|$?

71. The graph of a function f is illustrated in the figure.

(a) Draw the graph of $y = |f(x)|$.
(b) Draw the graph of $y = f(|x|)$.

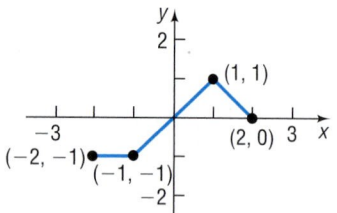

70. Exploration

(a) Use a graphing utility to graph $y = x + 1$ and $y = |x| + 1$.
(b) Graph $y = 4 - x^2$ and $y = 4 - |x|^2$.
(c) Graph $y = x^3 + x$ and $y = |x|^3 + |x|$.
(d) What do you conclude about the relationship between the graphs of $y = f(x)$ and $y = f(|x|)$?

72. The graph of a function f is illustrated in the figure.

(a) Draw the graph of $y = |f(x)|$.
(b) Draw the graph of $y = f(|x|)$.

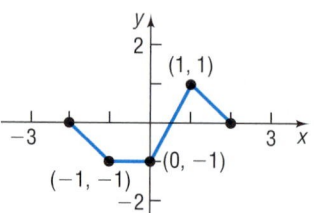

In Problems 73–78, complete the square of each quadratic expression. Then graph each function by hand using the technique of shifting. Verify your results using a graphing utility. (If necessary, refer to Section 1.3 to review completing the square.)

73. $f(x) = x^2 + 2x$

76. $f(x) = x^2 + 4x + 2$

74. $f(x) = x^2 - 6x$

77. $f(x) = x^2 + x + 1$

75. $f(x) = x^2 - 8x + 1$

78. $f(x) = x^2 - x + 1$

79. The equation $y = (x - c)^2$ defines a *family of parabolas*, one parabola for each value of c. On one set of coordinate axes, graph the members of the family for $c = 0$, $c = 3$, and $c = -2$.

80. Repeat Problem 79 for the family of parabolas $y = x^2 + c$.

81. Temperature Measurements   The relationship between the Celsius (°C) and Fahrenheit (°F) scales for measuring temperature is given by the equation

$$F = \frac{9}{5}C + 32$$

The relationship between the Celsius (°C) and Kelvin (K) scales is $K = C + 273$. Graph the equation $F = \frac{9}{5}C + 32$ using degrees Fahrenheit on the y-axis and degrees Celsius on the x-axis. Use the techniques introduced in this section to obtain the graph showing the relationship between Kelvin and Fahrenheit temperatures.

82. Period of a Pendulum   The period T (in seconds) of a simple pendulum is a function of its length l (in feet) defined by the equation

$$T = 2\pi \sqrt{\frac{l}{g}}$$

where $g \approx 32.2$ feet per second per second is the acceleration of gravity.

(a) Use a graphing utility to graph the function $T = T(l)$.
(b) Now graph the functions $T = T(l + 1), T = T(l + 2)$, and $T = T(l + 3)$.
(c) Discuss how adding to the length l changes the period T.
(d) Now graph the functions $T = T(2l), T = T(3l)$, and $T = T(4l)$.

(e) Discuss how multiplying the length l by factors of 2, 3, and 4 changes the period T.

83. Cigar Company Profits   The daily profits of a cigar company from selling x cigars are given by

$$p(x) = -0.05x^2 + 100x - 2000$$

The government wishes to impose a tax on cigars (sometimes called a *sin tax*) that gives the company the option of either paying a flat tax of $10,000 per day or a tax of 10% on profits. As chief financial officer (CFO) of the company, you need to decide which tax is the better option for the company.

(a) On the same screen, graph $Y_1 = p(x) - 10,000$ and $Y_2 = (1 - 0.10)p(x)$.
(b) Based on the graph, which option would you select? Why?
(c) Using the terminology learned in this section, describe each graph in terms of the graph of $p(x)$.
(d) Suppose that the government offered the options of a flat tax of $4800 or a tax of 10% on profits. Which would you select? Why?

PREPARING FOR THIS SECTION

Before getting started, review the following:

✓ Finding Values of a Function (Section 2.1, pp. 201–202)   ✓ Mixed Quotients (Section R.7, pp. 66–69)
✓ Domain of a Function (Section 2.1, pp. 203–204)   ✓ Adding Rational Expressions (Section R.7, pp. 62–66)

3.5   OPERATIONS ON FUNCTIONS; COMPOSITE FUNCTIONS

OBJECTIVES   **1** Form the Sum, Difference, Product, and Quotient of Two
Functions
2 Form the Composite Function and Find Its Domain

1   In this section, we introduce operations on functions. We shall see that functions, like numbers, can be added, subtracted, multiplied, and divided. For example, if $f(x) = x^2 + 9$ and $g(x) = 3x + 5$, then

$$f(x) + g(x) = (x^2 + 9) + (3x + 5) = x^2 + 3x + 14$$

The new function $y = x^2 + 3x + 14$ is called the *sum function f + g*.
  Similarly,

$$f(x) \cdot g(x) = (x^2 + 9)(3x + 5) = 3x^3 + 5x^2 + 27x + 45$$

The new function $y = 3x^3 + 5x^2 + 27x + 45$ is called the *product function
f · g*.
  More precise definitions are given next.

If f and g are functions:
The **sum $f + g$** is the function defined by

$$(f + g)(x) = f(x) + g(x)$$

  The domain of $f + g$ consists of the numbers x that are in the domains of both f and g.

The **difference $f - g$** is the function defined by

$$(f - g)(x) = f(x) - g(x)$$

  The domain of $f - g$ consists of the numbers x that are in the domains of both f and g.

The **product $f \cdot g$** is the function defined by

$$(f \cdot g)(x) = f(x) \cdot g(x)$$

  The domain of $f \cdot g$ consists of the numbers x that are in the domains of both f and g.

The **quotient** $\dfrac{f}{g}$ is the function defined by

$$\left(\frac{f}{g}\right)(x) = \frac{f(x)}{g(x)}, \qquad g(x) \neq 0$$

The domain of $\dfrac{f}{g}$ consists of the numbers x for which $g(x) \neq 0$ that are in the domains of both f and g.

EXAMPLE 1   **Operations on Functions**

Let f and g be two functions defined as

$$f(x) = \frac{1}{x + 2} \quad \text{and} \quad g(x) = \frac{x}{x - 1}$$

Find the following, and determine the domain in each case.

(a) $(f + g)(x)$ (b) $(f - g)(x)$ (c) $(f \cdot g)(x)$ (d) $\left(\dfrac{f}{g}\right)(x)$

Solution   The domain of f is $\{x \mid x \neq -2\}$ and the domain of g is $\{x \mid x \neq 1\}$.

(a) $(f + g)(x) = f(x) + g(x) = \dfrac{1}{x + 2} + \dfrac{x}{x - 1} = \dfrac{x - 1}{(x + 2)(x - 1)} + \dfrac{x(x + 2)}{(x + 2)(x - 1)} = \dfrac{x^2 + 3x - 1}{(x + 2)(x - 1)}$

The domain of $f + g$ consists of those numbers x that are in the domains of both f and g. Therefore, the domain of $f + g$ is $\{x \mid x \neq -2, x \neq 1\}$.

(b) $(f - g)(x) = f(x) - g(x) = \dfrac{1}{x + 2} - \dfrac{x}{x - 1} = \dfrac{x - 1}{(x + 2)(x - 1)} - \dfrac{x(x + 2)}{(x + 2)(x - 1)} = \dfrac{-(x^2 + x + 1)}{(x + 2)(x - 1)}$

The domain of $f - g$ consists of those numbers x that are in the domains of both f and g. Therefore, the domain of $f - g$ is $\{x \mid x \neq -2, x \neq 1\}$.

(c) $(f \cdot g)(x) = f(x) \cdot g(x) = \dfrac{1}{x + 2} \cdot \dfrac{x}{x - 1} = \dfrac{x}{(x + 2)(x - 1)}$

The domain of $f \cdot g$ consists of those numbers x that are in the domains of both f and g. Therefore, the domain of $f \cdot g$ is $\{x \mid x \neq -2, x \neq 1\}$.

(d) $\left(\dfrac{f}{g}\right)(x) = \dfrac{f(x)}{g(x)} = \dfrac{\dfrac{1}{x + 2}}{\dfrac{x}{x - 1}} = \dfrac{1}{x + 2} \cdot \dfrac{x - 1}{x} = \dfrac{x - 1}{x(x + 2)}$

The domain of $\dfrac{f}{g}$ consists of the numbers x for which $g(x) \neq 0$ that are in the domains of both f and g. Since $g(x) = 0$ when $x = 0$, we exclude 0 as well as -2 and 1. The domain of $\dfrac{f}{g}$ is $\{x \mid x \neq -2, x \neq 0, x \neq 1\}$. ■

NOW WORK PROBLEM 1.

In calculus, it is sometimes helpful to view a complicated function as the sum, difference, product, or quotient of simpler functions. For example,

$$F(x) = x^2 + \sqrt{x} \text{ is the sum of } f(x) = x^2 \text{ and } g(x) = \sqrt{x}.$$

$$H(x) = \frac{x^2 - 1}{x^2 + 1} \text{ is the quotient of } f(x) = x^2 - 1$$

$$\text{and } g(x) = x^2 + 1.$$

Composite Functions

2 Consider the function $y = (2x + 3)^2$. If we write $y = f(u) = u^2$ and $u = g(x) = 2x + 3$, then, by a substitution process, we can obtain the original function: $y = f(u) = f(g(x)) = (2x + 3)^2$. This process is called **composition.**

In general, suppose that f and g are two functions and that x is a number in the domain of g. By evaluating g at x, we get $g(x)$. If $g(x)$ is in the domain of f, then we may evaluate f at $g(x)$ and thereby obtain the expression $f(g(x))$. The correspondence from x to $f(g(x))$ is called a *composite function* $f \circ g$.

> Given two functions f and g, the **composite function,** denoted by $f \circ g$ (read as "f composed with g"), is defined by
>
> $$(f \circ g)(x) = f(g(x))$$
>
> The domain of $f \circ g$ is the set of all numbers x in the domain of g such that $g(x)$ is in the domain of f.

Look carefully at Figure 45. Only those x's in the domain of g for which $g(x)$ is in the domain of f can be in the domain of $f \circ g$. The reason is that if $g(x)$ is not in the domain of f then $f(g(x))$ is not defined. Because of this, the domain of $f \circ g$ is a subset of the domain of g; the range of $f \circ g$ is a subset of the range of f.

Figure 45

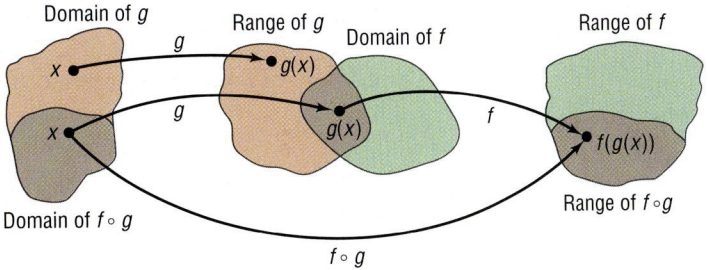

$f \circ g$

Figure 46 provides a second illustration of the definition. Notice that the "inside" function g in $f(g(x))$ is done first.

Figure 46

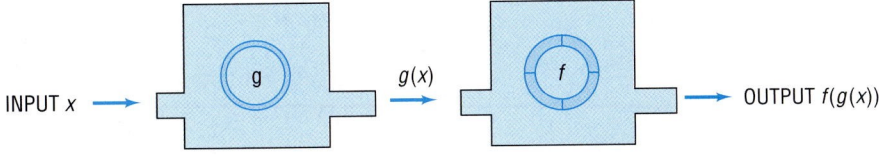

INPUT x → g → $g(x)$ → f → OUTPUT $f(g(x))$

Let's look at some examples.

EXAMPLE 2   **Evaluating a Composite Function**

Suppose that $f(x) = 2x^2 - 3$ and $g(x) = 4x$. Find:

(a) $(f \circ g)(1)$ (b) $(g \circ f)(1)$ (c) $(f \circ f)(-2)$ (d) $(g \circ g)(-1)$

Solution   (a) $(f \circ g)(1) = f(g(1)) = f(4) = 2 \cdot 16 - 3 = 29$

$$g(x) = 4x \qquad f(x) = 2x^2 - 3$$
$$g(1) = 4$$

(b) $(g \circ f)(1) = g(f(1)) = g(-1) = 4 \cdot (-1) = -4$

$$f(x) = 2x^2 - 3 \quad g(x) = 4x$$
$$f(1) = -1$$

(c) $(f \circ f)(-2) = f(f(-2)) = f(5) = 2 \cdot 25 - 3 = 47$

$$f(-2) = 2(-2)^2 - 3 = 5$$

Figure 47

(d) $(g \circ g)(-1) = g(g(-1)) = g(-4) = 4 \cdot (-4) = -16$

$$g(-1) = -4$$

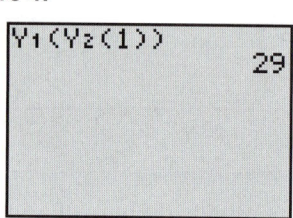

Graphing calculators can be used to evaluate composite functions.* Let $Y_1 = f(x) = 2x^2 - 3$ and $Y_2 = g(x) = 4x$. Then using a TI-83 graphing calculator, $(f \circ g)(1)$ would be found as shown in Figure 47. Notice that this is the result obtained in Example 2(a).

*NOW WORK PROBLEM **13**.*

EXAMPLE 3   **Finding a Composite Function**

Suppose that $f(x) = x^2 + 3x - 1$ and $g(x) = 2x + 3$. Find:

(a) $f \circ g$ (b) $g \circ f$

State the domain of each composite function.

Solution   The domain of f and the domain of g are all real numbers.

(a) $f \circ g = f(g(x)) = f(2x + 3) = (2x + 3)^2 + 3(2x + 3) - 1$

$$f(x) = x^2 + 3x - 1$$

$$= 4x^2 + 12x + 9 + 6x + 9 - 1 = 4x^2 + 18x + 17$$

Since the domains of both f and g are all real numbers, the domain of $f \circ g$ is all real numbers.

(b) $g \circ f = g(f(x)) = g(x^2 + 3x - 1) = 2(x^2 + 3x - 1) + 3$

$$g(x) = 2x + 3$$

$$= 2x^2 + 6x - 2 + 3 = 2x^2 + 6x + 1$$

Since the domains of both f and g are all real numbers, the domain of $f \circ g$ is all real numbers.

---

*Consult your owner's manual for the appropriate keystrokes.

Look back at Figure 45. In determining the domain of the composite function $(f \circ g)(x) = f(g(x))$, keep the following two thoughts in mind about the input x.

1. $g(x)$ must be defined so any x not in the domain of g must be excluded.
2. $f(g(x))$ must be defined so any x for which $g(x)$ is not in the domain of f must be excluded.

EXAMPLE 4 **Finding the Domain of $f \circ g$**

Find the domain of $(f \circ g)(x)$ if $f(x) = \dfrac{1}{x + 2}$ and $g(x) = \dfrac{4}{x - 1}$.

Solution For $(f \circ g)(x) = f(g(x))$, we first note that the domain of g is $\{x | x \neq 1\}$, so we exclude 1 from the domain of $f \circ g$. Next, we note that the domain of f is $\{x | x \neq -2\}$, which means that $g(x)$ cannot equal -2. We solve the equation $g(x) = -2$ to determine what values of x to exclude.

$$\frac{4}{x - 1} = -2 \qquad \textcolor{blue}{g(x) = -2.}$$
$$4 = -2(x - 1)$$
$$4 = -2x + 2$$
$$2x = -2$$
$$x = -1$$

We also exclude -1 from the domain of $f \circ g$. The domain of $f \circ g$ is $\{x | x \neq -1, x \neq 1\}$.

✔ **CHECK:** For $x = 1$, $g(x) = \dfrac{4}{x - 1}$ is not defined, so $(f \circ g)(x) = f(g(x))$ is not defined.

For $x = -1$, $g(-1) = \dfrac{4}{-2} = -2$, and $(f \circ g)(-1) = f(g(-1)) = f(-2)$ is not defined. ■ ■

▬▬ **NOW WORK PROBLEM 23.**

EXAMPLE 5 **Finding a Composite Function and Its Domain**

Suppose that $f(x) = \dfrac{1}{x + 2}$ and $g(x) = \dfrac{4}{x - 1}$. Find:

(a) $f \circ g$ (b) $f \circ f$

State the domain of each composite function.

Solution The domain of f is $\{x | x \neq -2\}$ and the domain of g is $\{x | x \neq 1\}$.

(a) $(f \circ g)(x) = f(g(x)) = f\left(\dfrac{4}{x - 1}\right) = \dfrac{1}{\dfrac{4}{x - 1} + 2} \underset{\textcolor{blue}{\text{Multiply by } \frac{x-1}{x-1}.}}{\uparrow} = \dfrac{x - 1}{4 + 2(x - 1)} = \dfrac{x - 1}{2x + 2} = \dfrac{x - 1}{2(x + 1)}$

In Example 4, we found the domain of $f \circ g$ to be $\{x \mid x \neq -1, x \neq 1\}$.

We could also find the domain of $f \circ g$ by first looking at the domain of g: $\{x \mid x \neq 1\}$. We exclude 1 from the domain of $f \circ g$ as a result. Then, we look at $f \circ g$ and notice that x cannot equal -1, since $x = -1$ results in division by 0. So we also exclude -1 from the domain of $f \circ g$. Therefore, the domain of $f \circ g$ is $\{x \mid x \neq -1, x \neq 1\}$.

(b) $(f \circ f)(x) = f(f(x)) = f\left(\dfrac{1}{x+2}\right) = \dfrac{1}{\dfrac{1}{x+2} + 2} = \dfrac{x+2}{1 + 2(x+2)} = \dfrac{x+2}{2x+5}$

The domain of $f \circ f$ consists of those x in the domain of $f(x \neq -2)$ for which

$$f(x) = \dfrac{1}{x+2} \neq -2 \qquad \dfrac{1}{x+2} = -2$$
$$1 = -2(x+2)$$
$$1 = -2x - 4$$
$$2x = -5$$
$$x = -\dfrac{5}{2}$$

or, equivalently,

$$x \neq -\dfrac{5}{2}$$

The domain of $f \circ f$ is $\left\{x \mid x \neq -\dfrac{5}{2}, x \neq -2\right\}$.

We could also find the domain of $f \circ f$ by recognizing that -2 is not in the domain of f and so should be excluded from $f \circ f$. Then, looking at $f \circ f$, we see that x cannot equal $-\dfrac{5}{2}$. Do you see why? Therefore, the domain of $f \circ f$ is $\left\{x \mid x \neq -\dfrac{5}{2}, x \neq -2\right\}$.

NOW WORK PROBLEMS 35 AND 37.

Look back at Example 3, which illustrates that, in general, $f \circ g \neq g \circ f$. Sometimes $f \circ g$ does equal $g \circ f$, as shown in the next example.

EXAMPLE 6   Showing That Two Composite Functions Are Equal

If $f(x) = 3x - 4$ and $g(x) = \dfrac{1}{3}(x+4)$, show that

$$(f \circ g)(x) = (g \circ f)(x) = x$$

for every x.

Solution
$$(f \circ g)(x) = f(g(x))$$
$$= f\left(\dfrac{x+4}{3}\right) \qquad g(x) = \dfrac{1}{3}(x+4) = \dfrac{x+4}{3}.$$
$$= 3\left(\dfrac{x+4}{3}\right) - 4 \qquad \text{Substitute } g(x) \text{ into the rule for } f, f(x) = 3x - 4.$$
$$= x + 4 - 4 = x$$

$$(g \circ f)(x) = g(f(x))$$
$$= g(3x - 4) \qquad f(x) = 3x - 4.$$
$$= \frac{1}{3}[(3x - 4) + 4] \qquad \text{Substitute } f(x) \text{ into the rule for } g, g(x) = \frac{1}{3}(x + 4).$$
$$= \frac{1}{3}(3x) = x$$

Thus, $(f \circ g)(x) = (g \circ f)(x) = x$. ∎

In Section 6.1, we shall see that there is an important relationship between functions f and g for which $(f \circ g)(x) = (g \circ f)(x) = x$.

 SEEING THE CONCEPT   Using a graphing calculator, let $Y_1 = f(x) = 3x - 4$, $Y_2 = g(x) = \frac{1}{3}(x + 4)$, $Y_3 = f \circ g$, and $Y_4 = g \circ f$. Using the viewing window $-3 \le x \le 3, -2 \le y \le 2$, graph only Y_3 and Y_4. What do you see? TRACE to verify that $Y_3 = Y_4$.

NOW WORK PROBLEM 47.

Calculus Application

Some techniques in calculus require that we be able to determine the components of a composite function. For example, the function $H(x) = \sqrt{x + 1}$ is the composition of the functions f and g, where $f(x) = \sqrt{x}$ and $g(x) = x + 1$, because $H(x) = (f \circ g)(x) = f(g(x)) = f(x + 1) = \sqrt{x + 1}$.

EXAMPLE 7   ### Finding the Components of a Composite Function

Find functions f and g such that $f \circ g = H$ if $H(x) = (x^2 + 1)^{50}$.

Solution   The function H takes $x^2 + 1$ and raises it to the power 50. A natural way to decompose H is to raise the function $g(x) = x^2 + 1$ to the power 50. If we let $f(x) = x^{50}$ and $g(x) = x^2 + 1$, then

$$(f \circ g)(x) = f(g(x))$$
$$= f(x^2 + 1)$$
$$= (x^2 + 1)^{50} = H(x)$$

Figure 48

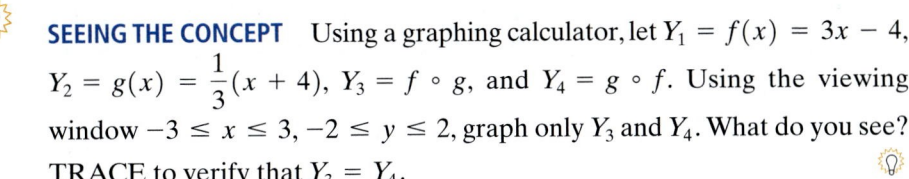

See Figure 48. ∎

Other functions f and g may be found for which $f \circ g = H$ in Example 7. For example, if $f(x) = x^2$ and $g(x) = (x^2 + 1)^{25}$, then

$$(f \circ g)(x) = f(g(x)) = f((x^2 + 1)^{25}) = [(x^2 + 1)^{25}]^2 = (x^2 + 1)^{50}$$

Although the functions f and g found as a solution to Example 7 are not unique, there is usually a "natural" selection for f and g that comes to mind first.

| EXAMPLE 8 | **Finding the Components of a Composite Function** |

Find functions f and g such that $f \circ g = H$ if $H(x) = \dfrac{1}{x + 1}$.

Solution Here H is the reciprocal of $g(x) = x + 1$. If we use the reciprocal function $f(x) = \dfrac{1}{x}$ with $g(x) = x + 1$, we find that

$$(f \circ g)(x) = f(g(x)) = f(x + 1) = \frac{1}{x + 1} = H(x) \quad \blacksquare$$

3.5 Concepts and Vocabulary

In Problems 1–3, fill in the blanks.

1. If the domain of f is all real numbers in the interval $[0, 7]$ and the domain of g is all real numbers in the interval $[-2, 5]$, the domain of $f + g$ is all real numbers in the interval _____.

2. The domain of $\dfrac{f}{g}$ consists of numbers x for which $g(x)$ _____ 0 that are in the domains of both _____ and _____.

3. If $f(x) = x + 1$ and $g(x) = x^3$, then _____ $= (x + 1)^3$.

In Problems 4–6, answer True or False to each statement.

4. $f(g(x)) = f(x) \cdot g(x)$.

5. The domain of $(f \cdot g)(x)$ consists of the numbers x that are in the domains of both f and g.

6. The domain of the composite function $(f \circ g)(x)$ is the same as the domain of $g(x)$.

7. Explain how you would find the domain of $(f \circ g)(x)$.

3.5 Exercises

In Problems 1–10, for the given functions f and g, find the following functions and state the domain of each.

(a) $f + g$ (b) $f - g$ (c) $f \cdot g$ (d) $\dfrac{f}{g}$

1. $f(x) = 3x + 4;\quad g(x) = 2x - 3$

2. $f(x) = 2x + 1;\quad g(x) = 3x - 2$

3. $f(x) = x - 1;\quad g(x) = 2x^2$

4. $f(x) = 2x^2 + 3;\quad g(x) = 4x^3 + 1$

5. $f(x) = \sqrt{x};\quad g(x) = 3x - 5$

6. $f(x) = |x|;\quad g(x) = x$

7. $f(x) = 1 + \dfrac{1}{x};\quad g(x) = \dfrac{1}{x}$

8. $f(x) = \sqrt{x - 2};\quad g(x) = \sqrt{4 - x}$

9. $f(x) = \dfrac{2x + 3}{3x - 2};\quad g(x) = \dfrac{4x}{3x - 2}$

10. $f(x) = \sqrt{x + 1};\quad g(x) = \dfrac{2}{x}$

11. Given $f(x) = 3x + 1$ and $(f + g)(x) = 6 - \dfrac{1}{2}x$, find the function g.

12. Given $f(x) = \dfrac{1}{x}$ and $\left(\dfrac{f}{g}\right)(x) = \dfrac{x + 1}{x^2 - x}$, find the function g.

In Problems 13–22, for the given functions f and g, find

(a) $(f \circ g)(4)$ (b) $(g \circ f)(2)$ (c) $(f \circ f)(1)$ (d) $(g \circ g)(0)$

Verify your results using a graphing utility.

13. $f(x) = 2x;\quad g(x) = 3x^2 + 1$

14. $f(x) = 3x + 2;\quad g(x) = 2x^2 - 1$

15. $f(x) = 4x^2 - 3$; $g(x) = 3 - \dfrac{1}{2}x^2$

16. $f(x) = 2x^2$; $g(x) = 1 - 3x^2$

17. $f(x) = \sqrt{x}$; $g(x) = 2x$

18. $f(x) = \sqrt{x + 1}$; $g(x) = 3x$

19. $f(x) = |x|$; $g(x) = \dfrac{1}{x^2 + 1}$

20. $f(x) = |x - 2|$; $g(x) = \dfrac{3}{x^2 + 2}$

21. $f(x) = \dfrac{3}{x + 1}$; $g(x) = \sqrt[3]{x}$

22. $f(x) = x^{3/2}$; $g(x) = \dfrac{2}{x + 1}$

In Problems 23–30, find the domain of the composite function $f \circ g$.

23. $f(x) = \dfrac{3}{x - 1}$; $g(x) = \dfrac{2}{x}$

24. $f(x) = \dfrac{1}{x + 3}$; $g(x) = \dfrac{-2}{x}$

25. $f(x) = \dfrac{x}{x - 1}$; $g(x) = \dfrac{-4}{x}$

26. $f(x) = \dfrac{x}{x + 3}$; $g(x) = \dfrac{2}{x}$

27. $f(x) = \sqrt{x}$; $g(x) = 2x + 3$

28. $f(x) = x - 2$; $g(x) = \sqrt{1 - x}$

29. $f(x) = x^2 + 1$; $g(x) = \sqrt{x - 1}$

30. $f(x) = x^2 + 4$; $g(x) = \sqrt{x - 2}$

In Problems 31–46, for the given functions f and g, find:

(a) $f \circ g$ *(b) $g \circ f$*   *(c) $f \circ f$*   *(d) $g \circ g$*

State the domain of each composite function.

31. $f(x) = 2x + 3$; $g(x) = 3x$

32. $f(x) = -x$; $g(x) = 2x - 4$

33. $f(x) = 3x + 1$; $g(x) = x^2$

34. $f(x) = x + 1$; $g(x) = x^2 + 4$

35. $f(x) = x^2$; $g(x) = x^2 + 4$

36. $f(x) = x^2 + 1$; $g(x) = 2x^2 + 3$

37. $f(x) = \dfrac{3}{x - 1}$; $g(x) = \dfrac{2}{x}$

38. $f(x) = \dfrac{1}{x + 3}$; $g(x) = \dfrac{-2}{x}$

39. $f(x) = \dfrac{x}{x - 1}$; $g(x) = \dfrac{-4}{x}$

40. $f(x) = \dfrac{x}{x + 3}$; $g(x) = \dfrac{2}{x}$

41. $f(x) = \sqrt{x}$; $g(x) = 2x + 3$

42. $f(x) = \sqrt{x - 2}$; $g(x) = 1 - 2x$

43. $f(x) = x^2 + 1$; $g(x) = \sqrt{x - 1}$

44. $f(x) = x^2 + 4$; $g(x) = \sqrt{x - 2}$

45. $f(x) = ax + b$; $g(x) = cx + d$

46. $f(x) = \dfrac{ax + b}{cx + d}$; $g(x) = mx$

In Problems 47–54, show that $(f \circ g)(x) = (g \circ f)(x) = x$.

47. $f(x) = 2x$; $g(x) = \dfrac{1}{2}x$

48. $f(x) = 4x$; $g(x) = \dfrac{1}{4}x$

49. $f(x) = x^3$; $g(x) = \sqrt[3]{x}$

50. $f(x) = x + 5$; $g(x) = x - 5$

51. $f(x) = 2x - 6$; $g(x) = \dfrac{1}{2}(x + 6)$

52. $f(x) = 4 - 3x$; $g(x) = \dfrac{1}{3}(4 - x)$

53. $f(x) = ax + b$; $g(x) = \dfrac{1}{a}(x - b), a \neq 0$

54. $f(x) = \dfrac{1}{x}$; $g(x) = \dfrac{1}{x}$

In Problems 55–60, find functions f and g so that $f \circ g = H$.

55. $H(x) = (2x + 3)^4$

56. $H(x) = (1 + x^2)^3$

57. $H(x) = \sqrt{x^2 + 1}$

58. $H(x) = \sqrt{1 - x^2}$

59. $H(x) = |2x + 1|$

60. $H(x) = |2x^2 + 3|$

61. If $f(x) = 2x^3 - 3x^2 + 4x - 1$ and $g(x) = 2$, find $(f \circ g)(x)$ and $(g \circ f)(x)$.

62. If $f(x) = \dfrac{x}{x - 1}$, find $(f \circ f)(x)$.

63. If $f(x) = 2x^2 + 5$ and $g(x) = 3x + a$, find a so that the graph of $f \circ g$ crosses the y-axis at 23.

64. If $f(x) = 3x^2 - 7$ and $g(x) = 2x + a$, find a so that the graph of $f \circ g$ crosses the y-axis at 68.

---

65. Surface Area of a Balloon The surface area S (in square meters) of a hot-air balloon is given by

$$S(r) = 4\pi r^2$$

where r is the radius of the balloon (in meters). If the radius r is increasing with time t (in seconds) according to the formula $r(t) = \dfrac{2}{3}t^3, t \geq 0$, find the surface area S of the balloon as a function of the time t.

66. Volume of a Balloon The volume V (in cubic meters) of the hot-air balloon described in Problem 65 is given by $V(r) = \dfrac{4}{3}\pi r^3$. If the radius r is the same function of t as in Problem 65, find the volume V as a function of the time t.

67. Automobile Production The number N of cars produced at a certain factory in 1 day after t hours of operation is given by $N(t) = 100t - 5t^2, 0 \leq t \leq 10$. If the cost C (in dollars) of producing N cars is $C(N) = 15,000 + 8000N$, find the cost C as a function of the time t of operation of the factory.

68. Environmental Concerns The spread of oil leaking from a tanker is in the shape of a circle. If the radius r (in feet) of the spread after t hours is $r(t) = 200\sqrt{t}$, find the area A of the oil slick as a function of the time t.

69. Production Cost The price p of a certain product and the quantity x sold obey the demand equation

$$p = -\dfrac{1}{4}x + 100, \qquad 0 \leq x \leq 400$$

Suppose that the cost C of producing x units is

$$C = \dfrac{\sqrt{x}}{25} + 600$$

Assuming that all items produced are sold, find the cost C as a function of the price p.

[**Hint:** Solve for x in the demand equation and then form the composite.]

70. Cost of a Commodity The price p of a certain commodity and the quantity x sold obey the demand equation

$$p = -\dfrac{1}{5}x + 200, \qquad 0 \leq x \leq 1000$$

Suppose that the cost C of producing x units is

$$C = \dfrac{\sqrt{x}}{10} + 400$$

Assuming that all items produced are sold, find the cost C as a function of the price p.

71. Volume of a Cylinder The volume V of a right circular cylinder of height h and radius r is $V = \pi r^2 h$. If the height is twice the radius, express the volume V as a function of r.

72. Volume of a Cone The volume V of a right circular cone is $V = \dfrac{1}{3}\pi r^2 h$. If the height is twice the radius, express the volume V as a function of r.

73. Economics The **participation rate** is the number of people in the labor force divided by the civilian population (excludes military). Let $L(x)$ represent the size of the labor force in year x and $P(x)$ represent the civilian population in year x. Determine a function that represents the participation rate R as a function of x.

74. Crimes Suppose that $V(x)$ represents the number of violent crimes committed in year x and $P(x)$ represents the number of property crimes committed in year x. Determine a function T that represents the number of violent crimes and property crimes in year x.

75. Healthcare Suppose that $P(x)$ represents the percentage of income spent on health care in year x and $I(x)$ represents income in year x. Determine a function H that represents total health care expenditures in year x.

76. Income Tax Suppose that $I(x)$ represents the income of an individual in year x before taxes and $T(x)$ represents the individual's tax bill in year x. Determine a function N that represents the individual's net income (income after taxes) in year x.

77. Foreign Exchange Traders often buy foreign currency in hope of making money when the currency's value changes. For example, on August 1, 2001, one U.S. dollar could purchase 1.136235 Euros and one Euro could purchase 109.846 Yen based on data obtained from Yahoo! Finance. Let $f(x)$ represent the number of Euros one can buy with x dollars and let $g(x)$ represent the number of Yen one can buy with x Euros.
(a) Find a function that relates dollars to Euros.
(b) Find a function that relates Euros to Yen.
(c) Use the results of parts (a) and (b) to find a function that relates dollars to Yen. That is, find $g(f(x))$.
(d) What is $g(f(1000))$?

78. If f and g are odd functions, show that the composite function $f \circ g$ is also odd.

79. If f is an odd function and g is an even function, show that the composite functions $f \circ g$ and $g \circ f$ are also even.

3.6 MATHEMATICAL MODELS; CONSTRUCTING FUNCTIONS

OBJECTIVES   ① Construct and Analyze Functions

① We have already seen examples of real-world problems that lead to linear functions (Section 2.2) and quadratic functions (Sections 2.3 and 2.4). Real-world problems may also lead to mathematical models that involve other types of functions.

EXAMPLE 1   ### Finding the Distance from the Origin to a Point on a Graph

Let $P = (x, y)$ be a point on the graph of $y = x^2 - 1$.

(a) Express the distance d from P to the origin O as a function of x.

(b) What is d if $x = 0$?

(c) What is d if $x = 1$?

(d) What is d if $x = \dfrac{\sqrt{2}}{2}$?

(e) Use a graphing utility to graph the function $d = d(x), x \geq 0$. Rounded to two decimal places, find the value of x at which d has a local minimum.

[This gives the point on the graph of $y = x^2 - 1$ closest to the origin.]

Solution   (a) Figure 49 illustrates the graph. The distance d from P to O is

Figure 49

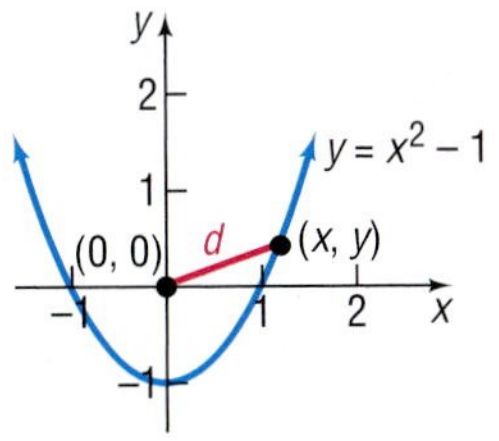

$$d = \sqrt{(x - 0)^2 + (y - 0)^2} = \sqrt{x^2 + y^2}$$

Since P is a point on the graph of $y = x^2 - 1$, we have

$$d(x) = \sqrt{x^2 + (x^2 - 1)^2} = \sqrt{x^4 - x^2 + 1}$$

We have expressed the distance d as a function of x.

(b) If $x = 0$, the distance d is

$$d(0) = \sqrt{1} = 1$$

(c) If $x = 1$, the distance d is

$$d(1) = \sqrt{1 - 1 + 1} = 1$$

(d) If $x = \dfrac{\sqrt{2}}{2}$, the distance d is

$$d\left(\frac{\sqrt{2}}{2}\right) = \sqrt{\left(\frac{\sqrt{2}}{2}\right)^4 - \left(\frac{\sqrt{2}}{2}\right)^2 + 1} = \sqrt{\frac{1}{4} - \frac{1}{2} + 1} = \frac{\sqrt{3}}{2}$$

Figure 50

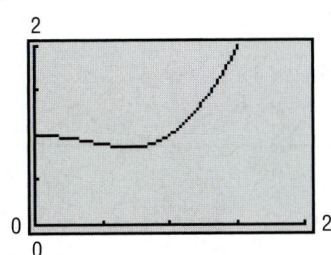

(e) Figure 50 shows the graph of $Y_1 = \sqrt{x^4 - x^2 + 1}$. Using the MINIMUM feature on a graphing utility, we find that when $x \approx 0.71$ the value of d is smallest ($d \approx 0.87$ rounded to two decimal places). By symmetry, it follows that when $x \approx -0.71$ the value of d is also a local minimum. ■

NOW WORK PROBLEM **1.**

Figure 51

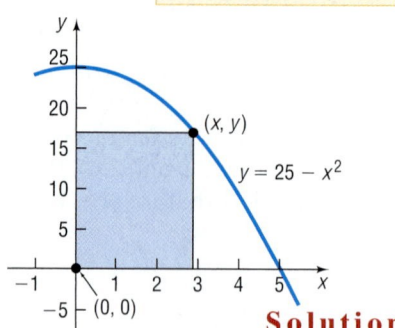

EXAMPLE 2   Area of a Rectangle

A rectangle has one corner on the graph of $y = 25 - x^2$, another at the origin, a third on the positive y-axis, and the fourth on the positive x-axis. See Figure 51.

(a) Express the area A of the rectangle as a function of x.

(b) What is the domain of A?

(c) Graph $A = A(x)$.

(d) For what value of x is the area largest?

Solution

(a) The area A of the rectangle is $A = xy$, where $y = 25 - x^2$. Substituting this expression for y, we obtain $A(x) = x(25 - x^2) = 25x - x^3$.

(b) Since x represents a side of the rectangle, we have $x > 0$. In addition, the area must be positive, so $y = 25 - x^2 > 0$, which implies that $x^2 < 25$, so $-5 < x < 5$. Combining these restrictions, we have the domain of A as $\{x | 0 < x < 5\}$ or $(0, 5)$ using interval notation.

(c) See Figure 52 for the graph of $A(x)$.

(d) Using MAXIMUM, we find the largest area is 48.11 occurring at $x = 2.89$, each rounded to two decimal places. See Figure 53.

Figure 52

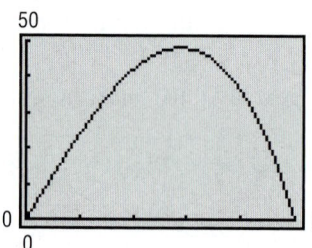

Figure 53

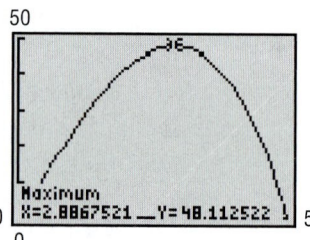

━━━ **NOW WORK PROBLEM 7.**

EXAMPLE 3   Getting from an Island to Town

An island is 2 miles from the nearest point P on a straight shoreline. A town is 12 miles down the shore from P.

(a) If a person can row a boat at an average speed of 3 miles per hour and the same person can walk 5 miles per hour, express the time T that it takes to go from the island to town as a function of the distance x from P to where the person lands the boat. See Figure 54.

Figure 54

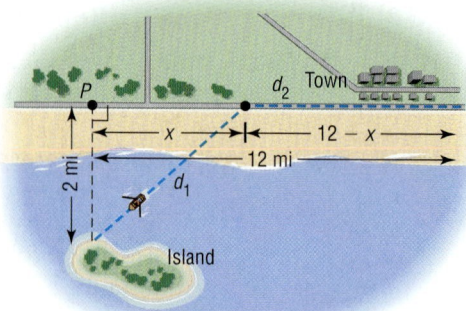

(b) What is the domain of T?

(c) How long will it take to travel from the island to town if the person lands the boat 4 miles from P?

(d) How long will it take if the person lands the boat 8 miles from P?

(e) Use a graphing utility to graph $T = T(x)$. What value of x results in the least time? What is the least time?

Solution (a) Figure 54 illustrates the situation. The distance d_1 from the island to the landing point is the hypotenuse of a right triangle. Using the Pythagorean Theorem, d_1 satisfies the equation

$$d_1^2 = 4 + x^2 \quad \text{or} \quad d_1 = \sqrt{4 + x^2}$$

Since the average speed of the boat is 3 miles per hour, the time t_1 that it takes to cover the distance d_1 satisfies

$$d_1 = 3t_1$$

Solving for t_1, we obtain

$$t_1 = \frac{d_1}{3} = \frac{\sqrt{4 + x^2}}{3}$$

The distance d_2 from the landing point to town is $12 - x$ and the time t_2 that it takes to cover this distance at an average walking speed of 5 miles per hour obeys the equation

$$d_2 = 5t_2$$

Solving for t_2, we obtain

$$t_2 = \frac{d_2}{5} = \frac{12 - x}{5}$$

The total time T of the trip is $t_1 + t_2$. Thus

$$T(x) = \frac{\sqrt{4 + x^2}}{3} + \frac{12 - x}{5}$$

(b) Since x equals the distance from P to where the boat lands, it follows that the domain of T is $0 \le x \le 12$.

(c) If the boat is landed 4 miles from P, then $x = 4$. The time T that the trip takes is

$$T(4) = \frac{\sqrt{20}}{3} + \frac{8}{5} \approx 3.09 \text{ hours}$$

(d) If the boat is landed 8 miles from P, then $x = 8$. The time T that the trip takes is

$$T(8) = \frac{\sqrt{68}}{3} + \frac{4}{5} \approx 3.55 \text{ hours}$$

(e) Figure 55(a) shows the graph of $T = T(x)$. Using MINIMUM, we find that the least time is 2.93 hours and occurs if $x = 1.50$ miles. See Figure 55(b). The boat should land about $12 - 1.5 = 10.5$ miles from town to minimize the travel time.

Figure 55

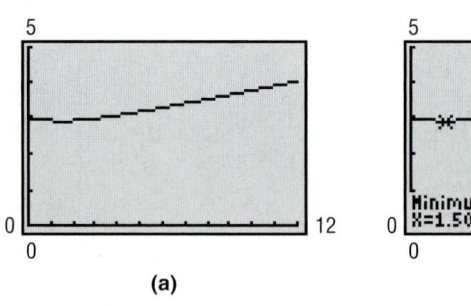

(a)   (b)

✏️ **NOW WORK PROBLEM 19.**

EXAMPLE 4   **Making a Playpen***

A manufacturer of children's playpens makes a square model that can be opened at one corner and attached at right angles to a wall or, perhaps, the side of a house. If each side is 3 feet in length, the open configuration doubles the available area in which the child can play from 9 square feet to 18 square feet. See Figure 56.

Figure 56

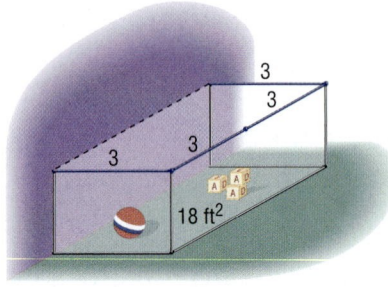

Now suppose that we place hinges at the outer corners to allow for a configuration like the one shown in Figure 57.

(a) Express the area A of this configuration as a function of the distance x between the two parallel sides.
(b) Find the domain of A.
(c) Find A if $x = 5$.
(d) Graph $A = A(x)$. For what value of x is the area largest? What is the maximum area?

Solution   (a) Refer to Figure 57. The area A that we seek consists of the area of a rectangle (with width 3 and length x) and the area of an isosceles triangle (with base x and two equal sides of length 3). The height h of the triangle may be found using the Pythagorean Theorem.

Figure 57

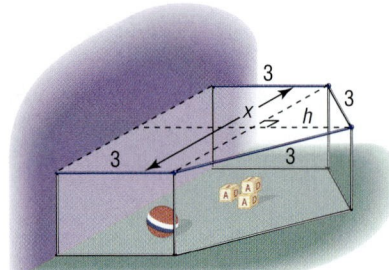

$$h^2 = 3^2 - \left(\frac{x}{2}\right)^2 = 9 - \frac{x^2}{4} = \frac{36 - x^2}{4}$$

$$h = \frac{1}{2}\sqrt{36 - x^2}$$

The area A enclosed by the playpen is

$$A = \text{area of rectangle} + \text{area of triangle} = 3x + \frac{1}{2}x\left(\frac{1}{2}\sqrt{36 - x^2}\right)$$

$$A(x) = 3x + \frac{x\sqrt{36 - x^2}}{4}$$

Now the area A is expressed as a function of x.

---

*Adapted from *Proceedings, Summer Conference for College Teachers on Applied Mathematics* (University of Missouri, Rolla), 1971.

(b) To find the domain of A, we note first that $x > 0$, since x is a length. Also, the expression under the radical must be positive, so

$$36 - x^2 > 0$$
$$x^2 < 36$$
$$-6 < x < 6$$

Combining these restrictions, we find that the domain of A is $0 < x < 6$, or $(0, 6)$, using interval notation.

(c) If $x = 5$, the area is

$$A(5) = 3(5) + \frac{5}{4}\sqrt{36 - (5)^2} \approx 19.15 \text{ square feet}$$

If the width of the playpen is 5 feet, its area is 19.15 square feet.

(d) See Figure 58. Using the MAXIMUM feature, the maximum area is about 19.82 square feet, obtained when x is about 5.58 feet. ■

Figure 58

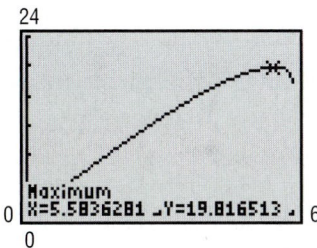

24

0

Maximum
X=5.5836281 Y=19.816513   6
0

EXAMPLE 5 **Filling a Swimming Pool**

Figure 59

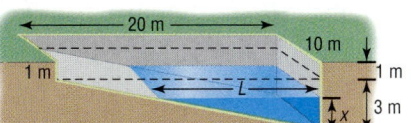

20 m

10 m

1 m

1 m

L

3 m

x

A rectangular swimming pool 20 meters long and 10 meters wide is 4 meters deep at one end and 1 meter deep at the other. Figure 59 illustrates a cross-sectional view of the pool. Water is being pumped into the pool to a height of 3 meters at the deep end.

(a) Find a function that expresses the volume V of water in the pool as a function of the height x of the water at the deep end.

(b) Find the volume when the height is 1 meter.

(c) Find the volume when the height is 2 meters.

(d) Use a graphing utility to graph the function $V = V(x)$. At what height is the volume 20 cubic meters? 100 cubic meters?

Solution (a) Let L denote the distance (in meters) measured at water level from the deep end to the short end. Notice that L and x form the sides of a triangle that is similar to the triangle whose sides are 20 meters by 3 meters. The variables L and x are related by the equation

$$\frac{L}{x} = \frac{20}{3} \quad \text{or} \quad L = \frac{20x}{3} \qquad 0 \le x \le 3$$

The volume V of water in the pool at any time is

$$V = \left(\begin{array}{c}\text{cross-sectional}\\\text{triangular area}\end{array}\right)(\text{width}) = \left(\frac{1}{2}Lx\right)(10) \quad \text{cubic meters}$$

Since $L = \dfrac{20x}{3}$, we have

$$V(x) = \left(\frac{1}{2} \cdot \frac{20x}{3} \cdot x\right)(10) = \frac{100}{3}x^2 \quad \text{cubic meters}$$

(b) When the height x of the water is 1 meter, the volume $V = V(x)$ is

$$V(1) = \frac{100}{3} \cdot 1^2 = 33\frac{1}{3} \quad \text{cubic meters}$$

Figure 60

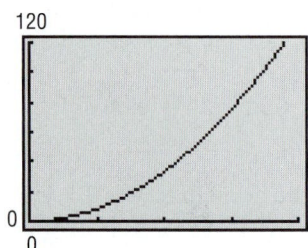

(c) When the height x of the water is 2 meters, the volume $V = V(x)$ is

$$V(2) = \frac{100}{3} \cdot 2^2 = \frac{400}{3} = 133\frac{1}{3} \quad \text{cubic meters}$$

(d) See Figure 60. When $x \approx 0.77$ meter, the volume is 20 cubic meters. When $x \approx 1.73$ meters, the volume is 100 cubic meters. ■

3.6 Exercises

1. Let $P = (x, y)$ be a point on the graph of $y = x^2 - 8$.
   (a) Express the distance d from P to the origin as a function of x.
   (b) What is d if $x = 0$?
   (c) What is d if $x = 1$?
   (d) Use a graphing utility to graph $d = d(x)$.
   (e) For what values of x is d smallest?

2. Let $P = (x, y)$ be a point on the graph of $y = x^2 - 8$.
   (a) Express the distance d from P to the point $(0, -1)$ as a function of x.
   (b) What is d if $x = 0$?
   (c) What is d if $x = -1$?
   (d) Use a graphing utility to graph $d = d(x)$.
   (e) For what values of x is d smallest?

3. Let $P = (x, y)$ be a point on the graph of $y = \sqrt{x}$.
   (a) Express the distance d from P to the point $(1, 0)$ as a function of x.
   (b) Use a graphing utility to graph $d = d(x)$.
   (c) For what values of x is d smallest?

4. Let $P = (x, y)$ be a point on the graph of $y = \frac{1}{x}$.
   (a) Express the distance d from P to the origin as a function of x.
   (b) Use a graphing utility to graph $d = d(x)$.
   (c) For what values of x is d smallest?

5. A right triangle has one vertex on the graph of $y = x^3, x > 0$, at (x, y), another at the origin, and the third on the positive y-axis at $(0, y)$, as shown in the figure. Express the area A of the triangle as a function of x.

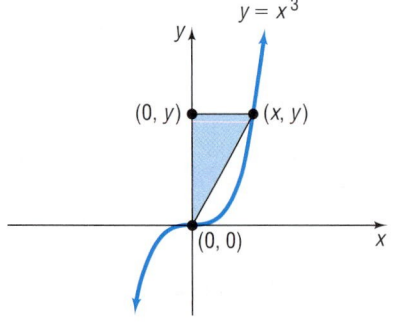

6. A right triangle has one vertex on the graph of $y = 9 - x^2, x > 0$, at (x, y), another at the origin, and the third on the positive x-axis at $(x, 0)$. Express the area A of the triangle as a function of x.

7. A rectangle has one corner on the graph of $y = 16 - x^2$, another at the origin, a third on the positive y-axis, and the fourth on the positive x-axis (see the figure).

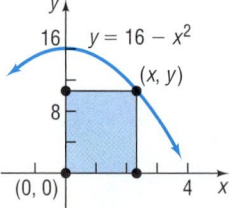

   (a) Express the area A of the rectangle as a function of x.
   (b) What is the domain of A?
   (c) Graph $A = A(x)$. For what value of x is A largest?

8. A rectangle is inscribed in a semicircle of radius 2 (see the figure). Let $P = (x, y)$ be the point in quadrant I that is a vertex of the rectangle and is on the circle.

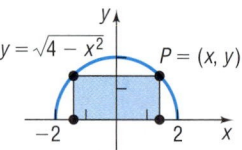

   (a) Express the area A of the rectangle as a function of x.
   (b) Express the perimeter p of the rectangle as a function of x.
   (c) Graph $A = A(x)$. For what value of x is A largest?
   (d) Graph $p = p(x)$. For what value of x is p largest?

9. A rectangle is inscribed in a circle of radius 2 (see the figure). Let $P = (x, y)$ be the point in quadrant I that is a vertex of the rectangle and is on the circle.

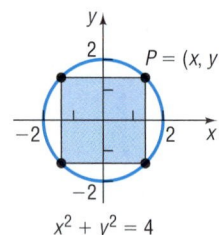

(a) Express the area A of the rectangle as a function of x.

(b) Express the perimeter p of the rectangle as a function of x.

(c) Graph $A = A(x)$. For what value of x is A largest?

(d) Graph $p = p(x)$. For what value of x is p largest?

10. A circle of radius r is inscribed in a square (see the figure).

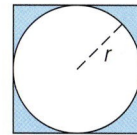

(a) Express the area A of the square as a function of the radius r of the circle.

(b) Express the perimeter p of the square as a function of r.

11. A wire 10 meters long is to be cut into two pieces. One piece will be shaped as a square, and the other piece will be shaped as a circle (see the figure).

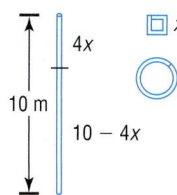

(a) Express the total area A enclosed by the pieces of wire as a function of the length x of a side of the square.

(b) What is the domain of A?

(c) Graph $A = A(x)$. For what value of x is A smallest?

12. A wire 10 meters long is to be cut into two pieces. One piece will be shaped as an equilateral triangle, and the other piece will be shaped as a circle.

(a) Express the total area A enclosed by the pieces of wire as a function of the length x of a side of the equilateral triangle.

(b) What is the domain of A?

(c) Graph $A = A(x)$. For what value of x is A smallest?

13. A wire of length x is bent into the shape of a circle.

(a) Express the circumference of the circle as a function of x.

(b) Express the area of the circle as a function of x.

14. A wire of length x is bent into the shape of a square.

(a) Express the perimeter of the square as a function of x.

(b) Express the area of the square as a function of x.

15. A semicircle of radius r is inscribed in a rectangle so that the diameter of the semicircle is the length of the rectangle (see the figure).

(a) Express the area A of the rectangle as a function of the radius r of the semicircle.

(b) Express the perimeter p of the rectangle as a function of r.

16. An equilateral triangle is inscribed in a circle of radius r. See the figure. Express the circumference C of the circle as a function of the length x of a side of the triangle.

[**Hint:** First show that $r^2 = \dfrac{x^2}{3}$.]

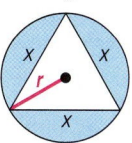

17. An equilateral triangle is inscribed in a circle of radius r. See the figure. Express the area A within the circle, but outside the triangle, as a function of the length x of a side of the triangle.

18. Water is poured into a container in the shape of a right circular cone with radius 4 feet and height 16 feet (see the figure). Express the volume V of the water in the cone as a function of the height h of the water.

[**Hint:** The volume V of a cone of radius r and height h is $V = \dfrac{1}{3}\pi r^2 h$.]

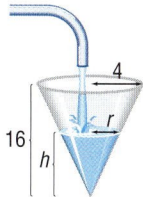

19. Installing Cable TV   MetroMedia Cable is asked to provide service to a customer whose house is located 2 miles from the road along which the cable is buried. The nearest connection box for the cable is located 5 miles down the road (see the figure).

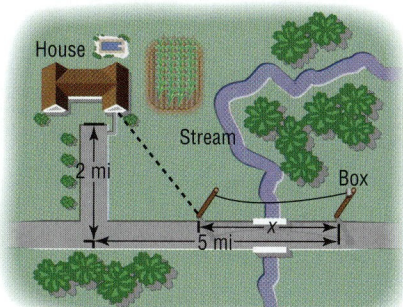

(a) If the installation cost is $100 per mile along the road and $140 per mile off the road, express the total cost C of installation as a function of the distance x (in miles) from the connection box to the point where the cable installation turns off the road. Give the domain.

(b) Compute the cost if $x = 1$ mile.

(c) Compute the cost if $x = 3$ miles.

(d) Graph the function $C = C(x)$.

(e) Create a TABLE with TblStart $= 0$ and ΔTbl $= 0.5$. To the nearest $\dfrac{1}{2}$ mile, which value of x results in the least cost?

(f) Using MINIMUM, what value of x results in the least cost?

20. Time Required to Go from an Island to a Town An island is 3 miles from the nearest point P on a straight shoreline. A town is located 20 miles down the shore from P.

(a) If a person has a boat that averages 12 miles per hour and the same person can run 5 miles per hour, express the time T that it takes to go from the island to town as a function of x, where x is the distance from P to where the person lands the boat. Give the domain.

(b) How long will it take to travel from the island to town if you land the boat 8 miles from P?

(c) How long will it take if you land the boat 12 miles from P?

(d) Graph the function $T = T(x)$.

(e) Create a TABLE with TblStart = 0 and ΔTbl = 1. To the nearest mile, determine which value of x results in the least time.

(f) Using MINIMUM, what value of x results in the least time?

 (g) The least time occurs by heading directly to town from the island. Explain why this solution makes sense.

21. Two cars leave an intersection at the same time. One is headed south at a constant speed of 30 miles per hour, the other is headed west at a constant speed of 40 miles per hour (see the figure). Express the distance d between the cars as a function of the time t.

[**Hint:** At $t = 0$ the cars leave the intersection.]

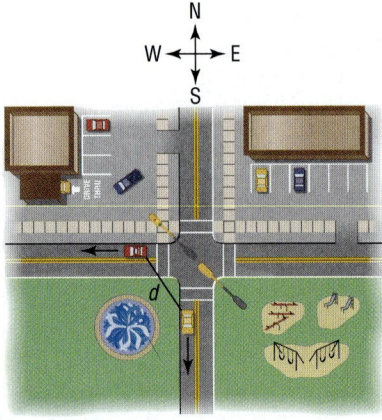

22. Two cars are approaching an intersection. One is 2 miles south of the intersection and is moving at a constant speed of 30 miles per hour. At the same time, the other car is 3 miles east of the intersection and is moving at a constant speed of 40 miles per hour.

(a) Express the distance d between the cars as a function of time t.

[**Hint:** At $t = 0$, the cars are 2 miles south and 3 miles east of the intersection, respectively.]

(b) Use a graphing utility to graph $d = d(t)$. For what value of t is d smallest?

23. Constructing an Open Box An open box with a square base is to be made from a square piece of cardboard 24

inches on a side by cutting out a square from each corner and turning up the sides (see the figure).

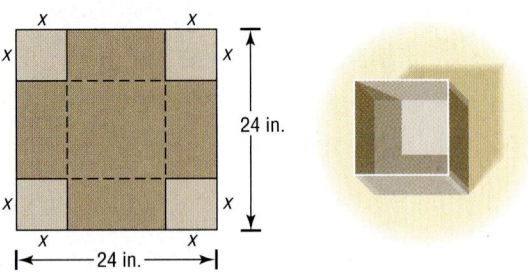

(a) Express the volume V of the box as a function of the length x of the side of the square cut from each corner.

(b) What is the volume if a 3-inch square is cut out?

(c) What is the volume if a 10-inch square is cut out?

(d) Graph $V = V(x)$. For what value of x is V largest?

24. Constructing an Open Box A open box with a square base is required to have a volume of 10 cubic feet.

(a) Express the amount A of material used to make such a box as a function of the length x of a side of the square base.

(b) How much material is required for a base 1 foot by 1 foot?

(c) How much material is required for a base 2 feet by 2 feet?

(d) Graph $A = A(x)$. For what value of x is A smallest?

25. Inscribing a Cylinder in a Sphere Inscribe a right circular cylinder of height h and radius r in a sphere of fixed radius R. See the illustration. Express the volume V of the cylinder as a function of h.

[**Hint:** $V = \pi r^2 h$. Note also the right triangle.]

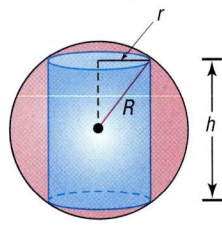

Sphere

26. Inscribing a Cylinder in a Cone Inscribe a right circular cylinder of height h and radius r in a cone of fixed radius R and fixed height H. See the illustration. Express the volume V of the cylinder as a function of r.

[**Hint:** $V = \pi r^2 h$. Note also the similar triangles.]

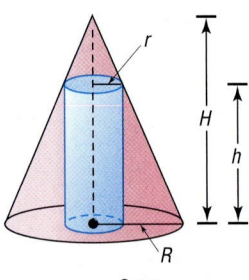

Cone

Chapter Review

Library of Functions

Linear function (p. 276)

$f(x) = mx + b$ Graph is a line with slope m and y-intercept b (see Figure 21)

Constant function (p. 277)

$f(x) = b$ Graph is a horizontal line with y-intercept b (see Figure 22)

Identity function (p. 277)

$f(x) = x$ Graph is a line with slope 1 and y-intercept 0 (see Figure 23)

Square function (p. 277)

$f(x) = x^2$ Graph is a parabola with vertex at $(0, 0)$ (see Figure 24)

Cube function (p. 278)

$f(x) = x^3$ See Figure 25.

Square root function (p.278)

$f(x) = \sqrt{x}$ See Figure 26.

Reciprocal function (p. 278)

$f(x) = \dfrac{1}{x}$ See Figure 27.

Absolute value function (p. 279)

$f(x) = |x|$ See Figure 28.

Things to Know

Average rate of change of a function (p. 264)   The average rate of change of f from c to x is

$$\frac{\Delta y}{\Delta x} = \frac{f(x) - f(c)}{x - c}, \qquad x \neq c$$

Increasing function (p. 266)   A function f is increasing on an open interval I if, for any choice of x_1 and x_2 in I, with $x_1 < x_2$, we have $f(x_1) < f(x_2)$.

Decreasing function (p. 266)   A function f is decreasing on an open interval I if, for any choice of x_1 and x_2 in I, with $x_1 < x_2$, we have $f(x_1) > f(x_2)$.

Constant function (p. 266)   A function f is constant on an open interval I if, for all choices of x in I, the values of $f(x)$ are equal.

Local maximum (p. 267)   A function f has a local maximum at c if there is an open interval I containing c so that, for all $x \neq c$ in I, $f(x) < f(c)$.

Local minimum (p. 268)   A function f has a local minimum at c if there is an open interval I containing c so that, for all $x \neq c$ in I, $f(x) > f(c)$.

Even function f (p. 269) $f(-x) = f(x)$ for every x in the domain ($-x$ must also be in the domain).

Odd function f (p. 269) $f(-x) = -f(x)$ for every x in the domain ($-x$ must also be in the domain).

Difference Quotient of f (p. 275) $\dfrac{f(x + h) - f(x)}{h}, \qquad h \neq 0$

Objectives

Section	You should be able to ...	Review Exercises
3.1	**1** Test an equation for symmetry with respect to (a) the x-axis, (b) the y-axis, and (c) the origin (p. 256)	3–10
	2 Know how to graph key equations (p. 258)	11, 12
3.2	**1** Find the average rate of change of a function (p. 263)	17–20, 70
	2 Use a graph to determine where a function is increasing, is decreasing, or is constant (p. 266)	1(b), 2(b)
	3 Use a graph to locate local maxima and minima (p. 267)	1(c), 2(c)

Review Exercises

Blue problem numbers indicate the authors' suggestions for use in a Practice Test.

In Problems 1 and 2, use the graph of the function f to find

(a) *The domain and range of f*
(b) *The intervals on which f is increasing, decreasing, or constant*
(c) *The local minima and local maxima*
(d) *Whether the graph is symmetric with respect to the x-axis, the y-axis, or the origin*
(e) *Whether the function is even, odd, or neither*
(f) *The intercepts, if any*

1.

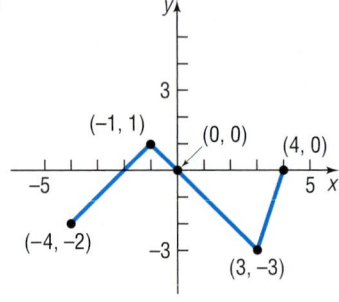

2.

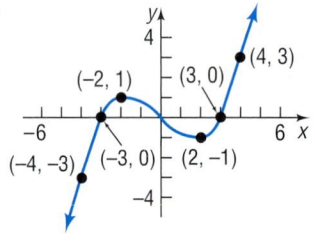

In Problems 3–10, test each equation for symmetry with respect to the x-axis, the y-axis, and the origin.

3. $2x = 3y^2$ **4.** $y = 5x$ **5.** $x^2 + 4y^2 = 16$ **6.** $9x^2 - y^2 = 9$

7. $y = x^4 + 2x^2 + 1$ **8.** $y = x^3 - x$ **9.** $x^2 + x + y^2 + 2y = 0$ **10.** $x^2 + 4x + y^2 - 2y = 0$

11. Sketch a graph of $y = x^3$.   **12.** Sketch a graph of $y = \sqrt{x}$.

In Problems 13–16:
(a) *Find the domain of each function.*   (b) *Locate any intercepts.*
(c) *Graph each function by hand.*   (d) *Based on the graph, find the range.*
(e) *Verify your results using a graphing utility.*

13. $f(x) = \begin{cases} 3x & -2 < x \le 1 \\ x + 1 & x > 1 \end{cases}$ **14.** $f(x) = \begin{cases} x - 1 & -3 < x < 0 \\ 3x - 1, & x \ge 0 \end{cases}$

15. $f(x) = \begin{cases} x & -4 \le x < 0 \\ 1 & x = 0 \\ 3x & x > 0 \end{cases}$ **16.** $f(x) = \begin{cases} x^2 & -2 \le x \le 2 \\ 2x - 1 & x > 2 \end{cases}$

In Problems 17–20, find the average rate of change from 2 to x for each function f. Be sure to simplify.

17. $f(x) = 2 - 5x$ **18.** $f(x) = 2x^2 + 7$ **19.** $f(x) = 3x - 4x^2$ **20.** $f(x) = x^2 - 3x + 2$

In Problems 21–28, determine (algebraically) whether the given function is even, odd, or neither.

21. $f(x) = x^3 - 4x$ **22.** $g(x) = \dfrac{4 + x^2}{1 + x^4}$ **23.** $h(x) = \dfrac{1}{x^4} + \dfrac{1}{x^2} + 1$ **24.** $F(x) = \sqrt{1 - x^3}$

25. $G(x) = 1 - x + x^3$ **26.** $H(x) = 1 + x + x^2$ **27.** $f(x) = \dfrac{x}{1 + x^2}$ **28.** $g(x) = \dfrac{1 + x^2}{x^3}$

In Problems 29–40, graph each function using transformations (shifting, compressing or stretching, and reflections). Identify any intercepts on the graph. State the domain and, based on the graph, find the range.

29. $F(x) = |x| - 4$ **30.** $f(x) = |x| + 4$ **31.** $g(x) = -2|x|$ **32.** $g(x) = \dfrac{1}{2}|x|$

33. $h(x) = \sqrt{x - 1}$ **34.** $h(x) = \sqrt{x} - 1$ **35.** $f(x) = \sqrt{1 - x}$ **36.** $f(x) = -\sqrt{x + 3}$

37. $h(x) = (x - 1)^2 + 2$ **38.** $h(x) = (x + 2)^2 - 3$ **39.** $g(x) = 3(x - 1)^3 + 1$ **40.** $g(x) = -2(x + 2)^3 - 8$

In Problems 41–44, use a graphing utility to graph each function over the indicated interval. Approximate any local maxima and local minima. Determine where the function is increasing and where it is decreasing.

41. $f(x) = 2x^3 - 5x + 1$ $(-3, 3)$ **42.** $f(x) = -x^3 + 3x - 5$ $(-3, 3)$

43. $f(x) = 2x^4 - 5x^3 + 2x + 1$ $(-2, 3)$ **44.** $f(x) = -x^4 + 3x^3 - 4x + 3$ $(-2, 3)$

In Problems 45–50, find $f + g$, $f - g$, $f \cdot g$, and $\dfrac{f}{g}$ for each pair of functions. State the domain of each.

45. $f(x) = 2 - x$; $g(x) = 3x + 1$ **46.** $f(x) = 2x - 1$; $g(x) = 2x + 1$

47. $f(x) = 3x^2 + x + 1$; $g(x) = 3x$ **48.** $f(x) = 3x$; $g(x) = 1 + x + x^2$

49. $f(x) = \dfrac{x + 1}{x - 1}$; $g(x) = \dfrac{1}{x}$ **50.** $f(x) = \dfrac{1}{x - 3}$; $g(x) = \dfrac{3}{x}$

In Problems 51–56, for the given functions f and g find:

(a) $(f \circ g)(2)$ *(b)* $(g \circ f)(-2)$ *(c)* $(f \circ f)(4)$ *(d)* $(g \circ g)(-1)$

Verify your results using a graphing utility.

51. $f(x) = 3x - 5$; $g(x) = 1 - 2x^2$ **52.** $f(x) = 4 - x$; $g(x) = 1 + x^2$

53. $f(x) = \sqrt{x + 2}$; $g(x) = 2x^2 + 1$ **54.** $f(x) = 1 - 3x^2$; $g(x) = \sqrt{4 - x}$

55. $f(x) = \dfrac{1}{x^2 + 4}$; $g(x) = 3x - 2$ **56.** $f(x) = \dfrac{2}{1 + 2x^2}$; $g(x) = 3x$

In Problems 57–62, find $f \circ g$, $g \circ f$, $f \circ f$, and $g \circ g$ for each pair of functions. State the domain of each composite function.

57. $f(x) = 2 - x$; $g(x) = 3x + 1$ **58.** $f(x) = 2x - 1$; $g(x) = 2x + 1$

59. $f(x) = 3x^2 + x + 1$; $g(x) = |3x|$ **60.** $f(x) = \sqrt{3x}$; $g(x) = 1 + x + x^2$

61. $f(x) = \dfrac{x + 1}{x - 1}$; $g(x) = \dfrac{1}{x}$ **62.** $f(x) = \sqrt{x - 3}$; $g(x) = \dfrac{3}{x}$

63. For the following graph of the function f draw the graph of:

  (a) $y = f(-x)$ (b) $y = -f(x)$ (c) $y = f(x + 2)$

  (d) $y = f(x) + 2$ (e) $y = 2f(x)$ (f) $y = f(3x)$

64. Repeat Problem 63 for the following graph of the function g.

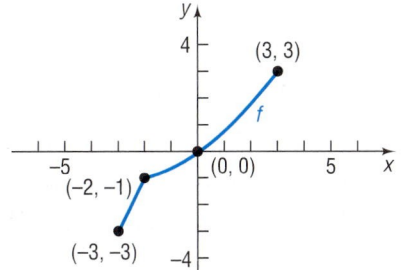

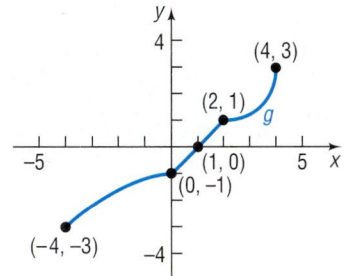

65. Find the point on the line $y = x$ that is closest to the point $(3, 1)$.

[**Hint:** Find the minimum value of the function $f(x) = d^2$, where d is the distance from $(3, 1)$ to a point on the line.]

66. Find the point on the line $y = x + 1$ that is closest to the point $(4, 1)$.

67. A rectangle has one vertex on the graph of $y = 10 - x^2$, $x > 0$, another at the origin, one on the positive x-axis, and one on the positive y-axis. Find the largest area A that can be enclosed by the rectangle.

68. **Constructing a Closed Box**   A closed box with a square base is required to have a volume of 10 cubic feet.
    (a) Express the amount A of material used to make such a box as a function of the length x of a side of the square base.
    (b) How much material is required for a base 1 foot by 1 foot?
    (c) How much material is required for a base 2 feet by 2 feet?
    (d) Graph $A = A(x)$. For what value of x is A smallest?

69. **Spheres**   The volume V of a sphere of radius r is $V = \frac{4}{3}\pi r^3$; the surface area S of this sphere is $S = 4\pi r^2$.

Express the volume V as a function of the surface area S. If the surface area doubles, how does the volume change?

70. **Productivity versus Earnings**   The following data represent the average hourly earnings and productivity (output per hour) of production workers for the years

1986–1995. Let productivity x be the independent variable and average hourly earnings y be the dependent variable.

Productivity	Average Hourly Earnings
94.2	8.76
94.1	8.98
94.6	9.28
95.3	9.66
96.1	10.01
96.7	10.32
100	10.57
100.2	10.83
100.7	11.12
100.8	11.44

Source: Bureau of Labor Statistics.

(a) Draw a scatter diagram of the data.
(b) Draw a line through the points $(94.2, 8.76)$ and $(96.7, 10.32)$ on the scatter diagram found in part (a).
(c) Find the average rate of change of hourly earnings for productivity from 94.2 to 96.7.
(d) Interpret the average rate of change found in part (c).
(e) Draw a line through the points $(96.7, 10.32)$ and $(100.8, 11.44)$ on the scatter diagram found in part (a).
(f) Find the average rate of change of hourly earnings for productivity from 96.7 to 100.8.
(g) Interpret the average rate of change found in part (f).
(h) What is happening to the average rate of change of hourly earnings as productivity increases?

Chapter Projects

1. **Oil Spill**   An oil tanker strikes a sand bar that rips a hole in the hull of the ship. Oil begins leaking out of the tanker

with the spilled oil forming a circle around the tanker. The radius of the circle is increasing at the rate of 2.2 feet per hour.
(a) Write the area of the circle as a function of the radius r.
(b) Write the radius of the circle as a function of time t.
(c) What is the radius of the circle after 2 hours? What is the radius of the circle after 2.5 hours?
(d) Use the result of part (c) to determine the area of the circle after 2 hours and 2.5 hours.
(e) Determine a function that represents area as a function of time t.
(f) Use the result of part (e) to determine the area of the circle after 2 hours and 2.5 hours.
(g) Compute the average rate of change of the area of the circle from 2 hours to 2.5 hours.
(h) Compute the average rate of change of the area of the circle from 3 hours to 3.5 hours.
(i) Based on the results obtained in parts (g) and (h), what is happening to the average rate of change of the area of the circle as time passes?

(j) If the oil tanker is 150 yards from shore, when will the oil spill first reach the shoreline? (1 yard = 3 feet)

(k) How long will it be until 6 miles of shoreline is contaminated with oil? (1 mile = 5280 feet)

2. **Economics** For a manufacturer of computer chips, it has been determined that the number Q of chips produced as a function of labor L, in hours, is given by $Q = 8L^{1/2}$. In addition, the cost C of production as a function of the number of chips produced per month is $C(Q) = 0.002q^3 + 0.5q^2 - q + 1000$.

(a) Determine cost as a function of labor.

(b) If the demand function for the chips is $Q(p) = 5000 - 20p$, where p is the price per chip, find the revenue R as a function of Q.
Hint: Revenue equals the number of units produced times the price per unit.

(c) Find revenue as a function of labor.

(d) Determine profit as a function of labor.

(e) Using a graphing utility, graph profit as a function of labor. Determine the implied domain of this function.

(f) Determine the profit-maximizing level of labor.

(g) Determine the maximum monthly profit of the firm.

(h) Determine the optimal (profit maximizing) level of output for the firm.

(i) Determine the optimal price per chip.

3. **Physics** Suppose a cannon ball is fired from a cannon whose muzzle is 10 feet above the ground at an inclination of 45° with an initial velocity of 250 ft/sec. From physics, the height $s(t)$ (in feet) at time t seconds after the cannon is fired is given by $s(t) = -16t^2 + 250t + 10$

(a) Either with a graphing utility or by hand, graph the function $s = s(t)$.

(b) The average rate of change of distance with respect to time is called average velocity. Determine the average velocity of the cannon ball from $t = 1$ sec to $t = 1.5$ sec.

(c) Determine the average velocity of the cannon ball from $t = 1$ sec to $t = 1.2$ sec.

(d) Determine the average velocity of the cannon ball from $t = 1$ sec to $t = 1.1$ sec.

(e) The average velocity represents the slope of the secant line joining the two points. For each of the average velocities found in parts (b), (c), and (d), sketch the graph of the secant lines on the graph of $s = s(t)$.

(f) What happens to the secant lines at t moves closer to 1?

(g) Compute the average rate of change from t_o to $t = 1$. Evaluate the average rate of change at $t_o = 1$. This is called the instantaneous rate of change of s with respect to t or instantaneous velocity.

(h) Graph the line whose slope is the instantaneous velocity at $t_o = 1$. This is the line tangent to s at $t = 1$.

Cumulative Review

1. Solve the following equation by factoring: $x^3 - 6x^2 + 8x = 0$. Verify your results using a graphing utility.

2. Solve the following inequality and draw a graph to illustrate the solution: $3x + 2 \leq 5x - 1$.

3. Find the center and radius of the circle $x^2 + 4x + y^2 - 2y - 4 = 0$. Graph the circle by hand.

4. For the equation $y = x^3 - 9x$, determine the intercepts and test for symmetry. Verify your results using a graphing utility.

5. Find an equation of the line perpendicular to $3x - 2y = 7$ that contains the point $(1, 5)$.

6. Determine whether the following graph represents a function:

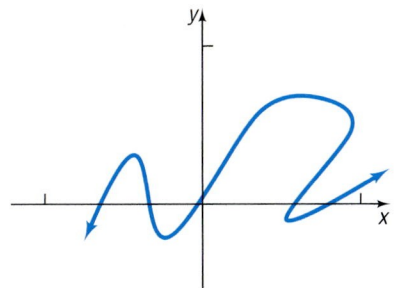

7. For the function $f(x) = x^2 + 5x - 2$, find

(a) $f(3)$ (b) $f(-x)$ (c) $-f(x)$

(d) $f(3x)$ (e) $\dfrac{f(x + h) - f(x)}{h}$, $h \neq 0$

8. Answer the following questions regarding the function $f(x) = \dfrac{x + 5}{x - 1}$.

(a) What is the domain of f?

(b) Is the point $(2, 6)$ on the graph of f?

(c) If $x = 3$, what is $f(x)$? What point is on the graph of f?

(d) If $f(x) = 9$, what is x? What point is on the graph of f?

9. Graph the function $f(x) = -3x + 7$

10. Graph $f(x) = 2x^2 - 4x + 1$ by hand by determining whether its graph opens up or down and by finding its vertex, axis of symmetry, y-intercept, and x-intercepts, if any. Verify your results using a graphing utility.

11. Find the average rate of change of $f(x) = x^2 + 3x + 1$ from 1 to x. Use this result to find the slope of the secant line containing $(1, f(1))$ and $(2, f(2))$.

12. In parts (a) to (f) use the graph below.

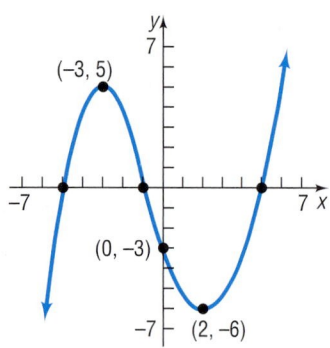

(a) Determine the intercepts.
(b) Based on the graph, tell whether the graph is symmetric with respect to the x-axis, the y-axis, and/or the origin.
(c) Based on the graph, tell whether the function is even, odd, or neither.
(d) List the intervals on which f is increasing. List the intervals on which f is decreasing.
(e) List the numbers, if any, at which f has a local maximum. What are these local maxima?
(f) List the numbers, if any, at which f has a local minimum. What are these local minima?

13. Use a graphing utility to graph $f(x) = -x^3 + 7x - 2$ for $-3 < x < 3$. Determine where f has a local maximum and where f has a local minimum. Determine where f is increasing and where it is decreasing.

14. Determine algebraically whether the function $f(x) = \dfrac{5x}{x^2 - 9}$ is even, odd, or neither.

15. For the function $f(x) = \begin{cases} 2x + 1, & -3 < x < 2 \\ -3x + 4, & x \geq 2 \end{cases}$

(a) Find the domain of f.
(b) Locate any intercepts.
(c) Graph the function by hand.
(d) Based on the graph, find the range.
(e) Verify your results using a graphing utility.

16. Graph the function $f(x) = -3(x + 1)^2 + 5$ using transformations.

17. Suppose that $f(x) = x^2 - 5x + 1$ and $g(x) = -4x - 7$.

(a) Find $f + g$ and state its domain.
(b) Find $\dfrac{f}{g}$ and state its domain.
(c) Find $(f \circ g)(2)$.
(d) Find $f \circ g$ and state its domain.

18. Suppose that $f(x) = \dfrac{3}{x - 1}$ and $g(x) = \dfrac{4}{x + 3}$. Find $f \circ g$ and state its domain.

19. Demand Equation The price p (in dollars) and the quantity x sold of a certain product obey the demand equation

$$p = -\frac{1}{10}x + 150, \qquad 0 \leq x \leq 1500$$

(a) Express the revenue R as a function of x.
(b) What is the revenue if 100 units are sold?
(c) What quantity x maximizes revenue? What is the maximum revenue?
(d) What price should the company charge to maximize revenue?

20. Long Distance Phone Company A phone company offers a deal by which a long distance phone call costs $0.99 for the first 20 minutes and $0.07 per minute thereafter. Write a piecewise-defined function for the cost C of making a phone call that lasts x minutes. Graph this function.

POLYNOMIAL AND RATIONAL FUNCTIONS

Weber on Vallas Wish List to Ease School Crowding

Vallas said that if a deal can be struck in the coming days, Weber could be reopened as an elementary school in the fall. Schools such as Burbank and other crowded institutions would benefit if such an arrangement were made, he said.

In recent years, Burbank teachers have held classes in the building's storage and loading areas. Tutoring sessions are often held in the foyer of a second-floor lavatory, and as many as four pupils are required to share a locker.

Enrollment is 1,072, nearly 300 more than the school was built to handle. Its heating system has been broken for some time, making it difficult to heat classrooms evenly. (Ray Quintanilla, *Chicago Tribune*.)

SEE CHAPTER PROJECT 1.

OUTLINE

For additional study help, go to

www.prenhall.com/sullivanegu3e

Materials include:

- Graphing Calculator Help
- Chapter Quiz
- Chapter Test
- PowerPoint Downloads
- Chapter Projects
- Student Tips

A Look Back, A Look Forward

In Chapter 2, we began our discussion of functions. We defined domain and range and independent and dependent variables; we found the value of a function and graphed functions. In Chapter 3, we continued our study of functions by listing the properties that a function might have, like being even or odd, and we created a library of functions, naming key functions and listing their properties, including their graphs.

In this chapter, we look at two general classes of functions: polynomial functions and rational functions, and examine their properties. Polynomial functions are arguably the simplest expressions in algebra. For this reason, they are often used to approximate other, more complicated, functions. Rational functions are simply ratios of polynomial functions.

PREPARING FOR THIS SECTION

Before getting started, review the following:

✓ Polynomials (Section R. 5, pp. 40–41)

✓ Intercepts (Section 1.2, pp. 106–107)

✓ Graphing Techniques: Transformations (Section 3.4, pp. 286–295)

4.1 POWER FUNCTIONS AND MODELS

OBJECTIVES   **1**   Graph Transformations of Power Functions

              2   Find the Power Function of Best Fit to Data

Power Functions

A *power function* is a function that is defined by a single monomial.

A **power function of degree *n*** is a function of the form

$$f(x) = ax^n \tag{1}$$

where a is a real number, $a \neq 0$, and $n > 0$ is an integer.

Examples of power functions are

$$f(x) = 3x \qquad f(x) = -5x^2 \qquad f(x) = 8x^3 \qquad f(x) = -5x^4$$

     degree 1   degree 2   degree 3   degree 4

In this section we discuss properties of power functions and graph transformations of power functions.

The graph of a power function of degree 1, $f(x) = ax$, is a line with slope a that passes through the origin. The graph of a power function of degree 2, $f(x) = ax^2$, is a parabola, with vertex at the origin, that opens up if $a > 0$ and down if $a < 0$.

If we know how to graph a power function of the form $f(x) = x^n$, then a compression or stretch and, perhaps, a reflection about the x-axis will enable us to obtain the graph of $g(x) = ax^n$ for any real number a. Consequently, we shall concentrate on graphing power functions of the form $f(x) = x^n$.

We begin with power functions of even degree of the form $f(x) = x^n$, $n \geq 2$ and *n even*.

 EXPLORATION   Using your graphing utility and the viewing window $-2 \leq x \leq 2$, $-4 \leq y \leq 16$, graph the function $Y_1 = f(x) = x^4$. On the same screen, graph $Y_2 = g(x) = x^8$. Now, also on the same screen, graph $Y_3 = h(x) = x^{12}$. What do you notice about the graphs as the magnitude of the exponent increases? Repeat this procedure for the viewing window $-1 \leq x \leq 1$, $0 \leq y \leq 1$. What do you notice? See Figures 1(a) and (b).

Figure 1

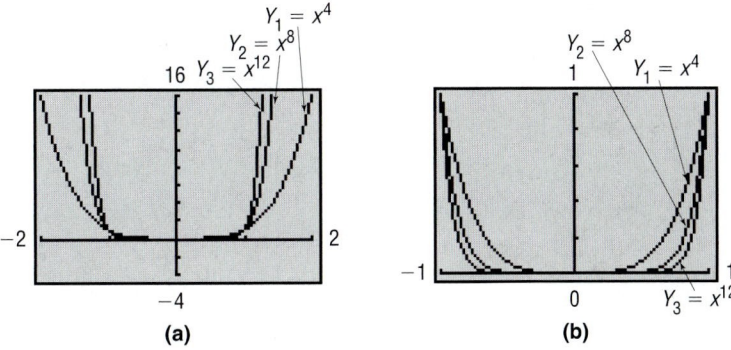

(a)   (b)

The domain of $f(x) = x^n$, $n \geq 2$ and *n* even is the set of all real numbers, and the range is the set of nonnegative real numbers. Such a power function is an even function (do you see why?), so its graph is symmetric with respect to the *y*-axis. Its graph always contains the origin $(0, 0)$ and the points $(-1, 1)$ and $(1, 1)$.

For large *n*, it appears that the graph coincides with the *x*-axis near the origin, but it does not; the graph actually touches the *x*-axis only at the origin. See Table 1, where $Y_2 = x^8$ and $Y_3 = x^{12}$. For *x* close to 0, the values of *y* are positive and close to 0. Also, for large *n*, it may appear that for $x < -1$ or for $x > 1$ the graph is vertical, but it is not; it is only increasing very rapidly. If you use TRACE along one of the graphs, these distinctions will be clear.

To summarize:

TABLE 1

X	Y2	Y3
-1	1	1
1	1	1
.5	.00391	2.4E-4
.1	1E-8	1E-12
.01	1E-16	1E-24
.001	1E-24	1E-36
0	0	0

Y2■X^8

> ### *Properties of Power Functions, $f(x) = x^n$, n is an Even Integer*
>
> 1. The graph is symmetric with respect to the *y*-axis, so *f* is even.
> 2. The domain is the set of all real numbers. The range is the set of nonnegative real numbers.
> 3. The graph always contains the points $(0, 0)$, $(1, 1)$, and $(-1, 1)$.
> 4. As the exponent *n* increases in magnitude, the graph becomes more vertical when $x < -1$ or $x > 1$; but for *x* near the origin, the graph tends to flatten out and lie closer to the *x*-axis.

Now we consider power functions of odd degree of the form $f(x) = x^n$, *n odd*.

 EXPLORATION   Using your graphing utility and the viewing window $-2 \leq x \leq 2$, $-16 \leq y \leq 16$, graph the function $Y_1 = f(x) = x^3$. On the same screen, graph $Y_2 = g(x) = x^7$ and $Y_3 = h(x) = x^{11}$. What do you notice about the graphs as the magnitude of the exponent increases? Repeat this procedure for the viewing window $-1 \leq x \leq 1$, $-1 \leq y \leq 1$. What do you notice? The graphs on your screen should look like Figures 2(a) and (b).

Figure 2

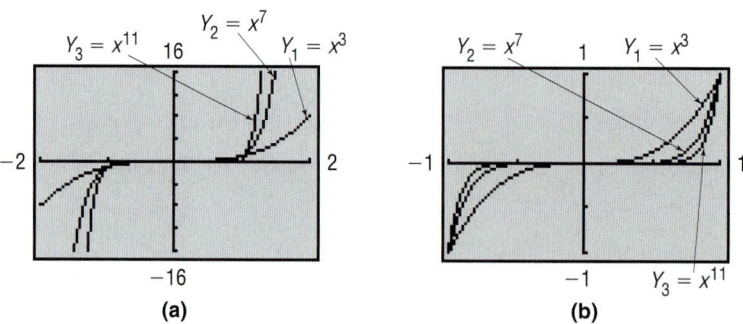

(a)   (b)

The domain and range of $f(x) = x^n$, $n \geq 3$ and n odd, are the set of real numbers. Such a power function is an odd function (do you see why?), so its graph is symmetric with respect to the origin. Its graph always contains the origin $(0, 0)$ and the points $(-1, -1)$ and $(1, 1)$.

It appears that the graph coincides with the x-axis near the origin, but it does not; the graph actually crosses the x-axis only at the origin. Also, it appears that as x increases the graph is vertical, but it is not; it is increasing very rapidly. TRACE along the graphs to verify these distinctions.

To summarize:

> **Properties of Power Functions, $f(x) = x^n$, n is an Odd Integer**
> 1. The graph is symmetric with respect to the origin, so f is odd.
> 2. The domain and range are the set of all real numbers.
> 3. The graph always contains the points $(0, 0)$, $(1, 1)$, and $(-1, -1)$.
> 4. As the exponent n increases in magnitude, the graph becomes more vertical when $x < -1$ or $x > 1$, but for x near the origin, the graph tends to flatten out and lie closer to the x-axis.

①   The methods of shifting, compression, stretching, and reflection studied in Section 3.4, when used with the facts just presented, will enable us to graph and analyze a variety of functions that are transformations of power functions.

EXAMPLE 1   **Graphing Transformations of Power Functions**

Graph: $f(x) = 1 - x^5$

Solution   Figure 3 shows the required stages.

Figure 3

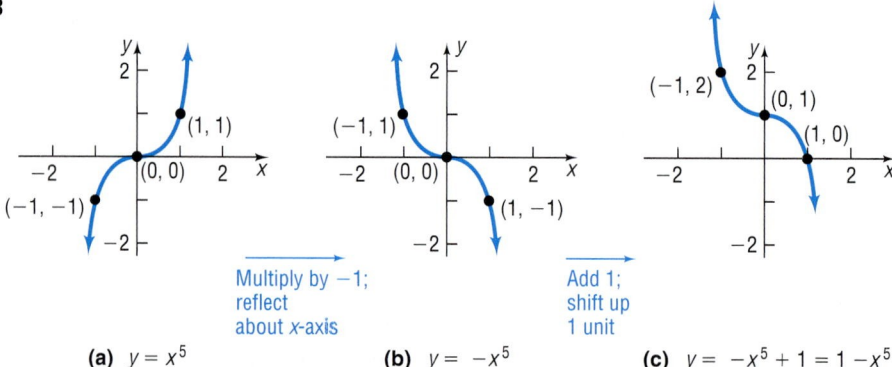

(a) $y = x^5$ **(b)** $y = -x^5$ **(c)** $y = -x^5 + 1 = 1 - x^5$

Multiply by -1; reflect about x-axis   Add 1; shift up 1 unit

✔ CHECK: Verify the result of Example 1 by graphing $Y_1 = f(x) = 1 - x^5$.

EXAMPLE 2   **Graphing Transformations of Power Functions**

Graph: $f(x) = \dfrac{1}{2}(x - 1)^4$

Solution   Figure 4 shows the required stages.

Figure 4

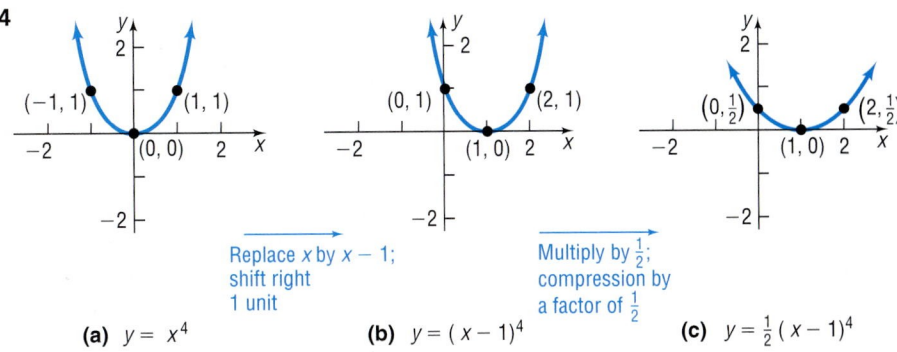

Replace x by $x - 1$;
shift right
1 unit

Multiply by $\frac{1}{2}$;
compression by
a factor of $\frac{1}{2}$

(a) $y = x^4$ **(b)** $y = (x - 1)^4$ **(c)** $y = \frac{1}{2}(x - 1)^4$

✔ CHECK: Verify the result of Example 2 by graphing
$$Y_1 = f(x) = \dfrac{1}{2}(x - 1)^4.$$

NOW WORK PROBLEM **1.**

Modeling Power Functions: Curve Fitting

 Figure 5 shows a scatter diagram that follows a power function.

Many situations lead to data that can be modeled using a power function.* One such situation involves falling objects.

Figure 5

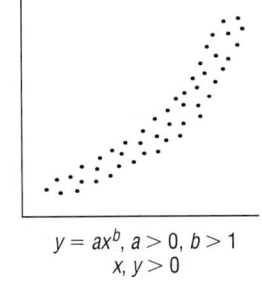

$y = ax^b,\ a > 0,\ b > 1$
$x,\ y > 0$

EXAMPLE 3   **Fitting Data to a Power Function**

Scott drops a ball from various heights and records the time that it takes for the ball to hit the ground. Using a motion detector connected to his graphing calculator, he collects the data shown in Table 2.

(a) Using a graphing utility, draw a scatter diagram using time t as the independent variable and distance s as the dependent variable.

(b) Using a graphing utility, find the power function of best fit $s = s(t)$ to these data.

(c) Graph the power function found in part (b) on the scatter diagram.

(d) Predict how long it will take the ball to fall 100 meters.

TABLE 2

Time t (seconds)	Distance s (meters)
1.528	11.46
2.015	19.99
3.852	72.41
4.154	84.45
4.625	104.23

* If the relation that exists between two variables, x and y, follows the function $y = ax^b$ where $a > 0, b > 0$ are real numbers, the function is called a **generalized power function**.

(e) Physics theory states the distance an object falls is directly proportional to the time squared. Rewrite the model found in part (b) so it is of the form $s = \frac{1}{2}gt^2$, where s is distance, g is acceleration due to gravity, and t is time. What is Scott's estimate of g? (It is known that the acceleration due to gravity is approximately 9.8 meters/sec².)

Solution

(a) After entering the data into the graphing utility, we obtain the scatter diagram shown in Figure 6.

(b) A graphing utility fits the data in Figure 6 to a power function of the form $y = ax^b$ by using the PoWeR REGression option. See Figure 7.* The power function of best fit is $s(t) = 4.93t^{1.993}$.

Figure 6

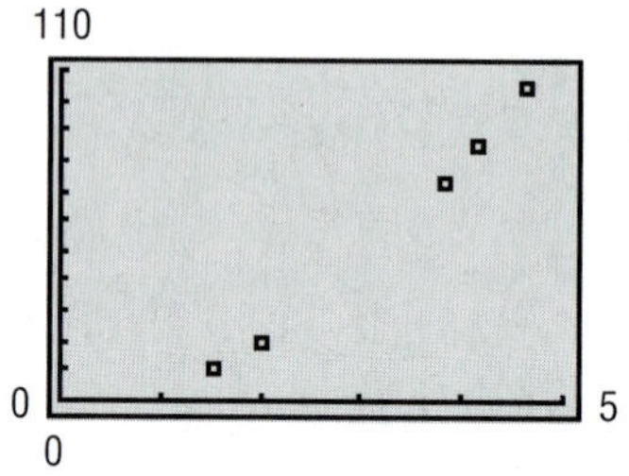

Figure 7

(c) Figure 8 shows the graph of the power function on the scatter diagram.

(d) If $s = 100$ meters, then $Y_1 = s(t) = 4.93t^{1.993}$ INTERSECTs $Y_2 = 100$ when $t = 4.53$ seconds, rounded to two decimal places. See Figure 9.

Figure 8

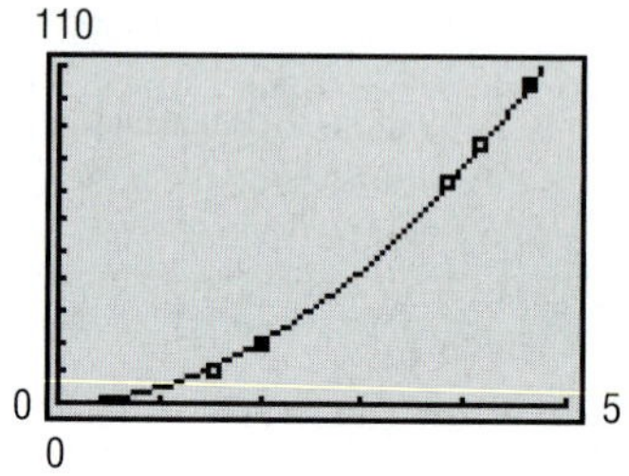

Figure 9

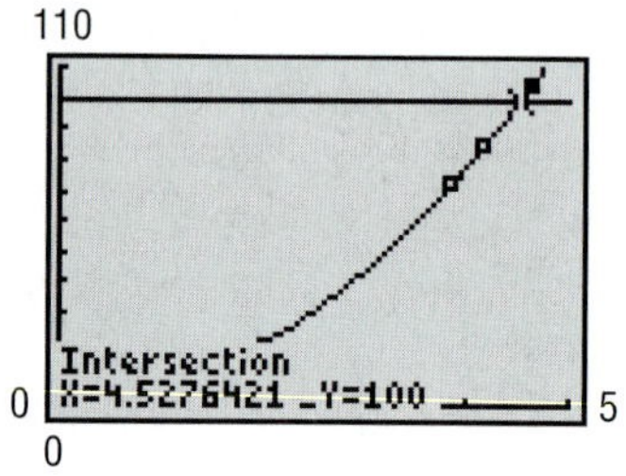

(e) The value of g obeys the equation

$$\frac{1}{2}g = 4.93$$

$$g = 9.86 \text{ meters/sec}^2$$

Scott's estimate for g is 9.86 meters/sec².

* For power functions, a value of r is given, since such functions can be expressed as linear functions using logarithms (see Chapter 6).

4.1 Concepts and Vocabulary

In Problems 1 and 2, fill in the blanks.

1. The degree of the power function $f(x) = -3x^5$ is _____ .

2. The graph of a power function $f(x) = x^n$, n odd, always contains the points _____ , _____ , and _____ .

In Problems 3 and 4, answer True or False to each statement.

3. The domain and the range of the power functions $f(x) = x^n$, n odd, are all real numbers.

4. A power function of degree n is a function of the form $f(x) = ax^n$, where n is an integer.

5. List four properties of power functions of even degree.

6. List four properties of power functions of odd degree.

4.1 Exercises

In Problems 1–16, use transformations of the graph of $y = x^4$ or $y = x^5$ to graph each function. Verify your results using a graphing utility.

1. $f(x) = (x + 1)^4$

2. $f(x) = (x - 2)^5$

3. $f(x) = x^5 - 3$

4. $f(x) = x^4 + 2$

5. $f(x) = \frac{1}{2}x^4$

6. $f(x) = 3x^5$

7. $f(x) = -x^5$

8. $f(x) = -x^4$

9. $f(x) = (x - 1)^5 + 2$

10. $f(x) = (x + 2)^4 - 3$

11. $f(x) = 2(x + 1)^4 + 1$

12. $f(x) = \frac{1}{2}(x - 1)^5 - 2$

13. $f(x) = 4 - (x - 2)^5$

14. $f(x) = 3 - (x + 2)^4$

15. $f(x) = -\frac{1}{2}(x - 2)^4 - 1$

16. $f(x) = 1 - 2(x + 1)^5$

17. CBL Experiment David is conducting an experiment to estimate the acceleration of an object due to gravity. David takes a ball and drops it from different heights and records the time it takes for the ball to hit the ground. Using an optic laser connected to a stopwatch in order to determine the time, he collects the following data:

Time (Seconds)	Distance (Feet)
1.003	16
1.365	30
1.769	50
2.093	70
2.238	80

(a) Draw a scatter diagram using time as the independent variable and distance as the dependent variable.
(b) Using a graphing utility, find the power function of best fit to these data.
(c) Graph the function found in part (b) on the scatter diagram.
(d) Use the function found in part (b) to predict the time that it will take an object to fall 100 feet.
(e) Physics theory states the distance an object falls is directly proportional to the time squared. Rewrite the model found in (b) so it is of the form $s = \frac{1}{2}gt^2$, where s is distance, g is the acceleration due to gravity, and t is time. What is David's estimate of g? (It is known that the acceleration due to gravity is approximately 32 feet/sec^2.)

18. CBL Experiment Paul, David's friend, doesn't believe David's estimate of acceleration due to gravity is correct. Paul repeats the experiment described in Problem 17 and obtains the following data:

Time (Seconds)	Distance (Feet)
0.7907	10
1.1160	20
1.4760	35
1.6780	45
1.9380	60

(a) Draw a scatter diagram using time as the independent variable and distance as the dependent variable.
(b) Using a graphing utility, find the power function of best fit to these data.
(c) Graph the function found in part (b) on the scatter diagram.
(d) Use the function found in part (b) to predict the time it will take an object to fall 100 feet.
(e) Use the function found in part (b) to estimate g. Why do you think this estimate is different than the one found in part (e) of Problem 17?

PREPARING FOR THIS SECTION

Before getting started, review the following:

✓ Polynomials (Section R.5, pp. 40–41)

✓ Using a Graphing Utility to Approximate Local Maxima and Local Minima (Section 3.2, pp. 267–269)

✓ Intercepts (Section 1.2, pp. 106–107)

4.2 POLYNOMIAL FUNCTIONS AND MODELS

OBJECTIVES
1. Identify Polynomial Functions and Their Degree
2. Identify the Zeros of a Polynomial Function and Its Multiplicity
3. Analyze the Graph of a Polynomial Function
4. Find the Cubic Function of Best Fit to Data

1 *Polynomial functions* are among the simplest expressions in algebra. They are easy to evaluate: only addition and repeated multiplication are required. Because of this, they are often used to approximate other, more complicated functions. In this section, we investigate properties of this important class of function.

> A **polynomial function** is a function of the form
>
> $$f(x) = a_n x^n + a_{n-1} x^{n-1} + \cdots + a_1 x + a_0 \qquad (1)$$
>
> where $a_n, a_{n-1}, \ldots, a_1, a_0$ are real numbers and n is a nonnegative integer. The domain consists of all real numbers.

A polynomial function is a function whose rule is given by a polynomial in one variable. The degree of a polynomial function is the degree of the polynomial in one variable, that is, the largest power of x that appears.

EXAMPLE 1 **Identifying Polynomial Functions**

Determine which of the following are polynomial functions. For those that are, state the degree; for those that are not, tell why not.

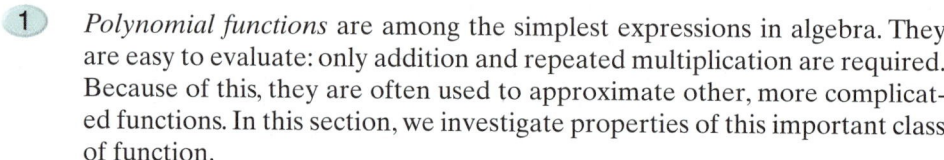

(a) $f(x) = 2 - 3x^4$ (b) $g(x) = \sqrt{x}$ (c) $h(x) = \dfrac{x^2 - 2}{x^3 - 1}$

(d) $F(x) = 0$ (e) $G(x) = 8$ (f) $H(x) = -2x^3(x - 1)^2$

Solution
(a) f is a polynomial function of degree 4.

(b) g is not a polynomial function. The variable x is raised to the $\dfrac{1}{2}$ power, which is not a nonnegative integer.

(c) h is not a polynomial function. It is the ratio of two polynomials, and the polynomial in the denominator is of positive degree.

(d) F is the zero polynomial function; it is not assigned a degree.

(e) G is a nonzero constant function; it is a polynomial function of degree 0 since $G(x) = 8 = 8x^0$.

(f) $H(x) = -2x^3(x-1)^2 = -2x^3(x^2 - 2x + 1) = -2x^5 + 4x^4 - 2x^3$.
So H is a polynomial function of degree 5. Do you see how to find the degree of H without multiplying out? ▇

NOW WORK PROBLEMS **1** AND **5**.

We have already discussed in detail polynomial functions of degrees 0, 1, and 2. See Table 3 for a summary of the properties of the graphs of these polynomial functions.

TABLE 3

Degree	Form	Name	Graph
No degree	$f(x) = 0$	Zero function	The x-axis
0	$f(x) = a_0, \quad a_0 \neq 0$	Constant function	Horizontal line with y-intercept a_0
1	$f(x) = a_1 x + a_0, \quad a_1 \neq 0$	Linear function	Nonvertical, nonhorizontal line with slope a_1 and y-intercept a_0
2	$f(x) = a_2 x^2 + a_1 x + a_0, \quad a_2 \neq 0$	Quadratic function	Parabola: Graph opens up if $a_2 > 0$; graph opens down if $a_2 < 0$; y-intercept a_0

Graphing Polynomial Functions

One of the objectives of this section is to analyze the graph of a polynomial function. If you take a course in calculus, you will learn that the graph of every polynomial function is both smooth and continuous. By **smooth,** we mean that the graph contains no sharp corners or cusps; by **continuous,** we mean that the graph has no gaps or holes and can be drawn without lifting pencil from paper. See Figures 10(a) and (b).

Figure 10

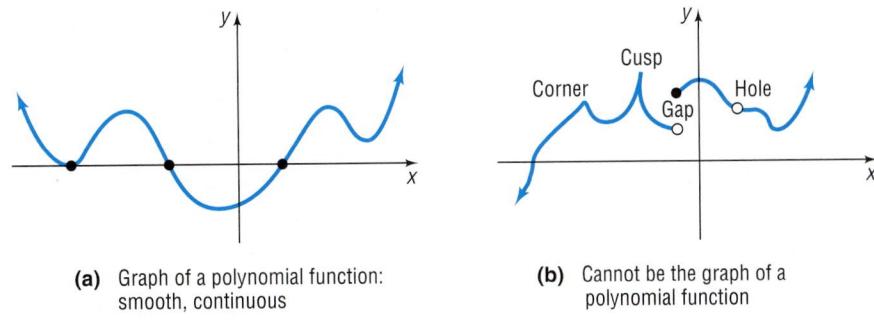

(a) Graph of a polynomial function: smooth, continuous

(b) Cannot be the graph of a polynomial function

Figure 11 shows the graph of a polynomial function with four x-intercepts. Notice that at the x-intercepts the graph must either cross the x-axis or touch the x-axis. Consequently, between consecutive x-intercepts the graph is either above the x-axis or below the x-axis. We will make use of this characteristic of the graph of a polynomial shortly.

Figure 11

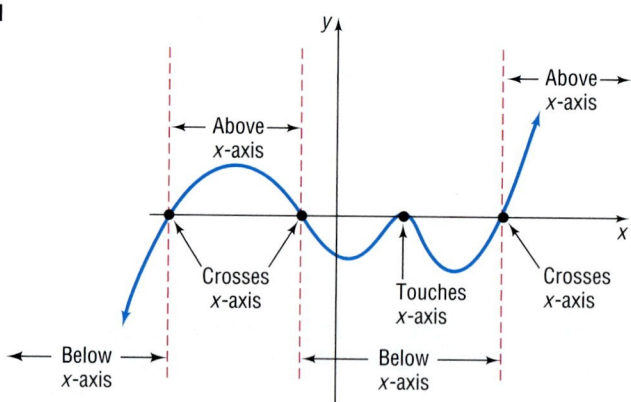

2. If a polynomial function f is factored completely, it is easy to solve the equation $f(x) = 0$ and locate the x-intercepts of the graph. For example, if $f(x) = (x - 1)^2(x + 3)$, then the solutions of the equation

$$f(x) = (x - 1)^2(x + 3) = 0$$

are identified as 1 and -3. Based on this result, we make the following observations:

> If f is a polynomial function and r is a real number for which $f(r) = 0$, then r is called a (real) **zero of** f, or **root of** f. If r is a (real) zero of f, then
>
> (a) r is an x-intercept of the graph of f.
> (b) $(x - r)$ is a factor of f.

EXAMPLE 2 **Finding a Polynomial from Its Zeros**

(a) Find a polynomial of degree 3 whose zeros are $-3, 0$, and 5.
(b) Graph the polynomial found in part (a) to verify your result.

Solution (a) If r is a zero of a polynomial f, then $x - r$ is a factor of f. This means that $x - (-3) = x + 3$, $x - 0 = x$, and $x - 5$ are factors of f. As a result, any polynomial of the form

$$f(x) = a(x + 3)(x - 0)(x - 5) = ax(x + 3)(x - 5)$$

where a is any nonzero real number, qualifies.

(b) The value of a causes a stretch, compression, or reflection, but does not affect the x-intercepts. We choose to graph f with $a = 1$.

$$f(x) = x(x + 3)(x - 5)$$

Figure 12 shows the graph of f. Notice that the x-intercepts are -3, 0, and 5.

Figure 12

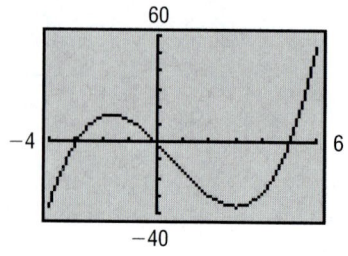

 SEEING THE CONCEPT Graph the function found in Example 2 for $a = 2$ and $a = -1$. Does the value of a affect the zeros of f? How does the value of a affect the graph of f?

NOW WORK PROBLEM 11.

If the same factor $x - r$ occurs more than once, then r is called a **repeated,** or **multiple, zero of f.** More precisely, we have the following definition.

If $(x - r)^m$ is a factor of a polynomial f and $(x - r)^{m+1}$ is not a factor of f, then r is called a **zero of multiplicity m of f.**

EXAMPLE 3 **Identifying Zeros and Their Multiplicities**

For the polynomial

$$f(x) = 5(x - 2)(x + 3)^2\left(x - \frac{1}{2}\right)^4$$

list the real zeros and identify their multiplicity.

Solution 2 is a zero of multiplicity 1.

-3 is a zero of multiplicity 2.

$\dfrac{1}{2}$ is a zero of multiplicity 4. ■

✏ **NOW WORK PROBLEM 19(a).**

In Example 3 notice that, if you add the multiplicities $(1 + 2 + 4 = 7)$, you obtain the degree of the polynomial.

Suppose that it is possible to factor completely a polynomial function and, as a result, locate all the x-intercepts of its graph (the real zeros of the function). The following example illustrates the role that the multiplicity of an x-intercept plays.

EXAMPLE 4 **Investigating the Role of Multiplicity**

For the polynomial $f(x) = x^2(x - 2)$:

(a) Find the x- and y-intercepts of the graph of f.

(b) Using a graphing utility, graph the polynomial.

(c) For each x-intercept, determine whether it is of odd or even multiplicity.

Solution (a) The y-intercept is $f(0) = 0^2(0 - 2) = 0$. The x-intercepts satisfy the equation

$$f(x) = x^2(x - 2) = 0$$

from which we find that

$$x^2 = 0 \qquad \text{or} \qquad x - 2 = 0$$
$$x = 0 \qquad\qquad\qquad x = 2$$

The x-intercepts are 0 and 2.

(b) See Figure 13 for the graph of f.

Figure 13

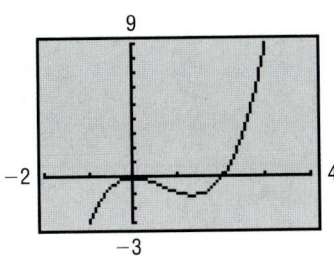

(c) We can see from the factored form of f that 0 is a zero or root of multiplicity 2, and 2 is a zero or root of multiplicity 1; so 0 is of even multiplicity and 2 is of odd multiplicity. ■

We can use a TABLE to further analyze the graph. See Table 4. The sign of $f(x)$ is the same on each side of the root 0, so the graph of f just touches the x-axis at 0 (a root of even multiplicity). The sign of $f(x)$ changes from one side of the root 2 to the other, so the graph of f crosses the x-axis at 2 (a root of odd multiplicity). These observations suggest the following result:

TABLE 4

X	Y1
-2	-16
-1	-3
0	0
1	-1
2	0
3	9
4	32

$Y_1 = X^2(X-2)$

If r is a Zero of Even Multiplicity

Sign of $f(x)$ does not change from one side of r to the other side of r.

Graph **touches** x-axis at r.

If r is a Zero of Odd Multiplicity

Sign of $f(x)$ changes from one side of r to the other side of r.

Graph **crosses** x-axis at r.

 **NOW WORK PROBLEM 19(b).**

Figure 14

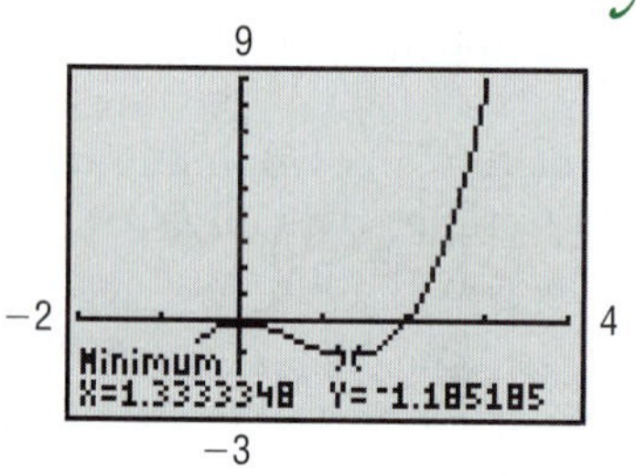

Minimum
X=1.3333348 Y=-1.185185

Look again at Figure 14. The graph of $f(x) = x^2(x - 2)$ has a local maximum at $x = 0$. The local maximum is 0. We can use a graphing utility to approximate the graph's local minimum in the interval $0 < x < 2$. After utilizing MINIMUM, we find that the graph's local minimum is -1.19 and it occurs at $x = 1.33$, each rounded to two decimal places. At points on the graph where the local extrema (local minima and local maxima) occur, the graph changes direction (that is, changes from an increasing function to a decreasing function, or vice versa). We call these points **turning points.** *

Look again at Figure 14. The graph of $f(x) = x^2(x - 2) = x^3 - 2x^2$, a polynomial of degree 3, has two turning points.

EXPLORATION Graph $Y_1 = x^3$, $Y_2 = x^3 - x$, and $Y_3 = x^3 + 3x^2 + 4$. How many turning points do you see? How does the number of turning points relate to the degree? Graph $Y_1 = x^4$, $Y_2 = x^4 - \frac{4}{3}x^3$, and $Y_3 = x^4 - 2x^2$. How many turning points do you see? How does the number of turning points compare to the degree?

The following theorem from calculus supplies the answer.

Theorem

If f is a polynomial function of degree n, then f has at most $n - 1$ turning points.

■

One last remark about Figure 14. Notice that the graph of $f(x) = x^2(x - 2)$ looks somewhat like the graph of $y = x^3$. In fact, for very large values of x, either positive or negative, there is little difference.

*Graphing utilities can be used to approximate turning points. Calculus is needed to find the exact location of turning points.

EXPLORATION   For each pair of functions Y_1 and Y_2 given in parts (a), (b), and (c) below, graph Y_1 and Y_2 on the same viewing window. Create a TABLE or TRACE for large positive and large negative values of x. What do you notice about the graphs of Y_1 and Y_2 as x becomes very large and positive or very large and negative?

(a) $Y_1 = x^2(x - 2);$ $Y_2 = x^3$
(b) $Y_1 = x^4 - 3x^3 + 7x - 3;$ $Y_2 = x^4$
(c) $Y_1 = -2x^3 + 4x^2 - 8x + 10;$ $Y_2 = -2x^3$

The behavior of the graph of a function for large values of x, either positive or negative, is referred to as its **end behavior.**

Theorem

> **End Behavior**
>
> For large values of x, either positive or negative, that is, for large $|x|$, the graph of the polynomial
>
> $$f(x) = a_n x^n + a_{n-1} x^{n-1} + \cdots + a_1 x + a_0$$
>
> resembles the graph of the power function
>
> $$y = a_n x^n$$

Look back at Figures 1 and 2 in Section 4.1. Based on the above theorem and the previous discussion on power functions, the end behavior of a polynomial can only be of four types. See Figure 15.

Figure 15
End Behavior

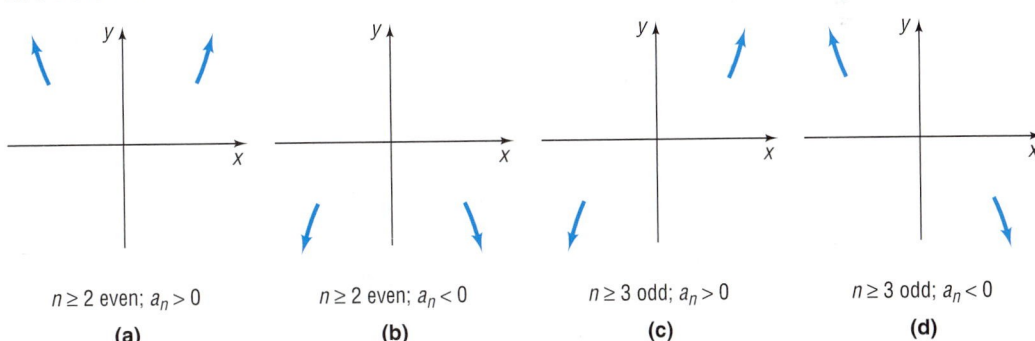

$n \geq 2$ even; $a_n > 0$	$n \geq 2$ even; $a_n < 0$	$n \geq 3$ odd; $a_n > 0$	$n \geq 3$ odd; $a_n < 0$
(a)	**(b)**	**(c)**	**(d)**

For example, consider the polynomial function $f(x) = -2x^4 + x^3 + 4x^2 - 7x + 1$. The graph of f will resemble the graph of the power function $y = -2x^4$ for large $|x|$. The graph of f will look like Figure 15(b) for large $|x|$.

NOW WORK PROBLEM **19(c).**

The following summarizes some features of the graph of a polynomial function.

SUMMARY **Graph of a Polynomial Function**

$$f(x) = a_n x^n + a_{n-1} x^{n-1} + \cdots + a_1 x + a_0, \quad a_n \neq 0$$

Degree of the polynomial f: $\quad n$

Maximum number of turning points: $\quad n - 1$

At zero of even multiplicity: $\quad$ The graph of f touches x-axis.

At zero of odd multiplicity: $\quad$ The graph of f crosses x-axis.

Between zeros, the graph of f is either above or below the x-axis.

End behavior: $\quad$ For large $|x|$, the graph of f behaves like the graph of $y = a_n x^n$.

EXAMPLE 5 **Analyzing the Graph of a Polynomial Function**

③ For the polynomial: $\quad f(x) = x^4 - 2x^3 - 8x^2$

(a) Find the x- and y-intercepts of the graph of f.

(b) Determine whether the graph crosses or touches the x-axis at each x-intercept.

(c) End behavior: find the power function that the graph of f resembles for large values of $|x|$.

(d) Use a graphing utility to graph f.

(e) Determine the number of turning points on the graph of f. Approximate the turning points, if any exist, rounded to two decimal places.

(f) Use the information obtained in parts (a) to (e) to draw a complete graph of f by hand.

Solution (a) The y-intercept is $f(0) = 0$. To find the x-intercepts, if any, we first factor f.

$$f(x) = x^4 - 2x^3 - 8x^2 = x^2(x^2 - 2x - 8) = x^2(x - 4)(x + 2)$$

We find the x-intercepts by solving the equation

$$f(x) = x^2(x - 4)(x + 2) = 0$$

So

$$x^2 = 0 \quad \text{or} \quad x - 4 = 0 \quad \text{or} \quad x + 2 = 0$$
$$x = 0 \qquad\qquad x = 4 \qquad\qquad x = -2$$

The x-intercepts are -2, 0, and 4.

(b) The x-intercept 0 is a zero of multiplicity 2, so the graph of f will touch the x-axis at 0; -2 and 4 are zeros of multiplicity 1, so the graph of f will cross the x-axis at -2 and 4.

(c) End behavior: the graph of f resembles that of the power function $y = x^4$ for large values of $|x|$.

(d) See Figure 16 for the graph of f.

(e) From the graph of f shown in Figure 16, we can see that f has three turning points. Using MAXIMUM, one turning point is at $(0, 0)$. Using MINIMUM, the two remaining turning points are at $(-1.39, -6.35)$ and $(2.89, -45.33)$, rounded to two decimal places.

Figure 16

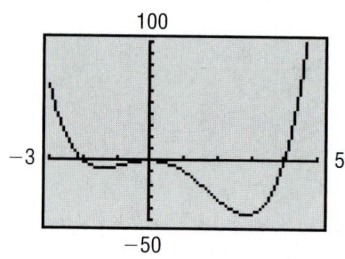

(f) Figure 17 shows a graph of f using the information obtained in parts (a) to (e).

Figure 17

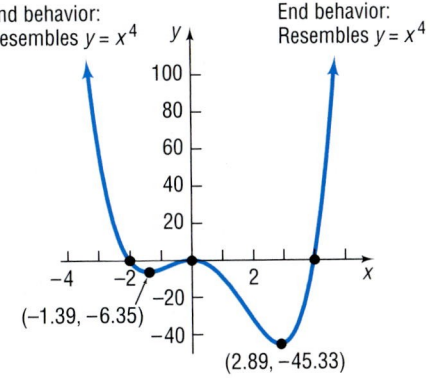

End behavior:
Resembles $y = x^4$

End behavior:
Resembles $y = x^4$

$(-1.39, -6.35)$

$(2.89, -45.33)$

N O W   W O R K   P R O B L E M   4 3 .

For polynomial functions that have noninteger coefficients and for polynomials that are not easily factored, we utilize the graphing utility early in the analysis of the graph. This is because the amount of information that can be obtained from algebraic analysis is limited.

EXAMPLE 6   **Analyzing the Graph of a Polynomial Function**

For the polynomial: $f(x) = x^3 + 2.48x^2 - 4.3155x + 1.484406$

(a) Use a graphing utility to graph f.

(b) Find the x- and y-intercepts of the graph of f.

(c) End behavior: find the power function that the graph of f resembles for large values of $|x|$.

(d) Determine the number of turning points on the graph of f. Approximate the turning points, if any exist, rounded to two decimal places.

(e) Use the information obtained in parts (a) to (d) to draw a complete graph of f by hand.

Figure 18

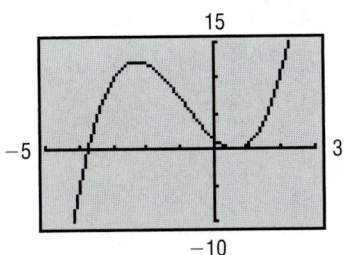

Solution

(a) See Figure 18 for the graph of f.

(b) The y-intercept is $f(0) = 1.484406$. In Example 5 we could easily factor $f(x)$ to find the x-intercepts. However, it is not readily apparent how $f(x)$ factors in this example. Therefore, we use a graphing utility's ZERO (or ROOT) feature and find the x-intercepts to be -3.74 and 0.63, rounded to two decimal places.

(c) The graph behaves like $y = x^3$ for large $|x|$.

(d) Since f is of degree 3, the graph can have at most two turning points. From the graph we see that it has two turning points: one between -3 and -2, the other between 0 and 1. Rounded to two decimal places, the local maximum is 12.36 and occurs and at $x = -2.28$; the local minimum is 0 and occurs at $x = 0.63$. The turning points are $(-2.28, 12.36)$ and $(0.63, 0)$.

Figure 19

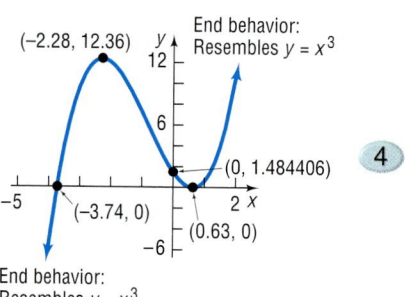

(e) Figure 19 shows a graph of f using the information obtained in parts (a) to (e).

NOW WORK PROBLEM 55.

Modeling Polynomial Functions: Curve Fitting

④ In Section 2.2, we found the line of best fit from data; in Section 2.4, we found the quadratic function of best fit; and in Section 4.1, we found the power function of best fit. It is also possible to find polynomial functions of best fit. However, most statisticians do not recommend finding polynomials of best fit of degree higher than 3.*

Data that follow a cubic relation should look like Figure 20(a) or (b).

Figure 20

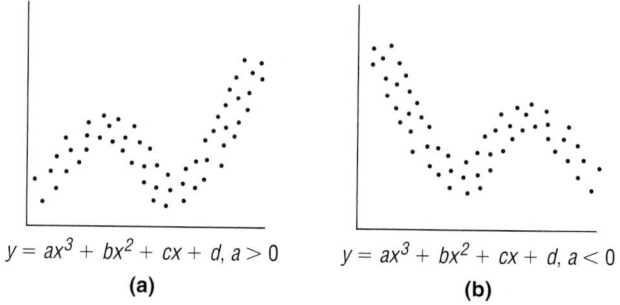

$$y = ax^3 + bx^2 + cx + d, a > 0$$
(a)

$$y = ax^3 + bx^2 + cx + d, a < 0$$
(b)

EXAMPLE 7 **A Cubic Function of Best Fit**

The data in Table 5 represent the average fuel consumption for cars (in gallons) in the United States for 1984–1997, where 1 represents 1984, 2 represents 1985, and so on.

(a) Draw a scatter diagram of the data using x as the independent variable and average fuel consumption C as the dependent variable. Comment on the type of relation that may exist between the two variables C and x.

(b) Using a graphing utility, find the cubic function of best fit $C = C(x)$ to these data.

(c) Graph the cubic function of best fit on your scatter diagram.

(d) Use the function found in part (b) to predict the average fuel consumption in 1998 ($x = 15$).

(e) The actual average fuel consumption in 1998 was 548 gallons. How well did the model predict average fuel consumption?

Solution (a) Figure 21 shows the scatter diagram.

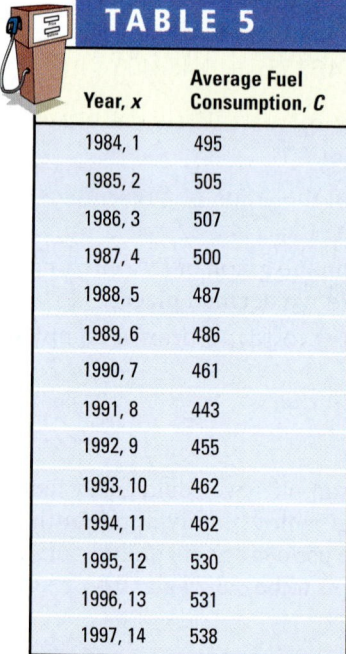

TABLE 5

Year, x	Average Fuel Consumption, C
1984, 1	495
1985, 2	505
1986, 3	507
1987, 4	500
1988, 5	487
1989, 6	486
1990, 7	461
1991, 8	443
1992, 9	455
1993, 10	462
1994, 11	462
1995, 12	530
1996, 13	531
1997, 14	538

Source: U.S. Federal Highway Administration

Figure 21

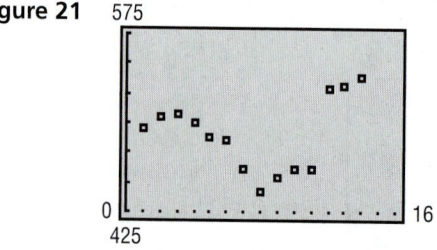

*Two points determine a unique linear function. Three noncollinear points determine a unique quadratic function. Four points determine a unique cubic function, and n points determine a unique polynomial of degree $n - 1$. Therefore, higher-degree polynomials will always "fit" data at least as well as lower-degree polynomials. However, higher-degree polynomials yield highly erratic predictions. Since the ultimate goal of curve fitting is not necessarily to find the model that best fits the data, but instead to explain relationships between two or more variables, polynomial models of degree 3 or less are usually used.

A cubic relation may exist between the two variables.

(b) Upon executing the CUBIC REGression program, we obtain the results shown in Figure 22. The output that the utility provides shows us the equation $y = ax^3 + bx^2 + cx + d$. The cubic function of best fit to the data is $C(x) = 0.2318x^3 - 3.6588x^2 + 10.2382x + 496.0669$.

(c) Figure 23 shows the graph of the cubic function of best fit on the scatter diagram. The function fits reasonably well.

Figure 22

Figure 23

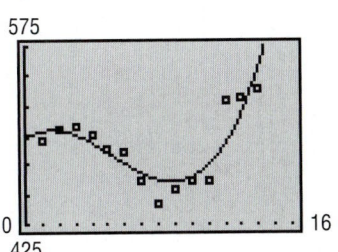

(d) We evaluate the function $C(x)$ for $x = 15$.

$$C(15) = 0.2318(15)^3 - 3.6588(15)^2 + 10.2382(15) + 496.0669 \approx 609$$

 (e) Comparing the predicted value of 609 gallons to the actual value of 538 gallons, we conclude that the function overestimates average fuel consumption.

4.2 Concepts and Vocabulary

In Problems 1–3, fill in the blanks.

1. The graph of every polynomial function is both _____ and _____.

2. A number r for which $f(r) = 0$ is call a(n) _____ of the function f.

3. If r is a zero of even multiplicity for a function f, the graph of f _____ the x-axis at r.

In Problems 4–6, answer True or False to each statement.

4. The graph of $f(x) = x^2(x - 3)(x + 4)$ has exactly three x-intercepts.

5. The x-intercepts of the graph of a polynomial function are called turning points.

6. End behavior: the graph of the function $f(x) = 3x^4 - 6x^2 + 2x + 5$ resembles $y = x^4$ for large values of $|x|$.

7. Can the graph of a polynomial function have no y-intercept? Can it have no x-intercepts? Explain.

8. The graph of a polynomial function is always smooth and continuous. Name a function from the Library of Functions that is smooth and not continuous. Name one that is continuous, but not smooth.

9. Which of the following statements are true regarding the graph of the cubic polynomial $f(x) = ax^3 + bx^2 + cx + d$? Give reasons for your conclusions.
   (a) It intersects the y-axis in one and only one point.
   (b) It intersects the x-axis in at most three points.
   (c) It intersects the x-axis at least once.
   (d) For $|x|$ very large, it behaves like the graph of $y = x^3$.
   (e) It is symmetric with respect to the origin.
   (f) It passes through the origin.

4.2 Exercises

In Problems 1–10, determine which functions are polynomial functions. For those that are, state the degree. For those that are not, tell why not.

1. $f(x) = 4x + x^3$

2. $f(x) = 5x^2 + 4x^4$

3. $g(x) = \dfrac{1 - x^2}{2}$

4. $h(x) = 3 - \dfrac{1}{2}x$

5. $f(x) = 1 - \dfrac{1}{x}$

6. $f(x) = x(x - 1)$

7. $g(x) = x^{3/2} - x^2 + 2$

8. $h(x) = \sqrt{x}(\sqrt{x} - 1)$

9. $F(x) = 5x^4 - \pi x^3 + \dfrac{1}{2}$

10. $F(x) = \dfrac{x^2 - 5}{x^3}$

In Problems 11–18, form a polynomial whose zeros and degree are given.

11. Zeros: $-1, 1, 3$; degree 3

12. Zeros: $-2, 2, 3$; degree 3

13. Zeros: $-3, 0, 4$; degree 3

14. Zeros: $-4, 0, 2$; degree 3

15. Zeros: $-4, -1, 2, 3$; degree 4

16. Zeros: $-3, -1, 2, 5$; degree 4

17. Zeros: -1, multiplicity 1; 3, multiplicity 2; degree 3

18. Zeros: -2, multiplicity 2; 4, multiplicity 1; degree 3

In Problems 19–30, for each polynomial function: (a) List each real zero and its multiplicity. (b) Determine whether the graph crosses or touches the x-axis at each x-intercept. (c) Find the power function that the graph of f resembles for large values of $|x|$.

19. $f(x) = 3(x - 7)(x + 3)^2$

20. $f(x) = 4(x + 4)(x + 3)^3$

21. $f(x) = 4(x^2 + 1)(x - 2)^3$

22. $f(x) = 2(x - 3)(x + 4)^3$

23. $f(x) = -2\left(x + \dfrac{1}{2}\right)^2(x^2 + 4)^2$

24. $f(x) = \left(x - \dfrac{1}{3}\right)^2(x - 1)^3$

25. $f(x) = (x - 5)^3(x + 4)^2$

26. $f(x) = (x + \sqrt{3})^2(x - 2)^4$

27. $f(x) = 3(x^2 + 8)(x^2 + 9)^2$

28. $f(x) = -2(x^2 + 3)^3$

29. $f(x) = -2x^2(x^2 - 2)$

30. $f(x) = 4x(x^2 - 3)$

In Problems 31–54, for each polynomial function f:

(a) *Find the x- and y-intercepts of the graph of f.*

(b) *Determine whether the graph crosses or touches the x-axis at each x-intercept.*

(c) *End behavior: Find the power function that the graph of f resembles for large values of $|x|$.*

(d) *Use a graphing utility to graph f.*

(e) *Determine the number of turning points on the graph of f. Approximate the turning points, if any exist, rounded to two decimal places.*

(f) *Use the information obtained in parts (a) to (e) to draw a complete graph of f by hand.*

31. $f(x) = (x - 1)^2$

32. $f(x) = (x - 2)^3$

33. $f(x) = x^2(x - 3)$

34. $f(x) = x(x + 2)^2$

35. $f(x) = 6x^3(x + 4)$

36. $f(x) = 5x(x - 1)^3$

37. $f(x) = -4x^2(x + 2)$

38. $f(x) = -\dfrac{1}{2}x^3(x + 4)$

39. $f(x) = (x + 1)(x - 2)(x + 4)$

40. $f(x) = (x - 1)(x + 4)(x - 3)$

41. $f(x) = 4x - x^3$

42. $f(x) = x - x^3$

43. $f(x) = x^2(x - 2)(x + 2)$

44. $f(x) = x^2(x - 3)(x + 4)$

45. $f(x) = (x + 1)^2(x - 2)^2$

46. $f(x) = (x + 1)^3(x - 3)$

47. $f(x) = x^2(x - 3)(x + 1)$

48. $f(x) = x^2(x - 3)(x - 1)$

49. $f(x) = (x + 2)^2(x - 4)^2$

50. $f(x) = (x - 2)^2(x + 2)(x + 4)$

51. $f(x) = x^2(x - 2)(x^2 + 3)$

52. $f(x) = x^2(x^2 + 1)(x + 4)$

53. $f(x) = -x^2(x^2 - 1)(x + 1)$

54. $f(x) = -x^2(x^2 - 4)(x - 5)$

In Problems 55–64, for each polynomial function f:

(a) *Use a graphing utility to graph f.*

(b) *Find the x- and y-intercepts of the graph of f.*

(c) *End behavior: find the power function that the graph of f resembles for large values of $|x|$.*

(d) *Determine the number of turning points on the graph of f. Approximate the turning points, if any exist, rounded to two decimal places.*

(e) *Use the information obtained in parts (a) to (d) to draw a complete graph of f by hand.*

55. $f(x) = x^3 + 0.2x^2 - 1.5876x - 0.31752$

56. $f(x) = x^3 - 0.8x^2 - 4.6656x + 3.73248$

57. $f(x) = x^3 + 2.56x^2 - 3.31x + 0.89$

58. $f(x) = x^3 - 2.91x^2 - 7.668x - 3.8151$

59. $f(x) = x^4 - 2.5x^2 + 0.5625$

60. $f(x) = x^4 - 18.5x^2 + 50.2619$

61. $f(x) = 2x^4 - \pi x^3 + \sqrt{5}x - 4$

62. $f(x) = -1.2x^4 + 0.5x^2 - \sqrt{3}x + 2$

63. $f(x) = -2x^5 - \sqrt{2}x^2 - x - \sqrt{2}$

64. $f(x) = \pi x^5 + \pi x^4 + \sqrt{3}x + 1$

In Problems 65–68, construct a polynomial function that might have this graph. (More than one answer may be possible.)

65.

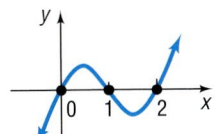

66.

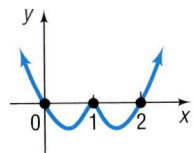

67.

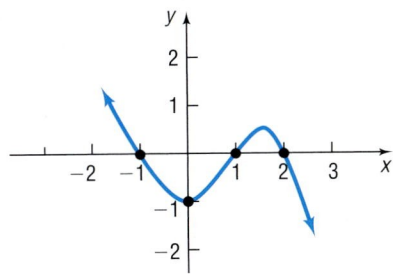

68.

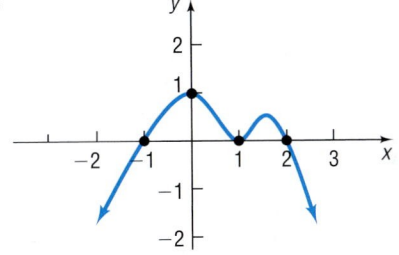

---

69. Cost of Manufacturing   The following data represent the cost C (in thousands of dollars) of manufacturing Chevy Cavaliers per hour and the number x of Cavaliers produced.

Number of Cavaliers Produced, x	Cost, C
0	10
1	23
2	31
3	38
4	43
5	50
6	59
7	70
8	85
9	105
10	135

(a) Draw a scatter diagram of the data using x as independent variable and C as the dependent variable. Comment on the type of relation that may exist between the two variables C and x.

(b) Find the average rate of change of cost from four to five Cavaliers.

(c) What is the average rate of change of cost from eight to nine Cavaliers?

(d) Use a graphing utility to find the cubic function of best fit $C = C(x)$.

(e) Graph the cubic function of best fit on the scatter diagram.

(f) Use the function found in part (d) to predict the cost of manufacturing 11 Cavaliers.

(g) Interpret the y-intercept.

70. Cost of Printing   The following data represent the weekly cost C (in thousands of dollars) of printing textbooks and the number x (in thousands) of texts printed.

Number of Text Books, x	Cost, C
0	100
5	128.1
10	144
13	153.5
17	161.2
18	162.6
20	166.3
23	178.9
25	190.2
27	221.8

(a) Draw a scatter diagram of the data using x as independent variable and C as the dependent variable. Comment on the type of relation that may exist between the two variables C and x.

(b) Find the average rate of change of cost from 10,000 to 13,000 textbooks.

(c) What is the average rate of change in the cost of producing from 18,000 to 20,000 textbooks?

(d) Use a graphing utility to find the cubic function of best fit $C = C(x)$.

(e) Graph the cubic function of best fit on the scatter diagram.

(f) Use the function found in part (d) to predict the cost of printing 22,000 texts per week.

(g) Interpret the y-intercept.

71. Motor Vehicle Thefts   The following data represent the number (in thousands) of motor vehicle thefts in the United States for the years 1987–1997, where $x = 1$ represents 1987, $x = 2$ represents 1988, and so on.

Year, x	Motor Vehicle Thefts, T
1987, 1	1289
1988, 2	1433
1989, 3	1565
1990, 4	1636
1991, 5	1662
1992, 6	1611
1993, 7	1563
1994, 8	1539
1995, 9	1472
1996, 10	1394
1997, 11	1354

Source: U.S. Federal Bureau of Investigation

(a) Draw a scatter diagram of the data using x as the independent variable and T as the dependent variable. Comment on the type of relation that may exist between the two variables T and x.

(b) Use a graphing utility to find the cubic function of best fit $T = T(x)$.

(c) Graph the cubic function of best fit on the scatter diagram.

(d) Use the function found in part (b) to predict the number of motor vehicle thefts in 1998.

(e) Check the prediction in part (d) against actual data. Do you think the function found in part (b) will be useful in predicting motor vehicle thefts in 2004?

72. Total Car Sales   The following data represent the total car sales S (used plus new car sales) in thousands of cars in the United States for the years 1990–1998, where $x = 1$ represents 1990, $x = 2$ represents 1991, and so on.

Year, x	Total Car Sales, S
1990, 1	46,830
1991, 2	45,465
1992, 3	45,163
1993, 4	46,575
1994, 5	49,132
1995, 6	50,353
1996, 7	49,355
1997, 8	48,542
1998, 9	48,372

Source: Statistical Abstract of the United States, 2000

(a) Draw a scatter diagram of the data using x as the independent variable and S as the dependent variable. Comment on the type of relation that may exist between the two variables S and x.

(b) Use a graphing utility to find the cubic function of best fit $S = S(x)$.

(c) Graph the cubic function of best fit on the scatter diagram.

(d) Use the function found in part (b) to predict total car sales in 1999.

(e) Check the prediction in part (d) against actual data. Do you think the function found in part (b) will be useful in predicting total car sales in 2004?

73. Write a few paragraphs that provide a general strategy for graphing a polynomial function. Be sure to mention the following: degree, intercepts, and turning points.

74. Make up a polynomial that has the following characteristics: crosses the x-axis at -1 and 4, touches the x-axis at 0 and 2, and is above the x-axis between 0 and 2. Give your polynomial to a fellow classmate and ask for a written critique of your polynomial.

75. Make up two polynomials, not of the same degree, with the following characteristics: crosses the x-axis at -2, touches the x-axis at 1, and is above the x-axis between -2 and 1. Give your polynomials to a fellow classmate and ask for a written critique of your polynomials.

PREPARING FOR THIS SECTION

Before getting started, review the following:

✓ Rational Expressions (Section R. 7, pp. 59–61)

✓ Graph of $f(x) = \dfrac{1}{x}$

(Section 3.1, Example 5, pp. 260–261)

✓ Polynomial Division (Section R. 5, pp. 45–48)

✓ Graphing Techniques: Transformations
(Section 3.4, pp. 286–295)

4.3   RATIONAL FUNCTIONS I

OBJECTIVES

1. Find the Domain of a Rational Function
2. Determine the Vertical Asymptotes of a Rational Function
3. Determine the Horizontal or Oblique Asymptotes of a Rational Function

Ratios of integers are called *rational numbers*. Similarly, ratios of polynomial functions are called *rational functions*.

A **rational function** is a function of the form

$$R(x) = \frac{p(x)}{q(x)}$$

where p and q are polynomial functions and q is not the zero polynomial. The domain consists of all real numbers except those for which the denominator q is 0.

EXAMPLE 1   **Finding the Domain of a Rational Function**

1.
(a) The domain of $R(x) = \dfrac{2x^2 - 4}{x + 5}$ consists of all real numbers x except -5.

(b) The domain of $R(x) = \dfrac{1}{x^2 - 4}$ consists of all real numbers x except -2 and 2.

(c) The domain of $R(x) = \dfrac{x^3}{x^2 + 1}$ consists of all real numbers.

(d) The domain of $R(x) = \dfrac{-x^2 + 2}{3}$ consists of all real numbers.

(e) The domain of $R(x) = \dfrac{x^2 - 1}{x - 1}$ consists of all real numbers x except 1.

It is important to observe that the functions

$$R(x) = \frac{x^2 - 1}{x - 1} \quad \text{and} \quad f(x) = x + 1$$

are not equal, since the domain of R is $\{x \mid x \neq 1\}$ and the domain of f is all real numbers.

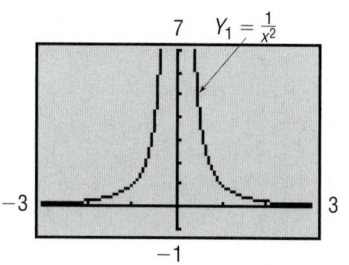 **NOW WORK PROBLEM 3.**

If $R(x) = \dfrac{p(x)}{q(x)}$ is a rational function and if p and q have no common factors, then the rational function R is said to be in **lowest terms.** For a rational function $R(x) = \dfrac{p(x)}{q(x)}$ in lowest terms, the zeros, if any, of the numerator are the x-intercepts of the graph of R and so will play a major role in the graph of R. The zeros of the denominator of R [that is, the numbers x, if any, for which $q(x) = 0$], although not in the domain of R, also play a major role in the graph of R. We will discuss this role shortly.

For now, let's look at two of the simplest rational functions. The first, $f(x) = \dfrac{1}{x}$ we have already discussed (refer to Example 5, page 260). The second, $H(x) = \dfrac{1}{x^2}$ we take up next

---

EXAMPLE 2 **Graphing** $y = \dfrac{1}{x^2}$

Analyze the graph of $H(x) = \dfrac{1}{x^2}$.

Solution The domain of $H(x) = \dfrac{1}{x^2}$ consists of all real numbers x except 0. The graph has no y-intercept, because x can never equal 0. The graph has no x-intercept because the equation $H(x) = 0$ has no solution. Therefore, the graph of H will not cross either coordinate axis.

Because

$$H(-x) = \frac{1}{(-x)^2} = \frac{1}{x^2} = H(x)$$

H is an even function, so its graph is symmetric with respect to the y-axis.

See Figure 24. Notice that the graph confirms the conclusions just reached. But what happens to the graph as the values of x get closer and closer to 0? We use a TABLE to answer the question. See Table 6. The first four rows show that as x approaches 0 the values of $H(x)$ become larger and larger positive numbers. When this happens, we say that $H(x)$ is **unbounded in the positive direction.** We symbolize this by writing $H(x) \to \infty$ [read as "$H(x)$ **approaches infinity**"]. In calculus, **limits** are used to convey these ideas. We use the symbolism $\lim\limits_{x \to 0} H(x) = \infty$, read as "the limit of $H(x)$ as x approaches 0 is infinity" to mean that $H(x) \to \infty$ as $x \to 0$.

Look at the last three rows of Table 6. As $x \to \infty$, the values of $H(x)$ approach 0 (the end behavior of the graph). This is symbolized in calculus by writing $\lim\limits_{x \to \infty} H(x) = 0$.

Figure 24

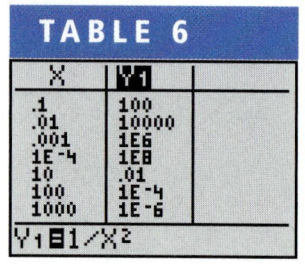

Figure 25 shows the graph of $y = \dfrac{1}{x^2}$ drawn by hand.

Figure 25

$H(x) = \dfrac{1}{x^2}$

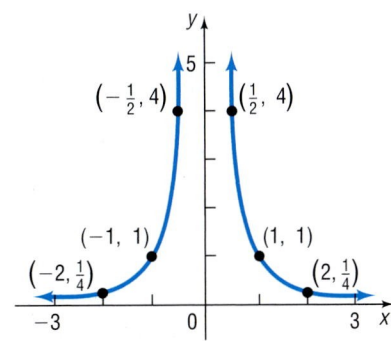

Sometimes transformations (shifting, compressing, stretching, and reflection) can be used to graph a rational function.

EXAMPLE 3   **Using Transformations to Graph a Rational Function**

Graph the rational function: $R(x) = \dfrac{1}{(x-2)^2} + 1$

Solution   First, we take note of the fact that the domain of R consists of all real numbers except $x = 2$. To graph R, we start with the graph of $y = \dfrac{1}{x^2}$. See Figure 26 for the stages.

Figure 26

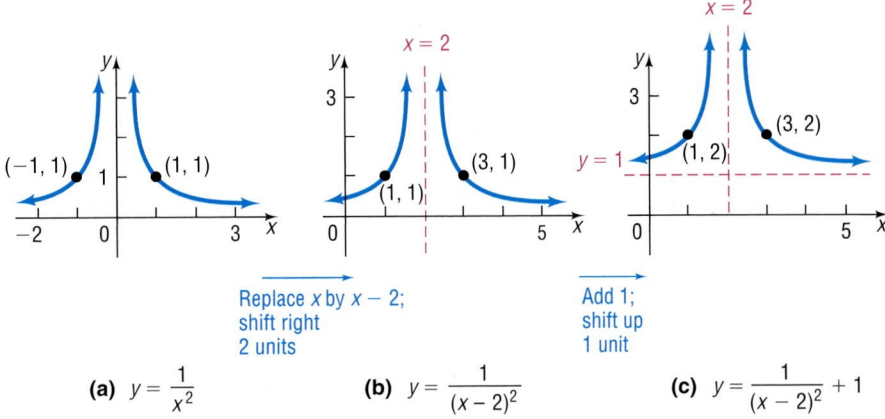

(a) $y = \dfrac{1}{x^2}$

Replace x by $x - 2$;
shift right
2 units

(b) $y = \dfrac{1}{(x-2)^2}$

Add 1;
shift up
1 unit

(c) $y = \dfrac{1}{(x-2)^2} + 1$

✔ **CHECK:**   Graph $Y_1 = \dfrac{1}{(x-2)^2} + 1$ using a graphing utility to verify the graph obtained in Figure 26(c).

NOW WORK PROBLEM 23.

Asymptotes

Notice that the y-axis in Figure 26(a) is transformed into the vertical line $x = 2$ in Figure 26(c) and the x-axis in Figure 26(a) is transformed into the horizontal line $y = 1$ in Figure 26(c). The **Exploration** that follows will help us analyze the role of these lines.

 EXPLORATION   Using a graphing utility and the TABLE feature, evaluate the function $H(x) = \dfrac{1}{(x-2)^2} + 1$ at $x = 10, 100, 1000,$ and 10,000. What happens to the values of H as x becomes unbounded in the positive direction, symbolized by $\lim\limits_{x \to \infty} H(x)$?

Evaluate H at $x = -10, -100, -1000,$ and $-10,000$. What happens to the values of H as x becomes unbounded in the negative direction, symbolized by $\lim\limits_{x \to -\infty} H(x)$?

Evaluate H at $x = 1.5, 1.9, 1.99, 1.999,$ and 1.9999. What happens to the values of H as x approaches $2, x < 2$, symbolized by $\lim\limits_{x \to 2^-} H(x)$?

Evaluate H at $x = 2.5, 2.1, 2.01, 2.001,$ and 2.0001. What happens to the values of H as x approaches $2, x > 2$, symbolized by $\lim\limits_{x \to 2^+} H(x)$?

RESULT   Table 7 shows the values of $Y_1 = H(x)$ as x approaches ∞. Notice that the values of H are approaching 1, so $\lim\limits_{x \to \infty} H(x) = 1$.

Table 8 shows the values of $Y_1 = H(x)$ as x approaches $-\infty$. Again the values of H are approaching 1, so $\lim\limits_{x \to -\infty} H(x) = 1$.

From Table 9 we see that, as x approaches $2, x < 2$, the values of H are increasing without bound, so $\lim\limits_{x \to 2^-} H(x) = \infty$.

Finally, Table 10 reveals that, as x approaches $2, x > 2$, the values of H are increasing without bound, so $\lim\limits_{x \to 2^+} H(x) = \infty$.

TABLE 7	
X	Y1
10	1.0156
100	1.0001
1000	1
10000	1
$Y_1 = 1/(X-2)^2+1$	

TABLE 8	
X	Y1
-10	1.0069
-100	1.0001
-1000	1
-10000	1
$Y_1 = 1/(X-2)^2+1$	

TABLE 9	
X	Y1
1.5	5
1.9	101
1.99	10001
1.999	1E6
1.9999	1E8
$Y_1 = 1/(X-2)^2+1$	

TABLE 10	
X	Y1
2.5	5
2.1	101
2.01	10001
2.001	1E6
2.0001	1E8
$Y_1 = 1/(X-2)^2+1$	

The results of the **Exploration** reveal an important characteristic of rational functions. The vertical line $x = 2$ and the horizontal line $y = 1$ are called *asymptotes* of the graph of H, which we define as follows:

Let R denote a function:

If, as $x \to -\infty$ or as $x \to \infty$, the values of $R(x)$ approach some fixed number L, then the line $y = L$ is a **horizontal asymptote** of the graph of R.

If, as x approaches some number c, the values $|R(x)| \to \infty$, then the line $x = c$ is a **vertical asymptote** of the graph of R.

Even though the asymptotes of a function are not part of the graph of the function, they provide information about how the graph looks. Figure 27 illustrates some of the possibilities.

Figure 27

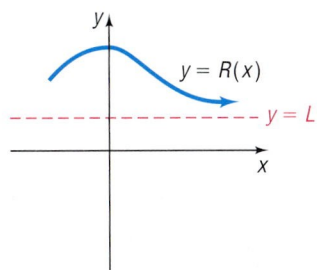

(a) End behavior:
As $x \to \infty$, the values of $R(x)$ approach L [$\lim\limits_{x \to \infty} R(x) = L$]. That is, the points on the graph of R are getting closer to the line $y = L$; $y = L$ is a horizontal asymptote.

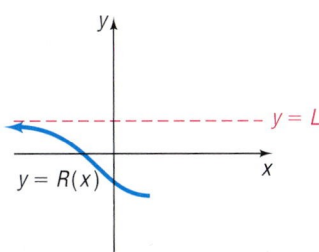

(b) End behavior:
As $x \to -\infty$, the values of $R(x)$ approach L [$\lim\limits_{x \to -\infty} R(x) = L$]. That is, the points on the graph of R are getting closer to the line $y = L$; $y = L$ is a horizontal asymptote.

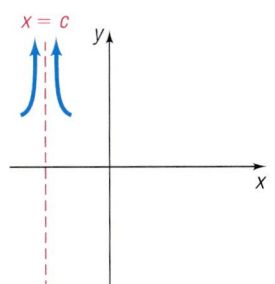

(c) As x approaches c, the values of $|R(x)| \to \infty$ [$\lim\limits_{x \to c^-} R(x) = \infty$; $\lim\limits_{x \to c^+} R(x) = \infty$]. That is, the points on the graph of R are getting closer to the line $x = c$; $x = c$ is a vertical asymptote.

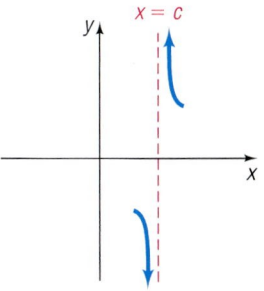

(d) As x approaches c, the values of $|R(x)| \to \infty$ [$\lim\limits_{x \to c^-} R(x) = -\infty$; $\lim\limits_{x \to c^+} R(x) = \infty$]. That is, the points on the graph of R are getting closer to the line $x = c$; $x = c$ is a vertical asymptote.

Figure 28
Oblique asymptote

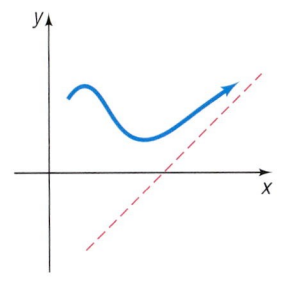

A horizontal asymptote, when it occurs, describes a certain behavior of the graph as $x \to \infty$ or as $x \to -\infty$, that is, its end behavior. The graph of a function may intersect a horizontal asymptote.

A vertical asymptote, when it occurs, describes a certain behavior of the graph when x is close to some number c. The graph of a rational function will never intersect one of its vertical asymptotes.

If an asymptote is neither horizontal nor vertical, it is called **oblique.** Figure 28 shows an oblique asymptote. An oblique asymptote, when it occurs, describes the end behavior of the graph. The graph of a function may intersect one of its oblique asymptotes.

Finding Asymptotes

The vertical asymptotes, if any, of a rational function $R(x) = \dfrac{p(x)}{q(x)}$, in lowest terms, are located at the zeros of the denominator of $q(x)$. Suppose that r is a zero so $x - r$ is a factor. Now, as x approaches r, symbolized as $x \to r$, the values of $x - r$ approach 0, causing the ratio to become unbounded, that is, causing $|R(x)| \to \infty$. Based on the definition, we conclude that the line $x = r$ is a vertical asymptote.

Theorem

Locating Vertical Asymptotes

A rational function $R(x) = \dfrac{p(x)}{q(x)}$, *in lowest terms,* will have a vertical asymptote $x = r$ if r is a real zero of the *denominator* q. That is, if $x - r$ is a factor of the denominator q of a rational function $R(x) = \dfrac{p(x)}{q(x)}$, in lowest terms, then R will have the vertical asymptote $x = r$.

> **WARNING:** If a rational function is not in lowest terms, an application of this theorem may result in an incorrect listing of vertical asymptotes.   ▪

EXAMPLE 4   **Finding Vertical Asymptotes**

Find the vertical asymptotes, if any, of the graph of each rational function.

(a) $R(x) = \dfrac{x}{x^2 - 4}$ (b) $F(x) = \dfrac{x + 3}{x - 1}$

(c) $H(x) = \dfrac{x^2}{x^2 + 1}$ (d) $G(x) = \dfrac{x^2 - 9}{x^2 + 4x - 21}$

Solution   (a) R is in lowest terms and the zeros of the denominator $x^2 - 4$ are -2 and 2. Hence, the lines $x = -2$ and $x = 2$ are the vertical asymptotes of the graph of R.

(b) F is in lowest terms and the only zero of the denominator is 1. Hence, the line $x = 1$ is the only vertical asymptote of the graph of F.

(c) H is in lowest terms and the denominator has no real zeros. Hence, the graph of H has no vertical asymptotes.

(d) Factor $G(x)$ to determine if it is in lowest terms.

$$G(x) = \frac{x^2 - 9}{x^2 + 4x - 21} = \frac{(x + 3)(x - 3)}{(x + 7)(x - 3)} = \frac{x + 3}{x + 7}, \qquad x \neq 3$$

The only zero of the denominator of $G(x)$ in lowest terms is -7. Hence, the line $x = -7$ is the only vertical asymptote of the graph of G.   ▪

 EXPLORATION   Graph each of the following rational functions:

$$R(x) = \frac{1}{x - 1} \qquad R(x) = \frac{1}{(x - 1)^2} \qquad R(x) = \frac{1}{(x - 1)^3} \qquad R(x) = \frac{1}{(x - 1)^4}$$

Each has the vertical asymptote $x = 1$. What happens to the value of $R(x)$ as x approaches 1 from the right side of the vertical asymptote; that is, what is $\lim\limits_{x \to 1^+} R(x)$? What happens to the value of $R(x)$ as x approaches 1 from the left side of the vertical asymptote; that is, what is $\lim\limits_{x \to 1^-} R(x)$? How does the multiplicity of the zero in the denominator affect the graph of R near its vertical asymptotes?   ▰

 ③   The procedure for finding horizontal and oblique asymptotes is somewhat more involved. To find such asymptotes, we need to know how the values of a function behave as $x \to -\infty$ or as $x \to \infty$.

If a rational function $R(x)$ is **proper,** that is, if the degree of the numerator is less than the degree of the denominator, then as $x \to -\infty$ or as $x \to \infty$, the values of $R(x)$ approach 0. Consequently, the line $y = 0$ (the x-axis) is a horizontal asymptote of the graph.

Theorem   If a rational function is proper, the line $y = 0$ is a horizontal asymptote of its graph.

EXAMPLE 5   Finding Horizontal Asymptotes

Find the horizontal asymptotes, if any, of the graph of

$$R(x) = \frac{x - 12}{4x^2 + x + 1}$$

Solution   The rational function R is proper, since the degree of the numerator, 1, is less than the degree of the denominator, 2. We conclude that the line $y = 0$ is a horizontal asymptote of the graph of R. ■

To see why $y = 0$ is a horizontal asymptote of the function R in Example 5, we need to investigate the behavior of R as $x \to -\infty$ and $x \to \infty$. When $|x|$ is unbounded, the numerator of R, which is $x - 12$, can be approximated by the power function $y = x$, while the denominator of R, which is $4x^2 + x + 1$, can be approximated by the power function $y = 4x^2$. Applying these ideas to $R(x)$, we find

$$R(x) = \frac{x - 12}{4x^2 + x + 1} \underset{\underset{\text{For } |x| \text{ unbounded}}{\uparrow}}{\approx} \frac{x}{4x^2} = \frac{1}{4x} \underset{\underset{\text{As } x \to -\infty \text{ or } x \to \infty}{\uparrow}}{\to} 0$$

This shows that the line $y = 0$ is a horizontal asymptote of the graph of R.

We verify these results in Tables 11(a) and (b). Notice, as $x \to -\infty$ (Table 11(a) or $x \to \infty$ (Table 11(b) respectively that the values of $R(x)$ approach 0.

TABLE 11

X	Y1		X	Y1
-10	-.0563		10	-.0049
-100	-.0028		100	.00219
-1000	-3E-4		1000	2.5E-4
-10000	-3E-5		10000	2.5E-5
-1E5	-3E-6		100000	2.5E-6
-1E6	-3E-7		1E6	2.5E-7
-1E7	-3E-8		1E7	2.5E-8

V1■(X-12)/(4X²+...	V1■(X-12)/(4X²+...
(a)	(b)

If a rational function $R(x) = \dfrac{p(x)}{q(x)}$ is **improper,** that is, if the degree of the numerator is greater than or equal to the degree of the denominator, we must use long division to write the rational function as the sum of a polynomial $f(x)$ plus a proper rational function $\dfrac{r(x)}{q(x)}$. That is, we write

$$R(x) = \frac{p(x)}{q(x)} = f(x) + \frac{r(x)}{q(x)}$$

where $f(x)$ is a polynomial and $\dfrac{r(x)}{q(x)}$ is a proper rational function. Since $\dfrac{r(x)}{q(x)}$ is proper, then $\dfrac{r(x)}{q(x)} \to 0$ as $x \to -\infty$ or as $x \to \infty$. As a result,

$$R(x) = \frac{p(x)}{q(x)} \to f(x), \qquad \text{as } x \to -\infty \text{ or as } x \to \infty$$

The possibilities are listed next.

1. If $f(x) = b$, a constant, then the line $y = b$ is a horizontal asymptote of the graph of R.
2. If $f(x) = ax + b$, $a \neq 0$, then the line $y = ax + b$ is an oblique asymptote of the graph of R.
3. In all other cases, the graph of R approaches the graph of f, and there are no horizontal or oblique asymptotes.

The following examples demonstrate these conclusions.

EXAMPLE 6   **Finding Horizontal or Oblique Asymptotes**

Find the horizontal or oblique asymptotes, if any, of the graph of

$$H(x) = \frac{3x^4 - x^2}{x^3 - x^2 + 1}$$

Solution   The rational function H is improper, since the degree of the numerator, 4, is larger than the degree of the denominator, 3. To find any horizontal or oblique asymptotes, we use long division.

$$
\begin{array}{r}
3x + 3 \\
x^3 - x^2 + 1 \overline{)3x^4 \qquad - x^2 \qquad\quad} \\
\underline{3x^4 - 3x^3 \qquad + 3x} \\
3x^3 - x^2 - 3x \\
\underline{3x^3 - 3x^2 \qquad + 3} \\
2x^2 - 3x - 3
\end{array}
$$

As a result,

$$H(x) = \frac{3x^4 - x^2}{x^3 - x^2 + 1} = 3x + 3 + \frac{2x^2 - 3x - 3}{x^3 - x^2 + 1}$$

As $x \to -\infty$ or as $x \to \infty$,

$$\frac{2x^2 - 3x - 3}{x^3 - x^2 + 1} \approx \frac{2x^2}{x^3} = \frac{2}{x} \to 0$$

As $x \to -\infty$ or as $x \to \infty$, we have $H(x) \to 3x + 3$. We conclude that the graph of the rational function H has an oblique asymptote $y = 3x + 3$. We verify these results in Tables 12(a) and (b) with $Y_1 = H(x)$ and $Y_2 = 3x + 3$. As $x \to -\infty$ or $x \to \infty$, the difference in the values between Y_1 and Y_2 is indistinguishable.

TABLE 12

X	Y1	Y2
-10	-27.21	-27
-100	-297	-297
-1000	-2997	-2997
-10000	-29997	-29997
-1E5	-3E5	-3E5
-1E6	-3E6	-3E6
-1E7	-3E7	-3E7

Y1 ◼(3X^4−X²)/(X...

(a)

X	Y1	Y2
10	33.185	33
100	303.02	303
1000	3003	3003
10000	30003	30003
100000	300003	300003
1E6	3E6	3E6
1E7	3E7	3E7

Y1 ◼(3X^4−X²)/(X...

(b)

EXAMPLE 7   **Finding Horizontal or Oblique Asymptotes**

Find the horizontal or oblique asymptotes, if any, of the graph of

$$R(x) = \frac{8x^2 - x + 2}{4x^2 - 1}$$

Solution   The rational function R is improper, since the degree of the numerator, 2, equals the degree of the denominator, 2. To find any horizontal or oblique asymptotes, we use long division.

$$\begin{array}{r} 2 \\ 4x^2 - 1 \overline{\smash{)}8x^2 - x + 2} \\ \underline{8x^2 - 2} \\ -x + 4 \end{array}$$

As a result,

$$R(x) = \frac{8x^2 - x + 2}{4x^2 - 1} = 2 + \frac{-x + 4}{4x^2 - 1}$$

Then, as $x \to -\infty$ or as $x \to \infty$,

$$\frac{-x + 4}{4x^2 - 1} \approx \frac{-x}{4x^2} = \frac{-1}{4x} \to 0$$

As $x \to -\infty$ or as $x \to \infty$, we have $R(x) \to 2$. We conclude that $y = 2$ is a horizontal asymptote of the graph.

✔ CHECK: Verify the results of Example 7 by creating a TABLE with $Y_1 = R(x)$ and $Y_2 = 2$.   ■ ■

   In Example 7, we note that the quotient 2 obtained by long division is the quotient of the leading coefficients of the numerator polynomial and the denominator polynomial $\left(\dfrac{8}{4}\right)$. This means that we can avoid the long division process for rational functions whose numerator and denominator *are of the same degree* and conclude that the quotient of the leading coefficients will give us the horizontal asymptote.

✏ NOW WORK PROBLEM 39.

EXAMPLE 8   **Finding Horizontal or Oblique Asymptotes**

Find the horizontal or oblique asymptotes, if any, of the graph of

$$G(x) = \frac{2x^5 - x^3 + 2}{x^3 - 1}$$

Solution   The rational function G is improper, since the degree of the numerator, 5, is larger than the degree of the denominator, 3. To find any horizontal or oblique asymptotes, we use long division.

$$
\begin{array}{r}
2x^2 - 1 \\
x^3 - 1 \overline{\smash{\big)}\ 2x^5 - x^3 + 2} \\
\underline{2x^5 - 2x^2} \\
-x^3 + 2x^2 + 2 \\
\underline{-x^3 + 1} \\
2x^2 + 1
\end{array}
$$

As a result,

$$
G(x) = \frac{2x^5 - x^3 + 2}{x^3 - 1} = 2x^2 - 1 + \frac{2x^2 + 1}{x^3 - 1}
$$

Then, as $x \to -\infty$ or as $x \to \infty$,

$$
\frac{2x^2 + 1}{x^3 - 1} \approx \frac{2x^2}{x^3} = \frac{2}{x} \to 0
$$

As $x \to -\infty$ or as $x \to \infty$, we have $G(x) \to 2x^2 - 1$. We conclude that, for large values of $|x|$, the graph of G approaches the graph of $y = 2x^2 - 1$. That is, the graph of G will look like the graph of $y = 2x^2 - 1$ as $x \to -\infty$ or $x \to \infty$. Since $y = 2x^2 - 1$ is not a linear function, G has no horizontal or oblique asymptotes. ■

We now summarize the procedure for finding horizontal and oblique asymptotes.

SUMMARY   **Finding Horizontal and Oblique Asymptotes of a Rational Function R**

Consider the rational function

$$
R(x) = \frac{p(x)}{q(x)} = \frac{a_n x^n + a_{n-1} x^{n-1} + \cdots + a_1 x + a_0}{b_m x^m + b_{m-1} x^{m-1} + \cdots + b_1 x + b_0}
$$

in which the degree of the numerator is n and the degree of the denominator is m.

1. If $n < m$, then R is a proper rational function, and the graph of R will have the horizontal asymptote $y = 0$ (the x-axis).

2. If $n \geq m$, then R is improper. Here long division is used.

   (a) If $n = m$, the quotient obtained will be the number $\dfrac{a_n}{b_m}$, and the line $y = \dfrac{a_n}{b_m}$ is a horizontal asymptote.

   (b) If $n = m + 1$, the quotient obtained is of the form $ax + b$ (a polynomial of degree 1), and the line $y = ax + b$ is an oblique asymptote.

   (c) If $n > m + 1$, the quotient obtained is a polynomial of degree 2 or higher, and R has neither a horizontal nor an oblique asymptote. In this case, for $|x|$ unbounded, the graph of R will behave like the graph of the quotient.

Note: Based on this analysis, the graph of a rational function either has one horizontal or one oblique asymptote or else has no horizontal and no oblique asymptote.

4.3 Concepts and Vocabulary

In Problems 1–3, fill in the blanks.

1. The line _____ is a horizontal asymptote of $R(x) = \dfrac{x^3 - 1}{x^3 + 1}$.

2. The line _____ is a vertical asymptote of $R(x) = \dfrac{x^3 - 1}{x^3 + 1}$.

3. For a rational function R, if the degree of the numerator is less than the degree of the denominator, then R is _____.

In Problems 4–6, answer True or False to each statement.

4. The domain of every rational function is all real numbers.

5. If an asymptote is neither horizontal nor vertical, it is called oblique.

6. If the degree of the numerator in a rational function equals the degree of the denominator, then the ratio of the leading coefficients gives rise to the horizontal asymptote.

7. Explain what the notation $\lim\limits_{x \to \infty} R(x) = L$ means.

8. The domain of a rational function and the location of the vertical asymptotes are related. Explain their relation. Do the values that are excluded from the domain of a rational function always give rise to vertical asymptotes? Why? Use examples to illustrate your point.

9. Can the graph of a rational function have both a horizontal and an oblique asymptote? Explain.

4.3 Exercises

In Problems 1–12, find the domain of each rational function.

1. $R(x) = \dfrac{4x}{x - 3}$

2. $R(x) = \dfrac{5x^2}{3 + x}$

3. $H(x) = \dfrac{-4x^2}{(x - 2)(x + 4)}$

4. $G(x) = \dfrac{6}{(x + 3)(4 - x)}$

5. $F(x) = \dfrac{3x(x - 1)}{2x^2 - 5x - 3}$

6. $Q(x) = \dfrac{-x(1 - x)}{3x^2 + 5x - 2}$

7. $R(x) = \dfrac{x}{x^3 - 8}$

8. $R(x) = \dfrac{x}{x^4 - 1}$

9. $H(x) = \dfrac{3x^2 + x}{x^2 + 4}$

10. $G(x) = \dfrac{x - 3}{x^4 + 1}$

11. $R(x) = \dfrac{3(x^2 - x - 6)}{4(x^2 - 9)}$

12. $F(x) = \dfrac{-2(x^2 - 4)}{3(x^2 + 4x + 4)}$

In Problems 13–22, use the graph shown to find:

    (a) *The domain and range of each function*   (b) *The intercepts, if any*
    (c) *Horizontal asymptotes, if any*   (d) *Vertical asymptotes, if any*
    (e) *Oblique asymptotes, if any*

13.

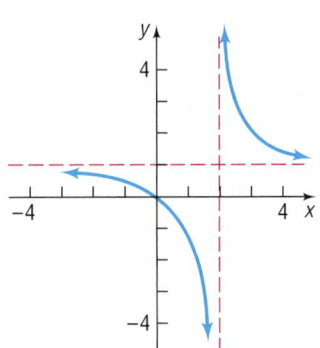

14.

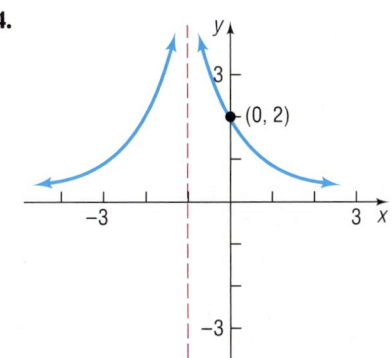

15.

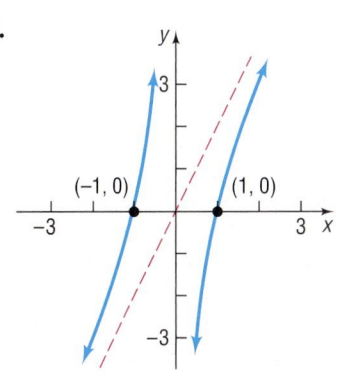

16.

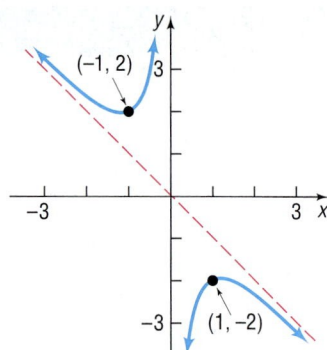

17.

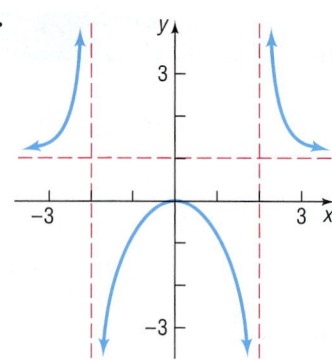

18.

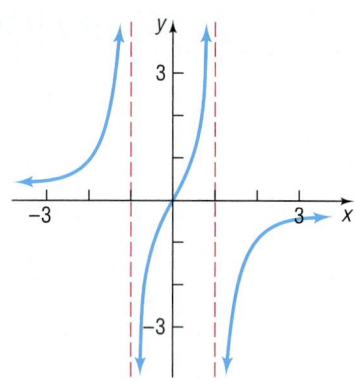

In Problems 19–22, assume that for any vertical asymptote $x = a$, a is an integer and for any horizontal asymtote $y = b$, b is an integer.

19.

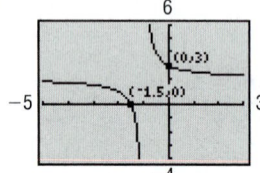

20.

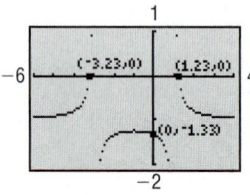

21.

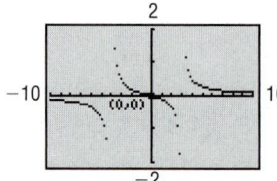

22.

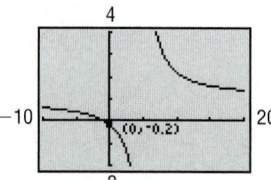

In Problems 23–34, graph each rational function using transformations. Verify your results using a graphing utility.

23. $R(x) = \dfrac{1}{(x-1)^2}$

24. $R(x) = \dfrac{3}{x}$

25. $H(x) = \dfrac{-2}{x+1}$

26. $G(x) = \dfrac{2}{(x+2)^2}$

27. $R(x) = \dfrac{1}{x^2 + 4x + 4}$

28. $R(x) = \dfrac{1}{x-1} + 1$

29. $F(x) = 1 - \dfrac{1}{x}$

30. $Q(x) = 1 + \dfrac{1}{x}$

31. $R(x) = \dfrac{x^2 - 4}{x^2}$

32. $R(x) = \dfrac{x-4}{x}$

33. $G(x) = 1 + \dfrac{2}{(x-3)^2}$

34. $F(x) = 2 - \dfrac{1}{x+1}$

In Problems 35–46, find the vertical, horizontal, and oblique asymptotes, if any, of each rational function without graphing. Verify your results using a graphing utility.

35. $R(x) = \dfrac{3x}{x+4}$

36. $R(x) = \dfrac{3x+5}{x-6}$

37. $H(x) = \dfrac{x^4 + 2x^2 + 1}{x^2 - x + 1}$

38. $G(x) = \dfrac{-x^2 + 1}{x+5}$

39. $T(x) = \dfrac{x^3}{x^4 - 1}$

40. $P(x) = \dfrac{4x^5}{x^3 - 1}$

41. $Q(x) = \dfrac{5 - x^2}{3x^4}$

42. $F(x) = \dfrac{-2x^2 + 1}{2x^3 + 4x^2}$

43. $R(x) = \dfrac{3x^4 + 4}{x^3 + 3x}$

44. $R(x) = \dfrac{6x^2 + x + 12}{3x^2 - 5x - 2}$

45. $G(x) = \dfrac{x^3 - 1}{x - x^2}$

46. $F(x) = \dfrac{x - 1}{x - x^3}$

47. If the graph of a rational function R has the vertical asymptote $x = 4$, then the factor $x - 4$ must be present in the denominator of R. Explain why.

48. If the graph of a rational function R has the horizontal asymptote $y = 2$, then the degree of the numerator of R equals the degree of the denominator of R. Explain why.

49. Make up a rational function that has $y = 2x + 1$ as an oblique asymptote. Explain the methodology that you used.

PREPARING FOR THIS SECTION

Before getting started, review the following:

✓ Intercepts (Section 1.2, pp. 106–107)

✓ Symmetry (Section 3.1, pp. 256–258)

✓ Even and Odd Functions (Section 3.2, pp. 269–272)

4.4 RATIONAL FUNCTIONS II: ANALYZING GRAPHS

OBJECTIVES

1. Analyze the Graph of a Rational Function
2. Solve Applied Problems Involving Rational Functions

1

Graphing utilities make the task of graphing rational functions less time consuming. However, the results of algebraic analysis must be taken into account before drawing conclusions based on the graph provided by the utility. We will use the information collected in the last section in conjunction with the graphing utility to analyze the graph of a rational function $R(x) = \dfrac{p(x)}{q(x)}$. The analysis will require the following steps:

Analyzing the Graph of a Rational Function

STEP 1: Find the domain of the rational function.

STEP 2: Write R in lowest terms.

STEP 3: Locate the intercepts of the graph. The x-intercepts, if any, of $R(x) = \dfrac{p(x)}{q(x)}$ in lowest terms satisfy the equation $p(x) = 0$. The y-intercept, if there is one, is $R(0)$.

STEP 4: Test for symmetry. Replace x by $-x$ in $R(x)$. If $R(-x) = R(x)$, there is symmetry with respect to the y-axis; if $R(-x) = -R(x)$, there is symmetry with respect to the origin.

STEP 5: Locate the vertical asymptotes. The vertical asymptotes, if any, of $R(x) = \dfrac{p(x)}{q(x)}$ in lowest terms are found by identifying the real zeros of $q(x)$. Each zero of the denominator gives rise to a vertical asymptote.

STEP 6: Locate the horizontal or oblique asymptotes, if any, using the procedure given in Section 4.3. Determine points, if any, at which the graph of R intersects these asymptotes.

STEP 7: Graph R using a graphing utility.

STEP 8: Use the results obtained in Steps 1 through 7 to graph R by hand.

EXAMPLE 1   **Analyzing the Graph of a Rational Function**

Analyze the graph of the rational function: $R(x) = \dfrac{x - 1}{x^2 - 4}$

Solution   First, we factor both the numerator and the denominator of R.

$$R(x) = \frac{x - 1}{(x + 2)(x - 2)}$$

STEP 1: The domain of R is $\{x | x \neq -2, x \neq 2\}$.

STEP 2: R is in lowest terms.

STEP 3: We locate the x-intercepts by finding the zeros of the numerator. By inspection, 1 is the only x-intercept. The y-intercept is $R(0) = \dfrac{1}{4}$.

STEP 4: Because

$$R(-x) = \frac{-x - 1}{x^2 - 4} = \frac{-(x + 1)}{x^2 - 4}$$

we conclude that R is neither even nor odd. There is no symmetry with respect to the y-axis or the origin.

STEP 5: We locate the vertical asymptotes by finding the zeros of the denominator. Since R is in lowest terms, the graph of R has two vertical asymptotes: the lines $x = -2$ and $x = 2$.

STEP 6: The degree of the numerator is less than the degree of the denominator, so R is proper and the line $y = 0$ (the x-axis) is a horizontal asymptote of the graph. To determine if the graph of R intersects the horizontal asymptote, we solve the equation $R(x) = 0$:

$$\frac{x - 1}{x^2 - 4} = 0$$

$$x - 1 = 0$$

$$x = 1$$

The only solution is $x = 1$, so the graph of R intersects the horizontal asymptote at $(1, 0)$.

STEP 7: The analysis just completed helps us to set the viewing window to obtain a complete graph. Figure 29(a) shows the graph of $R(x) = \dfrac{x - 1}{x^2 - 4}$ in connected mode, and Figure 29(b) shows it in dot mode. Notice in Figure 29(a) that the graph has vertical lines at $x = -2$ and $x = 2$. This is due to the fact that, when the graphing utility is in connected mode, it will "connect the dots" between consecutive pixels. We know that the graph of R does not cross the lines $x = -2$ and $x = 2$, since R is not defined at $x = -2$ or $x = 2$. When graphing rational functions, dot mode should be used if extraneous vertical lines appear. You should confirm that all the algebraic conclusions that we arrived at in Steps 1 through 6 are part of the graph. For example, the graph has vertical asymptotes at $x = -2$ and $x = 2$, and the graph has a horizontal asymptote at $y = 0$. The y-intercept is $\dfrac{1}{4}$ and the x-intercept is 1.

STEP 8: Using the information gathered in Steps 1 through 7, we obtain the graph of R shown in Figure 30.

Figure 29

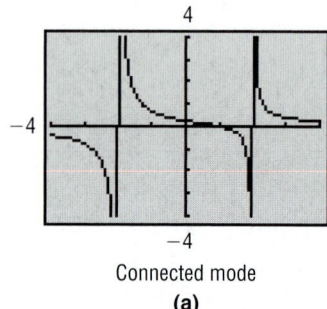

Connected mode
(a)

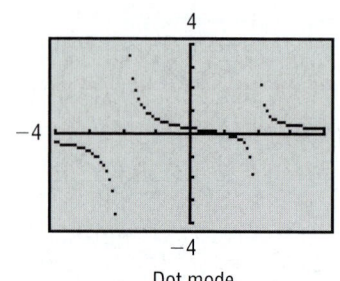

Dot mode
(b)

Figure 30

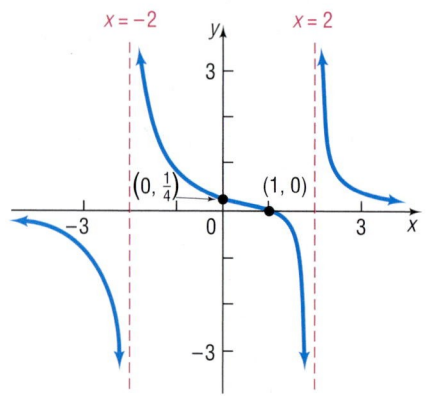

NOW WORK PROBLEM 1.

EXAMPLE 2	Analyzing the Graph of a Rational Function

Analyze the graph of the rational function: $\quad R(x) = \dfrac{x^2 - 1}{x}$

Solution
STEP 1: The domain of R is $\{x \mid x \neq 0\}$.

STEP 2: R is in lowest terms.

STEP 3: The graph has two x-intercepts: -1 and 1. There is no y-intercept, since x cannot equal 0.

STEP 4: Since $R(-x) = -R(x)$, the function is odd and the graph is symmetric with respect to the origin.

STEP 5: Since R is in lowest terms, the graph of $R(x)$ has the line $x = 0$ (the y-axis) as a vertical asymptote.

STEP 6: The rational function R is improper since the degree of the numerator, 2, is larger than the degree of the denominator, 1. To find any horizontal or oblique asymptotes, we use long division.

$$
\begin{array}{r}
x \\
x \overline{)\, x^2 - 1} \\
\underline{x^2 } \\
-1
\end{array}
$$

The quotient is x, so the line $y = x$ is an oblique asymptote of the graph. To determine whether the graph of R intersects the asymptote $y = x$, we solve the equation $R(x) = x$.

$$R(x) = \frac{x^2 - 1}{x} = x$$
$$x^2 - 1 = x^2$$
$$-1 = 0 \qquad \textcolor{teal}{\text{Impossible.}}$$

We conclude that the equation $\dfrac{x^2 - 1}{x} = x$ has no solution, so the graph of $R(x)$ does not intersect the line $y = x$.

STEP 7: See Figure 31. We see from the graph that there is no y-intercept and there are two x-intercepts, -1 and 1. The symmetry with respect to the origin is also evident. We can also see that there is a vertical asymptote at $x = 0$. Finally, it is not necessary to graph this function in dot mode since no extraneous vertical lines are present.

STEP 8: Using the information gathered in Steps 1 through 7, we obtain the graph of R shown in Figure 32. Notice how the oblique asymptote is used as a guide in graphing the rational function by hand.

Figure 31

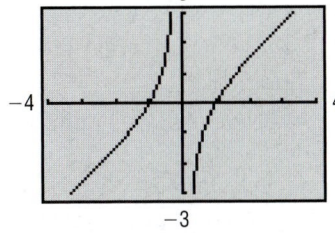

Figure 32

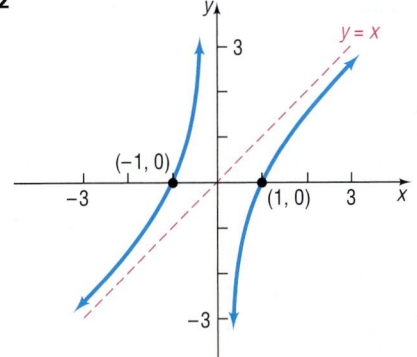

NOW WORK PROBLEM **9.**

EXAMPLE 3 **Analyzing the Graph of a Rational Function**

Analyze the graph of the rational function: $R(x) = \dfrac{3x^2 - 3x}{x^2 + x - 12}$

Solution We factor R to get

$$R(x) = \frac{3x(x - 1)}{(x + 4)(x - 3)}$$

STEP 1: The domain of R is $\{x \mid x \neq -4, x \neq 3\}$.

STEP 2: R is in lowest terms.

STEP 3: The graph has two x-intercepts: 0 and 1. The y-intercept is $R(0) = 0$.

STEP 4: Because

$$R(-x) = \frac{-3x(-x - 1)}{(-x + 4)(-x - 3)} = \frac{3x(x + 1)}{(x - 4)(x + 3)}$$

we conclude that R is neither even nor odd. There is no symmetry with respect to the y-axis or the origin.

STEP 5: Since R is in lowest terms, the graph of R has two vertical asymptotes: $x = -4$ and $x = 3$.

STEP 6: Since the degree of the numerator equals the degree of the denominator, the graph has a horizontal asymptote. To find it, we form the quotient of the leading coefficient of the numerator, 3, and the leading coefficient of the denominator, 1. The graph of R has the horizontal asymptote $y = 3$. To find out whether the graph of R intersects the asymptote, we solve the equation $R(x) = 3$.

$$R(x) = \frac{3x^2 - 3x}{x^2 + x - 12} = 3$$

$$3x^2 - 3x = 3x^2 + 3x - 36$$

$$-6x = -36$$

$$x = 6$$

The graph intersects the line $y = 3$ at $x = 6$, and $(6, 3)$ is a point on the graph of R.

STEP 7: Figure 33(a) shows the graph of R in connected mode. Notice the extraneous vertical lines at $x = -4$ and $x = 3$ (the vertical asymptotes). As a result, we also graph R in dot mode. See Figure 33(b).

STEP 8: Figure 33 does not display the graph between the two x-intercepts, 0 and 1. However, because the zeros in the numerator, 0 and 1, are of odd multiplicity (both are multiplicity 1), we know that the graph of R crosses the x-axis at 0 and 1. Therefore, the graph of R is above the x-axis for $0 < x < 1$. To see this part better, we graph R for $-1 \leq x \leq 2$ in Figure 34. Using MAXIMUM, we approximate the turning point to be $(0.52, 0.07)$, rounded to two decimal places.

Figure 33 also does not display the graph of R crossing the horizontal asymptote at $(6, 3)$. To see this part better, we graph R for $4 \leq x \leq 60$ in Figure 35. Using MINIMUM, we approximate the turning point to be $(11.48, 2.75)$, rounded to two decimal places.

Figure 33

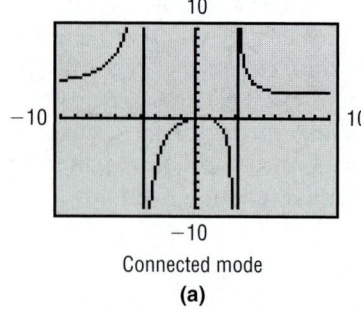

Connected mode
(a)

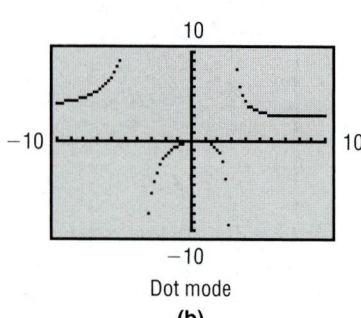

Dot mode
(b)

Figure 34

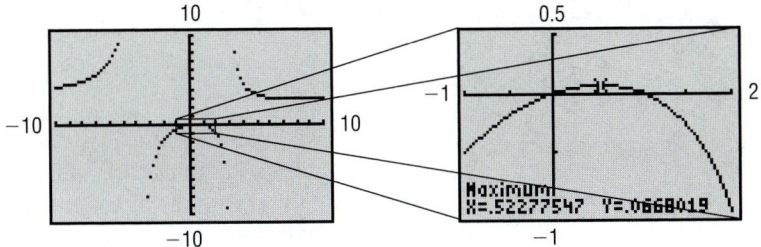

Figure 35

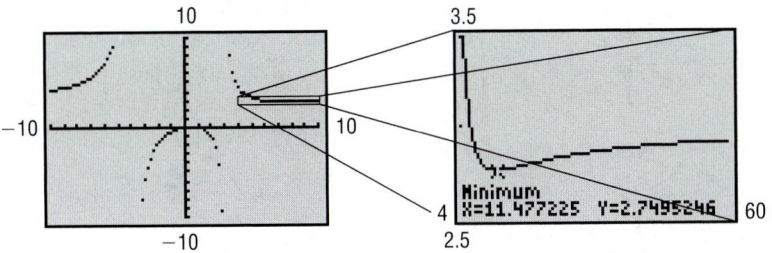

Using this information along with the information gathered in Steps 1 through 7, we obtain the graph of R shown in Figure 36.

Figure 36

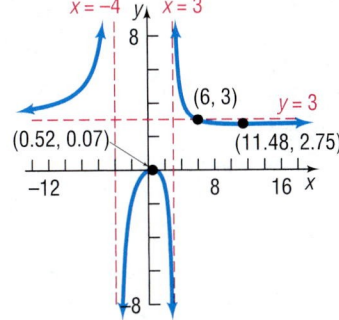

EXAMPLE 4 **Analyzing the Graph of a Rational Function with a Hole**

Analyze the graph of the rational function: $R(x) = \dfrac{2x^2 - 5x + 2}{x^2 - 4}$

Solution We factor R and obtain

$$R(x) = \frac{(2x - 1)(x - 2)}{(x + 2)(x - 2)}$$

STEP 1: The domain of R is $\{x \mid x \neq -2, x \neq 2\}$.
STEP 2: In lowest terms,

$$R(x) = \frac{2x - 1}{x + 2}, \qquad x \neq 2$$

STEP 3: The graph has one x-intercept: 0.5. The y-intercept is $R(0) = -0.5$.

STEP 4: Because

$$R(-x) = \frac{2x^2 + 5x + 2}{x^2 - 4}$$

we conclude that R is neither even nor odd. There is no symmetry with respect to the y-axis or the origin.

STEP 5: The graph has one vertical asymptote, $x = -2$, since $x + 2$ is the only factor of the denominator of $R(x)$ *in lowest terms*. However, the rational function is undefined at both $x = 2$ and $x = -2$.

STEP 6: Since the degree of the numerator equals the degree of the denominator, the graph has a horizontal asymptote. To find it, we form the quotient of the leading coefficient of the numerator, 2, and the leading coefficient of the denominator, 1. The graph of R has the horizontal asymptote $y = 2$. To find whether the graph of R intersects the asymptote, we solve the equation $R(x) = 2$.

$$R(x) = \frac{2x - 1}{x + 2} = 2$$

$$2x - 1 = 2(x + 2)$$

$$2x - 1 = 2x + 4$$

$$-1 = 4 \qquad \text{\textcolor{blue}{Impossible.}}$$

The graph does not intersect the line $y = 2$.

STEP 7: Figure 37 shows the graph of $R(x)$. Notice that the graph has one vertical asymptote at $x = -2$. Also, the function appears to be continuous at $x = 2$.

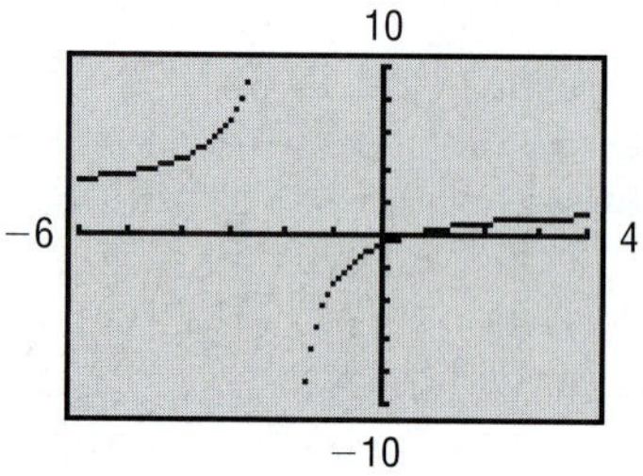

STEP 8: The analysis presented thus far does not explain the behavior of the graph at $x = 2$. We use the TABLE feature of our graphing utility to determine the behavior of the graph of R as x approaches 2. See Table 13. From the table, we conclude that the value of R approaches 0.75 as x approaches 2. This result is further verified by evaluating R in lowest terms at $x = 2$. We conclude that there is a hole in the graph at $(2, 0.75)$. Using the information gathered in Steps 1 through 7, we obtain the graph of R shown in Figure 38.

Figure 37

TABLE 13

X	Y1
1.99	.74687
1.999	.74969
1.9999	.74997
2	ERROR
2.0001	.75003
2.001	.75031
2.01	.75312

Y1 ◼ (2X²-5X+2)/(...

Figure 38

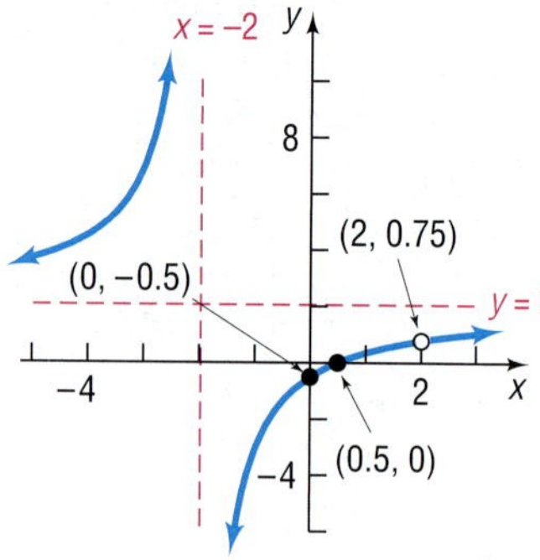

As Example 4 shows, the values excluded from the domain of a rational function give rise to either vertical asymptotes or holes.

 NOW WORK PROBLEM 27.

We now discuss the problem of finding a rational function from its graph.

EXAMPLE 5   **Constructing a Rational Function from Its Graph**

Make up a rational function that might have the graph shown in Figure 39.

Figure 39

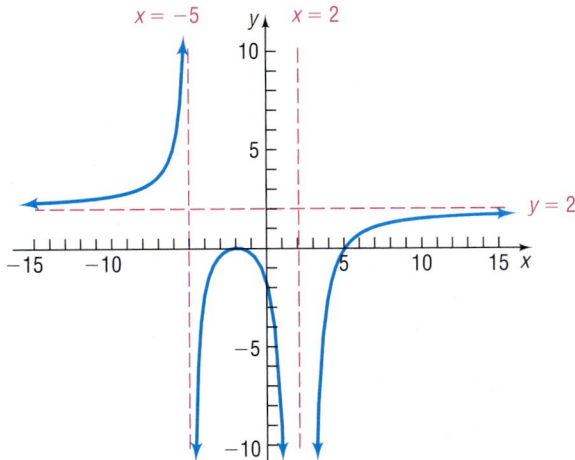

Solution   The numerator of a rational function $R(x) = \dfrac{p(x)}{q(x)}$ in lowest terms determines the x-intercepts of its graph. The graph shown in Figure 39 has x-intercepts -2 (even multiplicity; graph touches the x-axis) and 5 (odd multiplicity; graph crosses the x-axis). So, one possibility for the numerator is $p(x) = (x + 2)^2(x - 5)$.

The denominator of a rational function in lowest terms determines the vertical asymptotes of its graph. The vertical asymptotes of the graph are $x = -5$ and $x = 2$. Since $R(x)$ approaches ∞ from the left of $x = -5$ and $R(x)$ approaches $-\infty$ from the right of $x = -5$, we know that $(x + 5)$ is a factor of odd multiplicity in $q(x)$. Also, $R(x)$ approaches $-\infty$ from both sides of $x = 2$, so $(x - 2)$ is a factor of even multiplicity in $q(x)$. A possibility for the denominator is $q(x) = (x + 5)(x - 2)^2$.

Figure 40

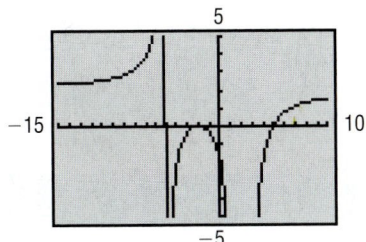

So far we have $R(x) = \dfrac{(x + 2)^2(x - 5)}{(x + 5)(x - 2)^2}$. However, the horizontal asymptote of the graph given in Figure 39 is $y = 2$, so we know that the degree of the numerator must equal the degree in the denominator and the quotient of leading coefficients must be $\dfrac{2}{1}$. This leads to $R(x) = \dfrac{2(x + 2)^2(x - 5)}{(x + 5)(x - 2)^2}$.
Figure 40 shows the graph of R drawn on a graphing utility. Since Figure 40 looks similar to Figure 39, we have found a rational function R for the graph in Figure 39. ■

NOW WORK PROBLEM **39.**

2   **Applications Involving Rational Functions**

EXAMPLE 6   **Finding the Least Cost of a Can**

Reynolds Metal Company manufactures aluminum cans in the shape of a cylinder with a capacity of 500 cubic centimeters $\left(\dfrac{1}{2} \text{ liter}\right)$. The top and bottom of the can are made of a special aluminum alloy that costs 0.05¢ per square centimeter. The sides of the can are made of material that costs 0.02¢ per square centimeter.

(a) Express the cost of material for the can as a function of the radius r of the can.
(b) Use a graphing utility to graph the function $C = C(r)$.
(c) What value of r will result in the least cost?
(d) What is this least cost?

Solution

(a) Figure 41 illustrates the situation. Notice that the material required to produce a cylindrical can of height h and radius r consists of a rectangle of area $2\pi rh$ and two circles, each of area πr^2. The total cost C (in cents) of manufacturing the can is therefore

Figure 41

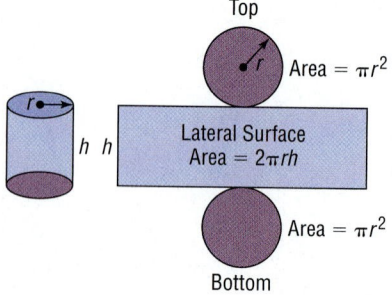

Top

Area $= \pi r^2$

Lateral Surface
Area $= 2\pi rh$

h h

Area $= \pi r^2$

Bottom

$$C = \text{Cost of top and bottom} + \text{Cost of side}$$
$$= \underbrace{(2\pi r^2 \text{ cm}^2)}_{\substack{\text{Total area} \\ \text{of top and} \\ \text{bottom}}} \underbrace{(0.05¢/\text{cm}^2)}_{\text{Cost/unit area}} + \underbrace{(2\pi rh \text{ cm}^2)}_{\substack{\text{Total area} \\ \text{of side}}} \underbrace{(0.02¢/\text{cm}^2)}_{\text{Cost/unit area}}$$
$$= 0.10\pi r^2 + 0.04\pi rh$$

But we have the additional restriction that the height h and radius r must be chosen so that the volume V of the can is 500 cubic centimeters. Since $V = \pi r^2 h$, we have

$$500 = \pi r^2 h \quad \text{or} \quad h = \frac{500}{\pi r^2}$$

Figure 42

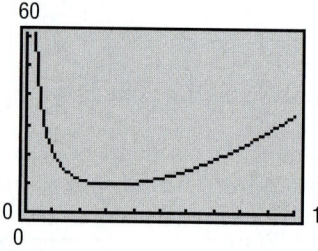

Substituting this expression for h, the cost C, in cents, as a function of the radius r is

$$C(r) = 0.10\pi r^2 + 0.04\pi r \frac{500}{\pi r^2} = 0.10\pi r^2 + \frac{20}{r} = \frac{0.10\pi r^3 + 20}{r}$$

(b) See Figure 42 for the graph of $C(r)$.
(c) Using the MINIMUM command, the cost is least for a radius of about 3.17 centimeters.
(d) The least cost is $C(3.17) \approx 9.47¢$. ∎

4.4 Concepts and Vocabulary

In Problem 1, fill in the blanks.

1. If the numerator and the denominator of a rational function have no common factors, the rational function is said to be in _____ .

In Problems 2 and 3, answer True or False to each statement.

2. The graph of a polynomial function sometimes has a hole.

3. The graph of a rational function never intersects a horizontal asymptote.

4. Under what conditions will a rational function have a hole in its graph? Give an example.

4.4 Exercises

In Problems 1–38, follow Steps 1 through 8 on page 355 to analyze the graph of each function.

1. $R(x) = \dfrac{x + 1}{x(x + 4)}$

2. $R(x) = \dfrac{x}{(x - 1)(x + 2)}$

3. $R(x) = \dfrac{3x + 3}{2x + 4}$

4. $R(x) = \dfrac{2x + 4}{x - 1}$

5. $R(x) = \dfrac{3}{x^2 - 4}$

6. $R(x) = \dfrac{6}{x^2 - x - 6}$

7. $P(x) = \dfrac{x^4 + x^2 + 1}{x^2 - 1}$

8. $Q(x) = \dfrac{x^4 - 1}{x^2 - 4}$

9. $H(x) = \dfrac{x^3 - 1}{x^2 - 9}$

10. $G(x) = \dfrac{x^3 + 1}{x^2 + 2x}$

11. $R(x) = \dfrac{x^2}{x^2 + x - 6}$

12. $R(x) = \dfrac{x^2 + x - 12}{x^2 - 4}$

13. $G(x) = \dfrac{x}{x^2 - 4}$

14. $G(x) = \dfrac{3x}{x^2 - 1}$

15. $R(x) = \dfrac{3}{(x - 1)(x^2 - 4)}$

16. $R(x) = \dfrac{-4}{(x + 1)(x^2 - 9)}$

17. $H(x) = \dfrac{4(x^2 - 1)}{x^4 - 16}$

18. $H(x) = \dfrac{x^2 + 4}{x^4 - 1}$

19. $F(x) = \dfrac{x^2 - 3x - 4}{x + 2}$

20. $F(x) = \dfrac{x^2 + 3x + 2}{x - 1}$

21. $R(x) = \dfrac{x^2 + x - 12}{x - 4}$

22. $R(x) = \dfrac{x^2 - x - 12}{x + 5}$

23. $F(x) = \dfrac{x^2 + x - 12}{x + 2}$

24. $G(x) = \dfrac{x^2 - x - 12}{x + 1}$

25. $R(x) = \dfrac{x(x - 1)^2}{(x + 3)^3}$

26. $R(x) = \dfrac{(x - 1)(x + 2)(x - 3)}{x(x - 4)^2}$

27. $R(x) = \dfrac{x^2 + x - 12}{x^2 - x - 6}$

28. $R(x) = \dfrac{x^2 + 3x - 10}{x^2 + 8x + 15}$

29. $R(x) = \dfrac{6x^2 - 7x - 3}{2x^2 - 7x + 6}$

30. $R(x) = \dfrac{8x^2 + 26x + 15}{2x^2 - x - 15}$

31. $R(x) = \dfrac{x^2 + 5x + 6}{x + 3}$

32. $R(x) = \dfrac{x^2 + x - 30}{x + 6}$

33. $f(x) = x + \dfrac{1}{x}$

34. $f(x) = 2x + \dfrac{9}{x}$

35. $f(x) = x^2 + \dfrac{1}{x}$

36. $f(x) = 2x^2 + \dfrac{9}{x}$

37. $f(x) = x + \dfrac{1}{x^3}$

38. $f(x) = 2x + \dfrac{9}{x^3}$

In Problems 39–42, make up a rational function that might have this graph. (More than one answer might be possible.)

39.

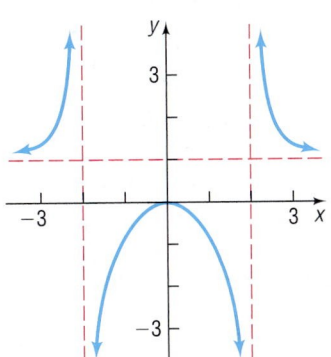

40.

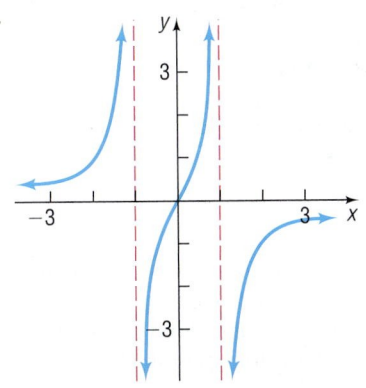

41.

42.

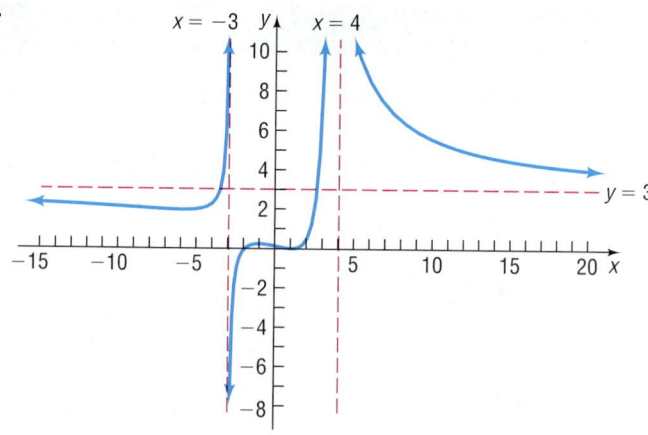

43. Make up a rational function that has the following characteristics: crosses the x-axis at 2; touches the x-axis at -1; one vertical asymptote at -5 and another at 6; and one horizontal asymptote, $y = 3$. Compare yours to a fellow classmate's. How do they differ? What are the similarities?

44. Make up a rational function that has the following characteristics: crosses the x-axis at 3; touches the x-axis at -2; one vertical asymptote, $x = 1$; and one horizontal asymptote, $y = 2$. Give your rational function to a fellow classmate and ask for a written critique of your rational function.

45. Gravity   In physics, it is established that the acceleration due to gravity g at a height h meters above sea level is given by

$$g(h) = \frac{3.99 \times 10^{14}}{(6.374 \times 10^6 + h)^2}$$

where 6.374×10^6 is the radius of Earth in meters.

(a) What is the acceleration due to gravity at sea level?
(b) The Sears Tower in Chicago, Illinois, is 443 meters tall. What is the acceleration due to gravity at the top of the Sears Tower?
(c) The peak of Mount Everest is 8848 meters above sea level. What is the acceleration due to gravity on the peak of Mount Everest?
(d) Find the horizontal asymptote of $g(h)$.
(e) Using your graphing utility, graph $g = g(h)$.
(f) Solve $g(h) = 0$. How do you interpret your answer?

46. Population Model   A rare species of insect was discovered in the Amazon Rain Forest. To protect the species, environmentalists declare the insect endangered and transplant the insects into a protected area. The population of the insect t months after being transplanted is given by

$$P(t) = \frac{50(1 + 0.5t)}{(2 + 0.01t)}$$

(a) How many insects were discovered? In other words, what was the population when $t = 0$?
(b) What will the population be after 5 years?
(c) Using your graphing utility, graph $P = P(t)$.

(d) Determine the horizontal asymptote of $P(t)$. What is the largest population that the protected area can sustain?
(e) TRACE $P(t)$ for large values of t to verify your answer to part (d).

47. Drug Concentration   The concentration C of a certain drug in a patient's bloodstream t hours after injection is given by

$$C(t) = \frac{t}{2t^2 + 1}$$

(a) Using your graphing utility, graph $C = C(t)$.
(b) Determine the time at which the concentration is highest.
(c) Find the horizontal asymptote of $C(t)$. What happens to the concentration of the drug as t increases?

48. Drug Concentration   The concentration C of a certain drug in a patient's bloodstream t minutes after injection is given by

$$C(t) = \frac{50t}{t^2 + 25}$$

(a) Using your graphing utility, graph $C = C(t)$.
(b) Determine the time at which the concentration is highest.
(c) Find the horizontal asymptote of $C(t)$. What happens to the concentration of the drug as t increases?

49. Average Cost   In Problem 69, Exercise 4.2, the cost function for manufacturing x Chevy Cavaliers per hour was found to be

$$C(x) = 0.2x^3 - 2.3x^2 + 14.3x + 10.2$$

Economists define the **average cost function** as

$$\bar{C}(x) = \frac{C(x)}{x}$$

(a) Find the average cost function.
(b) What is the average cost of producing six Cavaliers per hour?
(c) What is the average cost of producing nine Cavaliers per hour?
(d) Using your graphing utility, graph the average cost function.

(e) Using your graphing utility, find the number of Cavaliers that should be produced per hour to minimize average cost.

(f) What is the minimum average cost?

50. **Average Cost** In Problem 70, Exercise 4.2, the cost C (in thousands of dollars) for printing x thousand textbooks was found to be

$$C(x) = 0.015x^3 - 0.595x^2 + 9.15x + 98.43$$

(a) Find the average cost function (refer to Problem 51).

(b) What is the average cost of printing 13 thousand textbooks per week?

(c) What is the average cost of printing 25 thousand textbooks per week?

(d) Using your graphing utility, graph the average cost function.

(e) Using your graphing utility, find the number of textbooks that should be printed to minimize average cost.

(f) What is the minimum average cost?

51. **Minimizing Surface Area** United Parcel Service has contracted you to design a closed box with a square base that has a volume of 10,000 cubic inches. See the illustration.

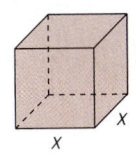

(a) Find a function for the surface areas of the box.

(b) Using a graphing utility, graph the function found in part (a).

(c) What is the minimum amount of cardboard that can be used to construct the box?

(d) What are the dimensions of the box that minimize surface area?

 (e) Why might UPS be interested in designing a box that minimizes surface area?

52. **Minimizing Surface Area** United Parcel Service has contracted you to design a closed box with a square

base that has a volume of 5000 cubic inches. See the illustration.

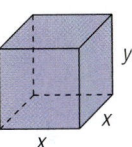

(a) Find a function for the surface areas of the box.

(b) Using a graphing utility, graph the function found in part (a).

(c) What is the minimum amount of cardboard that can be used to construct the box?

(d) What are the dimensions of the box that minimize surface area?

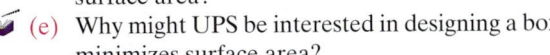 (e) Why might UPS be interested in designing a box that minimizes surface area?

53. **Cost of a Can** A can in the shape of a right circular cylinder is required to have a volume of 500 cubic centimeters. The top and bottom are made of material that costs 6¢ per square centimeter, while the sides are made of material that costs 4¢ per square centimeter.

(a) Express the total cost C of the material as a function of the radius r of the cylinder. (Refer to Figure 41.)

(b) Graph $C = C(r)$. For what value of r is the cost C least?

54. **Material Needed to Make a Drum** A steel drum in the shape of a right circular cylinder is required to have a volume of 100 cubic feet.

(a) Express the amount A of material required to make the drum as a function of the radius r of the cylinder.

(b) How much material is required if the drum is of radius 3 feet?

(c) Of radius 2 feet?

(d) Of radius 4 feet?

(e) Graph $A = A(r)$. For what value of r is A smallest?

55. Write a few paragraphs that provide a general strategy for graphing a rational function. Be sure to mention the following: proper, improper, intercepts, and asymptotes.

PREPARING FOR THIS SECTION

Before getting started, review the following:

✓ Solving Inequalities (Section 1.6, pp. 149–156)

✓ Rational Expressions (Section R.9, pp. 59–66)

✓ Quadratic Equations (Section 1.3, pp. 117–124)

4.5 POLYNOMIAL AND RATIONAL INEQUALITIES

OBJECTIVES

1 Solve Polynomial Inequalities Algebraically and Graphically

2 Solve Rational Inequalities Algebraically and Graphically

1 In this section we solve inequalities that involve polynomials of degree 2 and higher, as well as some that involve rational expressions. To solve such

inequalities, we use the information obtained in the previous three sections about the graph of polynomial and rational functions. The general idea follows:

Suppose that the polynomial or rational inequality is in one of the forms

$$f(x) > 0 \qquad f(x) \geq 0 \qquad f(x) < 0 \qquad f(x) \leq 0$$

Locate the zeros of f if f is a polynomial function, and locate the zeros of the numerator and the denominator if f is a rational function. If we use these zeros to divide the real number line into intervals, then we know that on each interval the graph of f is either above the x-axis $[f(x) > 0]$ or below the x-axis $[f(x) < 0]$. In other words, we have found the solution of the inequality.

The following steps provide more detail.

Steps for Solving Polynomial and Rational Inequalities Algebraically

STEP 1: Write the inequality so that a polynomial or rational expression f is on the left side and zero is on the right side in one of the following forms:

$$f(x) > 0 \qquad f(x) \geq 0 \qquad f(x) < 0 \qquad f(x) \leq 0$$

For rational expressions, be sure that the left side is written as a single quotient.

STEP 2: Determine the numbers at which the expression f on the left side equals zero and if the expression is rational, the numbers at which the expression f on the left side is undefined.

STEP 3: Use the numbers found in Step 2 to separate the real number line into intervals.

STEP 4: Select a **test number** in each interval and evaluate f at the test number.

(a) If the value of f is positive, then $f(x) > 0$ for all numbers x in the interval.

(b) If the value of f is negative, then $f(x) < 0$ for all numbers x in the interval.

If the inequality is not strict ($\geq$ or $\leq$), include the solutions of $f(x) = 0$ in the solution set, but be careful not to include values of x where the expression is undefined.

EXAMPLE 1 Solving Quadratic Inequalities

Solve the inequality $x^2 \leq 4x + 12$, and graph the solution set.

Algebraic Solution **STEP 1:** Rearrange the inequality so that 0 is on the right side.

$$x^2 \leq 4x + 12$$
$$x^2 - 4x - 12 \leq 0 \qquad \text{Subtract } 4x + 12 \text{ from both sides of the inequality.}$$

This inequality is equivalent to the one we wish to solve.

STEP 2: Find the zeros of $f(x) = x^2 - 4x - 12$ by solving the equation $x^2 - 4x - 12 = 0$.

$$x^2 - 4x - 12 = 0$$
$$(x + 2)(x - 6) = 0 \qquad \text{Factor.}$$
$$x = -2 \quad \text{or} \quad x = 6$$

STEP 3:   We use the zeros of f to separate the real number line into three intervals.

$$(-\infty, -2) \qquad (-2, 6) \qquad (6, \infty)$$

STEP 4:   We select a test number in each interval found in Step 3 and evaluate $f(x) = x^2 - 4x - 12$ at each test number to determine if $f(x)$ is positive or negative. See Table 14.

TABLE 14			
	$\xrightarrow{\hspace{1cm} \overset{-2}{\bullet} \hspace{2cm} \overset{6}{\bullet} \hspace{1cm}} x$		
Interval	$(-\infty, -2)$	$(-2, 6)$	$(6, \infty)$
Number Chosen	-3	0	7
Value of f	$f(-3) = 9$	$f(0) = -12$	$f(7) = 9$
Conclusion	Positive	Negative	Positive

Since we want to know where $f(x)$ is negative, we conclude that the solutions are all x such that $-2 < x < 6$. However, because the original inequality is not strict, numbers x that satisfy the equation $x^2 = 4x + 12$ are also solutions of the inequality $x^2 \leq 4x + 12$. Thus, we include -2 and 6. The solution set of the given inequality is $\{x | -2 \leq x \leq 6\}$ or, using interval notation, $[-2, 6]$.

Graphing Solution   We graph $Y_1 = x^2$ and $Y_2 = 4x + 12$ on the same screen. See Figure 43. Using the INTERSECT command, we find that Y_1 and Y_2 intersect at $x = -2$ and at $x = 6$. The graph of Y_1 is below that of Y_2, $Y_1 < Y_2$, between the points of intersection. Since the inequality is not strict, the solution set is $\{x | -2 \leq x \leq 6\}$ or, using interval notation, $[-2, 6]$.

Figure 43

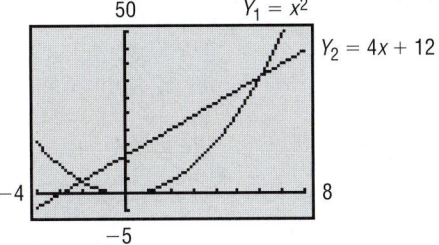

Figure 44
$-2 \leq x \leq 6$; $[-2, 6]$

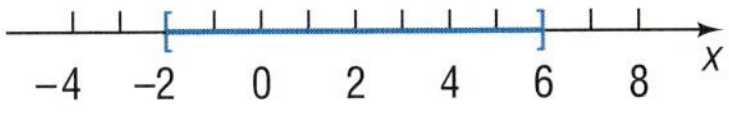

Figure 44 shows the graph of the solution set.

━ **NOW WORK PROBLEMS 3 AND 7.**

EXAMPLE 2   **Solving a Polynomial Inequality**

Solve the inequality $x^4 > x$, and graph the solution set.

Solution   **STEP 1:**   Rearrange the inequality so that 0 is on the right side.

$$x^4 > x$$
$$x^4 - x > 0 \qquad \text{\small Subtract } x \text{ from both sides of the inequality.}$$

This inequality is equivalent to the one we wish to solve.

Step 2: Find the zeros of $f(x) = x^4 - x$ by solving $x^4 - x = 0$.

$$x^4 - x = 0$$
$$x(x^3 - 1) = 0 \qquad \text{Factor out } x$$
$$x(x - 1)(x^2 + x + 1) = 0 \qquad \text{Factor the difference of two cubes.}$$
$$x = 0 \quad \text{or} \quad x - 1 = 0 \quad \text{or} \quad x^2 + x + 1 = 0 \qquad \text{Set each factor equal to zero and solve.}$$
$$x = 0 \quad \text{or} \quad x = 1$$

The equation $x^2 + x + 1 = 0$ has no real solutions. (Do you see why?)

Step 3: We use the zeros to separate the real number line into three intervals.

$$(-\infty, 0) \qquad (0, 1) \qquad (1, \infty)$$

Step 4: We select a test number in each interval found in Step 3 and evaluate $f(x) = x^4 - x$ at each test number to determine if $f(x)$ is positive or negative. See Table 15.

TABLE 15

	0		1	

Interval	$(-\infty, 0)$	$(0, 1)$	$(1, \infty)$
Number Chosen	-1	$\dfrac{1}{2}$	2
Value of f	$f(-1) = 2$	$f\left(\dfrac{1}{2}\right) = -\dfrac{7}{16}$	$f(2) = 14$
Conclusion	Positive	Negative	Positive

Since we want to know where $f(x)$ is positive, we conclude that the solutions are numbers x for which $x < 0$ or $x > 1$. Because the original inequality is strict, numbers x that satisfy the equation $x^4 = x$ are not solutions. The solution set of the given inequality is $\{x | x < 0 \text{ or } x > 1\}$ or, using interval notation, $(-\infty, 0)$ or $(1, \infty)$.

Graphing Solution We graph $Y_1 = x^4$ and $Y_2 = x$ on the same screen. See Figure 45. Using the INTERSECT command, we find that Y_1 and Y_2 intersect at $x = 0$ and at $x = 1$. The graph of Y_1 is above that of Y_2, $Y_1 > Y_2$, to the left of $x = 0$ and to the right of $x = 1$. Since the inequality is strict, the solution set is $\{x | x < 0 \text{ or } x > 1\}$ or, using interval notation, $(-\infty, 0)$ or $(1, \infty)$.

Figure 45

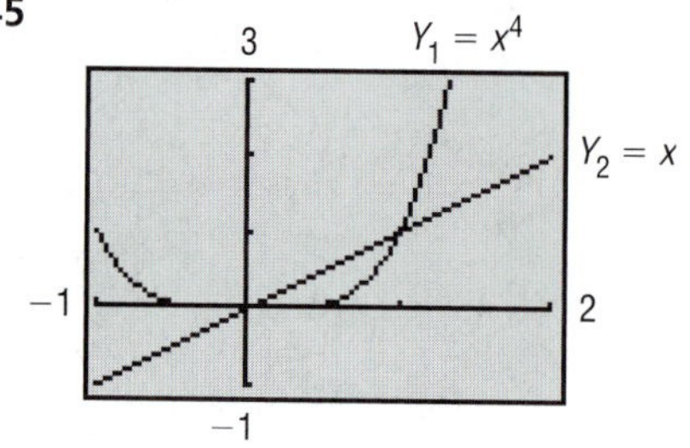

Figure 46
$x < 0$ or $x > 1$
$(-\infty, 0)$ or $(1, \infty)$

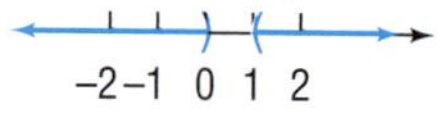

Figure 46 shows the graph of the solution set.

NOW WORK PROBLEM 19.

2   Let's solve a rational inequality.

EXAMPLE 3   Solving a Rational Inequality

Solve the inequality $\dfrac{4x + 5}{x + 2} \geq 3$, and graph the solution set.

Algebraic Solution   **STEP 1:**   We first note that the domain of the variable consists of all real numbers except -2. We rearrange terms so that 0 is on the right side.

$$\frac{4x + 5}{x + 2} - 3 \geq 0 \qquad \text{Subtract 3 from both sides of the inequality.}$$

STEP 2:   For $f(x) = \dfrac{4x + 5}{x + 2} - 3$, find the zeros of f and the values of x at which f is undefined. To find these numbers, we must express f as a single quotient.

$$f(x) = \frac{4x + 5}{x + 2} - 3 \qquad \text{Least Common Denominator: } x + 2.$$

$$= \frac{4x + 5}{x + 2} - 3\frac{x + 2}{x + 2} \qquad \text{Multiply } -3 \text{ by } \frac{x + 2}{x + 2}.$$

$$= \frac{4x + 5 - 3x - 6}{x + 2} \qquad \text{Write as a single quotient.}$$

$$= \frac{x - 1}{x + 2} \qquad \text{Combine like terms.}$$

The zero of f is 1. Also, f is undefined for $x = -2$.

STEP 3:   We use the numbers found in Step 2 to separate the real number line into three intervals.

$$(-\infty, -2) \qquad (-2, 1) \qquad (1, \infty)$$

STEP 4:   We select a test number in each interval found in Step 3 and evaluate $f(x) = \dfrac{4x + 5}{x + 2} - 3$ at each test number to determine if $f(x)$ is positive or negative. See Table 16.

TABLE 16

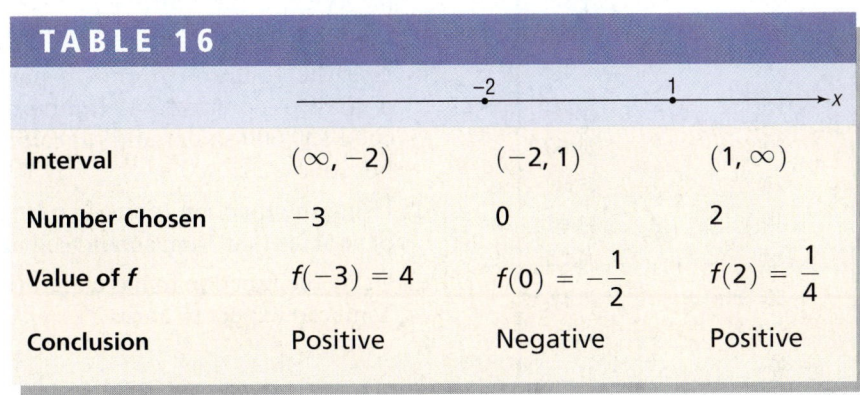

	$(\infty, -2)$	$(-2, 1)$	$(1, \infty)$
Interval	$(\infty, -2)$	$(-2, 1)$	$(1, \infty)$
Number Chosen	-3	0	2
Value of f	$f(-3) = 4$	$f(0) = -\dfrac{1}{2}$	$f(2) = \dfrac{1}{4}$
Conclusion	Positive	Negative	Positive

We want to know where $f(x)$ is positive. We conclude that the solutions are all x such that $x < -2$ or $x > 1$. Because the original inequality is not strict, numbers x that satisfy the equation $\dfrac{x-1}{x+2} = 0$ are also solutions of the inequality. Since $\dfrac{x-1}{x+2} = 0$ only if $x = 1$, we conclude that the solution set is $\{x | x < -2 \text{ or } x \geq 1\}$ or, using interval notation, $(-\infty, -2)$ or $[1, \infty)$.

Graphing Solution We first note that the domain of the variable consists of all real numbers except -2. We graph $Y_1 = \dfrac{4x+5}{x+2}$ and $Y_2 = 3$ on the same screen. See Figure 47. Using the INTERSECT command, we find that Y_1 and Y_2 intersect at $x = 1$. The graph of Y_1 is above that of Y_2, $Y_1 > Y_2$, to the left of $x = -2$ and to the right of $x = 1$. Since the inequality is not strict, the solution set is $\{x | x < -2 \text{ or } x \geq 1\}$. Using interval notation, the solution set is $(-\infty, -2)$ or $[1, \infty)$.

Figure 47

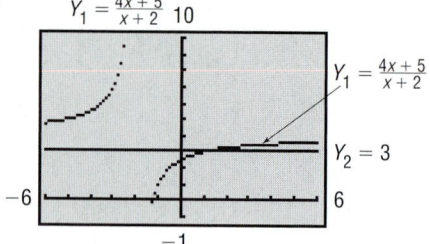

Figure 48

$x < -2$ or $x \geq 1$

$(-\infty, -2)$ or $[1, \infty)$

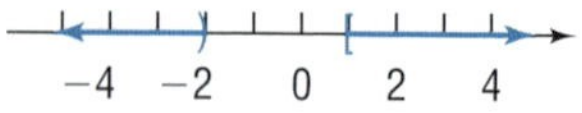

Figure 48 shows the graph of the solution set.

NOW WORK PROBLEM 37.

EXAMPLE 4 **Minimum Sales Requirements**

TABLE 17

Pounds of Cookies (in Hundreds), x	Profit, P
0	−20,000
50	−5990
75	412
120	10,932
200	26,583
270	36,948
340	44,381
420	49,638
525	49,225
610	44,381
700	34,220

Tami is considering leaving her $30,000 a year job and buying a cookie company. According to the financial records of the firm, the relationship between pounds of cookies sold and profit is exhibited by Table 17.

(a) Draw a scatter diagram of the data in Table 17 with the pounds of cookies sold as the independent variable.

(b) Use a graphing utility to find the quadratic function of best fit.

(c) Use the function found in part (b) to determine the number of pounds of cookies that Tami must sell in order for the profits to exceed $30,000 a year and therefore make it worthwhile for her to quit her job.

(d) Using the function found in part (b), determine the number of pounds of cookies that Tami should sell in order to maximize profits.

(e) Using the function found in part (b), determine the maximum profit that Tami can expect to earn.

Solution (a) Figure 49 shows the scatter diagram.

(b) Using a graphing utility, the quadratic function of best fit is

$$p(x) = -0.31x^2 + 295.86x - 20{,}042.52$$

See Figure 50.

(c) We want profit to exceed $30,000; therefore, we want to solve the inequality

$$-0.31x^2 + 295.86x - 20{,}042.52 > 30{,}000$$

We graph

$$Y_1 = p(x) = -0.31x^2 + 295.86x - 20{,}042.52 \quad \text{and} \quad Y_2 = 30{,}000$$

on the same screen. See Figure 51.

Figure 49

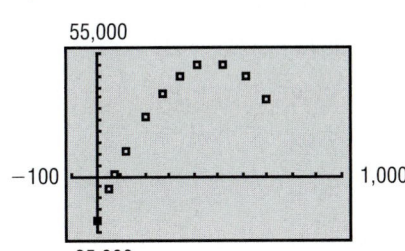

Figure 50

Figure 51

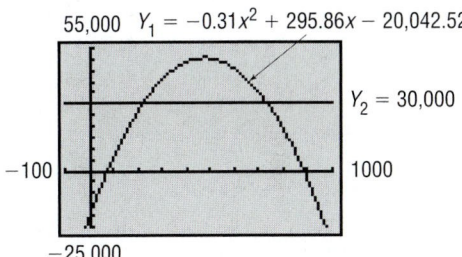

Using the INTERSECT command, we find that Y_1 and Y_2 intersect at $x = 220$ and at $x = 735$. The graph of Y_1 is above that of Y_2, $Y_1 > Y_2$, between the points of intersection. Since the inequality is strict, the solution set is $\{x | 220 < x < 735\}$.

(c) The function $p(x)$ given in part (b) is a quadratic function whose graph opens down. The vertex is therefore the highest point. The x-coordinate of the vertex is

$$x = -\frac{b}{2a} = -\frac{295.86}{2(-0.31)} = 477$$

To maximize profit, 477 hundred (47,700) pounds of cookies must be sold.

(d) The maximum profit is

$$p(477) = -0.31(477)^2 + 295.86(477) - 20{,}042.52 = \$50{,}549 \quad \blacksquare$$

4.5   Concepts and Vocabulary

In Problems 1 and 2, answer True or False to each statement.

1. A test number for the interval $-5 < x < -2$ is -1.

2. The inequality $x^2 + x + 4 > 0$ is true for all x.

3. The inequality $x^2 + 1 < -5$ has no solution. Explain why.

4. To solve polynomial inequalities, we locate the zeros and use these zeros to divide the real number line into intervals. We then know that on each interval the graph of the polynomial function is either above or below the x-axis. Explain why this is so.

5. To solve rational inequalities, we locate the zeros of the numerator and denominator. Then we use these zeros to divide the real number line into intervals. We know that on each interval the graph of the rational function is either above or below the x-axis. Explain why this is so.

4.5 Exercises

In Problems 1–58, solve each inequality algebraically. Verify your results using a graphing, utility.

1. $(x - 5)(x + 2) < 0$

2. $(x - 5)(x + 2) > 0$

3. $x^2 - 4x > 0$

4. $x^2 + 8x > 0$

5. $x^2 - 9 < 0$

6. $x^2 - 1 < 0$

7. $x^2 + x > 12$

8. $x^2 + 7x < -12$

9. $2x^2 < 5x + 3$

10. $6x^2 < 6 + 5x$

11. $x(x - 7) > 8$

12. $x(x + 1) > 20$

13. $4x^2 + 9 < 6x$

14. $25x^2 + 16 < 40x$

15. $6(x^2 - 1) > 5x$

16. $2(2x^2 - 3x) > -9$

17. $(x - 1)(x^2 + x + 1) > 0$

18. $(x + 2)(x^2 - x + 1) > 0$

19. $(x - 1)(x - 2)(x - 3) < 0$

20. $(x + 1)(x + 2)(x + 3) < 0$

21. $x^3 - 2x^2 - 3x > 0$

22. $x^3 + 2x^2 - 3x > 0$

23. $x^4 > x^2$

24. $x^4 < 4x^2$

25. $x^3 > x^2$

26. $x^3 < 3x^2$

27. $x^4 > 1$

28. $x^3 > 1$

29. $x^2 - 7x - 8 < 0$

30. $x^2 + 12x + 32 \geq 0$

31. $x^3 + x - 12 \geq 0$

32. $x^3 - 3x + 1 \leq 0$

33. $x^4 - 3x^2 - 4 > 0$

34. $x^4 - 5x^2 + 6 < 0$

35. $x^3 - 4 \geq 3x^2 + 5x - 3$

36. $x^4 - 4x \leq -x^2 + 2x + 1$

37. $\dfrac{x + 1}{x - 1} > 0$

38. $\dfrac{x - 3}{x + 1} > 0$

39. $\dfrac{(x - 1)(x + 1)}{x} < 0$

40. $\dfrac{(x - 3)(x + 2)}{x - 1} < 0$

41. $\dfrac{(x - 2)^2}{x^2 - 1} \geq 0$

42. $\dfrac{(x + 5)^2}{x^2 - 4} \geq 0$

43. $6x - 5 < \dfrac{6}{x}$

44. $x + \dfrac{12}{x} < 7$

45. $\dfrac{x + 4}{x - 2} \leq 1$

46. $\dfrac{x + 2}{x - 4} \geq 1$

47. $\dfrac{3x - 5}{x + 2} \leq 2$

48. $\dfrac{x - 4}{2x + 4} \geq 1$

49. $\dfrac{1}{x - 2} < \dfrac{2}{3x - 9}$

50. $\dfrac{5}{x - 3} > \dfrac{3}{x + 1}$

51. $\dfrac{2x + 5}{x + 1} > \dfrac{x + 1}{x - 1}$

52. $\dfrac{1}{x + 2} > \dfrac{3}{x + 1}$

53. $\dfrac{x^2(3 + x)(x + 4)}{(x + 5)(x - 1)} > 0$

54. $\dfrac{x(x^2 + 1)(x - 2)}{(x - 1)(x + 1)} > 0$

55. $\dfrac{2x^2 - x - 1}{x - 4} \leq 0$

56. $\dfrac{3x^2 + 2x - 1}{x + 2} > 0$

57. $\dfrac{x^2 + 3x - 1}{x + 3} > 0$

58. $\dfrac{x^2 - 5x + 3}{x - 5} < 0$

---

59. For what positive numbers will the cube of a number exceed four times its square?

60. For what positive numbers will the square of a number exceed twice the number?

61. What is the domain of the function $f(x) = \sqrt{x^2 - 16}$?

62. What is the domain of the function $f(x) = \sqrt{x^3 - 3x^2}$?

63. What is the domain of the function $R(x) = \sqrt{\dfrac{x - 2}{x + 4}}$?

64. What is the domain of in the function $R(x) = \sqrt{\dfrac{x - 1}{x + 4}}$?

65. Physics A ball is thrown vertically upward with an initial velocity of 80 feet per second. The distance *s* (in feet) of the ball from the ground after *t* seconds is $s = 80t - 16t^2$.

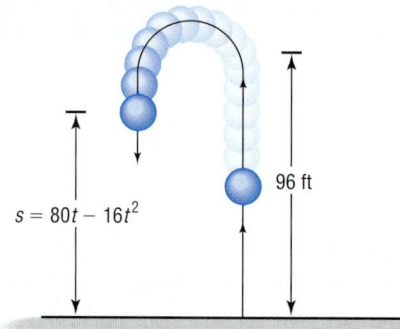

$s = 80t - 16t^2$

96 ft

(a) For what time interval is the ball more than 96 feet above the ground? (See the figure.)

(b) Using a graphing utility, graph the relation between s and t.

(c) What is the maximum height of the ball?

(d) After how many seconds does the ball reach the maximum height?

66. Physics A ball is thrown vertically upward with an initial velocity of 96 feet per second. The distance s (in feet) of the ball from the ground after t seconds is $s = 96t - 16t^2$.

(a) For what time interval is the ball more than 112 feet above the ground?

(b) Using a graphing utility, graph the relation between s and t.

(c) What is the maximum height of the ball?

(d) After how many seconds does the ball reach the maximum height?

67. Business The monthly revenue achieved by selling x wristwatches is figured to be $x(40 - 0.2x)$ dollars. The wholesale cost of each watch is $32.

(a) How many watches must be sold each month to achieve a profit (revenue − cost) of at least $50?

(b) Using a graphing utility, graph the revenue function.

(c) What is the maximum revenue that this firm could earn?

(d) How many wristwatches should the firm sell to maximize revenue?

(e) Using a graphing utility, graph the profit function.

(f) What is the maximum profit that this firm can earn?

(g) How many watches should the firm sell to maximize profit?

(h) Provide a reasonable explanation as to why the answers found in parts (d) and (g) differ. Is the shape of the revenue function reasonable in your opinion? Why?

68. Business The monthly revenue achieved by selling x boxes of candy is figured to be $x(5 - 0.05x)$ dollars. The wholesale cost of each box of candy is $1.50.

(a) How many boxes must be sold each month to achieve a profit of at least $60?

(b) Using a graphing utility, graph the revenue function.

(c) What is the maximum revenue that this firm could earn?

(d) How many boxes of candy should the firm sell to maximize revenue?

(e) Using a graphing utility, graph the profit function.

(f) What is the maximum profit that this firm can earn?

(g) How many boxes of candy should the firm sell to maximize profit?

(h) Provide a reasonable explanation as to why the answers found in parts (d) and (g) differ. Is the shape of the revenue function reasonable in your opinion? Why?

69. Cost of Manufacturing In Problem 49 of Section 4.4, a cubic function of best fit relating the cost C of manufacturing x Chevy Cavaliers per hour was given. Budget constraints will not allow Chevy to spend more than $97,000 per hour. Determine the number of Cavaliers that could be produced in an hour.

70. Cost of Printing In Problem 50 of Section 4.4, a cubic function of best fit relating the cost C of printing x textbooks in a week was given. Budget constraints will not allow the printer to spend more than $170,000 per week. Determine the number of textbooks that could be printed in a week.

71. Prove that if a, b are real numbers and $a \geq 0, b \geq 0$ then
$$a \leq b \quad \text{is equivalent to} \quad \sqrt{a} \leq \sqrt{b}$$
[**Hint:** $b - a = (\sqrt{b} - \sqrt{a})(\sqrt{b} + \sqrt{a})$].

72. Make up an inequality that has no solution. Make up one that has exactly one solution.

Chapter Review

Things to Know

Power function (pp. 325–326)

$f(x) = x^n, \quad n \geq 2$ even

Domain: all real numbers   Range: nonnegative real numbers

Passes through $(-1, 1), (0, 0), (1, 1)$

Even function

Decreasing on $(-\infty, 0)$, increasing on $(0, \infty)$

$f(x) = x^n, \quad n \geq 3$ odd

Domain: all real numbers   Range: all real numbers

Passes through $(-1, -1), (0, 0), (1, 1)$

Odd function

Increasing on $(-\infty, \infty)$

Polynomial function (p. 330)

$f(x) = a_n x^n + a_{n-1} x^{n-1} + \cdots$
$\quad + a_1 x + a_0, \quad a_n \neq 0$

Domain: all real numbers

At most $n - 1$ turning points

End behavior: Behaves like $y = a_n x^n$ for large $|x|$

Rational function (p. 343)

$R(x) = \dfrac{p(x)}{q(x)}$ p, q are polynomial functions   Domain: $\{x | q(x) \neq 0\}$

Vertical asymptotes: With $R(x)$ in lowest terms, if $q(r) = 0$, then $x = r$ is a vertical asymptote.

Horizontal or obligue asymptotes: See summary on page 352.

Objectives

Review Exercises

Blue problem numbers indicate the authors' suggestions for use in a Practice Test.

In Problems 1–4, determine which functions are polynomial functions. For those that are, state the degree. For those that are not, tell why not.

1. $f(x) = 4x^5 - 3x^2 + 5x - 2$

2. $f(x) = \dfrac{3x^5}{2x + 1}$

3. $f(x) = 3x^2 + 5x^{1/2} - 1$

4. $f(x) = 3$

In Problems 5–10, graph each function using transformations (shifting, compressing, stretching, and reflection). Show all the stages. Verify your results using a graphing utility.

5. $f(x) = (x + 2)^3$

6. $f(x) = -x^3 + 3$

7. $f(x) = -(x - 1)^4$

8. $f(x) = (x - 1)^4 - 2$

9. $f(x) = (x - 1)^4 + 2$

10. $f(x) = (1 - x)^3$

In Problems 11–18, for each polynomial function f:

    (a) *Find the x- and y-intercepts of the graph of f.*

    (b) *Determine whether the graph crosses or touches the x-axis at each x-intercept.*

    (c) *End behavior: Find the power function that the graph of f resembles for large values of $|x|$.*

    (d) *Use a graphing utility to graph f.*

    (e) *Determine the number of turning points on the graph of f. Approximate the turning points if any exist, rounded to two decimal places.*

    (f) *Use the information obtained in parts (a) to (e) to draw a complete graph of f by hand.*

11. $f(x) = x(x + 2)(x + 4)$

12. $f(x) = x(x - 2)(x - 4)$

13. $f(x) = (x - 2)^2(x + 4)$

14. $f(x) = (x - 2)(x + 4)^2$

15. $f(x) = x^3 - 4x^2$

16. $f(x) = x^3 + 4x$

17. $f(x) = (x - 1)^2(x + 3)(x + 1)$

18. $f(x) = (x - 4)(x + 2)^2(x - 2)$

In Problems 19–30, discuss each rational function following the eight steps on page 355.

19. $R(x) = \dfrac{2x - 6}{x}$

20. $R(x) = \dfrac{4 - x}{x}$

21. $H(x) = \dfrac{x + 2}{x(x - 2)}$

22. $H(x) = \dfrac{x}{x^2 - 1}$

23. $R(x) = \dfrac{x^2 + x - 6}{x^2 - x - 6}$

24. $R(x) = \dfrac{x^2 - 6x + 9}{x^2}$

25. $F(x) = \dfrac{x^3}{x^2 - 4}$

26. $F(x) = \dfrac{3x^3}{(x - 1)^2}$

27. $R(x) = \dfrac{2x^4}{(x - 1)^2}$

28. $R(x) = \dfrac{x^4}{x^2 - 9}$

29. $G(x) = \dfrac{x^2 - 4}{x^2 - x - 2}$

30. $F(x) = \dfrac{(x - 1)^2}{x^2 - 1}$

In Problems 31–40, solve each inequality (a) algebraically and (b) graphically.

31. $2x^2 + 5x - 12 < 0$

32. $3x^2 - 2x - 1 \geq 0$

33. $\dfrac{6}{x + 3} \geq 1$

34. $\dfrac{-2}{1 - 3x} < 1$

35. $\dfrac{2x - 6}{1 - x} < 2$

36. $\dfrac{3 - 2x}{2x + 5} \geq 2$

37. $\dfrac{(x - 2)(x - 1)}{x - 3} > 0$

38. $\dfrac{x + 1}{x(x - 5)} \leq 0$

39. $\dfrac{x^2 - 8x + 12}{x^2 - 16} > 0$

40. $\dfrac{x(x^2 + x - 2)}{x^2 + 9x + 20} \leq 0$

---

41. AIDS Cases in the United States   The following data represent the cumulative number of reported AIDS cases in the United States from 1990–1997.

Year, t	Number of AIDS Cases, A
1990, 1	193,878
1991, 2	251,638
1992, 3	326,648
1993, 4	399,613
1994, 5	457,280
1995, 6	528,215
1996, 7	594,760
1997, 8	653,253

Source: U.S. Center for Disease Control and Prevention

(a) Using a graphing utility, draw a scatter diagram of the data, treating the year t as the independent variable.

(b) Use a graphing utility to find the cubic function of best fit to these data.

(c) Draw the cubic function of best fit on the scatter diagram found in part (b).

(d) Use function found in (b) to predict the cumulative number of AIDS cases reported in the United States in 2000.

(e) Do you think the function given in part (b) will be useful in predicting the number of AIDS cases in 2010?

42. Making a Can   A can in the shape of a right circular cylinder is required to have a volume of 250 cubic centimeters.

(a) Express the amount A of material to make the drum as a function of the radius r of the cylinder.

(b) How much material is required if the drum is of radius 3 centimeters?

(c) Of radius 5 centimeters?

(d) Graph $A = A(r)$. For what value of r is A smallest?

43. Design a polynomial function with the following characteristics: degree 6; four real zeros, one of multiplicity 3; y-intercept 3; behaves like $y = -5x^6$ for large values of x. Is this polynomial unique? Compare your polynomial with those of other students. What terms will be the same as everyone else's? Add some more characteristics, such as symmetry or naming the real zeros. How does this modify the polynomial?

44. Design a rational function with the following characteristics: three real zeros, one of multiplicity 2; y-intercept 1; vertical asymptotes $x = -2$ and $x = 3$; oblique asymptote $y = 2x + 1$. Is this rational function unique? Compare yours with those of other students. What will be the same as everyone else's? Add some more characteristics, such as symmetry or naming the real zeros. How does this modify the rational function?

45. The illustration shows the graph of a polynomial function.

(a) Is the degree of the polynomial even or odd?

(b) Is the leading coefficient positive or negative?

(c) Is the function even, odd, or neither?

(d) Why is x^2 necessarily a factor of the polynomial?

(e) What is the minimum degree of the polynomial?

(f) Formulate five different polynomials whose graphs could look like the one shown. Compare yours to those of other students. What similarities do you see? What differences?

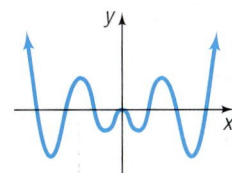

Chapter Projects

1. **Enrollment in Schools**   The following data represent the number of students enrolled in grades 9 through 12 in private institutions.

Year	Enrollment (000's)
1975	1300
1980	1339
1981	1400
1982	1400
1983	1400
1984	1400
1985	1362
1986	1336
1987	1247
1988	1206
1989	1193
1990	1137
1991	1125
1992	1163
1993	1191
1994	1236
1995	1260
1996	1297

(a) Using a graphing utility, draw a scatter diagram of the data.
(b) Using a graphing utility, find the cubic function of best fit.
(c) Using a graphing utility, graph the cubic function of best fit on your scatter diagram.
(d) Use the function found in part (b) to predict the number of students enrolled in private institutions for grades 9 through 12 in the year 2000.
(e) Compare the result found in part (d) with that of the U.S. National Center for Education Statistics, whose predictions can be found in the *Statistical Abstract of the United States*. What might account for any differences in your prediction versus that of the U.S. National Center for Education Statistics?
(f) Use the function found in part (b) to predict the year in which enrollment in private institutions for grades 9 through 12 will be 1,400,000. That is, solve $f(x) = 1400$.

2. **Finding a Polynomial Function**   Find a polynomial function of the form $P(x) = a_n x^n + a_{n-1} x^{n-1} + \cdots + a_0$ defined by the graph illustrated.

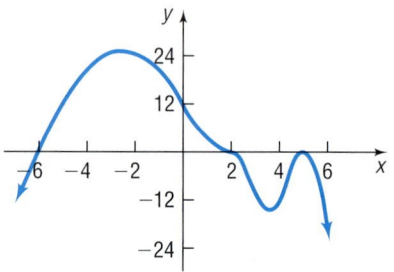

(a) How many zeros does the function have? What are their multiplicities?
(b) Construct a polynomial function whose zeros are those identified in part (a). What role does the multiplicity play in determining the factors? Graph the function using a graphing utility.
(c) Is the leading coefficient of the polynomial positive or negative? Why?
(d) What is the y-intercept of P?
(e) Construct a polynomial function whose zeros are those determined in part (a) and whose y-intercept is the one determined in part (d).
(f) Find the power function that the graph of the function resembles for large values of $|x|$.
(g) Find a second function that "fits" the graph. Be sure to check its graph on your graphing utility. What similarities do you find between the two functions? What differences do you find?
(h) Write a general polynomial in its factored form whose graph could look like the one given, based on the four points specified. Be sure to state any conditions on the factors, and give a general formula for the leading coefficient.

3. **Finding a Function from Its Graph**   Find a rational function of the form

$$R(x) = \frac{a_n x^n + a_{n-1} x^{n-1} + \cdots + a_0}{b_m x^m + b_{m-1} x^{m-1} + \cdots + b_0}$$

defined by the graph illustrated.

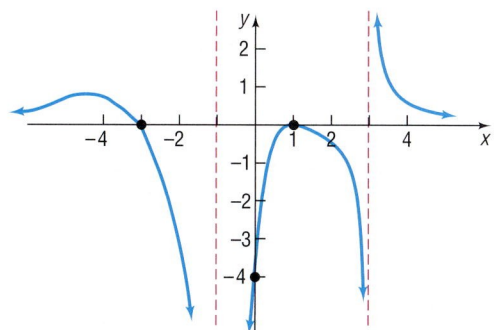

(a) There are two vertical asymptotes. How do they fit?

(b) There are two x-intercepts. Where do they show up in the function?

(c) Use a graphing utility to check what you have so far. Does it look like the original graph? Notice that 1 and -3 appear to be the only real zeros. What can you tell

about the multiplicity of each? Do you need to change the multiplicity of either zero in your solution?

(d) How can you account for the fact that the graph of the function approaches $-\infty$ on both sides of one vertical asymptote and at the other vertical asymptote it approaches $-\infty$ on one side and to ∞ on the other? Can you change the denominator in some way to account for this?

(e) Is the information that you have so far consistent with the horizontal asymptote $y = 0$? If not, how should you adjust your function so that it is consistent?

(f) Is the information that you have so far consistent with the y-intercept? If not, how should you adjust your function so that it is consistent?

(g) Are you finished? Have you checked your solution on a graphing utility? Are you satisfied that you have it right?

(h) Are there other functions whose graph might look like the one given? If so, what common characteristics do they have?

Cumulative Review

1. Find the exact solution(s) to the equation $4x^2 + 2x = 5$. Verify your results using a graphing utility

2. Find an equation for the line containing the points $(2, 3)$ and $(-3, 5)$.

3. Determine if the lines $3x - 4y = 5$ and $4x + 3y = 10$ are parallel, perpendicular, or neither.

4. Does the equation $3x^2 - 4y^2 = 10$ represent a function? Why or why not?

5. Find $g(4)$ if $g(t) = -16t^2 + 100t + 10$.

6. Graph the function $f(x) = x^2 - 4x + 1$ by determining whether the graph opens up or down and finding the vertex, intercepts, and axis of symmetry. Verify your results using a graphing utility.

7. Graph the function $f(x) = -2(x + 3)^2 + 5$ using transformations. Verify your results using a graphing utility.

8. Is the function $f(x) = x^3 - 4x$ even, odd, or neither? Is the graph of f symmetric with respect to the y-axis, the x-axis, and/or the origin?

9. Find the local maxima and the local minima of the function $f(x) = 0.2x^3 + x^2 + 1$. Determine the intervals where the function is increasing. Determine the intervals where the function is decreasing.

10. Suppose that $f(x) = \dfrac{1}{x + 2}$ and $g(x) = \dfrac{3}{x - 1}$. Find $f \circ g$ and state its domain.

11. For the polynomial function $f(x) = (x + 3)(x - 4)^2$:
    (a) Find the x- and y-intercepts of the graph of f.
    (b) Determine whether the graph crosses or touches the x-axis at each x-intercept.
    (c) End behavior: Find the power function that the graph of f resembles for large values of $|x|$.
    (d) Use a graphing utility to graph f.
    (e) Determine the number of turning points on the graph of f. Approximate the turning points, if any exist, rounded to two decimal places.
    (f) Use the information obtained in parts (a) to (e) to draw a complete graph of f by hand.

12. For the rational function $R(x) = \dfrac{x^2 - 1}{x^2 - 3x + 2}$:
    (a) What is the domain of R?
    (b) Determine the vertical asymptote(s).
    (c) Determine the horizontal or oblique asymptote, if any.
    (d) Determine the intercepts.
    (e) By hand, draw a complete graph of R.
    (f) Solve the inequality $R(x) \le 0$.

13. A stone is thrown into a pond. A series of concentric circles form. Suppose that the radius of each circle is increasing at a rate of 2 feet per second. Express the area of the circle as a function of time.

    Hint: This requires the use of a composite function.

THE ZEROS OF A POLYNOMIAL FUNCTION

Ragweed plus bad air will equal fall misery

A hot, dry, smoggy summer is adding up to an insufferable fall for allergy sufferers.

Experts say people with allergies to ragweed, who normally sneeze and wheeze through September, will be dabbing their eyes and nose well into October, even though pollen counts are actually no worse then usual.

The difference this year is that a summer of heavy smog has made the pollen in the air feel much thicker, making it more irritating.

"Smog plays an extremely important role in allergy suffering," says Frances Coates, owner of Ottawa-based Aerobiology Research, which measures pollen levels. "It holds particles in the air a lot longer."

Low rainfall delayed the ragweed season until the end of August. It usually hits early in the month.

SOURCE: Adapted by Louise Surette from Toronto Star, September 8, 2001.

SEE CHAPTER PROJECT 1.

 For additional study help, go to

www.prenhall.com/sullivanegu3e

Materials include:

- Graphing Calculator Help
- Chapter Quiz
- Chapter Test
- PowerPoint Downloads
- Chapter Projects
- Student Tips

A Look Back, A Look Forward

In Chapter 4, we studied polynomial functions and their graphs. The polynomial functions that we analyzed were already completely factored, which made identifying the zeros of the polynomial function straightforward. With the zeros (and their multiplicity) known, the graph of a polynomial function can be readily obtained. In this chapter, we introduce techniques that can be used to determine the zeros of poly-nomial functions that are not factored completely. We begin by developing techniques for identifying the real zeros of a polynomial function and continue by introducing a new number system, called the *complex number system*, concluding with a discussion of complex zeros. This chapter can be thought of as an introduction to an area of mathematics called the *theory of equations*.

PREPARING FOR THIS SECTION

Before getting started, review the following:

✓ Polynomial Division (Section R.5, pp. 45–48)

5.1   SYNTHETIC DIVISION*

OBJECTIVES   ① Use Synthetic Division

To find the quotient as well as the remainder when a polynomial function f of degree 1 or higher is divided by $g(x) = x - c$, a shortened version of long division, called **synthetic division**, makes the task simpler.

① To see how synthetic division works, we will use long division to divide the polynomial $f(x) = 2x^3 - x^2 + 3$ by $g(x) = x - 3$.

$$
\begin{array}{r}
2x^2 + 5x + 15 \quad \leftarrow \text{Quotient} \\
x - 3 \overline{)2x^3 - x^2 \qquad + 3} \\
\underline{2x^3 - 6x^2} \\
5x^2 \\
\underline{5x^2 - 15x} \\
15x + 3 \\
\underline{15x - 45} \\
48 \quad \leftarrow \text{Remainder}
\end{array}
$$

✔ CHECK: $(\text{Divisor}) \cdot (\text{Quotient}) + \text{Remainder}$

$$= (x - 3)(2x^2 + 5x + 15) + 48$$

$$= 2x^3 + 5x^2 + 15x - 6x^2 - 15x - 45 + 48$$

$$= 2x^3 - x^2 + 3$$ ∎

The process of synthetic division arises from rewriting the long division in a more compact form, using simpler notation. For example, in the long division above, the terms in blue are not really necessary because they are identical to the terms directly above them. With these terms removed, we have

---

*This section is not required for the discussion that follows, but its coverage does make the computations faster and easier.

$$
\begin{array}{r}
2x^2 + 5x\ \ + 15 \\
x - 3\overline{)2x^3 -\ \ x^2\qquad\quad + 3} \\
\underline{-\,6x^2} \\
5x^2 \\
\underline{-\,15x} \\
15x \\
\underline{-\,45} \\
48
\end{array}
$$

Most of the x's that appear in this process can also be removed, provided that we are careful about positioning each coefficient. In this regard, we will need to use 0 as the coefficient of x in the dividend, because that power of x is missing. Now we have

$$
\begin{array}{r}
2x^2 + 5x + 15 \\
x - 3\overline{)2\ \ -1\qquad 0\qquad 3} \\
\underline{-\,6} \\
5 \\
\underline{-\,15} \\
15 \\
\underline{-\,45} \\
48
\end{array}
$$

We can make this display more compact by moving the lines up until the numbers in color align horizontally.

$$
\begin{array}{rllll}
 & 2x^2 + 5x + 15 & & & \text{Row 1} \\
x - 3\overline{)2} & -1 & 0 & 3 & \text{Row 2} \\
\cline{1-4}
 & -6 & -15 & -45 & \text{Row 3} \\
\cline{1-4}
\bigcirc & 5 & 15 & 48 & \text{Row 4}
\end{array}
$$

Because the leading coefficient of the divisor is always 1, we know that the leading coefficient of the dividend will also be the leading coefficient of the quotient. So, we place the leading coefficient of the quotient, 2, in the circled position. Then the first three numbers in row 4 are precisely the coefficients of the quotient, and the last number in row 4 is the remainder. Now row 1 is not needed, so we compress the process to three rows, where the bottom row contains the coefficients of the quotient and the remainder.

$$
\begin{array}{rrrrl}
3\overline{)2} & -1 & 0 & 3 & \text{Row 1} \\
\underline{} & 6 & 15 & 45 & \text{Row 2 (add)} \\
2 & 5 & 15 & 48 & \text{Row 3}
\end{array}
$$

$$
\underbrace{2x^2 + 5x + 15}_{\text{Quotient}} \qquad \underset{\text{Remainder}}{48}
$$

Recall that the entries in row 3 are obtained by subtracting the entries in row 2 from those in row 1. Rather than subtracting the entries in row 2, we can change the sign of each entry and add. With this modification, our display will look like this:

$$
\begin{array}{rrrrl}
x - 3\overline{)2} & -1 & 0 & 3 & \text{Row 1} \\
\underline{} & 6 & 15 & 45 & \text{Row 2 (add)} \\
2 & 5 & 15 & 48 & \text{Row 3}
\end{array}
$$

Notice that the entries in row 2 are three times the prior entries in row 3. Our last modification to the display replaces the $x - 3$ by 3. The entries in row 3 give the quotient and the remainder, as shown next.

$$
\begin{array}{r|rrrr}
3) & 2 & -1 & 0 & 3 \\
   &   & 6 & 15 & 45 \\
\hline
   & 2 & 5 & 15 & 48
\end{array}
\quad
\begin{array}{l}
\text{Row 1} \\
\text{Row 2 (add)} \\
\text{Row 3}
\end{array}
$$

$$
\underbrace{2x^2 + 5x + 15}_{\text{Quotient}} \qquad \overset{\text{Remainder}}{48}
$$

Let's go through an example step by step.

EXAMPLE 1   Using Synthetic Division to Find the Quotient and Remainder

Use synthetic division to find the quotient $q(x)$ and remainder R when

$$f(x) = x^3 - 4x^2 - 5 \quad \text{is divided by} \quad g(x) = x - 3$$

Solution

STEP 1: Write the dividend in descending powers of x. Then copy the coefficients, remembering to insert a 0 for any missing powers of x.

$$
\begin{array}{cccc}
1 & -4 & 0 & -5
\end{array}
\qquad \text{Row 1}
$$

STEP 2: Insert the usual division symbol. In synthetic division, the divisor is of the form $x - c$, and c is the number placed to the left of the division symbol. Here, since the divisor is $x - 3$, we insert 3 to the left of the division symbol.

$$
3)\overline{\begin{array}{cccc} 1 & -4 & 0 & -5 \end{array}} \qquad \text{Row 1}
$$

STEP 3: Bring the 1 down two rows, and enter it in row 3.

$$
\begin{array}{r|rrrr}
3) & 1 & -4 & 0 & -5 \\
   & \downarrow & & & \\
\hline
   & 1 & & &
\end{array}
\quad
\begin{array}{l}
\text{Row 1} \\
\text{Row 2} \\
\text{Row 3}
\end{array}
$$

STEP 4: Multiply the latest entry in row 3 by 3, and place the result in row 2, one column over to the right.

$$
\begin{array}{r|rrrr}
3) & 1 & -4 & 0 & -5 \\
   &   & 3 & & \\
\hline
   & 1 & & &
\end{array}
\quad
\begin{array}{l}
\text{Row 1} \\
\text{Row 2} \\
\text{Row 3}
\end{array}
$$

STEP 5: Add the entry in row 2 to the entry above it in row 1, and enter the sum in row 3.

$$
\begin{array}{r|rrrr}
3) & 1 & -4 & 0 & -5 \\
   &   & 3 & & \\
\hline
   & 1 & -1 & &
\end{array}
\quad
\begin{array}{l}
\text{Row 1} \\
\text{Row 2} \\
\text{Row 3}
\end{array}
$$

STEP 6: Repeat Steps 4 and 5 until no more entries are available in row 1.

$$
\begin{array}{r|rrrr}
3) & 1 & -4 & 0 & -5 \\
   &   & 3 & -3 & -9 \\
\hline
   & 1 & -1 & -3 & -14
\end{array}
\quad
\begin{array}{l}
\text{Row 1} \\
\text{Row 2} \\
\text{Row 3}
\end{array}
$$

STEP 7: The final entry in row 3, the -14, is the remainder; the other entries in row 3 the $1, -1$, and -3, are the coefficients (in descending order) of a polynomial whose degree is 1 less than that of the dividend. This is the quotient. Do you see why? Thus,

$$q(x) = x^2 - x - 3 \qquad R = -14$$

✔ CHECK: (Divisor)(Quotient) + Remainder
$$= (x - 3)(x^2 - x - 3) + (-14)$$
$$= (x^3 - x^2 - 3x - 3x^2 + 3x + 9) + (-14)$$
$$= x^3 - 4x^2 - 5 = \text{Dividend}$$　■ ■

Let's do an example in which all seven steps are combined.

EXAMPLE 2　**Using Synthetic Division to Verify a Factor**

Use synthetic division to show that $x + 3$ is a factor of
$$f(x) = 2x^5 + 5x^4 - 2x^3 + 2x^2 - 2x + 3$$

Solution　The divisor is $x + 3 = x - (-3)$, so we place -3 to the left of the division symbol. Then the row 3 entries will be multiplied by -3, entered in row 2, and added to row 1.

$$
\begin{array}{r|rrrrrr}
-3) & 2 & 5 & -2 & 2 & -2 & 3 \\
    &   & -6 & 3 & -3 & 3 & -3 \\
\hline
    & 2 & -1 & 1 & -1 & 1 & 0
\end{array}
$$

Row 1
Row 2
Row 3

Because the remainder is 0, we have

(Divisor)(Quotient) + Remainder
$$= (x + 3)(2x^4 - x^3 + x^2 - x + 1) = 2x^5 + 5x^4 - 2x^3 + 2x^2 - 2x + 3$$
As we see, $x + 3$ is a factor of $2x^5 + 5x^4 - 2x^3 + 2x^2 - 2x + 3$.　■

As Example 2 illustrates, the remainder after division gives information about whether the divisor is, or is not, a factor. We shall have more to say about this in the next section.

✎━ **NOW WORK PROBLEMS 3 AND 13.**

5.1　Exercises

In Problems 1–12, use synthetic division to find the quotient $q(x)$ and remainder R when $f(x)$ is divided by $g(x)$.

1. $f(x) = x^3 - x^2 + 2x + 4$;　$g(x) = x - 2$
2. $f(x) = x^3 + 2x^2 - 3x + 1$;　$g(x) = x + 1$
3. $f(x) = 3x^3 + 2x^2 - x + 3$;　$g(x) = x - 3$
4. $f(x) = -4x^3 + 2x^2 - x + 1$;　$g(x) = x + 2$
5. $f(x) = x^5 - 4x^3 + x$;　$g(x) = x + 3$
6. $f(x) = x^4 + x^2 + 2$;　$g(x) = x - 2$
7. $f(x) = 4x^6 - 3x^4 + x^2 + 5$;　$g(x) = x - 1$
8. $f(x) = x^5 + 5x^3 - 10$;　$g(x) = x + 1$
9. $f(x) = 0.1x^3 + 0.2x$;　$g(x) = x + 1.1$
10. $f(x) = 0.1x^2 - 0.2$;　$g(x) = x + 2.1$
11. $f(x) = x^5 - 1$;　$g(x) = x - 1$
12. $f(x) = x^5 + 1$;　$g(x) = x + 1$

In Problems 13–22, use synthetic division to determine whether $x - c$ is a factor of $f(x)$.

13. $f(x) = 4x^3 - 3x^2 - 8x + 4$;　$x - 2$
14. $f(x) = -4x^3 + 5x^2 + 8$;　$x + 3$
15. $f(x) = 3x^4 - 6x^3 - 5x + 10$;　$x - 2$
16. $f(x) = 4x^4 - 15x^2 - 4$;　$x - 2$
17. $f(x) = 3x^6 + 82x^3 + 27$;　$x + 3$
18. $f(x) = 2x^6 - 18x^4 + x^2 - 9$;　$x + 3$
19. $f(x) = 4x^6 - 64x^4 + x^2 - 15$;　$x + 4$
20. $f(x) = x^6 - 16x^4 + x^2 - 16$;　$x + 4$
21. $f(x) = 2x^4 - x^3 + 2x - 1$;　$x - \dfrac{1}{2}$
22. $f(x) = 3x^4 + x^3 - 3x + 1$;　$x + \dfrac{1}{3}$

23. Find the sum of a, b, c, and d if
$$\frac{x^3 - 2x^2 + 3x + 5}{x + 2} = ax^2 + bx + c + \frac{d}{x + 2}$$

24. When dividing a polynomial by $x - c$, do you prefer to use long division or synthetic division? Does the value of c make a difference to you in choosing? Give reasons.

PREPARING FOR THIS SECTION

Before getting started, review the following:

✓ Polynomial Division (Section R.5, pp. 45–48)

✓ Polynomial Functions (Section 4.2, pp. 330–338)

✓ Quadratic Formula (Section 1.3, pp. 121–124)

✓ Factoring Polynomials (Section R.6, pp. 50–58)

5.2 THE REAL ZEROS OF A POLYNOMIAL FUNCTION

OBJECTIVES
1. Use the Remainder and Factor Theorems
2. Use the Rational Zeros Theorem to List the Potential Rational Zeros of a Polynomial Function
3. Find the Real Zeros of a Polynomial Function
4. Solve Polynomial Equations
5. Use the Theorem for Bounds on Zeros
6. Use the Intermediate Value Theorem

In this section, we discuss techniques that can be used to find the real zeros of a polynomial function. Recall that if r is a real zero of a polynomial function f then $f(r) = 0$; r is an x-intercept of the graph of f; and r is a solution of the equation $f(x) = 0$. For polynomial and rational functions, we have seen the importance of the zeros for graphing. In most cases, however, the zeros of a polynomial function are difficult to find using algebraic methods. No nice formulas like the quadratic formula are available to help us find zeros for polynomials of degree 3 or higher. Formulas do exist for solving any third- or fourth-degree polynomial equation, but they are somewhat complicated. No general formulas exist for polynomial equations of degree 5 or higher. Refer to the Historical Feature at the end of the section for more information.

Remainder and Factor Theorems

When we divide one polynomial (the dividend) by another (the divisor), we obtain a quotient polynomial and a remainder, the remainder being either the zero polynomial or a polynomial whose degree is less than the degree of the divisor. To check our work, we verify that

$$(\text{Quotient})(\text{Divisor}) + \text{Remainder} = \text{Dividend}$$

This checking routine is the basis for a famous theorem called the **division algorithm* for polynomials**, which we now state without proof.

Theorem

Division Algorithm for Polynomials

If $f(x)$ and $g(x)$ denote polynomial functions and if $g(x)$ is not the zero polynomial, then there are unique polynomial functions $q(x)$ and $r(x)$ such that

---

*A systematic process in which certain steps are repeated a finite number of times is called an **algorithm**. For example, long division is an algorithm.

$$\frac{f(x)}{g(x)} = q(x) + \frac{r(x)}{g(x)} \quad \text{or} \quad f(x) = q(x)g(x) + r(x) \quad (1)$$

dividend   quotient   divisor   remainder

where $r(x)$ is either the zero polynomial or a polynomial of degree less than that of $g(x)$.

In equation (1), $f(x)$ is the **dividend**, $g(x)$ is the **divisor**, $q(x)$ is the **quotient**, and $r(x)$ is the **remainder**.

If the divisor $g(x)$ is a first-degree polynomial of the form

$$g(x) = x - c, \qquad c \text{ a real number}$$

then the remainder $r(x)$ is either the zero polynomial or a polynomial of degree 0. As a result, for such divisors, the remainder is some number, say R, and we may write

$$f(x) = (x - c)q(x) + R \qquad (2)$$

This equation is an identity in x and is true for all real numbers x. Suppose that $x = c$. Then equation (2) becomes

$$f(c) = (c - c)q(c) + R$$
$$f(c) = R$$

Substitute $f(c)$ for R in equation (2) to obtain

$$f(x) = (x - c)q(x) + f(c) \qquad (3)$$

We have now proved the **Remainder Theorem**.

Remainder Theorem

Let f be a polynomial function. If $f(x)$ is divided by $x - c$, then the remainder is $f(c)$.

EXAMPLE 1   **Using the Remainder Theorem**

Find the remainder if $f(x) = x^3 - 4x^2 - 5$ is divided by

(a) $x - 3$ (b) $x + 2$

Solution   (a) We could use long division or synthetic division, but it is easier to use the Remainder Theorem, which says that the remainder is $f(3)$.

$$f(3) = (3)^3 - 4(3)^2 - 5 = 27 - 36 - 5 = -14$$

The remainder is -14.

(b) To find the remainder when $f(x)$ is divided by $x + 2 = x - (-2)$, we evaluate $f(-2)$.

$$f(-2) = (-2)^3 - 4(-2)^2 - 5 = -8 - 16 - 5 = -29$$

The remainder is -29.

Figure 1

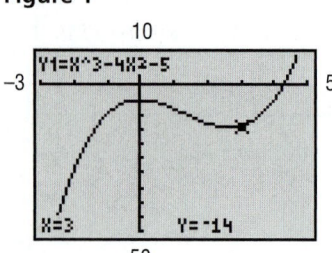

Compare the method used in Example 1(a) above with the method used in Example 1 on page 382 (synthetic division). Which method do you prefer? Give reasons.

COMMENT A graphing utility provides another way to find the value of a function, using the eVALUEate feature. Consult your manual for details. See Figure 1 for the results of Example 1(a). ∎

An important and useful consequence of the Remainder Theorem is the **Factor Theorem**.

Factor Theorem

Let f be a polynomial function. Then $x - c$ is a factor of $f(x)$ if and only if $f(c) = 0$.

∎

The Factor Theorem actually consists of two separate statements:

1. If $f(c) = 0$, then $x - c$ is a factor of $f(x)$.
2. If $x - c$ is a factor of $f(x)$, then $f(c) = 0$.

The proof requires two parts.

Proof

1. Suppose that $f(c) = 0$. Then, by equation (3), we have

$$f(x) = (x - c)q(x) + f(c)$$
$$= (x - c)q(x) + 0$$
$$= (x - c)q(x)$$

for some polynomial $q(x)$. That is, $x - c$ is a factor of $f(x)$.

2. Suppose that $x - c$ is a factor of $f(x)$. Then there is a polynomial function q such that

$$f(x) = (x - c)q(x)$$

Replacing x by c, we find that

$$f(c) = (c - c)q(c) = 0 \cdot q(c) = 0$$

This completes the proof. ∎

One use of the Factor Theorem is to determine whether a polynomial has a particular factor.

EXAMPLE 2 **Using the Factor Theorem**

Use the Factor Theorem to determine whether the function $f(x) = 2x^3 - x^2 + 2x - 3$ has the factor

(a) $x - 1$ (b) $x + 2$

Solution   The Factor Theorem states that if $f(c) = 0$ then $x - c$ is a factor.

(a) Because $x - 1$ is of the form $x - c$ with $c = 1$, we find the value of $f(1)$. We choose to use substitution.

$$f(1) = 2(1)^3 - (1)^2 + 2(1) - 3 = 2 - 1 + 2 - 3 = 0$$

See also Figure 2(a). By the Factor Theorem, $x - 1$ is a factor of $f(x)$.

(b) We first need to write $x + 2$ in the form $x - c$. Since $x + 2 = x - (-2)$, we find the value of $f(-2)$. See Figure 2(b). Because $f(-2) = -27 \neq 0$, we conclude from the Factor Theorem that $x - (-2) = x + 2$ is not a factor of $f(x)$.

Figure 2

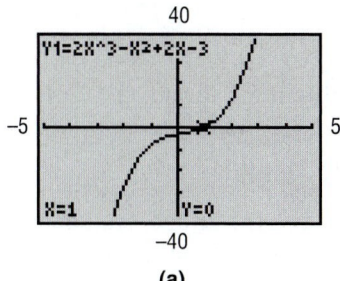

(a)

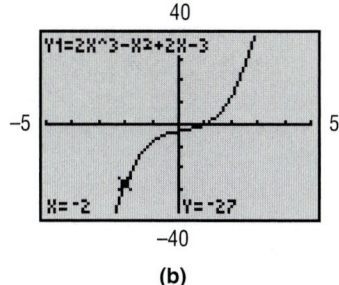

(b)

In Example 2, we found that $x - 1$ was a factor of f. To write f in factored form, we can use long division or synthetic division. Using synthetic division, we find that

$$
\begin{array}{r}
1\overline{)\,2 \quad -1 \quad 2 \quad -3} \\
\underline{2 \quad 1 \quad 3} \\
2 \quad 1 \quad 3 \quad 0
\end{array}
$$

The quotient is $q(x) = 2x^2 + x + 3$ with a remainder of 0, as expected. We can write f in factored form as

$$f(x) = 2x^3 - x^2 + 2x - 3 = (x - 1)(2x^2 + x + 3)$$

NOW WORK PROBLEM **1**.

The Number and Location of Real Zeros

The next theorem concerns the number of real zeros that a polynomial function may have. In counting the zeros of a polynomial, we count each zero as many times as its multiplicity.

Theorem   **Number of Real Zeros**

A polynomial function of degree $n, n \geq 1$, has at most n real zeros.

Proof   The proof is based on the Factor Theorem. If r is a zero of a polynomial function f, then $f(r) = 0$ and, hence, $x - r$ is a factor of $f(x)$. Each zero corresponds to a factor of degree 1. Because f cannot have more first-degree factors than its degree, the result follows.

Rational Zeros Theorem

② The next result, called the **Rational Zeros Theorem**, provides information about the rational zeros of a polynomial *with integer coefficients*.

Theorem

Rational Zeros Theorem

Let f be a polynomial function of degree 1 or higher of the form

$$f(x) = a_n x^n + a_{n-1} x^{n-1} + \cdots + a_1 x + a_0, \qquad a_n \neq 0, a_0 \neq 0$$

where each coefficient is an integer. If $\dfrac{p}{q}$, in lowest terms, is a rational zero of f, then p must be a factor of a_0, and q must be a factor of a_n.

EXAMPLE 3   **Listing Potential Rational Zeros**

List the potential rational zeros of

$$f(x) = 2x^3 + 11x^2 - 7x - 6$$

Solution   Because f has integer coefficients, we may use the Rational Zeros Theorem. First, we list all the integers p that are factors of $a_0 = -6$ and all the integers q that are factors of the leading coefficient $a_3 = 2$.

$$p: \quad \pm 1, \pm 2, \pm 3, \pm 6$$
$$q: \quad \pm 1, \pm 2$$

Now we form all possible ratios $\dfrac{p}{q}$.

$$\frac{p}{q}: \quad \pm 1, \pm 2, \pm 3, \pm 6, \pm \frac{1}{2}, \pm \frac{3}{2}$$

If f has a rational zero, it will be found in this list, which contains 12 possibilities.

 NOW WORK PROBLEM 11.

Be sure that you understand what the Rational Zeros Theorem says: For a polynomial with integer coefficients, *if* there is a rational zero, it is one of those listed. It may be the case that the function does not have any rational zeros.

The Rational Zeros Theorem provides a list of potential rational zeros of a function f. If we graph f, we can get a better sense of the location of the x-intercepts and test to see if they are rational. We can also use the potential rational zeros to select our initial viewing window to graph f and then adjust the window based on the results. The graphs shown throughout the text will be those obtained after setting the final viewing window.

EXAMPLE 4   **Finding the Rational Zeros of a Polynomial Function**

③ Continue working with Example 3 to find the rational zeros of

$$f(x) = 2x^3 + 11x^2 - 7x - 6$$

Solution   We gather all the information that we can about the zeros.

STEP 1:   There are at most three real zeros.
STEP 2:   We list the potential rational zeros obtained in Example 3:

$$\pm 1, \pm 2, \pm 3, \pm 6, \pm \frac{1}{2}, \pm \frac{3}{2}.$$

Figure 3

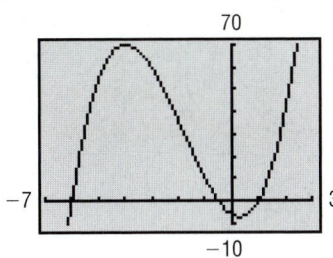

STEP 3: We could, of course, use the Factor Theorem to test each potential rational zero to see if the value of f there is zero. This is not very efficient. The graph of f will tell us approximately where the real zeros are. So we only need to test those rational zeros that are nearby. Figure 3 shows the graph of f. We see that f has three zeros: one near -6, one between -1 and 0, and one near 1. From our original list of potential rational zeros, we will test -6, using eVALUEate.

STEP 4: Because $f(-6) = 0$ we know that -6 is a zero and $x + 6$ is a factor of f. We can use long division or synthetic division to factor f.

$$f(x) = 2x^3 + 11x^2 - 7x - 6$$
$$= (x + 6)(2x^2 - x - 1)$$

Now any solution of the equation $2x^2 - x - 1 = 0$ will be a zero of f. Because of this, we call the equation $2x^2 - x - 1 = 0$ a **depressed equation** of f. Since the degree of the depressed equation of f is less than that of the original polynomial, we work with the depressed equation to find the zeros of f.

The depressed equation $2x^2 - x - 1 = 0$ is a quadratic equation with discriminant $b^2 - 4ac = (-1)^2 - 4(2)(-1) = 9 > 0$. The equation has two real solutions, which can be found by factoring.

$$2x^2 - x - 1 = (2x + 1)(x - 1) = 0$$
$$2x + 1 = 0 \quad \text{or} \quad x - 1 = 0$$
$$x = -\frac{1}{2} \quad \text{or} \quad x = 1$$

The zeros of f are -6, $-\dfrac{1}{2}$, and 1.

Because $f(x) = (x + 6)(2x^2 - x - 1)$, we completely factor f as follows:

$$f(x) = 2x^3 + 11x^2 - 7x - 6 = (x + 6)(2x^2 - x - 1) = (x + 6)(2x + 1)(x - 1)$$

Notice that the three zeros of f are among those given in the list of potential rational zeros in Example 3. ■

Steps for Finding the Real Zeros of a Polynomial Function

STEP 1: Use the degree of the polynomial to determine the maximum number of zeros.

STEP 2: If the polynomial has integer coefficients, use the Rational Zeros Theorem to identify those rational numbers that potentially can be zeros.

STEP 3: Using a graphing utility, graph the polynomial function.

STEP 4: (a) Use eVALUEate, substitution, synthetic division, or long division to test a potential rational zero based on the graph.

   (b) Each time that a zero (and thus a factor) is found, repeat Step 4 on the depressed equation. In attempting to find the zeros, remember to use (if possible) the factoring techniques that you already know (special products, factoring by grouping, and so on).

| **EXAMPLE 5** | **Finding the Real Zeros of a Polynomial Function** |

Find the real zeros of: $f(x) = x^5 - x^4 - 4x^3 + 8x^2 - 32x + 48$
Write f in factored form.

Solution   **STEP 1:** There are at most five real zeros.
    STEP 2: To obtain the list of potential rational zeros, we write the factors p
    of $a_0 = 48$ and the factors q of the leading coefficient $a_5 = 1$.

p: $\pm 1, \pm 2, \pm 3, \pm 4, \pm 6, \pm 8, \pm 12, \pm 16, \pm 24, \pm 48$
q: ± 1

The potential rational zeros consist of all possible quotients $\dfrac{p}{q}$:

$\dfrac{p}{q}$: $\pm 1, \pm 2, \pm 3, \pm 4, \pm 6, \pm 8, \pm 12, \pm 16, \pm 24, \pm 48$

Figure 4

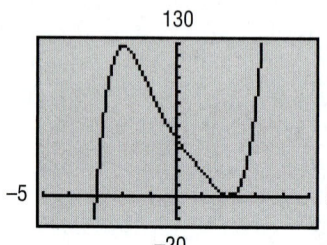

STEP 3: Figure 4 shows the graph of f. The graph has the characteristics that
    we expect of the given polynomial of degree 5: no more than four
    turning points, y-intercept 48, and it behaves like $y = x^5$ for large $|x|$.
STEP 4: Since -3 appears to be a zero and -3 is a potential rational zero,
    we eVALUEate f at -3 and find that $f(-3) = 0$. We use synthetic
    division to factor f.

$$
\begin{array}{r|rrrrr}
-3 & 1 & -1 & -4 & 8 & -32 & 48 \\
   &   & -3 & 12 & -24 & 48 & -48 \\
\hline
   & 1 & -4 & 8 & -16 & 16 & 0
\end{array}
$$

We can factor f as:

$$f(x) = x^5 - x^4 - 4x^3 + 8x^2 - 32x + 48 = (x + 3)(x^4 - 4x^3 + 8x^2 - 16x + 16)$$

We now work with the first depressed equation:

$$q_1(x) = x^4 - 4x^3 + 8x^2 - 16x + 16 = 0.$$

REPEAT STEP 4:   In looking back at Figure 4, it appears that 2 might be a zero
    of even multiplicity. We check the potential rational zero 2 and find
    that $f(2) = 0$. Using synthetic division,

$$
\begin{array}{r|rrrrr}
2 & 1 & -4 & 8 & -16 & 16 \\
  &   & 2 & -4 & 8 & -16 \\
\hline
  & 1 & -2 & 4 & -8 & 0
\end{array}
$$

Now we can factor f as:

$$f(x) = (x + 3)(x - 2)(x^3 - 2x^2 + 4x - 8)$$

REPEAT STEP 4:   The depressed equation $q_2(x) = x^3 - 2x^2 + 4x - 8 = 0$
    can be factored by grouping.

$$x^3 - 2x^2 + 4x - 8 = (x^3 - 2x^2) + (4x - 8) = x^2(x - 2) + 4(x - 2) = (x - 2)(x^2 + 4) = 0$$
$$x - 2 = 0 \quad \text{or} \quad x^2 + 4 = 0$$
$$x = 2$$

Since $x^2 + 4 = 0$ has no real solutions, the real zeros of f are -3 and
2, the latter being a zero of multiplicity 2. The factored form of f is

$$f(x) = x^5 - x^4 - 4x^3 + 8x^2 - 32x + 48$$
$$= (x + 3)(x - 2)^2(x^2 + 4)$$

■

▬▬▬ **NOW WORK PROBLEM 29.**

| EXAMPLE 6 | Solving a Polynomial Equation |

④ Solve the equation: $x^5 - x^4 - 4x^3 + 8x^2 - 32x + 48 = 0$

Solution The solutions of this equation are the zeros of the polynomial function

$$f(x) = x^5 - x^4 - 4x^3 + 8x^2 - 32x + 48$$

Using the result of Example 5, the real zeros are -3 and 2. These are the real solutions of the equation $x^5 - x^4 - 4x^3 + 8x^2 - 32x + 48 = 0$. ■

 **NOW WORK PROBLEM 53.**

In Example 5, the quadratic factor $x^2 + 4$ that appears in the factored form of $f(x)$ is called *irreducible*, because the polynomial $x^2 + 4$ cannot be factored over the real numbers. In general, we say that a quadratic factor $ax^2 + bx + c$ is **irreducible** if it cannot be factored over the real numbers, that is, if it is prime over the real numbers.

Refer back to Examples 4 and 5. The polynomial function of Example 4 has three real zeros, and its factored form contains three linear factors. The polynomial function of Example 5 has two distinct real zeros, and its factored form contains two distinct linear factors and one irreducible quadratic factor.

Theorem Every polynomial function (with real coefficients) can be uniquely factored into a product of linear factors and/or irreducible quadratic factors.

■

We shall prove this result in Section 5.4, and, in fact, we shall draw several additional conclusions about the zeros of a polynomial function. One conclusion is worth noting now. If a polynomial (with real coefficients) is of odd degree, then it must contain at least one linear factor. (Do you see why?) This means that it must have at least one real zero.

Corollary A polynomial function (with real coefficients) of odd degree has at least one real zero.

■

One challenge in using a graphing utility is to set the viewing window so that a complete graph is obtained. The next theorem is a tool that can be used to find bounds on the zeros. This will assure that the function does not have any zeros outside these bounds. Then using these bounds to set Xmin and Xmax assures that all the x-intercepts appear in the viewing window.

Bounds on Zeros

⑤ The search for the real zeros of a polynomial function can be reduced somewhat if *bounds* on the zeros are found. A number M is a **bound** on the zeros of a polynomial if every zero r lies between $-M$ and M, inclusive. That is, M is a bound to the zeros of a polynomial f if

$$-M \leq \text{any zero of } f \leq M$$

Theorem

Bounds on Zeros

Let f denote a polynomial function whose leading coefficient is 1.

$$f(x) = x^n + a_{n-1}x^{n-1} + \cdots + a_1x + a_0$$

A bound M on the zeros of f is the smaller of the two numbers

$$\text{Max}\{1, |a_0| + |a_1| + \cdots + |a_{n-1}|\}, \quad 1 + \text{Max}\{|a_0|, |a_1|, \ldots, |a_{n-1}|\} \qquad (4)$$

where Max{ } means "choose the largest entry in { }."

An example will help to make the theorem clear.

EXAMPLE 7   **Using the Theorem for Finding Bounds on Zeros**

Find a bound to the zeros of each polynomial.

(a) $f(x) = x^5 + 3x^3 - 9x^2 + 5$ (b) $g(x) = 4x^5 - 2x^3 + 2x^2 + 1$

Solution   (a) The leading coefficient of f is 1.

$$f(x) = x^5 + 3x^3 - 9x^2 + 5 \qquad a_4 = 0, a_3 = 3, a_2 = -9, a_1 = 0, a_0 = 5$$

We evaluate the expressions in equation (4).

$$\text{Max}\{1, |a_0| + |a_1| + \cdots + |a_{n-1}|\} = \text{Max}\{1, |5| + |0| + |-9| + |3| + |0|\}$$
$$= \text{Max}\{1, 17\} = 17$$
$$1 + \text{Max}\{|a_0|, |a_1|, \ldots, |a_{n-1}|\} = 1 + \text{Max}\{|5|, |0|, |-9|, |3|, |0|\}$$
$$= 1 + 9 = 10$$

The smaller of the two numbers, 10, is the bound. Every zero of f lies between -10 and 10.

(b) First we write g so that its leading coefficient is 1.

$$g(x) = 4x^5 - 2x^3 + 2x^2 + 1 = 4\left(x^5 - \frac{1}{2}x^3 + \frac{1}{2}x^2 + \frac{1}{4}\right)$$

Next we evaluate the two expressions in equation (4) with $a_4 = 0$, $a_3 = -\frac{1}{2}$, $a_2 = \frac{1}{2}$, $a_1 = 0$, and $a_0 = \frac{1}{4}$.

$$\text{Max}\{1, |a_0| + |a_1| + \cdots + |a_{n-1}|\} = \text{Max}\left\{1, \left|\frac{1}{4}\right| + |0| + \left|\frac{1}{2}\right| + \left|-\frac{1}{2}\right| + |0|\right\}$$
$$= \text{Max}\left\{1, \frac{5}{4}\right\} = \frac{5}{4}$$
$$1 + \text{Max}\{|a_0|, |a_1|, \ldots, |a_{n-1}|\} = 1 + \text{Max}\left\{\left|\frac{1}{4}\right|, |0|, \left|\frac{1}{2}\right|, \left|-\frac{1}{2}\right|, |0|\right\}$$
$$= 1 + \frac{1}{2} = \frac{3}{2}$$

The smaller of the two numbers, $\frac{5}{4}$, is the bound. Every zero of g lies between $-\frac{5}{4}$ and $\frac{5}{4}$.

| EXAMPLE 8 | **Obtaining Graphs Using Bounds on Zeros** |

Obtain a graph for each polynomial.

(a) $f(x) = x^5 + 3x^3 - 9x^2 + 5$ (b) $g(x) = 4x^5 - 2x^3 + 2x^2 + 1$

Solution   (a) Based on Example 7, every zero lies between -10 and 10. Using Xmin $= -10$ and Xmax $= 10$, we graph $Y_1 = f(x) = x^5 + 3x^3 - 9x^2 + 5$. Figure 5(a) shows the graph obtained using ZOOM-FIT. Figure 5(b) shows the graph after adjusting the viewing window to improve the graph.

Figure 5

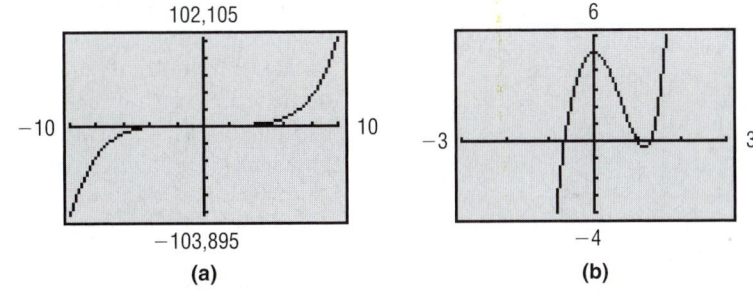

(a)   (b)

Figure 6

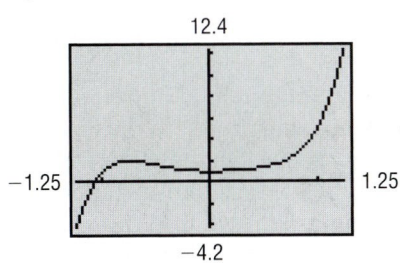

(b) Based on Example 7, every zero lies between $-\dfrac{5}{4}$ and $\dfrac{5}{4}$. Using Xmin $= -\dfrac{5}{4}$ and Xmax $= \dfrac{5}{4}$, we graph $Y_1 = g(x) = 4x^5 - 2x^3 + 2x^2 + 1$. Figure 6 shows the graph after using ZOOM-FIT. Here no adjustment of the viewing window is needed. ∎

━━ **NOW WORK PROBLEM 23.**

The next example shows how to proceed when some of the coefficients of the polynomial are not integers.

| EXAMPLE 9 | **Finding the Zeros of a Polynomial** |

Find all the real zeros of the polynomial function

$$f(x) = x^5 - 1.8x^4 - 17.79x^3 + 31.672x^2 + 37.95x - 8.7121$$

Solution   **STEP 1:** There are at most five real zeros.
STEP 2: Since there are noninteger coefficients, the Rational Zeros Theorem does not apply.
STEP 3: We determine the bounds of f. The leading coefficient of f is 1 with $a_4 = -1.8$, $a_3 = -17.79$, $a_2 = 31.672$, $a_1 = 37.95$, and $a_0 = -8.7121$. We evaluate the expressions using formula (4).

$$\text{Max}\{1, |-8.7121| + |37.95| + |31.672| + |-17.79| + |-1.8|\} = \text{Max}\{1, 97.9241\}$$
$$= 97.9241$$

$$1 + \text{Max}\{|-8.7121|, |37.95|, |31.672|, |-17.79|, |-1.8|\} = 1 + 37.95$$
$$= 38.95$$

Figure 7

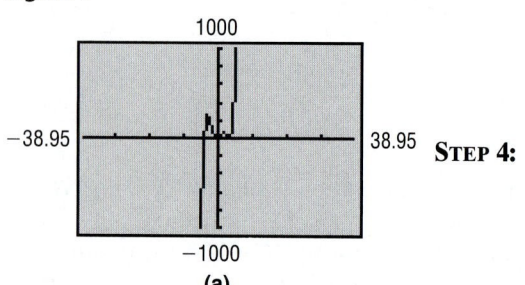

(a)

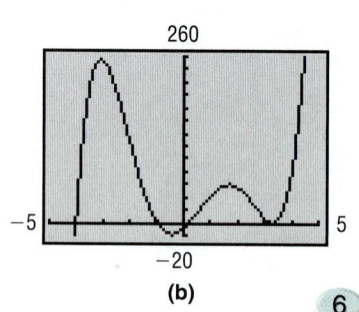

(b)

The smaller of the two numbers, 38.95, is the bound. Every real zero of f lies between -38.95 and 38.95. Figure 7(a) shows the graph of f with $X\text{min} = -38.95$ and $X\text{max} = 38.95$. Figure 7(b) shows a graph of f after adjusting the viewing window to improve the graph.

STEP 4: From Figure 7(b), we see that f appears to have four x-intercepts: one near -4, one near -1, one between 0 and 1, and one near 3. The x-intercept near 3 might be a zero of even multiplicity since the graph seems to touch the x-axis at that point.

We use the Factor Theorem to determine if -4 and -1 are zeros. Since $f(-4) = f(-1) = 0$, we know that -4 and -1 are zeros. Using ZERO (or ROOT), we find that the remaining zeros are 0.20 and 3.30, rounded to two decimal places.

There are no real zeros on the graph that have not already been identified. So either 3.30 is a zero of multiplicity 2 or there are two distinct zeros, each of which is 3.30, rounded to two decimal places. (Example 10 provides the answer.) ■

6 Intermediate Value Theorem

The Intermediate Value Theorem requires that the function be *continuous*. Although it requires calculus to explain the meaning precisely, the *idea* of a continuous function is easy to understand. Very basically, a function f is continuous when its graph can be drawn without lifting pencil from paper, that is, when the graph contains no "holes" or "jumps" or "gaps." For example, polynomial functions are continuous.

Intermediate Value Theorem

> Let f denote a continuous function. If $a < b$ and if $f(a)$ and $f(b)$ are of opposite sign, then f has at least one zero between a and b.

■

Although the proof of this result requires advanced methods in calculus, it is easy to "see" why the result is true. Look at Figure 8.

Figure 8
If $f(a) < 0$ and $f(b) > 0$ and if f is continuous, there is at least one zero between a and b.

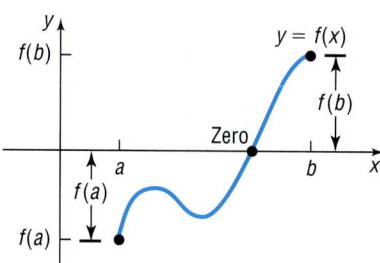

The Intermediate Value Theorem together with the TABLE feature of a graphing utility provides a basis for finding zeros.

EXAMPLE 10 **Using the Intermediate Value Theorem and a Graphing Utility to Locate Zeros**

Continue working with Example 9 to determine whether there is a repeated zero or two distinct zeros near 3.30.

Solution We use the TABLE feature of a graphing utility. See Table 1. Since $f(3.29) = 0.00956 > 0$ and $f(3.30) = -0.0001 < 0$, by the Intermediate Value Theorem there is a zero between 3.29 and 3.30. Similarly, since $f(3.3) = -0.0001 < 0$ and $f(3.31) = 0.0097 > 0$, there is another zero between 3.30 and 3.31. Now we know that the five zeros of f are distinct. ∎

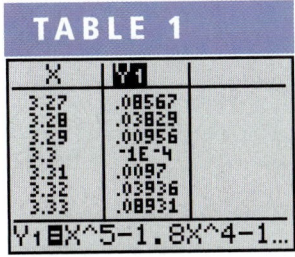

TABLE 1

X	Y1
3.27	.08567
3.28	.03829
3.29	.00956
3.3	-1E-4
3.31	.0097
3.32	.03936
3.33	.08931

Y1◻X^5-1.8X^4-1...

HISTORICAL FEATURE

Tartaglia (1500–1557)

Formulas for the solution of third- and fourth-degree polynomial equations exist, and, although not very practical, they do have an interesting history.

In the 1500s in Italy, mathematical contests were a popular pastime, and persons possessing methods for solving problems kept them secret. (Solutions that were published were already common knowledge.) Niccolo of Brescia (1500–1557), commonly referred to as Tartaglia ("the stammerer"), had the secret for solving cubic (third-degree) equations, which gave him a decided advantage in the contests. Girolamo Cardano (1501–1576) found out that Tartaglia had the secret and, being interested in cubics, he requested it from Tartaglia. The reluctant Tartaglia hesitated for some time, but finally, swearing Cardano to secrecy with midnight oaths by candlelight, told him the secret. Cardano then published the solution in his book *Ars Magna* (1545), giving Tartaglia the credit, but rather compromising the secrecy. Tartaglia exploded into bitter recriminations, and each wrote pamphlets that reflected on the other's mathematics, moral character, and ancestry. See the Historical Problems.

The quartic (fourth-degree) equation was solved by Cardano's student Lodovico Ferrari, and this solution also was included, with credit and this time with permission, in the *Ars Magna*.

Attempts were made to solve the fifth-degree equation in similar ways, all of which failed. In the early 1800s, P. Ruffini, Niels Abel, and Evariste Galois all found ways to show that it is not possible to solve fifth-degree equations by formula, but the proofs required the introduction of new methods. Galois' methods eventually developed into a large part of modern algebra.

HISTORICAL PROBLEMS

In Problems 1–8, develop the Tartaglia-Cardano solution of the cubic equation and show why it is not altogether practical.

1. Show that the general cubic equation $y^3 + by^2 + cy + d = 0$ can be transformed into an equation of the form $x^3 + px + q = 0$ by using the substitution $y = x - \dfrac{b}{3}$.

2. In the equation $x^3 + px + q = 0$, replace x by $H + K$. Let $3HK = -p$, and show that $H^3 + K^3 = -q$.
   [**Hint:** $3H^2K + 3HK^2 = 3HKx$.]

3. Based on Problem 2, we have the two equations
   $$3HK = -p \quad \text{and} \quad H^3 + K^3 = -q$$
   Solve for K in $3HK = -p$ and substitute into $H^3 + K^3 = -q$. Then show that
   $$H = \sqrt[3]{\dfrac{-q}{2} + \sqrt{\dfrac{q^2}{4} + \dfrac{p^3}{27}}}$$
   [**Hint:** Look for an equation that is quadratic in form.]

4. Use the solution for H from Problem 3 and the equation $H^3 + K^3 = -q$ to show that
   $$K = \sqrt[3]{\dfrac{-q}{2} - \sqrt{\dfrac{q^2}{4} + \dfrac{p^3}{27}}}$$

5. Use the results from Problems 2–4 to show that the solution of $x^3 + px + q = 0$ is
   $$x = \sqrt[3]{\dfrac{-q}{2} + \sqrt{\dfrac{q^2}{4} + \dfrac{p^3}{27}}} + \sqrt[3]{\dfrac{-q}{2} - \sqrt{\dfrac{q^2}{4} + \dfrac{p^3}{27}}}$$

6. Use the result of Problem 5 to solve the equation $x^3 - 6x - 9 = 0$.

7. Use a calculator and the result of Problem 5 to solve the equation $x^3 + 3x - 14 = 0$.

8. Use the methods of this chapter to solve the equation $x^3 + 3x - 14 = 0$.

5.2 Concepts and Vocabulary

In Problems 1–3, fill in the blanks.

1. In the process of polynomial division, $(\text{Divisor})(\text{Quotient}) + \underline{\hspace{1.5cm}} = \underline{\hspace{1.5cm}}$.

2. When a polynomial function f is divided by $x - c$, the remainder is $\underline{\hspace{1.5cm}}$.

3. If a function f whose domain is all real numbers is even and 4 is a zero of f, then $\underline{\hspace{1.5cm}}$ is also a zero.

In Problems 4–6, answer True or False to each statement.

4. Every polynomial function of degree 3 with real coefficients has exactly three real zeros.

5. The only potential rational zeros of $f(x) = 2x^5 - x^3 + x^2 - x + 1$ are $\pm 1, \pm 2$.

6. If f is a polynomial function of degree 4 and if $f(2) = 5$, then

$$\frac{f(x)}{x - 2} = p(x) + \frac{5}{x - 2}$$

where $p(x)$ is a polynomial of degree 3.

7. Is $\frac{1}{3}$ a zero of $f(x) = 2x^3 + 3x^2 - 6x + 7$? Explain.

8. A polynomial function of degree n can have at most n real zeros. In your own words, explain why this result is reasonable.

5.2 Exercises

In Problems 1–10, use the Factor Theorem to determine whether $x - c$ is a factor of f. If it is, write f in factored form, that is, write f in the form $f(x) = (x - c)(quotient)$.

1. $f(x) = 4x^3 - 3x^2 - 8x + 4$; $c = 3$

2. $f(x) = -4x^3 + 5x^2 + 8$; $c = -2$

3. $f(x) = 3x^4 - 6x^3 - 5x + 10$; $c = 1$

4. $f(x) = 4x^4 - 15x^2 - 4$; $c = 2$

5. $f(x) = 3x^6 + 2x^3 - 176$; $c = -2$

6. $f(x) = 2x^6 - 18x^4 + x^2 - 9$; $c = -3$

7. $f(x) = 4x^6 - 64x^4 + x^2 - 16$; $c = 4$

8. $f(x) = x^6 - 16x^4 + x^2 - 16$; $c = -4$

9. $f(x) = 2x^4 - x^3 + 2x - 1$; $c = -\dfrac{1}{2}$

10. $f(x) = 3x^4 + x^3 - 3x + 1$; $c = -\dfrac{1}{3}$

In Problems 11–22, tell the maximum number of real zeros that each polynomial function may have. Then list the potential rational zeros of each polynomial function. Do not attempt to find the zeros.

11. $f(x) = 3x^4 - 3x^3 + x^2 - x + 1$

12. $f(x) = x^5 - x^4 + 2x^2 + 3$

13. $f(x) = x^5 - 6x^2 + 9x - 3$

14. $f(x) = 2x^5 - x^4 - x^2 + 1$

15. $f(x) = -4x^3 - x^2 + x + 2$

16. $f(x) = 6x^4 - x^2 + 2$

17. $f(x) = 3x^4 - x^2 + 2$

18. $f(x) = -4x^3 + x^2 + x + 2$

19. $f(x) = 2x^5 - x^3 + 2x^2 + 4$

20. $f(x) = 3x^5 - x^2 + 2x + 3$

21. $f(x) = 6x^4 + 2x^3 - x^2 + 2$

22. $f(x) = -6x^3 - x^2 + x + 3$

In Problems 23–28, find the bounds to the zeros of each polynomial function. Use the bounds to obtain a complete graph of f.

23. $f(x) = 2x^3 + x^2 - 1$

24. $f(x) = 3x^3 - 2x^2 + x + 4$

25. $f(x) = x^3 - 5x^2 - 11x + 11$

26. $f(x) = 2x^3 - x^2 - 11x - 6$

27. $f(x) = x^4 + 3x^3 - 5x^2 + 9$

28. $f(x) = 4x^4 - 12x^3 + 27x^2 - 54x + 81$

In Problems 29–46, find the real zeros of f. Use the real zeros to factor f.

29. $f(x) = x^3 + 2x^2 - 5x - 6$

30. $f(x) = x^3 + 8x^2 + 11x - 20$

31. $f(x) = 2x^3 - 13x^2 + 24x - 9$

32. $f(x) = 2x^3 - 5x^2 - 4x + 12$

33. $f(x) = 3x^3 + 4x^2 + 4x + 1$

34. $f(x) = 3x^3 - 7x^2 + 12x - 28$

35. $f(x) = x^3 - 8x^2 + 17x - 6$

36. $f(x) = x^3 + 6x^2 + 6x - 4$

37. $f(x) = x^4 + x^3 - 3x^2 - x + 2$

38. $f(x) = x^4 - x^3 - 6x^2 + 4x + 8$

39. $f(x) = 2x^4 + 17x^3 + 35x^2 - 9x - 45$

40. $f(x) = 4x^4 - 15x^3 - 8x^2 + 15x + 4$

41. $f(x) = 2x^4 - 3x^3 - 21x^2 - 2x + 24$

42. $f(x) = 2x^4 + 11x^3 - 5x^2 - 43x + 35$

43. $f(x) = 4x^4 + 7x^2 - 2$

44. $f(x) = 4x^4 + 15x^2 - 4$

45. $f(x) = 4x^5 - 8x^4 - x + 2$

46. $f(x) = 4x^5 + 12x^4 - x - 3$

In Problems 47–52, find the real zeros of f rounded to two decimal places.

47. $f(x) = x^3 + 3.2x^2 - 16.83x - 5.31$

48. $f(x) = x^3 + 3.2x^2 - 7.25x - 6.3$

49. $f(x) = x^4 - 1.4x^3 - 33.71x^2 + 23.94x + 292.41$

50. $f(x) = x^4 + 1.2x^3 - 7.46x^2 - 4.692x + 15.2881$

51. $f(x) = x^3 + 19.5x^2 - 1021x + 1000.5$

52. $f(x) = x^3 + 42.2x^2 - 664.8x + 1490.4$

In Problems 53–62, find the real solutions of each equation.

53. $x^4 - x^3 + 2x^2 - 4x - 8 = 0$

54. $2x^3 + 3x^2 + 2x + 3 = 0$

55. $3x^3 + 4x^2 - 7x + 2 = 0$

56. $2x^3 - 3x^2 - 3x - 5 = 0$

57. $3x^3 - x^2 - 15x + 5 = 0$

58. $2x^3 - 11x^2 + 10x + 8 = 0$

59. $x^4 + 4x^3 + 2x^2 - x + 6 = 0$

60. $x^4 - 2x^3 + 10x^2 - 18x + 9 = 0$

61. $x^3 - \dfrac{2}{3}x^2 + \dfrac{8}{3}x + 1 = 0$

62. $x^3 - \dfrac{2}{3}x^2 + 3x - 2 = 0$

In Problems 63–68, use the Intermediate Value Theorem to show that each function has a zero in the given interval. Approximate the zero rounded to two decimal places.

63. $f(x) = 8x^4 - 2x^2 + 5x - 1;$ $[0, 1]$

64. $f(x) = x^4 + 8x^3 - x^2 + 2;$ $[-1, 0]$

65. $f(x) = 2x^3 + 6x^2 - 8x + 2;$ $[-5, -4]$

66. $f(x) = 3x^3 - 10x + 9;$ $[-3, -2]$

67. $f(x) = x^5 - x^4 + 7x^3 - 7x^2 - 18x + 18;$ $[1.4, 1.5]$

68. $f(x) = x^5 - 3x^4 - 2x^3 + 6x^2 + x + 2;$ $[1.7, 1.8]$

69. Cost of Manufacturing   In Problem 69 of Section 4.2, you found the cost function for manufacturing Chevy Cavaliers. Use the methods learned in this section to determine how many Cavaliers can be manufactured at a total cost of $3,000,000. In other words, solve the equation $C(x) = 3000$.

70. Cost of Printing   In Problem 70 of Section 4.2, you found the cost function for printing textbooks. Use the methods learned in this section to determine how many textbooks can be printed at a total cost of $200,000. In other words, solve the equation $C(x) = 200$.

71. Find k such that $f(x) = x^3 - kx^2 + kx + 2$ has the factor $x - 2$.

72. Find k such that $f(x) = x^4 - kx^3 + kx^2 + 1$ has the factor $x + 2$.

73. What is the remainder when

$$f(x) = 2x^{20} - 8x^{10} + x - 2$$

is divided by $x - 1$?

74. What is the remainder when

$$f(x) = -3x^{17} + x^9 - x^5 + 2x$$

is divided by $x + 1$?

75. Is $\dfrac{3}{5}$ a zero of $f(x) = 2x^6 - 5x^4 + x^3 - x + 1$? Explain.

76. Is $\dfrac{2}{3}$ a zero of $f(x) = x^7 + 6x^5 - x^4 + x + 2$? Explain.

77. What is the length of the edge of a cube if, after a slice 1 inch thick is cut from one side, the volume remaining is 294 cubic inches?

78. What is the length of the edge of a cube if its volume could be doubled by an increase of 6 centimeters in one edge, an increase of 12 centimeters in a second edge, and a decrease of 4 centimeters in the third edge?

PREPARING FOR THIS SECTION

Before getting started, review the following:

✓ Quadratic Formula (Section 1.3, pp. 121–124) ✓ Rationalizing Square Roots (Section R.8, pp. 74–75).

5.3 COMPLEX NUMBERS; QUADRATIC EQUATIONS WITH A NEGATIVE DISCRIMINANT

OBJECTIVES **1** Add, Subtract, Multiply, and Divide Complex Numbers

2 Solve Quadratic Equations with a Negative Discriminant

One property of a real number is that its square is nonnegative. For example, there is no real number x for which

$$x^2 = -1$$

To remedy this situation, we introduce a number called the **imaginary unit**, which we denote by i and whose square is -1; that is,

$$i^2 = -1$$

This should not surprise you. If our universe were to consist only of integers, there would be no number x for which $2x = 1$. This unfortunate circumstance was remedied by introducing numbers such as $\frac{1}{2}$ and $\frac{2}{3}$, the *rational numbers*. If our universe were to consist only of rational numbers, there would be no x whose square equals 2. That is, there would be no number x for which $x^2 = 2$. To remedy this, we introduced numbers such as $\sqrt{2}$ and $\sqrt[3]{5}$, the *irrational numbers*. The *real numbers*, you will recall, consist of the rational numbers and the irrational numbers. Now, if our universe were to consist only of real numbers, then there would be no number x whose square is -1. To remedy this, we introduce a number i, whose square is -1.

In the progression outlined, each time that we encountered a situation that was unsuitable, we introduced a new number system to remedy this situation. And each new number system contained the earlier number system as a subset. The number system that results from introducing the number i is called the **complex number system**.

Complex numbers are numbers of the form $a + bi$, where a and b are real numbers. The real number a is called the **real part** of the number $a + bi$; the real number b is called the **imaginary part** of $a + bi$.

For example, the complex number $-5 + 6i$ has the real part -5 and the imaginary part 6.

When a complex number is written in the form $a + bi$, where a and b are real numbers, we say it is in **standard form**. However, if the imaginary part of a complex number is negative, such as in the complex number $3 + (-2)i$, we agree to write it instead in the form $3 - 2i$.

Also, the complex number $a + 0i$ is usually written merely as a. This serves to remind us that the real numbers are a subset of the complex numbers. The complex number $0 + bi$ is usually written as bi. Sometimes the complex number bi is called a **pure imaginary number**.

Equality, addition, subtraction, and multiplication of complex numbers are defined so as to preserve the familiar rules of algebra for real numbers. Thus, two complex numbers are equal if and only if their real parts are equal and their imaginary parts are equal. That is,

Equality of Complex Numbers

$$a + bi = c + di \quad \text{if and only if } a = c \text{ and } b = d \qquad (1)$$

Two complex numbers are added by forming the complex number whose real part is the sum of the real parts and whose imaginary part is the sum of the imaginary parts. That is,

Sum of Complex Numbers

$$(a + bi) + (c + di) = (a + c) + (b + d)i \qquad (2)$$

To subtract two complex numbers, we use this rule:

Difference of Complex Numbers

$$(a + bi) - (c + di) = (a - c) + (b - d)i \qquad (3)$$

EXAMPLE 1   **Adding and Subtracting Complex Numbers**

Figure 9

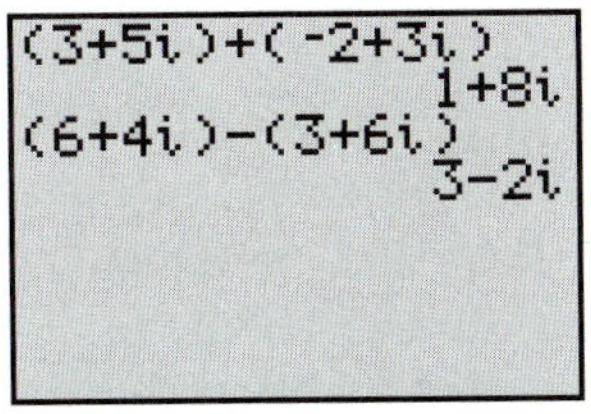

```
(3+5i)+(-2+3i)
              1+8i
(6+4i)-(3+6i)
              3-2i
```

(a) $(3 + 5i) + (-2 + 3i) = [3 + (-2)] + (5 + 3)i = 1 + 8i$

(b) $(6 + 4i) - (3 + 6i) = (6 - 3) + (4 - 6)i = 3 + (-2)i = 3 - 2i$ ◼

Some graphing calculators have the capability of handling complex numbers.* For example, Figure 9 shows the results of Example 1 using a TI-83 graphing calculator.

✏ **N O W W O R K P R O B L E M 5 .**

Products of complex numbers are calculated as illustrated in Example 2.

EXAMPLE 2   **Multiplying Complex Numbers**

$$
\begin{aligned}
(5 + 3i) \cdot (2 + 7i) &= 5 \cdot (2 + 7i) + 3i(2 + 7i) && \text{Distributive property} \\
&= 10 + 35i + 6i + 21i^2 && \text{Distributive property} \\
&= 10 + 41i + 21(-1) && i^2 = -1 \\
&= -11 + 41i
\end{aligned}
$$

◼

*Consult your user's manual for the appropriate keystrokes.

Based on the procedure of Example 2, we define the **product** of two complex numbers by the following formula:

Product of Complex Numbers

$$(a + bi) \cdot (c + di) = (ac - bd) + (ad + bc)i \qquad (4)$$

Do not bother to memorize formula (4). Instead, whenever it is necessary to multiply two complex numbers, follow the usual rules for multiplying two binomials, as in Example 2, remembering that $i^2 = -1$. For example,

$$(2i)(2i) = 4i^2 = -4$$
$$(2 + i)(1 - i) = 2 - 2i + i - i^2 = 3 - i$$

Figure 10

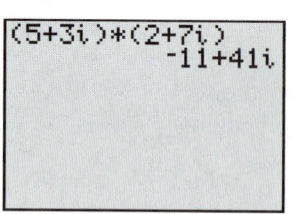

Graphing calculators may also be used to multiply complex numbers. Figure 10 shows the result obtained in Example 2 using a TI-83 graphing calculator.

NOW WORK PROBLEM 11.

Algebraic properties for addition and multiplication, such as the commutative, associative, and distributive properties, hold for complex numbers. However, the property that every nonzero complex number has a multiplicative inverse, or reciprocal, requires a closer look.

Conjugates

If $z = a + bi$ is a complex number, then its **conjugate**, denoted by $\bar{z}$, is defined as

---

$$\bar{z} = \overline{a + bi} = a - bi$$

For example, $\overline{2 + 3i} = 2 - 3i$ and $\overline{-6 - 2i} = -6 + 2i$.

EXAMPLE 3 | **Multiplying a Complex Number by Its Conjugate**

Find the product of the complex number $z = 3 + 4i$ and its conjugate $\bar{z}$.

Solution   Since $\bar{z} = 3 - 4i$, we have

$$z\bar{z} = (3 + 4i)(3 - 4i) = 9 + 12i - 12i - 16i^2 = 9 + 16 = 25 \quad \blacksquare$$

The result obtained in Example 3 has an important generalization.

Theorem   The product of a complex number and its conjugate is a nonnegative real number. That is, if $z = a + bi$, then

---

$$z\bar{z} = a^2 + b^2 \qquad (5)$$

Proof   If $z = a + bi$, then

$$z\bar{z} = (a + bi)(a - bi) = a^2 - (bi)^2 = a^2 - b^2 i^2 = a^2 + b^2 \quad \blacksquare$$

To express the reciprocal of a nonzero complex number z in standard form, multiply the numerator and denominator of $\dfrac{1}{z}$ by its conjugate $\bar{z}$. That is, if $z = a + bi$ is a nonzero complex number, then

$$\frac{1}{a + bi} = \frac{1}{z} = \frac{1}{z} \cdot \frac{\bar{z}}{\bar{z}} = \frac{\bar{z}}{z\bar{z}} \underset{\underset{\text{Use (5).}}{\uparrow}}{=} \frac{a - bi}{a^2 + b^2} = \frac{a}{a^2 + b^2} - \frac{b}{a^2 + b^2} i$$

EXAMPLE 4   **Writing the Reciprocal of a Complex Number in Standard Form**

Write $\dfrac{1}{3 + 4i}$ in standard form $a + bi$; that is, find the reciprocal of $3 + 4i$.

Solution   The idea is to multiply the numerator and denominator by the conjugate of $3 + 4i$, that is, the complex number $3 - 4i$. The result is

$$\frac{1}{3 + 4i} = \frac{1}{3 + 4i} \cdot \frac{3 - 4i}{3 - 4i} = \frac{3 - 4i}{9 + 16} = \frac{3}{25} - \frac{4}{25} i \quad \blacksquare$$

Figure 11

A graphing calculator can be used to verify the result of Example 4. See Figure 11.

To express the quotient of two complex numbers in standard form, we multiply the numerator and denominator of the quotient by the conjugate of the denominator.

EXAMPLE 5   **Writing the Quotient of Complex Numbers in Standard Form**

Write each of the following in standard form.

(a) $\dfrac{1 + 4i}{5 - 12i}$ (b) $\dfrac{2 - 3i}{4 - 3i}$

Solution   (a) $\dfrac{1 + 4i}{5 - 12i} = \dfrac{1 + 4i}{5 - 12i} \cdot \dfrac{5 + 12i}{5 + 12i} = \dfrac{5 + 20i + 12i + 48i^2}{25 + 144}$

$\qquad\qquad = \dfrac{-43 + 32i}{169} = \dfrac{-43}{169} + \dfrac{32}{169} i$

(b) $\dfrac{2 - 3i}{4 - 3i} = \dfrac{2 - 3i}{4 - 3i} \cdot \dfrac{4 + 3i}{4 + 3i} = \dfrac{8 - 12i + 6i - 9i^2}{16 + 9}$

$\qquad\qquad = \dfrac{17 - 6i}{25} = \dfrac{17}{25} - \dfrac{6}{25} i \quad \blacksquare$

NOW WORK PROBLEM **19**.

EXAMPLE 6   **Writing Other Expressions in Standard Form**

If $z = 2 - 3i$ and $w = 5 + 2i$, write each of the following expressions in standard form.

(a) $\dfrac{z}{w}$ (b) $\overline{z + w}$ (c) $z + \bar{z}$

Solution   (a) $\dfrac{z}{w} = \dfrac{z \cdot \bar{w}}{w \cdot \bar{w}} = \dfrac{(2 - 3i)(5 - 2i)}{(5 + 2i)(5 - 2i)} = \dfrac{10 - 15i - 4i + 6i^2}{25 + 4}$

$= \dfrac{4 - 19i}{29} = \dfrac{4}{29} - \dfrac{19}{29}i$

(b) $\overline{z + w} = \overline{(2 - 3i) + (5 + 2i)} = \overline{7 - i} = 7 + i$

(c) $z + \bar{z} = (2 - 3i) + (2 + 3i) = 4$

The conjugate of a complex number has certain general properties that we shall find useful later.

For a real number $a = a + 0i$, the conjugate is $\bar{a} = \overline{a + 0i} = a - 0i = a$. That is,

Theorem   The conjugate of a real number is the real number itself.

Other properties that are direct consequences of the definition of the conjugate are given next. In each statement, z and w represent complex numbers.

Theorem   The conjugate of the conjugate of a complex number is the complex number itself.

$$\left(\overline{\bar{z}} \right) = z \tag{6}$$

The conjugate of the sum of two complex numbers equals the sum of their conjugates.

$$\overline{z + w} = \bar{z} + \bar{w} \tag{7}$$

The conjugate of the product of two complex numbers equals the product of their conjugates.

$$\overline{z \cdot w} = \bar{z} \cdot \bar{w} \tag{8}$$

We leave the proofs of equations (6), (7), and (8) as exercises.

Powers of *i*

The **powers of *i*** follow a pattern that is useful to know.

$i^1 = i$ $i^5 = i^4 \cdot i = 1 \cdot i = i$

$i^2 = -1$ $i^6 = i^4 \cdot i^2 = -1$

$i^3 = i^2 \cdot i = -i$ $i^7 = i^4 \cdot i^3 = -i$

$i^4 = i^2 \cdot i^2 = (-1)(-1) = 1$ $i^8 = i^4 \cdot i^4 = 1$

And so on. The powers of *i* repeat with every fourth power.

EXAMPLE 7 **Evaluating Powers of i**

(a) $i^{27} = i^{24} \cdot i^3 = (i^4)^6 \cdot i^3 = 1^6 \cdot i^3 = -i$

(b) $i^{101} = i^{100} \cdot i^1 = (i^4)^{25} \cdot i = 1^{25} \cdot i = i$ ∎

EXAMPLE 8 **Writing the Power of a Complex Number in Standard Form**

Write $(2 + i)^3$ in standard form.

Solution We use the special product formula for $(x + a)^3$.

$$(x + a)^3 = x^3 + 3ax^2 + 3a^2x + a^3$$

Using this special product formula,

$$(2 + i)^3 = 2^3 + 3 \cdot i \cdot 2^2 + 3 \cdot i^2 \cdot 2 + i^3$$
$$= 8 + 12i + 6(-1) + (-i)$$
$$= 2 + 11i$$ ∎

✏️ **NOW WORK PROBLEMS 25 AND 33.**

Quadratic Equations with a Negative Discriminant

② Quadratic equations with a negative discriminant have no real number solution. However, if we extend our number system to allow complex numbers, quadratic equations will always have a solution. Since the solution to a quadratic equation involves the square root of the discriminant, we begin with a discussion of square roots of negative numbers.

If N is a positive real number, we define the **principal square root of** $-N$, denoted by $\sqrt{-N}$, as

$$\sqrt{-N} = \sqrt{N}i$$

where i is the imaginary unit and $i^2 = -1$.

EXAMPLE 9 **Evaluating the Square Root of a Negative Number**

(a) $\sqrt{-1} = \sqrt{1}i = i$ (b) $\sqrt{-4} = \sqrt{4}i = 2i$

(c) $\sqrt{-8} = \sqrt{8}i = 2\sqrt{2}i$ ∎

EXAMPLE 10 **Solving Equations**

Solve each equation in the complex number system.

(a) $x^2 = 4$ (b) $x^2 = -9$

Solution (a) $x^2 = 4$

$$x = \pm\sqrt{4} = \pm 2$$

The equation has two solutions, -2 and 2.

(b) $x^2 = -9$

$$x = \pm\sqrt{-9} = \pm\sqrt{9}i = \pm 3i$$

The equation has two solutions, $-3i$ and $3i$.   ■

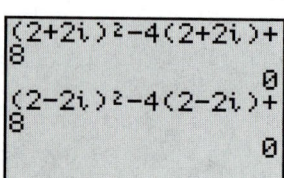 NOW WORK PROBLEM 45.

WARNING: When working with square roots of negative numbers, do not set the square root of a product equal to the product of the square roots (which can be done with positive numbers). To see why, look at this calculation: We know that $\sqrt{100} = 10$. However, it is also true that $100 = (-25)(-4)$, so

$$10 = \sqrt{100} = \sqrt{(-25)(-4)} \neq \sqrt{-25}\sqrt{-4} = (\sqrt{25}i)(\sqrt{4}i) = (5i)(2i) = 10i^2 = -10$$

↑
Here is the error.   ■

Because we have defined the square root of a negative number, we can now restate the quadratic formula without restriction.

Theorem

> In the complex number system, the solutions of the quadratic equation $ax^2 + bx + c = 0$, where a, b, and c are real numbers and $a \neq 0$, are given by the formula
>
> $$x = \frac{-b \pm \sqrt{b^2 - 4ac}}{2a} \qquad (9)$$

■

EXAMPLE 11   **Solving Quadratic Equations in the Complex Number System**

Solve the equation $x^2 - 4x + 8 = 0$ in the complex number system.

Solution   Here $a = 1, b = -4, c = 8$, and $b^2 - 4ac = 16 - 4(1)(8) = -16$. Using equation (9), we find that

$$x = \frac{-(-4) \pm \sqrt{-16}}{2(1)} = \frac{4 \pm \sqrt{16}i}{2} = \frac{4 \pm 4i}{2} = 2 \pm 2i$$

The equation has the solution set $\{2 - 2i, 2 + 2i\}$.

✔ CHECK:

$$2 + 2i: \quad (2 + 2i)^2 - 4(2 + 2i) + 8 = 4 + 8i + 4i^2 - 8 - 8i + 8$$
$$= 4 - 4 = 0$$
$$2 - 2i: \quad (2 - 2i)^2 - 4(2 - 2i) + 8 = 4 - 8i + 4i^2 - 8 + 8i + 8$$
$$= 4 - 4 = 0$$

■ ■

Figure 12

```
(2+2i)²-4(2+2i)+
8
              0
(2-2i)²-4(2-2i)+
8
              0
```

Figure 12 shows the check of the solution using a TI-83 graphing calculator.

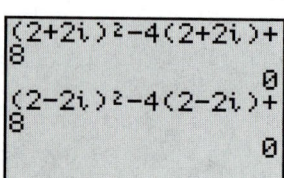 NOW WORK PROBLEM 51.

The discriminant $b^2 - 4ac$ of a quadratic equation still serves as a way to determine the character of the solutions.

Character of the Solutions of a Quadratic Equation

In the complex number system, consider a quadratic equation $ax^2 + bx + c = 0$ with real coefficients.

1. If $b^2 - 4ac > 0$, the equation has two unequal real solutions.
2. If $b^2 - 4ac = 0$, the equation has a repeated real solution, a double root.
3. If $b^2 - 4ac < 0$, the equation has two complex solutions that are not real. The solutions are conjugates of each other.

The third conclusion in the display is a consequence of the fact that if $b^2 - 4ac = -N < 0$ then, by the quadratic formula, the solutions are

$$x = \frac{-b + \sqrt{b^2 - 4ac}}{2a} = \frac{-b + \sqrt{-N}}{2a} = \frac{-b + \sqrt{N}i}{2a} = \frac{-b}{2a} + \frac{\sqrt{N}}{2a}i$$

and

$$x = \frac{-b - \sqrt{b^2 - 4ac}}{2a} = \frac{-b - \sqrt{-N}}{2a} = \frac{-b - \sqrt{N}i}{2a} = \frac{-b}{2a} - \frac{\sqrt{N}}{2a}i$$

which are conjugates of each other.

EXAMPLE 12 **Determining the Character of the Solutions of a Quadratic Equation**

Without solving, determine the character of the solutions of each equation.

(a) $3x^2 + 4x + 5 = 0$ (b) $2x^2 + 4x + 1 = 0$
(c) $9x^2 - 6x + 1 = 0$

Solution (a) Here $a = 3, b = 4$, and $c = 5$, so $b^2 - 4ac = 16 - 4(3)(5) = -44$. The solutions are complex numbers that are not real and are conjugates of each other.

(b) Here $a = 2, b = 4$, and $c = 1$, so $b^2 - 4ac = 16 - 8 = 8$. The solutions are two unequal real numbers.

(c) Here $a = 9, b = -6$, and $c = 1$, so $b^2 - 4ac = 36 - 4(9)(1) = 0$. The solution is a repeated real number, that is, a double root.   ■

═══ NOW WORK PROBLEM **65.**

5.3 Concepts and Vocabulary

In Problems 1 and 2, fill in the blanks.

1. In the complex number $5 + 2i$, the number 5 is called the _____ part; the number 2 is called the _____ part; the number i is called the _____ _____.

2. The equation $|x^2| = 4$ has four complex solutions: _____, _____, _____, and _____.

In Problems 3 and 4, answer True or False to each statement.

3. The conjugate of $2 + 5i$ is $-2 - 5i$.

4. All real numbers are complex numbers.

5.3 Exercises

In Problems 1–38, write each expression in the standard form a + bi. Verify your results using a graphing utility.

1. $(2 - 3i) + (6 + 8i)$
2. $(4 + 5i) + (-8 + 2i)$
3. $(-3 + 2i) - (4 - 4i)$
4. $(3 - 4i) - (-3 - 4i)$

5. $(2 - 5i) - (8 + 6i)$
6. $(-8 + 4i) - (2 - 2i)$
7. $3(2 - 6i)$
8. $-4(2 + 8i)$

9. $2i(2 - 3i)$
10. $3i(-3 + 4i)$
11. $(3 - 4i)(2 + i)$
12. $(5 + 3i)(2 - i)$

13. $(-6 + i)(-6 - i)$
14. $(-3 + i)(3 + i)$
15. $\dfrac{10}{3 - 4i}$
16. $\dfrac{13}{5 - 12i}$

17. $\dfrac{2 + i}{i}$
18. $\dfrac{2 - i}{-2i}$
19. $\dfrac{6 - i}{1 + i}$
20. $\dfrac{2 + 3i}{1 - i}$

21. $\left(\dfrac{1}{2} + \dfrac{\sqrt{3}}{2}i\right)^2$
22. $\left(\dfrac{\sqrt{3}}{2} - \dfrac{1}{2}i\right)^2$
23. $(1 + i)^2$
24. $(1 - i)^2$

25. i^{23}
26. i^{14}
27. i^{-15}
28. i^{-23}

29. $i^6 - 5$
30. $4 + i^3$
31. $6i^3 - 4i^5$
32. $4i^3 - 2i^2 + 1$

33. $(1 + i)^3$
34. $(3i)^4 + 1$
35. $i^7(1 + i^2)$
36. $2i^4(1 + i^2)$

37. $i^6 + i^4 + i^2 + 1$
38. $i^7 + i^5 + i^3 + i$

In Problems 39–44, perform the indicated operations and express your answer in the form a + bi.

39. $\sqrt{-4}$
40. $\sqrt{-9}$
41. $\sqrt{-25}$

42. $\sqrt{-64}$
43. $\sqrt{(3 + 4i)(4i - 3)}$
44. $\sqrt{(4 + 3i)(3i - 4)}$

In Problems 45–64, solve each equation in the complex number system. Check your results using a graphing utility.

45. $x^2 + 4 = 0$
46. $x^2 - 4 = 0$
47. $x^2 - 16 = 0$
48. $x^2 + 25 = 0$

49. $x^2 - 6x + 13 = 0$
50. $x^2 + 4x + 8 = 0$
51. $x^2 - 6x + 10 = 0$
52. $x^2 - 2x + 5 = 0$

53. $8x^2 - 4x + 1 = 0$
54. $10x^2 + 6x + 1 = 0$
55. $5x^2 + 2x + 1 = 0$
56. $13x^2 + 6x + 1 = 0$

57. $x^2 + x + 1 = 0$
58. $x^2 - x + 1 = 0$
59. $x^3 - 8 = 0$
60. $x^3 + 27 = 0$

61. $x^4 - 16 = 0$
62. $x^4 - 1 = 0$
63. $x^4 + 13x^2 + 36 = 0$
64. $x^4 + 3x^2 - 4 = 0$

In Problems 65–70, without solving, determine the character of the solutions of each equation. Verify your answer using a graphing utility.

65. $3x^2 - 3x + 4 = 0$
66. $2x^2 - 4x + 1 = 0$
67. $2x^2 + 3x - 4 = 0$

68. $x^2 + 2x + 6 = 0$
69. $9x^2 - 12x + 4 = 0$
70. $4x^2 + 12x + 9 = 0$

71. $2 + 3i$ is a solution of a quadratic equation with real coefficients. Find the other solution.

72. $4 - i$ is a solution of a quadratic equation with real coefficients. Find the other solution.

In Problems 73–76, $z = 3 - 4i$ and $w = 8 + 3i$. Write each expression in the standard form a + bi.

73. $z + \bar{z}$
74. $w - \bar{w}$
75. $z\bar{z}$
76. $\overline{z - w}$

---

77. Use $z = a + bi$ to show that $z + \bar{z} = 2a$ and that $z - \bar{z} = 2bi$.

78. Use $z = a + bi$ to show that $\left(\bar{\bar{z}}\right) = z$.

79. Use $z = a + bi$ and $w = c + di$ to show that $\overline{z + w} = \bar{z} + \bar{w}$.

80. Use $z = a + bi$ and $w = c + di$ to show that $\overline{z \cdot w} = \bar{z} \cdot \bar{w}$.

81. Explain to a friend how you would add two complex numbers and how you would multiply two complex numbers. Explain any differences in the two explanations.

82. Write a brief paragraph that compares the method used to rationalize denominators and the method used to write the quotient of two complex numbers in standard form.

83. Use an Internet search engine to investigate the origins of complex numbers. Write a paragraph describing what you find and present it to the class.

5.4 COMPLEX ZEROS; FUNDAMENTAL THEOREM OF ALGEBRA

OBJECTIVES
1. Utilize the Conjugate Pairs Theorem to Find the Complex Zeros of a Polynomial
2. Find a Polynomial Function with Specified Zeros
3. Find the Complex Zeros of a Polynomial Function

In Section 5.2, we found the **real** zeros of a polynomial function. In this section we will find the **complex** zeros of a polynomial function. Finding the complex zeros of a function requires finding all zeros of the form $a + bi$. These zeros will be real if $b = 0$.

A variable in the complex number system is referred to as a **complex variable**. A **complex polynomial function** f of degree n is a function of the form

$$f(x) = a_n x^n + a_{n-1} x^{n-1} + \cdots + a_1 x + a_0 \qquad (1)$$

where $a_n, a_{n-1}, \ldots, a_1, a_0$ are complex numbers, $a_n \neq 0$, n is a nonnegative integer, and x is a complex variable. As before, a_n is called the **leading coefficient** of f. A complex number r is called a (complex) **zero** of f if $f(r) = 0$.

We have learned that some quadratic equations have no real solutions, but that in the complex number system every quadratic equation has a solution, either real or complex. The next result, proved by Karl Friedrich Gauss (1777–1855) when he was 22 years old,* gives an extension to complex polynomials. In fact, this result is so important and useful that it has become known as the **Fundamental Theorem of Algebra**.

Fundamental Theorem of Algebra

Every complex polynomial function $f(x)$ of degree $n \geq 1$ has at least one complex zero.

We shall not prove this result, as the proof is beyond the scope of this book. However, using the Fundamental Theorem of Algebra and the Factor Theorem, we can prove the following result:

Theorem

Every complex polynomial function $f(x)$ of degree $n \geq 1$ can be factored into n linear factors (not necessarily distinct) of the form

$$f(x) = a_n(x - r_1)(x - r_2) \cdots \cdots (x - r_n) \qquad (2)$$

where $a_n, r_1, r_2, \ldots, r_n$ are complex numbers. That is, every complex polynomial function of degree $n \geq 1$ has exactly n (not necessarily distinct) zeros.

---

*In all, Gauss gave four different proofs of this theorem, the first one in 1799 being the subject of his doctoral dissertation.

Proof Let

$$f(x) = a_n x^n + a_{n-1} x^{n-1} + \cdots + a_1 x + a_0$$

By the Fundamental Theorem of Algebra, f has at least one zero, say r_1. Then, by the Factor Theorem, $x - r_1$ is a factor, and

$$f(x) = (x - r_1)q_1(x)$$

where $q_1(x)$ is a complex polynomial of degree $n - 1$ whose leading coefficient is a_n. Again by the Fundamental Theorem of Algebra, the complex polynomial $q_1(x)$ has at least one zero, say r_2. By the Factor Theorem, $q_1(x)$ has the factor $x - r_2$, so

$$q_1(x) = (x - r_2)q_2(x)$$

where $q_2(x)$ is a complex polynomial of degree $n - 2$ whose leading coefficient is a_n. Consequently,

$$f(x) = (x - r_1)(x - r_2)q_2(x)$$

Repeating this argument n times, we finally arrive at

$$f(x) = (x - r_1)(x - r_2) \cdot \ldots \cdot (x - r_n)q_n(x)$$

where $q_n(x)$ is a complex polynomial of degree $n - n = 0$ whose leading coefficient is a_n. Thus, $q_n(x) = a_n x^0 = a_n$, and so

$$f(x) = a_n(x - r_1)(x - r_2) \cdot \ldots \cdot (x - r_n)$$

We conclude that every complex polynomial function $f(x)$ of degree $n \geq 1$ has exactly n (not necessarily distinct) zeros. ◼

Complex Zeros of Polynomials with Real Coefficients

1 We can use the Fundamental Theorem of Algebra to obtain valuable information about the complex zeros of polynomials whose coefficients are real numbers.

Conjugate Pairs Theorem

> Let $f(x)$ be a polynomial whose coefficients are real numbers. If $r = a + bi$ is a zero of f, then the complex conjugate $\bar{r} = a - bi$ is also a zero of f.
>
> ◼

In other words, for polynomials whose coefficients are real numbers, the zeros occur in conjugate pairs.

Proof Let

$$f(x) = a_n x^n + a_{n-1} x^{n-1} + \cdots + a_1 x + a_0$$

where $a_n, a_{n-1}, \ldots, a_1, a_0$ are real numbers and $a_n \neq 0$. If $r = a + bi$ is a zero of f, then $f(r) = f(a + bi) = 0$, so

$$a_n r^n + a_{n-1} r^{n-1} + \cdots + a_1 r + a_0 = 0$$

We take the conjugate of both sides to get

$$\overline{a_n r^n + a_{n-1} r^{n-1} + \cdots + a_1 r + a_0} = \overline{0}$$

$$\overline{a_n r^n} + \overline{a_{n-1} r^{n-1}} + \cdots + \overline{a_1 r} + \overline{a_0} = \overline{0} \qquad \text{\color{teal}\textit{The conjugate of a sum equals}}$$
$$\color{teal}\textit{the sum of the conjugates (see}$$
$$\color{teal}\textit{Section 5.3).}$$

$$\overline{a_n}(\overline{r})^n + \overline{a_{n-1}}(\overline{r})^{n-1} + \cdots + \overline{a_1}\,\overline{r} + \overline{a_0} = \overline{0} \qquad \text{\color{teal}\textit{The conjugate of a product equals}}$$
$$\color{teal}\textit{the product of the conjugates.}$$

$$a_n(\overline{r})^n + a_{n-1}(\overline{r})^{n-1} + \cdots + a_1 \overline{r} + a_0 = 0 \qquad \text{\color{teal}\textit{The conjugate of a real number}}$$
$$\color{teal}\textit{equals the real number.}$$

This last equation states that $f(\overline{r}) = 0$; that is, $\overline{r} = a - bi$ is a zero of f.   ■

The value of this result should be clear. Once we know that, say, $3 + 4i$ is a zero of a polynomial with real coefficients, then we know that $3 - 4i$ is also a zero. This result has an important corollary.

Corollary

A polynomial f of odd degree with real coefficients has at least one real zero.

■

Proof   Because complex zeros occur as conjugate pairs in a polynomial with real coefficients, there will always be an even number of zeros that are not real numbers. Consequently, since f is of odd degree, one of its zeros has to be a real number.   ■

For example, the polynomial $f(x) = x^5 - 3x^4 + 4x^3 - 5$ has at least one zero that is a real number, since f is of degree 5 (odd) and has real coefficients.

EXAMPLE 1   **Using the Conjugate Pairs Theorem**

A polynomial f of degree 5 whose coefficients are real numbers has the zeros $1, 5i$, and $1 + i$. Find the remaining two zeros.

Solution   Since complex zeros appear as conjugate pairs, it follows that $-5i$, the conjugate of $5i$, and $1 - i$, the conjugate of $1 + i$, are the two remaining zeros.   ■

━ NOW WORK PROBLEM **1.**

EXAMPLE 2   **Finding a Polynomial Function Whose Zeros Are Given**

② (a) Find a polynomial f of degree 4 whose coefficients are real numbers and that has the zeros $1, 1$, and $-4 + i$.

(b) Graph the polynomial found in part (a) to verify your result.

Solution   (a) Since $-4 + i$ is a zero, by the Conjugate Pairs Theorem, $-4 - i$ must also be a zero of f. Because of the Factor Theorem, if $f(c) = 0$, then $x - c$ is a factor of $f(x)$. So we can now write f as

$$f(x) = a(x - 1)(x - 1)[x - (-4 + i)][x - (-4 - i)]$$

where a is any real number. If we let $a = 1$, we obtain

$$
\begin{aligned}
f(x) &= (x - 1)(x - 1)[x - (-4 + i)][x - (-4 - i)] \\
&= (x^2 - 2x + 1)[x^2 - (-4 + i)x - (-4 - i)x + (-4 + i)(-4 - i)] \\
&= (x^2 - 2x + 1)(x^2 + 4x - ix + 4x + ix + 16 + 4i - 4i - i^2) \\
&= (x^2 - 2x + 1)(x^2 + 8x + 17) \\
&= x^4 + 8x^3 + 17x^2 - 2x^3 - 16x^2 - 34x + x^2 + 8x + 17 \\
&= x^4 + 6x^3 + 2x^2 - 26x + 17
\end{aligned}
$$

Figure 13

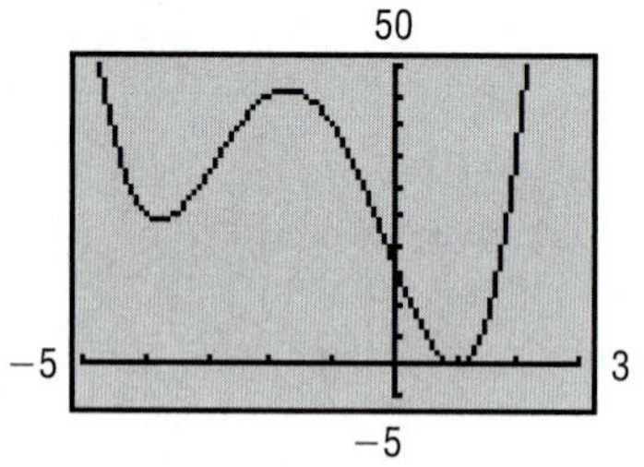

(b) A quick analysis of the polynomial f tells us what to expect:

> At most three turning points.
> For large $|x|$, the graph will behave like $y = x^4$.
> A repeated real zero at 1 so that the graph will touch the x-axis at 1.
> The only x-intercept is at 1.

Figure 13 shows the complete graph. (Do you see why? The graph has exactly three turning points and the degree of the polynomial is 4.) ■

EXPLORATION   Graph the function found in Example 2 for $a = 2$ and $a = -1$. Does the value of a affect the zeros of f? How does the value of a affect the graph of f?

Now we can prove the theorem we conjectured earlier in Section 5.2.

Theorem

Every polynomial function with real coefficients can be uniquely factored over the real numbers into a product of linear factors and/or irreducible quadratic factors.

■

Proof   Every complex polynomial f of degree n has exactly n zeros and can be factored into a product of n linear factors. If its coefficients are real, then those zeros that are complex numbers will always occur as conjugate pairs. As a result, if $r = a + bi$ is a complex zero, then so is $\bar{r} = a - bi$. Consequently, when the linear factors $x - r$ and $x - \bar{r}$ of f are multiplied, we have

$$(x - r)(x - \bar{r}) = x^2 - (r + \bar{r})x + r\bar{r} = x^2 - 2ax + a^2 + b^2$$

This second-degree polynomial has real coefficients and is irreducible (over the real numbers). Thus, the factors of f are either linear or irreducible quadratic factors. ■

EXAMPLE 3   **Finding the Complex Zeros of a Polynomial**

③ Find the complex zeros of the polynomial function

$$f(x) = 3x^4 + 5x^3 + 25x^2 + 45x - 18$$

Solution   **STEP 1:** The degree of f is 4. So f will have four complex zeros.

STEP 2: The Rational Zeros Theorem provides information about the potential rational zeros of polynomials with integer coefficients. For this polynomial (which has integer coefficients), the potential rational zeros are

$$\pm \frac{1}{3}, \pm \frac{2}{3}, \pm 1, \pm 2, \pm 3, \pm 6, \pm 9, \pm 18$$

Figure 14

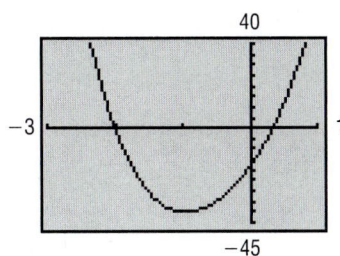

STEP 3: Figure 14 shows the graph of f. The graph has the characteristics that we expect of this polynomial of degree 4: It behaves like $y = 3x^4$ for large $|x|$ and has y-intercept -18. There are x-intercepts near -2 and between 0 and 1.

STEP 4: Because $f(-2) = 0$, we known that -2 is a zero and $x + 2$ is a factor of f. We can use long division or synthetic division to factor f:

$$f(x) = (x + 2)(3x^3 - x^2 + 27x - 9)$$

From the graph of f and the list of potential rational zeros, it appears that $\dfrac{1}{3}$ may also be a zero of f. Since $f\left(\dfrac{1}{3}\right) = 0$, we know that $\dfrac{1}{3}$ is a zero of f. We use synthetic division on the depressed equation of f to factor.

$$
\begin{array}{r|rrrr}
\frac{1}{3} & 3 & -1 & 27 & -9 \\
 & & 1 & 0 & 9 \\
\hline
 & 3 & 0 & 27 & 0
\end{array}
$$

Using the bottom row of the synthetic division, we find

$$f(x) = (x + 2)\left(x - \frac{1}{3}\right)(3x^2 + 27) = 3(x + 2)\left(x - \frac{1}{3}\right)(x^2 + 9)$$

The factor $x^2 + 9$ does not have any real zeros; its complex zeros are $\pm 3i$. The complex zeros of $f(x) = 3x^4 + 5x^3 + 25x^2 + 45x - 18$ are $-2, \dfrac{1}{3}, 3i, -3i$. ■

 **NOW WORK PROBLEM 27.**

5.4 Concepts and Vocabulary

In Problems 1 and 2, fill in the blanks.

1. Every polynomial function of odd degree with real coefficients will have at least _____ real zero(s).

2. If $3 + 4i$ is a zero of a polynomial function of degree 5 with real coefficients, then so is _____.

In Problems 3 and 4, answer True or False to each statement.

3. A polynomial function of degree n with real coefficients has exactly n complex zeros. At most n of them are real zeros.

4. A polynomial function of degree 4 with real coefficients could have $-3, 2 + i, 2 - i$, and $-3 + 5i$ as its zeros.

5. Suppose that $f(x)$ is a polynomial function of degree 4 whose coefficients are real numbers; three of its zeros are $2, 1 + 2i$, and $1 - 2i$. Explain why the remaining zero must be a real number.

6. Suppose that $f(x)$ is a polynomial function of degree 4 whose coefficients are real numbers; two of its zeros are -3 and $4 - i$. Explain why one of the remaining zeros must be a real number. Write down one of the missing zeros.

5.4 Exercises

In Problems 1–10, information is given about a polynomial $f(x)$ whose coefficients are real numbers. Find the remaining zeros of f.

1. Degree 3;   zeros: $3, 4 - i$

2. Degree 3;   zeros: $4, 3 + i$

3. Degree 4;   zeros: $i, 1 + i$

4. Degree 4;   zeros: $1, 2, 2 + i$

5. Degree 5;   zeros: $1, i, 2i$

6. Degree 5;   zeros: $0, 1, 2, i$

7. Degree 4;   zeros: $i, 2, -2$

8. Degree 4;   zeros: $2 - i, -i$

9. Degree 6;   zeros: $2, 2 + i, -3 - i, 0$

10. Degree 6;   zeros: $i, 3 - 2i, -2 + i$

In Problems 11–16, form a polynomial $f(x)$ with real coefficients having the given degree and zeros.

11. Degree 4;   zeros: $3 + 2i$; 4, multiplicity 2

12. Degree 4;   zeros: $i, 1 + 2i$

13. Degree 5;   zeros: 2, multiplicity 1; $-i$; $1 + i$

14. Degree 6;   zeros: $i, 4 - i; 2 + i$

15. Degree 4;   zeros: 3, multiplicity 2; $-i$

16. Degree 5;   zeros: 1, multiplicity 3; $1 + i$

In Problems 17–24, use the given zero to find the remaining zeros of each function. Graph the function to verify your results.

17. $f(x) = x^3 - 4x^2 + 4x - 16$;   zero: $2i$

18. $g(x) = x^3 + 3x^2 + 25x + 75$;   zero: $-5i$

19. $f(x) = 2x^4 + 5x^3 + 5x^2 + 20x - 12$;   zero: $-2i$

20. $h(x) = 3x^4 + 5x^3 + 25x^2 + 45x - 18$;   zero: $3i$

21. $h(x) = x^4 - 9x^3 + 21x^2 + 21x - 130$;   zero: $3 - 2i$

22. $f(x) = x^4 - 7x^3 + 14x^2 - 38x - 60$;   zero: $1 + 3i$

23. $h(x) = 3x^5 + 2x^4 + 15x^3 + 10x^2 - 528x - 352$;   zero: $-4i$

24. $g(x) = 2x^5 - 3x^4 - 5x^3 - 15x^2 - 207x + 108$;   zero: $3i$

In Problems 25–34, find the complex zeros of each polynomial function. Check your results by evaluating the function at each of the zeros.

25. $f(x) = x^3 - 1$

26. $f(x) = x^4 - 1$

27. $f(x) = x^3 - 8x^2 + 25x - 26$

28. $f(x) = x^3 + 13x^2 + 57x + 85$

29. $f(x) = x^4 + 5x^2 + 4$

30. $f(x) = x^4 + 13x^2 + 36$

31. $f(x) = x^4 + 2x^3 + 22x^2 + 50x - 75$

32. $f(x) = x^4 + 3x^3 - 19x^2 + 27x - 252$

33. $f(x) = 3x^4 - x^3 - 9x^2 + 159x - 52$

34. $f(x) = 2x^4 + x^3 - 35x^2 - 113x + 65$

In Problems 35 and 36, tell why the facts given are contradictory.

35. $f(x)$ is a polynomial of degree 3 whose coefficients are real numbers; its zeros are $4 + i, 4 - i$, and $2 + i$.

36. $f(x)$ is a polynomial of degree 3 whose coefficients are real numbers; its zeros are $2, 5$, and $3 + i$.

Chapter Review

Things To Know

Zeros of a polynomial f (p. 384)

Real or complex numbers for which $f(r) = 0$; when real these are the x-intercepts of the graph of f.

Remainder Theorem (p. 385)

If a polynomial $f(x)$ is divided by $x - c$, then the remainder is $f(c)$.

Factor Theorem (p. 386)

$x - c$ is a factor of a polynomial $f(x)$ if and only if $f(c) = 0$.

Rational Zeros Theorem (p. 388)

Let f be a polynomial function of degree 1 or higher of the form

$$f(x) = a_n x^n + a_{n-1} x^{n-1} + \cdots + a_1 x + a_0, \quad a_n \neq 0, a_0 \neq 0$$

where each coefficient is an integer. If $\dfrac{p}{q}$, in lowest terms, is a rational zero of f, then p must be a factor of a_0 and q must be a factor of a_n.

Intermediate Value Theorem (p. 394)

Let f denote a continuous function. If $a < b$ and $f(a)$ and $f(b)$ are of opposite sign, then f has at least one zero between a and b.

Quadratic equation and quadratic formula (p. 404)

If $ax^2 + bx + c = 0, a \neq 0$, then $x = \dfrac{-b \pm \sqrt{b^2 - 4ac}}{2a}$.

Discriminant (p. 405)

If $b^2 - 4ac > 0$, there are two distinct real solutions.
If $b^2 - 4ac = 0$, there is one repeated real solution.
If $b^2 - 4ac < 0$, there are two distinct complex solutions that are not real; the solutions are conjugates of each other.

Fundamental Theorem of Algebra (p. 407)

Every complex polynomial function $f(x)$ of degree $n \geq 1$ has at least one complex zero.

Conjugate Pairs Theorem (p. 408)

Let $f(x)$ be a polynomial whose coefficients are real numbers. If $r = a + bi$ is a zero of f, then its complex conjugate $\bar{r} = a - bi$ is also a zero of f.

Objectives

Section	You should be able to . . .	Problems
5.1	**1** Use synthetic division (p. 380)	1–4
5.2	**1** Use the Remainder and Factor Theorems (p. 384)	1–4
	2 Use the Rational Zeros Theorem to list the potential rational zeros of a polynomial function (p. 387)	5, 6
	3 Find the real zeros of a polynomial function (p. 388)	7–16
	4 Solve polynomial equations (p. 391)	17–20
	5 Use the theorem for Bounds on Zeros (p. 391)	21–24
	6 Use the Intermediate Value Theorem (p. 394)	25–28
5.3	**1** Add, subtract, multiply, and divide complex numbers (p. 399)	29–38
	2 Solve quadratic equations with a negative discriminant (p. 403)	43–50
5.4	**1** Utilize the Conjugate Pairs Theorem to find the complex zeros of a polynomial (p. 408)	39–42
	2 Find a polynomial function with specified zeros (p. 409)	39–42
	3 Find the complex zeros of a polynomial function (p. 410)	51–56

Review Exercises

Blue problem numbers indicate the authors' suggestions for use in a Practice Test.

In Problems 1–4, use synthetic division to find the quotient $q(x)$ and remainder R when $f(x)$ is divided by $g(x)$. Is g a factor of f?

1. $f(x) = 8x^3 - 3x^2 + x + 4$; $g(x) = x - 1$

2. $f(x) = 2x^3 + 8x^2 - 5x + 5$; $g(x) = x - 2$

3. $f(x) = x^4 - 2x^3 + 15x - 2$; $g(x) = x + 2$

4. $f(x) = x^4 - x^2 + 2x + 2$; $g(x) = x + 1$

In Problems 5 and 6, tell the maximum number of real zeros that each polynomial function may have. Then list the potential rational zeros of each polynomial function. Do not attempt to find the zeros.

5. $f(x) = 2x^8 - x^7 + 8x^4 - 2x^3 + x + 3$

6. $f(x) = -6x^5 + x^4 + 5x^3 + x + 1$

In Problems 7–12, find all the real zeros of each polynomial function.

7. $f(x) = x^3 - 3x^2 - 6x + 8$

8. $f(x) = x^3 - x^2 - 10x - 8$

9. $f(x) = 4x^3 + 4x^2 - 7x + 2$

10. $f(x) = 4x^3 - 4x^2 - 7x - 2$

11. $f(x) = x^4 - 4x^3 + 9x^2 - 20x + 20$

12. $f(x) = x^4 + 6x^3 + 11x^2 + 12x + 18$

In Problems 13–16, determine the real zeros of the polynomial function. Approximate all irrational zeros rounded to two decimal places.

13. $f(x) = 2x^3 - 11.84x^2 - 9.116x + 82.46$

14. $f(x) = 12x^3 + 39.8x^2 - 4.4x - 3.4$

15. $g(x) = 15x^4 - 21.5x^3 - 1718.3x^2 + 5308x + 3796.8$

16. $g(x) = 3x^4 + 67.93x^3 + 486.265x^2 + 1121.32x + 412.195$

In Problems 17–20, find the real solutions of each equation.

17. $2x^4 + 2x^3 - 11x^2 + x - 6 = 0$

18. $3x^4 + 3x^3 - 17x^2 + x - 6 = 0$

19. $2x^4 + 7x^3 + x^2 - 7x - 3 = 0$

20. $2x^4 + 7x^3 - 5x^2 - 28x - 12 = 0$

In Problems 21–24, find bounds to the zeros of each polynomial function. Obtain a complete graph of f.

21. $f(x) = x^3 - x^2 - 4x + 2$

22. $f(x) = x^3 + x^2 - 10x - 5$

23. $f(x) = 2x^3 - 7x^2 - 10x + 35$

24. $f(x) = 3x^3 - 7x^2 - 6x + 14$

In Problems 25–28, use the Intermediate Value Theorem to show that each polynomial has a zero in the given interval. Approximate the zero rounded to two decimal places.

25. $f(x) = 3x^3 - x - 1; \quad [0, 1]$

26. $f(x) = 2x^3 - x^2 - 3; \quad [1, 2]$

27. $f(x) = 8x^4 - 4x^3 - 2x - 1; \quad [0, 1]$

28. $f(x) = 3x^4 + 4x^3 - 8x - 2; \quad [1, 2]$

In Problems 29–38, write each expression in the standard form $a + bi$. Verify your results using a graphing utility.

29. $(6 + 3i) - (2 - 4i)$

30. $(8 - 3i) + (-6 + 2i)$

31. $4(3 - i) + 3(-5 + 2i)$

32. $2(1 + i) - 3(2 - 3i)$

33. $\dfrac{3}{3 + i}$

34. $\dfrac{4}{2 - i}$

35. i^{50}

36. i^{29}

37. $(2 + 3i)^3$

38. $(3 - 2i)^3$

In Problems 39–42, information is given about a complex polynomial $f(x)$ whose coefficients are real numbers. Find the remaining zeros of f. Write a polynomial function whose zeros are given.

39. Degree 3; zeros: $4 + i, 6$

40. Degree 3; zeros: $3 + 4i, 5$

41. Degree 4; zeros: $i, 1 + i$

42. Degree 4; zeros: $1, 2, 1 + i$

In Problems 43–56, solve each equation in the complex number system.

43. $x^2 + x + 1 = 0$

44. $x^2 - x + 1 = 0$

45. $2x^2 + x - 2 = 0$

46. $3x^2 - 2x - 1 = 0$

47. $x^2 + 3 = x$

48. $2x^2 + 1 = 2x$

49. $x(1 - x) = 6$

50. $x(1 + x) = 2$

51. $x^4 + 2x^2 - 8 = 0$

52. $x^4 + 8x^2 - 9 = 0$

53. $x^3 - x^2 - 8x + 12 = 0$

54. $x^3 - 3x^2 - 4x + 12 = 0$

55. $3x^4 - 4x^3 + 4x^2 - 4x + 1 = 0$

56. $x^4 + 4x^3 + 2x^2 - 8x - 8 = 0$

Chapter Projects

Date	Grains per cubic meter	Date	Grains per cubic meter
9/6	16	10/10	10
9/8	49	10/11	52
9/12	41	10/13	269
9/13	31	10/17	60
9/15	47	10/18	18
9/19	67	10/20	28
9/20	60	10/24	65
9/22	70	10/25	44
9/27	48	10/27	41
9/29	45	10/31	18
10/3	35	11/1	21
10/4	35	11/3	10
10/6	47		

1. **Weed Pollen**   Given in the table to the right is the weed pollen count for Dallas, Texas, from September 6, 2000, to November 3, 2000, the prime season for weed pollen.

Data Source: Nation Allergy Bureau at the American Academy of allergy, Asthma and Immunology website (*www.aaaai.org*)

(a) Let the independent variable D represent the date, where $D = 1$ on 9/6, $D = 3$ on 9/8, $D = 7$ on 9/12, ..., and $D = 59$ on 11/3. Let the dependent variable P represent the number of grains per cubic meter of weed pollen. Draw a scatter diagram of the data using your graphing utility and by hand on graph paper.

(b) In the hand drawn scatter diagram, sketch a smooth curve that "fits" the data. How many turning points are there? What does this tell you about the degree of the polynomial that could be fit to the data? If the ends of the graph on the left and on the right are extended downward, how many real zeros are indicated by the graph? How many complex? Why?

(c) Use your graphing utility to determine the fourth degree regression equation. Graph it on your graphing utility. Does it look like what you expected from your hand drawn graph? Explain your reasoning.

(d) Using your graphing utility, find the maximum and minimum values on the graph of the function found in part (c). What are the intercepts of the graph of the function?

(e) Using the D-intercepts found in part (d), factor the regression equation over the real numbers. Why does the factored form contain an irreducible quadratic factor?

(f) How well does a quadratic function fit the data? How well does a cubic function fit the data? Why will a cubic function always fit at least as well as a quadratic function? Why will a quartic function always fit at least as well as a cubic function?

(g) Investigate weed pollen counts at other times of the year. (Go to the National Allergy Bureau at www.aaaai.org.) Would the polynomial you found in part (c) work for other three month periods? Why or why not? What about for the same period of time in other years?

2. **Maclaurin Series** Polynomials are often used to approximate non-polynomial functions. In future chapters, several new functions will be introduced. These will be investigated by polynomial approximations in this project.

(a) Graph $Y_1 = \sin x$ on your graphing utility with $X\min = -2\pi$, $X\max = 2\pi$, $Y\min = -4$, $Y\max = 4$. (Make sure that your graphing utility is set to *radian* mode). Now graph $Y_2 = x$ on the same screen. Over what interval does Y_2 appear to equal Y_1?

(b) Clear the screen. Graph $Y_1 = \sin x$ and $Y_3 = x - \dfrac{x^3}{6}$ on your graphing utility with $X\min = -2\pi$, $X\max = 2\pi$, $Y\min = -4$, $Y\max = 4$. Over what interval does Y_3 appear to equal Y_1? Repeat this for $Y_4 = x - \dfrac{x^3}{6} + \dfrac{x^5}{120}$. Find the zeroes of Y_1, Y_2, Y_3, Y_4, and Y_5. Discuss how they are related.

(c) Evaluate $Y_1, Y_2, \ldots, Y_5$ for $x = \dfrac{\pi}{3}$, $x = \dfrac{\pi}{4}$, $x = \dfrac{\pi}{2}$. How well do each of these polynomials approximate $y = \sin x$ evaluated at each of these values?

(d) These polynomials are finite because they only have a finite number of terms. They are just portions of the **Maclaurin Series** for the function $y = \sin x$, which actually has an infinite number of terms. Based on the pattern given for Y_2 through Y_5, can you find a general formula for the series?

(e) Graph $Y_1 = e^x$ on your graphing utility with $X\min = -3.15$, $X\max = 3.15$, $Y\min = -3.1$, and $Y\max = 3.1$. Now graph the functions $Y_2 = 1$,

$$Y_3 = 1 + x, \quad Y_4 = 1 + x + \frac{x^2}{2}, \quad Y_5 = 1 + x + \frac{x^2}{2} + \frac{x^3}{6}$$

on the same window. Record the interval for which each function seems to approximate Y_1. Find the zeros of Y_1 through Y_5, if any, using your graphing utility. How do the zeros of Y_2 through Y_5 relate to the zeros of Y_1, if at all? Explain your results. Can you find a pattern to the polynomial series for $Y_1 = e^x$?

(f) Evaluate each of the polynomials for $x = 0$, $x = -1$, $x = -2$, $x = 1$, $x = 2$. How well does each polynomial approximate $y = e^x$ at each of those points?

3. (a) Solve $x^2 + 8x - 9 = 0$ by factoring. Find the sum and the product of the zeros.

(b) Suppose $x^2 + bx + c$ has zeros r_1 and r_2. Write the polynomial in factored form. Multiply the factored form out. What relationships do you find between this product and the coefficients in the original polynomial?

(c) Factor $f(x) = x^3 - x^2 - 10x - 8$ using the techniques of this chapter. Find the sum of the zeros and the product of the zeros. Now find the sum of all the double products of the zeros (i.e find $r_1 r_2 + r_1 r_3 + r_2 r_3$). What is the relationship between these three values and the coefficients of the polynomial?

(d) Suppose the monic (leading coefficient 1) cubic polynomial $f(x) = x^3 + bx^2 + cx + d$ has zeros r_1, r_2, and r_3. Write the polynomial in factored form and multiply the factors. Write the relationships between the coefficients of each form of the polynomial.

(e) Suppose the monic quartic polynomial function $f(x) = x^4 + bx^3 + cx^2 + dx + e$ has zeros r_1, r_2, r_3, and r_4. Repeat part (d) for this polynomial.

(f) What relationship do you find between the zeros of the polynomial function and the coefficients of the monic polynomial function? Can you state some general formulas? Will these formulas always work? Explain. Why do you think having formulas like this would be useful? Explain.

Cumulative Review

1. Evaluate each of the following expressions. In (c) and (d) use a calculator. If necessary, round answers to three decimal places.

   (a) 4^2 (b) 3^{-4} (c) 1.2^3 (d) $3^{3.2}$

2. Simplify: $\left(\dfrac{3x^2y}{2xy^5}\right)^{-2}$

3. Graph the function $f(x) = -2x + 5$.

4. Graph the function $f(x) = 2(x + 1)^2 - 3$ using transformations.

5. Use a graphing utility to approximate the local maxima and local minima of $f(x) = -2x^3 + 7x - 1$ on the interval $(-3, 3)$. Determine the intervals on which f is increasing. Determine the intervals on which f is decreasing.

6. Is the graph below the graph of a function?

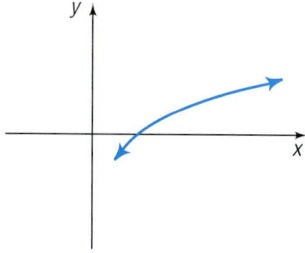

7. The graph of a function is given below. Use the graph to determine the domain and range of the function. Identify any intercepts.

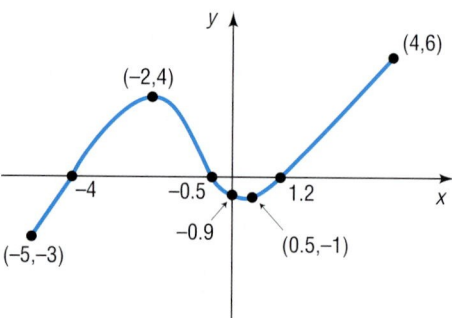

8. Suppose that $f(x) = x^2 + 3$ and $g(x) = \dfrac{3}{x - 4}$. Find $g \circ f$ and state its domain.

9. Analyze the rational function $R(x) = \dfrac{2x^2 - 5x - 3}{x^2 + x - 6}$ by following the steps listed on page 355. Solve the inequality $R(x) \geq 0$.

10. Find the real zeros of $f(x) = 3x^4 - 17x^3 - 6x^2 + 96x + 32$. Then draw a complete graph of f. Solve the inequality $f(x) > 0$.

EXPONENTIAL AND LOGARITHMIC FUNCTIONS

The McDonald's Scalding Coffee Case

April 3, 1996

There is a lot of hype about the McDonald's scalding coffee case. No one is in favor of frivolous cases or outlandish results; however, it is important to understand some points that were not reported in most of the stories about the case. McDonald's coffee was not only hot, it was scalding, capable of almost instantaneous destruction of skin, flesh and muscle.

Plaintiff's expert, a scholar in thermodynamics applied to human skin burns, testified that liquids, at 180 degrees, will cause a full thickness burn to human skin in two to seven seconds. Other testimony showed that as the temperature decreases toward 155 degrees, the extent of the burn relative to that temperature decreases exponentially. Thus, if (the) spill had involved coffee at 155 degrees, the liquid would have cooled and given her time to avoid a serious burn.

Miller, Norman, & Associates, Ltd., Attorneys, Moorhead, MN.

SEE CHAPTER PROJECT 1.

For additional
study help, go to

www.prenhall.com/sullivanegu3e

Materials include:
- Graphing Calculator Help
- Chapter Quiz
- Chapter Test
- PowerPoint Downloads
- Chapter Projects
- Student Tips

A Look Back, A Look Forward

Until now, our study of functions has concentrated on polynomial and rational functions. These functions belong to the class of **algebraic functions**, that is, functions that can be expressed in terms of sums, differences, products, quotients, powers, or roots of polynomials. Functions that are not algebraic are termed **transcendental** (they transcend, or go beyond, algebraic functions).

In this chapter, we study two transcendental functions: the *exponential* and *logarithmic* functions. The functions occur frequently in a wide variety of applications, such as biology, chemistry, economics, and psychology.

The chapter begins with a discussion of inverse functions.

PREPARING FOR THIS SECTION

Before getting started, review the following:

✓ Functions (Section 2.1, pp. 195–205)

✓ Increasing/Decreasing Functions (Section 3.2, pp. 266–267)

6.1 ONE-TO-ONE FUNCTIONS; INVERSE FUNCTIONS

OBJECTIVES   ① Determine the Inverse of a Function
                ② Obtain the Graph of the Inverse Function from the Graph of the Function
                ③ Find the Inverse Function f^{-1}

① In Section 2.1 we said that a function f can be thought of as a machine that receives as input a number, say x, from the domain, manipulates it, and outputs the value $f(x)$. The **inverse of f** receives as input a number $f(x)$, manipulates it, and outputs the value x.

EXAMPLE 1   **Finding the Inverse of a Function**

Find the inverse of the following functions.

(a) Let the domain of the function represent the employees of Yolanda's Preowned Car Mart and let the range represent their base salaries.

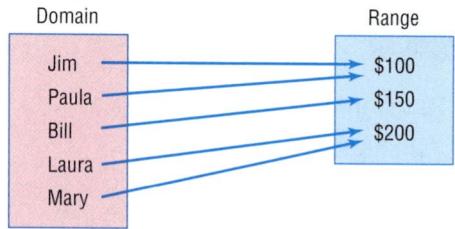

(b) Let the domain of the function represent the employees of Yolanda's Preowned Car Mart and let the range represent their spouse's names.

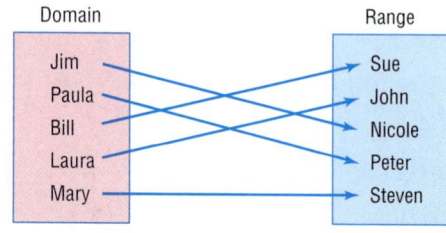

Solution   (a) The elements in the domain represent inputs to the function, and the elements in the range represent the outputs. To find the inverse, interchange the elements in the domain with the elements in the range. For example, the function receives as input Bill and outputs $150. So the inverse receives as input $150 and outputs Bill. The inverse of the given function takes the form

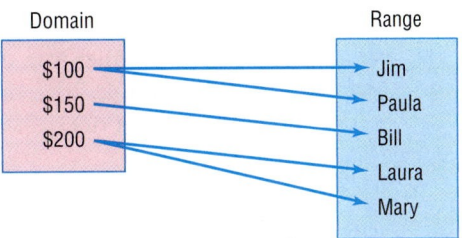

(b) The inverse of the given function is

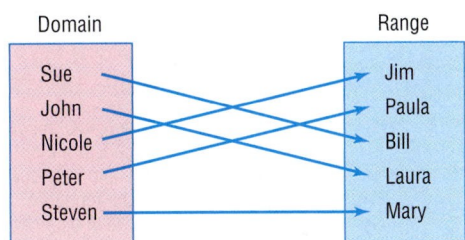

Notice that the inverse found in Example 1(b) is a function, since each element in the domain corresponds to exactly one element in the range. The inverse found in Example 1(a) is not a function, since each element in the domain does not correspond to exactly one element in the range.

If the function f is a set of ordered pairs (x, y), then the inverse of f is the set of ordered pairs (y, x).

EXAMPLE 2   **Finding the Inverse of a Function**

Find the inverse of the following functions:

(a) $\{(-3, -27), (-2, -8), (-1, -1), (0, 0), (1, 1), (2, 8), (3, 27)\}$
(b) $\{(-3, 9), (-2, 4), (-1, 1), (0, 0), (1, 1), (2, 4), (3, 9)\}$

Solution   (a) The inverse of the given function is found by interchanging the entries in each ordered pair and so is given by

$$\{(-27, -3), (-8, -2), (-1, -1), (0, 0), (1, 1), (8, 2), (27, 3)\}$$

(b) The inverse of the given function is

$$\{(9, -3), (4, -2), (1, -1), (0, 0), (1, 1), (4, 2), (9, 3)\}$$

The inverse obtained in the solution to Example 2(a) is a function, but the inverse obtained in Example 2(b) is not a function. So, sometimes the inverse of a function is a function, and sometimes it is not.

When the inverse of a function f is itself a function, then f is said to be a **one-to-one function**. That is, f is **one-to-one** if, for any choice of elements x_1 and x_2 in the domain of f, with $x_1 \neq x_2$, the corresponding values $f(x_1)$ and $f(x_2)$ are unequal, $f(x_1) \neq f(x_2)$.

In other words, a function f is one-to-one if no y in the range is the image of more than one x in the domain. A function is not one-to-one if two different elements in the domain correspond to the same element in the range. In Example 2(b), the elements -3 and 3 both correspond to 9, so the function is not one-to-one. Figure 1 illustrates the definition.

Figure 1

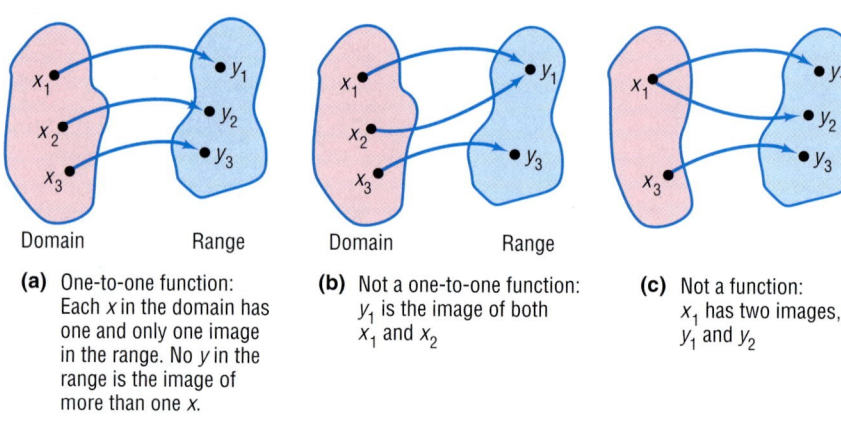

(a) One-to-one function:
Each x in the domain has one and only one image in the range. No y in the range is the image of more than one x.

(b) Not a one-to-one function: y_1 is the image of both x_1 and x_2

(c) Not a function: x_1 has two images, y_1 and y_2

NOW WORK PROBLEMS 1 AND 5.

If the graph of a function f is known, there is a simple test, called the **horizontal-line test**, to determine whether f is one-to-one.

Figure 2
$f(x_1) = f(x_2) = h$,
and $x_1 \neq x_2$;
f is not a
one-to-one function.

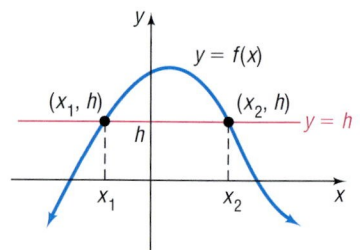

Theorem

Horizontal-line Test

If every horizontal line intersects the graph of a function f in at most one point, then f is one-to-one.

The reason that this test works can be seen in Figure 2, where the horizontal line $y = h$ intersects the graph at two distinct points, (x_1, h) and (x_2, h). Since h is the image of both x_1 and x_2, $x_1 \neq x_2$, f is not one-to-one. Based on Figure 2, we can state the horizontal-line test in another way: If the graph of any horizontal line intersects the graph of a function f at more than one point, then f is not one-to-one.

EXAMPLE 3   **Using the Horizontal-line Test**

For each function, use the graph to determine whether the function is one-to-one.

(a) $f(x) = x^2$ (b) $g(x) = x^3$

Solution   (a) Figure 3(a) illustrates the horizontal-line test for $f(x) = x^2$. The horizontal line $y = 1$ intersects the graph of f twice, at $(1, 1)$ and at $(-1, 1)$, so f is not one-to-one.

(b) Figure 3(b) illustrates the horizontal-line test for $g(x) = x^3$. Because every horizontal line will intersect the graph of g exactly once, it follows that g is one-to-one.

Figure 3

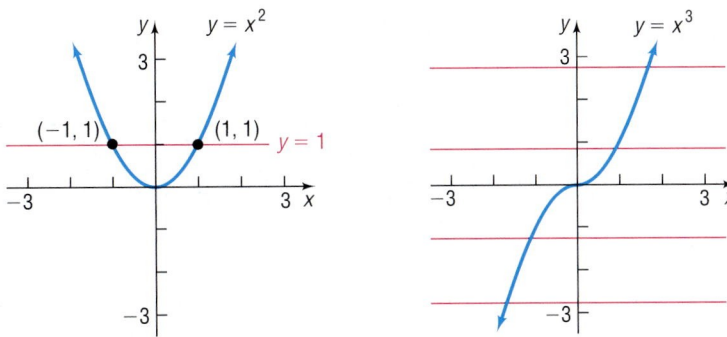

(a) A horizontal line intersects the graph twice; f is not one-to-one

(b) Every horizontal line intersects the graph exactly once; g is one-to-one

NOW WORK PROBLEM 9.

Let's look more closely at the one-to-one function $g(x) = x^3$. This function is an increasing function. Because an increasing (or decreasing) function will always have different y values for unequal x values, it follows that a function that is increasing (or decreasing) over its domain is also a one-to-one function.

Theorem

> A function that is increasing over its domain is a one-to-one function.
> A function that is decreasing over its domain is a one-to-one function.

Inverse Function of $y = f(x)$

If f is a one-to-one function, its inverse is a function. Then, to each x in the domain of f, there is exactly one y in the range (because f is a function); and to each y in the range of f, there is exactly one x in the domain (because f is one-to-one). The correspondence from the range of f back to the domain of f is called the **inverse function of f** and is denoted by the symbol f^{-1}. Figure 4 illustrates this definition.

Figure 4

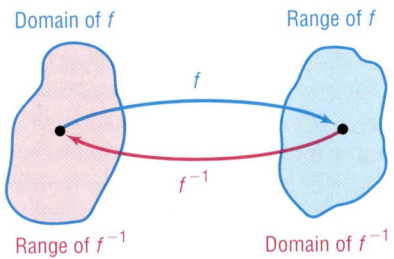

Domain of f

Range of f

Range of f^{-1}

Domain of f^{-1}

WARNING: Be careful! f^{-1} is a symbol for the inverse function of f. The -1 used in f^{-1} is not an exponent. That is, f^{-1} does *not* mean the reciprocal of f; $f^{-1}(x)$ is not equal to $\dfrac{1}{f(x)}$.

Two facts are now apparent about a function f and its inverse f^{-1}.

> Domain of f = Range of f^{-1} Range of f = Domain of f^{-1}

Look again at Figure 4 to visualize the relationship. If we start with x, apply f, and then apply f^{-1}, we get x back again. If we start with x, apply f^{-1}, and then apply f, we get the number x back again. To put it simply, what f does f^{-1} undoes, and vice versa.

$$\boxed{\text{Input } x} \xrightarrow{\text{Apply } f} \boxed{f(x)} \xrightarrow{\text{Apply } f^{-1}} \boxed{f^{-1}(f(x)) = x}$$

$$\boxed{\text{Input } x} \xrightarrow{\text{Apply } f^{-1}} \boxed{f^{-1}(x)} \xrightarrow{\text{Apply } f} \boxed{f(f^{-1}(x)) = x}$$

In other words,

$$f^{-1}(f(x)) = x \quad \text{and} \quad f(f^{-1}(x)) = x$$

Consider the function $f(x) = 2x$, which multiplies the argument x by 2. The inverse function f^{-1} undoes whatever f does. So the inverse function of f is $f^{-1}(x) = \dfrac{1}{2}x$, which divides the argument by 2. For example, $f(3) = 2(3) = 6$ and $f^{-1}(6) = \dfrac{1}{2}(6) = 3$, so f^{-1} undoes what f did. We can verify this by showing that

$$f^{-1}(f(x)) = f^{-1}(2x) = \frac{1}{2}(2x) = x \quad \text{and} \quad f(f^{-1}(x)) = f\left(\frac{1}{2}x\right) = 2\left(\frac{1}{2}x\right) = x$$

See Figure 5.

Figure 5

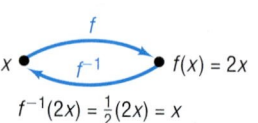

$f^{-1}(2x) = \frac{1}{2}(2x) = x$

EXAMPLE 4   **Verifying Inverse Functions**

(a) We verify that the inverse of $g(x) = x^3$ is $g^{-1}(x) = \sqrt[3]{x}$ by showing that

$$g^{-1}(g(x)) = g^{-1}(x^3) = \sqrt[3]{x^3} = x$$

and

$$g(g^{-1}(x)) = g(\sqrt[3]{x}) = (\sqrt[3]{x})^3 = x$$

(b) We verify that the inverse of $h(x) = 3x$ is $h^{-1}(x) = \dfrac{1}{3}x$ by showing that

$$h^{-1}(h(x)) = h^{-1}(3x) = \frac{1}{3}(3x) = x$$

and

$$h(h^{-1}(x)) = h\left(\frac{1}{3}x\right) = 3\left(\frac{1}{3}x\right) = x$$

(c) We verify that the inverse of $f(x) = 2x + 3$ is $f^{-1}(x) = \dfrac{1}{2}(x - 3)$ by showing that

$$f^{-1}(f(x)) = f^{-1}(2x + 3) = \frac{1}{2}[(2x + 3) - 3] = \frac{1}{2}(2x) = x$$

and

$$f(f^{-1}(x)) = f\left(\frac{1}{2}(x - 3)\right) = 2\left[\frac{1}{2}(x - 3)\right] + 3 = (x - 3) + 3 = x$$ ◼

EXPLORATION   Simultaneously graph $Y_1 = x$, $Y_2 = x^3$, and $Y_3 = \sqrt[3]{x}$ on a square screen with $-3 \leq x \leq 3$. What do you observe about the graphs of $Y_2 = x^3$, its inverse $Y_3 = \sqrt[3]{x}$, and the line $Y_1 = x$?

Repeat this experiment by simultaneously graphing $Y_1 = x$, $Y_2 = 2x + 3$, and $Y_3 = \frac{1}{2}(x - 3)$ on a square screen with $-6 \le x \le 3$. Do you see the symmetry of the graph of Y_2 and its inverse Y_3 with respect to the line $Y_1 = x$?

NOW WORK PROBLEM 21.

2 Geometric Interpretation

For the functions in Example 4(c), we list points on the graph of $f = Y_1$ and on the graph of $f^{-1} = Y_2$ in Table 1.

We notice that whenever (a, b) is on the graph of f then (b, a) is on the graph of f^{-1}. Figure 6 shows these points plotted. Also shown is the graph of $y = x$, which you should observe is a line of symmetry of the points.

Figure 7

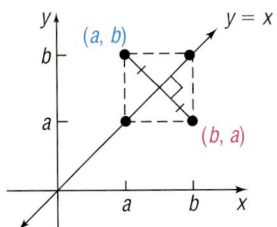

Figure 6

TABLE 1	
X | Y1	X | Y2
-5 | -7	-7 | -5
-4 | -5	-5 | -4
-3 | -3	-3 | -3
-2 | -1	-1 | -2
-1 | 1	1 | -1
0 | 3	3 | 0
1 | 5	5 | 1
Y₁⊟2X+3	Y₂⊟(1/2)(X−3)

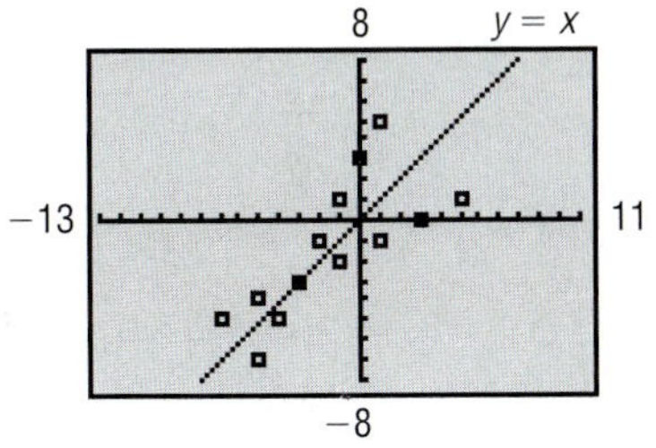

Suppose that (a, b) is a point on the graph of a one-to-one function f defined by $y = f(x)$. Then $b = f(a)$. This means that $a = f^{-1}(b)$, so (b, a) is a point on the graph of the inverse function f^{-1}. The relationship between the point (a, b) on f and the point (b, a) on f^{-1} is shown in Figure 7. The line segment containing (a, b) and (b, a) is perpendicular to the line $y = x$ and is bisected by the line $y = x$. (Do you see why?) It follows that the point (b, a) on f^{-1} is the reflection about the line $y = x$ of the point (a, b) on f.

Theorem

The graph of a function f and the graph of its inverse f^{-1} are symmetric with respect to the line $y = x$.

Figure 8 illustrates this result. Notice that, once the graph of f is known, the graph of f^{-1} may be obtained by reflecting the graph of f about the line $y = x$.

Figure 8

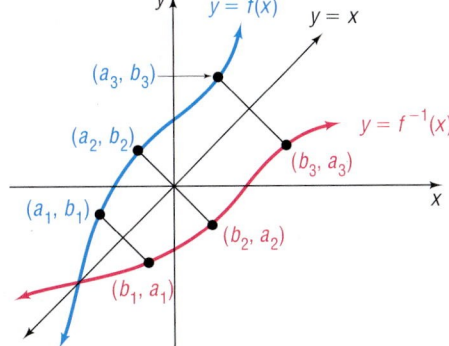

EXAMPLE 5 **Graphing the Inverse Function**

The graph in Figure 9(a) is that of a one-to-one function $y = f(x)$. Draw the graph of its inverse.

Solution We begin by adding the graph of $y = x$ to Figure 9(a). Since the points $(-2, -1)$, $(-1, 0)$, and $(2, 1)$ are on the graph of f, we know that the points $(-1, -2)$, $(0, -1)$, and $(1, 2)$ must be on the graph of f^{-1}. Keeping in mind that the graph of f^{-1} is the reflection about the line $y = x$ of the graph of f, we can draw f^{-1}. See Figure 9(b).

Figure 9

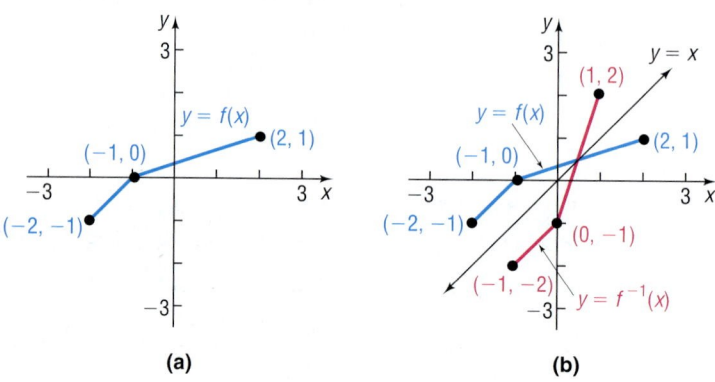

(a)   (b)

───── **NOW WORK PROBLEM 15.**

③ Finding the Inverse Function

The fact that the graph of a one-to-one function f and its inverse function f^{-1} are symmetric with respect to the line $y = x$ tells us more. It says that we can obtain f^{-1} by interchanging the roles of x and y in f. Look again at Figure 8. If f is defined by the equation

$$y = f(x)$$

then f^{-1} is defined by the equation

$$x = f(y)$$

The equation $x = f(y)$ defines f^{-1} *implicitly*. If we can solve this equation for y, we will have the *explicit* form of f^{-1}, that is,

$$y = f^{-1}(x)$$

Let's use this procedure to find the inverse of $f(x) = 2x + 3$. (Since f is a linear function and is increasing, we know that f is one-to-one and so has an inverse function.)

EXAMPLE 6 **Finding the Inverse Function**

Find the inverse of $f(x) = 2x + 3$. Also find the domain and range of f and f^{-1}. Graph f and f^{-1} on the same coordinate axes.

Solution In the equation $y = 2x + 3$, interchange the variables x and y. The result,

$$x = 2y + 3$$

is an equation that defines the inverse f^{-1} implicitly. To find the explicit form, we solve for y.

$$2y + 3 = x$$
$$2y = x - 3$$
$$y = \frac{1}{2}(x - 3)$$

The explicit form of the inverse f^{-1} is therefore

$$f^{-1}(x) = \frac{1}{2}(x - 3)$$

which we verified in Example 4(c).

Next we find

$$\text{Domain of } f = \text{Range of } f^{-1} = (-\infty, \infty)$$
$$\text{Range of } f = \text{Domain of } f^{-1} = (-\infty, \infty)$$

The graphs of $Y_1 = f(x) = 2x + 3$ and its inverse $Y_2 = f^{-1}(x) = \frac{1}{2}(x - 3)$ are shown in Figure 10. Note the symmetry of the graphs with respect to the line $Y_3 = x$.

Figure 10

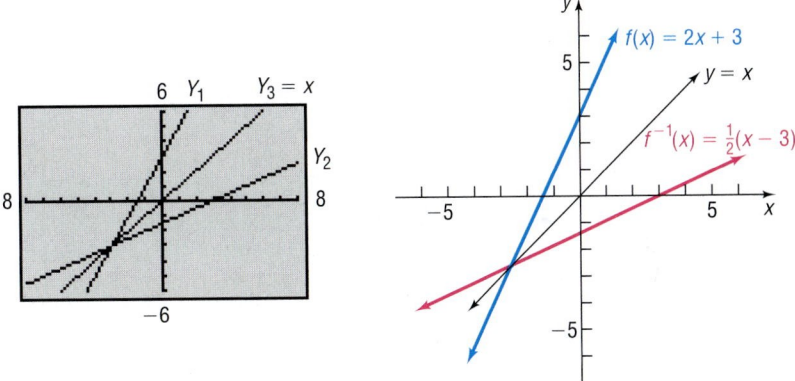

We outline next the steps to follow for finding the inverse of a one-to-one function.

Procedure for Finding the Inverse of a One-to-One Function

STEP 1: In $y = f(x)$, interchange the variables x and y to obtain

$$x = f(y)$$

This equation defines the inverse function f^{-1} implicitly.

STEP 2: If possible, solve the implicit equation for y in terms of x to obtain the explicit form of f^{-1}.

$$y = f^{-1}(x)$$

STEP 3: Check the result by showing that

$$f^{-1}(f(x)) = x \quad \text{and} \quad f(f^{-1}(x)) = x$$

EXAMPLE 7   **Finding the Inverse Function**

The function

$$f(x) = \frac{2x + 1}{x - 1}, \qquad x \neq 1$$

is one-to-one. Find its inverse and check the result.

Solution   **STEP 1:** Interchange the variables x and y in

$$y = \frac{2x + 1}{x - 1}$$

to obtain

$$x = \frac{2y + 1}{y - 1}$$

STEP 2: Solve for y.

$$x = \frac{2y + 1}{y - 1}$$

$x(y - 1) = 2y + 1$ Multiply both sides by y − 1.

$xy - x = 2y + 1$ Apply the distributive property.

$xy - 2y = x + 1$ Subtract 2y from both sides; add x to both sides.

$(x - 2)y = x + 1$ Factor.

$y = \dfrac{x + 1}{x - 2}$ Divide by x − 2.

The inverse is

$$f^{-1}(x) = \frac{x + 1}{x - 2}, \qquad x \neq 2 \qquad \text{Replace y by } f^{-1}(x).$$

STEP 3: ✔ CHECK:

$$f^{-1}(f(x)) = f^{-1}\!\left(\frac{2x + 1}{x - 1}\right) = \frac{\dfrac{2x + 1}{x - 1} + 1}{\dfrac{2x + 1}{x - 1} - 2} = \frac{2x + 1 + x - 1}{2x + 1 - 2(x - 1)} = \frac{3x}{3} = x$$

$$f(f^{-1}(x)) = f\!\left(\frac{x + 1}{x - 2}\right) = \frac{2\!\left(\dfrac{x + 1}{x - 2}\right) + 1}{\dfrac{x + 1}{x - 2} - 1} = \frac{2(x + 1) + x - 2}{x + 1 - (x - 2)} = \frac{3x}{3} = x$$

■ ■

🖳 **EXPLORATION**   In Example 7, we found that, if $f(x) = \dfrac{2x + 1}{x - 1}$, then $f^{-1}(x) = \dfrac{x + 1}{x - 2}$. Compare the vertical and horizontal asymptotes of f and f^{-1}. What did you find? Are you surprised?   ▧

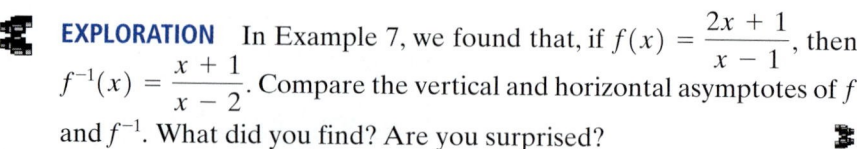

NOW WORK PROBLEM 33.

We said in Chapter 2 that finding the range of a function f is not easy. However, if f is one-to-one, we can find its range by finding the domain of the inverse function f^{-1}.

EXAMPLE 8   Finding the Range of a Function

Find the domain and range of

$$f(x) = \frac{2x + 1}{x - 1}$$

Solution   The domain of f is $\{x \mid x \neq 1\}$. To find the range of f, we first find the inverse f^{-1}. Based on Example 7, we have

$$f^{-1}(x) = \frac{x + 1}{x - 2}$$

The domain of f^{-1} is $\{x \mid x \neq 2\}$, so the range of f is $\{y \mid y \neq 2\}$. ■

━━ **NOW WORK PROBLEM 47.**

If a function is not one-to-one, then its inverse is not a function. Sometimes, though, an appropriate restriction on the domain of such a function will yield a new function that is one-to-one. Let's look at an example of this common practice.

EXAMPLE 9   Finding the Inverse of a Domain-restricted Function

Find the inverse of $y = f(x) = x^2$ if $x \geq 0$.

Solution   The function $y = x^2$ is not one-to-one. [Refer to Example 3(a).] However, if we restrict this function to only that part of its domain for which $x \geq 0$, as indicated, we have a new function that is increasing and therefore is one-to-one. As a result, the function defined by $y = f(x) = x^2$, $x \geq 0$, has an inverse function, f^{-1}.

We follow the steps given previously to find f^{-1}.

STEP 1:   In the equation $y = x^2$, $x \geq 0$, interchange the variables x and y. The result is

$$x = y^2, \qquad y \geq 0$$

This equation defines (implicitly) the inverse function.

STEP 2:   We solve for y to get the explicit form of the inverse. Since $y \geq 0$, only one solution for y is obtained.

$$y = \sqrt{x}$$

So $f^{-1}(x) = \sqrt{x}$.

STEP 3:   ✔ CHECK: $f^{-1}(f(x)) = f^{-1}(x^2) = \sqrt{x^2} = |x| = x$, since $x \geq 0$

$$f(f^{-1}(x)) = f(\sqrt{x}) = (\sqrt{x})^2 = x$$ ■

Figure 11 illustrates the graphs of $Y_1 = f(x) = x^2, x \geq 0$, and $Y_2 = f^{-1}(x) = \sqrt{x}$.

Figure 11

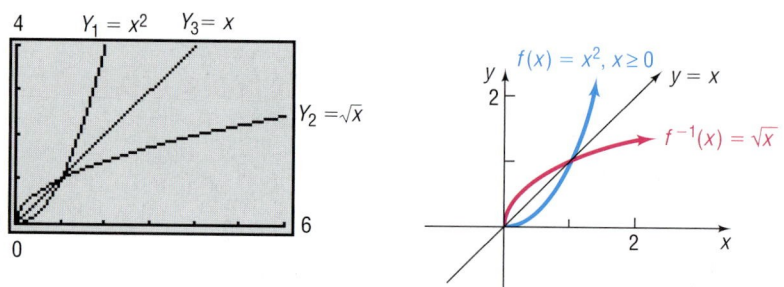

SUMMARY

1. If a function f is one-to-one, then it has an inverse function f^{-1}.
2. Domain of f = Range of f^{-1}; Range of f = Domain of f^{-1}.
3. To verify that f^{-1} is the inverse of f, show that $f^{-1}(f(x)) = x$ and $f(f^{-1}(x)) = x$.
4. The graphs of f and f^{-1} are symmetric with respect to the line $y = x$.
5. To find the range of a one-to-one function f, find the domain of the inverse function f^{-1}.

6.1 Concepts and Vocabulary

In Problems 1–3, fill in the blanks.

1. If every horizontal line intersects the graph of a function f at no more than one point, then f is a(n) _____ function.

2. If f^{-1} denotes the inverse of a function f, then the graphs of f and f^{-1} are symmetric with respect to the line _____.

3. If the domain of a one-to-one function f is $[4, \infty)$, then the range of its inverse, f^{-1}, is _____.

In Problems 4 and 5, answer True or False to each statement.

4. If f and g are inverse functions, then the domain of f is the same as the domain of g.

5. If f and g are inverse functions, then their graphs are symmetric with respect to the line $y = x$.

6. If a function f is even, can it be one-to-one? Explain.

7. Is every odd function one-to-one? Explain.

8. If the graph of a function and its inverse intersect, where must this necessarily occur? Can they intersect anywhere else? Must they intersect?

6.1 Exercises

In Problems 1–8, (a) find the inverse and (b) determine whether the inverse represents a function.

1.

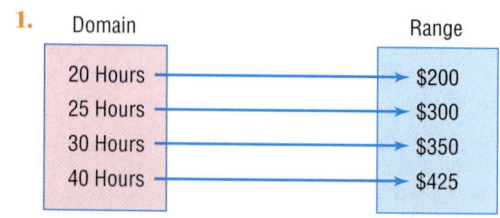

2.

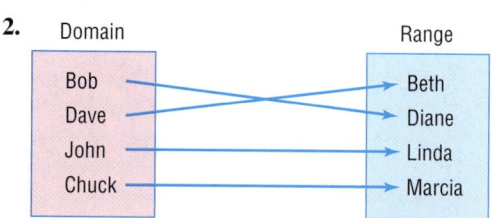

3.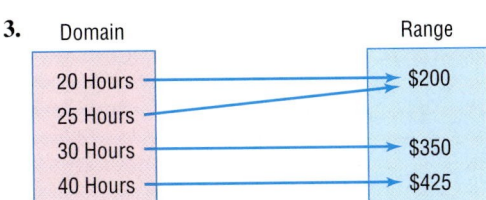

4.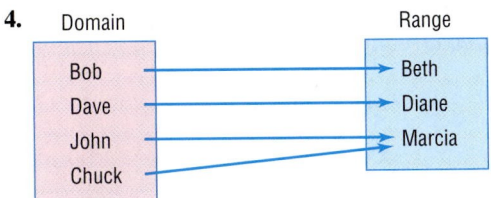

5. $\{(2,6),(-3,6),(4,9),(1,10)\}$

6. $\{(-2,5),(-1,3),(3,7),(4,12)\}$

7. $\{(0,0),(1,1),(2,16),(3,81)\}$

8. $\{(1,2),(2,8),(3,18),(4,32)\}$

In Problems 9–14, the graph of a function f is given. Use the horizontal line test to determine whether f is one-to-one.

9.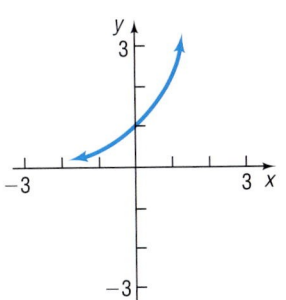

10.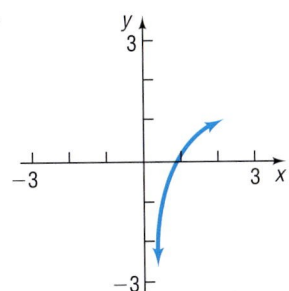

11.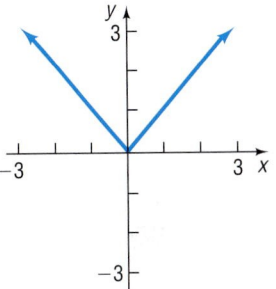

12.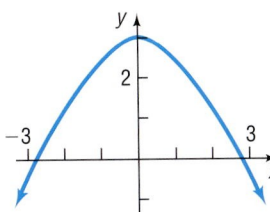

13.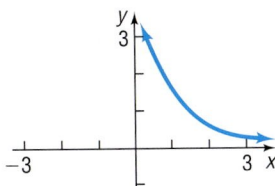

14.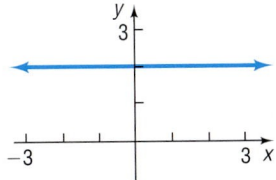

In Problems 15–20, the graph of a one-to-one function f is given. Draw the graph of the inverse function f^{-1}. For convenience (and as a hint), the graph of $y = x$ is also given.

15.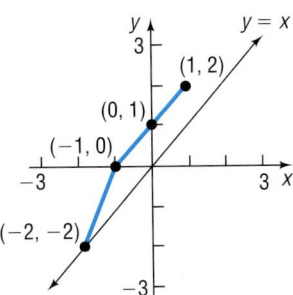

16.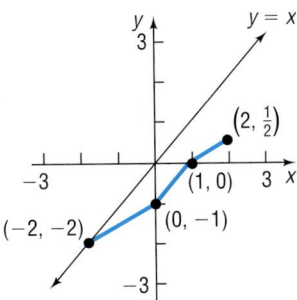

17.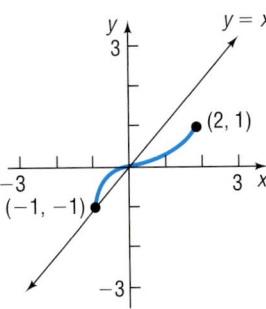

18.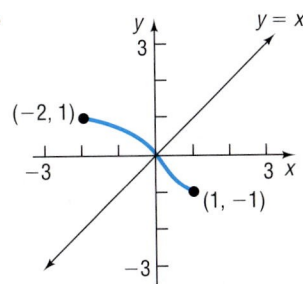

19.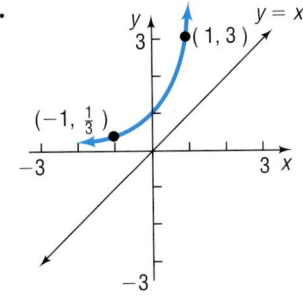

20.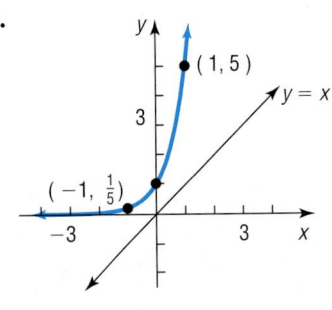

In Problems 21–30, verify that the functions f and g are inverses of each other by showing that $f(g(x)) = x$ and $g(f(x)) = x$. Using a graphing utility, simultaneously graph f, g, and $y = x$ on the same square screen.

21. $f(x) = 3x + 4$; $g(x) = \dfrac{1}{3}(x - 4)$

22. $f(x) = 3 - 2x$; $g(x) = -\dfrac{1}{2}(x - 3)$

23. $f(x) = 4x - 8$; $g(x) = \dfrac{x}{4} + 2$

24. $f(x) = 2x + 6$; $g(x) = \dfrac{1}{2}x - 3$

25. $f(x) = x^3 - 8$; $g(x) = \sqrt[3]{x + 8}$

26. $f(x) = (x - 2)^2$, $x \geq 2$; $g(x) = \sqrt{x} + 2$, $x \geq 0$

27. $f(x) = \dfrac{1}{x}$; $g(x) = \dfrac{1}{x}$

28. $f(x) = x$; $g(x) = x$

29. $f(x) = \dfrac{2x + 3}{x + 4}$; $g(x) = \dfrac{4x - 3}{2 - x}$

30. $f(x) = \dfrac{x - 5}{2x + 3}$; $g(x) = \dfrac{3x + 5}{1 - 2x}$

In Problems 31–42, the function f is one-to-one. Find its inverse and check your answer. State the domain and range of f and f^{-1}. By hand, graph f, f^{-1}, and $y = x$ on the same coordinate axes. Check your results using a graphing utility.

31. $f(x) = 3x$

32. $f(x) = -4x$

33. $f(x) = 4x + 2$

34. $f(x) = 1 - 3x$

35. $f(x) = x^3 - 1$

36. $f(x) = x^3 + 1$

37. $f(x) = x^2 + 4$, $x \geq 0$

38. $f(x) = x^2 + 9$, $x \geq 0$

39. $f(x) = \dfrac{4}{x}$

40. $f(x) = -\dfrac{3}{x}$

41. $f(x) = \dfrac{1}{x - 2}$

42. $f(x) = \dfrac{4}{x + 2}$

In Problems 43–54, the function f is one-to-one. Find its inverse and check your answer. State the domain of f and find its range using f^{-1}. Using a graphing utility, simultaneously graph f, f^{-1}, and $y = x$ on the same square screen.

43. $f(x) = \dfrac{2}{3 + x}$

44. $f(x) = \dfrac{4}{2 - x}$

45. $f(x) = (x + 2)^2$, $x \geq -2$

46. $f(x) = (x - 1)^2$, $x \geq 1$

47. $f(x) = \dfrac{2x}{x - 1}$

48. $f(x) = \dfrac{3x + 1}{x}$

49. $f(x) = \dfrac{3x + 4}{2x - 3}$

50. $f(x) = \dfrac{2x - 3}{x + 4}$

51. $f(x) = \dfrac{2x + 3}{x + 2}$

52. $f(x) = \dfrac{-3x - 4}{x - 2}$

53. $f(x) = 2\sqrt[3]{x}$

54. $f(x) = \dfrac{4}{\sqrt{x}}$

---

55. Find the inverse of the linear function

$$f(x) = mx + b, \quad m \neq 0$$

56. Find the inverse of the function

$$f(x) = \sqrt{r^2 - x^2}, \quad 0 \leq x \leq r$$

57. A function f has an inverse function. If the graph of f lies in quadrant I, in which quadrant does the graph of f^{-1} lie?

58. A function f has an inverse function. If the graph of f lies in quadrant II, in which quadrant does the graph of f^{-1} lie?

59. The function $f(x) = |x|$ is not one-to-one. Find a suitable restriction on the domain of f so that the new function that results is one-to-one. Then find the inverse of f.

60. The function $f(x) = x^4$ is not one-to-one. Find a suitable restriction on the domain of f so that the new function that results is one-to-one. Then find the inverse of f.

61. Temperature Conversion To convert from x degrees Celsius to y degrees Fahrenheit, we use the formula $y = f(x) = \dfrac{9}{5}x + 32$. To convert from x degrees Fahrenheit to y degrees Celsius, we use the formula $y = g(x) = \dfrac{5}{9}(x - 32)$. Show that f and g are inverse functions.

62. Demand for Corn The demand for corn obeys the equation $p(x) = 300 - 50x$, where p is the price per bushel (in dollars) and x is the number of bushels produced, in millions. Express the production amount x as a function of the price p.

63. Period of a Pendulum The period T (in seconds) of a simple pendulum is a function of its length l (in feet), given by $T(l) = 2\pi\sqrt{\dfrac{l}{g}}$, where $g \approx 32.2$ feet per second per second is the acceleration of gravity. Express the length l as a function of the period T.

64. Given

$$f(x) = \frac{ax + b}{cx + d}$$

find $f^{-1}(x)$. If $c \neq 0$, under what conditions on a, b, c, and d is $f = f^{-1}$?

 65. Can a one-to-one function and its inverse be equal? What must be true about the graph of f for this to happen? Give some examples to support your conclusion.

66. Draw the graph of a one-to-one function that contains the points $(-2, -3)$, $(0, 0)$, and $(1, 5)$. Now draw the graph of its inverse. Compare your graph to those of other students. Discuss any similarities. What differences do you see?

67. Give an example of a function whose domain is the set of real numbers and is neither increasing nor decreasing on its domain, but is one-to-one.

[**Hint:** Use a piecewise-defined function.]

PREPARING FOR THIS SECTION

Before getting started, review the following:

✓ Exponents (Section R.4, pp. 30–35 and Section R.9, pp. 79–82)

✓ Graphing Techniques: Transformations (Section 3.4, pp. 286–295)

✓ Solving Equations (Section 1.3, pp. 109–124, and Section 1.5, pp. 142–147)

✓ Horizontal Asymptotes (Section 4.3, pp. 346–351)

6.2 EXPONENTIAL FUNCTIONS

OBJECTIVES
1. Evaluate Exponential Functions
2. Graph Exponential Functions
3. Define the Number e
4. Solve Exponential Equations

1 In Section R.9, we give a definition for raising a real number a to a rational power. Based on that discussion, we gave meaning to expressions of the form

$$a^r$$

where the base a is a positive real number and the exponent r is a rational number.

 But what is the meaning of a^x, where the base a is a positive real number and the exponent x is an irrational number? Although a rigorous definition requires methods discussed in calculus, the basis for the definition is easy to follow: Select a rational number r that is formed by truncating (removing) all but a finite number of digits from the irrational number x. Then it is reasonable to expect that

$$a^x \approx a^r$$

For example, take the irrational number $\pi = 3.14159\ldots$. Then, an approximation to a^π is

$$a^\pi \approx a^{3.14}$$

where the digits after the hundredths position have been removed from the value for π. A better approximation would be

$$a^\pi \approx a^{3.14159}$$

where the digits after the hundred-thousandths position have been removed. Continuing in this way, we can obtain approximations to a^π to any desired degree of accuracy.

Graphing calculators can easily evaluate expressions of the form a^x as follows. Enter the base a, press the caret key $\boxed{^{}}$, enter the exponent x, and press $\boxed{\text{ENTER}}$.

EXAMPLE 1 **Using a Calculator to Evaluate Powers of 2**

Using a calculator, evaluate:

(a) $2^{1.4}$ (b) $2^{1.41}$ (c) $2^{1.414}$ (d) $2^{1.4142}$ (e) $2^{\sqrt{2}}$

Figure 12

```
2^1.4
         2.639015822
```

Solution Figure 12 shows the solution to part (a) using a TI-83 graphing calculator.

(a) $2^{1.4} \approx 2.639015822$

(b) $2^{1.41} \approx 2.657371628$

(c) $2^{1.414} \approx 2.66474965$

(d) $2^{1.4142} \approx 2.665119089$

(e) $2^{\sqrt{2}} \approx 2.665144143$

 NOW WORK PROBLEM **1**.

It can be shown that the familiar laws for rational exponents hold for real exponents.

Theorem

Laws of Exponents

If s, t, a, and b are real numbers with $a > 0$ and $b > 0$, then

$$a^s \cdot a^t = a^{s+t} \qquad (a^s)^t = a^{st} \qquad (ab)^s = a^s \cdot b^s$$

$$1^s = 1 \qquad a^{-s} = \frac{1}{a^s} = \left(\frac{1}{a}\right)^s \qquad a^0 = 1 \qquad (1)$$

We are now ready for the following definition:

An **exponential function** is a function of the form

$$f(x) = a^x$$

where a is a positive real number ($a > 0$) and $a \neq 1$. The domain of f is the set of all real numbers.

We exclude the base $a = 1$, because this function is simply the constant function $f(x) = 1^x = 1$. We also need to exclude bases that are negative, because, otherwise, we would have to exclude many values of x from the domain, such as $x = \frac{1}{2}$ and $x = \frac{3}{4}$. [Recall that $(-2)^{1/2} = \sqrt{-2}$, $(-3)^{3/4} = \sqrt[4]{(-3)^3} = \sqrt[4]{-27}$, and so on, are not defined in the system of real numbers.]

Some examples of exponential functions are

$$f(x) = 2^x, \qquad F(x) = \left(\frac{1}{3}\right)^x$$

Graphs of Exponential Functions

2 First, we graph the exponential function $f(x) = 2^x$.

| EXAMPLE 2 | **Graphing an Exponential Function** |

Graph the exponential function: $f(x) = 2^x$

Solution   The domain of $f(x) = 2^x$ consists of all real numbers. We begin by locating some points on the graph of $f(x) = 2^x$, as listed in Table 2.

Since $2^x > 0$ for all x, the range of f is $(0, \infty)$. From this, we conclude that the graph has no x-intercepts, and, in fact, the graph will lie above the x-axis for all x. As Table 2 indicates, the y-intercept is 1. Table 2 also indicates that as $x \to -\infty$ the value of $f(x) = 2^x$ gets closer and closer to 0. We conclude that the x-axis ($y = 0$) is a horizontal asymptote to the graph as $x \to -\infty$. This gives us the end behavior for x large and negative.

To determine the end behavior for x large and positive, look again at Table 2. As $x \to \infty$, $f(x) = 2^x$ grows very quickly, causing the graph of $f(x) = 2^x$ to rise very rapidly. It is apparent that f is an increasing function and hence is one-to-one.

Figure 13 shows the graph of $f(x) = 2^x$. Notice that all the conclusions given earlier are confirmed by the graph.   ■

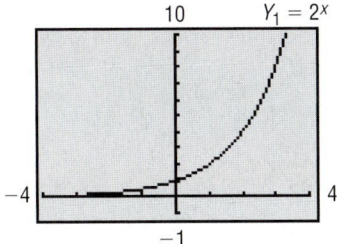

TABLE 2

Figure 13

As we shall see, graphs that look like the one in Figure 13 occur very frequently in a variety of situations. For example, look at the graph in Figure 14, which illustrates the closing price of a share of Harley Davidson stock. Investors might conclude from this graph that the price of Harley Davidson is *behaving exponentially*; that is, the graph exhibits rapid, or exponential, growth. We shall have more to say about situations that lead to exponential growth later in this chapter. For now, we continue to seek properties of the exponential functions.

Figure 14

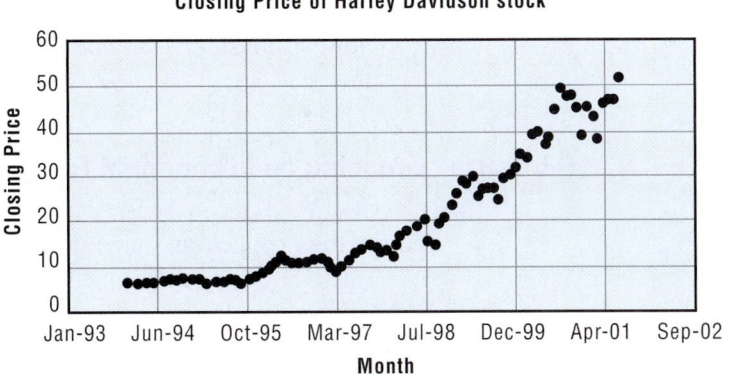

Closing Price of Harley Davidson stock

The graph of $f(x) = 2^x$ in Figure 13 is typical of all exponential functions that have a base larger than 1. Such functions are increasing functions and hence are one-to-one. Their graphs lie above the x-axis, pass through the point $(0, 1)$, and thereafter rise rapidly as $x \to \infty$. As $x \to -\infty$, the x-axis ($y = 0$) is a horizontal asymptote. There are no vertical asymptotes. Finally, the graphs are smooth and continuous, with no corners or gaps.

Figure 15 illustrates the graphs of two more exponential functions whose bases are larger than 1. Notice that for the larger base the graph is steeper when $x > 0$. Figure 16 shows that when $x < 0$ the graph of the equation with the larger base is closer to the x-axis.

Figure 15

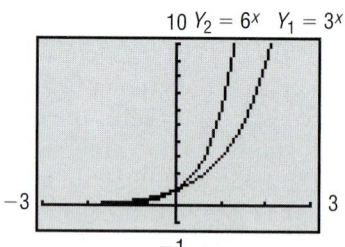

Figure 16

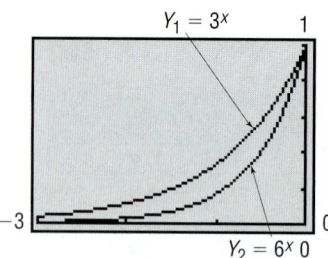

The following display summarizes the information that we have about $f(x) = a^x, a > 1$.

Figure 17
$f(x) = a^x, a > 1$

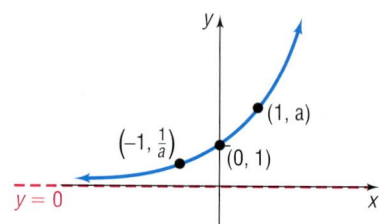

Properties of the Graph of an Exponential Function
$f(x) = a^x, a > 1$

1. The domain is all real numbers; the range is the set of positive real numbers.

2. There are no x-intercepts, the y-intercept is 1.

3. The x-axis ($y = 0$) is a horizontal asymptote as $x \to -\infty$.

4. $f(x) = a^x, a > 1$, is an increasing function and is one-to-one.

5. The graph of f contains the points $(0, 1)$, $(1, a)$, and $\left(-1, \dfrac{1}{a}\right)$.

6. The graph of f is smooth and continuous, with no corners or gaps. See Figure 17.

Now we consider $f(x) = a^x$ when $0 < a < 1$.

EXAMPLE 3   Graphing an Exponential Function

Graph the exponential function: $f(x) = \left(\dfrac{1}{2}\right)^x$

Solution   The domain of $f(x) = \left(\dfrac{1}{2}\right)^x$ consists of all real numbers. As before, we locate some points on the graph, as listed in Table 3. Since $\left(\dfrac{1}{2}\right)^x > 0$ for all x, the range of f is the interval $(0, \infty)$. The graph lies above the x-axis and so has no x-intercepts. The y-intercept is 1. As $x \to -\infty$, $f(x) = \left(\dfrac{1}{2}\right)^x$ grows very quickly. As $x \to \infty$, the values of $f(x)$ approach 0. The x-axis ($y = 0$) is a horizontal asymptote as $x \to \infty$. It is apparent that f is a decreasing function and hence is one-to-one. Figure 18 illustrates the graph.

TABLE 3

X	Y1
-10	1024
-3	8
-1	2
0	1
1	.5
3	.125
10	9.8E-4

Y1 ▤ (1/2)^X

Figure 18

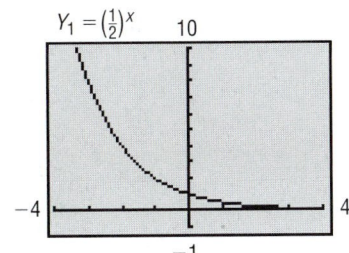

We could have obtained the graph of $y = \left(\dfrac{1}{2}\right)^x$ from the graph of $y = 2^x$ using transformations. If $f(x) = 2^x$, then $f(-x) = 2^{-x} = \dfrac{1}{2^x} = \left(\dfrac{1}{2}\right)^x$. The graph of $y = \left(\dfrac{1}{2}\right)^x = 2^{-x}$ is a reflection about the y-axis of the graph of $y = 2^x$. Compare Figures 13 and 18.

SEEING THE CONCEPT Using a graphing utility, simultaneously graph:

(a) $Y_1 = 3^x, Y_2 = \left(\dfrac{1}{3}\right)^x$ (b) $Y_1 = 6^x, Y_2 = \left(\dfrac{1}{6}\right)^x$

Conclude that the graph of $Y_2 = \left(\dfrac{1}{a}\right)^x$, for $a > 0$, is the reflection about the y-axis of the graph of $Y_1 = a^x$.

The graph of $f(x) = \left(\dfrac{1}{2}\right)^x$ in Figure 18 is typical of all exponential functions that have a base between 0 and 1. Such functions are decreasing and one-to-one. Their graphs lie above the x-axis and pass through the point $(0, 1)$. The graphs rise rapidly as $x \to -\infty$. As $x \to \infty$, the x-axis ($y = 0$) is a horizontal asymptote. There are no vertical asymptotes. Finally, the graphs are smooth and continuous, with no corners or gaps.

Figure 19 illustrates the graphs of two more exponential functions whose bases are between 0 and 1. Notice that the smaller base results in a graph that is steeper when $x < 0$. Figure 20 shows that when $x > 0$ the graph of the equation with the smaller base is closer to the x-axis.

Figure 19

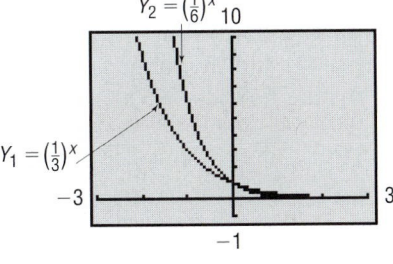

Figure 20

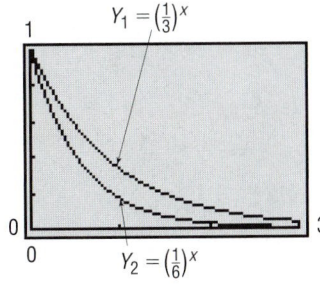

The following display summarizes the information that we have about the function $f(x) = a^x, 0 < a < 1$.

Figure 21
$f(x) = a^x, 0 < a < 1$

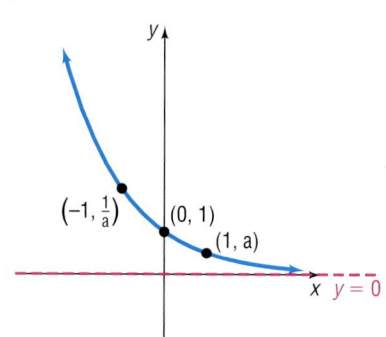

Properties of the Graph of an Exponential Function
$f(x) = a^x, 0 < a < 1$

1. The domain is all real numbers; the range is the set of positive real numbers.
2. There are no x-intercepts; the y-intercept is 1.
3. The x-axis ($y = 0$) is a horizontal asymptote as $x \to \infty$.
4. $f(x) = a^x, 0 < a < 1$, is a decreasing function and is one-to-one.
5. The graph of f contains the points $(0, 1)$, $(1, a)$, and $\left(-1, \dfrac{1}{a}\right)$.
6. The graph of f is smooth and continuous, with no corners or gaps. See Figure 21.

EXAMPLE 4   **Graphing Exponential Functions Using Transformations**

Graph $f(x) = 2^{-x} - 3$ and determine the domain, range, and horizontal asymptote of f.

Solution   We begin with the graph of $y = 2^x$. Figure 22 shows the stages.

Figure 22

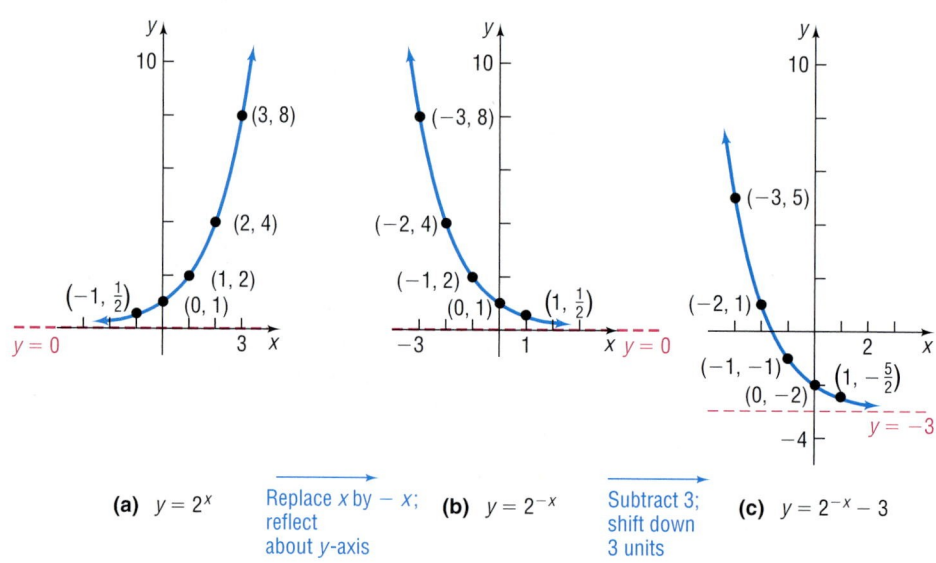

(a) $y = 2^x$ Replace x by $-x$; reflect about y-axis   (b) $y = 2^{-x}$ Subtract 3; shift down 3 units   (c) $y = 2^{-x} - 3$

As Figure 22(c) illustrates, the domain of $f(x) = 2^{-x} - 3$ is the interval $(-\infty, \infty)$ and the range is the interval $(-3, \infty)$. The horizontal asymptote of f is the line $y = -3$.

✔ CHECK: Graph $Y_1 = 2^{-x} - 3$ to verify the graph obtained in Figure 22(c).

━ NOW WORK PROBLEM **21.**

The Base e

③   As we shall see shortly, many problems that occur in nature require the use of an exponential function whose base is a certain irrational number, symbolized by the letter e.

Let's look at one way of arriving at this important number e.

The **number e** is defined as the number that the expression

$$\left(1 + \frac{1}{n}\right)^n \tag{2}$$

approaches as $n \to \infty$. In calculus, this is expressed using limit notation as

$$e = \lim_{n\to\infty}\left(1 + \frac{1}{n}\right)^n$$

Table 4 illustrates what happens to the defining expression (2) as n takes on increasingly large values. The last number in the right column in the table is correct to nine decimal places and is the same as the entry given for e on your calculator (if expressed correctly to nine decimal places).

TABLE 4

n	$\dfrac{1}{n}$	$1 + \dfrac{1}{n}$	$\left(1 + \dfrac{1}{n}\right)^n$
1	1	2	2
2	0.5	1.5	2.25
5	0.2	1.2	2.48832
10	0.1	1.1	2.59374246
100	0.01	1.01	2.704813829
1,000	0.001	1.001	2.716923932
10,000	0.0001	1.0001	2.718145927
100,000	0.00001	1.00001	2.718268237
1,000,000	0.000001	1.000001	2.718280469
1,000,000,000	10^{-9}	$1 + 10^{-9}$	2.718281827

TABLE 5

X	Y1
-2	.13534
-1	.36788
0	1
1	2.7183
2	7.3891

Y1 ▉ e^(X)

The exponential function $f(x) = e^x$, whose base is the number e, occurs with such frequency in applications that it is usually referred to as *the* exponential function. Graphing calculators have the key $\boxed{e^x}$ or $\boxed{\exp(x)}$, which may be used to evaluate the exponential function for a given value of x. Use your calculator to find e^x for $x = -2$, $x = -1$, $x = 0$, $x = 1$, and $x = 2$, as we have done to create Table 5. The graph of the exponential function $f(x) = e^x$ is given in Figure 23. Since $2 < e < 3$, the graph of $y = e^x$ is increasing and lies between the graphs of $y = 2^x$ and $y = 3^x$. [See Figure 23(c)].

Figure 23

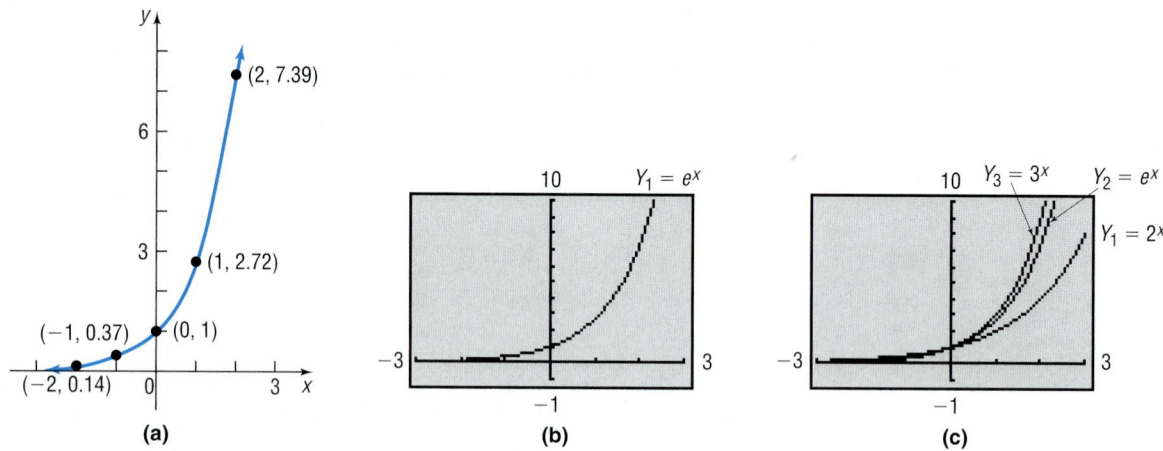

(a) (b) (c)

EXAMPLE 5 **Graphing Exponential Functions Using Transformations**

Graph $f(x) = -e^{x-3}$ and determine the domain, range, and horizontal asymptote of f.

Solution We begin with the graph of $y = e^x$. Figure 24 shows the stages.

Figure 24

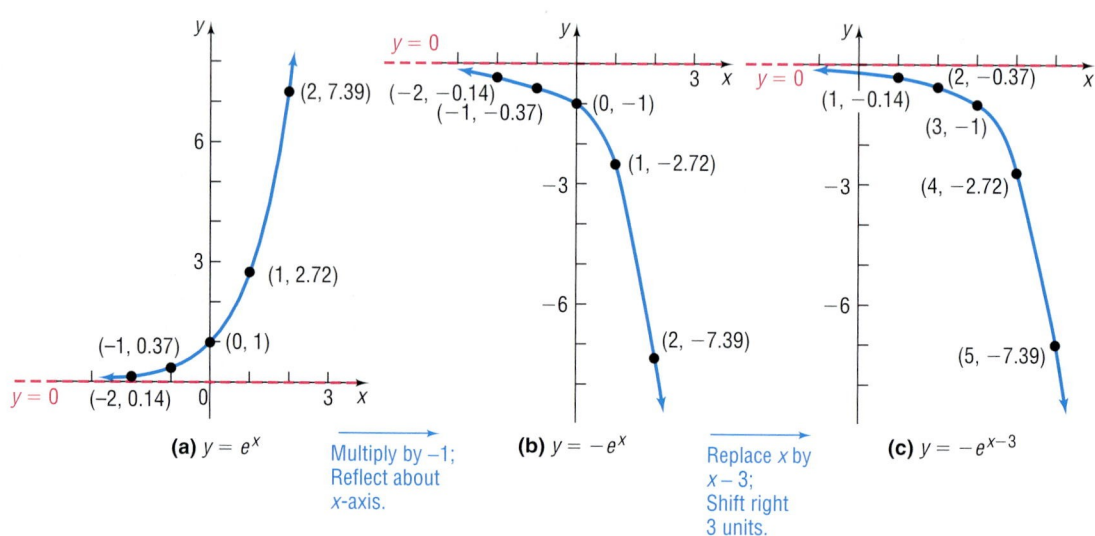

(a) $y = e^x$ Multiply by –1; Reflect about x-axis. **(b)** $y = -e^x$ Replace x by x – 3; Shift right 3 units. **(c)** $y = -e^{x-3}$

As Figure 24(c) illustrates, the domain of $f(x) = -e^{x-3}$ is the interval $(-\infty, \infty)$ and the range is the interval $(-\infty, 0)$. The horizontal asymptote is the line $y = 0$.

✔ CHECK: Graph $Y_1 = -e^{x-3}$ to verify the graph obtained in Figure 24(c). ■ ■

NOW WORK PROBLEM 29.

Exponential Equations

4 Equations that involve terms of the form $a^x, a > 0, a \neq 1$, are often referred to as **exponential equations**. Such equations can sometimes be solved by appropriately applying the Laws of Exponents and property (3).

$$\text{If} \quad a^u = a^v, \quad \text{then} \quad u = v \qquad (3)$$

Property (3) is a consequence of the fact that exponential functions are one-to-one. To use property (3), each side of the equality must be written with the same base.

EXAMPLE 6 | **Solving an Exponential Equation**

Solve: $3^{x+1} = 81$

Solution Since $81 = 3^4$, we can write the equation as

$$3^{x+1} = 81 = 3^4$$

Figure 25

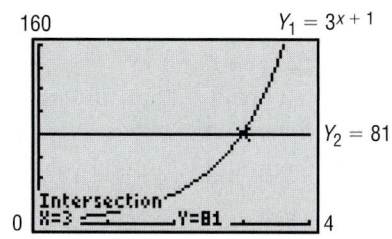

Now we have the same base, 3, on each side, so we can apply property (3) to obtain

$$x + 1 = 4$$
$$x = 3$$

✔ CHECK: We verify the solution by graphing $Y_1 = 3^{x+1}$ and $Y_2 = 81$ to determine where the graphs intersect. See Figure 25. The graphs intersect at $x = 3$. ∎

NOW WORK PROBLEM 43.

EXAMPLE 7 | **Solving an Exponential Equation**

Solve: $e^{-x^2} = (e^x)^2 \cdot \dfrac{1}{e^3}$

Solution We use Laws of Exponents first to get the base e on the right side.

$$(e^x)^2 \cdot \frac{1}{e^3} = e^{2x} \cdot e^{-3} = e^{2x-3}$$

As a result,

$$
\begin{array}{ll}
e^{-x^2} = e^{2x-3} & \\
-x^2 = 2x - 3 & \text{Apply Property (3).} \\
x^2 + 2x - 3 = 0 & \text{Place the quadratic equation in standard form.} \\
(x + 3)(x - 1) = 0 & \text{Factor.} \\
x = -3 \quad \text{or} \quad x = 1 & \text{Use the Zero-Product Property.}
\end{array}
$$

The solution set is $\{-3, 1\}$. ∎

You should verify these solutions using a graphing utility.

Many applications involve the exponential functions. Let's look at one.

EXAMPLE 8   **Exponential Probability**

Between 9:00 PM and 10:00 PM cars arrive at Burger King's drive-thru at the rate of 12 cars per hour (0.2 car per minute). The following formula from statistics can be used to determine the probability that a car will arrive within t minutes of 9:00 PM.

$$F(t) = 1 - e^{-0.2t}$$

(a) Determine the probability that a car will arrive within 5 minutes of 9 PM (that is, before 9:05 PM).

(b) Determine the probability that a car will arrive within 30 minutes of 9 PM (before 9:30 PM).

(c) Graph F using your graphing utility.

(d) What value does F approach as t becomes unbounded in the positive direction?

Solution   (a) The probability that a car will arrive within 5 minutes is found by evaluating $F(t)$ at $t = 5$.

$$F(5) = 1 - e^{-0.2(5)}$$

Figure 26

We evaluate this expression in Figure 26. We conclude that there is a 63% probability that a car will arrive within 5 minutes.

(b) The probability that a car will arrive within 30 minutes is found by evaluating $F(t)$ at $t = 30$.

$$F(30) = 1 - e^{-0.2(30)} \approx 0.9975$$

There is a 99.75% probability that a car will arrive within 30 minutes.

(c) See Figure 27 for the graph of F.

Figure 27

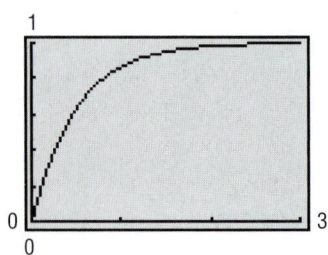

(d) As time passes, the probability that a car will arrive increases. The value that F approaches can be found by letting $t \to \infty$. Since $e^{-0.2t} = \dfrac{1}{e^{0.2t}}$, it follows that $e^{-0.2t} \to 0$ as $t \to \infty$. Thus, F approaches 1 as t gets large. ∎

NOW WORK PROBLEM **67.**

SUMMARY   **Properties of Exponential Functions**

$f(x) = a^x, \quad a > 1$
Domain: the interval $(-\infty, \infty)$; Range: the interval $(0, \infty)$
x-intercepts: none; y-intercept: 1; horizontal asymptote: x-axis ($y = 0$) as $x \to -\infty$; increasing; one-to-one; smooth; continuous
See Figure 17 for a typical graph.

$f(x) = a^x, \quad 0 < a < 1$
Domain: the interval $(-\infty, \infty)$; Range: the interval $(0, \infty)$
x-intercepts: none; y-intercept: 1; horizontal asymptote; x-axis ($y = 0$) as $x \to \infty$; decreasing; one-to-one; smooth; continuous
See Figure 21 for a typical graph.

If $a^u = a^v$, then $u = v$.

6.2 Concepts and Vocabulary

In Problems 1–3, fill in the blanks.

1. The graph of every exponential function $f(x) = a^x$, $a > 0$, $a \neq 1$, passes through three points: _____ , _____ , and _____ .

2. If the graph of every exponential function $f(x) = a^x$, $a > 0$, $a \neq 1$, is decreasing, then a must be less than _____ .

3. If $3^x = 3^4$, then $x =$ _____ .

In Problems 4–6, answer True or False to each statement.

4. The graph of an exponential function $f(x) = a^x$, $a < 0$, is decreasing.

5. The graphs of $y = 3^x$ and $y = \left(\dfrac{1}{3}\right)^x$ are identical.

6. The range of every exponential function $f(x) = a^x$, $a > 0$, $a \neq 1$, is all real numbers.

7. As the base a of an exponential function $f(x) = a^x$, $a > 1$, increases, what happens to the behavior of its graph for $x > 0$? What happens to the behavior of the graph for $x < 0$?

8. The graphs of $y = a^{-x}$ and $y = \left(\dfrac{1}{a}\right)^x$ are identical. Why?

6.2 Exercises

In Problems 1–10, approximate each number using a calculator. Express your answer rounded to three decimal places.

1. (a) $3^{2.2}$ (b) $3^{2.23}$ (c) $3^{2.236}$ (d) $3^{\sqrt{5}}$

2. (a) $5^{1.7}$ (b) $5^{1.73}$ (c) $5^{1.732}$ (d) $5^{\sqrt{3}}$

3. (a) $2^{3.14}$ (b) $2^{3.141}$ (c) $2^{3.1415}$ (d) 2^{π}

4. (a) $2^{2.7}$ (b) $2^{2.71}$ (c) $2^{2.718}$ (d) 2^{e}

5. (a) $3.1^{2.7}$ (b) $3.14^{2.71}$ (c) $3.141^{2.718}$ (d) π^{e}

6. (a) $2.7^{3.1}$ (b) $2.71^{3.14}$ (c) $2.718^{3.141}$ (d) e^{π}

7. $e^{1.2}$ **8.** $e^{-1.3}$ **9.** $e^{-0.85}$ **10.** $e^{2.1}$

In Problems 11–18, the graph of a transformed exponential function is given. Match each graph to one of the following functions without the aid of a graphing utility.

A. $y = 3^x$ **B.** $y = 3^{-x}$ **C.** $y = -3^x$ **D.** $y = -3^{-x}$

E. $y = 3^x - 1$ **F.** $y = 3^{x-1}$ **G.** $y = 3^{1-x}$ **H.** $y = 1 - 3^x$

11.

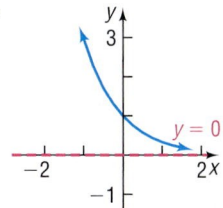

12.

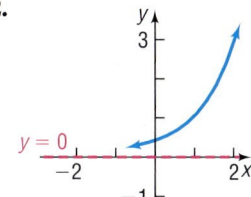

13.

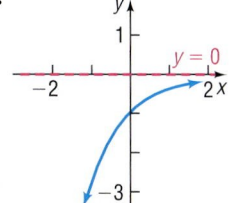

14.

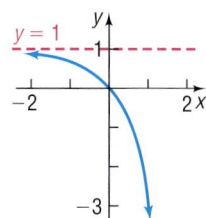

15.

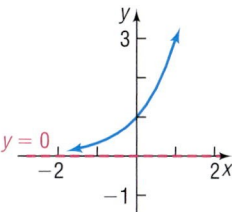

16.

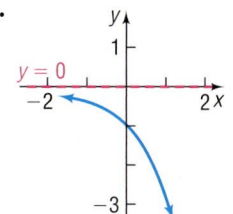

17.

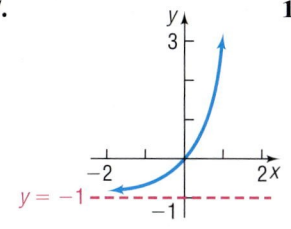

18.

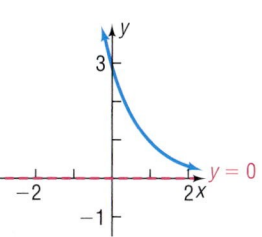

In Problems 19–36, use transformations to graph each function. Determine the domain, range, and horizontal asymptote of each function. Verify your results using a graphing utility.

19. $f(x) = 2^x + 1$

20. $f(x) = 2^{x+2}$

21. $f(x) = 3^{-x} - 2$

22. $f(x) = -3^x + 1$

23. $f(x) = 3(4^x)$

24. $f(x) = \frac{1}{2}(2^x)$

25. $f(x) = 3^{x/2}$

26. $f(x) = -2^{-x/3}$

27. $f(x) = 5 - 2(3^{x+1})$

28. $f(x) = 1 + 3^{x-4}$

29. $f(x) = e^{-x}$

30. $f(x) = -e^x$

31. $f(x) = e^{x+2}$

32. $f(x) = e^x - 1$

33. $f(x) = 5 + e^{-x}$

34. $f(x) = 9 + 3e^x$

35. $f(x) = 2 - e^{x/2}$

36. $f(x) = 7 - 3e^{-2x}$

In Problems 37–42, determine the exponential function whose graph is given.

37.

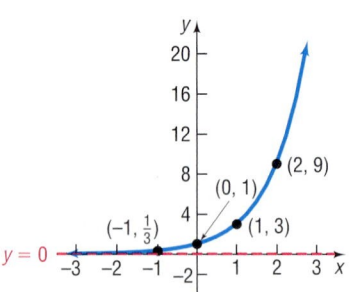

38.

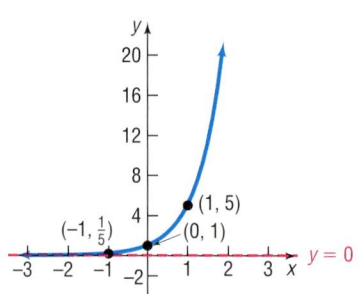

39.

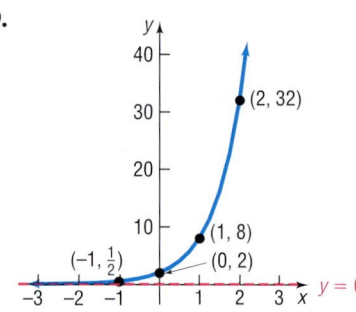

40.

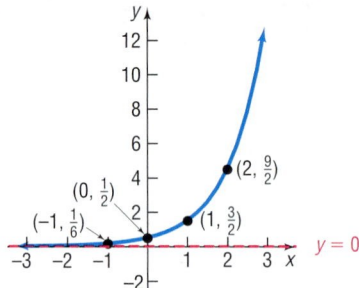

41.

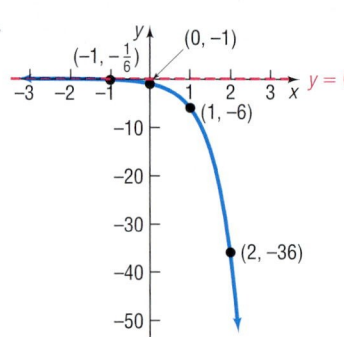

42.

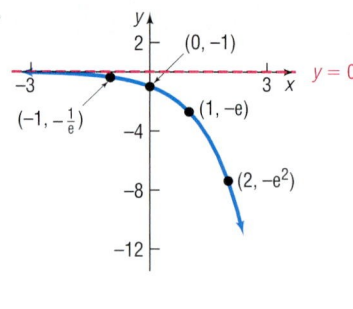

In Problems 43–56, solve each equation. Verify your solution using a graphing utility.

43. $2^{2x+1} = 4$

44. $5^{1-2x} = \frac{1}{5}$

45. $3^{x^3} = 9^x$

46. $4^{x^2} = 2^x$

47. $8^{x^2-2x} = \frac{1}{2}$

48. $9^{-x} = \frac{1}{3}$

49. $2^x \cdot 8^{-x} = 4^x$

50. $\left(\frac{1}{2}\right)^{1-x} = 4$

51. $\left(\frac{1}{5}\right)^{2-x} = 25$

52. $4^x - 2^x = 0$

53. $4^x = 8$

54. $9^{2x} = 27$

55. $e^{x^2} = (e^{3x}) \cdot \frac{1}{e^2}$

56. $(e^4)^x \cdot e^{x^2} = e^{12}$

57. If $4^x = 7$, what does 4^{-2x} equal?

58. If $2^x = 3$, what does 4^{-x} equal?

59. If $3^{-x} = 2$, what does 3^{2x} equal?

60. If $5^{-x} = 3$, what does 5^{3x} equal?

---

61. Optics If a single pane of glass obliterates 3% of the light passing through it, then the percent p of light that passes through n successive panes is given approximately by the function

$$p(n) = 100e^{-0.03n}$$

(a) What percent of light will pass through 10 panes?

(b) What percent of light will pass through 25 panes?

62. Atmospheric Pressure The atmospheric pressure p on a balloon or plane decreases with increasing height. This pressure, measured in millimeters of mercury, is related to the height h (in kilometers) above sea level by the function

$$p(h) = 760e^{-0.145h}$$

(a) Find the atmospheric pressure at a height of 2 kilometers (over a mile).

(b) What is it at a height of 10 kilometers (over 30,000 feet)?

63. Space Satellites The number of watts w provided by a space satellite's power supply over a period of d days is given by the function

$$w(d) = 50e^{-0.004d}$$

(a) How much power will be available after 30 days?
(b) How much power will be available after 1 year (365 days)?

64. Healing of Wounds   The normal healing of wounds can be modeled by an exponential function. If A_0 represents the original area of the wound and if A equals the area of the wound after n days, then the function

$$A(n) = A_0 e^{-0.35n}$$

describes the area of a wound on the nth day following an injury when no infection is present to retard the healing. Suppose that a wound initially had an area of 100 square millimeters.

(a) If healing is taking place, how large should the area of the wound be after 3 days?
(b) How large should it be after 10 days?

65. Drug Medication   The function

$$D(h) = 5e^{-0.4h}$$

can be used to find the number of milligrams D of a certain drug that is in a patient's bloodstream h hours after the drug has been administered. How many milligrams will be present after 1 hour? After 6 hours?

66. Spreading of Rumors   A model for the number of people N in a college community who have heard a certain rumor is

$$N = P(1 - e^{-0.15d})$$

where P is the total population of the community and d is the number of days that have elapsed since the rumor began. In a community of 1000 students, how many students will have heard the rumor after 3 days?

67. Exponential Probability   Between 12:00 PM and 1:00 PM, cars arrive at Citibank's drive-thru at the rate of 6 cars per hour (0.1 car per minute). The following formula from statistics can be used to determine the probability that a car will arrive within t minutes of 12:00 PM.

$$F(t) = 1 - e^{-0.1t}$$

(a) Determine the probability that a car will arrive within 10 minutes of 12:00 PM (that is, before 12:10 PM).
(b) Determine the probability that a car will arrive within 40 minutes of 12:00 PM (before 12:40 PM).
(c) Graph F using your graphing utility.
(d) What value does F approach as t becomes unbounded in the positive direction?

68. Exponential Probability   Between 5:00 PM and 6:00 PM, cars arrive at Jiffy Lube at the rate of 9 cars per hour (0.15 car per minute). The following formula from statistics can be used to determine the probability that a car will arrive within t minutes of 5:00 PM.

$$F(t) = 1 - e^{-0.15t}$$

(a) Determine the probability that a car will arrive within 15 minutes of 5:00 PM (that is, before 5:15 PM).
(b) Determine the probability that a car will arrive within 30 minutes of 5:00 PM (before 5:30 PM).
(c) Graph F using your graphing utility.
(d) What value does F approach as t becomes unbounded in the positive direction?

69. Poisson Probability   Between 5:00 PM and 6:00 PM, cars arrive at McDonald's drive-thru at the rate of 20 cars per hour. The following formula from statistics can be used to determine the probability that x cars will arrive between 5:00 PM and 6:00 PM.

$$P(x) = \frac{20^x e^{-20}}{x!}$$

where

$$x! = x \cdot (x - 1) \cdot (x - 2) \cdots \cdots 3 \cdot 2 \cdot 1$$

(a) Determine the probability that $x = 15$ cars will arrive between 5:00 PM and 6:00 PM.
(b) Determine the probability that $x = 20$ cars will arrive between 5:00 PM and 6:00 PM.

70. Poisson Probability   People enter a line for the *Demon Roller Coaster* at the rate of 4 per minute. The following formula from statistics can be used to determine the probability that x people will arrive within the next minute.

$$P(x) = \frac{4^x e^{-4}}{x!}$$

where

$$x! = x \cdot (x - 1) \cdot (x - 2) \cdots \cdots 3 \cdot 2 \cdot 1$$

(a) Determine the probability that $x = 5$ people will arrive within the next minute.
(b) Determine the probability that $x = 8$ people will arrive within the next minute.

71. Relative Humidity   The relative humidity is the ratio (expressed as a percent) of the amount of water vapor in the air to the maximum amount that it can hold at a specific temperature. The relative humidity, R, is found using the following formula:

$$R = 10^{\left(\frac{4221}{T+459.4} - \frac{4221}{D+459.4} + 2\right)}$$

where T is the air temperature (in °F) and D is the dew point temperature (in °F).

(a) Determine the relative humidity if the air temperature is 50° Fahrenheit and the dew point temperature is 41° Fahrenheit.

(b) Determine the relative humidity if the air temperature is 68° Fahrenheit and the dew point temperature is 59° Fahrenheit.

(c) What is the relative humidity if the air temperature and the dew point temperature are the same?

72. Learning Curve Suppose that a student has 500 vocabulary words to learn. If the student learns 15 words after 5 minutes, the function

$$L(t) = 500(1 - e^{-0.0061t})$$

approximates the number of words L that the student will learn after t minutes.

(a) How many words will the student learn after 30 minutes?

(b) How many words will the student learn after 60 minutes?

73. Current in an *RL* Circuit The equation governing the amount of current I (in amperes) after time t (in seconds) in a single RL circuit consisting of a resistance R (in ohms), an inductance L (in henrys), and an electromotive force E (in volts) is

$$I = \frac{E}{R}[1 - e^{-(R/L)t}]$$

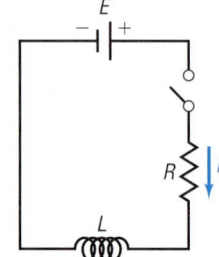

(a) If $E = 120$ volts, $R = 10$ ohms, and $L = 5$ henrys, how much current I_1 is flowing after 0.3 second? After 0.5 second? After 1 second?

(b) What is the maximum current?

(c) Graph this function $I = I_1(t)$, measuring I along the y-axis and t along the x-axis.

(d) If $E = 120$ volts, $R = 5$ ohms, and $L = 10$ henrys, how much current I_2 is flowing after 0.3 second? After 0.5 second? After 1 second?

(e) What is the maximum current?

(f) Graph this function $I = I_2(t)$ on the same viewing window as $I_1(t)$.

74. Current in an *RC* Circuit The equation governing the amount of current I (in amperes) after time t (in microseconds) in a single RC circuit consisting of a resistance R (in ohms), a capacitance C (in microfarads), and an electromotive force E (in volts) is

$$I = \frac{E}{R}e^{-t/(RC)}$$

(a) If $E = 120$ volts, $R = 2000$ ohms, and $C = 1.0$ microfarad, how much current I_1 is flowing initially $(t = 0)$? After 1000 microseconds? After 3000 microseconds?

(b) What is the maximum current?

(c) Graph this function $I = I_1(t)$, measuring I along the y-axis and t along the x-axis.

(d) If $E = 120$ volts, $R = 1000$ ohms, and $C = 2.0$ microfarads, how much current I_2 is flowing initially? After 1000 microseconds? After 3000 microseconds?

(e) What is the maximum current?

(f) Graph this function $I = I_2(t)$ on the same viewing window as $I_1(t)$.

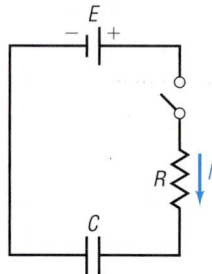

75. Another Formula for e Use a calculator to compute the values of

$$2 + \frac{1}{2!} + \frac{1}{3!} + \cdots + \frac{1}{n!}$$

for $n = 4, 6, 8,$ and 10. Compare each result with e.

[**Hint:** $1! = 1,\ 2! = 2 \cdot 1,\ 3! = 3 \cdot 2 \cdot 1,$
$n! = n(n - 1) \cdots \cdots (3)(2)(1)$]

76. Another Formula for e Use a calculator to compute the various values of the expression. Compare the values to e.

$$2 + \cfrac{1}{1 + \cfrac{1}{2 + \cfrac{2}{3 + \cfrac{3}{4 + \cfrac{4}{\text{etc.}}}}}}$$

77. Difference Quotient If $f(x) = a^x$, show that

$$\frac{f(x + h) - f(x)}{h} = a^x\left(\frac{a^h - 1}{h}\right)$$

78. If $f(x) = a^x$, show that $f(A + B) = f(A) \cdot f(B)$.

79. If $f(x) = a^x$, show that $f(-x) = \dfrac{1}{f(x)}$.

80. If $f(x) = a^x$, show that $f(\alpha x) = [f(x)]^\alpha$.

Problems 81 and 82 provide definitions for two other transcendental functions.

81. The **hyperbolic sine function**, designated by sinh x, is defined as

$$\sinh x = \frac{1}{2}(e^x - e^{-x})$$

(a) Show that $f(x) = \sinh x$ is an odd function.
(b) Graph $f(x) = \sinh x$ using a graphing utility.

82. The **hyperbolic cosine function**, designated by cosh x, is defined as

$$\cosh x = \frac{1}{2}(e^x + e^{-x})$$

(a) Show that $f(x) = \cosh x$ is an even function.
(b) Graph $f(x) = \cosh x$ using a graphing utility.
(c) Refer to Problem 81. Show that, for every x, $(\cosh x)^2 - (\sinh x)^2 = 1$.

83. Historical Problem Pierre de Fermat (1601–1665) conjectured that the function

$$f(x) = 2^{(2^x)} + 1$$

for $x = 1, 2, 3, \ldots$, would always have a value equal to a prime number. But Leonhard Euler (1707–1783) showed that this formula fails for $x = 5$. Use a calculator to determine the prime numbers produced by f for $x = 1, 2, 3, 4$. Then show that $f(5) = 641 \times 6{,}700{,}417$, which is not prime.

84. The bacteria in a 4-liter container double every minute. After 60 minutes the container is full. How long did it take to fill half the container?

85. Explain in your own words what the number e is. Provide at least two applications that require the use of this number.

86. Do you think that there is a power function that increases more rapidly than an exponential function whose base is greater than 1? Explain.

PREPARING FOR THIS SECTION

Before getting started, review the following:

✓ Solving Inequalities (Section 1.6, pp. 153–159)

✓ Polynomial and Rational Inequalities (Section 4.5, pp. 365–370)

6.3 LOGARITHMIC FUNCTIONS

OBJECTIVES
1. Change Exponential Expressions to Logarithmic Expressions
2. Change Logarithmic Expressions to Exponential Expressions
3. Evaluate Logarithmic Functions
4. Determine the Domain of a Logarithmic Function
5. Graph Logarithmic Functions
6. Solve Logarithmic Equations

Recall that a one-to-one function $y = f(x)$ has an inverse function that is defined (implicitly) by the equation $x = f(y)$. In particular, the exponential function $y = f(x) = a^x$, $a > 0$, $a \neq 1$, is one-to-one and hence has an inverse function that is defined implicitly by the equation

$$x = a^y, \qquad a > 0, \quad a \neq 1$$

This inverse function is so important that it is given a name, the *logarithmic function*.

> The **logarithmic function to the base a**, where $a > 0$ and $a \neq 1$, is denoted by $y = \log_a x$ (read as "y is the logarithm to the base a of x") and is defined by
>
> $$y = \log_a x \quad \text{if and only if} \quad x = a^y$$

The domain of the logarithmic function $y = \log_a x$ is $x > 0$.

EXAMPLE 1 **Relating Logarithms to Exponents**

(a) If $y = \log_3 x$, then $x = 3^y$. For example, $4 = \log_3 81$ is equivalent to $81 = 3^4$.

(b) If $y = \log_5 x$, then $x = 5^y$. For example, $-1 = \log_5\left(\dfrac{1}{5}\right)$ is equivalent to $\dfrac{1}{5} = 5^{-1}$. ∎

EXAMPLE 2 **Changing Exponential Expressions to Logarithmic Expressions**

① Change each exponential expression to an equivalent expression involving a logarithm.

(a) $1.2^3 = m$ (b) $e^b = 9$ (c) $a^4 = 24$

Solution We use the fact that $y = \log_a x$ and $x = a^y$, $a > 0, a \neq 1$, are equivalent.

(a) If $1.2^3 = m$, then $3 = \log_{1.2} m$.   (b) If $e^b = 9$, then $b = \log_e 9$.

(c) If $a^4 = 24$, then $4 = \log_a 24$. ∎

NOW WORK PROBLEM **1**.

EXAMPLE 3 **Changing Logarithmic Expressions to Exponential Expressions**

② Change each logarithmic expression to an equivalent expression involving an exponent.

(a) $\log_a 4 = 5$ (b) $\log_e b = -3$ (c) $\log_3 5 = c$

Solution (a) If $\log_a 4 = 5$, then $a^5 = 4$.   (b) If $\log_e b = -3$, then $e^{-3} = b$.

(c) If $\log_3 5 = c$, then $3^c = 5$. ∎

NOW WORK PROBLEM **13**.

③ To find the exact value of a logarithm, we write the logarithm in exponential notation and use the fact that if $a^u = a^v$, then $u = v$.

EXAMPLE 4 **Finding the Exact Value of a Logarithmic Expression**

Find the exact value of:

(a) $\log_2 16$ (b) $\log_3 \dfrac{1}{27}$

Solution (a) $y = \log_2 16$

$2^y = 16$ *Change to exponential form.*

$2^y = 2^4$ *$16 = 2^4$*

$y = 4$ *Equate exponents.*

Therefore, $\log_2 16 = 4$.

(b)
$$y = \log_3 \frac{1}{27}$$

$$3^y = \frac{1}{27} \qquad \textit{Change to exponential form.}$$

$$3^y = 3^{-3} \qquad \frac{1}{27} = \frac{1}{3^3} = 3^{-3}.$$

$$y = -3 \qquad \textit{Equate exponents.}$$

Therefore, $\log_3 \dfrac{1}{27} = -3$.

NOW WORK PROBLEM **25.**

Domain of a Logarithmic Function

④ The logarithmic function $y = \log_a x$ has been defined as the inverse of the exponential function $y = a^x$. That is, if $f(x) = a^x$, then $f^{-1}(x) = \log_a x$. Based on the discussion given in Section 6.1 on inverse functions, we know that, for a function f and its inverse f^{-1},

Domain of f^{-1} = Range of f and Range of f^{-1} = Domain of f

Consequently, it follows that

Domain of logarithmic function = Range of exponential function = $(0, \infty)$

Range of logarithmic function = Domain of exponential function = $(-\infty, \infty)$

In the next box, we summarize some properties of the logarithmic function:

$$y = \log_a x \quad (\text{defining equation:} \quad x = a^y)$$
$$\text{Domain:} \quad 0 < x < \infty \qquad \text{Range:} \quad -\infty < y < \infty$$

Notice that the domain of a logarithmic function consists of the *positive* real numbers. The argument of a logarithmic function must be greater than zero.

EXAMPLE 5

Finding the Domain of a Logarithmic Function

Find the domain of each logarithmic function.

(a) $F(x) = \log_2(x + 3)$ $\qquad$ (b) $g(x) = \log_5\left(\dfrac{1 + x}{1 - x}\right)$

(c) $h(x) = \log_{1/2}|x|$

Solution (a) The domain of F consists of all x for which $x + 3 > 0$, that is, $x > -3$ or, using interval notation, $(-3, \infty)$.

(b) The domain of g is restricted to

$$\frac{1 + x}{1 - x} > 0$$

Solving this inequality, we find that the domain of g consists of all x between -1 and 1, that is, $-1 < x < 1$, or using interval notation, $(-1, 1)$.

(c) Since $|x| > 0$, provided that $x \neq 0$, the domain of h consists of all real numbers except zero or, using interval notation, $(-\infty, 0)$ or $(0, \infty)$. ■

NOW WORK PROBLEMS **39** AND **45**.

Graphs of Logarithmic Functions

⑤ Since exponential functions and logarithmic functions are inverses of each other, the graph of a logarithmic function $y = \log_a x$ is the reflection about the line $y = x$ of the graph of the exponential function $y = a^x$, as shown in Figure 28.

Figure 28

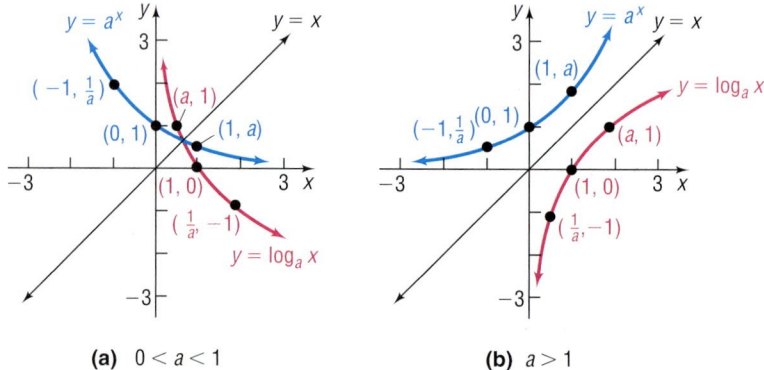

(a) $0 < a < 1$ (b) $a > 1$

Properties of the Graph of a Logarithmic Function $f(x) = \log_a x$

1. The domain is the set of positive real numbers; the range is all real numbers.
2. The x-intercept of the graph is 1. There is no y-intercept.
3. The y-axis ($x = 0$) is a vertical asymptote of the graph.
4. A logarithmic function is decreasing if $0 < a < 1$ and increasing if $a > 1$.
5. The graph of f contains the points $(1, 0)$, $(a, 1)$, and
   $$\left(\frac{1}{a}, -1\right).$$
6. The graph is smooth and continuous, with no corners or gaps.

If the base of a logarithmic function is the number e, then we have the **natural logarithm function**. This function occurs so frequently in applications that it is given a special symbol, **ln** (from the Latin, *logarithmus naturalis*). Thus,

$$y = \ln x \quad \text{if and only if} \quad x = e^y \tag{1}$$

Since $y = \ln x$ and the exponential function $y = e^x$ are inverse functions, we can obtain the graph of $y = \ln x$ by reflecting the graph of $y = e^x$ about the line $y = x$. See Figure 29.

Figure 29

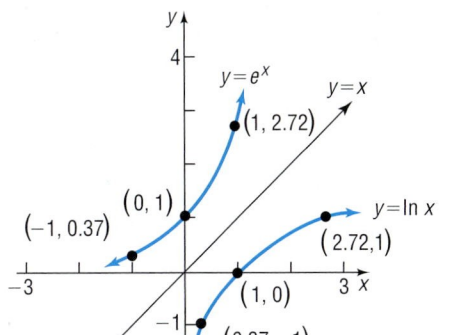

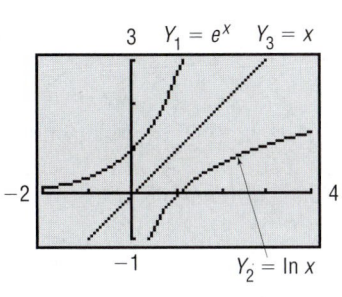

TABLE 6	
X	Y2
-1	ERROR
0.5	-.6931
1	0
2	.69315
2.7183	1
3	1.0986

Y2■ln(X)

Table 6 displays other points on the graph of $f(x) = \ln x$. Notice for $x \le 0$ that we obtain an error message. Do you recall why?

EXAMPLE 6   Graphing Logarithmic Functions Using Transformations

Graph $f(x) = -\ln(x + 2)$ by starting with the graph of $y = \ln x$. Determine the domain, range, and vertical asymptote.

Solution   The domain consists of all x for which

$$x + 2 > 0 \quad \text{or} \quad x > -2$$

To obtain the graph of $y = -\ln(x + 2)$, we use the steps illustrated in Figure 30.

Figure 30

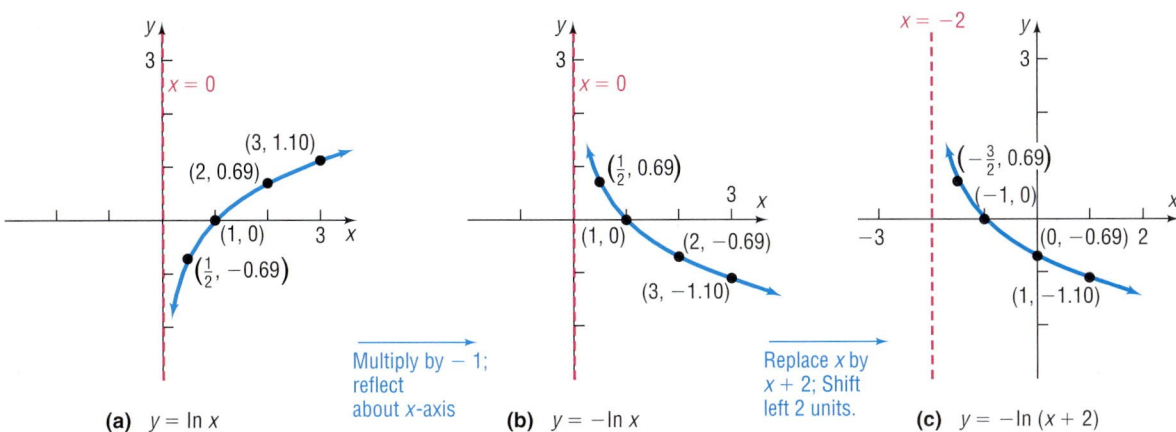

(a) $y = \ln x$

Multiply by -1;
reflect
about x-axis

(b) $y = -\ln x$

Replace x by
$x + 2$; Shift
left 2 units.

(c) $y = -\ln(x + 2)$

The range of $f(x) = -\ln(x + 2)$ is the interval $(-\infty, \infty)$, and the vertical asymptote is $x = -2$. [Do you see why? The original asymptote ($x = 0$) is shifted to the left 2 units.]

✔ CHECK: Graph $Y_1 = -\ln(x + 2)$ using a graphing utility to verify Figure 30(c).

NOW WORK PROBLEM **61.**

If the base of a logarithmic function is the number 10, then we have the **common logarithm function**. If the base a of the logarithmic function is not indicated, it is understood to be 10. Thus,

$$y = \log x \quad \text{if and only if} \quad x = 10^y$$

Since $y = \log x$ and the exponential function $y = 10^x$ are inverse functions, we can obtain the graph of $y = \log x$ by reflecting the graph of $y = 10^x$ about the line $y = x$. See Figure 31.

Figure 31

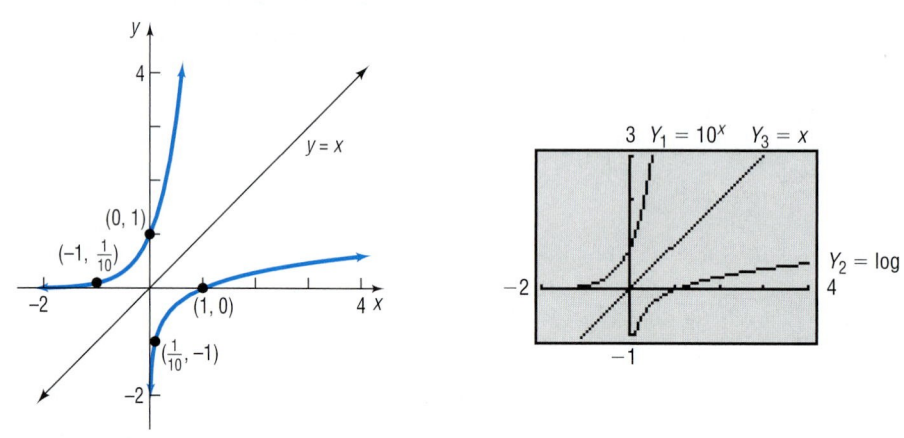

EXAMPLE 7   **Graphing Logarithmic Functions Using Transformations**

Graph $f(x) = 3 \log(x - 1)$. Determine the domain, range, and vertical asymptote of f.

Solution   The domain consists of all x for which

$$x - 1 > 0 \quad \text{or} \quad x > 1$$

To obtain the graph of $y = 3 \log(x - 1)$, we use the steps illustrated in Figure 32.

Figure 32

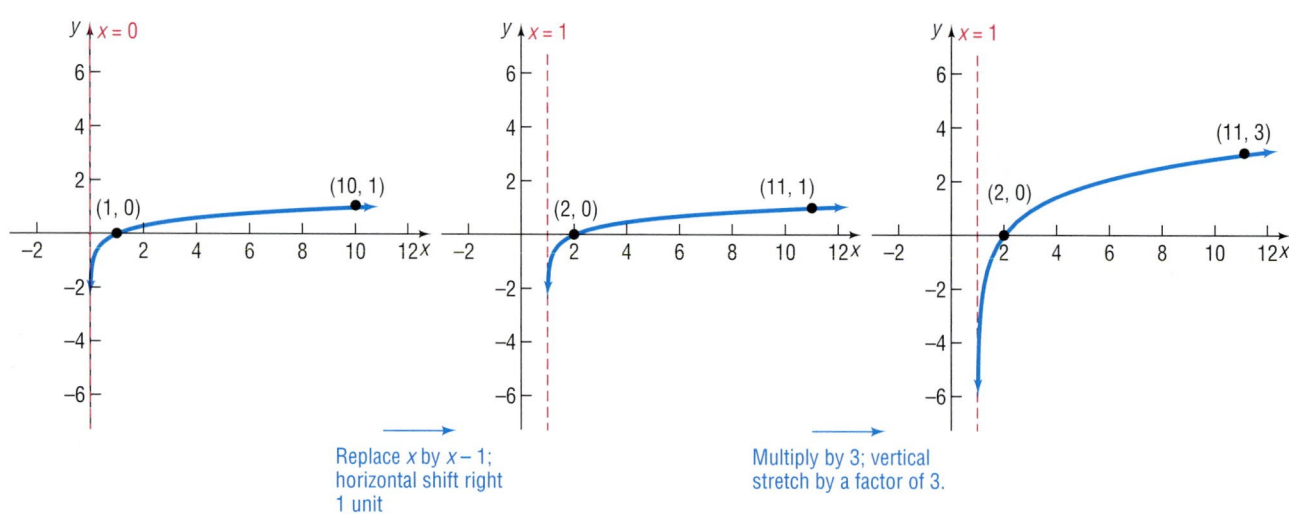

Replace x by $x - 1$; horizontal shift right 1 unit

Multiply by 3; vertical stretch by a factor of 3.

(a) $y = \log x$ **(b)** $y = \log(x - 1)$ **(c)** $y = 3 \log(x - 1)$

The range of $f(x) = 3\log(x - 1)$ is the interval $(-\infty, \infty)$, and the vertical asymptote is $x = 1$.

✔ CHECK: Graph $Y_1 = 3\log(x - 1)$ using a graphing utility to verify Figure 32(c). ■ ■

━━━ NOW WORK PROBLEM **75.**

Logarithmic Equations

6 Equations that contain logarithms are called **logarithmic equations**. Care must be taken when solving logarithmic equations algebraically. Be sure to check each apparent solution in the original equation and discard any that are extraneous. In the expression $\log_a M$, remember that a and M are positive and $a \neq 1$.

Some logarithmic equations can be solved by changing from a logarithmic expression to an exponential expression.

EXAMPLE 8 **Solving a Logarithmic Equation**

Solve: (a) $\log_3(4x - 7) = 2$ (b) $\log_x 64 = 2$

Solution (a) We can obtain an exact solution by changing the logarithm to exponential form.

$$\log_3(4x - 7) = 2$$
$$4x - 7 = 3^2 \quad \text{\textit{Change to exponential form.}}$$
$$4x - 7 = 9$$
$$4x = 16$$
$$x = 4$$

(b) We can obtain an exact solution by changing the logarithm to exponential form.

$$\log_x 64 = 2$$
$$x^2 = 64 \quad \text{\textit{Change to exponential form.}}$$
$$x = \pm\sqrt{64} = \pm 8$$

The base of a logarithm is always positive. As a result, we discard -8; the only solution is 8. ■

EXAMPLE 9 **Using Logarithms to Solve Exponential Equations**

Solve: $e^{2x} = 5$

Solution We can obtain an exact solution by changing the exponential equation to logarithmic form.

$$e^{2x} = 5$$
$$\ln 5 = 2x \quad \text{\textit{Change to a logarithmic expression using (1).}}$$
$$x = \frac{\ln 5}{2} \approx 0.805$$

■

━━━ NOW WORK PROBLEMS **85** AND **97.**

EXAMPLE 10 **Alcohol and Driving**

The concentration of alcohol in a person's blood is measurable. Recent medical research suggests that the risk R (given as a percent) of having an accident while driving a car can be modeled by the equation

$$R = 6e^{kx}$$

where x is the variable concentration of alcohol in the blood and k is a constant.

(a) Suppose that a concentration of alcohol in the blood of 0.04 results in a 10% risk ($R = 10$) of an accident. Find the constant k in the equation. Graph $R = 6e^{kx}$ using this value of k.

(b) Using this value of k, what is the risk if the concentration is 0.17?

(c) Using the same value of k, what concentration of alcohol corresponds to a risk of 100%?

(d) If the law asserts that anyone with a risk of having an accident of 20% or more should not have driving privileges, at what concentration of alcohol in the blood should a driver be arrested and charged with a DUI (Driving Under the Influence)?

Solution (a) For a concentration of alcohol in the blood of 0.04 and a risk of 10%, we let $x = 0.04$ and $R = 10$ in the equation and solve for k.

$$R = 6e^{kx}$$

$$10 = 6e^{k(0.04)} \qquad \text{\color{teal}{$R = 10$; $x = 0.04$.}}$$

$$\frac{10}{6} = e^{0.04k} \qquad \text{\color{teal}{Divide both sides by 6.}}$$

$$0.04k = \ln\frac{10}{6} = 0.5108256 \qquad \text{\color{teal}{Change to a logarithmic expression.}}$$

$$k = 12.77 \qquad \text{\color{teal}{Solve for k.}}$$

See Figure 33 for the graph of $R = 6e^{12.77x}$.

Figure 33

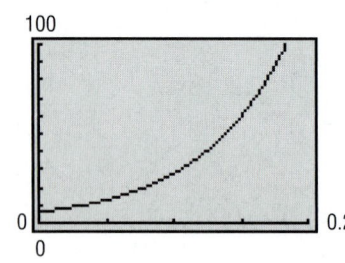

(b) Using $k = 12.77$ and $x = 0.17$ in the equation, we find the risk R to be

$$R = 6e^{kx} = 6e^{(12.77)(0.17)} = 52.6$$

For a concentration of alcohol in the blood of 0.17, the risk of an accident is about 52.6%. Verify this answer using eVALUEate.

(c) Using $k = 12.77$ and $R = 100$ in the equation, we find the concentration x of alcohol in the blood to be

$$R = 6e^{kx}$$

$$100 = 6e^{12.77x} \qquad \text{\color{teal}{$R = 100$; $k = 12.77$.}}$$

$$\frac{100}{6} = e^{12.77x} \qquad \text{\color{teal}{Divide both sides by 6.}}$$

$$12.77x = \ln\frac{100}{6} = 2.8134 \qquad \text{\color{teal}{Change to a logarithmic expression.}}$$

$$x = 0.22 \qquad \text{\color{teal}{Solve for x.}}$$

For a concentration of alcohol in the blood of 0.22, the risk of an accident is 100%. Verify this answer using a graphing utility.

(d) Using $k = 12.77$ and $R = 20$ in the equation, we find the concentration x of alcohol in the blood to be

$$R = 6e^{kx}$$

$$20 = 6e^{12.77x}$$

$$\frac{20}{6} = e^{12.77x}$$

$$12.77x = \ln \frac{20}{6} = 1.204$$

$$x = 0.094$$

A driver with a concentration of alcohol in the blood of 0.094 or more should be arrested and charged with DUI.

Note: Most states use 0.08 or 0.10 as the blood alcohol content at which a DUI citation is given.

SUMMARY Properties of the Logarithmic Function

$f(x) = \log_a x, \quad a > 1$
$(y = \log_a x \text{ means } x = a^y)$

Domain: the interval $(0, \infty)$; Range: the interval $(-\infty, \infty)$;
x-intercept: 1; y-intercept: none; vertical asymptote: $x = 0$ (y-axis);
increasing; one-to-one
See Figure 34(a) for a typical graph.

$f(x) = \log_a x, \quad 0 < a < 1$
$(y = \log_a x \text{ means } x = a^y)$

Domain: the interval $(0, \infty)$; Range: the interval $(-\infty, \infty)$;
x-intercept: 1; y-intercept: none; vertical asymptote: $x = 0$ (y-axis);
decreasing; one-to-one
See Figure 34(b) for a typical graph.

Figure 34

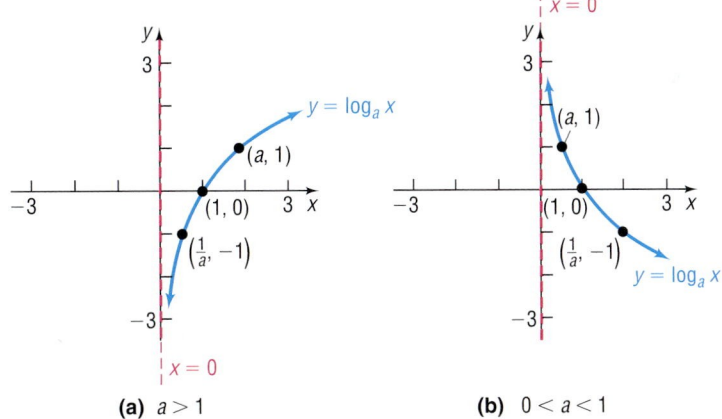

(a) $a > 1$ **(b)** $0 < a < 1$

6.3 Concepts and Vocabulary

In Problems 1–3, fill in the blanks.

1. The domain of the logarithmic function $f(x) = \log_a x$ is _____.

2. The graph of every logarithmic function $f(x) = \log_a x, a > 0, a \neq 1$, passes through three points: _____, _____, and _____.

3. If the graph of a logarithmic function $f(x) = \log_a x, a > 0, a \neq 1$, is increasing, then its base must be larger than _____.

In Problems 4 and 5, answer True or False to each statement.

4. If $y = \log_a x$, then $y = a^x$.

5. The graph of every logarithmic function $f(x) = \log_a x, a > 0, a \neq 1$, will contain the points $(1, 0), (a, 1)$ and $\left(\frac{1}{a}, -1\right)$.

6. In the definition of the logarithmic function, the base a is not allowed to equal 1. Why?

7. If the domain of a logarithmic function is the interval $(1, \infty)$, what is the range of its inverse?

6.3 Exercises

In Problems 1–12, change each exponential expression to an equivalent expression involving a logarithm.

1. $9 = 3^2$
2. $16 = 4^2$
3. $a^2 = 1.6$
4. $a^3 = 2.1$

5. $1.1^2 = M$
6. $2.2^3 = N$
7. $2^x = 7.2$
8. $3^x = 4.6$

9. $x^{\sqrt{2}} = \pi$
10. $x^\pi = e$
11. $e^x = 8$
12. $e^{2.2} = M$

In Problems 13–24, change each logarithmic expression to an equivalent expression involving an exponent.

13. $\log_2 8 = 3$
14. $\log_3\left(\frac{1}{9}\right) = -2$
15. $\log_a 3 = 6$
16. $\log_b 4 = 2$

17. $\log_3 2 = x$
18. $\log_2 6 = x$
19. $\log_2 M = 1.3$
20. $\log_3 N = 2.1$

21. $\log_{\sqrt{2}} \pi = x$
22. $\log_\pi x = \frac{1}{2}$
23. $\ln 4 = x$
24. $\ln x = 4$

In Problems 25–36, find the exact value of each logarithm without using a calculator.

25. $\log_2 1$
26. $\log_8 8$
27. $\log_5 25$
28. $\log_3\left(\frac{1}{9}\right)$

29. $\log_{1/2} 16$
30. $\log_{1/3} 9$
31. $\log_{10} \sqrt{10}$
32. $\log_5 \sqrt[3]{25}$

33. $\log_{\sqrt{2}} 4$
34. $\log_{\sqrt{3}} 9$
35. $\ln \sqrt{e}$
36. $\ln e^3$

In Problems 37–46, find the domain of each function.

37. $f(x) = \ln(x - 3)$
38. $g(x) = \ln(x - 1)$
39. $F(x) = \log_2 x^2$

40. $H(x) = \log_5 x^3$
41. $h(x) = \log_{1/2}(x^2 - 2x + 1)$
42. $G(x) = \log_{1/2}(x^2 - 1)$

43. $f(x) = \ln\left(\frac{1}{x + 1}\right)$
44. $g(x) = \ln\left(\frac{1}{x - 5}\right)$
45. $g(x) = \log_5\left(\frac{x + 1}{x}\right)$

46. $h(x) = \log_3\left(\frac{x}{x - 1}\right)$

In Problems 47–50, use a calculator to evaluate each expression. Round your answer to three decimal places.

47. $\ln \frac{5}{3}$
48. $\frac{\ln 5}{3}$
49. $\frac{\ln(10/3)}{0.04}$
50. $\frac{\ln(2/3)}{-0.1}$

51. Find a so that the graph of $f(x) = \log_a x$ contains the point $(2, 2)$.

52. Find a so that the graph of $f(x) = \log_a x$ contains the point $\left(\frac{1}{2}, -4\right)$.

In Problems 53–60, the graph of a logarithmic function is given. Match each graph to one of the following functions:

A. $y = \log_3 x$
B. $y = \log_3(-x)$
C. $y = -\log_3 x$
D. $y = -\log_3(-x)$

E. $y = \log_3 x - 1$
F. $y = \log_3(x - 1)$
G. $y = \log_3(1 - x)$
H. $y = 1 - \log_3 x$

53.

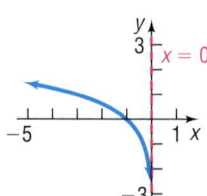

54.

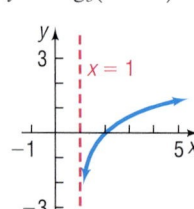

55.

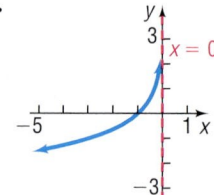

56.

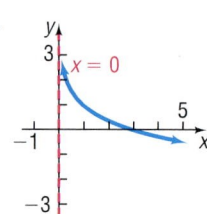

57.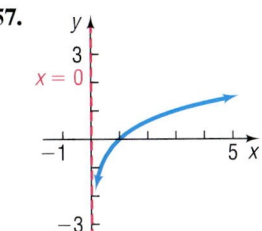

58.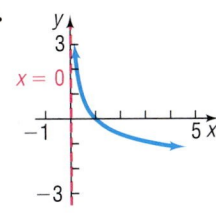

59.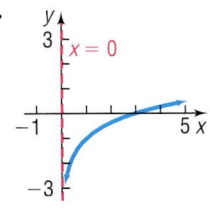

60.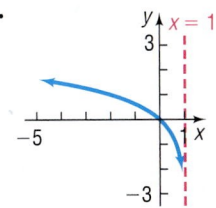

In Problems 61–84, use transformations to graph each function. Determine the domain, range, and vertical asymptote of each function. Verify your results using a graphing utility.

61. $f(x) = \ln(x + 4)$ **62.** $f(x) = \ln(x - 3)$ **63.** $f(x) = 2 + \ln x$ **64.** $f(x) = -\ln(-x)$

65. $g(x) = \ln(2x)$ **66.** $h(x) = \ln\left(\dfrac{1}{2}x\right)$ **67.** $f(x) = 3\ln x$ **68.** $f(x) = -2\ln x$

69. $g(x) = \ln(3 - x)$ **70.** $h(x) = \ln(4 - x)$ **71.** $f(x) = -\ln(x - 1)$ **72.** $f(x) = 2 - \ln x$
73. $f(x) = \log(x - 4)$ **74.** $f(x) = \log(x + 5)$ **75.** $h(x) = 4\log x$ **76.** $g(x) = -3\log x$
77. $F(x) = \log(2x)$ **78.** $G(x) = \log(5x)$ **79.** $f(x) = 2\log(x + 3)$ **80.** $f(x) = -3\log(x - 2)$
81. $h(x) = 3 + \log(x + 2)$ **82.** $g(x) = 2 - \log(x + 1)$ **83.** $F(x) = \log(2 - x)$ **84.** $G(x) = 3 + \log(4 - x)$

In Problems 85–104, solve each equation.

85. $\log_3 x = 2$ **86.** $\log_5 x = 3$ **87.** $\log_2(2x + 1) = 3$ **88.** $\log_3(3x - 2) = 2$

89. $\log_x 4 = 2$ **90.** $\log_x\left(\dfrac{1}{8}\right) = 3$ **91.** $\ln e^x = 5$ **92.** $\ln e^{-2x} = 8$

93. $\log_4 64 = x$ **94.** $\log_5 625 = x$ **95.** $\log_3 243 = 2x + 1$ **96.** $\log_6 36 = 5x + 3$

97. $e^{3x} = 10$ **98.** $e^{-2x} = \dfrac{1}{3}$ **99.** $e^{2x+5} = 8$ **100.** $e^{-2x+1} = 13$

101. $\log_3(x^2 + 1) = 2$ **102.** $\log_5(x^2 + x + 4) = 2$ **103.** $\log_2 8^x = -3$ **104.** $\log_3 3^x = -1$

In Problems 105–108, (a) graph each function and state its domain, range, and asymptote; (b) determine the inverse function; (c) use the graph obtained in (a) to graph the inverse and state its domain, range, and asymptote.

105. $f(x) = 2^x$ **106.** $f(x) = 5^x$ **107.** $f(x) = 2^{x+3}$ **108.** $f(x) = 5^x - 2$

---

109. Optics If a single pane of glass obliterates 10% of the light passing through it, then the percent P of light that passes through n successive panes is given approximately by the equation

$$P = 100e^{-0.1n}$$

(a) How many panes are necessary to block at least 50% of the light?

(b) How many panes are necessary to block at least 75% of the light?

110. Chemistry The pH of a chemical solution is given by the formula

$$pH = -\log_{10}[H^+]$$

where $[H^+]$ is the concentration of hydrogen ions in moles per liter. Values of pH range from 0 (acidic) to 14 (alkaline).

(a) What is the pH of a solution for which $[H^+]$ is 0.1?
(b) What is the pH of a solution for which $[H^+]$ is 0.01?
(c) What is the pH of a solution for which $[H^+]$ is 0.001?
(d) What happens to pH as the hydrogen ion concentration decreases?

(e) Determine the hydrogen ion concentration of an orange (pH = 3.5).
(f) Determine the hydrogen ion concentration of human blood (pH = 7.4).

111. Space Satellites The number of watts w provided by a space satellite's power supply after d days is given by the formula

$$w = 50e^{-0.004d}$$

(a) How long will it take for the available power to drop to 30 watts?
(b) How long will it take for the available power to drop to only 5 watts?

112. Healing of Wounds   The normal healing of wounds can be modeled by an exponential function. If A_0 represents the original area of the wound and if A equals the area of the wound after n days, then the formula

$$A = A_0 e^{-0.35n}$$

describes the area of a wound on the nth day following an injury when no infection is present to retard the healing. Suppose that a wound initially had an area of 100 square millimeters.
(a) If healing is taking place, how many days should pass before the wound is one-half its original size?
(b) How long before the wound is 10% of its original size?

113. Exponential Probability   Between 12:00 PM and 1:00 PM, cars arrive at Citibank's drive-thru at the rate of 6 cars per hour (0.1 car per minute). The following formula from statistics can be used to determine the probability that a car will arrive within t minutes of 12:00 PM.

$$F(t) = 1 - e^{-0.1t}$$

(a) Determine how many minutes are needed for the probability to reach 50%.
(b) Determine how many minutes are needed for the probability to reach 80%.
 (c) Is it possible for the probability to equal 100%? Explain.

114. Exponential Probability   Between 5:00 PM and 6:00 PM, cars arrive at Jiffy Lube at the rate of 9 cars per hour (0.15 car per minute). The following formula from statistics can be used to determine the probability that a car will arrive within t minutes of 5:00 PM.

$$F(t) = 1 - e^{-0.15t}$$

(a) Determine how many minutes are needed for the probability to reach 50%.
(b) Determine how many minutes are needed for the probability to reach 80%.

115. Drug Medication   The formula

$$D = 5e^{-0.4h}$$

can be used to find the number of milligrams D of a certain drug that is in a patient's bloodstream h hours after the drug has been administered. When the number of milligrams reaches 2, the drug is to be administered again. What is the time between injections?

116. Spreading of Rumors   A model for the number of people N in a college community who have heard a certain rumor is

$$N = P(1 - e^{-0.15d})$$

where P is the total population of the community and d is the number of days that have elapsed since the rumor began. In a community of 1000 students, how many days will elapse before 450 students have heard the rumor?

117. Current in an *RL* Circuit   The equation governing the amount of current I (in amperes) after time t (in seconds) in a simple RL circuit consisting of a resistance R (in ohms), an inductance L (in henrys), and an electromotive force E (in volts) is

$$I = \frac{E}{R}[1 - e^{-(R/L)t}]$$

If $E = 12$ volts, $R = 10$ ohms, and $L = 5$ henrys, how long does it take to obtain a current of 0.5 ampere? Of 1.0 ampere? Graph the equation.

118. Learning Curve   Psychologists sometimes use the function

$$L(t) = A(1 - e^{-kt})$$

to measure the amount L learned at time t. The number A represents the amount to be learned, and the number k measures the rate of learning. Suppose that a student has an amount A of 200 vocabulary words to learn. A psychologist determines that the student learned 20 vocabulary words after 5 minutes.
(a) Determine the rate of learning k.
(b) Approximately how many words will the student have learned after 10 minutes?
(c) After 15 minutes?
(d) How long does it take for the student to learn 180 words?

Loudness of Sound   *Problems 119–122 use the following discussion: The* **loudness** $L(x)$, *measured in decibels, of a sound of intensity* x, *measured in watts per square meter, is defined as* $L(x) = 10 \log \dfrac{x}{I_0}$, *where* $I_0 = 10^{-12}$ *watt per square meter is the least intense sound that a human ear can detect. Determine the loudness, in decibels, of each of the following sounds.*

119. Normal conversation: intensity of $x = 10^{-7}$ watt per square meter.

120. Heavy city traffic: intensity of $x = 10^{-3}$ watt per square meter.

121. Amplified rock music: intensity of 10^{-1} watt per square meter.

122. Diesel truck traveling 40 miles per hour 50 feet away: intensity 10 times that of a passenger car traveling 50 miles per hour 50 feet away whose loudness is 70 decibels.

Problems 123 and 124 use the following discussion: The **Richter scale** is one way of converting seismographic readings into numbers that provide an easy reference for measuring the magnitude M of an earthquake. All earthquakes are compared to a **zero-level earthquake** whose seismographic reading measures 0.001 millimeter at a distance of 100 kilometers from the epicenter. An earthquake whose seismographic reading measures x millimeters has **magnitude** M(x) given by

$$M(x) = \log\left(\frac{x}{x_0}\right)$$

where $x_0 = 10^{-3}$ is the reading of a zero-level earthquake the same distance from its epicenter. Determine the magnitude of the following earthquakes.

123. Magnitude of an Earthquake   Mexico City in 1985: seismographic reading of 125,892 millimeters 100 kilometers from the center.

124. Magnitude of an Earthquake   San Francisco in 1906: seismographic reading of 7943 millimeters 100 kilometers from the center.

125. Alcohol and Driving   The concentration of alcohol in a person's blood is measurable. Suppose that the risk R (given as a percent) of having an accident while driving a car can be modeled by the equation

$$R = 3e^{kx}$$

where x is the variable concentration of alcohol in the blood and k is a constant.
   (a) Suppose that a concentration of alcohol in the blood of 0.06 results in a 10% risk ($R = 10$) of an accident. Find the constant k in the equation.

   (b) Using this value of k, what is the risk if the concentration is 0.17?
   (c) Using the same value of k, what concentration of alcohol corresponds to a risk of 100%?
   (d) If the law asserts that anyone with a risk of having an accident of 15% or more should not have driving privileges, at what concentration of alcohol in the blood should a driver be arrested and charged with a DUI?
   (e) Compare this situation with that of Example 10. If you were a lawmaker, which situation would you support? Give your reasons.

126. Is there any function of the form $y = x^\alpha, 0 < \alpha < 1$, that increases more slowly than a logarithmic function whose base is greater than 1? Explain.

PREPARING FOR THIS SECTION

Before getting started, review the following:

✓ Inverse Functions (Section 6.1, pp. 421–422)

6.4 PROPERTIES OF LOGARITHMS

OBJECTIVES
   1 Work with the Properties of Logarithms
   2 Write a Logarithmic Expression as a Sum or Difference of Logarithms
   3 Write a Logarithmic Expression as a Single Logarithm
   4 Evaluate Logarithms Whose Base Is Neither 10 nor e
   5 Graph Logarithmic Functions Whose Base Is Neither 10 nor e

   1   Logarithms have some very useful properties that can be derived directly from the definition and the laws of exponents.

EXAMPLE 1   **Establishing Properties of Logarithms**

   (a) Show that $\log_a 1 = 0$.   (b) Show that $\log_a a = 1$.

Solution   (a) This fact was established when we graphed $y = \log_a x$ (see Figure 28). To show the result algebraically, let $y = \log_a 1$. Then

$$y = \log_a 1$$
$$a^y = 1 \qquad \text{Change to an exponent.}$$
$$a^y = a^0 \qquad a^0 = 1$$
$$y = 0 \qquad \text{Solve for } y.$$
$$\log_a 1 = 0 \qquad y = \log_a 1$$

(b) Let $y = \log_a a$. Then

$$y = \log_a a$$
$$a^y = a \qquad \text{Change to an exponent.}$$
$$a^y = a^1 \qquad a^1 = a$$
$$y = 1 \qquad \text{Solve for } y.$$
$$\log_a a = 1 \qquad y = \log_a a$$

To summarize:

$$\log_a 1 = 0 \qquad \log_a a = 1$$

Theorem

Properties of Logarithms

In the properties given next, M and a are positive real numbers, with $a \neq 1$, and r is any real number.

The number $\log_a M$ is the exponent to which a must be raised to obtain M. That is,

$$a^{\log_a M} = M \tag{1}$$

The logarithm to the base a of a raised to a power equals that power. That is,

$$\log_a a^r = r \tag{2}$$

The proof uses the fact that $y = a^x$ and $y = \log_a x$ are inverses.

Proof of Property (1) For inverse functions,

$$f(f^{-1}(x)) = x$$

Using $f(x) = a^x$ and $f^{-1}(x) = \log_a x$, we find

$$f(f^{-1}(x)) = f(\log_a x) = a^{\log_a x} = x$$

Now let $x = M$ to obtain $a^{\log_a M} = M$.

Proof of Property (2) For inverse functions,

$$f^{-1}(f(x)) = x$$

Using $f(x) = a^x$ and $f^{-1}(x) = \log_a x$, we find

$$f^{-1}(f(x)) = f^{-1}(a^x) = \log_a a^x = x$$

Now let $x = r$ to obtain $\log_a a^r = r$.

EXAMPLE 2   **Using Properties (1) and (2)**

(a) $2^{\log_2 \pi} = \pi$ (b) $\log_{0.2} 0.2^{-\sqrt{2}} = -\sqrt{2}$ (c) $\ln e^{kt} = kt$ ■

── NOW WORK PROBLEM **3.**

Other useful properties of logarithms are given next.

Theorem   **Properties of Logarithms**

In the following properties, M, N, and a are positive real numbers, with $a \neq 1$, and r is any real number.

The Log of a Product Equals the Sum of the Logs

$$\log_a(MN) = \log_a M + \log_a N \tag{3}$$

The Log of a Quotient Equals the Difference of the Logs

$$\log_a\!\left(\frac{M}{N}\right) = \log_a M - \log_a N \tag{4}$$

The Log of a Power Equals the Product of the Power and the Log

$$\log_a M^r = r \log_a M \tag{5}$$

We shall derive properties (3) and (5) and leave the derivation of property (4) as an exercise (see Problem 97).

Proof of Property (3)   Let $A = \log_a M$ and let $B = \log_a N$. These expressions are equivalent to the exponential expressions

$$a^A = M \quad \text{and} \quad a^B = N$$

Now

$$
\begin{aligned}
\log_a(MN) = \log_a(a^A a^B) &= \log_a a^{A+B} &&\text{Law of Exponents}\\
&= A + B &&\text{Property (2) of logarithms}\\
&= \log_a M + \log_a N &&\blacksquare
\end{aligned}
$$

Proof of Property (5)   Let $A = \log_a M$. This expression is equivalent to

$$a^A = M$$

Now

$$
\begin{aligned}
\log_a M^r = \log_a(a^A)^r &= \log_a a^{rA} &&\text{Law of Exponents}\\
&= rA &&\text{Property (2) of logarithms}\\
&= r \log_a M &&\blacksquare
\end{aligned}
$$

── NOW WORK PROBLEM **7.**

② Logarithms can be used to transform products into sums, quotients into differences, and powers into factors. Such transformations prove useful in certain types of calculus problems.

EXAMPLE 3 **Writing a Logarithmic Expression as a Sum of Logarithms**

Write $\log_a\left(x\sqrt{x^2 + 1}\right)$ as a sum of logarithms. Express all powers as factors.

Solution
$$
\begin{aligned}
\log_a\left(x\sqrt{x^2 + 1}\right) &= \log_a x + \log_a \sqrt{x^2 + 1} \qquad \text{Property (3)}\\
&= \log_a x + \log_a(x^2 + 1)^{1/2}\\
&= \log_a x + \frac{1}{2}\log_a(x^2 + 1) \qquad \text{Property (5)}
\end{aligned}
$$
◼

EXAMPLE 4 **Writing a Logarithmic Expression as a Difference of Logarithms**

Write

$$\ln\frac{x^2}{(x - 1)^3}$$

as a difference of logarithms. Express all powers as factors.

Solution $\ln\dfrac{x^2}{(x - 1)^3} = \ln x^2 - \ln(x - 1)^3 = 2\ln x - 3\ln(x - 1)$

         ↑ Property (4)   ↑ Property (5)
◼

EXAMPLE 5 **Writing a Logarithmic Expression as a Sum and Difference of Logarithms**

Write

$$\log_a\frac{x^3\sqrt{x^2 + 1}}{(x + 1)^4}$$

as a sum and difference of logarithms. Express all powers as factors.

Solution
$$
\begin{aligned}
\log_a\frac{x^3\sqrt{x^2 + 1}}{(x + 1)^4} &= \log_a\left(x^3\sqrt{x^2 + 1}\right) - \log_a(x + 1)^4 \qquad \text{Property (4)}\\
&= \log_a x^3 + \log_a \sqrt{x^2 + 1} - \log_a(x + 1)^4 \qquad \text{Property (3)}\\
&= \log_a x^3 + \log_a(x^2 + 1)^{1/2} - \log_a(x + 1)^4\\
&= 3\log_a x + \frac{1}{2}\log_a(x^2 + 1) - 4\log_a(x + 1) \qquad \text{Property (5)}
\end{aligned}
$$
◼

NOW WORK PROBLEM 39.

③ Another use of properties (3) through (5) is to write sums and/or differences of logarithms with the same base as a single logarithm. This skill will be needed to solve certain logarithmic equations discussed in the next section.

EXAMPLE 6 **Writing Expressions as a Single Logarithm**

Write each of the following as a single logarithm.

(a) $\log_a 7 + 4\log_a 3$ (b) $\dfrac{2}{3}\ln 8 - \ln(3^4 - 8)$

(c) $\log_a x + \log_a 9 + \log_a(x^2 + 1) - \log_a 5$

Solution (a) $\log_a 7 + 4\log_a 3 = \log_a 7 + \log_a 3^4$ Property (5)
$= \log_a 7 + \log_a 81$
$= \log_a(7 \cdot 81)$ Property (3)
$= \log_a 567$

(b) $\dfrac{2}{3}\ln 8 - \ln(3^4 - 8) = \ln 8^{2/3} - \ln(81 - 8)$ Property (5)
$= \ln 4 - \ln 73$ $8^{2/3} = \sqrt[3]{8^2} = \sqrt[3]{64} = 4$
$= \ln\left(\dfrac{4}{73}\right)$ Property (4)

(c) $\log_a x + \log_a 9 + \log_a(x^2 + 1) - \log_a 5 = \log_a(9x) + \log_a(x^2 + 1) - \log_a 5$
$= \log_a[9x(x^2 + 1)] - \log_a 5$
$= \log_a\left[\dfrac{9x(x^2 + 1)}{5}\right]$

WARNING: A common error made by some students is to express the logarithm of a sum as the sum of logarithms.

$\log_a(M + N)$ is not equal to $\log_a M + \log_a N$

Correct statement $\log_a(MN) = \log_a M + \log_a N$ Property (3)

Another common error is to express the difference of logarithms as the quotient of logarithms.

$\log_a M - \log_a N$ is not equal to $\dfrac{\log_a M}{\log_a N}$

Correct statement $\log_a M - \log_a N = \log_a\left(\dfrac{M}{N}\right)$ Property (4)

A third common error is to express a logarithm raised to a power as the product of the power times the logarithm.

$(\log_a M)^r$ is not equal to $r\log_a M$

Correct statement $\log_a M^r = r\log_a M$ Property (5)

✏ **NOW WORK PROBLEM 45.**

Two other properties of logarithms that we need to know are consequences of the fact that the logarithmic function $y = \log_a x$ is one-to-one.

Theorem In the following properties, M, N, and a are positive real numbers, with $a \neq 1$.

> If $M = N$, then $\log_a M = \log_a N$. (6)
> If $\log_a M = \log_a N$, then $M = N$. (7)

When property (6) is used, we start with the equation $M = N$ and say "take the logarithm of both sides" to obtain $\log_a M = \log_a N$.

Properties (6) and (7) are useful for solving exponential and logarithmic equations discussed in the next section.

Using a Calculator to Evaluate Logarithms with Bases Other Than 10 or e

④ Common logarithms, that is, logarithms to the base 10, were used to facilitate arithmetic computations before the widespread use of calculators. (See the Historical Feature at the end of this section.) Natural logarithms, that is, logarithms whose base is the number e, remain very important because they arise frequently in the study of natural phenomena.

Common logarithms are usually abbreviated by writing **log**, with the base understood to be 10 just as natural logarithms are abbreviated by **ln**, with the base understood to be e.

Most calculators have both $\boxed{\text{log}}$ and $\boxed{\text{ln}}$ keys to calculate the common logarithm and natural logarithm of a number. Let's look at an example to see how to approximate logarithms having a base other than 10 or e.

EXAMPLE 7   **Approximating Logarithms Whose Base Is Neither 10 nor e**

Approximate $\log_2 7$.
Round the answer to four decimal places.

Solution   Let $y = \log_2 7$. Then $2^y = 7$, so

$$2^y = 7$$

$$\ln 2^y = \ln 7 \qquad \text{Property (6)}$$

$$y \ln 2 = \ln 7 \qquad \text{Property (5)}$$

$$y = \frac{\ln 7}{\ln 2} \qquad \text{Solve for } y.$$

$$y \approx 2.8074 \qquad \text{Use your calculator (} \boxed{\text{ln}} \text{ key) and round to four decimal places.} \blacksquare$$

Example 7 shows how to approximate a logarithm whose base is 2 by changing to logarithms involving the base e. In general, we use the **Change-of-Base Formula**.

Theorem

Change-of-Base Formula

If $a \neq 1, b \neq 1$, and M are positive real numbers, then

$$\log_a M = \frac{\log_b M}{\log_b a} \qquad (8)$$

Proof   We derive this formula as follows: Let $y = \log_a M$. Then

$$a^y = M$$

$$\log_b a^y = \log_b M \qquad \text{\color{blue}Property (6)}$$

$$y \log_b a = \log_b M \qquad \text{\color{blue}Property (5)}$$

$$y = \frac{\log_b M}{\log_b a} \qquad \text{\color{blue}Solve for y.}$$

$$\log_a M = \frac{\log_b M}{\log_b a} \qquad \text{\color{blue}}y = \log_a M$$

Since calculators have only keys for $\boxed{\log}$ and $\boxed{\ln}$, in practice, the Change-of-Base Formula uses either $b = 10$ or $b = e$. Thus,

$$\log_a M = \frac{\log M}{\log a} \quad \text{and} \quad \log_a M = \frac{\ln M}{\ln a} \qquad (9)$$

EXAMPLE 8   **Using the Change-of-Base Formula**

Approximate:

(a) $\log_5 89$ (b) $\log_{\sqrt{2}} \sqrt{5}$

Round answers to four decimal places.

Solution   (a) $\log_5 89 = \dfrac{\log 89}{\log 5} \approx \dfrac{1.94939}{0.69897} = 2.7889$

or

$$\log_5 89 = \frac{\ln 89}{\ln 5} \approx \frac{4.4886}{1.6094} = 2.7889$$

(b) $\log_{\sqrt{2}} \sqrt{5} = \dfrac{\log \sqrt{5}}{\log \sqrt{2}} \approx 2.3219$

or

$$\log_{\sqrt{2}} \sqrt{5} = \frac{\ln \sqrt{5}}{\ln \sqrt{2}} \approx 2.3219$$

NOW WORK PROBLEMS **11** AND **61**.

⑤   We also use the Change-of-Base Formula to graph logarithmic functions whose base is neither 10 nor e.

EXAMPLE 9   **Graphing a Logarithmic Function Whose Base Is Neither 10 nor *e***

Use a graphing utility to graph $y = \log_2 x$.

Solution   Since graphing utilities only have logarithms with the base 10 or the base e, we need to use the Change-of-Base Formula to express $y = \log_2 x$ in terms

Figure 35

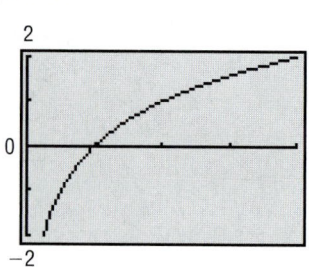

of logarithms with base 10 or base e. We can graph either $y = \dfrac{\ln x}{\ln 2}$ or $y = \dfrac{\log x}{\log 2}$ to obtain the graph of $y = \log_2 x$. See Figure 35.

✔ CHECK: Verify that $y = \dfrac{\ln x}{\ln 2}$ and $y = \dfrac{\log x}{\log 2}$ result in the same graph by graphing each on the same screen.   ■ ■

━━ **NOW WORK PROBLEM 69.**

SUMMARY	**Properties of Logarithms**

In the list that follows, $a > 0, a \neq 1$, and $b > 0, b \neq 1$; also, $M > 0$ and $N > 0$.

Definition

$y = \log_a x$ means $x = a^y$

Properties of logarithms

$\log_a 1 = 0; \quad \log_a a = 1$

$a^{\log_a M} = M; \quad \log_a a^r = r$

$\log_a(MN) = \log_a M + \log_a N$

$\log_a\left(\dfrac{M}{N}\right) = \log_a M - \log_a N$

$\log_a M^r = r \log_a M$

If $M = N$, then $\log_a M = \log_a N$.

If $\log_a M = \log_a N$, then $M = N$.

Change-of-Base Formula

$\log_a M = \dfrac{\log_b M}{\log_b a}$

HISTORICAL FEATURE

John Napier (1550–1617)

Logarithms were invented about 1590 by John Napier (1550–1617) and Joost Bürgi (1552–1632), working independently. Napier, whose work had the greater influence, was a Scottish lord, a secretive man whose neighbors were inclined to believe him to be in league with the devil. His approach to logarithms was very different from ours; it was based on the relationship between arithmetic and geometric sequences and not on the inverse function relationship of logarithms to exponential functions (described in Section 6.3). Napier's tables, published in 1614, listed what would now be called *natural logarithms* of sines and were rather difficult to use. A London professor, Henry Briggs, became interested in the tables and visited Napier. In their conversations, they developed the idea of common logarithms, which were published in 1617. Their importance for calculation was immediately recognized, and by 1650 they were being printed as far away as China. They remained an important calculation tool until the advent of the inexpensive handheld calculator about 1972, which has decreased their calculational, but not their theoretical, importance.

A side effect of the invention of logarithms was the popularization of the decimal system of notation for real numbers.

6.4 Concepts and Vocabulary

In Problems 1–3, fill in the blanks.

1. The logarithm of a product equals the _____ of the logarithms.

2. If $\log_8 M = \dfrac{\log_5 7}{\log_5 8}$, then $M = $ _____.

3. $\log_a M^r = $ _____.

In Problems 4–6, answer True or False to each statement.

4. $\ln(x + 3) - \ln(2x) = \dfrac{\ln(x + 3)}{\ln(2x)}$

5. $\log_2(3x^4) = 4\log_2(3x)$

6. $\log_2 16 = \dfrac{\ln 16}{\ln 2}$

7. Graph $Y_1 = \log(x^2)$ and $Y_2 = 2\log(x)$ on your graphing utility. Are they equivalent? What might account for any differences in the two functions?

6.4 Exercises

In Problems 1–16, use properties of logarithms to find the exact value of each expression. Do not use a calculator.

1. $\log_3 3^{71}$
2. $\log_2 2^{-13}$
3. $\ln e^{-4}$
4. $\ln e^{\sqrt{2}}$

5. $2^{\log_2 7}$
6. $e^{\ln 8}$
7. $\log_8 2 + \log_8 4$
8. $\log_6 9 + \log_6 4$

9. $\log_6 18 - \log_6 3$
10. $\log_8 16 - \log_8 2$
11. $\log_2 6 \cdot \log_6 4$
12. $\log_3 8 \cdot \log_8 9$

13. $3^{\log_3 5 - \log_3 4}$
14. $5^{\log_5 6 + \log_5 7}$
15. $e^{\log_{e^2} 16}$
16. $e^{\log_{e^2} 9}$

In Problems 17–24, suppose that $\ln 2 = a$ and $\ln 3 = b$. Use properties of logarithms to write each logarithm in terms of a and b.

17. $\ln 6$
18. $\ln \dfrac{2}{3}$
19. $\ln 1.5$
20. $\ln 0.5$

21. $\ln 8$
22. $\ln 27$
23. $\ln \sqrt[5]{6}$
24. $\ln \sqrt[4]{\dfrac{2}{3}}$

In Problems 25–44, write each expression as a sum and/or difference of logarithms. Express powers as factors.

25. $\log_5(25x)$
26. $\log_3 \dfrac{x}{9}$
27. $\log_2 z^3$
28. $\log_7(x^5)$

29. $\ln(ex)$
30. $\ln \dfrac{e}{x}$
31. $\ln(xe^x)$
32. $\ln \dfrac{x}{e^x}$

33. $\log_a(u^2 v^3)$
34. $\log_2\left(\dfrac{a}{b^2}\right)$
35. $\ln(x^2\sqrt{1 - x})$
36. $\ln\left(x\sqrt{1 + x^2}\right)$

37. $\log_2\left(\dfrac{x^3}{x - 3}\right)$
38. $\log_5\left(\dfrac{\sqrt[3]{x^2 + 1}}{x^2 - 1}\right)$
39. $\log\left[\dfrac{x(x + 2)}{(x + 3)^2}\right]$
40. $\log\left[\dfrac{x^3\sqrt{x + 1}}{(x - 2)^2}\right]$

41. $\ln\left[\dfrac{x^2 - x - 2}{(x + 4)^2}\right]^{1/3}$
42. $\ln\left[\dfrac{(x - 4)^2}{x^2 - 1}\right]^{2/3}$
43. $\ln\dfrac{5x\sqrt{1 - 3x}}{(x - 4)^3}$
44. $\ln\left[\dfrac{5x^2\sqrt[3]{1 - x}}{4(x + 1)^2}\right]$

In Problems 45–58, write each expression as a single logarithm.

45. $3 \log_5 u + 4 \log_5 v$

46. $2 \log_3 u - \log_3 v$

47. $\log_3 \sqrt{x} - \log_3 x^3$

48. $\log_2 \left(\dfrac{1}{x} \right) + \log_2 \left(\dfrac{1}{x^2} \right)$

49. $\log_4 (x^2 - 1) - 5 \log_4 (x + 1)$

50. $\log(x^2 + 3x + 2) - 2 \log(x + 1)$

51. $\ln \left(\dfrac{x}{x - 1} \right) + \ln \left(\dfrac{x + 1}{x} \right) - \ln(x^2 - 1)$

52. $\log \left(\dfrac{x^2 + 2x - 3}{x^2 - 4} \right) - \log \left(\dfrac{x^2 + 7x + 6}{x + 2} \right)$

53. $8 \log_2 \sqrt{3x - 2} - \log_2 \left(\dfrac{4}{x} \right) + \log_2 4$

54. $21 \log_3 \sqrt[3]{x} + \log_3 (9x^2) - \log_3 9$

55. $2 \log_a (5x^3) - \dfrac{1}{2} \log_a (2x + 3)$

56. $\dfrac{1}{3} \log(x^3 + 1) + \dfrac{1}{2} \log(x^2 + 1)$

57. $2 \log_2 (x + 1) - \log_2 (x + 3) - \log_2 (x - 1)$

58. $3 \log_5 (3x + 1) - 2 \log_5 (2x - 1) - \log_5 x$

59. Write the exponential model $y = ab^x$ as a linear model.
[**Hint:** Take the logarithm of both sides.]

60. Write the power model $y = ax^b$ as a linear model.

In Problems 61–68, use the Change-of-Base Formula and a calculator to evaluate each logarithm. Round your answer to three decimal places.

61. $\log_3 21$

62. $\log_5 18$

63. $\log_{1/3} 71$

64. $\log_{1/2} 15$

65. $\log_{\sqrt{2}} 7$

66. $\log_{\sqrt{5}} 8$

67. $\log_\pi e$

68. $\log_\pi \sqrt{2}$

In Problems 69–74, graph each function using a graphing utility and the Change-of-Base Formula.

69. $y = \log_4 x$

70. $y = \log_5 x$

71. $y = \log_2 (x + 2)$

72. $y = \log_4 (x - 3)$

73. $y = \log_{x-1} (x + 1)$

74. $y = \log_{x+2} (x - 2)$

In Problems 75–84, express y as a function of x. The constant C is a positive number.

75. $\ln y = \ln x + \ln C$

76. $\ln y = \ln(x + C)$

77. $\ln y = \ln x + \ln(x + 1) + \ln C$

78. $\ln y = 2 \ln x - \ln(x + 1) + \ln C$

79. $\ln y = 3x + \ln C$

80. $\ln y = -2x + \ln C$

81. $\ln(y - 3) = -4x + \ln C$

82. $\ln(y + 4) = 5x + \ln C$

83. $3 \ln y = \dfrac{1}{2} \ln(2x + 1) - \dfrac{1}{3} \ln(x + 4) + \ln C$

84. $2 \ln y = -\dfrac{1}{2} \ln x + \dfrac{1}{3} \ln(x^2 + 1) + \ln C$

85. Find the value of $\log_2 3 \cdot \log_3 4 \cdot \log_4 5 \cdot \log_5 6 \cdot \log_6 7 \cdot \log_7 8$.

86. Find the value of $\log_2 4 \cdot \log_4 6 \cdot \log_6 8$.

87. Find the value of $\log_2 3 \cdot \log_3 4 \cdot \ldots \cdot \log_n (n + 1) \cdot \log_{n+1} 2$.

88. Find the value of $\log_2 2 \cdot \log_2 4 \cdot \ldots \cdot \log_2 2^n$.

89. Show that $\log_a \left(x + \sqrt{x^2 - 1} \right) + \log_a \left(x - \sqrt{x^2 - 1} \right) = 0$.

90. Show that $\log_a (\sqrt{x} + \sqrt{x - 1}) + \log_a (\sqrt{x} - \sqrt{x - 1}) = 0$.

91. Show that $\ln(1 + e^{2x}) = 2x + \ln(1 + e^{-2x})$.

92. Difference Quotient If $f(x) = \log_a x$, show that $\dfrac{f(x + h) - f(x)}{h} = \log_a \left(1 + \dfrac{h}{x} \right)^{1/h}, h \neq 0$.

93. If $f(x) = \log_a x$, show that $-f(x) = \log_{1/a} x$.

94. If $f(x) = \log_a x$, show that $f(AB) = f(A) + f(B)$.

95. If $f(x) = \log_a x$, show that $f \left(\dfrac{1}{x} \right) = -f(x)$.

96. If $f(x) = \log_a x$, show that $f(x^\alpha) = \alpha f(x)$.

97. Show that $\log_a \left(\dfrac{M}{N} \right) = \log_a M - \log_a N$, where a, M, and N are positive real numbers, with $a \neq 1$.

98. Show that $\log_a \left(\dfrac{1}{N} \right) = -\log_a N$, where a and N are positive real numbers, with $a \neq 1$.

PREPARING FOR THIS SECTION

Before getting started, review the following:

✓ Solving Equations Using a Graphing Utility; Solving Linear and Quadratic Equations (Section 1.3, pp. 109–124)

6.5 LOGARITHMIC AND EXPONENTIAL EQUATIONS

OBJECTIVES
1. Solve Logarithmic Equations Using the Properties of Logarithms
2. Solve Exponential Equations
3. Solve Logarithmic and Exponential Equations Using a Graphing Utility

Logarithmic Equations

1. In Section 6.3 we solved logarithmic equations by changing a logarithm to exponential form. Often, however, some manipulation of the equation (usually using the properties of logarithms) is required before we can change to exponential form.

Our practice will be to solve equations, whenever possible, by finding exact solutions using algebraic methods. In such cases, we will also verify the solution obtained by using a graphing utility. When algebraic methods cannot be used, approximate solutions will be obtained using a graphing utility. The reader is encouraged to pay particular attention to the form of equations for which exact solutions are possible.

EXAMPLE 1   **Solving a Logarithmic Equation**

Solve: $2 \log_5 x = \log_5 9$

Solution   Because each logarithm is to the same base, 5, we can obtain an exact solution as follows:

$$2 \log_5 x = \log_5 9$$
$$\log_5 x^2 = \log_5 9 \qquad \text{\small $\log_a M^r = r \log_a M$}$$
$$x^2 = 9 \qquad \text{\small If $\log_a M = \log_a N$, then $M = N$}$$
$$x = 3 \quad \text{or} \quad \cancel{x = -3} \qquad \text{\small Recall that logarithms of negative numbers are not defined, so, in the expression $2 \log_5 x$, x must be positive. Therefore -3 is extraneous and we discard it.}$$

Figure 36

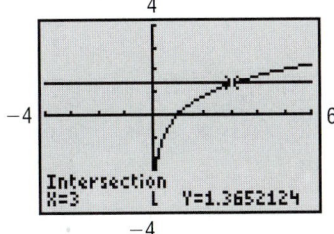

The equation has only one solution, 3.

✔ CHECK: To verify that 3 is the only solution using a graphing utility, graph $Y_1 = 2 \log_5 x = \dfrac{2 \log x}{\log 5}$ and $Y_2 = \log_5 9 = \dfrac{\log 9}{\log 5}$, and determine the point of intersection. See Figure 36.

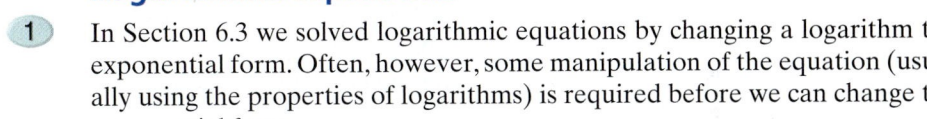 NOW WORK PROBLEM **5.**

EXAMPLE 2 **Solving a Logarithmic Equation**

Solve: $\log_4(x + 3) + \log_4(2 - x) = 1$

Solution To obtain an exact solution, we need to express the left side as a single logarithm. Then we will change the expression to exponential form.

$$\log_4(x + 3) + \log_4(2 - x) = 1$$
$$\log_4[(x + 3)(2 - x)] = 1 \qquad \log_a M + \log_a N = \log_a(MN)$$
$$(x + 3)(2 - x) = 4^1 = 4 \qquad \text{Change to an exponential expression.}$$
$$-x^2 - x + 6 = 4 \qquad \text{Simplify.}$$
$$x^2 + x - 2 = 0 \qquad \text{Place the quadratic equation in standard form.}$$
$$(x + 2)(x - 1) = 0 \qquad \text{Factor.}$$
$$x = -2 \quad \text{or} \quad x = 1 \qquad \text{Zero-Product Property}$$

Since the arguments of each logarithmic expression in the equation are positive for both $x = -2$ and $x = 1$, neither is extraneous. The solution set is $\{-2, 1\}$.

You should verify that both of these are solutions using a graphing utility.

═══ **NOW WORK PROBLEM 9.**

Exponential Equations

2 In Sections 6.2 and 6.3 we solved certain exponential equations by expressing each side of the equation with the same base. However, many exponential equations cannot be rewritten so each side has the same base. In such cases, sometimes properties of logarithms along with algebraic techniques can be used to obtain a solution.

EXAMPLE 3 **Solving an Exponential Equation**

Solve: $4^x - 2^x - 12 = 0$

Solution We note that $4^x = (2^2)^x = 2^{2x} = (2^x)^2$, so the equation is actually quadratic in form, and we can rewrite it as

$$(2^x)^2 - 2^x - 12 = 0 \qquad \text{Let } u = 2^x; \text{then } u^2 - u - 12 = 0.$$

Now we can factor as usual.

$$(2^x - 4)(2^x + 3) = 0 \qquad (u - 4)(u + 3) = 0$$
$$2^x - 4 = 0 \quad \text{or} \quad 2^x + 3 = 0 \qquad u - 4 = 0 \quad \text{or} \quad u + 3 = 0$$
$$2^x = 4 \qquad\qquad 2^x = -3 \qquad u = 2^x = 4 \qquad u = 2^x = -3$$

The equation on the left has the solution $x = 2$, since $2^x = 4 = 2^2$; the equation on the right has no solution, since $2^x > 0$ for all x. The only solution is 2.

In the preceding example, we were able to write the exponential expression using the same base after utilizing some algebra, obtaining an exact solution to the equation. When this is not possible, logarithms can sometimes be used to obtain the solution.

EXAMPLE 4 **Solving an Exponential Equation**

Solve: $2^x = 5$.

Solution Since 5 cannot be written as an integral power of 2, we write the exponential equation as the equivalent logarithmic equation.

$$2^x = 5$$

$$x = \log_2 5 \underset{\uparrow}{=} \frac{\ln 5}{\ln 2}$$

Change-of-Base Formula (9), Section 6.4

Alternatively, we can solve the equation $2^x = 5$ by taking the natural logarithm (or common logarithm) of each side. Taking the natural logarithm,

$$2^x = 5$$

$$\ln 2^x = \ln 5 \qquad \text{If } M = N, \text{ then } \ln M = \ln N$$

$$x \ln 2 = \ln 5 \qquad \ln M^r = r \ln M$$

$$x = \frac{\ln 5}{\ln 2} \qquad \text{Solve for x.}$$

Using a calculator, the solution, rounded to three decimal places, is

$$x = \frac{\ln 5}{\ln 2} \approx 2.322$$

━━ NOW WORK PROBLEM **17.**

EXAMPLE 5 **Solving an Exponential Equation**

Solve: $8 \cdot 3^x = 5$

Solution

$$8 \cdot 3^x = 5$$

$$3^x = \frac{5}{8} \qquad \text{Solve for } 3^x.$$

$$x = \log_3\left(\frac{5}{8}\right) = \frac{\ln\left(\frac{5}{8}\right)}{\ln 3} \qquad \text{Solve for x.}$$

The solution, rounded to three decimal places, is

$$x = \frac{\ln\left(\frac{5}{8}\right)}{\ln 3} \approx -0.428$$

EXAMPLE 6 **Solving an Exponential Equation**

Solve: $5^{x-2} = 3^{3x+2}$

Solution Because the bases are different, we first apply Property (6), Section 6.4 (take the natural logarithm of each side) and then use appropriate properties of logarithms. The result is an equation in x that we can solve.

$$5^{x-2} = 3^{3x+2}$$

$$\ln 5^{x-2} = \ln 3^{3x+2} \qquad \text{If } M = N, \ln M = \ln N$$

$$(x-2)\ln 5 = (3x+2)\ln 3 \qquad \ln M^r = r\ln M$$

$$(\ln 5)x - 2\ln 5 = (3\ln 3)x + 2\ln 3 \qquad \text{Distribute.}$$

$$(\ln 5)x - (3\ln 3)x = 2\ln 3 + 2\ln 5 \qquad \begin{array}{l}\text{Place terms involving } x \text{ on}\\ \text{the left.}\end{array}$$

$$(\ln 5 - 3\ln 3)x = 2(\ln 3 + \ln 5) \qquad \text{Factor.}$$

$$x = \frac{2(\ln 3 + \ln 5)}{\ln 5 - 3\ln 3} \approx -3.212 \qquad \text{Solve for } x.$$

━━ **NOW WORK PROBLEM 25.**

Graphing Utility Solutions

③ The techniques introduced in this section apply only to certain types of logarithmic and exponential equations. Solutions for other types are usually studied in calculus, using numerical methods. However, we can use a graphing utility to approximate the solution.

EXAMPLE 7   **Solving Equations Using a Graphing Utility**

Solve: $\log_3 x + \log_4 x = 4$
Express the solution(s) rounded to two decimal places.

Solution   The solution is found by graphing

Figure 37

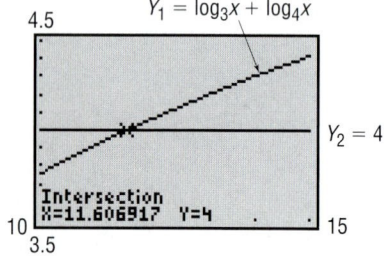

$$Y_1 = \log_3 x + \log_4 x = \frac{\log x}{\log 3} + \frac{\log x}{\log 4} \quad \text{and} \quad Y_2 = 4$$

(Remember that you must use the Change-of-Base Formula to graph Y_1.) Y_1 is an increasing function (do you know why?), and so there is only one point of intersection for Y_1 and Y_2. Figure 37 shows the graphs of Y_1 and Y_2. Using the INTERSECT command, the solution is 11.61, rounded to two decimal places.

EXPLORATION   Can you discover an algebraic solution to Example 7? [**Hint:** Factor $\log x$ from Y_1.]

EXAMPLE 8   **Solving Equations Using a Graphing Utility**

Solve: $x + e^x = 2$
Express the solution(s) rounded to two decimal places.

Figure 38

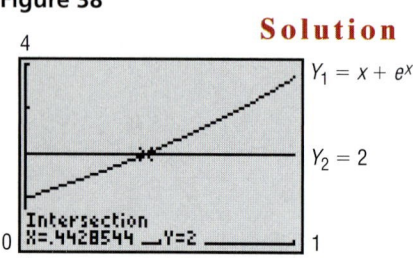

Solution   The solution is found by graphing $Y_1 = x + e^x$ and $Y_2 = 2$. Y_1 is an increasing function (do you know why?), and so there is only one point of intersection for Y_1 and Y_2. Figure 38 shows the graphs of Y_1 and Y_2. Using the INTERSECT command, the solution is 0.44 rounded to two decimal places.

━━ **NOW WORK PROBLEM 49.**

6.5 Exercises

In Problems 1–44, solve each equation. Verify your solution using a graphing utility.

1. $\log_4(x + 2) = \log_4 8$

2. $\log_5(2x + 3) = \log_5 3$

3. $\frac{1}{2}\log_3 x = 2\log_3 2$

4. $-2\log_4 x = \log_4 9$

5. $2\log_5 x = 3\log_5 4$

6. $3\log_2 x = -\log_2 27$

7. $3\log_2(x - 1) + \log_2 4 = 5$

8. $2\log_3(x + 4) - \log_3 9 = 2$

9. $\log x + \log(x + 15) = 2$

10. $\log_4 x + \log_4(x - 3) = 1$

11. $\ln x + \ln(x + 2) = 4$

12. $\ln(x + 1) - \ln x = 2$

13. $2^{2x} + 2^x - 12 = 0$

14. $3^{2x} + 3^x - 2 = 0$

15. $3^{2x} + 3^{x+1} - 4 = 0$

16. $2^{2x} + 2^{x+2} - 12 = 0$

17. $2^x = 10$

18. $3^x = 14$

19. $8^{-x} = 1.2$

20. $2^{-x} = 1.5$

21. $3^{1-2x} = 4^x$

22. $2^{x+1} = 5^{1-2x}$

23. $\left(\frac{3}{5}\right)^x = 7^{1-x}$

24. $\left(\frac{4}{3}\right)^{1-x} = 5^x$

25. $1.2^x = (0.5)^{-x}$

26. $0.3^{1+x} = 1.7^{2x-1}$

27. $\pi^{1-x} = e^x$

28. $e^{x+3} = \pi^x$

29. $5(2^{3x}) = 8$

30. $0.3(4^{0.2x}) = 0.2$

31. $\log_a(x - 1) - \log_a(x + 6) = \log_a(x - 2) - \log_a(x + 3)$

32. $\log_a x + \log_a(x - 2) = \log_a(x + 4)$

33. $\log_{1/3}(x^2 + x) - \log_{1/3}(x^2 - x) = -1$

34. $\log_4(x^2 - 9) - \log_4(x + 3) = 3$

35. $\log_2(x + 1) - \log_4 x = 1$

36. $\log_2(3x + 2) - \log_4 x = 3$

[**Hint:** Change $\log_4 x^2$ to base 2.]

37. $\log_{16} x + \log_4 x + \log_2 x = 7$

38. $\log_9 x + 3\log_3 x = 14$

39. $(\sqrt[3]{2})^{2-x} = 2^{x^2}$

40. $\log_2 x^{\log_2 x} = 4$

41. $\frac{e^x + e^{-x}}{2} = 1$

42. $\frac{e^x + e^{-x}}{2} = 3$

43. $\frac{e^x - e^{-x}}{2} = 2$

44. $\frac{e^x - e^{-x}}{2} = -2$

[**Hint:** Multiply each side by e^x.]

In Problems 45–60, use a graphing utility to solve each equation. Express your answer rounded to two decimal places.

45. $\log_5 x + \log_3 x = 1$

46. $\log_2 x + \log_6 x = 3$

47. $\log_5(x + 1) - \log_4(x - 2) = 1$

48. $\log_2(x - 1) - \log_6(x + 2) = 2$

49. $e^x = -x$

50. $e^{2x} = x + 2$

51. $e^x = x^2$

52. $e^x = x^3$

53. $\ln x = -x$

54. $\ln(2x) = -x + 2$

55. $\ln x = x^3 - 1$

56. $\ln x = -x^2$

57. $e^x + \ln x = 4$

58. $e^x - \ln x = 4$

59. $e^{-x} = \ln x$

60. $e^{-x} = -\ln x$

PREPARING FOR THIS SECTION

Before getting started, review the following:

✓ Simple Interest (Section 1.4, pp. 130–131)

6.6 COMPOUND INTEREST

OBJECTIVES

1 Determine the Future Value of a Lump Sum of Money

2 Calculate Effective Rates of Return

3 Determine the Present Value of a Lump Sum of Money

4 Determine the Time Required to Double a Lump Sum of Money

1 Interest is money paid for the use of money. The total amount borrowed (whether by an individual from a bank in the form of a loan or by a bank from an individual in the form of a savings account) is called the **principal**. The **rate of interest**, expressed as a percent, is the amount charged for the use of the principal for a given period of time, usually on a yearly (that is, per annum) basis.

Simple Interest Formula

If a principal of P dollars is borrowed for a period of t years at a per annum interest rate r, expressed as a decimal, the interest I charged is

$$I = Prt \tag{1}$$

Interest charged according to formula (1) is called **simple interest**.

In working with problems involving interest, we use the term **payment period** as follows:

Annually	Once per year	Monthly	12 times per year
Semiannually	Twice per year	Daily	365 times per year*
Quarterly	Four times per year		

When the interest due at the end of a payment period is added to the principal so that the interest computed at the end of the next payment period is based on this new principal amount (old principal + interest), the interest is said to have been **compounded. Compound interest** is interest paid on previously earned interest.

EXAMPLE 1 **Computing Compound Interest**

A credit union pays interest of 8% per annum compounded quarterly on a certain savings plan. If $1000 is deposited in such a plan and the interest is left to accumulate, how much is in the account after 1 year?

Solution We use the simple interest formula, $I = Prt$. The principal P is $1000 and the rate of interest is 8% = 0.08. After the first quarter of a year, the time t is $\frac{1}{4}$ year, so the interest earned is

$$I = Prt = (\$1000)(0.08)\left(\frac{1}{4}\right) = \$20$$

The new principal is $P + I = \$1000 + \$20 = \$1020$. At the end of the second quarter, the interest on this principal is

$$I = (\$1020)(0.08)\left(\frac{1}{4}\right) = \$20.40$$

*Most banks use a 360-day "year." Why do you think they do?

At the end of the third quarter, the interest on the new principal of $1020 + $20.40 = $1040.40 is

$$I = (\$1040.40)(0.08)\left(\frac{1}{4}\right) = \$20.81$$

Finally, after the fourth quarter, the interest is

$$I = (\$1061.21)(0.08)\left(\frac{1}{4}\right) = \$21.22$$

After 1 year the account contains $1082.43. ■

The pattern of the calculations performed in Example 1 leads to a general formula for compound interest. To fix our ideas, let P represent the principal to be invested at a per annum interest rate r that is compounded n times per year. (For computing purposes, r is expressed as a decimal.) The interest earned after each compounding period is the principal times $\frac{r}{n}$. The amount A after one compounding period is

$$A = P + P\left(\frac{r}{n}\right) = P\left(1 + \frac{r}{n}\right)$$

After two compounding periods, the amount A, based on the new principal $P\left(1 + \frac{r}{n}\right)$, is

$$A = \underbrace{P\left(1 + \frac{r}{n}\right)}_{\substack{\text{New} \\ \text{principal}}} + \underbrace{P\left(1 + \frac{r}{n}\right)\left(\frac{r}{n}\right)}_{\substack{\text{Interest on} \\ \text{new principal}}} = P\left(1 + \frac{r}{n}\right)\left(1 + \frac{r}{n}\right) = P\left(1 + \frac{r}{n}\right)^2$$

After three compounding periods,

$$A = P\left(1 + \frac{r}{n}\right)^2 + P\left(1 + \frac{r}{n}\right)^2\left(\frac{r}{n}\right) = P\left(1 + \frac{r}{n}\right)^2\left(1 + \frac{r}{n}\right) = P\left(1 + \frac{r}{n}\right)^3$$

Continuing this way, after n compounding periods (1 year),

$$A = P\left(1 + \frac{r}{n}\right)^n$$

Because t years will contain $n \cdot t$ compounding periods, after t years we have

$$A = P\left(1 + \frac{r}{n}\right)^{nt}$$

Theorem

Compound Interest Formula

The amount A after t years due to a principal P invested at an annual interest rate r compounded n times per year is

$$A = P\left(1 + \frac{r}{n}\right)^{nt} \tag{2}$$

For example, to rework Example 1, we would use $P = \$1000$, $r = 0.08$, $n = 4$ (quarterly compounding), and $t = 1$ year to obtain

$$A = 1000\left(1 + \frac{0.08}{4}\right)^{4(1)} = \$1082.43$$

In equation (2), the amount A is typically referred to as the **accumulated value or future value** of the account, while P is called the **present value**.

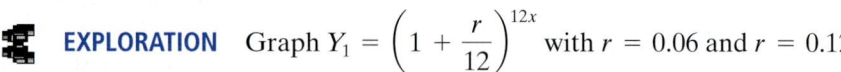 **EXPLORATION** Graph $Y_1 = \left(1 + \dfrac{r}{12}\right)^{12x}$ with $r = 0.06$ and $r = 0.12$

for $0 \le x \le 30$. What is the future value of $1 in 30 years when the interest rate per annum is $r = 0.06$ (6%)? What is the future value of $1 in 30 years when the interest rate per annum is $r = 0.12$ (12%)? Does doubling the interest rate double the future value?

Note: In using your calculator, be sure to use stored values, rather than approximations, in order to avoid round-off errors. At the final step, round money to the nearest cent.

NOW WORK PROBLEM 1.

EXAMPLE 2 **Comparing Investments Using Different Compounding Periods**

Investing $1000 at an annual rate of 10% compounded annually, quarterly, monthly, and daily will yield the following amounts after 1 year:

Annual compounding: $A = P(1 + r)$
$= (\$1000)(1 + 0.10) = \$1100.00$

Quarterly compounding: $A = P\left(1 + \dfrac{r}{4}\right)^4$
$= (\$1000)(1 + 0.025)^4 = \$1103.81$

Monthly compounding: $A = P\left(1 + \dfrac{r}{12}\right)^{12}$
$= (\$1000)(1 + 0.00833)^{12} = \$1104.71$

Daily compounding: $A = P\left(1 + \dfrac{r}{365}\right)^{365}$
$= (\$1000)(1 + 0.000274)^{365} = \$1105.16$ ■

From Example 2 we can see that the effect of compounding more frequently is that the amount after 1 year is higher: $1000 compounded 4 times a year at 10% results in $1103.81; $1000 compounded 12 times a year at 10% results in $1104.71; and $1000 compounded 365 times a year at 10% results in $1105.16. This leads to the following question: What would happen to the amount after 1 year if the number of times that the interest is compounded were increased without bound?

Let's find the answer. Suppose that P is the principal, r is the per annum interest rate, and n is the number of times that the interest is compounded each year. The amount after 1 year is

$$A = P\left(1 + \frac{r}{n}\right)^n$$

Now suppose that the number n of times that the interest is compounded per year gets larger and larger; that is, suppose that $n \rightarrow \infty$. Then

$$A = P\left(1 + \frac{r}{n}\right)^n = P\left[1 + \frac{1}{\frac{n}{r}}\right]^n = P\left(\left[1 + \frac{1}{\frac{n}{r}}\right]^{\frac{n}{r}}\right)^r = P\left[\left(1 + \frac{1}{h}\right)^h\right]^r \qquad (3)$$

$$h = \frac{n}{r}$$

In (3), as $n \rightarrow \infty$, then $h = \dfrac{n}{r} \rightarrow \infty$, and the expression in brackets equals e. [Refer to (2) on p. 437]. That is, $A \rightarrow Pe^r$.

Table 7 compares $\left(1 + \dfrac{r}{n}\right)^n$, for large values of n, to e^r for $r = 0.05$, $r = 0.10$, $r = 0.15$, and $r = 1$. The larger that n gets, the closer $\left(1 + \dfrac{r}{n}\right)^n$ gets to e^r. No matter how frequent the compounding, the amount after 1 year has the definite ceiling Pe^r.

TABLE 7

	$\left(1 + \dfrac{r}{n}\right)^n$			
	$n = 100$	$n = 1000$	$n = 10{,}000$	e^r
$r = 0.05$	1.0512579	1.0512698	1.051271	1.0512711
$r = 0.10$	1.1051157	1.1051654	1.1051703	1.1051709
$r = 0.15$	1.1617037	1.1618212	1.1618329	1.1618342
$r = 1$	2.7048138	2.7169239	2.7181459	2.7182818

When interest is compounded so that the amount after 1 year is Pe^r, we say the interest is **compounded continuously**.

Theorem

Continuous Compounding

The amount A after t years due to a principal P invested at an annual interest rate r compounded continuously is

$$A = Pe^{rt} \qquad (4)$$

EXAMPLE 3   **Using Continuous Compounding**

The amount A that results from investing a principal P of $1000 at an annual rate r of 10% compounded continuously for a time t of 1 year is

$$A = \$1000e^{0.10(1)} = (\$1000)(1.10517) = \$1105.17$$

━ **NOW WORK PROBLEM 9.**

② The **effective rate of interest** is the equivalent annual simple rate of interest that would yield the same amount as compounding after 1 year. For example, based on Example 3, a principal of $1000 will result in $1105.17 at a rate of 10% compounded continuously. To get this same amount using a simple rate of interest would require that interest of $1105.17 − $1000.00 = $105.17 be earned on the principal. Since $105.17 is 10.517% of $1000, a simple rate of interest of 10.517% is needed to equal 10% compounded continuously. The effective rate of interest of 10% compounded continuously is 10.517%.

Based on the results of Examples 2 and 3, we find the following comparisons:

	Annual Rate	Effective Rate
Annual compounding	10%	10%
Quarterly compounding	10%	10.381%
Monthly compounding	10%	10.471%
Daily compounding	10%	10.516%
Continuous compounding	10%	10.517%

EXAMPLE 4   Computing the Value of an IRA

On January 2, 2002, $2000 is placed in an Individual Retirement Account (IRA) that will pay interest of 10% per annum compounded continuously.
(a) What will the IRA be worth on January 1, 2022?
(b) What is the effective rate of interest?

Solution   (a) The amount A after 20 years is

$$A = Pe^{rt} = \$2000e^{(0.10)(20)} = \$14{,}778.11$$

(b) First, we compute the interest earned on $2000 at $r = 10\%$ compounded continuously for 1 year.

$$A = 2000e^{0.10(1)}$$
$$= \$2210.34$$

So, the interest earned is $2210.34 − $2000.00 = $210.34. Use the simple interest formula, $I = Prt$, with $I = \$210.34$, $P = \$2000$, and $t = 1$, and solve for r, the effective rate of interest.

$$\$210.34 = \$2000 \cdot r \cdot 1$$
$$r = \frac{\$210.34}{\$2000} = 0.10517$$

The effective rate of interest is 10.517%. ∎

 NOW WORK PROBLEM **21.**

EXPLORATION How long will it be until $A = \$4000$? $\$6000$? [**Hint:** Graph $Y_1 = 2000e^{0.1x}$ and $Y_2 = 4000$. Use INTERSECT to find x.]

③

Time is money

When people engaged in finance speak of the "time value of money," they are usually referring to the *present value* of money. The **present value** of A dollars to be received at a future date is the principal that you would need to invest now so that it would grow to A dollars in the specified time period. The present value of money to be received at a future date is always less than the amount to be received, since the amount to be received will equal the present value (money invested now) *plus* the interest accrued over the time period.

We use the compound interest formula (2) to get a formula for present value. If P is the present value of A dollars to be received after t years at a per annum interest rate r compounded n times per year, then, by formula (2),

$$A = P\left(1 + \frac{r}{n}\right)^{nt}$$

To solve for P, we divide both sides by $\left(1 + \frac{r}{n}\right)^{nt}$, and the result is

$$\frac{A}{\left(1 + \frac{r}{n}\right)^{nt}} = P \quad \text{or} \quad P = A\left(1 + \frac{r}{n}\right)^{-nt}$$

Theorem

Present Value Formulas

The present value P of A dollars to be received after t years, assuming a per annum interest rate r compounded n times per year, is

$$P = A\left(1 + \frac{r}{n}\right)^{-nt} \tag{5}$$

If the interest is compounded continuously, then

$$P = Ae^{-rt} \tag{6}$$

To prove (6), solve formula (4) for P.

EXAMPLE 5 **Computing the Value of a Zero-Coupon Bond**

A zero-coupon (noninterest-bearing) bond can be redeemed in 10 years for $\$1000$. How much should you be willing to pay for it now if you want a return of:

(a) 8% compounded monthly?   (b) 7% compounded continuously?

Solution (a) We are seeking the present value of $1000. We use formula (5) with $A = \$1000, n = 12, r = 0.08$, and $t = 10$.

$$P = A\left(1 + \frac{r}{n}\right)^{-nt}$$

$$= \$1000\left(1 + \frac{0.08}{12}\right)^{-12(10)}$$

$$= \$450.52$$

For a return of 8% compounded monthly, you should pay $450.52 for the bond.

(b) Here we use formula (6) with $A = \$1000, r = 0.07$, and $t = 10$.

$$P = Ae^{-rt}$$

$$= \$1000e^{-(0.07)(10)}$$

$$= \$496.59$$

For a return of 7% compounded continuously, you should pay $496.59 for the bond.

 **NOW WORK PROBLEM 11.**

| **EXAMPLE 6** | **Rate of Interest Required to Double an Investment** |

④ What annual rate of interest compounded annually should you seek if you want to double your investment in 5 years?

Algebraic Solution If P is the principal and we want P to double, the amount A will be $2P$. We use the compound interest formula with $n = 1$ and $t = 5$ to find r.

$$2P = P(1 + r)^5$$
$$2 = (1 + r)^5$$
$$1 + r = \sqrt[5]{2}$$
$$r = \sqrt[5]{2} - 1 = 1.148698 - 1 = 0.148698$$

The annual rate of interest needed to double the principal in 5 years is 14.87%.

Graphing Solution We solve the equation

$$2 = (1 + r)^5$$

for r by graphing the two functions $Y_1 = 2$ and $Y_2 = (1 + x)^5$. The x-coordinate of their point of intersection is the rate r that we seek. See Figure 39.

Figure 39

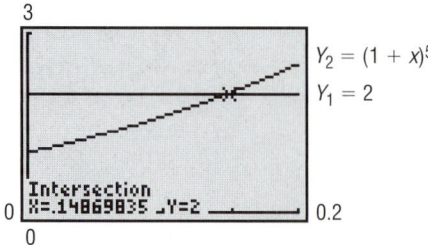

Using the INTERSECT command, we find that the point of intersection of Y_1 and Y_2 is $(0.14869835, 2)$, so $r = 0.148698$ as before. ■

NOW WORK PROBLEM **23.**

EXAMPLE 7   **Doubling Time for an Investment**

How long will it take for an investment to double in value if it earns 5% compounded continuously?

Algebraic Solution   If P is the initial investment and we want P to double, the amount A will be $2P$. We use formula (4) for continuously compounded interest with $r = 0.05$. Then

$$A = Pe^{rt}$$
$$2P = Pe^{0.05t} \qquad \text{Let } A = 2P$$
$$2 = e^{0.05t} \qquad \text{Divide both sides by } P$$
$$0.05t = \ln 2 \qquad \text{Rewrite as a logarithm}$$
$$t = \frac{\ln 2}{0.05} = 13.86 \qquad \text{Solve for } t.$$

It will take about 14 years to double the investment.

Graphing Solution   We solve the equation

$$2 = e^{0.05t}$$

for t by graphing the two functions $Y_1 = 2$ and $Y_2 = e^{0.05x}$. Their point of intersection is $(13.86, 2)$. See Figure 40.

Figure 40

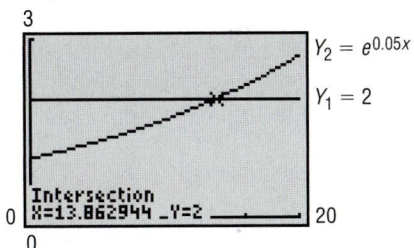

NOW WORK PROBLEM **29.**

6.6   Exercises

In Problems 1–10, find the amount that results from each investment.

1. $100 invested at 4% compounded quarterly after a period of 2 years

2. $50 invested at 6% compounded monthly after a period of 3 years

3. $500 invested at 8% compounded quarterly after a period of $2\frac{1}{2}$ years

4. $300 invested at 12% compounded monthly after a period of $1\frac{1}{2}$ years

5. $600 invested at 5% compounded daily after a period of 3 years

6. $700 invested at 6% compounded daily after a period of 2 years

7. $10 invested at 11% compounded continuously after a period of 2 years

8. $40 invested at 7% compounded continuously after a period of 3 years

9. $100 invested at 10% compounded continuously after a period of $2\frac{1}{4}$ years

10. $100 invested at 12% compounded continuously after a period of $3\frac{3}{4}$ years

In Problems 11–20, find the principal needed now to get each amount; that is, find the present value.

11. To get $100 after 2 years at 6% compounded monthly

12. To get $75 after 3 years at 8% compounded quarterly

13. To get $1000 after $2\frac{1}{2}$ years at 6% compounded daily

14. To get $800 after $3\frac{1}{2}$ years at 7% compounded monthly

15. To get $600 after 2 years at 4% compounded quarterly

16. To get $300 after 4 years at 3% compounded daily

17. To get $80 after $3\frac{1}{4}$ years at 9% compounded continuously

18. To get $800 after $2\frac{1}{2}$ years at 8% compounded continuously

19. To get $400 after 1 year at 10% compounded continuously

20. To get $1000 after 1 year at 12% compounded continuously

21. Find the effective rate of interest for $5\frac{1}{4}$% compounded quarterly.

22. What interest rate compounded quarterly will give an effective interest rate of 7%?

23. What annual rate of interest is required to double an investment in 3 years? Verify your answer using a graphing utility.

24. What annual rate of interest is required to double an investment in 10 years? Verify your answer using a graphing utility.

In Problems 25–28, which of the two rates would yield the larger amount in 1 year?

[**Hint:** Start with a principal of $10,000 in each instance.]

25. 6% compounded quarterly or $6\frac{1}{4}$% compounded annually

26. 9% compounded quarterly or $9\frac{1}{4}$% compounded annually

27. 9% compounded monthly or 8.8% compounded daily

28. 8% compounded semiannually or 7.9% compounded daily

---

29. How long does it take for an investment to double in value if it is invested at 8% per annum compounded monthly? Compounded continuously? Verify your answer using a graphing utility.

30. How long does it take for an investment to double in value if it is invested at 10% per annum compounded monthly? Compounded continuously? Verify your answer using a graphing utility.

31. If Tanisha has $100 to invest at 8% per annum compounded monthly, how long will it be before she has $150? If the compounding is continuous, how long will it be?

32. If Angela has $100 to invest at 10% per annum compounded monthly, how long will it be before she has $175? If the compounding is continuous, how long will it be?

33. How many years will it take for an initial investment of $10,000 to grow to $25,000? Assume a rate of interest of 6% compounded continuously.

34. How many years will it take for an initial investment of $25,000 to grow to $80,000? Assume a rate of interest of 7% compounded continuously.

35. What will a $90,000 house cost 5 years from now if the inflation rate over that period averages 3% compounded annually?

36. Sears charges 1.25% per month on the unpaid balance for customers with charge accounts (interest is compounded monthly). A customer charges $200 and does not pay her bill for 6 months. What is the bill at that time?

37. Jerome will be buying a used car for $15,000 in 3 years. How much money should he ask his parents for now so that, if he invests it at 5% compounded continuously, he will have enough to buy the car?

38. John will require $3000 in 6 months to pay off a loan that has no prepayment privileges. If he has the $3000 now, how much of it should he save in an account paying 3% compounded monthly so that in 6 months he will have exactly $3000?

39. George is contemplating the purchase of 100 shares of a stock selling for $15 per share. The stock pays no dividends. The history of the stock indicates that it should grow at an annual rate of 15% per year. How much will the 100 shares of stock be worth in 5 years?

40. Tracy is contemplating the purchase of 100 shares of a stock selling for $15 per share. The stock pays no dividends. Her broker says that the stock will be worth $20 per share in 2 years. What is the annual rate of return on this investment?

41. A business purchased for $650,000 in 1997 is sold in 2000 for $850,000. What is the annual rate of return for this investment?

42. Tanya has just inherited a diamond ring appraised at $5000. If diamonds have appreciated in value at an annual rate of 8%, what was the value of the ring 10 years ago when the ring was purchased?

43. Jim places $1000 in a bank account that pays 5.6% compounded continuously. After 1 year, will he have enough money to buy a computer system that costs $1060? If another bank will pay Jim 5.9% compounded monthly, is this a better deal?

44. On January 1, Kim places $1000 in a certificate of deposit that pays 6.8% compounded continuously and matures in 3 months. Then Kim places the $1000 and the interest in a passbook account that pays 5.25% compounded monthly. How much does Kim have in the passbook account on May 1?

45. Will invests $2000 in a bond trust that pays 9% interest compounded semiannually. His friend Henry invests $2000 in a certificate of deposit (CD) that pays $8\frac{1}{2}$% compounded continuously. Who has more money after 20 years, Will or Henry?

46. Suppose that April has access to an investment that will pay 10% interest compounded continuously. Which is better: To be given $1000 now so that she can take advantage of this investment opportunity or to be given $1325 after 3 years?

47. Colleen and Bill have just purchased a house for $150,000, with the seller holding a second mortgage of $50,000. They promise to pay the seller $50,000 plus all accrued interest 5 years from now. The seller offers them three interest options on the second mortgage:
   (a) Simple interest at 12% per annum
   (b) $11\frac{1}{2}$% interest compounded monthly
   (c) $11\frac{1}{4}$% interest compounded continuously

   Which option is best; that is, which results in the least interest on the loan?

48. The First National Bank advertises that it pays interest on savings accounts at the rate of 4.25% compounded daily. Find the effective rate if the bank uses (a) 360 days or (b) 365 days in determining the daily rate.

Problems 49–52 involve zero-coupon bonds. A zero-coupon bond is a bond that is sold now at a discount and will pay its face value at some time when it matures; no interest payments are made.

49. A zero-coupon bond can be redeemed in 20 years for $10,000. How much should you be willing to pay for it now if you want a return of:
   (a) 10% compounded monthly?
   (b) 10% compounded continuously?

50. A child's grandparents are considering buying a $40,000 face value zero-coupon bond at birth so that she will have enough money for her college education 17 years later. If they want a rate of return of 8% compounded annually, what should they pay for the bond?

51. How much should a $10,000 face value zero-coupon bond, maturing in 10 years, be sold for now if its rate of return is to be 8% compounded annually?

52. If Pat pays $12,485.52 for a $25,000 face value zero-coupon bond that matures in 8 years, what is his annual rate of return?

53. Explain in your own words what the term *compound interest* means. What does *continuous compounding* mean?

54. Explain in your own words the meaning of *present value*.

55. Time to Double or Triple an Investment The formula

$$y = \frac{\ln m}{n \ln\left(1 + \frac{r}{n}\right)}$$

can be used to find the number of years *y* required to multiply an investment *m* times when *r* is the per annum interest rate compounded *n* times a year.

   (a) How many years will it take to double the value of an IRA that compounds annually at the rate of 12%?
   (b) How many years will it take to triple the value of a savings account that compounds quarterly at an annual rate of 6%?
   (c) Give a derivation of this formula.

56. Time to Reach an Investment Goal The formula

$$y = \frac{\ln A - \ln P}{r}$$

can be used to find the number of years *y* required for an investment *P* to grow to a value *A* when compounded continuously at an annual rate *r*.

   (a) How long will it take to increase an initial investment of $1000 to $8000 at an annual rate of 10%?
   (b) What annual rate is required to increase the value of a $2000 IRA to $30,000 in 35 years?
   (c) Give a derivation of this formula.

57. Critical Thinking You have just contracted to buy a house and will seek financing in the amount of $100,000. You go to several banks. Bank 1 will lend you $100,000 at the rate of 8.75% amortized over 30 years with a loan origination fee of 1.75%. Bank 2 will lend you $100,000 at the rate of 8.375% amortized over 15 years with a loan origination fee of 1.5%. Bank 3 will lend you $100,000 at the rate of 9.125% amortized over 30 years with no loan origination fee. Bank 4 will lend you $100,000 at the rate of 8.625% amortized over 15 years with no loan origination fee. Which loan would you take? Why? Be sure to have sound reasons for your choice. Use the information in the table to assist you. If the amount of the monthly payment does not matter to you, which loan would you take? Again, have sound reasons for your choice. Compare your final decision with others in the class. Discuss.

	Monthly Payment	Loan Origination Fee
Bank 1	$786.70	$1,750.00
Bank 2	$977.42	$1,500.00
Bank 3	$813.63	$0.00
Bank 4	$990.68	$0.00

6.7 GROWTH AND DECAY

OBJECTIVES

1. Find Equations of Populations That Obey the Law of Uninhibited Growth
2. Find Equations of Populations That Obey the Law of Decay
3. Use Newton's Law of Cooling
4. Use Logistic Growth Models

1. Many natural phenomena have been found to follow the law that an amount A varies with time t according to

$$A = A_0 e^{kt} \tag{1}$$

Here A_0 is the original amount ($t = 0$) and $k \neq 0$ is a constant.

If $k > 0$, then equation (1) states that the amount A is increasing over time; if $k < 0$, the amount A is decreasing over time. In either case, when an amount A varies over time according to equation (1), it is said to follow the **exponential law** or the **law of uninhibited growth** ($k > 0$) **or decay** ($k < 0$). See Figure 41.

Figure 41

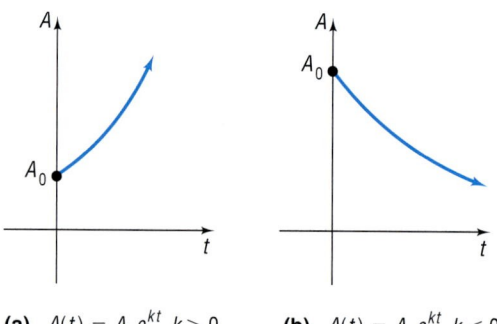

(a) $A(t) = A_0 e^{kt}, k > 0$ (b) $A(t) = A_0 e^{kt}, k < 0$

For example, we saw in Section 6.6 that continuously compounded interest follows the law of uninhibited growth. In this section we shall look at three additional phenomena that follow the exponential law.

Uninhibited Growth

Cell division is the growth process of many living organisms, such as amoebas, plants, and human skin cells. Based on an ideal situation in which no cells die and no by-products are produced, the number of cells present at a given time follows the law of uninhibited growth. Actually, however, after enough time has passed, growth at an exponential rate will cease due to the influence of factors such as lack of living space and dwindling food supply. The law of uninhibited growth accurately reflects only the early stages of the cell division process.

The cell division process begins with a culture containing N_0 cells. Each cell in the culture grows for a certain period of time and then divides into two identical cells. We assume that the time needed for each cell to divide in two is constant and does not change as the number of cells increases. These new cells then grow, and eventually each divides in two, and so on.

A model that gives the number N of cells in the culture after a time t has passed (in the early stages of growth) is given next.

Uninhibited Growth of Cells

$$N(t) = N_0 e^{kt}, \qquad k > 0 \qquad\qquad (2)$$

where N_0 is the initial number of cells and k is a positive constant that represents the growth rate of the cells.

In using formula (2) to model the growth of cells, we are using a function that yields positive real numbers, even though we are counting the number of cells, which must be an integer. This is a common practice in many applications.

EXAMPLE 1   **Bacterial Growth**

A colony of bacteria grows according to the law of uninhibited growth according to the function $N(t) = 100e^{0.045t}$, where N is measured in grams and t is measured in days.

(a) Determine the initial amount of bacteria.

(b) What is the growth rate of the bacteria?

(c) Graph the function using a graphing utility.

(d) What is the population after 5 days?

(e) How long will it take for the population to reach 140 grams?

(f) What is the doubling time for the population?

Solution   (a) The initial amount of bacteria, N_0, is obtained when $t = 0$, so
$N_0 = N(0) = 100e^{0.045(0)} = 100$ grams.

(b) Compare $N(t) = 100e^{0.045t}$ to $N(t) = 100e^{kt}$. The value of k, 0.045, indicates a growth rate of 4.5%.

(c) Figure 42 shows the graph of $N(t) = 100e^{0.045t}$.

(d) The population after 5 days is $N(5) = 100e^{0.045(5)} = 125.2$ grams.

(e) We solve the equation $N(t) = 140$.

$$100e^{0.045t} = 140$$
$$e^{0.045t} = 1.4 \qquad \text{Divide both sides of the equation by 100.}$$
$$0.045t = \ln 1.4 \qquad \text{Rewrite as a logarithm.}$$
$$t = \frac{\ln 1.4}{0.045} \qquad \text{Divide both sides of the equation by 0.045.}$$
$$\approx 7.5 \text{ days}$$

Figure 42

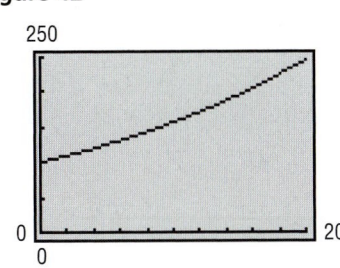

(f) The population doubles when $N(t) = 200$ grams, so we find the doubling time by solving the equation $200 = 100e^{0.045t}$ for t.

$$200 = 100e^{0.045t}$$

$$2 = e^{0.045t} \qquad \text{Divide both sides of the equation by 100.}$$

$$\ln 2 = 0.045t \qquad \text{Rewrite as a logarithm.}$$

$$t = \frac{\ln 2}{0.045} \qquad \text{Divide both sides of the equation by 0.045.}$$

$$\approx 15.4 \text{ days}$$

NOW WORK PROBLEM 1.

EXAMPLE 2 **Bacterial Growth**

A colony of bacteria increases according to the law of uninhibited growth.

(a) If the number of bacteria doubles in 3 hours, find the function that gives the number of cells in the culture.

(b) How long will it take for the size of the colony to triple?

(c) How long will it take for the population to double a second time (that is, increase four times)?

Solution (a) Using formula (2), the number N of cells at a time t is

$$N(t) = N_0 e^{kt}$$

where N_0 is the initial number of bacteria present and k is a positive number. We first seek the number k. The number of cells doubles in 3 hours, so we have

$$N(3) = 2N_0$$

But $N(3) = N_0 e^{k(3)}$, so

$$N_0 e^{k(3)} = 2N_0$$

$$e^{3k} = 2 \qquad \text{Divide both sides by } N_0.$$

$$3k = \ln 2 \qquad \text{Write the exponential equation as a logarithm.}$$

$$k = \frac{1}{3}\ln 2 \approx \frac{1}{3}(0.6931) = 0.2310$$

Formula (2) for this growth process is therefore

$$N(t) = N_0 e^{\left(\frac{1}{3}\ln 2\right)t}$$

(b) The time t needed for the size of the colony to triple requires that $N = 3N_0$. We substitute $3N_0$ for N to get

$$3N_0 = N_0 e^{\left(\frac{1}{3}\ln 2\right)t}$$

$$3 = e^{\left(\frac{1}{3}\ln 2\right)t}$$

$$\left(\frac{1}{3}\ln 2\right)t = \ln 3$$

$$t = \frac{3\ln 3}{\ln 2} \approx 4.755 \text{ hours}$$

It will take about 4.755 hours or 4 hours, 45 minutes for the size of the colony to triple.

 (c) If a population doubles in 3 hours, it will double a second time in 3 more hours, for a total time of 6 hours. ■

Radioactive Decay

 Radioactive materials follow the law of uninhibited decay. The amount A of a radioactive material present at time t is given by the following model:

> **Uninhibited Radioactive Decay**
>
> $$A(t) = A_0 e^{kt}, \qquad k < 0 \tag{3}$$
>
> where A_0 is the original amount of radioactive material and k is a negative number that represents the rate of decay.

All radioactive substances have a specific **half-life**, which is the time required for half of the radioactive substance to decay. In **carbon dating**, we use the fact that all living organisms contain two kinds of carbon, carbon 12 (a stable carbon) and carbon 14 (a radioactive carbon, with a half-life of 5600 years). While an organism is living, the ratio of carbon 12 to carbon 14 is constant. But when an organism dies, the original amount of carbon 12 present remains unchanged, whereas the amount of carbon 14 begins to decrease. This change in the amount of carbon 14 present relative to the amount of carbon 12 present makes it possible to calculate when an organism died.

EXAMPLE 3 **Estimating the Age of Ancient Tools**

Traces of burned wood along with ancient stone tools in an archaeological dig in Chile were found to contain approximately 1.67% of the original amount of carbon 14.

(a) If the half-life of carbon 14 is 5600 years, approximately when was the tree cut and burned?

(b) Using a graphing utility, graph the relation between the percentage of carbon 14 remaining and time.

(c) Determine the time that elapses until half of the carbon 14 remains. This answer should equal the half-life of carbon 14.

(d) Use a graphing utility to verify the answer found in part (a).

Solution (a) Using formula (3), the amount A of carbon 14 present at time t is

$$A(t) = A_0 e^{kt}$$

where A_0 is the original amount of carbon 14 present and k is a negative number. We first seek the number k. To find it, we use the fact that after 5600 years half of the original amount of carbon 14 remains, so $A(5600) = \dfrac{1}{2} A_0$. Thus,

$$\frac{1}{2}A_0 = A_0 e^{k(5600)}$$

$$\frac{1}{2} = e^{5600k} \qquad \text{Divide both sides of the equation by } A_0.$$

$$5600k = \ln\frac{1}{2} \qquad \text{Rewrite as a logarithm.}$$

$$k = \frac{1}{5600}\ln\frac{1}{2} \approx -0.000124$$

Formula (3) therefore becomes

$$A(t) = A_0 e^{\frac{\ln\frac{1}{2}}{5600}t}$$

If the amount A of carbon 14 now present is 1.67% of the original amount, it follows that

$$0.0167A_0 = A_0 e^{\frac{\ln\frac{1}{2}}{5600}t}$$

$$0.0167 = e^{\frac{\ln\frac{1}{2}}{5600}t} \qquad \text{Divide both sides of the equation by } A_0.$$

$$\frac{\ln\frac{1}{2}}{5600}t = \ln 0.0167 \qquad \text{Rewrite as a logarithm.}$$

$$t = \frac{5600}{\ln\frac{1}{2}}\ln 0.0167 \approx 33{,}062 \text{ years}$$

The tree was cut and burned about 33,062 years ago. Some archaeologists use this conclusion to argue that humans lived in the Americas 33,000 years ago, much earlier than is generally accepted.

Figure 43

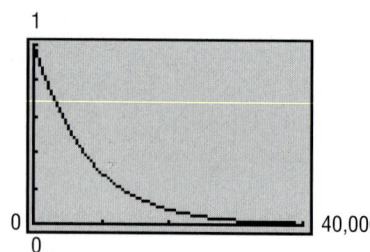

(b) Figure 43 shows the graph of $y = e^{\frac{\ln\frac{1}{2}}{5600}x}$, where y is the fraction of carbon 14 present and x is the time.

(c) By graphing $Y_1 = 0.5$ and $Y_2 = e^{\frac{\ln\frac{1}{2}}{5600}x}$, where x is time, and using INTERSECT, we find that it takes 5600 years until half the carbon 14 remains. The half-life of carbon 14 is 5600 years.

(d) By graphing $Y_1 = 0.0167$ and $Y_2 = e^{\frac{\ln\frac{1}{2}}{5600}x}$, where x is time, and using INTERSECT, we find that it takes 33,062 years until 1.67% of the carbon 14 remains.

NOW WORK PROBLEM 3.

Newton's Law of Cooling

③ **Newton's Law of Cooling*** states that the temperature of a heated object decreases exponentially over time toward the temperature of the surrounding medium. That is, the temperature u of a heated object at a given time t can be modeled by the following function:

Newton's Law of Cooling

$$u(t) = T + (u_0 - T)e^{kt} \qquad k < 0 \qquad (4)$$

where T is the constant temperature of the surrounding medium, u_0 is the initial temperature of the heated object, and k is a negative constant.

EXAMPLE 4 **Using Newton's Law of Cooling**

An object is heated to 100°C (degrees Celsius) and is then allowed to cool in a room whose air temperature is 30°C.

(a) If the temperature of the object is 80°C after 5 minutes, when will its temperature be 50°C?

(b) Using a graphing utility, graph the relation found between the temperature and time.

(c) Using a graphing utility, verify that after 18.6 minutes the temperature is 50°C.

(d) Using a graphing utility, determine the elapsed time before the object is 35°C.

(e) What do you notice about the temperature as time passes?

Solution (a) Using formula (4) with $T = 30$ and $u_0 = 100$, the temperature (in degrees Celsius) of the object at time t (in minutes) is

$$u(t) = 30 + (100 - 30)e^{kt} = 30 + 70e^{kt} \qquad (5)$$

where k is a negative constant. To find k, we use the fact that $u = 80$ when $t = 5$. Then

$$80 = 30 + 70e^{k(5)}$$

$$50 = 70e^{5k}$$

$$e^{5k} = \frac{50}{70}$$

$$5k = \ln\frac{5}{7}$$

$$k = \frac{1}{5}\ln\frac{5}{7} \approx -0.0673$$

Formula (4) therefore becomes

$$u(t) = 30 + 70e^{\frac{\ln\frac{5}{7}}{5}t}$$

*Named after Sir Isaac Newton (1642–1727), one of the cofounders of calculus.

We want to find t when $u = 50°C$, so

$$50 = 30 + 70e^{\frac{\ln\frac{5}{7}}{5}t}$$

$$20 = 70e^{\frac{\ln\frac{5}{7}}{5}t}$$

$$e^{\frac{\ln\frac{5}{7}}{5}t} = \frac{20}{70}$$

$$\frac{\ln\frac{5}{7}}{5}t = \ln\frac{2}{7}$$

$$t = \frac{5}{\ln\frac{5}{7}}\ln\frac{2}{7} \approx 18.6 \text{ minutes}$$

The temperature of the object will be 50°C after about 18.6 minutes or 18 minutes, 37 seconds.

Figure 44

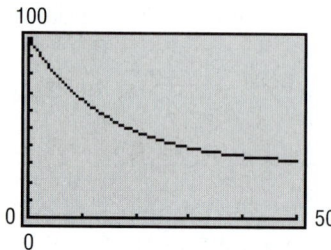

(b) Figure 44 shows the graph of $y = 30 + 70e^{\frac{\ln\frac{5}{7}}{5}x}$, where y is the temperature and x is the time.

(c) By graphing $Y_1 = 50$ and $Y_2 = 30 + 70e^{\frac{\ln\frac{5}{7}}{5}x}$, where x is time, and using INTERSECT, we find that it takes $x = 18.6$ minutes (18 minutes, 37 seconds) for the temperature to cool to 50°C.

(d) By graphing $Y_1 = 35$ and $Y_2 = 30 + 70e^{\frac{\ln\frac{5}{7}}{5}x}$, where x is time, and using INTERSECT, we find that it takes $x = 39.22$ minutes (39 minutes, 13 seconds) for the temperature to cool to 35°C.

(e) As x increases, the value of $e^{\frac{\ln\frac{5}{7}}{5}x}$ approaches zero, so the value of y approaches 30°C. The temperature of the object approaches 30°C. ∎

Logistic Models

④ The exponential growth model $A(t) = A_0e^{kt}, k > 0$, assumes uninhibited growth, meaning that the value of the function grows without limit. Recall we stated that cell division could be modeled using this function, assuming that no cells die and no by-products are produced. However, cell division would eventually be limited by factors such as living space and food supply. The **logistic growth model** is an exponential function that can model situations where the growth of the dependent variable is limited.

Other situations that lead to a logistic growth model include population growth and the sales of a product due to advertising. See Problems 21 through 25. The logistic growth model is given next.

Logistic Growth Model

$$P(t) = \frac{c}{1 + ae^{-bt}}$$

where a, b, and c are constants with $c > 0$ and $b > 0$.

The number c is called the **carrying capacity** because the value $P(t)$ approaches c as t approaches infinity; that is, $\lim_{t \to \infty} P(t) = c$. The number b is the growth rate.

EXAMPLE 5   **Fruit Fly Population**

Fruit flies are placed in a half-pint milk bottle with a banana (for food) and yeast plants (for food and to provide a stimulus to lay eggs). Suppose that the fruit fly population after t days is given by

$$P(t) = \frac{230}{1 + 56.5e^{-0.37t}}$$

(a) State the carrying capacity and the growth rate.

(b) Determine the initial population.

(c) Use a graphing utility to graph $P(t)$.

(d) What is the population after 5 days?

(e) How long does it take for the population to reach 180?

(f) How long does it take for the population to reach one-half of the carrying capacity?

Solution (a) As $t \to \infty$, $e^{-0.37t} \to 0$ and $P(t) \to 230/1$. The carrying capacity of the half-pint bottle is 230 fruit flies. The growth rate is $0.37 = 37\%$.

(b) To find the initial number of fruit flies in the half-pint bottle, we evaluate $P(0)$.

$$P(0) = \frac{230}{1 + 56.5e^{-0.37(0)}}$$

$$= \frac{230}{1 + 56.5}$$

$$= 4$$

So initially there were four fruit flies in the half-pint bottle.

Figure 45

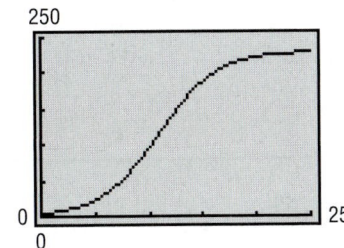

(c) See Figure 45 for the graph of $P(t)$.

(d) To find the number of fruit flies in the half-pint bottle after 5 days, we evaluate $P(5)$.

$$P(5) = \frac{230}{1 + 56.5e^{-0.37(5)}} \approx 23 \text{ fruit flies}$$

After 5 days, there are 23 fruit flies in the bottle.

(e) To determine when the population of fruit flies will be 180, we solve the equation

$$\frac{230}{1 + 56.5e^{-0.37t}} = 180$$

$$230 = 180(1 + 56.5e^{-0.37t})$$

$$1.2778 = 1 + 56.5e^{-0.37t} \qquad \text{Divide both sides by 180.}$$

$$0.2778 = 56.5e^{-0.37t} \qquad \text{Subtract 1 from both sides.}$$

$$0.0049 = e^{-0.37t} \qquad \text{Divide both sides by 56.5.}$$

$$\ln(0.0049) = -0.37t \qquad \text{Rewrite as a logarithmic expression.}$$

$$t \approx 14.4 \text{ days} \qquad \text{Divide both sides by } -0.37.$$

It will take approximately 14.4 days (14 days, 9 hours) for the population to reach 180 fruit flies.

We could also solve this problem by graphing $Y_1 = \dfrac{230}{1 + 56.5e^{-0.37t}}$ and $Y_2 = 180$, using INTERSECT to find the solution shown in Figure 46.

Figure 46

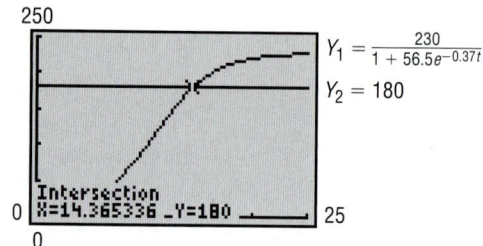

(f) One-half of the carrying capacity is 115 fruit flies. We solve $P(t) = 115$ by graphing $Y_1 = \dfrac{230}{1 + 56.5e^{-0.37t}}$ and $Y_2 = 115$ and using INTERSECT. See Figure 47. The population will reach one-half of the carrying capacity in about 10.9 days (10 days, 22 hours).

Figure 47

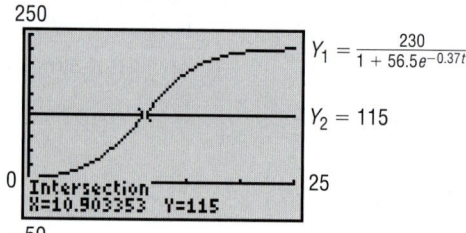

    Look back at Figure 47. Notice the point where the graph reaches 115 fruit flies (one-half of the carrying capacity); the graph changes from being curved upward to being curved downward. Using the language of calculus, we say the graph changes from increasing at an increasing rate to increasing at a decreasing rate. For any logistic growth function, when the population reaches one-half the carrying capacity, the population growth starts to slow down.

EXPLORATION   On the same viewing rectangle, graph $Y_1 = \dfrac{500}{1 + 24e^{-0.03t}}$ and $Y_2 = \dfrac{500}{1 + 24e^{-0.08t}}$. What effect does the growth rate b have on the logistic growth function?

6.7 Exercises

1. **Growth of an Insect Population** The size P of a certain insect population at time t (in days) obeys the function $P(t) = 500e^{0.02t}$.
   (a) Determine the number of insects at $t = 0$ days.
   (b) What is the growth rate of the insect population?
   (c) Graph the function using a graphing utility.
   (d) What is the population after 10 days?
   (e) When will the insect population reach 800?
   (f) When will the insect population double?

2. **Growth of Bacteria** The number N of bacteria present in a culture at time t (in hours) obeys the function $N(t) = 1000e^{0.01t}$.
   (a) Determine the number of bacteria at $t = 0$ hours.
   (b) What is the growth rate of the bacteria?
   (c) Graph the function using a graphing utility.
   (d) What is the population after 4 hours?
   (e) When will the number of bacteria reach 1700?
   (f) When will the number of bacteria double?

3. **Radioactive Decay** Strontium 90 is a radioactive material that decays according to the function $A(t) = A_0 e^{-0.0244t}$, where A_0 is the initial amount present and A is the amount present at time t (in years). Assume that a scientist has a sample of 500 grams of strontium 90.
   (a) What is the decay rate of strontium 90?
   (b) Graph the function using a graphing utility.
   (c) How much strontium 90 is left after 10 years?
   (d) When will 400 grams of strontium 90 be left?
   (e) What is the half-life of strontium 90?

4. **Radioactive Decay** Iodine 131 is a radioactive material that decays according to the function $A(t) = A_0 e^{-0.087t}$ where A_0 is the initial amount present and A is the amount present at time t (in days). Assume that a scientist has a sample of 100 grams of iodine 131.
   (a) What is the decay rate of iodine 131?
   (b) Graph the function using a graphing utility.
   (c) How much iodine 131 is left after 9 days?
   (d) When will 70 grams of iodine 131 be left?
   (e) What is the half-life of iodine 131?

5. **Growth of a Colony of Mosquitoes** The population of a colony of mosquitoes obeys the law of uninhibited growth. If there are 1000 mosquitoes initially and there are 1800 after 1 day, what is the size of the colony after 3 days? How long is it until there are 10,000 mosquitoes?

6. **Bacterial Growth** A culture of bacteria obeys the law of uninhibited growth. If 500 bacteria are present initially and there are 800 after 1 hour, how many will be present in the culture after 5 hours? How long is it until there are 20,000 bacteria?

7. **Population Growth** The population of a southern city follows the exponential law. If the population doubled in size over an 18-month period and the current population is 10,000, what will the population be 2 years from now?

8. **Population Decline** The population of a midwestern city follows the exponential law. If the population decreased from 900,000 to 800,000 from 1993 to 1995, what will the population be in 1997?

9. **Radioactive Decay** The half-life of radium is 1690 years. If 10 grams is present now, how much will be present in 50 years?

10. **Radioactive Decay** The half-life of radioactive potassium is 1.3 billion years. If 10 grams is present now, how much will be present in 100 years? In 1000 years?

11. **Estimating the Age of a Tree** A piece of charcoal is found to contain 30% of the carbon 14 that it originally had.
    (a) When did the tree from which the charcoal came die? Use 5600 years as the half-life of carbon 14.
    (b) Using a graphing utility, graph the relation between the percentage of carbon 14 remaining and time.
    (c) Using INTERSECT, determine the time that elapses until half of the carbon 14 remains.
    (d) Verify the answer found in part (a).

12. **Estimating the Age of a Fossil** A fossilized leaf contains 70% of its normal amount of carbon 14.
    (a) How old is the fossil?
    (b) Using a graphing utility, graph the relation between the percentage of carbon 14 remaining and time.
    (c) Using INTERSECT, determine the time that elapses until half of the carbon 14 remains.
    (d) Verify the answer found in part (a).

13. **Cooling Time of a Pizza** A pizza baked at 450°F is removed from the oven at 5:00 PM into a room that is a constant 70°F. After 5 minutes, the pizza is at 300°F.
    (a) At what time can you begin eating the pizza if you want its temperature to be 135°F?
    (b) Using a graphing utility, graph the relation between temperature and time.
    (c) Using INTERSECT, determine the time that needs to elapse before the pizza is 160°F.
    (d) TRACE the function for large values of time. What do you notice about y, the temperature?

14. **Newton's Law of Cooling** A thermometer reading 72°F is placed in a refrigerator where the temperature is a constant 38°F.

(a) If the thermometer reads 60°F after 2 minutes, what will it read after 7 minutes?

(b) How long will it take before the thermometer reads 39°F?

(c) Using a graphing utility, graph the relation between temperature and time.

(d) Using INTERSECT, determine the time needed to elapse before the thermometer reads 45°F.

(e) TRACE the function for large values of time. What do you notice about y, the temperature?

15. Newton's Law of Cooling   A thermometer reading 8°C is brought into a room with a constant temperature of 35°C.

(a) If the thermometer reads 15°C after 3 minutes, what will it read after being in the room for 5 minutes? For 10 minutes?

(b) Graph the relation between temperature and time. TRACE to verify that your answers are correct.

[**Hint:** You need to construct a formula similar to equation (4).]

16. Thawing Time of a Steak   A frozen steak has a temperature of 28°F. It is placed in a room with a constant temperature of 70°F. After 10 minutes, the temperature of the steak has risen to 35°F. What will the temperature of the steak be after 30 minutes? How long will it take the steak to thaw to a temperature of 45°F? [See the hint given for Problem 15.] Graph the relation between temperature and time. TRACE to verify that your answer is correct.

17. Decomposition of Salt in Water   Salt (NaCl) decomposes in water into sodium (NA$^+$) and chloride (Cl$^-$) ions according to the law of uninhibited decay. If the initial amount of salt is 25 kilograms and, after 10 hours, 15 kilograms of salt is left, how much salt is left after 1 day? How long does it take until $\frac{1}{2}$ kilogram of salt is left?

18. Voltage of a Conductor   The voltage of a certain conductor decreases over time according to the law of uninhibited decay. If the initial voltage is 40 volts, and 2 seconds later it is 10 volts, what is the voltage after 5 seconds?

19. Radioactivity from Chernobyl   After the release of radioactive material into the atmosphere from a nuclear power plant at Chernobyl (Ukraine) in 1986, the hay in Austria was contaminated by iodine 131 (half-life 8 days). If it is all right to feed the hay to cows when 10% of the iodine 131 remains, how long do the farmers need to wait to use this hay?

20. Pig Roasts   The hotel Bora-Bora is having a pig roast. At noon, the chef put the pig in a large earthen oven. The pig's original temperature was 75°F. At 2:00 PM the chef checked the pig's temperature and was upset because it had reached only 100°F. If the oven's temperature remains a constant 325°F, at what time may the hotel serve its guests, assuming that pork is done when it reaches 175°F?

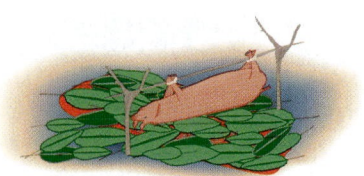

21. Proportion of the Population That Owns a VCR   The logistic growth model

$$P(t) = \frac{0.9}{1 + 6e^{-0.32t}}$$

relates the proportion of U.S. households that own a VCR to the year. Let $t = 0$ represent 1984, $t = 1$ represent 1985, and so on.

(a) Determine the maximum percentage of households that will own a VCR.

(b) What percentage of households owned a VCR in 1984 ($t = 0$)?

(c) Use a graphing utility to graph $P(t)$.

(d) What percentage of households owned a VCR in 1999 ($t = 15$)?

(e) When will 0.8 (80%) of U.S. households own a VCR?

(f) How long will it be before 0.45 (45%) of the population owns a VCR?

22. Market Penetration of Intel's Coprocessor   The logistic growth model

$$P(t) = \frac{0.90}{1 + 3.5e^{-0.339t}}$$

relates the proportion of new personal computers (PCs) sold at Best Buy that have Intel's latest coprocessor t months after it has been introduced.

(a) Determine the maximum percentage of PCs sold at Best Buy that will have Intel's latest coprocessor.

(b) What percentage of computers sold at Best Buy will have Intel's latest coprocessor when it is first introduced ($t = 0$)?

(c) Use a graphing utility to graph $P(t)$.

(d) What percentage of PCs sold will have Intel's latest coprocessor $t = 4$ months after it is introduced?

(e) When will 0.75 (75%) of PCs sold by Best Buy have Intel's latest coprocessor?

(f) How long will it be before 0.45 (45%) of the PCs sold by Best Buy have Intel's latest coprocessor?

23. Population of a Bacteria Culture   The logistic growth model

$$P(t) = \frac{1000}{1 + 32.33e^{-0.439t}}$$

represents the population (in grams) of a bacterium after t hours.

(a) Determine the carrying capacity of the environment. What is the growth rate of the bacteria?

(b) Determine the initial population size.

(c) Use a graphing utility to graph $P(t)$.
(d) What is the population after 9 hours?
(e) When will the population be 700 grams?
(f) How long does it take for the population to reach one-half of the carrying capacity?

24. **Population of a Endangered Species**   Often environmentalists will capture an endangered species and transport the species to a controlled environment where the species can produce offspring and regenerate its population. Suppose that six American bald eagles are captured, transported to Montana, and set free. Based on experience, the environmentalists expect the population to grow according to the model

$$P(t) = \frac{500}{1 + 83.33e^{-0.162t}}$$

where t is measured in years.
(a) Determine the carrying capacity of the environment. What is the growth rate of the bald eagle?
(b) Use a graphing utility to graph $P(t)$.
(c) What is the population after 3 years?
(d) When will the population be 300 eagles?
(e) How long does it take for the population to reach one-half of the carrying capacity?

25. **The *Challenger* Disaster***   After the *Challenger* disaster in 1986, a study of the 23 launches that preceded the fatal flight was made. A mathematical model was developed involving the relationship between the Fahrenheit temperature x around the O-rings and the number y of eroded or leaky primary O-rings. The model stated that

$$y = \frac{6}{1 + e^{-(5.085 - 0.1156x)}}$$

where the number 6 indicates the 6 primary O-rings on the spacecraft.
(a) What is the predicted number of eroded or leaky primary O-rings at a temperature of 100°F?
(b) What is the predicted number of eroded or leaky primary O-rings at a temperature of 60°F?
(c) What is the predicted number of eroded or leaky primary O-rings at a temperature of 30°F?
(d) Graph the equation. At what temperature is the predicted number of eroded or leaky O-rings 1? 3? 5?

*Linda Tappin, "Analyzing Data Relating to the *Challenger* Disaster," *Mathematics Teacher*, Vol. 87, No. 6, September 1994, pp. 423–426.

PREPARING FOR THIS SECTION

Before getting started, review the following:

✓ Linear Models (Section 2.2, pp. 217–221)

✓ Power Models (Section 4.1, pp. 327–328)

✓ Quadratic Models (Section 2.4, pp. 242–244)

✓ Cubic Models (Section 4.2, pp. 338–339)

6.8   EXPONENTIAL, LOGARITHMIC, AND LOGISTIC MODELS

OBJECTIVES
1 Use a Graphing Utility to Fit an Exponential Function to Data
2 Use a Graphing Utility to Fit a Logarithmic Function to Data
3 Use a Graphing Utility to Fit a Logistic Function to Data

In Section 2.2 we discussed how to find the linear function of best fit ($y = ax + b$), in Section 2.4 we discussed how to find the quadratic function of best fit ($y = ax^2 + bx + c$), and in Chapter 4 we discussed how to find the power function of best fit ($y = ax^b$) and the cubic function of best fit ($y = ax^3 + bx^2 + cx + d$).

In this section we will discuss how to use a graphing utility to find equations of best fit that describe the relation between two variables when the relation is thought to be exponential ($y = ab^x$), logarithmic ($y = a + b \ln x$), or logistic $\left(y = \dfrac{c}{1 + ae^{-bx}} \right)$. As before, we draw a scatter diagram of the data to help determine the appropriate model to use.

Figure 48 shows scatter diagrams that will typically be observed for the three models. Below each scatter diagram are any restrictions on the values of the parameters.

Figure 48

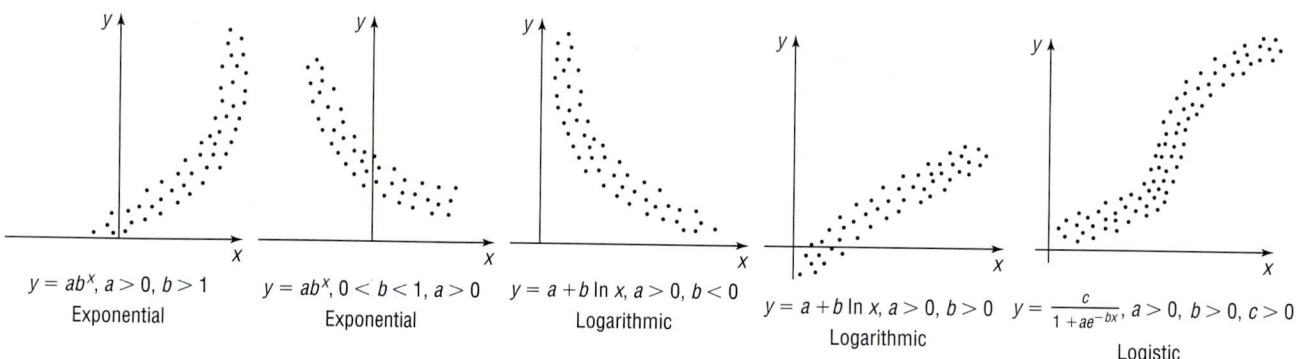

$y = ab^x, a > 0, b > 1$
Exponential

$y = ab^x, 0 < b < 1, a > 0$
Exponential

$y = a + b \ln x, a > 0, b < 0$
Logarithmic

$y = a + b \ln x, a > 0, b > 0$
Logarithmic

$y = \dfrac{c}{1 + ae^{-bx}}, a > 0, b > 0, c > 0$
Logistic

Most graphing utilities have REGression options that fit data to a specific type of curve. Once the data have been entered and a scatter diagram obtained, the type of curve that you want to fit to the data is selected. Then that REGression option is used to obtain the curve of "best fit" of the type selected.

The correlation coefficient r will appear only if the model can be written as a linear expression. Thus, r will appear for the linear, power, exponential, and logarithmic models, since these models can be written as a linear expression (see Problems 59 and 60 in Exercise 6.4). Remember, the closer $|r|$ is to 1, the better the fit.

Let's look at some examples.

Exponential Models

 We saw in Section 6.6 that the future value of money behaves exponentially, and we saw in Section 6.7 that growth and decay models also behave exponentially. The next example shows how data can lead to an exponential model.

EXAMPLE 1 **Fitting an Exponential Function to Data**

Beth is interested in finding a function that explains the closing price of Harley Davidson stock at the end of each month. She obtains the following data:

Year, x	Closing Price, y
1987 (x = 1)	0.392
1988 (x = 2)	0.7652
1989 (x = 3)	1.1835
1990 (x = 4)	1.1609
1991 (x = 5)	2.6988
1992 (x = 6)	4.5381
1993 (x = 7)	5.3379
1994 (x = 8)	6.8032
1995 (x = 9)	7.0328
1996 (x = 10)	11.5585
1997 (x = 11)	13.4799
1998 (x = 12)	23.5424
1999 (x = 13)	31.9342
2000 (x = 14)	39.7277

Source: http://finance.yahoo.com

(a) Using a graphing utility, draw a scatter diagram with year as the independent variable.

(b) Using a graphing utility, fit an exponential function to the data.

(c) Express the function found in part (b) in the form $A = A_0 e^{kt}$.

(d) Graph the exponential function found in part (b) or (c) on the scatter diagram.

(e) Using the solution to part (b) or (c), predict the closing price of Harley Davidson stock at the end of 2001.

(f) Interpret the value of k found in part (c).

Solution

(a) Enter the data into the graphing utility, letting 1 represent 1987, 2 represent 1998, and so on. We obtain the scatter diagram shown in Figure 49.

Figure 49

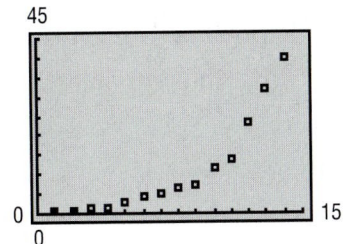

(b) A graphing utility fits the data in Figure 49 to an exponential function of the form $y = ab^x$ by using the EXPonential REGression option. See Figure 50. Thus, $y = ab^x = 0.394503(1.40276)^x$.

(c) To express $y = ab^x$ in the form $A = A_0 e^{kt}$, where $x = t$ and $y = A$, we proceed as follows:

$$ab^x = A_0 e^{kt}, \quad x = t$$

When $x = t = 0$, we find $a = A_0$. This leads to

$$a = A_0, \qquad b^x = e^{kt}$$
$$b^x = (e^k)^t$$
$$b = e^k \qquad x = t$$

Since $y = ab^x = 0.394503(1.40276)^x$, we find that $a = 0.394503$ and $b = 1.40276$, so

$$a = A_0 = 0.394503 \quad \text{and} \quad b = 1.40276 = e^k$$

We want to find k, so we rewrite $1.40276 = e^k$ as a logarithm and obtain

$$k = \ln(1.40276) \approx 0.33844$$

As a result, $A = A_0 e^{kt} = 0.394503 e^{0.33844t}$.

Figure 50

ExpReg
y=a*b^x
a=.394502732
b=1.402763529

Figure 51

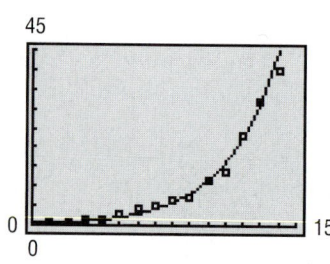

(d) See Figure 51 for the graph of the exponential function of best fit.

(e) Let $t = 15$ (end of 2001) in the function found in part (c). The predicted closing price of Harley Davidson stock at the end of 2001 is

$$A = 0.394503e^{0.33844(15)} = \$63.21$$

(f) The value of k represents the annual interest rate compounded continuously.

$$A = A_0 e^{kt} = 0.394503e^{0.33844t}$$
$$= Pe^{rt} \qquad \text{Equation (4), Section 6.6}$$

The price of Harley Davidson stock has grown at an annual rate of 33.844% (compounded continuously) between 1987 and 2000. ■

 **NOW WORK PROBLEM 1.**

Logarithmic Models

2 Many relations between variables do not follow an exponential model, but, instead, the independent variable is related to the dependent variable using a logarithmic model.

EXAMPLE 2 **Fitting a Logarithmic Function to Data**

Jodi, a meteorologist, is interested in finding a function that explains the relation between the height of a weather balloon (in kilometers) and the atmospheric pressure (measured in millimeters of mercury) on the balloon. She collects the data shown in the table.

Atmospheric Pressure, p	Height, h
760	0
740	0.184
725	0.328
700	0.565
650	1.079
630	1.291
600	1.634
580	1.862
550	2.235

(a) Using a graphing utility, draw a scatter diagram of the data with atmospheric pressure as the independent variable.

(b) Using a graphing utility, fit a logarithmic function to the data.

(c) Draw the logarithmic function found in part (b) on the scatter diagram.

(d) Use the function found in part (b) to predict the height of the weather balloon if the atmospheric pressure is 560 millimeters of mercury.

Solution (a) After entering the data into the graphing utility, we obtain the scatter diagram shown in Figure 52.

(b) A graphing utility fits the data in Figure 52 to a logarithmic function of the form $y = a + b \ln x$ by using the Logarithm REGression option. See Figure 53. The logarithmic function of best fit to the data is

$$h(p) = 45.7863 - 6.9025 \ln p$$

where h is the height of the weather balloon and p is the atmospheric pressure. Notice that $|r|$ is close to 1, indicating a good fit.

(c) Figure 54 shows the graph of $h(p) = 45.7863 - 6.9025 \ln p$ on the scatter diagram.

(d) Using the function found in part (b), Jodi predicts the height of the weather balloon when the atmospheric pressure is 560 to be

$$h(560) = 45.7863 - 6.9025 \ln 560$$

$$\approx 2.108 \text{ kilometers}$$

Figure 52

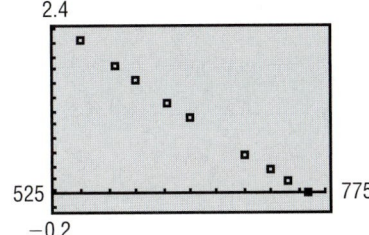

Figure 53

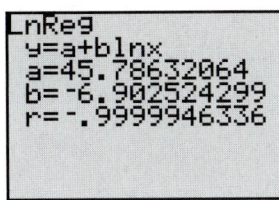

Figure 54

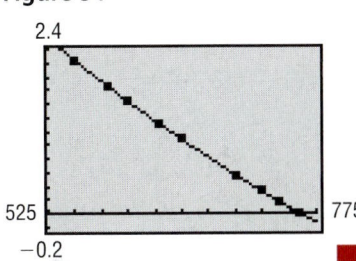

NOW WORK PROBLEM 7.

Logistic Models

3 Logistic growth models can be used to model situations where the value of the dependent variable is limited. Many real-world situations conform to this scenario. For example, the population of the human race is limited by the availability of natural resources such as food and shelter. When the value of the dependent variable is limited, a logistic growth model is often appropriate.

EXAMPLE 3 **Fitting a Logistic Function to Data**

The data in the table on page 498, obtained from Tor Carlson (Über Geschwindigkeit und Grösse der Hefevermehrung in Würze, *Biochemishe Zeitschrift*, Bd. 57, pp. 313–334, 1913), represent the amount of yeast biomass after t hours in a culture.

(a) Using a graphing utility, draw a scatter diagram of the data with time as the independent variable.

(b) Using a graphing utility, fit a logistic function to the data.

(c) Using a graphing utility, graph the function found in part (b) on the scatter diagram.

Time (in hours)	Yeast Biomass
0	9.6
1	18.3
2	29.0
3	47.2
4	71.1
5	119.1
6	174.6
7	257.3
8	350.7
9	441.0
10	513.3
11	559.7
12	594.8
13	629.4
14	640.8
15	651.1
16	655.9
17	659.6
18	661.8

(d) What is the predicted carrying capacity of the culture?
(e) Use the function found in part (b) to predict the population of the culture at $t = 19$ hours.

Solution (a) See Figure 55 for a scatter diagram of the data.

Figure 55

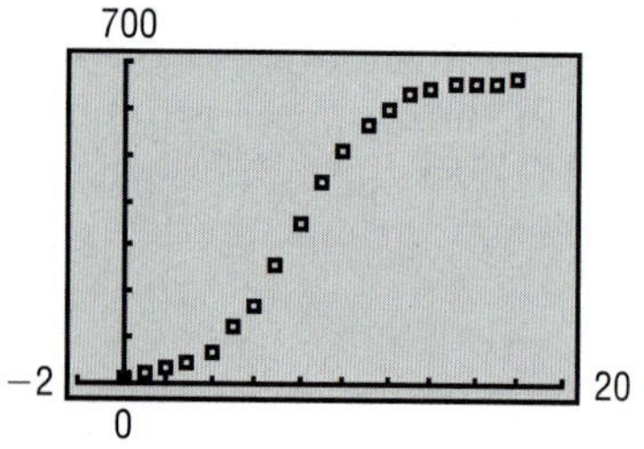

(b) A graphing utility fits a logistic growth model of the form $y = \dfrac{c}{1 + ae^{-bx}}$ by using the LOGISTIC regression option. See Figure 56. The logistic function of best fit to the data is

$$y = \frac{663.0}{1 + 71.6e^{-0.5470x}}$$

where y is the amount of yeast biomass in the culture and x is the time.
(c) See Figure 57 for the graph of the logistic function of best fit.
(d) Based on the logistic growth function found in part (b), the carrying capacity of the culture is 663.
(e) Using the logistic growth function found in part (b), the predicted amount of yeast biomass at $t = 19$ hours is

$$y = \frac{663.0}{1 + 71.6e^{-0.5470(19)}} = 661.5$$

Figure 56

Figure 57

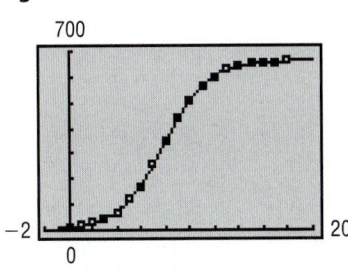

NOW WORK PROBLEM **9.**

6.8 Exercises

1. Biology   A strain of E-coli Beu 397-recA441 is placed into a petri dish at 30° Celsius and allowed to grow. The following data are collected. Theory states that the number of bacteria in the petri dish will initially grow according to the law of uninhibited growth. The population is measured using an optical device in which the amount of light that passes through the petri dish is measured.

Time (hours), x	Population, y
0	0.09
2.5	0.18
3.5	0.26
4.5	0.35
6	0.50

Source: Dr. Polly Lavery, Joliet Junior College

(a) Draw a scatter diagram treating time as the predictor variable.
(b) Using a graphing utility, fit an exponential function to the data.
(c) Express the function found in part (b) in the form $N(t) = N_0 e^{kt}$.
(d) Graph the exponential function found in part (b) or (c) on the scatter diagram.
(e) Use the exponential function from part (b) or (c) to predict the population at $x = 7$ hours.
(f) Use the exponential function from part (b) or (c) to predict when the population will reach 0.75.

2. Biology   A strain of E-coli SC18del-recA718 is placed into a petri dish at 30° Celsius and allowed to grow. The following data are collected. Theory states that the number of bacteria in the petri dish will initially grow according to the law of uninhibited growth. The population is measured using an optical device in which the amount of light that passes through the petri dish is measured.

Time (hours), x	Population, y
2.5	0.175
3.5	0.38
4.5	0.63
4.75	0.76
5.25	1.20

Source: Dr. Polly Lavery, Joliet Junior College

(a) Draw a scatter diagram treating time as the predictor variable.
(b) Using a graphing utility, fit an exponential function to the data.
(c) Express the function found in part (b) in the form $N(t) = N_0 e^{kt}$.
(d) Graph the exponential function found in part (b) or (c) on the scatter diagram.
(e) Use the exponential function from part (b) or (c) to predict the population at $x = 6$ hours.
(f) Use the exponential function from part (b) or (c) to predict when the population will reach 2.1.

3. Chemistry   A chemist has a 100-gram sample of a radioactive material. He records the amount of radioactive material every week for 6 weeks and obtains the following data:

Week	Weight (in Grams)
0	100.0
1	88.3
2	75.9
3	69.4
4	59.1
5	51.8
6	45.5

(a) Using a graphing utility, draw a scatter diagram with week as the independent variable.
(b) Using a graphing utility, fit an exponential function to the data.
(c) Express the function found in part (b) in the form $A(t) = A_0 e^{kt}$.
(d) Graph the exponential function found in part (b) or (c) on the scatter diagram.
(e) From the result found in part (b), determine the half-life of the radioactive material.
(f) How much radioactive material will be left after 50 weeks?
(g) When will there be 20 grams of radioactive material?

4. Chemistry A chemist has a 1000-gram sample of a radioactive material. She records the amount of radioactive material remaining in the sample every day for a week and obtains the following data:

Day	Weight (in Grams)
0	1000.0
1	897.1
2	802.5
3	719.8
4	651.1
5	583.4
6	521.7
7	468.3

(a) Using a graphing utility, draw a scatter diagram with day as the independent variable.
(b) Using a graphing utility, fit an exponential function to the data.
(c) Express the function found in part (b) in the form $A(t) = A_0 e^{kt}$.
(d) Graph the exponential function found in part (b) or (c) on the scatter diagram.
(e) From the result found in part (b), find the half-life of the radioactive material.
(f) How much radioactive material will be left after 20 days?
(g) When will there be 200 grams of radioactive material?

5. Finance The following data represent the amount of money an investor has in an investment account each year for 10 years. She wishes to determine the average annual rate of return on her investment.
(a) Using a graphing utility, draw a scatter diagram with time as the independent variable and the value of the account as the dependent variable.
(b) Using a graphing utility, fit an exponential function to the data.

Year	Value of Account
1991	$10,000
1992	$10,573
1993	$11,260
1994	$11,733
1995	$12,424
1996	$13,269
1997	$13,968
1998	$14,823
1999	$15,297
2000	$16,539

(c) Based on the answer in part (b), what was the average annual rate of return from this account over the past 10 years?
(d) If the investor plans on retiring in 2021, what will the predicted value of this account be?
(e) When will the account be worth $50,000?

6. Finance The following data show the amount of money an investor has in an investment account each year for 7 years. He wishes to determine the average annual rate of return on his investment.

Year	Value of Account
1994	$20,000
1995	$21,516
1996	$23,355
1997	$24,885
1998	$27,434
1999	$30,053
2000	$32,622

(a) Using a graphing utility, draw a scatter diagram with time as the independent variable and the value of the account as the dependent variable.
(b) Using a graphing utility, fit an exponential function to the data.
(c) Based on the answer to part (b), what was the average annual rate of return from this account over the past 7 years?
(d) If the investor plans on retiring in 2020, what will the predicted value of this account be?
(e) When will the account be worth $80,000?

7. Economics and Marketing The following data represent the price and quantity demanded in 2000 for IBM personal computers at Best Buy.

Price ($/Computer)	Quantity Demanded
2300	152
2000	159
1700	164
1500	171
1300	176
1200	180
1000	189

(a) Using a graphing utility, draw a scatter diagram of the data with price as the dependent variable.
(b) Using a graphing utility, fit a logarithmic function to the data.
(c) Using a graphing utility, draw the logarithmic function found in part (b) on the scatter diagram.
(d) Use the function found in part (b) to predict the number of IBM personal computers that would be demanded if the price were $1650.

8. Economics and Marketing The following data represent the price and quantity supplied in 2000 for IBM personal computers.

Price ($/Computer)	Quantity Supplied
2300	180
2000	173
1700	160
1500	150
1300	137
1200	130
1000	113

(a) Using a graphing utility, draw a scatter diagram of the data with price as the dependent variable.
(b) Using a graphing utility, fit a logarithmic function to the data.
(c) Using a graphing utility, draw the logarithmic function found in part (b) on the scatter diagram.
(d) Use the function found in part (b) to predict the number of IBM personal computers that would be supplied if the price were $1650.

9. Population Model The following data obtained from the U.S. Census Bureau represent the population of the United States. An ecologist is interested in finding a function that describes the population of the United States.

Year	Population
1900	76,212,168
1910	92,228,496
1920	106,021,537
1930	123,202,624
1940	132,164,569
1950	151,325,798
1960	179,323,175
1970	203,302,031
1980	226,542,203
1990	248,709,873
2000	281,421,906

(a) Using a graphing utility, draw a scatter diagram of the data using the year as the independent variable and population as the dependent variable.

(b) Using a graphing utility, fit a logistic function to the data.
(c) Using a graphing utility, draw the function found in part (b) on the scatter diagram.
(d) Based on the function found in part (b), what is the carrying capacity of the United States?
(e) Use the function found in part (b) to predict the population of the United States in 2001.
(f) When will the United States population be 300,000,000?

10. Population Model The following data obtained from the U.S. Census Bureau represent the world population. An ecologist is interested in finding a function that describes the world population.

Year	Population (in Billions)
1993	5.531
1994	5.611
1995	5.691
1996	5.769
1997	5.847
1998	5.925
1999	6.003
2000	6.080
2001	6.157

(a) Using a graphing utility, draw a scatter diagram of the data using year as the independent variable and population as the dependent variable.
(b) Using a graphing utility, fit a logistic function to the data.
(c) Using a graphing utility, draw the function found in part (b) on the scatter diagram.
(d) Based on the function found in part (b), what is the carrying capacity of the world?
(e) Use the function found in part (b) to predict the population of the world in 2002.
(f) When will world population be 7 billion?

11. Population Model The data on page 502 obtained from the U.S. Census Bureau represent the population of Illinois. An urban economist is interested in finding a model that describes the population of Illinois.
(a) Using a graphing utility, draw a scatter diagram of the data using year as the independent variable and population as the dependent variable.
(b) Using a graphing utility, fit a logistic function to the data.
(c) Using a graphing utility, draw the function found in part (b) on the scatter diagram.
(d) Based on the function found in part (b), what is the carrying capacity of Illinois?

Year	Population
1900	4,821,550
1910	5,638,591
1920	6,485,280
1930	7,630,654
1940	7,897,241
1950	8,712,176
1960	10,081,158
1970	11,110,285
1980	11,427,409
1990	11,430,602
2000	12,419,293

Year	Population
1900	6,302,115
1910	7,665,111
1920	8,720,017
1930	9,631,350
1940	9,900,180
1950	10,498,012
1960	11,319,366
1970	11,800,766
1980	11,864,720
1990	11,881,643
2000	12,281,054

(e) Use the function found in part (b) to predict the population of Illinois in 2010.

12. **Population Model** The following data obtained from the U.S. Census Bureau represent the population of Pennsylvania. An urban economist is interested in finding a model that describes the population of Pennsylvania.

(a) Using a graphing utility, draw a scatter diagram of the data using year as the independent variable and population as the dependent variable.
(b) Using a graphing utility, fit a logistic function to the data.
(c) Using a graphing utility, draw the function found in part (b) on the scatter diagram.
(d) Based on the function found in part (b), what is the carrying capacity of Pennsylvania?
(e) Use the function found in part (b) to predict the population of Pennsylvania in 2010.

Chapter Review

Things To Know

One-to-one function f (p. 420)

A function whose inverse is also a function.
For any choice of elements x_1, x_2 in the domain of f, if $x_1 \neq x_2$, then $f(x_1) \neq f(x_2)$.

Horizontal-line test (p. 420)

If every horizontal line intersects the graph of a function f in at most one point, then f is one-to-one.

Inverse function f^{-1} **of** f (pp. 421–428)

Domain of f = Range of f^{-1}; Range of f = Domain of f^{-1}.
$f^{-1}(f(x)) = x$ and $f(f^{-1}(x)) = x$.
Graphs of f and f^{-1} are symmetric with respect to the line $y = x$.

Properties of the exponential function (pp. 434 and 436)

$f(x) = a^x$, $a > 1$

Domain: the interval $(-\infty, \infty)$;
Range: the interval $(0, \infty)$;
x-intercepts: none; y-intercept: 1;
horizontal asymptote: x-axis ($y = 0$) as $x \rightarrow -\infty$;
increasing; one-to-one; smooth; continuous
See Figure 17 for a typical graph.

$f(x) = a^x$, $0 < a < 1$

Domain: the interval $(-\infty, \infty)$;
Range: the interval $(0, \infty)$;
x-intercepts: none; y-intercept: 1;
horizontal asymptote: x-axis ($y = 0$) as $x \rightarrow \infty$;
decreasing; one-to-one; smooth; continuous
See Figure 21 for a typical graph.

Number e (p. 437)   Value approached by the expression $\left(1 + \dfrac{1}{n}\right)^n$ as $n \to \infty$; that is,

$$\lim_{n \to \infty}\left(1 + \frac{1}{n}\right)^n = e.$$

Property of exponents (p. 439)   If $a^u = a^v$, then $u = v$.

Properties of the logarithmic
functions (p. 448)
$f(x) = \log_a x, \quad a > 1$ Domain: the interval $(0, \infty)$;
$(y = \log_a x$ means $x = a^y)$ Range: the interval $(-\infty, \infty)$;
                                         x-intercept: 1; y-intercept: none;
                                         vertical asymptote: $x = 0$ (y-axis);
                                         increasing; one-to-one; smooth; continuous
                                         See Figure 28(b) for a typical graph.

$f(x) = \log_a x, \quad 0 < a < 1$ Domain: the interval $(0, \infty)$;
$(y = \log_a x$ means $x = a^y)$ Range: the interval $(-\infty, \infty)$;
                                          x-intercept: 1; y-intercept: none;
                                         vertical asymptote: $x = 0$ (y-axis);
                                         decreasing; one-to-one; smooth; continuous
                                         See Figure 28(b) for a typical graph.

Natural logarithm (p. 448) $y = \ln x$ means $x = e^y$.

Properties of logarithms (pp. 458–459) $\log_a 1 = 0$ $\log_a a = 1$ $a^{\log_a M} = M$ $\log_a a^r = r$

$$\log_a(MN) = \log_a M + \log_a N \qquad \log_a\!\left(\frac{M}{N}\right) = \log_a M - \log_a N$$

$$\log_a M^r = r \log_a M$$

(p. 461)   If $M = N$, then $\log_a M = \log_a N$.
                     If $\log_a M = \log_a N$, then $M = N$.

▨ Formulas

Change-of-Base Formula (p. 462) $\log_a M = \dfrac{\log_b M}{\log_b a}$

Compound Interest Formula (p. 473) $A = P\left(1 + \dfrac{r}{n}\right)^{nt}$

Continuous compounding (p. 475) $A = Pe^{rt}$

Present Value Formula (p. 477) $P = A\left(1 + \dfrac{r}{n}\right)^{-nt}$ or $P = Ae^{-rt}$

Growth and Decay (p. 482) $A(t) = A_0 e^{kt}$

Newton's Law of Cooling (p. 487) $u(t) = T + (u_0 - T)e^{kt}, \quad k < 0$

Logistic Growth Model (p. 489) $P(t) = \dfrac{c}{1 + ae^{-bt}}$

▨ Objectives

Section	You should be able to . . .	Review Exercises
6.1	① Determine the inverse of a function (p. 418)	1, 2
	② Obtain the graph of the inverse function from the graph of a function (p. 423)	3, 4
	③ Find the inverse function f^{-1} (p. 424)	5–10
6.2	① Evaluate exponential functions (p. 431)	11, 12
	② Graph exponential functions (p. 433)	43–47, 49, 51

▓ Review Exercises

Blue problem numbers indicate the authors' suggestions for use in a Practice Test.

In Problems 1 and 2, (a) find the inverse of the function, and (b) determine whether the inverse represents a function.

1. $\{(1, 2), (3, 5), (5, 8), (6, 10)\}$

2. $\{(-1, 4), (0, 2), (1, 4), (3, 7)\}$

In Problems 3 and 4, the graph of a one-to-one function is given. Draw the graph of the inverse function f^{-1}. For convenience (and as a hint), the graph of $y = x$ is also given.

3.

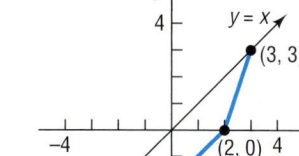

4.

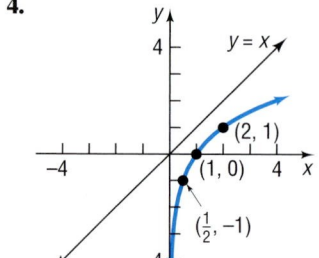

In Problems 5–10, the function f is one-to-one. Find the inverse of each function and check your answer. Find the domain and range of f and f^{-1}. Use a graphing utility to simultaneously graph f, f^{-1}, and y = x on the same square screen.

5. $f(x) = \dfrac{2x + 3}{5x - 2}$

6. $f(x) = \dfrac{2 - x}{3 + x}$

7. $f(x) = \dfrac{1}{x - 1}$

8. $f(x) = \sqrt{x - 2}$

9. $f(x) = \dfrac{3}{x^{1/3}}$

10. $f(x) = x^{1/3} + 1$

In Problems 11 and 12, suppose that $f(x) = 3^x$ and $g(x) = \log_3 x$.

11. Evaluate the following: (a) $f(4)$ (b) $g(9)$ (c) $f(-2)$ (d) $g\left(\dfrac{1}{27}\right)$

12. Evaluate the following: (a) $f(1)$ (b) $g(81)$ (c) $f(-4)$ (d) $g\left(\dfrac{1}{243}\right)$

In Problems 13 and 14, convert each exponential expression to an equivalent expression involving a logarithm. In Problems 15 and 16, convert each logarithmic expression to an equivalent expression involving an exponent.

13. $5^2 = z$

14. $a^5 = m$

15. $\log_5 u = 13$

16. $\log_a 4 = 3$

In Problems 17–20, find the domain of each logarithmic function.

17. $\log(3x - 2)$

18. $\log_5(2x + 1)$

19. $\log_2(x^2 - 3x + 2)$

20. $\ln(x^2 - 9)$

In Problems 21–26, evaluate each expression. Do not use a graphing utility.

21. $\log_2\left(\dfrac{1}{8}\right)$

22. $\log_3 81$

23. $\ln e^{\sqrt{2}}$

24. $e^{\ln 0.1}$

25. $2^{\log_2 0.4}$

26. $\log_2 2^{\sqrt{3}}$

In Problems 27–32, write each expression as the sum and/or difference of logarithms. Express powers as factors.

27. $\log_3\left(\dfrac{uv^2}{w}\right)$

28. $\log_2(a^2\sqrt{b})^4$

29. $\log(x^2\sqrt{x^3 + 1})$

30. $\log_5\left(\dfrac{x^2 + 2x + 1}{x^2}\right)$

31. $\ln\left(\dfrac{x\sqrt[3]{x^2 + 1}}{x - 3}\right)$

32. $\ln\left(\dfrac{2x + 3}{x^2 - 3x + 2}\right)^2$

In Problems 33–38, write each expression as a single logarithm.

33. $3\log_4 x^2 + \dfrac{1}{2}\log_4 \sqrt{x}$

34. $-2\log_3\left(\dfrac{1}{x}\right) + \dfrac{1}{3}\log_3 \sqrt{x}$

35. $\ln\left(\dfrac{x - 1}{x}\right) + \ln\left(\dfrac{x}{x + 1}\right) - \ln(x^2 - 1)$

36. $\log(x^2 - 9) - \log(x^2 + 7x + 12)$

37. $2\log 2 + 3\log x - \dfrac{1}{2}[\log(x + 3) + \log(x - 2)]$

38. $\dfrac{1}{2}\ln(x^2 + 1) - 4\ln\dfrac{1}{2} - \dfrac{1}{2}[\ln(x - 4) + \ln x]$

In Problems 39 and 40, use the Change-of-Base Formula and a calculator to evaluate each logarithm. Round your answer to three decimal places.

39. $\log_4 19$

40. $\log_2 21$

In Problems 41 and 42, graph each function using a graphing utility and the Change-of-Base Formula.

41. $y = \log_3 x$

42. $y = \log_7 x$

In Problems 43–52, use transformations to graph each function. Determine the domain, range, and any asymptotes. Verify your results using a graphing utility.

43. $f(x) = 2^{x-3}$

44. $f(x) = -2^x + 3$

45. $f(x) = \dfrac{1}{2}(3^{-x})$

46. $f(x) = 1 + 3^{2x}$

47. $f(x) = 1 - e^x$

48. $f(x) = 3 + \ln x$

49. $f(x) = 3e^x$

50. $f(x) = \dfrac{1}{2} \ln x$

51. $f(x) = 3 - e^{-x}$

52. $f(x) = 4 - \ln(-x)$

In Problems 53–72, solve each equation. Verify your result using a graphing utility.

53. $4^{1-2x} = 2$

54. $8^{6+3x} = 4$

55. $3^{x^2+x} = \sqrt{3}$

56. $4^{x-x^2} = \dfrac{1}{2}$

57. $\log_x 64 = -3$

58. $\log_{\sqrt{2}} x = -6$

59. $5^x = 3^{x+2}$

60. $5^{x+2} = 7^{x-2}$

61. $9^{2x} = 27^{3x-4}$

62. $25^{2x} = 5^{x^2-12}$

63. $\log_3 \sqrt{x-2} = 2$

64. $2^{x+1} \cdot 8^{-x} = 4$

65. $8 = 4^{x^2} \cdot 2^{5x}$

66. $2^x \cdot 5 = 10^x$

67. $\log_6(x+3) + \log_6(x+4) = 1$

68. $\log_{10}(7x - 12) = 2 \log_{10} x$

69. $e^{1-x} = 5$

70. $e^{1-2x} = 4$

71. $2^{3x} = 3^{2x+1}$

72. $2^{x^3} = 3^{x^2}$

In Problems 73 and 74, use the following result: If x is the atmospheric pressure (measured in millimeters of mercury), then the formula for the altitude h(x) (measured in meters above sea level) is

$$h(x) = (30T + 8000) \log\left(\frac{P_0}{x}\right)$$

where T is the temperature (in degrees Celsius) and P_0 is the atmospheric pressure at sea level, which is approximately 760 millimeters of mercury.

73. Finding the Altitude of an Airplane At what height is a Piper Cub whose instruments record an outside temperature of 0°C and a barometric pressure of 300 millimeters of mercury?

74. Finding the Height of a Mountain How high is a mountain if instruments placed on its peak record a temperature of 5°C and a barometric pressure of 500 millimeters of mercury?

---

75. Amplifying Sound An amplifier's power output P (in watts) is related to its decibel voltage gain d by the formula $P = 25e^{0.1d}$.

(a) Find the power output for a decibel voltage gain of 4 decibels.
(b) For a power output of 50 watts, what is the decibel voltage gain?

76. Limiting Magnitude of a Telescope A telescope is limited in its usefulness by the brightness of the star it is aimed at and by the diameter of its lens. One measure of a star's brightness is its *magnitude*: the dimmer the star, the larger its magnitude. A formula for the limiting magnitude L of a telescope, that is, the magnitude of the dimmest star that it can be used to view, is given by

$$L = 9 + 5.1 \log d$$

where d is the diameter (in inches) of the lens.
(a) What is the limiting magnitude of a 3.5-inch telescope?
(b) What diameter is required to view a star of magnitude 14?

77. Salvage Value The number of years n for a piece of machinery to depreciate to a known salvage value can be found using the formula

$$n = \frac{\log s - \log i}{\log(1 - d)}$$

where s is the salvage value of the machinery, i is its initial value, and d is the annual rate of depreciation.
(a) How many years will it take for a piece of machinery to decline in value from $90,000 to $10,000 if the annual rate of depreciation is 0.20 (20%)?
(b) How many years will it take for a piece of machinery to lose half of its value if the annual rate of depreciation is 15%?

78. Funding a College Education A child's grandparents purchase a $10,000 bond fund that matures in 18 years to be used for her college education. The bond fund pays 4% interest compounded semiannually. How much will the bond fund be worth at maturity? What is the effective rate of interest? How long would it take the bond to double in value under these terms?

79. Funding a College Education A child's grandparents wish to purchase a bond fund that matures in 18 years to be used for her college education. The bond fund pays 4% interest compounded semiannually. How much should they purchase so that the bond fund will be worth $85,000 at maturity?

80. Funding an IRA First Colonial Bankshares Corporation advertised the following IRA investment plans.

Target IRA Plans

For each $5000 Maturity Value Desired	
Deposit:	**At a Term of:**
$620.17	20 Years
$1045.02	15 Years
$1760.92	10 Years
$2967.26	5 Years

(a) Assuming continuous compounding, what was the annual rate of interest that they offered?

(b) First Colonial Bankshares claims that $4000 invested today will have a value of over $32,000 in 20 years. Use the answer found in part (a) to find the actual value of $4000 in 20 years. Assume continuous compounding.

81. Estimating the Date That a Prehistoric Man Died The bones of a prehistoric man found in the desert of New Mexico contain approximately 5% of the original amount of carbon 14. If the half-life of carbon 14 is 5600 years, approximately how long ago did the man die?

82. Temperature of a Skillet A skillet is removed from an oven whose temperature is 450°F and placed in a room whose temperature is 70°F. After 5 minutes, the temperature of the skillet is 400°F. How long will it be until its temperature is 150°F?

83. World Population According to the U.S. Census Bureau, the growth rate of the world's population in 1997 was $k = 1.33\% = 0.0133$. The population of the world in 1997 was 5,840,445,216. Letting $t = 0$ represent 1997, use the uninhibited growth model to predict the world's population in the year 2000.

84. Radioactive Decay The half-life of radioactive cobalt is 5.27 years. If 100 grams of radioactive cobalt is present now, how much will be present in 20 years? In 40 years?

85. Logistic Growth The logistic growth model

$$P(t) = \frac{0.8}{1 + 1.67e^{-0.16t}}$$

represents the proportion of new computers sold that utilize the Microsoft Windows 2000 operating system. Let $t = 0$ represent 2000, $t = 1$ represent 2001, and so on.

(a) What proportion of new computers sold in 2000 utilized Windows 2000?

(b) Determine the maximum proportion of new computers sold that will utilize Windows 2000.

(c) Using a graphing utility, graph $P(t)$.

(d) When will 75% of new computers sold utilize Windows 2000?

86. CBL Experiment The following data were collected by placing a temperature probe in a portable heater, removing the probe, and then recording temperature over time.

According to Newton's Law of Cooling, these data should follow an exponential model.

(a) Using a graphing utility, draw a scatter diagram for the data.

(b) Using a graphing utility, fit an exponential function to the data.

(c) Graph the exponential function found in part (b) on the scatter diagram.

(d) Predict how long it will take for the probe to reach a temperature of 110°F.

Time	Temperature (F°)
0	165.07
1	164.77
2	163.99
3	163.22
4	162.82
5	161.96
6	161.20
7	160.45
8	159.35
9	158.61
10	157.89
11	156.83
12	156.11
13	155.08
14	154.40
15	153.72

87. Wind Chill Factor The following data represent the wind speed (mph) and wind chill factor at an air temperature of 15°F.

Wind Speed (mph)	Wind Chill Factor
5	12
10	−3
15	−11
20	−17
25	−22
30	−25
35	−27

Source: Information Please Almanac

(a) Using a graphing utility, draw a scatter diagram with wind speed as the independent variable.

(b) Using a graphing utility, fit a logarithmic function to the data.

(c) Using a graphing utility, draw the logarithmic function found in part (b) on the scatter diagram.

(d) Use the function found in part (b) to predict the wind chill factor if the air temperature is 15°F and the wind speed is 23 mph.

88. Spreading of a Disease Jack and Diane live in a small town of 50 people. Unfortunately, Jack and Diane both have a cold. Those who come in contact with someone who has this cold will themselves catch the cold. The following data represent the number of people in the small town who have caught the cold after t days.

Days, t	Number of People with Cold, C
0	2
1	4
2	8
3	14
4	22
5	30
6	37
7	42
8	44

(a) Using a graphing utility, draw a scatter diagram of the data. Comment on the type of relation that appears to exist between the days and number of people with a cold.
(b) Using a graphing utility, fit a logistic function to the data.
(c) Graph the function found in part (b) on the scatter diagram.
(d) According to the function found in part (b), what is the maximum number of people who will catch the cold? In reality, what is the maximum number of people who could catch the cold?
(e) Sometime between the second and third day, 10 people in the town had a cold. According to the model found in part (b), when did 10 people have a cold?
(f) How long will it take for 46 people to catch the cold?

Chapter Projects

1. **Hot Coffee** A fast-food restaurant wants a special container to hold coffee. The restaurant wishes the container to quickly cool the coffee from 200°F to 130°F and keep the liquid between 110° and 130°F as long as possible. The restaurant has three containers to select from.

(1) The CentiKeeper Company has a container that reduces the temperature of a liquid from 200°F to 100°F in 30 minutes by maintaining a constant temperature of 70°F.
(2) The TempControl Company has a container that reduces the temperature of a liquid from 200°F to 110°F in 25 minutes by maintaining a constant temperature of 60°F.
(3) The Hot'n'Cold Company has a container that reduces the temperature of a liquid from 200°F to 120°F in 20 minutes by maintaining a constant temperature of 65°F.

You need to recommend which container the restaurant should purchase.

(a) Use Newton's Law of Cooling to find a function relating the temperature of the liquid over time.
(b) Graph each function using a graphing utility.
(c) How long does it take each container to lower the coffee temperature from 200°F to 130°F?
(d) How long will the coffee temperature remain between 110°F and 130°F? This temperature is considered the optimal drinking temperature.
(e) Which company would you recommend to the restaurant? Why?
(f) How might the cost of the container affect your decision?

2. **Depreciation of a New Car** In buying a new car, one consideration might be how well the price of the car holds up over time. Different makes of cars have different depreciation rates. We will discuss one way of computing the depreciation rate for a car. Look up the value of your favorite car for the past 5 years in the NADA ("blue book") available at your local bank, public library, or via the Internet. Let $t = 0$ represent the cost of this car if you purchased it new, $t = 1$ represent the cost of the same car when it is one year old, and so on. Fill in the following table:

Year	Value
$t = 0$	
$t = 1$	
$t = 2$	
$t = 3$	
$t = 4$	
$t = 5$	

(a) Using a graphing utility, draw a scatter diagram with time as the independent variable and value as the dependent variable.

(b) Find the exponential function of best fit.

(c) Express this function in the form $A = A_0e^{rt}$.

(d) What is the value of A_0? Compare this value to the purchase price of the vehicle.

(e) What is the value of r, the depreciation rate?

(f) Compare your value of r to others in your class. Which car has the lowest rate of depreciation? The highest?

(g) How might depreciation factor into your decision regarding a new car purchase?

3. **CBL Experiment** The following data were collected by placing a temperature probe in a cup of hot water,

removing the probe, and then recording temperature over time. According to Newton's Law of Cooling, these data should follow an exponential model.

(a) Using a graphing utility, draw a scatter diagram for the data.

(b) Using a graphing utility, fit an exponential function to the data.

(c) Graph the exponential function found in part (b) on the scatter diagram.

(d) Predict how long it will take for the water to reach a temperature of 110°F.

(e) Write the model found in part (b) in the form of equation (4) on page 487.

Time	Temperature (F°)	Time	Temperature (F°)
0	175.69	13	148.50
1	173.52	14	146.84
2	171.21	15	145.33
3	169.07	16	143.83
4	166.59	17	142.38
5	164.21	18	141.22
6	161.89	19	140.09
7	159.66	20	138.69
8	157.86	21	137.59
9	155.75	22	136.78
10	153.70	23	135.70
11	151.93	24	134.91
12	150.08	25	133.86

Cumulative Review

1. Is the following graph the graph of a function? If it is, is the function one-to-one?

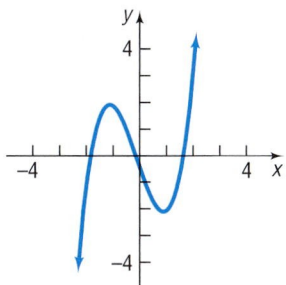

2. For the function $f(x) = 2x^2 - 3x + 1$, find the following:
   (a) $f(3)$ (b) $f(-x)$ (c) $f(x + h)$

3. Determine which of the following points are on the graph of $x^2 + y^2 = 1$.
   (a) $\left(\dfrac{1}{2}, \dfrac{1}{2}\right)$ (b) $\left(\dfrac{1}{2}, \dfrac{\sqrt{3}}{2}\right)$

4. Solve the equation $3(x - 2) = 4(x + 5)$.

5. Graph the line $2x - 4y = 16$.

6. Graph the quadratic function $f(x) = -x^2 + 2x - 3$ by determining whether its graph opens up or down and by finding its vertex, axis of symmetry, y-intercept, and x-intercept(s), if any. Verify your results using a graphing utility.

7. Determine the quadratic function whose graph is given in the figure below.

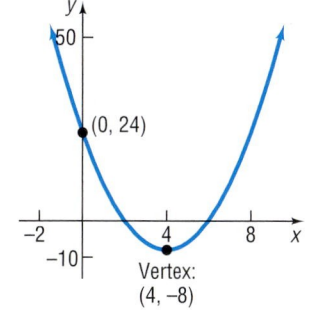

Vertex:
(4, −8)

8. Graph $f(x) = 3(x + 1)^3 - 2$ using transformations.

9. Given that $f(x) = x^2 + 2$ and $g(x) = \dfrac{2}{x - 3}$, find $f(g(x))$ and state its domain. What is $f(g(3))$?

10. For the polynomial function
$f(x) = 4x^3 + 9x^2 - 30x - 8$:
(a) Find the real zeros of f.
(b) Determine the intercepts of the graph of f.
(c) Use a graphing utility to approximate the local maxima and local minima.

(d) By hand, draw a complete graph of f. Be sure to label the intercepts and turning points.

11. For the function $g(x) = 3^x + 2$:
(a) Graph g using transformations. State the domain, range, and horizontal asymptote of g.
(b) Determine the inverse of g. State the domain, range, and vertical asymptote of g^{-1}.
(c) On the same graph as g, graph g^{-1}.

12. Solve the equation $4^{x-3} = 8^{2x}$.

13. Solve the equation $\log_3(x + 1) + \log_3(2x - 3) = 1$.

SYSTEMS OF EQUATIONS AND INEQUALITIES

Educational attainment

Changes in educational attainment over time indicate changes in the demand for skills and knowledge in the work force, as well as cultural evolution. An increase in the overall level of educational attainment can reflect the increasing emphasis society places on completing high school and college. Completing high school or college is an important educational accomplishment that yields many benefits such as better job opportunities and higher earnings.

- The educational attainment of 25- to 29-year-olds increased between 1971 and 1997. The percentage of students completing high school rose from 78 to 87 percent; the percentage of high school completers with some college rose from 44 to 65 percent; and the percentage of high school completers with 4 or more years of college rose from 22 to 32 percent.

SOURCE: National Center for Education Statistics, The Condition of Education, 1998.

SEE CHAPTER PROJECT 1.

OUTLINE

 For additional study help, go to

www.prenhall.com/sullivanegu3e

Materials include:

- Graphing Calculator Help
- Chapter Quiz
- Chapter Test
- PowerPoint Downloads
- Chapter Projects
- Student Tips

A Look Back, A Look Forward

In Chapters 1, 4, and 5, we solved equations and inequalities involving a single variable. In this chapter we take up the problem of solving equations and inequalities containing two or more variables. There are various ways to solve such problems:

The *method of substitution* for solving equations in several unknowns goes back to ancient times.

The *method of elimination*, although it had existed for centuries, was put into systematic order by Karl Friedrich Gauss (1777–1855) and by Camille Jordan (1838–1922). This method is now used for solving large systems by computer.

The theory of *matrices* was developed in 1857 by Arthur Cayley (1821–1895), although only later were matrices used as we use them in this chapter. Matrices have become a very flexible instrument, useful in almost all areas of mathematics.

The method of *determinants* was invented by Takakazu Seki Kōwa (1642–1708) in 1683 in Japan and by Gottfried Wilhelm von Leibniz (1646–1716) in 1693 in Germany. Both used them only in relation to linear equations. *Cramer's Rule* is named after Gabriel Cramer (1704–1752) of Switzerland, who popularized the use of determinants for solving linear systems.

Section 7.6 introduces *linear programming*, a modern application of linear inequalities to certain types of problems. This topic is particularly useful for students interested in operations research.

Section 7.7, on *partial fraction decomposition*, provides an application of systems of equations. This particular application is one that is used in integral calculus.

PREPARING FOR THIS SECTION

Before getting started, review the following:

✓ Solving Linear Equations (Section 1.3, pp. 115–117)   ✓ Lines (Section 1.7, pp. 163–173)

7.1   SYSTEMS OF LINEAR EQUATIONS: TWO EQUATIONS CONTAINING TWO VARIABLES

OBJECTIVES
1. Solve Systems of Equations Using a Graphing Utility
2. Solve Systems of Equations by Substitution
3. Solve Systems of Equations by Elimination
4. Identify Inconsistent Systems
5. Express the Solutions of a System of Dependent Equations

We begin with an example.

EXAMPLE 1   **Movie Theater Ticket Sales**

A movie theater sells tickets for $8.00 each, with seniors receiving a discount of $2.00. One evening the theater took in $3580 in revenue. If x represents the number of tickets sold at $8.00 and y the number of tickets sold at the discounted price of $6.00, write an equation that relates these variables.

Solution   Each nondiscounted ticket brings in $8.00, so x tickets will bring in $8x$ dollars. Similarly, y discounted tickets bring in $6y$ dollars. Since the total brought in is $3580, we must have

$$8x + 6y = 3580$$

In Example 1, suppose that we also know that 525 tickets were sold that evening. Then we have another equation relating the variables x and y.

$$x + y = 525$$

The two equations

$$8x + 6y = 3580$$
$$x + y = 525$$

form a *system* of equations.

In general, a **system of equations** is a collection of two or more equations, each containing one or more variables. Example 2 gives some examples of systems of equations.

EXAMPLE 2 **Examples of Systems of Equations**

(a) $\begin{cases} 2x + \ y = \ \ 5 \\ -4x + 6y = -2 \end{cases}$ (1) Two equations containing two variables, x and y
(2)

(b) $\begin{cases} x + y^2 = 5 \\ 2x + y \ = 4 \end{cases}$ (1) Two equations containing two variables, x and y
(2)

(c) $\begin{cases} x + \ y + \ z = 6 \\ 3x - 2y + 4z = 9 \\ x - \ y - \ z = 0 \end{cases}$ (1) Three equations containing three variables, x, y, and z
(2)
(3)

(d) $\begin{cases} x + y + z = 5 \\ x - y = 2 \end{cases}$ (1) Two equations containing three variables, x, y, and z
(2)

(e) $\begin{cases} x + y + \ z = 6 \\ 2x \ + 2z = 4 \\ y + \ z = 2 \\ x \ = 4 \end{cases}$ (1) Four equations containing three variables, x, y, and z
(2)
(3)
(4)

We use a brace, as shown, to remind us that we are dealing with a system of equations. We also will find it convenient to number each equation in the system.

A **solution** of a system of equations consists of values for the variables that are solutions of each equation of the system. To **solve** a system of equations means to find all solutions of the system.

For example, $x = 2, y = 1$ is a solution of the system in Example 2(a), because

$$\begin{cases} 2x + \ y = \ \ 5 \\ -4x + 6y = -2 \end{cases} \text{(1)} \qquad \begin{cases} 2(2) + \ \ 1 \ = \ \ 4 + 1 = \ \ 5 \\ -4(2) + 6(1) = -8 + 6 = -2 \end{cases}$$

A solution of the system in Example 2(b) is $x = 1, y = 2$, because

$$\begin{cases} x + y^2 = 5 \\ 2x + y \ = 4 \end{cases} \text{(1)} \qquad \begin{cases} 1 \ + 2^2 = 1 + 4 = 5 \\ 2(1) + 2 \ = 2 + 2 = 4 \end{cases}$$

Another solution of the system in Example 2(b) is $x = \dfrac{11}{4}, y = -\dfrac{3}{2}$, which you can check for yourself.

A solution of the system in Example 2(c) is $x = 3, y = 2, z = 1$, because

$$\begin{cases} x + y + z = 6 & \text{(1)} \\ 3x - 2y + 4z = 9 & \text{(2)} \\ x - y - z = 0 & \text{(3)} \end{cases} \qquad \begin{cases} 3 + 2 + 1 = 6 & \text{(1)} \\ 3(3) - 2(2) + 4(1) = 9 - 4 + 4 = 9 & \text{(2)} \\ 3 - 2 - 1 = 0 & \text{(3)} \end{cases}$$

Note that $x = 3, y = 3, z = 0$ is not a solution of the system in Example 2(c).

$$\begin{cases} x + y + z = 6 & \text{(1)} \\ 3x - 2y + 4z = 9 & \text{(2)} \\ x - y - z = 0 & \text{(3)} \end{cases} \qquad \begin{cases} 3 + 3 + 0 = 6 & \text{(1)} \\ 3(3) - 2(3) + 4(0) = 3 \neq 9 & \text{(2)} \\ 3 - 3 - 0 = 0 & \text{(3)} \end{cases}$$

Although these values satisfy equations (1) and (3), they do not satisfy equation (2). Any solution of the system must satisfy *each* equation of the system.

NOW WORK PROBLEM 3.

When a system of equations has at least one solution, it is said to be **consistent**; otherwise, it is called **inconsistent**.

An equation in n variables is said to be **linear** if it is equivalent to an equation of the form

$$a_1x_1 + a_2x_2 + \cdots + a_nx_n = b$$

where $x_1, x_2, \ldots, x_n$ are n distinct variables, $a_1, a_2, \ldots, a_n, b$ are constants, and at least one of the a's is not 0.

Some examples of linear equations are

$$2x + 3y = 2 \qquad 5x - 2y + 3z = 10 \qquad 8x + 8y - 2z + 5w = 0$$

If each equation in a system of equations is linear, then we have a **system of linear equations**. The systems in Examples 2(a), (c), (d), and (e) are linear, whereas the system in Example 2(b) is nonlinear. In this chapter we shall solve linear systems in Sections 7.1–7.4. We discuss nonlinear systems in a later chapter.

Two Linear Equations Containing Two Variables

We can view the problem of solving a system of two linear equations containing two variables as a geometry problem. The graph of each equation in such a system is a line. So, a system of two equations containing two variables represents a pair of lines. The lines either (1) intersect or (2) are parallel or (3) are **coincident** (that is, identical).

1. If the lines intersect, then the system of equations has one solution, given by the point of intersection. The system is **consistent** and the equations are **independent**. See Figure 1(a).

2. If the lines are parallel, then the system of equations has no solution, because the lines never intersect. The system is **inconsistent**. See Figure 1(b).

3. If the lines are coincident, then the system of equations has infinitely many solutions, represented by the totality of points on the line. The system is **consistent** and the equations are **dependent**. See Figure 1(c).

Figure 1

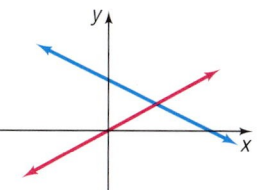

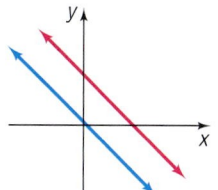

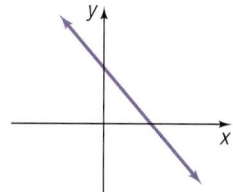

(a) Intersecting lines; system has one solution

(b) Parallel lines; system has no solution

(c) Coincident lines; system has infinitely many solutions

EXAMPLE 3 **Solving a System of Linear Equations Using a Graphing Utility**

① Solve: $\begin{cases} 2x + y = 5 & (1) \\ -4x + 6y = 12 & (2) \end{cases}$

Solution First, we solve each equation for y. This is equivalent to writing each equation in slope–intercept form. Equation (1) in slope–intercept form is $Y_1 = -2x + 5$. Equation (2) in slope–intercept form is $Y_2 = \dfrac{2}{3}x + 2$. Figure 2 shows the graphs using a graphing utility. From the graph in Figure 2, we see that the lines intersect, so the system is consistent and the equations are independent. Using INTERSECT, we obtain the solution $(1.125, 2.75)$. ∎

Figure 2

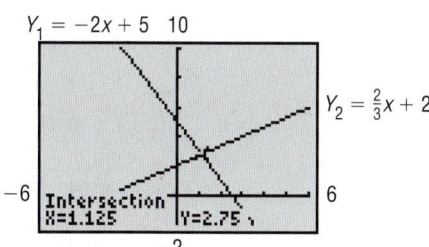

Sometimes, in order to obtain exact solutions, we must use algebraic methods. The first algebraic method that we discuss is the *method of substitution*.

Method of Substitution

② We illustrate the **method of substitution** by solving the system given in Example 4.

EXAMPLE 4 **Solving a System of Linear Equations by Substitution**

Solve: $\begin{cases} 2x + y = 5 & (1) \\ -4x + 6y = 12 & (2) \end{cases}$

Solution We solve the first equation for y, obtaining

$$y = -2x + 5 \qquad \text{Subtract 2x from each side of (1).}$$

We substitute this result for y in the second equation. The result is an equation containing just the variable x, which we can then solve for.

$$-4x + 6y = 12 \qquad (2)$$
$$-4x + 6(-2x + 5) = 12 \qquad \text{Substitute } y = -2x + 5 \text{ in (2).}$$
$$-4x - 12x + 30 = 12 \qquad \text{Remove parentheses.}$$
$$-16x = -18 \qquad \text{Combine like terms and subtract } 30 \text{ from both sides.}$$
$$x = \frac{-18}{-16} = \frac{9}{8} \qquad \text{Divide each side by } -16.$$

Once we know that $x = \dfrac{9}{8}$, we can easily find the value of y by **back-substitution**, that is, by substituting $\dfrac{9}{8}$ for x in one of the original equations. We use the first one.

$$2x + y = 5 \qquad \text{(1)}$$

$$2\left(\frac{9}{8}\right) + y = 5 \qquad \textcolor{blue}{\text{Substitute } x = \frac{9}{8} \text{ in (1).}}$$

$$\frac{9}{4} + y = 5 \qquad \textcolor{blue}{\text{Simplify.}}$$

$$y = 5 - \frac{9}{4} \qquad \textcolor{blue}{\text{Subtract } \frac{9}{4} \text{ from both sides.}}$$

$$= \frac{20}{4} - \frac{9}{4} = \frac{11}{4}$$

The solution of the system is $x = \dfrac{9}{8} = 1.125$, $y = \dfrac{11}{4} = 2.75$.

✔ CHECK:
$$\begin{cases} 2x + y = 5: & 2\left(\dfrac{9}{8}\right) + \dfrac{11}{4} = \dfrac{9}{4} + \dfrac{11}{4} = \dfrac{20}{4} = 5 \\[2mm] -4x + 6y = 12: & -4\left(\dfrac{9}{8}\right) + 6\left(\dfrac{11}{4}\right) = -\dfrac{9}{2} + \dfrac{33}{2} = \dfrac{24}{2} = 12 \end{cases}$$

The method used to solve the system in Example 4 is called **substitution**. The steps to be used are outlined next.

Steps for Solving by Substitution

STEP 1: Pick one of the equations and solve for one of the variables in terms of the remaining variables.

STEP 2: Substitute the result in the remaining equations.

STEP 3: If one equation in one variable results, solve this equation. Otherwise repeat Steps 1 and 2 until a single equation with one variable remains.

STEP 4: Find the values of the remaining variables by back-substitution.

STEP 5: Check the solution found.

EXAMPLE 5 Solving a System of Linear Equations by Substitution

Solve: $\begin{cases} 3x - 2y = 5 & \text{(1)} \\ 5x - y = 6 & \text{(2)} \end{cases}$

Solution **STEP 1:** After looking at the two equations, we conclude that it is easier to solve for the variable y in equation (2).

$$5x - y = 6 \qquad \text{(2)}$$

$$5x - y - 6 = 0 \qquad \textcolor{blue}{\text{Subtract 6 from both sides.}}$$

$$5x - 6 = y \qquad \textcolor{blue}{\text{Add } y \text{ to both sides.}}$$

STEP 2: We substitute this result into equation (1).

$$3x - 2y = 5 \quad \text{(1)}$$

$$3x - 2(5x - 6) = 5 \quad \text{Substitute } y = 5x - 6 \text{ in (1).}$$

STEP 3: We solve the resulting equation for x.

$$-7x + 12 = 5 \quad \text{Remove parentheses and combine like terms.}$$

$$-7x = -7 \quad \text{Subtract 12 from both sides.}$$

$$x = 1 \quad \text{Divide both sides by } -7.$$

Because we now have one solution, $x = 1$, we proceed to Step 4.

STEP 4: Knowing $x = 1$, we can find y from the equation

$$y = 5x - 6 = 5(1) - 6 = -1$$

$$\uparrow$$
$$x = 1$$

STEP 5: ✔ CHECK:
$$\begin{cases} 3(1) - 2(-1) = 3 + 2 = 5 \\ 5(1) - (-1) = 5 + 1 = 6 \end{cases}$$

The solution of the system is $x = 1$, $y = -1$.

✎ **NOW USE SUBSTITUTION TO WORK PROBLEM 11.**

Method of Elimination

③ A second method for solving a system of linear equations is the *method of elimination*. This method is usually preferred over substitution if substitution leads to fractions or if the system contains more than two variables. Elimination also provides the necessary motivation for solving systems using matrices (the subject of Section 7.3).

The idea behind the method of elimination is to replace the original system of equations by an equivalent system so that adding two of the equations eliminates a variable. The rules for obtaining equivalent equations are the same as those studied earlier. However, we may also interchange any two equations of the system and/or replace any equation in the system by the sum (or difference) of that equation and any other equation in the system.

> **Rules for Obtaining an Equivalent System of Equations**
>
> 1. Interchange any two equations in the system.
> 2. Multiply (or divide) each side of an equation by the same nonzero constant.
> 3. Replace any equation in the system by the sum (or difference) of that equation and a nonzero multiple of any other equation in the system.

An example will give you the idea. As you work through the example, pay particular attention to the pattern being followed.

EXAMPLE 6 **Solving a System of Linear Equations by Elimination**

Solve: $\begin{cases} 2x + 3y = 1 \quad \text{(1)} \\ -x + y = -3 \quad \text{(2)} \end{cases}$

Solution We multiply each side of equation (2) by 2 so that the coefficients of x in the two equations are negatives of one another. The result is the equivalent system

$$\begin{cases} 2x + 3y = 1 & (1) \\ -2x + 2y = -6 & (2) \end{cases}$$

If we now replace equation (2) of this system by the sum of the two equations, we obtain an equation containing just the variable y, which we can solve for.

$$\begin{cases} 2x + 3y = 1 & (1) \\ \underline{-2x + 2y = -6} & (2) \\ 5y = -5 & \text{Add (1) and (2).} \\ y = -1 & \text{Solve for } y. \end{cases}$$

We back-substitute this value for y in equation (1) and simplify to get

$$\begin{aligned} 2x + 3y &= 1 & (1) \\ 2x + 3(-1) &= 1 & \text{Substitute } y = -1 \text{ in (1).} \\ 2x &= 4 & \text{Simplify.} \\ x &= 2 & \text{Solve for } x. \end{aligned}$$

The solution of the original system is $x = 2$, $y = -1$. We leave it to you to check the solution. ∎

The procedure used in Example 6 is called the **method of elimination**. Notice the pattern of the solution. First, we eliminated the variable x from the second equation. Then we back-substituted; that is, we substituted the value found for y back into the first equation to find x.

Let's return to the movie theater example (Example 1).

EXAMPLE 7 | **Movie Theater Ticket Sales**

A movie theater sells tickets for $8.00 each, with seniors receiving a discount of $2.00. One evening the theater sold 525 tickets and took in $3580 in revenue. How many of each type of ticket were sold?

Solution If x represents the number of tickets sold at $8.00 and y the number of tickets sold at the discounted price of $6.00, then the given information results in the system of equations

$$\begin{cases} 8x + 6y = 3580 & (1) \\ x + y = 525 & (2) \end{cases}$$

We use elimination and multiply equation (2) by -6 and then add the equations

$$\begin{cases} 8x + 6y = 3580 \\ \underline{-6x - 6y = -3150} \\ 2x = 430 \qquad \text{Add the equations.} \\ x = 215 \end{cases}$$

Since $x + y = 525$, then $y = 525 - x = 525 - 215 = 310$. Thus, 215 nondiscounted tickets and 310 senior discount tickets were sold. ∎

NOW USE ELIMINATION TO WORK PROBLEM **11.**

④ The previous examples dealt with consistent systems of equations that had one solution. The next two examples deal with two other possibilities that may occur, the first being a system that has no solution.

EXAMPLE 8   An Inconsistent System of Linear Equations

Solve: $\begin{cases} 2x + y = 5 & \text{(1)} \\ 4x + 2y = 8 & \text{(2)} \end{cases}$

Solution   We choose to use the method of substitution and solve equation (1) for y.

$$2x + y = 5 \qquad \text{(1)}$$
$$y = -2x + 5 \qquad \text{Subtract 2x from each side.}$$

Now substitute $y = -2x + 5$ for y in equation (2) and solve for x.

$$4x + 2y = 8 \qquad \text{(2)}$$
$$4x + 2(-2x + 5) = 8 \qquad \text{Substitute } y = -2x + 5 \text{ in (2).}$$
$$4x - 4x + 10 = 8 \qquad \text{Remove parentheses.}$$
$$0 \cdot x = -2 \qquad \text{Subtract 10 from both sides.}$$

This equation has no solution. We conclude that the system itself has no solution and is therefore inconsistent. ∎

Figure 3

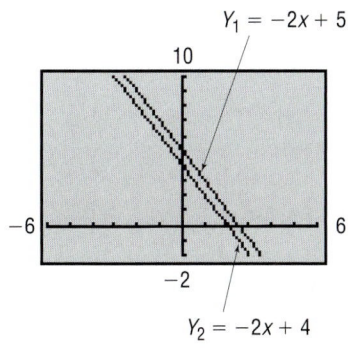

$Y_1 = -2x + 5$

$Y_2 = -2x + 4$

Figure 3 illustrates the pair of lines whose equations form the system in Example 8. Notice that the graphs of the two equations are lines, each with slope -2; one has a y-intercept of 5, the other a y-intercept of 4. The lines are parallel and have no point of intersection. This geometric statement is equivalent to the algebraic statement that the system has no solution.

⑤ The next example is an illustration of a system with infinitely many solutions.

EXAMPLE 9   Solving a System of Dependent Equations

Solve: $\begin{cases} 2x + y = 4 & \text{(1)} \\ -6x - 3y = -12 & \text{(2)} \end{cases}$

Solution   We choose to use the method of elimination.

$$\begin{cases} 2x + y = 4 & \text{(1)} \\ -6x - 3y = -12 & \text{(2)} \end{cases}$$

$$\begin{cases} 6x + 3y = 12 & \text{(1)} \quad \text{Multiply each side of equation (1) by 3.} \\ -6x - 3y = -12 & \text{(2)} \end{cases}$$

$$\begin{cases} 6x + 3y = 12 & \text{(1)} \\ 0 = 0 & \text{(2)} \quad \text{Replace equation (2) by the sum of equations (1) and (2).} \end{cases}$$

The original system is equivalent to a system containing one equation, so the equations are dependent. This means that any values of x and y for which $6x + 3y = 12$ or, equivalently, $2x + y = 4$ are solutions. For example, $x = 2$, $y = 0$; $x = 0$, $y = 4$; $x = -2$, $y = 8$; $x = 4$, $y = -4$; and so on, are solutions. There are, in fact, infinitely many values of x and y for which

$2x + y = 4$, so the original system has infinitely many solutions. We will write the solutions of the original systems either as

$$y = -2x + 4$$

where x can be any real number, or as

$$x = -\frac{1}{2}y + 2$$

where y can be any real number. ∎

Figure 4
$Y_1 = Y_2 = -2x + 4$

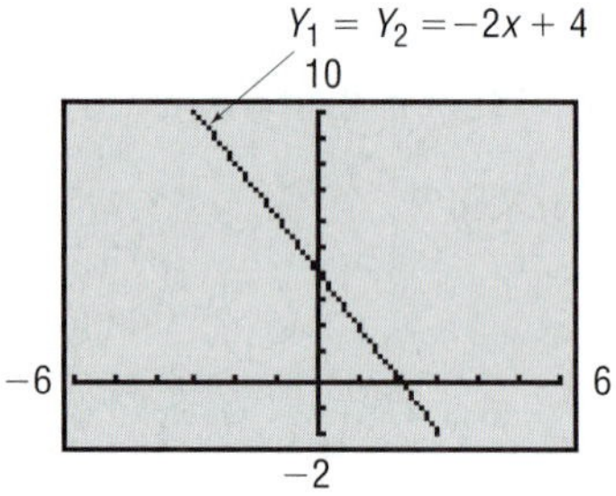

Figure 4 illustrates the situation presented in Example 9. Notice that the graphs of the two equations are lines, each with slope -2 and each with y-intercept 4. The lines are coincident. Notice also that equation (2) in the original system is just -3 times equation (1), indicating that the two equations are dependent.

For the system in Example 9, we can write down some of the infinite number of solutions by assigning values to x and then finding $y = -2x + 4$.

If $x = -2$, then $y = 8$.
If $x = 0$, then $y = 4$.
If $x = 2$, then $y = 0$.

The pairs (x, y) are points on the line in Figure 4.

✏ **NOW WORK PROBLEMS 17 AND 21.**

7.1 Concepts and Vocabulary

In Problems 1 and 2, fill in the blanks.

1. If a system of equations has no solution, it is said to be _____.

2. If a system of equations has infinitely many solutions, the system is said to be _____ and the equations are _____.

In Problems 3 and 4, answer True or False to each statement.

3. A system of two linear equations containing two variables always has at least one solution.

4. A solution of a system of equations consists of values for the variables that are solutions of each equation of the system.

5. Explain why replacing any equation in the system by the sum of that equation and a nonzero multiple of any other equation in the system results in an equivalent system of equations.

6. In this section, we presented two algebraic methods for solving a system of linear equations. Are there any circumstances where one method is preferable to the other? What are these circumstances?

7.1 Exercises

In Problems 1–8, verify that the values of the variables listed are solutions of the system of equations.

1. $\begin{cases} 2x - y = 5 \\ 5x + 2y = 8 \end{cases}$

$x = 2, y = -1$

2. $\begin{cases} 3x + 2y = 2 \\ x - 7y = -30 \end{cases}$

$x = -2, y = 4$

3. $\begin{cases} 3x + 4y = 4 \\ \frac{1}{2}x - 3y = -\frac{1}{2} \end{cases}$

$x = 2, y = \frac{1}{2}$

4. $\begin{cases} 2x + \frac{1}{2}y = 0 \\ 3x - 4y = -\frac{19}{2} \end{cases}$

$x = -\frac{1}{2}, y = 2$

5. $\begin{cases} x - y = 3 \\ \dfrac{1}{2}x + y = 3 \end{cases}$

$x = 4, y = 1$

6. $\begin{cases} x - y = 3 \\ -3x + y = 1 \end{cases}$

$x = -2, y = -5$

7. $\begin{cases} 3x + 3y + 2z = 4 \\ x - y - z = 0 \\ 2y - 3z = -8 \end{cases}$

$x = 1, y = -1, z = 2$

8. $\begin{cases} 4x - z = 7 \\ 8x + 5y - z = 0 \\ -x - y + 5z = 6 \end{cases}$

$x = 2, y = -3, z = 1$

In Problems 9–32, solve each system of equations. If the system has no solution, say that it is inconsistent. Use either substitution or elimination. Verify your solution using a graphing utility.

9. $\begin{cases} x + y = 8 \\ x - y = 4 \end{cases}$

10. $\begin{cases} x + 2y = 5 \\ x + y = 3 \end{cases}$

11. $\begin{cases} 5x - y = 13 \\ 2x + 3y = 12 \end{cases}$

12. $\begin{cases} x + 3y = 5 \\ 2x - 3y = -8 \end{cases}$

13. $\begin{cases} 3x = 24 \\ x + 2y = 0 \end{cases}$

14. $\begin{cases} 4x + 5y = -3 \\ -2y = -4 \end{cases}$

15. $\begin{cases} 3x - 6y = 2 \\ 5x + 4y = 1 \end{cases}$

16. $\begin{cases} 2x + 4y = \dfrac{2}{3} \\ 3x - 5y = -10 \end{cases}$

17. $\begin{cases} 2x + y = 1 \\ 4x + 2y = 3 \end{cases}$

18. $\begin{cases} x - y = 5 \\ -3x + 3y = 2 \end{cases}$

19. $\begin{cases} 2x - y = 0 \\ 3x + 2y = 7 \end{cases}$

20. $\begin{cases} 3x + 3y = -1 \\ 4x + y = \dfrac{8}{3} \end{cases}$

21. $\begin{cases} x + 2y = 4 \\ 2x + 4y = 8 \end{cases}$

22. $\begin{cases} 3x - y = 7 \\ 9x - 3y = 21 \end{cases}$

23. $\begin{cases} 2x - 3y = -1 \\ 10x + y = 11 \end{cases}$

24. $\begin{cases} 3x - 2y = 0 \\ 5x + 10y = 4 \end{cases}$

25. $\begin{cases} 2x + 3y = 6 \\ x - y = \dfrac{1}{2} \end{cases}$

26. $\begin{cases} \dfrac{1}{2}x + y = -2 \\ x - 2y = 8 \end{cases}$

27. $\begin{cases} \dfrac{1}{2}x + \dfrac{1}{3}y = 3 \\ \dfrac{1}{4}x - \dfrac{2}{3}y = -1 \end{cases}$

28. $\begin{cases} \dfrac{1}{3}x - \dfrac{3}{2}y = -5 \\ \dfrac{3}{4}x + \dfrac{1}{3}y = 11 \end{cases}$

29. $\begin{cases} 3x - 5y = 3 \\ 15x + 5y = 21 \end{cases}$

30. $\begin{cases} 2x - y = -1 \\ x + \dfrac{1}{2}y = \dfrac{3}{2} \end{cases}$

31. $\begin{cases} \dfrac{1}{x} + \dfrac{1}{y} = 8 \\ \dfrac{3}{x} - \dfrac{5}{y} = 0 \end{cases}$

32. $\begin{cases} \dfrac{4}{x} - \dfrac{3}{y} = 0 \\ \dfrac{6}{x} + \dfrac{3}{2y} = 2 \end{cases}$

[**Hint:** Let $u = \dfrac{1}{x}$ and $v = \dfrac{1}{y}$, and solve for u and v.

Then $x = \dfrac{1}{u}$ and $y = \dfrac{1}{v}$.]

In Problems 33–38, use a graphing utility to solve each system of equations. Express the solution rounded to two decimal places.

33. $\begin{cases} y = \sqrt{2}x - 20\sqrt{7} \\ y = -0.1x + 20 \end{cases}$

34. $\begin{cases} y = -\sqrt{3}x + 100 \\ y = 0.2x + \sqrt{19} \end{cases}$

35. $\begin{cases} \sqrt{2}x + \sqrt{3}y + \sqrt{6} = 0 \\ \sqrt{3}x - \sqrt{2}y + 60 = 0 \end{cases}$

36. $\begin{cases} \sqrt{5}x - \sqrt{6}y + 60 = 0 \\ 0.2x + 0.3y + \sqrt{5} = 0 \end{cases}$

37. $\begin{cases} \sqrt{3}x + \sqrt{2}y = \sqrt{0.3} \\ 100x - 95y = 20 \end{cases}$

38. $\begin{cases} \sqrt{6}x - \sqrt{5}y + \sqrt{1.1} = 0 \\ y = -0.2x + 0.1 \end{cases}$

---

39. Supply and Demand Suppose that the quantity supplied Q_s and quantity demanded Q_d of T-shirts at a concert is given by the following equations:

$$Q_s = -200 + 50p$$
$$Q_d = 1000 - 25p$$

where p is the price. The equilibrium price of a market is defined as the price at which quantity supplied equals quantity demanded $(Q_s = Q_d)$. Find the equilibrium price for T-shirts at this concert. What is the equilibrium quantity?

40. Supply and Demand Suppose that the quantity supplied Q_s and quantity demanded Q_d of hot dogs at a baseball game is given by the following equations:

$$Q_s = -2000 + 3000p$$
$$Q_d = 10{,}000 - 1000p$$

where p is the price. The equilibrium price of a market is defined as the price at which quantity supplied equals quantity demanded $(Q_s = Q_d)$. Find the equilibrium price for hot dogs at the baseball game. What is the equilibrium quantity?

41. The perimeter of a rectangular floor is 90 feet. Find the dimensions of the floor if the length is twice the width.

42. The length of fence required to enclose a rectangular field is 3000 meters. What are the dimensions of the field if it is known that the difference between its length and width is 50 meters?

43. Cost of Fast Food   Four large cheeseburgers and two chocolate shakes cost a total of $7.90. Two shakes cost 15¢ more than one cheeseburger. What is the cost of a cheeseburger? A shake?

44. Movie Theater Tickets   A movie theater charges $9.00 for adults and $7.00 for senior citizens. On a day when 325 people paid an admission, the total receipts were $2495. How many who paid were adults? How many were seniors?

45. Mixing Nuts   A store sells cashews for $5.00 per pound and peanuts for $1.50 per pound. The manager decides to mix 30 pounds of peanuts with some cashews and sell the mixture for $3.00 per pound. How many pounds of cashews should be mixed with the peanuts so that the mixture will produce the same revenue as would selling the nuts separately?

46. Financial Planning   A recently retired couple need $12,000 per year to supplement their Social Security. They have $150,000 to invest to obtain this income. They have decided on two investment options: AA bonds yielding 10% per annum and a Bank Certificate yielding 5%.
(a) How much should be invested in each to realize exactly $12,000?
(b) If, after two years, the couple requires $14,000 per year in income, how should they reallocate their investment to achieve the new amount?

47. Computing Wind Speed   With a tail wind, a small Piper aircraft can fly 600 miles in 3 hours. Against this same wind, the Piper can fly the same distance in 4 hours. Find the average wind speed and the average airspeed of the Piper.

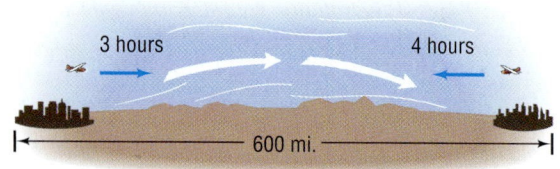

48. Computing Wind Speed   The average airspeed of a single-engine aircraft is 150 miles per hour. If the aircraft flew the same distance in 2 hours with the wind as it flew in 3 hours against the wind, what was the wind speed?

49. Restaurant Management   A restaurant manager wants to purchase 200 sets of dishes. One design costs $25 per set, while another costs $45 per set. If she only has $7400 to spend, how many of each design should be ordered?

50. Cost of Fast Food   One group of people purchased 10 hot dogs and 5 soft drinks at a cost of $12.50. A second bought 7 hot dogs and 4 soft drinks at a cost of $9.00. What is the cost of a single hot dog? A single soft drink?

We paid $12.50.
How much is one hot dog?
How much is one cola?

We paid $9.00.
How much is one hot dog?
How much is one cola?

51. Computing a Refund   The grocery store we use does not mark prices on its goods. My wife went to this store, bought three 1-pound packages of bacon and two cartons of eggs, and paid a total of $7.45. Not knowing that she went to the store, I also went to the same store, purchased two 1-pound packages of bacon and three cartons of eggs, and paid a total of $6.45. Now we want to return two 1-pound packages of bacon and two cartons of eggs. How much will be refunded?

52. Finding the Current of a Stream   Pamela requires 3 hours to swim 15 miles downstream on the Illinois River. The return trip upstream takes 5 hours. Find Pamela's average speed in still water. How fast is the current? (Assume that Pamela's speed is the same in each direction.)

53. Pharmacy   A doctor's prescription calls for a daily intake of liquid containing 40 mg of vitamin C and 30 mg of vitamin D. Your pharmacy stocks two liquids that can be used: one contains 20% vitamin C and 30% vitamin D, the other 40% vitamin C and 20% vitamin D. How many milligrams of each liquid should be mixed to fill the prescription?

54. Pharmacy   A doctor's prescription calls for the creation of pills that contain 12 units of vitamin B_{12} and 12 units of vitamin E. Your pharmacy stocks two powders that can be used to make these pills: one contains 20% vitamin B_{12} and 30% vitamin E, the other 40% vitamin B_{12} and 20% vitamin E. How many units of each powder should be mixed in each pill?

*The point at which a company's profits equal zero is called the company's **break-even point**. For Problems 55 and 56, let R represent a company's revenue, let C represent the company's costs, and let x represent the number of units produced and sold each day. Find the firm's break-even point; that is, find x so that R = C.*

55. $R = 8x$
  $C = 4.5x + 17,500$

56. $R = 12x$
  $C = 10x + 15,000$

57. Make up a system of two linear equations containing two variables that has:
 (a) No solution
 (b) Exactly one solution
 (c) Infinitely many solutions
 Give the three systems to a friend to solve and critique.

58. Write a brief paragraph outlining your strategy for solving a system of two linear equations containing two variables.

59. Do you prefer the method of substitution or the method of elimination for solving a system of two linear equations containing two variables? Give reasons.

7.2 SYSTEMS OF LINEAR EQUATIONS: THREE EQUATIONS CONTAINING THREE VARIABLES

OBJECTIVES
 1 Solve Systems of Three Equations Containing Three Variables
 2 Identify Inconsistent Systems
 3 Express the Solutions of a System of Dependent Equations

Just as with a system of two linear equations containing two variables, a system of three linear equations containing three variables also has either (1) exactly one solution (a consistent system with independent equations), or (2) no solution (an inconsistent system), or (3) infinitely many solutions (a consistent system with dependent equations).

We can view the problem of solving a system of three linear equations containing three variables as a geometry problem. The graph of each equation in such a system is a plane in space. A system of three linear equations containing three variables represents three planes in space. Figure 5 illustrates some of the possibilities.

Figure 5

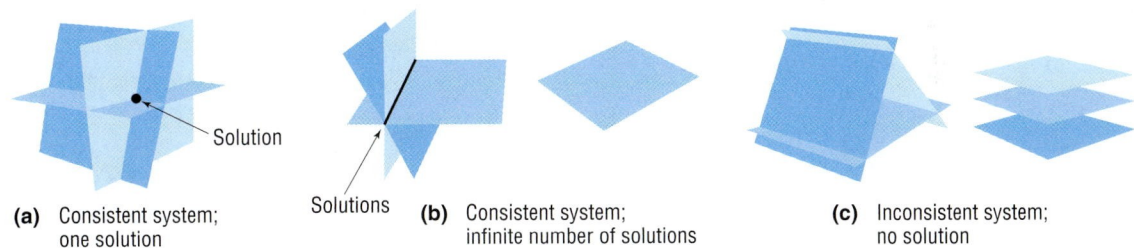

(a) Consistent system; one solution

Solutions **(b)** Consistent system; infinite number of solutions

(c) Inconsistent system; no solution

Recall that a **solution** to a system of equations consists of values for the variables that are solutions of each equation of the system. For example, $x = 3, y = -1, z = -5$ is a solution to the system of equations

$$\begin{cases} x + y + z = -3 \\ 2x - 3y + 6z = -21 \\ -3x + 5y = -14 \end{cases}$$

(1) $3 + (-1) + (-5) = -3$
(2) $2(3) - 3(-1) + 6(-5) = 6 + 3 - 30 = -21$
(3) $-3(3) + 5(-1) = -9 - 5 = -14$

because these values of the variables are solutions of each equation.

Typically, when solving a system of three linear equations containing three variables, we use the method of elimination. Recall that the idea behind the method of elimination is to form equivalent equations so that adding two of the equations eliminates a variable. We list the rules for obtaining an equivalent system of equations as a reminder.

> **Rules for Obtaining an Equivalent System of Equations**
> 1. Interchange any two equations in the system.
> 2. Multiply (or divide) each side of an equation by the same nonzero constant.
> 3. Replace any equation in the system by the sum (or difference) of that equation and a nonzero multiple of any other equation in the system.

1   Let's see how elimination works on a system of three equations containing three variables.

EXAMPLE 1   **Solving a System of Three Linear Equations with Three Variables**

Use the method of elimination to solve the system of equations.

$$\begin{cases} x + y - z = -1 & (1) \\ 4x - 3y + 2z = 16 & (2) \\ 2x - 2y - 3z = 5 & (3) \end{cases}$$

Solution   For a system of three equations, we attempt to eliminate one variable at a time, using pairs of equations until an equation with a single variable remains. Our plan of attack on this system will be to first eliminate the variable x from equations (2) and (3).

We begin by multiplying each side of equation (1) by -4 and adding the result to equation (2). (Do you see why? The coefficients on x are now opposites of each other.) We also multiply equation (1) by -2 and add the result to equation (3). Notice that these two procedures result in the removal of the x-variable from equations (2) and (3).

$$\begin{array}{l} x + y - z = -1 \;\; (1) \quad \text{Multiply by } -4 \\ 4x - 3y + 2z = 16 \;\; (2) \end{array}$$

$$\begin{array}{r} -4x - 4y + 4z = 4 \;\; (1) \\ 4x - 3y + 2z = 16 \;\; (2) \\ \hline -7y + 6z = 20 \quad \text{Add.} \end{array}$$

$$\begin{array}{l} x + y - z = -1 \;\; (1) \quad \text{Multiply by } -2 \\ 2x - 2y - 3z = 5 \;\; (3) \end{array}$$

$$\begin{array}{r} -2x - 2y + 2z = 2 \;\; (1) \\ 2x - 2y - 3z = 5 \;\; (3) \\ \hline -4y - z = 7 \quad \text{Add.} \end{array}$$

$$\begin{cases} x + y - z = -1 & (1) \\ -7y + 6z = 20 & (2) \\ -4y - z = 7 & (3) \end{cases}$$

We now concentrate on equations (2) and (3), treating them as a system of two equations containing two variables. It is easiest to eliminate z. We multiply each side of equation (3) by 6 and add equations (2) and (3). The result is the new equation (3).

$$\begin{array}{l} -7y + 6z = 20 \;\; (2) \\ -4y - z = 7 \;\; (3) \quad \text{Multiply by 6} \end{array}$$

$$\begin{array}{r} -7y + 6z = 20 \;\; (2) \\ -24y - 6z = 42 \;\; (3) \\ \hline -31y = 62 \quad \text{Add.} \end{array}$$

$$\begin{cases} x + y - z = -1 & (1) \\ -7y + 6z = 20 & (2) \\ -31y = 62 & (3) \end{cases}$$

We now solve equation (3) for y by dividing both sides of the equation by -31.

$$\begin{cases} x + y - z = -1 & (1) \\ -7y + 6z = 20 & (2) \\ y = -2 & (3) \end{cases}$$

Back-substitute $y = -2$ in equation (2) and solve for z.

$$-7y + 6z = 20 \qquad (2)$$

$$-7(-2) + 6z = 20 \qquad \text{Substitute } y = -2 \text{ in (2).}$$

$$6z = 6 \qquad \text{Subtract 14 from both sides of the equation.}$$

$$z = 1 \qquad \text{Divide both sides of the equation by 6.}$$

Finally, we back-substitute $y = -2$ and $z = 1$ in equation (1) and solve for x.

$$x + y - z = -1 \qquad (1)$$

$$x + (-2) - 1 = -1 \qquad \text{Substitute } y = -2 \text{ and } z = 1 \text{ in (1).}$$

$$x - 3 = -1 \qquad \text{Simplify.}$$

$$x = 2 \qquad \text{Add 3 to both sides.}$$

The solution of the original system is $x = 2$, $y = -2$, $z = 1$. You should verify this solution.　■

Look back over the solution given in Example 1. Note the pattern of removing one of the variables from two of the equations, followed by solving this system of two equations and two unknowns. Although which variables to remove is your choice, the methodology remains the same for all systems.

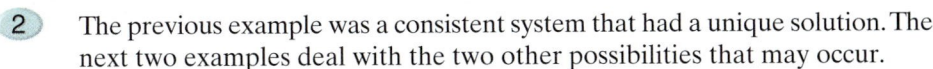 **NOW WORK PROBLEM 5.**

2 The previous example was a consistent system that had a unique solution. The next two examples deal with the two other possibilities that may occur.

EXAMPLE 2   **An Inconsistent System of Linear Equations**

Solve: $\begin{cases} 2x + y - z = -2 & (1) \\ x + 2y - z = -9 & (2) \\ x - 4y + z = 1 & (3) \end{cases}$

Solution   Our plan of attack is the same as in Example 1. However, in this system, it seems easiest to eliminate the variable z first. Do you see why?

Multiply each side of equation (1) by -1 and add the result to equation (2). Add equations (2) and (3).

$$\begin{array}{ll} -2x - y + z = 2 & (1) \text{ Multiply by } -1 \\ x + 2y - z = -9 & (2) \\ \hline -x + y = -7 & \text{Add.} \end{array}$$

$$\begin{array}{ll} x + 2y - z = -9 & (2) \\ x - 4y + z = 1 & (3) \\ \hline 2x - 2y = -8 & \text{Add.} \end{array}$$

$\begin{cases} 2x + y - z = -2 & (1) \\ -x + y = -7 & (2) \\ 2x - 2y = -8 & (3) \end{cases}$

We now concentrate on equations (2) and (3), treating them as a system of two equations containing two variables. Multiply each side of equation (2) by 2 and add the result to equation (3).

$$\begin{array}{l} -x + y = -7 \quad (2) \text{ Multiply by 2} \\ 2x - 2y = -8 \quad (3) \end{array} \qquad \begin{array}{l} -2x + 2y = -14 \quad (2) \\ 2x - 2y = -8 \quad (3) \\ \hline 0 = -22 \quad \text{Add.} \end{array} \qquad \begin{cases} 2x + y - z = -2 & (1) \\ -x + y = -7 & (2) \\ 0 = -22 & (3) \end{cases}$$

Equation (3) has no solution and the system is inconsistent.　■

3 Now let's look at a system of dependent equations.

EXAMPLE 3 **Solving a System of Dependent Equations**

Solve: $\begin{cases} x - 2y - z = 8 & (1) \\ 2x - 3y + z = 23 & (2) \\ 4x - 5y + 5z = 53 & (3) \end{cases}$

Solution Multiply each side of equation (1) by -2 and add the result to equation (2). Also, multiply each side of equation (1) by -4 and add the result to equation (3).

$\begin{array}{l} x - 2y - z = 8 \quad (1) \\ 2x - 3y + z = 23 \quad (2) \end{array}$ Multiply by -2 $\begin{array}{l} -2x + 4y + 2z = -16 \quad (1) \\ \underline{2x - 3y + z = 23} \quad (2) \\ \quad\quad y + 3z = 7 \quad \text{Add.} \end{array}$

$\begin{array}{l} x - 2y - z = 8 \quad (1) \\ 4x - 5y + 5z = 53 \quad (3) \end{array}$ Multiply by -4 $\begin{array}{l} -4x + 8y + 4z = -32 \quad (1) \\ \underline{4x - 5y + 5z = 53} \quad (2) \\ \quad\quad 3y + 9z = 21 \quad \text{Add.} \end{array}$

$\begin{cases} x - 2y - z = 8 & (1) \\ y + 3z = 7 & (2) \\ 3y + 9z = 21 & (3) \end{cases}$

Treat equations (2) and (3) as a system of two equations containing two variables, and eliminate the y-variable by multiplying each side of equation (2) by -3 and adding the result to equation (3).

$\begin{array}{l} y + 3z = 7 \\ 3y + 9z = 21 \end{array}$ Multiply by -3. $\begin{array}{l} -3y - 9z = -21 \\ \underline{3y + 9z = 21} \quad \text{Add.} \\ \quad\quad\quad 0 = 0 \end{array}$ $\begin{cases} x - 2y - z = 8 & (1) \\ y + 3z = 7 & (2) \\ 0 = 0 & (3) \end{cases}$

The original system is equivalent to a system containing two equations, so the equations are dependent and the system has infinitely many solutions. If we let z represent any real number, then, solving equation (2) for y, we determine that $y = -3z + 7$. Substitute this expression into equation (1) to determine x in terms of z.

$$\begin{array}{ll} x - 2y - z = 8 & (1) \\ x - 2(-3z + 7) - z = 8 & \text{Substitute } y = -3z + 7 \text{ in (1).} \\ x + 6z - 14 - z = 8 & \text{Remove parentheses.} \\ x + 5z = 22 & \text{Combine like terms.} \\ x = -5z + 22 & \text{Solve for } x. \end{array}$$

We will write the solution to the system as

$$\begin{cases} x = -5z + 22 \\ y = -3z + 7 \end{cases}$$

where z can be any real number.

To find specific solutions to the system, choose any value of z and use the equations $x = -5z + 22$ and $y = -3z + 7$ to determine x and y. For example, if $z = 0$, then $x = 22$ and $y = 7$, and if $z = 1$, then $x = 17$ and $y = 4$. ∎

NOW WORK PROBLEM 7.

Two points in the Cartesian plane determine a unique line. Given three noncollinear points, we can find the (unique) quadratic function whose graph contains these three points.

EXAMPLE 4   Curve Fitting

Find real numbers $a, b,$ and c so that the graph of the quadratic function $y = ax^2 + bx + c$ contains the points $(-1, -4), (1, 6),$ and $(3, 0)$.

Solution   We require that the three points satisfy the equation $y = ax^2 + bx + c$.

For the point $(-1, -4)$ we have: $-4 = a(-1)^2 + b(-1) + c$ $-4 = a - b + c$

For the point $(1, 6)$ we have: $6 = a(1)^2 + b(1) + c$ $6 = a + b + c$

For the point $(3, 0)$ we have: $0 = a(3)^2 + b(3) + c$ $0 = 9a + 3b + c$

We wish to determine $a, b,$ and c such that each equation is satisfied. That is, we want to solve the following system of three equations containing three variables:

$$\begin{cases} a - b + c = -4 & (1) \\ a + b + c = 6 & (2) \\ 9a + 3b + c = 0 & (3) \end{cases}$$

Figure 6

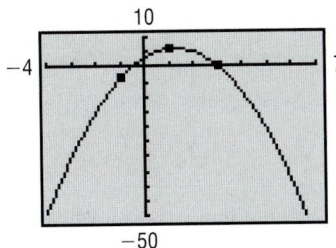

Solving this system of equations, we obtain $a = -2, b = 5,$ and $c = 3$. So the quadratic function whose graph contains the points $(-1, -4), (1, 6),$ and $(3, 0)$ is

$$y = -2x^2 + 5x + 3 \qquad y = ax^2 + bx + c, \quad a = -2, b = 5, c = 3$$

Figure 6 shows the graph of the function along with the three points.   ∎

7.2   Exercises

In Problems 1 and 2, verify that the values of the variables listed are solutions of the system of equations.

1. $\begin{cases} 3x + 3y + 2z = 4 \\ x - y - z = 0 \\ 2y - 3z = -8 \end{cases}$

$x = 1, y = -1, z = 2$

2. $\begin{cases} 4x - z = 7 \\ 8x + 5y - z = 0 \\ -x - y + 5z = 6 \end{cases}$

$x = 2, y = -3, z = 1$

In Problems 3–16, solve each system of equations. If the system has no solution, say that it is inconsistent.

3. $\begin{cases} x - y = 6 \\ 2x - 3z = 16 \\ 2y + z = 4 \end{cases}$

4. $\begin{cases} 2x + y = -4 \\ -2y + 4z = 0 \\ 3x - 2z = -11 \end{cases}$

5. $\begin{cases} x - 2y + 3z = 7 \\ 2x + y + z = 4 \\ -3x + 2y - 2z = -10 \end{cases}$

6. $\begin{cases} 2x + y - 3z = 0 \\ -2x + 2y + z = -7 \\ 3x - 4y - 3z = 7 \end{cases}$

7. $\begin{cases} x - y - z = 1 \\ 2x + 3y + z = 2 \\ 3x + 2y = 0 \end{cases}$

8. $\begin{cases} 2x - 3y - z = 0 \\ -x + 2y + z = 5 \\ 3x - 4y - z = 1 \end{cases}$

9. $\begin{cases} x - y - z = 1 \\ -x + 2y - 3z = -4 \\ 3x - 2y - 7z = 0 \end{cases}$

10. $\begin{cases} 2x - 3y - z = 0 \\ 3x + 2y + 2z = 2 \\ x + 5y + 3z = 2 \end{cases}$

11. $\begin{cases} 2x - 2y + 3z = 6 \\ 4x - 3y + 2z = 0 \\ -2x + 3y - 7z = 1 \end{cases}$

12. $\begin{cases} 3x - 2y + 2z = 6 \\ 7x - 3y + 2z = -1 \\ 2x - 3y + 4z = 0 \end{cases}$

13. $\begin{cases} x + y - z = 6 \\ 3x - 2y + z = -5 \\ x + 3y - 2z = 14 \end{cases}$

14. $\begin{cases} x - y + z = -4 \\ 2x - 3y + 4z = -15 \\ 5x + y - 2z = 12 \end{cases}$

15. $\begin{cases} x + 2y - z = -3 \\ 2x - 4y + z = -7 \\ -2x + 2y - 3z = 4 \end{cases}$

16. $\begin{cases} x + 4y - 3z = -8 \\ 3x - y + 3z = 12 \\ x + y + 6z = 1 \end{cases}$

---

17. Curve Fitting Find real numbers $a, b,$ and c so that the graph of the function $y = ax^2 + bx + c$ contains the points $(-1, 4), (2, 3),$ and $(0, 1)$.

18. Curve Fitting Find real numbers $a, b,$ and c so that the graph of the function $y = ax^2 + bx + c$ contains the points $(-1, -2), (1, -4),$ and $(2, 4)$.

19. Electricity: Kirchhoff's Rules An application of Kirchhoff's Rules to the circuit shown results in the following system of equations:

$$\begin{cases} I_2 = I_1 + I_3 \\ 5 - 3I_1 - 5I_2 = 0 \\ 10 - 5I_2 - 7I_3 = 0 \end{cases}$$

Find the currents $I_1, I_2,$ and I_3.*

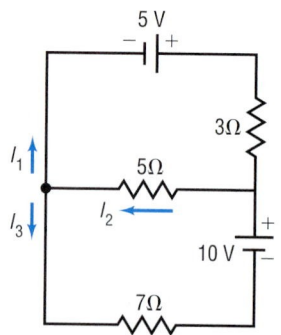

20. Electricity: Kirchhoff's Rules An application of Kirchhoff's Rules to the circuit shown results in the following system of equations:

$$\begin{cases} I_3 = I_1 + I_2 \\ 8 = 4I_3 + 6I_2 \\ 8I_1 = 4 + 6I_2 \end{cases}$$

Find the currents $I_1, I_2,$ and I_3.†

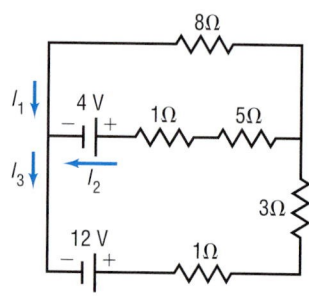

21. Theater Revenues A Broadway theater has 500 seats, divided into orchestra, main, and balcony seating. Orchestra seats sell for $50, main seats for $35, and balcony seats for $25. If all the seats are sold, the gross revenue to the theater is $17,100. If all the main and balcony seats are sold, but only half the orchestra seats are sold, the gross revenue is $14,600. How many are there of each kind of seat?

22. Theater Revenues A movie theater charges $8.00 for adults, $4.50 for children, and $6.00 for senior citizens. One day the theater sold 405 tickets and collected $2320 in receipts. There were twice as many children's tickets sold as adult tickets. How many adults, children, and senior citizens went to the theater that day?

23. Nutrition A dietitian wishes a patient to have a meal that has 66 grams of protein, 94.5 grams of carbohydrates, and 910 milligrams of calcium. The hospital food service tells the dietitian that the dinner for today is chicken, corn, and 2% milk. Each serving of chicken has 30 grams of protein, 35 grams of carbohydrates, and 200 milligrams of calcium. Each serving of corn has 3 grams of protein, 16 grams of carbohydrates, and 10 milligrams of calcium. Each glass of 2% milk has 9 grams of protein, 13 grams of carbohydrates, and 300 milligrams of calcium. How many servings of each food should the dietitian provide for the patient?

24. Investments Kelly has $20,000 to invest. As her financial planner, you recommend that she diversify into three investments: Treasury bills that yield 5% simple interest, Treasury bonds that yield 7% simple interest, and corporate bonds that yield 10% simple interest. Kelly wishes to earn $1390 per year in income. Also, Kelly wants her investment in Treasury bills to be $3000 more than her investment in corporate bonds. How much money should Kelly place in each investment?

25. Prices of Fast Food One group of customers bought 8 deluxe hamburgers, 6 orders of large fries, and 6 large colas for $26.10. A second group ordered 10 deluxe hamburgers, 6 large fries, and 8 large colas and paid $31.60. Is there sufficient information to determine the price of each food item? If not, construct a table showing the various possibilities. Assume that the hamburgers cost between $1.75 and $2.25, the fries between $0.75 and $1.00, and the colas between $0.60 and $0.90.

---

*Source: Based on Raymond Serway, *Physics*, 3rd ed. (Philadelphia: Saunders, 1990), Prob. 26, p. 790.
†Source: Ibid., Prob. 27, p. 790.

26. **Prices of Fast Food** Use the information given in Problem 25, and suppose that a third group purchased 3 deluxe hamburgers, 2 large fries, and 4 large colas for $10.95. Now is there sufficient information to determine the price of each food item? If so, determine each price.

27. **Painting a House** Three painters, Beth, Bill, and Edie, working together, can paint the exterior of a home in 10 hours. Bill and Edie together have painted a similar house in 15 hours. One day, all three worked on this same kind of house for 4 hours, after which Edie left. Beth and Bill required 8 more hours to finish. Assuming no gain or loss in efficiency, how long should it take each person to complete such a job alone?

PREPARING FOR THIS SECTION

Before getting started on this section, review the following:

✓ Solving Linear Equations (Section 1.3, pp. 115–117) ✓ Lines (Section 1.7, pp. 163–173)

7.3 SYSTEMS OF LINEAR EQUATIONS: MATRICES

OBJECTIVES
1. Write the Augmented Matrix of a System of Linear Equations
2. Write the System from the Augmented Matrix
3. Perform Row Operations on a Matrix
4. Solve Systems of Linear Equations Using Matrices
5. Use a Graphing Utility to Solve a System of Linear Equations

The systematic approach of the method of elimination for solving a system of linear equations provides another method of solution that involves a simplified notation.

Consider the following system of linear equations:

$$\begin{cases} x + 4y = 14 \\ 3x - 2y = 0 \end{cases}$$

If we choose not to write the symbols used for the variables, we can represent this system as

$$\left[\begin{array}{cc|c} 1 & 4 & 14 \\ 3 & -2 & 0 \end{array} \right]$$

where it is understood that the first column represents the coefficients of the variable x, the second column the coefficients of y, and the third column the constants on the right side of the equal signs. The vertical line serves as a reminder of the equal signs. The large square brackets are the traditional symbols used to denote a *matrix* in algebra.

A **matrix** is defined as a rectangular array of numbers,

$$
\begin{array}{c}
\\
\text{Row 1} \\
\text{Row 2} \\
\vdots \\
\text{Row } i \\
\vdots \\
\text{Row } m
\end{array}
\begin{bmatrix}
a_{11} & a_{12} & \cdots & a_{1j} & \cdots & a_{1n} \\
a_{21} & a_{22} & \cdots & a_{2j} & \cdots & a_{2n} \\
\vdots & \vdots & & \vdots & & \vdots \\
a_{i1} & a_{i2} & \cdots & a_{ij} & \cdots & a_{in} \\
\vdots & \vdots & & \vdots & & \vdots \\
a_{m1} & a_{m2} & \cdots & a_{mj} & \cdots & a_{mn}
\end{bmatrix}
\quad (1)
$$

with column headers: Column 1, Column 2, Column j, Column n.

Each number a_{ij} of the matrix has two indexes: the **row index** i and the **column index** j. The matrix shown in display (1) has m rows and n columns. The numbers a_{ij} are usually referred to as the **entries** of the matrix. For example, a_{23} refers to the entry in the second row, third column.

Now we will use matrix notation to represent a system of linear equations. The matrices used to represent systems of linear equations are called **augmented matrices**. In writing the augmented matrix of a system, the variables of each equation must be on the left side of the equal sign and the constants on the right side. A variable that does not appear in an equation has a coefficient of 0.

EXAMPLE 1   Writing the Augmented Matrix of a System of Linear Equations

Write the augmented matrix of each system of equations.

(a) $\begin{cases} 3x - 4y = -6 & (1) \\ 2x - 3y = -5 & (2) \end{cases}$
(b) $\begin{cases} 2x - y + z = 0 & (1) \\ x + z - 1 = 0 & (2) \\ x + 2y - 8 = 0 & (3) \end{cases}$

Solution (a) The augmented matrix is

$$
\begin{bmatrix}
3 & -4 & | & -6 \\
2 & -3 & | & -5
\end{bmatrix}
$$

(b) Care must be taken that the system be written so that the coefficients of all variables are present (if any variable is missing, its coefficient is 0). Also, all constants must be to the right of the equal sign. We need to rearrange the given system as follows:

$$
\begin{cases}
2x - y + z = 0 & (1) \\
x + z - 1 = 0 & (2) \\
x + 2y - 8 = 0 & (3)
\end{cases}
$$

$$
\begin{cases}
2x - y + z = 0 & (1) \\
x + 0 \cdot y + z = 1 & (2) \\
x + 2y + 0 \cdot z = 8 & (3)
\end{cases}
$$

The augmented matrix is

$$
\begin{bmatrix}
2 & -1 & 1 & | & 0 \\
1 & 0 & 1 & | & 1 \\
1 & 2 & 0 & | & 8
\end{bmatrix}
$$

If we do not include the constants to the right of the equal sign, that is, to the right of the vertical bar in the augmented matrix of a system of equations, the resulting matrix is called the **coefficient matrix** of the system. For the systems discussed in Example 1, the coefficient matrices are

$$\begin{bmatrix} 3 & -4 \\ 2 & -3 \end{bmatrix} \text{ and } \begin{bmatrix} 2 & -1 & 1 \\ 1 & 0 & 1 \\ 1 & 2 & 0 \end{bmatrix}$$

 **NOW WORK PROBLEM 3.**

EXAMPLE 2

Writing the System of Linear Equations from the Augmented Matrix

② Write the system of linear equations corresponding to each augmented matrix.

(a) $\begin{bmatrix} 5 & 2 & | & 13 \\ -3 & 1 & | & -10 \end{bmatrix}$ (b) $\begin{bmatrix} 3 & -1 & -1 & | & 7 \\ 2 & 0 & 2 & | & 8 \\ 0 & 1 & 1 & | & 0 \end{bmatrix}$

Solution (a) The matrix has two rows and so represents a system of two equations. The two columns to the left of the vertical bar indicate that the system has two variables. If x and y are used to denote these variables, the system of equations is

$$\begin{cases} 5x + 2y = 13 & (1) \\ -3x + y = -10 & (2) \end{cases}$$

(b) Since the augmented matrix has three rows, it represents a system of three equations. Since there are three columns to the left of the vertical bar, the system contains three variables. If x, y, and z are the three variables, the system of equations is

$$\begin{cases} 3x - y - z = 7 & (1) \\ 2x + 2z = 8 & (2) \\ y + z = 0 & (3) \end{cases}$$ ■

Row Operations on a Matrix

③ **Row operations** on a matrix are used to solve systems of equations when the system is written as an augmented matrix. There are three basic row operations.

> ***Row Operations***
>
>   1. Interchange any two rows.
>   2. Replace a row by a nonzero multiple of that row.
>   3. Replace a row by the sum of that row and a constant nonzero multiple of some other row.

These three row operations correspond to the three rules given earlier for obtaining an equivalent system of equations. When a row operation is performed on a matrix, the resulting matrix represents a system of equations equivalent to the system represented by the original matrix.

For example, consider the augmented matrix

$$\begin{bmatrix} 1 & 2 & | & 3 \\ 4 & -1 & | & 2 \end{bmatrix}$$

Suppose that we want to apply a row operation to this matrix that results in a matrix whose entry in row 2, column 1 is a 0. The row operation to use is

> Multiply each entry in row 1 by -4 and add the result to the corresponding entries in row 2.   (2)

If we use R_2 to represent the new entries in row 2 and we use r_1 and r_2 to represent the original entries in rows 1 and 2, respectively, then we can represent the row operation in statement (2) by

$$R_2 = -4r_1 + r_2$$

Then

$$\begin{bmatrix} 1 & 2 & | & 3 \\ 4 & -1 & | & 2 \end{bmatrix} \xrightarrow[R_2 = -4r_1 + r_2]{} \begin{bmatrix} 1 & 2 & | & 3 \\ -4(1) + 4 & -4(2) + (-1) & | & -4(3) + 2 \end{bmatrix} = \begin{bmatrix} 1 & 2 & | & 3 \\ 0 & -9 & | & -10 \end{bmatrix}$$

As desired, we now have the entry 0 in row 2, column 1.

EXAMPLE 3   Applying a Row Operation to an Augmented Matrix

Apply the row operation $R_2 = -3r_1 + r_2$ to the augmented matrix

$$\begin{bmatrix} 1 & -2 & | & 2 \\ 3 & -5 & | & 9 \end{bmatrix}$$

Solution   The row operation $R_2 = -3r_1 + r_2$ tells us that the entries in row 2 are to be replaced by the entries obtained after multiplying each entry in row 1 by -3 and adding the result to the corresponding entries in row 2. Thus,

$$\begin{bmatrix} 1 & -2 & | & 2 \\ 3 & -5 & | & 9 \end{bmatrix} \xrightarrow[R_2 = -3r_1 + r_2]{} \begin{bmatrix} 1 & -2 & | & 2 \\ -3(1) + 3 & (-3)(-2) + (-5) & | & -3(2) + 9 \end{bmatrix} = \begin{bmatrix} 1 & -2 & | & 2 \\ 0 & 1 & | & 3 \end{bmatrix}$$

NOW WORK PROBLEM **13**.

EXAMPLE 4   Finding a Particular Row Operation

Using the augmented matrix

$$\begin{bmatrix} 1 & -2 & | & 2 \\ 0 & 1 & | & 3 \end{bmatrix}$$

find a row operation that will result in this augmented matrix having a 0 in row 1, column 2.

Solution   We want a 0 in row 1, column 2. This result can be accomplished by multiplying row 2 by 2 and adding the result to row 1. That is, we apply the row operation $R_1 = 2r_2 + r_1$.

$$\begin{bmatrix} 1 & -2 & | & 2 \\ 0 & 1 & | & 3 \end{bmatrix} \rightarrow \begin{bmatrix} 2(0) + 1 & 2(1) + (-2) & | & 2(3) + 2 \\ 0 & 1 & | & 3 \end{bmatrix} = \begin{bmatrix} 1 & 0 & | & 8 \\ 0 & 1 & | & 3 \end{bmatrix}$$

$$\uparrow$$
$$R_1 = 2r_2 + r_1$$

A word about the notation that we have introduced. A row operation such as $R_1 = 2r_2 + r_1$ changes the entries in row 1. Note also that for this type of row operation we change the entries in a given row by multiplying the entries in some other row by an appropriate nonzero number and adding the results to the original entries of the row to be changed.

Solving a System of Linear Equations Using Matrices

④   To solve a system of linear equations using matrices, we use row operations on the augmented matrix of the system to obtain a matrix that is in *row echelon form*.

> A matrix is in **row echelon form** when
>
> 1. The entry in row 1, column 1 is a 1, and 0's appear below it.
>
> 2. The first nonzero entry in each row after the first row is a 1, 0's appear below it, and it appears to the right of the first nonzero entry in any row above.
>
> 3. Any rows that contain all 0's to the left of the vertical bar appear at the bottom.

For example, for a system of three equations containing three variables with a unique solution, the augmented matrix is in row echelon form if it is of the form

$$\begin{bmatrix} 1 & a & b & | & d \\ 0 & 1 & c & | & e \\ 0 & 0 & 1 & | & f \end{bmatrix}$$

where a, b, c, d, e, and f are real numbers. The last row of the augmented matrix states that $z = f$. We can then determine the value of y using back-substitution with $z = f$, since row 2 represents the equation $y + cz = e$. Finally, x is determined using back-substitution again.

Two advantages of solving a system of equations by writing the augmented matrix in row echelon form are the following:

1. The process is algorithmic; that is, it consists of repetitive steps that can be programmed on a computer.

2. The process works on any system of linear equations, no matter how many equations or variables are present.

The next example shows how to write a matrix in row echelon form.

EXAMPLE 5   **Solving a System of Linear Equations Using Matrices (Row Echelon Form)**

Solve: $\begin{cases} 2x + 2y = 6 & (1) \\ x + y + z = 1 & (2) \\ 3x + 4y - z = 13 & (3) \end{cases}$

Solution   First, we write the augmented matrix that represents this system.

$$\left[\begin{array}{ccc|c} 2 & 2 & 0 & 6 \\ 1 & 1 & 1 & 1 \\ 3 & 4 & -1 & 13 \end{array} \right]$$

The first step requires getting the entry 1 in row 1, column 1. An interchange of rows 1 and 2 is the easiest way to do this. [Note that this is equivalent to interchanging equations (1) and (2) of the system.]

$$\left[\begin{array}{ccc|c} 1 & 1 & 1 & 1 \\ 2 & 2 & 0 & 6 \\ 3 & 4 & -1 & 13 \end{array} \right]$$

Next, we want a 0 in row 2, column 1 and a 0 in row 3, column 1. We use the row operations $R_2 = -2r_1 + r_2$ and $R_3 = -3r_1 + r_3$ to accomplish this. Notice that row 1 is unchanged using these row operations. Also, do you see that performing these row operations simultaneously is the same as doing one followed by the other?

$$\left[\begin{array}{ccc|c} 1 & 1 & 1 & 1 \\ 2 & 2 & 0 & 6 \\ 3 & 4 & -1 & 13 \end{array} \right] \rightarrow \left[\begin{array}{ccc|c} 1 & 1 & 1 & 1 \\ 0 & 0 & -2 & 4 \\ 0 & 1 & -4 & 10 \end{array} \right]$$

$$\uparrow \quad \begin{array}{l} R_2 = -2r_1 + r_2 \\ R_3 = -3r_1 + r_3 \end{array}$$

Now we want the entry 1 in row 2, column 2. Interchanging rows 2 and 3 will accomplish this.

$$\left[\begin{array}{ccc|c} 1 & 1 & 1 & 1 \\ 0 & 0 & -2 & 4 \\ 0 & 1 & -4 & 10 \end{array} \right] \rightarrow \left[\begin{array}{ccc|c} 1 & 1 & 1 & 1 \\ 0 & 1 & -4 & 10 \\ 0 & 0 & -2 & 4 \end{array} \right]$$

Finally, we want a 1 in row 3, column 3. To obtain it, we use the row operation $R_3 = -\dfrac{1}{2}r_3$. The result is

$$\left[\begin{array}{ccc|c} 1 & 1 & 1 & 1 \\ 0 & 1 & -4 & 10 \\ 0 & 0 & -2 & 4 \end{array} \right] \rightarrow \left[\begin{array}{ccc|c} 1 & 1 & 1 & 1 \\ 0 & 1 & -4 & 10 \\ 0 & 0 & 1 & -2 \end{array} \right]$$

$$\uparrow \quad R_3 = -\frac{1}{2}r_3$$

This matrix is the row echelon form of the augmented matrix. The third row of this matrix represents the equation $z = -2$. Using $z = -2$, we back-substitute into the equation $y - 4z = 10$ (from the second row) and obtain

$$\begin{aligned} y - 4z &= 10 \\ y - 4(-2) &= 10 \qquad z = -2 \\ y &= 2 \qquad \text{Solve for } y. \end{aligned}$$

Finally, we back-substitute $y = 2$ and $z = -2$ into the equation $x + y + z = 1$ (from the first row) and obtain

$$x + y + z = 1$$
$$x + 2 + (-2) = 1 \qquad \text{\textcolor{blue}{$y = 2, z = -2$}}$$
$$x = 1 \qquad \text{\textcolor{blue}{Solve for x.}}$$

The solution of the system is $x = 1, y = 2, z = -2$. ■

The steps that we used to solve the system of linear equations in Example 5 can be summarized as follows:

> **Matrix Method for Solving a System of Linear Equations (Row Echelon Form)**
>
> **STEP 1:** Write the augmented matrix that represents the system.
>
> **STEP 2:** Perform row operations that place the entry 1 in row 1, column 1.
>
> **STEP 3:** Perform row operations that leave the entry 1 in row 1, column 1 unchanged, while causing 0's to appear below it in column 1.
>
> **STEP 4:** Perform row operations that place the entry 1 in row 2, column 2, but leave the entries in columns to the left unchanged. If it is impossible to place a 1 in row 2, column 2, then proceed to place a 1 in row 2, column 3. Once a 1 is in place, perform row operations to place 0's below it.
>
> [If any rows are obtained that contain only 0's on the left side of the vertical bar, place such rows at the bottom of the matrix.]
>
> **STEP 5:** Now repeat Step 4, placing a 1 in the next row, but one column to the right. Continue until the bottom row or the vertical bar is reached.
>
> **STEP 6:** The matrix that results is the row echelon form of the augmented matrix. Analyze the system of equations corresponding to it to solve the original system.

In the next example, we solve a system of linear equations using these steps. In addition, we solve the system using a graphing utility.

EXAMPLE 6 **Solving a System of Linear Equations Using Matrices (Row Echelon Form)**

Solve: $\begin{cases} x - y + z = 8 & \text{\textcolor{blue}{(1)}} \\ 2x + 3y - z = -2 & \text{\textcolor{blue}{(2)}} \\ 3x - 2y - 9z = 9 & \text{\textcolor{blue}{(3)}} \end{cases}$

Algebraic Solution **STEP 1:** The augmented matrix of the system is

$$\begin{bmatrix} 1 & -1 & 1 & | & 8 \\ 2 & 3 & -1 & | & -2 \\ 3 & -2 & -9 & | & 9 \end{bmatrix}$$

STEP 2: Because the entry 1 is already present in row 1, column 1, we can go to step 3.

STEP 3: Perform the row operations $R_2 = -2r_1 + r_2$ and $R_3 = -3r_1 + r_3$. Each of these leaves the entry 1 in row 1, column 1 unchanged, while causing 0's to appear under it.

$$\begin{bmatrix} 1 & -1 & 1 & | & 8 \\ 2 & 3 & -1 & | & -2 \\ 3 & -2 & -9 & | & 9 \end{bmatrix} \rightarrow \begin{bmatrix} 1 & -1 & 1 & | & 8 \\ 0 & 5 & -3 & | & -18 \\ 0 & 1 & -12 & | & -15 \end{bmatrix}$$
$$\begin{array}{l} R_2 = -2r_1 + r_2 \\ R_3 = -3r_1 + r_3 \end{array}$$

STEP 4: The easiest way to obtain the entry 1 in row 2, column 2 without altering column 1 is to interchange rows 2 and 3 (another way would be to multiply row 2 by $\dfrac{1}{5}$, but this introduces fractions).

$$\begin{bmatrix} 1 & -1 & 1 & | & 8 \\ 0 & 1 & -12 & | & -15 \\ 0 & 5 & -3 & | & -18 \end{bmatrix}$$

To get a 0 under the 1 in row 2, column 2, perform the row operation $R_3 = -5r_2 + r_3$.

$$\begin{bmatrix} 1 & -1 & 1 & | & 8 \\ 0 & 1 & -12 & | & -15 \\ 0 & 5 & -3 & | & -18 \end{bmatrix} \rightarrow \begin{bmatrix} 1 & -1 & 1 & | & 8 \\ 0 & 1 & -12 & | & -15 \\ 0 & 0 & 57 & | & 57 \end{bmatrix}$$
$$R_3 = -5r_2 + r_3$$

STEP 5: Continuing, we obtain a 1 in row 3, column 3 by using $R_3 = \dfrac{1}{57}r_3$.

$$\begin{bmatrix} 1 & -1 & 1 & | & 8 \\ 0 & 1 & -12 & | & -15 \\ 0 & 0 & 57 & | & 57 \end{bmatrix} \rightarrow \begin{bmatrix} 1 & -1 & 1 & | & 8 \\ 0 & 1 & -12 & | & -15 \\ 0 & 0 & 1 & | & 1 \end{bmatrix}$$
$$R_3 = \dfrac{1}{57}r_3$$

STEP 6: The matrix on the right is the row echelon form of the augmented matrix. The system of equations represented by the matrix in row echelon form is

$$\begin{cases} x - y + z = 8 & (1) \\ y - 12z = -15 & (2) \\ z = 1 & (3) \end{cases}$$

Using $z = 1$, we back-substitute to get

$$\begin{cases} x - y + 1 = 8 & (1) \\ y - 12(1) = -15 & (2) \end{cases} \xrightarrow{\text{Simplify.}} \begin{cases} x - y = 7 & (1) \\ y = -3 & (2) \end{cases}$$

We get $y = -3$, and back-substituting into $x - y = 7$, we find that $x = 4$. The solution of the system is $x = 4$, $y = -3$, $z = 1$.

⑤ **Graphing Solution** The augmented matrix of the system is

$$\begin{bmatrix} 1 & -1 & 1 & | & 8 \\ 2 & 3 & -1 & | & -2 \\ 3 & -2 & -9 & | & 9 \end{bmatrix}$$

We enter this matrix into our graphing utility and name it A. See Figure 7(a). Using the REF (Row Echelon Form) command on matrix A, we obtain the results shown in Figure 7(b). Since the entire matrix does not fit on the screen, we need to scroll right to see the rest of it. See Figure 7(c).

Figure 7

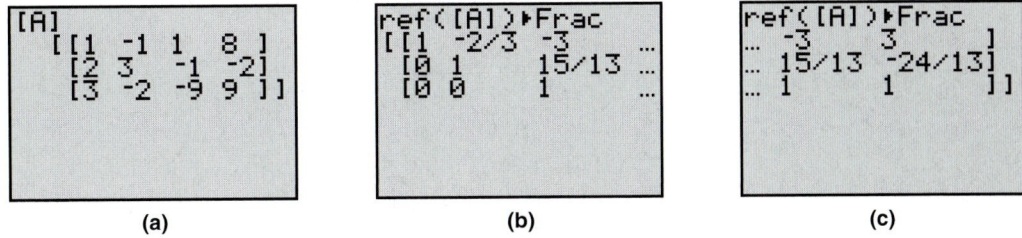

(a) (b) (c)

The system of equations represented by the matrix in row echelon form is

$$\begin{cases} x - \dfrac{2}{3}y - 3z = 3 & \text{(1)} \\[2mm] y + \dfrac{15}{13}z = -\dfrac{24}{13} & \text{(2)} \\[2mm] z = 1 & \text{(3)} \end{cases}$$

Using $z = 1$, we back-substitute to get

$$\begin{cases} x - \dfrac{2}{3}y - 3(1) = 3 & \text{(1)} \\[2mm] y + \dfrac{15}{13}(1) = -\dfrac{24}{13} & \text{(2)} \end{cases} \xrightarrow[\text{Simplify.}]{} \begin{cases} x - \dfrac{2}{3}y = 6 & \text{(1)} \\[2mm] y = -\dfrac{39}{13} = -3 & \text{(2)} \end{cases}$$

Solving the second equation for y, we find that $y = -3$. Back-substituting $y = -3$ into $x - \dfrac{2}{3}y = 6$, we find that $x = 4$. The solution of the system is $x = 4, y = -3, z = 1$. ∎

Notice that the row echelon form of the augmented matrix using the graphing utility differs from the row echelon form in our algebraic solution, yet both matrices provide the same solution! This is because the two solutions used different row operations to obtain the row echelon form. In all likelihood, the two solutions parted ways in Step 4 of the algebraic solution, where we avoided introducing fractions by interchanging rows 2 and 3.

Sometimes it is advantageous to write a matrix in **reduced row echelon form**. In this form, row operations are used to obtain entries that are 0 above (as well as below) the leading 1 in a row. For example, the row echelon form obtained in the algebraic solution to Example 6 is

$$\begin{bmatrix} 1 & -1 & 1 & | & 8 \\ 0 & 1 & -12 & | & -15 \\ 0 & 0 & 1 & | & 1 \end{bmatrix}$$

To write this matrix in reduced row echelon form, we proceed as follows:

$$\begin{bmatrix} 1 & -1 & 1 & | & 8 \\ 0 & 1 & -12 & | & -15 \\ 0 & 0 & 1 & | & 1 \end{bmatrix} \underset{\substack{\uparrow \\ R_1 = r_2 + r_1}}{\rightarrow} \begin{bmatrix} 1 & 0 & -11 & | & -7 \\ 0 & 1 & -12 & | & -15 \\ 0 & 0 & 1 & | & 1 \end{bmatrix} \underset{\substack{\uparrow \\ R_1 = 11r_3 + r_1 \\ R_2 = 12r_3 + r_2}}{\rightarrow} \begin{bmatrix} 1 & 0 & 0 & | & 4 \\ 0 & 1 & 0 & | & -3 \\ 0 & 0 & 1 & | & 1 \end{bmatrix}$$

The matrix is now written in reduced row echelon form. The advantage of writing the matrix in this form is that the solution to the system, $x = 4$, $y = -3, z = 1$, is readily found, without the need to back-substitute. Another advantage will be seen in Section 7.5, where the inverse of a matrix is discussed.

Most graphing utilities also have the ability to put a matrix in reduced row echelon form. Figure 8 shows the reduced row echelon form of the augmented matrix from Example 6 using the RREF command on a TI-83 graphing calculator.

For the remaining examples in this section, we will only provide algebraic solutions to the systems. The reader is encouraged to verify the results using a graphing utility.

Figure 8

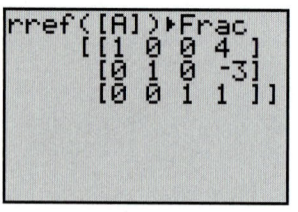

$\blacktriangleright$ **NOW WORK PROBLEMS 33 AND 43.**

The matrix method for solving a system of linear equations also identifies systems that have infinitely many solutions and systems that are inconsistent. Let's see how.

EXAMPLE 7 **Solving a System of Linear Equations Using Matrices**

Solve: $\begin{cases} 6x - y - z = 4 & \text{(1)} \\ -12x + 2y + 2z = -8 & \text{(2)} \\ 5x + y - z = 3 & \text{(3)} \end{cases}$

Solution We start with the augmented matrix of the system.

$$\begin{bmatrix} 6 & -1 & -1 & | & 4 \\ -12 & 2 & 2 & | & -8 \\ 5 & 1 & -1 & | & 3 \end{bmatrix} \underset{\substack{\uparrow \\ R_1 = -1r_3 + r_1}}{\rightarrow} \begin{bmatrix} 1 & -2 & 0 & | & 1 \\ -12 & 2 & 2 & | & -8 \\ 5 & 1 & -1 & | & 3 \end{bmatrix} \underset{\substack{\uparrow \\ R_2 = 12r_1 + r_2 \\ R_3 = -5r_1 + r_3}}{\rightarrow} \begin{bmatrix} 1 & -2 & 0 & | & 1 \\ 0 & -22 & 2 & | & 4 \\ 0 & 11 & -1 & | & -2 \end{bmatrix}$$

Obtaining a 1 in row 2, column 2 without altering column 1 can be accomplished by $R_2 = -\dfrac{1}{22}r_2$ or by $R_3 = \dfrac{1}{11}r_3$ and interchanging rows 2 and 3 or by $R_2 = \dfrac{23}{11}r_3 + r_2$. We shall use the first of these.

$$\begin{bmatrix} 1 & -2 & 0 & | & 1 \\ 0 & -22 & 2 & | & 4 \\ 0 & 11 & -1 & | & -2 \end{bmatrix} \rightarrow \begin{bmatrix} 1 & -2 & 0 & | & 1 \\ 0 & 1 & -\dfrac{1}{11} & | & -\dfrac{2}{11} \\ 0 & 11 & -1 & | & -2 \end{bmatrix} \rightarrow \begin{bmatrix} 1 & -2 & 0 & | & 1 \\ 0 & 1 & -\dfrac{1}{11} & | & -\dfrac{2}{11} \\ 0 & 0 & 0 & | & 0 \end{bmatrix}$$

$$R_2 = -\dfrac{1}{22}r_2 \qquad\qquad R_3 = -11r_2 + r_3$$

This matrix is in row echelon form. Because the bottom row consists entirely of 0's, the system actually consists of only two equations.

$$\begin{cases} x - 2y = 1 & (1) \\ y - \dfrac{1}{11}z = -\dfrac{2}{11} & (2) \end{cases}$$

We back-substitute the solution for y from the second equation, $y = \dfrac{1}{11}z - \dfrac{2}{11}$, into the first equation to get

$$x = 2y + 1 = 2\left(\dfrac{1}{11}z - \dfrac{2}{11}\right) + 1 = \dfrac{2}{11}z + \dfrac{7}{11}$$

The original system is equivalent to the system

$$\begin{cases} x = \dfrac{2}{11}z + \dfrac{7}{11} & (1) \\ y = \dfrac{1}{11}z - \dfrac{2}{11} & (2) \end{cases}$$

where z can be any real number.

Let's look at the situation. The original system of three equations is equivalent to a system containing two equations. This means that any values of x, y, z that satisfy both

$$x = \dfrac{2}{11}z + \dfrac{7}{11} \quad \text{and} \quad y = \dfrac{1}{11}z - \dfrac{2}{11}$$

will be solutions. For example, $z = 0, x = \dfrac{7}{11}, y = -\dfrac{2}{11}$; $z = 1, x = \dfrac{9}{11}$, $y = -\dfrac{1}{11}$; and $z = -1, x = \dfrac{5}{11}, y = -\dfrac{3}{11}$ are some of the solutions of the original system. There are, in fact, infinitely many values of $x, y,$ and z

for which the two equations are satisfied. That is, the original system has infinitely many solutions. We will write the solution of the original system as

$$\begin{cases} x = \dfrac{2}{11}z + \dfrac{7}{11} \\[2mm] y = \dfrac{1}{11}z - \dfrac{2}{11} \end{cases}$$

where z can be any real number.   ◼

We can also find the solution by writing the augmented matrix in reduced row echelon form. Starting with the row echelon form, we have

$$\begin{bmatrix} 1 & -2 & 0 & \bigm| & 1 \\ 0 & 1 & -\dfrac{1}{11} & \bigm| & -\dfrac{2}{11} \\ 0 & 0 & 0 & \bigm| & 0 \end{bmatrix} \xrightarrow[\;R_1 = 2r_2 + r_1\;]{} \begin{bmatrix} 1 & 0 & -\dfrac{2}{11} & \bigm| & \dfrac{7}{11} \\ 0 & 1 & -\dfrac{1}{11} & \bigm| & -\dfrac{2}{11} \\ 0 & 0 & 0 & \bigm| & 0 \end{bmatrix}$$

The matrix on the right is in reduced row echelon form. The corresponding system of equations is

$$\begin{cases} x - \dfrac{2}{11}z = \dfrac{7}{11} & (1) \\[2mm] y - \dfrac{1}{11}z = -\dfrac{2}{11} & (2) \end{cases}$$

or, equivalently,

$$\begin{cases} x = \dfrac{2}{11}z + \dfrac{7}{11} & (1) \\[2mm] y = \dfrac{1}{11}z - \dfrac{2}{11} & (2) \end{cases}$$

where z can be any real number.

—— **NOW WORK PROBLEM 49.**

EXAMPLE 8   Solving a System of Linear Equations Using Matrices

Solve: $\begin{cases} x + y + z = 6 \\ 2x - y - z = 3 \\ x + 2y + 2z = 0 \end{cases}$

Solution   We proceed as follows, beginning with the augmented matrix.

$$\begin{bmatrix} 1 & 1 & 1 & \bigm| & 6 \\ 2 & -1 & -1 & \bigm| & 3 \\ 1 & 2 & 2 & \bigm| & 0 \end{bmatrix} \xrightarrow[\substack{R_2 = -2r_1 + r_2 \\ R_3 = -1r_1 + r_3}]{} \begin{bmatrix} 1 & 1 & 1 & \bigm| & 6 \\ 0 & -3 & -3 & \bigm| & -9 \\ 0 & 1 & 1 & \bigm| & -6 \end{bmatrix} \xrightarrow[\text{Interchange rows 2 and 3.}]{} \begin{bmatrix} 1 & 1 & 1 & \bigm| & 6 \\ 0 & 1 & 1 & \bigm| & -6 \\ 0 & -3 & -3 & \bigm| & -9 \end{bmatrix} \xrightarrow[\;R_3 = 3r_2 + r_3\;]{} \begin{bmatrix} 1 & 1 & 1 & \bigm| & 6 \\ 0 & 1 & 1 & \bigm| & -6 \\ 0 & 0 & 0 & \bigm| & -27 \end{bmatrix}$$

This matrix is in row echelon form. The bottom row is equivalent to the equation

$$0x + 0y + 0z = -27$$

which has no solution. Hence, the original system is inconsistent. ■

━━ **NOW WORK PROBLEM 23.**

The matrix method is especially effective for systems of equations for which the number of equations and the number of variables are unequal. Here, too, such a system is either inconsistent or consistent. If it is consistent, it will have either exactly one solution or infinitely many solutions.

Let's look at a system of four equations containing three variables.

EXAMPLE 9 **Solving a System of Linear Equations Using Matrices**

Solve: $\begin{cases} x - 2y + z = 0 & (1) \\ 2x + 2y - 3z = -3 & (2) \\ y - z = -1 & (3) \\ -x + 4y + 2z = 13 & (4) \end{cases}$

Solution We proceed as follows, beginning with the augmented matrix.

$$\begin{bmatrix} 1 & -2 & 1 & | & 0 \\ 2 & 2 & -3 & | & -3 \\ 0 & 1 & -1 & | & -1 \\ -1 & 4 & 2 & | & 13 \end{bmatrix} \rightarrow \begin{bmatrix} 1 & -2 & 1 & | & 0 \\ 0 & 6 & -5 & | & -3 \\ 0 & 1 & -1 & | & -1 \\ 0 & 2 & 3 & | & 13 \end{bmatrix} \rightarrow \begin{bmatrix} 1 & -2 & 1 & | & 0 \\ 0 & 1 & -1 & | & -1 \\ 0 & 6 & -5 & | & -3 \\ 0 & 2 & 3 & | & 13 \end{bmatrix}$$

$R_2 = -2r_1 + r_2$ Interchange rows 2 and 3.
$R_4 = r_1 + r_4$

$$\rightarrow \begin{bmatrix} 1 & -2 & 1 & | & 0 \\ 0 & 1 & -1 & | & -1 \\ 0 & 0 & 1 & | & 3 \\ 0 & 0 & 5 & | & 15 \end{bmatrix} \rightarrow \begin{bmatrix} 1 & -2 & 1 & | & 0 \\ 0 & 1 & -1 & | & -1 \\ 0 & 0 & 1 & | & 3 \\ 0 & 0 & 0 & | & 0 \end{bmatrix}$$

$R_3 = -6r_2 + r_3$ $R_4 = -5r_3 + r_4$
$R_4 = -2r_2 + r_4$

We could stop here, since the matrix is in row echelon form, and back-substitute $z = 3$ to find x and y. Or we can continue to obtain the reduced row echelon form.

$$\rightarrow \begin{bmatrix} 1 & 0 & -1 & | & -2 \\ 0 & 1 & -1 & | & -1 \\ 0 & 0 & 1 & | & 3 \\ 0 & 0 & 0 & | & 0 \end{bmatrix} \rightarrow \begin{bmatrix} 1 & 0 & 0 & | & 1 \\ 0 & 1 & 0 & | & 2 \\ 0 & 0 & 1 & | & 3 \\ 0 & 0 & 0 & | & 0 \end{bmatrix}$$

$R_1 = 2r_2 + r_1$ $R_1 = r_3 + r_1$
                           $R_2 = r_3 + r_2$

The matrix is now in reduced row echelon form, and we can see that the solution is $x = 1, y = 2, z = 3$. ■

━━ **NOW WORK PROBLEM 65.**

EXAMPLE 10 **Nutrition**

A dietitian at Cook County Hospital wants a patient to have a meal that has 65 grams of protein, 95 grams of carbohydrates, and 905 milligrams of calcium. The hospital food service tells the dietitian that the dinner for today is chicken a la king, baked potatoes, and 2% milk. Each serving of chicken a la king has 30 grams of protein, 35 grams of carbohydrates, and 200 milligrams of calcium. Each serving of baked potatoes contains 4 grams of protein, 33 grams of carbohydrates, and 10 milligrams of calcium. Each glass of 2% milk contains 9 grams of protein, 13 grams of carbohydrates, and 300 milligrams of calcium. How many servings of each food should the dietitian provide for the patient?

Solution Let c, p, and m represent the number of servings of chicken a la king, baked potatoes, and milk, respectively. The dietitian wants the patient to have 65 grams of protein. Each serving of chicken a la king has 30 grams of protein, so c servings will have $30c$ grams of protein. Each serving of baked potatoes contains 4 grams of protein, so p potatoes will have $4p$ grams of protein. Finally, each glass of milk has 9 grams of protein, so m glasses of milk will have $9m$ grams of protein. The same logic will result in equations for carbohydrates and calcium, and we have the following system of equations:

$$\begin{cases} 30c + 4p + 9m = 65 & \text{protein equation} \\ 35c + 33p + 13m = 95 & \text{carbohydrate equation} \\ 200c + 10p + 300m = 905 & \text{calcium equation} \end{cases}$$

We begin with the augmented matrix and proceed as follows:

$$\begin{bmatrix} 30 & 4 & 9 & | & 65 \\ 35 & 33 & 13 & | & 95 \\ 200 & 10 & 300 & | & 905 \end{bmatrix} \rightarrow \begin{bmatrix} 1 & \frac{2}{15} & \frac{3}{10} & | & \frac{13}{6} \\ 35 & 33 & 13 & | & 95 \\ 200 & 10 & 300 & | & 905 \end{bmatrix} \rightarrow \begin{bmatrix} 1 & \frac{2}{15} & \frac{3}{10} & | & \frac{13}{6} \\ 0 & \frac{85}{3} & \frac{5}{2} & | & \frac{115}{6} \\ 0 & -\frac{50}{3} & 240 & | & \frac{1415}{3} \end{bmatrix}$$

$$R_1 = \left(\frac{1}{30}\right)r_1 \qquad R_2 = -35r_1 + r_2$$
$$R_3 = -200r_1 + r_3$$

$$\rightarrow \begin{bmatrix} 1 & \frac{2}{15} & \frac{3}{10} & | & \frac{13}{6} \\ 0 & 1 & \frac{3}{34} & | & \frac{23}{34} \\ 0 & -\frac{50}{3} & 240 & | & \frac{1415}{3} \end{bmatrix} \rightarrow \begin{bmatrix} 1 & \frac{2}{15} & \frac{3}{10} & | & \frac{13}{6} \\ 0 & 1 & \frac{3}{34} & | & \frac{23}{34} \\ 0 & 0 & \frac{4105}{17} & | & \frac{8210}{17} \end{bmatrix} \rightarrow \begin{bmatrix} 1 & \frac{2}{15} & \frac{3}{10} & | & \frac{13}{6} \\ 0 & 1 & \frac{3}{34} & | & \frac{23}{34} \\ 0 & 0 & 1 & | & 2 \end{bmatrix}$$

$$R_2 = \left(\frac{3}{85}\right)r_2 \qquad R_3 = \left(\frac{50}{3}\right)r_2 + r_3 \qquad R_3 = \left(\frac{17}{4105}\right)r_3$$

The matrix is now in row echelon form. The final matrix represents the system

$$\begin{cases} c + \dfrac{2}{15}p + \dfrac{3}{10}m = \dfrac{13}{6} & \text{(1)} \\[2mm] p + \dfrac{3}{34}m = \dfrac{23}{34} & \text{(2)} \\[2mm] m = 2 & \text{(3)} \end{cases}$$

From (3), we determine that 2 glasses of milk should be served. Back-substitute $m = 2$ into equation (2) to find that $p = \dfrac{1}{2}$, so $\dfrac{1}{2}$ of a baked potato should be served. Back-substitute these values into equation (1) and find that $c = 1.5$, so 1.5 servings of chicken a la king should be given to the patient to meet the dietary requirements. ∎

7.3 Concepts and Vocabulary

In Problems 1 and 2, fill in the blanks.

1. An m by n rectangular array of numbers is called a(n) _____ .

2. The matrix used to represent a system of linear equations is called a(n) _____ matrix.

In Problems 3 and 4, answer True or False for each statement.

3. The augmented matrix of a system of two equations containing three variables has two rows and four columns.

4. The matrix $\begin{bmatrix} 1 & 3 & -2 \\ 0 & 1 & 5 \\ 0 & 0 & 4 \end{bmatrix}$ is in row echelon form.

5. Where would the entry a_{46} be located in a matrix?

6. What does it mean for a matrix to be in reduced row echelon form?

7.3 Exercises

In Problems 1–12, write the augmented matrix of the given system of equations.

1. $\begin{cases} x - 5y = 5 \\ 4x + 3y = 6 \end{cases}$

2. $\begin{cases} 3x + 4y = 7 \\ 4x - 2y = 5 \end{cases}$

3. $\begin{cases} 2x + 3y - 6 = 0 \\ 4x - 6y + 2 = 0 \end{cases}$

4. $\begin{cases} 9x - y = 0 \\ 3x - y - 4 = 0 \end{cases}$

5. $\begin{cases} 0.01x - 0.03y = 0.06 \\ 0.13x + 0.10y = 0.20 \end{cases}$

6. $\begin{cases} \dfrac{4}{3}x - \dfrac{3}{2}y = \dfrac{3}{4} \\[2mm] -\dfrac{1}{4}x + \dfrac{1}{3}y = \dfrac{2}{3} \end{cases}$

7. $\begin{cases} x - y + z = 10 \\ 3x + 3y = 5 \\ x + y + 2z = 2 \end{cases}$

8. $\begin{cases} 5x - y - z = 0 \\ x + y = 5 \\ 2x - 3z = 2 \end{cases}$

9. $\begin{cases} x + y - z = 2 \\ 3x - 2y = 2 \\ 5x + 3y - z = 1 \end{cases}$

10. $\begin{cases} 2x + 3y - 4z = 0 \\ x - 5z + 2 = 0 \\ x + 2y - 3z = -2 \end{cases}$

11. $\begin{cases} x - y - z = 10 \\ 2x + y + 2z = -1 \\ -3x + 4y = 5 \\ 4x - 5y + z = 0 \end{cases}$

12. $\begin{cases} x - y + 2z - w = 5 \\ x + 3y - 4z + 2w = 2 \\ 3x - y - 5z - w = -1 \end{cases}$

In Problems 13–20, perform each row operation on the given augmented matrix.

13. $\begin{bmatrix} 1 & -3 & | & -2 \\ 2 & -5 & | & 5 \end{bmatrix}$ (a) $R_2 = -2r_1 + r_2$

14. $\begin{bmatrix} 1 & -3 & | & -3 \\ 2 & -5 & | & -4 \end{bmatrix}$ (a) $R_2 = -2r_1 + r_2$

15. $\begin{bmatrix} 1 & -3 & 4 & | & 3 \\ 2 & -5 & 6 & | & 6 \\ -3 & 3 & 4 & | & 6 \end{bmatrix}$ (a) $R_2 = -2r_1 + r_2$ (b) $R_3 = 3r_1 + r_3$

16. $\begin{bmatrix} 1 & -3 & 3 & | & -5 \\ 2 & -5 & -3 & | & -5 \\ -3 & -2 & 4 & | & 6 \end{bmatrix}$ (a) $R_2 = -2r_1 + r_2$ (b) $R_3 = 3r_1 + r_3$

17. $\begin{bmatrix} 1 & -3 & 2 & | & -6 \\ 2 & -5 & 3 & | & -4 \\ -3 & -6 & 4 & | & 6 \end{bmatrix}$ (a) $R_2 = -2r_1 + r_2$ (b) $R_3 = 3r_1 + r_3$

18. $\begin{bmatrix} 1 & -3 & -4 & | & -6 \\ 2 & -5 & 6 & | & -6 \\ -3 & 1 & 4 & | & 6 \end{bmatrix}$ (a) $R_2 = -2r_1 + r_2$ (b) $R_3 = 3r_1 + r_3$

19. $\begin{bmatrix} 1 & -3 & 1 & | & -2 \\ 2 & -5 & 6 & | & -2 \\ -3 & 1 & 4 & | & 6 \end{bmatrix}$ (a) $R_2 = -2r_1 + r_2$ (b) $R_3 = 3r_1 + r_3$

20. $\begin{bmatrix} 1 & -3 & -1 & | & 2 \\ 2 & -5 & 2 & | & 6 \\ -3 & -6 & 4 & | & 6 \end{bmatrix}$ (a) $R_2 = -2r_1 + r_2$ (b) $R_3 = 3r_1 + r_3$

In Problems 21–32, the reduced row echelon form of a system of linear equations is given. Write the system of equations corresponding to the given matrix. Use x, y; or x, y, z; or x_1, x_2, x_3, x_4 as variables. Determine whether the system is consistent or inconsistent. If it is consistent, give the solution.

21. $\begin{bmatrix} 1 & 0 & | & 5 \\ 0 & 1 & | & -1 \end{bmatrix}$

22. $\begin{bmatrix} 1 & 0 & | & -4 \\ 0 & 1 & | & 0 \end{bmatrix}$

23. $\begin{bmatrix} 1 & 0 & 0 & | & 1 \\ 0 & 1 & 0 & | & 2 \\ 0 & 0 & 0 & | & 3 \end{bmatrix}$

24. $\begin{bmatrix} 1 & 0 & 0 & | & 0 \\ 0 & 1 & 0 & | & 0 \\ 0 & 0 & 0 & | & 2 \end{bmatrix}$

25. $\begin{bmatrix} 1 & 0 & 2 & | & -1 \\ 0 & 1 & -4 & | & -2 \\ 0 & 0 & 0 & | & 0 \end{bmatrix}$

26. $\begin{bmatrix} 1 & 0 & 4 & | & 4 \\ 0 & 1 & 3 & | & 2 \\ 0 & 0 & 0 & | & 0 \end{bmatrix}$

27. $\begin{bmatrix} 1 & 0 & 0 & 0 & | & 1 \\ 0 & 1 & 0 & 1 & | & 2 \\ 0 & 0 & 1 & 2 & | & 3 \end{bmatrix}$

28. $\begin{bmatrix} 1 & 0 & 0 & 0 & | & 1 \\ 0 & 1 & 0 & 2 & | & 2 \\ 0 & 0 & 1 & 3 & | & 0 \end{bmatrix}$

29. $\begin{bmatrix} 1 & 0 & 0 & 4 & | & 2 \\ 0 & 1 & 1 & 3 & | & 3 \\ 0 & 0 & 0 & 0 & | & 0 \end{bmatrix}$

30. $\begin{bmatrix} 1 & 0 & 0 & 0 & | & 1 \\ 0 & 1 & 0 & 0 & | & 2 \\ 0 & 0 & 1 & 2 & | & 3 \end{bmatrix}$

31. $\begin{bmatrix} 1 & 0 & 0 & 1 & | & -2 \\ 0 & 1 & 0 & 2 & | & 2 \\ 0 & 0 & 1 & -1 & | & 0 \\ 0 & 0 & 0 & 0 & | & 0 \end{bmatrix}$

32. $\begin{bmatrix} 1 & 0 & 0 & 0 & | & 1 \\ 0 & 1 & 0 & 0 & | & 2 \\ 0 & 0 & 1 & 0 & | & 3 \\ 0 & 0 & 0 & 1 & | & 0 \end{bmatrix}$

In Problems 33–68, solve each system of equations using matrices (row operations). If the system has no solution, say that it is inconsistent.

33. $\begin{cases} x + y = 8 \\ x - y = 4 \end{cases}$

34. $\begin{cases} x + 2y = 5 \\ x + y = 3 \end{cases}$

35. $\begin{cases} 2x - 4y = -2 \\ 3x + 2y = 3 \end{cases}$

36. $\begin{cases} 3x + 3y = 3 \\ 4x + 2y = \dfrac{8}{3} \end{cases}$

37. $\begin{cases} x + 2y = 4 \\ 2x + 4y = 8 \end{cases}$

38. $\begin{cases} 3x - y = 7 \\ 9x - 3y = 21 \end{cases}$

39. $\begin{cases} 2x + 3y = 6 \\ x - y = \dfrac{1}{2} \end{cases}$

40. $\begin{cases} \dfrac{1}{2}x + y = -2 \\ x - 2y = 8 \end{cases}$

41. $\begin{cases} 3x - 5y = 3 \\ 15x + 5y = 21 \end{cases}$

42. $\begin{cases} 2x - y = -1 \\ x + \dfrac{1}{2}y = \dfrac{3}{2} \end{cases}$

43. $\begin{cases} x - y = 6 \\ 2x - 3z = 16 \\ 2y + z = 4 \end{cases}$

44. $\begin{cases} 2x + y = -4 \\ -2y + 4z = 0 \\ 3x - 2z = -11 \end{cases}$

45. $\begin{cases} x - 2y + 3z = 7 \\ 2x + y + z = 4 \\ -3x + 2y - 2z = -10 \end{cases}$

46. $\begin{cases} 2x + y - 3z = 0 \\ -2x + 2y + z = -7 \\ 3x - 4y - 3z = 7 \end{cases}$

47. $\begin{cases} 2x - 2y - 2z = 2 \\ 2x + 3y + z = 2 \\ 3x + 2y = 0 \end{cases}$

48. $\begin{cases} 2x - 3y - z = 0 \\ -x + 2y + z = 5 \\ 3x - 4y - z = 1 \end{cases}$

49. $\begin{cases} -x + y + z = -1 \\ -x + 2y - 3z = -4 \\ 3x - 2y - 7z = 0 \end{cases}$

50. $\begin{cases} 2x - 3y - z = 0 \\ 3x + 2y + 2z = 2 \\ x + 5y + 3z = 2 \end{cases}$

51. $\begin{cases} 2x - 2y + 3z = 6 \\ 4x - 3y + 2z = 0 \\ -2x + 3y - 7z = 1 \end{cases}$

52. $\begin{cases} 3x - 2y + 2z = 6 \\ 7x - 3y + 2z = -1 \\ 2x - 3y + 4z = 0 \end{cases}$

53. $\begin{cases} x + y - z = 6 \\ 3x - 2y + z = -5 \\ x + 3y - 2z = 14 \end{cases}$

54. $\begin{cases} x - y + z = -4 \\ 2x - 3y + 4z = -15 \\ 5x + y - 2z = 12 \end{cases}$

55. $\begin{cases} x + 2y - z = -3 \\ 2x - 4y + z = -7 \\ -2x + 2y - 3z = 4 \end{cases}$

56. $\begin{cases} x + 4y - 3z = -8 \\ 3x - y + 3z = 12 \\ x + y + 6z = 1 \end{cases}$

57. $\begin{cases} 3x + y - z = \dfrac{2}{3} \\ 2x - y + z = 1 \\ 4x + 2y = \dfrac{8}{3} \end{cases}$

58. $\begin{cases} x + y = 1 \\ 2x - y + z = 1 \\ x + 2y + z = \dfrac{8}{3} \end{cases}$

59. $\begin{cases} x + y + z + w = 4 \\ 2x - y + z = 0 \\ 3x + 2y + z - w = 6 \\ x - 2y - 2z + 2w = -1 \end{cases}$

60. $\begin{cases} x + y + z + w = 4 \\ -x + 2y + z = 0 \\ 2x + 3y + z - w = 6 \\ -2x + y - 2z + 2w = -1 \end{cases}$

61. $\begin{cases} x + 2y + z = 1 \\ 2x - y + 2z = 2 \\ 3x + y + 3z = 3 \end{cases}$

62. $\begin{cases} x + 2y - z = 3 \\ 2x - y + 2z = 6 \\ x - 3y + 3z = 4 \end{cases}$

63. $\begin{cases} x - y + z = 5 \\ 3x + 2y - 2z = 0 \end{cases}$

64. $\begin{cases} 2x + y - z = 4 \\ -x + y + 3z = 1 \end{cases}$

65. $\begin{cases} 2x + 3y - z = 3 \\ x - y - z = 0 \\ -x + y + z = 0 \\ x + y + 3z = 5 \end{cases}$

66. $\begin{cases} x - 3y + z = 1 \\ 2x - y - 4z = 0 \\ x - 3y + 2z = 1 \\ x - 2y = 5 \end{cases}$

67. $\begin{cases} 4x + y + z - w = 4 \\ x - y + 2z + 3w = 3 \end{cases}$

68. $\begin{cases} -4x + y = 5 \\ 2x - y + z - w = 5 \\ z + w = 4 \end{cases}$

69. Curve Fitting Find the function $y = ax^2 + bx + c$ whose graph contains the points $(1, 2), (-2, -7)$, and $(2, -3)$.

70. Curve Fitting Find the function $y = ax^2 + bx + c$ whose graph contains the points $(1, -1), (3, -1)$, and $(-2, 14)$.

71. Curve Fitting Find the function $f(x) = ax^3 + bx^2 + cx + d$ for which $f(-3) = -112, f(-1) = -2, f(1) = 4$, and $f(2) = 13$.

72. Curve Fitting Find the function $f(x) = ax^3 + bx^2 + cx + d$ for which $f(-2) = -10, f(-1) = 3, f(1) = 5$, and $f(3) = 15$.

73. Nutrition A dietitian at Palos Community Hospital wants a patient to have a meal that has 78 grams of protein, 59 grams of carbohydrates, and 75 milligrams of vitamin A. The hospital food service tells the dietitian that the dinner for today is salmon steak, baked eggs, and acorn squash. Each serving of salmon steak has 30 grams of protein, 20 grams of carbohydrates, and 2 milligrams of vitamin A. Each serving of baked eggs contains 15 grams of protein, 2 grams of carbohydrates, and 20 milligrams of vitamin A. Each serving of acorn squash contains 3 grams of protein, 25 grams of carbohydrates, and 32 milligrams of vitamin A. How many servings of each food should the dietitian provide for the patient?

74. Nutrition A dietitian at General Hospital wants a patient to have a meal that has 47 grams of protein, 58 grams of carbohydrates, and 630 milligrams of calcium. The hospital food service tells the dietitian that the dinner for today is pork chops, corn on the cob, and 2% milk. Each serving of pork chops has 23 grams of protein, 0 grams of carbohydrates, and 10 milligrams of calcium. Each serving of corn on the cob contains 3 grams of protein, 16 grams of carbohydrates, and 10 milligrams of calcium. Each glass of 2% milk contains 9 grams of protein, 13 grams of carbohydrates, and 300 milligrams of calcium. How many servings of each food should the dietitian provide for the patient?

75. Financial Planning Carletta has $10,000 to invest. As her financial consultant, you recommend that she invest in Treasury bills that yield 6%, Treasury bonds that yield 7%, and corporate bonds that yield 8%. Carletta wants to have an annual income of $680, and the amount invested in corporate bonds must be half that invested in Treasury bills. Find the amount in each investment.

76. Financial Planning John has $20,000 to invest. As his financial consultant, you recommend that he invest in Treasury bills that yield 5%, Treasury bonds that yield 7%, and corporate bonds that yield 9%. John wants to have an annual income of $1280, and the amount invested in Treasury bills must be two times the amount invested in corporate bonds. Find the amount in each investment.

77. Production To manufacture an automobile requires painting, drying, and polishing. Epsilon Motor Company produces three types of cars: the Delta, the Beta, and the Sigma. Each Delta requires 10 hours for painting, 3 hours for drying, and 2 hours for polishing. A Beta requires 16 hours of painting, 5 hours of drying, and 3 hours of polishing, while a Sigma requires 8 hours for painting, 2 hours for drying, and 1 hour for polishing. If the company has 240 hours for painting, 69 hours for drying, and 41 hours for polishing per month, how many of each type of car are produced?

78. Production A Florida juice company completes the preparation of its products by sterilizing, filling, and labeling bottles. Each case of orange juice requires 9 minutes for sterilizing, 6 minutes for filling, and 1 minute for labeling. Each case of grapefruit juice requires 10 minutes for sterilizing, 4 minutes for filling, and 2 minutes for labeling. Each case of tomato juice requires 12 minutes for sterilizing, 4 minutes for filling, and 1 minute for labeling. If the company runs the sterilizing machine for 398 minutes, the filling machine for 164 minutes, and the labeling machine for 58 minutes, how many cases of each type of juice are prepared?

79. Electricity: Kirchhoff's Rules An application of Kirchhoff's Rules to the circuit shown results in the following system of equations:

$$\begin{cases} -4 + 8 - 2I_2 = 0 \\ 8 = 5I_4 + I_1 \\ 4 = 3I_3 + I_1 \\ I_3 + I_4 = I_1 \end{cases}$$

Find the currents I_1, I_2, I_3, and I_4.*

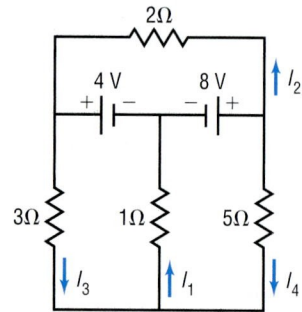

80. Electricity: Kirchhoff's Rules An application of Kirchhoff's Rules to the circuit shown results in the following system of equations:

$$\begin{cases} I_1 = I_3 + I_2 \\ 24 - 6I_1 - 3I_3 = 0 \\ 12 + 24 - 6I_1 - 6I_2 = 0 \end{cases}$$

Find the currents I_1, I_2, and I_3.†

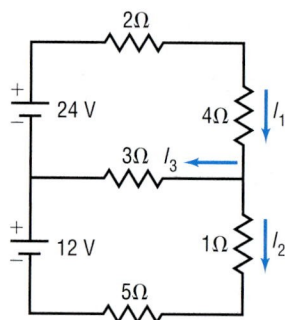

81. Financial Planning Three retired couples each require an additional annual income of $2000 per year. As their financial consultant, you recommend that they invest some money in Treasury bills that yield 7%, some money in corporate bonds that yield 9%, and some money in junk bonds that yield 11%. Prepare a table for each couple showing the various ways that their goals can be achieved:
(a) If the first couple has $20,000 to invest.
(b) If the second couple has $25,000 to invest.
(c) If the third couple has $30,000 to invest.
(d) What advice would you give each couple regarding the amount to invest and the choices available?

[**Hint:** Higher yields generally carry more risk.]

82. Financial Planning A young couple has $25,000 to invest. As their financial consultant, you recommend that they invest some money in Treasury bills that yield 7%, some money in corporate bonds that yield 9%, and some money in junk bonds that yield 11%. Prepare a table showing the various ways this couple can achieve the following goals:
(a) The couple wants $1500 per year in income.
(b) The couple wants $2000 per year in income.
(c) The couple wants $2500 per year in income.
(d) What advice would you give this couple regarding the income that they require and the choices available?

[**Hint:** Higher yields generally carry more risk.]

83. Pharmacy A doctor's prescription calls for a daily intake of liquid containing 40 mg of vitamin C and 30 mg of vitamin D. Your pharmacy stocks three liquids that can be used: one contains 20% vitamin C and 30% vitamin D; a second, 40% vitamin C and 20% vitamin D; and a third, 30% vitamin C and 50% vitamin D. Create a table showing the possible combinations that could be used to fill the prescription.

84. Pharmacy A doctor's prescription calls for the creation of pills that contain 12 units of vitamin B_{12} and 12 units of vitamin E. Your pharmacy stocks three powders that can be used to make these pills: one contains 20% vitamin B_{12} and 30% vitamin E; a second, 40% vitamin B_{12} and 20% vitamin E; and a third, 30% vitamin B_{12} and 40% vitamin E. Create a table showing the possible combinations of each powder that could be mixed in each pill.

---

*Source: Based on Raymond Serway, *Physics*, 3rd ed. (Philadelphia: Saunders, 1990), Prob. 34, p. 79.
†Source: Ibid., Prob. 38, p. 79.

85. Write a brief paragraph or two that outlines your strategy for solving a system of linear equations using matrices.

86. When solving a system of linear equations using matrices, do you prefer to place the augmented matrix in row echelon form or in reduced row echelon form? Give reasons for your choice.

87. Make up a system of three linear equations containing three variables that has:
(a) No solution
(b) Exactly one solution
(c) Infinitely many solutions
Give the three systems to a friend to solve and critique.

7.4 SYSTEMS OF LINEAR EQUATIONS: DETERMINANTS

OBJECTIVES
1. Evaluate 2 by 2 Determinants
2. Use Cramer's Rule to Solve a System of Two Equations, Two Variables
3. Evaluate 3 by 3 Determinants
4. Use Cramer's Rule to Solve a System of Three Equations, Three Variables
5. Know Properties of Determinants

1. In the preceding section, we described a method for using matrices to solve a system of linear equations. This section deals with yet another method for solving systems of linear equations; however, it can be used only when the number of equations equals the number of variables. Although the method will work for any system (provided that the number of equations equals the number of variables), it is most often used for systems of two equations containing two variables or three equations containing three variables. This method, called *Cramer's Rule*, is based on the concept of a *determinant*.

2 by 2 Determinants

If $a, b, c,$ and d are four real numbers, the symbol

$$D = \begin{vmatrix} a & b \\ c & d \end{vmatrix}$$

is called a **2 by 2 determinant**. Its value is the number $ad - bc$; that is,

$$D = \begin{vmatrix} a & b \\ c & d \end{vmatrix} = ad - bc \qquad (1)$$

A device that may be helpful for remembering the value of a 2 by 2 determinant is the following:

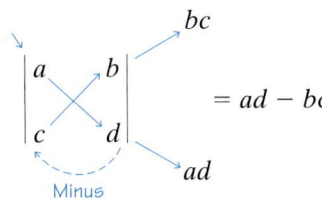

| EXAMPLE 1 | Evaluating a 2 × 2 Determinant |

$$\begin{vmatrix} 3 & -2 \\ 6 & 1 \end{vmatrix} = (3)(1) - (6)(-2) = 3 - (-12) = 15$$

Graphing utilities can also be used to evaluate determinants.

| EXAMPLE 2 | Using a Graphing Utility to Evaluate a 2 × 2 Determinant |

Use a graphing utility to evaluate the determinant from Example 1: $\begin{vmatrix} 3 & -2 \\ 6 & 1 \end{vmatrix}$.

Solution First, we enter the matrix whose entries are those of the determinant into the graphing utility and name it A. Using the determinant command, we obtain the result shown in Figure 9.

Figure 9

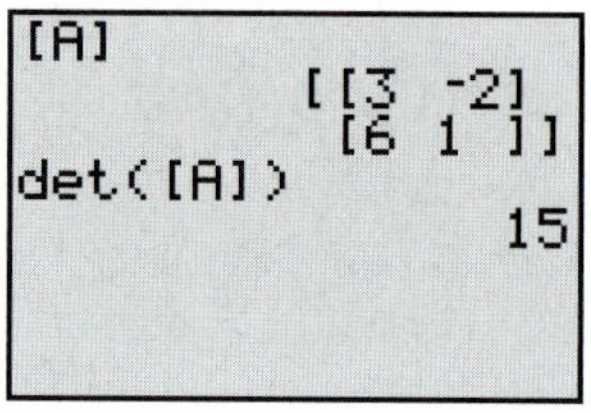

NOW WORK PROBLEM 3.

Cramer's Rule

2 Let's now see the role that a 2 by 2 determinant plays in the solution of a system of two equations containing two variables. Consider the system

$$\begin{cases} ax + by = s & (1) \\ cx + dy = t & (2) \end{cases} \qquad (2)$$

We shall use the method of elimination to solve this system.

Provided $d \neq 0$ and $b \neq 0$, this system is equivalent to the system

$$\begin{cases} adx + bdy = sd & (1) \text{ Multiply by } d. \\ bcx + bdy = tb & (2) \text{ Multiply by } b. \end{cases}$$

On subtracting the second equation from the first equation, we get

$$\begin{cases} (ad - bc)x + 0 \cdot y = sd - tb & (1) \\ bcx \quad + bdy = tb & (2) \end{cases}$$

Now, the first equation can be rewritten using determinant notation.

$$\begin{vmatrix} a & b \\ c & d \end{vmatrix} x = \begin{vmatrix} s & b \\ t & d \end{vmatrix}$$

If $D = \begin{vmatrix} a & b \\ c & d \end{vmatrix} = ad - bc \neq 0$, we can solve for x to get

$$x = \frac{\begin{vmatrix} s & b \\ t & d \end{vmatrix}}{\begin{vmatrix} a & b \\ c & d \end{vmatrix}} = \frac{\begin{vmatrix} s & b \\ t & d \end{vmatrix}}{D} \qquad (3)$$

Return now to the original system (2). Provided that $a \neq 0$ and $c \neq 0$, the system is equivalent to

$$\begin{cases} acx + bcy = cs & \text{(1) Multiply by } c. \\ acx + ady = at & \text{(2) Multiply by } a. \end{cases}$$

On subtracting the first equation from the second equation, we get

$$\begin{cases} acx + \quad bcy \quad = \quad cs & \text{(1)} \\ 0 \cdot x + (ad - bc)y = at - cs & \text{(2)} \end{cases}$$

The second equation can now be rewritten using determinant notation.

$$\begin{vmatrix} a & b \\ c & d \end{vmatrix} y = \begin{vmatrix} a & s \\ c & t \end{vmatrix}$$

If $D = \begin{vmatrix} a & b \\ c & d \end{vmatrix} = ad - bc \neq 0$, we can solve for y to get

$$y = \frac{\begin{vmatrix} a & s \\ c & t \end{vmatrix}}{\begin{vmatrix} a & b \\ c & d \end{vmatrix}} = \frac{\begin{vmatrix} a & s \\ c & t \end{vmatrix}}{D} \qquad (4)$$

Equations (3) and (4) lead us to the following result, called **Cramer's Rule**.

Theorem

Cramer's Rule for Two Equations Containing Two Variables

The solution to the system of equations

$$\begin{cases} ax + by = s & \text{(1)} \\ cx + dy = t & \text{(2)} \end{cases} \qquad (5)$$

is given by

$$x = \frac{\begin{vmatrix} s & b \\ t & d \end{vmatrix}}{\begin{vmatrix} a & b \\ c & d \end{vmatrix}}, \qquad y = \frac{\begin{vmatrix} a & s \\ c & t \end{vmatrix}}{\begin{vmatrix} a & b \\ c & d \end{vmatrix}} \qquad (6)$$

provided that

$$D = \begin{vmatrix} a & b \\ c & d \end{vmatrix} = ad - bc \neq 0$$

In the derivation given for Cramer's Rule above, we assumed that none of the numbers a, b, c, and d was 0. In Problem 58 you will be asked to complete the proof under the less stringent condition that $D = ad - bc \neq 0$.

Now look carefully at the pattern in Cramer's Rule. The denominator in the solution (6) is the determinant of the coefficients of the variables.

$$\begin{cases} ax + by = s \\ cx + dy = t \end{cases} \qquad D = \begin{vmatrix} a & b \\ c & d \end{vmatrix}$$

In the solution for x, the numerator is the determinant, denoted by D_x, formed by replacing the entries in the first column (the coefficients of x) in D by the constants on the right side of the equal sign.

$$D_x = \begin{vmatrix} s & b \\ t & d \end{vmatrix}$$

In the solution for y, the numerator is the determinant, denoted by D_y, formed by replacing the entries in the second column (the coefficients of y) in D by the constants on the right side of the equal sign.

$$D_y = \begin{vmatrix} a & s \\ c & t \end{vmatrix}$$

Cramer's Rule then states that, if $D \neq 0$,

$$x = \frac{D_x}{D}, \qquad y = \frac{D_y}{D} \qquad\qquad (7)$$

EXAMPLE 3   Solving a System of Linear Equations Using Determinants

Use Cramer's Rule, if applicable, to solve the system

$$\begin{cases} 3x - 2y = 4 & (1) \\ 6x + y = 13 & (2) \end{cases}$$

Algebraic Solution   The determinant D of the coefficients of the variables is

$$D = \begin{vmatrix} 3 & -2 \\ 6 & 1 \end{vmatrix} = (3)(1) - (6)(-2) = 15$$

Because $D \neq 0$, Cramer's Rule (7) can be used.

$$x = \frac{D_x}{D} = \frac{\begin{vmatrix} 4 & -2 \\ 13 & 1 \end{vmatrix}}{15} = \frac{30}{15} = 2 \qquad y = \frac{D_y}{D} = \frac{\begin{vmatrix} 3 & 4 \\ 6 & 13 \end{vmatrix}}{15} = \frac{15}{15} = 1$$

The solution is $x = 2$, $y = 1$.

Graphing Solution   We enter the coefficient matrix into our graphing utility. Call it A and evaluate $\det[A]$. Since $\det[A] \neq 0$, we can use Cramer's Rule. We enter the matrices D_x and D_y into our graphing utility and call them B and C,

Figure 10

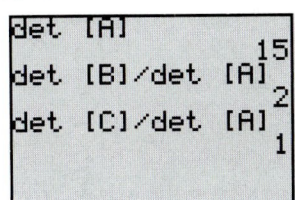

respectively. Finally, we find x by calculating $\dfrac{\det[B]}{\det[A]}$ and y by calculating $\dfrac{\det[C]}{\det[A]}$. The results are shown in Figure 10.

The solution is $x = 2$, $y = 1$. ∎

In attempting to use Cramer's Rule, if the determinant D of the coefficients of the variables is found to equal 0 (so that Cramer's Rule is not applicable), then the system either is inconsistent or has infinitely many solutions.

✏ **NOW WORK PROBLEM 11.**

3 by 3 Determinants

③ In order to use Cramer's Rule to solve a system of three equations containing three variables, we need to define a 3 by 3 determinant.

A **3 by 3 determinant** is symbolized by

$$\begin{vmatrix} a_{11} & a_{12} & a_{13} \\ a_{21} & a_{22} & a_{23} \\ a_{31} & a_{32} & a_{33} \end{vmatrix} \tag{8}$$

in which $a_{11}, a_{12}, \dots,$ are real numbers.

As with matrices, we use a double subscript to identify an entry by indicating its row and column numbers. For example, the entry a_{23} is in row 2, column 3.

The value of a 3 by 3 determinant may be defined in terms of 2 by 2 determinants by the following formula:

$$\begin{vmatrix} a_{11} & a_{12} & a_{13} \\ a_{21} & a_{22} & a_{23} \\ a_{31} & a_{32} & a_{33} \end{vmatrix} = a_{11} \begin{vmatrix} a_{22} & a_{23} \\ a_{32} & a_{33} \end{vmatrix} \overset{\text{Minus}}{-} a_{12} \begin{vmatrix} a_{21} & a_{23} \\ a_{31} & a_{33} \end{vmatrix} + a_{13} \begin{vmatrix} a_{21} & a_{22} \\ a_{31} & a_{32} \end{vmatrix} \tag{9}$$

2 by 2 determinant left after removing row and column containing a_{11} 2 by 2 determinant left after removing row and column containing a_{12} 2 by 2 determinant left after removing row and column containing a_{13}

The 2 by 2 determinants shown in formula (9) are called **minors** of the 3 by 3 determinant. For an n by n determinant, the **minor** M_{ij} of entry a_{ij} is the determinant resulting from removing the ith row and jth column.

EXAMPLE 4 **Finding Minors of a 3 by 3 Determinant**

For the determinant $A = \begin{vmatrix} 2 & -1 & 3 \\ -2 & 5 & 1 \\ 0 & 6 & -9 \end{vmatrix}$, find: (a) M_{12} (b) M_{23}

Solution   (a) M_{12} is the determinant that results from removing the first row and second column from A.

$$A = \begin{vmatrix} 2 & -1 & 3 \\ -2 & 5 & 1 \\ 0 & 6 & -9 \end{vmatrix} \qquad M_{12} = \begin{vmatrix} -2 & 1 \\ 0 & -9 \end{vmatrix} = (-2)(-9) - (0)(1) = 18$$

(b) M_{23} is the determinant that results from removing the second row and third column from A.

$$A = \begin{vmatrix} 2 & -1 & 3 \\ -2 & 5 & 1 \\ 0 & 6 & -9 \end{vmatrix} \qquad M_{23} = \begin{vmatrix} 2 & -1 \\ 0 & 6 \end{vmatrix} = (2)(6) - (0)(-1) = 12$$

■

Referring back to formula (9), we see that each entry a_{ij} is multiplied by its minor, but sometimes this term is added and other times subtracted. To determine whether to add or subtract a term, we must consider the *cofactor*.

For an n by n determinant A, the **cofactor** of entry a_{ij}, denoted by A_{ij} is given by

$$A_{ij} = (-1)^{i+j} M_{ij}$$

where M_{ij} is the minor of entry a_{ij}.

The exponent of $(-1)^{i+j}$ is the sum of the row and column of the entry a_{ij} so if $i + j$ is even, $(-1)^{i+j}$ will equal 1, and if $i + j$ is odd, $(-1)^{i+j}$ will equal -1.

To find the value of a determinant, multiply each entry in any row or column by its cofactor and sum the results, This process is referred to as *expanding across a row or column*. For example, the value of the 3 by 3 determinant in formula (9) was found by expanding across row 1.

If we choose to expand down column 2, we obtain

$$\begin{vmatrix} a_{11} & a_{12} & a_{13} \\ a_{21} & a_{22} & a_{23} \\ a_{31} & a_{32} & a_{33} \end{vmatrix} = (-1)^{1+2} a_{12} \begin{vmatrix} a_{21} & a_{23} \\ a_{31} & a_{33} \end{vmatrix} + (-1)^{2+2} a_{22} \begin{vmatrix} a_{11} & a_{13} \\ a_{31} & a_{33} \end{vmatrix} + (-1)^{3+2} a_{32} \begin{vmatrix} a_{11} & a_{13} \\ a_{21} & a_{23} \end{vmatrix}$$

Expand down column 2.

If we choose to expand across row 3, we obtain

$$\begin{vmatrix} a_{11} & a_{12} & a_{13} \\ a_{21} & a_{22} & a_{23} \\ a_{31} & a_{32} & a_{33} \end{vmatrix} = (-1)^{3+1} a_{31} \begin{vmatrix} a_{12} & a_{13} \\ a_{22} & a_{23} \end{vmatrix} + (-1)^{3+2} a_{32} \begin{vmatrix} a_{11} & a_{13} \\ a_{21} & a_{23} \end{vmatrix} + (-1)^{3+3} a_{33} \begin{vmatrix} a_{11} & a_{12} \\ a_{21} & a_{22} \end{vmatrix}$$

Expand across row 3.

It can be shown that the value of a determinant does not depend on the choice of the row or column used in the expansion. However, expanding across a row or column that has an entry equal to 0 reduces the amount of work needed to compute the value of the determinant.

<table>
<tr><td>**EXAMPLE 5**</td><td>**Evaluating a 3 × 3 Determinant**</td></tr>
</table>

Find the value of the 3 by 3 determinant: $\begin{vmatrix} 3 & 4 & -1 \\ 4 & 6 & 2 \\ 8 & -2 & 3 \end{vmatrix}$

Solution We choose to expand across row 1.

$$\begin{vmatrix} 3 & 4 & -1 \\ 4 & 6 & 2 \\ 8 & -2 & 3 \end{vmatrix} = (-1)^{1+1}3\begin{vmatrix} 6 & 2 \\ -2 & 3 \end{vmatrix} + (-1)^{1+2}4\begin{vmatrix} 4 & 2 \\ 8 & 3 \end{vmatrix} + (-1)^{1+3}(-1)\begin{vmatrix} 4 & 6 \\ 8 & -2 \end{vmatrix}$$

$$= 3(18 + 4) - 4(12 - 16) + (-1)(-8 - 48)$$

$$= 3(22) - 4(-4) + (-1)(-56)$$

$$= 66 + 16 + 56 = 138 \qquad \blacksquare$$

We could also find the value of the 3 by 3 determinant in Example 5 by expanding down column 3.

$$\begin{vmatrix} 3 & 4 & -1 \\ 4 & 6 & 2 \\ 8 & -2 & 3 \end{vmatrix} = (-1)^{1+3}(-1)\begin{vmatrix} 4 & 6 \\ 8 & -2 \end{vmatrix} + (-1)^{2+3}2\begin{vmatrix} 3 & 4 \\ 8 & -2 \end{vmatrix} + (-1)^{3+3}3\begin{vmatrix} 3 & 4 \\ 4 & 6 \end{vmatrix}$$

$$= -1(-8 - 48) - 2(-6 - 32) + 3(18 - 16)$$

$$= 56 + 76 + 6 = 138$$

Evaluating 3 × 3 determinants on a graphing utility follows the same procedure as evaluating 2 × 2 determinants.

✏️➤ **NOW WORK PROBLEM 7.**

Systems of Three Equations Containing Three Variables

④ Consider the following system of three equations containing three variables.

$$\begin{cases} a_{11}x + a_{12}y + a_{13}z = c_1 \\ a_{21}x + a_{22}y + a_{23}z = c_2 \\ a_{31}x + a_{32}y + a_{33}z = c_3 \end{cases} \qquad (10)$$

It can be shown that if the determinant D of the coefficients of the variables is not 0, that is, if

$$D = \begin{vmatrix} a_{11} & a_{12} & a_{13} \\ a_{21} & a_{22} & a_{23} \\ a_{31} & a_{32} & a_{33} \end{vmatrix} \neq 0$$

then the unique solution of system (10) is given by

Cramer's Rule for Three Equations Containing Three Variables

$$x = \frac{D_x}{D} \qquad y = \frac{D_y}{D} \qquad z = \frac{D_z}{D}$$

where

$$D_x = \begin{vmatrix} c_1 & a_{12} & a_{13} \\ c_2 & a_{22} & a_{23} \\ c_3 & a_{32} & a_{33} \end{vmatrix} \qquad D_y = \begin{vmatrix} a_{11} & c_1 & a_{13} \\ a_{21} & c_2 & a_{23} \\ a_{31} & c_3 & a_{33} \end{vmatrix} \qquad D_z = \begin{vmatrix} a_{11} & a_{12} & c_1 \\ a_{21} & a_{22} & c_2 \\ a_{31} & a_{32} & c_3 \end{vmatrix}$$

The similarity of this pattern and the pattern observed earlier for a system of two equations containing two variables should be apparent.

EXAMPLE 6   **Using Cramer's Rule**

Use Cramer's Rule, if applicable, to solve the following system:

$$\begin{cases} 2x + y - z = 3 & (1) \\ -x + 2y + 4z = -3 & (2) \\ x - 2y - 3z = 4 & (3) \end{cases}$$

Solution   The value of the determinant D of the coefficients of the variables is

$$D = \begin{vmatrix} 2 & 1 & -1 \\ -1 & 2 & 4 \\ 1 & -2 & -3 \end{vmatrix} = (-1)^{1+1}2 \begin{vmatrix} 2 & 4 \\ -2 & -3 \end{vmatrix} + (-1)^{1+2}1 \begin{vmatrix} -1 & 4 \\ 1 & -3 \end{vmatrix} + (-1)^{1+3}(-1) \begin{vmatrix} -1 & 2 \\ 1 & -2 \end{vmatrix}$$

$$= 2(2) - 1(-1) + (-1)(0)$$

$$= 4 + 1 = 5$$

Because $D \neq 0$, we proceed to find the values of $D_x, D_y,$ and D_z.

$$D_x = \begin{vmatrix} 3 & 1 & -1 \\ -3 & 2 & 4 \\ 4 & -2 & -3 \end{vmatrix} = (-1)^{1+1}3 \begin{vmatrix} 2 & 4 \\ -2 & -3 \end{vmatrix} + (-1)^{1+2}1 \begin{vmatrix} -3 & 4 \\ 4 & -3 \end{vmatrix} + (-1)^{1+3}(-1) \begin{vmatrix} -3 & 2 \\ 4 & -2 \end{vmatrix}$$

$$= 3(2) - 1(-7) + (-1)(-2) = 15$$

$$D_y = \begin{vmatrix} 2 & 3 & -1 \\ -1 & -3 & 4 \\ 1 & 4 & -3 \end{vmatrix} = (-1)^{1+1}2 \begin{vmatrix} -3 & 4 \\ 4 & -3 \end{vmatrix} + (-1)^{1+2}3 \begin{vmatrix} -1 & 4 \\ 1 & -3 \end{vmatrix} + (-1)^{1+3}(-1) \begin{vmatrix} -1 & -3 \\ 1 & 4 \end{vmatrix}$$

$$= 2(-7) - 3(-1) + (-1)(-1)$$

$$= -14 + 3 + 1 = -10$$

$$D_z = \begin{vmatrix} 2 & 1 & 3 \\ -1 & 2 & -3 \\ 1 & -2 & 4 \end{vmatrix} = (-1)^{1+1}2 \begin{vmatrix} 2 & -3 \\ -2 & 4 \end{vmatrix} + (-1)^{1+2}1 \begin{vmatrix} -1 & -3 \\ 1 & 4 \end{vmatrix} + (-1)^{1+3}3 \begin{vmatrix} -1 & 2 \\ 1 & -2 \end{vmatrix}$$

$$= 2(2) - 1(-1) + 3(0) = 5$$

As a result,

$$x = \frac{D_x}{D} = \frac{15}{5} = 3 \qquad y = \frac{D_y}{D} = \frac{-10}{5} = -2 \qquad z = \frac{D_z}{D} = \frac{5}{5} = 1$$

The solution is $x = 3$, $y = -2$, $z = 1$.   ■

If the determinant of the coefficients of the variables of a system of three linear equations containing three variables is 0, then Cramer's Rule is not applicable. In such a case, the system either is inconsistent or has infinitely many solutions.

Solving systems of three equations containing three variables using Cramer's Rule on a graphing utility follows the same procedure as that for solving systems of two equations containing two variables.

✎ **NOW WORK PROBLEM 29.**

Properties of Determinants

⑤ Determinants have several properties that are sometimes helpful for obtaining their value. We list some of them here.

Theorem   The value of a determinant changes sign if any two rows
(or any two columns) are interchanged.   (11)

■

Proof for 2 by 2 Determinants

$$\begin{vmatrix} a & b \\ c & d \end{vmatrix} = ad - bc \quad \text{and} \quad \begin{vmatrix} c & d \\ a & b \end{vmatrix} = bc - ad = -(ad - bc)$$

■

EXAMPLE 7   **Demonstrating Theorem (11)**

$$\begin{vmatrix} 3 & 4 \\ 1 & 2 \end{vmatrix} = 6 - 4 = 2 \qquad \begin{vmatrix} 1 & 2 \\ 3 & 4 \end{vmatrix} = 4 - 6 = -2$$

■

Theorem   If all the entries in any row (or any column) equal 0,
the value of the determinant is 0.   (12)

■

Proof   Merely expand across the row (or down the column) containing the 0's.   ■

Theorem   If any two rows (or any two columns) of a determinant have corresponding
entries that are equal, the value of the determinant is 0.   (13)

■

You are asked to prove this result for a 3 by 3 determinant in which the entries in column 1 equal the entries in column 3 in Problem 61.

EXAMPLE 8   **Demonstrating Theorem (13)**

$$\begin{vmatrix} 1 & 2 & 3 \\ 1 & 2 & 3 \\ 4 & 5 & 6 \end{vmatrix} = (-1)^{1+1}1\begin{vmatrix} 2 & 3 \\ 5 & 6 \end{vmatrix} + (-1)^{1+2}2\begin{vmatrix} 1 & 3 \\ 4 & 6 \end{vmatrix} + (-1)^{1+3}3\begin{vmatrix} 1 & 2 \\ 4 & 5 \end{vmatrix}$$

$$= 1(-3) - 2(-6) + 3(-3) = -3 + 12 - 9 = 0$$

Theorem   If any row (or any column) of a determinant is multiplied by a non-zero number k, the value of the determinant is also changed by a factor of k.   (14)

You are asked to prove this result for a 3 by 3 determinant using row 2 in Problem 60.

EXAMPLE 9   **Demonstrating Theorem (14)**

$$\begin{vmatrix} 1 & 2 \\ 4 & 6 \end{vmatrix} = 6 - 8 = -2$$

$$\begin{vmatrix} k & 2k \\ 4 & 6 \end{vmatrix} = 6k - 8k = -2k = k(-2) = k\begin{vmatrix} 1 & 2 \\ 4 & 6 \end{vmatrix}$$

Theorem   If the entries of any row (or any column) of a determinant are multiplied by a nonzero number k and the result is added to the corresponding entries of another row (or column), the value of the determinant remains unchanged.   (15)

In Problem 62, you are asked to prove this result for a 3 by 3 determinant using rows 1 and 2.

EXAMPLE 10   **Demonstrating Theorem (15)**

$$\begin{vmatrix} 3 & 4 \\ 5 & 2 \end{vmatrix} = \begin{vmatrix} -7 & 0 \\ 5 & 2 \end{vmatrix} = -14$$

Multiply row 2 by -2 and add to row 1.

7.4 Concepts and Vocabulary

In Problems 1 and 2, fill in the blanks.

1. Cramer's Rule uses _____ to solve a system of linear equations.

2. $D = \begin{vmatrix} a & b \\ c & d \end{vmatrix} = $ _____ .

In Problems 3 and 4, answer True or False for each statement.

3. A 3 by 3 determinant can never equal 0.

4. The value of a determinant remains unchanged if any two rows or any two columns are interchanged.

5. Why can't Cramer's Rule be used if the determinant of the coefficient matrix is zero? If the determinant of the coefficient matrix is zero, how many solutions might the system have?

7.4 Exercises

In Problems 1–10, find the value of each determinant (a) by hand and (b) by using a graphing utility.

1. $\begin{vmatrix} 3 & 1 \\ 4 & 2 \end{vmatrix}$
  2. $\begin{vmatrix} 6 & 1 \\ 5 & 2 \end{vmatrix}$
  3. $\begin{vmatrix} 6 & 4 \\ -1 & 3 \end{vmatrix}$
  4. $\begin{vmatrix} 8 & -3 \\ 4 & 2 \end{vmatrix}$
  5. $\begin{vmatrix} -3 & -1 \\ 4 & 2 \end{vmatrix}$

6. $\begin{vmatrix} -4 & 2 \\ -5 & 3 \end{vmatrix}$
  7. $\begin{vmatrix} 3 & 4 & 2 \\ 1 & -1 & 5 \\ 1 & 2 & -2 \end{vmatrix}$
  8. $\begin{vmatrix} 1 & 3 & -2 \\ 6 & 1 & -5 \\ 8 & 2 & 3 \end{vmatrix}$
  9. $\begin{vmatrix} 4 & -1 & 2 \\ 6 & -1 & 0 \\ 1 & -3 & 4 \end{vmatrix}$
  10. $\begin{vmatrix} 3 & -9 & 4 \\ 1 & 4 & 0 \\ 8 & -3 & 1 \end{vmatrix}$

In Problems 11–40, solve each system of equations using Cramer's Rule, if it is applicable, (a) by hand and (b) by using a graphing utility. If Cramer's Rule is not applicable, say so.

11. $\begin{cases} x + y = 8 \\ x - y = 4 \end{cases}$
  12. $\begin{cases} x + 2y = 5 \\ x - y = 3 \end{cases}$
  13. $\begin{cases} 5x - y = 13 \\ 2x + 3y = 12 \end{cases}$
  14. $\begin{cases} x + 3y = 5 \\ 2x - 3y = -8 \end{cases}$

15. $\begin{cases} 3x = 24 \\ x + 2y = 0 \end{cases}$
  16. $\begin{cases} 4x + 5y = -3 \\ -2y = -4 \end{cases}$
  17. $\begin{cases} 3x - 6y = 24 \\ 5x + 4y = 12 \end{cases}$
  18. $\begin{cases} 2x + 4y = 16 \\ 3x - 5y = -9 \end{cases}$

19. $\begin{cases} 3x - 2y = 4 \\ 6x - 4y = 0 \end{cases}$
  20. $\begin{cases} -x + 2y = 5 \\ 4x - 8y = 6 \end{cases}$
  21. $\begin{cases} 2x - 4y = -2 \\ 3x + 2y = 3 \end{cases}$
  22. $\begin{cases} 3x + 3y = 3 \\ 4x + 2y = \dfrac{8}{3} \end{cases}$

23. $\begin{cases} 2x - 3y = -1 \\ 10x + 10y = 5 \end{cases}$
  24. $\begin{cases} 3x - 2y = 0 \\ 5x + 10y = 4 \end{cases}$
  25. $\begin{cases} 2x + 3y = 6 \\ x - y = \dfrac{1}{2} \end{cases}$
  26. $\begin{cases} \dfrac{1}{2}x + y = -2 \\ x - 2y = 8 \end{cases}$

27. $\begin{cases} 3x - 5y = 3 \\ 15x + 5y = 21 \end{cases}$
  28. $\begin{cases} 2x - y = -1 \\ x + \dfrac{1}{2}y = \dfrac{3}{2} \end{cases}$
  29. $\begin{cases} x + y - z = 6 \\ 3x - 2y + z = -5 \\ x + 3y - 2z = 14 \end{cases}$
  30. $\begin{cases} x - y + z = -4 \\ 2x - 3y + 4z = -15 \\ 5x + y - 2z = 12 \end{cases}$

31. $\begin{cases} x + 2y - z = -3 \\ 2x - 4y + z = -7 \\ -2x + 2y - 3z = 4 \end{cases}$
  32. $\begin{cases} x + 4y - 3z = -8 \\ 3x - y + 3z = 12 \\ x + y + 6z = 1 \end{cases}$
  33. $\begin{cases} x - 2y + 3z = 1 \\ 3x + y - 2z = 0 \\ 2x - 4y + 6z = 2 \end{cases}$
  34. $\begin{cases} x - y + 2z = 5 \\ 3x + 2y = 4 \\ -2x + 2y - 4z = -10 \end{cases}$

35. $\begin{cases} x + 2y - z = 0 \\ 2x - 4y + z = 0 \\ -2x + 2y - 3z = 0 \end{cases}$
  36. $\begin{cases} x + 4y - 3z = 0 \\ 3x - y + 3z = 0 \\ x + y + 6z = 0 \end{cases}$
  37. $\begin{cases} x - 2y + 3z = 0 \\ 3x + y - 2z = 0 \\ 2x - 4y + 6z = 0 \end{cases}$
  38. $\begin{cases} x - y + 2z = 0 \\ 3x + 2y = 0 \\ -2x + 2y - 4z = 0 \end{cases}$

39. $\begin{cases} \dfrac{1}{x} + \dfrac{1}{y} = 8 \\ \dfrac{3}{x} - \dfrac{5}{y} = 0 \end{cases}$

40. $\begin{cases} \dfrac{4}{x} - \dfrac{3}{y} = 0 \\ \dfrac{6}{x} + \dfrac{3}{2y} = 2 \end{cases}$

[**Hint:** Let $u = \dfrac{1}{x}$ and $v = \dfrac{1}{y}$ and solve for u and v.]

In Problems 41–46, solve for x.

41. $\begin{vmatrix} x & x \\ 4 & 3 \end{vmatrix} = 5$

42. $\begin{vmatrix} x & 1 \\ 3 & x \end{vmatrix} = -2$

43. $\begin{vmatrix} x & 1 & 1 \\ 4 & 3 & 2 \\ -1 & 2 & 5 \end{vmatrix} = 2$

44. $\begin{vmatrix} 3 & 2 & 4 \\ 1 & x & 5 \\ 0 & 1 & -2 \end{vmatrix} = 0$

45. $\begin{vmatrix} x & 2 & 3 \\ 1 & x & 0 \\ 6 & 1 & -2 \end{vmatrix} = 7$

46. $\begin{vmatrix} x & 1 & 2 \\ 1 & x & 3 \\ 0 & 1 & 2 \end{vmatrix} = -4x$

In Problems 47–54, use properties of determinants to find the value of each determinant if it is known that

$$\begin{vmatrix} x & y & z \\ u & v & w \\ 1 & 2 & 3 \end{vmatrix} = 4$$

47. $\begin{vmatrix} 1 & 2 & 3 \\ u & v & w \\ x & y & z \end{vmatrix}$

48. $\begin{vmatrix} x & y & z \\ u & v & w \\ 2 & 4 & 6 \end{vmatrix}$

49. $\begin{vmatrix} x & y & z \\ -3 & -6 & -9 \\ u & v & w \end{vmatrix}$

50. $\begin{vmatrix} 1 & 2 & 3 \\ x - u & y - v & z - w \\ u & v & w \end{vmatrix}$

51. $\begin{vmatrix} 1 & 2 & 3 \\ x - 3 & y - 6 & z - 9 \\ 2u & 2v & 2w \end{vmatrix}$

52. $\begin{vmatrix} x & y & z - x \\ u & v & w - u \\ 1 & 2 & 2 \end{vmatrix}$

53. $\begin{vmatrix} 1 & 2 & 3 \\ 2x & 2y & 2z \\ u - 1 & v - 2 & w - 3 \end{vmatrix}$

54. $\begin{vmatrix} x + 3 & y + 6 & z + 9 \\ 3u - 1 & 3v - 2 & 3w - 3 \\ 1 & 2 & 3 \end{vmatrix}$

---

55. Geometry: Equation of a Line An equation of the line containing the two points (x_1, y_1) and (x_2, y_2) may be expressed as the determinant

$$\begin{vmatrix} x & y & 1 \\ x_1 & y_1 & 1 \\ x_2 & y_2 & 1 \end{vmatrix} = 0$$

Prove this result by expanding the determinant and comparing the result to the two-point form of the equation of a line.

56. Geometry: Collinear Points Using the result obtained in Problem 55, show that three distinct points (x_1, y_1), (x_2, y_2), and (x_3, y_3) are collinear (lie on the same line) if and only if

$$\begin{vmatrix} x_1 & y_1 & 1 \\ x_2 & y_2 & 1 \\ x_3 & y_3 & 1 \end{vmatrix} = 0$$

57. Show that $\begin{vmatrix} x^2 & x & 1 \\ y^2 & y & 1 \\ z^2 & z & 1 \end{vmatrix} = (y - z)(x - y)(x - z)$.

58. Complete the proof of Cramer's Rule for two equations containing two variables.

[**Hint:** In system (5), page 549, if $a = 0$, then $b \neq 0$ and $c \neq 0$, since $D = -bc \neq 0$. Now show that equation (6) provides a solution of the system when $a = 0$. There are then three remaining cases: $b = 0$, $c = 0$, and $d = 0$.]

59. Interchange columns 1 and 3 of a 3 by 3 determinant. Show that the value of the new determinant is -1 times the value of the original determinant.

60. Multiply each entry in row 2 of a 3 by 3 determinant by the number $k, k \neq 0$. Show that the value of the new determinant is k times the value of the original determinant.

61. Prove that a 3 by 3 determinant in which the entries in column 1 equal those in column 3 has the value 0.

62. Prove that if row 2 of a 3 by 3 determinant is multiplied by $k, k \neq 0$, and the result is added to the entries in row 1, then there is no change in the value of the determinant.

7.5 MATRIX ALGEBRA

OBJECTIVES

1. Find the Sum and Difference of Two Matrices
2. Find Scalar Multiples of a Matrix
3. Find the Product of Two Matrices
4. Find the Inverse of a Matrix
5. Solve Systems of Equations Using Inverse Matrices

In Section 7.3, we defined a matrix as an array of real numbers and used an augmented matrix to represent a system of linear equations. There is, however, a branch of mathematics, called **linear algebra**, that deals with matrices in such a way that an algebra of matrices is permitted. In this section, we provide a survey of how this **matrix algebra** is developed.

Before getting started, we restate the definition of a matrix.

A **matrix** is defined as a rectangular array of numbers:

$$
\begin{array}{ccccc}
 & \text{Column 1} & \text{Column 2} & \text{Column } j & \text{Column } n \\
\text{Row 1} & \begin{bmatrix} a_{11} & a_{12} & \cdots & a_{1j} & \cdots & a_{1n} \\
\text{Row 2} & a_{21} & a_{22} & \cdots & a_{2j} & \cdots & a_{2n} \\
\vdots & \vdots & \vdots & & \vdots & & \vdots \\
\text{Row } i & a_{i1} & a_{i2} & \cdots & a_{ij} & \cdots & a_{in} \\
\vdots & \vdots & \vdots & & \vdots & & \vdots \\
\text{Row } m & a_{m1} & a_{m2} & \cdots & a_{mj} & \cdots & a_{mn} \end{bmatrix}
\end{array}
$$

Each number a_{ij} of the matrix has two indexes: the **row index** i and the **column index** j. The matrix shown above has m rows and n columns. The $m \cdot n$ numbers a_{ij} are usually referred to as the **entries** of the matrix. For example, a_{23} refers to the entry in the second row, third column.

Let's begin with an example that illustrates how matrices can be used to conveniently represent an array of information.

EXAMPLE 1 **Arranging Data in a Matrix**

In a survey of 900 people, the following information was obtained:

200 males	Thought federal defense spending was too high
150 males	Thought federal defense spending was too low
45 males	Had no opinion
315 females	Thought federal defense spending was too high
125 females	Thought federal defense spending was too low
65 females	Had no opinion

We can arrange the above data in a rectangular array as follows:

	Too High	Too Low	No Opinion
Male	200	150	45
Female	315	125	65

or as the matrix

$$\begin{bmatrix} 200 & 150 & 45 \\ 315 & 125 & 65 \end{bmatrix}$$

This matrix has two rows (representing males and females) and three columns (representing "too high," "too low," and "no opinion"). ▪

The matrix we developed in Example 1 has 2 rows and 3 columns. In general, a matrix with m rows and n columns is called an **m by n matrix**. The matrix we developed in Example 1 is a 2 by 3 matrix and contains $2 \cdot 3 = 6$ entries. An m by n matrix will contain $m \cdot n$ entries.

If an m by n matrix has the same number of rows as columns, that is, if $m = n$, then the matrix is referred to as a **square matrix**.

EXAMPLE 2   **Examples of Matrices**

(a) $\begin{bmatrix} 5 & 0 \\ -6 & 1 \end{bmatrix}$ A 2 by 2 square matrix   (b) $\begin{bmatrix} 1 & 0 & 3 \end{bmatrix}$ A 1 by 3 matrix

(c) $\begin{bmatrix} 6 & -2 & 4 \\ 4 & 3 & 5 \\ 8 & 0 & 1 \end{bmatrix}$ A 3 by 3 square matrix

▪

The Sum and Difference of Two Matrices

① We begin our discussion of matrix algebra by first defining what is meant by two matrices being equal and then defining the operations of addition and subtraction. It is important to note that these definitions require each matrix to have the same number of rows *and* the same number of columns as a prerequisite for equality and for addition and subtraction.

We usually represent matrices by capital letters, such as A, B, C, and so on.

Two m by n matrices A and B are said to be **equal**, written as

$$A = B$$

provided that each entry a_{ij} in A is equal to the corresponding entry b_{ij} in B.

For example,

$$\begin{bmatrix} 2 & 1 \\ 0.5 & -1 \end{bmatrix} = \begin{bmatrix} \sqrt{4} & 1 \\ \frac{1}{2} & -1 \end{bmatrix} \quad \text{and} \quad \begin{bmatrix} 3 & 2 & 1 \\ 0 & 1 & -2 \end{bmatrix} = \begin{bmatrix} \sqrt{9} & \sqrt{4} & 1 \\ 0 & 1 & \sqrt[3]{-8} \end{bmatrix}$$

$$\begin{bmatrix} 4 & 1 \\ 6 & 1 \end{bmatrix} \neq \begin{bmatrix} 4 & 0 \\ 6 & 1 \end{bmatrix}$$ Because the entries in row 1, column 2 are not equal

$$\begin{bmatrix} 4 & 1 & 2 \\ 6 & 1 & 2 \end{bmatrix} \neq \begin{bmatrix} 4 & 1 & 2 & 3 \\ 6 & 1 & 2 & 4 \end{bmatrix}$$ Because the matrix on the left is 2 by 3 and the matrix on the right is 2 by 4

Suppose that A and B represent two m by n matrices. We define their **sum $A + B$** to be the m by n matrix formed by adding the corresponding entries a_{ij} of A and b_{ij} of B. The **difference $A - B$** is defined as the m by n matrix formed by subtracting the entries b_{ij} in B from the corresponding entries a_{ij} in A. Addition and subtraction of matrices are allowed only for matrices having the same number m of rows and the same number n of columns. For example, a 2 by 3 matrix and a 2 by 4 matrix cannot be added or subtracted.

Graphing utilities make the sometimes tedious process of matrix algebra easy. Let's compare how a graphing utility adds and subtracts matrices with doing it by hand.

EXAMPLE 3   **Adding and Subtracting Matrices**

Suppose that

$$A = \begin{bmatrix} 2 & 4 & 8 & -3 \\ 0 & 1 & 2 & 3 \end{bmatrix} \quad \text{and} \quad B = \begin{bmatrix} -3 & 4 & 0 & 1 \\ 6 & 8 & 2 & 0 \end{bmatrix}$$

Find: (a) $A + B$ (b) $A - B$

Algebraic Solution   (a) $A + B = \begin{bmatrix} 2 & 4 & 8 & -3 \\ 0 & 1 & 2 & 3 \end{bmatrix} + \begin{bmatrix} -3 & 4 & 0 & 1 \\ 6 & 8 & 2 & 0 \end{bmatrix}$

$$= \begin{bmatrix} 2 + (-3) & 4 + 4 & 8 + 0 & -3 + 1 \\ 0 + 6 & 1 + 8 & 2 + 2 & 3 + 0 \end{bmatrix} \qquad \text{Add corresponding entries.}$$

$$= \begin{bmatrix} -1 & 8 & 8 & -2 \\ 6 & 9 & 4 & 3 \end{bmatrix}$$

(b) $A - B = \begin{bmatrix} 2 & 4 & 8 & -3 \\ 0 & 1 & 2 & 3 \end{bmatrix} - \begin{bmatrix} -3 & 4 & 0 & 1 \\ 6 & 8 & 2 & 0 \end{bmatrix}$

$$= \begin{bmatrix} 2 - (-3) & 4 - 4 & 8 - 0 & -3 - 1 \\ 0 - 6 & 1 - 8 & 2 - 2 & 3 - 0 \end{bmatrix} \qquad \text{Subtract corresponding entries.}$$

$$= \begin{bmatrix} 5 & 0 & 8 & -4 \\ -6 & -7 & 0 & 3 \end{bmatrix}$$

Graphing Solution   Enter the matrices into a graphing utility. Name them $[A]$ and $[B]$. Figure 11 shows the results of adding and subtracting $[A]$ and $[B]$.

Figure 11

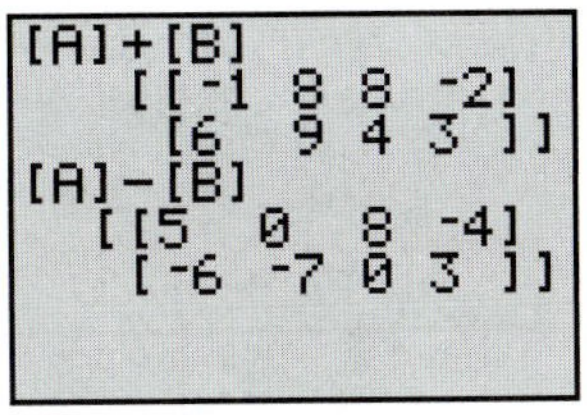

```
[A]+[B]
    [[-1 8 8 -2]
     [6  9 4  3 ]]
[A]-[B]
    [[5  0  8 -4]
     [-6 -7 0  3 ]]
```

NOW WORK PROBLEM **1**.

Many of the algebraic properties of sums of real numbers are also true for sums of matrices. Suppose that A, B, and C are m by n matrices. Then matrix addition is **commutative**. That is,

Commutative Property

$$A + B = B + A$$

Matrix addition is also **associative**. That is,

Associative Property

$$(A + B) + C = A + (B + C)$$

Although we shall not prove these results, the proofs, as the following example illustrates, are based on the commutative and associative properties for real numbers.

EXAMPLE 4   **Demonstrating the Commutative Property**

$$\begin{bmatrix} 2 & 3 & -1 \\ 4 & 0 & 7 \end{bmatrix} + \begin{bmatrix} -1 & 2 & 1 \\ 5 & -3 & 4 \end{bmatrix} = \begin{bmatrix} 2 + (-1) & 3 + 2 & -1 + 1 \\ 4 + 5 & 0 + (-3) & 7 + 4 \end{bmatrix}$$

$$= \begin{bmatrix} -1 + 2 & 2 + 3 & 1 + (-1) \\ 5 + 4 & -3 + 0 & 4 + 7 \end{bmatrix}$$

$$= \begin{bmatrix} -1 & 2 & 1 \\ 5 & -3 & 4 \end{bmatrix} + \begin{bmatrix} 2 & 3 & -1 \\ 4 & 0 & 7 \end{bmatrix}$$

A matrix whose entries are all equal to 0 is called a **zero matrix**. Each of the following matrices is a zero matrix.

$$\begin{bmatrix} 0 & 0 \\ 0 & 0 \end{bmatrix}$$ 2 by 2 square zero matrix $$\begin{bmatrix} 0 & 0 & 0 \\ 0 & 0 & 0 \end{bmatrix}$$ 2 by 3 zero matrix $$\begin{bmatrix} 0 & 0 & 0 \end{bmatrix}$$ 1 by 3 zero matrix

Zero matrices have properties similar to the real number 0. If A is an m by n matrix and 0 is an m by n zero matrix, then

$$A + 0 = A$$

In other words, the zero matrix is the additive identity in matrix algebra.

Scalar Multiples of a Matrix

② We can also multiply a matrix by a real number. If k is a real number and A is an m by n matrix, the matrix kA is the m by n matrix formed by multiplying each entry a_{ij} in A by k. The number k is sometimes referred to as a **scalar**, and the matrix kA is called a **scalar multiple** of A.

EXAMPLE 5   **Operations Using Matrices**

Suppose that

$$A = \begin{bmatrix} 3 & 1 & 5 \\ -2 & 0 & 6 \end{bmatrix} \qquad B = \begin{bmatrix} 4 & 1 & 0 \\ 8 & 1 & -3 \end{bmatrix} \qquad C = \begin{bmatrix} 9 & 0 \\ -3 & 6 \end{bmatrix}$$

Find: (a) $4A$ (b) $\frac{1}{3}C$ (c) $3A - 2B$

Algebraic Solution

(a) $4A = 4\begin{bmatrix} 3 & 1 & 5 \\ -2 & 0 & 6 \end{bmatrix} = \begin{bmatrix} 4 \cdot 3 & 4 \cdot 1 & 4 \cdot 5 \\ 4(-2) & 4 \cdot 0 & 4 \cdot 6 \end{bmatrix} = \begin{bmatrix} 12 & 4 & 20 \\ -8 & 0 & 24 \end{bmatrix}$

(b) $\frac{1}{3}C = \frac{1}{3}\begin{bmatrix} 9 & 0 \\ -3 & 6 \end{bmatrix} = \begin{bmatrix} \frac{1}{3} \cdot 9 & \frac{1}{3} \cdot 0 \\ \frac{1}{3}(-3) & \frac{1}{3} \cdot 6 \end{bmatrix} = \begin{bmatrix} 3 & 0 \\ -1 & 2 \end{bmatrix}$

(c) $3A - 2B = 3\begin{bmatrix} 3 & 1 & 5 \\ -2 & 0 & 6 \end{bmatrix} - 2\begin{bmatrix} 4 & 1 & 0 \\ 8 & 1 & -3 \end{bmatrix}$

$= \begin{bmatrix} 3 \cdot 3 & 3 \cdot 1 & 3 \cdot 5 \\ 3(-2) & 3 \cdot 0 & 3 \cdot 6 \end{bmatrix} - \begin{bmatrix} 2 \cdot 4 & 2 \cdot 1 & 2 \cdot 0 \\ 2 \cdot 8 & 2 \cdot 1 & 2(-3) \end{bmatrix}$

$= \begin{bmatrix} 9 & 3 & 15 \\ -6 & 0 & 18 \end{bmatrix} - \begin{bmatrix} 8 & 2 & 0 \\ 16 & 2 & -6 \end{bmatrix}$

$= \begin{bmatrix} 9 - 8 & 3 - 2 & 15 - 0 \\ -6 - 16 & 0 - 2 & 18 - (-6) \end{bmatrix}$

$= \begin{bmatrix} 1 & 1 & 15 \\ -22 & -2 & 24 \end{bmatrix}$

Graphing Solution   Enter the matrices $[A], [B]$, and $[C]$ into a graphing utility. Figure 12 shows the required computations.

Figure 12

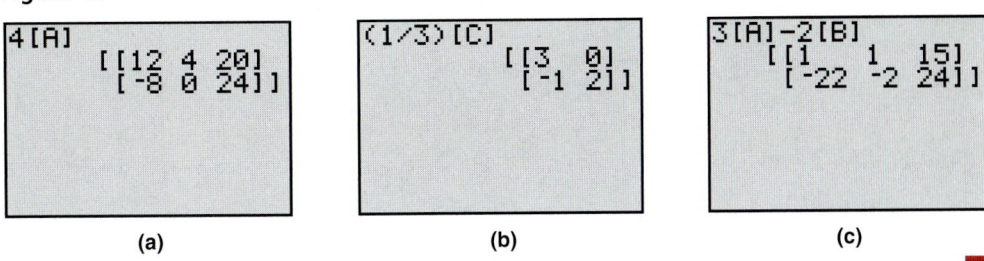

(a)   (b)   (c)

✏ NOW WORK PROBLEM 5.

We list next some of the algebraic properties of scalar multiplication. Let h and k be real numbers, and let A and B be m by n matrices. Then

Properties of Scalar Multiplication

$$k(hA) = (kh)A$$
$$(k + h)A = kA + hA$$
$$k(A + B) = kA + kB$$

The Product of Two Matrices

3   Unlike the straightforward definition for adding two matrices, the definition for multiplying two matrices is not what we might expect. In preparation for this definition, we need the following definitions:

A **row vector** R is a 1 by n matrix

$$R = [r_1 \quad r_2 \quad \cdots \quad r_n]$$

A **column vector** C is an n by 1 matrix

$$C = \begin{bmatrix} c_1 \\ c_2 \\ \vdots \\ c_n \end{bmatrix}$$

The **product** RC of R times C is defined as the number

$$RC = [r_1 \; r_2 \cdots r_n] \begin{bmatrix} c_1 \\ c_2 \\ \vdots \\ c_n \end{bmatrix} = r_1 c_1 + r_2 c_2 + \cdots + r_n c_n$$

Notice that a row vector and a column vector can be multiplied only if they contain the same number of entries.

EXAMPLE 6   **The Product of a Row Vector by a Column Vector**

If $R = [3 \quad -5 \quad 2]$ and $C = \begin{bmatrix} 3 \\ 4 \\ -5 \end{bmatrix}$, then

$$RC = [3 \quad -5 \quad 2] \begin{bmatrix} 3 \\ 4 \\ -5 \end{bmatrix} = 3 \cdot 3 + (-5)4 + 2(-5) = 9 - 20 - 10 = -21$$

■

Let's look at an application of the product of a row vector by a column vector.

EXAMPLE 7   **Using Matrices to Compute Revenue**

A clothing store sells men's shirts for $40, silk ties for $20, and wool suits for $375. Last month, the store had sales consisting of 100 shirts, 200 ties, and 50 suits. What was the total revenue due to these sales?

Solution   We set up a row vector R to represent the prices of each item and a column vector C to represent the corresponding number of items sold. Then

$$\begin{array}{cc} \text{Prices} & \text{Number} \\ \text{Shirts Ties Suits} & \text{sold} \end{array}$$

$$R = \begin{bmatrix} 40 & 20 & 375 \end{bmatrix} \qquad C = \begin{bmatrix} 100 \\ 200 \\ 50 \end{bmatrix} \begin{array}{l} \text{Shirts} \\ \text{Ties} \\ \text{Suits} \end{array}$$

The total revenue obtained is the product RC. That is,

$$RC = \begin{bmatrix} 40 & 20 & 375 \end{bmatrix} \begin{bmatrix} 100 \\ 200 \\ 50 \end{bmatrix}$$

$$= \underbrace{40 \cdot 100}_{\text{Shirt revenue}} + \underbrace{20 \cdot 200}_{\text{Tie revenue}} + \underbrace{375 \cdot 50}_{\text{Suit revenue}} = \underbrace{\$26{,}750}_{\text{Total revenue}}$$

The definition for multiplying two matrices is based on the definition of a row vector times a column vector.

> Let A denote an m by r matrix, and let B denote an r by n matrix. The **product** AB is defined as the m by n matrix whose entry in row i, column j is the product of the ith row of A and the jth column of B.

The definition of the product AB of two matrices A and B, in this order, requires that the number of columns of A equal the number of rows of B; otherwise, no product is defined.

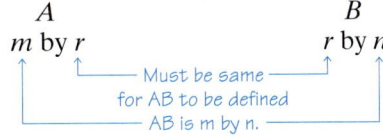

An example will help to clarify the definition.

EXAMPLE 8   **Multiplying Two Matrices**

Find the product AB if

$$A = \begin{bmatrix} 2 & 4 & -1 \\ 5 & 8 & 0 \end{bmatrix} \quad \text{and} \quad B = \begin{bmatrix} 2 & 5 & 1 & 4 \\ 4 & 8 & 0 & 6 \\ -3 & 1 & -2 & -1 \end{bmatrix}$$

Solution   First, we note that A is 2 by 3 and B is 3 by 4, so the product AB is defined and will be a 2 by 4 matrix.

Algebraic Solution   Suppose that we want the entry in row 2, column 3 of AB. To find it, we find the product of the row vector from row 2 of A and the column vector from column 3 of B.

$$\underset{\text{Row 2 of } A}{[5 \quad 8 \quad 0]} \overset{\text{Column 3 of } B}{\begin{bmatrix} 1 \\ 0 \\ -2 \end{bmatrix}} = 5 \cdot 1 + 8 \cdot 0 + 0(-2) = 5$$

So far, we have

$$AB = \begin{bmatrix} \underline{\quad} & \underline{\quad} & \overset{\text{Column 3}}{\underset{5}{\downarrow}} & \underline{\quad} \\ \underline{\quad} & \underline{\quad} & 5 & \underline{\quad} \end{bmatrix} \quad \leftarrow \text{Row 2}$$

Now, to find the entry in row 1, column 4 of AB, we find the product of row 1 of A and column 4 of B.

$$\underset{\text{Row 1 of } A}{[2 \quad 4 \quad -1]} \overset{\text{Column 4 of } B}{\begin{bmatrix} 4 \\ 6 \\ -1 \end{bmatrix}} = 2 \cdot 4 + 4 \cdot 6 + (-1)(-1) = 33$$

Continuing in this fashion, we find AB.

$$AB = \begin{bmatrix} 2 & 4 & -1 \\ 5 & 8 & 0 \end{bmatrix} \begin{bmatrix} 2 & 5 & 1 & 4 \\ 4 & 8 & 0 & 6 \\ -3 & 1 & -2 & -1 \end{bmatrix}$$

$$= \begin{bmatrix} \begin{matrix} \text{Row 1 of } A \\ \text{times} \\ \text{column 1 of } B \end{matrix} & \begin{matrix} \text{Row 1 of } A \\ \text{times} \\ \text{column 2 of } B \end{matrix} & \begin{matrix} \text{Row 1 of } A \\ \text{times} \\ \text{column 3 of } B \end{matrix} & \begin{matrix} \text{Row 1 of } A \\ \text{times} \\ \text{column 4 of } B \end{matrix} \\ \begin{matrix} \text{Row 2 of } A \\ \text{times} \\ \text{column 1 of } B \end{matrix} & \begin{matrix} \text{Row 2 of } A \\ \text{times} \\ \text{column 2 of } B \end{matrix} & \begin{matrix} \text{Row 2 of } A \\ \text{times} \\ \text{column 3 of } B \end{matrix} & \begin{matrix} \text{Row 2 of } A \\ \text{times} \\ \text{column 4 of } B \end{matrix} \end{bmatrix}$$

$$= \begin{bmatrix} 2 \cdot 2 + 4 \cdot 4 + (-1)(-3) & 2 \cdot 5 + 4 \cdot 8 + (-1)1 & 2 \cdot 1 + 4 \cdot 0 + (-1)(-2) & 33\,(\text{from earlier}) \\ 5 \cdot 2 + 8 \cdot 4 + 0(-3) & 5 \cdot 5 + 8 \cdot 8 + 0 \cdot 1 & 5\,(\text{from earlier}) & 5 \cdot 4 + 8 \cdot 6 + 0(-1) \end{bmatrix}$$

$$= \begin{bmatrix} 23 & 41 & 4 & 33 \\ 42 & 89 & 5 & 68 \end{bmatrix}$$

Graphing Solution   Enter the matrices A and B into a graphing utility. Figure 13 shows the product AB.

Figure 13

```
[A] [B]
  [[23 41 4 33]
   [42 89 5 68]]
```

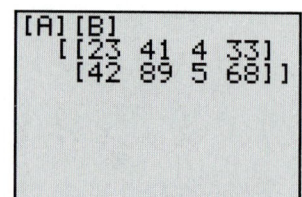

NOW WORK PROBLEM **17.**

Notice that for the matrices given in Example 8 the product BA is not defined, because B is 3 by 4 and A is 2 by 3. Try calculating BA on a graphing utility. What do you notice?

Another result that can occur when multiplying two matrices is illustrated in the next example.*

EXAMPLE 9   **Multiplying Two Matrices**

If

$$A = \begin{bmatrix} 2 & 1 & 3 \\ 1 & -1 & 0 \end{bmatrix} \quad \text{and} \quad B = \begin{bmatrix} 1 & 0 \\ 2 & 1 \\ 3 & 2 \end{bmatrix}$$

find: (a) AB (b) BA

Solution   (a) $AB = \underset{\text{2 by 3}}{\begin{bmatrix} 2 & 1 & 3 \\ 1 & -1 & 0 \end{bmatrix}} \underset{\text{3 by 2}}{\begin{bmatrix} 1 & 0 \\ 2 & 1 \\ 3 & 2 \end{bmatrix}} = \underset{\text{2 by 2}}{\begin{bmatrix} 13 & 7 \\ -1 & -1 \end{bmatrix}}$

(b) $BA = \underset{\text{3 by 2}}{\begin{bmatrix} 1 & 0 \\ 2 & 1 \\ 3 & 2 \end{bmatrix}} \underset{\text{2 by 3}}{\begin{bmatrix} 2 & 1 & 3 \\ 1 & -1 & 0 \end{bmatrix}} = \underset{\text{3 by 3}}{\begin{bmatrix} 2 & 1 & 3 \\ 5 & 1 & 6 \\ 8 & 1 & 9 \end{bmatrix}}$

Notice in Example 9 that AB is 2 by 2 and BA is 3 by 3. It is possible for both AB and BA to be defined, yet be unequal. In fact, even if A and B are both n by n matrices, so that AB and BA are each defined and n by n, AB and BA will usually be unequal.

EXAMPLE 10   **Multiplying Two Square Matrices**

If

$$A = \begin{bmatrix} 2 & 1 \\ 0 & 4 \end{bmatrix} \quad \text{and} \quad B = \begin{bmatrix} -3 & 1 \\ 1 & 2 \end{bmatrix}$$

find: (a) AB (b) BA

Solution   (a) $AB = \begin{bmatrix} 2 & 1 \\ 0 & 4 \end{bmatrix} \begin{bmatrix} -3 & 1 \\ 1 & 2 \end{bmatrix} = \begin{bmatrix} -5 & 4 \\ 4 & 8 \end{bmatrix}$

(b) $BA = \begin{bmatrix} -3 & 1 \\ 1 & 2 \end{bmatrix} \begin{bmatrix} 2 & 1 \\ 0 & 4 \end{bmatrix} = \begin{bmatrix} -6 & 1 \\ 2 & 9 \end{bmatrix}$

The preceding examples demonstrate that an important property of real numbers, the commutative property of multiplication, is not shared by matrices. Thus, in general:

---

*For most of the examples that follow, we will multiply matrices by hand. You should verify each result using a graphing utility.

| Theorem | Matrix multiplication is not commutative. |

NOW WORK PROBLEMS **7** AND **9.**

Next we give two of the properties of real numbers that are shared by matrices. Assuming that each product and sum is defined, we have the following:

Associative Property

$$A(BC) = (AB)C$$

Distributive Property

$$A(B + C) = AB + AC$$

The Identity Matrix

For an n by n square matrix, the entries located in row i, column i, $1 \leq i \leq n$, are called the **diagonal entries**. An n by n square matrix whose diagonal entries are 1's, while all other entries are 0's, is called the **identity matrix I_n**. For example,

$$I_2 = \begin{bmatrix} 1 & 0 \\ 0 & 1 \end{bmatrix} \qquad I_3 = \begin{bmatrix} 1 & 0 & 0 \\ 0 & 1 & 0 \\ 0 & 0 & 1 \end{bmatrix}$$

and so on.

EXAMPLE 11   **Multiplication with an Identity Matrix**

Let

$$A = \begin{bmatrix} -1 & 2 & 0 \\ 0 & 1 & 3 \end{bmatrix} \quad \text{and} \quad B = \begin{bmatrix} 3 & 2 \\ 4 & 6 \\ 5 & 2 \end{bmatrix}$$

Find: (a) AI_3 (a) I_2A (c) BI_2

Solution   (a) $AI_3 = \begin{bmatrix} -1 & 2 & 0 \\ 0 & 1 & 3 \end{bmatrix} \begin{bmatrix} 1 & 0 & 0 \\ 0 & 1 & 0 \\ 0 & 0 & 1 \end{bmatrix} = \begin{bmatrix} -1 & 2 & 0 \\ 0 & 1 & 3 \end{bmatrix} = A$

(b) $I_2A = \begin{bmatrix} 1 & 0 \\ 0 & 1 \end{bmatrix} \begin{bmatrix} -1 & 2 & 0 \\ 0 & 1 & 3 \end{bmatrix} = \begin{bmatrix} -1 & 2 & 0 \\ 0 & 1 & 3 \end{bmatrix} = A$

(c) $BI_2 = \begin{bmatrix} 3 & 2 \\ 4 & 6 \\ 5 & 2 \end{bmatrix} \begin{bmatrix} 1 & 0 \\ 0 & 1 \end{bmatrix} = \begin{bmatrix} 3 & 2 \\ 4 & 6 \\ 5 & 2 \end{bmatrix} = B$

Example 11 demonstrates the following property:

Identity Property

If A is an m by n matrix, then

$$I_m A = A \quad \text{and} \quad A I_n = A$$

If A is an n by n square matrix, then $A I_n = I_n A = A$.

An identity matrix has properties analogous to those of the real number 1. In other words, the identity matrix is a multiplicative identity in matrix algebra.

The Inverse of a Matrix

④ Let A be a square n by n matrix. If there exists an n by n matrix A^{-1}, read "A inverse," for which

$$AA^{-1} = A^{-1}A = I_n$$

then A^{-1} is called the **inverse** of the matrix A.

As we shall soon see, not every square matrix has an inverse. When a matrix A does have an inverse A^{-1}, then A is said to be **nonsingular**. If a matrix A has no inverse, it is called **singular**.*

EXAMPLE 12   **Multiplying a Matrix by Its Inverse**

Show that the inverse of

$$A = \begin{bmatrix} 3 & 1 \\ 2 & 1 \end{bmatrix} \quad \text{is} \quad A^{-1} = \begin{bmatrix} 1 & -1 \\ -2 & 3 \end{bmatrix}$$

Solution   We need to show that $AA^{-1} = A^{-1}A = I_2$.

$$AA^{-1} = \begin{bmatrix} 3 & 1 \\ 2 & 1 \end{bmatrix}\begin{bmatrix} 1 & -1 \\ -2 & 3 \end{bmatrix} = \begin{bmatrix} 1 & 0 \\ 0 & 1 \end{bmatrix} = I_2$$

$$A^{-1}A = \begin{bmatrix} 1 & -1 \\ -2 & 3 \end{bmatrix}\begin{bmatrix} 3 & 1 \\ 2 & 1 \end{bmatrix} = \begin{bmatrix} 1 & 0 \\ 0 & 1 \end{bmatrix} = I_2$$

We now show one way to find the inverse of

$$A = \begin{bmatrix} 3 & 1 \\ 2 & 1 \end{bmatrix}$$

Suppose that A^{-1} is given by

$$A^{-1} = \begin{bmatrix} x & y \\ z & w \end{bmatrix} \tag{1}$$

---

*If the determinant of A is zero, then A is singular. (Refer to Section 7.4.)

where x, y, z, and w are four variables. Based on the definition of an inverse, if, indeed, A has an inverse, we have

$$AA^{-1} = I_2$$

$$\begin{bmatrix} 3 & 1 \\ 2 & 1 \end{bmatrix} \begin{bmatrix} x & y \\ z & w \end{bmatrix} = \begin{bmatrix} 1 & 0 \\ 0 & 1 \end{bmatrix}$$

$$\begin{bmatrix} 3x + z & 3y + w \\ 2x + z & 2y + w \end{bmatrix} = \begin{bmatrix} 1 & 0 \\ 0 & 1 \end{bmatrix}$$

Because corresponding entries must be equal, it follows that this matrix equation is equivalent to four ordinary equations.

$$\begin{cases} 3x + z = 1 \\ 2x + z = 0 \end{cases} \qquad \begin{cases} 3y + w = 0 \\ 2y + w = 1 \end{cases}$$

The augmented matrix of each system is

$$\left[\begin{array}{cc|c} 3 & 1 & 1 \\ 2 & 1 & 0 \end{array} \right] \qquad \left[\begin{array}{cc|c} 3 & 1 & 0 \\ 2 & 1 & 1 \end{array} \right] \qquad (2)$$

The usual procedure would be to transform each augmented matrix into reduced row echelon form. Notice, though, that the left sides of the augmented matrices are equal, so the same row operations (see Section 7.3) can be used to reduce each one. Thus, we find it more efficient to combine the two augmented matrices (2) into a single matrix, as shown next, and then transform it into reduced row echelon form.

$$\left[\begin{array}{cc|cc} 3 & 1 & 1 & 0 \\ 2 & 1 & 0 & 1 \end{array} \right]$$

Now we attempt to transform the left side into an identity matrix.

$$\left[\begin{array}{cc|cc} 3 & 1 & 1 & 0 \\ 2 & 1 & 0 & 1 \end{array} \right] \underset{\underset{R_1 = -1r_2 + r_1}{\uparrow}}{\rightarrow} \left[\begin{array}{cc|cc} 1 & 0 & 1 & -1 \\ 2 & 1 & 0 & 1 \end{array} \right]$$

$$\underset{\underset{R_2 = -2r_1 + r_2}{\uparrow}}{\rightarrow} \left[\begin{array}{cc|cc} 1 & 0 & 1 & -1 \\ 0 & 1 & -2 & 3 \end{array} \right] \qquad (3)$$

Matrix (3) is in reduced row echelon form. Now we reverse the earlier step of combining the two augmented matrices in (2) and write the single matrix (3) as two augmented matrices.

$$\left[\begin{array}{cc|c} 1 & 0 & 1 \\ 0 & 1 & -2 \end{array} \right] \quad \text{and} \quad \left[\begin{array}{cc|c} 1 & 0 & -1 \\ 0 & 1 & 3 \end{array} \right]$$

We conclude from these matrices that $x = 1$, $z = -2$, and $y = -1$, $w = 3$. Substituting these values into matrix (1), we find that

$$A^{-1} = \begin{bmatrix} 1 & -1 \\ -2 & 3 \end{bmatrix}$$

Notice in display (3) that the 2 by 2 matrix to the right of the vertical bar is, in fact, the inverse of A. Also notice that the identity matrix I_2 is the matrix that appears to the left of the vertical bar. These observations and the procedures followed above will work in general.

> ***Procedure for Finding the Inverse of a Nonsingular Matrix***
>
> To find the inverse of an n by n nonsingular matrix A, proceed as follows:
>
> **STEP 1:** Form the matrix $[A|I_n]$.
> **STEP 2:** Transform the matrix $[A|I_n]$ into reduced row echelon form.
> **STEP 3:** The reduced row echelon form of $[A|I_n]$ will contain the identity matrix I_n on the left of the vertical bar; the n by n matrix on the right of the vertical bar is the inverse of A.

In other words, if A is nonsingular, we begin with the matrix $[A|I_n]$ and, after transforming it into reduced row echelon form, we end up with the matrix $[I_n|A^{-1}]$.

Let's look at another example.

EXAMPLE 13   Finding the Inverse of a Matrix

The matrix

$$A = \begin{bmatrix} 1 & 1 & 0 \\ -1 & 3 & 4 \\ 0 & 4 & 3 \end{bmatrix}$$

is nonsingular. Find its inverse.

Algebraic Solution   First, we form the matrix

$$[A|I_3] = \left[\begin{array}{ccc|ccc} 1 & 1 & 0 & 1 & 0 & 0 \\ -1 & 3 & 4 & 0 & 1 & 0 \\ 0 & 4 & 3 & 0 & 0 & 1 \end{array}\right]$$

Next, we use row operations to transform $[A|I_3]$ into reduced row echelon form.

$$\left[\begin{array}{ccc|ccc} 1 & 1 & 0 & 1 & 0 & 0 \\ -1 & 3 & 4 & 0 & 1 & 0 \\ 0 & 4 & 3 & 0 & 0 & 1 \end{array}\right] \rightarrow \left[\begin{array}{ccc|ccc} 1 & 1 & 0 & 1 & 0 & 0 \\ 0 & 4 & 4 & 1 & 1 & 0 \\ 0 & 4 & 3 & 0 & 0 & 1 \end{array}\right] \rightarrow \left[\begin{array}{ccc|ccc} 1 & 1 & 0 & 1 & 0 & 0 \\ 0 & 1 & 1 & \frac{1}{4} & \frac{1}{4} & 0 \\ 0 & 4 & 3 & 0 & 0 & 1 \end{array}\right]$$

$$R_2 = r_1 + r_2 \qquad\qquad R_2 = \frac{1}{4}r_2$$

$$\rightarrow \left[\begin{array}{ccc|ccc} 1 & 0 & -1 & \frac{3}{4} & -\frac{1}{4} & 0 \\ 0 & 1 & 1 & \frac{1}{4} & \frac{1}{4} & 0 \\ 0 & 0 & -1 & -1 & -1 & 1 \end{array}\right] \rightarrow \left[\begin{array}{ccc|ccc} 1 & 0 & -1 & \frac{3}{4} & -\frac{1}{4} & 0 \\ 0 & 1 & 1 & \frac{1}{4} & \frac{1}{4} & 0 \\ 0 & 0 & 1 & 1 & 1 & -1 \end{array}\right]$$

$$R_1 = -1r_2 + r_1 \qquad\qquad R_3 = -1r_3$$
$$R_3 = -4r_2 + r_3$$

$$\rightarrow \left[\begin{array}{ccc|ccc} 1 & 0 & 0 & \frac{7}{4} & \frac{3}{4} & -1 \\ 0 & 1 & 0 & -\frac{3}{4} & -\frac{3}{4} & 1 \\ 0 & 0 & 1 & 1 & 1 & -1 \end{array}\right]$$

$$R_1 = r_3 + r_1$$
$$R_2 = -1r_3 + r_2$$

The matrix $[A|I_3]$ is now in reduced row echelon form, and the identity matrix I_3 is on the left of the vertical bar. Hence, the inverse of A is

$$A^{-1} = \begin{bmatrix} \dfrac{7}{4} & \dfrac{3}{4} & -1 \\[2mm] -\dfrac{3}{4} & -\dfrac{3}{4} & 1 \\[2mm] 1 & 1 & -1 \end{bmatrix}$$

Graphing Solution Enter the matrix A into a graphing utility. Figure 14 shows A^{-1}.

Figure 14

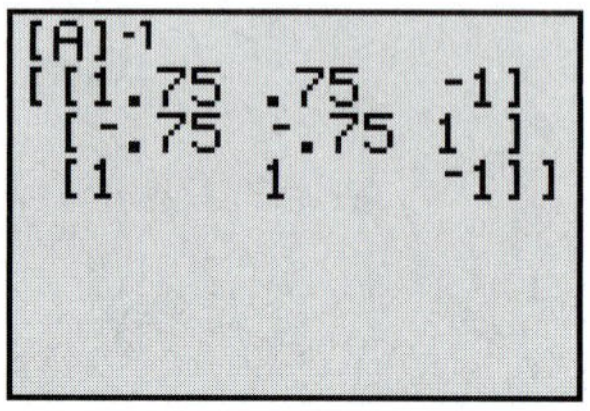

```
[A]-1
[[1.75 .75  -1]
 [-.75 -.75 1 ]
 [1    1    -1]]
```

You can (and should) verify that this is the correct inverse by showing that $AA^{-1} = A^{-1}A = I_3$.

━━━ **NOW WORK PROBLEM 23.**

If transforming the matrix $[A|I_n]$ into reduced row echelon form does not result in the identity matrix I_n to the left of the vertical bar, then A is singular and has no inverse. The next example demonstrates such a matrix.

EXAMPLE 14 **Showing That a Matrix Has No Inverse**

Show that the following matrix has no inverse.

$$A = \begin{bmatrix} 4 & 6 \\ 2 & 3 \end{bmatrix}$$

Algebraic Solution Proceeding as in Example 13, we form the matrix

$$[A|I_2] = \begin{bmatrix} 4 & 6 & | & 1 & 0 \\ 2 & 3 & | & 0 & 1 \end{bmatrix}$$

Then we use row operations to transform $[A|I_2]$ into reduced row echelon form.

$$[A|I_2] = \begin{bmatrix} 4 & 6 & | & 1 & 0 \\ 2 & 3 & | & 0 & 1 \end{bmatrix} \rightarrow \begin{bmatrix} 1 & \dfrac{3}{2} & | & \dfrac{1}{4} & 0 \\[2mm] 2 & 3 & | & 0 & 1 \end{bmatrix} \rightarrow \begin{bmatrix} 1 & \dfrac{3}{2} & | & \dfrac{1}{4} & 0 \\[2mm] 0 & 0 & | & -\dfrac{1}{2} & 1 \end{bmatrix}$$

$$R_1 = \frac{1}{4}r_1 \qquad\qquad R_2 = -2r_1 + r_2$$

The matrix $[A|I_2]$ is sufficiently reduced for us to see that the identity matrix cannot appear to the left of the vertical bar. We conclude that A is singular and so has no inverse.

Graphing Solution   Enter the matrix A. Figure 15 shows the result when we try to find its inverse. The ERRor comes about because A is singular.

Figure 15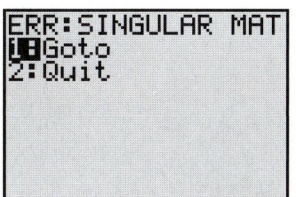

 SEEING THE CONCEPT   Compute the determinant of A in Example 14 using a graphing utility. What is the result? Are you surprised?

 **NOW WORK PROBLEM 51.**

Solving Systems of Linear Equations

⑤   Inverse matrices can be used to solve systems of equations in which the number of equations is the same as the number of variables.

EXAMPLE 15   ### Using the Inverse Matrix to Solve a System of Linear Equations

Solve the system of equations:
$$\begin{cases} x + y = 3 \\ -x + 3y + 4z = -3 \\ 4y + 3z = 2 \end{cases}$$

Solution   If we let

$$A = \begin{bmatrix} 1 & 1 & 0 \\ -1 & 3 & 4 \\ 0 & 4 & 3 \end{bmatrix} \qquad X = \begin{bmatrix} x \\ y \\ z \end{bmatrix} \qquad B = \begin{bmatrix} 3 \\ -3 \\ 2 \end{bmatrix}$$

then the original system of equations can be written compactly as the matrix equation

$$AX = B \qquad (4)$$

We know from Example 13 that the matrix A has the inverse A^{-1}, so we multiply each side of equation (4) by A^{-1}.

$$AX = B$$
$$A^{-1}(AX) = A^{-1}B \qquad \text{Multiply both sides by } A^{-1}.$$
$$(A^{-1}A)X = A^{-1}B \qquad \text{Associative property of multiplication}$$
$$I_3 X = A^{-1}B \qquad \text{Definition of inverse matrix}$$
$$X = A^{-1}B \qquad \text{Property of identity matrix} \qquad (5)$$

Now we use (5) to find $X = \begin{bmatrix} x \\ y \\ z \end{bmatrix}$.

Algebraic Solution

$$X = \begin{bmatrix} x \\ y \\ z \end{bmatrix} = A^{-1}B = \begin{bmatrix} \dfrac{7}{4} & \dfrac{3}{4} & -1 \\ -\dfrac{3}{4} & -\dfrac{3}{4} & 1 \\ 1 & 1 & -1 \end{bmatrix} \begin{bmatrix} 3 \\ -3 \\ 2 \end{bmatrix} = \begin{bmatrix} 1 \\ 2 \\ -2 \end{bmatrix}$$

Example 13

Thus, $x = 1$, $y = 2$, $z = -2$.

Graphing Solution   Enter the matrices A and B into a graphing utility. Figure 16 shows the solution to the system of equations.

Figure 16

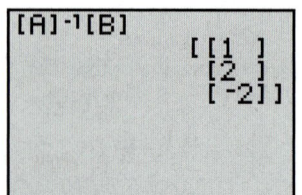

```
[A]⁻¹[B]
            [[1 ]
             [2 ]
             [-2]]
```

The method used in Example 15 to solve a system of equations is particularly useful when it is necessary to solve several systems of equations in which the constants appearing to the right of the equal signs change, while the coefficients of the variables on the left side remain the same. See Problems 31–50 for some illustrations. Be careful; this method can only be used if the inverse exists. If it does not exist, row reduction must be used since the system is either inconsistent or dependent.

HISTORICAL FEATURE

Arthur Cayley (1821–1895)

Matrices were invented in 1857 by Arthur Cayley (1821–1895) as a way of efficiently computing the result of substituting one linear system into another (see Historical Problem 2). The resulting system had incredible richness, in the sense that a very wide variety of mathematical systems could be mimicked by the matrices. Cayley and his friend James J. Sylvester (1814–1897) spent much of the rest of their lives elaborating the theory. The torch was then passed to Georg Frobenius (1849–1917), whose deep investigations established a central place for matrices in modern mathematics. In 1924, rather to the surprise of physicists, it was found that matrices (with complex numbers in them) were exactly the right tool for describing the behavior of atomic systems. Today, matrices are used in a wide variety of applications.

HISTORICAL PROBLEMS

1. *Matrices and Complex Numbers* Frobenius emphasized in his research how matrices could be used to mimic other mathematical systems. Here, we mimic the behavior of complex numbers using matrices. Mathematicians call such a relationship an *isomorphism*.

   Complex number ⟷ Matrix

   $$a + bi \longleftrightarrow \begin{bmatrix} a & b \\ -b & a \end{bmatrix}$$

   Note that the complex number can be read off the top line of the matrix. Thus,

   $$2 + 3i \longleftrightarrow \begin{bmatrix} 2 & 3 \\ -3 & 2 \end{bmatrix} \quad \text{and} \quad \begin{bmatrix} 4 & -2 \\ 2 & 4 \end{bmatrix} \longleftrightarrow 4 - 2i$$

   (a) Find the matrices corresponding to $2 - 5i$ and $1 + 3i$.

   (b) Multiply the two matrices.

   (continued)

(c) Find the corresponding complex number for the matrix found in part (b).

(d) Multiply $2 - 5i$ and $1 + 3i$. The result should be the same as that found in part (c).

The process also works for addition and subtraction. Try it for yourself.

2. *Cayley's Definition of Matrix Multiplication* Cayley invented matrix multiplication to simplify the following problem:

$$\begin{cases} u = ar + bs \\ v = cr + ds \end{cases} \qquad \begin{cases} x = ku + lv \\ y = mu + nv \end{cases}$$

(a) Find x and y in terms of r and s by substituting u and v from the first system of equations into the second system of equations.

(b) Use the result of part (a) to find the 2 by 2 matrix A in

$$\begin{bmatrix} x \\ y \end{bmatrix} = A \begin{bmatrix} r \\ s \end{bmatrix}$$

(c) Now look at the following way to do it. Write the equations in matrix form.

$$\begin{bmatrix} u \\ v \end{bmatrix} = \begin{bmatrix} a & b \\ c & d \end{bmatrix} \begin{bmatrix} r \\ s \end{bmatrix} \qquad \begin{bmatrix} x \\ y \end{bmatrix} = \begin{bmatrix} k & l \\ m & n \end{bmatrix} \begin{bmatrix} u \\ v \end{bmatrix}$$

So

$$\begin{bmatrix} x \\ y \end{bmatrix} = \begin{bmatrix} k & l \\ m & n \end{bmatrix} \begin{bmatrix} a & b \\ c & d \end{bmatrix} \begin{bmatrix} r \\ s \end{bmatrix}$$

Do you see how Cayley defined matrix multiplication?

7.5 Concepts and Vocabulary

In Problems 1–3, fill in the blanks.

1. A matrix B, for which $AB = I_n$, the identity matrix, is called the _____ of A.

2. A matrix that has the same number of rows as columns is called a(n) _____ matrix.

3. In the algebra of matrices, the matrix that has properties similar to the number 1 is called the _____ matrix.

In Problems 4–6, answer True or False for each statement.

4. Every square matrix has an inverse.

5. Matrix multiplication is commutative.

6. Any pair of matrices can be multiplied.

7. Under what circumstances can two matrices be multiplied?

8. What does it mean if a matrix is singular? If the coefficient matrix in a system of equations is singular, how many solutions might the system have?

7.5 Exercises

In Problems 1–16, use the following matrices to compute the given expression (a) by hand and (b) by using a graphing utility.

$$A = \begin{bmatrix} 0 & 3 & -5 \\ 1 & 2 & 6 \end{bmatrix} \qquad B = \begin{bmatrix} 4 & 1 & 0 \\ -2 & 3 & -2 \end{bmatrix} \qquad C = \begin{bmatrix} 4 & 1 \\ 6 & 2 \\ -2 & 3 \end{bmatrix}$$

1. $A + B$
2. $A - B$
3. $4A$
4. $-3B$
5. $3A - 2B$
6. $2A + 4B$
7. AC
8. BC
9. CA
10. CB
11. $C(A + B)$
12. $(A + B)C$
13. $AC - 3I_2$
14. $CA + 5I_3$
15. $CA - CB$
16. $AC + BC$

In Problems 17–20, compute each product (a) by hand and (b) by using a graphing utility.

17. $\begin{bmatrix} 2 & -2 \\ 1 & 0 \end{bmatrix} \begin{bmatrix} 2 & 1 & 4 & 6 \\ 3 & -1 & 3 & 2 \end{bmatrix}$

18. $\begin{bmatrix} 4 & 1 \\ 2 & 1 \end{bmatrix} \begin{bmatrix} -6 & 6 & 1 & 0 \\ 2 & 5 & 4 & -1 \end{bmatrix}$

19. $\begin{bmatrix} 1 & 0 & 1 \\ 2 & 4 & 1 \\ 3 & 6 & 1 \end{bmatrix} \begin{bmatrix} 1 & 3 \\ 6 & 2 \\ 8 & -1 \end{bmatrix}$

20. $\begin{bmatrix} 4 & -2 & 3 \\ 0 & 1 & 2 \\ -1 & 0 & 1 \end{bmatrix} \begin{bmatrix} 2 & 6 \\ 1 & -1 \\ 0 & 2 \end{bmatrix}$

In Problems 21–30, each matrix is nonsingular. Find the inverse of each matrix. Be sure to check your answer using a graphing utility (when possible).

21. $\begin{bmatrix} 2 & 1 \\ 1 & 1 \end{bmatrix}$ **22.** $\begin{bmatrix} 3 & -1 \\ -2 & 1 \end{bmatrix}$ ✎ **23.** $\begin{bmatrix} 6 & 5 \\ 2 & 2 \end{bmatrix}$ **24.** $\begin{bmatrix} -4 & 1 \\ 6 & -2 \end{bmatrix}$ **25.** $\begin{bmatrix} 2 & 1 \\ a & a \end{bmatrix}, \; a \neq 0$

26. $\begin{bmatrix} b & 3 \\ b & 2 \end{bmatrix}, \; b \neq 0$ **27.** $\begin{bmatrix} 1 & -1 & 1 \\ 0 & -2 & 1 \\ -2 & -3 & 0 \end{bmatrix}$ **28.** $\begin{bmatrix} 1 & 0 & 2 \\ -1 & 2 & 3 \\ 1 & -1 & 0 \end{bmatrix}$ **29.** $\begin{bmatrix} 1 & 1 & 1 \\ 3 & 2 & -1 \\ 3 & 1 & 2 \end{bmatrix}$ **30.** $\begin{bmatrix} 3 & 3 & 1 \\ 1 & 2 & 1 \\ 2 & -1 & 1 \end{bmatrix}$

In Problems 31–50, use the inverses found in Problems 21–30, to solve each system of equations by hand.

31. $\begin{cases} 2x + y = 8 \\ x + y = 5 \end{cases}$ **32.** $\begin{cases} 3x - y = 8 \\ -2x + y = 4 \end{cases}$ **33.** $\begin{cases} 2x + y = 0 \\ x + y = 5 \end{cases}$ **34.** $\begin{cases} 3x - y = 4 \\ -2x + y = 5 \end{cases}$

35. $\begin{cases} 6x + 5y = 7 \\ 2x + 2y = 2 \end{cases}$ **36.** $\begin{cases} -4x + y = 0 \\ 6x - 2y = 14 \end{cases}$ **37.** $\begin{cases} 6x + 5y = 13 \\ 2x + 2y = 5 \end{cases}$ **38.** $\begin{cases} -4x + y = 5 \\ 6x - 2y = -9 \end{cases}$

39. $\begin{cases} 2x + y = -3 \\ ax + ay = -a \end{cases}, \; a \neq 0$ **40.** $\begin{cases} bx + 3y = 2b + 3 \\ bx + 2y = 2b + 2 \end{cases}, \; b \neq 0$ **41.** $\begin{cases} 2x + y = \dfrac{7}{a} \\ ax + ay = 5 \end{cases}, \; a \neq 0$

42. $\begin{cases} bx + 3y = 14 \\ bx + 2y = 10 \end{cases}, \; b \neq 0$ **43.** $\begin{cases} x - y + z = 0 \\ -2y + z = -1 \\ -2x - 3y = -5 \end{cases}$ **44.** $\begin{cases} x + 2z = 6 \\ -x + 2y + 3z = -5 \\ x - y = 6 \end{cases}$

45. $\begin{cases} x - y + z = 2 \\ -2y + z = 2 \\ -2x - 3y = \dfrac{1}{2} \end{cases}$ **46.** $\begin{cases} x + 2z = 2 \\ -x + 2y + 3z = -\dfrac{3}{2} \\ x - y = 2 \end{cases}$ **47.** $\begin{cases} x + y + z = 9 \\ 3x + 2y - z = 8 \\ 3x + y + 2z = 1 \end{cases}$

48. $\begin{cases} 3x + 3y + z = 8 \\ x + 2y + z = 5 \\ 2x - y + z = 4 \end{cases}$ **49.** $\begin{cases} x + y + z = 2 \\ 3x + 2y - z = \dfrac{7}{3} \\ 3x + y + 2z = \dfrac{10}{3} \end{cases}$ **50.** $\begin{cases} 3x + 3y + z = 1 \\ x + 2y + z = 0 \\ 2x - y + z = 4 \end{cases}$

In Problems 51–56, by hand, show that each matrix has no inverse. Verify your result using a graphing utility.

✎ **51.** $\begin{bmatrix} 4 & 2 \\ 2 & 1 \end{bmatrix}$ **52.** $\begin{bmatrix} -3 & \dfrac{1}{2} \\ 6 & -1 \end{bmatrix}$ **53.** $\begin{bmatrix} 15 & 3 \\ 10 & 2 \end{bmatrix}$

54. $\begin{bmatrix} -3 & 0 \\ 4 & 0 \end{bmatrix}$ **55.** $\begin{bmatrix} -3 & 1 & -1 \\ 1 & -4 & -7 \\ 1 & 2 & 5 \end{bmatrix}$ **56.** $\begin{bmatrix} 1 & 1 & -3 \\ 2 & -4 & 1 \\ -5 & 7 & 1 \end{bmatrix}$

In Problems 57–60, use a graphing utility to find the inverse, if it exists, of each matrix. Round answers to two decimal places.

57. $\begin{bmatrix} 25 & 61 & -12 \\ 18 & -2 & 4 \\ 8 & 35 & 21 \end{bmatrix}$ **58.** $\begin{bmatrix} 18 & -3 & 4 \\ 6 & -20 & 14 \\ 10 & 25 & -15 \end{bmatrix}$ **59.** $\begin{bmatrix} 44 & 21 & 18 & 6 \\ -2 & 10 & 15 & 5 \\ 21 & 12 & -12 & 4 \\ -8 & -16 & 4 & 9 \end{bmatrix}$ **60.** $\begin{bmatrix} 16 & 22 & -3 & 5 \\ 21 & -17 & 4 & 8 \\ 2 & 8 & 27 & 20 \\ 5 & 15 & -3 & -10 \end{bmatrix}$

In Problems 61–64, use the idea behind Example 15 with a graphing utility to solve the following systems of equations. Round answers to two decimal places.

61. $\begin{cases} 25x + 61y - 12z = 10 \\ 18x - 12y + 7z = -9 \\ 3x + 4y - z = 12 \end{cases}$ **62.** $\begin{cases} 25x + 61y - 12z = 15 \\ 18x - 12y + 7z = -3 \\ 3x + 4y - z = 12 \end{cases}$ **63.** $\begin{cases} 25x + 61y - 12z = 21 \\ 18x - 12y + 7z = 7 \\ 3x + 4y - z = -2 \end{cases}$ **64.** $\begin{cases} 25x + 61y - 12z = 25 \\ 18x - 12y + 7z = 10 \\ 3x + 4y - z = -4 \end{cases}$

---

65. Computing the Cost of Production   The Acme Steel Company is a producer of stainless steel and aluminum containers. On a certain day, the following stainless steel containers were manufactured: 500 with 10-gallon capacity, 350 with 5-gallon capacity, and 400 with 1-gallon capacity. On the same day, the following aluminum containers were manufactured: 700 with 10-gallon capacity, 500 with 5-gallon capacity, and 850 with 1-gallon capacity.

(a) Find a 2 by 3 matrix representing the above data. Find a 3 by 2 matrix to represent the same data.
(b) If the amount of material used in the 10-gallon containers is 15 pounds, the amount used in the 5-gallon containers is 8 pounds, and the amount used in the 1-gallon containers is 3 pounds, find a 3 by 1 matrix representing the amount of material.
(c) Multiply the 2 by 3 matrix found in part (a) and the 3 by 1 matrix found in part (b) to get a 2 by 1 matrix showing the day's usage of material.
(d) If stainless steel costs Acme $0.10 per pound and aluminum costs $0.05 per pound, find a 1 by 2 matrix representing cost.
(e) Multiply the matrices found in parts (c) and (d) to determine what the total cost of the day's production was.

66. Computing Profit   Rizza Ford has two locations, one in the city and the other in the suburbs. In January, the city location sold 400 subcompacts, 250 intermediate-size cars, and 50 station wagons; in February, it sold 350 subcompacts, 100 intermediates, and 30 station wagons. At the suburban location in January, 450 subcompacts, 200 intermediates, and 140 station wagons were sold. In February, the suburban location sold 350 subcompacts, 300 intermediates, and 100 station wagons.

(a) Find 2 by 3 matrices that summarize the sales data for each location for January and February (one matrix for each month).
(b) Use matrix addition to obtain total sales for the two-month period.
(c) The profit on each kind of car is $100 per subcompact, $150 per intermediate, and $200 per station wagon. Find a 3 by 1 matrix representing this profit.
(d) Multiply the matrices found in parts (b) and (c) to get a 2 by 1 matrix showing the profit at each location.

67. Make up a situation different from any found in the text that can be represented by a matrix.

PREPARING FOR THIS SECTION

Before getting started, review the following:

✓ Solving Inequalities (Section 1.6, pp. 149–155)   ✓ Lines (Section 1.7, pp. 163–172)

7.6   SYSTEMS OF LINEAR INEQUALITIES; LINEAR PROGRAMMING

OBJECTIVES
1 Graph a Linear Inequality by Hand
2 Graph a Linear Inequality Using a Graphing Utility
3 Graph a System of Linear Inequalities
4 Set up a Linear Programming Problem
5 Solve a Linear Programming Problem

In Chapter 1, we discussed inequalities in one variable. In this section, we discuss linear inequalities in two variables.

Linear inequalities in two variables are inequalities in one of the forms

$$Ax + By < C \quad Ax + By > C \quad Ax + By \leq C \quad Ax + By \geq C$$

where A and B are not both zero.

Figure 17

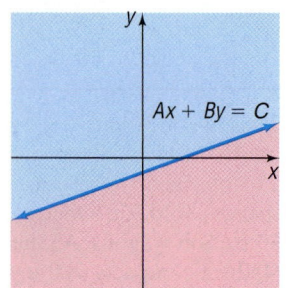

The graph of the corresponding equation of a linear inequality is a line, which separates the xy-plane into two regions, called **half-planes**. See Figure 17.

As shown, $Ax + By = C$ is the equation of the boundary line and it divides the plane into two half-planes, one for which $Ax + By < C$ and the other for which $Ax + By > C$.

1

A linear inequality in two variables x and y is **satisfied** by an ordered pair (a, b) if, when x is replaced by a and y by b, a true statement results. A **graph of a linear inequality in two variables** x and y consists of all points (x, y) whose coordinates satisfy the inequality.

Let's look at an example.

EXAMPLE 1   **Graphing a Linear Inequality by Hand**

Graph the linear inequality: $3x + y \le 6$

Solution   We begin by graphing the linear equation

$$3x + y = 6$$

formed by replacing (for now) the $\le$ symbol with an $=$ sign. The graph of the linear equation is a line. See Figure 18(a). This line is part of the graph of the inequality that we seek because the inequality is nonstrict. (Do you see why? We are seeking points for which $3x + y$ is less than *or equal to* 6.)

Figure 18

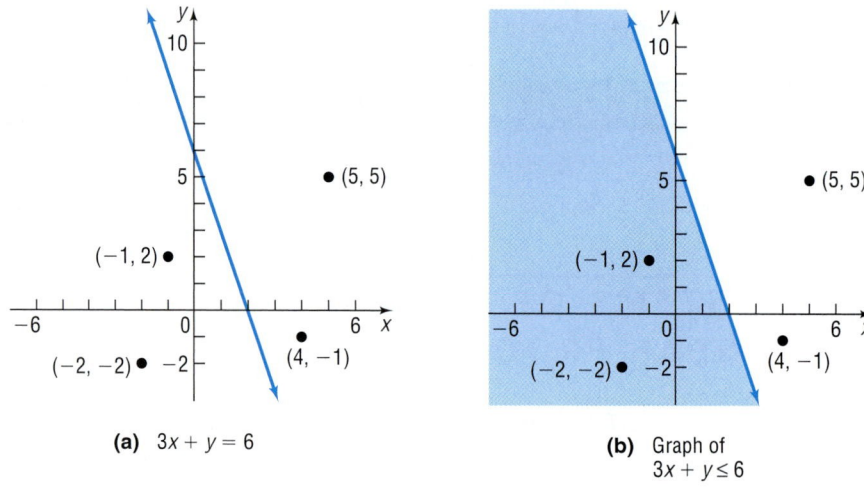

(a) $3x + y = 6$

(b) Graph of $3x + y \le 6$

Now let's test a few randomly selected points to see whether they belong to the graph of the inequality.

	$3x + y \le 6$	**Conclusion**
$(4, -1)$	$3(4) + (-1) = 11 > 6$	Does not belong to graph
$(5, 5)$	$3(5) + 5 = 20 > 6$	Does not belong to graph
$(-1, 2)$	$3(-1) + 2 = -1 \le 6$	Belongs to graph
$(-2, -2)$	$3(-2) + (-2) = -8 \le 6$	Belongs to graph

Look again at Figure 18(a). Notice that the two points that belong to the graph both lie on the same side of the line, and the two points that do not belong to the graph lie on the opposite side. As it turns out, this is always the case. Thus, the graph we seek consists of all points that lie on the same side of the line as do $(-1, 2)$ and $(-2, -2)$, that is, the shaded region in Figure 18(b).   ■

The graph of any linear inequality in two variables may be obtained by graphing the equation corresponding to the inequality, using dashes if the inequality is strict ($<$ or $>$) and a solid line if it is nonstrict ($\leq$ or $\geq$). This graph will separate the xy-plane into two half-planes. In each half-plane either all points satisfy the inequality or no points satisfy the inequality. So the use of a single test point is all that is required to determine whether the points of that half-plane are part of the graph or not. The steps to follow are given next.

Steps for Graphing an Inequality by Hand

STEP 1: Replace the inequality symbol by an equal sign and graph the resulting equation. If the inequality is strict, use dashes; if it is nonstrict, use a solid mark. This graph separates the xy-plane into two half-planes.

STEP 2: Select a test point P in one of the half-planes.

(a) If the coordinates of P satisfy the inequality, then so do all the points in that half-plane. Indicate this by shading the half-plane.

(b) If the coordinates of P do not satisfy the inequality, then none of the points in that half-plane do, so shade the opposite half-plane.

Graphing utilities can also be used to graph linear inequalities. The steps to follow to graph an inequality using a graphing utility are given next.

Steps for Graphing an Inequality Using a Graphing Utility

STEP 1: Replace the inequality symbol by an equal sign and graph the resulting equation.

STEP 2: Select a test point P in one of the half-planes.

(a) Use a graphing utility to determine if the test point P satisfies the inequality. If the test point satisfies the inequality, then so do all the points in this half-plane. Indicate this by using the graphing utility to shade the half-plane.

(b) If the coordinates of P do not satisfy the inequality, then none of the points in that half-plane do, so shade the opposite half-plane.

EXAMPLE 2   **Graphing a Linear Inequality Using a Graphing Utility**

Use a graphing utility to graph $3x + y \leq 6$.

Solution **Step 1:** We begin by graphing the equation $3x + y = 6$ ($Y_1 = -3x + 6$). See Figure 19.

Figure 19

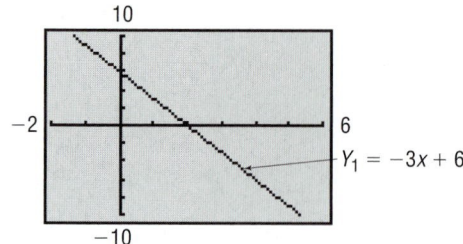

$Y_1 = -3x + 6$

Step 2: Select a test point in one of the half-planes and determine whether it satisfies the inequality. To test the point $(-1, 2)$, for example, enter $3(-1) + 2 \leq 6$. See Figure 20(a). The 1 that appears indicates that the statement entered (the inequality) is true. When the point $(5, 5)$ is tested, a 0 appears, indicating that the statement entered is false. Thus, $(-1, 2)$ is a part of the graph of the inequality and $(5, 5)$ is not, so we shade the half-plane below Y_1. Figure 20(b) shows the graph of the inequality on a TI-83.*

Figure 20

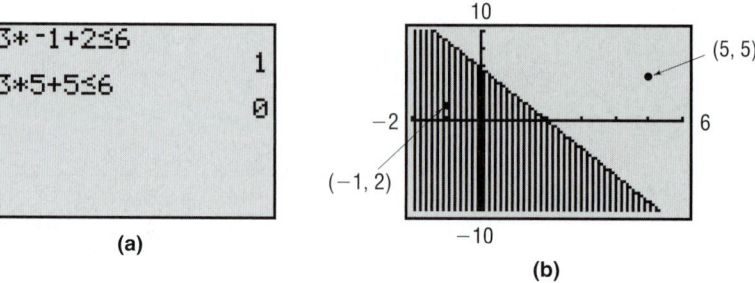

(a)

(b)

NOW WORK PROBLEM **5.**

EXAMPLE 3 **Graphing Linear Inequalities by Hand**

Graph: (a) $y < 2$ (b) $y \geq 2x$

Solution (a) The graph of the equation $y = 2$ is a horizontal line and is not part of the graph of the inequality, so we use a dashed line. Since $(0, 0)$ satisfies the inequality, the graph consists of the half-plane below the line $y = 2$. See Figure 21.

(b) The graph of the equation $y = 2x$ is a line and is part of the graph of the inequality, so we use a solid line. Using $(3, 0)$ as a test point, we find that it does not satisfy the inequality $[0 < 2 \cdot 3]$. Points in the half-plane on the opposite side of $y = 2x$ satisfy the inequality. See Figure 22.

*Consult your owner's manual for shading techniques.

Figure 21

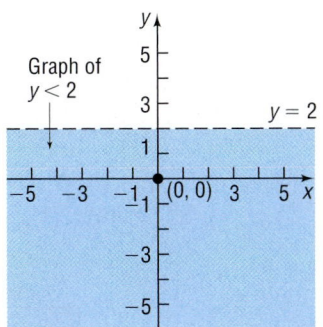

Figure 22

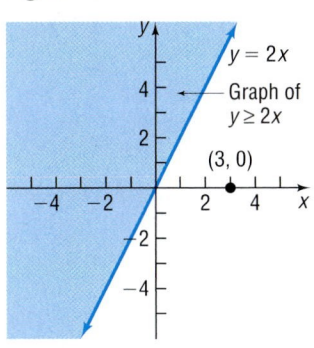

NOW WORK PROBLEM **3.**

Systems of Linear Inequalities in Two Variables

③ The **graph of a system of linear inequalities** in two variables x and y is the set of all points (x, y) that simultaneously satisfies *each* of the inequalities in the system. The graph of a system of linear inequalities can be obtained by graphing each inequality individually and then determining where, if at all, they intersect.

EXAMPLE 4 **Graphing a System of Linear Inequalities by Hand**

Graph the system: $\begin{cases} x + y \geq 2 \\ 2x - y \leq 4 \end{cases}$

Solution First, we graph the inequality $x + y \geq 2$ as the shaded region in Figure 23(a). Next, we graph the inequality $2x - y \leq 4$ as the shaded region in Figure 23(b). Now superimpose the two graphs, as shown in Figure 23(c). The points that are in both shaded regions [the overlapping, purple region in Figure 23(c)] are the solutions we seek to the system, because they simultaneously satisfy each linear inequality.

Figure 23

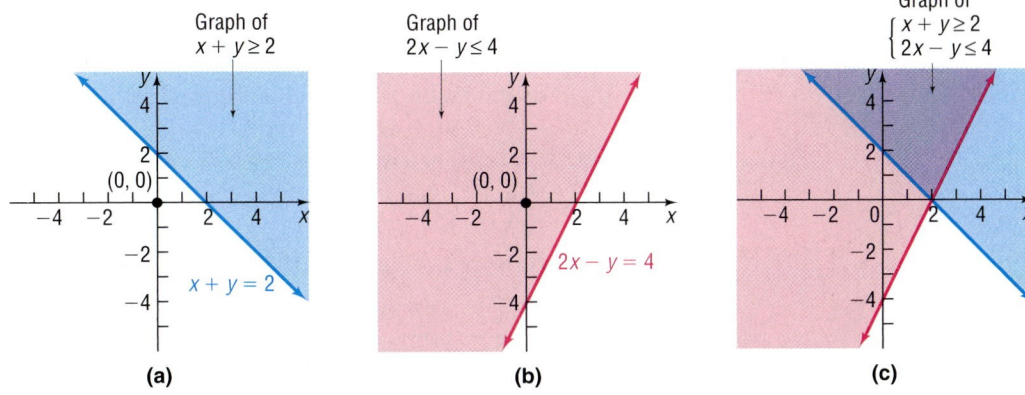

(a)   (b)   (c)

EXAMPLE 5   **Graphing a System of Linear Inequalities Using a Graphing Utility**

Graph the system: $\begin{cases} x + y \geq 2 \\ 2x - y \leq 4 \end{cases}$

Solution   First, we graph the lines $x + y = 2$ ($Y_1 = -x + 2$) and $2x - y = 4$ ($Y_2 = 2x - 4$). See Figure 24.

Figure 24

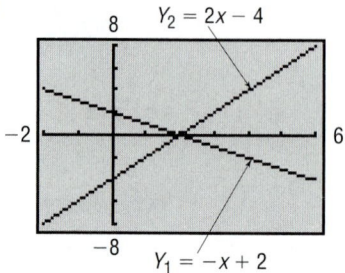

Notice that the graphs divide the viewing window into four regions. We select a test point for each region and determine whether the point makes *both* inequalities true. We choose to test $(0, 0)$, $(2, 3)$, $(4, 0)$, and $(2, -2)$. Figure 25(a) shows that $(2, 3)$ is the only point for which both inequalities are true. We obtain the graph shown in Figure 25(b).

Figure 25

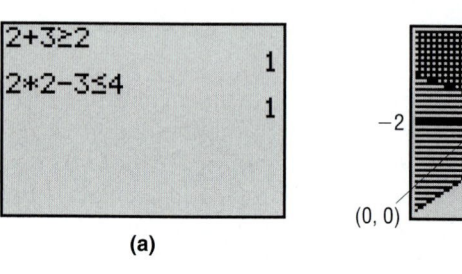

(a)

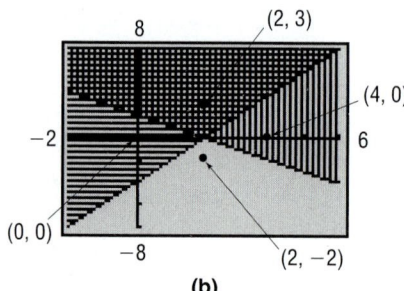

(b)

Rather than testing four points, we could just test the point $(0, 0)$ on each inequality. For example, $(0, 0)$ does not satisfy $x + y \geq 2$, so we shade above the line $x + y = 2$. In addition, $(0, 0)$ does satisfy $2x - y \leq 4$, so we shade above the line $2x - y = 4$. The intersection of the shaded regions gives us the result presented in Figure 25(b).

 **NOW WORK PROBLEM 11.**

EXAMPLE 6   **Graphing a System of Linear Inequalities by Hand**

Graph the system: $\begin{cases} x + y \leq 2 \\ x + y \geq 0 \end{cases}$

Solution   See Figure 26. The overlapping purple-shaded region between the two boundary lines is the graph of the system.

Figure 26 $x + y = 0$ $x + y = 2$

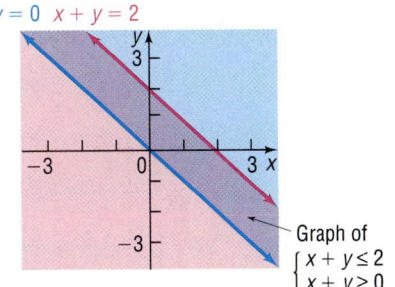

Graph of
$\begin{cases} x + y \leq 2 \\ x + y \geq 0 \end{cases}$

NOW WORK PROBLEM **15.**

EXAMPLE 7   **Graphing a System of Linear Inequalities by Hand**

Graph the system: $\begin{cases} 2x - y \geq 0 \\ 2x - y \geq 2 \end{cases}$

Solution   See Figure 27. The overlapping purple-shaded region is the graph of the system. Note that the graph of the system is identical to the graph of the single inequality $2x - y \geq 2$.

Figure 27

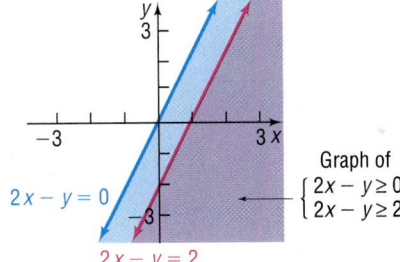

$2x - y = 0$

$2x - y = 2$

Graph of
$\begin{cases} 2x - y \geq 0 \\ 2x - y \geq 2 \end{cases}$

EXAMPLE 8   **Graphing a System of Linear Inequalities by Hand**

Graph the system: $\begin{cases} x + 2y \leq 2 \\ x + 2y \geq 6 \end{cases}$

Solution   See Figure 28. Because no overlapping region results, there are no points in the xy-plane that simultaneously satisfy each inequality. Hence, the system has no solution.

Figure 28

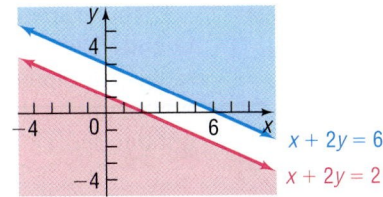

$x + 2y = 6$

$x + 2y = 2$

EXAMPLE 9 **Graphing a System of Four Linear Inequalities**

Graph the system:
$$\begin{cases} x + y \leq 3 \\ 2x + y \leq 4 \\ x \geq 0 \\ y \geq 0 \end{cases}$$

Solution The two inequalities $x \geq 0$ and $y \geq 0$ require that the graph be in quadrant I. We set our viewing window accordingly. Figure 29 shows the graph of $x + y = 3$ ($Y_1 = -x + 3$) and $2x + y = 4$ ($Y_2 = -2x + 4$). The graphs divide the viewing window into four regions.

Figure 29

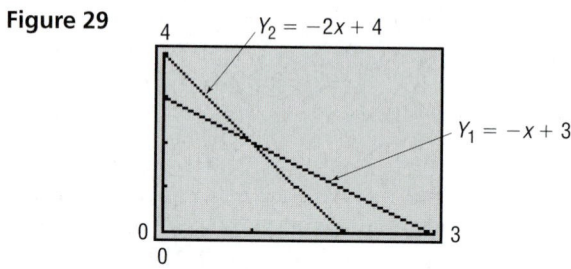

The graph of the inequality $x + y \leq 3$ consists of the half-plane below Y_1, and the graph of $2x + y \leq 4$ consists of the half-plane below Y_2, so we shade accordingly. See Figures 30(a) and (b).

Figure 30

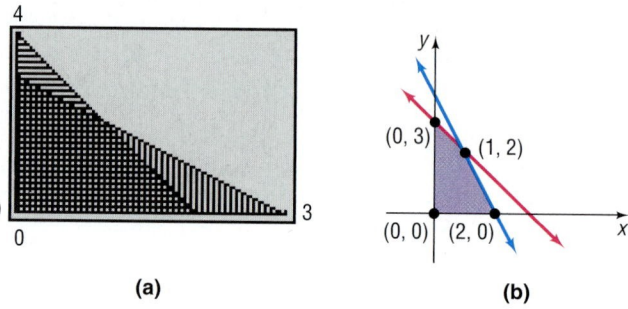

(a)   (b)

Notice in Figure 30(b) that those points belonging to the graph that are also points of intersection of boundary lines have been plotted. Such points are referred to as **vertices** or **corner points** of the graph. The corner points $(0, 0)$, $(0, 3)$, $(1, 2)$ and $(2, 0)$ are found by solving the system of equations formed by the two boundary lines. So the corner points are found by solving the following systems:

$$\begin{cases} x = 0 \\ y = 0 \end{cases} \qquad \begin{cases} x + y = 3 \\ x = 0 \end{cases} \qquad \begin{cases} x + y = 3 \\ 2x + y = 4 \end{cases} \qquad \begin{cases} 2x + y = 4 \\ y = 0 \end{cases}$$

Corner point: $(0, 0)$ Corner point: $(0, 3)$ Corner point: $(1, 2)$ Corner point: $(2, 0)$

The graph of the system of linear inequalities in Figure 30 is said to be **bounded**, because it can be contained within some circle of sufficiently large radius. A graph that cannot be contained in any circle is said to be **unbounded**. For example, the graph of the system of linear inequalities in Figure 27 is unbounded, since it extends indefinitely in a particular direction.

NOW WORK PROBLEM 23.

Linear Programming

4 Historically, linear programming evolved as a technique for solving problems involving resource allocation of goods and materials for the U.S. Air Force during World War II. Today, linear programming techniques are used to solve a wide variety of problems, such as optimizing airline scheduling and establishing telephone lines. Although most practical linear programming problems involve systems of several hundred linear inequalities containing several hundred variables, we will limit our discussion to problems containing only two variables, because we can solve such problems using graphing techniques.*

EXAMPLE 10　**Financial Planning**

A retired couple has up to $25,000 to invest. As their financial adviser, you recommend that they place at least $15,000 in Treasury bills yielding 6% and at most $5000 in corporate bonds yielding 9%. How much money should be placed in each investment so that income is maximized?

The problem given here is typical of a *linear programming problem*. The problem requires that a certain linear expression, the income, be maximized. If I represents income, x the amount invested in Treasury bills at 6%, and y the amount invested in corporate bonds at 9%, then

$$I = 0.06x + 0.09y$$

We shall assume x and y are in thousands of dollars.

The linear expression $I = 0.06x + 0.09y$ is called the **objective function**. Furthermore, the problem requires that the maximum income be achieved under certain conditions or **constraints**, each of which is a linear inequality involving the variables.

The linear programming problem given in Example 10 may be restated as

Maximize:　$I = 0.06x + 0.09y$

subject to the conditions that

$$x \geq 0, \quad y \geq 0$$
$$x + y \leq 25 \quad \text{Up to \$25,000 to invest}$$
$$x \geq 15 \quad \text{At least \$15,000 in Treasury bills}$$
$$y \leq 5 \quad \text{At most \$5000 in corporate bonds}$$

In general, every linear programming problem has two components:

1. A linear objective function that is to be maximized or minimized
2. A collection of linear inequalities that must be satisfied simultaneously

---

A **linear programming problem** in two variables x and y consists of maximizing (or minimizing) a linear objective function

$$z = Ax + By$$

where A and B are real numbers, not both 0, subject to certain conditions, or constraints, expressible as linear inequalities in x and y.

---

*The **simplex method** is a way to solve linear programming problems involving many inequalities and variables. This method was developed by George Dantzig in 1946 and is particularly well suited for computerization. In 1984, Narendra Karmarkar of Bell Laboratories discovered a way of solving large linear programming problems that improves on the simplex method.

To maximize (or minimize) the quantity $z = Ax + By$, we need to identify points (x, y) that make the expression for z the largest (or smallest) possible. But not all points (x, y) are eligible; only those that also satisfy each linear inequality (constraint) can be used. We refer to each point (x, y) that satisfies the system of linear inequalities (the constraints) as a **feasible point**. In a linear programming problem, we seek the feasible point(s) that maximizes (or minimizes) the objective function.

Let's look again at the linear programming problem in Example 10.

EXAMPLE 11 **Analyzing a Linear Programming Problem**

Consider the following linear programming problem:

$$\text{Maximize:} \quad I = 0.06x + 0.09y$$

subject to the conditions that

$$x \geq 0, \quad y \geq 0$$
$$x + y \leq 25$$
$$x \geq 15$$
$$y \leq 5$$

Graph the constraints. Then graph the objective function for $I = 0, 0.9, 1.35, 1.65,$ and 1.8.

Solution   Figure 31 shows the graph of the constraints. We superimpose on this graph the graph of the objective function for the given values of I.

For $I = 0$, the objective function is the line $0 = 0.06x + 0.09y$.

For $I = 0.9$, the objective function is the line $0.9 = 0.06x + 0.09y$.

For $I = 1.35$, the objective function is the line $1.35 = 0.06x + 0.09y$.

For $I = 1.65$, the objective function is the line $1.65 = 0.06x + 0.09y$.

For $I = 1.8$, the objective function is the line $1.8 = 0.06x + 0.09y$.

Figure 31

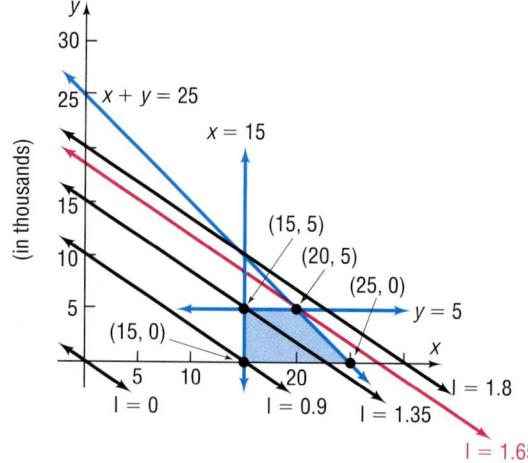

A **solution** to a linear programming problem consists of the feasible point(s) that maximizes (or minimizes) the objective function, together with the corresponding value of the objective function.

⑤   One condition for a linear programming problem in two variables to have a solution is that the graph of the feasible points be bounded. (Refer to page 584.)

If none of the feasible points maximizes (or minimizes) the objective function or if there are no feasible points, then the linear programming problem has no solution.

Consider the linear programming problem stated in Example 11, and look again at Figure 31. The feasible points are the points that lie in the shaded region. For example, $(20, 3)$ is a feasible point, as is $(15, 5)$, $(20, 5)$, $(18, 4)$, and so on. To find the solution of the problem requires that we find a feasible point (x, y) that makes $I = 0.06x + 0.09y$ as large as possible. Notice that, as I increases in value from $I = 0$ to $I = 0.9$ to $I = 1.35$ to $I = 1.65$ to $I = 1.8$, we obtain a collection of parallel lines. Furthermore, notice that the largest value of I that can be obtained while feasible points are present is $I = 1.65$, which corresponds to the line $1.65 = 0.06x + 0.09y$. Any larger value of I results in a line that does not pass through any feasible points. Finally, notice that the feasible point that yields $I = 1.65$ is the point $(20, 5)$, a corner point. These observations form the basis of the following result, which we state without proof.

Theorem   **Location of the Solution of a Linear Programming Problem**

If a linear programming problem has a solution, it is located at a corner point of the graph of the feasible points.

If a linear programming problem has multiple solutions, at least one of them is located at a corner point of the graph of the feasible points.

In either case, the corresponding value of the objective function is unique.

We shall not consider here linear programming problems that have no solution. As a result, we can outline the procedure for solving a linear programming problem as follows:

Procedure for Solving a Linear Programming Problem

STEP 1:   Write an expression to the quantity to be maximized (or minimized). This expression is the objective function.

STEP 2:   Write all the constraints as a system of linear inequalities and graph the system.

STEP 3:   List the corner points of the graph of the feasible points.

STEP 4:   List the corresponding values of the objective function at each corner point. The largest (or smallest) of these is the solution.

EXAMPLE 12   **Solving a Minimum Linear Programming Problem**

Minimize the expression

$$z = 2x + 3y$$

subject to the constraints

$$y \leq 5 \qquad x \leq 6 \qquad x + y \geq 2 \qquad x \geq 0 \qquad y \geq 0$$

Solution The objective function is $z = 2x + 3y$. We seek the smallest value of z that can occur if x and y are solutions of the system of linear inequalities

$$\begin{cases} y \leq 5 \\ x \leq 6 \\ x + y \geq 2 \\ x \geq 0 \\ y \geq 0 \end{cases}$$

The graph of this system (the feasible points) is shown as the shaded region in Figure 32. We have also plotted the corner points. Table 1 lists the corner points and the corresponding values of the objective function. From the table, we can see that the minimum value of z is 4, and it occurs at the point $(2, 0)$.

Figure 32

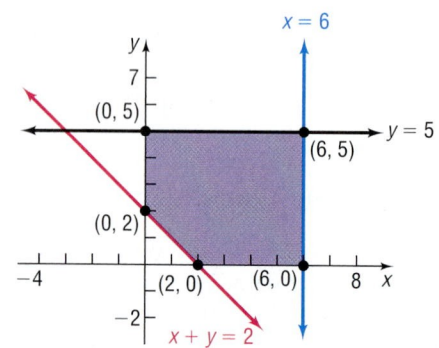

TABLE 1	
Corner Point (x, y)	**Value of the Objective Function** $z = 2x + 3y$
$(0, 2)$	$z = 2(0) + 3(2) = 6$
$(0, 5)$	$z = 2(0) + 3(5) = 15$
$(6, 5)$	$z = 2(6) + 3(5) = 27$
$(6, 0)$	$z = 2(6) + 3(0) = 12$
$(2, 0)$	$z = 2(2) + 3(0) = 4$

NOW WORK PROBLEMS 37 AND 43.

EXAMPLE 13 **Maximizing Profit**

At the end of every month, after filling orders for its regular customers, a coffee company has some pure Colombian coffee and some special-blend coffee remaining. The practice of the company has been to package a mixture of the two coffees into 1-pound packages as follows: a low-grade mixture containing 4 ounces of Colombian coffee and 12 ounces of special-blend coffee and a high-grade mixture containing 8 ounces of Colombian and 8 ounces of special-blend coffee. A profit of $0.30 per package is made on the low-grade mixture, whereas a profit of $0.40 per package is made on the high-grade mixture. This month, 120 pounds of special-blend coffee and 100 pounds of pure Colombian coffee remain. How many packages of each mixture should be prepared to achieve a maximum profit? Assume that all packages prepared can be sold.

Solution We begin by assigning symbols for the two variables.

x = Number of packages of the low-grade mixture
y = Number of packages of the high-grade mixture

If P denotes the profit, then

$$P = \$0.30x + \$0.40y$$

This expression is the objective function. We seek to maximize P subject to certain constraints on x and y. Because x and y represent numbers of packages, the only meaningful values for x and y are nonnegative integers. Thus, we have the two constraints

$$x \geq 0 \qquad y \geq 0 \qquad \text{Nonnegative constraints}$$

We also have only so much of each type of coffee available. For example, the total amount of Colombian coffee used in the two mixtures cannot exceed 100 pounds, or 1600 ounces. Because we use 4 ounces in each low-grade package and 8 ounces in each high-grade package, we are led to the constraint

$$4x + 8y \leq 1600 \qquad \text{Colombian coffee constraint}$$

Similarly, the supply of 120 pounds, or 1920 ounces, of special-blend coffee leads to the constraint

$$12x + 8y \leq 1920 \qquad \text{Special-blend coffee constraint}$$

The linear programming problem may be stated as

$$\text{Maximize:} \quad P = 0.3x + 0.4y$$

subject to the constraints

$$x \geq 0 \qquad y \geq 0 \qquad 4x + 8y \leq 1600 \qquad 12x + 8y \leq 1920$$

The graph of the constraints (the feasible points) is illustrated in Figure 33. We list the corner points and evaluate the objective function at each one. In Table 2, we can see that the maximum profit, $84, is achieved with 40 packages of the low-grade mixture and 180 packages of the high-grade mixture.

Figure 33

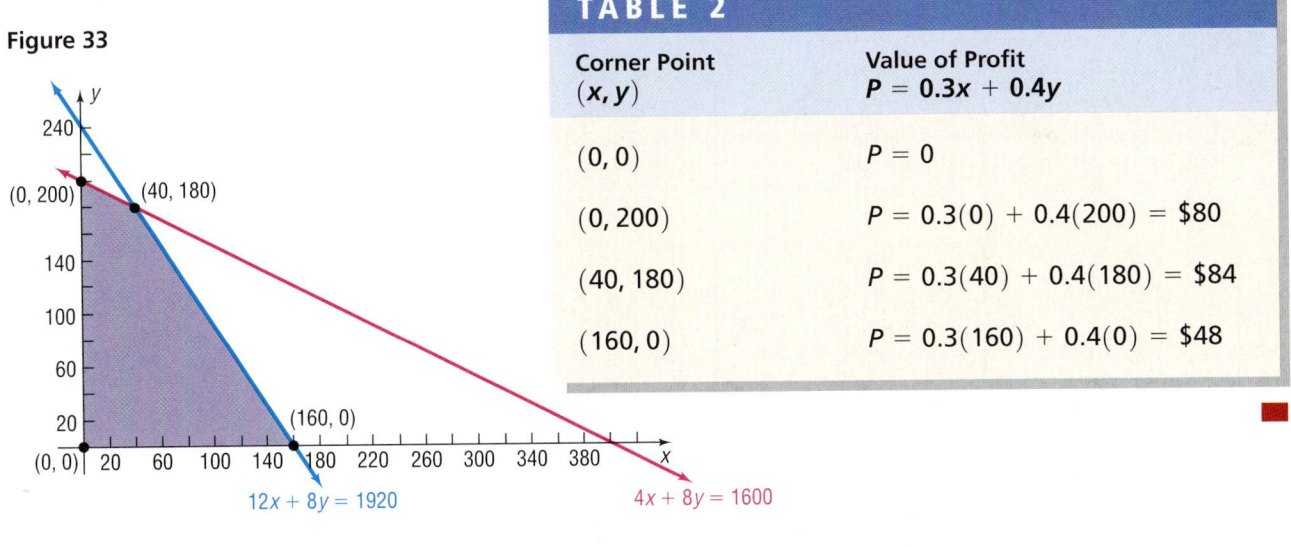

TABLE 2	
Corner Point (x, y)	**Value of Profit** $P = 0.3x + 0.4y$
$(0, 0)$	$P = 0$
$(0, 200)$	$P = 0.3(0) + 0.4(200) = \$80$
$(40, 180)$	$P = 0.3(40) + 0.4(180) = \$84$
$(160, 0)$	$P = 0.3(160) + 0.4(0) = \$48$

NOW WORK PROBLEM **51.**

7.6 Concepts and Vocabulary

In Problems 1–3, fill in the blanks.

1. The graph of a linear inequality is called a(n) _____ .

2. A linear programming problem requires that a linear expression, called the _____ , be maximized or minimized.

3. Each point that satisfies the constraints of a linear programming problem is called a(n) _____ .

In Problems 4–6, answer True or False for each statement.

4. The graph of a linear inequality is a line.

5. The graph of a system of linear inequalities is sometimes unbounded.

6. If a linear programming problem has a solution, it is located at a corner point of the graph of the feasible points.

7. If an ordered pair does not satisfy a linear inequality, then we shade the side opposite the point. Explain why this works.

7.6 Exercises

In Problems 1–8, graph each inequality (a) by hand and (b) using a graphing utility.

1. $x \geq 0$ **2.** $y \geq 0$ **3.** $x \geq 4$ **4.** $y \leq 2$

5. $x + y > 1$ **6.** $x + y \leq 9$ **7.** $2x + y \geq 6$ **8.** $3x + 2y \leq 6$

In Problems 9–20, graph each system of inequalities by hand.

9. $\begin{cases} x + y \leq 2 \\ 2x + y \geq 4 \end{cases}$ **10.** $\begin{cases} 3x - y \geq 6 \\ x + 2y \leq 2 \end{cases}$ **11.** $\begin{cases} 2x - y \leq 4 \\ 3x + 2y \geq -6 \end{cases}$ **12.** $\begin{cases} 4x - 5y \leq 0 \\ 2x - y \geq 2 \end{cases}$

13. $\begin{cases} 2x - 3y \leq 0 \\ 3x + 2y \leq 6 \end{cases}$ **14.** $\begin{cases} 4x - y \geq 2 \\ x + 2y \geq 2 \end{cases}$ **15.** $\begin{cases} x - 2y \leq 6 \\ 2x - 4y \geq 0 \end{cases}$ **16.** $\begin{cases} x + 4y \leq 8 \\ x + 4y \geq 4 \end{cases}$

17. $\begin{cases} 2x + y \geq -2 \\ 2x + y \geq 2 \end{cases}$ **18.** $\begin{cases} x - 4y \leq 4 \\ x - 4y \geq 0 \end{cases}$ **19.** $\begin{cases} 2x + 3y \geq 6 \\ 2x + 3y \leq 0 \end{cases}$ **20.** $\begin{cases} 2x + y \geq 0 \\ 2x + y \geq 2 \end{cases}$

In Problems 21–30, graph each system of linear inequalities by hand. Tell whether the graph is bounded or unbounded, and label the corner points.

21. $\begin{cases} x \geq 0 \\ y \geq 0 \\ 2x + y \leq 6 \\ x + 2y \leq 6 \end{cases}$ **22.** $\begin{cases} x \geq 0 \\ y \geq 0 \\ x + y \geq 4 \\ 2x + 3y \geq 6 \end{cases}$ **23.** $\begin{cases} x \geq 0 \\ y \geq 0 \\ x + y \geq 2 \\ 2x + y \geq 4 \end{cases}$ **24.** $\begin{cases} x \geq 0 \\ y \geq 0 \\ 3x + y \leq 6 \\ 2x + y \leq 2 \end{cases}$ **25.** $\begin{cases} x \geq 0 \\ y \geq 0 \\ x + y \geq 2 \\ 2x + 3y \leq 12 \\ 3x + y \leq 12 \end{cases}$

26. $\begin{cases} x \geq 0 \\ y \geq 0 \\ x + y \geq 2 \\ x + y \leq 10 \\ 2x + y \leq 3 \end{cases}$ **27.** $\begin{cases} x \geq 0 \\ y \geq 0 \\ x + y \geq 2 \\ x + y \leq 8 \\ 2x + y \leq 10 \end{cases}$ **28.** $\begin{cases} x \geq 0 \\ y \geq 0 \\ x + y \geq 2 \\ x + y \leq 8 \\ x + 2y \geq 1 \end{cases}$ **29.** $\begin{cases} x \geq 0 \\ y \geq 0 \\ x + 2y \geq 1 \\ x + 2y \leq 10 \end{cases}$ **30.** $\begin{cases} x \geq 0 \\ y \geq 0 \\ x + 2y \geq 1 \\ x + 2y \leq 10 \\ x + y \geq 2 \\ x + y \leq 8 \end{cases}$

In Problems 31–34, write a system of linear inequalities that has the given graph.

31.

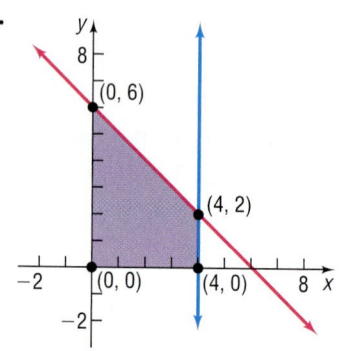

32.

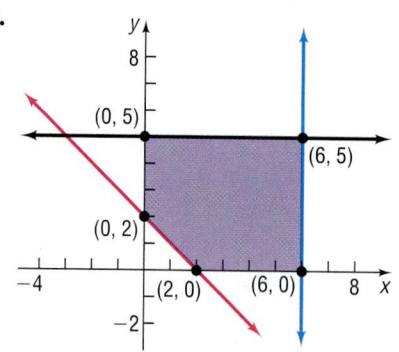

33.

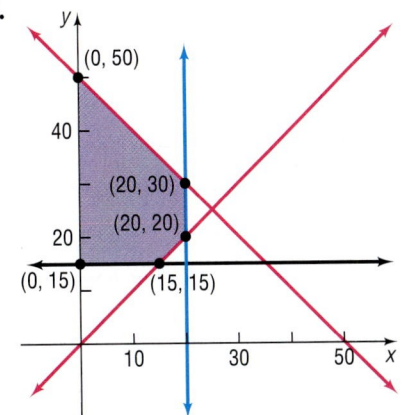

34.

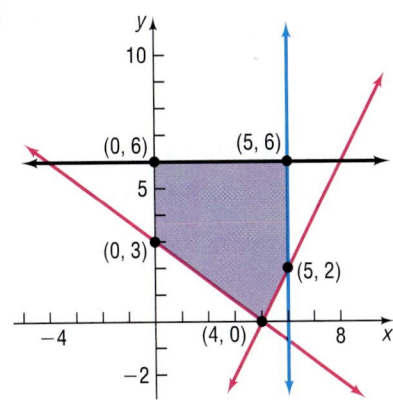

In Problems 35–40, find the maximum and minimum value of the given objective function of a linear programming problem. The figure illustrates the graph of the feasible points.

35. $z = x + y$ **36.** $z = 2x + 3y$ **37.** $z = x + 10y$

38. $z = 10x + y$ **39.** $z = 5x + 7y$ **40.** $z = 7x + 5y$

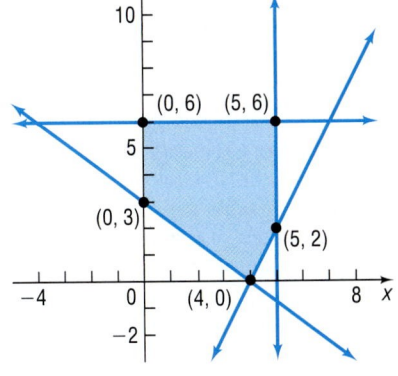

In Problems 41–50, solve each linear programming problem.

41. Maximize $z = 2x + y$ subject to $x \geq 0$, $y \geq 0$, $x + y \leq 6$, $x + y \geq 1$

42. Maximize $z = x + 3y$ subject to $x \geq 0$, $y \geq 0$, $x + y \geq 3$, $x \leq 5$, $y \leq 7$

43. Minimize $z = 2x + 5y$ subject to $x \geq 0$, $y \geq 0$, $x + y \geq 2$, $x \leq 5$, $y \leq 3$

44. Minimize $z = 3x + 4y$ subject to $x \geq 0$, $y \geq 0$, $2x + 3y \geq 6$, $x + y \leq 8$

45. Maximize $z = 3x + 5y$ subject to $x \geq 0$, $y \geq 0$, $x + y \geq 2$, $2x + 3y \leq 12$, $3x + 2y \leq 12$

46. Maximize $z = 5x + 3y$ subject to $x \geq 0$, $y \geq 0$, $x + y \geq 2$, $x + y \leq 8$, $2x + y \leq 10$

47. Minimize $z = 5x + 4y$ subject to $x \geq 0$, $y \geq 0$, $x + y \geq 2$, $2x + 3y \leq 12$, $3x + y \leq 12$

48. Minimize $z = 2x + 3y$ subject to $x \geq 0$, $y \geq 0$, $x + y \geq 3$, $x + y \leq 9$, $x + 3y \geq 6$

49. Maximize $z = 5x + 2y$ subject to $x \geq 0$, $y \geq 0$, $x + y \leq 10$, $2x + y \geq 10$, $x + 2y \geq 10$

50. Maximize $z = 2x + 4y$ subject to $x \geq 0$, $y \geq 0$, $2x + y \geq 4$, $x + y \leq 9$

51. Maximizing Profit A manufacturer of skis produces two types: downhill and cross country. Use the following table to determine how many of each kind of ski should be produced to achieve a maximum profit. What is the maximum profit? What would the maximum profit be if the maximum time available for manufacturing is increased to 48 hours?

	Downhill	Cross Country	Maximum Time Available
Manufacturing time per ski	2 hours	1 hour	40 hours
Finishing time per ski	1 hour	1 hour	32 hours
Profit per ski	$70	$50	

52. Farm Management A farmer has 70 acres of land available for planting either soybeans or wheat. The cost of preparing the soil, the workdays required, and the expected profit per acre planted for each type of crop are given in the following table:

	Soybeans	Wheat
Preparation cost per acre	$60	$30
Workdays required per acre	3	4
Profit per acre	$180	$100

The farmer cannot spend more than $1800 in preparation costs nor use more than a total of 120 workdays. How many acres of each crop should be planted in order to maximize the profit? What is the maximum profit? What is the maximum profit if the farmer is willing to spend no more than $2400 on preparation?

53. Farm Management A small farm in Illinois has 100 acres of land available on which to grow corn and soybeans. The following table shows the cultivation cost per acre, the labor cost per acre, and the expected profit per acre. The column on the right shows the amount of money available for each of these expenses. Find the number of acres of each crop that should be planted in order to maximize profit.

	Soybeans	Corn	Money Available
Cultivation cost per acre	$40	$60	$1800
Labor cost per acre	$60	$60	$2400
Profit per acre	$200	$250	

54. Dietary Requirements A certain diet requires at least 60 units of carbohydrates, 45 units of protein, and 30 units of fat each day. Each ounce of Supplement A provides 5 units of carbohydrates, 3 units of protein, and 4 units of fat. Each ounce of Supplement B provides 2 units of carbohydrates, 2 units of protein, and 1 unit of fat. If Supplement A costs $1.50 per ounce and Supplement B costs $1.00 per ounce, how many ounces of each supplement should be taken daily to minimize the cost of the diet?

55. Production Scheduling In a factory, machine 1 produces 8-inch pliers at the rate of 60 units per hour and 6-inch pliers at the rate of 70 units per hour. Machine 2 produces 8-inch pliers at the rate of 40 units per hour and 6-inch pliers at the rate of 20 units per hour. It costs $50 per hour to operate machine 1, while machine 2 costs $30 per hour to operate. The production schedule requires that at least 240 units of 8-inch pliers and at least 140 units of 6-inch pliers must be produced during each 10-hour day. Which combination of machines will cost the least money to operate?

56. Farm Management An owner of a fruit orchard hires a crew of workers to prune at least 25 of his 50 fruit trees. Each newer tree requires 1 hour to prune, while each older tree needs 1.5 hours. The crew contracts to work for at least 30 hours and charge $15 for each newer tree and $20 for each older tree. To minimize his cost, how many of each kind of tree will the orchard owner have pruned? What will be the cost?

57. Managing a Meat Market A meat market combines ground beef and ground pork in a single package for meat loaf. The ground beef is 75% lean (75% beef, 25% fat) and costs the market $0.75 per pound. The ground pork is 60% lean and costs the market $0.45 per pound. The meat loaf must be at least 70% lean. If the market wants to use at least 50 pounds of its available pork, but no more than 200 pounds of its available ground beef, how much ground beef should be mixed with ground pork so that the cost is minimized?

58. Return on Investment An investment broker is instructed by her client to invest up to $20,000, some in a junk bond yielding 9% per annum and some in Treasury bills yielding 7% per annum. The client wants to invest at least $8000 in T-bills and no more than $12,000 in the junk bond.
(a) How much should the broker recommend that the client place in each investment to maximize income if the client insists that the amount invested in T-bills must equal or exceed the amount placed in junk bonds?
(b) How much should the broker recommend that the client place in each investment to maximize income if the client insists that the amount invested in T-bills must not exceed the amount placed in junk bonds?

59. Maximizing Profit on Ice Skates A factory manufactures two kinds of ice skates: racing skates and figure skates. The racing skates require 6 work-hours in the fabrication department, whereas the figure skates require 4 work-hours there. The racing skates require 1 work-hour in the finishing department, whereas the figure skates require 2 work-hours there. The fabricating department has available at most 120 work-hours per day, and the finishing department has no more than 40 work-hours per day available. If the profit on each racing skate is $10 and the profit on each figure skate is $12, how many of each should be manufactured each day to maximize profit? (Assume that all skates made are sold.)

60. Financial Planning A retired couple has up to $50,000 to place in fixed-income securities. Their financial advisor suggests two securities to them: one is a AAA bond that yields 8% per annum; the other is a Certificate of Deposit (CD) that yields 4%. After careful consideration of the alternatives, the couple decides to place at most $20,000 in the AAA bond and at least $15,000 in the CD. They also instruct the financial adviser to place at least as much in the CD as in the AAA bond. How should the financial adviser proceed to maximize the return on their investment?

61. Product Design An entrepreneur is having a design group produce at least six samples of a new kind of fastener that he wants to market. It costs $9.00 to produce each metal fastener and $4.00 to produce each plastic fastener. He wants to have at least two of each version of the fastener and needs to have all the samples 24 hours from now. It takes 4 hours to produce each metal sample and 2 hours to produce each plastic sample. To minimize the cost of the samples, how many of each kind should the entrepreneur order? What will be the cost of the samples?

62. Animal Nutrition Kevin's dog Amadeus likes two kinds of canned dog food. Gourmet Dog costs 40 cents a can and has 20 units of a vitamin complex; the calorie content is 75 calories. Chow Hound costs 32 cents a can and has 35 units of vitamins and 50 calories. Kevin likes Amadeus to have at least 1175 units of vitamins a month and at least 2375 calories during the same time period. Kevin has space to store only 60 cans of dog food at a time. How much of each kind of dog food should Kevin buy once a month in order to minimize his cost?

63. Airline Revenue An airline has two classes of service: first class and coach. Management's experience has been that each aircraft should have at least 8 but not more than 16 first-class seats and at least 80 but not more than 120 coach seats.

(a) If management decides that the ratio of first-class to coach seats should never exceed 1 : 12, with how many of each type seat should an aircraft be configured to maximize revenue?

(b) If management decides that the ratio of first-class to coach seats should never exceed 1 : 8, with how many of each type seat should an aircraft be configured to maximize revenue?

(c) If you were management, what would you do?

[**Hint:** Assume that the airline charges $C for a coach seat and $F for a first-class seat; $C > 0, F > 0$.]

64. Minimizing Cost A firm that specializes in raising frying chickens supplements the regular chicken feed with four vitamins. The owner wants the supplemental food to contain at least 50 units of vitamin I, 90 units of vitamin II, 60 units of vitamin III, and 100 units of vitamin IV per 100 ounces of feed. Two supplements are available: supplement A, which contains 5 units of vitamin I, 25 units of vitamin II, 10 units of vitamin III, and 35 units of vitamin IV per ounce; and supplement B, which contains 25 units of vitamin I, 10 units of vitamin II, 10 units of vitamin III, and 20 units of vitamin IV per ounce. If supplement A costs $0.06 per ounce and supplement B costs $0.08 per ounce, how much of each supplement should the manager of the farm buy to add to each 100 ounces of feed in order to keep the total cost at a minimum, while still meeting the owner's vitamin specifications?

65. Explain in your own words what a linear programming problem is and how it can be solved.

PREPARING FOR THIS SECTION

Before getting started, review the following:

✓ Identity (Section 1.3, p. 110)

✓ Proper and Improper Rational Functions (Section 4.3, p. 348)

✓ Factoring Polynomials (Section R.6, pp. 50–58)

✓ Fundamental Theorem of Algebra (Section 5.4, p. 407)

7.7 PARTIAL FRACTION DECOMPOSITION

OBJECTIVES

1 Decompose $\dfrac{P}{Q}$, Where Q Has Only Nonrepeated Linear Factors

2 Decompose $\dfrac{P}{Q}$, Where Q Has Repeated Linear Factors

3 Decompose $\dfrac{P}{Q}$, Where Q Has Nonrepeated Irreducible Quadratic Factors

4 Decompose $\dfrac{P}{Q}$, Where Q Has Repeated Irreducible Quadratic Factors

Consider the problem of adding two fractions:

$$\frac{3}{x + 4} \quad \text{and} \quad \frac{2}{x - 3}$$

The result is

$$\frac{3}{x + 4} + \frac{2}{x - 3} = \frac{3(x - 3) + 2(x + 4)}{(x + 4)(x - 3)} = \frac{5x - 1}{x^2 + x - 12}$$

 The reverse procedure, of starting with the rational expression $\frac{5x - 1}{x^2 + x - 12}$ and writing it as the sum (or difference) of the two simpler fractions $\frac{3}{x + 4}$ and $\frac{2}{x - 3}$, is referred to as **partial fraction decomposition**, and the two simpler fractions are called **partial fractions**. Decomposing a rational expression into a sum of partial fractions is important in solving certain types of calculus problems. This section presents a systematic way to decompose rational expressions.

We begin by recalling that a rational expression is the ratio of two polynomials, say, P and $Q \neq 0$. We assume that P and Q have no common factors.

Recall also that a rational expression $\frac{P}{Q}$ is called **proper** if the degree of the polynomial in the numerator is less than the degree of the polynomial in the denominator. Otherwise, the rational expression is termed **improper**.

Because any improper rational expression can be reduced by long division to a mixed form consisting of the sum of a polynomial and a proper rational expression, we shall restrict the discussion that follows to proper rational expressions.

The partial fraction decomposition of the rational expression $\frac{P}{Q}$ depends on the factors of the denominator Q. Recall (from Section 5.2) that any polynomial whose coefficients are real numbers can be factored (over the real numbers) into products of linear and/or irreducible quadratic factors. Thus, the denominator Q of the rational expression $\frac{P}{Q}$ will contain only factors of one or both of the following types:

1. *Linear factors* of the form $x - a$, where a is a real number.
2. *Irreducible quadratic factors* of the form $ax^2 + bx + c$, where $a, b,$ and c are real numbers, $a \neq 0$, and $b^2 - 4ac < 0$ (which guarantees that $ax^2 + bx + c$ cannot be written as the product of two linear factors with real coefficients).

1 As it turns out, there are four cases to be examined. We begin with the case for which Q has only nonrepeated linear factors.

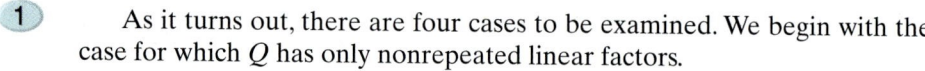

Case 1: Q has only nonrepeated linear factors.

Under the assumption that Q has only nonrepeated linear factors, the polynomial Q has the form

$$Q(x) = (x - a_1)(x - a_2) \cdots \cdot (x - a_n)$$

where none of the numbers $a_1, a_2, \ldots, a_n$ are equal. In this case, the partial fraction decomposition of $\dfrac{P}{Q}$ is of the form

$$\frac{P(x)}{Q(x)} = \frac{A_1}{x - a_1} + \frac{A_2}{x - a_2} + \cdots + \frac{A_n}{x - a_n} \qquad (1)$$

where the numbers $A_1, A_2, \ldots, A_n$ are to be determined.

We show how to find these numbers in the example that follows.

EXAMPLE 1   **Nonrepeated Linear Factors**

Write the partial fraction decomposition of $\dfrac{x}{x^2 - 5x + 6}$.

Solution   First, we factor the denominator,

$$x^2 - 5x + 6 = (x - 2)(x - 3)$$

and conclude that the denominator contains only nonrepeated linear factors. Then we decompose the rational expression according to equation (1):

$$\frac{x}{x^2 - 5x + 6} = \frac{A}{x - 2} + \frac{B}{x - 3} \qquad (2)$$

where A and B are to be determined. To find A and B, we clear the fractions by multiplying each side by $(x - 2)(x - 3) = x^2 - 5x + 6$. The result is

$$x = A(x - 3) + B(x - 2) \qquad (3)$$

or

$$x = (A + B)x + (-3A - 2B)$$

This equation is an identity in x. Thus, we may equate the coefficients of like powers of x to get

$$\begin{cases} 1 = A + B & \text{Equate coefficients of } x\text{: } 1x = (A + B)x. \\ 0 = -3A - 2B & \text{Equate coefficients of } x^0, \text{ the constants:} \\ & 0x^0 = (-3A - 2B)x^0. \end{cases}$$

This system of two equations containing two variables, A and B, can be solved using whatever method you wish. Solving it, we get

$$A = -2 \qquad B = 3$$

From equation (2), the partial fraction decomposition is

$$\frac{x}{x^2 - 5x + 6} = \frac{-2}{x - 2} + \frac{3}{x - 3}$$

✔ CHECK: The decomposition can be checked by adding the fractions.

$$\frac{-2}{x-2} + \frac{3}{x-3} = \frac{-2(x-3) + 3(x-2)}{(x-2)(x-3)} = \frac{x}{(x-2)(x-3)}$$

$$= \frac{x}{x^2 - 5x + 6}$$

■ ■

▬▬▬▬▬ **NOW WORK PROBLEM 9.**

The numbers to be found in the partial fraction decomposition can sometimes be found more readily by using suitable choices for x (which may include complex numbers) in the identity obtained after fractions have been cleared. In Example 1, the identity after clearing fractions, equation (3), is

$$x = A(x-3) + B(x-2)$$

If we let $x = 2$ in this expression, the term containing B drops out, leaving $2 = A(-1)$, or $A = -2$. Similarly, if we let $x = 3$, the term containing A drops out, leaving $3 = B$. As before, $A = -2$ and $B = 3$.

We use this method in the next example.

②

Case 2: Q has repeated linear factors.

If the polynomial Q has a repeated factor, say $(x - a)^n, n \geq 2$ an integer, then, in the partial fraction decomposition of $\dfrac{P}{Q}$, we allow for the terms

$$\frac{A_1}{x-a} + \frac{A_2}{(x-a)^2} + \cdots + \frac{A_n}{(x-a)^n}$$

where the numbers $A_1, A_2, \ldots, A_n$ are to be determined.

EXAMPLE 2   **Repeated Linear Factors**

Write the partial fraction decomposition of $\dfrac{x+2}{x^3 - 2x^2 + x}$.

Solution   First, we factor the denominator,

$$x^3 - 2x^2 + x = x(x^2 - 2x + 1) = x(x-1)^2$$

and find that the denominator has the nonrepeated linear factor x and the repeated linear factor $(x-1)^2$. By Case 1, we must allow for the term $\dfrac{A}{x}$ in the decomposition; and, by Case 2, we must allow for the terms $\dfrac{B}{x-1} + \dfrac{C}{(x-1)^2}$ in the decomposition.

We write

$$\frac{x+2}{x^3 - 2x^2 + x} = \frac{A}{x} + \frac{B}{x-1} + \frac{C}{(x-1)^2} \tag{4}$$

Again, we clear fractions by multiplying each side by $x^3 - 2x^2 + x = x(x - 1)^2$. The result is the identity

$$x + 2 = A(x - 1)^2 + Bx(x - 1) + Cx \tag{5}$$

If we let $x = 0$ in this expression, the terms containing B and C drop out, leaving $2 = A(-1)^2$, or $A = 2$. Similarly, if we let $x = 1$, the terms containing A and B drop out, leaving $3 = C$. Thus, equation (5) becomes

$$x + 2 = 2(x - 1)^2 + Bx(x - 1) + 3x$$

Now let $x = 2$ (any choice other than 0 or 1 will work as well). The result is

$$4 = 2(1)^2 + B(2)(1) + 3(2)$$
$$2B = 4 - 2 - 6 = -4$$
$$B = -2$$

Thus, we have $A = 2$, $B = -2$, and $C = 3$.

From equation (4), the partial fraction decomposition is

$$\frac{x + 2}{x^3 - 2x^2 + x} = \frac{2}{x} + \frac{-2}{x - 1} + \frac{3}{(x - 1)^2}$$

■

EXAMPLE 3   **Repeated Linear Factors**

Write the partial fraction decomposition of $\dfrac{x^3 - 8}{x^2(x - 1)^3}$.

Solution   The denominator contains the repeated linear factor x^2 and the repeated linear factor $(x - 1)^3$. Thus, the partial fraction decomposition takes the form

$$\frac{x^3 - 8}{x^2(x - 1)^3} = \frac{A}{x} + \frac{B}{x^2} + \frac{C}{x - 1} + \frac{D}{(x - 1)^2} + \frac{E}{(x - 1)^3} \tag{6}$$

As before, we clear fractions and obtain the identity

$$x^3 - 8 = Ax(x - 1)^3 + B(x - 1)^3 + Cx^2(x - 1)^2 + Dx^2(x - 1) + Ex^2 \tag{7}$$

Let $x = 0$. (Do you see why this choice was made?) Then

$$-8 = B(-1)$$
$$B = 8$$

Now let $x = 1$ in equation (7). Then

$$-7 = E$$

Use $B = 8$ and $E = -7$ in equation (7) and collect like terms.

$$x^3 - 8 = Ax(x - 1)^3 + 8(x - 1)^3$$
$$+ Cx^2(x - 1)^2 + Dx^2(x - 1) - 7x^2$$
$$x^3 - 8 - 8(x^3 - 3x^2 + 3x - 1) + 7x^2 = Ax(x - 1)^3 + Cx^2(x - 1)^2 + Dx^2(x - 1)$$
$$-7x^3 + 31x^2 - 24x = x(x - 1)[A(x - 1)^2 + Cx(x - 1) + Dx]$$
$$x(x - 1)(-7x + 24) = x(x - 1)[A(x - 1)^2 + Cx(x - 1) + Dx]$$
$$-7x + 24 = A(x - 1)^2 + Cx(x - 1) + Dx \qquad (8)$$

We now work with equation (8). Let $x = 0$. Then

$$24 = A$$

Now let $x = 1$ in equation (8). Then

$$17 = D$$

Use $A = 24$ and $D = 17$ in equation (8) and collect like terms.

$$-7x + 24 = 24(x - 1)^2 + Cx(x - 1) + 17x$$
$$-24x^2 + 48x - 24 - 17x - 7x + 24 = Cx(x - 1)$$
$$-24x^2 + 24x = Cx(x - 1)$$
$$-24x(x - 1) = Cx(x - 1)$$
$$-24 = C$$

We now know all the numbers $A, B, C, D,$ and E, so, from equation (6), we have the decomposition

$$\frac{x^3 - 8}{x^2(x - 1)^3} = \frac{24}{x} + \frac{8}{x^2} + \frac{-24}{x - 1} + \frac{17}{(x - 1)^2} + \frac{-7}{(x - 1)^3} \qquad \blacksquare$$

The method employed in Example 3, although somewhat tedious, is still preferable to solving the system of five equations containing five variables that the expansion of equation (6) leads to.

✏ **NOW WORK PROBLEM 15.**

③ The final two cases involve irreducible quadratic factors. A quadratic factor is irreducible if it cannot be factored into linear factors with real coefficients. A quadratic expression $ax^2 + bx + c$ is irreducible whenever $b^2 - 4ac < 0$. For example, $x^2 + x + 1$ and $x^2 + 4$ are irreducible.

Case 3: Q contains a nonrepeated irreducible quadratic factor.

If Q contains a nonrepeated irreducible quadratic factor of the form $ax^2 + bx + c$, then, in the partial fraction decomposition of $\dfrac{P}{Q}$, allow for the term

$$\frac{Ax + B}{ax^2 + bx + c}$$

where the numbers A and B are to be determined.

EXAMPLE 4   **Nonrepeated Irreducible Quadratic Factor**

Write the partial factor decomposition of $\dfrac{3x - 5}{x^3 - 1}$.

Solution   We factor the denominator,

$$x^3 - 1 = (x - 1)(x^2 + x + 1)$$

and find that it has a nonrepeated linear factor $x - 1$ and a nonrepeated irreducible quadratic factor $x^2 + x + 1$. Thus, we allow for the term $\dfrac{A}{x - 1}$ by Case 1, and we allow for the term $\dfrac{Bx + C}{x^2 + x + 1}$ by Case 3. Hence, we write

$$\frac{3x - 5}{x^3 - 1} = \frac{A}{x - 1} + \frac{Bx + C}{x^2 + x + 1} \qquad (9)$$

We clear fractions by multiplying each side of equation (9) by $x^3 - 1 = (x - 1)(x^2 + x + 1)$ to obtain the identity

$$3x - 5 = A(x^2 + x + 1) + (Bx + C)(x - 1) \qquad (10)$$

Now let $x = 1$. Then equation (10) gives $-2 = A(3)$, or $A = -\dfrac{2}{3}$. We use this value of A in equation (10) and simplify.

$$3x - 5 = -\frac{2}{3}(x^2 + x + 1) + (Bx + C)(x - 1)$$

$$3(3x - 5) = -2(x^2 + x + 1) + 3(Bx + C)(x - 1) \qquad \text{\color{blue}Multiply each side by 3.}$$
$$9x - 15 = -2x^2 - 2x - 2 + 3(Bx + C)(x - 1)$$
$$2x^2 + 11x - 13 = 3(Bx + C)(x - 1) \qquad \text{\color{blue}Collect terms.}$$
$$(2x + 13)(x - 1) = 3(Bx + C)(x - 1) \qquad \text{\color{blue}Factor the left side.}$$
$$2x + 13 = 3Bx + 3C$$
$$2 = 3B \quad \text{and} \quad 13 = 3C \qquad \text{\color{blue}Equate coefficients.}$$
$$B = \frac{2}{3} \qquad\qquad C = \frac{13}{3}$$

Thus, from equation (9), we see that

$$\frac{3x - 5}{x^3 - 1} = \frac{-\dfrac{2}{3}}{x - 1} + \frac{\dfrac{2}{3}x + \dfrac{13}{3}}{x^2 + x + 1}$$

━ **NOW WORK PROBLEM 17.**

④   **Case 4: Q contains repeated irreducible quadratic factors.**

If the polynomial Q contains a repeated irreducible quadratic factor $(ax^2 + bx + c)^n, n \geq 2, n$ an integer, then, in the partial fraction decomposition of $\dfrac{P}{Q}$, allow for the terms

$$\frac{A_1x + B_1}{ax^2 + bx + c} + \frac{A_2x + B_2}{(ax^2 + bx + c)^2} + \cdots + \frac{A_nx + B_n}{(ax^2 + bx + c)^n}$$

where the numbers $A_1, B_1, A_2, B_2, \ldots, A_n, B_n$ are to be determined.

EXAMPLE 5	**Repeated Irreducible Quadratic Factor**

Write the partial fraction decomposition of $\dfrac{x^3 + x^2}{(x^2 + 4)^2}$.

Solution The denominator contains the repeated irreducible quadratic factor $(x^2 + 4)^2$, so we write

$$\frac{x^3 + x^2}{(x^2 + 4)^2} = \frac{Ax + B}{x^2 + 4} + \frac{Cx + D}{(x^2 + 4)^2} \qquad (11)$$

We clear fractions to obtain

$$x^3 + x^2 = (Ax + B)(x^2 + 4) + Cx + D$$

Collecting like terms yields

$$x^3 + x^2 = Ax^3 + Bx^2 + (4A + C)x + D + 4B$$

Equating coefficients, we arrive at the system

$$\begin{cases} A = 1 \\ B = 1 \\ 4A + C = 0 \\ D + 4B = 0 \end{cases}$$

The solution is $A = 1, B = 1, C = -4, D = -4$. Hence, from equation (11),

$$\frac{x^3 + x^2}{(x^2 + 4)^2} = \frac{x + 1}{x^2 + 4} + \frac{-4x - 4}{(x^2 + 4)^2}$$

NOW WORK PROBLEM **31.**

7.7 Exercises

In Problems 1–8, tell whether the given rational expression is proper or improper. If improper, rewrite it as the sum of a polynomial and a proper rational expression.

1. $\dfrac{x}{x^2 - 1}$
2. $\dfrac{5x + 2}{x^3 - 1}$
3. $\dfrac{x^2 + 5}{x^2 - 4}$
4. $\dfrac{3x^2 - 2}{x^2 - 1}$

5. $\dfrac{5x^3 + 2x - 1}{x^2 - 4}$
6. $\dfrac{3x^4 + x^2 - 2}{x^3 + 8}$
7. $\dfrac{x(x - 1)}{(x + 4)(x - 3)}$
8. $\dfrac{2x(x^2 + 4)}{x^2 + 1}$

In Problems 9–42, write the partial fraction decomposition of each rational expression.

9. $\dfrac{4}{x(x - 1)}$
10. $\dfrac{3x}{(x + 2)(x - 1)}$
11. $\dfrac{1}{x(x^2 + 1)}$
12. $\dfrac{1}{(x + 1)(x^2 + 4)}$

13. $\dfrac{x}{(x - 1)(x - 2)}$
14. $\dfrac{3x}{(x + 2)(x - 4)}$
15. $\dfrac{x^2}{(x - 1)^2(x + 1)}$
16. $\dfrac{x + 1}{x^2(x - 2)}$

17. $\dfrac{1}{x^3 - 8}$
18. $\dfrac{2x + 4}{x^3 - 1}$
19. $\dfrac{x^2}{(x - 1)^2(x + 1)^2}$
20. $\dfrac{x + 1}{x^2(x - 2)^2}$

21. $\dfrac{x - 3}{(x + 2)(x + 1)^2}$
22. $\dfrac{x^2 + x}{(x + 2)(x - 1)^2}$
23. $\dfrac{x + 4}{x^2(x^2 + 4)}$
24. $\dfrac{10x^2 + 2x}{(x - 1)^2(x^2 + 2)}$

25. $\dfrac{x^2 + 2x + 3}{(x + 1)(x^2 + 2x + 4)}$

26. $\dfrac{x^2 - 11x - 18}{x(x^2 + 3x + 3)}$

27. $\dfrac{x}{(3x - 2)(2x + 1)}$

28. $\dfrac{1}{(2x + 3)(4x - 1)}$

29. $\dfrac{x}{x^2 + 2x - 3}$

30. $\dfrac{x^2 - x - 8}{(x + 1)(x^2 + 5x + 6)}$

31. $\dfrac{x^2 + 2x + 3}{(x^2 + 4)^2}$

32. $\dfrac{x^3 + 1}{(x^2 + 16)^2}$

33. $\dfrac{7x + 3}{x^3 - 2x^2 - 3x}$

34. $\dfrac{x^5 + 1}{x^6 - x^4}$

35. $\dfrac{x^2}{x^3 - 4x^2 + 5x - 2}$

36. $\dfrac{x^2 + 1}{x^3 + x^2 - 5x + 3}$

37. $\dfrac{x^3}{(x^2 + 16)^3}$

38. $\dfrac{x^2}{(x^2 + 4)^3}$

39. $\dfrac{4}{2x^2 - 5x - 3}$

40. $\dfrac{4x}{2x^2 + 3x - 2}$

41. $\dfrac{2x + 3}{x^4 - 9x^2}$

42. $\dfrac{x^2 + 9}{x^4 - 2x^2 - 8}$

Chapter Review

Things To Know

Systems of equations (page 513)

Systems with no solutions are inconsistent.   Systems with a solution are consistent.

Consistent systems of linear equations have either a unique solution or an infinite number of solutions.

Determinants and Cramer's Rule (pp. 549 and 554)

Matrix (page 530 and 559)   Rectangular array of numbers, called entries

m by n matrix (page 560)   Matrix with m rows and n columns

Identity matrix I (page 568)   Square matrix whose diagonal entries are 1's, while all other entries are 0's

Inverse of a matrix (page 569) A^{-1} is the inverse of A if $AA^{-1} = A^{-1}A = I$

Nonsingular matrix (page 569)   A matrix that has an inverse

Linear programming problem (page 585)

Maximize (or minimize) a linear objective function, $z = Ax + By$, subject to certain conditions, or constraints, expressible as linear inequalities in x and y. A feasible point (x, y) is a point that satisfies the constraints of a linear programming problem.

Location of solution (page 587)

If a linear programming problem has a solution, it is located at a corner point of the graph of the feasible points.

If a linear programming problem has multiple solutions, at least one of them is located at a corner point of the graph of the feasible points.

In either case, the corresponding value of the objective function is unique.

Objectives

Section		You should be able to …	Review Exercises
7.1	1	Solve systems of equations using a graphing utility (p. 515)	1–14
	2	Solve systems of equations by substitution (p. 515)	1–14
	3	Solve systems of equations by elimination (p. 517)	1–14
	4	Identify inconsistent systems (p. 519)	10, 13, 82
	5	Express the solution of a system of dependent equations (p. 519)	9, 14, 81
7.2	1	Solve systems of three equations containing three variables (p. 524)	15–18, 83, 84
	2	Identify inconsistent systems (p. 525)	18
	3	Express the solutions of a system of dependent systems (p. 526)	17
7.3	1	Write the augmented matrix of a system of linear equations (p. 530)	35–44
	2	Write the system from the augmented matrix (p. 531)	19, 20

▰ Review Exercises

Blue problem numbers indicate the authors' suggestions for use in a Practice Test.

In Problems 1–18, solve each system of equations algebraically using the method of substitution or the method of elimination. If the system has no solution, say that it is inconsistent. Verify your result using a graphing utility.

1. $\begin{cases} 2x - y = 5 \\ 5x + 2y = 8 \end{cases}$

2. $\begin{cases} 2x + 3y = 2 \\ 7x - y = 3 \end{cases}$

3. $\begin{cases} 3x - 4y = 4 \\ x - 3y = \dfrac{1}{2} \end{cases}$

4. $\begin{cases} 2x + y = 0 \\ 5x - 4y = -\dfrac{13}{2} \end{cases}$

5. $\begin{cases} x - 2y - 4 = 0 \\ 3x + 2y - 4 = 0 \end{cases}$

6. $\begin{cases} x - 3y + 5 = 0 \\ 2x + 3y - 5 = 0 \end{cases}$

7. $\begin{cases} y = 2x - 5 \\ x = 3y + 4 \end{cases}$

8. $\begin{cases} x = 5y + 2 \\ y = 5x + 2 \end{cases}$

9. $\begin{cases} x - 3y + 4 = 0 \\ \dfrac{1}{2}x - \dfrac{3}{2}y + \dfrac{4}{3} = 0 \end{cases}$

10. $\begin{cases} x + \dfrac{1}{4}y = 2 \\ y + 4x + 2 = 0 \end{cases}$

11. $\begin{cases} 2x + 3y - 13 = 0 \\ 3x - 2y = 0 \end{cases}$

12. $\begin{cases} 4x + 5y = 21 \\ 5x + 6y = 42 \end{cases}$

13. $\begin{cases} 3x - 2y = 8 \\ x - \dfrac{2}{3}y = 12 \end{cases}$

14. $\begin{cases} 2x + 5y = 10 \\ 4x + 10y = 20 \end{cases}$

15. $\begin{cases} x + 2y - z = 6 \\ 2x - y + 3z = -13 \\ 3x - 2y + 3z = -16 \end{cases}$

16. $\begin{cases} x + 5y - z = 2 \\ 2x + y + z = 7 \\ x - y + 2z = 11 \end{cases}$

17. $\begin{cases} 2x - 4y + z = -15 \\ x + 2y - 4z = 27 \\ 5x - 6y - 2z = -3 \end{cases}$

18. $\begin{cases} x - 4y + 3z = 15 \\ -3x + y - 5z = -5 \\ -7x - 5y - 9z = 10 \end{cases}$

In Problems 19 and 20, write the system of equations corresponding to the given augmented matrix.

19. $\begin{bmatrix} 3 & 2 & | & 8 \\ 1 & 4 & | & -1 \end{bmatrix}$

20. $\begin{bmatrix} 1 & 2 & 5 & | & -2 \\ 5 & 0 & -3 & | & 8 \\ 2 & -1 & 0 & | & 0 \end{bmatrix}$

In Problems 21–28, use the following matrices to compute each expression. Verify your result using a graphing utility.

$$A = \begin{bmatrix} 1 & 0 \\ 2 & 4 \\ -1 & 2 \end{bmatrix} \quad B = \begin{bmatrix} 4 & -3 & 0 \\ 1 & 1 & -2 \end{bmatrix} \quad C = \begin{bmatrix} 3 & -4 \\ 1 & 5 \\ 5 & -2 \end{bmatrix}$$

21. $A + C$

22. $A - C$

23. $6A$

24. $-4B$

25. AB

26. BA

27. CB

28. BC

In Problems 29–34, find the inverse of each matrix algebraically, if there is one. If there is not an inverse, say that the matrix is singular. Verify your result using a graphing utility.

29. $\begin{bmatrix} 4 & 6 \\ 1 & 3 \end{bmatrix}$

30. $\begin{bmatrix} -3 & 2 \\ 1 & -2 \end{bmatrix}$

31. $\begin{bmatrix} 1 & 3 & 3 \\ 1 & 2 & 1 \\ 1 & -1 & 2 \end{bmatrix}$

32. $\begin{bmatrix} 3 & 1 & 2 \\ 3 & 2 & -1 \\ 1 & 1 & 1 \end{bmatrix}$

33. $\begin{bmatrix} 4 & -8 \\ -1 & 2 \end{bmatrix}$

34. $\begin{bmatrix} -3 & 1 \\ -6 & 2 \end{bmatrix}$

In Problems 35–44, solve each system of equations algebraically using matrices. If the system has no solution, say that it is inconsistent. Verify your result using a graphing utility.

35. $\begin{cases} 3x - 2y = 1 \\ 10x + 10y = 5 \end{cases}$

36. $\begin{cases} 3x + 2y = 6 \\ x - y = -\dfrac{1}{2} \end{cases}$

37. $\begin{cases} 5x + 6y - 3z = 6 \\ 4x - 7y - 2z = -3 \\ 3x + y - 7z = 1 \end{cases}$

38. $\begin{cases} 2x + y + z = 5 \\ 4x - y - 3z = 1 \\ 8x + y - z = 5 \end{cases}$

39. $\begin{cases} x - 2z = 1 \\ 2x + 3y = -3 \\ 4x - 3y - 4z = 3 \end{cases}$

40. $\begin{cases} x + 2y - z = 2 \\ 2x - 2y + z = -1 \\ 6x + 4y + 3z = 5 \end{cases}$

41. $\begin{cases} x - y + z = 0 \\ x - y - 5z - 6 = 0 \\ 2x - 2y + z - 1 = 0 \end{cases}$

42. $\begin{cases} 4x - 3y + 5z = 0 \\ 2x + 4y - 3z = 0 \\ 6x + 2y + z = 0 \end{cases}$

43. $\begin{cases} x - y - z - t = 1 \\ 2x + y - z + 2t = 3 \\ x - 2y - 2z - 3t = 0 \\ 3x - 4y + z + 5t = -3 \end{cases}$

44. $\begin{cases} x - 3y + 3z - t = 4 \\ x + 2y - z = -3 \\ x + 3z + 2t = 3 \\ x + y + 5z = 6 \end{cases}$

In Problems 45–50, find the value of each determinant algebraically. Verify your result using a graphing utility.

45. $\begin{vmatrix} 3 & 4 \\ 1 & 3 \end{vmatrix}$

46. $\begin{vmatrix} -4 & 0 \\ 1 & 3 \end{vmatrix}$

47. $\begin{vmatrix} 1 & 4 & 0 \\ -1 & 2 & 6 \\ 4 & 1 & 3 \end{vmatrix}$

48. $\begin{vmatrix} 2 & 3 & 10 \\ 0 & 1 & 5 \\ -1 & 2 & 3 \end{vmatrix}$

49. $\begin{vmatrix} 2 & 1 & -3 \\ 5 & 0 & 1 \\ 2 & 6 & 0 \end{vmatrix}$

50. $\begin{vmatrix} -2 & 1 & 0 \\ 1 & 2 & 3 \\ -1 & 4 & 2 \end{vmatrix}$

In Problems 51–56, use Cramer's Rule, if applicable, to solve each system.

51. $\begin{cases} x - 2y = 4 \\ 3x + 2y = 4 \end{cases}$

52. $\begin{cases} x - 3y = -5 \\ 2x + 3y = 5 \end{cases}$

53. $\begin{cases} 2x + 3y - 13 = 0 \\ 3x - 2y = 0 \end{cases}$

54. $\begin{cases} 3x - 4y - 12 = 0 \\ 5x + 2y + 6 = 0 \end{cases}$

55. $\begin{cases} x + 2y - z = 6 \\ 2x - y + 3z = -13 \\ 3x - 2y + 3z = -16 \end{cases}$

56. $\begin{cases} x - y + z = 8 \\ 2x + 3y - z = -2 \\ 3x - y - 9z = 9 \end{cases}$

In Problems 57 and 58, use properties of determinants to find the value of each determinant if it is known that

$$\begin{vmatrix} x & y \\ a & b \end{vmatrix} = 8$$

57. $\begin{vmatrix} 2x & y \\ 2a & b \end{vmatrix}$

58. $\begin{vmatrix} y & x \\ b & a \end{vmatrix}$

In Problems 59 and 60, graph each inequality (a) by hand and (b) using a graphing utility.

59. $3x + 4y \leq 12$

60. $2x - 3y \geq 6$

In Problems 61–66, graph each system of inequalities. Tell whether the graph is bounded or unbounded, and label the corner points.

61. $\begin{cases} -2x + y \leq 2 \\ x + y \geq 2 \end{cases}$

62. $\begin{cases} x - 2y \leq 6 \\ 2x + y \geq 2 \end{cases}$

63. $\begin{cases} x \geq 0 \\ y \geq 0 \\ x + y \leq 4 \\ 2x + 3y \leq 6 \end{cases}$

64. $\begin{cases} x \geq 0 \\ y \geq 0 \\ 3x + y \geq 6 \\ 2x + y \geq 2 \end{cases}$

65. $\begin{cases} x \geq 0 \\ y \geq 0 \\ 2x + y \leq 8 \\ x + 2y \geq 2 \end{cases}$

66. $\begin{cases} x \geq 0 \\ y \geq 0 \\ 3x + y \leq 9 \\ 2x + 3y \geq 6 \end{cases}$

In Problems 67–70, solve each linear programming problem.

67. Maximize $z = 3x + 4y$ subject to $x \geq 0$, $y \geq 0$, $3x + 2y \geq 6$, $x + y \leq 8$

68. Maximize $z = 2x + 4y$ subject to $x \geq 0$, $y \geq 0$, $x + y \leq 6$, $x \geq 2$

69. Minimize $z = 3x + 5y$ subject to $x \geq 0$, $y \geq 0$, $x + y \geq 1$, $3x + 2y \leq 12$, $x + 3y \leq 12$

70. Minimize $z = 3x + y$ subject to $x \geq 0$, $y \geq 0$, $x \leq 8$, $y \leq 6$, $2x + y \geq 4$

In Problems 71–80, write the partial fraction decomposition of each rational expression.

71. $\dfrac{6}{x(x - 4)}$

72. $\dfrac{x}{(x + 2)(x - 3)}$

73. $\dfrac{x - 4}{x^2(x - 1)}$

74. $\dfrac{2x - 6}{(x - 2)^2(x - 1)}$

75. $\dfrac{x}{(x^2 + 9)(x + 1)}$

76. $\dfrac{3x}{(x - 2)(x^2 + 1)}$

77. $\dfrac{x^3}{(x^2 + 4)^2}$

78. $\dfrac{x^3 + 1}{(x^2 + 16)^2}$

79. $\dfrac{x^2}{(x^2 + 1)(x^2 - 1)}$

80. $\dfrac{4}{(x^2 + 4)(x^2 - 1)}$

81. Find A such that the system of equations has infinitely many solutions.

$$\begin{cases} 2x + 5y = 5 \\ 4x + 10y = A \end{cases}$$

82. Find A such that the system in Problem 81 is inconsistent.

83. **Curve Fitting**   Find the quadratic function $y = ax^2 + bx + c$ that passes through the three points $(0, 1)$, $(1, 0)$, and $(-2, 1)$.

84. Curve Fitting   Find the general equation of the circle that passes through the three points $(0, 1)$, $(1, 0)$, and $(-2, 1)$.

[**Hint:** The general equation of a circle is $x^2 + y^2 + Dx + Ey + F = 0$.]

85. Blending Coffee   A coffee distributor is blending a new coffee that will cost $3.90 per pound. It will consist of a blend of $3.00 per pound coffee and $6.00 per pound coffee. What amounts of each type of coffee should be mixed to achieve the desired blend?

[**Hint:** Assume that the weight of the blended coffee is 100 pounds.]

$3.00/lb $3.90/lb $6.00/lb

86. Farming   A 1000-acre farm in Illinois is used to raise corn and soy beans. The cost per acre for raising corn is $65 and the cost per acre for soy beans is $45. If $54,325 has been budgeted for costs and all the acreage is to be used, how many acres should be allocated for each crop?

87. Cookie Orders   A cookie company makes three kinds of cookies, oatmeal raisin, chocolate chip, and shortbread, packaged in small, medium, and large boxes. The small box contains 1 dozen oatmeal raisin and 1 dozen chocolate chip; the medium box has 2 dozen oatmeal raisin, 1 dozen chocolate chip, and 1 dozen shortbread; the large box contains 2 dozen oatmeal raisin, 2 dozen chocolate chip, and 3 dozen shortbread. If you require exactly 15 dozen oatmeal raisin, 10 dozen chocolate chip, and 11 dozen shortbread, how many of each size box should you buy?

88. Mixed Nuts   A store that specializes in selling nuts has 72 pounds of cashews and 120 pounds of peanuts available. These are to be mixed in 12-ounce packages as follows: a lower-priced package containing 8 ounces of peanuts and 4 ounces of cashews and a quality package containing 6 ounces of peanuts and 6 ounces of cashews.
  (a) Use x to denote the number of lower-priced packages, and use y to denote the number of quality packages. Write a system of linear inequalities that describes the possible number of each kind of package.
  (b) Graph the system and label the corner points.

89. Determining the Speed of the Current of the Aguarice River   On a recent trip to the Cuyabeno Wildlife Reserve in the Amazon region of Ecuador, a 100-kilometer trip by speedboat was taken down the Aguarico River from Chiritza to the Flotel Orellana. As I watched the Amazon unfold, I wondered how fast the speedboat was going and how fast the current of the white-water Aguarico River was. I timed the trip downstream at 2.5 hours and the return trip at 3 hours. What were the two speeds?

90. Finding the Speed of the Jet Stream   On a flight between Midway Airport in Chicago and Ft. Lauderdale, Florida, a Boeing 737 jet maintains an airspeed of 475 miles per hour. If the trip from Chicago to Ft. Lauderdale takes 2 hours, 30 minutes and the return flight takes 2 hours, 50 minutes, what is the speed of the jet stream? (Assume that the speed of the jet stream remains constant at the various altitudes of the plane and that the plane flies with the jet stream one way and against it the other way.)

91. Constant Rate Jobs   If Bruce and Bryce work together for 1 hour and 20 minutes, they will finish a certain job. If Bryce and Marty work together for 1 hour and 36 minutes, the same job can be finished. If Marty and Bruce work together, they can complete this job in 2 hours and 40 minutes. How long will it take each of them working alone to finish the job?

92. Maximizing Profit on Figurines   A factory manufactures two kinds of ceramic figurines: a dancing girl and a mermaid. Each requires three processes: molding, painting, and glazing. The daily labor available for molding is no more than 90 work-hours, labor available for painting does not exceed 120 work-hours, and labor available for glazing is no more than 60 work-hours. The dancing girl requires 3 work-hours for molding, 6 work-hours for painting, and 2 work-hours for glazing. The mermaid requires 3 work-hours for molding, 4 work-hours for painting, and 3 work-hours for glazing. If the profit on each figurine is $25 for dancing girls and $30 for mermaids, how many of each should be produced each day to maximize profit? If management decides to produce the number of each figurine that maximizes profit, determine which of these processes has excess work-hours assigned to it.

93. Minimizing Production Cost   A factory produces gasoline engines and diesel engines. Each week the factory is obligated to deliver at least 20 gasoline engines and at least 15 diesel engines. Due to physical limitations, however, the factory cannot make more than 60 gasoline engines nor more than 40 diesel engines. Finally, to prevent layoffs, a total of at least 50 engines must be produced. If gasoline engines cost $450 each to produce and diesel engines cost $550 each to produce, how many of each should be produced per week to minimize the cost? What is the excess capacity of the factory; that is, how many of each kind of engine are being produced in excess of the number that the factory is obligated to deliver?

94. Describe four ways of solving a system of three linear equations containing three variables. Which method do you prefer? Why?

Chapter Projects

1. Markov Chains A **Markov chain** (or process) is one in which future outcomes are determined by a current state. Future outcomes are based on probabilities. The probability of moving to a certain state depends only on the state previously occupied and does not vary with time. An example of a Markov chain would be the maximum education achieved by children based on the highest education attained of their parents, where the states are (1) earned college degree, (2) high-school diploma only, (3) elementary school only. If p_{ij} is the probability of moving from state i to state j, then the **transition matrix** is the $m \times m$ matrix.

$$P = \begin{bmatrix} p_{11} & p_{12} & \cdots & p_{1m} \\ \vdots & \vdots & & \vdots \\ p_{m1} & p_{m2} & \cdots & p_{mm} \end{bmatrix}$$

The following table represents the probabilities of the highest educational level of children based on the highest educational level of their parents. For example, the table shows that the probability p_{21} is 40% that parents with a high-school education (row 2) will have children with a college education (column 1).

(a) Convert the percentages to decimals.

(b) What is the transition matrix?

(c) Sum across the rows. What do you notice? Why do you think that you obtained this result?

(d) If P is the transition matrix of a Markov chain, then the (i, j)th entry of P^n (nth power of P) gives the probability of passing from state i to state j in n stages. What is the probability that a grandchild of a college graduate is a college graduate?

(e) What is the probability that the grandchild of a high school graduate finishes college?

(f) The row vector $v^{(0)} = [0.252 \ 0.582 \ 0.166]$ represents the proportion of the U.S. population that has college, high school, and elementary school, respectively, as the highest educational level in 1999.* In a Markov chain the probability distribution $v^{(k)}$ after k stages is $v^{(k)} = v^{(0)}P^k$, where P^k is the kth power of the transition matrix. What will be the distribution of highest educational attainment of the grandchildren of the current population?

(g) Calculate $P^3, P^4, P^5, \ldots$. Continue until the matrix does not change. This is called the long-run distribution. What is the long-run distribution of highest educational attainment of the population?

2. Using Matrices to Find the Line of Best Fit Recall, in Section 2.2, that we found the line of best fit for a set of data that appeared to be linearly related using a graphing utility. The equation was of the form $y = ax + b$, where a was the slope and b was the y-intercept. We now use matrices to find the line of best fit for the following data:

x	3	4	5	6	7	8	9
y	4	6	7	10	12	14	16

We can think of these points as being values of x and y that satisfy the equation

$$b + ax = y$$

*Source: U.S. Census Bureau.

Highest Educational Level of Parents	Maximum Education That Children Achieve		
	College	**High School**	**Elementary**
College	80%	18%	2%
High school	40%	50%	10%
Elementary	20%	60%	20%

where a is the slope and b is the y-intercept. The ordered pairs (x, y) yield a system of equations as follows:

$$\begin{cases} b + 3a = 4 \\ b + 4a = 6 \\ \vdots \qquad \vdots \\ b + 9a = 16 \end{cases}$$

which can be written in matrix form as follows:

$$\begin{bmatrix} 1 & 3 \\ 1 & 4 \\ 1 & 5 \\ 1 & 6 \\ 1 & 7 \\ 1 & 8 \\ 1 & 9 \end{bmatrix} \begin{bmatrix} b \\ a \end{bmatrix} = \begin{bmatrix} 4 \\ 6 \\ 7 \\ 10 \\ 12 \\ 14 \\ 16 \end{bmatrix}$$

$$AB = Y$$

It can be shown that the solution $\boldsymbol{B} = \begin{bmatrix} b \\ a \end{bmatrix}$ to this system is given by

$$\boldsymbol{B} = (\boldsymbol{A}^T\boldsymbol{A})^{-1}\boldsymbol{A}^T\boldsymbol{Y}$$

where $\boldsymbol{A}^T$ is found by putting the elements in row 1 of $\boldsymbol{A}$ in column 1 of $\boldsymbol{A}^T$, the elements in row 2 of $\boldsymbol{A}$ are inserted in column 2 of $\boldsymbol{A}^T$, and so on. In other words, the rows of $\boldsymbol{A}$ become the columns of $\boldsymbol{A}^T$.

(a) Find $\boldsymbol{A}^T$.

(b) Find $\boldsymbol{B} = \begin{bmatrix} b \\ a \end{bmatrix}$.

(c) What is the line of best fit using the matrix method described above?

(d) Verify your results by using a graphing utility to find the line of best fit.

3. **CBL Experiment**   Two students walk toward each other at a *constant* rate. Plotting distance against time on the same viewing window for each student results in a system of linear equations. By finding the intersection point of the two graphs the time and location where the students pass each other is determined. (Activity 4, Real-World Math with the CBL System, 1994.)

Cumulative Review

In Problems 1–6, solve each equation algebraically. Verify your results using a graphing utility.

1. $2x^2 - x = -3$

2. $\sqrt{3x + 1} = 4$

3. $2x^3 - 3x^2 - 8x - 3 = 0$

4. $3^x = 9^{x+1}$

5. $\log_3(x - 1) + \log_3(2x + 1) = 2$

6. $3^x = e$

7. Determine whether the function $g(x) = \dfrac{2x^3}{x^4 + 1}$ is even, odd, or neither. Is the graph of g symmetric with respect to the x-axis, y-axis, or origin?

8. Find the center and radius of the circle $x^2 + y^2 - 2x + 4y - 11 = 0$. Graph the circle by hand.

9. Graph $f(x) = 3^{x-2} + 1$ using transformations. What is the domain, range, and horizontal asymptote of f?

10. The function $f(x) = \dfrac{5}{x + 2}$ is one-to-one. Find f^{-1}. Find the domain and range of f and the domain and range of f^{-1}.

11. Depreciation of the Chevy Camaro The following data represent the price of used Chevy Camaros on December 21, 2000, by age. For example, a 2000 Camaro is considered to be 1-year old. Treat the age of the car as the independent variable and the price as the dependent variable.

Age, a (yr)	Price, P ($)
2	15,900
5	10,988
2	16,980
5	9995
4	11,995
5	10,995
1	20,365
2	16,463
6	10,824
1	19,995
1	18,650
4	10,488

Source: www.onlineauto.com

(a) Does the relation defined by the set of ordered pairs (a, P) represent a function?
(b) Draw a scatter diagram of the data.
(c) Using a graphing utility, find the line of best fit relating age and price.
(d) Interpret the slope of the line of best fit.
(e) Express the relationship found in part (c) using function notation.
(f) Predict the price of a Camaro that is 7 years old.

TRIGONOMETRIC FUNCTIONS

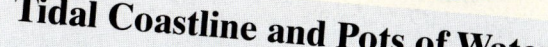

Tidal Coastline and Pots of Water

In Florida, they post times of tides coming in and going out very precisely, like 11:23 A.M. How can they be so precise?

There is more to tides, the rise and fall of ocean waters, than the gravitational pull of the moon and sun.

These are the chief factors, of course. And because the movement of the Earth, sun and moon in relationship to each other are known precisely, the rhythm of tides rising and falling along coastlines is easy to predict.

Yet the time and heights of high and low tide may vary along different stretches of the same coast though they're reacting to similar driving forces and pressures.

Historic observation makes possible the exact timing of high tides and low tides along a particular section of coast for a month, a year, or far into the future.

The reason for the difference is oscillation. Think of pots and pans filled with varying levels of water on a table, says Charles O'Reilly, Chief of Tidal Analysis for the Geological Survey of Canada's hydrographic service in Dartmouth, N.S. Then kick the table.

"You'll notice the water in the pots and pans will slosh differently. That's their natural oscillation," he said. "If you kick the table rhythmically, you'll find each pot continues to slosh differently because it has its own rhythm.

"Now, if you join those pots and pans together, that's sort of like a coastal ocean. They're all feeling the same 'kick,' but they are all responding differently. In order to predict a tide, you have to measure them for some period of time."

SOURCE: *Toronto Star*, June 13, 2001, p. GT02

SEE CHAPTER PROJECT 1.

OUTLINE

 For additional study help, go to

www.prenhall.com/sullivanegu3e

Materials include:

- Graphing Calculator Help
- Chapter Quiz
- Chapter Test
- PowerPoint Downloads
- Chapter Projects
- Student Tips

A Look Back, A Look Forward

In Chapter 2, we began our discussion of functions. We defined domain and range and independent and dependent variables; we found the value of a function and graphed functions. In Chapter 3, we continued our study of functions by listing properties that a function might have, like being even or odd, and we created a library of functions, naming key functions and listing their properties, including the graph.

In this chapter we define the trigonometric functions, six functions that have a wide application. We shall talk about their domain and range and how to find values, graph them, and develop a list of their properties.

There are two widely accepted approaches to the development of the trigonometric functions: one uses right triangles; the other uses circles, especially the unit circle. In this book, we develop the trigonometric functions using right triangles. In Section 8.5, we introduce trigonometric functions using the unit circle and show that this approach leads to the definition using right triangles.

PREPARING FOR THIS SECTION

Before getting started, review the following:

✓ Circumference and Area of a Circle
  (Section R.3, p. 27)

✓ Rectangular Coordinates
  (Section 1.1, pp. 90–91)

8.1 ANGLES AND THEIR MEASURE

OBJECTIVES
1. Convert between Degrees, Minutes, Seconds, and Decimal Forms for Angles
2. Find the Arc Length of a Circle
3. Convert from Degrees to Radians
4. Convert from Radians to Degrees
5. Find the Area of a Sector of a Circle
6. Find the Linear Speed of an Object Traveling in Circular Motion

Figure 1

A **ray**, or **half-line**, is that portion of a line that starts at a point V on the line and extends indefinitely in one direction. The starting point V of a ray is called its **vertex**. See Figure 1.

If two rays are drawn with a common vertex, they form an **angle**. We call one of the rays of an angle the **initial side** and the other the **terminal side**. The angle that is formed is identified by showing the direction and amount of rotation from the initial side to the terminal side. If the rotation is in the counterclockwise direction, the angle is **positive**; if the rotation is clockwise, the angle is **negative**. See Figure 2. Lowercase Greek letters, such as α (alpha), β (beta), γ (gamma), and θ (theta), will be used to denote angles. Notice in Figure 2(a) that the angle α is positive because the direction of the rotation from the initial side to the terminal side is counterclockwise. The angle β in Figure 2(b) is negative because the rotation is clockwise. The angle γ in Figure 2(c) is positive. Notice that the angle α in Figure 2(a) and the angle γ in Figure 2(c) have the same initial side and the same terminal side. However, α and γ are unequal, because the amount of rotation required to go from the initial side to the terminal side is greater for angle γ than for angle α.

Figure 2

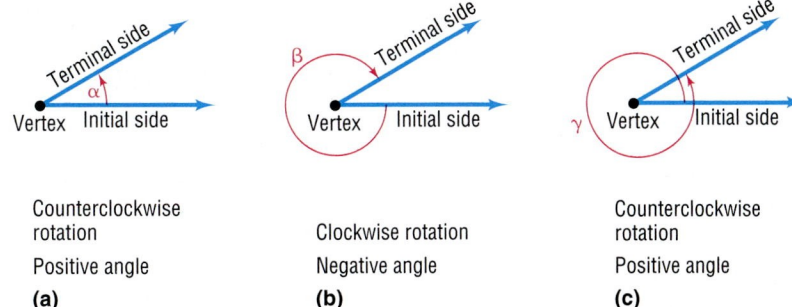

Counterclockwise
rotation

Positive angle

(a)

Clockwise rotation

Negative angle

(b)

Counterclockwise
rotation

Positive angle

(c)

An angle θ is said to be in **standard position** if its vertex is at the origin of a rectangular coordinate system and its initial side coincides with the positive x-axis. See Figure 3.

Figure 3

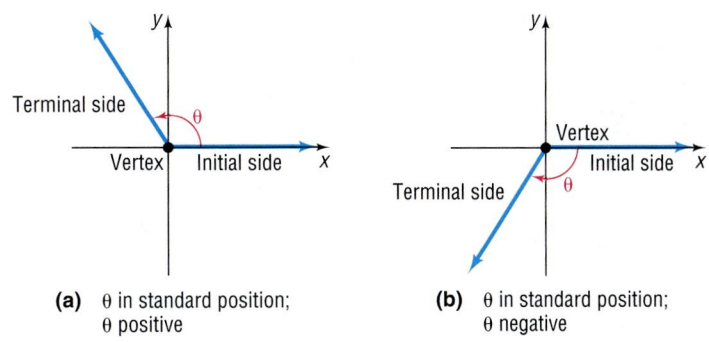

(a) θ in standard position;
θ positive

(b) θ in standard position;
θ negative

When an angle θ is in standard position, the terminal side will lie either in a quadrant, in which case we say that θ **lies in that quadrant**, or on the x-axis or the y-axis, in which case we say that θ is a **quadrantal angle**. For example, the angle θ in Figure 4(a) lies in quadrant II, the angle θ in Figure 4(b) lies in quadrant IV, and the angle θ in Figure 4(c) is a quadrantal angle.

Figure 4

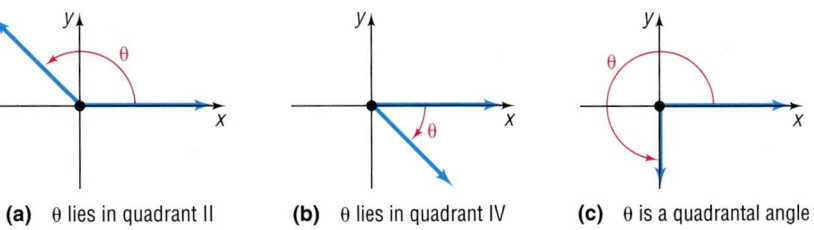

(a) θ lies in quadrant II

(b) θ lies in quadrant IV

(c) θ is a quadrantal angle

We measure angles by determining the amount of rotation needed for the initial side to become coincident with the terminal side. The two commonly used measures for angles are *degrees* and *radians*.

Degrees

The angle formed by rotating the initial side exactly once in the counterclockwise direction until it coincides with itself (1 revolution) is said to measure 360 degrees, abbreviated 360°. **One degree, 1°,** is $\dfrac{1}{360}$ revolution. A **right angle** is an angle that measures 90°, or $\dfrac{1}{4}$ revolution; a **straight angle** is an angle that measures 180°, or $\frac{1}{2}$ revolution. See Figure 5. As Figure 5(b) shows, it is customary to indicate a right angle by using the symbol ⌐.

Figure 5

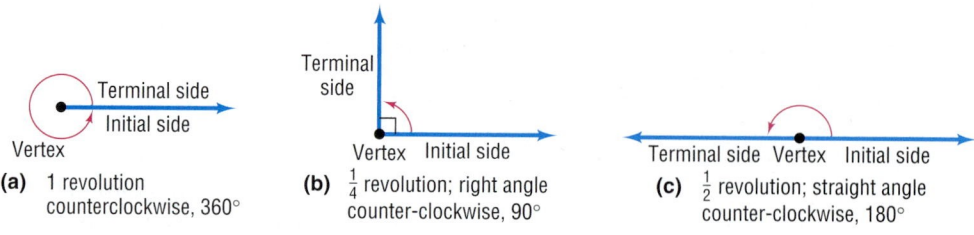

(a) 1 revolution counterclockwise, 360°

(b) $\frac{1}{4}$ revolution; right angle counter-clockwise, 90°

(c) $\frac{1}{2}$ revolution; straight angle counter-clockwise, 180°

It is also customary to refer to an angle that measures θ degrees as an angle of θ degrees.

EXAMPLE 1 **Drawing an Angle**

Draw each angle.

(a) 45°   (b) −90°   (c) 225°   (d) 405°

Solution (a) An angle of 45° is $\frac{1}{2}$ of a right angle. See Figure 6.

(b) An angle of −90° is $\frac{1}{4}$ revolution in the clockwise direction. See Figure 7.

Figure 6

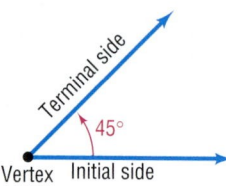

Figure 7

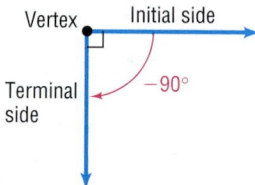

(c) An angle of 225° consists of a rotation through 180° followed by a rotation through 45°. See Figure 8.

(d) An angle of 405° consists of 1 revolution (360°) followed by a rotation through 45°. See Figure 9.

Figure 8

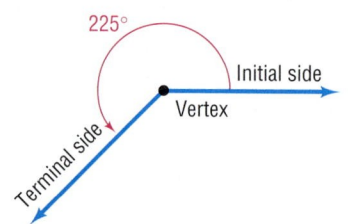

Figure 9

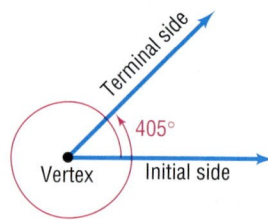

NOW WORK PROBLEM **1.**

1   Although subdivisions of a degree may be obtained by using decimals, we also may use the notion of *minutes* and *seconds*. **One minute,** denoted by **1′,** is defined as $\dfrac{1}{60}$ degree. **One second,** denoted by **1″,** is defined as $\dfrac{1}{60}$ minute or, equivalently, $\dfrac{1}{3600}$ degree. An angle of, say, 30 degrees, 40 minutes, 10 seconds is written compactly as 30°40′10″. To summarize:

$$
\begin{array}{c}
1 \text{ counterclockwise revolution} = 360° \\
1° = 60' \qquad 1' = 60''
\end{array}
\tag{1}
$$

It is sometimes necessary to convert from the degree, minute, second notation (D°M′S″) to a decimal form, and vice versa. Check your calculator; it should be capable of doing the conversion for you.

Before getting started you must set the mode to degrees, because there are two common ways to measure angles: degree mode and radian mode. (We will define radians shortly.) Usually, a menu is used to change from one mode to another. Check your owner's manual to find out how your particular calculator works.

Now let's see how to convert by hand from the degree, minute, second notation (D°M′S″) to a decimal form, and vice versa, by looking at some examples: $15°30' = 15.5°$ because $30' = \dfrac{1}{2}^{\circ} = 0.5°$

$32.25° = 32°15'$ because $0.25° = \dfrac{1}{4}^{\circ} = \dfrac{1}{4}(60') = 15'.$

EXAMPLE 2   **Converting between Degrees, Minutes, Seconds, and Decimal Forms by Hand**

(a) Convert 50°6′21″ to a decimal in degrees.

(b) Convert 21.256° to the D°M′S″ form.

Algebraic Solution   (a) Because $1' = \dfrac{1}{60}^{\circ}$ and $1'' = \dfrac{1}{60}' = \left(\dfrac{1}{60} \cdot \dfrac{1}{60}\right)^{\circ}$, we convert as follows:

$$
50°6'21'' = 50° + 6' + 21'' = 50° + 6 \cdot \dfrac{1}{60}^{\circ} + 21 \cdot \dfrac{1}{60} \cdot \dfrac{1}{60}^{\circ}
$$

$$
\approx 50° + 0.1° + 0.005833°
$$

$$
= 50.105833°
$$

(b) We start with the decimal part of 21.256°, that is, 0.256°.

$$
0.256° = (0.256)(1°) = (0.256)(60') = 15.36'
$$

$$
\uparrow
$$
$$
1° = 60'
$$

Now we work with the decimal part of 15.36′, that is, 0.36′.

$$
0.36' = (0.36)(1') = (0.36)(60'') = 21.6'' \approx 22''
$$

$$
\uparrow
$$
$$
1' = 60''
$$

Thus,

$$
21.256° = 21° + 0.256° = 21° + 15.36' = 21° + 15' + 0.36'
$$

$$
= 21° + 15' + 21.6'' \approx 21°15'22''
$$

Graphing Solution   (a) Figure 10 shows the solution using a TI-83 graphing calculator.   (b) Figure 11 shows the solution using a TI-83 graphing calculator.

Figure 10

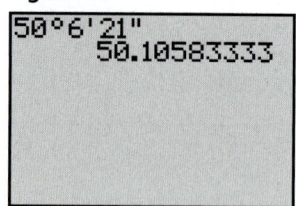

Figure 11

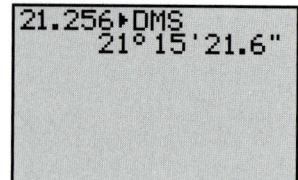

   **NOW WORK PROBLEMS 69 AND 75.**

    In many applications, such as describing the exact location of a star or the precise position of a boat at sea, angles measured in degrees, minutes, and even seconds are used. For calculation purposes, these are transformed to decimal form. In other applications, especially those in calculus, angles are measured using *radians*.

Radians

A **central angle** is an angle whose vertex is at the center of a circle. The rays of a central angle subtend (intersect) an arc on the circle. If the radius of the circle is r and the length of the arc subtended by the central angle is also r, then the measure of the angle is **1 radian.** See Figure 12(a).

For a circle of radius 1, the rays of a central angle with measure 1 radian would subtend an arc of length 1. For a circle of radius 3, the rays of a central angle with measure 1 radian would subtend an arc of length 3. See Figure 12(b).

Figure 12

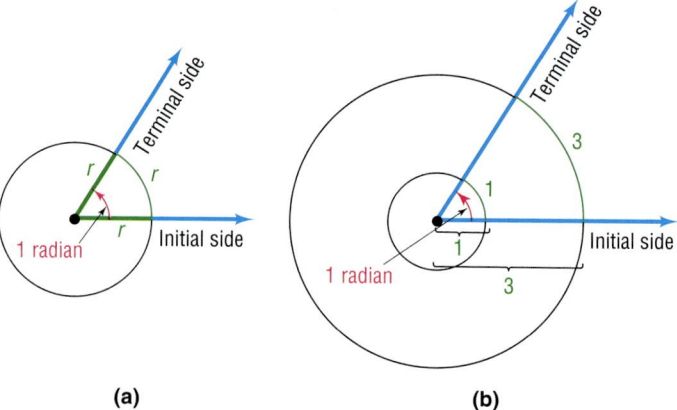

(a)   (b)

Figure 13

$$\frac{\theta}{\theta_1} = \frac{s}{s_1}$$

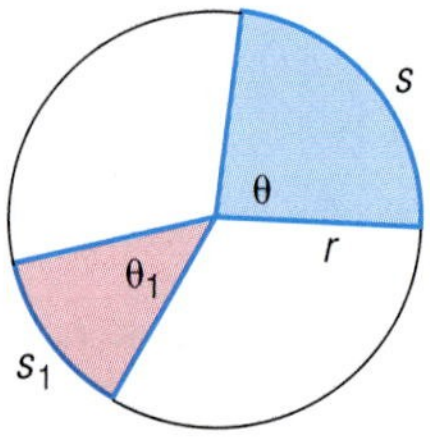

2   Now consider a circle of radius r and two central angles, θ and θ_1, measured in radians. Suppose that these central angles subtend arcs of lengths s and s_1, respectively, as shown in Figure 13. From geometry, we know that the ratio of the measures of the angles equals the ratio of the corresponding lengths of the arcs subtended by these angles; that is,

$$\frac{\theta}{\theta_1} = \frac{s}{s_1} \tag{2}$$

Suppose that $\theta_1 = 1$ radian. Refer again to Figure 12(a). The amount of arc s_1 subtended by the central angle $\theta_1 = 1$ radian equals the radius r of the circle. Then $s_1 = r$, so formula (2) reduces to

$$\frac{\theta}{1} = \frac{s}{r} \quad \text{or} \quad s = r\theta \tag{3}$$

Theorem

Arc Length

For a circle of radius r, a central angle of θ radians subtends an arc whose length s is

$$s = r\theta \tag{4}$$

Note: Formulas must be consistent with regard to the units used. In equation (4), we write

$$s = r\theta$$

To see the units, however, we must go back to equation (3) and write

$$\frac{\theta \text{ radians}}{1 \text{ radian}} = \frac{s \text{ length units}}{r \text{ length units}}$$

$$s \text{ length units} = r \text{ length units} \frac{\theta \text{ radians}}{1 \text{ radian}}$$

Since the radians cancel, we are left with

$$s \text{ length units} = (r \text{ length units})\theta \qquad s = r\theta$$

where θ appears to be "dimensionless" but, in fact, is measured in radians. So, in using the formula $s = r\theta$, the dimension for θ is radians, and any convenient unit of length (such as inches or meters) may be used for s and r.

EXAMPLE 3

Finding the Length of Arc of a Circle

Find the length of the arc of a circle of radius 2 meters subtended by a central angle of 0.25 radian.

Solution

We use equation (4) with $r = 2$ meters and $\theta = 0.25$. The length s of the arc is

$$s = r\theta = 2(0.25) = 0.5 \text{ meter}$$

 **NOW WORK PROBLEM 37.**

Figure 14
1 revolution = 2π radians

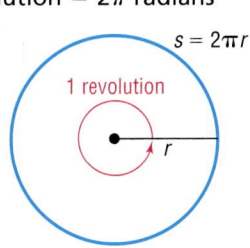

Relationship Between Degrees And Radians

Consider a circle of radius r. A central angle of 1 revolution will subtend an arc equal to the circumference of the circle (Figure 14). Because the circumference of a circle equals $2\pi r$, we use $s = 2\pi r$ in equation (4) to find that, for an angle θ of 1 revolution,

$$s = r\theta$$
$$2\pi r = r\theta \qquad \theta = 1 \text{ revolution}; s = 2\pi r$$
$$\theta = 2\pi \text{ radians} \qquad \text{Solve for } \theta$$

From this we have,

$$1 \text{ revolution} = 2\pi \text{ radians} \qquad (5)$$

so

$$360° = 2\pi \text{ radians}$$

or

$$180° = \pi \text{ radians} \qquad (6)$$

Divide both sides of equation (6) by 180. Then

$$1 \text{ degree} = \frac{\pi}{180} \text{radian}$$

Divide both sides of (6) by π. Then

$$\frac{180}{\pi} \text{degrees} = 1 \text{ radian}$$

We have the following two conversion formulas:

$$1 \text{ degree} = \frac{\pi}{180} \text{radian} \qquad 1 \text{ radian} = \frac{180}{\pi} \text{degrees} \qquad (7)$$

EXAMPLE 4 Converting from Degrees to Radians

③ Convert each angle in degrees to radians.

(a) 60° (b) 150° (c) −45° (d) 90°

Solution (a) $60° = 60 \cdot 1 \text{ degree} = 60 \cdot \frac{\pi}{180} \text{radian} = \frac{\pi}{3} \text{radians}$

(b) $150° = 150 \cdot \frac{\pi}{180} \text{radian} = \frac{5\pi}{6} \text{radians}$

(c) $-45° = -45 \cdot \frac{\pi}{180} \text{radian} = -\frac{\pi}{4} \text{radian}$

(d) $90° = 90 \cdot \frac{\pi}{180} \text{radian} = \frac{\pi}{2} \text{radians}$ ∎

Example 4 illustrates that angles that are fractions of a revolution are expressed in radian measure as fractional multiples of π, rather than as decimals. For example, a right angle, as in Example 4(d), is left in the form $\frac{\pi}{2}$ radians, which is exact, rather than using the approximation $\frac{\pi}{2} \approx \frac{3.1416}{2} = 1.5708$ radians.

NOW WORK PROBLEM 13.

EXAMPLE 5   **Converting Radians to Degrees**

   Convert each angle in radians to degrees.

(a) $\dfrac{\pi}{6}$ radian   (b) $\dfrac{3\pi}{2}$ radians   (c) $-\dfrac{3\pi}{4}$ radians   (d) $\dfrac{7\pi}{3}$ radians

Solution   (a) $\dfrac{\pi}{6}$ radian $= \dfrac{\pi}{6} \cdot 1$ radian $= \dfrac{\pi}{6} \cdot \dfrac{180}{\pi}$ degrees $= 30°$

(b) $\dfrac{3\pi}{2}$ radians $= \dfrac{3\pi}{2} \cdot \dfrac{180}{\pi}$ degrees $= 270°$

(c) $-\dfrac{3\pi}{4}$ radians $= -\dfrac{3\pi}{4} \cdot \dfrac{180}{\pi}$ degrees $= -135°$

(d) $\dfrac{7\pi}{3}$ radians $= \dfrac{7\pi}{3} \cdot \dfrac{180}{\pi}$ degrees $= 420°$ ■

✏ ─ **NOW WORK PROBLEM 25.**

Table 1 lists the degree and radian measures of some commonly encountered angles. You should learn to feel equally comfortable using degree or radian measure for these angles.

TABLE 1									
Degrees	0°	30°	45°	60°	90°	120°	135°	150°	180°
Radians	0	$\dfrac{\pi}{6}$	$\dfrac{\pi}{4}$	$\dfrac{\pi}{3}$	$\dfrac{\pi}{2}$	$\dfrac{2\pi}{3}$	$\dfrac{3\pi}{4}$	$\dfrac{5\pi}{6}$	π
Degrees		210°	225°	240°	270°	300°	315°	330°	360°
Radians		$\dfrac{7\pi}{6}$	$\dfrac{5\pi}{4}$	$\dfrac{4\pi}{3}$	$\dfrac{3\pi}{2}$	$\dfrac{5\pi}{3}$	$\dfrac{7\pi}{4}$	$\dfrac{11\pi}{6}$	2π

EXAMPLE 6   **Finding the Distance between Two Cities**

See Figure 15(a). The latitude of a location L is the angle formed by a ray drawn from the center of Earth to the Equator and a ray drawn from the center of Earth to L. See Figure 15(b). Glasgow, Montana, is due north of Albuquerque, New Mexico. Find the distance between Glasgow (48°9′ north latitude) and Albuquerque (35°5′ north latitude). Assume that the radius of Earth is 3960 miles.

Figure 15

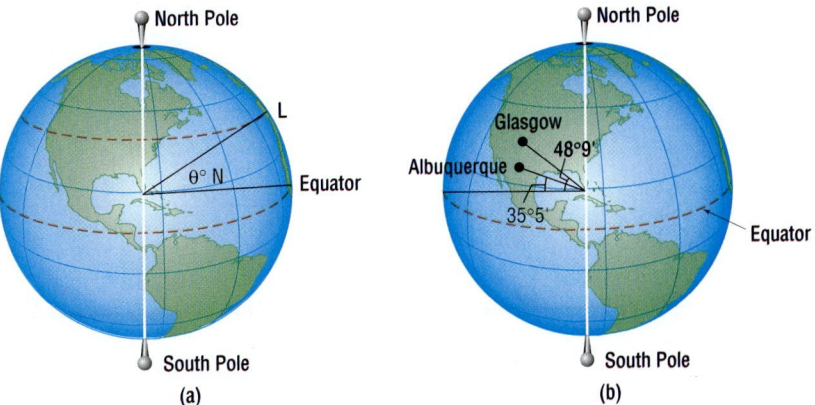

(a)   (b)

Solution   The measure of the central angle between the two cities is $48°9' - 35°5' = 13°4'$. We use equation (4), $s = r\theta$, but first we must convert the angle of $13°4'$ to radians.

$$\theta = 13°4' \approx 13.0667° = 13.0667 \cdot \frac{\pi}{180}\,\text{radian} \approx 0.228\,\text{radian}$$

We use $\theta = 0.228$ radian and $r = 3960$ miles in equation (4). The distance between the two cities is

$$s = r\theta = 3960 \cdot 0.228 \approx 903\,\text{miles}$$

When an angle is measured in degrees, the degree symbol will always be shown. However, when an angle is measured in radians, we will follow the usual practice and omit the word *radians*. So, if the measure of an angle is given as $\frac{\pi}{6}$ it is understood to mean $\frac{\pi}{6}$ radian.

NOW WORK PROBLEM 91.

Area of a Sector

Figure 16

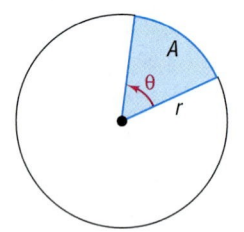

5   Consider a circle of radius r. Suppose that θ, measured in radians, is a central angle of this circle. See Figure 16. We seek a formula for the area A of the sector formed by the angle θ (shown in blue).

Now consider a circle of radius r and two central angles θ and θ_1, both measured in radians. See Figure 17. From geometry, we know the ratio of the measures of the angles equals the ratio of the corresponding areas of the sectors formed by these angles. That is,

Figure 17

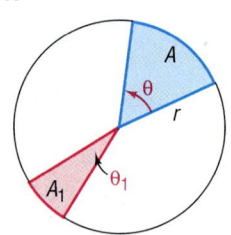

$$\frac{\theta}{\theta_1} = \frac{A}{A_1}$$

Suppose that $\theta_1 = 2\pi$ radians. Then A_1 = area of the circle = πr^2. Solving for A, we find

$$A = A_1 \frac{\theta}{\theta_1} = \pi r^2 \frac{\theta}{2\pi} = \frac{1}{2}r^2\theta$$

Theorem

Area of a Sector

The area A of the sector of a circle of radius r formed by a central angle of θ radians is

$$A = \frac{1}{2}r^2\theta \qquad (8)$$

EXAMPLE 7   **Finding the Area of a Sector of a Circle**

Find the area of the sector of a circle of radius 2 feet formed by an angle of $30°$. Round the answer to two decimal places.

Solution   We use equation (8) with $r = 2$ feet and $\theta = 30° = \dfrac{\pi}{6}$ radian. [Remember, in equation (8), θ must be in radians.] The area A of the sector is

$$A = \frac{1}{2}r^2\theta = \frac{1}{2}(2)^2\frac{\pi}{6} = \frac{\pi}{3}\text{ square feet} \approx 1.05 \text{ square feet}$$

rounded to two decimal places.

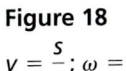 **NOW WORK PROBLEM 45.**

Circular Motion

⑥   We have already defined the average speed of an object as the distance traveled divided by the elapsed time. Suppose that an object moves around a circle of radius r at a constant speed. If s is the distance traveled in time t around this circle, then the **linear speed** v of the object is defined as

$$v = \frac{s}{t} \tag{9}$$

Figure 18

$v = \dfrac{s}{t}; \omega = \dfrac{\theta}{t}$

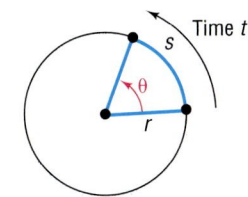

Time t

As this object travels around the circle, suppose that θ (measured in radians) is the central angle swept out in time t. See Figure 18. Then the **angular speed** ω (the Greek letter omega) of this object is the angle (measured in radians) swept out divided by the elapsed time; that is,

$$\omega = \frac{\theta}{t} \tag{10}$$

Angular speed is the way the turning rate of an engine is described. For example, an engine idling at 900 rpm (revolutions per minute) is one that rotates at an angular speed of

$$900\,\frac{\text{revolutions}}{\text{minute}} = 900\,\frac{\cancel{\text{revolutions}}}{\text{minute}} \cdot 2\pi\,\frac{\text{radians}}{\cancel{\text{revolution}}} = 1800\pi\,\frac{\text{radians}}{\text{minute}}$$

There is an important relationship between linear speed and angular speed:

$$\text{linear speed} = v = \underset{\underset{(9)}{\uparrow}}{\frac{s}{t}} = \underset{\underset{s\,=\,r\theta}{\uparrow}}{\frac{r\theta}{t}} = r\left(\frac{\theta}{t}\right)$$

Then, using equation (10), we obtain

$$v = r\omega \tag{11}$$

where ω is measured in radians per unit time.

When using equation (11), remember that $v = \dfrac{s}{t}$ (the linear speed) has the dimensions of length per unit of time (such as feet per second or miles per hour), r (the radius of the circular motion) has the same length dimension as s, and ω (the angular speed) has the dimensions of radians per unit of time. If the angular speed is given in terms of *revolutions* per unit of time (as is often the case), be sure to convert it to *radians* per unit of time before attempting to use equation (11).

EXAMPLE 8   **Finding Linear Speed**

A child is spinning a rock at the end of a 2-foot rope at the rate of 180 revolutions per minute (rpm). Find the linear speed of the rock when it is released.

Figure 19

Solution   Look at Figure 19. The rock is moving around a circle of radius $r = 2$ feet. The angular speed ω of the rock is

$$\omega = 180\frac{\text{revolutions}}{\text{minute}} = 180\frac{\cancel{\text{revolutions}}}{\text{minute}} \cdot 2\pi\frac{\text{radians}}{\cancel{\text{revolution}}} = 360\pi\frac{\text{radians}}{\text{minute}}$$

From equation (11), the linear speed v of the rock is

$$v = r\omega = 2\text{ feet} \cdot 360\pi\frac{\text{radians}}{\text{minute}} = 720\pi\frac{\text{feet}}{\text{minute}} \approx 2262\frac{\text{feet}}{\text{minute}}$$

The linear speed of the rock when it is released is 2262 ft/min $\approx$ 25.7 mi/hr. ∎

 **NOW WORK PROBLEM 87.**

8.1 Concepts and Vocabulary

In Problems 1–3, fill in the blanks.

1. An angle θ is in _____ _____ if its vertex is at the origin of a rectangular coordinate system and its initial side coincides with the positive x-axis.

2. On a circle of radius r, a central angle of θ radians subtends an arc of length $s =$ _____; the area of the sector formed by this angle θ is $A =$ _____.

3. An object travels around a circle of radius r with constant speed. If s is the distance traveled in time t around the circle and θ is the central angle (in radians) swept out in time t, then the linear speed of the object is $v =$ _____ and the angular speed of the object is $\omega =$ _____.

In Problems 4–6, answer True or False to each statement.

4. $\pi = 180$.

5. $180° = \pi$ radians

6. On the unit circle, if s is the length of the arc subtended by a central angle θ, measured in radians, then $s = \theta$.

7. What is 1 radian?

8. Which angle has the larger measure: 1 degree or 1 radian? Or are they equal?

9. Explain the difference between linear speed and angular speed.

8.1 Exercises

In Problems 1–12, draw each angle.

1. 30° **2.** 60° **3.** 135° **4.** −120° **5.** 450° **6.** 540°

7. $\dfrac{3\pi}{4}$ **8.** $\dfrac{4\pi}{3}$ **9.** $-\dfrac{\pi}{6}$ **10.** $-\dfrac{2\pi}{3}$ **11.** $\dfrac{16\pi}{3}$ **12.** $\dfrac{21\pi}{4}$

In Problems 13–24, convert each angle in degrees to radians. Express your answer as a multiple of π.

13. 30° **14.** 120° **15.** 240° **16.** 330° **17.** −60° **18.** −30°

19. 180° **20.** 270° **21.** −135° **22.** −225° **23.** −90° **24.** −180°

In Problems 25–36, convert each angle in radians to degrees.

25. $\dfrac{\pi}{3}$ **26.** $\dfrac{5\pi}{6}$ **27.** $-\dfrac{5\pi}{4}$ **28.** $-\dfrac{2\pi}{3}$ **29.** $\dfrac{\pi}{2}$ **30.** 4π

31. $\dfrac{\pi}{12}$ **32.** $\dfrac{5\pi}{12}$ **33.** $-\dfrac{\pi}{2}$ **34.** $-\pi$ **35.** $-\dfrac{\pi}{6}$ **36.** $-\dfrac{3\pi}{4}$

In Problems 37–44, s denotes the length of the arc of a circle of radius r subtended by the central angle θ. Find the missing quantity. Round answers to three decimal places, if necessary.

37. $r = 10$ meters, $\theta = \dfrac{1}{2}$ radian, $s = ?$

38. $r = 6$ feet, $\theta = 2$ radians, $s = ?$

39. $\theta = \dfrac{1}{3}$ radian, $s = 2$ feet, $r = ?$

40. $\theta = \dfrac{1}{4}$ radian, $s = 6$ centimeters, $r = ?$

41. $r = 5$ miles, $s = 3$ miles, $\theta = ?$

42. $r = 6$ meters, $s = 8$ meters, $\theta = ?$

43. $r = 2$ inches, $\theta = 30°$, $s = ?$

44. $r = 3$ meters, $\theta = 120°$, $s = ?$

In Problems 45–52, A denotes the area of the sector of a circle of radius r formed by the central angle θ. Find the missing quantity. Round answers to three decimal places, if necessary.

45. $r = 10$ meters, $\theta = \dfrac{1}{2}$ radian, $A = ?$

46. $r = 6$ feet, $\theta = 2$ radians, $A = ?$

47. $\theta = \dfrac{1}{3}$ radian, $A = 2$ square feet, $r = ?$

48. $\theta = \dfrac{1}{4}$ radian, $A = 6$ square centimeters, $r = ?$

49. $r = 5$ miles, $A = 3$ square miles, $\theta = ?$

50. $r = 6$ meters, $A = 8$ square meters, $\theta = ?$

51. $r = 2$ inches, $\theta = 30°$, $A = ?$

52. $r = 3$ meters, $\theta = 120°$, $A = ?$

In Problems 53–56, find the length s and area A. Round answers to three decimal places.

53. 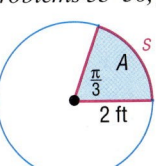 **54.** 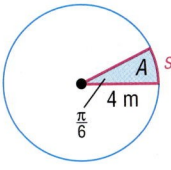 **55.** 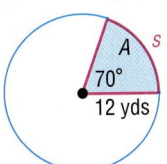 **56.**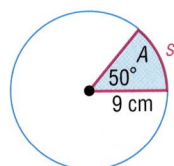

In Problems 57–62, convert each angle in degrees to radians. Express your answer in decimal form, rounded to two decimal places.

57. 17° **58.** 73° **59.** −40° **60.** −51° **61.** 125° **62.** 350°

In Problems 63–68, convert each angle in radians to degrees. Express your answer in decimal form, rounded to two decimal places.

63. 3.14 **64.** 0.75 **65.** 2 **66.** 3 **67.** 6.32 **68.** $\sqrt{2}$

In Problems 69–74, convert each angle to a decimal in degrees. Round your answer to two decimal places.

69. 40°10′25″ **70.** 61°42′21″ **71.** 1°2′3″ **72.** 73°40′40″ **73.** 9°9′9″ **74.** 98°22′45″

In Problems 75–80, convert each angle to D°M′S″ form. Round your answer to the nearest second.

75. 40.32° **76.** 61.24° **77.** 18.255° **78.** 29.411° **79.** 19.99° **80.** 44.01°

81. Minute Hand of a Clock The minute hand of a clock is 6 inches long. How far does the tip of the minute hand move in 15 minutes? How far does it move in 25 minutes?

82. Movement of a Pendulum A pendulum swings through an angle of 20° each second. If the pendulum is 40 inches long, how far does its tip move each second?

83. Area of a Sector Find the area of the sector of a circle of radius 4 meters formed by an angle of 45°. Round the answer to two decimal places.

84. Area of a Sector Find the area of the sector of a circle of radius 3 centimeters formed by an angle of 60°. Round the answer to two decimal places.

85. Watering a Lawn A water sprinkler sprays water over a distance of 30 feet while rotating through an angle of 135°. What area of lawn receives water?

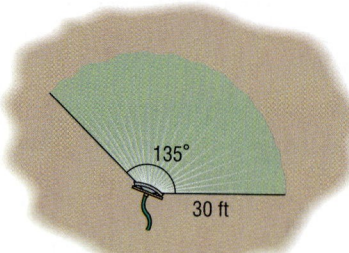

135°
30 ft

86. Designing a Water Sprinkler An engineer is asked to design a water sprinkler that will cover a field of 100 square yards that is in the shape of a sector of a circle of radius 50 yards. Through what angle should the sprinkler rotate?

87. Motion on a Circle An object is traveling around a circle with a radius of 5 centimeters. If in 20 seconds a central angle of $\frac{1}{3}$ radian is swept out, what is the angular speed of the object? What is its linear speed?

88. Motion on a Circle An object is traveling around a circle with a radius of 2 meters. If in 20 seconds the object travels 5 meters, what is its angular speed? What is its linear speed?

89. Bicycle Wheels The diameter of each wheel of a bicycle is 26 inches. If you are traveling at a speed of 35 miles per hour on this bicycle, through how many revolutions per minute are the wheels turning?

90. Car Wheels The radius of each wheel of a car is 15 inches. If the wheels are turning at the rate of 3 revolutions per second, how fast is the car moving? Express your answer in inches per second and in miles per hour.

In Problems 91–94, the latitude of a location L is the angle formed by a ray drawn from the center of Earth to the Equator and a ray drawn from the center of Earth to L. See the figure.

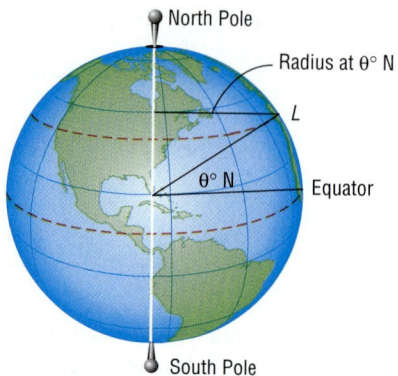

North Pole
Radius at θ° N
L
θ° N
Equator
South Pole

91. Distance between Cities Memphis, Tennessee, is due north of New Orleans, Louisiana. Find the distance between Memphis (35°9′ north latitude) and New Orleans (29°57′ north latitude). Assume that the radius of Earth is 3960 miles.

92. Distance between Cities Charleston, West Virginia, is due north of Jacksonville, Florida. Find the distance between Charleston (38°21′ north latitude) and Jacksonville (30°20′ north latitude). Assume that the radius of Earth is 3960 miles.

93. Linear Speed on Earth Earth rotates on an axis through its poles. The distance from the axis to a location on Earth 30° north latitude is about 3429.5 miles. Therefore, a location on Earth at 30° north latitude is spinning on a circle of radius 3429.5 miles. Compute the linear speed on the surface of Earth at 30° north latitude.

94. Linear Speed on Earth Earth rotates on an axis through its poles. The distance from the axis to a location on Earth 40° north latitude is about 3033.5 miles. Therefore, a location on Earth at 40° north latitude is spinning on a circle of radius 3033.5 miles. Compute the linear speed on the surface of Earth at 40° north latitude.

95. Speed of the Moon The mean distance of the Moon from Earth is 2.39×10^5 miles. Assuming that the orbit of the Moon around Earth is circular and that 1 revolution takes 27.3 days, find the linear speed of the Moon. Express your answer in miles per hour.

96. Speed of Earth The mean distance of Earth from the Sun is 9.29×10^7 miles. Assuming that the orbit of Earth around the Sun is circular and that 1 revolution takes 365 days, find the linear speed of Earth. Express your answer in miles per hour.

97. Pulleys Two pulleys, one with radius 2 inches and the other with radius 8 inches, are connected by a belt. (See the figure on page 623.) If the 2-inch pulley is caused to rotate at 3 revolutions per minute, determine the revolutions per minute of the 8-inch pulley.

[**Hint:** The linear speeds of the pulleys, that is, the speed of the belt, are the same.]

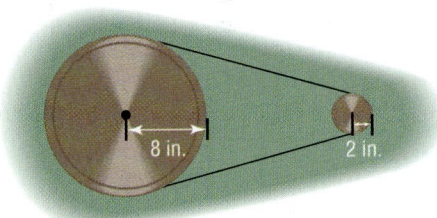

98. Ferris Wheels A neighborhood carnival has a Ferris wheel whose radius is 30 feet. You measure the time it takes for one revolution to be 70 seconds. What is the linear speed (in feet per second) of this Ferris wheel? What is the angular speed in radians per second?

99. Computing the Speed of a River Current To approximate the speed of the current of a river, a circular paddle wheel with radius 4 feet is lowered into the water. If the current causes the wheel to rotate at a speed of 10 revolutions per minute, what is the speed of the current? Express your answer in miles per hour.

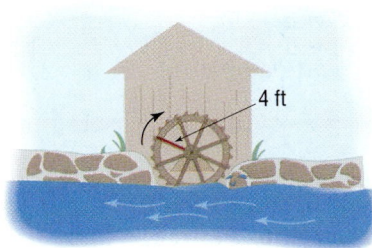

100. Spin Balancing Tires A spin balancer rotates the wheel of a car at 480 revolutions per minute. If the diameter of the wheel is 26 inches, what road speed is being tested? Express your answer in miles per hour. At how many revolutions per minute should the balancer be set to test a road speed of 80 miles per hour?

101. The Cable Cars of San Francisco At the Cable Car Museum you can see the four cable lines that are used to pull cable cars up and down the hills of San Francisco. Each cable travels at a speed of 9.55 miles per hour, caused by a rotating wheel whose diameter is 8.5 feet. How fast is the wheel rotating? Express your answer in revolutions per minute.

102. Difference in Time of Sunrise Naples, Florida, is approximately 90 miles due west of Ft. Lauderdale. How much sooner would a person in Ft. Lauderdale first see the rising Sun than a person in Naples?

[**Hint:** Consult the figure. When a person at Q sees the first rays of the Sun, a person at P is still in the dark. The person at P sees the first rays after Earth has rotated so that P is at the location Q. Now use the fact that at the

latitude of Ft. Lauderdale in 24 hours a length of arc of $2\pi(3559)$ miles is subtended.]

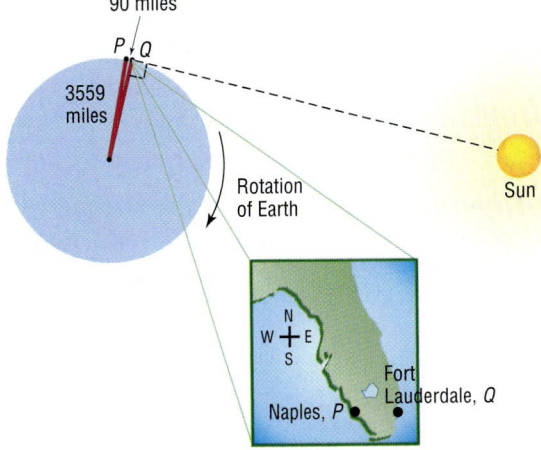

103. Keeping Up with the Sun How fast would you have to travel on the surface of Earth at the equator to keep up with the Sun (that is, so that the Sun would appear to remain in the same position in the sky)?

104. Nautical Miles A **nautical mile** equals the length of arc subtended by a central angle of 1 minute on a great circle* on the surface of Earth. (See the figure.) If the radius of Earth is taken as 3960 miles, express 1 nautical mile in terms of ordinary, or **statute,** miles.

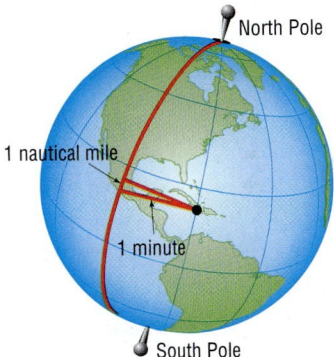

105. Pulleys Two pulleys, one with radius r_1 and the other with radius r_2, are connected by a belt. The pulley with radius r_1 rotates at ω_1 revolutions per minute, whereas the pulley with radius r_2 rotates at ω_2 revolutions per minute. Show that $\dfrac{r_1}{r_2} = \dfrac{\omega_2}{\omega_1}$.

106. Do you prefer to measure angles using degrees or radians? Provide justification and a rationale for your choice.

107. Discuss why ships and airplanes use nautical miles to measure distance. Explain the difference between a nautical mile and a statute mile.

108. Investigate the way that speed bicycles work. In particular, explain the differences and similarities between 5-speed and 9-speed derailleurs. Be sure to include a discussion of linear speed and angular speed.

*Any circle drawn on the surface of Earth that divides Earth into two equal hemispheres.

PREPARING FOR THIS SECTION

Before getting started, review the following:

✓ Pythagorean Theorem (Section R.3, pp. 25–26)   ✓ Functions (Section 2.1, pp. 196–207)

8.2 RIGHT TRIANGLE TRIGONOMETRY

OBJECTIVES

1. Find the Value of Trigonometric Functions of Acute Angles
2. Use the Fundamental Identities
3. Find the Remaining Trigonometric Functions Given the Value of One of Them
4. Use the Complementary Angle Theorem

A triangle in which one angle is a right angle (90°) is called a **right triangle.** Recall that the side opposite the right angle is called the **hypotenuse,** and the remaining two sides are called the **legs** of the triangle. In Figure 20 we have labeled the hypotenuse as c to indicate that its length is c units, and, in a like manner, we have labeled the legs as a and b. Because the triangle is a right triangle, the Pythagorean Theorem tells us that

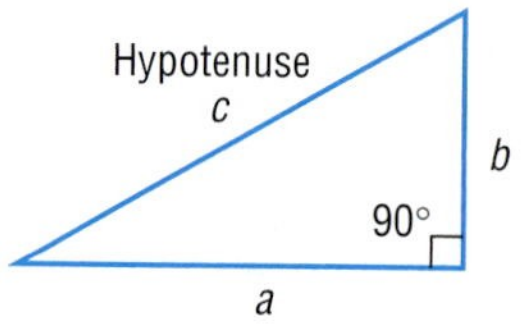

$$c^2 = a^2 + b^2$$

Now, suppose that θ is an **acute angle;** that is, $0° < \theta < 90°$ (if θ is measured in degrees) and $0 < \theta < \dfrac{\pi}{2}$ (if θ is measured in radians). See Figure 21(a). Using this acute angle θ, we can form a right triangle, like the one illustrated in Figure 21(b), with hypotenuse of length c and legs of lengths a and b. Using the three sides of this triangle, we can form exactly six ratios:

$$\frac{b}{c}, \quad \frac{a}{c}, \quad \frac{b}{a}, \quad \frac{c}{b}, \quad \frac{c}{a}, \quad \frac{a}{b}$$

Figure 20

(triangle figure with Hypotenuse c, legs a and b, and the right angle $90°$)

Figure 21

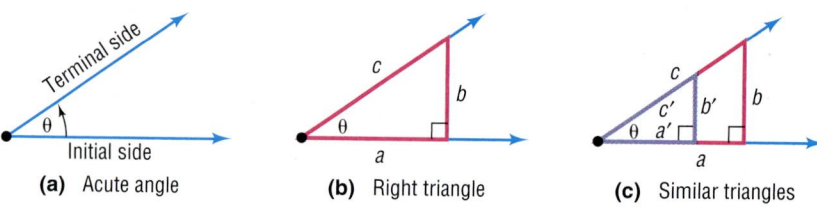

(a) Acute angle   **(b)** Right triangle   **(c)** Similar triangles

In fact, these ratios depend only on the size of the angle θ and not on the triangle formed. To see why, look at Figure 21(c). Any two right triangles formed using the angle θ will be similar and, hence, corresponding ratios will be equal. As a result,

$$\frac{b}{c} = \frac{b'}{c'} \qquad \frac{a}{c} = \frac{a'}{c'} \qquad \frac{b}{a} = \frac{b'}{a'} \qquad \frac{c}{b} = \frac{c'}{b'} \qquad \frac{c}{a} = \frac{c'}{a'} \qquad \frac{a}{b} = \frac{a'}{b'}$$

Because the ratios depend only on the angle θ and not on the triangle itself, we give each ratio a name that involves θ: sine of θ, cosine of θ, tangent of θ, cosecant of θ, secant of θ, and cotangent of θ.

The six ratios of a right triangle are called **trigonometric functions of acute angles** and are defined as follows:

Function Name	Abbreviation	Value
sine of θ	$\sin \theta$	$\dfrac{b}{c}$
cosine of θ	$\cos \theta$	$\dfrac{a}{c}$
tangent of θ	$\tan \theta$	$\dfrac{b}{a}$
cosecant of θ	$\csc \theta$	$\dfrac{c}{b}$
secant of θ	$\sec \theta$	$\dfrac{c}{a}$
cotangent of θ	$\cot \theta$	$\dfrac{a}{b}$

Figure 22

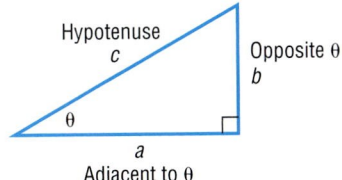

Hypotenuse
c

Opposite θ
b

θ

a
Adjacent to θ

As an aid to remembering these definitions, it may be helpful to refer to the lengths of the sides of the triangle by the names *hypotenuse (c)*, *opposite (b)*, and *adjacent (a)*. See Figure 22. In terms of these names, we have the following ratios:

$$\sin \theta = \frac{\text{opposite}}{\text{hypotenuse}} = \frac{b}{c} \qquad \cos \theta = \frac{\text{adjacent}}{\text{hypotenuse}} = \frac{a}{c} \qquad \tan \theta = \frac{\text{opposite}}{\text{adjacent}} = \frac{b}{a}$$

$$\csc \theta = \frac{\text{hypotenuse}}{\text{opposite}} = \frac{c}{b} \qquad \sec \theta = \frac{\text{hypotenuse}}{\text{adjacent}} = \frac{c}{a} \qquad \cot \theta = \frac{\text{adjacent}}{\text{opposite}} = \frac{a}{b} \tag{1}$$

Notice that each of the trigonometric functions of the acute angle θ is positive.

EXAMPLE 1 **Finding the Value of Trigonometric Functions**

Figure 23

5

Opposite θ

θ

3

Find the value of each of the six trigonometric functions of the angle θ in Figure 23.

Solution We see in Figure 23 that the two given sides of the triangle are

$$c = \text{hypotenuse} = 5 \qquad a = \text{adjacent} = 3$$

To find the length of the opposite side, we use the Pythagorean Theorem.

$$(\text{adjacent})^2 + (\text{opposite})^2 = (\text{hypotenuse})^2$$
$$3^2 + (\text{opposite})^2 = 5^2$$
$$(\text{opposite})^2 = 25 - 9 = 16$$
$$\text{opposite} = 4$$

Now that we know the lengths of the three sides, we use the ratios in (1) to find the value of each of the six trigonometric functions:

$$\sin \theta = \frac{\text{opposite}}{\text{hypotenuse}} = \frac{4}{5} \qquad \cos \theta = \frac{\text{adjacent}}{\text{hypotenuse}} = \frac{3}{5} \qquad \tan \theta = \frac{\text{opposite}}{\text{adjacent}} = \frac{4}{3}$$

$$\csc \theta = \frac{\text{hypotenuse}}{\text{opposite}} = \frac{5}{4} \qquad \sec \theta = \frac{\text{hypotenuse}}{\text{adjacent}} = \frac{5}{3} \qquad \cot \theta = \frac{\text{adjacent}}{\text{opposite}} = \frac{3}{4}$$

NOW WORK PROBLEM 1.

Fundamental Identities

You may have observed some relationships that exist among the six trigonometric functions of acute angles. For example, the **reciprocal identities** are

Reciprocal Identities

$$\csc \theta = \frac{1}{\sin \theta} \qquad \sec \theta = \frac{1}{\cos \theta} \qquad \cot \theta = \frac{1}{\tan \theta} \qquad (2)$$

Two other fundamental identities that are easy to see are the **quotient identities.**

Quotient Identities

$$\tan \theta = \frac{\sin \theta}{\cos \theta} \qquad \cot \theta = \frac{\cos \theta}{\sin \theta} \qquad (3)$$

If $\sin \theta$ and $\cos \theta$ are known, formulas (2) and (3) make it easy to find the values of the remaining trigonometric functions.

EXAMPLE 2 **Finding the Value of the Remaining Trigonometric Functions, Given $\sin \theta$ and $\cos \theta$**

Given $\sin \theta = \dfrac{\sqrt{5}}{5}$ and $\cos \theta = \dfrac{2\sqrt{5}}{5}$, find the value of each of the four remaining trigonometric functions of θ.

Solution Based on formula (3), we have

$$\tan \theta = \frac{\sin \theta}{\cos \theta} = \frac{\dfrac{\sqrt{5}}{5}}{\dfrac{2\sqrt{5}}{5}} = \frac{1}{2}$$

Then we use the reciprocal identities from formula (2) to get

$$\csc \theta = \frac{1}{\sin \theta} = \frac{1}{\dfrac{\sqrt{5}}{5}} = \frac{5}{\sqrt{5}} = \sqrt{5} \qquad \sec \theta = \frac{1}{\cos \theta} = \frac{1}{\dfrac{2\sqrt{5}}{5}} = \frac{5}{2\sqrt{5}} = \frac{\sqrt{5}}{2} \qquad \cot \theta = \frac{1}{\tan \theta} = \frac{1}{\dfrac{1}{2}} = 2$$

NOW WORK PROBLEM 11.

Figure 24
$a^2 + b^2 = c^2$

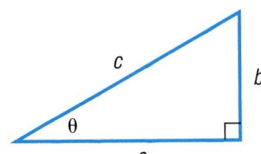

Refer now to the triangle in Figure 24. The Pythagorean Theorem states that $a^2 + b^2 = c^2$, which we write as

$$b^2 + a^2 = c^2$$

Dividing each side by c^2, we get

$$\frac{b^2}{c^2} + \frac{a^2}{c^2} = 1 \quad \text{or} \quad \left(\frac{b}{c}\right)^2 + \left(\frac{a}{c}\right)^2 = 1$$

In terms of trigonometric functions of the angle θ, this equation states that

$$(\sin \theta)^2 + (\cos \theta)^2 = 1 \qquad (4)$$

Equation (4) is, in fact, an identity, since the equation is true for any acute angle θ.

It is customary to write $\sin^2 \theta$ instead of $(\sin \theta)^2$, $\cos^2 \theta$ instead of $(\cos \theta)^2$, and so on. With this notation, we can rewrite equation (4) as

$$\sin^2 \theta + \cos^2 \theta = 1 \qquad (5)$$

Another identity can be obtained from equation (5) by dividing each side by $\cos^2 \theta$.

$$\frac{\sin^2 \theta}{\cos^2 \theta} + 1 = \frac{1}{\cos^2 \theta}$$

Now use formulas (2) and (3) to get

$$\tan^2 \theta + 1 = \sec^2 \theta \qquad (6)$$

Similarly, by dividing each side of equation (5) by $\sin^2 \theta$, we get

$$1 + \cot^2 \theta = \csc^2 \theta \qquad (7)$$

Collectively, the identities in equations (5), (6), and (7) are referred to as the **Pythagorean identities.**

Let's pause here to summarize the fundamental identities.

Fundamental Identities

$$\tan \theta = \frac{\sin \theta}{\cos \theta} \qquad \cot \theta = \frac{\cos \theta}{\sin \theta}$$

$$\csc \theta = \frac{1}{\sin \theta} \qquad \sec \theta = \frac{1}{\cos \theta} \qquad \cot \theta = \frac{1}{\tan \theta}$$

$$\sin^2 \theta + \cos^2 \theta = 1 \qquad \tan^2 \theta + 1 = \sec^2 \theta \qquad 1 + \cot^2 \theta = \csc^2 \theta$$

EXAMPLE 3 **Finding the Exact Value of a Trigonometric Expression Using Identities**

Find the exact value of each expression. Do not use a calculator.

(a) $\tan 20° - \dfrac{\sin 20°}{\cos 20°}$ (b) $\sin^2 \dfrac{\pi}{12} + \dfrac{1}{\sec^2 \dfrac{\pi}{12}}$

Solution (a) $\tan 20° - \underset{\underset{\frac{\sin \theta}{\cos \theta} = \tan \theta}{\uparrow}}{\dfrac{\sin 20°}{\cos 20°}} = \tan 20° - \tan 20° = 0$

(b) $\sin^2\dfrac{\pi}{12} + \dfrac{1}{\sec^2\dfrac{\pi}{12}} = \sin^2\dfrac{\pi}{12} + \cos^2\dfrac{\pi}{12} = 1$

$\uparrow$ $\uparrow$

$\cos\theta = \dfrac{1}{\sec\theta}$ $\sin^2\theta + \cos^2\theta = 1$

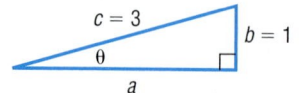 **NOW WORK PROBLEM 29.**

③   Once the value of one trigonometric function is known, it is possible to find the value of each of the remaining five trigonometric functions.

EXAMPLE 4   **Finding the Value of the Remaining Trigonometric Functions, Given $\sin\theta$, θ Acute**

Given that $\sin\theta = \dfrac{1}{3}$ and θ is an acute angle, find the exact value of each of the remaining five trigonometric functions of θ.

Solution   We solve this problem in two ways: The first way uses the definition of the trigonometric functions; the second method uses the fundamental identities.

Solution 1
Using the Definition   We draw a right triangle with acute angle θ, opposite side of length $b = 1$, and hypotenuse of length $c = 3$ $\left(\text{because } \sin\theta = \dfrac{1}{3} = \dfrac{b}{c}\right)$. See Figure 25. The adjacent side a can be found by using the Pythagorean Theorem.

Figure 25

$c = 3$ $b = 1$
θ
a

$$a^2 + 1^2 = 3^2$$
$$a^2 + 1 = 9$$
$$a^2 = 8$$
$$a = 2\sqrt{2}$$

Now the definitions given in equation (1) can be used to find the value of each of the remaining five trigonometric functions. (Refer back to the method used in Example 1.) Using $a = 2\sqrt{2}$, $b = 1$, and $c = 3$, we have

$$\cos\theta = \frac{a}{c} = \frac{2\sqrt{2}}{3} \qquad\qquad \tan\theta = \frac{b}{a} = \frac{1}{2\sqrt{2}} = \frac{\sqrt{2}}{4}$$

$$\csc\theta = \frac{c}{b} = \frac{3}{1} = 3 \qquad \sec\theta = \frac{c}{a} = \frac{3}{2\sqrt{2}} = \frac{3\sqrt{2}}{4} \qquad \cot\theta = \frac{a}{b} = \frac{2\sqrt{2}}{1} = 2\sqrt{2}$$

Solution 2
Using Identities   We begin by seeking $\cos\theta$, which can be found by using the Pythagorean identity from equation (5).

$$\sin^2\theta + \cos^2\theta = 1$$
$$\frac{1}{9} + \cos^2\theta = 1 \qquad\qquad \sin\theta = \frac{1}{3}$$
$$\cos^2\theta = 1 - \frac{1}{9} = \frac{8}{9}$$

Because $\cos\theta > 0$ for an acute angle θ, we have

$$\cos\theta = \sqrt{\frac{8}{9}} = \frac{2\sqrt{2}}{3}$$

Now we know that $\sin\theta = \dfrac{1}{3}$ and $\cos\theta = \dfrac{2\sqrt{2}}{3}$, so we can proceed as we did in Example 2.

$$\tan\theta = \frac{\sin\theta}{\cos\theta} = \frac{\dfrac{1}{3}}{\dfrac{2\sqrt{2}}{3}} = \frac{1}{2\sqrt{2}} = \frac{\sqrt{2}}{4} \qquad \cot\theta = \frac{1}{\tan\theta} = \frac{1}{\dfrac{\sqrt{2}}{4}} = \frac{4}{\sqrt{2}} = 2\sqrt{2}$$

$$\sec\theta = \frac{1}{\cos\theta} = \frac{1}{\dfrac{2\sqrt{2}}{3}} = \frac{3}{2\sqrt{2}} = \frac{3\sqrt{2}}{4} \qquad \csc\theta = \frac{1}{\sin\theta} = \frac{1}{\dfrac{1}{3}} = 3$$

Finding the Values of the Trigonometric Functions When One Is Known

Given the value of one trigonometric function of an acute angle θ, the exact value of each of the remaining five trigonometric functions of θ can be found in either of two ways.

Method 1 Using the Definition

STEP 1: Draw a right triangle showing the acute angle θ.

STEP 2: Two of the sides can then be assigned values based on the given trigonometric function.

STEP 3: Find the length of the third side by using the Pythagorean Theorem.

STEP 4: Use the definitions in equation (1) to find the value of each of the remaining trigonometric functions.

Method 2 Using Identities

Use appropriately selected identities to find the value of each of the remaining trigonometric functions.

EXAMPLE 5   **Given One Value of a Trigonometric Function, Find the Remaining Ones**

Given $\tan\theta = \dfrac{1}{2}$, θ an acute angle, find the exact value of each of the remaining five trigonometric functions of θ.

Solution 1
Using the Definition

Figure 26 shows a right triangle with acute angle θ, where

$$\tan\theta = \frac{1}{2} = \frac{\text{opposite}}{\text{adjacent}} = \frac{b}{a}$$

Choose $b = 1$ and $a = 2$. The hypotenuse c can be found by using the Pythagorean Theorem.

$$c^2 = a^2 + b^2 = 2^2 + 1^2 = 5$$
$$c = \sqrt{5}$$

Figure 26
$\tan\theta = \dfrac{1}{2}$

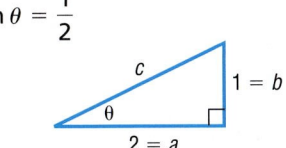

Now apply the definitions using $a = 2$, $b = 1$, and $c = \sqrt{5}$.

$$\sin\theta = \frac{b}{c} = \frac{1}{\sqrt{5}} = \frac{\sqrt{5}}{5} \qquad \cos\theta = \frac{a}{c} = \frac{2}{\sqrt{5}} = \frac{2\sqrt{5}}{5}$$

$$\csc\theta = \frac{c}{b} = \frac{\sqrt{5}}{1} = \sqrt{5} \qquad \sec\theta = \frac{c}{a} = \frac{\sqrt{5}}{2} \qquad \cot\theta = \frac{a}{b} = \frac{2}{1} = 2$$

Solution 2
Using Identities

We use the Pythagorean identity that involves $\tan\theta$:

$$\tan^2\theta + 1 = \sec^2\theta$$

$$\left(\frac{1}{2}\right)^2 + 1 = \sec^2\theta \qquad\qquad tan\,\theta = \frac{1}{2}$$

$$\sec^2\theta = \frac{1}{4} + 1 = \frac{5}{4} \qquad\qquad \textit{Proceed to solve for sec }\theta.$$

$$\sec\theta = \frac{\sqrt{5}}{2} \qquad\qquad sec\,\theta > 0,\, \theta \textit{ acute}$$

Now

$$\cos\theta = \frac{1}{\sec\theta} = \frac{1}{\dfrac{\sqrt{5}}{2}} = \frac{2}{\sqrt{5}} = \frac{2\sqrt{5}}{5}$$

$$\tan\theta = \frac{\sin\theta}{\cos\theta}, \quad \text{so} \quad \sin\theta = (\tan\theta)(\cos\theta) = \frac{1}{2}\cdot\frac{2\sqrt{5}}{5} = \frac{\sqrt{5}}{5}$$

$$\csc\theta = \frac{1}{\sin\theta} = \frac{1}{\dfrac{\sqrt{5}}{5}} = \sqrt{5}$$

$$\cot\theta = \frac{1}{\tan\theta} = \frac{1}{\dfrac{1}{2}} = 2$$

 NOW WORK PROBLEM 15.

Complementary Angles; Cofunctions

④ Two acute angles are called **complementary** if their sum is a right angle. Because the sum of the angles of any triangle is 180°, it follows that, for a right triangle, the two acute angles are complementary.

Refer now to Figure 27; we have labeled the angle opposite side b as β and the angle opposite side a as α. Notice that side a is adjacent to angle β and is opposite angle α. Similarly, side b is opposite angle β and is adjacent to angle α. As a result,

Figure 27

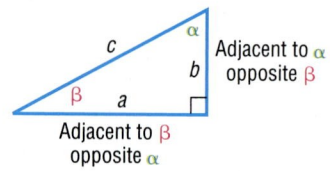

$$\sin\beta = \frac{b}{c} = \cos\alpha \qquad \cos\beta = \frac{a}{c} = \sin\alpha \qquad \tan\beta = \frac{b}{a} = \cot\alpha$$

$$\hspace{11cm}(8)$$

$$\csc\beta = \frac{c}{b} = \sec\alpha \qquad \sec\beta = \frac{c}{a} = \csc\alpha \qquad \cot\beta = \frac{a}{b} = \tan\alpha$$

Because of these relationships, the functions sine and cosine, tangent and cotangent, and secant and cosecant are called **cofunctions** of each other. The identities (8) may be expressed in words as follows:

Theorem   **Complementary Angle Theorem**

Cofunctions of complementary angles are equal.

Here are examples of this theorem.

Complementary angles	Complementary angles	Complementary angles
$\sin 30° = \cos 60°$	$\tan 40° = \cot 50°$	$\sec 80° = \csc 10°$
Cofunctions	Cofunctions	Cofunctions

If an angle θ is measured in degrees, we will use the degree symbol when writing a trigonometric function of θ, as, for example, in $\sin 30°$ and $\tan 45°$. If an angle θ is measured in radians, then no symbol is used when writing a trigonometric function of θ, as, for example, in $\cos \pi$ and $\sec \dfrac{\pi}{3}$.

If θ is an acute angle measured in degrees, the angle $90° - \theta$ (or $\dfrac{\pi}{2} - \theta$, if θ is in radians) is the angle complementary to θ. Table 2 restates the preceding theorem on cofunctions.

TABLE 2

θ (Degrees)	θ (Radians)
$\sin \theta = \cos(90° - \theta)$	$\sin \theta = \cos\left(\dfrac{\pi}{2} - \theta\right)$
$\cos \theta = \sin(90° - \theta)$	$\cos \theta = \sin\left(\dfrac{\pi}{2} - \theta\right)$
$\tan \theta = \cot(90° - \theta)$	$\tan \theta = \cot\left(\dfrac{\pi}{2} - \theta\right)$
$\csc \theta = \sec(90° - \theta)$	$\csc \theta = \sec\left(\dfrac{\pi}{2} - \theta\right)$
$\sec \theta = \csc(90° - \theta)$	$\sec \theta = \csc\left(\dfrac{\pi}{2} - \theta\right)$
$\cot \theta = \tan(90° - \theta)$	$\cot \theta = \tan\left(\dfrac{\pi}{2} - \theta\right)$

Although the angle θ in Table 2 is acute, we will see later (Section 9.4) that these results are valid for any angle θ.

EXAMPLE 6   **Using the Complementary Angle Theorem**

(a) $\sin 62° = \cos(90° - 62°) = \cos 28°$

(b) $\tan \dfrac{\pi}{12} = \cot\left(\dfrac{\pi}{2} - \dfrac{\pi}{12}\right) = \cot \dfrac{5\pi}{12}$

(c) $\cos \dfrac{\pi}{4} = \sin\left(\dfrac{\pi}{2} - \dfrac{\pi}{4}\right) = \sin \dfrac{\pi}{4}$

(d) $\csc \dfrac{\pi}{6} = \sec\left(\dfrac{\pi}{2} - \dfrac{\pi}{6}\right) = \sec \dfrac{\pi}{3}$

EXAMPLE 7   **Using the Complementary Angle Theorem**

Find the exact value of each expression. Do not use a calculator.

(a) $\sec 28° - \csc 62°$ (b) $\dfrac{\sin 35°}{\cos 55°}$

Solution   (a) $\sec 28° - \csc 62° = \csc(90° - 28°) - \csc 62°$
$$= \csc 62° - \csc 62° = 0$$

(b) $\dfrac{\sin 35°}{\cos 55°} = \dfrac{\cos(90° - 35°)}{\cos 55°} = \dfrac{\cos 55°}{\cos 55°} = 1$

NOW WORK PROBLEMS 33 AND 47.

HISTORICAL FEATURE

The name *sine* for the sine function is due to a medieval confusion. The name comes from the Sanskrit word *jīva* (meaning chord), first used in India by Araybhata the Elder (AD 510). He really meant half-chord, but abbreviated it. This was brought into Arabic as *jība,* which was meaningless. Because the proper Arabic word *jaib* would be written the same way (short vowels are not written out in Arabic), *jība* was pronounced as *jaib,* which meant bosom or hollow, and *jaib* remains as the Arabic word for sine to this day. Scholars translating the Arabic works into Latin found that the word *sinus* also meant bosom or hollow, and from *sinus* we get the word *sine.*

The name *tangent,* due to Thomas Finck (1583), can be understood by looking at Figure 28. The line segment $\overline{DC}$ is tangent to the circle at C. If $d(O, B) = d(O, C) = 1$, then the length of the line segment $\overline{DC}$ is

$$d(D, C) = \frac{d(D, C)}{1} = \frac{d(D, C)}{d(O, C)} = \tan \alpha$$

The old name for the tangent is *umbra versa* (meaning turned shadow), referring to the use of the tangent in solving height problems with shadows.

Figure 28

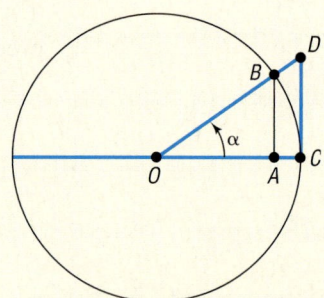

The names of the cofunctions came about as follows. If α and β are complementary angles, then $\cos \alpha = \sin \beta$. Because β is the complement of α, it was natural to write the cosine of α as *sin co* α. Probably for reasons involving ease of pronunciation, the *co* migrated to the front, and then cosine received a three-letter abbreviation to match sin, sec, and tan. The two other cofunctions were similarly treated, except that the long forms *cotan* and *cosec* survive to this day in some countries.

8.2 Concepts and Vocabulary

In Problems 1–3, fill in the blanks.

1. Two acute angles whose sum is a right angle are called _____.

2. The sine and _____ functions are cofunctions.

3. $\tan 28° = \cot$ _____.

In Problems 4–6, answer True or False to each statement.

4. $1 + \tan^2 \theta = \csc^2 \theta$.

5. If θ is an acute angle and $\sec \theta = 3$, then $\cos \theta = \dfrac{1}{3}$.

6. $\tan \dfrac{\pi}{5} = \cot \dfrac{4\pi}{5}$.

7. Write down the Reciprocal Identities.

8. Explain the Complementary Angle Theorem. Give an example that uses it.

9. If you know the value of $\tan\theta$, θ an acute angle, how would you find $\sec\theta$?

10. Name the two functions that are reciprocals and cofunctions of each other.

8.2 Exercises

In Problems 1–10, find the value of the six trigonometric functions of the angle θ in each figure.

1.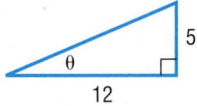

2.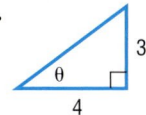

3.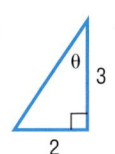

4.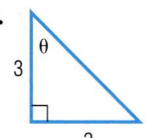

5.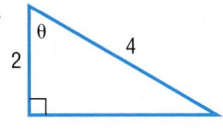

6.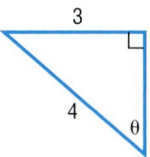

7.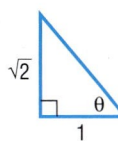

8.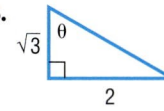

9.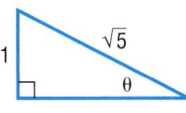

10.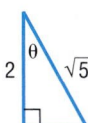

In Problems 11–14, use identities to find the exact value of each of the four remaining trigonometric functions of the acute angle θ.

11. $\sin\theta = \dfrac{1}{2}, \quad \cos\theta = \dfrac{\sqrt{3}}{2}$

12. $\sin\theta = \dfrac{\sqrt{3}}{2}, \quad \cos\theta = \dfrac{1}{2}$

13. $\sin\theta = \dfrac{2}{3}, \quad \cos\theta = \dfrac{\sqrt{5}}{3}$

14. $\sin\theta = \dfrac{1}{3}, \quad \cos\theta = \dfrac{2\sqrt{2}}{3}$

In Problems 15–26, use the definition or identities to find the exact value of each of the remaining five trigonometric functions of the acute angle θ.

15. $\sin\theta = \dfrac{\sqrt{2}}{2}$

16. $\cos\theta = \dfrac{\sqrt{2}}{2}$

17. $\cos\theta = \dfrac{1}{3}$

18. $\sin\theta = \dfrac{\sqrt{3}}{4}$

19. $\tan\theta = \dfrac{1}{2}$

20. $\cot\theta = \dfrac{1}{2}$

21. $\sec\theta = 3$

22. $\csc\theta = 5$

23. $\tan\theta = \sqrt{2}$

24. $\sec\theta = \dfrac{5}{3}$

25. $\csc\theta = 2$

26. $\cot\theta = 2$

In Problems 27–44, use Fundamental Identities and/or the Complementary Angle Theorem to find the exact value of each expression. Do not use a calculator.

27. $\sin^2 20° + \cos^2 20°$

28. $\sec^2 28° - \tan^2 28°$

29. $\sin 80° \csc 80°$

30. $\tan 10° \cot 10°$

31. $\tan 50° - \dfrac{\sin 50°}{\cos 50°}$

32. $\cot 25° - \dfrac{\cos 25°}{\sin 25°}$

33. $\sin 38° - \cos 52°$

34. $\tan 12° - \cot 78°$

35. $\dfrac{\cos 10°}{\sin 80°}$

36. $\dfrac{\cos 40°}{\sin 50°}$

37. $1 - \cos^2 20° - \cos^2 70°$

38. $1 + \tan^2 5° - \csc^2 85°$

39. $\tan 20° - \dfrac{\cos 70°}{\cos 20°}$

40. $\cot 40° - \dfrac{\sin 50°}{\sin 40°}$

41. $\tan 35° \cdot \sec 55° \cdot \cos 35°$

42. $\cot 25° \cdot \csc 65° \cdot \sin 25°$

43. $\cos 35° \sin 55° + \cos 55° \sin 35°$

44. $\sec 35° \csc 55° - \tan 35° \cot 55°$

45. Given $\sin 30° = \dfrac{1}{2}$, use trigonometric identities to find the exact value of

(a) $\cos 60°$ (b) $\cos^2 30°$

(c) $\csc \dfrac{\pi}{6}$ (d) $\sec \dfrac{\pi}{3}$

46. Given $\sin 60° = \dfrac{\sqrt{3}}{2}$, use trigonometric identities to find the exact value of

(a) $\cos 30°$ (b) $\cos^2 60°$

(c) $\sec \dfrac{\pi}{6}$ (d) $\csc \dfrac{\pi}{3}$

47. Given $\tan \theta = 4$, use trigonometric identities to find the exact value of

(a) $\sec^2 \theta$ (b) $\cot \theta$

(c) $\cot\left(\dfrac{\pi}{2} - \theta\right)$ (d) $\csc^2 \theta$

48. Given $\sec \theta = 3$, use trigonometric identities to find the exact value of

(a) $\cos \theta$ (b) $\tan^2 \theta$
(c) $\csc(90° - \theta)$ (d) $\sin^2 \theta$

49. Given $\csc \theta = 4$, use trigonometric identities to find the exact value of

(a) $\sin \theta$ (b) $\cot^2 \theta$
(c) $\sec(90° - \theta)$ (d) $\sec^2 \theta$

50. Given $\cot \theta = 2$, use trigonometric identities to find the exact value of

(a) $\tan \theta$ (b) $\csc^2 \theta$

(c) $\tan\left(\dfrac{\pi}{2} - \theta\right)$ (d) $\sec^2 \theta$

51. Given the approximation $\sin 38° \approx 0.62$, use trigonometric identities to find the approximate value of

(a) $\cos 38°$ (b) $\tan 38°$
(c) $\cot 38°$ (d) $\sec 38°$
(e) $\csc 38°$ (f) $\sin 52°$
(g) $\cos 52°$ (h) $\tan 52°$

52. Given the approximation $\cos 21° \approx 0.93$, use trigonometric identities to find the approximate value of

(a) $\sin 21°$ (b) $\tan 21°$
(c) $\cot 21°$ (d) $\sec 21°$
(e) $\csc 21°$ (f) $\sin 69°$
(g) $\cos 69°$ (h) $\tan 69°$

53. If $\sin \theta = 0.3$, find the exact value of $\sin \theta + \cos\left(\dfrac{\pi}{2} - \theta\right)$.

54. If $\tan \theta = 4$, find the exact value of $\tan \theta + \tan\left(\dfrac{\pi}{2} - \theta\right)$.

55. Find an acute angle θ that satisfies the equation $\sin \theta = \cos(2\theta + 30°)$.

56. Find an acute angle θ that satisfies the equation $\tan \theta = \cot(\theta + 45°)$.

57. Calculating the Time of a Trip   From a parking lot you want to walk to a house on the ocean. The house is located 1500 feet down a paved path that parallels the beach, which is 500 feet wide. Along the path you can walk 300

feet per minute, but in the sand on the beach you can only walk 100 feet per minute. See the illustration.

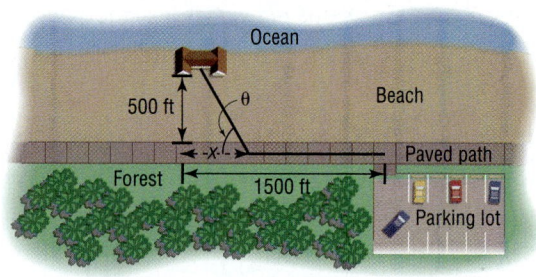

(a) Calculate the time T if you walk 1500 feet along the paved path and then 500 feet in the sand to the house.
(b) Calculate the time T if you walk in the sand directly toward the ocean for 500 feet and then turn left and walk along the beach for 1500 feet to the house.
(c) Express the time T to get from the parking lot to the beachhouse as a function of the angle θ shown in the illustration.
(d) Calculate the time T if you walk directly from the parking lot to the house.

[**Hint:** $\tan \theta = 500/1500$]

(e) Calculate the time T if you walk 1000 feet along the paved path and then walk directly to the house.
(f) Use a graphing utility to graph $T = T(\theta)$. For what angle θ is T least? What is x for this angle? What is the minimum time?
(g) Explain why $\tan \theta = \dfrac{1}{3}$ gives the smallest angle θ that is possible.

58. Carrying a Ladder around a Corner   A ladder of length L is carried horizontally around a corner from a hall 3 feet wide into a hall 4 feet wide. See the illustration. Find the length L of the ladder as a function of the angle θ shown in the illustration.

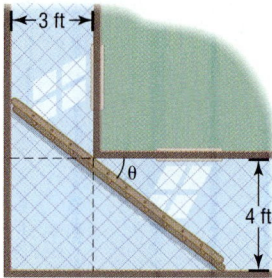

59. Suppose that the angle θ is a central angle of a circle of radius 1 (see the figure). Show that

(a) Angle $OAC = \dfrac{\theta}{2}$

(b) $|CD| = \sin \theta$ and $|OD| = \cos \theta$

(c) $\tan \dfrac{\theta}{2} = \dfrac{\sin \theta}{1 + \cos \theta}$

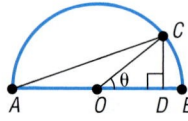

60. Show that the area A of an isosceles triangle is $A = a^2 \sin\theta \cos\theta$, where a is the length of one of the two equal sides and θ is the measure of one of the two equal angles (see the figure).

61. Let $n \geq 1$ be any real number and let θ be any angle for which $0 < n\theta < \dfrac{\pi}{2}$. Then we can draw a triangle with the angles θ and $n\theta$ and included side of length 1 (do you see why?) and place it on the unit circle as illustrated. Now, drop the perpendicular from C to $D = (x, 0)$ and show that

$$x = \frac{\tan(n\theta)}{\tan\theta + \tan(n\theta)}$$

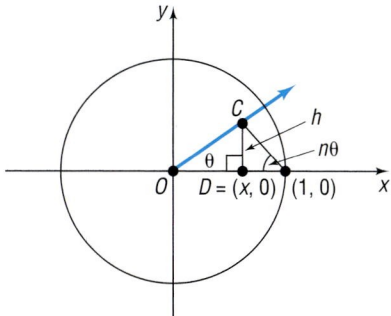

62. Refer to the figure. The smaller circle, whose radius is a, is tangent to the larger circle, whose radius is b. The ray $\overrightarrow{OA}$ contains a diameter of each circle, and the ray $\overrightarrow{OB}$ is tangent to each circle. Show that

$$\cos\theta = \frac{\sqrt{ab}}{\dfrac{a+b}{2}}$$

(This shows that $\cos\theta$ equals the ratio of the geometric mean of a and b to the arithmetic mean of a and b.)

[**Hint:** First show that $\sin\theta = \dfrac{b-a}{b+a}$.]

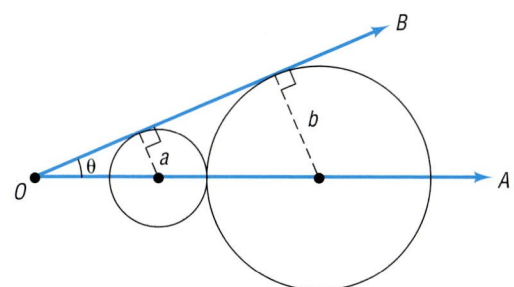

63. Refer to the figure. If $|OA| = 1$, show that

(a) Area $\triangle OAC = \dfrac{1}{2}\sin\alpha\cos\alpha$

(b) Area $\triangle OCB = \dfrac{1}{2}|OB|^2 \sin\beta\cos\beta$

(c) Area $\triangle OAB = \dfrac{1}{2}|OB|\sin(\alpha + \beta)$

(d) $|OB| = \dfrac{\cos\alpha}{\cos\beta}$

(e) $\sin(\alpha + \beta) = \sin\alpha\cos\beta + \cos\alpha\sin\beta$

[**Hint:** Area $\triangle OAB =$ Area $\triangle OAC +$ Area $\triangle OCB$]

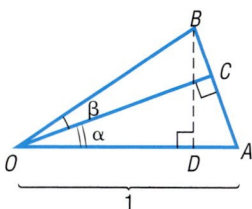

64. Refer to the figure, where a unit circle is drawn. The line DB is tangent to the circle.

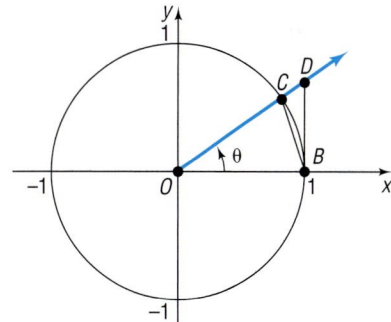

(a) Express the area of $\triangle OBC$ in terms of $\sin\theta$ and $\cos\theta$.
[**Hint:** Use the altitude from C to the base $\overline{OB} = 1$.]

(b) Express the area of $\triangle OBD$ in terms of $\sin\theta$ and $\cos\theta$.

(c) The area of the sector of the circle OBC is $\dfrac{1}{2}\theta$, where θ is measured in radians. Use the results of parts (a) and (b) and the fact that

$$\text{Area } \triangle OBC < \text{Area } \overset{\frown}{OBC} < \text{Area } \triangle OBD$$

to show that

$$1 < \frac{\theta}{\sin\theta} < \frac{1}{\cos\theta}$$

65. If $\cos\alpha = \tan\beta$ and $\cos\beta = \tan\alpha$, where α and β are acute angles, show that

$$\sin\alpha = \sin\beta = \sqrt{\frac{3 - \sqrt{5}}{2}}$$

66. If θ is an acute angle, explain why $\sec\theta > 1$.

67. If θ is an acute angle, explain why $0 < \sin\theta < 1$.

68. How would you explain the meaning of the sine function to a fellow student who has just completed college algebra?

8.3 COMPUTING THE VALUES OF TRIGONOMETRIC FUNCTIONS OF GIVEN ANGLES

OBJECTIVES

1. Find the Exact Value of the Trigonometric Functions of $\frac{\pi}{4} = 45°$

2. Find the Exact Value of the Trigonometric Functions of $\frac{\pi}{6} = 30°$ and $\frac{\pi}{3} = 60°$

3. Use a Calculator to Approximate the Value of the Trigonometric Functions of Acute Angles

In the previous section, we developed ways to find the value of each trigonometric function of an acute angle when one of the functions is known. In this section, we discuss the problem of finding the value of each trigonometric function of an acute angle when the angle is given.

For three special acute angles, we can use some results from plane geometry to find the exact value of each of the six trigonometric functions.

1 Trigonometric Functions of $\frac{\pi}{4} = 45°$

EXAMPLE 1 **Finding the Exact Value of the Trigonometric Functions of $\frac{\pi}{4} = 45°$**

Find the exact value of the six trigonometric functions of $\frac{\pi}{4} = 45°$.

Figure 29

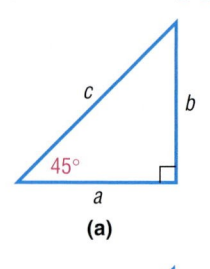

(a)

(b)

Solution Form the right triangle in Figure 29(a), in which one of the angles is $\frac{\pi}{4} = 45°$. It follows that the other acute angle is also $\frac{\pi}{4} = 45°$, and hence the triangle is isosceles. As a result, side a and side b are equal in length. Because the values of the trigonometric functions of an angle depend only on the angle and not on the size of the triangle, we may choose to use the triangle for which

$$a = b = 1$$

Then, by the Pythagorean Theorem,

$$c^2 = a^2 + b^2 = 1 + 1 = 2$$
$$c = \sqrt{2}$$

As a result, we have the triangle in Figure 29(b), from which we find

$$\sin \frac{\pi}{4} = \sin 45° = \frac{b}{c} = \frac{1}{\sqrt{2}} = \frac{\sqrt{2}}{2} \qquad \cos \frac{\pi}{4} = \cos 45° = \frac{a}{c} = \frac{1}{\sqrt{2}} = \frac{\sqrt{2}}{2}$$

Using Quotient and Reciprocal Identities, we find

$$\tan \frac{\pi}{4} = \tan 45° = \frac{\sin 45°}{\cos 45°} = \frac{\frac{\sqrt{2}}{2}}{\frac{\sqrt{2}}{2}} = 1 \qquad \cot \frac{\pi}{4} = \cot 45° = \frac{1}{\tan 45°} = \frac{1}{1} = 1$$

$$\sec \frac{\pi}{4} = \sec 45° = \frac{1}{\cos 45°} = \frac{1}{\frac{1}{\sqrt{2}}} = \sqrt{2} \qquad \csc \frac{\pi}{4} = \csc 45° = \frac{1}{\sin 45°} = \frac{1}{\frac{1}{\sqrt{2}}} = \sqrt{2}$$

EXAMPLE 2 **Finding the Exact Value of a Trigonometric Expression**

Find the exact value of each expression.

(a) $(\sin 45°)(\tan 45°)$ (b) $\left(\sec \dfrac{\pi}{4}\right)\left(\cot \dfrac{\pi}{4}\right)$

Solution We use the results obtained in Example 1.

(a) $(\sin 45°)(\tan 45°) = \dfrac{\sqrt{2}}{2} \cdot 1 = \dfrac{\sqrt{2}}{2}$

(b) $\left(\sec \dfrac{\pi}{4}\right)\left(\cot \dfrac{\pi}{4}\right) = \sqrt{2} \cdot 1 = \sqrt{2}$

◾

NOW WORK PROBLEMS **1** AND **13**.

2 **Trigonometric Functions of** $\dfrac{\pi}{6} = 30°$ **and** $\dfrac{\pi}{3} = 60°$

EXAMPLE 3 **Finding the Exact Value of the Trigonometric Functions**

of $\dfrac{\pi}{6} = 30°$ **and** $\dfrac{\pi}{3} = 60°$

Find the exact value of the six trigonometric functions of $\dfrac{\pi}{6} = 30°$ and $\dfrac{\pi}{3} = 60°$.

Figure 30

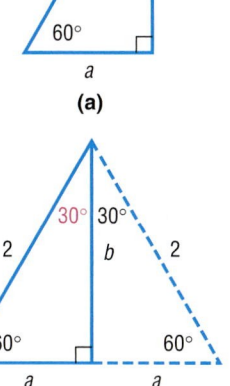

(a)

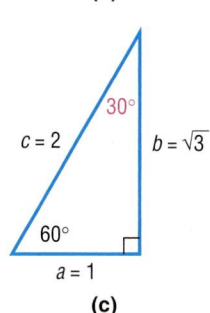

(b)

Solution Form a right triangle in which one of the angles is $\dfrac{\pi}{6} = 30°$. It then follows that the other angle is $\dfrac{\pi}{3} = 60°$. Figure 30(a) illustrates such a triangle with hypotenuse of length 2. Our problem is to determine a and b.

We begin by placing next to the triangle in Figure 30(a) another triangle congruent to the first, as shown in Figure 30(b). Notice that we now have a triangle whose angles are each 60°. This triangle is therefore equilateral, so each side is of length 2. In particular, the base is $2a = 2$, so $a = 1$. By the Pythagorean Theorem, b satisfies the equation $a^2 + b^2 = c^2$, so we have

$$a^2 + b^2 = c^2$$
$$1^2 + b^2 = 2^2 \qquad {\color{blue} a = 1, c = 2}$$
$$b^2 = 4 - 1 = 3$$
$$b = \sqrt{3}$$

Using the triangle in Figure 30(c) and the fact that $\dfrac{\pi}{6} = 30°$ and $\dfrac{\pi}{3} = 60°$ are complementary angles, we find

$$\sin \dfrac{\pi}{6} = \sin 30° = \dfrac{\text{opposite}}{\text{hypotenuse}} = \dfrac{1}{2} \qquad\qquad \cos \dfrac{\pi}{3} = \cos 60° = \dfrac{1}{2}$$

$$\cos \dfrac{\pi}{6} = \cos 30° = \dfrac{\text{adjacent}}{\text{hypotenuse}} = \dfrac{\sqrt{3}}{2} \qquad\qquad \sin \dfrac{\pi}{3} = \sin 60° = \dfrac{\sqrt{3}}{2}$$

$$\tan \dfrac{\pi}{6} = \tan 30° = \dfrac{\sin 30°}{\cos 30°} = \dfrac{\dfrac{1}{2}}{\dfrac{\sqrt{3}}{2}} = \dfrac{1}{\sqrt{3}} = \dfrac{\sqrt{3}}{3} \qquad\qquad \cot \dfrac{\pi}{3} = \cot 60° = \dfrac{\sqrt{3}}{3}$$

(c)

$$\csc \frac{\pi}{6} = \csc 30° = \frac{1}{\sin 30°} = \frac{1}{\frac{1}{2}} = 2 \qquad\qquad \sec \frac{\pi}{3} = \sec 60° = 2$$

$$\sec \frac{\pi}{6} = \sec 30° = \frac{1}{\cos 30°} = \frac{1}{\frac{\sqrt{3}}{2}} = \frac{2}{\sqrt{3}} = \frac{2\sqrt{3}}{3} \qquad\qquad \csc \frac{\pi}{3} = \csc 60° = \frac{2\sqrt{3}}{3}$$

$$\cot \frac{\pi}{6} = \cot 30° = \frac{1}{\tan 30°} = \frac{1}{\frac{\sqrt{3}}{3}} = \frac{3}{\sqrt{3}} = \sqrt{3} \qquad\qquad \tan \frac{\pi}{3} = \tan 60° = \sqrt{3}$$ ■

Table 3 summarizes the information just derived for the angles $\frac{\pi}{6} = 30°$, $\frac{\pi}{4} = 45°$, and $\frac{\pi}{3} = 60°$. Until you memorize the entries in Table 3, you should draw the appropriate triangle to determine the values given in the table.

TABLE 3

θ (Radians)	θ (Degrees)	sin θ	cos θ	tan θ	csc θ	sec θ	cot θ
$\frac{\pi}{6}$	30°	$\frac{1}{2}$	$\frac{\sqrt{3}}{2}$	$\frac{\sqrt{3}}{3}$	2	$\frac{2\sqrt{3}}{3}$	$\sqrt{3}$
$\frac{\pi}{4}$	45°	$\frac{\sqrt{2}}{2}$	$\frac{\sqrt{2}}{2}$	1	$\sqrt{2}$	$\sqrt{2}$	1
$\frac{\pi}{3}$	60°	$\frac{\sqrt{3}}{2}$	$\frac{1}{2}$	$\sqrt{3}$	$\frac{2\sqrt{3}}{3}$	2	$\frac{\sqrt{3}}{3}$

EXAMPLE 4 **Finding the Exact Value of a Trigonometric Expression**

Find the exact value of each expression.

(a) $\sin 45°\cos 30°$ (b) $\tan \frac{\pi}{4} - \sin \frac{\pi}{3}$ (c) $\tan^2 \frac{\pi}{6} + \sin^2 \frac{\pi}{4}$

Solution   (a) $\sin 45° \cos 30° = \frac{\sqrt{2}}{2} \cdot \frac{\sqrt{3}}{2} = \frac{\sqrt{6}}{4}$

(b) $\tan \frac{\pi}{4} - \sin \frac{\pi}{3} = 1 - \frac{\sqrt{3}}{2} = \frac{2 - \sqrt{3}}{2}$

(c) $\tan^2 \frac{\pi}{6} + \sin^2 \frac{\pi}{4} = \left(\frac{\sqrt{3}}{3}\right)^2 + \left(\frac{\sqrt{2}}{2}\right)^2 = \frac{1}{3} + \frac{1}{2} = \frac{5}{6}$ ■

NOW WORK PROBLEMS **5** AND **15**.

The exact values of the trigonometric functions for the angles $\frac{\pi}{6} = 30°$, $\frac{\pi}{4} = 45°$, and $\frac{\pi}{3} = 60°$ are relatively easy to calculate, because the triangles that contain such angles have "nice" geometric features. For most other angles, we can only approximate the value of each trigonometric function. To do this, we will need a calculator.

Using a Calculator to Find Values of Trigonometric Functions

③ Before getting started, you must first decide whether to enter the angle in the calculator using radians or degrees and then set the calculator to the correct MODE. (Check your instruction manual to find out how your calculator handles degrees and radians.) Your calculator has the keys marked $\boxed{\sin}$, $\boxed{\cos}$, and $\boxed{\tan}$. To find the values of the remaining three trigonometric functions (secant, cosecant, and cotangent), we use the reciprocal identities.

$$\sec \theta = \frac{1}{\cos \theta} \qquad \csc \theta = \frac{1}{\sin \theta} \qquad \cot \theta = \frac{1}{\tan \theta}$$

EXAMPLE 5   **Using a Calculator to Approximate the Value of Trigonometric Functions**

Use a calculator to find the approximate value of:

(a) $\cos 48°$ (b) $\csc 21°$ (c) $\tan \dfrac{\pi}{12}$

Express your answer rounded to two decimal places.

Solution   (a) First, we set the MODE to receive degrees. See Figure 31(a). Figure 31(b) shows the solution using a TI-83 graphing calculator.

Figure 31

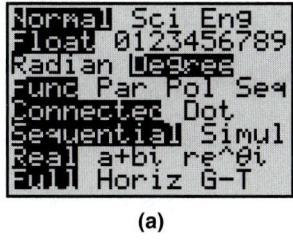

(a)   (b)

Figure 32

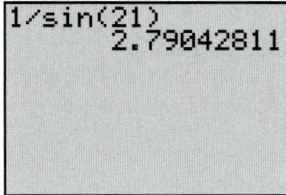

Rounded to two decimal places,

$$\cos 48° = 0.67$$

(b) Most calculators do not have a csc key. The manufacturers assume the user knows some trigonometry. To find the value of csc 21°, we use the fact that $\csc 21° = \dfrac{1}{\sin 21°}$. Figure 32 shows the solution using a TI-83 graphing calculator. Rounded to two decimal places,

$$\csc 21° = 2.79$$

Figure 33

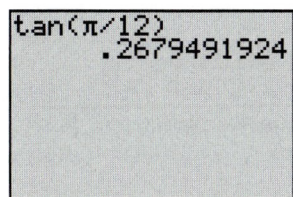

(c) Set the MODE to receive radians. Figure 33 shows the solution using a TI-83 graphing calculator. Rounded to two decimal places,

$$\tan \frac{\pi}{12} = 0.27$$

■

✏ **NOW WORK PROBLEM 25.**

EXAMPLE 6

Constructing a Rain Gutter

Figure 34

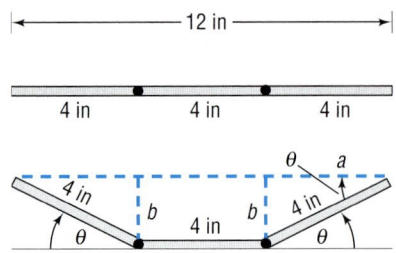

A rain gutter is to be constructed of aluminum sheets 12 inches wide. After marking off a length of 4 inches from each edge, this length is bent up at an angle θ. See Figure 34.

(a) Express the area A of the opening as a function of θ.

  [**Hint:** Let b denote the vertical height of the bend.]

(b) Find the area A of the opening for $\theta = 30°, \theta = 45°, \theta = 60°,$ and $\theta = 75°$.

(c) Graph $A = A(\theta)$. Find the angle θ that makes A largest. (This bend will allow the most water to flow through the gutter.)

Solution

(a) Look again at Figure 34. The area A of the opening is the sum of the areas of two congruent right triangles and one rectangle. Look at Figure 35, showing the triangle in Figure 34 redrawn. We see that

Figure 35

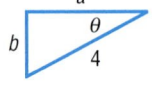

$$\cos \theta = \frac{a}{4} \quad \text{so} \quad a = 4 \cos \theta \qquad \sin \theta = \frac{b}{4} \quad \text{so} \quad b = 4 \sin \theta$$

The area of the triangle is

$$\text{area} = \frac{1}{2}(\text{base})(\text{height}) = \frac{1}{2}ab = \frac{1}{2}(4 \cos \theta)(4 \sin \theta) = 8 \sin \theta \cos \theta$$

So the area of the two triangles is $16 \sin \theta \cos \theta$.

  The rectangle has length 4 and height b, so its area is

$$4b = 4(4 \sin \theta) = 16 \sin \theta$$

The area A of the opening is

$$A = \text{area of the two triangles } + \text{ area of the rectangle}$$
$$A(\theta) = 16 \sin \theta \cos \theta + 16 \sin \theta = 16 \sin \theta(\cos \theta + 1)$$

(b) For $\theta = 30°$: $\qquad A(30°) = 16 \sin 30°(\cos 30° + 1)$

$$= 16\left(\frac{1}{2}\right)\left(\frac{\sqrt{3}}{2} + 1\right) = 4\sqrt{3} + 8$$

The area of the opening for $\theta = 30°$ is about 14.9 square inches.

For $\theta = 45°$: $\qquad A(45°) = 16 \sin 45°(\cos 45° + 1)$

$$= 16\left(\frac{\sqrt{2}}{2}\right)\left(\frac{\sqrt{2}}{2} + 1\right) = 8 + 8\sqrt{2}$$

The area of the opening for $\theta = 45°$ is about 19.3 square inches.

For $\theta = 60°$: $\qquad A(60°) = 16 \sin 60°(\cos 60° + 1)$

$$= 16\left(\frac{\sqrt{3}}{2}\right)\left(\frac{1}{2} + 1\right) = 12\sqrt{3}$$

The area of the opening for $\theta = 60°$ is about 20.8 square inches.

For $\theta = 75°$: $\qquad A(75°) = 16 \sin 75°(\cos 75° + 1) \approx 19.5$

The area of the opening for $\theta = 75°$ is about 19.5 square inches.

Figure 36

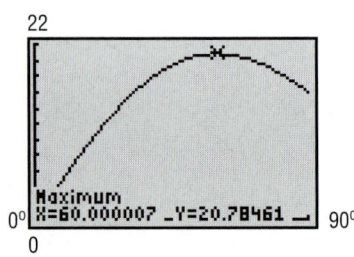

(c) Figure 36 shows the graph of $A = A(\theta)$. Using MAXIMUM, the angle θ that makes A largest is 60°.

8.3 Concepts and Vocabulary

In Problems 1 and 2, fill in the blanks.

1. $\tan \dfrac{\pi}{4} + \sin 30° = $ _____.

2. Using a calculator, $\sin 2 = $ _____ , rounded to two decimal places.

In Problems 3 and 4, answer True or False to each statement.

3. Exact values can be found for the trigonometric functions of $60°$.

4. Exact values can be found for the sine of any angle.

5. Explain how you remember that $\sin 45° = \dfrac{\sqrt{2}}{2}$.

6. Explain how you remember that $\sin \dfrac{\pi}{6} = \dfrac{1}{2}$.

8.3 Exercises

1. Write down the exact value of each of the six trigonometric functions of $45°$.

2. Write down the exact value of each of the six trigonometric functions of $30°$ and of $60°$.

In Problems 3–12, find the exact value of each expression if $\theta = 60°$. Do not use a calculator.

3. $\sin \theta$	**4.** $\cos \theta$	**5.** $\sin \dfrac{\theta}{2}$	**6.** $\cos \dfrac{\theta}{2}$	**7.** $(\sin \theta)^2$
8. $(\cos \theta)^2$	**9.** $2 \sin \theta$	**10.** $2 \cos \theta$	**11.** $\dfrac{\sin \theta}{2}$	**12.** $\dfrac{\cos \theta}{2}$

In Problems 13–24, find the exact value of each expression. Do not use a calculator.

13. $4 \cos 45° - 2 \sin 45°$ **14.** $2 \sin 45° + 4 \cos 30°$ **15.** $6 \tan 45° - 8 \cos 60°$

16. $\sin 30° \cdot \tan 60°$ **17.** $\sec \dfrac{\pi}{4} + 2 \csc \dfrac{\pi}{3}$ **18.** $\tan \dfrac{\pi}{4} + \cot \dfrac{\pi}{4}$

19. $\sec^2 \dfrac{\pi}{6} - 4$ **20.** $4 + \tan^2 \dfrac{\pi}{3}$ **21.** $\sin^2 30° + \cos^2 60°$

22. $\sec^2 60° - \tan^2 45°$ **23.** $1 - \cos^2 30° - \cos^2 60°$ **24.** $1 + \tan^2 30° - \csc^2 45°$

In Problems 25–42, use a calculator to find the approximate value of each expression. Round the answer to two decimal places.

25. $\sin 28°$	**26.** $\cos 14°$	**27.** $\tan 21°$	**28.** $\cot 70°$	**29.** $\sec 41°$	**30.** $\csc 55°$
31. $\sin \dfrac{\pi}{10}$	**32.** $\cos \dfrac{\pi}{8}$	**33.** $\tan \dfrac{5\pi}{12}$	**34.** $\cot \dfrac{\pi}{18}$	**35.** $\sec \dfrac{\pi}{12}$	**36.** $\csc \dfrac{5\pi}{13}$
37. $\sin 1$	**38.** $\tan 1$	**39.** $\sin 1°$	**40.** $\tan 1°$	**41.** $\tan 0.3$	**42.** $\tan 0.1$

Projectile Motion The path of a projectile fired at an inclination θ to the horizontal with initial speed v_0 is a parabola (see the figure). The range R of the projectile, that is, the horizontal distance that the projectile travels, is found by using the formula

$$R = \dfrac{2v_0^2 \sin \theta \cos \theta}{g}$$

where $g \approx 32.2$ feet per second per second ≈ 9.8 meters per second per second is the acceleration due to gravity.

The maximum height H of the projectile is

$$H = \dfrac{v_0^2 \sin^2 \theta}{2g}$$

v_0 = Initial speed

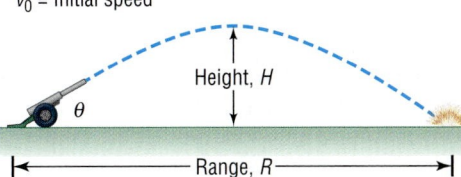

Height, H

Range, R

In Problems 43–46, find the range R and maximum height H. Round answers to two decimal places.

43. The projectile is fired at an angle of 45° to the horizontal with an initial speed of 100 feet per second.

44. The projectile is fired at an angle of 30° to the horizontal with an initial speed of 150 meters per second.

45. The projectile is fired at an angle of 25° to the horizontal with an initial speed of 500 meters per second.

46. The projectile is fired at an angle of 50° to the horizontal with an initial speed of 200 feet per second.

47. Inclined Plane   If friction is ignored, the time t (in seconds) required for a block to slide down an inclined plane (see the figure) is given by the formula

$$t = \sqrt{\frac{2a \sin \theta}{g \cos \theta}}$$

where a is the length (in feet) of the base and $g \approx 32$ feet per second per second is the acceleration of gravity. How long does it take a block to slide down an inclined plane with base $a = 10$ feet when

(a) $\theta = 30°$?   (b) $\theta = 45°$?   (c) $\theta = 60°$?

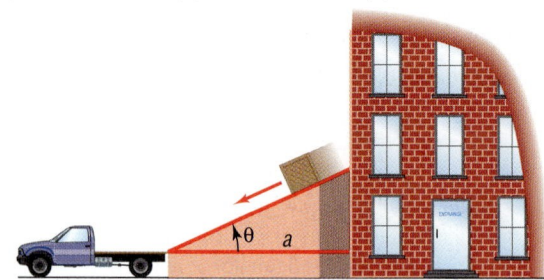

48. Piston Engines   In a certain piston engine, the distance x (in meters) from the center of the drive shaft to the head of the piston is given by

$$x = \cos \theta + \sqrt{16 + 0.5(2 \cos^2 \theta - 1)}$$

where θ is the angle between the crank and the path of the piston head (see the figure). Find x when $\theta = 30°$ and when $\theta = 45°$.

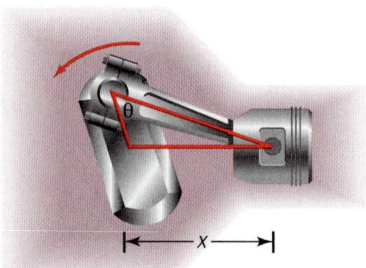

49. Calculating the Time of a Trip   Two oceanfront homes are located 8 miles apart on a straight stretch of beach, each a distance of 1 mile from a paved road that parallels the ocean. Sally can jog 8 miles per hour along the paved road, but only 3 miles per hour in the sand on the beach. Because of a river between the two houses, it is necessary to jog on the sand to the road, continue on the road, and then jog on the sand to get from one house to the other. See the illustration.

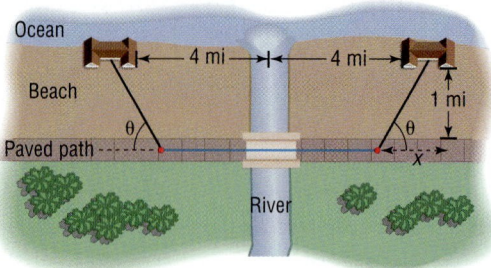

(a) Express the time T to get from one house to the other as a function of the angle θ shown in the illustration.

(b) Calculate the time T for $\theta = 30°$. How long is Sally on the paved road?

(c) Calculate the time T for $\theta = 45°$. How long is Sally on the paved road?

(d) Calculate the time T for $\theta = 60°$. How long is Sally on the paved road?

(e) Calculate the time T for $\theta = 90°$. Describe the path taken.

(f) Calculate the time T for $\tan \theta = \dfrac{1}{4}$. Describe the path taken. Explain why θ must be larger than 14°.

(g) Use a graphing utility to graph $T = T(\theta)$. What angle θ results in the least time? What is the least time? How long is Sally on the paved road?

50. Designing Fine Decorative Pieces   A designer of decorative art plans to market solid gold spheres encased in clear crystal cones. Each sphere is of fixed radius R and will be enclosed in a cone of height h and radius r. See the illustration. Many cones can be used to enclose the sphere, each having a different slant angle θ.

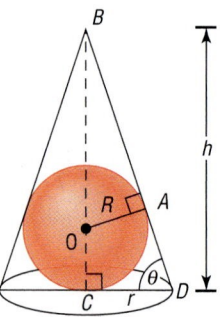

(a) Express the volume V of the cone as a function of the slant angle θ of the cone.

 [**Hint:** The volume V of a cone of height h and radius r is $V = \dfrac{1}{3}\pi r^2 h$.]

(b) What volume V is required to enclose a sphere of radius 2 centimeters in a cone whose slant angle θ is 30°? 45°? 60°?

(c) What slant angle θ should be used for the volume V of the cone to be a minimum? (This choice minimizes the amount of crystal required and gives maximum emphasis to the gold sphere.)

51. Use a calculator set in radian mode to complete the following table. What can you conclude about the ratio $\dfrac{\sin\theta}{\theta}$ as θ approaches 0?

θ	0.5	0.4	0.2	0.1	0.01	0.001	0.0001	0.00001
$\sin\theta$								
$\dfrac{\sin\theta}{\theta}$								

52. Use a calculator set in radian mode to complete the following table. What can you conclude about the ratio $\dfrac{\cos\theta - 1}{\theta}$ as θ approaches 0?

θ	0.5	0.4	0.2	0.1	0.01	0.001	0.0001	0.00001
$\cos\theta - 1$								
$\dfrac{\cos\theta - 1}{\theta}$								

53. Find the exact value of $\tan 1° \cdot \tan 2° \cdot \tan 3° \cdot \ldots \cdot \tan 89°$.

54. Find the exact value of $\cot 1° \cdot \cot 2° \cdot \cot 3° \cdot \ldots \cdot \cot 89°$.

55. Find the exact value of $\cos 1° \cdot \cos 2° \cdot \ldots \cdot \cos 45° \cdot \csc 46° \cdot \ldots \cdot \csc 89°$.

56. Find the exact value of $\sin 1° \cdot \sin 2° \cdot \ldots \cdot \sin 45° \cdot \sec 46° \cdot \ldots \cdot \sec 89°$.

57. Write a brief paragraph that explains how to quickly compute the trigonometric functions of $30°, 45°$, and $60°$.

8.4 TRIGONOMETRIC FUNCTIONS OF GENERAL ANGLES

OBJECTIVES
1. Find the Exact Value of the Trigonometric Functions for General Angles
2. Use Coterminal Angles to Find the Exact Value of a Trigonometric Function
3. Determine the Sign of the Trigonometric Functions of an Angle in a Given Quadrant
4. Find the Reference Angle of a General Angle
5. Use the Theorem on Reference Angles
6. Find the Exact Value of Trigonometric Functions of an Angle Given One of them and the Quadrant of the Angle

Figure 37

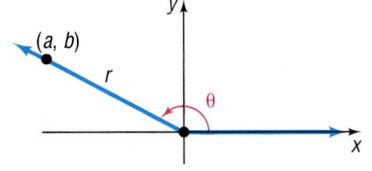

To extend the definitions of the trigonometric functions to include angles that are not acute, we employ a rectangular coordinate system and place the angle in the standard position so that its vertex is at the origin and its initial side is along the positive *x*-axis. See Figure 37.

Let θ be any angle in standard position, and let (a, b) denote the coordinates of any point, except the origin $(0, 0)$, on the terminal side of θ. If $r = \sqrt{a^2 + b^2}$ denotes the distance from $(0, 0)$ to (a, b), then the **six trigonometric functions of θ** are defined as the ratios

$$\sin\theta = \frac{b}{r} \qquad \cos\theta = \frac{a}{r} \qquad \tan\theta = \frac{b}{a}$$

$$\csc\theta = \frac{r}{b} \qquad \sec\theta = \frac{r}{a} \qquad \cot\theta = \frac{a}{b}$$

provided no denominator equals 0. If a denominator equals 0, that trigonometric function of the angle θ is not defined.

Figure 38

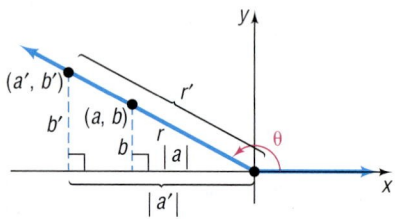

Notice in the preceding definitions that if $a = 0$, that is, the point $(0, b)$ is on the y-axis, then the tangent function and the secant function are undefined. Also, if $b = 0$, that is, the point $(a, 0)$ is on the x-axis, then the cosecant function and the cotangent function are undefined.

By constructing similar triangles, you should be convinced that the values of the six trigonometric functions of an angle θ do not depend on the selection of the point (a, b) on the terminal side of θ, but rather depend only on the angle θ itself. See Figure 38 for an illustration when θ lies in quadrant II.

Figure 39

$$\sin\theta = \frac{b}{r} = \frac{\text{opposite}}{\text{hypotenuse}}$$

$$\cos\theta = \frac{a}{r} = \frac{\text{adjacent}}{\text{hypotenuse}} \text{ and so on.}$$

Since the triangles are similar, the ratio $\dfrac{b}{r}$ equals the ratio $\dfrac{b'}{r'}$, the common value being $\sin\theta$. Also, the ratio $\dfrac{|a|}{r}$ equals the ratio $\dfrac{|a'|}{r'}$, so $\dfrac{a}{r} = \dfrac{a'}{r'}$, the common value being $\cos\theta$. And so on.

Also, observe that if θ is acute these definitions reduce to the right triangle definitions given in Section 8.2, as illustrated in Figure 39.

Finally, from the definition of the six trigonometric functions of a general angle, we see that the Quotient and Reciprocal Identities hold. Also, using $r^2 = a^2 + b^2$ and dividing each side by r^2, we can derive the Pythagorean Identities for general angles.

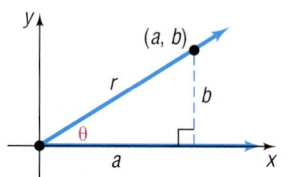

EXAMPLE 1   **Finding the Exact Value of the Six Trigonometric Functions of θ Given a Point on the Terminal Side**

① Find the exact value of each of the six trigonometric functions of a positive angle θ if $(4, -3)$ is a point on its terminal side.

Solution   Figure 40 illustrates the situation. For the point $(a, b) = (4, -3)$, we have $a = 4$ and $b = -3$. Then $r = \sqrt{a^2 + b^2} = \sqrt{16 + 9} = 5$, so that

$$\sin\theta = \frac{b}{r} = -\frac{3}{5} \qquad \cos\theta = \frac{a}{r} = \frac{4}{5} \qquad \tan\theta = \frac{b}{a} = -\frac{3}{4}$$

$$\csc\theta = \frac{r}{b} = -\frac{5}{3} \qquad \sec\theta = \frac{r}{a} = \frac{5}{4} \qquad \cot\theta = \frac{a}{b} = -\frac{4}{3}$$

Figure 40

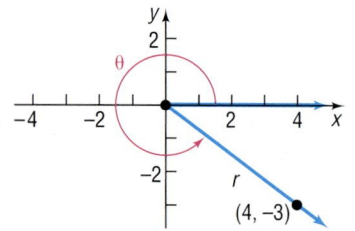

NOW WORK PROBLEM **1.**

In the next example, we find the exact value of each of the six trigonometric functions at the quadrantal angles $0, \dfrac{\pi}{2}, \pi,$ and $\dfrac{3\pi}{2}$.

EXAMPLE 2 Finding the Exact Value of the Six Trigonometric Functions of Quadrantal Angles

Find the exact value of each of the six trigonometric functions of

(a) $\theta = 0 = 0°$ (b) $\theta = \dfrac{\pi}{2} = 90°$

(c) $\theta = \pi = 180°$ (d) $\theta = \dfrac{3\pi}{2} = 270°$

Solution

Figure 41
$\theta = 0 = 0°$

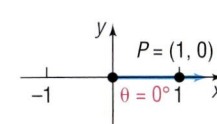

(a) The point $P = (1, 0)$ is on the terminal side of $\theta = 0 = 0°$ and is a distance of $r = 1$ unit from the origin. See Figure 41. Then,

$$\sin 0 = \sin 0° = \frac{0}{1} = 0 \qquad \cos 0 = \cos 0° = \frac{1}{1} = 1$$

$$\tan 0 = \tan 0° = \frac{0}{1} = 0 \qquad \sec 0 = \sec 0° = \frac{1}{1} = 1$$

Since the y-coordinate of P is 0, $\csc 0$ and $\cot 0$ are not defined.

(b) The point $P = (0, 1)$ is on the terminal side of $\theta = \dfrac{\pi}{2} = 90°$ and is a distance of $r = 1$ unit from the origin. See Figure 42. Then

Figure 42
$\theta = \dfrac{\pi}{2} = 90°$

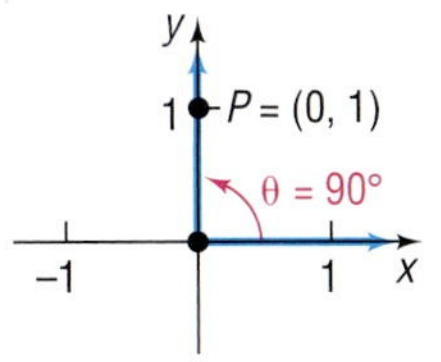

$$\sin \frac{\pi}{2} = \sin 90° = \frac{1}{1} = 1 \qquad \cos \frac{\pi}{2} = \cos 90° = \frac{0}{1} = 0$$

$$\csc \frac{\pi}{2} = \csc 90° = \frac{1}{1} = 1 \qquad \cot \frac{\pi}{2} = \cot 90° = \frac{0}{1} = 0$$

Since the x-coordinate of P is 0, $\tan \dfrac{\pi}{2}$ and $\sec \dfrac{\pi}{2}$ are not defined.

(c) The point $P = (-1, 0)$ is on the terminal side of $\theta = \pi = 180°$ and is a distance of $r = 1$ unit from the origin. See Figure 43. Then,

Figure 43
$\theta = \pi = 180°$

$$\sin \pi = \sin 180° = \frac{0}{1} = 0 \qquad \cos \pi = \cos 180° = \frac{-1}{1} = -1$$

$$\tan \pi = \tan 180° = \frac{0}{-1} = 0 \qquad \sec \pi = \sec 180° = \frac{1}{-1} = -1$$

Since the y-coordinate of P is 0, $\csc \pi$ and $\cot \pi$ are not defined.

Figure 44
$\theta = \dfrac{3\pi}{2} = 270°$

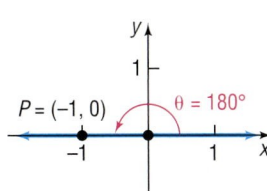

(d) The point $P = (0, -1)$ is on the terminal side of $\theta = \dfrac{3\pi}{2} = 270°$ and is a distance of $r = 1$ unit from the origin. See Figure 44. Then,

$$\sin \frac{3\pi}{2} = \sin 270° = \frac{-1}{1} = -1 \qquad \cos \frac{3\pi}{2} = \cos 270° = \frac{0}{1} = 0$$

$$\csc \frac{3\pi}{2} = \csc 270° = \frac{1}{-1} = -1 \qquad \cot \frac{3\pi}{2} = \cot 270° = \frac{0}{-1} = 0$$

Since the x-coordinate of P is 0, $\tan \dfrac{3\pi}{2}$ and $\sec \dfrac{3\pi}{2}$ are not defined.

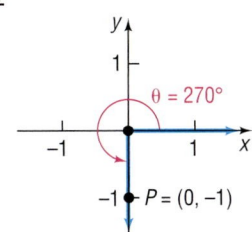

Table 4 summarizes the values of the trigonometric functions found in Example 2.

TABLE 4

θ (Radians)	θ (Degrees)	sin θ	cos θ	tan θ	csc θ	sec θ	cot θ
0	0°	0	1	0	Not defined	1	Not defined
$\dfrac{\pi}{2}$	90°	1	0	Not defined	1	Not defined	0
π	180°	0	−1	0	Not defined	−1	Not defined
$\dfrac{3\pi}{2}$	270°	−1	0	Not defined	−1	Not defined	0

There is no need to memorize Table 4. To find the value of a trigonometric function of a quadrantal angle, draw the angle and apply the definition as we did in Example 2.

Coterminal Angles

Two angles in standard position are said to be **coterminal** if they have the same terminal side.

See Figure 45.

Figure 45

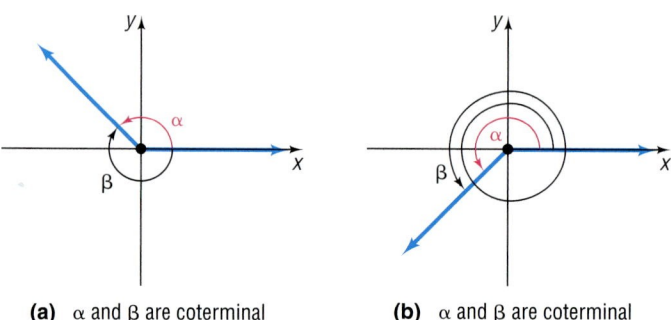

(a) α and β are coterminal   **(b)** α and β are coterminal

In general, if θ is an angle measured in degrees, then $\theta \pm 360°k$, where k is any integer, is an angle coterminal with θ. If θ is measured in radians, then $\theta \pm 2\pi k$, where k is any integer, is an angle coterminal with θ.

② Because coterminal angles have the same terminal side, it follows that the values of the trigonometric functions of coterminal angles are equal. We use this fact in the next example.

EXAMPLE 3 ### Using the Coterminal Angle to Find the Exact Value of a Trigonometric Function

Find the exact value of each of the following:

(a) $\sin 390°$ (b) $\cos 420°$ (c) $\tan \dfrac{9\pi}{4}$ (d) $\sec\left(-\dfrac{7\pi}{4}\right)$ (e) $\csc(-270°)$

Solution (a) It is best to sketch the angle first. See Figure 46. The angle 390° is coterminal with 30°.

$$\sin 390° = \sin(360° + 30°)$$
$$= \sin 30° = \frac{1}{2}$$

(b) See Figure 47. The angle 420° is coterminal with 60°.

$$\cos 420° = \cos(360° + 60°)$$
$$= \cos 60° = \frac{1}{2}$$

Figure 46

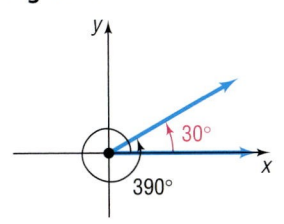

Figure 47

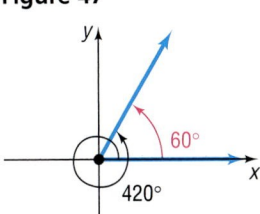

Figure 48

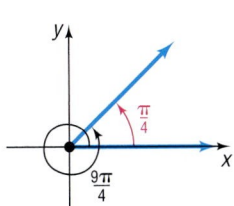

(c) The angle $\dfrac{9\pi}{4}$ is coterminal with $\dfrac{\pi}{4}\left(\dfrac{9\pi}{4} = \dfrac{8\pi}{4} + \dfrac{\pi}{4} = 2\pi + \dfrac{\pi}{4}\right)$. See Figure 48.

$$\tan\frac{9\pi}{4} = \tan\left(2\pi + \frac{\pi}{4}\right)$$
$$= \tan\frac{\pi}{4} = 1$$

(d) See Figure 49. The angle $-\dfrac{7\pi}{4}$ is coterminal with $\dfrac{\pi}{4}$.

$$\sec\left(-\frac{7\pi}{4}\right) = \sec\left(-2\pi + \frac{\pi}{4}\right) = \sec\frac{\pi}{4} = \sqrt{2}$$

Figure 49

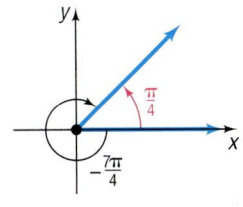

(e) See Figure 50. The angle $-270°$ is coterminal with 90°.

$$\csc(-270°) = \csc(-360° + 90°)$$
$$= \csc 90° = 1$$

Figure 50

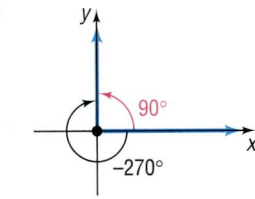

As Example 3 illustrates, the value of a trigonometric function of any angle is equal to the value of the same trigonometric function of an angle θ coterminal to it where $0° \leq \theta < 360°$ (or $0 \leq \theta < 2\pi$). Because the angles θ and $\theta \pm 360°k$ (or $\theta \pm 2\pi k$), where k is any integer, are coterminal, and because the values of the trigonometric functions are equal for coterminal angles, it follows that

$$\begin{aligned}
\sin(\theta \pm 360°k) &= \sin\theta & \sin(\theta \pm 2\pi k) &= \sin\theta \\
\cos(\theta \pm 360°k) &= \cos\theta & \cos(\theta \pm 2\pi k) &= \cos\theta \\
\tan(\theta \pm 360°k) &= \tan\theta & \tan(\theta \pm 2\pi k) &= \tan\theta \\
\csc(\theta \pm 360°k) &= \csc\theta & \csc(\theta \pm 2\pi k) &= \csc\theta \\
\sec(\theta \pm 360°k) &= \sec\theta & \sec(\theta \pm 2\pi k) &= \sec\theta \\
\cot(\theta \pm 360°k) &= \cot\theta & \cot(\theta \pm 2\pi k) &= \cot\theta
\end{aligned} \tag{1}$$

These formulas show that the values of the trigonometric functions repeat themselves every 360° (or 2π radians).

NOW WORK PROBLEM 37.

The Signs of the Trigonometric Functions

③ If θ is not a quadrantal angle, then it will lie in a particular quadrant. In such a case, the signs of the x-coordinate and the y-coordinate of a point (a, b) on the terminal side of θ are known. Because $r = \sqrt{a^2 + b^2} > 0$, it follows that the signs of the trigonometric functions of an angle θ can be found if we know in which quadrant θ lies.

For example, if θ lies in quadrant II, as shown in Figure 51, then a point (a, b) on the terminal side of θ has a negative x-coordinate and a positive y-coordinate. Thus,

Figure 51
θ in quadrant II; $a < 0$, $b > 0$, $r > 0$

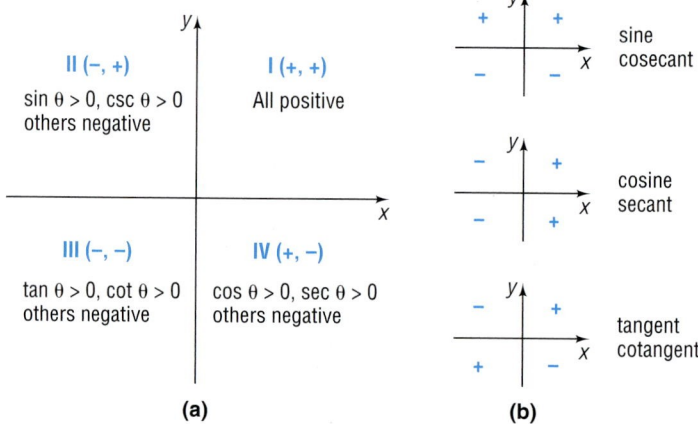

$$\sin \theta = \frac{b}{r} > 0 \qquad \cos \theta = \frac{a}{r} < 0 \qquad \tan \theta = \frac{b}{a} < 0$$

$$\csc \theta = \frac{r}{b} > 0 \qquad \sec \theta = \frac{r}{a} < 0 \qquad \cot \theta = \frac{a}{b} < 0$$

Table 5 lists the signs of the six trigonometric functions for each quadrant. See also Figure 52.

TABLE 5

Quadrant of θ	$\sin \theta$, $\csc \theta$	$\cos \theta$, $\sec \theta$	$\tan \theta$, $\cot \theta$
I	Positive	Positive	Positive
II	Positive	Negative	Negative
III	Negative	Negative	Positive
IV	Negative	Positive	Negative

Figure 52

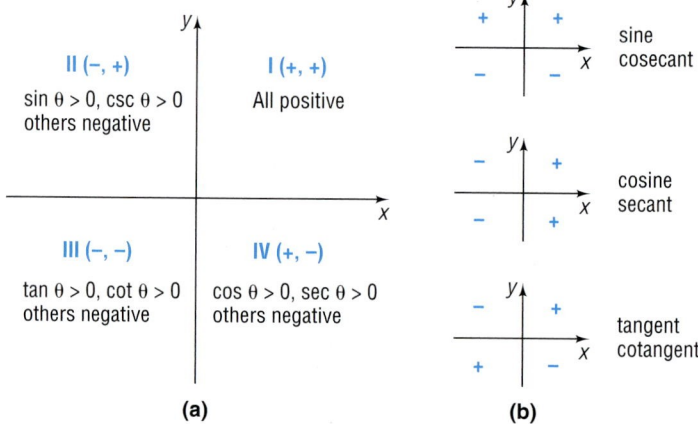

(a)

(b)

EXAMPLE 4 **Finding the Quadrant in Which an Angle Lies**

If $\sin \theta < 0$ and $\cos \theta < 0$, name the quadrant in which the angle θ lies.

Solution If $\sin \theta < 0$, then θ lies in quadrant III or IV. If $\cos \theta < 0$, then θ lies in quadrant II or III. Therefore, θ lies in quadrant III. ∎

NOW WORK PROBLEM **11**.

Reference Angle

4 Once we know in which quadrant an angle lies, we know the sign of each trigonometric function of this angle. The use of a certain reference angle may help us to evaluate the trigonometric functions of such an angle.

> Let θ denote a nonacute angle that lies in a quadrant. The acute angle formed by the terminal side of θ and either the positive x-axis or the negative x-axis is called the **reference angle** for θ.

Figure 53 illustrates the reference angle for some general angles θ. Note that a reference angle is always an acute angle, that is, an angle whose measure is between $0°$ and $90°$.

Figure 53

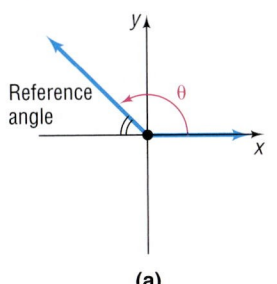

(a)

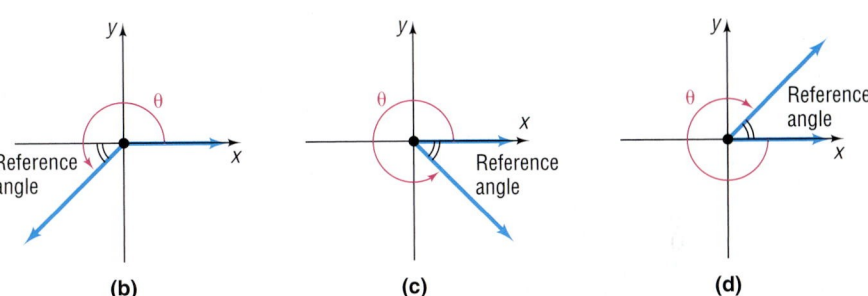

(b)　(c)　(d)

Although formulas can be given for calculating reference angles, usually it is easier to find the reference angle for a given angle by making a quick sketch of the angle.

EXAMPLE 5　Finding Reference Angles

Find the reference angle for each of the following angles:

(a) $150°$　(b) $-45°$　(c) $\dfrac{9\pi}{4}$　(d) $-\dfrac{5\pi}{6}$

Solution　(a) Refer to Figure 54. The reference angle for $150°$ is $30°$.

(b) Refer to Figure 55. The reference angle for $-45°$ is $45°$.

Figure 54

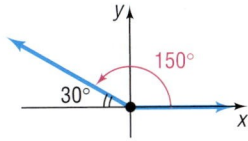

Figure 55

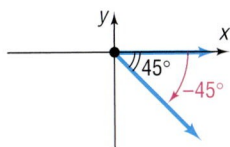

(c) Refer to Figure 56. The reference angle for $\dfrac{9\pi}{4}$ is $\dfrac{\pi}{4}$.

(d) Refer to Figure 57. The reference angle for $-\dfrac{5\pi}{6}$ is $\dfrac{\pi}{6}$.

Figure 56

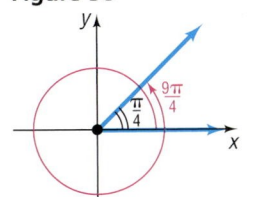

Figure 57

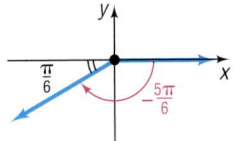

NOW WORK PROBLEM 19.

⑤   The advantage of using reference angles is that, except for the correct sign, the values of the trigonometric functions of a general angle θ equal the values of the trigonometric functions of its reference angle.

Theorem   **Reference Angles**

If θ is an angle that lies in a quadrant and if α is its reference angle, then

$$\begin{array}{lll} \sin\theta = \pm\sin\alpha & \cos\theta = \pm\cos\alpha & \tan\theta = \pm\tan\alpha \\ \csc\theta = \pm\csc\alpha & \sec\theta = \pm\sec\alpha & \cot\theta = \pm\cot\alpha \end{array} \qquad (2)$$

where the $+$ or $-$ sign depends on the quadrant in which θ lies.

Figure 58
$\sin\theta = b/r$, $\sin\alpha = b/r$,
$\cos\theta = a/r$, $\cos\alpha = |a|/r$

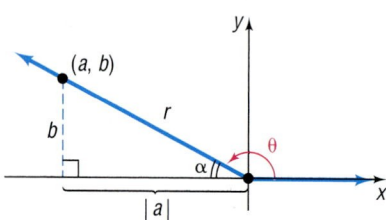

For example, suppose that θ lies in quadrant II and α is its reference angle. See Figure 58. If (a, b) is a point on the terminal side of θ and if $r = \sqrt{a^2 + b^2}$, we have

$$\sin\theta = \frac{b}{r} = \sin\alpha \qquad \cos\theta = \frac{a}{r} = \frac{-|a|}{r} = -\cos\alpha$$

and so on.

The next example illustrates how the theorem on reference angles is used.

EXAMPLE 6   **Using the Reference Angle to Find the Exact Value of a Trigonometric Function**

Find the exact value of each of the following trigonometric functions using reference angles.

(a) $\sin 135°$ (b) $\cos 240°$ (c) $\cos\dfrac{5\pi}{6}$ (d) $\tan\left(-\dfrac{\pi}{3}\right)$

Solution   (a) Refer to Figure 59. The angle $135°$ is in quadrant II, where the sine function is positive. The reference angle for $135°$ is $45°$.

$$\sin 135° = \sin 45° = \frac{\sqrt{2}}{2}$$

Figure 59

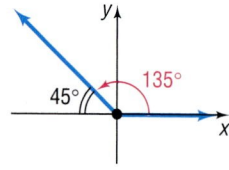

(b) Refer to Figure 60. The angle $240°$ is in quadrant III, where the cosine function is negative. The reference angle for $240°$ is $60°$.

$$\cos 240° = -\cos 60° = -\frac{1}{2}$$

Figure 60

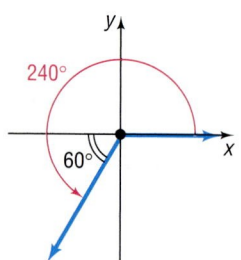

(c) Refer to Figure 61. The angle $\dfrac{5\pi}{6}$ is in quadrant II, where the cosine function is negative. The reference angle for $\dfrac{5\pi}{6}$ is $\dfrac{\pi}{6}$.

$$\cos\dfrac{5\pi}{6} = -\cos\dfrac{\pi}{6} = -\dfrac{\sqrt{3}}{2}$$

Figure 61

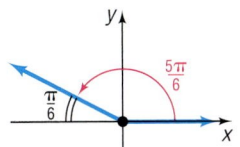

(d) Refer to Figure 62. The angle $-\dfrac{\pi}{3}$ is in quadrant IV, where the tangent function is negative. The reference angle for $-\dfrac{\pi}{3}$ is $\dfrac{\pi}{3}$.

$$\tan\left(-\dfrac{\pi}{3}\right) = -\tan\dfrac{\pi}{3} = -\sqrt{3}$$

Figure 62

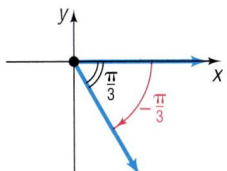

NOW WORK PROBLEM **53.**

EXAMPLE 7 | **Finding the Exact Value of Trigonometric Functions**

⑥ Given that $\cos\theta = -\dfrac{2}{3}$, $\dfrac{\pi}{2} < \theta < \pi$, find the exact value of each of the remaining trigonometric functions.

Solution   The angle θ lies in quadrant II, so we know that $\sin\theta$ and $\csc\theta$ are positive, whereas the other trigonometric functions are negative. If α is the reference angle for θ, then $\cos\alpha = \dfrac{2}{3}$. The values of the remaining trigonometric functions of the angle α can be found by drawing the appropriate triangle and using the Pythagorean Theorem. We use Figure 63 to obtain

Figure 63
$$\cos\alpha = \dfrac{2}{3}$$

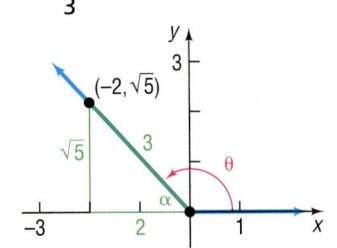

$$\sin\alpha = \dfrac{\sqrt{5}}{3} \qquad \cos\alpha = \dfrac{2}{3} \qquad \tan\alpha = \dfrac{\sqrt{5}}{2}$$

$$\csc\alpha = \dfrac{3}{\sqrt{5}} = \dfrac{3\sqrt{5}}{5} \qquad \sec\alpha = \dfrac{3}{2} \qquad \cot\alpha = \dfrac{2}{\sqrt{5}} = \dfrac{2\sqrt{5}}{5}$$

Now we assign the appropriate signs to each of these values to find the values of the trigonometric functions of θ.

$$\sin\theta = \dfrac{\sqrt{5}}{3} \qquad \cos\theta = -\dfrac{2}{3} \qquad \tan\theta = -\dfrac{\sqrt{5}}{2}$$

$$\csc\theta = \dfrac{3\sqrt{5}}{5} \qquad \sec\theta = -\dfrac{3}{2} \qquad \cot\theta = -\dfrac{2\sqrt{5}}{5}$$

NOW WORK PROBLEM **79.**

EXAMPLE 8 | **Finding the Exact Value of Trigonometric Functions**

If $\tan\theta = -4$ and $\sin\theta < 0$, find the exact value of each of the remaining trigonometric functions of θ.

Solution   Since $\tan\theta = -4 < 0$ and $\sin\theta < 0$, it follows that θ lies in quadrant IV. If α is the reference angle for θ, then $\tan\alpha = 4$. See Figure 64. Then

Figure 64
$\tan\alpha = 4$

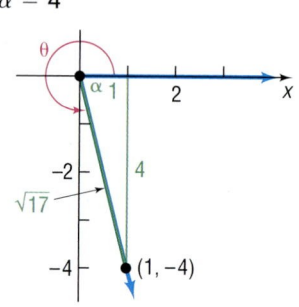

$$\sin\alpha = \frac{4}{\sqrt{17}} = \frac{4\sqrt{17}}{17} \qquad \cos\alpha = \frac{1}{\sqrt{17}} = \frac{\sqrt{17}}{17} \qquad \tan\alpha = \frac{4}{1} = 4$$

$$\csc\alpha = \frac{\sqrt{17}}{4} \qquad \sec\alpha = \frac{\sqrt{17}}{1} = \sqrt{17} \qquad \cot\alpha = \frac{1}{4}$$

We assign the appropriate sign to each of these to obtain the values of the trigonometric functions of θ.

$$\sin\theta = -\frac{4\sqrt{17}}{17} \qquad \cos\theta = \frac{\sqrt{17}}{17} \qquad \tan\theta = -4$$

$$\csc\theta = -\frac{\sqrt{17}}{4} \qquad \sec\theta = \sqrt{17} \qquad \cot\theta = -\frac{1}{4}$$

■

━ **NOW WORK PROBLEM 89.**

SUMMARY

To find the values of the trigonometric functions of a general angle:

STEP 1: If the angle θ is a quadrantal angle, draw the angle, pick a point on its terminal side, and apply the definitions of the trigonometric functions.

STEP 2: If the angle θ lies in a quadrant, determine the correct signs of the trigonometric functions in that quadrant and determine the reference angle α for θ. Now express each trigonometric function of θ in terms of the same value (except for the sign) of the trigonometric function of α, an acute angle. Finally, apply the correct sign to each function.

8.4   Concepts and Vocabulary

In Problems 1–3, fill in the blanks.

1. For an angle θ that lies in quadrant III, the trigonometric functions _____ and _____ are positive.

2. Two angles in standard position that have the same terminal side are _____.

3. The reference angle of 240° is _____.

In Problems 4–6, answer True or False to each statement.

4. $\sin 182° = \cos 2°$.

5. $\tan\dfrac{\pi}{2}$ is not defined.

6. The reference angle is always an acute angle.

7. Explain how you would find the reference angle of 240°.

8. In which quadrants is the cosine function positive?

9. If $0 \le \theta < 2\pi$, for what angles θ, if any, is $\tan\theta$ undefined?

10. Find an acute angle that is coterminal with −130°.

8.4 Exercises

In Problems 1–10, a point on the terminal side of an angle θ is given. Find the exact value of each of the six trigonometric functions of θ.

1. $(-3, 4)$
2. $(5, -12)$
3. $(2, -3)$
4. $(-1, -2)$
5. $(-3, -3)$
6. $(2, -2)$
7. $\left(\dfrac{\sqrt{3}}{2}, \dfrac{1}{2}\right)$
8. $\left(-\dfrac{1}{2}, \dfrac{\sqrt{3}}{2}\right)$
9. $\left(\dfrac{\sqrt{2}}{2}, -\dfrac{\sqrt{2}}{2}\right)$
10. $\left(-\dfrac{\sqrt{2}}{2}, -\dfrac{\sqrt{2}}{2}\right)$

In Problems 11–18, name the quadrant in which the angle θ lies.

11. $\sin \theta > 0$, $\cos \theta < 0$
12. $\sin \theta < 0$, $\cos \theta > 0$
13. $\sin \theta < 0$, $\tan \theta < 0$
14. $\cos \theta > 0$, $\tan \theta > 0$
15. $\cos \theta > 0$, $\cot \theta < 0$
16. $\sin \theta < 0$, $\cot \theta > 0$
17. $\sec \theta < 0$, $\tan \theta > 0$
18. $\csc \theta > 0$, $\cot \theta < 0$

In Problems 19–36, find the reference angle of each angle.

19. $-30°$
20. $60°$
21. $120°$
22. $300°$
23. $210°$
24. $330°$
25. $\dfrac{5\pi}{4}$
26. $\dfrac{5\pi}{6}$
27. $\dfrac{8\pi}{3}$
28. $\dfrac{7\pi}{4}$
29. $-135°$
30. $-240°$
31. $-\dfrac{2\pi}{3}$
32. $-\dfrac{7\pi}{6}$
33. $440°$
34. $490°$
35. $-\dfrac{3\pi}{4}$
36. $-\dfrac{11\pi}{6}$

In Problems 37–78, find the exact value of each expression. Do not use a calculator.

37. $\sin 405°$
38. $\cos 420°$
39. $\tan 405°$
40. $\sin 390°$
41. $\csc 450°$
42. $\sec 540°$
43. $\cot 390°$
44. $\sec 420°$
45. $\cos \dfrac{33\pi}{4}$
46. $\sin \dfrac{9\pi}{4}$
47. $\tan (21\pi)$
48. $\csc \dfrac{9\pi}{2}$
49. $\sec \dfrac{17\pi}{4}$
50. $\cot \dfrac{17\pi}{4}$
51. $\tan \dfrac{19\pi}{6}$
52. $\sec \dfrac{25\pi}{6}$
53. $\sin 150°$
54. $\cos 210°$
55. $\cos 315°$
56. $\sin 120°$
57. $\sec 240°$
58. $\csc 300°$
59. $\cot 330°$
60. $\tan 225°$
61. $\sin \dfrac{3\pi}{4}$
62. $\cos \dfrac{2\pi}{3}$
63. $\cot \dfrac{7\pi}{6}$
64. $\csc \dfrac{7\pi}{4}$
65. $\cos(-60°)$
66. $\tan(-120°)$
67. $\sin\left(-\dfrac{2\pi}{3}\right)$
68. $\cot\left(-\dfrac{\pi}{6}\right)$
69. $\tan \dfrac{14\pi}{3}$
70. $\sec \dfrac{11\pi}{4}$
71. $\csc(-315°)$
72. $\sec(-225°)$
73. $\sin(8\pi)$
74. $\cos(-2\pi)$
75. $\tan(7\pi)$
76. $\cot(5\pi)$
77. $\sec(-3\pi)$
78. $\csc\left(-\dfrac{5\pi}{2}\right)$

In Problems 79–96, find the exact value of each of the remaining trigonometric functions of θ.

79. $\sin \theta = \dfrac{12}{13}$, θ in quadrant II
80. $\cos \theta = \dfrac{3}{5}$, θ in quadrant IV
81. $\cos \theta = -\dfrac{4}{5}$, θ in quadrant III
82. $\sin \theta = -\dfrac{5}{13}$, θ in quadrant III
83. $\sin \theta = \dfrac{5}{13}$, $90° < \theta < 180°$
84. $\cos \theta = \dfrac{4}{5}$, $270° < \theta < 360°$
85. $\cos \theta = -\dfrac{1}{3}$, $180° < \theta < 270°$
86. $\sin \theta = -\dfrac{2}{3}$, $180° < \theta < 270°$
87. $\sin \theta = \dfrac{2}{3}$, $\tan \theta < 0$
88. $\cos \theta = -\dfrac{1}{4}$, $\tan \theta > 0$
89. $\sec \theta = 2$, $\sin \theta < 0$
90. $\csc \theta = 3$, $\cot \theta < 0$
91. $\tan \theta = \dfrac{3}{4}$, $\sin \theta < 0$
92. $\cot \theta = \dfrac{4}{3}$, $\cos \theta < 0$
93. $\tan \theta = -\dfrac{1}{3}$, $\sin \theta > 0$
94. $\sec \theta = -2$, $\tan \theta > 0$
95. $\csc \theta = -2$, $\tan \theta > 0$
96. $\cot \theta = -2$, $\sec \theta > 0$

97. Find the exact value of $\sin 45° + \sin 135° + \sin 225° + \sin 315°$.

98. Find the exact value of $\tan 60° + \tan 150°$.

99. If $\sin \theta = 0.2$, find $\sin(\theta + \pi)$.

100. If $\cos \theta = 0.4$, find $\cos(\theta + \pi)$.

101. If $\tan \theta = 3$, find $\tan(\theta + \pi)$.

102. If $\cot \theta = -2$, find $\cot(\theta + \pi)$.

103. If $\sin \theta = \dfrac{1}{5}$, find $\csc \theta$.

104. If $\cos \theta = \dfrac{2}{3}$, find $\sec \theta$.

105. Find the exact value of

$$\sin 1° + \sin 2° + \sin 3° + \cdots + \sin 358° + \sin 359°$$

106. Find the exact value of

$$\cos 1° + \cos 2° + \cos 3° + \cdots + \cos 358° + \cos 359°$$

---

107. Projectile Motion   An object is propelled upward at an angle θ, $45° < \theta < 90°$, to the horizontal with an initial velocity of v_0 feet per second from the base of a plane that makes an angle of $45°$ with the horizontal. See the illustration. If air resistance is ignored, the distance R that it travels up the inclined plane is given by

$$R = \frac{v_0^2 \sqrt{2}}{32}[\sin(2\theta) - \cos(2\theta) - 1]$$

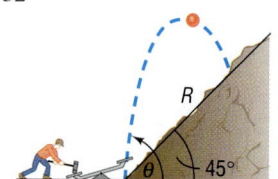

(a) Find the distance R that the object travels along the inclined plane if the initial velocity is 32 feet per second and $\theta = 60°$.

(b) Use a graphing utility to graph $R = R(\theta)$ if the initial velocity is 32 feet per second.

(c) What value of θ makes R largest?

108. Give three examples that demonstrate how to use the theorem on reference angles.

109. Write a brief paragraph that explains how to quickly compute the value of the trigonometric functions of $0°$, $90°$, $180°$, and $270°$.

PREPARING FOR THIS SECTION

Before getting started, review the following:

✓ Unit Circle (Section 1.8, p. 180)

✓ Symmetry (Section 3.1, pp. 256–258)

✓ Functions (Section 2.1, pp. 196–204)

✓ Even and Odd Functions (Section 3.2, pp. 269–272)

8.5   PROPERTIES OF THE TRIGONOMETRIC FUNCTIONS; UNIT CIRCLE APPROACH

OBJECTIVES

1. Find the Exact Value of the Trigonometric Functions Using the Unit Circle
2. Know the Domain and Range of the Trigonometric Functions
3. Use the Periodic Properties to Find the Exact Value of the Trigonometric Functions
4. Use Even-Odd Properties to Find the Exact Value of the Trigonometric Functions

In this section, we develop important properties of the trigonometric functions. We begin by introducing the trigonometric functions using the unit circle. This approach will lead to the definition given earlier of the trigonometric functions of a general angle.

The Unit Circle

1. Recall that the unit circle is a circle whose radius is 1 and whose center is at the origin of a rectangular coordinate system. Also recall that any circle of radius r has circumference of length $2\pi r$. Therefore, the unit circle (radius = 1) has a circumference of length 2π. In other words, for 1 revolution around the unit circle the length of arc is 2π units.

The following discussion sets the stage for defining the trigonometric functions.

Let $t \geq 0$ be any real number and let s be the distance from the origin to t on the real number line. See the red portion of Figure 65(a). Now look at the unit circle in Figure 65(a). Beginning at the the point $(1, 0)$ on the unit circle, travel $s = t$ units in the counterclockwise direction along the circle to arrive at the point $P = (a, b)$. In this sense, the length $s = t$ units is being **wrapped** around the unit circle.

Figure 65

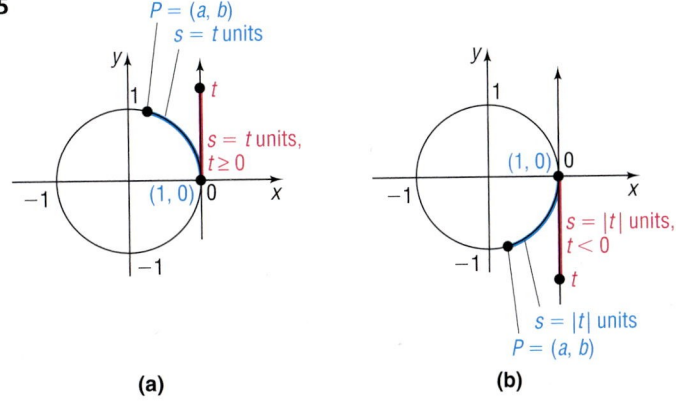

(a)   (b)

If $t < 0$, we begin at the point $(1, 0)$ on the unit circle and travel $s = |t|$ units in the clockwise direction to arrive at the point $P = (a, b)$. See Figure 65(b).

If $t > 2\pi$ or if $t < -2\pi$, it will be necessary to travel around the unit circle more than once before arriving at point P. Do you see why?

Let's describe this process another way. Picture a string of length $s = |t|$ units being wrapped around a circle of radius 1 unit. We start wrapping the string around the circle at the point $(1, 0)$. If $t \geq 0$, we wrap the string in the counterclockwise direction; if $t < 0$, we wrap the string in the clockwise direction. The point $P = (a, b)$ is the point where the string ends.

This discussion tells us that, for any real number t, we can locate a unique point $P = (a, b)$ on the unit circle. This point is called **the point P on the unit circle that corresponds to t.** This is the important idea here. No matter what real number t is chosen, there is a unique point P on the unit circle corresponding to it. We use the coordinates of the point $P = (a, b)$ on the unit circle corresponding to the real number t to define the **six trigonometric functions of t.**

Let t be a real number and let $P = (a, b)$ be the point on the unit circle that corresponds to t.

The **sine function** associates with t the y-coordinate of P and is denoted by

$$\sin t = b$$

The **cosine function** associates with t the x-coordinate of P and is denoted by

$$\cos t = a$$

If $a \neq 0$, the **tangent function** is defined as

$$\tan t = \frac{b}{a}$$

If $b \neq 0$, the **cosecant function** is defined as

$$\csc t = \frac{1}{b}$$

If $a \neq 0$, the **secant function** is defined as

$$\sec t = \frac{1}{a}$$

If $b \neq 0$, the **cotangent function** is defined as

$$\cot t = \frac{a}{b}$$

Once again, notice in these definitions that if $a = 0$ [that is, the point $P = (0, b)$ is on the y-axis] the tangent function and the secant function are undefined. Also, if $b = 0$ [that is, the point $P = (a, 0)$ is on the x-axis] the cosecant function and the cotangent function are undefined.

Because we use the unit circle in these definitions of the trigonometric functions, they are also sometimes referred to as **circular functions**.

EXAMPLE 1

Finding the Value of the Trigonometric Functions Using a Point on the Unit Circle

Figure 66

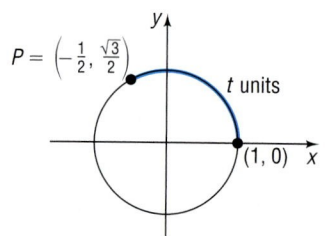

Find the value of $\sin t$, $\cos t$, $\tan t$, $\csc t$, $\sec t$, and $\cot t$ if $P = \left(-\frac{1}{2}, \frac{\sqrt{3}}{2}\right)$ is the point on the unit circle that corresponds to the real number t.

Solution See Figure 66. We follow the definition of the six trigonometric functions using $P = \left(-\frac{1}{2}, \frac{\sqrt{3}}{2}\right) = (a, b)$. Then, with $a = -\frac{1}{2}$, $b = \frac{\sqrt{3}}{2}$, we have

$$\sin t = b = \frac{\sqrt{3}}{2} \qquad\qquad \cos t = a = -\frac{1}{2} \qquad\qquad \tan t = \frac{b}{a} = \frac{\dfrac{\sqrt{3}}{2}}{-\dfrac{1}{2}} = -\sqrt{3}$$

$$\csc t = \frac{1}{b} = \frac{1}{\dfrac{\sqrt{3}}{2}} = \frac{2}{\sqrt{3}} = \frac{2\sqrt{3}}{3} \qquad \sec t = \frac{1}{a} = \frac{1}{-\dfrac{1}{2}} = -2 \qquad \cot t = \frac{a}{b} = \frac{-\dfrac{1}{2}}{\dfrac{\sqrt{3}}{2}} = -\frac{1}{\sqrt{3}} = -\frac{\sqrt{3}}{3}$$

NOW WORK PROBLEM 1.

Trigonometric Functions of Angles

Let $P = (a, b)$ be the point on the unit circle corresponding to the real number t. See Figure 67(a). Let θ be the angle in standard position, measured in radians, whose terminal side is the ray from the origin through P. See Figure 67(b). Since the unit circle has radius 1 unit, from the formula for arc length, $s = r\theta$, we find that

$$s = r\theta = \theta$$
$$\uparrow$$
$$r = 1$$

So, if $s = |t|$ units, then $\theta = t$ radians. See Figures 67(c) and (d).

Figure 67

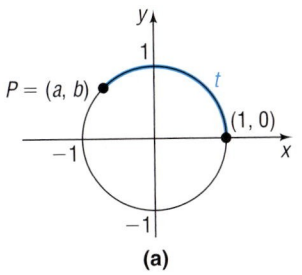
(a)

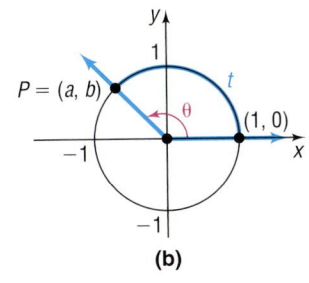
(b)

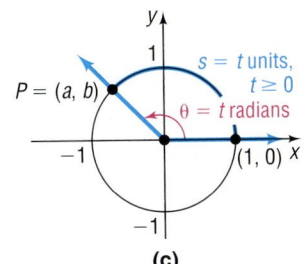
(c)

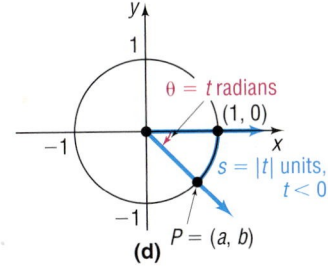
(d)

The point $P = (a, b)$ on the unit circle that corresponds to the real number t is the point P on the terminal side of the angle $\theta = t$ radians. As a result, we can say that

$$\sin t = \sin \theta$$
$$\uparrow \qquad\quad \uparrow$$
Real number $\theta = t$ radians

and so on. We can now define the trigonometric functions of the angle θ.

> If $\theta = t$ radians, the six **trigonometric functions of the angle θ** are defined as
>
> | | | |
> |---|---|---|
> | $\sin \theta = \sin t$ | $\cos \theta = \cos t$ | $\tan \theta = \tan t$ |
> | $\csc \theta = \csc t$ | $\sec \theta = \sec t$ | $\cot \theta = \cot t$ |

Even though the distinction between trigonometric functions of real numbers and trigonometric functions of angles is important, it is customary to refer to trigonometric functions of real numbers and trigonometric functions of angles collectively as *the trigonometric functions*. We will follow this practice from now on.

Since the values of the trigonometric functions of an angle θ are determined by the coordinates of the point $P = (a, b)$ on the unit circle corresponding to θ, the units used to measure the angle θ are irrelevant. For example, it does not matter whether we write $\theta = \dfrac{\pi}{2}$ radians or $\theta = 90°$. The point on the unit circle corresponding to this angle is $P = (0, 1)$. Hence,

$$\sin \frac{\pi}{2} = \sin 90° = 1 \quad \text{and} \quad \cos \frac{\pi}{2} = \cos 90° = 0$$

To find the exact value of a trigonometric function of an angle θ requires that we locate the corresponding point $P^* = (a^*, b^*)$ on the unit circle. In fact, though, any circle whose center is at the origin can be used.

Let θ be any nonquadrantal angle placed in standard position. Let $P = (a, b)$ be the point on the circle $x^2 + y^2 = r^2$ that corresponds to θ. See Figure 68.

Notice that the triangles OA^*P^* and OAP are similar so that the ratios of the corresponding sides are equal.

Figure 68

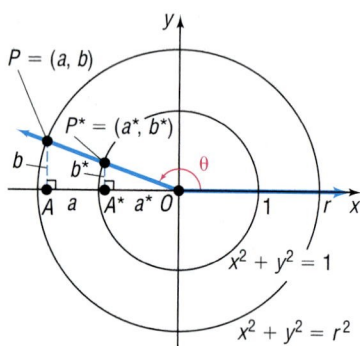

$$\frac{b^*}{1} = \frac{b}{r} \qquad \frac{a^*}{1} = \frac{a}{r} \qquad \frac{b^*}{a^*} = \frac{b}{a}$$

$$\frac{1}{b^*} = \frac{r}{b} \qquad \frac{1}{a^*} = \frac{r}{a} \qquad \frac{a^*}{b^*} = \frac{a}{b}$$

These results lead us to formulate the following theorem:

Theorem

For an angle θ in standard position, let $P = (a, b)$ be any point on the terminal side of θ that is also on the circle $x^2 + y^2 = r^2$. Then

$\sin \theta = \dfrac{b}{r}$	$\cos \theta = \dfrac{a}{r}$	$\tan \theta = \dfrac{b}{a}, \quad a \neq 0$
$\csc \theta = \dfrac{r}{b}, \quad b \neq 0$	$\sec \theta = \dfrac{r}{a}, \quad a \neq 0$	$\cot \theta = \dfrac{a}{b}, \quad b \neq 0$

This result coincides with the definition given in Section 8.4 for the six trigonometric functions of a general angle θ.

NOW WORK PROBLEM 7.

Domain and Range of the Trigonometric Functions

Figure 69

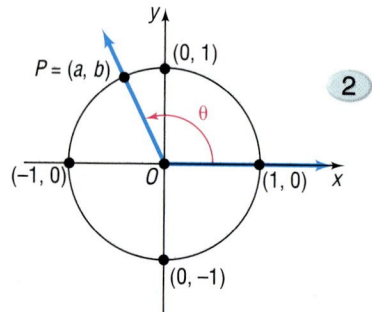

Let θ be an angle in standard position, and let $P = (a, b)$ be the point on the unit circle that corresponds to θ. See Figure 69. Then, by definition,

$\sin \theta = b$	$\cos \theta = a$	$\tan \theta = \dfrac{b}{a}, \quad a \neq 0$
$\csc \theta = \dfrac{1}{b}, \quad b \neq 0$	$\sec \theta = \dfrac{1}{a}, \quad a \neq 0$	$\cot \theta = \dfrac{a}{b}, \quad b \neq 0$

For $\sin\theta$ and $\cos\theta$, θ can be any angle, so it follows that the domain of the sine function and cosine function is the set of all real numbers.

> The domain of the sine function is the set of all real numbers.
>
> The domain of the cosine function is the set of all real numbers.

If $a = 0$, then the tangent function and the secant function are not defined. That is, for the tangent function and secant function, the x-coordinate of $P = (a, b)$ cannot be 0. On the unit circle, there are two such points $(0, 1)$ and $(0, -1)$. These two points correspond to the angles $\frac{\pi}{2}\,(90°)$ and $\frac{3\pi}{2}\,(270°)$ or, more generally, to any angle that is an odd multiple of $\frac{\pi}{2}\,(90°)$, such as $\frac{\pi}{2}\,(90°)$, $\frac{3\pi}{2}\,(270°)$, $\frac{5\pi}{2}\,(450°)$, $-\frac{\pi}{2}\,(-90°)$, and $-\frac{3\pi}{2}\,(-270°)$. Such angles must therefore be excluded from the domain of the tangent function and secant function.

> The domain of the tangent function is the set of all real numbers, except odd multiples of $\frac{\pi}{2}\,(90°)$.
>
> The domain of the secant function is the set of all real numbers, except odd multiples of $\frac{\pi}{2}\,(90°)$.

If $b = 0$, then the cotangent function and the cosecant function are not defined. For the cotangent function and cosecant function, the y-coordinate of $P = (a, b)$ cannot be 0. On the unit circle, there are two such points, $(1, 0)$ and $(-1, 0)$. These two points correspond to the angles $0\,(0°)$ and $\pi\,(180°)$ or, more generally, to any angle that is an integral multiple of $\pi\,(180°)$, such as $0\,(0°)$, $\pi\,(180°)$, $2\pi\,(360°)$, $3\pi\,(540°)$, and $-\pi\,(-180°)$. Such angles must therefore be excluded from the domain of the cotangent function and cosecant function.

> The domain of the cotangent function is the set of all real numbers, except integral multiples of $\pi\,(180°)$.
>
> The domain of the cosecant function is the set of all real numbers, except integral multiples of $\pi\,(180°)$.

Next, we determine the range of each of the six trigonometric functions. Refer again to Figure 69. Let $P = (a, b)$ be the point on the unit circle that corresponds to the angle θ. It follows that $-1 \le a \le 1$ and $-1 \le b \le 1$. Since $\sin\theta = b$ and $\cos\theta = a$, we have

> $$-1 \le \sin\theta \le 1 \quad \text{and} \quad -1 \le \cos\theta \le 1$$

The range of both the sine function and the cosine function consists of all real numbers between -1 and 1, inclusive. Using absolute value notation, we have $|\sin\theta| \le 1$ and $|\cos\theta| \le 1$.

If θ is not a multiple of π (180°), then $\csc \theta = \dfrac{1}{b}$. Since $b = \sin \theta$ and $|b| = |\sin \theta| \leq 1$, it follows that $|\csc \theta| = \dfrac{1}{|\sin \theta|} = \dfrac{1}{|b|} \geq 1$. The range of the cosecant function consists of all real numbers less than or equal to -1 or greater than or equal to 1. That is,

$$\csc \theta \leq -1 \quad \text{or} \quad \csc \theta \geq 1$$

Using absolute value notation, we have $|\csc \theta| \geq 1$.

If θ is not an odd multiple of $\dfrac{\pi}{2}$ (90°), then, by definition, $\sec \theta = \dfrac{1}{a}$. Since $a = \cos \theta$ and $|a| = |\cos \theta| \leq 1$, it follows that $|\sec \theta| = \dfrac{1}{|\cos \theta|} = \dfrac{1}{|a|} \geq 1$. The range of the secant function consists of all real numbers less than or equal to -1 or greater than or equal to 1. That is,

$$\sec \theta \leq -1 \quad \text{or} \quad \sec \theta \geq 1$$

Using absolute value notation, we have $|\sec \theta| \geq 1$.

The range of both the tangent function and the cotangent function consists of all real numbers. That is,

$$-\infty < \tan \theta < \infty \quad \text{and} \quad -\infty < \cot \theta < \infty$$

You are asked to prove this in Problems 81 and 82.

Table 6 summarizes these results.

TABLE 6

Function	Symbol	Domain	Range
sine	$f(\theta) = \sin \theta$	All real numbers	All real numbers from -1 to 1, inclusive
cosine	$f(\theta) = \cos \theta$	All real numbers	All real numbers from -1 to 1, inclusive
tangent	$f(\theta) = \tan \theta$	All real numbers, except odd multiples of $\dfrac{\pi}{2}$ (90°)	All real numbers
cosecant	$f(\theta) = \csc \theta$	All real numbers, except integral multiples of π (180°)	All real numbers greater than or equal to 1 or less than or equal to -1
secant	$f(\theta) = \sec \theta$	All real numbers, except odd multiples of $\dfrac{\pi}{2}$ (90°)	All real numbers greater than or equal to 1 or less than or equal to -1
cotangent	$f(\theta) = \cot \theta$	All real numbers, except integral multiples of π (180°)	All real numbers

NOW WORK PROBLEMS 53 AND 57.

Period of the Trigonometric Functions

Figure 70

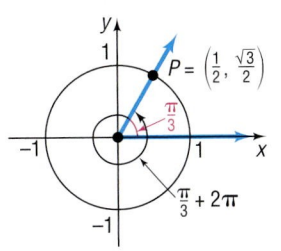

3 Look at Figure 70. This figure shows that for an angle of $\frac{\pi}{3}$ radians the corresponding point P on the unit circle is $\left(\frac{1}{2}, \frac{\sqrt{3}}{2}\right)$. Notice that for an angle of $\frac{\pi}{3} + 2\pi$ radians the corresponding point P on the unit circle is also $\left(\frac{1}{2}, \frac{\sqrt{3}}{2}\right)$. Thus,

$$\sin\frac{\pi}{3} = \frac{\sqrt{3}}{2} \quad \text{and} \quad \sin\left(\frac{\pi}{3} + 2\pi\right) = \frac{\sqrt{3}}{2}$$

$$\cos\frac{\pi}{3} = \frac{1}{2} \quad \text{and} \quad \cos\left(\frac{\pi}{3} + 2\pi\right) = \frac{1}{2}$$

This example illustrates a more general situation. For a given angle θ, measured in radians, suppose that we know the corresponding point $P = (a, b)$ on the unit circle. Now add 2π to θ. The point on the unit circle corresponding to $\theta + 2\pi$ is identical to the point P on the unit circle corresponding to θ. See Figure 71. The values of the trigonometric functions of $\theta + 2\pi$ are equal to the values of the corresponding trigonometric functions of θ.

Figure 71

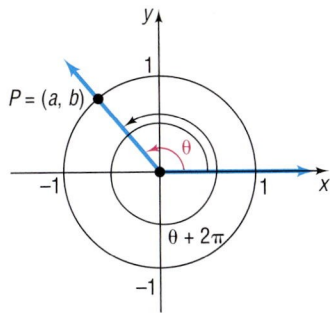

If we add (or subtract) integral multiples of 2π to θ, the trigonometric values remain unchanged. That is, for all θ,

$$\sin(\theta + 2\pi k) = \sin\theta \quad \cos(\theta + 2\pi k) = \cos\theta \qquad (1)$$
$$\text{where } k \text{ is any integer}$$

Functions that exhibit this kind of behavior are called *periodic functions*.

A function f is called **periodic** if there is a positive number p such that, whenever θ is in the domain of f, so is $\theta + p$, and

$$f(\theta + p) = f(\theta)$$

If there is a smallest such number p, this smallest value is called the **(fundamental) period** of f.

Based on equation (1), the sine and cosine functions are periodic. In fact, the sine and cosine functions have period 2π. You are asked to prove this fact in Problems 83 and 84. The secant and cosecant functions are also periodic with period 2π; and the tangent and cotangent functions are periodic with period π. You are asked to prove these statements in Problems 85 through 88.

These facts are summarized as follows:

Periodic Properties

$$\sin(\theta + 2\pi) = \sin\theta \quad \cos(\theta + 2\pi) = \cos\theta \quad \tan(\theta + \pi) = \tan\theta$$
$$\csc(\theta + 2\pi) = \csc\theta \quad \sec(\theta + 2\pi) = \sec\theta \quad \cot(\theta + \pi) = \cot\theta$$

Because the sine, cosine, secant, and cosecant functions have period 2π, once we know their values for $0 \le \theta < 2\pi$, we know all their values; similarly, since the tangent and cotangent functions have period π, once we know their values for $0 \le \theta < \pi$, we know all their values.

EXAMPLE 2 **Using Periodic Properties to Find Exact Values**

Find the exact value of: (a) $\sin 420°$ (b) $\tan \dfrac{5\pi}{4}$ (c) $\cos \dfrac{11\pi}{4}$

Solution (a) $\sin 420° = \sin(360° + 60°) = \sin 60° = \dfrac{\sqrt{3}}{2}$

(b) $\tan \dfrac{5\pi}{4} = \tan\left(\pi + \dfrac{\pi}{4}\right) = \tan \dfrac{\pi}{4} = 1$

(c) $\cos \dfrac{11\pi}{4} = \cos\left(\dfrac{3\pi}{4} + \dfrac{8\pi}{4}\right) = \cos\left(\dfrac{3\pi}{4} + 2\pi\right) = \cos\dfrac{3\pi}{4} = -\dfrac{\sqrt{2}}{2}$ ∎

NOW WORK PROBLEMS 13 AND 71.

The periodic properties of the trigonometric functions will be very helpful to us when we study their graphs in the next section.

Even–Odd Properties

④ Recall that a function f is even if $f(-\theta) = f(\theta)$ for all θ in the domain of f; a function f is odd if $f(-\theta) = -f(\theta)$ for all θ in the domain of f. We will now show that the trigonometric functions sine, tangent, cotangent, and cosecant are odd functions, whereas the functions cosine and secant are even functions.

Theorem **Even–Odd Properties**

$$\sin(-\theta) = -\sin\theta \qquad \cos(-\theta) = \cos\theta \qquad \tan(-\theta) = -\tan\theta$$
$$\csc(-\theta) = -\csc\theta \qquad \sec(-\theta) = \sec\theta \qquad \cot(-\theta) = -\cot\theta$$

Proof Let $P = (a, b)$ be the point on the unit circle that corresponds to the angle θ. See Figure 72. The point Q on the unit circle that corresponds to the angle $-\theta$ will have coordinates $(a, -b)$. Using the definition for the trigonometric functions, we have

$$\sin\theta = b \qquad \cos\theta = a \qquad \sin(-\theta) = -b \qquad \cos(-\theta) = a$$

so

$$\sin(-\theta) = -b = -\sin\theta \qquad \cos(-\theta) = a = \cos\theta$$

Now, using these results and some of the Fundamental Identities, we have

$$\tan(-\theta) = \frac{\sin(-\theta)}{\cos(-\theta)} = \frac{-\sin\theta}{\cos\theta} = -\tan\theta$$

$$\cot(-\theta) = \frac{1}{\tan(-\theta)} = \frac{1}{-\tan\theta} = -\cot\theta$$

Figure 72

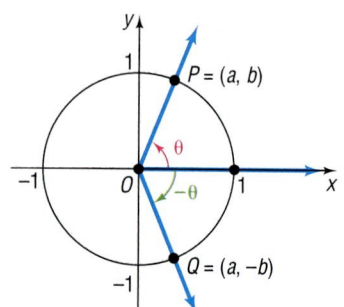

$$\sec(-\theta) = \frac{1}{\cos(-\theta)} = \frac{1}{\cos\theta} = \sec\theta$$

$$\csc(-\theta) = \frac{1}{\sin(-\theta)} = \frac{1}{-\sin\theta} = -\csc\theta$$

EXAMPLE 3   **Finding Exact Values Using Even–Odd Properties**

Find the exact value of:

(a) $\sin(-45°)$ (b) $\cos(-\pi)$ (c) $\cot\left(-\dfrac{3\pi}{2}\right)$ (d) $\tan\left(-\dfrac{37\pi}{4}\right)$

Solution   (a) $\underset{\substack{\uparrow \\ \textit{Odd function}}}{\sin(-45°) = -\sin 45°} = -\dfrac{\sqrt{2}}{2}$ (b) $\underset{\substack{\uparrow \\ \textit{Even function}}}{\cos(-\pi) = \cos\pi} = -1$

(c) $\underset{\substack{\uparrow \\ \textit{Odd function}}}{\cot\left(-\dfrac{3\pi}{2}\right) = -\cot\dfrac{3\pi}{2}} = 0$

(d) $\underset{\substack{\uparrow \\ \textit{Odd function}}}{\tan\left(-\dfrac{37\pi}{4}\right) = -\tan\dfrac{37\pi}{4}} = \underset{\substack{\uparrow \\ \textit{Period is } \pi}}{-\tan\left(\dfrac{\pi}{4} + 9\pi\right) = -\tan\dfrac{\pi}{4}} = -1$

SEEING THE CONCEPT   To see that the cosine function is even, use a graphing utility to graph $Y_1 = \cos x$. Do you see the symmetry? Why would you conjecture that the cosine function is even? Clear the screen. Now graph $Y_1 = \sin x$. Do you see the symmetry? Do you conjecture that the sine function is odd?

NOW WORK PROBLEMS 29 AND 65.

8.5   Concepts and Vocabulary

In Problems 1–3, fill in the blanks.

1. The sine, cosine, cosecant, and secant functions have period _____; the tangent and cotangent functions have period _____.

2. The domain of the tangent function is _____.

3. The range of the sine function is _____.

In Problems 4–6, answer True or False to each statement.

4. The only even trigonometric functions are the cosine and secant functions.

5. All the trigonometric functions are periodic, with period 2π.

6. The range of the secant function is the set of positive real numbers.

7. Explain how you would find the value of $\sin 390°$ using periodic properties.

8. Explain how you would find the value of $\cos(-45°)$ using even–odd properties.

8.5 Exercises

In Problems 1–6, the point P on the unit circle that corresponds to a real number t is given. Find $\sin t$, $\cos t$, $\tan t$, $\csc t$, $\sec t$, and $\cot t$.

1. $\left(\dfrac{\sqrt{3}}{2}, -\dfrac{1}{2}\right)$

2. $\left(-\dfrac{\sqrt{3}}{2}, -\dfrac{1}{2}\right)$

3. $\left(-\dfrac{\sqrt{2}}{2}, -\dfrac{\sqrt{2}}{2}\right)$

4. $\left(\dfrac{\sqrt{2}}{2}, -\dfrac{\sqrt{2}}{2}\right)$

5. $\left(\dfrac{\sqrt{5}}{3}, \dfrac{2}{3}\right)$

6. $\left(-\dfrac{\sqrt{5}}{5}, \dfrac{2\sqrt{5}}{5}\right)$

In Problems 7–12, the point P on the circle $x^2 + y^2 = r^2$ that is also on the terminal side of an angle θ in standard position is given. Find $\sin \theta$, $\cos \theta$, $\tan \theta$, $\csc \theta$, $\sec \theta$, and $\cot \theta$.

7. $(3, -4)$ **8.** $(4, -3)$ **9.** $(-2, 3)$ **10.** $(2, -4)$ **11.** $(-1, -1)$ **12.** $(-3, 1)$

In Problems 13–28, use the fact that the trigonometric functions are periodic to find the exact value of each expression. Do not use a calculator.

13. $\sin 405°$

14. $\cos 420°$

15. $\tan 405°$

16. $\sin 390°$

17. $\csc 450°$

18. $\sec 540°$

19. $\cot 390°$

20. $\sec 420°$

21. $\cos \dfrac{33\pi}{4}$

22. $\sin \dfrac{9\pi}{4}$

23. $\tan(21\pi)$

24. $\csc \dfrac{9\pi}{2}$

25. $\sec \dfrac{17\pi}{4}$

26. $\cot \dfrac{17\pi}{4}$

27. $\tan \dfrac{19\pi}{6}$

28. $\sec \dfrac{25\pi}{6}$

In Problems 29–46, use the even-odd properties to find the exact value of each expression. Do not use a calculator.

29. $\sin(-60°)$

30. $\cos(-30°)$

31. $\tan(-30°)$

32. $\sin(-135°)$

33. $\sec(-60°)$

34. $\csc(-30°)$

35. $\sin(-90°)$

36. $\cos(-270°)$

37. $\tan\left(-\dfrac{\pi}{4}\right)$

38. $\sin(-\pi)$

39. $\cos\left(-\dfrac{\pi}{4}\right)$

40. $\sin\left(-\dfrac{\pi}{3}\right)$

41. $\tan(-\pi)$

42. $\sin\left(-\dfrac{3\pi}{2}\right)$

43. $\csc\left(-\dfrac{\pi}{4}\right)$

44. $\sec(-\pi)$

45. $\sec\left(-\dfrac{\pi}{6}\right)$

46. $\csc\left(-\dfrac{\pi}{3}\right)$

In Problems 47–52, find the exact value of each expression. Do not use a calculator.

47. $\sin(-\pi) + \cos(5\pi)$

48. $\tan\left(-\dfrac{5\pi}{6}\right) - \cot\dfrac{7\pi}{2}$

49. $\sec(-\pi) + \csc\left(-\dfrac{\pi}{2}\right)$

50. $\tan(-6\pi) + \cos\dfrac{9\pi}{4}$

51. $\sin\left(-\dfrac{9\pi}{4}\right) - \tan\left(-\dfrac{9\pi}{4}\right)$

52. $\cos\left(-\dfrac{17\pi}{4}\right) - \sin\left(-\dfrac{3\pi}{2}\right)$

53. What is the domain of the sine function?

54. What is the domain of the cosine function?

55. For what numbers θ is $f(\theta) = \tan \theta$ not defined?

56. For what numbers θ is $f(\theta) = \cot \theta$ not defined?

57. For what numbers θ is $f(\theta) = \sec \theta$ not defined?

58. For what numbers θ is $f(\theta) = \csc \theta$ not defined?

59. What is the range of the sine function?

60. What is the range of the cosine function?

61. What is the range of the tangent function?

62. What is the range of the cotangent function?

63. What is the range of the secant function?

64. What is the range of the cosecant function?

65. Is the sine function even, odd, or neither? Is its graph symmetric? With respect to what?

66. Is the cosine function even, odd, or neither? Is its graph symmetric? With respect to what?

67. Is the tangent function even, odd, or neither? Is its graph symmetric? With respect to what?

68. Is the cotangent function even, odd, or neither? Is its graph symmetric? With respect to what?

69. Is the secant function even, odd, or neither? Is its graph symmetric? With respect to what?

70. Is the cosecant function even, odd, or neither? Is its graph symmetric? With respect to what?

71. If $\sin \theta = 0.3$, find the value of

$$\sin \theta + \sin(\theta + 2\pi) + \sin(\theta + 4\pi)$$

72. If $\cos \theta = 0.2$, find the value of

$$\cos \theta + \cos(\theta + 2\pi) + \cos(\theta + 4\pi)$$

73. If $\tan \theta = 3$, find the value of

$$\tan \theta + \tan(\theta + \pi) + \tan(\theta + 2\pi).$$

74. If $\cot \theta = -2$, find the value of

$$\cot \theta + \cot(\theta - \pi) + \cot(\theta - 2\pi).$$

In Problems 75–80, use the periodic and even-odd properties.

75. If $f(x) = \sin x$ and $f(a) = \dfrac{1}{3}$, find the exact value of:
(a) $f(-a)$ (b) $f(a) + f(a + 2\pi) + f(a + 4\pi)$

76. If $f(x) = \cos x$ and $f(a) = \dfrac{1}{4}$, find the exact value of:
(a) $f(-a)$ (b) $f(a) + f(a + 2\pi) + f(a - 2\pi)$

77. If $f(x) = \tan x$ and $f(a) = 2$, find the exact value of:
(a) $f(-a)$ (b) $f(a) + f(a + \pi) + f(a + 2\pi)$

78. If $f(x) = \cot x$ and $f(a) = -3$, find the exact value of:
(a) $f(-a)$ (b) $f(a) + f(a + \pi) + f(a + 4\pi)$

79. If $f(x) = \sec x$ and $f(a) = -4$, find the exact value of:
(a) $f(-a)$ (b) $f(a) + f(a + 2\pi) + f(a + 4\pi)$

80. If $f(x) = \csc x$ and $f(a) = 2$, find the exact value of:
(a) $f(-a)$ (b) $f(a) + f(a + 2\pi) + f(a + 4\pi)$

81. Show that the range of the tangent function is the set of all real numbers.

82. Show that the range of the cotangent function is the set of all real numbers.

83. Show that the period of $f(\theta) = \sin \theta$ is 2π.

[**Hint:** Assume that $0 < p < 2\pi$ exists so that $\sin(\theta + p) = \sin \theta$ for all θ. Let $\theta = 0$ to find p. Then let $\theta = \dfrac{\pi}{2}$ to obtain a contradiction.]

84. Show that the period of $f(\theta) = \cos \theta$ is 2π.

85. Show that the period of $f(\theta) = \sec \theta$ is 2π.

86. Show that the period of $f(\theta) = \csc \theta$ is 2π.

87. Show that the period of $f(\theta) = \tan \theta$ is π.

88. Show that the period of $f(\theta) = \cot \theta$ is π.

89. If $\theta\,(0 < \theta < \pi)$ is the angle between a horizontal ray directed to the right (say, the positive x-axis) and a non-horizontal, nonvertical line L, show that the slope m of L equals $\tan \theta$. The angle θ is called the **inclination** of L.
[**Hint:** See the illustration, where we have drawn the line L^* parallel to L and passing through the origin. Use the fact that L^* intersects the unit circle at the point $(\cos \theta, \sin \theta)$.]

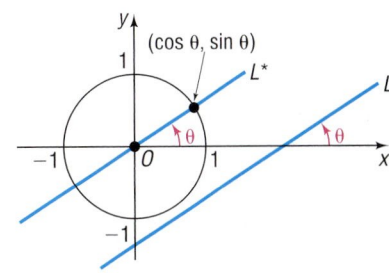

90. Write down five properties of the tangent function. Explain the meaning of each.

91. Describe your understanding of the meaning of a periodic function.

PREPARING FOR THIS SECTION

Before getting started, review the following:

✓ Graphing Techniques: Transformations (Section 3.4, pp. 286–295)

8.6 GRAPHS OF THE SINE AND COSINE FUNCTIONS

OBJECTIVES

  1 Graph Transformations of the Sine Function

  2 Graph Transformations of the Cosine Function

  3 Determine the Amplitude and Period of Sinusoidal Functions

  4 Graph Sinusoidal Functions: $y = A\sin(\omega x)$

  5 Find an Equation for a Sinusoidal Graph

Since we want to graph the trigonometric functions in the *xy*-plane, we shall use the traditional symbols x for the independent variable (or argument) and y for the dependent variable (or value at x) for each function. So we write the six trigonometric functions as

$$
\begin{array}{lll}
y = f(x) = \sin x & y = f(x) = \cos x & y = f(x) = \tan x \\
y = f(x) = \csc x & y = f(x) = \sec x & y = f(x) = \cot x
\end{array}
$$

 Here the independent variable x represents an angle, measured in radians. In calculus, x will usually be treated as a real number. As we said earlier, these are equivalent ways of viewing x.

The Graph of $y = \sin x$

Since the sine function has period 2π, we need to graph $y = \sin x$ only on the interval $[0, 2\pi]$. The remainder of the graph will consist of repetitions of this portion of the graph.

We begin by constructing Table 7, which lists some points on the graph of $y = \sin x$, $0 \leq x \leq 2\pi$. As the table shows, the graph of $y = \sin x$, $0 \leq x \leq 2\pi$, begins at the origin. As x increases from 0 to $\dfrac{\pi}{2}$, the value of $y = \sin x$ increases from 0 to 1; as x increases from $\dfrac{\pi}{2}$ to π to $\dfrac{3\pi}{2}$, the value of y decreases from 1 to 0 to -1; as x increases from $\dfrac{3\pi}{2}$ to 2π, the value of y increases from -1 to 0. If we plot the points listed in Table 7 and connect them with a smooth curve, we obtain the graph shown in Figure 73.

TABLE 7

x	$y = \sin x$	(x, y)
0	0	$(0, 0)$
$\dfrac{\pi}{6}$	$\dfrac{1}{2}$	$\left(\dfrac{\pi}{6}, \dfrac{1}{2}\right)$
$\dfrac{\pi}{2}$	1	$\left(\dfrac{\pi}{2}, 1\right)$
$\dfrac{5\pi}{6}$	$\dfrac{1}{2}$	$\left(\dfrac{5\pi}{6}, \dfrac{1}{2}\right)$
π	0	$(\pi, 0)$
$\dfrac{7\pi}{6}$	$-\dfrac{1}{2}$	$\left(\dfrac{7\pi}{6}, -\dfrac{1}{2}\right)$
$\dfrac{3\pi}{2}$	-1	$\left(\dfrac{3\pi}{2}, -1\right)$
$\dfrac{11\pi}{6}$	$-\dfrac{1}{2}$	$\left(\dfrac{11\pi}{6}, -\dfrac{1}{2}\right)$
2π	0	$(2\pi, 0)$

Figure 73

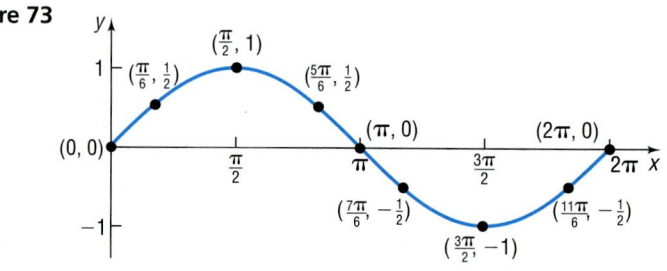

The graph in Figure 73 is one period, or **cycle**, of the graph of $y = \sin x$. To obtain a more complete graph of $y = \sin x$, we repeat this period in each direction, as shown in Figure 74(a). Figure 74(b) shows the graph on a TI-83 graphing calculator.

Figure 74
$y = \sin x, \; -\infty < x < \infty$

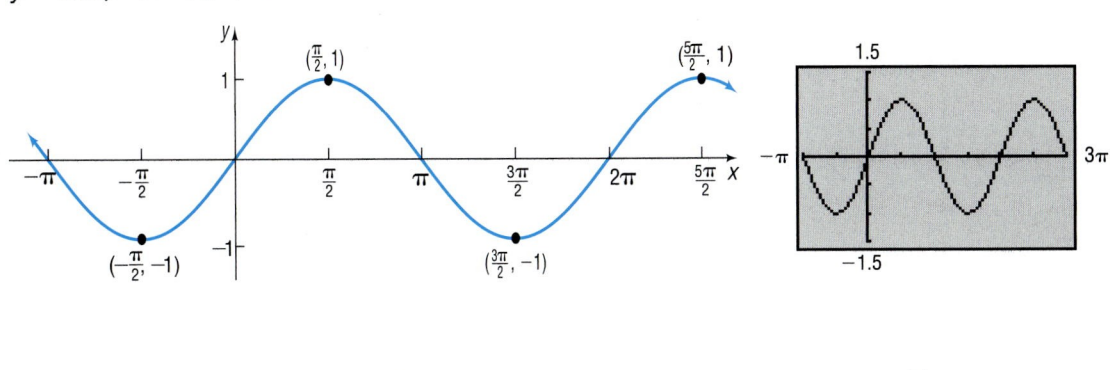

(a)   (b)

The graph of $y = \sin x$ illustrates some of the facts that we already know about the sine function.

Properties of the Sine Function

1. The domain is the set of all real numbers.
2. The range consists of all real numbers from -1 to 1, inclusive.
3. The sine function is an odd function, as the symmetry of the graph with respect to the origin indicates.
4. The sine function is periodic, with period 2π.
5. The x-intercepts are $\ldots, -2\pi, -\pi, \; 0, \; \pi, 2\pi, 3\pi, \ldots;$ the y-intercept is 0.
6. The maximum value is 1 and occurs at $x = \ldots, -\dfrac{3\pi}{2}, \dfrac{\pi}{2}, \dfrac{5\pi}{2},$ $\dfrac{9\pi}{2}, \ldots;$ the minimum value is -1 and occurs at $x = \ldots, -\dfrac{\pi}{2},$ $\dfrac{3\pi}{2}, \dfrac{7\pi}{2}, \dfrac{11\pi}{2}, \ldots.$

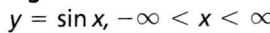 **NOW WORK PROBLEMS 1, 3, AND 5.**

①   The graphing techniques introduced in Chapter 3 may be used to graph functions that are transformations of the sine function (refer to Section 3.4).

EXAMPLE 1   **Graphing Variations of $y = \sin x$ Using Transformations**

Use the graph of $y = \sin x$ to graph $y = \sin\left(x - \dfrac{\pi}{4}\right)$.

Solution   Figure 75 illustrates the steps.

Figure 75

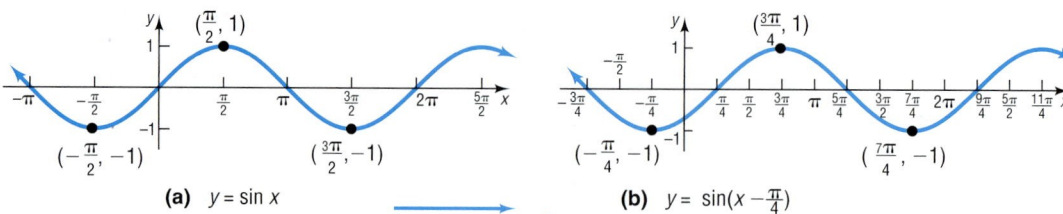

(a) $y = \sin x$

(b) $y = \sin(x - \frac{\pi}{4})$

Replace x by $x - \frac{\pi}{4}$; horizontal shift to the right $\frac{\pi}{4}$ units.

✔ CHECK: Figure 76 shows the graph of $Y_1 = \sin\left(x - \dfrac{\pi}{4}\right)$.

Figure 76

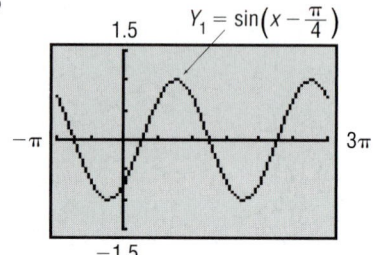

$Y_1 = \sin\left(x - \frac{\pi}{4}\right)$

EXAMPLE 2 **Graphing Variations of $y = \sin x$ Using Transformations**

Use the graph of $y = \sin x$ to graph $y = -\sin x + 2$.

Figure 77

Solution Figure 77 illustrates the steps.

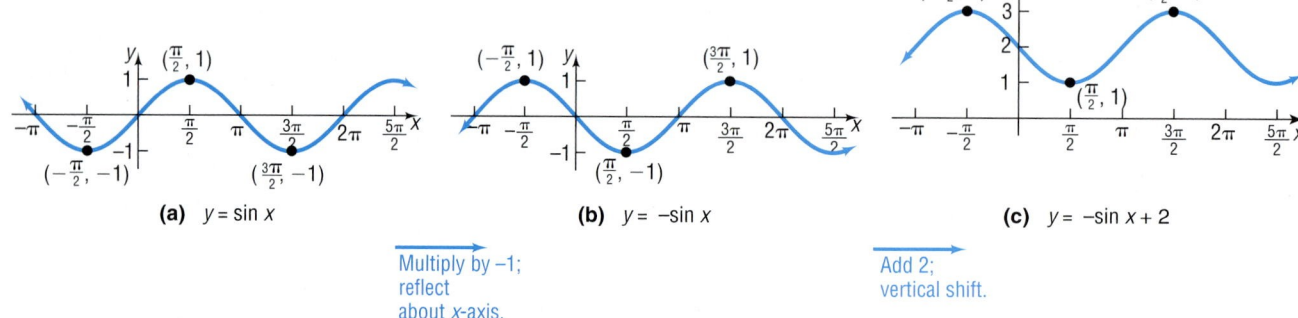

(a) $y = \sin x$

(b) $y = -\sin x$

(c) $y = -\sin x + 2$

Multiply by –1; reflect about x-axis.

Add 2; vertical shift.

✔ CHECK: Figure 78 shows the graph of $Y_1 = -\sin x + 2$.

Figure 78

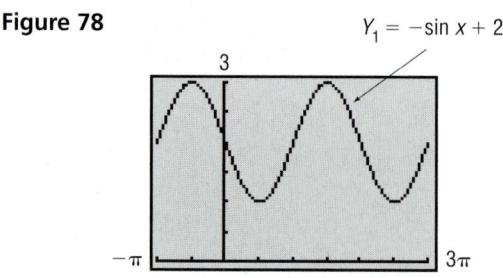

$Y_1 = -\sin x + 2$

NOW WORK PROBLEM **17.**

TABLE 8

x	$y = \cos x$	(x, y)
0	1	$(0, 1)$
$\dfrac{\pi}{3}$	$\dfrac{1}{2}$	$\left(\dfrac{\pi}{3}, \dfrac{1}{2}\right)$
$\dfrac{\pi}{2}$	0	$\left(\dfrac{\pi}{2}, 0\right)$
$\dfrac{2\pi}{3}$	$-\dfrac{1}{2}$	$\left(\dfrac{2\pi}{3}, -\dfrac{1}{2}\right)$
π	-1	$(\pi, -1)$
$\dfrac{4\pi}{3}$	$-\dfrac{1}{2}$	$\left(\dfrac{4\pi}{3}, -\dfrac{1}{2}\right)$
$\dfrac{3\pi}{2}$	0	$\left(\dfrac{3\pi}{2}, 0\right)$
$\dfrac{5\pi}{3}$	$\dfrac{1}{2}$	$\left(\dfrac{5\pi}{3}, \dfrac{1}{2}\right)$
2π	1	$(2\pi, 1)$

The Graph of $y = \cos x$

The cosine function also has period 2π. We proceed as we did with the sine function by constructing Table 8, which lists some points on the graph of $y = \cos x$, $0 \le x \le 2\pi$. As the table shows, the graph of $y = \cos x$, $0 \le x \le 2\pi$, begins at the point $(0, 1)$. As x increases from 0 to $\dfrac{\pi}{2}$ to π, the value of y decreases from 1 to 0 to -1; as x increases from π to $\dfrac{3\pi}{2}$ to 2π, the value of y increases from -1 to 0 to 1. As before, we plot the points in Table 8 to get one period or cycle of the graph. See Figure 79.

Figure 79
$y = \cos x, 0 \le x \le 2\pi$

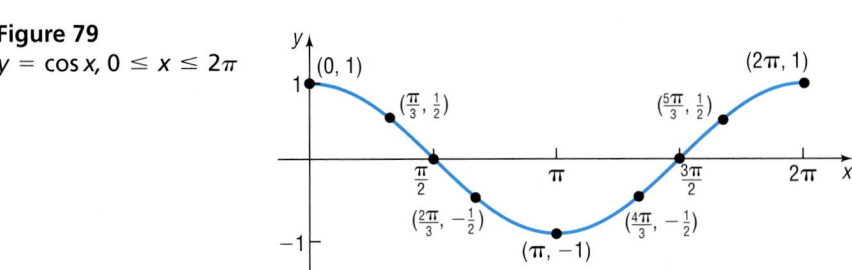

A more complete graph of $y = \cos x$ is obtained by repeating this period in each direction, as shown in Figure 80(a). Figure 80(b) shows the graph on a TI-83 graphing calculator.

Figure 80
$y = \cos x, -\infty < x < \infty$

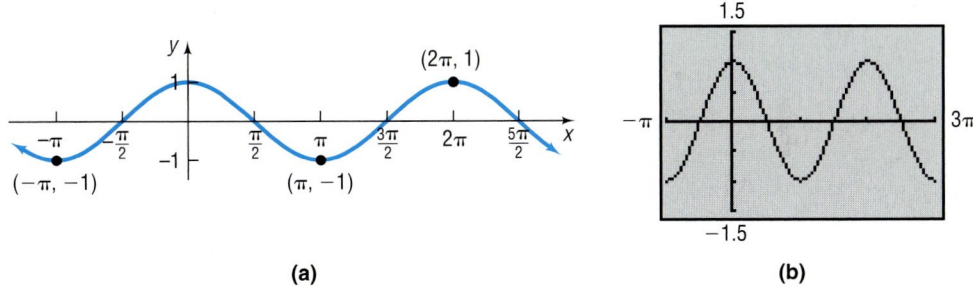

(a)   (b)

The graph of $y = \cos x$ illustrates some of the facts that we already know about the cosine function.

> ### Properties of the Cosine Function
>
> 1. The domain is the set of all real numbers.
> 2. The range consists of all real numbers from -1 to 1, inclusive.
> 3. The cosine function is an even function, as the symmetry of the graph with respect to the y-axis indicates.
> 4. The cosine function is periodic, with period 2π.
> 5. The x-intercepts are $\ldots, -\dfrac{3\pi}{2}, -\dfrac{\pi}{2}, \dfrac{\pi}{2}, \dfrac{3\pi}{2}, \dfrac{5\pi}{2}, \ldots$; the y-intercept is 1.
> 6. The maximum value is 1 and occurs at $x = \ldots, -2\pi, 0, 2\pi, 4\pi, 6\pi, \ldots$; the minimum value is -1 and occurs at $x = \ldots, -\pi, \pi, 3\pi, 5\pi, \ldots$.

(2) Again, the graphing techniques from Chapter 3 may be used to graph transformations of the cosine function.

EXAMPLE 3 Graphing Variations of $y = \cos x$ Using Transformations

Use the graph of $y = \cos x$ to graph $y = 2 \cos x$.

Solution Figure 81 illustrates the steps.

Figure 81

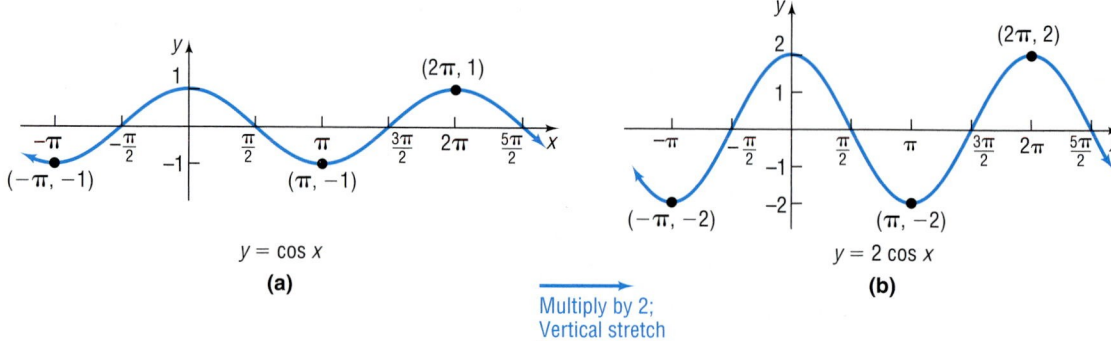

$y = \cos x$
(a)

Multiply by 2;
Vertical stretch
by a factor of 2.

$y = 2 \cos x$
(b)

✔ CHECK: Figure 82 shows the graph of $Y_1 = 2 \cos x$.

Figure 82 $Y_1 = 2 \cos x$

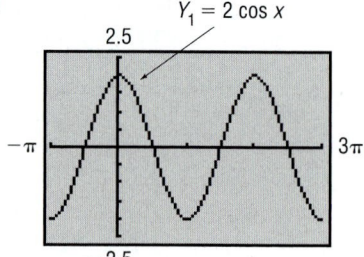

EXAMPLE 4 Graphing Variations of $y = \cos x$ Using Transformations

Use the graph of $y = \cos x$ to graph $y = \cos(3x)$.

Solution Figure 83 illustrates the graph, which is a horizontal compression of the graph of $y = \cos x$. (Multiply each x-coordinate by $\frac{1}{3}$.) Notice that, due to this compression, the period of $y = \cos(3x)$ is $\frac{2\pi}{3}$, whereas the period of $y = \cos x$ is 2π.

Figure 83

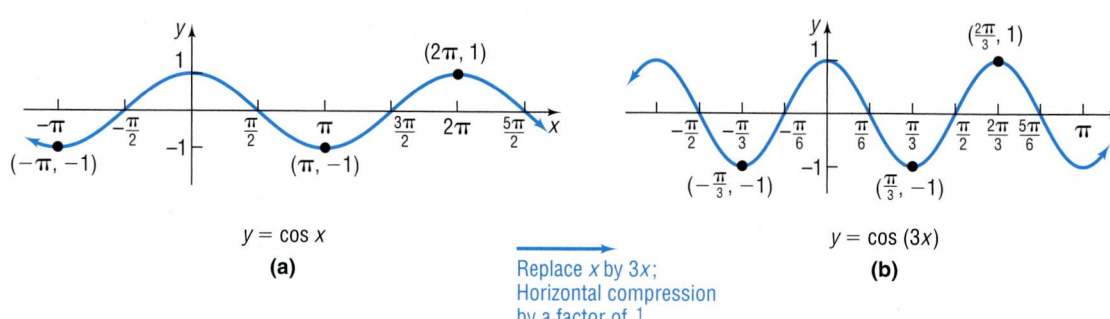

$y = \cos x$
(a)

Replace x by $3x$;
Horizontal compression
by a factor of $\frac{1}{3}$.

$y = \cos (3x)$
(b)

✔ CHECK: Figure 84 shows the graph of $Y_1 = \cos(3x)$.

Figure 84

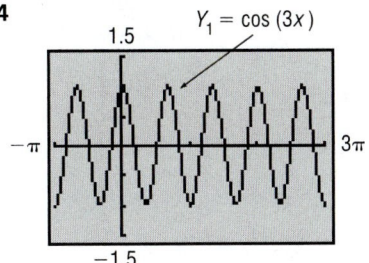

 SEEING THE CONCEPT Graph $Y_1 = \cos(3x)$ with $X\text{min} = 0$, $X\text{max} = \dfrac{2\pi}{3}$ and $X\text{scl} = \dfrac{\pi}{6}$ to verify that the period is $\dfrac{2\pi}{3}$.

━ **NOW WORK PROBLEM 25.**

Sinusoidal Graphs

The graph of $y = \cos x$, when compared to the graph of $y = \sin x$, suggests that the graph of $y = \sin x$ is the same as the graph of $y = \cos\left(x - \dfrac{\pi}{2}\right)$. See Figure 85.

Figure 85

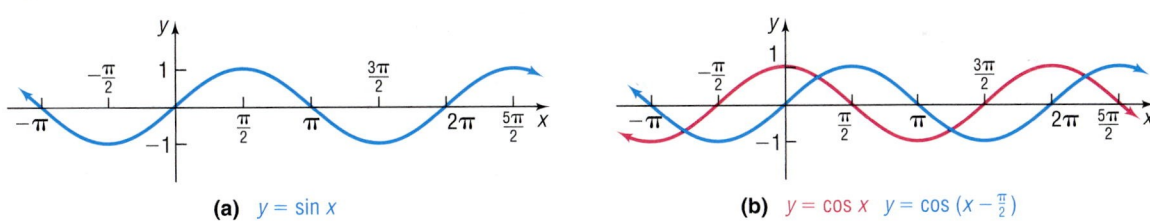

(a) $y = \sin x$ **(b)** $y = \cos x$ $y = \cos\left(x - \frac{\pi}{2}\right)$

Based on Figure 85, we conjecture that

$$\sin x = \cos\left(x - \frac{\pi}{2}\right)$$

(We shall prove this fact in Chapter 9.) Because of this similarity, the graphs of sine functions and cosine functions are referred to as **sinusoidal graphs**.

 SEEING THE CONCEPT Graph $Y_1 = \sin x$ and $Y_2 = \cos\left(x - \dfrac{\pi}{2}\right)$. How many graphs do you see?

Let's look at some general properties of sinusoidal graphs.

3   In Example 3 we obtained the graph of $y = 2 \cos x$, which we reproduce in Figure 86. Notice that the values of $y = 2 \cos x$ lie between -2 and 2, inclusive.

Figure 86
$y = 2 \cos x$

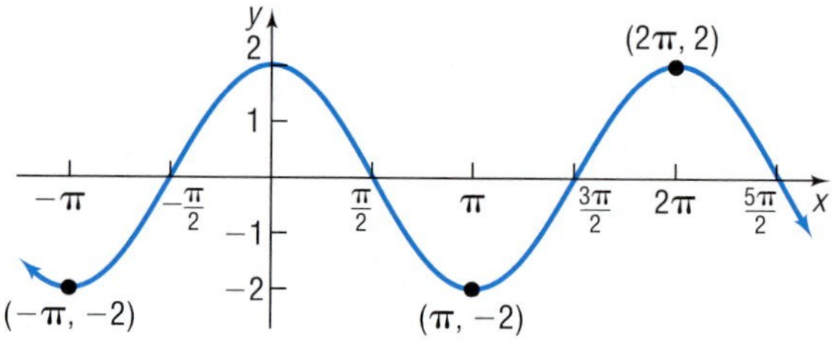

In general, the values of the functions $y = A \sin x$ and $y = A \cos x$, where $A \neq 0$, will always satisfy the inequalities

$$-|A| \leq A \sin x \leq |A| \quad \text{and} \quad -|A| \leq A \cos x \leq |A|$$

respectively. The number $|A|$ is called the **amplitude** of $y = A \sin x$ or $y = A \cos x$. See Figure 87.

Figure 87

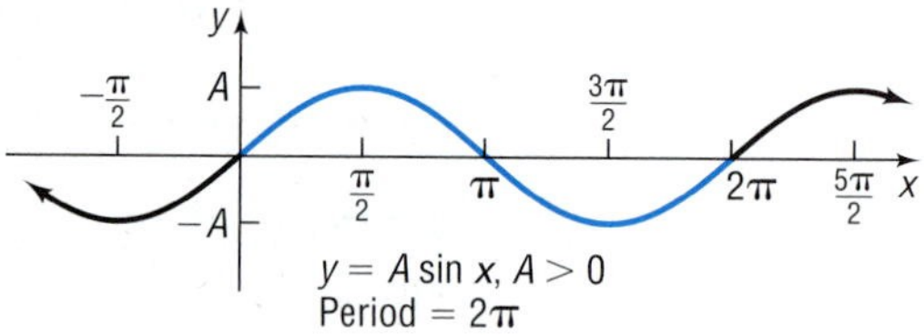

In Example 4, we obtained the graph of $y = \cos(3x)$, which we reproduce in Figure 88. Notice that the period of this function is $\dfrac{2\pi}{3}$.

Figure 88
$y = \cos(3x)$

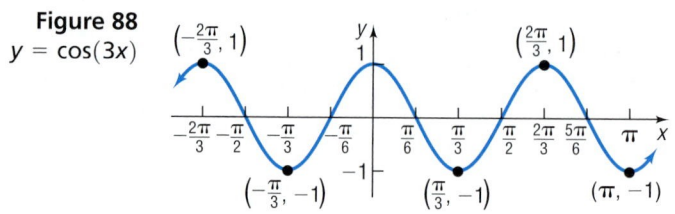

In general, if $\omega > 0$, the functions $y = \sin(\omega x)$ and $y = \cos(\omega x)$ will have period $T = \dfrac{2\pi}{\omega}$. To see why, recall that the graph of $y = \sin(\omega x)$ is obtained from the graph of $y = \sin x$ by performing a horizontal compression or stretch by a factor $\dfrac{1}{\omega}$. This horizontal compression replaces the interval $[0, 2\pi]$, which contains one period of the graph of $y = \sin x$, by the interval $\left[0, \dfrac{2\pi}{\omega}\right]$, which contains one period of the graph of $y = \sin(\omega x)$. The period of the functions $y = \sin(\omega x)$ and $y = \cos(\omega x)$, $\omega > 0$, is $\dfrac{2\pi}{\omega}$.

For example, for the function $y = \cos(3x)$, graphed in Figure 88, $\omega = 3$, so the period is $\dfrac{2\pi}{\omega} = \dfrac{2\pi}{3}$.

One period of the graph of $y = \sin(\omega x)$ or $y = \cos(\omega x)$ is called a **cycle**. Figure 89 illustrates the general situation. The blue portion of the graph is one cycle.

Figure 89

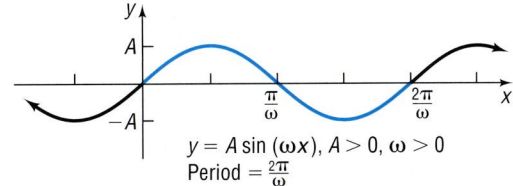

$y = A \sin(\omega x), A > 0, \omega > 0$
Period $= \frac{2\pi}{\omega}$

If $\omega < 0$ in $y = \sin(\omega x)$ or $y = \cos(\omega x)$, we use the even-odd properties of the sine and cosine functions as follows:

$$\sin(-\omega x) = -\sin(\omega x) \quad \text{and} \quad \cos(-\omega x) = \cos(\omega x)$$

This gives us an equivalent form in which the coefficient of x is positive. For example,

$$\sin(-2x) = -\sin(2x) \quad \text{and} \quad \cos(-\pi x) = \cos(\pi x)$$

Theorem

If $\omega > 0$, the amplitude and period of $y = A \sin(\omega x)$ and $y = A \cos(\omega x)$ are given by

$$\text{Amplitude} = |A| \qquad \text{Period} = T = \frac{2\pi}{\omega} \qquad (1)$$

EXAMPLE 5 **Finding the Amplitude and Period of a Sinusoidal Function**

Determine the amplitude and period of $y = 3 \sin(4x)$.

Solution Comparing $y = 3 \sin(4x)$ to $y = A \sin(\omega x)$, we find that $A = 3$ and $\omega = 4$. From equation (1),

$$\text{Amplitude} = |A| = 3 \qquad \text{Period} = T = \frac{2\pi}{\omega} = \frac{2\pi}{4} = \frac{\pi}{2}$$

 **NOW WORK PROBLEM 33.**

④ Earlier, we graphed sine and cosine functions using transformations. We now introduce another method that can be used to graph these functions.

Figure 90 shows one cycle of the graphs of $y = \sin x$ and $y = \cos x$ on the interval $[0, 2\pi]$. Notice that each graph consists of four parts corresponding to the four subintervals:

$$\left[0, \frac{\pi}{2}\right], \quad \left[\frac{\pi}{2}, \pi\right], \quad \left[\pi, \frac{3\pi}{2}\right], \quad \left[\frac{3\pi}{2}, 2\pi\right]$$

Each of these subintervals is of length $\dfrac{\pi}{2}$ (the period 2π divided by 4) and the endpoints of these intervals give rise to five key points, as shown in Figure 90.

Figure 90

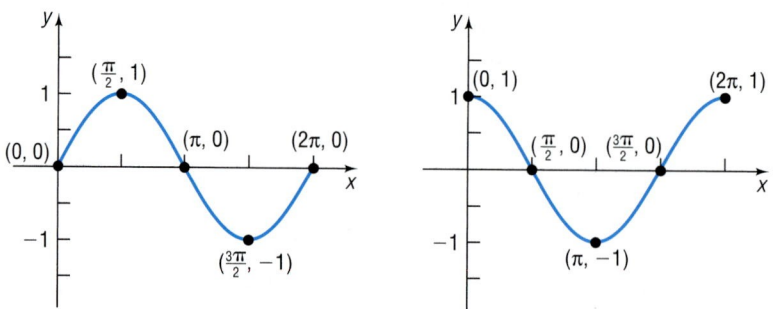

When graphing a sinusoidal function of the form $y = A\sin(\omega x)$ or $y = A\cos(\omega x)$ by hand, we use the amplitude to determine the maximum and minimum values of the function. The period is used to divide the x-axis into four subintervals. The endpoints of the subintervals give rise to five key points on the graph, which are used to sketch one cycle. Finally, extend the graph in either direction to make it complete.

To graph a sinusoidal function using a graphing utility, we use the amplitude to set Ymin and Ymax and use the period to set Xmin and Xmax.

Let's look at an example.

EXAMPLE 6   **Graphing a Sinusoidal Function**

Graph: $y = 3\sin(4x)$

Solution   From Example 5, the amplitude is 3 and the period is $\dfrac{\pi}{2}$. The graph of $y = 3\sin(4x)$ will lie between -3 and 3 on the y-axis. One cycle will begin at $x = 0$ and end at $x = \dfrac{\pi}{2}$.

We divide the interval $\left[0, \dfrac{\pi}{2}\right]$ into four subintervals, each of length $\dfrac{\pi}{2} \div 4 = \dfrac{\pi}{8}$:

$$\left[0, \dfrac{\pi}{8}\right], \quad \left[\dfrac{\pi}{8}, \dfrac{\pi}{4}\right], \quad \left[\dfrac{\pi}{4}, \dfrac{3\pi}{8}\right], \quad \left[\dfrac{3\pi}{8}, \dfrac{\pi}{2}\right]$$

The endpoints of these intervals give rise to five key points on the graph:

$$(0, 0), \quad \left(\dfrac{\pi}{8}, 3\right), \quad \left(\dfrac{\pi}{4}, 0\right), \quad \left(\dfrac{3\pi}{8}, -3\right), \quad \left(\dfrac{\pi}{2}, 0\right)$$

We plot these five points and fill in the graph of the sine curve as shown in Figure 91(a). If we extend the graph in either direction, we obtain the complete graph shown in Figure 91(b).

Figure 91

Figure 91(c) shows the graph using a graphing utility.

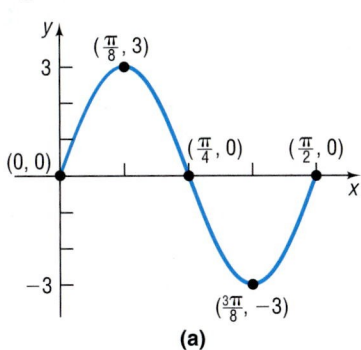

(a)

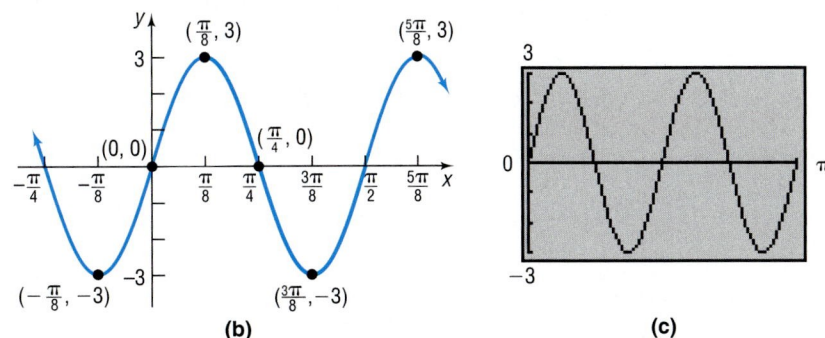

(b)   (c)

✔ CHECK: Graph $y = 3\sin(4x)$ by hand using transformations. Which graphing method do you prefer?

NOW WORK PROBLEM **39.**

EXAMPLE 7

Finding the Amplitude and Period of a Sinusoidal Function and Graphing It

Determine the amplitude and period of $y = -4\cos(\pi x)$, and graph the function.

Solution   Comparing $y = -4\cos(\pi x)$ with $y = A\cos(\omega x)$, we find that $A = -4$ and $\omega = \pi$. The amplitude is $|A| = |-4| = 4$, and the period is $T = \dfrac{2\pi}{\omega} = \dfrac{2\pi}{\pi} = 2$.

The graph of $y = -4\cos(\pi x)$ will lie between -4 and 4 on the y-axis. One cycle will begin at $x = 0$ and end at $x = 2$. We divide the interval $[0, 2]$ into four subintervals, each of length $2 \div 4 = \dfrac{1}{2}$:

$$\left[0, \frac{1}{2}\right], \quad \left[\frac{1}{2}, 1\right], \quad \left[1, \frac{3}{2}\right], \quad \left[\frac{3}{2}, 2\right].$$

The five key points on the graph are

$$(0, -4), \quad \left(\frac{1}{2}, 0\right), \quad (1, 4), \quad \left(\frac{3}{2}, 0\right), \quad (2, -4).$$

We plot these five points and fill in the graph of the cosine function as shown in Figure 92(a). Extending the graph in either direction, we obtain Figure 92(b). Figure 92(c) shows the graph using a graphing utility.

Figure 92

$y = -4\cos(\pi x)$

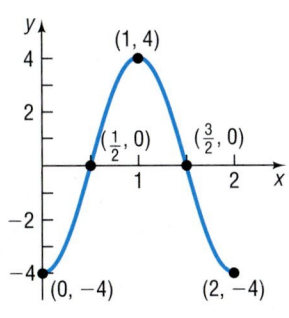

(a)

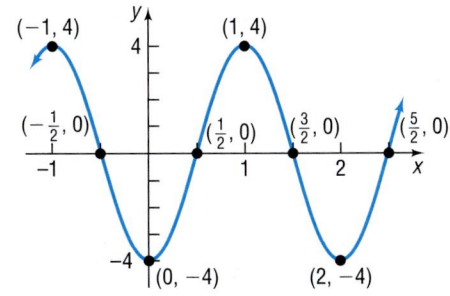

(b)

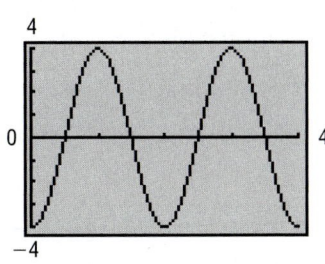

(c)

✔ CHECK: Graph $y = -4\cos(\pi x)$ by hand using transformations. Which graphing method do you prefer? ■ ■

EXAMPLE 8   **Finding the Amplitude and Period of a Sinusoidal Function and Graphing It**

Determine the amplitude and period of $y = 2\sin\left(-\dfrac{\pi}{2}x\right)$, and graph the function.

Solution   Since the sine function is odd, we use the equivalent form:

$$y = -2\sin\left(\frac{\pi}{2}x\right)$$

Comparing $y = -2\sin\left(\dfrac{\pi}{2}x\right)$ to $y = A\sin(\omega x)$, we find that $A = -2$ and $\omega = \dfrac{\pi}{2}$. The amplitude is $|A| = 2$, and the period is $T = \dfrac{2\pi}{\omega} = \dfrac{2\pi}{\dfrac{\pi}{2}} = 4$.

The graph of $y = -2\sin\left(\dfrac{\pi}{2}x\right)$ will lie between -2 and 2 on the y-axis. One cycle will begin at $x = 0$ and end at $x = 4$. We divide the interval $[0, 4]$ into four subintervals, each of length $4 \div 4 = 1$:

$$[0, 1], \quad [1, 2], \quad [2, 3], \quad [3, 4]$$

The five key points on the graph are

$$(0, 0), \quad (1, -2), \quad (2, 0), \quad (3, 2), \quad (4, 0)$$

We plot these five points and fill in the graph of the sine function as shown in Figure 93(a). Extending the graph in either direction, we obtain Figure 93(b).

Figure 93(c) shows the graph using a graphing utility.

Figure 93

$$y = 2\sin\left(-\frac{\pi}{2}x\right)$$

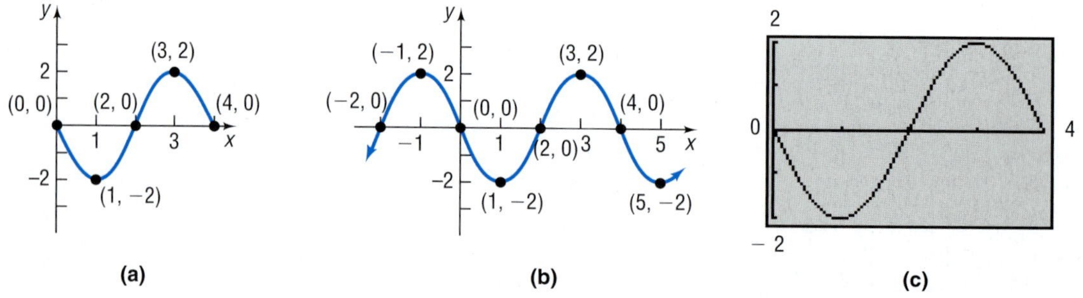

(a)

(b)

(c)

✔ CHECK: Graph $y = 2\sin\left(-\dfrac{\pi}{2}x\right)$ by hand using transformations. Which graphing method do you prefer? ■ ■

NOW WORK PROBLEM **53.**

⑤ We can also use the ideas of amplitude and period to identify a sinusoidal function when its graph is given.

EXAMPLE 9 Finding an Equation for a Sinusoidal Graph

Find an equation for the graph shown in Figure 94.

Figure 94

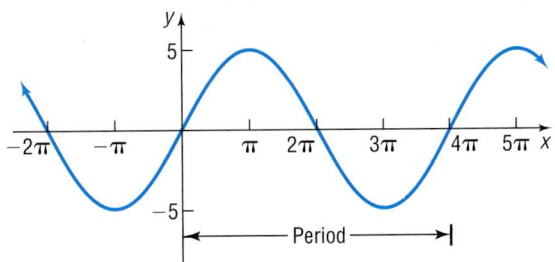

Solution This graph can be viewed as the graph of a sine function* with amplitude $A = 5$. The period T is observed to be 4π. By equation (1),

$$T = \frac{2\pi}{\omega}$$

$$4\pi = \frac{2\pi}{\omega}$$

$$\omega = \frac{2\pi}{4\pi} = \frac{1}{2}$$

The sine function whose graph is given in Figure 94 is

$$y = A\sin(\omega x) = 5\sin\left(\frac{1}{2}x\right)$$

✔ CHECK: Graph $Y_1 = 5\sin\left(\frac{1}{2}x\right)$ and compare the result with Figure 94. ∎

EXAMPLE 10 Finding an Equation for a Sinusoidal Graph

Find an equation for the graph shown in Figure 95.

Figure 95

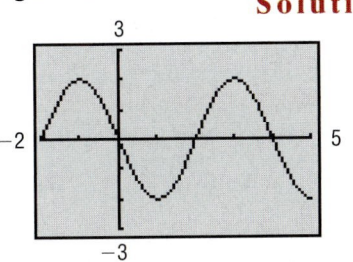

Solution The graph is sinusoidal, with amplitude $A = 2$. The period is 4, so $\dfrac{2\pi}{\omega} = 4$ or $\omega = \dfrac{\pi}{2}$. Since the graph passes through the origin, it is easiest to view the equation as a sine function, but notice that the graph is actually the reflection of a sine function about the x-axis (since the graph is decreasing near the origin). Thus, we have

$$y = -A\sin(\omega x) = -2\sin\left(\frac{\pi}{2}x\right)$$

*The equation could also be viewed as a cosine function with a horizontal shift, but viewing it as a sine function is easier.

✔ CHECK: Graph $Y_1 = -2\sin\left(\dfrac{\pi}{2}x\right)$ and compare the result with Figure 95.

■ ■

NOW WORK PROBLEMS 63 AND 67.

8.6 Concepts and Vocabulary

In Problems 1–3, fill in the blanks.

1. The maximum value of $y = \sin x$ is _____ and occurs at $x =$ _____.

2. The function $y = A\sin(\omega x)$ has amplitude 3 and period 2; then $A =$ _____ and $\omega =$ _____.

3. The function $y = 3\cos(6x)$ has amplitude _____ and period _____.

In Problems 4–6, answer True or False to each statement.

4. The graphs of $y = \sin x$ and $y = \cos x$ are identical except for a horizontal shift.

5. For $y = 2\sin(\pi x)$, the amplitude is 2 and the period is $\dfrac{\pi}{2}$.

6. The graph of the sine function has infinitely many x-intercepts.

7. Draw a quick sketch of $y = \sin x$. Be sure to label at least five points.

8. Explain how you would scale the x-axis and the y-axis before graphing $y = 3\cos(\pi x)$.

9. Explain the term *amplitude* as it relates to the graph of a sinusoidal function.

8.6 Exercises

In Problems 1–10, if necessary, refer to the graphs to answer each question.

1. What is the y-intercept of $y = \sin x$?

2. What is the y-intercept of $y = \cos x$?

3. For what numbers x, $-\pi \le x \le \pi$, is the graph of $y = \sin x$ increasing?

4. For what numbers x, $-\pi \le x \le \pi$, is the graph of $y = \cos x$ decreasing?

5. What is the largest value of $y = \sin x$?

6. What is the smallest value of $y = \cos x$?

7. For what numbers x, $0 \le x \le 2\pi$, does $\sin x = 0$?

8. For what numbers x, $0 \le x \le 2\pi$, does $\cos x = 0$?

9. For what numbers x, $-2\pi \le x \le 2\pi$, does $\sin x = 1$? What about $\sin x = -1$?

10. For what numbers x, $-2\pi \le x \le 2\pi$, does $\cos x = 1$? What about $\cos x = -1$?

In Problems 11 and 12, match the graph to a function. Three answers are possible.

A. $y = -\sin x$

B. $y = -\cos x$

C. $y = \sin\left(x - \dfrac{\pi}{2}\right)$

D. $y = -\cos\left(x - \dfrac{\pi}{2}\right)$

E. $y = \sin(x + \pi)$

F. $y = \cos(x + \pi)$

11.

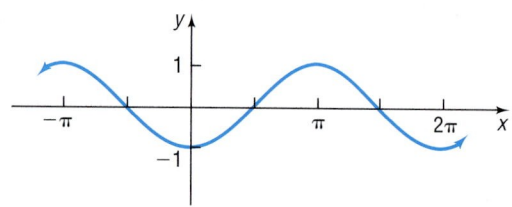

12.

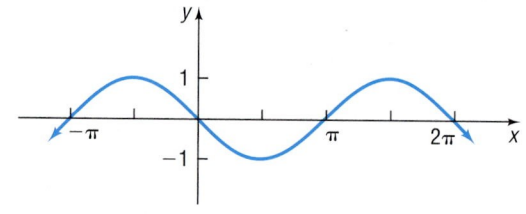

In Problems 13–28, use transformations to graph each function.

13. $y = 3 \sin x$

14. $y = 4 \cos x$

15. $y = \cos\left(x + \dfrac{\pi}{4}\right)$

16. $y = \sin(x - \pi)$

17. $y = \sin x - 1$

18. $y = \cos x + 1$

19. $y = -2 \sin x$

20. $y = -3 \cos x$

21. $y = \sin(\pi x)$

22. $y = \cos\left(\dfrac{\pi}{2} x\right)$

23. $y = 2 \sin x + 2$

24. $y = 3 \cos x + 3$

25. $y = -2 \cos\left(x - \dfrac{\pi}{2}\right)$

26. $y = -3 \sin\left(x + \dfrac{\pi}{2}\right)$

27. $y = 3 \sin(\pi - x)$

28. $y = 2 \cos(\pi - x)$

In Problems 29–38, determine the amplitude and period of each function without graphing.

29. $y = 2 \sin x$

30. $y = 3 \cos x$

31. $y = -4 \cos(2x)$

32. $y = -\sin\left(\dfrac{1}{2} x\right)$

33. $y = 6 \sin(\pi x)$

34. $y = -3 \cos(3x)$

35. $y = -\dfrac{1}{2}\cos\left(\dfrac{3}{2} x\right)$

36. $y = \dfrac{4}{3}\sin\left(\dfrac{2}{3} x\right)$

37. $y = \dfrac{5}{3}\sin\left(-\dfrac{2\pi}{3} x\right)$

38. $y = \dfrac{9}{5}\cos\left(-\dfrac{3\pi}{2} x\right)$

In Problems 39–48, match the given function to one of the graphs (A)–(J).

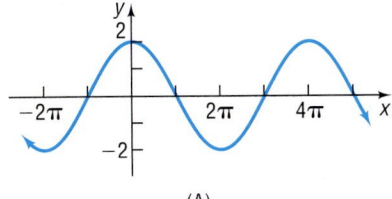

(A)

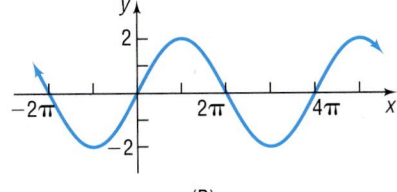

(B)

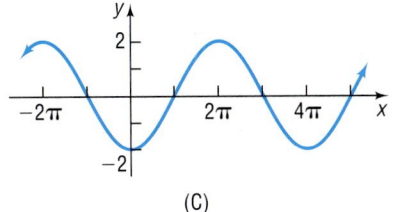

(C)

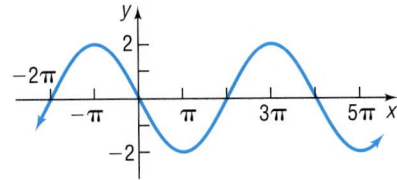

(D)

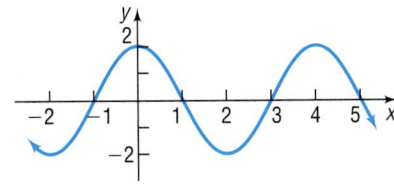

(E)

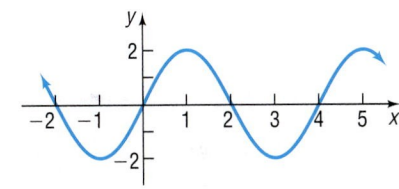

(F)

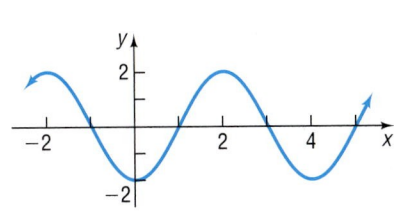

(G)

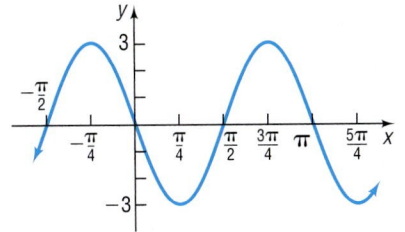

(H)

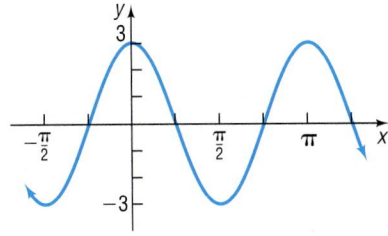

(I)

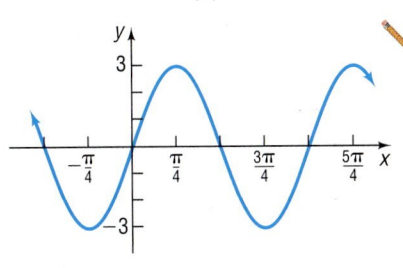

(J)

39. $y = 2 \sin\left(\dfrac{\pi}{2} x\right)$

40. $y = 2 \cos\left(\dfrac{\pi}{2} x\right)$

41. $y = 2 \cos\left(\dfrac{1}{2} x\right)$

42. $y = 3 \cos(2x)$

43. $y = -3 \sin(2x)$

44. $y = 2 \sin\left(\dfrac{1}{2} x\right)$

45. $y = -2 \cos\left(\dfrac{1}{2} x\right)$

46. $y = -2 \cos\left(\dfrac{\pi}{2} x\right)$

47. $y = 3 \sin(2x)$

48. $y = -2 \sin\left(\dfrac{1}{2} x\right)$

In Problems 49–52, match the given function to one of the graphs (A)–(D).

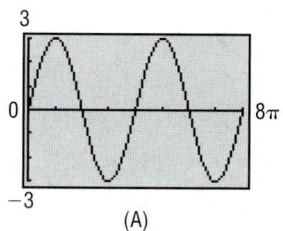
(A)

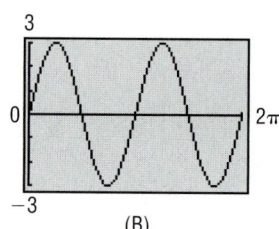
(B)

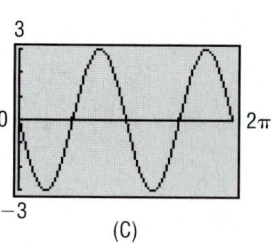
(C)

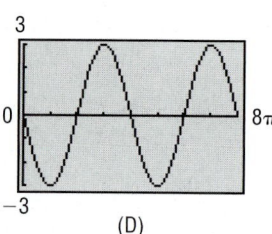
(D)

49. $y = 3\sin\left(\dfrac{1}{2}x\right)$ **50.** $y = -3\sin(2x)$ **51.** $y = 3\sin(2x)$ **52.** $y = -3\sin\left(\dfrac{1}{2}x\right)$

In Problems 53–62, graph each sinusoidal function.

53. $y = 5\sin(4x)$ **54.** $y = 4\cos(6x)$ **55.** $y = 5\cos(\pi x)$ **56.** $y = 2\sin(\pi x)$

57. $y = -2\cos(2\pi x)$ **58.** $y = -5\cos(2\pi x)$ **59.** $y = -4\sin\left(\dfrac{1}{2}x\right)$ **60.** $y = -2\cos\left(\dfrac{1}{2}x\right)$

61. $y = \dfrac{3}{2}\sin\left(-\dfrac{2}{3}x\right)$ **62.** $y = \dfrac{4}{3}\cos\left(-\dfrac{1}{3}x\right)$

In Problems 63–66, write the equation of a sine function that has the given characteristics.

63. Amplitude: 3   **64.** Amplitude: 2   **65.** Amplitude: 3   **66.** Amplitude: 4
     Period: π Period: 4π Period: 2   Period: 1

In Problems 67–78, find an equation for each graph.

67.

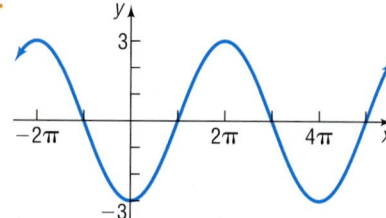

68.

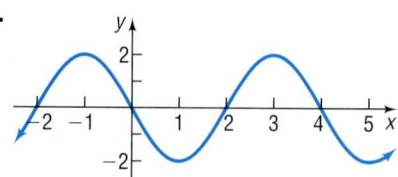

69.

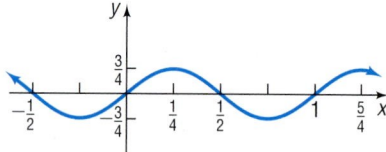

70.

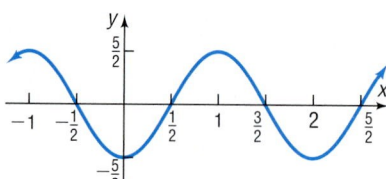

71.

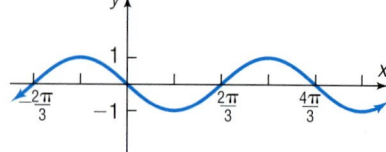

72.

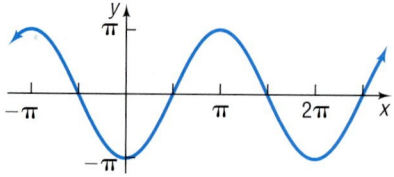

73.

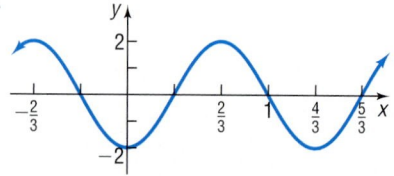

74.

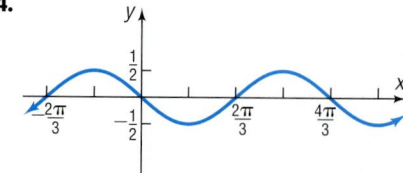

75. 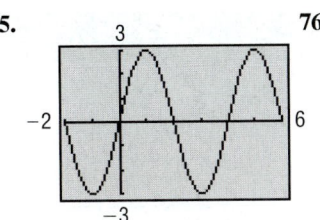   **76.**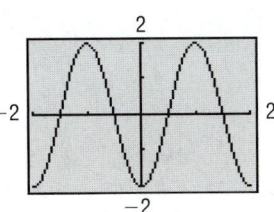

77. 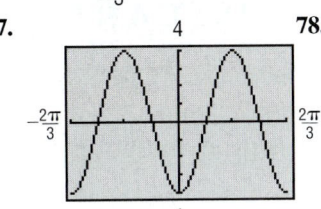   **78.**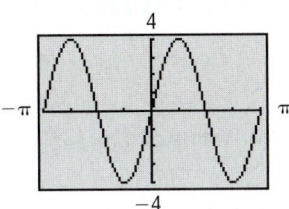

79. Alternating Current (ac) Circuits   The current I, in amperes, flowing through an ac (alternating current) circuit at time t is

$$I = 220 \sin(60\pi t), \qquad t \geq 0$$

What is the period? What is the amplitude? Graph this function over two periods.

80. Alternating Current (ac) Circuits   The current I, in amperes, flowing through an ac (alternating current) circuit at time t is

$$I = 120 \sin(30\pi t), \qquad t \geq 0$$

What is the period? What is the amplitude? Graph this function over two periods.

81. Alternating Current (ac) Generators   The voltage V produced by an ac generator is

$$V = 220 \sin(120\pi t)$$

(a)   What is the amplitude? What is the period?
(b)   Graph V over two periods, beginning at $t = 0$.
(c)   If a resistance of $R = 10$ ohms is present, what is the current I?
   [**Hint:** Use Ohm's Law, $V = IR$.]
(d)   What is the amplitude and period of the current I?
(e)   Graph I over two periods, beginning at $t = 0$.

82. Alternating Current (ac) Generators   The voltage V produced by an ac generator is

$$V = 120 \sin(120\pi t)$$

(a)   What is the amplitude? What is the period?
(b)   Graph V over two periods, beginning at $t = 0$.
(c)   If a resistance of $R = 20$ ohms is present, what is the current I?
   [**Hint:** Use Ohm's Law, $V = IR$.]
(d)   What is the amplitude and period of the current I?
(e)   Graph I over two periods, beginning at $t = 0$.

83. Alternating Current (ac) Generators   The voltage V produced by an ac generator is sinusoidal. As a function of time, the voltage V is

$$V = V_0 \sin(2\pi f t)$$

where f is the **frequency**, the number of complete oscillations (cycles) per second. [In the United States and Canada, f is 60 hertz (Hz).] The **power** P delivered to a resistance R at any time t is defined as

$$P = \frac{V^2}{R}$$

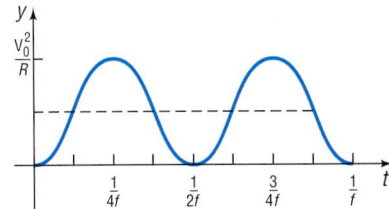

Power in an ac generator

(a)   Show that $P = \dfrac{V_0^2}{R} \sin^2(2\pi f t)$.
(b)   The graph of P is shown in the figure. Express P as a sinusoidal function.
(c)   Deduce that

$$\sin^2(2\pi f t) = \frac{1}{2}\big[1 - \cos(4\pi f t)\big]$$

84. Biorhythms   In the theory of biorhythms, a sine function of the form

$$P = 50 \sin(\omega t) + 50$$

is used to measure the percent P of a person's potential at time t, where t is measured in days and $t = 0$ is the person's birthday. Three characteristics are commonly measured:

   Physical potential: period of 23 days
   Emotional potential: period of 28 days
   Intellectual potential: period of 33 days

(a)   Find ω for each characteristic.
(b)   Use a graphing utility to graph all three functions.
(c)   Is there a time t when all three characteristics have 100% potential? When is it?
(d)   Suppose that you are 20 years old today ($t = 7305$ days). Describe your physical, emotional, and intellectual potential for the next 30 days.

85. Graph $y = |\cos x|, -2\pi \leq x \leq 2\pi$.

86. Graph $y = |\sin x|, -2\pi \leq x \leq 2\pi$.

87. Explain how the amplitude and period of a sinusoidal graph are used to establish the scale on each coordinate axis.

88. Find an application in your major field that leads to a sinusoidal graph. Write a paper about your findings.

PREPARING FOR THIS SECTION

Before getting started, review the following:

✓ Vertical Asymptotes (Section 4.3, pp. 346–348)

8.7 GRAPHS OF THE TANGENT, COTANGENT, COSECANT, AND SECANT FUNCTIONS

OBJECTIVES

1 Graph Transformations of the Tangent Function and Cotangent Function

2 Graph Transformations of the Cosecant Function and Secant Function

The Graphs of $y = \tan x$ and $y = \cot x$

1 Because the tangent function has period π, we only need to determine the graph over some interval of length π. The rest of the graph will consist of repetitions of that graph. Because the tangent function is not defined at $\ldots, -\dfrac{3\pi}{2}, -\dfrac{\pi}{2}, \dfrac{\pi}{2}, \dfrac{3\pi}{2}, \ldots$, we will concentrate on the interval $\left(-\dfrac{\pi}{2}, \dfrac{\pi}{2}\right)$, of length π, and construct Table 9, which lists some points on the graph of $y = \tan x$, $-\dfrac{\pi}{2} < x < \dfrac{\pi}{2}$. We plot the points in the table and connect them with a smooth curve. See Figure 96 for a partial graph of $y = \tan x$, where $-\dfrac{\pi}{3} \le x \le \dfrac{\pi}{3}$.

TABLE 9

x	$y = \tan x$	(x, y)
$-\dfrac{\pi}{3}$	$-\sqrt{3} \approx -1.73$	$\left(-\dfrac{\pi}{3}, -\sqrt{3}\right)$
$-\dfrac{\pi}{4}$	-1	$\left(-\dfrac{\pi}{4}, -1\right)$
$-\dfrac{\pi}{6}$	$-\dfrac{\sqrt{3}}{3} \approx -0.58$	$\left(-\dfrac{\pi}{6}, -\dfrac{\sqrt{3}}{3}\right)$
0	0	$(0, 0)$
$\dfrac{\pi}{6}$	$\dfrac{\sqrt{3}}{3} \approx 0.58$	$\left(\dfrac{\pi}{6}, \dfrac{\sqrt{3}}{3}\right)$
$\dfrac{\pi}{4}$	1	$\left(\dfrac{\pi}{4}, 1\right)$
$\dfrac{\pi}{3}$	$\sqrt{3} \approx 1.73$	$\left(\dfrac{\pi}{3}, \sqrt{3}\right)$

Figure 96

$y = \tan x$, $-\dfrac{\pi}{3} \le x \le \dfrac{\pi}{3}$

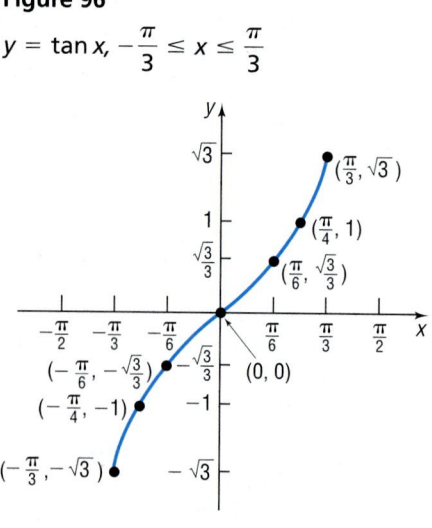

To complete one period of the graph of $y = \tan x$, we need to investigate the behavior of the function as x approaches $-\dfrac{\pi}{2}$ and $\dfrac{\pi}{2}$. We must be careful, though, because $y = \tan x$ is not defined at these numbers. To determine this behavior, we use the identity

$$\tan x = \frac{\sin x}{\cos x}$$

If x is close to $\dfrac{\pi}{2} \approx 1.5708$, but remains less than $\dfrac{\pi}{2}$, then $\sin x$ will be close to 1 and $\cos x$ will be positive and close to 0. (Refer back to the graphs of the sine function and the cosine function.) Hence, the ratio $\dfrac{\sin x}{\cos x}$ will be positive and large. In fact, the closer x gets to $\dfrac{\pi}{2}$, the closer $\sin x$ gets to 1 and $\cos x$ gets to 0, so $\tan x$ approaches ∞ $\left(\lim\limits_{x \to \frac{\pi}{2}^{-}} \tan x = \infty \right)$. In other words, the vertical line $x = \dfrac{\pi}{2}$ is a vertical asymptote to the graph of $y = \tan x$. See Table 10.

If x is close to $-\dfrac{\pi}{2}$, but remains greater than $-\dfrac{\pi}{2}$, then $\sin x$ will be close to -1 and $\cos x$ will be positive and close to 0. Hence, the ratio $\dfrac{\sin x}{\cos x}$ approaches $-\infty$ $\left(\lim\limits_{x \to -\frac{\pi}{2}^{+}} \tan x = -\infty \right)$. In other words, the vertical line $x = -\dfrac{\pi}{2}$ is also a vertical asymptote to the graph.

With these observations, we can complete one period of the graph. We obtain the complete graph of $y = \tan x$ by repeating this period, as shown in Figure 97(a).

Figure 97(b) shows the graph of $y = \tan x$, $-\infty < x < \infty$ using a graphing utility. Notice we used dot mode when graphing $y = \tan x$. Do you know why?

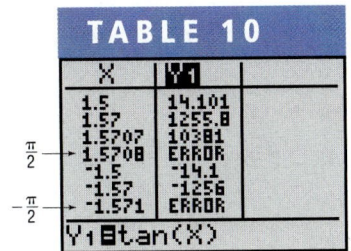

TABLE 10

X	Y1
1.5	14.101
1.57	1255.8
1.5707	10381
1.5708	ERROR
-1.5	-14.1
-1.57	-1256
-1.571	ERROR

$Y_1 \boxminus \tan(X)$

Figure 97
$y = \tan x$, $-\infty < x < \infty$, x

not equal to odd multiples of $\dfrac{\pi}{2}$

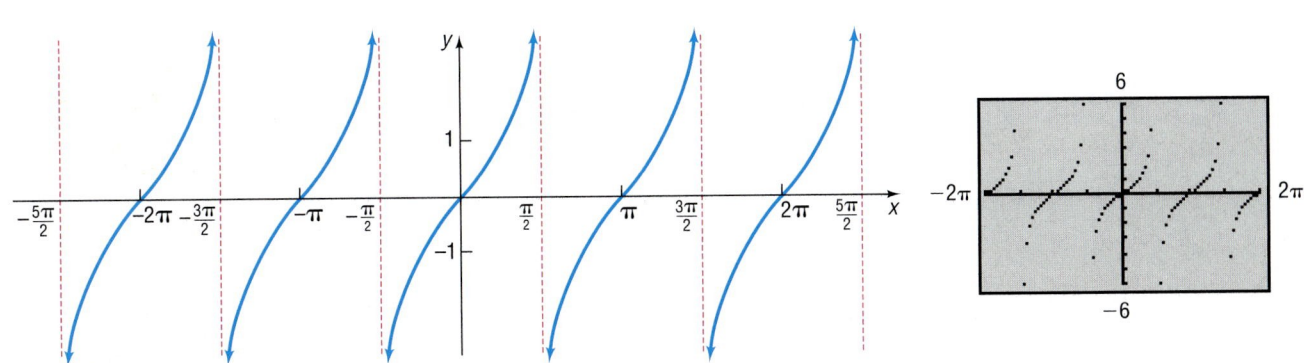

(a)   (b)

The graph of $y = \tan x$ illustrates some facts that we already know about the tangent function.

> **Properties of the Tangent Function**
>
> 1. The domain is the set of all real numbers, except odd multiples of $\frac{\pi}{2}$.
> 2. The range consists of all real numbers.
> 3. The tangent function is an odd function, as the symmetry of the graph with respect to the origin indicates.
> 4. The tangent function is periodic, with period π.
> 5. The x-intercepts are $\ldots, -2\pi, -\pi, 0, \pi, 2\pi, 3\pi, \ldots$; the y-intercept is 0.
> 6. Vertical asymptotes occur at $x = \ldots, -\frac{3\pi}{2}, -\frac{\pi}{2}, \frac{\pi}{2}, \frac{3\pi}{2}, \ldots$.

✏️ **NOW WORK PROBLEMS 1 AND 9.**

EXAMPLE 1   **Graphing Variations of $y = \tan x$ Using Transformations**

Graph: $y = 2 \tan x$

Solution   We start with the graph of $y = \tan x$ and vertically stretch it by a factor of 2. See Figure 98.

Figure 98

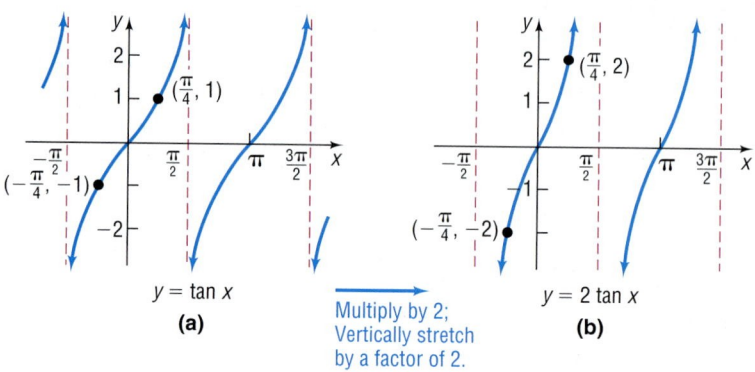

$y = \tan x$

(a)

Multiply by 2; Vertically stretch by a factor of 2.

$y = 2 \tan x$

(b)

Figure 99 shows the graph using a graphing utility.

Figure 99

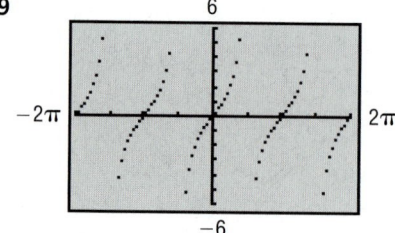

EXAMPLE 2   **Graphing Variations of $y = \tan x$ Using Transformations**

Graph: $y = -\tan\left(x + \frac{\pi}{4}\right)$

Solution   We start with the graph of $y = \tan x$. See Figure 100.

Figure 100

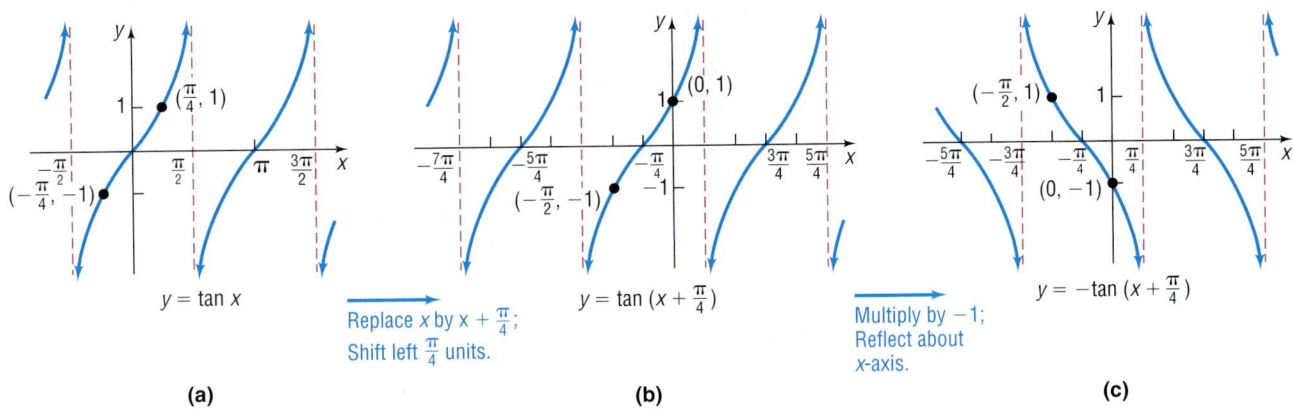

(a)

Replace x by $x + \frac{\pi}{4}$;
Shift left $\frac{\pi}{4}$ units.

$y = \tan\left(x + \frac{\pi}{4}\right)$

(b)

Multiply by -1;
Reflect about
x-axis.

$y = -\tan\left(x + \frac{\pi}{4}\right)$

(c)

✔ CHECK: Graph $Y_1 = -\tan\left(x + \dfrac{\pi}{4}\right)$ and compare the result to Figure 100(c). ■ ■

NOW WORK PROBLEM **19**.

TABLE 11

x	$y = \cot x$	(x, y)
$\dfrac{\pi}{6}$	$\sqrt{3}$	$\left(\dfrac{\pi}{6}, \sqrt{3}\right)$
$\dfrac{\pi}{4}$	1	$\left(\dfrac{\pi}{4}, 1\right)$
$\dfrac{\pi}{3}$	$\dfrac{\sqrt{3}}{3}$	$\left(\dfrac{\pi}{3}, \dfrac{\sqrt{3}}{3}\right)$
$\dfrac{\pi}{2}$	0	$\left(\dfrac{\pi}{2}, 0\right)$
$\dfrac{2\pi}{3}$	$-\dfrac{\sqrt{3}}{3}$	$\left(\dfrac{2\pi}{3}, -\dfrac{\sqrt{3}}{3}\right)$
$\dfrac{3\pi}{4}$	-1	$\left(\dfrac{3\pi}{4}, -1\right)$
$\dfrac{5\pi}{6}$	$-\sqrt{3}$	$\left(\dfrac{5\pi}{6}, -\sqrt{3}\right)$

We obtain the graph of $y = \cot x$ as we did the graph of $y = \tan x$. The period of $y = \cot x$ is π. Because the cotangent function is not defined for integral multiples of π, we will concentrate on the interval $(0, \pi)$. Table 11 lists some points on the graph of $y = \cot x, 0 < x < \pi$. As x approaches 0, but remains greater than 0, the value of $\cos x$ will be close to 1 and the value of $\sin x$ will be positive and close to 0. Hence, the ratio $\dfrac{\cos x}{\sin x} = \cot x$ will be positive and large; so as x approaches 0, with $x > 0$, $\cot x$ approaches ∞ $\left(\lim\limits_{x \to 0^+} \cot x = \infty\right)$. Similarly, as x approaches π, but remains less than π, the value of $\cos x$ will be close to -1, and the value of $\sin x$ will be positive and close to 0. Hence, the ratio $\dfrac{\cos x}{\sin x} = \cot x$ will be negative and will approach $-\infty$ as x approaches π $\left(\lim\limits_{x \to \pi^-} \cot x = -\infty\right)$. Figure 101 shows the graph.

Figure 101
$y = \cot x, -\infty < x < \infty,$
x not equal to integral
multiples of $\pi, -\infty < y < \infty$

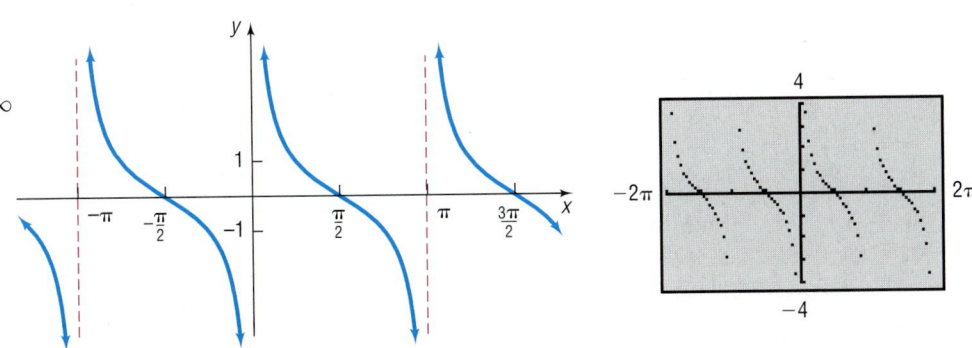

NOW WORK PROBLEM **25**.

The Graphs of $y = \csc x$ and $y = \sec x$

2 The cosecant and secant functions, sometimes referred to as **reciprocal functions**, are graphed by making use of the reciprocal identities

$$\csc x = \frac{1}{\sin x} \quad \text{and} \quad \sec x = \frac{1}{\cos x}$$

For example, the value of the cosecant function $y = \csc x$ at a given number x equals the reciprocal of the corresponding value of the sine function, provided that the value of the sine function is not 0. If the value of $\sin x$ is 0, then, at such numbers x, (integral multiples of π) the cosecant function is not defined. In fact, the graph of the cosecant function has vertical asymptotes at integral multiples of π. Figure 102 shows the graph.

Figure 102

$y = \csc x$, $-\infty < x < \infty$, x not equal to integral multiples of π, $|y| \geq 1$

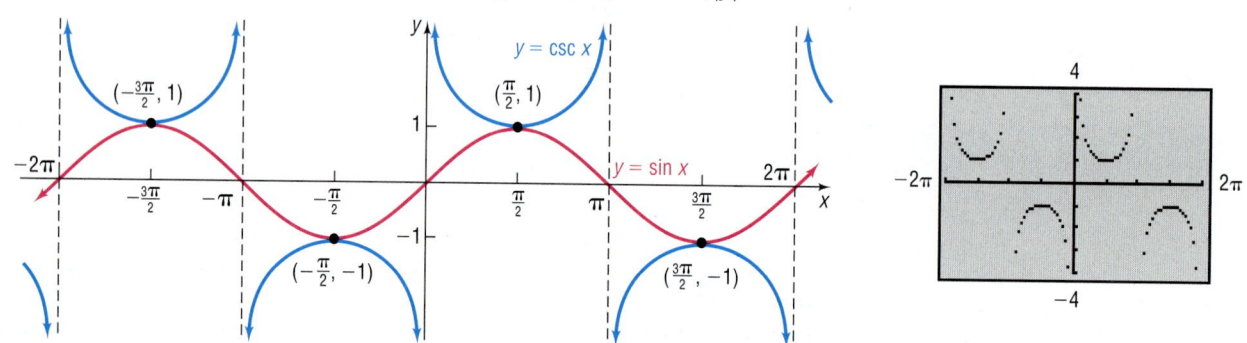

EXAMPLE 3   Graphing Variations of $y = \csc x$ Using Transformations

Graph: $y = 2 \csc\left(x - \frac{\pi}{2}\right)$

Solution   Figure 103 shows the required steps.

Figure 103

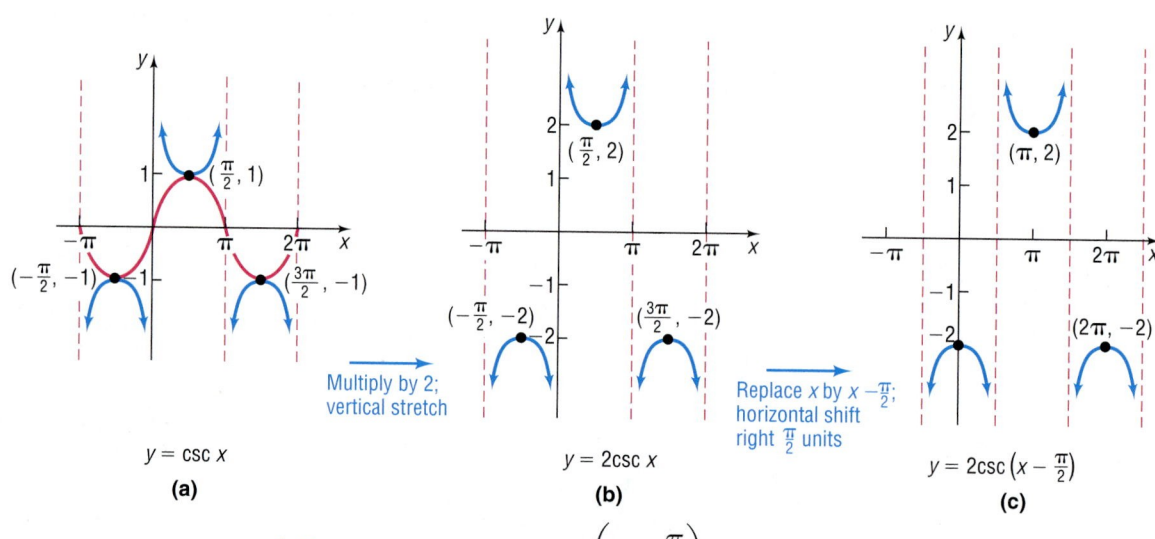

Multiply by 2; vertical stretch

Replace x by $x - \frac{\pi}{2}$; horizontal shift right $\frac{\pi}{2}$ units

$y = \csc x$
(a)

$y = 2\csc x$
(b)

$y = 2\csc\left(x - \frac{\pi}{2}\right)$
(c)

✔ CHECK: Graph $Y_1 = 2 \csc\left(x - \frac{\pi}{2}\right)$ and compare the result with Figure 103(c).

■ ■

NOW WORK PROBLEM **31.**

Using the idea of reciprocals, we can similarly obtain the graph of $y = \sec x$. See Figure 104.

Figure 104

$y = \sec x$, $-\infty < x < \infty$, x not equal to odd multiples of $\dfrac{\pi}{2}$, $|y| \geq 1$

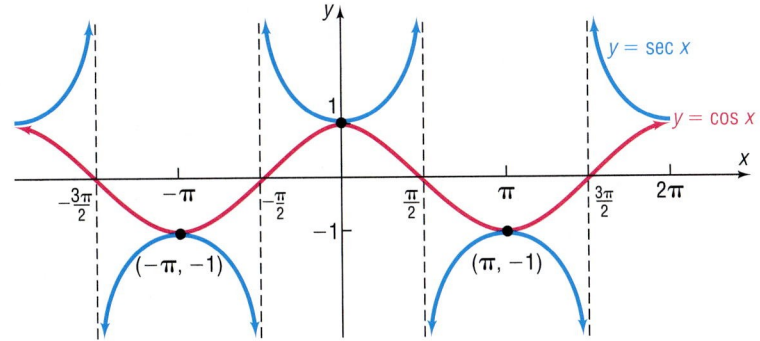
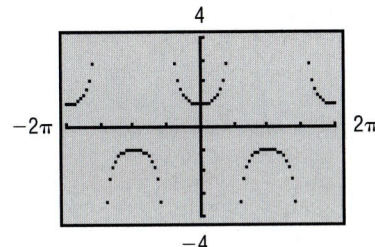

8.7 Concepts and Vocabulary

In Problems 1–3, fill in the blanks.

1. The graph of $y = \tan x$ is symmetric with respect to the _____ and has vertical asymptotes at _____.
2. The graph of $y = \sec x$ is symmetric with respect to the _____ and has vertical asymptotes at _____.
3. It is easiest to graph $y = \sec x$ by first sketching the graph of _____.

In Problem 4, answer True of False.

4. The graphs of $y = \tan x$, $y = \cot x$, $y = \sec x$, and $y = \csc x$ each have infinitely many vertical asymptotes.

5. Explain how you would graph $y = \csc x$ using the idea of a reciprocal function.
6. Sketch the graph of $y = \tan x$.

8.7 Exercises

In Problems 1–10, if necessary, refer to the graphs to answer each question.

1. What is the y-intercept of $y = \tan x$?
2. What is the y-intercept of $y = \cot x$?

3. What is the y-intercept of $y = \sec x$?
4. What is the y-intercept of $y = \csc x$?

5. For what numbers x, $-2\pi \leq x \leq 2\pi$, does $\sec x = 1$? What about $\sec x = -1$?
6. For what numbers x, $-2\pi \leq x \leq 2\pi$, does $\csc x = 1$? What about $\csc x = -1$?

7. For what numbers x, $-2\pi \leq x \leq 2\pi$, does the graph of $y = \sec x$ have vertical asymptotes?
8. For what numbers x, $-2\pi \leq x \leq 2\pi$, does the graph of $y = \csc x$ have vertical asymptotes?

9. For what numbers x, $-2\pi \leq x \leq 2\pi$, does the graph of $y = \tan x$ have vertical asymptotes?
10. For what numbers x, $-2\pi \leq x \leq 2\pi$, does the graph of $y = \cot x$ have vertical asymptotes?

In Problems 11–14, match each function to its graph.

A. $y = -\tan x$ B. $y = \tan\left(x + \dfrac{\pi}{2}\right)$ C. $y = \tan(x + \pi)$ D. $y = -\tan\left(x - \dfrac{\pi}{2}\right)$

11.

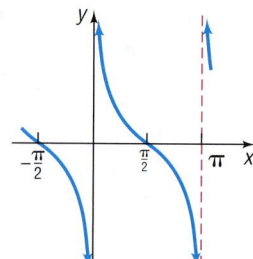

12.

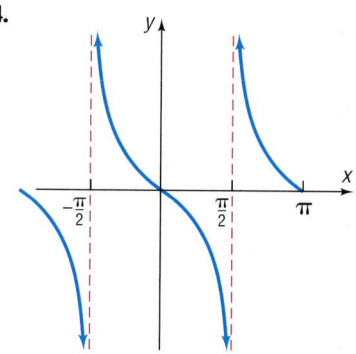

13.

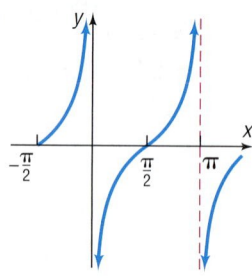

14.

In Problems 15–34, use tranformations to graph each function. Verify your result using a graphing utility.

15. $y = -\sec x$ **16.** $y = -\cot x$ **17.** $y = \sec\left(x - \dfrac{\pi}{2}\right)$ **18.** $y = \csc(x - \pi)$

19. $y = \tan(x - \pi)$ **20.** $y = \cot(x - \pi)$ **21.** $y = 3\tan(2x)$ **22.** $y = 4\tan\left(\dfrac{1}{2}x\right)$

23. $y = \sec(2x)$ **24.** $y = \csc\left(\dfrac{1}{2}x\right)$ **25.** $y = \cot(\pi x)$ **26.** $y = \cot(2x)$

27. $y = -3\tan(4x)$ **28.** $y = -3\tan(2x)$ **29.** $y = 2\sec\left(\dfrac{1}{2}x\right)$ **30.** $y = 2\sec(3x)$

31. $y = -3\csc\left(x + \dfrac{\pi}{4}\right)$ **32.** $y = -2\tan\left(x + \dfrac{\pi}{4}\right)$ **33.** $y = \dfrac{1}{2}\cot\left(x - \dfrac{\pi}{4}\right)$ **34.** $y = 3\sec\left(x + \dfrac{\pi}{2}\right)$

---

35. Carrying a Ladder around a Corner A ladder of length L is carried horizontally around a corner from a hall 3 feet wide into a hall 4 feet wide. See the illustration below.

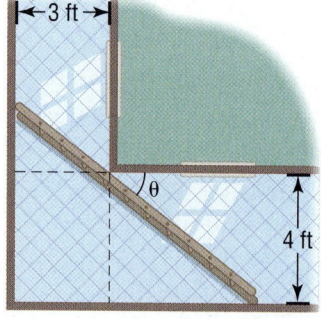

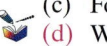

(a) Show that the length L of the ladder as a function of the angle θ is

$$L(\theta) = 3\sec\theta + 4\csc\theta$$

(b) Use a graphing utility to graph $L, 0 < \theta < \dfrac{\pi}{2}$.

(c) For what value of θ is L the least?

(d) What is the length of the longest ladder that can be carried around the corner? Why is this also the least value of L?

36. Exploration Graph

$$y = \tan x \quad \text{and} \quad y = -\cot\left(x + \dfrac{\pi}{2}\right)$$

Do you think that $\tan x = -\cot\left(x + \dfrac{\pi}{2}\right)$?

8.8 PHASE SHIFT; SINUSOIDAL CURVE FITTING

OBJECTIVES
1. Determine the Phase Shift of a Sinusoidal Function
2. Graph Sinusoidal Functions: $y = A \sin(\omega x - \phi)$
3. Find a Sinusoidal Function from Data

Phase Shift

1

We have seen that the graph of $y = A \sin(\omega x)$, $\omega > 0$, has amplitude $|A|$ and period $T = \dfrac{2\pi}{\omega}$. One cycle can be drawn as x varies from 0 to $\dfrac{2\pi}{\omega}$ or, equivalently, as ωx varies from 0 to 2π. See Figure 105.

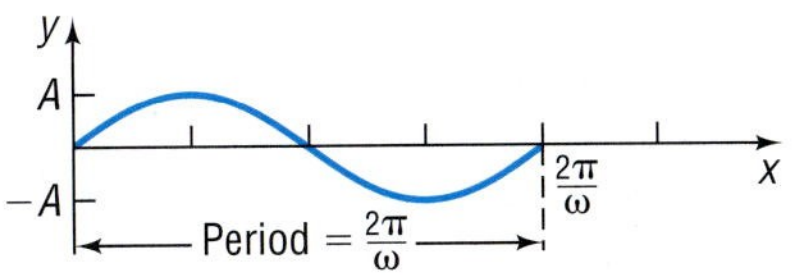

We now want to discuss the graph of

$$y = A \sin(\omega x - \phi) = A \sin\left[\omega\left(x - \frac{\phi}{\omega}\right)\right]$$

Figure 105
One cycle
$y = A \sin(\omega x)$, $A > 0$, $\omega > 0$

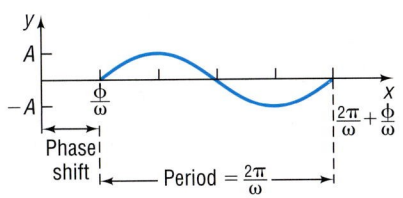

where $\omega > 0$ and ϕ (the Greek letter phi) are real numbers. The graph will be a sine curve of amplitude $|A|$. As $\omega x - \phi$ varies from 0 to 2π, one period will be traced out. This period will begin when

$$\omega x - \phi = 0 \quad \text{or} \quad x = \frac{\phi}{\omega}$$

and will end when

$$\omega x - \phi = 2\pi \quad \text{or} \quad x = \frac{2\pi}{\omega} + \frac{\phi}{\omega}$$

Figure 106
One cycle
$y = A \sin(\omega x - \phi)$, $A > 0$, $\omega > 0$, $\phi > 0$

See Figure 106.

We see that the graph of $y = A \sin(\omega x - \phi) = A \sin\left[\omega\left(x - \dfrac{\phi}{\omega}\right)\right]$ is the same as the graph of $y = A \sin(\omega x)$, except that it has been shifted $\dfrac{\phi}{\omega}$ units (to the right if $\phi > 0$ and to the left if $\phi < 0$). This number $\dfrac{\phi}{\omega}$ is called the **phase shift** of the graph of $y = A \sin(\omega x - \phi)$.

For the graphs of $y = A \sin(\omega x - \phi)$ or $y = A \cos(\omega x - \phi)$, $\omega > 0$,

$$\text{Amplitude} = |A| \qquad \text{Period} = T = \frac{2\pi}{\omega} \qquad \text{Phase shift} = \frac{\phi}{\omega}$$

The phase shift is to the left if $\phi < 0$ and to the right if $\phi > 0$.

EXAMPLE 1   **Finding the Amplitude, Period, and Phase Shift of a Sinusoidal Function and Graphing It**

2

Find the amplitude, period, and phase shift of $y = 3 \sin(2x - \pi)$, and graph the function.

Solution Comparing

$$y = 3\sin(2x - \pi) = 3\sin\left[2\left(x - \frac{\pi}{2}\right)\right]$$

to

$$y = A\sin(\omega x - \phi) = A\sin\left[\omega\left(x - \frac{\phi}{\omega}\right)\right]$$

we find that $A = 3$, $\omega = 2$, and $\phi = \pi$. The graph is a sine curve with amplitude $|A| = 3$, period $T = \dfrac{2\pi}{\omega} = \dfrac{2\pi}{2} = \pi$, and phase shift $= \dfrac{\phi}{\omega} = \dfrac{\pi}{2}$.

The graph of $y = 3\sin(2x - \pi)$ will lie between -3 and 3 on the y-axis. One cycle will begin at $x = \dfrac{\phi}{\omega} = \dfrac{\pi}{2}$ and end at $x = \dfrac{2\pi}{\omega} + \dfrac{\phi}{\omega} = \pi + \dfrac{\pi}{2} = \dfrac{3\pi}{2}$. We divide the interval $\left[\dfrac{\pi}{2}, \dfrac{3\pi}{2}\right]$ into four subintervals, each of length $\pi \div 4 = \dfrac{\pi}{4}$:

$$\left[\frac{\pi}{2}, \frac{3\pi}{4}\right], \quad \left[\frac{3\pi}{4}, \pi\right], \quad \left[\pi, \frac{5\pi}{4}\right], \quad \left[\frac{5\pi}{4}, \frac{3\pi}{2}\right]$$

The five key points on the graph are

$$\left(\frac{\pi}{2}, 0\right), \quad \left(\frac{3\pi}{4}, 3\right), \quad (\pi, 0), \quad \left(\frac{5\pi}{4}, -3\right), \quad \left(\frac{3\pi}{2}, 0\right)$$

We plot these five points and fill in the graph of the sine function as shown in Figure 107(a). Extending the graph in either direction, we obtain Figure 107(b).

Figure 107

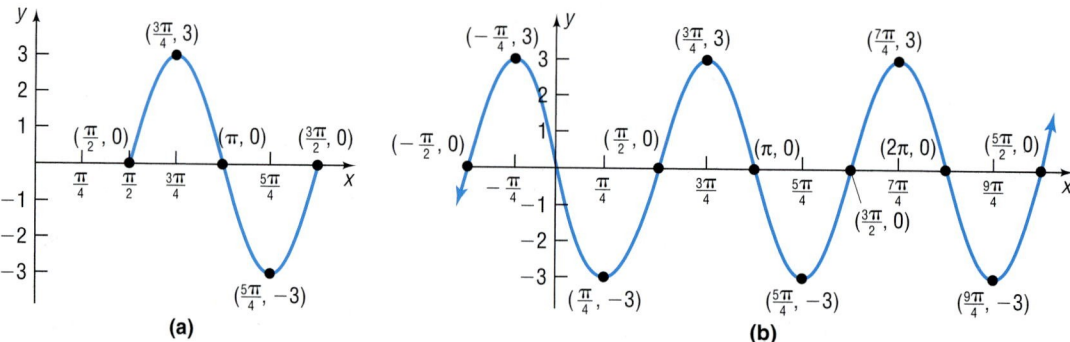

(a)

(b)

The graph of $y = 3\sin(2x - \pi) = 3\sin\left[2\left(x - \dfrac{\pi}{2}\right)\right]$ may also be obtained using transformations. See Figure 108.

Figure 108

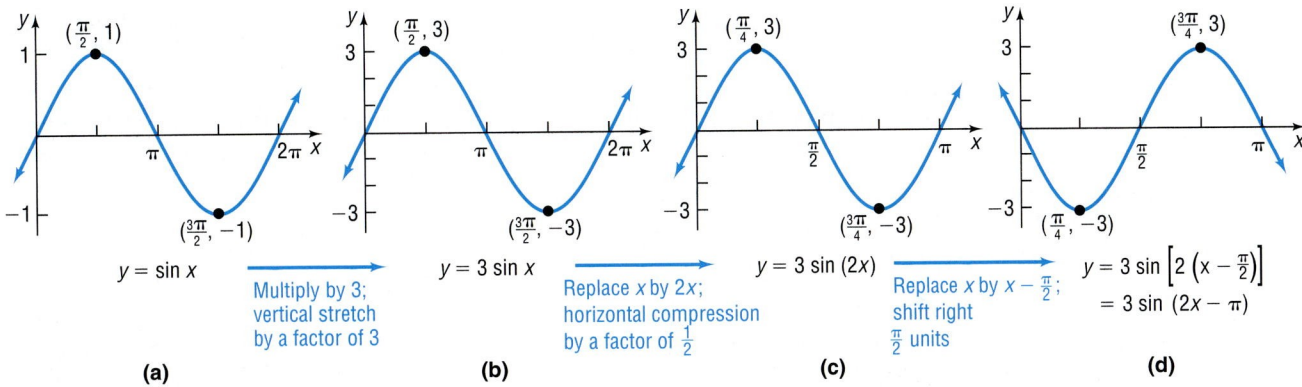

(a)   (b)   (c)   (d)

✔ CHECK: Figure 109 shows the graph of $Y_1 = 3\sin(2x - \pi)$ using a graphing utility.

Figure 109

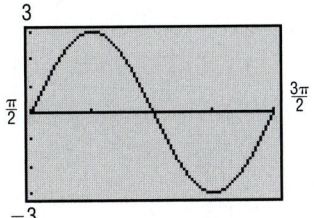

EXAMPLE 2   **Finding the Amplitude, Period, and Phase Shift of a Sinusoidal Function and Graphing It**

Find the amplitude, period, and phase shift of $y = 2\cos(4x + 3\pi)$, and graph the function.

Solution   Comparing

$$y = 2\cos(4x + 3\pi) = 2\cos\left[4\left(x + \frac{3\pi}{4}\right)\right]$$

to

$$y = A\cos(\omega x - \phi) = A\cos\left[\omega\left(x - \frac{\phi}{\omega}\right)\right]$$

we see that $A = 2$, $\omega = 4$, and $\phi = -3\pi$. The graph is a cosine curve with amplitude $|A| = 2$, period $T = \dfrac{2\pi}{\omega} = \dfrac{2\pi}{4} = \dfrac{\pi}{2}$, and phase shift $= \dfrac{\phi}{\omega} = -\dfrac{3\pi}{4}$.

The graph of $y = 2\cos(4x + 3\pi)$ will lie between -2 and 2 on the y-axis. One cycle will begin at $x = \dfrac{\phi}{\omega} = -\dfrac{3\pi}{4}$ and end at $x = \dfrac{2\pi}{\omega} + \dfrac{\phi}{\omega} = \dfrac{\pi}{2} + \left(-\dfrac{3\pi}{4}\right) = -\dfrac{\pi}{4}$. We divide the interval $\left[-\dfrac{3\pi}{4}, -\dfrac{\pi}{4}\right]$ into four subintervals, each of the length $\dfrac{\pi}{2} \div 4 = \dfrac{\pi}{8}$:

$$\left[-\frac{3\pi}{4}, -\frac{5\pi}{8}\right], \quad \left[-\frac{5\pi}{8}, -\frac{\pi}{2}\right], \quad \left[-\frac{\pi}{2}, -\frac{3\pi}{8}\right], \quad \left[-\frac{3\pi}{8}, -\frac{\pi}{4}\right]$$

The five key points on the graph are

$$\left(-\frac{3\pi}{4}, 2\right), \quad \left(-\frac{5\pi}{8}, 0\right), \quad \left(-\frac{\pi}{2}, -2\right), \quad \left(-\frac{3\pi}{8}, 0\right), \quad \left(-\frac{\pi}{4}, 2\right)$$

We plot these five points and fill in the graph of the cosine function as shown in Figure 110(a). Extending the graph in either direction, we obtain Figure 110(b).

Figure 110

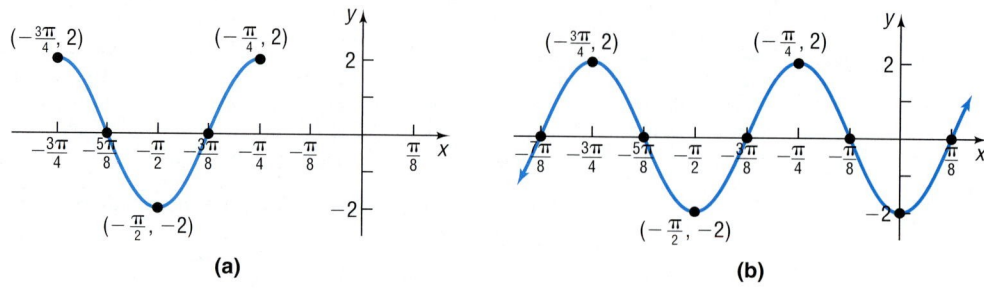

(a)　　　　　　　　　　(b)

The graph of $y = 2\cos(4x + 3\pi) = 2\cos\left[4\left(x + \frac{3\pi}{4}\right)\right]$ may also be obtained using transformations. See Figure 111.

Figure 111

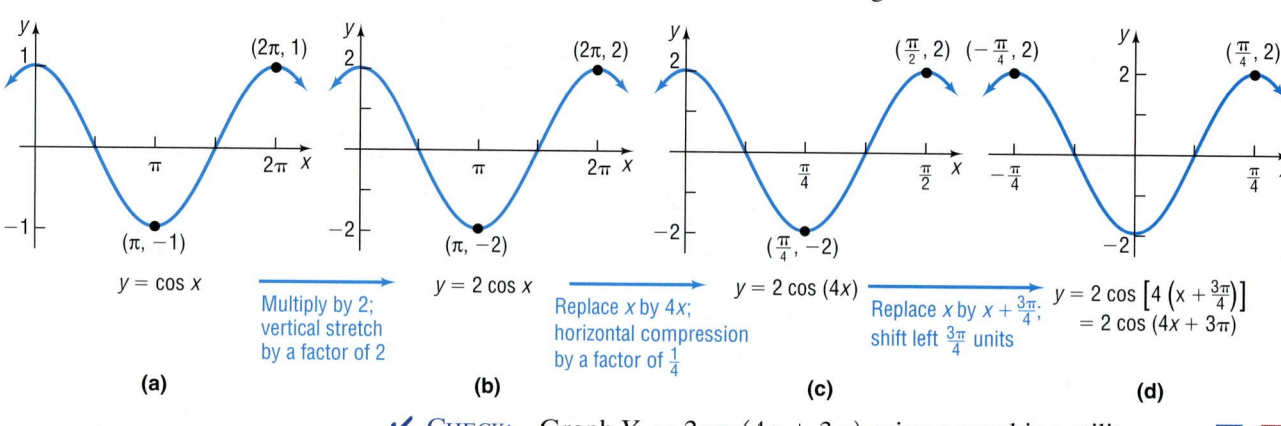

(a)　　　　　　(b)　　　　　　(c)　　　　　　(d)

✔ CHECK: Graph $Y_1 = 2\cos(4x + 3\pi)$ using a graphing utility. ■ ■

NOW WORK PROBLEM **1**.

SUMMARY　Steps for Graphing Sinusoidal Functions

To graph sinusoidal functions of the form $y = A\sin(\omega x - \phi)$ or $y = A\cos(\omega x - \phi)$:

STEP 1: Determine the amplitude $|A|$ and period $T = \dfrac{2\pi}{\omega}$.

STEP 2: Determine the starting point of one cycle of the graph, $\dfrac{\phi}{\omega}$.

STEP 3: Determine the ending point of one cycle of the graph, $\dfrac{2\pi}{\omega} + \dfrac{\phi}{\omega}$.

STEP 4: Divide the interval $\left[\dfrac{\phi}{\omega}, \dfrac{2\pi}{\omega} + \dfrac{\phi}{\omega}\right]$ into four subintervals, each of length $\dfrac{2\pi}{\omega} \div 4$.

STEP 5: Use the endpoints of the subintervals to find the five key points on the graph.

STEP 6: Fill in one cycle of the graph.

STEP 7: Extend the graph in each direction to make it complete.

Finding Sinusoidal Functions from Data

3   Scatter diagrams of data sometimes take the form of a sinusoidal function. Let's look at an example.

The data given in Table 12 represent the average monthly temperatures in Denver, Colorado. Since the data represent average monthly temperatures collected over many years, the data will not vary much from year to year and so will essentially repeat each year. In other words, the data are periodic. Figure 112 shows the scatter diagram of these data repeated over two years, where $x = 1$ represents January, $x = 2$ represents February, and so on.

Figure 112

TABLE 12		
	Month, x	**Average Monthly Temperature, °F**
	January, 1	29.7
	February, 2	33.4
	March, 3	39.0
	April, 4	48.2
	May, 5	57.2
	June, 6	66.9
	July, 7	73.5
	August, 8	71.4
	September, 9	62.3
	October, 10	51.4
	November, 11	39.0
	December, 12	31.0

Source: U.S. National Oceanic and Atmospheric Administration

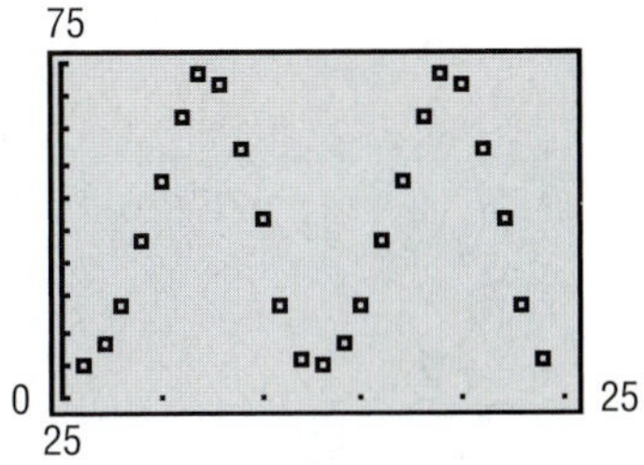

Notice that the scatter diagram looks like the graph of a sinusoidal function. We choose to fit the data to a sine function of the form

$$y = A \sin(\omega x - \phi) + B$$

where $A, B, \omega,$ and ϕ are constants.

EXAMPLE 3   **Finding a Sinusoidal Function from Temperature Data**

Fit a sine function to the data in Table 12.

Figure 113

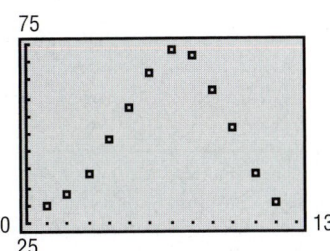

Solution   We begin with a scatter diagram of the data for one year. See Figure 113. The data will be fitted to a sine function of the form

$$y = A \sin(\omega x - \phi) + B$$

STEP 1:   To find the amplitude A, we compute

$$\text{Amplitude} = \frac{\text{largest data value} - \text{smallest data value}}{2}$$

$$= \frac{73.5 - 29.7}{2} = 21.9$$

To see the remaining steps in this process, we superimpose the graph of the function $y = 21.9 \sin x$, where x represents months, on the scatter diagram. Figure 114 shows the two graphs.

To fit the data, the graph needs to be shifted vertically, shifted horizontally, and stretched horizontally.

Figure 114

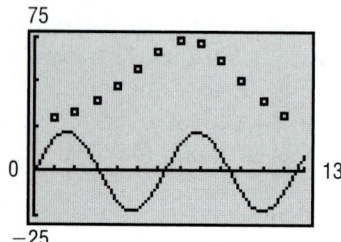

STEP 2: We determine the vertical shift by finding the average of the highest and lowest data value.

$$\text{Vertical shift} = \frac{73.5 + 29.7}{2} = 51.6$$

Now we superimpose the graph of $y = 21.9 \sin x + 51.6$ on the scatter diagram. See Figure 115.

We see that the graph needs to be shifted horizontally and stretched horizontally.

Figure 115

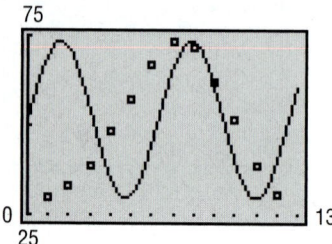

STEP 3: It is easier to find the horizontal stretch factor first. Since the temperatures repeat every 12 months, the period of the function is $T = 12$. Since $T = \dfrac{2\pi}{\omega} = 12$,

$$\omega = \frac{2\pi}{12} = \frac{\pi}{6}$$

Now we superimpose the graph of $y = 21.9 \sin\left(\dfrac{\pi}{6} x\right) + 51.6$ on the scatter diagram. See Figure 116.

We see that the graph still needs to be shifted horizontally.

Figure 116

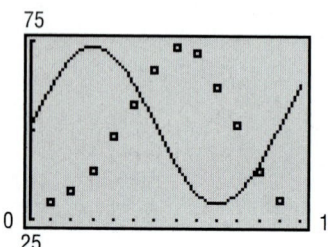

STEP 4: To determine the horizontal shift, we solve the equation

$$y = 21.9 \sin\left(\frac{\pi}{6} x - \phi\right) + 51.6$$

for ϕ by letting $y = 29.7$ and $x = 1$ (the average temperature in Denver in January).*

$$29.7 = 21.9 \sin\left(\frac{\pi}{6} \cdot 1 - \phi\right) + 51.6$$

$$-21.9 = 21.9 \sin\left(\frac{\pi}{6} - \phi\right) \qquad \text{Subtract 51.6 from both sides of the equation.}$$

$$-1 = \sin\left(\frac{\pi}{6} - \phi\right) \qquad \text{Divide both sides of the equation by 21.9.}$$

$$\frac{\pi}{6} - \phi = -\frac{\pi}{2} \qquad \sin\theta = -1 \text{ when } \theta = -\frac{\pi}{2}.$$

$$\phi = \frac{2\pi}{3} \qquad \text{Solve for } \phi.$$

The sine function that fits the data is

$$y = 21.9 \sin\left(\frac{\pi}{6} x - \frac{2\pi}{3}\right) + 51.6$$

---

*The data point selected here to find ϕ is arbitrary. Selecting a different data point will usually result in a different value for ϕ. To maintain consistency, we will always choose the data point for which y is smallest (in this case, January gives the lowest temperature).

Figure 117

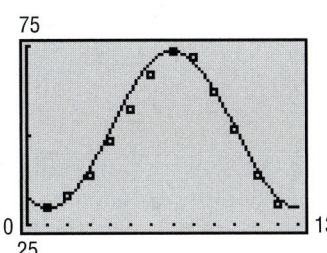

The graph of $y = 21.9 \sin\left(\dfrac{\pi}{6}x - \dfrac{2\pi}{3}\right) + 51.6$ and the scatter diagram of the data are shown in Figure 117.   ▪

The steps to fit a sine function

$$y = A \sin(\omega x - \phi) + B$$

to sinusoidal data follow:

Steps for Fitting Data to a Sine Function $y = A \sin(\omega x - \phi) + B$

STEP 1:   Determine A, the amplitude of the function.

$$\text{Amplitude} = \frac{\text{largest data value} - \text{smallest data value}}{2}$$

STEP 2:   Determine B, the vertical shift of the function.

$$\text{Vertical shift} = \frac{\text{largest data value} + \text{smallest data value}}{2}$$

STEP 3:   Determine ω. Since the period T, the time it takes for the data to repeat, is $T = \dfrac{2\pi}{\omega}$, we have

$$\omega = \frac{2\pi}{T}$$

STEP 4:   Determine the horizontal shift of the function by solving the equation

$$y = A \sin(\omega x - \phi) + B$$

for ϕ by choosing an ordered pair (x, y) from the data. Since answers will vary depending on the ordered pair selected, we will always choose the ordered pair for which y is smallest in order to maintain consistency.

✏️ **NOW WORK PROBLEM 19(a)–(c).**

Certain graphing utilities (such as a TI-83 and TI-86) have the capability of finding the sine function of best fit for sinusoidal data. At least four data points are required for this process.

EXAMPLE 4   **Finding the Sine Function of Best Fit**

Use a graphing utility to find the sine function of best fit for the data in Table 12. Graph this function with the scatter diagram of the data.

Figure 118

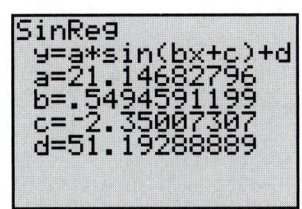

Solution   Enter the data from Table 12 and execute the SINe REGression program. The result is shown in Figure 118.

The output that the utility provides shows us the equation

$$y = a \sin(bx + c) + d$$

The sinusoidal function of best fit is

$$y = 21.15 \sin(0.55x - 2.35) + 51.19$$

Figure 119

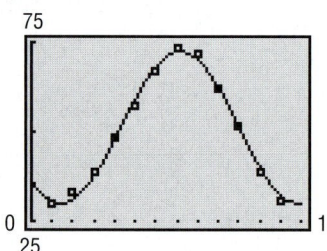

where *x* represents the month and *y* represents the average temperature.

Figure 119 shows the graph of the sinusoidal function of best fit on the scatter diagram.

▬ **NOW WORK PROBLEM 19(d) AND (e).**

Since the number of hours of sunlight in a day cycles annually, the number of hours of sunlight in a day for a given location can be modeled by a sinusoidal function.

The longest day of the year (in terms of hours of sunlight) occurs on the day of the summer solstice. The summer solstice is the time when the sun is farthest north (for locations in the northern hemisphere). In 2001, the summer solstice occurred on June 21 (the 172nd day of the year) at 3.38 AM EDT. The shortest day of the year occurs on the day of the winter solstice. The winter solstice is the time when the sun is farthest south (again, for locations in the northern hemisphere). In 2001, the winter solstice occurred on December 21 (the 355th day of the year) at 2:22 PM (EST).

EXAMPLE 5   **Finding a Sinusoidal Function for Hours of Daylight**

According to the *Old Farmer's Almanac*, the number of hours of sunlight in Boston on the summer solstice is 15.283 and the number of hours of sunlight on the winter solstice is 9.067.

(a) Find a sinusoidal function of the form $y = A \sin(\omega x - \phi) + B$ that fits the data.*

(b) Use the function found in part (a) to predict the number of hours of sunlight on April 1, the 91st day of the year.

(c) Draw a graph of the function found in part (a).

(d) Look up the number of hours of sunlight for April 1 in the *Old Farmer's Almanac* and compare the actual hours of daylight to the results found in part (b).

Solution   (a) **STEP 1:** Amplitude = $\dfrac{\text{largest data value} - \text{smallest data value}}{2}$

$$= \frac{15.283 - 9.067}{2} = 3.108$$

STEP 2:   Vertical shift = $\dfrac{\text{largest data value} + \text{smallest data value}}{2}$

$$= \frac{15.283 + 9.067}{2} = 12.175$$

STEP 3:   The data repeat every 365 days. Since $T = \dfrac{2\pi}{\omega} = 365$, we find

$$\omega = \frac{2\pi}{365}$$

So far, we have $y = 3.108 \sin\left(\dfrac{2\pi}{365} x - \phi\right) + 12.175$.

---

*Notice that only two data points are given, so a graphing utility cannot be used to find the sine function of best fit.

STEP 4: To determine the horizontal shift, we solve the equation

$$y = 3.108 \sin\left(\frac{2\pi}{365}x - \phi\right) + 12.175$$

for ϕ by letting $y = 9.067$ and $x = 355$ (the number of hours of daylight in Boston on December 21).

$$9.067 = 3.108 \sin\left(\frac{2\pi}{365} \cdot 355 - \phi\right) + 12.175$$

$$-3.108 = 3.108 \sin\left(\frac{2\pi}{365} \cdot 355 - \phi\right) \qquad \text{Subtract 12.175 from both sides of the equation.}$$

$$-1 = \sin\left(\frac{2\pi}{365} \cdot 355 - \phi\right) \qquad \text{Divide both sides of the equation by 3.108.}$$

$$\frac{2\pi}{365} \cdot 355 - \phi = -\frac{\pi}{2} \qquad \text{$\sin\theta = -1$ when $\theta = -\frac{\pi}{2}$.}$$

$$\phi = \frac{357}{146}\pi \qquad \text{Solve for ϕ.}$$

The function that provides the number of hours of daylight in Boston for any day, x, is given by

$$y = 3.108 \sin\left(\frac{2\pi}{365}x - \frac{357}{146}\pi\right) + 12.175$$

(b) To predict the number of hours of daylight on April 1, we let $x = 91$ in the function found in part (a) and obtain

$$y = 3.108 \sin\left(\frac{2\pi}{365} \cdot 91 - \frac{357}{146}\pi\right) + 12.175$$

$$\approx 12.69$$

So we predict that there will be about 12.69 hours of sunlight on April 1 in Boston.

(c) The graph of the function found in part (a) is given in Figure 120.

(d) According to the *Old Farmer's Almanac*, there will be 12 hours 43 minutes of sunlight on April 1 in Boston. Our prediction of 12.69 hours converts to 12 hours 41 minutes. ▪

Figure 120

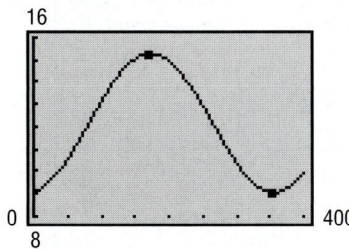

 **NOW WORK PROBLEM 25.**

8.8 Concepts and Vocabulary

In Problem 1, fill in the blank.

1. For the graph of $y = A\sin(\omega x - \phi)$, the number $\dfrac{\phi}{\omega}$ is called the _____.

In Problem 2, answer True or False.

2. Only two data points are required by a graphing utility to find the sine function of best fit.

8.8 Exercises

In Problems 1–12, find the amplitude, period, and phase shift of each function. Graph each function by hand. Show at least one period. Verify the result using a graphing utility.

1. $y = 4\sin(2x - \pi)$

2. $y = 3\sin(3x - \pi)$

3. $y = 2\cos\left(3x + \dfrac{\pi}{2}\right)$

4. $y = 3\cos(2x + \pi)$

5. $y = -3\sin\left(2x + \dfrac{\pi}{2}\right)$

6. $y = -2\cos\left(2x - \dfrac{\pi}{2}\right)$

7. $y = 4\sin(\pi x + 2)$

8. $y = 2\cos(2\pi x + 4)$

9. $y = 3\cos(\pi x - 2)$

10. $y = 2\cos(2\pi x - 4)$

11. $y = 3\sin\left(-2x + \dfrac{\pi}{2}\right)$

12. $y = 3\cos\left(-2x + \dfrac{\pi}{2}\right)$

In Problems 13–16, write the equation of a sine function that has the given characteristics.

13. Amplitude: 2

Period: π

Phase shift: $\dfrac{1}{2}$

14. Amplitude: 3

Period: $\dfrac{\pi}{2}$

Phase shift: 2

15. Amplitude: 3

Period: 3π

Phase shift: $-\dfrac{1}{3}$

16. Amplitude: 2

Period: π

Phase shift: -2

---

17. Alternating Current (ac) Circuits The current I, in amperes, flowing through an ac (alternating current) circuit at time t is

$$I = 120\sin\left(30\pi t - \dfrac{\pi}{3}\right), \qquad t \geq 0$$

What is the period? What is the amplitude? What is the phase shift? Graph this function over two periods.

18. Alternating Current (ac) Circuits The current I, in amperes, flowing through an ac (alternating current) circuit at time t is

$$I = 220\sin\left(60\pi t - \dfrac{\pi}{6}\right), \qquad t \geq 0$$

What is the period? What is the amplitude? What is the phase shift? Graph this function over two periods.

19. Monthly Temperature The following data represent the average monthly temperatures for Juneau, Alaska.

Month, x	Average Monthly Temperature, °F
January, 1	24.2
February, 2	28.4
March, 3	32.7
April, 4	39.7
May, 5	47.0
June, 6	53.0
July, 7	56.0
August, 8	55.0
September, 9	49.4
October, 10	42.2
November, 11	32.0
December, 12	27.1

Source: U.S. National Oceanic and Atmospheric Administration.

(a) Use a graphing utility to draw a scatter diagram of the data for one period.

(b) By hand, find a sinusoidal function of the form $y = A\sin(\omega x - \phi) + B$ that fits the data.

(c) Draw the sinusoidal function found in part (b) on the scatter diagram.

(d) Use a graphing utility to find the sinusoidal function of best fit.

(e) Draw the sinusoidal function of best fit on the scatter diagram.

20. Monthly Temperature The following data represent the average monthly temperatures for Washington, D.C.

Month, x	Average Monthly Temperature, °F
January, 1	34.6
February, 2	37.5
March, 3	47.2
April, 4	56.5
May, 5	66.4
June, 6	75.6
July, 7	80.0
August, 8	78.5
September, 9	71.3
October, 10	59.7
November, 11	49.8
December, 12	39.4

Source: U.S. National Oceanic and Atmospheric Administration.

(a) Use a graphing utility to draw a scatter diagram of the data for one period.

(b) By hand, find a sinusoidal function of the form
$y = A \sin(\omega x - \phi) + B$ that fits the data.
(c) Draw the sinusoidal function found in part (b) on the scatter diagram.
(d) Use a graphing utility to find the sinusoidal function of best fit.
(e) Graph the sinusoidal function of best fit on the scatter diagram.

21. Monthly Temperature   The following data represent the average monthly temperatures for Indianapolis, Indiana.

Month, x	Average Monthly Temperature, °F
January, 1	25.5
February, 2	29.6
March, 3	41.4
April, 4	52.4
May, 5	62.8
June, 6	71.9
July, 7	75.4
August, 8	73.2
September, 9	66.6
October, 10	54.7
November, 11	43.0
December, 12	30.9

Source: U.S. National Oceanic and Atmospheric Administration.

(a) Use a graphing utility to draw a scatter diagram of the data for one period.
(b) By hand, find a sinusoidal function of the form
$y = A \sin(\omega x - \phi) + B$ that fits the data.
(c) Draw the sinusoidal function found in part (b) on the scatter diagram.
(d) Use a graphing utility to find the sinusoidal function of best fit.
(e) Graph the sinusoidal function of best fit on the scatter diagram.

22. Monthly Temperature   The data at the top of the right column represent the average monthly temperatures for Baltimore, Maryland.

(a) Use a graphing utility to draw a scatter diagram of the data for one period.
(b) By hand, find a sinusoidal function of the form
$y = A \sin(\omega x - \phi) + B$ that fits the data.
(c) Draw the sinusoidal function found in part (b) on the scatter diagram.
(d) Use a graphing utility to find the sinusoidal function of best fit.
(e) Graph the sinusoidal function of best fit on the scatter diagram.

Month, x	Average Monthly Temperature, °F
January, 1	31.8
February, 2	34.8
March, 3	44.1
April, 4	53.4
May, 5	63.4
June, 6	72.5
July, 7	77.0
August, 8	75.6
September, 9	68.5
October, 10	56.6
November, 11	46.8
December, 12	36.7

Source: U.S. National Oceanic and Atmospheric Administration.

23. Tides   Suppose that the length of time between consecutive high tides is approximately 12.5 hours. According to the National Oceanic and Atmospheric Administration, on Saturday, June 28, 1997, in Savannah, Georgia, high tide occurred at 3:38 AM (3.6333 hours) and low tide occurred at 10:08 AM (10.1333 hours). Water heights are measured as the amounts above or below the mean lower low water. The height of the water at high tide was 8.2 feet and the height of the water at low tide was −0.6 foot.
(a) Approximately when will the next high tide occur?
(b) Find a sinusoidal function of the form
$y = A \sin(\omega x - \phi) + B$ that fits the data.
(c) Draw a graph of the function found in part (b).
(d) Use the function found in part (b) to predict the height of the water at the next high tide.

24. Tides   Suppose that the length of time between consecutive high tides is approximately 12.5 hours. According to the National Oceanic and Atmospheric Administration, on Saturday, June 28, 1997, in Juneau, Alaska, high tide occurred at 8:11 AM (8.1833 hours) and low tide occurred at 2:14 PM (14.2333 hours). Water heights are measured as the amounts above or below the mean lower low water. The height of the water at high tide was 13.2 feet and the height of the water at low tide was 2.2 feet.
(a) Approximately when will the next high tide occur?
(b) Find a sinusoidal function of the form
$y = A \sin(\omega x - \phi) + B$ that fits the data.
(c) Draw a graph of the function found in part (b).
(d) Use the function found in part (b) to predict the height of the water at the next high tide.

25. Hours of Daylight   According to the *Old Farmer's Almanac*, in Miami, Florida, the number of hours of sunlight on the summer solstice is 12.75 and the number of hours of sunlight on the winter solstice is 10.583.
(a) Find a sinusoidal function of the form
$y = A \sin(\omega x - \phi) + B$ that fits the data.

(b) Draw a graph of the function found in part (a).

(c) Use the function found in part (a) to predict the number of hours of sunlight on April 1, the 91st day of the year.

 (d) Look up the number of hours of sunlight for April 1 in the *Old Farmer's Almanac* and compare the actual hours of daylight to the results found in part (c).

26. **Hours of Daylight** According to the *Old Farmer's Almanac*, in Detroit, Michigan, the number of hours of sunlight on the summer solstice is 13.65 and the number of hours of sunlight on the winter solstice is 9.067.

(a) Find a sinusoidal function of the form $y = A \sin(\omega x - \phi) + B$ that fits the data.

(b) Draw a graph of the function found in part (a).

(c) Use the function found in part (a) to predict the number of hours of sunlight on April 1, the 91st day of the year.

(d) Look up the number of hours of sunlight for April 1 in the *Old Farmer's Almanac* and compare the actual hours of daylight to the results found in part (c).

27. **Hours of Daylight** According to the *Old Farmer's Almanac*, in Anchorage, Alaska, the number of hours of sunlight on the summer solstice is 16.233 and the number of hours of sunlight on the winter solstice is 5.45.

(a) Find a sinusoidal function of the form $y = A \sin(\omega x - \phi) + B$ that fits the data.

(b) Draw a graph of the function found in part (a).

(c) Use the function found in part (a) to predict the number of hours of sunlight on April 1, the 91st day of the year.

(d) Look up the number of hours of sunlight for April 1 in the *Old Farmer's Almanac* and compare the actual hours of daylight to the results found in part (c).

28. **Hours of Daylight** According to the *Old Farmer's Almanac*, in Honolulu, Hawaii, the number of hours of sunlight on the summer solstice is 12.767 and the number of hours of sunlight on the winter solstice is 10.783.

(a) Find a sinusoidal function of the form $y = A \sin(\omega x - \phi) + B$ that fits the data.

(b) Draw a graph of the function found in part (a).

(c) Use the function found in part (a) to predict the number of hours of sunlight on April 1, the 91st day of the year.

(d) Look up the number of hours of sunlight for April 1 in the *Old Farmer's Almanac* and compare the actual hours of daylight to the results found in part (c).

Chapter Review

Things To Know

Definitions

Angle in standard position (p. 611)	Vertex is at the origin; initial side is along the positive *x*-axis
Degree (1°) (p. 611)	$1° = \dfrac{1}{360}$ revolution
Radian (1) (p. 614)	The measure of a central angle of a circle whose rays subtend an arc whose length is the radius of the circle
Acute angle (p. 624)	An angle θ whose measure is $0° < \theta < 90°$ $\left(\text{or } 0 < \theta < \dfrac{\pi}{2}\right)$
Trigonometric functions (p. 644)	$P = (a, b)$ is the point on the terminal side of θ a distance r from the origin:

$$\sin\theta = \frac{b}{r} \qquad \cos\theta = \frac{a}{r} \qquad \tan\theta = \frac{b}{a}, \quad a \neq 0$$

$$\csc\theta = \frac{r}{b}, \quad b \neq 0 \qquad \sec\theta = \frac{r}{a}, \quad a \neq 0 \qquad \cot\theta = \frac{a}{b}, \quad b \neq 0$$

Complementary angles (p. 630)	Two acute angles whose sum is $90°\left(\dfrac{\pi}{2}\right)$
Cofunction (p. 630)	The following pairs of functions are cofunctions of each other: sine and cosine; tangent and cotangent; secant and cosecant
Reference angle of θ (p. 649)	The acute angle formed by the terminal side of θ and either the positive or negative *x*-axis
Periodic function (p. 661)	$f(\theta + p) = f(\theta)$, for all θ, $p > 0$, where the smallest such p is the fundamental period

Formulas

1 revolution $= 360°$ (p. 611)

$\qquad\qquad = 2\pi$ radians (p. 616)

$s = r\theta$ (p. 615)

θ is measured in radians; s is the length of arc subtended by the central angle θ of the circle of radius r; A is the area of the sector

$A = \dfrac{1}{2}r^2\theta$ (p. 618)

$v = r\omega$ (p. 619)

v is the linear speed along the circle of radius r; ω is the angular speed (measured in radians per unit time)

TABLE OF VALUES

θ (Radians)	θ (Degrees)	$\sin\theta$	$\cos\theta$	$\tan\theta$	$\csc\theta$	$\sec\theta$	$\cot\theta$
0	0°	0	1	0	Not defined	1	Not defined
$\dfrac{\pi}{6}$	30°	$\dfrac{1}{2}$	$\dfrac{\sqrt{3}}{2}$	$\dfrac{\sqrt{3}}{3}$	2	$\dfrac{2\sqrt{3}}{3}$	$\sqrt{3}$
$\dfrac{\pi}{4}$	45°	$\dfrac{\sqrt{2}}{2}$	$\dfrac{\sqrt{2}}{2}$	1	$\sqrt{2}$	$\sqrt{2}$	1
$\dfrac{\pi}{3}$	60°	$\dfrac{\sqrt{3}}{2}$	$\dfrac{1}{2}$	$\sqrt{3}$	$\dfrac{2\sqrt{3}}{3}$	2	$\dfrac{\sqrt{3}}{3}$
$\dfrac{\pi}{2}$	90°	1	0	Not defined	1	Not defined	0
π	180°	0	−1	0	Not defined	−1	Not defined
$\dfrac{3\pi}{2}$	270°	−1	0	Not defined	−1	Not defined	0

Fundamental Identities (p. 627)

$$\tan\theta = \frac{\sin\theta}{\cos\theta}, \quad \cot\theta = \frac{\cos\theta}{\sin\theta}$$

$$\cot\theta = \frac{1}{\tan\theta}, \quad \sec\theta = \frac{1}{\cos\theta}, \quad \csc\theta = \frac{1}{\sin\theta}$$

$$\sin^2\theta + \cos^2\theta = 1, \quad \tan^2\theta + 1 = \sec^2\theta, \quad 1 + \cot^2\theta = \csc^2\theta$$

Properties of the Trigonometric Functions

$y = \sin x$ Domain: $-\infty < x < \infty$

(p. 667)   Range: $-1 \le y \le 1$

              Periodic: period $= 2\pi(360°)$

              Odd function

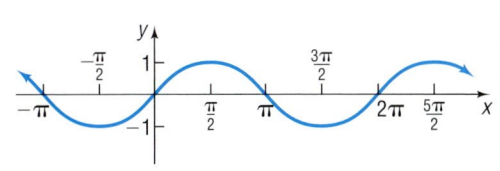

$y = \cos x$ Domain: $-\infty < x < \infty$

(p. 669)   Range: $-1 \le y \le 1$

              Periodic: period $= 2\pi(360°)$

              Even function

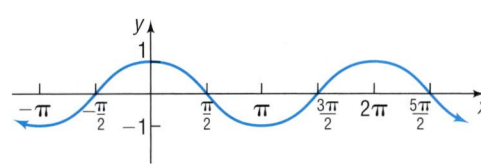

$y = \tan x$
(p. 684)
Domain: $-\infty < x < \infty$, except odd multiples of $\frac{\pi}{2}(90°)$
Range: $-\infty < y < \infty$
Periodic: period $= \pi(180°)$
Odd function

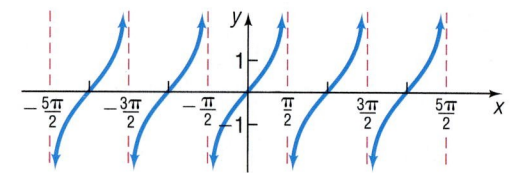

$y = \cot x$
(p. 685)
Domain: $-\infty < x < \infty$, except integral multiples of $\pi(180°)$
Range: $-\infty < y < \infty$
Periodic: period $= \pi(180°)$
Odd function

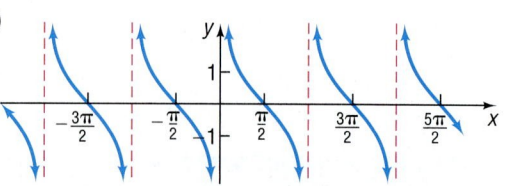

$y = \csc x$
(p. 686)
Domain: $-\infty < x < \infty$, except integral multiples of $\pi(180°)$
Range: $|y| \geq 1$
Periodic: period $= 2\pi(360°)$
Odd function

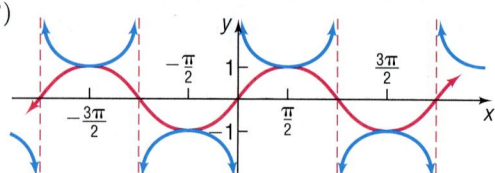

$y = \sec x$
(p. 687)
Domain: $-\infty < x < \infty$, except odd multiples of $\frac{\pi}{2}(90°)$
Range: $|y| \geq 1$
Periodic: period $= 2\pi(360°)$
Even function

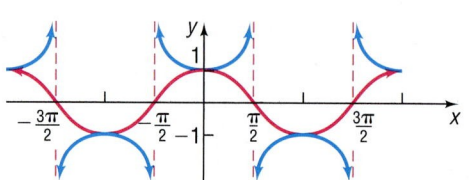

Sinusoidal graphs (pp. 673, 689)

$y = A \sin(\omega x), \quad \omega > 0$

$y = A \cos(\omega x), \quad \omega > 0$

$y = A \sin(\omega x - \phi) = A \sin\left[\omega\left(x - \frac{\phi}{\omega}\right)\right]$

$y = A \cos(\omega x - \phi) = A \cos\left[\omega\left(x - \frac{\phi}{\omega}\right)\right]$

$\text{Period} = \dfrac{2\pi}{\omega}$

$\text{Amplitude} = |A|$

$\text{Phase shift} = \dfrac{\phi}{\omega}$

Objectives

Section		You should be able to:	Review Exercises
8.1	1	Convert between degrees, minutes, seconds, and decimal forms for angles (p. 613)	82
	2	Find the arc length of a circle (p. 614)	83–84
	3	Convert from degrees to radians (p. 616)	1–4
	4	Convert from radians to degrees (p. 617)	5–8
	5	Find the area of a sector of a circle (p. 618)	83
	6	Find the linear speed of an object traveling in circular motion (p. 619)	85–88
8.2	1	Find the value of trigonometric functions of acute angles (p. 625)	75
	2	Use the fundamental identities (p. 626)	21–24
	3	Find the remaining trigonometric functions given the value of one of them (p. 628)	31–32
	4	Use the complementary angle theorem (p. 630)	25–26

Review Exercises

Blue problem numbers indicate the authors' suggestions for use in a Practice Test.

In Problems 1–4, convert each angle in degrees to radians. Express your answer as a multiple of π.

1. 135° **2.** 210° **3.** 18° **4.** 15°

In Problems 5–8, convert each angle in radians to degrees.

5. $\dfrac{3\pi}{4}$ **6.** $\dfrac{2\pi}{3}$ **7.** $-\dfrac{5\pi}{2}$ **8.** $-\dfrac{3\pi}{2}$

In Problems 9–30, find the exact value of each expression. Do not use a calculator.

9. $\tan\dfrac{\pi}{4} - \sin\dfrac{\pi}{6}$

10. $\cos\dfrac{\pi}{3} + \sin\dfrac{\pi}{4}$

11. $3\sin 45° - 4\tan\dfrac{\pi}{6}$

12. $4\cos 60° + 3\tan\dfrac{\pi}{3}$

13. $6\cos\dfrac{3\pi}{4} + 2\tan\left(-\dfrac{\pi}{3}\right)$

14. $3\sin\dfrac{2\pi}{3} - 4\cos\dfrac{5\pi}{2}$

15. $\sec\left(-\dfrac{\pi}{3}\right) - \cot\left(-\dfrac{5\pi}{4}\right)$

16. $4\csc\dfrac{3\pi}{4} - \cot\left(-\dfrac{\pi}{4}\right)$

17. $\tan \pi + \sin \pi$

18. $\cos\dfrac{\pi}{2} - \csc\left(-\dfrac{\pi}{2}\right)$

19. $\cos 540° - \tan(-45°)$

20. $\sin 630° + \cos(-180°)$

21. $\sin^2 20° + \dfrac{1}{\sec^2 20°}$

22. $\dfrac{1}{\cos^2 40°} - \dfrac{1}{\cot^2 40°}$

23. $\sec 50° \cos 50°$

24. $\tan 10° \cot 10°$

25. $\dfrac{\sin 50°}{\cos 40°}$

26. $\dfrac{\tan 20°}{\cot 70°}$

27. $\dfrac{\sin(-40°)}{\cos 50°}$

28. $\tan(-20°)\cot 20°$

29. $\sin 400° \sec(-50°)$

30. $\cot 200° \cot(-70°)$

In Problems 31–46, find the exact value of each of the remaining trigonometric functions.

31. $\sin\theta = \dfrac{4}{5}$, θ acute

32. $\cos\theta = \dfrac{3}{5}$, θ acute

33. $\tan\theta = \dfrac{12}{5}$, $\sin\theta < 0$

34. $\cot\theta = \dfrac{12}{5}$, $\cos\theta < 0$

35. $\sec\theta = -\dfrac{5}{4}$, $\tan\theta < 0$

36. $\csc\theta = -\dfrac{5}{3}$, $\cot\theta < 0$

37. $\sin\theta = \dfrac{12}{13}$, θ in quadrant II

38. $\cos\theta = -\dfrac{3}{5}$, θ in quadrant III

39. $\sin\theta = -\dfrac{5}{13}$, $\dfrac{3\pi}{2} < \theta < 2\pi$

40. $\cos\theta = \dfrac{12}{13}$, $\dfrac{3\pi}{2} < \theta < 2\pi$

41. $\tan\theta = \dfrac{1}{3}$, $180° < \theta < 270°$

42. $\tan\theta = -\dfrac{2}{3}$, $90° < \theta < 180°$

43. $\sec\theta = 3$, $\dfrac{3\pi}{2} < \theta < 2\pi$

44. $\csc\theta = -4$, $\pi < \theta < \dfrac{3\pi}{2}$

45. $\cot\theta = -2$, $\dfrac{\pi}{2} < \theta < \pi$

46. $\tan\theta = -2$, $\dfrac{3\pi}{2} < \theta < 2\pi$

In Problems 47–58, graph each function. Each graph should contain at least one period.

47. $y = 2\sin(4x)$

48. $y = -3\cos(2x)$

49. $y = -2\cos\left(x + \dfrac{\pi}{2}\right)$:

50. $y = 3\sin(x - \pi)$

51. $y = \tan(x + \pi)$

52. $y = -\tan\left(x - \dfrac{\pi}{2}\right)$

53. $y = -2\tan(3x)$

54. $y = 4\tan(2x)$

55. $y = \cot\left(x + \dfrac{\pi}{4}\right)$

56. $y = -4\cot(2x)$

57. $y = \sec\left(x - \dfrac{\pi}{4}\right)$

58. $y = \csc\left(x + \dfrac{\pi}{4}\right)$

In Problems 59–62, determine the amplitude and period of each function without graphing.

59. $y = 4\cos x$

60. $y = \sin(2x)$

61. $y = -8\sin\left(\dfrac{\pi}{2}x\right)$

62. $y = -2\cos(3\pi x)$

In Problems 63–70, find the amplitude, period, and phase shift of each function. Graph each function. Show at least one period.

63. $y = 4\sin(3x)$

64. $y = 2\cos\left(\dfrac{1}{3}x\right)$

65. $y = 2\sin(2x - \pi)$

66. $y = -\cos\left(\dfrac{1}{2}x + \dfrac{\pi}{2}\right)$

67. $y = \dfrac{1}{2}\sin\left(\dfrac{3}{2}x - \pi\right)$

68. $y = \dfrac{3}{2}\cos(6x + 3\pi)$

69. $y = -\dfrac{2}{3}\cos(\pi x - 6)$

70. $y = -7\sin\left(\dfrac{\pi}{3}x + \dfrac{4}{3}\right)$

In Problems 71–74, find a function whose graph is given.

71.

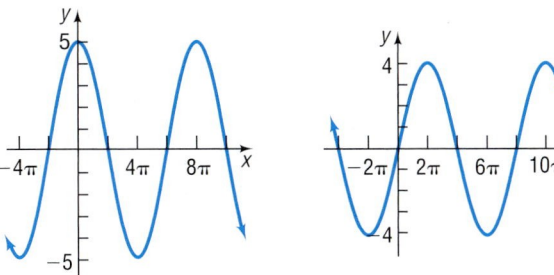

72.

73.

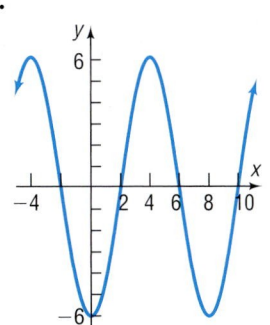

74.

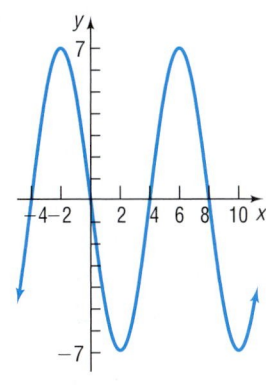

75. Find the value of each of the six trigonometric functions of the angle θ in the illustration.

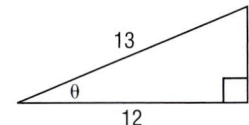

76. Use a calculator to approximate $\sec 10°$. Round the answer to two decimal places.

77. Find the exact value of each of the six trigonometric functions of an angle θ if $(3, -4)$ is a point on the terminal side of θ.

78. Name the quadrant θ lies in if $\cos \theta > 0$ and $\tan \theta < 0$.

79. Find the reference angle of $-\dfrac{4\pi}{5}$.

80. Find the exact value of $\sin t, \cos t$, and $\tan t$ if $P = \left(-\dfrac{3}{5}, \dfrac{4}{5}\right)$

is the point on the unit circle that corresponds to t.

81. Find the domain and range of the secant function.

82. (a) Convert the angle $32°20'35''$ to a decimal in degrees. Round the answer to two decimal places.
(b) Convert the angle $63.18°$ to $D°M'S''$ form. Express the answer to the nearest second.

83. Find the length of arc subtended by a central angle of $30°$ on a circle of radius 2 feet. What is the area of the sector?

84. The minute hand of a clock is 8 inches long. How far does the tip of the minute hand move in 30 minutes? How far does it move in 20 minutes?

85. Angular Speed of a Race Car A race car is driven around a circular track at a constant speed of 180 miles per hour. If the diameter of the track is $\dfrac{1}{2}$ mile, what is the angular speed of the car? Express your answer in revolutions per hour (which is equivalent to laps per hour).

86. Merry-Go-Rounds A neighborhood carnival has a merry-go-round whose radius is 25 feet. If the time for one revolution is 30 seconds, how fast is the merry-go-round going?

87. Lighthouse Beacons The Montauk Point Lighthouse on Long Island has dual beams (two light sources opposite each other). Ships at sea observe a blinking light every 5 seconds. What rotation speed is required to do this?

88. Spin Balancing Tires The radius of each wheel of a car is 16 inches. At how many revolutions per minute should a spin balancer be set to balance the tires at a speed of 90 miles per hour? Is the setting different for a wheel of radius 14 inches? If so, what is this setting?

89. Alternating Voltage The electromotive force E, in volts, in a certain ac circuit obeys the equation

$$E = 120 \sin(120\pi t), \quad t \geq 0$$

where t is measured in seconds.
(a) What is the maximum value of E?
(b) What is the period?
(c) Graph this function over two periods.

90. Alternating Current The current I, in amperes, flowing through an ac (alternating current) circuit at time t is

$$I = 220 \sin\left(30\pi t + \frac{\pi}{6}\right), \quad t \geq 0$$

(a) What is the period?
(b) What is the amplitude?
(c) What is the phase shift?
(d) Graph this function over two periods.

91. Monthly Temperature The following data represent the average monthly temperatures for Phoenix, Arizona.

Month, m	Average Monthly Temperature, T
January, 1	51
February, 2	55
March, 3	63
April, 4	67
May, 5	77
June, 6	86
July, 7	90
August, 8	90
September, 9	84
October, 10	71
November, 11	59
December, 12	52

Source: U.S. National Oceanic and Atmospheric Administration.

(a) Use a graphing utility to draw a scatter diagram of the data for one period.

(b) By hand, find a sinusoidal function of the form $y = A \sin(\omega x - \phi) + B$ that fits the data.

(c) Draw the sinusoidal function found in part (b) on the scatter diagram.

(d) Use a graphing utility to find the sinusoidal function of best fit.

(e) Graph the sinusoidal function of best fit on the scatter diagram.

92. Monthly Temperature The following data represent the average monthly temperatures for Chicago, Illinois.

Month, m	Average Monthly Temperature, T
January, 1	25
February, 2	28
March, 3	36
April, 4	48
May, 5	61
June, 6	72
July, 7	74
August, 8	75
September, 9	66
October, 10	55
November, 11	39
December, 12	28

Source: U.S. National Oceanic and Atmospheric Administration.

(a) Use a graphing utility to draw a scatter diagram of the data for one period.

(b) By hand, find a sinusoidal function of the form $y = A \sin(\omega x - \phi) + B$ that fits the data.

(c) Draw the sinusoidal function found in part (b) on the scatter diagram.

(d) Use a graphing utility to find the sinusoidal function of best fit.

(e) Graph the sinusoidal function of best fit on the scatter diagram.

93. Hours of Daylight According to the *Old Farmer's Almanac*, in Las Vegas, Nevada, the number of hours of sunlight on the summer solstice is 13.367 and the number of hours of sunlight on the winter solstice is 9.667.

(a) Find a sinusoidal function of the form
$$y = A \sin(\omega x - \phi) + B$$
that fits the data.

(b) Draw a graph of the function found in part (a).

(c) Use the function found in part (a) to predict the number of hours of sunlight on April 1, the 91st day of the year.

(d) Look up the number of hours of sunlight for April 1 in the *Old Farmer's Almanac* and compare the actual hours of daylight to the results found in part (c).

94. Hours of Daylight According to the *Old Farmer's Almanac*, in Seattle, Washington, the number of hours of sunlight on the summer solstice is 13.967 and the number of hours of sunlight on the winter solstice is 8.417.

(a) Find a sinusoidal function of the form
$$y = A \sin(\omega x - \phi) + B$$
that fits the data.

(b) Draw a graph of the function found in part (a).

(c) Use the function found in part (a) to predict the number of hours of sunlight on April 1, the 91st day of the year.

(d) Look up the number of hours of sunlight for April 1 in the *Old Farmer's Almanac* and compare the actual hours of daylight to the results found in part (c).

Chapter Projects

1. **Tides** A partial tide table for September, 2001, for Sabine Pass along the Texas Gulf Coast is given in the table on page 707:

(a) On September 15, when was the tide high? This is called "high tide". On September 19, when was the tide low? This is called "low tide". Most days will have two low tides and two high tides.

(b) Why do you think there is a negative height for the low tide on September 14? What is the tide height measured against?

(c) On your graphing utility, draw a scatter diagram for the data in the table. Let T (time) be the independent variable, with $T = 0$ being 12:00 A.M. on September 1, $T = 24$ being 12:00 A.M. on September 2, etc. Remember that there are 60 minutes in an hour. Let H be the height in feet when converting the times. Also, make sure your graphing utility is in radian mode.

(d) What shape does the data take? What is the period of the data? What is the amplitude? Is the amplitude constant? Explain.

Sept	High Tide Time	High Tide Ht(ft)	High Tide Time	High Tide Ht(ft)	Low Tide Time	Low Tide Ht(ft)	Low Tide Time	Low Tide Ht(ft)	Sun/Moon phase Rise/Set
F 14	03:08a	2.4	11:12a	2.2	08:14a	2.0	07:19p	−0.1	7:00a/7:23p
S 15	03:33a	2.4	12:56p	2.2	08:15a	1.9	08:13p	0.0	7:00a/7:22p
S 16	03:57a	2.3	02:17p	2.3	08:45a	1.6	09:05p	0.3	7:01a/7:20p
M17	04:20a	2.2	03:33p	2.3	09:24a	1.4	09:54p	0.5	7:01a/7:19p
T18	04:41a	2.2	04:47p	2.3	10:08a	1.0	10:43p	1.0	7:02a/7:08p
W19	05:01a	2.0	06:04p	2.3	10:54a	0.7	11:32p	1.4	7:02a/7:17p
T20	05:20a	2.0	07:27p	2.3	11:44a	0.4			7:03a/7:15p

Date from: *www.harbortides.com*

(e) Using Steps 1–4, given on page 695, fit a sine curve to the data. Let the amplitude be the average of the amplitudes that you found in part (c), unless the amplitude was constant. Is there a vertical shift? Is there a phase shift?

(f) Using your graphing utility, find the sinusoidal function of best fit. How does it compare to your equation?

(g) Using the equation found in part (e) and using the sinusoidal equation of best fit found in part (f), predict the high tides and the low tides on September 21.

(h) Looking at the times of day that the low tides occur, what do you think causes the low tides to vary so much each day? Explain. Does this seem to have the same type of effect on the high tides? Explain.

2. **Identifying Mountain Peaks in Hawaii**   Suppose that you are standing on the southeastern shore of Oahu and you see three mountain peaks on the horizon. You want to determine which mountains are visible from Oahu. The possible mountain peaks that can be seen from Oahu and the height (above sea level) of their peaks are:

Island	Distance (miles)	Mountain	Height (feet)
Lanai	65	Lanaihale	3,370
Maui	110	Haleakala	10,023
Hawaii	190	Mauna Kea	13,796
Molokai	40	Kamakou	4,961

(a) To determine which of these mountain peaks would be visible from Oahu, consider that you are standing on the shore and looking "straight out" so that your line of sight is tangent to the surface of Earth at the point you're standing. Make a sketch of the right triangle formed by your sight line, the radius from the center of Earth to the point you are standing, and the line from the center of Earth through Lanai.

(b) Assuming that the radius of Earth is 3960 miles, determine the angle formed at the center of Earth.

(c) Determine the length of the hypotenuse of the triangle. Is Lanaihale visible from Oahu?

(d) Repeat parts (a)–(d) for the other three islands.

(e) Which three mountains are visible from Oahu?

3. **CBL Experiment**   Using a CBL, the microphone probe and a tuning fork, record the amplitude, frequency, and period of the sound from the graph of the sound created by the tuning fork over time. Repeat the experiment for different tuning forks.

Cumulative Review

1. Find the real solutions, if any, of the equation
   $2x^2 + x - 1 = 0$.

2. Find an equation for the line with slope -3, containing the point $(-2, 5)$.

3. Find an equation for the circle of radius 4 and center at the point $(0, -2)$.

4. Graph the equation $2x - 3y = 12$.

5. Graph the equation $x^2 + y^2 - 2x + 4y - 4 = 0$.

6. Use transformations to graph the function
   $y = (x - 3)^2 + 2$.

7. Sketch a graph of each of the following functions. Label at least three points on each graph.
   (a) $y = x^2$
   (b) $y = x^3$
   (c) $y = e^x$
   (d) $y = \ln x$
   (e) $y = \sin x$
   (f) $y = \tan x$

8. Find the inverse function of $f(x) = 3x - 2$.

9. Find the exact value of $(\sin 14°)^2 + (\sin 76°)^2 - 3$.

10. Graph $y = 3 \sin(2x)$.

11. Find the exact value of $\tan \dfrac{\pi}{4} - 3 \cos \dfrac{\pi}{6} + \csc \dfrac{\pi}{6}$.

12. Find an exponential function for the following graph. Express your answer in the form $y = Ab^x$

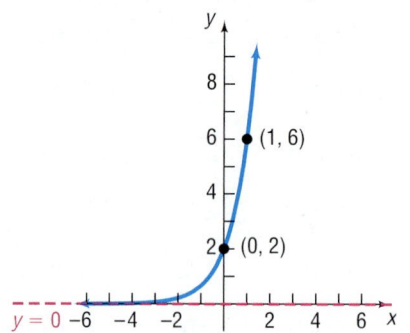

13. Find a sinusoidal function for the following graph.

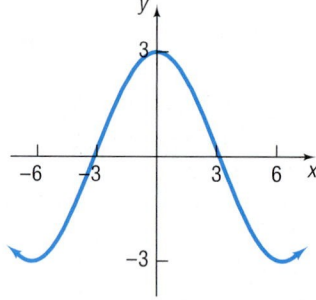

ANALYTIC TRIGONOMETRY

Tremor Brought First Hint of Doom

Underwater earthquakes that triggered the devastating tsunami in PNG would have been felt by villagers about 30 minutes before the waves struck, scientists said yesterday. "The tremor was felt by coastal residents who may not have realised its significance or did not have time to retreat," said associate Professor Ted Bryant, a geoscientist at the University of Wollongong.

The tsunami would have sounded like a fleet of bombers as it crashed into a 30-kilometre stretch of coast. "The tsunami would have been caused by a rapid uplift or drop of the sea floor," said an applied mathematician and cosmologist from Monash University, Professor Joe Monaghan, who is one of Australia's leading experts on tsunamis . . . .

Tsunamis, ridges of water hundreds of kilometres long and stretching from front to back for several kilometres, line up parallel to the beach. "We are talking about a huge volume of water moving very fast-300 kilometres per hour would be typical," Professor Monaghan said.

(SOURCE: Peter Spinks, The Age, Tuesday, July 21, 1998.)

SEE CHAPTER PROJECT 1.

OUTLINE

 For additional study help, go to

www.prenhall.com/sullivanegu3e

Materials include:

- Graphing Calculator Help
- Chapter Quiz
- Chapter Test
- PowerPoint Downloads
- Chapter Projects
- Student Tips

A Look Back, A Look Forward

In Chapter 6, we defined inverse functions and developed their properties, particularly the relationship between the domain and range of a function and its inverse. We learned that the graph of a function and its inverse are symmetric with respect to the line $y = x$. We continued in Chapter 6 by defining the exponential function and the inverse of the exponential function, the logarithmic function. In the first two sections of this chapter, we define the six inverse trigonometric functions and investigate their properties.

In Chapter 8, we derived several identities involving the trigonometric functions. In Sections 9.3 through 9.6 of this chapter, we continue the derivation of identities. These identities play an important role in calculus, the physical and life sciences, and economics, where they are used to simplify complicated expressions. The last two sections of this chapter deal with equations that contain trigonometric functions.

PREPARING FOR THIS SECTION

Before getting started, review the following:

✓ Inverse Functions (Section 6.1, pp. 418–424)

✓ Composite Functions (Section 3.5, pp. 301–306)

✓ Definition of the Trigonometric Functions (Section 8.4, p. 644)

✓ Values of the Trigonometric Functions of Certain Angles (Section 8.3, p. 638, and Section 8.4, p. 646)

✓ Domain and Range of the Sine, Cosine, and Tangent Functions (Section 8.5, pp. 658–660)

✓ Graphs of the Sine, Cosine, and Tangent Functions (Section 8.6, pp. 666–671, and Section 8.7, pp. 682–685)

9.1 THE INVERSE SINE, COSINE, AND TANGENT FUNCTIONS

OBJECTIVES   1 Find the Exact Value of the Inverse Sine, Cosine, and Tangent Functions

2 Find an Approximate Value of the Inverse Sine, Cosine, and Tangent Functions

In Section 6.1 we discussed inverse functions, and we noted that if a function is one-to-one it will have an inverse function. We also observed that if a function is not one-to-one it may be possible to restrict its domain in some suitable manner so that the restricted function is one-to-one.

Next, we review some properties of a function f and its inverse function f^{-1}.

1. $f^{-1}(f(x)) = x$ for every x in the domain of f and $f(f^{-1}(x)) = x$ for every x in the domain of f^{-1}.

2. Domain of f = range of f^{-1} and range of f = domain of f^{-1}.

3. The graph of f and the graph of f^{-1} are symmetric with respect to the line $y = x$.

4. If a function $y = f(x)$ has an inverse function, the equation of the inverse function is $x = f(y)$. The solution of this equation is $y = f^{-1}(x)$.

The Inverse Sine Function

In Figure 1, we reproduce the graph of $y = \sin x$. Because every horizontal line $y = b$, where b is between -1 and 1, intersects the graph of $y = \sin x$ infinitely many times, it follows from the horizontal-line test that the function $y = \sin x$ is not one-to-one.

Figure 1
$y = \sin x$,
$-\infty < x < \infty, -1 \leq y \leq 1$

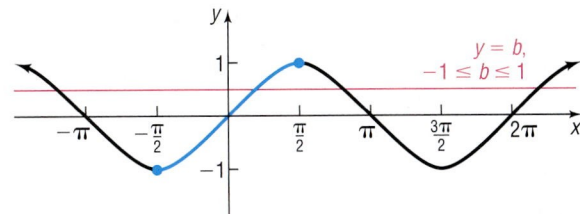

However, if we restrict the domain of $y = \sin x$ to the interval $\left[-\dfrac{\pi}{2}, \dfrac{\pi}{2}\right]$, the restricted function

$$y = \sin x, \qquad -\frac{\pi}{2} \leq x \leq \frac{\pi}{2}$$

is one-to-one and, hence, will have an inverse function.* See Figure 2.

An equation for the inverse of $y = f(x) = \sin x$ is obtained by interchanging x and y. The implicit form of the inverse function is $x = \sin y$, $-\dfrac{\pi}{2} \leq y \leq \dfrac{\pi}{2}$. The explicit form is called the **inverse sine** of x and is symbolized by $y = f^{-1}(x) = \sin^{-1} x$.

Figure 2

$y = \sin x$,

$-\dfrac{\pi}{2} \leq x \leq \dfrac{\pi}{2}, -1 \leq y \leq 1$

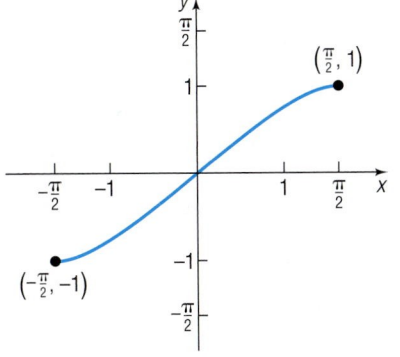

> $$y = \sin^{-1} x \quad \text{means} \quad x = \sin y \qquad (1)$$
>
> $$\text{where} \quad -1 \leq x \leq 1 \quad \text{and} \quad -\frac{\pi}{2} \leq y \leq \frac{\pi}{2}$$

Because $y = \sin^{-1} x$ means $x = \sin y$, we read $y = \sin^{-1} x$ as "y is the angle or real number whose sine equals x." Alternatively, we can say that "y is the inverse sine of x." Be careful about the notation used. The superscript -1 that appears in $y = \sin^{-1} x$ is not an exponent, but is reminiscent of the symbolism f^{-1} used to denote the inverse function of f. [To avoid this notation, some books use the notation $y = \arcsin x$ instead of $y = \sin^{-1} x$.]

The inverse of a function f receives as input an element from the range of f and returns as output an element in the domain of f. The restricted sine function, $y = f(x) = \sin x$, receives as input an angle or real number x in the interval $\left[-\dfrac{\pi}{2}, \dfrac{\pi}{2}\right]$ and outputs a real number in the interval $[-1, 1]$. Therefore, the inverse sine function receives as input a real number in the

---

*Although there are many other ways to restrict the domain and obtain a one-to-one function, mathematicians have agreed on a consistent use of the interval $\left[-\dfrac{\pi}{2}, \dfrac{\pi}{2}\right]$ in order to define the inverse of $y = \sin x$.

interval $[-1, 1]$ and outputs an angle or real number in the interval $\left[-\dfrac{\pi}{2}, \dfrac{\pi}{2}\right]$.

Since the domain of f = range of f^{-1} and the range of f = domain of f^{-1}, the domain of the inverse sine function, $y = f^{-1}(x) = \sin^{-1} x$, is $[-1, 1]$ or $-1 \le x \le 1$, and the range of the inverse sine function is $\left[-\dfrac{\pi}{2}, \dfrac{\pi}{2}\right]$ or $-\dfrac{\pi}{2} \le y \le \dfrac{\pi}{2}$. The graph of the inverse sine function can be obtained by reflecting the restricted portion of the graph of $y = f(x) = \sin x$ about the line $y = x$, as shown in Figure 3(a). Figure 3(b) shows the graph using a graphing utility.

Figure 3
$y = \sin^{-1} x,$
$-1 \le x \le 1, -\dfrac{\pi}{2} \le y \le \dfrac{\pi}{2}$

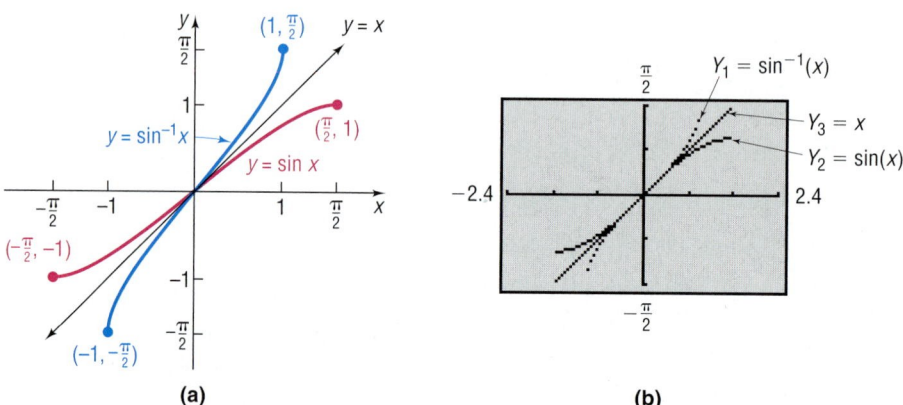

(a)   (b)

When we discussed functions and their inverses in Section 6.1, we found that $f^{-1}(f(x)) = x$ and $f(f^{-1}(x)) = x$. In terms of the sine function and its inverse, these properties are of the form

$$f^{-1}(f(x)) = \sin^{-1}(\sin x) = x, \quad \text{where } -\dfrac{\pi}{2} \le x \le \dfrac{\pi}{2} \qquad \text{(2a)}$$
$$f(f^{-1}(x)) = \sin(\sin^{-1} x) = x, \quad \text{where } -1 \le x \le 1 \qquad \text{(2b)}$$

For example, because $\dfrac{\pi}{8}$ lies in the interval $\left[-\dfrac{\pi}{2}, \dfrac{\pi}{2}\right]$, the restricted domain of the sine function, we have

$$\sin^{-1}\left[\sin\left(\dfrac{\pi}{8}\right)\right] = \dfrac{\pi}{8}$$

Also, because 0.8 lies in the interval $[-1, 1]$, the domain of the inverse sine function, we have

$$\sin[\sin^{-1}(0.8)] = 0.8$$

Figure 4

See Figure 4 for these calculations on a graphing calculator.

However, because $\dfrac{5\pi}{8}$ is not in the interval $\left[-\dfrac{\pi}{2}, \dfrac{\pi}{2}\right]$

$$\sin^{-1}\left[\sin\left(\dfrac{5\pi}{8}\right)\right] \ne \dfrac{5\pi}{8}$$

See Figure 5.

Also, because 1.8 is not in the interval $[-1, 1]$,

$$\sin[\sin^{-1}(1.8)] \neq 1.8$$

See Figure 6. Can you explain why the error appears?

Figure 5

Figure 6

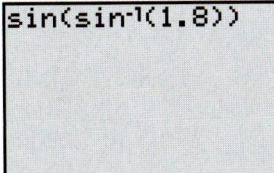

①   For some numbers x it is possible to find the exact value of $y = \sin^{-1} x$.

EXAMPLE 1   **Finding the Exact Value of an Inverse Sine Function**

Find the exact value of: $\sin^{-1} 1$

Solution   Let $\theta = \sin^{-1} 1$. We seek the angle θ, $-\dfrac{\pi}{2} \leq \theta \leq \dfrac{\pi}{2}$, whose sine equals 1.

$$\theta = \sin^{-1} 1, \qquad -\frac{\pi}{2} \leq \theta \leq \frac{\pi}{2}$$

$$\sin \theta = 1, \qquad -\frac{\pi}{2} \leq \theta \leq \frac{\pi}{2} \qquad \text{\color{blue}{By definition of } } y = \sin^{-1} x.$$

Now look at Table 1 and Figure 7.

TABLE 1

θ	$\sin \theta$
$-\dfrac{\pi}{2}$	-1
$-\dfrac{\pi}{3}$	$-\dfrac{\sqrt{3}}{2}$
$-\dfrac{\pi}{4}$	$-\dfrac{\sqrt{2}}{2}$
$-\dfrac{\pi}{6}$	$-\dfrac{1}{2}$
0	0
$\dfrac{\pi}{6}$	$\dfrac{1}{2}$
$\dfrac{\pi}{4}$	$\dfrac{\sqrt{2}}{2}$
$\dfrac{\pi}{3}$	$\dfrac{\sqrt{3}}{2}$
$\dfrac{\pi}{2}$	1

Figure 7

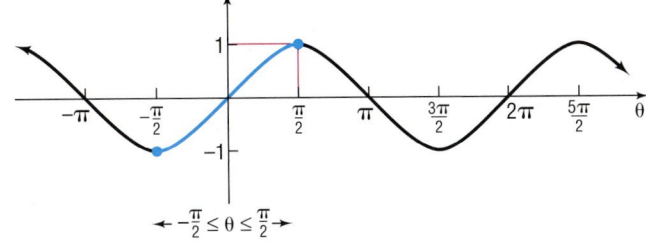

We see that the only angle θ within the interval $\left[-\dfrac{\pi}{2}, \dfrac{\pi}{2}\right]$ whose sine is

1 is $\dfrac{\pi}{2}$. [Note that $\sin \dfrac{5\pi}{2}$ also equals 1, but $\dfrac{5\pi}{2}$ lies outside the interval

$\left[-\dfrac{\pi}{2}, \dfrac{\pi}{2}\right]$ and hence is not admissible.] Since $\sin \dfrac{\pi}{2} = 1$ and $\dfrac{\pi}{2}$ is in the

interval $\left[-\dfrac{\pi}{2}, \dfrac{\pi}{2}\right]$, we conclude that

$$\sin^{-1} 1 = \frac{\pi}{2}$$

Figure 8

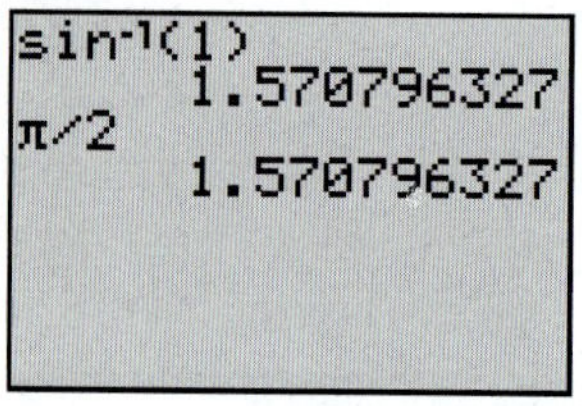

```
sin⁻¹(1)
          1.570796327
π/2
          1.570796327
```

✔ CHECK: We can verify the solution by evaluating $\sin^{-1} 1$ with our graphing calculator in radian mode. See Figure 8.

For the remainder of the section, the reader is encouraged to verify the solutions obtained using a graphing utility.

➤ NOW WORK PROBLEM 1.

EXAMPLE 2   **Finding the Exact Value of an Inverse Sine Function**

Find the exact value of: $\sin^{-1}\left(-\dfrac{1}{2}\right)$

Solution   Let $\theta = \sin^{-1}\left(-\dfrac{1}{2}\right)$. We seek the angle θ, $-\dfrac{\pi}{2} \leq \theta \leq \dfrac{\pi}{2}$, whose sine equals $-\dfrac{1}{2}$.

$$\theta = \sin^{-1}\left(-\frac{1}{2}\right), \qquad -\frac{\pi}{2} \leq \theta \leq \frac{\pi}{2}$$

$$\sin \theta = -\frac{1}{2}, \qquad -\frac{\pi}{2} \leq \theta \leq \frac{\pi}{2}$$

(Refer to Table 1 and Figure 7, if necessary.) The only angle within the interval $\left[-\dfrac{\pi}{2}, \dfrac{\pi}{2}\right]$ whose sine is $-\dfrac{1}{2}$ is $-\dfrac{\pi}{6}$. So, since $\sin\left(-\dfrac{\pi}{6}\right) = -\dfrac{1}{2}$ and $-\dfrac{\pi}{6}$ is in the interval $\left[-\dfrac{\pi}{2}, \dfrac{\pi}{2}\right]$, we conclude that

$$\sin^{-1}\left(-\frac{1}{2}\right) = -\frac{\pi}{6}$$

➤ NOW WORK PROBLEM 7.

② For most numbers x, the value $y = \sin^{-1} x$ must be approximated.

EXAMPLE 3   **Finding an Approximate Value of an Inverse Sine Function**

Find an approximate value of:

(a) $\sin^{-1}\dfrac{1}{3}$ (b) $\sin^{-1}\left(-\dfrac{1}{4}\right)$

Express the answer in radians rounded to two decimal places.

Solution   Because we want the angle measured in radians, we first set the mode to radians.

(a) Figure 9(a) shows the solution using a TI-83 graphing calculator.

(b) Figure 9(b) shows the solution using a TI-83 graphing calculator.

Figure 9(a)

Figure 9(b)

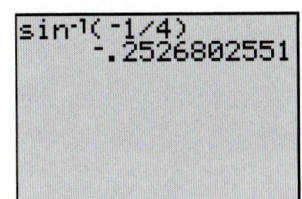

We have $\sin^{-1}\dfrac{1}{3} = 0.34$, rounded to two decimal places.

We have $\sin^{-1}\left(-\dfrac{1}{4}\right) = -0.25$, rounded to two decimal places.

━━━━━━━━━━ **NOW WORK PROBLEM 13.**

The Inverse Cosine Function

In Figure 10 we reproduce the graph of $y = \cos x$. Because every horizontal line $y = b$, where b is between -1 and 1, intersects the graph of $y = \cos x$ infinitely many times, it follows that the cosine function is not one-to-one.

Figure 10
$y = \cos x$,
$-\infty < x < \infty, -1 \le y \le 1$

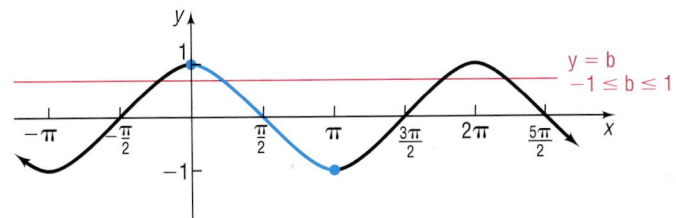

However, if we restrict the domain of $y = \cos x$ to the interval $[0, \pi]$, the restricted function

$$y = \cos x, \qquad 0 \le x \le \pi$$

Figure 11
$y = \cos x, 0 \le x \le \pi, -1 \le y \le 1$

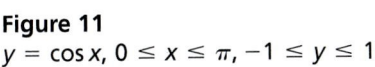
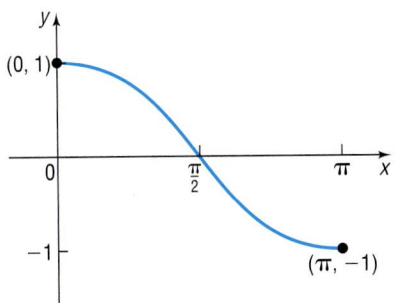

is one-to-one and hence will have an inverse function.* See Figure 11.

An equation for the inverse of $y = f(x) = \cos x$ is obtained by interchanging x and y. The implicit form of the inverse function is $x = \cos y$, $0 \le y \le \pi$. The explicit form is called the **inverse cosine** of x and is symbolized by $y = f^{-1}(x) = \cos^{-1} x$ (or by $y = \arccos x$).

$$y = \cos^{-1} x \quad \text{means} \quad x = \cos y \qquad (3)$$
$$\text{where} \quad -1 \le x \le 1 \quad \text{and} \quad 0 \le y \le \pi$$

Here, y is the angle whose cosine is x. The domain of the function $y = \cos^{-1} x$ is $-1 \le x \le 1$, and its range is $0 \le y \le \pi$. (Do you know why?) The graph of $y = \cos^{-1} x$ can be obtained by reflecting the restricted portion of the graph of $y = \cos x$ about the line $y = x$, as shown in Figure 12(a). Figure 12(b) shows the graph using a graphing utility.

*This is the generally accepted restriction.

Figure 12
$y = \cos^{-1} x,$
$-1 \leq x \leq 1, 0 \leq y \leq \pi$

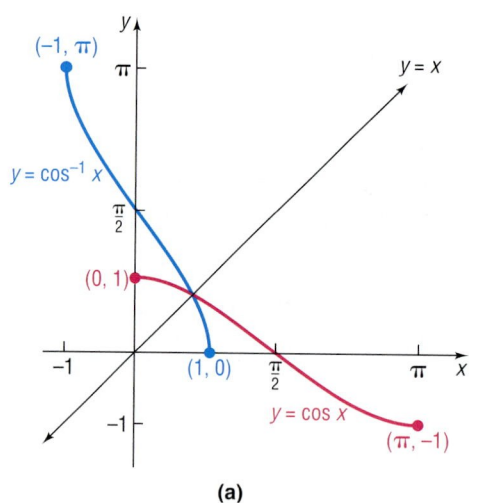

(a)

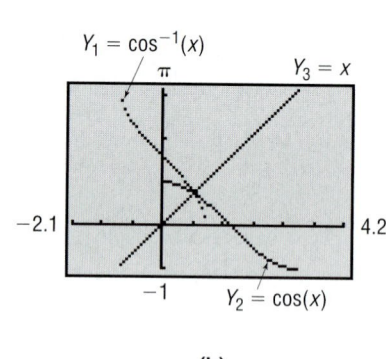

(b)

EXAMPLE 4 **Finding the Exact Value of an Inverse Cosine Function**

Find the exact value of: $\cos^{-1} 0$

Solution Let $\theta = \cos^{-1} 0$. We seek the angle θ, $0 \leq \theta \leq \pi$, whose cosine equals 0.

$$\theta = \cos^{-1} 0, \qquad 0 \leq \theta \leq \pi$$
$$\cos \theta = 0, \qquad 0 \leq \theta \leq \pi$$

Look at Table 2 and Figure 13.

Figure 13

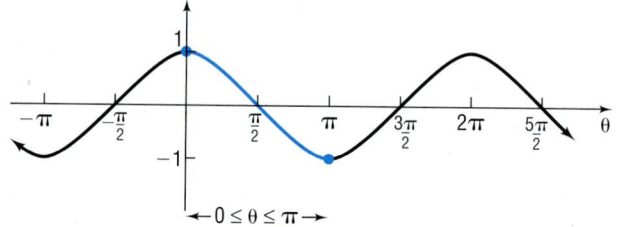

We see that the only angle θ within the interval $[0, \pi]$ whose cosine is 0 is $\dfrac{\pi}{2}$. [Note that $\cos \dfrac{3\pi}{2}$ also equals 0, but $\dfrac{3\pi}{2}$ lies outside the interval $[0, \pi]$ and hence is not admissible.] Since $\cos \dfrac{\pi}{2} = 0$ and $\dfrac{\pi}{2}$ is in the interval $[0, \pi]$, we conclude that

$$\cos^{-1} 0 = \frac{\pi}{2}$$

TABLE 2	
θ	$\cos \theta$
0	1
$\dfrac{\pi}{6}$	$\dfrac{\sqrt{3}}{2}$
$\dfrac{\pi}{4}$	$\dfrac{\sqrt{2}}{2}$
$\dfrac{\pi}{3}$	$\dfrac{1}{2}$
$\dfrac{\pi}{2}$	0
$\dfrac{2\pi}{3}$	$-\dfrac{1}{2}$
$\dfrac{3\pi}{4}$	$-\dfrac{\sqrt{2}}{2}$
$\dfrac{5\pi}{6}$	$-\dfrac{\sqrt{3}}{2}$
π	-1

EXAMPLE 5 **Finding the Exact Value of an Inverse Cosine Function**

Find the exact value of: $\cos^{-1} \dfrac{\sqrt{2}}{2}$

Solution   Let $\theta = \cos^{-1}\dfrac{\sqrt{2}}{2}$. We seek the angle θ, $0 \le \theta \le \pi$, whose cosine equals $\dfrac{\sqrt{2}}{2}$.

$$\theta = \cos^{-1}\frac{\sqrt{2}}{2}, \qquad 0 \le \theta \le \pi$$

$$\cos\theta = \frac{\sqrt{2}}{2}, \qquad 0 \le \theta \le \pi$$

Look at Table 2 and Figure 14.

Figure 14

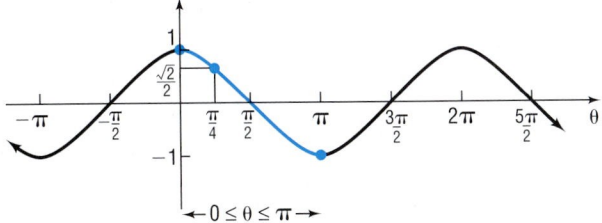

We see that the only angle θ within the interval $[0, \pi]$, whose cosine is $\dfrac{\sqrt{2}}{2}$ is $\dfrac{\pi}{4}$. Since $\cos\dfrac{\pi}{4} = \dfrac{\sqrt{2}}{2}$ and $\dfrac{\pi}{4}$ is in the interval $[0, \pi]$, we conclude that

$$\cos^{-1}\frac{\sqrt{2}}{2} = \frac{\pi}{4}$$

━ **NOW WORK PROBLEM 11.**

For the cosine function $f(x) = \cos x$ and its inverse $f^{-1}(x) = \cos^{-1} x$, the following properties hold:

$$f^{-1}(f(x)) = \cos^{-1}(\cos x) = x, \quad \text{where } 0 \le x \le \pi \qquad \text{(4a)}$$
$$f(f^{-1}(x)) = \cos(\cos^{-1} x) = x, \quad \text{where } -1 \le x \le 1 \qquad \text{(4b)}$$

EXAMPLE 6   **Finding the Exact Value of a Composite Function**

Find the exact value of: (a) $\cos^{-1}\left[\cos\left(\dfrac{\pi}{12}\right)\right]$ (b) $\cos[\cos^{-1}(-0.4)]$

Solution   (a) $\cos^{-1}\left[\cos\left(\dfrac{\pi}{12}\right)\right] = \dfrac{\pi}{12}$ By Property (4a).

(b) $\cos[\cos^{-1}(-0.4)] = -0.4$ By Property (4b).

━ **NOW WORK PROBLEM 27.**

The Inverse Tangent Function

In Figure 15 we reproduce the graph of $y = \tan x$. Because every horizontal line intersects the graph infinitely many times, it follows that the tangent function is not one-to-one.

Figure 15

$y = \tan x, -\infty < x < \infty,$
x not equal to odd multiples
of $\dfrac{\pi}{2}, -\infty < y < \infty$

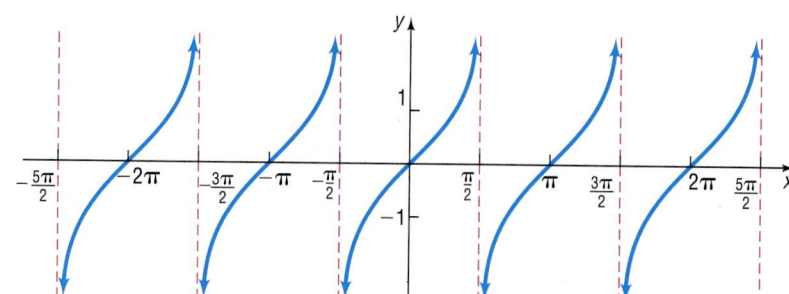

However, if we restrict the domain of $y = \tan x$ to the interval $\left(-\dfrac{\pi}{2}, \dfrac{\pi}{2}\right)$, the restricted function

$$y = \tan x, \qquad -\dfrac{\pi}{2} < x < \dfrac{\pi}{2}$$

is one-to-one and hence has an inverse function.* See Figure 16.

An equation for the inverse of $y = f(x) = \tan x$ is obtained by interchanging x and y. The implicit form of the inverse function is $x = \tan y$, $-\dfrac{\pi}{2} < y < \dfrac{\pi}{2}$. The explicit form is called the **inverse tangent** of x and is symbolized by $y = f^{-1}(x) = \tan^{-1} x$ (or by $y = \arctan x$).

Figure 16

$y = \tan x,$

$-\dfrac{\pi}{2} < x < \dfrac{\pi}{2}, -\infty < y < \infty$

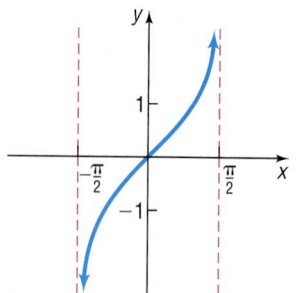

$$y = \tan^{-1} x \quad \text{means} \quad x = \tan y \qquad (5)$$
$$\text{where} \quad -\infty < x < \infty \quad \text{and} \quad -\dfrac{\pi}{2} < y < \dfrac{\pi}{2}$$

Here, y is the angle whose tangent is x. The domain of the function $y = \tan^{-1} x$ is $-\infty < x < \infty$, and its range is $-\dfrac{\pi}{2} < y < \dfrac{\pi}{2}$. The graph of $y = \tan^{-1} x$ can be obtained by reflecting the restricted portion of the graph of $y = \tan x$ about the line $y = x$, as shown in Figure 17(a). Figure 17(b) shows the graph using a graphing utitity.

Figure 17
$y = \tan^{-1} x,$
$-\infty < x < \infty, -\dfrac{\pi}{2} < y < \dfrac{\pi}{2}$

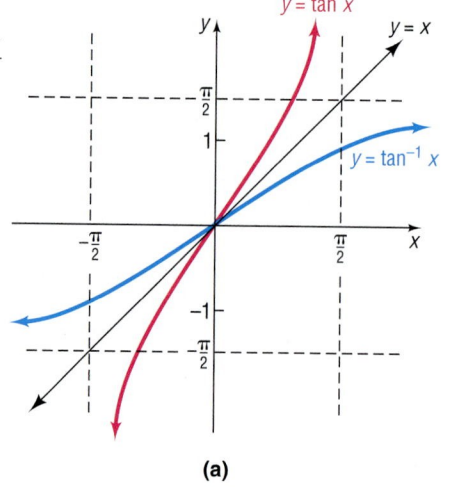

(a)

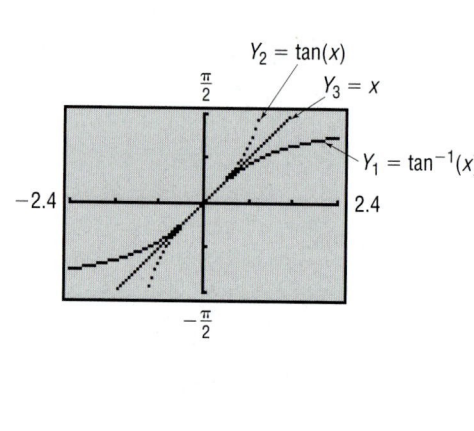

(b)

*This is the generally accepted restriction.

For the tangent function and its inverse, the following properties hold:

$$f^{-1}(f(x)) = \tan^{-1}(\tan x) = x, \quad \text{where } -\frac{\pi}{2} < x < \frac{\pi}{2}$$

$$f(f^{-1}(x)) = \tan(\tan^{-1} x) = x, \quad \text{where } -\infty < x < \infty$$

EXAMPLE 7   **Finding the Exact Value of an Inverse Tangent Function**

Find the exact value of: $\tan^{-1} 1$

Solution   Let $\theta = \tan^{-1} 1$. We seek the angle θ, $-\frac{\pi}{2} < \theta < \frac{\pi}{2}$, whose tangent equals 1.

$$\theta = \tan^{-1} 1, \quad -\frac{\pi}{2} < \theta < \frac{\pi}{2}$$

$$\tan \theta = 1, \quad -\frac{\pi}{2} < \theta < \frac{\pi}{2}$$

Look at Table 3 or Figure 16. The only angle θ within the interval $\left(-\frac{\pi}{2}, \frac{\pi}{2}\right)$ whose tangent is 1 is $\frac{\pi}{4}$. Since $\tan \frac{\pi}{4} = 1$ and $\frac{\pi}{4}$ is in the interval $\left(-\frac{\pi}{2}, \frac{\pi}{2}\right)$, we conclude that

$$\tan^{-1} 1 = \frac{\pi}{4}$$

TABLE 3

θ	$\tan \theta$
$-\dfrac{\pi}{2}$	Undefined
$-\dfrac{\pi}{3}$	$-\sqrt{3}$
$-\dfrac{\pi}{4}$	-1
$-\dfrac{\pi}{6}$	$-\dfrac{\sqrt{3}}{3}$
0	0
$\dfrac{\pi}{6}$	$\dfrac{\sqrt{3}}{3}$
$\dfrac{\pi}{4}$	1
$\dfrac{\pi}{3}$	$\sqrt{3}$
$\dfrac{\pi}{2}$	Undefined

EXAMPLE 8   **Finding the Exact Value of an Inverse Tangent Function**

Find the exact value of: $\tan^{-1}(-\sqrt{3})$

Solution   Let $\theta = \tan^{-1}(-\sqrt{3})$. We seek the angle θ, $-\frac{\pi}{2} < \theta < \frac{\pi}{2}$, whose tangent equals $-\sqrt{3}$.

$$\theta = \tan^{-1}(-\sqrt{3}), \quad -\frac{\pi}{2} < \theta < \frac{\pi}{2}$$

$$\tan \theta = -\sqrt{3}, \quad -\frac{\pi}{2} < \theta < \frac{\pi}{2}$$

Look at Table 3 or Figure 16 if necessary. The only angle θ within the interval $\left(-\frac{\pi}{2}, \frac{\pi}{2}\right)$ whose tangent is $-\sqrt{3}$ is $-\frac{\pi}{3}$. Since $\tan\left(-\frac{\pi}{3}\right) = -\sqrt{3}$ and $-\frac{\pi}{3}$ is in the interval $\left(-\frac{\pi}{2}, \frac{\pi}{2}\right)$, we conclude that

$$\tan^{-1}(-\sqrt{3}) = -\frac{\pi}{3}$$

NOW WORK PROBLEM 5.

9.1   Concepts and Vocabulary

In Problems 1–3, fill in the blanks.

1. $y = \sin^{-1} x$ means _____ , where $-1 \le x \le 1$ and $-\dfrac{\pi}{2} \le y \le \dfrac{\pi}{2}$.

2. The value of $\sin^{-1}\left[\cos \dfrac{\pi}{2}\right]$ is _____ .

3. $\cos^{-1}\left[\cos \dfrac{\pi}{5}\right] = $ _____ .

In Problems 4–6, answer True or False to each statement.

4. The domain of $y = \sin^{-1} x$ is $-\dfrac{\pi}{2} \le x \le \dfrac{\pi}{2}$.

5. $\cos(\sin^{-1} 0) = 1$ and $\sin(\cos^{-1} 0) = 1$.

6. $y = \tan^{-1} x$ means $x = \tan y$, where $-\infty < x < \infty$ and $-\dfrac{\pi}{2} < y < \dfrac{\pi}{2}$.

7. How would you quickly sketch the graph of $y = \cos^{-1} x$?
8. What restrictions are on x if $\sin^{-1}(\sin x) = x$?
9. What restrictions are on x if $\sin(\sin^{-1} x) = x$?
10. Explain how you would find the exact value of $\tan^{-1}(-1)$.

9.1   Exercises

In Problems 1–12, find the exact value of each expression. Verify your results using a graphing utility.

1. $\sin^{-1} 0$
2. $\cos^{-1} 1$
3. $\sin^{-1}(-1)$
4. $\cos^{-1}(-1)$

5. $\tan^{-1} 0$
6. $\tan^{-1}(-1)$
7. $\sin^{-1} \dfrac{\sqrt{2}}{2}$
8. $\tan^{-1} \dfrac{\sqrt{3}}{3}$

9. $\tan^{-1} \sqrt{3}$
10. $\sin^{-1}\left(-\dfrac{\sqrt{3}}{2}\right)$
11. $\cos^{-1}\left(-\dfrac{\sqrt{3}}{2}\right)$
12. $\sin^{-1}\left(-\dfrac{\sqrt{2}}{2}\right)$

In Problems 13–24, use a calculator to find the value of each expression rounded to two decimal places.

13. $\sin^{-1} 0.1$
14. $\cos^{-1} 0.6$
15. $\tan^{-1} 5$
16. $\tan^{-1} 0.2$

17. $\cos^{-1} \dfrac{7}{8}$
18. $\sin^{-1} \dfrac{1}{8}$
19. $\tan^{-1}(-0.4)$
20. $\tan^{-1}(-3)$

21. $\sin^{-1}(-0.12)$
22. $\cos^{-1}(-0.44)$
23. $\cos^{-1} \dfrac{\sqrt{2}}{3}$
24. $\sin^{-1} \dfrac{\sqrt{3}}{5}$

In Problems 25–32, find the exact value of the expression. Do not use a calculator.

25. $\sin[\sin^{-1}(0.54)]$
26. $\tan[\tan^{-1}(7.4)]$
27. $\cos^{-1}\left[\cos\left(\dfrac{4\pi}{5}\right)\right]$
28. $\sin^{-1}\left[\sin\left(-\dfrac{\pi}{10}\right)\right]$

29. $\tan[\tan^{-1}(-3.5)]$
30. $\cos[\cos^{-1}(-0.05)]$
31. $\sin^{-1}\left[\sin\left(-\dfrac{3\pi}{7}\right)\right]$
32. $\tan^{-1}\left[\tan\left(\dfrac{2\pi}{5}\right)\right]$

In Problems 33–44, do not use a calculator.

33. Does $\sin^{-1}\left[\sin\left(-\dfrac{\pi}{6}\right)\right] = -\dfrac{\pi}{6}$? Why or why not?

34. Does $\sin^{-1}\left[\sin\left(\dfrac{2\pi}{3}\right)\right] = \dfrac{2\pi}{3}$? Why or why not?

35. Does $\sin[\sin^{-1}(2)] = 2$? Why or why not?

36. Does $\sin\left[\sin^{-1}\left(-\dfrac{1}{2}\right)\right] = -\dfrac{1}{2}$? Why or why not?

37. Does $\cos^{-1}\left[\cos\left(-\dfrac{\pi}{6}\right)\right] = -\dfrac{\pi}{6}$? Why or why not?

38. Does $\cos^{-1}\left[\cos\left(\dfrac{2\pi}{3}\right)\right] = \dfrac{2\pi}{3}$? Why or why not?

39. Does $\cos\left[\cos^{-1}\left(-\dfrac{1}{2}\right)\right] = -\dfrac{1}{2}$? Why or why not?

40. Does $\cos[\cos^{-1}(2)] = 2$? Why or why not?

41. Does $\tan^{-1}\left[\tan\left(-\dfrac{\pi}{3}\right)\right] = -\dfrac{\pi}{3}$? Why or why not?

42. Does $\tan^{-1}\left[\tan\left(\dfrac{2\pi}{3}\right)\right] = \dfrac{2\pi}{3}$? Why or why not?

43. Does $\tan[\tan^{-1}(2)] = 2$? Why or why not?

44. Does $\tan\left[\tan^{-1}\left(-\dfrac{1}{2}\right)\right] = -\dfrac{1}{2}$? Why or why not?

In Problems 45–50, use the following: The formula

$$D = 24\left[1 - \frac{\cos^{-1}(\tan i \tan \theta)}{\pi}\right]$$

can be used to approximate the number of hours of daylight when the declination of the Sun is $i°$ at a location $\theta°$ north latitude for any date between the vernal equinox and autumnal equinox. The declination of the Sun is defined as the angle i between the equatorial plane and any ray of light from the Sun. The latitude of a location is the angle θ formed by a ray from the center of Earth to the Equator and a ray drawn from the center of Earth to the location. See the figure. To use the formula, $\cos^{-1}(\tan i \tan \theta)$ must be expressed in radians.

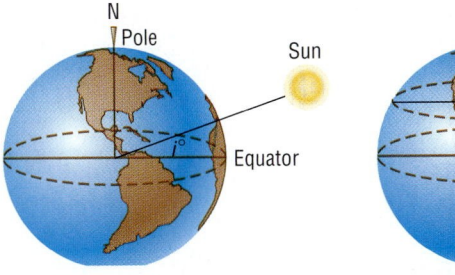 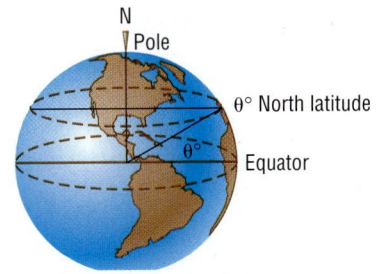

45. Approximate the number of hours of daylight in Houston, Texas (29°45′ north latitude), for the following dates:
  (a) Summer solstice ($i = 23.5°$)
  (b) Vernal equinox ($i = 0°$)
  (c) July 4 ($i = 22°48′$)

46. Approximate the number of hours of daylight in New York, New York (40°45′ north latitude), for the following dates:
  (a) Summer solstice ($i = 23.5°$)
  (b) Vernal equinox ($i = 0°$)
  (c) July 4 ($i = 22°48′$)

47. Approximate the number of hours of daylight in Honolulu, Hawaii (21°18′ north latitude), for the following dates:
  (a) Summer solstice ($i = 23.5°$)
  (b) Vernal equinox ($i = 0°$)
  (c) July 4 ($i = 22°48′$)

48. Approximate the number of hours of daylight in Anchorage, Alaska (61°10′ north latitude), for the following dates:
  (a) Summer solstice ($i = 23.5°$)
  (b) Vernal equinox ($i = 0°$)
  (c) July 4 ($i = 22°48′$)

49. Approximate the number of hours of daylight at the Equator (0° north latitude) for the following dates:
    (a) Summer solstice ($i = 23.5°$)
    (b) Vernal equinox ($i = 0°$)
    (c) July 4 ($i = 22°48'$)
    (d) What do you conclude about the number of hours of daylight throughout the year for a location at the Equator?

50. Approximate the number of hours of daylight for any location that is 66°30' north latitude for the following dates:
    (a) Summer solstice ($i = 23.5°$)
    (b) Vernal equinox ($i = 0°$)
    (c) July 4 ($i = 22°48'$)
    (d) The number of hours of daylight on the winter solstice may be found by computing the number of hours of daylight on the summer solstice and subtracting this result from 24 hours, due to the symmetry of the orbital path of Earth around the Sun. Compute the number of hours of daylight for this location on the winter solstice. What do you conclude about daylight for a location at 66°30' north latitude?

51. **Being the First to See the Rising Sun** Cadillac Mountain, elevation 1530 feet, is located in Acadia National Park, Maine, and is the highest peak on the east coast of the United States. It is said that a person standing on the summit will be the first person in the United States to see the rays of the rising Sun. How much sooner would a person atop Cadillac Mountain see the first rays than a person standing below, at sea level?

[**Hint:** Consult the figure. When the person at D sees the first rays of the Sun, the person at P does not. The person at P sees the first rays of the Sun only after Earth has rotated so that P is at location Q. Compute the length of the arc subtended by the central angle θ. Then use the fact that, at the latitude of Cadillac Mountain, in 24 hours a length of $2\pi (2710)$ miles is subtended, and find the time it takes to subtend this length.]

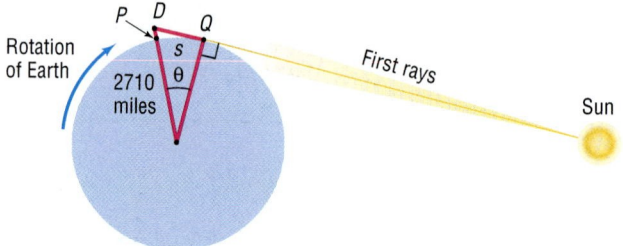

PREPARING FOR THIS SECTION

Before getting started, review the following:

✓ Finding Exact Values Given the Value of a Trigonometric Function and the Quadrant of the Angle (Section 8.4, pp. 651–652)

✓ Graphs of the Secant, Cosecant, and Cotangent Functions (Section 8.7, pp. 685–687)

✓ Domain and Range of the Secant, Cosecant, and Cotangent Functions (Section 8.5, pp. 658–660)

9.2 THE INVERSE TRIGONOMETRIC FUNCTIONS [CONTINUED]

OBJECTIVES

1 Find the Exact Value of Expressions Involving the Inverse Sine, Cosine, and Tangent Functions

2 Find the Exact Value of the Inverse Secant, Cosecant, and Cotangent Functions

3 Use a Calculator to Evaluate $\sec^{-1} x$, $\csc^{-1} x$, and $\cot^{-1} x$

1 In this section we continue our discussion of the inverse trigonometric functions.

EXAMPLE 1   **Finding the Exact Value of Expressions Involving Inverse Trigonometric Functions**

Find the exact value of: $\sin^{-1}\left(\sin \dfrac{5\pi}{4}\right)$

Solution $\sin^{-1}\left(\sin\dfrac{5\pi}{4}\right) = \sin^{-1}\left(-\dfrac{\sqrt{2}}{2}\right) = -\dfrac{\pi}{4}$

Figure 18

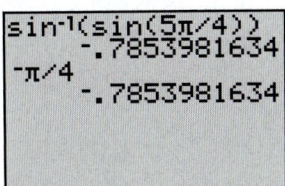

✔ CHECK: We can verify the solution by evaluating $\sin^{-1}\left(\sin\dfrac{5\pi}{4}\right)$ with our graphing calculator in radian mode. See Figure 18. ■ ■

Notice in the solution to Example 1 that we did not use Property (2a), page 712. This is because the argument of the sine function is not in the interval $\left[-\dfrac{\pi}{2}, \dfrac{\pi}{2}\right]$, as required. If we use the fact that

$$\sin\dfrac{5\pi}{4} = -\sin\dfrac{\pi}{4} \underset{\substack{\uparrow \\ y = \sin x \text{ is odd}}}{=} \sin\left(-\dfrac{\pi}{4}\right)$$

then we can use Property (2a):

$$\sin^{-1}\left(\sin\dfrac{5\pi}{4}\right) = \sin^{-1}\left[\sin\left(-\dfrac{\pi}{4}\right)\right] \underset{\substack{\uparrow \\ \text{Property (2a)}}}{=} -\dfrac{\pi}{4}$$

For the remainder of the section, the reader is encouraged to verify the solutions obtained using a graphing utility.

✎ **NOW WORK PROBLEM 13.**

EXAMPLE 2 **Finding the Exact Value of Expressions Involving Inverse Trigonometric Functions**

Find the exact value of: $\sin\left(\tan^{-1}\dfrac{1}{2}\right)$

Solution Let $\theta = \tan^{-1}\dfrac{1}{2}$. Then $\tan\theta = \dfrac{1}{2}$, where $-\dfrac{\pi}{2} < \theta < \dfrac{\pi}{2}$. Because $\tan\theta > 0$, it follows that $0 < \theta < \dfrac{\pi}{2}$, so θ lies in quadrant I. Now, in Figure 19 we draw a triangle in quadrant I depicting $\tan\theta = \dfrac{1}{2}$. The hypotenuse of this triangle is found to be $\sqrt{5}$. Then $\sin\theta = \dfrac{1}{\sqrt{5}}$, and

$$\sin\left(\tan^{-1}\dfrac{1}{2}\right) = \sin\theta = \dfrac{1}{\sqrt{5}} = \dfrac{\sqrt{5}}{5}$$ ■

Figure 19

$\tan\theta = \dfrac{1}{2}$

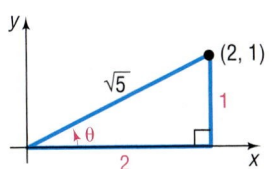

EXAMPLE 3 **Finding the Exact Value of Expressions Involving Inverse Trigonometric Functions**

Find the exact value of: $\cos\left[\sin^{-1}\left(-\dfrac{1}{3}\right)\right]$

Figure 20

$\sin \theta = -\dfrac{1}{3}$

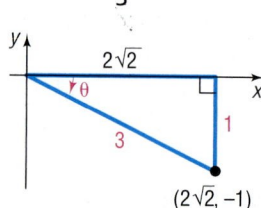

$(2\sqrt{2}, -1)$

Solution Let $\theta = \sin^{-1}\left(-\dfrac{1}{3}\right)$. Then $\sin \theta = -\dfrac{1}{3}$ and $-\dfrac{\pi}{2} \leq \theta \leq \dfrac{\pi}{2}$. Because

$\sin \theta < 0$, it follows that $-\dfrac{\pi}{2} \leq \theta < 0$, so θ lies in quadrant IV. Figure 20

illustrates $\sin \theta = -\dfrac{1}{3}$ for θ in quadrant IV. Then

$$\cos\left[\sin^{-1}\left(-\dfrac{1}{3}\right)\right] = \cos \theta = \dfrac{2\sqrt{2}}{3}$$ ■

EXAMPLE 4 **Finding the Exact Value of Expressions Involving Inverse Trigonometric Functions**

Find the exact value of: $\tan\left[\cos^{-1}\left(-\dfrac{1}{3}\right)\right]$

Figure 21

$\cos \theta = -\dfrac{1}{3}$

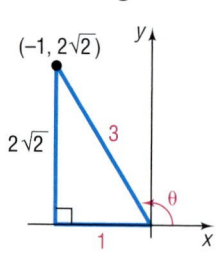

Solution Let $\theta = \cos^{-1}\left(-\dfrac{1}{3}\right)$. Then $\cos \theta = -\dfrac{1}{3}$ and $0 \leq \theta \leq \pi$. Because $\cos \theta < 0$,

it follows that $\dfrac{\pi}{2} < \theta \leq \pi$, so θ lies in quadrant II. Figure 21 illustrates

$\cos \theta = -\dfrac{1}{3}$ for θ in quadrant II. Then

$$\tan\left[\cos^{-1}\left(-\dfrac{1}{3}\right)\right] = \tan \theta = \dfrac{2\sqrt{2}}{-1} = -2\sqrt{2}$$ ■

NOW WORK PROBLEMS 1 AND 19.

The Remaining Inverse Trigonometric Functions

2 The inverse secant, inverse cosecant, and inverse cotangent functions are defined as follows:

$$y = \sec^{-1} x \quad \text{means} \quad x = \sec y \qquad (1)$$

$$\text{where} \quad |x| \geq 1 \quad \text{and} \quad 0 \leq y \leq \pi, \quad y \neq \dfrac{\pi}{2} \, ^*$$

$$y = \csc^{-1} x \quad \text{means} \quad x = \csc y \qquad (2)$$

$$\text{where} \quad |x| \geq 1 \quad \text{and} \quad -\dfrac{\pi}{2} \leq y \leq \dfrac{\pi}{2}, \quad y \neq 0 \, ^\dagger$$

$$y = \cot^{-1} x \quad \text{means} \quad x = \cot y \qquad (3)$$

$$\text{where} \quad -\infty < x < \infty \quad \text{and} \quad 0 < y < \pi$$

You are encouraged to review the graphs of the cotangent, cosecant, and secant functions in Figures 101 (p. 685), 102 (p. 686), and 104 (p. 687) in Section 8.7 to help you to see the basis for these definitions.

*Most books use this definition. A few use the restriction $0 \leq y < \dfrac{\pi}{2}, \pi \leq y < \dfrac{3\pi}{2}$.

†Most books use this definition. A few use the restriction $-\pi < y \leq -\dfrac{\pi}{2}, 0 < y \leq \dfrac{\pi}{2}$.

EXAMPLE 5 **Finding the Exact Value of an Inverse Cosecant Function**

Find the exact value of: $\csc^{-1} 2$

Solution Let $\theta = \csc^{-1} 2$. We seek the angle θ, $-\dfrac{\pi}{2} \leq \theta \leq \dfrac{\pi}{2}, \theta \neq 0$, whose cosecant equals 2.

$$\theta = \csc^{-1} 2, \qquad -\frac{\pi}{2} \leq \theta \leq \frac{\pi}{2}, \quad \theta \neq 0$$

$$\csc \theta = 2, \qquad -\frac{\pi}{2} \leq \theta \leq \frac{\pi}{2}, \quad \theta \neq 0$$

The only angle θ in the interval $-\dfrac{\pi}{2} \leq \theta \leq \dfrac{\pi}{2}, \theta \neq 0$, whose cosecant is 2 is $\dfrac{\pi}{6}$, so $\csc^{-1} 2 = \dfrac{\pi}{6}$. ■

NOW WORK PROBLEM 31.

③ Most calculators do not have keys for evaluating the inverse cotangent, cosecant, and secant functions. The easiest way to evaluate them is to convert to an inverse trigonometric function whose range is the same as the one to be evaluated. In this regard, notice that $y = \cot^{-1} x$ and $y = \sec^{-1} x$ (except where undefined) each have the same range as $y = \cos^{-1} x$; also, $y = \csc^{-1} x$, except where undefined, has the same range as $y = \sin^{-1} x$.

EXAMPLE 6 **Approximating the Value of Inverse Trigonometric Functions**

Use a calculator to approximate each expression in radians rounded to two decimal places.

(a) $\sec^{-1} 3$ (b) $\csc^{-1}(-4)$ (c) $\cot^{-1} \dfrac{1}{2}$ (d) $\cot^{-1}(-2)$

Solution First, set your calculator to radian mode.

(a) Let $\theta = \sec^{-1} 3$. Then $\sec \theta = 3$ and $0 \leq \theta \leq \pi, \theta \neq \dfrac{\pi}{2}$. Since $\sec \theta = \dfrac{1}{\cos \theta} = 3$, we have $\cos \theta = \dfrac{1}{3}$, so that $\theta = \cos^{-1}\left(\dfrac{1}{3}\right)$. Therefore,

$$\sec^{-1} 3 = \theta = \cos^{-1} \frac{1}{3} \approx 1.23$$

↑ Use a calculator.

(b) Let $\theta = \csc^{-1}(-4)$. Then $\csc \theta = -4, -\dfrac{\pi}{2} \leq \theta \leq \dfrac{\pi}{2}, \theta \neq 0$. Since $\sin \theta = -\dfrac{1}{4}$ and $\theta = \sin^{-1}\left(-\dfrac{1}{4}\right)$, we have

$$\csc^{-1}(-4) = \theta = \sin^{-1}\left(-\frac{1}{4}\right) \approx -0.25$$

Figure 22
$\cot \theta = \dfrac{1}{2}, 0 < \theta < \pi$

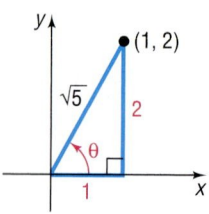

Figure 23
$\cot \theta = -2, 0 < \theta < \pi$

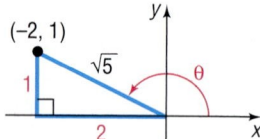

(c) Let $\theta = \cot^{-1} \dfrac{1}{2}$. Then $\cot \theta = \dfrac{1}{2}, 0 < \theta < \pi$. From these facts we know that θ lies in quadrant I. We draw Figure 22 to help us to find $\cos \theta$. Then $\cos \theta = \dfrac{1}{\sqrt{5}}$, so $\theta = \cos^{-1}\left(\dfrac{1}{\sqrt{5}}\right)$, and

$$\cot^{-1} \dfrac{1}{2} = \theta = \cos^{-1}\left(\dfrac{1}{\sqrt{5}}\right) \approx 1.11$$

(d) Let $\theta = \cot^{-1}(-2)$. Then $\cot \theta = -2, 0 < \theta < \pi$. From these facts we know that θ lies in quadrant II. We draw Figure 23 to help us to find $\cos \theta$. Then $\cos \theta = -\dfrac{2}{\sqrt{5}}$, so $\theta = \cos^{-1}\left(-\dfrac{2}{\sqrt{5}}\right)$, and

$$\cot^{-1}(-2) = \theta = \cos^{-1}\left(-\dfrac{2}{\sqrt{5}}\right) \approx 2.68$$

NOW WORK PROBLEM 37.

9.2 Concepts and Vocabulary

In Problems 1 and 2, fill in the blanks.

1. $y = \sec^{-1} x$ means _____, where $|x|$ _____ and _____ $\le y \le$ _____, $y \ne \dfrac{\pi}{2}$.

2. $\cos(\tan^{-1} 1) =$ _____.

In Problems 3–5, answer True or False to each statement.

3. It is impossible to obtain exact values for the inverse secant function.

4. $\csc^{-1} 0.5$ is not defined.

5. The domain of the inverse cotangent function is the set of real numbers.

6. Explain how you would evaluate $\sec^{-1} 2$ using a calculator.

7. Explain how you would evaluate $\cot^{-1}(-4)$ using a calculator.

9.2 Exercises

In Problems 1–28, find the exact value of each expression. Verify your results using a graphing utility.

1. $\cos\left(\sin^{-1} \dfrac{\sqrt{2}}{2}\right)$

2. $\sin\left(\cos^{-1} \dfrac{1}{2}\right)$

3. $\tan\left[\cos^{-1}\left(-\dfrac{\sqrt{3}}{2}\right)\right]$

4. $\tan\left[\sin^{-1}\left(-\dfrac{1}{2}\right)\right]$

5. $\sec\left(\cos^{-1} \dfrac{1}{2}\right)$

6. $\cot\left[\sin^{-1}\left(-\dfrac{1}{2}\right)\right]$

7. $\csc(\tan^{-1} 1)$

8. $\sec(\tan^{-1} \sqrt{3})$

9. $\sin[\tan^{-1}(-1)]$

10. $\cos\left[\sin^{-1}\left(-\dfrac{\sqrt{3}}{2}\right)\right]$

11. $\sec\left[\sin^{-1}\left(-\dfrac{1}{2}\right)\right]$

12. $\csc\left[\cos^{-1}\left(-\dfrac{\sqrt{3}}{2}\right)\right]$

13. $\cos^{-1}\left(\cos \dfrac{5\pi}{4}\right)$

14. $\tan^{-1}\left(\tan \dfrac{2\pi}{3}\right)$

15. $\sin^{-1}\left[\sin\left(-\dfrac{7\pi}{6}\right)\right]$

16. $\cos^{-1}\left[\cos\left(-\dfrac{\pi}{3}\right)\right]$

17. $\tan\left(\sin^{-1} \dfrac{1}{3}\right)$

18. $\tan\left(\cos^{-1} \dfrac{1}{3}\right)$

19. $\sec\left(\tan^{-1} \dfrac{1}{2}\right)$

20. $\cos\left(\sin^{-1} \dfrac{\sqrt{2}}{3}\right)$

21. $\cot\left[\sin^{-1}\left(-\dfrac{\sqrt{2}}{3}\right)\right]$ **22.** $\csc[\tan^{-1}(-2)]$ **23.** $\sin[\tan^{-1}(-3)]$ **24.** $\cot\left[\cos^{-1}\left(-\dfrac{\sqrt{3}}{3}\right)\right]$

25. $\sec\left(\sin^{-1}\dfrac{2\sqrt{5}}{5}\right)$ **26.** $\csc\left(\tan^{-1}\dfrac{1}{2}\right)$ **27.** $\sin^{-1}\left(\cos\dfrac{3\pi}{4}\right)$ **28.** $\cos^{-1}\left(\sin\dfrac{7\pi}{6}\right)$

In Problems 29–36, find the exact value of each expression. Verify your results using a graphing utility.

29. $\cot^{-1}\sqrt{3}$ **30.** $\cot^{-1}1$ **31.** $\csc^{-1}(-1)$ **32.** $\csc^{-1}\sqrt{2}$

33. $\sec^{-1}\dfrac{2\sqrt{3}}{3}$ **34.** $\sec^{-1}(-2)$ **35.** $\cot^{-1}\left(-\dfrac{\sqrt{3}}{3}\right)$ **36.** $\csc^{-1}\left(-\dfrac{2\sqrt{3}}{3}\right)$

In Problems 37–48, use a calculator to find the value of each expression rounded to two decimal places.

37. $\sec^{-1}4$ **38.** $\csc^{-1}5$ **39.** $\cot^{-1}2$ **40.** $\sec^{-1}(-3)$

41. $\csc^{-1}(-3)$ **42.** $\cot^{-1}\left(-\dfrac{1}{2}\right)$ **43.** $\cot^{-1}\sqrt{5}$ **44.** $\cot^{-1}(-8.1)$

45. $\csc^{-1}\left(-\dfrac{3}{2}\right)$ **46.** $\sec^{-1}\left(-\dfrac{4}{3}\right)$ **47.** $\cot^{-1}\left(-\dfrac{3}{2}\right)$ **48.** $\cot^{-1}(-\sqrt{10})$

---

49. Using a graphing utility, graph $y = \cot^{-1}x$.

50. Using a graphing utility, graph $y = \sec^{-1}x$.

51. Using a graphing utility, graph $y = \csc^{-1}x$.

52. Explain in your own words how you would use your calculator to find the value of $\cot^{-1}10$.

53. Consult three books on calculus and write down the definition in each of $y = \sec^{-1}x$ and $y = \csc^{-1}x$. Compare these with the definition given in this book.

PREPARING FOR THIS SECTION

Before getting started, review the following:

✓ Fundamental Identities (Section 8.2, p. 627)

9.3 TRIGONOMETRIC IDENTITIES

OBJECTIVES   **1**   Establish Identities

We saw in the previous chapter that the trigonometric functions lend themselves to a wide variety of identities. Before establishing some additional identities, let's review the definition of an *identity*.

> Two functions f and g are said to be **identically equal** if
>
> $$f(x) = g(x)$$
>
> for every value of x for which both functions are defined. Such an equation is referred to as an **identity**. An equation that is not an identity is called a **conditional equation**.

For example, the following are identities:

$$(x + 1)^2 = x^2 + 2x + 1 \qquad \sin^2 x + \cos^2 x = 1 \qquad \csc x = \frac{1}{\sin x}$$

The following are conditional equations:

$$2x + 5 = 0 \qquad \text{True only if } x = -\frac{5}{2}.$$

$$\sin x = 0 \qquad \text{True only if } x = k\pi, k \text{ an integer.}$$

$$\sin x = \cos x \qquad \text{True only if } x = \frac{\pi}{4} + k\pi, k \text{ an integer.}$$

The following boxes summarize the trigonometric identities that we have established thus far.

Quotient Identities

$$\tan \theta = \frac{\sin \theta}{\cos \theta} \qquad \cot \theta = \frac{\cos \theta}{\sin \theta}$$

Reciprocal Identities

$$\csc \theta = \frac{1}{\sin \theta} \qquad \sec \theta = \frac{1}{\cos \theta} \qquad \cot \theta = \frac{1}{\tan \theta}$$

Pythagorean Identities

$$\sin^2 \theta + \cos^2 \theta = 1 \qquad \tan^2 \theta + 1 = \sec^2 \theta$$
$$1 + \cot^2 \theta = \csc^2 \theta$$

Even–Odd Identities

$$\sin(-\theta) = -\sin \theta \qquad \cos(-\theta) = \cos \theta \qquad \tan(-\theta) = -\tan \theta$$
$$\csc(-\theta) = -\csc \theta \qquad \sec(-\theta) = \sec \theta \qquad \cot(-\theta) = -\cot \theta$$

This list of identities comprises what we shall refer to as the **basic trigonometric identities**. These identities should not merely be memorized, but should be *known* (just as you know your name rather than have it memorized). In fact, minor variations of a basic identity are often used. For example, we might want to use $\sin^2 \theta = 1 - \cos^2 \theta$ or $\cos^2 \theta = 1 - \sin^2 \theta$ instead of $\sin^2 \theta + \cos^2 \theta = 1$. For this reason, among others, you need to know these relationships and be quite comfortable with variations of them.

① In the examples that follow, the directions will read "Establish the identity . . . ." As you will see, this is accomplished by starting with one side of the given equation (usually the one containing the more complicated expression) and, using appropriate basic identities and algebraic manipulations, arriving at the other side. The selection of appropriate basic identities to obtain the desired result is learned only through experience and lots of practice.

EXAMPLE 1	Establishing an Identity

Establish the identity: $\csc\theta \cdot \tan\theta = \sec\theta$

Solution We start with the left side, because it contains the more complicated expression, and apply a reciprocal identity and a quotient identity.

$$\csc\theta \cdot \tan\theta = \frac{1}{\sin\theta} \cdot \frac{\sin\theta}{\cos\theta} = \frac{1}{\cos\theta} = \sec\theta$$

Having arrived at the right side, the identity is established. ■

COMMENT: A graphing utility can be used to provide evidence of an identity. For example, if we graph $Y_1 = \csc\theta \cdot \tan\theta$ and $Y_2 = \sec\theta$, the graphs appear to be the same. This provides evidence that $Y_1 = Y_2$. However, it does not prove their equality. A graphing utility *cannot be used to establish an identity*—identities must be established algebraically. ■

✏ **NOW WORK PROBLEM 1.**

EXAMPLE 2	Establishing an Identity

Establish the identity: $\sin^2(-\theta) + \cos^2(-\theta) = 1$

Solution We begin with the left side and apply even–odd identities.

$$\begin{aligned} \sin^2(-\theta) + \cos^2(-\theta) &= [\sin(-\theta)]^2 + [\cos(-\theta)]^2 \\ &= (-\sin\theta)^2 + (\cos\theta)^2 && \text{Even–odd identities.} \\ &= (\sin\theta)^2 + (\cos\theta)^2 \\ &= 1 && \text{Pythagorean Identity.} \end{aligned}$$
■

EXAMPLE 3	Establishing an Identity

Establish the identity: $\dfrac{\sin^2(-\theta) - \cos^2(-\theta)}{\sin(-\theta) - \cos(-\theta)} = \cos\theta - \sin\theta$

Solution We begin with two observations: The left side appears to contain the more complicated expression. Also, the left side contains expressions with the argument $-\theta$, whereas the right side contains expressions with the argument θ. We decide, therefore, to start with the left side and apply even–odd identities.

$$\begin{aligned} \frac{\sin^2(-\theta) - \cos^2(-\theta)}{\sin(-\theta) - \cos(-\theta)} &= \frac{[\sin(-\theta)]^2 - [\cos(-\theta)]^2}{\sin(-\theta) - \cos(-\theta)} \\ &= \frac{(-\sin\theta)^2 - (\cos\theta)^2}{-\sin\theta - \cos\theta} && \text{Even–odd identities.} \\ &= \frac{(\sin\theta)^2 - (\cos\theta)^2}{-\sin\theta - \cos\theta} && \text{Simplify.} \\ &= \frac{(\sin\theta - \cos\theta)(\sin\theta + \cos\theta)}{-(\sin\theta + \cos\theta)} && \text{Factor.} \\ &= \cos\theta - \sin\theta && \text{Cancel and simplify.} \end{aligned}$$
■

EXAMPLE 4 **Establishing an Identity**

Establish the identity: $\dfrac{1 + \tan \theta}{1 + \cot \theta} = \tan \theta$

Solution $\dfrac{1 + \tan \theta}{1 + \cot \theta} = \dfrac{1 + \tan \theta}{1 + \dfrac{1}{\tan \theta}} = \dfrac{1 + \tan \theta}{\dfrac{\tan \theta + 1}{\tan \theta}} = \dfrac{\tan \theta \, (1 + \tan \theta)}{\tan \theta + 1} = \tan \theta$

NOW WORK PROBLEM **9.**

When sums or differences of quotients appear, it is usually best to rewrite them as a single quotient, especially if the other side of the identity consists of only one term.

EXAMPLE 5 **Establishing an Identity**

Establish the identity: $\dfrac{\sin \theta}{1 + \cos \theta} + \dfrac{1 + \cos \theta}{\sin \theta} = 2 \csc \theta$

Solution The left side is more complicated, so we start with it and proceed to add.

$$\dfrac{\sin \theta}{1 + \cos \theta} + \dfrac{1 + \cos \theta}{\sin \theta} = \dfrac{\sin^2 \theta + (1 + \cos \theta)^2}{(1 + \cos \theta)(\sin \theta)} \qquad \text{Add the quotients.}$$

$$= \dfrac{\sin^2 \theta + 1 + 2 \cos \theta + \cos^2 \theta}{(1 + \cos \theta)(\sin \theta)} \qquad \begin{array}{l}\text{Remove parentheses} \\ \text{in numerator.}\end{array}$$

$$= \dfrac{(\sin^2 \theta + \cos^2 \theta) + 1 + 2 \cos \theta}{(1 + \cos \theta)(\sin \theta)} \qquad \text{Regroup.}$$

$$= \dfrac{2 + 2 \cos \theta}{(1 + \cos \theta)(\sin \theta)} \qquad \text{Pythagorean Identity.}$$

$$= \dfrac{2(1 + \cos \theta)}{(1 + \cos \theta)(\sin \theta)} \qquad \text{Factor and cancel.}$$

$$= \dfrac{2}{\sin \theta}$$

$$= 2 \csc \theta \qquad \text{Reciprocal Identity.}$$

NOW WORK PROBLEM **31.**

Sometimes it helps to write one side in terms of sines and cosines only.

EXAMPLE 6 **Establishing an Identity**

Establish the identity: $\dfrac{\tan \theta + \cot \theta}{\sec \theta \csc \theta} = 1$

Solution

$$\frac{\tan\theta + \cot\theta}{\sec\theta\csc\theta} \underset{\uparrow}{=} \frac{\dfrac{\sin\theta}{\cos\theta} + \dfrac{\cos\theta}{\sin\theta}}{\dfrac{1}{\cos\theta}\dfrac{1}{\sin\theta}} \underset{\uparrow}{=} \frac{\dfrac{\sin^2\theta + \cos^2\theta}{\cos\theta\sin\theta}}{\dfrac{1}{\cos\theta\sin\theta}}$$

Change to sines and cosines.   Add the quotients in the numerator.

$$\underset{\uparrow}{=} \frac{1}{\cos\theta\sin\theta} \cdot \frac{\cos\theta\sin\theta}{1} = 1$$

Divide quotient; $\sin^2\theta + \cos^2\theta = 1$

∎

NOW WORK PROBLEM **51.**

Sometimes, multiplying the numerator and denominator by an appropriate factor will result in a simplification.

EXAMPLE 7 **Establishing an Identity**

Establish the identity: $\dfrac{1 - \sin\theta}{\cos\theta} = \dfrac{\cos\theta}{1 + \sin\theta}$

Solution We start with the left side and multiply the numerator and the denominator by $1 + \sin\theta$. (Alternatively, we could multiply the numerator and denominator of the right side by $1 - \sin\theta$.)

$$\frac{1 - \sin\theta}{\cos\theta} = \frac{1 - \sin\theta}{\cos\theta} \cdot \frac{1 + \sin\theta}{1 + \sin\theta}$$ Multiply numerator and denominator by $1 + \sin\theta$.

$$= \frac{1 - \sin^2\theta}{\cos\theta(1 + \sin\theta)}$$

$$= \frac{\cos^2\theta}{\cos\theta(1 + \sin\theta)}$$ $1 - \sin^2\theta = \cos^2\theta$.

$$= \frac{\cos\theta}{1 + \sin\theta}$$ Cancel.

∎

NOW WORK PROBLEM **35.**

EXAMPLE 8 **Establishing an Identity Involving Inverse Trigonometric Functions**

Show that $\sin(\tan^{-1}v) = \dfrac{v}{\sqrt{1 + v^2}}$.

Solution Let $\theta = \tan^{-1}v$ so that $\tan\theta = v$, $-\dfrac{\pi}{2} < \theta < \dfrac{\pi}{2}$. As a result, we know that $\sec\theta > 0$.

$$\sin(\tan^{-1}v) = \sin\theta = \sin\theta \cdot \frac{\cos\theta}{\cos\theta} \underset{\uparrow}{=} \tan\theta\cos\theta = \frac{\tan\theta}{\sec\theta} \underset{\uparrow}{=} \frac{\tan\theta}{\sqrt{1 + \tan^2\theta}} = \frac{v}{\sqrt{1 + v^2}}$$

$\dfrac{\sin\theta}{\cos\theta} = \tan\theta$ $\sec^2\theta = 1 + \tan^2\theta$
$\sec\theta > 0$

∎

NOW WORK PROBLEM **81.**

Although a lot of practice is the only real way to learn how to establish identities, the following guidelines should prove helpful.

> **Guidelines for Establishing Identities**
>
> 1. It is almost always preferable to start with the side containing the more complicated expression.
> 2. Rewrite sums or differences of quotients as a single quotient.
> 3. Sometimes, rewriting one side in terms of sines and cosines only will help.
> 4. Always keep your goal in mind. As you manipulate one side of the expression, you must keep in mind the form of the expression on the other side.

WARNING: Be careful not to handle identities to be established as if they were conditional equations. You *cannot* establish an identity by such methods as adding the same expression to each side and obtaining a true statement. This practice is not allowed, because the original statement is precisely the one that you are trying to establish. You do not know until it has been established that it is, in fact, true. ■

9.3 Concepts and Vocabulary

In Problems 1–3, fill in the blanks.

1. Suppose that f and g are two functions with the same domain. If $f(x) = g(x)$ for every x in the domain, the equation is called a(n) _____ . Otherwise, it is called a(n) _____ equation.

2. $\tan^2 \theta - \sec^2 \theta =$ _____ .

3. $\cos(-\theta) - \cos \theta =$ _____ .

In Problems 4–6, answer True or False to each statement.

4. $\sin(-\theta) + \sin \theta = 0$ for any value of θ.

5. In establishing an identity, it is often easiest to just multiply both sides by a well-chosen nonzero expression involving the variable.

6. $[\tan \theta][\cos \theta] = \sin \theta$.

7. Write down the three Pythagorean Identities.

8. Why do you think it is usually preferable to start with the side containing the more complicated expression when establishing an identity?

9. Make up an identity that is not a Fundamental Identity.

9.3 Exercises

In Problems 1–80, establish each identity.

1. $\csc \theta \cdot \cos \theta = \cot \theta$
2. $\sec \theta \cdot \sin \theta = \tan \theta$
3. $1 + \tan^2(-\theta) = \sec^2 \theta$
4. $1 + \cot^2(-\theta) = \csc^2 \theta$
5. $\cos \theta(\tan \theta + \cot \theta) = \csc \theta$
6. $\sin \theta(\cot \theta + \tan \theta) = \sec \theta$
7. $\tan \theta \cot \theta - \cos^2 \theta = \sin^2 \theta$
8. $\sin \theta \csc \theta - \cos^2 \theta = \sin^2 \theta$
9. $(\sec \theta - 1)(\sec \theta + 1) = \tan^2 \theta$
10. $(\csc \theta - 1)(\csc \theta + 1) = \cot^2 \theta$
11. $(\sec \theta + \tan \theta)(\sec \theta - \tan \theta) = 1$
12. $(\csc \theta + \cot \theta)(\csc \theta - \cot \theta) = 1$
13. $\cos^2 \theta(1 + \tan^2 \theta) = 1$
14. $(1 - \cos^2 \theta)(1 + \cot^2 \theta) = 1$
15. $(\sin \theta + \cos \theta)^2 + (\sin \theta - \cos \theta)^2 = 2$
16. $\tan^2 \theta \cos^2 \theta + \cot^2 \theta \sin^2 \theta = 1$
17. $\sec^4 \theta - \sec^2 \theta = \tan^4 \theta + \tan^2 \theta$
18. $\csc^4 \theta - \csc^2 \theta = \cot^4 \theta + \cot^2 \theta$

19. $\sec \theta - \tan \theta = \dfrac{\cos \theta}{1 + \sin \theta}$

20. $\csc \theta - \cot \theta = \dfrac{\sin \theta}{1 + \cos \theta}$

21. $3 \sin^2 \theta + 4 \cos^2 \theta = 3 + \cos^2 \theta$

22. $9 \sec^2 \theta - 5 \tan^2 \theta = 5 + 4 \sec^2 \theta$

23. $1 - \dfrac{\cos^2 \theta}{1 + \sin \theta} = \sin \theta$

24. $1 - \dfrac{\sin^2 \theta}{1 - \cos \theta} = -\cos \theta$

25. $\dfrac{1 + \tan \theta}{1 - \tan \theta} = \dfrac{\cot \theta + 1}{\cot \theta - 1}$

26. $\dfrac{\csc \theta - 1}{\csc \theta + 1} = \dfrac{1 - \sin \theta}{1 + \sin \theta}$

27. $\dfrac{\sec \theta}{\csc \theta} + \dfrac{\sin \theta}{\cos \theta} = 2 \tan \theta$

28. $\dfrac{\csc \theta - 1}{\cot \theta} = \dfrac{\cot \theta}{\csc \theta + 1}$

29. $\dfrac{1 + \sin \theta}{1 - \sin \theta} = \dfrac{\csc \theta + 1}{\csc \theta - 1}$

30. $\dfrac{\cos \theta + 1}{\cos \theta - 1} = \dfrac{1 + \sec \theta}{1 - \sec \theta}$

31. $\dfrac{1 - \sin \theta}{\cos \theta} + \dfrac{\cos \theta}{1 - \sin \theta} = 2 \sec \theta$

32. $\dfrac{\cos \theta}{1 + \sin \theta} + \dfrac{1 + \sin \theta}{\cos \theta} = 2 \sec \theta$

33. $\dfrac{\sin \theta}{\sin \theta - \cos \theta} = \dfrac{1}{1 - \cot \theta}$

34. $1 - \dfrac{\sin^2 \theta}{1 + \cos \theta} = \cos \theta$

35. $\dfrac{1 - \sin \theta}{1 + \sin \theta} = (\sec \theta - \tan \theta)^2$

36. $\dfrac{1 - \cos \theta}{1 + \cos \theta} = (\csc \theta - \cot \theta)^2$

37. $\dfrac{\cos \theta}{1 - \tan \theta} + \dfrac{\sin \theta}{1 - \cot \theta} = \sin \theta + \cos \theta$

38. $\dfrac{\cot \theta}{1 - \tan \theta} + \dfrac{\tan \theta}{1 - \cot \theta} = 1 + \tan \theta + \cot \theta$

39. $\tan \theta + \dfrac{\cos \theta}{1 + \sin \theta} = \sec \theta$

40. $\dfrac{\sin \theta \cos \theta}{\cos^2 \theta - \sin^2 \theta} = \dfrac{\tan \theta}{1 - \tan^2 \theta}$

41. $\dfrac{\tan \theta + \sec \theta - 1}{\tan \theta - \sec \theta + 1} = \tan \theta + \sec \theta$

42. $\dfrac{\sin \theta - \cos \theta + 1}{\sin \theta + \cos \theta - 1} = \dfrac{\sin \theta + 1}{\cos \theta}$

43. $\dfrac{\tan \theta - \cot \theta}{\tan \theta + \cot \theta} = \sin^2 \theta - \cos^2 \theta$

44. $\dfrac{\sec \theta - \cos \theta}{\sec \theta + \cos \theta} = \dfrac{\sin^2 \theta}{1 + \cos^2 \theta}$

45. $\dfrac{\tan \theta - \cot \theta}{\tan \theta + \cot \theta} + 1 = 2 \sin^2 \theta$

46. $\dfrac{\tan \theta - \cot \theta}{\tan \theta + \cot \theta} + 2 \cos^2 \theta = 1$

47. $\dfrac{\sec \theta + \tan \theta}{\cot \theta + \cos \theta} = \tan \theta \sec \theta$

48. $\dfrac{\sec \theta}{1 + \sec \theta} = \dfrac{1 - \cos \theta}{\sin^2 \theta}$

49. $\dfrac{1 - \tan^2 \theta}{1 + \tan^2 \theta} + 1 = 2 \cos^2 \theta$

50. $\dfrac{1 - \cot^2 \theta}{1 + \cot^2 \theta} + 2 \cos^2 \theta = 1$

51. $\dfrac{\sec \theta - \csc \theta}{\sec \theta \csc \theta} = \sin \theta - \cos \theta$

52. $\dfrac{\sin^2 \theta - \tan \theta}{\cos^2 \theta - \cot \theta} = \tan^2 \theta$

53. $\sec \theta - \cos \theta - \sin \theta \tan \theta = 0$

54. $\tan \theta + \cot \theta - \sec \theta \csc \theta = 0$

55. $\dfrac{1}{1 - \sin \theta} + \dfrac{1}{1 + \sin \theta} = 2 \sec^2 \theta$

56. $\dfrac{1 + \sin \theta}{1 - \sin \theta} - \dfrac{1 - \sin \theta}{1 + \sin \theta} = 4 \tan \theta \sec \theta$

57. $\dfrac{\sec \theta}{1 - \sin \theta} = \dfrac{1 + \sin \theta}{\cos^3 \theta}$

58. $\dfrac{1 - \sin \theta}{1 + \sin \theta} = (\sec \theta - \tan \theta)^2$

59. $\dfrac{(\sec \theta - \tan \theta)^2 + 1}{\csc \theta (\sec \theta - \tan \theta)} = 2 \tan \theta$

60. $\dfrac{\sec^2 \theta - \tan^2 \theta + \tan \theta}{\sec \theta} = \sin \theta + \cos \theta$

61. $\dfrac{\sin \theta + \cos \theta}{\cos \theta} - \dfrac{\sin \theta - \cos \theta}{\sin \theta} = \sec \theta \csc \theta$

62. $\dfrac{\sin \theta + \cos \theta}{\sin \theta} - \dfrac{\cos \theta - \sin \theta}{\cos \theta} = \sec \theta \csc \theta$

63. $\dfrac{\sin^3 \theta + \cos^3 \theta}{\sin \theta + \cos \theta} = 1 - \sin \theta \cos \theta$

64. $\dfrac{\sin^3 \theta + \cos^3 \theta}{1 - 2 \cos^2 \theta} = \dfrac{\sec \theta - \sin \theta}{\tan \theta - 1}$

65. $\dfrac{\cos^2 \theta - \sin^2 \theta}{1 - \tan^2 \theta} = \cos^2 \theta$

66. $\dfrac{\cos \theta + \sin \theta - \sin^3 \theta}{\sin \theta} = \cot \theta + \cos^2 \theta$

67. $\dfrac{(2 \cos^2 \theta - 1)^2}{\cos^4 \theta - \sin^4 \theta} = 1 - 2 \sin^2 \theta$

68. $\dfrac{1 - 2 \cos^2 \theta}{\sin \theta \cos \theta} = \tan \theta - \cot \theta$

69. $\dfrac{1 + \sin \theta + \cos \theta}{1 + \sin \theta - \cos \theta} = \dfrac{1 + \cos \theta}{\sin \theta}$

70. $\dfrac{1 + \cos \theta + \sin \theta}{1 + \cos \theta - \sin \theta} = \sec \theta + \tan \theta$

71. $(a \sin \theta + b \cos \theta)^2 + (a \cos \theta - b \sin \theta)^2 = a^2 + b^2$

72. $(2a \sin \theta \cos \theta)^2 + a^2 (\cos^2 \theta - \sin^2 \theta)^2 = a^2$

73. $\dfrac{\tan \alpha + \tan \beta}{\cot \alpha + \cot \beta} = \tan \alpha \tan \beta$

74. $(\tan \alpha + \tan \beta)(1 - \cot \alpha \cot \beta) + (\cot \alpha + \cot \beta)(1 - \tan \alpha \tan \beta) = 0$

75. $(\sin \alpha + \cos \beta)^2 + (\cos \beta + \sin \alpha)(\cos \beta - \sin \alpha) = 2 \cos \beta (\sin \alpha + \cos \beta)$

76. $(\sin \alpha - \cos \beta)^2 + (\cos \beta + \sin \alpha)(\cos \beta - \sin \alpha) = -2 \cos \beta (\sin \alpha - \cos \beta)$

77. $\ln|\sec\theta| = -\ln|\cos\theta|$

78. $\ln|\tan\theta| = \ln|\sin\theta| - \ln|\cos\theta|$

79. $\ln|1 + \cos\theta| + \ln|1 - \cos\theta| = 2\ln|\sin\theta|$

80. $\ln|\sec\theta + \tan\theta| + \ln|\sec\theta - \tan\theta| = 0$

81. Show that $\sec(\tan^{-1}v) = \sqrt{1 + v^2}$.

82. Show that $\tan(\sin^{-1}v) = \dfrac{v}{\sqrt{1 - v^2}}$.

83. Show that $\tan(\cos^{-1}v) = \dfrac{\sqrt{1 - v^2}}{v}$.

84. Show that $\sin(\cos^{-1}v) = \sqrt{1 - v^2}$.

85. Show that $\cos(\sin^{-1}v) = \sqrt{1 - v^2}$.

86. Show that $\cos(\tan^{-1}v) = \dfrac{1}{\sqrt{1 + v^2}}$.

87. Write a few paragraphs outlining your strategy for establishing identities.

PREPARING FOR THIS SECTION

Before getting started, review the following:

✓ Distance Formula (Section 1.1, pp. 93–95)

✓ Values of the Trigonometric Functions of Certain Angles (Section 8.3, p. 638, and Section 8.4, p. 646)

9.4 SUM AND DIFFERENCE FORMULAS

OBJECTIVES

1. Use Sum and Difference Formulas to Find Exact Values
2. Use Sum and Difference Formulas to Establish Identities
3. Use Sum and Difference Formulas Involving Inverse Trigonometric Functions

In this section, we continue our derivation of trigonometric identities by obtaining formulas that involve the sum or difference of two angles, such as $\cos(\alpha + \beta)$, $\cos(\alpha - \beta)$, or $\sin(\alpha + \beta)$. These formulas are referred to as the **sum and difference formulas**. We begin with the formulas for $\cos(\alpha + \beta)$ and $\cos(\alpha - \beta)$.

Theorem

Sum and Difference Formulas for Cosines

$$\cos(\alpha + \beta) = \cos\alpha\cos\beta - \sin\alpha\sin\beta \qquad (1)$$

$$\cos(\alpha - \beta) = \cos\alpha\cos\beta + \sin\alpha\sin\beta \qquad (2)$$

In words, formula (1) states that the cosine of the sum of two angles equals the cosine of the first angle times the cosine of the second angle minus the sine of the first angle times the sine of the second angle.

Proof We will prove formula (2) first. Although this formula is true for all numbers α and β, we shall assume in our proof that $0 < \beta < \alpha < 2\pi$. We begin with the unit circle and place the angles α and β in standard position, as shown in Figure 24(a). The point P_1 lies on the terminal side of β, so its coordinates are $(\cos\beta, \sin\beta)$; and the point P_2 lies on the terminal side of α so its coordinates are $(\cos\alpha, \sin\alpha)$.

Figure 24

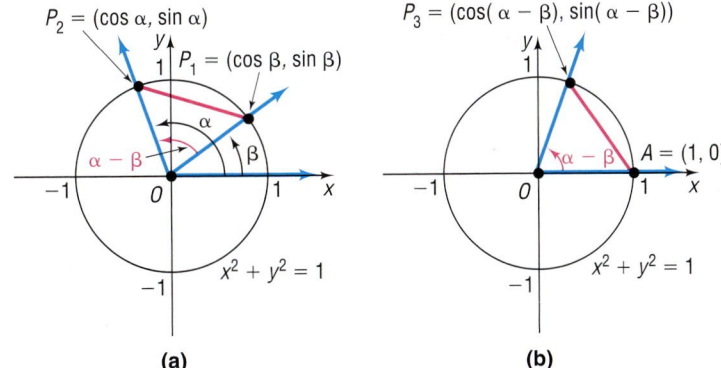

(a)

(b)

Now, place the angle $\alpha - \beta$ in standard position, as shown in Figure 24(b). The point A has coordinates $(1, 0)$, and the point P_3 is on the terminal side of the angle $\alpha - \beta$, so its coordinates are $(\cos(\alpha - \beta), \sin(\alpha - \beta))$.

Looking at triangle OP_1P_2 in Figure 24(a) and triangle OAP_3 in Figure 24(b), we see that these triangles are congruent. (Do you see why? Two sides and the included angle, $\alpha - \beta$, are equal.) As a result, the unknown side of each triangle must equal; that is,

$$d(A, P_3) = d(P_1, P_2)$$

Using the distance formula, we find that

$$\sqrt{[\cos(\alpha - \beta) - 1]^2 + [\sin(\alpha - \beta) - 0]^2} = \sqrt{(\cos \alpha - \cos \beta)^2 + (\sin \alpha - \sin \beta)^2} \quad \textcolor{gray}{d(A, P_3) = d(P_1, P_2).}$$

$$[\cos(\alpha - \beta) - 1]^2 + \sin^2(\alpha - \beta) = (\cos \alpha - \cos \beta)^2 + (\sin \alpha - \sin \beta)^2 \quad \textcolor{gray}{\text{Square both sides.}}$$

$$\cos^2(\alpha - \beta) - 2\cos(\alpha - \beta) + 1 + \sin^2(\alpha - \beta) = \cos^2 \alpha - 2\cos \alpha \cos \beta + \cos^2 \beta \quad \textcolor{gray}{\text{Multiply out the squared terms.}}$$

$$+ \sin^2 \alpha - 2 \sin \alpha \sin \beta + \sin^2 \beta$$

$$2 - 2\cos(\alpha - \beta) = 2 - 2\cos \alpha \cos \beta - 2 \sin \alpha \sin \beta \quad \textcolor{gray}{\text{Apply a Pythagorean Identity (3 times).}}$$

$$-2\cos(\alpha - \beta) = -2\cos \alpha \cos \beta - 2 \sin \alpha \sin \beta \quad \textcolor{gray}{\text{Subtract 2 from each side.}}$$

$$\cos(\alpha - \beta) = \cos \alpha \cos \beta + \sin \alpha \sin \beta \quad \textcolor{gray}{\text{Divide each side by } -2.}$$

which is formula (2).

The proof of formula (1) follows from formula (2) and the Even–Odd Identities. We use the fact that $\alpha + \beta = \alpha - (-\beta)$. Then

$$\cos(\alpha + \beta) = \cos[\alpha - (-\beta)]$$

$$= \cos \alpha \cos(-\beta) + \sin \alpha \sin(-\beta) \quad \textcolor{gray}{\text{Use formula (2).}}$$

$$= \cos \alpha \cos \beta - \sin \alpha \sin \beta \quad \textcolor{gray}{\text{Even–odd Identities}} \quad ∎$$

1 One use of formulas (1) and (2) is to obtain the exact value of the cosine of an angle that can be expressed as the sum or difference of angles whose sine and cosine are known exactly.

EXAMPLE 1   **Using the Sum Formula to Find Exact Values**

Find the exact value of $\cos 75°$.

Solution   Since $75° = 45° + 30°$, we use formula (1) to obtain

$$\cos 75° = \cos(45° + 30°) = \cos 45° \cos 30° - \sin 45° \sin 30°$$

↑
Formula (1)

$$= \frac{\sqrt{2}}{2} \cdot \frac{\sqrt{3}}{2} - \frac{\sqrt{2}}{2} \cdot \frac{1}{2} = \frac{1}{4}(\sqrt{6} - \sqrt{2})$$

■

EXAMPLE 2   **Using the Difference Formula to Find Exact Values**

Find the exact value of $\cos \dfrac{\pi}{12}$.

Solution

$$\cos \frac{\pi}{12} = \cos\left(\frac{3\pi}{12} - \frac{2\pi}{12} \right) = \cos\left(\frac{\pi}{4} - \frac{\pi}{6} \right)$$

$$= \cos \frac{\pi}{4} \cos \frac{\pi}{6} + \sin \frac{\pi}{4} \sin \frac{\pi}{6} \qquad \text{Use formula (2).}$$

$$= \frac{\sqrt{2}}{2} \cdot \frac{\sqrt{3}}{2} + \frac{\sqrt{2}}{2} \cdot \frac{1}{2} = \frac{1}{4}(\sqrt{6} + \sqrt{2})$$

■

➤ **NOW WORK PROBLEM 3.**

② Another use of formulas (1) and (2) is to establish other identities. One important pair of identities is given next.

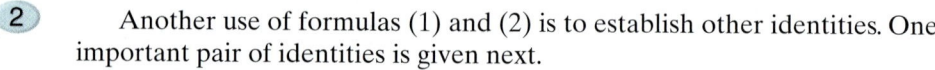

$$\cos\left(\frac{\pi}{2} - \theta \right) = \sin \theta \qquad \text{(3a)}$$

$$\sin\left(\frac{\pi}{2} - \theta \right) = \cos \theta \qquad \text{(3b)}$$

SEEING THE CONCEPT   Graph $Y_1 = \cos\left(\dfrac{\pi}{2} - \theta \right)$ and $Y_2 = \sin \theta$ in radian mode on the same screen. Does this demonstrate the result 3(a)? How would you demonstrate the result 3(b)?

Proof   To prove formula (3a), we use the formula for $\cos(\alpha - \beta)$ with $\alpha = \dfrac{\pi}{2}$ and $\beta = \theta$.

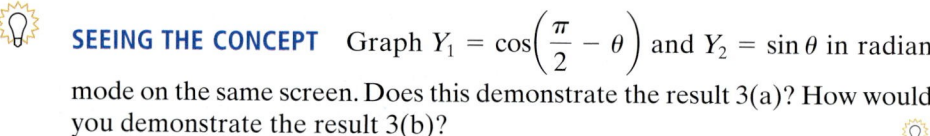

$$\cos\left(\frac{\pi}{2} - \theta \right) = \cos \frac{\pi}{2} \cos \theta + \sin \frac{\pi}{2} \sin \theta$$

$$= 0 \cdot \cos \theta + 1 \cdot \sin \theta$$

$$= \sin \theta$$

To prove formula (3b), we make use of the identity (3a) just established.

$$\sin\left(\frac{\pi}{2} - \theta\right) \underset{\substack{\uparrow \\ \text{Use (3a).}}}{=} \cos\left[\frac{\pi}{2} - \left(\frac{\pi}{2} - \theta\right)\right] = \cos\theta$$ ■

Formulas (3a) and (3b) should look familiar. They are the basis for the theorem stated in Chapter 8: Cofunctions of complementary angles are equal.

Also, since

$$\cos\left(\frac{\pi}{2} - \theta\right) = \cos\left[-\left(\theta - \frac{\pi}{2}\right)\right] \underset{\substack{\uparrow \\ \text{Even Property of Cosine}}}{=} \cos\left(\theta - \frac{\pi}{2}\right)$$

and

$$\cos\left(\frac{\pi}{2} - \theta\right) \underset{\substack{\uparrow \\ 3(a)}}{=} \sin\theta$$

it follows that $\cos\left(\theta - \frac{\pi}{2}\right) = \sin\theta$. The graphs of $y = \cos\left(\theta - \frac{\pi}{2}\right)$ and $y = \sin\theta$ are identical, a fact that we conjectured earlier in Section 8.6.

Formulas For $\sin(\alpha + \beta)$ and $\sin(\alpha - \beta)$

Having established the identities in formulas (3a) and (3b), we now can derive the sum and difference formulas for $\sin(\alpha + \beta)$ and $\sin(\alpha - \beta)$.

Proof

$$
\begin{aligned}
\sin(\alpha + \beta) &= \cos\left[\frac{\pi}{2} - (\alpha + \beta)\right] && \text{Formula (3a)}\\
&= \cos\left[\left(\frac{\pi}{2} - \alpha\right) - \beta\right]\\
&= \cos\left(\frac{\pi}{2} - \alpha\right)\cos\beta + \sin\left(\frac{\pi}{2} - \alpha\right)\sin\beta && \text{Formula (2)}\\
&= \sin\alpha\cos\beta + \cos\alpha\sin\beta && \text{Formulas (3a) and (3b)}\\
\sin(\alpha - \beta) &= \sin[\alpha + (-\beta)]\\
&= \sin\alpha\cos(-\beta) + \cos\alpha\sin(-\beta) && \text{Use the sum formula for}\\
& && \text{sine just obtained.}\\
&= \sin\alpha\cos\beta + \cos\alpha(-\sin\beta) && \text{Even–odd Identities}\\
&= \sin\alpha\cos\beta - \cos\alpha\sin\beta
\end{aligned}
$$
■

Theorem

Sum and Difference Formulas for Sines

$$\sin(\alpha + \beta) = \sin\alpha\cos\beta + \cos\alpha\sin\beta \qquad (4)$$
$$\sin(\alpha - \beta) = \sin\alpha\cos\beta - \cos\alpha\sin\beta \qquad (5)$$

In words, formula (4) states that the sine of the sum of two angles equals the sine of the first angle times the cosine of the second angle plus the cosine of the first angle times the sine of the second angle.

EXAMPLE 3 **Using the Sum Formula to Find Exact Values**

Find the exact value of $\sin\dfrac{7\pi}{12}$.

Solution
$$\sin\frac{7\pi}{12} = \sin\left(\frac{3\pi}{12} + \frac{4\pi}{12}\right) = \sin\left(\frac{\pi}{4} + \frac{\pi}{3}\right)$$

$$= \sin\frac{\pi}{4}\cos\frac{\pi}{3} + \cos\frac{\pi}{4}\sin\frac{\pi}{3} \qquad \text{Formula (4)}$$

$$= \frac{\sqrt{2}}{2}\cdot\frac{1}{2} + \frac{\sqrt{2}}{2}\cdot\frac{\sqrt{3}}{2} = \frac{1}{4}(\sqrt{2} + \sqrt{6}) \qquad ■$$

NOW WORK PROBLEM 9.

EXAMPLE 4 **Using the Difference Formula to Find Exact Values**

Find the exact value of $\sin 80° \cos 20° - \cos 80° \sin 20°$.

Solution The form of the expression $\sin 80° \cos 20° - \cos 80° \sin 20°$ is that of the right side of the formula (5) for $\sin(\alpha - \beta)$ with $\alpha = 80°$ and $\beta = 20°$. Thus,

$$\sin 80° \cos 20° - \cos 80° \sin 20° = \sin(80° - 20°) = \sin 60° = \frac{\sqrt{3}}{2} \qquad ■$$

NOW WORK PROBLEMS 15 AND 19.

EXAMPLE 5 **Finding Exact Values**

If it is known that $\sin\alpha = \dfrac{4}{5}, \dfrac{\pi}{2} < \alpha < \pi$, and that

$\sin\beta = -\dfrac{2}{\sqrt{5}} = -\dfrac{2\sqrt{5}}{5}, \pi < \beta < \dfrac{3\pi}{2}$, find the exact value of

(a) $\cos\alpha$ (b) $\cos\beta$ (c) $\cos(\alpha + \beta)$ (d) $\sin(\alpha + \beta)$

Solution (a) Since $\sin\alpha = \dfrac{4}{5} = \dfrac{b}{r}$ and $\dfrac{\pi}{2} < \alpha < \pi$, we let $b = 4$ and $r = 5$ and place α in quadrant II. See Figure 25. Since $(a, 4)$ is in quadrant II, we have $a < 0$. The distance from $(a, 4)$ to $(0, 0)$ is 5, so

$$a^2 + 4^2 = 5^2,$$
$$a^2 + 16 = 25$$
$$a^2 = 25 - 16 = 9$$
$$a = -3 \quad a < 0$$

Figure 25

Given $\sin\alpha = \dfrac{4}{5}, \dfrac{\pi}{2} < \alpha < \pi$

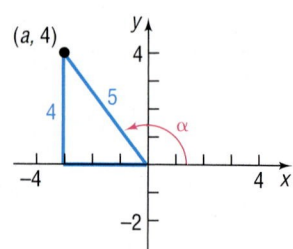

Then,

$$\cos\alpha = \frac{a}{r} = -\frac{3}{5}$$

Alternatively, we can find $\cos \alpha$ using identities, as follows:

$$\cos \alpha = -\sqrt{1 - \sin^2 \alpha} = -\sqrt{1 - \frac{16}{25}} = -\sqrt{\frac{9}{25}} = -\frac{3}{5}$$

↑
α in quadrant II,
$\cos \alpha < 0$

(b) Since $\sin \beta = \dfrac{-2}{\sqrt{5}} = \dfrac{b}{r}$ and $\pi < \beta < \dfrac{3\pi}{2}$, we let $b = -2$ and $r = \sqrt{5}$ and place β in quadrant III. See Figure 26. Since $(a, -2)$ is in quadrant III, we have $a < 0$. The distance from $(a, -2)$ to $(0, 0)$ is $\sqrt{5}$, so

$$a^2 + 4 = (\sqrt{5})^2,$$
$$a^2 = 5 - 4 = 1$$
$$a = -1 \quad a < 0$$

Figure 26

Given $\sin \beta = \dfrac{-2}{\sqrt{5}}, \pi < \beta < \dfrac{3\pi}{2}$

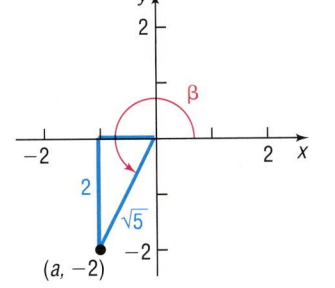

Then,

$$\cos \beta = \frac{a}{r} = -\frac{1}{\sqrt{5}} = -\frac{\sqrt{5}}{5}$$

Alternatively, we can find $\cos \beta$ using identities, as follows:

$$\cos \beta = -\sqrt{1 - \sin^2 \beta} = -\sqrt{1 - \frac{4}{5}} = -\sqrt{\frac{1}{5}} = -\frac{\sqrt{5}}{5}$$

(c) Using the results found in parts (a) and (b) and formula (1), we have

$$\cos(\alpha + \beta) = \cos \alpha \cos \beta - \sin \alpha \sin \beta$$
$$= -\frac{3}{5}\left(-\frac{\sqrt{5}}{5}\right) - \frac{4}{5}\left(-\frac{2\sqrt{5}}{5}\right) = \frac{11\sqrt{5}}{25}$$

(d) $\sin(\alpha + \beta) = \sin \alpha \cos \beta + \cos \alpha \sin \beta$
$$= \frac{4}{5}\left(-\frac{\sqrt{5}}{5}\right) + \left(-\frac{3}{5}\right)\left(-\frac{2\sqrt{5}}{5}\right) = \frac{2\sqrt{5}}{25}$$

■

NOW WORK PROBLEM **23 (a), (b), AND (c).**

EXAMPLE 6 **Establishing an Identity**

Establish the identity: $\dfrac{\cos(\alpha - \beta)}{\sin \alpha \sin \beta} = \cot \alpha \cot \beta + 1$

Solution

$$\frac{\cos(\alpha - \beta)}{\sin \alpha \sin \beta} = \frac{\cos \alpha \cos \beta + \sin \alpha \sin \beta}{\sin \alpha \sin \beta}$$
$$= \frac{\cos \alpha \cos \beta}{\sin \alpha \sin \beta} + \frac{\sin \alpha \sin \beta}{\sin \alpha \sin \beta}$$
$$= \frac{\cos \alpha}{\sin \alpha} \frac{\cos \beta}{\sin \beta} + 1$$
$$= \cot \alpha \cot \beta + 1$$

■

NOW WORK PROBLEMS **31 AND 43.**

Formulas for $\tan(\alpha + \beta)$ and $\tan(\alpha - \beta)$

We use the identity $\tan\theta = \dfrac{\sin\theta}{\cos\theta}$ and the sum formulas for $\sin(\alpha + \beta)$ and $\cos(\alpha + \beta)$ to derive a formula for $\tan(\alpha + \beta)$.

Proof

$$\tan(\alpha + \beta) = \frac{\sin(\alpha + \beta)}{\cos(\alpha + \beta)} = \frac{\sin\alpha\cos\beta + \cos\alpha\sin\beta}{\cos\alpha\cos\beta - \sin\alpha\sin\beta}$$

Now we divide the numerator and denominator by $\cos\alpha\cos\beta$.

$$\tan(\alpha + \beta) = \frac{\dfrac{\sin\alpha\cos\beta + \cos\alpha\sin\beta}{\cos\alpha\cos\beta}}{\dfrac{\cos\alpha\cos\beta - \sin\alpha\sin\beta}{\cos\alpha\cos\beta}} = \frac{\dfrac{\sin\alpha\cos\beta}{\cos\alpha\cos\beta} + \dfrac{\cos\alpha\sin\beta}{\cos\alpha\cos\beta}}{\dfrac{\cos\alpha\cos\beta}{\cos\alpha\cos\beta} - \dfrac{\sin\alpha\sin\beta}{\cos\alpha\cos\beta}}$$

$$= \frac{\dfrac{\sin\alpha}{\cos\alpha} + \dfrac{\sin\beta}{\cos\beta}}{1 - \dfrac{\sin\alpha\sin\beta}{\cos\alpha\cos\beta}} = \frac{\tan\alpha + \tan\beta}{1 - \tan\alpha\tan\beta}$$

We use the sum formula for $\tan(\alpha + \beta)$ and even–odd properties to get the difference formula.

$$\tan(\alpha - \beta) = \tan[\alpha + (-\beta)] = \frac{\tan\alpha + \tan(-\beta)}{1 - \tan\alpha\tan(-\beta)} = \frac{\tan\alpha - \tan\beta}{1 + \tan\alpha\tan\beta}$$

We have proved the following results:

Theorem

> **Sum and Difference Formulas for Tangents**
>
> $$\tan(\alpha + \beta) = \frac{\tan\alpha + \tan\beta}{1 - \tan\alpha\tan\beta} \tag{6}$$
>
> $$\tan(\alpha - \beta) = \frac{\tan\alpha - \tan\beta}{1 + \tan\alpha\tan\beta} \tag{7}$$

In words, formula (6) states that the tangent of the sum of two angles equals the tangent of the first angle plus the tangent of the second angle, all divided by 1 minus their product.

NOW WORK PROBLEM 23(d).

EXAMPLE 7   **Establishing an Identity**

Prove the identity: $\tan(\theta + \pi) = \tan \theta$

Solution $\tan(\theta + \pi) = \dfrac{\tan \theta + \tan \pi}{1 - \tan \theta \tan \pi} = \dfrac{\tan \theta + 0}{1 - \tan \theta \cdot 0} = \tan \theta$ ∎

The result obtained in Example 7 verifies that the tangent function is periodic with period π, a fact that we mentioned earlier.

WARNING: Be careful when using formulas (6) and (7). These formulas can be used only for angles α and β for which $\tan \alpha$ and $\tan \beta$ are defined, that is, all angles except odd multiples of $\dfrac{\pi}{2}$. ∎

EXAMPLE 8   **Establishing an Identity**

Prove the identity: $\tan\left(\theta + \dfrac{\pi}{2}\right) = -\cot \theta$

Solution   We cannot use formula (6), since $\tan \dfrac{\pi}{2}$ is not defined. Instead, we proceed as follows:

$$\tan\left(\theta + \dfrac{\pi}{2}\right) = \dfrac{\sin\left(\theta + \dfrac{\pi}{2}\right)}{\cos\left(\theta + \dfrac{\pi}{2}\right)} = \dfrac{\sin \theta \cos \dfrac{\pi}{2} + \cos \theta \sin \dfrac{\pi}{2}}{\cos \theta \cos \dfrac{\pi}{2} - \sin \theta \sin \dfrac{\pi}{2}}$$

$$= \dfrac{(\sin \theta)(0) + (\cos \theta)(1)}{(\cos \theta)(0) - (\sin \theta)(1)} = \dfrac{\cos \theta}{-\sin \theta} = -\cot \theta$$ ∎

EXAMPLE 9   **Finding the Exact Value of Expressions Involving Inverse Trigonometric Functions**

③   Find the exact value of: $\sin\left(\cos^{-1}\dfrac{1}{2} + \sin^{-1}\dfrac{3}{5}\right)$

Solution   We seek the sine of the sum of two angles, $\alpha = \cos^{-1}\dfrac{1}{2}$ and $\beta = \sin^{-1}\dfrac{3}{5}$. Then

$$\cos \alpha = \dfrac{1}{2}, \quad 0 \le \alpha \le \pi, \quad \text{and} \quad \sin \beta = \dfrac{3}{5}, \quad -\dfrac{\pi}{2} \le \beta \le \dfrac{\pi}{2}$$

We use Pythagorean Identities to obtain $\sin \alpha$ and $\cos \beta$. Since $\sin \alpha > 0$ and $\cos \beta > 0$ (do you know why?), we find

$$\sin \alpha = \sqrt{1 - \cos^2 \alpha} = \sqrt{1 - \dfrac{1}{4}} = \sqrt{\dfrac{3}{4}} = \dfrac{\sqrt{3}}{2}$$

$$\cos \beta = \sqrt{1 - \sin^2 \beta} = \sqrt{1 - \dfrac{9}{25}} = \sqrt{\dfrac{16}{25}} = \dfrac{4}{5}$$

As a result,

$$\sin\left(\cos^{-1}\frac{1}{2} + \sin^{-1}\frac{3}{5}\right) = \sin(\alpha + \beta) = \sin\alpha\cos\beta + \cos\alpha\sin\beta$$

$$= \frac{\sqrt{3}}{2}\cdot\frac{4}{5} + \frac{1}{2}\cdot\frac{3}{5} = \frac{4\sqrt{3} + 3}{10} \quad\blacksquare$$

NOW WORK PROBLEM **59**.

EXAMPLE 10 — Writing a Trigonometric Expression as an Algebraic Expression

Write $\sin(\sin^{-1}u + \cos^{-1}v)$ as an algebraic expression containing u and v (that is, without any trigonometric functions).

Solution Let $\alpha = \sin^{-1}u$ and $\beta = \cos^{-1}v$. Then

$$\sin\alpha = u, \quad -\frac{\pi}{2} \le \alpha \le \frac{\pi}{2} \quad \text{and} \quad \cos\beta = v, \quad 0 \le \beta \le \pi$$

Since $-\dfrac{\pi}{2} \le \alpha \le \dfrac{\pi}{2}$, we know that $\cos\alpha \ge 0$. As a result,

$$\cos\alpha = \sqrt{1 - \sin^2\alpha} = \sqrt{1 - u^2}$$

Similarly, since $0 \le \beta \le \pi$, we know that $\sin\beta \ge 0$. As a result,

$$\sin\beta = \sqrt{1 - \cos^2\beta} = \sqrt{1 - v^2}$$

Now

$$\sin(\sin^{-1}u + \cos^{-1}v) = \sin(\alpha + \beta) = \sin\alpha\cos\beta + \cos\alpha\sin\beta$$

$$= uv + \sqrt{1 - u^2}\sqrt{1 - v^2} \quad\blacksquare$$

NOW WORK PROBLEM **69**.

SUMMARY

The following box summarizes the sum and difference formulas.

Sum and Difference Formulas

$$\cos(\alpha + \beta) = \cos\alpha\cos\beta - \sin\alpha\sin\beta \qquad \cos(\alpha - \beta) = \cos\alpha\cos\beta + \sin\alpha\sin\beta$$

$$\sin(\alpha + \beta) = \sin\alpha\cos\beta + \cos\alpha\sin\beta \qquad \sin(\alpha - \beta) = \sin\alpha\cos\beta - \cos\alpha\sin\beta$$

$$\tan(\alpha + \beta) = \frac{\tan\alpha + \tan\beta}{1 - \tan\alpha\tan\beta} \qquad \tan(\alpha - \beta) = \frac{\tan\alpha - \tan\beta}{1 + \tan\alpha\tan\beta}$$

9.4 Concepts and Vocabulary

In Problems 1 and 2, fill in the blanks.

1. $\cos(\alpha + \beta) = \cos \alpha \cos \beta \underline{\hspace{1.5cm}} \sin \alpha \sin \beta.$

2. $\sin(\alpha - \beta) = \sin \alpha \cos \beta \underline{\hspace{1.5cm}} \cos \alpha \sin \beta.$

In Problems 3–5, answer True or False to each statement.

3. $\sin(\alpha + \beta) = \sin \alpha + \sin \beta + 2 \sin \alpha \sin \beta.$

4. $\tan 75° = \tan 30° + \tan 45°.$

5. $\cos\left(\dfrac{\pi}{2} - \theta\right) = \cos \theta.$

6. Use a sum formula to find the exact value of $\sin 75°$.

7. Use a difference formula to find the exact value of $\sin 75°$.

8. Find the exact value of $\cos 65° \cos 25° - \sin 65° \sin 25°$.

9.4 Exercises

In Problems 1–12, find the exact value of each trigonometric function.

1. $\sin \dfrac{5\pi}{12}$

2. $\sin \dfrac{\pi}{12}$

3. $\cos \dfrac{7\pi}{12}$

4. $\tan \dfrac{7\pi}{12}$

5. $\cos 165°$

6. $\sin 105°$

7. $\tan 15°$

8. $\tan 195°$

9. $\sin \dfrac{17\pi}{12}$

10. $\tan \dfrac{19\pi}{12}$

11. $\sec\left(-\dfrac{\pi}{12}\right)$

12. $\cot\left(-\dfrac{5\pi}{12}\right)$

In Problems 13–22, find the exact value of each expression.

13. $\sin 20° \cos 10° + \cos 20° \sin 10°$

14. $\sin 20° \cos 80° - \cos 20° \sin 80°$

15. $\cos 70° \cos 20° - \sin 70° \sin 20°$

16. $\cos 40° \cos 10° + \sin 40° \sin 10°$

17. $\dfrac{\tan 20° + \tan 25°}{1 - \tan 20° \tan 25°}$

18. $\dfrac{\tan 40° - \tan 10°}{1 + \tan 40° \tan 10°}$

19. $\sin \dfrac{\pi}{12} \cos \dfrac{7\pi}{12} - \cos \dfrac{\pi}{12} \sin \dfrac{7\pi}{12}$

20. $\cos \dfrac{5\pi}{12} \cos \dfrac{7\pi}{12} - \sin \dfrac{5\pi}{12} \sin \dfrac{7\pi}{12}$

21. $\cos \dfrac{\pi}{12} \cos \dfrac{5\pi}{12} + \sin \dfrac{5\pi}{12} \sin \dfrac{\pi}{12}$

22. $\sin \dfrac{\pi}{18} \cos \dfrac{5\pi}{18} + \cos \dfrac{\pi}{18} \sin \dfrac{5\pi}{18}$

In Problems 23–28, find the exact value of each of the following under the given conditions:

(a) $\sin(\alpha + \beta)$ (b) $\cos(\alpha + \beta)$ (c) $\sin(\alpha - \beta)$ (d) $\tan(\alpha - \beta)$

23. $\sin \alpha = \dfrac{3}{5}, \quad 0 < \alpha < \dfrac{\pi}{2}; \quad \cos \beta = \dfrac{2\sqrt{5}}{5}, \quad -\dfrac{\pi}{2} < \beta < 0$

24. $\cos \alpha = \dfrac{\sqrt{5}}{5}, \quad 0 < \alpha < \dfrac{\pi}{2}; \quad \sin \beta = -\dfrac{4}{5}, \quad -\dfrac{\pi}{2} < \beta < 0$

25. $\tan \alpha = -\dfrac{4}{3}, \quad \dfrac{\pi}{2} < \alpha < \pi; \quad \cos \beta = \dfrac{1}{2}, \quad 0 < \beta < \dfrac{\pi}{2}$

26. $\tan \alpha = \dfrac{5}{12}, \quad \pi < \alpha < \dfrac{3\pi}{2}; \quad \sin \beta = -\dfrac{1}{2}, \quad \pi < \beta < \dfrac{3\pi}{2}$

27. $\sin \alpha = \dfrac{5}{13}, \quad -\dfrac{3\pi}{2} < \alpha < -\pi; \quad \tan \beta = -\sqrt{3}, \quad \dfrac{\pi}{2} < \beta < \pi$

28. $\cos \alpha = \dfrac{1}{2}, \quad -\dfrac{\pi}{2} < \alpha < 0; \quad \sin \beta = \dfrac{1}{3}, \quad 0 < \beta < \dfrac{\pi}{2}$

29. If $\sin \theta = \dfrac{1}{3}, \theta$ in quadrant II, find the exact value of:

(a) $\cos \theta$

(b) $\sin\left(\theta + \dfrac{\pi}{6}\right)$

(c) $\cos\left(\theta - \dfrac{\pi}{3}\right)$

(d) $\tan\left(\theta + \dfrac{\pi}{4}\right)$

30. If $\cos \theta = \dfrac{1}{4}, \theta$ in quadrant IV, find the exact value of:

(a) $\sin \theta$

(b) $\sin\left(\theta - \dfrac{\pi}{6}\right)$

(c) $\cos\left(\theta + \dfrac{\pi}{3}\right)$

(d) $\tan\left(\theta - \dfrac{\pi}{4}\right)$

In Problems 31–56, establish each identity.

31. $\sin\left(\dfrac{\pi}{2} + \theta\right) = \cos \theta$

32. $\cos\left(\dfrac{\pi}{2} + \theta\right) = -\sin \theta$

33. $\sin(\pi - \theta) = \sin \theta$

34. $\cos(\pi - \theta) = -\cos \theta$

35. $\sin(\pi + \theta) = -\sin \theta$

36. $\cos(\pi + \theta) = -\cos \theta$

37. $\tan(\pi - \theta) = -\tan \theta$

38. $\tan(2\pi - \theta) = -\tan \theta$

39. $\sin\left(\dfrac{3\pi}{2} + \theta\right) = -\cos \theta$

40. $\cos\left(\dfrac{3\pi}{2} + \theta\right) = \sin \theta$

41. $\sin(\alpha + \beta) + \sin(\alpha - \beta) = 2 \sin \alpha \cos \beta$

42. $\cos(\alpha + \beta) + \cos(\alpha - \beta) = 2 \cos \alpha \cos \beta$

43. $\dfrac{\sin(\alpha + \beta)}{\sin \alpha \cos \beta} = 1 + \cot \alpha \tan \beta$

44. $\dfrac{\sin(\alpha + \beta)}{\cos \alpha \cos \beta} = \tan \alpha + \tan \beta$

45. $\dfrac{\cos(\alpha + \beta)}{\cos \alpha \cos \beta} = 1 - \tan \alpha \tan \beta$

46. $\dfrac{\cos(\alpha - \beta)}{\sin \alpha \cos \beta} = \cot \alpha + \tan \beta$

47. $\dfrac{\sin(\alpha + \beta)}{\sin(\alpha - \beta)} = \dfrac{\tan \alpha + \tan \beta}{\tan \alpha - \tan \beta}$

48. $\dfrac{\cos(\alpha + \beta)}{\cos(\alpha - \beta)} = \dfrac{1 - \tan \alpha \tan \beta}{1 + \tan \alpha \tan \beta}$

49. $\cot(\alpha + \beta) = \dfrac{\cot \alpha \cot \beta - 1}{\cot \beta + \cot \alpha}$

50. $\cot(\alpha - \beta) = \dfrac{\cot \alpha \cot \beta + 1}{\cot \beta - \cot \alpha}$

51. $\sec(\alpha + \beta) = \dfrac{\csc \alpha \csc \beta}{\cot \alpha \cot \beta - 1}$

52. $\sec(\alpha - \beta) = \dfrac{\sec \alpha \sec \beta}{1 + \tan \alpha \tan \beta}$

53. $\sin(\alpha - \beta) \sin(\alpha + \beta) = \sin^2 \alpha - \sin^2 \beta$

54. $\cos(\alpha - \beta) \cos(\alpha + \beta) = \cos^2 \alpha - \sin^2 \beta$

55. $\sin(\theta + k\pi) = (-1)^k \sin \theta, \quad k$ any integer

56. $\cos(\theta + k\pi) = (-1)^k \cos \theta, \quad k$ any integer

In Problems 57–68, find the exact value of each expression.

57. $\sin\left(\sin^{-1}\dfrac{1}{2} + \cos^{-1} 0\right)$

58. $\sin\left(\sin^{-1}\dfrac{\sqrt{3}}{2} + \cos^{-1} 1\right)$

59. $\sin\left[\sin^{-1}\dfrac{3}{5} - \cos^{-1}\left(-\dfrac{4}{5}\right)\right]$

60. $\sin\left[\sin^{-1}\left(-\dfrac{4}{5}\right) - \tan^{-1}\dfrac{3}{4}\right]$

61. $\cos\left(\tan^{-1}\dfrac{4}{3} + \cos^{-1}\dfrac{5}{13}\right)$

62. $\cos\left[\tan^{-1}\dfrac{5}{12} - \sin^{-1}\left(-\dfrac{3}{5}\right)\right]$

63. $\cos\left(\sin^{-1}\dfrac{5}{13} - \tan^{-1}\dfrac{3}{4}\right)$

64. $\cos\left(\tan^{-1}\dfrac{4}{3} + \cos^{-1}\dfrac{12}{13}\right)$

65. $\tan\left(\sin^{-1}\dfrac{3}{5} + \dfrac{\pi}{6}\right)$

66. $\tan\left(\dfrac{\pi}{4} - \cos^{-1}\dfrac{3}{5}\right)$

67. $\tan\left(\sin^{-1}\dfrac{4}{5} + \cos^{-1} 1\right)$

68. $\tan\left(\cos^{-1}\dfrac{4}{5} + \sin^{-1} 1\right)$

In Problems 69–74, write each trigonometric expression as an algebraic expression containing u and v.

69. $\cos(\cos^{-1} u + \sin^{-1} v)$

70. $\sin(\sin^{-1} u - \cos^{-1} v)$

71. $\sin(\tan^{-1} u - \sin^{-1} v)$

72. $\cos(\tan^{-1} u + \tan^{-1} v)$

73. $\tan(\sin^{-1} u - \cos^{-1} v)$

74. $\sec(\tan^{-1} u + \cos^{-1} v)$

75. Show that $\sin^{-1} v + \cos^{-1} v = \dfrac{\pi}{2}$.

76. Show that $\tan^{-1} v + \cot^{-1} v = \dfrac{\pi}{2}$.

77. Show that $\tan^{-1}\left(\dfrac{1}{v}\right) = \dfrac{\pi}{2} - \tan^{-1} v$, if $v > 0$.

78. Show that $\cot^{-1} e^v = \tan^{-1} e^{-v}$.

79. Show that $\sin(\sin^{-1} v + \cos^{-1} v) = 1$.

80. Show that $\cos(\sin^{-1} v + \cos^{-1} v) = 0$.

81. Calculus Show that the difference quotient for $f(x) = \sin x$ is given by

$$\frac{f(x+h) - f(x)}{h} = \frac{\sin(x+h) - \sin x}{h}$$

$$= \cos x \cdot \frac{\sin h}{h} - \sin x \cdot \frac{1 - \cos h}{h}$$

82. Calculus Show that the difference quotient for $f(x) = \cos x$ is given by

$$\frac{f(x+h) - f(x)}{h} = \frac{\cos(x+h) - \cos x}{h}$$

$$= -\sin x \cdot \frac{\sin h}{h} - \cos x \cdot \frac{1 - \cos h}{h}$$

83. Explain why formula (7) cannot be used to show that

$$\tan\left(\frac{\pi}{2} - \theta\right) = \cot \theta$$

Establish this identity by using formulas (3a) and (3b).

84. If $\tan \alpha = x + 1$ and $\tan \beta = x - 1$, show that $2 \cot(\alpha - \beta) = x^2$.

85. Geometry: Angle between Two Lines Let L_1 and L_2 denote two nonvertical intersecting lines, and let θ denote the acute angle between L_1 and L_2 (see the figure). Show that

$$\tan \theta = \frac{m_2 - m_1}{1 + m_1 m_2}$$

where m_1 and m_2 are the slopes of L_1 and L_2, respectively.

[**Hint:** Use the facts that $\tan \theta_1 = m_1$ and $\tan \theta_2 = m_2$.]

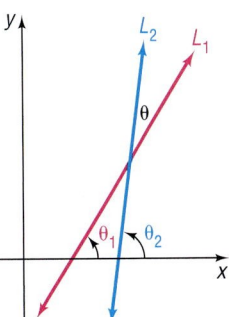

86. If $\alpha + \beta + \gamma = 180°$ and

$$\cot \theta = \cot \alpha + \cot \beta + \cot \gamma, \quad 0 < \theta < 90°$$

show that

$$\sin^3 \theta = \sin(\alpha - \theta) \sin(\beta - \theta) \sin(\gamma - \theta)$$

87. Discuss the following derivation:

$$\tan\left(\theta + \frac{\pi}{2}\right) = \frac{\tan \theta + \tan \frac{\pi}{2}}{1 - \tan \theta \tan \frac{\pi}{2}}$$

$$= \frac{\dfrac{\tan \theta}{\tan \dfrac{\pi}{2}} + 1}{\dfrac{1}{\tan \dfrac{\pi}{2}} - \tan \theta} = \frac{0 + 1}{0 - \tan \theta}$$

$$= \frac{1}{-\tan \theta} = -\cot \theta$$

Can you justify each step?

9.5 DOUBLE-ANGLE AND HALF-ANGLE FORMULAS

OBJECTIVES

1. Use Double-Angle Formulas to Find Exact Values
2. Use Double-Angle and Half-Angle Formulas to Establish Identities
3. Use Half-Angle Formulas to Find Exact Values

In this section we derive formulas for $\sin(2\theta)$, $\cos(2\theta)$, $\sin\left(\dfrac{1}{2}\theta\right)$, and $\cos\left(\dfrac{1}{2}\theta\right)$ in terms of $\sin \theta$ and $\cos \theta$. They are easily derived using the sum formulas.

Double-Angle Formulas

In the sum formulas for $\sin(\alpha + \beta)$ and $\cos(\alpha + \beta)$, let $\alpha = \beta = \theta$. Then

$$\sin(\alpha + \beta) = \sin \alpha \cos \beta + \cos \alpha \sin \beta$$

$$\sin(\theta + \theta) = \sin \theta \cos \theta + \cos \theta \sin \theta$$

$$\sin(2\theta) = 2 \sin \theta \cos \theta \tag{1}$$

and

$$\cos(\alpha + \beta) = \cos\alpha\cos\beta - \sin\alpha\sin\beta$$
$$\cos(\theta + \theta) = \cos\theta\cos\theta - \sin\theta\sin\theta$$
$$\cos(2\theta) = \cos^2\theta - \sin^2\theta \tag{2}$$

An application of the Pythagorean Identity $\sin^2\theta + \cos^2\theta = 1$ results in two other ways to write formula (2) for $\cos(2\theta)$.

$$\cos(2\theta) = \cos^2\theta - \sin^2\theta = (1 - \sin^2\theta) - \sin^2\theta = 1 - 2\sin^2\theta$$

and

$$\cos(2\theta) = \cos^2\theta - \sin^2\theta = \cos^2\theta - (1 - \cos^2\theta) = 2\cos^2\theta - 1$$

We have established the following **double-angle formulas:**

Theorem

Double-Angle Formulas

$$\sin(2\theta) = 2\sin\theta\cos\theta \tag{1}$$
$$\cos(2\theta) = \cos^2\theta - \sin^2\theta \tag{2}$$
$$\cos(2\theta) = 1 - 2\sin^2\theta \tag{3}$$
$$\cos(2\theta) = 2\cos^2\theta - 1 \tag{4}$$

EXAMPLE 1   **Finding Exact Values Using the Double-Angle Formulas**

 If $\sin\theta = \dfrac{3}{5}, \dfrac{\pi}{2} < \theta < \pi$, find the exact value of:

(a) $\sin(2\theta)$ (b) $\cos(2\theta)$

Solution   (a) Because $\sin(2\theta) = 2\sin\theta\cos\theta$ and we already know that $\sin\theta = \dfrac{3}{5}$, we only need to find $\cos\theta$. Since $\sin\theta = \dfrac{3}{5} = \dfrac{b}{r}, \dfrac{\pi}{2} < \theta < \pi$, we let $b = 3$ and $r = 5$, and place θ in quadrant II. See Figure 27. The point $(a, 3)$ is in quadrant II, so $a < 0$. The distance from $(a, 3)$ to $(0, 0)$ is 5, so

$$a^2 + 3^2 = 5^2,$$
$$a^2 + 9 = 25$$
$$a^2 = 25 - 9 = 16$$
$$a = -4 \quad \textcolor{blue}{a < 0}$$

Figure 27

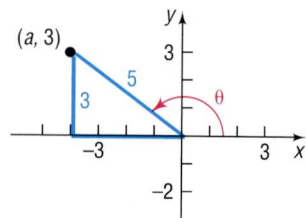

We find that $\cos\theta = \dfrac{a}{r} = -\dfrac{4}{5}$. Now we use formula (1) to obtain

$$\sin(2\theta) = 2\sin\theta\cos\theta = 2\left(\frac{3}{5}\right)\left(-\frac{4}{5}\right) = -\frac{24}{25}$$

(b) Because we are given $\sin \theta = \dfrac{3}{5}$, it is easiest to use formula (3) to get $\cos(2\theta)$.

$$\cos(2\theta) = 1 - 2\sin^2 \theta = 1 - 2\left(\frac{9}{25}\right) = 1 - \frac{18}{25} = \frac{7}{25} \quad \blacksquare$$

WARNING: In finding $\cos(2\theta)$ in Example 1(b), we chose to use a version of the double-angle formula, formula (3). Note that we are unable to use the Pythagorean Identity $\cos(2\theta) = \pm\sqrt{1 - \sin^2(2\theta)}$, with $\sin(2\theta) = -\dfrac{24}{25}$, because we have no way of knowing which sign to choose. $\quad \blacksquare$

✏️ **NOW WORK PROBLEMS 1(a) AND (b).**

EXAMPLE 2 **Establishing Identities**

② (a) Develop a formula for $\tan(2\theta)$ in terms of $\tan \theta$.
(b) Develop a formula for $\sin(3\theta)$ in terms of $\sin \theta$ and $\cos \theta$.

Solution (a) In the sum formula for $\tan(\alpha + \beta)$, let $\alpha = \beta = \theta$. Then

$$\tan(\alpha + \beta) = \frac{\tan \alpha + \tan \beta}{1 - \tan \alpha \tan \beta}$$

$$\tan(\theta + \theta) = \frac{\tan \theta + \tan \theta}{1 - \tan \theta \tan \theta}$$

$$\boxed{\tan(2\theta) = \frac{2\tan \theta}{1 - \tan^2 \theta}} \qquad (5)$$

(b) To get a formula for $\sin(3\theta)$, we use the sum formula and write 3θ as $2\theta + \theta$.

$$\sin(3\theta) = \sin(2\theta + \theta) = \sin(2\theta)\cos \theta + \cos(2\theta)\sin \theta$$

Now use the double-angle formulas to get

$$\sin(3\theta) = (2\sin \theta \cos \theta)(\cos \theta) + (\cos^2 \theta - \sin^2 \theta)(\sin \theta)$$
$$= 2\sin \theta \cos^2 \theta + \sin \theta \cos^2 \theta - \sin^3 \theta$$
$$= 3\sin \theta \cos^2 \theta - \sin^3 \theta \qquad \blacksquare$$

The formula obtained in Example 2(b) can also be written as

$$\sin(3\theta) = 3\sin \theta \cos^2 \theta - \sin^3 \theta = 3\sin \theta(1 - \sin^2 \theta) - \sin^3 \theta$$
$$= 3\sin \theta - 4\sin^3 \theta$$

That is, $\sin(3\theta)$ is a third-degree polynomial in the variable $\sin \theta$. In fact, $\sin(n\theta)$, n a positive odd integer, can always be written as a polynomial of degree n in the variable $\sin \theta$.*

✏️ **NOW WORK PROBLEM 47.**

---

*Due to the work done by P.L. Chebyshëv, these polynomials are sometimes called *Chebyshëv polynomials*.

Other Variations of the Double-Angle Formulas

By rearranging the double-angle formulas (3) and (4), we obtain other formulas that we will use later in this section.

We begin with formula (3) and proceed to solve for $\sin^2 \theta$.

$$\cos(2\theta) = 1 - 2\sin^2 \theta$$

$$2\sin^2 \theta = 1 - \cos(2\theta)$$

$$\sin^2 \theta = \frac{1 - \cos(2\theta)}{2} \tag{6}$$

Similarly, using formula (4), we proceed to solve for $\cos^2 \theta$.

$$\cos(2\theta) = 2\cos^2 \theta - 1$$

$$2\cos^2 \theta = 1 + \cos(2\theta)$$

$$\cos^2 \theta = \frac{1 + \cos(2\theta)}{2} \tag{7}$$

Formulas (6) and (7) can be used to develop a formula for $\tan^2 \theta$.

$$\tan^2 \theta = \frac{\sin^2 \theta}{\cos^2 \theta} = \frac{\dfrac{1 - \cos(2\theta)}{2}}{\dfrac{1 + \cos(2\theta)}{2}}$$

$$\tan^2 \theta = \frac{1 - \cos(2\theta)}{1 + \cos(2\theta)} \tag{8}$$

 Formulas (6) through (8) do not have to be memorized since their derivations are so straightforward.

Formulas (6) and (7) are important in calculus. The next example illustrates a problem that arises in calculus requiring the use of formula (7).

EXAMPLE 3   **Establishing an Identity**

Write an equivalent expression for $\cos^4 \theta$ that does not involve any powers of sine or cosine greater than 1.

Solution   The idea here is to apply formula (7) twice.

$$\cos^4 \theta = (\cos^2 \theta)^2 = \left(\frac{1 + \cos(2\theta)}{2} \right)^2 \qquad \text{Formula (7)}$$

$$= \frac{1}{4}[1 + 2\cos(2\theta) + \cos^2(2\theta)]$$

$$= \frac{1}{4} + \frac{1}{2}\cos(2\theta) + \frac{1}{4}\cos^2(2\theta)$$

$$= \frac{1}{4} + \frac{1}{2}\cos(2\theta) + \frac{1}{4}\left\{\frac{1 + \cos[2(2\theta)]}{2}\right\} \qquad \text{Formula (7)}$$

$$= \frac{1}{4} + \frac{1}{2}\cos(2\theta) + \frac{1}{8}[1 + \cos(4\theta)]$$

$$= \frac{3}{8} + \frac{1}{2}\cos(2\theta) + \frac{1}{8}\cos(4\theta)$$

 **NOW WORK PROBLEM 23.**

Identities, such as the double-angle formulas, can sometimes be used to rewrite expressions in a more suitable form. Let's look at an example.

EXAMPLE 4 **Projectile Motion**

An object is propelled upward at an angle θ to the horizontal with an initial velocity of v_0 feet per second. See Figure 28. If air resistance is ignored, the **range** R, the horizontal distance that the object travels, is given by

Figure 28

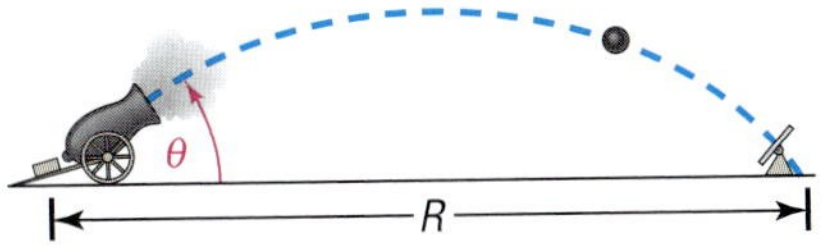

$$R = \frac{1}{16}v_0^2 \sin\theta \cos\theta$$

(a) Show that $R = \frac{1}{32}v_0^2 \sin(2\theta)$.

(b) Find the angle θ for which R is a maximum.

Solution (a) We rewrite the given expression for the range using the double-angle formula $\sin(2\theta) = 2\sin\theta\cos\theta$. Then

$$R = \frac{1}{16}v_0^2 \sin\theta\cos\theta = \frac{1}{16}v_0^2 \frac{2\sin\theta\cos\theta}{2} = \frac{1}{32}v_0^2 \sin(2\theta)$$

(b) In this form, the largest value for the range R can be found. For a fixed initial speed v_0, the angle θ of inclination to the horizontal determines the value of R. Since the largest value of a sine function is 1, occurring when the argument 2θ is 90°, it follows that for maximum R we must have

$$2\theta = 90°$$

$$\theta = 45°$$

An inclination to the horizontal of 45° results in maximum range.

✔ CHECK: Graph $Y_1 = \frac{1}{16}v_0^2 \sin\theta\cos\theta$ with $v_0 = 1$, in degree mode. Use MAXIMUM to determine the angle θ that maximizes the range R.

Half-Angle Formulas

Another important use of formulas (6) through (8) is to prove the **half-angle formulas**. In formulas (6) through (8), let $\theta = \frac{\alpha}{2}$. Then

$$\sin^2\frac{\alpha}{2} = \frac{1 - \cos\alpha}{2} \qquad \cos^2\frac{\alpha}{2} = \frac{1 + \cos\alpha}{2} \qquad \tan^2\frac{\alpha}{2} = \frac{1 - \cos\alpha}{1 + \cos\alpha} \qquad (9)$$

 Note: The identities in box (9) will prove useful in integral calculus.

If we solve for the trigonometric functions on the left sides of equations (9), we obtain the half-angle formulas.

Theorem   **Half-Angle Formulas**

$$\sin\frac{\alpha}{2} = \pm\sqrt{\frac{1 - \cos\alpha}{2}} \tag{10a}$$

$$\cos\frac{\alpha}{2} = \pm\sqrt{\frac{1 + \cos\alpha}{2}} \tag{10b}$$

$$\tan\frac{\alpha}{2} = \pm\sqrt{\frac{1 - \cos\alpha}{1 + \cos\alpha}} \tag{10c}$$

where the $+$ or $-$ sign is determined by the quadrant of the angle $\dfrac{\alpha}{2}$.

We use the half-angle formulas in the next example.

EXAMPLE 5   **Finding Exact Values Using Half-Angle Formulas**

3   Find the exact value of:

(a) $\cos 15°$ (b) $\sin(-15°)$

Solution   (a) Because $15° = \dfrac{30°}{2}$, we can use the half-angle formula for $\cos\dfrac{\alpha}{2}$ with $\alpha = 30°$. Also, because $15°$ is in quadrant I, $\cos 15° > 0$, so we choose the $+$ sign in using formula (10b):

$$\cos 15° = \cos\frac{30°}{2} = \sqrt{\frac{1 + \cos 30°}{2}}$$

$$= \sqrt{\frac{1 + \dfrac{\sqrt{3}}{2}}{2}} = \sqrt{\frac{2 + \sqrt{3}}{4}} = \frac{\sqrt{2 + \sqrt{3}}}{2}$$

(b) We use the fact that $\sin(-15°) = -\sin 15°$ and then apply formula (10a).

$$\sin(-15°) = -\sin\frac{30°}{2} = -\sqrt{\frac{1 - \cos 30°}{2}}$$

$$= -\sqrt{\frac{1 - \dfrac{\sqrt{3}}{2}}{2}} = -\sqrt{\frac{2 - \sqrt{3}}{4}} = -\frac{\sqrt{2 - \sqrt{3}}}{2}$$

It is interesting to compare the answer found in Example 5(a) with the answer to Example 2 of Section 9.4. There we calculated

$$\cos\frac{\pi}{12} = \cos 15° = \frac{1}{4}(\sqrt{6} + \sqrt{2})$$

Based on this and the result of Example 5(a), we conclude that

$$\frac{1}{4}(\sqrt{6} + \sqrt{2}) \quad \text{and} \quad \frac{\sqrt{2 + \sqrt{3}}}{2}$$

are equal. (Since each expression is positive, you can verify this equality by squaring each expression.) Two very different looking, yet correct, answers can be obtained, depending on the approach taken to solve a problem.

NOW WORK PROBLEM 13.

EXAMPLE 6   **Finding Exact Values Using Half-Angle Formulas**

If $\cos \alpha = -\dfrac{3}{5}, \pi < \alpha < \dfrac{3\pi}{2}$, find the exact value of:

(a) $\sin \dfrac{\alpha}{2}$ (b) $\cos \dfrac{\alpha}{2}$ (c) $\tan \dfrac{\alpha}{2}$

Solution   First, we observe that if $\pi < \alpha < \dfrac{3\pi}{2}$, then $\dfrac{\pi}{2} < \dfrac{\alpha}{2} < \dfrac{3\pi}{4}$. As a result, $\dfrac{\alpha}{2}$ lies in quadrant II.

(a) Because $\dfrac{\alpha}{2}$ lies in quadrant II, $\sin \dfrac{\alpha}{2} > 0$, so we use the $+$ sign in formula (10a) to get

$$\sin \frac{\alpha}{2} = \sqrt{\frac{1 - \cos \alpha}{2}} = \sqrt{\frac{1 - \left(-\dfrac{3}{5}\right)}{2}} = \sqrt{\frac{\dfrac{8}{5}}{2}}$$

$$= \sqrt{\frac{4}{5}} = \frac{2}{\sqrt{5}} = \frac{2\sqrt{5}}{5}$$

(b) Because $\dfrac{\alpha}{2}$ lies in quadrant II, $\cos \dfrac{\alpha}{2} < 0$, so we use the $-$ sign in formula (10b) to get

$$\cos \frac{\alpha}{2} = -\sqrt{\frac{1 + \cos \alpha}{2}} = -\sqrt{\frac{1 + \left(-\dfrac{3}{5}\right)}{2}} = -\sqrt{\frac{\dfrac{2}{5}}{2}}$$

$$= -\frac{1}{\sqrt{5}} = -\frac{\sqrt{5}}{5}$$

(c) Because $\dfrac{\alpha}{2}$ lies in quadrant II, $\tan \dfrac{\alpha}{2} < 0$, so we use the $-$ sign in formula (10c) to get

$$\tan \frac{\alpha}{2} = -\sqrt{\frac{1 - \cos \alpha}{1 + \cos \alpha}} = -\sqrt{\frac{1 - \left(-\dfrac{3}{5}\right)}{1 + \left(-\dfrac{3}{5}\right)}} = -\sqrt{\frac{\dfrac{8}{5}}{\dfrac{2}{5}}} = -2$$

Another way to solve Example 6(c) is to use the solutions found in parts (a) and (b).

$$\tan\frac{\alpha}{2} = \frac{\sin\frac{\alpha}{2}}{\cos\frac{\alpha}{2}} = \frac{\frac{2\sqrt{5}}{5}}{-\frac{\sqrt{5}}{5}} = -2$$

NOW WORK PROBLEMS 1(c) AND (d).

There is a formula for $\tan\frac{\alpha}{2}$ that does not contain $+$ and $-$ signs, making it more useful than formula 10(c). Because

$$1 - \cos\alpha = 2\sin^2\frac{\alpha}{2} \qquad \text{Formula (9)}$$

and

$$\sin\alpha = \sin\left[2\left(\frac{\alpha}{2}\right)\right] = 2\sin\frac{\alpha}{2}\cos\frac{\alpha}{2} \qquad \text{Double-angle formula}$$

we have

$$\frac{1 - \cos\alpha}{\sin\alpha} = \frac{2\sin^2\frac{\alpha}{2}}{2\sin\frac{\alpha}{2}\cos\frac{\alpha}{2}} = \frac{\sin\frac{\alpha}{2}}{\cos\frac{\alpha}{2}} = \tan\frac{\alpha}{2}$$

Since it also can be shown that

$$\frac{1 - \cos\alpha}{\sin\alpha} = \frac{\sin\alpha}{1 + \cos\alpha}$$

we have the following two half-angle formulas:

Half-Angle Formulas for $\tan\dfrac{\alpha}{2}$

$$\tan\frac{\alpha}{2} = \frac{1 - \cos\alpha}{\sin\alpha} = \frac{\sin\alpha}{1 + \cos\alpha} \tag{11}$$

With this formula, the solution to Example 6(c) can be given as

$$\cos\alpha = -\frac{3}{5}$$

$$\sin\alpha = -\sqrt{1 - \cos^2\alpha} = -\sqrt{1 - \frac{9}{25}} = -\sqrt{\frac{16}{25}} = -\frac{4}{5}$$

Then, by equation (11),

$$\tan\frac{\alpha}{2} = \frac{1 - \cos\alpha}{\sin\alpha} = \frac{1 - \left(-\frac{3}{5}\right)}{-\frac{4}{5}} = \frac{\frac{8}{5}}{-\frac{4}{5}} = -2$$

9.5 Concepts and Vocabulary

In Problems 1–3, fill in the blanks.

1. $\cos(2\theta) = \cos^2\theta - \underline{\hspace{1cm}} = \underline{\hspace{1cm}} - 1 = 1 - \underline{\hspace{1cm}}$.

2. $\sin^2\dfrac{\theta}{2} = \dfrac{\underline{\hspace{0.5cm}}}{2}$.

3. $\tan\dfrac{\theta}{2} = \dfrac{1 - \cos\theta}{\underline{\hspace{0.5cm}}}$

In Problems 4 and 5, answer True or False to each statement.

4. $\cos(2\theta)$ has three equivalent forms: $\cos^2\theta - \sin^2\theta$; $1 - 2\sin^2\theta$; and $2\cos^2\theta - 1$.

5. $\sin(2\theta)$ has two equivalent forms: $2\sin\theta\cos\theta$ and $\sin^2\theta - \cos^2\theta$.

6. If you are given the value of $\cos\theta$ and want the exact value of $\cos(2\theta)$, what form of the double-angle formula for $\cos(2\theta)$ is most efficient to use?

7. Use a half-angle formula to find the exact value of $\sin 15°$.

8. Use a difference formula to find the exact value of $\sin 15°$.

9. Show that the answers found in Problems 7 and 8 are the same.

9.5 Exercises

In Problems 1–12, use the information given about the angle $\theta, 0 \leq \theta < 2\pi$, to find the exact value of

(a) $\sin(2\theta)$ (b) $\cos(2\theta)$ (c) $\sin\dfrac{\theta}{2}$ (d) $\cos\dfrac{\theta}{2}$

1. $\sin\theta = \dfrac{3}{5}, \quad 0 < \theta < \dfrac{\pi}{2}$

2. $\cos\theta = \dfrac{3}{5}, \quad 0 < \theta < \dfrac{\pi}{2}$

3. $\tan\theta = \dfrac{4}{3}, \quad \pi < \theta < \dfrac{3\pi}{2}$

4. $\tan\theta = \dfrac{1}{2}, \quad \pi < \theta < \dfrac{3\pi}{2}$

5. $\cos\theta = -\dfrac{\sqrt{6}}{3}, \quad \dfrac{\pi}{2} < \theta < \pi$

6. $\sin\theta = -\dfrac{\sqrt{3}}{3}, \quad \dfrac{3\pi}{2} < \theta < 2\pi$

7. $\sec\theta = 3, \quad \sin\theta > 0$

8. $\csc\theta = -\sqrt{5}, \quad \cos\theta < 0$

9. $\cot\theta = -2, \quad \sec\theta < 0$

10. $\sec\theta = 2, \quad \csc\theta < 0$

11. $\tan\theta = -3, \quad \sin\theta < 0$

12. $\cot\theta = 3, \quad \cos\theta < 0$

In Problems 13–22, use the half-angle formulas to find the exact value of each trigonometric function.

13. $\sin 22.5°$

14. $\cos 22.5°$

15. $\tan\dfrac{7\pi}{8}$

16. $\tan\dfrac{9\pi}{8}$

17. $\cos 165°$

18. $\sin 195°$

19. $\sec\dfrac{15\pi}{8}$

20. $\csc\dfrac{7\pi}{8}$

21. $\sin\left(-\dfrac{\pi}{8}\right)$

22. $\cos\left(-\dfrac{3\pi}{8}\right)$

23. Show that $\sin^4\theta = \dfrac{3}{8} - \dfrac{1}{2}\cos(2\theta) + \dfrac{1}{8}\cos(4\theta)$.

24. Develop a formula for $\cos(3\theta)$ as a third-degree polynomial in the variable $\cos\theta$.

25. Show that $\sin(4\theta) = (\cos\theta)(4\sin\theta - 8\sin^3\theta)$.

26. Develop a formula for $\cos(4\theta)$ as a fourth-degree polynomial in the variable $\cos\theta$.

27. Find an expression for $\sin(5\theta)$ as a fifth-degree polynomial in the variable $\sin\theta$.

28. Find an expression for $\cos(5\theta)$ as a fifth-degree polynomial in the variable $\cos\theta$.

In Problems 29–50, establish each identity.

29. $\cos^4\theta - \sin^4\theta = \cos(2\theta)$

30. $\dfrac{\cot\theta - \tan\theta}{\cot\theta + \tan\theta} = \cos(2\theta)$

31. $\cot(2\theta) = \dfrac{\cot^2\theta - 1}{2\cot\theta}$

32. $\cot(2\theta) = \dfrac{1}{2}(\cot\theta - \tan\theta)$

33. $\sec(2\theta) = \dfrac{\sec^2\theta}{2 - \sec^2\theta}$

34. $\csc(2\theta) = \dfrac{1}{2}\sec\theta\csc\theta$

35. $\cos^2(2\theta) - \sin^2(2\theta) = \cos(4\theta)$

36. $(4\sin\theta\cos\theta)(1 - 2\sin^2\theta) = \sin(4\theta)$

37. $\dfrac{\cos(2\theta)}{1 + \sin(2\theta)} = \dfrac{\cot\theta - 1}{\cot\theta + 1}$

38. $\sin^2\theta\cos^2\theta = \dfrac{1}{8}[1 - \cos(4\theta)]$

39. $\sec^2\dfrac{\theta}{2} = \dfrac{2}{1 + \cos\theta}$

40. $\csc^2\dfrac{\theta}{2} = \dfrac{2}{1 - \cos\theta}$

41. $\cot^2\dfrac{\theta}{2} = \dfrac{\sec\theta + 1}{\sec\theta - 1}$

42. $\tan\dfrac{\theta}{2} = \csc\theta - \cot\theta$

43. $\cos\theta = \dfrac{1 - \tan^2\dfrac{\theta}{2}}{1 + \tan^2\dfrac{\theta}{2}}$

44. $1 - \dfrac{1}{2}\sin(2\theta) = \dfrac{\sin^3\theta + \cos^3\theta}{\sin\theta + \cos\theta}$

45. $\dfrac{\sin(3\theta)}{\sin\theta} - \dfrac{\cos(3\theta)}{\cos\theta} = 2$

46. $\dfrac{\cos\theta + \sin\theta}{\cos\theta - \sin\theta} - \dfrac{\cos\theta - \sin\theta}{\cos\theta + \sin\theta} = 2\tan(2\theta)$

47. $\tan(3\theta) = \dfrac{3\tan\theta - \tan^3\theta}{1 - 3\tan^2\theta}$

48. $\tan\theta + \tan(\theta + 120°) + \tan(\theta + 240°) = 3\tan(3\theta)$

49. $\ln|\sin\theta| = \dfrac{1}{2}(\ln|1 - \cos(2\theta)| - \ln 2)$

50. $\ln|\cos\theta| = \dfrac{1}{2}(\ln|1 + \cos(2\theta)| - \ln 2)$

In Problems 51–62, find the exact value of each expression.

51. $\sin\left(2\sin^{-1}\dfrac{1}{2}\right)$

52. $\sin\left[2\sin^{-1}\dfrac{\sqrt{3}}{2}\right]$

53. $\cos\left(2\sin^{-1}\dfrac{3}{5}\right)$

54. $\cos\left(2\cos^{-1}\dfrac{4}{5}\right)$

55. $\tan\left[2\cos^{-1}\left(-\dfrac{3}{5}\right)\right]$

56. $\tan\left(2\tan^{-1}\dfrac{3}{4}\right)$

57. $\sin\left(2\cos^{-1}\dfrac{4}{5}\right)$

58. $\cos\left[2\tan^{-1}\left(-\dfrac{4}{3}\right)\right]$

59. $\sin^2\left(\dfrac{1}{2}\cos^{-1}\dfrac{3}{5}\right)$

60. $\cos^2\left(\dfrac{1}{2}\sin^{-1}\dfrac{3}{5}\right)$

61. $\sec\left(2\tan^{-1}\dfrac{3}{4}\right)$

62. $\csc\left[2\sin^{-1}\left(-\dfrac{3}{5}\right)\right]$

63. If $x = 2\tan\theta$, express $\sin(2\theta)$ as a function of x.

64. If $x = 2\tan\theta$, express $\cos(2\theta)$ as a function of x.

65. Find the value of the number C:

$$\dfrac{1}{2}\sin^2 x + C = -\dfrac{1}{4}\cos(2x)$$

66. Find the value of the number C:

$$\dfrac{1}{2}\cos^2 x + C = \dfrac{1}{4}\cos(2x)$$

67. If $z = \tan\dfrac{\alpha}{2}$, show that $\sin\alpha = \dfrac{2z}{1 + z^2}$.

68. If $z = \tan\dfrac{\alpha}{2}$, show that $\cos\alpha = \dfrac{1 - z^2}{1 + z^2}$.

69. Area of an Isosceles Triangle Show that the area A of an isosceles triangle whose equal sides are of length s and θ is the angle between them is

$$\dfrac{1}{2}s^2\sin\theta$$

[**Hint:** See the illustration. The height h bisects the angle θ and is the perpendicular bisector of the base.]

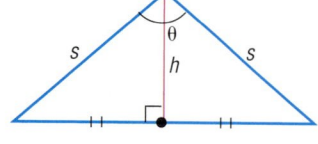

70. Geometry A rectangle is inscribed in a semicircle of radius 1. See the illustration.

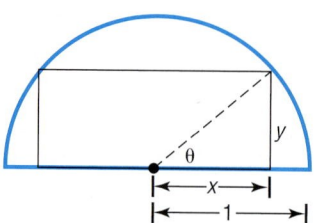

(a) Express the area A of the rectangle as a function of the angle θ shown in the illustration.

(b) Show that $A = \sin(2\theta)$.

(c) Find the angle θ that results in the largest area A.

(d) Find the dimensions of this largest rectangle.

71. Graph $f(x) = \sin^2 x = \dfrac{1 - \cos(2x)}{2}$ for $0 \leq x \leq 2\pi$ by using transformations.

72. Repeat Problem 71 for $g(x) = \cos^2 x$.

73. Use the fact that

$$\cos\frac{\pi}{12} = \frac{1}{4}(\sqrt{6} + \sqrt{2})$$

to find $\sin\dfrac{\pi}{24}$ and $\cos\dfrac{\pi}{24}$.

74. Show that

$$\cos\frac{\pi}{8} = \frac{\sqrt{2 + \sqrt{2}}}{2}$$

and use it to find $\sin\dfrac{\pi}{16}$ and $\cos\dfrac{\pi}{16}$.

75. Show that

$$\sin^3\theta + \sin^3(\theta + 120°) + \sin^3(\theta + 240°) = -\frac{3}{4}\sin(3\theta)$$

76. If $\tan\theta = a\tan\dfrac{\theta}{3}$, express $\tan\dfrac{\theta}{3}$ in terms of a.

77. Projectile Motion An object is propelled upward at an angle θ, $45° < \theta < 90°$, to the horizontal with an initial velocity of v_0 feet per second from the base of a plane that makes an angle of $45°$ with the horizontal. See the illustration. If air resistance is ignored, the distance R that it travels up the inclined plane is given by

$$R = \frac{v_0^2\sqrt{2}}{16}\cos\theta(\sin\theta - \cos\theta)$$

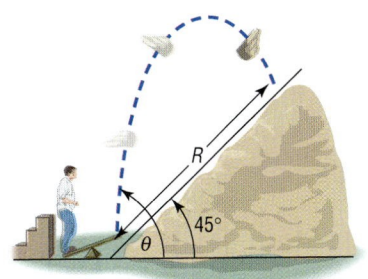

(a) Show that

$$R = \frac{v_0^2\sqrt{2}}{32}[\sin(2\theta) - \cos(2\theta) - 1]$$

(b) Graph $R = R(\theta)$. (Use $v_0 = 32$ feet per second.)

(c) What value of θ makes R the largest? (Use $v_0 = 32$ feet per second.)

78. Sawtooth Curve An oscilloscope often displays a sawtooth curve. This curve can be approximated by sinusoidal curves of varying periods and amplitudes. A first approximation to the sawtooth curve is given by

$$y = \frac{1}{2}\sin(2\pi x) + \frac{1}{4}\sin(4\pi x)$$

Show that $y = \sin(2\pi x)\cos^2(\pi x)$.

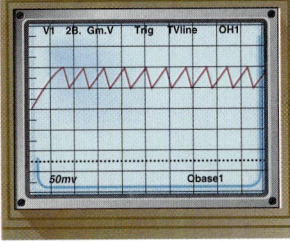

79. Go to the library and research Chebyshëv polynomials. Write a report on your findings.

9.6 PRODUCT-TO-SUM AND SUM-TO-PRODUCT FORMULAS

OBJECTIVES

1 Express Products as Sums

2 Express Sums as Products

1 The sum and difference formulas can be used to derive formulas for writing the products of sines and/or cosines as sums or differences. These identities are usually called the **Product-to-Sum Formulas.**

Theorem **Product-to-Sum Formulas**

$$\sin\alpha\sin\beta = \frac{1}{2}[\cos(\alpha - \beta) - \cos(\alpha + \beta)] \qquad (1)$$

$$\cos\alpha\cos\beta = \frac{1}{2}[\cos(\alpha - \beta) + \cos(\alpha + \beta)] \qquad (2)$$

$$\sin\alpha\cos\beta = \frac{1}{2}[\sin(\alpha + \beta) + \sin(\alpha - \beta)] \qquad (3)$$

These formulas do not have to be memorized. Instead, you should remember how they are derived. Then, when you want to use them, either look them up or derive them, as needed.

To derive formulas (1) and (2), write down the sum and difference formulas for the cosine:

$$\cos(\alpha - \beta) = \cos\alpha\cos\beta + \sin\alpha\sin\beta \qquad (4)$$
$$\cos(\alpha + \beta) = \cos\alpha\cos\beta - \sin\alpha\sin\beta \qquad (5)$$

Subtract equation (5) from equation (4) to get

$$\cos(\alpha - \beta) - \cos(\alpha + \beta) = 2\sin\alpha\sin\beta$$

from which

$$\sin\alpha\sin\beta = \frac{1}{2}[\cos(\alpha - \beta) - \cos(\alpha + \beta)]$$

Now, add equations (4) and (5) to get

$$\cos(\alpha - \beta) + \cos(\alpha + \beta) = 2\cos\alpha\cos\beta$$

from which

$$\cos\alpha\cos\beta = \frac{1}{2}[\cos(\alpha - \beta) + \cos(\alpha + \beta)]$$

To derive Product-to-Sum Formula (3), use the Sum and Difference Formulas for sine in a similar way. (You are asked to do this in Problem 41.)

EXAMPLE 1   **Expressing Products as Sums**

Express each of the following products as a sum containing only sines or cosines.

(a) $\sin(6\theta)\sin(4\theta)$ (b) $\cos(3\theta)\cos\theta$ (c) $\sin(3\theta)\cos(5\theta)$

Solution   (a) We use formula (1) to get

$$\sin(6\theta)\sin(4\theta) = \frac{1}{2}[\cos(6\theta - 4\theta) - \cos(6\theta + 4\theta)]$$

$$= \frac{1}{2}[\cos(2\theta) - \cos(10\theta)]$$

(b) We use formula (2) to get

$$\cos(3\theta)\cos\theta = \frac{1}{2}[\cos(3\theta - \theta) + \cos(3\theta + \theta)]$$

$$= \frac{1}{2}[\cos(2\theta) + \cos(4\theta)]$$

(c) We use formula (3) to get

$$\sin(3\theta)\cos(5\theta) = \frac{1}{2}[\sin(3\theta + 5\theta) + \sin(3\theta - 5\theta)]$$

$$= \frac{1}{2}[\sin(8\theta) + \sin(-2\theta)] = \frac{1}{2}[\sin(8\theta) - \sin(2\theta)] \quad \blacksquare$$

NOW WORK PROBLEM 1.

②   The **Sum-to-Product Formulas** are given next.

Theorem   **Sum-to-Product Formulas**

$$\sin \alpha + \sin \beta = 2 \sin \frac{\alpha + \beta}{2} \cos \frac{\alpha - \beta}{2} \tag{6}$$

$$\sin \alpha - \sin \beta = 2 \sin \frac{\alpha - \beta}{2} \cos \frac{\alpha + \beta}{2} \tag{7}$$

$$\cos \alpha + \cos \beta = 2 \cos \frac{\alpha + \beta}{2} \cos \frac{\alpha - \beta}{2} \tag{8}$$

$$\cos \alpha - \cos \beta = -2 \sin \frac{\alpha + \beta}{2} \sin \frac{\alpha - \beta}{2} \tag{9}$$

We will derive formula (6) and leave the derivations of formulas (7) through (9) as exercises (see Problems 42 through 44).

Proof

$$2 \sin \frac{\alpha + \beta}{2} \cos \frac{\alpha - \beta}{2} = 2 \cdot \frac{1}{2} \left[\sin \left(\frac{\alpha + \beta}{2} + \frac{\alpha - \beta}{2} \right) + \sin \left(\frac{\alpha + \beta}{2} - \frac{\alpha - \beta}{2} \right) \right]$$

$$\uparrow$$
Product-to-Sum Formula (3)

$$= \sin \frac{2\alpha}{2} + \sin \frac{2\beta}{2} = \sin \alpha + \sin \beta$$

EXAMPLE 2   **Expressing Sums (or Differences) as a Product**

Express each sum or difference as a product of sines and/or cosines.

(a) $\sin(5\theta) - \sin(3\theta)$ (b) $\cos(3\theta) + \cos(2\theta)$

Solution   (a) We use formula (7) to get

$$\sin(5\theta) - \sin(3\theta) = 2 \sin \frac{5\theta - 3\theta}{2} \cos \frac{5\theta + 3\theta}{2}$$

$$= 2 \sin \theta \cos(4\theta)$$

(b) $\cos(3\theta) + \cos(2\theta) = 2 \cos \dfrac{3\theta + 2\theta}{2} \cos \dfrac{3\theta - 2\theta}{2}$ Formula (8)

$$= 2 \cos \frac{5\theta}{2} \cos \frac{\theta}{2}$$

NOW WORK PROBLEM 11.

9.6 Exercises

In Problems 1–10, express each product as a sum containing only sines or cosines.

1. $\sin(4\theta) \sin(2\theta)$ **2.** $\cos(4\theta) \cos(2\theta)$ **3.** $\sin(4\theta) \cos(2\theta)$ **4.** $\sin(3\theta) \sin(5\theta)$

5. $\cos(3\theta)\cos(5\theta)$

6. $\sin(4\theta)\cos(6\theta)$

7. $\sin\theta\sin(2\theta)$

8. $\cos(3\theta)\cos(4\theta)$

9. $\sin\dfrac{3\theta}{2}\cos\dfrac{\theta}{2}$

10. $\sin\dfrac{\theta}{2}\cos\dfrac{5\theta}{2}$

In Problems 11–18, express each sum or difference as a product of sines and/or cosines.

11. $\sin(4\theta) - \sin(2\theta)$

12. $\sin(4\theta) + \sin(2\theta)$

13. $\cos(2\theta) + \cos(4\theta)$

14. $\cos(5\theta) - \cos(3\theta)$

15. $\sin\theta + \sin(3\theta)$

16. $\cos\theta + \cos(3\theta)$

17. $\cos\dfrac{\theta}{2} - \cos\dfrac{3\theta}{2}$

18. $\sin\dfrac{\theta}{2} - \sin\dfrac{3\theta}{2}$

In Problems 19–36, establish each identity.

19. $\dfrac{\sin\theta + \sin(3\theta)}{2\sin(2\theta)} = \cos\theta$

20. $\dfrac{\cos\theta + \cos(3\theta)}{2\cos(2\theta)} = \cos\theta$

21. $\dfrac{\sin(4\theta) + \sin(2\theta)}{\cos(4\theta) + \cos(2\theta)} = \tan(3\theta)$

22. $\dfrac{\cos\theta - \cos(3\theta)}{\sin(3\theta) - \sin\theta} = \tan(2\theta)$

23. $\dfrac{\cos\theta - \cos(3\theta)}{\sin\theta + \sin(3\theta)} = \tan\theta$

24. $\dfrac{\cos\theta - \cos(5\theta)}{\sin\theta + \sin(5\theta)} = \tan(2\theta)$

25. $\sin\theta[\sin\theta + \sin(3\theta)] = \cos\theta[\cos\theta - \cos(3\theta)]$

26. $\sin\theta[\sin(3\theta) + \sin(5\theta)] = \cos\theta[\cos(3\theta) - \cos(5\theta)]$

27. $\dfrac{\sin(4\theta) + \sin(8\theta)}{\cos(4\theta) + \cos(8\theta)} = \tan(6\theta)$

28. $\dfrac{\sin(4\theta) - \sin(8\theta)}{\cos(4\theta) - \cos(8\theta)} = -\cot(6\theta)$

29. $\dfrac{\sin(4\theta) + \sin(8\theta)}{\sin(4\theta) - \sin(8\theta)} = -\dfrac{\tan(6\theta)}{\tan(2\theta)}$

30. $\dfrac{\cos(4\theta) - \cos(8\theta)}{\cos(4\theta) + \cos(8\theta)} = \tan(2\theta)\tan(6\theta)$

31. $\dfrac{\sin\alpha + \sin\beta}{\sin\alpha - \sin\beta} = \tan\dfrac{\alpha+\beta}{2}\cot\dfrac{\alpha-\beta}{2}$

32. $\dfrac{\cos\alpha + \cos\beta}{\cos\alpha - \cos\beta} = -\cot\dfrac{\alpha+\beta}{2}\cot\dfrac{\alpha-\beta}{2}$

33. $\dfrac{\sin\alpha + \sin\beta}{\cos\alpha + \cos\beta} = \tan\dfrac{\alpha+\beta}{2}$

34. $\dfrac{\sin\alpha - \sin\beta}{\cos\alpha - \cos\beta} = -\cot\dfrac{\alpha+\beta}{2}$

35. $1 + \cos(2\theta) + \cos(4\theta) + \cos(6\theta) = 4\cos\theta\cos(2\theta)\cos(3\theta)$

36. $1 - \cos(2\theta) + \cos(4\theta) - \cos(6\theta) = 4\sin\theta\cos(2\theta)\sin(3\theta)$

---

37. Touch-Tone Phones   On a Touch-Tone phone, each button produces a unique sound. The sound produced is the sum of two tones, given by

$$y = \sin(2\pi l t) \quad \text{and} \quad y = \sin(2\pi h t)$$

where l and h are the low and high frequencies (cycles per second) shown in the illustration. For example, if you

Touch-Tone phone

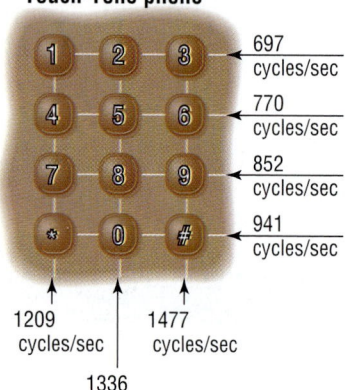

697 cycles/sec
770 cycles/sec
852 cycles/sec
941 cycles/sec

1209 cycles/sec   1477 cycles/sec

1336 cycles/sec

touch 7, the low frequency is $l = 852$ cycles per second

$$y = \sin[2\pi(852)t] + \sin[2\pi(1209)t]$$

and the high frequency is $h = 1209$ cycles per second. The sound emitted by touching 7 is

(a) Write this sound as a product of sines and/or cosines.
(b) Determine the maximum value of y.
(c) Graph the sound emitted by touching 7.

38. Touch-Tone Phones

(a) Write the sound emitted by touching the # key as a product of sines and/or cosines.
(b) Determine the maximum value of y.
(c) Graph the sound emitted by touching the # key.

39. If $\alpha + \beta + \gamma = \pi$, show that

$$\sin(2\alpha) + \sin(2\beta) + \sin(2\gamma) = 4\sin\alpha\sin\beta\sin\gamma$$

40. If $\alpha + \beta + \gamma = \pi$, show that

$$\tan\alpha + \tan\beta + \tan\gamma = \tan\alpha\tan\beta\tan\gamma$$

41. Derive formula (3).

42. Derive formula (7).

43. Derive formula (8).

44. Derive formula (9).

PREPARING FOR THIS SECTION

Before getting started, review the following:

✓ Solving Equations (Section 1.3, pp. 109–113)

✓ Values of the Trigonometric Functions of Certain Angles (Section 8.3, p. 638, and Section 8.4, p. 646)

9.7 TRIGONOMETRIC EQUATIONS (I)

OBJECTIVES   **1** Solve Equations Involving a Single Trigonometric Function

1 The previous four sections of this chapter were devoted to trigonometric identities, that is, equations involving trigonometric functions that are satisfied by every value in the domain of the variable. In the remaining two sections, we discuss **trigonometric equations**, that is, equations involving trigonometric functions that are satisfied only by some values of the variable (or, possibly, are not satisfied by any values of the variable). The values that satisfy the equation are called **solutions** of the equation.

EXAMPLE 1 **Checking Whether a Given Number Is a Solution of a Trigonometric Equation**

Determine whether $\theta = \dfrac{\pi}{4}$ is a solution of the equation $\sin \theta = \dfrac{1}{2}$. Is $\theta = \dfrac{\pi}{6}$ a solution?

Solution Replace θ by $\dfrac{\pi}{4}$ in the given equation. The result is

$$\sin \frac{\pi}{4} = \frac{\sqrt{2}}{2} \neq \frac{1}{2}$$

We conclude that $\dfrac{\pi}{4}$ is not a solution.

Next, replace θ by $\dfrac{\pi}{6}$ in the equation. The result is

$$\sin \frac{\pi}{6} = \frac{1}{2}$$

We conclude that $\dfrac{\pi}{6}$ is a solution of the given equation. ■

Figure 29

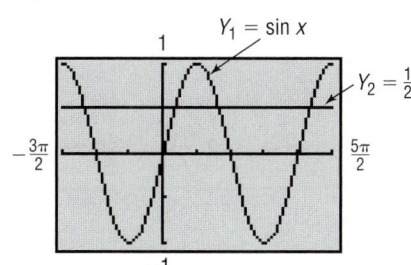

The equation given in Example 1 has other solutions besides $\theta = \dfrac{\pi}{6}$. For example, $\theta = \dfrac{5\pi}{6}$ is also a solution, as is $\theta = \dfrac{13\pi}{6}$. (You should check this for yourself.) In fact, the equation has an infinite number of solutions due to the periodicity of the sine function. See Figure 29.

As before, our practice will be to solve equations, whenever possible, by finding exact solutions. In such cases, we will also verify the solution obtained by using a graphing utility. When traditional methods cannot be used, approximate solutions will be obtained using a graphing utility. The reader is encouraged to pay particular attention to the form of equations for which exact solutions are possible.

Unless the domain of the variable is restricted, we need to find *all* the solutions of a trigonometric equation. As the next example illustrates, finding all the solutions can be accomplished by first finding solutions over an interval whose length equals the period of the function and then adding multiples of that period to the solutions found. Let's look at some examples.

EXAMPLE 2   **Finding All the Solutions of a Trigonometric Equation**

Solve the equation: $\cos \theta = \dfrac{1}{2}$

Give a general formula for all the solutions. List six solutions.

Figure 30

$\cos \theta = \dfrac{1}{2}$

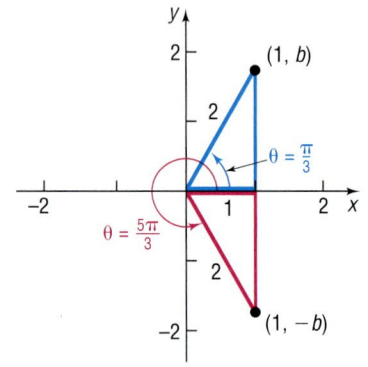

Solution   The period of the cosine function is 2π. In the interval $[0, 2\pi)$, the two angles θ for which $\cos \theta = \dfrac{1}{2}$ are $\theta = \dfrac{\pi}{3}$ and $\theta = \dfrac{5\pi}{3}$. See Figure 30. Because the cosine function has period 2π, all the solutions of $\cos \theta = \dfrac{1}{2}$ may be given by the general formula

$$\theta = \frac{\pi}{3} + 2k\pi \quad \text{or} \quad \theta = \frac{5\pi}{3} + 2k\pi \qquad \textit{k any integer}$$

Some of the solutions are

$$\underbrace{\frac{\pi}{3}, \frac{5\pi}{3},}_{k = 0} \quad \underbrace{\frac{7\pi}{3}, \frac{11\pi}{3},}_{k = 1} \quad \underbrace{\frac{13\pi}{3}, \frac{17\pi}{3},}_{k = 2} \quad \text{and so on}$$

Figure 31

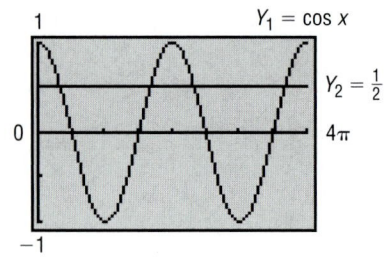

✔ CHECK: We can verify the solutions by graphing $Y_1 = \cos x$ and $Y_2 = \dfrac{1}{2}$ to determine where the graphs intersect. (Be sure to graph in radian mode.) See Figure 31. The graph of Y_1 intersects the graph of Y_2 at $x = 1.05\left(\approx \dfrac{\pi}{3}\right), 5.24\left(\approx \dfrac{5\pi}{3}\right), 7.33\left(\approx \dfrac{7\pi}{3}\right)$, and $11.52\left(\approx \dfrac{11\pi}{3}\right)$, rounded to two decimal places.   ■ ■

NOW WORK PROBLEM 1.

In most of our work, we shall be interested only in finding solutions of trigonometric equations for $0 \le \theta < 2\pi$.

EXAMPLE 3   **Solving a Linear Trigonometric Equation**

Solve the equation: $2 \sin \theta + \sqrt{3} = 0, \quad 0 \le \theta < 2\pi$

Solution   We solve the equation for $\sin \theta$.

$$2 \sin \theta + \sqrt{3} = 0$$
$$2 \sin \theta = -\sqrt{3} \qquad \text{Subtract } \sqrt{3} \text{ from both sides.}$$
$$\sin \theta = -\frac{\sqrt{3}}{2} \qquad \text{Divide both sides by 2.}$$

The period of the sine function is 2π. In the interval $[0, 2\pi)$, the two angles θ for which $\sin\theta = -\dfrac{\sqrt{3}}{2}$ are $\theta = \dfrac{4\pi}{3}$ and $\theta = \dfrac{5\pi}{3}$.

━━━ **NOW WORK PROBLEM 11.**

EXAMPLE 4 | **Solving a Trigonometric Equation**

Solve the equation: $\sin(2\theta) = \dfrac{1}{2}, \quad 0 \le \theta < 2\pi$

Solution

Figure 32

$\sin(2\theta) = \dfrac{1}{2}, 0 \le \theta < 2\pi$

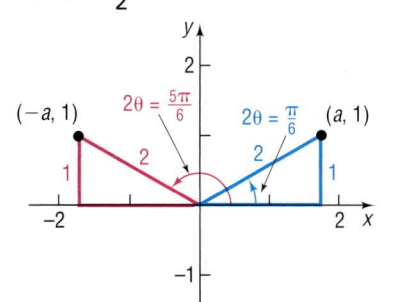

The period of the sine function is 2π. In the interval $[0, 2\pi)$, the sine function has a value $\dfrac{1}{2}$ at $\dfrac{\pi}{6}$ and $\dfrac{5\pi}{6}$. See Figure 32. Because the argument is 2θ in the equation $\sin(2\theta) = \dfrac{1}{2}$, we have

$$2\theta = \dfrac{\pi}{6} + 2k\pi \quad \text{or} \quad 2\theta = \dfrac{5\pi}{6} + 2k\pi \qquad \text{\textit{k} any integer}$$

$$\theta = \dfrac{\pi}{12} + k\pi \qquad\qquad \theta = \dfrac{5\pi}{12} + k\pi \qquad \text{Divide by 2.}$$

Then

$$\theta = \dfrac{\pi}{12} + (-1)\pi = -\dfrac{11\pi}{12} \qquad k = -1 \qquad \theta = \dfrac{5\pi}{12} + (-1)\pi = -\dfrac{7\pi}{12}$$

$$\theta = \dfrac{\pi}{12} + (0)\pi = \dfrac{\pi}{12} \qquad\quad k = 0 \qquad\quad \theta = \dfrac{5\pi}{12} + (0)\pi = \dfrac{5\pi}{12}$$

$$\theta = \dfrac{\pi}{12} + (1)\pi = \dfrac{13\pi}{12} \qquad k = 1 \qquad\quad \theta = \dfrac{5\pi}{12} + (1)\pi = \dfrac{17\pi}{12}$$

$$\theta = \dfrac{\pi}{12} + (2)\pi = \dfrac{25\pi}{12} \qquad k = 2 \qquad\quad \theta = \dfrac{5\pi}{12} + (2)\pi = \dfrac{29\pi}{12}$$

In the interval $[0, 2\pi)$, the solutions of $\sin(2\theta) = \dfrac{1}{2}$ are $\theta = \dfrac{\pi}{12}$, $\theta = \dfrac{13\pi}{12}, \theta = \dfrac{5\pi}{12}$, and $\theta = \dfrac{17\pi}{12}$.

✔ CHECK: Verify these solutions by graphing $Y_1 = \sin(2x)$ and $Y_2 = \dfrac{1}{2}$ for $0 \le x \le 2\pi$. ■ ■

WARNING: In solving a trigonometric equation for $\theta, 0 \le \theta < 2\pi$, in which the argument is not θ (as in Example 4), you must write down all the solutions first and then list those that are in the interval $[0, 2\pi)$. Otherwise, solutions may be lost. For example, in solving $\sin(2\theta) = \dfrac{1}{2}$, if you merely write the solutions $2\theta = \dfrac{\pi}{6}$ and $2\theta = \dfrac{5\pi}{6}$, you will find only $\theta = \dfrac{\pi}{12}$ and $\theta = \dfrac{5\pi}{12}$ and miss the other solutions. ■

━━━ **NOW WORK PROBLEM 17.**

EXAMPLE 5 **Solving a Trigonometric Equation**

Solve the equation: $\tan\left(\theta - \dfrac{\pi}{2}\right) = 1, \quad 0 \le \theta < 2\pi$

Solution The period of the tangent function is π. In the interval $[0, \pi)$, the tangent function has the value 1 when the argument is $\dfrac{\pi}{4}$. Because the argument is $\theta - \dfrac{\pi}{2}$ in the given equation, we have

$$\theta - \frac{\pi}{2} = \frac{\pi}{4} + k\pi \qquad \textcolor{blue}{k \text{ any integer}}$$

$$\theta = \frac{3\pi}{4} + k\pi$$

In the interval $[0, 2\pi)$, $\theta = \dfrac{3\pi}{4}$ and $\theta = \dfrac{3\pi}{4} + \pi = \dfrac{7\pi}{4}$ are the only solutions.

✔ CHECK: Verify these solutions using a graphing utility. ▪ ▪

The next example illustrates how to solve trigonometric equations using a calculator. Remember that the function keys on a calculator will only give values consistent with the definition of the function.

EXAMPLE 6 **Solving a Trigonometric Equation with a Calculator**

Use a calculator to solve the equation: $\sin\theta = 0.3, \quad 0 \le \theta < 2\pi$ Express any solutions in radians, rounded to two decimal places.

Solution To solve $\sin\theta = 0.3$ on a calculator, first set the mode to radians. Then use the $\boxed{\sin^{-1}}$ key to obtain

$$\theta = \sin^{-1}(0.3) \approx 0.304692654$$

Rounded to two decimal places, $\theta = \sin^{-1}(0.3) = 0.30$ radian. Because of the definition of $y = \sin^{-1}x$, the angle θ that we obtain is the angle $-\dfrac{\pi}{2} \le \theta \le \dfrac{\pi}{2}$ for which $\sin\theta = 0.3$. Another angle for which $\sin\theta = 0.3$ is $\pi - 0.30$. See Figure 33. The angle $\pi - 0.30$ is the angle in quadrant II, where $\sin\theta = 0.3$. The solutions for $\sin\theta = 0.3, 0 \le \theta < 2\pi$, are

$$\theta = 0.30 \text{ radian} \quad \text{and} \quad \theta = \pi - 0.30 \approx 2.84 \text{ radians} \quad ▪$$

Figure 33
$\sin\theta = 0.3$

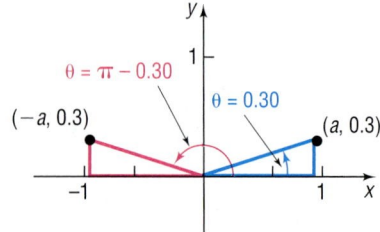

A second method for solving $\sin\theta = 0.3, 0 \le \theta < 2\pi$ would be to graph $Y_1 = \sin x$ and $Y_2 = 0.3$ for $0 \le x < 2\pi$ and find their point(s) of intersection. Try this method for yourself to verify the results obtained in Example 6.

WARNING: Example 6 illustrates that caution must be exercised when solving trigonometric equations on a calculator. Remember that the calculator supplies an angle only within the restrictions of the definition of the inverse trigonometric function. To find the remaining solutions, you must identify other quadrants, if any, in which the angle may be located. ▪

✏ **NOW WORK PROBLEM 35.**

9.7 Concepts and Vocabulary

In Problems 1 and 2, fill in the blanks.

1. Two solutions of the equation $\sin \theta = \dfrac{1}{2}$ are _____ and _____.

2. All the solutions of the equation $\sin \theta = \dfrac{1}{2}$ are _____.

In Problems 3–5, answer True or False to each statement.

3. Most trigonometric equations have unique solutions.

4. The equation $\tan^{-1} \theta = \dfrac{\pi}{2}$ has no solution.

5. The equation $\sin \theta = 2$ has a solution that can be found using a graphing calculator.

9.7 Exercises

In Problems 1–10, solve each equation. Give a general formula for all the solutions. List six solutions.

1. $\sin \theta = \dfrac{1}{2}$ **2.** $\tan \theta = 1$ **3.** $\tan \theta = -\dfrac{\sqrt{3}}{3}$ **4.** $\cos \theta = -\dfrac{\sqrt{3}}{2}$ **5.** $\cos \theta = 0$

6. $\sin \theta = \dfrac{\sqrt{2}}{2}$ **7.** $\cos(2\theta) = -\dfrac{1}{2}$ **8.** $\sin(2\theta) = -1$ **9.** $\sin \dfrac{\theta}{2} = -\dfrac{\sqrt{3}}{2}$ **10.** $\tan \dfrac{\theta}{2} = -1$

In Problems 11–34, solve each equation on the interval $0 \le \theta < 2\pi$.

11. $2 \sin \theta + 3 = 2$ **12.** $1 - \cos \theta = \dfrac{1}{2}$ **13.** $4 \cos^2 \theta = 1$ **14.** $\tan^2 \theta = \dfrac{1}{3}$

15. $2 \sin^2 \theta - 1 = 0$ **16.** $4 \cos^2 \theta - 3 = 0$ **17.** $\sin(3\theta) = -1$ **18.** $\tan \dfrac{\theta}{2} = \sqrt{3}$

19. $\cos(2\theta) = -\dfrac{1}{2}$ **20.** $\tan(2\theta) = -1$ **21.** $\sec \dfrac{3\theta}{2} = -2$ **22.** $\cot \dfrac{2\theta}{3} = -\sqrt{3}$

23. $\cos\left(2\theta - \dfrac{\pi}{2}\right) = -1$ **24.** $\sin\left(3\theta + \dfrac{\pi}{18}\right) = 1$ **25.** $\tan\left(\dfrac{\theta}{2} + \dfrac{\pi}{3}\right) = 1$ **26.** $\cos\left(\dfrac{\theta}{3} - \dfrac{\pi}{4}\right) = \dfrac{1}{2}$

27. $2 \sin \theta + 1 = 0$ **28.** $\cos \theta + 1 = 0$ **29.** $\tan \theta + 1 = 0$ **30.** $\sqrt{3} \cot \theta + 1 = 0$

31. $4 \sec \theta + 6 = -2$ **32.** $5 \csc \theta - 3 = 2$ **33.** $3\sqrt{2} \cos \theta + 2 = -1$ **34.** $4 \sin \theta + 3\sqrt{3} = \sqrt{3}$

In Problems 35–42, use a calculator to solve each equation on the interval $0 \le \theta < 2\pi$. Round answers to two decimal places.

35. $\sin \theta = 0.4$ **36.** $\cos \theta = 0.6$ **37.** $\tan \theta = 5$ **38.** $\cot \theta = 2$

39. $\cos \theta = -0.9$ **40.** $\sin \theta = -0.2$ **41.** $\sec \theta = -4$ **42.** $\csc \theta = -3$

The following discussion of **Snell's Law of Refraction** (named after Willebrord Snell, 1580–1626) is needed for Problems 43–49. *Light, sound, and other waves travel at different speeds, depending on the media (air, water, wood, and so on) through which they pass. Suppose that light travels from a point A in one medium, where its speed is v_1, to a point B in another medium, where its speed is v_2. Refer to the figure (page 764), where the angle θ_1 is called the **angle of incidence** and the angle θ_2 is the **angle of refraction**. Snell's Law,* which can be proved using calculus, states that*

$$\frac{\sin \theta_1}{\sin \theta_2} = \frac{v_1}{v_2}$$

*Because this law was also deduced by René Descartes in France, it is also known as Descartes' Law.

*The ratio v_1/v_2 is called the **index of refraction**. Some values are given in the following table.*

Medium	Index of Refraction*
Water	1.33
Ethyl alcohol	1.36
Carbon bisulfide	1.63
Air (1 atm and 20°C)	1.0003
Methylene iodide	1.74
Fused quartz	1.46
Glass, crown	1.52
Glass, dense flint	1.66
Sodium chloride	1.53

SOME INDEXES OF REFRACTION

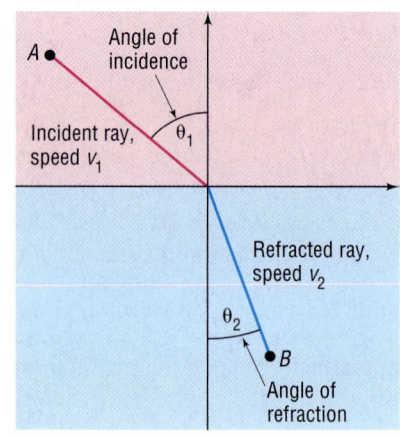

*For light of wavelength 589 nanometers, measured with respect to a vacuum. The index with respect to air is negligibly different in most cases.

43. The index of refraction of light in passing from a vacuum into water is 1.33. If the angle of incidence is 40°, determine the angle of refraction.

44. The index of refraction of light in passing from a vacuum into dense glass is 1.66. If the angle of incidence is 50°, determine the angle of refraction.

45. Ptolemy, who lived in the city of Alexandria in Egypt during the second century AD, gave the measured values in the table below for the angle of incidence θ_1 and the angle of refraction θ_2 for a light beam passing from air into water. Do these values agree with Snell's Law? If so, what index of refraction results? (These data are interesting as the oldest recorded physical measurements.)*

θ_1	θ_2	θ_1	θ_2
10°	7°45′	50°	35°0′
20°	15°30′	60°	40°30′
30°	22°30′	70°	45°30′
40°	29°0′	80°	50°0′

46. The speed of yellow sodium light (wavelength of 589 nanometers) in a certain liquid is measured to be 1.92×10^8 meters per second. What is the index of refraction of this liquid, with respect to air, for sodium light?[†]

[**Hint:** The speed of light in air is approximately 2.99×10^8 meters per second.]

47. A beam of light with a wavelength of 589 nanometers traveling in air makes an angle of incidence of 40° on a slab of transparent material, and the refracted beam makes an angle of refraction of 26°. Find the index of refraction of the material.[†]

48. A light ray with a wavelength of 589 nanometers (produced by a sodium lamp) traveling through air makes an angle of incidence of 30° on a smooth, flat slab of crown glass. Find the angle of refraction.[†]

49. A light beam passes through a thick slab of material whose index of refraction is n_2. Show that the emerging beam is parallel to the incident beam.[†]

50. Explain in your own words how you would use your calculator to solve the equation $\sin x = 0.3, 0 \le x < 2\pi$. How would you modify your approach in order to solve the equation $\cot x = 5, 0 < x < 2\pi$?

*Adapted from Halliday and Resnick, *Physics, Parts 1 & 2*, 3rd ed. New York: Wiley, 1978, p. 953.
[†]Adapted from Serway, *Physics*, 3rd ed. Philadelphia: W. B. Saunders, p. 805.

PREPARING FOR THIS SECTION

Before getting started, review the following:

✓ Solving Quadratic Equations by Factoring
(Section 1.3, pp. 117–119)

✓ The Quadratic Formula (Section 1.3, p. 122)

✓ Solving Equations Using a Graphing Utility
(Section 1.3, pp. 113–115)

✓ Equations Quadratic in Form (Section 1.5, pp. 145–146)

9.8 TRIGONOMETRIC EQUATIONS (II)

OBJECTIVES

1. Solve Trigonometric Equations Quadratic in Form
2. Solve Trigonometric Equations Using Identities
3. Solve Trigonometric Equations Linear in Sine and Cosine
4. Solve Trigonometric Equations Using a Graphing Utility

1. In this section we continue our study of trigonometric equations. Many trigonometric equations can be solved by applying techniques that we already know, such as applying the quadratic formula (if the equation is a second-degree polynomial) or factoring.

EXAMPLE 1 | Solving a Trigonometric Equation Quadratic in Form

Solve the equation: $2 \sin^2 \theta - 3 \sin \theta + 1 = 0, \quad 0 \le \theta < 2\pi$

Solution The equation that we wish to solve is a quadratic equation (in $\sin \theta$) that can be factored.

$$2 \sin^2 \theta - 3 \sin \theta + 1 = 0 \qquad 2x^2 - 3x + 1 = 0, \quad x = \sin \theta$$
$$(2 \sin \theta - 1)(\sin \theta - 1) = 0 \qquad (2x - 1)(x - 1) = 0$$
$$2 \sin \theta - 1 = 0 \quad \text{or} \quad \sin \theta - 1 = 0$$
$$\sin \theta = \frac{1}{2} \qquad \qquad \sin \theta = 1$$

Solving each equation in the interval $[0, 2\pi)$, we obtain

$$\theta = \frac{\pi}{6}, \qquad \theta = \frac{5\pi}{6}, \qquad \theta = \frac{\pi}{2}$$

━━━ **NOW WORK PROBLEM 3.**

2. When a trigonometric equation contains more than one trigonometric function, identities sometimes can be used to obtain an equivalent equation that contains only one trigonometric function.

EXAMPLE 2 | Solving a Trigonometric Equation Using Identities

Solve the equation: $3 \cos \theta + 3 = 2 \sin^2 \theta, \quad 0 \le \theta < 2\pi$

Solution The equation in its present form contains sines and cosines. However, a form of the Pythagorean Identity can be used to transform the equation into an equivalent expression containing only cosines.

$$3 \cos \theta + 3 = 2 \sin^2 \theta$$

$$3 \cos \theta + 3 = 2(1 - \cos^2 \theta) \qquad {\scriptstyle \sin^2 \theta \,=\, 1\, -\, \cos^2 \theta}$$

$$3 \cos \theta + 3 = 2 - 2 \cos^2 \theta$$

$$2 \cos^2 \theta + 3 \cos \theta + 1 = 0 \qquad\qquad {\scriptstyle \text{Quadratic in } \cos \theta}$$

$$(2 \cos \theta + 1)(\cos \theta + 1) = 0 \qquad\qquad {\scriptstyle \text{Factor.}}$$

$$2 \cos \theta + 1 = 0 \quad \text{or} \quad \cos \theta + 1 = 0$$

$$\cos \theta = -\frac{1}{2} \qquad\qquad \cos \theta = -1$$

Solving each equation in the interval $[0, 2\pi)$, we obtain

$$\theta = \frac{2\pi}{3}, \qquad \theta = \frac{4\pi}{3}, \qquad \theta = \pi$$

✔ CHECK: Graph $Y_1 = 3 \cos x + 3$ and $Y_2 = 2 \sin^2 x$, $0 \le x \le 2\pi$, and find the points of intersection. How close are your approximate solutions to the exact ones found in this example? ■ ■

EXAMPLE 3	Solving a Trigonometric Equation Using Identities

Solve the equation: $\cos(2\theta) + 3 = 5 \cos \theta, \quad 0 \le \theta < 2\pi$

Solution First, we observe that the given equation contains two cosine functions, but with different arguments, θ and 2θ. We use the Double-Angle Formula $\cos(2\theta) = 2 \cos^2 \theta - 1$ to obtain an equivalent equation containing only $\cos \theta$.

$$\cos(2\theta) + 3 = 5 \cos \theta$$

$$(2 \cos^2 \theta - 1) + 3 = 5 \cos \theta$$

$$2 \cos^2 \theta - 5 \cos \theta + 2 = 0$$

$$(\cos \theta - 2)(2 \cos \theta - 1) = 0$$

$$\cos \theta = 2 \quad \text{or} \quad \cos \theta = \frac{1}{2}$$

For any angle θ, $-1 \le \cos \theta \le 1$; therefore, the equation $\cos \theta = 2$ has no solution. The solutions of $\cos \theta = \frac{1}{2}$, $0 \le \theta < 2\pi$, are

$$\theta = \frac{\pi}{3}, \qquad \theta = \frac{5\pi}{3}$$

✔ CHECK: Graph $Y_1 = \cos(2x) + 3$ and $Y_2 = 5 \cos x$, $0 \le x \le 2\pi$, and find the points of intersection. ■ ■

✏ **NOW WORK PROBLEM 19.**

EXAMPLE 4	Solving a Trigonometric Equation Using Identities

Solve the equation: $\cos^2 \theta + \sin \theta = 2, \quad 0 \le \theta < 2\pi$

Solution   This equation involves two trigonometric functions, sine and cosine. Since it is easier to work with only one, we use a form of the Pythagorean Identity, $\sin^2\theta + \cos^2\theta = 1$ to rewrite the equation.

$$\cos^2\theta + \sin\theta = 2$$

$$(1 - \sin^2\theta) + \sin\theta = 2 \qquad \cos^2\theta = 1 - \sin^2\theta$$

$$\sin^2\theta - \sin\theta + 1 = 0$$

This is a quadratic equation in $\sin\theta$. The discriminant is $b^2 - 4ac = 1 - 4 = -3 < 0$. Therefore, the equation has no real solution.

✔ CHECK: Graph $Y_1 = \cos^2 x + \sin x$ and $Y_2 = 2$. See Figure 34. The two graphs do not intersect, so the equation $Y_1 = Y_2$ has no real solution.   ■ ■

Figure 34

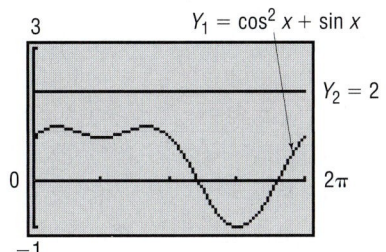

$Y_1 = \cos^2 x + \sin x$

$Y_2 = 2$

| EXAMPLE 5 | **Solving a Trigonometric Equation Using Identities** |

Solve the equation: $\sin\theta\cos\theta = -\dfrac{1}{2}, \quad 0 \le \theta < 2\pi$

Solution   The left side of the given equation is in the form of the Double-Angle Formula $2\sin\theta\cos\theta = \sin(2\theta)$, except for a factor of 2. We multiply each side by 2.

$$\sin\theta\cos\theta = -\frac{1}{2}$$

$$2\sin\theta\cos\theta = -1 \qquad \text{Multiply each side by 2.}$$

$$\sin(2\theta) = -1 \qquad \text{Double-Angle Formula}$$

The argument here is 2θ. So we need to write all the solutions of this equation and then list those that are in the interval $[0, 2\pi)$.

$$2\theta = \frac{3\pi}{2} + 2k\pi \qquad \text{k any integer}$$

$$\theta = \frac{3\pi}{4} + k\pi$$

$$\theta = \frac{3\pi}{4} + (-1)\pi = -\frac{\pi}{4}, \quad \theta = \frac{3\pi}{4} + (0)\pi = \frac{3\pi}{4}, \quad \theta = \frac{3\pi}{4} + (1)\pi = \frac{7\pi}{4}, \quad \theta = \frac{3\pi}{4} + (2)\pi = \frac{11\pi}{4}$$
$$\uparrow \qquad\qquad \uparrow \qquad\qquad \uparrow \qquad\qquad \uparrow$$
$$k = -1 \qquad\quad k = 0 \qquad\quad k = 1 \qquad\quad k = 2$$

The solutions in the interval $[0, 2\pi)$ are

$$\theta = \frac{3\pi}{4}, \qquad \theta = \frac{7\pi}{4} \qquad\qquad ■$$

③   Sometimes it is necessary to square both sides of an equation in order to obtain expressions that allow the use of identities. Remember, however, that when squaring both sides, extraneous solutions may be introduced. As a result, apparent solutions must be checked.

| EXAMPLE 6 | **Other Methods for Solving a Trigonometric Equation** |

Solve the equation: $\sin\theta + \cos\theta = 1,\quad 0 \le \theta < 2\pi$

Solution A   Attempts to use available identities do not lead to equations that are easy to solve. (Try it yourself.) Given the form of this equation, we decide to square each side.

$$\sin\theta + \cos\theta = 1$$
$$(\sin\theta + \cos\theta)^2 = 1 \qquad \textit{Square each side.}$$
$$\sin^2\theta + 2\sin\theta\cos\theta + \cos^2\theta = 1 \qquad \textit{Remove parentheses.}$$
$$2\sin\theta\cos\theta = 0 \qquad \textit{$\sin^2\theta + \cos^2\theta = 1$}$$
$$\sin\theta\cos\theta = 0$$

Setting each factor equal to zero, we obtain

$$\sin\theta = 0 \quad \text{or} \quad \cos\theta = 0$$

The apparent solutions are

$$\theta = 0, \qquad \theta = \pi, \qquad \theta = \frac{\pi}{2}, \qquad \theta = \frac{3\pi}{2}$$

Because we squared both sides of the original equation, we must check these apparent solutions to see if any are extraneous.

$$\theta = 0: \quad \sin 0 + \cos 0 = 0 + 1 = 1 \qquad \textit{A solution}$$
$$\theta = \pi: \quad \sin\pi + \cos\pi = 0 + (-1) = -1 \qquad \textit{Not a solution}$$
$$\theta = \frac{\pi}{2}: \quad \sin\frac{\pi}{2} + \cos\frac{\pi}{2} = 1 + 0 = 1 \qquad \textit{A solution}$$
$$\theta = \frac{3\pi}{2}: \quad \sin\frac{3\pi}{2} + \cos\frac{3\pi}{2} = -1 + 0 = -1 \qquad \textit{Not a solution}$$

Therefore, $\theta = \dfrac{3\pi}{2}$ and $\theta = \pi$ are extraneous. The only solutions are $\theta = 0$ and $\theta = \dfrac{\pi}{2}$. ■

We can solve the equation given in Example 6 in another way.

Solution B   We start with the equation

$$\sin\theta + \cos\theta = 1$$

and divide each side by $\sqrt{2}$. (The reason for this choice will become apparent shortly.) Then

$$\frac{1}{\sqrt{2}}\sin\theta + \frac{1}{\sqrt{2}}\cos\theta = \frac{1}{\sqrt{2}}$$

The left side now resembles the formula for the sine of the sum of two angles, one of which is θ. The other angle is unknown (call it ϕ.) Then

$$\sin(\theta + \phi) = \sin\theta\cos\phi + \cos\theta\sin\phi = \frac{1}{\sqrt{2}} = \frac{\sqrt{2}}{2} \qquad (1)$$

where

$$\cos\phi = \frac{1}{\sqrt{2}} = \frac{\sqrt{2}}{2}, \qquad \sin\phi = \frac{1}{\sqrt{2}} = \frac{\sqrt{2}}{2}, \qquad 0 \le \phi < 2\pi$$

The angle ϕ is therefore $\dfrac{\pi}{4}$. As a result, equation (1) becomes

$$\sin\left(\theta + \frac{\pi}{4}\right) = \frac{\sqrt{2}}{2}$$

Figure 35

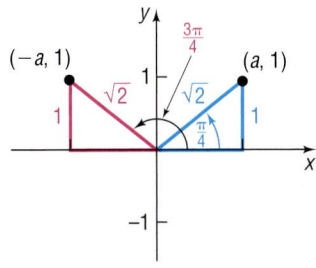

There are two angles whose sine is $\dfrac{\sqrt{2}}{2}$: $\dfrac{\pi}{4}$ and $\dfrac{3\pi}{4}$. See Figure 35. As a result,

$$\theta + \frac{\pi}{4} = \frac{\pi}{4} \quad \text{or} \quad \theta + \frac{\pi}{4} = \frac{3\pi}{4}$$
$$\theta = 0 \qquad\qquad \theta = \frac{\pi}{2}$$

These solutions agree with the solutions found earlier. ■

This second method of solution can be used to solve any linear equation in the variables $\sin\theta$ and $\cos\theta$.

EXAMPLE 7 **Solving a Trigonometric Equation Linear in $\sin\theta$ and $\cos\theta$**

Solve:

$$a\sin\theta + b\cos\theta = c, \qquad 0 \le \theta < 2\pi \tag{2}$$

where a, b, and c are constants and either $a \ne 0$ or $b \ne 0$.

Solution We divide each side of equation (2) by $\sqrt{a^2 + b^2}$. Then

$$\frac{a}{\sqrt{a^2 + b^2}}\sin\theta + \frac{b}{\sqrt{a^2 + b^2}}\cos\theta = \frac{c}{\sqrt{a^2 + b^2}} \tag{3}$$

There is a unique angle ϕ, $0 \le \phi < 2\pi$, for which

$$\cos\phi = \frac{a}{\sqrt{a^2 + b^2}} \quad \text{and} \quad \sin\phi = \frac{b}{\sqrt{a^2 + b^2}} \tag{4}$$

Figure 36

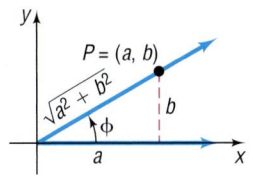

(see Figure 36). Equation (3) may be written as

$$\sin\theta\cos\phi + \cos\theta\sin\phi = \frac{c}{\sqrt{a^2 + b^2}}$$

or, equivalently,

$$\sin(\theta + \phi) = \frac{c}{\sqrt{a^2 + b^2}} \tag{5}$$

where ϕ satisfies equations (4).

If $|c| > \sqrt{a^2 + b^2}$, then $\sin(\theta + \phi) > 1$ or $\sin(\theta + \phi) < -1$, and equation (5) has no solution.

If $|c| \le \sqrt{a^2 + b^2}$, then the solutions of equation (5) are

$$\theta + \phi = \sin^{-1}\frac{c}{\sqrt{a^2 + b^2}} \quad \text{or} \quad \theta + \phi = \pi - \sin^{-1}\frac{c}{\sqrt{a^2 + b^2}}$$

Because the angle ϕ is determined by equations (4), these are the solutions to equation (2). ■

✏ **NOW WORK PROBLEM 33.**

⚠ ④ ## Graphing Utility Solutions

The techniques introduced in this section apply only to certain types of trigonometric equations. Solutions for other types are usually studied in calculus, using numerical methods. In the next example, we show how a graphing utility may be used to obtain solutions.

EXAMPLE 8   ### Solving Trigonometric Equations Using a Graphing Utility

Solve: $5 \sin x + x = 3$
Express the solution(s) rounded to two decimal places.

Solution   This type of trigonometric equation cannot be solved by previous methods. A graphing utility, though, can be used here. The solutions of this equation are the same as the points of intersection of the graphs of $Y_1 = 5 \sin x + x$ and $Y_2 = 3$. See Figure 37.

Figure 37

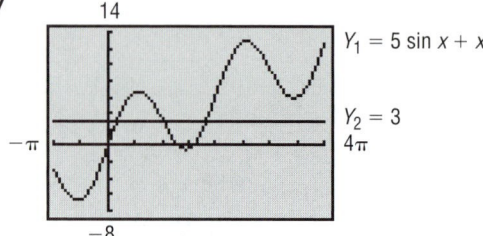

There are three points of intersection; the x-coordinates are the solutions that we seek. Using INTERSECT, we find

$$x = 0.52, \qquad x = 3.18, \qquad x = 5.71$$

rounded to two decimal places.

━━✏ **NOW WORK PROBLEM 45.**

9.8 Exercises

In Problems 1–38, solve each equation on the interval $0 \le \theta < 2\pi$.

1. $2 \cos^2 \theta + \cos \theta = 0$

2. $\sin^2 \theta - 1 = 0$

3. $2 \sin^2 \theta - \sin \theta - 1 = 0$

4. $2 \cos^2 \theta + \cos \theta - 1 = 0$

5. $(\tan \theta - 1)(\sec \theta - 1) = 0$

6. $(\cot \theta + 1)\left(\csc \theta - \dfrac{1}{2} \right) = 0$

7. $\sin^2 \theta - \cos^2 \theta = 1 + \cos \theta$

8. $\cos^2 \theta - \sin^2 \theta + \sin \theta = 0$

9. $\sin^2 \theta = 6(\cos \theta + 1)$

10. $2 \sin^2 \theta = 3(1 - \cos \theta)$

11. $\cos(2\theta) + 6 \sin^2 \theta = 4$

12. $\cos(2\theta) = 2 - 2 \sin^2 \theta$

13. $\cos \theta = \sin \theta$

14. $\cos \theta + \sin \theta = 0$

15. $\tan \theta = 2 \sin \theta$

16. $\sin(2\theta) = \cos \theta$

17. $\sin \theta = \csc \theta$

18. $\tan \theta = \cot \theta$

19. $\cos(2\theta) = \cos \theta$

20. $\sin(2\theta) \sin \theta = \cos \theta$

21. $\sin(2\theta) + \sin(4\theta) = 0$

22. $\cos(2\theta) + \cos(4\theta) = 0$

23. $\cos(4\theta) - \cos(6\theta) = 0$

24. $\sin(4\theta) - \sin(6\theta) = 0$

25. $1 + \sin \theta = 2 \cos^2 \theta$

26. $\sin^2 \theta = 2 \cos \theta + 2$

27. $\tan^2 \theta = \dfrac{3}{2} \sec \theta$

28. $\csc^2 \theta = \cot \theta + 1$

29. $3 - \sin \theta = \cos(2\theta)$

30. $\cos(2\theta) + 5 \cos \theta + 3 = 0$

31. $\sec^2 \theta + \tan \theta = 0$

32. $\sec \theta = \tan \theta + \cot \theta$

33. $\sin \theta - \sqrt{3} \cos \theta = 1$

34. $\sqrt{3}\sin\theta + \cos\theta = 1$

35. $\tan(2\theta) + 2\sin\theta = 0$

36. $\tan(2\theta) + 2\cos\theta = 0$

37. $\sin\theta + \cos\theta = \sqrt{2}$

38. $\sin\theta + \cos\theta = -\sqrt{2}$

In Problems 39–44, solve each equation for x, $-\pi \le x \le \pi$. Express the solution(s) rounded to two decimal places.

39. Solve the equation $\cos x = e^x$ by graphing $Y_1 = \cos x$ and $Y_2 = e^x$ and finding their point(s) of intersection.

40. Solve the equation $\cos x = e^x$ by graphing $Y_1 = \cos x - e^x$ and finding the x-intercept(s).

41. Solve the equation $2\sin x = 0.7x$ by graphing $Y_1 = 2\sin x$ and $Y_2 = 0.7x$ and finding their point(s) of intersection.

42. Solve the equation $2\sin x = 0.7x$ by graphing $Y_1 = 2\sin x - 0.7x$ and finding the x-intercept(s).

43. Solve the equation $\cos x = x^2$ by graphing $Y_1 = \cos x$ and $Y_2 = x^2$ and finding their point(s) of intersection.

44. Solve the equation $\cos x = x^2$ by graphing $Y_1 = \cos x - x^2$ and finding the x-intercept(s).

In Problems 45–56, use a graphing utility to solve each equation. Express the solution(s) rounded to two decimal places.

45. $x + 5\cos x = 0$

46. $x - 4\sin x = 0$

47. $22x - 17\sin x = 3$

48. $19x + 8\cos x = 2$

49. $\sin x + \cos x = x$

50. $\sin x - \cos x = x$

51. $x^2 - 2\cos x = 0$

52. $x^2 + 3\sin x = 0$

53. $x^2 - 2\sin(2x) = 3x$

54. $x^2 = x + 3\cos(2x)$

55. $6\sin x - e^x = 2$, $x > 0$

56. $4\cos(3x) - e^x = 1$, $x > 0$

---

57. Constructing a Rain Gutter   A rain gutter is to be constructed of aluminum sheets 12 inches wide. After marking off a length of 4 inches from each edge, this length is bent up at an angle θ. See the illustration. The area A of the opening as a function of θ is given by

$$A(\theta) = 16\sin\theta(\cos\theta + 1),   0° < \theta < 90°$$

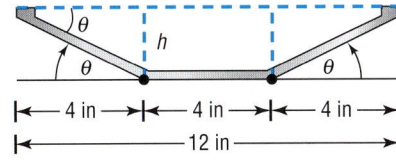

(a) In calculus, you will be asked to find the angle θ that maximizes A by solving the equation

$$\cos(2\theta) + \cos\theta = 0,   0° < \theta < 90°$$

Solve this equation for θ by using the Double-Angle Formula.

(b) Solve the equation for θ by writing the sum of the two cosines as a product.

(c) What is the maximum area A of the opening?

(d) Graph A, $0° \le \theta \le 90°$, and find the angle θ that maximizes the area A. Also find the maximum area. Compare the results to the answers found earlier.

58. Projectile Motion   An object is propelled upward at an angle θ, $45° < \theta < 90°$, to the horizontal with an initial velocity of v_0 feet per second from the base of a plane that makes an angle of $45°$ with the horizontal. See the illustration. If air resistance is ignored, the distance R that it travels up the inclined plane is given by

$$R = \frac{v_0^2\sqrt{2}}{32}[\sin(2\theta) - \cos(2\theta) - 1]$$

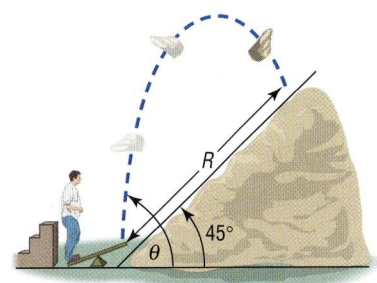

(a) In calculus, you will be asked to find the angle θ that maximizes R by solving the equation

$$\sin(2\theta) + \cos(2\theta) = 0$$

Solve this equation for θ by squaring both sides of the equation.

(b) Solve this equation for θ by dividing each side by $\cos(2\theta)$.

(c) What is the maximum distance R if $v_0 = 32$ feet per second?

(d) Graph R, $45° \le \theta \le 90°$, and find the angle θ that maximizes the distance R. Also find the maximum distance. Use $v_0 = 32$ feet per second. Compare the results with the answers found earlier.

59. Heat Transfer   In the study of heat transfer, the equation $x + \tan x = 0$ occurs. Graph $Y_1 = -x$ and $Y_2 = \tan x$ for $x \ge 0$. Conclude that there are an infinite number of points of intersection of these two graphs. Now find the first two positive solutions of $x + \tan x = 0$ rounded to two decimal places.

60. Carrying a Ladder around a Corner A ladder of length L is carried horizontally around a corner from a hall 3 feet wide into a hall 4 feet wide. See the illustration.

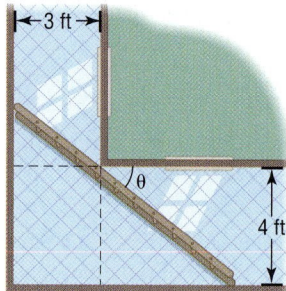

(a) Express L as a function of θ.
(b) In calculus, you will be asked to find the length of the longest ladder that can turn the corner by solving the equation

$$3 \sec \theta \tan \theta - 4 \csc \theta \cot \theta = 0, \quad 0° < \theta < 90°$$

Solve this equation for θ.
(c) What is the length of the longest ladder that can be carried around the corner?
(d) Graph L, $0° \le \theta \le 90°$, and find the angle θ that minimizes the length L.
(e) Compare the result with the one found in part (b). Explain why the two answers are the same.

61. Projectile Motion The horizontal distance that a projectile will travel in the air is given by the equation

$$R = \frac{v_0^2 \sin(2\theta)}{g}$$

where v_0 is the initial velocity of the projectile, θ is the angle of elevation, and g is acceleration due to gravity (9.8 meters per second squared).
(a) If you can throw a baseball with an initial speed of 34.8 meters per second, at what angle of elevation θ should you direct the throw so that the ball travels a distance of 107 meters before striking the ground?
(b) Determine the maximum distance that you can throw the ball.
(c) Graph R, with $v_0 = 34.8$ meters per second.
(d) Verify the results obtained in parts (a) and (b) using ZERO or ROOT.

62. Projectile Motion Refer to Problem 61.
(a) If you can throw a baseball with an initial speed of 40 meters per second, at what angle of elevation θ should you direct the throw so that the ball travels a distance of 110 meters before striking the ground?
(b) Determine the maximum distance that you can throw the ball.
(c) Graph R, with $v_0 = 40$ meters per second.
(d) Verify the results obtained in parts (a) and (b) using ZERO or ROOT.

Chapter Review

Things To Know

Definitions of the six inverse trigonometric functions

$y = \sin^{-1} x$	means	$x = \sin y$	where	$-1 \le x \le 1,\quad -\dfrac{\pi}{2} \le y \le \dfrac{\pi}{2}$	(p. 711)
$y = \cos^{-1} x$	means	$x = \cos y$	where	$-1 \le x \le 1,\quad 0 \le y \le \pi$	(p. 715)
$y = \tan^{-1} x$	means	$x = \tan y$	where	$-\infty < x < \infty,\quad -\dfrac{\pi}{2} < y < \dfrac{\pi}{2}$	(p. 718)
$y = \sec^{-1} x$	means	$x = \sec y$	where	$|x| \ge 1,\quad 0 \le y \le \pi,\quad y \ne \dfrac{\pi}{2}$	(p. 724)
$y = \csc^{-1} x$	means	$x = \csc y$	where	$|x| \ge 1,\quad -\dfrac{\pi}{2} \le y \le \dfrac{\pi}{2},\quad y \ne 0$	(p. 724)
$y = \cot^{-1} x$	means	$x = \cot y$	where	$-\infty < x < \infty,\quad 0 < y < \pi$	(p. 724)

Sum and Difference Formulas (pp. 734, 737, and 740)

$\cos(\alpha + \beta) = \cos \alpha \cos \beta - \sin \alpha \sin \beta$ $\cos(\alpha - \beta) = \cos \alpha \cos \beta + \sin \alpha \sin \beta$

$\sin(\alpha + \beta) = \sin \alpha \cos \beta + \cos \alpha \sin \beta$ $\sin(\alpha - \beta) = \sin \alpha \cos \beta - \cos \alpha \sin \beta$

$\tan(\alpha + \beta) = \dfrac{\tan \alpha + \tan \beta}{1 - \tan \alpha \tan \beta}$ $\tan(\alpha - \beta) = \dfrac{\tan \alpha - \tan \beta}{1 + \tan \alpha \tan \beta}$

Double-Angle Formulas (pp. 746 and 747)

$\sin(2\theta) = 2 \sin \theta \cos \theta$ $\cos(2\theta) = \cos^2 \theta - \sin^2 \theta$ $\cos(2\theta) = 1 - 2 \sin^2 \theta$

$\cos(2\theta) = 2 \cos^2 \theta - 1$ $\tan(2\theta) = \dfrac{2 \tan \theta}{1 - \tan^2 \theta}$

Half-Angle Formulas (pp. 749, 750, and 752)

$$\sin^2\frac{\alpha}{2} = \frac{1 - \cos\alpha}{2} \qquad\qquad \cos^2\frac{\alpha}{2} = \frac{1 + \cos\alpha}{2} \qquad\qquad \tan^2\frac{\alpha}{2} = \frac{1 - \cos\alpha}{1 + \cos\alpha}$$

$$\sin\frac{\alpha}{2} = \pm\sqrt{\frac{1 - \cos\alpha}{2}} \qquad\qquad \cos\frac{\alpha}{2} = \pm\sqrt{\frac{1 + \cos\alpha}{2}} \qquad\qquad \tan\frac{\alpha}{2} = \pm\sqrt{\frac{1 - \cos\alpha}{1 + \cos\alpha}} = \frac{1 - \cos\alpha}{\sin\alpha} = \frac{\sin\alpha}{1 + \cos\alpha}$$

where the $+$ or $-$ sign is determined by the quadrant of $\dfrac{\alpha}{2}$

Product-to-Sum Formulas (p. 755)

$$\sin\alpha\sin\beta = \frac{1}{2}[\cos(\alpha - \beta) - \cos(\alpha + \beta)]$$

$$\cos\alpha\cos\beta = \frac{1}{2}[\cos(\alpha - \beta) + \cos(\alpha + \beta)]$$

$$\sin\alpha\cos\beta = \frac{1}{2}[\sin(\alpha + \beta) + \sin(\alpha - \beta)]$$

Sum-to-Product Formulas (p. 757)

$$\sin\alpha + \sin\beta = 2\sin\frac{\alpha + \beta}{2}\cos\frac{\alpha - \beta}{2} \qquad\qquad \sin\alpha - \sin\beta = 2\sin\frac{\alpha - \beta}{2}\cos\frac{\alpha + \beta}{2}$$

$$\cos\alpha + \cos\beta = 2\cos\frac{\alpha + \beta}{2}\cos\frac{\alpha - \beta}{2} \qquad\qquad \cos\alpha - \cos\beta = -2\sin\frac{\alpha + \beta}{2}\sin\frac{\alpha - \beta}{2}$$

Objectives

Section		You should be able to:	Review Exercises
9.1	1	Find the exact value of the inverse sine, cosine, and tangent functions (p. 713)	1–6
	2	Find an approximate value of the inverse sine, cosine, and tangent functions (p. 714)	101–104
9.2	1	Find the exact value of expressions involving the inverse sine, cosine and tangent functions (p. 722)	9–20
	2	Find the exact value of the inverse secant, cosecant, and cotangent functions (p. 724)	7–8
	3	Use a calculator to evaluate $\sec^{-1}x$, $\csc^{-1}x$, and $\cot^{-1}x$ (p. 725)	105–106
9.3	1	Establish identities (p. 728)	21–38
9.4	1	Use sum and difference formulas to find exact values (p. 735)	53–56, 59–60, 61–70(a)–(d)
	2	Use sum and difference formulas to establish identities (p. 736)	39–42
	3	Use sum and difference formulas involving inverse trigonometric functions (p. 741)	71–74
9.5	1	Use double-angle formulas to find exact values (p. 746)	61–70(e)–(f)
	2	Use double-angle and half-angle formulas to establish identities (p. 747)	43–48
	3	Use half-angle formulas to find exact values (p. 750)	61–70(g)–(h)
9.6	1	Express products as sums (p. 755)	49–50
	2	Express sums as products (p. 757)	51–52
9.7	1	Solve equations involving a single trigonometric function (p. 759)	77–86
9.8	1	Solve trigonometric equations quadratic in form (p. 765)	93–94
	2	Solve trigonometric equations using identities (p. 765)	87–92, 95–98
	3	Solve trigonometric equations linear in sine and cosine (p. 767)	99–100
	4	Solve trigonometric equations using a graphing utility (p. 770)	107–112

Review Exercises

Blue problem numbers indicate the authors' suggestions for use in a Practice Test.

In Problems 1–20, find the exact value of each expression. Do not use a calculator.

1. $\sin^{-1} 1$

2. $\cos^{-1} 0$

3. $\tan^{-1} 1$

4. $\sin^{-1}\left(-\dfrac{1}{2}\right)$

5. $\cos^{-1}\left(-\dfrac{\sqrt{3}}{2}\right)$

6. $\tan^{-1}(-\sqrt{3})$

7. $\sec^{-1}\sqrt{2}$

8. $\cot^{-1}(-1)$

9. $\tan\left[\sin^{-1}\left(-\dfrac{\sqrt{3}}{2}\right)\right]$

10. $\tan\left[\cos^{-1}\left(-\dfrac{1}{2}\right)\right]$

11. $\sec\left(\tan^{-1}\dfrac{\sqrt{3}}{3}\right)$

12. $\csc\left(\sin^{-1}\dfrac{\sqrt{3}}{2}\right)$

13. $\sin\left(\tan^{-1}\dfrac{3}{4}\right)$

14. $\cos\left(\sin^{-1}\dfrac{3}{5}\right)$

15. $\tan\left[\sin^{-1}\left(-\dfrac{4}{5}\right)\right]$

16. $\tan\left[\cos^{-1}\left(-\dfrac{3}{5}\right)\right]$

17. $\sin^{-1}\left(\cos\dfrac{2\pi}{3}\right)$

18. $\cos^{-1}\left(\tan\dfrac{3\pi}{4}\right)$

19. $\tan^{-1}\left(\tan\dfrac{7\pi}{4}\right)$

20. $\cos^{-1}\left(\cos\dfrac{7\pi}{6}\right)$

In Problems 21–52, establish each identity.

21. $\tan\theta\cot\theta - \sin^2\theta = \cos^2\theta$

22. $\sin\theta\csc\theta - \sin^2\theta = \cos^2\theta$

23. $\cos^2\theta(1 + \tan^2\theta) = 1$

24. $(1 - \cos^2\theta)(1 + \cot^2\theta) = 1$

25. $4\cos^2\theta + 3\sin^2\theta = 3 + \cos^2\theta$

26. $4\sin^2\theta + 2\cos^2\theta = 4 - 2\cos^2\theta$

27. $\dfrac{1 - \cos\theta}{\sin\theta} + \dfrac{\sin\theta}{1 - \cos\theta} = 2\csc\theta$

28. $\dfrac{\sin\theta}{1 + \cos\theta} + \dfrac{1 + \cos\theta}{\sin\theta} = 2\csc\theta$

29. $\dfrac{\cos\theta}{\cos\theta - \sin\theta} = \dfrac{1}{1 - \tan\theta}$

30. $1 - \dfrac{\cos^2\theta}{1 + \sin\theta} = \sin\theta$

31. $\dfrac{\csc\theta}{1 + \csc\theta} = \dfrac{1 - \sin\theta}{\cos^2\theta}$

32. $\dfrac{1 + \sec\theta}{\sec\theta} = \dfrac{\sin^2\theta}{1 - \cos\theta}$

33. $\csc\theta - \sin\theta = \cos\theta\cot\theta$

34. $\dfrac{\csc\theta}{1 - \cos\theta} = \dfrac{1 + \cos\theta}{\sin^3\theta}$

35. $\dfrac{1 - \sin\theta}{\sec\theta} = \dfrac{\cos^3\theta}{1 + \sin\theta}$

36. $\dfrac{1 - \cos\theta}{1 + \cos\theta} = (\csc\theta - \cot\theta)^2$

37. $\dfrac{1 - 2\sin^2\theta}{\sin\theta\cos\theta} = \cot\theta - \tan\theta$

38. $\dfrac{(2\sin^2\theta - 1)^2}{\sin^4\theta - \cos^4\theta} = 1 - 2\cos^2\theta$

39. $\dfrac{\cos(\alpha + \beta)}{\cos\alpha\sin\beta} = \cot\beta - \tan\alpha$

40. $\dfrac{\sin(\alpha - \beta)}{\sin\alpha\cos\beta} = 1 - \cot\alpha\tan\beta$

41. $\dfrac{\cos(\alpha - \beta)}{\cos\alpha\cos\beta} = 1 + \tan\alpha\tan\beta$

42. $\dfrac{\cos(\alpha + \beta)}{\sin\alpha\cos\beta} = \cot\alpha - \tan\beta$

43. $(1 + \cos\theta)\left(\tan\dfrac{\theta}{2}\right) = \sin\theta$

44. $\sin\theta\tan\dfrac{\theta}{2} = 1 - \cos\theta$

45. $2\cot\theta\cot(2\theta) = \cot^2\theta - 1$

46. $2\sin(2\theta)(1 - 2\sin^2\theta) = \sin(4\theta)$

47. $1 - 8\sin^2\theta\cos^2\theta = \cos(4\theta)$

48. $\dfrac{\sin(3\theta)\cos\theta - \sin\theta\cos(3\theta)}{\sin(2\theta)} = 1$

49. $\dfrac{\sin(2\theta) + \sin(4\theta)}{\cos(2\theta) + \cos(4\theta)} = \tan(3\theta)$

50. $\dfrac{\sin(2\theta) + \sin(4\theta)}{\sin(2\theta) - \sin(4\theta)} + \dfrac{\tan(3\theta)}{\tan\theta} = 0$

51. $\dfrac{\cos(2\theta) - \cos(4\theta)}{\cos(2\theta) + \cos(4\theta)} - \tan\theta\tan(3\theta) = 0$

52. $\cos(2\theta) - \cos(10\theta) = \tan(4\theta)[\sin(2\theta) + \sin(10\theta)]$

In Problems 53–60, find the exact value of each expression.

53. $\sin 165°$

54. $\tan 105°$

55. $\cos\dfrac{5\pi}{12}$

56. $\sin\left(-\dfrac{\pi}{12}\right)$

57. $\cos 80°\cos 20° + \sin 80°\sin 20°$

58. $\sin 70°\cos 40° - \cos 70°\sin 40°$

59. $\tan\dfrac{\pi}{8}$

60. $\sin\dfrac{5\pi}{8}$

In Problems 61–70, use the information given about the angles α and β to find the exact value of:

(a) $\sin(\alpha + \beta)$

(b) $\cos(\alpha + \beta)$

(c) $\sin(\alpha - \beta)$

(d) $\tan(\alpha + \beta)$

(e) $\sin(2\alpha)$

(f) $\cos(2\beta)$

(g) $\sin\dfrac{\beta}{2}$

(h) $\cos\dfrac{\alpha}{2}$

61. $\sin\alpha = \dfrac{4}{5}, \quad 0 < \alpha < \dfrac{\pi}{2}; \quad \sin\beta = \dfrac{5}{13}, \quad \dfrac{\pi}{2} < \beta < \pi$

62. $\cos\alpha = \dfrac{4}{5}, \quad 0 < \alpha < \dfrac{\pi}{2}; \quad \cos\beta = \dfrac{5}{13}, \quad -\dfrac{\pi}{2} < \beta < 0$

63. $\sin \alpha = -\dfrac{3}{5}, \quad \pi < \alpha < \dfrac{3\pi}{2}; \quad \cos \beta = \dfrac{12}{13}, \quad \dfrac{3\pi}{2} < \beta < 2\pi$

64. $\sin \alpha = -\dfrac{4}{5}, \quad -\dfrac{\pi}{2} < \alpha < 0; \quad \cos \beta = -\dfrac{5}{13}, \quad \dfrac{\pi}{2} < \beta < \pi$

65. $\tan \alpha = \dfrac{3}{4}, \quad \pi < \alpha < \dfrac{3\pi}{2}; \quad \tan \beta = \dfrac{12}{5}, \quad 0 < \beta < \dfrac{\pi}{2}$

66. $\tan \alpha = -\dfrac{4}{3}, \quad \dfrac{\pi}{2} < \alpha < \pi; \quad \cot \beta = \dfrac{12}{5}, \quad \pi < \beta < \dfrac{3\pi}{2}$

67. $\sec \alpha = 2, \quad -\dfrac{\pi}{2} < \alpha < 0; \quad \sec \beta = 3, \quad \dfrac{3\pi}{2} < \beta < 2\pi$

68. $\csc \alpha = 2, \quad \dfrac{\pi}{2} < \alpha < \pi; \quad \sec \beta = -3, \quad \dfrac{\pi}{2} < \beta < \pi$

69. $\sin \alpha = -\dfrac{2}{3}, \quad \pi < \alpha < \dfrac{3\pi}{2}; \quad \cos \beta = -\dfrac{2}{3}, \quad \pi < \beta < \dfrac{3\pi}{2}$

70. $\tan \alpha = -2, \quad \dfrac{\pi}{2} < \alpha < \pi; \quad \cot \beta = -2, \quad \dfrac{\pi}{2} < \beta < \pi$

In Problems 71–76, find the exact value of each expression.

71. $\cos\left(\sin^{-1} \dfrac{3}{5} - \cos^{-1} \dfrac{1}{2} \right)$

72. $\sin\left(\cos^{-1} \dfrac{5}{13} - \cos^{-1} \dfrac{4}{5} \right)$

73. $\tan\left[\sin^{-1}\left(-\dfrac{1}{2}\right) - \tan^{-1} \dfrac{3}{4} \right]$

74. $\cos\left[\tan^{-1}(-1) + \cos^{-1}\left(-\dfrac{4}{5}\right) \right]$

75. $\sin\left[2 \cos^{-1}\left(-\dfrac{3}{5}\right) \right]$

76. $\cos\left(2 \tan^{-1} \dfrac{4}{3} \right)$

In Problems 77–100, solve each equation on the interval $0 \le \theta < 2\pi$.

77. $\cos \theta = \dfrac{1}{2}$

78. $\sin \theta = -\dfrac{\sqrt{3}}{2}$

79. $2 \cos \theta + \sqrt{2} = 0$

80. $\tan \theta + \sqrt{3} = 0$

81. $\sin(2\theta) + 1 = 0$

82. $\cos(2\theta) = 0$

83. $\tan(2\theta) = 0$

84. $\sin(3\theta) = 1$

85. $\sec^2 \theta = 4$

86. $\csc^2 \theta = 1$

87. $\sin \theta = \tan \theta$

88. $\cos \theta = \sec \theta$

89. $\sin \theta + \sin(2\theta) = 0$

90. $\cos(2\theta) = \sin \theta$

91. $\sin(2\theta) - \cos \theta - 2 \sin \theta + 1 = 0$

92. $\sin(2\theta) - \sin \theta - 2 \cos \theta + 1 = 0$

93. $2 \sin^2 \theta - 3 \sin \theta + 1 = 0$

94. $2 \cos^2 \theta + \cos \theta - 1 = 0$

95. $4 \sin^2 \theta = 1 + 4 \cos \theta$

96. $8 - 12 \sin^2 \theta = 4 \cos^2 \theta$

97. $\sin(2\theta) = \sqrt{2} \cos \theta$

98. $1 + \sqrt{3} \cos \theta + \cos(2\theta) = 0$

99. $\sin \theta - \cos \theta = 1$

100. $\sin \theta - \sqrt{3} \cos \theta = 2$

In Problems 101–106, use a calculator to find an approximate value for each expression, rounded to two decimal places.

101. $\sin^{-1} 0.7$

102. $\cos^{-1} \dfrac{4}{5}$

103. $\tan^{-1}(-2)$

104. $\cos^{-1}(-0.2)$

105. $\sec^{-1} 3$

106. $\cot^{-1}(-4)$

In Problems 107–112, use a graphing utility to solve each equation on the interval $0 \le x \le 2\pi$. Approximate any solutions rounded to two decimal places.

107. $2x = 5 \cos x$

108. $2x = 5 \sin x$

109. $2 \sin x + 3 \cos x = 4x$

110. $3 \cos x + x = \sin x$

111. $\sin x = \ln x$

112. $\sin x = e^{-x}$

Chapter Projects

1. **Waves**   A stretched string that is attached at both ends, pulled in a direction perpendicular to the string, and released, has motion that is described as wave motion. If we assume no friction and a length such that there are no "echoes" (that is, the wave doesn't bounce back), the transverse motion (motion perpendicular to the string) can be described by the equation

$$y = y_m \sin(kx - \omega t)$$

where y_m is the amplitude measured in meters

k and ω are constants

The height of the sound wave depends on the distance x from one endpoint of the string and time t. Thus, a typical wave has horizontal and vertical motion over time.

(a) What is the amplitude of the wave $y = 0.00421 \sin(68.3x - 2.68t)$?

(b) The value of ω is the angular frequency measured in radians per second. What is the angular frequency of the wave given in part (a)?

(c) The frequency f is the number of vibrations per second (hertz) made by the wave as it passes a certain point. Its value is found using the formula $f = \dfrac{\omega}{2\pi}$. What is the frequency of the wave given in part (a)?

(d) The wavelength, λ, of a wave is the shortest distance at which the wave pattern repeats itself for a constant t. Thus, $\lambda = \dfrac{2\pi}{k}$. What is the wavelength of the wave given in part (a)?

(e) Graph the height of the string a distance $x = 1$ meter from an endpoint.

(f) If two waves travel simultaneously along the same stretched string, the vertical displacement of the string when both waves act is $y = y_1 + y_2$, where y_1 is the vertical displacement of the first wave and y_2 is the vertical displacement of the second wave. This result is called the Principle of Superposition and was analyzed by the French mathematician Jean Baptiste Fourier (1768–1830). When two waves travel along

the same string, one wave will differ from the other wave by a phase constant ϕ.

$$y_1 = y_m \sin(kx - \omega t)$$
$$y_2 = y_m \sin(kx - \omega t + \phi)$$

assuming that each wave has the same amplitude. Write $y_1 + y_2$ as a product using the Sum-to-Product formulas.

(g) Suppose that two waves are moving in the same direction along a stretched string. The amplitude of each string is 0.0045 meter and the phase difference between them is 2.5 radians. The wavelength, λ, of each wave is 0.09 meter and the frequency, f, is 2.3 hertz. Find y_1, y_2 and $y_1 + y_2$.

(h) Using a graphing utility, graph y_1, y_2 and $y_1 + y_2$ on the same viewing window.

(i) Redo parts (g) and (h) with the phase difference between the waves being 0.4 radian.

(j) What effect does the phase difference have on the amplitude of $y_1 + y_2$?

2. **Jacob's Field**   In Jacob's Field, home of the Cleveland Indians, the distance from home plate to the outfield fence varies from 325 feet (first and third base lines) to 410 feet (dead center field). We wish to estimate the minimum distance from home plate to the highest point of Jacob's Field and from the base of the outfield fence to the highest point of the stadium so that we can estimate the minimum distance that a ball must travel to be hit out of Jacob's Field.

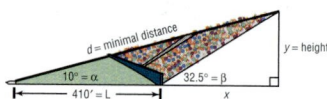

(a) Sketch a picture showing home plate, the fence, and the highest point of the stadium. Draw lines from home plate to the highest point and from the base of the fence to the highest point.

(b) Using a transit, the angle of elevation from home plate to the highest point of the stadium in dead center field is 10° and the angle of elevation from the base of the fence to the highest point of the stadium in dead center field is 32.5°. Use the distance from home plate to the base of the wall in dead center field to compute the minimum distance from home plate to the highest point of the stadium and the distance from the base of the outfield wall to the highest point of the stadium.

(c) Develop a general formula for computing the distance from home plate to the highest point of the stadium and the distance from the base of the outfield wall to the highest point of the stadium. Let α and β denote the angles of elevation to the top of the roof from home plate and from the base of the outfield fence, respectively. Let L be the distance from home plate to the outfield fence.

(d) Write each formula found in part (c) in the form

$$\text{Height} = \frac{L \sin\alpha \sin\beta}{\sin(\beta - \alpha)} \qquad \text{Distance} = \frac{L \sin\beta}{\sin(\beta - \alpha)}$$

3. Suppose $f(x) = \sin x$.

(a) Build a table of values for $f(x)$ where $x = 0, \dfrac{\pi}{6}, \dfrac{\pi}{4},$ $\dfrac{\pi}{3}, \dfrac{\pi}{2}, \dfrac{2\pi}{3}, \dfrac{3\pi}{4}, \dfrac{5\pi}{6}, \pi, \dfrac{7\pi}{6}, \dfrac{5\pi}{4}, \dfrac{4\pi}{3}, \dfrac{3\pi}{2}, \dfrac{5\pi}{3}, \dfrac{7\pi}{4},$ $\dfrac{11\pi}{6}, 2\pi$. Use exact values.

(b) Find the **first differences** for each consecutive pair of values in part (a). That is, evaluate $g(x_i) = \dfrac{\Delta f(x_i)}{\Delta x_i} = \dfrac{f(x_{i+1}) - f(x_i)}{x_{i+1} - x_i}$, where $x_1 = 0$, $x_2 = \dfrac{\pi}{6}, \ldots,$ $x_{17} = 2\pi$. Use your calculator to approximate each value rounded to three decimal places.

(c) Plot the points $(x_i, g(x_i))$ for $i = 1, \ldots, 16$ on a scatter diagram. What shape does the set of points give? What function does this resemble? Fit a sine curve of best fit to the points. How does that relate to your guess?

(d) Find the first differences for each consecutive pair of values in part (b). That is, evaluate $h(x_i) = \dfrac{\Delta g(x_i)}{\Delta x_i} = \dfrac{g(x_{i+1}) - g(x_i)}{x_{i+1} - x_i}$ where $x_1 = 0$, $x_2 = \dfrac{\pi}{6}, \ldots, x_{16} = \dfrac{11\pi}{6}$. This is the set of **second differences** of $f(x)$. Use your calculator to approximate each value rounded to three decimal places. Plot the points $(x_i, h(x_i))$ for $i = 1, \ldots, 15$ on a scatter diagram. What shape does the set of points give? What function does this resemble? Fit a sine curve of best fit to the points. How does that relate to your guess?

(e) Find the first differences for each consecutive pair of values in part (d). That is, evaluate $k(x_i) = \dfrac{\Delta h(x_i)}{\Delta x_i}$ $= \dfrac{h(x_{i+1}) - h(x_i)}{x_{i+1} - x_i}$, where $x_1 = 0$, $x_2 = \dfrac{\pi}{6}, \ldots, x_{15}$ $= \dfrac{7\pi}{4}$. This is the set of **third differences** of $f(x)$. Use your calculator to approximate each value rounded to three decimal places. Plot the points $(x_i, k(x_i))$ for $i = 1, \ldots, 14$ on a scatter diagram. What shape does the set of points give? What function does this resemble? Fit a sine curve of best fit to the points. How does that relate to your guess?

(f) Find the first differences for each consecutive pair of values in part (e). That is, evaluate $m(x_i) = \dfrac{\Delta k(x_i)}{\Delta x_i}$ $= \dfrac{k(x_{i+1}) - k(x_i)}{x_{i+1} - x_i}$, where $x_1 = 0$, $x_2 = \dfrac{\pi}{6}, \ldots,$ $x_{14} = \dfrac{5\pi}{3}$. This is the set of **fourth differences** of $f(x)$.

Use your calculator to approximate each value rounded to three decimal places. Plot the points $(x_i, m(x_i))$ for $i = 1, \ldots, 13$ on a scatter diagram. What shape does the set of points give? What function does this resemble? Fit a sine curve of best fit to the points. How does that relate to your guess?

(g) What pattern do you notice about the curves that you found? What happened in part (f)? Can you make a generalization about what happened as you computed the differences? Explain your answers.

Cumulative Review

1. Find the real solutions, if any, of the equation $3x^2 + x - 1 = 0$.

2. Find an equation for the line containing the points $(-2, 5)$ and $(4, -1)$. What is the distance between these points? What is their midpoint?

3. Test the equation $3x + y^2 = 9$ for symmetry with respect to the x-axis, the y-axis, and the origin. List the intercepts.

4. Use transformations to graph the equation $y = |x - 3| + 2$.

5. Use transformations to graph the equation $y = 3e^x - 2$.

6. Use transformations to graph the equation $y = \cos\left(x - \dfrac{\pi}{2}\right) - 1$.

7. Sketch a graph of each of the following functions. Label at least three points on each graph. Name the inverse function of each one and show its graph.

(a) $y = x^3$

(b) $y = e^x$

(c) $y = \sin x$, $-\dfrac{\pi}{2} \le x \le \dfrac{\pi}{2}$

(d) $y = \cos x$, $0 \le x \le \pi$

8. If $\sin \theta = -\dfrac{1}{3}$ and $\pi < \theta < \dfrac{3\pi}{2}$, find the exact value of:

(a) $\cos \theta$ (b) $\tan \theta$ (c) $\sin(2\theta)$

(d) $\cos(2\theta)$ (e) $\sin\left(\dfrac{1}{2}\theta\right)$ (f) $\cos\left(\dfrac{1}{2}\theta\right)$

9. Find the exact value of $\cos(\tan^{-1} 2)$.

10. If $\sin \alpha = \dfrac{1}{3}, \dfrac{\pi}{2} < \alpha < \pi$, and $\cos \beta = -\dfrac{1}{3}, \pi < \beta < \dfrac{3\pi}{2}$,

find the exact value of

(a) $\cos \alpha$ (b) $\sin \beta$ (c) $\cos(2\alpha)$

(d) $\cos(\alpha + \beta)$ (e) $\sin \dfrac{\beta}{2}$

11. For the triangle shown, find h if $\theta = 20°$.

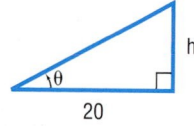

20

12. For the function $f(x) = 2x^5 - x^4 - 4x^3 + 2x^2 + 2x - 1$
   (a) Find the real zeros and their multiplicity.
   (b) Find the intercepts.
   (c) Find the power function that the graph of f resembles for large $|x|$.
   (d) Graph f using a graphing utility.
   (e) Approximate the turning points, if any exist.
   (f) Use the information obtained in parts (a)–(e) to sketch a graph of f by hand.

APPLICATIONS OF TRIGONOMETRIC FUNCTIONS

New maps pinpoint Lewis and Clark's journey through Missouri

KANSAS CITY, Mo.—Nearly two centuries ago, Congress commissioned Meriwether Lewis and William Clark to explore trade routes to the West.

Their long-documented journey took them through Missouri, but their exact route has been up for debate.

Now, modern computer technology combined with 19th century land surveys may provide a precise picture. The latest computer-generated maps of Lewis and Clark's expedition were unveiled here this week by Missouri Secretary of State Matt Blunt.

The maps combine the terrain of the early 19th century with contemporary geographical markers. They are the most precise to date of Lewis and Clark's journeys through Missouri in 1804 and 1806, said the project's lead researcher, Jim Harlan.

"We've known what they've done, but not with this much certainty," said Harlan, geographic resource project director at the University of Missouri-Columbia. "I think we need more information and less speculation. That's what this is all about."

SOURCE: Sophia Maines, "New Maps Pinpoint Lewis and Clark's Journey Through Missouri," *The Kansas City Star*, August 1, 2001 Distributed by Knight Ridder/Tribune Information Services.

SEE CHAPTER PROJECT 1.

For additional study help, go to

www.prenhall.com/sullivanegu3e

Materials include:

- Graphing Calculator Help
- Chapter Quiz
- Chapter Test
- PowerPoint Downloads
- Chapter Projects
- Student Tips

A Look Back, A Look Forward

In Chapter 8, we defined the six trigonometric functions using right triangles and then extended the definition to include general angles. In particular, we learned to evaluate the trigonometric functions. We also learned how to graph sinusoidal functions.

In this chapter, we use the trigonometric functions to solve applied problems. The first four sections deal with applications involving right triangles and *oblique*

triangles, triangles that do not have a right angle. To solve problems involving oblique triangles, we will develop the Law of Sines and the Law of Cosines. We will also develop formulas for finding the area of a triangle.

The final section deals with applications of sinusoidal functions involving simple harmonic motion and damped motion.

PREPARING FOR THIS SECTION

Before getting started, review the following:

✓ Pythagorean Theorem (Section R.3, pp. 25–26)

✓ Trigonometric Equations (I) (Section 9.7, pp. 759–762)

✓ Complementary Angle Theorem (Section 8.2, pp. 630–632)

10.1 APPLICATIONS INVOLVING RIGHT TRIANGLES

OBJECTIVES   1 Solve Right Triangles
2 Solve Applied Problems

Solving Right Triangles

1   In the discussion that follows, we will always label a right triangle so that side a is opposite angle α, side b is opposite angle β, and side c is the hypotenuse, as shown in Figure 1. **To solve a right triangle** means to find the missing lengths of its sides and the measurements of its angles. We shall follow the practice of expressing the lengths of the sides rounded to two decimal places and of expressing angles in degrees rounded to one decimal place. (Be sure that your calculator is in degree mode.)

To solve a right triangle, we need to know one of the acute angles α or β and a side, or else two sides. Then we make use of the Pythagorean Theorem and the fact that the sum of the angles of a triangle is 180°. The sum of the angles α and β in a right triangle is therefore 90°.

Figure 1

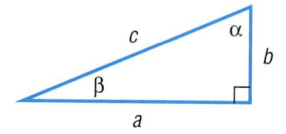

For the right triangle shown in Figure 1, we have

$$c^2 = a^2 + b^2, \qquad \alpha + \beta = 90°$$

Figure 2

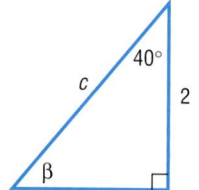

EXAMPLE 1   Solving a Right Triangle

Use Figure 2. If $b = 2$ and $\alpha = 40°$, find a, c, and β.

Solution   Since $\alpha = 40°$ and $\alpha + \beta = 90°$, we find that $\beta = 50°$. To find the sides a and c, we use the facts that

$$\tan 40° = \frac{a}{2} \quad \text{and} \quad \cos 40° = \frac{2}{c}$$

Now solve for a and c.

$$a = 2 \tan 40° \approx 1.68 \quad \text{and} \quad c = \frac{2}{\cos 40°} \approx 2.61$$

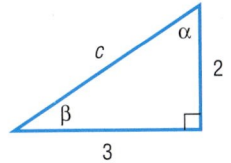 **NOW WORK PROBLEM 1.**

EXAMPLE 2 **Solving a Right Triangle**

Use Figure 3. If $a = 3$ and $b = 2$, find c, α, and β.

Figure 3

Solution Since $a = 3$ and $b = 2$, then, by the Pythagorean Theorem, we have

$$c^2 = a^2 + b^2 = 3^2 + 2^2 = 9 + 4 = 13$$
$$c = \sqrt{13} \approx 3.61$$

To find angle α, we use the fact that

$$\tan \alpha = \frac{3}{2} \quad \text{so} \quad \alpha = \tan^{-1} \frac{3}{2}$$

Set the mode on your calculator to degrees. Then, rounded to one decimal place, we find that $\alpha = 56.3°$. Since $\alpha + \beta = 90°$, we find that $\beta = 33.7°$.

Note: To avoid round-off errors when using a calculator, we will store unrounded values in memory for use in subsequent calculations.

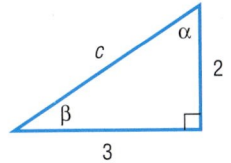 **NOW WORK PROBLEM 11.**

Applications

② One common use for trigonometry is to measure heights and distances that are either awkward or impossible to measure by ordinary means.

EXAMPLE 3 **Finding the Width of a River**

A surveyor can measure the width of a river by setting up a transit* at a point C on one side of the river and taking a sighting of a point A on the other side. Refer to Figure 4. After turning through an angle of $90°$ at C, the surveyor walks a distance of 200 meters to point B. Using the transit at B, the angle β is measured and found to be $20°$. What is the width of the river rounded to the nearest meter?

Figure 4

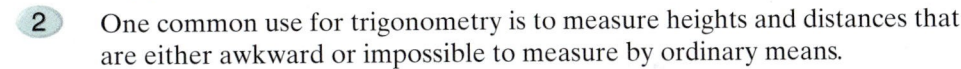

Solution We seek the length of side b. We know a and β, so we use the fact that

$$\tan \beta = \frac{b}{a}$$

to get

$$\tan 20° = \frac{b}{200}$$
$$b = 200 \tan 20° \approx 72.79 \text{ meters}$$

The width of the river is 73 meters, rounded to the nearest meter.

 **NOW WORK PROBLEM 21.**

*An instrument used in surveying to measure angles.

EXAMPLE 4 **Finding the Inclination of a Mountain Trail**

Figure 5

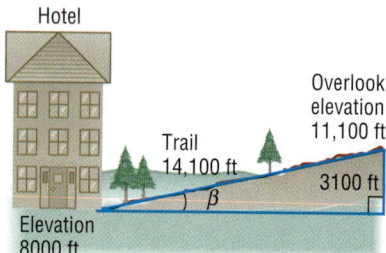

Hotel
Trail 14,100 ft
Overlook elevation 11,100 ft
3100 ft
β
Elevation 8000 ft

A straight trail leads from the Alpine Hotel, elevation 8000 feet, to a scenic overlook, elevation 11,100 feet. The length of the trail is 14,100 feet. What is the inclination of the trail? That is, what is the angle β in Figure 5?

Solution As Figure 5 illustrates, the angle β obeys the equation

$$\sin \beta = \frac{3100}{14,100}$$

Using a calculator,

$$\beta = \sin^{-1} \frac{3100}{14,100} \approx 12.7°$$

The inclination of the trail is approximately 12.7°. ∎

Vertical heights can sometimes be measured using either the angle of elevation or the angle of depression. If a person is looking up at an object, the acute angle measured from the horizontal to a line-of-sight observation of the object is called the **angle of elevation**. See Figure 6(a).

If a person is standing on a cliff looking down at an object, the acute angle made by the line-of-sight observation of the object and the horizontal is called the **angle of depression**. See Figure 6(b).

Figure 6

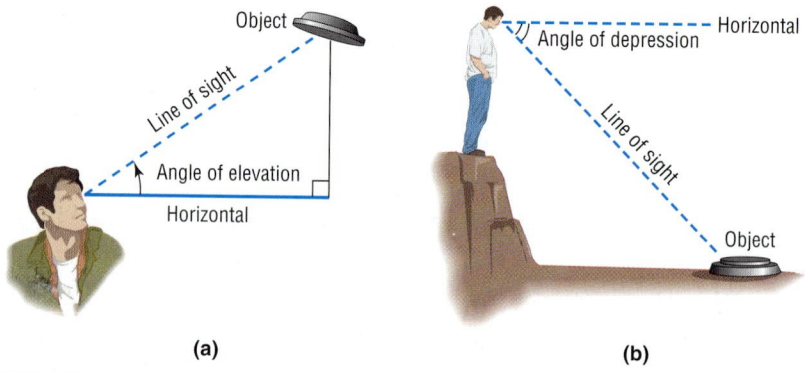

Object
Line of sight
Angle of elevation
Horizontal
(a)

Horizontal
Angle of depression
Line of sight
Object
(b)

EXAMPLE 5 **Finding the Height of a Cloud**

Meteorologists find the height of a cloud using an instrument called a **ceilometer**. A ceilometer consists of a **light projector** that directs a vertical light beam up to the cloud base and a **light detector** that scans the cloud to detect the light beam. See Figure 7(a). On December 1, 2001, at Midway Airport in Chicago, a ceilometer with a base of 300 feet was employed to find the height of the cloud cover. If the angle of elevation of the light detector is 75°, what is the height of the cloud cover?

Figure 7

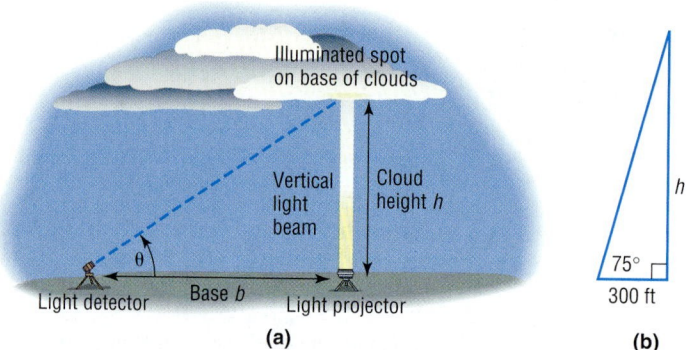

Illuminated spot on base of clouds
Vertical light beam
Cloud height h
θ
Light detector
Base b
Light projector
(a)

h
75°
300 ft
(b)

Solution Figure 7(b) illustrates the situation. To find the height h, we use the fact that $\tan 75° = \dfrac{h}{300}$, so

$$h = 300 \tan 75° \approx 1120 \text{ feet}$$

The ceiling (height to the base of the cloud cover) is approximately 1120 feet. ▪

 NOW WORK PROBLEM **23.**

The idea behind Example 5 can also be used to find the height of an object with a base that is not accessible to the horizontal.

EXAMPLE 6 **Finding the Height of a Statue on a Building**

Adorning the top of the Board of Trade building in Chicago is a statue of Ceres, the Greek goddess of wheat. From street level, two observations are taken 400 feet from the center of the building. The angle of elevation to the base of the statue is found to be 45°; the angle of elevation to the top of the statue is 47.2°. See Figure 8(a). What is the height of the statue?

Figure 8

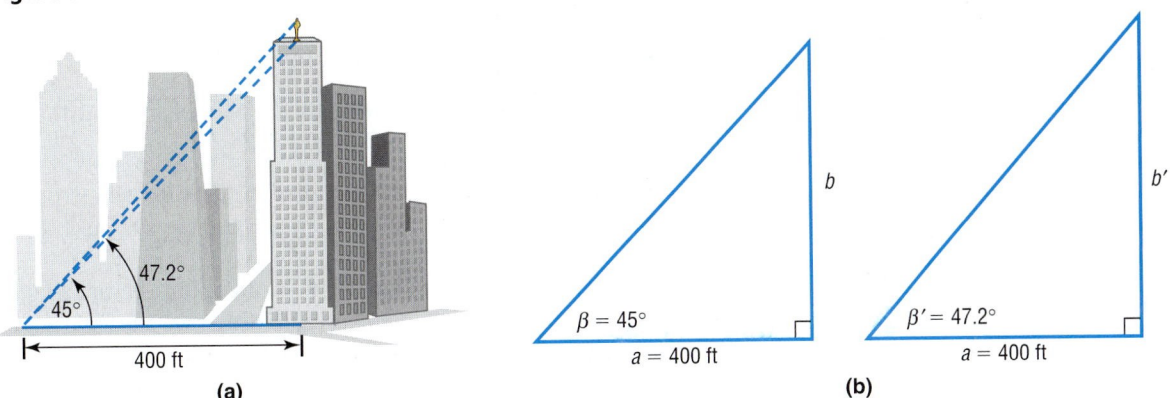

(a)   (b)

Solution Figure 8(b) shows two triangles that replicate Figure 8(a). The height of the statue of Ceres will be $b' - b$. To find b and b', we refer to Figure 8(b).

$$\tan 45° = \frac{b}{400} \qquad\qquad \tan 47.2° = \frac{b'}{400}$$

$$b = 400 \tan 45° = 400 \qquad\qquad b' = 400 \tan 47.2° \approx 432$$

The height of the statue is approximately $432 - 400 = 32$ feet. ▪

 NOW WORK PROBLEM **31.**

EXAMPLE 7 **The Gibb's Hill Lighthouse, Southampton, Bermuda**

In operation since 1846, the Gibb's Hill Lighthouse stands 117 feet high on a hill 245 feet high, so its beam of light is 362 feet above sea level. A brochure states that the light can be seen on the horizon about 26 miles distant. Verify the accuracy of this statement.

Solution Figure 9 illustrates the situation. The central angle θ, positioned at the center of Earth, radius 3960 miles, obeys the equation

Figure 9

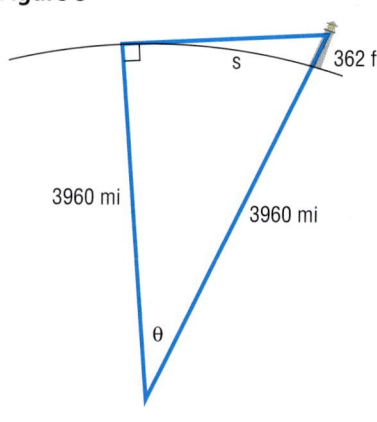

$$\cos\theta = \frac{3960}{3960 + \dfrac{362}{5280}} \approx 0.999982687 \qquad \text{1 mile = 5280 feet}$$

Solving for θ, we find

$$\theta \approx 0.33715° \approx 20.23'$$

The brochure does not indicate whether the distance is measured in nautical miles or statute miles. Let's calculate both distances.

The distance s in nautical miles (refer to Problem 104, p. 623) is the measure of angle θ in minutes, so $s = 20.23$ nautical miles.

The distance s in statute miles is given by the formula $s = r\theta$, where θ is measured in radians. Then, since

$$\theta = 20.23' \overset{\uparrow}{=} 0.33715° \overset{\uparrow}{=} 0.00588 \text{ radian}$$
$$\qquad\quad 1' = \frac{1}{60}° \qquad 1° = \frac{\pi}{180}\text{ radian}$$

we find that

$$s = r\theta = (3960)(0.00588) = 23.3 \text{ miles}$$

In either case, it would seem that the brochure overstated the distance somewhat. ∎

In navigation and surveying, the **direction** or **bearing** from a point O to a point P equals the acute angle θ between the ray OP and the vertical line through O, the north–south line.

Figure 10

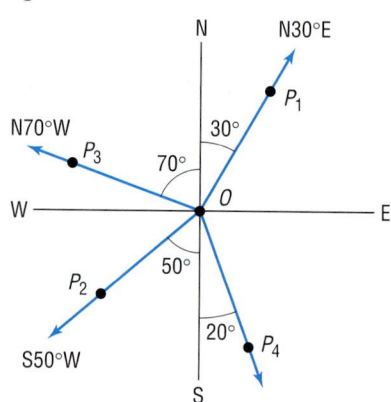

Figure 10 illustrates some bearings. Notice that the bearing from O to P_1 is denoted by the symbolism N30°E, indicating that the bearing is 30° east of north. In writing the bearing from O to P, the direction north or south always appears first, followed by an acute angle, followed by east or west. In Figure 10, the bearing from O to P_2 is S50°W, and from O to P_3 it is N70°W.

EXAMPLE 8 Finding the Bearing of an Object

In Figure 10, what is the bearing from O to an object at P_4?

Solution The acute angle between the ray OP_4 and the north–south line through O is given as 20°. The bearing from O to P_4 is S20°E. ∎

EXAMPLE 9 Finding the Bearing of an Airplane

A Boeing 777 aircraft takes off from O'Hare Airport on runway 2 LEFT which has a bearing of N20°E.* After flying for 1 mile, the pilot of the air craft requests permission to turn 90° and head toward the northwest. The request is granted. After the plane goes 2 miles in this direction, what bearing should the control tower use to locate the aircraft?

---

*In air navigation, the term **azimuth** is employed to denote the positive angle measured clockwise from the north (N) to a ray OP. In Figure 10, the azimuth from O to P_1 is 30°; the azimuth from O to P_2 is 230°; the azimuth from O to P_3 is 290°. In naming runways, the units digit is left off the azimuth. Runway 2 LEFT means the left runway with a direction of azimuth 20° (bearing N20°E). Runway 23 is the runway with azimuth 230° and bearing S50°W.

Figure 11

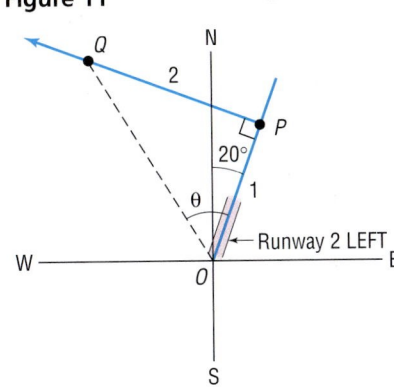

Solution   Figure 11 illustrates the situation. After flying 1 mile from the airport O (the control tower), the aircraft is at P. After turning $90°$ toward the northwest and flying 2 miles, the aircraft is at the point Q. In triangle OPQ, the angle θ obeys the equation

$$\tan \theta = \frac{2}{1} = 2 \quad \text{so} \quad \theta = \tan^{-1} 2 \approx 63.4°$$

The acute angle between the north-south line and the ray OQ is $63.4° - 20° = 43.4°$. The bearing of the aircraft from O to Q is N43.4°W. ■

NOW WORK PROBLEM 39.

10.1 Concepts and Vocabulary

In Problems 1 and 2, fill in the blanks.

1. When you look up at an object, the acute angle measured from the horizontal to a line-of-sight observation of the object is called the _____ _____ _____.

2. When you look down at an object, the acute angle described in Problem 1 is called the _____ _____ _____.

In Problems 3 and 4, answer True or False to each statement.

3. In a right triangle, if two sides are known, we can solve the triangle.

4. In a right triangle, if we know the two acute angles, we can solve the triangle.

5. Explain how you would measure the width of the Grand Canyon from a point on its ridge.

6. Explain how you would measure the height of a TV tower that is on the roof of a tall building.

10.1 Exercises

In Problems 1–14, use the right triangle shown in the margin. Then, using the given information, solve the triangle.

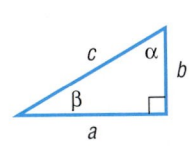

1. $b = 5$, $\beta = 20°$; find $a, c,$ and α
2. $b = 4$, $\beta = 10°$; find $a, c,$ and α
3. $a = 6$, $\beta = 40°$; find $b, c,$ and α
4. $a = 7$, $\beta = 50°$; find $b, c,$ and α
5. $b = 4$, $\alpha = 10°$; find $a, c,$ and β
6. $b = 6$, $\alpha = 20°$; find $a, c,$ and β
7. $a = 5$, $\alpha = 25°$; find $b, c,$ and β
8. $a = 6$, $\alpha = 40°$; find $b, c,$ and β
9. $c = 9$, $\beta = 20°$; find $b, a,$ and α
10. $c = 10$, $\alpha = 40°$; find $b, a,$ and β
11. $a = 5$, $b = 3$; find $c, \alpha,$ and β
12. $a = 2$, $b = 8$; find $c, \alpha,$ and β
13. $a = 2$, $c = 5$; find $b, \alpha,$ and β
14. $b = 4$, $c = 6$; find $a, \alpha,$ and β

15. **Geometry**   A right triangle has a hypotenuse of length 8 inches. If one angle is $35°$, find the length of each leg.

16. **Geometry**   A right triangle has a hypotenuse of length 10 centimeters. If one angle is $40°$, find the length of each leg.

17. **Geometry**   A right triangle contains a $25°$ angle. If one leg is of length 5 inches, what is the length of the hypotenuse?

[**Hint:** Two answers are possible.]

18. Geometry A right triangle contains an angle of $\pi/8$ radian. If one leg is of length 3 meters, what is the length of the hypotenuse?

[**Hint:** Two answers are possible.]

19. Geometry The hypotenuse of a right triangle is 5 inches. If one leg is 2 inches, find the degree measure of each angle.

20. Geometry The hypotenuse of a right triangle is 3 feet. If one leg is 1 foot, find the degree measure of each angle.

21. Finding the Width of a Gorge Find the distance from A to C across the gorge illustrated in the figure.

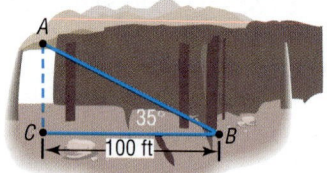

22. Finding the Distance across a Pond Find the distance from A to C across the pond illustrated in the figure.

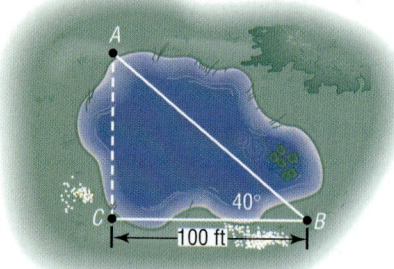

23. The Eiffel Tower The tallest tower built before the era of television masts, the Eiffel Tower was completed on March 31, 1889. Find the height of the Eiffel Tower (before a television mast was added to the top) using the information given in the illustration.

24. Finding the Distance of a Ship from Shore A ship, offshore from a vertical cliff known to be 100 feet in height, takes a sighting of the top of the cliff. If the angle of elevation is found to be 25°, how far offshore is the ship?

25. Finding the Distance to a Plateau Suppose that you are headed toward a plateau 50 meters high. If the angle of elevation to the top of the plateau is 20°, how far are you from the base of the plateau?

26. Statue of Liberty A ship is just offshore of New York City. A sighting is taken of the Statue of Liberty, which is about 305 feet tall. If the angle of elevation to the top of the statue is 20°, how far is the ship from the base of the statue?

27. Finding the Reach of a Ladder A 22-foot extension ladder leaning against a building makes a 70° angle with the ground. How far up the building does the ladder touch?

28. Finding the Height of a Building To measure the height of a building, two sightings are taken a distance of 50 feet apart. If the first angle of elevation is 40° and the second is 32°, what is the height of the building?

29. Directing a Laser Beam A laser beam is to be directed through a small hole in the center of a circle of radius 10 feet. The origin of the beam is 35 feet from the circle (see the figure). At what angle of elevation should the beam be aimed to ensure that it goes through the hole?

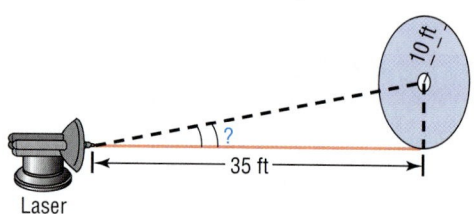

30. Finding the Angle of Elevation of the Sun At 10 AM on April 26, 2000, a building 300 feet high casts a shadow 50 feet long. What is the angle of elevation of the Sun?

31. Mt. Rushmore To measure the height of Lincoln's caricature on Mt. Rushmore, two sightings 800 feet from the base of the mountain are taken. If the angle of elevation to the bottom of Lincoln's face is 32° and the angle of elevation to the top is 35°, what is the height of Lincoln's face?

32. Finding the Distance between Two Objects A blimp, suspended in the air at a height of 500 feet, lies directly over a line from Soldier Field to the Adler Planetarium on Lake Michigan (see the figure). If the angle of depression from the blimp to the stadium is 32° and from the blimp to the planetarium is 23°, find the distance between Soldier Field and the Adler Planetarium.

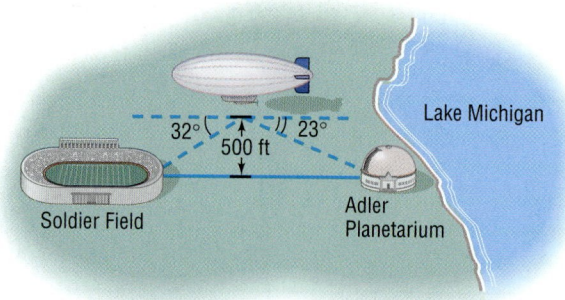

33. Finding the Length of a Guy Wire A radio transmission tower is 200 feet high. How long should a guy wire be if it is to be attached to the tower 10 feet from the top and is to make an angle of 21° with the ground?

34. Finding the Height of a Tower   A guy wire 80 feet long is attached to the top of a radio transmission tower, making an angle of 25° with the ground. How high is the tower?

35. Washington Monument   The angle of elevation to the top of the Washington Monument is 35.1° at the instant it casts a shadow 789 feet long. Use this information to calculate the height of the monument.

36. Finding the Length of a Mountain Trail   A straight trail with an inclination of 17° leads from a hotel at an elevation of 9000 feet to a mountain lake at an elevation of 11,200 feet. What is the length of the trail?

37. Finding the Speed of a Truck   A state trooper is hidden 30 feet from a highway. One second after a truck passes, the angle θ between the highway and the line of observation from the patrol car to the truck is measured. See the illustration.

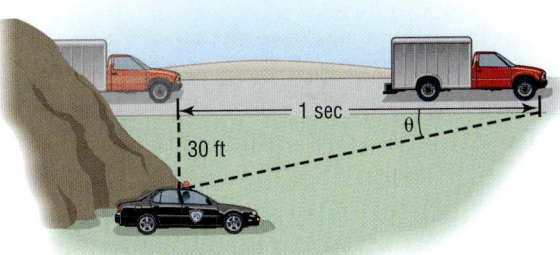

(a) If the angle measures 15°, how fast is the truck traveling? Express the answer in feet per second and in miles per hour.

(b) If the angle measures 20°, how fast is the truck traveling? Express the answer in feet per second and in miles per hour.

(c) If the speed limit is 55 miles per hour and a speeding ticket is issued for speeds of 5 miles per hour or more over the limit, for what angles should the trooper issue a ticket?

38. Security   A security camera in a neighborhood bank is mounted on a wall 9 feet above the floor. What angle of depression should be used if the camera is to be directed to a spot 6 feet above the floor and 12 feet from the wall?

39. Finding the Bearing of an Aircraft   A DC-9 aircraft leaves Midway Airport from runway 4 RIGHT, whose bearing is N40°E. After flying for 1/2 mile, the pilot requests permission to turn 90° and head toward the southeast. The permission is granted. After the airplane goes 1 mile in this direction, what bearing should the control tower use to locate the aircraft?

40. Finding the Bearing of a Ship   A ship leaves the port of Miami with a bearing of S80°E and a speed of 15 knots. After 1 hour, the ship turns 90° toward the south. After 2 hours, maintaining the same speed, what is the bearing to the ship from port?

41. Shooting Free Throws in Basketball   The eyes of a basketball player are 6 feet above the floor. The player is at the free-throw line, which is 15 feet from the center of the basket rim (see the figure). What is the angle of elevation from the player's eyes to the center of the rim?

[**Hint:** The rim is 10 feet above the floor.]

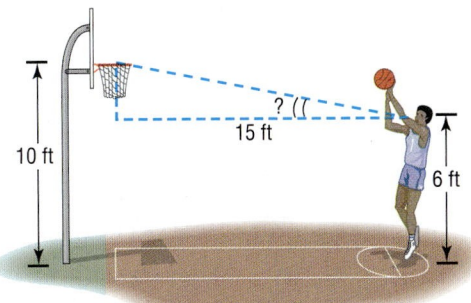

42. Finding the Pitch of a Roof   A carpenter is preparing to put a roof on a garage that is 20 feet by 40 feet by 20 feet. A steel support beam 46 feet in length is positioned in the center of the garage. To support the roof, another beam will be attached to the top of the center beam (see the figure). At what angle of elevation is the new beam? In other words, what is the pitch of the roof?

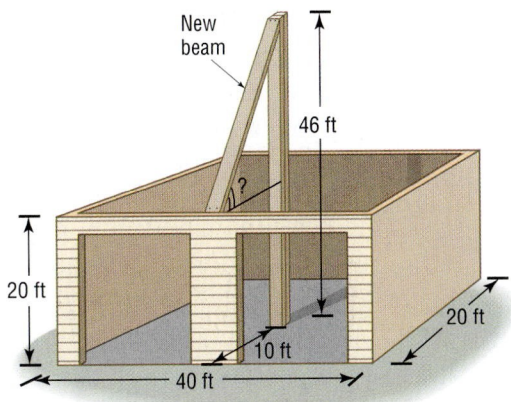

43. Constructing a Highway   A highway whose primary directions are north–south is being constructed along the west coast of Florida. Near Naples, a bay obstructs the straight path of the road. Since the cost of a bridge is prohibitive, engineers decide to go around the bay. The illustration shows the path that they decide on and the measurements taken. What is the length of highway needed to go around the bay?

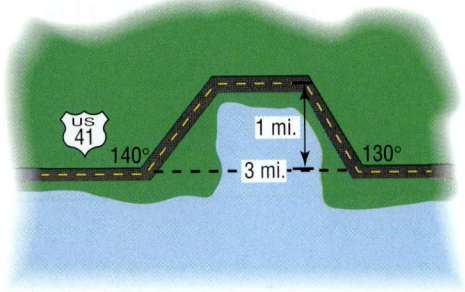

44. Surveillance Satellites   A surveillance satellite circles Earth at a height of h miles above the surface. Suppose that d is the distance, in miles, on the surface of Earth that can be observed from the satellite. See the illustration.
(a) Find an equation that relates the central angle θ to the height h.
(b) Find an equation that relates the observable distance d and θ.
(c) Find an equation that relates d and h.
(d) If d is to be 2500 miles, how high must the satellite orbit above Earth?
(e) If the satellite orbits at a height of 300 miles, what distance d on the surface can be observed?

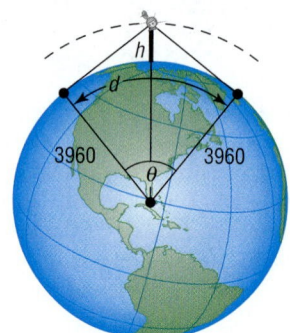

45. Photography   A camera is mounted on a tripod 4 feet high at a distance of 10 feet from George, who is 6 feet tall. See the illustration. If the camera lens has angles of depression and elevation of 20°, will George's feet and head be seen by the lens? If not, how far back will the camera need to be moved to include George's feet and head?

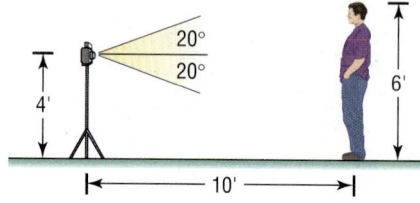

46. Construction   A ramp for wheelchair accessibility is to be constructed with an angle of elevation of 15° and a final height of 5 feet. How long is the ramp?

47. Geometry   A rectangle is inscribed in a semicircle of radius 1. See the illustration.

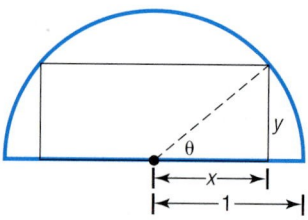

(a) Express the area A of the rectangle as a function of the angle θ shown in the illustration.
(b) Show that $A = \sin(2\theta)$.
(c) Find the angle θ that results in the largest area A.
(d) Find the dimensions of this largest rectangle.

48. Area of an Isosceles Triangle   Show that the area A of an isosceles triangle, whose equal sides are of length s and the angle between them is θ, is

$$A = \frac{1}{2}s^2 \sin \theta$$

[**Hint:** See the illustration. The height h bisects the angle θ and is the perpendicular bisector of the base.]

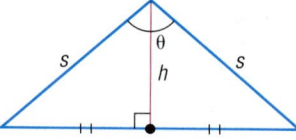

49. The Gibb's Hill Lighthouse, Southampton, Bermuda   In operation since 1846, the Gibb's Hill Lighthouse stands 117 feet high on a hill 245 feet high, so its beam of light is 362 feet above sea level. A brochure states that ships 40 miles away can see the light and planes flying at 10,000 feet can see it 120 miles away. Verify the accuracy of these statements. What assumption did the brochure make about the height of the ship?

PREPARING FOR THIS SECTION

Before getting started, review the following:

✓ Trigonometric Equations (I) (Section 9.7, pp. 759–762)   ✓ Difference Formulas for Sine (Section 9.4, p. 737)

10.2   THE LAW OF SINES

OBJECTIVES   1 Solve SAA or ASA Triangles
2 Solve SSA Triangles
3 Solve Applied Problems Using the Law of Sines

If none of the angles of a triangle is a right angle, the triangle is called **oblique**. An oblique triangle will have either three acute angles or two acute angles and one obtuse angle (an angle between 90° and 180°). See Figure 12.

Figure 12

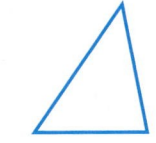 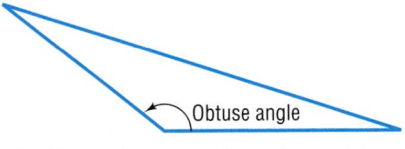

(a) All angles are acute　　**(b)** Two acute angles and one obtuse angle

Figure 13

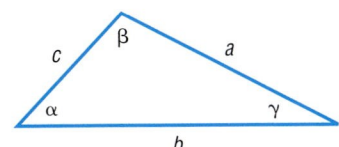

In the discussion that follows, we will always label an oblique triangle so that side a is opposite angle α, side b is opposite angle β, and side c is opposite angle γ, as shown in Figure 13.

To **solve an oblique triangle** means to find the lengths of its sides and the measurements of its angles. To do this, we need to know the length of one side along with: (1) two angles or (2) the other two sides or (3) one angle and one other side. Knowing three angles of a triangle determines a family of *similar triangles,* that is, triangles that have the same shape but different sizes. There are four possibilities to consider:

> **CASE 1:** One side and two angles are known (ASA or SAA).
> **CASE 2:** Two sides and the angle opposite one of them are known (SSA).
> **CASE 3:** Two sides and the included angle are known (SAS).
> **CASE 4:** Three sides are known (SSS).

Figure 14 illustrates the four cases.

Figure 14

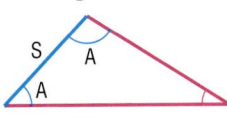 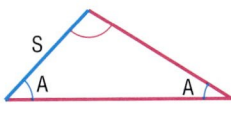 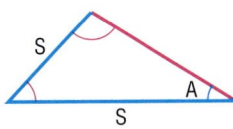 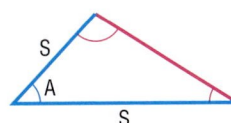 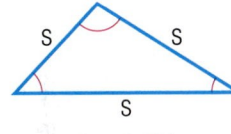

Case 1: ASA　　　Case 1: SAA　　　Case 2: SSA　　　Case 3: SAS　　　Case 4: SSS

The **Law of Sines** is used to solve triangles for which Case 1 or 2 holds. Cases 3 and 4 are considered when we study the Law of Cosines in Section 10.3.

Theorem

Law of Sines

For a triangle with sides a, b, c and opposite angles α, β, γ, respectively,

$$\frac{\sin \alpha}{a} = \frac{\sin \beta}{b} = \frac{\sin \gamma}{c} \tag{1}$$

Proof To prove the Law of Sines, we construct an altitude of length h from one of the vertices of a triangle. Figure 15(a) shows h for a triangle with three acute angles, and Figure 15(b) shows h for a triangle with an obtuse angle. In each case, the altitude is drawn from the vertex at β. Using either illustration, we have

$$\sin \gamma = \frac{h}{a}$$

Figure 15

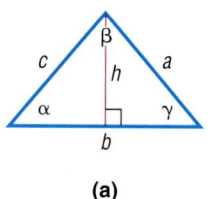

(a)

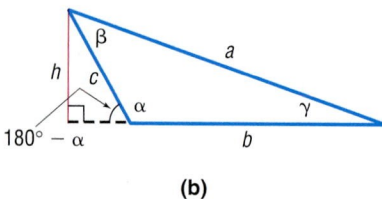

(b)

from which

$$h = a \sin \gamma \qquad (2)$$

From Figure 15(a), it also follows that

$$\sin \alpha = \frac{h}{c}$$

from which

$$h = c \sin \alpha \qquad (3)$$

From Figure 15(b), it follows that

$$\sin(180° - \alpha) = \sin \alpha = \frac{h}{c}$$

Difference formula

which again gives

$$h = c \sin \alpha$$

So, whether the triangle has three acute angles or has two acute angles and one obtuse angle, equations (2) and (3) hold. As a result, we may equate the expressions for h in equations (2) and (3) to get

$$a \sin \gamma = c \sin \alpha$$

from which

$$\frac{\sin \alpha}{a} = \frac{\sin \gamma}{c} \qquad (4)$$

In a similar manner, by constructing the altitude h' from the vertex of angle α as shown in Figure 16, we can show that

Figure 16

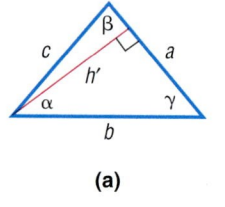

(a)

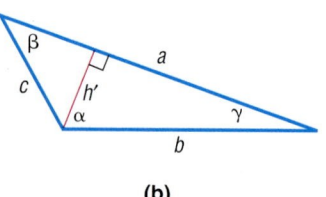

(b)

$$\sin \beta = \frac{h'}{c} \quad \text{and} \quad \sin \gamma = \frac{h'}{b}$$

Equating the expressions for h', we find that

$$h' = c \sin \beta = b \sin \gamma$$

from which

$$\frac{\sin \beta}{b} = \frac{\sin \gamma}{c} \qquad (5)$$

When equations (4) and (5) are combined, we have equation (1), the Law of Sines. ∎

In applying the Law of Sines to solve triangles, we use the fact that the sum of the angles of any triangle equals 180°; that is,

$$\boxed{\alpha + \beta + \gamma = 180° \qquad (6)}$$

1   Our first two examples show how to solve a triangle when one side and two angles are known (Case 1: SAA or ASA).

| EXAMPLE 1 | **Using the Law of Sines to Solve a SAA Triangle** |

Solve the triangle: $\alpha = 40°, \beta = 60°, a = 4$

Solution   Figure 17 shows the triangle that we want to solve. The third angle γ is found using equation (6).

Figure 17

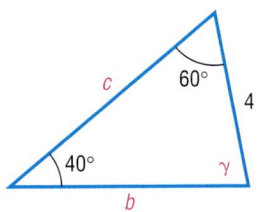

$$\alpha + \beta + \gamma = 180°$$
$$40° + 60° + \gamma = 180°$$
$$\gamma = 80°$$

Now we use the Law of Sines (twice) to find the unknown sides b and c.

$$\frac{\sin \alpha}{a} = \frac{\sin \beta}{b} \qquad \frac{\sin \alpha}{a} = \frac{\sin \gamma}{c}$$

Because $a = 4, \alpha = 40°, \beta = 60°$, and $\gamma = 80°$, we have

$$\frac{\sin 40°}{4} = \frac{\sin 60°}{b} \qquad \frac{\sin 40°}{4} = \frac{\sin 80°}{c}$$

Solving for b and c, we find that

$$b = \frac{4 \sin 60°}{\sin 40°} \approx 5.39 \qquad c = \frac{4 \sin 80°}{\sin 40°} \approx 6.13 \qquad ■$$

Notice in Example 1 that we found b and c by working with the given side a. This is better than finding b first and working with a rounded value of b to find c.

✏ **NOW WORK PROBLEM 1.**

| EXAMPLE 2 | **Using the Law of Sines to Solve an ASA Triangle** |

Solve the triangle: $\alpha = 35°, \beta = 15°, c = 5$

Solution   Figure 18 illustrates the triangle that we want to solve. Because we know two angles ($\alpha = 35°$ and $\beta = 15°$), we find the third angle using equation (6).

Figure 18

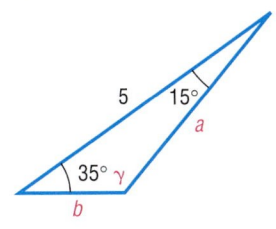

$$\alpha + \beta + \gamma = 180°$$
$$35° + 15° + \gamma = 180°$$
$$\gamma = 130°$$

Now we know the three angles and one side ($c = 5$) of the triangle. To find the remaining two sides a and b, we use the Law of Sines (twice).

$$\frac{\sin \alpha}{a} = \frac{\sin \gamma}{c} \qquad \frac{\sin \beta}{b} = \frac{\sin \gamma}{c}$$

$$\frac{\sin 35°}{a} = \frac{\sin 130°}{5} \qquad \frac{\sin 15°}{b} = \frac{\sin 130°}{5}$$

$$a = \frac{5 \sin 35°}{\sin 130°} \approx 3.74 \qquad b = \frac{5 \sin 15°}{\sin 130°} \approx 1.69 \qquad ■$$

✏ **NOW WORK PROBLEM 15.**

The Ambiguous Case

(2) Case 2 (SSA), which applies to triangles for which two sides and the angle opposite one of them are known, is referred to as the **ambiguous case**, because the known information may result in one triangle, two triangles, or no triangle at all. Suppose that we are given sides a and b and angle α, as illustrated in Figure 19. The key to determining the possible triangles, if any, that may be formed from the given information lies primarily with the height h and the fact that $h = b \sin \alpha$.

Figure 19

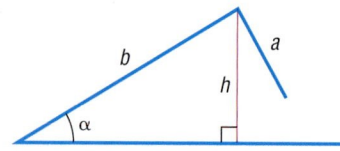

No Triangle If $a < h = b \sin \alpha$, then side a is not sufficiently long to form a triangle. See Figure 20.

One Right Triangle If $a = h = b \sin \alpha$, then side a is just long enough to form a right triangle. See Figure 21.

Figure 20
$a < h = b \sin \alpha$

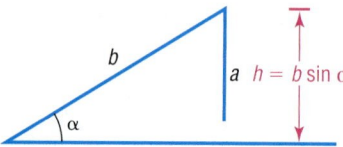

Figure 21
$a = b \sin \alpha$

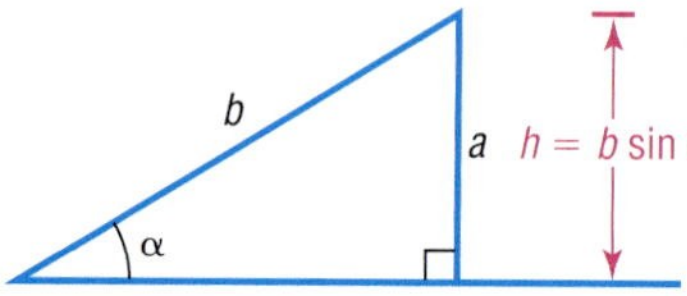

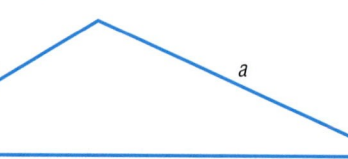

Two Triangles If $a < b$ and $h = b \sin \alpha < a$, then two distinct triangles can be formed from the given information. See Figure 22.

One Triangle If $a \geq b$, then only one triangle can be formed. See Figure 23.

Figure 22
$b \sin \alpha < a$ and $a < b$

Figure 23
$a \geq b$

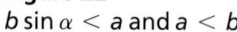
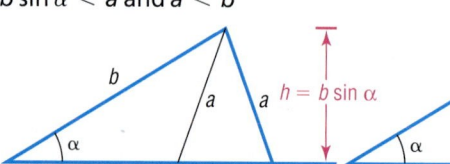

Fortunately, we do not have to rely on an illustration to draw the correct conclusion in the ambiguous case. The Law of Sines will lead us to the correct determination. Let's see how.

EXAMPLE 3

Using the Law of Sines to Solve a SSA Triangle (One Solution)

Solve the triangle: $a = 3, b = 2, \alpha = 40°$

Solution

See Figure 24(a). Because $a = 3, b = 2$, and $\alpha = 40°$ are known, we use the Law of Sines to find the angle β.

Figure 24(a)

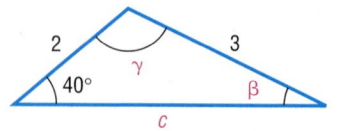

$$\frac{\sin \alpha}{a} = \frac{\sin \beta}{b}$$

Then

$$\frac{\sin 40°}{3} = \frac{\sin \beta}{2}$$

$$\sin \beta = \frac{2 \sin 40°}{3} \approx 0.43$$

There are two angles β, $0° < \beta < 180°$, for which $\sin \beta \approx 0.43$.

$$\beta_1 \approx 25.4° \quad \text{and} \quad \beta_2 \approx 154.6°$$

Note: Here we computed β using the stored value of $\sin \beta$. If you use the rounded value, $\sin \beta \approx 0.43$, you will obtain slightly different results.

The second possibility, $\beta_2 \approx 154.6°$, is ruled out, because $\alpha = 40°$, making $\alpha + \beta_2 \approx 194.6° > 180°$. Now, using $\beta_1 \approx 25.4°$, we find that

$$\gamma = 180° - \alpha - \beta_1 \approx 180° - 40° - 25.4° = 114.6°$$

The third side c may now be determined using the Law of Sines.

$$\frac{\sin \alpha}{a} = \frac{\sin \gamma}{c}$$

$$\frac{\sin 40°}{3} = \frac{\sin 114.6°}{c}$$

$$c = \frac{3 \sin 114.6°}{\sin 40°} \approx 4.24$$

Figure 24(b)

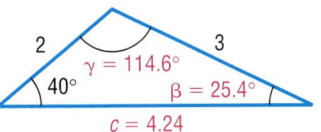

Figure 24(b) illustrates the solved triangle. ■

EXAMPLE 4   **Using the Law of Sines to Solve a SSA Triangle (Two Solutions)**

Solve the triangle: $a = 6, b = 8, \alpha = 35°$

Solution   See Figure 25(a). Because $a = 6, b = 8$, and $\alpha = 35°$ are known, we use the Law of Sines to find the angle β.

Figure 25(a)

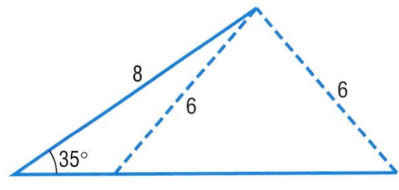

$$\frac{\sin \alpha}{a} = \frac{\sin \beta}{b}$$

Then

$$\frac{\sin 35°}{6} = \frac{\sin \beta}{8}$$

$$\sin \beta = \frac{8 \sin 35°}{6} \approx 0.76$$

$$\beta_1 \approx 49.9° \quad \text{or} \quad \beta_2 \approx 130.1°$$

For both choices of β, we have $\alpha + \beta < 180°$. There are two triangles one containing the angle $\beta_1 \approx 49.9°$ and the other containing the angle $\beta_2 \approx 130.1°$. The third angle γ is either

$$\gamma_1 = 180° - \alpha - \beta_1 \approx 95.1° \quad \text{or} \quad \gamma_2 = 180° - \alpha - \beta_2 \approx 14.9°$$

$$\uparrow \qquad\qquad\qquad\qquad\qquad\qquad \uparrow$$
$$\alpha = 35° \qquad\qquad\qquad\qquad\qquad \alpha = 35°$$
$$\beta_1 = 49.9° \qquad\qquad\qquad\qquad\; \beta_2 = 130.1°$$

Figure 25(b)

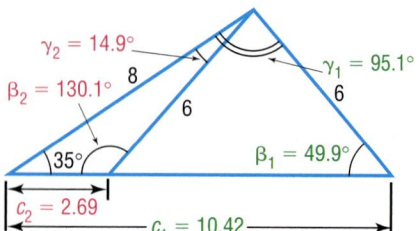

The third side c obeys the Law of Sines, so we have

$$\frac{\sin \alpha}{a} = \frac{\sin \gamma_1}{c_1} \qquad\qquad \frac{\sin \alpha}{a} = \frac{\sin \gamma_2}{c_2}$$

$$\frac{\sin 35°}{6} = \frac{\sin 95.1°}{c_1} \qquad\qquad \frac{\sin 35°}{6} = \frac{\sin 14.9°}{c_2}$$

$$c_1 = \frac{6 \sin 95.1°}{\sin 35°} \approx 10.42 \qquad\qquad c_2 = \frac{6 \sin 14.9°}{\sin 35°} \approx 2.69$$

The two solved triangles are illustrated in Figure 25(b). ∎

EXAMPLE 5 **Using the Law of Sines to Solve a SSA Triangle (No Solution)**

Solve the triangle: $a = 2, c = 1, \gamma = 50°$

Solution Because $a = 2, c = 1$, and $\gamma = 50°$ are known, we use the Law of Sines to find the angle α.

$$\frac{\sin \alpha}{a} = \frac{\sin \gamma}{c}$$

$$\frac{\sin \alpha}{2} = \frac{\sin 50°}{1}$$

$$\sin \alpha = 2 \sin 50° \approx 1.53$$

Figure 26

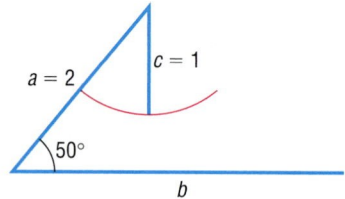

Since there is no angle α for which $\sin \alpha > 1$, there can be no triangle with the given measurements. Figure 26 illustrates the measurements given. Notice that, no matter how we attempt to position side c, it will never touch side b to form a triangle. ∎

NOW WORK PROBLEMS 17 AND 23.

Applications

③ The Law of Sines is particularly useful for solving certain applied problems.

EXAMPLE 6 **Finding the Height of a Mountain**

To measure the height of a mountain, a surveyor takes two sightings of the peak at a distance 900 meters apart on a direct line to the mountain.* See Figure 27(a). The first observation results in an angle of elevation of 47°, whereas the second results in an angle of elevation of 35°. If the transit is 2 meters high, what is the height h of the mountain?

Figure 27

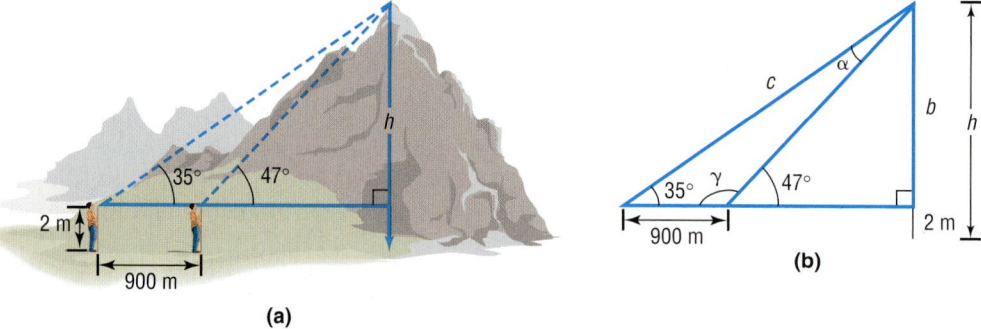

(a)

(b)

*For simplicity, we assume that these sightings are at the same level.

Solution Figure 27(b) shows the triangles that replicate the illustration in Figure 27(a). Since $\gamma + 47° = 180°$, we find that $\gamma = 133°$. Also, since $\alpha + \beta + \gamma = 180°$, we find that $\alpha = 180° - 35° - 133° = 12°$. We use the Law of Sines to find c.

$$\frac{\sin \alpha}{a} = \frac{\sin \gamma}{c} \qquad \textcolor{blue}{\alpha = 12°, \gamma = 133°, a = 900}$$

$$c = \frac{900 \sin 133°}{\sin 12°} = 3165.86$$

Using the larger right triangle, we have

$$\sin 35° = \frac{b}{c} \qquad \textcolor{blue}{c = 3165.86}$$

$$b = 3165.86 \sin 35° = 1815.86 \approx 1816 \text{ meters}$$

The height of the peak from ground level is approximately $1816 + 2 = 1818$ meters.

NOW WORK PROBLEM **31.**

EXAMPLE 7 **Rescue at Sea**

Coast Guard Station Zulu is located 120 miles due west of Station X-ray. A ship at sea sends an SOS call that is received by each station. The call to Station Zulu indicates that the bearing of the ship from Zulu is N40°E (40° east of north). The call to Station X-ray indicates that the bearing of the ship from X-ray is N30°W (30° west of north).

(a) How far is each station from the ship?

(b) If a helicopter capable of flying 200 miles per hour is dispatched from the nearest station to the ship, how long will it take to reach the ship?

Solution (a) Figure 28 illustrates the situation. The angle γ is found to be

$$\gamma = 180° - 50° - 60° = 70°$$

Figure 28

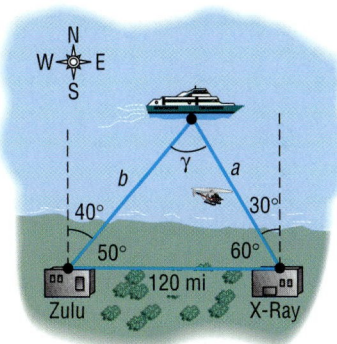

The Law of Sines can now be used to find the two distances a and b that we seek.

$$\frac{\sin 50°}{a} = \frac{\sin 70°}{120}$$

$$a = \frac{120 \sin 50°}{\sin 70°} \approx 97.82 \text{ miles}$$

$$\frac{\sin 60°}{b} = \frac{\sin 70°}{120}$$

$$b = \frac{120 \sin 60°}{\sin 70°} \approx 110.59 \text{ miles}$$

Station Zulu is about 111 miles from the ship, and Station X-ray is about 98 miles from the ship.

(b) The time t needed for the helicopter to reach the ship from Station X-ray is found by using the formula

$$(\text{Velocity}, v)(\text{Time}, t) = \text{Distance}, a$$

Then

$$t = \frac{a}{v} = \frac{97.82}{200} \approx 0.49 \text{ hour} \approx 29 \text{ minutes}$$

It will take about 29 minutes for the helicopter to reach the ship.   ■

 **NOW WORK PROBLEM 29.**

10.2 Concepts and Vocabulary

In Problems 1 and 2, fill in the blanks.

1. If none of the angles of a triangle is a right angle, the triangle is called _____.

2. For a triangle with sides a, b, c and opposite angles α, β, γ, the Law of Sines states that _____.

In Problems 3–5, answer True or False to each statement.

3. An oblique triangle in which two sides and an angle are given always results in at least one triangle.

4. The sum of the angles of any triangle equals 180°.

5. The ambiguous case refers to the fact that, when two sides and the angle opposite one of them is given, sometimes the Law of Sines cannot be used.

6. What do you do first if you are asked to solve a triangle and are given one side and two angles?

7. What do you do first if you are asked to solve a triangle and are given two sides and the angle opposite one of them?

10.2 Exercises

In Problems 1–8 solve each triangle.

1.
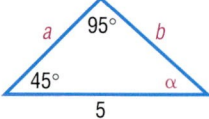

2.
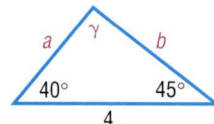

3.
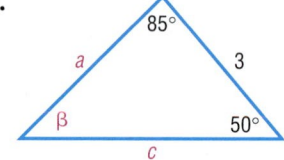

4.
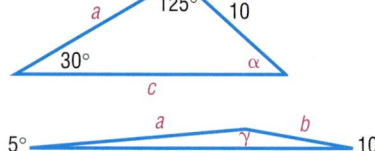

5.

6.
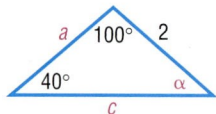

7.

8.
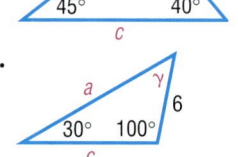

In Problems 9–14 solve each triangle.

9. $\alpha = 40°$, $\beta = 20°$, $a = 2$

10. $\alpha = 50°$, $\gamma = 20°$, $a = 3$

11. $\beta = 70°$, $\gamma = 10°$, $b = 5$

12. $\alpha = 70°$, $\beta = 60°$, $c = 4$

13. $\alpha = 110°$, $\gamma = 30°$, $c = 3$

14. $\beta = 10°$, $\gamma = 100°$, $b = 2$

15. $\alpha = 40°$, $\beta = 40°$, $c = 2$

16. $\beta = 20°$, $\gamma = 70°$, $a = 1$

In Problems 17–28, two sides and an angle are given. Determine whether the given information results in one triangle, two triangles, or no triangle at all. Solve any resulting triangle(s).

17. $a = 3$, $b = 2$, $\alpha = 50°$

18. $b = 4$, $c = 3$, $\beta = 40°$

19. $b = 5$, $c = 3$, $\beta = 100°$

20. $a = 2$, $c = 1$, $\alpha = 120°$

21. $a = 4$, $b = 5$, $\alpha = 60°$

22. $b = 2$, $c = 3$, $\beta = 40°$

23. $b = 4$, $c = 6$, $\beta = 20°$

24. $a = 3$, $b = 7$, $\alpha = 70°$

25. $a = 2$, $c = 1$, $\gamma = 100°$

26. $b = 4$, $c = 5$, $\beta = 95°$

27. $a = 2$, $c = 1$, $\gamma = 25°$

28. $b = 4$, $c = 5$, $\beta = 40°$

29. Rescue at Sea   Coast Guard Station Able is located 150 miles due south of Station Baker. A ship at sea sends an SOS call that is received by each station. The call to Station Able indicates that the ship is located N55°E; the call to Station Baker indicates that the ship is located S60°E.
(a) How far is each station from the ship?
(b) If a helicopter capable of flying 200 miles per hour is dispatched from the nearest station to the ship, how long will it take to reach the ship?

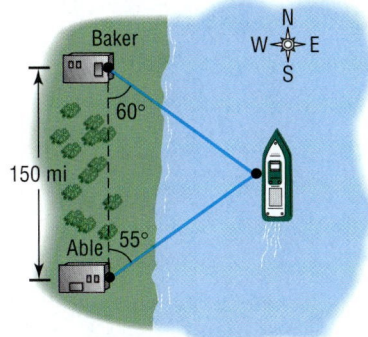

30. Surveying   Consult the figure. To find the distance from the house at *A* to the house at *B*, a surveyor measures the angle *BAC* to be 40° and then walks off a distance of 100 feet to *C* and measures the angle *ACB* to be 50°. What is the distance from *A* to *B*?

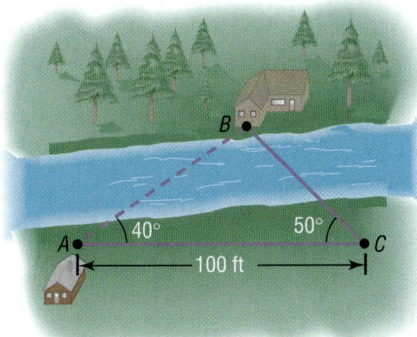

31. Finding the Length of a Ski Lift   Consult the figure. To find the length of the span of a proposed ski lift from *A* to *B*, a surveyor measures the angle *DAB* to be 25° and then walks off a distance of 1000 feet to *C* and measures the angle *ACB* to be 15°. What is the distance from *A* to *B*?

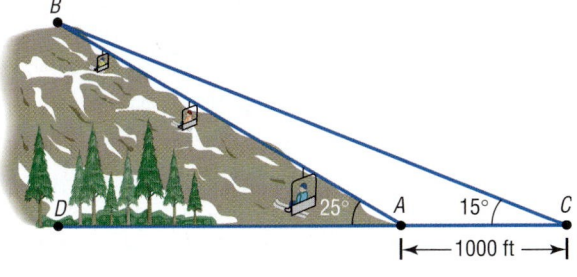

32. Finding the Height of a Mountain   Use the illustration in Problem 31 to find the height *BD* of the mountain at *B*.

33. Finding the Height of an Airplane   An aircraft is spotted by two observers who are 1000 feet apart. As the airplane passes over the line joining them, each observer takes a sighting of the angle of elevation to the plane, as indicated in the figure. How high is the airplane?

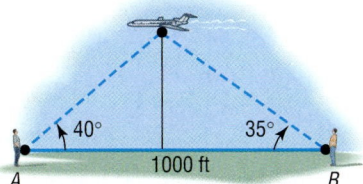

34. Finding the Height of the Bridge over the Royal Gorge   The highest bridge in the world is the bridge over the Royal Gorge of the Arkansas River in Colorado.* Sightings to the same point at water level directly under the bridge are taken from each side of the 880-foot-long bridge, as indicated in the figure. How high is the bridge?

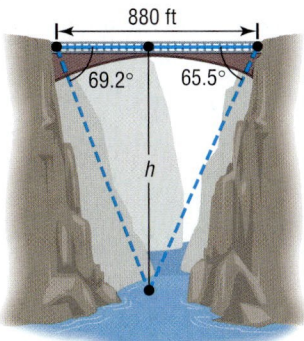

35. Navigation   An airplane flies from city *A* to city *B*, a distance of 150 miles, and then turns through an angle of 40° and heads toward city *C*, as shown in the figure.
(a) If the distance between cities *A* and *C* is 300 miles, how far is it from city *B* to city *C*?
(b) Through what angle should the pilot turn at city *C* to return to city *A*?

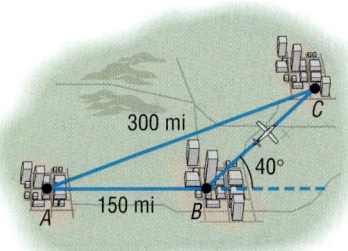

Source: Guinness Book of World Records.

36. Time Lost due to a Navigation Error   In attempting to fly from city A to city B, an aircraft followed a course that was 10° in error, as indicated in the figure. After flying a distance of 50 miles, the pilot corrected the course by turning at point C and flying 70 miles farther. If the constant speed of the aircraft was 250 miles per hour, how much time was lost due to the error?

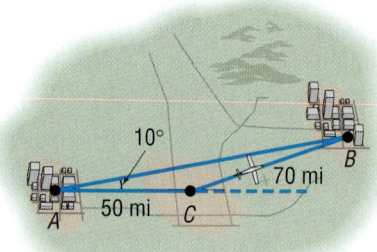

37. Finding the Lean of the Leaning Tower of Pisa   The famous Leaning Tower of Pisa was originally 184.5 feet high. At a distance of 123 feet from the base of the tower, the angle of elevation to the top of the tower is found to be 60°. Find the angle CAB indicated in the figure. Also, find the perpendicular distance from C to AB.

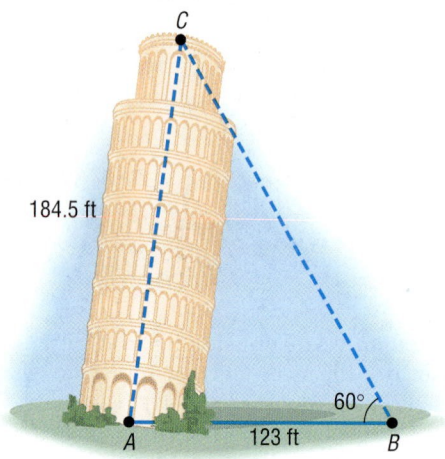

38. Crankshafts on Cars   On a certain automobile, the crankshaft is 3 inches long and the connecting rod is 9 inches long (see the figure). At the time when the angle OPA is 15°, how far is the piston (P) from the center (O) of the crankshaft?

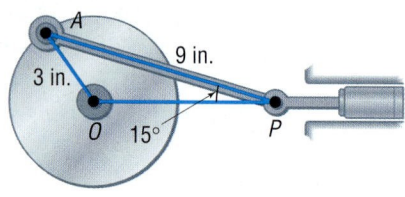

39. Constructing a Highway   U.S. 41, a highway whose primary directions are north–south, is being constructed along the west coast of Florida. Near Naples, a bay obstructs the straight path of the road. Since the cost of a bridge is prohibitive, engineers decide to go around the bay. The illustration shows the path that they decide on and the measurements taken. What is the length of highway needed to go around the bay?

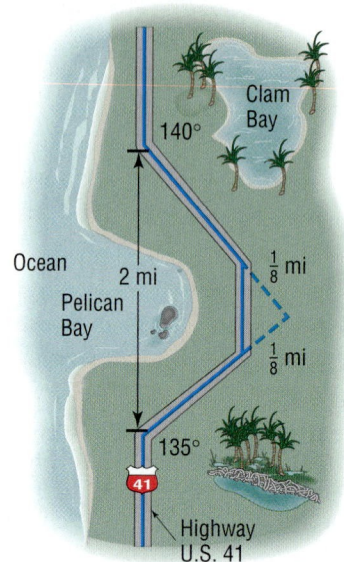

40. Determining Distances at Sea   The navigator of a ship at sea spots two lighthouses that she knows to be 3 miles apart along a straight seashore. She determines that the angles formed between two line-of-sight observations of the lighthouses and the line from the ship directly to shore are 15° and 35°. See the illustration.
(a) How far is the ship from lighthouse A?
(b) How far is the ship from lighthouse B?
(c) How far is the ship from shore?

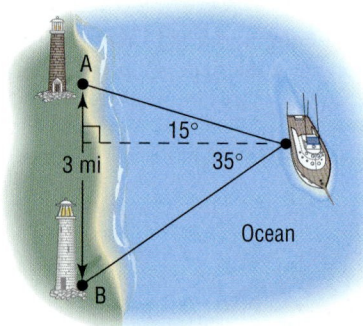

41. Calculating Distance at Sea   The navigator of a ship at sea has the harbor in sight at which the ship is to dock. She spots a lighthouse that she knows is 1 mile up the coast from the mouth of the harbor, and she measures the angle between the line-of-sight observations of the harbor and lighthouse to be 20°. With the ship heading directly toward the harbor, she repeats this measurement after 5 minutes of traveling at 12 miles per hour. If the new angle is 30°, how far is the ship from the harbor?

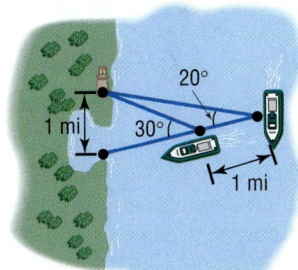

42. Finding Distances A forest ranger is walking on a path inclined at 5° to the horizontal directly toward a 100-foot-tall fire observation tower. The angle of elevation from the path to the top of the tower is 40°. How far is the ranger from the tower at this time?

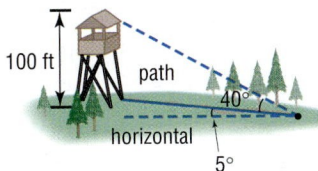

43. Great Pyramid of Cheops One of the original Seven Wonders of the World, the Great Pyramid of Cheops was built about 2580 BC. Its original height was 480 feet 11 inches, but due to the loss of its topmost stones, it is now shorter.* Find the current height of the Great Pyramid, using the information given in the illustration.

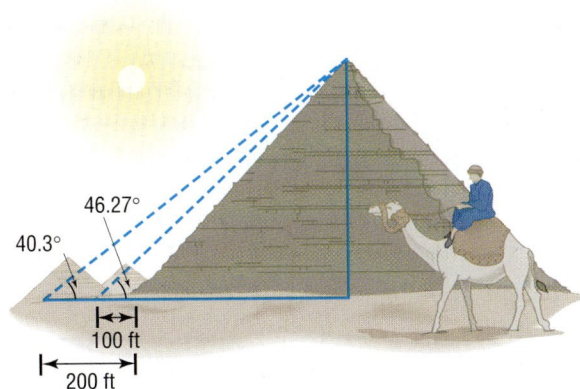

44. Determining the Height of an Aircraft Two sensors are spaced 700 feet apart along the approach to a small airport. When an aircraft is nearing the airport, the angle of elevation from the first sensor to the aircraft is 20°, and from the second sensor to the aircraft it is 15°. Determine how high the aircraft is at this time.

Source: Guinness Book of World Records.

45. Landscaping Pat needs to determine the height of a tree before cutting it down to be sure that it will not fall on a nearby fence. The angle of elevation of the tree from one position on a flat path from the tree is 30°, and from a second position 40 feet farther along this path it is 20°. What is the height of the tree?

46. Construction A loading ramp 10 feet long that makes an angle of 18° with the horizontal is to be replaced by one that makes an angle of 12° with the horizontal. How long is the new ramp?

47. Finding the Height of a Helicopter Two observers simultaneously measure the angle of elevation of a helicopter. One angle is measured as 25°, the other as 40° (see the figure). If the observers are 100 feet apart and the helicopter lies over the line joining them, how high is the helicopter?

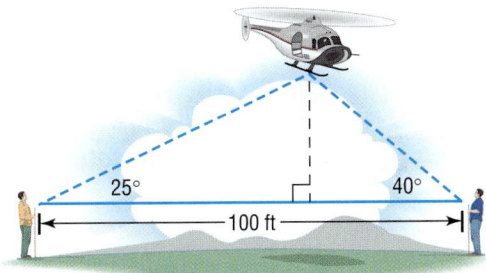

48. Mollweide's Formula For any triangle, Mollweide's Formula (named after Karl Mollweide, 1774–1825) states that

$$\frac{a+b}{c} = \frac{\cos\left[\frac{1}{2}(\alpha - \beta)\right]}{\sin\left(\frac{1}{2}\gamma\right)}$$

Derive it.

[**Hint:** Use the Law of Sines and then a sum-to-product formula. Notice that this formula involves all six parts of a triangle. As a result, it is sometimes used to check the solution of a triangle.]

49. Mollweide's Formula Another form of Mollweide's Formula is

$$\frac{a-b}{c} = \frac{\sin\left[\frac{1}{2}(\alpha - \beta)\right]}{\cos\left(\frac{1}{2}\gamma\right)}$$

Derive it.

50. For any triangle, derive the formula

$$a = b \cos \gamma + c \cos \beta$$

[**Hint:** Use the fact that $\sin \alpha = \sin(180° - \beta - \gamma)$.]

51. Law of Tangents   For any triangle, derive the Law of Tangents.

$$\frac{a - b}{a + b} = \frac{\tan\left[\frac{1}{2}(\alpha - \beta)\right]}{\tan\left[\frac{1}{2}(\alpha + \beta)\right]}$$

[**Hint:** Use Mollweide's Formula.]

52. Circumscribing a Triangle   Show that

$$\frac{\sin \alpha}{a} = \frac{\sin \beta}{b} = \frac{\sin \gamma}{c} = \frac{1}{2r}$$

where r is the radius of the circle circumscribing the triangle ABC whose sides are a, b, and c, as shown in the figure.

[**Hint:** Draw the diameter AB'. Then β = angle ABC = angle $AB'C$, and angle ACB' = 90°.]

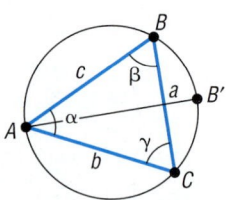

 53. Make up three problems involving oblique triangles. One should result in one triangle, the second in two triangles, and the third in no triangle.

PREPARING FOR THIS SECTION

Before getting started, review the following:

✓ Trigonometric Equations (I) (Section 9.7, pp. 759–762)   ✓ The Distance Formula (Section 1.1, pp. 93–95)

10.3   THE LAW OF COSINES

OBJECTIVES   **1** Solve SAS Triangles

2 Solve SSS Triangles

3 Solve Applied Problems Using the Law of Cosines

In Section 10.2, we used the Law of Sines to solve Case 1 (SAA or ASA) and Case 2 (SSA) of an oblique triangle. In this section, we derive the Law of Cosines and use it to solve the remaining cases, 3 and 4.

> **CASE 3:**   Two sides and the included angle are known (SAS).
>
> **CASE 4:**   Three sides are known (SSS).

Theorem   **Law of Cosines**

For a triangle with sides a, b, c and opposite angles α, β, γ, respectively,

$$c^2 = a^2 + b^2 - 2ab \cos \gamma \qquad (1)$$
$$b^2 = a^2 + c^2 - 2ac \cos \beta \qquad (2)$$
$$a^2 = b^2 + c^2 - 2bc \cos \alpha \qquad (3)$$

Proof   We will prove only formula (1) here. Formulas (2) and (3) may be proved using the same argument.

Figure 29

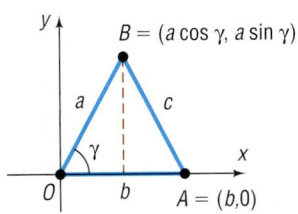

(a) Angle γ is acute

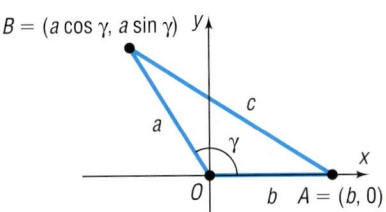

(b) Angle γ is obtuse

We begin by strategically placing a triangle on a rectangular coordinate system so that the vertex of angle γ is at the origin and side b lies along the positive x-axis. Regardless of whether γ is acute, as in Figure 29(a), or obtuse, as in Figure 29(b), the vertex B has coordinates $(a \cos \gamma, a \sin \gamma)$. Vertex A has coordinates $(b, 0)$.

We can now use the distance formula to compute c^2.

$$c^2 = (b - a \cos \gamma)^2 + (0 - a \sin \gamma)^2$$

$$= b^2 - 2ab \cos \gamma + a^2 \cos^2 \gamma + a^2 \sin^2 \gamma$$

$$= b^2 - 2ab \cos \gamma + a^2(\cos^2 \gamma + \sin^2 \gamma)$$

$$= a^2 + b^2 - 2ab \cos \gamma$$

Each of formulas (1), (2), and (3) may be stated in words as follows:

Theorem

Law of Cosines

The square of one side of a triangle equals the sum of the squares of the other two sides minus twice their product times the cosine of their included angle.

Observe that if the triangle is a right triangle (so that, say, $\gamma = 90°$) then formula (1) becomes the familiar Pythagorean Theorem: $c^2 = a^2 + b^2$. Thus, the Pythagorean Theorem is a special case of the Law of Cosines.

① Let's see how to use the Law of Cosines to solve Case 3 (SAS), which applies to triangles for which two sides and the included angle are known.

EXAMPLE 1   Using the Law of Cosines to Solve a SAS Triangle

Solve the triangle: $a = 2, b = 3, \gamma = 60°$.

Figure 30

Solution   See Figure 30. The Law of Cosines makes it easy to find the third side, c.

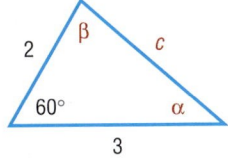

$$c^2 = a^2 + b^2 - 2ab \cos \gamma$$

$$= 4 + 9 - 2 \cdot 2 \cdot 3 \cdot \cos 60°$$

$$= 13 - \left(12 \cdot \frac{1}{2}\right) = 7$$

$$c = \sqrt{7}$$

Side c is of length $\sqrt{7}$. To find the angles α and β, we may use either the Law of Sines or the Law of Cosines. It is preferable to use the Law of Cosines, since it will lead to an equation with one solution. Using the Law of Sines would lead to an equation with two solutions that would need to be checked to determine which solution fits the given data. We choose to use formulas (2) and (3) of the Law of Cosines to find α and β.

For α:

$$a^2 = b^2 + c^2 - 2bc \cos \alpha$$
$$2bc \cos \alpha = b^2 + c^2 - a^2$$
$$\cos \alpha = \frac{b^2 + c^2 - a^2}{2bc} = \frac{9 + 7 - 4}{2 \cdot 3\sqrt{7}} = \frac{12}{6\sqrt{7}} = \frac{2\sqrt{7}}{7}$$
$$\alpha \approx 40.9°$$

For β:

$$b^2 = a^2 + c^2 - 2ac \cos \beta$$
$$\cos \beta = \frac{a^2 + c^2 - b^2}{2ac} = \frac{4 + 7 - 9}{4\sqrt{7}} = \frac{1}{2\sqrt{7}} = \frac{\sqrt{7}}{14}$$
$$\beta \approx 79.1°$$

Notice that $\alpha + \beta + \gamma = 40.9° + 79.1° + 60° = 180°$, as required.   ■

NOW WORK PROBLEM **1.**

②   The next example illustrates how the Law of Cosines is used when three sides of a triangle are known, Case 4 (SSS).

EXAMPLE 2   Using the Law of Cosines to Solve a SSS Triangle

Solve the triangle: $a = 4, b = 3, c = 6$.

Solution   See Figure 31. To find the angles α, β, and γ, we proceed as we did in the latter part of the solution to Example 1.

Figure 31

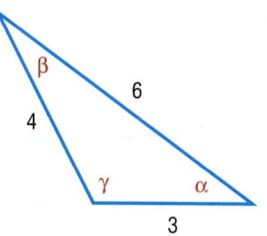

For α:

$$\cos \alpha = \frac{b^2 + c^2 - a^2}{2bc} = \frac{9 + 36 - 16}{2 \cdot 3 \cdot 6} = \frac{29}{36}$$
$$\alpha \approx 36.3°$$

For β:

$$\cos \beta = \frac{a^2 + c^2 - b^2}{2ac} = \frac{16 + 36 - 9}{2 \cdot 4 \cdot 6} = \frac{43}{48}$$
$$\beta \approx 26.4°$$

Since we know α and β,

$$\gamma = 180° - \alpha - \beta \approx 180° - 36.3° - 26.4° = 117.3°$$ ■

NOW WORK PROBLEM **7.**

EXAMPLE 3   Correcting a Navigational Error

③   A motorized sailboat leaves Naples, Florida, bound for Key West, 150 miles away. Maintaining a constant speed of 15 miles per hour, but encountering heavy crosswinds and strong currents, the crew finds, after 4 hours, that the sailboat is off course by 20°.

(a) How far is the sailboat from Key West at this time?

(b) Through what angle should the sailboat turn to correct its course?

(c) How much time has been added to the trip because of this? (Assume that the speed remains at 15 miles per hour.)

Solution   See Figure 32. With a speed of 15 miles per hour, the sailboat has gone 60 miles after 4 hours. We seek the distance x of the sailboat from Key West. We also seek the angle θ that the sailboat should turn through to correct its course.

Figure 32

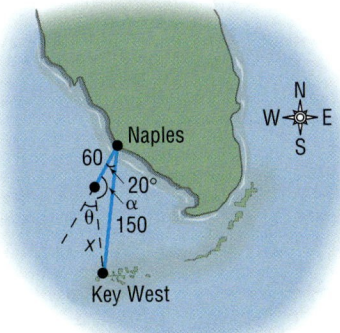

(a) To find x, we use the Law of Cosines, since we know two sides and the included angle.

$$x^2 = 150^2 + 60^2 - 2(150)(60)\cos 20° \approx 9186$$
$$x \approx 95.8$$

The sailboat is about 96 miles from Key West.

(b) We now know three sides of the triangle, so we can use the Law of Cosines again to find the angle α opposite the side of length 150 miles.

$$150^2 = 96^2 + 60^2 - 2(96)(60)\cos \alpha$$
$$9684 = -11{,}520 \cos \alpha$$
$$\cos \alpha \approx -0.8406$$
$$\alpha \approx 147.2°$$

The sailboat should turn through an angle of

$$\theta = 180° - \alpha \approx 180° - 147.2° = 32.8°$$

The sailboat should turn through an angle of about 33° to correct its course.

(c) The total length of the trip is now $60 + 96 = 156$ miles. The extra 6 miles will only require about 0.4 hour or 24 minutes more if the speed of 15 miles per hour is maintained. ■

 **NOW WORK PROBLEM 27.**

HISTORICAL FEATURE

The Law of Sines was known vaguely long before it was explicitly stated by Nasîr Eddîn (about AD 1250). Ptolemy (about AD 150) was aware of it in a form using a chord function instead of the sine function. But it was first clearly stated in Europe by Regiomontanus, writing in 1464.

The Law of Cosines appears first in Euclid's *Elements* (Book II), but in a well-disguised form in which squares built on the sides of triangles are added and a rectangle representing the cosine term is subtracted. It was thus known to all mathematicians because of their familiarity with Euclid's work. An early modern form of the Law of Cosines, that for finding the angle when the sides are known, was stated by François Viète (in 1593).

The Law of Tangents (see Problem 51 of Exercise 10.2) has become obsolete. In the past it was used in place of the Law of Cosines, because the Law of Cosines was very inconvenient for calculation with logarithms or slide rules. Mixing of addition and multiplication is now very easy on a calculator, however, and the Law of Tangents has been shelved along with the slide rule.

10.3 Concepts and Vocabulary

In Problems 1–3, fill in the blanks.

1. If three sides of a triangle are given, the Law of _____ is used to solve the triangle.

2. If one side and two angles of a triangle are given, the Law of _____ is used to solve the triangle.

3. If two sides and the included angle of a triangle are given, the Law of _____ is used to solve the triangle.

In Problems 4–6, answer True or False to each statement.

4. Given only the three sides of a triangle, there is insufficient information to solve the triangle.

5. Given two sides and the included angle, the first thing to do to solve the triangle is to use the Law of Sines.

6. A special case of the Law of Cosines is the Pythagorean Theorem.

7. What do you do first if you are asked to solve a triangle and are given two sides and the included angle?

8. What do you do first if you are asked to solve a triangle and are given three sides?

9. Make up an applied problem that requires using the Law of Cosines.

10.3 Exercises

In Problems 1–8, solve each triangle.

1.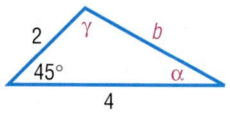

2.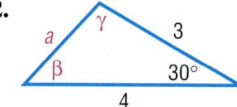

3.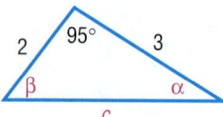

4.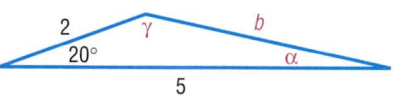

5.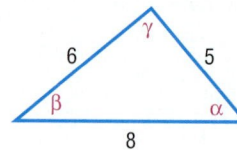

6.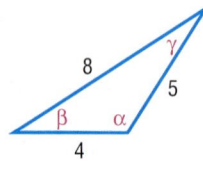

7.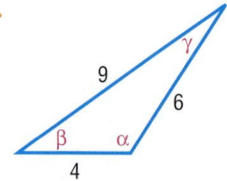

8.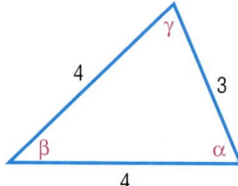

In Problems 9–24, solve each triangle.

9. $a = 3, \quad b = 4, \quad \gamma = 40°$

10. $a = 2, \quad c = 1, \quad \beta = 10°$

11. $b = 1, \quad c = 3, \quad \alpha = 80°$

12. $a = 6, \quad b = 4, \quad \gamma = 60°$

13. $a = 3, \quad c = 2, \quad \beta = 110°$

14. $b = 4, \quad c = 1, \quad \alpha = 120°$

15. $a = 2, \quad b = 2, \quad \gamma = 50°$

16. $a = 3, \quad c = 2, \quad \beta = 90°$

17. $a = 12, \quad b = 13, \quad c = 5$

18. $a = 4, \quad b = 5, \quad c = 3$

19. $a = 2, \quad b = 2, \quad c = 2$

20. $a = 3, \quad b = 3, \quad c = 2$

21. $a = 5, \quad b = 8, \quad c = 9$

22. $a = 4, \quad b = 3, \quad c = 6$

23. $a = 10, \quad b = 8, \quad c = 5$

24. $a = 9, \quad b = 7, \quad c = 10$

---

25. Surveying Consult the figure. To find the distance from the house at A to the house at B, a surveyor measures the angle ACB, which is found to be 70°, and then walks off the distance to each house, 50 feet and 70 feet, respectively. How far apart are the houses?

(a) How far is it from Ft. Myers to Orlando?
(b) Through what angle should the pilot turn at Orlando to return to Ft. Myers?

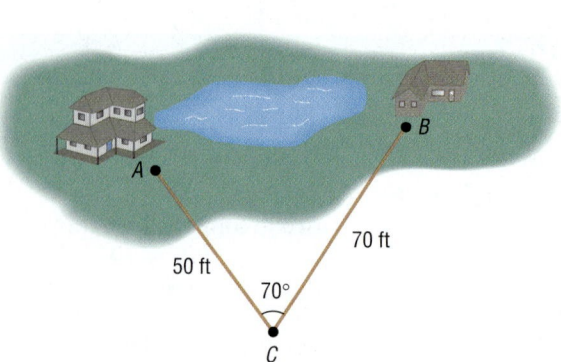

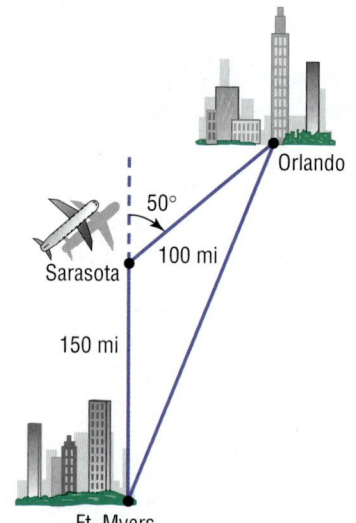

26. Navigation An airplane flies from Ft. Myers to Sarasota, a distance of 150 miles, and then turns through an angle of 50° and flies to Orlando, a distance of 100 miles (see the figure).

27. **Revising a Flight Plan** In attempting to fly from Chicago to Louisville, a distance of 330 miles, a pilot inadvertently took a course that was 10° in error, as indicated in the figure.
    (a) If the aircraft maintains an average speed of 220 miles per hour and if the error in direction is discovered after 15 minutes, through what angle should the pilot turn to head toward Louisville?
    (b) What new average speed should the pilot maintain so that the total time of the trip is 90 minutes?

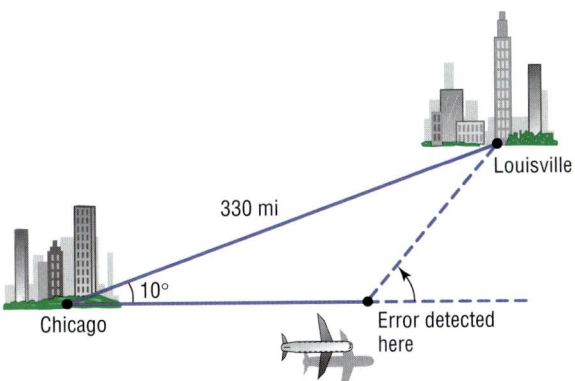

28. **Avoiding a Tropical Storm** A cruise ship maintains an average speed of 15 knots in going from San Juan, Puerto Rico, to Barbados, West Indies, a distance of 600 nautical miles. To avoid a tropical storm, the captain heads out of San Juan in a direction of 20° off a direct heading to Barbados. The captain maintains the 15-knot speed for 10 hours, after which time the path to Barbados becomes clear of storms.
    (a) Through what angle should the captain turn to head directly to Barbados?
    (b) Once the turn is made, how long will it be before the ship reaches Barbados if the same 15-knot speed is maintained?

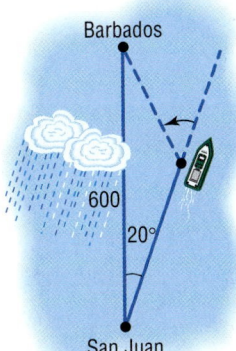

29. **Major League Baseball Field** A Major League baseball diamond is actually a square 90 feet on a side. The pitching rubber is located 60.5 feet from home plate on a line joining home plate and second base.

(a) How far is it from the pitching rubber to first base?
(b) How far is it from the pitching rubber to second base?
(c) If a pitcher faces home plate, through what angle does he need to turn to face first base?

30. **Little League Baseball Field** According to Little League baseball official regulations, the diamond is a square 60 feet on a side. The pitching rubber is located 46 feet from home plate on a line joining home plate and second base.
    (a) How far is it from the pitching rubber to first base?
    (b) How far is it from the pitching rubber to second base?
    (c) If a pitcher faces home plate, through what angle does he need to turn to face first base?

31. **Finding the Length of a Guy Wire** The height of a radio tower is 500 feet, and the ground on one side of the tower slopes upward at an angle of 10° (see the figure).
    (a) How long should a guy wire be if it is to connect to the top of the tower and be secured at a point on the sloped side 100 feet from the base of the tower?
    (b) How long should a second guy wire be if it is to connect to the middle of the tower and be secured at a point 100 feet from the base on the flat side?

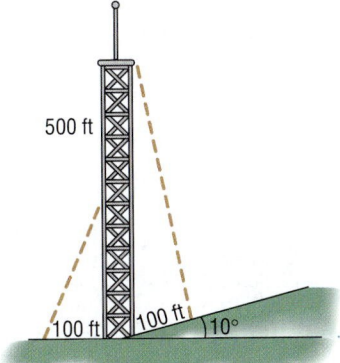

32. **Finding the Length of a Guy Wire** A radio tower 500 feet high is located on the side of a hill with an inclination to the horizontal of 5° (see the figure). How long should two guy wires be if they are to connect to the top of the tower and be secured at two points 100 feet directly above and directly below the base of the tower?

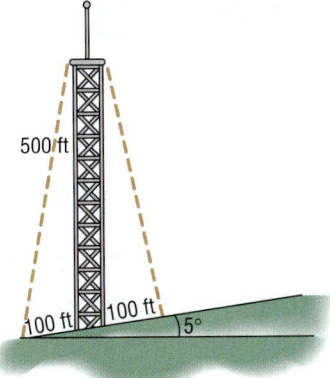

33. Wrigley Field, Home of the Chicago Cubs   The distance from home plate to the fence in dead center in Wrigley Field is 400 feet (see the figure). How far is it from the fence in dead center to third base?

400 ft

90 ft   90 ft

34. Little League Baseball   The distance from home plate to the fence in dead center at the Oak Lawn Little League field is 280 feet. How far is it from the fence in dead center to third base?

[**Hint:** The distance between the bases in Little League in 60 feet.]

35. Rods and Pistons   Rod OA (see the figure) rotates about the fixed point O so that point A travels on a circle of radius r. Connected to point A is another rod AB of length $L > 2r$, and point B is connected to a piston. Show that the distance x between point O and point B is given by

$$x = r \cos \theta + \sqrt{r^2 \cos^2 \theta + L^2 - r^2}$$

where θ is the angle of rotation of rod OA.

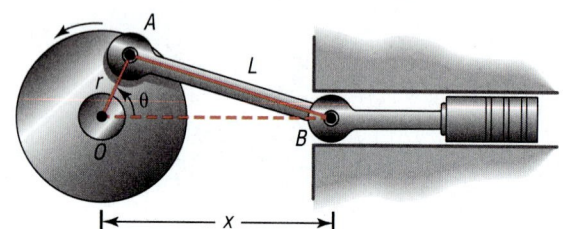

A

L

r

θ

O

B

x

36. Geometry   Show that the length d of a chord of a circle of radius r is given by the formula

$$d = 2r \sin \frac{\theta}{2}$$

where θ is the central angle formed by the radii to the ends of the chord (see the figure). Use this result to derive the fact that $\sin \theta < \theta$, where $\theta > 0$ is measure in radians.

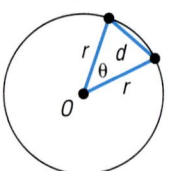

r   d

θ

O   r

37. For any triangle, show that

$$\cos \frac{\gamma}{2} = \sqrt{\frac{s(s-c)}{ab}}$$

where $s = \frac{1}{2}(a + b + c)$.

[**Hint:** Use a Half-angle Formula and the Law of Cosines.]

38. For any triangle show that

$$\sin \frac{\gamma}{2} = \sqrt{\frac{(s-a)(s-b)}{ab}}$$

where $s = \frac{1}{2}(a + b + c)$.

39. Use the Law of Cosines to prove the identity

$$\frac{\cos \alpha}{a} + \frac{\cos \beta}{b} + \frac{\cos \gamma}{c} = \frac{a^2 + b^2 + c^2}{2abc}$$

 40. Write down your strategy for solving an oblique triangle.

PREPARING FOR THIS SECTION

Before getting started, review the following:

✓ Geometry Review (Section R.3, pp. 27–28)

10.4   AREA OF A TRIANGLE

OBJECTIVES   **1** Find the Area of SAS Triangles

2 Find the Area of SSS Triangles

In this section, we will derive several formulas for calculating the area A of a triangle. The most familiar of these is the following:

Theorem

The area A of a triangle is

$$A = \frac{1}{2}bh \tag{1}$$

where b is the base and h is an altitude drawn to that base.

■

Figure 33

$A = \frac{1}{2}bh$

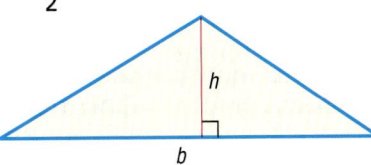

Proof The derivation of this formula is rather easy once a rectangle of base b and height h is constructed around the triangle. See Figures 33 and 34.

Triangles 1 and 2 in Figure 34 are equal in area, as are triangles 3 and 4. Consequently, the area of the triangle with base b and altitude h is exactly half the area of the rectangle, which is bh. ■

Figure 34

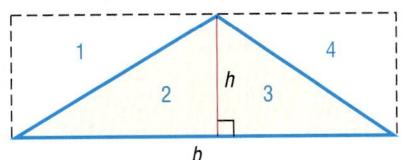

If the base b and altitude h to that base are known, then we can find the area of such a triangle using formula (1). Usually, though, the information required to use formula (1) is not given. Suppose, for example, that we know two sides a and b and the included angle γ (see Figure 35). Then the altitude h can be found by noting that

$$\frac{h}{a} = \sin \gamma$$

Figure 35

$h = a \sin \gamma$

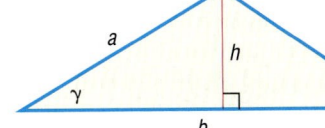

so that

$$h = a \sin \gamma$$

Using this fact in formula (1) produces

$$A = \frac{1}{2}bh = \frac{1}{2}b(a \sin \gamma) = \frac{1}{2}ab \sin \gamma$$

We now have the formula

$$A = \frac{1}{2}ab \sin \gamma \tag{2}$$

By dropping altitudes from the other two vertices of the triangle, we obtain the following corresponding formulas:

$$A = \frac{1}{2}bc \sin \alpha \tag{3}$$

$$A = \frac{1}{2}ac \sin \beta \tag{4}$$

It is easiest to remember these formulas using the following wording:

Theorem

The area A of a triangle equals one-half the product of two of its sides and the sine of their included angle.

■

EXAMPLE 1   **Finding the Area of a SAS Triangle**

① Find the area A of the triangle for which $a = 8$, $b = 6$, and $\gamma = 30°$.

Figure 36

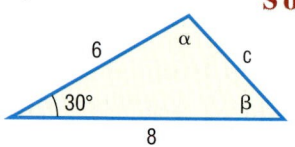

Solution   See Figure 36. We use formula (2) to get

$$A = \frac{1}{2}ab \sin \gamma = \frac{1}{2} \cdot 8 \cdot 6 \sin 30° = 12$$

➥ **NOW WORK PROBLEM 1.**

② If the three sides of a triangle are known, another formula, called **Heron's Formula** (named after Heron of Alexandria), can be used to find the area of a triangle.

Theorem

> **Heron's Formula**
>
> The area A of a triangle with sides a, b, and c is
>
> $$A = \sqrt{s(s - a)(s - b)(s - c)} \qquad (5)$$
>
> where $s = \frac{1}{2}(a + b + c)$.

Proof   The proof we shall give uses formula (2) on page 807.

$$A = \frac{1}{2}ab \sin \gamma$$

$$A = \frac{1}{2}ab\sqrt{1 - \cos^2 \gamma} \qquad \qquad \sin \gamma = \sqrt{1 - \cos^2 \gamma}, 0° < \gamma < 180°$$

$$A = \frac{1}{2}ab\sqrt{1 - \left(\frac{a^2 + b^2 - c^2}{2ab}\right)^2} \qquad \text{Use the Law of Cosines}$$

$$A = \frac{1}{2}ab\sqrt{\frac{4a^2b^2 - (a^2 + b^2 - c^2)^2}{4a^2b^2}} \qquad \text{Write the expression under the radical as a single quotient}$$

$$A = \frac{1}{4}\sqrt{[2ab + (a^2 + b^2 - c^2)][2ab - (a^2 + b^2 - c^2)]} \qquad \text{Simply; factor}$$

$$A = \frac{1}{4}\sqrt{(a^2 + 2ab + b^2 - c^2)(c^2 - (a^2 - 2ab + b^2))} \qquad \text{Rearrange terms}$$

$$A = \frac{1}{4}\sqrt{[(a + b)^2 - c^2][c^2 - (a - b)^2]} \qquad \text{Factor}$$

$$A = \sqrt{\frac{1}{16}[(a + b) + c][(a + b) - c][c + (a - b)][c - (a - b)]}$$ Difference of two squares; $\frac{1}{4} = \sqrt{\frac{1}{16}}$

$$A = \sqrt{\frac{1}{2}(a + b + c) \cdot \frac{1}{2}(b + c - a) \cdot \frac{1}{2}(a + c - b) \cdot \frac{1}{2}(a + b - c)}$$ Simplify; Rearrange terms

$$A = \sqrt{\frac{1}{2}(a + b + c) \cdot \left[\frac{1}{2}(a + b + c) - a\right] \cdot \left[\frac{1}{2}(a + b + c) - b\right] \cdot \left[\frac{1}{2}(a + b + c) - c\right]}$$

If we let $s = \frac{1}{2}(a + b + c)$, then we have $A = \sqrt{s \cdot (s - a) \cdot (s - b) \cdot (s - c)}$. ∎

EXAMPLE 2 **Finding the Area of a SSS Triangle**

Find the area of a triangle whose sides are 4, 5, and 7.

Solution We let $a = 4$, $b = 5$, and $c = 7$. Then

$$s = \frac{1}{2}(a + b + c) = \frac{1}{2}(4 + 5 + 7) = 8$$

Heron's Formula then gives the area A as

$$A = \sqrt{s(s - a)(s - b)(s - c)} = \sqrt{8 \cdot 4 \cdot 3 \cdot 1} = \sqrt{96} = 4\sqrt{6}$$ ∎

NOW WORK PROBLEM **7**.

HISTORICAL FEATURE

Heron's formula (also known as *Hero's Formula*) is due to Heron of Alexandria (first century AD), who had, besides his mathematical talents, a good deal of engineering skills. In various temples his mechanical devices produced effects that seemed supernatural, and visitors presumably were thus influenced to generosity. Heron's book *Metrica*, on making such devices, has survived and was discovered in 1896 in the city of Constantinople.

Heron's Formulas for the area of a triangle caused some mild discomfort in Greek mathematics, because a product with two factors was an area, while one with three factors was a volume, but four factors seemed contradictory in Heron's time.

10.4 Concepts and Vocabulary

In Problem 1, fill in the blank.

1. If three sides of a triangle are given, _____ Formula is used to find the area of the triangle.

In Problems 2 and 3, answer True or False to each statement.

2. No formula exists for finding the area of a triangle when only three sides are given.

3. Given two sides and the included angle, there is a formula that can be used to find the area of the triangle.

4. What do you do first if you are asked to find the area of a triangle and are given two sides and the included angle?

5. What do you do first if you are asked to find the area of a triangle and are given three sides?

10.4 Exercises

In Problems 1–8, find the area of each triangle. Round answers to two decimal places.

1.
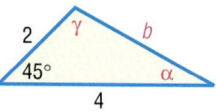

2.
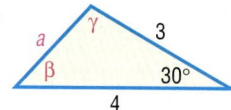

3.
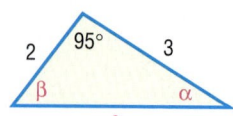

4.
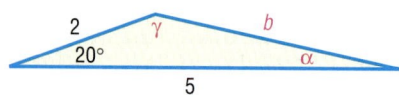

5.
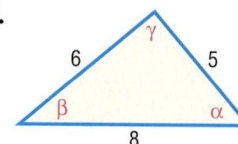

6.
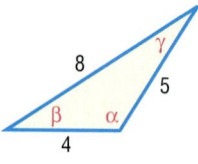

7.
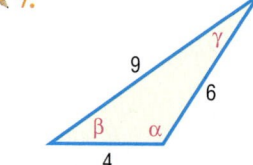

8.
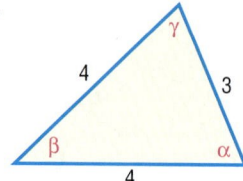

In Problems 9–24, find the area of each triangle. Round answers to two decimal places.

9. $a = 3$, $b = 4$, $\gamma = 40°$

10. $a = 2$, $c = 1$, $\beta = 10°$

11. $b = 1$, $c = 3$, $\alpha = 80°$

12. $a = 6$, $b = 4$, $\gamma = 60°$

13. $a = 3$, $c = 2$, $\beta = 110°$

14. $b = 4$, $c = 1$, $\alpha = 120°$

15. $a = 2$, $b = 2$, $\gamma = 50°$

16. $a = 3$, $c = 2$, $\beta = 90°$

17. $a = 12$, $b = 13$, $c = 5$

18. $a = 4$, $b = 5$, $c = 3$

19. $a = 2$, $b = 2$, $c = 2$

20. $a = 3$, $b = 3$, $c = 2$

21. $a = 5$, $b = 8$, $c = 9$

22. $a = 4$, $b = 3$, $c = 6$

23. $a = 10$, $b = 8$, $c = 5$

24. $a = 9$, $b = 7$, $c = 10$

---

25. Area of a Segment   Find the area of the segment (shaded in blue in the figure) of a circle whose radius is 8 feet, formed by a central angle of 70°.

[**Hint:** Subtract the area of the triangle from the area of the sector to obtain the area of the segment.]

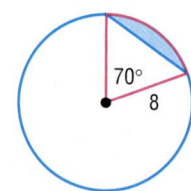

26. Area of a Segment   Find the area of the segment of a circle whose radius is 5 inches, formed by a central angle of 40°.

27. Cost of a Triangular Lot   The dimensions of a triangular lot are 100 feet by 50 feet by 75 feet. If the price of such land is $3 per square foot, how much does the lot cost?

28. Amount of Materials to Make a Tent   A cone-shaped tent is made from a circular piece of canvas 24 feet in diameter by removing a sector with central angle 100° and connecting the ends. What is the surface area of the tent?

29. Computing Areas   Find the area of the shaded region enclosed in a semicircle of diameter 8 centimeters. The length of the chord AB is 6 centimeters.

[**Hint:** Triangle ABC is a right triangle.]

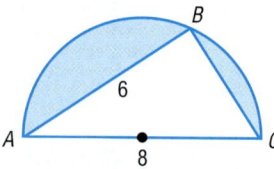

30. Computing Areas   Find the area of the shaded region enclosed in a semicircle of diameter 10 inches. The length of the chord AB is 8 inches.

[**Hint:** Triangle ABC is a right triangle.]

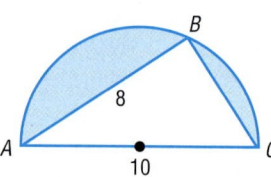

31. Area of a Triangle Prove that the area A of a triangle is given by the formula

$$A = \frac{a^2 \sin \beta \sin \gamma}{2 \sin \alpha}$$

32. Area of a Triangle Prove the two other forms of the formula given in Problem 31.

$$A = \frac{b^2 \sin \alpha \sin \gamma}{2 \sin \beta} \quad \text{and} \quad A = \frac{c^2 \sin \alpha \sin \beta}{2 \sin \gamma}$$

In Problems 33–38, use the results of Problem 31 or 32 to find the area of each triangle. Round answers to two decimal places.

33. $\alpha = 40°$, $\beta = 20°$, $a = 2$

34. $\alpha = 50°$, $\gamma = 20°$, $a = 3$

35. $\beta = 70°$, $\gamma = 10°$, $b = 5$

36. $\alpha = 70°$, $\beta = 60°$, $c = 4$

37. $\alpha = 110°$, $\gamma = 30°$, $c = 3$

38. $\beta = 10°$, $\gamma = 100°$, $b = 2$

---

39. Geometry Consult the figure, which shows a circle of radius r with center at O. Find the area A of the shaded region as a function of the central angle θ.

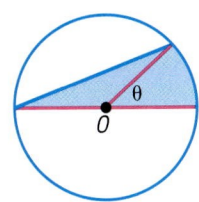

40. Approximating the Area of a Lake To approximate the area of a lake, a surveyor walks around the perimeter of the lake, taking the measurements shown in the illustration. Using this technique, what is the approximate area of the lake?

[**Hint:** Use the Law of Cosines on the three triangles shown and then find the sum of their areas.]

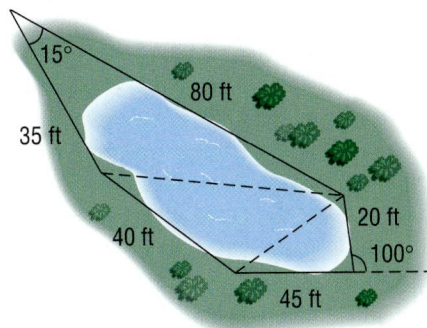

41. The Cow Problem* A cow is tethered to one corner of a square barn, 10 feet by 10 feet, with a rope 100 feet long. What is the maximum grazing area for the cow?

[**Hint:** See the illustration.]

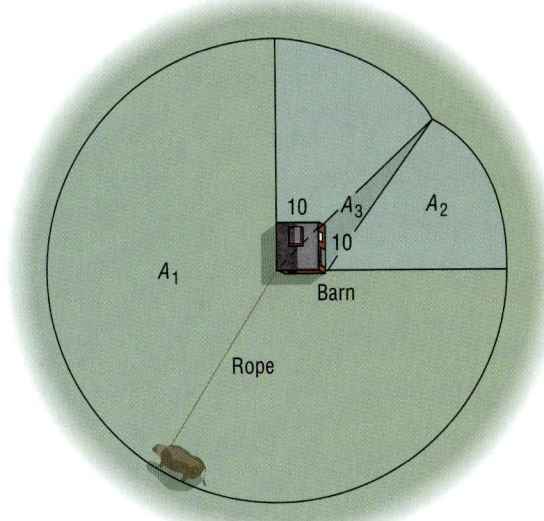

42. Another Cow Problem If the barn in Problem 41 is rectangular, 10 feet by 20 feet, what is the maximum grazing area for the cow?

43. If h_1, h_2, and h_3 are the altitudes dropped from A, B, and C, respectively, in a triangle (see the figure), show that

$$\frac{1}{h_1} + \frac{1}{h_2} + \frac{1}{h_3} = \frac{s}{K}$$

where K is the area of the triangle and $s = \frac{1}{2}(a + b + c)$.

[**Hint:** $h_1 = \frac{2K}{a}$.]

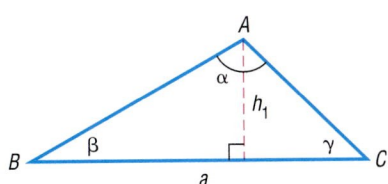

---

*Suggested by Professor Teddy Koukounas of SUNY at Old Westbury, who learned of it from an old farmer in Virginia. Solution provided by Professor Kathleen Miranda of SUNY at Old Westbury.

44. Show that a formula for the altitude h from a vertex to the opposite side a of a triangle is

$$h = \frac{a \sin \beta \sin \gamma}{\sin \alpha}$$

Inscribed Circle   *For Problems 45–48, the lines that bisect each angle of a triangle meet in a single point O, and the perpendicular distance r from O to each side of the triangle is the same. The circle with center at O and radius r is called the **inscribed circle** of the triangle (see the figure).*

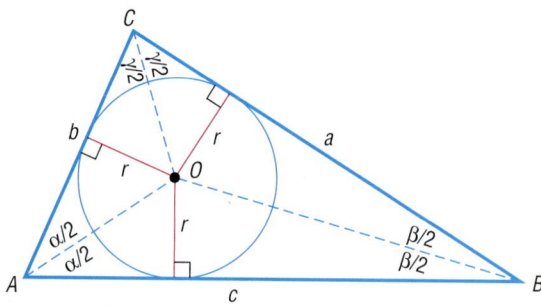

45. Apply Problem 44 to triangle OAB to show that

$$r = \frac{c \sin \dfrac{\alpha}{2} \sin \dfrac{\beta}{2}}{\cos \dfrac{\gamma}{2}}$$

46. Use the results of Problem 45 (above) and Problem 38 in Section 10.3 to show that

$$\cot \frac{\gamma}{2} = \frac{s - c}{r}$$

47. Show that

$$\cot \frac{\alpha}{2} + \cot \frac{\beta}{2} + \cot \frac{\gamma}{2} = \frac{s}{r}$$

48. Show that the area K of triangle ABC is $K = rs$. Then show that

$$r = \sqrt{\frac{(s - a)(s - b)(s - c)}{s}}$$

where $s = \dfrac{1}{2}(a + b + c)$.

PREPARING FOR THIS SECTION

Before getting started, review the following:

✓ Sinusoidal Graphs (Section 8.6, pp. 671–678)

10.5 SIMPLE HARMONIC MOTION; DAMPED MOTION; COMBINING WAVES

OBJECTIVES
1. Find an Equation for an Object in Simple Harmonic Motion
2. Analyze Simple Harmonic Motion
3. Analyze an Object in Damped Motion
4. Graph the Sum of Two Functions

Simple Harmonic Motion

Many physical phenomena can be described as simple harmonic motion. Radio and television waves, light waves, sound waves, and water waves exhibit motion that is simple harmonic.

The swinging of a pendulum, the vibrations of a tuning fork, and the bobbing of a weight attached to a coiled spring are examples of vibrational motion. In this type of motion, an object swings back and forth over the same path. In each illustration in Figure 37, the point B is the **equilibrium (rest) position** of the vibrating object. The **amplitude** of vibration is the distance from the object's rest position to its point of greatest displacement (either point A or point C in Figure 37). The **period** of a vibrating object is the time required to complete one vibration, that is, the time it takes to go from, say, point A through B to C and back to A.

Figure 37

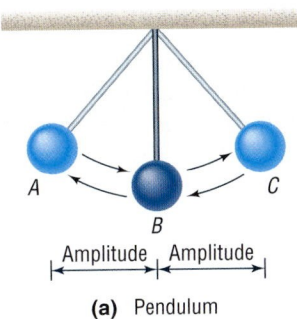

(a) Pendulum

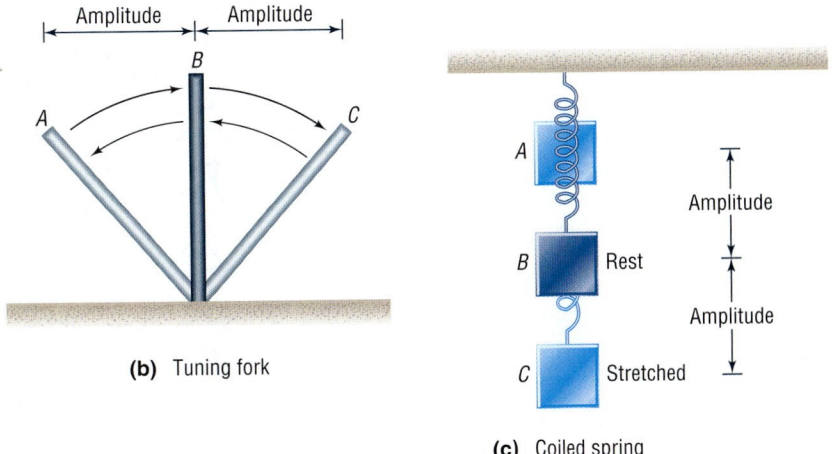

(b) Tuning fork

(c) Coiled spring

Simple harmonic motion is a special kind of vibrational motion in which the acceleration a of the object is directly proportional to the negative of its displacement d from its rest position. That is, $a = -kd, k > 0$.

For example, when the mass hanging from the spring in Figure 37(c) is pulled down from its rest position B to the point C, the force of the spring tries to restore the mass to its rest position. Assuming that there is no frictional force* to retard the motion, the amplitude will remain constant. The force increases in direct proportion to the distance that the mass is pulled from its rest position. Since the force increases directly, the acceleration of the mass of the object must do likewise, because (by Newton's Second Law of Motion) force is directly proportional to acceleration. Thus, the acceleration of the object varies directly with its displacement, and the motion is an example of simple harmonic motion.

Simple harmonic motion is related to circular motion. To see this relationship, consider a circle of radius a, with center at $(0, 0)$. See Figure 38. Suppose that an object initially placed at $(a, 0)$ moves counterclockwise around the circle at constant angular speed ω. Suppose further that after time t has elapsed the object is at the point $P = (x, y)$ on the circle. The angle θ, in radians, swept out by the ray $\overrightarrow{OP}$ in this time t is

$$\theta = \omega t$$

The coordinates of the point P at time t are

$$x = a \cos \theta = a \cos(\omega t)$$
$$y = a \sin \theta = a \sin(\omega t)$$

Figure 38

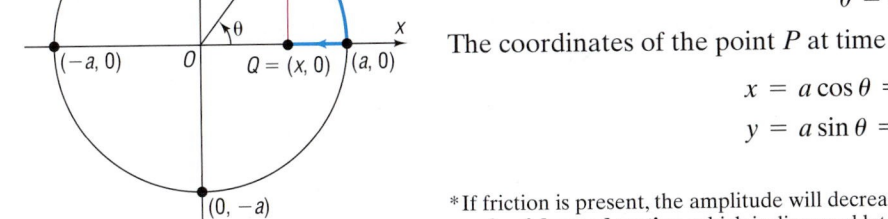

*If friction is present, the amplitude will decrease with time to 0. This type of motion is an example of **damped motion**, which is discussed later in this section.

Corresponding to each position $P = (x, y)$ of the object moving about the circle, there is the point $Q = (x, 0)$, called the **projection of P on the x-axis.** As P moves around the circle at a constant rate, the point Q moves back and forth between the points $(a, 0)$ and $(-a, 0)$ along the x-axis with a motion that is simple harmonic. Similarly, for each point P there is a point $Q' = (0, y)$, called the **projection of P on the y-axis.** As P moves around the circle, the point Q' moves back and forth between the points $(0, a)$ and $(0, -a)$ on the y-axis with a motion that is simple harmonic. Simple harmonic motion can be described as the projection of constant circular motion on a coordinate axis.

To put it another way, again consider a mass hanging from a spring where the mass is pulled down from its rest position to the point C and then released. See Figure 39(a). The graph shown in Figure 39(b) describes the displacement d of the object from its rest position as a function of time t, assuming that no frictional force is present.

Figure 39

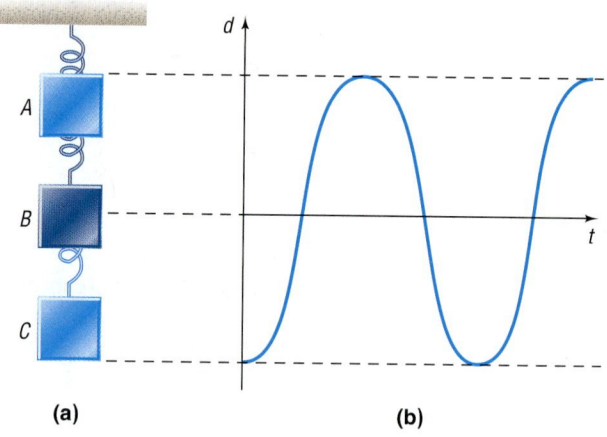

(a)   (b)

Theorem   **Simple Harmonic Motion**

An object that moves on a coordinate axis so that its distance d from the origin at time t is given by either

$$d = a \cos(\omega t) \quad \text{or} \quad d = a \sin(\omega t)$$

where a and $\omega > 0$ are constants, moves with simple harmonic motion. The motion has amplitude $|a|$ and period $T = \dfrac{2\pi}{\omega}$.

The **frequency** f of an object in simple harmonic motion is the number of oscillations per unit time. Since the period is the time required for one oscillation, it follows that frequency is the reciprocal of the period; that is,

$$f = \frac{\omega}{2\pi}, \quad \omega > 0$$

EXAMPLE 1 Finding an Equation for an Object in Harmonic Motion

① Suppose that an object attached to a coiled spring is pulled down a distance of 5 inches from its rest position and then released. If the time for one oscillation is 3 seconds, write an equation that relates the displacement d of the object from its rest position after time t (in seconds). Assume no friction.

Figure 40

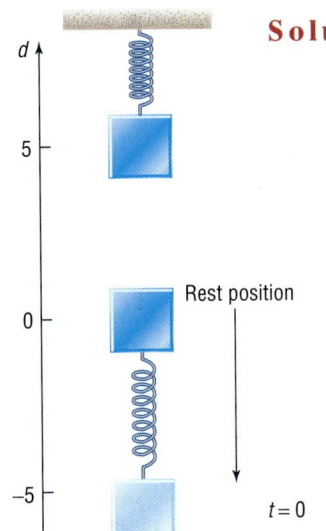

Solution The motion of the object is simple harmonic. See Figure 40. When the object is released $(t = 0)$, the displacement of the object from the rest position is -5 units (since the object was pulled down). Because $d = -5$ when $t = 0$, it is easier to use the cosine function*

$$d = a\cos(\omega t)$$

to describe the motion. Now the amplitude is $|-5| = 5$ and the period is 3, so

$$a = -5 \quad \text{and} \quad \frac{2\pi}{\omega} = \text{period} = 3, \quad \omega = \frac{2\pi}{3}$$

An equation of the motion of the object is

$$d = -5\cos\left(\frac{2\pi}{3}t\right)$$

Note: In the solution to Example 1, we let $a = -5$, since the initial motion is down. If the initial direction were up, we would let $a = 5$.

━ **NOW WORK PROBLEM 1.**

EXAMPLE 2 Analyzing the Motion of an Object

② Suppose that the displacement d (in meters) of an object at time t (in seconds) satisfies the equation

$$d = 10\sin(5t)$$

(a) Describe the motion of the object.
(b) What is the maximum displacement from its resting position?
(c) What is the time required for one oscillation?
(d) What is the frequency?

Solution We observe that the given equation is of the form

$$d = a\sin(\omega t) \qquad \textcolor{teal}{d = 10 \sin(5t)}$$

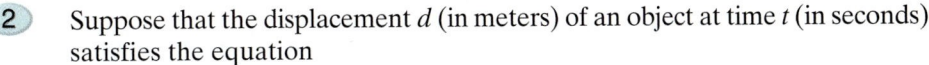

where $a = 10$ and $\omega = 5$.

(a) The motion is simple harmonic.
(b) The maximum displacement of the object from its resting position is the amplitude: $|a| = 10$ meters.
(c) The time required for one oscillation is the period:

$$\text{Period} = \frac{2\pi}{\omega} = \frac{2\pi}{5} \text{ seconds}$$

(d) The frequency is the reciprocal of the period. Thus,

$$\text{Frequency} = f = \frac{5}{2\pi} \text{ oscillations per second}$$

━ **NOW WORK PROBLEM 9.**

*No phase shift is required if a cosine function is used.

Damped Motion

③ Most physical phenomena are affected by friction or other resistive forces. These forces remove energy from a moving system and thereby damp its motion. For example, when a mass hanging from a spring is pulled down a distance a and released, the friction in the spring causes the distance that the mass moves from its at-rest position to decrease over time. Thus, the amplitude of any real oscillating spring or swinging pendulum decreases with time due to air resistance, friction, and so forth. See Figure 41.

Figure 41

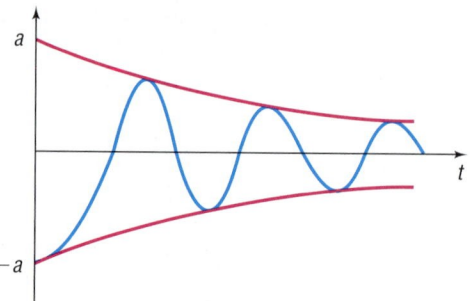

A function that describes this phenomenon maintains a sinusoidal component, but the amplitude of this component will decrease with time in order to account for the damping effect. In addition, the period of the oscillating component will be affected by the damping. The next result, from physics, describes damped motion.

Theorem

Damped Motion

The displacement d of an oscillating object from its at-rest position at time t is given by

$$d(t) = ae^{-bt/2m} \cos\left(\sqrt{\omega^2 - \frac{b^2}{4m^2}}\, t \right)$$

where b is a **damping factor** (most physics texts call this a **damping coefficient**) and m is the mass of the oscillating object.

Notice for $b = 0$ (zero damping) that we have the formula for simple harmonic motion with amplitude $|a|$ and period $\dfrac{2\pi}{\omega}$.

Figure 42

EXAMPLE 3 **Analyzing the Motion of a Pendulum with Damped Motion**

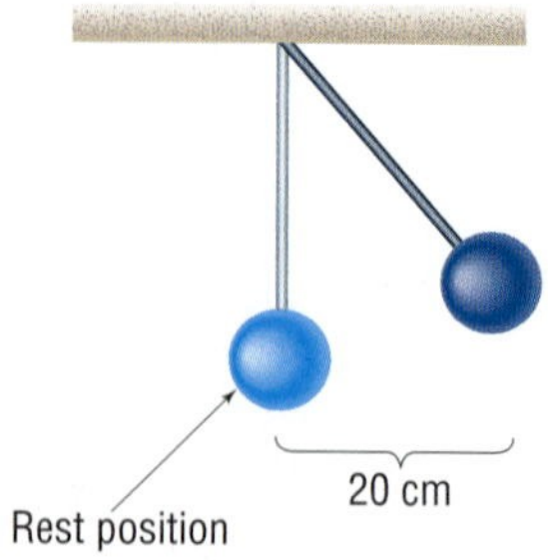

Suppose that a simple pendulum with a bob of mass 10 grams and a damping factor of 0.8 gram/second is pulled 20 centimeters from its at-rest position and released, see Figure 42. The period of the pendulum without the damping effect is 4 seconds.

(a) Find an equation that describes the position of the pendulum bob.

(b) Using a graphing utility, graph the function found in part (a).

(c) Determine the maximum displacement of the bob after the first oscillation.

20 cm

Rest position

(d) What happens to the displacement of the bob as time increases without bound?

Solution (a) We have $m = 10, a = 20,$ and $b = 0.8$. Since the period of the pendulum under simple harmonic motion is 4 seconds, we have

$$4 = \frac{2\pi}{\omega}$$

$$\omega = \frac{2\pi}{4} = \frac{\pi}{2}$$

Substituting these values into the equation for damped motion, we obtain

$$d = 20e^{-0.8t/2(10)} \cos\left(\sqrt{\left(\frac{\pi}{2}\right)^2 - \frac{0.8^2}{4(10)^2}}\, t\right)$$

$$d = 20e^{-0.8t/20} \cos\left(\sqrt{\frac{\pi^2}{4} - \frac{0.64}{400}}\, t\right)$$

(b) See Figure 43.

(c) See Figure 44. After the first oscillation, the maximum displacement is approximately 17.05 centimeters.

Figure 43

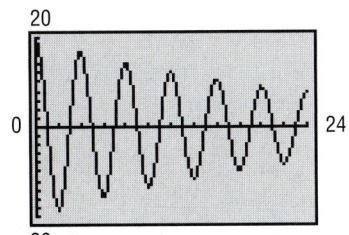

Figure 44

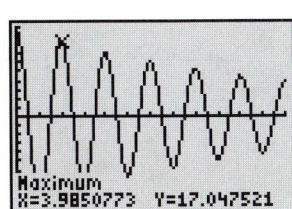

 (d) As t increases without bound $e^{-0.8t/20} \to 0$, so the displacement of the bob approaches zero. As a result, the pendulum will eventually come to rest.

 **NOW WORK PROBLEM 17.**

Combining Waves

④ Many physical and biological applications require the graph of the sum of two functions, such as

$$f(x) = x + \sin x \quad \text{or} \quad g(x) = \sin x + \cos(2x)$$

For example, if two tones are emitted, the sound produced is the sum of the waves produced by the two tones. See Problem 39 for an explanation of Touch-Tone phones.

To graph the sum of two (or more) functions, we can use the method of adding y-coordinates described next.

EXAMPLE 4 **Graphing the Sum of Two Functions**

Use the method of adding y-coordinates to graph $f(x) = x + \sin x$.

Solution First, we graph the component functions,

$$y = f_1(x) = x \qquad y = f_2(x) = \sin x$$

in the same coordinate system. See Figure 45(a). Now, select several values of x, say, $x = 0$, $x = \dfrac{\pi}{2}$, $x = \pi$, $x = \dfrac{3\pi}{2}$, and $x = 2\pi$, at which we compute $f(x) = f_1(x) + f_2(x)$. Table 1 shows the computation. We plot these points and connect them to get the graph, as shown in Figure 45(b).

TABLE 1

x	0	$\dfrac{\pi}{2}$	π	$\dfrac{3\pi}{2}$	2π
$y = f_1(x) = x$	0	$\dfrac{\pi}{2}$	π	$\dfrac{3\pi}{2}$	2π
$y = f_2(x) = \sin x$	0	1	0	-1	0
$f(x) = x + \sin x$	0	$\dfrac{\pi}{2} + 1 \approx 2.57$	π	$\dfrac{3\pi}{2} - 1 \approx 3.71$	2π
Point on Graph of f	$(0, 0)$	$\left(\dfrac{\pi}{2}, 2.57\right)$	(π, π)	$\left(\dfrac{3\pi}{2}, 3.71\right)$	$(2\pi, 2\pi)$

Figure 45

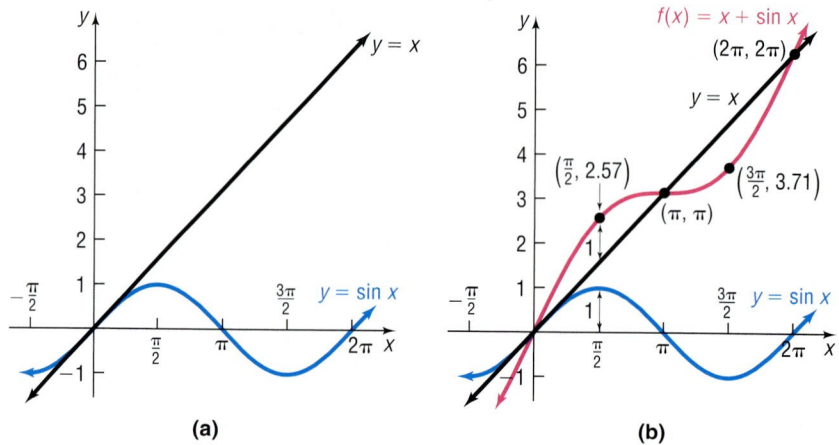

(a)　　　　(b)

Note that the graph of $f(x) = x + \sin x$ intersects the line $y = x$ whenever $\sin x = 0$. Also, notice that the graph of f is not periodic.

✔ CHECK:　Graph $Y_1 = x + \sin x$ and compare the result with Figure 45(b). Use INTERSECT to verify that the graphs intersect when $\sin x = 0$. ■ ■

The next example shows a periodic graph of the sum of two functions.

EXAMPLE 5　Graphing the Sum of two Sinusoidal Functions

Use the method of adding y-coordinates to graph

$$f(x) = \sin x + \cos(2x)$$

Solution   Table 2 shows the steps for computing several points on the graph of f. Figure 46 illustrates the graphs of the component functions, $y = f_1(x) = \sin x$ and $y = f_2(x) = \cos(2x)$, and the graph of $f(x) = \sin x + \cos(2x)$, which is shown in red.

TABLE 2

x	$-\dfrac{\pi}{2}$	0	$\dfrac{\pi}{2}$	π	$\dfrac{3\pi}{2}$	2π
$y = f_1(x) = \sin x$	-1	0	1	0	-1	0
$y = f_2(x) = \cos(2x)$	-1	1	-1	1	-1	1
$f(x) = \sin x + \cos(2x)$	-2	1	0	1	-2	1
Point on Graph of f	$\left(-\dfrac{\pi}{2}, -2\right)$	$(0, 1)$	$\left(\dfrac{\pi}{2}, 0\right)$	$(\pi, 1)$	$\left(\dfrac{3\pi}{2}, -2\right)$	$(2\pi, 1)$

Figure 46
$f(x) = \sin x + \cos(2x)$

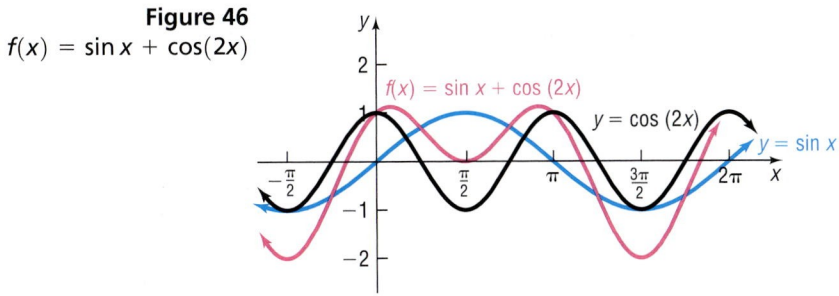

✔ CHECK: Graph $Y_1 = \sin x + \cos(2x)$ and compare the result with Figure 46.

━ NOW WORK PROBLEM **29**.

10.5 Concepts and Vocabulary

In Problems 1 and 2, fill in the blanks.

1. The motion of an object obeys the equation $d = 4 \cos(6t)$. Such motion is described as _____ _____ . The number 4 is called the _____ .

2. When a mass hanging from a spring is pulled down and then released, the motion is called _____ _____ _____ if there is no frictional force to retard the motion, and the motion is called _____ _____ if there is friction.

In Problem 3, answer True or False to the statement.

3. If the distance d of an object from the origin at time t is given by a sinusoidal graph, the motion of the object is simple harmonic motion.

10.5 Exercises

In Problems 1–4, an object attached to a coiled spring is pulled down a distance a from its rest position and then released. Assuming that the motion is simple harmonic with period T, write an equation that relates the displacement d of the object from its rest position after t seconds. Also assume that the positive direction of the motion is up.

1. $a = 5$; $T = 2$ seconds

2. $a = 10$; $T = 3$ seconds

3. $a = 6$; $T = \pi$ seconds

4. $a = 4$; $T = \dfrac{\pi}{2}$ seconds

5. Rework Problem 1 under the same conditions except that, at time $t = 0$, the object is at its resting position and moving down.

6. Rework Problem 2 under the same conditions except that, at time $t = 0$, the object is at its resting position and moving down.

7. Rework Problem 3 under the same conditions except that, at time $t = 0$, the object is at its resting position and moving down.

8. Rework Problem 4 under the same conditions except that, at time $t = 0$, the object is at its resting position and moving down.

In Problems 9–16, the displacement d (in meters) of an object at time t (in seconds) is given.
  (a) *Describe the motion of the object.*
  (b) *What is the maximum displacement from its resting position?*
  (c) *What is the time required for one oscillation?*
  (d) *What is the frequency?*

9. $d = 5\sin(3t)$

10. $d = 4\sin(2t)$

11. $d = 6\cos(\pi t)$

12. $d = 5\cos\left(\dfrac{\pi}{2}t\right)$

13. $d = -3\sin\left(\dfrac{1}{2}t\right)$

14. $d = -2\cos(2t)$

15. $d = 6 + 2\cos(2\pi t)$

16. $d = 4 + 3\sin(\pi t)$

In Problems 17–22, an object of mass m attached to a coiled spring with damping factor b is pulled down a distance a from its rest position and then released. Assume that the positive direction of the motion is up and the period of the first oscillation is T.
  (a) *Write an equation that relates the distance d of the object from its rest position after t seconds.*
  (b) *Graph the equation found in part (a) for 5 oscillations using a graphing utility.*

17. $m = 25$ grams; $a = 10$ centimeters; $b = 0.7$ gram/second; $T = 5$ seconds

18. $m = 20$ grams; $a = 15$ centimeters; $b = 0.75$ gram/second; $T = 6$ seconds

19. $m = 30$ grams; $a = 18$ centimeters; $b = 0.6$ gram/second; $T = 4$ seconds

20. $m = 15$ grams; $a = 16$ centimeters; $b = 0.65$ gram/second; $T = 5$ seconds

21. $m = 10$ grams; $a = 5$ centimeters; $b = 0.8$ gram/second; $T = 3$ seconds

22. $m = 10$ grams; $a = 5$ centimeters; $b = 0.7$ gram/second; $T = 3$ seconds

In Problems 23–28, the distance d (in meters) of the bob of a pendulum of mass m (in kilograms) from its rest position at time t (in seconds) is given.
  (a) *Describe the motion of the object. Be sure to give the mass and damping factor.*
  (b) *What is the initial displacement of the bob? That is, what is the displacement at $t = 0$?*
  (c) *Graph the motion using a graphing utility.*
  (d) *What is the maximum displacement after the first oscillation?*
  (e) *What happens to the displacement of the bob as time increases without bound?*

23. $d = -20e^{-0.7t/40}\cos\left(\sqrt{\left(\dfrac{2\pi}{5}\right)^2 - \dfrac{0.49}{1600}}\,t\right)$

24. $d = -20e^{-0.8t/40}\cos\left(\sqrt{\left(\dfrac{2\pi}{5}\right)^2 - \dfrac{0.64}{1600}}\,t\right)$

25. $d = -30e^{-0.6t/80}\cos\left(\sqrt{\left(\dfrac{2\pi}{7}\right)^2 - \dfrac{0.36}{6400}}\,t\right)$

26. $d = -30e^{-0.5t/70}\cos\left(\sqrt{\left(\dfrac{\pi}{2}\right)^2 - \dfrac{0.25}{4900}}\,t\right)$

27. $d = -15e^{-0.9t/30}\cos\left(\sqrt{\left(\dfrac{\pi}{3}\right)^2 - \dfrac{0.81}{900}}\,t\right)$

28. $d = -10e^{-0.8t/50}\cos\left(\sqrt{\left(\dfrac{2\pi}{3}\right)^2 - \dfrac{0.64}{2500}}\,t\right)$

In Problem 29–36, use the method of adding y-coordinates to graph each function. Verify your result using a graphing utility.

29. $f(x) = x + \cos x$

30. $f(x) = x + \cos(2x)$

31. $f(x) = x - \sin x$

32. $f(x) = x - \cos x$

33. $f(x) = \sin x + \cos x$

34. $f(x) = \sin(2x) + \cos x$

35. $g(x) = \sin x + \sin(2x)$

36. $g(x) = \cos(2x) + \cos x$

---

37. Charging a Capacitor If a charged capacitor is connected to a coil by closing a switch (see the figure), energy is transferred to the coil and then back to the capacitor in an oscillatory motion. The voltage V (in volts) across the capacitor will gradually diminish to 0 with time t (in seconds).

(a) By hand, graph the equation relating V and t:
$$V(t) = e^{-t/3}\cos(\pi t), \quad 0 \le t \le 3$$

(b) At what times t will the graph of V touch the graph of $y = e^{-t/3}$? When does V touch the graph of $y = -e^{-t/3}$?

(c) When will the voltage V be between -0.4 and 0.4 volt?

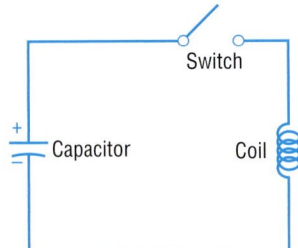

38. The Sawtooth Curve An oscilloscope often displays a *sawtooth curve*. This curve can be approximated by sinusoidal curves of varying periods and amplitudes.

(a) Use a graphing utility to graph the following function, which can be used to approximate the sawtooth curve.
$$f(x) = \frac{1}{2}\sin(2\pi x) + \frac{1}{4}\sin(4\pi x), \quad 0 \le x \le 2$$

(b) A better approximation to the sawtooth curve is given by
$$f(x) = \frac{1}{2}\sin(2\pi x) + \frac{1}{4}\sin(4\pi x) + \frac{1}{8}\sin(8\pi x)$$
Use a graphing utility to graph this function for $0 \le x \le 4$ and compare the result to the graph obtained in part (a).

(c) A third and even better approximation to the sawtooth curve is given by
$$f(x) = \frac{1}{2}\sin(2\pi x) + \frac{1}{4}\sin(4\pi x) + \frac{1}{8}\sin(8\pi x) + \frac{1}{16}\sin(16\pi x)$$
Use a graphing utility to graph this function for $0 \le x \le 4$ and compare the result to the graphs obtained in parts (a) and (b).

(d) What do you think the next approximation to the sawtooth curve is?

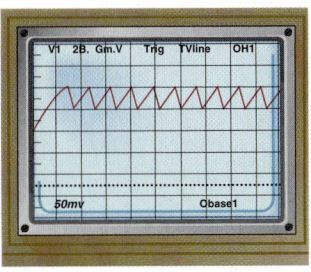

39. Touch-Tone Phones On a Touch-Tone phone, each button produces a unique sound. The sound produced is the sum of two tones, given by
$$y = \sin(2\pi l t) \quad \text{and} \quad y = \sin(2\pi h t)$$
where l and h are the low and high frequencies (cycles per second) shown on the illustration. For example, if you touch 7, the low frequency is $l = 852$ cycles per second and the high frequency is $h = 1209$ cycles per second. The sound emitted by touching 7 is
$$y = \sin[2\pi(852)t] + \sin[2\pi(1209)t]$$
Use a graphing utility to graph the sound emitted by touching 7.

Touch-Tone phone

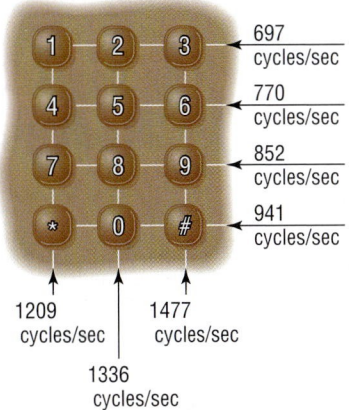

40. Use a graphing utility to graph the sound emitted by the * key on a Touch-Tone phone. See Problem 39.

41. Use a graphing utility to graph the function $f(x) = \dfrac{\sin x}{x}, x > 0$. Based on the graph, what do you conjecture about the value of $\dfrac{\sin x}{x}$ for x close to 0?

42. Use a graphing utility to graph $y = x \sin x$, $y = x^2 \sin x$, and $y = x^3 \sin x$ for $x > 0$. What patterns do you observe?

43. Use a graphing utility to graph $y = \dfrac{1}{x}\sin x$, $y = \dfrac{1}{x^2}\sin x$, and $y = \dfrac{1}{x^3}\sin x$ for $x > 0$. What patterns do you observe?

44. CBL Experiment Pendulum motion is analyzed to estimate simple harmonic motion. A plot is generated with the position of the pendulum over time. The graph is used to find a sinusoidal curve of the form $y = A\cos B(x - C) + D$. Determine the amplitude, period and frequency. (Activity 16, Real-World Math with the CBL System.)

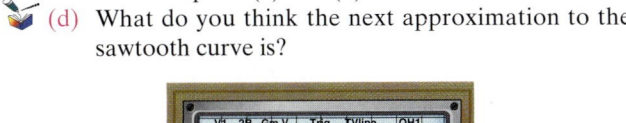

45. CBL Experiment The sound from a tuning fork is collected over time. The amplitude, frequency, and period of the graph are determined. A model of the form $y = A \cos B(x - C)$ is fitted to the data. (Activity 23, Real-World Math with the CBL System.)

46. How would you explain to a friend what simple harmonic motion is? How would you explain damped motion?

Chapter Review

Things To Know

Formulas

Law of Sines (p. 789)

$$\frac{\sin \alpha}{a} = \frac{\sin \beta}{b} = \frac{\sin \gamma}{c}$$

Law of Cosines (p. 800)

$$c^2 = a^2 + b^2 - 2ab \cos \gamma$$
$$b^2 = a^2 + c^2 - 2ac \cos \beta$$
$$a^2 = b^2 + c^2 - 2bc \cos \alpha$$

Area of a triangle (pp. 807–808)

$$A = \frac{1}{2}bh$$

$$A = \frac{1}{2}ab \sin \gamma$$

$$A = \frac{1}{2}bc \sin \alpha$$

$$A = \frac{1}{2}ac \sin \beta$$

$$A = \sqrt{s(s - a)(s - b)(s - c)}, \quad \text{where} \quad s = \frac{1}{2}(a + b + c)$$

Objectives

Section		You should be able to:	Review Exercises
10.1	1	Solve right triangles (p. 780)	1–4
	2	Solve applied problems using right triangle trigonometry (p. 781)	35–40
10.2	1	Solve SAA or ASA triangles (p. 790)	5–6, 22
	2	Solve SSA triangles (p. 792)	7–10, 12, 17–18, 21
	3	Solve applied problems using the Law of Sines (p. 794)	41, 43, 44
10.3	1	Solve SAS triangles (p. 801)	11, 15–16, 23–24
	2	Solve SSS triangles (p. 802)	13–14, 19–20
	3	Solve applied problems using the Law of Cosines (p. 802)	42, 45, 46, 47
10.4	1	Find the area of SAS triangles (p. 808)	25–28, 47–49
	2	Find the area of SSS triangles (p. 808)	29–32
10.5	1	Find an equation for an object in simple harmonic motion (p. 815)	57–58
	2	Analyze simple harmonic motion (p. 815)	53–58
	3	Analyze an object in damped motion (p. 816)	59–62
	4	Graph the sum of two functions (p. 817)	63, 64

Review Exercises

Blue problem numbers indicate the authors' suggestions for use in a Practice Test.

In Problems 1–4, solve each triangle.

1.

2.

3.

4.

In Problems 5–24, find the remaining angle(s) and side(s) of each triangle, if it (they) exists. If no triangle exists, say "No triangle."

5. $\alpha = 50°$, $\beta = 30°$, $a = 1$
6. $\alpha = 10°$, $\gamma = 40°$, $c = 2$
7. $\alpha = 100°$, $a = 5$, $c = 2$

8. $a = 2$, $c = 5$, $\alpha = 60°$
9. $a = 3$, $c = 1$, $\gamma = 110°$
10. $a = 3$, $c = 1$, $\gamma = 20°$

11. $a = 3$, $c = 1$, $\beta = 100°$
12. $a = 3$, $b = 5$, $\beta = 80°$
13. $a = 2$, $b = 3$, $c = 1$

14. $a = 10$, $b = 7$, $c = 8$
15. $a = 1$, $b = 3$, $\gamma = 40°$
16. $a = 4$, $b = 1$, $\gamma = 100°$

17. $a = 5$, $b = 3$, $\alpha = 80°$
18. $a = 2$, $b = 3$, $\alpha = 20°$
19. $a = 1$, $b = \dfrac{1}{2}$, $c = \dfrac{4}{3}$

20. $a = 3$, $b = 2$, $c = 2$
21. $a = 3$, $\alpha = 10°$, $b = 4$
22. $a = 4$, $\alpha = 20°$, $\beta = 100°$

23. $c = 5$, $b = 4$, $\alpha = 70°$
24. $a = 1$, $b = 2$, $\gamma = 60°$

In Problems 25–34, find the area of each triangle.

25. $a = 2$, $b = 3$, $\gamma = 40°$
26. $b = 5$, $c = 5$, $\alpha = 20°$
27. $b = 4$, $c = 10$, $\alpha = 70°$

28. $a = 2$, $b = 1$, $\gamma = 100°$
29. $a = 4$, $b = 3$, $c = 5$
30. $a = 10$, $b = 7$, $c = 8$

31. $a = 4$, $b = 2$, $c = 5$
32. $a = 3$, $b = 2$, $c = 2$
33. $\alpha = 50°$, $\beta = 30°$, $a = 1$

34. $\alpha = 10°$, $\gamma = 40°$, $c = 3$

35. **Measuring the Length of a Lake** From a stationary hot-air balloon 500 feet above the ground, two sightings of a lake are made (see the figure). How long is the lake?

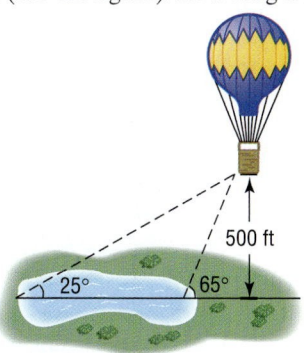

36. **Finding the Speed of a Glider** From a glider 200 feet above the ground, two sightings of a stationary object directly in front are taken 1 minute apart (see the figure). What is the speed of the glider?

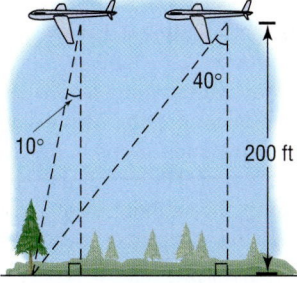

37. **Finding the Width of a River** Find the distance from A to C across the river illustrated in the figure.

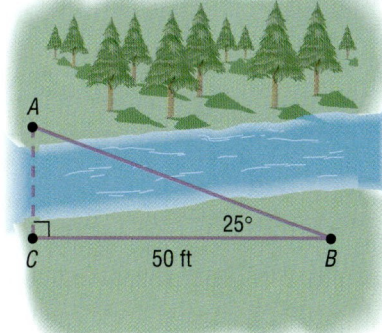

38. **Finding the Height of a Building** Find the height of the building shown in the figure.

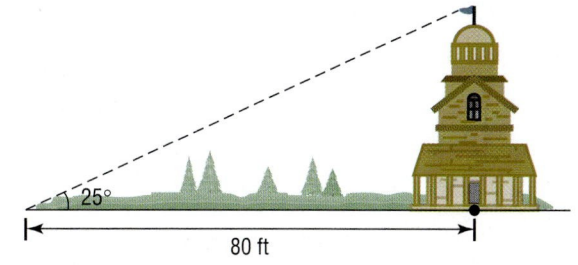

39. Finding the Distance to Shore The Sears Tower in Chicago is 1454 feet tall and is situated about 1 mile inland from the shore of Lake Michigan, as indicated in the figure. An observer in a pleasure boat on the lake directly in front of the Sears Tower looks at the top of the tower and measures the angle of elevation as 5°. How far offshore is the boat?

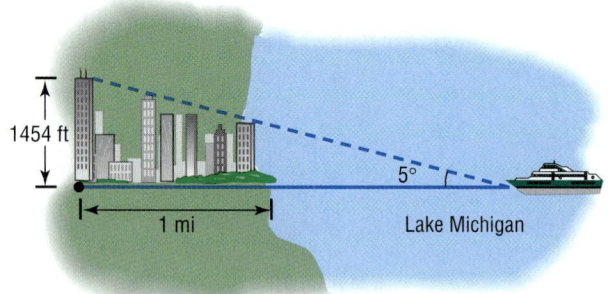

40. Finding the Grade of a Mountain Trail A straight trail with a uniform inclination leads from a hotel, elevation 5000 feet, to a lake in a valley, elevation 4100 feet. The length of the trail is 4100 feet. What is the inclination of the trail?

41. Navigation An airplane flies from city A to city B, a distance of 100 miles, and then turns through an angle of 20° and heads toward city C, as indicated in the figure. If the distance from A to C is 300 miles, how far is it from city B to city C?

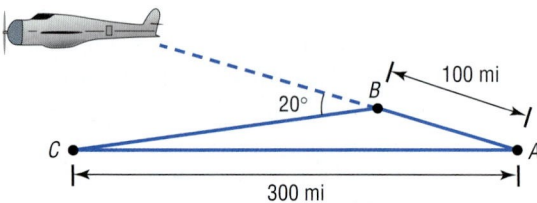

42. Correcting a Navigation Error Two cities A and B are 300 miles apart. In flying from city A to city B, a pilot inadvertently took a course that was 5° in error.
(a) If the error was discovered after flying 10 minutes at a constant speed of 420 miles per hour, through what angle should the pilot turn to correct the course? (Consult the figure.)
(b) What new constant speed should be maintained so that no time is lost due to the error? (Assume that the speed would have been a constant 420 miles per hour if no error had occurred.)

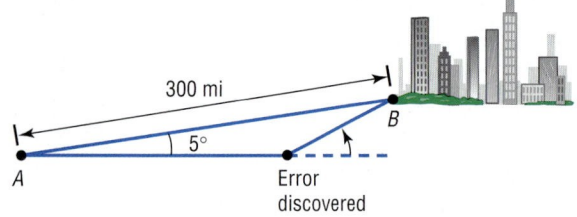

43. Determining Distances at Sea Rebecca, the navigator of a ship at sea, spots two lighthouses that she knows to be 2 miles apart along a straight shoreline. She determines that the angles formed between two line-of-sight observations of the lighthouses and the line from the ship directly to shore are 12° and 30°. See the illustration.
(a) How far is the ship from lighthouse A?
(b) How far is the ship from lighthouse B?
(c) How far is the ship from shore?

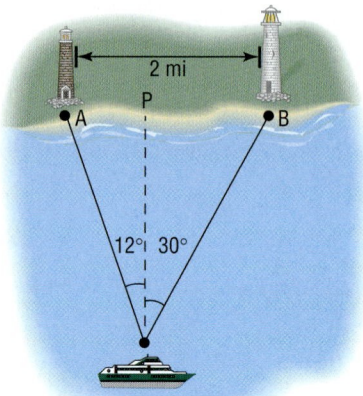

44. Constructing a Highway A highway whose primary directions are north–south is being constructed along the west coast of Florida. Near Naples, a bay obstructs the straight path of the road. Since the cost of a bridge is prohibitive, engineers decide to go around the bay. The illustration shows the path that they decide on and the measurements taken. What is the length of highway needed to go around the bay?

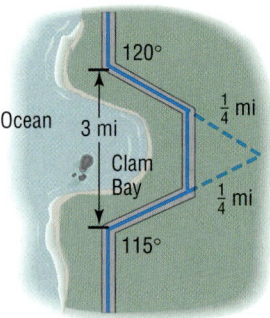

45. Correcting a Navigational Error A sailboat leaves St. Thomas bound for an island in the British West Indies, 200 miles away. Maintaining a constant speed of 18 miles per hour, but encountering heavy crosswinds and strong currents, the crew finds after 4 hours that the sailboat is off course by 15°.
(a) How far is the sailboat from the island at this time?
(b) Through what angle should the sailboat turn to correct its course?

(c) How much time has been added to the trip because of this? (Assume that the speed remains at 18 miles per hour.)

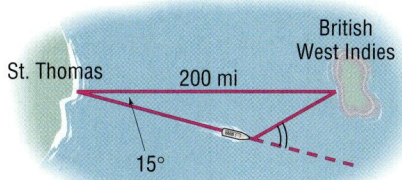

46. Surveying Two homes are located on opposite sides of a small hill. See the illustration. To measure the distance between them, a surveyor walks a distance of 50 feet from house A to point C, uses a transit to measure the angle ACB, which is found to be $80°$, and then walks to house B, a distance of 60 feet. How far apart are the houses?

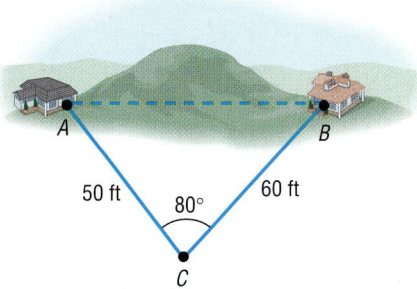

47. Approximating the Area of a Lake To approximate the area of a lake, Cindy walks around the perimeter of the lake, taking the measurements shown in the illustration. Using this technique, what is the approximate area of the lake?

[**Hint:** Use the Law of Cosines on the three triangles shown and then find the sum of their areas.]

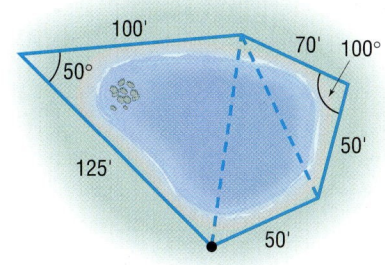

48. Calculating the Cost of Land The irregular parcel of land shown in the Figure is being sold for $100 per square foot. What is the cost of this parcel?

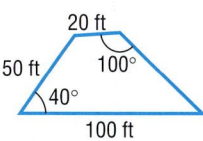

49. Area of a Segment Find the area of the segment of a circle whose radius is 6 inches formed by a central angle of $50°$.

50. Finding the Bearing of a Ship The *Majesty* leaves the Port at Boston for Bermuda with a bearing of S80°E at an average speed of 10 knots. After 1 hour, the ship turns $90°$ toward the southwest. After 2 hours at an average speed of 20 knots, what is the bearing of the ship from Boston?

51. The drive wheel of an engine is 13 inches in diameter, and the pulley on the rotary pump is 5 inches in diameter. If the shafts of the drive wheel and the pulley are 2 feet apart, what length of belt is required to join them as shown in the figure?

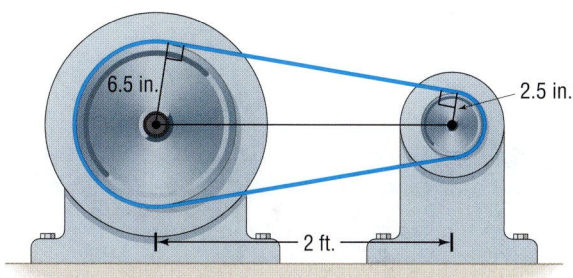

52. Rework Problem 51 if the belt is crossed, as shown in the figure.

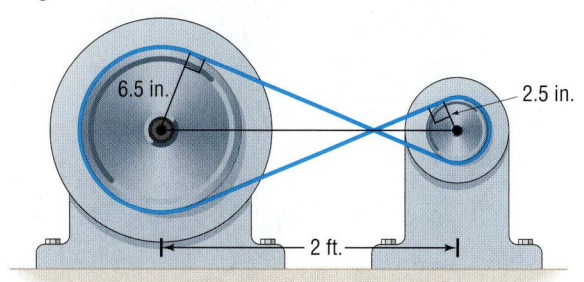

In Problems 53–56, the distance d (in feet) that an object travels in time t (in seconds) is given.
  (a) Describe the motion of the object.
  (b) What is the maximum displacement from its resting position?
  (c) What is the time required for one oscillation?
  (d) What is the frequency?

53. $d = 6\sin(2t)$ **54.** $d = 2\cos(4t)$ **55.** $d = -2\cos(\pi t)$ **56.** $d = -3\sin\left[\dfrac{\pi}{2}t\right]$

In Problems 57 and 58, an object attached to a coiled spring is pulled down a distance a from its rest position and then released. Assuming that the motion is simple harmonic with period T, write an equation that relates the displacement d of the object from its rest position after t seconds. Also assume that the positive direction of the motion is up.

57. $a = 4$, $T = 3$ seconds

58. $a = 5$, $T = 2$ seconds

In Problems 59 and 60, an object of mass m attached to a coiled spring with damping factor b is pulled down a distance a from its rest position and then released. Assume that the positive direction of the motion is up and the period of the first oscillation is T.

   (a) *Write an equation that relates the distance d of the object from its rest position after t seconds.*

   (b) *Graph the equation found in part (a) for 5 oscillations.*

59. $m = 40$ grams; $a = 15$ centimeters; $b = 0.75$ gram/second; $T = 5$ seconds

60. $m = 25$ grams; $a = 13$ centimeters; $b = 0.65$ gram/second; $T = 4$ seconds

In Problems 61 and 62, the distance d (in meters) of the bob of a pendulum of mass m (in kilograms) from its rest position at time t (in seconds) is given.

   (a) *Describe the motion of the object.*

   (b) *What is the initial displacement of the bob? That is, what is the displacement at $t = 0$?*

   (c) *Graph the motion using a graphing utility.*

   (d) *What is the maximum displacement after the first oscillation?*

   (e) *What happens to the displacement of the bob as time increases without bound?*

61. $d = -15e^{-0.6t/40} \cos\left(\sqrt{\left(\dfrac{2\pi}{5}\right)^2 - \dfrac{0.36}{1600}} \, t \right)$

62. $d = -20e^{-0.5t/60} \cos\left(\sqrt{\left(\dfrac{2\pi}{3}\right)^2 - \dfrac{0.25}{3600}} \, t \right)$

In Problems 63 and 64, use the method of adding y-coordinates to graph each function. Verify your result using a graphing utility.

63. $y = 2 \sin x + \cos(2x)$

64. $y = 2 \cos(2x) + \sin \dfrac{x}{2}$

Chapter Projects

1. **A.** When the distance between two locations on the surface of the earth is small we can compute the distance in statutory miles. Using this assumption, we can use the Law of Sines and the Law of Cosines to approximate distances and angles. However, if you look at a globe, you notice that the Earth is a sphere, so as the distance between two points on its surface increases, the linear distance is less accurate because of curvature. Under this circumstance, we need to take into account the curvature of the Earth when using the Law of Sines and the Law of Cosines.

(a) Draw a spherical triangle and label each vertex by $A, B,$ and C. Then connect each vertex by a radius to the center O of the sphere. Now, draw tangent lines to the sides a and b of the triangle that go through C. Extend the lines OA and OB to intersect the tangent lines at P and Q, respectively. See the diagram. List the plane right triangles. Determine the measures of the central angles.

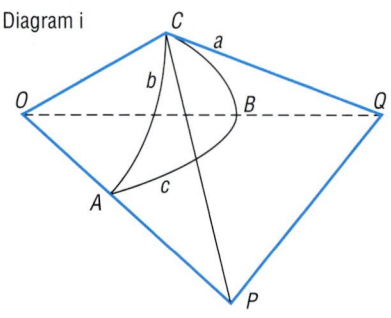

Diagram i

(b) Apply the Law of Cosines to triangles OPQ and CPQ to find two expressions for the length of PQ.

(c) Subtract the expressions in part (b) from each other. Solve for the term containing $\cos C$.

(d) Use the Pythagorean Theorem to find another value for $OQ^2 - CQ^2$ and $OP^2 - CP^2$. Now solve for $\cos C$.

(e) Replacing the ratios in part (d) by the cosines of the sides of the spherical triangle, you should now have the Law of Cosines for spherical triangles:

$$\cos C = \cos A \cos B + \sin A \sin B \cos C.$$

B. Lewis and Clark followed several rivers in their trek from what is now Great Falls, MT, to the Pacific coast. First, they went down the Missouri and Jefferson Rivers from Great Falls to Lemhi, ID. Because the two cities are on different longitudes and different latitudes, we must account for the curvature of the earth when computing the distance that they traveled. Assume that the radius of the earth is 3960 miles.

(a) Great Falls is at approximately 47.5°N and 111.3°W. Lemhi is at approximately 45.0°N and 113.5°W. (We will assume that the rivers flow straight from Great Falls to Lemhi on the surface of the earth.) This line is called a geodesic line. Apply the Law of Cosines for a spherical triangle to find the angle between Great Falls and Lemhi. (The central angles are found by using the differences in the latitudes and longitudes of the towns. See the diagram.) Then find the length of the arc joining the two towns. (Recall $s = r\theta$)

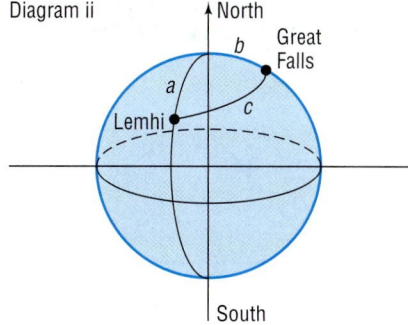

(b) From Lemhi, they went up the Bitteroot River and the Snake River to what is now Lewiston and Clarkston on the border of Idaho and Washington. Although this is not really a side to a triangle, we will make a side that goes from Lemhi to Lewiston/Clarkston. If Lewiston and Clarkston is at about 46.5°N 117.0°W, find the distance from Lemhi using the Law of Cosines for a spherical triangle and the arc length.

(c) How far did the explorers travel just to get that far?

(d) Draw a plane triangle connecting the three towns. If the distance from Lewiston to Great Falls is 282 miles and the angle at Great Falls is 42° and the angle at Lewiston is 48.5°, find the distance from Great Falls to Lemhi and from Lemhi to Lewiston. How do these distances compare with the ones computed in parts (a) and (b)?

For spherical Law of Cosines: *Mathematics from the Birth of Numbers* by Jan Gullberg. WW Norton & Co Publishers. 1996. Pg. 491–494.

For Lewis and Clark Expedition: *American Journey: The Quest for Liberty to 1877, Texas Edition*. Prentice Hall. 1992. Pg. 345

For map coordinates: *National Geographic Atlas of the World*, published by National Geographic Society, 1981. Pg. 74–75

2. **Leaning Tower of Pisa**   PISA, Italy—Workers began removing two sets of steel suspenders attached to the leaning Tower of Pisa on Tuesday, in one of the final phases of a bold plan to partially straighten the famously titled monument.

The 340 foot-long cables had been recurred to the tower in 1998 as a precaution in case it needed to be yanked back up while the soil under its foundation was being excavated. Anchored to giant winches dug into the ground about 100 yards from the tower, the suspenders haven't been needed.

Removal of the suspenders will be completed in time for an inauguration ceremony of the newly straightened tower scheduled for June 16, said Paolo Heiniger, who overseas the project.

The tower was closed to the public more than a decade ago, when officials feared it was beginning to lean so much that it might topple over. Work to stop the tower's increasing tilt has taken far longer than planned, but officials expect it to be open to tourists in the fall.

When work began the tower leaned 6 degrees, or 13 feet, off the perpendicular on its south side. By removing a small amount of soil, the tower has settled better and now leans about 16 inches less—nearly the tilt it had 300 years ago.

The decrease in lean isn't enough for the naked eye to detect but sufficient to stabilize the monument, experts have said.

Use the fact that the tower was 184.5 feet tall when it stood upright to answer the following questions:

(a) The article states that when work began, the tower leaned 6°, or 13 feet, off the perpendicular. Draw a sketch and label the two measurements.

(b) How high is the tower with a lean of 6°?

(c) The article goes on to say that the tower now leans 16 inches less. Draw a new sketch that shows this measurement.

(d) What is the degree measure of this lean?

(e) What is the height of the tower with this lean?

(f) Comment on the fact that $\sin 4° = \dfrac{13}{184.5} = 0.07046$, while $\sin 6° = 0.10453$. Can you reconcile this seeming discrepancy?

(g) Investigate further and write a report on your findings.

Source: *Naples Daily News*, Wednesday, May 16, 2001.

3. **Locating Lost Treasure**   While Scuba diving off Wreck Hill in Bermuda, a group of five entrepreneurs discovered a treasure map in a small watertight cask on a pirate schooner that had sunk in 1747. The map directed them to an area of Bermuda now known as The Flatts.

The directions on the map read as follows:

(1) From the tallest palm tree, sight the highest hill. Drop your eyes vertically until you sight the base of the hill.

(2) Turn 40° clockwise from that line and walk 70 paces to the big red rock.

(3) From the red rock walk 50 paces back to the sight line between the palm tree and the hill. Dig there.

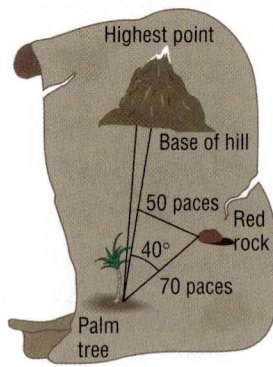

Upon reaching The Flatts, the five entrepreneurs believed that they had found the red rock and the highest hill in the vicinity, but the "tallest palm tree" had long since fallen and disintegrated. It occurred to them that the treasure must be located on a circle with radius 50 "paces" centered around the red rock, but they decided against digging a trench 942 feet in circumference, especially since they had no assurance that the treasure was still there. (They had decided that a "pace" must be about a yard.)

(a) Determine a plan to locate the position of the lost palm tree.

(b) One solution follows: From the location of the palm tree, turn 40° counter-clockwise from the rock to the hill, then go about 50 yards to the circle traced about the rock. Verify this solution.

(c) This location did not yield any treasure. Find the other solution and the treasure. Where is the treasure? How far is it from the palm tree?

Cumulative Review

1. Find the real solutions, if any, of the equation $3x^2 + 1 = 4x$.

2. Find an equation for the circle with center at the point $(-5, 1)$ and radius 3. Graph this circle.

3. What is the domain of the function $f(x) = \sqrt{x^2 - 3x - 4}$?

4. Graph the function $y = 3 \sin(\pi x)$.

5. Graph the function $y = -2 \cos(2x - \pi)$.

6. If $\tan \theta = -2$ and $\dfrac{3\pi}{2} < \theta < 2\pi$, find the exact value of:

(a) $\sin \theta$ (b) $\cos \theta$ (c) $\sin(2\theta)$

(d) $\cos(2\theta)$ (e) $\sin\left(\dfrac{1}{2}\theta\right)$ (f) $\cos\left(\dfrac{1}{2}\theta\right)$

7. Use a graphing utility to graph each of the following functions on the interval $[0, 4]$:

(a) $y = e^x$ (b) $y = \sin x$

(c) $y = e^x \sin x$ (d) $y = 2x + \sin x$

8. Sketch the graph of each of the following functions:

(a) $y = x$ (b) $y = x^2$ (c) $y = \sqrt{x}$

(d) $y = x^3$ (e) $y = e^x$ (f) $y = \ln x$

(g) $y = \sin x$ (h) $y = \cos x$ (i) $y = \tan x$

9. Solve the triangle:

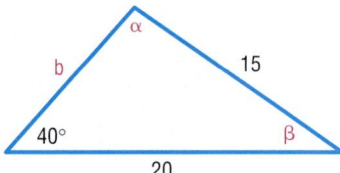

10. In the complex number system, solve the equation:
$$3x^5 - 10x^4 + 21x^3 - 42x^2 + 36x - 8 = 0$$

11. Analyze the graph of the rational function
$$R(x) = \frac{2x^2 - 7x - 4}{x^2 + 2x - 15}$$

12. Solve $3^x = 12$. Round your answer to two decimal places.

13. Solve $\log_3(x + 8) + \log_3 x = 2$.

POLAR COORDINATES; VECTORS

Multifractals and the Market

An extensive mathematical basis already exists for fractals and multifractals. Fractal patterns appear not just in the price changes of securities but in the distribution of galaxies throughout the cosmos, in the shape of coastlines and in the decorative designs generated by innumerable computer programs.

A fractal is a geometric shape that can be separated into parts, each of which is a reduced scale version of the whole. In finance, this concept is not a rootless abstraction but a theoretical reformulation of a down-to-earth bit of market folklore—namely, that movements of stock or currency all look alike when a market chart is enlarged or reduced so that it fits the same time and price scale. An observer then cannot tell which of the data concern prices that change from week to week, day to day or hour to hour. This quality defines charts as fractal curves and makes available many powerful tools of mathematical and computer analysis.

SOURCE: Benoit Mandelbrot, Scientific American, February, 1999.

SEE CHAPTER PROJECT 1.

OUTLINE

For additional study help, go to

www.prenhall.com/sullivanegu3e

Materials include:

- Graphing Calculator Help
- Chapter Quiz
- Chapter Test
- PowerPoint Downloads
- Chapter Projects
- Student Tips

A Look Back, A Look Forward

This chapter is in two parts: Polar Coordinates, Sections 11.1–11.3, and Vectors, Sections 11.4–11.5. They are independent of each other and may be covered in any order.

Sections 11.1–11.3: In Chapter 1 we introduced rectangular coordinates (x, y) and discussed the graph of an equation in two variables involving x and y. In Sections 11.1 and 11.2, we introduce an alternative to rectangular coordinates, polar coordinates, and discuss graphing equations that involve polar coordinates. In

Section 6.2, we discussed raising a real number to a real power. In Section 11.3 we extend this idea by raising a complex number to a real power. As it turns out, polar coordinates are useful for the discussion.

Sections 11.4–11.5: We saw in Chapter 1 that often we are required to solve an equation to obtain a solution to applied problems. In the last two sections of this chapter, we develop the notion of a vector, which is the basis for solving certain types of applied problems, particularly in physics and engineering.

PREPARING FOR THIS SECTION

Before getting started, review the following:

✓ Rectangular Coordinates (Section 1.1, pp. 90–91)

✓ Definitions of the Sine and Cosine Functions (Section 8.4, p. 644)

✓ Inverse Tangent Function (Section 9.1, pp. 717–719)

✓ Completing the Square (Section 1.3, pp. 120–121)

11.1 POLAR COORDINATES

OBJECTIVES

1. Plot Points Using Polar Coordinates
2. Convert from Polar Coordinates to Rectangular Coordinates
3. Convert from Rectangular Coordinates to Polar Coordinates

So far, we have always used a system of rectangular coordinates to plot points in the plane. Now we are ready to describe another system called *polar coordinates*. As we shall soon see, in many instances polar coordinates offer certain advantages over rectangular coordinates.

In a rectangular coordinate system, you will recall, a point in the plane is represented by an ordered pair of numbers (x, y), where x and y equal the signed distance of the point from the y-axis and x-axis, respectively. In a polar coordinate system, we select a point, called the **pole,** and then a ray with vertex at the pole, called the **polar axis.** Comparing the rectangular and polar coordinate systems, we see (in Figure 1) that the origin in rectangular coordinates coincides with the pole in polar coordinates, and the positive x-axis in rectangular coordinates coincides with the polar axis in polar coordinates.

A point P in a polar coordinate system is represented by an ordered pair of numbers (r, θ). If $r > 0$, then r is the distance of the point from the pole, while θ is the angle (in degrees or radians) formed by the polar axis and a ray from the pole through the point. We call the ordered pair (r, θ) the **polar coordinates** of the point. See Figure 2.

As an example, suppose that the polar coordinates of a point P are $\left(2, \dfrac{\pi}{4}\right)$. We locate P by first drawing an angle of $\dfrac{\pi}{4}$ radian, placing its vertex at the pole and its initial side along the polar axis. Then we go out a distance of 2 units along the terminal side of the angle to reach the point P. See Figure 3.

Figure 1

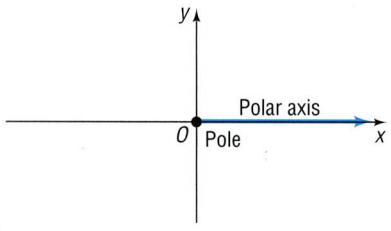

Figure 2

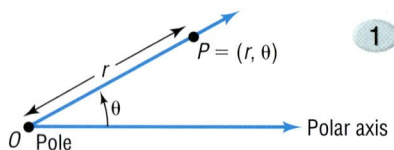

Figure 3

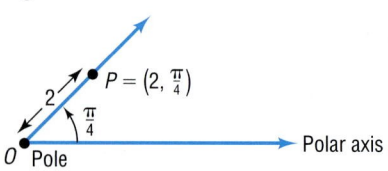

▱▱▱ **NOW WORK PROBLEM 9.**

Recall that an angle measured counterclockwise is positive, whereas an angle measured clockwise is negative. This convention has some interesting consequences relating to polar coordinates. Let's see what these consequences are.

EXAMPLE 1   **Finding Several Polar Coordinates of a Single Point**

Consider again the point P with polar coordinates $\left(2, \dfrac{\pi}{4}\right)$, as shown in Figure 4(a). Because $\dfrac{\pi}{4}, \dfrac{9\pi}{4}$, and $-\dfrac{7\pi}{4}$ all have the same terminal side, we also could have located this point P by using the polar coordinates $\left(2, \dfrac{9\pi}{4}\right)$ or $\left(2, -\dfrac{7\pi}{4}\right)$, as shown in Figures 4(b) and (c).

Figure 4

Figure 5

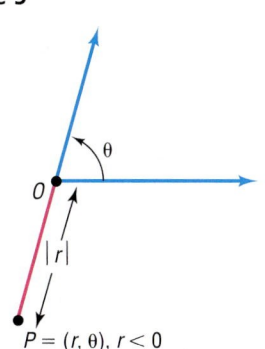

$P = (r, \theta), r < 0$

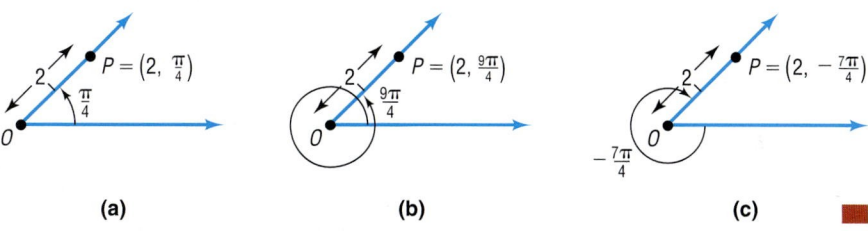

(a)   (b)   (c)

In using polar coordinates (r, θ), it is possible for the first entry r to be negative. When this happens, instead of the point being on the terminal side of θ, it is on the ray from the pole extending in the direction *opposite* the terminal side of θ at a distance $|r|$ from the pole. See Figure 5 for an illustration.

EXAMPLE 2   **Polar Coordinates $(r, \theta), r < 0$**

Consider again the point P with polar coordinates $\left(2, \dfrac{\pi}{4}\right)$, as shown in Figure 6(a). This same point P can be assigned the polar coordinates $\left(-2, \dfrac{5\pi}{4}\right)$, as indicated in Figure 6(b). To locate the point $\left(-2, \dfrac{5\pi}{4}\right)$, we use the ray in the opposite direction of $\dfrac{5\pi}{4}$ and go out 2 units along that ray to find the point P.

Figure 6

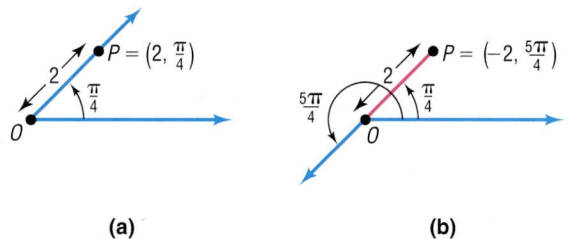

(a)   (b)

These examples show a major difference between rectangular coordinates and polar coordinates. In the former, each point has exactly one pair of rectangular coordinates; in the latter, a point can have infinitely many pairs of polar coordinates.

EXAMPLE 3   **Plotting Points Using Polar Coordinates**

Plot the points with the following polar coordinates:

(a) $\left(3, \dfrac{5\pi}{3}\right)$ (b) $\left(2, -\dfrac{\pi}{4}\right)$ (c) $(3, 0)$ (d) $\left(-2, \dfrac{\pi}{4}\right)$

Solution   Figure 7 shows the points.

Figure 7

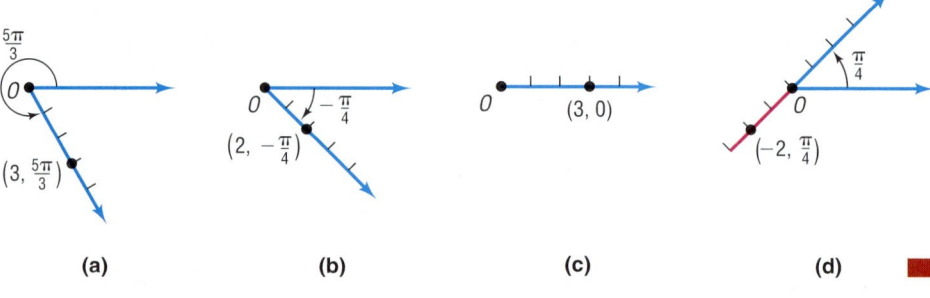

(a)   (b)   (c)   (d)

▸ **NOW WORK PROBLEMS 1 AND 17.**

EXAMPLE 4   **Finding Other Polar Coordinates of a Given Point**

Plot the point P with polar coordinates $\left(3, \dfrac{\pi}{6}\right)$, and find other polar coordinates (r, θ) of this same point for which:

(a) $r > 0$, $2\pi \le \theta < 4\pi$ (b) $r < 0$, $0 \le \theta < 2\pi$
(c) $r > 0$, $-2\pi \le \theta < 0$

Solution   The point $\left(3, \dfrac{\pi}{6}\right)$ is plotted in Figure 8.

Figure 8

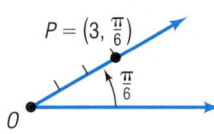

(a) We add 1 revolution (2π radians) to the angle $\dfrac{\pi}{6}$ to get
$P = \left(3, \dfrac{\pi}{6} + 2\pi\right) = \left(3, \dfrac{13\pi}{6}\right)$. See Figure 9(a).

(b) We add $\dfrac{1}{2}$ revolution (π radians) to the angle $\dfrac{\pi}{6}$ and replace 3 by -3 to
get $P = \left(-3, \dfrac{\pi}{6} + \pi\right) = \left(-3, \dfrac{7\pi}{6}\right)$. See Figure 9(b).

(c) We subtract 2π from the angle $\dfrac{\pi}{6}$ to get $P = \left(3, \dfrac{\pi}{6} - 2\pi\right) = \left(3, -\dfrac{11\pi}{6}\right)$.
See Figure 9(c).

Figure 9

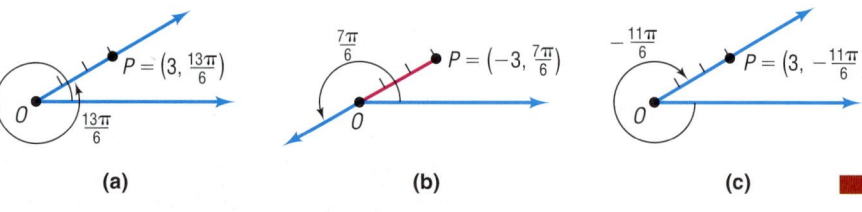

(a)   (b)   (c)

▸ **NOW WORK PROBLEM 21.**

SUMMARY

A point with polar coordinates (r, θ) also can be represented by either of the following:

$$(r, \theta + 2k\pi) \quad \text{or} \quad (-r, \theta + \pi + 2k\pi), \quad k \text{ any integer}$$

The polar coordinates of the pole are $(0, \theta)$, where θ can be any angle.

Conversion from Polar Coordinates to Rectangular Coordinates, and Vice Versa

2 It is sometimes convenient and, indeed, necessary to be able to convert coordinates or equations in rectangular form to polar form, and vice versa. To do this, we recall that the origin in rectangular coordinates is the pole in polar coordinates and that the positive x-axis in rectangular coordinates is the polar axis in polar coordinates.

Theorem

Conversion from Polar Coordinates to Rectangular Coordinates

If P is a point with polar coordinates (r, θ), the rectangular coordinates (x, y) of P are given by

$$x = r \cos \theta \qquad y = r \sin \theta \tag{1}$$

Figure 10

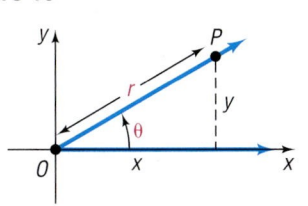

Proof Suppose that P has the polar coordinates (r, θ). We seek the rectangular coordinates (x, y) of P. Refer to Figure 10.

If $r = 0$, then, regardless of θ, the point P is the pole, for which the rectangular coordinates are $(0, 0)$. Formula (1) is valid for $r = 0$.

If $r > 0$, the point P is on the terminal side of θ, and $r = d(O, P) = \sqrt{x^2 + y^2}$. Since

$$\cos \theta = \frac{x}{r} \qquad \sin \theta = \frac{y}{r}$$

we have

$$x = r \cos \theta \qquad y = r \sin \theta$$

If $r < 0$, then the point $P = (r, \theta)$ can be represented as $(-r, \pi + \theta)$, where $-r > 0$. Since

$$\cos(\pi + \theta) = -\cos \theta = \frac{x}{-r} \qquad \sin(\pi + \theta) = -\sin \theta = \frac{y}{-r}$$

we have

$$x = r \cos \theta \qquad y = r \sin \theta \qquad \blacksquare$$

EXAMPLE 5 **Converting from Polar Coordinates to Rectangular Coordinates**

Find the rectangular coordinates of the points with the following polar coordinates:

(a) $\left(6, \dfrac{\pi}{6} \right)$ (b) $\left(-4, -\dfrac{\pi}{4} \right)$

Solution   We use formula (1): $x = r \cos \theta$ and $y = r \sin \theta$.

Figure 11

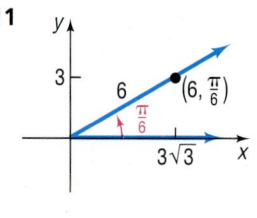

(a)

(a) Figure 11(a) shows $\left(6, \dfrac{\pi}{6}\right)$ plotted. With $r = 6$ and $\theta = \dfrac{\pi}{6}$, we have

$$x = r \cos \theta = 6 \cos \frac{\pi}{6} = 6 \cdot \frac{\sqrt{3}}{2} = 3\sqrt{3}$$

$$y = r \sin \theta = 6 \sin \frac{\pi}{6} = 6 \cdot \frac{1}{2} = 3$$

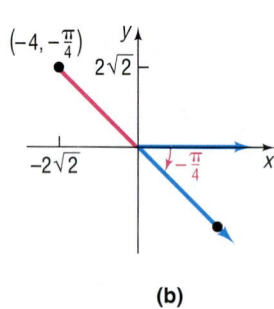

(b)

The rectangular coordinates of the point $\left(6, \dfrac{\pi}{6}\right)$ are $(3\sqrt{3}, 3)$.

(b) Figure 11(b) shows $\left(-4, -\dfrac{\pi}{4}\right)$ plotted. With $r = -4$ and $\theta = -\dfrac{\pi}{4}$, we have

$$x = r \cos \theta = -4 \cos\left(-\frac{\pi}{4}\right) = -4 \cdot \frac{\sqrt{2}}{2} = -2\sqrt{2}$$

$$y = r \sin \theta = -4 \sin\left(-\frac{\pi}{4}\right) = -4\left(-\frac{\sqrt{2}}{2}\right) = 2\sqrt{2}$$

The rectangular coordinates of the point $\left(-4, -\dfrac{\pi}{4}\right)$ are $(-2\sqrt{2}, 2\sqrt{2})$.   ■

Figure 12

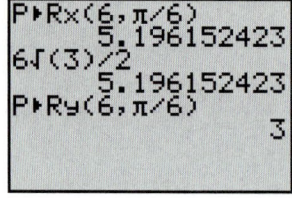

Most graphing calculators have the capability of converting from polar coordinates to rectangular coordinates. Consult you owner's manual for the proper keystrokes. Figure 12 verifies the results obtained in Example 5(a) using a TI-83. Note that the calculator is in radian mode.

━ **NOW WORK PROBLEMS 29 AND 41.**

3   Converting from rectangular coordinates (x, y) to polar coordinates (r, θ) is a little more complicated. Notice that we begin each example by plotting the given rectangular coordinates.

EXAMPLE 6   ### Converting from Rectangular Coordinates to Polar Coordinates

Find polar coordinates of a point whose rectangular coordinates are $(0, 3)$.

Figure 13

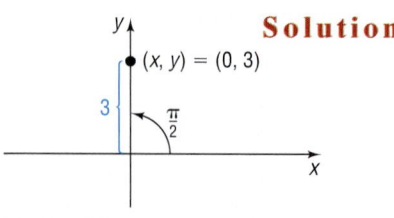

Solution   See Figure 13. The point $(0, 3)$ lies on the y-axis a distance of 3 units from the origin (pole), so $r = 3$. A ray with vertex at the pole through $(0, 3)$ forms an angle $\theta = \dfrac{\pi}{2}$ with the polar axis. Polar coordinates for this point can be given by $\left(3, \dfrac{\pi}{2}\right)$.   ■

Figure 14

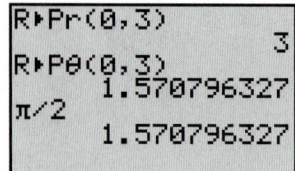

Most graphing calculators have the capability of converting from rectangular coordinates to polar coordinates. Consult you owner's manual for the proper keystrokes. Figure 14 verifies the results obtained in Example 6 using a TI-83. Note that the calculator is in radian mode.

Figure 15 shows polar coordinates of points that lie on either the x-axis or the y-axis. In each illustration, $a > 0$.

Figure 15

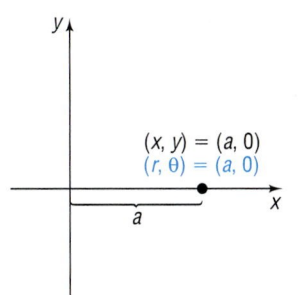

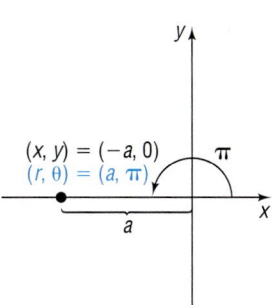

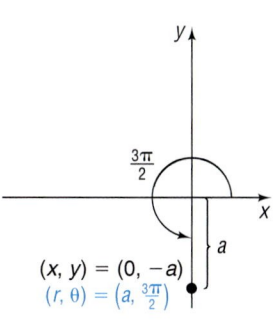

(a) $(x, y) = (a, 0)$, $a > 0$ **(b)** $(x, y) = (0, a)$, $a > 0$ **(c)** $(x, y) = (-a, 0)$, $a > 0$ **(d)** $(x, y) = (0, -a)$, $a > 0$

NOW WORK PROBLEM **45.**

EXAMPLE 7

Converting from Rectangular Coordinates to Polar Coordinates

Find polar coordinates of a point whose rectangular coordinates are:

(a) $(2, -2)$ (b) $(-1, -\sqrt{3})$

Solution (a) See Figure 16(a). The distance r from the origin to the point $(2, -2)$ is

$$r = \sqrt{x^2 + y^2} = \sqrt{(2)^2 + (-2)^2} = \sqrt{8} = 2\sqrt{2}$$

To find θ, we use the reference angle α. Then

$$\alpha = \tan^{-1}\left|\frac{y}{x}\right| = \tan^{-1}\left|\frac{-2}{2}\right| = \tan^{-1}\frac{2}{2} = \tan^{-1} 1 = \frac{\pi}{4}$$

Figure 16

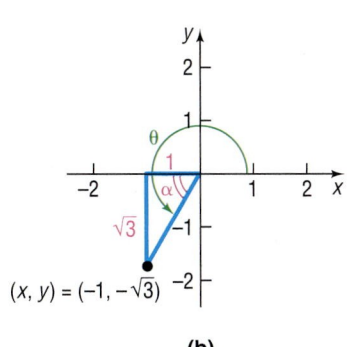

(a)

Using Figure 16(a), $\theta = -\dfrac{\pi}{4}$, and a set of polar coordinates for this point is $\left(2\sqrt{2}, -\dfrac{\pi}{4}\right)$. Other possible representations include $\left(2\sqrt{2}, \dfrac{7\pi}{4}\right)$ and $\left(-2\sqrt{2}, \dfrac{3\pi}{4}\right)$.

(b) See Figure 16(b). The distance r from the origin to the point $(-1, -\sqrt{3})$ is

$$r = \sqrt{x^2 + y^2} = \sqrt{(-1)^2 + (-\sqrt{3})^2} = \sqrt{4} = 2$$

To find θ, we use the reference angle α. Then

$$\alpha = \tan^{-1}\left|\frac{y}{x}\right| = \tan^{-1}\left|\frac{-\sqrt{3}}{-1}\right| = \tan^{-1}\sqrt{3} = \frac{\pi}{3}$$

(b)

Using Figure 16(b),

$$\theta = \pi + \alpha = \pi + \frac{\pi}{3} = \frac{4\pi}{3}$$

and a set of polar coordinates is $\left(2, \dfrac{4\pi}{3}\right)$. Other possible representations include $\left(-2, \dfrac{\pi}{3}\right)$ and $\left(2, -\dfrac{2\pi}{3}\right)$.

Figure 17 shows how to find polar coordinates of a point that lies in a quadrant when its rectangular coordinates (x, y) are given.

Figure 17

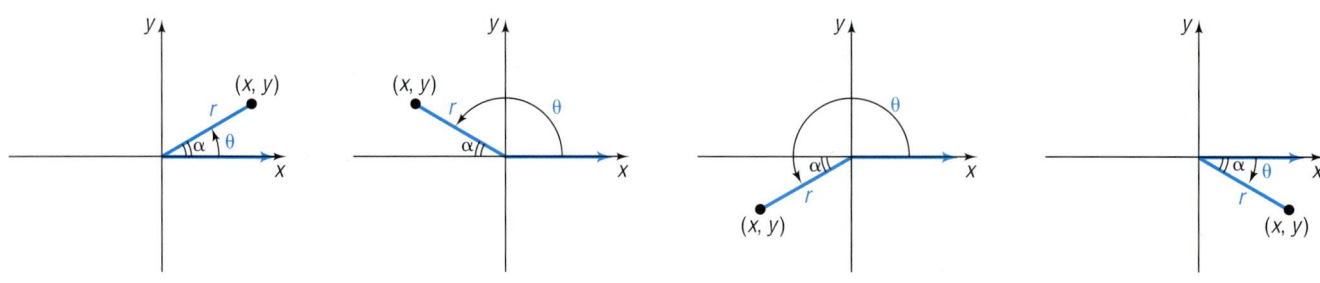

(a) $r = \sqrt{x^2 + y^2}$
$\theta = \alpha = \tan^{-1} \left|\frac{y}{x}\right|$

(b) $r = \sqrt{x^2 + y^2}$
$\theta = \pi - \alpha = \pi - \tan^{-1} \left|\frac{y}{x}\right|$

(c) $r = \sqrt{x^2 + y^2}$
$\theta = \pi + \alpha = \pi + \tan^{-1} \left|\frac{y}{x}\right|$

(d) $r = \sqrt{x^2 + y^2}$
$\theta = -\alpha = -\tan^{-1} \left|\frac{y}{x}\right|$

Based on the preceding discussion, we have the formulas

$$r^2 = x^2 + y^2 \qquad \tan \theta = \frac{y}{x} \qquad \text{if } x \neq 0 \qquad (2)$$

To use formula (2) effectively, follow these steps:

> ### Steps for Converting from Rectangular to Polar Coordinates
>
> **STEP 1:** Always plot the point (x, y) first, as we did in Examples 6 and 7.
> **STEP 2:** To find r, compute the distance from the origin to (x, y).
>
> **STEP 3:** To find θ, it is best to compute the reference angle α of θ,
> $$\alpha = \tan^{-1} \left|\frac{y}{x}\right|, \text{ if } x \neq 0, \text{ and then use your illustration to find } \theta.$$

Look again at Figure 17 and Example 7.

NOW WORK PROBLEM 49.

Formulas (1) and (2) may also be used to transform equations.

EXAMPLE 8   **Transforming an Equation from Polar to Rectangular Form**

Transform the equation $r = 4 \sin \theta$ from polar coordinates to rectangular coordinates, and identify the graph.

Solution   If we multiply each side by r, it will be easier to apply formulas (1) and (2).

$$r = 4 \sin \theta$$
$$r^2 = 4r \sin \theta \qquad \text{Multiply each side by } r.$$
$$x^2 + y^2 = 4y \qquad r^2 = x^2 + y^2; \quad y = r \sin \theta$$

This is the equation of a circle; we proceed to complete the square to obtain the standard form of the equation.

$$x^2 + (y^2 - 4y) = 0 \quad \textit{General form}$$

$$x^2 + (y^2 - 4y + 4) = 4 \quad \textit{Complete the square in y.}$$

$$x^2 + (y - 2)^2 = 4 \quad \textit{Standard form}$$

The center of the circle is at $(0, 2)$, and its radius is 2. ■

━━━━━━━━━ **NOW WORK PROBLEM 65.**

EXAMPLE 9   **Transforming an Equation from Rectangular to Polar Form**

Transform the equation $4xy = 9$ from rectangular coordinates to polar coordinates.

Solution   We use formula (1).

$$4xy = 9$$

$$4(r\cos\theta)(r\sin\theta) = 9 \quad x = r\cos\theta, y = r\sin\theta$$

$$4r^2\cos\theta\sin\theta = 9$$

$$2r^2(2\sin\theta\cos\theta) = 9$$

$$2r^2\sin(2\theta) = 9 \quad \textit{Double-angle formula} \quad ■$$

11.1 Concepts and Vocabulary

In Problems 1–3, fill in the blanks.

1. In polar coordinates, the origin is called the _____ and the positive x-axis is referred to as the _____ _____ _____.

2. Another representation in polar coordinates for the point $\left(2, \dfrac{\pi}{3}\right)$ is $\left(\underline{\hspace{1cm}}, \dfrac{4\pi}{3}\right)$.

3. The polar coordinates $\left(-2, \dfrac{\pi}{6}\right)$ are represented in rectangular coordinates by $(\underline{\hspace{1cm}}, \underline{\hspace{1cm}})$.

In Problems 4–6, answer True or False to each statement.

4. The polar coordinates of a point are unique.

5. The rectangular coordinates of a point are unique.

6. In (r, θ), the number r can be negative.

7. In converting from polar coordinates to rectangular coordinates, what formulas will you use?

8. Explain how you proceed to convert from rectangular coordinates to polar coordinates.

9. Is the street system in your town based on a rectangular coordinate system, a polar coordinate system, or some other system? Explain.

11.1 Exercises

In Problems 1–8, match each point in polar coordinates with either A, B, C, or D on the graph.

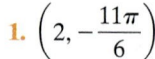

1. $\left(2, -\dfrac{11\pi}{6}\right)$

2. $\left(-2, -\dfrac{\pi}{6}\right)$

3. $\left(-2, \dfrac{\pi}{6}\right)$

4. $\left(2, \dfrac{7\pi}{6}\right)$

5. $\left(2, \dfrac{5\pi}{6}\right)$

6. $\left(-2, \dfrac{5\pi}{6}\right)$

7. $\left(-2, \dfrac{7\pi}{6}\right)$

8. $\left(2, \dfrac{11\pi}{6}\right)$

In Problems 9–20, plot each point given in polar coordinates.

9. $(3, 90°)$

10. $(4, 270°)$

11. $(-2, 0)$

12. $(-3, \pi)$

13. $\left(6, \dfrac{\pi}{6}\right)$

14. $\left(5, \dfrac{5\pi}{3}\right)$

15. $(-2, 135°)$

16. $(-3, 120°)$

17. $\left(-1, -\dfrac{\pi}{3}\right)$

18. $\left(-3, -\dfrac{3\pi}{4}\right)$

19. $(-2, -\pi)$

20. $\left(-3, -\dfrac{\pi}{2}\right)$

In Problems 21–28, plot each point given in polar coordinates, and find other polar coordinates (r, θ) of the point for which:

(a) $r > 0$, $-2\pi \le \theta < 0$ (b) $r < 0$, $0 \le \theta < 2\pi$ (c) $r > 0$, $2\pi \le \theta < 4\pi$

21. $\left(5, \dfrac{2\pi}{3}\right)$

22. $\left(4, \dfrac{3\pi}{4}\right)$

23. $(-2, 3\pi)$

24. $(-3, 4\pi)$

25. $\left(1, \dfrac{\pi}{2}\right)$

26. $(2, \pi)$

27. $\left(-3, -\dfrac{\pi}{4}\right)$

28. $\left(-2, -\dfrac{2\pi}{3}\right)$

In Problems 29–44, polar coordinates of a point are given. Find the rectangular coordinates of each point. Verify your results using a graphing utility.

29. $\left(3, \dfrac{\pi}{2}\right)$

30. $\left(4, \dfrac{3\pi}{2}\right)$

31. $(-2, 0)$

32. $(-3, \pi)$

33. $(6, 150°)$

34. $(5, 300°)$

35. $\left(-2, \dfrac{3\pi}{4}\right)$

36. $\left(-2, \dfrac{2\pi}{3}\right)$

37. $\left(-1, -\dfrac{\pi}{3}\right)$

38. $\left(-3, -\dfrac{3\pi}{4}\right)$

39. $(-2, -180°)$

40. $(-3, -90°)$

41. $(7.5, 110°)$

42. $(-3.1, 182°)$

43. $(6.3, 3.8)$

44. $(8.1, 5.2)$

In Problems 45–56, the rectangular coordinates of a point are given. Find polar coordinates for each point. Verify your results using a graphing utility.

45. $(3, 0)$

46. $(0, 2)$

47. $(-1, 0)$

48. $(0, -2)$

49. $(1, -1)$

50. $(-3, 3)$

51. $(\sqrt{3}, 1)$

52. $(-2, -2\sqrt{3})$

53. $(1.3, -2.1)$

54. $(-0.8, -2.1)$

55. $(8.3, 4.2)$

56. $(-2.3, 0.2)$

In Problems 57–64, the letters x and y represent rectangular coordinates. Write each equation using polar coordinates (r, θ).

57. $2x^2 + 2y^2 = 3$

58. $x^2 + y^2 = x$

59. $x^2 = 4y$

60. $y^2 = 2x$

61. $2xy = 1$

62. $4x^2 y = 1$

63. $x = 4$

64. $y = -3$

In Problems 65–72, the letters r and θ represent polar coordinates. Write each equation using rectangular coordinates (x, y).

65. $r = \cos \theta$

66. $r = \sin \theta + 1$

67. $r^2 = \cos \theta$

68. $r = \sin \theta - \cos \theta$

69. $r = 2$

70. $r = 4$

71. $r = \dfrac{4}{1 - \cos \theta}$

72. $r = \dfrac{3}{3 - \cos \theta}$

73. Show that the formula for the distance d between two points $P_1 = (r_1, \theta_1)$ and $P_2 = (r_2, \theta_2)$ is

$$d = \sqrt{r_1^2 + r_2^2 - 2r_1 r_2 \cos(\theta_2 - \theta_1)}$$

PREPARING FOR THIS SECTION

Before getting started, review the following:

✓ Introduction to Graphing Equations
(Section 1.2, pp. 100–102)

✓ Even-Odd Properties of Trigonometric Functions
(Section 8.5, pp. 662–663)

✓ Circles (Section 1.8, pp. 180–184)

✓ Difference Formulas (Section 9.4, pp. 734 and 737)

✓ Value of the Sine and Cosine Functions at Certain Angles
(Section 8.3, p. 638, and Section 8.4, p. 646)

11.2 POLAR EQUATIONS AND GRAPHS

OBJECTIVES

1. Graph and Identify Polar Equations by Converting to Rectangular Equations
2. Graph Polar Equations Using a Graphing Utility
3. Test Polar Equations for Symmetry
4. Graph Polar Equations by Plotting Points

Just as a rectangular grid may be used to plot points given by rectangular coordinates, as in Figure 18(a), we can use a grid consisting of concentric circles (with centers at the pole) and rays (with vertices at the pole) to plot points given by polar coordinates, as shown in Figure 18(b). We shall use such **polar grids** to graph *polar equations*.

Figure 18

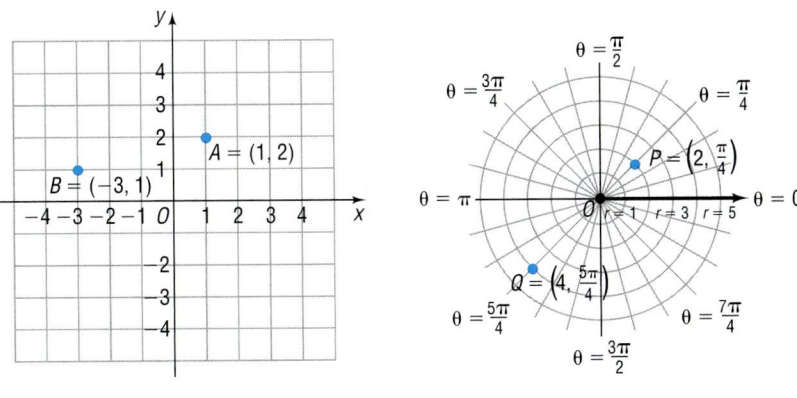

(a) Rectangular grid   **(b)** Polar grid

An equation whose variables are polar coordinates is called a **polar equation**. The **graph of a polar equation** consists of all points whose polar coordinates satisfy the equation.

1.   One method that we can use to graph a polar equation is to convert the equation to rectangular coordinates. In the discussion that follows, (x, y) represent the rectangular coordinates of a point P, and (r, θ) represent polar coordinates of the point P.

EXAMPLE 1 **Identifying and Graphing a Polar Equation by Hand (Circle)**

Identify and graph the equation: $r = 3$

Solution We convert the polar equation to a rectangular equation.

$$r = 3$$
$$r^2 = 9 \qquad \text{\color{blue}Square both sides.}$$
$$x^2 + y^2 = 9 \qquad \text{\color{blue}} r^2 = x^2 + y^2$$

The graph of $r = 3$ is a circle, with center at the pole and radius 3. See Figure 19.

Figure 19
$r = 3$ or $x^2 + y^2 = 9$

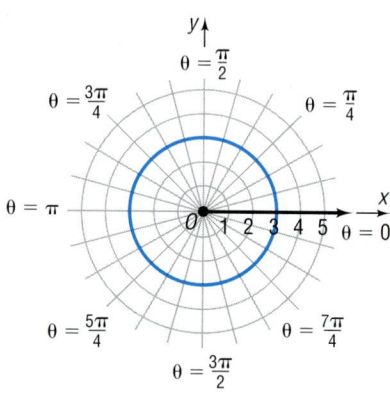

━━━ **NOW WORK PROBLEM 1.**

EXAMPLE 2 **Identifying and Graphing a Polar Equation by Hand (Line)**

Identify and graph the equation: $\theta = \dfrac{\pi}{4}$

Figure 20
$\theta = \dfrac{\pi}{4}$ or $y = x$

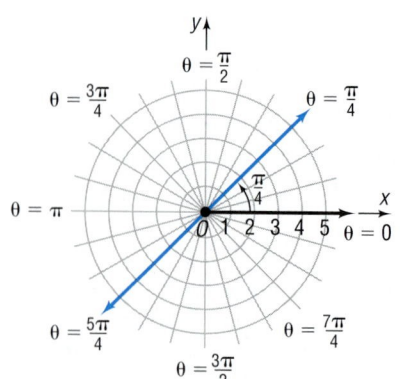

Solution We convert the polar equation to a rectangular equation.

$$\theta = \frac{\pi}{4}$$
$$\tan \theta = \tan \frac{\pi}{4} = 1$$
$$\frac{y}{x} = 1 \qquad \text{\color{blue}} \tan \theta = \frac{y}{x}$$
$$y = x$$

The graph of $\theta = \dfrac{\pi}{4}$ is a line passing through the pole making an angle of $\dfrac{\pi}{4}$ with the polar axis. See Figure 20.

━━━ **NOW WORK PROBLEM 3.**

| EXAMPLE 3 | **Identifying and Graphing a Polar Equation by Hand (Horizontal Line)** |

Identify and graph the equation: $r \sin \theta = 2$

Solution   Since $y = r \sin \theta$, we can write the equation as

$$y = 2$$

We conclude that the graph of $r \sin \theta = 2$ is a horizontal line 2 units above the pole. See Figure 21.

Figure 21
$r \sin \theta = 2$ or $y = 2$

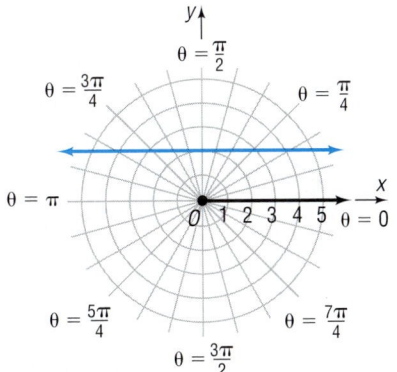

2   A second method we can use to graph a polar equation is to graph the equation using a graphing utility.

Most graphing utilities require the following steps in order to obtain the graph of an equation:

> *Graphing a Polar Equation Using a Graphing Utility*
>
> **STEP 1:** Solve the equation for r in terms of θ.
>
> **STEP 2:** Select the viewing window in POLar mode. In addition to setting Xmin, Xmax, Xscl, and so forth, the viewing window in polar mode requires setting minimum and maximum values for θ and an increment setting for θ (θstep). Finally, a square screen and radian measure should be used.
>
> **STEP 3:** Enter the expression involving θ that you found in Step 1. (Consult your manual for the correct way to enter the expression.)
>
> **STEP 4:** Graph.

| EXAMPLE 4 | **Graphing a Polar Equation Using a Graphing Utility** |

Use a graphing utility to graph the polar equation $r \sin \theta = 2$.

Solution   **STEP 1:** We solve the equation for r in terms of θ.

$$r \sin \theta = 2$$

$$r = \frac{2}{\sin \theta}$$

Step 2: From the polar mode, select a square viewing window. We will use the one given next.

$$\theta\text{min} = 0 \qquad X\text{min} = -9 \qquad Y\text{min} = -6$$
$$\theta\text{max} = 2\pi \qquad X\text{max} = 9 \qquad Y\text{max} = 6$$
$$\theta\text{step} = \frac{\pi}{24} \qquad X\text{scl} = 1 \qquad Y\text{scl} = 1$$

θstep determines the number of points the graphing utility will plot. For example, if θstep is $\frac{\pi}{24}$, then the graphing utility will evaluate r at $\theta = 0(\theta\text{min}), \frac{\pi}{24}, \frac{2\pi}{24}, \frac{3\pi}{24}$, and so forth, up to $2\pi(\theta\text{max})$.

The smaller θstep, the more points the graphing utility will plot. The student is encouraged to experiment with different values for θmin, θmax, and θstep to see how the graph is affected.

Figure 22

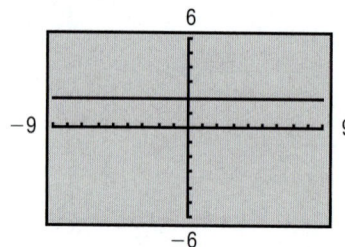

Step 3: Enter the expression $\dfrac{2}{\sin \theta}$ after the prompt $r = \quad$.

Step 4: Graph.

The graph is shown in Figure 22.   ■

EXAMPLE 5   **Identifying and Graphing a Polar Equation (Vertical Line)**

Identify and graph the equation: $r \cos \theta = -3$

Solution   Since $x = r \cos \theta$, we can write the equation as

$$x = -3$$

We conclude that the graph of $r \cos \theta = -3$ is a vertical line 3 units to the left of the pole. Figure 23(a) shows the graph drawn by hand. Figure 23(b) shows the graph using a graphing utility with $\theta\text{min} = 0, \theta\text{max} = 2\pi$, and $\theta\text{step} = \dfrac{\pi}{24}$.

Figure 23
$r \cos \theta = -3$ or $x = -3$

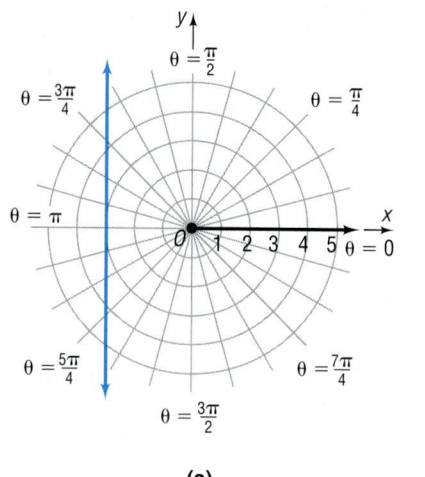

(a)

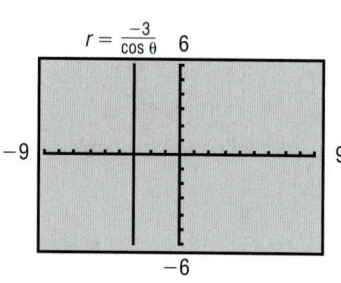

(b)   ■

Based on Examples 3, 4, and 5, we are led to the following results. (The proofs are left as exercises.)

Theorem

Let a be a nonzero real number. Then the graph of the equation

$$r \sin \theta = a$$

is a horizontal line a units above the pole if $a > 0$ and $|a|$ units below the pole if $a < 0$.

The graph of the equation

$$r \cos \theta = a$$

is a vertical line a units to the right of the pole if $a > 0$ and $|a|$ units to the left of the pole if $a < 0$.

NOW WORK PROBLEM 7.

EXAMPLE 6   **Identifying and Graphing a Polar Equation (Circle)**

Identify and graph the equation: $r = 4 \sin \theta$

Solution   To transform the equation to rectangular coordinates, we multiply each side by r.

$$r^2 = 4r \sin \theta$$

Now we use the facts that $r^2 = x^2 + y^2$ and $y = r \sin \theta$. Then

$$x^2 + y^2 = 4y$$
$$x^2 + (y^2 - 4y) = 0$$
$$x^2 + (y^2 - 4y + 4) = 4 \qquad \text{\textcolor{teal}{Complete the square in y.}}$$
$$x^2 + (y - 2)^2 = 4 \qquad \text{\textcolor{teal}{Standard equation of a circle}}$$

This is the equation of a circle with center at $(0, 2)$ in rectangular coordinates and radius 2. Figure 24(a) shows the graph drawn by hand. Figure 24(b) shows the graph using a graphing utility with $\theta\text{min} = 0$, $\theta\text{max} = 2\pi$, and $\theta\text{step} = \dfrac{\pi}{24}$.

Figure 24
$r = 4 \sin \theta$ or $x^2 + (y - 2)^2 = 4$

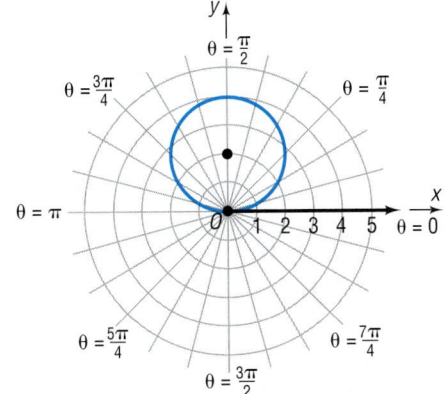

(a)

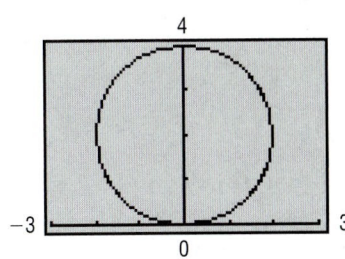

(b)

| EXAMPLE 7 | **Identifying and Graphing a Polar Equation (Circle)** |

Identify and graph the equation: $r = -2 \cos \theta$

Solution We proceed as in Example 6.

$$r^2 = -2r \cos \theta \qquad \text{Multiply both sides by } r.$$

$$x^2 + y^2 = -2x \qquad r^2 = x^2 + y^2; \quad x = r \cos \theta$$

$$x^2 + 2x + y^2 = 0$$

$$(x^2 + 2x + 1) + y^2 = 1 \qquad \text{Complete the square in } x.$$

$$(x + 1)^2 + y^2 = 1 \qquad \text{Standard equation of a circle}$$

This is the equation of a circle with center at $(-1, 0)$ in rectangular coordinates and radius 1. Figure 25(a) shows the graph drawn by hand. Figure 25(b) shows the graph using a graphing utility with $\theta \text{min} = 0$, $\theta \text{max} = 2\pi$, and $\theta \text{step} = \dfrac{\pi}{24}$.

Figure 25
$r = -2 \cos \theta$ or $(x + 1)^2 + y^2 = 1$

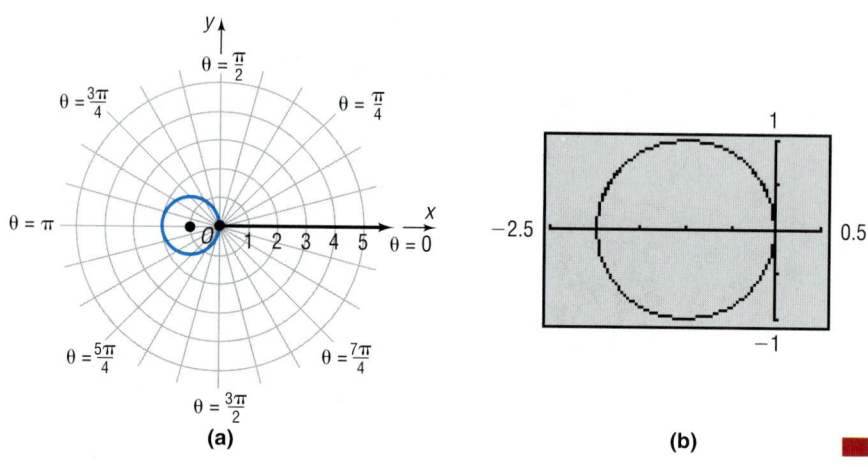

(a)

(b)

EXPLORATION Using a square screen, graph $r_1 = \sin \theta$, $r_2 = 2 \sin \theta$, and $r_3 = 3 \sin \theta$. Do you see the pattern? Clear the screen and graph $r_1 = -\sin \theta$, $r_2 = -2 \sin \theta$, and $r_3 = -3 \sin \theta$. Do you see the pattern? Clear the screen and graph $r_1 = \cos \theta$, $r_2 = 2 \cos \theta$, and $r_3 = 3 \cos \theta$. Do you see the pattern? Clear the screen and graph $r_1 = -\cos \theta$, $r_2 = -2 \cos \theta$, and $r_3 = -3 \cos \theta$. Do you see the pattern?

Based on Examples 6 and 7 and the preceding Exploration, we are led to the following results. (The proofs are left as exercises.)

Theorem

Let a be a positive real number. Then,

Equation	Description
(a) $r = 2a \sin \theta$	Circle: radius a; center at $(0, a)$ in rectangular coordinates
(b) $r = -2a \sin \theta$	Circle: radius a; center at $(0, -a)$ in rectangular coordinates
(c) $r = 2a \cos \theta$	Circle: radius a; center at $(a, 0)$ in rectangular coordinates
(d) $r = -2a \cos \theta$	Circle: radius a; center at $(-a, 0)$ in rectangular coordinates

Each circle passes through the pole.

 **NOW WORK PROBLEM 9.**

The method of converting a polar equation to an identifiable rectangular equation in order to obtain the graph is not always helpful, nor is it always necessary. Usually, we set up a table that lists several points on the graph. By checking for symmetry, it may be possible to reduce the number of points needed to draw the graph.

Symmetry

In polar coordinates, the points (r, θ) and $(r, -\theta)$ are symmetric with respect to the polar axis (and to the x-axis). See Figure 26(a). The points (r, θ) and $(r, \pi - \theta)$ are symmetric with respect to the line $\theta = \dfrac{\pi}{2}$ (the y-axis). See Figure 26(b). The points (r, θ) and $(-r, \theta)$ are symmetric with respect to the pole (the origin). See Figure 26(c).

Figure 26

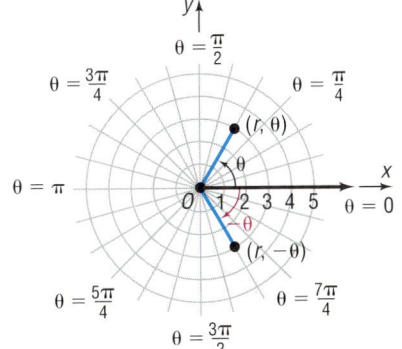

(a) Points symmetric with respect to the polar axis

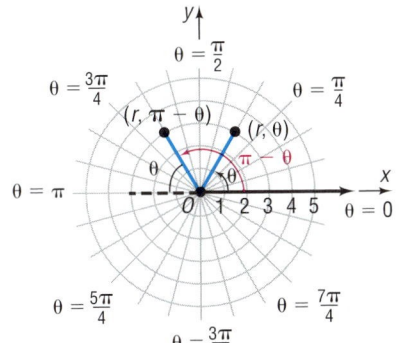

(b) Points symmetric with respect to the line $\theta = \dfrac{\pi}{2}$

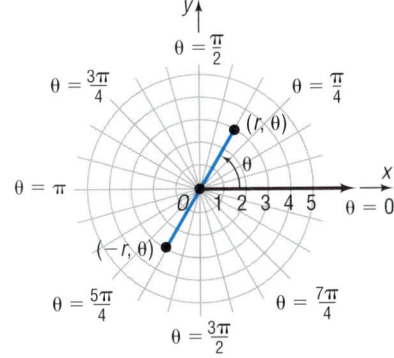

(c) Points symmetric with respect to the pole

The following tests are a consequence of these observations.

Theorem

Tests for Symmetry

Symmetry with Respect to the Polar Axis (x-Axis)

In a polar equation, replace θ by $-\theta$. If an equivalent equation results, the graph is symmetric with respect to the polar axis.

Symmetry with Respect to the Line $\theta = \dfrac{\pi}{2}$ (y-Axis)

In a polar equation, replace θ by $\pi - \theta$. If an equivalent equation results, the graph is symmetric with respect to the line $\theta = \dfrac{\pi}{2}$.

Symmetry with Respect to the Pole (Origin)

In a polar equation, replace r by $-r$. If an equivalent equation results, the graph is symmetric with respect to the pole.

The three tests for symmetry given here are *sufficient* conditions for symmetry, but they are not *necessary* conditions. That is, an equation may fail these tests and still have a graph that is symmetric with respect to the polar axis, the line $\theta = \dfrac{\pi}{2}$, or the pole. For example, the graph of $r = \sin(2\theta)$ turns out to be symmetric with respect to the polar axis, the line $\theta = \dfrac{\pi}{2}$, and the pole, but all three tests given here fail. See also Problems 71, 72, and 73.

EXAMPLE 8 **Graphing a Polar Equation (Cardioid)**

Graph the equation: $r = 1 - \sin\theta$

Solution We check for symmetry first.

TABLE 1

θ	$r = 1 - \sin\theta$
$-\dfrac{\pi}{2}$	$1 - (-1) = 2$
$-\dfrac{\pi}{3}$	$1 - \left(-\dfrac{\sqrt{3}}{2}\right) \approx 1.87$
$-\dfrac{\pi}{6}$	$1 - \left(-\dfrac{1}{2}\right) = \dfrac{3}{2}$
0	$1 - 0 = 1$
$\dfrac{\pi}{6}$	$1 - \dfrac{1}{2} = \dfrac{1}{2}$
$\dfrac{\pi}{3}$	$1 - \dfrac{\sqrt{3}}{2} \approx 0.13$
$\dfrac{\pi}{2}$	$1 - 1 = 0$

Polar Axis: Replace θ by $-\theta$. The result is

$$r = 1 - \sin(-\theta) = 1 + \sin\theta$$

The test fails, so the graph may or may not be symmetric with respect to the polar axis.

The Line $\theta = \dfrac{\pi}{2}$: Replace θ by $\pi - \theta$. The result is

$$r = 1 - \sin(\pi - \theta) = 1 - (\sin\pi\cos\theta - \cos\pi\sin\theta)$$
$$= 1 - [0\cdot\cos\theta - (-1)\sin\theta] = 1 - \sin\theta$$

The test is satisfied, so the graph is symmetric with respect to the line $\theta = \dfrac{\pi}{2}$.

The Pole: Replace r by $-r$. Then the result is $-r = 1 - \sin\theta$, so $r = -1 + \sin\theta$. The test fails, so the graph may or may not be symmetric with respect to the pole.

Next, we identify points on the graph by assigning values to the angle θ and calculating the corresponding values of r. Due to the symmetry with respect to the line $\theta = \dfrac{\pi}{2}$, we only need to assign values to θ from $-\dfrac{\pi}{2}$ to $\dfrac{\pi}{2}$, as given in Table 1.

Now we plot the points (r, θ) from Table 1 and trace out the graph, beginning at the point $\left(2, -\dfrac{\pi}{2}\right)$ and ending at the point $\left(0, \dfrac{\pi}{2}\right)$. Then we reflect this portion of the graph about the line $\theta = \dfrac{\pi}{2}$ (the y-axis) to obtain the complete graph. Figure 27(a) shows the graph drawn by hand. Figure 27(b) shows the graph using a graphing utility with θmin $= 0$, θmax $= 2\pi$, and θstep $= \dfrac{\pi}{24}$.

Figure 27
$r = 1 - \sin\theta$

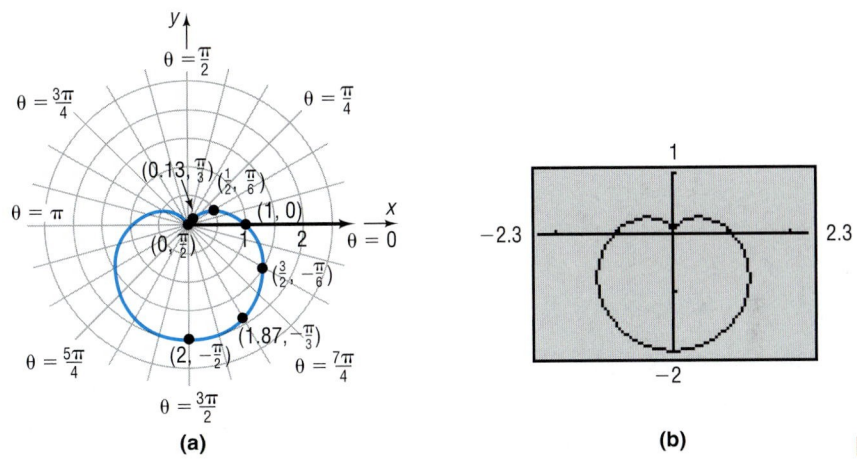

(a)   (b)

EXPLORATION   Graph $r_1 = 1 + \sin\theta$. Clear the screen and graph $r_1 = 1 - \cos\theta$. Clear the screen and graph $r_1 = 1 + \cos\theta$. Do you see a pattern?

The curve in Figure 27 is an example of a *cardioid* (a heart-shaped curve).

Cardioids are characterized by equations of the form

$$r = a(1 + \cos\theta) \qquad r = a(1 + \sin\theta)$$
$$r = a(1 - \cos\theta) \qquad r = a(1 - \sin\theta)$$

where $a > 0$. The graph of a cardioid passes through the pole.

 **NOW WORK PROBLEM 31.**

EXAMPLE 9   Graphing a Polar Equation (Limaçon without Inner Loop)

Graph the equation: $r = 3 + 2\cos\theta$

Solution   We check for symmetry first.

Polar Axis: Replace θ by $-\theta$. The result is

$$r = 3 + 2\cos(-\theta) = 3 + 2\cos\theta$$

The test is satisfied, so the graph is symmetric with respect to the polar axis.

TABLE 2	
θ	$r = 3 + 2\cos\theta$
0	$3 + 2(1) = 5$
$\dfrac{\pi}{6}$	$3 + 2\left(\dfrac{\sqrt{3}}{2}\right) \approx 4.73$
$\dfrac{\pi}{3}$	$3 + 2\left(\dfrac{1}{2}\right) = 4$
$\dfrac{\pi}{2}$	$3 + 2(0) = 3$
$\dfrac{2\pi}{3}$	$3 + 2\left(-\dfrac{1}{2}\right) = 2$
$\dfrac{5\pi}{6}$	$3 + 2\left(-\dfrac{\sqrt{3}}{2}\right) \approx 1.27$
π	$3 + 2(-1) = 1$

The Line $\theta = \dfrac{\pi}{2}$:   Replace θ by $\pi - \theta$. The result is

$$r = 3 + 2\cos(\pi - \theta) = 3 + 2(\cos\pi\cos\theta + \sin\pi\sin\theta)$$
$$= 3 - 2\cos\theta$$

The test fails, so the graph may or may not be symmetric with respect to the line $\theta = \dfrac{\pi}{2}$.

The Pole: Replace r by $-r$. The test fails, so the graph may or may not be symmetric with respect to the pole.

Next, we identify points on the graph by assigning values to the angle θ and calculating the corresponding values of r. Due to the symmetry with respect to the polar axis, we only need to assign values to θ from 0 to π, as given in Table 2.

Now we plot the points (r, θ) from Table 2 and trace out the graph, beginning at the point $(5, 0)$ and ending at the point $(1, \pi)$. Then we reflect this portion of the graph about the polar axis (the x-axis) to obtain the complete graph. Figure 28(a) shows the graph drawn by hand. Figure 28(b) shows the graph using a graphing utility with $\theta\text{min} = 0$, $\theta\text{max} = 2\pi$, and $\theta\text{step} = \dfrac{\pi}{24}$.

Figure 28
$r = 3 + 2\cos\theta$

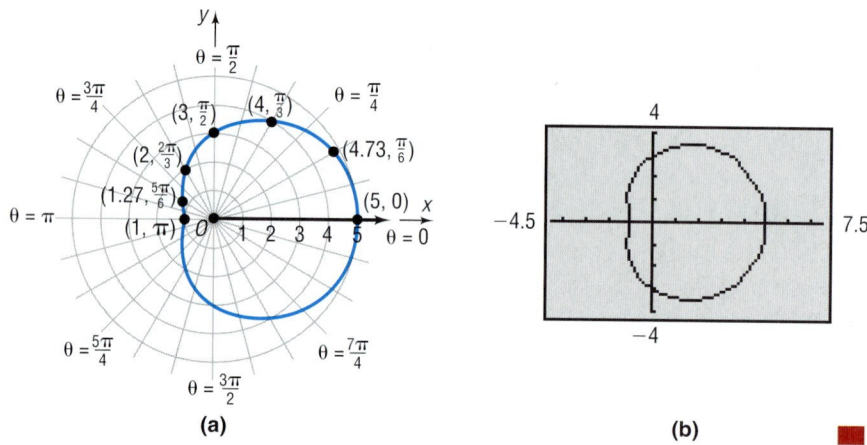

(a)   (b)

EXPLORATION   Graph $r_1 = 3 - 2\cos\theta$. Clear the screen and graph $r_1 = 3 + 2\sin\theta$. Clear the screen and graph $r_1 = 3 - 2\sin\theta$. Do you see a pattern?

The curve in Figure 28 is an example of a limaçon (the French word for *snail*) without an inner loop.

Limaçons without an inner loop are characterized by equations of the form

$r = a + b\cos\theta$	$r = a + b\sin\theta$
$r = a - b\cos\theta$	$r = a - b\sin\theta$

where $a > 0$, $b > 0$, and $a > b$. The graph of a limaçon without an inner loop does not pass through the pole.

NOW WORK PROBLEM **37.**

EXAMPLE 10 Graphing a Polar Equation (Limaçon with Inner Loop)

Graph the equation: $r = 1 + 2\cos\theta$

Solution First, we check for symmetry.

Polar Axis: Replace θ by $-\theta$. The result is

$$r = 1 + 2\cos(-\theta) = 1 + 2\cos\theta$$

The test is satisfied, so the graph is symmetric with respect to the polar axis.

The Line $\theta = \dfrac{\pi}{2}$: Replace θ by $\pi - \theta$. The result is

$$r = 1 + 2\cos(\pi - \theta) = 1 + 2(\cos\pi\cos\theta + \sin\pi\sin\theta)$$
$$= 1 - 2\cos\theta$$

The test fails, so the graph may or may not be symmetric with respect to the line $\theta = \dfrac{\pi}{2}$.

The Pole: Replace r by $-r$. The test fails, so the graph may or may not be symmetric with respect to the pole.

Next, we identify points on the graph of $r = 1 + 2\cos\theta$ by assigning values to the angle θ and calculating the corresponding values of r. Due to the symmetry with respect to the polar axis, we only need to assign values to θ from 0 to π, as given in Table 3.

Now we plot the points (r, θ) from Table 3, beginning at $(3, 0)$ and ending at $(-1, \pi)$. See Figure 29(a). Finally, we reflect this portion of the graph about the polar axis (the x-axis) to obtain the complete graph. See Figure 29(b). Figure 29(c) shows the graph using a graphing utility with $\theta\text{min} = 0$, $\theta\text{max} = 2\pi$, and $\theta\text{step} = \dfrac{\pi}{24}$.

TABLE 3	
θ	$r = 1 + 2\cos\theta$
0	$1 + 2(1) = 3$
$\dfrac{\pi}{6}$	$1 + 2\left(\dfrac{\sqrt{3}}{2}\right) \approx 2.73$
$\dfrac{\pi}{3}$	$1 + 2\left(\dfrac{1}{2}\right) = 2$
$\dfrac{\pi}{2}$	$1 + 2(0) = 1$
$\dfrac{2\pi}{3}$	$1 + 2\left(-\dfrac{1}{2}\right) = 0$
$\dfrac{5\pi}{6}$	$1 + 2\left(-\dfrac{\sqrt{3}}{2}\right) \approx -0.73$
π	$1 + 2(-1) = -1$

Figure 29

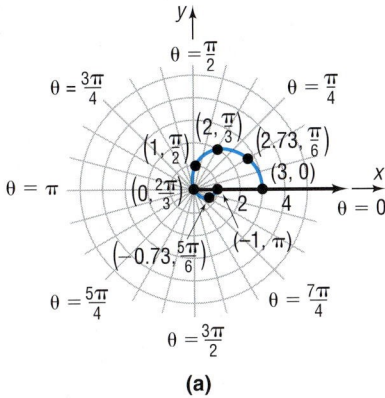

(a)

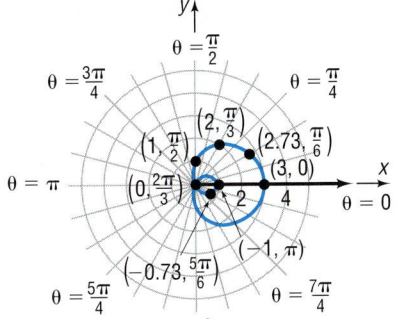

(b) $r = 1 + 2\cos\theta$

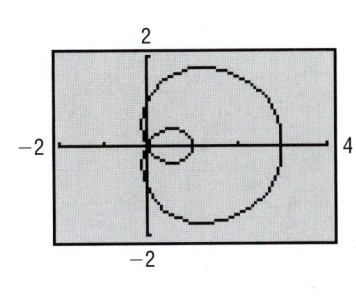

(c)

EXPLORATION Graph $r_1 = 1 - 2\cos\theta$. Clear the screen and graph $r_1 = 1 + 2\sin\theta$. Clear the screen and graph $r_1 = 1 - 2\sin\theta$. Do you see a pattern?

The curve in Figure 29(b) or 29(c) is an example of a limaçon with an inner loop.

Limaçons with an inner loop are characterized by equations of the form

$$r = a + b \cos \theta \qquad r = a + b \sin \theta$$
$$r = a - b \cos \theta \qquad r = a - b \sin \theta$$

where $a > 0, b > 0$, and $a < b$. The graph of a limaçon with an inner loop will pass through the pole twice.

✏ **NOW WORK PROBLEM 39.**

EXAMPLE 11 **Graphing a Polar Equation (Rose)**

Graph the equation: $r = 2 \cos(2\theta)$

Solution We check for symmetry.

Polar Axis: If we replace θ by $-\theta$, the result is

$$r = 2 \cos[2(-\theta)] = 2 \cos(2\theta)$$

The test is satisfied, so the graph is symmetric with respect to the polar axis.

The Line $\theta = \dfrac{\pi}{2}$: If we replace θ by $\pi - \theta$, we obtain

$$r = 2 \cos[2(\pi - \theta)] = 2 \cos(2\pi - 2\theta) = 2 \cos(2\theta)$$

The test is satisfied, so the graph is symmetric with respect to the line $\theta = \dfrac{\pi}{2}$.

The Pole: Since the graph is symmetric with respect to both the polar axis and the line $\theta = \dfrac{\pi}{2}$, it must be symmetric with respect to the pole.

Next, we construct Table 4. Due to the symmetry with respect to the polar axis, the line $\theta = \dfrac{\pi}{2}$, and the pole, we consider only values of θ from 0 to $\dfrac{\pi}{2}$.

We plot and connect these points in Figure 30(a). Finally, because of symmetry, we reflect this portion of the graph first about the polar axis (the x-axis) and then about the line $\theta = \dfrac{\pi}{2}$ (the y-axis) to obtain the complete

TABLE 4

θ	$r = 2 \cos(2\theta)$
0	$2(1) = 2$
$\dfrac{\pi}{6}$	$2\left(\dfrac{1}{2}\right) = 1$
$\dfrac{\pi}{4}$	$2(0) = 0$
$\dfrac{\pi}{3}$	$2\left(-\dfrac{1}{2}\right) = -1$
$\dfrac{\pi}{2}$	$2(-1) = -2$

Figure 30

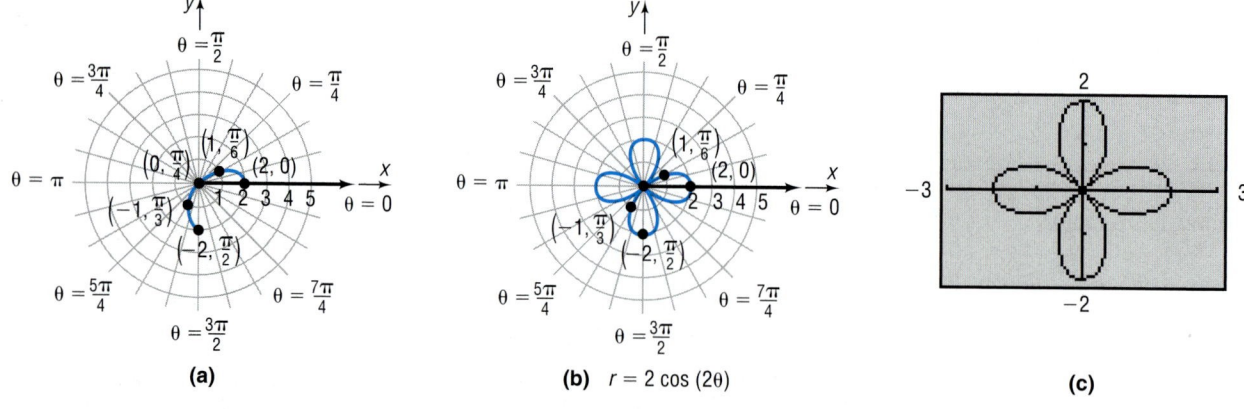

(a)

(b) $r = 2 \cos(2\theta)$

(c)

graph. See Figure 30(b). Figure 30(c) shows the graph using a graphing utility with $\theta\text{min} = 0$, $\theta\text{max} = 2\pi$, and $\theta\text{step} = \dfrac{\pi}{24}$. ■

 EXPLORATION Graph $r = 2\cos(4\theta)$; clear the screen and graph $r = 2\cos(6\theta)$. How many petals did each of these graphs have?

Clear the screen and graph, in order, each on a clear screen, $r = 2\cos(3\theta)$, $r = 2\cos(5\theta)$, and $r = 2\cos(7\theta)$. What do you notice about the number of petals?

The curve in Figure 30(b) or (c) is called a *rose* with four petals.

> **Rose** curves are characterized by equations of the form
>
> $$r = a\cos(n\theta), \qquad r = a\sin(n\theta), \qquad a \neq 0$$
>
> and have graphs that are rose shaped. If $n \neq 0$ is even, the rose has $2n$ petals; if $n \neq \pm 1$ is odd, the rose has n petals.

NOW WORK PROBLEM 43.

EXAMPLE 12 **Graphing a Polar Equation (Lemniscate)**

Graph the equation: $r^2 = 4\sin(2\theta)$

Solution   We leave it to you to verify that the graph is symmetric with respect to the pole. Table 5 lists points on the graph for values of $\theta = 0$ through $\theta = \dfrac{\pi}{2}$. Note that there are no points on the graph for $\dfrac{\pi}{2} < \theta < \pi$ (quadrant II), since $\sin(2\theta) < 0$ for such values. The points from Table 5 where $r \geq 0$ are plotted in Figure 31(a). The remaining points on the graph may be obtained by using symmetry. Figure 31(b) shows the final graph drawn by hand. Figure 31(c) shows the graph using a graphing utility with $\theta\text{min} = 0$, $\theta\text{max} = 2\pi$, and $\theta\text{step} = \dfrac{\pi}{24}$.

TABLE 5

θ	$r^2 = 4\sin(2\theta)$	r
0	$4(0) = 0$	0
$\dfrac{\pi}{6}$	$4\left(\dfrac{\sqrt{3}}{2}\right) = 2\sqrt{3}$	± 1.9
$\dfrac{\pi}{4}$	$4(1) = 4$	± 2
$\dfrac{\pi}{3}$	$4\left(\dfrac{\sqrt{3}}{2}\right) = 2\sqrt{3}$	± 1.9
$\dfrac{\pi}{2}$	$4(0) = 0$	0

Figure 31

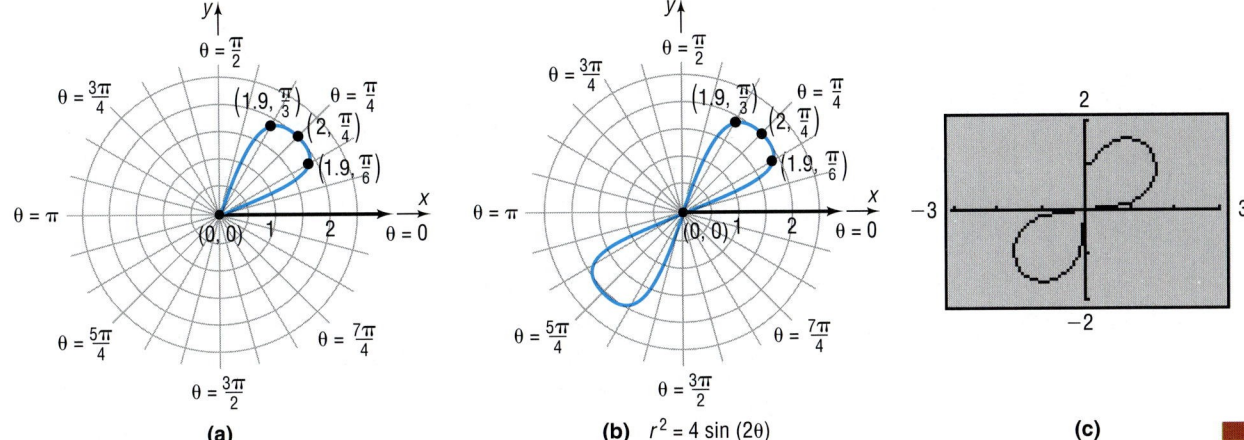

(a)

(b) $r^2 = 4\sin(2\theta)$

(c)

The curve in Figure 31(b) or (c) is an example of a *lemniscate*.

Lemniscates are characterized by equations of the form

$$r^2 = a^2 \sin(2\theta) \qquad r^2 = a^2 \cos(2\theta)$$

where $a \neq 0$, and have graphs that are propeller shaped.

— NOW WORK PROBLEM 47.

EXAMPLE 13   Graphing a Polar Equation (Spiral)

Graph the equation: $r = e^{\theta/5}$

Solution   The tests for symmetry with respect to the pole, the polar axis, and the line $\theta = \dfrac{\pi}{2}$ fail. Furthermore, there is no number θ for which $r = 0$, so the graph does not pass through the pole. We observe that r is positive for all θ, r increases as θ increases, $r \to 0$ as $\theta \to -\infty$, and $r \to \infty$ as $\theta \to \infty$. With the help of a calculator, we obtain the values in Table 6. See Figure 32(a) for the graph drawn by hand. Figure 32(b) shows the graph using a graphing utility with θmin $= -4\pi$, θmax $= 3\pi$, and θstep $= \dfrac{\pi}{24}$.

TABLE 6	
θ	$r = e^{\theta/5}$
$-\dfrac{3\pi}{2}$	0.39
$-\pi$	0.53
$-\dfrac{\pi}{2}$	0.73
$-\dfrac{\pi}{4}$	0.85
0	1
$\dfrac{\pi}{4}$	1.17
$\dfrac{\pi}{2}$	1.37
π	1.87
$\dfrac{3\pi}{2}$	2.57
2π	3.51

Figure 32
$r = e^{\theta/5}$

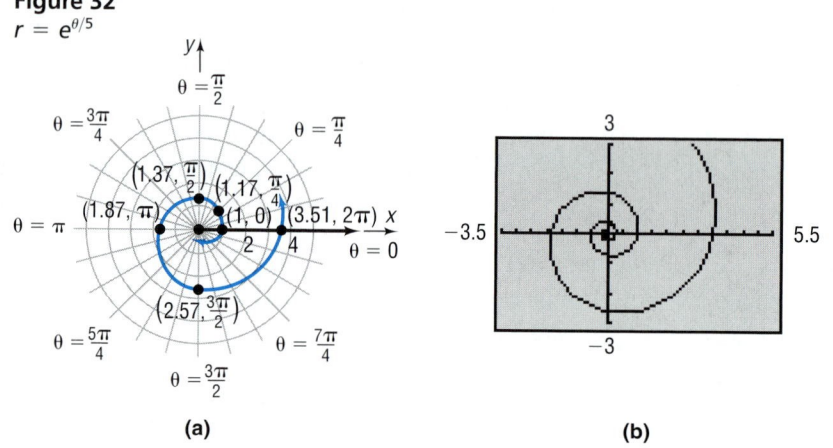

(a)   (b)

The curve in Figure 32 is called a **logarithmic spiral**, since its equation may be written as $\theta = 5 \ln r$ and it spirals infinitely both toward the pole and away from it.

Classification of Polar Equations

The equations of some lines and circles in polar coordinates and their corresponding equations in rectangular coordinates are given in Table 7 on pages 853–854. Also included are the names and the graphs of a few of the more frequently encountered polar equations.

TABLE 7

Lines

Description	Line passing through the pole making an angle α with the polar axis	Vertical line	Horizontal line
Rectangular equation	$y = (\tan \alpha)x$	$x = a$	$y = b$
Polar equation	$\theta = \alpha$	$r\cos \theta = a$	$r\sin \theta = b$
Typical graph	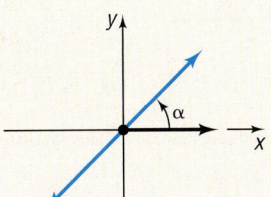	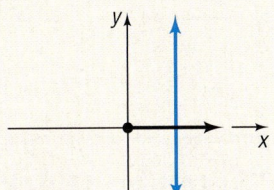	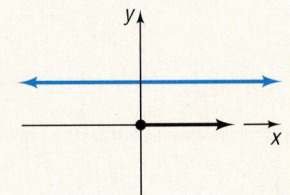

Circles

Description	Center at the pole, radius a	Passing through the pole, tangent to the y-axis center on the polar x-axis, radius a	Passing through the pole, tangent to the x-axis, center on the y-axis, radius a
Rectangular equation	$x^2 + y^2 = a^2, \quad a > 0$	$x^2 + y^2 = \pm 2ax, \quad a > 0$	$x^2 + y^2 = \pm 2ay, \quad a > 0$
Polar equation	$r = a, \quad a > 0$	$r = \pm 2a\cos \theta, \quad a > 0$	$r = \pm 2a\sin \theta, \quad a > 0$
Typical graph	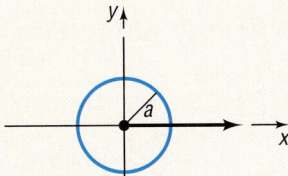	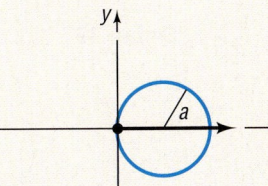	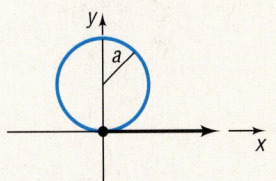

Other Equations

Name	Cardioid	Limaçon without inner loop	Limaçon with inner loop
Polar equations	$r = a \pm a\cos \theta, \quad a > 0$ $r = a \pm a\sin \theta, \quad a > 0$	$r = a \pm b\cos \theta, \quad 0 < b < a$ $r = a \pm b\sin \theta, \quad 0 < b < a$	$r = a \pm b\cos \theta, \quad 0 < a < b$ $r = a \pm b\sin \theta, \quad 0 < a < b$
Typical graph	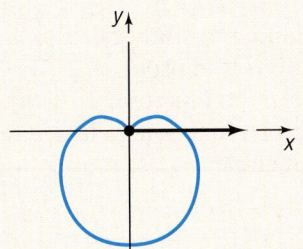	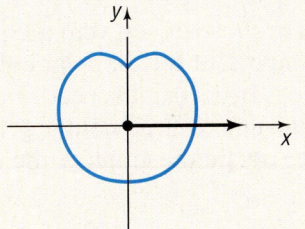	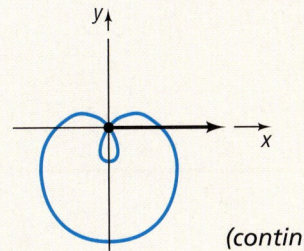

(continued)

TABLE 7 (continued)

Other Equations

Name	Lemniscate	Rose with three petals	Rose with four petals
Polar equations	$r^2 = a^2 \cos(2\theta)$, $a > 0$ $r^2 = a^2 \sin(2\theta)$, $a > 0$	$r = a \sin(3\theta)$, $a > 0$ $r = a \cos(3\theta)$, $a > 0$	$r = a \sin(2\theta)$, $a > 0$ $r = a \cos(2\theta)$, $a > 0$
Typical graph	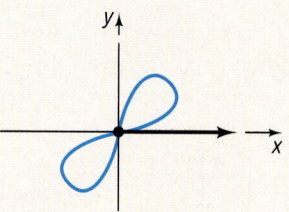	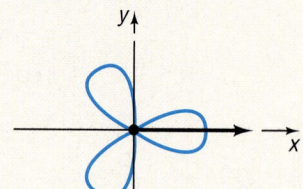	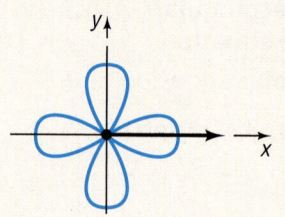

Sketching Quickly

If a polar equation only involves a sine (or cosine) function, you can quickly obtain a sketch of its graph by making use of Table 7, periodicity, and a short table.

EXAMPLE 14   Sketching the Graph of a Polar Equation Quickly by Hand

Graph the equation: $r = 2 + 2\sin\theta$

Solution   We recognize the polar equation: Its graph is a cardioid. The period of $\sin\theta$ is 2π, so we form a table using $0 \le \theta \le 2\pi$, compute r, plot the points (r, θ), and sketch the graph of a cardioid as θ varies from 0 to 2π. See Table 8 and Figure 33.

TABLE 8

θ	$r = 2 + 2\sin\theta$
0	$2 + 2(0) = 2$
$\dfrac{\pi}{2}$	$2 + 2(1) = 4$
π	$2 + 2(0) = 2$
$\dfrac{3\pi}{2}$	$2 + 2(-1) = 0$
2π	$2 + 2(0) = 2$

Figure 33

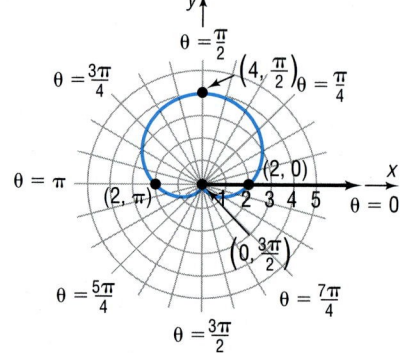

Calculus Comment

For those of you who are planning to study calculus, a comment about one important role of polar equations is in order.

In rectangular coordinates, the equation $x^2 + y^2 = 1$, whose graph is the unit circle, is not the graph of a function. In fact, it requires two functions to obtain the graph of the unit circle:

$$y_1 = \sqrt{1 - x^2} \qquad \text{Upper semicircle}$$

$$y_2 = -\sqrt{1 - x^2} \qquad \text{Lower semicircle}$$

In polar coordinates, the equation $r = 1$, whose graph is also the unit circle, does define a function. That is, for each choice of θ there is only one corresponding value of r, namely, $r = 1$. Since many uses of calculus require that functions be used, the opportunity to express nonfunctions in rectangular coordinates as functions in polar coordinates becomes extremely useful.

Note also that the vertical-line test for functions is valid only for equations in rectangular coordinates.

HISTORICAL FEATURE

Jakob Bernoulli
1654–1705

Polar coordinates seem to have been invented by Jakob Bernoulli (1654–1705) in about 1691, although, as with most such ideas, earlier traces of the notion exist. Early users of calculus remained committed to rectangular coordinates, and polar coordinates did not become widely used until the early 1800s. Even then, it was mostly geometers who used them for describing odd curves. Finally, about the mid-1800s, applied mathematicians realized the tremendous simplification that polar coordinates make possible in the description of objects with circular or cylindrical symmetry. From then on their use became widespread.

11.2 Concepts and Vocabulary

In Problems 1–3, fill in the blanks.

1. An equation whose variables are polar coordinates is called a _____ _____.

2. Using polar coordinates (r, θ), the circle $x^2 + y^2 = 2x$ takes the form _____.

3. A polar equation is symmetric with respect to the pole if an equivalent equation results when r is replaced by _____.

In Problems 4–6, answer True or False to each statement.

4. The tests for symmetry in polar coordinates are necessary, but not sufficient.

5. The graph of a cardioid never passes through the pole.

6. All polar equations have a symmetric feature.

7. What is the graph of $r = 4$?

8. What is the graph of $\theta = \dfrac{\pi}{6}$?

9. What is the shape of a cardioid?

10. What is the shape of a lemniscate?

11.2 Exercises

In Problems 1–16, transform each polar equation to an equation in rectangular coordinates. Then identify and graph the equation. Verify your graph using a graphing utility.

1. $r = 4$
2. $r = 2$
3. $\theta = \dfrac{\pi}{3}$
4. $\theta = -\dfrac{\pi}{4}$

5. $r \sin \theta = 4$
6. $r \cos \theta = 4$
7. $r \cos \theta = -2$
8. $r \sin \theta = -2$

9. $r = 2 \cos \theta$
10. $r = 2 \sin \theta$
11. $r = -4 \sin \theta$
12. $r = -4 \cos \theta$

13. $r \sec \theta = 4$
14. $r \csc \theta = 8$
15. $r \csc \theta = -2$
16. $r \sec \theta = -4$

In Problems 17–24, match each of the graphs (A) through (H) to one of the following polar equations.

17. $r = 2$

18. $\theta = \dfrac{\pi}{4}$

19. $r = 2\cos\theta$

20. $r\cos\theta = 2$

21. $r = 1 + \cos\theta$

22. $r = 2\sin\theta$

23. $\theta = \dfrac{3\pi}{4}$

24. $r\sin\theta = 2$

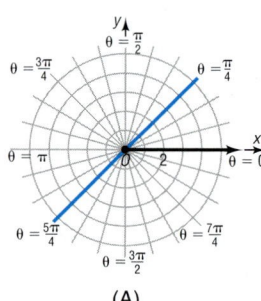
(A)

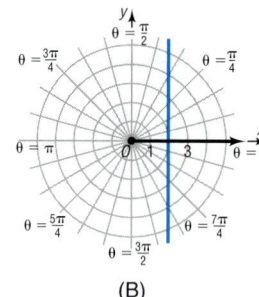
(B)

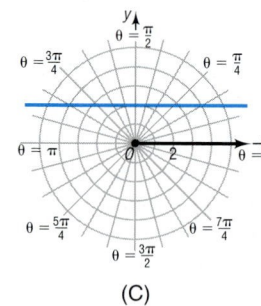
(C)

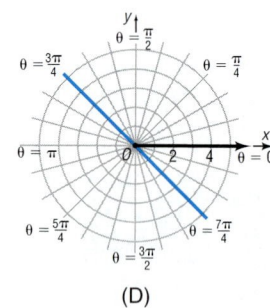
(D)

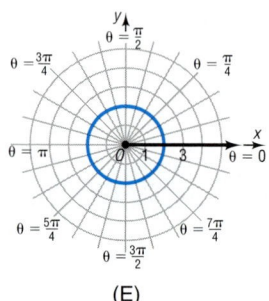
(E)

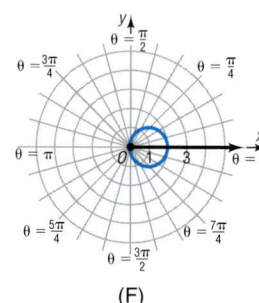
(F)

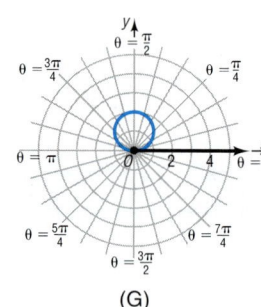
(G)

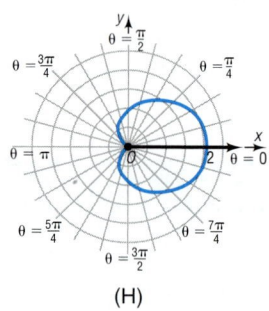
(H)

In Problems 25–30, match each of the graphs (A) through (F) to one of the following polar equations.

25. $r = 4$

26. $r = 3\cos\theta$

27. $r = 3\sin\theta$

28. $r\sin\theta = 3$

29. $r\cos\theta = 3$

30. $r = 2 + \sin\theta$

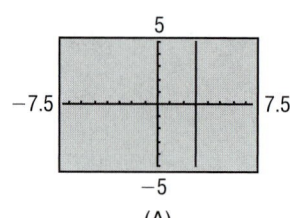
(A)

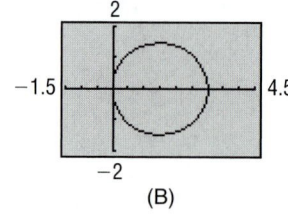
(B)

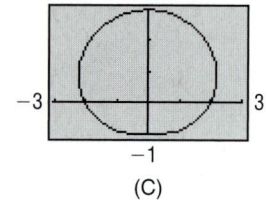
(C)

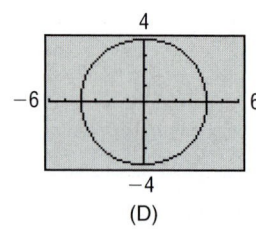
(D)

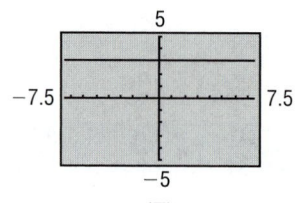
(E)

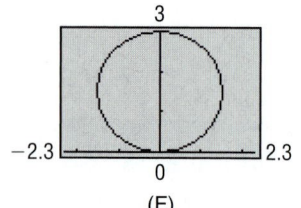
(F)

In Problems 31–54, identify and graph each polar equation. Verify your graph using a graphing utility.

31. $r = 2 + 2\cos\theta$

32. $r = 1 + \sin\theta$

33. $r = 3 - 3\sin\theta$

34. $r = 2 - 2\cos\theta$

35. $r = 2 + \sin\theta$

36. $r = 2 - \cos\theta$

37. $r = 4 - 2\cos\theta$

38. $r = 4 + 2\sin\theta$

39. $r = 1 + 2\sin\theta$

40. $r = 1 - 2\sin\theta$

41. $r = 2 - 3\cos\theta$

42. $r = 2 + 4\cos\theta$

43. $r = 3\cos(2\theta)$

44. $r = 2\sin(3\theta)$

45. $r = 4\sin(5\theta)$

46. $r = 3\cos(4\theta)$

47. $r^2 = 9\cos(2\theta)$ **48.** $r^2 = \sin(2\theta)$ **49.** $r = 2^\theta$ **50.** $r = 3^\theta$

51. $r = 1 - \cos\theta$ **52.** $r = 3 + \cos\theta$ **53.** $r = 1 - 3\cos\theta$ **54.** $r = 4\cos(3\theta)$

In Problems 55–64, graph each polar equation. Verify your graph using a graphing utility.

55. $r = \dfrac{2}{1 - \cos\theta}$ (parabola)

56. $r = \dfrac{2}{1 - 2\cos\theta}$ (hyperbola)

57. $r = \dfrac{1}{3 - 2\cos\theta}$ (ellipse)

58. $r = \dfrac{1}{1 - \cos\theta}$ (parabola)

59. $r = \theta, \quad \theta \geq 0$ (spiral of Archimedes)

60. $r = \dfrac{3}{\theta}$ (reciprocal spiral)

61. $r = \csc\theta - 2, \quad 0 < \theta < \pi$ (conchoid)

62. $r = \sin\theta\tan\theta$ (cissoid)

63. $r = \tan\theta, \quad -\dfrac{\pi}{2} < \theta < \dfrac{\pi}{2},$ (kappa curve)

64. $r = \cos\dfrac{\theta}{2}$

---

65. Show that the graph of the equation $r\sin\theta = a$ is a horizontal line a units above the pole if $a > 0$ and $|a|$ units below the pole if $a < 0$.

66. Show that the graph of the equation $r\cos\theta = a$ is a vertical line a units to the right of the pole if $a > 0$ and $|a|$ units to the left of the pole if $a < 0$.

67. Show that the graph of the equation $r = 2a\sin\theta, a > 0$, is a circle of radius a with center at $(0, a)$ in rectangular coordinates.

68. Show that the graph of the equation $r = -2a\sin\theta, a > 0$, is a circle of radius a with center at $(0, -a)$ in rectangular coordinates.

69. Show that the graph of the equation $r = 2a\cos\theta, a > 0$, is a circle of radius a with center at $(a, 0)$ in rectangular coordinates.

70. Show that the graph of the equation $r = -2a\cos\theta$, $a > 0$, is a circle of radius a with center at $(-a, 0)$ in rectangular coordinates.

71. Explain why the following test for symmetry is valid: Replace r by $-r$ and θ by $-\theta$ in a polar equation. If an equivalent equation results, the graph is symmetric with respect to the line $\theta = \dfrac{\pi}{2}$ (y-axis).

 (a) Show that the test on page 846 fails for $r^2 = \cos\theta$, but this new test works.

 (b) Show that the test on page 846 works for $r^2 = \sin\theta$, yet this new test fails.

72. Develop a new test for symmetry with respect to the pole.

 (a) Find a polar equation for which this new test fails, yet the test on page 846 works.

 (b) Find a polar equation for which the test on page 846 fails yet the new test works.

73. Write down two different tests for symmetry with respect to the polar axis. Find examples in which one test works and the other fails. Which test do you prefer to use? Justify your answer.

PREPARING FOR THIS SECTION

Before getting started, review the following:

✓ Complex Numbers (Section 5.3, pp. 398–403)

✓ Definitions of the Sine and Cosine Functions (Section 8.4, p. 644)

✓ Value of the Sine and Cosine Functions at Certain Angles (Section 8.3, p. 638, and Section 8.4, p. 646)

✓ Sum and Difference Formulas for Sine and Cosine (Section 9.4, pp. 734 and 737)

11.3 THE COMPLEX PLANE; DE MOIVRE'S THEOREM

OBJECTIVES

 1. Convert a Complex Number from Rectangular Form to Polar Form

 2. Plot Points in the Complex Plane

 3. Find Products and Quotients of Complex Numbers in Polar Form

 4. Use De Moivre's Theorem

 5. Find Complex Roots

When we first introduced complex numbers, we were not prepared to give a geometric interpretation of a complex number. Now we are ready. Although we could give several interpretations, the one that follows is the easiest to understand.

A complex number $z = x + yi$ can be interpreted geometrically as the point (x, y) in the xy-plane. Each point in the plane corresponds to a complex number and, conversely, each complex number corresponds to a point in the plane. We shall refer to the collection of such points as the **complex plane**. The x-axis will be referred to as the **real axis**, because any point that lies on the real axis is of the form $z = x + 0i = x$, a real number. The y-axis is called the **imaginary axis**, because any point that lies on it is of the form $z = 0 + yi = yi$, a pure imaginary number. See Figure 34.

Figure 34
Complex plane

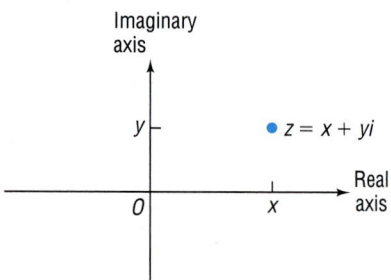

Let $z = x + yi$ be a complex number. The **magnitude** or **modulus** of z, denoted by $|z|$, is defined as the distance from the origin to the point (x, y). That is,

$$|z| = \sqrt{x^2 + y^2} \qquad (1)$$

Figure 35

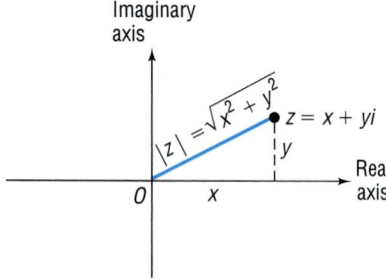

See Figure 35 for an illustration.

This definition for $|z|$ is consistent with the definition for the absolute value of a real number: If $z = x + yi$ is real, then $z = x + 0i$ and

$$|z| = \sqrt{x^2 + 0^2} = \sqrt{x^2} = |x|$$

For this reason, the magnitude of z is sometimes called the absolute value of z.

Recall (Section 5.3) that if $z = x + yi$ then its **conjugate**, denoted by $\bar{z}$, is $\bar{z} = x - yi$. Because $z\bar{z} = x^2 + y^2$, it follows from equation (1) that the magnitude of z can be written as

$$|z| = \sqrt{z\bar{z}} \qquad (2)$$

Polar Form of a Complex Number

When a complex number is written in the standard form $z = x + yi$, we say that it is in **rectangular**, or **Cartesian, form** because (x, y) are the rectangular coordinates of the corresponding point in the complex plane. Suppose that (r, θ) are the polar coordinates of this point. Then

$$x = r \cos \theta \qquad y = r \sin \theta \qquad (3)$$

Figure 36

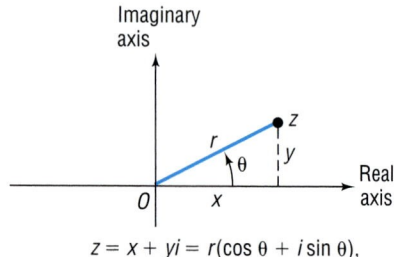

$z = x + yi = r(\cos \theta + i \sin \theta)$,
$r \geq 0, 0 \leq \theta < 2\pi$

If $r \geq 0$ and $0 \leq \theta < 2\pi$, the complex number $z = x + yi$ may be written in **polar form** as

$$z = x + yi = (r \cos \theta) + (r \sin \theta)i = r(\cos \theta + i \sin \theta) \qquad (4)$$

See Figure 36.

If $z = r(\cos \theta + i \sin \theta)$ is the polar form of a complex number, the angle $\theta, 0 \leq \theta < 2\pi$, is called the **argument of z**.

Also, because $r \geq 0$, we have $r = \sqrt{x^2 + y^2}$. From equation (1) it follows that the magnitude of $z = r(\cos \theta + i \sin \theta)$ is

$$|z| = r$$

EXAMPLE 1 | **Plotting a Point in the Complex Plane and Writing a Complex Number in Polar Form**

② Plot the point corresponding to $z = \sqrt{3} - i$ in the complex plane, and write an expression for z in polar form.

Solution The point corresponding to $z = \sqrt{3} - i$ has the rectangular coordinates $(\sqrt{3}, -1)$. The point, located in quadrant IV, is plotted in Figure 37. Because $x = \sqrt{3}$ and $y = -1$, it follows that

$$r = \sqrt{x^2 + y^2} = \sqrt{(\sqrt{3})^2 + (-1)^2} = \sqrt{4} = 2$$

and

$$\sin \theta = \frac{y}{r} = \frac{-1}{2}, \qquad \cos \theta = \frac{x}{r} = \frac{\sqrt{3}}{2}, \qquad 0 \leq \theta < 2\pi$$

Then $\theta = \dfrac{11\pi}{6}$ and $r = 2$, so the polar form of $z = \sqrt{3} - i$ is

$$z = r(\cos \theta + i \sin \theta) = 2\left(\cos \frac{11\pi}{6} + i \sin \frac{11\pi}{6} \right)$$

Figure 37

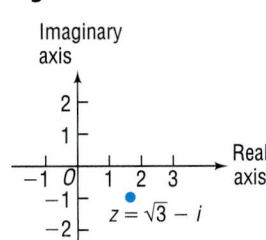

➤ NOW WORK PROBLEM 1.

EXAMPLE 2 | **Plotting a Point in the Complex Plane and Converting from Polar to Rectangular Form**

Plot the point corresponding to $z = 2(\cos 30° + i \sin 30°)$ in the complex plane, and write an expression for z in rectangular form.

Solution To plot the complex number $z = 2(\cos 30° + i \sin 30°)$, we plot the point whose polar coordinates are $(r, \theta) = (2, 30°)$, as shown in Figure 38. In rectangular form,

$$z = 2(\cos 30° + i \sin 30°) = 2\left(\frac{\sqrt{3}}{2} + \frac{1}{2}i \right) = \sqrt{3} + i$$

Figure 38

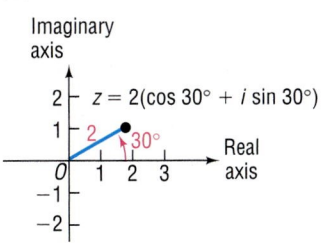

➤ NOW WORK PROBLEM 13.

③ The polar form of a complex number provides an alternative method for finding products and quotients of complex numbers.

Theorem   Let $z_1 = r_1(\cos \theta_1 + i \sin \theta_1)$ and $z_2 = r_2(\cos \theta_2 + i \sin \theta_2)$ be two complex numbers. Then

$$z_1 z_2 = r_1 r_2 [\cos(\theta_1 + \theta_2) + i \sin(\theta_1 + \theta_2)] \qquad (5)$$

If $z_2 \neq 0$, then

$$\frac{z_1}{z_2} = \frac{r_1}{r_2} [\cos(\theta_1 - \theta_2) + i \sin(\theta_1 - \theta_2)] \qquad (6)$$

Proof   We will prove formula (5). The proof of formula (6) is left as an exercise (see Problem 56).

$$\begin{aligned} z_1 z_2 &= [r_1(\cos \theta_1 + i \sin \theta_1)][r_2(\cos \theta_2 + i \sin \theta_2)] \\ &= r_1 r_2 [(\cos \theta_1 + i \sin \theta_1)(\cos \theta_2 + i \sin \theta_2)] \\ &= r_1 r_2 [(\cos \theta_1 \cos \theta_2 - \sin \theta_1 \sin \theta_2) + i(\sin \theta_1 \cos \theta_2 + \cos \theta_1 \sin \theta_2)] \\ &= r_1 r_2 [\cos(\theta_1 + \theta_2) + i \sin(\theta_1 + \theta_2)] \end{aligned}$$

Because the magnitude of a complex number z is r and its argument is θ, when $z = r(\cos \theta + i \sin \theta)$, we can restate this theorem as follows:

Theorem   The magnitude of the product (quotient) of two complex numbers equals the product (quotient) of their magnitudes; the argument of the product (quotient) of two complex numbers is determined by the sum (difference) of their arguments.

Let's look at an example of how this theorem can be used.

EXAMPLE 3   **Finding Products and Quotients of Complex Numbers in Polar Form**

If $z = 3(\cos 20° + i \sin 20°)$ and $w = 5(\cos 100° + i \sin 100°)$, find the following (leave your answers in polar form):

(a) zw (b) $\dfrac{z}{w}$

Solution   (a) $\begin{aligned} zw &= [3(\cos 20° + i \sin 20°)][5(\cos 100° + i \sin 100°)] \\ &= (3 \cdot 5)[\cos(20° + 100°) + i \sin(20° + 100°)] \\ &= 15(\cos 120° + i \sin 120°) \end{aligned}$

(b) $\begin{aligned} \frac{z}{w} &= \frac{3(\cos 20° + i \sin 20°)}{5(\cos 100° + i \sin 100°)} \\ &= \frac{3}{5}[\cos(20° - 100°) + i \sin(20° - 100°)] \\ &= \frac{3}{5}[\cos(-80°) + i \sin(-80°)] \\ &= \frac{3}{5}(\cos 280° + i \sin 280°) \quad \text{Argument must lie between } 0° \text{ and } 360°. \end{aligned}$

▸ **NOW WORK PROBLEM 23.**

De Moivre's Theorem

De Moivre's Theorem, stated by Abraham De Moivre (1667–1754) in 1730, but already known to many people by 1710, is important for the following reason: The fundamental processes of algebra are the four operations of addition, subtraction, multiplication, and division, together with powers and the extraction of roots. De Moivre's Theorem allows these latter fundamental algebraic operations to be applied to complex numbers.

De Moivre's Theorem, in its most basic form, is a formula for raising a complex number z to the power n, where $n \geq 1$ is a positive integer. Let's see if we can guess the form of the result.

Let $z = r(\cos \theta + i \sin \theta)$ be a complex number. Then, based on equation (5), we have

$$n = 2: \quad z^2 = r^2[\cos(2\theta) + i \sin(2\theta)] \qquad \text{Equation (5)}$$

$$n = 3: \quad z^3 = z^2 \cdot z$$
$$= \{r^2[\cos(2\theta) + i \sin(2\theta)]\}[r(\cos \theta + i \sin \theta)]$$
$$= r^3[\cos(3\theta) + i \sin(3\theta)] \qquad \text{Equation (5)}$$

$$n = 4: \quad z^4 = z^3 \cdot z$$
$$= \{r^3[\cos(3\theta) + i \sin(3\theta)]\}[r(\cos \theta + i \sin \theta)]$$
$$= r^4[\cos(4\theta) + i \sin(4\theta)] \qquad \text{Equation (5)}$$

The pattern should now be clear.

Theorem

De Moivre's Theorem

If $z = r(\cos \theta + i \sin \theta)$ is a complex number, then

$$z^n = r^n[\cos(n\theta) + i \sin(n\theta)] \qquad (7)$$

where $n \geq 1$ is a positive integer.

We will not prove De Moivre's Theorem because the proof requires mathematical induction (which is not discussed until Section 13.4).

Let's look at some examples.

EXAMPLE 4   **Using De Moivre's Theorem**

Write $[2(\cos 20° + i \sin 20°)]^3$ in the standard form $a + bi$.

Solution
$$[2(\cos 20° + i \sin 20°)]^3 = 2^3[\cos(3 \cdot 20°) + i \sin(3 \cdot 20°)]$$
$$= 8(\cos 60° + i \sin 60°)$$
$$= 8\left(\frac{1}{2} + \frac{\sqrt{3}}{2}i\right) = 4 + 4\sqrt{3}i$$

NOW WORK PROBLEM 31.

| EXAMPLE 5 | **Using De Moivre's Theorem** |

Write $(1 + i)^5$ in the standard form $a + bi$.

Algebraic Solution To apply De Moivre's Theorem, we must first write the complex number in polar form. Since the magnitude of $1 + i$ is $\sqrt{1^2 + 1^2} = \sqrt{2}$, we begin by writing

$$1 + i = \sqrt{2}\left(\frac{1}{\sqrt{2}} + \frac{1}{\sqrt{2}}i\right) = \sqrt{2}\left(\cos\frac{\pi}{4} + i\sin\frac{\pi}{4}\right)$$

Now

$$\begin{aligned}
(1 + i)^5 &= \left[\sqrt{2}\left(\cos\frac{\pi}{4} + i\sin\frac{\pi}{4}\right)\right]^5 \\
&= (\sqrt{2})^5\left[\cos\left(5\cdot\frac{\pi}{4}\right) + i\sin\left(5\cdot\frac{\pi}{4}\right)\right] \\
&= 4\sqrt{2}\left(\cos\frac{5\pi}{4} + i\sin\frac{5\pi}{4}\right) \\
&= 4\sqrt{2}\left[-\frac{1}{\sqrt{2}} + \left(-\frac{1}{\sqrt{2}}\right)i\right] = -4 - 4i
\end{aligned}$$

Graphing Solution Using a TI-83 graphing calculator, we obtain the solution shown in Figure 39.

Figure 39

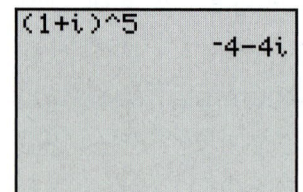

Complex Roots

⑤ Let w be a given complex number, and let $n \geq 2$ denote a positive integer. Any complex number z that satisfies the equation

$$z^n = w$$

is called a **complex nth root** of w. In keeping with previous usage, if $n = 2$, the solutions of the equation $z^2 = w$ are called **complex square roots** of w, and if $n = 3$, the solutions of the equation $z^3 = w$ are called **complex cube roots** of w.

Theorem **Finding Complex Roots**

Let $w = r(\cos\theta_0 + i\sin\theta_0)$ be a complex number and let $n \geq 2$ be an integer. If $w \neq 0$, there are n distinct complex roots of w, given by the formula

$$z_k = \sqrt[n]{r}\left[\cos\left(\frac{\theta_0}{n} + \frac{2k\pi}{n}\right) + i\sin\left(\frac{\theta_0}{n} + \frac{2k\pi}{n}\right)\right] \qquad (8)$$

where $k = 0, 1, 2, \ldots, n - 1$.

Proof (Outline) We will not prove this result in its entirety. Instead, we shall show only that each z_k in equation (8) satisfies the equation $z_k^n = w$, proving that each z_k is a complex nth root of w.

$$z_k^n = \left\{ \sqrt[n]{r}\left[\cos\left(\frac{\theta_0}{n} + \frac{2k\pi}{n}\right) + i\sin\left(\frac{\theta_0}{n} + \frac{2k\pi}{n}\right)\right]\right\}^n$$

$$= \left(\sqrt[n]{r}\right)^n\left\{\cos\left[n\left(\frac{\theta_0}{n} + \frac{2k\pi}{n}\right)\right] + i\sin\left[n\left(\frac{\theta_0}{n} + \frac{2k\pi}{n}\right)\right]\right\} \qquad \text{De Moivre's Theorem}$$

$$= r\left[\cos(\theta_0 + 2k\pi) + i\sin(\theta_0 + 2k\pi)\right]$$

$$= r\left(\cos\theta_0 + i\sin\theta_0\right) = w \qquad \text{Periodic Property}$$

So, each z_k, $k = 0, 1, \ldots, n - 1$, is a complex nth root of w. To complete the proof, we would need to show that each z_k, $k = 0, 1, \ldots, n - 1$, is, in fact, distinct and that there are no complex nth roots of w other than those given by equation (8). ◼

EXAMPLE 6 **Finding Complex Cube Roots**

Find the complex cube roots of $-1 + \sqrt{3}i$. Leave your answers in polar form, with θ in degrees.

Solution First, we express $-1 + \sqrt{3}i$ in polar form using degrees.

$$-1 + \sqrt{3}i = 2\left(-\frac{1}{2} + \frac{\sqrt{3}}{2}i\right) = 2(\cos 120° + i\sin 120°)$$

So, $r = 2$ and $\theta_0 = 120°$. The three complex cube roots of $-1 + \sqrt{3}i = 2(\cos 120° + i\sin 120°)$ are

$$z_k = \sqrt[3]{2}\left[\cos\left(\frac{120°}{3} + \frac{360°k}{3}\right) + i\sin\left(\frac{120°}{3} + \frac{360°k}{3}\right)\right], \qquad k = 0, 1, 2$$

$$= \sqrt[3]{2}\left[\cos(40° + 120°k) + i\sin(40° + 120°k)\right], \qquad k = 0, 1, 2$$

So,

$$z_0 = \sqrt[3]{2}\left[\cos(40° + 120° \cdot 0) + i\sin(40° + 120° \cdot 0)\right] = \sqrt[3]{2}\left(\cos 40° + i\sin 40°\right)$$

$$z_1 = \sqrt[3]{2}\left[\cos(40° + 120° \cdot 1) + i\sin(40° + 120° \cdot 1)\right] = \sqrt[3]{2}\left(\cos 160° + i\sin 160°\right)$$

$$z_2 = \sqrt[3]{2}\left[\cos(40° + 120° \cdot 2) + i\sin(40° + 120° \cdot 2)\right] = \sqrt[3]{2}\left(\cos 280° + i\sin 280°\right) ▪$$

WARNING: Most graphing utilities will only provide the answer z_0 to the calculation $(-1 + \sqrt{3}i)^{\wedge}\left(\frac{1}{3}\right)$. The following paragraph explains how to obtain z_1 and z_2 from z_0. ◼

Notice that each of the three complex roots of $-1 + \sqrt{3}i$ has the same magnitude, $\sqrt[3]{2}$. This means that the points corresponding to each cube root lie the same distance from the origin; that is, the three points lie on a circle with center at the origin and radius $\sqrt[3]{2}$. Furthermore, the arguments of these cube roots are $40°$, $160°$, and $280°$, the difference of consecutive pairs being

$120° = \dfrac{360°}{3}$. This means that the three points are equally spaced on the circle, as shown in Figure 40. These results are not coincidental. In fact, you are asked to show that these results hold for complex nth roots in Problems 53 through 55.

Figure 40

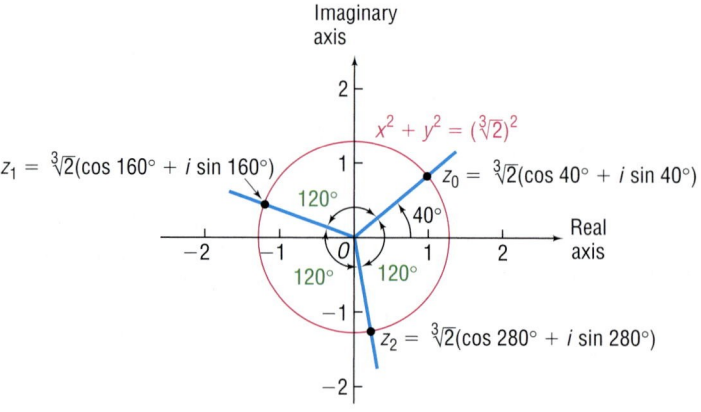

NOW WORK PROBLEM 43.

HISTORICAL FEATURE

John Wallis

The Babylonians, Greeks, and Arabs considered square roots of negative quantities to be impossible and equations with complex solutions to be unsolvable. The first hint that there was some connection between real solutions of equations and complex numbers came when Girolamo Cardano (1501–1576) and Tartaglia (1499–1557) found *real* roots of cubic equations by taking cube roots of *complex* quantities. For centuries thereafter, mathematicians worked with complex numbers without much belief in their actual existence. In 1673, John Wallis appears to have been the first to suggest the graphical representation of complex numbers, a truly significant idea that was not pursued further until about 1800. Several people, including Karl Friedrich Gauss (1777–1855), then rediscovered the idea, and the graphical representation helped to establish complex numbers as equal members of the number family. In practical applications, complex numbers have found their greatest uses in the area of alternating current, where they are a commonplace tool, and in the area of subatomic physics.

HISTORICAL PROBLEMS

1. The quadratic formula will work perfectly well if the coefficients are complex numbers. Solve the following using De Moivre's Theorem where necessary.
   [**Hint:** The answers are "nice."]

   (a) $z^2 - (2 + 5i)z - 3 + 5i = 0$
   (b) $z^2 - (1 + i)z - 2 - i = 0$

11.3 Concepts and Vocabulary

In Problems 1–3, fill in the blanks.

1. When a complex number z is written in the polar form $z = r(\cos\theta + i\sin\theta)$, the nonnegative number r is the _____ or _____ of z, and the angle θ, $0 \le \theta < 2\pi$, is the _____ of z.

2. _____ Theorem can be used to raise a complex number to a power.

3. A complex number will, in general, have _____ cube roots.

In Problems 4–6, answer True or False to each statement.

4. De Moivre's Theorem is useful for raising a complex number to a positive integer power.

5. Using De Moivre's Theorem, the square of a complex number will have two answers.

6. The polar form of a complex number is unique.

7. Explain how to multiply two complex numbers given in polar form.

8. If $z = r(\cos\theta + i\sin\theta)$, use De Moivre's Theorem to write z^2.

9. If $z = r(\cos\theta + i\sin\theta)$, write down the two complex square roots of z.

11.3 Exercises

In Problems 1–12, plot each complex number in the complex plane and write it in polar form. Express the argument in degrees.

1. $1 + i$ **2.** $-1 + i$ **3.** $\sqrt{3} - i$ **4.** $1 - \sqrt{3}i$

5. $-3i$ **6.** -2 **7.** $4 - 4i$ **8.** $9\sqrt{3} + 9i$

9. $3 - 4i$ **10.** $2 + \sqrt{3}i$ **11.** $-2 + 3i$ **12.** $\sqrt{5} - i$

In Problems 13–22, write each complex number in rectangular form.

13. $2(\cos 120° + i\sin 120°)$ **14.** $3(\cos 210° + i\sin 210°)$ **15.** $4\left(\cos\dfrac{7\pi}{4} + i\sin\dfrac{7\pi}{4}\right)$

16. $2\left(\cos\dfrac{5\pi}{6} + i\sin\dfrac{5\pi}{6}\right)$ **17.** $3\left(\cos\dfrac{3\pi}{2} + i\sin\dfrac{3\pi}{2}\right)$ **18.** $4\left(\cos\dfrac{\pi}{2} + i\sin\dfrac{\pi}{2}\right)$

19. $0.2(\cos 100° + i\sin 100°)$ **20.** $0.4(\cos 200° + i\sin 200°)$ **21.** $2\left(\cos\dfrac{\pi}{18} + i\sin\dfrac{\pi}{18}\right)$

22. $3\left(\cos\dfrac{\pi}{10} + i\sin\dfrac{\pi}{10}\right)$

In Problems 23–30, find zw and $\dfrac{z}{w}$. Leave your answers in polar form.

23. $z = 2(\cos 40° + i\sin 40°)$
$\quad w = 4(\cos 20° + i\sin 20°)$

24. $z = \cos 120° + i\sin 120°$
$\quad w = \cos 100° + i\sin 100°$

25. $z = 3(\cos 130° + i\sin 130°)$
$\quad w = 4(\cos 270° + i\sin 270°)$

26. $z = 2(\cos 80° + i\sin 80°)$
$\quad w = 6(\cos 200° + i\sin 200°)$

27. $z = 2\left(\cos\dfrac{\pi}{8} + i\sin\dfrac{\pi}{8}\right)$
$\quad w = 2\left(\cos\dfrac{\pi}{10} + i\sin\dfrac{\pi}{10}\right)$

28. $z = 4\left(\cos\dfrac{3\pi}{8} + i\sin\dfrac{3\pi}{8}\right)$
$\quad w = 2\left(\cos\dfrac{9\pi}{16} + i\sin\dfrac{9\pi}{16}\right)$

29. $z = 2 + 2i$
$\quad w = \sqrt{3} - i$

30. $z = 1 - i$
$\quad w = 1 - \sqrt{3}i$

In Problems 31–42, write each expression in the standard form $a + bi$. Verify your answer using a graphing utility.

31. $[4(\cos 40° + i\sin 40°)]^3$ **32.** $[3(\cos 80° + i\sin 80°)]^3$ **33.** $\left[2\left(\cos\dfrac{\pi}{10} + i\sin\dfrac{\pi}{10}\right)\right]^5$

34. $\left[\sqrt{2}\left(\cos\dfrac{5\pi}{16} + i\sin\dfrac{5\pi}{16}\right)\right]^4$ **35.** $[\sqrt{3}(\cos 10° + i\sin 10°)]^6$ **36.** $\left[\dfrac{1}{2}(\cos 72° + i\sin 72°)\right]^5$

37. $\left[\sqrt{5}\left(\cos\dfrac{3\pi}{16} + i\sin\dfrac{3\pi}{16}\right)\right]^4$ **38.** $\left[\sqrt{3}\left(\cos\dfrac{5\pi}{18} + i\sin\dfrac{5\pi}{18}\right)\right]^6$ **39.** $(1 - i)^5$

40. $(\sqrt{3} - i)^6$ **41.** $(\sqrt{2} - i)^6$ **42.** $(1 - \sqrt{5}i)^8$

In Problems 43–50, find all the complex roots. Leave your answers in polar form with the argument in degrees.

43. The complex cube roots of $1 + i$

44. The complex fourth roots of $\sqrt{3} - i$

45. The complex fourth roots of $4 - 4\sqrt{3}i$

46. The complex cube roots of $-8 - 8i$

47. The complex fourth roots of $-16i$

48. The complex cube roots of -8

49. The complex fifth roots of i

50. The complex fifth roots of $-i$

---

51. Find the four complex fourth roots of unity (1) and plot each.

52. Find the six complex sixth roots of unity (1) and plot each.

53. Show that each complex nth root of a nonzero complex number w has the same magnitude.

54. Use the result of Problem 53 to draw the conclusion that each complex nth root lies on a circle with center at the origin. What is the radius of this circle?

55. Refer to Problem 54. Show that the complex nth roots of a nonzero complex number w are equally spaced on the circle.

56. Prove formula (6).

PREPARING FOR THIS SECTION

Before getting started, review the following:

✓ Rectangular Coordinates (Section 1.1, pp. 90–91)

✓ Pythagorean Theorem (Section R.3, p. 25)

11.4 VECTORS

OBJECTIVES

1. Graph Vectors
2. Find a Position Vector
3. Add and Subtract Vectors
4. Find a Scalar Product and the Magnitude of a Vector
5. Find a Unit Vector
6. Find a Vector from Its Direction and Magnitude
7. Work with Objects in Static Equilibrium

In simple terms, a **vector** (derived from the Latin *vehere*, meaning "to carry") is a quantity that has both magnitude and direction. It is customary to represent a vector by using an arrow. The length of the arrow represents the **magnitude** of the vector, and the arrowhead indicates the **direction** of the vector.

Many quantities in physics can be represented by vectors. For example, the velocity of an aircraft can be represented by an arrow that points in the direction of movement; the length of the arrow represents speed. If the aircraft speeds up, we lengthen the arrow; if the aircraft changes direction, we introduce an arrow in the new direction. See Figure 41. Based on this representation, it is not surprising that vectors and directed line segments are somehow related.

Figure 41

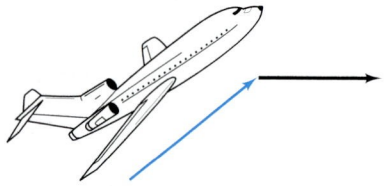

Geometric Vectors

If P and Q are two distinct points in the *xy*-plane, there is exactly one line containing both P and Q [Figure 42(a)]. The points on that part of the line that joins P to Q, including P and Q, form what is called the **line segment** $\overline{PQ}$ [Figure 42(b)]. If we order the points so that they proceed from P to Q, we have a **directed line segment** from P to Q, or a **geometric vector**, which we

Figure 42

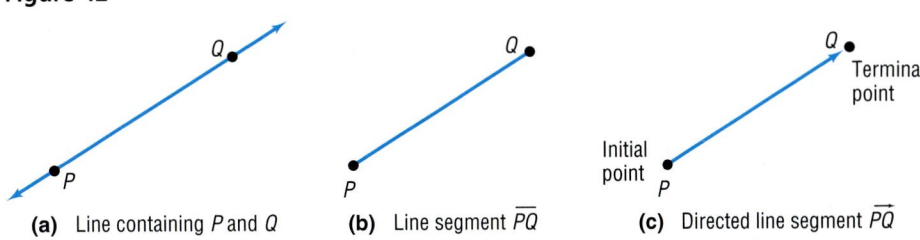

(a) Line containing P and Q (b) Line segment $\overline{PQ}$ (c) Directed line segment $\overrightarrow{PQ}$

denote by $\overrightarrow{PQ}$. In a directed line segment $\overrightarrow{PQ}$, we call P the **initial point** and Q the **terminal point**, as indicated in Figure 42(c).

The magnitude of the directed line segment $\overrightarrow{PQ}$ is the distance from the point P to the point Q; that is, it is the length of the line segment. The direction of $\overrightarrow{PQ}$ is from P to Q. If a vector $\mathbf{v}^*$ has the same magnitude and the same direction as the directed line segment $\overrightarrow{PQ}$, we write

$$\mathbf{v} = \overrightarrow{PQ}$$

The vector $\mathbf{v}$ whose magnitude is 0 is called the **zero vector, 0**. The zero vector is assigned no direction.

Two vectors $\mathbf{v}$ and $\mathbf{w}$ are **equal**, written

$$\mathbf{v} = \mathbf{w}$$

if they have the same magnitude and the same direction.

For example, the vectors shown in Figure 43 have the same magnitude and the same direction, so they are equal, even though they have different initial points and different terminal points. As a result, we find it useful to think of a vector simply as an arrow, keeping in mind that two arrows (vectors) are equal if they have the same direction and the same magnitude (length).

Figure 43

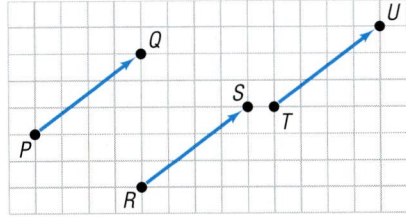

Adding Vectors

Figure 44

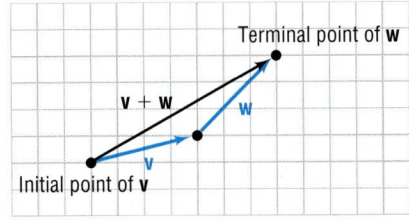

The **sum $\mathbf{v} + \mathbf{w}$** of two vectors is defined as follows: We position the vectors $\mathbf{v}$ and $\mathbf{w}$ so that the terminal point of $\mathbf{v}$ coincides with the initial point of $\mathbf{w}$, as shown in Figure 44. The vector $\mathbf{v} + \mathbf{w}$ is then the unique vector whose initial point coincides with the initial point of $\mathbf{v}$ and whose terminal point coincides with the terminal point of $\mathbf{w}$.

Vector addition is **commutative**. That is, if $\mathbf{v}$ and $\mathbf{w}$ are any two vectors, then

$$\mathbf{v} + \mathbf{w} = \mathbf{w} + \mathbf{v}$$

Figure 45 illustrates this fact. (Observe that the commutative property is another way of saying that opposite sides of a parallelogram are equal and parallel.)

Vector addition is also **associative**. That is, if $\mathbf{u}, \mathbf{v}$, and $\mathbf{w}$ are vectors, then

$$\mathbf{u} + (\mathbf{v} + \mathbf{w}) = (\mathbf{u} + \mathbf{v}) + \mathbf{w}$$

Figure 45

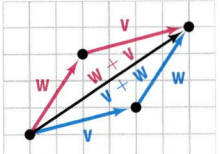

*Boldface letters will be used to denote vectors, in order to distinguish them from numbers. For handwritten work, an arrow is placed over the letter to signify a vector.

Figure 46
(u + v) + w = u + (v + w)

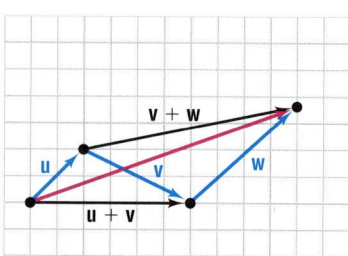

Figure 47

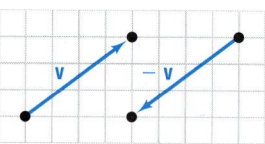

Figure 48

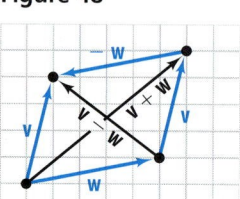

Figure 46 illustrates the associative property for vectors.

The zero vector has the property that

$$\mathbf{v} + \mathbf{0} = \mathbf{0} + \mathbf{v} = \mathbf{v}$$

for any vector **v**.

If **v** is a vector, then −**v** is the vector having the same magnitude as **v**, but whose direction is opposite to **v**, as shown in Figure 47.

Furthermore,

$$\mathbf{v} + (-\mathbf{v}) = \mathbf{0}$$

If **v** and **w** are two vectors, we define the **difference v − w** as

$$\mathbf{v} - \mathbf{w} = \mathbf{v} + (-\mathbf{w})$$

Figure 48 illustrates the relationships among **v**, **w**, **v** + **w**, and **v** − **w**.

Multiplying Vectors by Numbers

When dealing with vectors, we refer to real numbers as **scalars**. Scalars are quantities that have only magnitude. Examples from physics of scalar quantities are temperature, speed, and time. We now define how to multiply a vector by a scalar.

> If α is a scalar and **v** is a vector, the **scalar product** $\alpha\mathbf{v}$ is defined as follows:
>
> 1. If $\alpha > 0$, the product $\alpha\mathbf{v}$ is the vector whose magnitude is α times the magnitude of **v** and whose direction is the same as **v**.
> 2. If $\alpha < 0$, the product $\alpha\mathbf{v}$ is the vector whose magnitude is $|\alpha|$ times the magnitude of **v** and whose direction is opposite that of **v**.
> 3. If $\alpha = 0$ or if $\mathbf{v} = \mathbf{0}$, then $\alpha\mathbf{v} = \mathbf{0}$.

Figure 49

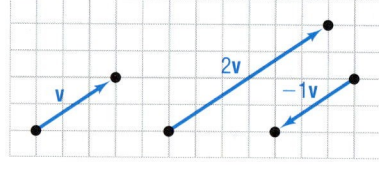

See Figure 49 for some illustrations.

For example, if **a** is the acceleration of an object of mass m due to a force **F** being exerted on it, then, by Newton's second law of motion, **F** = m**a**. Here, m**a** is the product of the scalar m and the vector **a**.

Scalar products have the following properties:

$$0\mathbf{v} = \mathbf{0} \qquad 1\mathbf{v} = \mathbf{v} \qquad -1\mathbf{v} = -\mathbf{v}$$
$$(\alpha + \beta)\mathbf{v} = \alpha\mathbf{v} + \beta\mathbf{v} \qquad \alpha(\mathbf{v} + \mathbf{w}) = \alpha\mathbf{v} + \alpha\mathbf{w}$$
$$\alpha(\beta\mathbf{v}) = (\alpha\beta)\mathbf{v}$$

EXAMPLE 1 **Graphing Vectors**

Figure 50

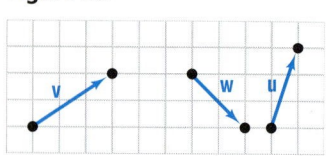

Use the vectors illustrated in Figure 50 to graph each of the following vectors:

(a) **v** − **w**   (b) 2**v** + 3**w**   (c) 2**v** − **w** + **u**

Solution Figure 51 illustrates each graph.

Figure 51

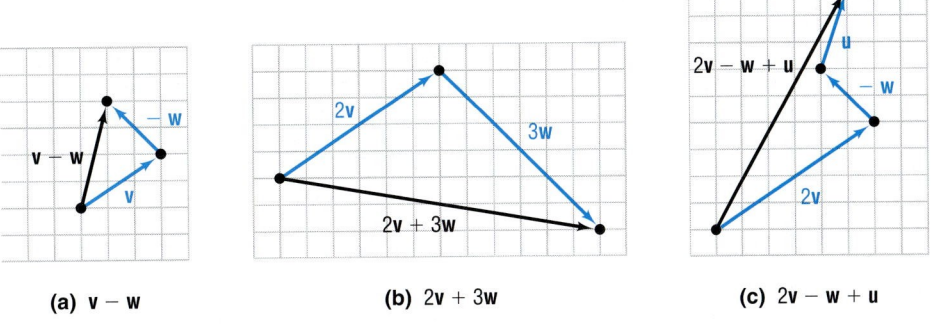

(a) **v** − **w**   (b) 2**v** + 3**w**   (c) 2**v** − **w** + **u**

NOW WORK PROBLEMS **1** AND **3**.

Magnitudes of Vectors

If **v** is a vector, we use the symbol $\|\mathbf{v}\|$ to represent the **magnitude** of **v**. Since $\|\mathbf{v}\|$ equals the length of a directed line segment, it follows that $\|\mathbf{v}\|$ has the following properties:

Theorem

> **Properties of** $\|\mathbf{v}\|$
>
> If **v** is a vector and if α is a scalar, then
>
> (a) $\|\mathbf{v}\| \geq 0$ (b) $\|\mathbf{v}\| = 0$ if and only if $\mathbf{v} = \mathbf{0}$
>
> (c) $\|-\mathbf{v}\| = \|\mathbf{v}\|$ (d) $\|\alpha\mathbf{v}\| = |\alpha|\|\mathbf{v}\|$

Property (a) is a consequence of the fact that distance is a nonnegative number. Property (b) follows, because the length of the directed line segment $\overrightarrow{PQ}$ is positive unless P and Q are the same point, in which case the length is 0. Property (c) follows because the length of the line segment $\overline{PQ}$ equals the length of the line segment $\overline{QP}$. Property (d) is a direct consequence of the definition of a scalar product.

> A vector **u** for which $\|\mathbf{u}\| = 1$ is called a **unit vector**.

To compute the magnitude and direction of a vector, we need an algebraic way of representing vectors.

Algebraic Vectors

② An **algebraic vector v** is represented as

$$\mathbf{v} = \langle a, b \rangle$$

where a and b are real numbers (scalars) called the **components** of the vector **v**.

We use a rectangular coordinate system to represent algebraic vectors in the plane. If $\mathbf{v} = \langle a, b \rangle$ is an algebraic vector whose initial point is at the origin, then **v** is called a **position vector**. See Figure 52. Notice that the terminal point of the position vector $\mathbf{v} = \langle a, b \rangle$ is $P = (a, b)$.

Figure 52

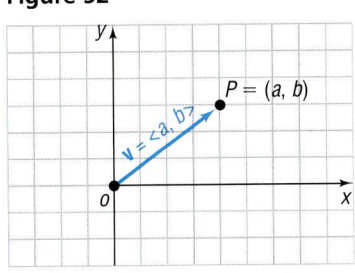

The next result states that any vector whose initial point is not at the origin is equal to a unique position vector.

Theorem

Suppose that $\mathbf{v}$ is a vector with initial point $P_1 = (x_1, y_1)$, not necessarily the origin, and terminal point $P_2 = (x_2, y_2)$. If $\mathbf{v} = \overrightarrow{P_1 P_2}$, then $\mathbf{v}$ is equal to the position vector

$$\mathbf{v} = \langle x_2 - x_1, y_2 - y_1 \rangle \qquad (1)$$

To see why this is true, look at Figure 53. Triangle OPA and triangle $P_1 P_2 Q$ are congruent. (Do you see why? The line segments have the same magnitude, so $d(O, P) = d(P_1, P_2)$; and they have the same direction, so $\angle POA = \angle P_2 P_1 Q$. Since the triangles are right triangles, we have angle-side-angle.) It follows that corresponding sides are equal. As a result, $x_2 - x_1 = a$ and $y_2 - y_1 = b$, so $\mathbf{v}$ may be written as

$$\mathbf{v} = \langle a, b \rangle = \langle x_2 - x_1, y_2 - y_1 \rangle$$

Figure 53

$\langle a, b \rangle = \langle x_2 - x_1, y_2 - y_1 \rangle$

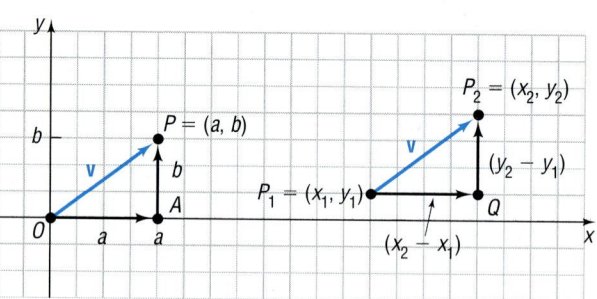

Because of this result, we can replace any algebraic vector by a unique position vector, and vice versa. This flexibility is one of the main reasons for the wide use of vectors. Unless otherwise specified, from now on the term *vector* will mean the unique position vector equal to it.

EXAMPLE 2

Finding a Position Vector

Figure 54

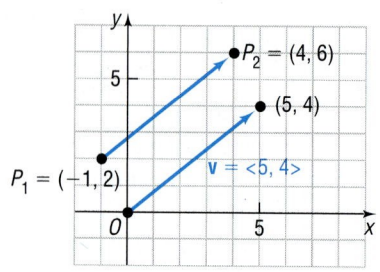

Find the position vector of the vector $\mathbf{v} = \overrightarrow{P_1 P_2}$ if $P_1 = (-1, 2)$ and $P_2 = (4, 6)$.

Solution   By equation (1), the position vector equal to $\mathbf{v}$ is

$$\mathbf{v} = \langle 4 - (-1), 6 - 2 \rangle = \langle 5, 4 \rangle$$

See Figure 54.

Two position vectors $\mathbf{v}$ and $\mathbf{w}$ are equal if and only if the terminal point of $\mathbf{v}$ is the same as the terminal point of $\mathbf{w}$. This leads to the following result:

Theorem

Equality of Vectors

Two vectors **v** and **w** are equal if and only if their corresponding components are equal. That is,

> If $\mathbf{v} = \langle a_1, b_1 \rangle$ and $\mathbf{w} = \langle a_2, b_2 \rangle$
> then $\mathbf{v} = \mathbf{w}$ if and only if $a_1 = a_2$ and $b_1 = b_2$.

Figure 55

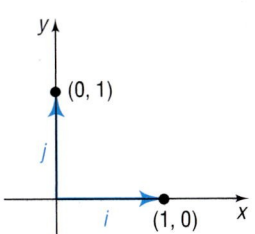

We now present an alternative representation of a vector in the plane that is common in the physical sciences. Let **i** denote the unit vector whose direction is along the positive x-axis; let **j** denote the unit vector whose direction is along the positive y-axis. Then $\mathbf{i} = \langle 1, 0 \rangle$ and $\mathbf{j} = \langle 0, 1 \rangle$, as shown in Figure 55. Any vector $\mathbf{v} = \langle a, b \rangle$ can be written using the unit vectors **i** and **j** as follows:

$$\mathbf{v} = \langle a, b \rangle = a\langle 1, 0 \rangle + b\langle 0, 1 \rangle = a\mathbf{i} + b\mathbf{j}$$

We call a and b the **horizontal** and **vertical components** of **v**, respectively.

↪ **NOW WORK PROBLEM 21.**

We define addition, subtraction, scalar product, and magnitude in terms of the components of a vector.

> Let $\mathbf{v} = a_1\mathbf{i} + b_1\mathbf{j} = \langle a_1, b_1 \rangle$ and $\mathbf{w} = a_2\mathbf{i} + b_2\mathbf{j} = \langle a_2, b_2 \rangle$ be two vectors, and let α be a scalar. Then
>
> $$\mathbf{v} + \mathbf{w} = (a_1 + a_2)\mathbf{i} + (b_1 + b_2)\mathbf{j} = \langle a_1 + a_2, b_1 + b_2 \rangle \qquad (2)$$
> $$\mathbf{v} - \mathbf{w} = (a_1 - a_2)\mathbf{i} + (b_1 - b_2)\mathbf{j} = \langle a_1 - a_2, b_1 - b_2 \rangle \qquad (3)$$
> $$\alpha\mathbf{v} = (\alpha a_1)\mathbf{i} + (\alpha b_1)\mathbf{j} = \langle \alpha a_1, \alpha b_1 \rangle \qquad (4)$$
> $$\|\mathbf{v}\| = \sqrt{a_1^2 + b_1^2} \qquad (5)$$

These definitions are compatible with the geometric definitions given earlier in this section. See Figure 56.

Figure 56

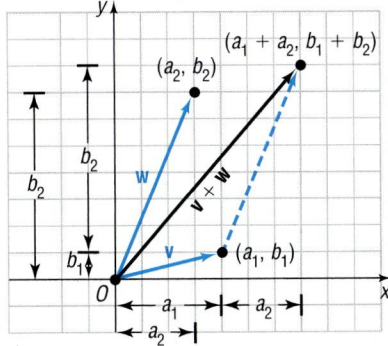

(a) Illustration of property (2)

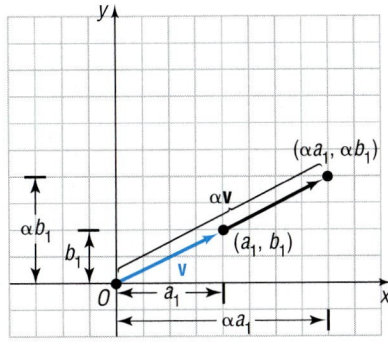

(b) Illustration of property (4), $\alpha > 0$

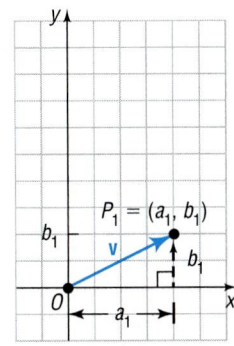

(c) Illustration of property (5):
$\| \mathbf{v} \| =$ Distance from O to P_1
$\| \mathbf{v} \| = \sqrt{a_1^2 + b_1^2}$

To add two vectors, add corresponding components. To subtract two vectors, subtract corresponding components.

EXAMPLE 3   Adding and Subtracting Vectors

③   If $\mathbf{v} = 2\mathbf{i} + 3\mathbf{j} = \langle 2, 3 \rangle$ and $\mathbf{w} = 3\mathbf{i} - 4\mathbf{j} = \langle 3, -4 \rangle$, find:

(a) $\mathbf{v} + \mathbf{w}$ (b) $\mathbf{v} - \mathbf{w}$

Solution   (a) $\mathbf{v} + \mathbf{w} = (2\mathbf{i} + 3\mathbf{j}) + (3\mathbf{i} - 4\mathbf{j}) = (2 + 3)\mathbf{i} + (3 - 4)\mathbf{j} = 5\mathbf{i} - \mathbf{j}$
or
$\mathbf{v} + \mathbf{w} = \langle 2, 3 \rangle + \langle 3, -4 \rangle = \langle 2 + 3, 3 + (-4) \rangle = \langle 5, -1 \rangle$

(b) $\mathbf{v} - \mathbf{w} = (2\mathbf{i} + 3\mathbf{j}) - (3\mathbf{i} - 4\mathbf{j}) = (2 - 3)\mathbf{i} + [3 - (-4)]\mathbf{j} = -\mathbf{i} + 7\mathbf{j}$
or
$\mathbf{v} - \mathbf{w} = \langle 2, 3 \rangle - \langle 3, -4 \rangle = \langle 2 - 3, 3 - (-4) \rangle = \langle -1, 7 \rangle$ ■

EXAMPLE 4   Finding Scalar Products and Magnitudes

④   If $\mathbf{v} = 2\mathbf{i} + 3\mathbf{j} = \langle 2, 3 \rangle$ and $\mathbf{w} = 3\mathbf{i} - 4\mathbf{j} = \langle 3, -4 \rangle$, find:

(a) $3\mathbf{v}$ (b) $2\mathbf{v} - 3\mathbf{w}$ (c) $\|\mathbf{v}\|$

Solution   (a) $3\mathbf{v} = 3(2\mathbf{i} + 3\mathbf{j}) = 6\mathbf{i} + 9\mathbf{j}$
or
$3\mathbf{v} = 3\langle 2, 3 \rangle = \langle 6, 9 \rangle$

(b) $2\mathbf{v} - 3\mathbf{w} = 2(2\mathbf{i} + 3\mathbf{j}) - 3(3\mathbf{i} - 4\mathbf{j}) = 4\mathbf{i} + 6\mathbf{j} - 9\mathbf{i} + 12\mathbf{j}$
$= -5\mathbf{i} + 18\mathbf{j}$
or
$2\mathbf{v} - 3\mathbf{w} = 2\langle 2, 3 \rangle - 3\langle 3, -4 \rangle = \langle 4, 6 \rangle - \langle 9, -12 \rangle$
$= \langle 4 - 9, 6 - (-12) \rangle = \langle -5, 18 \rangle$

(c) $\|\mathbf{v}\| = \|2\mathbf{i} + 3\mathbf{j}\| = \sqrt{2^2 + 3^2} = \sqrt{13}$ ■

✏ **NOW WORK PROBLEMS 27 AND 33.**

For the remainder of the section, we will express a vector $\mathbf{v}$ in the form $a\mathbf{i} + b\mathbf{j}$.

⑤   Recall that a unit vector $\mathbf{u}$ is a vector for which $\|\mathbf{u}\| = 1$. In many applications, it is useful to be able to find a unit vector $\mathbf{u}$ that has the same direction as a given vector $\mathbf{v}$.

Theorem

Unit Vector in the Direction of v

For any nonzero vector $\mathbf{v}$, the vector

$$\mathbf{u} = \frac{\mathbf{v}}{\|\mathbf{v}\|}$$

is a unit vector that has the same direction as $\mathbf{v}$.

Proof Let $\mathbf{v} = a\mathbf{i} + b\mathbf{j}$. Then $\|\mathbf{v}\| = \sqrt{a^2 + b^2}$ and

$$\mathbf{u} = \frac{\mathbf{v}}{\|\mathbf{v}\|} = \frac{a\mathbf{i} + b\mathbf{j}}{\sqrt{a^2 + b^2}} = \frac{a}{\sqrt{a^2 + b^2}}\mathbf{i} + \frac{b}{\sqrt{a^2 + b^2}}\mathbf{j}$$

The vector $\mathbf{u}$ is in the same direction as $\mathbf{v}$, since $\|\mathbf{v}\| > 0$. Furthermore,

$$\|\mathbf{u}\| = \sqrt{\frac{a^2}{a^2 + b^2} + \frac{b^2}{a^2 + b^2}} = \sqrt{\frac{a^2 + b^2}{a^2 + b^2}} = 1$$

Thus, $\mathbf{u}$ is a unit vector in the direction of $\mathbf{v}$.

As a consequence of this theorem, if $\mathbf{u}$ is a unit vector in the same direction as a vector $\mathbf{v}$, then $\mathbf{v}$ may be expressed as

$$\mathbf{v} = \|\mathbf{v}\|\mathbf{u} \tag{6}$$

This way of expressing a vector is useful in many applications.

EXAMPLE 5 **Finding a Unit Vector**

Find a unit vector in the same direction as $\mathbf{v} = 4\mathbf{i} - 3\mathbf{j}$.

Solution We find $\|\mathbf{v}\|$ first.

$$\|\mathbf{v}\| = \|4\mathbf{i} - 3\mathbf{j}\| = \sqrt{16 + 9} = 5$$

Now we multiply $\mathbf{v}$ by the scalar $\dfrac{1}{\|\mathbf{v}\|} = \dfrac{1}{5}$. A unit vector in the same direction as $\mathbf{v}$ is

$$\frac{\mathbf{v}}{\|\mathbf{v}\|} = \frac{4\mathbf{i} - 3\mathbf{j}}{5} = \frac{4}{5}\mathbf{i} - \frac{3}{5}\mathbf{j}$$

✔ CHECK: This vector is, in fact, a unit vector because

$$\left(\frac{4}{5}\right)^2 + \left(-\frac{3}{5}\right)^2 = \frac{16}{25} + \frac{9}{25} = \frac{25}{25} = 1$$

━ NOW WORK PROBLEM **43**.

Writing a Vector in Terms of Its Magnitude and Direction

⑥ If a vector represents the speed and direction of an object, it is called a **velocity vector**. If a vector represents the direction and amount of a force acting on an object, it is called a **force vector**. In many applications, a vector is described in terms of its magnitude and direction, rather than in terms of its components. For example, a ball thrown with an initial speed of 25 miles per hour at an angle 30° to the horizontal is a velocity vector.

Suppose that we are given the magnitude $\|\mathbf{v}\|$ of a nonzero vector $\mathbf{v}$ and the angle $\alpha, 0° \leq \alpha < 360°$, between $\mathbf{v}$ and $\mathbf{i}$. To express $\mathbf{v}$ in terms of $\|\mathbf{v}\|$ and α, we first find the unit vector $\mathbf{u}$ having the same direction as $\mathbf{v}$.

$$\mathbf{u} = \frac{\mathbf{v}}{\|\mathbf{v}\|} \quad \text{or} \quad \mathbf{v} = \|\mathbf{v}\|\mathbf{u} \tag{7}$$

Figure 57

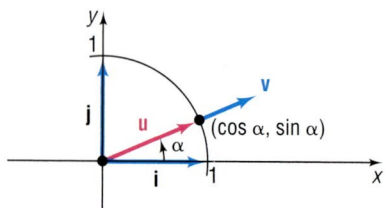

Look at Figure 57. The coordinates of the terminal point of $\mathbf{u}$ are $(\cos \alpha, \sin \alpha)$. Then $\mathbf{u} = \cos \alpha\mathbf{i} + \sin \alpha\mathbf{j}$ and, from (7),

$$\mathbf{v} = \|\mathbf{v}\|(\cos \alpha\mathbf{i} + \sin \alpha\mathbf{j}) \tag{8}$$

where α is the angle between $\mathbf{v}$ and $\mathbf{i}$.

EXAMPLE 6   Writing a Vector When Its Magnitude and Direction Are Given

A ball is thrown with an initial speed of 25 miles per hour in a direction that makes an angle of $30°$ with the positive x-axis. Express the velocity vector $\mathbf{v}$ in terms of $\mathbf{i}$ and $\mathbf{j}$. What is the initial speed in the horizontal direction? What is the initial speed in the vertical direction?

Solution   The magnitude of $\mathbf{v}$ is $\|\mathbf{v}\| = 25$ miles per hour, and the angle between the direction of $\mathbf{v}$ and $\mathbf{i}$, the positive x-axis, is $\alpha = 30°$. By equation (8),

$$\mathbf{v} = \|\mathbf{v}\|(\cos \alpha\mathbf{i} + \sin \alpha\mathbf{j}) = 25(\cos 30°\mathbf{i} + \sin 30°\mathbf{j}) = 25\left(\frac{\sqrt{3}}{2}\mathbf{i} + \frac{1}{2}\mathbf{j}\right) = \frac{25\sqrt{3}}{2}\mathbf{i} + \frac{25}{2}\mathbf{j}$$

The initial speed of the ball in the horizontal direction is the horizontal component of $\mathbf{v}$, $\frac{25\sqrt{3}}{2} \approx 21.65$ miles per hour. The initial speed in the vertical direction is the vertical component of $\mathbf{v}$, $\frac{25}{2} = 12.5$ miles per hour.

Application: Static Equilibrium

Because forces can be represented by vectors, two forces "combine" the way that vectors "add." If $\mathbf{F}_1$ and $\mathbf{F}_2$ are two forces simultaneously acting on an object, the vector sum $\mathbf{F}_1 + \mathbf{F}_2$ is the **resultant force**. The resultant force produces the same effect on the object as that obtained when the two forces $\mathbf{F}_1$ and $\mathbf{F}_2$ act on the object. See Figure 58. An application of this concept is *static equilibrium*. An object is said to be in **static equilibrium** if (1) the object is at rest and (2) the sum of all forces acting on the object is zero, that is, if the resultant force is 0.

Figure 58

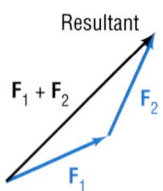

EXAMPLE 7   An Object in Static Equilibrium

A box of supplies that weighs 1200 pounds is suspended by two cables attached to the ceiling, as shown in Figure 59. What is the tension in the two cables?

Figure 59

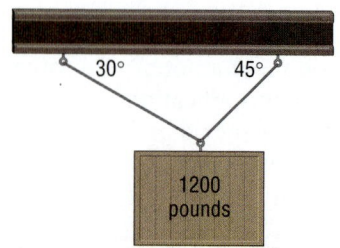

Solution   We draw a force diagram with the vectors drawn as shown in Figure 60. The tensions in the cables are the magnitudes $\|\mathbf{F}_1\|$ and $\|\mathbf{F}_2\|$ of the force vectors $\mathbf{F}_1$ and $\mathbf{F}_2$. The magnitude of the force vector $\mathbf{F}_3$ equals 1200 pounds, the weight of the box. Now write each force vector in terms of the unit vectors $\mathbf{i}$ and $\mathbf{j}$. For $\mathbf{F}_1$ and $\mathbf{F}_2$, we use equation (8). Remember that α is the angle between the vector and the positive x-axis.

Figure 60

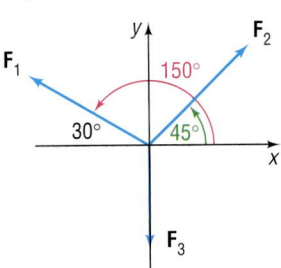

$$\mathbf{F}_1 = \|\mathbf{F}_1\|(\cos 150°\mathbf{i} + \sin 150°\mathbf{j}) = \|\mathbf{F}_1\|\left(-\frac{\sqrt{3}}{2}\mathbf{i} + \frac{1}{2}\mathbf{j}\right) = -\frac{\sqrt{3}}{2}\|\mathbf{F}_1\|\mathbf{i} + \frac{1}{2}\|\mathbf{F}_1\|\mathbf{j}$$

$$\mathbf{F}_2 = \|\mathbf{F}_2\|(\cos 45°\mathbf{i} + \sin 45°\mathbf{j}) = \|\mathbf{F}_2\|\left(\frac{\sqrt{2}}{2}\mathbf{i} + \frac{\sqrt{2}}{2}\mathbf{j}\right) = \frac{\sqrt{2}}{2}\|\mathbf{F}_2\|\mathbf{i} + \frac{\sqrt{2}}{2}\|\mathbf{F}_2\|\mathbf{j}$$

$$\mathbf{F}_3 = -1200\mathbf{j}$$

For static equilibrium, the sum of the force vectors must equal zero.

$$\mathbf{F}_1 + \mathbf{F}_2 + \mathbf{F}_3 = -\frac{\sqrt{3}}{2}\|\mathbf{F}_1\|\mathbf{i} + \frac{1}{2}\|\mathbf{F}_1\|\mathbf{j} + \frac{\sqrt{2}}{2}\|\mathbf{F}_2\|\mathbf{i} + \frac{\sqrt{2}}{2}\|\mathbf{F}_2\|\mathbf{j} - 1200\mathbf{j} = \mathbf{0}$$

The **i** component and **j** component will each equal zero. This results in the two equations

$$-\frac{\sqrt{3}}{2}\|\mathbf{F}_1\| + \frac{\sqrt{2}}{2}\|\mathbf{F}_2\| = 0 \tag{9}$$

$$\frac{1}{2}\|\mathbf{F}_1\| + \frac{\sqrt{2}}{2}\|\mathbf{F}_2\| - 1200 = 0 \tag{10}$$

We solve equation (9) for $\|\mathbf{F}_2\|$ and obtain

$$\|\mathbf{F}_2\| = \frac{\sqrt{3}}{\sqrt{2}}\|\mathbf{F}_1\| \tag{11}$$

Substituting into equation (10) and solving for $\|\mathbf{F}_1\|$, we obtain

$$\frac{1}{2}\|\mathbf{F}_1\| + \frac{\sqrt{2}}{2}\left(\frac{\sqrt{3}}{\sqrt{2}}\|\mathbf{F}_1\|\right) - 1200 = 0$$

$$\frac{1}{2}\|\mathbf{F}_1\| + \frac{\sqrt{3}}{2}\|\mathbf{F}_1\| - 1200 = 0$$

$$\frac{1 + \sqrt{3}}{2}\|\mathbf{F}_1\| = 1200$$

$$\|\mathbf{F}_1\| = \frac{2400}{1 + \sqrt{3}} \approx 878.5 \text{ pounds}$$

Substituting this value into equation (11) yields $\|\mathbf{F}_2\|$.

$$\|\mathbf{F}_2\| = \frac{\sqrt{3}}{\sqrt{2}}\|\mathbf{F}_1\| = \frac{\sqrt{3}}{\sqrt{2}}\frac{2400}{1 + \sqrt{3}} \approx 1075.9 \text{ pounds}$$

The left cable has tension of approximately 878.5 pounds and the right cable has tension of approximately 1075.9 pounds. ■

HISTORICAL FEATURE

Josiah Gibbs
1839–1903

The history of vectors is surprisingly complicated for such a natural concept. In the *xy*-plane, complex numbers do a good job of imitating vectors. About 1840, mathematicians became interested in finding a system that would do for three dimensions what the complex numbers do for two dimensions. Hermann Grassmann (1809–1877), in Germany, and William Rowan Hamilton (1805–1865), in Ireland, both attempted to find solutions.

Hamilton's system was the *quaternions*, which are best thought of as a real number plus a vector, and do for four dimensions what complex numbers do for two dimensions. In this system the order of multiplication matters; that is, **ab** ≠ **ba**. Also, two products of vectors emerged, the scalar (or dot) product and the vector (or cross) product.

Grassmann's abstract style, although easily read today, was almost impenetrable during the previous century, and only a few of his ideas were appreciated. Among those few were the same scalar and vector products that Hamilton had found.

About 1880, the American physicist Josiah Willard Gibbs (1839–1903) worked out an algebra involving only the simplest concepts: the vectors and the two products. He then added some calculus, and the resulting system was simple, flexible, and well adapted to expressing a large number of physical laws. This system remains in use essentially unchanged. Hamilton's and Grassmann's more extensive systems each gave birth to much interesting mathematics, but little of this mathematics is seen at elementary levels.

11.4 Concepts and Vocabulary

In Problems 1–3, fill in the blanks.

1. A vector whose magnitude is 1 is called a(n) _____ vector.
2. The product of a vector by a number is called a _____ product.
3. If $\mathbf{v} = a\mathbf{i} + b\mathbf{j}$, then a is called the _____ component of $\mathbf{v}$ and b is the _____ component of $\mathbf{v}$.

In Problems 4–6, answer True or False to each statement.

4. Vectors are quantities that have magnitude and direction.
5. Force is a physical example of a vector.
6. Mass is a physical example of a vector.

7. Explain in words how to add two vectors drawn on a piece of paper.

11.4 Exercises

In Problems 1–8, use the vectors in the figure at the right to graph each of the following vectors.

1. $\mathbf{v} + \mathbf{w}$
2. $\mathbf{u} + \mathbf{v}$
3. $3\mathbf{v}$
4. $4\mathbf{w}$
5. $\mathbf{v} - \mathbf{w}$
6. $\mathbf{u} - \mathbf{v}$
7. $3\mathbf{v} + \mathbf{u} - 2\mathbf{w}$
8. $2\mathbf{u} - 3\mathbf{v} + \mathbf{w}$

In Problems 9–16, use the figure at the right. Determine whether the given statement is true or false.

9. $\mathbf{A} + \mathbf{B} = \mathbf{F}$
10. $\mathbf{K} + \mathbf{G} = \mathbf{F}$
11. $\mathbf{C} = \mathbf{D} - \mathbf{E} + \mathbf{F}$
12. $\mathbf{G} + \mathbf{H} + \mathbf{E} = \mathbf{D}$
13. $\mathbf{E} + \mathbf{D} = \mathbf{G} + \mathbf{H}$
14. $\mathbf{H} - \mathbf{C} = \mathbf{G} - \mathbf{F}$
15. $\mathbf{A} + \mathbf{B} + \mathbf{K} + \mathbf{G} = 0$
16. $\mathbf{A} + \mathbf{B} + \mathbf{C} + \mathbf{H} + \mathbf{G} = 0$

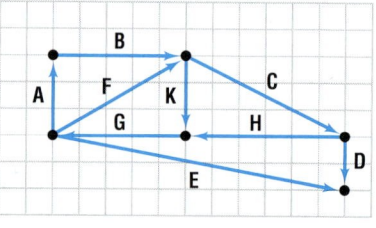

17. If $\|\mathbf{v}\| = 4$, what is $\|3\mathbf{v}\|$?
18. If $\|\mathbf{v}\| = 2$, what is $\|-4\mathbf{v}\|$?

In Problems 19–26, the vector $\mathbf{v}$ has initial point P and terminal point Q. Write $\mathbf{v}$ in the form $a\mathbf{i} + b\mathbf{j}$, that is, find its position vector.

19. $P = (0,0); \quad Q = (3,4)$
20. $P = (0,0); \quad Q = (-3,-5)$
21. $P = (3,2); \quad Q = (5,6)$
22. $P = (-3,2); \quad Q = (6,5)$
23. $P = (-2,-1); \quad Q = (6,-2)$
24. $P = (-1,4); \quad Q = (6,2)$
25. $P = (1,0); \quad Q = (0,1)$
26. $P = (1,1); \quad Q = (2,2)$

In Problems 27–32, find $\|\mathbf{v}\|$.

27. $\mathbf{v} = 3\mathbf{i} - 4\mathbf{j}$
28. $\mathbf{v} = -5\mathbf{i} + 12\mathbf{j}$
29. $\mathbf{v} = \mathbf{i} - \mathbf{j}$
30. $\mathbf{v} = -\mathbf{i} - \mathbf{j}$
31. $\mathbf{v} = -2\mathbf{i} + 3\mathbf{j}$
32. $\mathbf{v} = 6\mathbf{i} + 2\mathbf{j}$

In Problems 33–38, find each quantity if $\mathbf{v} = 3\mathbf{i} - 5\mathbf{j}$ and $\mathbf{w} = -2\mathbf{i} + 3\mathbf{j}$.

33. $2\mathbf{v} + 3\mathbf{w}$ **34.** $3\mathbf{v} - 2\mathbf{w}$ **35.** $\|\mathbf{v} - \mathbf{w}\|$

36. $\|\mathbf{v} + \mathbf{w}\|$ **37.** $\|\mathbf{v}\| - \|\mathbf{w}\|$ **38.** $\|\mathbf{v}\| + \|\mathbf{w}\|$

In Problems 39–44, find the unit vector having the same direction as $\mathbf{v}$.

39. $\mathbf{v} = 5\mathbf{i}$ **40.** $\mathbf{v} = -3\mathbf{j}$ **41.** $\mathbf{v} = 3\mathbf{i} - 4\mathbf{j}$

42. $\mathbf{v} = -5\mathbf{i} + 12\mathbf{j}$ **43.** $\mathbf{v} = \mathbf{i} - \mathbf{j}$ **44.** $\mathbf{v} = 2\mathbf{i} - \mathbf{j}$

45. Find a vector $\mathbf{v}$ whose magnitude is 4 and whose component in the $\mathbf{i}$ direction is twice the component in the $\mathbf{j}$ direction.

46. Find a vector $\mathbf{v}$ whose magnitude is 3 and whose component in the $\mathbf{i}$ direction is equal to the component in the $\mathbf{j}$ direction.

47. If $\mathbf{v} = 2\mathbf{i} - \mathbf{j}$ and $\mathbf{w} = x\mathbf{i} + 3\mathbf{j}$, find all numbers x for which $\|\mathbf{v} + \mathbf{w}\| = 5$.

48. If $P = (-3, 1)$ and $Q = (x, 4)$, find all numbers x such that the vector represented by $\overrightarrow{PQ}$ has length 5.

In Problems 49–54, write the vector $\mathbf{v}$ in the form $a\mathbf{i} + b\mathbf{j}$, given its magnitude $\|\mathbf{v}\|$ and the angle α it makes with the positive x-axis.

49. $\|\mathbf{v}\| = 5, \quad \alpha = 60°$ **50.** $\|\mathbf{v}\| = 8, \quad \alpha = 45°$ **51.** $\|\mathbf{v}\| = 14, \quad \alpha = 120°$

52. $\|\mathbf{v}\| = 3, \quad \alpha = 240°$ **53.** $\|\mathbf{v}\| = 25, \quad \alpha = 330°$ **54.** $\|\mathbf{v}\| = 15, \quad \alpha = 315°$

---

55. A child pulls a wagon with a force of 40 pounds. The handle of the wagon makes an angle of 30° with the ground. Express the force vector $\mathbf{F}$ in terms of $\mathbf{i}$ and $\mathbf{j}$.

56. A man pushes a wheelbarrow up an incline of 20° with a force of 100 pounds. Express the force vector $\mathbf{F}$ in terms of $\mathbf{i}$ and $\mathbf{j}$.

57. Resultant Force Two forces of magnitude 40 newtons (N) and 60 newtons act on an object at angles of 30° and $-45°$ with the positive x-axis as shown in the figure. Find the direction and magnitude of the resultant force; that is, find $\mathbf{F}_1 + \mathbf{F}_2$.

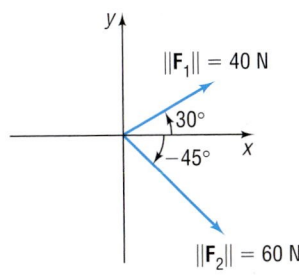

58. Resultant Force Two forces of magnitude 30 newtons (N) and 70 newtons act on an object at angles of 45° and 120° with the positive x-axis as shown in the figure. Find the direction and magnitude of the resultant force; that is, find $\mathbf{F}_1 + \mathbf{F}_2$.

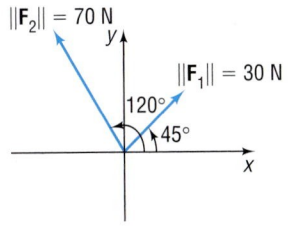

59. Static Equilibrium A weight of 1000 pounds is suspended from two cables as shown in the figure. What is the tension of the two cables?

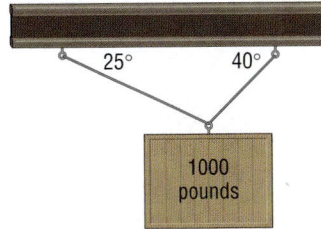

60. Static Equilibrium A weight of 800 pounds is suspended from two cables as shown in the figure. What is the tension of the two cables?

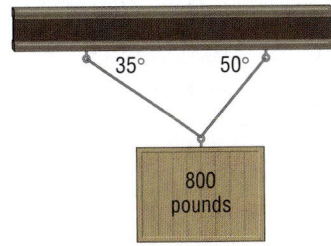

61. Static Equilibrium A tightrope walker located at a certain point deflects the rope as indicated in the figure. If the weight of the tightrope walker is 150 pounds, how much tension is in each part of the rope?

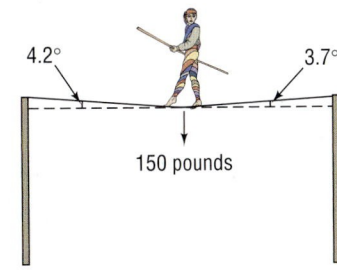

62. Static Equilibrium Repeat Problem 61 if the left angle is 3.8°, the right angle is 2.6°, and the weight of the tightrope walker is 135 pounds.

63. Show on the following graph the force needed for the object at P to be in static equilibrium.

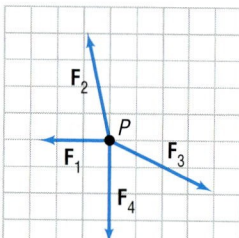

64. Explain in your own words what a vector is. Give an example of a vector.

65. Write a brief paragraph comparing the algebra of complex numbers and the algebra of vectors.

PREPARING FOR THIS SECTION

Before getting started, review the following:

✓ Law of Cosines (Section 10.3, p. 800)

11.5 THE DOT PRODUCT

OBJECTIVES

1. Find the Dot Product of Two Vectors
2. Find the Angle between Two Vectors
3. Determine Whether Two Vectors Are Parallel
4. Determine Whether Two Vectors Are Orthogonal
5. Decompose a Vector into Two Orthogonal Vectors
6. Compute Work

① The definition for a product of two vectors is somewhat unexpected. However, such a product has meaning in many geometric and physical applications.

If $\mathbf{v} = a_1\mathbf{i} + b_1\mathbf{j}$ and $\mathbf{w} = a_2\mathbf{i} + b_2\mathbf{j}$ are two vectors, the **dot product** $\mathbf{v} \cdot \mathbf{w}$ is defined as

$$\mathbf{v} \cdot \mathbf{w} = a_1a_2 + b_1b_2 \tag{1}$$

EXAMPLE 1   Finding Dot Products

If $\mathbf{v} = 2\mathbf{i} - 3\mathbf{j}$ and $\mathbf{w} = 5\mathbf{i} + 3\mathbf{j}$, find:

(a) $\mathbf{v} \cdot \mathbf{w}$ (b) $\mathbf{w} \cdot \mathbf{v}$ (c) $\mathbf{v} \cdot \mathbf{v}$

(d) $\mathbf{w} \cdot \mathbf{w}$ (e) $\|\mathbf{v}\|$ (f) $\|\mathbf{w}\|$

Solution   (a) $\mathbf{v} \cdot \mathbf{w} = 2(5) + (-3)3 = 1$ (b) $\mathbf{w} \cdot \mathbf{v} = 5(2) + 3(-3) = 1$

(c) $\mathbf{v} \cdot \mathbf{v} = 2(2) + (-3)(-3) = 13$ (d) $\mathbf{w} \cdot \mathbf{w} = 5(5) + 3(3) = 34$

(e) $\|\mathbf{v}\| = \sqrt{2^2 + (-3)^2} = \sqrt{13}$ (f) $\|\mathbf{w}\| = \sqrt{5^2 + 3^2} = \sqrt{34}$ ■

Since the dot product $\mathbf{v} \cdot \mathbf{w}$ of two vectors $\mathbf{v}$ and $\mathbf{w}$ is a real number (scalar), we sometimes refer to it as the **scalar product**.

Properties

The results obtained in Example 1 suggest some general properties.

Theorem

Properties of the Dot Product

If $\mathbf{u}, \mathbf{v}$, and $\mathbf{w}$ are vectors, then

Commutative Property

$$\mathbf{u} \cdot \mathbf{v} = \mathbf{v} \cdot \mathbf{u} \tag{2}$$

Distributive Property

$$\mathbf{u} \cdot (\mathbf{v} + \mathbf{w}) = \mathbf{u} \cdot \mathbf{v} + \mathbf{u} \cdot \mathbf{w} \tag{3}$$

$$\mathbf{v} \cdot \mathbf{v} = \|\mathbf{v}\|^2 \tag{4}$$

$$\mathbf{0} \cdot \mathbf{v} = 0 \tag{5}$$

Proof   We will prove properties (2) and (4) here and leave properties (3) and (5) as exercises (see Problems 33 and 34).

To prove property (2), we let $\mathbf{u} = a_1\mathbf{i} + b_1\mathbf{j}$ and $\mathbf{v} = a_2\mathbf{i} + b_2\mathbf{j}$. Then

$$\mathbf{u} \cdot \mathbf{v} = a_1 a_2 + b_1 b_2 = a_2 a_1 + b_2 b_1 = \mathbf{v} \cdot \mathbf{u}$$

To prove property (4), we let $\mathbf{v} = a\mathbf{i} + b\mathbf{j}$. Then

$$\mathbf{v} \cdot \mathbf{v} = a^2 + b^2 = \|\mathbf{v}\|^2$$

One use of the dot product is to calculate the angle between two vectors.

Angle Between Vectors

Let $\mathbf{u}$ and $\mathbf{v}$ be two vectors with the same initial point A. Then the vectors $\mathbf{u}$, $\mathbf{v}$, and $\mathbf{u} - \mathbf{v}$ form a triangle. The angle θ at vertex A of the triangle is the angle between the vectors $\mathbf{u}$ and $\mathbf{v}$. See Figure 61. We wish to find a formula for calculating the angle θ.

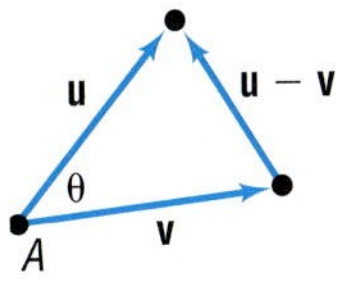

The sides of the triangle have lengths $\|\mathbf{v}\|$, $\|\mathbf{u}\|$, and $\|\mathbf{u} - \mathbf{v}\|$, and θ is the included angle between the sides of length $\|\mathbf{v}\|$ and $\|\mathbf{u}\|$. The Law of Cosines (Section 10.3) can be used to find the cosine of the included angle.

$$\|\mathbf{u} - \mathbf{v}\|^2 = \|\mathbf{u}\|^2 + \|\mathbf{v}\|^2 - 2\|\mathbf{u}\|\|\mathbf{v}\| \cos \theta$$

Now we use property (4) to rewrite this equation in terms of dot products.

$$(\mathbf{u} - \mathbf{v}) \cdot (\mathbf{u} - \mathbf{v}) = \mathbf{u} \cdot \mathbf{u} + \mathbf{v} \cdot \mathbf{v} - 2\|\mathbf{u}\|\|\mathbf{v}\| \cos \theta \tag{6}$$

Then we apply the distributive property (3) twice on the left side of (6) to obtain

$$\begin{aligned}
(\mathbf{u} - \mathbf{v}) \cdot (\mathbf{u} - \mathbf{v}) &= \mathbf{u} \cdot (\mathbf{u} - \mathbf{v}) - \mathbf{v} \cdot (\mathbf{u} - \mathbf{v}) \\
&= \mathbf{u} \cdot \mathbf{u} - \mathbf{u} \cdot \mathbf{v} - \mathbf{v} \cdot \mathbf{u} + \mathbf{v} \cdot \mathbf{v} \\
&= \mathbf{u} \cdot \mathbf{u} + \mathbf{v} \cdot \mathbf{v} - 2\mathbf{u} \cdot \mathbf{v} \tag{7}
\end{aligned}$$

↑
Property (2)

Figure 61

Combining equations (6) and (7), we have

$$\mathbf{u} \cdot \mathbf{u} + \mathbf{v} \cdot \mathbf{v} - 2\mathbf{u} \cdot \mathbf{v} = \mathbf{u} \cdot \mathbf{u} + \mathbf{v} \cdot \mathbf{v} - 2\|\mathbf{u}\|\|\mathbf{v}\| \cos \theta$$

$$\mathbf{u} \cdot \mathbf{v} = \|\mathbf{u}\|\|\mathbf{v}\| \cos \theta$$

We have proved the following result:

Theorem

Angle between Vectors

If $\mathbf{u}$ and $\mathbf{v}$ are two nonzero vectors, the angle θ, $0 \le \theta \le \pi$, between $\mathbf{u}$ and $\mathbf{v}$ is determined by the formula

$$\cos \theta = \frac{\mathbf{u} \cdot \mathbf{v}}{\|\mathbf{u}\|\|\mathbf{v}\|} \tag{8}$$

EXAMPLE 2   **Finding the Angle θ between Two Vectors**

Find the angle θ between $\mathbf{u} = 4\mathbf{i} - 3\mathbf{j}$ and $\mathbf{v} = 2\mathbf{i} + 5\mathbf{j}$.

Solution   We compute the quantities $\mathbf{u} \cdot \mathbf{v}$, $\|\mathbf{u}\|$, and $\|\mathbf{v}\|$.

$$\mathbf{u} \cdot \mathbf{v} = 4(2) + (-3)(5) = -7$$

$$\|\mathbf{u}\| = \sqrt{4^2 + (-3)^2} = 5$$

$$\|\mathbf{v}\| = \sqrt{2^2 + 5^2} = \sqrt{29}$$

By formula (8), if θ is the angle between $\mathbf{u}$ and $\mathbf{v}$, then

$$\cos \theta = \frac{\mathbf{u} \cdot \mathbf{v}}{\|\mathbf{u}\|\|\mathbf{v}\|} = \frac{-7}{5\sqrt{29}} \approx -0.26$$

We find that $\theta \approx 105°$. See Figure 62.

Figure 62

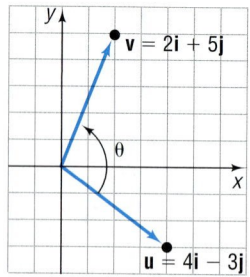

NOW WORK PROBLEMS **1(a)** AND **(b)**.

EXAMPLE 3   **Finding the Actual Speed and Direction of an Aircraft**

A Boeing 737 aircraft maintains a constant airspeed of 500 miles per hour in the direction due south. The velocity of the jet stream is 80 miles per hour in a northeasterly direction. Find the actual speed and direction of the aircraft relative to the ground.

Solution   We set up a coordinate system in which north (N) is along the positive y-axis. See Figure 63. Let

$$\mathbf{v}_a = \text{velocity of aircraft relative to the air} = -500\mathbf{j}$$

$$\mathbf{v}_g = \text{velocity of aircraft relative to ground}$$

$$\mathbf{v}_w = \text{velocity of jet stream}$$

The velocity of the jet stream $\mathbf{v}_w$ has magnitude 80 and direction NE (northeast), so $\alpha = 45°$. We express $\mathbf{v}_w$ in terms of $\mathbf{i}$ and $\mathbf{j}$ as

$$\mathbf{v}_w = 80(\cos 45°\mathbf{i} + \sin 45°\mathbf{j}) = 80\left(\frac{\sqrt{2}}{2}\mathbf{i} + \frac{\sqrt{2}}{2}\mathbf{j}\right) = 40\sqrt{2}(\mathbf{i} + \mathbf{j})$$

Figure 63

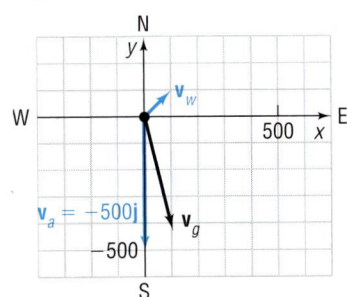

The velocity of the aircraft relative to the ground is

$$\mathbf{v}_g = \mathbf{v}_a + \mathbf{v}_w = -500\mathbf{j} + 40\sqrt{2}(\mathbf{i} + \mathbf{j}) = 40\sqrt{2}\mathbf{i} + (40\sqrt{2} - 500)\mathbf{j}$$

The actual speed of the aircraft is

$$\|\mathbf{v}_g\| = \sqrt{(40\sqrt{2})^2 + (40\sqrt{2} - 500)^2} \approx 447 \text{ miles per hour}$$

The angle θ between $\mathbf{v}_g$ and the vector $\mathbf{v}_a = -500\mathbf{j}$ (the velocity of the aircraft relative to the air) is determined by the equation

$$\cos\theta = \frac{\mathbf{v}_g \cdot \mathbf{v}_a}{\|\mathbf{v}_g\|\|\mathbf{v}_a\|} = \frac{(40\sqrt{2} - 500)(-500)}{(447)(500)} \approx 0.9920$$

$$\theta \approx 7.3°$$

The direction of the aircraft relative to the ground is approximately S7.3°E (about 7.3° east of south).

 **NOW WORK PROBLEM 19.**

Parallel and Orthogonal Vectors

3   Two vectors $\mathbf{v}$ and $\mathbf{w}$ are said to be **parallel** if there is a nonzero scalar α so that $\mathbf{v} = \alpha\mathbf{w}$. In this case, the angle θ between $\mathbf{v}$ and $\mathbf{w}$ is 0 or π.

EXAMPLE 4   **Determining Whether Vectors Are Parallel**

The vectors $\mathbf{v} = 3\mathbf{i} - \mathbf{j}$ and $\mathbf{w} = 6\mathbf{i} - 2\mathbf{j}$ are parallel, since $\mathbf{v} = \dfrac{1}{2}\mathbf{w}$. Furthermore, since

$$\cos\theta = \frac{\mathbf{v} \cdot \mathbf{w}}{\|\mathbf{v}\|\|\mathbf{w}\|} = \frac{18 + 2}{\sqrt{10}\sqrt{40}} = \frac{20}{\sqrt{400}} = 1$$

the angle θ between $\mathbf{v}$ and $\mathbf{w}$ is 0.

4   If the angle θ between two nonzero vectors $\mathbf{v}$ and $\mathbf{w}$ is $\dfrac{\pi}{2}$, the vectors $\mathbf{v}$ and $\mathbf{w}$ are called **orthogonal**.* See Figure 64.

It follows from formula (8) that if $\mathbf{v}$ and $\mathbf{w}$ are orthogonal then $\mathbf{v} \cdot \mathbf{w} = 0$, since $\cos\dfrac{\pi}{2} = 0$.

On the other hand, if $\mathbf{v} \cdot \mathbf{w} = 0$, then either $\mathbf{v} = 0$ or $\mathbf{w} = 0$ or $\cos\theta = 0$. In the latter case, $\theta = \dfrac{\pi}{2}$, and $\mathbf{v}$ and $\mathbf{w}$ are orthogonal. If $\mathbf{v}$ or $\mathbf{w}$ is the zero vector, then, since the zero vector has no specific direction, we adopt the convention that the zero vector is orthogonal to every vector.

Figure 64
v is orthogonal to w

Theorem   Two vectors $\mathbf{v}$ and $\mathbf{w}$ are orthogonal if and only if

$$\mathbf{v} \cdot \mathbf{w} = 0$$

*Orthogonal, perpendicular, and normal are all terms that mean "meet at a right angle." It is customary to refer to two vectors as being orthogonal, two lines as being perpendicular, and a line and a plane or a vector and a plane as being normal.

Figure 65

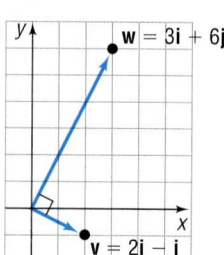

EXAMPLE 5   **Determining Whether Two Vectors Are Orthogonal**

The vectors
$$\mathbf{v} = 2\mathbf{i} - \mathbf{j} \quad \text{and} \quad \mathbf{w} = 3\mathbf{i} + 6\mathbf{j}$$
are orthogonal, since
$$\mathbf{v} \cdot \mathbf{w} = 6 - 6 = 0$$
See Figure 65.

━ NOW WORK PROBLEM 1(c).

Projection of a Vector onto Another Vector

⑤   In many physical applications, it is necessary to find "how much" of a vector is applied in a given direction. Look at Figure 66. The force $\mathbf{F}$ due to gravity is pulling straight down (toward the center of Earth) on the block. To study the effect of gravity on the block, it is necessary to determine how much of $\mathbf{F}$ is actually pushing the block down the incline ($\mathbf{F}_1$) and how much is pressing the block against the incline ($\mathbf{F}_2$), at a right angle to the incline. Knowing the **decomposition** of $\mathbf{F}$ often will allow us to determine when friction is overcome and the block will slide down the incline.

Figure 66

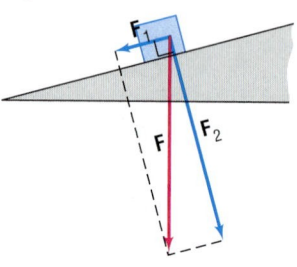

Suppose that $\mathbf{v}$ and $\mathbf{w}$ are two nonzero vectors with the same initial point P. We seek to decompose $\mathbf{v}$ into two vectors: $\mathbf{v}_1$, which is parallel to $\mathbf{w}$, and $\mathbf{v}_2$, which is orthogonal to $\mathbf{w}$. See Figure 67(a) and (b). The vector $\mathbf{v}_1$ is called the **vector projection of v onto w**.

The vector $\mathbf{v}_1$ is obtained as follows: From the terminal point of $\mathbf{v}$, drop a perpendicular to the line containing $\mathbf{w}$. The vector $\mathbf{v}_1$ is the vector from P to the foot of this perpendicular. The vector $\mathbf{v}_2$ is given by $\mathbf{v}_2 = \mathbf{v} - \mathbf{v}_1$. Note that $\mathbf{v} = \mathbf{v}_1 + \mathbf{v}_2$, $\mathbf{v}_1$ is parallel to $\mathbf{w}$, and $\mathbf{v}_2$ is orthogonal to $\mathbf{w}$. This is the decomposition of $\mathbf{v}$ that we wanted.

Figure 67

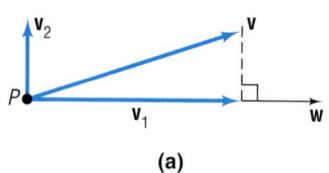

(a)

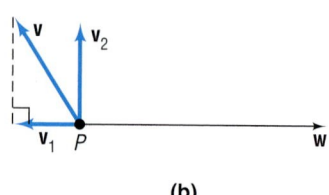

(b)

Now we seek a formula for $\mathbf{v}_1$ that is based on a knowledge of the vectors $\mathbf{v}$ and $\mathbf{w}$. Since $\mathbf{v} = \mathbf{v}_1 + \mathbf{v}_2$, we have
$$\mathbf{v} \cdot \mathbf{w} = (\mathbf{v}_1 + \mathbf{v}_2) \cdot \mathbf{w} = \mathbf{v}_1 \cdot \mathbf{w} + \mathbf{v}_2 \cdot \mathbf{w} \tag{9}$$
Since $\mathbf{v}_2$ is orthogonal to $\mathbf{w}$, we have $\mathbf{v}_2 \cdot \mathbf{w} = 0$. Since $\mathbf{v}_1$ is parallel to $\mathbf{w}$, we have $\mathbf{v}_1 = \alpha \mathbf{w}$ for some scalar α. Equation (9) can be written as
$$\mathbf{v} \cdot \mathbf{w} = \alpha \mathbf{w} \cdot \mathbf{w} = \alpha \|\mathbf{w}\|^2$$
$$\alpha = \frac{\mathbf{v} \cdot \mathbf{w}}{\|\mathbf{w}\|^2}$$
Then
$$\mathbf{v}_1 = \alpha \mathbf{w} = \frac{\mathbf{v} \cdot \mathbf{w}}{\|\mathbf{w}\|^2} \mathbf{w}$$

Theorem   If $\mathbf{v}$ and $\mathbf{w}$ are two nonzero vectors, the vector projection of $\mathbf{v}$ onto $\mathbf{w}$ is
$$\mathbf{v}_1 = \frac{\mathbf{v} \cdot \mathbf{w}}{\|\mathbf{w}\|^2} \mathbf{w} \tag{10}$$

The decomposition of $\mathbf{v}$ into $\mathbf{v}_1$ and $\mathbf{v}_2$, where $\mathbf{v}_1$ is parallel to $\mathbf{w}$ and $\mathbf{v}_2$ is perpendicular to $\mathbf{w}$, is
$$\mathbf{v}_1 = \frac{\mathbf{v} \cdot \mathbf{w}}{\|\mathbf{w}\|^2} \mathbf{w} \qquad \mathbf{v}_2 = \mathbf{v} - \mathbf{v}_1 \tag{11}$$

EXAMPLE 6   Decomposing a Vector into Two Orthogonal Vectors

Find the vector projection of $\mathbf{v} = \mathbf{i} + 3\mathbf{j}$ onto $\mathbf{w} = \mathbf{i} + \mathbf{j}$. Decompose $\mathbf{v}$ into two vectors $\mathbf{v}_1$ and $\mathbf{v}_2$, where $\mathbf{v}_1$ is parallel to $\mathbf{w}$ and $\mathbf{v}_2$ is orthogonal to $\mathbf{w}$.

Figure 68

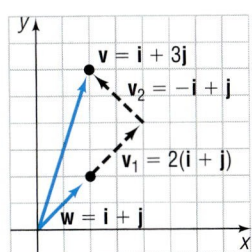

Solution   We use formulas (10) and (11).

$$\mathbf{v}_1 = \frac{\mathbf{v} \cdot \mathbf{w}}{\|\mathbf{w}\|^2}\mathbf{w} = \frac{1 + 3}{(\sqrt{2})^2}\mathbf{w} = 2\mathbf{w} = 2(\mathbf{i} + \mathbf{j})$$

$$\mathbf{v}_2 = \mathbf{v} - \mathbf{v}_1 = (\mathbf{i} + 3\mathbf{j}) - 2(\mathbf{i} + \mathbf{j}) = -\mathbf{i} + \mathbf{j}$$

See Figure 68.   ∎

✎ **NOW WORK PROBLEM 13.**

Work Done by a Constant Force

⑥   In elementary physics, the **work** W done by a constant force $\mathbf{F}$ in moving an object from a point A to a point B is defined as

$$W = (\text{magnitude of force})(\text{distance}) = \|\mathbf{F}\|\|\overrightarrow{AB}\|$$

Figure 69

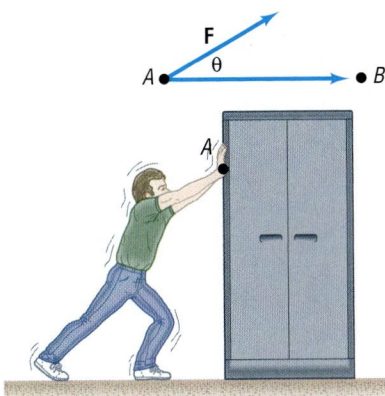

Work is commonly measured in foot-pounds or in Newton-meters (joules).

In this definition, it is assumed that the force $\mathbf{F}$ is applied along the line of motion. If the constant force $\mathbf{F}$ is not along the line of motion, but, instead, is at an angle θ to the direction of motion, as illustrated in Figure 69, then the **work** W **done by** $\mathbf{F}$ in moving an object from A to B is defined as

$$W = \mathbf{F} \cdot \overrightarrow{AB} \tag{12}$$

This definition is compatible with the force times distance definition given above, since

$$W = (\text{amount of force in the direction of } \overrightarrow{AB})(\text{distance})$$

$$= \|\text{projection of } \mathbf{F} \text{ on } \overrightarrow{AB}\|\|\overrightarrow{AB}\| = \frac{\mathbf{F} \cdot \overrightarrow{AB}}{\|\overrightarrow{AB}\|^2}\|\overrightarrow{AB}\|\|\overrightarrow{AB}\| = \mathbf{F} \cdot \overrightarrow{AB}$$

EXAMPLE 7   Computing Work

Figure 70(a) shows a girl pulling a wagon with a force of 50 pounds. How much work is done in moving the wagon 100 feet if the handle makes an angle of $30°$ with the ground?

Figure 70

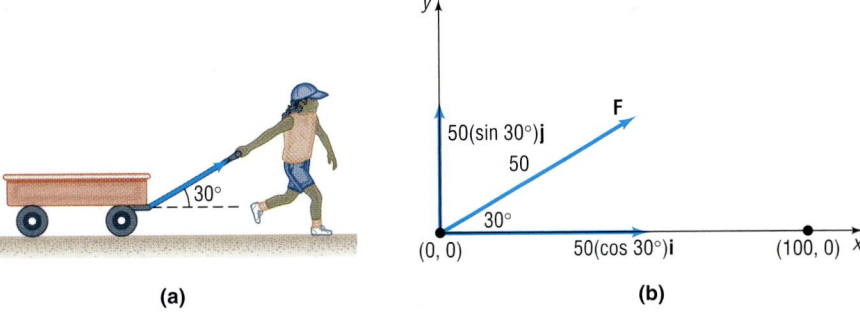

(a)   (b)

Solution   We position the vectors in a coordinate system in such a way that the wagon is moved from $(0, 0)$ to $(100, 0)$. The motion is from $A = (0, 0)$ to $B = (100, 0)$, so $\overrightarrow{AB} = 100\mathbf{i}$. The force vector $\mathbf{F}$, as shown in Figure 70(b), is

$$\mathbf{F} = 50(\cos 30°\mathbf{i} + \sin 30°\mathbf{j}) = 50\left(\frac{\sqrt{3}}{2}\mathbf{i} + \frac{1}{2}\mathbf{j}\right) = 25(\sqrt{3}\mathbf{i} + \mathbf{j})$$

By formula (12), the work done is

$$W = \mathbf{F} \cdot \overrightarrow{AB} = 25(\sqrt{3}\mathbf{i} + \mathbf{j}) \cdot 100\mathbf{i} = 2500\sqrt{3} \text{ foot-pounds} \quad \blacksquare$$

─── **NOW WORK PROBLEM 29.**

HISTORICAL FEATURE

1. We stated in an earlier Historical Feature that complex numbers were used as vectors in the plane before the general notion of a vector was clarified. Suppose that we make the correspondence

$$\text{Vector} \leftrightarrow \text{Complex number}$$
$$a\mathbf{i} + b\mathbf{j} \leftrightarrow a + bi$$
$$c\mathbf{i} + d\mathbf{j} \leftrightarrow c + di$$

Show that

$$(a\mathbf{i} + b\mathbf{j}) \cdot (c\mathbf{i} + d\mathbf{j}) = \text{real part}[\overline{(a + bi)}(c + di)]$$

This is how the dot product was found originally. The imaginary part is also interesting. It is a determinant (see Section 7.4) and represents the area of the parallelogram whose edges are the vectors. This is close to some of Hermann Grassmann's ideas and is also connected with the scalar triple product of three-dimensional vectors.

11.5 Concepts and Vocabulary

In Problems 1–3, fill in the blanks.

1. If the angle between two vectors is $\dfrac{\pi}{2}$, the dot product of the two vectors equals _____.

2. If $\mathbf{v} \cdot \mathbf{w} = 0$, then the two vectors $\mathbf{v}$ and $\mathbf{w}$ are _____.

3. If $\mathbf{v} = 3\mathbf{w}$, then the two vectors $\mathbf{v}$ and $\mathbf{w}$ are _____.

In Problems 4–6, answer True or False to each statement.

4. If $\mathbf{v}$ and $\mathbf{w}$ are parallel vectors, then $\mathbf{v} \cdot \mathbf{w} = 0$.

5. Given two nonzero vectors $\mathbf{v}$ and $\mathbf{w}$, it is always possible to decompose $\mathbf{v}$ into two vectors, one parallel to $\mathbf{w}$ and the other perpendicular to $\mathbf{w}$.

6. Work is a physical example of a vector.

7. What does the word orthogonal mean?

11.5 Exercises

In Problems 1–10, (a) find the dot product $\mathbf{v} \cdot \mathbf{w}$; (b) find the angle between $\mathbf{v}$ and $\mathbf{w}$; (c) state whether the vectors are parallel, orthogonal, or neither.

1. $\mathbf{v} = \mathbf{i} - \mathbf{j}, \quad \mathbf{w} = \mathbf{i} + \mathbf{j}$
2. $\mathbf{v} = \mathbf{i} + \mathbf{j}, \quad \mathbf{w} = -\mathbf{i} + \mathbf{j}$
3. $\mathbf{v} = 2\mathbf{i} + \mathbf{j}, \quad \mathbf{w} = \mathbf{i} + 2\mathbf{j}$
4. $\mathbf{v} = 2\mathbf{i} + 2\mathbf{j}, \quad \mathbf{w} = \mathbf{i} + 2\mathbf{j}$
5. $\mathbf{v} = \sqrt{3}\mathbf{i} - \mathbf{j}, \quad \mathbf{w} = \mathbf{i} + \mathbf{j}$
6. $\mathbf{v} = \mathbf{i} + \sqrt{3}\mathbf{j}, \quad \mathbf{w} = \mathbf{i} - \mathbf{j}$
7. $\mathbf{v} = 3\mathbf{i} + 4\mathbf{j}, \quad \mathbf{w} = 4\mathbf{i} + 3\mathbf{j}$
8. $\mathbf{v} = 3\mathbf{i} - 4\mathbf{j}, \quad \mathbf{w} = 4\mathbf{i} - 3\mathbf{j}$
9. $\mathbf{v} = 4\mathbf{i}, \quad \mathbf{w} = \mathbf{j}$
10. $\mathbf{v} = \mathbf{i}, \quad \mathbf{w} = -3\mathbf{j}$

11. Find a so that the vectors $\mathbf{v} = \mathbf{i} - a\mathbf{j}$ and $\mathbf{w} = 2\mathbf{i} + 3\mathbf{j}$ are orthogonal.

12. Find b so that the vectors $\mathbf{v} = \mathbf{i} + \mathbf{j}$ and $\mathbf{w} = \mathbf{i} + b\mathbf{j}$ are orthogonal.

In Problems 13–18, decompose $\mathbf{v}$ into two vectors $\mathbf{v}_1$ and $\mathbf{v}_2$, where $\mathbf{v}_1$ is parallel to $\mathbf{w}$ and $\mathbf{v}_2$ is orthogonal to $\mathbf{w}$.

13. $\mathbf{v} = 2\mathbf{i} - 3\mathbf{j}, \quad \mathbf{w} = \mathbf{i} - \mathbf{j}$
14. $\mathbf{v} = -3\mathbf{i} + 2\mathbf{j}, \quad \mathbf{w} = 2\mathbf{i} + \mathbf{j}$
15. $\mathbf{v} = \mathbf{i} - \mathbf{j}, \quad \mathbf{w} = \mathbf{i} + 2\mathbf{j}$
16. $\mathbf{v} = 2\mathbf{i} - \mathbf{j}, \quad \mathbf{w} = \mathbf{i} - 2\mathbf{j}$
17. $\mathbf{v} = 3\mathbf{i} + \mathbf{j}, \quad \mathbf{w} = -2\mathbf{i} - \mathbf{j}$
18. $\mathbf{v} = \mathbf{i} - 3\mathbf{j}, \quad \mathbf{w} = 4\mathbf{i} - \mathbf{j}$

19. Finding the Actual Speed and Direction of an Aircraft
A DC-10 jumbo jet maintains an airspeed of 550 miles per hour in a southwesterly direction. The velocity of the jet stream is a constant 80 miles per hour from the west. Find the actual speed and direction of the aircraft.

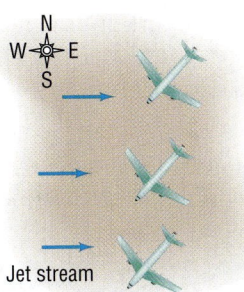

Jet stream

20. Finding the Correct Compass Heading The pilot of an aircraft wishes to head directly east, but is faced with a wind speed of 40 miles per hour from the northwest. If the pilot maintains an airspeed of 250 miles per hour, what compass heading should be maintained? What is the actual speed of the aircraft?

21. Correct Direction for Crossing a River A river has a constant current of 3 kilometers per hour. At what angle to a boat dock should a motorboat, capable of maintaining a constant speed of 20 kilometers per hour, be headed in order to reach a point directly opposite the dock? If the river is $\frac{1}{2}$ kilometer wide, how long will it take to cross?

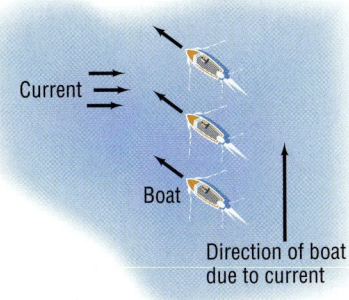

Current

Boat

Direction of boat
due to current

22. Correct Direction for Crossing a River Repeat Problem 21 if the current is 5 kilometers per hour.

23. Braking Load A Toyota Sienna with a gross weight of 5300 pounds is parked on a street with a slope of 8°. See the figure. Find the force required to keep the Sienna from rolling down the hill. What is the force perpendicular to the hill?

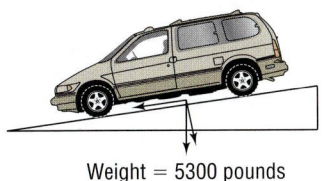

Weight = 5300 pounds

24. Braking Load A Pontiac Bonneville with a gross weight of 4500 pounds is parked on a street with a slope of 10°. Find the force required to keep the Bonneville from rolling down the hill. What is the force perpendicular to the hill?

25. Ground Speed and Direction of an Airplane An airplane has an airspeed of 500 kilometers per hour bearing N45°E. The wind velocity is 60 kilometers per hour in the direction N30°W. Find the resultant vector representing the path of the plane relative to the ground. What is the ground speed of the plane? What is its direction?

26. Ground Speed and Direction of an Airplane An airplane has an airspeed of 600 kilometers per hour bearing S30°E. The wind velocity is 40 kilometers per hour in the direction S45°E. Find the resultant vector representing the path of the plane relative to the ground. What is the ground speed of the plane? What is its direction?

27. Crossing a River A small motorboat in still water maintains a speed of 20 miles per hour. In heading directly across a river (that is, perpendicular to the current) whose current is 3 miles per hour, find a vector representing the speed and direction of the motorboat. What is the true speed of the motorboat? What is its direction?

28. Crossing a River A small motorboat in still water maintains a speed of 10 miles per hour. In heading directly across a river (that is, perpendicular to the current) whose current is 4 miles per hour, find a vector representing the speed and direction of the motorboat. What is the true speed of the motorboat? What is its direction?

29. Computing Work Find the work done by a force of 3 pounds acting in the direction 60° to the horizontal in moving an object 2 feet from $(0, 0)$ to $(2, 0)$.

30. Computing Work Find the work done by a force of 1 pound acting in the direction 45° to the horizontal in moving an object 5 feet from $(0, 0)$ to $(5, 0)$.

31. Computing Work A wagon is pulled horizontally by exerting a force of 20 pounds on the handle at an angle of 30° with the horizontal. How much work is done in moving the wagon 100 feet?

32. Find the acute angle that a constant unit force vector makes with the positive x-axis if the work done by the force in moving a particle from $(0, 0)$ to $(4, 0)$ equals 2.

33. Prove the distributive property:
$$\mathbf{u} \cdot (\mathbf{v} + \mathbf{w}) = \mathbf{u} \cdot \mathbf{v} + \mathbf{u} \cdot \mathbf{w}$$

34. Prove property (5), $\mathbf{0} \cdot \mathbf{v} = 0$.

35. If $\mathbf{v}$ is a unit vector and the angle between $\mathbf{v}$ and $\mathbf{i}$ is α, show that $\mathbf{v} = \cos \alpha \mathbf{i} + \sin \alpha \mathbf{j}$.

36. Suppose that $\mathbf{v}$ and $\mathbf{w}$ are unit vectors. If the angle between $\mathbf{v}$ and $\mathbf{i}$ is α and if the angle between $\mathbf{w}$ and $\mathbf{i}$ is β, use the idea of the dot product $\mathbf{v} \cdot \mathbf{w}$ to prove that
$$\cos(\alpha - \beta) = \cos \alpha \cos \beta + \sin \alpha \sin \beta$$

37. Show that the projection of $\mathbf{v}$ onto $\mathbf{i}$ is $(\mathbf{v} \cdot \mathbf{i})\mathbf{i}$. In fact, show that we can always write a vector $\mathbf{v}$ as
$$\mathbf{v} = (\mathbf{v} \cdot \mathbf{i})\mathbf{i} + (\mathbf{v} \cdot \mathbf{j})\mathbf{j}$$

38. (a) If $\mathbf{u}$ and $\mathbf{v}$ have the same magnitude, show that $\mathbf{u} + \mathbf{v}$ and $\mathbf{u} - \mathbf{v}$ are orthogonal.

(b) Use this to prove that an angle inscribed in a semi-circle is a right angle (see the figure).

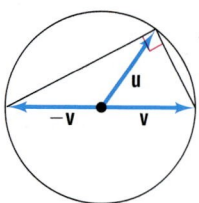

39. Let $\mathbf{v}$ and $\mathbf{w}$ denote two nonzero vectors. Show that the vector $\mathbf{v} - \alpha\mathbf{w}$ is orthogonal to $\mathbf{w}$ if $\alpha = \dfrac{\mathbf{v} \cdot \mathbf{w}}{\|\mathbf{w}\|^2}$.

40. Let $\mathbf{v}$ and $\mathbf{w}$ denote two nonzero vectors. Show that the vectors $\|\mathbf{w}\|\mathbf{v} + \|\mathbf{v}\|\mathbf{w}$ and $\|\mathbf{w}\|\mathbf{v} - \|\mathbf{v}\|\mathbf{w}$ are orthogonal.

41. In the definition of work given in this section, what is the work done if $\mathbf{F}$ is orthogonal to $\overrightarrow{AB}$?

42. Prove the **polarization identity**,
$$\|\mathbf{u} + \mathbf{v}\|^2 - \|\mathbf{u} - \mathbf{v}\|^2 = 4(\mathbf{u} \cdot \mathbf{v})$$

 43. Make up an application different from any found in the text that requires the dot product.

Chapter Review

Things To Know

Relationship between polar coordinates (r, θ) and rectangular coordinates (x, y) (pp. 833 and 836)	$x = r\cos\theta, y = r\sin\theta$ $r^2 = x^2 + y^2, \tan\theta = \dfrac{y}{x}, \quad x \neq 0$
Polar form of a complex number (p. 858)	If $z = x + yi$, then $z = r(\cos\theta + i\sin\theta)$, where $r = |z| = \sqrt{x^2 + y^2}$, $\sin\theta = \dfrac{y}{r}$, $\cos\theta = \dfrac{x}{r}$, $\quad 0 \leq \theta < 2\pi$
De Moivre's Theorem (p. 861)	If $z = r(\cos\theta + i\sin\theta)$, then $z^n = r^n[\cos(n\theta) + i\sin(n\theta)]$, where $n \geq 1$ is a positive integer
nth root of a complex number $z = r(\cos\theta_0 + i\sin\theta_0)$ (p. 862)	$\sqrt[n]{z} = \sqrt[n]{r}\left[\cos\left(\dfrac{\theta_0}{n} + \dfrac{2k\pi}{n}\right) + i\sin\left(\dfrac{\theta_0}{n} + \dfrac{2k\pi}{n}\right)\right], \quad k = 0, 1, \ldots, n - 1,$ where $n \geq 2$ is an integer.
Vector (p. 866)	Quantity having magnitude and direction; equivalent to a directed line segment $\overrightarrow{PQ}$
Position vector (p. 869)	Vector whose initial point is at the origin
Unit vector (pp. 869 and 872)	Vector whose magnitude is 1
Dot product (p. 878)	If $\mathbf{v} = a_1\mathbf{i} + b_1\mathbf{j}$ and $\mathbf{w} = a_2\mathbf{i} + b_2\mathbf{j}$, then $\mathbf{v} \cdot \mathbf{w} = a_1a_2 + b_1b_2$.
Angle θ between two nonzero vectors $\mathbf{u}$ and $\mathbf{v}$ (p. 880)	$\cos\theta = \dfrac{\mathbf{u} \cdot \mathbf{v}}{\|\mathbf{u}\|\|\mathbf{v}\|}$

Objectives

Review Exercises

Blue problem numbers indicate the authors' suggestions for use in a Practice Test.

In Problems 1–6, plot each point given in polar coordinates, and find its rectangular coordinates.

1. $\left(3, \dfrac{\pi}{6}\right)$ **2.** $\left(4, \dfrac{2\pi}{3}\right)$ **3.** $\left(-2, \dfrac{4\pi}{3}\right)$

4. $\left(-1, \dfrac{5\pi}{4}\right)$ **5.** $\left(-3, -\dfrac{\pi}{2}\right)$ **6.** $\left(-4, -\dfrac{\pi}{4}\right)$

In Problems 7–12, the rectangular coordinates of a point are given. Find two pairs of polar coordinates (r, θ) for each point, one with $r > 0$ and the other with $r < 0$. Express θ in radians.

7. $(-3, 3)$ **8.** $(1, -1)$ **9.** $(0, -2)$ **10.** $(2, 0)$ **11.** $(3, 4)$ **12.** $(-5, 12)$

In Problems 13–18, the letters r and θ represent polar coordinates. Write each polar equation as an equation in rectangular coordinates (x, y). Identify the equation and graph it.

13. $r = 2 \sin \theta$ **14.** $3r = \sin \theta$ **15.** $r = 5$

16. $\theta = \dfrac{\pi}{4}$ **17.** $r \cos \theta + 3r \sin \theta = 6$ **18.** $r^2 + 4r \sin \theta - 8r \cos \theta = 5$

In Problems 19–22, graph each polar equation using a graphing utility.

19. $r = 2 + \sin \theta$ **20.** $r = \sin^2 \theta$ **21.** $r \tan \theta = 2$ **22.** $(r + 5) \sin^2 \theta = 2$

In Problems 23–28, sketch the graph of each polar equation. Be sure to test for symmetry. Verify your graph using a graphing utility.

23. $r = 4 \cos \theta$ **24.** $r = 3 \sin \theta$ **25.** $r = 3 - 3 \sin \theta$

26. $r = 2 + \cos \theta$ **27.** $r = 4 - \cos \theta$ **28.** $r = 1 - 2 \sin \theta$

In Problems 29–32, write each complex number in polar form. Express each argument in degrees.

29. $-1 - i$ **30.** $-\sqrt{3} + i$ **31.** $4 - 3i$ **32.** $3 - 2i$

In Problems 33–38, write each complex number in the standard form a + bi and plot each in the complex plane.

33. $2(\cos 150° + i \sin 150°)$

34. $3(\cos 60° + i \sin 60°)$

35. $3\left(\cos \dfrac{2\pi}{3} + i \sin \dfrac{2\pi}{3}\right)$

36. $4\left(\cos \dfrac{3\pi}{4} + i \sin \dfrac{3\pi}{4}\right)$

37. $0.1(\cos 350° + i \sin 350°)$

38. $0.5(\cos 160° + i \sin 160°)$

In Problems 39–44, find zw and $\dfrac{z}{w}$. Leave your answers in polar form.

39. $z = \cos 80° + i \sin 80°$
  $w = \cos 50° + i \sin 50°$

40. $z = \cos 205° + i \sin 205°$
  $w = \cos 85° + i \sin 85°$

41. $z = 3\left(\cos \dfrac{9\pi}{5} + i \sin \dfrac{9\pi}{5}\right)$
  $w = 2\left(\cos \dfrac{\pi}{5} + i \sin \dfrac{\pi}{5}\right)$

42. $z = 2\left(\cos \dfrac{5\pi}{3} + i \sin \dfrac{5\pi}{3}\right)$
  $w = 3\left(\cos \dfrac{\pi}{3} + i \sin \dfrac{\pi}{3}\right)$

43. $z = 5(\cos 10° + i \sin 10°)$
  $w = \cos 355° + i \sin 355°$

44. $z = 4(\cos 50° + i \sin 50°)$
  $w = \cos 340° + i \sin 340°$

In Problems 45–52, write each expression in the standard form a + bi.

45. $[3(\cos 20° + i \sin 20°)]^3$

46. $[2(\cos 50° + i \sin 50°)]^3$

47. $\left[\sqrt{2}\left(\cos \dfrac{5\pi}{8} + i \sin \dfrac{5\pi}{8}\right)\right]^4$

48. $\left[2\left(\cos \dfrac{5\pi}{16} + i \sin \dfrac{5\pi}{16}\right)\right]^4$

49. $(1 - \sqrt{3}i)^6$

50. $(2 - 2i)^8$

51. $(3 + 4i)^4$

52. $(1 - 2i)^4$

53. Find all the complex cube roots of 27.

54. Find all the complex fourth roots of -16.

In Problems 55–58, use the figure to graph each of the following:

55. $\mathbf{u} + \mathbf{v}$

56. $\mathbf{v} + \mathbf{w}$

57. $2\mathbf{u} + 3\mathbf{v}$

58. $5\mathbf{v} - 2\mathbf{w}$

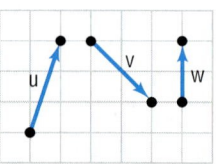

In Problems 59–62, the vector $\mathbf{v}$ is represented by the directed line segment $\overrightarrow{PQ}$. Write $\mathbf{v}$ in the form $a\mathbf{i} + b\mathbf{j}$ and find $\|\mathbf{v}\|$.

59. $P = (1, -2); \quad Q = (3, -6)$

60. $P = (-3, 1); \quad Q = (4, -2)$

61. $P = (0, -2); \quad Q = (-1, 1)$

62. $P = (3, -4); \quad Q = (-2, 0)$

In Problems 63–72, use the vectors $\mathbf{v} = -2\mathbf{i} + \mathbf{j}$ and $\mathbf{w} = 4\mathbf{i} - 3\mathbf{j}$ to find:

63. $\mathbf{v} + \mathbf{w}$

64. $\mathbf{v} - \mathbf{w}$

65. $4\mathbf{v} - 3\mathbf{w}$

66. $-\mathbf{v} + 2\mathbf{w}$

67. $\|\mathbf{v}\|$

68. $\|\mathbf{v} + \mathbf{w}\|$

69. $\|\mathbf{v}\| + \|\mathbf{w}\|$

70. $\|2\mathbf{v}\| - 3\|\mathbf{w}\|$

71. Find a unit vector in the same direction as $\mathbf{v}$.

72. Find a unit vector in the opposite direction of $\mathbf{w}$.

73. Find the vector $\mathbf{v}$ with magnitude 3 if the angle between $\mathbf{v}$ and $\mathbf{i}$ is $60°$.

74. Find the vector $\mathbf{v}$ with magnitude 5 if the angle between $\mathbf{v}$ and $\mathbf{i}$ is $150°$.

In Problems 75–78, find the dot product $\mathbf{v} \cdot \mathbf{w}$ and the angle between $\mathbf{v}$ and $\mathbf{w}$.

75. $\mathbf{v} = -2\mathbf{i} + \mathbf{j}, \quad \mathbf{w} = 4\mathbf{i} - 3\mathbf{j}$

76. $\mathbf{v} = 3\mathbf{i} - \mathbf{j}, \quad \mathbf{w} = \mathbf{i} + \mathbf{j}$

77. $\mathbf{v} = \mathbf{i} - 3\mathbf{j}, \quad \mathbf{w} = -\mathbf{i} + \mathbf{j}$

78. $\mathbf{v} = \mathbf{i} + 4\mathbf{j}, \quad \mathbf{w} = 3\mathbf{i} - 2\mathbf{j}$

In Problems 79–84, determine whether $\mathbf{v}$ and $\mathbf{w}$ are parallel, orthogonal, or neither.

79. $\mathbf{v} = 2\mathbf{i} + 3\mathbf{j}; \quad \mathbf{w} = -4\mathbf{i} - 6\mathbf{j}$

80. $\mathbf{v} = -2\mathbf{i} - \mathbf{j}; \quad \mathbf{w} = 2\mathbf{i} + \mathbf{j}$

81. $\mathbf{v} = 3\mathbf{i} - 4\mathbf{j}; \quad \mathbf{w} = -3\mathbf{i} + 4\mathbf{j}$

82. $\mathbf{v} = -2\mathbf{i} + 2\mathbf{j}; \quad \mathbf{w} = -3\mathbf{i} + 2\mathbf{j}$

83. $\mathbf{v} = 3\mathbf{i} - 2\mathbf{j}; \quad \mathbf{w} = 4\mathbf{i} + 6\mathbf{j}$

84. $\mathbf{v} = -4\mathbf{i} + 2\mathbf{j}; \quad \mathbf{w} = 2\mathbf{i} + 4\mathbf{j}$

In Problems 85 and 86, decompose **v** into two vectors, one parallel to **w** and the other orthogonal to **w**.

85. $\mathbf{v} = 2\mathbf{i} + \mathbf{j}; \quad \mathbf{w} = -4\mathbf{i} + 3\mathbf{j}$

86. $\mathbf{v} = -3\mathbf{i} + 2\mathbf{j}; \quad \mathbf{w} = -2\mathbf{i} + \mathbf{j}$

87. Find the vector projection of $\mathbf{v} = 2\mathbf{i} + 3\mathbf{j}$ onto $\mathbf{w} = 3\mathbf{i} + \mathbf{j}$.

88. Find the vector projection of $\mathbf{v} = -\mathbf{i} + 2\mathbf{j}$ onto $\mathbf{w} = 3\mathbf{i} - \mathbf{j}$.

89. Actual Speed and Direction of a Swimmer A swimmer can maintain a constant speed of 5 miles per hour. If the swimmer heads directly across a river that has a current moving at the rate of 2 miles per hour, what is the actual speed of the swimmer? (See the figure.) If the river is 1 mile wide, how far downstream will the swimmer end up from the point directly across the river from the starting point?

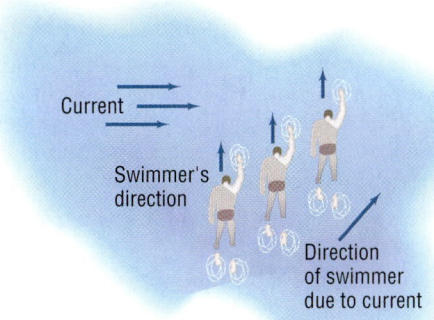

90. Actual Speed and Direction of an Airplane An airplane has an airspeed of 500 kilometers per hour in a northerly direction. The wind velocity is 60 kilometers per hour in a southeasterly direction. Find the actual speed and direction of the plane relative to the ground.

91. Static Equilibrium A weight of 2000 pounds is suspended from two cables as shown in the figure. What are the tensions of each cable?

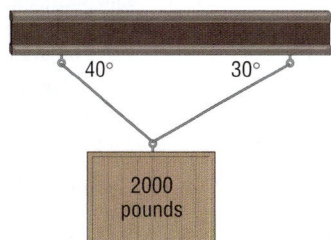

92. Actual Speed and Distance of a Motorboat A small motorboat is moving at a true speed of 11 miles per hour in a southerly direction. The current is known to be from the northeast at 3 miles per hour. What is the speed of the motorboat relative to the water? In what direction does the compass indicate that the boat is headed?

93. Computing Work Find the work done by a force of 5 pounds acting in the direction $60°$ to the horizontal in moving an object 20 feet from $(0, 0)$ to $(20, 0)$.

Chapter Projects

1. Mandelbrot Sets

(a) Let $z = x + yi$ be a complex number. We can plot complex numbers using a coordinate system called the complex plane. The x-axis will be referred to as the real axis, because any point that lies on the real axis is of the form $z = x + 0i = x$, a real number.

The y-axis is called the imaginary axis, because any point that lies on it is of the form $z = 0 + yi = yi$, a pure imaginary number. To plot the complex number $z = x + yi$, plot the ordered pair (x, y) where x is the signed distance from the imaginary axis and y is the signed distance from the real axis. Draw a complex plane and plot the points $z_1 = 3 + 4i$, $z_2 = -2 + i, z_3 = 0 - 2i, z_4 = -2$.

(b) Consider the expression $a_n = (a_{n-1})^2 + z$, where z is some complex number (called the **seed**) and $a_0 = z$. Compute $a_1(= a_0^2 + z)$, $a_2(= a_1^2 + z)$, $a_3(= a_2^2 + z)$, a_4, a_5, a_6 for the following seeds: $z_1 = 0.1 - 0.4i, z_2 = 0.5 + 0.8i, z_3 = -0.9 + 0.7i$, $z_4 = -1.1 + 0.1i, z_5 = 0 - 1.3i$, and $z_6 = 1 + 1i$.

(c) The dark portion of the graph, shown on page 890, represents the set of all values $z = x + yi$ that are in the Mandelbrot Set. Determine which complex numbers in (b) are in this set by plotting them on the graph. Do the complex numbers that are not in the Mandelbrot Set have any common characteristics regarding the values of a_6 found in (b)?

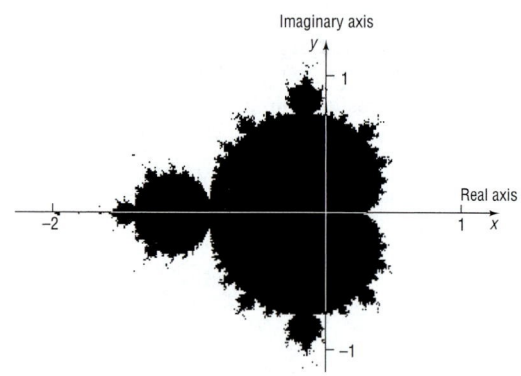

Imaginary axis

Real axis

(d) Compute $|z| = \sqrt{x^2 + y^2}$ for each of the complex numbers in (b). Now compute $|a_6|$ for each of the complex numbers in (b). For which complex numbers is $|a_6| \geq |z|$ and $|z| > 2$? Conclude that the criterion for a complex number to be in the Mandelbrot Set is that $|a_n| \geq |z|$ and $|z| > 2$.

2. **Compound Interest** In Example 13 on page 852, we graphed a logarithmic spiral. If we write the logarithmic spiral in the form $A = Pe^{r\theta}$ we have the formula for continuously compounded interest where A is the amount resulting from investing a principal P at an annual interest rate r for θ years. Set your graphing calculator to radian and polar coordinate mode.

(a) Graph the amount A that results from a $100 deposit earning 5% compounded continuously using θstep = 0.1.

(b) Using TRACE, determine the amount of time it takes (to the nearest tenth of a year) to double your investment.

(c) Using TRACE, determine the amount of time it takes (to the nearest tenth of a year) to triple your investment.

(d) Graph the amount A that results from a $100 deposit earning 10% compounded continuously using θstep = 0.1.

(e) Using TRACE, determine the amount of time it takes (to the nearest tenth of a year) to double your investment.

(f) Using TRACE, determine the amount of time it takes (to the nearest tenth of a year) to triple your investment.

(g) Does doubling the interest rate reduce the time it takes the investment to double by half? Support your answer.

3. **A.** The equation $e^{i\pi} + 1 = 0$ is sometimes called **Euler's equation**, named after Leonhard Euler, a Swiss mathematician from the 18th century. Euler was instrumental in establishing a connection between complex numbers and trigonometric functions. One of his identities is $e^{ix} = \cos x + i \sin x$.

(a) Verify Euler's equation, using the Euler identity stated above.

(b) Verify the other two Euler identities:

$$\sin x = \frac{e^{ix} - e^{-ix}}{2i} \quad \text{and} \quad \cos x = \frac{e^{ix} + e^{-ix}}{2}.$$

(c) Use the Euler identities to write $\sin(1 + i)$ in the standard form $a + bi$. Round to three decimal places.

(d) Prove the theorems for the product and quotient of two complex numbers using Euler's identities.

B. Gauss established a proof of the Fundamental Theorem of Algebra (see Section 5.4) in his doctoral dissertation (at the age of 20). In it, he graphed complex numbers as was shown in Section 11.3 to help establish the existence of complex solutions.

(a) Solve $ix^3 + 8 = 0$ graphically by answering the following:

1. Make the substitution $x = u + iv$ and simplify.

2. Equate the real parts and the imaginary parts on both sides of the equation.

3. Graph each real equation resulting in part 2 on the same set of axes. (Use your graphing calculator.)

4. Find the points of intersection (u, v) by using the INTERSECT feature on your calculator. What are the solutions of the equation? How many should there be? Is that the number you found? Are they complex (but not real) or real? Do they make sense?

(b) Solve the equation by using DeMoivre's Theorem. Do the solutions that you found in part (a) match those that you found here?

Source: Carl B. Boyer, a History of Mathematics, 2nd edition. John Wiley & sons, 1989. Part a–pg. 520; Part b–pgs 560, 561, 596.

Cumulative Review

1. Find the real solutions, if any, of the equation: $e^{x^2-9} = 1$.

2. Find an equation for the line containing the origin that makes an angle of $30°$ with the positive x-axis.

3. Find an equation for the circle with center at the point $(0, 1)$ and radius 3. Graph this circle.

4. What is the domain of the function $f(x) = \ln(1 - 2x)$?

5. Test the equation $x^2 + y^3 = 2x^4$ for symmetry with respect to the x-axis, the y-axis, and the origin.

6. Graph the function $y = |\ln x|$.

7. Graph the function $y = |\sin x|$.

8. Graph the function $y = \sin |x|$.

9. Find the exact value of $\sin^{-1}\left(-\frac{1}{2}\right)$.

10. Graph the equations $x = 3$ and $y = 4$ using the same set of rectangular coordinates.

11. Graph the equations $r = 2$ and $\theta = \dfrac{\pi}{3}$ using the same set of polar coordinates.

ANALYTIC GEOMETRY

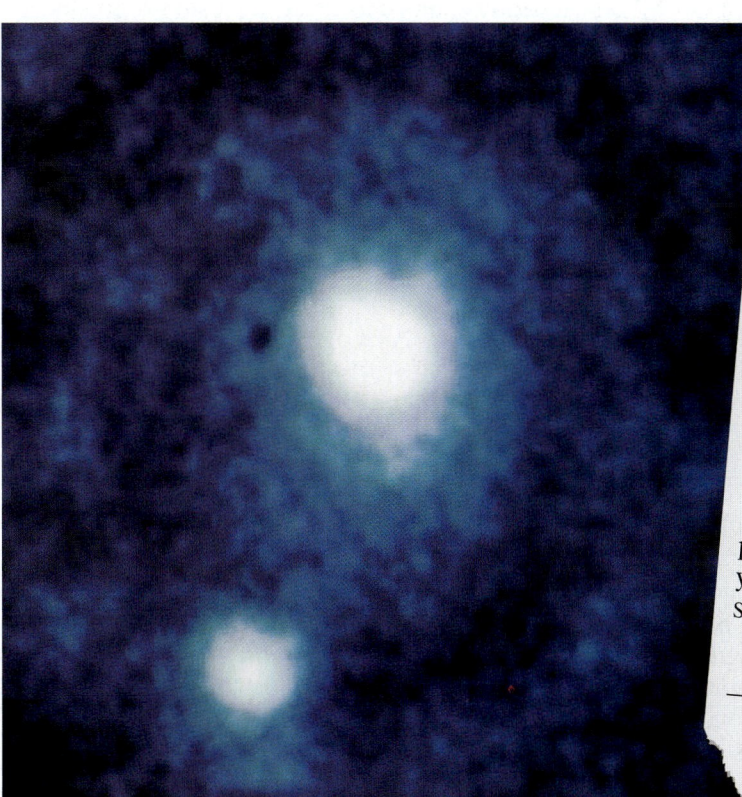

Pluto resumes 'farthest orbit'

WASHINGTON (AP)—Mere days after surviving attacks on its status as a planet, diminutive Pluto is resuming its traditional spot farthest from the sun.

Pluto was on course to cross the orbit of Neptune at 5:08 a.m. Thursday, NASA reported.

Normally the most distant planet from the sun. Pluto has a highly elliptical orbit that occasionally brings it inside the orbit of Neptune.

That last took place on Feb. 7, 1979, and since then Neptune had been the most distant planet.

Now Pluto once again becomes the farthest planet from the sun, where it will remain for 228 years. It takes pluto 248 years to circle the sun.

SOURCE: *Naples Daily News*, February 11, 1999.

SEE CHAPTER PROJECT 1.

OUTLINE

 For additional study help, go to

www.prenhall.com/sullivanegu3e

Materials include:

- Graphing Calculator Help
- Chapter Quiz
- Chapter Test
- PowerPoint Downloads
- Chapter Projects
- Student Tips

A Look Back, A Look Forward

In Chapter 1, we introduced rectangular coordinates and showed how geometry problems can be solved algebraically. We defined a circle geometrically and then used the distance formula and rectangular coordinates to obtain an equation for a circle. In this chapter we give geometric definitions for the conics and use the distance formula and rectangular coordinates to obtain their equations.

Historically, Apollonius (200 B.C.) was among the first to study *conics* and discover some of their interesting properties. Today, conics are still studied because of their many uses. *Paraboloids of revolution* (parabolas rotated about their axes of symmetry) are used as signal collectors (the satellite dishes used with radar and cable TV, for example), as solar energy col-

lectors, and as reflectors (telescopes, light projection, and so on). The planets circle the Sun in approximately *elliptical* orbits. Elliptical surfaces can be used to reflect signals such as light and sound from one place to another. And *hyperbolas* can be used to determine the positions of ships at sea.

The Greeks used the methods of Euclidean geometry to study conics. However, we shall use the more powerful methods of analytic geometry, bringing to bear both algebra and geometry, for our study of conics.

The chapter concludes with sections on equations of conics in polar coordinates, plane curves and parametric equations, and systems of nonlinear equations.

12.1 CONICS

OBJECTIVES   ①   Know the Names of the Conics

Figure 1

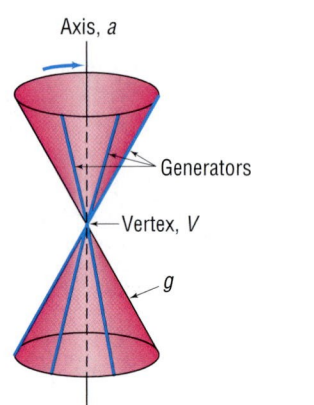

Axis, *a*

Generators

Vertex, *V*

g

①   The word *conic* derives from the word *cone*, which is a geometric figure that can be constructed in the following way: Let *a* and *g* be two distinct lines that intersect at a point *V*. Keep the line *a* fixed. Now rotate the line *g* about *a* while maintaining the same angle between *a* and *g*. The collection of points swept out (generated) by the line *g* is called a **(right circular) cone**. See Figure 1. The fixed line *a* is called the **axis** of the cone; the point *V* is called its **vertex**; the lines that pass through *V* and make the same angle with *a* as *g* are called **generators** of the cone. Each generator is a line that lies entirely on the cone. The cone consists of two parts, called **nappes**, that intersect at the vertex.

Conics, an abbreviation for **conic sections**, are curves that result from the intersection of a (right circular) cone and a plane. The conics we shall study arise when the plane does not contain the vertex, as shown in Figure 2. These conics are **circles** when the plane is perpendicular to the axis of the

Figure 2

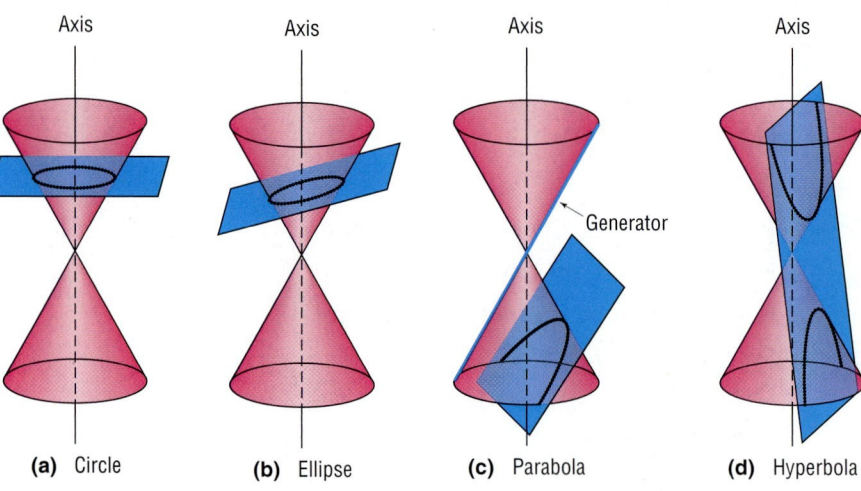

Axis	Axis	Axis	Axis
(a) Circle	**(b)** Ellipse	**(c)** Parabola	**(d)** Hyperbola

Generator

cone and intersects each generator; **ellipses** when the plane is tilted slightly so that it intersects each generator, but intersects only one nappe of the cone; **parabolas** when the plane is tilted farther so that it is parallel to one (and only one) generator and intersects only one nappe of the cone; and **hyperbolas** when the plane intersects both nappes.

If the plane does contain the vertex, the intersection of the plane and the cone is a point, a line, or a pair of intersecting lines. These are usually called **degenerate conics**.

PREPARING FOR THIS SECTION

Before getting started, review the following:

✓ Distance Formula (Section 1.1, p. 93)

✓ Symmetry (Section 3.1, pp. 256–258)

✓ Square Root Method (Section 1.3, pp. 119–120)

✓ Completing the Square (Section 1.3, pp. 120–121)

✓ Graphing Techniques: Transformations (Section 3.4, pp. 286–295)

12.2 THE PARABOLA

OBJECTIVES

1. Find the Equation of a Parabola
2. Graph Parabolas
3. Discuss the Equation of a Parabola
4. Work with Parabolas with Vertex at (h, k)
5. Solve Applied Problems Involving Parabolas

We stated earlier (Section 2.3) that the graph of a quadratic function is a parabola. In this section, we begin with a geometric definition of a parabola and use it to obtain an equation.

> A **parabola** is the collection of all points P in the plane that are the same distance from a fixed point F as they are from a fixed line D. The point F is called the **focus** of the parabola, and the line D is its **directrix**. As a result, a parabola is the set of points P for which
>
> $$d(F, P) = d(P, D) \tag{1}$$

Figure 3

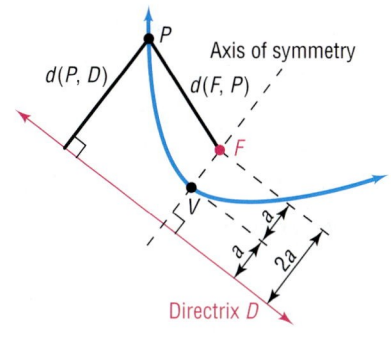

① Figure 3 shows a parabola. The line through the focus F and perpendicular to the directrix D is called the **axis of symmetry** of the parabola. The point of intersection of the parabola with its axis of symmetry is called the **vertex** V.

Because the vertex V lies on the parabola, it must satisfy equation (1): $d(F, V) = d(V, D)$. The vertex is midway between the focus and the directrix. We shall let a equal the distance $d(F, V)$ from F to V. Now we are ready to derive an equation for a parabola. To do this, we use a rectangular system of coordinates positioned so that the vertex V, focus F, and directrix D of the parabola are conveniently located. If we choose to locate the vertex V at the origin $(0, 0)$, then we can conveniently position the focus F on either the x-axis or the y-axis.

Figure 4
$y^2 = 4ax$

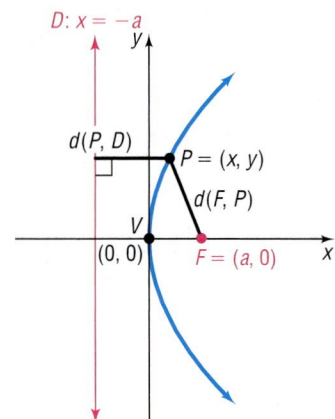

First, we consider the case where the focus F is on the positive x-axis, as shown in Figure 4. Because the distance from F to V is a, the coordinates of F will be $(a, 0)$ with $a > 0$. Similarly, because the distance from V to the directrix D is also a and because D must be perpendicular to the x-axis (since the x-axis is the axis of symmetry), the equation of the directrix D must be $x = -a$. Now, if $P = (x, y)$ is any point on the parabola, then P must obey equation (1):

$$d(F, P) = d(P, D)$$

So we have

$$\sqrt{(x - a)^2 + y^2} = |x + a| \qquad \text{Use the distance formula.}$$
$$(x - a)^2 + y^2 = (x + a)^2 \qquad \text{Square both sides.}$$
$$x^2 - 2ax + a^2 + y^2 = x^2 + 2ax + a^2 \qquad \text{Remove parentheses.}$$
$$y^2 = 4ax \qquad \text{Simplify.}$$

Theorem

Equation of a Parabola; Vertex at $(0, 0)$, Focus at $(a, 0)$, $a > 0$

The equation of a parabola with vertex at $(0, 0)$, focus at $(a, 0)$, and directrix $x = -a, a > 0$, is

$$y^2 = 4ax \qquad (2)$$

EXAMPLE 1 **Finding the Equation of a Parabola and Graphing It by Hand**

Find an equation of the parabola with vertex at $(0, 0)$ and focus at $(3, 0)$. Graph the equation by hand.

Solution The distance from the vertex $(0, 0)$ to the focus $(3, 0)$ is $a = 3$. Based on equation (2), the equation of this parabola is

$$y^2 = 4ax$$
$$y^2 = 12x \qquad a = 3$$

Figure 5
$y^2 = 12x$

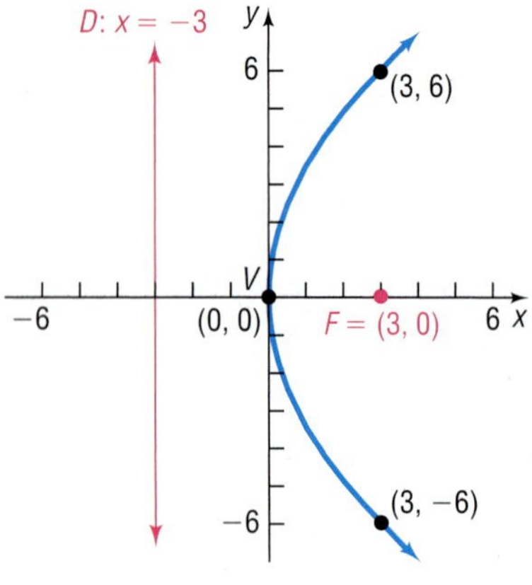

To graph this parabola by hand, it is helpful to plot the two points on the graph above and below the focus. To locate them, we let $x = 3$. Then

$$y^2 = 12x = 12(3) = 36$$
$$y = \pm 6 \qquad \text{Solve for } y.$$

The points on the parabola above and below the focus are $(3, 6)$ and $(3, -6)$. These points help in graphing the parabola because they determine the "opening." See Figure 5.

In general, the points on a parabola $y^2 = 4ax$ that lie above and below the focus $(a, 0)$ are each at a distance $2a$ from the focus. This follows from the fact that if $x = a$ then $y^2 = 4ax = 4a^2$ so $y = \pm 2a$. The line segment joining these two points is called the **latus rectum**; its length is $4a$.

NOW WORK PROBLEM 15.

EXAMPLE 2 **Graphing a Parabola Using a Graphing Utility**

Graph the parabola $y^2 = 12x$.

Figure 6
$y^2 = 12x$

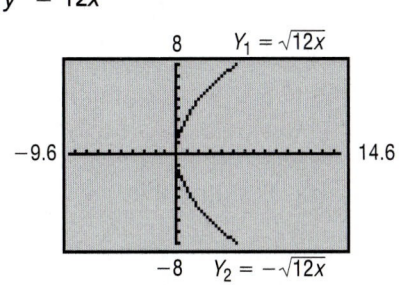

Solution To graph the parabola $y^2 = 12x$, we need to graph the two functions $Y_1 = \sqrt{12x}$ and $Y_2 = -\sqrt{12x}$ on a square screen. Figure 6 shows the graph of $y^2 = 12x$. Notice that the graph fails the vertical line test, so $y^2 = 12x$ is not a function. ∎

By reversing the steps we used to obtain equation (2), it follows that the graph of an equation of the form of equation (2), $y^2 = 4ax$, is a parabola; its vertex is at $(0, 0)$, its focus is at $(a, 0)$, its directrix is the line $x = -a$, and its axis of symmetry is the x-axis.

3 For the remainder of this section, the direction "Discuss the equation" will mean to find the vertex, focus, and directrix of the parabola and graph it.

EXAMPLE 3 **Discussing the Equation of a Parabola**

Discuss the equation: $y^2 = 8x$

Solution The equation $y^2 = 8x$ is of the form $y^2 = 4ax$, where $4a = 8$ so that $a = 2$. Consequently, the graph of the equation is a parabola with vertex at $(0, 0)$ and focus on the positive x-axis at $(2, 0)$. The directrix is the vertical line $x = -2$. The two points defining the latus rectum are obtained by letting $x = 2$. Then $y^2 = 16$ so $y = \pm 4$. See Figure 7(a) for the graph drawn by hand. Figure 7(b) shows the graph obtained using a graphing utility.

Figure 7
$y^2 = 8x$

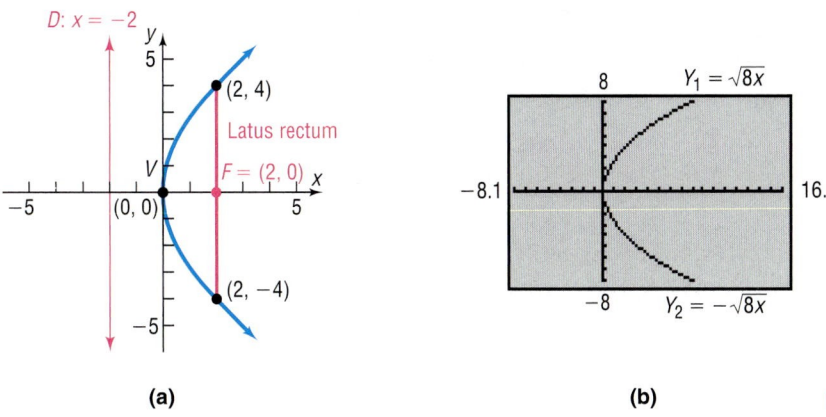

(a)   (b)

Recall that we arrived at equation (2) after placing the focus on the positive x-axis. If the focus is placed on the negative x-axis, positive y-axis, or negative y-axis, a different form of the equation for the parabola results. The four forms of the equation of a parabola with vertex at $(0, 0)$ and focus on a coordinate axis a distance a from $(0, 0)$ are given in Table 1, and their graphs are given in Figure 8. Notice that each graph is symmetric with respect to its axis of symmetry.

TABLE 1	Equations of a Parabola: Vertex at $(0,0)$; Focus on an Axis; $a > 0$				
Vertex	**Focus**	**Directrix**	**Equation**	**Description**
$(0,0)$	$(a, 0)$	$x = -a$	$y^2 = 4ax$	Parabola, axis of symmetry is the x-axis, opens to right
$(0,0)$	$(-a, 0)$	$x = a$	$y^2 = -4ax$	Parabola, axis of symmetry is the x-axis, opens to left
$(0,0)$	$(0, a)$	$y = -a$	$x^2 = 4ay$	Parabola, axis of symmetry is the y-axis, opens up
$(0,0)$	$(0, -a)$	$y = a$	$x^2 = -4ay$	Parabola, axis of symmetry is the y-axis, opens down

Figure 8

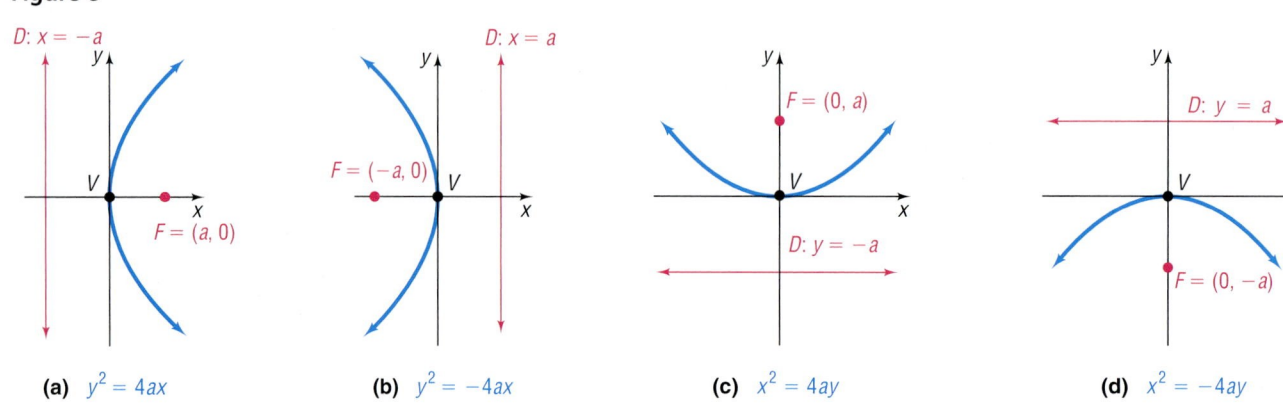

(a) $y^2 = 4ax$ (b) $y^2 = -4ax$ (c) $x^2 = 4ay$ (d) $x^2 = -4ay$

EXAMPLE 4   **Discussing the Equation of a Parabola**

Discuss the equation: $x^2 = -12y$

Solution   The equation $x^2 = -12y$ is of the form $x^2 = -4ay$, with $a = 3$. Consequently, the graph of the equation is a parabola with vertex at $(0,0)$, focus at $(0, -3)$, and directrix the line $y = 3$. The parabola opens down, and its axis of symmetry is the y-axis. To obtain the points defining the latus rectum, let $y = -3$. Then $x^2 = 36$ so $x = \pm 6$. See Figure 9(a) for the graph drawn by hand. Figure 9(b) shows the graph obtained using a graphing utility.

Figure 9
$x^2 = -12y$

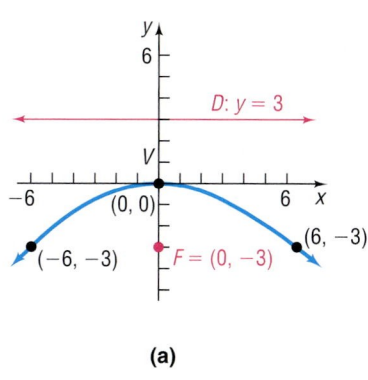

(a)

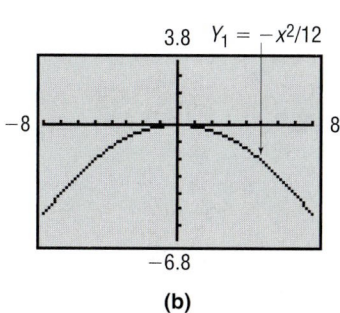

(b)

NOW WORK PROBLEM **33.**

EXAMPLE 5   Finding the Equation of a Parabola

Find the equation of the parabola with focus at $(0, 4)$ and directrix the line $y = -4$. Graph the equation by hand.

Solution   A parabola whose focus is at $(0, 4)$ and whose directrix is the horizontal line $y = -4$ will have its vertex at $(0, 0)$. (Do you see why? The vertex is midway between the focus and the directrix.) Since the focus is on the positive y-axis at $(0, 4)$, the equation of this parabola is of the form $x^2 = 4ay$, with $a = 4$;

$$x^2 = 4ay = 4(4)y = 16y$$
$$\uparrow$$
$$a = 4$$

Figure 10 shows the graph of $x^2 = 16y$.

Figure 10
$x^2 = 16y$

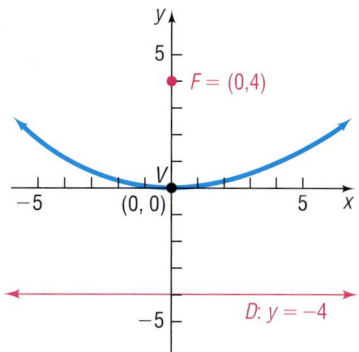

EXAMPLE 6   Finding the Equation of a Parabola

Find the equation of a parabola with vertex at $(0, 0)$ if its axis of symmetry is the x-axis and its graph contains the point $\left(-\dfrac{1}{2}, 2\right)$. Find its focus and directrix, and graph the equation by hand.

Solution   The vertex is at the origin, the axis of symmetry is the x-axis, and the graph contains a point in the second quadrant, so the parabola opens to the left. We see from Table 1 that the form of the equation is

$$y^2 = -4ax$$

Figure 11
$y^2 = -8x$

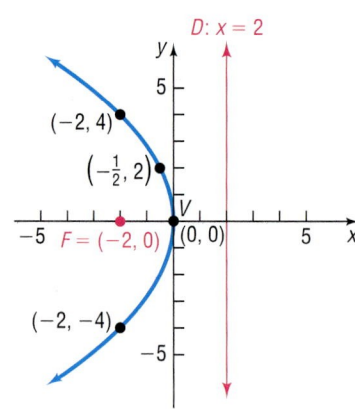

Because the point $\left(-\frac{1}{2}, 2\right)$ is on the parabola, the coordinates $x = -\frac{1}{2}, y = 2$ must satisfy the equation. Substituting $x = -\frac{1}{2}$ and $y = 2$ into the equation, we find that

$$4 = -4a\left(-\frac{1}{2}\right) \qquad y^2 = -4ax, x = -\frac{1}{2}, y = 2$$

$$a = 2$$

The equation of the parabola is

$$y^2 = -4(2)x = -8x$$

The focus is at $(-2, 0)$ and the directrix is the line $x = 2$. Letting $x = -2$, we find $y^2 = 16$ so $y = \pm 4$. The points $(-2, 4)$ and $(-2, -4)$ define the latus rectum. See Figure 11.

NOW WORK PROBLEM 25.

Vertex at (h, k)

4 If a parabola with vertex at the origin and axis of symmetry along a coordinate axis is shifted horizontally h units and then vertically k units, the result is a parabola with vertex at (h, k) and axis of symmetry parallel to a coordinate axis. The equations of such parabolas have the same forms as those in Table 1, but with x replaced by $x - h$ (the horizontal shift) and y replaced by $y - k$ (the vertical shift). Table 2 gives the forms of the equations of such parabolas. Figure 12(a)–(d) illustrates the graphs for $h > 0, k > 0$.

Figure 12

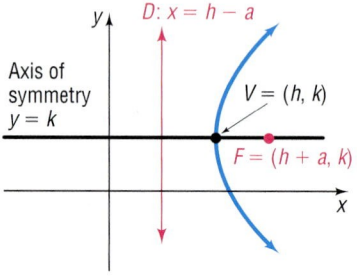

(a) $(y - k)^2 = 4a(x - h)$

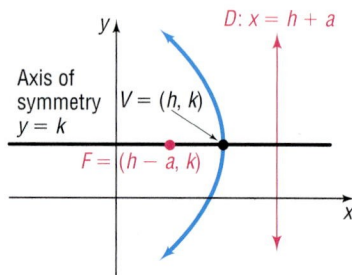

(b) $(y - k)^2 = -4a(x - h)$

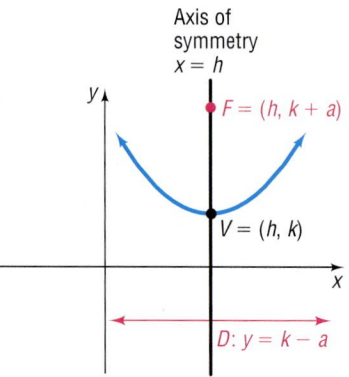

(c) $(x - h)^2 = 4a(y - k)$

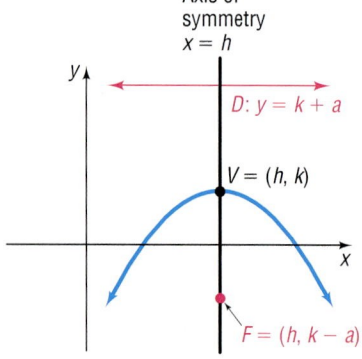

(d) $(x - h)^2 = -4a(y - k)$

TABLE 2 Parabolas with Vertex at (h, k); Axis of Symmetry Parallel to a Coordinate Axis, $a > 0$				
Vertex	**Focus**	**Directrix**	**Equation**	**Description**
(h, k)	$(h + a, k)$	$x = h - a$	$(y - k)^2 = 4a(x - h)$	Parabola, axis of symmetry parallel to x-axis, opens to right
(h, k)	$(h - a, k)$	$x = h + a$	$(y - k)^2 = -4a(x - h)$	Parabola, axis of symmetry parallel to x-axis, opens to left
(h, k)	$(h, k + a)$	$y = k - a$	$(x - h)^2 = 4a(y - k)$	Parabola, axis of symmetry parallel to y-axis, opens up
(h, k)	$(h, k - a)$	$y = k + a$	$(x - h)^2 = -4a(y - k)$	Parabola, axis of symmetry parallel to y-axis, opens down

EXAMPLE 7   Finding the Equation of a Parabola, Vertex Not at Origin

Find an equation of the parabola with vertex at $(-2, 3)$ and focus at $(0, 3)$. Graph the equation by hand.

Figure 13
$(y - 3)^2 = 8(x + 2)$

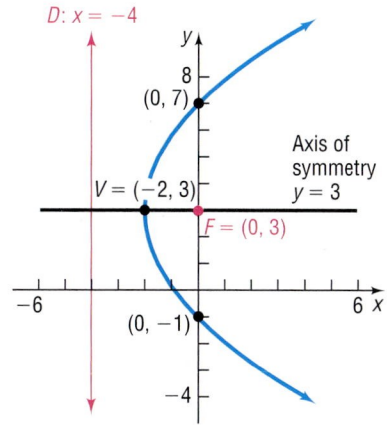

Solution   The vertex $(-2, 3)$ and focus $(0, 3)$ both lie on the horizontal line $y = 3$ (the axis of symmetry). The distance a from the vertex $(-2, 3)$ to the focus $(0, 3)$ is $a = 2$. Also, because the focus lies to the right of the vertex, we know that the parabola opens to the right. Consequently, the form of the equation is

$$(y - k)^2 = 4a(x - h)$$

where $(h, k) = (-2, 3)$ and $a = 2$. Therefore, the equation is

$$(y - 3)^2 = 4 \cdot 2[x - (-2)]$$
$$(y - 3)^2 = 8(x + 2)$$

If $x = 0$, then $(y - 3)^2 = 16$. Then, $y - 3 = \pm 4$ so $y = -1$ or $y = 7$. The points $(0, -1)$ and $(0, 7)$ define the latus rectum; the line $x = -4$ is the directrix. See Figure 13.   ■

➤ **NOW WORK PROBLEM 23.**

EXAMPLE 8   Using a Graphing Utility to Graph a Parabola, Vertex Not at Origin

Figure 14

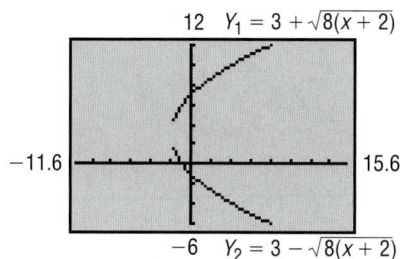

Using a graphing utility, graph the equation $(y - 3)^2 = 8(x + 2)$.

Solution   First, we must solve the equation for y.

$$(y - 3)^2 = 8(x + 2)$$
$$y - 3 = \pm\sqrt{8(x + 2)} \qquad \text{Use the Square Root Method.}$$
$$y = 3 \pm \sqrt{8(x + 2)} \qquad \text{Add 3 to both sides.}$$

Figure 14 shows the graphs of the equations $Y_1 = 3 + \sqrt{8(x + 2)}$ and $Y_2 = 3 - \sqrt{8(x + 2)}$.   ■

➤ **NOW WORK PROBLEM 39.**

Polynomial equations define parabolas whenever they involve two variables that are quadratic in one variable and linear in the other. To discuss this type of equation, we first complete the square of the variable that is quadratic.

EXAMPLE 9 | **Discussing the Equation of a Parabola**

Discuss the equation: $x^2 + 4x - 4y = 0$

Figure 15
$x^2 + 4x - 4y = 0$

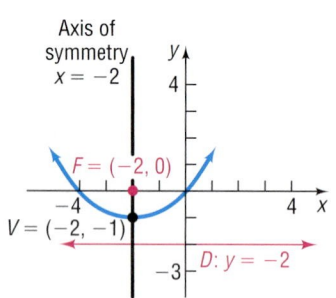

Solution | To discuss the equation $x^2 + 4x - 4y = 0$, we complete the square involving the variable x.

$$x^2 + 4x - 4y = 0$$
$$x^2 + 4x = 4y \qquad \text{Isolate the terms involving } x \text{ on the left side.}$$
$$x^2 + 4x + 4 = 4y + 4 \qquad \text{Complete the square on the left side.}$$
$$(x + 2)^2 = 4(y + 1) \qquad \text{Factor.}$$

This equation is of the form $(x - h)^2 = 4a(y - k)$, with $h = -2, k = -1$, and $a = 1$. The graph is a parabola with vertex at $(h, k) = (-2, -1)$ that opens up. The focus is at $(-2, 0)$, and the directrix is the line $y = -2$. See Figure 15.

NOW WORK PROBLEM **41.**

5

Parabolas find their way into many applications. For example, as we discussed in Section 2.3, suspension bridges have cables in the shape of a parabola. Another property of parabolas that is used in applications is their reflecting property.

Reflecting Property

Suppose that a mirror is shaped like a **paraboloid of revolution**, a surface formed by rotating a parabola about its axis of symmetry. If a light (or any other emitting source) is placed at the focus of the parabola, all the rays emanating from the light will reflect off the mirror in lines parallel to the axis of symmetry. This principle is used in the design of searchlights, flashlights, certain automobile headlights, and other such devices. See Figure 16.

Conversely, suppose that rays of light (or other signals) emanate from a distant source so that they are essentially parallel. When these rays strike the surface of a parabolic mirror whose axis of symmetry is parallel to these rays, they are reflected to a single point at the focus. This principle is used in the design of some solar energy devices, satellite dishes, and the mirrors used in some types of telescopes. See Figure 17.

Figure 16
Searchlight

Figure 17
Telescope

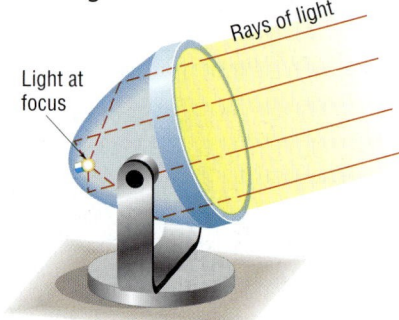

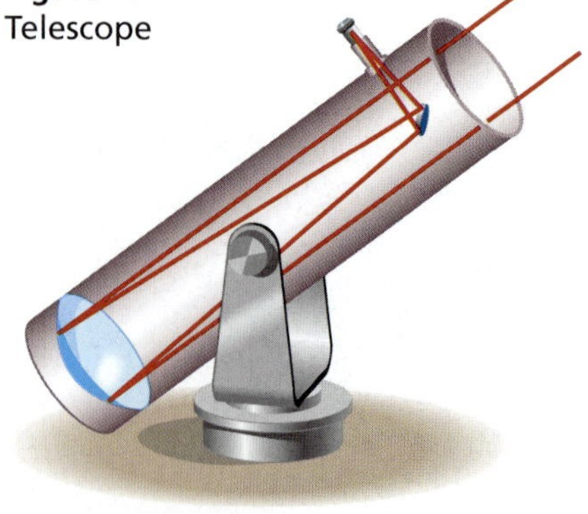

EXAMPLE 10   **Satellite Dish**

A satellite dish is shaped like a paraboloid of revolution. The signals that emanate from a satellite strike the surface of the dish and are reflected to a single point, where the receiver is located. If the dish is 8 feet across at its opening and is 3 feet deep at its center, at what position should the receiver be placed?

Solution   Figure 18(a) shows the satellite dish. We draw the parabola used to form the dish on a rectangular coordinate system so that the vertex of the parabola is at the origin and its focus is on the positive *y*-axis. See Figure 18(b).

Figure 18

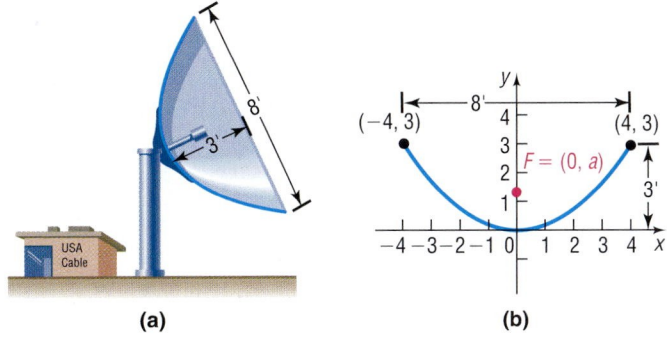

(a)   (b)

The form of the equation of the parabola is

$$x^2 = 4ay$$

and its focus is at $(0, a)$. Since $(4, 3)$ is a point on the graph, we have

$$4^2 = 4a(3)$$

$$a = \frac{4}{3}$$

The receiver should be located $1\frac{1}{3}$ feet from the base of the dish, along its axis of symmetry.   ■

 **NOW WORK PROBLEM 57.**

12.2 Concepts and Vocabulary

In Problems 1–3, fill in the blanks.

1. A(n) _____ is the collection of all points in the plane such that the distance from each point to a fixed point equals its distance to a fixed line.

2. The surface formed by rotating a parabola about its axis of symmetry is called a _____ _____ _____.

3. The line segment joining the two points on a parabola above and below its focus is called the _____ _____.

In Problems 4–6, answer True or False to each statement.

4. The vertex of a parabola is a point on the parabola that also is on its axis of symmetry.

5. If a light is placed at the focus of a parabola, all the rays reflected off the parabola will be parallel to the axis of symmetry.

6. The graph of a quadratic function is a parabola.

7. Write down the four equations that are parabolas with vertex at the origin and axis along a coordinate axis.

8. Draw a parabola, labeling its vertex, axis of symmetry, focus, and directrix.

12.2 Exercises

In Problems 1–8, the graph of a parabola is given. Match each graph to its equation.

A. $y^2 = 4x$ C. $y^2 = -4x$ E. $(y - 1)^2 = 4(x - 1)$ G. $(y - 1)^2 = -4(x - 1)$

B. $x^2 = 4y$ D. $x^2 = -4y$ F. $(x + 1)^2 = 4(y + 1)$ H. $(x + 1)^2 = -4(y + 1)$

1.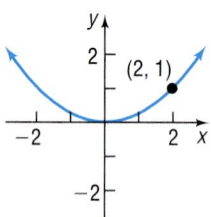

2.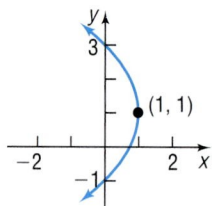

3.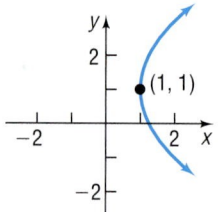

4.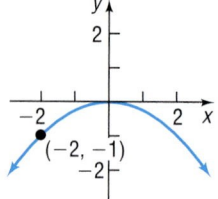

5.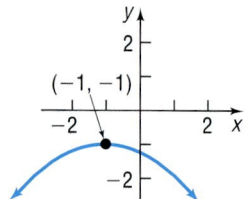

6.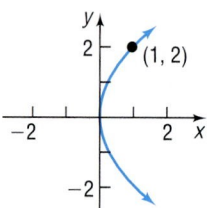

7.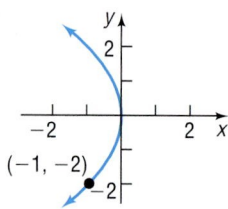

8.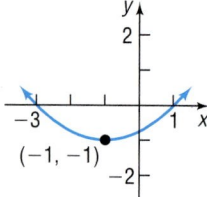

In Problems 9–14, the graph of a parabola is given. Match each graph to its equation.

A. $x^2 = 6y$ B. $y^2 = 6x$ C. $y^2 = -6x$

D. $(x + 2)^2 = -6(y - 2)$ E. $(y - 2)^2 = 6(x + 2)$ F. $(x + 2)^2 = 6(y - 2)$

9.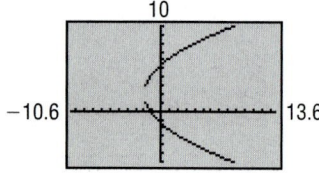

10.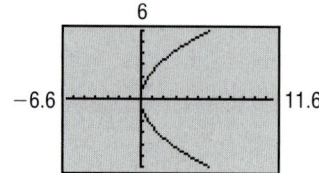

11.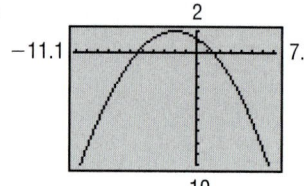

12.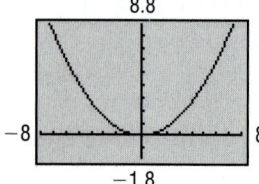

13.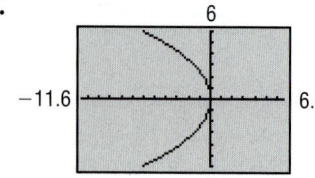

14.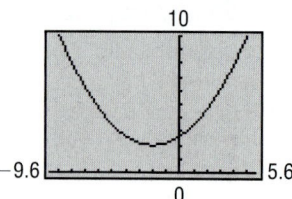

In Problems 15–30, find the equation of the parabola described. Find the two points that define the latus rectum, and graph the equation by hand.

15. Focus at $(4, 0)$;   vertex at $(0, 0)$

16. Focus at $(0, 2)$;   vertex at $(0, 0)$

17. Focus at $(0, -3)$;   vertex at $(0, 0)$

18. Focus at $(-4, 0)$;   vertex at $(0, 0)$

19. Focus at $(-2, 0)$;   directrix the line $x = 2$

20. Focus at $(0, -1)$;   directrix the line $y = 1$

21. Directrix the line $y = -\dfrac{1}{2}$;   vertex at $(0, 0)$

22. Directrix the line $x = -\dfrac{1}{2}$;   vertex at $(0, 0)$

23. Vertex at $(2, -3)$;   focus at $(2, -5)$

24. Vertex at $(4, -2)$;   focus at $(6, -2)$

25. Vertex at $(0, 0)$;   axis of symmetry the y-axis; containing the point $(2, 3)$

26. Vertex at $(0, 0)$;   axis of symmetry the x-axis; containing the point $(2, 3)$

27. Focus at $(-3, 4)$;   directrix the line $y = 2$

28. Focus at $(2, 4)$;   directrix the line $x = -4$

29. Focus at $(-3, -2)$;   directrix the line $x = 1$

30. Focus at $(-4, 4)$;   directrix the line $y = -2$

In Problems 31–48, find the vertex, focus, and directrix of each parabola. Graph the equation (a) by hand and (b) by using a graphing utility.

31. $x^2 = 4y$

32. $y^2 = 8x$

33. $y^2 = -16x$

34. $x^2 = -4y$

35. $(y - 2)^2 = 8(x + 1)$

36. $(x + 4)^2 = 16(y + 2)$

37. $(x - 3)^2 = -(y + 1)$

38. $(y + 1)^2 = -4(x - 2)$

39. $(y + 3)^2 = 8(x - 2)$

40. $(x - 2)^2 = 4(y - 3)$

41. $y^2 - 4y + 4x + 4 = 0$

42. $x^2 + 6x - 4y + 1 = 0$

43. $x^2 + 8x = 4y - 8$

44. $y^2 - 2y = 8x - 1$

45. $y^2 + 2y - x = 0$

46. $x^2 - 4x = 2y$

47. $x^2 - 4x = y + 4$

48. $y^2 + 12y = -x + 1$

In Problems 49–56, write an equation for each parabola.

49.
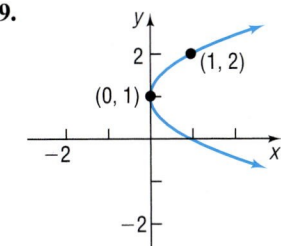

50.
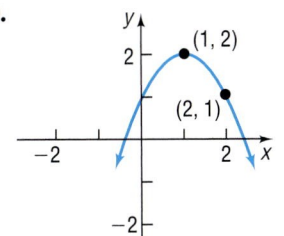

51.
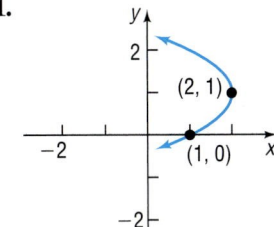

52.
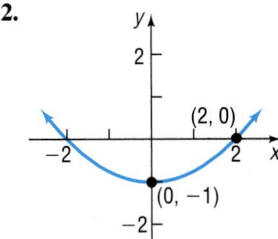

53.
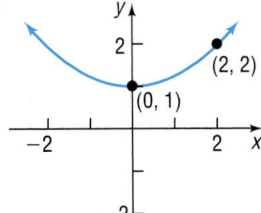

54.
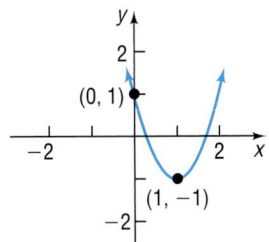

55.

56.
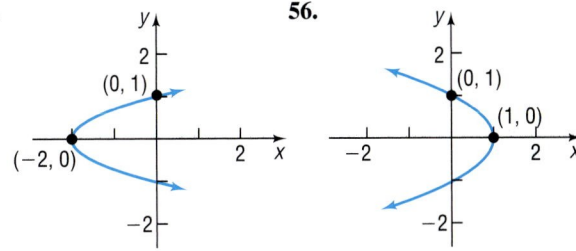

57. Satellite Dish   A satellite dish is shaped like a paraboloid of revolution. The signals that emanate from a satellite strike the surface of the dish and are reflected to a single point, where the receiver is located. If the dish is 10 feet across at its opening and is 4 feet deep at its center, at what position should the receiver be placed?

58. Constructing a TV Dish   A cable TV receiving dish is in the shape of a paraboloid of revolution. Find the location

of the receiver, which is placed at the focus, if the dish is 6 feet across at its opening and 2 feet deep.

59. Constructing a Flashlight   The reflector of a flashlight is in the shape of a paraboloid of revolution. Its diameter is 4 inches and its depth is 1 inch. How far from the vertex should the light bulb be placed so that the rays will be reflected parallel to the axis?

60. Constructing a Headlight A sealed-beam headlight is in the shape of a paraboloid of revolution. The bulb, which is placed at the focus, is 1 inch from the vertex. If the depth is to be 2 inches, what is the diameter of the headlight at its opening?

61. Suspension Bridge The cables of a suspension bridge are in the shape of a parabola, as shown in the figure. The towers supporting the cable are 600 feet apart and 80 feet high. If the cables touch the road surface midway between the towers, what is the height of the cable at a point 150 feet from the center of the bridge?

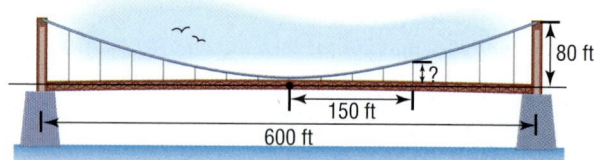

62. Suspension Bridge The cables of a suspension bridge are in the shape of a parabola. The towers supporting the cable are 400 feet apart and 100 feet high. If the cables are at a height of 10 feet midway between the towers, what is the height of the cable at a point 50 feet from the center of the bridge?

63. Searchlight A searchlight is shaped like a paraboloid of revolution. If the light source is located 2 feet from the base along the axis of symmetry and the opening is 5 feet across, how deep should the searchlight be?

64. Searchlight A searchlight is shaped like a paraboloid of revolution. If the light source is located 2 feet from the base along the axis of symmetry and the depth of the searchlight is 4 feet, what should the width of the opening be?

65. Solar Heat A mirror is shaped like a paraboloid of revolution and will be used to concentrate the rays of the sun at its focus, creating a heat source. (See the figure.) If the mirror is 20 feet across at its opening and is 6 feet deep, where will the heat source be concentrated?

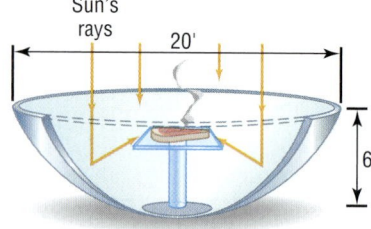

66. Reflecting Telescope A reflecting telescope contains a mirror shaped like a paraboloid of revolution. If the mirror is 4 inches across at its opening and is 3 feet deep, where will the collected light be concentrated?

67. Parabolic Arch Bridge A bridge is built in the shape of a parabolic arch. The bridge has a span of 120 feet and a maximum height of 25 feet. See the illustration. Choose a suitable rectangular coordinate system and find the height of the arch at distances of 10, 30, and 50 feet from the center.

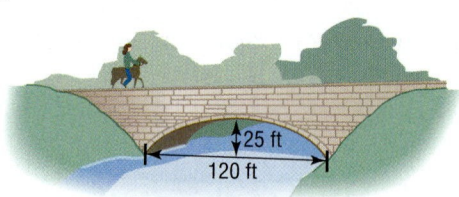

68. Parabolic Arch Bridge A bridge is to be built in the shape of a parabolic arch and is to have a span of 100 feet. The height of the arch a distance of 40 feet from the center is to be 10 feet. Find the height of the arch at its center.

69. Show that an equation of the form

$$Ax^2 + Ey = 0, \qquad A \neq 0, E \neq 0$$

is the equation of a parabola with vertex at $(0,0)$ and axis of symmetry the y-axis. Find its focus and directrix.

70. Show that an equation of the form

$$Cy^2 + Dx = 0, \quad C \neq 0, D \neq 0$$

is the equation of a parabola with vertex at $(0,0)$ and axis of symmetry the x-axis. Find its focus and directrix.

71. Show that the graph of an equation of the form

$$Ax^2 + Dx + Ey + F = 0, \quad A \neq 0$$

(a) Is a parabola if $E \neq 0$.
(b) Is a vertical line if $E = 0$ and $D^2 - 4AF = 0$.
(c) Is two vertical lines if $E = 0$ and $D^2 - 4AF > 0$.
(d) Contains no points if $E = 0$ and $D^2 - 4AF < 0$.

72. Show that the graph of an equation of the form

$$Cy^2 + Dx + Ey + F = 0, \quad C \neq 0$$

(a) Is a parabola if $D \neq 0$.
(b) Is a horizontal line if $D = 0$ and $E^2 - 4CF = 0$.
(c) Is two horizontal lines if $D = 0$ and $E^2 - 4CF > 0$.
(d) Contains no points if $D = 0$ and $E^2 - 4CF < 0$.

PREPARING FOR THIS SECTION

Before getting started, review the following:

✓ Distance Formula (Section 1.1, p. 93)

✓ Completing the Square (Section 1.3, pp. 120–121)

✓ Intercepts (Section 1.2, pp. 106–107)

✓ Square Root Method (Section 1.3, 119–120)

✓ Symmetry (Section 3.1, pp. 256–258)

✓ Circles (Section 1.8, pp. 180–184)

✓ Graphing Techniques: Transformations (Section 3.4, pp. 286–295)

12.3 THE ELLIPSE

OBJECTIVES
1. Find the Equation of an Ellipse
2. Graph Ellipses
3. Discuss the Equation of an Ellipse
4. Work with Ellipses with Center at (h, k)
5. Solve Applied Problems Involving Ellipses

> An **ellipse** is the collection of all points in the plane the sum of whose distances from two fixed points, called the **foci**, is a constant.

Figure 19

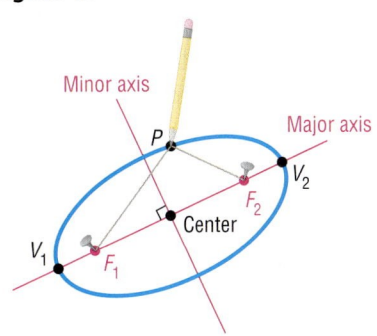

The definition actually contains within it a physical means for drawing an ellipse. Find a piece of string (the length of this string is the constant referred to in the definition). Then take two thumbtacks (the foci) and stick them on a piece of cardboard so that the distance between them is less than the length of the string. Now attach the ends of the string to the thumbtacks and, using the point of a pencil, pull the string taut. See Figure 19. Keeping the string taut, rotate the pencil around the two thumbtacks. The pencil traces out an ellipse, as shown in Figure 19.

In Figure 19, the foci are labeled F_1 and F_2. The line containing the foci is called the **major axis**. The midpoint of the line segment joining the foci is called the **center** of the ellipse. The line through the center and perpendicular to the major axis is called the **minor axis**.

The two points of intersection of the ellipse and the major axis are the **vertices**, V_1 and V_2, of the ellipse. The distance from one vertex to the other is called the **length of the major axis**. The ellipse is symmetric with respect to its major axis and with respect to its minor axis.

Figure 20
$$d(F_1, P) + d(F_2, P) = 2a$$

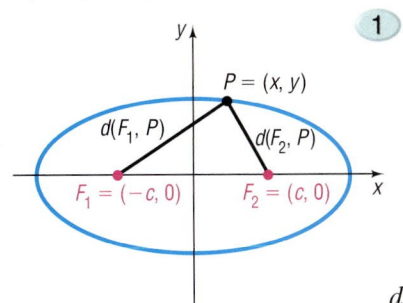

With these ideas in mind, we are now ready to find the equation of an ellipse in a rectangular coordinate system. First, we place the center of the ellipse at the origin. Second, we position the ellipse so that its major axis coincides with a coordinate axis. Suppose that the major axis coincides with the x-axis, as shown in Figure 20. If c is the distance from the center to a focus, then one focus will be at $F_1 = (-c, 0)$ and the other at $F_2 = (c, 0)$. As we shall see, it is convenient to let $2a$ denote the constant distance referred to in the definition. Then, if $P = (x, y)$ is any point on the ellipse, we have

$$d(F_1, P) + d(F_2, P) = 2a$$
Sum of the distances from P to the foci equals a constant, $2a$.

$$\sqrt{(x + c)^2 + y^2} + \sqrt{(x - c)^2 + y^2} = 2a$$
Use the distance formula.

$$\sqrt{(x + c)^2 + y^2} = 2a - \sqrt{(x - c)^2 + y^2}$$
Isolate one radical.

$$(x + c)^2 + y^2 = 4a^2 - 4a\sqrt{(x - c)^2 + y^2} \\ + (x - c)^2 + y^2$$
Square both sides.

$$x^2 + 2cx + c^2 + y^2 = 4a^2 - 4a\sqrt{(x - c)^2 + y^2} \\ + x^2 - 2cx + c^2 + y^2$$
Remove parentheses.

$$4cx - 4a^2 = -4a\sqrt{(x - c)^2 + y^2}$$
Simplify; Isolate the radical.

$$cx - a^2 = -a\sqrt{(x - c)^2 + y^2}$$
Divide each side by 4.

$$(cx - a^2)^2 = a^2[(x - c)^2 + y^2]$$
Square both sides again.

$$c^2x^2 - 2a^2cx + a^4 = a^2(x^2 - 2cx + c^2 + y^2) \qquad \text{Remove parentheses.}$$
$$(c^2 - a^2)x^2 - a^2y^2 = a^2c^2 - a^4 \qquad \text{Rearrange the terms.}$$
$$(a^2 - c^2)x^2 + a^2y^2 = a^2(a^2 - c^2) \qquad \text{Multiply each side by } -1; \qquad (1)$$
$$\text{factor } a^2 \text{ on the right side.}$$

To obtain points on the ellipse off the x-axis, it must be that $a > c$. To see why, look again at Figure 20.

$$d(F_1, P) + d(F_2, P) > d(F_1, F_2) \qquad \text{The sum of the lengths of two sides of a}$$
$$\text{triangle is greater than the length of the third side.}$$
$$2a > 2c \qquad d(F_1, P) + d(F_2, P) = 2a; d(F_1, F_2) = 2c.$$
$$a > c$$

Since $a > c$, we also have $a^2 > c^2$, so $a^2 - c^2 > 0$. Let $b^2 = a^2 - c^2, b > 0$. Then $a > b$ and equation (1) can be written as

$$b^2x^2 + a^2y^2 = a^2b^2$$
$$\frac{x^2}{a^2} + \frac{y^2}{b^2} = 1 \qquad \text{Divide each side by } a^2 b^2.$$

Theorem

Equation of an Ellipse; Center at $(0,0)$; Foci at $(\pm c, 0)$; Major Axis along the x-Axis

An equation of the ellipse with center at $(0,0)$, foci at $(-c, 0)$ and $(c, 0)$, and vertices at $(-a, 0)$ and $(a, 0)$ is

$$\frac{x^2}{a^2} + \frac{y^2}{b^2} = 1, \qquad \text{where } a > b > 0 \text{ and } b^2 = a^2 - c^2 \quad (2)$$

The major axis is the x-axis.

② As you can verify, the ellipse defined by equation (2) is symmetric with respect to the x-axis, y-axis, and origin.

To find the vertices of the ellipse defined by equation (2), let $y = 0$. The vertices satisfy the equation $\frac{x^2}{a^2} = 1$, the solutions of which are $x = \pm a$. Consequently, the vertices of the ellipse given by equation (2) are $V_1 = (-a, 0)$ and $V_2 = (a, 0)$. The y-intercepts of the ellipse, found by letting $x = 0$, have coordinates $(0, -b)$ and $(0, b)$. These four intercepts, $(a, 0)$, $(-a, 0)$, $(0, b)$, and $(0, -b)$, are used to graph the ellipse. See Figure 21.

Notice in Figure 21 the right triangle formed with the points $(0,0)$, $(c, 0)$, and $(0, b)$. Because $b^2 = a^2 - c^2$ (or $b^2 + c^2 = a^2$), the distance from the focus at $(c, 0)$ to the point $(0, b)$ is a.

Figure 21

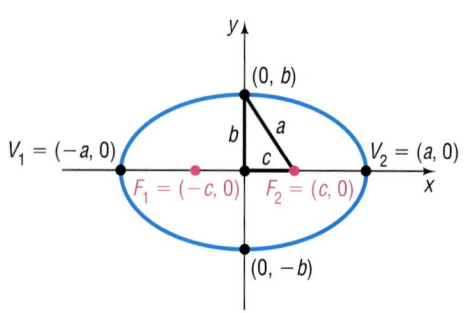

EXAMPLE 1 **Finding an Equation of an Ellipse**

Find an equation of the ellipse with center at the origin, one focus at $(3, 0)$, and a vertex at $(-4, 0)$. Graph the equation by hand.

Figure 22
$$\frac{x^2}{16} + \frac{y^2}{7} = 1$$

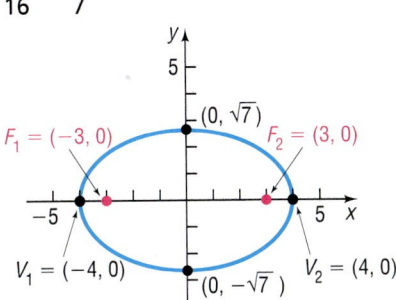

Solution The ellipse has its center at the origin and, since the given focus and vertex lie on the x-axis, the major axis is the x-axis. The distance from the center, $(0, 0)$, to one of the foci, $(3, 0)$, is $c = 3$. The distance from the center, $(0, 0)$, to one of the vertices, $(-4, 0)$, is $a = 4$. From equation (2), it follows that

$$b^2 = a^2 - c^2 = 16 - 9 = 7$$

so an equation of the ellipse is

$$\frac{x^2}{16} + \frac{y^2}{7} = 1$$

Figure 22 shows the graph drawn by hand. ■

Notice in Figure 22 how we used the intercepts of the equation to graph the ellipse. Following this practice will make it easier for you to obtain an accurate graph of an ellipse when graphing by hand. It also tells you how to set the initial viewing window when using a graphing utility.

EXAMPLE 2 **Graphing an Ellipse Using a Graphing Utility**

Use a graphing utility to graph the ellipse $\frac{x^2}{16} + \frac{y^2}{7} = 1$.

Solution First, we must solve $\frac{x^2}{16} + \frac{y^2}{7} = 1$ for y.

Figure 23

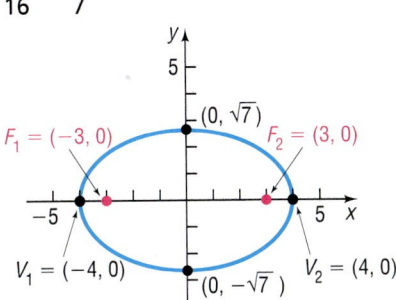

$$\frac{y^2}{7} = 1 - \frac{x^2}{16} \qquad \text{Subtract } \frac{x^2}{16} \text{ from each side.}$$

$$y^2 = 7\left(1 - \frac{x^2}{16}\right) \qquad \text{Multiply both sides by 7.}$$

$$y = \pm\sqrt{7\left(1 - \frac{x^2}{16}\right)} \qquad \text{Apply the Square Root Method.}$$

Figure 23* shows the graphs of $Y_1 = \sqrt{7\left(1 - \frac{x^2}{16}\right)}$ and $Y_2 = -\sqrt{7\left(1 - \frac{x^2}{16}\right)}$.

Notice in Figure 23 that we used a square screen. As with circles and parabolas, this is done to avoid a distorted view of the graph.

 NOW WORK PROBLEM 19.

An equation of the form of equation (2), with $a > b$, is the equation of an ellipse with center at the origin, foci on the x-axis at $(-c, 0)$ and $(c, 0)$, where $c^2 = a^2 - b^2$, and major axis along the x-axis.

③ For the remainder of this section, the direction "Discuss the equation" will mean to find the center, major axis, foci, and vertices of the ellipse and graph it.

---

*The initial viewing window selected was $X\text{min} = -4$, $X\text{max} = 4$, $Y\text{min} = -3$, $Y\text{max} = 3$. Then we used the ZOOM-SQUARE option to obtain the window shown.

EXAMPLE 3 **Discussing the Equation of an Ellipse**

Discuss the equation: $\dfrac{x^2}{25} + \dfrac{y^2}{9} = 1$

Solution The given equation is of the form of equation (2), with $a^2 = 25$ and $b^2 = 9$. The equation is that of an ellipse with center $(0,0)$ and major axis along the x-axis. The vertices are at $(\pm a, 0) = (\pm 5, 0)$. Because $b^2 = a^2 - c^2$, we find that

$$c^2 = a^2 - b^2 = 25 - 9 = 16$$

The foci are at $(\pm c, 0) = (\pm 4, 0)$. Figure 24(a) shows the graph drawn by hand. Figure 24(b) shows the graph obtained using a graphing utility.

Figure 24
$\dfrac{x^2}{25} + \dfrac{y^2}{9} = 1$

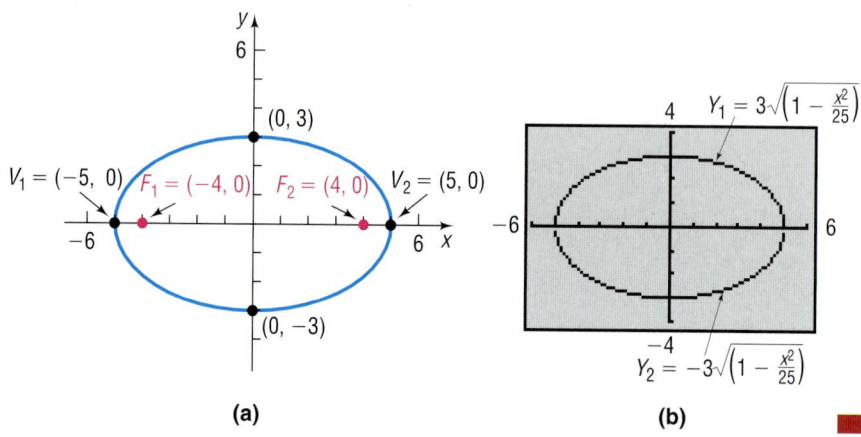

(a)   (b)

NOW WORK PROBLEM 9.

If the major axis of an ellipse with center at $(0,0)$ lies on the y-axis, then the foci are at $(0, -c)$ and $(0, c)$. Using the same steps as before, the definition of an ellipse leads to the following result:

Theorem **Equation of an Ellipse; Center at $(0,0)$; Foci at $(0, \pm c)$; Major Axis along the y-Axis**

An equation of the ellipse with center at $(0,0)$, foci at $(0, -c)$ and $(0, c)$, and vertices at $(0, -a)$ and $(0, a)$ is

$$\dfrac{x^2}{b^2} + \dfrac{y^2}{a^2} = 1, \qquad \text{where } a > b > 0 \text{ and } b^2 = a^2 - c^2 \quad (3)$$

The major axis is the y-axis.

Figure 25

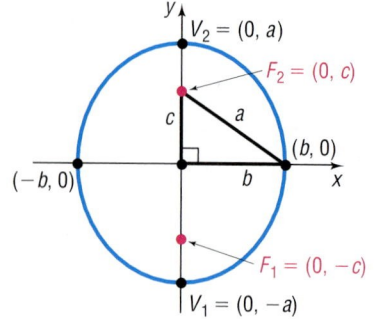

Figure 25 illustrates the graph of such an ellipse. Again, notice the right triangle with the points at $(0,0)$, $(b, 0)$, and $(0, c)$.

Look closely at equations (2) and (3). Although they may look alike, there is a difference! In equation (2), the larger number, a^2, is in the denominator of the x^2-term, so the major axis of the ellipse is along the x-axis.

In equation (3), the larger number, a^2, is in the denominator of the y^2-term, so the major axis is along the y-axis.

EXAMPLE 4 **Discussing the Equation of an Ellipse**

Discuss the equation: $9x^2 + y^2 = 9$

Solution To put the equation in proper form, we divide each side by 9.

$$x^2 + \frac{y^2}{9} = 1$$

The larger number, 9, is in the denominator of the y^2-term so, based on equation (3), this is the equation of an ellipse with center at the origin and major axis along the y-axis. Also, we conclude that $a^2 = 9, b^2 = 1$, and $c^2 = a^2 - b^2 = 9 - 1 = 8$. The vertices are at $(0, \pm a) = (0, \pm 3)$, and the foci are at $(0, \pm c) = (0, \pm 2\sqrt{2})$. Figure 26(a) shows the graph drawn by hand. Figure 26(b) shows the graph obtained using a graphing utility.

Figure 26
$$x^2 + \frac{y^2}{9} = 1$$

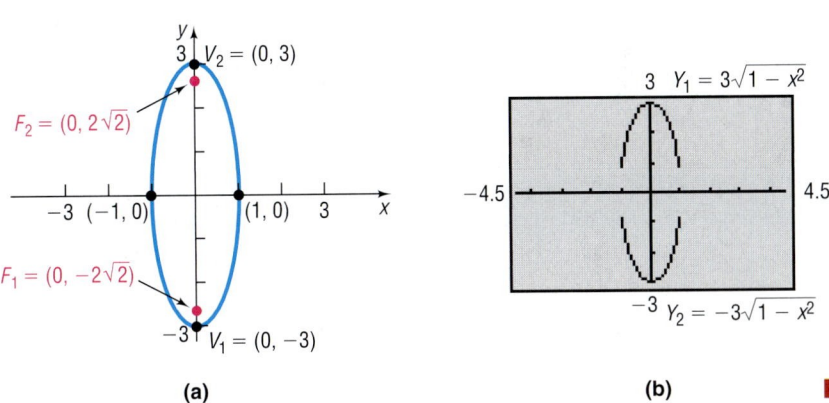

(a)

(b)

NOW WORK PROBLEM **13**.

EXAMPLE 5 **Finding an Equation of an Ellipse**

Find an equation of the ellipse having one focus at $(0, 2)$ and vertices at $(0, -3)$ and $(0, 3)$. Graph the equation by hand.

Figure 27
$$\frac{x^2}{5} + \frac{y^2}{9} = 1$$

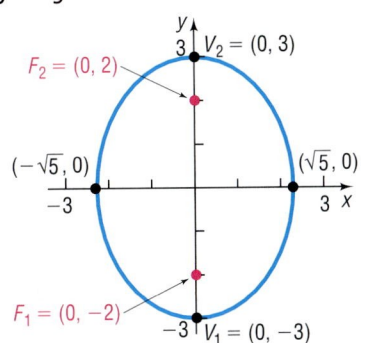

Solution Because the vertices are at $(0, -3)$ and $(0, 3)$, the center of this ellipse is at their midpoint, the origin. Also, its major axis lies on the y-axis. The distance from the center, $(0, 0)$, to one of the foci, $(0, 2)$, is $c = 2$. The distance from the center, $(0, 0)$, to one of the vertices, $(0, 3)$, is $a = 3$. So $b^2 = a^2 - c^2 = 9 - 4 = 5$. The form of the equation of this ellipse is given by equation (3).

$$\frac{x^2}{b^2} + \frac{y^2}{a^2} = 1$$

$$\frac{x^2}{5} + \frac{y^2}{9} = 1$$

Figure 27 shows the graph.

NOW WORK PROBLEM **21**.

The circle may be considered a special kind of ellipse. To see why, let $a = b$ in equation (2) or (3). Then

$$\frac{x^2}{a^2} + \frac{y^2}{a^2} = 1$$

$$x^2 + y^2 = a^2$$

This is the equation of a circle with center at the origin and radius a. The value of c is

$$c^2 = a^2 - b^2 = 0$$

We conclude that the closer the two foci of an ellipse are to the center, the more the ellipse will look like a circle.

Center at (h, k)

④ If an ellipse with center at the origin and major axis coinciding with a coordinate axis is shifted horizontally h units and then vertically k units, the result is an ellipse with center at (h, k) and major axis parallel to a coordinate axis. The equations of such ellipses have the same forms as those given in equations (2) and (3), except that x is replaced by $x - h$ (the horizontal shift) and y is replaced by $y - k$ (the vertical shift). Table 3 gives the forms of the equations of such ellipses, and Figure 28 shows their graphs.

TABLE 3   Ellipses with Center at (h, k) and Major Axis Parallel to a Coordinate Axis

Center	Major Axis	Foci	Vertices	Equation
(h, k)	Parallel to x-axis	$(h + c, k)$	$(h + a, k)$	$\dfrac{(x - h)^2}{a^2} + \dfrac{(y - k)^2}{b^2} = 1,$
		$(h - c, k)$	$(h - a, k)$	$a > b$ and $b^2 = a^2 - c^2$
(h, k)	Parallel to y-axis	$(h, k + c)$	$(h, k + a)$	$\dfrac{(x - h)^2}{b^2} + \dfrac{(y - k)^2}{a^2} = 1,$
		$(h, k - c)$	$(h, k - a)$	$a > b$ and $b^2 = a^2 - c^2$

Figure 28

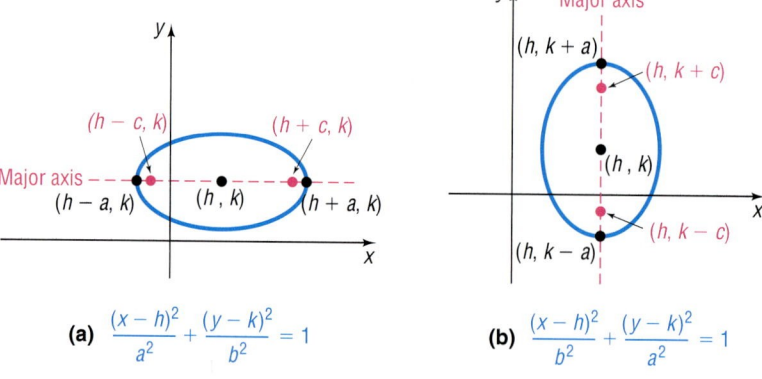

(a) $\dfrac{(x - h)^2}{a^2} + \dfrac{(y - k)^2}{b^2} = 1$

(b) $\dfrac{(x - h)^2}{b^2} + \dfrac{(y - k)^2}{a^2} = 1$

EXAMPLE 6   **Finding an Equation of an Ellipse, Center Not at the Origin**

Find an equation for the ellipse with center at $(2, -3)$, one focus at $(3, -3)$, and one vertex at $(5, -3)$. Graph the equation by hand.

Solution   The center is at $(h, k) = (2, -3)$, so $h = 2$ and $k = -3$. Since the center, focus, and vertex all lie on the line $y = -3$, the major axis is parallel to the x-axis. The distance from the center $(2, -3)$ to a focus $(3, -3)$ is $c = 1$; the distance from the center $(2, -3)$ to a vertex $(5, -3)$ is $a = 3$. Then, $b^2 = a^2 - c^2 = 9 - 1 = 8$. The form of the equation is

$$\frac{(x - h)^2}{a^2} + \frac{(y - k)^2}{b^2} = 1, \quad \text{where } h = 2, k = -3, a = 3, b = 2\sqrt{2}$$

$$\frac{(x - 2)^2}{9} + \frac{(y + 3)^2}{8} = 1$$

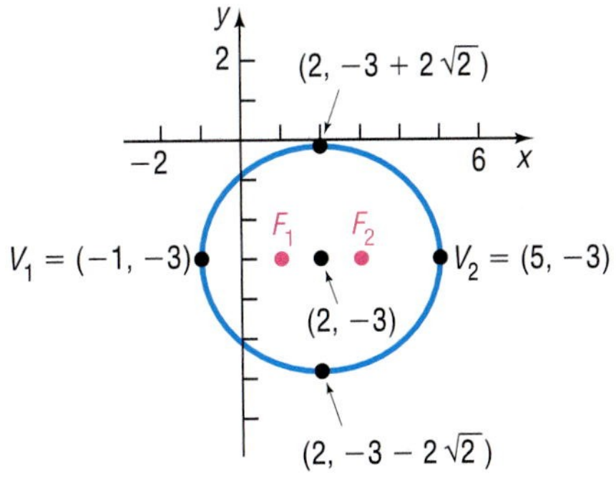

To graph the equation, we use the center $(h, k) = (2, -3)$ to locate the vertices. The major axis is parallel to the x-axis, so the vertices are $a = 3$ units left and right of the center $(2, -3)$. Therefore, the vertices are

$$V_1 = (2 - 3, -3) = (-1, -3) \quad \text{and} \quad V_2 = (2 + 3, -3) = (5, -3)$$

Since $c = 1$ and the major axis is parallel to the x-axis, the foci are 1 unit left and right of the center. Therefore, the foci are

$$F_1 = (2 - 1, -3) = (1, -3) \quad \text{and} \quad F_2 = (2 + 1, -3) = (3, -3)$$

Finally, we use the value of $b = 2\sqrt{2}$ to find the two points above and below the center.

$$(2, -3 - 2\sqrt{2}) \quad \text{and} \quad (2, -3 + 2\sqrt{2})$$

Figure 29 shows the graph.

Figure 29
$$\frac{(x - 2)^2}{9} + \frac{(y + 3)^2}{8} = 1$$

NOW WORK PROBLEM 45.

EXAMPLE 7   **Using a Graphing Utility to Graph an Ellipse, Center Not at the Origin**

Using a graphing utility, graph the ellipse: $\dfrac{(x - 2)^2}{9} + \dfrac{(y + 3)^2}{8} = 1$

Solution   First, we must solve the equation $\dfrac{(x - 2)^2}{9} + \dfrac{(y + 3)^2}{8} = 1$ for y.

$$\frac{(y + 3)^2}{8} = 1 - \frac{(x - 2)^2}{9} \qquad \text{Subtract } \frac{(x-2)^2}{9} \text{ from each side.}$$

$$(y + 3)^2 = 8\left[1 - \frac{(x - 2)^2}{9}\right] \qquad \text{Multiply each side by 8.}$$

$$y + 3 = \pm\sqrt{8\left[1 - \frac{(x - 2)^2}{9}\right]} \qquad \text{Apply the Square Root Method.}$$

$$y = -3 \pm \sqrt{8\left[1 - \frac{(x - 2)^2}{9}\right]} \qquad \text{Subtract 3 from each side.}$$

Figure 30

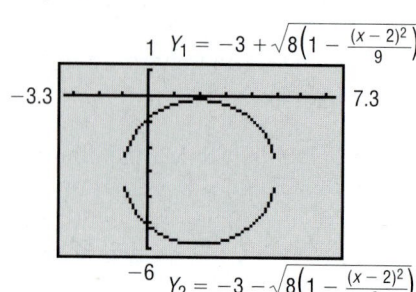

Figure 30 shows the graphs of $Y_1 = -3 + \sqrt{8\left[1 - \dfrac{(x-2)^2}{9}\right]}$ and $Y_2 = -3 - \sqrt{8\left[1 - \dfrac{(x-2)^2}{9}\right]}$.

━━━━ **NOW WORK PROBLEM 33.**

EXAMPLE 8 **Discussing the Equation of an Ellipse**

Discuss the equation: $4x^2 + y^2 - 8x + 4y + 4 = 0$

Solution We proceed to complete the squares in x and in y.

$$4x^2 + y^2 - 8x + 4y + 4 = 0$$

$$4x^2 - 8x + y^2 + 4y = -4 \qquad \text{Group like variables; place the constant on the right side.}$$

$$4(x^2 - 2x) + (y^2 + 4y) = -4 \qquad \text{Factor out 4 from the first two terms.}$$

$$4(x^2 - 2x + 1) + (y^2 + 4y + 4) = -4 + 4 + 4 \qquad \text{Complete each square.}$$

$$4(x - 1)^2 + (y + 2)^2 = 4 \qquad \text{Factor.}$$

$$(x - 1)^2 + \frac{(y + 2)^2}{4} = 1 \qquad \text{Divide each side by 4.}$$

This is the equation of an ellipse with center at $(1, -2)$ and major axis parallel to the y-axis. Since $a^2 = 4$ and $b^2 = 1$, we have $c^2 = a^2 - b^2 = 4 - 1 = 3$. The vertices are at $(h, k \pm a) = (1, -2 \pm 2)$ or $(1, 0)$ and $(1, -4)$. The foci are at $(h, k \pm c) = (1, -2 \pm \sqrt{3})$ or $(1, -2 - \sqrt{3})$ and $(1, -2 + \sqrt{3})$. Figure 31(a) shows the graph drawn by hand. Figure 31(b) shows the graph obtained using a graphing utility.

Figure 31

$$(x - 1)^2 + \frac{(y + 2)^2}{4} = 1$$

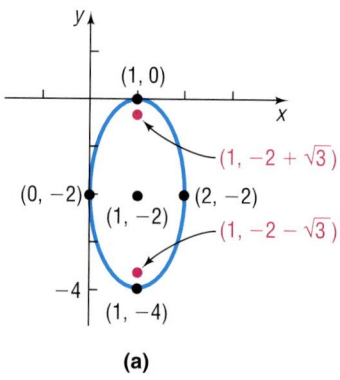

(a)

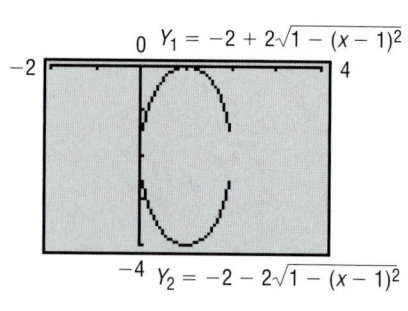

(b)

━━━━ **NOW WORK PROBLEM 37.**

Applications

5   Ellipses are found in many applications in science and engineering. For example, the orbits of the planets around the Sun are elliptical, with the Sun's position at a focus. See Figure 32.

Figure 32

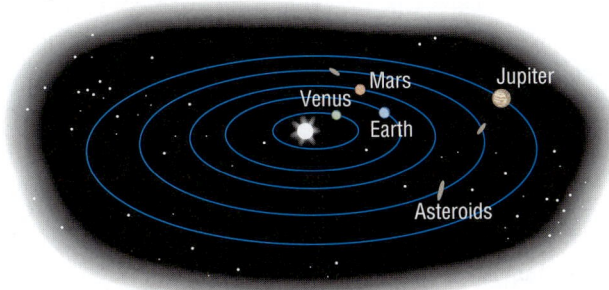

Stone and concrete bridges are often shaped as semielliptical arches. Elliptical gears are used in machinery when a variable rate of motion is required.

Ellipses also have an interesting reflection property. If a source of light (or sound) is placed at one focus, the waves transmitted by the source will reflect off the ellipse and concentrate at the other focus. This is the principle behind *whispering galleries*, which are rooms designed with elliptical ceilings. A person standing at one focus of the ellipse can whisper and be heard by a person standing at the other focus, because all the sound waves that reach the ceiling are reflected to the other person.

EXAMPLE 9 **Whispering Galleries**

Figure 33 shows the specifications for an elliptical ceiling in a hall designed to be a whispering gallery. In a whispering gallery, a person standing at one focus of the ellipse can whisper and be heard by another person standing at the other focus, because all the sound waves that reach the ceiling from one focus are reflected to the other focus. Where are the foci located in the hall?

Solution   We set up a rectangular coordinate system so that the center of the ellipse is at the origin and the major axis is along the x-axis. See Figure 34. The equation of the ellipse is

$$\frac{x^2}{a^2} + \frac{y^2}{b^2} = 1$$

where $a = 25$ and $b = 20$.

Figure 33

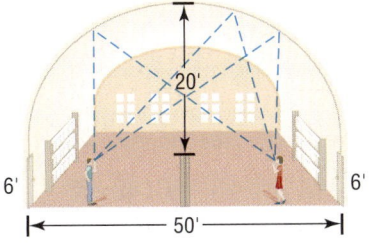

Figure 34

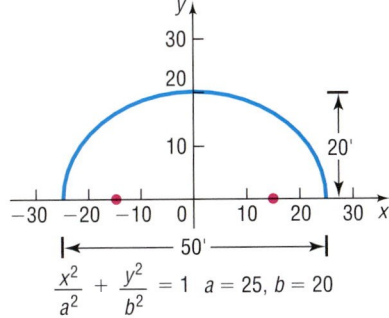

$$\frac{x^2}{a^2} + \frac{y^2}{b^2} = 1 \quad a = 25, b = 20$$

Then, since

$$c^2 = a^2 - b^2 = 25^2 - 20^2 = 625 - 400 = 225$$

we have $c = 15$. The foci are located 15 feet from the center of the ellipse along the major axis.

12.3 Concepts and Vocabulary

In Problems 1–3, fill in the blanks.

1. A(n) _____ is the collection of all points in the plane the sum of whose distances from two fixed points is a constant.

2. For an ellipse, the foci lie on a line called the _____ axis.

3. For the ellipse $\dfrac{x^2}{4} + \dfrac{y^2}{25} = 1$, the vertices are the points _____ and _____.

In Problems 4–6, answer True or False to each statement.

4. The foci, vertices, and center of an ellipse lie on a line called the axis of symmetry.

5. If the center of an ellipse is at the origin and the foci lie on the y-axis, the ellipse is symmetric with respect to the x-axis, the y-axis, and the origin.

6. A circle is a certain type of ellipse.

7. Explain how to draw an ellipse using two thumbtacks and a piece of string.

8. Make up an equation for an ellipse with center at the origin and foci on the x-axis.

9. Make up an equation for an ellipse with center at the origin and foci on the y-axis.

12.3 Exercises

In Problems 1–4, the graph of an ellipse is given. Match each graph to its equation.

A. $\dfrac{x^2}{4} + y^2 = 1$ B. $x^2 + \dfrac{y^2}{4} = 1$ C. $\dfrac{x^2}{16} + \dfrac{y^2}{4} = 1$ D. $\dfrac{x^2}{4} + \dfrac{y^2}{16} = 1$

1.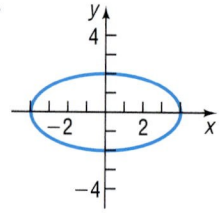

2.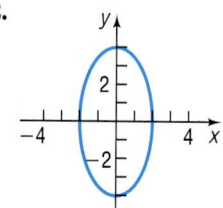

3.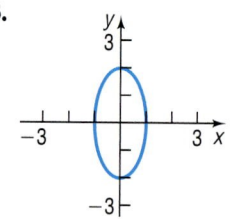

4.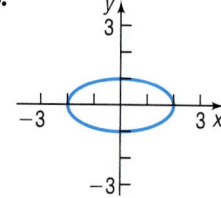

In Problems 5–8, the graph of an ellipse is given. Match each graph to its equation.

A. $\dfrac{(x+1)^2}{4} + \dfrac{(y-1)^2}{9} = 1$ B. $\dfrac{(x-1)^2}{4} + \dfrac{(y+1)^2}{9} = 1$

C. $\dfrac{(x-1)^2}{9} + \dfrac{(y+1)^2}{4} = 1$ D. $\dfrac{(x+1)^2}{9} + \dfrac{(y-1)^2}{4} = 1$

5. 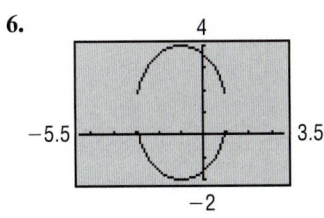   **6.** 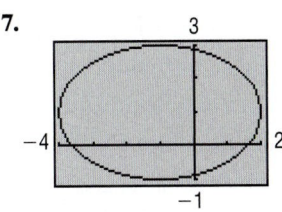   **7.** 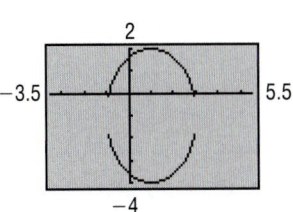   **8.**

In Problems 9–18, find the vertices and foci of each ellipse. Graph each equation (a) by hand and (b) by using a graphing utility.

9. $\dfrac{x^2}{25} + \dfrac{y^2}{4} = 1$ **10.** $\dfrac{x^2}{9} + \dfrac{y^2}{4} = 1$ **11.** $\dfrac{x^2}{9} + \dfrac{y^2}{25} = 1$ **12.** $x^2 + \dfrac{y^2}{16} = 1$

13. $4x^2 + y^2 = 16$ **14.** $x^2 + 9y^2 = 18$ **15.** $4y^2 + x^2 = 8$ **16.** $4y^2 + 9x^2 = 36$

17. $x^2 + y^2 = 16$ **18.** $x^2 + y^2 = 4$

In Problems 19–28, find an equation for each ellipse. Graph the equation by hand.

19. Center at $(0,0)$;   focus at $(3,0)$;   vertex at $(5,0)$
20. Center at $(0,0)$;   focus at $(-1,0)$;   vertex at $(3,0)$
21. Center at $(0,0)$;   focus at $(0,-4)$;   vertex at $(0,5)$
22. Center at $(0,0)$;   focus at $(0,1)$;   vertex at $(0,-2)$
23. Foci at $(\pm 2,0)$;   length of the major axis is 6
24. Focus at $(0,-4)$;   vertices at $(0,\pm 8)$
25. Foci at $(0,\pm 3)$; x-intercepts are ± 2
26. Foci at $(0,\pm 2)$;   length of the major axis is 8
27. Center at $(0,0)$;   vertex at $(0,4)$; $b = 1$
28. Vertices at $(\pm 5,0)$; $c = 2$

In Problems 29–32, write an equation for each ellipse.

29. 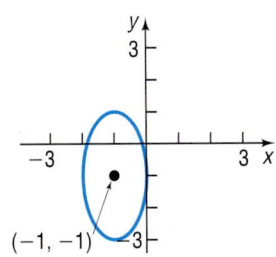   **30.** 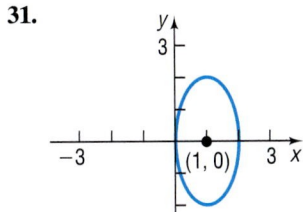   **31.** 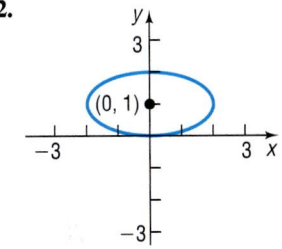   **32.**

In Problems 33–44, discuss each equation, that is, find the center, foci, and vertices of each ellipse. Graph each equation (a) by hand and (b) by using a graphing utility.

33. $\dfrac{(x-3)^2}{4} + \dfrac{(y+1)^2}{9} = 1$ **34.** $\dfrac{(x+4)^2}{9} + \dfrac{(y+2)^2}{4} = 1$

35. $(x+5)^2 + 4(y-4)^2 = 16$ **36.** $9(x-3)^2 + (y+2)^2 = 18$

37. $x^2 + 4x + 4y^2 - 8y + 4 = 0$ **38.** $x^2 + 3y^2 - 12y + 9 = 0$

39. $2x^2 + 3y^2 - 8x + 6y + 5 = 0$ **40.** $4x^2 + 3y^2 + 8x - 6y = 5$

41. $9x^2 + 4y^2 - 18x + 16y - 11 = 0$ **42.** $x^2 + 9y^2 + 6x - 18y + 9 = 0$

43. $4x^2 + y^2 + 4y = 0$ **44.** $9x^2 + y^2 - 18x = 0$

In Problems 45–54, find an equation for each ellipse. Graph the equation by hand.

45. Center at $(2,-2)$;   vertex at $(7,-2)$;   focus at $(4,-2)$
46. Center at $(-3,1)$;   vertex at $(-3,3)$;   focus at $(-3,0)$
47. Vertices at $(4,3)$ and $(4,9)$;   focus at $(4,8)$
48. Foci at $(1,2)$ and $(-3,2)$;   vertex at $(-4,2)$
49. Foci at $(5,1)$ and $(-1,1)$;   length of the major axis is 8
50. Vertices at $(2,5)$ and $(2,-1)$; $c = 2$
51. Center at $(1,2)$;   focus at $(4,2)$;   contains the point $(1,3)$
52. Center at $(1,2)$;   focus at $(1,4)$;   contains the point $(2,2)$
53. Center at $(1,2)$;   vertex at $(4,2)$;   contains the point $(1,3)$
54. Center at $(1,2)$;   vertex at $(1,4)$;   contains the point $(2,2)$

In Problems 55–58, graph each function by hand. Use a graphing utility to verify your results.

[**Hint:** Notice that each function is half an ellipse.]

55. $f(x) = \sqrt{16 - 4x^2}$ **56.** $f(x) = \sqrt{9 - 9x^2}$ **57.** $f(x) = -\sqrt{64 - 16x^2}$ **58.** $f(x) = -\sqrt{4 - 4x^2}$

59. Semielliptical Arch Bridge   An arch in the shape of the upper half of an ellipse is used to support a bridge that is to span a river 20 meters wide. The center of the arch is 6 meters above the center of the river (see the figure). Write an equation for the ellipse in which the *x*-axis coincides with the water level and the *y*-axis passes through the center of the arch.

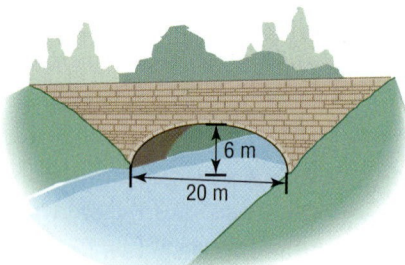

60. Semielliptical Arch Bridge   The arch of a bridge is a semiellipse with a horizontal major axis. The span is 30 feet, and the top of the arch is 10 feet above the major axis. The roadway is horizontal and is 2 feet above the top of the arch. Find the vertical distance from the roadway to the arch at 5-foot intervals along the roadway.

61. Whispering Gallery   A hall 100 feet in length is to be designed as a whispering gallery. If the foci are located 25 feet from the center, how high will the ceiling be at the center?

62. Whispering Gallery   Jim, standing at one focus of a whispering gallery, is 6 feet from the nearest wall. His friend is standing at the other focus, 100 feet away. What is the length of this whispering gallery? How high is its elliptical ceiling at the center?

63. Semielliptical Arch Bridge   A bridge is built in the shape of a semielliptical arch. The bridge has a span of 120 feet and a maximum height of 25 feet. Choose a suitable rectangular coordinate system and find the height of the arch at distances of 10, 30, and 50 feet from the center.

64. Semielliptical Arch Bridge   A bridge is to be built in the shape of a semielliptical arch and is to have a span of 100 feet. The height of the arch, at a distance of 40 feet from the center, is to be 10 feet. Find the height of the arch at its center.

65. Semielliptical Arch   An arch in the form of half an ellipse is 40 feet wide and 15 feet high at the center. Find the height of the arch at intervals of 10 feet along its width.

66. Semielliptical Arch Bridge   An arch for a bridge over a highway is in the form of half an ellipse. The top of the arch is 20 feet above the ground level (the major axis). The highway has four lanes, each 12 feet wide; a center safety strip 8 feet wide; and two side strips, each 4 feet wide. What should the span of the bridge be (the length of its major axis) if the height 28 feet from the center is to be 13 feet?

*In Problems 67–70, use the fact that the orbit of a planet about the Sun is an ellipse, with the Sun at one focus. The **aphelion** of a planet is its greatest distance from the Sun, and the **perihelion** is its shortest distance. The **mean distance** of a planet from the Sun is the length of the semimajor axis of the elliptical orbit. See the illustration.*

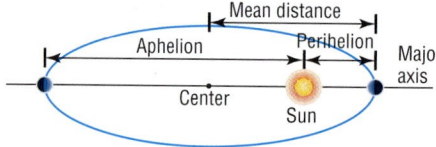

67. Earth   The mean distance of Earth from the Sun is 93 million miles. If the aphelion of Earth is 94.5 million miles, what is the perihelion? Write an equation for the orbit of Earth around the Sun.

68. Mars   The mean distance of Mars from the Sun is 142 million miles. If the perihelion of Mars is 128.5 million miles, what is the aphelion? Write an equation for the orbit of Mars about the Sun.

69. Jupiter   The aphelion of Jupiter is 507 million miles. If the distance from the Sun to the center of its elliptical orbit is 23.2 million miles, what is the perihelion? What is the mean distance? Write an equation for the orbit of Jupiter around the Sun.

70. Pluto   The perihelion of Pluto is 4551 million miles, and the distance of the Sun from the center of its elliptical orbit is 897.5 million miles. Find the aphelion of Pluto. What is the mean distance of Pluto from the Sun? Write an equation for the orbit of Pluto about the Sun.

71. Racetrack Design Consult the figure. A racetrack is in the shape of an ellipse, 100 feet long and 50 feet wide. What is the width 10 feet from a vertex?

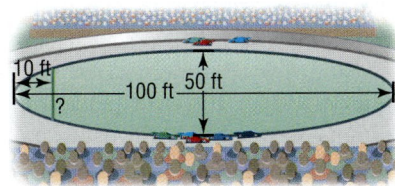

72. Racetrack Design A racetrack is in the shape of an ellipse 80 feet long and 40 feet wide. What is the width 10 feet from a vertex?

73. Show that an equation of the form

$$Ax^2 + Cy^2 + F = 0, \qquad A \neq 0, C \neq 0, F \neq 0$$

where A and C are of the same sign and F is of opposite sign,

(a) Is the equation of an ellipse with center at $(0,0)$ if $A \neq C$.

(b) Is the equation of a circle with center $(0,0)$ if $A = C$.

74. Show that the graph of an equation of the form

$$Ax^2 + Cy^2 + Dx + Ey + F = 0, \qquad A \neq 0, C \neq 0$$

where A and C are of the same sign,

(a) Is an ellipse if $\dfrac{D^2}{4A} + \dfrac{E^2}{4C} - F$ is the same sign as A.

(b) Is a point if $\dfrac{D^2}{4A} + \dfrac{E^2}{4C} - F = 0$.

(c) Contains no points if $\dfrac{D^2}{4A} + \dfrac{E^2}{4C} - F$ is of opposite sign to A.

75. The **eccentricity** e of an ellipse is defined as the number $\dfrac{c}{a}$, where a and c are the numbers given in equation (2). Because $a > c$, it follows that $e < 1$. Write a brief paragraph about the general shape of each of the following ellipses. Be sure to justify your conclusions.

(a) Eccentricity close to 0

(b) Eccentricity = 0.5

(c) Eccentricity close to 1

PREPARING FOR THIS SECTION

Before getting started, review the following:

✓ Distance Formula (Section 1.1, p. 93)

✓ Completing the Square (Section 1.3, pp. 120–121)

✓ Symmetry (Section 3.1, pp. 256–258)

✓ Asymptotes (Section 4.3, pp. 346–352)

✓ Graphing Techniques: Transformations (Section 3.4, pp. 286–295)

✓ Square Root Method (Section 1.3, pp. 119–120)

12.4 THE HYPERBOLA

OBJECTIVES

1. Find the Equation of a Hyperbola
2. Graph Hyperbolas
3. Discuss the Equation of a Hyperbola
4. Find the Asymptotes of a Hyperbola
5. Work with Hyperbolas with Center at (h, k)
6. Solve Applied Problems Involving Hyperbolas

A **hyperbola** is the collection of all points in the plane the difference of whose distances from two fixed points, called the **foci**, is a constant.

1   Figure 35 illustrates a hyperbola with foci F_1 and F_2. The line containing the foci is called the **transverse axis**. The midpoint of the line segment joining the foci is called the **center** of the hyperbola. The line through the center

Figure 35

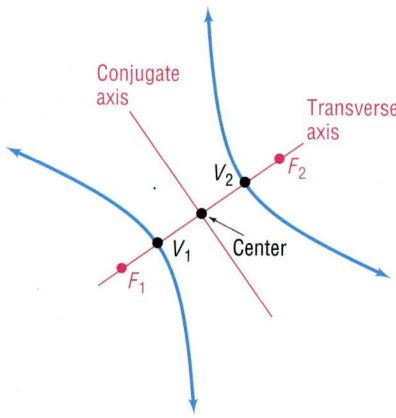

Figure 36
$$d(F_1, P) - d(F_2, P) = \pm 2a$$

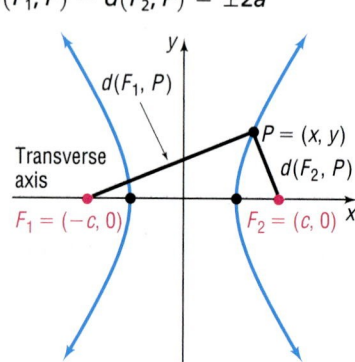

and perpendicular to the transverse axis is called the **conjugate axis**. The hyperbola consists of two separate curves, called **branches**, that are symmetric with respect to the transverse axis, conjugate axis, and center. The two points of intersection of the hyperbola and the transverse axis are the **vertices**, V_1 and V_2, of the hyperbola.

With these ideas in mind, we are now ready to find the equation of a hyperbola in a rectangular coordinate system. First, we place the center at the origin. Next, we position the hyperbola so that its transverse axis coincides with a coordinate axis. Suppose that the transverse axis coincides with the x-axis, as shown in Figure 36.

If c is the distance from the center to a focus, then one focus will be at $F_1 = (-c, 0)$ and the other at $F_2 = (c, 0)$. Now we let the constant difference of the distances from any point $P = (x, y)$ on the hyperbola to the foci F_1 and F_2 be denoted by $\pm 2a$. (If P is on the right branch, the $+$ sign is used; if P is on the left branch, the $-$ sign is used.) The coordinates of P must satisfy the equation

$$d(F_1, P) - d(F_2, P) = \pm 2a \qquad \text{\small Difference of the distances from } P \text{ to the foci equals } \pm 2a.$$

$$\sqrt{(x + c)^2 + y^2} - \sqrt{(x - c)^2 + y^2} = \pm 2a \qquad \text{\small Use the distance formula.}$$

$$\sqrt{(x + c)^2 + y^2} = \pm 2a + \sqrt{(x - c)^2 + y^2} \qquad \text{\small Isolate one radical.}$$

$$(x + c)^2 + y^2 = 4a^2 \pm 4a\sqrt{(x - c)^2 + y^2} \qquad \text{\small Square both sides.}$$
$$+ (x - c)^2 + y^2$$

Next, we remove the parentheses.

$$x^2 + 2cx + c^2 + y^2 = 4a^2 \pm 4a\sqrt{(x - c)^2 + y^2} + x^2 - 2cx + c^2 + y^2$$

$$4cx - 4a^2 = \pm 4a\sqrt{(x - c)^2 + y^2} \qquad \text{\small Simplify; isolate the radical.}$$

$$cx - a^2 = \pm a\sqrt{(x - c)^2 + y^2} \qquad \text{\small Divide each side by 4.}$$

$$(cx - a^2)^2 = a^2[(x - c)^2 + y^2] \qquad \text{\small Square both sides.}$$

$$c^2x^2 - 2ca^2x + a^4 = a^2(x^2 - 2cx + c^2 + y^2) \qquad \text{\small Simplify.}$$

$$c^2x^2 + a^4 = a^2x^2 + a^2c^2 + a^2y^2 \qquad \text{\small Remove parentheses and simplify.}$$

$$(c^2 - a^2)x^2 - a^2y^2 = a^2c^2 - a^4 \qquad \text{\small Rearrange terms.}$$

$$(c^2 - a^2)x^2 - a^2y^2 = a^2(c^2 - a^2) \qquad \text{\small Factor } a^2 \text{ on the right side.} \qquad (1)$$

To obtain points on the hyperbola off the x-axis, it must be that $a < c$. To see why, look again at Figure 36.

$$d(F_1, P) < d(F_2, P) + d(F_1, F_2) \qquad \text{\small Use triangle } F_1PF_2.$$
$$d(F_1, P) - d(F_2, P) < d(F_1, F_2) \qquad \text{\small } P \text{ is on the right branch, so}$$
$$\text{\small } d(F_1, P) - d(F_2, P) = 2a.$$

$$2a < 2c$$

$$a < c$$

Since $a < c$, we also have $a^2 < c^2$, so $c^2 - a^2 > 0$. Let $b^2 = c^2 - a^2, b > 0$. Then equation (1) can be written as

$$b^2x^2 - a^2y^2 = a^2b^2$$

$$\frac{x^2}{a^2} - \frac{y^2}{b^2} = 1 \qquad \text{Divide each side by } a^2b^2.$$

To find the vertices of the hyperbola defined by this equation, let $y = 0$. The vertices satisfy the equation $\dfrac{x^2}{a^2} = 1$, the solutions of which are $x = \pm a$. Consequently, the vertices of the hyperbola are $V_1 = (-a, 0)$ and $V_2 = (a, 0)$. Notice that the distance from the center $(0, 0)$ to either vertex is a.

Theorem

> **Equation of a Hyperbola; Center at $(0, 0)$; Foci at $(\pm c, 0)$; Vertices at $(\pm a, 0)$; Transverse Axis along the x-Axis**
>
> An equation of the hyperbola with center at $(0, 0)$, foci at $(-c, 0)$ and $(c, 0)$, and vertices at $(-a, 0)$ and $(a, 0)$ is
>
> $$\boxed{\frac{x^2}{a^2} - \frac{y^2}{b^2} = 1, \qquad \text{where } b^2 = c^2 - a^2} \qquad (2)$$
>
> The transverse axis is the x-axis.

Figure 37
$\dfrac{x^2}{a^2} - \dfrac{y^2}{b^2} = 1, b^2 = c^2 - a^2$

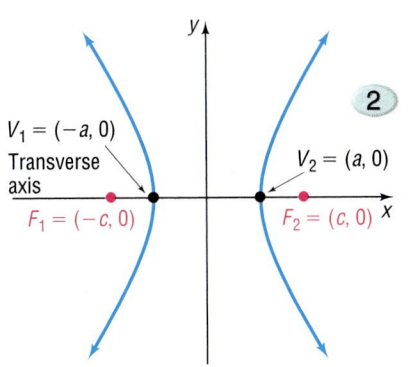

As you can verify, the hyperbola defined by equation (2) is symmetric with respect to the x-axis, y-axis, and origin. To find the y-intercepts, if any, let $x = 0$ in equation (2). This results in the equation $\dfrac{y^2}{b^2} = -1$, which has no real solution. We conclude that the hyperbola defined by equation (2) has no y-intercepts. In fact, since $\dfrac{x^2}{a^2} - 1 = \dfrac{y^2}{b^2} \geq 0$, it follows that $\dfrac{x^2}{a^2} \geq 1$. There are no points on the graph for $-a < x < a$. See Figure 37.

EXAMPLE 1   **Finding and Graphing an Equation of a Hyperbola**

Find an equation of the hyperbola with center at the origin, one focus at $(3, 0)$, and one vertex at $(-2, 0)$. Graph the equation by hand.

Solution   The hyperbola has its center at the origin, and the transverse axis coincides with the x-axis. One focus is at $(c, 0) = (3, 0)$, so $c = 3$. One vertex is at $(-a, 0) = (-2, 0)$, so $a = 2$. From equation (2), it follows that $b^2 = c^2 - a^2 = 9 - 4 = 5$, so an equation of the hyperbola is

$$\frac{x^2}{4} - \frac{y^2}{5} = 1$$

To graph a hyperbola, it is helpful to locate and plot other points on the graph. For example, to find the points above and below the foci, we let $x = \pm 3$. Then

$$\frac{x^2}{4} - \frac{y^2}{5} = 1$$

$$\frac{(\pm 3)^2}{4} - \frac{y^2}{5} = 1 \qquad {\color{blue} x = \pm 3}$$

$$\frac{9}{4} - \frac{y^2}{5} = 1$$

$$\frac{y^2}{5} = \frac{5}{4}$$

$$y^2 = \frac{25}{4}$$

$$y = \pm \frac{5}{2}$$

Figure 38

$\dfrac{x^2}{4} - \dfrac{y^2}{5} = 1$

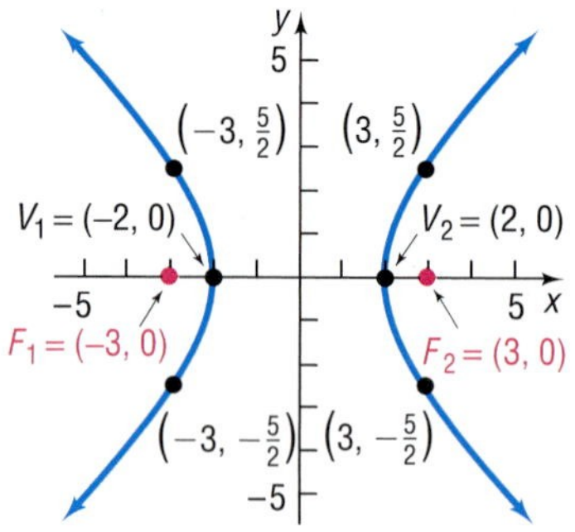

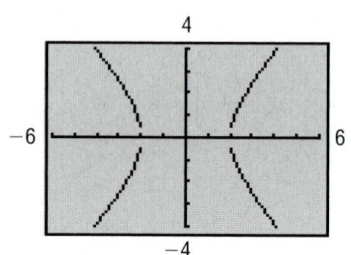

The points above and below the foci are $\left(\pm 3, \dfrac{5}{2} \right)$ and $\left(\pm 3, -\dfrac{5}{2} \right)$. These points help because they determine the "opening" of the hyperbola. See Figure 38. ■

EXAMPLE 2   Using a Graphing Utility to Graph a Hyperbola

Using a graphing utility, graph the hyperbola: $\dfrac{x^2}{4} - \dfrac{y^2}{5} = 1$

Figure 39

Solution   To graph the hyperbola $\dfrac{x^2}{4} - \dfrac{y^2}{5} = 1$, we need to graph the two functions $Y_1 = \sqrt{5}\sqrt{\dfrac{x^2}{4} - 1}$ and $Y_2 = -\sqrt{5}\sqrt{\dfrac{x^2}{4} - 1}$. As with graphing circles, parabolas, and ellipses on a graphing utility, we use a square screen setting so that the graph is not distorted. Figure 39 shows the graph of the hyperbola. ■

✎ **NOW WORK PROBLEM 9.**

An equation of the form of equation (2) is the equation of a hyperbola with center at the origin; foci on the x-axis at $(-c, 0)$ and $(c, 0)$, where $c^2 = a^2 + b^2$; and transverse axis along the x-axis.

③   For the remainder of this section, the direction "Discuss the equation" will mean to find the center, transverse axis, vertices, and foci of the hyperbola and graph it.

EXAMPLE 3   Discussing the Equation of a Hyperbola

Discuss the equation: $\dfrac{x^2}{16} - \dfrac{y^2}{4} = 1$

Solution   The given equation is of the form of equation (2), with $a^2 = 16$ and $b^2 = 4$. The graph of the equation is a hyperbola with center at $(0, 0)$ and transverse axis along the x-axis. Also, we know that $c^2 = a^2 + b^2 = 16 + 4 = 20$. The vertices are at $(\pm a, 0) = (\pm 4, 0)$, and the foci are at $(\pm c, 0) = (\pm 2\sqrt{5}, 0)$.

To locate the points on the graph above and below the foci, we let $x = \pm 2\sqrt{5}$. Then

$$\frac{x^2}{16} - \frac{y^2}{4} = 1$$

$$\frac{(\pm 2\sqrt{5})^2}{16} - \frac{y^2}{4} = 1 \qquad x = \pm 2\sqrt{5}$$

$$\frac{20}{16} - \frac{y^2}{4} = 1$$

$$\frac{5}{4} - \frac{y^2}{4} = 1$$

$$\frac{y^2}{4} = \frac{1}{4}$$

$$y = \pm 1$$

The points above and below the foci are $(\pm 2\sqrt{5}, 1)$ and $(\pm 2\sqrt{5}, -1)$. See Figure 40(a) for the graph drawn by hand. Figure 40(b) shows the graph obtained using a graphing utility.

Figure 40
$$\frac{x^2}{16} - \frac{y^2}{4} = 1$$

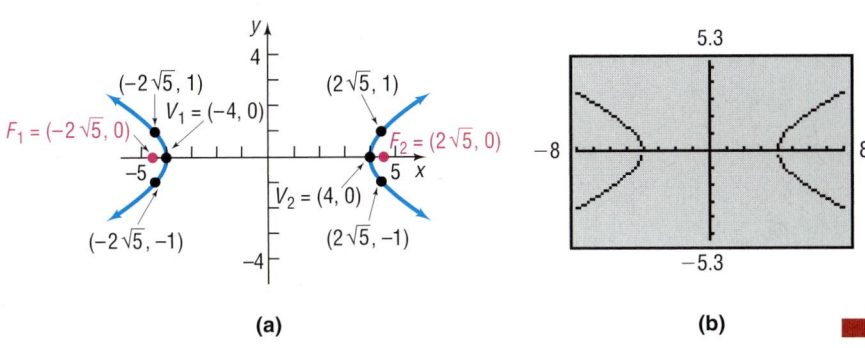

(a)   (b)

NOW WORK PROBLEM **19.**

The next result gives the form of the equation of a hyperbola with center at the origin and transverse axis along the y-axis.

Theorem

Equation of a Hyperbola; Center at $(0, 0)$; Foci at $(0, \pm c)$; Vertices at $(0, \pm a)$; Transverse Axis along the y-Axis

An equation of the hyperbola with center at $(0, 0)$, foci at $(0, -c)$ and $(0, c)$, and vertices at $(0, -a)$ and $(0, a)$ is

$$\frac{y^2}{a^2} - \frac{x^2}{b^2} = 1, \qquad \text{where } b^2 = c^2 - a^2 \qquad (3)$$

The transverse axis is the y-axis.

Figure 41
$$\frac{y^2}{a^2} - \frac{x^2}{b^2} = 1, b^2 = c^2 - a^2$$

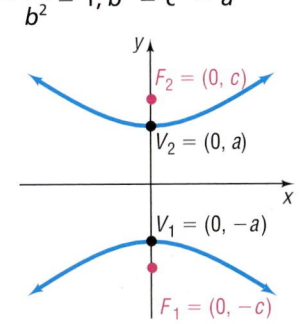

Figure 41 shows the graph of a typical hyperbola defined by equation (3).

An equation of the form of equation (2), $\dfrac{x^2}{a^2} - \dfrac{y^2}{b^2} = 1$, is the equation of a hyperbola with center at the origin, foci on the x-axis at $(-c, 0)$ and $(c, 0)$, where $c^2 = a^2 + b^2$, and transverse axis along the x-axis.

An equation of the form of equation (3), $\dfrac{y^2}{a^2} - \dfrac{x^2}{b^2} = 1$, is the equation of a hyperbola with center at the origin, foci on the y-axis at $(0, -c)$ and $(0, c)$, where $c^2 = a^2 + b^2$, and transverse axis along the y-axis.

Notice the difference in the forms of equations (2) and (3). When the y^2-term is subtracted from the x^2-term, the transverse axis is the x-axis. When the x^2-term is subtracted from the y^2-term, the transverse axis is the y-axis.

EXAMPLE 4   Discussing the Equation of a Hyperbola

Discuss the equation: $y^2 - 4x^2 = 4$

Solution   To put the equation in proper form, we divide each side by 4:

$$\frac{y^2}{4} - x^2 = 1$$

Since the x^2-term is subtracted from the y^2-term, the equation is that of a hyperbola with center at the origin and transverse axis along the y-axis. Also, comparing the above equation to equation (3), we find $a^2 = 4$, $b^2 = 1$, and $c^2 = a^2 + b^2 = 5$. The vertices are at $(0, \pm a) = (0, \pm 2)$, and the foci are at $(0, \pm c) = (0, \pm\sqrt{5})$.

To locate other points on the graph, we let $x = \pm 2$. Then

$$\begin{aligned} y^2 - 4x^2 &= 4 \\ y^2 - 4(\pm 2)^2 &= 4 \qquad \color{blue}{x = \pm 2} \\ y^2 - 16 &= 4 \\ y^2 &= 20 \\ y &= \pm 2\sqrt{5} \end{aligned}$$

Four other points on the graph are $(\pm 2, 2\sqrt{5})$ and $(\pm 2, -2\sqrt{5})$. See Figure 42(a) for the graph drawn by hand. Figure 42(b) shows the graph obtained using a graphing utility.

Figure 42

$$\frac{y^2}{4} - x^2 = 1$$

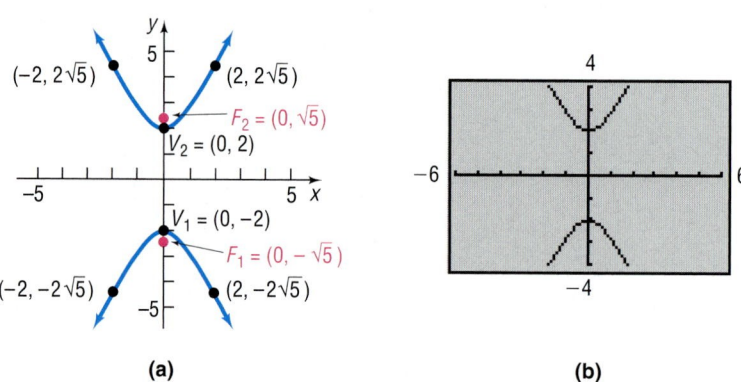

(a)

(b)

EXAMPLE 5   Finding an Equation of a Hyperbola

Find an equation of the hyperbola having one vertex at $(0, 2)$ and foci at $(0, -3)$ and $(0, 3)$. Graph the equation by hand.

Figure 43

$$\frac{y^2}{4} - \frac{x^2}{5} = 1$$

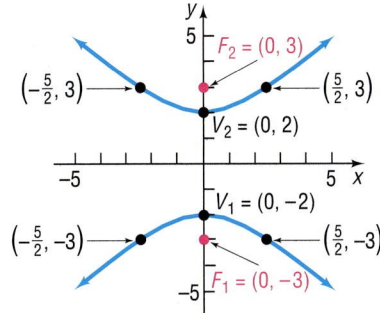

Solution Since the foci are at $(0, -3)$ and $(0, 3)$, the center of the hyperbola is at their midpoint, the origin. Also, the transverse axis is along the y-axis. The given information also reveals that $c = 3, a = 2$, and $b^2 = c^2 - a^2 = 9 - 4 = 5$. The form of the equation of the hyperbola is given by equation (3):

$$\frac{y^2}{a^2} - \frac{x^2}{b^2} = 1$$

$$\frac{y^2}{4} - \frac{x^2}{5} = 1$$

Let $y = \pm 3$ to obtain points on the graph across from the foci. See Figure 43. ∎

NOW WORK PROBLEM 13.

Look at the equations of the hyperbolas in Examples 3 and 5. For the hyperbola in Example 3, $a^2 = 16$ and $b^2 = 4$, so $a > b$; for the hyperbola in Example 5, $a^2 = 4$ and $b^2 = 5$, so $a < b$. We conclude that, for hyperbolas, there are no requirements involving the relative sizes of a and b. Contrast this situation to the case of an ellipse, in which the relative sizes of a and b dictate which axis is the major axis. Hyperbolas have another feature to distinguish them from ellipses and parabolas: Hyperbolas have asymptotes.

Asymptotes

④ Recall from Section 4.3 that a horizontal or oblique asymptote of a graph is a line with the property that the distance from the line to points on the graph approaches 0 as $x \to -\infty$ or as $x \to \infty$. The asymptotes provide information about the end behavior of the graph of a hyperbola.

Theorem

> **Asymptotes of a Hyperbola**
>
> The hyperbola $\dfrac{x^2}{a^2} - \dfrac{y^2}{b^2} = 1$ has the two oblique asymptotes
>
> $$y = \frac{b}{a}x \quad \text{and} \quad y = -\frac{b}{a}x \qquad (4)$$

Proof We begin by solving for y in the equation of the hyperbola.

$$\frac{x^2}{a^2} - \frac{y^2}{b^2} = 1$$

$$\frac{y^2}{b^2} = \frac{x^2}{a^2} - 1$$

$$y^2 = b^2\left(\frac{x^2}{a^2} - 1\right)$$

Since $x \neq 0$, we can rearrange the right side in the form

$$y^2 = \frac{b^2 x^2}{a^2}\left(1 - \frac{a^2}{x^2}\right)$$

$$y = \pm \frac{bx}{a}\sqrt{1 - \frac{a^2}{x^2}}$$

Now, as $x \to -\infty$ or as $x \to \infty$, the term $\dfrac{a^2}{x^2}$ approaches 0, so the expression under the radical approaches 1. Thus, as $x \to -\infty$ or as $x \to \infty$, the value of y approaches $\pm \dfrac{bx}{a}$, that is, the graph of the hyperbola approaches the lines

$$y = -\frac{b}{a}x \quad \text{and} \quad y = \frac{b}{a}x$$

These lines are oblique asymptotes of the hyperbola.

The asymptotes of a hyperbola are not part of the hyperbola, but they do serve as a guide for graphing a hyperbola. For example, suppose that we want to graph the equation

$$\frac{x^2}{a^2} - \frac{y^2}{b^2} = 1$$

We begin by plotting the vertices $(-a, 0)$ and $(a, 0)$. Then we plot the points $(0, -b)$ and $(0, b)$ and use these four points to construct a rectangle, as shown in Figure 44. The diagonals of this rectangle have slopes $\dfrac{b}{a}$ and $-\dfrac{b}{a}$, and their extensions are the asymptotes $y = \dfrac{b}{a}x$ and $y = -\dfrac{b}{a}x$ of the hyperbola. If we graph the asymptotes, we can use them to establish the "opening" of the hyperbola and avoid plotting other points.

Figure 44
$\dfrac{x^2}{a^2} - \dfrac{y^2}{b^2} = 1$

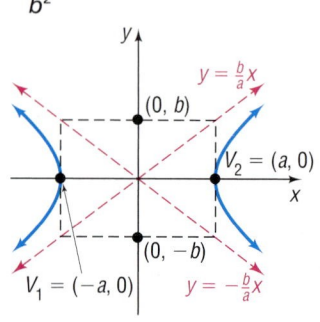

Theorem

Asymptotes of a Hyperbola

The hyperbola $\dfrac{y^2}{a^2} - \dfrac{x^2}{b^2} = 1$ has the two oblique asymptotes

$$y = \frac{a}{b}x \quad \text{and} \quad y = -\frac{a}{b}x \qquad (5)$$

You are asked to prove this result in Problem 64.

For the remainder of this section, the direction "Discuss the equation" will mean to find the center, transverse axis, vertices, foci, and asymptotes of the hyperbola and graph it.

EXAMPLE 6   **Discussing the Equation of a Hyperbola**

Discuss the equation: $9x^2 - 4y^2 = 36$

Solution   Divide each side of the equation by 36 to put the equation in proper form.

$$\frac{x^2}{4} - \frac{y^2}{9} = 1$$

We now proceed to analyze the equation. The center of the hyperbola is the origin. Since the x^2-term is first in the equation, we know that the transverse axis is along the x-axis and the vertices and foci will lie on the x-axis. Using equation (2), we find $a^2 = 4$, $b^2 = 9$, and $c^2 = a^2 + b^2 = 13$. The vertices are $a = 2$ units left and right of the center at $(\pm a, 0) = (\pm 2, 0)$; the foci are $c = \sqrt{13}$ units left and right of the center at $(\pm c, 0) = (\pm \sqrt{13}, 0)$; and the asymptotes have the equations

$$y = \frac{b}{a}x = \frac{3}{2}x \quad \text{and} \quad y = -\frac{b}{a}x = -\frac{3}{2}x$$

To graph the hyperbola by hand, form the rectangle containing the points $(\pm a, 0)$ and $(0, \pm b)$, that is, $(-2, 0)$, $(2, 0)$, $(0, -3)$, and $(0, 3)$. The extensions of the diagonals of this rectangle are the asymptotes. See Figure 45(a) for the graph drawn by hand. Figure 45(b) shows the graph obtained using a graphing utility.

Figure 45

$$\frac{x^2}{4} - \frac{y^2}{9} = 1$$

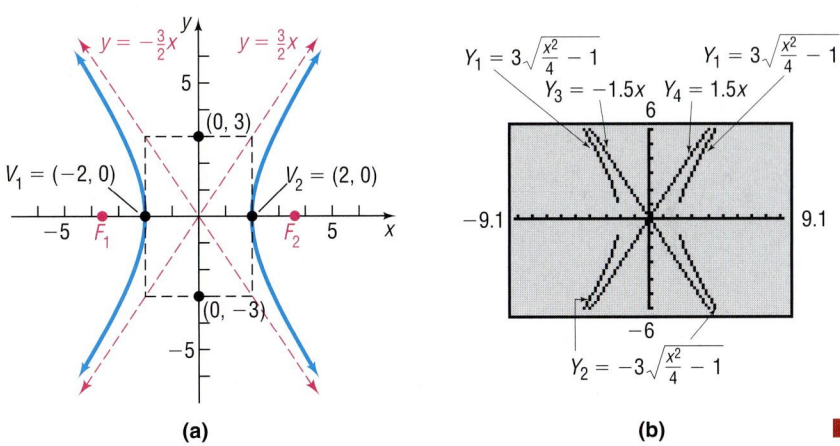

(a)   (b)

  NOW WORK PROBLEM 21.

SEEING THE CONCEPT Refer to Figure 45(b). Create a TABLE using Y_1 and Y_4 with $x = 10$, 100, 1000, and 10,000. Compare the values of Y_1 and Y_4. Repeat for Y_1 and Y_3, Y_2 and Y_3, and Y_2 and Y_4.

Center at (h, k)

⑤ If a hyperbola with center at the origin and transverse axis coinciding with a coordinate axis is shifted horizontally h units and then vertically k units, the result is a hyperbola with center at (h, k) and transverse axis parallel to a coordinate axis. The equations of such hyperbolas have the same forms as those given in equations (2) and (3), except that x is replaced by $x - h$ (the horizontal shift) and y is replaced by $y - k$ (the vertical shift). Table 4 gives the forms of the equations of such hyperbolas. See Figure 46 for the graphs.

TABLE 4 Hyperbolas With Center at (h, k), and Transverse Axis Parallel to a Coordinate Axis

Center	Transverse Axis	Foci	Vertices	Equation	Asymptotes
(h, k)	Parallel to x-axis	$(h \pm c, k)$	$(h \pm a, k)$	$\dfrac{(x-h)^2}{a^2} - \dfrac{(y-k)^2}{b^2} = 1$, $\quad b^2 = c^2 - a^2$	$y - k = \pm\dfrac{b}{a}(x-h)$
(h, k)	Parallel to y-axis	$(h, k \pm c)$	$(h, k \pm a)$	$\dfrac{(y-k)^2}{a^2} - \dfrac{(x-h)^2}{b^2} = 1$, $\quad b^2 = c^2 - a^2$	$y - k = \pm\dfrac{a}{b}(x-h)$

Figure 46

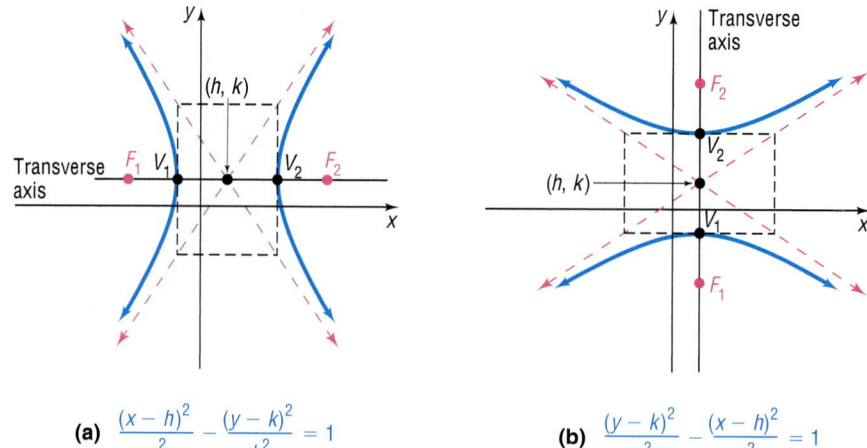

(a) $\dfrac{(x-h)^2}{a^2} - \dfrac{(y-k)^2}{b^2} = 1$

(b) $\dfrac{(y-k)^2}{a^2} - \dfrac{(x-h)^2}{b^2} = 1$

| EXAMPLE 7 | **Finding an Equation of a Hyperbola, Center Not at the Origin** |

Find an equation for the hyperbola with center at $(1, -2)$, one focus at $(4, -2)$, and one vertex at $(3, -2)$. Graph the equation by hand.

Figure 47
$\dfrac{(x-1)^2}{4} - \dfrac{(y+2)^2}{5} = 1$

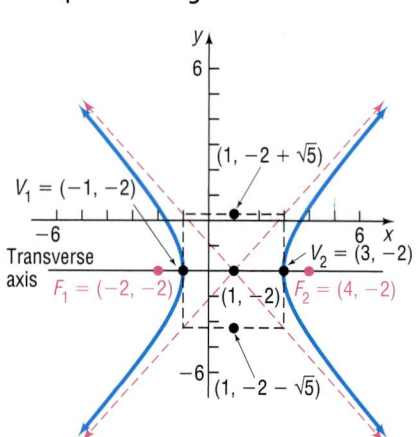

Solution The center is at $(h, k) = (1, -2)$, so $h = 1$ and $k = -2$. Since the center, focus, and vertex all lie on the line $y = -2$, the transverse axis is parallel to the x-axis. The distance from the center $(1, -2)$ to the focus $(4, -2)$ is $c = 3$; the distance from the center $(1, -2)$ to the vertex $(3, -2)$ is $a = 2$. Thus, $b^2 = c^2 - a^2 = 9 - 4 = 5$. The equation is

$$\frac{(x-h)^2}{a^2} - \frac{(y-k)^2}{b^2} = 1$$

$$\frac{(x-1)^2}{4} - \frac{(y+2)^2}{5} = 1$$

See Figure 47.

✎ **NOW WORK PROBLEM 31.**

| EXAMPLE 8 | **Discussing the Equation of a Hyperbola** |

Discuss the equation: $-x^2 + 4y^2 - 2x - 16y + 11 = 0$

Solution We complete the squares in x and in y.

$$-x^2 + 4y^2 - 2x - 16y + 11 = 0$$

$$-(x^2 + 2x) + 4(y^2 - 4y) = -11 \qquad \text{\textit{Group terms.}}$$

$$-(x^2 + 2x + 1) + 4(y^2 - 4y + 4) = -11 - 1 + 16 \qquad \text{\textit{Complete each square.}}$$

$$-(x+1)^2 + 4(y-2)^2 = 4$$

$$(y-2)^2 - \frac{(x+1)^2}{4} = 1 \qquad \text{\textit{Divide each side by 4.}}$$

This is the equation of a hyperbola with center at $(-1, 2)$ and transverse axis parallel to the y-axis. Also, $a^2 = 1$ and $b^2 = 4$, so $c^2 = a^2 + b^2 = 5$. Since the transverse axis is parallel to the y-axis, the vertices and foci are located a and c units above and below the center, respectively. The vertices are at $(h, k \pm a) = (-1, 2 \pm 1)$, or $(-1, 1)$ and $(-1, 3)$. The foci are at $(h, k \pm c) = (-1, 2 \pm \sqrt{5})$. The asymptotes are $y - 2 = \frac{1}{2}(x + 1)$ and $y - 2 = -\frac{1}{2}(x + 1)$. Figure 48(a) shows the graph drawn by hand. Figure 48(b) shows the graph obtained using a graphing utility.

Figure 48

$$(y - 2)^2 - \frac{(x + 1)^2}{4} = 1$$

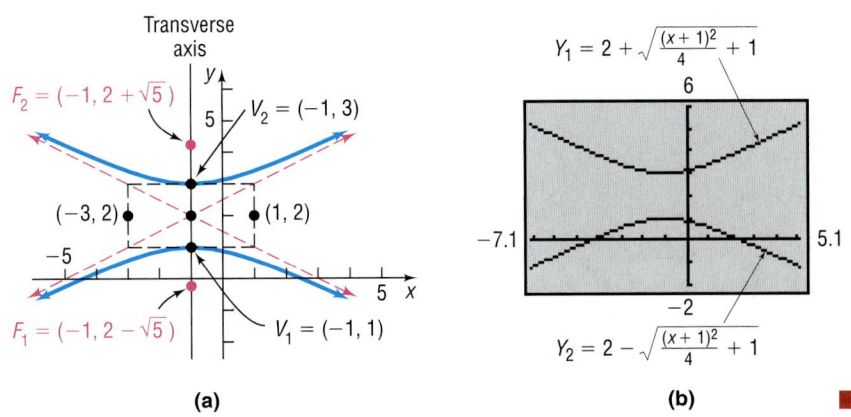

(a)

(b)

NOW WORK PROBLEM **45.**

Applications

Figure 49

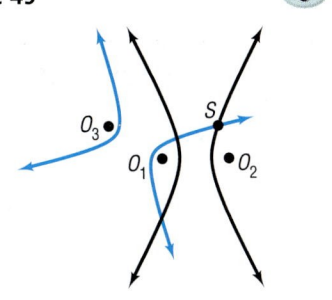

6

See Figure 49. Suppose that a gun is fired from an unknown source S. An observer at O_1 hears the report (sound of gun shot) 1 second after another observer at O_2. Because sound travels at about 1100 feet per second, it follows that the point S must be 1100 feet closer to O_2 than to O_1. S lies on one branch of a hyperbola with foci at O_1 and O_2. (Do you see why? The difference of the distances from S to O_1 and from S to O_2 is the constant 1100.) If a third observer at O_3 hears the same report 2 seconds after O_1 hears it, then S will lie on a branch of a second hyperbola with foci at O_1 and O_3. The intersection of the two hyperbolas will pinpoint the location of S.

Loran

Figure 50

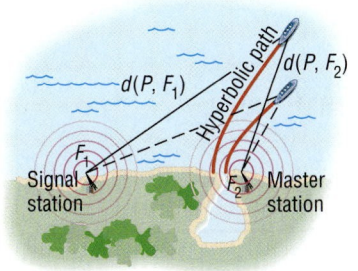

$d(P, F_1) - d(P, F_2) = $ constant

In the LOng RAnge Navigation system (LORAN), a master radio sending station and a secondary sending station emit signals that can be received by a ship at sea. See Figure 50. Because a ship monitoring the two signals will usually be nearer to one of the two stations, there will be a difference in the distance that the two signals travel, which will register as a slight time difference between the signals. As long as the time difference remains constant, the difference of the two distances will also be constant. If the ship follows a path corresponding to the fixed time difference, it will follow the path of a hyperbola whose foci are located at the positions of the two sending stations. So for each time difference a different hyperbolic path results, each bringing the ship to a different shore location. Navigation charts show the various hyperbolic paths corresponding to different time differences.

EXAMPLE 9   LORAN

Two LORAN stations are positioned 250 miles apart along a straight shore.

(a) A ship records a time difference of 0.00054 second between the LORAN signals. Set up an appropriate rectangular coordinate system to determine where the ship would reach shore if it were to follow the hyperbola corresponding to this time difference.

(b) If the ship wants to enter a harbor located between the two stations 25 miles from the master station, what time difference should it be looking for?

(c) If the ship is 80 miles offshore when the desired time difference is obtained, what is the approximate location of the ship?

[*Note:* The speed of each radio signal is 186,000 miles per second.]

Solution

Figure 51

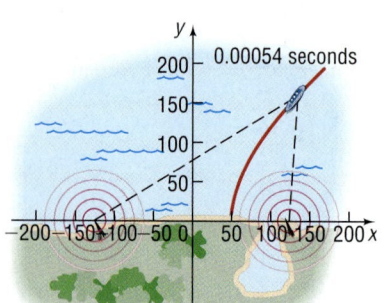

(a) We set up a rectangular coordinate system so that the two stations lie on the x-axis and the origin is midway between them. See Figure 51. The ship lies on a hyperbola whose foci are the locations of the two stations. The reason for this is that the constant time difference of the signals from each station results in a constant difference in the distance of the ship from each station. Since the time difference is 0.00054 second and the speed of the signal is 186,000 miles per second, the difference of the distances from the ship to each station (foci) is

$$\text{Distance} = \text{Speed} \times \text{Time} = 186{,}000 \times 0.00054 \approx 100 \text{ miles}$$

The difference of the distances from the ship to each station, 100, equals $2a$, so $a = 50$ and the vertex of the corresponding hyperbola is at $(50, 0)$. Since the focus is at $(125, 0)$, following this hyperbola the ship would reach shore 75 miles from the master station.

(b) To reach shore 25 miles from the master station, the ship would follow a hyperbola with vertex at $(100, 0)$. For this hyperbola, $a = 100$, so the constant difference of the distances from the ship to each station is $2a = 200$. The time difference that the ship should look for is

$$\text{Time} = \frac{\text{Distance}}{\text{Speed}} = \frac{200}{186{,}000} \approx 0.001075 \text{ second}$$

(c) To find the approximate location of the ship, we need to find the equation of the hyperbola with vertex at $(100, 0)$ and a focus at $(125, 0)$. The form of the equation of this hyperbola is

$$\frac{x^2}{a^2} - \frac{y^2}{b^2} = 1$$

where $a = 100$. Since $c = 125$, we have

$$b^2 = c^2 - a^2 = 125^2 - 100^2 = 5625$$

The equation of the hyperbola is

$$\frac{x^2}{100^2} - \frac{y^2}{5625} = 1$$

Figure 52

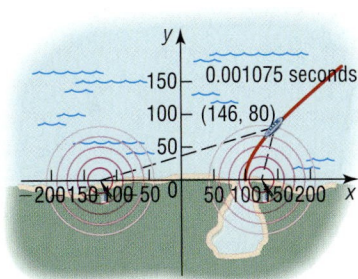

Since the ship is 80 miles from shore, we use $y = 80$ in the equation and solve for x.

$$\frac{x^2}{100^2} - \frac{80^2}{5625} = 1$$

$$\frac{x^2}{100^2} = 1 + \frac{80^2}{5625} \approx 2.14$$

$$x^2 \approx 100^2(2.14)$$

$$x \approx 146$$

The ship is at the position $(146, 80)$. See Figure 52.

⟶ **NOW WORK PROBLEM 57.**

12.4 Concepts and Vocabulary

In Problems 1–3, fill in the blanks.

1. A(n) _____ is the collection of points in the plane the difference of whose distances from two fixed points is a constant.

2. For a hyperbola, the foci lie on a line called the _____ _____.

3. The asymptotes of the hyperbola $\dfrac{x^2}{4} - \dfrac{y^2}{9} = 1$ are _____ and _____.

In Problems 4–6, answer True or False to each statement.

4. The foci of a hyperbola lie on a line called the axis of symmetry.

5. Hyperbolas always have asymptotes.

6. A hyperbola will never intersect its transverse axis.

7. Explain how the asymptotes of a hyperbola are helpful in obtaining its graph.

8. Make up an equation for a hyperbola with center at the origin and transverse axis along the x-axis.

9. Make up an equation for a hyperbola with center at the origin and transverse axis along the y-axis.

12.4 Exercises

In Problems 1–4, the graph of a hyperbola is given. Match each graph to its equation.

A. $\dfrac{x^2}{4} - y^2 = 1$ **B.** $x^2 - \dfrac{y^2}{4} = 1$ **C.** $\dfrac{y^2}{4} - x^2 = 1$ **D.** $y^2 - \dfrac{x^2}{4} = 1$

1.

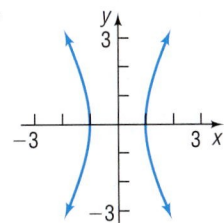

2.

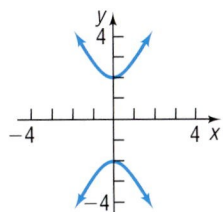

3.

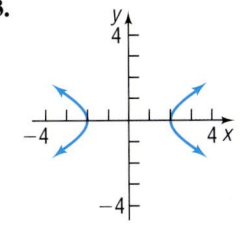

4.

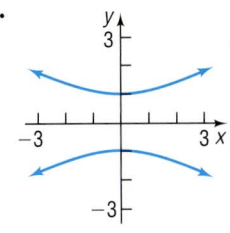

In Problems 5–8, the graph of a hyperbola is given. Match each graph to its equation.

A. $\dfrac{x^2}{16} - \dfrac{y^2}{9} = 1$ B. $\dfrac{x^2}{9} - \dfrac{y^2}{16} = 1$ C. $\dfrac{y^2}{16} - \dfrac{x^2}{9} = 1$ D. $\dfrac{y^2}{9} - \dfrac{x^2}{16} = 1$

5.

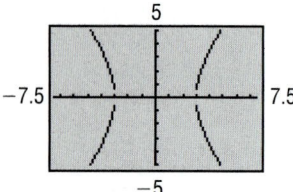

6.

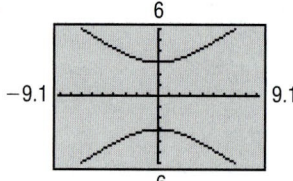

7.

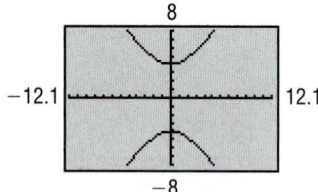

8.

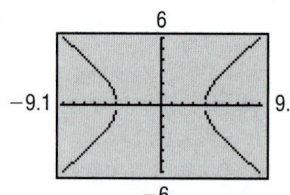

In Problems 9–18, find an equation for the hyperbola described. Graph the equation by hand.

9. Center at $(0,0)$; focus at $(3,0)$; vertex at $(1,0)$

10. Center at $(0,0)$; focus at $(0,5)$; vertex at $(0,3)$

11. Center at $(0,0)$; focus at $(0,-6)$; vertex at $(0,4)$

12. Center at $(0,0)$; focus at $(-3,0)$; vertex at $(2,0)$

13. Foci at $(-5,0)$ and $(5,0)$; vertex at $(3,0)$

14. Focus at $(0,6)$; vertices at $(0,-2)$ and $(0,2)$

15. Vertices at $(0,-6)$ and $(0,6)$; asymptote the line $y = 2x$

16. Vertices at $(-4,0)$ and $(4,0)$; asymptote the line $y = 2x$

17. Foci at $(-4,0)$ and $(4,0)$; asymptote the line $y = -x$

18. Foci at $(0,-2)$ and $(0,2)$; asymptote the line $y = -x$

In Problems 19–26, find the center, transverse axis, vertices, foci, and asymptotes. Graph each equation (a) by hand and (b) by using a graphing utility.

19. $\dfrac{x^2}{25} - \dfrac{y^2}{9} = 1$

20. $\dfrac{y^2}{16} - \dfrac{x^2}{4} = 1$

21. $4x^2 - y^2 = 16$

22. $4y^2 - x^2 = 16$

23. $y^2 - 9x^2 = 9$

24. $x^2 - y^2 = 4$

25. $y^2 - x^2 = 25$

26. $2x^2 - y^2 = 4$

In Problems 27–30, write an equation for each hyperbola.

27.

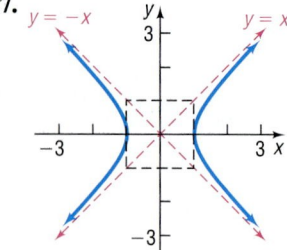

28.

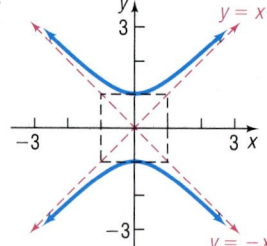

29.

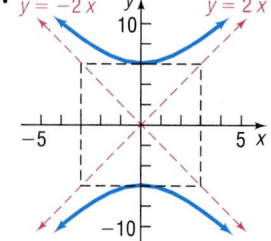

30.

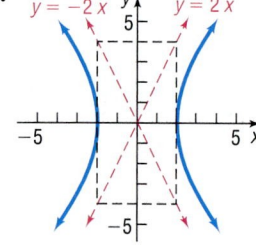

In Problems 31–38, find an equation for the hyperbola described. Graph the equation by hand.

31. Center at $(4,-1)$; focus at $(7,-1)$; vertex at $(6,-1)$

32. Center at $(-3,1)$; focus at $(-3,6)$; vertex at $(-3,4)$

33. Center at $(-3,-4)$; focus at $(-3,-8)$; vertex at $(-3,-2)$

34. Center at $(1,4)$; focus at $(-2,4)$; vertex at $(0,4)$

35. Foci at $(3,7)$ and $(7,7)$; vertex at $(6,7)$

36. Focus at $(-4,0)$; vertices at $(-4,4)$ and $(-4,2)$

37. Vertices at $(-1,-1)$ and $(3,-1)$; asymptote the line
$$y + 1 = \frac{3}{2}(x - 1)$$

38. Vertices at $(1,-3)$ and $(1,1)$; asymptote the line
$$y + 1 = \frac{3}{2}(x - 1)$$

In Problems 39–52, find the center, transverse axis, vertices, foci, and asymptotes. Graph each equation (a) by hand and (b) by using a graphing utility.

39. $\dfrac{(x-2)^2}{4} - \dfrac{(y+3)^2}{9} = 1$

40. $\dfrac{(y+3)^2}{4} - \dfrac{(x-2)^2}{9} = 1$

41. $(y-2)^2 - 4(x+2)^2 = 4$

42. $(x+4)^2 - 9(y-3)^2 = 9$

43. $(x+1)^2 - (y+2)^2 = 4$

44. $(y-3)^2 - (x+2)^2 = 4$

45. $x^2 - y^2 - 2x - 2y - 1 = 0$

46. $y^2 - x^2 - 4y + 4x - 1 = 0$

47. $y^2 - 4x^2 - 4y - 8x - 4 = 0$

48. $2x^2 - y^2 + 4x + 4y - 4 = 0$

49. $4x^2 - y^2 - 24x - 4y + 16 = 0$

50. $2y^2 - x^2 + 2x + 8y + 3 = 0$

51. $y^2 - 4x^2 - 16x - 2y - 19 = 0$

52. $x^2 - 3y^2 + 8x - 6y + 4 = 0$

In Problems 53–56, graph each function.

[**Hint:** Notice that each function is half a hyperbola.]

53. $f(x) = \sqrt{16 + 4x^2}$

54. $f(x) = -\sqrt{9 + 9x^2}$

55. $f(x) = -\sqrt{-25 + x^2}$

56. $f(x) = \sqrt{-1 + x^2}$

---

57. LORAN Two LORAN stations are positioned 200 miles apart along a straight shore.

(a) A ship records a time difference of 0.00038 second between the LORAN signals. Set up an appropriate rectangular coordinate system to determine where the ship would reach shore if it were to follow the hyperbola corresponding to this time difference.

(b) If the ship wants to enter a harbor located between the two stations 20 miles from the master station, what time difference should it be looking for?

(c) If the ship is 50 miles offshore when the desired time difference is obtained, what is the approximate location of the ship?

[*Note:* The speed of each radio signal is 186,000 miles per second.]

58. LORAN Two LORAN stations are positioned 100 miles apart along a straight shore.

(a) A ship records a time difference of 0.00032 second between the LORAN signals. Set up an appropriate rectangular coordinate system to determine where the ship would reach shore if it were to follow the hyperbola corresponding to this time difference.

(b) If the ship wants to enter a harbor located between the two stations 10 miles from the master station, what time difference should it be looking for?

(c) If the ship is 20 miles offshore when the desired time difference is obtained, what is the approximate location of the ship?

[*Note:* The speed of each radio signal is 186,000 miles per second.]

59. Calibrating Instruments In a test of their recording devices, a team of seismologists positioned two of the devices 2000 feet apart, with the device at point A to the west of the device at point B. At a point between the devices and 200 feet from point B, a small amount of explosive was detonated and a note made of the time at which the sound reached each device. A second explosion is to be carried out at a point directly north of point B.

(a) How far north should the site of the second explosion be chosen so that the measured time difference recorded by the devices for the second detonation is the same as that recorded for the first detonation?

(b) Explain why this experiment can be used to calibrate the instruments.

60. Explain in your own words the LORAN system of navigation.

61. The **eccentricity** e of a hyperbola is defined as the number $\dfrac{c}{a}$, where a and c are the numbers given in equation (2). Because $c > a$, it follows that $e > 1$. Describe the general shape of a hyperbola whose eccentricity is close to 1. What is the shape if e is very large?

62. A hyperbola for which $a = b$ is called an **equilateral hyperbola**. Find the eccentricity e of an equilateral hyperbola.

[*Note:* The eccentricity of a hyperbola is defined in Problem 61.]

63. Two hyperbolas that have the same set of asymptotes are called **conjugate**. Show that the hyperbolas

$$\frac{x^2}{4} - y^2 = 1 \quad \text{and} \quad y^2 - \frac{x^2}{4} = 1$$

are conjugate. Graph each hyperbola on the same set of coordinate axes.

64. Prove that the hyperbola

$$\frac{y^2}{a^2} - \frac{x^2}{b^2} = 1$$

has the two oblique asymptotes

$$y = \frac{a}{b}x \quad \text{and} \quad y = -\frac{a}{b}x$$

65. Show that the graph of an equation of the form

$$Ax^2 + Cy^2 + F = 0, \qquad A \neq 0, C \neq 0, F \neq 0$$

where A and C are of opposite sign, is a hyperbola with center at $(0, 0)$.

66. Show that the graph of an equation of the form

$$Ax^2 + Cy^2 + Dx + Ey + F = 0, \qquad A \neq 0, C \neq 0$$

where A and C are of opposite sign,

(a) Is a hyperbola if $\dfrac{D^2}{4A} + \dfrac{E^2}{4C} - F \neq 0$.

(b) Is two intersecting lines if

$$\frac{D^2}{4A} + \frac{E^2}{4C} - F = 0$$

PREPARING FOR THIS SECTION

Before getting started, review the following:

✓ Sum Formulas for Sine and Cosine
(Section 9.4, pp. 734 and 737)

✓ Half-Angle Formulas for Sine and Cosine
(Section 9.5, p. 750)

✓ Double-Angle Formulas for Sine and Cosine
(Section 9.5, p. 746)

12.5 ROTATION OF AXES; GENERAL FORM OF A CONIC

OBJECTIVES

  1 Identify a Conic

  2 Use a Rotation of Axes to Transform Equations

  3 Discuss an Equation Using a Rotation of Axes

  4 Identify Conics without a Rotation of Axes

In this section, we show that the graph of a general second-degree polynomial containing two variables x and y, that is, an equation of the form

$$Ax^2 + Bxy + Cy^2 + Dx + Ey + F = 0 \tag{1}$$

where A, B, and C are not simultaneously 0, is a conic. We shall not concern ourselves here with the degenerate cases of equation (1), such as $x^2 + y^2 = 0$, whose graph is a single point $(0, 0)$; or $x^2 + 3y^2 + 3 = 0$, whose graph contains no points; or $x^2 - 4y^2 = 0$, whose graph is two lines, $x - 2y = 0$ and $x + 2y = 0$.

We begin with the case where $B = 0$. In this case, the term containing xy is not present, so equation (1) has the form

$$Ax^2 + Cy^2 + Dx + Ey + F = 0$$

where either $A \neq 0$ or $C \neq 0$.

 We have already discussed the procedure for identifying the graph of this kind of equation; we complete the squares of the quadratic expressions in x or y, or both. Once this has been done, the conic can be identified by comparing it to one of the forms studied in Sections 12.2 through 12.4.

In fact, though, we can identify the conic directly from the equation without completing the squares.

Theorem

> ## Identifying Conics without Completing the Squares
>
> Excluding degenerate cases, the equation
>
> $$Ax^2 + Cy^2 + Dx + Ey + F = 0 \qquad (2)$$
>
> where A and C cannot both equal zero:
>
> (a) Defines a parabola if $AC = 0$.
> (b) Defines an ellipse (or a circle) if $AC > 0$.
> (c) Defines a hyperbola if $AC < 0$.

Proof

(a) If $AC = 0$, then either $A = 0$ or $C = 0$, but not both, so the form of equation (2) is either

$$Ax^2 + Dx + Ey + F = 0, \qquad A \neq 0$$

or

$$Cy^2 + Dx + Ey + F = 0, \qquad C \neq 0$$

Using the results of Problems 71 and 72 in Exercise 12.2, it follows that, except for the degenerate cases, the equation is a parabola.

(b) If $AC > 0$, then A and C are of the same sign. Using the results of Problems 73 and 74 in Exercise 12.3, except for the degenerate cases, the equation is an ellipse if $A \neq C$ or a circle if $A = C$.

(c) If $AC < 0$, then A and C are of opposite sign. Using the results of Problems 65 and 66 in Exercise 12.4, except for the degenerate cases, the equation is a hyperbola.

We will not be concerned with the degenerate cases of equation (2). However, in practice, you should be alert to the possibility of degeneracy.

EXAMPLE 1 | **Identifying a Conic without Completing the Squares**

Identify each equation without completing the squares.

(a) $3x^2 + 6y^2 + 6x - 12y = 0$ (b) $2x^2 - 3y^2 + 6y + 4 = 0$
(c) $y^2 - 2x + 4 = 0$

Solution (a) We compare the given equation to equation (2) and conclude that $A = 3$ and $C = 6$. Since $AC = 18 > 0$, the equation is an ellipse.

(b) Here, $A = 2$ and $C = -3$, so $AC = -6 < 0$. The equation is a hyperbola.

(c) Here, $A = 0$ and $C = 1$, so $AC = 0$. The equation is a parabola.

NOW WORK PROBLEM 1.

Although we can now identify the type of conic represented by any equation of the form of equation (2) without completing the squares, we will still need to complete the squares if we desire additional information about a conic.

Now we turn our attention to equations of the form of equation (1), where $B \neq 0$. To discuss this case, we first need to investigate a new procedure: *rotation of axes*.

Rotation of Axes

Figure 53

2

In a **rotation of axes**, the origin remains fixed while the x-axis and y-axis are rotated through an angle θ to a new position; the new positions of the x- and y-axes are denoted by x' and y', respectively, as shown in Figure 53(a).

Now look at Figure 53(b). There the point P has the coordinates (x, y) relative to the xy-plane, while the same point P has coordinates (x', y') relative to the $x'y'$-plane. We seek relationships that will enable us to express x and y in terms of x', y', and θ.

As Figure 53(b) shows, r denotes the distance from the origin O to the point P, and α denotes the angle between the positive x'-axis and the ray from O through P. Then, using the definitions of sine and cosine, we have

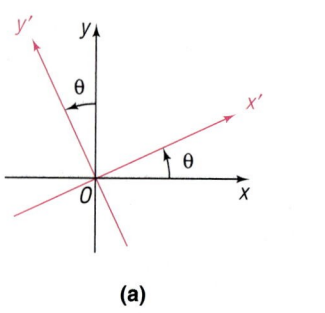

(a)

$$x' = r\cos\alpha \qquad y' = r\sin\alpha \qquad (3)$$
$$x = r\cos(\theta + \alpha) \qquad y = r\sin(\theta + \alpha) \qquad (4)$$

Now

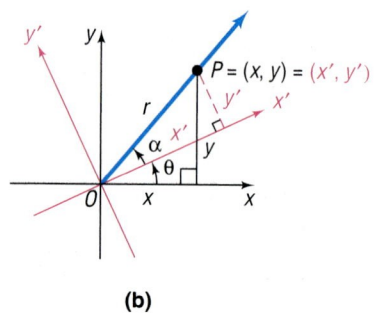

(b)

$$
\begin{aligned}
x &= r\cos(\theta + \alpha) \\
&= r(\cos\theta\cos\alpha - \sin\theta\sin\alpha) &&\text{\color{teal}Sum formula for cosine}\\
&= (r\cos\alpha)(\cos\theta) - (r\sin\alpha)(\sin\theta) \\
&= x'\cos\theta - y'\sin\theta &&\text{\color{teal}By equation (3)}
\end{aligned}
$$

Similarly,

$$
\begin{aligned}
y &= r\sin(\theta + \alpha) \\
&= r(\sin\theta\cos\alpha + \cos\theta\sin\alpha) \\
&= x'\sin\theta + y'\cos\theta
\end{aligned}
$$

Theorem

Rotation Formulas

If the x- and y-axes are rotated through an angle θ, the coordinates (x, y) of a point P relative to the xy-plane and the coordinates (x', y') of the same point relative to the new x'- and y'-axes are related by the formulas

$$x = x'\cos\theta - y'\sin\theta \qquad y = x'\sin\theta + y'\cos\theta \qquad (5)$$

EXAMPLE 2   Rotating Axes

Express the equation $xy = 1$ in terms of new $x'y'$-coordinates by rotating the axes through a 45° angle. Discuss the new equation.

Solution   Let $\theta = 45°$ in equation (5). Then

$$x = x'\cos 45° - y'\sin 45° = x'\frac{\sqrt{2}}{2} - y'\frac{\sqrt{2}}{2} = \frac{\sqrt{2}}{2}(x' - y')$$

$$y = x'\sin 45° + y'\cos 45° = x'\frac{\sqrt{2}}{2} + y'\frac{\sqrt{2}}{2} = \frac{\sqrt{2}}{2}(x' + y')$$

Substituting these expressions for x and y in $xy = 1$ gives

$$\left[\frac{\sqrt{2}}{2}(x' - y')\right]\left[\frac{\sqrt{2}}{2}(x' + y')\right] = 1$$

$$\frac{1}{2}(x'^2 - y'^2) = 1$$

$$\frac{x'^2}{2} - \frac{y'^2}{2} = 1$$

This is the equation of a hyperbola with center at $(0, 0)$ and transverse axis along the x'-axis. The vertices are at $(\pm\sqrt{2}, 0)$ on the x'-axis; the asymptotes are $y' = x'$ and $y' = -x'$ (which correspond to the original x- and y-axes). See Figure 54 for the graph drawn by hand. ◼

Figure 54

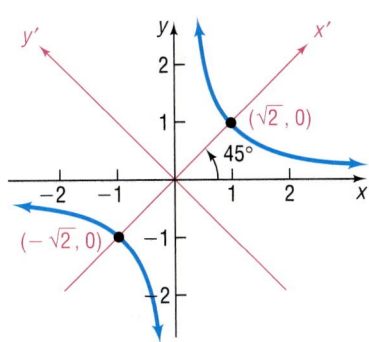

As Example 2 illustrates, a rotation of axes through an appropriate angle can transform a second-degree equation in x and y containing an xy-term into one in x' and y' in which no $x'y'$-term appears. In fact, we will show that a rotation of axes through an appropriate angle will transform any equation of the form of equation (1) into an equation in x' and y' without an $x'y'$-term.

To find the formula for choosing an appropriate angle θ through which to rotate the axes, we begin with equation (1),

$$Ax^2 + Bxy + Cy^2 + Dx + Ey + F = 0, \qquad B \neq 0$$

Next we rotate through an angle θ using rotation formulas (5).

$$A(x' \cos\theta - y' \sin\theta)^2 + B(x' \cos\theta - y' \sin\theta)(x' \sin\theta + y' \cos\theta)$$
$$+ C(x' \sin\theta + y' \cos\theta)^2 + D(x' \cos\theta - y' \sin\theta)$$
$$+ E(x' \sin\theta + y' \cos\theta) + F = 0$$

By expanding and collecting like terms, we obtain

$$(A\cos^2\theta + B\sin\theta\cos\theta + C\sin^2\theta)x'^2 + [B(\cos^2\theta - \sin^2\theta) + 2(C - A)(\sin\theta\cos\theta)]x'y'$$
$$+ (A\sin^2\theta - B\sin\theta\cos\theta + C\cos^2\theta)y'^2$$
$$+ (D\cos\theta + E\sin\theta)x'$$
$$+ (-D\sin\theta + E\cos\theta)y' + F = 0 \qquad (6)$$

In equation (6), the coefficient of $x'y'$ is

$$2(C - A)(\sin\theta\cos\theta) + B(\cos^2\theta - \sin^2\theta)$$

Since we want to eliminate the $x'y'$-term, we select an angle θ so that

$$2(C - A)(\sin\theta\cos\theta) + B(\cos^2\theta - \sin^2\theta) = 0$$
$$(C - A)\sin(2\theta) + B\cos(2\theta) = 0 \quad \text{Double-angle formulas}$$
$$B\cos(2\theta) = (A - C)\sin(2\theta)$$
$$\cot(2\theta) = \frac{A - C}{B}, \qquad B \neq 0$$

Theorem

To transform the equation

$$Ax^2 + Bxy + Cy^2 + Dx + Ey + F = 0, \qquad B \neq 0$$

into an equation in x' and y' without an $x'y'$-term, rotate the axes through an angle θ that satisfies the equation

$$\cot(2\theta) = \frac{A - C}{B} \tag{7}$$

Equation (7) has an infinite number of solutions for θ. We shall adopt the convention of choosing the acute angle θ that satisfies (7). Then we have the following two possibilities:

If $\cot(2\theta) \geq 0$, then $0° < 2\theta \leq 90°$ so that $0° < \theta \leq 45°$.

If $\cot(2\theta) < 0$, then $90° < 2\theta < 180°$ so that $45° < \theta < 90°$.

Each of these results in a counterclockwise rotation of the axes through an acute angle θ.*

WARNING Be careful if you use a calculator to solve equation (7).

1. If $\cot(2\theta) = 0$, then $2\theta = 90°$ and $\theta = 45°$.

2. If $\cot(2\theta) \neq 0$, first find $\cos(2\theta)$. Then use the inverse cosine function key(s) to obtain $2\theta, 0° < 2\theta < 180°$. Finally, divide by 2 to obtain the correct acute angle θ.

EXAMPLE 3 **Discussing an Equation Using a Rotation of Axes**

Discuss the equation: $x^2 + \sqrt{3}xy + 2y^2 - 10 = 0$

Solution Since an xy-term is present, we must rotate the axes. Using $A = 1, B = \sqrt{3}$, and $C = 2$ in equation (7), the appropriate acute angle θ through which to rotate the axes satisfies the equation

$$\cot(2\theta) = \frac{A - C}{B} = \frac{-1}{\sqrt{3}} = -\frac{\sqrt{3}}{3}, \qquad 0° < 2\theta < 180°$$

Since $\cot(2\theta) = -\dfrac{\sqrt{3}}{3}$, we find $2\theta = 120°$, so $\theta = 60°$. Using $\theta = 60°$ in rotation formulas (5), we find

$$x = x' \cos 60° - y' \sin 60° = \frac{1}{2}x' - \frac{\sqrt{3}}{2}y' = \frac{1}{2}(x' - \sqrt{3}y')$$

$$y = x' \sin 60° + y' \cos 60° = \frac{\sqrt{3}}{2}x' + \frac{1}{2}y' = \frac{1}{2}(\sqrt{3}x' + y')$$

---

*Any rotation (clockwise or counterclockwise) through an angle θ that satisfies $\cot(2\theta) = \dfrac{A - C}{B}$ will eliminate the $x'y'$-term. However, the final form of the transformed equation may be different (but equivalent), depending on the angle chosen.

Substituting these values into the original equation and simplifying, we have

$$x^2 + \sqrt{3}xy + 2y^2 - 10 = 0$$

$$\frac{1}{4}(x' - \sqrt{3}y')^2 + \sqrt{3}\left[\frac{1}{2}(x' - \sqrt{3}y')\right]\left[\frac{1}{2}(\sqrt{3}x' + y')\right] + 2\left[\frac{1}{4}(\sqrt{3}x' + y')^2\right] = 10$$

Multiply both sides by 4 and expand to obtain

$$x'^2 - 2\sqrt{3}x'y' + 3y'^2 + \sqrt{3}(\sqrt{3}x'^2 - 2x'y' - \sqrt{3}y'^2) + 2(3x'^2 + 2\sqrt{3}x'y' + y'^2) = 40$$

$$10x'^2 + 2y'^2 = 40$$

$$\frac{x'^2}{4} + \frac{y'^2}{20} = 1$$

Figure 55

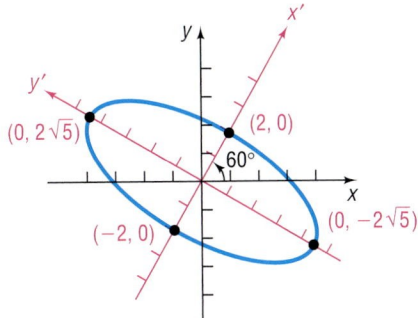

This is the equation of an ellipse with center at $(0,0)$ and major axis along the y'-axis. The vertices are at $(0, \pm 2\sqrt{5})$ on the y'-axis. See Figure 55 for the graph drawn by hand.

To graph the equation $x^2 + \sqrt{3}xy + 2y^2 - 10 = 0$ using a graphing utility, we need to solve the equation for y. Rearranging the terms we observe the equation is quadratic in the variable y: $2y^2 + \sqrt{3}xy + (x^2 - 10) = 0$. We can solve the equation for y using the quadratic formula with $a = 2$, $b = \sqrt{3}x$, and $c = x^2 - 10$.

$$Y_1 = \frac{-\sqrt{3}x + \sqrt{(\sqrt{3}x)^2 - 4(2)(x^2 - 10)}}{2(2)} = \frac{-\sqrt{3}x + \sqrt{-5x^2 + 80}}{4}$$

and

$$Y_2 = \frac{-\sqrt{3}x - \sqrt{(\sqrt{3}x)^2 - 4(2)(x^2 - 10)}}{2(2)} = \frac{-\sqrt{3}x - \sqrt{-5x^2 + 80}}{4}$$

Figure 56 shows the graph of Y_1 and Y_2.

Figure 56

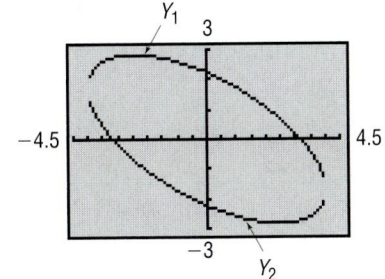

━━━━━━━ **NOW WORK PROBLEM 21.**

In Example 3, the acute angle θ through which to rotate the axes was easy to find because of the numbers that we used in the given equation. In general, the equation $\cot(2\theta) = \dfrac{A - C}{B}$ will not have such a "nice" solution. As the next example shows, we can still find the appropriate rotation formulas without using a calculator approximation by applying half-angle formulas.

EXAMPLE 4   **Discussing an Equation Using a Rotation of Axes**

Discuss the equation: $4x^2 - 4xy + y^2 + 5\sqrt{5}x + 5 = 0$

Solution   Letting $A = 4$, $B = -4$, and $C = 1$ in equation (7), the appropriate angle θ through which to rotate the axes satisfies

$$\cot(2\theta) = \frac{A - C}{B} = \frac{3}{-4} = -\frac{3}{4}$$

In order to use rotation formulas (5), we need to know the values of $\sin\theta$ and $\cos\theta$. Since we seek an acute angle θ, we know that $\sin\theta > 0$ and $\cos\theta > 0$. We use the half-angle formulas in the form

$$\sin \theta = \sqrt{\frac{1 - \cos(2\theta)}{2}} \qquad \cos \theta = \sqrt{\frac{1 + \cos(2\theta)}{2}}$$

Now we need to find the value of $\cos(2\theta)$. Since $\cot(2\theta) = -\dfrac{3}{4}$ and $90° < 2\theta < 180°$ (Do you know why?), it follows that $\cos(2\theta) = -\dfrac{3}{5}$. Then

$$\sin \theta = \sqrt{\frac{1 - \cos(2\theta)}{2}} = \sqrt{\frac{1 - \left(-\dfrac{3}{5}\right)}{2}} = \sqrt{\frac{4}{5}} = \frac{2}{\sqrt{5}} = \frac{2\sqrt{5}}{5}$$

$$\cos \theta = \sqrt{\frac{1 + \cos(2\theta)}{2}} = \sqrt{\frac{1 + \left(-\dfrac{3}{5}\right)}{2}} = \sqrt{\frac{1}{5}} = \frac{1}{\sqrt{5}} = \frac{\sqrt{5}}{5}$$

With these values, the rotation formulas (5) are

$$x = \frac{\sqrt{5}}{5}x' - \frac{2\sqrt{5}}{5}y' = \frac{\sqrt{5}}{5}(x' - 2y')$$

$$y = \frac{2\sqrt{5}}{5}x' + \frac{\sqrt{5}}{5}y' = \frac{\sqrt{5}}{5}(2x' + y')$$

Substituting these values in the original equation and simplifying, we obtain

$$4x^2 - 4xy + y^2 + 5\sqrt{5}x + 5 = 0$$

$$4\left[\frac{\sqrt{5}}{5}(x' - 2y')\right]^2 - 4\left[\frac{\sqrt{5}}{5}(x' - 2y')\right]\left[\frac{\sqrt{5}}{5}(2x' + y')\right]$$

$$+ \left[\frac{\sqrt{5}}{5}(2x' + y')\right]^2 + 5\sqrt{5}\left[\frac{\sqrt{5}}{5}(x' - 2y')\right] = -5$$

Multiply both sides by 5 and expand to obtain

$$4(x'^2 - 4x'y' + 4y'^2) - 4(2x'^2 - 3x'y' - 2y'^2)$$

$$+ 4x'^2 + 4x'y' + y'^2 + 25(x' - 2y') = -25$$

$$25y'^2 - 50y' + 25x' = -25 \qquad \textcolor{purple}{\text{Combine like terms.}}$$

$$y'^2 - 2y' + x' = -1 \qquad \textcolor{purple}{\text{Divide by 25.}}$$

$$y'^2 - 2y' + 1 = -x' \qquad \textcolor{purple}{\text{Complete the square in } y'.}$$

$$(y' - 1)^2 = -x'$$

Figure 57

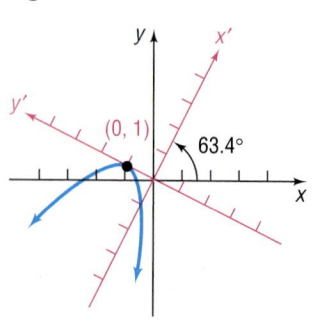

This is the equation of a parabola with vertex at $(0, 1)$ in the $x'y'$-plane. The axis of symmetry is parallel to the x'-axis. Using a calculator to solve $\sin \theta = \dfrac{2\sqrt{5}}{5}$, we find that $\theta \approx 63.4°$. See Figure 57 for the graph drawn by hand.

To graph the equation $4x^2 - 4xy + y^2 + 5\sqrt{5}x + 5 = 0$ using a graphing utility, we need to solve the equation for y. Rearranging the terms we observe the equation is quadratic in the variable y: $y^2 - 4xy + (4x^2 + 5\sqrt{5}x + 5) = 0$. We can solve the equation for y using the quadratic formula with $a = 1, b = -4x$, and $c = 4x^2 + 5\sqrt{5}x + 5$.

Figure 58

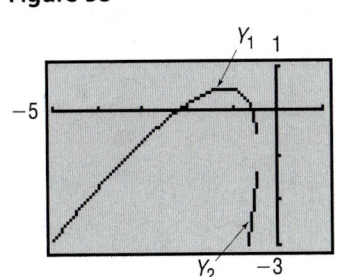

$$Y_1 = \frac{-(-4x) + \sqrt{(-4x)^2 - 4(1)(4x^2 + 5\sqrt{5}x + 5)}}{2(1)} = 2x + \sqrt{-5(\sqrt{5}x + 1)}$$

$$Y_2 = \frac{-(-4x) - \sqrt{(-4x)^2 - 4(1)(4x^2 + 5\sqrt{5}x + 5)}}{2(1)} = 2x - \sqrt{-5(\sqrt{5}x + 1)}$$

Figure 58 shows the graph of Y_1 and Y_2.

➤ **NOW WORK PROBLEM 27.**

Identifying Conics without a Rotation of Axes

④ Suppose that we are required only to identify (rather than discuss) an equation of the form

$$Ax^2 + Bxy + Cy^2 + Dx + Ey + F = 0, \qquad B \neq 0 \qquad (8)$$

If we apply rotation formulas (5) to this equation, we obtain an equation of the form

$$A'x'^2 + B'x'y' + C'y'^2 + D'x' + E'y' + F' = 0 \qquad (9)$$

where $A', B', C', D', E',$ and F' can be expressed in terms of $A, B, C, D, E,$ $F,$ and the angle θ of rotation (see Problem 43). It can be shown that the value of $B^2 - 4AC$ in equation (8) and the value of $B'^2 - 4A'C'$ in equation (9) are equal no matter what angle θ of rotation is chosen (see Problem 45). In particular, if the angle θ of rotation satisfies equation (7), then $B' = 0$ in equation (9), and $B^2 - 4AC = -4A'C'$. Since equation (9) then has the form of equation (2),

$$A'x'^2 + C'y'^2 + D'x' + E'y' + F' = 0$$

we can identify it without completing the squares, as we did in the beginning of this section. In fact, now we can identify the conic described by any equation of the form of equation (8) without a rotation of axes.

Theorem | **Identifying Conics without a Rotation of Axes**

Except for degenerate cases, the equation

$$\boxed{Ax^2 + Bxy + Cy^2 + Dx + Ey + F = 0}$$

(a) Defines a parabola if $B^2 - 4AC = 0$.
(b) Defines an ellipse (or a circle) if $B^2 - 4AC < 0$.
(c) Defines a hyperbola if $B^2 - 4AC > 0$.

You are asked to prove this theorem in Problem 46.

EXAMPLE 5 | **Identifying a Conic without a Rotation of Axes**

Identify the equation: $8x^2 - 12xy + 17y^2 - 4\sqrt{5}x - 2\sqrt{5}y - 15 = 0$

Solution Here $A = 8,$ $B = -12,$ and $C = 17,$ so $B^2 - 4AC = -400.$ Since $B^2 - 4AC < 0,$ the equation defines an ellipse.

➤ **NOW WORK PROBLEM 33.**

12.5 Concepts and Vocabulary

In Problems 1–3, fill in the blanks.

1. To transform the equation

$$Ax^2 + Bxy + Cy^2 + Dx + Ey + F = 0, \qquad B \neq 0$$

   into one in x' and y' without an $x'y'$-term, rotate the axes through an acute angle θ that satisfies the equation _____ .

2. Identify the conic $x^2 - 2y^2 + x - y - 18 = 0$: _____ .
3. Identify the conic $x^2 + 2xy + 3y^2 - 2x + 4y + 10 = 0$: _____ .

In Problems 4–6, answer True or False to each statement.

4. The equation $ax^2 + 6y^2 - 12y = 0$ defines an ellipse if $a > 0$.
5. The equation $3x^2 + bxy + 12y^2 = 10$ defines a parabola if $b = -12$.
6. To eliminate the xy-term from the equation $x^2 - 2xy + y^2 - 2x + 3y + 5 = 0$, rotate the axes through an angle θ, where $\cot \theta = B^2 - 4AC$.

7. Write down an equation of the form $Ax^2 + Bxy + Cy^2 = 10$ that is a parabola.
8. Write down an equation of the form $Ax^2 + Bxy + Cy^2 = 10$ that is an ellipse.
9. Write down an equation of the form $Ax^2 + Bxy + Cy^2 = 10$ that is a hyperbola.

12.5 Exercises

In Problems 1–10, identify each equation without completing the squares.

1. $x^2 + 4x + y + 3 = 0$
2. $2y^2 - 3y + 3x = 0$
3. $6x^2 + 3y^2 - 12x + 6y = 0$
4. $2x^2 + y^2 - 8x + 4y + 2 = 0$
5. $3x^2 - 2y^2 + 6x + 4 = 0$
6. $4x^2 - 3y^2 - 8x + 6y + 1 = 0$
7. $2y^2 - x^2 - y + x = 0$
8. $y^2 - 8x^2 - 2x - y = 0$
9. $x^2 + y^2 - 8x + 4y = 0$
10. $2x^2 + 2y^2 - 8x + 8y = 0$

In Problems 11–20, determine the appropriate rotation formulas to use so that the new equation contains no xy-term.

11. $x^2 + 4xy + y^2 - 3 = 0$
12. $x^2 - 4xy + y^2 - 3 = 0$
13. $5x^2 + 6xy + 5y^2 - 8 = 0$
14. $3x^2 - 10xy + 3y^2 - 32 = 0$
15. $13x^2 - 6\sqrt{3}xy + 7y^2 - 16 = 0$
16. $11x^2 + 10\sqrt{3}xy + y^2 - 4 = 0$
17. $4x^2 - 4xy + y^2 - 8\sqrt{5}x - 16\sqrt{5}y = 0$
18. $x^2 + 4xy + 4y^2 + 5\sqrt{5}y + 5 = 0$
19. $25x^2 - 36xy + 40y^2 - 12\sqrt{13}x - 8\sqrt{13}y = 0$
20. $34x^2 - 24xy + 41y^2 - 25 = 0$

In Problems 21–32, rotate the axes so that the new equation contains no xy-term. Discuss and graph the new equation by hand. Refer to Problems 11–20 for Problems 21–30.

21. $x^2 + 4xy + y^2 - 3 = 0$
22. $x^2 - 4xy + y^2 - 3 = 0$
23. $5x^2 + 6xy + 5y^2 - 8 = 0$
24. $3x^2 - 10xy + 3y^2 - 32 = 0$
25. $13x^2 - 6\sqrt{3}xy + 7y^2 - 16 = 0$
26. $11x^2 + 10\sqrt{3}xy + y^2 - 4 = 0$
27. $4x^2 - 4xy + y^2 - 8\sqrt{5}x - 16\sqrt{5}y = 0$
28. $x^2 + 4xy + 4y^2 + 5\sqrt{5}y + 5 = 0$
29. $25x^2 - 36xy + 40y^2 - 12\sqrt{13}x - 8\sqrt{13}y = 0$
30. $34x^2 - 24xy + 41y^2 - 25 = 0$
31. $16x^2 + 24xy + 9y^2 - 130x + 90y = 0$
32. $16x^2 + 24xy + 9y^2 - 60x + 80y = 0$

In Problems 33–42, identify each equation without applying a rotation of axes.

33. $x^2 + 3xy - 2y^2 + 3x + 2y + 5 = 0$
34. $2x^2 - 3xy + 4y^2 + 2x + 3y - 5 = 0$
35. $x^2 - 7xy + 3y^2 - y - 10 = 0$
36. $2x^2 - 3xy + 2y^2 - 4x - 2 = 0$
37. $9x^2 + 12xy + 4y^2 - x - y = 0$
38. $10x^2 + 12xy + 4y^2 - x - y + 10 = 0$
39. $10x^2 - 12xy + 4y^2 - x - y - 10 = 0$
40. $4x^2 + 12xy + 9y^2 - x - y = 0$
41. $3x^2 - 2xy + y^2 + 4x + 2y - 1 = 0$
42. $3x^2 + 2xy + y^2 + 4x - 2y + 10 = 0$

In Problems 43–46, apply rotation formulas (5) to

$$Ax^2 + Bxy + Cy^2 + Dx + Ey + F = 0$$

to obtain the equation

$$A'x'^2 + B'x'y' + C'y'^2 + D'x' + E'y' + F' = 0$$

43. Express A', B', C', D', E', and F' in terms of A, B, C, D, E, F, and the angle θ of rotation.

[**Hint:** Refer to Equation (6)].

44. Show that $A + C = A' + C'$, and thus show that $A + C$ is **invariant;** that is, its value does not change under a rotation of axes.

45. Refer to Problem 44. Show that $B^2 - 4AC$ is invariant.

46. Prove that, except for degenerate cases, the equation

$$Ax^2 + Bxy + Cy^2 + Dx + Ey + F = 0$$

(a) Defines a parabola if $B^2 - 4AC = 0$.
(b) Defines an ellipse (or a circle) if $B^2 - 4AC < 0$.
(c) Defines a hyperbola if $B^2 - 4AC > 0$.

47. Use rotation formulas (5) to show that distance is invariant under a rotation of axes. That is, show that the distance from $P_1 = (x_1, y_1)$ to $P_2 = (x_2, y_2)$ in the xy-plane equals the distance from $P_1 = (x_1', y_1')$ to $P_2 = (x_2', y_2')$ in the $x'y'$-plane.

48. Show that the graph of the equation $x^{\frac{1}{2}} + y^{\frac{1}{2}} = a^{\frac{1}{2}}$ is part of the graph of a parabola.

49. Formulate a strategy for discussing and graphing an equation of the form

$$Ax^2 + Cy^2 + Dx + Ey + F = 0$$

How does your strategy change if the equation is of the form

$$Ax^2 + Bxy + Cy^2 + Dx + Ey + F = 0$$

PREPARING FOR THIS SECTION

Before getting started, review the following:

✓ Polar Coordinates (Section 11.1, pp. 830–837)

12.6 POLAR EQUATIONS OF CONICS

OBJECTIVES

1. Discuss and Graph Polar Equations of Conics
2. Convert a Polar Equation of a Conic to a Rectangular Equation

1. In Sections 12.2 through 12.4, we gave separate definitions for the parabola, ellipse, and hyperbola based on geometric properties and the distance formula. In this section, we present an alternative definition that simultaneously defines all these conics. As we shall see, this approach is well suited to polar coordinate representation. (Refer to Section 11.1.)

Let D denote a fixed line called the **directrix;** let F denote a fixed point called the **focus,** which is not on D; and let e be a fixed positive number called the **eccentricity.** A **conic** is the set of points P in the plane such that the ratio of the distance from F to P to the distance from D to P equals e. That is, a conic is the collection of points P for which

$$\frac{d(F, P)}{d(D, P)} = e \qquad (1)$$

If $e = 1$, the conic is a **parabola.**
If $e < 1$, the conic is an **ellipse.**

If $e > 1$, the conic is a **hyperbola.**

Observe that if $e = 1$ the definition of a parabola in equation (1) is exactly the same as the definition used earlier in Section 12.2.

In the case of an ellipse, the **major axis** is a line through the focus perpendicular to the directrix. In the case of a hyperbola, the **transverse axis** is a line through the focus perpendicular to the directrix. For both an ellipse and a hyperbola, the eccentricity e satisfies

$$e = \frac{c}{a} \tag{2}$$

where c is the distance from the center to the focus and a is the distance from the center to a vertex.

Just as we did earlier using rectangular coordinates, we derive equations for the conics in polar coordinates by choosing a convenient position for the focus F and the directrix D. The focus F is positioned at the pole, and the directrix D is either parallel to the polar axis or perpendicular to it.

Suppose that we start with the directrix D perpendicular to the polar axis at a distance p units to the left of the pole (the focus F). See Figure 59.

If $P = (r, \theta)$ is any point on the conic, then, by equation (1),

Figure 59

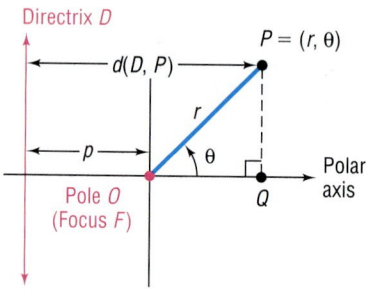

$$\frac{d(F, P)}{d(D, P)} = e \quad \text{or} \quad d(F, P) = e \cdot d(D, P) \tag{3}$$

Now we use the point Q obtained by dropping the perpendicular from P to the polar axis to calculate $d(D, P)$.

$$d(D, P) = p + d(O, Q) = p + r \cos \theta$$

Using this expression and the fact that $d(F, P) = d(O, P) = r$ in equation (3), we get

$$d(F, P) = e \cdot d(D, P)$$
$$r = e(p + r \cos \theta)$$
$$r = ep + er \cos \theta$$
$$r - er \cos \theta = ep$$
$$r(1 - e \cos \theta) = ep$$
$$r = \frac{ep}{1 - e \cos \theta}$$

Theorem

Polar Equation of a Conic; Focus at Pole; Directrix Perpendicular to Polar Axis a Distance p to the Left of the Pole

The polar equation of a conic with focus at the pole and directrix perpendicular to the polar axis at a distance p to the left of the pole is

$$r = \frac{ep}{1 - e \cos \theta} \tag{4}$$

where e is the eccentricity of the conic.

| EXAMPLE 1 | **Discussing and Graphing the Polar Equation of a Conic** |

Discuss and graph the equation: $r = \dfrac{4}{2 - \cos\theta}$

Solution   The given equation is not quite in the form of equation (4), since the first term in the denominator is 2 instead of 1. We divide the numerator and denominator by 2 to obtain

$$r = \dfrac{2}{1 - \dfrac{1}{2}\cos\theta} \qquad r = \dfrac{ep}{1 - e\cos\theta}$$

This equation is in the form of equation (4), with

$$e = \dfrac{1}{2} \quad \text{and} \quad ep = \dfrac{1}{2}p = 2, \quad p = 4$$

We conclude that the conic is an ellipse, since $e = \dfrac{1}{2} < 1$. One focus is at the pole, and the directrix is perpendicular to the polar axis, a distance of $p = 4$ units to the left of the pole. It follows that the major axis is along the polar axis. To find the vertices, we let $\theta = 0$ and $\theta = \pi$. The vertices of the ellipse are $(4, 0)$ and $\left(\dfrac{4}{3}, \pi\right)$. The midpoint of the vertices, $\left(\dfrac{4}{3}, 0\right)$ in polar coordinates, is the center of the ellipse. [Do you see why? The vertices $(4, 0)$ and $\left(\dfrac{4}{3}, \pi\right)$ in polar coordinates are $(4, 0)$ and $\left(-\dfrac{4}{3}, 0\right)$ in rectangular coordinates. The midpoint in rectangular coordinates is $\left(\dfrac{4}{3}, 0\right)$, which is also $\left(\dfrac{4}{3}, 0\right)$ in polar coordinates.] Then a = distance from the center to a vertex = $\dfrac{8}{3}$. Using $a = \dfrac{8}{3}$ and $e = \dfrac{1}{2}$ in equation (2), $e = \dfrac{c}{a}$, we find $c = \dfrac{4}{3}$.

Finally, using $a = \dfrac{8}{3}$ and $c = \dfrac{4}{3}$ in $b^2 = a^2 - c^2$, we have

$$b^2 = a^2 - c^2 = \dfrac{64}{9} - \dfrac{16}{9} = \dfrac{48}{9}$$

$$b = \dfrac{4\sqrt{3}}{3}$$

Figure 60(a) shows the graph drawn by hand.

Figures 60(b) shows the graph of the equation obtained using a graphing utility in POLar mode with θmin $= 0$, θmax $= 2\pi$ and θstep $= \dfrac{\pi}{24}$. ■

Figure 60

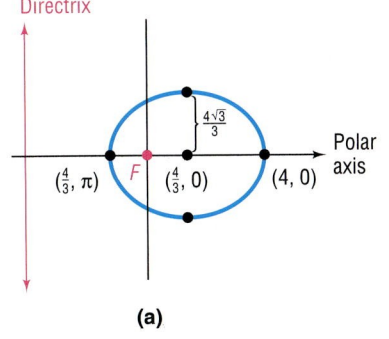

(a)

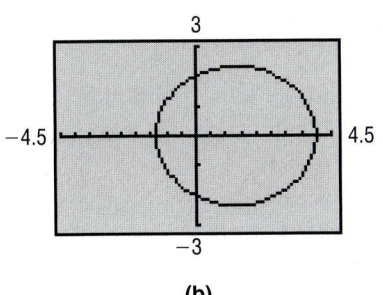

(b)

EXPLORATION   Graph $r_1 = \dfrac{4}{2 + \cos\theta}$ and compare the result with Figure 60(b). What do you conclude? Clear the screen and graph $r_1 = \dfrac{4}{2 - \sin\theta}$ and then $r_1 = \dfrac{4}{2 + \sin\theta}$. Compare each of these graphs with Figure 60(b). What do you conclude?

— **NOW WORK PROBLEM 5.**

Equation (4) was obtained under the assumption that the directrix was perpendicular to the polar axis at a distance p units to the left of the pole. A similar derivation (see Problem 37), in which the directrix is perpendicular to the polar axis at a distance p units to the right of the pole, results in the equation

$$r = \frac{ep}{1 + e \cos \theta}$$

In Problems 38 and 39 you are asked to derive the polar equations of conics with focus at the pole and directrix parallel to the polar axis. Table 5 summarizes the polar equations of conics.

TABLE 5 Polar Equations of Conics (Focus at the Pole, Eccentricity e)

Equation	Description
(a) $r = \dfrac{ep}{1 - e \cos \theta}$	Directrix is perpendicular to the polar axis at a distance p units to the left of the pole.
(b) $r = \dfrac{ep}{1 + e \cos \theta}$	Directrix is perpendicular to the polar axis at a distance p units to the right of the pole.
(c) $r = \dfrac{ep}{1 + e \sin \theta}$	Directrix is parallel to the polar axis at a distance p units above the pole.
(d) $r = \dfrac{ep}{1 - e \sin \theta}$	Directrix is parallel to the polar axis at a distance p units below the pole.

Eccentricity

If $e = 1$, the conic is a parabola; the axis of symmetry is perpendicular to the directrix.

If $e < 1$, the conic is an ellipse; the major axis is perpendicular to the directrix.

If $e > 1$, the conic is a hyperbola; the transverse axis is perpendicular to the directrix.

EXAMPLE 2 **Discussing and Graphing the Polar Equation of a Conic**

Discuss and graph the equation: $r = \dfrac{6}{3 + 3 \sin \theta}$

Solution To place the equation in proper form, we divide the numerator and denominator by 3 to get

$$r = \frac{2}{1 + \sin \theta}$$

Figure 61

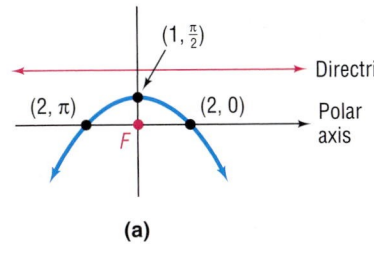

(a)

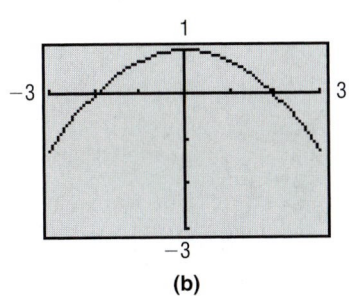

(b)

Referring to Table 5, we conclude that this equation is in the form of equation (c) with

$$e = 1 \quad \text{and} \quad ep = 2$$

$$p = 2$$

The conic is a parabola with focus at the pole. The directrix is parallel to the polar axis at a distance 2 units above the pole; the axis of symmetry is perpendicular to the polar axis. The vertex of the parabola is at $\left(1, \dfrac{\pi}{2}\right)$. (Do you see why?) See Figure 61(a) for the graph drawn by hand. Notice that we plotted two additional points, $(2, 0)$ and $(2, \pi)$, to assist in graphing.

Figure 61(b) shows the graph of the equation using a graphing utility in POLar mode with θmin $= 0$, θmax $= 2\pi$, and θstep $= \dfrac{\pi}{24}$. ■

━━━━ **NOW WORK PROBLEM 7.**

EXAMPLE 3   **Discussing and Graphing the Polar Equation of a Conic**

Discuss and graph the equation: $r = \dfrac{3}{1 + 3\cos\theta}$

Solution   This equation is in the form of equation (b) in Table 5. We conclude that

$$e = 3 \quad \text{and} \quad ep = 3p = 3$$

$$p = 1$$

This is the equation of a hyperbola with a focus at the pole. The directrix is perpendicular to the polar axis, 1 unit to the right of the pole. The transverse axis is along the polar axis. To find the vertices, we let $\theta = 0$ and $\theta = \pi$. The vertices are $\left(\dfrac{3}{4}, 0\right)$ and $\left(-\dfrac{3}{2}, \pi\right)$. The center, which is at the midpoint of $\left(\dfrac{3}{4}, 0\right)$ and $\left(-\dfrac{3}{2}, \pi\right)$, is $\left(\dfrac{9}{8}, 0\right)$. Then $c =$ distance from the center to a focus $= \dfrac{9}{8}$. Since $e = 3$, it follows from equation (2), $e = \dfrac{c}{a}$, that $a = \dfrac{3}{8}$. Finally, using $a = \dfrac{3}{8}$ and $c = \dfrac{9}{8}$ in $b^2 = c^2 - a^2$, we find

$$b^2 = c^2 - a^2 = \frac{81}{64} - \frac{9}{64} = \frac{72}{64} = \frac{9}{8}$$

$$b = \frac{3}{2\sqrt{2}} = \frac{3\sqrt{2}}{4}$$

Figure 62(a) shows the graph drawn by hand. Notice that we plotted two additional points, $\left(3, \frac{\pi}{2}\right)$ and $\left(3, \frac{3\pi}{2}\right)$, on the left branch and used symmetry to obtain the right branch. The asymptotes of this hyperbola were found in the usual way by constructing the rectangle shown.

Figures 62(b) and (c) show the graph of the equation using a graphing utility in POLar mode with θmin $= 0$, θmax $= 2\pi$, and θstep $= \frac{\pi}{24}$, using both dot mode and connected mode. Notice the extraneous asymptotes in the connected mode.

Figure 62

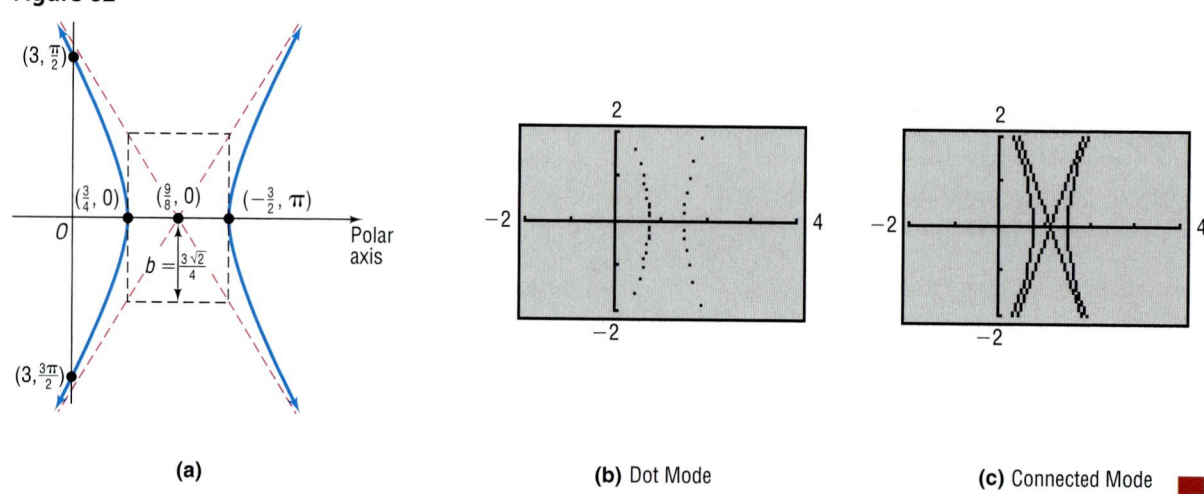

(a)   (b) Dot Mode   (c) Connected Mode

NOW WORK PROBLEM 11.

EXAMPLE 4   **Converting a Polar Equation to a Rectangular Equation**

2   Convert the polar equation

$$r = \frac{1}{3 - 3\cos\theta}$$

to a rectangular equation.

Solution   The strategy here is first to rearrange the equation and square each side before using the transformation equations.

$$r = \frac{1}{3 - 3\cos\theta}$$

$$3r - 3r\cos\theta = 1$$

$$3r = 1 + 3r\cos\theta \qquad \textit{Rearrange the equation.}$$

$$9r^2 = (1 + 3r\cos\theta)^2 \qquad \textit{Square each side.}$$

$$9(x^2 + y^2) = (1 + 3x)^2 \qquad \textit{x}^2 + y^2 = r^2; x = r\cos\theta$$

$$9x^2 + 9y^2 = 9x^2 + 6x + 1$$

$$9y^2 = 6x + 1$$

This is the equation of a parabola in rectangular coordinates.

NOW WORK PROBLEM 19.

12.6 Concepts and Vocabulary

In Problems 1 and 2, fill in the blanks.

1. The polar equation $r = \dfrac{8}{4 - 2\sin\theta}$ is a conic whose eccentricity is _____ . It is a(n) _____ whose directrix is _____ to the polar axis at a distance _____ units _____ the pole.

2. The eccentricity e of a parabola is _____ , of an ellipse it is _____ , and of a hyperbola it is _____ .

In Problems 3 and 4, answer True or False to each statement.

3. If (r, θ) are polar coordinates, the equation $r = \dfrac{2}{2 + 3\sin\theta}$ defines a hyperbola.

4. The eccentricity of any parabola is 1.

5. Write down an equation of the form $r = \dfrac{3e}{1 + e\cos\theta}$ that is a parabola.

6. Write down an equation of the form $r = \dfrac{3e}{1 + e\cos\theta}$ that is an ellipse.

7. Write down an equation of the form $r = \dfrac{3e}{1 + e\cos\theta}$ that is a hyperbola.

12.6 Exercises

In Problems 1–6, identify the conic that each polar equation represents. Also, give the position of the directrix.

1. $r = \dfrac{1}{1 + \cos\theta}$

2. $r = \dfrac{3}{1 - \sin\theta}$

3. $r = \dfrac{4}{2 - 3\sin\theta}$

4. $r = \dfrac{2}{1 + 2\cos\theta}$

5. $r = \dfrac{3}{4 - 2\cos\theta}$

6. $r = \dfrac{6}{8 + 2\sin\theta}$

In Problems 7–18, discuss each equation and graph it by hand. Verify the graph using a graphing utility.

7. $r = \dfrac{1}{1 + \cos\theta}$

8. $r = \dfrac{3}{1 - \sin\theta}$

9. $r = \dfrac{8}{4 + 3\sin\theta}$

10. $r = \dfrac{10}{5 + 4\cos\theta}$

11. $r = \dfrac{9}{3 - 6\cos\theta}$

12. $r = \dfrac{12}{4 + 8\sin\theta}$

13. $r = \dfrac{8}{2 - \sin\theta}$

14. $r = \dfrac{8}{2 + 4\cos\theta}$

15. $r(3 - 2\sin\theta) = 6$

16. $r(2 - \cos\theta) = 2$

17. $r = \dfrac{6\sec\theta}{2\sec\theta - 1}$

18. $r = \dfrac{3\csc\theta}{\csc\theta - 1}$

In Problems 19–30, convert each polar equation to a rectangular equation.

19. $r = \dfrac{1}{1 + \cos\theta}$

20. $r = \dfrac{3}{1 - \sin\theta}$

21. $r = \dfrac{8}{4 + 3\sin\theta}$

22. $r = \dfrac{10}{5 + 4\cos\theta}$

23. $r = \dfrac{9}{3 - 6\cos\theta}$

24. $r = \dfrac{12}{4 + 8\sin\theta}$

25. $r = \dfrac{8}{2 - \sin\theta}$

26. $r = \dfrac{8}{2 + 4\cos\theta}$

27. $r(3 - 2\sin\theta) = 6$

28. $r(2 - \cos\theta) = 2$

29. $r = \dfrac{6\sec\theta}{2\sec\theta - 1}$

30. $r = \dfrac{3\csc\theta}{\csc\theta - 1}$

In Problems 31–36, find a polar equation for each conic. For each, a focus is at the pole.

31. $e = 1$; directrix is parallel to the polar axis 1 unit above the pole

32. $e = 1$; directrix is parallel to the polar axis 2 units below the pole

33. $e = \dfrac{4}{5}$; directrix is perpendicular to the polar axis 3 units to the left of the pole

34. $e = \dfrac{2}{3}$; directrix is parallel to the polar axis 3 units above the pole

35. $e = 6$; directrix is parallel to the polar axis 2 units below the pole

36. $e = 5$; directrix is perpendicular to the polar axis 5 units to the right of the pole

37. Derive equation (b) in Table 5:

$$r = \frac{ep}{1 + e\cos\theta}$$

38. Derive equation (c) in Table 5:

$$r = \frac{ep}{1 + e\sin\theta}$$

39. Derive equation (d) in Table 5:

$$r = \frac{ep}{1 - e\sin\theta}$$

40. Orbit of Mercury   The planet Mercury travels around the Sun in an elliptical orbit given approximately by

$$r = \frac{(3.442)10^7}{1 - 0.206\cos\theta}$$

where r is measured in miles and the Sun is at the pole. Find the distance from Mercury to the Sun at *aphelion* (greatest distance from the Sun) and at *perihelion* (shortest distance from the Sun). See the figure. Use the aphelion and perihelion to graph the orbit of Mercury using a graphing utility.

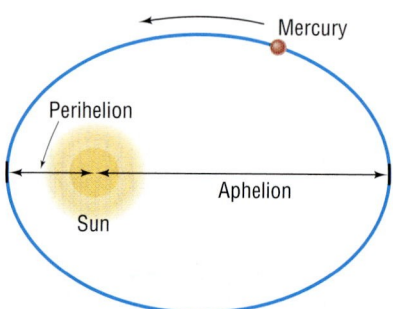

PREPARING FOR THIS SECTION

Before getting started, review the following:

✓ Amplitude and Period of Sinusoidal Graphs (Section 8.6, p. 673)

12.7 PLANE CURVES AND PARAMETRIC EQUATIONS

OBJECTIVES

1. Graph Parametric Equations by Hand
2. Graph Parametric Equations Using a Graphing Utility
3. Find a Rectangular Equation for a Curve Defined Parametrically
4. Use Time as a Parameter in Parametric Equations
5. Find Parametric Equations for Curves Defined by Rectangular Equations

Equations of the form $y = f(x)$, where f is a function, have graphs that are intersected no more than once by any vertical line. The graphs of many of the conics and certain other, more complicated graphs do not have this characteristic. Yet each graph, like the graph of a function, is a collection of points (x, y) in the xy-plane; that is, each is a *plane curve*. In this section, we discuss another way of representing such graphs.

Let $x = f(t)$ and $y = g(t)$, where f and g are two functions whose common domain is some interval I. The collection of points defined by

$$(x, y) = (f(t), g(t))$$

is called a **plane curve.** The equations

$$x = f(t) \qquad y = g(t)$$

where t is in I, are called **parametric equations** of the curve. The variable t is called a **parameter.**

Parametric equations are particularly useful in describing movement along a curve. Suppose that a curve is defined by the parametric equations

$$x = f(t), \qquad y = g(t), \qquad a \le t \le b$$

Figure 63

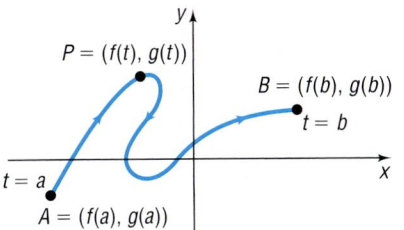

where f and g are each defined over the interval $a \le t \le b$. For a given value of t, we can find the value of $x = f(t)$ and $y = g(t)$, obtaining a point (x, y) on the curve. In fact, as t varies over the interval from $t = a$ to $t = b$, successive values of t give rise to a directed movement along the curve; that is, the curve is traced out in a certain direction by the corresponding succession of points (x, y). See Figure 63. The arrows show the direction, or **orientation**, along the curve as t varies from a to b.

EXAMPLE 1   Discussing a Curve Defined by Parametric Equations

Discuss the curve defined by the parametric equations

$$x = 3t^2, \qquad y = 2t, \qquad -2 \le t \le 2 \tag{1}$$

Solution   For each number $t, -2 \le t \le 2$, there corresponds a number x and a number y. For example, when $t = -2$, then $x = 12$ and $y = -4$. When $t = 0$, then $x = 0$ and $y = 0$. Indeed, we can set up a table listing various choices of the parameter t and the corresponding values for x and y, as shown in Table 6. Plotting these points and connecting them with a smooth curve leads to Figure 64. The arrows in Figure 64 are used to indicate the orientation.

TABLE 6

t	x	y	(x, y)
−2	12	−4	(12, −4)
−1	3	−2	(3, −2)
0	0	0	(0, 0)
1	3	2	(3, 2)
2	12	4	(12, 4)

Figure 64

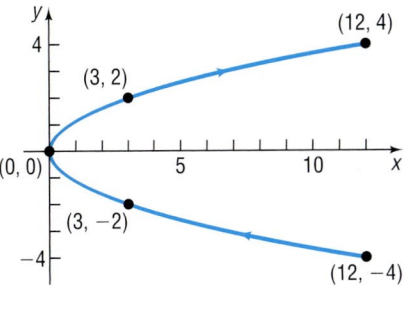

2   Most graphing utilities have the capability of graphing parametric equations. The following steps are usually required in order to obtain the graph of parametric equations. Check your owner's manual to see how yours works.

> ### Graphing Parametric Equations Using a Graphing Utility
>
> **STEP 1:** Set the mode to PARametric. Enter $x(t)$ and $y(t)$.
>
> **STEP 2:** Select the viewing window. In addition to setting Xmin, Xmax, Xscl, and so on, the viewing window in parametric mode requires setting minimum and maximum values for the parameter t and an increment setting for t (Tstep).
>
> **STEP 3:** Graph.

EXAMPLE 2   ### Graphing a Curve Defined by Parametric Equations Using a Graphing Utility

Graph the curve defined by the parametric equations

$$x = 3t^2 \qquad y = 2t \qquad -2 \le t \le 2$$

Solution   **STEP 1:** Enter the equations $x(t) = 3t^2$, $y(t) = 2t$ with the graphing utility in PARametric mode.

STEP 2: Select the viewing window. The interval I is $-2 \le t \le 2$, so we select the following square viewing window:

$$
\begin{array}{lll}
T\text{min} = -2 & X\text{min} = 0 & Y\text{min} = -5 \\
T\text{max} = 2 & X\text{max} = 15 & Y\text{max} = 5 \\
T\text{step} = 0.1 & X\text{scl} = 1 & Y\text{scl} = 1
\end{array}
$$

We choose Tmin $= -2$ and Tmax $= 2$ because $-2 \le t \le 2$. Finally, the choice for Tstep will determine the number of points the graphing utility will plot. For example, with Tstep at 0.1, the graphing utility will evaluate x and y at $t = -2, -1.9, -1.8$, and so on. The smaller the Tstep, the more points the graphing utility will plot. The reader is encouraged to experiment with different values of Tstep to see how the graph is affected.

STEP 3: Graph. Notice the direction the graph is drawn. This direction shows the orientation of the curve.

The graph shown in Figure 65 is complete.   ■

Figure 65

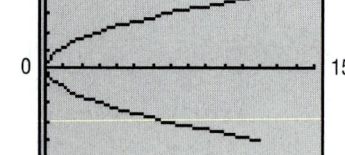

EXPLORATION   Graph the following parametric equations using a graphing utility with Xmin $= 0$, Xmax $= 15$, Ymin $= -5$, Ymax $= 5$, and Tstep $= 0.1$:

1. $x = \dfrac{3t^2}{4}, y = t, -4 \le t \le 4$

2. $x = 3t^2 + 12t + 12, y = 2t + 4, -4 \le t \le 0$

3. $x = 3t^{\frac{2}{3}}, y = 2\sqrt[3]{t}, -8 \le t \le 8$

Compare these graphs to the graph in Figure 65. Conclude that parametric equations defining a curve are not unique, that is, different parametric equations can represent the same graph.

③   The curve given in Examples 1 and 2 should be familiar. To identify it accurately, we find the corresponding rectangular equation by eliminating the parameter t from the parametric equations (1) given in Example 1:

$$x = 3t^2 \qquad y = 2t, \qquad -2 \le t \le 2$$

Noting that we can readily solve for t in $y = 2t$, obtaining $t = \dfrac{y}{2}$, we substitute this expression in the other equation:

$$x = 3t^2 = 3\left(\frac{y}{2}\right)^2 = \frac{3y^2}{4}$$
$$\underset{t = \frac{y}{2}}{\uparrow}$$

This equation, $x = \dfrac{3y^2}{4}$, is the equation of a parabola with vertex at $(0, 0)$ and axis of symmetry along the x-axis.

⚙ EXPLORATION   In FUNCtion mode graph $x = \dfrac{3y^2}{4}\left(Y_1 = \sqrt{\dfrac{4x}{3}} \text{ and }\right.$

$\left. Y_2 = -\sqrt{\dfrac{4x}{3}}\right)$ with $X\text{min} = 0, X\text{max} = 15, Y\text{min} = -5, Y\text{max} = 5.$

Compare this graph with Figure 65. Why do the graphs differ?   **⚙**

Note that the parameterized curve defined by equation (1) and shown in Figure 64 (or 65) is only a part of the parabola $x = \dfrac{3y^2}{4}$. The graph of the rectangular equation obtained by eliminating the parameter will, in general, contain more points than the original parameterized curve. Care must therefore be taken when a parameterized curve is sketched by hand after eliminating the parameter. Even so, the process of eliminating the parameter t of a parameterized curve in order to identify it accurately is sometimes a better approach than merely plotting points. However, the elimination process sometimes requires a little ingenuity.

EXAMPLE 3   Finding the Rectangular Equation of a Curve Defined Parametrically

Find the rectangular equation of the curve whose parametric equations are

$$x = a\cos t \qquad y = a\sin t$$

where $a > 0$ is a constant. By hand, graph this curve, indicating its orientation.

Solution   The presence of sines and cosines in the parametric equations suggests that we use a Pythagorean identity. In fact, since

$$\cos t = \frac{x}{a} \qquad \sin t = \frac{y}{a}$$

we find that

$$\cos^2 t + \sin^2 t = 1$$

$$\left(\frac{x}{a}\right)^2 + \left(\frac{y}{a}\right)^2 = 1$$

$$x^2 + y^2 = a^2$$

Figure 66

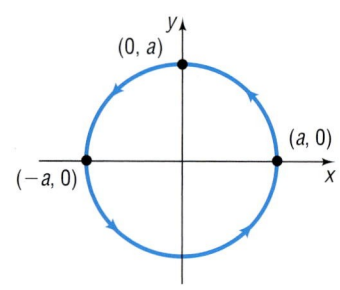

The curve is a circle with center at $(0,0)$ and radius a. As the parameter t increases, say from $t = 0$ [the point $(a, 0)$] to $t = \dfrac{\pi}{2}$ [the point $(0, a)$] to $t = \pi$ [the point $(-a, 0)$], we see that the corresponding points are traced in a counterclockwise direction around the circle. The orientation is as indicated in Figure 66. ▪

NOW WORK PROBLEMS 1 AND 13.

Let's discuss the curve in Example 3 further. The domain of each parametric equation is $-\infty < t < \infty$. Thus, the graph in Figure 66 is actually being repeated each time that t increases by 2π.

If we wanted the curve to consist of exactly 1 revolution in the counterclockwise direction, we could write

$$x = a \cos t, \qquad y = a \sin t, \qquad 0 \le t \le 2\pi$$

This curve starts at $t = 0$ [the point $(a, 0)$] and, proceeding counterclockwise around the circle, ends at $t = 2\pi$ [also the point $(a, 0)$].

If we wanted the curve to consist of exactly three revolutions in the counterclockwise direction, we could write

$$x = a \cos t, \qquad y = a \sin t, \qquad -2\pi \le t \le 4\pi$$

or

$$x = a \cos t, \qquad y = a \sin t, \qquad 0 \le t \le 6\pi$$

or

$$x = a \cos t, \qquad y = a \sin t, \qquad 2\pi \le t \le 8\pi$$

EXAMPLE 4   **Describing Parametric Equations**

Find rectangular equations for and graph the following curves defined by parametric equations.

(a) $x = a \cos t, \qquad y = a \sin t, \qquad 0 \le t \le \pi, \qquad a > 0$
(b) $x = -a \sin t, \qquad y = -a \cos t, \qquad 0 \le t \le \pi, \qquad a > 0$

Solution   (a) We eliminate the parameter t using a Pythagorean identity.

$$\left(\frac{x}{a}\right)^2 + \left(\frac{y}{a}\right)^2 = \cos^2 t + \sin^2 t = 1$$

$$x^2 + y^2 = a^2$$

The curve defined by these parametric equations is a circle, with radius a and center at $(0, 0)$. The circle begins at the point $(a, 0), t = 0$; passes through the point $(0, a), t = \dfrac{\pi}{2}$; and ends at the point $(-a, 0), t = \pi$.

Figure 67

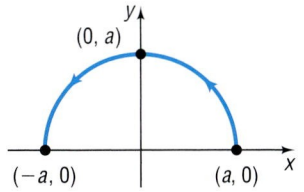

Figure 68

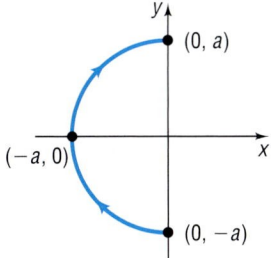

The parametric equations define an upper semicircle of radius a with a counterclockwise orientation. See Figure 67. The rectangular equation is

$$y = a\sqrt{1 - \left(\frac{x}{a}\right)^2}, \qquad -a \le x \le a$$

(b) We eliminate the parameter t using a Pythagorean identity.

$$\left(\frac{x}{-a}\right)^2 + \left(\frac{y}{-a}\right)^2 = \sin^2 t + \cos^2 t = 1$$

$$x^2 + y^2 = a^2$$

The curve defined by these parametric equations is a circle, with radius a and center at $(0, 0)$. The circle begins at the point $(0, -a)$, $t = 0$; passes through the point $(-a, 0)$, $t = \frac{\pi}{2}$; and ends at the point $(0, a)$, $t = \pi$. The parametric equations define a left semicircle of radius a with a clockwise orientation. See Figure 68. The rectangular equation is

$$x = -a\sqrt{1 - \left(\frac{y}{a}\right)^2}, \qquad -a \le y \le a$$

Example 4 illustrates the versatility of parametric equations for replacing complicated rectangular equations, while providing additional information about orientation. These characteristics make parametric equations very useful in applications, such as projectile motion.

 SEEING THE CONCEPT Graph $x = \cos t$, $y = \sin t$ for $0 \le t \le 2\pi$. Compare to Figure 66. Graph $x = \cos t$, $y = \sin t$ for $0 \le t \le \pi$. Compare to Figure 67. Graph $x = -\sin t$, $y = -\cos t$ for $0 \le t \le \pi$. Compare to Figure 68.

Time as a Parameter: Projectile Motion; Simulated Motion

 If we think of the parameter t as time, then the parametric equations $x = f(t)$ and $y = g(t)$ of a curve C specify how the x- and y-coordinates of a moving point vary with time.

For example, we can use parametric equations to describe the motion of an object, sometimes referred to as **curvilinear motion**. Using parametric equations, we can specify not only where the object travels, that is, its location (x, y), but also when it gets there, that is, the time t.

When an object is propelled upward at an inclination θ to the horizontal with initial speed v_0, the resulting motion is called **projectile motion.** See Figure 69(a) on page 954.

 In calculus it is shown that the parametric equations of the path of a projectile fired at an inclination θ to the horizontal, with an initial speed v_0, from a height h above the horizontal are

$$x = (v_0 \cos \theta)t \qquad y = -\frac{1}{2}gt^2 + (v_0 \sin \theta)t + h \qquad (2)$$

where t is the time and g is the constant acceleration due to gravity (approximately 32 ft/sec/sec or 9.8 m/sec/sec). See Figure 69(b) on page 954.

Figure 69

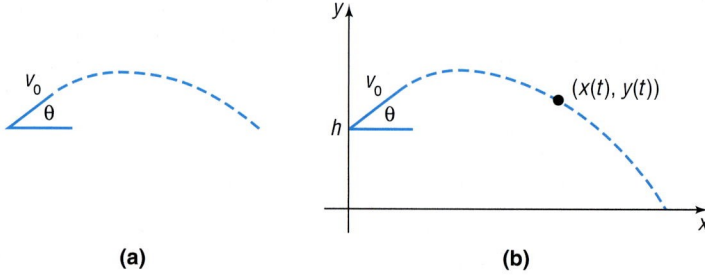

(a)   (b)

EXAMPLE 5   Projectile Motion

Figure 70

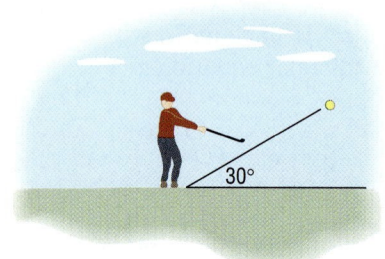

Suppose that Jim hit a golf ball with an initial velocity of 150 feet per second at an angle of 30° to the horizontal. See Figure 70.

(a) Find parametric equations that describe the position of the ball as a function of time.

(b) How long is the golf ball in the air?

(c) When is the ball at its maximum height? Determine the maximum height of the ball.

(d) Determine the distance that the ball traveled.

(e) Using a graphing utility, simulate the motion of the golf ball by simultaneously graphing the equations found in part (a).

Solution   (a) We have $v_0 = 150, \theta = 30°, h = 0$ (the ball is on the ground), and $g = 32$ (since units are in feet and seconds). Substituting these values into equations (2), we find that

$$x = (v_0 \cos \theta)t = (150 \cos 30°)t = 75\sqrt{3}t$$

$$y = -\frac{1}{2}gt^2 + (v_0 \sin \theta)t + h = -\frac{1}{2}(32)t^2 + (150 \sin 30°)t + 0$$

$$= -16t^2 + 75t$$

(b) To determine the length of time that the ball is in the air, we solve the equation $y = 0$.

$$-16t^2 + 75t = 0$$

$$t(-16t + 75) = 0$$

$$t = 0 \sec \quad \text{or} \quad t = \frac{75}{16} = 4.6875 \sec$$

The ball will strike the ground after 4.6875 seconds.

(c) Notice that the height y of the ball is a quadratic function of t, so the maximum height of the ball can be found by determining the vertex of $y = -16t^2 + 75t$. The value of t at the vertex is

$$t = \frac{-b}{2a} = \frac{-75}{-32} = 2.34375 \sec$$

The ball is at its maximum height after 2.34375 seconds. The maximum height of the ball is found by evaluating the function y at $t = 2.34375$ seconds.

Maximum height $= -16(2.34375)^2 + (75)2.34375 \approx 87.89$ feet

Figure 71

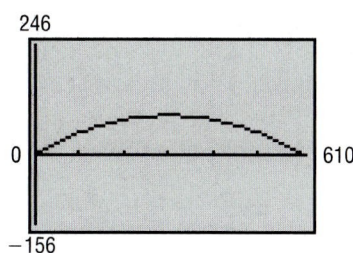

(d) Since the ball is in the air for 4.6875 seconds, the horizontal distance that the ball travels is

$$x = (75\sqrt{3})4.6875 \approx 608.92 \text{ feet}$$

(e) We enter the equations from part (a) into a graphing utility with $T\text{min} = 0$, $T\text{max} = 4.7$, and $T\text{step} = 0.1$. We use ZOOM-SQUARE to avoid any distortion to the angle of elevation. See Figure 71.

 EXPLORATION Simulate the motion of a ball thrown straight up with an initial speed of 100 feet per second from a height of 5 feet above the ground. Use PARametric mode with $T\text{min} = 0$, $T\text{max} = 6.5$, $T\text{step} = 0.1$, $X\text{min} = 0$, $X\text{max} = 5$, $Y\text{min} = 0$, and $Y\text{max} = 180$. What happens to the speed with which the graph is drawn as the ball goes up and then comes back down? How do you interpret this physically? Repeat the experiment using other values for $T\text{step}$. How does this affect the experiment?

[**Hint:** In the projectile motion equations, let $\theta = 90°$, $v_0 = 100$, $h = 5$, and $g = 32$. We use $x = 3$ instead of $x = 0$ to see the vertical motion better.]

RESULT See Figure 72. In Figure 72(a) the ball is going up. In Figure 72(b) the ball is near its highest point. Finally, in Figure 72(c) the ball is coming back down.

Notice that, as the ball goes up, its speed decreases, until at the highest point it is zero. Then the speed increases as the ball comes back down.

Figure 72

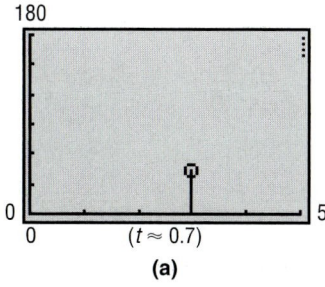

(a)

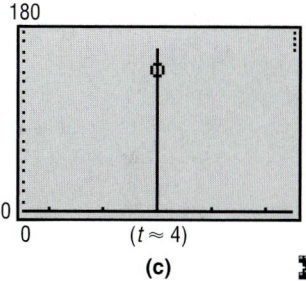

(b)

(c)

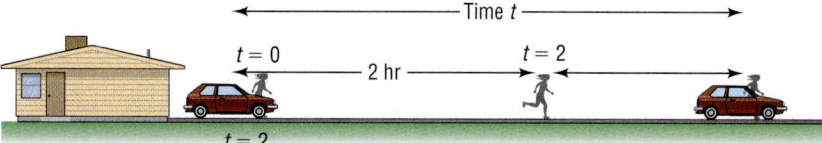 **NOW WORK PROBLEM 27.**

A graphing utility can be used to simulate other kinds of motion as well. Let's work again Example 7 from Section 1.4.

EXAMPLE 6 **Simulating Motion**

Tanya, who is a long distance runner, runs at an average velocity of 8 miles per hour. Two hours after Tanya leaves your house, you leave in your Honda and follow the same route. If your average velocity is 40 miles per hour, how long will it be before you catch up to Tanya? See Figure 73. Use a simulation of the two motions to verify the answer.

Figure 73

Time t

$t = 0$

2 hr

$t = 2$

$t = 2$

Solution We begin with two sets of parametric equations: one to describe Tanya's motion, the other to describe the motion of the Honda. We choose time $t = 0$ to be when Tanya leaves the house. If we choose $y_1 = 2$ as Tanya's path, then we can use $y_2 = 4$ as the parallel path of the Honda. The horizontal distances traversed in time t (Distance = Velocity × Time) are

$$\text{Tanya:} \quad x_1 = 8t \qquad \text{Honda:} \quad x_2 = 40(t - 2)$$

The Honda catches up to Tanya when $x_1 = x_2$.

$$8t = 40(t - 2)$$
$$8t = 40t - 80$$
$$-32t = -80$$
$$t = \frac{-80}{-32} = 2.5$$

The Honda catches up to Tanya 2.5 hours after Tanya leaves the house.

In PARametric mode with Tstep = 0.01, we simultaneously graph

$$\text{Tanya:} \quad x_1 = 8t \qquad\qquad \text{Honda:} \quad x_2 = 40(t - 2)$$
$$y_1 = 2 \qquad\qquad\qquad\qquad y_2 = 4$$

for $0 \leq t \leq 3$.

Figure 74 shows the relative position of Tanya and the Honda for $t = 0, t = 2, t = 2.25, t = 2.5$, and $t = 2.75$.

Figure 74

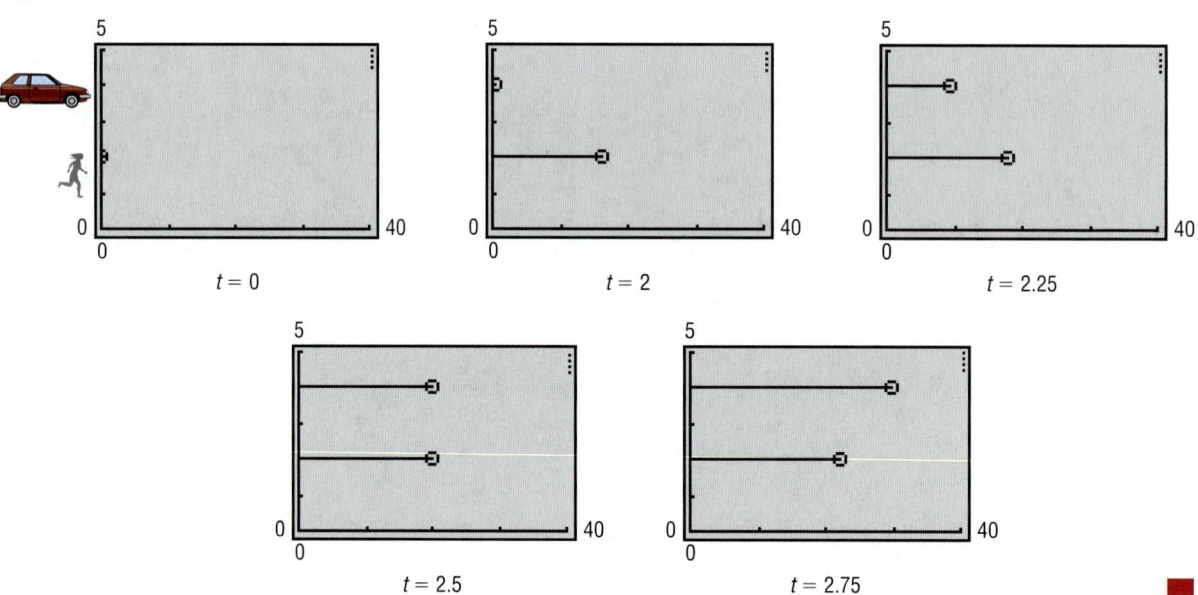

$t = 0$ $\qquad\qquad$ $t = 2$ $\qquad\qquad$ $t = 2.25$

$t = 2.5$ $\qquad\qquad$ $t = 2.75$

Finding Parametric Equations

We now take up the question of how to find parametric equations of a given curve.

⑤ If a curve is defined by the equation $y = f(x)$, where f is a function, one way of finding parametric equations is to let $x = t$. Then $y = f(t)$ and

$$x = t, \quad y = f(t), \qquad t \text{ in the domain of } f$$

are parametric equations of the curve.

EXAMPLE 7   **Finding Parametric Equations for a Curve Defined by a Rectangular Equation**

Find parametric equations for the equation $y = x^2 - 4$.

Solution   Let $x = t$. Then the parametric equations are

$$x = t, \quad y = t^2 - 4, \qquad -\infty < t < \infty$$

Another less obvious approach to Example 7 is to let $x = t^3$. Then the parametric equations become

$$x = t^3, \quad y = t^6 - 4, \qquad -\infty < t < \infty$$

Care must be taken when using this approach, since the substitution for x must be a function that allows x to take on all the values stipulated by the domain of f. For example, letting $x = t^2$ so that $y = t^4 - 4$ does not result in equivalent parametric equations for $y = x^2 - 4$, since only points for which $x \geq 0$ are obtained.

EXAMPLE 8   **Finding Parametric Equations for an Object in Motion**

Find parametric equations for the ellipse

$$x^2 + \frac{y^2}{9} = 1$$

where the parameter t is time (in seconds) and

(a) The motion around the ellipse is clockwise, begins at the point $(0, 3)$, and requires 1 second for a complete revolution.
(b) The motion around the ellipse is counterclockwise, begins at the point $(1, 0)$, and requires 2 seconds for a complete revolution.

Solution   (a) See Figure 75. Since the motion begins at the point $(0, 3)$, we want $x = 0$ and $y = 3$ when $t = 0$. Furthermore, since the given equation is an ellipse, we begin by letting

Figure 75

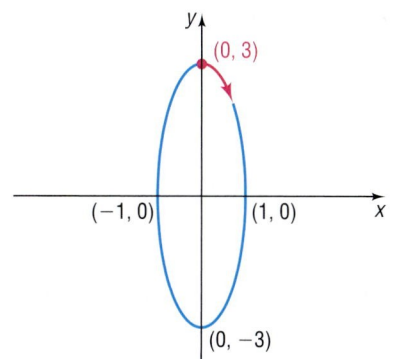

$$x = \sin(\omega t) \qquad \frac{y}{3} = \cos(\omega t)$$

for some constant ω. These parametric equations satisfy the equation of the ellipse. Furthermore, with this choice, when $t = 0$, we have $x = 0$ and $y = 3$.

For the motion to be clockwise, the motion will have to begin with the value of x increasing and y decreasing as t increases. This requires that $\omega > 0$. [Do you know why? If $\omega > 0$, then $x = \sin(\omega t)$ is increasing when $t > 0$ is near zero and $y = 3\cos(\omega t)$ is decreasing when $t > 0$ is near zero]. See the red part of the graph in Figure 75.

Finally, since 1 revolution requires 1 second, the period $\dfrac{2\pi}{\omega} = 1$, so $\omega = 2\pi$. Parametric equations that satisfy the conditions stipulated are

$$x = \sin(2\pi t), \qquad y = 3\cos(2\pi t), \qquad 0 \leq t \leq 1 \qquad (3)$$

Figure 76

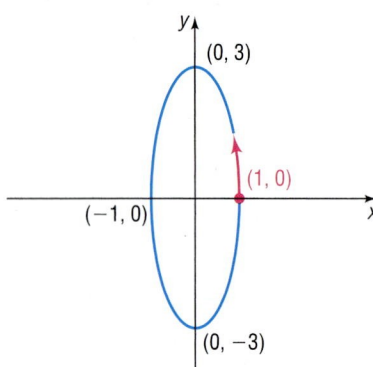

(b) See Figure 76. Since the motion begins at the point $(1, 0)$, we want $x = 1$ and $y = 0$ when $t = 0$. Furthermore, since the given equation is an ellipse, we begin by letting

$$x = \cos(\omega t) \qquad \frac{y}{3} = \sin(\omega t)$$

for some constant ω. These parametric equations satisfy the equation of the ellipse. Furthermore, with this choice, when $t = 0$, we have $x = 1$ and $y = 0$.

For the motion to be counterclockwise, the motion will have to begin with the value of x decreasing and y increasing as t increases. This requires that $\omega > 0$. [Do you know why?] Finally, since 1 revolution requires 2 seconds, the period is $\dfrac{2\pi}{\omega} = 2$, so $\omega = \pi$. The parametric equations that satisfy the conditions stipulated are

$$x = \cos(\pi t), \qquad y = 3\sin(\pi t), \qquad 0 \le t \le 2 \qquad (4) \quad \blacksquare$$

Either of equations (3) or (4) can serve as parametric equations for the ellipse $x^2 + \dfrac{y^2}{9} = 1$ given in Example 8. The direction of the motion, the beginning point, and the time for 1 revolution merely serve to help us arrive at a particular parametric representation.

 **NOW WORK PROBLEM 43.**

The Cycloid

Suppose that a circle of radius a rolls along a horizontal line without slipping. As the circle rolls along the line, a point P on the circle will trace out a curve called a **cycloid** (see Figure 77). We now seek parametric equations* for a cycloid.

Figure 77

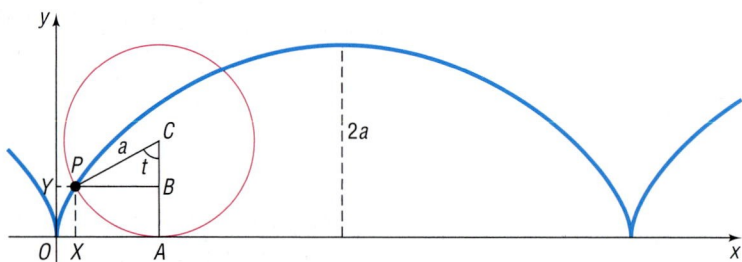

We begin with a circle of radius a and take the fixed line on which the circle rolls as the x-axis. Let the origin be one of the points at which the point P comes in contact with the x-axis. Figure 77 illustrates the position of this point P after the circle has rolled somewhat. The angle t (in radians) measures the angle through which the circle has rolled.

*Any attempt to derive the rectangular equation of a cycloid would soon demonstrate how complicated the task is.

Since we require no slippage, it follows that

$$\text{Arc } AP = d(O, A)$$

The length of the arc AP is given by $s = r\theta$, where $r = a$ and $\theta = t$ radians. Then,

$$at = d(O, A) \qquad s = r\theta, \text{ where } r = a \text{ and } \theta = t$$

The x-coordinate of the point P is

$$d(O, X) = d(O, A) - d(X, A) = at - a\sin t = a(t - \sin t)$$

The y-coordinate of the point P is equal to

$$d(O, Y) = d(A, C) - d(B, C) = a - a\cos t = a(1 - \cos t)$$

The parametric equations of the cycloid are

$$x = a(t - \sin t) \qquad y = a(1 - \cos t) \tag{5}$$

 EXPLORATION   Graph $x = t - \sin t$, $y = 1 - \cos t$, $0 \le t \le 3\pi$, using your graphing utility with Tstep $= \dfrac{\pi}{36}$ and a square screen. Compare your results with Figure 77.

Applications to Mechanics

If a is negative in equation (5), we obtain an inverted cycloid, as shown in Figure 78(a). The inverted cycloid occurs as a result of some remarkable applications in the field of mechanics. We shall mention two of them: the *brachistochrone* and the *tautochrone*.*

Figure 78

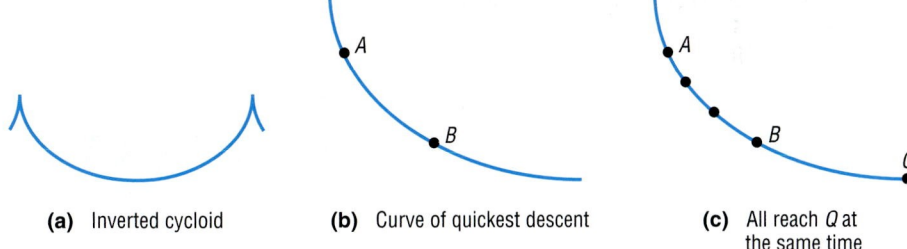

(a) Inverted cycloid   **(b)** Curve of quickest descent   **(c)** All reach Q at
the same time

 The **brachistochrone** is the curve of quickest descent. If a particle is constrained to follow some path from one point A to a lower point B (not on the same vertical line) and is acted on only by gravity, the time needed to make the descent is least if the path is an inverted cycloid. See Figure 78(b). This remarkable discovery, which is attributed to many famous mathematicians (including Johann Bernoulli and Blaise Pascal), was a significant step in creating the branch of mathematics known as the *calculus of variations*.

To define the **tautochrone**, let Q be the lowest point on an inverted cycloid. If several particles placed at various positions on an inverted cycloid simultaneously begin to slide down the cycloid, they will reach the point Q at the same time, as indicated in Figure 78(c). The tautochrone property of

*In Greek, *brachistochrone* means "the shortest time" and *tautochrone* means "equal time."

Figure 79

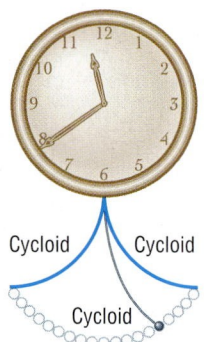

Cycloid   Cycloid

Cycloid

the cycloid was used by Christiaan Huygens (1629–1695), the Dutch mathematician, physicist, and astronomer, to construct a pendulum clock with a bob that swings along a cycloid (see Figure 79). In Huygen's clock, the bob was made to swing along a cycloid by suspending the bob on a thin wire constrained by two plates shaped like cycloids. In a clock of this design, the period of the pendulum is independent of its amplitude.

12.7 Concepts and Vocabulary

In Problems 1–3, fill in the blanks.

1. Let $x = f(t)$ and $y = g(t)$, where f and g are two functions whose common domain is some interval I. The collection of points defined by $(x, y) = (f(t), g(t))$ is called a _____ _____. The variable t is called a _____.

2. The parametric equations $x = 2 \sin t$, $y = 3 \cos t$ define a(n) _____.

3. If a circle rolls along a horizontal line without slippage, a point P on the circle will trace out a curve called a _____.

In Problems 4 and 5, answer True or False to each statement.

4. Parametric equations defining a curve are unique.

5. Curves defined using parametric equations have an orientation.

6. Make up parametric equations that define an ellipse.

7. Explain what a brachistochrone is.

12.7 Exercises

In Problems 1–20, graph by hand the curve whose parametric equations are given and show its orientation. Verify the graph using a graphing utility. Find the rectangular equation of each curve.

1. $x = 3t + 2$, $y = t + 1$; $0 \leq t \leq 4$

2. $x = t - 3$, $y = 2t + 4$; $0 \leq t \leq 2$

3. $x = t + 2$, $y = \sqrt{t}$; $t \geq 0$

4. $x = \sqrt{2t}$, $y = 4t$; $t \geq 0$

5. $x = t^2 + 4$, $y = t^2 - 4$; $-\infty < t < \infty$

6. $x = \sqrt{t} + 4$, $y = \sqrt{t} - 4$; $t \geq 0$

7. $x = 3t^2$, $y = t + 1$; $-\infty < t < \infty$

8. $x = 2t - 4$, $y = 4t^2$; $-\infty < t < \infty$

9. $x = 2e^t$, $y = 1 + e^t$; $t \geq 0$

10. $x = e^t$, $y = e^{-t}$; $t \geq 0$

11. $x = \sqrt{t}$, $y = t^{\frac{3}{2}}$; $t \geq 0$

12. $x = t^{\frac{3}{2}} + 1$, $y = \sqrt{t}$; $t \geq 0$

13. $x = 2 \cos t$, $y = 3 \sin t$; $0 \leq t \leq 2\pi$

14. $x = 2 \cos t$, $y = 3 \sin t$; $0 \leq t \leq \pi$

15. $x = 2 \cos t$, $y = 3 \sin t$; $-\pi \leq t \leq 0$

16. $x = 2 \cos t$, $y = \sin t$; $0 \leq t \leq \dfrac{\pi}{2}$

17. $x = \sec t$, $y = \tan t$; $0 \leq t \leq \dfrac{\pi}{4}$

18. $x = \csc t$, $y = \cot t$; $\dfrac{\pi}{4} \leq t \leq \dfrac{\pi}{2}$

19. $x = \sin^2 t$, $y = \cos^2 t$; $0 \leq t \leq 2\pi$

20. $x = t^2$, $y = \ln t$; $t > 0$

21. Projectile Motion Bob throws a ball straight up with an initial speed of 50 feet per second from a height of 6 feet.
(a) Find parametric equations that describe the motion of the ball as a function of time.
(b) How long is the ball in the air?
(c) When is the ball at its maximum height? Determine the maximum height of the ball.
(d) Simulate the motion of the ball by simultaneously graphing the equations found in part (a).

22. Projectile Motion Alice throws a ball straight up with an initial speed of 40 feet per second from a height of 5 feet.
(a) Find parametric equations that describe the motion of the ball as a function of time.
(b) How long is the ball in the air?
(c) When is the ball at its maximum height? Determine the maximum height of the ball.
(d) Simulate the motion of the ball by simultaneously graphing the equations found in part (a).

23. Catching a Train Bill's train leaves at 8:06 AM and accelerates at the rate of 2 meters per second per second. Bill, who can run 5 meters per second, arrives at the train station 5 seconds after the train has left.
(a) Find parametric equations that describe the motion of the train and Bill as a function of time.

   [**Hint:** The position s at time t of an object having acceleration a is $s = \frac{1}{2}at^2$.]

(b) Determine algebraically whether Bill will catch the train. If so, when?
(c) Simulate the motion of the train and Bill by simultaneously graphing the equations found in part (a).

24. Catching a Bus Jodi's bus leaves at 5:30 pm and accelerates at the rate of 3 meters per second per second. Jodi, who can run 5 meters per second, arrives at the bus station 2 seconds after the bus has left.
(a) Find parametric equations that describe the motion of the bus and Jodi as a function of time.

   [**Hint:** The position s at time t of an object having acceleration a is $s = \frac{1}{2}at^2$.]

(b) Determine algebraically whether Jodi will catch the bus. If so, when?
(c) Simulate the motion of the bus and Jodi by simultaneously graphing the equations found in part (a).

25. Projectile Motion Nolan Ryan throws a baseball with an initial speed of 145 feet per second at an angle of 20° to the horizontal. The ball leaves Nolan Ryan's hand at a height of 5 feet.
(a) Find parametric equations that describe the position of the ball as a function of time.
(b) How long is the ball in the air?
(c) When is the ball at its maximum height? Determine the maximum height of the ball.
(d) Determine the distance that the ball traveled.
(e) Using a graphing utility, simultaneously graph the equations found in part (a).

26. Projectile Motion Mark McGwire hit a baseball with an initial speed of 180 feet per second at an angle of 40° to the horizontal. The ball was hit at a height of 3 feet off the ground.
(a) Find parametric equations that describe the position of the ball as a function of time.
(b) How long is the ball in the air?
(c) When is the ball at its maximum height? Determine the maximum height of the ball.
(d) Determine the distance that the ball traveled.
(e) Using a graphing utility, simultaneously graph the equations found in part (a).

27. Projectile Motion Suppose that Adam throws a tennis ball off a cliff 300 meters high with an initial speed of 40 meters per second at an angle of 45° to the horizontal.
(a) Find parametric equations that describe the position of the ball as a function of time.
(b) How long is the ball in the air?
(c) When is the ball at its maximum height? Determine the maximum height of the ball.
(d) Determine the distance that the ball traveled.
(e) Using a graphing utility, simultaneously graph the equations found in part (a).

28. Projectile Motion Suppose that Adam throws a tennis ball off a cliff 300 meters high with an initial speed of 40 meters per second at an angle of 45° to the horizontal on the Moon (gravity on the Moon is one-sixth of that on Earth).
(a) Find parametric equations that describe the position of the ball as a function of time.
(b) How long is the ball in the air?
(c) When is the ball at its maximum height? Determine the maximum height of the ball.
(d) Determine the distance that the ball traveled.
(e) Using a graphing utility, simultaneously graph the equations found in part (a).

29. Uniform Motion A Toyota Paseo (traveling east at 40 mph) and Pontiac Bonneville (traveling north at 30 mph) are heading toward the same intersection. The Paseo is 5 miles from the intersection when the Bonneville is 4 miles from the intersection. See the figure.

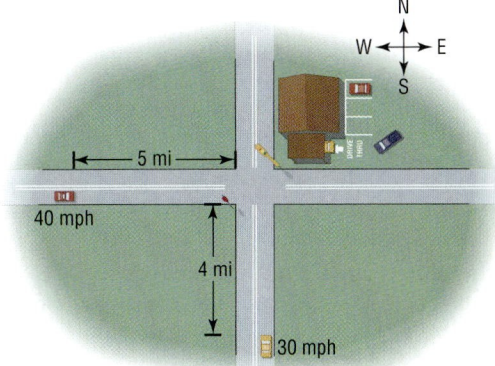

(a) Find parametric equations that describe the motion of the Paseo and Bonneville.
(b) Find a formula for the distance between the cars as a function of time.

(c) Graph the function in part (b) using a graphing utility.

(d) What is the minimum distance between the cars? When are the cars closest?

(e) Simulate the motion of the cars by simultaneously graphing the equations found in part (a).

30. Uniform Motion   A Cessna (heading south at 120 mph) and a Boeing 747 (heading west at 600 mph) are flying toward the same point at the same altitude. The Cessna is 100 miles from the point where the flight patterns intersect and the 747 is 550 miles from this intersection point. See the figure.

(a) Find parametric equations that describe the motion of the Cessna and 747.

(b) Find a formula for the distance between the planes as a function of time.

(c) Graph the function in part (b) using a graphing utility.

(d) What is the minimum distance between the planes? When are the planes closest?

(e) Simulate the motion of the planes by simultaneously graphing the equations found in part (a).

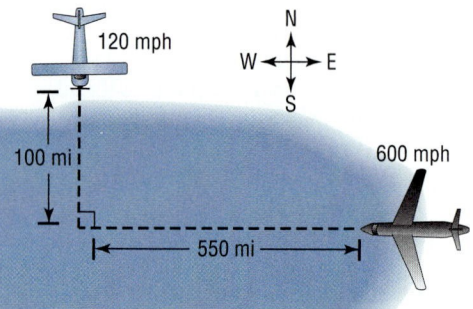

In Problems 31–38, find two different parametric equations for each rectangular equation.

31. $y = 4x - 1$

32. $y = -8x + 3$

33. $y = x^2 + 1$

34. $y = -2x^2 + 1$

35. $y = x^3$

36. $y = x^4 + 1$

37. $x = y^{3/2}$

38. $x = \sqrt{y}$

In Problems 39–42, find parametric equations that define the curve shown.

39.

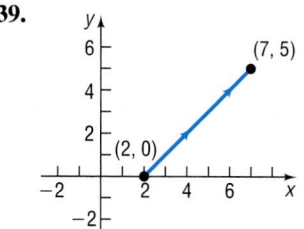

40.

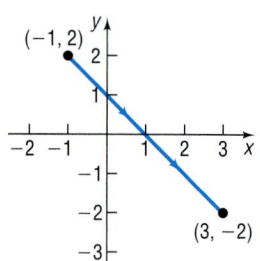

41.

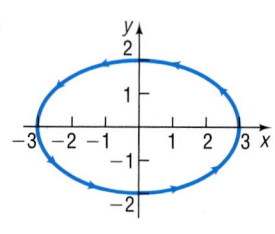

42.

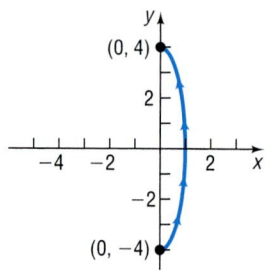

In Problems 43–46, find parametric equations for an object that moves along the ellipse $\dfrac{x^2}{4} + \dfrac{y^2}{9} = 1$ with the motion described.

43. The motion begins at $(2, 0)$, is clockwise, and requires 2 seconds for a complete revolution.

44. The motion begins at $(0, 3)$, is counterclockwise, and requires 1 second for a complete revolution.

45. The motion begins at $(0, 3)$, is clockwise, and requires 1 second for a complete revolution.

46. The motion begins at $(2, 0)$, is counterclockwise, and requires 3 seconds for a complete revolution.

In Problems 47 and 48, the parametric equations of four curves are given. By hand, graph each of them, indicating the orientation.

47. C_1: $x = t$, $y = t^2$; $-4 \le t \le 4$
C_2: $x = \cos t$, $y = 1 - \sin^2 t$; $0 \le t \le \pi$
C_3: $x = e^t$, $y = e^{2t}$; $0 \le t \le \ln 4$
C_4: $x = \sqrt{t}$, $y = t$; $0 \le t \le 16$

48. C_1: $x = t$, $y = \sqrt{1 - t^2}$; $-1 \le t \le 1$
C_2: $x = \sin t$, $y = \cos t$; $0 \le t \le 2\pi$
C_3: $x = \cos t$, $y = \sin t$; $0 \le t \le 2\pi$
C_4: $x = \sqrt{1 - t^2}$, $y = t$; $-1 \le t \le 1$

49. Show that the parametric equations for a line passing through the points (x_1, y_1) and (x_2, y_2) are

$$x = (x_2 - x_1)t + x_1$$
$$y = (y_2 - y_1)t + y_1, \quad -\infty < t < \infty$$

What is the orientation of this line?

50. Projectile Motion The position of a projectile fired with an initial velocity v_0 feet per second and at an angle θ to the horizontal at the end of t seconds is given by the parametric equations

$$x = (v_0 \cos \theta)t \qquad y = (v_0 \sin \theta)t - 16t^2$$

See the following illustration.

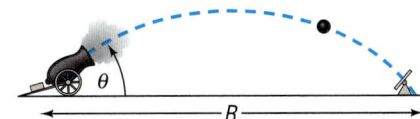

(a) Obtain the rectangular equation of the trajectory and identify the curve.

(b) Show that the projectile hits the ground $(y = 0)$ when $t = \dfrac{1}{16} v_0 \sin \theta$.

(c) How far has the projectile traveled (horizontally) when it strikes the ground? In other words, find the range R.

(d) Find the time t when $x = y$. Then find the horizontal distance x and the vertical distance y traveled by the projectile in this time. Then compute $\sqrt{x^2 + y^2}$. This is the distance R, the range, that the projectile travels up a plane inclined at $45°$ to the horizontal $(x = y)$. See the following illustration. (See also Problem 107 in Exercise 8.4.)

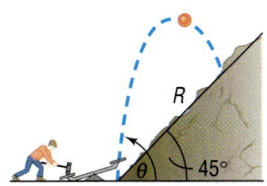

In Problems 51–54, use a graphing utility to graph the curve defined by the given parametric equations.

51. $x = t \sin t, \quad y = t \cos t$

52. $x = \sin t + \cos t, \quad y = \sin t - \cos t$

53. $x = 4 \sin t - 2 \sin(2t)$
$y = 4 \cos t - 2 \cos(2t)$

54. $x = 4 \sin t + 2 \sin(2t)$
$y = 4 \cos t + 2 \cos(2t)$

---

55. Hypocycloid The hypocycloid is a curve defined by the parametric equations

$$x(t) = \cos^3 t, \qquad y(t) = \sin^3 t, \qquad 0 \le t \le 2\pi$$

(a) Graph the hypocycloid using a graphing utility.
(b) Find rectangular equations of the hypocycloid.

56. In Problem 55, we graphed the hypocycloid. Now graph the rectangular equations of the hypocycloid. Did you obtain a complete graph? If not, experiment until you do.

57. Look up the curves called *hypocycloid* and *epicycloid*. Write a report on what you find. Be sure to draw comparisons with the cycloid.

PREPARING FOR THIS SECTION

Before getting started, review the following:

✓ Lines (Section 1.7, pp. 163–175)

✓ Circles (Section 1.8, pp. 180–184)

✓ Systems of Linear Equations: Two Equations Containing Two Variables (Section 7.1, pp. 512–520)

12.8 SYSTEMS OF NONLINEAR EQUATIONS

OBJECTIVES
1. Solve a System of Nonlinear Equations Using Substitution
2. Solve a System of Nonlinear Equations Using Elimination

1. In Section 7.1 we observed that the solution to a system of linear equations could be found geometrically by determining the point of intersection of the equations in the system. Similarly, when solving systems of nonlinear equations, the solution(s) also represent the point(s) of intersection of the equations.

There is no general methodology for solving a system of nonlinear equations by hand. There are times when substitution is best; other times, elimination is best; and there are times when neither of these methods works. Experience and a certain degree of imagination are your allies here.

Before we begin, two comments are in order:

1. If the system contains two variables, then graph them. By graphing each equation in the system, we can get an idea of how many solutions a system has and approximately where they are located.

2. Extraneous solutions can creep in when solving nonlinear systems algebraically, so it is imperative that all apparent solutions be checked.

EXAMPLE 1   Solving a System of Nonlinear Equations

Solve the following system of equations:

$$\begin{cases} 3x - y = -2 & \text{(1) A line} \\ 2x^2 - y = 0 & \text{(2) A parabola} \end{cases}$$

Algebraic Solution Using Substitution

First, we notice that the system contains two variables and that we know how to graph each equation by hand. See Figure 80. The system apparently has two solutions.

We will use substitution to solve the system. Equation (1) is easily solved for y.

$$3x - y = -2 \qquad \text{(1)}$$
$$y = 3x + 2$$

We substitute this expression for y in equation (2). The result is an equation containing just the variable x, which we can then solve.

$$2x^2 - y = 0 \qquad \text{(2)}$$
$$2x^2 - (3x + 2) = 0 \qquad \text{Substitute } y = 3x + 2 \text{ in (2).}$$
$$2x^2 - 3x - 2 = 0 \qquad \text{Remove parentheses.}$$
$$(2x + 1)(x - 2) = 0 \qquad \text{Factor.}$$
$$2x + 1 = 0 \quad \text{or} \quad x - 2 = 0 \qquad \text{Apply the Zero-Product Property.}$$
$$x = -\frac{1}{2} \quad \text{or} \quad x = 2$$

Using these values for x in $y = 3x + 2$, we find

$$y = 3\left(-\frac{1}{2}\right) + 2 = \frac{1}{2} \quad \text{or} \quad y = 3(2) + 2 = 8$$

The apparent solutions are $x = -\frac{1}{2}, y = \frac{1}{2}$ and $x = 2, y = 8$.

Figure 80

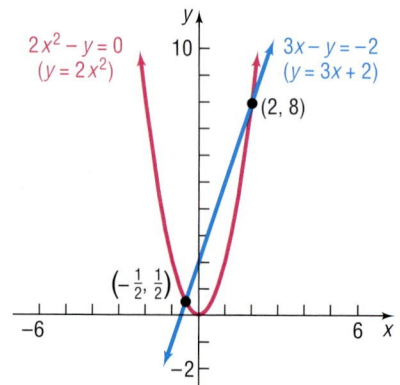

✔ CHECK: For $x = -\frac{1}{2}, y = \frac{1}{2}$:

$$\begin{cases} 3\left(-\frac{1}{2}\right) - \frac{1}{2} = -\frac{3}{2} - \frac{1}{2} = -2 & \text{(1)} \\ 2\left(-\frac{1}{2}\right)^2 - \frac{1}{2} = 2\left(\frac{1}{4}\right) - \frac{1}{2} = 0 & \text{(2)} \end{cases}$$

For $x = 2$, $y = 8$:

$$\begin{cases} 3(2) - 8 = \quad 6 - 8 = -2 & \text{(1)} \\ 2(2)^2 - 8 = 2(4) - 8 = \quad 0 & \text{(2)} \end{cases}$$

Each solution checks. Now we know that the graphs in Figure 80 intersect at $\left(-\dfrac{1}{2}, \dfrac{1}{2}\right)$ and at $(2, 8)$.

Graphing Solution We use a graphing utility to graph $Y_1 = 3x + 2$ and $Y_2 = 2x^2$. From Figure 81 we see that the system apparently has two solutions. Using INTERSECT, the solutions to the system of equations are $(-0.5, 0.5)$ and $(2, 8)$.

Figure 81

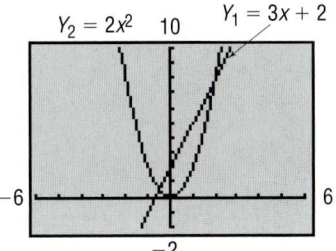

NOW WORK PROBLEM **11** USING SUBSTITUTION.

2 Our next example illustrates how the method of elimination works for nonlinear systems.

EXAMPLE 2 Solving a System of Nonlinear Equations

Solve: $\begin{cases} x^2 + y^2 = 13 & \text{(1) } A\ circle \\ x^2 - y \ = \ 7 & \text{(2) } A\ parabola \end{cases}$

Algebraic Solution First, we graph each equation, as shown in Figure 82. Based on the graph, we
Using Elimination expect four solutions. By subtracting equation (2) from equation (1), the variable x can be eliminated.

$$\begin{cases} x^2 + y^2 = 13 \\ x^2 - y \ = 7 \quad \textit{Subtract} \\ \hline \quad\ y^2 + y = 6 \end{cases}$$

This quadratic equation in y is easily solved by factoring.

$$y^2 + y - 6 = 0$$
$$(y + 3)(y - 2) = 0$$
$$y = -3 \quad \text{or} \quad y = 2$$

We use these values for y in equation (2) to find x.

If $y = 2$, then $x^2 = y + 7 = 9$ so $x = 3$ or -3.

If $y = -3$, then $x^2 = y + 7 = 4$ so $x = 2$ or -2.

We have four solutions: $x = 3$, $y = 2$; $x = -3$, $y = 2$; $x = 2$, $y = -3$; and $x = -2$, $y = -3$.

You should verify that, in fact, these four solutions also satisfy equation (1), so all four are solutions of the system. The four points, $(3, 2)$, $(-3, 2)$,

Figure 82

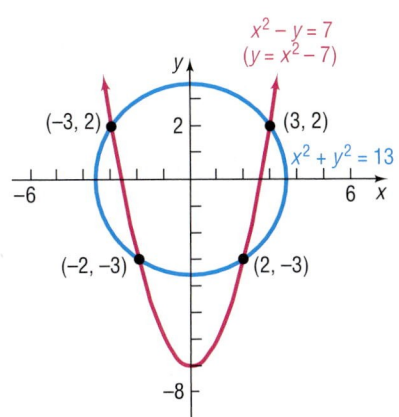

$(2, -3)$, and $(-2, -3)$, are the points of intersection of the graphs. Look again at Figure 82.

Graphing Solution We use a graphing utility to graph $x^2 + y^2 = 13$ and $x^2 - y = 7$. (Remember that to graph $x^2 + y^2 = 13$ requires two functions, $Y_1 = \sqrt{13 - x^2}$ and $Y_2 = -\sqrt{13 - x^2}$, and a square screen.) From Figure 83 we see that the system apparently has four solutions. Using INTERSECT, the solutions to the system of equations are $(-3, 2)$, $(3, 2)$, $(-2, -3)$, and $(2, -3)$.

Figure 83

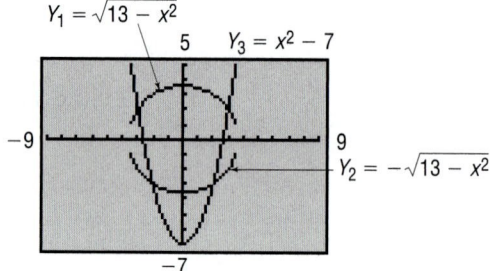

NOW WORK PROBLEM **9** USING ELIMINATION.

EXAMPLE 3   **Solving a System of Nonlinear Equations**

Solve:
$$\begin{cases} x^2 + x + y^2 - 3y + 2 = 0 & (1) \\ x + 1 + \dfrac{y^2 - y}{x} = 0 & (2) \end{cases}$$

Algebraic Solution Using Elimination First, we multiply equation (2) by x to eliminate the fraction. The result is an equivalent system because x cannot be 0. [Look at equation (2) to see why].

$$\begin{cases} x^2 + x + y^2 - 3y + 2 = 0 & (1) \\ x^2 + x + y^2 - y = 0 & (2) \end{cases}$$

Now subtract equation (2) from equation (1) to eliminate x. The result is

$$-2y + 2 = 0$$
$$y = 1 \qquad \text{Solve for } y.$$

To find x, we back-substitute $y = 1$ in equation (1):

$$x^2 + x + y^2 - 3y + 2 = 0 \qquad (1)$$
$$x^2 + x + 1 - 3 + 2 = 0 \qquad \text{Substitute } y = 1 \text{ in (1)}.$$
$$x^2 + x = 0 \qquad \text{Simplify.}$$
$$x(x + 1) = 0 \qquad \text{Factor.}$$
$$x = 0 \quad \text{or} \quad x = -1 \qquad \text{Apply the Zero-Product Property.}$$

Because x cannot be 0, the value $x = 0$ is extraneous, and we discard it. The solution is $x = -1$, $y = 1$.

✔ CHECK: We now check $x = -1$, $y = 1$:

$$\begin{cases} (-1)^2 + (-1) + 1^2 - 3(1) + 2 = 1 - 1 + 1 - 3 + 2 = 0 & (1) \\ -1 + 1 + \dfrac{1^2 - 1}{-1} = 0 + \dfrac{0}{-1} = 0 & (2) \end{cases}$$

Graphing Solution First, we multiply equation (2) by x to eliminate the fraction. The result is an equivalent system because x cannot be 0 [look at equation (2) to see why]:

$$\begin{cases} x^2 + x + y^2 - 3y + 2 = 0 & (1) \\ x + y^2 - y = 0 & (2) \end{cases}$$

We need to solve each equation for y. First, we solve equation (1) for y:

$$x^2 + x + y^2 - 3y + 2 = 0$$

$$y^2 - 3y = -x^2 - x - 2 \qquad \text{\color{blue}Rearrange so terms involving }y\text{ are on left side.}$$

$$y^2 - 3y + \frac{9}{4} = -x^2 - x - 2 + \frac{9}{4} \qquad \text{\color{blue}Complete the square involving }y.$$

$$\left(y - \frac{3}{2}\right)^2 = -x^2 - x + \frac{1}{4}$$

$$y - \frac{3}{2} = \pm\sqrt{-x^2 - x + \frac{1}{4}} \qquad \text{\color{blue}Square Root Method.}$$

$$y = \frac{3}{2} \pm \sqrt{-x^2 - x + \frac{1}{4}} \qquad \text{\color{blue}Solve for }y.$$

Now we solve equation (2) for y:

$$x^2 + x + y^2 - y = 0$$

$$y^2 - y = -x^2 - x \qquad \text{\color{blue}Rearrange so terms involving }y\text{ are on left side.}$$

$$y^2 - y + \frac{1}{4} = -x^2 - x + \frac{1}{4} \qquad \text{\color{blue}Complete the square involving }y.$$

$$\left(y - \frac{1}{2}\right)^2 = -x^2 - x + \frac{1}{4}$$

$$y - \frac{1}{2} = \pm\sqrt{-x^2 - x + \frac{1}{4}} \qquad \text{\color{blue}Square Root Method.}$$

$$y = \frac{1}{2} \pm \sqrt{-x^2 - x + \frac{1}{4}} \qquad \text{\color{blue}Solve for }y.$$

Figure 84

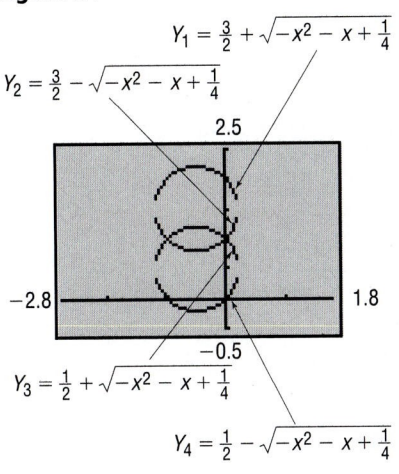

$Y_1 = \frac{3}{2} + \sqrt{-x^2 - x + \frac{1}{4}}$

$Y_2 = \frac{3}{2} - \sqrt{-x^2 - x + \frac{1}{4}}$

2.5

−2.8   1.8

−0.5

$Y_3 = \frac{1}{2} + \sqrt{-x^2 - x + \frac{1}{4}}$

$Y_4 = \frac{1}{2} - \sqrt{-x^2 - x + \frac{1}{4}}$

Now graph each equation using a graphing utility. See Figure 84.

Using INTERSECT, the points of intersection are $(-1, 1)$ and $(0, 1)$. Since $x \neq 0$ [look back at the original equation (2)], the graph of Y_3 has a hole at the point $(0, 1)$ and Y_4 has a hole at $(0, 0)$. The value $x = 0$ is extraneous, and we discard it. The only solution is $x = -1$ and $y = 1$. ∎

━━ **NOW WORK PROBLEMS 25 AND 45.**

EXAMPLE 4 **Solving a System of Nonlinear Equations**

Solve: $\begin{cases} x^2 - y^2 = 4 & \text{\color{blue}(1) A hyperbola} \\ y = x^2 & \text{\color{blue}(2) A parabola} \end{cases}$

Algebraic Solution Either substitution or elimination can be used here. We use substitution and replace x^2 by y in equation (1). The result is

$$y - y^2 = 4$$

$$y^2 - y + 4 = 0$$

This is a quadratic equation whose discriminant is $(-1)^2 - 4 \cdot 1 \cdot 4 = 1 - 4 \cdot 4 = -15 < 0$. The equation has no real solutions, so the system is inconsistent. The graphs of these two equations do not intersect. See Figure 85.

Figure 85

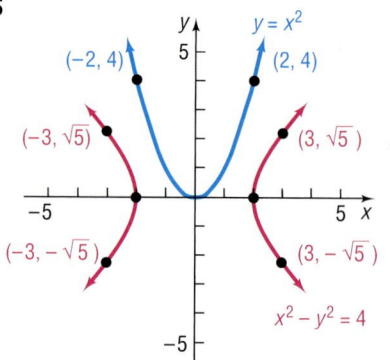

Graphing Solution   We graph $Y_1 = x^2$ and $x^2 - y^2 = 4$ in Figure 86. You will need to graph $x^2 - y^2 = 4$ as two functions:

$$Y_2 = \sqrt{x^2 - 4} \qquad \text{and} \qquad Y_3 = -\sqrt{x^2 - 4}$$

From Figure 86 we see that the graphs of these two equations do not intersect. The system is inconsistent.

Figure 86

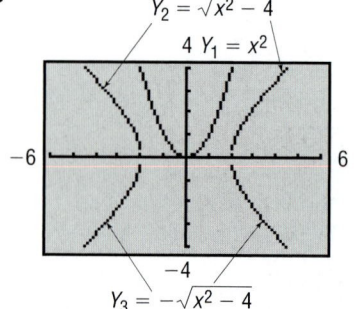

EXAMPLE 5   **Solving a System of Nonlinear Equations**

Solve: $\begin{cases} 3xy - 2y^2 = -2 & (1) \\ 9x^2 + 4y^2 = 10 & (2) \end{cases}$

Algebraic Solution   We multiply equation (1) by 2 and add the result to equation (2) to eliminate the y^2 terms.

$$\begin{cases} 6xy - 4y^2 = -4 & (1) \\ \underline{9x^2 + 4y^2 = 10} & (2) \ \text{Add.} \end{cases}$$
$$9x^2 + 6xy = 6$$
$$3x^2 + 2xy = 2 \qquad \text{Divide each side by 3.}$$

Since $x \neq 0$ (do you see why?), we can solve for y in this equation to get

$$y = \frac{2 - 3x^2}{2x}, \qquad x \neq 0 \qquad (1)$$

Now substitute for y in equation (2) of the system.

$$9x^2 + 4y^2 = 10 \qquad (2)$$

$$9x^2 + 4\left(\frac{2 - 3x^2}{2x}\right)^2 = 10 \qquad \text{Substitute } y = \frac{2 - 3x^2}{2x} \text{ in (2).}$$

$$9x^2 + \frac{4 - 12x^2 + 9x^4}{x^2} = 10$$

$$9x^4 + 4 - 12x^2 + 9x^4 = 10x^2 \qquad \text{Multiply both sides by } x^2.$$

$$18x^4 - 22x^2 + 4 = 0 \qquad \text{Subtract } 10x^2 \text{ from both sides.}$$

$$9x^4 - 11x^2 + 2 = 0 \qquad \text{Divide both sides by 2.}$$

This quadratic equation (in x^2) can be factored:

$$(9x^2 - 2)(x^2 - 1) = 0$$

$$9x^2 - 2 = 0 \qquad\qquad \text{or} \qquad x^2 - 1 = 0$$

$$x^2 = \frac{2}{9} \qquad\qquad\qquad\qquad x^2 = 1$$

$$x = \pm\sqrt{\frac{2}{9}} = \pm\frac{\sqrt{2}}{3} \qquad\qquad\qquad x = \pm 1$$

To find y, we use equation (1):

If $x = \dfrac{\sqrt{2}}{3}$: $\qquad y = \dfrac{2 - 3x^2}{2x} = \dfrac{2 - \dfrac{2}{3}}{2\left(\dfrac{\sqrt{2}}{3}\right)} = \dfrac{4}{2\sqrt{2}} = \sqrt{2}$

If $x = -\dfrac{\sqrt{2}}{3}$: $\quad y = \dfrac{2 - 3x^2}{2x} = \dfrac{2 - \dfrac{2}{3}}{2\left(-\dfrac{\sqrt{2}}{3}\right)} = \dfrac{4}{-2\sqrt{2}} = -\sqrt{2}$

If $x = 1$: $\qquad\quad y = \dfrac{2 - 3x^2}{2x} = \dfrac{2 - 3(1)^2}{2} = -\dfrac{1}{2}$

If $x = -1$: $\qquad\; y = \dfrac{2 - 3x^2}{2x} = \dfrac{2 - 3(-1)^2}{-2} = \dfrac{1}{2}$

The system has four solutions. Check them for yourself.

Graphing Solution To graph $3xy - 2y^2 = -2$, we need to solve for y. In this instance, it is easier to view the equation as a quadratic equation in the variable y.

$$3xy - 2y^2 = -2$$

$$2y^2 - 3xy - 2 = 0 \qquad \text{Place in standard form.}$$

$$y = \frac{-(-3x) \pm \sqrt{(-3x)^2 - 4(2)(-2)}}{2(2)} \qquad \begin{array}{l}\text{Use the quadratic}\\\text{formula with}\\ a = 2, b = -3x,\\ c = -2.\end{array}$$

$$y = \frac{3x \pm \sqrt{9x^2 + 16}}{4} \qquad \text{Simplify.}$$

Using a graphing utility, we graph $Y_1 = \dfrac{3x + \sqrt{9x^2 + 16}}{4}$ and $Y_2 = \dfrac{3x - \sqrt{9x^2 + 16}}{4}$. From (2), we graph $Y_3 = \dfrac{\sqrt{10 - 9x^2}}{2}$ and $Y_4 = \dfrac{-\sqrt{10 - 9x^2}}{2}$. See Figure 87.

Figure 87

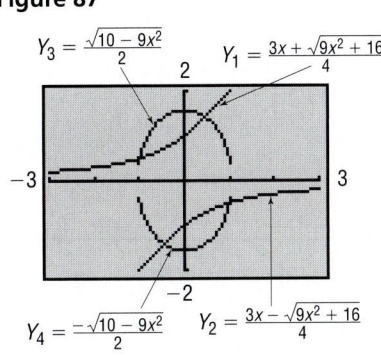

$Y_3 = \dfrac{\sqrt{10 - 9x^2}}{2}$ $\qquad Y_1 = \dfrac{3x + \sqrt{9x^2 + 16}}{4}$

$Y_4 = \dfrac{-\sqrt{10 - 9x^2}}{2}$ $\qquad Y_2 = \dfrac{3x - \sqrt{9x^2 + 16}}{4}$

Using INTERSECT, the solutions to the system of equations are $(-1, 0.5)$, $(0.47, 1.41)$, $(1, -0.5)$, and $(-0.47, -1.41)$, each rounded to two decimal places. ■

✏ **N O W   W O R K   P R O B L E M   43.**

EXAMPLE 6 | Running a Long Distance Race

In a 50-mile race, the winner crosses the finish line 1 mile ahead of the second-place runner and 4 miles ahead of the third-place runner. Assuming that each runner maintains a constant speed throughout the race, by how many miles does the second-place runner beat the third place runner?

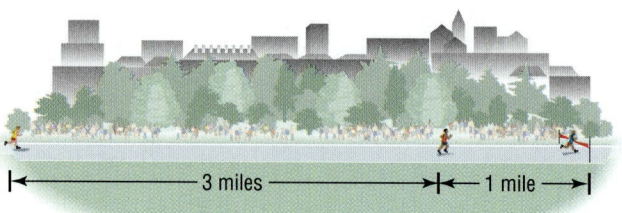

|←——————— 3 miles ———————→|←— 1 mile —→|

Solution   Let v_1, v_2, v_3 denote the speeds of the first-, second-, and third-place runners, respectively. Let t_1 and t_2 denote the times (in hours) required for the first-place runner and second-place runner to finish the race. Then we have the system of equations

$$
\begin{cases}
50 = v_1 t_1 & \text{(1) First-place runner goes 50 miles in } t_1 \\
49 = v_2 t_1 & \text{(2) Second-place runner goes 49 miles in } t_1 \\
46 = v_3 t_1 & \text{(3) Third-place runner goes 46 miles in } t_1 \\
50 = v_2 t_2 & \text{(4) Second-place runner goes 50 miles in } t_2
\end{cases}
$$

We seek the distance of the third-place runner from the finish at time t_2. At time t_2, the third-place runner has gone a distance of $v_3 t_2$ miles, so the distance remaining is $50 - v_3 t_2$. Now

$$
50 - v_3 t_2 = 50 - v_3 \left(t_1 \cdot \frac{t_2}{t_1} \right)
$$

$$
= 50 - (v_3 t_1) \cdot \frac{t_2}{t_1}
$$

$$
= 50 - 46 \cdot \frac{\dfrac{50}{v_2}}{\dfrac{50}{v_1}} \qquad \left\{ \begin{array}{l} \text{From (3), } v_3 t_1 = 46; \\[4pt] \text{from (4), } t_2 = \dfrac{50}{v_2}; \\[4pt] \text{from (1), } t_1 = \dfrac{50}{v_1}. \end{array} \right.
$$

$$
= 50 - 46 \cdot \frac{v_1}{v_2}
$$

$$
= 50 - 46 \cdot \frac{50}{49} \qquad \text{Form the quotient of (1) and (2).}
$$

$$
\approx 3.06 \text{ miles}
$$

■

HISTORICAL FEATURE

In the beginning of this section, we said that imagination and experience are important in solving simultaneous nonlinear equations. Indeed, these kinds of problems lead into some of the deepest and most difficult parts of modern mathematics. Look again at the graphs in Examples 1 and 2 of this section (Figures 80 and 82). We see that Example 1 has two solutions, and Example 2 has four solutions. We might conjecture that the number of solutions is equal to the product of the degrees of the equations involved. This conjecture was indeed made by Etienne Bezout (1730–1783), but working out the details took about 150 years. It turns out that, to arrive at the correct number of intersections, we must count not only the complex number intersections, but also those intersections that, in a certain sense, lie at infinity. For example, a parabola and a line lying on the axis of the parabola intersect at the vertex and at infinity. This topic is part of the study of algebraic geometry.

HISTORICAL PROBLEMS

A papyrus dating back to 1950 BC contains the following problem: "A given surface area of 100 units of area shall be represented as the sum of two squares whose sides are to each other as $1 : \frac{3}{4}$." Solve for the sides by solving the system of equations

$$\begin{cases} x^2 + y^2 = 100 \\ x = \frac{3}{4}y \end{cases}$$

12.8 Exercises

In Problems 1–20, graph each equation of the system by hand. Then solve the system algebraically to find the points of intersection. Verify your results using a graphing utility.

1. $\begin{cases} y = x^2 + 1 \\ y = x + 1 \end{cases}$

2. $\begin{cases} y = x^2 + 1 \\ y = 4x + 1 \end{cases}$

3. $\begin{cases} y = \sqrt{36 - x^2} \\ y = 8 - x \end{cases}$

4. $\begin{cases} y = \sqrt{4 - x^2} \\ y = 2x + 4 \end{cases}$

5. $\begin{cases} y = \sqrt{x} \\ y = 2 - x \end{cases}$

6. $\begin{cases} y = \sqrt{x} \\ y = 6 - x \end{cases}$

7. $\begin{cases} x = 2y \\ x = y^2 - 2y \end{cases}$

8. $\begin{cases} y = x - 1 \\ y = x^2 - 6x + 9 \end{cases}$

9. $\begin{cases} x^2 + y^2 = 4 \\ x^2 + 2x + y^2 = 0 \end{cases}$

10. $\begin{cases} x^2 + y^2 = 8 \\ x^2 + y^2 + 4y = 0 \end{cases}$

11. $\begin{cases} y = 3x - 5 \\ x^2 + y^2 = 5 \end{cases}$

12. $\begin{cases} x^2 + y^2 = 10 \\ y = x + 2 \end{cases}$

13. $\begin{cases} x^2 + y^2 = 4 \\ y^2 - x = 4 \end{cases}$

14. $\begin{cases} x^2 + y^2 = 16 \\ x^2 - 2y = 8 \end{cases}$

15. $\begin{cases} xy = 4 \\ x^2 + y^2 = 8 \end{cases}$

16. $\begin{cases} x^2 = y \\ xy = 1 \end{cases}$

17. $\begin{cases} x^2 + y^2 = 4 \\ y = x^2 - 9 \end{cases}$

18. $\begin{cases} xy = 1 \\ y = 2x + 1 \end{cases}$

19. $\begin{cases} y = x^2 - 4 \\ y = 6x - 13 \end{cases}$

20. $\begin{cases} x^2 + y^2 = 10 \\ xy = 3 \end{cases}$

In Problems 21–50, solve each system algebraically. Use any method you wish.

21. $\begin{cases} 2x^2 + y^2 = 18 \\ xy = 4 \end{cases}$

22. $\begin{cases} x^2 - y^2 = 21 \\ x + y = 7 \end{cases}$

23. $\begin{cases} y = 2x + 1 \\ 2x^2 + y^2 = 1 \end{cases}$

24. $\begin{cases} x^2 - 4y^2 = 16 \\ 2y - x = 2 \end{cases}$

25. $\begin{cases} x + y + 1 = 0 \\ x^2 + y^2 + 6y - x = -5 \end{cases}$

26. $\begin{cases} 2x^2 - xy + y^2 = 8 \\ xy = 4 \end{cases}$

27. $\begin{cases} 4x^2 - 3xy + 9y^2 = 15 \\ 2x + 3y = 5 \end{cases}$

28. $\begin{cases} 2y^2 - 3xy + 6y + 2x + 4 = 0 \\ 2x - 3y + 4 = 0 \end{cases}$

29. $\begin{cases} x^2 - 4y^2 + 7 = 0 \\ 3x^2 + y^2 = 31 \end{cases}$

30. $\begin{cases} 3x^2 - 2y^2 + 5 = 0 \\ 2x^2 - y^2 + 2 = 0 \end{cases}$

31. $\begin{cases} 7x^2 - 3y^2 + 5 = 0 \\ 3x^2 + 5y^2 = 12 \end{cases}$

32. $\begin{cases} x^2 - 3y^2 + 1 = 0 \\ 2x^2 - 7y^2 + 5 = 0 \end{cases}$

33. $\begin{cases} x^2 + 2xy = 10 \\ 3x^2 - xy = 2 \end{cases}$

34. $\begin{cases} 5xy + 13y^2 + 36 = 0 \\ xy + 7y^2 = 6 \end{cases}$

35. $\begin{cases} 2x^2 + y^2 = 2 \\ x^2 - 2y^2 + 8 = 0 \end{cases}$

36. $\begin{cases} y^2 - x^2 + 4 = 0 \\ 2x^2 + 3y^2 = 6 \end{cases}$

37. $\begin{cases} x^2 + 2y^2 = 16 \\ 4x^2 - y^2 = 24 \end{cases}$

38. $\begin{cases} 4x^2 + 3y^2 = 4 \\ 2x^2 - 6y^2 = -3 \end{cases}$

39. $\begin{cases} \dfrac{5}{x^2} - \dfrac{2}{y^2} + 3 = 0 \\ \dfrac{3}{x^2} + \dfrac{1}{y^2} = 7 \end{cases}$

40. $\begin{cases} \dfrac{2}{x^2} - \dfrac{3}{y^2} + 1 = 0 \\ \dfrac{6}{x^2} - \dfrac{7}{y^2} + 2 = 0 \end{cases}$

41. $\begin{cases} \dfrac{1}{x^4} + \dfrac{6}{y^4} = 6 \\ \dfrac{2}{x^4} - \dfrac{2}{y^4} = 19 \end{cases}$

42. $\begin{cases} \dfrac{1}{x^4} - \dfrac{1}{y^4} = 1 \\ \dfrac{1}{x^4} + \dfrac{1}{y^4} = 4 \end{cases}$

43. $\begin{cases} x^2 - 3xy + 2y^2 = 0 \\ x^2 + xy = 6 \end{cases}$

44. $\begin{cases} x^2 - xy - 2y^2 = 0 \\ xy + x + 6 = 0 \end{cases}$

45. $\begin{cases} y^2 + y + x^2 - x - 2 = 0 \\ y + 1 + \dfrac{x - 2}{y} = 0 \end{cases}$

46. $\begin{cases} x^3 - 2x^2 + y^2 + 3y - 4 = 0 \\ x - 2 + \dfrac{y^2 - y}{x^2} = 0 \end{cases}$

47. $\begin{cases} \log_x y = 3 \\ \log_x(4y) = 5 \end{cases}$

48. $\begin{cases} \log_x(2y) = 3 \\ \log_x(4y) = 2 \end{cases}$

49. $\begin{cases} \ln x = 4 \ln y \\ \log_3 x = 2 + 2 \log_3 y \end{cases}$

50. $\begin{cases} \ln x = 5 \ln y \\ \log_2 x = 3 + 2 \log_2 y \end{cases}$

In Problems 51–56, graph each equation by hand and find the point(s) of intersection, if any.

51. The line $x + 2y = 0$ and the circle $(x - 1)^2 + (y - 1)^2 = 5$

52. The line $x + 2y + 6 = 0$ and the circle $(x + 1)^2 + (y + 1)^2 = 5$

53. The circle $(x - 1)^2 + (y + 2)^2 = 4$ and the parabola $y^2 + 4y - x + 1 = 0$

54. The circle $(x + 2)^2 + (y - 1)^2 = 4$ and the parabola $y^2 - 2y - x - 5 = 0$

55. The graph of $y = \dfrac{4}{x - 3}$ and the circle $x^2 - 6x + y^2 + 1 = 0$

56. The graph of $y = \dfrac{4}{x + 2}$ and the circle $x^2 + 4x + y^2 - 4 = 0$

In Problems 57–64, use a graphing utility to solve each system of equations. Express the solution(s) rounded to two decimal places.

57. $\begin{cases} y = x^{2/3} \\ y = e^{-x} \end{cases}$

58. $\begin{cases} y = x^{3/2} \\ y = e^{-x} \end{cases}$

59. $\begin{cases} x^2 + y^3 = 2 \\ x^3 y = 4 \end{cases}$

60. $\begin{cases} x^3 + y^2 = 2 \\ x^2 y = 4 \end{cases}$

61. $\begin{cases} x^4 + y^4 = 12 \\ xy^2 = 2 \end{cases}$

62. $\begin{cases} x^4 + y^4 = 6 \\ xy = 1 \end{cases}$

63. $\begin{cases} xy = 2 \\ y = \ln x \end{cases}$

64. $\begin{cases} x^2 + y^2 = 4 \\ y = \ln x \end{cases}$

65. The difference of two numbers is 2 and the sum of their squares is 10. Find the numbers.

66. The sum of two numbers is 7 and the difference of their squares is 21. Find the numbers.

67. The product of two numbers is 4 and the sum of their squares is 8. Find the numbers.

68. The product of two numbers is 10 and the difference of their squares is 21. Find the numbers.

69. The difference of two numbers is the same as their product, and the sum of their reciprocals is 5. Find the numbers.

70. The sum of two numbers is the same as their product, and the difference of their reciprocals is 3. Find the numbers.

71. The ratio of a to b is $\dfrac{2}{3}$. The sum of a and b is 10. What is the ratio of $a + b$ to $b - a$?

72. The ratio of a to b is $4:3$. The sum of a and b is 14. What is the ratio of $a - b$ to $a + b$?

73. Geometry The perimeter of a rectangle is 16 inches and its area is 15 square inches. What are its dimensions?

74. Geometry An area of 52 square feet is to be enclosed by two squares whose sides are in the ratio of $2:3$. Find the sides of the squares.

75. Geometry Two circles have circumferences that add up to 12π centimeters and areas that add up to 20π square centimeters. Find the radius of each circle.

76. Geometry The altitude of an isosceles triangle drawn to its base is 3 centimeters, and its perimeter is 18 centimeters. Find the length of its base.

77. The Tortoise and the Hare   In a 21-meter race between a tortoise and a hare, the tortoise leaves 9 minutes before the hare. The hare, by running at an average speed of 0.5 meter per hour faster than the tortoise, crosses the finish line 3 minutes before the tortoise. What are the average speeds of the tortoise and the hare?

21 meters

78. Running a Race   In a 1-mile race, the winner crosses the finish line 10 feet ahead of the second-place runner and 20 feet ahead of the third-place runner. Assuming that each runner maintains a constant speed throughout the race, by how many feet does the second-place runner beat the third-place runner?

79. Constructing a Box   A rectangular piece of cardboard, whose area is 216 square centimeters, is made into an open box by cutting a 2-centimeter square from each corner and turning up the sides. See the figure. If the box is to have a volume of 224 cubic centimeters, what size cardboard should you start with?

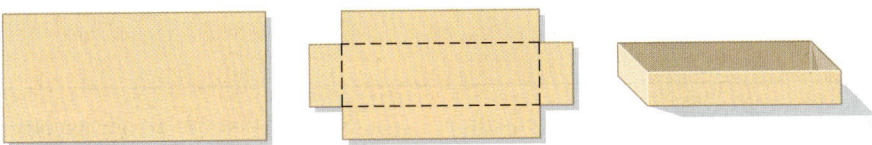

80. Constructing a Cylindrical Tube   A rectangular piece of cardboard, whose area is 216 square centimeters, is made into a cylindrical tube by joining together two sides of the rectangle. See the figure. If the tube is to have a volume of 224 cubic centimeters, what size cardboard should you start with?

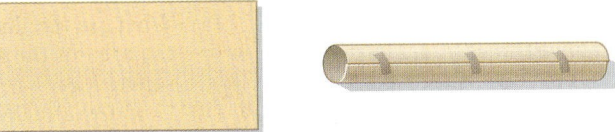

81. Fencing   A farmer has 300 feet of fence available to enclose a 4500 square foot region in the shape of adjoining squares, with sides of length x and y. See the figure. Find x and y.

82. Bending Wire   A wire 60 feet long is cut into two pieces. Is it possible to bend one piece into the shape of a square and the other into the shape of a circle so that the total area enclosed by the two pieces is 100 square feet? If this is possible, find the length of the side of the square and the radius of the circle.

83. Geometry   Find formulas for the length l and width w of a rectangle in terms of its area A and perimeter P.

84. Geometry   Find formulas for the base b and one of the equal sides l of an isosceles triangle in terms of its altitude h and perimeter P.

85. Descartes's Method of Equal Roots   Descartes's method for finding tangents depends on the idea that, for many graphs, the tangent line at a given point is the *unique* line that intersects the graph at that point only. We will apply his method to find an equation of the tangent line to the parabola $y = x^2$ at the point $(2, 4)$. See the figure on page 974. First, we know that the equation of the tangent line must be in the form $y = mx + b$. Using the fact that the point $(2, 4)$ is on the line, we can solve for b in terms of m and get the equation $y = mx + (4 - 2m)$. Now we want $(2, 4)$ to be the *unique* solution to the system

$$\begin{cases} y = x^2 \\ y = mx + 4 - 2m \end{cases}$$

From this system, we get $x^2 = mx + 4 - 2m$ or $x^2 - mx + (2m - 4) = 0$. By using the quadratic formula, we get

$$x = \frac{m \pm \sqrt{m^2 - 4(2m - 4)}}{2}$$

To obtain a unique solution for x, the two roots must be equal; in other words, the discriminant $m^2 - 4(2m - 4)$ must be 0. Complete the work to get m, and write an equation of the tangent line.

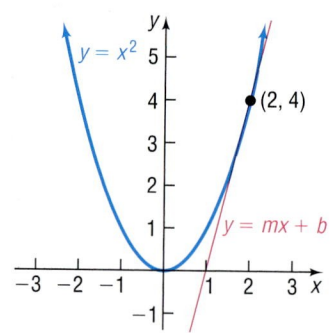

In Problems 86–92, use Descartes's method from Problem 85 to find the equation of the line tangent to each graph at the given point.

86. $x^2 + y^2 = 10$;   at $(1, 3)$

87. $y = x^2 + 2$;   at $(1, 3)$

88. $x^2 + y = 5$;   at $(-2, 1)$

89. $2x^2 + 3y^2 = 14$;   at $(1, 2)$

90. $3x^2 + y^2 = 7$;   at $(-1, 2)$

91. $x^2 - y^2 = 3$;   at $(2, 1)$

92. $2y^2 - x^2 = 14$;   at $(2, 3)$

93. If r_1 and r_2 are two solutions of a quadratic equation $ax^2 + bx + c = 0$, then it can be shown that

$$r_1 + r_2 = -\frac{b}{a} \quad \text{and} \quad r_1 r_2 = \frac{c}{a}$$

Solve this system of equations for r_1 and r_2.

94. A circle and a line intersect at most twice. A circle and a parabola intersect at most four times. Deduce that a circle and the graph of a polynomial of degree 3 intersect at most six times. What do you conjecture about a polynomial of degree 4? What about a polynomial of degree n? Can you explain your conclusions using an algebraic argument?

95. Suppose that you are the manager of a sheet metal shop. A customer asks you to manufacture 10,000 boxes, each box being open on top. The boxes are required to have a square base and a 9-cubic-foot capacity. You construct the boxes by cutting out a square from each corner of a square piece of sheet metal and folding along the edges.

(a) What are the dimensions of the square to be cut if the area of the square piece of sheet metal is 100 square feet?

(b) Could you make the box using a smaller piece of sheet metal? Make a list of the dimensions of the box for various pieces of sheet metal.

Chapter Review

Things To Know

Equations

Parabola	See Tables 1 and 2 (pp. 896 and 899).
Ellipse	See Table 3 (p. 910).
Hyperbola	See Table 4 (p. 925).

General equation of a conic (p. 939) $Ax^2 + Bxy + Cy^2 + Dx + Ey + F = 0$

Parabola if $B^2 - 4AC = 0$
Ellipse (or circle) if $B^2 - 4AC < 0$
Hyperbola if $B^2 - 4AC > 0$

Conic in polar coordinates (p. 941) $\dfrac{d(F, P)}{d(P, D)} = e$

Parabola if $e = 1$
Ellipse if $e < 1$
Hyperbola if $e > 1$

Polar equations of a conic with focus at the pole   See Table 5 (p. 944).

Parametric equations of a curve (p. 949) $x = f(t), y = g(t), t$ is the parameter

Definitions

Parabola (p. 893)   Set of points P in the plane for which $d(F, P) = d(P, D)$, where F is the focus and D is the directrix

Ellipse (p. 905)   Set of points P in the plane, the sum of whose distances from two fixed points (the foci) is a constant

Hyperbola (p. 917)   Set of points P in the plane, the difference of whose distances from two fixed points (the foci) is a constant

Conic in polar coordinates (p. 941) $\dfrac{d(F, P)}{d(P, D)} = e$ Parabola if $e = 1$
Ellipse if $e < 1$
Hyperbola if $e > 1$

Formulas

Rotation formulas (p. 934) $x = x' \cos\theta - y' \sin\theta$
$y = x' \sin\theta + y' \cos\theta$

Angle θ of rotation that eliminates the $x'y'$-term (p. 936) $\cot(2\theta) = \dfrac{A - C}{B}, \quad 0° < \theta < 90°$

Objectives

Section	You should be able to:	Review Exercises
12.1	1 Know the names of the conics (p. 892)	1–36
12.2	1 Find the equation of a parabola (p. 893)	21, 24
	2 Graph parabolas (p. 894)	21, 24
	3 Discuss the equation of a parabola (p. 895)	1, 2
	4 Work with parabolas with vertex at (h, k) (p. 898)	7, 11, 12, 17, 18, 27, 30
	5 Solve applied problems involving parabolas (p. 900)	87, 88
12.3	1 Find the equation of an ellipse (p. 905)	22, 25
	2 Graph ellipses (p. 906)	22, 25
	3 Discuss the equation of an ellipse (p. 907)	5, 6, 10
	4 Work with ellipses with center at (h, k) (p. 910)	14–16, 19, 28, 31
	5 Solve applied problems involving ellipses (p. 913)	89, 90
12.4	1 Find the equation of a hyperbola (p. 917)	23, 26
	2 Graph hyperbolas (p. 919)	23, 26
	3 Discuss the equation of a hyperbola (p. 920)	3, 4, 8, 9
	4 Find the asymptotes of a hyperbola (p. 923)	3, 4, 8, 9
	5 Work with hyperbolas with center at (h, k) (p. 925)	13, 20, 29, 32–36
	6 Solve applied problems involving hyperbolas (p. 927)	91
12.5	1 Identify a conic (p. 932)	37–40
	2 Use a rotation of axes to transform equations (p. 934)	47–52
	3 Discuss an equation using a rotation of axes (p. 936)	47–52
	4 Identify conics without a rotation of axes (p. 939)	41–46
12.6	1 Discuss and graph polar equations of conics (p. 941)	53–58
	2 Convert a polar equation of a conic to a rectangular equation (p. 946)	59–62
12.7	1 Graph parametric equations by hand (p. 949)	63–68
	2 Graph parametric equations using a graphing utility (p. 950)	63–68
	3 Find a rectangular equation for a curve defined parametrically (p. 951)	63–68

▦ Review Exercises

Blue problem numbers indicate the authors' suggestions for use in a Practice Test.

In Problems 1–20, identify each equation. If it is a parabola, gives its vertex, focus, and directrix; if it is an ellipse, give its center, vertices, and foci; if it is a hyperbola, give its center, vertices, foci, and asymptotes.

1. $y^2 = -16x$

2. $16x^2 = y$

3. $\dfrac{x^2}{25} - y^2 = 1$

4. $\dfrac{y^2}{25} - x^2 = 1$

5. $\dfrac{y^2}{25} + \dfrac{x^2}{16} = 1$

6. $\dfrac{x^2}{9} + \dfrac{y^2}{16} = 1$

7. $x^2 + 4y = 4$

8. $3y^2 - x^2 = 9$

9. $4x^2 - y^2 = 8$

10. $9x^2 + 4y^2 = 36$

11. $x^2 - 4x = 2y$

12. $2y^2 - 4y = x - 2$

13. $y^2 - 4y - 4x^2 + 8x = 4$

14. $4x^2 + y^2 + 8x - 4y + 4 = 0$

15. $4x^2 + 9y^2 - 16x - 18y = 11$

16. $4x^2 + 9y^2 - 16x + 18y = 11$

17. $4x^2 - 16x + 16y + 32 = 0$

18. $4y^2 + 3x - 16y + 19 = 0$

19. $9x^2 + 4y^2 - 18x + 8y = 23$

20. $x^2 - y^2 - 2x - 2y = 1$

In Problems 21–36, obtain an equation of the conic described. Graph the equation by hand.

21. Parabola; focus at $(-2, 0)$; directrix the line $x = 2$

22. Ellipse; center at $(0, 0)$; focus at $(0, 3)$; vertex at $(0, 5)$

23. Hyperbola; center at $(0, 0)$; focus at $(0, 4)$; vertex at $(0, -2)$

24. Parabola; vertex at $(0, 0)$; directrix the line $y = -3$

25. Ellipse; foci at $(-3, 0)$ and $(3, 0)$; vertex at $(4, 0)$

26. Hyperbola; vertices at $(-2, 0)$ and $(2, 0)$; focus at $(4, 0)$

27. Parabola; vertex at $(2, -3)$; focus at $(2, -4)$

28. Ellipse; center at $(-1, 2)$; focus at $(0, 2)$; vertex at $(2, 2)$

29. Hyperbola; center at $(-2, -3)$; focus at $(-4, -3)$; vertex at $(-3, -3)$

30. Parabola; focus at $(3, 6)$; directrix the line $y = 8$

31. Ellipse; foci at $(-4, 2)$ and $(-4, 8)$; vertex at $(-4, 10)$

32. Hyperbola; vertices at $(-3, 3)$ and $(5, 3)$; focus at $(7, 3)$

33. Center at $(-1, 2)$; $a = 3$; $c = 4$; transverse axis parallel to the x-axis

34. Center at $(4, -2)$; $a = 1$; $c = 4$; transverse axis parallel to the y-axis

35. Vertices at $(0, 1)$ and $(6, 1)$; asymptote the line $3y + 2x = 9$

36. Vertices at $(4, 0)$ and $(4, 4)$; asymptote the line $y + 2x = 10$

In Problems 37–46, identify each conic without completing the squares and without applying a rotation of axes.

37. $y^2 + 4x + 3y - 8 = 0$

38. $2x^2 - y + 8x = 0$

39. $x^2 + 2y^2 + 4x - 8y + 2 = 0$

40. $x^2 - 8y^2 - x - 2y = 0$

41. $9x^2 - 12xy + 4y^2 + 8x + 12y = 0$

42. $4x^2 + 4xy + y^2 - 8\sqrt{5}x + 16\sqrt{5}y = 0$

43. $4x^2 + 10xy + 4y^2 - 9 = 0$

44. $4x^2 - 10xy + 4y^2 - 9 = 0$

45. $x^2 - 2xy + 3y^2 + 2x + 4y - 1 = 0$

46. $4x^2 + 12xy - 10y^2 + x + y - 10 = 0$

In Problems 47–52, rotate the axes so that the new equation contains no xy-term. Discuss and graph the new equation.

47. $2x^2 + 5xy + 2y^2 - \dfrac{9}{2} = 0$

48. $2x^2 - 5xy + 2y^2 - \dfrac{9}{2} = 0$

49. $6x^2 + 4xy + 9y^2 - 20 = 0$

50. $x^2 + 4xy + 4y^2 + 16\sqrt{5}x - 8\sqrt{5}y = 0$

51. $4x^2 - 12xy + 9y^2 + 12x + 8y = 0$

52. $9x^2 - 24xy + 16y^2 + 80x + 60y = 0$

In Problems 53–58, identify the conic that each polar equation represents and graph it.

53. $r = \dfrac{4}{1 - \cos\theta}$

54. $r = \dfrac{6}{1 + \sin\theta}$

55. $r = \dfrac{6}{2 - \sin\theta}$

56. $r = \dfrac{2}{3 + 2\cos\theta}$

57. $r = \dfrac{8}{4 + 8\cos\theta}$

58. $r = \dfrac{10}{5 + 20\sin\theta}$

In Problems 59–62, convert each polar equation to a rectangular equation.

59. $r = \dfrac{4}{1 - \cos\theta}$

60. $r = \dfrac{6}{2 - \sin\theta}$

61. $r = \dfrac{8}{4 + 8\cos\theta}$

62. $r = \dfrac{2}{3 + 2\cos\theta}$

In Problems 63–68, by hand graph the curve whose parametric equations are given and show its orientation. Find the rectangular equation of each curve. Verify your results using a graphing utility.

63. $x = 4t - 2, \quad y = 1 - t; \quad -\infty < t < \infty$

64. $x = 2t^2 + 6, \quad y = 5 - t; \quad -\infty < t < \infty$

65. $x = 3\sin t, \quad y = 4\cos t + 2; \quad 0 \le t \le 2\pi$

66. $x = \ln t, y = t^3; \quad t > 0$

67. $x = \sec^2 t, \quad y = \tan^2 t; \quad 0 \le t \le \dfrac{\pi}{4}$

68. $x = t^{\frac{3}{2}}, \quad y = 2t + 4; \quad t \ge 0$

In Problems 69 and 70, find two different parametric equations for each rectangular equation.

69. $y = -2x + 4$

70. $y = 2x^2 - 8$

In Problems 71 and 72, find parametric equations for an object that moves along the ellipse $\dfrac{x^2}{16} + \dfrac{y^2}{9} = 1$ with the motion described.

71. The motion begins at $(4, 0)$, is counterclockwise, and requires 4 seconds for a complete revolution.

72. The motion begins at $(0, 3)$, is clockwise, and requires 5 seconds for a complete revolution.

In Problems 73–82, solve each system of equations algebraically. Verify your result using a graphing utility.

73. $\begin{cases} 2x + y + 3 = 0 \\ x^2 + y^2 = 5 \end{cases}$

74. $\begin{cases} x^2 + y^2 = 16 \\ 2x - y^2 = -8 \end{cases}$

75. $\begin{cases} 2xy + y^2 = 10 \\ 3y^2 - xy = 2 \end{cases}$

76. $\begin{cases} 3x^2 - y^2 = 1 \\ 7x^2 - 2y^2 - 5 = 0 \end{cases}$

77. $\begin{cases} x^2 + y^2 = 6y \\ x^2 = 3y \end{cases}$

78. $\begin{cases} 2x^2 + y^2 = 9 \\ x^2 + y^2 = 9 \end{cases}$

79. $\begin{cases} 3x^2 + 4xy + 5y^2 = 8 \\ x^2 + 3xy + 2y^2 = 0 \end{cases}$

80. $\begin{cases} 3x^2 + 2xy - 2y^2 = 6 \\ xy - 2y^2 + 4 = 0 \end{cases}$

81. $\begin{cases} x^2 - 3x + y^2 + y = -2 \\ \dfrac{x^2 - x}{y} + y + 1 = 0 \end{cases}$

82. $\begin{cases} x^2 + x + y^2 = y + 2 \\ x + 1 = \dfrac{2 - y}{x} \end{cases}$

---

83. Find an equation of the hyperbola whose foci are the vertices of the ellipse $4x^2 + 9y^2 = 36$ and whose vertices are the foci of this ellipse.

84. Find an equation of the ellipse whose foci are the vertices of the hyperbola $x^2 - 4y^2 = 16$ and whose vertices are the foci of this hyperbola.

85. Describe the collection of points in a plane so that the distance from each point to the point $(3, 0)$ is three-fourths of its distance from the line $x = \dfrac{16}{3}$.

86. Describe the collection of points in a plane so that the distance from each point to the point $(5, 0)$ is five-fourths of its distance from the line $x = \dfrac{16}{5}$.

87. Mirrors A mirror is shaped like a paraboloid of revolution. If a light source is located 1 foot from the base along the axis of symmetry and the opening is 2 feet across, how deep should the mirror be?

88. Parabolic Arch Bridge A bridge is built in the shape of a parabolic arch. The bridge has a span of 60 feet and a maximum height of 20 feet. Find the height of the arch at distances of 5, 10, and 20 feet from the center.

89. Semi-elliptical Arch Bridge A bridge is built in the shape of a semi-elliptical arch. The bridge has a span of 60 feet and a maximum height of 20 feet. Find the height of the arch at distances of 5, 10, and 20 feet from the center.

90. Whispering Galleries The figure shows the specifications for an elliptical ceiling in a hall designed to be a whispering gallery. Where are the foci located in the hall?

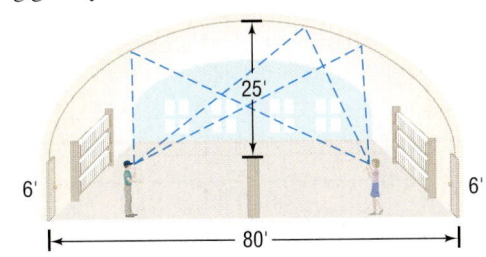

91. LORAN   Two LORAN stations are positioned 150 miles apart along a straight shore.
  (a)   A ship records a time difference of 0.00032 second between the LORAN signals. Set up an appropriate rectangular coordinate system to determine where the ship would reach shore if it were to follow the hyperbola corresponding to this time difference.
  (b)   If the ship wants to enter a harbor located between the two stations 15 miles from the master station, what time difference should it be looking for?
  (c)   If the ship is 20 miles offshore when the desired time difference is obtained, what is the approximate location of the ship?
    [*Note:* The speed of each radio signal is 186,000 miles per second.]

92. Uniform Motion   Mary's train leaves at 7:15 AM and accelerates at the rate of 3 meters per second per second. Mary, who can run 6 meters per second, arrives at the train station 2 seconds after the train has left.
  (a)   Find parametric equations that describe the motion of the train and Mary as a function of time.
    [**Hint:** The position s at time t of an object having acceleration a is $s = \dfrac{1}{2}at^2$.]

  (b)   Determine algebraically whether Mary will catch the train. If so, when?
  (c)   Simulate the motion of the train and Mary by simultaneously graphing the equations found in part (a).

93. Projectile Motion   Drew Bledsoe throws a football with an initial speed of 100 feet per second at an angle of 35° to the horizontal. The ball leaves Drew Bledsoe's hand at a height of 6 feet.
  (a)   Find parametric equations that describe the position of the ball as a function of time.
  (b)   How long is the ball in the air?
  (c)   When is the ball at its maximum height? Determine the maximum height of the ball.
  (d)   Determine the distance that the ball travels.
  (e)   Using a graphing utility, simultaneously graph the equations found in part (a).

94.   Formulate a strategy for discussing and graphing an equation of the form

$$Ax^2 + Bxy + Cy^2 + Dx + Ey + F = 0$$

Chapter Projects

1. The Orbits of Neptune and Pluto   The orbit of a planet about the Sun is an ellipse, with the Sun at one focus. The **aphelion** of a planet is its greatest distance from the Sun and the **perihelion** is its shortest distance. The **mean distance** of a planet from the Sun is the length of the semimajor axis of the elliptical orbit. See the illustration.

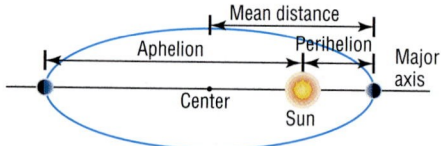

  (a)   The aphelion of Neptune is 4532.2×10^6 km and its perihelion is 4458.0×10^6 km. Find the equation for the orbit of Neptune around the Sun.
  (b)   The aphelion of Pluto is 7381.2×10^6 km and its perihelion is 4445.8×10^6 km. Find the equation for the orbit of Pluto around the Sun.
  (c)   Graph the orbits of Pluto and Neptune on a graphing utility. The orbits of the planets do not intersect! But the orbits do intersect. What is the explanation?
  (d)   The graphs of the orbits have the same center, so their foci lie in different locations. To see an accurate representation, the location of the Sun (a focus) needs to be the same for both graphs. This can be accomplished by shifting Pluto's orbit to the left. The shift amount is equal to Pluto's distance from the center [in the graph in part (c)] to the Sun minus Neptune's distance from the center to the Sun. Find the new equation representing the orbit of Pluto.
  (e)   Graph the equation for the orbit of Pluto found in part (d) along with the equation of the orbit of Neptune. Do you see that Pluto's orbit is sometimes inside Neptune's?
  (f)   Find the point(s) of intersection of the two orbits.
  (g)   Do you think two planets ever collide?

2. Constructing a Bridge Over the East River   A new bridge is to be constructed over the East River in New York City. The space between the supports needs to be 1050 feet; the height at the center of the arch needs to be 350 feet. Two structural possibilities exist: the support could be in the shape of a parabola or the support could be in the shape of a semiellipse.

An empty tanker needs a 280 foot clearance to pass beneath the bridge. The width of the channel for each of the two plans must be determined to verify that the tanker can pass through the bridge.

(a) Determine the equation of a parabola with these characteristics.

[**Hint:** Place the vertex of the parabola at the origin to simplify calculations.]

(b) How wide is the channel that the tanker can pass through?

(c) Determine the equation of a semiellipse with these characteristics.

[**Hint:** Place the center of the semiellipse at the origin to simplify calculations.]

(d) How wide is the channel that the tanker can pass through?

(e) If the river were to flood and rise 10 feet, how would the clearances of the two bridges be affected? Does this affect your decision as to which design to choose? Why?

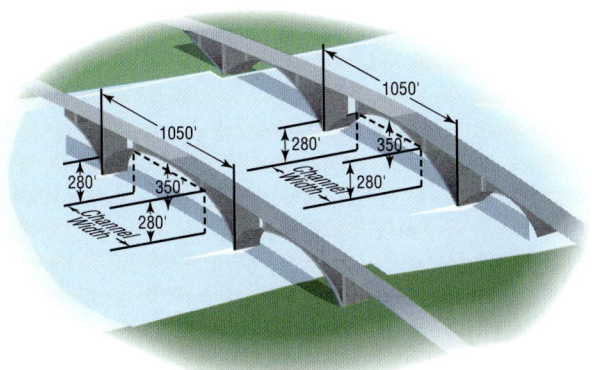

3. **Systems of Parametric Equations**   Consider the following systems of parametric equations:

I. $x_1 = 4t - 2$, $y_1 = 1 - t$, $-\infty < t < \infty$

$x_2 = \sec^2 t$, $y_2 = \tan^2 t$, $0 \le t \le \dfrac{\pi}{4}$

II. $x_1 = \ln t$, $y_2 = t^3$, $t > 0$

$x_2 = t^{\frac{3}{2}}$, $y_2 = 2t + 4$, $t \ge 0$

III. $x_1 = 3 \sin t$, $y_1 = 4 \cos t + 2$, $0 \le t \le 2\pi$

$x_2 = 2 \cos t$, $y_2 = 4 \sin t$, $0 \le t \le 2\pi$

(a) For system I, set $x_1 = x_2$ and $y_1 = y_2$ and solve each equation for t. If you can solve the resulting equations algebraically, do so. If they cannot be solved algebraically, solve them graphically, using your graphing calculator. Remember that the value of t must be the same for both the x and y equations in order to have a solution for the system.

(b) Now, graph the system of parametric equations using your graphing calculator and find the point(s) of intersection, if there are any. (You will need to use the TRACE feature to do this. Make sure the same value of t gives any points of intersection of each curve.) What did you notice?

(c) Did any solutions you found in part (b) match any of those that you found in part (a)? Why or why not? Explain.

(d) Convert the parametric equations in system I to rectangular coordinates and state the domain and range for each equation. Find the solution to the system either algebraically or graphically. How does this solution compare to what you found in part (c)?

(e) Repeat parts (a)–(d) for System II.

(f) Repeat parts (a)–(d) for System III.

(g) Which method is more efficient—solving in the parametric form or solving in rectangular form? Does this depend on the equations? What must you watch for when solving systems of parametric equations? Explain.

Cumulative Review

1. Find all the solutions of the equation $\sin(2\theta) = 0.5$.

2. Find a polar equation for the line containing the origin that makes an angle of $30°$ with the positive x-axis.

3. Find a polar equation for the circle with center at the point $(0, 4)$ and radius 4. Graph this circle.

4. What is the domain of the function $f(x) = \dfrac{3}{\sin x + \cos x}$?

5. Solve the system of linear equations: $\begin{cases} 2x - 3y = 1 \\ x + 4y = 6 \end{cases}$

6. Solve the system of nonlinear equations: $\begin{cases} x + y = 6 \\ y = x^2 \end{cases}$

7. Graph the inequality $x + y \le 6$.

8. For what numbers x is $6 - x \ge x^2$?

9. Solve the equation $\cot(2\theta) = 1$, where $0° < \theta < 90°$.

10. Find an equation for each of the following graphs:

(a) Line:

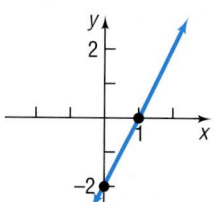

(b) Circle:

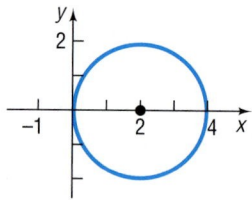

(c) Ellipse:

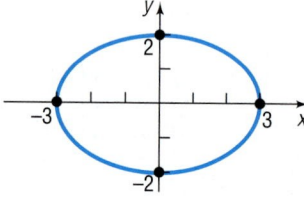

(d) Parabola:

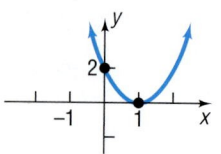

(e) Hyperbola:

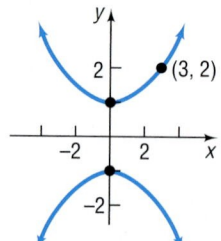

(f) Exponential:

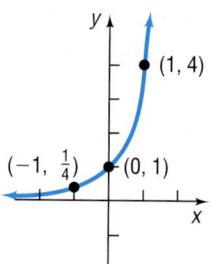

SEQUENCES; INDUCTION; THE BINOMIAL THEOREM

Will Humans Overwhelm the Earth?

Two hundred years ago the Rev. Thomas Robert Malthus, an English economist and mathematician, anonymously published an essay predicting that the world's burgeoning population would overwhelm the Earth's capacity to sustain it.

Malthus's gloomy forecast, called 'An Essay on the Principle of Population As It Affects the Future Improvement of Society,' was condemned by Karl Marx, Friedrich Engels and many other theorists, and it was still striking sparks last week at a meeting in Philadelphia of the American Anthropological Society. Despite continuing controversy, it was clear that Malthus's conjectures are far from dead.

Among the scores of special conferences organized for the 5,000 participating anthropologists, many touched directly and indirectly on the Malthusian dilemma: Although global food supplies increase arithmetically, the population increases geometrically—a vastly faster rate. The consequence Malthus believed, was that poverty and the misery it imposes will inevitably increase unless the population increase is curbed. ("Will Humans Overwhelm the Earth? The Debate continues," Malcolm W. Browne, *New York Times*, December 8, 1998.)

SEE CHAPTER PROJECT 1.

For additional study help, go to

www.prenhall.com/sullivanegu3e

Materials include:

- Graphing Calculator Help
- Chapter Quiz
- Chapter Test
- PowerPoint Downloads
- Chapter Projects
- Student Tips

A Look Back, A Look Forward

This chapter may be divided into three independent parts: Sections 13.1–13.3, Section 13.4, and Section 13.5.

In Chapter 2, we defined a function and its domain, which was usually some set of real numbers. In Sections 13.1–13.3, we discuss sequences, which are functions whose domain is the set of positive integers.

Throughout this text, where it seemed appropriate, we have given proofs of many of the results. In Section 13.4, Mathematical Induction, a technique for proving theorems involving natural numbers is discussed.

In Chapter R, we gave formulas for expanding $(x + a)^2$ and $(x + a)^3$. In Section 13.5, we discuss the Binomial Theorem, a formula for the expansion of $(x + a)^n$, where n is a positive integer.

The topics introduced in this chapter are covered in more detail in courses titled *Discrete Mathematics*. Applications of these topics can be found in the fields of computer science, engineering, business and economics, the social sciences, and the physical and biological sciences.

PREPARING FOR THIS SECTION

Before getting started, review the following:

✓ Evaluating Functions (Section 2.1, pp. 201–202)

✓ Compound Interest (Section 6.6, pp. 471–475)

✓ Integer Exponents (Section R.4, pp. 30–35)

13.1 SEQUENCES

OBJECTIVES
1. Write the First Several Terms of a Sequence
2. Write the Terms of a Sequence Defined by a Recursive Formula
3. Use Summation Notation
4. Find the Sum of a Sequence by Hand and by Using a Graphing Utility
5. Solve Annuity and Amortization Problems

A **sequence** is a function whose domain is the set of positive integers.

Because a sequence is a function, it will have a graph. In Figure 1(a), we have the graph of the function $f(x) = \dfrac{1}{x}, x > 0$. If all the points on this graph were removed except those whose x-coordinates are positive integers, that is, if all points were removed except $(1, 1)$, $\left(2, \dfrac{1}{2}\right)$, $\left(3, \dfrac{1}{3}\right)$, and so on, the remaining points would be the graph of the sequence $f(n) = \dfrac{1}{n}$, as shown in Figure 1(b). Notice that we use n to represent the independent variable in a sequence. This serves to remind us that n is a positive integer.

Figure 1

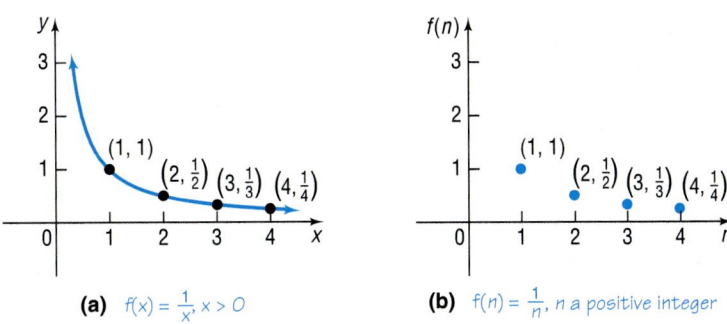

(a) $f(x) = \frac{1}{x}, x > 0$ **(b)** $f(n) = \frac{1}{n}, n$ a positive integer

A sequence may be represented by listing its values in order. For example, the sequence whose graph is given in Figure 1(b) might be represented as

$$f(1), f(2), f(3), f(4), \ldots \quad \text{or} \quad 1, \frac{1}{2}, \frac{1}{3}, \frac{1}{4}, \ldots$$

The list never ends, as the ellipsis indicates. The numbers in this ordered list are called the **terms** of the sequence.

In dealing with sequences, we use subscripted letters, such as a_1, to represent the first term, a_2 for the second term, a_3 for the third term, and so on. For the sequence $f(n) = \dfrac{1}{n}$, we write

$$a_1 = f(1) = 1, \quad a_2 = f(2) = \frac{1}{2}, \quad a_3 = f(3) = \frac{1}{3}, \quad a_4 = f(4) = \frac{1}{4}, \ldots, \quad a_n = f(n) = \frac{1}{n}, \ldots$$

In other words, we do not use the traditional function notation $f(n)$ for sequences. For this particular sequence, we have a rule for the nth term, which is $a_n = \dfrac{1}{n}$, so it is easy to find any term of the sequence.

When a formula for the nth term (sometimes called the **general term**) of a sequence is known, rather than write out the terms of the sequence, we may represent the entire sequence by placing braces around the formula for the nth term. For example, the sequence whose nth term is $b_n = \left(\dfrac{1}{2}\right)^n$ may be represented as

$$\{b_n\} = \left\{\left(\frac{1}{2}\right)^n\right\}$$

or by

$$b_1 = \frac{1}{2}, \quad b_2 = \frac{1}{4}, \quad b_3 = \frac{1}{8}, \ldots, \quad b_n = \left(\frac{1}{2}\right)^n, \ldots$$

EXAMPLE 1   **Writing the First Several Terms of a Sequence**

Write down the first six terms of the following sequence and graph it.

$$\{a_n\} = \left\{\frac{n-1}{n}\right\}$$

Figure 2

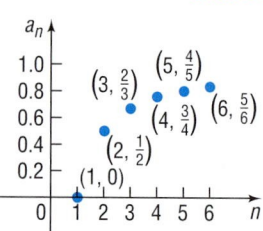

Solution   The first six terms of the sequence are

$$a_1 = 0, \quad a_2 = \frac{1}{2}, \quad a_3 = \frac{2}{3}, \quad a_4 = \frac{3}{4}, \quad a_5 = \frac{4}{5}, \quad a_6 = \frac{5}{6}$$

See Figure 2 for the graph. ■

Graphing utilities can be used to write the terms of a sequence and graph them, as the following example illustrates.

EXAMPLE 2   **Using a Graphing Utility to Write the First Several Terms of a Sequence**

Use a graphing utility to write the first six terms of the following sequence and graph it.

$$\{a_n\} = \left\{ \frac{n-1}{n} \right\}$$

Solution   Figure 3 shows the sequence generated on a TI-83 graphing calculator. We can see the first few terms of the sequence on the screen. You need to press the right arrow key to scroll right in order to see the remaining terms of the sequence.

Figure 3

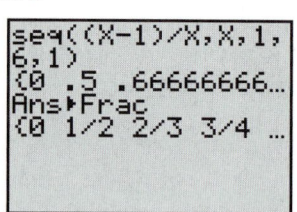

We could also obtain the terms of the sequence using the TABLE feature. First, put the graphing utility in SEQuence mode. Using $Y =$, enter the formula for the sequence into the graphing utility. See Figure 4. Set up the table with TblStart $= 1$ and ΔTbl $= 1$. See Table 1. Finally, we can graph the sequence. See Figure 5. Notice that the first term of the sequence is not visible since it lies on the x-axis. TRACEing the graph will allow you to determine the terms of the sequence.

Figure 4

TABLE 1

n	$u(n)$
1	0
2	.5
3	.66667
4	.75
5	.8
6	.83333
7	.85714

$u(n)⊟(n-1)/n$

Figure 5

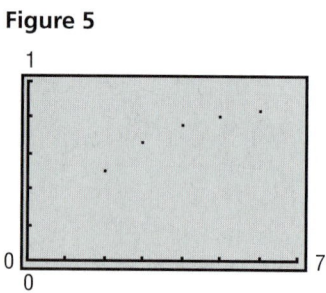

We will usually provide solutions done by hand. The reader is encouraged to verify solutions using a graphing utility.

EXAMPLE 3   **Writing the First Several Terms of a Sequence**

Write down the first six terms of the following sequence and graph it by hand.

$$\{b_n\} = \left\{ (-1)^{n-1} \left(\frac{2}{n} \right) \right\}$$

Solution   The first six terms of the sequence are

$$b_1 = 2, \quad b_2 = -1, \quad b_3 = \frac{2}{3}, \quad b_4 = -\frac{1}{2}, \quad b_5 = \frac{2}{5}, \quad b_6 = -\frac{1}{3}$$

See Figure 6 for the graph.

Figure 6

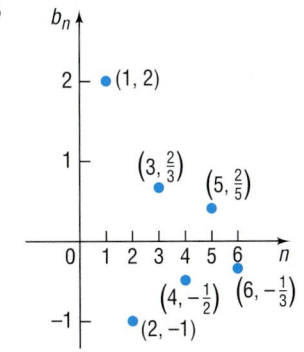

EXAMPLE 4   **Writing the First Several Terms of a Sequence**

Write down the first six terms of the following sequence and graph it.

$$\{c_n\} = \begin{cases} n & \text{if } n \text{ is even} \\ \dfrac{1}{n} & \text{if } n \text{ is odd} \end{cases}$$

Solution   The first six terms of the sequence are

$$c_1 = 1, \quad c_2 = 2, \quad c_3 = \frac{1}{3}, \quad c_4 = 4, \quad c_5 = \frac{1}{5}, \quad c_6 = 6$$

See Figure 7 for the graph.

Figure 7

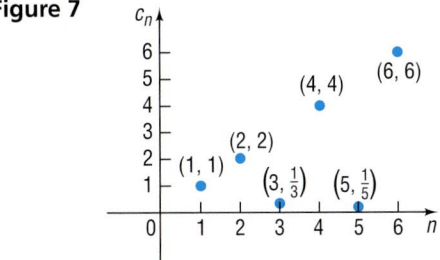

NOW WORK PROBLEMS **3** AND **5.**

Sometimes a sequence is indicated by an observed pattern in the first few terms that makes it possible to infer the makeup of the nth term. In the example that follows, a sufficient number of terms of the sequence are given so that a natural choice for the nth term is suggested.

EXAMPLE 5   **Determining a Sequence from a Pattern**

(a) $e, \dfrac{e^2}{2}, \dfrac{e^3}{3}, \dfrac{e^4}{4}, \ldots$ $a_n = \dfrac{e^n}{n}$

(b) $1, \dfrac{1}{3}, \dfrac{1}{9}, \dfrac{1}{27}, \ldots$ $b_n = \dfrac{1}{3^{n-1}}$

(c) $1, 3, 5, 7, \ldots$ $c_n = 2n - 1$

(d) $1, 4, 9, 16, 25, \ldots$ $d_n = n^2$

(e) $1, -\dfrac{1}{2}, \dfrac{1}{3}, -\dfrac{1}{4}, \dfrac{1}{5}, \ldots$ $e_n = (-1)^{n+1}\left(\dfrac{1}{n}\right)$ ∎

Notice in the sequence $\{e_n\}$ in Example 5(e) that the signs of the terms **alternate**. When this occurs, we use factors such as $(-1)^{n+1}$, which equals 1 if n is odd and -1 if n is even, or $(-1)^n$, which equals -1 if n is odd and 1 if n is even.

✏ **NOW WORK PROBLEM 13.**

The Factorial Symbol

If $n \geq 0$ is an integer, the **factorial symbol** $n!$ is defined as follows:

$$0! = 1 \qquad 1! = 1$$
$$n! = n(n - 1) \cdot \ldots \cdot 3 \cdot 2 \cdot 1 \qquad \text{if } n \geq 2$$

For example, $2! = 2 \cdot 1 = 2$, $3! = 3 \cdot 2 \cdot 1 = 6$, $4! = 4 \cdot 3 \cdot 2 \cdot 1 = 24$, and so on. Table 2 lists the values of $n!$ for $0 \leq n \leq 6$.

Because

$$n! = n\underbrace{(n - 1)(n - 2) \cdot \ldots \cdot 3 \cdot 2 \cdot 1}_{(n-1)!}$$

we can use the formula

TABLE 2

n	0	1	2	3	4	5	6
$n!$	1	1	2	6	24	120	720

$$n! = n(n - 1)!$$

to find successive factorials. For example, because $6! = 720$, we have

$$7! = 7 \cdot 6! = 7(720) = 5040$$

and

$$8! = 8 \cdot 7! = 8(5040) = 40{,}320$$

COMMENT   Your calculator has a factorial key. Use it to see how fast factorials increase in value. Find the value of 69!. What happens when you try to find 70!? In fact, 70! is larger than 10^{100} (a **googol**), the largest number most calculators can display.   ∎

Recursive Formulas

2 A second way of defining a sequence is to assign a value to the first (or the first few) term(s) and specify the nth term by a formula or equation that involves one or more of the terms preceding it. Sequences defined this way are said to be defined **recursively**, and the rule or formula is called a **recursive formula**.

EXAMPLE 6   **Writing the Terms of a Recursively Defined Sequence**

Write down the first five terms of the following recursively defined sequence.

$$s_1 = 1, \qquad s_n = 4s_{n-1}$$

Solution   The first term is given as $s_1 = 1$. To get the second term, we use $n = 2$ in the formula to get $s_2 = 4s_1 = 4 \cdot 1 = 4$. To get the third term, we use $n = 3$ in the formula to get $s_3 = 4s_2 = 4 \cdot 4 = 16$. To get a new term requires that we know the value of the preceding term. The first five terms are

$$s_1 = 1$$
$$s_2 = 4 \cdot 1 = 4$$
$$s_3 = 4 \cdot 4 = 16$$
$$s_4 = 4 \cdot 16 = 64$$
$$s_5 = 4 \cdot 64 = 256$$

Graphing utilities can be used to generate recursively defined sequences.

EXAMPLE 7   **Using a Graphing Utility to Write the Terms of a Recursively Defined Sequence**

Use a graphing utility to write down the first five terms of the following recursively defined sequence.

$$s_1 = 1 \qquad s_2 = 4s_{n-1}$$

Solution   First, put the graphing utility into SEQuence mode. Using $Y =$, enter the recursive formula into the graphing utility. See Figure 8(a). Next, set up the viewing window to generate the desired sequence. Finally, graph the recursion relation and use TRACE to determine the terms in the sequence. See Figure 8(b) For example, we see that the fourth term of the sequence is 64. Table 3 also shows the terms of the sequence.

Figure 8

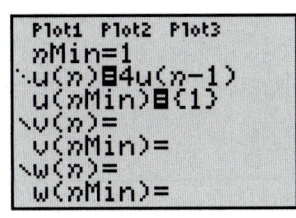

(a)

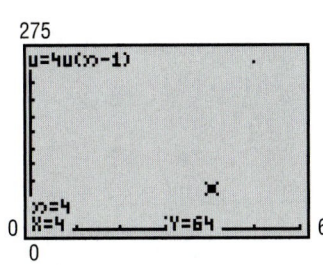

(b)

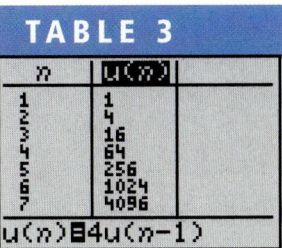

EXAMPLE 8   **Writing the Terms of a Recursively Defined Sequence**

Write down the first five terms of the following recursively defined sequence.

$$u_1 = 1, \qquad u_2 = 1, \qquad u_{n+2} = u_n + u_{n+1}$$

Solution   We are given the first two terms. To get the third term requires that we know each of the previous two terms. Thus,

$$u_1 = 1$$
$$u_2 = 1$$
$$u_3 = u_1 + u_2 = 1 + 1 = 2$$
$$u_4 = u_2 + u_3 = 1 + 2 = 3$$
$$u_5 = u_3 + u_4 = 2 + 3 = 5$$ ■

The sequence defined in Example 8 is called a **Fibonacci sequence**, and the terms of this sequence are called **Fibonacci numbers**. These numbers appear in a wide variety of applications (see Problems 75–78).

✏ **NOW WORK PROBLEMS 21 AND 29.**

Summation Notation

③ It is often important to be able to find the sum of the first n terms of a sequence $\{a_n\}$, that is,

$$a_1 + a_2 + a_3 + \cdots + a_n \tag{1}$$

Rather than write down all these terms, we introduce a more concise way to express the sum, called **summation notation**. Using summation notation, we would write the sum (1) as

$$a_1 + a_2 + a_3 + \cdots + a_n = \sum_{k=1}^{n} a_k$$

The symbol Σ (a stylized version of the Greek letter sigma, which is an S in our alphabet) is simply an instruction to sum, or add up, the terms. The integer k is called the **index** of the sum; it tells you where to start the sum and where to end it. The expression

$$\sum_{k=1}^{n} a_k \tag{2}$$

is an instruction to add the terms a_k of the sequence $\{a_n\}$ from $k = 1$ through $k = n$. We read expression (2) as "the sum of a_k from $k = 1$ to $k = n$."

EXAMPLE 9   **Expanding Summation Notation**

Write out each sum.

(a) $\displaystyle\sum_{k=1}^{10} \frac{1}{k}$ (b) $\displaystyle\sum_{k=1}^{n} k!$

Solution   (a) $\displaystyle\sum_{k=1}^{10} \frac{1}{k} = 1 + \frac{1}{2} + \frac{1}{3} + \cdots + \frac{1}{10}$ (b) $\displaystyle\sum_{k=1}^{n} k! = 1! + 2! + \cdots + n!$ ■

EXAMPLE 10	Writing a Sum in Summation Notation

Express each sum using summation notation.

(a) $1^2 + 2^2 + 3^2 + \cdots + 9^2$

(b) $1 + \dfrac{1}{2} + \dfrac{1}{4} + \dfrac{1}{8} + \cdots + \dfrac{1}{2^{n-1}}$

Solution   (a) The sum $1^2 + 2^2 + 3^2 + \cdots + 9^2$ has 9 terms, each of the form k^2, and starts at $k = 1$ and ends at $k = 9$:

$$1^2 + 2^2 + 3^2 + \cdots + 9^2 = \sum_{k=1}^{9} k^2$$

(b) The sum

$$1 + \frac{1}{2} + \frac{1}{4} + \frac{1}{8} + \cdots + \frac{1}{2^{n-1}}$$

has n terms, each of the form $\dfrac{1}{2^{k-1}}$ (do you see why?), and starts at $k = 1$ and ends at $k = n$:

$$1 + \frac{1}{2} + \frac{1}{4} + \frac{1}{8} + \cdots + \frac{1}{2^{n-1}} = \sum_{k=1}^{n} \frac{1}{2^{k-1}}$$ ■

The index of summation need not always begin at 1 or end at n; for example, we could have expressed the sum in Example 10(b) as

$$\sum_{k=0}^{n-1} \frac{1}{2^k} = 1 + \frac{1}{2} + \frac{1}{4} + \cdots + \frac{1}{2^{n-1}}$$

Letters other than k may be used as the index. For example,

$$\sum_{j=1}^{n} j! \quad \text{and} \quad \sum_{i=1}^{n} i!$$

each represent the same sum as the one given in Example 9(b).

 ✏ **NOW WORK PROBLEMS 37 AND 47.**

Adding the First n Terms of a Sequence

④ Next, we list some properties of sequences using summation notation. These properties are useful for adding the terms of a sequence.

Theorem **Properties of Sequences**

If $\{a_n\}$ and $\{b_n\}$ are two sequences and c is a real number, then:

1. $\displaystyle\sum_{k=1}^{n} c = \underbrace{c + c + \cdots + c}_{n \text{ terms}} = cn$

2. $\displaystyle\sum_{k=1}^{n} (ca_k) = ca_1 + ca_2 + \cdots + ca_n = c(a_1 + a_2 + \cdots + a_n) = c\sum_{k=1}^{n} a_k$

3. $\displaystyle\sum_{k=1}^{n} (a_k + b_k) = \sum_{k=1}^{n} a_k + \sum_{k=1}^{n} b_k$

4. $\displaystyle\sum_{k=1}^{n} (a_k - b_k) = \sum_{k=1}^{n} a_k - \sum_{k=1}^{n} b_k$

5. $\displaystyle\sum_{k=1}^{n} a_k = \sum_{k=1}^{j} a_k + \sum_{k=j+1}^{n} a_k,$ where $1 < j < n$

6. $\displaystyle\sum_{k=1}^{n} k = 1 + 2 + 3 + \cdots + n = \frac{n(n+1)}{2}$

7. $\displaystyle\sum_{k=1}^{n} k^2 = 1^2 + 2^2 + 3^2 + \cdots + n^2 = \frac{n(n+1)(2n+1)}{6}$

8. $\displaystyle\sum_{k=1}^{n} k^3 = 1^3 + 2^3 + 3^3 + \cdots + n^3 = \left(\frac{n(n+1)}{2}\right)^2$

The proofs of properties 1 and 2 are shown and the proofs of 3 through 5 are based on properties of real numbers. The proofs of 7 and 8 require mathematical induction, which is discussed in Section 13.4. See Problem 79 for a derivation of 6.

EXAMPLE 11 **Finding the Sum of a Sequence**

Find the sum of each sequence.

(a) $\displaystyle\sum_{k=1}^{5} (3k)$ (b) $\displaystyle\sum_{k=1}^{3} (k^3 + 1)$ (c) $\displaystyle\sum_{k=1}^{4} (k^2 - 7k + 2)$

Solution (a) $\displaystyle\sum_{k=1}^{5} (3k) = 3\sum_{k=1}^{5} k = 3\left(\frac{5(5+1)}{2}\right) = 3(15) = 45$

  Property 2   Property 6

(b) $\displaystyle\sum_{k=1}^{3} (k^3 + 1) = \sum_{k=1}^{3} k^3 + \sum_{k=1}^{3} 1$ Property 3

$\displaystyle\qquad\qquad\quad = \left(\frac{3(3+1)}{2}\right)^2 + 1(3)$ Properties 1, 8

$\displaystyle\qquad\qquad\quad = 36 + 3$

$\displaystyle\qquad\qquad\quad = 39$

(c) $\displaystyle\sum_{k=1}^{4} (k^2 - 7k + 2) = \sum_{k=1}^{4} k^2 - \sum_{k=1}^{4} (7k) + \sum_{k=1}^{4} 2$ Properties 3, 4

$$= \sum_{k=1}^{4} k^2 - 7\sum_{k=1}^{4} k + \sum_{k=1}^{4} 2 \qquad \text{\color{blue}Property 2}$$

$$= \frac{4(4+1)(2\cdot 4 + 1)}{6} - 7\left(\frac{4(4+1)}{2}\right) + 2(4)$$

Properties 1, 6, 7

$$= 30 - 70 + 8$$
$$= -32$$

■

NOW WORK PROBLEM **59(a)**.

EXAMPLE 12	**Using a Graphing Utility to Find the Sum of a Sequence**

Using a graphing utility, find the sum of each sequence.

(a) $\displaystyle\sum_{k=1}^{5} 3k$ (b) $\displaystyle\sum_{k=1}^{3} (k^3 + 1)$ (c) $\displaystyle\sum_{k=1}^{4} (k^2 - 7k + 2)$

Solution   (a) Figure 9 shows the solution using a TI-83 graphing calculator.

$$\sum_{k=1}^{5} 3k = 45$$

(b) Figure 10 shows the solution using a TI-83 graphing calculator.

$$\sum_{k=1}^{3} (k^3 + 1) = 39$$

(c) Figure 11 shows the solution using a TI-83 graphing calculator.

$$\sum_{k=1}^{4} (k^2 - 7k + 2) = -32$$

Figure 9

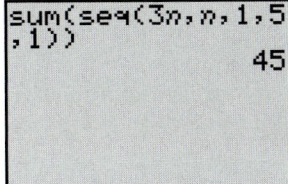

Figure 10

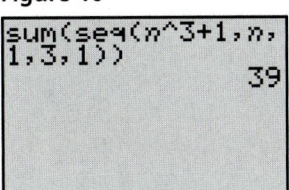

Figure 11

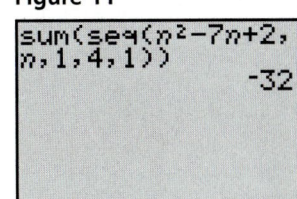

■

NOW WORK PROBLEM **59(b)**.

Annuities; Amortization

⑤   In Section 6.6 we developed the compound interest formula, which gives the future value when a fixed amount of money is deposited in an account that pays interest compounded periodically. Often, though, money is invested in small amounts at periodic intervals. An **annuity** is a sequence of equal periodic deposits. The periodic deposits may be made annually, quarterly, monthly, or daily.

When deposits are made at the same time that the interest is credited, the annuity is called **ordinary**. We will only deal with ordinary annuities here. The **amount of an annuity** is the sum of all deposits made plus all interest paid.

Suppose that the initial amount deposited in an annuity is M, the periodic deposit is P, and the per annum rate of interest is $r\%$ (expressed as a decimal) compounded N times per year. The periodic deposit is made at the same time that the interest is credited, so N deposits are made per year. The amount A_n of the annuity after n deposits will equal the amount of the annuity after $n - 1$ deposits, A_{n-1}, plus the interest earned on this amount plus the periodic deposit P. That is,

$$A_n = A_{n-1} + \frac{r}{N}A_{n-1} + P = \left(1 + \frac{r}{N}\right)A_{n-1} + P$$

Amount   Amount   Interest   Periodic

after   in previous   earned   deposit

n deposits   period

We have established the following result:

Theorem

Annuity Formula

If $A_0 = M$ represents the initial amount deposited in an annuity that earns $r\%$ per annum compounded N times per year, and if P is the periodic deposit made at each payment period, then the amount A_n of the annuity after n deposits is given by the recursive sequence

$$A_0 = M, \qquad A_n = \left(1 + \frac{r}{N}\right)A_{n-1} + P, \qquad n \geq 1 \qquad (3)$$

Formula (3) may be explained as follows: the money in the account initially, A_0, is M; the money in the account after $n - 1$ payments (A_{n-1}) earns interest $\dfrac{r}{N}$ during the nth period; so when the periodic payment of P dollars is added, the amount after n payments, A_n, is obtained.

EXAMPLE 13   Saving for Spring Break

A trip to Cancun during spring break will cost $450 and full payment is due March 2. To have the money, a student, on September 1, deposits $100 in a savings account that pays 4% per annum compounded monthly. On the first of each month, the student deposits $50 in this account.

(a) Find a recursive sequence that explains how much is in the account after n months.

(b) Use the TABLE feature to list the amounts of the annuity for the first 6 months.

(c) After the deposit on March 1 is made, is there enough in the account to pay for the Cancun trip?

(d) If the student deposits $60 each month, will there be enough for the trip?

Solution (a) The initial amount deposited in the account is $A_0 = \$100$. The monthly deposit is $P = \$50$, and the per annum rate of interest is $r = 0.04$ compounded $N = 12$ times per year. The amount A_n in the account after n monthly deposits is given by the recursive sequence

$$A_0 = 100, \qquad A_n = \left(1 + \frac{r}{N}\right)A_{n-1} + P = \left(1 + \frac{0.04}{12}\right)A_{n-1} + 50$$

(b) In SEQuence mode on a TI-83, enter the sequence $\{A_n\}$ and create Table 4. On September 1 ($n = 0$), there is $100 in the account. After the first payment on October 1, the value of the account is $150.33. After the second payment on November 1, the value of the account is $200.83. After the third payment on December 1, the value of the account is $251.50, and so on.

(c) On March 1 ($n = 6$), there is only $404.53, not enough to pay for the trip to Cancun.

(d) If the periodic deposit, P, is $60, then on March 1, there is $465.03 in the account, enough for the trip. See Table 5. ■

TABLE 4

n	$u(n)$
0	100
1	150.33
2	200.83
3	251.5
4	302.34
5	353.35
6	404.53

$u(n) = (1 + .04/12)...$

TABLE 5

n	$u(n)$
0	100
1	160.33
2	220.87
3	281.6
4	342.54
5	403.68
6	465.03

$u(n) = (1 + .04/12)...$

Amortization

Recursive sequences can also be used to compute information about loans. When equal periodic payments are made to pay off a loan, the loan is said to be **amortized**.

Theorem **Amortization Formula**

If $\$B$ is borrowed at an interest rate of $r\%$ (expressed as a decimal) per annum compounded monthly, the balance A_n due after n monthly payments of $\$P$ is given by the recursive sequence

$$A_0 = B, \qquad A_n = \left(1 + \frac{r}{12}\right)A_{n-1} - P, \qquad n \geq 1 \quad (4)$$

Formula (4) may be explained as follows: The initial loan balance is $\$B$. The balance due A_n after n payments will equal the balance due previously, A_{n-1}, plus the interest charged on that amount reduced by the periodic payment P.

EXAMPLE 14 **Mortgage Payments**

John and Wanda borrowed $180,000 at 7% per annum compounded monthly for 30 years to purchase a home. Their monthly payment is determined to be $1197.54.

(a) Find a recursive formula that represents their balance after each payment of $1197.54 has been made.

(b) Determine their balance after the first payment is made.

(c) When will their balance be below $170,000?

Solution (a) We use formula (4) with $A_0 = 180,000$, $r = 0.07$, and $P = \$1197.54$. Then

$$A_0 = 180,000 \qquad A_n = \left(1 + \frac{0.07}{12}\right)A_{n-1} - 1197.54$$

(b) In SEQuence mode on a TI-83, enter the sequence $\{A_n\}$ and create Table 6. After the first payment is made, the balance is $A_1 = \$179,852$.

(c) Scroll down until the balance is below $170,000. See Table 7. After the fifty-eighth payment is made ($n = 58$), the balance is below $170,000.

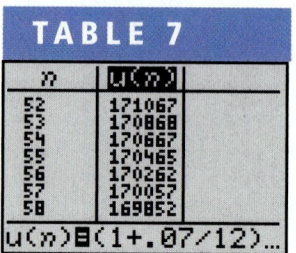

TABLE 6	
n	u(n)
0	180000
1	179852
2	179704
3	179555
4	179405
5	179254
6	179102
u(n)8(1+.07/12)...	

TABLE 7	
n	u(n)
52	171067
53	170868
54	170667
55	170465
56	170262
57	170057
58	169852
u(n)8(1+.07/12)...	

NOW WORK PROBLEM 67.

13.1 Concepts and Vocabulary

In Problems 1–3, fill in the blanks.

1. A(n) _____ is a function whose domain is the set of positive integers.

2. For the sequence $\{s_n\} = \{4n - 1\}$, the first term is $s_1 =$ _____ and the fourth term is $s_4 =$ _____.

3. $\displaystyle\sum_{k=1}^{4} (2k) =$ _____.

In Problems 4–6, answer True or False to each statement.

4. Sequences are sometimes defined recursively.

5. A sequence is a function.

6. $\displaystyle\sum_{k=1}^{2} k = 3$

7. Explain how a sequence and a function are related.

13.1 Exercises

In Problems 1–12, write down the first five terms of each sequence by hand. Verify your results using a graphing utility.

1. $\{n\}$

2. $\{n^2 + 1\}$

3. $\left\{\dfrac{n}{n+2}\right\}$

4. $\left\{\dfrac{2n+1}{2n}\right\}$

5. $\{(-1)^{n+1}n^2\}$

6. $\left\{(-1)^{n-1}\left(\dfrac{n}{2n-1}\right)\right\}$

7. $\left\{\dfrac{2^n}{3^n+1}\right\}$

8. $\left\{\left(\dfrac{4}{3}\right)^n\right\}$

9. $\left\{\dfrac{(-1)^n}{(n+1)(n+2)}\right\}$

10. $\left\{\dfrac{3^n}{n}\right\}$

11. $\left\{\dfrac{n}{e^n}\right\}$

12. $\left\{\dfrac{n^2}{2^n}\right\}$

In Problems 13–20, the given pattern continues. Write down the nth term of each sequence suggested by the pattern.

13. $\dfrac{1}{2}, \dfrac{2}{3}, \dfrac{3}{4}, \dfrac{4}{5}, \cdots$

14 $\dfrac{1}{1 \cdot 2}, \dfrac{1}{2 \cdot 3}, \dfrac{1}{3 \cdot 4}, \dfrac{1}{4 \cdot 5}, \cdots$

15 $1, \dfrac{1}{2}, \dfrac{1}{4}, \dfrac{1}{8}, \cdots$

16. $\dfrac{2}{3}, \dfrac{4}{9}, \dfrac{8}{27}, \dfrac{16}{81}, \cdots$

17. $1, -1, 1, -1, 1, -1, \ldots$

18. $1, \dfrac{1}{2}, 3, \dfrac{1}{4}, 5, \dfrac{1}{6}, 7, \dfrac{1}{8}, \cdots$

19. $1, -2, 3, -4, 5, -6, \ldots$

20. $2, -4, 6, -8, 10, \ldots$

In Problems 21–34, a sequence is defined recursively. Write the first five terms by hand. Use a graphing utility to verify your results.

21. $a_1 = 2; \quad a_n = 3 + a_{n-1}$

22. $a_1 = 3; \quad a_n = 4 - a_{n-1}$

23. $a_1 = -2; \quad a_n = n + a_{n-1}$

24. $a_1 = 1; \quad a_n = n - a_{n-1}$

25. $a_1 = 5; \quad a_n = 2a_{n-1}$

26. $a_1 = 2; \quad a_n = -a_{n-1}$

27. $a_1 = 3; \quad a_n = \dfrac{a_{n-1}}{n}$

28. $a_1 = -2; \quad a_n = n + 3a_{n-1}$

29. $a_1 = 1; \quad a_2 = 2; \quad a_n = a_{n-1} \cdot a_{n-2}$

30. $a_1 = -1; \quad a_2 = 1; \quad a_n = a_{n-2} + na_{n-1}$

31. $a_1 = A; \quad a_n = a_{n-1} + d$

32. $a_1 = A; \quad a_n = ra_{n-1}, \quad r \neq 0$

33. $a_1 = \sqrt{2}; \quad a_n = \sqrt{2 + a_{n-1}}$

34. $a_1 = \sqrt{2}; \quad a_n = \sqrt{\dfrac{a_{n-1}}{2}}$

In Problems 35–44, write out each sum.

35. $\displaystyle\sum_{k=1}^{5} (k + 2)$

36. $\displaystyle\sum_{k=1}^{4} (2k + 1)$

37. $\displaystyle\sum_{k=1}^{8} \dfrac{k^2}{2}$

38. $\displaystyle\sum_{k=1}^{7} (k + 1)^2$

39. $\displaystyle\sum_{k=0}^{n} \dfrac{1}{3^k}$

40. $\displaystyle\sum_{k=0}^{n} \left(\dfrac{3}{2}\right)^k$

41. $\displaystyle\sum_{k=0}^{n-1} \dfrac{1}{3^{k+1}}$

42. $\displaystyle\sum_{k=0}^{n-1} (2k + 1)$

43. $\displaystyle\sum_{k=2}^{n} (-1)^k \ln k$

44. $\displaystyle\sum_{k=3}^{n} (-1)^{k+1} 2^k$

In Problems 45–54, express each sum using summation notation.

45. $1 + 2 + 3 + \cdots + 20$

46. $1^3 + 2^3 + 3^3 + \cdots + 8^3$

47. $\dfrac{1}{2} + \dfrac{2}{3} + \dfrac{3}{4} + \cdots + \dfrac{13}{13 + 1}$

48. $1 + 3 + 5 + 7 + \cdots + [2(12) - 1]$

49. $1 - \dfrac{1}{3} + \dfrac{1}{9} - \dfrac{1}{27} + \cdots + (-1)^6 \left(\dfrac{1}{3^6}\right)$

50. $\dfrac{2}{3} - \dfrac{4}{9} + \dfrac{8}{27} - \cdots + (-1)^{11+1} \left(\dfrac{2}{3}\right)^{11}$

51. $3 + \dfrac{3^2}{2} + \dfrac{3^3}{3} + \cdots + \dfrac{3^n}{n}$

52. $\dfrac{1}{e} + \dfrac{2}{e^2} + \dfrac{3}{e^3} + \cdots + \dfrac{n}{e^n}$

53. $a + (a + d) + (a + 2d) + \cdots + (a + nd)$

54. $a + ar + ar^2 + \cdots + ar^{n-1}$

In Problems 55–66, find the sum of each sequence (a) by hand and (b) by using a graphing utility.

55. $\displaystyle\sum_{k=1}^{10} 5$

56. $\displaystyle\sum_{k=1}^{20} 8$

57. $\displaystyle\sum_{k=1}^{6} k$

58. $\displaystyle\sum_{k=1}^{4} (-k)$

59. $\displaystyle\sum_{k=1}^{5} (5k + 3)$

60. $\displaystyle\sum_{k=1}^{6} (3k - 7)$

61. $\displaystyle\sum_{k=1}^{3} (k^2 + 4)$

62. $\displaystyle\sum_{k=0}^{4} (k^2 - 4)$

63. $\displaystyle\sum_{k=1}^{6} (-1)^k 2^k$

64. $\displaystyle\sum_{k=1}^{4} (-1)^k 3^k$

65. $\displaystyle\sum_{k=1}^{4} (k^3 - 1)$

66. $\displaystyle\sum_{k=0}^{3} (k^3 + 2)$

---

67. Credit Card Debt John has a balance of $3000 on his Discover card that charges 1% interest per month on any unpaid balance. John can afford to pay $100 toward the balance each month. His balance each month after making a $100 payment is given by the recursively defined sequence

$$B_0 = \$3000, \quad B_n = 1.01B_{n-1} - 100$$

(a) Determine John's balance after making the first payment. That is, determine B_1.

(b) Using a graphing utility, determine when John's balance will be below $2000. How many payments of $100 have been made?

(c) Using a graphing utility, determine when John will pay off the balance. What is the total of all the payments?

(d) What was John's interest expense?

68. Car Loans   Phil bought a car by taking out a loan for $18,500 at 0.5% interest per month. Phil's normal monthly payment is $434.47 per month, but he decides that he can afford to pay $100 extra toward the balance each month. His balance each month is given by the recursively defined sequence

$$B_0 = \$18,500, \quad B_n = 1.005B_{n-1} - 534.47$$

(a) Determine Phil's balance after making the first payment. That is, determine B_1.
(b) Using a graphing utility, determine when Phil's balance will be below $10,000. How many payments of $534.47 have been made?
(c) Using a graphing utility, determine when Phil will pay off the balance. What is the total of all the payments?
(d) What was Phil's interest expense?

69. Trout Population   A pond currently has 2000 trout in it. A fish hatchery decides to add an additional 20 trout each month. In addition, it is known that the trout population is growing 3% per month. The size of the population after n months is given by the recursively defined sequence

$$p_0 = 2000, \quad p_n = 1.03p_{n-1} + 20$$

(a) How many trout are in the pond at the end of the second month? That is, what is p_2?
(b) Using a graphing utility, determine how long it will be before the trout population reaches 5000.

70. Environmental Control   The Environmental Protection Agency (EPA) determines that Maple Lake has 250 tons of pollutants as a result of industrial waste and that 10% of the pollutant present is neutralized by solar oxidation every year. The EPA imposes new pollution control laws that result in 15 tons of new pollutant entering the lake each year. The amount of pollutant in the lake at the end of each year is given by the recursively defined sequence

$$p_0 = 250, \quad p_n = 0.9p_{n-1} + 15$$

(a) Determine the amount of pollutant in the lake at the end of the second year. That is, determine p_2.
(b) Using a graphing utility, provide pollutant amounts for the next 20 years.
(c) What is the equilibrium level of pollution in Maple Lake? That is, what is $\lim_{n\to\infty} p_n$?

71. Roth IRA   On January 1, 1999, Bob decides to place $500 at the end of each quarter into a Roth Individual Retirement Account.
(a) Find a recursive formula that represents Bob's balance at the end of each quarter if the rate of return is assumed to be 8% per annum compounded quarterly.
(b) How long will it be before the value of the account exceeds $100,000?
(c) What will be the value of the account in 25 years when Bob retires?

72. Education IRA   On January 1, 1999, John's parents decide to place $45 at the end of each month into an Education IRA.

(a) Find a recursive formula that represents the balance at the end of each month if the rate of return is assumed to be 6% per annum compounded monthly.
(b) How long will it be before the value of the account exceeds $4000?
(c) What will be the value of the account in 16 years when John goes to college?

73. Home Loan   Bill and Laura borrowed $150,000 at 6% per annum compounded monthly for 30 years to purchase a home. Their monthly payment is determined to be $899.33.
(a) Find a recursive formula for their balance after each monthly payment has been made.
(b) Determine Bill and Laura's balance after the first payment.
(c) Using a graphing utility, create a table showing Bill and Laura's balance after each monthly payment.
(d) Using a graphing utility, determine when Bill and Laura's balance will be below $140,000.
(e) Using a graphing utility, determine when Bill and Laura will pay off the balance.
(f) Determine Bill and Laura's interest expense when the loan is paid.
(g) Suppose that Bill and Laura decide to pay an additional $100 each month on their loan. Answer parts (a) to (f) under this scenario.
(h) Is it worthwhile for Bill and Laura to pay the additional $100? Explain.

74. Home Loan   Jodi and Jeff borrowed $120,000 at 6.5% per annum compounded monthly for 30 years to purchase a home. Their monthly payment is determined to be $758.48.
(a) Find a recursive formula for their balance after each monthly payment has been made.
(b) Determine Jodi and Jeff's balance after the first payment.
(c) Using a graphing utility, create a table showing Jodi and Jeff's balance after each monthly payment.
(d) Using a graphing utility, determine when Jodi and Jeff's balance will be below $100,000.
(e) Using a graphing utility, determine when Jodi and Jeff will pay off the balance.
(f) Determine Jodi and Jeff's interest expense when the loan is paid.
(g) Suppose that Jodi and Jeff decide to pay an additional $100 each month on their loan. Answer parts (a) to (f) under this scenario.
(h) Is it worthwhile for Jodi and Jeff to pay the additional $100? Explain.

75. Growth of a Rabbit Colony   A colony of rabbits begins with one pair of mature rabbits, which will produce a pair of offspring (one male, one female) each month. Assume that all rabbits mature in 1 month and produce a pair of offspring (one male, one female) after 2 months. If no rabbits ever die, how many pairs of mature rabbits are there after 7 months?

[**Hint:** A Fibonacci sequence models this colony. Do you see why?]

1 mature pair
1 mature pair
2 mature pairs
3 mature pairs

76. Fibonacci Sequence   Let

$$u_n = \frac{(1 + \sqrt{5})^n - (1 - \sqrt{5})^n}{2^n\sqrt{5}}$$

define the nth term of a sequence.

(a) Show that $u_1 = 1$ and $u_2 = 1$.
(b) Show that $u_{n+2} = u_{n+1} + u_n$.
(c) Draw the conclusion that $\{u_n\}$ is a Fibonacci sequence.

77. Pascal's Triangle   Divide the triangular array shown (called Pascal's triangle) using diagonal lines as indicated. Find the sum of the numbers in each of these diagonal rows. Do you recognize this sequence?

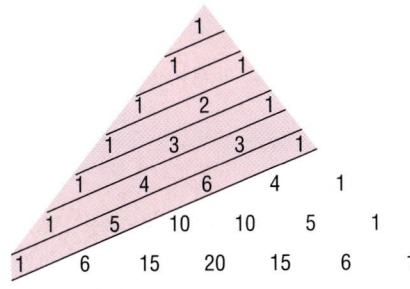

78. Fibonacci Sequence   Use the result of Problem 76 to do the following problems:

(a) Write the first 10 terms of the Fibonacci sequence.

(b) Compute the ratio $\dfrac{u_{n+1}}{u_n}$ for the first 10 terms.

(c) As n gets large, what number does the ratio approach? This number is referred to as the **golden ratio**. Rectangles whose sides are in this ratio were considered pleasing to the eye by the Greeks. For example, the facade of the Parthenon was constructed using the golden ratio.

(d) Compute the ratio $\dfrac{u_n}{u_{n+1}}$ for the first 10 terms.

(e) As n gets large, what number does the ratio approach? This number is also referred to as the **golden ratio**. This ratio is believed to have been used in the construction of the Great Pyramid in Egypt. The ratio equals the sum of the areas of the four face triangles divided by the total surface area of the Great Pyramid.

79. Show that

$$1 + 2 + \cdots + (n - 1) + n = \frac{n(n + 1)}{2}$$

[**Hint:** Let

$$S = 1 + 2 + \cdots + (n - 1) + n$$
$$S = n + (n - 1) + (n - 2) + \cdots + 1$$

Add these equations. Then

$$2S = \underbrace{[1 + n] + [2 + (n - 1)] + \cdots + [n + 1]}_{n \text{ terms in brackets}}$$

Now complete the derivation.]

80. Investigate various applications that lead to a Fibonacci sequence, such as art, architecture, or financial markets. Write an essay on these applications.

13.2   ARITHMETIC SEQUENCES

OBJECTIVES   **1** Determine If a Sequence Is Arithmetic
2 Find a Formula for an Arithmetic Sequence
3 Find the Sum of an Arithmetic Sequence

1   When the difference between successive terms of a sequence is always the same number, the sequence is called **arithmetic**. An **arithmetic sequence*** may be defined recursively as $a_1 = a$, $a_n - a_{n-1} = d$, or as

$$a_1 = a, \quad a_n = a_{n-1} + d \tag{1}$$

where $a = a_1$ and d are real numbers. The number a is the first term, and the number d is called the **common difference**.

*Sometimes called an **arithmetic progression**.

The terms of an arithmetic sequence with first term a and common difference d follow the pattern

$$a, \quad a + d, \quad a + 2d, \quad a + 3d, \ldots$$

EXAMPLE 1 **Determining If a Sequence Is Arithmetic**

The sequence

$$4, \quad 7, \quad 10, \quad 13, \ldots$$

is arithmetic since the difference of successive terms is 3. The first term is 4, and the common difference is 3. ∎

EXAMPLE 2 **Determining If a Sequence Is Arithmetic**

Show that the following sequence is arithmetic. Find the first term and the common difference.

$$\{s_n\} = \{3n + 5\}$$

Solution The first term is $s_1 = 3 \cdot 1 + 5 = 8$. The nth and $(n - 1)$st terms of the sequence $\{s_n\}$ are

$$s_n = 3n + 5 \quad \text{and} \quad s_{n-1} = 3(n - 1) + 5 = 3n + 2$$

Their difference is

$$s_n - s_{n-1} = (3n + 5) - (3n + 2) = 5 - 2 = 3$$

Since the difference of two successive terms is constant, the sequence is arithmetic and the common difference is 3. ∎

EXAMPLE 3 **Determining If a Sequence Is Arithmetic**

Show that the sequence $\{t_n\} = \{4 - n\}$ is arithmetic. Find the first term and the common difference.

Solution The first term is $t_1 = 4 - 1 = 3$. The nth and $(n - 1)$st terms are

$$t_n = 4 - n \quad \text{and} \quad t_{n-1} = 4 - (n - 1) = 5 - n$$

Their difference is

$$t_n - t_{n-1} = (4 - n) - (5 - n) = 4 - 5 = -1$$

Since the difference of two successive terms is constant, $\{t_n\}$ is an arithmetic sequence whose common difference is -1. ∎

✏️ **NOW WORK PROBLEM 3.**

② Suppose that a is the first term of an arithmetic sequence whose common difference is d. We seek a formula for the nth term, a_n. To see the pattern, we write down the first few terms.

$$a_1 = a$$
$$a_2 = a_1 + d = a + 1 \cdot d$$
$$a_3 = a_2 + d = (a + d) + d = a + 2 \cdot d$$
$$a_4 = a_3 + d = (a + 2 \cdot d) + d = a + 3 \cdot d$$
$$a_5 = a_4 + d = (a + 3 \cdot d) + d = a + 4 \cdot d$$
$$\vdots$$
$$a_n = a_{n-1} + d = [a + (n - 2)d] + d = a + (n - 1)d$$

We are led to the following result:

Theorem

nth Term of an Arithmetic Sequence

For an arithmetic sequence $\{a_n\}$ whose first term is a and whose common difference is d, the nth term is determined by the formula

$$a_n = a + (n - 1)d \tag{2}$$

EXAMPLE 4   **Finding a Particular Term of an Arithmetic Sequence**

Find the thirteenth term of the arithmetic sequence: $2, 6, 10, 14, 18, \ldots$

Solution   The first term of this arithmetic sequence is $a = 2$, and the common difference is 4. By formula (2), the nth term is

$$a_n = 2 + (n - 1)4$$

Hence, the thirteenth term is

$$a_{13} = 2 + 12 \cdot 4 = 50$$

EXPLORATION   Use a graphing utility to find the thirteenth term of the sequence given in Example 4. Use it to find the twentieth and fiftieth terms.

EXAMPLE 5   **Finding a Recursive Formula for an Arithmetic Sequence**

The eighth term of an arithmetic sequence is 75, and the twentieth term is 39. Find the first term and the common difference. Give a recursive formula for the sequence.

Solution   By formula (2), we know that $a_n = a + (n - 1)d$. As a result,

$$\begin{cases} a_8 = a + 7d = 75 \\ a_{20} = a + 19d = 39 \end{cases}$$

This is a system of two linear equations containing two variables, a and d, which we can solve by elimination. Subtracting the second equation from the first equation, we get

$$-12d = 36$$
$$d = -3$$

With $d = -3$, we find that $a = 75 - 7d = 75 - 7(-3) = 96$. The first term is $a = 96$, and the common difference is $d = -3$. A recursive formula for this sequence is found using formula (1).

$$a_1 = 96, \quad a_n = a_{n-1} - 3$$

Based on formula (2), a formula for the nth term of the sequence $\{a_n\}$ in Example 5 is

$$a_n = a + (n - 1)d = 96 + (n - 1)(-3) = 99 - 3n$$

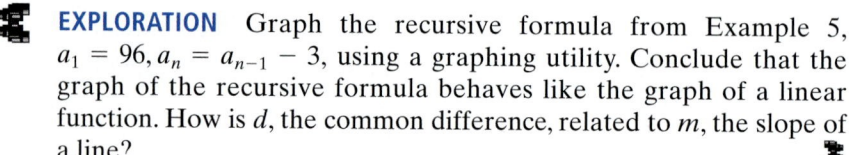

NOW WORK PROBLEMS 19 AND 25.

EXPLORATION Graph the recursive formula from Example 5, $a_1 = 96, a_n = a_{n-1} - 3$, using a graphing utility. Conclude that the graph of the recursive formula behaves like the graph of a linear function. How is d, the common difference, related to m, the slope of a line?

Adding the First n Terms of an Arithmetic Sequence

③ The next result gives a formula for finding the sum of the first n terms of an arithmetic sequence.

Theorem

Sum of n Terms of an Arithmetic Sequence

Let $\{a_n\}$ be an arithmetic sequence with first term a and common difference d. The sum S_n of the first n terms of $\{a_n\}$ is

$$S_n = \frac{n}{2}[2a + (n - 1)d] = \frac{n}{2}(a + a_n) \qquad (3)$$

Proof

$$\begin{aligned}
S_n &= a_1 + a_2 + a_3 + \cdots + a_n && \textcolor{teal}{\text{Sum of first } n \text{ terms}} \\
&= a + (a + d) + (a + 2d) + \cdots + [a + (n - 1)d] && \textcolor{teal}{\text{Formula (2)}} \\
&= \underbrace{(a + a + \cdots + a)}_{n \text{ terms}} + [d + 2d + \cdots + (n - 1)d] && \textcolor{teal}{\text{Rearrange terms}} \\
&= na + d[1 + 2 + \cdots + (n - 1)] \\
&= na + d\left[\frac{(n - 1)n}{2}\right] && \textcolor{teal}{\text{Property 6, Section 13.1}} \\
&= na + \frac{n}{2}(n - 1)d \\
&= \frac{n}{2}[2a + (n - 1)d] && \textcolor{teal}{\text{Factor out } \frac{n}{2}} && (4) \\
&= \frac{n}{2}[a + a + (n - 1)d] \\
&= \frac{n}{2}(a + a_n) && \textcolor{teal}{\text{Formula (2)}} && (5)
\end{aligned}$$

Formula (3) provides two ways to find the sum of the first n terms of an arithmetic sequence. Notice that formula (4) involves the first term and common difference, whereas formula (5) involves the first term and the nth term. Use whichever form is easier.

| EXAMPLE 6 | Finding the Sum of n Terms of an Arithmetic Sequence |

Find the sum S_n of the first n terms of the sequence $\{3n + 5\}$; that is, find

$$8 + 11 + 14 + \cdots + (3n + 5)$$

Solution The sequence $\{3n + 5\}$ is an arithmetic sequence with first term $a = 8$ and the nth term $(3n + 5)$. To find the sum S_n, we use formula (3), as given in (5).

$$S_n = \frac{n}{2}(a + a_n) = \frac{n}{2}[8 + (3n + 5)] = \frac{n}{2}(3n + 13)$$ ∎

✎ NOW WORK PROBLEM **33.**

| EXAMPLE 7 | Using a Graphing Utility to Find the Sum of 20 Terms of an Arithmetic Sequence |

Use a graphing utility to find the sum S_n of the first 20 terms of the sequence $\{9.5n + 2.6\}$.

Solution Figure 12 shows the results obtained using a TI-83 graphing calculator.

Figure 12

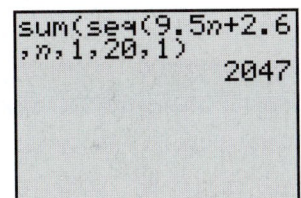

```
sum(seq(9.5n+2.6
,n,1,20,1)
              2047
```

The sum of the first 20 terms of the sequence $\{9.5n + 2.6\}$ is 2047. ∎

✎ WORK EXAMPLE **7** USING FORMULA **(3).**

✎ NOW WORK PROBLEM **41.**

| EXAMPLE 8 | Creating a Floor Design |

A ceramic tile floor is designed in the shape of a trapezoid 20 feet wide at the base and 10 feet wide at the top. See Figure 13. The tiles, 12 inches by 12 inches, are to be placed so that each successive row contains one less tile than the preceding row. How many tiles will be required?

Figure 13

Solution   The bottom row requires 20 tiles and the top row, 10 tiles. Since each successive row requires one less tile, the total number of tiles required is

$$S = 20 + 19 + 18 + \cdots + 11 + 10$$

This is the sum of an arithmetic sequence; the common difference is -1. The number of terms to be added is $n = 11$, with the first term $a = 20$ and the last term $a_{11} = 10$. The sum S is

$$S = \frac{n}{2}(a + a_{11}) = \frac{11}{2}(20 + 10) = 165$$

In all, 165 tiles will be required.   ■

13.2 Concepts and Vocabulary

In Problem 1, fill in the blank.

1. In a(n) _____ sequence, the difference between successive terms is a constant.

In Problems 2 and 3, answer True or False to each statement.

2. In an arithmetic sequence the sum of the first and last terms equals twice the sum of all the terms.

3. In an arithmetic sequence the difference between the first and the last term is the common difference.

4. Given a sequence, how do you determine if it is arithmetic?

13.2 Exercises

In Problems 1–10, an arithmetic sequence is given. Find the common difference and write out the first four terms.

1. $\{n + 4\}$
2. $\{n - 5\}$
3. $\{2n - 5\}$
4. $\{3n + 1\}$
5. $\{6 - 2n\}$

6. $\{4 - 2n\}$
7. $\left\{\dfrac{1}{2} - \dfrac{1}{3}n\right\}$
8. $\left\{\dfrac{2}{3} + \dfrac{n}{4}\right\}$
9. $\{\ln 3^n\}$
10. $\{e^{\ln n}\}$

In Problems 11–18, find the nth term of the arithmetic sequence whose initial term a and common difference d are given. What is the fifth term?

11. $a = 2; \quad d = 3$
12. $a = -2; \quad d = 4$
13. $a = 5; \quad d = -3$
14. $a = 6; \quad d = -2$

15. $a = 0; \quad d = \dfrac{1}{2}$
16. $a = 1; \quad d = -\dfrac{1}{3}$
17. $a = \sqrt{2}; \quad d = \sqrt{2}$
18. $a = 0; \quad d = \pi$

In Problems 19–24, find the indicated term in each arithmetic sequence.

19. 12th term of $2, 4, 6, \ldots$
20. 8th term of $-1, 1, 3, \ldots$

21. 10th term of $1, -2, -5, \ldots$
22. 9th term of $5, 0, -5, \ldots$

23. 8th term of $a, a + b, a + 2b, \ldots$
24. 7th term of $2\sqrt{5}, 4\sqrt{5}, 6\sqrt{5}, \ldots$

In Problems 25–32, find the first term and the common difference of the arithmetic sequence described. Give a recursive formula for the sequence.

25. 8th term is 8; 20th term is 44
26. 4th term is 3; 20th term is 35

27. 9th term is -5; 15th term is 31
28. 8th term is 4; 18th term is -96

29. 15th term is 0; 40th term is -50
30. 5th term is -2; 13th term is 30

31. 14th term is -1; 18th term is -9
32. 12th term is 4; 18th term is 28

In Problems 33–40, find the sum.

33. $1 + 3 + 5 + \cdots + (2n - 1)$

34. $2 + 4 + 6 + \cdots + 2n$

35. $7 + 12 + 17 + \cdots + (2 + 5n)$

36. $-1 + 3 + 7 + \cdots + (4n - 5)$

37. $2 + 4 + 6 + \cdots + 70$

38. $1 + 3 + 5 + \cdots + 59$

39. $5 + 9 + 13 + \cdots + 49$

40. $2 + 5 + 8 + \cdots + 41$

For Problems 41–46, use a graphing utility to find the sum of each sequence.

41. $\{3.45n + 4.12\}, \quad n = 20$

42. $\{2.67n - 1.23\}, \quad n = 25$

43. $2.8 + 5.2 + 7.6 + \cdots + 36.4$

44. $5.4 + 7.3 + 9.2 + \cdots + 32$

45. $4.9 + 7.48 + 10.06 + \cdots + 66.82$

46. $3.71 + 6.9 + 10.09 + \cdots + 80.27$

---

47. Find x so that $x + 3, 2x + 1$, and $5x + 2$ are consecutive terms of an arithmetic sequence.

48. Find x so that $2x, 3x + 2$, and $5x + 3$ are consecutive terms of an arithmetic sequence.

49. **Drury Lane Theater** The Drury Lane Theater has 25 seats in the first row and 30 rows in all. Each successive row contains one additional seat. How many seats are in the theater?

50. **Football Stadium** The corner section of a football stadium has 15 seats in the first row and 40 rows in all. Each successive row contains two additional seats. How many seats are in this section?

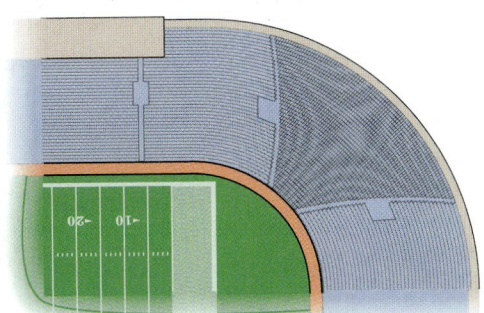

51. **Creating a Mosaic** A mosaic is designed in the shape of an equilateral triangle, 20 feet on each side. Each tile in the mosaic is in the shape of an equilateral triangle, 12 inches to a side. The tiles are to alternate in color as shown in the illustration. How many tiles of each color will be required?

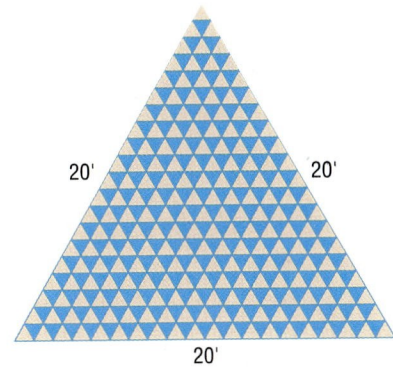

52. **Constructing a Brick Staircase** A brick staircase has a total of 30 steps. The bottom step requires 100 bricks. Each successive step requires two less bricks than the prior step.
    (a) How many bricks are required for the top step?
    (b) How many bricks are required to build the staircase?

53. **Stadium Construction** How many rows are in the corner section of a stadium containing 2040 seats if the first row has 10 seats and each successive row has 4 additional seats?

54. **Salary** Suppose that you just received a job offer with a starting salary of $35,000 per year and a guaranteed raise of $1400 per year. How many years will it take before your aggregate salary is $280,000?

    [**Hint:** Your aggregate salary after two years is $35,000 + ($35,000 + $1400).]

55. Make up an arithmetic sequence. Give it to a friend and ask for its twentieth term.

13.3 GEOMETRIC SEQUENCES; GEOMETRIC SERIES

OBJECTIVES
1. Determine If a Sequence Is Geometric
2. Find a Formula for a Geometric Sequence
3. Find the Sum of a Geometric Sequence
4. Find the Sum of a Geometric Series

1 When the ratio of successive terms of a sequence is always the same non-zero number, the sequence is called **geometric**. A **geometric sequence*** may

*Sometimes called a **geometric progression**.

be defined recursively as $a_1 = a, \dfrac{a_n}{a_{n-1}} = r$, or as

$$a_1 = a, \quad a_n = ra_{n-1} \tag{1}$$

where $a_1 = a$ and $r \neq 0$ are real numbers. The number a is the first term, and the nonzero number r is called the **common ratio**.

The terms of a geometric sequence with first term a and common ratio r follow the pattern

$$a, \quad ar, \quad ar^2, \quad ar^3, \ldots$$

EXAMPLE 1   Determining If a Sequence Is Geometric

The sequence

$$2, 6, 18, 54, 162, \ldots$$

is geometric since the ratio of successive terms is $3\left(\dfrac{6}{2} = \dfrac{18}{6} = \cdots = 3\right)$. The first term is 2, and the common ratio is 3. ∎

EXAMPLE 2   Determining If a Sequence Is Geometric

Show that the following sequence is geometric. Find the first term and the common ratio.

$$\{s_n\} = 2^{-n}$$

Solution   The first term is $s_1 = 2^{-1} = \dfrac{1}{2}$. The nth and $(n-1)$st terms of the sequence $\{s_n\}$ are

$$s_n = 2^{-n} \quad \text{and} \quad s_{n-1} = 2^{-(n-1)}$$

Their ratio is

$$\frac{s_n}{s_{n-1}} = \frac{2^{-n}}{2^{-(n-1)}} = 2^{-n+(n-1)} = 2^{-1} = \frac{1}{2}$$

Because the ratio of successive terms is a nonzero constant, the sequence $\{s_n\}$ is geometric with common ratio $\dfrac{1}{2}$. ∎

EXAMPLE 3   Determining If a Sequence Is Geometric

Show that the following sequence is geometric. Find the first term and the common ratio.

$$\{t_n\} = \{4^n\}$$

Solution   The first term is $t_1 = 4^1 = 4$. The nth and $(n-1)$st terms are

$$t_n = 4^n \quad \text{and} \quad t_{n-1} = 4^{n-1}$$

Their ratio is

$$\frac{t_n}{t_{n-1}} = \frac{4^n}{4^{n-1}} = 4^{n-(n-1)} = 4$$

The sequence $\{t_n\}$ is a geometric sequence with common ratio 4. ■

NOW WORK PROBLEM **3**.

2 Suppose that a is the first term of a geometric sequence with common ratio $r \neq 0$. We seek a formula for the nth term a_n. To see the pattern, we write down the first few terms:

$$a_1 = 1a = ar^0$$
$$a_2 = ra_1 = ar^1$$
$$a_3 = ra_2 = r(ar) = ar^2$$
$$a_4 = ra_3 = r(ar^2) = ar^3$$
$$a_5 = ra_4 = r(ar^3) = ar^4$$
$$\vdots$$
$$a_n = ra_{n-1} = r(ar^{n-2}) = ar^{n-1}$$

We are led to the following result:

Theorem

nth Term of a Geometric Sequence

For a geometric sequence $\{a_n\}$ whose first term is a and whose common ratio is r, the nth term is determined by the formula

$$a_n = ar^{n-1}, \qquad r \neq 0 \tag{2}$$

■

EXAMPLE 4 **Finding a Particular Term of a Geometric Sequence**

(a) Find the ninth term of the geometric sequence: $\quad 10, 9, \dfrac{81}{10}, \dfrac{729}{100} \cdots$

(b) Find a recursive formula for this sequence.

Solution (a) The first term of this geometric sequence is $a = 10$ and the common ratio is $\dfrac{9}{10}$. (Use $\dfrac{9}{10}$, or $\dfrac{81/10}{9} = \dfrac{9}{10}$, or any two successive terms.) By formula (2), the nth term is

$$a_n = 10\left(\frac{9}{10}\right)^{n-1}$$

The ninth term is

$$a_9 = 10\left(\frac{9}{10}\right)^{9-1} = 10\left(\frac{9}{10}\right)^8 = 4.3046721$$

(b) The first term in the sequence is 10 and the common ratio is $r = \dfrac{9}{10}$.

Using formula (1), the recursive formula is $a_1 = 10, a_n = \dfrac{9}{10}a_{n-1}$.

EXPLORATION   Use a graphing utility to find the ninth term of the sequence given in Example 4. Use it to find the twentieth and fiftieth terms. Now use a graphing utility to graph the recursive formula found in Example 4(b). Conclude that the graph of the recursive formula behaves like the graph of an exponential function. How is r, the common ratio, related to a, the base of the exponential function $y = a^x$?

NOW WORK PROBLEMS 25 AND 33.

Adding the First n Terms of a Geometric Sequence

③ The next result gives us a formula for finding the sum of the first n terms of a geometric sequence.

Theorem

Sum of n Terms of a Geometric Sequence

Let $\{a_n\}$ be a geometric sequence with first term a and common ratio r where $r \neq 0, r \neq 1$. The sum S_n of the first n terms of $\{a_n\}$ is

$$S_n = a\frac{1 - r^n}{1 - r}, \qquad r \neq 0, 1 \tag{3}$$

Proof   The sum S_n of the first n terms of $\{a_n\} = \{ar^{n-1}\}$ is

$$S_n = a + ar + \cdots + ar^{n-1} \tag{4}$$

Multiply each side by r to obtain

$$rS_n = ar + ar^2 + \cdots + ar^n \tag{5}$$

Now, subtract (5) from (4). The result is

$$S_n - rS_n = a - ar^n$$
$$(1 - r)S_n = a(1 - r^n)$$

Since $r \neq 1$, we can solve for S_n.

$$S_n = a\frac{1 - r^n}{1 - r}$$

EXAMPLE 5   **Finding the Sum of n Terms of a Geometric Sequence**

Find the sum S_n of the first n terms of the sequence $\left\{ \left(\dfrac{1}{2}\right)^n \right\}$; that is, find

$$\frac{1}{2} + \frac{1}{4} + \frac{1}{8} + \cdots + \left(\frac{1}{2}\right)^n$$

Solution The sequence $\left\{\left(\dfrac{1}{2}\right)^n\right\}$ is a geometric sequence with $a = \dfrac{1}{2}$ and $r = \dfrac{1}{2}$. The sum S_n that we seek is the sum of the first n terms of the sequence, so we use formula (3) to get

$$S_n = \sum_{k=1}^{n}\left(\frac{1}{2}\right)^k = \frac{1}{2} + \frac{1}{4} + \frac{1}{8} + \cdots + \left(\frac{1}{2}\right)^n$$

$$= \frac{1}{2}\left[\frac{1 - \left(\dfrac{1}{2}\right)^n}{1 - \dfrac{1}{2}}\right] \qquad \text{Formula (3)}$$

$$= \frac{1}{2}\left[\frac{1 - \left(\dfrac{1}{2}\right)^n}{\dfrac{1}{2}}\right]$$

$$= 1 - \left(\frac{1}{2}\right)^n$$

NOW WORK PROBLEM 39.

EXAMPLE 6 **Using a Graphing Utility to Find the Sum of a Geometric Sequence**

Use a graphing utility to find the sum of the first 15 terms of the sequence $\left\{\left(\dfrac{1}{3}\right)^n\right\}$; that is, find

$$S_{15} = \frac{1}{3} + \frac{1}{9} + \frac{1}{27} + \cdots + \left(\frac{1}{3}\right)^{15}$$

Solution Figure 14 shows the result obtained using a TI-83 graphing calculator.

Figure 14

```
sum(seq((1/3)^n,
n,1,15,1))
          .4999999652
```

The sum of the first 15 terms of the sequence $\left\{\left(\dfrac{1}{3}\right)^n\right\}$ is 0.4999999652.

NOW WORK PROBLEM 45.

Geometric Series

An infinite sum of the form

$$a + ar + ar^2 + \cdots + ar^{n-1} + \cdots$$

with first term a and common ratio r, is called an **infinite geometric series** and is denoted by

$$\sum_{k=1}^{\infty} ar^{k-1}$$

④ Based on formula (3), the sum S_n of the first n terms of a geometric series is

$$S_n = a\frac{1 - r^n}{1 - r} = \frac{a}{1 - r} - \frac{ar^n}{1 - r} \tag{6}$$

If this finite sum S_n approaches a number L as $n \to \infty$, then we call L the **sum of the infinite geometric series**, and we write

$$L = \sum_{k=1}^{\infty} ar^{k-1}$$

Theorem

Sum of an Infinite Geometric Series

If $|r| < 1$, the sum of the infinite geometric series $\displaystyle\sum_{k=1}^{\infty} ar^{k-1}$ is

$$\sum_{k=1}^{\infty} ar^{k-1} = \frac{a}{1 - r} \tag{7}$$

Intuitive Proof Since $|r| < 1$, it follows that $|r^n|$ approaches 0 as $n \to \infty$.

Then, based on formula (6), the sum S_n approaches $\dfrac{a}{1 - r}$ as $n \to \infty$. ∎

EXAMPLE 7 **Finding the Sum of a Geometric Series**

Find the sum of the geometric series: $2 + \dfrac{4}{3} + \dfrac{8}{9} + \cdots$

Solution The first term is $a = 2$, and the common ratio is

$$r = \frac{\dfrac{4}{3}}{2} = \frac{4}{6} = \frac{2}{3}$$

Since $|r| < 1$, we use formula (7) to find that

$$2 + \frac{4}{3} + \frac{8}{9} + \cdots = \frac{2}{1 - \frac{2}{3}} = 6$$

 **NOW WORK PROBLEM 51.**

 EXPLORATION Use a graphing utility to graph $U_n = 2\left(\frac{2}{3}\right)^{n-1}$ in sequence mode. TRACE the graph for large values of n. What happens to the value of U_n as n increases without bound? What can you conclude about $\displaystyle\sum_{n=1}^{\infty} 2\left(\frac{2}{3}\right)^{(n-1)}$?

EXAMPLE 8 **Repeating Decimals**

Show that the repeating decimal $0.999\ldots$ equals 1.

Solution

$$0.999\ldots = \frac{9}{10} + \frac{9}{100} + \frac{9}{1000} + \cdots$$

The decimal, $0.999\ldots$ is a geometric series with first term $\frac{9}{10}$ and common ratio $\frac{1}{10}$. Using formula (7), we find

$$0.999\ldots = \frac{\dfrac{9}{10}}{1 - \dfrac{1}{10}} = \frac{\dfrac{9}{10}}{\dfrac{9}{10}} = 1$$

Figure 15

EXAMPLE 9 **Pendulum Swings**

Initially, a pendulum swings through an arc of 18 inches. See Figure 15. On each successive swing, the length of the arc is 0.98 of the previous length.

(a) What is the length of the arc after 10 swings?
(b) On which swing is the length of the arc first less than 12 inches?
(c) After 15 swings, what total distance will the pendulum have swung?
(d) When it stops, what total distance will the pendulum have swung?

Solution (a) The length of the first swing is 18 inches.
The length of the second swing is 0.98(18) inches.
The length of the third swing is $0.98(0.98)(18) = 0.98^2(18)$ inches.

The length of the arc of the tenth swing is

$$(0.98)^9(18) = 15.007 \text{ inches}$$

(b) The length of the arc of the nth swing is $(0.98)^{n-1}(18)$. For this to be exactly 12 inches requires that

$$(0.98)^{n-1}(18) = 12$$

$$(0.98)^{n-1} = \frac{12}{18} = \frac{2}{3} \qquad\qquad \textit{Divide both sides by 18.}$$

$$n - 1 = \log_{0.98}\left(\frac{2}{3}\right) \qquad\qquad \textit{Express as a logarithm}$$

$$n = 1 + \frac{\ln\left(\dfrac{2}{3}\right)}{\ln 0.98} \approx 1 + 20.07 \approx 21.07 \qquad \textit{Solve for n; Change of Base Formula}$$

The length of the arc of the pendulum exceeds 12 inches on the twenty-first swing and is first less than 12 inches on the twenty-second swing.

(c) After 15 swings, the pendulum will have swung the following total distance L:

$$L = \underset{\text{1st}}{18} + \underset{\text{2nd}}{0.98(18)} + \underset{\text{3rd}}{(0.98)^2(18)} + \underset{\text{4th}}{(0.98)^3(18)} + \cdots + \underset{\text{15th}}{(0.98)^{14}(18)}$$

This is the sum of a geometric sequence. The common ratio is 0.98; the first term is 18. The sum has 15 terms, so

$$L = 18\frac{1 - 0.98^{15}}{1 - 0.98} \approx 18(13.07) \approx 235.29 \text{ inches}$$

The pendulum will have swung through 235.29 inches after 15 swings.

(d) When the pendulum stops, it will have swung the following total distance T:

$$T = 18 + 0.98(18) + (0.98)^2(18) + (0.98)^3(18) + \cdots$$

This is the sum of a geometric series. The common ratio is $r = 0.98$; the first term is $a = 18$. The sum is

$$T = \frac{a}{1 - r} = \frac{18}{1 - 0.98} = 900$$

The pendulum will have swung a total of 900 inches when it finally stops.

HISTORICAL FEATURE

Fibonacci

Sequences are among the oldest objects of mathematical investigation, having been studied for over 3500 years. After the initial steps, however, little progress was made until about 1600.

Arithmetic and geometric sequences appear in the Rhind papyrus, a mathematical text containing 85 problems copied around 1650 BC by the Egyptian scribe Ahmes from an earlier work (see Historical Problem 1). Fibonacci (AD 1220) wrote about problems similar to those found in the Rhind papyrus, leading one to suspect that Fibonacci may have had material available that is now lost. This material would have been in the non-Euclidean Greek tradition of

Heron (about AD 75) and Diophantus (about AD 250). One problem, again modified slightly, is still with us in the familiar puzzle rhyme "As I was going to St. Ives ..." (see Historical Problem 2).

The Rhind papyrus indicates that the Egyptians knew how to add up the terms of an arithmetic or geometric sequence, as did the Babylonians. The rule for summing up a geometric sequence is found in Euclid's *Elements* (book IX, 35, 36), where, like all Euclid's algebra, it is presented in a geometric form.

Investigations of other kinds of sequences began in the 1500s, when algebra became sufficiently developed to handle the more complicated problems. The development of calculus in the 1600s added a powerful new tool, especially for finding the sum of infinite series, and the subject continues to flourish today.

HISTORICAL PROBLEMS

1. *Arithmetic sequence problem from the Rhind papyrus (statement modified slightly for clarity)* One hundred loaves of bread are to be divided among five people so that the amounts that they receive form an arithmetic sequence. The first two together receive one-seventh of what the last three receive. How many does each receive? [*Partial answer:* First person receives $1\frac{2}{3}$ loaves.]

2. The following old English children's rhyme resembles one of the Rhind papyrus problems.

   As I was going to St. Ives
   I met a man with seven wives
   Each wife had seven sacks

   Each sack had seven cats
   Each cat had seven kits [kittens]
   Kits, cats, sacks, wives
   How many were going to St. Ives?

   (a) Assuming that the speaker and the cat fanciers met by traveling in opposite directions, what is the answer?

   (b) How many kittens are being transported?

   (c) Kits, cats, sacks, wives; how many?

   [**Hint:** It is easier to include the man, find the sum with the formula, and then subtract 1 for the man.]

13.3 Concepts and Vocabulary

In Problems 1 and 2, fill in the blanks.

1. In a(n) _____ sequence the ratio of successive terms is a constant.

2. If $|r| < 1$, the sum of the geometric series $\sum\limits_{k=1}^{\infty} ar^{k-1}$ is _____.

In Problems 3–5, answer True or False to each statement.

3. A geometric sequence may be defined recursively.

4. In a geometric sequence the common ratio is a positive number.

5. For a geometric sequence with first term a and common ratio r, where $r \neq 0, r \neq 1$, the sum of the first n terms is
   $$S_n = a \cdot \frac{1 - r^n}{1 - r}.$$

6. How do you determine if a sequence might be geometric?

7. How do you determine if a geometric series has a sum?

13.3 Exercises

In Problems 1–10, a geometric sequence is given. Find the common ratio and write out the first four terms.

1. $\{3^n\}$

2. $\{(-5)^n\}$

3. $\left\{-3\left(\dfrac{1}{2}\right)^n\right\}$

4. $\left\{\left(\dfrac{5}{2}\right)^n\right\}$

5. $\left\{\dfrac{2^{n-1}}{4}\right\}$

6. $\left\{\dfrac{3^n}{9}\right\}$

7. $\{2^{n/3}\}$

8. $\{3^{2n}\}$

9. $\left\{\dfrac{3^{n-1}}{2^n}\right\}$

10. $\left\{\dfrac{2^n}{3^{n-1}}\right\}$

In Problems 11–24, determine whether the given sequence is arithmetic, geometric, or neither. If the sequence is arithmetic, find the common difference; if it is geometric, find the common ratio.

11. $\{n+2\}$

12. $\{2n-5\}$

13. $\{4n^2\}$

14. $\{5n^2+1\}$

15. $\left\{3-\dfrac{2}{3}n\right\}$

16. $\left\{8-\dfrac{3}{4}n\right\}$

17. $1,3,6,10,\ldots$

18. $2,4,6,8,\ldots$

19. $\left\{\left(\dfrac{2}{3}\right)^n\right\}$

20. $\left\{\left(\dfrac{5}{4}\right)^n\right\}$

21. $-1,-2,-4,-8,\ldots$

22. $1,1,2,3,5,8,\ldots$

23. $\{3^{n/2}\}$

24. $\{(-1)^n\}$

In Problems 25–32, find the fifth term and the nth term of the geometric sequence whose initial term a and common ratio r are given.

25. $a=2;\quad r=3$

26. $a=-2;\quad r=4$

27. $a=5;\quad r=-1$

28. $a=6;\quad r=-2$

29. $a=0;\quad r=\dfrac{1}{2}$

30. $a=1;\quad r=-\dfrac{1}{3}$

31. $a=\sqrt{2};\quad r=\sqrt{2}$

32. $a=0;\quad r=\dfrac{1}{\pi}$

In Problems 33–38, find the indicated term of each geometric sequence.

33. 7th term of $1,\dfrac{1}{2},\dfrac{1}{4},\ldots$

34. 8th term of $1,3,9,\ldots$

35. 9th term of $1,-1,1,\ldots$

36. 10th term of $-1,2,-4,\ldots$

37. 8th term of $0.4,0.04,0.004,\ldots$

38. 7th term of $0.1,1.0,10.0,\ldots$

In Problems 39–44, find the sum.

39. $\dfrac{1}{4}+\dfrac{2}{4}+\dfrac{2^2}{4}+\dfrac{2^3}{4}+\cdots+\dfrac{2^{n-1}}{4}$

40. $\dfrac{3}{9}+\dfrac{3^2}{9}+\dfrac{3^3}{9}+\cdots+\dfrac{3^n}{9}$

41. $\displaystyle\sum_{k=1}^{n}\left(\dfrac{2}{3}\right)^k$

42. $\displaystyle\sum_{k=1}^{n}4\cdot3^{k-1}$

43. $-1-2-4-8-\cdots-(2^{n-1})$

44. $2+\dfrac{6}{5}+\dfrac{18}{25}+\cdots+2\left(\dfrac{3}{5}\right)^n$

For Problems 45–50, use a graphing utility to find the sum of each geometric sequence.

45. $\dfrac{1}{4}+\dfrac{2}{4}+\dfrac{2^2}{4}+\dfrac{2^3}{4}+\cdots+\dfrac{2^{14}}{4}$

46. $\dfrac{3}{9}+\dfrac{3^2}{9}+\dfrac{3^3}{9}+\cdots+\dfrac{3^{15}}{9}$

47. $\displaystyle\sum_{n=1}^{15}\left(\dfrac{2}{3}\right)^n$

48. $\displaystyle\sum_{n=1}^{15}4\cdot3^{n-1}$

49. $-1-2-4-8-\cdots-2^{14}$

50. $2+\dfrac{6}{5}+\dfrac{18}{25}+\cdots+2\left(\dfrac{3}{5}\right)^{15}$

In Problems 51–60, find the sum of each infinite geometric series.

51. $1+\dfrac{1}{3}+\dfrac{1}{9}+\cdots$

52. $2+\dfrac{4}{3}+\dfrac{8}{9}+\cdots$

53. $8+4+2+\cdots$

54. $6+2+\dfrac{2}{3}+\cdots$

55. $2-\dfrac{1}{2}+\dfrac{1}{8}-\dfrac{1}{32}+\cdots$

56. $1-\dfrac{3}{4}+\dfrac{9}{16}-\dfrac{27}{64}+\cdots$

57. $\displaystyle\sum_{k=1}^{\infty}5\left(\dfrac{1}{4}\right)^{k-1}$

58. $\displaystyle\sum_{k=1}^{\infty}8\left(\dfrac{1}{3}\right)^{k-1}$

59. $\displaystyle\sum_{k=1}^{\infty}6\left(-\dfrac{2}{3}\right)^{k-1}$

60. $\displaystyle\sum_{k=1}^{\infty}4\left(-\dfrac{1}{2}\right)^{k-1}$

61. Find x so that $x, x + 2,$ and $x + 3$ are consecutive terms of a geometric sequence.

62. Find x so that $x - 1, x,$ and $x + 2$ are consecutive terms of a geometric sequence.

63. Salary Increases   Suppose that you have just been hired at an annual salary of $18,000 and expect to receive annual increases of 5%. What will your salary be when you begin your fifth year?

64. Equipment Depreciation   A new piece of equipment cost a company $15,000. Each year, for tax purposes, the company depreciates the value by 15%. What value should the company give the equipment after 5 years?

65. Pendulum Swings   Initially, a pendulum swings through an arc of 2 feet. On each successive swing, the length of the arc is 0.9 of the previous length.
  (a) What is the length of the arc after 10 swings?
  (b) On which swing is the length of the arc first less than 1 foot?
  (c) After 15 swings, what total length will the pendulum have swung?
  (d) When it stops, what total length will the pendulum have swung?

66. Bouncing Balls   A ball is dropped from a height of 30 feet. Each time it strikes the ground, it bounces up to 0.8 of the previous height.

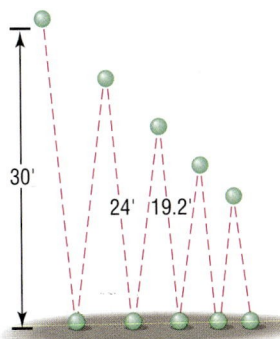

  (a) What height will the ball bounce up to after it strikes the ground for the third time?
  (b) How high will it bounce after it strikes the ground for the nth time?
  (c) How many times does the ball need to strike the ground before its bounce is less than 6 inches?
  (d) What total distance does the ball travel before it stops bouncing?

67. Critical Thinking   You are interviewing for a job and receive two offers:
  A: $20,000 to start, with guaranteed annual increases of 6% for the first 5 years
  B: $22,000 to start, with guaranteed annual increases of 3% for the first 5 years
Which offer is best if your goal is to be making as much as possible after 5 years? Which is best if your goal is to make as much money as possible over the contract (5 years)?

68. Critical Thinking   Which of the following choices, A or B, results in more money?
  A: To receive $1000 on day 1, $999 on day 2, $998 on day 3, with the process to end after 1000 days
  B: To receive $1 on day 1, $2 on day 2, $4 on day 3, for 19 days

69. Critical Thinking   You have just signed a 7-year professional football league contract with a beginning salary of $2,000,000 per year. Management gives you the following options with regard to your salary over the 7 years.
  1. A bonus of $100,000 each year
  2. An annual increase of 4.5% per year beginning after 1 year
  3. An annual increase of $95,000 per year beginning after 1 year
Which option provides the most money over the 7-year period? Which the least? Which would you choose? Why?

70. A Rich Man's Promise   A rich man promises to give you $1000 on September 1, 2001. Each day thereafter he will give you $\dfrac{9}{10}$ of what he gave you the previous day. What is the first date on which the amount you receive is less than 1¢? How much have you received when this happens?

71. Grains of Wheat on a Chess Board   In an old fable, a commoner who had just saved the king's life was told he could ask the king for any just reward. Being a shrewd man, the commoner said, "A simple wish, sire. Place one grain of wheat on the first square of a chessboard, two grains on the second square, four grains on the third square, continuing until you have filled the board. This is all I seek." Compute the total number of grains needed to do this to see why the request, seemingly simple, could not be granted. (A chessboard consists of $8 \times 8 = 64$ squares.)

72. Look at the figure below. What fraction of the square is eventually shaded if the indicated shading process continues indefinitely?

73. Multiplier   Suppose that, throughout the U.S. economy, individuals spend 90% of every additional dollar that they earn. Economists would say that an individual's **marginal propensity to consume** is 0.90. For example, if Jane earns an additional dollar, she will spend $0.9(1) = \$0.90$ of it. The individual that earns \$0.90 (from Jane) will spend 90% of it or \$0.81. This process of spending continues and results in an infinite geometric series as follows:

$$1, 0.90, 0.90^2, 0.90^3, 0.90^4, \ldots$$

The sum of this infinite geometric series is called the **multiplier**. What is the multiplier if individuals spend 90% of every additional dollar that they earn?

74. Multiplier   Refer to Problem 73. Suppose that the marginal propensity to consume throughout the U.S. economy is 0.95. What is the multiplier for the U.S. economy?

75. Stock Price   One method of pricing a stock is to discount the stream of future dividends of the stock. Suppose that a stock pays $\$P$ per year in dividends and, historically, the dividend has been increased $i\%$ per year. If you desire an annual rate of return of $r\%$, this method of pricing a stock states that the price that you should pay is the present value of an infinite stream of payments:

$$\text{Price} = P + P\frac{1+i}{1+r} + P\left(\frac{1+i}{1+r}\right)^2 + P\left(\frac{1+i}{1+r}\right)^3 + \ldots$$

The price of the stock is the sum of an infinite geometric series. Suppose that a stock pays an annual dividend of \$4.00 and, historically, the dividend has been increased 3% per year. You desire an annual rate of return of 9%. What is the most you should pay for the stock?

76. Stock Price   Refer to Problem 75. Suppose that a stock pays an annual dividend of \$2.50 and, historically, the dividend has increased 4% per year. You desire an annual rate of return of 11%. What is the most that you should pay for the stock?

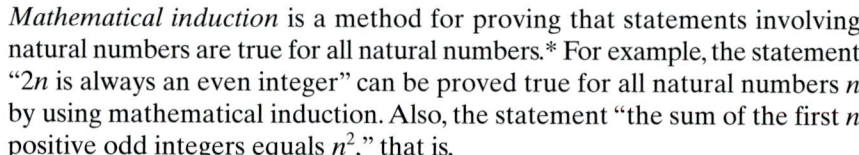 **77.** Can a sequence be both arithmetic and geometric? Give reasons for your answer.

78. Make up a geometric sequence. Give it to a friend and ask for its 20th term.

79. Make up two infinite geometric series, one that has a sum and one that does not. Give them to a friend and ask for the sum of each series.

13.4 MATHEMATICAL INDUCTION

OBJECTIVES   1   Prove Statements Using Mathematical Induction

1   *Mathematical induction* is a method for proving that statements involving natural numbers are true for all natural numbers.* For example, the statement "2*n* is always an even integer" can be proved true for all natural numbers *n* by using mathematical induction. Also, the statement "the sum of the first *n* positive odd integers equals n^2," that is,

$$1 + 3 + 5 + \cdots + (2n - 1) = n^2 \tag{1}$$

can be proved for all natural numbers *n* by using mathematical induction.

Before stating the method of mathematical induction, let's try to gain a sense of the power of the method. We shall use the statement in equation (1) for this purpose by restating it for various values of $n = 1, 2, 3, \ldots$.

$n = 1$ The sum of the first positive odd integer is 1^2; $1 = 1^2$.

$n = 2$ The sum of the first 2 positive odd integers is 2^2; $1 + 3 = 4 = 2^2$.

$n = 3$ The sum of the first 3 positive odd integers is 3^2; $1 + 3 + 5 = 9 = 3^2$.

$n = 4$ The sum of the first 4 positive odd integers is 4^2; $1 + 3 + 5 + 7 = 16 = 4^2$.

Although from this pattern we might conjecture that statement (1) is true for any choice of *n*, can we really be sure that it does not fail for some choice of *n*? The method of proof by mathematical induction will, in fact, prove that the statement is true for all *n*.

---

*Recall that the natural numbers are the numbers $1, 2, 3, 4, \ldots$. In other words, the terms *natural numbers* and *positive integers* are synonymous.

Theorem

The Principle of Mathematical Induction

Suppose that the following two conditions are satisfied with regard to a statement about natural numbers:

CONDITION I: The statement is true for the natural number 1.

CONDITION II: If the statement is true for some natural number k, it is also true for the next natural number $k + 1$.

Then the statement is true for all natural numbers.

We shall not prove this principle. However, we can provide a physical interpretation that will help us to see why the principle works. Think of a collection of natural numbers obeying a statement as a collection of infinitely many dominoes (see Figure 16).

Figure 16

Now, suppose that we are told two facts:

1. The first domino is pushed over.
2. If one domino falls over, say the kth domino, then so will the next one, the $(k + 1)$st domino.

Is it safe to conclude that *all* the dominoes fall over? The answer is yes, because if the first one falls (Condition I), then the second one does also (by Condition II); and if the second one falls, then so does the third (by Condition II); and so on.

Now let's prove some statements about natural numbers using mathematical induction.

EXAMPLE 1   **Using Mathematical Induction**

Show that the following statement is true for all natural numbers n.

$$1 + 3 + 5 + \cdots + (2n - 1) = n^2 \qquad (2)$$

Solution   We need to show first that statement (2) holds for $n = 1$. Because $1 = 1^2$, statement (2) is true for $n = 1$. Condition I holds.

Next, we need to show that Condition II holds. Suppose that we know for some k that

$$1 + 3 + \cdots + (2k - 1) = k^2 \qquad (3)$$

We wish to show that, based on equation (3), statement (2) holds for $k + 1$. We look at the sum of the first $k + 1$ positive odd integers to determine whether this sum equals $(k + 1)^2$.

$$1 + 3 + \cdots + (2k - 1) + (2k + 1) = \underbrace{[1 + 3 + \cdots + (2k - 1)]}_{= \, k^2 \text{ by equation (3)}} + (2k + 1)$$
$$= k^2 + (2k + 1)$$
$$= k^2 + 2k + 1 = (k + 1)^2$$

Conditions I and II are satisfied; by the Principle of Mathematical Induction, statement (2) is true for all natural numbers n.

EXAMPLE 2   Using Mathematical Induction

Show that the following statement is true for all natural numbers n.

$$2^n > n$$

Solution   First, we show that the statement $2^n > n$ holds when $n = 1$. Because $2^1 = 2 > 1$, the inequality is true for $n = 1$. Condition I holds.

Next, we assume, for some natural number k, that $2^k > k$. We wish to show that the formula holds for $k + 1$; that is, we wish to show that $2^{k+1} > k + 1$. Now,

$$2^{k+1} = 2 \cdot 2^k > 2 \cdot k = k + k \geq k + 1$$

$$\underset{\substack{\uparrow \\ \text{We know that} \\ 2^k > k.}}{} \qquad \underset{\substack{\uparrow \\ k \geq 1}}{}$$

If $2^k > k$, then $2^{k+1} > k + 1$, so Condition II of the Principle of Mathematical Induction is satisfied. The statement $2^n > n$ is true for all natural numbers n.   ∎

EXAMPLE 3   Using Mathematical Induction

Show that the following formula is true for all natural numbers n.

$$1 + 2 + 3 + \cdots + n = \frac{n(n + 1)}{2} \tag{4}$$

Solution   First, we show that formula (4) is true when $n = 1$. Because

$$\frac{1(1 + 1)}{2} = \frac{1(2)}{2} = 1$$

Condition I of the Principle of Mathematical Induction holds.

Next, we assume that formula (4) holds for some k, and we determine whether the formula then holds for $k + 1$. We assume that

$$1 + 2 + 3 + \cdots + k = \frac{k(k + 1)}{2} \qquad \text{for some } k \tag{5}$$

Now we need to show that

$$1 + 2 + 3 + \cdots + k + (k + 1) = \frac{(k + 1)(k + 1 + 1)}{2} = \frac{(k + 1)(k + 2)}{2}$$

We do this as follows:

$$1 + 2 + 3 + \cdots + k + (k + 1) = \underbrace{[1 + 2 + 3 + \cdots + k]}_{\substack{= \frac{k(k+1)}{2} \quad \text{by equation (5)}}} + (k + 1)$$

$$= \frac{k(k + 1)}{2} + (k + 1)$$

$$= \frac{k^2 + k + 2k + 2}{2}$$

$$= \frac{k^2 + 3k + 2}{2} = \frac{(k + 1)(k + 2)}{2}$$

Condition II also holds. As a result, formula (4) is true for all natural numbers n.   ∎

═══ NOW WORK PROBLEM **1**.

EXAMPLE 4   **Using Mathematical Induction**

Show that $3^n - 1$ is divisible by 2 for all natural numbers n.

Solution   First, we show that the statement is true when $n = 1$. Because $3^1 - 1 = 3 - 1 = 2$ is divisible by 2, the statement is true when $n = 1$. Condition I is satisfied.

Next, we assume that the statement holds for some k, and we determine whether the statement then holds for $k + 1$. We assume that $3^k - 1$ is divisible by 2 for some k. We need to show that $3^{k+1} - 1$ is divisible by 2. Now

$$3^{k+1} - 1 = 3^{k+1} - 3^k + 3^k - 1 \qquad \textcolor{teal}{\textit{Subtract and add } 3^k}$$
$$= 3^k(3 - 1) + (3^k - 1) = 3^k \cdot 2 + (3^k - 1)$$

Because $3^k \cdot 2$ is divisible by 2 and $3^k - 1$ is divisible by 2, it follows that $3^k \cdot 2 + (3^k - 1) = 3^{k+1} - 1$ is divisible by 2. Condition II is also satisfied. As a result, the statement "$3^n - 1$ is divisible by 2" is true for all natural numbers n. ■

WARNING: The conclusion that a statement involving natural numbers is true for all natural numbers is made only after *both* Conditions I and II of the Principle of Mathematical Induction have been satisfied. Problem 27 demonstrates a statement for which only Condition I holds, but the statement is not true for all natural numbers. Problem 28 demonstrates a statement for which only Condition II holds, but the statement is *not* true for any natural number. ■

13.4 Concepts and Vocabulary

In Problem 1, fill in the blanks.

1. _____ _____ can sometimes be used to prove theorems involving natural numbers.

13.4 Exercises

In Problems 1–26, use the Principle of Mathematical Induction to show that the given statement is true for all natural numbers n.

1. $2 + 4 + 6 + \cdots + 2n = n(n + 1)$

2. $1 + 5 + 9 + \cdots + (4n - 3) = n(2n - 1)$

3. $3 + 4 + 5 + \cdots + (n + 2) = \dfrac{1}{2}n(n + 5)$

4. $3 + 5 + 7 + \cdots + (2n + 1) = n(n + 2)$

5. $2 + 5 + 8 + \cdots + (3n - 1) = \dfrac{1}{2}n(3n + 1)$

6. $1 + 4 + 7 + \cdots + (3n - 2) = \dfrac{1}{2}n(3n - 1)$

7. $1 + 2 + 2^2 + \cdots + 2^{n-1} = 2^n - 1$

8. $1 + 3 + 3^2 + \cdots + 3^{n-1} = \dfrac{1}{2}(3^n - 1)$

9. $1 + 4 + 4^2 + \cdots + 4^{n-1} = \dfrac{1}{3}(4^n - 1)$

10. $1 + 5 + 5^2 + \cdots + 5^{n-1} = \dfrac{1}{4}(5^n - 1)$

11. $\dfrac{1}{1 \cdot 2} + \dfrac{1}{2 \cdot 3} + \dfrac{1}{3 \cdot 4} + \cdots + \dfrac{1}{n(n + 1)} = \dfrac{n}{n + 1}$

12. $\dfrac{1}{1 \cdot 3} + \dfrac{1}{3 \cdot 5} + \dfrac{1}{5 \cdot 7} + \cdots + \dfrac{1}{(2n - 1)(2n + 1)} = \dfrac{n}{2n + 1}$

13. $1^2 + 2^2 + 3^2 + \cdots + n^2 = \dfrac{1}{6}n(n + 1)(2n + 1)$

14. $1^3 + 2^3 + 3^3 + \cdots + n^3 = \dfrac{1}{4}n^2(n + 1)^2$

15. $4 + 3 + 2 + \cdots + (5 - n) = \dfrac{1}{2}n(9 - n)$

16. $-2 - 3 - 4 - \cdots - (n + 1) = -\dfrac{1}{2}n(n + 3)$

17. $1 \cdot 2 + 2 \cdot 3 + 3 \cdot 4 + \cdots + n(n + 1) = \dfrac{1}{3}n(n + 1)(n + 2)$

18. $1 \cdot 2 + 3 \cdot 4 + 5 \cdot 6 + \cdots + (2n - 1)(2n) = \dfrac{1}{3}n(n + 1)(4n - 1)$

19. $n^2 + n$ is divisible by 2.

20. $n^3 + 2n$ is divisible by 3.

21. $n^2 - n + 2$ is divisible by 2.

22. $n(n + 1)(n + 2)$ is divisible by 6.

23. If $x > 1$, then $x^n > 1$.

24. If $0 < x < 1$, then $0 < x^n < 1$.

25. $a - b$ is a factor of $a^n - b^n$.

26. $a + b$ is a factor of $a^{2n+1} + b^{2n+1}$.

[**Hint:** $a^{k+1} - b^{k+1} = a(a^k - b^k) + b^k(a - b)$]

---

27. Show that the statement "$n^2 - n + 41$ is a prime number" is true for $n = 1$, but is not true for $n = 41$.

28. Show that the formula

$$2 + 4 + 6 + \cdots + 2n = n^2 + n + 2$$

obeys Condition II of the Principle of Mathematical Induction. That is, show that if the formula is true for some k it is also true for $k + 1$. Then show that the formula is false for $n = 1$ (or for any other choice of n).

29. Use mathematical induction to prove that if $r \neq 1$ then

$$a + ar + ar^2 + \cdots + ar^{n-1} = a\frac{1 - r^n}{1 - r}$$

30. Use mathematical induction to prove that

$$a + (a + d) + (a + 2d) + \cdots$$
$$+ [a + (n - 1)d] = na + d\frac{n(n - 1)}{2}$$

31. Extended Principle of Mathematical Induction The Extended Principle of Mathematical Induction states that if Conditions I and II hold, that is,

(I)   A statement is true for a natural number j.

(II) If the statement is true for some natural number $k \geq j$, then it is also true for the next natural number $k + 1$.

then the statement is true for all natural numbers $\geq j$.

Use the Extended Principle of Mathematical Induction to show that the number of diagonals in a convex polygon of n sides is $\dfrac{1}{2}n(n - 3)$.

[**Hint:** Begin by showing that the result is true when $n = 4$ (Condition I).]

32. Geometry Use the Extended Principle of Mathematical Induction to show that the sum of the interior angles of a convex polygon of n sides equals $(n - 2) \cdot 180°$.

  33. How would you explain to a friend the Principle of Mathematical Induction?

13.5 THE BINOMIAL THEOREM

OBJECTIVES   **1** Evaluate a Binomial Coefficient

2 Expand a Binomial

Formulas have been given for expanding $(x + a)^n$ for $n = 2$ and $n = 3$. The *Binomial Theorem** is a formula for the expansion of $(x + a)^n$ for any positive integer n. If $n = 1, 2, 3,$ and 4, the expansion of $(x + a)^n$ is straightforward.

$$(x + a)^1 = x + a$$
Two terms, beginning with x^1 and ending with a^1

$$(x + a)^2 = x^2 + 2ax + a^2$$
Three terms, beginning with x^2 and ending with a^2

$$(x + a)^3 = x^3 + 3ax^2 + 3a^2x + a^3$$
Four terms, beginning with x^3 and ending with a^3

$$(x + a)^4 = x^4 + 4ax^3 + 6a^2x^2 + 4a^3x + a^4$$
Five terms, beginning with x^4 and ending with a^4

*The name *binomial* is derived from the fact that $x + a$ is a binomial, that is, it contains two terms.

Notice that each expansion of $(x + a)^n$ begins with x^n and ends with a^n. As you read from left to right, the powers of x are decreasing, while the powers of a are increasing. Also, the number of terms that appears equals $n + 1$. Notice, too, that the degree of each monomial in the expansion equals n. For example, in the expansion of $(x + a)^3$, each monomial $(x^3, 3ax^2, 3a^2x, a^3)$ is of degree 3. As a result, we might conjecture that the expansion of $(x + a)^n$ would look like this:

$$(x + a)^n = x^n + \_\_ ax^{n-1} + \_\_ a^2x^{n-2} + \cdots + \_\_ a^{n-1}x + a^n$$

where the blanks are numbers to be found. This is, in fact, the case, as we shall see shortly.

First, we need to introduce a symbol.

The Symbol $\dbinom{n}{j}$

1   We define the symbol $\dbinom{n}{j}$, read "n taken j at a time," as follows:

If j and n are integers with $0 \leq j \leq n$, the symbol $\dbinom{n}{j}$ is defined as

$$\binom{n}{j} = \frac{n!}{j!(n - j)!} \tag{1}$$

COMMENT   On a graphing calculator, the symbol $\dbinom{n}{j}$ may be denoted by the key $\boxed{\text{nCr}}$.   ∎

EXAMPLE 1   Evaluating $\dbinom{n}{j}$

Find:

(a) $\dbinom{3}{1}$ (b) $\dbinom{4}{2}$ (c) $\dbinom{8}{7}$ (d) $\dbinom{65}{15}$

Solution   (a) $\dbinom{3}{1} = \dfrac{3!}{1!(3 - 1)!} = \dfrac{3!}{1!2!} = \dfrac{3 \cdot 2 \cdot 1}{1(2 \cdot 1)} = \dfrac{6}{2} = 3$

(b) $\dbinom{4}{2} = \dfrac{4!}{2!(4 - 2)!} = \dfrac{4!}{2!2!} = \dfrac{4 \cdot 3 \cdot 2 \cdot 1}{(2 \cdot 1)(2 \cdot 1)} = \dfrac{24}{4} = 6$

(c) $\dbinom{8}{7} = \dfrac{8!}{7!(8 - 7)!} = \dfrac{8!}{7!1!} = \dfrac{8 \cdot \cancel{7!}}{\underset{\underset{8! \,=\, 8 \cdot 7!}{\uparrow}}{\cancel{7!} \cdot 1!}} = \dfrac{8}{1} = 8$

Figure 17

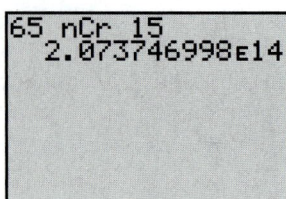

(d) Figure 17 shows the solution using a TI-83 graphing calculator:
$\binom{65}{15} = 2.073746998 \times 10^{14}$.

NOW WORK PROBLEM **1**.

Four useful formulas involving the symbol $\binom{n}{j}$ are

$$\binom{n}{0} = 1 \qquad \binom{n}{1} = n \qquad \binom{n}{n-1} = n \qquad \binom{n}{n} = 1$$

Proof

$$\binom{n}{0} = \frac{n!}{0!(n-0)!} = \frac{n!}{0!n!} = \frac{1}{1} = 1$$

$$\binom{n}{1} = \frac{n!}{1!(n-1)!} = \frac{n!}{(n-1)!} = \frac{n(n-1)!}{(n-1)!} = n$$

You are asked to show the remaining two formulas in Problem 41.

Suppose that we arrange the various values of the symbol $\binom{n}{j}$ in a triangular display, as shown next and in Figure 18.

$$\binom{0}{0}$$

$$\binom{1}{0} \quad \binom{1}{1}$$

$$\binom{2}{0} \quad \binom{2}{1} \quad \binom{2}{2}$$

$$\binom{3}{0} \quad \binom{3}{1} \quad \binom{3}{2} \quad \binom{3}{3}$$

$$\binom{4}{0} \quad \binom{4}{1} \quad \binom{4}{2} \quad \binom{4}{3} \quad \binom{4}{4}$$

$$\binom{5}{0} \quad \binom{5}{1} \quad \binom{5}{2} \quad \binom{5}{3} \quad \binom{5}{4} \quad \binom{5}{5}$$

Figure 18
Pascal triangle

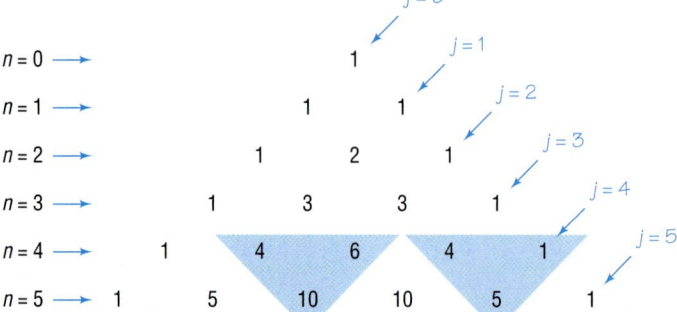

This display is called the **Pascal triangle**, named after Blaise Pascal (1623–1662), a French mathematician.

The Pascal triangle has 1's down the sides. To get any other entry, merely add the two nearest entries in the row above it. The shaded triangles in Figure 18 illustrate this feature of the Pascal triangle. Based on this feature, the row corresponding to $n = 6$ is found as follows:

$$n = 5 \rightarrow \quad 1 \quad 5 \quad 10 \quad 10 \quad 5 \quad 1$$
$$n = 6 \rightarrow \quad 1 \quad 6 \quad 15 \quad 20 \quad 15 \quad 6 \quad 1$$

Later we shall prove that this addition always works (see the theorem on page 1023).

Although the Pascal triangle provides an interesting and organized display of the symbol $\binom{n}{j}$, in practice it is not all that helpful. For example, if you wanted to know the value of $\binom{12}{5}$, you would need to produce 12 rows of the triangle before seeing the answer. It is much faster to use the definition (1).

Binomial Theorem

2 Now we are ready to state the **Binomial Theorem**. You are asked to prove this result in Problem 47.

Theorem

Binomial Theorem

Let x and a be real numbers. For any positive integer n, we have

$$(x + a)^n = \binom{n}{0}x^n + \binom{n}{1}ax^{n-1} + \cdots + \binom{n}{j}a^j x^{n-j} + \cdots + \binom{n}{n}a^n$$

$$= \sum_{j=0}^{n}\binom{n}{j}x^{n-j}a^j \qquad (2)$$

Now you know why we needed to introduce the symbol $\binom{n}{j}$; these symbols are the numerical coefficients that appear in the expansion of $(x + a)^n$. Because of this, the symbol $\binom{n}{j}$ is called the **binomial coefficient**.

EXAMPLE 2   **Expanding a Binomial**

Use the Binomial Theorem to expand $(x + 2)^5$.

Solution   In the Binomial Theorem, let $a = 2$ and $n = 5$. Then

$$(x + 2)^5 = \binom{5}{0}x^5 + \binom{5}{1}2x^4 + \binom{5}{2}2^2x^3 + \binom{5}{3}2^3x^2 + \binom{5}{4}2^4x + \binom{5}{5}2^5$$

Use equation (2).

$$= 1 \cdot x^5 + 5 \cdot 2x^4 + 10 \cdot 4x^3 + 10 \cdot 8x^2 + 5 \cdot 16x + 1 \cdot 32$$

Use row $n = 5$ of the Pascal triangle or formula (1) for $\binom{n}{j}$.

$$= x^5 + 10x^4 + 40x^3 + 80x^2 + 80x + 32 \qquad \blacksquare$$

EXAMPLE 3 **Expanding a Binomial**

Expand $(2y - 3)^4$ using the Binomial Theorem.

Solution First, we rewrite the expression $(2y - 3)^4$ as $[2y + (-3)]^4$. Now we use the Binomial Theorem with $n = 4$, $x = 2y$, and $a = -3$.

$$[2y + (-3)]^4 = \binom{4}{0}(2y)^4 + \binom{4}{1}(-3)(2y)^3 + \binom{4}{2}(-3)^2(2y)^2$$

$$+ \binom{4}{3}(-3)^3(2y) + \binom{4}{4}(-3)^4$$

$$= 1 \cdot 16y^4 + 4(-3)8y^3 + 6 \cdot 9 \cdot 4y^2 + 4(-27)2y + 1 \cdot 81$$

Use row $n = 4$ of the Pascal triangle or formula (1) for $\binom{n}{j}$.

$$= 16y^4 - 96y^3 + 216y^2 - 216y + 81$$

In this expansion, note that the signs alternate due to the fact that $a = -3 < 0$. $\qquad \blacksquare$

═══ NOW WORK PROBLEM **17**.

EXAMPLE 4 **Finding a Particular Coefficient in a Binomial Expansion**

Find the coefficient of y^8 in the expansion of $(2y + 3)^{10}$.

Solution We write out the expansion using the Binomial Theorem.

$$(2y + 3)^{10} = \binom{10}{0}(2y)^{10} + \binom{10}{1}(2y)^9(3)^1 + \binom{10}{2}(2y)^8(3)^2 + \binom{10}{3}(2y)^7(3)^3$$

$$+ \binom{10}{4}(2y)^6(3)^4 + \cdots + \binom{10}{9}(2y)(3)^9 + \binom{10}{10}(3)^{10}$$

From the third term in the expansion, the coefficient of y^8 is

$$\binom{10}{2}(2)^8(3)^2 = \frac{10!}{2!8!} \cdot 2^8 \cdot 9 = \frac{10 \cdot 9 \cdot 8!}{2 \cdot 8!} \cdot 2^8 \cdot 9 = 103,680 \qquad \blacksquare$$

As this solution demonstrates, we can use the Binomial Theorem to find a particular term in an expansion without writing the entire expansion.

Based on the expansion of $(x + a)^n$, the term containing x^j is

$$\binom{n}{n - j} a^{n-j} x^j \tag{3}$$

For example, we can solve Example 4 by using formula (3) with $n = 10$, $a = 3$, $x = 2y$, and $j = 8$. Then the term containing y^8 is

$$\binom{10}{10 - 8} 3^{10-8} (2y)^8 = \binom{10}{2} \cdot 3^2 \cdot 2^8 \cdot y^8 = \frac{10!}{2!8!} \cdot 9 \cdot 2^8 y^8$$

$$= \frac{10 \cdot 9 \cdot 8!}{2 \cdot 8!} \cdot 9 \cdot 2^8 y^8 = 103{,}680 y^8$$

EXAMPLE 5   Finding a Particular Term in a Binomial Expansion

Find the sixth term in the expansion of $(x + 2)^9$.

Solution A   We expand using the Binomial Theorem until the sixth term is reached.

$$(x + 2)^9 = \binom{9}{0} x^9 + \binom{9}{1} x^8 \cdot 2 + \binom{9}{2} x^7 \cdot 2^2 + \binom{9}{3} x^6 \cdot 2^3 + \binom{9}{4} x^5 \cdot 2^4$$

$$+ \binom{9}{5} x^4 \cdot 2^5 + \cdots$$

The sixth term is

$$\binom{9}{5} x^4 \cdot 2^5 = \frac{9!}{5!4!} \cdot x^4 \cdot 32 = 4032 x^4$$

Solution B   The sixth term in the expansion of $(x + 2)^9$, which has 10 terms total, contains x^4. (Do you see why?) By formula (3), the sixth term is

$$\binom{9}{9 - 4} 2^{9-4} x^4 = \binom{9}{5} 2^5 x^4 = \frac{9!}{5!4!} \cdot 32 x^4 = 4032 x^4 \qquad ■$$

✏️ ━━━ **NOW WORK PROBLEMS 25 AND 31.**

Next we show that the *triangular addition* feature of the Pascal triangle illustrated in Figure 18 always works.

Theorem   If n and j are integers with $1 \leq j \leq n$, then

$$\binom{n}{j - 1} + \binom{n}{j} = \binom{n + 1}{j} \tag{4}$$

Proof

$$\binom{n}{j-1} + \binom{n}{j} = \frac{n!}{(j-1)![n-(j-1)]!} + \frac{n!}{j!(n-j)!}$$

$$= \frac{n!}{(j-1)!(n-j+1)!} + \frac{n!}{j!(n-j)!}$$

$$= \frac{jn!}{j(j-1)!(n-j+1)!} + \frac{(n-j+1)n!}{j!(n-j+1)(n-j)!}$$

$$= \frac{jn!}{j!(n-j+1)!} + \frac{(n-j+1)n!}{j!(n-j+1)!}$$

$$= \frac{jn! + (n-j+1)n!}{j!(n-j+1)!}$$

$$= \frac{n!(j+n-j+1)}{j!(n-j+1)!}$$

$$= \frac{n!(n+1)}{j!(n-j+1)!} = \frac{(n+1)!}{j![(n+1)-j]!} = \binom{n+1}{j}$$

Multiply the first term by $\frac{j}{j}$ and the second term by $\frac{n-j+1}{n-j+1}$.

Now the denominators are equal

HISTORICAL FEATURE

Omar Khayyám (1050–1123)

The case $n = 2$ of the Binomial Theorem, $(a + b)^2$, was known to Euclid in 300 BC, but the general law seems to have been discovered by the Persian mathematician and astronomer Omar Khayyám (1050–1123), who is also well known as the author of the *Rubáiyát*, a collection of four-line poems making observations on the human condition. Omar Khayyám did not state the Binomial Theorem explicitly, but he claimed to have a method for extracting third, fourth, fifth roots, and so on. A little study shows that one must know the Binomial Theorem to create such a method.

The heart of the Binomial Theorem is the formula for the numerical coefficients, and, as we saw, they can be written out in a symmetric triangular form. The Pascal triangle appears first in the books of Yang Hui (about 1270) and Chu Shih-chieh (1303). Pascal's name is attached to the triangle because of the many applications he made of it, especially to counting and probability. In establishing these results, he was one of the earliest users of mathematical induction.

Many people worked on the proof of the Binomial Theorem, which was finally completed for all n (including complex numbers) by Niels Abel (1802–1829).

13.5 Concepts and Vocabulary

In Problems 1 and 2, fill in the blanks.

1. The _____ _____ is a triangular display of the binomial coefficients.

2. $\binom{6}{2} = $ _____ .

In Problems 3 and 4, answer True or False to each statement.

3. $\binom{n}{j} = \dfrac{j!}{(n-j)!n!}$

4. What do you do first if asked to expand $(x + 5)^4$?

5. Write down the first four rows of the Pascal triangle.

13.5 Exercises

In Problems 1–12, evaluate each expression by hand. Use a graphing utility to verify your answers.

1. $\dbinom{5}{3}$ **2.** $\dbinom{7}{3}$ **3.** $\dbinom{7}{5}$ **4.** $\dbinom{9}{7}$

5. $\dbinom{50}{49}$ **6.** $\dbinom{100}{98}$ **7.** $\dbinom{1000}{1000}$ **8.** $\dbinom{1000}{0}$

9. $\dbinom{55}{23}$ **10.** $\dbinom{60}{20}$ **11.** $\dbinom{47}{25}$ **12.** $\dbinom{37}{19}$

In Problems 13–24, expand each expression using the Binomial Theorem.

13. $(x + 1)^5$ **14.** $(x - 1)^5$ **15.** $(x - 2)^6$ **16.** $(x + 3)^5$

17. $(3x + 1)^4$ **18.** $(2x + 3)^5$ **19.** $(x^2 + y^2)^5$ **20.** $(x^2 - y^2)^6$

21. $(\sqrt{x} + \sqrt{2})^6$ **22.** $(\sqrt{x} - \sqrt{3})^4$ **23.** $(ax + by)^5$ **24.** $(ax - by)^4$

In Problems 25–38, use the Binomial Theorem to find the indicated coefficient or term.

25. The coefficient of x^6 in the expansion of $(x + 3)^{10}$

26. The coefficient of x^3 in the expansion of $(x - 3)^{10}$

27. The coefficient of x^7 in the expansion of $(2x - 1)^{12}$

28. The coefficient of x^3 in the expansion of $(2x + 1)^{12}$

29. The coefficient of x^7 in the expansion of $(2x + 3)^9$

30. The coefficient of x^2 in the expansion of $(2x - 3)^9$

31. The fifth term in the expansion of $(x + 3)^7$

32. The third term in the expansion of $(x - 3)^7$

33. The third term in the expansion of $(3x - 2)^9$

34. The sixth term in the expansion of $(3x + 2)^8$

35. The coefficient of x^0 in the expansion of $\left(x^2 + \dfrac{1}{x} \right)^{12}$

36. The coefficient of x^0 in the expansion of $\left(x - \dfrac{1}{x^2} \right)^9$

37. The coefficient of x^4 in the expansion of $\left(x - \dfrac{2}{\sqrt{x}} \right)^{10}$

38. The coefficient of x^2 in the expansion of $\left(\sqrt{x} + \dfrac{3}{\sqrt{x}} \right)^8$

39. Use the Binomial Theorem to find the numerical value of $(1.001)^5$ correct to five decimal places.

[**Hint:** $(1.001)^5 = (1 + 10^{-3})^5$]

40. Use the Binomial Theorem to find the numerical value of $(0.998)^6$ correct to five decimal places.

41. Show that $\dbinom{n}{n - 1} = n$ and $\dbinom{n}{n} = 1$.

42. Show that if n and j are integers with $0 \le j \le n$ then

$$\binom{n}{j} = \binom{n}{n - j}$$

Conclude that the Pascal triangle is symmetric with respect to a vertical line drawn from the topmost entry.

43. If n is a positive integer, show that

$$\binom{n}{0} + \binom{n}{1} + \cdots + \binom{n}{n} = 2^n$$

[**Hint:** $2^n = (1 + 1)^n$; now use the Binomial Theorem.]

44. If n is a positive integer, show that

$$\binom{n}{0} - \binom{n}{1} + \binom{n}{2} - \cdots + (-1)^n \binom{n}{n} = 0$$

45. $\dbinom{5}{0}\left(\dfrac{1}{4}\right)^5 + \dbinom{5}{1}\left(\dfrac{1}{4}\right)^4\left(\dfrac{3}{4}\right) + \dbinom{5}{2}\left(\dfrac{1}{4}\right)^3\left(\dfrac{3}{4}\right)^2$

$+ \dbinom{5}{3}\left(\dfrac{1}{4}\right)^2\left(\dfrac{3}{4}\right)^3 + \dbinom{5}{4}\left(\dfrac{1}{4}\right)\left(\dfrac{3}{4}\right)^4 + \dbinom{5}{5}\left(\dfrac{3}{4}\right)^5 = ?$

46. Stirling's formula An approximation for $n!$, when n is large, is given by

$$n! \approx \sqrt{2n\pi}\left(\dfrac{n}{e}\right)^n\left(1 + \dfrac{1}{12n - 1}\right)$$

Calculate 12!, 20!, and 25! on your calculator. Then use Stirling's formula to approximate 12!, 20!, and 25!.

47. Prove the Binomial Theorem.

[**Hint:** Use Mathematical Induction.]

Chapter Review

Things To Know

Sequence (p. 982)

A function whose domain is the set of positive integers.

Factorials (p. 986)

$0! = 1, 1! = 1, n! = n(n - 1) \cdot \ldots \cdot 3 \cdot 2 \cdot 1$ if $n \geq 2$

Arithmetic sequence
(pp. 997 and 999)

$a_1 = a, a_n = a_{n-1} + d$, where a = first term, d = common difference
$a_n = a + (n - 1)d$

Sum of the first n terms
of an arithmetic sequence (p. 1000)

$S_n = \dfrac{n}{2}[2a + (n - 1)d] = \dfrac{n}{2}(a + a_n)$

Geometric sequence
(pp. 1004 and 1005)

$a_1 = a, \quad a_n = ra_{n-1}$, where a = first term, r = common ratio
$a_n = ar^{n-1}, \quad r \neq 0$

Sum of the first n terms of
a geometric sequence (p. 1006)

$S_n = a\dfrac{1 - r^n}{1 - r}, \quad r \neq 0, 1$

Infinite geometric series (p. 1008)

$a + ar + \cdots + ar^{n-1} + \cdots = \displaystyle\sum_{k=1}^{\infty} ar^{k-1}$

Sum of an infinite
geometric series (p. 1008)

$\displaystyle\sum_{k=1}^{\infty} ar^{k-1} = \dfrac{a}{1 - r}, \quad |r| < 1$

Principle of Mathematical
Induction (p. 1015)

Suppose the following two conditions are satisfied.
Condition I: The statement is true for the natural number 1.
Condition II: If the statement is true for some natural number k, it is also true
for $k + 1$.
Then the statement is true for all natural numbers n.

Binomial coefficient (p. 1019)

$\dbinom{n}{j} = \dfrac{n!}{j!(n - j)!}$

Pascal triangle (p. 1020)

See Figure 18.

Binomial Theorem (p. 1021)

$(x + a)^n = \dbinom{n}{0}x^n + \dbinom{n}{1}ax^{n-1} + \cdots + \dbinom{n}{j}a^j x^{n-j} + \cdots + \dbinom{n}{n}a^n$

Objectives

Section		You should be able to:	Review Exercises
13.1	1	Write the first several terms of a sequence (p. 983)	1–4
	2	Write the terms of a sequence defined by a recursive formula (p. 987)	5–8
	3	Use summation notation (p. 988)	9–12
	4	Find the sum of a sequence by hand and by using a graphing utility (p. 989)	25–30
	5	Solve annuity and amortization problems (p. 991)	66, 67
13.2	1	Determine if a sequence is arithmetic (p. 997)	13–24
	2	Find a formula for an arithmetic sequence (p. 998)	31, 32, 37–40, 63, 64
	3	Find the sum of an arithmetic sequence (p. 1000)	13, 14, 19, 20, 63, 64
13.3	1	Determine if a sequence is geometric (p. 1003)	13–24
	2	Find a formula for a geometric sequence (p. 1005)	17, 18, 21, 22, 33–36

▩ Review Exercises

Blue problem numbers indicate the authors' suggestions for use in a Practice Test.

In Problems 1–8, write down the first five terms of each sequence.

1. $\left\{(-1)^n\left(\dfrac{n+3}{n+2}\right)\right\}$ **2.** $\{(-1)^{n+1}(2n+3)\}$ **3.** $\left\{\dfrac{2^n}{n^2}\right\}$ **4.** $\left\{\dfrac{e^n}{n}\right\}$

5. $a_1 = 3; \quad a_n = \dfrac{2}{3}a_{n-1}$ **6.** $a_1 = 4; \quad a_n = -\dfrac{1}{4}a_{n-1}$ **7.** $a_1 = 2; \quad a_n = 2 - a_{n-1}$ **8.** $a_1 = -3; \quad a_n = 4 + a_{n-1}$

In Problems 9 and 10, write out each sum.

9. $\displaystyle\sum_{k=1}^{4}(4k+2)$ **10.** $\displaystyle\sum_{k=1}^{3}(3-k^2)$

In Problems 11 and 12, express each sum using summation notation.

11. $1 - \dfrac{1}{2} + \dfrac{1}{3} - \dfrac{1}{4} + \cdots + \dfrac{1}{13}$ **12.** $2 + \dfrac{2^2}{3} + \dfrac{2^3}{3^2} + \cdots + \dfrac{2^{n+1}}{3^n}$

In Problems 13–24, determine whether the given sequence is arithmetic, geometric, or neither. If the sequence is arithmetic, find the common difference and the sum of the first n terms. If the sequence is geometric, find the common ratio and the sum of the first n terms.

13. $\{n+5\}$ **14.** $\{4n+3\}$ **15.** $\{2n^3\}$ **16.** $\{2n^2 - 1\}$

17. $\{2^{3n}\}$ **18.** $\{3^{2n}\}$ **19.** $0, 4, 8, 12, \ldots$ **20.** $1, -3, -7, -11, \ldots$

21. $3, \dfrac{3}{2}, \dfrac{3}{4}, \dfrac{3}{8}, \dfrac{3}{16}, \cdots$ **22.** $5, -\dfrac{5}{3}, \dfrac{5}{9}, -\dfrac{5}{27}, \dfrac{5}{81}, \cdots$ **23.** $\dfrac{2}{3}, \dfrac{3}{4}, \dfrac{4}{5}, \dfrac{5}{6}, \cdots$ **24.** $\dfrac{3}{2}, \dfrac{5}{4}, \dfrac{7}{6}, \dfrac{9}{8}, \dfrac{11}{10}, \cdots$

In Problems 25–30, evaluate each sum (a) by hand and (b) by using a graphing utility.

25. $\displaystyle\sum_{k=1}^{5}(k^2 + 12)$ **26.** $\displaystyle\sum_{k=1}^{3}(k+2)^2$ **27.** $\displaystyle\sum_{k=1}^{10}(3k-9)$

28. $\displaystyle\sum_{k=1}^{9}(-2k+8)$ **29.** $\displaystyle\sum_{k=1}^{7}\left(\dfrac{1}{3}\right)^k$ **30.** $\displaystyle\sum_{k=1}^{10}(-2)^k$

In Problems 31–36, find the indicated term in each sequence (a) by hand and (b) by using a graphing utility.
[**Hint:** Find the general term first.]

31. 9th term of $3, 7, 11, 15, \ldots$ **32.** 8th term of $1, -1, -3, -5, \ldots$ **33.** 11th term of $1, \dfrac{1}{10}, \dfrac{1}{100}, \cdots$

34. 11th term of $1, 2, 4, 8, \ldots$ **35.** 9th term of $\sqrt{2}, 2\sqrt{2}, 3\sqrt{2}, \ldots$ **36.** 9th term of $\sqrt{2}, 2, 2^{3/2}, \ldots$

In Problems 37–40, find a general formula for each arithmetic sequence.

37. 7th term is 31; 20th term is 96 **38.** 8th term is -20; 17th term is -47

39. 10th term is 0; 18th term is 8 **40.** 12th term is 30; 22nd term is 50

In Problems 41–46, find the sum of each infinite geometric series.

41. $3 + 1 + \dfrac{1}{3} + \dfrac{1}{9} + \cdots$ **42.** $2 + 1 + \dfrac{1}{2} + \dfrac{1}{4} + \cdots$ **43.** $2 - 1 + \dfrac{1}{2} - \dfrac{1}{4} + \cdots$

44. $6 - 4 + \dfrac{8}{3} - \dfrac{16}{9} + \cdots$

45. $\displaystyle\sum_{k=1}^{\infty} 4\left(\dfrac{1}{2}\right)^{k-1}$

46. $\displaystyle\sum_{k=1}^{\infty} 3\left(-\dfrac{3}{4}\right)^{k-1}$

In Problems 47–52, use the Principle of Mathematical Induction to show that the given statement is true for all natural numbers.

47. $3 + 6 + 9 + \cdots + 3n = \dfrac{3n}{2}(n + 1)$

48. $2 + 6 + 10 + \cdots + (4n - 2) = 2n^2$

49. $2 + 6 + 18 + \cdots + 2 \cdot 3^{n-1} = 3^n - 1$

50. $3 + 6 + 12 + \cdots + 3 \cdot 2^{n-1} = 3(2^n - 1)$

51. $1^2 + 4^2 + 7^2 + \cdots + (3n - 2)^2 = \dfrac{1}{2}n(6n^2 - 3n - 1)$

52. $1 \cdot 3 + 2 \cdot 4 + 3 \cdot 5 + \cdots + n(n + 2) = \dfrac{n}{6}(n + 1)(2n + 7)$

In Problems 53 and 54, evaluate each binomial coefficient.

53. $\dbinom{5}{2}$

54. $\dbinom{8}{6}$

In Problems 55–58, expand each expression using the Binomial Theorem.

55. $(x + 2)^5$

56. $(x - 3)^4$

57. $(2x + 3)^5$

58. $(3x - 4)^4$

---

59. Find the coefficient of x^7 in the expansion of $(x + 2)^9$.

60. Find the coefficient of x^3 in the expansion of $(x - 3)^8$.

61. Find the coefficient of x^2 in the expansion of $(2x + 1)^7$.

62. Find the coefficient of x^6 in the expansion of $(2x + 1)^8$.

63. Constructing a Brick Staircase A brick staircase has a total of 25 steps. The bottom step requires 80 bricks. Each successive step requires three less bricks than the prior step.
(a) How many bricks are required for the top step?
(b) How many bricks are required to build the staircase?

64. Creating a Floor Design A mosaic tile floor is designed in the shape of a trapezoid 30 feet wide at the base and 15 feet wide at the top. The tiles, 12 inches by 12 inches, are to be placed so that each successive row contains one less tile than the row below. How many tiles will be required?

[**Hint:** Refer to Figure 13 on page 1001.]

65. Bouncing Balls A ball is dropped from a height of 20 feet. Each time it strikes the ground, it bounces up to three-quarters of the previous height.
(a) What height will the ball bounce up to after it strikes the ground for the third time?
(b) How high will it bounce after it strikes the ground for the nth time?
(c) How many times does the ball need to strike the ground before its bounce is less than 6 inches?
(d) What total distance does the ball travel before it stops bouncing?

66. Home Loan Mike and Yola borrowed $190,000 at 6.75% per annum compounded monthly for 30 years to purchase a home. Their monthly payment is determined to be $1232.34.
(a) Find a recursive formula for their balance after each monthly payment has been made.

(b) Determine Mike and Yola's balance after the first payment.
(c) Using a graphing utility, create a table showing Mike and Yola's balance after each monthly payment.
(d) Using a graphing utility, determine when Mike and Yola's balance will be below $100,000.
(e) Using a graphing utility, determine when Mike and Yola will pay off the balance.
(f) Determine Mike and Yola's interest expense when the loan is paid.
(g) Suppose that Mike and Yola decide to pay an additional $100 each month on their loan. Answer parts (a) to (f) under this scenario.

67. Credit Card Debt Beth just charged $5000 on a VISA card that charges 1.5% interest per month on any unpaid balance. She can afford to pay $100 toward the balance each month. Her balance at the beginning of each month after making a $100 payment is given by the recursively defined sequence

$$b_1 = \$5000, \quad b_n = 1.015b_{n-1} - 100$$

(a) Determine Beth's balance after making the first payment; that is, determine b_2, the balance at the beginning of the second month.
(b) Using a graphing utility, graph the recursively defined sequence.
(c) Using a graphing utility, determine when Beth's balance will be below $4000. How many payments of $100 have been made?
(d) Using a graphing utility, determine the number of payments it will take until Beth pays off the balance. What is the total of all the payments?
(e) What was Beth's interest expense?

Chapter Projects

1. **Population Growth**   The size of the population of the United States essentially depends on its current population, the birth and death rates of the population, and immigration. Suppose that b represents the birth rate of the U.S. population and d represents the death rate of the U.S. population. Then $r = b - d$ represents the growth rate of the population, where r varies from year to year. The U.S. population after n years can be modeled using the recursive function

$$p_n = (1 + r)p_{n-1} + I$$

where I represents net immigration into the United States.

(a) Using data from the National Center for Health Statistics *http://www.fedstats.gov*, determine the birth and death rates for all races for the most recent year that data are available. Birth rates are given as the number of live births per 1000 population, while death rates are given as the number of deaths per 100,000 population. Each one must be computed as the number of births (deaths) per individual. For example, in 1990, the birth rate was 16.7 per 1000 and the death rate was 863.8 per 100,000,   so $b = 16.7/1000 = 0.0167$,   while $d = \dfrac{863.8}{100,000} = 0.008638$.

Next, using data from the Immigration and Naturalization Service *http://www.fedstats.gov*, determine the immigration to the United States for the same year as the one used to obtain b and d in part (a).

(b) Determine the value of r, the growth rate of the population.

(c) Find a recursive formula for the population of the United States.

(d) Use the recursive formula to predict the population of the United States in the following year. In other words, if data are available up to the year 1994, predict the U.S. population in 1995.

(e) Compare your prediction to actual data.

(f) Do you think the recursive formula found in part (c) will be useful in predicting future populations? Why or why not?

2. **Economics**   In many situations, the amount of output a producer manufactures is a function of the price of the product in the prior time period. This is especially true in areas where the producer's output decisions must be made in advance of the date of sale, such as agricultural output, where seeds planted today are sold well into the future and planting decisions must be made based on today's price. We can represent the quantity supplied by the equation $Q_{st} = -a + bP_{t-1}$ (i.e., quantity supplied in time period t depends on the price in period $t - 1$). Since consumers base their decision on today's price, demand is given by the equation $Q_{dt} = c - dP_t$. Equilibrium occurs when $Q_{st} = Q_{dt}$ or $-a + bP_{t-1} = c - dP_t$. Solving for P_t, we find the recursion formula

$$P_t = \frac{a + c - bP_{t-1}}{d} \tag{1}$$

If $b < d$, the system will converge to an equilibrium price, a price where quantity supplied equals quantity demanded and $P_{t-1} = P_t$.

Consider the following model:

$$Q_{st} = -3 + 2P_{t-1} \quad Q_{dt} = 18 - 3P_t \quad P_0 = \$2 \text{ (initial price)}$$

The value of b is 2, while the value of d is 3. Therefore, the price will converge to an equilibrium price. To analyze this model, put your graphing utility in WEB format and SEQuence mode.

(a) Find a recursive formula for the model using formula (1).

(b) Graph the recursive formula in WEB format. When graphing in WEB format, the x-axis represents P_{t-1} and the y-axis represents P_t.

(c) TRACE the recursive function. What is the price after one time period $(n = 1)$? Determine Q_{st} and Q_{dt} for this price. They should be equal! This is due to the fact the price will adjust so that $Q_{st} = Q_{dt}$. So, when the price was $2, quantity supplied was such that the quantity demanded was larger than the amount being supplied. This caused competition among buyers, which caused the price to increase. Now TRACE to the next time period. Explain what happened in the markets from $t = 1$ to $t = 2$.

(d) Find the equilibrium point (if any) for each by TRACEing the function until the price is correct to the nearest penny.

(e) How many time periods does it take until this equilibrium is reached?

(f) Verify that this price is an equilibrium by finding Q_{st} and Q_{dt}. What is the equilibrium quantity supplied and demanded?

3. **Standardized Tests**   On some types of tests, like tests that are given to elementary school children or on IQ tests for older students or on some tests for pre-employment, sequences of numbers are given, and the person being tested is asked to find the next number in the sequence.

Consider the sequence of numbers: $1, 2, 4, 7, 11, \ldots$

(a) Find the next three numbers in the sequence.
(b) Is the sequence arithmetic? Why or why not? Is the sequence geometric? Why or why not?
(c) Draw a scatter diagram of the sequence by plotting the following points:

$$(1,1), (2,2), (3,4), (4,7), (5,11)$$

(d) Find the line of best fit using your graphing utility. Graph the graph. Does the graph pass through any of the sequence points? What are the next three terms in the sequence using that equation? Evaluate the regression equation for the values $x = 1, 2, 3, 4, 5$. Find

$$\sum_{i=1}^{5} (y_{r_i} - y_i)$$

where y_{r_i} is the regression value for x_i and y_i is the sequence value for x_i. What do you find? What does this result mean?

(e) Find the quadratic function of best fit using your graphing utility. Graph the function. Does the graph pass through any of the sequence points? What are the next three terms in the sequence using that function? Evaluate the function for the values $x = 1, 2, 3, 4, 5$. Find

$$\sum_{i=1}^{5} (y_{r_i} - y_i)$$

where y_{r_i} is the regression value for x_i and y_i is the sequence value for x_i. What do you find?

(f) Repeat part (e) for a cubic, a fourth degree polynomial and an exponential function of best fit.
(g) Do any of the functions from parts (d), (e), or (f) provide the same three terms that you stated in part (a)? Why or why not?
(d) Could a logarithmic or sinusoidal regression fit the sequence? Can you tell right away? Why or why not?

Cumulative Review

1. Find all the solutions, real and complex, of the equation $|x^2| = 9$.

2. (a) Graph the circle $x^2 + y^2 = 100$ and the parabola $y = 3x^2$.

   (b) Solve the system of equations: $\begin{cases} x^2 + y^2 = 100 \\ y = 3x^2 \end{cases}$

   (c) Where do the circle and the parabola intersect?

3. Solve the equation $2e^x = 5$.

4. Solve the equation $2\sin^2 x - \sin x - 3 = 0, 0 < x < 2\pi$.

5. Find the exact value of $\cos^{-1}(-0.5)$.

6. If $\sin \theta = \dfrac{1}{4}$ and θ is in the second quadrant, find:

   (a) $\cos \theta$ (b) $\tan \theta$
   (c) $\sin(2\theta)$ (d) $\cos(2\theta)$
   (e) $\sin\left(\dfrac{1}{2}\theta\right)$

7. Find the equation of an ellipse with center at the origin, a focus at $(0, 3)$, and a vertex at $(0, 4)$. What are the parametric equations of this ellipse?

8. Find the equation of a parabola with vertex at $(-1, 2)$ and focus at $(-1, 3)$

9. Find the polar equation of a circle with center at $(0, 4)$ that passes through the pole. What is its rectangular equation?

COUNTING AND PROBABILITY

The Two-Children Problem

PROBLEM: A woman and a man (unrelated) each have two children. At least one of the woman's children is a boy, and the man's older child is a boy. Do the chances that the woman has two boys equal the chances that the man has two boys?

The above problem was posed to Marilyn vos Savant in her column *Ask Marilyn*. Her original answer, based on theoretical probabilities, was that the chances that the woman has two boys are 1 in 3 and the chances that the man has two boys are 1 in 2. This is found by looking at the sample space of two-child families: BB, BG, GB, GG. In the case of the man, we know that his older child is a boy and the sample space reduces to BB and GB. Hence, the probability that he has two boys is 1 out of 2. In the case of the woman, since we only know that she has at least one boy, the sample space reduces to BB, BG, and GB. Thus, her chances of having two boys is 1 out of 3.

The answer about the woman's chances created quite a bit of controversy resulting in many letters which challenged the correctness of her answer (*Parade*, July 27, 1997). Marilyn proposed that readers with exactly two children and at least one boy write in and tell the sex of both their children.

SEE CHAPTER PROJECT 1.

For additional study help, go to

www.prenhall.com/sullivanegu3e

Materials include:

- Graphing Calculator Help
- Chapter Quiz
- Chapter Test
- PowerPoint Downloads
- Chapter Projects
- Student Tips

14.1 SETS AND COUNTING

OBJECTIVES
1. Find All the Subsets of a Set
2. Find the Intersection and Union of Sets
3. Find the Complement of a Set
4. Count the Number of Elements in a Set

Sets

A **set** is a well-defined collection of distinct objects. The objects of a set are called its **elements**. By **well-defined**, we mean that there is a rule that enables us to determine whether a given object is an element of the set. If a set has no elements, it is called the **empty set**, or **null set**, and is denoted by the symbol ∅.

Because the elements of a set are distinct, we never repeat elements. For example, we would never write $\{1, 2, 3, 2\}$; the correct listing is $\{1, 2, 3\}$. Because a set is a collection, the order in which the elements are listed is immaterial. $\{1, 2, 3\}$, $\{1, 3, 2\}$, $\{2, 1, 3\}$, and so on, all represent the same set.

EXAMPLE 1 | **Writing the Elements of a Set**

Write the set consisting of the possible results (outcomes) from tossing a coin twice. Use H for *heads* and T for *tails*.

Solution In tossing a coin twice, we can get heads each time, HH; or heads the first time and tails the second, HT; or tails the first time and heads the second, TH; or tails each time, TT. Because no other possibilities exist, the set of outcomes is

$$\{HH, HT, TH, TT\}$$

① If two sets A and B have precisely the same elements, we say that A and B are **equal** and write $A = B$.

If each element of a set A is also an element of a set B, we say that A is a **subset** of B and write $A \subseteq B$.

If $A \subseteq B$ and $A \neq B$, then we say that A is a **proper subset** of B and write $A \subset B$.

If $A \subseteq B$, every element in set A is also in set B, but B may or may not have additional elements. If $A \subset B$, every element in A is also in B, and B has at least one element not found in A.

Finally, we agree that the empty set is a subset of every set; that is,

$$\varnothing \subseteq A, \quad \text{for any set } A$$

EXAMPLE 2 | **Finding All the Subsets of a Set**

Write down all the subsets of the set $\{a, b, c\}$.

Solution To organize our work, we write down all the subsets with no elements, then those with one element, then those with two elements, and finally those with three elements. These will give us all the subsets. Do you see why?

0 Elements	1 Element	2 Elements	3 Elements
∅	$\{a\}, \{b\}, \{c\}$	$\{a, b\}, \{b, c\}, \{a, c\}$	$\{a, b, c\}$

 ━ NOW WORK PROBLEM **21.**

2 If A and B are sets, the **intersection** of A with B, denoted $A \cap B$, is the set consisting of elements that belong to both A and B. The **union** of A with B, denoted $A \cup B$, is the set consisting of elements that belong to either A or B, or both.

EXAMPLE 3   Finding the Intersection and Union of Sets

Let $A = \{1, 3, 5, 8\}$, $B = \{3, 5, 7\}$, and $C = \{2, 4, 6, 8\}$. Find:

(a) $A \cap B$ (b) $A \cup B$ (c) $B \cap (A \cup C)$

Solution (a) $A \cap B = \{1, 3, 5, 8\} \cap \{3, 5, 7\} = \{3, 5\}$
(b) $A \cup B = \{1, 3, 5, 8\} \cup \{3, 5, 7\} = \{1, 3, 5, 7, 8\}$
(c) $B \cap (A \cup C) = \{3, 5, 7\} \cap [\{1, 3, 5, 8\} \cup \{2, 4, 6, 8\}]$
                  $= \{3, 5, 7\} \cap \{1, 2, 3, 4, 5, 6, 8\} = \{3, 5\}$ ■

 NOW WORK PROBLEM **5**.

3 Usually, in working with sets, we designate a **universal set** U, the set consisting of all the elements that we wish to consider. Once a universal set has been designated, we can consider elements of the universal set not found in a given set.

If A is a set, the **complement** of A, denoted $\overline{A}$, is the set consisting of all the elements in the universal set that are not in A.

Note: Some books use the notation A' for the complement of A.

EXAMPLE 4   Finding the Complement of a Set

If the universal set is $U = \{1, 2, 3, 4, 5, 6, 7, 8, 9\}$, and if $A = \{1, 3, 5, 7, 9\}$, then $\overline{A} = \{2, 4, 6, 8\}$. ■

It follows that $A \cup \overline{A} = U$ and $A \cap \overline{A} = \varnothing$. Do you see why?

 NOW WORK PROBLEM **13**.

Figure 1

Universal set

It is often helpful to draw pictures of sets. Such pictures, called **Venn diagrams**, represent sets as circles enclosed in a rectangle, which represents the universal set. Such diagrams often help us to visualize various relationships among sets. See Figure 1.

If we know that $A \subset B$, we might use the Venn diagram in Figure 2(a). If we know that A and B have no elements in common, that is, if $A \cap B = \varnothing$, we might use the Venn diagram in Figure 2(b). The sets A and B in Figure 2(b) are said to be **disjoint**.

Figure 2

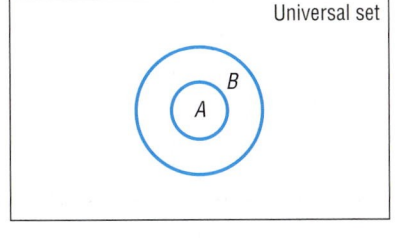

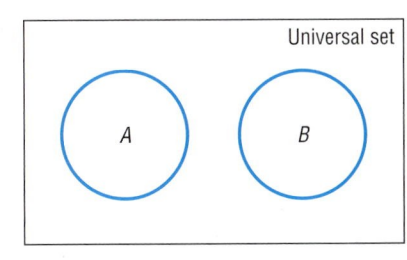

(a) $A \subset B$
proper subset

(b) $A \cap B = \varnothing$
disjoint sets

Figures 3(a), 3(b), and 3(c) use Venn diagrams to illustrate the definitions of intersection, union, and complement, respectively.

Figure 3

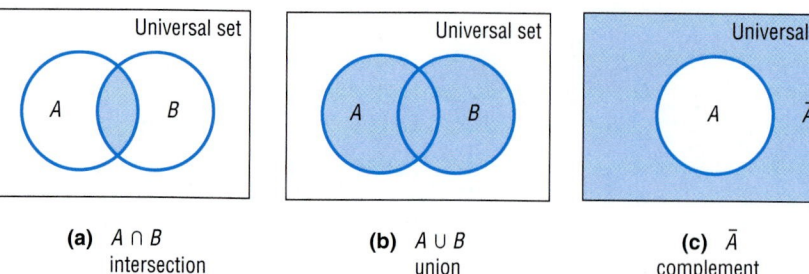

(a) $A \cap B$
intersection

(b) $A \cup B$
union

(c) $\bar{A}$
complement

Counting

4

As you count the number of students in a classroom or the number of pennies in your pocket, what you are really doing is matching, on a one-to-one basis, each object to be counted with the set of counting numbers $1, 2, 3, \ldots, n$, for some number n. If a set A matched up in this fashion with the set $\{1, 2, \ldots, 25\}$, you would conclude that there are 25 elements in the set A. We use the notation $n(A) = 25$ to indicate that there are 25 elements in the set A.

Because the empty set has no elements, we write

$$n(\varnothing) = 0$$

If the number of elements in a set is a nonnegative integer, we say that the set is **finite**. Otherwise, it is **infinite**. We shall concern ourselves only with finite sets.

Look again at Example 2. A set with 3 elements has $2^3 = 8$ subsets. This result can be generalized.

> If A is a set with n elements, then A has 2^n subsets.

For example, the set $\{a, b, c, d, e\}$ has $2^5 = 32$ subsets.

EXAMPLE 5 **Analyzing Survey Data**

In a survey of 100 college students, 35 were registered in College Algebra, 52 were registered in Computer Science I, and 18 were registered in both courses.

(a) How many students were registered in College Algebra or Computer Science I?

(b) How many were registered in neither course?

Solution (a) First, let $A =$ set of students in College Algebra
$B =$ set of students in Computer Science I
Then the given information tells us that

$$n(A) = 35 \qquad n(B) = 52 \qquad n(A \cap B) = 18$$

Refer to Figure 4. Since $n(A \cap B) = 18$, we know that the common part of the circles representing set A and set B has 18 elements. In addition, we know that the remaining portion of the circle representing set A will have $35 - 18 = 17$ elements. Similarly, we know that the remaining por-

Figure 4

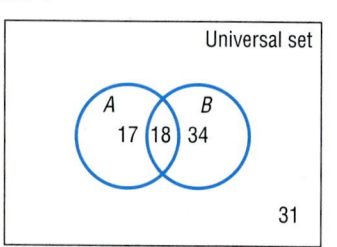

tion of the circle representing set B has $52 - 18 = 34$ elements. We conclude that $17 + 18 + 34 = 69$ students were registered in College Algebra or Computer Science I.

(b) Since 100 students were surveyed, it follows that $100 - 69 = 31$ were registered in neither course.

NOW WORK PROBLEM **35.**

The solution to Example 5 contains the basis for a general counting formula. If we count the elements in each of two sets A and B, we necessarily count twice any elements that are in both A and B, that is, those elements in $A \cap B$. To count correctly the elements that are in A or B, that is, to find $n(A \cup B)$, we need to subtract those in $A \cap B$ from $n(A) + n(B)$.

Theorem

Counting Formula

If A and B are finite sets, then

$$n(A \cup B) = n(A) + n(B) - n(A \cap B) \qquad (1)$$

Refer back to Example 5. Using (1), we have

$$n(A \cup B) = n(A) + n(B) - n(A \cap B)$$
$$= 35 + 52 - 18$$
$$= 69$$

There are 69 students registered in College Algebra or Computer Science I.

A special case of the counting formula (1) occurs if A and B have no elements in common. In this case, $A \cap B = \emptyset$, so $n(A \cap B) = 0$.

Theorem

Addition Principle of Counting

If two sets A and B have no elements in common, that is

$$\text{if } A \cap B = \emptyset, \text{ then } n(A \cup B) = n(A) + n(B) \qquad (2)$$

We can generalize formula (2).

Theorem

General Addition Principle of Counting

If, for n sets $A_1, A_2, \ldots, A_n$, no two have elements in common, then

$$n(A_1 \cup A_2 \cup \cdots \cup A_n) = n(A_1) + n(A_2) + \cdots + n(A_n) \qquad (3)$$

EXAMPLE 6	Counting

In 2000 there were 88,496 full-time law-enforcement officers in the United States federal government. Table 1 lists the type of law-enforcement officer and the corresponding number of full-time officers. No officer is classified as more than one type of officer.

TABLE 1

Type of Officer	Number of Full-Time Officers
Criminal	36,439
Corrections	16,082
Noncriminal	11,884
Court operations	3,599
Security/protection	3,060
Police response and patrol	16,788
Other	644

Source: Bureau of Justice Statistics

(a) How many full-time law-enforcement officers in the United States federal government were criminal officers or corrections officers?

(b) How many full-time law-enforcement officers in the United States federal government were criminal officers, corrections officers, or noncriminal officers?

Solution   Let A represent the set of criminal officers, B represent the set of corrections officers, and C represent the set of noncriminal officers. No two of the sets A, B, and C have elements in common since a single officer cannot be classified as more than one type of officer.

(a) Using formula (2), we have

$$n(A \cup B) = n(A) + n(B) = 36{,}439 + 16{,}082 = 52{,}521$$

There were 52,521 officers that were criminal or corrections officers.

(b) Using formula (3), we have

$$n(A \cup B \cup C) = n(A) + n(B) + n(C) = 36{,}439 + 16{,}082 + 11{,}884 = 64{,}405$$

There were 64,405 officers that were criminal, corrections, or noncriminal officers.

NOW WORK PROBLEM 39.

14.1 Concepts and Vocabulary

In Problems 1 and 2, fill in the blanks.

1. The _____ of A and B consists of all elements in either A or B or both.

2. The _____ of A with B consists of all elements in both A and B.

In Problems 3 and 4, answer True or False to each statement.

3. The intersection of two sets is always a subset of their union.

4. If A is a set, the complement of A is the set of all the elements in the universal set that are not in A.

5. Under what circumstances is a set A a proper subset of a set B?

6. What does it mean if two sets are disjoint?

14.1 Exercises

In Problems 1–10, use $A = \{1, 3, 5, 7, 9\}$, $B = \{1, 5, 6, 7\}$, and $C = \{1, 2, 4, 6, 8, 9\}$ to find each set.

1. $A \cup B$
2. $A \cup C$
3. $A \cap B$
4. $A \cap C$
5. $(A \cup B) \cap C$
6. $(A \cap C) \cup (B \cap C)$
7. $(A \cap B) \cup C$
8. $(A \cup B) \cup C$
9. $(A \cup C) \cap (B \cup C)$
10. $(A \cap B) \cap C$

In Problems 11–20, use $U = universal\ set = \{0, 1, 2, 3, 4, 5, 6, 7, 8, 9\}$, $A = \{1, 3, 4, 5, 9\}$, $B = \{2, 4, 6, 7, 8\}$, and $C = \{1, 3, 4, 6\}$ to find each set.

11. $\overline{A}$
12. $\overline{C}$
13. $\overline{A \cap B}$
14. $\overline{B \cup C}$
15. $\overline{A} \cup \overline{B}$
16. $\overline{B} \cap \overline{C}$
17. $\overline{A} \cap \overline{C}$
18. $\overline{\overline{B} \cup C}$
19. $\overline{A \cup B \cup C}$
20. $\overline{A \cap B \cap C}$

21. Write down all the subsets of $\{a, b, c, d\}$.

22. Write down all the subsets of $\{a, b, c, d, e\}$.

23. If $n(A) = 15$, $n(B) = 20$, and $n(A \cap B) = 10$, find $n(A \cup B)$.

24. If $n(A) = 20$, $n(B) = 40$, and $n(A \cup B) = 35$, find $n(A \cap B)$.

25. If $n(A \cup B) = 50$, $n(A \cap B) = 10$, and $n(B) = 20$, find $n(A)$.

26. If $n(A \cup B) = 60$, $n(A \cap B) = 40$, and $n(A) = n(B)$, find $n(A)$.

In Problems 27–34, use the information given in the figure.

27. How many are in set A?

28. How many are in set B?

29. How many are in A or B?

30. How many are in A and B?

31. How many are in A but not C?

32. How many are not in A?

33. How many are in A and B and C?

34. How many are in A or B or C?

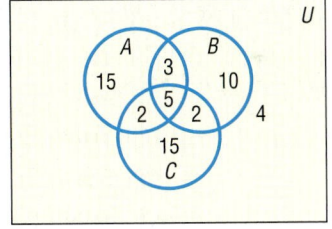

35. Analyzing Survey Data   In a consumer survey of 500 people, 200 indicated that they would be buying a major appliance within the next month; 150 indicated that they would buy a car, and 25 said that they would purchase both a major appliance and a car. How many will purchase neither? How many will purchase only a car?

36. Analyzing Survey Data   In a student survey, 200 indicated that they would attend Summer Session I and 150 indicated Summer Session II. If 75 students plan to attend both summer sessions and 275 indicated that they would attend neither session, how many students participated in the survey?

37. Analyzing Survey Data   In a survey of 100 investors in the stock market,

    50 owned shares in IBM
    40 owned shares in AT&T
    45 owned shares in GE

    20 owned shares in both IBM and GE
    15 owned shares in both AT&T and GE
    20 owned shares in both IBM and AT&T
    5 owned shares in all three

(a) How many of the investors surveyed did not have shares in any of the three companies?

(b) How many owned just IBM shares?

(c) How many owned just GE shares?

(d) How many owned neither IBM nor GE?

(e) How many owned either IBM or AT&T but no GE?

38. Classifying Blood Types   Human blood is classified as either Rh+ or Rh−. Blood is also classified by type: A, if it contains an A antigen; B, if it contains a B antigen; AB, if it contains both A and B antigens; and O, if it contains neither antigen. Draw a Venn diagram illustrating the various blood types. Based on this classification, how many different kinds of blood are there?

39. The following data represent the marital status of males 18 years old and older in March 1999.

Marital Status	Number (in thousands)
Married	58,986
Widowed	2,542
Divorced	8,543
Never married	25,782

Source: Current Population Survey

(a) Determine the number of males 18 years old and older who are married or widowed.
(b) Determine the number of males 18 years old and older who are widowed or divorced.
(c) Determine the number of males 18 years old and older who are married, widowed, or divorced.

40. The following data represent the marital status of females 18 years old and older in March 1999.

Marital Status	Number (in thousands)
Married	59,918
Widowed	10,944
Divorced	11,141
Never married	21,865

Source: Current Population Survey

(a) Determine the number of females 18 years old and older who are married or widowed.
(b) Determine the number of females 18 years old and older who are widowed or divorced.
(c) Determine the number of females 18 years old and older who are married, widowed, or divorced.

41. Make up a problem different from any found in the text that requires the addition principle of counting to solve. Give it to a friend to solve and critique.

42. Investigate the notion of counting as it relates to infinite sets. Write an essay on your findings.

PREPARING FOR THIS SECTION

Before getting started, review the following:

✓ Factorial (Section 13.1, p. 986)

14.2 PERMUTATIONS AND COMBINATIONS

OBJECTIVES

1. Solve Counting Problems Using the Multiplication Principle
2. Solve Counting Problems Using Permutations
3. Solve Counting Problems Using Combinations
4. Solve Counting Problems Using Permutations Involving *n* Non-Distinct Objects

1   Counting plays a major role in many diverse areas, such as probability, statistics, and computer science; counting techniques are a part of a branch of mathematics called **combinatorics**. In this section we shall look at special types of counting problems and develop general formulas for solving them.
  We begin with an example that will demonstrate a general counting principle.

EXAMPLE 1   Counting the Number of Possible Meals

The fixed-price dinner at Mabenka Restaurant provides the following choices:

  Appetizer: soup or salad
  Entree: baked chicken, broiled beef patty, baby beef liver, or roast beef au jus
  Dessert: ice cream or cheese cake

How many different meals can be ordered?

Solution   Ordering such a meal requires three separate decisions:

Choose an Appetizer   **Choose an Entree**   **Choose a Dessert**

2 choices   4 choices   2 choices

Look at the **tree diagram** in Figure 5. We see that, for each choice of appetizer, there are 4 choices of entrees. And for each of these $2 \cdot 4 = 8$ choices, there are 2 choices for dessert. A total of

$$2 \cdot 4 \cdot 2 = 16$$

different meals can be ordered.

Figure 5

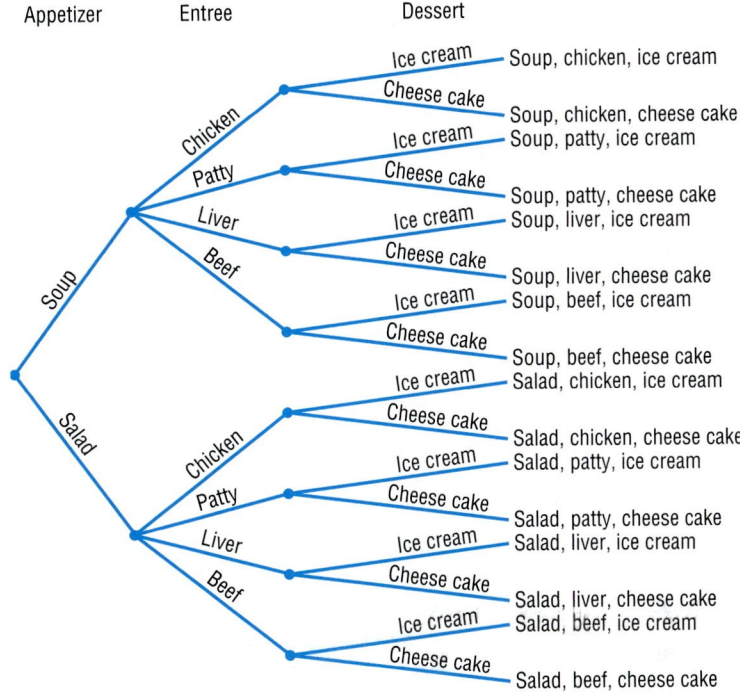

Theorem

Multiplication Principle of Counting

If a task consists of a sequence of choices in which there are p selections for the first choice, q selections for the second choice, r selections for the third choice, and so on, then the task of making these selections can be done in

$$p \cdot q \cdot r \cdot \ldots$$

different ways.

EXAMPLE 2   **Forming Codes**

How many two-symbol codewords can be formed if the first symbol is a letter (uppercase) and the second symbol is a digit?

Solution   It sometimes helps to begin by listing some of the possibilities. The code consists of a letter (uppercase) followed by a digit, so some possibilities are A1, A2, B3, X0, and so on. The task consists of making two selections: the first selection requires choosing an uppercase letter (26 choices) and the second task requires choosing a digit (10 choices). By the Multiplication Principle, there are

$$26 \cdot 10 = 260$$

different codewords of the type described.

NOW WORK PROBLEM 25.

Permutations

② We begin with the definition

A **permutation** is an ordered arrangement of r objects chosen from n objects.

We discuss three types of permutations:

1. The n objects are distinct (different), and repetition is allowed in the selection of r of them. [Distinct, with repetition]
2. The n objects are distinct (different), and repetition is not allowed in the selection of r of them, where $r \leq n$. [Distinct, without repetition]
3. The n objects are not distinct, and we use all of them in the arrangement. [Not distinct]

We take up the first two types here and deal with the third type at the end of this section.

The first type of permutation is handled using the Multiplication Principle.

EXAMPLE 3 | **Counting Airport Codes [Permutation: Distinct, with Repetition]**

The International Airline Transportation Association (IATA) assigns three-letter codes to represent airport locations. For example, the airport code for Ft. Lauderdale, Florida is FLL. Notice that repetition is allowed in forming this code. How many airport codes are possible?

Solution We are choosing 3 letters from 26 letters and arranging them in order. In the ordered arrangement a letter may be repeated. This is an example of a permutation with repetition in which 3 objects are chosen from 26 distinct objects.

The task of counting the number of such arrangements consists of making three selections. Each selection requires choosing a letter of the alphabet (26 choices). By the Multiplication Principle, there are

$$26 \cdot 26 \cdot 26 = 17,576$$

different airport codes.

The solution given to Example 3 can be generalized.

Theorem | **Permutations: Distinct Objects, With Repetition**

The number of ordered arrangements of r objects chosen from n objects, in which the n objects are distinct and repetition is allowed, is n^r.

NOW WORK PROBLEM 29.

We begin the discussion of permutations in which the objects are distinct and repetition is not allowed with an example.

EXAMPLE 4

Forming Codes [Permutation: Distinct, without Repetition]

Suppose that we wish to establish a three-letter code using any of the 26 uppercase letters of the alphabet, but we require that no letter be used more than once. How many different three-letter codes are there?

Solution Some of the possibilities are: ABC, ABD, ABZ, ACB, CBA, and so on. The task consists of making three selections. The first selection requires choosing from 26 letters. Because no letter can be used more than once, the second selection requires choosing from 25 letters. The third selection requires choosing from 24 letters. (Do you see why?) By the Multiplication Principle, there are

$$26 \cdot 25 \cdot 24 = 15,600$$

different three-letter codes with no letter repeated.

For the second type of permutation, we introduce the following symbol.

The symbol $P(n, r)$ represents the number of ordered arrangements of r objects chosen from n distinct objects, where $r \leq n$ and repetition is not allowed.

For example, the question posed in Example 4 asks for the number of ways that the 26 letters of the alphabet can be arranged in order using three nonrepeated letters. The answer is

$$P(26, 3) = 26 \cdot 25 \cdot 24 = 15,600$$

EXAMPLE 5

Lining Up People

In how many ways can 5 people be lined up?

Solution The 5 people are distinct. Once a person is in line, that person will not be repeated elsewhere in the line; and, in lining up people, order is important. We have a permutation of 5 objects taken 5 at a time. We can line up 5 people in

$$P(5, 5) = \underbrace{5 \cdot 4 \cdot 3 \cdot 2 \cdot 1}_{5 \text{ factors}} = 120 \text{ ways}$$

NOW WORK PROBLEM 31.

To arrive at a formula for $P(n, r)$, we note that the task of obtaining an ordered arrangement of n objects in which only $r \leq n$ of them are used, without repeating any of them, requires making r selections. For the first selection, there are n choices; for the second selection, there are $n - 1$ choices; for the third selection, there are $n - 2$ choices; ...; for the rth selection, there are $n - (r - 1)$ choices. By the Multiplication Principle, we have

$$
\begin{array}{cccc}
& \text{1st} \quad \text{2nd} \quad\;\; \text{3rd} & & r\text{th} \\
P(n,r) &= n\cdot(n-1)\cdot(n-2)\cdot\ldots\cdot[n-(r-1)] & & \\
&= n\cdot(n-1)\cdot(n-2)\cdot\ldots\cdot(n-r+1) & &
\end{array}
$$

This formula for $P(n,r)$ can be compactly written using factorial notation.*

$$
\begin{aligned}
P(n,r) &= n\cdot(n-1)\cdot(n-2)\cdot\ldots\cdot(n-r+1) \\
&= n\cdot(n-1)\cdot(n-2)\cdot\ldots\cdot(n-r+1)\cdot\frac{(n-r)\cdot\ldots\cdot 3\cdot 2\cdot 1}{(n-r)\cdot\ldots\cdot 3\cdot 2\cdot 1} = \frac{n!}{(n-r)!}
\end{aligned}
$$

Theorem

Permutations of r Objects Chosen from n Distinct Objects without Repetition

The number of arrangements of n objects using $r \le n$ of them, in which

1. the n objects are distinct,
2. once an object is used it cannot be repeated, and
3. order is important,

is given by the formula

$$
P(n,r) = \frac{n!}{(n-r)!} \tag{1}
$$

EXAMPLE 6   Computing Permutations

Evaluate: (a) $P(7,3)$ (b) $P(6,1)$ (c) $P(52,5)$

Solution   We shall work parts (a) and (b) in two ways.

(a) $P(7,3) = \underbrace{7\cdot 6\cdot 5}_{3\text{ factors}} = 210$

or

$$
P(7,3) = \frac{7!}{(7-3)!} = \frac{7!}{4!} = \frac{7\cdot 6\cdot 5\cdot \cancel{4!}}{\cancel{4!}} = 210
$$

(b) $P(6,1) = \underbrace{6}_{1\text{ factor}} = 6$

Figure 6

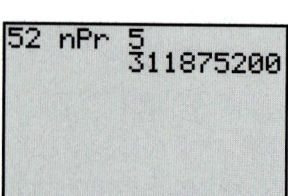
```
52 nPr 5
          311875200
```

or

$$
P(6,1) = \frac{6!}{(6-1)!} = \frac{6!}{5!} = \frac{6\cdot \cancel{5!}}{\cancel{5!}} = 6
$$

(c) Figure 6 shows the solution using a TI-83 graphing calculator: $P(52,5) = 311{,}875{,}200$.

━ NOW WORK PROBLEM **1.**

*Recall that $0! = 1$, $1! = 1$, $2! = 2\cdot 1,\ldots, n! = n(n-1)\cdot\ldots\cdot 3\cdot 2\cdot 1$.

EXAMPLE 7   **The Birthday Problem**

All we know about Shannon, Patrick, and Ryan is that they have different birthdays. If we listed all the possible ways this could occur, how many would there be? Assume that there are 365 days in a year.

Solution   This is an example of a permutation in which 3 birthdays are selected from a possible 365 days, and no birthday may repeat itself. The number of ways that this can occur is

$$P(365, 3) = \frac{365!}{(365 - 3)!} = \frac{365 \cdot 364 \cdot 363 \cdot \cancel{362!}}{\cancel{362!}} = 365 \cdot 364 \cdot 363 = 48{,}228{,}180$$

There are 48,228,180 ways in a group of three people that each has a different birthday. ∎

━━━ NOW WORK PROBLEM **47**.

Combinations

③   In a permutation, order is important; for example, the arrangements ABC, CAB, BAC,... are considered different arrangements of the letters A, B, and C. In many situations, though, order is unimportant. For example, in the card game of poker, the order in which the cards are received does not matter; it is the *combination* of the cards that matters.

> A **combination** is an arrangement, without regard to order, of r objects selected from n distinct objects without repetition, where $r \leq n$. The symbol $C(n, r)$ represents the number of combinations of n distinct objects using r of them.

EXAMPLE 8   **Listing Combinations**

List all the combinations of the 4 objects a, b, c, d taken 2 at a time. What is $C(4, 2)$?

Solution   One combination of a, b, c, d taken 2 at a time is

$$ab$$

We exclude ba from the list because order is not important in a combination. The list of all such combinations (convince yourself of this) is

$$ab, \quad ac, \quad ad, \quad bc, \quad bd, \quad cd$$

so,

$$C(4, 2) = 6 \qquad ∎$$

We can find a formula for $C(n, r)$ by noting that the only difference between a permutation of type 2 and a combination is that we disregard order in combinations. To determine $C(n, r)$, we need only eliminate from the formula for $P(n, r)$ the number of permutations that were simply rearrangements of a given set of r objects. This can be determined from the formula for

$P(n, r)$ by calculating $P(r, r) = r!$. So, if we divide $P(n, r)$ by $r!$, we will have the desired formula for $C(n, r)$:

$$C(n, r) = \frac{P(n, r)}{r!} = \underset{\underset{\text{Use formula (1).}}{\uparrow}}{\frac{n!/(n-r)!}{r!}} = \frac{n!}{(n-r)!r!}$$

We have proved the following result:

Theorem

Number of Combinations of n Distinct Objects Taken r at a Time

The number of arrangements of n objects using $r \le n$ of them, in which

1. the n objects are distinct,
2. once an object is used, it cannot be repeated, and
3. order is not important, is given by the formula

$$C(n, r) = \frac{n!}{(n-r)!r!} \tag{2}$$

Based on formula (2), we discover that the symbol $C(n, r)$ and the symbol $\binom{n}{r}$ for the binomial coefficients are, in fact, the same. The Pascal triangle (see Section 13.5) can be used to find the value of $C(n, r)$. However, because it is more practical and convenient, we will use formula (2) instead.

EXAMPLE 9   **Using Formula (2)**

Use formula (2) to find the value of each expression.

(a) $C(3, 1)$ (b) $C(6, 3)$ (c) $C(n, n)$ (d) $C(n, 0)$ (e) $C(52, 5)$

Solution   (a) $C(3, 1) = \dfrac{3!}{(3-1)!1!} = \dfrac{3!}{2!1!} = \dfrac{3 \cdot 2 \cdot 1}{2 \cdot 1 \cdot 1} = 3$

(b) $C(6, 3) = \dfrac{6!}{(6-3)!3!} = \dfrac{6 \cdot 5 \cdot 4 \cdot 3!}{3! \cdot 3!} = \dfrac{6 \cdot 5 \cdot 4}{6} = 20$

Figure 7

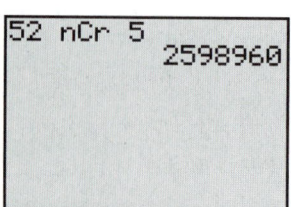

```
52 nCr 5
           2598960
```

(c) $C(n, n) = \dfrac{n!}{(n-n)!n!} = \dfrac{n!}{0!n!} = \dfrac{1}{1} = 1$

(d) $C(n, 0) = \dfrac{n!}{(n-0)!0!} = \dfrac{n!}{n!0!} = \dfrac{1}{1} = 1$

(e) Figure 7 shows the solution using a TI-83 graphing calculator: $C(52, 5) = 2,598,960$.

✎━━━ **NOW WORK PROBLEM 9.**

EXAMPLE 10   **Forming Committees**

How many different committees of 3 people can be formed from a pool of 7 people?

Solution   The 7 people are distinct. More important, though, is the observation that the order of being selected for a committee is not significant. The problem asks for the number of combinations of 7 objects taken 3 at a time.

$$C(7,3) = \frac{7!}{4!3!} = \frac{7 \cdot 6 \cdot 5 \cdot 4!}{4!3!} = \frac{7 \cdot 6 \cdot 5}{6} = 35$$

EXAMPLE 11   **Forming Committees**

In how many ways can a committee consisting of 2 faculty members and 3 students be formed if 6 faculty members and 10 students are eligible to serve on the committee?

Solution   The problem can be separated into two parts: the number of ways that the faculty members can be chosen, $C(6,2)$, and the number of ways that the student members can be chosen, $C(10,3)$. By the Multiplication Principle, the committee can be formed in

$$C(6,2) \cdot C(10,3) = \frac{6!}{4!2!} \cdot \frac{10!}{7!3!} = \frac{6 \cdot 5 \cdot 4!}{4!2!} \cdot \frac{10 \cdot 9 \cdot 8 \cdot 7!}{7!3!}$$

$$= \frac{30}{2} \cdot \frac{720}{6} = 1800 \text{ ways}$$

NOW WORK PROBLEM **49**.

Permutations Involving *n* Objects That Are Not Distinct

④   We begin with an example.

EXAMPLE 12   **Forming Different Words**

How many different words (real or imaginary) can be formed using all the letters in the word REARRANGE?

Solution   Each word formed will have 9 letters: 3 R's, 2 A's, 2 E's, 1 N, and 1 G. To construct each word, we need to fill in 9 positions with the 9 letters:

$$\overline{1} \; \overline{2} \; \overline{3} \; \overline{4} \; \overline{5} \; \overline{6} \; \overline{7} \; \overline{8} \; \overline{9}$$

The process of forming a word consists of five tasks:

Task 1:   Choose the positions for the 3 R's.
Task 2:   Choose the positions for the 2 A's.
Task 3:   Choose the positions for the 2 E's.
Task 4:   Choose the position for the 1 N.
Task 5:   Choose the position for the 1 G.

Task 1 can be done in $C(9,3)$ ways. There then remain 6 positions to be filled, so Task 2 can be done in $C(6,2)$ ways. There remain 4 positions to be filled, so Task 3 can be done in $C(4,2)$ ways. There remain 2 positions to be filled, so Task 4 can be done in $C(2,1)$ ways. The last position can be filled in $C(1,1)$ way. Using the Multiplication Principle, the number of possible words that can be formed is

$$C(9,3) \cdot C(6,2) \cdot C(4,2) \cdot C(2,1) \cdot C(1,1) = \frac{9!}{3! \cdot 6!} \frac{6!}{2! \cdot 4!} \frac{4!}{2! \cdot 2!} \frac{2!}{1! \cdot 1!} \frac{1!}{0! \cdot 1!}$$

$$= \frac{9!}{3! \cdot 2! \cdot 2! \cdot 1! \cdot 1!} = 15,120$$

The form of the answer to Example 12 is suggestive of a general result. Had the letters in REARRANGE each been different, there would have been $P(9, 9) = 9!$ possible words formed. This is the numerator of the answer. The presence of 3 R's, 2 A's, and 2 E's reduces the number of different words, as the entries in the denominator illustrate. We are led to the following result:

Theorem | **Permutations Involving n Objects That Are Not Distinct**

The number of permutations of n objects of which n_1 are of one kind, n_2 are of a second kind, ..., and n_k are of a kth kind is given by

$$\frac{n!}{n_1! \cdot n_2! \cdot \ldots \cdot n_k!} \qquad (3)$$

where $n = n_1 + n_2 + \cdots + n_k$.

EXAMPLE 13 | **Arranging Flags**

How many different vertical arrangements are there of 8 flags if 4 are white, 3 are blue, and 1 is red?

Solution We seek the number of permutations of 8 objects, of which 4 are of one kind, 3 of a second kind, and 1 of a third kind. Using formula (3), we find that there are

$$\frac{8!}{4! \cdot 3! \cdot 1!} = \frac{8 \cdot 7 \cdot 6 \cdot 5 \cdot 4!}{4! \cdot 3! \cdot 1!} = 280 \text{ different arrangements}$$

 **NOW WORK PROBLEM 51.**

14.2 Concepts and Vocabulary

In Problems 1 and 2, fill in the blanks.

1. A(n) _____ is an ordered arrangement of r objects chosen from n objects.

2. A(n) _____ is an arrangement of r objects chosen from n distinct objects, without repetition and without regard to order.

In Problems 3 and 4, answer True or False to each statement.

3. In a combination problem, order is not important.

4. In a permutation problem, once an object is used, it cannot be repeated.

5. State the three types of permutations. How are they different? How are they the same?

14.2 Exercises

In Problems 1–8, find the value of each permutation.

1. $P(6, 2)$
2. $P(7, 2)$
3. $P(4, 4)$
4. $P(8, 8)$
5. $P(7, 0)$
6. $P(9, 0)$
7. $P(8, 4)$
8. $P(8, 3)$

In Problems 9–16, use formula (2) to find the value of each combination.

9. $C(8, 2)$ **10.** $C(8, 6)$ **11.** $C(7, 4)$ **12.** $C(6, 2)$

13. $C(15, 15)$ **14.** $C(18, 1)$ **15.** $C(26, 13)$ **16.** $C(18, 9)$

---

17. List all the ordered arrangements of 5 objects *a, b, c, d,* and *e* choosing 3 at a time without repetition. What is $P(5, 3)$?

18. List all the ordered arrangements of 5 objects *a, b, c, d,* and *e* choosing 2 at a time without repetition. What is $P(5, 2)$?

19. List all the ordered arrangements of 4 objects 1, 2, 3, and 4 choosing 3 at a time without repetition. What is $P(4, 3)$?

20. List all the ordered arrangements of 6 objects 1, 2, 3, 4, 5, and 6 choosing 3 at a time without repetition. What is $P(6, 3)$?

21. List all the combinations of 5 objects *a, b, c, d,* and *e* taken 3 at a time. What is $C(5, 3)$?

22. List all the combinations of 5 objects *a, b, c, d,* and *e* taken 2 at a time. What is $C(5, 2)$?

23. List all the combinations of 4 objects 1, 2, 3, and 4 taken 3 at a time. What is $C(4, 3)$?

24. List all the combinations of 6 objects 1, 2, 3, 4, 5, and 6 taken 3 at a time. What is $C(6, 3)$?

25. Shirts and Ties   A man has 5 shirts and 3 ties. How many different shirt and tie arrangements can he wear?

26. Blouses and Skirts   A woman has 3 blouses and 5 skirts. How many different outfits can she wear?

27. Forming Codes   How many two-letter codes can be formed using the letters *A, B, C,* and *D*? Repeated letters are allowed.

28. Forming Codes   How many two-letter codes can be formed using the letters *A, B, C, D,* and *E*? Repeated letters are allowed.

29. Forming Numbers   How many three-digit numbers can be formed using the digits 0 and 1? Repeated digits are allowed.

30. Forming Numbers   How many three-digit numbers can be formed using the digits 0, 1, 2, 3, 4, 5, 6, 7, 8, and 9? Repeated digits are allowed.

31. Lining People Up   In how many ways can 4 people be lined up?

32. Stacking Boxes   In how many ways can 5 different boxes be stacked?

33. Forming Codes   How many different three-letter codes are there if only the letters *A, B, C, D,* and *E* can be used and no letter can be used more than once?

34. Forming Codes   How many different four-letter codes are there if only the letters *A, B, C, D, E,* and *F* can be used and no letter can be used more than once?

35. Stocks on the NYSE   Companies whose stocks are listed on the New York Stock Exchange (NYSE) have their company name represented by either 1, 2, or 3 letters (repetition of letters is allowed). What is the maximum number of companies that can be listed on the NYSE?

36. Stocks on the NASDAQ   Companies whose stocks are listed on the NASDAQ stock exchange have their company name represented by either 4 or 5 letters (repetition

of letters is allowed). What is the maximum number of companies that can be listed on the NASDAQ?

37. Establishing Committees   In how many ways can a committee of 4 students be formed from a pool of 7 students?

38. Establishing Committees   In how many ways can a committee of 3 professors be formed from a department having 8 professors?

39. Possible Answers on a True/False Test   How many arrangements of answers are possible for a true/false test with 10 questions?

40. Possible Answers on a Multiple-choice Test   How many arrangements of answers are possible in a multiple-choice test with 5 questions, each of which has 4 possible answers?

41. Four-Digit Numbers   How many four-digit numbers can be formed using the digits 0, 1, 2, 3, 4, 5, 6, 7, 8, and 9 if the first digit cannot be 0? Repeated digits are allowed.

42. Five-Digit Numbers   How many five-digit numbers can be formed using the digits 0, 1, 2, 3, 4, 5, 6, 7, 8, and 9 if the first digit cannot be 0 or 1? Repeated digits are allowed.

43. Arranging Books   Five different mathematics books are to be arranged on a student's desk. How many arrangements are possible?

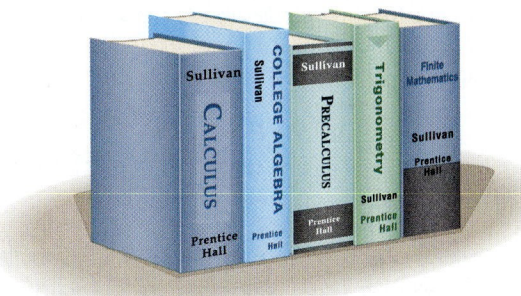

44. Forming License Plate Numbers   How many different license plate numbers can be made using 2 letters followed by 4 digits selected from the digits 0 through 9, if
(a) Letters and digits may be repeated?
(b) Letters may be repeated, but digits may not be repeated?
(c) Neither letters nor digits may be repeated?

45. Stock Portfolios   As a financial planner, you are asked to select one stock each from the following groups: 8 DOW stocks, 15 NASDAQ stocks, and 4 global stocks. How many different portfolios are possible?

46. Combination Locks   A combination lock displays 50 numbers. To open it, you turn to a number, then rotate clockwise to a second number, and then counterclockwise to the third number. How many different lock combinations are there?

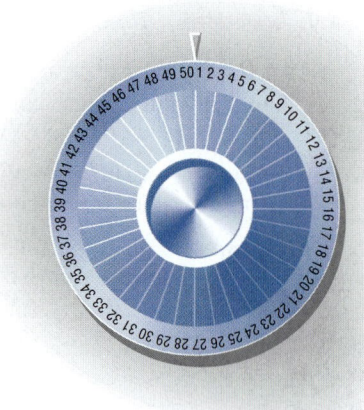

47. Birthday Problem In how many ways can 2 people each have different birthdays? Assume that there are 365 days in a year.

48. Birthday Problem In how many ways can 5 people each have different birthdays? Assume that there are 365 days in a year.

49. Forming a Committee A student dance committee is to be formed consisting of 2 boys and 3 girls. If the membership is to be chosen from 4 boys and 8 girls, how many different committees are possible?

50. Forming a Committee The student relations committee of a college consists of 2 administrators, 3 faculty members, and 5 students. Four administrators, 8 faculty members, and 20 students are eligible to serve. How many different committees are possible?

51. Forming Words How many different 9-letter words (real or imaginary) can be formed from the letters in the word ECONOMICS?

52. Forming Words How many different 11-letter words (real or imaginary) can be formed from the letters in the word MATHEMATICS?

53. Selecting Objects An urn contains 7 white balls and 3 red balls. Three balls are selected. In how many ways can the 3 balls be drawn from the total of 10 balls:
(a) If 2 balls are white and 1 is red?
(b) If all 3 balls are white?
(c) If all 3 balls are red?

54. Selecting Objects An urn contains 15 red balls and 10 white balls. Five balls are selected. In how many ways can the 5 balls be drawn from the total of 25 balls:
(a) If all 5 balls are red?
(b) If 3 balls are red and 2 are white?
(c) If at least 4 are red balls?

55. Senate Committees The U.S. Senate has 100 members. Suppose that it is desired to place each senator on exactly 1 of 7 possible committees. The first committee has 22 members, the second has 13, the third has 10, the fourth has 5, the fifth has 16, and the sixth and seventh have 17 apiece. In how many ways can these committees be formed?

56. Football Teams A defensive football squad consists of 25 players. Of these, 10 are linemen, 10 are linebackers, and 5 are safeties. How many different teams of 5 linemen, 3 linebackers, and 3 safeties can be formed?

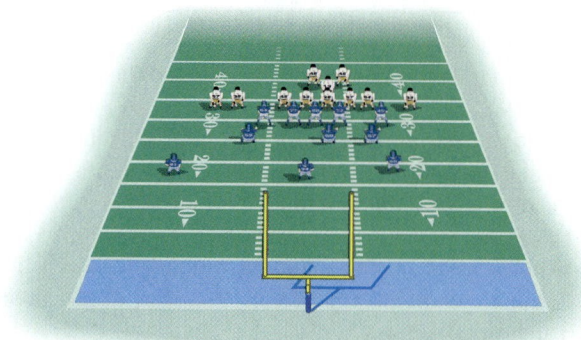

57. Baseball In the American Baseball League, a designated hitter may be used. How many batting orders is it possible for a manager to use? (There are 9 regular players on a team.)

58. Baseball In the National Baseball League, the pitcher usually bats ninth. If this is the case, how many batting orders is it possible for a manager to use?

59. Baseball Teams A baseball team has 15 members. Four of the players are pitchers, and the remaining 11 members can play any position. How many different teams of 9 players can be formed?

60. World Series In the World Series the American League team (A) and the National League team (N) play until one team wins four games. If the sequence of winners is designated by letters (for example, $NAAAA$ means that the National League team won the first game and the American League won the next four), how many different sequences are possible?

61. Basketball Teams A basketball team has 6 players who only play guard (2 of 5 starting positions). How many different teams are possible, assuming that the remaining 3 positions are filled and it is not possible to distinguish a left guard from a right guard?

62. Basketball Teams On a basketball team of 12 players, 2 only play center, 3 only play guard, and the rest play forward (5 players on a team: 2 forwards, 2 guards, and 1 center). How many different teams are possible, assuming that it is not possible to distinguish left and right guards and left and right forwards?

63. Make up a problem different from any found in the text that requires the Multiplication Principle of counting to solve. Give it to a friend to solve and critique.

64. Make up a problem different from any found in the text that requires a permutation to solve. Give it to a friend to solve and critique.

65. Make up a problem different from any found in the text that requires a combination to solve. Give it to a friend to solve and critique.

66. Explain the difference between a permutation and a combination. Give an example to illustrate your explanation.

14.3 PROBABILITY OF EQUALLY LIKELY OUTCOMES

OBJECTIVES
1. Construct Probability Models
2. Compute Probabilities of Equally Likely Outcomes
3. Utilize the Addition Rule to Find Probabilities
4. Utilize the Complement Rule to Find Probabilities
5. Compute Probabilities Using Permutations and Combinations

Probability is an area of mathematics that deals with experiments that yield random results, yet admit a certain regularity. Such experiments do not always produce the same result or outcome, so the result of any one trial of the experiment is not known. However, the results of the experiment over a long period do produce regular patterns that enable us to predict with remarkable accuracy.

EXAMPLE 1

Tossing a Fair Coin

In tossing a fair coin, we know that the outcome is either heads or tails. On any particular throw, we cannot know, ahead of time what will happen, but, if we toss the coin many times, we observe that the number of times that heads comes up is approximately equal to the number of times that we get tails.

It seems reasonable, therefore, to assign a probability of $\frac{1}{2}$ that heads comes up and a probability of $\frac{1}{2}$ that tails comes up.   ∎

Probability Models

1. The discussion in Example 1 constitutes the construction of a **probability model** for the experiment of tossing a fair coin once. A probability model has two components: a sample space and an assignment of probabilities. A **sample space** S is a set whose elements represent all the possibilities that can occur as a result of the experiment. Each element of S is called an **outcome**. To each outcome, we assign a number, called the **probability** of that outcome, which has two properties:

1. Each probability is nonnegative.
2. The sum of all the probabilities equals 1.

If a probability model has the sample space

$$S = \{e_1, e_2, \ldots, e_n\}$$

where $e_1, e_2, \ldots, e_n$ are the possible outcomes, and if $P(e_1), P(e_2), \ldots, P(e_n)$ denote the respective probabilities of these outcomes, then

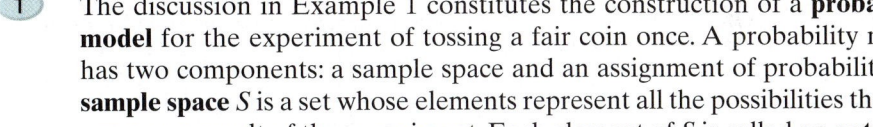

$$P(e_1) \geq 0, P(e_2) \geq 0, \ldots, P(e_n) \geq 0 \qquad (1)$$

$$\sum_{i=1}^{n} P(e_i) = P(e_1) + P(e_2) + \cdots + P(e_n) = 1 \qquad (2)$$

EXAMPLE 2   Determining Probability Models

In a bag of M&Ms the candies are colored red, green, blue, brown, yellow, and orange. Suppose that a candy is drawn from the bag and the color is recorded. The sample space of this experiment is {red, green, blue, brown, yellow, orange}. Determine which of the following are probability models.

(a)

Outcome	Probability
{red}	0.3
{green}	0.15
{blue}	0
{brown}	0.15
{yellow}	0.2
{orange}	0.2

(b)

Outcome	Probability
{red}	0.1
{green}	0.1
{blue}	0.1
{brown}	0.4
{yellow}	0.2
{orange}	0.3

(c)

Outcome	Probability
{red}	0.3
{green}	−0.3
{blue}	0.2
{brown}	0.4
{yellow}	0.2
{orange}	0.2

(d)

Outcome	Probability
{red}	0
{green}	0
{blue}	0
{brown}	0
{yellow}	1
{orange}	0

Solution   (a) This model is a probability model since all the outcomes have probabilities that are nonnegative and the sum of the probabilities is 1.

(b) This model is not a probability model because the sum of the probabilities is not 1.

(c) This model is not a probability model because $P(\text{green})$ is less than 0. Recall, all probabilities must be nonnegative.

(d) This model is a probability model because all the outcomes have probabilities that are nonnegative, and the sum of the probabilities is 1. Notice that $P(\text{yellow}) = 1$, meaning that this outcome will occur with 100% certainty each time that the experiment is repeated. This means that the entire bag of M&Ms has yellow candies. ■

 **NOW WORK PROBLEM 3.**

Let's look at an example of constructing a probability model.

EXAMPLE 3   Constructing a Probability Model

Figure 8

An experiment consists of rolling a fair die once.* Construct a probability model for this experiment.

Solution   A sample space S consists of all the possibilities that can occur. Because rolling the die will result in one of six faces showing, the sample space S consists of

$$S = \{1, 2, 3, 4, 5, 6\}$$

*A die is a cube with each face having either 1, 2, 3, 4, 5, or 6 dots on it. See Figure 8.

Because the die is fair, one face is no more likely to occur than another. As a result, our assignment of probabilities is

$$P(1) = \frac{1}{6} \qquad P(2) = \frac{1}{6}$$

$$P(3) = \frac{1}{6} \qquad P(4) = \frac{1}{6}$$

$$P(5) = \frac{1}{6} \qquad P(6) = \frac{1}{6}$$

Now suppose that a die is loaded so that the probability assignments are

$$P(1) = 0, \quad P(2) = 0, \quad P(3) = \frac{1}{3}, \quad P(4) = \frac{2}{3}, \quad P(5) = 0, \quad P(6) = 0$$

This assignment would be made if the die were loaded so that only a 3 or 4 could occur and the 4 is twice as likely as the 3 to occur. This assignment is consistent with the definition, since each assignment is nonnegative and the sum of all the probability assignments equals 1.

✏️ **N O W   W O R K   P R O B L E M   19 .**

EXAMPLE 4   **Constructing a Probability Model**

An experiment consists of tossing a coin. The coin is weighted so that heads (H) is three times as likely to occur as tails (T). Construct a probability model for this experiment.

Solution   The sample space S is $S = \{H, T\}$. If x denotes the probability that a tail occurs, then

$$P(T) = x \quad \text{and} \quad P(H) = 3x$$

Since the sum of the probabilities of the possible outcomes must equal 1, we have

$$P(T) + P(H) = x + 3x = 1$$
$$4x = 1$$
$$x = \frac{1}{4}$$

Thus, we assign the probabilities

$$P(T) = \frac{1}{4} \qquad P(H) = \frac{3}{4}$$

✏️ **N O W   W O R K   P R O B L E M   23 .**

In working with probability models, the term **event** is used to describe a set of possible outcomes of the experiment. Thus, an event E is some subset of the sample space S. The **probability of an event** $E, E \neq \varnothing$, denoted by $P(E)$, is defined as the sum of the probabilities of the outcomes in E. We can also think of the probability of an event E as the likelihood that the event E occurs. If $E = \varnothing$, then $P(E) = 0$; if $E = S$, then $P(E) = P(S) = 1$.

Equally Likely Outcomes

  When the same probability is assigned to each outcome of the sample space, the experiment is said to have **equally likely outcomes**.

Theorem

Probability for Equally Likely Outcomes

If an experiment has n equally likely outcomes and if the number of ways that an event E can occur is m, then the probability of E is

$$P(E) = \frac{\text{Number of ways that } E \text{ can occur}}{\text{Number of all logical possibilities}} = \frac{m}{n} \qquad (3)$$

Thus, if S is the sample space of this experiment, then

$$P(E) = \frac{n(E)}{n(S)} \qquad (4)$$

EXAMPLE 5

Calculating Probabilities of Equally Likely Events

Calculate the probability that in a 3-child family there are 2 boys and 1 girl. Assume equally likely outcomes.

Figure 9

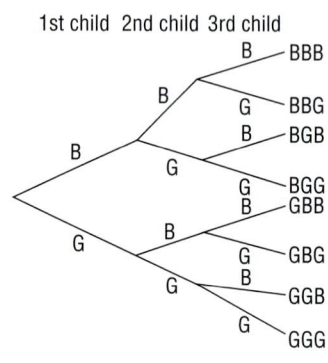

1st child 2nd child 3rd child

Solution We begin by constructing a tree diagram to help in listing the possible outcomes of the experiment. See Figure 9, where B stands for boy and G for girl. The sample space S of this experiment is

$$S = \{\text{BBB, BBG, BGB, BGG, GBB, GBG, GGB, GGG}\}$$

so $n(S) = 8$.

We wish to know the probability of the event E: "having two boys and one girl." From Figure 9, we conclude that $E = \{\text{BBG, BGB, GBB}\}$, so $n(E) = 3$. Since the outcomes are equally likely, the probability of E is

$$P(E) = \frac{n(E)}{n(S)} = \frac{3}{8}$$

NOW WORK PROBLEM 33.

Compound Probabilities

So far, we have calculated probabilities of single events. We will now compute probabilities of multiple events, called **compound probabilities**.

EXAMPLE 6

Computing Compound Probabilities

Consider the experiment of rolling a single fair die. Let E represent the event "roll an odd number" and let F represent the event "roll a 1 or 2."

(a) Write the event E and F.   (b) Write the event E or F.

(c) Compute $P(E)$ and $P(F)$.   (d) Compute $P(E \cap F)$.

(e) Compute $P(E \cup F)$.

Solution   The sample space S of the experiment is $\{1, 2, 3, 4, 5, 6\}$, so $n(S) = 6$. Since the die is fair, the outcomes are equally likely. The event E: "roll an odd number" is $\{1, 3, 5\}$, and the event F: "roll a 1 or 2" is $\{1, 2\}$, so $n(E) = 3$ and $n(F) = 2$.

(a) The word *and* in probability means the intersection of two events. The event E and F is

$$E \cap F = \{1, 3, 5\} \cap \{1, 2\} = \{1\}, \qquad n(E \cap F) = 1$$

(b) The word *or* in probability means the union of the two events. The event E or F is

$$E \cup F = \{1, 3, 5\} \cup \{1, 2\} = \{1, 2, 3, 5\}, \qquad n(E \cup F) = 4$$

(c) We use formula (4).

$$P(E) = \frac{n(E)}{n(S)} = \frac{3}{6} = \frac{1}{2}, \qquad P(F) = \frac{n(F)}{n(S)} = \frac{2}{6} = \frac{1}{3}$$

(d) $P(E \cap F) = \dfrac{n(E \cap F)}{n(S)} = \dfrac{1}{6}$

(e) $P(E \cup F) = \dfrac{n(E \cup F)}{n(S)} = \dfrac{4}{6} = \dfrac{2}{3}$ ■

3   The **Addition Rule** can be used to find the probability of the union of two events.

Theorem   **Addition Rule**

For any two events E and F,

$$P(E \cup F) = P(E) + P(F) - P(E \cap F) \qquad (5)$$

For example, we can use the Addition Rule to find $P(E \cup F)$ in Example 6(e). Then

$$P(E \cup F) = P(E) + P(F) - P(E \cap F) = \frac{1}{2} + \frac{1}{3} - \frac{1}{6} = \frac{3}{6} + \frac{2}{6} - \frac{1}{6} = \frac{4}{6} = \frac{2}{3}$$

as before.

EXAMPLE 7   **Computing Probabilities of Compound Events Using the Addition Rule**

If $P(E) = 0.2$, $P(F) = 0.3$, and $P(E \cap F) = 0.1$, find $P(E \cup F)$.

Solution   We use the Addition Rule, formula (5).

$$P(E \cup F) = P(E) + P(F) - P(E \cap F) = 0.2 + 0.3 - 0.1 = 0.4$$ ■

A Venn diagram can sometimes be used to obtain probabilities. To construct a Venn diagram representing the information in Example 7, we draw two sets E and F. We begin with the fact that $P(E \cap F) = 0.1$. See Figure 10(a). Then, since $P(E) = 0.2$ and $P(F) = 0.3$, we fill in E with $0.2 - 0.1 = 0.1$ and F with $0.3 - 0.1 = 0.2$. See Figure 10(b). Since $P(S) = 1$, we complete the diagram by inserting $1 - [0.1 + 0.1 + 0.2] = 0.6$. See Figure 10(c). Now it is easy to see, for example, that the probability of F, but not E, is 0.2. Also, the probability of neither E nor F is 0.6.

Figure 10

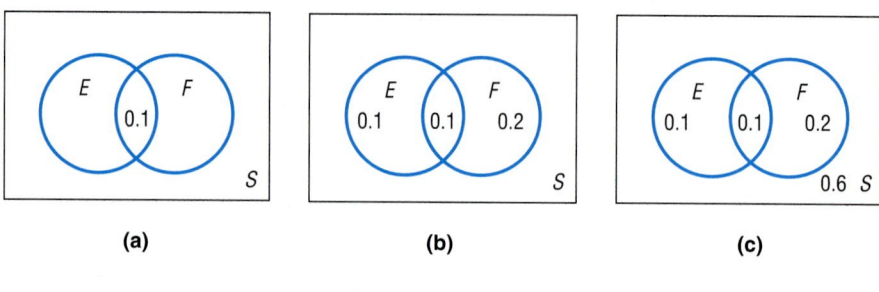

(a)   (b)   (c)

NOW WORK PROBLEM **41.**

If events E and F are disjoint so that $E \cap F = \varnothing$, we say they are **mutually exclusive**. In this case, $P(E \cap F) = 0$, and the Addition Rule takes the following form:

Theorem

Mutually Exclusive Events

If E and F are **mutually exclusive events**, then

$$P(E \cup F) = P(E) + P(F) \qquad (6)$$

EXAMPLE 8   **Computing Compound Probabilities of Mutually Exclusive Events**

If $P(E) = 0.4$ and $P(F) = 0.25$, and E and F are mutually exclusive, find $P(E \cup F)$.

Solution   Since E and F are mutually exclusive, we use formula (6).

$$P(E \cup F) = P(E) + P(F) = 0.4 + 0.25 = 0.65$$

NOW WORK PROBLEM **43.**

Complements

④ Recall, if A is a set, the complement of A, denoted $\overline{A}$, is the set of all elements in the universal set U not in A. We similarly define the complement of an event.

Complement of an Event

Let S denote the sample space of an experiment and let E denote an event. The **complement of E**, denoted $\overline{E}$, is the set of all outcomes in the sample space S that are not outcomes in the event E.

The complement of an event E, that is, $\overline{E}$, in a sample space S has the following two properties:

$$E \cap \overline{E} = \emptyset \qquad E \cup \overline{E} = S$$

Since E and $\overline{E}$ are mutually exclusive, it follows from (6) that

$$P(E \cup \overline{E}) = P(S) = 1 \qquad P(E) + P(\overline{E}) = 1 \qquad P(\overline{E}) = 1 - P(E)$$

We have the following result.

Theorem

> **Computing Probabilities of Complementary Events**
>
> If E represents any event and $\overline{E}$ represents the complement of E, then
>
> $$P(\overline{E}) = 1 - P(E) \qquad (7)$$

EXAMPLE 9 **Computing Probabilities Using Complements**

On the local news the weather reporter stated that the probability of rain is 40%. What is the probability that it will not rain?

Solution The complement of the event "rain" is "no rain." Thus,

$$P(\text{no rain}) = 1 - P(\text{rain}) = 1 - 0.4 = 0.6$$

There is a 60% chance of no rain tomorrow.

━━━ **NOW WORK PROBLEM 47.**

EXAMPLE 10 **Birthday Problem**

What is the probability that in a group of 10 people at least 2 people have the same birthday? Assume that there are 365 days in a year.

Solution We assume that a person is as likely to be born on one day as another, so we have equally likely outcomes.

We first determine the number of outcomes in the sample space S. There are 365 possibilities for each person's birthday. Since there are 10 people in the group, there are 365^{10} possibilities for the birthdays. [For one person in the group, there are 365 days on which his or her birthday can fall; for two people, there are $(365)(365) = 365^2$ pairs of days; and, in general, using the Multiplication Principle, for n people there are 365^n possibilities.] So

$$n(S) = 365^{10}$$

We wish to find the probability of the event E: "at least two people have the same birthday." It is difficult to count the elements in this set; it is much easier to count the elements of the complementary event $\overline{E}$: "no two people have the same birthday."

We find $n(\overline{E})$ as follows: Choose one person at random. There are 365 possibilities for his or her birthday. Choose a second person. There are 364 possibilities for this birthday, if no two people are to have the same birthday. Choose a third person. There are 363 possibilities left for this birthday. Finally, we arrive at the tenth person. There are 356 possibilities left for this birthday. By the Multiplication Principle, the total number of possibilities is

$$n(\overline{E}) = 365 \cdot 364 \cdot 363 \cdot \ldots \cdot 356$$

Hence, the probability of event $\bar{E}$ is

$$P(\bar{E}) = \frac{n(\bar{E})}{n(S)} = \frac{365 \cdot 364 \cdot 363 \cdot \ldots \cdot 356}{365^{10}} \approx 0.883$$

The probability of two or more people in a group of 10 people having the same birthday is then

$$P(E) = 1 - P(\bar{E}) = 1 - 0.883 = 0.117 \quad \blacksquare$$

The birthday problem can be solved for any group size. The following table gives the probabilities for two or more people having the same birthday for various group sizes. Notice that the probability is greater than $\frac{1}{2}$ for any group of 23 or more people.

						Number of People											
5	**10**	**15**	**20**	**21**	**22**	**23**	**24**	**25**	**30**	**40**	**50**	**60**	**70**	**80**	**90**		
Probability That Two or More Have the Same Birthday																	
0.027	0.117	0.253	0.411	0.444	0.476	0.507	0.538	0.569	0.706	0.891	0.970	0.994	0.99916	0.99991	0.99999		

NOW WORK PROBLEM **65**.

5   Probabilities Involving Combinations and Permutations

EXAMPLE 11   Computing Probabilities

Because of a mistake in packaging, 5 defective phones were packaged with 15 good ones. All phones look alike and have equal probability of being chosen. Three phones are selected.

(a) What is the probability that all 3 are defective?
(b) What is the probability that exactly 2 are defective?
(c) What is the probability that at least 2 are defective?

Solution   The sample space S consists of the number of ways that 3 objects can be selected from 20 objects, that is, the number of combinations of 20 things taken 3 at a time.

$$n(S) = C(20, 3) = \frac{20!}{17! \cdot 3!} = \frac{20 \cdot 19 \cdot 18}{6} = 1140$$

Each of these outcomes is equally likely to occur.

(a) If E is the event "3 are defective," the number of elements in E is the number of ways the 3 defective phones can be chosen from the 5 defective phones: $C(5, 3) = 10$. Thus, the probability of E is

$$P(E) = \frac{n(E)}{n(S)} = \frac{C(5, 3)}{C(20, 3)} = \frac{10}{1140} \approx 0.0088$$

(b) If F is the event "exactly 2 are defective" and 3 phones are selected, the number of elements in F is the number of ways to select 2 defective phones from the 5 defective phones and 1 good phone from the 15 good ones. The first of these can be done in $C(5, 2)$ ways and the second in $C(15, 1)$ ways. By the Multiplication Principle, the event F can occur in

$$C(5, 2) \cdot C(15, 1) = \frac{5!}{3! \cdot 2!} \cdot \frac{15!}{14! \cdot 1!} = 10 \cdot 15 = 150 \text{ ways}$$

The probability of F is therefore

$$P(F) = \frac{n(F)}{n(S)} = \frac{C(5, 2) \cdot C(15, 1)}{C(20, 3)} = \frac{150}{1140} \approx 0.1316$$

(c) The event G, "at least two are defective," when 3 are chosen is equivalent to requiring that either exactly 2 defective are chosen or exactly 3 defective are chosen. That is, $G = E \cup F$. Since E and F are mutually exclusive (it is not possible to select 2 defective phones and, at the same time, select 3 defective phones), we find that

$$P(G) = P(E) + P(F) = \frac{C(5, 2)}{C(20, 3)} + \frac{C(5, 2) \cdot C(15, 1)}{C(20, 3)} \approx 0.0088 + 0.1316 = 0.1404$$

■

NOW WORK PROBLEM 73.

EXAMPLE 12   Tossing a Coin

A fair coin is tossed 6 times.

(a) What is the probability of obtaining exactly 5 heads and 1 tail?
(b) What is the probability of obtaining between 4 and 6 heads, inclusive?

Solution   The number of elements in the sample space S is found using the Multiplication Principle. Each toss results in a head (H) or a tail (T). Since the coin is tossed 6 times, we have

$$n(S) = \underbrace{2 \cdot 2 \cdot \ldots \cdot 2}_{6 \text{ tosses}} = 2^6 = 64$$

The outcomes are equally likely since the coin is fair.

(a) Any sequence that contains 5 heads and 1 tail is determined once the position of the 5 heads (or 1 tail) is known. The number of ways that we can position 5 heads in a sequence of 6 slots is $C(6, 5) = 6$. The probability of the event E, exactly 5 heads and 1 tail, is

$$P(E) = \frac{n(E)}{n(S)} = \frac{C(6, 5)}{2^6} = \frac{6}{64} \approx 0.0938$$

(b) Let F be the event "between 4 and 6 heads, inclusive." To obtain between 4 and 6 heads is equivalent to the event "either 4 heads or 5 heads or

6 heads." Since these are mutually exclusive (it is impossible to obtain both 4 heads and 5 heads when tossing a coin 6 times), we have

$$P(F) = P(4 \text{ heads or } 5 \text{ heads or } 6 \text{ heads})$$
$$= P(4 \text{ heads}) + P(5 \text{ heads}) + P(6 \text{ heads})$$

We proceed as in part (a).

$$P(F) = \frac{C(6,4)}{2^6} + \frac{C(6,5)}{2^6} + \frac{C(6,6)}{2^6} = \frac{15}{64} + \frac{6}{64} + \frac{1}{64} = \frac{22}{64} \approx 0.3438$$

HISTORICAL FEATURE

Blaise Pascal
(1623–1662)

Set theory, counting, and probability first took form as a systematic theory in an exchange of letters (1654) between Pierre de Fermat (1601–1665) and Blaise Pascal (1623–1662). They discussed the problem of how to divide the stakes in a game that is interrupted before completion, knowing how many points each player needs to win. Fermat solved the problem by listing all possibilities and counting the favorable ones, whereas Pascal made use of the triangle that now bears his name. As mentioned in the text, the entries in Pascal's triangle are equivalent to $C(n, r)$. This recognition of the role of $C(n, r)$ in counting is the foundation of all further developments.

The first book on probability, the work of Christian Huygens (1629–1695), appeared in 1657. In it, the notion of mathematical expectation is explored. This allows the calculation of the profit or loss that a gambler might expect, knowing the probabilities involved in the game (see the Historical Problems that follow).

Although Girolamo Cardano (1501–1576) wrote a treatise on probability, it was not published until 1663 in Cardano's collected works, and this was too late to have any effect on the development of the theory.

In 1713, the posthumously published *Ars Conjectandi* of Jakob Bernoulli (1654–1705) gave the theory the form it would have until 1900. In the current century, both combinatorics (counting) and probability have undergone rapid development due to the use of computers.

A final comment about notation. The notations $C(n, r)$ and $P(n, r)$ are variants of a form of notation developed in England after 1830. The notation $\binom{n}{r}$ for $C(n, r)$ goes back to Leonhard Euler (1707–1783), but is now losing ground because it has no clearly related symbolism of the same type for permutations. The set symbols $\cup$ and $\cap$ were introduced by Giuseppe Peano (1858–1932) in 1888 in a slightly different context. The inclusion symbol $\subset$ was introduced by E. Schroeder (1841–1902) about 1890. The treatment of set theory in the text is due to George Boole (1815–1864), who wrote $A + B$ for $A \cup B$ and AB for $A \cap B$ (statisticians still use AB for $A \cap B$).

HISTORICAL PROBLEMS

1. *The Problem Discussed by Fermat and Pascal* A game between two equally skilled players, A and B, is interrupted when A needs 2 points to win and B needs 3 points. In what proportion would the stakes be divided?

   [*Note:* If each play results in 1 point for either player, at most four more plays will decide the game.]

   (a) *Fermat's solution* List all possible outcomes that will end the game to form the sample space (for example, *ABA, ABBB*, etc.). Determine the probabilities for A to win and B to win then determine how the stakes should be divided.

   (b) *Pascal's solution* Use combinations to determine the number of ways that the 2 points needed for A to win could occur in four plays. Then use combinations to determine the number of ways that the 3 points needed for B to win could occur. This is trickier than it looks, since A can win with 2 points in either two plays, three plays, or four

   plays. Compute the probabilities and compare with the results in part (a).

2. *Huygen's Mathematical Expectation* In a game with n possible outcomes with probabilities $p_1, p_2, \ldots, p_n$, suppose that the *net* winnings are $w_1, w_2, \ldots, w_n$, respectively. Then the mathematical expectation is

   $$E = p_1 w_1 + p_2 w_2 + \cdots + p_n w_n$$

   The number E represents the profit or loss per game in the long run. The following problems are a modification of those of Huygens.

   (a) A fair die is tossed. A gambler wins \$3 if he throws a 6 and \$6 if he throws a 5. What is his expectation? [*Note:* $w_1 = w_2 = w_3 = w_4 = 0$]

   (b) A gambler plays the same game as in part (a), but now the gambler must pay \$1 to play. This means that $w_5 = \$5$, $w_6 = \$2$, and $w_1 = w_2 = w_3 = w_4 = -\$1$. What is the expectation?

14.3 Concepts and Vocabulary

In Problems 1 and 2, fill in the blanks.

1. When the same probability is assigned to each outcome of a sample space, the experiment is said to have _____ _____ outcomes.

2. The _____ of an event *E* is the set of all outcomes in the sample space *S* that are not outcomes in the event *E*.

In Problems 3 and 4, answer True or False to each statement.

3. The probability of an event can never equal 0.

4. In a probability model, the sum of all probabilities is 1.

5. The probability of an outcome has two properties. What are they?

14.3 Exercises

1. In a probability model, which of the following numbers could be the probability of an outcome:

   0, 0.01, 0.35, −0.4, 1, 1.4?

2. In a probability model, which of the following numbers could be the probability of an outcome:

   $1.5, \quad \dfrac{1}{2}, \quad \dfrac{3}{4}, \quad \dfrac{2}{3}, \quad 0, \quad -\dfrac{1}{4}?$

3. Determine whether the following is a probability model.

Outcome	Probability
{1}	0.2
{2}	0.3
{3}	0.1
{4}	0.4

4. Determine whether the following is a probability model.

Outcome	Probability
{Melody}	0.4
{Bob}	0.3
{Faye}	0.1
{Patricia}	0.2

5. Determine whether the following is a probability model.

Outcome	Probability
{Linda}	0.3
{Jean}	0.2
{Grant}	0.1
{Elena}	0.3

6. Determine whether the following is a probability model.

Outcome	Probability
{Dave}	0.3
{Joanne}	0.2
{Nelson}	0.1
{Rich}	0.5
{Judy}	−0.1

In Problems 7–12, construct a probability model for each experiment.

7. Tossing a fair coin twice

8. Tossing two fair coins once

9. Tossing two fair coins, then a fair die

10. Tossing a fair coin, a fair die, and then a fair coin

11. Tossing three fair coins once

12. Tossing one fair coin three times

In Problems 13–18, use the following spinners to construct a probability model for each experiment.

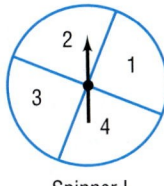
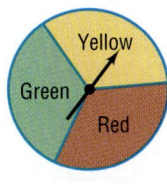
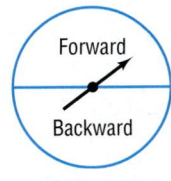

Spinner I   Spinner II   Spinner III

13. Spin spinner I, then spinner II. What is the probability of getting a 2 or a 4, followed by Red?

14. Spin spinner III, then spinner II. What is the probability of getting Forward, followed by Yellow or Green?

15. Spin spinner I, then II, then III. What is the probability of getting a 1, followed by Red or Green, followed by Backward?

16. Spin spinner II, then I, then III. What is the probability of getting Yellow, followed by a 2 or a 4, followed by Forward?

17. Spin spinner I twice, then spinner II. What is the probability of getting a 2, followed by a 2 or a 4, followed by Red or Green?

18. Spin spinner III, then spinner I twice. What is the probability of getting Forward, followed by a 1 or a 3, followed by a 2 or a 4?

In Problems 19–22, consider the experiment of tossing a coin twice. The table lists six possible assignments of probabilities for this experiment. Using this table, answer the following questions.

19. Which of the assignments of probabilities are consistent with the definition of a probability model?

20. Which of the assignments of probabilities should be used if the coin is known to be fair?

21. Which of the assignments of probabilities should be used if the coin is known to always come up tails?

22. Which of the assignments of probabilities should be used if tails is twice as likely as heads to occur?

	Sample Space			
Assignments	HH	HT	TH	TT
A	$\dfrac{1}{4}$	$\dfrac{1}{4}$	$\dfrac{1}{4}$	$\dfrac{1}{4}$
B	0	0	0	1
C	$\dfrac{3}{16}$	$\dfrac{5}{16}$	$\dfrac{5}{16}$	$\dfrac{3}{16}$
D	$\dfrac{1}{2}$	$\dfrac{1}{2}$	$-\dfrac{1}{2}$	$\dfrac{1}{2}$
E	$\dfrac{1}{4}$	$\dfrac{1}{4}$	$\dfrac{1}{4}$	$\dfrac{1}{8}$
F	$\dfrac{1}{9}$	$\dfrac{2}{9}$	$\dfrac{2}{9}$	$\dfrac{4}{9}$

23. Assigning Probabilities A coin is weighted so that heads is four times as likely as tails to occur. What probability should we assign to heads? to tails?

24. Assigning Probabilities A coin is weighted so that tails is twice as likely as heads to occur. What probability should we assign to heads? to tails?

25. Assigning Probabilities A die is weighted so that an odd-numbered face is twice as likely as an even-numbered face. What probability should we assign to each face?

26. Assigning Probabilities A die is weighted so that a six cannot appear. The other faces occur with the same probability. What probability should we assign to each face?

For Problems 27–30, let the sample space be $S = \{1, 2, 3, 4, 5, 6, 7, 8, 9, 10\}$. Suppose that the outcomes are equally likely.

27. Compute the probability of the event $E = \{1, 2, 3\}$.

28. Compute the probability of the event $F = \{3, 5, 9, 10\}$.

29. Compute the probability of the event E: "an even number."

30. Compute the probability of the event F: "an odd number."

For Problems 31 and 32, an urn contains 5 white marbles, 10 green marbles, 8 yellow marbles, and 7 black marbles.

31. If one marble is selected, determine the probability that it is white.

32. If one marble is selected, determine the probability that it is black.

In Problems 33–36, assume equally likely outcomes.

33. Determine the probability of having 3 boys in a 3-child family.

34. Determine the probability of having 3 girls in a 3-child family.

35. Determine the probability of having 1 girl and 3 boys in a 4-child family.

36. Determine the probability of having 2 girls and 2 boys in a 4-child family.

For Problems 37–40, two fair dice are rolled.

37. Determine the probability that the sum of the two dice is 7.

38. Determine the probability that the sum of the two dice is 11.

39. Determine the probability that the sum of the two dice is 3.

40. Determine the probability that the sum of the two dice is 12.

In Problems 41–44, find the probability of the indicated event if $P(A) = 0.25$ and $P(B) = 0.45$.

41. $P(A \cup B)$ if $P(A \cap B) = 0.15$

42. $P(A \cap B)$ if $P(A \cup B) = 0.6$

43. $P(A \cup B)$ if A, B are mutually exclusive

44. $P(A \cap B)$ if A, B are mutually exclusive

45. If $P(A) = 0.60$, $P(A \cup B) = 0.85$, and $P(A \cap B) = 0.05$, find $P(B)$.

46. If $P(B) = 0.30$, $P(A \cup B) = 0.65$, and $P(A \cap B) = 0.15$, find $P(A)$.

47. According to the Federal Bureau of Investigation, in 1997 there was a 25.3% probability of theft involving a motor vehicle. If a victim of theft is randomly selected, what is the probability that he or she was not a victim of motor vehicle theft?

48. According to the Federal Bureau of Investigation, in 1997 there was a 5.6% probability of theft involving a bicycle. If a victim of theft is randomly selected, what is the probability that he or she was not a victim of bicycle theft?

49. In Chicago, there is a 30% probability that Memorial Day will have a high temperature in the 70s. What is the probability that next Memorial Day will not have a high temperature in the 70s in Chicago?

50. In Chicago, there is a 4% probability that Memorial Day will have a low temperature in the 30s. What is the probability that next Memorial Day will not have a low temperature in the 30s in Chicago?

For Problems 51–54, a golf ball is selected at random from a container. If the container has 9 white balls, 8 green balls, and 3 orange balls, find the probability of each event.

51. The golf ball is white or green.

52. The golf ball is white or orange.

53. The golf ball is not white.

54. The golf ball is not green.

55. On "The Price is Right" there is a game in which a bag is filled with 3 strike chips and 5 numbers. Let's say that the numbers in the bag are 0, 1, 3, 6, and 9. What is the probability of selecting a strike chip or the number 1?

56. Another game on the "Price is Right" requires the contestant to spin a wheel with numbers 5, 10, 15, 20, ..., 100. What is the probability that the contestant spins 100 or 30?

Problems 57–60 are based on a consumer survey of annual incomes in 100 households. The following table gives the data.

Income	$0–9999	$10,000–19,999	$20,000–29,999	$30,000–39,999	$40,000 or more
Number of households	5	35	30	20	10

57. What is the probability that a household has an annual income of $30,000 or more?

58. What is the probability that a household has an annual income between $10,000 and $29,999, inclusive?

59. What is the probability that a household has an annual income of less than $20,000?

60. What is the probability that a household has an annual income of $20,000 or more?

61. Surveys In a survey about the number of TV sets in a house, the following probability table was constructed:

Number of TV sets	0	1	2	3	4 or more
Probability	0.05	0.24	0.33	0.21	0.17

Find the probability of a house having:
(a) 1 or 2 TV sets
(b) 1 or more TV sets
(c) 3 or fewer TV sets
(d) 3 or more TV sets
(e) Less than 2 TV sets
(f) Less than 1 TV set
(g) 1, 2, or 3 TV sets
(h) 2 or more TV sets

62. Checkout Lines Through observation it has been determined that the probability for a given number of people waiting in line at the "5 items or less" checkout register of a supermarket is:

Number waiting in line	0	1	2	3	4 or more
Probability	0.10	0.15	0.20	0.24	0.31

Find the probability of:
(a) At most 2 people in line
(b) At least 2 people in line
(c) At least 1 person in line

63. In a certain College Algebra class, there are 18 freshmen and 15 sophomores. Of the 18 freshmen, 10 are male, and of the 15 sophomores, 8 are male. Find the probability that a randomly selected student is:
(a) A freshman or female
(b) A sophomore or male

64. The faculty of the mathematics department at Joliet Junior College is composed of 4 females and 9 males. Of the 4 females, 2 are under the age of 40, and 3 of the males are under age 40. Find the probability that a randomly selected faculty member is:
(a) Female or under age 40
(b) Male or over age 40

65. Birthday Problem   What is the probability that at least 2 people have the same birthday in a group of 12 people? Assume that there are 365 days in a year.

66. Birthday Problem   What is the probability that at least 2 people have the same birthday in a group of 35 people? Assume that there are 365 days in a year.

67. Winning a Lottery   In a certain lottery, there are ten balls, numbered 1, 2, 3, 4, 5, 6, 7, 8, 9, 10. Of these, five are drawn in order. If you pick five numbers that match those drawn in the correct order, you win $1,000,000. What is the probability of winning such a lottery?

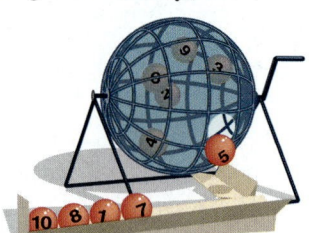

68. A committee of 6 people is to be chosen at random from a group of 14 people consisting of 2 supervisors, 5 skilled laborers, and 7 unskilled laborers. What is the probability that the committee chosen consists of 2 skilled and 4 unskilled laborers?

69. A fair coin is tossed 5 times.
(a) Find the probability that exactly 3 heads appear.
(b) Find the probability that no heads appear.

70. A fair coin is tossed 4 times.
(a) Find the probability that exactly 1 tail appears.
(b) Find the probability that no more than 1 tail appears.

71. A pair of fair dice is tossed 3 times.
(a) Find the probability that the sum of 7 appears 3 times.
(b) Find the probability that a sum of 7 or 11 appears at least twice.

72. A pair of fair dice is tossed 5 times.
(a) Find the probability that the sum is never 2.
(b) Find the probability that the sum is never 7.

73. Through a mix-up on the production line, 5 defective TVs were shipped out with 25 good ones. If 5 are selected at random, what is the probability that all 5 are defective? What is the probability that at least 2 of them are defective?

74. In a shipment of 50 transformers, 10 are known to be defective. If 30 transformers are picked at random, what is the probability that all 30 are nondefective? Assume that all transformers look alike and have an equal probability of being chosen.

75. In a promotion, 50 silver dollars are placed in a bag, one of which is valued at more than $10,000. The winner of the promotion is given the opportunity to reach into the bag, while blindfolded, and pull out 5 coins. What is the probability that 1 of the 5 coins is the one valued at more than $10,000?

14.4 OBTAINING PROBABILITIES FROM DATA

OBJECTIVES   **1** Compute Probabilities from Data
2 Simulate Experiments to Approximate Probabilities

1   In Section 14.3, probabilities were computed by counting the number of equally likely ways that an event E could occur and dividing this result by the number of possible outcomes of the experiment. Thus, we obtained the probability of an event without actually conducting the experiment. For example, in Example 5, the probability of the event E, "having two boys and one girl," was determined to be $\dfrac{3}{8}$ without actually observing three-child families.

A second method for computing probabilities relies on the performance of the experiment or collection of data. The probability of an event E is then computed by determining the *relative frequency* of the event. The **relative**

frequency of an event is found by dividing the number of elements in an event or category by the total number of items or repetitions of the experiment. Probabilities assigned in this way are called **empirical probabilities**.

EXAMPLE 1	Computing Empirical Probabilities from Data

The data in Table 2 represent the daily highs in Chicago from November 22 to November 30 for the years 1872 through 1998.

    Construct a probability model for the daily high temperatures for a day between November 22 and 30 in Chicago.

Solution   There are a total of $1 + 8 + 13 + \cdots + 16 = 1143$ days of data. To construct a probability model, we compute the relative frequency of each category of data. The relative frequency of the first category, $5°$ to $9°F$, is found by dividing the number of days with a high temperature between $5°$ and $9°F$ by the total number of days. The relative frequency of this category, and therefore the probability, is $\dfrac{1}{1143} \approx 0.0009 = 0.09\%$. Following this procedure for the remaining categories of temperature, we obtain Table 3.

TABLE 2

Temperature (°F)	Frequency
5–9	1
10–14	8
15–19	13
20–24	28
25–29	77
30–34	145
35–39	221
40–44	252
45–49	151
50–54	114
55–59	73
60–64	44
65–69	16

Source: Chicago Tribune

TABLE 3

Temperature (°F)	Frequency	Probability %
5–9	1	0.09
10–14	8	0.70
15–19	13	1.14
20–24	28	2.45
25–29	77	6.74
30–34	145	12.69
35–39	221	19.34
40–44	252	22.05
45–49	151	13.21
50–54	114	9.97
55–59	73	6.39
60–64	44	3.85
65–69	16	1.40

From Table 3, we see the probability that a randomly selected day in Chicago from November 22 to November 30 has a high temperature between $40°$ and $44°F$ is 22.05%.

NOW WORK PROBLEM 1.

EXAMPLE 2	Determining the Probability of an Event from an Experiment

A mathematics professor at Joliet Junior College randomly selects 40 currently enrolled students and finds that 25 of them pay their own tuition. Find the probability that a randomly selected student pays his or her own tuition.

Solution   Let E represent the event "pays own tuition." The experiment involves asking students, "Do you pay your own tuition?" The experiment is repeated 40 times and the frequency of "yes" responses is 25, so

$$P(E) = \frac{25}{40} = \frac{5}{8} = 0.625$$

There is a 62.5% probability that a randomly selected student pays his or her own tuition. ∎

SEEING THE CONCEPT   Perform the preceding experiment yourself by asking 10 students, "Do you pay your own tuition?" Repeat the experiment three times. Are your probabilities the same? Why might they be different?

Simulation

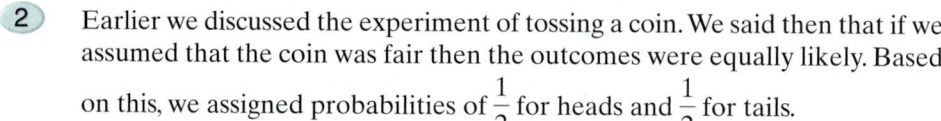

Earlier we discussed the experiment of tossing a coin. We said then that if we assumed that the coin was fair then the outcomes were equally likely. Based on this, we assigned probabilities of $\frac{1}{2}$ for heads and $\frac{1}{2}$ for tails.

We could also assign probabilities by actually tossing the coin, say 100 times, and recording the frequencies of heads and tails. Perhaps we obtain 47 heads and 53 tails. Then, based on the experiment, we would use empirical probability and assign $P(H) = 0.47$ and $P(T) = 0.53$.

Instead of actually physically tossing the coin, we could use simulation; that is, we could use a graphing utility to replicate the experiment of tossing a coin.

EXPLORATION   (a) Simulate tossing a coin 100 times. What percent of the time do you obtain heads? (b) Simulate tossing the coin 100 times again. Are the results the same as the first 100 flips? (c) Repeat the simulation by tossing the coin 250 times. Does the percent of heads get closer to $\frac{1}{2}$?

Figure 11

```
randInt(1,2,100)
→L₁
{2 2 2 2 2 2 1 ...
```

RESULT   (a) Using the randInt* feature on a TI-83, we can simulate flipping a coin 100 times by letting 1 represent heads and 2 represent tails and then storing the random integers in L1. See Figure 11. Figure 12 shows the number of heads and the number of tails that result from flipping the coin 100 times. Based on this simulation, we would assign probabilities as

$$P(1) = P(\text{heads}) = \frac{48}{100} = 0.48 \quad \text{and} \quad P(2) = P(\text{tails}) = \frac{52}{100} = 0.52$$

(b) Figure 13 shows the number of heads and the number of tails that result from a second simulation of flipping the coin 100 times. Based on this simulation, we would assign probabilities as

$$P(1) = P(\text{heads}) = \frac{47}{100} = 0.47 \quad \text{and} \quad P(2) = P(\text{tails}) = \frac{53}{100} = 0.53$$

---

*The randInt feature involves a mathematical formula that uses a seed number to generate a sequence of random integers. Consult your owner's manual for setting the seed so that the same random numbers are not generated each time that you repeat this exploration.

Since repetitions of an experiment (simulation) do not necessarily result in the same outcomes, the empirical probabilities from two experiments will generally be different.

(c) Figure 14 shows the number of heads and the number of tails that result from a third simulation of flipping the coin 250 times. Based on this simulation, we would assign probabilities as

$$P(1) = P(\text{heads}) = \frac{124}{250} = 0.496 \quad \text{and} \quad P(2) = P(\text{tails}) = \frac{126}{250} = 0.504$$

Figure 12	**Figure 13**	**Figure 14**

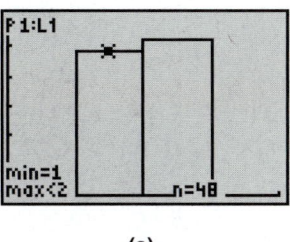

(a)

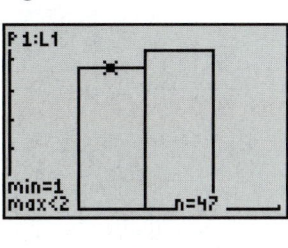

(a)

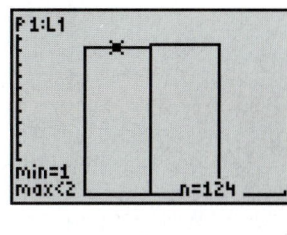

(a)

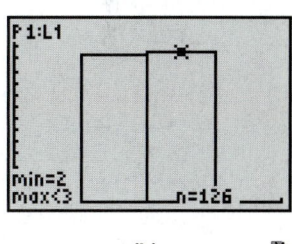

(b)

(b)

(b)

Notice that the result of flipping the coin more often gives probability assignments that get closer to the probabilities obtained using the method of equally likely outcomes. This demonstrates the **Law of Large Numbers**, which states that the more times an experiment involving equally likely outcomes is performed, the closer the empirical probability will come to the probability predicted by equally likely outcomes.

14.4 Exercises

1. **Causes of Death** The data to the right represent the causes of death for 15 to 24 year olds in 1998.
   (a) Construct a probability model for these data.
   (b) What is the probability that the cause of death for a randomly selected 15 to 24 year old was suicide?
   (c) What is the probability that the cause of death for a randomly selected 15 to 24 year old was suicide or malignant neoplasms?
   (d) What is the probability that the cause of death for a randomly selected 15 to 24 year old was not suicide nor malignant neoplasms?

Cause of Death	Number
Accidents and adverse effects	13,349
Homicide and legal intervention	5,506
Suicide	4,135
Malignant neoplasms	1,699
Diseases of heart	1,057
Human immunodeficiency virus infection	194
Congenital anomalies	450
Chronic obstructive pulmonary diseases	239
Pneumonia and influenza	215
Cerebrovascular diseases	176
All other causes	3,605

Source: National Center for Health Statistics, 1998

2. Earthquakes The following data represent the number of earthquakes worldwide whose magnitude on the Richter scale was less than 8.0 in 1998.

Magnitude	Number
0–0.9	2,389
1.0–1.9	752
2.0–2.9	3,851
3.0–3.9	5,639
4.0–4.9	6,943
5.0–5.9	832
6.0–6.9	113
7.0–7.9	10

Source: National Earthquake Information Center

(a) Construct a probability model for these data.
(b) What is the probability that a randomly selected earthquake in 1998 was 3.0–3.9 on the Richter scale?
(c) What is the probability that a randomly selected earthquake in 1998 was 3.0–3.9 or 5.0–5.9 on the Richter scale?
(d) What is the probability that a randomly selected earthquake in 1998 was not 3.0–3.9 on the Richter scale?

3. College Survey In a national survey conducted in 1995 in order to determine their health-risk behaviors, college students were asked, "How often do you wear a seat belt when riding in a car driven by someone else?" The frequencies are displayed next.

Response	Number
Never	125
Rarely	324
Sometimes	552
Most of the time	1,257
Always	2,518

(a) Construct a probability model for these data.
(b) What is the probability that a randomly selected college student never wears a seat belt when riding in a car driven by someone else?
(c) What is the probability that a randomly selected college student never or rarely wears a seat belt when riding in a car driven by someone else?

4. College Survey In a national survey conducted in 1995 in order to determine their health-risk behaviors, college students were asked, "How often do you wear a seat belt when driving a car?" The frequencies are displayed next.

Response	Number
Never	118
Rarely	249
Sometimes	345
Most of the time	716
Always	3093

(a) Construct a probability model for these data.

(b) What is the probability that a randomly selected college student never wears a seat belt when driving a car?
(c) What is the probability that a randomly selected college student never or rarely wears a seat belt when driving a car?

5. Birth Weight The following data represent the birth weight of all babies born in the United States in 1998.

Weight (in grams)	Number
Less than 500	5,950
500–999	22,471
1,000–1,499	28,555
1,500–1,999	58,921
2,000–2,499	182,311
2,500–2,999	649,658
3,000–3,499	1,457,401
3,500–3,999	1,135,572
4,000–4,499	335,087
4,500–4,999	54,809
5,000 or more	6,200
Total	**3,936,935**

Source: National Vital Statistics Report, Vol. 48, No. 3 March 28, 2000

(a) Construct a probability model for these data.
(b) What is the probability a that randomly selected baby born in 1998 weighs 3500–3999 grams?
(c) What is the probability that a randomly selected baby born in 1998 weighs 3500–4499 grams?
(d) What is the probability that a randomly selected baby born in 1998 does not weigh 5000 grams or more?

6. Multiple Births The following data represent the number of live multiple delivery births (three or more babies) in 1996 for women 15 to 49 years old.

Age	Number of Multiple Births
15–19	77
20–24	372
25–29	1,455
30–34	2,546
35–39	1,242
40–44	210
45–49	34
Total	**5,936**

Source: Time Almanac, 2000

(a) Construct a probability model for these data.
(b) What is the probability that a randomly selected multiple delivery in 1996 is a mother 40–44 years old?
(c) What is the probability that a randomly selected multiple delivery in 1996 is a mother 40–49 years old?
(d) What is the probability that a randomly selected multiple delivery in 1996 is not a mother 45–49 years old?

7. While golfing, you had the good fortune of finding 10 golf balls, 4 of which were Titleists. Based on this, what is the probability that a golfer plays with Titleist golf balls?

8. In a recent survey, 200 people were asked if they favor a shopping mall in their neighborhood. Of the 200 people questioned, 120 were in favor of the shopping mall. What is the probability that a randomly selected individual in this neighborhood would favor the shopping mall?

9. In a survey of 50 families with 3 children, it was determined that 20 of them had 2 boys and 1 girl. Based on the results of this survey, what is the probability of having 2 boys and 1 girl?

10. Compaq computer just received a shipment of 500 hard disk drives, 12 of which were defective. What is the probability that a randomly selected disk drive is defective?

11. Of the 28,538 tornadoes that occurred between 1916 and 1985 in the United States, 3262 of them occurred between the hours of 5 and 6 PM. What is the probability that a tornado will occur between 5 and 6 PM? What is the probability that tornado will not occur between 5 and 6 PM? (*Source:* U.S. Tornadoes Part 1, Dr. T. Fujita)

12. Of the 28,538 tornadoes that occurred between 1916 and 1985 in the United States, 324 of them occurred between the hours of 5 and 6 AM. What is the probability that a tornado will occur between 5 and 6 AM? What is the probability that tornado will not occur between 5 and 6 AM? (*Source:* U.S. Tornadoes Part 1, Dr. T. Fujita)

13. In Chicago, during the month of June, 11.5 days are partly cloudy, 7.3 days are clear, and 11.2 days are cloudy. What is the probability that a randomly selected day in Chicago in June is cloudy? What is the probability that a randomly selected day is clear? What is the probability that a randomly selected day is not clear?

14. In Chicago, during the month of July, 12.1 days are partly cloudy, 8.0 days are clear, and 10.3 days are cloudy. What is the probability that a randomly selected day in Chicago in July is cloudy? What is the probability that a randomly selected day is clear? What is the probability that a randomly selected day is not clear?

15. In Chicago, during the month of June, 6.4 days are thunderstorm days. What is the probability that a randomly selected day in June will be a thunderstorm day in Chicago? What is the probability that a randomly selected day in June will not be a thunderstorm day in Chicago?

16. In Chicago, during the month of July, 6.0 days are thunderstorm days. What is the probability that a randomly selected day in July will be a thunderstorm day in Chicago? What is the probability that randomly selected day in July will not be a thunderstorm day in Chicago?

17. On September 8, 1998, Mark McGwire hit his sixty-second home run of the season. Of the 62 home runs he hit, 26 went to left field, 21 went to left center, 12 went to center field, 3 went to right center field, and 0 went to right field.
    (a) Construct a probability model for Mark McGwire's home runs.
    (b) What is the probability that a randomly selected home run was hit to left field?
    (c) What is the probability that a randomly selected home run was hit to center field?
    (d) What is the probability that a randomly selected home run was hit to right field?
    (e) Is it impossible for Mark McGwire to hit a home run to right field? Explain.

18. Conduct a survey in your school by randomly asking 50 students whether they drive to school. Based on the results of the survey, determine the probability that a randomly selected student drives to school.

19. **Simulation**   Use a graphing utility to simulate rolling a six-sided die 100 times.
    (a) Use the results to compute the probability of obtaining a 1.
    (b) Repeat the simulation. Compute the probability of obtaining a 1.
    (c) Simulate rolling a six-sided die 500 times. Compute the probability of obtaining a 1.
    (d) Which simulation resulted in the closest estimate to the probability that would be obtained using equally likely outcomes?

Chapter Review

Things To Know

Set (p. 1032)		Well-defined collection of distinct objects, called elements
Null set (p. 1032)	$\varnothing$	Set that has no elements
Equality (p. 1032)	$A = B$	A and B have the same elements
Subset (p. 1032)	$A \subseteq B$	Each element of A is also an element of B.
Intersection (p. 1033)	$A \cap B$	Set consisting of elements that belong to both A and B
Union (p. 1033)	$A \cup B$	Set consisting of elements that belong to either A or B, or both
Universal set (p. 1033)	U	Set consisting of all the elements that we wish to consider
Complement (p. 1033)	$\overline{A}$	Set consisting of elements of the universal set that are not in A
Finite set (p. 1034)		The number of elements in the set is a nonnegative integer
Infinite set (p. 1034)		A set that is not finite

Counting formula (p. 1035) $n(A \cup B) = n(A) + n(B) - n(A \cap B)$

Addition Principle (p. 1035)   If $A \cap B = \emptyset$, then $n(A \cup B) = n(A) + n(B)$.

Multiplication Principle (p. 1039)   If a task consists of a sequence of choices in which there are p selections for the first choice, q selections for the second choice, and so on, then the task of making these selections can be done in $p \cdot q \cdot \ldots$ different ways.

Permutation (p. 1040)   An ordered arrangement of r objects chosen from n objects.

Permutation: Distinct, with Repetition (p. 1040) n^r The n objects are distinct (different) and repetition is allowed in the selection of r of them.

Permutation: Distinct, without Repetition (p. 1041) $P(n,r) = n(n-1) \cdot \ldots \cdot [n - (r-1)]$ An ordered arrangement of n distinct objects without repetition

$$= \frac{n!}{(n-r)!}$$

Combination (p. 1043) $C(n,r) = \dfrac{P(n,r)}{r!}$ An arrangement, without regard to order, of n distinct objects without repetition

$$= \frac{n!}{(n-r)!\,r!}$$

Permutation: Not distinct, with repetition (p. 1046) $\dfrac{n!}{n_1!\, n_2! \cdots n_k!}$ The number of permutations of n objects of which n_1 are of one kind, n_2 are of a second kind, $\ldots$, and n_k are of a kth kind, where $n = n_1 + n_2 + \cdots + n_k$

Sample space (p. 1049)   Set whose elements represent all the logical possibilities that can occur as a result of an experiment

Probability (p. 1049)   A nonnegative number assigned to each outcome of a sample space; the sum of all the probabilities of the outcomes equals 1.

Equally likely outcomes (p. 1052) $P(E) = \dfrac{n(E)}{n(S)}$ The same probability is assigned to each outcome.

Addition Rule (p. 1053) $P(E \cup F) = P(E) + P(F) - P(E \cap F)$

Complement of an event (p. 1055) $P(\overline{E}) = 1 - P(E)$

■ Objectives

Section		You should be able to:	Review Exercises
14.1	1	Find all the subsets of a set (p. 1032)	1, 2
	2	Find the intersection and union of sets (p. 1033)	3–10
	3	Find the complement of a set (p. 1033)	7–10
	4	Count the number of elements in a set (p. 1034)	11–18
14.2	1	Solve counting problems using the Multiplication Principle (p. 1038)	23–26, 32–36
	2	Solve counting problems using permutations (p. 1040)	19, 20, 27, 28
	3	Solve counting problems using combinations (p. 1043)	21, 22, 27, 29–31, 39–40
	4	Solve counting problems using permutations involving n nondistinct objects (p. 1045)	37, 38
14.3	1	Construct probability models (p. 1049)	41(a), 42(a)
	2	Compute probabilities of equally likely outcomes (p. 1052)	43, 46–50
	3	Utilize the Addition Rule to find probabilities (p. 1053)	41(c), 42(c), 52
	4	Utilize the Complement Rule to find probabilities (p. 1054)	41(d), 42(d), 43(c), 44(b), 45(b), 46, 50
	5	Compute probabilities using permutations and combinations (p. 1056)	50, 51, 53, 54
14.4	1	Compute probabilities from data (p. 1062)	41(b), 42(b), 44(a), 45(a)
	2	Simulate experiments to approximate probabilities (p. 1064)	51(c), 55

Review Exercises

Blue problem numbers indicate the authors' suggestions for use in a Practice Test.

1. Write down all the subsets of the set {Dave, Joanne, Erica}.

2. Write down all the subsets of the set {Green, Blue, Red}.

In Problems 3–10, use U = universal set = {1, 2, 3, 4, 5, 6, 7, 8, 9}, A = {1, 3, 5, 7}, B = {3, 5, 6, 7, 8}, and C = {2, 3, 7, 8, 9} to find each set.

3. $A \cup B$
4. $B \cup C$
5. $A \cap C$
6. $A \cap B$

7. $\overline{A} \cup \overline{B}$
8. $\overline{B} \cap \overline{C}$
9. $\overline{B \cap C}$
10. $\overline{A \cup B}$

11. If $n(A) = 8$, $n(B) = 12$, and $n(A \cap B) = 3$, find $n(A \cup B)$.

12. If $n(A) = 12$, $n(A \cup B) = 30$, and $n(A \cap B) = 6$, find $n(B)$.

In Problems 13–18, use the information supplied in the figure:

13. How many are in A?

14. How many are in A or B?

15. How many are in A and C?

16. How many are not in B?

17. How many are in neither A nor C?

18. How many are in B but not in C?

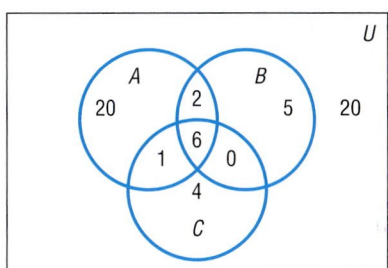

In Problems 19–22, compute the given expression.

19. $P(8, 3)$
20. $P(7, 3)$
21. $C(8, 3)$
22. $C(7, 3)$

---

23. A clothing store sells pure wool and polyester–wool suits. Each suit comes in 3 colors and 10 sizes. How many suits are required for a complete assortment?

24. In connecting a certain electrical device, 5 wires are to be connected to 5 different terminals. How many different wirings are possible if 1 wire is connected to each terminal?

25. **Baseball** On a given day, the American Baseball League schedules 7 games. How many different outcomes are possible, assuming that each game is played to completion?

26. **Baseball** On a given day, the National Baseball League schedules 6 games. How many different outcomes are possible, assuming that each game is played to completion?

27. If 4 people enter a bus having 9 vacant seats, in how many ways can they be seated?

28. How many different arrangements are there of the letters in the word ROSE?

29. In how many ways can a squad of 4 relay runners be chosen from a track team of 8 runners?

30. A professor has 10 similar problems to put on a test with 3 problems. How many different tests can she design?

31. **Baseball** In how many ways can 2 teams from 14 teams in the American League be chosen without regard to which team is at home?

32. **Arranging Books on a Shelf** There are 5 different French books and 5 different Spanish books. How many ways are there to arrange them on a shelf if:
    (a) Books of the same language must be grouped together, French on the left, Spanish on the right?
    (b) French and Spanish books must alternate in the grouping, beginning with a French book?

33. **Telephone Numbers** Using the digits 0, 1, 2, . . . , 9, how many 7-digit numbers can be formed if the first digit cannot be 0 or 9 and if the last digit is greater than or equal to 2 and less than or equal to 3? Repeated digits are allowed.

34. **Home Choices** A contractor constructs homes with 5 different choices of exterior finish, 3 different roof arrangements, and 4 different window designs. How many different types of homes can be built?

35. **License Plate Possibilities** A license plate consists of 1 letter, excluding O and I, followed by a 4-digit number that cannot have a 0 in the lead position. How many different plates are possible?

36. Using the digits 0 and 1, how many different numbers consisting of 8 digits can be formed?

37. **Forming Different Words** How many different words can be formed using all the letters in the word MISSING?

38. **Arranging Flags** How many different vertical arrangements are there of 10 flags if 4 are white, 3 are blue, 2 are green, and 1 is red?

39. **Forming Committees** A group of 9 people is going to be formed into committees of 4, 3, and 2 people. How many committees can be formed if:
    (a) A person can serve on any number of committees?
    (b) No person can serve on more than one committee?

40. **Forming Committees** A group consists of 5 men and 8 women. A committee of 4 is to be formed from this group, and policy dictates that at least 1 woman be on this committee.
    (a) How many committees can be formed that contain exactly 1 man?

(b) How many committees can be formed that contain exactly 2 women?

(c) How many committees can be formed that contain at least 1 man?

41. Vehicle Fatalities   The frequency distribution represents the number of drivers in fatal crashes in 1996 by age for males aged 20 to 84 years old.

Age	Number of Drivers
20–24	6,148
25–29	5,073
30–34	4,834
35–39	4,414
40–44	3,563
45–49	2,935
50–54	2,164
55–59	1,655
60–64	1,398
65–69	1,154
70–74	1,055
75–79	894
80–84	684

Source: National Highway Traffic Safety Administration

(a) Construct a probability model for these data.

(b) What is the probability that a randomly selected male driver fatality is 25–29 years old?

(c) What is the probability that a randomly selected male driver fatality is 20–29 years old?

(d) What is the probability that a randomly selected male driver fatality is not 20–24 years old?

42. Vehicle Fatalities   The frequency distribution represents the number of drivers in fatal crashes in 1996 by age for females aged 20 to 84 years old.

Age	Number of Drivers
20–24	1,747
25–29	1,558
30–34	1,561
35–39	1,503
40–44	1,180
45–49	957
50–54	752
55–59	522
60–64	498
65–69	491
70–74	550
75–79	485
80–84	314

Source: National Highway Traffic Safety Administration

(a) Construct a probability model for these data.

(b) What is the probability that a randomly selected female driver fatality is 25–29 years old?

(c) What is the probability that a randomly selected female driver fatality is 20–29 years old?

(d) What is the probability that a randomly selected female driver fatality is not 20–24 years old?

43. Birthday Problem   For this problem, assume that the year has 365 days.

(a) How many ways can 18 people have different birthdays?

(b) What is the probability that nobody has the same birthday in a group of 18 people?

(c) What is the probability in a group of 18 people that at least 2 people have the same birthday?

44. Death Rates   According to the U.S. National Center for Health Statistics, 32.1% of all deaths in 1994 were due to heart disease.

(a) What is the probability that a randomly selected death in 1994 was due to heart disease?

(b) What is the probability that a randomly selected death in 1994 was not due to heart disease?

45. Unemployment   According to the U.S. Bureau of Labor Statistics, 5.4% of the U.S. labor force was unemployed in 1996.

(a) What is the probability that a randomly selected member of the U.S. labor force was unemployed in 1996?

(b) What is the probability that a randomly selected member of the U.S. labor force was not unemployed in 1996?

46. From a box containing three 40-watt bulbs, six 60-watt bulbs, and eleven 75-watt bulbs, a bulb is drawn at random. What is the probability that the bulb is 40 watts? What is the probability that it is not a 75-watt bulb?

47. You have four $1 bills, three $5 bills, and two $10 bills in your wallet. If you pick a bill at random, what is the probability that it will be a $1 bill?

48. Each of the letters in the word ROSE is written on an index card and the cards are then shuffled. What is the probability that, when the cards are dealt out, they spell the word ROSE?

49. Each of the numbers, 1, 2, ..., 100 is written on an index card and the cards are then shuffled. If a card is selected at random, what is the probability that the number on the card is divisible by 5? What is the probability that the card selected is either a 1 or names a prime number?

50. **Computing Probabilities** Because of a mistake in packaging, a case of 12 bottles of red wine contained 5 Merlot and 7 Cabernet, each without labels. All the bottles look alike and have equal probability of being chosen. Three bottles are selected.
    (a) What is the probability that all 3 are Merlot?
    (b) What is the probability that exactly 2 are Merlot?
    (c) What is the probability that none is a Merlot?

51. **Tossing a Coin** A fair coin is tossed 10 times.
    (a) What is the probability of obtaining exactly 5 heads?
    (b) What is the probability of obtaining all heads?
    (c) Use a graphing utility to simulate this experiment 100 times. What is the empirical probability of obtaining exactly 5 heads? All heads?

52. At the Milex tune-up and brake repair shop, the manager has found that a car will require a tune-up with a probability of 0.6, a brake job with a probability of 0.1, and both with a probability of 0.02.
    (a) What is the probability that a car requires either a tune-up or a brake job?
    (b) What is the probability that a car requires a tune-up but not a brake job?
    (c) What is the probability that a car requires neither type of repair?

53. **Selecting a Jury** The grade appeal process at a university requires that a jury be structured by randomly selecting 5 individuals from a pool of 8 students and 10 faculty.
    (a) What is the probability a selecting a jury of all students?
    (b) What is the probability a selecting a jury of all faculty?
    (c) What is the probability a selecting a jury of 2 students and 3 faculty?

54. **Selecting a Committee** Suppose that there are 55 Democrats and 45 Republicans in the U.S. Senate. A committee of 7 senators is to be formed by randomly selecting members of the Senate.
    (a) What is the probability that the committee is comprised of all Democrats?
    (b) What is the probability that the committee is comprised of all Republicans?
    (c) What is the probability that the committee is comprised of 3 Democrats and 4 Republicans?

55. **Simulation** Use a graphing calculator or statistical software to simulate rolling a six-sided die 100 times.
    (a) Use the results of the simulation to compute the probability of obtaining a 1.
    (b) Repeat the simulation. Compute the probability of rolling a 1.
    (c) Simulate rolling a six-sided die 500 times. Compute the probability of rolling a 1.
    (d) Which simulation resulted in the closest estimate to the probability that would be obtained using the equally-likely outcomes method?

Chapter Projects

1. **Simulation** In the Winter 1998 edition of *Eightysomething!*, Mike Koehler uses simulation to calculate the following probabilities: "A woman and man (unrelated) each have two children. At least one of the woman's children is a boy, and the man's older child is a boy. Do the chances that the woman has two boys equal the chances that the man has two boys?" Perform a simulation to answer the question.

2. **Surveys** Conduct a survey of 25 students in your school. Ask each student the following questions:
    (a) Do you drive yourself to school?
    (b) Do you pay your own tuition?
    (c) Do you study (i) 0–5 hours per week, (ii) 6–10 hours per week, (iii) 11–15 hours per week, (iv) 16–20 hours per week, (v) more than 20 hours per week?
    (d) Compute the probability that a student (1) drives himself or herself to school, (2) pays his or her own tuition, (3) studies 6–10 hours per week.
    (e) Compare your results with other students in your class. Why might your probabilities differ from the probabilities obtained by other students in the class?

3. **Law of Large Numbers** We can demonstrate the Law of Large Numbers using a graphing calculator such as the TI-83. We will simulate flipping a fair coin 200 times.
    (a) In L_1, store the integers 1–200 using the command $seq(n,n,1,200) \rightarrow L_1$.
    (b) In L_2, store the results of flipping a coin 200 times using the command $randInt(0,1,200) \rightarrow L_2$. Note that the 0 represents flipping a tail and 1 represents flipping a head.

(c) In L_3, store the cumulative number of heads flipped using the command $cumSum(L_2) \rightarrow L_3$.

(d) L_4 will represent the proportion of heads thrown and is found using the command $L_3/L_1 \rightarrow L_4$.

(e) Graph the line $Y_1 = 0.5$ and the data in L_4 versus the data in L_1 using the viewing window:

$$X\min = 0, \quad X\max = 200, \quad Y\min = 0, \quad Y\max = 1$$

(f) What do you observe about the graph of the data as the number of flips increases?

Cumulative Review

1. Solve $3x^2 - 2x = -1$.

2. Graph $f(x) = x^2 + 4x - 5$ by determining whether the graph opens up or down and by finding the vertex, axis of symmetry, and intercepts.

3. Graph $f(x) = 2(x + 1)^2 - 4$ using transformations.

4. Solve $|x - 4| \leq 0.01$

5. Find the complex zeros of
$f(x) = 5x^4 - 9x^3 - 7x^2 - 31x - 6$.

6. Graph $g(x) = 3^{x-1} + 5$ using transformations. Determine the domain, range, and horizontal asymptote of g.

7. What is the exact value of $\log_3 9^x$?

8. Solve algebraically $\log_2(3x - 2) + \log_2 x = 4$.

9. Graph $y = 3\sin(2x + \pi)$.

10. Solve the following triangle and determine its area.

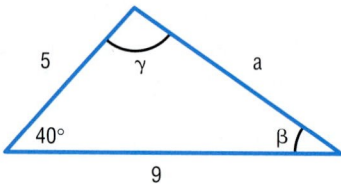

11. Solve the system:
$$\begin{cases} x - 2y + z = 15 \\ 3x + y - 3z = -8 \\ -2x + 4y - z = -27 \end{cases}$$

12. What is the 33rd term in the sequence $-3, 1, 5, 9, \ldots$? What is the sum of the first 20 terms?

C H A P T E R R Review

Historical Problems *(page 15)*

1. (a) 1,20 **(b)** 2,50

R.1 Concepts and Vocabulary *(page 15)*

1. Rational **2.** 31 **3.** Distributive **4.** T **5.** F **6.** F **7.** Natural Numbers, Integers, Rational Numbers, Irrational Numbers
8. $(2 + 4) \cdot 4 + 3 \cdot 4$ **9.** If the product of two number is zero, at least one of the numbers must equal zero.
$\quad\quad 6 \quad \cdot 4 + 3 \cdot 4$ Parentheses first $\quad\quad$ Symbolically, if $a \cdot b = 0$, either $a = 0$ or $b = 0$ or both equal zero.
$\quad\quad 24 + 12$ Multiply before adding
$\quad\quad\quad 36$

R.1 Exercises *(page 16)*

1. (a) $\{2, 5\}$ **(b)** $\{-6, 2, 5\}$ **(c)** $\left\{-6, \dfrac{1}{2}, -1.333..., 2, 5\right\}$ **(d)** $\{\pi\}$ **(e)** $\left\{-6, \dfrac{1}{2}, -1.333..., \pi, 2, 5\right\}$ **3. (a)** $\{1\}$ **(b)** $\{0, 1\}$
(c) $\left\{0, 1, \dfrac{1}{2}, \dfrac{1}{3}, \dfrac{1}{4}\right\}$ **(d)** None **(e)** $\left\{0, 1, \dfrac{1}{2}, \dfrac{1}{3}, \dfrac{1}{4}\right\}$ **5. (a)** None **(b)** None **(c)** None **(d)** $\left\{\sqrt{2}, \pi, \sqrt{2} + 1, \pi + \dfrac{1}{2}\right\}$
(e) $\left\{\sqrt{2}, \pi, \sqrt{2} + 1, \pi + \dfrac{1}{2}\right\}$ **7. (a)** 18.953 **(b)** 18.952 **9. (a)** 28.653 **(b)** 28.653 **11. (a)** 0.063 **(b)** 0.062
13. (a) 9.999 **(b)** 9.998 **15. (a)** 0.429 **(b)** 0.428 **17. (a)** 34.733 **(b)** 34.733 **19.** $3 + 2 = 5$ **21.** $x + 2 = 3 \cdot 4$ **23.** $3 \cdot y = 1 + 2$
25. $x - 2 = 6$ **27.** $\dfrac{x}{2} = 6$ **29.** 7 **31.** 6 **33.** 1 **35.** $\dfrac{13}{3}$ **37.** -11 **39.** 11 **41.** -4 **43.** 1 **45.** 6 **47.** $\dfrac{2}{7}$ **49.** $\dfrac{4}{45}$ **51.** $\dfrac{23}{20}$ **53.** $\dfrac{79}{30}$
55. $\dfrac{13}{36}$ **57.** $-\dfrac{16}{45}$ **59.** $\dfrac{1}{60}$ **61.** $\dfrac{15}{22}$ **63.** $6x + 24$ **65.** $x^2 - 4x$ **67.** $x^2 + 6x + 8$ **69.** $x^2 - x - 2$ **71.** $x^2 - 10x + 16$
73. $x^2 - 4$ **75.** $2x + 3x = (2 + 3)x = 5x$ **77.** $2(3 \cdot 4) = 2 \cdot 12 = 24; (2 \cdot 3) \cdot (2 \cdot 4) = 6 \cdot 8 = 48$ **79.** No; $2 - 3 \neq 3 - 2$
81. No; $\dfrac{2}{3} \neq \dfrac{3}{2}$ **83.** Symmetric property **85.** No; no **87.** 1

R.2 Concepts and Vocabulary *(page 22)*

1. Variable **2.** Origin **3.** Strict **4.** T **5.** F **6.** F **7.** $8 - 5$ is positive. **8.** ⟶$\!\!\!\!$ **9.** $\{x | x \neq 3\}$
$\quad\quad\quad\quad\quad\quad\quad\quad\quad\quad\quad\quad\quad\quad\quad\quad\quad\quad\quad\quad\quad\quad\quad -3$

R.2 Exercises *(page 22)*

1. ⟵—•——•——••—••——————•—⟶ **3.** $>$ **5.** $>$ **7.** $>$ **9.** $=$ **11.** $<$ **13.** $x > 0$ **15.** $x < 2$ **17.** $x \leq 1$
$\quad -2.5 \quad -1 \quad 0 \;\; \frac{3}{4} \; 1 \quad\quad \frac{5}{2}$ **19.** ⟶ **21.** ⟶ **23.** 1 **25.** 2 **27.** 6 **29.** 4 **31.** -28 **33.** $\dfrac{4}{5}$
$\quad\quad\quad\quad\quad 0.25 \quad\quad\quad\quad\quad\quad\quad\quad\quad\quad\quad -2 \quad\quad\quad\quad -1$
35. 0 **37.** 1 **39.** 5 **41.** 1 **43.** 22 **45.** 2 **47.** $x = 0$ **49.** $x = 3$ **51.** None **53.** $x = 1, x = 0, x = -1$ **55.** $\{x | x \neq 5\}$
57. $\{x | x \neq -4\}$ **59.** 0°C **61.** 25°C **63.** $A = lw$ **65.** $C = \pi d$ **67.** $A = \dfrac{\sqrt{3}}{4}x^2$ **69.** $V = \dfrac{4}{3}\pi r^3$ **71.** $V = x^3$
73. (a) \$6000 **(b)** \$8000 **75. (a)** $2 \leq 5$ **(b)** $6 > 5$ **77. (a)** Yes **(b)** No **79.** No; $\dfrac{1}{3}$ is larger; $0.000333...$ **81.** No

R.3 Concepts and Vocabulary *(page 28)*

1. Right; hypotenuse **2.** $A = \dfrac{1}{2}bh$ **3.** $C = 2\pi r$ **4.** T **5.** T **6.** F **7.** $A = \pi r^2$ **8.** $V = \dfrac{4}{3}\pi r^3$

R.3 Exercises *(page 28)*

1. 13 **3.** 26 **5.** 25 **7.** Right triangle; 5 **9.** Not a right triangle **11.** Right triangle; 25 **13.** Not a right triangle **15.** 8 in^2 **17.** 4 in^2
19. $A = 25\pi$ m^2; $C = 10\pi$ m **21.** 224 ft^3 **23.** $V = \dfrac{256}{3}\pi$ cm^3; $S = 64\pi$ cm^2 **25.** 648π in^3 **27.** π square units **29.** 2π square units
31. $\dfrac{16}{3}\pi \approx 16.8$ ft **33.** 64 ft^2 **35.** $24 + 2\pi \approx 30.28$ ft^2; $16 + 2\pi \approx 22.28$ ft **37.** About 5.477 mi **39.** About 12.247 mi; about 15.000 mi

R.4 Concepts and Vocabulary *(page 37)*

1. Base; exponent or power   **2.** a^{mn} **3.** 1.2345678×10^3 **4.** F   **5.** F   **6.** F   **7.** a multiplied by itself n times   **8.** 8.76×10^{-3} **9.** 4.25

R.4 Exercises *(page 38)*

1. 16   **3.** $\dfrac{1}{16}$ **5.** $-\dfrac{1}{16}$ **7.** $\dfrac{1}{8}$ **9.** $\dfrac{1}{4}$ **11.** $\dfrac{1}{9}$ **13.** $\dfrac{81}{64}$ **15.** $\dfrac{27}{8}$ **17.** $\dfrac{81}{2}$ **19.** $\dfrac{4}{81}$ **21.** $\dfrac{1}{12}$ **23.** $-\dfrac{2}{3}$ **25.** y^2 **27.** $\dfrac{x}{y^2}$ **29.** $\dfrac{1}{64x^6}$ **31.** $-\dfrac{4}{x}$

33. 3   **35.** $\dfrac{1}{x^3 y}$ **37.** $\dfrac{1}{xy}$ **39.** $\dfrac{y}{x}$ **41.** $\dfrac{25x^2}{16y^2}$ **43.** $\dfrac{1}{x^2 y^2}$ **45.** $\dfrac{1}{x^3 y^3}$ **47.** $-\dfrac{8x^3}{9yz^2}$ **49.** $\dfrac{y^3}{x^8}$ **51.** $\dfrac{16x^2}{9y^2}$ **53.** $\dfrac{1}{x^3 y}$ **55.** $\dfrac{y^2}{x^2}$ **57.** $10; 0$ **59.** 81

61. 304,006.671   **63.** 0.004   **65.** 481.890   **67.** 0.000   **69.** 4.542×10^2 **71.** 1.3×10^{-2} **73.** 3.2155×10^4 **75.** 4.23×10^{-4} **77.** 61,500
79. 0.001214   **81.** 110,000,000   **83.** 0.081   **85.** 400,000,000 m   **87.** 0.0000005 m   **89.** 5×10^{-4} in.   **91.** 1.92×10^9 barrels
93. 5.865696×10^{12} mi

R.5 Concepts and Vocabulary *(page 49)*

1. $4; 8$ **2.** $-25; x^2 - 5x + 12$ **3.** $x^3 - 8$ **4.** F   **5.** T   **6.** F   **7.** $2x^4 + 5$ **8.** No; x has a non-integer exponent, $\dfrac{1}{2}$.

R.5 Exercises *(page 49)*

1. Monomial; Variable: x; Coefficient: 2; Degree: 3   **3.** Not a monomial   **5.** Monomial; Variables: x, y; Coefficient: -2; Degree: 3
7. Not a monomial   **9.** Not a monomial   **11.** Yes; 2   **13.** Yes; 0   **15.** No   **17.** Yes; 3   **19.** No   **21.** $x^2 + 7x + 2$ **23.** $x^3 - 4x^2 + 9x + 7$
25. $6x^5 + 5x^4 + 3x^2 + x$ **27.** $7x^2 - x - 7$ **29.** $-2x^3 + 18x^2 - 18$ **31.** $2x^2 - 4x + 6$ **33.** $15y^2 - 27y + 30$ **35.** $x^3 + x^2 - 4x$
37. $-8x^5 - 10x^2$ **39.** $x^3 + 3x^2 - 2x - 4$ **41.** $x^2 + 6x + 8$ **43.** $2x^2 + 9x + 10$ **45.** $x^2 - 2x - 8$ **47.** $x^2 - 5x + 6$
49. $2x^2 - x - 6$ **51.** $-2x^2 + 11x - 12$ **53.** $2x^2 + 8x + 8$ **55.** $x^2 - xy - 2y^2$ **57.** $-6x^2 - 13xy - 6y^2$ **59.** $x^2 - 49$ **61.** $4x^2 - 9$
63. $x^2 + 8x + 16$ **65.** $x^2 - 8x + 16$ **67.** $9x^2 - 16$ **69.** $4x^2 - 12x + 9$ **71.** $x^2 - y^2$ **73.** $9x^2 - y^2$ **75.** $x^2 + 2xy + y^2$
77. $x^2 - 4xy + 4y^2$ **79.** $x^3 - 6x^2 + 12x - 8$ **81.** $8x^3 + 12x^2 + 6x + 1$ **83.** $4x^2 - 3x + 1$; remainder 1
85. $4x^2 - 11x + 23$; remainder -45 **87.** $4x - 3$; remainder $x + 1$ **89.** $4x - 3$; remainder $-7x + 7$ **91.** 2; remainder $-3x^2 + x + 3$
93. $2x - \dfrac{5}{2}$; remainder $\dfrac{3}{2}x + \dfrac{7}{2}$ **95.** $-4x^2 - 3x - 3$; remainder -7 **97.** $x^2 - x - 1$; remainder $2x + 2$ **99.** $x^2 + ax + a^2$; remainder 0
101. -9

R.6 Concepts and Vocabulary *(page 58)*

1. $3x(x - 2)(x + 2)$ **2.** Prime   **3.** T   **4.** F   **5.** $x^2 + 2x + 1$ **6.** It means the polynomial is written as a product of prime polynomials.

R.6 Exercises *(page 58)*

1. $3(x + 2)$ **3.** $a(x^2 + 1)$ **5.** $x(x^2 + x + 1)$ **7.** $2x(x - 1)$ **9.** $3xy(x - 2y + 4)$ **11.** $(x - 1)(x + 1)$ **13.** $(2x - 1)(2x + 1)$
15. $(x - 4)(x + 4)$ **17.** $(5x - 2)(5x + 2)$ **19.** $(x + 1)^2$ **21.** $(x + 2)^2$ **23.** $(x - 5)^2$ **25.** $(2x + 1)^2$ **27.** $(4x + 1)^2$
29. $(x - 3)(x^2 + 3x + 9)$ **31.** $(x + 3)(x^2 - 3x + 9)$ **33.** $(2x + 3)(4x^2 - 6x + 9)$ **35.** $(x + 2)(x + 3)$ **37.** $(x + 6)(x + 1)$
39. $(x + 5)(x + 2)$ **41.** $(x - 8)(x - 2)$ **43.** $(x - 8)(x + 1)$ **45.** $(x + 8)(x - 1)$ **47.** $(x + 2)(2x + 3)$ **49.** $(x - 2)(2x + 1)$
51. $(2x + 3)(3x + 2)$ **53.** $(3x + 1)(x + 1)$ **55.** $(z + 1)(2z + 3)$ **57.** $(x + 2)(3x - 4)$ **59.** $(x - 2)(3x + 4)$
61. $(x + 4)(3x + 2)$ **63.** $(x + 4)(3x - 2)$ **65.** $(x - 6)(x + 6)$ **67.** $2(1 + 2x)(1 - 2x)$ **69.** $(x + 2)(x + 5)$ **71.** $(x - 7)(x - 3)$
73. $4(x^2 - 2x + 8)$ **75.** Prime   **77.** $-(x - 5)(x + 3)$ **79.** $3(x + 2)(x - 6)$ **81.** $y^2(y + 5)(y + 6)$ **83.** $(2x + 3)^2$
85. $2(3x + 1)(x + 1)$ **87.** $(x - 3)(x + 3)(x^2 + 9)$ **89.** $(x - 1)^2(x^2 + x + 1)^2$ **91.** $x^5(x - 1)(x + 1)$ **93.** $(4x + 3)^2$
95. $-(4x - 5)(4x + 1)$ **97.** $(2y - 5)(2y - 3)$ **99.** $-(3x - 1)(3x + 1)(x^2 + 1)$ **101.** $(x + 3)(x - 6)$ **103.** $(x + 2)(x - 3)$
105. $(3x - 5)(9x^2 - 3x + 7)$ **107.** $(x + 5)(3x + 11)$ **109.** $(x - 1)(x + 1)(x + 2)$ **111.** $(x - 1)(x + 1)(x^2 - x + 1)$
113. The integer pairs whose product is 4 are: $1, 4; -1, -4; 2, 2; -2, -2$; none of these have sums equal to 0, the coefficient of the middle term.

R.7 Concepts and Vocabulary *(page 69)*

1. Reduced to lowest terms or simplified   **2.** Least common multiple   **3.** T   **4.** F   **5.** A quotient of two polynomials.

R.7 Exercises *(page 69)*

1. $\dfrac{3}{x - 3}$ **3.** $\dfrac{x}{3}$ **5.** $\dfrac{4x}{2x - 1}$ **7.** $\dfrac{y + 5}{2(y + 1)}$ **9.** $\dfrac{x + 5}{x - 1}$ **11.** $-(x + 7)$ **13.** $\dfrac{3}{5x(x - 2)}$ **15.** $\dfrac{2x}{x + 4}$ **17.** $\dfrac{8}{3x}$ **19.** $\dfrac{x - 3}{x + 7}$

21. $\dfrac{4x}{(x - 2)(x - 3)}$ **23.** $\dfrac{4}{5(x - 1)}$ **25.** $-\dfrac{(x - 4)^2}{4x}$ **27.** $\dfrac{(x + 3)^2}{(x - 3)^2}$ **29.** $\dfrac{(x - 4)(x + 3)}{(x - 1)(2x + 1)}$ **31.** $\dfrac{x + 5}{2}$ **33.** $\dfrac{(x - 2)(x + 2)}{2x - 3}$

35. $\dfrac{3x-2}{x-3}$ **37.** $\dfrac{x+9}{2x-1}$ **39.** $\dfrac{4-x}{x-2}$ **41.** $\dfrac{2(x+5)}{(x-1)(x+2)}$ **43.** $\dfrac{3x^2-2x-3}{(x+1)(x-1)}$ **45.** $\dfrac{-11x-2}{(x+2)(x-2)}$ **47.** $\dfrac{2(x^2-2)}{x(x-2)(x+2)}$

49. $(x-2)(x+2)(x+1)$ **51.** $x(x-1)(x+1)$ **53.** $x^3(2x-1)^2$ **55.** $x(x-1)^2(x+1)(x^2+x+1)$ **57.** $\dfrac{5x}{(x-6)(x-1)(x+4)}$

59. $\dfrac{2(2x^2+5x-2)}{(x-2)(x+2)(x+3)}$ **61.** $\dfrac{5x+1}{(x-1)^2(x+1)^2}$ **63.** $\dfrac{-x^2+3x+13}{(x-2)(x+1)(x+4)}$ **65.** $\dfrac{x^3-2x^2+4x+3}{x^2(x+1)(x-1)}$ **67.** $\dfrac{-1}{x(x+h)}$ **69.** $\dfrac{x+1}{x-1}$

71. $\dfrac{(x-1)(x+1)}{x^2+1}$ **73.** $\dfrac{2(5x-1)}{(x-2)(x+1)^2}$ **75.** $\dfrac{-2x(x^2-2)}{(x+2)(x^2-x-3)}$ **77.** $\dfrac{-1}{x-1}$ **79.** $f=\dfrac{R_1\cdot R_2}{(n-1)(R_1+R_2)};\dfrac{2}{15}$ m

R.8 Concepts and Vocabulary *(page 77)*

1. Radical sign **2.** Index; radicand **3.** F **4.** T **5.** T **6.** F **7.** $\sqrt[3]{27}$

R.8 Exercises *(page 78)*

1. 5 **3.** 3 **5.** -4 **7.** $\dfrac{1}{3}$ **9.** $5x^2$ **11.** $2(1+x)$ **13.** $2\sqrt{2}$ **15.** $5\sqrt{2}$ **17.** $2\sqrt[3]{2}$ **19.** $-2\sqrt[3]{2}$ **21.** $\dfrac{5}{3}|x|$ **23.** x^3y^2 **25.** $6\sqrt{x}$ **27.** $6x\sqrt{x}$

29. 1 **31.** $\dfrac{4y^2}{3x}$ **33.** $15\sqrt[3]{3}$ **35.** $\dfrac{1}{x(2x+3)}$ **37.** 1 **39.** $7\sqrt{2}$ **41.** $\sqrt{2}$ **43.** $-\sqrt[3]{2}$ **45.** $(2x-15)\sqrt{2x}$ **47.** $(-x-5y)\sqrt[3]{2xy}$

49. $36\sqrt{2}$ **51.** $3-4\sqrt{3}$ **53.** $42+9\sqrt{7}$ **55.** $3-2\sqrt{2}$ **57.** $1-3\sqrt[3]{4}+3\sqrt[3]{2}$ **59.** $4x+4\sqrt{x}-15$ **61.** $\dfrac{-x^2}{\sqrt{1-x^2}}$ **63.** $\dfrac{2\sqrt{5}}{5}$

65. $\dfrac{4\sqrt{6}}{3}$ **67.** $\dfrac{\sqrt{x}}{x}$ **69.** $\dfrac{15-3\sqrt{2}}{23}$ **71.** $\dfrac{4-\sqrt{7}}{3}$ **73.** $\dfrac{15-2\sqrt{5}}{41}$ **75.** $5-2\sqrt{6}$ **77.** $\dfrac{\sqrt{x}-2}{x-4},x\neq4$ **79.** 1.41 **81.** 1.59

83. 4.89 **85.** 2.15 **87. (a)** About 15,660.4 gal **(b)** About 390.7 gal **89.** $2\sqrt{2}\pi\approx8.89$ s **91.** $\dfrac{\pi\sqrt{3}}{6}\approx0.91$ s

R.9 Concepts and Vocabulary *(page 83)*

1. $\sqrt[n]{a};\sqrt[n]{a}$ **2.** T **3.** T **4.** T **5.** T

R.9 Exercises *(page 83)*

1. 4 **3.** 9 **5.** $\dfrac{1}{8}$ **7.** $\dfrac{1}{27}$ **9.** $\dfrac{27}{8}$ **11.** $\dfrac{27}{8}$ **13.** 8 **15.** 8 **17.** 27 **19.** $\dfrac{1}{5}$ **21.** 9 **23.** $\dfrac{1}{7}$ **25.** 2 **27.** 3 **29.** $\dfrac{1}{2}$ **31.** $6^{2/3}$ **33.** $2^{5/6}$

35. $x^{1/2}$ **37.** $x^{7/4}$ **39.** x **41.** x^2y^4 **43.** $x^{4/3}y^{5/3}$ **45.** $\dfrac{8x^{3/2}}{y^{1/4}}$ **47.** $\dfrac{x^{11}}{y^3}$ **49.** $\dfrac{3x+2}{(1+x)^{1/2}}$ **51.** $\dfrac{x(3x^2+2)}{(x^2+1)^{1/2}}$ **53.** $\dfrac{22x+5}{10\sqrt{x}-5\sqrt{4x+3}}$

55. $\dfrac{2+x}{2(1+x)^{3/2}}$ **57.** $\dfrac{4-x}{(x+4)^{3/2}}$ **59.** $\dfrac{1}{x^2(x^2-1)^{1/2}}$ **61.** $\dfrac{1-3x^2}{2\sqrt{x}(1+x^2)^2}$ **63.** $\dfrac{1}{2}(5x+2)(x+1)^{1/2}$ **65.** $2x^{1/2}(3x-4)(x+1)$

67. $(x^2+4)^{1/3}(11x^2+12)$ **69.** $(3x+5)^{1/3}(2x+3)^{1/2}(17x+27)$ **71.** $\dfrac{3(x+2)}{2x^{1/2}}$

Review Exercises *(page 86)*

1. (a) None **(b)** $\{-10\}$ **(c)** $\left\{-10,0.65,1.343434...,\dfrac{1}{9}\right\}$ **(d)** $\{\sqrt{7}\}$ **(e)** $\left\{-10,0.65,1.343434...,\sqrt{7},\dfrac{1}{9}\right\}$ **3.** 14 **5.** $\dfrac{15}{22}$ **7.** $4x-12$

9. **11.** 5 **13.** -10 **15.** $\{x\,|\,x\neq6\}$ **17.** 4 **19.** $4\sqrt{x}(x+1)$ **21.** $\dfrac{125}{x^2y}$ **23.** $-5;3$ **25.** $2x^4-2x^3+x^2+5x+3$

27. $6x^2-7x-5$ **29.** $16x^2-1$ **31.** x^3-7x-6 **33.** $3x^2+8x+25$; remainder 79 **35.** $-3x^2+4$; remainder -2

37. $x^4-x^3+x^2-x+1$; remainder 0 **39.** $(x+7)(x-2)$ **41.** $(3x+2)(2x-3)$ **43.** $3(x+2)(x-7)$ **45.** $(2x+1)(4x^2-2x+1)$

47. $(2x+3)(x-1)(x+1)$ **49.** $(5x-2)(5x+2)$ **51.** Prime **53.** $(x+4)^2$ **55.** $\dfrac{2x+7}{x-2}$ **57.** $\dfrac{3(3x-1)}{(x+3)(3x+1)}$

59. $\dfrac{4x}{(x+1)(x-1)}$ **61.** $\dfrac{x^2+17x+2}{(x-2)(x+2)^2}$ **63.** $\dfrac{x-1}{x-3}$ **65.** $\dfrac{3x}{5y^2}$ **67.** $3xy^4\sqrt[3]{x}$ **69.** $\dfrac{4\sqrt{5}}{5}$ **71.** $-2(1+\sqrt{2})$ **73.** $-\dfrac{3+\sqrt{5}}{2}$

75. $\dfrac{2(1+x^2)}{(2+x^2)^{1/2}}$ **77.** $\dfrac{x(3x+16)}{2(x+4)^{3/2}}$ **79.** 2.81421906×10^8 **81.** Yes **83.** \$0.35 per share **85.** 216 ft^2; 84 ft **87.** Yes; about 229.2 mi

C H A P T E R 1   Graphs

1.1 Concepts and Vocabulary *(page 97)*

1. Abscissa; ordinate   **2.** Quadrants   **3.** Midpoint   **4.** F   **5.** F   **6.** T   **7.** Determining Xmin, Xmax, Xscl, Ymin, Ymax, Yscl

1.1 Exercises *(page 97)*

1. **(a)** Quadrant II   **(b)** Positive x-axis
(c) Quadrant III   **(d)** Quadrant I
(e) Negative y-axis **(f)** Quadrant IV

3. The points will be on a vertical line that is 2 units to the right of the y-axis

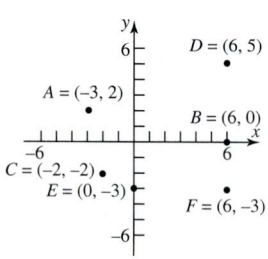

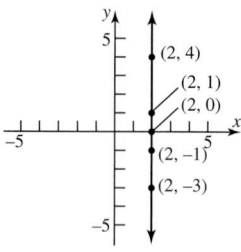

5. $(-1, 4)$; Quadrant II   **7.** $(3, 1)$; Quadrant I   **9.** Xmin $= -11$, Xmax $= 5$, Xscl $= 1$, Ymin $= -3$, Ymax $= 6$, Yscl $= 1$
11. Xmin $= -30$, Xmax $= 50$, Xscl $= 10$, Ymin $= -90$, Ymax $= 50$, Yscl $= 10$ **13.** Xmin $= -10$, Xmax $= 110$, Xscl $= 10$,
Ymin $= -10$, Ymax $= 160$, Yscl $= 10$ **15.** Xmin $= -6$, Xmax $= 6$, Xscl $= 2$, Ymin $= -4$, Ymax $= 4$, Yscl $= 2$
17. Xmin $= -6$, Xmax $= 6$, Xscl $= 2$, Ymin $= -1$, Ymax $= 3$, Yscl $= 1$ **19.** Xmin $= 3$, Xmax $= 9$, Xscl $= 1$, Ymin $= 2$,
Ymax $= 10$, Yscl $= 2$ **21.** $\sqrt{5}$ **23.** $\sqrt{10}$ **25.** $2\sqrt{5}$ **27.** $\sqrt{85}$ **29.** $\sqrt{53}$ **31.** $\sqrt{6.89} \approx 2.62$ **33.** $\sqrt{a^2 + b^2}$ **35.** $4\sqrt{10}$ **37.** $2\sqrt{65}$

39. $d(A, B) = \sqrt{13}$
$d(B, C) = \sqrt{13}$
$d(A, C) = \sqrt{26}$
$(\sqrt{13})^2 + (\sqrt{13})^2 = (\sqrt{26})^2$
Area $= \dfrac{13}{2}$ square units

41. $d(A, B) = \sqrt{130}$
$d(B, C) = \sqrt{26}$
$d(A, C) = 2\sqrt{26}$
$(\sqrt{26})^2 + (2\sqrt{26})^2 = (\sqrt{130})^2$
Area $= 26$ square units

43. $d(A, B) = 4$
$d(A, C) = 5$
$d(B, C) = \sqrt{41}$
$4^2 + 5^2 = 16 + 25 = (\sqrt{41})^2$
Area $= 10$ square units

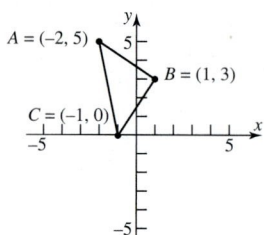

45. $(2, 2)$; $(2, -4)$ **47.** $(0, 0)$; $(8, 0)$ **49.** $(4, 3)$ **51.** $\left(\dfrac{3}{2}, 1\right)$ **53.** $(5, -1)$ **55.** $(1.05, 0.7)$ **57.** $\left(\dfrac{a}{2}, \dfrac{b}{2}\right)$ **59.** $\sqrt{17}$; $2\sqrt{5}$; $\sqrt{29}$

61. $d(P_1, P_2) = 6$; $d(P_2, P_3) = 4$; $d(P_1, P_3) = 2\sqrt{13}$; right triangle   **63.** $d(P_1, P_2) = 2\sqrt{17}$; $d(P_2, P_3) = \sqrt{34}$; $d(P_1, P_3) = \sqrt{34}$;
isosceles right triangle   **65.** $90\sqrt{2} \approx 127.28$ ft   **67.** **(a)** $(90, 0)$, $(90, 90)$, $(0, 90)$ **(b)** $5\sqrt{2161} \approx 232.43$ ft   **(c)** $30\sqrt{149} \approx 366.20$ ft
69. $d = 50t$

1.2 Concepts and Vocabulary *(page 108)*

1. Intercepts   **2.** Zeros; roots   **3.** T   **4.** F   **5.** A complete graph presents enough of the illustration so that a viewer of the graph can
visualize the rest of the graph as an obvious continuation.   **6.** Xmin $= -10$, Xmax $= 10$; Xscl $= 1$, Ymin $= -10$, Ymax $= 10$, Yscl $= 1$
7. Answers will vary. One example is shown in Exercise 14 on page 108.

1.2 Exercises *(page 108)*

1. $(0, 0)$ is on the graph.   **3.** $(0, 3)$ is on the graph.   **5.** $(0, 2)$ and $(\sqrt{2}, \sqrt{2})$ are on the graph.   **7.** $(-1, 0)$, $(1, 0)$

9. $\left(-\dfrac{\pi}{2}, 0\right)$, $(0, 1)$, $\left(\dfrac{\pi}{2}, 0\right)$ **11.** $(0, 0)$ **13.** $(-4, 0)$, $(-1, 0)$, $(0, -3)$, $(4, 0)$ **15.** $(-1.5, 0)$, $(0, -2)$, $(1.5, 0)$ **17.** None   **19.** $-\dfrac{2}{5}$
21. $2a + 3b = 6$

23. x-intercept: -2;
y-intercept: 2

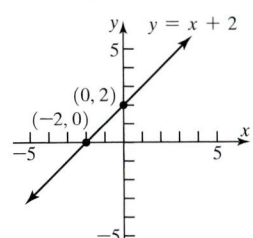

25. x-intercept: -4;
y-intercept: 8

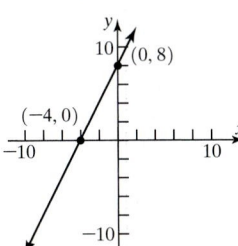

27. x-intercepts: $-1, 1$;
y-intercept: -1

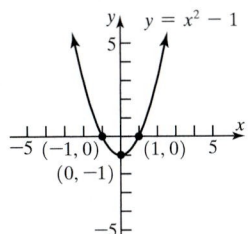

29. x-intercepts: $-2, 2$;
y-intercept: 4

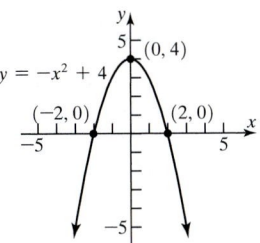

31. x-intercept: 3;
y-intercept: 2

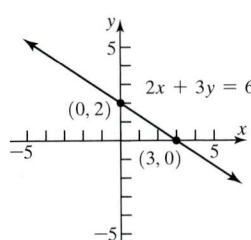

33. x-intercepts: $-2, 2$;
y-intercept: 9

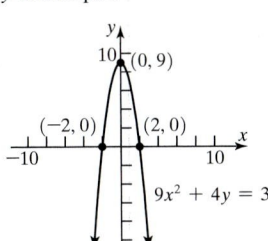

35. x-intercept: 6.5;
y-intercept: -13

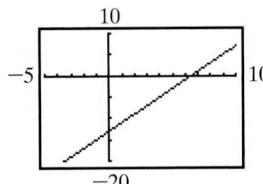

37. x-intercepts: $-2.74, 2.74$;
y-intercept: -15

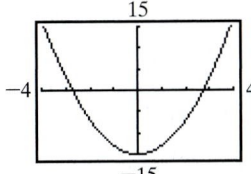

39. x-intercept: 14.33;
y-intercept: -21.5

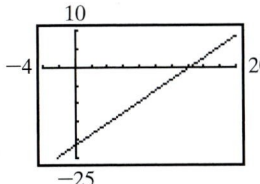

41. x-intercepts: $-2.72, 2.72$;
y-intercept: 12.33

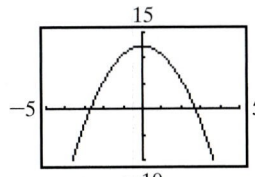

1.3 Concepts and Vocabulary *(page 125)*

1. Repeated; multiplicity two   **2.** Discriminant; negative   **3.** First; second   **4.** T   **5.** F   **6.** T
7. Factoring, completing the square, quadratic formula, using a graphing utility. Only using a graphing utility results in non-exact solutions.
8. It tells you the nature of the solution.

1.3 Exercises *(page 125)*

1. $-2.21, 0.54, 1.68$ **3.** $-1.55, 1.15$ **5.** $-1.12, 0.36$ **7.** $-2.69, -0.49, 1.51$ **9.** $-2.86, -1.34, 0.20, 1.00$ **11.** No real solutions

13. $\left\{\dfrac{4}{3}\right\}$ **15.** $\{3\}$ **17.** $\{-2\}$ **19.** $\left\{-\dfrac{4}{3}\right\}$ **21.** $\{-18\}$ **23.** $\{-4\}$ **25.** $\{2\}$ **27.** $\{0.5\}$ **29.** $\left\{\dfrac{46}{5}\right\}$ **31.** $\{3\}$ **33.** $\{2\}$ **35.** $\{-1\}$

37. $\{0, 9\}$ **39.** $\{-5, 5\}$ **41.** $\{-3, 2\}$ **43.** $\left\{-\dfrac{1}{2}, 3\right\}$ **45.** $\{-4, 4\}$ **47.** $\{-6, -2\}$ **49.** $\left\{\dfrac{3}{2}\right\}$ **51.** $\left\{-\dfrac{5}{2}, 3\right\}$ **53.** $\left\{-\dfrac{3}{4}, 2\right\}$

55. $\{-5, 0, 4\}$ **57.** $\{-1, 1\}$ **59.** $\{-2, 2, 3\}$ **61.** $\{-5, 5\}$ **63.** $\{-1, 3\}$ **65.** $\{-3, 0\}$ **67.** 16   **69.** $\dfrac{1}{16}$ **71.** $\{-7, 3\}$ **73.** $\left\{-\dfrac{1}{4}, \dfrac{3}{4}\right\}$

75. $\left\{\dfrac{-1 - \sqrt{7}}{6}, \dfrac{-1 + \sqrt{7}}{6}\right\}$ **77.** $\{2 - \sqrt{2}, 2 + \sqrt{2}\}$ **79.** $\{2 - \sqrt{5}, 2 + \sqrt{5}\}$ **81.** $\left\{1, \dfrac{3}{2}\right\}$ **83.** No real solution

85. $\left\{\dfrac{-1 - \sqrt{5}}{4}, \dfrac{-1 + \sqrt{5}}{4}\right\}$ **87.** $\left\{\dfrac{9 - \sqrt{113}}{8}, \dfrac{9 + \sqrt{113}}{8}\right\}$ **89.** $\left\{\dfrac{1}{3}\right\}$ **91.** $\left\{-\dfrac{2}{3}, 1\right\}$ **93.** $\left\{\dfrac{1 - \sqrt{33}}{8}, \dfrac{1 + \sqrt{33}}{8}\right\}$ **95.** $\{-\sqrt{5}, \sqrt{5}\}$

97. $\left\{\dfrac{1}{4}\right\}$ **99.** $\left\{-\dfrac{3}{5}, \dfrac{5}{2}\right\}$ **101.** $\left\{-\dfrac{1}{2}, \dfrac{2}{3}\right\}$ **103.** $\left\{\dfrac{-\sqrt{2} + 2}{2}, \dfrac{-\sqrt{2} - 2}{2}\right\}$ **105.** $\left\{\dfrac{-1 - \sqrt{17}}{2}, \dfrac{-1 + \sqrt{17}}{2}\right\}$ **107.** No real solution

109. Repeated real solution   **111.** Two unequal real solutions   **113.** $x = \dfrac{b + c}{a}$ **115.** $x = \dfrac{abc}{a + b}$ **117.** 3   **119.** $R = \dfrac{R_1 R_2}{R_1 + R_2}$

121. $R = \dfrac{mv^2}{F}$

1.4 Concepts and Vocabulary *(page 137)*

1. Mathematical modeling   **2.** Interest   **3.** Uniform motion   **4.** T   **5.** T   **6.** $100 - x$

1.4 Exercises *(page 137)*

1. $A = \pi r^2$; r = Radius, A = Area   **3.** $A = s^2$; A = Area, s = Length of a side   **5.** $F = ma$; F = Force, m = Mass, a = Acceleration
7. $W = Fd$; W = Work, F = Force, d = Distance   **9.** $C = 150x$; C = Total cost in dollars, x = Number of dishwashers
11. \$11,000 will be invested in bonds and \$9000 in CDs.   **13.** David will receive \$400,000, Paige \$300,000, and Dan \$200,000.
15. The regular hourly rate is \$8.50   **17.** The Bears got 5 touchdowns.   **19.** The length is 19 ft; the width is 11 ft
21. Brooke needs a score of 85.   **23.** The original price was \$147,058.82; purchasing the model saves \$22,058.82.
25. The bookstore paid \$44.80.   **27.** Invest \$31,250 in bonds and \$18,750 in CDs.   **29.** \$11,600 was loaned out at 8%.
31. Mix 75 lb of Earl Gray tea with 25 lb of Orange Pekoe tea.   **33.** Mix 40 lb of cashews with the peanuts.
35. Add $\dfrac{20}{3} \approx 6.67$ oz of pure water.   **37.** The Metra commuter averages 30 mph; the Amtrak averages 80 mph.
39. The speed of the current is $\dfrac{16}{7} \approx 2.29$ mph.   **41.** Working together, it takes 12 min.   **43.** Start the auxiliary pump at 9:45 A.M.
45. The dimensions are 11 ft by 13 ft.   **47.** The dimensions are 5 m by 8 m.   **49.** The speed of the current is 5 mph.
51. (a) The dimensions are 10 ft by 5 ft.   **(b)** The area is 50 sq ft.   **(c)** The dimensions would be 7.5 ft by 7.5 ft.
(d) The area would be 56.25 sq ft.   **53.** The border will be approximately 2.71 ft wide.   **55.** The border will be approximately 2.56 ft wide.
57. Add $\dfrac{2}{3}$ gal of water.   **59.** 5 lb must be added.   **61.** The dimensions should be approximately 11.55 cm by 6.55 cm by 3 cm.
63. 40 g of 12 karat gold should be mixed with 20 g of pure gold.   **65.** Mike passes Dan $\dfrac{1}{3}$ mi from the start, 2 min from the time
Mike started to race.   **67.** 60 minutes   **69.** The most you can invest in the CD is \$66,666.67.   **71.** The average speed is 49.5 mph.
73. Set the original price at \$40. At 50% off, there will be no profit at all.

1.5 Concepts and Vocabulary *(page 147)*

1. $-a$ **2.** Extraneous   **3.** -2 and 2   **4.** T   **5.** T   **6.** F   **7.** Squaring both sides
9. The absolute value of a real number is always greater than or equal to zero.

1.5 Exercises *(page 147)*

1. $\{9\}$ **3.** $\{22\}$ **5.** $\{1\}$ **7.** No real solution   **9.** $\{-13\}$ **11.** $\{0, 36\}$ **13.** $\{3\}$ **15.** $\{2\}$ **17.** $\left\{-\dfrac{8}{5}\right\}$ **19.** $\{8\}$ **21.** $\{-1, 3\}$
23. $\{1, 5\}$ **25.** $\{5\}$ **27.** $\{2\}$ **29.** $\{-4, 4\}$ **31.** $\{-2, 2\}$ **33.** $\{-2, -1, 1, 2\}$ **35.** $\{-1, 1\}$ **37.** $\{-2, 1\}$ **39.** $\{-6, -5\}$ **41.** $\left\{-\dfrac{1}{3}\right\}$
43. $\left\{-\dfrac{3}{2}, 2\right\}$ **45.** $\{0, 16\}$ **47.** $\{16\}$ **49.** $\{1\}$ **51.** $\left\{\left(\dfrac{9 - \sqrt{17}}{8}\right)^4, \left(\dfrac{9 + \sqrt{17}}{8}\right)^4\right\}$ **53.** $\{\sqrt{2}, \sqrt{3}\}$ **55.** $\left\{-2, -\dfrac{1}{2}\right\}$ **57.** $\left\{-\dfrac{3}{2}, \dfrac{1}{3}\right\}$
59. $\left\{-\dfrac{1}{8}, 27\right\}$ **61.** $\left\{-2, -\dfrac{4}{5}\right\}$ **63.** $\{-6, 6\}$ **65.** $\{-4, 1\}$ **67.** $\left\{-1, \dfrac{3}{2}\right\}$ **69.** $\{-4, 4\}$ **71.** $\left\{-\dfrac{1}{2}, \dfrac{1}{2}\right\}$ **73.** $\left\{-\dfrac{27}{2}, \dfrac{27}{2}\right\}$ **75.** $\left\{-\dfrac{36}{5}, \dfrac{24}{5}\right\}$
77. No real solution   **79.** $\{-3, 3\}$ **81.** $\{-1, 3\}$ **83.** $\{-2, -1, 0, 1\}$ **85.** $\{0.34, 11.66\}$ **87.** $\{-1.03, 1.03\}$ **89.** $\{-1.85, 0.17\}$
91. $\left\{\dfrac{3}{2}, 5\right\}$ **93.** The distance is approximately 229.94 ft.

1.6 Concepts and Vocabulary *(page 159)*

1. Negative   **2.** Closed interval   **3.** Multiplication properties   **4.** T   **5.** T   **6.** F   **7.** The absolute value of a real number x is always
greater than or equal to zero.   **8.** Because $x^2 + 1$ is greater than or equal to 1 for all real numbers x.

1.6 Exercises *(page 160)*

1. $[0, 2]$; $0 \le x \le 2$ **3.** $(-1, 2)$; $-1 < x < 2$ **5.** $[0, 3)$; $0 \le x < 3$ **7. (a)** $6 < 8$ **(b)** $-2 < 0$ **(c)** $9 < 15$ **(d)** $-6 > -10$
9. (a) $7 > 0$ **(b)** $-1 > -8$ **(c)** $12 > -9$ **(d)** $-8 < 6$ **11. (a)** $2x + 4 < 5$ **(b)** $2x - 4 < -3$ **(c)** $6x + 3 < 6$ **(d)** $-4x - 2 > -4$
13. $[0, 4]$

15. $[4, 6)$

17. $[4, \infty)$

19. $(-\infty, -4)$

21. $2 \le x \le 5$

23. $-3 < x < -2$

25. $x \ge 4$

27. $x < -3$

29. $<$ **31.** $>$ **33.** $\ge$ **35.** $<$ **37.** $\le$ **39.** $<$ **41.** $\ge$

43. $\{x|x < 4\}; (-\infty, 4)$

4

45. $\{x|x \geq -1\}; [-1, \infty)$

-1

47. $\{x|x > 3\}; (3, \infty)$

3

49. $\{x|x \geq 2\}; [2, \infty)$

2

51. $\{x|x > -7\}; (-7, \infty)$

-7

53. $\left\{x \middle| x \leq \dfrac{2}{3}\right\}; \left(-\infty, \dfrac{2}{3}\right]$

$\dfrac{2}{3}$

55. $\{x|x < -20\}; (-\infty, -20)$

-20

57. $\left\{x \middle| x \geq \dfrac{4}{3}\right\}; \left[\dfrac{4}{3}, \infty\right)$

$\dfrac{4}{3}$

59. $\{x|3 \leq x \leq 5\}; [3, 5]$

3   5

61. $\left\{x \middle| \dfrac{2}{3} \leq x \leq 3\right\}; \left[\dfrac{2}{3}, 3\right]$

$\dfrac{2}{3}$ 3

63. $\left\{x \middle| -\dfrac{11}{2} < x < \dfrac{1}{2}\right\}; \left(-\dfrac{11}{2}, \dfrac{1}{2}\right)$

$-\dfrac{11}{2}$ $\dfrac{1}{2}$

65. $\{x|-6 < x < 0\}; (-6, 0)$

-6   0

67. $\{x|x < -5\}; (-\infty, -5)$

-5

69. $\{x|x \geq -1\}; [-1, \infty)$

-1

71. $\left\{x \middle| \dfrac{1}{2} \leq x < \dfrac{5}{4}\right\}; \left[\dfrac{1}{2}, \dfrac{5}{4}\right)$

$\dfrac{1}{2}$ $\dfrac{5}{4}$

73. $\{x|-6 < x < 6\}; (-6, 6)$

-6   6

75. $\{x|x < -4 \text{ or } x > 4\}$

$(-\infty, -4) \text{ or } (4, \infty)$

-4   -4

77. $\{x|-4 < x < 4\}; (-4, 4)$

-4   4

79. $\{x|x < -4 \text{ or } x > 4\};$

$(-\infty, -4) \text{ or } (4, \infty)$

-4   4

81. $\{x|1 < x < 3\}; (1, 3)$

1   3

83. $\left\{t \middle| -\dfrac{2}{3} \leq t \leq 2\right\}; \left[-\dfrac{2}{3}, 2\right]$

$-\dfrac{2}{3}$ 2

85. $\{x|x \leq 1 \text{ or } x \geq 5\};$

$(-\infty, 1] \text{ or } [5, \infty)$

1   5

87. $\left\{x \middle| -1 < x < \dfrac{3}{2}\right\}; \left(-1, \dfrac{3}{2}\right)$

-1 $\dfrac{3}{2}$

89. $\{x|x < -1 \text{ or } x > 2\};$

$(-\infty, -1) \text{ or } (2, \infty)$

-1   2

91. No real solution; $\varnothing$

0

93. $\left|x - 2\right| < \dfrac{1}{2}; \left\{x \middle| \dfrac{3}{2} < x < \dfrac{5}{2}\right\}$

95. $|x + 3| > 2; \{x|x < -5 \text{ or } x > -1\}$ **97.** $21 < \text{age} < 30$ **99.** $|x - 98.6| \geq 1.5; \{x|x \leq 97.1 \text{ or } x \geq 100.1\}$

101. (a) Male ≥ 73.4 **(b)** Female ≥ 79.7 **(c)** A female can expect to live at least 6.3 years longer.   **103.** The agent's commission ranges from $45,000 to $95,000, inclusive. As a percent of selling price, the commission ranges from 5% to approximately 8.6%, inclusive.

105. The amount withheld varies from $72.14 to $93.14, inclusive.

107. The usage varies from approximately 675.43 to 2500.86 kilowatt-hours, inclusive.

109. The dealer's cost varies from $7457.63 to $7857.14, inclusive.   **111.** You need at least a 74 on the last test.

113. The amount of gasoline ranged from 12 to 20 gal, inclusive.

115. $\dfrac{a + b}{2} - a = \dfrac{a + b - 2a}{2} = \dfrac{b - a}{2} > 0;$ therefore, $a < \dfrac{a + b}{2}.$

$b - \dfrac{a + b}{2} = \dfrac{2b - a - b}{2} = \dfrac{b - a}{2} > 0;$ therefore, $b > \dfrac{a + b}{2}.$

117. $(\sqrt{ab})^2 - a^2 = ab - a^2 = a(b - a) > 0;$ thus, $(\sqrt{ab})^2 > a^2$ and $\sqrt{ab} > a$

$b^2 - (\sqrt{ab})^2 = b^2 - ab = b(b - a) > 0;$ thus $b^2 > (\sqrt{ab})^2$ and $b > \sqrt{ab}$

119. $h = \dfrac{1}{\dfrac{1}{2}\left(\dfrac{1}{a} + \dfrac{1}{b}\right)} = \dfrac{2}{\dfrac{b}{ab} + \dfrac{a}{ab}} = \dfrac{2ab}{a + b}$

$h - a = \dfrac{2ab}{a + b} - a = \dfrac{ab - a^2}{a + b} = \dfrac{a(b - a)}{a + b} > 0;$ thus, $h > a.$

$b - h = b - \dfrac{2ab}{a + b} = \dfrac{b^2 - ab}{a + b} = \dfrac{b(b - a)}{a + b} > 0;$ thus $h < b.$

1.7 Concepts and Vocabulary *(page 176)*

1. Undefined; zero   **2.** $m_1 = m_2$; y-intercepts; $m_1 = -\dfrac{1}{m_2}$ **3.** $y = b$; y-intercept   **4.** T   **5.** F   **6.** F   **7.** No, consider vertical lines

8. No; no   **9.** They are the same line.   **10.** They are the same line.   **11.** No   **12.** No

1.7 Exercises *(page 176)*

1. (a) $\dfrac{1}{2}$ **(b)** If x increases by 2 units, y will increase by 1 unit. **3. (a)** $-\dfrac{1}{3}$ **(b)** If x increases by 3 units, y will decrease by 1 unit.

5. Slope $= -\dfrac{3}{2}$ **7.** Slope $= -\dfrac{1}{2}$ **9.** Slope $= 0$ **11.** Slope undefined

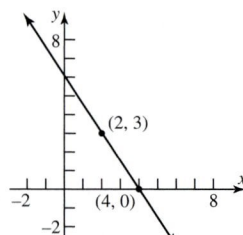 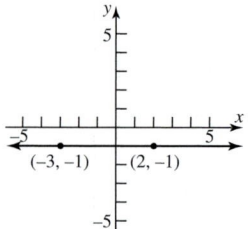

13.   **15.**   **17.**   **19.**

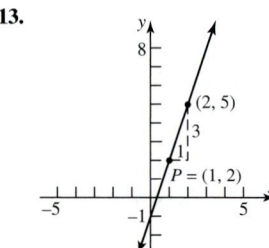 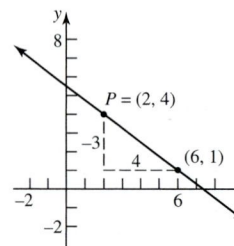 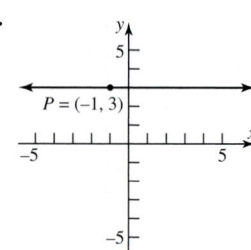 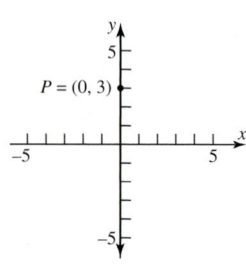

21. $(2, 6); (3, 10); (4, 14)$ **23.** $(4, -7); (6, -10); (8, -13)$ **25.** $(-1, -5); (0, -7); (1, -9)$ **27.** $x - 2y = 0$ or $y = \dfrac{1}{2}x$

29. $x + y = 2$ or $y = -x + 2$ **31.** $2x - y = 3$ or $y = 2x - 3$ **33.** $x + 2y = 5$ or $y = -\dfrac{1}{2}x + \dfrac{5}{2}$ **35.** $3x - y = -9$ or $y = 3x + 9$

37. $2x + 3y = -1$ or $y = -\dfrac{2}{3}x - \dfrac{1}{3}$ **39.** $x - 2y = -5$ or $y = \dfrac{1}{2}x + \dfrac{5}{2}$ **41.** $3x + y = 3$ or $y = -3x + 3$

43. $x - 2y = 2$ or $y = \dfrac{1}{2}x - 1$ **45.** $x = 2$; no slope-intercept form   **47.** $2x - y = -4$ or $y = 2x + 4$ **49.** $2x - y = 0$ or $y = 2x$

51. $x = 4$; no slope–intercept form   **53.** $2x + y = 0$ or $y = -2x$ **55.** $x - 2y = -3$ or $y = \dfrac{1}{2}x + \dfrac{3}{2}$ **57.** $y = 4$

59. Slope $= 2$; y-intercept $= 3$ **61.** $y = 2x - 2$; Slope $= 2$; y-intercept $= -2$ **63.** Slope $= \dfrac{1}{2}$; y-intercept $= 2$

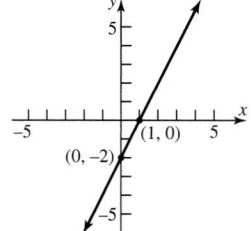 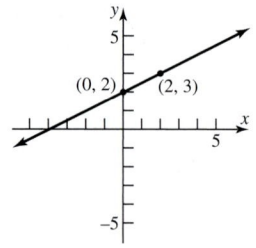

65. $y = -\dfrac{1}{2}x + 2$; Slope $= -\dfrac{1}{2}$;   **67.** $y = \dfrac{2}{3}x - 2$; Slope $= \dfrac{2}{3}$;   **69.** $y = -x + 1$; Slope $= -1$;

y-intercept $= 2$ y-intercept $= -2$ y-intercept $= 1$

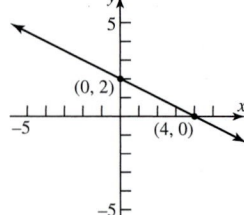 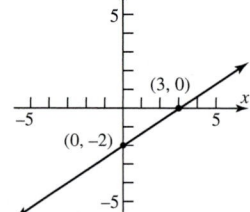 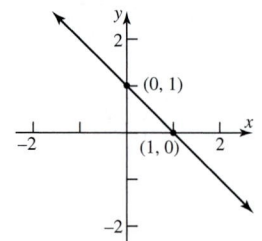

71. Slope undefined; no y-intercept

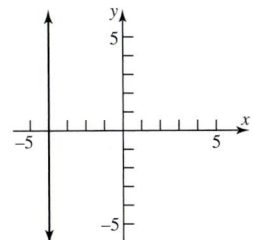

73. Slope $= 0$; y-intercept $= 5$

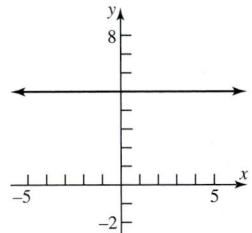

75. $y = x$; Slope $= 1$; y-intercept $= 0$

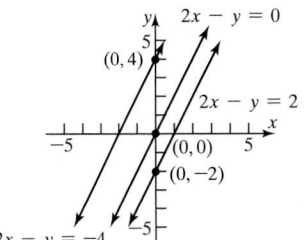

77. $y = \dfrac{3}{2}x$; Slope $= \dfrac{3}{2}$; y-intercept $= 0$

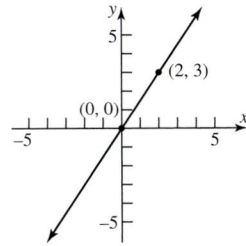

79. $y = 0$

81. (b)   **83.** (d)

85. $x - y = -2$ or $y = x + 2$

87. $x + 3y = 3$ or $y = -\dfrac{1}{3}x + 1$

89. $C = 0.07x + 29$; $36.70; $45.10

91. $^{\circ}\mathrm{C} = \dfrac{5}{9}(^{\circ}\mathrm{F} - 32)$; approximately $21.11^{\circ}\mathrm{C}$

93. (a) $P = 0.5x - 100$
(b) $400   (c) $2400

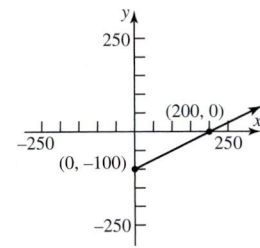

95. $C = 0.06543x + 5.65$; $25.28; $54.72

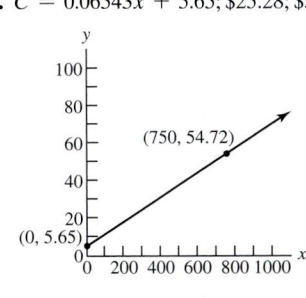

97. All have the same slope, 2;
the lines are parallel.

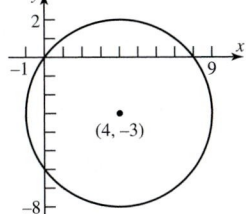

99. (b), (c), (e), (g)   **101.** (c)

1.8 Concepts and Vocabulary *(page 184)*

1. Add; $\dfrac{25}{4}$ **2.** Radius   **3.** T   **4.** F   **7.** 3

1.8 Exercises *(page 184)*

1. Center $(2, 1)$; Radius 2; $(x - 2)^2 + (y - 1)^2 = 4$ **3.** Center $\left(\dfrac{5}{2}, 2\right)$; Radius $\dfrac{3}{2}$; $\left(x - \dfrac{5}{2}\right)^2 + (y - 2)^2 = \dfrac{9}{4}$

5. $x^2 + y^2 = 4$;
$x^2 + y^2 - 4 = 0$

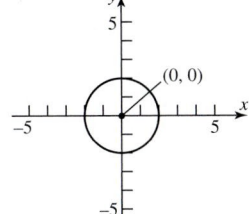

7. $x^2 + (y - 2)^2 = 4$;
$x^2 + y^2 - 4y = 0$

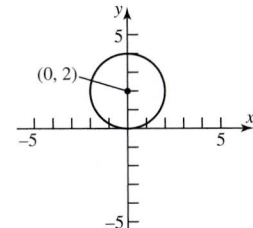

9. $(x - 4)^2 + (y + 3)^2 = 25$;
$x^2 + y^2 - 8x + 6y = 0$

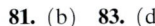

11. $(x + 2)^2 + (y - 1)^2 = 16$;
$x^2 + y^2 + 4x - 2y - 11 = 0$

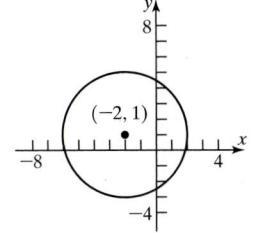

13. $\left(x - \dfrac{1}{2}\right)^2 + y^2 = \dfrac{1}{4}$;

$x^2 + y^2 - x = 0$

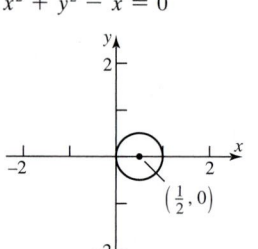

15. $(h, k) = (0, 0); r = 5$

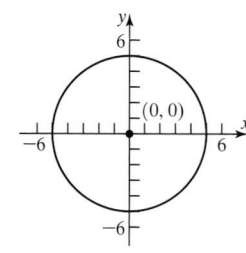

17. $(h, k) = (2, 0); r = 2$

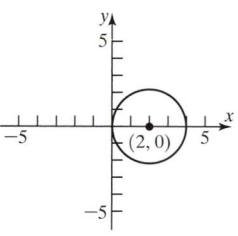

19. $(h, k) = (-2, 2); r = 3$

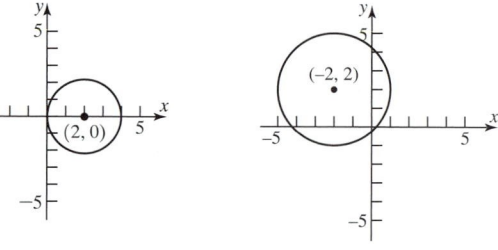

21. $(h, k) = \left(\dfrac{1}{2}, -1\right); r = \dfrac{1}{2}$

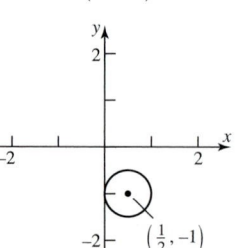

23. $(h, k) = (3, -2); r = 5$

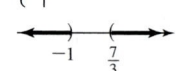

25. $x^2 + y^2 - 13 = 0$

27. $x^2 + y^2 - 4x - 6y + 4 = 0$

29. $x^2 + y^2 + 2x - 6y + 5 = 0$

31. (c)   **33.** (b)

35. $(x + 3)^2 + (y - 1)^2 = 16$

37. $(x - 2)^2 + (y - 2)^2 = 9$

39. (b), (c), (e), (g)

41. $x^2 + y^2 + 2x + 4y - 4168.16 = 0$

43. $\sqrt{2}x + 4y - 9\sqrt{2} = 0$ **45.** $(1, 0)$ **47.** $y = 2$

Review Exercises *(page 189)*

1. $\{-18\}$ **3.** $\{6\}$ **5.** $\left\{\dfrac{1}{5}\right\}$ **7.** $\{6\}$ **9.** No real solution   **11.** $\left\{\dfrac{11}{8}\right\}$ **13.** $\left\{-2, \dfrac{3}{2}\right\}$ **15.** $\left\{\dfrac{1 - \sqrt{13}}{4}, \dfrac{1 + \sqrt{13}}{4}\right\}$ **17.** $\{-3, 3\}$

19. No real solution   **21.** $\{-2, -1, 1, 2\}$ **23.** $\{2\}$ **25.** $\{0\}$ **27.** $\left\{\dfrac{\sqrt{5}}{2}\right\}$ **29.** $\left\{-\dfrac{1}{8}, 1\right\}$ **31.** $\left\{-1, \dfrac{1}{2}\right\}$ **33.** $\left\{-\dfrac{9}{5}\right\}$ **35.** $\{-5, 2\}$

37. $\left\{-\dfrac{5}{3}, 3\right\}$ **39.** $-2.49, 0.66, 1.83$ **41.** $-1.14, 1.64$

43. $\{x | x \geq 14\}; [14, \infty)$

14

45. $\left\{x \left| -\dfrac{31}{2} \leq x \leq \dfrac{33}{2}\right.\right\}; \left[-\dfrac{31}{2}, \dfrac{33}{2}\right]$

$-\dfrac{31}{2}$ $\dfrac{33}{2}$

47. $\{x | -23 < x < -7\}; (-23, -7)$

-23 -7

49. $\left\{x \left| -\dfrac{3}{2} < x < -\dfrac{7}{6}\right.\right\}; \left(-\dfrac{3}{2}, -\dfrac{7}{6}\right)$

$-\dfrac{3}{2}$ $\dfrac{7}{6}$

51. $\{x | x \leq -2 \text{ or } x \geq 7\}; (-\infty, -2] \text{ or } [7, \infty)$

-2 7

53. $\left\{x \left| 0 \leq x \leq \dfrac{4}{3}\right.\right\}; \left[0, \dfrac{4}{3}\right]$

0 $\dfrac{4}{3}$

55. $\left\{x \left| x < -1 \text{ or } x > \dfrac{7}{3}\right.\right\}; (-\infty, -1) \text{ or } \left(\dfrac{7}{3}, \infty\right)$

-1 $\dfrac{7}{3}$

57.

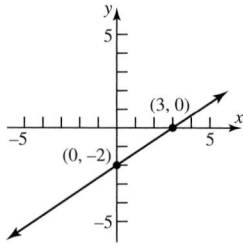

59.

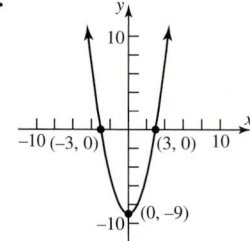

61.

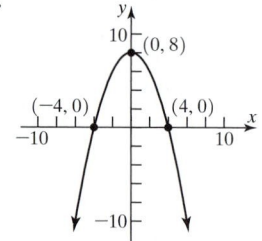

63. Center $(1, -2)$; radius $= 3$

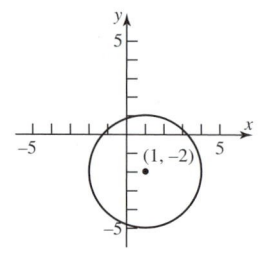

65. Center $(1, -2)$; radius $= \sqrt{5}$

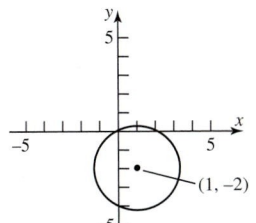

67. $2x + y = 5$ or $y = -2x + 5$

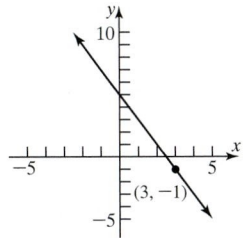

69. $x = -3$; no slope-intercept form

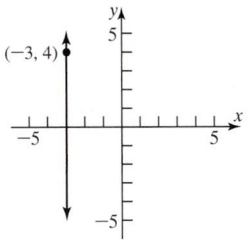

71. $x + 5y = -10$ or $y = -\dfrac{1}{5}x - 2$

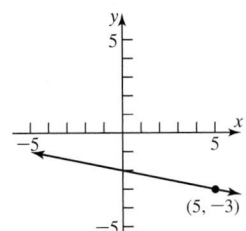

73. $2x - 3y = -19$ or $y = \dfrac{2}{3}x + \dfrac{19}{3}$

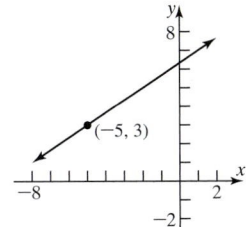

75. $-x + y = -7$ or $y = x - 7$

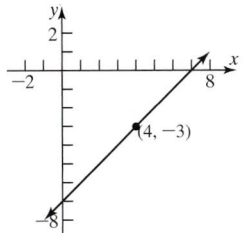

77. Slope $= \dfrac{1}{5}$; distance $= 2\sqrt{26}$; midpoint $= (2, 3)$ **79.** $d(A, B) = \sqrt{13}$; $d(B, C) = \sqrt{13}$ **81.** $m_{AB} = -1$; $m_{BC} = -1$

83. Center $(1, -2)$; radius $= 4\sqrt{2}$; $x^2 + y^2 - 2x + 4y - 27 = 0$ **85.** $P = 2l + 2w$ **87.** The interest is $630.
89. Mix 90 cm³ of 15% HCl to obtain 150 cm³ of 25% HCl.   **91.** Add 256 oz of water.   **93.** The search plane can go as far as 616 mi.

95. The helicopter will reach the life raft in $\dfrac{30}{19}$ hr, a little less than 1 hr, 35 min.   **97.** It takes Clarissa 10 days by herself.

99. The freight train is $\dfrac{572}{3} \approx 190.67$ ft long.   **101.** It will take the smaller pump another 2 hr.

103. 36 seniors went on the trip; each one paid $13.40.   **105.** 5, 12   **109.** (a) No   (b) Todd wins again.   (c) Todd wins by $\dfrac{1}{4}$ m.

(d) Todd should line up 5.26 m behind the start line.   **(e)** Yes

C H A P T E R   2   Linear and Quadratic Functions

2.1 Concepts and Vocabulary *(page 209)*

1. Independent; dependent   **2.** Vertical   **3.** Range   **4.** F   **5.** T   **6.** T   **8.** No limit; 1   **9.** Yes

2.1 Exercises *(page 210)*

1. Function; Domain: {Dad, Colleen, Kaleigh, Marissa}, Range: {January 8, March 15, September 17}   **3.** Not a function   **5.** Not a function
7. Function; Domain: {1, 2, 3, 4}, Range: {3}   **9.** Not a function   **11.** Function; Domain: {−2, −1, 0, 1}, Range: {0, 1, 4}
13. (a) 5   (b) 7   (c) 3   (d) $-2x + 5$ (e) $-2x - 5$ (f) $2x + 7$ (g) $4x + 5$ (h) $2x + 2h + 5$ **15.** (a) −4   (b) 1   (c) −3
(d) $3x^2 - 2x - 4$ (e) $-3x^2 - 2x + 4$ (f) $3x^2 + 8x + 1$ (g) $12x^2 + 4x - 4$ (h) $3x^2 + 6xh + 3h^2 + 2x + 2h - 4$
17. (a) 0   (b) $\dfrac{1}{2}$ (c) $-\dfrac{1}{2}$ (d) $\dfrac{-x}{x^2 + 1}$ (e) $\dfrac{-x}{x^2 + 1}$ (f) $\dfrac{x + 1}{x^2 + 2x + 2}$ (g) $\dfrac{2x}{4x^2 + 1}$ (h) $\dfrac{x + h}{x^2 + 2xh + h^2 + 1}$
19. (a) 4   (b) 5   (c) 5   (d) $|x| + 4$ (e) $-|x| - 4$ (f) $|x + 1| + 4$ (g) $2|x| + 4$ (h) $|x + h| + 4$ **21.** Function   **23.** Function
25. Not a function   **27.** Not a function   **29.** Function   **31.** Not a function   **33.** All real numbers   **35.** All real numbers
37. $\{x | x \neq -4, x \neq 4\}$ **39.** $\{x | x \neq 0\}$ **41.** $\{x | x \geq 4\}$ **43.** $\{x | x > 9\}$ **45.** $\{x | x > 1\}$
47. (a) $f(0) = 3$; $f(-6) = -3$ (b) $f(6) = 0$; $f(11) = 1$ (c) Positive   (d) Negative   (e) $-3, 6,$ and 10   (f) $-3 < x < 6$; $10 < x \leq 11$
(g) $\{x | -6 \leq x \leq 11\}$ (h) $\{y | -3 \leq y \leq 4\}$ (i) $-3, 6, 10$ (j) 3   (k) 3 times   (l) Once   (m) $0, 4$ (n) $-5, 8$ **49.** Not a function
51. Function   (a) Domain: $\{x | -\pi \leq x \leq \pi\}$; Range: $\{y | -1 \leq y \leq 1\}$ (b) $\left(-\dfrac{\pi}{2}, 0\right), \left(\dfrac{\pi}{2}, 0\right), (0, 1)$
53. Not a function   **55.** Function   (a) Domain: $\{x | x > 0\}$; Range: all real numbers   (b) $(1, 0)$
57. Function   (a) Domain: all real numbers; Range: $\{y | y \leq 2\}$ (b) $(-3, 0), (3, 0), (0, 2)$
59. Function   (a) Domain: all real numbers; Range: $\{y | y \geq -3\}$ (b) $(1, 0), (3, 0), (0, 9)$
61. (a) Yes   (b) $f(-2) = 9$; $(-2, 9)$ (c) $0, \dfrac{1}{2}$; $(0, -1), \left(\dfrac{1}{2}, -1\right)$ (d) All real numbers   (e) $-\dfrac{1}{2}, 1$ (f) −1

63. (a) No **(b)** $f(4) = -3; (4, -3)$ **(c)** $14; (14, 2)$ **(d)** $\{x | x \neq 6\}$ **(e)** -2 **(f)** $-\dfrac{1}{3}$ **65. (a)** Yes **(b)** $f(2) = \dfrac{8}{17}; \left(2, \dfrac{8}{17}\right)$

(c) $-1, 1; (-1, 1), (1, 1)$ **(d)** All real numbers **(e)** 0 **(f)** 0 **67.** $C = -3$ **69.** $A = -4$ **71.** $A = 8$; undefined at $x = 3$ **73.** 4

75. $2x + h - 1$ **77.** $3x^2 + 3xh + h^2$ **79. (a)** III **(b)** IV **(c)** I **(d)** V **(e)** II

81.

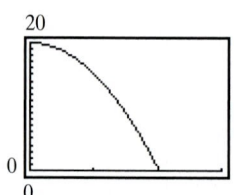

83. (a) 2 hr elapsed during which Kevin was between 0 and 3 mi from home.
(b) 0.5 hr elapsed during which Kevin was 3 mi from home.
(c) 0.3 hr elapsed during which Kevin was between 0 and 3 mi from home.
(d) 0.2 hr elapsed during which Kevin was 0 mi from home.
(e) 0.9 hr elapsed during which Kevin was between 0 and 2.8 mi from home.
(f) 0.3 hr elapsed during which Kevin was 2.8 mi from home.
(g) 1.1 hr elapsed during which Kevin was between 0 and 2.8 mi from home.
(h) 3 mi
(i) 2 times

85. (a)

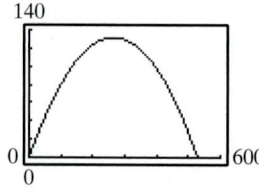

87. $A(x) = \dfrac{1}{2}x^2$

89. $G(x) = 10x$

(b) About 15.1 m, 14.07 m, 12.94 m, 11.72 m
(c) After about 1.01 sec, 1.43 sec, 1.75 sec
(d) After about 2.02 sec

91. (a) About 81.07 ft **(b)** About 129.59 ft
(c) About 26.63 ft
(d)

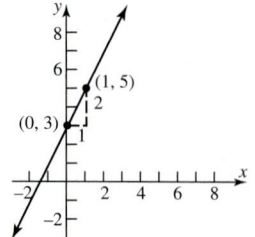

93. (a) $222 **(b)** $225 **(c)** $220 **(d)** $230
(e)

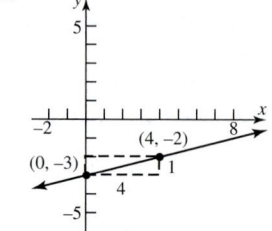

(f) 600 mph

(e) $\dfrac{4225 \pm 65\sqrt{1345}}{16} \approx 115.07$ feet and 413.05 ft

(f) 275 ft; maximum height shown in the table is 131.8 ft
(g) 264 ft; about 132.03 ft
(h) $\{x | 0 \leq x \leq 528.125\}$

95. Only $h(x) = 2x$ **97.** No $f(x)$ has a domain of all real numbers, while $g(x)$ has a domain of $\{x | x \neq -1\}$.

2.2 Concepts and Vocabulary *(page 222)*

1. Slope; y-intercept **2.** Scatter diagram **3.** $y = kx$ **4.** F **5.** T **6.** T **7.** $y = 1.5x + 3.5; 1$ **8.** No linear relation
9. Average $= k \cdot$ Time

2.2 Exercises *(page 223)*

1.

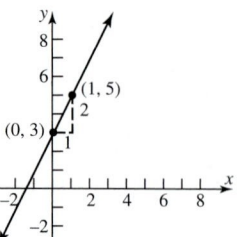

3.

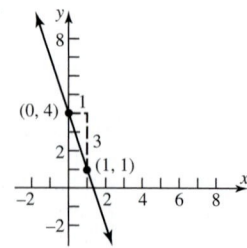

5.

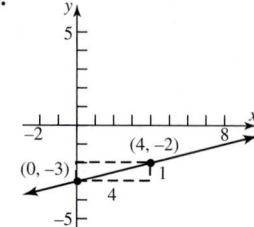

7.

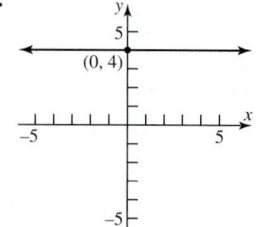

9. (a) $V(x) = -1000x + 3000$

(b)

(c) $1000

11. (a) $C(x) = 90x + 1800$

(b)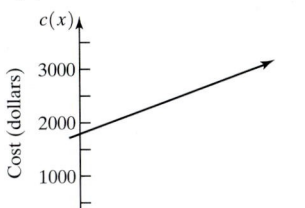

(c) $3060

13. Linear relation   **15.** Linear relation   **17.** Nonlinear relation

19. (a)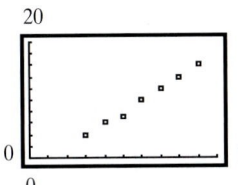

(b) Answers will vary. Using $(4, 6)$ and $(8, 14)$: $y = 2x - 2$

(c)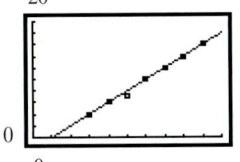

(d) $y = 2.0357x - 2.3571$

(e)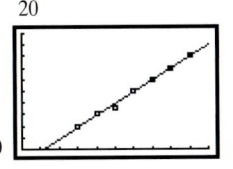

21. (a)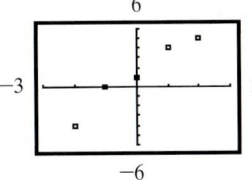

(b) Answers will vary. Using $(-2, -4)$ and $(1, 4)$: $y = \dfrac{8}{3}x + \dfrac{4}{3}$

(c)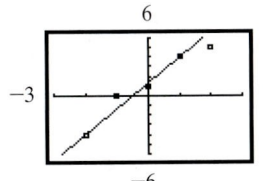

(d) $y = 2.2x + 1.2$

(e)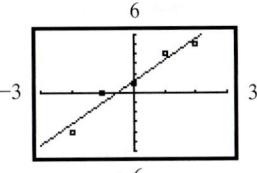

23. (a)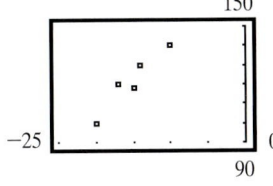

(b) Answers will vary. Using $(-20, 100)$ and $(-15, 118)$: $y = \dfrac{18}{5}x + 172$

(c)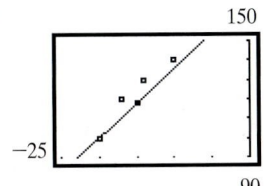

(d) $y = 3.8613x + 180.2920$

(e)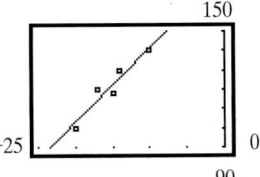

25. (a)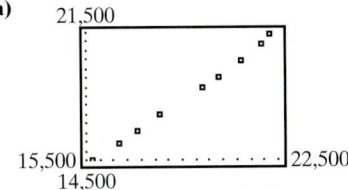

(b) $C(I) = 1.0374I - 1897.5071$

(c) If income increases by $1, then consumption increases by about $1.04.

(d) About $20,408

27. (a)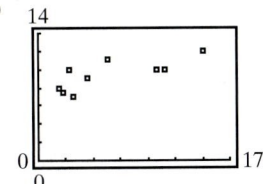

(b) $L(G) = 0.0261G + 7.8738$

(c) If gestation period increases by 1 day, then life expectancy increases by about 0.0261 years.

(d) About 10.2 years

29. (a) No   **(b)**

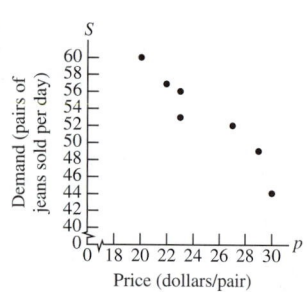

(c) $D = -1.3355p + 86.1974$
(d) If the price increases \$1, the quantity sold per day
decreases by about 1.34 pairs of jeans.
(e) $D(p) = -1.3355p + 86.1974$
(f) $\{p|p > 0\}$
(g) $D(28) \approx 48.80$; about 49 pairs

31. $P(B) = 0.00649B$; \$941.05   **33.** $R(g) = 1.32g$; \$13.86   **35.** 144 ft; 2 s

2.3 Concepts and Vocabulary *(page 234)*

1. Parabola   **2.** Axis of symmetry   **3.** T   **4.** T   **5.** T   **6.** $b^2 - 4ac < 0$

2.3 Exercises *(page 235)*

1. C   **3.** F   **5.** G   **7.** H   **9.** B   **11.** D

13.
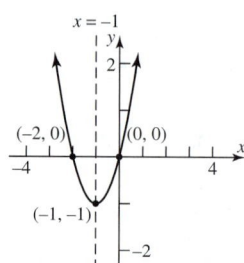

15.
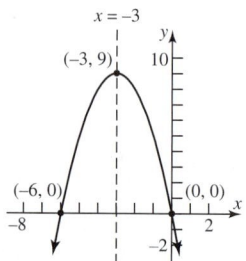

17.
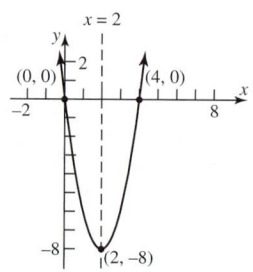

19.
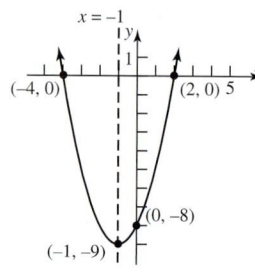

21.
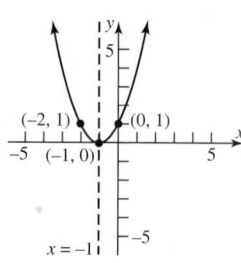

23.
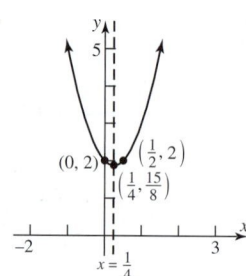

25.
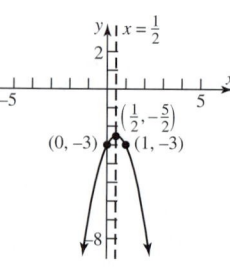

27.
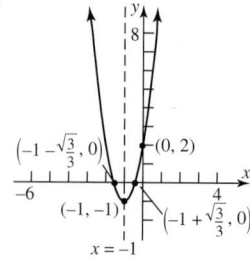

29.
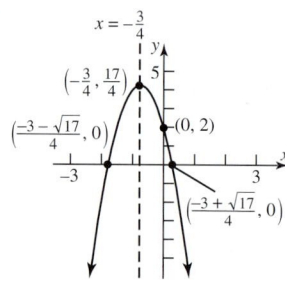

31. $f(x) = (x + 1)^2 - 2 = x^2 + 2x - 1$
33. $f(x) = -(x + 3)^2 + 5 = -x^2 - 6x - 4$
35. $f(x) = 2(x - 1)^2 - 3 = 2x^2 - 4x - 1$

37. (a) $a = 1: f(x) = (x + 3)(x - 1) = x^2 + 2x - 3$
   $a = 2: f(x) = 2(x + 3)(x - 1) = 2x^2 + 4x - 6$
   $a = -2: f(x) = -2(x + 3)(x - 1) = -2x^2 - 4x + 6$
   $a = 5: f(x) = 5(x + 3)(x - 1) = 5x^2 + 10x - 15$
(b) The value of a does not affect the intercepts.

(c) The value of a does not affect the axis of symmetry. It is $x = -1$
   for all values of a.
(d) The value of a does not affect the x-coordinate of the vertex.
   However, the y-coordinate of the vertex is multiplied by a.
(e) They are the same.

39. $a = 6, b = 0, c = 2$ **41.** $R(x) = -\dfrac{1}{6}x^2 + 100x$ **43.** $R(x) = -\dfrac{1}{5}x^2 + 20x$ **45.** $A(x) = -x^2 + 200x$

2.4 Concepts and Vocabulary *(page 244)*

1. $a > 0; -\dfrac{b}{2a}$ **2.** $a < 0; f\left(-\dfrac{b}{2a}\right)$ **3.** T **4.** F **5.** Quantity demanded is 0 (the calculator is too expensive).

2.4 Exercises *(page 244)*

1. Minimum value; 2 **3.** Maximum value; 4 **5.** Minimum value; −18 **7.** Minimum value; −21
9. Maximum value; 21 **11.** Maximum value; 13 **13.** \$500; \$1,000,000 **15.** 40; \$400

17. (a) $R(x) = -\dfrac{1}{6}x^2 + 100x$ **19. (a)** $R(x) = -\dfrac{1}{5}x^2 + 20x$ **21. (a)** $A(x) = -x^2 + 200x$ **23.** 2,000,000 m²

   (b) \$13,333   **(b)** \$255   **(b)** A is largest when
   (c) 300; \$15,000   **(c)** 50; \$500 $x = 100$ yd.
   (d) \$50   **(d)** \$10   **(c)** 10,000 yd²

25. (a) $\dfrac{625}{16} \approx 39$ ft **(b)** $\dfrac{7025}{32} \approx 219.5$ ft **(c)** 170 ft **27.** 18.75 m

   (d)   **(f)** When the height is 100 ft, the projectile **29.** 3 in.
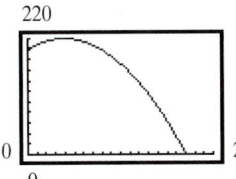   is about 135.7 ft from the cliff.   **31.** $\dfrac{200}{\pi} \approx 63.7$ m by 100 m

33. (a)   **(b)** $I(x) = -47.71x^2 + 4262.66x - 42,777.73$
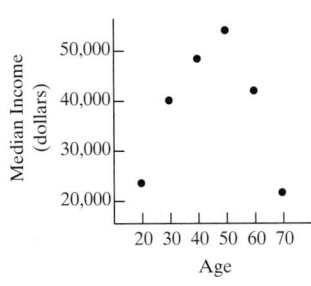   **(c)** About 45 years old
   (d) \$52,434
   (e)
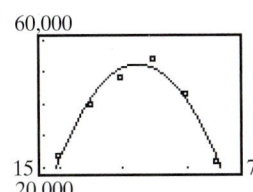

35. (a) Quadratic, $a < 0$.
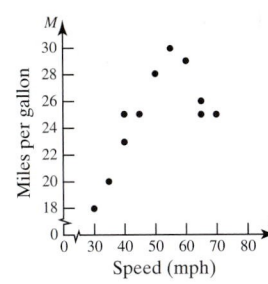   **(b)** $M(s) = -0.0175s^2 + 1.93s - 25.34$ **37.** $x = \dfrac{a}{2}$ **39.** $\dfrac{38}{3}$ **41.** $\dfrac{248}{3}$
   (c) Approximately 55.1 mph
   (d) Approximately 26.8 mpg
   (e)
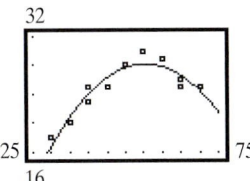

Review Exercises *(page 249)*

1. $f(x) = -2x + 3$ **3.** $A = 11$ **5.** b, c, d

7. (a) $f(-x) = \dfrac{-3x}{x^2 - 4}$ **(b)** $-f(x) = \dfrac{-3x}{x^2 - 4}$ **(c)** $f(x + 2) = \dfrac{3x + 6}{x^2 + 4x}$ **(d)** $f(x - 2) = \dfrac{3x - 6}{x^2 - 4x}$

9. (a) $f(-x) = \sqrt{x^2 - 4}$ **(b)** $-f(x) = -\sqrt{x^2 - 4}$ **(c)** $f(x + 2) = \sqrt{x^2 + 4x}$ **(d)** $f(x - 2) = \sqrt{x^2 - 4x}$

11. (a) $f(-x) = \dfrac{x^2 - 4}{x^2}$ **(b)** $-f(x) = -\dfrac{x^2 - 4}{x^2}$ **(c)** $f(x + 2) = \dfrac{x^2 + 4x}{x^2 + 4x + 4}$ **(d)** $f(x - 2) = \dfrac{x^2 - 4x}{x^2 - 4x + 4}$

13. $\{x | x \neq -3, x \neq 3\}$ **15.** $\{x | x \leq 2\}$ **17.** $\{x | x > 0\}$ **19.** $\{x | x \neq -3, x \neq 1\}$

21.

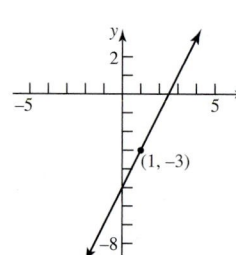

23.

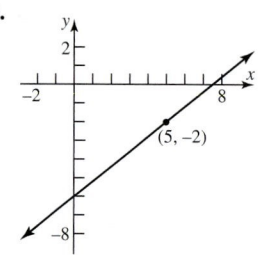

25.

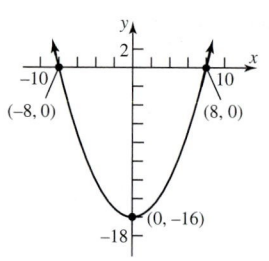

27.

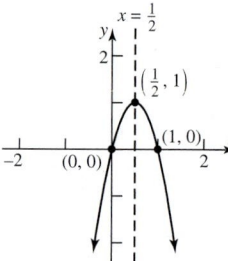

29.

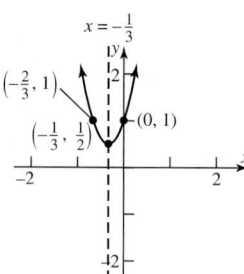

31.

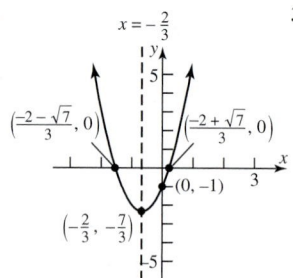

33.

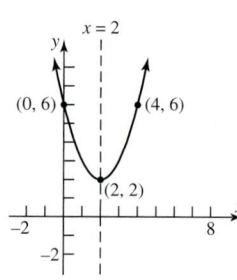

35. Minimum value; 1
37. Maximum value; 12
39. Maximum value; 16

41. (a) Yes
(b)

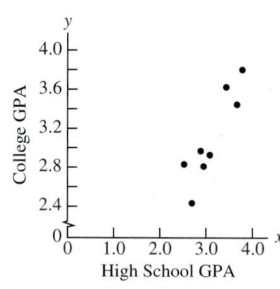

(c) $G = 0.964x + 0.072$
(d) As high school GPA increases 1 point, college GPA increases by 0.964 point.
(e) $G(x) = 0.964x + 0.072$
(f) $\{x \mid 0 \le x \le 4\}$
(g) 3.19

43. $R(g) = 1.18g$; $13.22
45. 2 by 8 ft
47. (a) 63
(b) $151.90

49. (a) Quadratic, $a < 0$.

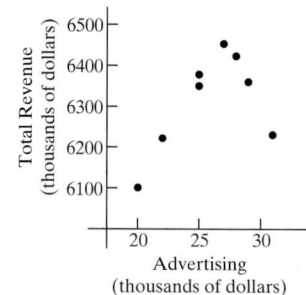

(b) $R(A) = -7.760A^2 + 411.875A + 942.721$
(c) About $26.5 thousand
(d) $6408 thousand
(e)

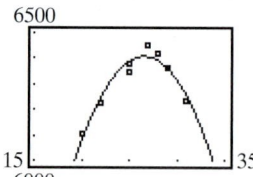

Cumulative Review Exercises *(page 253)*

1. $5\sqrt{2}$ **2.** $(-2, -1)$ and $(2, 3)$ are on the graph. **3.** x-intercepts: $-5, \dfrac{1}{3}$; y-intercept: -5 **4.** $-1.10, 0.26, 1.48, 2.36$

5. $\left\{ x \mid x \ge -\dfrac{3}{5} \right\}$ or $\left[-\dfrac{3}{5}, \infty \right)$

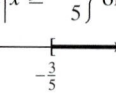

7. $y = -2x + 2$

8. $y = -\dfrac{1}{2}x + \dfrac{13}{2}$

9. $(x - 2)^2 + (y + 4)^2 = 25$

6. $\left\{ x \mid -\dfrac{1}{3} < x < 3 \right\}$ or $\left(-\dfrac{1}{3}, 3 \right)$

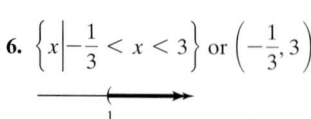

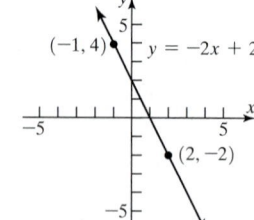

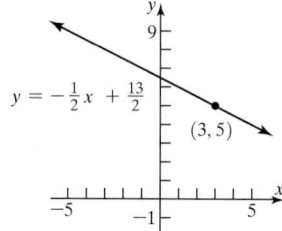

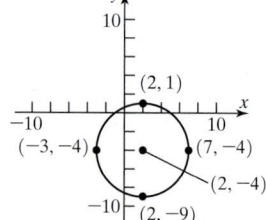

10. Yes, a function   **11. (a)** -3 **(b)** $x^2 - 4x - 2$ **(c)** $x^2 + 4x + 1$ **(d)** $-x^2 + 4x - 1$ **(e)** $x^2 - 3$ **(f)** $2x + h - 4$
12. $\{z | z \neq -1, z \neq 7\}$ **13.** Yes, a function   **14. (a)** No   **(b)** $-1; (-2, -1)$ is on the graph.   **(c)** $-8; (-8, 2)$ is on the graph.

15.

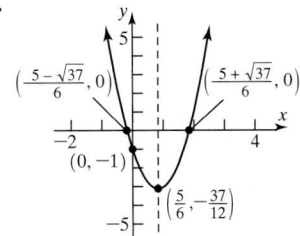

16. Minimum; -11 **17. (a)** $R(x) = -\dfrac{1}{4}x^2 + 25x$ **(b)** \$400   **(c)** $50; \$625$ **(d)** \$12.50

C H A P T E R   3   Functions and Their Graphs

3.1 Concepts and Vocabulary *(page 262)*

1. y-axis   **2.** origin   **3.** T   **4.** T   **5.** y-axis, x-axis, origin   **6.** $(-1, -2)$ **7.** -6

3.1 Exercises *(page 262)*

1. (a) $(3, -4)$ **(b)** $(-3, 4)$ **(c)** $(-3, -4)$ **3. (a)** $(-2, -1)$ **(b)** $(2, 1)$ **(c)** $(2, -1)$ **5. (a)** $(1, -1)$ **(b)** $(-1, 1)$ **(c)** $(-1, -1)$
7. (a) $(-3, 4)$ **(b)** $(3, -4)$ **(c)** $(3, 4)$ **9. (a)** $(0, 3)$ **(b)** $(0, -3)$ **(c)** $(0, 3)$ **11.** x-axis, y-axis, and origin
13. y-axis   **15.** x-axis   **17.** No symmetry

19.

21.

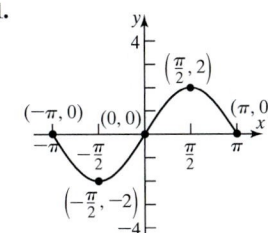

23. Symmetric with respect to the y-axis
25. Symmetric with respect to the origin
27. Symmetric with respect to the y-axis
29. No symmetry
31. No symmetry
33. Symmetric with respect to the origin

35.

37.

3.2 Concepts and Vocabulary *(page 272)*

1. Slope   **2.** Increasing   **3.** Even; odd   **4.** T   **5.** T   **6.** F   **9.** No

3.2 Exercises *(page 273)*

1. Yes   **3.** No   **5.** $(-8, -2); (0, 2); (5, 10)$ **7.** Yes; $f(2) = 10$ **9.** -2 and $2; f(-2) = 6$ and $f(2) = 10$
11. (a) $(-2, 0), (0, 3), (2, 0)$ **(b)** Domain $\{x | -4 \leq x \leq 4\}$ or $[-4, 4]$; Range: $\{y | 0 \leq y \leq 3\}$ or $[0, 3]$
   (c) Increasing on $(-2, 0)$ and $(2, 4)$; Decreasing on $(-4, -2)$ and $(0, 2)$.   **(d)** Even
13. (a) $(0, 1)$ **(b)** Domain: all real numbers or $(-\infty, \infty)$; Range: $\{y | y > 0\}$ or $(0, \infty)$.   **(c)** Increasing on $(-\infty, \infty)$ **(d)** Neither
15. (a) $(-\pi, 0), (0, 0), (\pi, 0)$ **(b)** Domain: $\{x | -\pi \leq x \leq \pi\}$ or $[-\pi, \pi]$; Range: $\{y | -1 \leq y \leq 1\}$ or $[-1, 1]$.
   (c) Increasing on $\left(-\dfrac{\pi}{2}, \dfrac{\pi}{2}\right)$; Decreasing on $\left(-\pi, -\dfrac{\pi}{2}\right)$ and on $\left(\dfrac{\pi}{2}, \pi\right)$ **(d)** Odd
17. (a) $\left(0, \dfrac{1}{2}\right), \left(\dfrac{1}{2}, 0\right), \left(\dfrac{5}{2}, 0\right)$ **(b)** Domain: $\{x | -3 \leq x \leq 3\}$ or $[-3, 3]$; Range: $\{y | -1 \leq y \leq 2\}$ or $[-1, 2]$
   (c) Increasing on $(2, 3)$; Decreasing on $(-1, 1)$; Constant on $(-3, -1)$ and on $(1, 2)$.   **(d)** Neither
19. (a) $(-2, 0), (0, 2), (2, 0)$ **(b)** Domain: $\{x | -4 \leq x \leq 4\}$ or $[-4, 4]$; Range: $\{y | 0 \leq y \leq 2\}$ or $[0, 2]$
   (c) Increasing on $(-2, 0)$ and on $(2, 4)$; Decreasing on $(-4, -2)$ and on $(0, 2)$.   **(d)** Even

21. (a) $0; f(0) = 3$ **(b)** -2 and $2; f(-2) = 0$ and $f(2) = 0$ **23. (a)** $\dfrac{\pi}{2}; f\left(\dfrac{\pi}{2}\right) = 1$ **(b)** $-\dfrac{\pi}{2}; f\left(-\dfrac{\pi}{2}\right) = -1$

25. (a) 5 **(b)** 5 **(c)** $y = 5x$
(d)

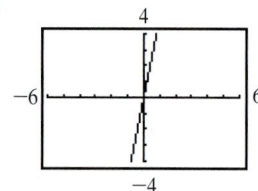

27. (a) -3 **(b)** -3 **(c)** $y = -3x + 1$
(d)

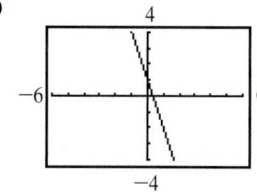

29. (a) $x - 1$ **(b)** 1 **(c)** $y = x - 2$
(d)

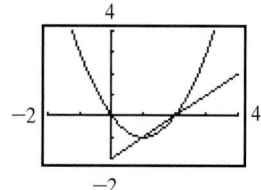

31. (a) $x(x + 1)$ **(b)** 6 **(c)** $y = 6x - 6$
(d)

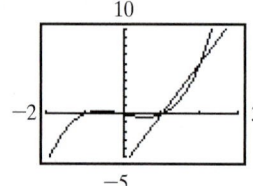

33. (a) $-\dfrac{1}{x + 1}$ **(b)** $-\dfrac{1}{3}$
(c) $y = -\dfrac{1}{3}x + \dfrac{4}{3}$
(d)

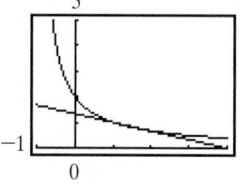

35. (a) $\dfrac{1}{\sqrt{x} + 1}$ **(b)** $\dfrac{1}{\sqrt{2} + 1}$
(c) $y = \dfrac{1}{\sqrt{2} + 1}x - \dfrac{1}{\sqrt{2} + 1} + 1$
(d)

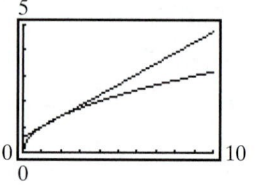

37. Odd   **39.** Even   **41.** Odd   **43.** Neither   **45.** Even   **47.** Odd   **49.** At most one

51.

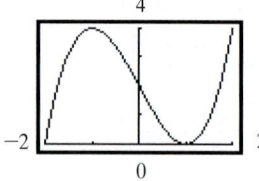

Increasing: $(-2, -1), (1, 2)$
Decreasing: $(-1, 1)$
Local maximum: $(-1, 4)$
Local minimum: $(1, 0)$

53.

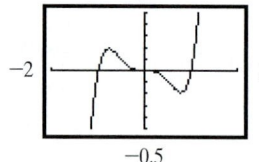

Increasing: $(-2, -0.77), (0.77, 2)$
Decreasing: $(-0.77, 0.77)$
Local maximum: $(-0.77, 0.19)$
Local minimum: $(0.77, -0.19)$

55.

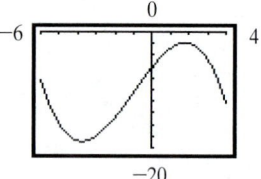

Increasing: $(-3.77, 1.77)$
Decreasing: $(-6, -3.77), (1.77, 4)$
Local maximum: $(1.77, -1.91)$
Local minimum: $(-3.77, -18.89)$

57.

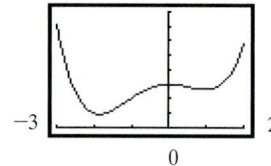

Increasing: $(-1.87, 0), (0.97, 2)$
Decreasing: $(-3, -1.87), (0, 0.97)$
Local maximum: $(0, 3)$
Local minima: $(-1.87, 0.95), (0.97, 2.65)$

59. (a) 2
(b) $2; 2; 2; m_{\text{sec}} = 2$
(c) $y = 2x + 5$
(d)

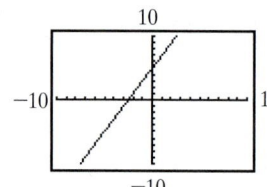

61. (a) $2x + h + 2$
(b) $4.5; 4.1; 4.01; m_{\text{sec}} = 4$
(c) $y = 4.01x - 1.01$
(d)

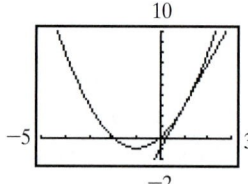

63. (a) $4x + 2h - 3$ **(b)** $2; 1.2; 1.02; m_{\text{sec}} = 1$

(c) $y = 1.02x - 1.02$

(d)

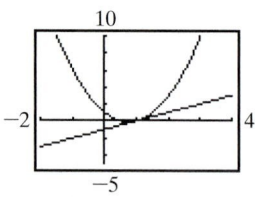

65. (a) $-\dfrac{1}{(x + h) \cdot x}$ **(b)** $-\dfrac{2}{3}; -\dfrac{10}{11}; -\dfrac{100}{101}; m_{\text{sec}} = -1$

(c) $y = -\dfrac{100}{101}x + \dfrac{201}{101}$

(d)

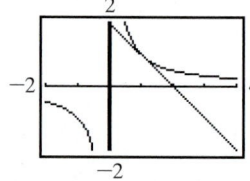

67.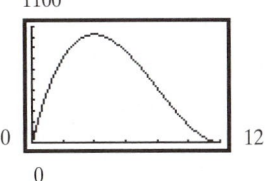

The volume is largest at $x = 4$ inches.

69. (a)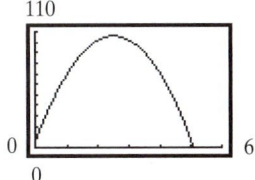

(b) 2.5 seconds   **(c)** 106 feet

71. (a), (b), (e)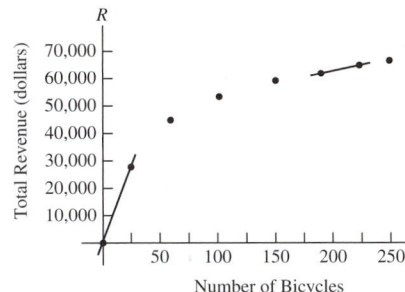

(c) 1120 dollars/bicycle

(d) For each additional bicycle sold between 0 and 25 bicycles, total revenue increases, on average, by \$1120.

(f) 75 dollars/bicycle

(g) For each additional bicycle sold between 190 and 223 bicycles, total revenue increases, on average, by \$75.

73. (a), (b), (e)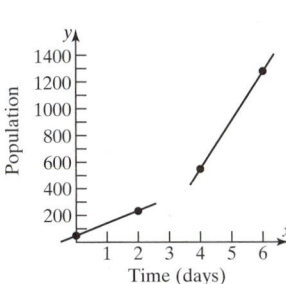

(c) 92 bacteria/day

(d) The population is increasing at an average rate of 92 bacteria per day between day 0 and day 2.

(f) 366.5 bacteria/day

(g) The population is increasing at an average rate of 366.5 bacteria per day between day 4 and day 6.

(h) The average rate of change is increasing.

3.3 Concepts and Vocabulary *(page 283)*

1. Less than   **2.** Piecewise-defined   **3.** T   **4.** F   **5.** Answers will vary. See Figure 24 on page 277 for a graph.

3.3 Exercises *(page 283)*

1. C   **3.** E   **5.** B   **7.** F

9.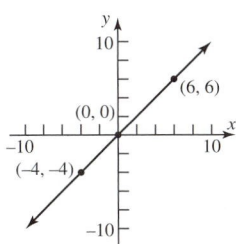

11.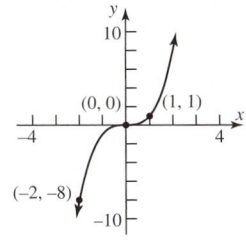

13.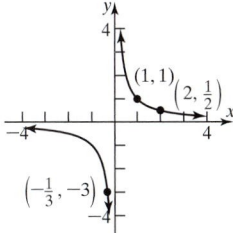

15. (a) 4   **(b)** 2   **(c)** 5   **17. (a)** 2   **(b)** 3   **(c)** -4

19. (a) All real numbers

(b) $(0, 1)$

(c)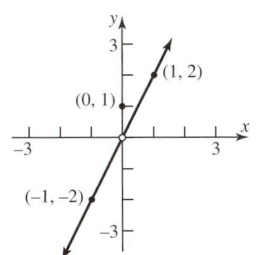

(d) $\{y | y \neq 0\}; (-\infty, 0)$ or $(0, \infty)$

21. (a) All real numbers

(b) $(0, 3)$

(c)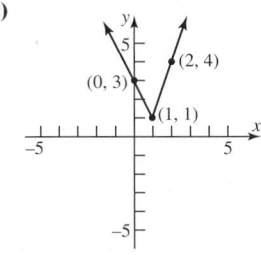

(d) $\{y | y \geq 1\}; [1, \infty)$

23. (a) $\{x | x \geq -2\}; [-2, \infty)$

(b) $(0, 3), (2, 0)$

(c)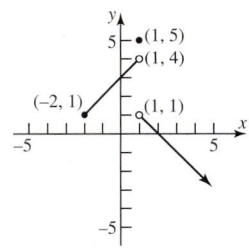

(d) $\{y | y < 4, y = 5\}; (-\infty, 4)$ or $\{5\}$

25. (a) All real numbers

(b) $(-1, 0), (0, 0)$

(c)

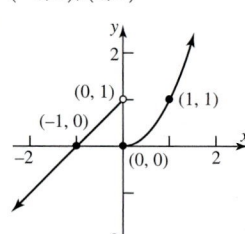

(d) All real numbers

27. (a) $\{x | x \geq -2\}; [-2, \infty)$

(b) $(0, 1)$

(c)

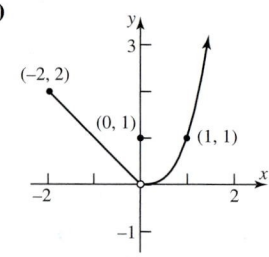

(d) $\{y | y > 0\}; (0, \infty)$

29. (a) All real numbers

(b) $(x, 0)$ for $0 \leq x < 1$

(c)

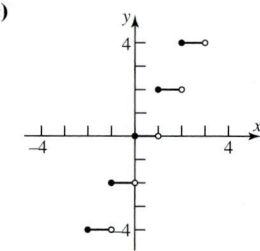

(d) Set of even integers

31. $f(x) = \begin{cases} -x & \text{if } -1 \leq x \leq 0 \\ \frac{1}{2}x & \text{if } 0 < x \leq 2 \end{cases}$ (Other answers are possible.)

33. $f(x) = \begin{cases} -x & \text{if } x \leq 0 \\ -x + 2 & \text{if } 0 < x \leq 2 \end{cases}$ (Other answers are possible.)

35. (a)

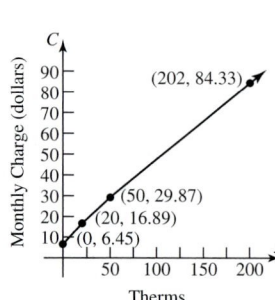

(b) $C = 50 + 0.4(x - 100)$

(c) $C = 170 + 0.25(x - 400)$

37. $f(x) = \begin{cases} x & \text{if} & 0 \leq x < 10 \\ 10 & \text{if} & 10 \leq x < 500 \\ 30 & \text{if} & 500 \leq x < 1000 \\ 50 & \text{if} & 1000 \leq x < 1500 \\ 70 & \text{if} & 1500 \leq x \end{cases}$

39. For schedule X: $f(x) = \begin{cases} 0.15x & \text{if} & 0 \leq x \leq 27{,}050 \\ 4057.50 + 0.28(x - 27{,}050) & \text{if} & 27{,}050 < x \leq 65{,}550 \\ 14{,}837.50 + 0.31(x - 65{,}550) & \text{if} & 65{,}550 < x \leq 136{,}750 \\ 36{,}909.50 + 0.36(x - 136{,}750) & \text{if} & 136{,}750 < x \leq 297{,}300 \\ 94{,}707.50 + 0.396(x - 297{,}300) & \text{if} & x > 297{,}300 \end{cases}$

For schedule Y-1: $f(x) = \begin{cases} 0.15x & \text{if} & x \leq 45{,}200 \\ 6780 + 0.28(x - 45{,}200) & \text{if} & 45{,}200 < x \leq 109{,}250 \\ 24{,}714 + 0.31(x - 109{,}250) & \text{if} & 109{,}250 < x \leq 166{,}450 \\ 42{,}446 + 0.36(x - 166{,}450) & \text{if} & 166{,}450 < x \leq 297{,}300 \\ 89{,}552 + 0.396(x - 297{,}300) & \text{if} & x > 297{,}300 \end{cases}$

41. (a) \$25.54 **(b)** \$84.33 **(c)** $C = \begin{cases} 6.45 + 0.5221x & \text{if } 0 \leq x \leq 20 \\ 8.24 + 0.4326x & \text{if } 20 < x \leq 50 \\ 11.955 + 0.3583x & \text{if } x > 50 \end{cases}$ **(d)**

43. Each graph is that of $y = x^2$, but shifted vertically. If $y = x^2 + k, k > 0$, the shift is up k units; if $y = x^2 + k, k < 0$, the shift is down $|k|$ units. **45.** Each graph is that of $y = |x|$, but either compressed or stretched. If $y = k|x|$ and $k > 1$, the graph is stretched vertically; if $y = k|x|, 0 < k < 1$, the graph is compressed vertically. **47.** The graph of $y = f(-x)$ is the reflection about the y-axis of the graph of $y = f(x)$. **49.** They are all $\cup$-shaped and open upward. All three go through the points $(-1, 1)$, $(0, 0)$ and $(1, 1)$. As the exponent increases, the steepness of the curve increases (except between -1 and 1).

3.4 Concepts and Vocabulary *(page 295)*

1. Horizontal; right   **2.** y **3.** Compression; 3   **4.** T   **5.** F   **6.** T   **8.** $4f(x)$ is a vertical stretch; $f(4x)$ is a horizontal compression.

3.4 Exercises *(page 296)*

1. B   **3.** H   **5.** I   **7.** L   **9.** F   **11.** G   **13.** C   **15.** B   **17.** $y = (x - 4)^3$ **19.** $y = x^3 + 4$ **21.** $y = -x^3$ **23.** $y = 4x^3$
25. (1) $y = \sqrt{x} + 2$; (2) $y = -(\sqrt{x} + 2)$; (3) $y = -(\sqrt{-x} + 2)$ **27.** (1) $y = -\sqrt{x}$; (2) $y = -\sqrt{x} + 2$; (3) $y = -\sqrt{x + 3} + 2$
29. (c)   **31.** (c)

33.
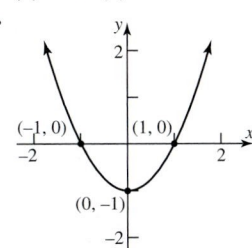

35.
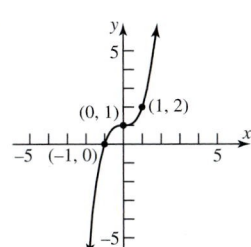

37.
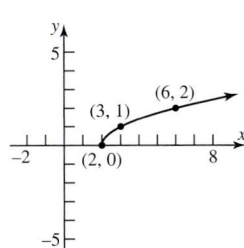

39.
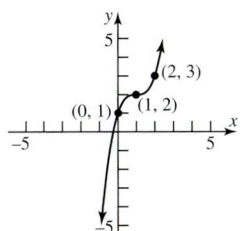

41.
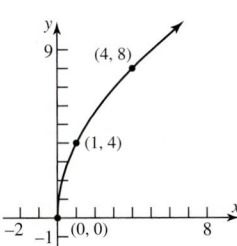

43.
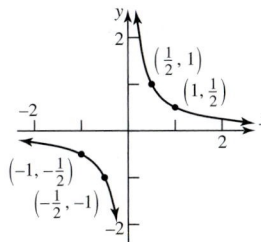

45.
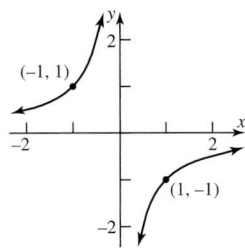

47.
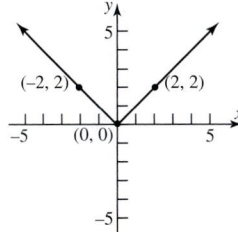

49.
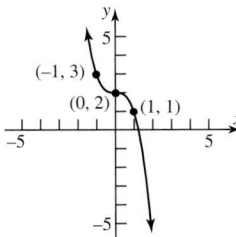

51.
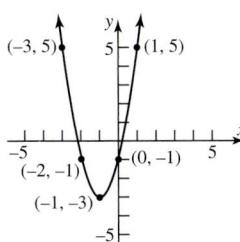

53.
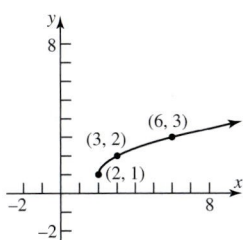

55.
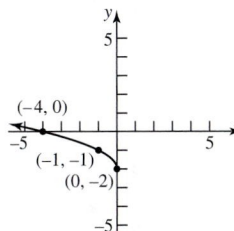

57.
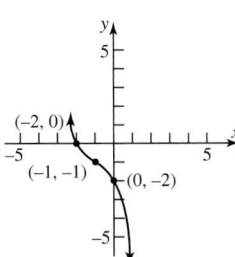

59.
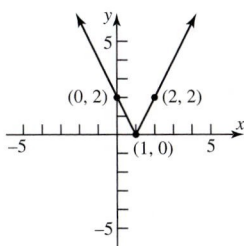

61.
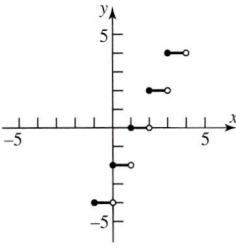

63. (a) $F(x) = f(x) + 3$ **(b)** $G(x) = f(x + 2)$ **(c)** $P(x) = -f(x)$ **(d)** $H(x) = f(x + 1) - 2$

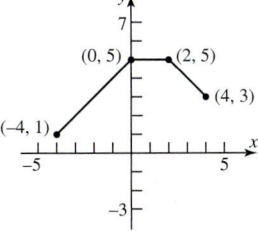

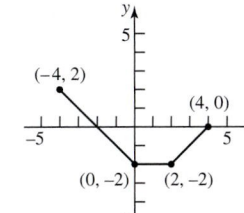

(e) $Q(x) = \dfrac{1}{2}f(x)$

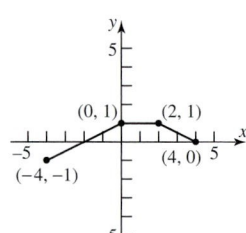

(f) $g(x) = f(-x)$

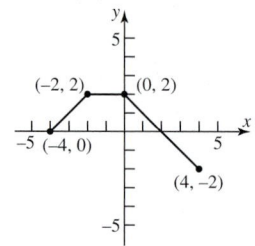

(g) $h(x) = f(2x)$

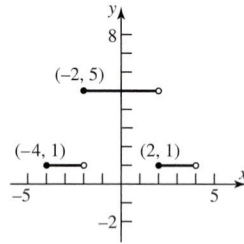

65. (a) $F(x) = f(x) + 3$

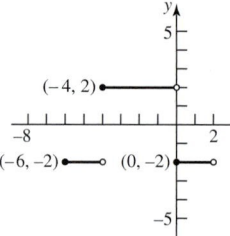

(b) $G(x) = f(x + 2)$

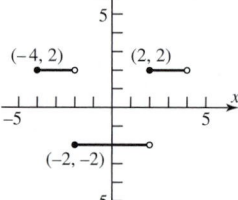

(c) $P(x) = -f(x)$

(d) $H(x) = f(x + 1) - 2$

(e) $Q(x) = \dfrac{1}{2}f(x)$

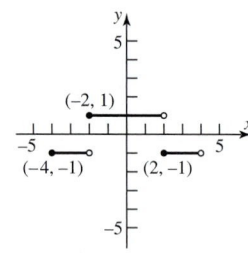

(f) $g(x) = f(-x)$

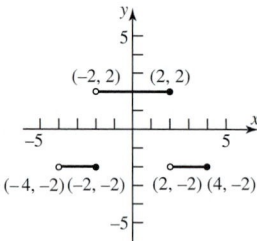

(g) $h(x) = f(2x)$

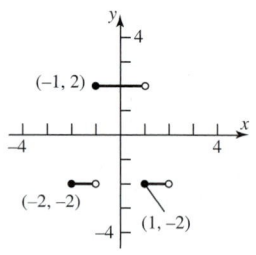

67. (a) $F(x) = f(x) + 3$

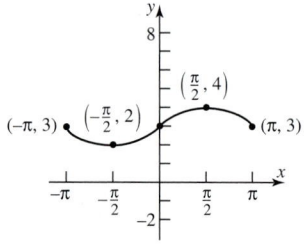

(b) $G(x) = f(x + 2)$

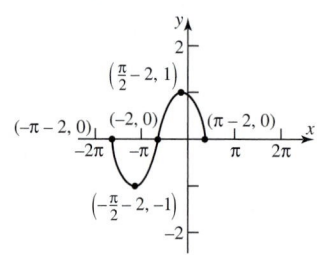

(c) $P(x) = -f(x)$

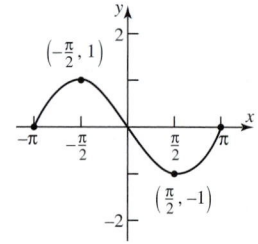

(d) $H(x) = f(x + 1) - 2$

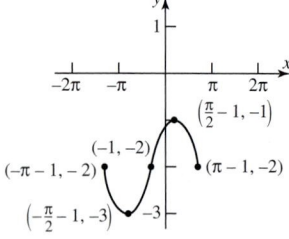

(e) $Q(x) = \dfrac{1}{2}f(x)$

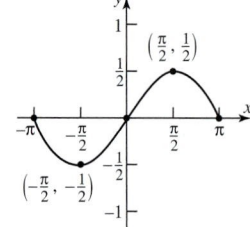

(f) $g(x) = f(-x)$

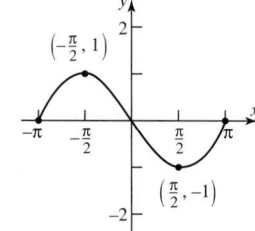

(g) $h(x) = f(2x)$

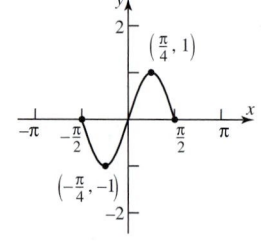

69. (a) $y = |x + 1|$

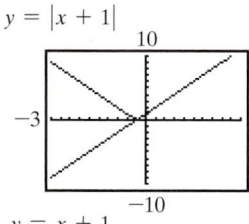

$y = x + 1$

(b) $y = |4 - x^2|$

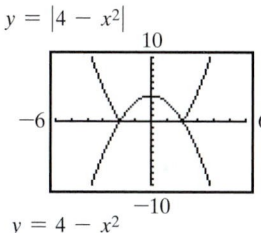

$y = 4 - x^2$

(c) $y = |x^3 + x|$

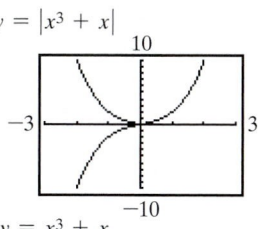

$y = x^3 + x$

(d) Any part of the graph of $y = f(x)$ that lies below the x-axis is reflected about the x-axis to obtain the graph of $y = |f(x)|$.

71. (a)

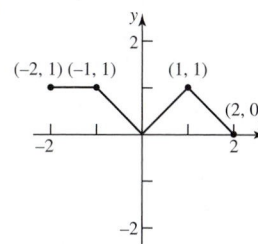

(b)

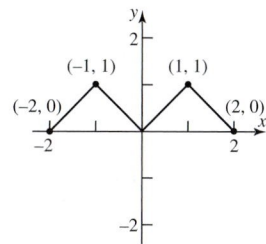

73. $f(x) = (x + 1)^2 - 1$

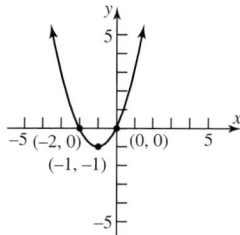

75. $f(x) = (x - 4)^2 - 15$

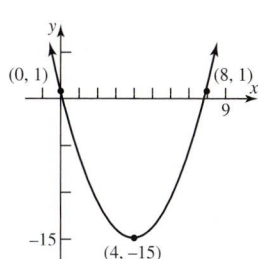

77. $f(x) = \left(x + \dfrac{1}{2}\right)^2 + \dfrac{3}{4}$

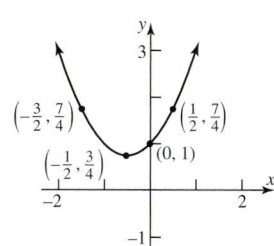

79.

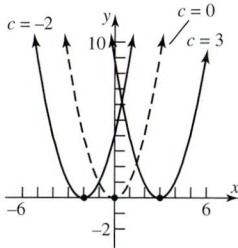

81.

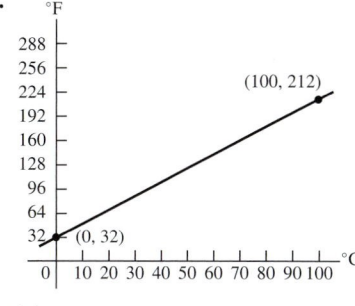

83. (a)

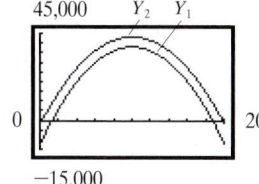

(b) 10% tax
(c) Y_1 is the graph of $p(x)$ shifted down vertically 10,000 units.
  Y_2 is the graph of $p(x)$ vertically compressed by a factor of 0.9.
(d) 10% tax

3.5 Concepts and Vocabulary *(page 306)*

1. $[0, 5]$ **2.** $\neq; f; g$ **3.** $g(f(x))$ **4.** F   **5.** T   **6.** F

3.5 Exercises *(page 306)*

1. (a) $(f + g)(x) = 5x + 1$; All real numbers   **(b)** $(f - g)(x) = x + 7$; All real numbers
(c) $(f \cdot g)(x) = 6x^2 - x - 12$; All real numbers   **(d)** $\left(\dfrac{f}{g}\right)(x) = \dfrac{3x + 4}{2x - 3}; \left\{x \middle| x \neq \dfrac{3}{2}\right\}$

3. (a) $(f + g)(x) = 2x^2 + x - 1$; All real numbers   **(b)** $(f - g)(x) = -2x^2 + x - 1$; All real numbers

(c) $(f \cdot g)(x) = 2x^3 - 2x^2$; All real numbers   **(d)** $\left(\dfrac{f}{g}\right)(x) = \dfrac{x-1}{2x^2}$; $\{x | x \neq 0\}$ **5. (a)** $(f + g)(x) = \sqrt{x} + 3x - 5$; $\{x | x \geq 0\}$

(b) $(f - g)(x) = \sqrt{x} - 3x + 5$; $\{x | x \geq 0\}$ **(c)** $(f \cdot g)(x) = 3x\sqrt{x} - 5\sqrt{x}$; $\{x | x \geq 0\}$ **(d)** $\left(\dfrac{f}{g}\right)(x) = \dfrac{\sqrt{x}}{3x - 5}$; $\left\{x | x \geq 0, x \neq \dfrac{5}{3}\right\}$

7. (a) $(f + g)(x) = 1 + \dfrac{2}{x}$; $\{x | x \neq 0\}$ **(b)** $(f - g)(x) = 1$; $\{x | x \neq 0\}$ **(c)** $(f \cdot g)(x) = \dfrac{1}{x} + \dfrac{1}{x^2}$; $\{x | x \neq 0\}$

(d) $\left(\dfrac{f}{g}\right)(x) = x + 1$; $\{x | x \neq 0\}$ **9. (a)** $(f + g)(x) = \dfrac{6x + 3}{3x - 2}$; $\left\{x | x \neq \dfrac{2}{3}\right\}$ **(b)** $(f - g)(x) = \dfrac{-2x + 3}{3x - 2}$; $\left\{x | x \neq \dfrac{2}{3}\right\}$

(c) $(f \cdot g)(x) = \dfrac{8x^2 + 12x}{(3x - 2)^2}$; $\left\{x | x \neq \dfrac{2}{3}\right\}$ **(d)** $\left(\dfrac{f}{g}\right)(x) = \dfrac{2x + 3}{4x}$; $\left\{x | x \neq 0, x \neq \dfrac{2}{3}\right\}$ **11.** $g(x) = 5 - \dfrac{7}{2}x$ **13. (a)** 98   **(b)** 49   **(c)** 4

(d) 4   **15. (a)** 97   **(b)** $-\dfrac{163}{2}$ **(c)** 1   **(d)** $-\dfrac{3}{2}$ **17. (a)** $2\sqrt{2}$ **(b)** $2\sqrt{2}$ **(c)** 1   **(d)** 0   **19. (a)** $\dfrac{1}{17}$ **(b)** $\dfrac{1}{5}$ **(c)** 1   **(d)** $\dfrac{1}{2}$

21. (a) $\dfrac{3}{\sqrt[3]{4} + 1}$ **(b)** 1   **(c)** $\dfrac{6}{5}$ **(d)** 0   **23.** $\{x | x \neq 0, x \neq 2\}$ **25.** $\{x | x \neq -4, x \neq 0\}$ **27.** $\left\{x | x \geq -\dfrac{3}{2}\right\}$ **29.** $\{x | x \geq 1\}$

31. (a) $(f \circ g)(x) = 6x + 3$; All real numbers   **(b)** $(g \circ f)(x) = 6x + 9$; All real numbers   **(c)** $(f \circ f)(x) = 4x + 9$; All real numbers

(d) $(g \circ g)(x) = 9x$; All real numbers   **33. (a)** $(f \circ g)(x) = 3x^2 + 1$; All real numbers   **(b)** $(g \circ f)(x) = 9x^2 + 6x + 1$; All real numbers

(c) $(f \circ f)(x) = 9x + 4$; All real numbers   **(d)** $(g \circ g)(x) = x^4$; All real numbers   **35. (a)** $(f \circ g)(x) = x^4 + 8x^2 + 16$; All real numbers

(b) $(g \circ f)(x) = x^4 + 4$; All real numbers   **(c)** $(f \circ f)(x) = x^4$; All real numbers   **(d)** $(g \circ g)(x) = x^4 + 8x^2 + 20$; All real numbers

37. (a) $(f \circ g)(x) = \dfrac{3x}{2 - x}$; $\{x | x \neq 0, x \neq 2\}$ **(b)** $(g \circ f)(x) = \dfrac{2(x - 1)}{3}$; $\{x | x \neq 1\}$ **(c)** $(f \circ f)(x) = \dfrac{3(x - 1)}{4 - x}$; $\{x | x \neq 1, x \neq 4\}$

(d) $(g \circ g)(x) = x$; $\{x | x \neq 0\}$ **39. (a)** $(f \circ g)(x) = \dfrac{4}{4 + x}$; $\{x | x \neq -4, x \neq 0\}$ **(b)** $(g \circ f)(x) = -\dfrac{4(x - 1)}{x}$; $\{x | x \neq 0, x \neq 1\}$

(c) $(f \circ f)(x) = x$; $\{x | x \neq 1\}$ **(d)** $(g \circ g)(x) = x$; $\{x | x \neq 0\}$ **41. (a)** $(f \circ g)(x) = \sqrt{2x + 3}$; $\left\{x | x \geq -\dfrac{3}{2}\right\}$

(b) $(g \circ f)(x) = 2\sqrt{x} + 3$; $\{x | x \geq 0\}$ **(c)** $(f \circ f)(x) = \sqrt[4]{x}$; $\{x | x \geq 0\}$ **(d)** $(g \circ g)(x) = 4x + 9$; All real numbers

43. (a) $(f \circ g)(x) = x$; $\{x | x \geq 1\}$ **(b)** $(g \circ f)(x) = |x|$; All real numbers   **(c)** $(f \circ f)(x) = x^4 + 2x^2 + 2$; All real numbers

(d) $(g \circ g)(x) = \sqrt{\sqrt{x - 1} - 1}$; $\{x | x \geq 2\}$ **45. (a)** $(f \circ g)(x) = acx + ad + b$; All real numbers   **(b)** $(g \circ f)(x) = acx + bc + d$;

All real numbers   **(c)** $(f \circ f)(x) = a^2x + ab + b$; All real numbers   **(d)** $(g \circ g)(x) = c^2x + cd + d$; All real numbers

47. $(f \circ g)(x) = f(g(x)) = f\left(\dfrac{1}{2}x\right) = 2\left(\dfrac{1}{2}x\right) = x$; $(g \circ f)(x) = g(f(x)) = g(2x) = \dfrac{1}{2}(2x) = x$

49. $(f \circ g)(x) = f(g(x)) = f(\sqrt[3]{x}) = (\sqrt[3]{x})^3 = x$; $(g \circ f)(x) = g(f(x)) = g(x^3) = \sqrt[3]{x^3} = x$

51. $(f \circ g)(x) = f(g(x)) = f\left(\dfrac{1}{2}(x + 6)\right) = 2\left[\dfrac{1}{2}(x + 6)\right] - 6 = x + 6 - 6 = x$;

$(g \circ f)(x) = g(f(x)) = g(2x - 6) = \dfrac{1}{2}(2x - 6 + 6) = x$

53. $(f \circ g)(x) = f(g(x)) = f\left(\dfrac{1}{a}(x - b)\right) = a\left[\dfrac{1}{a}(x - b)\right] + b = x$; $(g \circ f)(x) = g(f(x)) = g(ax + b) = \dfrac{1}{a}(ax + b - b) = x$

55. $f(x) = x^4$; $g(x) = 2x + 3$ (Other answers are possible.)   **57.** $f(x) = \sqrt{x}$; $g(x) = x^2 + 1$ (Other answers are possible.)

59. $f(x) = |x|$; $g(x) = 2x + 1$ (Other answers are possible.)   **61.** $(f \circ g)(x) = 11$; $(g \circ f)(x) = 2$ **63.** $-3, 3$ **65.** $S(r(t)) = \dfrac{16}{9}\pi t^6$

67. $C(N(t)) = 15{,}000 + 800{,}000t - 40{,}000t^2$ **69.** $C = \dfrac{2\sqrt{100 - p}}{25} + 600, 0 \leq p \leq 100$ **71.** $V(r) = 2\pi r^3$ **73.** $R(x) = \dfrac{L(x)}{P(x)}$

75. $H(x) = P(x) \cdot I(x)$ **77. (a)** $f(x) = 1.136235x$ **(b)** $g(x) = 109.846x$ **(c)** $g(f(x)) = g(1.136235x) = 124.81087x$

(d) 124,810.87 Yen

3.6 Exercises *(page 314)*

1. (a) $d(x) = \sqrt{x^4 - 15x^2 + 64}$ **(b)** $d(0) = 8$ **3. (a)** $d(x) = \sqrt{x^2 - x + 1}$ **5.** $A(x) = \dfrac{1}{2}x^4$

(c) $d(1) = \sqrt{50} \approx 7.07$ **(b)**

(d)

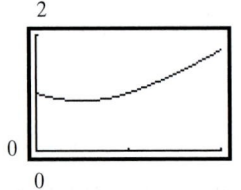

(e) d is smallest when $x \approx -2.74$ or $x \approx 2.74$.

(c) d is smallest when $x = 0.50$.

7. (a) $A(x) = x(16 - x^2)$

(b) Domain: $\{x \mid 0 < x < 4\}$

(c) The area is largest when $x \approx 2.31$.

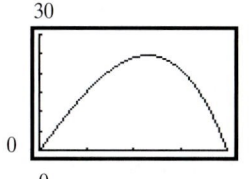

9. (a) $A(x) = 4x\sqrt{4 - x^2}$ **(b)** $p(x) = 4x + 4\sqrt{4 - x^2}$

(c) The area is largest when $x \approx 1.41$.   **(d)** The perimeter is largest when $x \approx 1.41$.

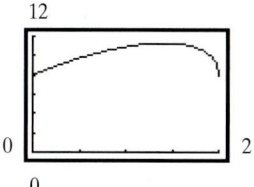

11. (a) $A(x) = x^2 + \dfrac{(5 - 2x)^2}{\pi}$

(b) Domain: $\{x \mid 0 < x < 2.5\}$

(c) The area is smallest when $x \approx 1.40$ meters.

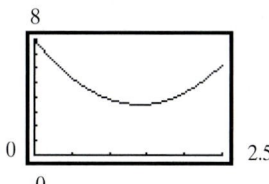

13. (a) $C(x) = x$ **(b)** $A(x) = \dfrac{x^2}{4\pi}$

15. (a) $A(r) = 2r^2$ **(b)** $p(r) = 6r$

17. $A(x) = \left(\dfrac{\pi}{3} - \dfrac{\sqrt{3}}{4}\right)x^2$

19. (a) $C(x) = 100x + 140\sqrt{x^2 - 10x + 29}$;

Domain: $\{x \mid 0 \le x \le 5\}$

(b) $C(1) = \$726.10$ **(c)** $C(3) = \$695.98$

(d)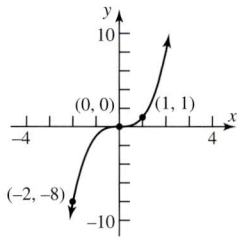

(e) $x \approx 3.0$ miles

(f) $x \approx 2.96$ miles

21. $d = 50t$

23. (a) $V(x) = x(24 - 2x)^2$

(b) 972 in^3 **(c)** 160 in^3

(d) The volume is largest when $x = 4$.

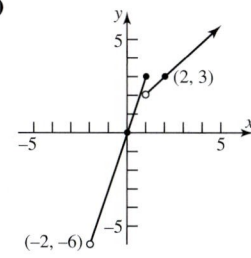

25. $V(h) = \pi h\left(R^2 - \dfrac{h^2}{4}\right)$

Review Exercises (page 318)

1. (a) Domain: $\{x \mid -4 \le x \le 4\}$ or $[-4, 4]$; Range: $\{y \mid -3 \le y \le 1\}$ or $[-3, 1]$ **(b)** Increasing on $(-4, -1)$ and $(3, 4)$; Decreasing on $(-1, 3)$
(c) Local maximum is 1 and occurs at $x = -1$; Local minimum is -3 and occurs at $x = 3$ **(d)** No symmetry   **(e)** Neither
(f) x-intercepts: $-2, 0, 4$; y-intercept: 0   **3.** x-axis   **5.** x-axis, y-axis, origin   **7.** y-axis   **9.** No symmetry

11.

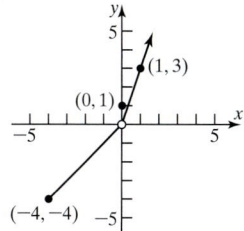

13. (a) $\{x \mid x > -2\}$; $(-2, \infty)$
(b) $(0, 0)$
(c)

(d) $\{y \mid y > -6\}$; $(-6, \infty)$

15. (a) $\{x \mid x \ge -4\}$; $[-4, \infty)$
(b) $(0, 1)$
(c)

(d) $\{y \mid y \ge -4, y \ne 0\}$; $[-4, 0)$ or $(0, \infty)$

17. -5 **19.** $-4x - 5$ **21.** Odd   **23.** Even   **25.** Neither   **27.** Odd

29.

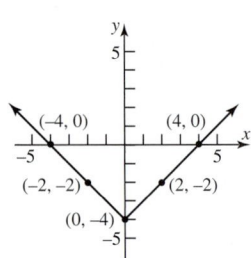

Intercepts: $(-4, 0), (4, 0), (0, -4)$
Domain: all real numbers
Range: $\{y | y \geq -4\}$

31.

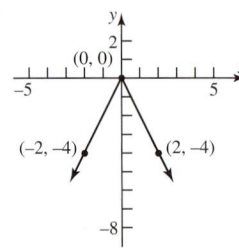

Intercept: $(0, 0)$
Domain: all real numbers
Range: $\{y | y \leq 0\}$

33.

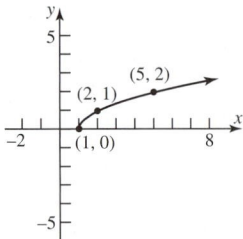

Intercept: $(1, 0)$
Domain: $\{x | x \geq 1\}$
Range: $\{y | y \geq 0\}$

35.

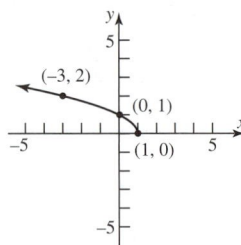

Intercepts: $(0, 1), (1, 0)$
Domain: $\{x | x \leq 1\}$
Range: $\{y | y \geq 0\}$

37.

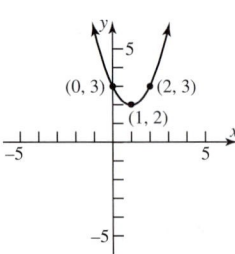

Intercept: $(0, 3)$

Domain: all real numbers
Range: $\{y | y \geq 2\}$

39.

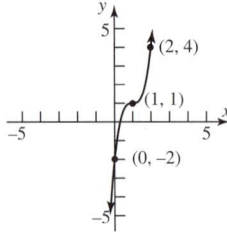

Intercepts: $\left(\sqrt[3]{-\dfrac{1}{3}} + 1, 0 \right), (0, -2)$

Domain: all real numbers
Range: all real numbers

41.

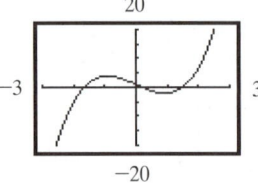

Local maximum: $(-0.91, 4.04)$
Local minimum: $(0.91, -2.04)$
Increasing: $(-3, -0.91); (0.91, 3)$
Decreasing: $(-0.91, 0.91)$

43.

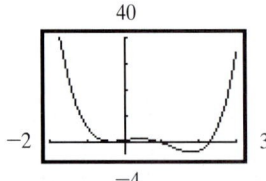

Local maximum: $(0.41, 1.53)$
Local minimum: $(-0.34, 0.54); (1.80, -3.56)$
Increasing: $(-0.34, 0.41); (1.80, 3)$
Decreasing: $(-2, -0.34); (0.41, 1.80)$

45. $(f + g)(x) = 2x + 3$; Domain: all real numbers
$(f - g)(x) = -4x + 1$; Domain: all real numbers
$(f \cdot g)(x) = -3x^2 + 5x + 2$; Domain: all real numbers
$\left(\dfrac{f}{g}\right)(x) = \dfrac{2 - x}{3x + 1}$; Domain: $\left\{ x \Big| x \neq -\dfrac{1}{3} \right\}$

47. $(f + g)(x) = 3x^2 + 4x + 1$; Domain: all real numbers
$(f - g)(x) = 3x^2 - 2x + 1$; Domain: all real numbers
$(f \cdot g)(x) = 9x^3 + 3x^2 + 3x$; Domain: all real numbers
$\left(\dfrac{f}{g}\right)(x) = \dfrac{3x^2 + x + 1}{3x}$; Domain: $\{x | x \neq 0\}$

49. $(f + g)(x) = \dfrac{x^2 + 2x - 1}{x(x - 1)}$; Domain: $\{x | x \neq 0, 1\}$

$(f - g)(x) = \dfrac{x^2 + 1}{x(x - 1)}$; Domain: $\{x | x \neq 0, 1\}$

$(f \cdot g)(x) = \dfrac{x + 1}{x(x - 1)}$; Domain: $\{x | x \neq 0, 1\}$

$\left(\dfrac{f}{g}\right)(x) = \dfrac{x(x + 1)}{x - 1}$; Domain: $\{x | x \neq 0, 1\}$

51. (a) -26 (b) -241 (c) 16 (d) -1

53. (a) $\sqrt{11}$ (b) 1 (c) $\sqrt{\sqrt{6} + 2}$ (d) 19

55. (a) $\dfrac{1}{20}$ (b) $-\dfrac{13}{8}$ (c) $\dfrac{400}{1601}$ (d) -17

57. $(f \circ g)(x) = 1 - 3x$; all real numbers; $(g \circ f)(x) = 7 - 3x$; all real numbers;
$(f \circ f)(x) = x$; all real numbers; $(g \circ g)(x) = 9x + 4$; all real numbers

59. $(f \circ g)(x) = 27x^2 + 3|x| + 1$; all real numbers; $(g \circ f)(x) = 3|3x^2 + x + 1|$; all real numbers;
$(f \circ f)(x) = 3(3x^2 + x + 1)^2 + 3x^2 + x + 2$; all real numbers; $(g \circ g)(x) = 9|x|$; all real numbers

61. $(f \circ g)(x) = \dfrac{1 + x}{1 - x}$; $\{x | x \neq 0, x \neq 1\}$; $(g \circ f)(x) = \dfrac{x - 1}{x + 1}$; $\{x | x \neq -1, x \neq 1\}$; $(f \circ f)(x) = x$; $\{x | x \neq 1\}$; $(g \circ g)(x) = x$; $\{x | x \neq 0\}$

63. (a)

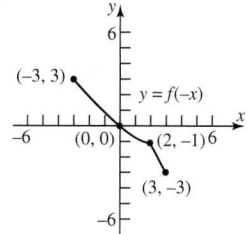

(b)

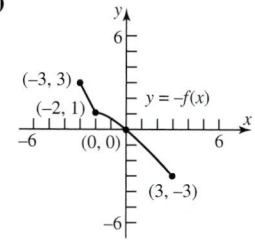

(c)

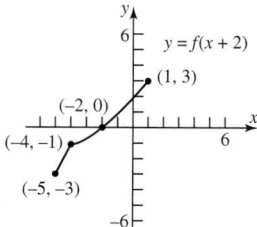

(d)

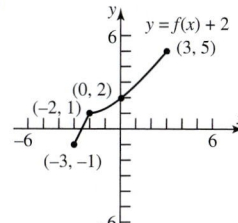

(e)

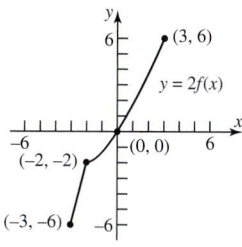

(f)

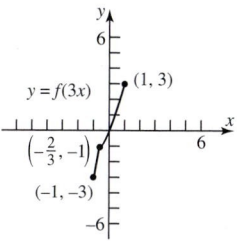

65. $(2, 2)$ **67.** $\dfrac{20\sqrt{30}}{9} \approx 12.17$ square units   **69.** $V(S) = \dfrac{S}{6}\sqrt{\dfrac{S}{\pi}}$; If the surface area doubles, the volume increases by a factor of $2\sqrt{2}$.

Cumulative Review Exercises *(page 321)*

1. $0, 2, 4$ **2.** $\left\{ x \,\middle|\, x \ge \dfrac{3}{2} \right\}; \left[\dfrac{3}{2}, \infty\right)$ **3.** Center: $(-2, 1)$; Radius: 3   **4.** x-intercepts: $-3, 0, 3$

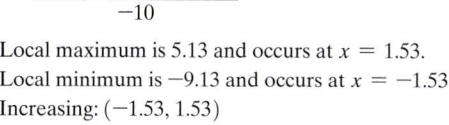

y-intercept: 0
Symmetric with respect to the origin

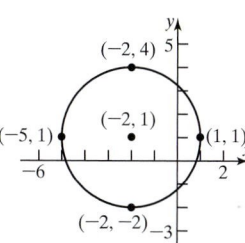

5. $y = -\dfrac{2}{3}x + \dfrac{17}{3}$

6. Not a function

7. (a) 22 **(b)** $x^2 - 5x - 2$
(c) $-x^2 - 5x + 2$
(d) $9x^2 + 15x - 2$
(e) $2x + h + 5$

8. (a) $\{x \mid x \ne 1\}$ **(b)** No, $(2, 7)$ is on the graph.   **(c)** 4; $(3, 4)$ is on the graph.   **(d)** $\dfrac{7}{4}; \left(\dfrac{7}{4}, 9\right)$ is on the graph.

9.

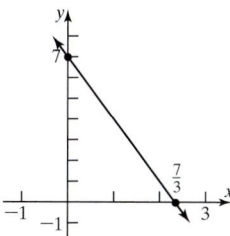

10.

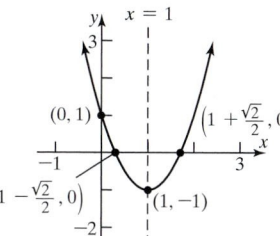

11. $x + 4; m_{\sec} = 6$

12. (a) x-intercepts: $-5, -1, 5$; y-intercept: -1
(b) No symmetry
(c) Neither
(d) Increasing: $(-\infty, -3)$ and $(2, \infty)$
  Decreasing: $(-3, 2)$
(e) Local maximum is 5 and occurs at $x = -3$.
(f) Local minimum is -6 and occurs at $x = 2$.

13.

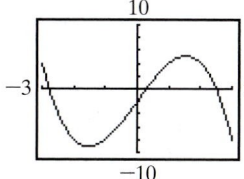

Local maximum is 5.13 and occurs at $x = 1.53$.
Local minimum is -9.13 and occurs at $x = -1.53$.
Increasing: $(-1.53, 1.53)$
Decreasing: $(-3, -1.53)$ and $(1.53, 3)$

14. Odd
15. (a) Domain: $\{x \mid -3 < x\}$ or $(-3, \infty)$

(b) x-intercept: $-\dfrac{1}{2}$; y-intercept: 1

(c)

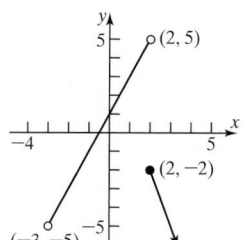

(d) Range: $\{y \mid y < 5\}$ or $(-\infty, 5)$

16.

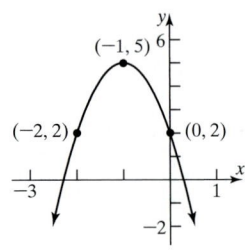

17. (a) $(f + g)(x) = x^2 - 9x - 6$; Domain: all real numbers

(b) $\left(\dfrac{f}{g}\right)(x) = \dfrac{x^2 - 5x + 1}{-4x - 7}$; Domain: $\left\{x \,\middle|\, x \neq -\dfrac{7}{4}\right\}$

(c) 301

(d) $(f \circ g)(x) = 16x^2 + 76x + 85$; Domain: all real numbers

18. $(f \circ g)(x) = -\dfrac{3(x + 3)}{x - 1}$; Domain: $\{x \,|\, x \neq -3, 1\}$

19. (a) $R(x) = -\dfrac{1}{10}x^2 + 150x$

(b) \$14,000 **(c)** 750; \$56,250 **(d)** \$75

20. $C(x) = \begin{cases} 0.99 & \text{if } 0 \leq x \leq 20 \\ 0.99 + 0.07(x - 20) & \text{if } 20 < x \end{cases}$

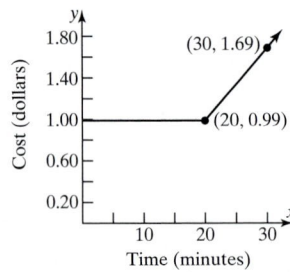

C H A P T E R 4 Polynomial and Rational Functions

4.1 Concepts and Vocabulary *(page 328)*

1. 5 **2.** $(-1, -1); (0, 0); (1, 1)$ **3.** T **4.** F

4.1 Exercises *(page 329)*

1.

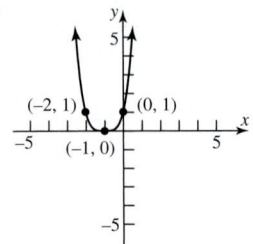

3.

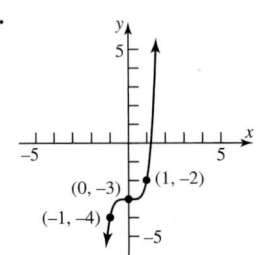

5.

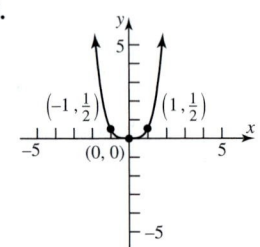

7.

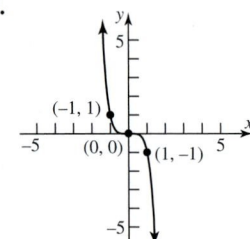

9.

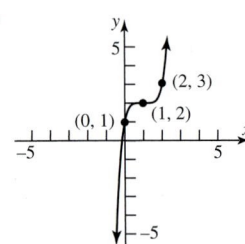

11.

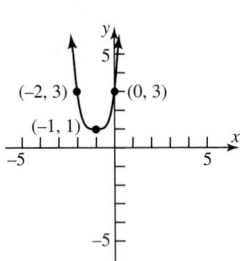

13.

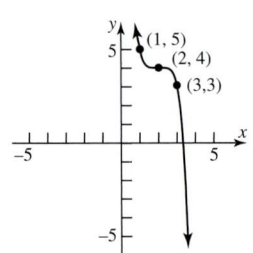

15.

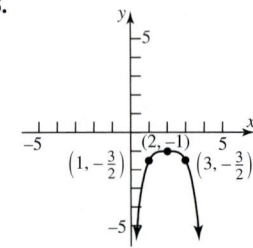

17. (a)

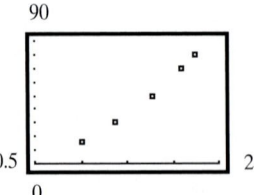

(b) $y = 15.9727x^{2.0018}$ **(c)**

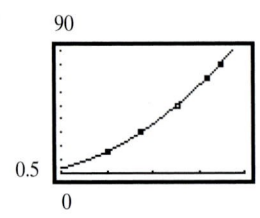

(d) $t \approx 2.5$ sec

(e) $s = \dfrac{1}{2} \cdot 31.9454t^2$,

$g \approx 31.9454$ ft/sec^2

4.2 Concepts and Vocabulary (page 339)

1. Smooth; continuous   **2.** Zero   **3.** Touches   **4.** T   **5.** F   **6.** F   **7.** No; yes   **8.** $f(x) = \text{int}(x); f(x) = |x|$ **9.** (a), (b), (c)

4.2 Exercises (page 340)

1. Yes; degree 3   **3.** Yes; degree 2   **5.** No; x is raised to the -1 power.   **7.** No; x is raised to the $\frac{3}{2}$ power.   **9.** Yes; degree 4

11. $f(x) = x^3 - 3x^2 - x + 3$ for $a = 1$ **13.** $f(x) = x^3 - x^2 - 12x$ for $a = 1$ **15.** $f(x) = x^4 - 15x^2 + 10x + 24$ for $a = 1$
17. $f(x) = x^3 - 5x^2 + 3x + 9$ for $a = 1$ **19.** **(a)** 7, multiplicity 1; -3, multiplicity 2
(b) graph touches the x-axis at -3 and crosses it at 7   **(c)** $y = 3x^3$ **21.** **(a)** 2, multiplicity 3   **(b)** graph crosses the x-axis at 2

(c) $y = 4x^5$ **23.** **(a)** $-\frac{1}{2}$, multiplicity 2   **(b)** graph touches the x-axis at $-\frac{1}{2}$ **(c)** $y = -2x^6$

25. **(a)** 5, multiplicity 3; -4, multiplicity 2   **(b)** graph touches the x-axis at -4 and crosses it at 5   **(c)** $y = x^5$
27. **(a)** No real zeros   **(b)** graph neither crosses nor touches the x-axis   **(c)** $y = 3x^6$
29. **(a)** 0, multiplicity 2; $-\sqrt{2}, \sqrt{2}$, multiplicity 1   **(b)** graph touches the x-axis at 0 and crosses at $-\sqrt{2}$ and $\sqrt{2}$ **(c)** $y = -2x^4$

31. **(a)** x-intercept: 1; y-intercept: 1
(b) 1: Touches
(c) $y = x^2$
(d)
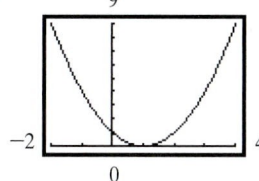
(e) 1; $(1, 0)$
(f)
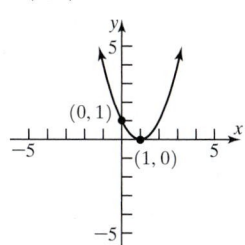

33. **(a)** x-intercepts: 0, 3; y-intercept: 0
(b) 0: Touches; 3: Crosses
(c) $y = x^3$
(d)
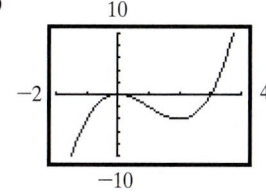
(e) 2; $(2, -4), (0, 0)$
(f)
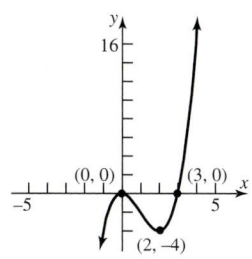

35. **(a)** x-intercepts: -4, 0; y-intercept: 0
(b) -4, 0: Crosses
(c) $y = 6x^4$
(d)
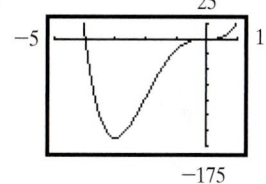
(e) 1; $(-3, -162)$
(f)
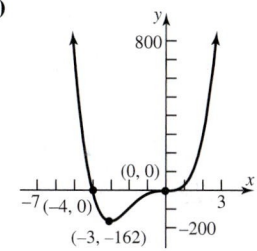

37. **(a)** x-intercepts: -2, 0; y-intercept: 0
(b) -2: Crosses; 0: Touches
(c) $y = -4x^3$
(d)
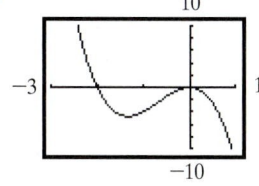
(e) 2; $(-1.33, -4.74), (0, 0)$
(f)
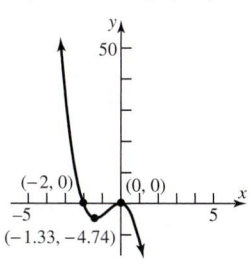

39. **(a)** x-intercepts: $-4, -1, 2$; y-intercept: -8
(b) $-4, -1, 2$: Crosses
(c) $y = x^3$
(d)
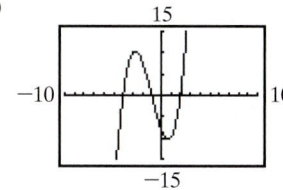
(e) 2; $(-2.73, 10.39); (0.73, -10.39)$
(f)
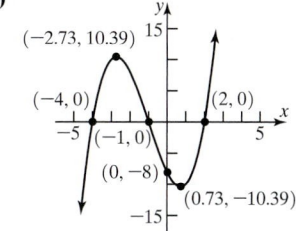

41. **(a)** x-intercepts: -2, 0, 2; y-intercept: 0
(b) -2, 0, 2: Crosses
(c) $y = -x^3$
(d)
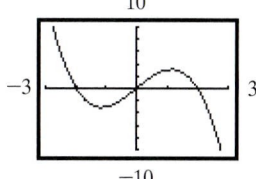
(e) 2; $(-1.15, -3.08), (1.15, 3.08)$
(f)
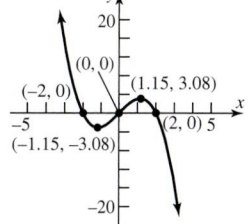

43. (a) *x*-intercepts: $-2, 0, 2$; *y*-intercept: 0
(b) $-2, 2$: Crosses; 0: Touches
(c) $y = x^4$
(d)

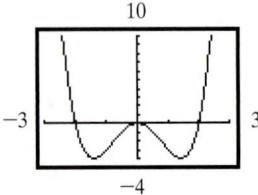

(e) $3; (-1.41, -4), (0, 0), (1.41, -4)$
(f)

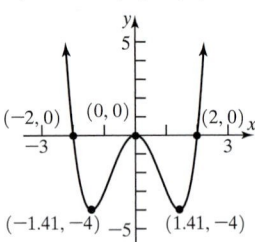

45. (a) *x*-intercepts: $-1, 2$; *y*-intercept: 4
(b) $-1, 2$: Touches
(c) $y = x^4$
(d)

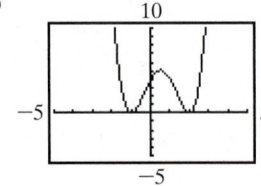

(e) $3; (-1, 0), (0.5, 5.06); (2, 0)$
(f)

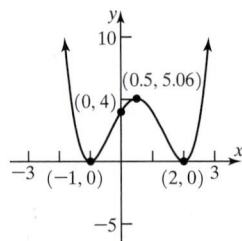

47. (a) *x*-intercepts: $-1, 0, 3$; *y*-intercept: 0
(b) $-1, 3$: Crosses; 0: Touches
(c) $y = x^4$
(d)

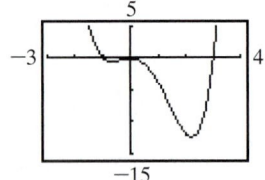

(e) $3; (-0.69, -0.54), (0, 0), (2.19, -12.39)$
(f)

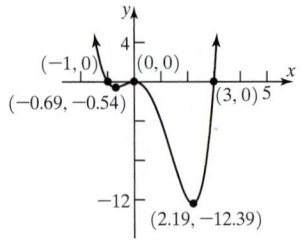

49. (a) *x*-intercepts: $-2, 4$; *y*-intercept: 64
(b) $-2, 4$: Touches
(c) $y = x^4$
(d)

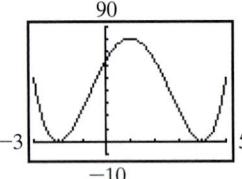

(e) $3; (-2, 0), (1, 81), (4, 0)$
(f)

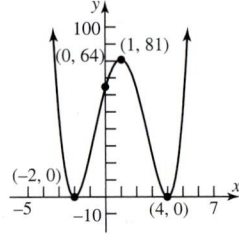

51. (a) *x*-intercepts: $0, 2$; *y*-intercept: 0
(b) 0: Touches; 2: Crosses
(c) $y = x^5$
(d)

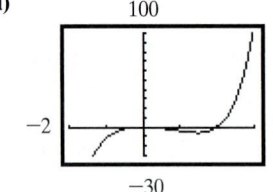

(e) $2; (0, 0), (1.48, -5.91)$
(f)

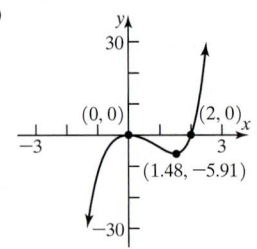

53. (a) *x*-intercepts: $-1, 0, 1$; *y*-intercept: 0
(b) $-1, 0$: Touches; 1: Crosses
(c) $y = -x^5$
(d)

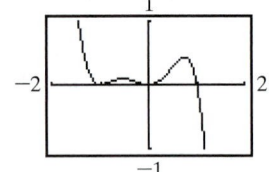

(e) $4; (-1, 0), (-0.54, 0.10), (0, 0),$
$(0.74, 0.43)$
(f)

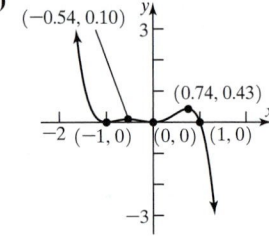

55. (a)

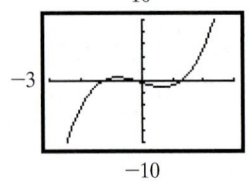

(b) *x*-intercepts: $-1.26, -0.20, 1.26$; *y*-intercept: -0.31752
(c) $y = x^3$
(d) $2; (-0.80, 0.57), (0.66, -0.99)$ **(e)**

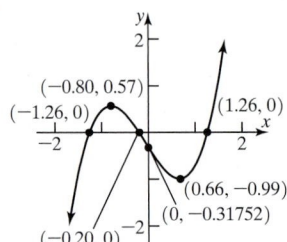

57. (a)

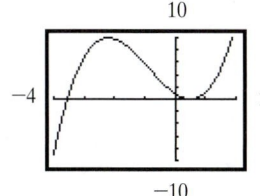

(b) x-intercepts: $-3.56, 0.50$;
   y-intercept: 0.89
(c) $y = x^3$
(d) $2; (-2.21, 9.91), (0.50, 0)$
(e)

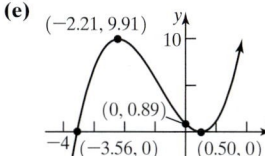

59. (a)

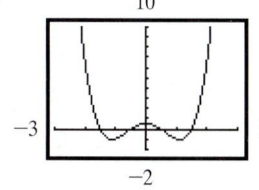

(b) x-intercepts: $-1.50, -0.50, 0.50, 1.50$;
   y-intercept: 0.5625
(c) $y = x^4$
(d) $3; (-1.12, -1), (0, 0.5625), (1.12, -1)$
(e)

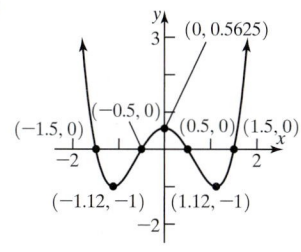

61. (a)

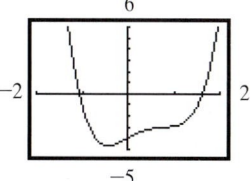

(b) x-intercepts: $-1.07, 1.62$;
   y-intercept: -4
(c) $y = 2x^4$
(d) $1; (-0.42, -4.64)$
(e)

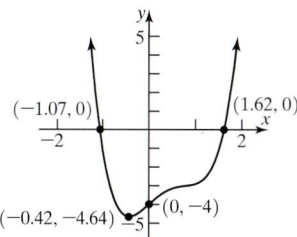

63. (a)

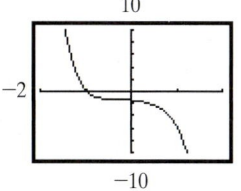

(b) x-intercept: -0.98; y-intercept: $-\sqrt{2}$
(c) $y = -2x^5$
(d) 0 **(e)**

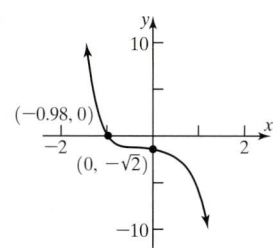

65. Answers will vary. One possibility is $f(x) = x(x - 1)(x - 2)$.

67. Answers will vary. One possibility is $f(x) = -\dfrac{1}{2}(x + 1)(x - 1)(x - 2)$.

69. (a) Cubic, $a > 0$

(b) \$7000 per car   **(c)** \$20,000 per car

(d) $C(x) = 0.2156x^3 - 2.3473x^2 + 14.3275x + 10.2238$

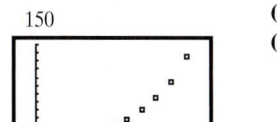

(e)

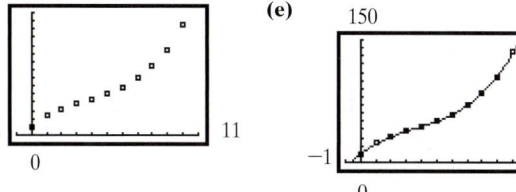

(f) About \$171,000
(g) Fixed costs of about \$10,200

71. (a) Cubic, $a > 0$

(b) $T(x) = 1.5243x^3 - 39.8089x^2 + 282.2881x + 1035.5$

(c)

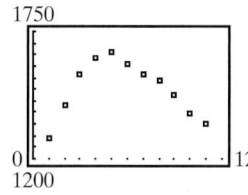

(d) 1324

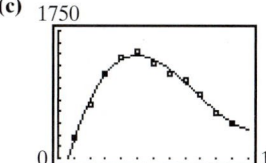

4.3 Concepts and Vocabulary *(page 353)*

1. $y = 1$ **2.** $x = -1$ **3.** Proper   **4.** F   **5.** T   **6.** T   **7.** The value of R approaches L as x increases.   **9.** No

4.3 Exercises *(page 353)*

1. $\{x \mid x \neq 3\}$ **3.** $\{x \mid x \neq -4, x \neq 2\}$ **5.** $\left\{x \mid x \neq -\dfrac{1}{2}, x \neq 3\right\}$ **7.** $\{x \mid x \neq 2\}$ **9.** All real numbers   **11.** $\{x \mid x \neq -3, x \neq 3\}$

13. (a) Domain: $\{x \mid x \neq 2\}$; Range: $\{y \mid y \neq 1\}$ **(b)** $(0, 0)$ **(c)** $y = 1$ **(d)** $x = 2$ **(e)** None

15. (a) Domain: $\{x \mid x \neq 0\}$; Range: all real numbers   **(b)** $(-1, 0), (1, 0)$ **(c)** None   **(d)** $x = 0$ **(e)** $y = 2x$

17. (a) Domain: $\{x \mid x \neq -2, x \neq 2\}$; Range: $\{y \mid y \leq 0 \text{ or } y > 1\}$ **(b)** $(0, 0)$ **(c)** $y = 1$ **(d)** $x = -2, x = 2$ **(e)** None

19. (a) Domain: $\{x \mid x \neq -1\}$; Range: $\{y \mid y \neq 2\}$ **(b)** $(-1.5, 0), (0, 3)$ **(c)** $y = 2$ **(d)** $x = -1$ **(e)** None

21. (a) Domain: $\{x \mid x \neq -4, x \neq 3\}$; Range: all real numbers   **(b)** $(0, 0)$ **(c)** $y = 0$ **(d)** $x = -4; x = 3$ **(e)** None

23. 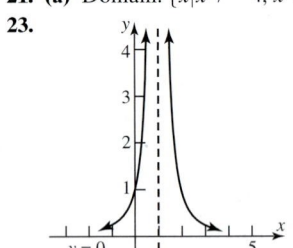   **25.** 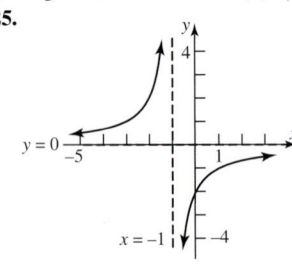   **27.** 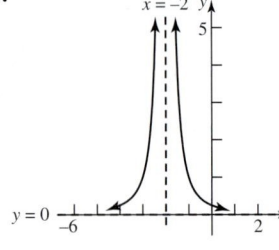   **29.**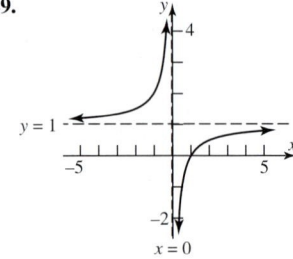

31. 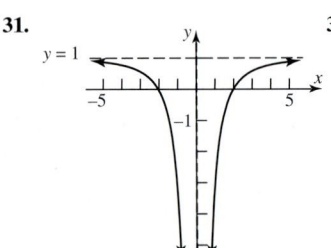   **33.**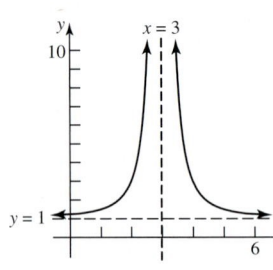

35. Horizontal asymptote: $y = 3$; vertical asymptote: $x = -4$

37. No asymptotes

39. Horizontal asymptote: $y = 0$; vertical asymptotes: $x = 1, x = -1$

41. Horizontal asymptote: $y = 0$; vertical asymptote: $x = 0$

43. Oblique asymptote: $y = 3x$; vertical asymptote: $x = 0$

45. Oblique asymptote: $y = -(x + 1)$; vertical asymptote: $x = 0$

4.4 Concepts and Vocabulary *(page 362)*

1. In lowest terms   **2.** F   **3.** F

4.4 Exercises *(page 363)*

1. 1. Domain: $\{x \mid x \neq 0, x \neq -4\}$
   2. $R(x)$ is in lowest terms.
   3. x-intercept: -1; no y-intercept
   4. No symmetry
   5. Vertical asymptotes: $x = 0, x = -4$
   6. Horizontal asymptote: $y = 0$, intersected at $(-1, 0)$

7.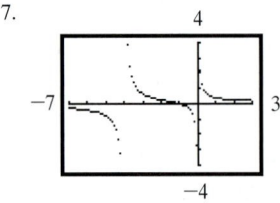

8.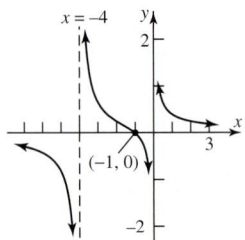

3. 1. Domain: $\{x \mid x \neq -2\}$

   2. $R(x) = \dfrac{3(x + 1)}{2(x + 2)}$

   3. x-intercept: -1; y-intercept: $\dfrac{3}{4}$

   4. No symmetry

   5. Vertical asymptote: $x = -2$

   6. Horizontal asymptote: $y = \dfrac{3}{2}$, not intersected

7.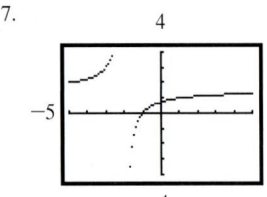

8.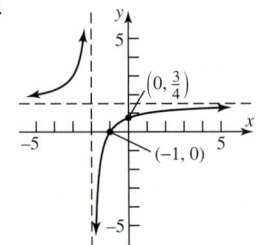

5. 1. Domain: $\{x|x \neq -2, x \neq 2\}$

2. $R(x) = \dfrac{3}{(x+2)(x-2)}$

3. No x-intercept; y-intercept: $-\dfrac{3}{4}$

4. Symmetric with respect to the y-axis

5. Vertical asymptotes: $x = 2, x = -2$

6. Horizontal asymptote: $y = 0$, not intersected

7.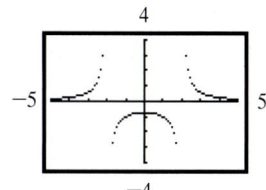

8.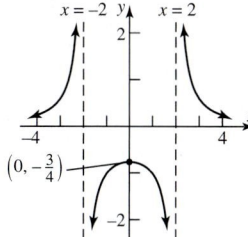

7. 1. Domain: $\{x|x \neq -1, x \neq 1\}$

2. $P(x) = \dfrac{x^4 + x^2 + 1}{(x+1)(x-1)}$

3. No x-intercept; y-intercept: -1

4. Symmetric with respect to the y-axis

5. Vertical asymptotes: $x = -1, x = 1$

6. No horizontal or oblique asymptotes

7.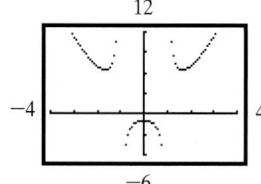

8.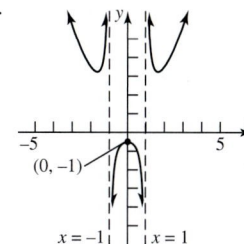

9. 1. Domain: $\{x|x \neq -3, x \neq 3\}$

2. $H(x) = \dfrac{(x-1)(x^2+x+1)}{(x+3)(x-3)}$

3. x-intercept: 1; y-intercept: $\dfrac{1}{9}$

4. No symmetry

5. Vertical asymptotes: $x = 3, x = -3$

6. Oblique asymptote: $y = x$, intersected at $\left(\dfrac{1}{9}, \dfrac{1}{9}\right)$

7.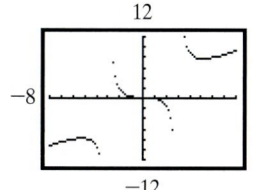

8.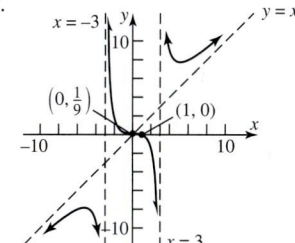

11. 1. Domain: $\{x \neq -3, x \neq 2\}$

2. $R(x) = \dfrac{x^2}{(x+3)(x-2)}$

3. Intercept: $(0, 0)$

4. No symmetry

5. Vertical asymptotes: $x = 2, x = -3$

6. Horizontal asymptote: $y = 1$, intersected at $(6, 1)$

7.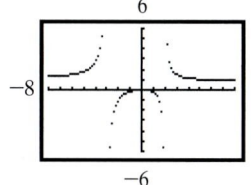

8.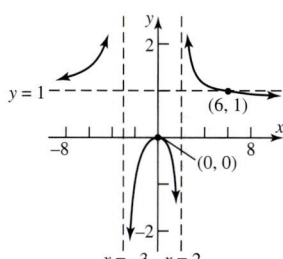

13. 1. Domain: $\{x|x \neq -2, x \neq 2\}$

2. $G(x) = \dfrac{x}{(x+2)(x-2)}$

3. Intercept: $(0, 0)$

4. Symmetry with respect to the origin

5. Vertical asymptotes: $x = -2, x = 2$

6. Horizontal asymptote: $y = 0$, intersected at $(0, 0)$

7.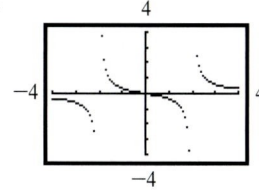

8.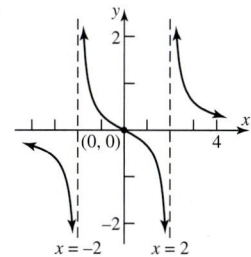

15. 1. Domain: $\{x|x \neq 1, x \neq -2, x \neq 2\}$

2. $R(x) = \dfrac{3}{(x-1)(x-2)(x+2)}$

3. No x-intercept; y-intercept: $\dfrac{3}{4}$

4. No symmetry

5. Vertical asymptotes: $x = -2, x = 1, x = 2$

6. Horizontal asymptote: $y = 0$, not intersected

7.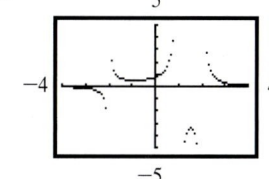

8.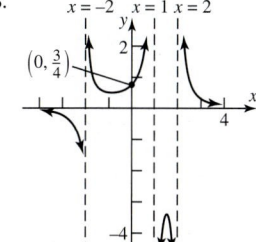

17. 1. Domain: $\{x \mid x \neq -2, x \neq 2\}$

2. $H(x) = \dfrac{4(x + 1)(x - 1)}{(x^2 + 4)(x + 2)(x - 2)}$

3. x-intercepts: $-1, 1$; y-intercept: $\dfrac{1}{4}$

4. Symmetry with respect to the y-axis

5. Vertical asymptotes: $x = -2, x = 2$

6. Horizontal asymptote: $y = 0$,
   intersected at $(-1, 0)$ and $(1, 0)$

7.
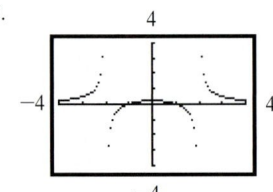

8.
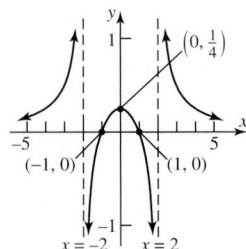

19. 1. Domain: $\{x \mid x \neq -2\}$

2. $F(x) = \dfrac{(x - 4)(x + 1)}{x + 2}$

3. x-intercepts: $-1, 4$; y-intercept: -2

4. No symmetry

5. Vertical asymptote: $x = -2$

6. Oblique asymptote: $y = x - 5$, not intersected

7.
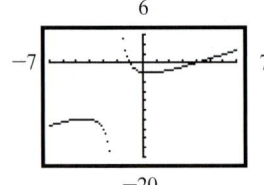

8.
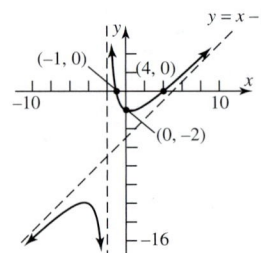

21. 1. Domain: $\{x \mid x \neq 4\}$

2. $R(x) = \dfrac{(x + 4)(x - 3)}{x - 4}$

3. x-intercepts: $-4, 3$; y-intercept: 3

4. No symmetry

5. Vertical asymptote: $x = 4$

6. Oblique asymptote: $y = x + 5$, not intersected

7.
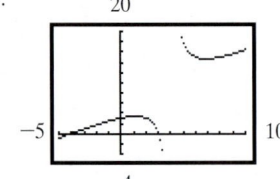

8.
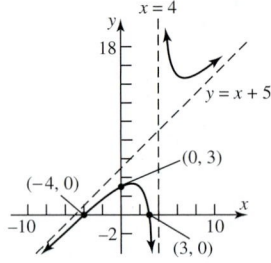

23. 1. Domain: $\{x \mid x \neq -2\}$

2. $F(x) = \dfrac{(x + 4)(x - 3)}{x + 2}$

3. x-intercepts: $-4, 3$; y-intercept: -6

4. No symmetry

5. Vertical asymptote: $x = -2$

6. Oblique asymptote: $y = x - 1$, not intersected

7.
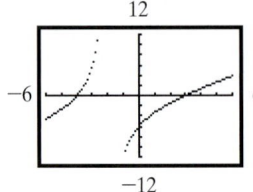

8.
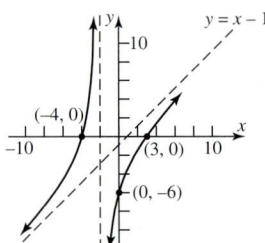

25. 1. Domain: $\{x \mid x \neq -3\}$

2. $R(x)$ is in lowest terms.

3. x-intercepts: $0, 1$; y-intercept: 0

4. No symmetry

5. Vertical asymptote: $x = -3$

6. Horizontal asymptote: $y = 1$, not intersected

7.
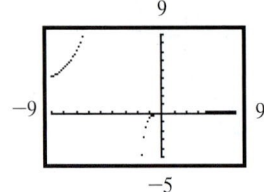

8.
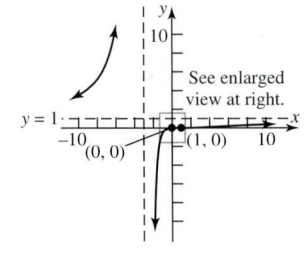

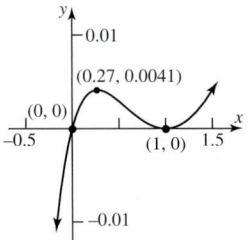

Enlarged view

27. 1. Domain: $\{x | x \neq -2, x \neq 3\}$

2. $R(x) = \dfrac{x+4}{x+2}$

3. x-intercept: -4; y-intercept: 2

4. No symmetry

5. Vertical asymptote: $x = -2$; hole at $\left(3, \dfrac{7}{5}\right)$

6. Horizontal asymptote: $y = 1$, not intersected

7.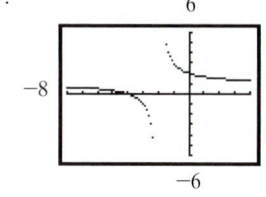

8.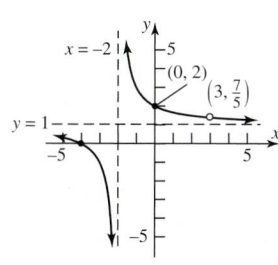

29. 1. Domain: $\left\{x \middle| x \neq \dfrac{3}{2}, x \neq 2\right\}$

2. $R(x) = \dfrac{3x+1}{x-2}$

3. x-intercept: $-\dfrac{1}{3}$; y-intercept: $-\dfrac{1}{2}$

4. No symmetry

5. Vertical asymptote: $x = 2$; hole at $\left(\dfrac{3}{2}, -11\right)$

6. Horizontal asymptote: $y = 3$, not intersected

7.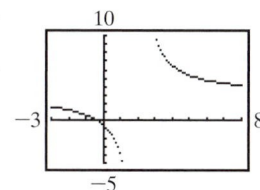

8.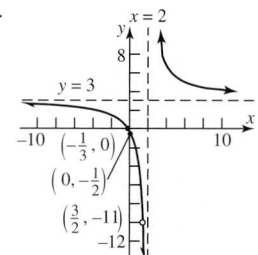

31. 1. Domain: $\{x | x \neq -3\}$

2. $R(x) = x + 2$

3. x-intercept: -2; y-intercept: 2

4. No symmetry

5. Vertical asymptote: none; hole at $(-3, -1)$

6. Oblique asymptote: $y = x + 2$

7.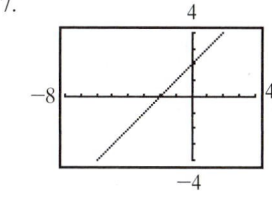

8.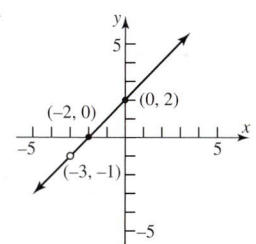

33. 1. Domain: $\{x | x \neq 0\}$

2. $f(x) = \dfrac{x^2 + 1}{x}$

3. No x-intercepts; no y-intercepts

4. Symmetric with respect to the origin

5. Vertical asymptote: $x = 0$

6. Oblique asymptotes: $y = x$, not intersected

7.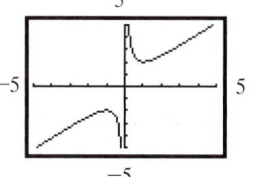

8.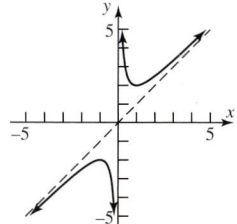

35. 1. Domain: $\{x | x \neq 0\}$

2. $f(x) = \dfrac{x^3 + 1}{x} = \dfrac{(x+1)(x^2 - x + 1)}{x}$

3. x-intercept: -1; no y-intercepts

4. No symmetry

5. Vertical asymptote: $x = 0$

6. No horizontal or oblique asymptotes

7.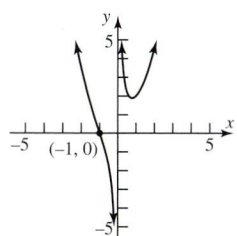

37. 1. Domain: $\{x | x \neq 0\}$

2. $f(x) = \dfrac{x^4 + 1}{x^3}$

3. No x-intercepts; no y-intercepts

4. Symmetric with respect to the origin

5. Vertical asymptote: $x = 0$

6. Oblique asymptote: $y = x$, not intersected

7.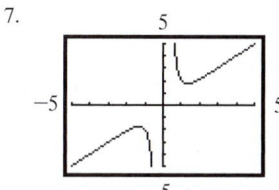

8.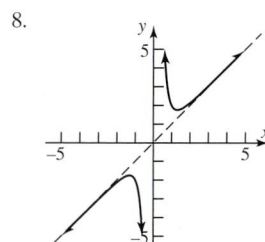

39. One possibility: $R(x) = \dfrac{x^2}{x^2 - 4}$ **41.** One possibility: $R(x) = \dfrac{(x-1)(x-3)(x^2 + \frac{4}{3})}{(x+1)^2(x-2)^2}$

45. (a) ≈ 9.8208 m/sec² **(b)** ≈ 9.8195 m/sec²
(c) ≈ 9.7936 m/sec² **(d)** h-axis
(e)

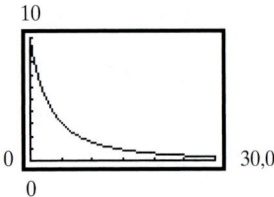

(f) $g(h)$ is never equal to 0, but $g(h) \to 0$ as $h \to \infty$. The further away from sea level you get, the lower the acceleration due to gravity.

47. (a)

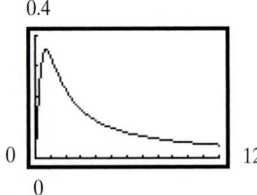

(b) 0.71 hr
(c) $C(t) = 0$; As t increases, the concentration decreases to near zero.

49. (a) $\overline{C}(x) = \dfrac{0.2x^3 - 2.3x^2 + 14.3x + 10.2}{x}$

(b) $\overline{C}(6) = \$9400$ per car
(c) $\overline{C}(9) \approx \$10,933$ per car
(d)

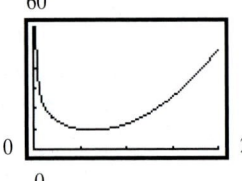

(e) 6.38 cars
(f) $9366 per car

51. (a) $S(x) = 2x^2 + \dfrac{40,000}{x}$

(b)

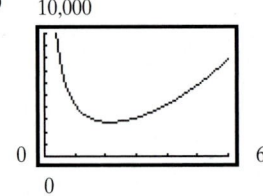

(c) 2784.95 in²
(d) 21.54 in. × 21.54 in. × 21.54 in.

53. (a) $C(r) = 12\pi r^2 + \dfrac{4000}{r}$ **(b)** 3.76 cm

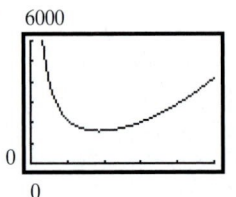

C is smallest when $r \approx 3.76$ cm.

4.5 Concepts and Vocabulary *(page 371)*

1. F **2.** T **3.** Because $x^2 + 1 > 0$ for all x.

4.5 Exercises *(page 372)*

1. $\{x|-2 < x < 5\}; (-2, 5)$ **3.** $\{x|x < 0 \text{ or } x > 4\}; (-\infty, 0) \text{ or } (4, \infty)$ **5.** $\{x|-3 < x < 3\}; (-3, 3)$ **7.** $\{x|x < -4 \text{ or } x > 3\};$
$(-\infty, -4) \text{ or } (3, \infty)$ **9.** $\left\{x\left|-\dfrac{1}{2} < x < 3\right.\right\}; \left(-\dfrac{1}{2}, 3\right)$ **11.** $\{x|x < -1 \text{ or } x > 8\}; (-\infty, -1) \text{ or } (8, \infty)$ **13.** No real solution

15. $\left\{x\left|x < -\dfrac{2}{3} \text{ or } x > \dfrac{3}{2}\right.\right\}; \left(-\infty, -\dfrac{2}{3}\right) \text{ or } \left(\dfrac{3}{2}, \infty\right)$ **17.** $\{x|x > 1\}; (1, \infty)$ **19.** $\{x|x < 1 \text{ or } 2 < x < 3\}; (-\infty, 1) \text{ or } (2, 3)$

21. $\{x|-1 < x < 0 \text{ or } x > 3\}; (-1, 0) \text{ or } (3, \infty)$ **23.** $\{x|x < -1 \text{ or } x > 1\}; (-\infty, -1) \text{ or } (1, \infty)$ **25.** $\{x|x > 1\}; (1, \infty)$
27. $\{x|x < -1 \text{ or } x > 1\}; (-\infty, -1) \text{ or } (1, \infty)$ **29.** $\{x|-1 < x < 8\}; (-1, 8)$ **31.** $\{x|x \geq 2.14\}; [2.14, \infty)$ **33.** $\{x|x < -2 \text{ or } x > 2\};$
$(-\infty, -2) \text{ or } (2, \infty)$ **35.** $\{x|-1 \leq x \leq 2 - \sqrt{5} \text{ or } x \geq 2 + \sqrt{5}\}; [-1, 2 - \sqrt{5}] \text{ or } [2 + \sqrt{5}, \infty)$ **37.** $\{x|x < -1 \text{ or } x > 1\};$
$(-\infty, -1) \text{ or } (1, \infty)$ **39.** $\{x|x < -1 \text{ or } 0 < x < 1\}; (-\infty, -1) \text{ or } (0, 1)$ **41.** $\{x|x < -1 \text{ or } x > 1\}; (-\infty, -1) \text{ or } (1, \infty)$

43. $\left\{x\left|x < -\dfrac{2}{3} \text{ or } 0 < x < \dfrac{3}{2}\right.\right\}; \left(-\infty, -\dfrac{2}{3}\right) \text{ or } \left(0, \dfrac{3}{2}\right)$ **45.** $\{x|x < 2\}; (-\infty, 2)$ **47.** $\{x|-2 < x \leq 9\}; (-2, 9]$

49. $\{x|x < 2 \text{ or } 3 < x < 5\}; (-\infty, 2) \text{ or } (3, 5)$ **51.** $\{x|x < -3 \text{ or } -1 < x < 1 \text{ or } x > 2\}; (-\infty, -3) \text{ or } (-1, 1) \text{ or } (2, \infty)$

53. $\{x|x < -5 \text{ or } -4 < x < -3 \text{ or } x > 1\}; (-\infty, -5) \text{ or } (-4, -3) \text{ or } (1, \infty)$ **55.** $\left\{x\left|x \leq -\dfrac{1}{2} \text{ or } 1 \leq x < 4\right.\right\}; \left(-\infty, -\dfrac{1}{2}\right] \text{ or } [1, 4)$

57. $\left\{x\left|\dfrac{-3-\sqrt{13}}{2}<x<-3 \text{ or } x>\dfrac{-3+\sqrt{13}}{2}\right.\right\}; \left(\dfrac{-3-\sqrt{13}}{2},-3\right)$ or $\left(\dfrac{-3+\sqrt{13}}{2},\infty\right)$

59. $\{x|x>4\}; (4,\infty)$ **61.** $\{x|x\le -4 \text{ or } x\ge 4\}; (-\infty,-4] \text{ or } [4,\infty)$ **63.** $\{x|x<-4 \text{ or } x\ge 2\}; (-\infty,-4) \text{ or } [2,\infty)$

65. (a) The ball is more than 96 feet above the ground from time t between 2 and 3 seconds, $2<t<3$.   **(b)** 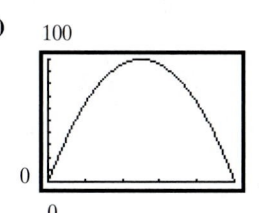   **(c)** 100 ft   **(d)** 2.5 sec

67. (a) For a profit of at least \$50, between 8 and 32 watches must be sold, $8\le x\le 32$.   **(b)** 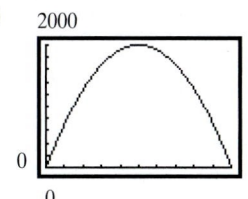   **(c)** \$2000   **(d)** 100   **(e)** 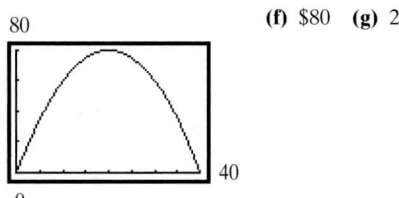   **(f)** \$80   **(g)** 20

69. Chevy can produce at most 8 Cavaliers in an hour, assuming that cars cannot be partially completed in an hour.

Review Exercises *(page 374)*

1. Polynomial of degree 5   **3.** Not a polynomial

5.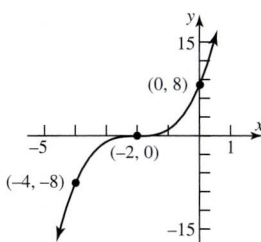

7.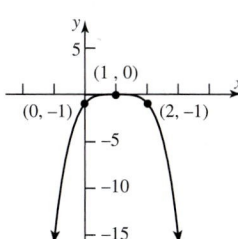

9.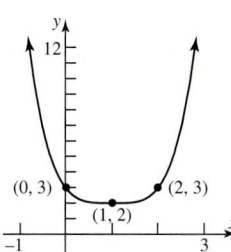

11. (a) x-intercepts: $-4, -2, 0$; y-intercept: 0
(b) $-4, -2, 0$: Crosses
(c) $y=x^3$
(d)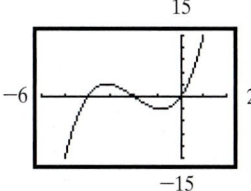
(e) $2; (-3.15, 3.08), (-0.85, -3.08)$
(f)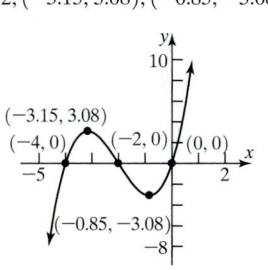

13. (a) x-intercepts: $-4, 2$; y-intercept: 16
(b) -4: Crosses; 2: Touches
(c) $y=x^3$
(d)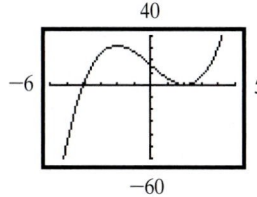
(e) $2; (-2, 32), (2, 0)$
(f)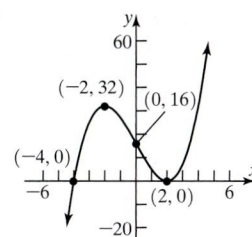

15. $f(x)=x^3-4x^2=x^2(x-4)$
(a) x-intercepts: 0, 4; y-intercept: 0
(b) 0: Touches; 4: Crosses
(c) $y=x^3$
(d)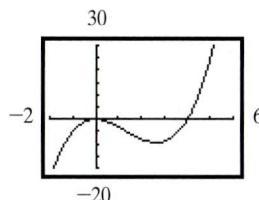
(e) $2; (0, 0), (2.67, -9.48)$
(f)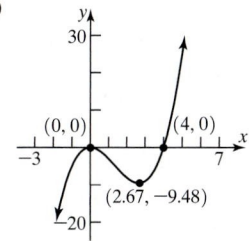

17. **(a)** x-intercepts: $-3, -1, 1$; y-intercept: 3   **(b)** $-3, -1$: Crosses; 1: Touches   **(c)** $y = x^4$
   (d)   **(e)** $3; (-2.28, -9.91), (-0.22, 3.23), (1, 0)$
   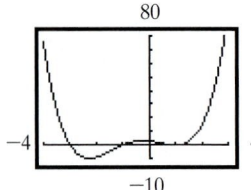   **(f)**

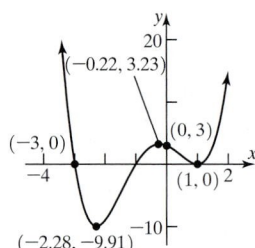

19. 1. Domain: $\{x | x \neq 0\}$
   2. $R(x)$ is in lowest terms.
   3. x-intercept: 3; no y-intercept
   4. No symmetry
   5. Vertical asymptote: $x = 0$
   6. Horizontal asymptote: $y = 2$, not intersected

7.
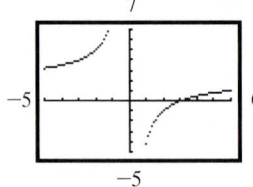

8.
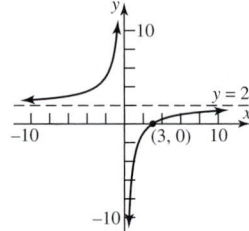

21. 1. Domain: $\{x | x \neq 0, x \neq 2\}$
   2. $H(x)$ is in lowest terms
   3. x-intercept: -2; no y-intercept
   4. No symmetry
   5. Vertical asymptotes: $x = 0, x = 2$
   6. Horizontal asymptote: $y = 0$, intersected at $(-2, 0)$

7.
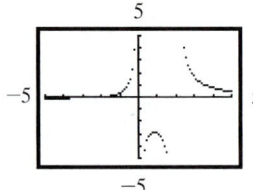

8.
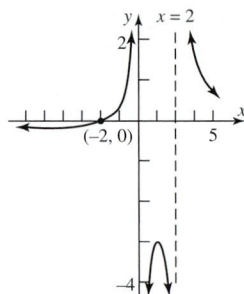

23. 1. Domain: $\{x | x \neq -2, x \neq 3\}$
   2. $R(x) = \dfrac{(x + 3)(x - 2)}{(x - 3)(x + 2)}$
   3. x-intercepts: $-3, 2$; y-intercept: 1
   4. No symmetry
   5. Vertical asymptotes: $x = -2, x = 3$
   6. Horizontal asymptote: $y = 1$, intersected at $(0, 1)$

7.
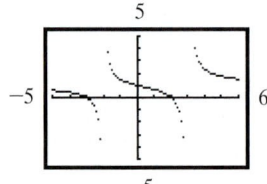

8.
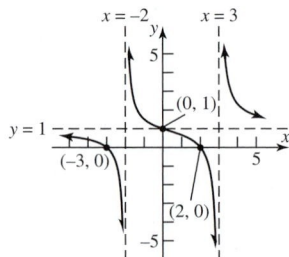

25. 1. Domain: $\{x | x \neq -2, x \neq 2\}$
   2. $F(x) = \dfrac{x^3}{(x + 2)(x - 2)}$
   3. Intercept: $(0, 0)$
   4. Symmetric with respect to the origin
   5. Vertical asymptotes: $x = -2, x = 2$
   6. Oblique asymptote: $y = x$, intersected at $(0, 0)$

7.
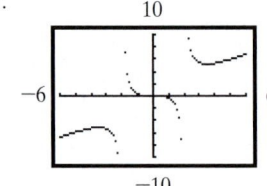

8.
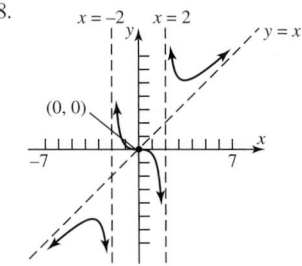

27. 1. Domain: $\{x|x \neq 1\}$
2. $R(x)$ is in lowest terms
3. Intercept: $(0, 0)$
4. No symmetry
5. Vertical asymptote: $x = 1$
6. No oblique or horizontal asymptote

7.
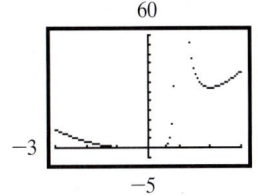

8.
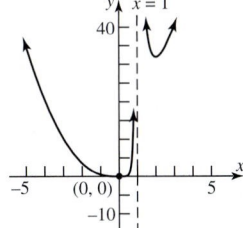

29. 1. Domain: $\{x|x \neq -1, x \neq 2\}$
2. $G(x) = \dfrac{x + 2}{x + 1}$
3. x-intercept: -2; y-intercept: 2
4. No symmetry
5. Vertical asymptote: $x = -1$; hole at $\left(2, \dfrac{4}{3}\right)$
6. Horizontal asymptote: $y = 1$, not intersected

7.
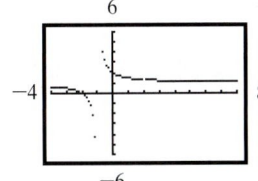

8.
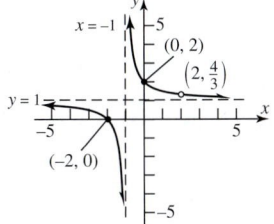

31. $\left\{x\left|-4 < x < \dfrac{3}{2}\right.\right\}; \left(-4, \dfrac{3}{2}\right)$ **33.** $\{x|-3 < x \leq 3\}; (-3, 3]$ **35.** $\{x|x < 1 \text{ or } x > 2\}; (-\infty, 1) \text{ or } (2, \infty)$

37. $\{x|1 < x < 2 \text{ or } x > 3\}; (1, 2) \text{ or } (3, \infty)$ **39.** $\{x|x < -4 \text{ or } 2 < x < 4 \text{ or } x > 6\}; (-\infty, -4) \text{ or } (2, 4) \text{ or } (6, \infty)$

41. **(a)**
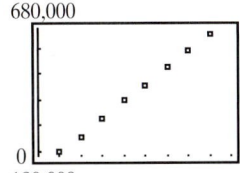

(b) $A(t) = -212.0076t^3 + 2429.132t^2 + 59{,}568.8539t + 130{,}003.1429$
(c)
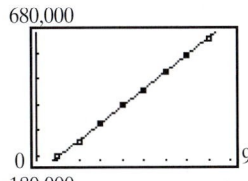

(d) About 797,000

Cumulative Review Exercises *(page 377)*

1. $\left\{\dfrac{-1 - \sqrt{21}}{4}, \dfrac{-1 + \sqrt{21}}{4}\right\}$ **2.** $y = -\dfrac{2}{5}x + \dfrac{19}{5}$ **3.** Perpendicular   **4.** No   **5.** 154

6.
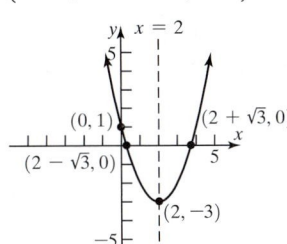

7.
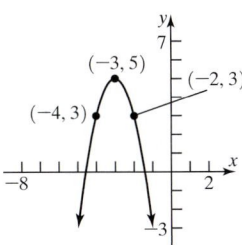

8. Odd; symmetric with respect to the origin
9. Local maximum is 4.70 and occurs at $x \approx -3.33$
Local minimum is 1 and occurs at $x = 0$
Increasing: $(-\infty, -3.33)$ or $(0, \infty)$;
Decreasing: $(-3.33, 0)$
10. $f(g(x)) = \dfrac{x - 1}{1 + 2x}$;
Domain of $f \circ g$: $\left\{x\left|x \neq -\dfrac{1}{2}, x \neq 1\right.\right\}$

11. **(a)** x-intercepts: $-3, 4$; y-intercept: 48   **(b)** -3: Crosses; 4: Touches   **(c)** $y = x^3$
(d)
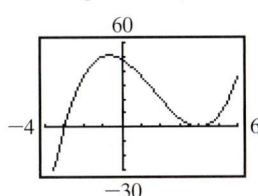

(e) $2; (-0.67, 50.81), (4, 0)$ **(f)**
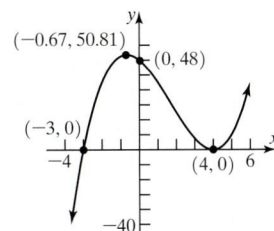

12. (a) $\{x|x \neq 1, x \neq 2\}$ **(b)** $x = 2$ **(c)** $y = 1$ **(d)** x-intercept: -1; y-intercept: $-\dfrac{1}{2}$ **13.** $A(t) = 4\pi t^2$

(e)

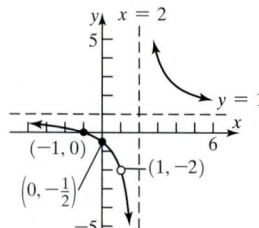

(f) $\{x|-1 \leq x < 1 \text{ or } 1 < x < 2\}; [-1, 1) \text{ or } (1, 2)$

C H A P T E R   5   The Zeros of a Polynomial Function

5.1 Exercises *(page 383)*

1. $q(x) = x^2 + x + 4; R = 12$ **3.** $q(x) = 3x^2 + 11x + 32; R = 99$ **5.** $q(x) = x^4 - 3x^3 + 5x^2 - 15x + 46; R = -138$
7. $q(x) = 4x^5 + 4x^4 + x^3 + x^2 + 2x + 2; R = 7$ **9.** $q(x) = 0.1x^2 - 0.11x + 0.321; R = -0.3531$
11. $q(x) = x^4 + x^3 + x^2 + x + 1; R = 0$ **13.** No **15.** Yes **17.** Yes **19.** No **21.** Yes **23.** $a + b + c + d = -9$

Historical Problems *(page 395)*

1.
$$\left(x - \frac{b}{3}\right)^3 + b\left(x - \frac{b}{3}\right)^2 + c\left(x - \frac{b}{3}\right) + d = 0$$

$$x^3 - bx^2 + \frac{b^2x}{3} - \frac{b^3}{27} + bx^2 - \frac{2b^2x}{3} + \frac{b^3}{9} + cx - \frac{bc}{3} + d = 0$$

$$x^3 + \left(c - \frac{b^2}{3}\right)x + \left(\frac{2b^3}{27} - \frac{bc}{3} + d\right) = 0$$

Let $p = c - \dfrac{b^2}{3}$ and $q = \dfrac{2b^3}{27} - \dfrac{bc}{3} + d$. Then $x^3 + px + q = 0$.

3.
$$3HK = -p$$

$$K = -\frac{p}{3H}$$

$$H^3 + \left(-\frac{p}{3H}\right)^3 = -q$$

$$H^3 - \frac{p^3}{27H^3} = -q$$

$$27H^6 - p^3 = -27qH^3$$

$$27H^6 + 27qH^3 - p^3 = 0$$

$$H^3 = \frac{-27q \pm \sqrt{(27q)^2 - 4(27)(-p^3)}}{2 \cdot 27}$$

$$H^3 = \frac{-q}{2} \pm \sqrt{\frac{27^2q^2}{2^2(27^2)} + \frac{4(27)p^3}{2^2(27^2)}}$$

$$H^3 = \frac{-q}{2} \pm \sqrt{\frac{q^2}{4} + \frac{p^3}{27}}$$ Choose the positive root for now.

$$H = \sqrt[3]{\frac{-q}{2} + \sqrt{\frac{q^2}{4} + \frac{p^3}{27}}}$$

5. $x = H + K$

$$x = \sqrt[3]{\frac{-q}{2} + \sqrt{\frac{q^2}{4} + \frac{p^3}{27}}} + \sqrt[3]{\frac{-q}{2} - \sqrt{\frac{q^2}{4} + \frac{p^3}{27}}}$$ (Note that if we had used the negative root in 3, the result would be the same.)

5.2 Concepts and Vocabulary *(page 396)*

1. Remainder; dividend **2.** $f(c)$ **3.** -4 **4.** F **5.** F **6.** T **7.** No, because 3 is not a factor of 2.

5.2 Exercises *(page 396)*

1. No; $f(3) = 61$ **3.** No; $f(1) = 2$ **5.** Yes; $f(x) = (x + 2)(3x^5 - 6x^4 + 12x^3 - 22x^2 + 44x - 88)$

7. Yes; $f(x) = (x - 4)(4x^5 + 16x^4 + x + 4)$ **9.** No; $f\left(-\dfrac{1}{2}\right) = -\dfrac{7}{4}$ **11.** $4; \pm 1, \pm\dfrac{1}{3}$ **13.** $5; \pm 1, \pm 3$ **15.** $3; \pm 1, \pm 2, \pm\dfrac{1}{4}, \pm\dfrac{1}{2}$

17. $4; \pm 1, \pm 2, \pm\dfrac{1}{3}, \pm\dfrac{2}{3}$ **19.** $5; \pm 1, \pm 2, \pm 4, \pm\dfrac{1}{2}$ **21.** $4; \pm 1, \pm 2, \pm\dfrac{1}{6}, \pm\dfrac{1}{3}, \pm\dfrac{1}{2}, \pm\dfrac{2}{3}$

23. -1 and 1 **25.** -12 and 12 **27.** -10 and 10

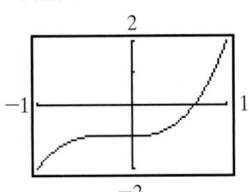
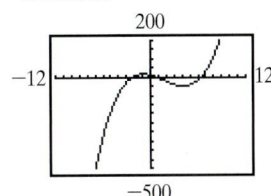
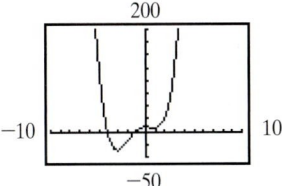

29. $-3, -1, 2; f(x) = (x + 3)(x + 1)(x - 2)$ **31.** $\dfrac{1}{2}, 3, 3; f(x) = (2x - 1)(x - 3)^2$ **33.** $-\dfrac{1}{3}; f(x) = (3x + 1)(x^2 + x + 1)$

35. $3, \dfrac{5 + \sqrt{17}}{2}, \dfrac{5 - \sqrt{17}}{2}; f(x) = (x - 3)\left(x - \left(\dfrac{5 + \sqrt{17}}{2}\right)\right)\left(x - \left(\dfrac{5 - \sqrt{17}}{2}\right)\right)$ **37.** $-2, -1, 1, 1; f(x) = (x + 2)(x + 1)(x - 1)^2$

39. $-5, -3, -\dfrac{3}{2}, 1; f(x) = (x + 5)(x + 3)(2x + 3)(x - 1)$ **41.** $-2, -\dfrac{3}{2}, 1, 4; f(x) = (x + 2)(2x + 3)(x - 1)(x - 4)$

43. $-\dfrac{1}{2}, \dfrac{1}{2}; f(x) = (2x + 1)(2x - 1)(x^2 + 2)$ **45.** $\dfrac{\sqrt{2}}{2}, -\dfrac{\sqrt{2}}{2}, 2; f(x) = (x - 2)(2x - \sqrt{2})(2x + \sqrt{2})\left(x^2 + \dfrac{1}{2}\right)$ **47.** $-5.9, -0.3, 3$

49. $-3.8, 4.5$ **51.** $-43.5, 1, 23$ **53.** $\{-1, 2\}$ **55.** $\left\{\dfrac{2}{3}, -1 + \sqrt{2}, -1 - \sqrt{2}\right\}$ **57.** $\left\{\dfrac{1}{3}, \sqrt{5}, -\sqrt{5}\right\}$ **59.** $\{-3, -2\}$ **61.** $-\dfrac{1}{3}$

63. $f(0) = -1; f(1) = 10;$ Zero: 0.22 **65.** $f(-5) = -58; f(-4) = 2;$ Zero: -4.05 **67.** $f(1.4) = -0.17536; f(1.5) = 1.40625;$ Zero: 1.41

69. ≈ 27 Cavaliers **71.** $k = 5$ **73.** -7 **75.** No, $\dfrac{1}{2}$ and 1 are the only possible positive rational zeros. **77.** 7 in.

5.3 Concepts and Vocabulary *(page 405)*

1. Real; imaginary; imaginary unit **2.** $-2, 2, -2i, 2i$ **3.** F **4.** T

5.3 Exercises *(page 406)*

1. $8 + 5i$ **3.** $-7 + 6i$ **5.** $-6 - 11i$ **7.** $6 - 18i$ **9.** $6 + 4i$ **11.** $10 - 5i$ **13.** 37 **15.** $\dfrac{6}{5} + \dfrac{8}{5}i$ **17.** $1 - 2i$ **19.** $\dfrac{5}{2} - \dfrac{7}{2}i$

21. $-\dfrac{1}{2} + \dfrac{\sqrt{3}}{2}i$ **23.** $2i$ **25.** $-i$ **27.** i **29.** -6 **31.** $-10i$ **33.** $-2 + 2i$ **35.** 0 **37.** 0 **39.** $2i$ **41.** $5i$ **43.** $5i$ **45.** $\{-2i, 2i\}$

47. $\{-4, 4\}$ **49.** $\{3 - 2i, 3 + 2i\}$ **51.** $\{3 - i, 3 + i\}$ **53.** $\left\{\dfrac{1}{4} - \dfrac{1}{4}i, \dfrac{1}{4} + \dfrac{1}{4}i\right\}$ **55.** $\left\{-\dfrac{1}{5} - \dfrac{2}{5}i, -\dfrac{1}{5} + \dfrac{2}{5}i\right\}$

57. $\left\{-\dfrac{1}{2} - \dfrac{\sqrt{3}}{2}i, -\dfrac{1}{2} + \dfrac{\sqrt{3}}{2}i\right\}$ **59.** $\{2, -1 - \sqrt{3}i, -1 + \sqrt{3}i\}$ **61.** $\{-2, 2, -2i, 2i\}$ **63.** $\{-3i, -2i, 2i, 3i\}$ **65.** Two complex solutions

67. Two unequal real solutions **69.** A repeated real solution **71.** $2 - 3i$ **73.** 6 **75.** 25

77. $z + \overline{z} = (a + bi) + (a - bi) = 2a; z - \overline{z} = (a + bi) - (a - bi) = 2bi$

79. $\overline{z + w} = \overline{(a + bi) + (c + di)} = \overline{(a + c) + (b + d)i} = (a + c) - (b + d)i = (a - bi) + (c - di) = \overline{z} + \overline{w}$

5.4 Concepts and Vocabulary *(page 411)*

1. One **2.** $3 - 4i$ **3.** T **4.** F **5.** If it were complex, $a + bi$, then the conjugate, $a - bi$, would also be a zero. This would give us 5 zeros, which is more than the degree. **6.** Because $4 + i$ must be a zero, there is only one remaining unknown zero. It must be real because nonreal complex zeros occur in conjugate pairs when the coefficients are real.

5.4 Exercises *(page 411)*

1. $4 + i$ **3.** $-i, 1 - i$ **5.** $-i, -2i$ **7.** $-i$ **9.** $2 - i, -3 + i$ **11.** $f(x) = x^4 - 14x^3 + 77x^2 - 200x + 208; a = 1$

13. $f(x) = x^5 - 4x^4 + 7x^3 - 8x^2 + 6x - 4; a = 1$ **15.** $f(x) = x^4 - 6x^3 + 10x^2 - 6x + 9; a = 1$ **17.** $-2i, 4$ **19.** $2i, -3, \dfrac{1}{2}$

21. $3 + 2i, -2, 5$ **23.** $4i, -\sqrt{11}, \sqrt{11}, -\dfrac{2}{3}$ **25.** $1, -\dfrac{1}{2} - \dfrac{\sqrt{3}}{2}i, -\dfrac{1}{2} + \dfrac{\sqrt{3}}{2}i$ **27.** $2, 3 - 2i, 3 + 2i$ **29.** $-i, i, -2i, 2i$ **31.** $-3, 1, -5i, 5i$

33. $-4, \dfrac{1}{3}, 2 - 3i, 2 + 3i$ **35.** Zeros that are complex numbers must occur in conjugate pairs; or a polynomial with real coefficients of odd degree must have at least one real zero.

Review Exercises *(page 413)*

1. $q(x) = 8x^2 + 5x + 6; R = 10; g$ is not a factor of f.   **3.** $q(x) = x^3 - 4x^2 + 8x - 1; R = 0; g$ is a factor of f.   **5.** $8; \pm\dfrac{1}{2}, \pm1, \pm\dfrac{3}{2}, \pm3$

7. $-2, 1, 4$ **9.** $\dfrac{1}{2}$, multiplicity 2; -2 **11.** 2, multiplicity 2   **13.** $-2.5, 3.1, 5.32$ **15.** $-11.3, -0.6, 4, 9.33$ **17.** $\{-3, 2\}$ **19.** $\left\{-3, -1, -\dfrac{1}{2}, 1\right\}$

21. -5 and 5 **23.** $-\dfrac{37}{2}$ and $\dfrac{37}{2}$ **25.** $f(0) = -1, f(1) = 1; 0.85$

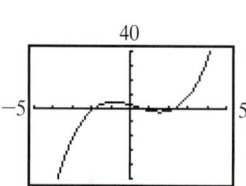

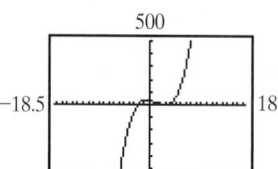

27. $f(0) = -1; f(1) = 1; 0.94$

29. $4 + 7i$ **31.** $-3 + 2i$ **33.** $\dfrac{9}{10} - \dfrac{3}{10}i$ **35.** -1

37. $-46 + 9i$ **39.** $4 - i; f(x) = x^3 - 14x^2 + 65x - 102$

41. $-i, 1 - i; f(x) = x^4 - 2x^3 + 3x^2 - 2x + 2$

43. $\left\{-\dfrac{1}{2} - \dfrac{\sqrt{3}}{2}i, -\dfrac{1}{2} + \dfrac{\sqrt{3}}{2}i\right\}$ **45.** $\left\{\dfrac{-1 - \sqrt{17}}{4}, \dfrac{-1 + \sqrt{17}}{4}\right\}$

47. $\left\{\dfrac{1}{2} - \dfrac{\sqrt{11}}{2}i, \dfrac{1}{2} + \dfrac{\sqrt{11}}{2}i\right\}$ **49.** $\left\{\dfrac{1}{2} - \dfrac{\sqrt{23}}{2}i, \dfrac{1}{2} + \dfrac{\sqrt{23}}{2}i\right\}$ **51.** $\{-\sqrt{2}, \sqrt{2}, -2i, 2i\}$ **53.** $\{-3, 2\}$ **55.** $\left\{\dfrac{1}{3}, 1, -i, i\right\}$

Cumulative Review Exercises *(page 416)*

1. (a) 16   **(b)** $\dfrac{1}{81}$ **(c)** 1.728   **(d)** 33.635   **2.** $\dfrac{4y^8}{9x^2}$

3.

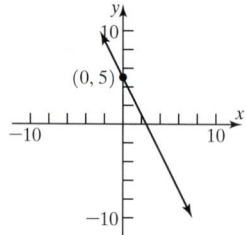

4.

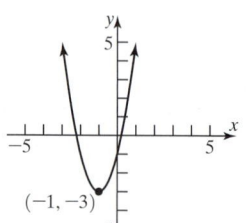

5. Local minimum -6.04 occurs at $x = -1.08$
Local maximum 4.04 occurs at $x = 1.08$
Increasing: $(-1.08, 1.08)$
Decreasing: $(-3, -1.08), (1.08, 3)$

6. Function

7. Domain: $[-5, 4]$; Range: $[-3, 6]$;
x-intercepts: $-4, -0.5, 1.2$;
y-intercept: -0.9

8. $g \circ f = g(f(x)) = \dfrac{3}{x^2 - 1}$; Domain: $\{x | x \neq -1, x \neq 1\}$

9. 1. Domain: $\{x | x \neq -3, x \neq 2\}$

2. $R(x) = \dfrac{(2x + 1)(x - 3)}{(x + 3)(x - 2)}$

3. x-intercepts: $-\dfrac{1}{2}, 3$; y-intercept: $\dfrac{1}{2}$

4. No symmetry

5. Vertical asymptotes: $x = -3, x = 2$

6. Horizontal asymptote: $y = 2$, intersected at $\left(\dfrac{9}{7}, 2\right)$

$R(x) \geq 0$ on the interval $(-\infty, -3)$ or $\left[-\dfrac{1}{2}, 2\right)$ or $[3, \infty)$

7.

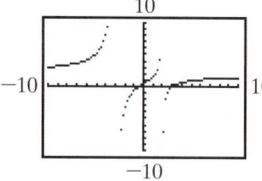

8.

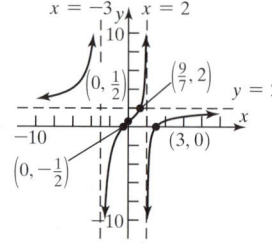

10. $-2, -\dfrac{1}{3}, 4$ (multiplicity 2);

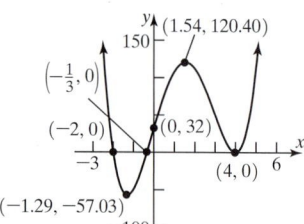

$; f(x) > 0$ on the interval $(-\infty, -2)$ or $\left(-\dfrac{1}{3}, 4\right)$ or $(4, \infty)$

C H A P T E R   6   Exponential and Logarithmic Functions

6.1 Concepts and Vocabulary *(page 428)*

1. One-to-one   **2.** $y = x$ **3.** $[4, \infty)$ **4.** F   **5.** T   **6.** Yes, if the domain is $\{x \mid x = 0\}$.   **7.** No   **8.** On the line $y = x$. No. No.

6.1 Exercises *(page 428)*

1. (a)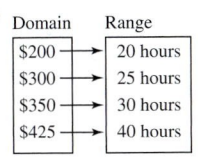

(b) Inverse is a function

3. (a)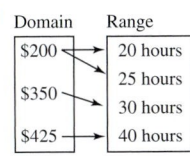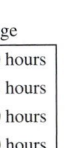

(b) Inverse is not a function

5. (a) $\{(6, 2), (6, -3), (9, 4), (10, 1)\}$
  (b) Inverse is not a function
7. (a) $\{(0, 0), (1, 1), (16, 2), (81, 3)\}$
  (b) Inverse is a function
9. One-to-one   **11.** Not one-to-one   **13.** One-to-one

15.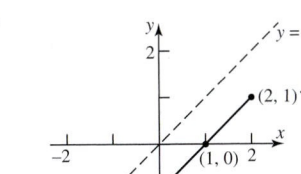

17.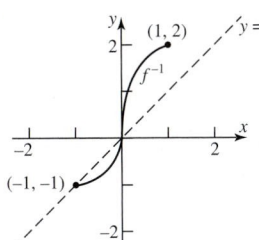

19.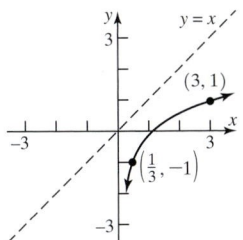

21. $f(g(x)) = f\left(\dfrac{1}{3}(x - 4)\right) = 3\left[\dfrac{1}{3}(x - 4)\right] + 4 = x$

  $g(f(x)) = g(3x + 4) = \dfrac{1}{3}[(3x + 4) - 4] = x$

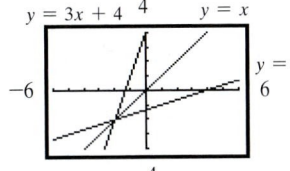

23. $f(g(x)) = 4\left[\dfrac{x}{4} + 2\right] - 8 = x$

  $g(f(x)) = \dfrac{4x - 8}{4} + 2 = x$

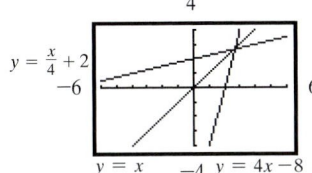

25. $f(g(x)) = (\sqrt[3]{x + 8})^3 - 8 = x$
  $g(f(x)) = \sqrt[3]{(x^3 - 8)} + 8 = x$

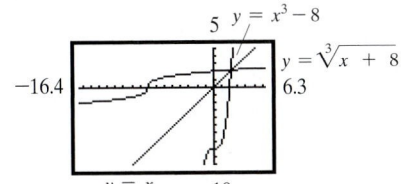

27. $f(g(x)) = \dfrac{1}{\left(\dfrac{1}{x}\right)} = x$

  $g(f(x)) = \dfrac{1}{\left(\dfrac{1}{x}\right)} = x$

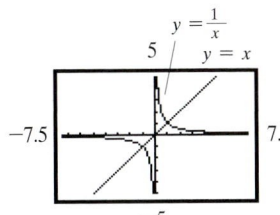

29. $f(g(x)) = \dfrac{2\left(\dfrac{4x-3}{2-x}\right)+3}{\dfrac{4x-3}{2-x}+4} = x$

$g(f(x)) = \dfrac{4\left(\dfrac{2x+3}{x+4}\right)-3}{2-\dfrac{2x+3}{x+4}} = x$

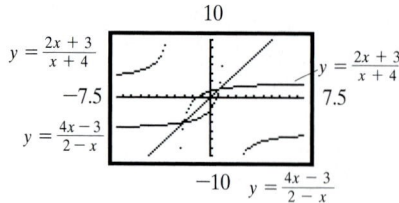

$y = \dfrac{2x+3}{x+4}$ $y = \dfrac{2x+3}{x+4}$

$y = \dfrac{4x-3}{2-x}$ $y = \dfrac{4x-3}{2-x}$

31. $f^{-1}(x) = \dfrac{1}{3}x$

$f(f^{-1}(x)) = 3\left(\dfrac{1}{3}x\right) = x$

$f^{-1}(f(x)) = \dfrac{1}{3}(3x) = x$

Domain f = Range $f^{-1} = (-\infty, \infty)$
Range f = Domain $f^{-1} = (-\infty, \infty)$

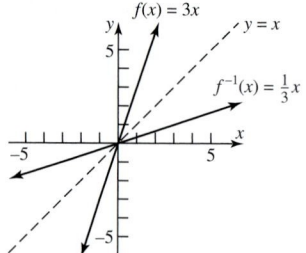

33. $f^{-1}(x) = \dfrac{x}{4} - \dfrac{1}{2}$

$f(f^{-1}(x)) = 4\left(\dfrac{x}{4} - \dfrac{1}{2}\right) + 2 = x$

$f^{-1}(f(x)) = \dfrac{4x+2}{4} - \dfrac{1}{2} = x$

Domain f = Range $f^{-1} = (-\infty, \infty)$
Range f = Domain $f^{-1} = (-\infty, \infty)$

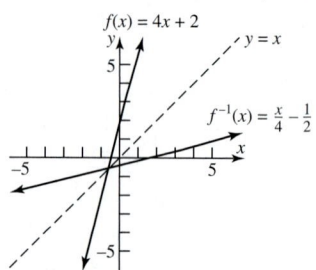

35. $f^{-1}(x) = \sqrt[3]{x+1}$

$f(f^{-1}(x)) = (\sqrt[3]{x+1})^3 - 1 = x$

$f^{-1}(f(x)) = \sqrt[3]{(x^3-1)+1} = x$

Domain f = Range $f^{-1} = (-\infty, \infty)$
Range f = Domain $f^{-1} = (-\infty, \infty)$

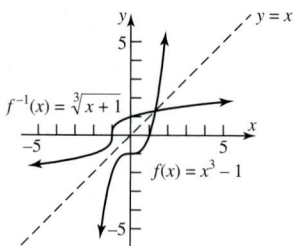

37. $f^{-1}(x) = \sqrt{x-4}$

$f(f^{-1}(x)) = (\sqrt{x-4})^2 + 4 = x$

$f^{-1}(f(x)) = \sqrt{(x^2+4)-4} = \sqrt{x^2} = |x|$
$\qquad\qquad = x$

Domain f = Range $f^{-1} = [0, \infty)$
Range f = Domain $f^{-1} = [4, \infty)$

39. $f^{-1}(x) = \dfrac{4}{x}$

$f(f^{-1}(x)) = \dfrac{4}{\dfrac{4}{x}} = x$

$f^{-1}(f(x)) = \dfrac{4}{\dfrac{4}{x}} = x$

Domain f = Range f^{-1} = all real numbers except 0
Range f = Domain f^{-1} = all real numbers except 0

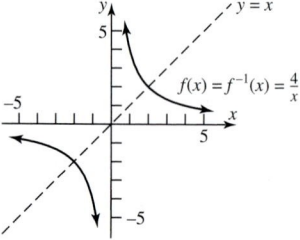

41. $f^{-1}(x) = \dfrac{2x + 1}{x}$

$$f(f^{-1}(x)) = \dfrac{1}{\dfrac{2x + 1}{x} - 2} = x$$

$$f^{-1}(f(x)) = \dfrac{2\left(\dfrac{1}{x - 2}\right) + 1}{\dfrac{1}{x - 2}} = x$$

Domain f = Range f^{-1} = all real numbers except 2
Range f = Domain f^{-1} = all real numbers except 0

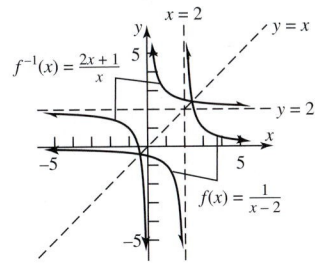

43. $f^{-1}(x) = \dfrac{2 - 3x}{x}$

$$f(f^{-1}(x)) = \dfrac{2}{3 + \dfrac{2 - 3x}{x}} = x$$

$$f^{-1}(f(x)) = \dfrac{2 - 3\left(\dfrac{2}{3 + x}\right)}{\dfrac{2}{3 + x}} = x$$

Domain f = Range f^{-1} = all real numbers except -3
Range f = Domain f^{-1} = all real numbers except 0

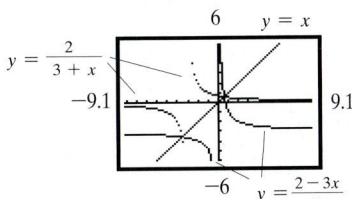

45. $f^{-1}(x) = \sqrt{x} - 2$
$f(f^{-1}(x)) = (\sqrt{x} - 2 + 2)^2 = x$
$f^{-1}(f(x)) = \sqrt{(x + 2)^2} - 2 = |x + 2| - 2 = x, x \geq -2$
Domain f = Range f^{-1} = $[-2, \infty)$
Range f = Domain f^{-1} = $[0, \infty)$

$y = (x + 2)^2, x \geq -2$ ⟶ 6
-9.1 ⟶ $y = \sqrt{x} - 2$ ⟶ 9.1
-6

47. $f^{-1}(x) = \dfrac{x}{x - 2}$

$$f(f^{-1}(x)) = \dfrac{2\left(\dfrac{x}{x - 2}\right)}{\dfrac{x}{x - 2} - 1} = x$$

$$f^{-1}(f(x)) = \dfrac{\dfrac{2x}{x - 1}}{\dfrac{2x}{x - 1} - 2} = x$$

Domain f = Range f^{-1} = all real numbers except 1
Range f = Domain f^{-1} = all real numbers except 2

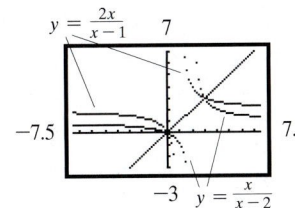

49. $f^{-1}(x) = \dfrac{3x + 4}{2x - 3}$

$$f(f^{-1}(x)) = \frac{3\left(\dfrac{3x + 4}{2x - 3}\right) + 4}{2\left(\dfrac{3x + 4}{2x - 3}\right) - 3} = x$$

$$f^{-1}(f(x)) = \frac{3\left(\dfrac{3x + 4}{2x - 3}\right) + 4}{2\left(\dfrac{3x + 4}{2x - 3}\right) - 3} = x$$

Domain f = Range f^{-1} = all real numbers except $\dfrac{3}{2}$

Range f = Domain f^{-1} = all real numbers except $\dfrac{3}{2}$

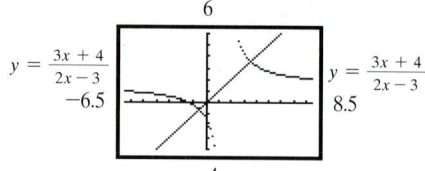

51. $f^{-1}(x) = \dfrac{-2x + 3}{x - 2}$

$$f(f^{-1}(x)) = \frac{2\left(\dfrac{-2x + 3}{x - 2}\right) + 3}{\dfrac{-2x + 3}{x - 2} + 2} = x$$

$$f^{-1}(f(x)) = \frac{-2\left(\dfrac{2x + 3}{x + 2}\right) + 3}{\dfrac{2x + 3}{x + 2} - 2} = x$$

Domain f = Range f^{-1} = all real numbers except -2

Range f = Domain f^{-1} = all real numbers except 2

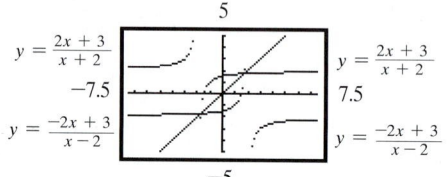

53. $f^{-1}(x) = \dfrac{x^3}{8}$

$$f(f^{-1}(x)) = 2\sqrt[3]{\dfrac{x^3}{8}} = x$$

$$f^{-1}(f(x)) = \dfrac{(2\sqrt[3]{x})^3}{8} = x$$

Domain f = Range f^{-1} = $(-\infty, \infty)$

Range f = Domain f^{-1} = $(-\infty, \infty)$

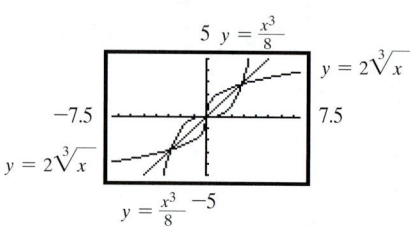

55. $f^{-1}(x) = \dfrac{1}{m}(x - b), m \neq 0$

57. Quadrant I

59. Possible answer: $f(x) = |x|, x \geq 0$, is one-to-one; $f^{-1}(x) = x, x \geq 0$

61. $f(g(x)) = \dfrac{9}{5}\left[\dfrac{5}{9}(x - 32)\right] + 32 = x$; $g(f(x)) = \dfrac{5}{9}\left[\left(\dfrac{9}{5}x + 32\right) - 32\right] = x$

63. $l(T) = \dfrac{gT^2}{4\pi^2}, T > 0$

65. Yes, the graph must be symmetric about the line $y = x$; $y = \dfrac{1}{x}$ is an example.

67. $y = \begin{cases} \dfrac{1}{x}, & x < 0 \\ x, & x \geq 0 \end{cases}$ is neither an increasing nor a decreasing function on its domain but is one-to-one.

6.2 Concepts and Vocabulary *(page 441)*

1. $\left(-1, \dfrac{1}{a}\right), (0, 1), (1, a)$ **2.** 1 **3.** 4 **4.** F **5.** F **6.** F **7.** Becomes steeper; Lies closer to the x-axis

8. Because $y = a^{-x} = (a^{-1})^x = \left(\dfrac{1}{a}\right)^x$.

6.2 Exercises *(page 441)*

1. (a) 11.212 **(b)** 11.587 **(c)** 11.664 **(d)** 11.665 **3. (a)** 8.815 **(b)** 8.821 **(c)** 8.824 **(d)** 8.825
5. (a) 21.217 **(b)** 22.217 **(c)** 22.440 **(d)** 22.459 **7.** 3.320 **9.** 0.427 **11.** B **13.** D **15.** A **17.** E

19.

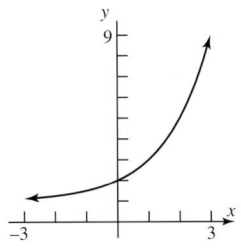

Domain: $(-\infty, \infty)$ or
all real numbers
Range: $(1, \infty)$ or $\{y|y > 1\}$
Horizontal asymptote: $y = 1$

21.

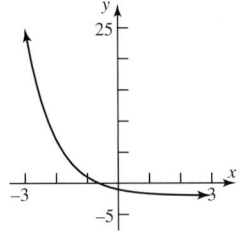

Domain: $(-\infty, \infty)$ or
all real numbers
Range: $(-2, \infty)$ or $\{y|y > -2\}$
Horizontal asymptote: $y = -2$

23.

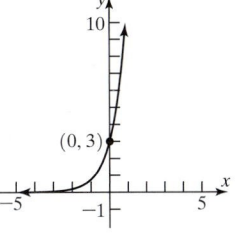

Domain: $(-\infty, \infty)$ or
all real numbers
Range: $(0, \infty)$ or $\{y|y > 0\}$
Horizontal asymptote: $y = 0$

25.

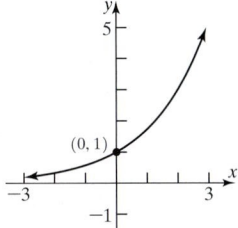

Domain: $(-\infty, \infty)$ or
all real numbers
Range: $(0, \infty)$ or $\{y|y > 0\}$
Horizontal asymptote: $y = 0$

27.

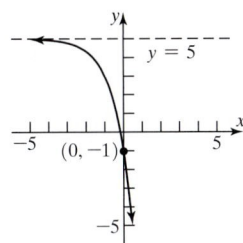

Domain: $(-\infty, \infty)$ or
all real numbers
Range: $(-\infty, 5)$ or $\{y|y < 5\}$
Horizontal asymptote: $y = 5$

29.

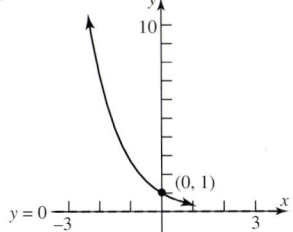

Domain: $(-\infty, \infty)$ or
all real numbers
Range: $(0, \infty)$ or $\{y|y > 0\}$
Horizontal asymptote: $y = 0$

31.

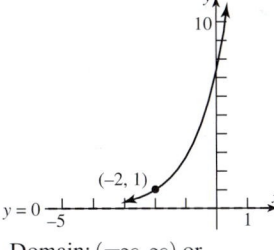

Domain: $(-\infty, \infty)$ or
all real numbers
Range: $(0, \infty)$ or $\{y|y > 0\}$
Horizontal asymptote: $y = 0$

33.

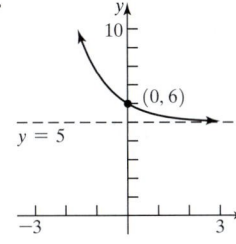

Domain: $(-\infty, \infty)$ or
all real numbers
Range: $(5, \infty)$ or $\{y|y > 5\}$
Horizontal asymptote: $y = 5$

35.

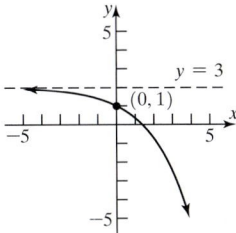

Domain: $(-\infty, \infty)$ or all real numbers
Range: $(-\infty, 2)$ or $\{y|y < 2\}$
Horizontal asymptote: $y = 2$

37. $f(x) = 3^x$ **39.** $f(x) = 2(4^x)$ **41.** $f(x) = -6^x$

43. $\dfrac{1}{2}$ **45.** $\{-\sqrt{2}, 0, \sqrt{2}\}$ **47.** $\left\{1 - \dfrac{\sqrt{6}}{3}, 1 + \dfrac{\sqrt{6}}{3}\right\}$

49. 0 **51.** 4 **53.** $\dfrac{3}{2}$ **55.** $\{1, 2\}$ **57.** $\dfrac{1}{49}$ **59.** $\dfrac{1}{4}$

61. (a) 74% (b) 47% **63.** (a) 44 watts (b) 11.6 watts

65. 3.35 milligrams; 0.45 milligrams

67. (a) 0.632 (b) 0.982 (c) (d) 1

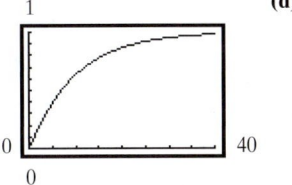

69. (a) 0.052 (b) 0.089 **71.** (a) 70.95% (b) 72.62% (c) 100%

73. (a) 5.414 amperes, 7.585 amperes, 10.376 amperes (b) 12 amperes
(d) 3.343 amperes, 5.309 amperes, 9.443 amperes (e) 24 amperes
(c), (f)

75. $n = 4: 2.7083; n = 6: 2.7181; n = 8: 2.7182788; n = 10: 2.7182818$

77. $\dfrac{f(x+h)-f(x)}{h} = \dfrac{a^{x+h}-a^x}{h} = \dfrac{a^x a^h - a^x}{h} = \dfrac{a^x(a^h-1)}{h}$ **79.** $f(-x) = a^{-x} = \dfrac{1}{a^x} = \dfrac{1}{f(x)}$

81. (a) $f(-x) = \dfrac{1}{2}(e^{-x} - e^{-(-x)}) = \dfrac{1}{2}(e^{-x} - e^x)$ **(b)**

$\qquad\qquad = -\dfrac{1}{2}(e^x - e^{-x}) = -f(x)$

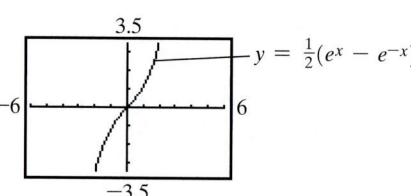

83. $f(1) = 5, f(2) = 17, f(3) = 257, f(4) = 65{,}537, f(5) = 4{,}294{,}967{,}297 = 641 \times 6{,}700{,}417$

6.3 Concepts and Vocabulary *(page 453)*

1. $\{x | x > 0\}$ or $(0, \infty)$ **2.** $(1, 0), (a, 1), \left(\dfrac{1}{a}, -1\right)$ **3.** 1   **4.** F   **5.** T

6. Because $y = \log_1 x$ means $1^y = 1 = x$, which cannot be true for $x \neq 1$.   **7.** $(1, \infty)$

6.3 Exercises *(page 454)*

1. $2 = \log_3 9$ **3.** $2 = \log_a 1.6$ **5.** $2 = \log_{1.1} M$ **7.** $x = \log_2 7.2$ **9.** $\sqrt{2} = \log_x \pi$ **11.** $x = \ln 8$ **13.** $2^3 = 8$ **15.** $a^6 = 3$ **17.** $3^x = 2$

19. $2^{1.3} = M$ **21.** $(\sqrt{2})^x = \pi$ **23.** $e^x = 4$ **25.** 0   **27.** 2   **29.** -4 **31.** $\dfrac{1}{2}$ **33.** 4   **35.** $\dfrac{1}{2}$ **37.** $\{x | x > 3\}; (3, \infty)$

39. $\{x | x \neq 0\}; (-\infty, 0)$ or $(0, \infty)$ **41.** $\{x | x \neq 1\}; (-\infty, 1)$ or $(1, \infty)$ **43.** $\{x | x > -1\}; (-1, \infty)$

45. $\{x | x < -1$ or $x > 0\}; (-\infty, -1)$ or $(0, \infty)$ **47.** 0.511   **49.** 30.099   **51.** $\sqrt{2}$ **53.** B   **55.** D   **57.** A   **59.** E

61.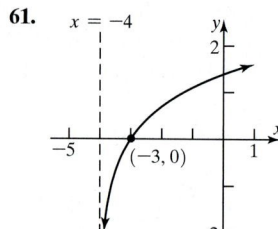

Domain: $(-4, \infty)$
Range: $(-\infty, \infty)$
Vertical asymptote: $x = -4$

63.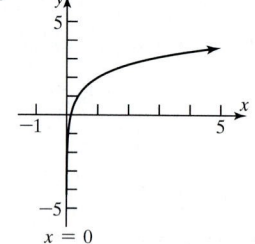

Domain: $(0, \infty)$
Range: $(-\infty, \infty)$
Vertical asymptote: $x = 0$

65.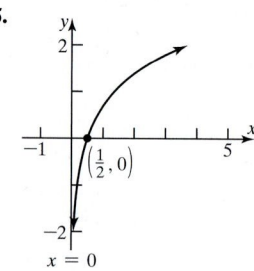

Domain: $(0, \infty)$
Range: $(-\infty, \infty)$
Vertical asymptote: $x = 0$

67.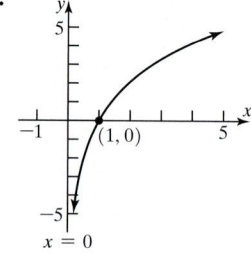

Domain: $(0, \infty)$
Range: $(-\infty, \infty)$
Vertical asymptote: $x = 0$

69.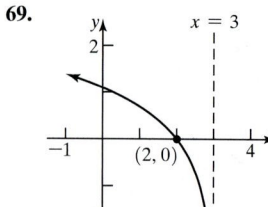

Domain: $(-\infty, 3)$
Range: $(-\infty, \infty)$
Vertical asymptote: $x = 3$

71.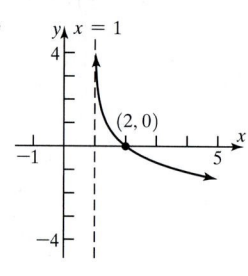

Domain: $(1, \infty)$
Range: $(-\infty, \infty)$
Vertical asymptote: $x = 1$

73.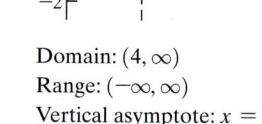

Domain: $(4, \infty)$
Range: $(-\infty, \infty)$
Vertical asymptote: $x = 4$

75.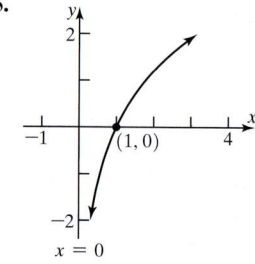

Domain: $(0, \infty)$
Range: $(-\infty, \infty)$
Vertical asymptote: $x = 0$

77.

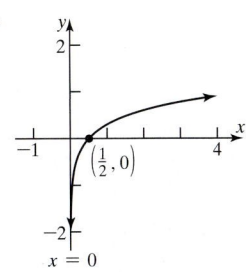

Domain: $(0, \infty)$
Range: $(-\infty, \infty)$
Vertical asymptote: $x = 0$

79.

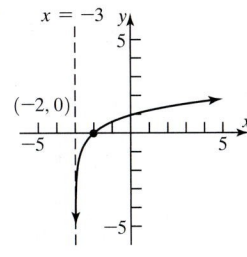

Domain: $(-3, \infty)$
Range: $(-\infty, \infty)$
Vertical asymptote: $x = -3$

81.

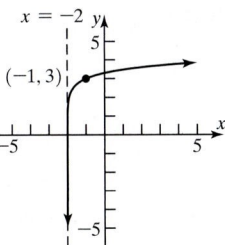

Domain: $(-2, \infty)$
Range: $(-\infty, \infty)$
Vertical asymptote: $x = -2$

83.

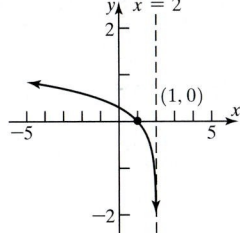

Domain: $(-\infty, 2)$
Range: $(-\infty, \infty)$
Vertical asymptote: $x = 2$

85. 9 **87.** $\dfrac{7}{2}$ **89.** 2 **91.** 5 **93.** 3 **95.** 2 **97.** $\dfrac{\ln 10}{3}$ **99.** $\dfrac{\ln 8 - 5}{2}$ **101.** $\{-2\sqrt{2}, 2\sqrt{2}\}$ **103.** -1

105. (a)

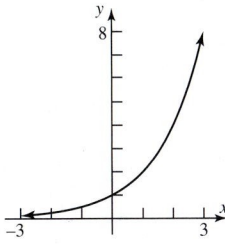

Domain: $(-\infty, \infty)$
Range: $(0, \infty)$
Horizontal asymptote: $y = 0$

(b) $f^{-1}(x) = \log_2 x$

(c)

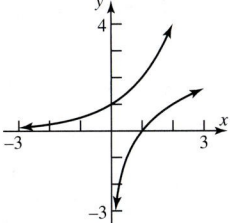

Domain of f^{-1} = Range of f = $(0, \infty)$
Range of f^{-1} = Domain of f = $(-\infty, \infty)$
Vertical asymptote of f^{-1}: $x = 0$

107. (a)

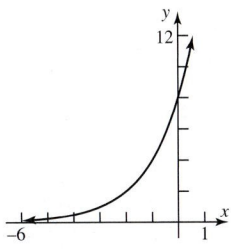

Domain: $(-\infty, \infty)$
Range: $(0, \infty)$
Horizontal asymptote: $y = 0$

(b) $f^{-1}(x) = \log_2 x - 3$

(c)

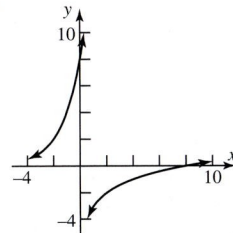

Domain of f^{-1} = Range of f = $(0, \infty)$
Range of f^{-1} = Domain of f = $(-\infty, \infty)$
Vertical asymptote of f^{-1}: $x = 0$

109. (a) $n \approx 6.93$ so 7 panes are necessary **(b)** $n \approx 13.86$ so 14 panes are necessary
111. (a) $d \approx 127.7$ so it takes about 128 days **(b)** $d \approx 575.6$ so it takes about 576 days
113. (a) 6.93 min **(b)** 16.09 min **(c)** No, since $F(t)$ can never equal one.
115. $h \approx 2.29$ so the time between injections is about 2 hours, 17 minutes
117. 0.2695 seconds
0.8959 seconds

119. 50 decibels
121. 110 decibels
123. 8.1
125. (a) $k = 20.07$
  (b) 91%
  (c) 0.175
  (d) 0.08

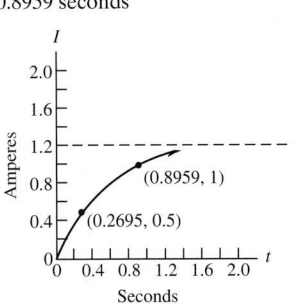

6.4 Concepts and Vocabulary *(page 465)*

1. Sum **2.** 7 **3.** $r \log_a M$ **4.** F **5.** F **6.** T

6.4 Exercises *(page 465)*

1. 71 **3.** -4 **5.** 7 **7.** 1 **9.** 1 **11.** 2 **13.** $\dfrac{5}{4}$ **15.** 4 **17.** $a + b$ **19.** $b - a$ **21.** $3a$ **23.** $\dfrac{1}{5}(a + b)$ **25.** $2 + \log_5 x$ **27.** $3 \log_2 z$

29. $1 + \ln x$ **31.** $\ln x + x$ **33.** $2 \log_a u + 3 \log_a v$ **35.** $2 \ln x + \dfrac{1}{2}\ln(1 - x)$ **37.** $3 \log_2 x - \log_2(x - 3)$

39. $\log x + \log(x + 2) - 2 \log(x + 3)$ **41.** $\dfrac{1}{3}\ln(x - 2) + \dfrac{1}{3}\ln(x + 1) - \dfrac{2}{3}\ln(x + 4)$ **43.** $\ln 5 + \ln x + \dfrac{1}{2}\ln(1 - 3x) - 3 \ln(x - 4)$

45. $\log_5 u^3 v^4$ **47.** $-\dfrac{5}{2}\log_3 x$ **49.** $\log_4 \dfrac{x - 1}{(x + 1)^4}$ **51.** $-2 \ln(x - 1)$ **53.** $\log_2[x(3x - 2)^4]$ **55.** $\log_a\left(\dfrac{25x^6}{\sqrt{2x + 3}}\right)$

57. $\log_2 \dfrac{(x + 1)^2}{(x + 3)(x - 1)}$ **59.** $\log y = \log a + x \log b$ **61.** 2.771 **63.** -3.880 **65.** 5.615 **67.** 0.874

69. $y = \dfrac{\log x}{\log 4}$

71. $y = \dfrac{\log(x + 2)}{\log 2}$

73. $y = \dfrac{\log(x + 1)}{\log(x - 1)}$

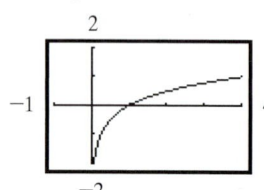

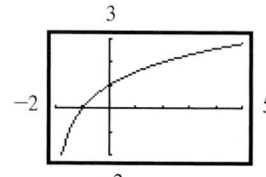

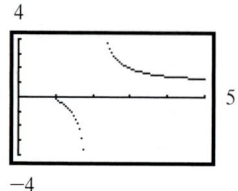

75. $y = Cx$ **77.** $y = Cx(x + 1)$ **79.** $y = Ce^{3x}$ **81.** $y = Ce^{-4x} + 3$ **83.** $y = \dfrac{\sqrt[3]{C}(2x + 1)^{1/6}}{(x + 4)^{1/9}}$ **85.** 3 **87.** 1

89. $\log_a(x + \sqrt{x^2 - 1}) + \log_a(x - \sqrt{x^2 - 1}) = \log_a[(x + \sqrt{x^2 - 1})(x - \sqrt{x^2 - 1})] = \log_a[x^2 - (x^2 - 1)] = \log_a 1 = 0$

91. $\ln(1 + e^{2x}) = \ln(e^{2x}(e^{-2x} + 1)) = \ln e^{2x} + \ln(e^{-2x} + 1) = 2x + \ln(1 + e^{-2x})$

93. $y = f(x) = \log_a x; a^y = x$ implies $a^y = \left(\dfrac{1}{a}\right)^{-y} = x$, so $-y = \log_{1/a} x = -f(x)$.

95. $f(x) = \log_a x; f\left(\dfrac{1}{x}\right) = \log_a \dfrac{1}{x} = \log_a 1 - \log_a x = -f(x)$

97. $\log_a \dfrac{M}{N} = \log_a (M \cdot N^{-1}) = \log_a M + \log_a N^{-1} = \log_a M - \log_a N,$

since $a^{\log_a N^{-1}} = N^{-1}$ implies $a^{-\log_a N^{-1}} = N$, i.e., $\log_a N = -\log_a N^{-1}$

6.5 Exercises *(page 471)*

1. 6 **3.** 16 **5.** 8 **7.** 3 **9.** 5 **11.** $-1 + \sqrt{1 + e^4} \approx 6.456$ **13.** $\dfrac{\ln 3}{\ln 2} \approx 1.585$ **15.** 0 **17.** $\dfrac{\ln 10}{\ln 2} \approx 3.322$

19. $-\dfrac{\ln 1.2}{\ln 8} \approx -0.088$ **21.** $\dfrac{\ln 3}{2 \ln 3 + \ln 4} \approx 0.307$ **23.** $\dfrac{\ln 7}{\ln 0.6 + \ln 7} \approx 1.356$ **25.** 0 **27.** $\dfrac{\ln \pi}{1 + \ln \pi} \approx 0.534$ **29.** $\dfrac{\ln 1.6}{3 \ln 2} \approx 0.226$

31. $\dfrac{9}{2}$ **33.** 2 **35.** 1 **37.** 16 **39.** $-1, \dfrac{2}{3}$ **41.** 0 **43.** $\ln(2 + \sqrt{5}) \approx 1.444$ **45.** 1.92 **47.** 2.79 **49.** -0.57 **51.** -0.70 **53.** 0.57

55. 0.39, 1.00 **57.** 1.32 **59.** 1.31

6.6 Exercises *(page 479)*

1. $108.29 **3.** $609.50 **5.** $697.09 **7.** $12.46 **9.** $125.23 **11.** $88.72 **13.** $860.72 **15.** $554.09 **17.** $59.71 **19.** $361.93 **21.** 5.35%

23. 26% **25.** $6\dfrac{1}{4}$% compounded annually **27.** 9% compounded monthly **29.** 104.32 months; 103.97 months **31.** 61.02 months; 60.82 months

33. 15.27 years **35.** $104,335 **37.** $12,910.62 **39.** About $30.17 per share or $3017 **41.** 9.35% **43.** Not quite. Jim will have $1057.60. The second bank gives a better deal, since Jim will have $1060.62 after 1 year. **45.** Will has $11,632.73; Henry has $10,947.89.
47. **(a)** Interest is $30,000 **(b)** Interest is $38,613.59 **(c)** Interest is $37,752.73. Simple interest at 12% is best.
49. **(a)** $1364.62 **(b)** $1353.35 **51.** $4631.93

55. (a) 6.12 years   **(b)** 18.45 years   **(c)** $mP = P\left(1 + \dfrac{r}{n}\right)^{nt}$

$$m = \left(1 + \dfrac{r}{n}\right)^{nt}$$

$$\ln m = \ln\left(1 + \dfrac{r}{n}\right)^{nt} = nt \ln\left(1 + \dfrac{r}{n}\right)$$

$$t = \dfrac{\ln m}{n \ln\left(1 + \dfrac{r}{n}\right)}$$

6.7 Exercises *(page 491)*

1. (a) 500 insects
   (b) $0.02 = 2\%$
   (c)
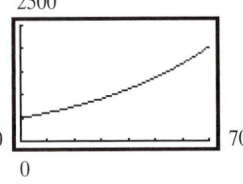
   (d) $\approx$ 611 insects
   (e) After about 23.5 days
   (f) After about 34.7 days

3. (a) $-0.0244 = -2.44\%$
   (b)
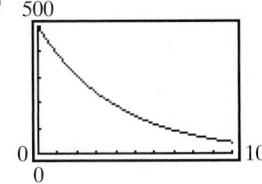
   (c) About 391.7 grams
   (d) After about 9.1 years
   (e) 28.4 years

5. 5832; 3.9 days
7. 25,198
9. 9.797 grams
11. (a) 9727 years ago
   (b)
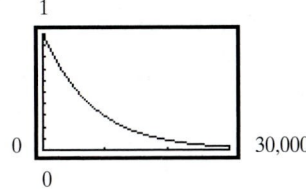
   (c) 5600 years

13. (a) 5:18 PM
   (b)
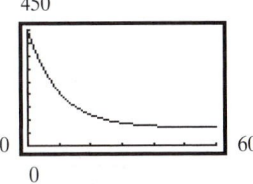
   (c) About 14.3 minutes
   (d) The temperature of the pizza approaches 70°F.

15. (a) 18.63°C; 25.1°C
   (b)
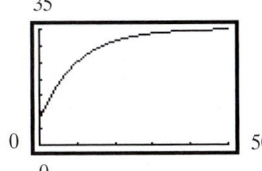

17. 7.34 kg; 76.6 hours
19. 26.6 days

21. (a) 90% **(b)** 12.86%
   (c)
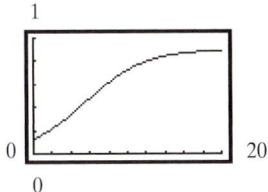
   (d) 85.77%   **(e)** 1996
   (f) About 5.6 years

23. (a) 1000 g, 43.9%   **(b)** 30 g
   (c)
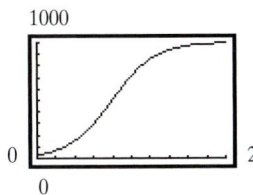
   (d) 616.6 g
   (e) After 9.85 hours
   (f) About 7.9 hours

25. (a) 9.23×10^{-3}, or about 0
   (b) 0.81, or about 1
   (c) 5.01, or about 5
   (d) 57.91°, 43.99°, 30.07°
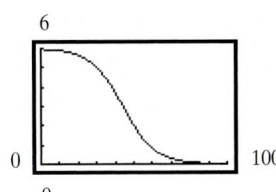

6.8 Exercises *(page 499)*

1. (a)

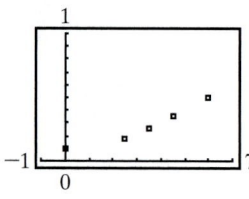

(b) $y = 0.0903(1.3384)^x$
(c) $N(t) = 0.0903e^{0.2915t}$
(d)

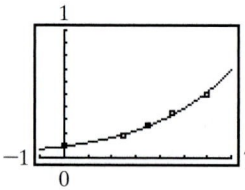

(e) 0.69
(f) After about 7.26 hours

3. (a)

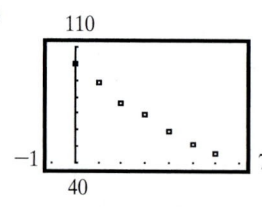

(b) $y = 100.326(0.8769)^x$
(c) $A = 100.326e^{-0.1314t}$
(d)

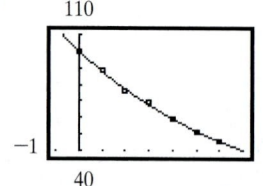

(e) 5.3 weeks **(f)** 0.14 grams
(g) After about 12.3 weeks

5. (a)

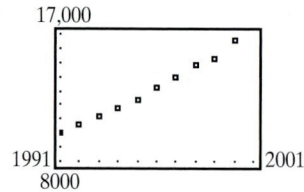

(b) value $= 2.7018 \times 10^{-44}(1.056554737)^{\text{year}}$
(c) 5.66%
(d) $52,166
(e) In 2029

7. (a)

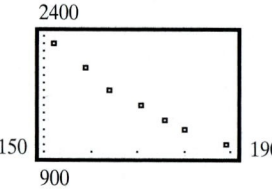

(b) $y = 32,741.02 - 6070.96 \ln x$

(c)

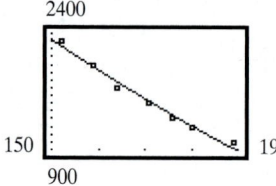

(d) Approximately 168 computers

9. (a)

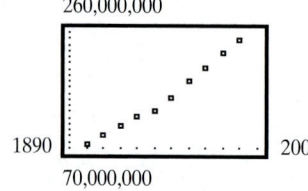

(b) $y = \dfrac{799,475,916.5}{1 + 1.56344 \times 10^{14}e^{-0.0160x}}$

(c)

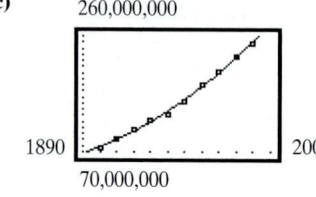

(d) 799,475,917
(e) 283,391,335 **(f)** 2007

11. (a)

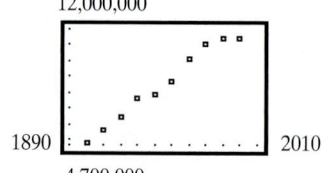

(b) $y = \dfrac{14,471,245.24}{1 + 3.860 \times 10^{20}\, e^{0.0246x}}$

(c)

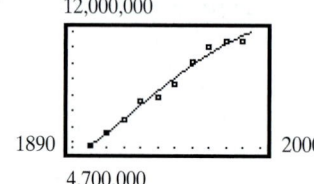

(d) 14,471,245
(e) Approximately 12,750,816

Review Exercises *(page 504)*

1. (a) $\{(2, 1), (5, 3), (8, 5), (10, 6)\}$ **(b)** Inverse is a function **3.**

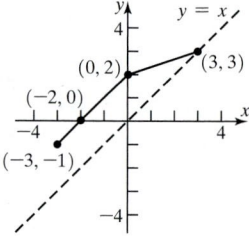

5. $f^{-1}(x) = \dfrac{2x + 3}{5x - 2}$

$$f(f^{-1}(x)) = \dfrac{2\left(\dfrac{2x + 3}{5x - 2}\right) + 3}{5\left(\dfrac{2x + 3}{5x - 2}\right) - 2} = x$$

$$f^{-1}(f(x)) = \dfrac{2\left(\dfrac{2x + 3}{5x - 2}\right) + 3}{5\left(\dfrac{2x + 3}{5x - 2}\right) - 2} = x$$

Domain f = Range f^{-1} = all real numbers except $\dfrac{2}{5}$

Range f = Domain f^{-1} = all real numbers except $\dfrac{2}{5}$

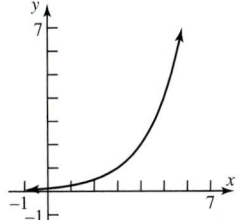

7. $f^{-1}(x) = \dfrac{x + 1}{x}$

$$f(f^{-1}(x)) = \dfrac{1}{\dfrac{x + 1}{x} - 1} = x$$

$$f^{-1}(f(x)) = \dfrac{\dfrac{1}{x - 1} + 1}{\dfrac{1}{x - 1}} = x$$

Domain f = Range f^{-1} = all real numbers except 1
Range f = Domain f^{-1} = all real numbers except 0

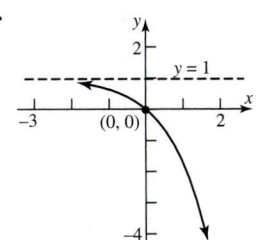

9. $f^{-1}(x) = \dfrac{27}{x^3}$

$$f(f^{-1}(x)) = \dfrac{3}{\left(\dfrac{27}{x^3}\right)^{1/3}} = x$$

$$f^{-1}(f(x)) = \dfrac{27}{\left(\dfrac{3}{x^{1/3}}\right)^3} = x$$

Domain f = Range f^{-1} = all real numbers except 0
Range f = Domain f^{-1} = all real numbers except 0

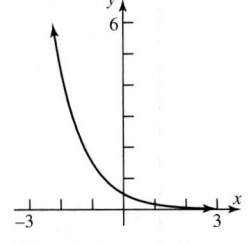

11. (a) 81 **(b)** 2 **(c)** $\dfrac{1}{9}$ **(d)** -3 **13.** $\log_5 z = 2$ **15.** $5^{13} = u$

17. $\left\{x \mid x > \dfrac{2}{3}\right\}; \left(\dfrac{2}{3}, \infty\right)$ **19.** $\left\{x \mid x < 1 \text{ or } x > 2\right\}; (-\infty, 1) \text{ or } (2, \infty)$

21. -3 **23.** $\sqrt{2}$ **25.** 0.4 **27.** $\log_3 u + 2 \log_3 v - \log_3 w$

29. $2 \log x + \dfrac{1}{2} \log(x^3 + 1)$ **31.** $\ln x + \dfrac{1}{3} \ln(x^2 + 1) - \ln(x - 3)$

33. $\dfrac{25}{4} \log_4 x$ **35.** $-2 \ln(x + 1)$ **37.** $\log\left(\dfrac{4x^3}{[(x + 3)(x - 2)]^{1/2}}\right)$

39. 2.124 **41.**

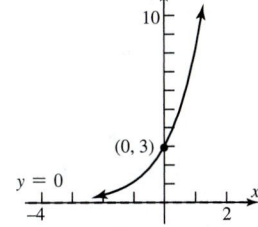

43.

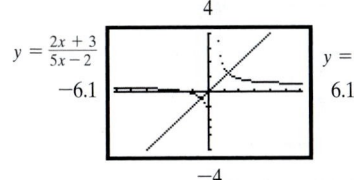

Domain: $(-\infty, \infty)$
Range: $(0, \infty)$
Horizontal asymptote: $y = 0$

45.

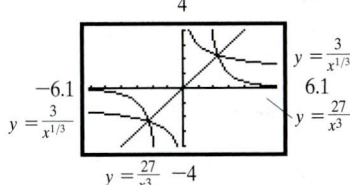

Domain: $(-\infty, \infty)$
Range: $(0, \infty)$
Horizontal asymptote: $y = 0$

47.

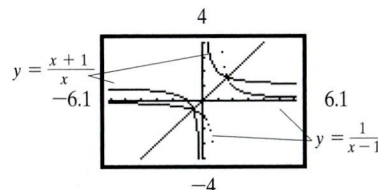

Domain: $(-\infty, \infty)$
Range: $(-\infty, 1)$
Horizontal asymptote: $y = 1$

49.

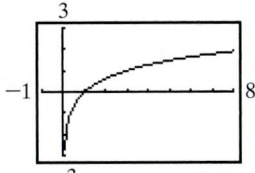

Domain: $(-\infty, \infty)$
Range: $(0, \infty)$
Horizontal asymptote: $y = 0$

51.

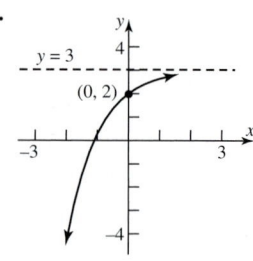

Domain: $(-\infty, \infty)$
Range: $(-\infty, 3)$
Horizontal asymptote: $y = 3$

53. $\left\{\dfrac{1}{4}\right\}$ **55.** $\left\{\dfrac{-1 - \sqrt{3}}{2}, \dfrac{-1 + \sqrt{3}}{2}\right\}$ **57.** $\left\{\dfrac{1}{4}\right\}$ **59.** $\left\{\dfrac{2 \ln 3}{\ln 5 - \ln 3} \approx 4.301\right\}$

61. $\left\{\dfrac{12}{5}\right\}$ **63.** $\{83\}$ **65.** $\left\{\dfrac{1}{2}, -3\right\}$ **67.** $\{-1\}$ **69.** $\{1 - \ln 5 \approx -0.609\}$

71. $\left\{\dfrac{\ln 3}{3 \ln 2 - 2 \ln 3} \approx -9.327\right\}$ **73.** 3229.5 meters

75. (a) 37.3 watts **(b)** 6.9 decibels   **77. (a)** 9.85 years   **(b)** 4.27 years

79. \$41,668.97   **81.** 24,203 years ago   **83.** 6,078,190,457

85. (a) 0.3   **(b)** 0.8
(c)

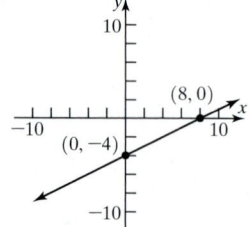

(d) About 20.1 years

87. (a)

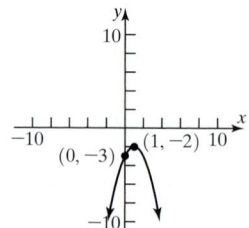

(b) Wind chill $= 44.198 - 20.331 \ln$ (wind speed)
(c)

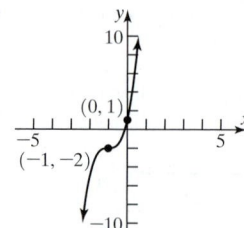

(d) Approximately -20 degrees Fahrenheit

Cumulative Review Exercises *(page 509)*

1. Yes; no   **2. (a)** 10   **(b)** $2x^2 + 3x + 1$ **(c)** $2x^2 + 4xh + 2h^2 - 3x - 3h + 1$ **3.** $\left(\dfrac{1}{2}, \dfrac{\sqrt{3}}{2}\right)$ is on the graph   **4.** -26

5.

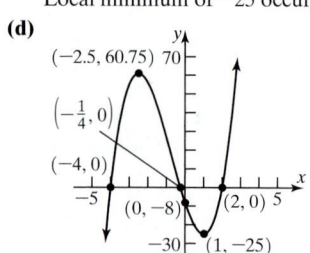

6.

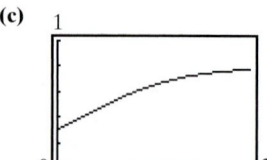

7. $f(x) = 2(x - 4)^2 - 8 = 2x^2 - 16x + 24$

8.

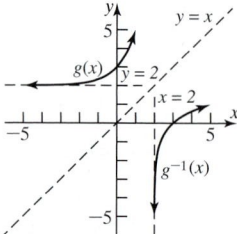

9. $f(g(x)) = \dfrac{4}{(x - 3)^2} + 2$; domain: $\{x | x \neq 3\}$

10. (a) Zeros: $-4, -\dfrac{1}{4}, 2$

(b) x-intercepts: $-4, -\dfrac{1}{4}, 2$; y-intercept: -8

(c) Local maximum of 60.75 occurs at $x = -2.5$;
Local minimum of -25 occurs at $x = 1$

(d)

11. (a), (c)

Domain g = Range $g^{-1} = (-\infty, \infty)$
Range g = Domain $g^{-1} = (2, \infty)$
(b) $g^{-1}(x) = \log_3(x - 2)$

12. $-\dfrac{3}{2}$ **13.** 2

C H A P T E R   7   Systems of Equations and Inequalities

7.1 Concepts and Vocabulary *(page 520)*

1. Inconsistent   **2.** Consistent; dependent   **3.** F   **4.** T

7.1 Exercises *(page 520)*

1. $\begin{cases} 2(2) - (-1) = 5 \\ 5(2) + 2(-1) = 8 \end{cases}$ **3.** $\begin{cases} 3(2) - 4\left(\dfrac{1}{2}\right) = 4 \\ \dfrac{1}{2}(2) - 3\left(\dfrac{1}{2}\right) = -\dfrac{1}{2} \end{cases}$ **5.** $\begin{cases} 4 - 1 = 3 \\ \dfrac{1}{2}(4) + 1 = 3 \end{cases}$ **7.** $\begin{cases} 3(1) + 3(-1) + 2(2) = 4 \\ 1 - (-1) - 2 = 0 \\ 2(-1) - 3(2) = -8 \end{cases}$ **9.** $x = 6, y = 2$

11. $x = 3, y = 2$ **13.** $x = 8, y = -4$ **15.** $x = \dfrac{1}{3}, y = -\dfrac{1}{6}$ **17.** Inconsistent   **19.** $x = 1, y = 2$ **21.** $x = 4 - 2y$, y is any real number

23. $x = 1, y = 1$ **25.** $x = \dfrac{3}{2}, y = 1$ **27.** $x = 4, y = 3$ **29.** $x = \dfrac{4}{3}, y = \dfrac{1}{5}$ **31.** $x = \dfrac{1}{5}, y = \dfrac{1}{3}$ **33.** $x = 48.15, y = 15.18$

35. $x = -21.48, y = 16.12$ **37.** $x = 0.26, y = 0.07$ **39.** $p = \$16, q = 600$ T-shirts   **41.** Length 30 feet; width 15 feet
43. Cheeseburger $1.55; shake $0.85   **45.** 22.5 pounds   **47.** Average wind speed 25 mph; average airspeed 175 mph
49. 80 $25 sets and 120 $45 sets   **51.** $5.56   **53.** Mix 50 mg of first liquid with 75 mg of second.   **55.** 5000

7.2 Exercises *(page 527)*

1. $3(1) + 3(-1) + 2(2) = 4, 1 - (-1) - 2 = 0,$ and $2(-1) - 3(2) = -8$ **3.** $x = 8, y = 2, z = 0$ **5.** $x = 2, y = -1, z = 1$
7. Inconsistent   **9.** $x = 5z - 2, y = 4z - 3$, where z is any real number   **11.** Inconsistent   **13.** $x = 1, y = 3, z = -2$

15. $x = -3, y = \dfrac{1}{2}, z = 1$ **17.** $a = \dfrac{4}{3}, b = -\dfrac{5}{3}, c = 1$ **19.** $I_1 = \dfrac{10}{71}, I_2 = \dfrac{65}{71}, I_3 = \dfrac{55}{71}$

21. 100 orchestra, 210 main, and 190 balcony seats   **23.** 1.5 chicken, 1 corn, 2 milk

25. If $x =$ price of hamburgers, $y =$ price of fries, $z =$ price of colas, then $x = 2.75 - z, y = \dfrac{41}{60} + \dfrac{1}{3}z, \$0.60 \leq z \leq \$0.90.$

There is not sufficient information:

x	$2.13	$2.01	$1.86
y	$0.89	$0.93	$0.98
z	$0.62	$0.74	$0.89

27. It will take Beth 30 hr, Bill 24 hr, and Edie 40 hr.

7.3 Concepts and Vocabulary *(page 543)*

1. Matrix   **2.** Augmented   **3.** T   **4.** F   **5.** Fourth row, sixth column

7.3 Exercises *(page 543)*

1. $\begin{bmatrix} 1 & -5 & | & 5 \\ 4 & 3 & | & 6 \end{bmatrix}$ **3.** $\begin{bmatrix} 2 & 3 & | & 6 \\ 4 & -6 & | & -2 \end{bmatrix}$ **5.** $\begin{bmatrix} 0.01 & -0.03 & | & 0.06 \\ 0.13 & 0.10 & | & 0.20 \end{bmatrix}$ **7.** $\begin{bmatrix} 1 & -1 & 1 & | & 10 \\ 3 & 3 & 0 & | & 5 \\ 1 & 1 & 2 & | & 2 \end{bmatrix}$ **9.** $\begin{bmatrix} 1 & 1 & -1 & | & 2 \\ 3 & -2 & 0 & | & 2 \\ 5 & 3 & -1 & | & 1 \end{bmatrix}$

11. $\begin{bmatrix} 1 & -1 & -1 & | & 10 \\ 2 & 1 & 2 & | & -1 \\ -3 & 4 & 0 & | & 5 \\ 4 & -5 & 1 & | & 0 \end{bmatrix}$ **13.** $\begin{bmatrix} 1 & -3 & | & -2 \\ 0 & 1 & | & 9 \end{bmatrix}$ **15. (a)** $\begin{bmatrix} 1 & -3 & 4 & | & 3 \\ 0 & 1 & -2 & | & 0 \\ -3 & 3 & 4 & | & 6 \end{bmatrix}$ **(b)** $\begin{bmatrix} 1 & -3 & 4 & | & 3 \\ 2 & -5 & 6 & | & 6 \\ 0 & -6 & 16 & | & 15 \end{bmatrix}$

17. (a) $\begin{bmatrix} 1 & -3 & 2 & | & -6 \\ 0 & 1 & -1 & | & 8 \\ -3 & -6 & 4 & | & 6 \end{bmatrix}$ **(b)** $\begin{bmatrix} 1 & -3 & 2 & | & -6 \\ 2 & -5 & 3 & | & -4 \\ 0 & -15 & 10 & | & -12 \end{bmatrix}$ **19. (a)** $\begin{bmatrix} 1 & -3 & 1 & | & -2 \\ 0 & 1 & 4 & | & 2 \\ -3 & 1 & 4 & | & 6 \end{bmatrix}$ **(b)** $\begin{bmatrix} 1 & -3 & 1 & | & -2 \\ 2 & -5 & 6 & | & -2 \\ 0 & -8 & 7 & | & 0 \end{bmatrix}$

21. $\begin{cases} x = 5 \\ y = -1 \end{cases}$

consistent; $x = 5, y = -1$

23. $\begin{cases} x = 1 \\ y = 2 \\ 0 = 3 \end{cases}$

inconsistent

25. $\begin{cases} x + 2z = -1 \\ y - 4z = -2 \\ 0 = 0 \end{cases}$

consistent;
$x = -1 - 2z,$
$y = -2 + 4z,$
z is any real number

27. $\begin{cases} x_1 = 1 \\ x_2 + x_4 = 2 \\ x_3 + 2x_4 = 3 \end{cases}$

consistent;
$x_1 = 1, x_2 = 2 - x_4,$
$x_3 = 3 - 2x_4,$
x_4 is any real number

29. $\begin{cases} x_1 + 4x_4 = 2 \\ x_2 + x_3 + 3x_4 = 3 \\ 0 = 0 \end{cases}$

consistent;
$x_1 = 2 - 4x_4,$
$x_2 = 3 - x_3 - 3x_4,$
x_3, x_4 are any real numbers

31. $\begin{cases} x_1 + x_4 = -2 \\ x_2 + 2x_4 = 2 \\ x_3 - x_4 = 0 \end{cases}$

consistent;
$x_1 = -2 - x_4,$
$x_2 = 2 - 2x_4,$
$x_3 = x_4,$
x_4 is any real number

33. $x = 6, y = 2$ **35.** $x = \dfrac{1}{2}, y = \dfrac{3}{4}$

37. $x = 4 - 2y$, y is any real number   **39.** $x = \dfrac{3}{2}, y = 1$

41. $x = \dfrac{4}{3}, y = \dfrac{1}{5}$ **43.** $x = 8, y = 2, z = 0$ **45.** $x = 2, y = -1, z = 1$

47. Inconsistent   **49.** $x = 5z - 2, y = 4z - 3$, where z is any real number

51. Inconsistent   **53.** $x = 1, y = 3, z = -2$ **55.** $x = -3, y = \dfrac{1}{2}, z = 1$ **57.** $x = \dfrac{1}{3}, y = \dfrac{2}{3}, z = 1$ **59.** $x = 1, y = 2, z = 0, w = 1$

61. $y = 0, z = 1 - x$, x is any real number   **63.** $x = 2, y = z - 3$, z is any real number   **65.** $x = \dfrac{13}{9}, y = \dfrac{7}{18}, z = \dfrac{19}{18}$

67. $x = \dfrac{7}{5} - \dfrac{3}{5}z - \dfrac{2}{5}w, y = -\dfrac{8}{5} + \dfrac{7}{5}z + \dfrac{13}{5}w$, where z and w are any real numbers   **69.** $y = -2x^2 + x + 3$

71. $f(x) = 3x^3 - 4x^2 + 5$ **73.** 1.5 salmon steak, 2 baked eggs, 1 acorn squash

75. \$4000 in Treasury bills, \$4000 in Treasury bonds, \$2000 in corporate bonds   **77.** 8 Deltas, 5 Betas, 10 Sigmas

79. $I_1 = \dfrac{44}{23}, I_2 = 2, I_3 = \dfrac{16}{23}, I_4 = \dfrac{28}{23}$

81. (a)

Amount Invested At

7%	9%	11%
0	10,000	10,000
1000	8000	11,000
2000	6000	12,000
3000	4000	13,000
4000	2000	14,000
5000	0	15,000

(b)

Amount Invested At

7%	9%	11%
12,500	12,500	0
14,500	8500	2000
16,500	4500	4000
18,750	0	6250

(c) All the money invested at 7% provides \$2100, more than what is required.

83.

First Liquid	Second Liquid	Third Liquid
50 mg	75 mg	0 mg
36 mg	76 mg	8 mg
22 mg	77 mg	16 mg
8 mg	78 mg	24 mg

7.4 Concepts and Vocabulary *(page 557)*

1. Determinants   **2.** $ad - bc$ **3.** F   **4.** F

7.4 Exercises *(page 557)*

1. 2   **3.** 22   **5.** -2 **7.** 10   **9.** -26 **11.** $x = 6, y = 2$ **13.** $x = 3, y = 2$ **15.** $x = 8, y = -4$ **17.** $x = 4, y = -2$

19. Not applicable   **21.** $x = \dfrac{1}{2}, y = \dfrac{3}{4}$ **23.** $x = \dfrac{1}{10}, y = \dfrac{2}{5}$ **25.** $x = \dfrac{3}{2}, y = 1$ **27.** $x = \dfrac{4}{3}, y = \dfrac{1}{5}$ **29.** $x = 1, y = 3, z = -2$

31. $x = -3, y = \dfrac{1}{2}, z = 1$ **33.** Not applicable   **35.** $x = 0, y = 0, z = 0$ **37.** Not applicable   **39.** $x = \dfrac{1}{5}, y = \dfrac{1}{3}$ **41.** -5 **43.** $\dfrac{13}{11}$

45. 0 or -9 **47.** -4 **49.** 12   **51.** 8   **53.** 8

55. $(y_1 - y_2)x - (x_1 - x_2)y + (x_1y_2 - x_2y_1) = 0$

$\qquad (y_1 - y_2)x + (x_2 - x_1)y = x_2y_1 - x_1y_2$

$\qquad (x_2 - x_1)y - (x_2 - x_1)y_1 = (y_2 - y_1)x + x_2y_1 - x_1y_2 - (x_2 - x_1)y_1$

$\qquad (x_2 - x_1)(y - y_1) = (y_2 - y_1)x - (y_2 - y_1)x_1$

$\qquad y - y_1 = \dfrac{y_2 - y_1}{x_2 - x_1}(x - x_1)$

57. $\begin{vmatrix} x^2 & x & 1 \\ y^2 & y & 1 \\ z^2 & z & 1 \end{vmatrix} = x^2 \begin{vmatrix} y & 1 \\ z & 1 \end{vmatrix} - x \begin{vmatrix} y^2 & 1 \\ z^2 & 1 \end{vmatrix} + \begin{vmatrix} y^2 & y \\ z^2 & z \end{vmatrix} = x^2(y - z) - x(y^2 - z^2) + yz(y - z)$

$\qquad = (y - z)[x^2 - x(y + z) + yz] = (y - z)[(x^2 - xy) - (xz - yz)] = (y - z)[x(x - y) - z(x - y)]$

$\qquad = (y - z)(x - y)(x - z)$

59. $\begin{vmatrix} a_{13} & a_{12} & a_{11} \\ a_{23} & a_{22} & a_{21} \\ a_{33} & a_{32} & a_{31} \end{vmatrix} = a_{13}(a_{22}a_{31} - a_{32}a_{21}) - a_{12}(a_{23}a_{31} - a_{33}a_{21}) + a_{11}(a_{23}a_{32} - a_{33}a_{22})$

$$= -a_{11}(a_{22}a_{33} - a_{32}a_{23}) + a_{12}(a_{21}a_{33} - a_{31}a_{23}) - a_{13}(a_{21}a_{32} - a_{31}a_{22}) = -\begin{vmatrix} a_{11} & a_{12} & a_{13} \\ a_{21} & a_{22} & a_{23} \\ a_{31} & a_{32} & a_{33} \end{vmatrix}$$

61. $\begin{vmatrix} a_{11} & a_{12} & a_{11} \\ a_{21} & a_{22} & a_{21} \\ a_{31} & a_{32} & a_{31} \end{vmatrix} = a_{11}(a_{22}a_{31} - a_{32}a_{21}) - a_{12}(a_{21}a_{31} - a_{31}a_{21}) + a_{11}(a_{21}a_{32} - a_{31}a_{22})$

$$= a_{11}a_{22}a_{31} - a_{11}a_{32}a_{21} - a_{12}(0) + a_{11}a_{21}a_{32} - a_{11}a_{31}a_{22} = 0$$

Historical Problems

1. (a) $2 - 5i \longleftrightarrow \begin{bmatrix} 2 & -5 \\ 5 & 2 \end{bmatrix}, 1 + 3i \longleftrightarrow \begin{bmatrix} 1 & 3 \\ -3 & 1 \end{bmatrix}$ **(b)** $\begin{bmatrix} 2 & -5 \\ 5 & 2 \end{bmatrix}\begin{bmatrix} 1 & 3 \\ -3 & 1 \end{bmatrix} = \begin{bmatrix} 17 & 1 \\ -1 & 17 \end{bmatrix}$ **(c)** $17 + i$ **(d)** $17 + i$

7.5 Concepts and Vocabulary *(page 575)*

1. Inverse **2.** Square **3.** Identity **4.** F **5.** F **6.** F **7.** Number of columns in 1st matrix equals the number of rows in 2nd matrix.
8. It doesn't have an inverse. Either no solution or infinitely many solutions.

7.5 Exercises *(page 575)*

1. $\begin{bmatrix} 4 & 4 & -5 \\ -1 & 5 & 4 \end{bmatrix}$ **3.** $\begin{bmatrix} 0 & 12 & -20 \\ 4 & 8 & 24 \end{bmatrix}$ **5.** $\begin{bmatrix} -8 & 7 & -15 \\ 7 & 0 & 22 \end{bmatrix}$ **7.** $\begin{bmatrix} 28 & -9 \\ 4 & 23 \end{bmatrix}$ **9.** $\begin{bmatrix} 1 & 14 & -14 \\ 2 & 22 & -18 \\ 3 & 0 & 28 \end{bmatrix}$ **11.** $\begin{bmatrix} 15 & 21 & -16 \\ 22 & 34 & -22 \\ -11 & 7 & 22 \end{bmatrix}$

13. $\begin{bmatrix} 25 & -9 \\ 4 & 20 \end{bmatrix}$ **15.** $\begin{bmatrix} -13 & 7 & -12 \\ -18 & 10 & -14 \\ 17 & -7 & 34 \end{bmatrix}$ **17.** $\begin{bmatrix} -2 & 4 & 2 & 8 \\ 2 & 1 & 4 & 6 \end{bmatrix}$ **19.** $\begin{bmatrix} 9 & 2 \\ 34 & 13 \\ 47 & 20 \end{bmatrix}$ **21.** $\begin{bmatrix} 1 & -1 \\ -1 & 2 \end{bmatrix}$ **23.** $\begin{bmatrix} 1 & -\frac{5}{2} \\ -1 & 3 \end{bmatrix}$

25. $\begin{bmatrix} 1 & -\frac{1}{a} \\ -1 & \frac{2}{a} \end{bmatrix}$ **27.** $\begin{bmatrix} 3 & -3 & 1 \\ -2 & 2 & -1 \\ -4 & 5 & -2 \end{bmatrix}$ **29.** $\begin{bmatrix} -\frac{5}{7} & \frac{1}{7} & \frac{3}{7} \\ \frac{9}{7} & \frac{1}{7} & -\frac{4}{7} \\ \frac{3}{7} & -\frac{2}{7} & \frac{1}{7} \end{bmatrix}$ **31.** $x = 3, y = 2$ **33.** $x = -5, y = 10$ **35.** $x = 2, y = -1$

37. $x = \frac{1}{2}, y = 2$ **39.** $x = -2, y = 1$ **41.** $x = \frac{2}{a}, y = \frac{3}{a}$ **43.** $x = -2, y = 3, z = 5$ **45.** $x = \frac{1}{2}, y = -\frac{1}{2}, z = 1$

47. $x = -\frac{34}{7}, y = \frac{85}{7}, z = \frac{12}{7}$ **49.** $x = \frac{1}{3}, y = 1, z = \frac{2}{3}$ **51.** $\begin{bmatrix} 4 & 2 & | & 1 & 0 \\ 2 & 1 & | & 0 & 1 \end{bmatrix} \rightarrow \begin{bmatrix} 1 & \frac{1}{2} & | & \frac{1}{4} & 0 \\ 2 & 1 & | & 0 & 1 \end{bmatrix} \rightarrow \begin{bmatrix} 1 & \frac{1}{2} & | & \frac{1}{4} & 0 \\ 0 & 0 & | & -\frac{1}{2} & 1 \end{bmatrix}$

53. $\begin{bmatrix} 15 & 3 & | & 1 & 0 \\ 10 & 2 & | & 0 & 1 \end{bmatrix} \rightarrow \begin{bmatrix} 1 & \frac{1}{5} & | & \frac{1}{15} & 0 \\ 10 & 2 & | & 0 & 1 \end{bmatrix} \rightarrow \begin{bmatrix} 1 & \frac{1}{5} & | & \frac{1}{15} & 0 \\ 0 & 0 & | & -\frac{2}{3} & 1 \end{bmatrix}$

55. $\begin{bmatrix} -3 & 1 & -1 & | & 1 & 0 & 0 \\ 1 & -4 & -7 & | & 0 & 1 & 0 \\ 1 & 2 & 5 & | & 0 & 0 & 1 \end{bmatrix} \rightarrow \begin{bmatrix} 1 & 2 & 5 & | & 0 & 0 & 1 \\ 1 & -4 & -7 & | & 0 & 1 & 0 \\ -3 & 1 & -1 & | & 1 & 0 & 0 \end{bmatrix} \rightarrow \begin{bmatrix} 1 & 2 & 5 & | & 0 & 0 & 1 \\ 0 & -6 & -12 & | & 0 & 1 & -1 \\ 0 & 7 & 14 & | & 1 & 0 & 3 \end{bmatrix} \rightarrow \begin{bmatrix} 1 & 2 & 5 & | & 0 & 0 & 1 \\ 0 & 1 & 2 & | & 0 & -\frac{1}{6} & \frac{1}{6} \\ 0 & 1 & 2 & | & \frac{1}{7} & 0 & \frac{3}{7} \end{bmatrix}$

$\rightarrow \begin{bmatrix} 1 & 2 & 5 & | & 0 & 0 & 1 \\ 0 & 1 & 2 & | & 0 & -\frac{1}{6} & \frac{1}{6} \\ 0 & 0 & 0 & | & \frac{1}{7} & \frac{1}{6} & \frac{11}{42} \end{bmatrix}$

57. $\begin{bmatrix} 0.01 & 0.05 & -0.01 \\ 0.01 & -0.02 & 0.01 \\ -0.02 & 0.01 & 0.03 \end{bmatrix}$ **59.** $\begin{bmatrix} 0.02 & -0.04 & -0.01 & 0.01 \\ -0.02 & 0.05 & 0.03 & -0.03 \\ 0.02 & 0.01 & -0.04 & 0.00 \\ -0.02 & 0.06 & 0.07 & 0.06 \end{bmatrix}$

61. $x = 4.57, y = -6.44, z = -24.07$ **63.** $x = -1.19, y = 2.46, z = 8.27$

65. (a) $\begin{bmatrix} 500 & 350 & 400 \\ 700 & 500 & 850 \end{bmatrix}; \begin{bmatrix} 500 & 700 \\ 350 & 500 \\ 400 & 850 \end{bmatrix}$ **(b)** $\begin{bmatrix} 15 \\ 8 \\ 3 \end{bmatrix}$ **(c)** $\begin{bmatrix} 11{,}500 \\ 17{,}050 \end{bmatrix}$ **(d)** $\begin{bmatrix} 0.10 & 0.05 \end{bmatrix}$ **(e)** $2002.50

7.6 Concepts and Vocabulary *(page 589)*

1. Half-plane **2.** Objective function **3.** Feasible point **4.** F **5.** T **6.** T

7.6 Exercises *(page 590)*

1. (a) 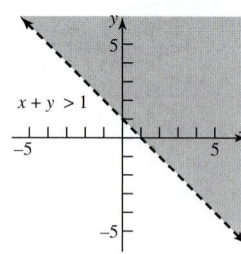 **(b)**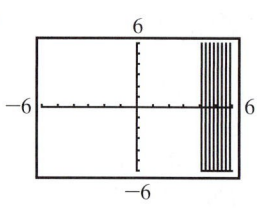

3. (a) 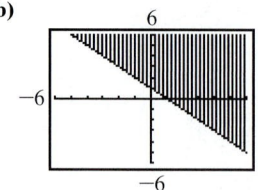 **(b)**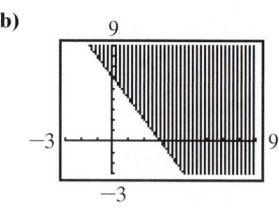

5. (a) 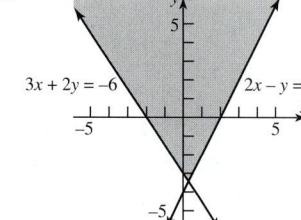 **(b)**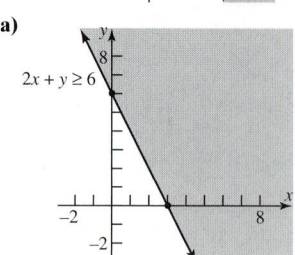

7. (a) 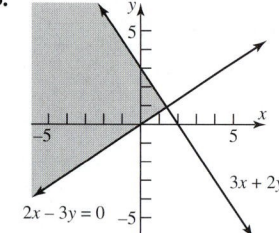 **(b)**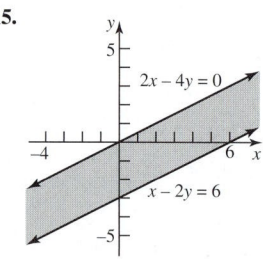

9.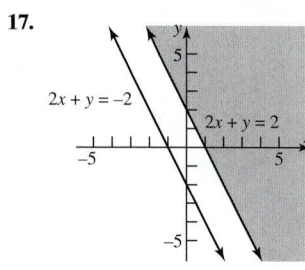

11.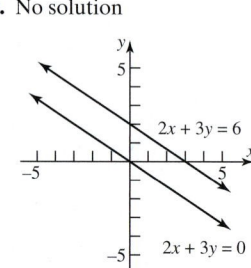

13.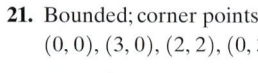

15.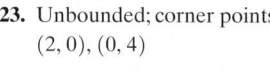

17.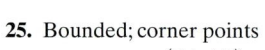

19. No solution

21. Bounded; corner points $(0, 0), (3, 0), (2, 2), (0, 3)$

23. Unbounded; corner points $(2, 0), (0, 4)$

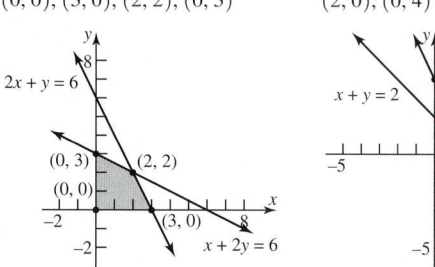

25. Bounded; corner points
$(2, 0), (4, 0), \left(\frac{24}{7}, \frac{12}{7}\right), (0, 4), (0, 2)$

27. Bounded; corner points
$(2, 0), (5, 0), (2, 6), (0, 8), (0, 2)$

29. Bounded; corner points
$(1, 0), (10, 0), (0, 5), \left(0, \frac{1}{2}\right)$

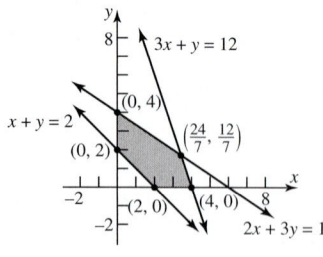

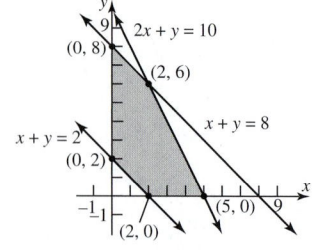

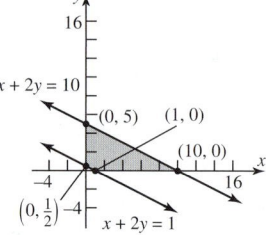

31. $\begin{cases} x \leq 4 \\ x + y \leq 6 \\ x \geq 0 \\ y \geq 0 \end{cases}$

33. $\begin{cases} x \leq 20 \\ y \geq 15 \\ x + y \leq 50 \\ x - y \leq 0 \\ x \geq 0 \end{cases}$

35. Maximum value is 11; minimum value is 3   **37.** Maximum value is 65; minimum value is 4

39. Maximum value is 67; minimum value is 20

41. The maximum value of z is 12, and it occurs at the point $(6, 0)$.

43. The minimum value of z is 4, and it occurs at the point $(2, 0)$.   **45.** The maximum value of z is 20, and it occurs at the point $(0, 4)$.

47. The minimum value of z is 8, and it occurs at the point $(0, 2)$.   **49.** The maximum value of z is 50, and it occurs at the point $(10, 0)$.
51. 8 downhill, 24 cross-country; 1760; 1920 **53.** 30 acres of soybeans and 10 acres of corn.
55. $\frac{1}{2}$ hour on machine 1; $5\frac{1}{4}$ hours on machine 2   **57.** 100 pounds of ground beef and 50 pounds of pork   **59.** 10 racing skates, 15 figure skates
61. 2 metal samples, 4 plastic samples; $34   **63. (a)** 10 first class, 120 coach   **(b)** 15 first class, 120 coach

7.7 Exercises *(page 600)*

1. Proper   **3.** Improper; $1 + \dfrac{9}{x^2 - 4}$ **5.** Improper; $5x + \dfrac{22x - 1}{x^2 - 4}$ **7.** Improper; $1 + \dfrac{-2(x - 6)}{(x + 4)(x - 3)}$ **9.** $\dfrac{-4}{x} + \dfrac{4}{x - 1}$

11. $\dfrac{1}{x} + \dfrac{-x}{x^2 + 1}$ **13.** $\dfrac{-1}{x - 1} + \dfrac{2}{x - 2}$ **15.** $\dfrac{\frac{1}{4}}{x + 1} + \dfrac{\frac{3}{4}}{x - 1} + \dfrac{\frac{1}{2}}{(x - 1)^2}$ **17.** $\dfrac{\frac{1}{12}}{x - 2} + \dfrac{-\frac{1}{12}(x + 4)}{x^2 + 2x + 4}$

19. $\dfrac{\frac{1}{4}}{x - 1} + \dfrac{\frac{1}{4}}{(x - 1)^2} + \dfrac{-\frac{1}{4}}{x + 1} + \dfrac{\frac{1}{4}}{(x + 1)^2}$ **21.** $\dfrac{-5}{x + 2} + \dfrac{5}{x + 1} + \dfrac{-4}{(x + 1)^2}$ **23.** $\dfrac{\frac{1}{4}}{x} + \dfrac{1}{x^2} + \dfrac{-\frac{1}{4}(x + 4)}{x^2 + 4}$

25. $\dfrac{\frac{2}{3}}{x + 1} + \dfrac{\frac{1}{3}(x + 1)}{x^2 + 2x + 4}$ **27.** $\dfrac{\frac{2}{7}}{3x - 2} + \dfrac{\frac{1}{7}}{2x + 1}$ **29.** $\dfrac{\frac{3}{4}}{x + 3} + \dfrac{\frac{1}{4}}{x - 1}$ **31.** $\dfrac{1}{x^2 + 4} + \dfrac{2x - 1}{(x^2 + 4)^2}$ **33.** $\dfrac{-1}{x} + \dfrac{2}{x - 3} + \dfrac{-1}{x + 1}$

35. $\dfrac{4}{x - 2} + \dfrac{-3}{x - 1} + \dfrac{-1}{(x - 1)^2}$ **37.** $\dfrac{x}{(x^2 + 16)^2} + \dfrac{-16x}{(x^2 + 16)^3}$ **39.** $\dfrac{-\frac{8}{7}}{2x + 1} + \dfrac{\frac{4}{7}}{x - 3}$ **41.** $\dfrac{-\frac{2}{9}}{x} + \dfrac{-\frac{1}{3}}{x^2} + \dfrac{\frac{1}{6}}{x - 3} + \dfrac{\frac{1}{18}}{x + 3}$

Review Exercises *(page 602)*

1. $x = 2, y = -1$ **3.** $x = 2, y = \dfrac{1}{2}$ **5.** $x = 2, y = -1$ **7.** $x = \dfrac{11}{5}, y = -\dfrac{3}{5}$ **9.** Inconsistent   **11.** $x = 2, y = 3$ **13.** Inconsistent

15. $x = -1, y = 2, z = -3$ **17.** $x = \dfrac{7}{4}z + \dfrac{39}{4}, y = \dfrac{9}{8}z + \dfrac{69}{8}, z$ is any real number   **19.** $\begin{cases} 3x + 2y = 8 \\ x + 4y = -1 \end{cases}$ **21.** $\begin{bmatrix} 4 & -4 \\ 3 & 9 \\ 4 & 0 \end{bmatrix}$ **23.** $\begin{bmatrix} 6 & 0 \\ 12 & 24 \\ -6 & 12 \end{bmatrix}$

25. $\begin{bmatrix} 4 & -3 & 0 \\ 12 & -2 & -8 \\ -2 & 5 & -4 \end{bmatrix}$ **27.** $\begin{bmatrix} 8 & -13 & 8 \\ 9 & 2 & -10 \\ 18 & -17 & 4 \end{bmatrix}$ **29.** $\begin{bmatrix} \frac{1}{2} & -1 \\ -\frac{1}{6} & \frac{2}{3} \end{bmatrix}$ **31.** $\begin{bmatrix} -\frac{5}{7} & \frac{9}{7} & \frac{3}{7} \\ \frac{1}{7} & \frac{1}{7} & -\frac{2}{7} \\ \frac{3}{7} & -\frac{4}{7} & \frac{1}{7} \end{bmatrix}$ **33.** Singular   **35.** $x = \dfrac{2}{5}, y = \dfrac{1}{10}$

37. $x = \dfrac{1}{2}, y = \dfrac{2}{3}, z = \dfrac{1}{6}$ **39.** $x = -\dfrac{1}{2}, y = -\dfrac{2}{3}, z = -\dfrac{3}{4}$ **41.** $z = -1, x = y + 1, y$ is any real number   **43.** $x = 4, y = 2, z = 3, t = -2$

45. 5   **47.** 108   **49.** -100 **51.** $x = 2, y = -1$ **53.** $x = 2, y = 3$ **55.** $x = -1, y = 2, z = -3$ **57.** 16

59. (a)

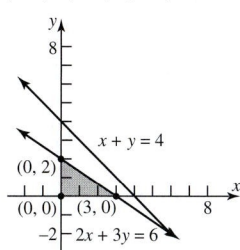

(b)

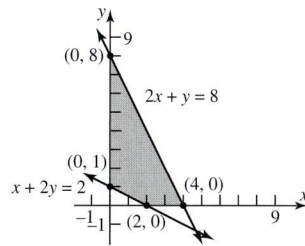

61. Unbounded; corner point $(0, 2)$

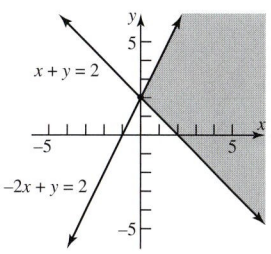

63. Bounded; corner points
$(0, 0), (0, 2), (3, 0)$

65. Bounded; corner points
$(0, 1), (0, 8), (4, 0), (2, 0)$

67. The maximum value is 32 when $x = 0$ and $y = 8$.
69. The minimum value is 3 when $x = 1$ and $y = 0$.

71. $\dfrac{-\frac{3}{2}}{x} + \dfrac{\frac{3}{2}}{x - 4}$ **73.** $\dfrac{-3}{x - 1} + \dfrac{3}{x} + \dfrac{4}{x^2}$

75. $\dfrac{-\frac{1}{10}}{x + 1} + \dfrac{\frac{1}{10}x + \frac{9}{10}}{x^2 + 9}$ **77.** $\dfrac{x}{x^2 + 4} + \dfrac{-4x}{(x^2 + 4)^2}$

79. $\dfrac{\frac{1}{2}}{x^2 + 1} + \dfrac{\frac{1}{4}}{x - 1} + \dfrac{-\frac{1}{4}}{x + 1}$

81. 10   **83.** $y = -\dfrac{1}{3}x^2 - \dfrac{2}{3}x + 1$ **85.** 70 pounds of $3 coffee and 30 pounds of $6 coffee   **87.** 1 small, 5 medium, 2 large

89. Speedboat: 36.7 km/hr; Aguarico River: 3.3 km/hr   **91.** Bruce: 4 hours; Bryce: 2 hours; Marty: 8 hours

93. 35 gasoline engines, 15 diesel engines; 15 gasoline engines, 0 diesel engines

Cumulative Review Exercises *(page 607)*

1. $\left\{\dfrac{1}{4} - \dfrac{\sqrt{23}}{4}i, \dfrac{1}{4} + \dfrac{\sqrt{23}}{4}i\right\}$ **2.** $\{5\}$ **3.** $\left\{-1, -\dfrac{1}{2}, 3\right\}$ **4.** $\{-2\}$ **5.** $\left\{\dfrac{5}{2}\right\}$ **6.** $\left\{\dfrac{1}{\ln 3}\right\}$ **7.** Odd; symmetric with respect to origin

8. Center: $(1, -2)$; radius $= 4$

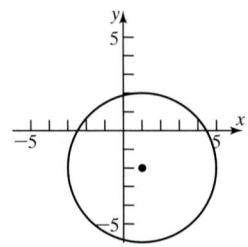

9. Domain: all real numbers
Range: $\{y | y > 1\}$
Horizontal asymptote: $y = 1$

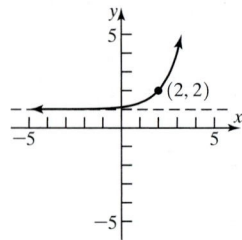

10. $f^{-1}(x) = \dfrac{5}{x} - 2$
Domain of f: $\{x | x \neq -2\}$
Range of f: $\{y | y \neq 0\}$
Domain of f^{-1}: $\{x | x \neq 0\}$
Range of f^{-1}: $\{y | y \neq -2\}$

11. (a) No   **(b)**

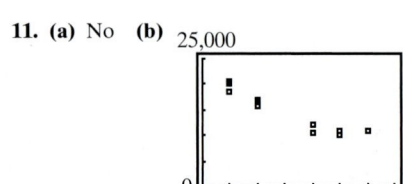

(c) $P = -2054.6a + 20,976.2$
(d) If the age of the car increases by 1 year, the price decreases by $2054.60.
(e) $P(a) = -2054.6a + 20,976.2$
(f) $6594

C H A P T E R   8   Trigonometric Functions

8.1 Concepts and Vocabulary *(page 620)*

1. Standard position   **2.** $r\theta; \dfrac{1}{2}r^2\theta$ **3.** $\dfrac{s}{t}; \dfrac{\theta}{t}$ **4.** F   **5.** T   **6.** T   **7.** On a circle of radius r, if the length of arc subtended by a central angle is also r, then the measure of the angle is 1 radian.   **8.** 1 radian is larger.   **9.** When an object travels around a circle, linear speed measures the length of the arc traveled per unit time; angular speed measures the angle swept out per unit time.

8.1 Exercises *(page 621)*

1.   **3.** 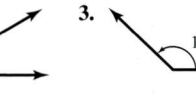   **5.** 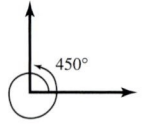   **7.** 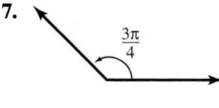   **9.** 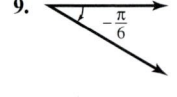   **11.**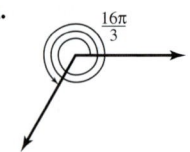

13. $\dfrac{\pi}{6}$ **15.** $\dfrac{4\pi}{3}$ **17.** $-\dfrac{\pi}{3}$ **19.** π **21.** $-\dfrac{3\pi}{4}$ **23.** $-\dfrac{\pi}{2}$ **25.** $60°$ **27.** $-225°$ **29.** $90°$ **31.** $15°$ **33.** $-90°$ **35.** $-30°$ **37.** 5 m   **39.** 6 ft

41. 0.6 radian   **43.** $\dfrac{\pi}{3} \approx 1.047$ in.   **45.** 25 m²   **47.** $2\sqrt{3} \approx 3.464$ ft   **49.** 0.24 radian   **51.** $\dfrac{\pi}{3} \approx 1.047$ in²   **53.** $s = 2.094$ ft; $A = 2.094$ ft²

55. $s = 14.661$ yd; $A = 87.965$ yd²   **57.** 0.30   **59.** -0.70 **61.** 2.18   **63.** $179.91°$ **65.** $114.59°$ **67.** $362.11°$ **69.** $40.17°$ **71.** $1.03°$

73. $9.15°$ **75.** $40°19'12''$ **77.** $18°15'18''$ **79.** $19°59'24''$ **81.** $3\pi \approx 9.42$ in.; $5\pi \approx 15.71$ in.   **83.** $2\pi \approx 6.28$ m²   **85.** $\dfrac{675\pi}{2} \approx 1060.29$ ft²

87. $\omega = \dfrac{1}{60}$ radian/sec; $v = \dfrac{1}{12}$ cm/sec   **89.** Approximately 452.49 rpm   **91.** Approximately 359.40 mi   **93.** Approximately 897.84 mi/hr

95. Approximately 2291.94 mi/hr   **97.** $\dfrac{3}{4}$ rpm   **99.** Approximately 2.86 mi/hr   **101.** Approximately 31.47 rpm

103. Approximately 1036.73 mi/hr   **105.** $v_1 = r_1\omega_1, v_2 = r_2\omega_2$, and $v_1 = v_2$ so $r_1\omega_1 = r_2\omega_2 \Rightarrow \dfrac{r_1}{r_2} = \dfrac{\omega_2}{\omega_1}$.

8.2 Concepts and Vocabulary *(page 632)*

1. Complementary **2.** Cosine **3.** $62°$ **4.** F **5.** T **6.** F **7.** $\csc \theta = \dfrac{1}{\sin \theta}$; $\sec \theta = \dfrac{1}{\cos \theta}$; $\cot \theta = \dfrac{1}{\tan \theta}$ **10.** Tangent, cotangent

8.2 Exercises *(page 633)*

1. $\sin \theta = \dfrac{5}{13}$; $\cos \theta = \dfrac{12}{13}$; $\tan \theta = \dfrac{5}{12}$; $\csc \theta = \dfrac{13}{5}$; $\sec \theta = \dfrac{13}{12}$; $\cot \theta = \dfrac{12}{5}$ **3.** $\sin \theta = \dfrac{2\sqrt{13}}{13}$; $\cos \theta = \dfrac{3\sqrt{13}}{13}$; $\tan \theta = \dfrac{2}{3}$; $\csc \theta = \dfrac{\sqrt{13}}{2}$;

$\sec \theta = \dfrac{\sqrt{13}}{3}$; $\cot \theta = \dfrac{3}{2}$ **5.** $\sin \theta = \dfrac{\sqrt{3}}{2}$; $\cos \theta = \dfrac{1}{2}$; $\tan \theta = \sqrt{3}$; $\csc \theta = \dfrac{2\sqrt{3}}{3}$; $\sec \theta = 2$; $\cot \theta = \dfrac{\sqrt{3}}{3}$ **7.** $\sin \theta = \dfrac{\sqrt{6}}{3}$; $\cos \theta = \dfrac{\sqrt{3}}{3}$;

$\tan \theta = \sqrt{2}$; $\csc \theta = \dfrac{\sqrt{6}}{2}$; $\sec \theta = \sqrt{3}$; $\cot \theta = \dfrac{\sqrt{2}}{2}$ **9.** $\sin \theta = \dfrac{\sqrt{5}}{5}$; $\cos \theta = \dfrac{2\sqrt{5}}{5}$; $\tan \theta = \dfrac{1}{2}$; $\csc \theta = \sqrt{5}$; $\sec \theta = \dfrac{\sqrt{5}}{2}$; $\cot \theta = 2$

11. $\tan \theta = \dfrac{\sqrt{3}}{3}$; $\csc \theta = 2$; $\sec \theta = \dfrac{2\sqrt{3}}{3}$; $\cot \theta = \sqrt{3}$ **13.** $\tan \theta = \dfrac{2\sqrt{5}}{5}$; $\csc \theta = \dfrac{3}{2}$; $\sec \theta = \dfrac{3\sqrt{5}}{5}$; $\cot \theta = \dfrac{\sqrt{5}}{2}$

15. $\cos \theta = \dfrac{\sqrt{2}}{2}$; $\tan \theta = 1$; $\csc \theta = \sqrt{2}$; $\sec \theta = \sqrt{2}$; $\cot \theta = 1$ **17.** $\sin \theta = \dfrac{2\sqrt{2}}{3}$; $\tan \theta = 2\sqrt{2}$; $\csc \theta = \dfrac{3\sqrt{2}}{4}$; $\sec \theta = 3$; $\cot \theta = \dfrac{\sqrt{2}}{4}$

19. $\sin \theta = \dfrac{\sqrt{5}}{5}$; $\cos \theta = \dfrac{2\sqrt{5}}{5}$; $\csc \theta = \sqrt{5}$; $\sec \theta = \dfrac{\sqrt{5}}{2}$; $\cot \theta = 2$ **21.** $\sin \theta = \dfrac{2\sqrt{2}}{3}$; $\cos \theta = \dfrac{1}{3}$; $\tan \theta = 2\sqrt{2}$; $\csc \theta = \dfrac{3\sqrt{2}}{4}$; $\cot \theta = \dfrac{\sqrt{2}}{4}$

23. $\sin \theta = \dfrac{\sqrt{6}}{3}$; $\cos \theta = \dfrac{\sqrt{3}}{3}$; $\csc \theta = \dfrac{\sqrt{6}}{2}$; $\sec \theta = \sqrt{3}$; $\cot \theta = \dfrac{\sqrt{2}}{2}$ **25.** $\sin \theta = \dfrac{1}{2}$; $\cos \theta = \dfrac{\sqrt{3}}{2}$; $\tan \theta = \dfrac{\sqrt{3}}{3}$; $\sec \theta = \dfrac{2\sqrt{3}}{3}$; $\cot \theta = \sqrt{3}$

27. 1 **29.** 1 **31.** 0 **33.** 0 **35.** 1 **37.** 0 **39.** 0 **41.** 1 **43.** 1 **45. (a)** $\dfrac{1}{2}$ **(b)** $\dfrac{3}{4}$ **(c)** 2 **(d)** 2 **47. (a)** 17 **(b)** $\dfrac{1}{4}$ **(c)** 4 **(d)** $\dfrac{17}{16}$

49. (a) $\dfrac{1}{4}$ **(b)** 15 **(c)** 4 **(d)** $\dfrac{16}{15}$ **51. (a)** 0.78 **(b)** 0.79 **(c)** 1.27 **(d)** 1.28 **(e)** 1.61 **(f)** 0.78 **(g)** 0.62 **(h)** 1.27 **53.** 0.6 **55.** $20°$

57. (a) 10 min **(b)** 20 min **(c)** $T(\theta) = 5\left(1 - \dfrac{1}{3 \tan \theta} + \dfrac{1}{\sin \theta}\right)$

(d) Approximately 15.81 min **(e)** Approximately 10.40 min

(f)

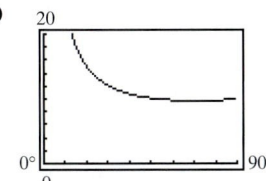

Approximately $70.53°$
Approximately 176.78 ft
Approximately 9.71 min

59. (a) $|OA| = |OC| = 1$; angle $OAC = $ angle OCA
angle $OAC + $ angle $OCA + (180° - \theta) = 180°$

$2(\text{angle } OAC) - \theta = 0$; angle $OAC = \dfrac{\theta}{2}$

(b) $\sin \theta = \dfrac{|CD|}{|OC|} = |CD|$; $\cos \theta = \dfrac{|OD|}{|OC|} = |OD|$

(c) $\tan \dfrac{\theta}{2} = \dfrac{|CD|}{|AD|} = \dfrac{\sin \theta}{1 + |OD|} = \dfrac{\sin \theta}{1 + \cos \theta}$

61. $h = x \tan \theta$ and $h = (1 - x) \tan(n\theta)$; thus, $x \tan \theta = (1 - x) \tan(n\theta)$ so $x = \dfrac{\tan(n\theta)}{\tan \theta + \tan(n\theta)}$.

63. (a) Area $\triangle OAC = \dfrac{1}{2}|AC||OC| = \dfrac{1}{2} \cdot \dfrac{|AC|}{1} \cdot \dfrac{|OC|}{1} = \dfrac{1}{2} \sin \alpha \cos \alpha$

(b) Area $\triangle OCB = \dfrac{1}{2}|BC||OC| = \dfrac{1}{2}|OB|^2 \dfrac{|BC|}{|OB|} \cdot \dfrac{|OC|}{|OB|} = \dfrac{1}{2}|OB|^2 \sin \beta \cos \beta$

(c) Area $\triangle OAB = \dfrac{1}{2}|BD||OA| = \dfrac{1}{2}|OB|\dfrac{|BD|}{|OB|} = \dfrac{1}{2}|OB| \sin(\alpha + \beta)$ **(d)** $\dfrac{\cos \alpha}{\cos \beta} = \dfrac{\dfrac{|OC|}{1}}{\dfrac{|OC|}{|OB|}} = |OB|$

(e) Area $\triangle OAB = $ area $\triangle OAC + $ area $\triangle OCB$

$\dfrac{1}{2}|OB|\sin(\alpha + \beta) = \dfrac{1}{2} \sin \alpha \cos \alpha + \dfrac{1}{2}|OB|^2 \sin \beta \cos \beta$

$\sin(\alpha + \beta) = \dfrac{\sin \alpha \cos \alpha + |OB|^2 \sin \beta \cos \beta}{|OB|}$

$\sin(\alpha + \beta) = \dfrac{\sin \alpha(|OB|\cos \beta) + |OB|^2 \sin \beta\left(\dfrac{\cos \alpha}{|OB|}\right)}{|OB|}$

$\sin(\alpha + \beta) = \sin \alpha \cos \beta + \cos \alpha \sin \beta$

65. $\sin \alpha = \tan \alpha \cos \alpha = \cos \beta \cos \alpha = \cos \beta \tan \beta = \sin \beta$;

$$\sin^2\alpha + \cos^2\alpha = 1$$
$$\sin^2\alpha + \tan^2\beta = 1$$
$$\sin^2\alpha + \frac{\sin^2\beta}{\cos^2\beta} = 1$$
$$\sin^2\alpha + \frac{\sin^2\alpha}{1-\sin^2\alpha} = 1$$
$$\sin^2\alpha - \sin^4\alpha + \sin^2\alpha = 1 - \sin^2\alpha$$
$$\sin^4\alpha - 3\sin^2\alpha + 1 = 0$$
$$\sin^2\alpha = \frac{3 \pm \sqrt{5}}{2}$$
$$\sin^2\alpha = \frac{3 - \sqrt{5}}{2}$$
$$\sin\alpha = \sqrt{\frac{3 - \sqrt{5}}{2}}$$

Note that $\dfrac{3 + \sqrt{5}}{2} > 1$.

67. Assume (a, b) is a point on the terminal side of θ. Since θ is acute, $a > 0$ and $b > 0$. Since $a^2 + b^2 = c^2$, then $0 < b^2 < c^2$, so $0 < b < c$.

Thus, $0 < \dfrac{b}{c} < 1$ and $0 < \sin\theta < 1$.

8.3 Concepts and Vocabulary *(page 641)*

1. $\dfrac{3}{2}$ **2.** 0.91   **3.** T   **4.** F

8.3 Exercises *(page 641)*

1. $\sin 45° = \dfrac{\sqrt{2}}{2}$; $\cos 45° = \dfrac{\sqrt{2}}{2}$; $\tan 45° = 1$; $\csc 45° = \sqrt{2}$; $\sec 45° = \sqrt{2}$; $\cot 45° = 1$ **3.** $\dfrac{\sqrt{3}}{2}$ **5.** $\dfrac{1}{2}$ **7.** $\dfrac{3}{4}$ **9.** $\sqrt{3}$ **11.** $\dfrac{\sqrt{3}}{4}$ **13.** $\sqrt{2}$

15. 2   **17.** $\sqrt{2} + \dfrac{4\sqrt{3}}{3}$ **19.** $-\dfrac{8}{3}$ **21.** $\dfrac{1}{2}$ **23.** 0   **25.** 0.47   **27.** 0.38   **29.** 1.33   **31.** 0.31   **33.** 3.73   **35.** 1.04   **37.** 0.84   **39.** 0.02

41. 0.31   **43.** $R \approx 310.56$ ft; $H \approx 77.64$ ft   **45.** $R \approx 19{,}541.95$ m; $H \approx 2278.14$ m   **47. (a)** 0.60 sec   **(b)** 0.79 sec   **(c)** 1.04 sec

49. (a) $T(\theta) = 1 + \dfrac{2}{3\sin\theta} - \dfrac{1}{4\tan\theta}$

(b) 1.90 hr; 0.57 hr
(c) 1.69 hr; 0.75 hr
(d) 1.63 hr; 0.86 hr
(e) 1.67 hr   **(f)** 2.75 hr

(g)

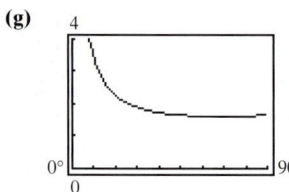

67.98°; 1.62 hr; 0.90 hr

51.

θ	0.5	0.4	0.2	0.1	0.01	0.001	0.0001	0.00001
$\sin\theta$	0.4794	0.3894	0.1987	0.0998	0.0100	0.0010	0.0001	0.00001
$\dfrac{\sin\theta}{\theta}$	0.9589	0.9735	0.9933	0.9983	1.0000	1.0000	1.0000	1.0000

$\dfrac{(\sin\theta)}{\theta}$ approaches 1 as $\theta \to 0$.

53. 1   **55.** $\dfrac{\sqrt{2}}{2}$

8.4 Concepts and Vocabulary *(page 652)*

1. Tangent; cotangent   **2.** Coterminal   **3.** 60°   **4.** F   **5.** T   **6.** T   **8.** I, IV   **9.** $\dfrac{\pi}{2}, \dfrac{3\pi}{2}$ **10.** 50°

8.4 Exercises *(page 653)*

1. $\sin\theta = \dfrac{4}{5}$; $\cos\theta = -\dfrac{3}{5}$; $\tan\theta = -\dfrac{4}{3}$; $\csc\theta = \dfrac{5}{4}$; $\sec\theta = -\dfrac{5}{3}$; $\cot\theta = -\dfrac{3}{4}$ **3.** $\sin\theta = -\dfrac{3\sqrt{13}}{13}$; $\cos\theta = \dfrac{2\sqrt{13}}{13}$; $\tan\theta = -\dfrac{3}{2}$;

$\csc\theta = -\dfrac{\sqrt{13}}{3}$; $\sec\theta = \dfrac{\sqrt{13}}{2}$; $\cot\theta = -\dfrac{2}{3}$ **5.** $\sin\theta = -\dfrac{\sqrt{2}}{2}$; $\cos\theta = -\dfrac{\sqrt{2}}{2}$; $\tan\theta = 1$; $\csc\theta = -\sqrt{2}$; $\sec\theta = -\sqrt{2}$; $\cot\theta = 1$

7. $\sin\theta = \dfrac{1}{2}$; $\cos\theta = \dfrac{\sqrt{3}}{2}$; $\tan\theta = \dfrac{\sqrt{3}}{3}$; $\csc\theta = 2$; $\sec\theta = \dfrac{2\sqrt{3}}{3}$; $\cot\theta = \sqrt{3}$ **9.** $\sin\theta = -\dfrac{\sqrt{2}}{2}$; $\cos\theta = \dfrac{\sqrt{2}}{2}$; $\tan\theta = -1$; $\csc\theta = -\sqrt{2}$;

$\sec \theta = \sqrt{2}$; $\cot \theta = -1$ **11.** II   **13.** IV   **15.** IV   **17.** III   **19.** $30°$ **21.** $60°$ **23.** $30°$ **25.** $\dfrac{\pi}{4}$ **27.** $\dfrac{\pi}{3}$ **29.** $45°$ **31.** $\dfrac{\pi}{3}$ **33.** $80°$

35. $\dfrac{\pi}{4}$ **37.** $\dfrac{\sqrt{2}}{2}$ **39.** 1   **41.** 1   **43.** $\sqrt{3}$ **45.** $\dfrac{\sqrt{2}}{2}$ **47.** 0   **49.** $\sqrt{2}$ **51.** $\dfrac{\sqrt{3}}{3}$ **53.** $\dfrac{1}{2}$ **55.** $\dfrac{\sqrt{2}}{2}$ **57.** -2 **59.** $-\sqrt{3}$ **61.** $\dfrac{\sqrt{2}}{2}$ **63.** $\sqrt{3}$

65. $\dfrac{1}{2}$ **67.** $-\dfrac{\sqrt{3}}{2}$ **69.** $-\sqrt{3}$ **71.** $\sqrt{2}$ **73.** 0   **75.** 0   **77.** -1 **79.** $\cos \theta = -\dfrac{5}{13}$; $\tan \theta = -\dfrac{12}{5}$; $\csc \theta = \dfrac{13}{12}$; $\sec \theta = -\dfrac{13}{5}$; $\cot \theta = -\dfrac{5}{12}$

81. $\sin \theta = -\dfrac{3}{5}$; $\tan \theta = \dfrac{3}{4}$; $\csc \theta = -\dfrac{5}{3}$; $\sec \theta = -\dfrac{5}{4}$; $\cot \theta = \dfrac{4}{3}$ **83.** $\cos \theta = -\dfrac{12}{13}$; $\tan \theta = -\dfrac{5}{12}$; $\csc \theta = \dfrac{13}{5}$; $\sec \theta = -\dfrac{13}{12}$; $\cot \theta = -\dfrac{12}{5}$

85. $\sin \theta = -\dfrac{2\sqrt{2}}{3}$; $\tan \theta = 2\sqrt{2}$; $\csc \theta = -\dfrac{3\sqrt{2}}{4}$; $\sec \theta = -3$; $\cot \theta = \dfrac{\sqrt{2}}{4}$ **87.** $\cos \theta = -\dfrac{\sqrt{5}}{3}$; $\tan \theta = -\dfrac{2\sqrt{5}}{5}$; $\csc \theta = \dfrac{3}{2}$; $\sec \theta = -\dfrac{3\sqrt{5}}{5}$;

$\cot \theta = -\dfrac{\sqrt{5}}{2}$ **89.** $\sin \theta = -\dfrac{\sqrt{3}}{2}$; $\cos \theta = \dfrac{1}{2}$; $\tan \theta = -\sqrt{3}$; $\csc \theta = -\dfrac{2\sqrt{3}}{3}$; $\cot \theta = -\dfrac{\sqrt{3}}{3}$ **91.** $\sin \theta = -\dfrac{3}{5}$; $\cos \theta = -\dfrac{4}{5}$; $\csc \theta = -\dfrac{5}{3}$;

$\sec \theta = -\dfrac{5}{4}$; $\cot \theta = \dfrac{4}{3}$ **93.** $\sin \theta = \dfrac{\sqrt{10}}{10}$; $\cos \theta = -\dfrac{3\sqrt{10}}{10}$; $\csc \theta = \sqrt{10}$; $\sec \theta = -\dfrac{\sqrt{10}}{3}$; $\cot \theta = -3$ **95.** $\sin \theta = -\dfrac{1}{2}$; $\cos \theta = -\dfrac{\sqrt{3}}{2}$;

$\tan \theta = \dfrac{\sqrt{3}}{3}$; $\sec \theta = -\dfrac{2\sqrt{3}}{3}$; $\cot \theta = \sqrt{3}$ **97.** 0   **99.** -0.2 **101.** 3   **103.** 5   **105.** 0   **107. (a)** Approximately 16.56 ft

(b)   **(c)** $67.50°$

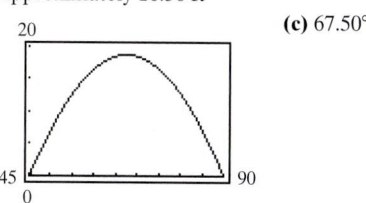

8.5 Concepts and Vocabulary *(page 663)*

1. 2π; π **2.** All real numbers except odd multiples of $\dfrac{\pi}{2}$ **3.** All real numbers from -1 to 1 inclusive   **4.** T   **5.** F   **6.** F

8.5 Exercises *(page 664)*

1. $\sin t = -\dfrac{1}{2}$; $\cos t = \dfrac{\sqrt{3}}{2}$; $\tan t = -\dfrac{\sqrt{3}}{3}$; $\csc t = -2$; $\sec t = \dfrac{2\sqrt{3}}{3}$; $\cot t = -\sqrt{3}$ **3.** $\sin t = -\dfrac{\sqrt{2}}{2}$; $\cos t = -\dfrac{\sqrt{2}}{2}$; $\tan t = 1$;

$\csc t = -\sqrt{2}$; $\sec t = -\sqrt{2}$; $\cot t = 1$ **5.** $\sin t = \dfrac{2}{3}$; $\cos t = \dfrac{\sqrt{5}}{3}$; $\tan t = \dfrac{2\sqrt{5}}{5}$; $\csc t = \dfrac{3}{2}$; $\sec t = \dfrac{3\sqrt{5}}{5}$; $\cot t = \dfrac{\sqrt{5}}{2}$

7. $\sin \theta = -\dfrac{4}{5}$; $\cos \theta = \dfrac{3}{5}$; $\tan \theta = -\dfrac{4}{3}$; $\csc \theta = -\dfrac{5}{4}$; $\sec \theta = \dfrac{5}{3}$; $\cot \theta = -\dfrac{3}{4}$ **9.** $\sin \theta = \dfrac{3\sqrt{13}}{13}$; $\cos \theta = -\dfrac{2\sqrt{13}}{13}$; $\tan \theta = -\dfrac{3}{2}$;

$\csc \theta = \dfrac{\sqrt{13}}{3}$; $\sec \theta = -\dfrac{\sqrt{13}}{2}$; $\cot \theta = -\dfrac{2}{3}$ **11.** $\sin \theta = -\dfrac{\sqrt{2}}{2}$; $\cos \theta = -\dfrac{\sqrt{2}}{2}$; $\tan \theta = 1$; $\csc \theta = -\sqrt{2}$; $\sec \theta = -\sqrt{2}$; $\cot \theta = 1$

13. $\dfrac{\sqrt{2}}{2}$ **15.** 1   **17.** 1   **19.** $\sqrt{3}$ **21.** $\dfrac{\sqrt{2}}{2}$ **23.** 0   **25.** $\sqrt{2}$ **27.** $\dfrac{\sqrt{3}}{3}$ **29.** $-\dfrac{\sqrt{3}}{2}$ **31.** $-\dfrac{\sqrt{3}}{3}$ **33.** 2   **35.** -1 **37.** -1 **39.** $\dfrac{\sqrt{2}}{2}$ **41.** 0

43. $-\sqrt{2}$ **45.** $\dfrac{2\sqrt{3}}{3}$ **47.** -1 **49.** -2 **51.** $\dfrac{2-\sqrt{2}}{2}$ **53.** All real numbers   **55.** Odd multiples of $\dfrac{\pi}{2}$ **57.** Odd multiples of $\dfrac{\pi}{2}$

59. $[-1, 1]$ **61.** $(-\infty, \infty)$ **63.** $(-\infty, -1]$ or $[1, \infty)$ **65.** Odd; yes; origin   **67.** Odd; yes; origin   **69.** Even; yes; y-axis   **71.** 0.9   **73.** 9

75. (a) $-\dfrac{1}{3}$ **(b)** 1   **77. (a)** -2 **(b)** 6   **79. (a)** -4 **(b)** -12

81. Let $P = (x, y)$ be the point on the unit circle that corresponds to t. Consider the equation $\tan t = \dfrac{y}{x} = a$. Then $y = ax$. But

$x^2 + y^2 = 1$ so that $x^2 + a^2 x^2 = 1$. Thus, $x = \pm \dfrac{1}{\sqrt{1 + a^2}}$ and $y = \pm \dfrac{a}{\sqrt{1 + a^2}}$; that is, for any real number a, there is a point

$P = (x, y)$ on the unit circle for which $\tan t = a$. In other words, $-\infty < \tan t < \infty$, and the range of the tangent function is the set of all real numbers.

83. Suppose there is a number p, $0 < p < 2\pi$, for which $\sin(\theta + p) = \sin \theta$ for all θ. If $\theta = 0$, then $\sin(0 + p) = \sin p = \sin 0 = 0$;

so that $p = \pi$. If $\theta = \dfrac{\pi}{2}$, then $\sin\left(\dfrac{\pi}{2} + p\right) = \sin\left(\dfrac{\pi}{2}\right)$. But $p = \pi$. Thus, $\sin\left(\dfrac{3\pi}{2}\right) = -1 = \sin\left(\dfrac{\pi}{2}\right) = 1$. This is impossible.

Therefore, the smallest positive number p for which $\sin(\theta + p) = \sin \theta$ for all θ is $p = 2\pi$.

85. $\sec \theta = \dfrac{1}{\cos \theta}$; since $\cos \theta$ has period 2π, so does $\sec \theta$.

87. If $P = (a, b)$ is the point on the unit circle corresponding to θ, then $Q = (-a, -b)$ is the point on the unit circle corresponding to

$\theta + \pi$. Thus, $\tan(\theta + \pi) = \dfrac{-b}{-a} = \dfrac{b}{a} = \tan\theta$. If there exists a number $p, 0 < p < \pi$, for which $\tan(\theta + p) = \tan\theta$ for all θ, then when

$\theta = 0, \tan(p) = \tan 0 = 0$. But this means that p is a multiple of π. Since no multiple of π exists in the interval $(0, \pi)$, this is impossible. Therefore, the fundamental period of $f(\theta) = \tan\theta$ is π.

89. $m = \dfrac{\sin\theta - 0}{\cos\theta - 0} = \dfrac{\sin\theta}{\cos\theta} = \tan\theta$

8.6 Concepts and Vocabulary *(page 678)*

1. $1; \ldots -\dfrac{3\pi}{2}, \dfrac{\pi}{2}, \dfrac{5\pi}{2}, \ldots$ **2.** $\pm 3; \pi$ **3.** $3; \dfrac{\pi}{3}$ **4.** T   **5.** F   **6.** T

8.6 Exercises *(page 678)*

1. 0   **3.** $-\dfrac{\pi}{2} \le x \le \dfrac{\pi}{2}$ **5.** 1   **7.** $0, \pi, 2\pi$ **9.** $\sin x = 1$ for $x = -\dfrac{3\pi}{2}, \dfrac{\pi}{2}$; $\sin x = -1$ for $x = -\dfrac{\pi}{2}, \dfrac{3\pi}{2}$ **11.** B, C, F

13.
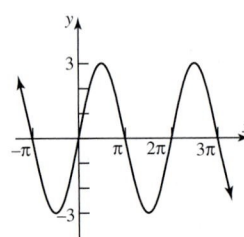

15.
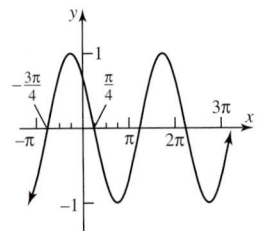

17.
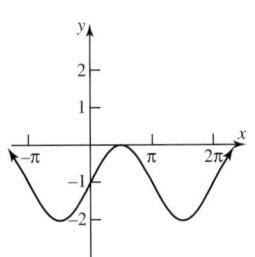

19.
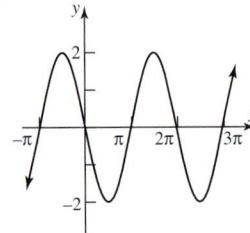

21.
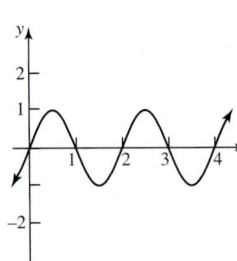

23.
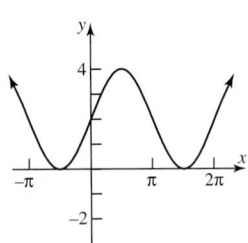

25.
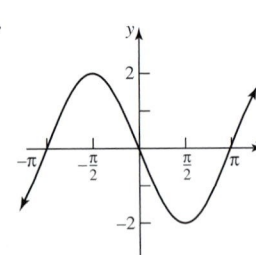

27.
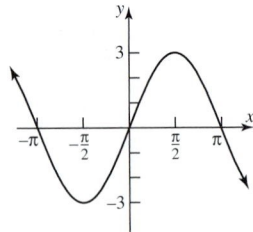

29. Amplitude = 2; period = 2π **31.** Amplitude = 4; period = π **33.** Amplitude = 6; period = 2

35. Amplitude = $\dfrac{1}{2}$; period = $\dfrac{4\pi}{3}$ **37.** Amplitude = $\dfrac{5}{3}$; period = 3   **39.** F   **41.** A   **43.** H   **45.** C   **47.** J   **49.** A   **51.** B

53.
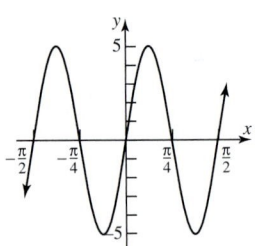

55.
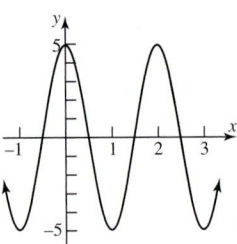

57.
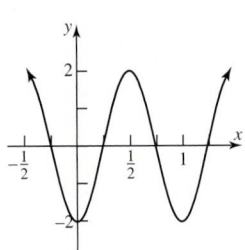

59.
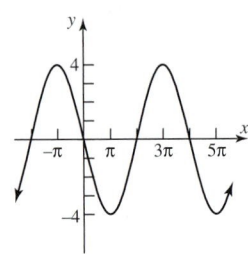

61.
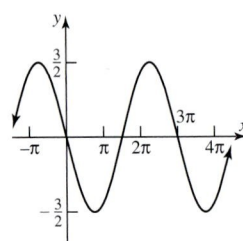

63. $y = \pm 3\sin(2x)$ **65.** $y = \pm 3\sin(\pi x)$

67. $y = -3\cos\left(\dfrac{1}{2}x\right)$ **69.** $y = \dfrac{3}{4}\sin(2\pi x)$ **71.** $y = -\sin\left(\dfrac{3}{2}x\right)$

73. $y = -2\cos\left(\dfrac{3\pi}{2}x\right)$ **75.** $y = 3\sin\left(\dfrac{\pi}{2}x\right)$ **77.** $y = -4\cos(3x)$

79. Period $= \dfrac{1}{30}$, amplitude $= 220$

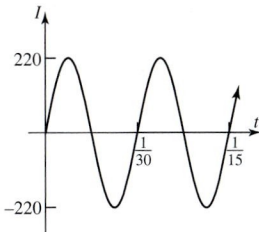

81. (a) Amplitude $= 220$, period $= \dfrac{1}{60}$

(b), (e)

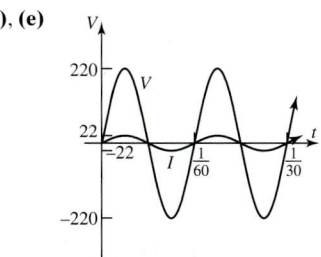

(c) $I = 22 \sin(120\pi t)$

(d) Amplitude $= 22$, period $= \dfrac{1}{60}$

83. (a) $P = \dfrac{[V_0 \sin(2\pi ft)]^2}{R} = \dfrac{V_0^2}{R} \sin^2[2\pi ft]$

(b) Since the graph of P has amplitude $\dfrac{V_0^2}{2R}$,

period $\dfrac{1}{2f}$, and is of the form $y = A\cos(\omega t) + B$,

then $A = -\dfrac{V_0^2}{2R}$ and $B = \dfrac{V_0^2}{2R}$. Since $\dfrac{1}{2f} = \dfrac{2\pi}{\omega}$,

then $\omega = 4\pi f$. Therefore,

$P = -\dfrac{V_0^2}{2R} \cos(4\pi ft) + \dfrac{V_0^2}{2R} = \dfrac{V_0^2}{2R}[1 - \cos(4\pi ft)]$.

85.

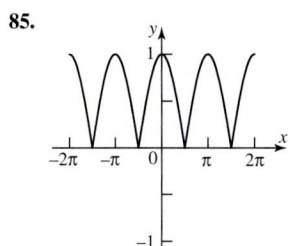

8.7 Concepts and Vocabulary *(page 687)*

1. Origin; odd multiples of $\dfrac{\pi}{2}$ **2.** y-axis; odd multiples of $\dfrac{\pi}{2}$ **3.** $y = \cos x$ **4.** T

8.7 Exercises *(page 687)*

1. 0   **3.** 1   **5.** $\sec x = 1$ for $x = -2\pi, 0, 2\pi$; $\sec x = -1$ for $x = -\pi, \pi$ **7.** $-\dfrac{3\pi}{2}, -\dfrac{\pi}{2}, \dfrac{\pi}{2}, \dfrac{3\pi}{2}$ **9.** $-\dfrac{3\pi}{2}, -\dfrac{\pi}{2}, \dfrac{\pi}{2}, \dfrac{3\pi}{2}$ **11.** D   **13.** B

15.

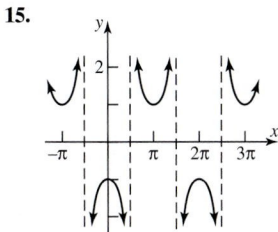

17.

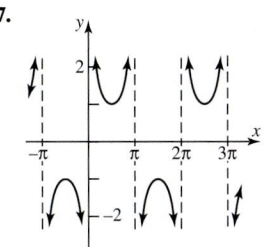

19.

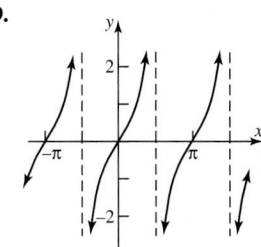

21.

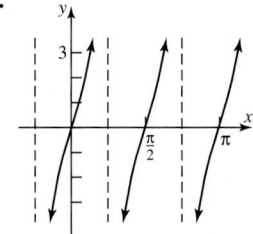

23.

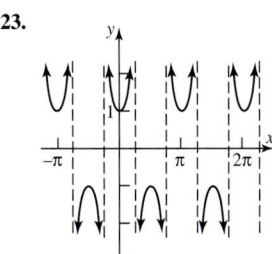

25.

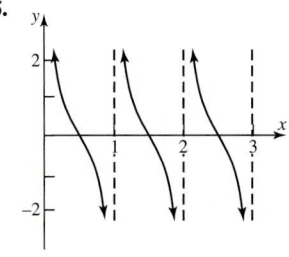

27.

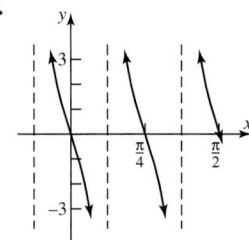

29.

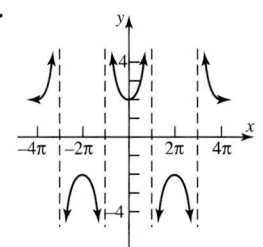

31.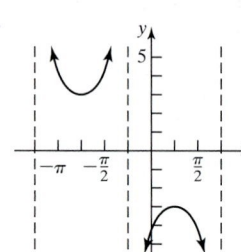

33.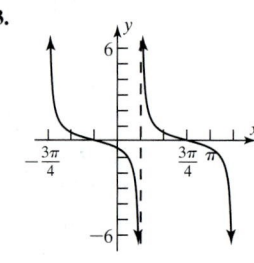

35. (a) $L(\theta) = \dfrac{3}{\cos \theta} + \dfrac{4}{\sin \theta} = 3 \sec \theta + 4 \csc \theta$

(b)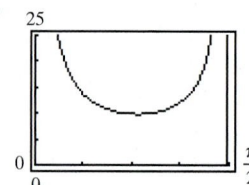

(c) 0.83
(d) 9.87 ft

8.8 Concepts and Vocabulary *(page 697)*

1. Phase shift   **2.** F

8.8 Exercises *(page 698)*

1. Amplitude $= 4$
Period $= \pi$
Phase shift $= \dfrac{\pi}{2}$

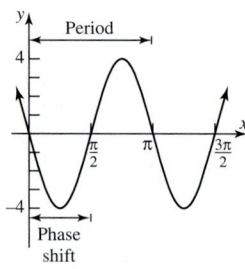

3. Amplitude $= 2$
Period $= \dfrac{2\pi}{3}$
Phase shift $= -\dfrac{\pi}{6}$

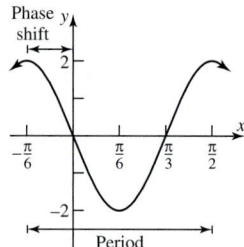

5. Amplitude $= 3$
Period $= \pi$
Phase shift $= -\dfrac{\pi}{4}$

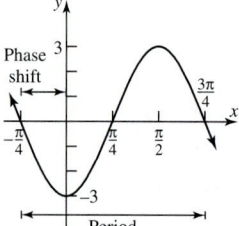

7. Amplitude $= 4$
Period $= 2$
Phase shift $= -\dfrac{2}{\pi}$

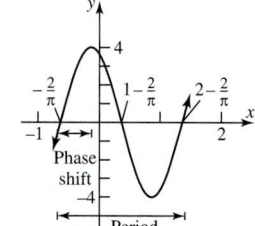

9. Amplitude $= 3$
Period $= 2$
Phase shift $= \dfrac{2}{\pi}$

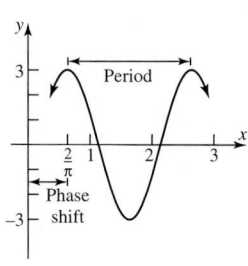

11. Amplitude $= 3$
Period $= \pi$
Phase shift $= \dfrac{\pi}{4}$

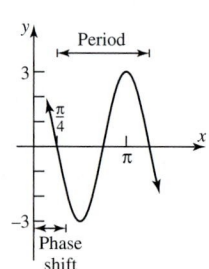

13. $y = \pm 2 \sin\left[2\left(x - \dfrac{1}{2} \right) \right]$ or $y = \pm 2 \sin(2x - 1)$

15. $y = \pm 3 \sin\left[\dfrac{2}{3}\left(x + \dfrac{1}{3} \right) \right]$ or $y = \pm 3 \sin\left(\dfrac{2}{3}x + \dfrac{2}{9} \right)$

17. Period $= \dfrac{1}{15}$; amplitude $= 120$; phase shift $= \dfrac{1}{90}$

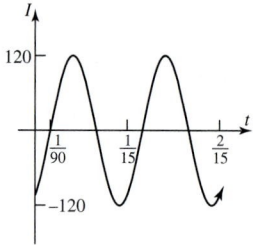

19. (a)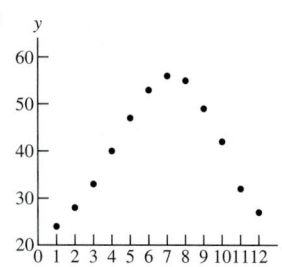

(b) $y = 15.9 \sin\left(\dfrac{\pi}{6}x - \dfrac{2\pi}{3} \right) + 40.1$

(c)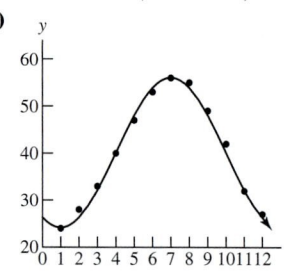

(d) $y = 15.62 \sin(0.517x - 2.096) + 40.377$

(e)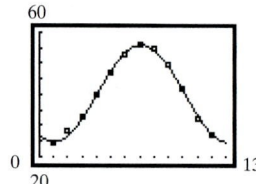

21. (a)

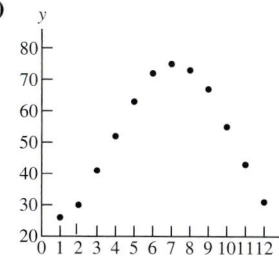

(b) $y = 24.95 \sin\left(\dfrac{\pi}{6} x - \dfrac{2\pi}{3}\right) + 50.45$

(c)

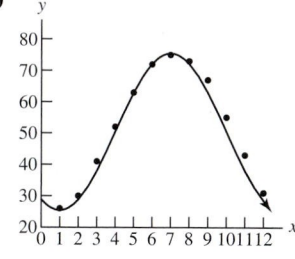

(d) $y = 25.693 \sin(0.476x - 1.814) + 49.854$

(e)

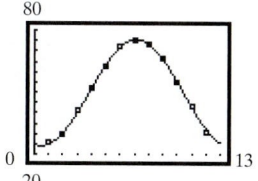

23. (a) 4:08 P.M.

(b) $y = 4.4 \sin\left(\dfrac{4\pi}{25}x - 6.66\right) + 3.8$

(c)

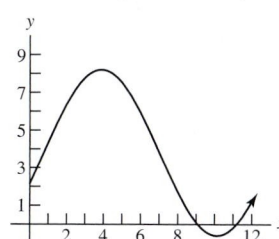

(d) 8.2 ft

25. (a) $y = 1.0835 \sin\left(\dfrac{2\pi}{365} x - 2.45\pi\right) + 11.6665$

(b)

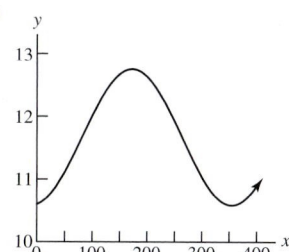

(c) 11.83 hr

27. (a) $y = 5.3915 \sin\left(\dfrac{2\pi}{365} x - 2.45\pi\right) + 10.8415$

(b)

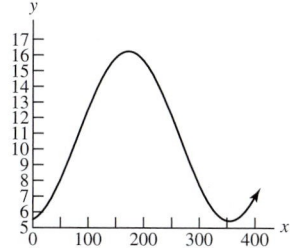

(c) 11.66 hr

Review Exercises *(page 703)*

1. $\dfrac{3\pi}{4}$ **3.** $\dfrac{\pi}{10}$ **5.** $135°$ **7.** $-450°$ **9.** $\dfrac{1}{2}$ **11.** $\dfrac{3\sqrt{2}}{2} - \dfrac{4\sqrt{3}}{3}$ **13.** $-3\sqrt{2} - 2\sqrt{3}$ **15.** 3   **17.** 0   **19.** 0   **21.** 1   **23.** 1   **25.** 1   **27.** -1

29. 1   **31.** $\cos\theta = \dfrac{3}{5}; \tan\theta = \dfrac{4}{3}; \csc\theta = \dfrac{5}{4}; \sec\theta = \dfrac{5}{3}; \cot\theta = \dfrac{3}{4}$ **33.** $\sin\theta = -\dfrac{12}{13}; \cos\theta = -\dfrac{5}{13}; \csc\theta = -\dfrac{13}{12}; \sec\theta = -\dfrac{13}{5}; \cot\theta = \dfrac{5}{12}$

35. $\sin\theta = \dfrac{3}{5}; \cos\theta = -\dfrac{4}{5}; \tan\theta = -\dfrac{3}{4}; \csc\theta = \dfrac{5}{3}; \cot\theta = -\dfrac{4}{3}$ **37.** $\cos\theta = -\dfrac{5}{13}; \tan\theta = -\dfrac{12}{5}; \csc\theta = \dfrac{13}{12}; \sec\theta = -\dfrac{13}{5}; \cot\theta = -\dfrac{5}{12}$

39. $\cos\theta = \dfrac{12}{13}; \tan\theta = -\dfrac{5}{12}; \csc\theta = -\dfrac{13}{5}; \sec\theta = \dfrac{13}{12}; \cot\theta = -\dfrac{12}{5}$ **41.** $\sin\theta = -\dfrac{\sqrt{10}}{10}; \cos\theta = -\dfrac{3\sqrt{10}}{10}; \csc\theta = -\sqrt{10};$

$\sec\theta = -\dfrac{\sqrt{10}}{3}; \cot\theta = 3$ **43.** $\sin\theta = -\dfrac{2\sqrt{2}}{3}; \cos\theta = \dfrac{1}{3}; \tan\theta = -2\sqrt{2}; \csc\theta = -\dfrac{3\sqrt{2}}{4}; \cot\theta = -\dfrac{\sqrt{2}}{4}$

45. $\sin\theta = \dfrac{\sqrt{5}}{5}; \cos\theta = -\dfrac{2\sqrt{5}}{5}; \tan\theta = -\dfrac{1}{2}; \csc\theta = \sqrt{5}; \sec\theta = -\dfrac{\sqrt{5}}{2}$

47.

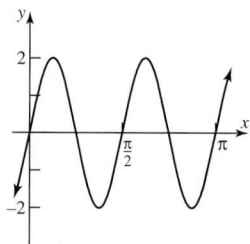

49.

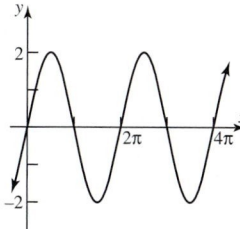

51.

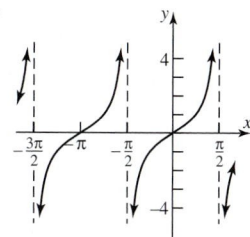

53.

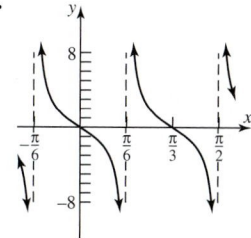

55.

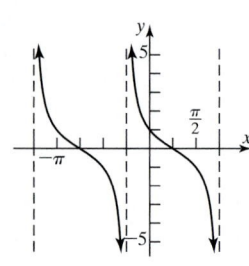

57.

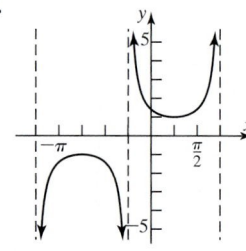

59. Amplitude $= 4$; period $= 2\pi$
61. Amplitude $= 8$; period $= 4$

63. Amplitude $= 4$

Period $= \dfrac{2\pi}{3}$

Phase shift $= 0$

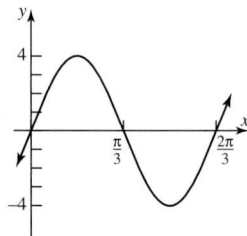

65. Amplitude $= 2$

Period $= \pi$

Phase Shift $= \dfrac{\pi}{2}$

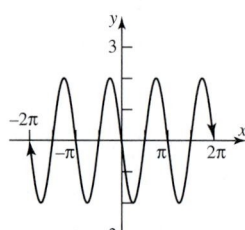

67. Amplitude $= \dfrac{1}{2}$

Period $= \dfrac{4\pi}{3}$

Phase shift $= \dfrac{2\pi}{3}$

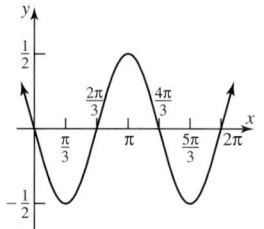

69. Amplitude $= \dfrac{2}{3}$

Period $= 2$

Phase shift $= \dfrac{6}{\pi}$

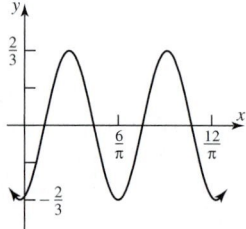

71. $y = 5\cos\dfrac{x}{4}$ **73.** $y = -6\cos\left(\dfrac{\pi}{4}x\right)$ **75.** $\sin\theta = \dfrac{5}{13}, \cos\theta = \dfrac{12}{13}, \tan\theta = \dfrac{5}{12}, \csc\theta = \dfrac{13}{5}, \sec\theta = \dfrac{13}{12}, \cot\theta = \dfrac{12}{5}$

77. $\sin\theta = -\dfrac{4}{5}, \cos\theta = \dfrac{3}{5}, \tan\theta = -\dfrac{4}{3}, \csc\theta = -\dfrac{5}{4}, \sec\theta = \dfrac{5}{3}, \cot\theta = -\dfrac{3}{4}$ **79.** $\dfrac{\pi}{5}$

81. Domain: $\left\{x \middle| x \neq \text{odd multiple of } \dfrac{\pi}{2}\right\}$; range: $\{y \mid \lvert y \rvert \geq 1\}$ **83.** $\dfrac{\pi}{3} \approx 1.05$ ft; $\dfrac{\pi}{3} \approx 1.05$ ft^2 **85.** Approximately 114.59 revolutions/hr

87. 0.1 revolution/sec $= \dfrac{\pi}{5}$ radian/sec

89. (a) 120   **(b)** $\dfrac{1}{60}$ **(c)**

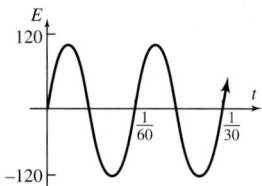

91. (a)

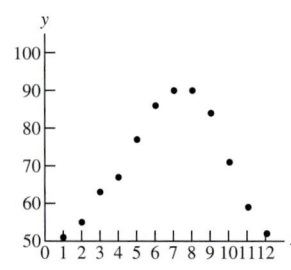

(b) $y = 19.5\sin\left(\dfrac{\pi}{6}x - \dfrac{2\pi}{3}\right) + 70.5$

(c)

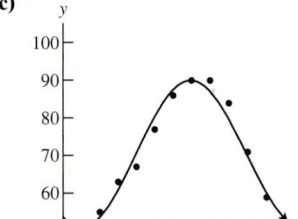

(d) $y = 19.52\sin(0.54x - 2.28) + 71.01$
(e)

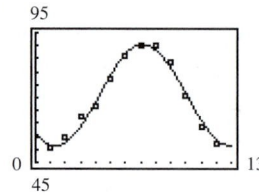

93. (a) $y = 1.85 \sin\left(\dfrac{2\pi}{365}x - 2.45\pi\right) + 11.517$ **(b)**

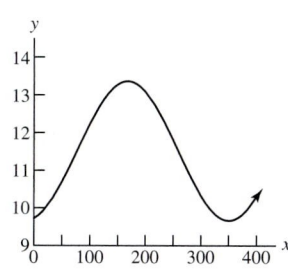

(c) 11.80 hr

Cumulative Review Exercises *(page 707)*

1. $\left\{-1, \dfrac{1}{2}\right\}$ **2.** $y - 5 = -3(x + 2)$ or $y = -3x - 1$ **3.** $x^2 + (y + 2)^2 = 16$

4.

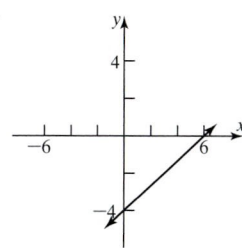

5.

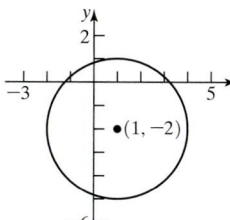

6.

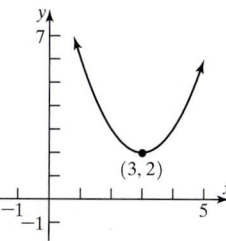

7. (a)

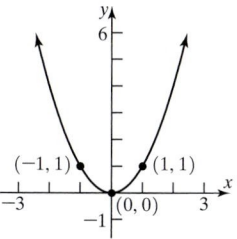

(b)

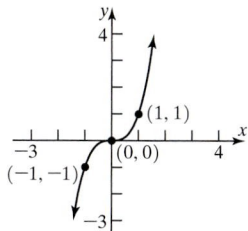

(c)

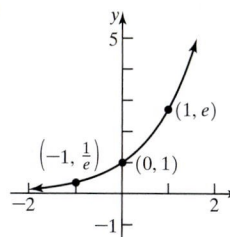

(d)

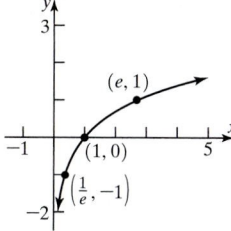

(e)

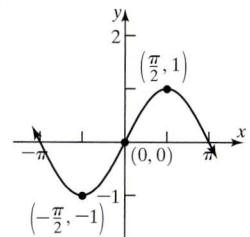

(f)

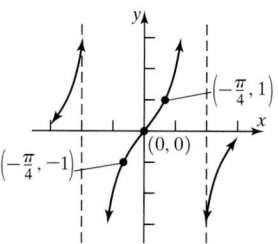

8. $f^{-1}(x) = \dfrac{1}{3}(x + 2)$ **9.** -2

10.

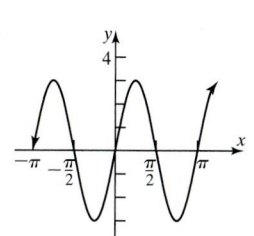

11. $3 - \dfrac{3\sqrt{3}}{2}$

12. $y = 2(3^x)$

13. $y = 3 \cos\left(\dfrac{\pi}{6}x\right)$

C H A P T E R   9   Analytic Trigonometry

9.1 Concepts and Vocabulary *(page 720)*

1. $x = \sin y$ **2.** 0 **3.** $\dfrac{\pi}{5}$ **4.** F **5.** T **6.** T **8.** $-\dfrac{\pi}{2} \le x \le \dfrac{\pi}{2}$ **9.** $-1 \le x \le 1$

9.1 Exercises *(page 720)*

1. 0 **3.** $-\dfrac{\pi}{2}$ **5.** 0 **7.** $\dfrac{\pi}{4}$ **9.** $\dfrac{\pi}{3}$ **11.** $\dfrac{5\pi}{6}$ **13.** 0.10 **15.** 1.37 **17.** 0.51 **19.** -0.38 **21.** -0.12 **23.** 1.08 **25.** 0.54 **27.** $\dfrac{4\pi}{5}$

29. -3.5 **31.** $-\dfrac{3\pi}{7}$ **33.** Yes; $-\dfrac{\pi}{6}$ lies in the interval $\left[-\dfrac{\pi}{2}, \dfrac{\pi}{2}\right]$. **35.** No; 2 is not in the domain of $\sin^{-1} x$.

37. No; $-\dfrac{\pi}{6}$ does not lie in the interval $[0, \pi]$. **39.** Yes, $-\dfrac{1}{2}$ is in the domain of $\cos^{-1} x$. **41.** Yes; $-\dfrac{\pi}{3}$ lies in the interval $\left(-\dfrac{\pi}{2}, \dfrac{\pi}{2}\right)$.

43. Yes; 2 is in the domain of $\tan^{-1} x$. **45. (a)** 13.92 hr or 13 hr, 55 min **(b)** 12 hr **(c)** 13.85 hr or 13 hr, 51 min

47. (a) 13.3 hr or 13 hr, 18 min **(b)** 12 hr **(c)** 13.26 hr or 13 hr, 15 min **49. (a)** 12 hr **(b)** 12 hr **(c)** 12 hr **(d)** It's 12 hr.

51. 3.35 min

9.2 Concepts and Vocabulary *(page 726)*

1. $x = \sec y$; ≥ 1; 0; π **2.** $\dfrac{\sqrt{2}}{2}$ **3.** F **4.** T **5.** T

9.2 Exercises *(page 726)*

1. $\dfrac{\sqrt{2}}{2}$ **3.** $-\dfrac{\sqrt{3}}{3}$ **5.** 2 **7.** $\sqrt{2}$ **9.** $-\dfrac{\sqrt{2}}{2}$ **11.** $\dfrac{2\sqrt{3}}{3}$ **13.** $\dfrac{3\pi}{4}$ **15.** $\dfrac{\pi}{6}$ **17.** $\dfrac{\sqrt{2}}{4}$ **19.** $\dfrac{\sqrt{5}}{2}$ **21.** $-\dfrac{\sqrt{14}}{2}$ **23.** $-\dfrac{3\sqrt{10}}{10}$ **25.** $\sqrt{5}$ **27.** $-\dfrac{\pi}{4}$

29. $\dfrac{\pi}{6}$ **31.** $-\dfrac{\pi}{2}$ **33.** $\dfrac{\pi}{6}$ **35.** $\dfrac{2\pi}{3}$ **37.** 1.32 **39.** 0.46 **41.** -0.34 **43.** 0.42 **45.** -0.73 **47.** 2.55

49.

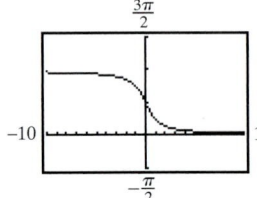

51.

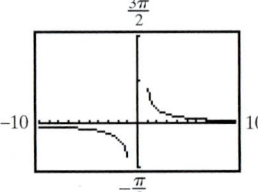

9.3 Concepts and Vocabulary *(page 732)*

1. Identity; conditional **2.** -1 **3.** 0 **4.** T **5.** F **6.** T **7.** $\sin^2\theta + \cos^2\theta = 1$; $\tan^2\theta + 1 = \sec^2\theta$; $1 + \cot^2\theta = \csc^2\theta$

9.3 Exercises *(page 732)*

1. $\csc\theta \cdot \cos\theta = \dfrac{1}{\sin\theta} \cdot \cos\theta = \dfrac{\cos\theta}{\sin\theta} = \cot\theta$ **3.** $1 + \tan^2(-\theta) = 1 + (-\tan\theta)^2 = 1 + \tan^2\theta = \sec^2\theta$

5. $\cos\theta(\tan\theta + \cot\theta) = \cos\theta\left(\dfrac{\sin\theta}{\cos\theta} + \dfrac{\cos\theta}{\sin\theta}\right) = \cos\theta\left(\dfrac{\sin^2\theta + \cos^2\theta}{\cos\theta\sin\theta}\right) = \cos\theta\left(\dfrac{1}{\cos\theta\sin\theta}\right) = \dfrac{1}{\sin\theta} = \csc\theta$

7. $\tan\theta\cot\theta - \cos^2\theta = \dfrac{\sin\theta}{\cos\theta} \cdot \dfrac{\cos\theta}{\sin\theta} - \cos^2\theta = 1 - \cos^2\theta = \sin^2\theta$ **9.** $(\sec\theta - 1)(\sec\theta + 1) = \sec^2\theta - 1 = \tan^2\theta$

11. $(\sec\theta + \tan\theta)(\sec\theta - \tan\theta) = \sec^2\theta - \tan^2\theta = 1$

13. $\cos^2\theta(1 + \tan^2\theta) = \cos^2\theta + \cos^2\theta\tan^2\theta = \cos^2\theta + \cos^2\theta \cdot \dfrac{\sin^2\theta}{\cos^2\theta} = \cos^2\theta + \sin^2\theta = 1$

15. $(\sin\theta + \cos\theta)^2 + (\sin\theta - \cos\theta)^2 = \sin^2\theta + 2\sin\theta\cos\theta + \cos^2\theta + \sin^2\theta - 2\sin\theta\cos\theta + \cos^2\theta = \sin^2\theta + \cos^2\theta + \sin^2\theta + \cos^2\theta$
$= 1 + 1 = 2$

17. $\sec^4\theta - \sec^2\theta = \sec^2\theta(\sec^2\theta - 1) = (1 + \tan^2\theta)\tan^2\theta = \tan^4\theta + \tan^2\theta$

19. $\sec\theta - \tan\theta = \dfrac{1}{\cos\theta} - \dfrac{\sin\theta}{\cos\theta} = \dfrac{1 - \sin\theta}{\cos\theta} \cdot \dfrac{1 + \sin\theta}{1 + \sin\theta} = \dfrac{1 - \sin^2\theta}{\cos\theta(1 + \sin\theta)} = \dfrac{\cos^2\theta}{\cos\theta(1 + \sin\theta)} = \dfrac{\cos\theta}{1 + \sin\theta}$

21. $3\sin^2\theta + 4\cos^2\theta = 3\sin^2\theta + 3\cos^2\theta + \cos^2\theta = 3(\sin^2\theta + \cos^2\theta) + \cos^2\theta = 3 + \cos^2\theta$

23. $1 - \dfrac{\cos^2\theta}{1 + \sin\theta} = 1 - \dfrac{1 - \sin^2\theta}{1 + \sin\theta} = 1 - \dfrac{(1 + \sin\theta)(1 - \sin\theta)}{1 + \sin\theta} = 1 - (1 - \sin\theta) = \sin\theta$

25. $\dfrac{1 + \tan\theta}{1 - \tan\theta} = \dfrac{1 + \dfrac{1}{\cot\theta}}{1 - \dfrac{1}{\cot\theta}} = \dfrac{\dfrac{\cot\theta + 1}{\cot\theta}}{\dfrac{\cot\theta - 1}{\cot\theta}} = \dfrac{\cot\theta + 1}{\cot\theta - 1}$ **27.** $\dfrac{\sec\theta}{\csc\theta} + \dfrac{\sin\theta}{\cos\theta} = \dfrac{\dfrac{1}{\cos\theta}}{\dfrac{1}{\sin\theta}} + \tan\theta = \dfrac{\sin\theta}{\cos\theta} + \tan\theta = \tan\theta + \tan\theta = 2\tan\theta$

29. $\dfrac{1 + \sin\theta}{1 - \sin\theta} = \dfrac{1 + \dfrac{1}{\csc\theta}}{1 - \dfrac{1}{\csc\theta}} = \dfrac{\dfrac{\csc\theta + 1}{\csc\theta}}{\dfrac{\csc\theta - 1}{\csc\theta}} = \dfrac{\csc\theta + 1}{\csc\theta - 1}$

31. $\dfrac{1 - \sin\theta}{\cos\theta} + \dfrac{\cos\theta}{1 - \sin\theta} = \dfrac{(1 - \sin\theta)^2 + \cos^2\theta}{\cos\theta(1 - \sin\theta)} = \dfrac{1 - 2\sin\theta + \sin^2\theta + \cos^2\theta}{\cos\theta(1 - \sin\theta)} = \dfrac{2 - 2\sin\theta}{\cos\theta(1 - \sin\theta)} = \dfrac{2(1 - \sin\theta)}{\cos\theta(1 - \sin\theta)} = \dfrac{2}{\cos\theta}$
$= 2\sec\theta$

33. $\dfrac{\sin\theta}{\sin\theta - \cos\theta} = \dfrac{1}{\dfrac{\sin\theta - \cos\theta}{\sin\theta}} = \dfrac{1}{1 - \dfrac{\cos\theta}{\sin\theta}} = \dfrac{1}{1 - \cot\theta}$

35. $(\sec\theta - \tan\theta)^2 = \sec^2\theta - 2\sec\theta\tan\theta + \tan^2\theta = \dfrac{1}{\cos^2\theta} - \dfrac{2\sin\theta}{\cos^2\theta} + \dfrac{\sin^2\theta}{\cos^2\theta} = \dfrac{1 - 2\sin\theta + \sin^2\theta}{\cos^2\theta} = \dfrac{(1-\sin\theta)^2}{1 - \sin^2\theta}$

$= \dfrac{(1-\sin\theta)^2}{(1-\sin\theta)(1+\sin\theta)} = \dfrac{1-\sin\theta}{1+\sin\theta}$

37. $\dfrac{\cos\theta}{1-\tan\theta} + \dfrac{\sin\theta}{1-\cot\theta} = \dfrac{\cos\theta}{1 - \dfrac{\sin\theta}{\cos\theta}} + \dfrac{\sin\theta}{1 - \dfrac{\cos\theta}{\sin\theta}} = \dfrac{\cos\theta}{\dfrac{\cos\theta - \sin\theta}{\cos\theta}} + \dfrac{\sin\theta}{\dfrac{\sin\theta - \cos\theta}{\sin\theta}} = \dfrac{\cos^2\theta}{\cos\theta - \sin\theta} + \dfrac{\sin^2\theta}{\sin\theta - \cos\theta}$

$= \dfrac{\cos^2\theta - \sin^2\theta}{\cos\theta - \sin\theta} = \dfrac{(\cos\theta - \sin\theta)(\cos\theta + \sin\theta)}{\cos\theta - \sin\theta} = \sin\theta + \cos\theta$

39. $\tan\theta + \dfrac{\cos\theta}{1+\sin\theta} = \dfrac{\sin\theta}{\cos\theta} + \dfrac{\cos\theta}{1+\sin\theta} = \dfrac{\sin\theta(1+\sin\theta) + \cos^2\theta}{\cos\theta(1+\sin\theta)} = \dfrac{\sin\theta + \sin^2\theta + \cos^2\theta}{\cos\theta(1+\sin\theta)} = \dfrac{\sin\theta + 1}{\cos\theta(1+\sin\theta)} = \dfrac{1}{\cos\theta} = \sec\theta$

41. $\dfrac{\tan\theta + \sec\theta - 1}{\tan\theta - \sec\theta + 1} = \dfrac{\tan\theta + (\sec\theta - 1)}{\tan\theta - (\sec\theta - 1)} \cdot \dfrac{\tan\theta + (\sec\theta - 1)}{\tan\theta + (\sec\theta - 1)} = \dfrac{\tan^2\theta + 2\tan\theta(\sec\theta - 1) + \sec^2\theta - 2\sec\theta + 1}{\tan^2\theta - (\sec^2\theta - 2\sec\theta + 1)}$

$= \dfrac{\sec^2\theta - 1 + 2\tan\theta(\sec\theta - 1) + \sec^2\theta - 2\sec\theta + 1}{\sec^2\theta - 1 - \sec^2\theta + 2\sec\theta - 1} = \dfrac{2\sec^2\theta - 2\sec\theta + 2\tan\theta(\sec\theta - 1)}{-2 + 2\sec\theta}$

$= \dfrac{2\sec\theta(\sec\theta - 1) + 2\tan\theta(\sec\theta - 1)}{2(\sec\theta - 1)} \cdot = \dfrac{2(\sec\theta - 1)(\sec\theta + \tan\theta)}{2(\sec\theta - 1)} = \tan\theta + \sec\theta$

43. $\dfrac{\tan\theta - \cot\theta}{\tan\theta + \cot\theta} = \dfrac{\dfrac{\sin\theta}{\cos\theta} - \dfrac{\cos\theta}{\sin\theta}}{\dfrac{\sin\theta}{\cos\theta} + \dfrac{\cos\theta}{\sin\theta}} = \dfrac{\dfrac{\sin^2\theta - \cos^2\theta}{\cos\theta\sin\theta}}{\dfrac{\sin^2\theta + \cos^2\theta}{\cos\theta\sin\theta}} = \dfrac{\sin^2\theta - \cos^2\theta}{1} = \sin^2\theta - \cos^2\theta$

45. $\dfrac{\tan\theta - \cot\theta}{\tan\theta + \cot\theta} + 1 = \dfrac{\dfrac{\sin\theta}{\cos\theta} - \dfrac{\cos\theta}{\sin\theta}}{\dfrac{\sin\theta}{\cos\theta} + \dfrac{\cos\theta}{\sin\theta}} + 1 = \dfrac{\dfrac{\sin^2\theta - \cos^2\theta}{\cos\theta\sin\theta}}{\dfrac{\sin^2\theta + \cos^2\theta}{\cos\theta\sin\theta}} + 1 = \sin^2\theta - \cos^2\theta + 1 = \sin^2\theta + (1 - \cos^2\theta) = 2\sin^2\theta$

47. $\dfrac{\sec\theta + \tan\theta}{\cot\theta + \cos\theta} = \dfrac{\dfrac{1}{\cos\theta} + \dfrac{\sin\theta}{\cos\theta}}{\dfrac{\cos\theta}{\sin\theta} + \cos\theta} = \dfrac{\dfrac{1+\sin\theta}{\cos\theta}}{\dfrac{\cos\theta + \cos\theta\sin\theta}{\sin\theta}} = \dfrac{1+\sin\theta}{\cos\theta} \cdot \dfrac{\sin\theta}{\cos\theta(1+\sin\theta)} = \dfrac{\sin\theta}{\cos\theta} \cdot \dfrac{1}{\cos\theta} = \tan\theta\sec\theta$

49. $\dfrac{1 - \tan^2\theta}{1 + \tan^2\theta} + 1 = \dfrac{1 - \tan^2\theta}{\sec^2\theta} + 1 = \dfrac{1}{\sec^2\theta} - \dfrac{\tan^2\theta}{\sec^2\theta} + 1 = \cos^2\theta - \dfrac{\dfrac{\sin^2\theta}{\cos^2\theta}}{\dfrac{1}{\cos^2\theta}} + 1 = \cos^2\theta - \sin^2\theta + 1$

$= \cos^2\theta + (1 - \sin^2\theta) = 2\cos^2\theta$

51. $\dfrac{\sec\theta - \csc\theta}{\sec\theta\csc\theta} = \dfrac{\sec\theta}{\sec\theta\csc\theta} - \dfrac{\csc\theta}{\sec\theta\csc\theta} = \dfrac{1}{\csc\theta} - \dfrac{1}{\sec\theta} = \sin\theta - \cos\theta$

53. $\sec\theta - \cos\theta - \sin\theta\tan\theta = \left(\dfrac{1}{\cos\theta} - \cos\theta\right) - \sin\theta \cdot \dfrac{\sin\theta}{\cos\theta} = \dfrac{1 - \cos^2\theta}{\cos\theta} - \dfrac{\sin^2\theta}{\cos\theta} = \dfrac{\sin^2\theta}{\cos\theta} - \dfrac{\sin^2\theta}{\cos\theta} = 0$

55. $\dfrac{1}{1-\sin\theta} + \dfrac{1}{1+\sin\theta} = \dfrac{1+\sin\theta + 1 - \sin\theta}{(1+\sin\theta)(1-\sin\theta)} = \dfrac{2}{1-\sin^2\theta} = \dfrac{2}{\cos^2\theta} = 2\sec^2\theta$

57. $\dfrac{\sec\theta}{1-\sin\theta} = \dfrac{\sec\theta}{1-\sin\theta} \cdot \dfrac{1+\sin\theta}{1+\sin\theta} = \dfrac{\sec\theta(1+\sin\theta)}{1-\sin^2\theta} = \dfrac{\sec\theta(1+\sin\theta)}{\cos^2\theta} = \dfrac{1+\sin\theta}{\cos^3\theta}$

59. $\dfrac{(\sec\theta - \tan\theta)^2 + 1}{\csc\theta(\sec\theta - \tan\theta)} = \dfrac{\sec^2\theta - 2\sec\theta\tan\theta + \tan^2\theta + 1}{\dfrac{1}{\sin\theta}\left(\dfrac{1}{\cos\theta} - \dfrac{\sin\theta}{\cos\theta}\right)} = \dfrac{2\sec^2\theta - 2\sec\theta\tan\theta}{\dfrac{1}{\sin\theta}\left(\dfrac{1-\sin\theta}{\cos\theta}\right)} = \dfrac{\dfrac{2}{\cos^2\theta} - \dfrac{2\sin\theta}{\cos^2\theta}}{\dfrac{1-\sin\theta}{\sin\theta\cos\theta}} = \dfrac{2 - 2\sin\theta}{\cos^2\theta} \cdot \dfrac{\sin\theta\cos\theta}{1-\sin\theta}$

$= \dfrac{2(1-\sin\theta)}{\cos\theta} \cdot \dfrac{\sin\theta}{1-\sin\theta} = \dfrac{2\sin\theta}{\cos\theta} = 2\tan\theta$

61. $\dfrac{\sin\theta + \cos\theta}{\cos\theta} - \dfrac{\sin\theta - \cos\theta}{\sin\theta} = \dfrac{\sin\theta}{\cos\theta} + 1 - 1 + \dfrac{\cos\theta}{\sin\theta} = \dfrac{\sin^2\theta + \cos^2\theta}{\cos\theta\sin\theta} = \dfrac{1}{\cos\theta\sin\theta} = \sec\theta\csc\theta$

63. $\dfrac{\sin^3 \theta + \cos^3 \theta}{\sin \theta + \cos \theta} = \dfrac{(\sin \theta + \cos \theta)(\sin^2 \theta - \sin \theta \cos \theta + \cos^2 \theta)}{\sin \theta + \cos \theta} = \sin^2 \theta + \cos^2 \theta - \sin \theta \cos \theta = 1 - \sin \theta \cos \theta$

65. $\dfrac{\cos^2 \theta - \sin^2 \theta}{1 - \tan^2 \theta} = \dfrac{\cos^2 \theta - \sin^2 \theta}{1 - \dfrac{\sin^2 \theta}{\cos^2 \theta}} = \dfrac{\cos^2 \theta - \sin^2 \theta}{\dfrac{\cos^2 \theta - \sin^2 \theta}{\cos^2 \theta}} = \cos^2 \theta$

67. $\dfrac{(2\cos^2 \theta - 1)^2}{\cos^4 \theta - \sin^4 \theta} = \dfrac{[2\cos^2 \theta - (\sin^2 \theta + \cos^2 \theta)]^2}{(\cos^2 \theta - \sin^2 \theta)(\cos^2 \theta + \sin^2 \theta)} = \dfrac{(\cos^2 \theta - \sin^2 \theta)^2}{\cos^2 \theta - \sin^2 \theta} = \cos^2 \theta - \sin^2 \theta = (1 - \sin^2 \theta) - \sin^2 \theta = 1 - 2\sin^2 \theta$

69. $\dfrac{1 + \sin \theta + \cos \theta}{1 + \sin \theta - \cos \theta} = \dfrac{(1 + \sin \theta) + \cos \theta}{(1 + \sin \theta) - \cos \theta} \cdot \dfrac{(1 + \sin \theta) + \cos \theta}{(1 + \sin \theta) + \cos \theta} = \dfrac{1 + 2\sin \theta + \sin^2 \theta + 2(1 + \sin \theta)(\cos \theta) + \cos^2 \theta}{1 + 2\sin \theta + \sin^2 \theta - \cos^2 \theta}$

$= \dfrac{1 + 2\sin \theta + \sin^2 \theta + 2(1 + \sin \theta)(\cos \theta) + (1 - \sin^2 \theta)}{1 + 2\sin \theta + \sin^2 \theta - (1 - \sin^2 \theta)} = \dfrac{2 + 2\sin \theta + 2(1 + \sin \theta)(\cos \theta)}{2\sin \theta + 2\sin^2 \theta}$

$= \dfrac{2(1 + \sin \theta) + 2(1 + \sin \theta)(\cos \theta)}{2\sin \theta(1 + \sin \theta)} = \dfrac{2(1 + \sin \theta)(1 + \cos \theta)}{2\sin \theta(1 + \sin \theta)} = \dfrac{1 + \cos \theta}{\sin \theta}$

71. $(a \sin \theta + b \cos \theta)^2 + (a \cos \theta - b \sin \theta)^2 = a^2 \sin^2 \theta + 2ab \sin \theta \cos \theta + b^2 \cos^2 \theta + a^2 \cos^2 \theta - 2ab \sin \theta \cos \theta + b^2 \sin^2 \theta$
$= a^2(\sin^2 \theta + \cos^2 \theta) + b^2(\cos^2 \theta + \sin^2 \theta) = a^2 + b^2$

73. $\dfrac{\tan \alpha + \tan \beta}{\cot \alpha + \cot \beta} = \dfrac{\tan \alpha + \tan \beta}{\dfrac{1}{\tan \alpha} + \dfrac{1}{\tan \beta}} = \dfrac{\tan \alpha + \tan \beta}{\dfrac{\tan \beta + \tan \alpha}{\tan \alpha \tan \beta}} = (\tan \alpha + \tan \beta) \cdot \dfrac{\tan \alpha \tan \beta}{\tan \alpha + \tan \beta} = \tan \alpha \tan \beta$

75. $(\sin \alpha + \cos \beta)^2 + (\cos \beta + \sin \alpha)(\cos \beta - \sin \alpha) = (\sin^2 \alpha + 2\sin \alpha \cos \beta + \cos^2 \beta) + (\cos^2 \beta - \sin^2 \alpha)$
$= 2\cos^2 \beta + 2\sin \alpha \cos \beta = 2\cos \beta(\cos \beta + \sin \alpha)$

77. $\ln|\sec \theta| = \ln|\cos \theta|^{-1} = -\ln|\cos \theta|$

79. $\ln|1 + \cos \theta| + \ln|1 - \cos \theta| = \ln(|1 + \cos \theta||1 - \cos \theta|) = \ln|1 - \cos^2 \theta| = \ln|\sin^2 \theta| = 2\ln|\sin \theta|$

81. Let $\theta = \tan^{-1} v$. Then $\tan \theta = v, -\dfrac{\pi}{2} < \theta < \dfrac{\pi}{2}$, so $\sec \theta > 0$ and $\tan^2 \theta + 1 = \sec^2 \theta$. Thus, $\sec(\tan^{-1} v) = \sec \theta = \sqrt{1 + v^2}$.

83. Let $\theta = \cos^{-1} v$. Then $\cos \theta = v, 0 \le \theta \le \pi$, so $\sin \theta \ge 0$ and $\sin \theta = \sqrt{1 - \cos^2 \theta} = \sqrt{1 - v^2}$.

Thus, $\tan(\cos^{-1} v) = \tan \theta = \dfrac{\sin \theta}{\cos \theta} = \dfrac{\sqrt{1 - v^2}}{v}$.

85. Let $\theta = \sin^{-1} v$. Then $\sin \theta = v, -\dfrac{\pi}{2} \le \theta \le \dfrac{\pi}{2}$, so $\cos \theta \ge 0$ and $\cos \theta = \sqrt{1 - \sin^2 \theta} = \sqrt{1 - v^2}$.

Thus, $\cos(\sin^{-1} v) = \cos \theta = \sqrt{1 - v^2}$.

9.4 Concepts and Vocabulary *(page 743)*

1. $-$ **2.** $-$ **3.** F   **4.** F   **5.** F   **6.** $\dfrac{1}{4}(\sqrt{6} + \sqrt{2})$ **7.** $\dfrac{1}{4}(\sqrt{2} + \sqrt{6})$ **8.** 0

9.4 Exercises *(page 743)*

1. $\dfrac{1}{4}(\sqrt{6} + \sqrt{2})$ **3.** $\dfrac{1}{4}(\sqrt{2} - \sqrt{6})$ **5.** $-\dfrac{1}{4}(\sqrt{2} + \sqrt{6})$ **7.** $2 - \sqrt{3}$ **9.** $-\dfrac{1}{4}(\sqrt{6} + \sqrt{2})$ **11.** $\sqrt{6} - \sqrt{2}$ **13.** $\dfrac{1}{2}$ **15.** 0   **17.** 1

19. -1 **21.** $\dfrac{1}{2}$ **23.** (a) $\dfrac{2\sqrt{5}}{25}$ (b) $\dfrac{11\sqrt{5}}{25}$ (c) $\dfrac{2\sqrt{5}}{5}$ (d) 2   **25.** (a) $\dfrac{4 - 3\sqrt{3}}{10}$ (b) $\dfrac{-3 - 4\sqrt{3}}{10}$ (c) $\dfrac{4 + 3\sqrt{3}}{10}$ (d) $\dfrac{25\sqrt{3} + 48}{39}$

27. (a) $-\dfrac{5 + 12\sqrt{3}}{26}$ (b) $\dfrac{12 - 5\sqrt{3}}{26}$ (c) $-\dfrac{5 - 12\sqrt{3}}{26}$ (d) $\dfrac{-240 + 169\sqrt{3}}{69}$ **29.** (a) $-\dfrac{2\sqrt{2}}{3}$ (b) $\dfrac{-2\sqrt{2} + \sqrt{3}}{6}$ (c) $\dfrac{-2\sqrt{2} + \sqrt{3}}{6}$

(d) $\dfrac{9 - 4\sqrt{2}}{7}$ **31.** $\sin\left(\dfrac{\pi}{2} + \theta\right) = \sin \dfrac{\pi}{2} \cos \theta + \cos \dfrac{\pi}{2} \sin \theta = 1 \cdot \cos \theta + 0 \cdot \sin \theta = \cos \theta$

33. $\sin(\pi - \theta) = \sin \pi \cos \theta - \cos \pi \sin \theta = 0 \cdot \cos \theta - (-1)\sin \theta = \sin \theta$

35. $\sin(\pi + \theta) = \sin \pi \cos \theta + \cos \pi \sin \theta = 0 \cdot \cos \theta + (-1)\sin \theta = -\sin \theta$

37. $\tan(\pi - \theta) = \dfrac{\tan \pi - \tan \theta}{1 + \tan \pi \tan \theta} = \dfrac{0 - \tan \theta}{1 + 0 \cdot \tan \theta} = -\tan \theta$

39. $\sin\left(\dfrac{3\pi}{2} + \theta\right) = \sin \dfrac{3\pi}{2} \cos \theta + \cos \dfrac{3\pi}{2} \sin \theta = (-1)\cos \theta + 0 \cdot \sin \theta = -\cos \theta$

41. $\sin(\alpha + \beta) + \sin(\alpha - \beta) = \sin \alpha \cos \beta + \cos \alpha \sin \beta + \sin \alpha \cos \beta - \cos \alpha \sin \beta = 2\sin \alpha \cos \beta$

43. $\dfrac{\sin(\alpha + \beta)}{\sin \alpha \cos \beta} = \dfrac{\sin \alpha \cos \beta + \cos \alpha \sin \beta}{\sin \alpha \cos \beta} = \dfrac{\sin \alpha \cos \beta}{\sin \alpha \cos \beta} + \dfrac{\cos \alpha \sin \beta}{\sin \alpha \cos \beta} = 1 + \cot \alpha \tan \beta$

45. $\dfrac{\cos(\alpha+\beta)}{\cos\alpha\cos\beta} = \dfrac{\cos\alpha\cos\beta-\sin\alpha\sin\beta}{\cos\alpha\cos\beta} = \dfrac{\cos\alpha\cos\beta}{\cos\alpha\cos\beta} - \dfrac{\sin\alpha\sin\beta}{\cos\alpha\cos\beta} = 1-\tan\alpha\tan\beta$

47. $\dfrac{\sin(\alpha+\beta)}{\sin(\alpha-\beta)} = \dfrac{\sin\alpha\cos\beta+\cos\alpha\sin\beta}{\sin\alpha\cos\beta-\cos\alpha\sin\beta} = \dfrac{\dfrac{\sin\alpha\cos\beta+\cos\alpha\sin\beta}{\cos\alpha\cos\beta}}{\dfrac{\sin\alpha\cos\beta-\cos\alpha\sin\beta}{\cos\alpha\cos\beta}} = \dfrac{\dfrac{\sin\alpha\cos\beta}{\cos\alpha\cos\beta}+\dfrac{\cos\alpha\sin\beta}{\cos\alpha\cos\beta}}{\dfrac{\sin\alpha\cos\beta}{\cos\alpha\cos\beta}-\dfrac{\cos\alpha\sin\beta}{\cos\alpha\cos\beta}} = \dfrac{\tan\alpha+\tan\beta}{\tan\alpha-\tan\beta}$

49. $\cot(\alpha+\beta) = \dfrac{\cos(\alpha+\beta)}{\sin(\alpha+\beta)} = \dfrac{\cos\alpha\cos\beta-\sin\alpha\sin\beta}{\sin\alpha\cos\beta+\cos\alpha\sin\beta} = \dfrac{\dfrac{\cos\alpha\cos\beta-\sin\alpha\sin\beta}{\sin\alpha\sin\beta}}{\dfrac{\sin\alpha\cos\beta+\cos\alpha\sin\beta}{\sin\alpha\sin\beta}} = \dfrac{\dfrac{\cos\alpha\cos\beta}{\sin\alpha\sin\beta}-\dfrac{\sin\alpha\sin\beta}{\sin\alpha\sin\beta}}{\dfrac{\sin\alpha\cos\beta}{\sin\alpha\sin\beta}+\dfrac{\cos\alpha\sin\beta}{\sin\alpha\sin\beta}} = \dfrac{\cot\alpha\cot\beta-1}{\cot\beta+\cot\alpha}$

51. $\sec(\alpha+\beta) = \dfrac{1}{\cos(\alpha+\beta)} = \dfrac{1}{\cos\alpha\cos\beta-\sin\alpha\sin\beta} = \dfrac{\dfrac{1}{\sin\alpha\sin\beta}}{\dfrac{\cos\alpha\cos\beta-\sin\alpha\sin\beta}{\sin\alpha\sin\beta}} = \dfrac{\dfrac{1}{\sin\alpha}\cdot\dfrac{1}{\sin\beta}}{\dfrac{\cos\alpha\cos\beta}{\sin\alpha\sin\beta}-\dfrac{\sin\alpha\sin\beta}{\sin\alpha\sin\beta}} = \dfrac{\csc\alpha\csc\beta}{\cot\alpha\cot\beta-1}$

53. $\sin(\alpha-\beta)\sin(\alpha+\beta) = (\sin\alpha\cos\beta-\cos\alpha\sin\beta)(\sin\alpha\cos\beta+\cos\alpha\sin\beta) = \sin^2\alpha\cos^2\beta-\cos^2\alpha\sin^2\beta$
$= (\sin^2\alpha)(1-\sin^2\beta)-(1-\sin^2\alpha)(\sin^2\beta) = \sin^2\alpha-\sin^2\beta$

55. $\sin(\theta+k\pi) = \sin\theta\cos k\pi+\cos\theta\sin k\pi = (\sin\theta)(-1)^k+(\cos\theta)(0) = (-1)^k\sin\theta, k$ any integer

57. $\dfrac{\sqrt3}{2}$ **59.** $-\dfrac{24}{25}$ **61.** $-\dfrac{33}{65}$ **63.** $\dfrac{63}{65}$ **65.** $\dfrac{48+25\sqrt3}{39}$ **67.** $\dfrac{4}{3}$ **69.** $u\sqrt{1-v^2}-v\sqrt{1-u^2}$

71. $\dfrac{u\sqrt{1-v^2}-v}{\sqrt{1+u^2}}$ **73.** $\dfrac{uv-\sqrt{1-u^2}\sqrt{1-v^2}}{v\sqrt{1-u^2}+u\sqrt{1-v^2}}$

75. Let $\alpha=\sin^{-1}v$ and $\beta=\cos^{-1}v$. Then $\sin\alpha=\cos\beta=v$, and since $\sin\alpha=\cos\left(\dfrac{\pi}{2}-\alpha\right)$, $\cos\left(\dfrac{\pi}{2}-\alpha\right)=\cos\beta$.

If $v\ge0$, then $0\le\alpha\le\dfrac{\pi}{2}$, so that $\left(\dfrac{\pi}{2}-\alpha\right)$ and β both lie in $\left[0,\dfrac{\pi}{2}\right]$. If $v<0$, then $-\dfrac{\pi}{2}\le\alpha<0$, so that $\left(\dfrac{\pi}{2}-\alpha\right)$ and β both lie in

$\left(\dfrac{\pi}{2},\pi\right]$. Either way, $\cos\left(\dfrac{\pi}{2}-\alpha\right)=\cos\beta$ implies $\dfrac{\pi}{2}-\alpha=\beta$, or $\alpha+\beta=\dfrac{\pi}{2}$.

77. Let $\alpha=\tan^{-1}\dfrac{1}{v}$, and $\beta=\tan^{-1}v$. Because $v\ne0$, $\alpha,\beta\ne0$. Then $\tan\alpha=\dfrac{1}{v}=\dfrac{1}{\tan\beta}=\cot\beta$, and since

$\tan\alpha=\cot\left(\dfrac{\pi}{2}-\alpha\right)$, $\cot\left(\dfrac{\pi}{2}-\alpha\right)=\cot\beta$. Because $v>0$, $0<\alpha<\dfrac{\pi}{2}$ and so $\left(\dfrac{\pi}{2}-\alpha\right)$ and β both lie in $\left(0,\dfrac{\pi}{2}\right)$.

Thus $\cot\left(\dfrac{\pi}{2}-\alpha\right)=\cot\beta$ implies $\dfrac{\pi}{2}-\alpha=\beta$, so $\alpha=\dfrac{\pi}{2}-\beta$, or $\tan^{-1}\dfrac{1}{v}=\dfrac{\pi}{2}-\tan^{-1}v$.

79. $\sin(\sin^{-1}v+\cos^{-1}v) = \sin(\sin^{-1}v)\cos(\cos^{-1}v)+\cos(\sin^{-1}v)\sin(\cos^{-1}v) = (v)(v)+\sqrt{1-v^2}\sqrt{1-v^2} = v^2+1-v^2 = 1$

81. $\dfrac{\sin(x+h)-\sin x}{h} = \dfrac{\sin x\cos h+\cos x\sin h-\sin x}{h} = \dfrac{\cos x\sin h-\sin x(1-\cos h)}{h} = \cos x\cdot\dfrac{\sin h}{h}-\sin x\cdot\dfrac{1-\cos h}{h}$

83. $\tan\dfrac{\pi}{2}$ is not defined; $\tan\left(\dfrac{\pi}{2}-\theta\right) = \dfrac{\sin\left(\dfrac{\pi}{2}-\theta\right)}{\cos\left(\dfrac{\pi}{2}-\theta\right)} = \dfrac{\cos\theta}{\sin\theta} = \cot\theta$ **85.** $\tan\theta=\tan(\theta_2-\theta_1) = \dfrac{\tan\theta_2-\tan\theta_1}{1+\tan\theta_1\tan\theta_2} = \dfrac{m_2-m_1}{1+m_1m_2}$

87. No; $\tan\dfrac{\pi}{2}$ is undefined.

9.5 Concepts and Vocabulary *(page 753)*

1. $\sin^2\theta; 2\cos^2\theta; 2\sin^2\theta$ **2.** $1-\cos\theta$ **3.** $\sin\theta$ **4.** T   **5.** F   **6.** $2\cos^2\theta-1$

7. $\sin\dfrac{30°}{2} = \dfrac{\sqrt{2-\sqrt3}}{2}$ **8.** $\sin(45°-30°) = \dfrac{1}{4}(\sqrt6-\sqrt2)$

9.5 Exercises *(page 753)*

1. (a) $\dfrac{24}{25}$ (b) $\dfrac{7}{25}$ (c) $\dfrac{\sqrt{10}}{10}$ (d) $\dfrac{3\sqrt{10}}{10}$ **3.** (a) $\dfrac{24}{25}$ (b) $-\dfrac{7}{25}$ (c) $\dfrac{2\sqrt5}{5}$ (d) $-\dfrac{\sqrt5}{5}$ **5.** (a) $-\dfrac{2\sqrt2}{3}$ (b) $\dfrac{1}{3}$ (c) $\sqrt{\dfrac{3+\sqrt6}{6}}$

(d) $\sqrt{\dfrac{3-\sqrt6}{6}}$ **7.** (a) $\dfrac{4\sqrt2}{9}$ (b) $-\dfrac{7}{9}$ (c) $\dfrac{\sqrt3}{3}$ (d) $\dfrac{\sqrt6}{3}$ **9.** (a) $-\dfrac{4}{5}$ (b) $\dfrac{3}{5}$ (c) $\sqrt{\dfrac{5+2\sqrt5}{10}}$ (d) $\sqrt{\dfrac{5-2\sqrt5}{10}}$

11. (a) $-\dfrac{3}{5}$ (b) $-\dfrac{4}{5}$ (c) $\dfrac{1}{2}\sqrt{\dfrac{10-\sqrt{10}}{5}}$ (d) $-\dfrac{1}{2}\sqrt{\dfrac{10+\sqrt{10}}{5}}$ **13.** $\dfrac{\sqrt{2-\sqrt2}}{2}$ **15.** $1-\sqrt2$ **17.** $-\dfrac{\sqrt{2+\sqrt3}}{2}$

19. $\dfrac{2}{\sqrt{2+\sqrt{2}}} = (2-\sqrt{2})\sqrt{2+\sqrt{2}}$ **21.** $-\dfrac{\sqrt{2-\sqrt{2}}}{2}$

23. $\sin^4\theta = (\sin^2\theta)^2 = \left(\dfrac{1-\cos(2\theta)}{2}\right)^2 = \dfrac{1}{4}(1 - 2\cos(2\theta) + \cos^2(2\theta)) = \dfrac{1}{4} - \dfrac{1}{2}\cos(2\theta) + \dfrac{1}{4}\cos^2(2\theta)$

$= \dfrac{1}{4} - \dfrac{1}{2}\cos(2\theta) + \dfrac{1}{4}\left(\dfrac{1+\cos(4\theta)}{2}\right) = \dfrac{1}{4} - \dfrac{1}{2}\cos(2\theta) + \dfrac{1}{8} + \dfrac{1}{8}\cos(4\theta) = \dfrac{3}{8} - \dfrac{1}{2}\cos(2\theta) + \dfrac{1}{8}\cos(4\theta)$

25. $\sin(4\theta) = \sin[2(2\theta)] = 2\sin(2\theta)\cos(2\theta) = (4\sin\theta\cos\theta)(1 - 2\sin^2\theta) = 4\sin\theta\cos\theta - 8\sin^3\theta\cos\theta = (\cos\theta)(4\sin\theta - 8\sin^3\theta)$

27. $\sin(5\theta) = 16\sin^5\theta - 20\sin^3\theta + 5\sin\theta$ **29.** $\cos^4\theta - \sin^4\theta = (\cos^2\theta + \sin^2\theta)(\cos^2\theta - \sin^2\theta) = \cos(2\theta)$

31. $\cot(2\theta) = \dfrac{1}{\tan(2\theta)} = \dfrac{1}{\dfrac{2\tan\theta}{1-\tan^2\theta}} = \dfrac{1-\tan^2\theta}{2\tan\theta} = \dfrac{1-\dfrac{1}{\cot^2\theta}}{2\left(\dfrac{1}{\cot\theta}\right)} = \dfrac{\dfrac{\cot^2\theta-1}{\cot^2\theta}}{\dfrac{2}{\cot\theta}} = \dfrac{\cot^2\theta-1}{\cot^2\theta}\cdot\dfrac{\cot\theta}{2} = \dfrac{\cot^2\theta-1}{2\cot\theta}$

33. $\sec(2\theta) = \dfrac{1}{\cos(2\theta)} = \dfrac{1}{2\cos^2\theta-1} = \dfrac{1}{\dfrac{2}{\sec^2\theta}-1} = \dfrac{1}{\dfrac{2-\sec^2\theta}{\sec^2\theta}} = \dfrac{\sec^2\theta}{2-\sec^2\theta}$

35. $\cos^2(2\theta) - \sin^2(2\theta) = \cos[2(2\theta)] = \cos(4\theta)$

37. $\dfrac{\cos(2\theta)}{1+\sin(2\theta)} = \dfrac{\cos^2\theta - \sin^2\theta}{1+2\sin\theta\cos\theta} = \dfrac{(\cos\theta-\sin\theta)(\cos\theta+\sin\theta)}{\sin^2\theta+\cos^2\theta+2\sin\theta\cos\theta} = \dfrac{(\cos\theta-\sin\theta)(\cos\theta+\sin\theta)}{(\sin\theta+\cos\theta)(\sin\theta+\cos\theta)} = \dfrac{\cos\theta-\sin\theta}{\cos\theta+\sin\theta}$

$= \dfrac{\dfrac{\cos\theta-\sin\theta}{\sin\theta}}{\dfrac{\cos\theta+\sin\theta}{\sin\theta}} = \dfrac{\dfrac{\cos\theta}{\sin\theta}-\dfrac{\sin\theta}{\sin\theta}}{\dfrac{\cos\theta}{\sin\theta}+\dfrac{\sin\theta}{\sin\theta}} = \dfrac{\cot\theta-1}{\cot\theta+1}$

39. $\sec^2\dfrac{\theta}{2} = \dfrac{1}{\cos^2\left(\dfrac{\theta}{2}\right)} = \dfrac{1}{\dfrac{1+\cos\theta}{2}} = \dfrac{2}{1+\cos\theta}$

41. $\cot^2\dfrac{\theta}{2} = \dfrac{1}{\tan^2\left(\dfrac{\theta}{2}\right)} = \dfrac{1}{\dfrac{1-\cos\theta}{1+\cos\theta}} = \dfrac{1+\cos\theta}{1-\cos\theta} = \dfrac{1+\dfrac{1}{\sec\theta}}{1-\dfrac{1}{\sec\theta}} = \dfrac{\dfrac{\sec\theta+1}{\sec\theta}}{\dfrac{\sec\theta-1}{\sec\theta}} = \dfrac{\sec\theta+1}{\sec\theta}\cdot\dfrac{\sec\theta}{\sec\theta-1} = \dfrac{\sec\theta+1}{\sec\theta-1}$

43. $\dfrac{1-\tan^2\left(\dfrac{\theta}{2}\right)}{1+\tan^2\left(\dfrac{\theta}{2}\right)} = \dfrac{1-\dfrac{1-\cos\theta}{1+\cos\theta}}{1+\dfrac{1-\cos\theta}{1+\cos\theta}} = \dfrac{\dfrac{1+\cos\theta-(1-\cos\theta)}{1+\cos\theta}}{\dfrac{1+\cos\theta+1-\cos\theta}{1+\cos\theta}} = \dfrac{2\cos\theta}{1+\cos\theta}\cdot\dfrac{1+\cos\theta}{2} = \cos\theta$

45. $\dfrac{\sin(3\theta)}{\sin\theta} - \dfrac{\cos(3\theta)}{\cos\theta} = \dfrac{\sin(3\theta)\cos\theta - \cos(3\theta)\sin\theta}{\sin\theta\cos\theta} = \dfrac{\sin(3\theta-\theta)}{\dfrac{1}{2}(2\sin\theta\cos\theta)} = \dfrac{2\sin(2\theta)}{\sin(2\theta)} = 2$

47. $\tan(3\theta) = \tan(\theta + 2\theta) = \dfrac{\tan\theta+\tan(2\theta)}{1-\tan\theta\tan(2\theta)} = \dfrac{\tan\theta+\dfrac{2\tan\theta}{1-\tan^2\theta}}{1-\dfrac{\tan\theta(2\tan\theta)}{1-\tan^2\theta}} = \dfrac{\tan\theta-\tan^3\theta+2\tan\theta}{1-\tan^2\theta-2\tan^2\theta} = \dfrac{3\tan\theta-\tan^3\theta}{1-3\tan^2\theta}$

49. $\dfrac{1}{2}(\ln|1-\cos(2\theta)| - \ln 2) = \ln\left(\dfrac{|1-\cos(2\theta)|}{2}\right)^{1/2} = \ln|\sin^2\theta|^{1/2} = \ln|\sin\theta|$

51. $\dfrac{\sqrt{3}}{2}$ **53.** $\dfrac{7}{25}$ **55.** $\dfrac{24}{7}$ **57.** $\dfrac{24}{25}$ **59.** $\dfrac{1}{5}$ **61.** $\dfrac{25}{7}$ **63.** $\sin(2\theta) = \dfrac{4x}{4+x^2}$ **65.** $-\dfrac{1}{4}$

67. $\dfrac{2z}{1+z^2} = \dfrac{2\tan\left(\dfrac{\alpha}{2}\right)}{1+\tan^2\left(\dfrac{\alpha}{2}\right)} = \dfrac{2\tan\left(\dfrac{\alpha}{2}\right)}{\sec^2\left(\dfrac{\alpha}{2}\right)} = \dfrac{\dfrac{2\sin\left(\dfrac{\alpha}{2}\right)}{\cos\left(\dfrac{\alpha}{2}\right)}}{\dfrac{1}{\cos^2\left(\dfrac{\alpha}{2}\right)}} = 2\sin\left(\dfrac{\alpha}{2}\right)\cos\left(\dfrac{\alpha}{2}\right) = \sin\left(2\cdot\dfrac{\alpha}{2}\right) = \sin\alpha$

69. $A = \dfrac{1}{2}h(\text{base}) = h\left(\dfrac{1}{2}\,\text{base}\right) = s\cos\dfrac{\theta}{2}\cdot s\sin\dfrac{\theta}{2} = \dfrac{1}{2}s^2\sin\theta$

71.

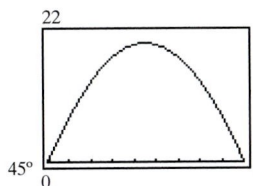

73. $\sin\dfrac{\pi}{24} = \dfrac{\sqrt{2}}{4}\sqrt{4 - \sqrt{6} - \sqrt{2}}$; $\cos\dfrac{\pi}{24} = \dfrac{\sqrt{2}}{4}\sqrt{4 + \sqrt{6} + \sqrt{2}}$

75. $\sin^3\theta + \sin^3(\theta + 120°) + \sin^3(\theta + 240°) = \sin^3\theta + (\sin\theta\cos 120° + \cos\theta\sin 120°)^3 + (\sin\theta\cos 240° + \cos\theta\sin 240°)^3$

$\displaystyle = \sin^3\theta + \left(-\frac{1}{2}\sin\theta + \frac{\sqrt{3}}{2}\cos\theta\right)^3 + \left(-\frac{1}{2}\sin\theta - \frac{\sqrt{3}}{2}\cos\theta\right)^3$

$\displaystyle = \sin^3\theta + \frac{1}{8}(3\sqrt{3}\cos^3\theta - 9\cos^2\theta\sin\theta + 3\sqrt{3}\cos\theta\sin^2\theta - \sin^3\theta) - \frac{1}{8}(\sin^3\theta + 3\sqrt{3}\sin^2\theta\cos\theta + 9\sin\theta\cos^2\theta + 3\sqrt{3}\cos^3\theta)$

$\displaystyle = \frac{3}{4}\sin^3\theta - \frac{9}{4}\cos^2\theta\sin\theta = \frac{3}{4}[\sin^3\theta - 3\sin\theta(1 - \sin^2\theta)] = \frac{3}{4}(4\sin^3\theta - 3\sin\theta) = -\frac{3}{4}\sin(3\theta)$ (from Example 2b)

77. (a) $R = \dfrac{v_0^2\sqrt{2}}{16}(\sin\theta\cos\theta - \cos^2\theta)$

$\displaystyle = \frac{v_0^2\sqrt{2}}{16}\left(\frac{1}{2}\sin(2\theta) - \frac{1 + \cos(2\theta)}{2}\right)$

$\displaystyle = \frac{v_0^2\sqrt{2}}{32}[\sin(2\theta) - \cos(2\theta) - 1]$

(b)

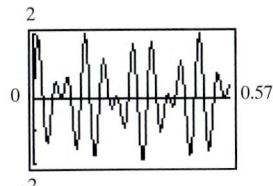

(c) $\theta = 67.5°$ makes R largest

9.6 Exercises *(page 757)*

1. $\dfrac{1}{2}[\cos(2\theta) - \cos(6\theta)]$ **3.** $\dfrac{1}{2}[\sin(6\theta) + \sin(2\theta)]$ **5.** $\dfrac{1}{2}[\cos(2\theta) + \cos(8\theta)]$ **7.** $\dfrac{1}{2}[\cos\theta - \cos(3\theta)]$ **9.** $\dfrac{1}{2}[\sin(2\theta) + \sin\theta]$

11. $2\sin\theta\cos(3\theta)$ **13.** $2\cos(3\theta)\cos\theta$ **15.** $2\sin(2\theta)\cos\theta$ **17.** $2\sin\theta\sin\dfrac{\theta}{2}$ **19.** $\dfrac{\sin\theta + \sin(3\theta)}{2\sin(2\theta)} = \dfrac{2\sin(2\theta)\cos\theta}{2\sin(2\theta)} = \cos\theta$

21. $\dfrac{\sin(4\theta) + \sin(2\theta)}{\cos(4\theta) + \cos(2\theta)} = \dfrac{2\sin(3\theta)\cos\theta}{2\cos(3\theta)\cos\theta} = \dfrac{\sin(3\theta)}{\cos(3\theta)} = \tan(3\theta)$ **23.** $\dfrac{\cos\theta - \cos(3\theta)}{\sin\theta + \sin(3\theta)} = \dfrac{2\sin(2\theta)\sin\theta}{2\sin(2\theta)\cos\theta} = \dfrac{\sin\theta}{\cos\theta} = \tan\theta$

25. $\sin\theta[\sin\theta + \sin(3\theta)] = \sin\theta[2\sin(2\theta)\cos\theta] = \cos\theta[2\sin(2\theta)\sin\theta] = \cos\theta\left[2\cdot\dfrac{1}{2}[\cos\theta - \cos(3\theta)]\right] = \cos\theta[\cos\theta - \cos(3\theta)]$

27. $\dfrac{\sin(4\theta) + \sin(8\theta)}{\cos(4\theta) + \cos(8\theta)} = \dfrac{2\sin(6\theta)\cos(2\theta)}{2\cos(6\theta)\cos(2\theta)} = \dfrac{\sin(6\theta)}{\cos(6\theta)} = \tan(6\theta)$

29. $\dfrac{\sin(4\theta) + \sin(8\theta)}{\sin(4\theta) - \sin(8\theta)} = \dfrac{2\sin(6\theta)\cos(-2\theta)}{2\sin(-2\theta)\cos(6\theta)} = \dfrac{\sin(6\theta)}{\cos(6\theta)}\cdot\dfrac{\cos(2\theta)}{-\sin(2\theta)} = \tan(6\theta)[-\cot(2\theta)] = -\dfrac{\tan(6\theta)}{\tan(2\theta)}$

31. $\dfrac{\sin\alpha + \sin\beta}{\sin\alpha - \sin\beta} = \dfrac{2\sin\dfrac{\alpha+\beta}{2}\cos\dfrac{\alpha-\beta}{2}}{2\sin\dfrac{\alpha-\beta}{2}\cos\dfrac{\alpha+\beta}{2}} = \dfrac{\sin\dfrac{\alpha+\beta}{2}}{\cos\dfrac{\alpha+\beta}{2}}\cdot\dfrac{\cos\dfrac{\alpha-\beta}{2}}{\sin\dfrac{\alpha-\beta}{2}} = \tan\dfrac{\alpha+\beta}{2}\cot\dfrac{\alpha-\beta}{2}$

33. $\dfrac{\sin\alpha + \sin\beta}{\cos\alpha + \cos\beta} = \dfrac{2\sin\dfrac{\alpha+\beta}{2}\cos\dfrac{\alpha-\beta}{2}}{2\cos\dfrac{\alpha+\beta}{2}\cos\dfrac{\alpha-\beta}{2}} = \dfrac{\sin\dfrac{\alpha+\beta}{2}}{\cos\dfrac{\alpha+\beta}{2}} = \tan\dfrac{\alpha+\beta}{2}$

35. $1 + \cos(2\theta) + \cos(4\theta) + \cos(6\theta) = [1 + \cos(6\theta)] + [\cos(2\theta) + \cos(4\theta)] = 2\cos^2(3\theta) + 2\cos(3\theta)\cos(-\theta)$

$= 2\cos(3\theta)[\cos(3\theta) + \cos\theta] = 2\cos(3\theta)[2\cos(2\theta)\cos\theta] = 4\cos\theta\cos(2\theta)\cos(3\theta)$

37. (a) $y = 2\sin(2061\pi t)\cos(357\pi t)$ **(b)** $y_{\max} = 2$ **(c)**

39. $\sin(2\alpha) + \sin(2\beta) + \sin(2\gamma) = 2\sin(\alpha + \beta)\cos(\alpha - \beta) + \sin(2\gamma) = 2\sin(\alpha + \beta)\cos(\alpha - \beta) + 2\sin\gamma\cos\gamma$

$= 2\sin(\pi - \gamma)\cos(\alpha - \beta) + 2\sin\gamma\cos\gamma = 2\sin\gamma\cos(\alpha - \beta) + 2\sin\gamma\cos\gamma = 2\sin\gamma[\cos(\alpha - \beta) + \cos\gamma]$

$= 2\sin\gamma\left(2\cos\dfrac{\alpha - \beta + \gamma}{2}\cos\dfrac{\alpha - \beta - \gamma}{2}\right) = 4\sin\gamma\cos\dfrac{\pi - 2\beta}{2}\cos\dfrac{2\alpha - \pi}{2} = 4\sin\gamma\cos\left(\dfrac{\pi}{2} - \beta\right)\cos\left(\alpha - \dfrac{\pi}{2}\right)$

$= 4\sin\gamma\sin\beta\sin\alpha$

41. $\qquad\qquad \sin(\alpha - \beta) = \sin\alpha\cos\beta - \cos\alpha\sin\beta$

$\qquad\qquad \sin(\alpha + \beta) = \sin\alpha\cos\beta + \cos\alpha\sin\beta$

$\sin(\alpha - \beta) + \sin(\alpha + \beta) = 2\sin\alpha\cos\beta$

$\qquad\qquad \sin\alpha\cos\beta = \dfrac{1}{2}[\sin(\alpha + \beta) + \sin(\alpha - \beta)]$

43. $2\cos\dfrac{\alpha + \beta}{2}\cos\dfrac{\alpha - \beta}{2} = 2\cdot\dfrac{1}{2}\left[\cos\left(\dfrac{\alpha + \beta}{2} + \dfrac{\alpha - \beta}{2}\right) + \cos\left(\dfrac{\alpha + \beta}{2} - \dfrac{\alpha - \beta}{2}\right)\right] = \cos\dfrac{2\alpha}{2} + \cos\dfrac{2\beta}{2} = \cos\alpha + \cos\beta$

9.7 Concepts and Vocabulary *(page 763)*

1. $\left\{\dfrac{\pi}{6}, \dfrac{5\pi}{6}\right\}$ **2.** $\left\{\theta \,\middle|\, \theta = \dfrac{\pi}{6} + 2\pi k, \theta = \dfrac{5\pi}{6} + 2\pi k,\, k \text{ any integer}\right\}$ **3.** F   **4.** T   **5.** F

9.7 Exercises *(page 763)*

1. $\left\{\theta \,\middle|\, \theta = \dfrac{\pi}{6} + 2k\pi, \theta = \dfrac{5\pi}{6} + 2k\pi\right\}; \dfrac{\pi}{6}, \dfrac{5\pi}{6}, \dfrac{13\pi}{6}, \dfrac{17\pi}{6}, \dfrac{25\pi}{6}, \dfrac{29\pi}{6}$ **3.** $\left\{\theta \,\middle|\, \theta = \dfrac{5\pi}{6} + k\pi\right\}; \dfrac{5\pi}{6}, \dfrac{11\pi}{6}, \dfrac{17\pi}{6}, \dfrac{23\pi}{6}, \dfrac{29\pi}{6}, \dfrac{35\pi}{6}$

5. $\left\{\theta \,\middle|\, \theta = \dfrac{\pi}{2} + 2k\pi, \theta = \dfrac{3\pi}{2} + 2k\pi\right\}; \dfrac{\pi}{2}, \dfrac{3\pi}{2}, \dfrac{5\pi}{2}, \dfrac{7\pi}{2}, \dfrac{9\pi}{2}, \dfrac{11\pi}{2}$ **7.** $\left\{\theta \,\middle|\, \theta = \dfrac{\pi}{3} + k\pi, \theta = \dfrac{2\pi}{3} + k\pi\right\}; \dfrac{\pi}{3}, \dfrac{2\pi}{3}, \dfrac{4\pi}{3}, \dfrac{5\pi}{3}, \dfrac{7\pi}{3}, \dfrac{8\pi}{3}$

9. $\left\{\theta \,\middle|\, \theta = \dfrac{8\pi}{3} + 4k\pi, \theta = \dfrac{10\pi}{3} + 4k\pi\right\}; \dfrac{8\pi}{3}, \dfrac{10\pi}{3}, \dfrac{20\pi}{3}, \dfrac{22\pi}{3}, \dfrac{32\pi}{3}, \dfrac{34\pi}{3}$ **11.** $\left\{\dfrac{7\pi}{6}, \dfrac{11\pi}{6}\right\}$ **13.** $\left\{\dfrac{\pi}{3}, \dfrac{2\pi}{3}, \dfrac{4\pi}{3}, \dfrac{5\pi}{3}\right\}$ **15.** $\left\{\dfrac{\pi}{4}, \dfrac{3\pi}{4}, \dfrac{5\pi}{4}, \dfrac{7\pi}{4}\right\}$

17. $\left\{\dfrac{\pi}{2}, \dfrac{7\pi}{6}, \dfrac{11\pi}{6}\right\}$ **19.** $\left\{\dfrac{\pi}{3}, \dfrac{2\pi}{3}, \dfrac{4\pi}{3}, \dfrac{5\pi}{3}\right\}$ **21.** $\left\{\dfrac{4\pi}{9}, \dfrac{8\pi}{9}, \dfrac{16\pi}{9}\right\}$ **23.** $\left\{\dfrac{3\pi}{4}, \dfrac{7\pi}{4}\right\}$ **25.** $\left\{\dfrac{11\pi}{6}\right\}$ **27.** $\left\{\dfrac{7\pi}{6}, \dfrac{11\pi}{6}\right\}$ **29.** $\left\{\dfrac{3\pi}{4}, \dfrac{7\pi}{4}\right\}$

31. $\left\{\dfrac{2\pi}{3}, \dfrac{4\pi}{3}\right\}$ **33.** $\left\{\dfrac{3\pi}{4}, \dfrac{5\pi}{4}\right\}$ **35.** $\{0.41, 2.73\}$ **37.** $\{1.37, 4.51\}$ **39.** $\{2.69, 3.59\}$ **41.** $\{1.82, 4.46\}$ **43.** $28.90°$

45. Yes; it varies from 1.28 to 1.34   **47.** 1.47

49. If θ is the original angle of incidence and ϕ is the angle of refraction, then $\dfrac{\sin\theta}{\sin\phi} = n_2$. The angle of incidence of the emerging

beam is also ϕ, and the index of refraction is $\dfrac{1}{n_2}$. Thus, θ is the angle of refraction of the emerging beam.

9.8 Exercises *(page 770)*

1. $\left\{\dfrac{\pi}{2}, \dfrac{2\pi}{3}, \dfrac{4\pi}{3}, \dfrac{3\pi}{2}\right\}$ **3.** $\left\{\dfrac{\pi}{2}, \dfrac{7\pi}{6}, \dfrac{11\pi}{6}\right\}$ **5.** $\left\{0, \dfrac{\pi}{4}, \dfrac{5\pi}{4}\right\}$ **7.** $\left\{\dfrac{\pi}{2}, \dfrac{2\pi}{3}, \dfrac{4\pi}{3}, \dfrac{3\pi}{2}\right\}$ **9.** $\{\pi\}$ **11.** $\left\{\dfrac{\pi}{3}, \dfrac{2\pi}{3}, \dfrac{4\pi}{3}, \dfrac{5\pi}{3}\right\}$ **13.** $\left\{\dfrac{\pi}{4}, \dfrac{5\pi}{4}\right\}$

15. $\left\{0, \dfrac{\pi}{3}, \pi, \dfrac{5\pi}{3}\right\}$ **17.** $\left\{\dfrac{\pi}{2}, \dfrac{3\pi}{2}\right\}$ **19.** $\left\{0, \dfrac{2\pi}{3}, \dfrac{4\pi}{3}\right\}$ **21.** $\left\{0, \dfrac{\pi}{3}, \dfrac{\pi}{2}, \dfrac{2\pi}{3}, \pi, \dfrac{4\pi}{3}, \dfrac{3\pi}{2}, \dfrac{5\pi}{3}\right\}$ **23.** $\left\{0, \dfrac{\pi}{5}, \dfrac{2\pi}{5}, \dfrac{3\pi}{5}, \dfrac{4\pi}{5}, \pi, \dfrac{6\pi}{5}, \dfrac{7\pi}{5}, \dfrac{8\pi}{5}, \dfrac{9\pi}{5}\right\}$

25. $\left\{\dfrac{\pi}{6}, \dfrac{5\pi}{6}, \dfrac{3\pi}{2}\right\}$ **27.** $\left\{\dfrac{\pi}{3}, \dfrac{5\pi}{3}\right\}$ **29.** No real solutions   **31.** No real solutions   **33.** $\left\{\dfrac{\pi}{2}, \dfrac{7\pi}{6}\right\}$ **35.** $\left\{0, \dfrac{\pi}{3}, \pi, \dfrac{5\pi}{3}\right\}$ **37.** $\left\{\dfrac{\pi}{4}\right\}$

39. $\{-1.29, 0\}$ **41.** $\{-2.24, 0, 2.24\}$ **43.** $\{-0.82, 0.82\}$

45. $\{-1.31, 1.98, 3.84\}$ **47.** $\{0.52\}$ **49.** $\{1.26\}$ **51.** $\{-1.02, 1.02\}$ **53.** $\{0, 2.15\}$ **55.** $\{0.76, 1.35\}$

57. (a) $60°$ **(b)** $60°$ **(c)** $A(60°) = 12\sqrt{3} \approx 20.78 \text{ in}^2$ **(d)**   **59.** $2.03, 4.91$

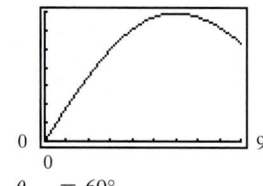

$\theta_{\max} = 60°$

Maximum Area $\approx 20.78 \text{ in}^2$

61. (a) $\approx 29.99°$ or $\approx 60.01°$ **(b)** ≈ 123.56 m   **(c)**

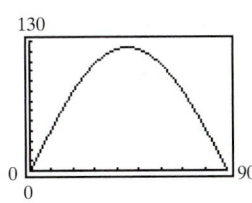

Review Exercises *(page 774)*

1. $\dfrac{\pi}{2}$ **3.** $\dfrac{\pi}{4}$ **5.** $\dfrac{5\pi}{6}$ **7.** $\dfrac{\pi}{4}$ **9.** $-\sqrt{3}$ **11.** $\dfrac{2\sqrt{3}}{3}$ **13.** $\dfrac{3}{5}$ **15.** $-\dfrac{4}{3}$ **17.** $-\dfrac{\pi}{6}$ **19.** $-\dfrac{\pi}{4}$ **21.** $\tan\theta\cot\theta - \sin^2\theta = 1 - \sin^2\theta = \cos^2\theta$

23. $\cos^2\theta(1 + \tan^2\theta) = \cos^2\theta\sec^2\theta = 1$ **25.** $4\cos^2\theta + 3\sin^2\theta = \cos^2\theta + 3(\cos^2\theta + \sin^2\theta) = 3 + \cos^2\theta$

27. $\dfrac{1 - \cos\theta}{\sin\theta} + \dfrac{\sin\theta}{1 - \cos\theta} = \dfrac{(1 - \cos\theta)^2 + \sin^2\theta}{\sin\theta(1 - \cos\theta)} = \dfrac{1 - 2\cos\theta + \cos^2\theta + \sin^2\theta}{\sin\theta(1 - \cos\theta)} = \dfrac{2(1 - \cos\theta)}{\sin\theta(1 - \cos\theta)} = 2\csc\theta$

29. $\dfrac{\cos\theta}{\cos\theta - \sin\theta} = \dfrac{\dfrac{\cos\theta}{\cos\theta}}{\dfrac{\cos\theta - \sin\theta}{\cos\theta}} = \dfrac{1}{1 - \dfrac{\sin\theta}{\cos\theta}} = \dfrac{1}{1 - \tan\theta}$

31. $\dfrac{\csc\theta}{1 + \csc\theta} = \dfrac{\dfrac{1}{\sin\theta}}{1 + \dfrac{1}{\sin\theta}} = \dfrac{1}{1 + \sin\theta} = \dfrac{1}{1 + \sin\theta}\cdot\dfrac{1 - \sin\theta}{1 - \sin\theta} = \dfrac{1 - \sin\theta}{1 - \sin^2\theta} = \dfrac{1 - \sin\theta}{\cos^2\theta}$

33. $\csc\theta - \sin\theta = \dfrac{1}{\sin\theta} - \sin\theta = \dfrac{1 - \sin^2\theta}{\sin\theta} = \dfrac{\cos^2\theta}{\sin\theta} = \cos\theta\cdot\dfrac{\cos\theta}{\sin\theta} = \cos\theta\cot\theta$

35. $\dfrac{1 - \sin\theta}{\sec\theta} = \cos\theta(1 - \sin\theta)\cdot\dfrac{1 + \sin\theta}{1 + \sin\theta} = \dfrac{\cos\theta(1 - \sin^2\theta)}{1 + \sin\theta} = \dfrac{\cos^3\theta}{1 + \sin\theta}$

37. $\cot\theta - \tan\theta = \dfrac{\cos\theta}{\sin\theta} - \dfrac{\sin\theta}{\cos\theta} = \dfrac{\cos^2\theta - \sin^2\theta}{\sin\theta\cos\theta} = \dfrac{1 - 2\sin^2\theta}{\sin\theta\cos\theta}$

39. $\dfrac{\cos(\alpha + \beta)}{\cos\alpha\sin\beta} = \dfrac{\cos\alpha\cos\beta - \sin\alpha\sin\beta}{\cos\alpha\sin\beta} = \dfrac{\cos\alpha\cos\beta}{\cos\alpha\sin\beta} - \dfrac{\sin\alpha\sin\beta}{\cos\alpha\sin\beta} = \cot\beta - \tan\alpha$

41. $\dfrac{\cos(\alpha - \beta)}{\cos\alpha\cos\beta} = \dfrac{\cos\alpha\cos\beta + \sin\alpha\sin\beta}{\cos\alpha\cos\beta} = \dfrac{\cos\alpha\cos\beta}{\cos\alpha\cos\beta} + \dfrac{\sin\alpha\sin\beta}{\cos\alpha\cos\beta} = 1 + \tan\alpha\tan\beta$

43. $(1 + \cos\theta)\left(\tan\dfrac{\theta}{2}\right) = \left(2\cos^2\dfrac{\theta}{2}\right)\dfrac{\sin\left(\dfrac{\theta}{2}\right)}{\cos\left(\dfrac{\theta}{2}\right)} = 2\sin\dfrac{\theta}{2}\cos\dfrac{\theta}{2} = \sin\theta$

45. $2\cot\theta\cot 2\theta = 2\left(\dfrac{\cos\theta}{\sin\theta}\right)\left(\dfrac{\cos 2\theta}{\sin 2\theta}\right) = \dfrac{2\cos\theta(\cos^2\theta - \sin^2\theta)}{2\sin^2\theta\cos\theta} = \dfrac{\cos^2\theta - \sin^2\theta}{\sin^2\theta} = \cot^2\theta - 1$

47. $1 - 8\sin^2\theta\cos^2\theta = 1 - 2(2\sin\theta\cos\theta)^2 = 1 - 2\sin^2(2\theta) = \cos(4\theta)$ **49.** $\dfrac{\sin(2\theta) + \sin(4\theta)}{\cos(2\theta) + \cos(4\theta)} = \dfrac{2\sin(3\theta)\cos(-\theta)}{2\cos(3\theta)\cos(-\theta)} = \tan(3\theta)$

51. $\dfrac{\cos(2\theta) - \cos(4\theta)}{\cos(2\theta) + \cos(4\theta)} - \tan\theta\tan(3\theta) = \dfrac{-2\sin(3\theta)\sin(-\theta)}{2\cos(3\theta)\cos(-\theta)} - \tan\theta\tan(3\theta) = \tan(3\theta)\tan\theta - \tan\theta\tan(3\theta) = 0$

53. $\dfrac{1}{4}(\sqrt{6} - \sqrt{2})$ **55.** $\dfrac{1}{4}(\sqrt{6} - \sqrt{2})$ **57.** $\dfrac{1}{2}$ **59.** $\sqrt{2} - 1$ **61. (a)** $-\dfrac{33}{65}$ **(b)** $-\dfrac{56}{65}$ **(c)** $-\dfrac{63}{65}$ **(d)** $\dfrac{33}{56}$ **(e)** $\dfrac{24}{25}$ **(f)** $\dfrac{119}{169}$ **(g)** $\dfrac{5\sqrt{26}}{26}$

(h) $\dfrac{2\sqrt{5}}{5}$ **63. (a)** $-\dfrac{16}{65}$ **(b)** $-\dfrac{63}{65}$ **(c)** $-\dfrac{56}{65}$ **(d)** $\dfrac{16}{63}$ **(e)** $\dfrac{24}{25}$ **(f)** $-\dfrac{119}{169}$ **(g)** $\dfrac{\sqrt{26}}{26}$ **(h)** $-\dfrac{\sqrt{10}}{10}$ **65. (a)** $-\dfrac{63}{65}$ **(b)** $\dfrac{16}{65}$ **(c)** $\dfrac{33}{65}$

(d) $-\dfrac{63}{16}$ **(e)** $\dfrac{24}{25}$ **(f)** $-\dfrac{119}{169}$ **(g)** $\dfrac{2\sqrt{13}}{13}$ **(h)** $-\dfrac{\sqrt{10}}{10}$ **67. (a)** $\dfrac{-\sqrt{3} - 2\sqrt{2}}{6}$ **(b)** $\dfrac{1 - 2\sqrt{6}}{6}$ **(c)** $\dfrac{-\sqrt{3} + 2\sqrt{2}}{6}$ **(d)** $\dfrac{8\sqrt{2} + 9\sqrt{3}}{23}$

(e) $-\dfrac{\sqrt{3}}{2}$ **(f)** $-\dfrac{7}{9}$ **(g)** $\dfrac{\sqrt{3}}{3}$ **(h)** $\dfrac{\sqrt{3}}{2}$ **69. (a)** 1 **(b)** 0 **(c)** $-\dfrac{1}{9}$ **(d)** Not defined   **(e)** $\dfrac{4\sqrt{5}}{9}$ **(f)** $-\dfrac{1}{9}$ **(g)** $\dfrac{\sqrt{30}}{6}$

(h) $-\dfrac{\sqrt{6}\sqrt{3 - \sqrt{5}}}{6}$ **71.** $\dfrac{4 + 3\sqrt{3}}{10}$ **73.** $-\dfrac{48 + 25\sqrt{3}}{39}$ **75.** $-\dfrac{24}{25}$ **77.** $\left\{\dfrac{\pi}{3}, \dfrac{5\pi}{3}\right\}$ **79.** $\left\{\dfrac{3\pi}{4}, \dfrac{5\pi}{4}\right\}$ **81.** $\left\{\dfrac{3\pi}{4}, \dfrac{7\pi}{4}\right\}$ **83.** $\left\{0, \dfrac{\pi}{2}, \pi, \dfrac{3\pi}{2}\right\}$

85. $\left\{\dfrac{\pi}{3}, \dfrac{2\pi}{3}, \dfrac{4\pi}{3}, \dfrac{5\pi}{3}\right\}$ **87.** $\{0, \pi\}$ **89.** $\left\{0, \dfrac{2\pi}{3}, \pi, \dfrac{4\pi}{3}\right\}$ **91.** $\left\{0, \dfrac{\pi}{6}, \dfrac{5\pi}{6}\right\}$ **93.** $\left\{\dfrac{\pi}{6}, \dfrac{\pi}{2}, \dfrac{5\pi}{6}\right\}$ **95.** $\left\{\dfrac{\pi}{3}, \dfrac{5\pi}{3}\right\}$ **97.** $\left\{\dfrac{\pi}{4}, \dfrac{\pi}{2}, \dfrac{3\pi}{4}, \dfrac{3\pi}{2}\right\}$

99. $\left\{\dfrac{\pi}{2}, \pi\right\}$ **101.** 0.78 **103.** -1.11 **105.** 1.23 **107.** $\{1.11\}$ **109.** $\{0.87\}$ **111.** $\{2.22\}$

Cumulative Review Exercises *(page 777)*

1. $\left\{\dfrac{-1 - \sqrt{13}}{6}, \dfrac{-1 + \sqrt{13}}{6}\right\}$ **2.** $y + 1 = -1(x - 4)$ or $x + y = 3; 6\sqrt{2}; (1, 2)$ **3.** *x*-axis symmetry; $(0, -3), (0, 3), (3, 0)$

4. 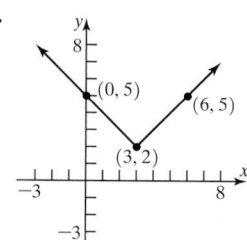   **5.** 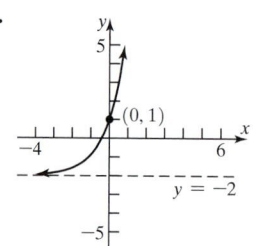   **6.**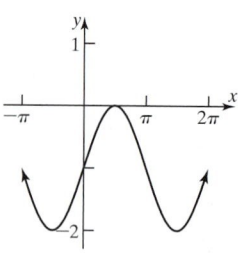

7. (a) 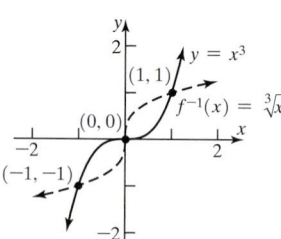   **(b)** 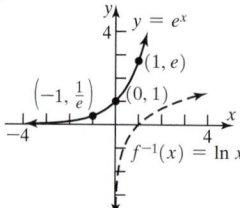   **(c)** 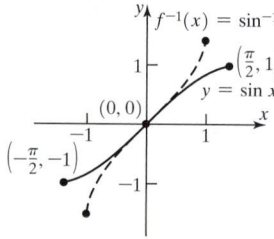   **(d)**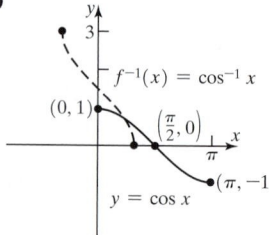

8. (a) $-\dfrac{2\sqrt{2}}{3}$ **(b)** $\dfrac{\sqrt{2}}{4}$ **(c)** $\dfrac{4\sqrt{2}}{9}$ **(d)** $\dfrac{7}{9}$ **(e)** $\sqrt{\dfrac{3 + 2\sqrt{2}}{6}}$ **(f)** $\sqrt{\dfrac{3 - 2\sqrt{2}}{6}}$ **9.** $\dfrac{\sqrt{5}}{5}$

10. (a) $-\dfrac{2\sqrt{2}}{3}$ **(b)** $-\dfrac{2\sqrt{2}}{3}$ **(c)** $\dfrac{7}{9}$ **(d)** $\dfrac{4\sqrt{2}}{9}$ **(e)** $\dfrac{\sqrt{6}}{3}$ **11.** 7.28

12. (a) $f(x) = (2x - 1)(x - 1)^2(x + 1)^2; \dfrac{1}{2}$ multiplicity one, 1 and -1 multiplicity two

(b) $(0, -1); \left(\dfrac{1}{2}, 0\right); (-1, 0); (1, 0)$ **(c)** $y = 2x^5$

(d) 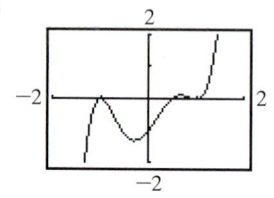   **(e)** Minimum $(-0.29, -1.33)$; maxima: $(-1, 0), (0.69, 0.10)$

(f)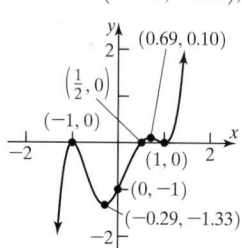

C H A P T E R   1 0   Applications of Trigonometric Functions

10.1 Concepts and Vocabulary *(page 785)*

1. Angle of elevation   **2.** Angle of depression   **3.** T   **4.** F

10.1 Exercises *(page 785)*

1. $a \approx 13.74, c \approx 14.62, \alpha = 70°$ **3.** $b \approx 5.03, c \approx 7.83, \alpha = 50°$ **5.** $a \approx 0.71, c \approx 4.06, \beta = 80°$ **7.** $b \approx 10.72, c \approx 11.83, \beta = 65°$
9. $b \approx 3.08, a \approx 8.46, \alpha = 70°$ **11.** $c \approx 5.83, \alpha \approx 59.0°, \beta = 31.0°$ **13.** $b \approx 4.58, \alpha \approx 23.6°, \beta = 66.4°$ **15.** 4.59 in., 6.55 in.
17. 5.52 in. or 11.83 in.   **19.** 23.6° and 66.4°   **21.** 70.02 ft   **23.** 985.91 ft   **25.** 137.37 m   **27.** 20.67 ft   **29.** 15.9°   **31.** 60.27 ft
33. 530.18 ft   **35.** 554.52 ft   **37. (a)** 111.96 ft/sec or 76.3 mi/hr   **(b)** 82.42 ft/sec or 56.2 mi/hr   **(c)** 18.8° or less   **39.** S76.6°E
41. 14.9°   **43.** 3.83 mi   **45.** No; Move the tripod back about 1 ft.

47. (a) $A(\theta) = 2 \sin \theta \cos \theta$ **(b)** From double-angle formula, since $2 \sin \theta \cos \theta = \sin(2\theta)$ **(c)** $\theta = 45°$ **(d)** $\dfrac{\sqrt{2}}{2}$ by $\sqrt{2}$

10.2 Concepts and Vocabulary *(page 796)*

1. Oblique   **2.** $\dfrac{\sin \alpha}{a} = \dfrac{\sin \beta}{b} = \dfrac{\sin \gamma}{c}$ **3.** F   **4.** T   **5.** F

10.2 Exercises *(page 796)*

1. $a \approx 3.23, b \approx 3.55, \alpha = 40°$ **3.** $a \approx 3.25, c \approx 4.23, \beta = 45°$ **5.** $\gamma = 95°, c \approx 9.86, a \approx 6.36$ **7.** $\alpha = 40°, a = 2, c \approx 3.06$
9. $\gamma = 120°, b \approx 1.06, c \approx 2.69$ **11.** $\alpha = 100°, a \approx 5.24, c \approx 0.92$ **13.** $\beta = 40°, a \approx 5.64, b \approx 3.86$ **15.** $\gamma = 100°, a \approx 1.31, b \approx 1.31$
17. One triangle; $\beta \approx 30.7°, \gamma \approx 99.3°, c \approx 3.86$ **19.** One triangle; $\gamma \approx 36.2°, \alpha \approx 43.8°, a \approx 3.51$ **21.** No triangle
23. Two triangles; $\gamma_1 \approx 30.9°, \alpha_1 \approx 129.1°, a_1 \approx 9.07$ or $\gamma_2 \approx 149.1°, \alpha_2 \approx 10.9°, a_2 \approx 2.20$ **25.** No triangle   **27.** Two triangles; $a_1 \approx 57.7°$,
$\beta_1 \approx 97.3°, b_1 \approx 2.35$ or $\alpha_2 \approx 122.3°, \beta_2 \approx 32.7°, b_2 \approx 1.28$ **29. (a)** Station Able is about 143.33 mi from the ship; Station Baker is about
135.58 mi from the ship.   **(b)** Approximately 41 min   **31.** 1490.48 ft   **33.** 381.69 ft   **35. (a)** 169.18 mi   **(b)** 161.3°   **37.** 84.7°; 183.72 ft
39. 2.64 mi   **41.** 1.88 mi or 1.53 mi   **43.** 449.36 ft   **45.** 39.39 ft   **47.** 29.97 ft

49. $\dfrac{a - b}{c} = \dfrac{a}{c} - \dfrac{b}{c} = \dfrac{\sin \alpha}{\sin \gamma} - \dfrac{\sin \beta}{\sin \gamma} = \dfrac{\sin \alpha - \sin \beta}{\sin \gamma} = \dfrac{2 \sin\left(\dfrac{\alpha - \beta}{2}\right) \cos\left(\dfrac{\alpha + \beta}{2}\right)}{2 \sin \dfrac{\gamma}{2} \cos \dfrac{\gamma}{2}} = \dfrac{\sin\left(\dfrac{\alpha - \beta}{2}\right) \cos\left(\dfrac{\pi}{2} - \dfrac{\gamma}{2}\right)}{\sin \dfrac{\gamma}{2} \cos \dfrac{\gamma}{2}} = \dfrac{\sin\left(\dfrac{\alpha - \beta}{2}\right)}{\cos \dfrac{\gamma}{2}}$

51. $\dfrac{a - b}{a + b} = \dfrac{\dfrac{a - b}{c}}{\dfrac{a + b}{c}} = \dfrac{\dfrac{\sin\left[\dfrac{1}{2}(\alpha - \beta)\right]}{\cos \dfrac{\gamma}{2}}}{\dfrac{\cos\left[\dfrac{1}{2}(\alpha - \beta)\right]}{\sin \dfrac{\gamma}{2}}} = \dfrac{\tan\left[\dfrac{1}{2}(\alpha - \beta)\right]}{\cot \dfrac{\gamma}{2}} = \dfrac{\tan\left[\dfrac{1}{2}(\alpha - \beta)\right]}{\tan\left(\dfrac{\pi}{2} - \dfrac{\gamma}{2}\right)} = \dfrac{\tan\left[\dfrac{1}{2}(\alpha - \beta)\right]}{\tan\left[\dfrac{1}{2}(\alpha + \beta)\right]}$

10.3 Concepts and Vocabulary *(page 803)*

1. Cosines   **2.** Sines   **3.** Cosines   **4.** F   **5.** F   **6.** T

10.3 Exercises *(page 804)*

1. $b \approx 2.95, \alpha \approx 28.7°, \gamma \approx 106.3°$ **3.** $c \approx 3.75, \alpha \approx 32.1°, \beta \approx 52.9°$ **5.** $\alpha \approx 48.5°, \beta \approx 38.6°, \gamma \approx 92.9°$
7. $\alpha \approx 127.2°, \beta \approx 32.1°, \gamma \approx 20.7°$ **9.** $c \approx 2.57, \alpha \approx 48.6°, \beta \approx 91.4°$ **11.** $a \approx 2.99, \beta \approx 19.2°, \gamma \approx 80.8°$
13. $b \approx 4.14, \alpha \approx 43.0°, \gamma \approx 27.0°$ **15.** $c \approx 1.69, \alpha \approx 65.0°, \beta \approx 65.0°$ **17.** $\alpha \approx 67.4°, \beta \approx 90°, \gamma \approx 22.6°$
19. $\alpha = 60°, \beta = 60°, \gamma = 60°$ **21.** $\alpha \approx 33.6°, \beta \approx 62.2°, \gamma \approx 84.3°$ **23.** $\alpha \approx 97.9°, \beta \approx 52.4°, \gamma \approx 29.7°$ **25.** 70.75 ft
27. (a) 12.0°   **(b)** 220.8 mph   **29. (a)** 63.7 ft   **(b)** 66.8 ft   **(c)** 92.8°   **31. (a)** 492.6 ft   **(b)** 269.3 ft   **33.** 342.33 ft

35. Using the Law of Cosines:
$L^2 = x^2 + r^2 - 2rx \cos \theta$
$x^2 - 2rx \cos \theta + r^2 - L^2 = 0$
Then, using the quadratic formula:
$x = r \cos \theta + \sqrt{r^2 \cos^2 \theta + L^2 - r^2}$

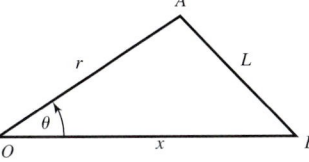

37. $\cos \dfrac{\gamma}{2} = \sqrt{\dfrac{1 + \cos \gamma}{2}} = \sqrt{\dfrac{1 + \dfrac{a^2 + b^2 - c^2}{2ab}}{2}} = \sqrt{\dfrac{2ab + a^2 + b^2 - c^2}{4ab}} = \sqrt{\dfrac{(a + b)^2 - c^2}{4ab}} = \sqrt{\dfrac{(a + b + c)(a + b - c)}{4ab}}$

$= \sqrt{\dfrac{2s(2s - 2c)}{4ab}} = \sqrt{\dfrac{s(s - c)}{ab}}$

39. $\dfrac{\cos \alpha}{a} + \dfrac{\cos \beta}{b} + \dfrac{\cos \gamma}{c} = \dfrac{b^2 + c^2 - a^2}{2abc} + \dfrac{a^2 + c^2 - b^2}{2abc} + \dfrac{a^2 + b^2 - c^2}{2abc} = \dfrac{b^2 + c^2 - a^2 + a^2 + c^2 - b^2 + a^2 + b^2 - c^2}{2abc}$

$= \dfrac{a^2 + b^2 + c^2}{2abc}$

10.4 Concepts and Vocabulary *(page 809)*

1. Heron's   **2.** F   **3.** T

10.4 Exercises *(page 810)*

1. 2.83   **3.** 2.99   **5.** 14.98   **7.** 9.56   **9.** 3.86   **11.** 1.48   **13.** 2.82   **15.** 1.53   **17.** 30   **19.** 1.73   **21.** 19.90   **23.** 19.81   **25.** 9.03 sq ft

27. \$5446.38   **29.** 9.26 sq cm   **31.** $A = \dfrac{1}{2} ab \sin \gamma = \dfrac{1}{2} a \sin \gamma \left(\dfrac{a \sin \beta}{\sin \alpha} \right) = \dfrac{a^2 \sin \beta \sin \gamma}{2 \sin \alpha}$ **33.** 0.92   **35.** 2.27   **37.** 5.44

39. $A = \dfrac{1}{2} r^2 (\theta + \sin \theta)$ **41.** 31,145.15 sq ft

43. $h_1 = 2\dfrac{K}{a}, h_2 = 2\dfrac{K}{b}, h_3 = 2\dfrac{K}{c}.$ Then $\dfrac{1}{h_1} + \dfrac{1}{h_2} + \dfrac{1}{h_3} = \dfrac{a}{2K} + \dfrac{b}{2K} + \dfrac{c}{2K} = \dfrac{a + b + c}{2K} = \dfrac{2s}{2K} = \dfrac{s}{K}.$

45. Angle AOB measures $180° - \left(\dfrac{\alpha}{2} + \dfrac{\beta}{2} \right) = 180° - \dfrac{1}{2}(180° - \gamma) = 90° + \dfrac{\gamma}{2},$ and $\sin\left(90° + \dfrac{\gamma}{2} \right) = \cos\left(-\dfrac{\gamma}{2} \right) = \cos \dfrac{\gamma}{2}$ since cosine is

an even function. Therefore, $r = \dfrac{c \sin \dfrac{\alpha}{2} \sin \dfrac{\beta}{2}}{\sin\left(90° + \dfrac{\gamma}{2} \right)} = \dfrac{c \sin \dfrac{\alpha}{2} \sin \dfrac{\beta}{2}}{\cos \dfrac{\gamma}{2}}.$

47. $\cot \dfrac{\alpha}{2} + \cot \dfrac{\beta}{2} + \cot \dfrac{\gamma}{2} = \dfrac{s - a}{r} + \dfrac{s - b}{r} + \dfrac{s - c}{r} = \dfrac{3s - (a + b + c)}{r} = \dfrac{3s - 2s}{r} = \dfrac{s}{r}$

10.5 Concepts and Vocabulary *(page 819)*

1. Simple harmonic; amplitude   **2.** Simple harmonic; damped   **3.** T

10.5 Exercises *(page 820)*

1. $d = -5 \cos(\pi t)$ **3.** $d = -6 \cos(2t)$ **5.** $d = -5 \sin(\pi t)$ **7.** $d = -6 \sin(2t)$ **9. (a)** Simple harmonic   **(b)** 5 m

(c) $\dfrac{2\pi}{3}$ sec   **(d)** $\dfrac{3}{2\pi}$ oscillation/sec   **11. (a)** Simple harmonic   **(b)** 6 m   **(c)** 2 sec   **(d)** $\dfrac{1}{2}$ oscillation/sec   **13. (a)** Simple harmonic

(b) 3 m   **(c)** 4π sec   **(d)** $\dfrac{1}{4\pi}$ oscillation/sec   **15. (a)** Simple harmonic   **(b)** 2 m   **(c)** 1 sec   **(d)** 1 oscillation/sec

17. (a) $d = -10e^{-0.7t/50} \cos\left(\sqrt{\dfrac{4\pi^2}{25} - \dfrac{0.49}{2500}}\, t \right)$ **19. (a)** $d = -18e^{-0.6t/60} \cos\left(\sqrt{\dfrac{\pi^2}{4} - \dfrac{0.36}{3600}}\, t \right)$ **21. (a)** $d = -5e^{-0.8t/20} \cos\left(\sqrt{\dfrac{4\pi^2}{9} - \dfrac{0.64}{400}} \right)$

(b) [graph: range 0 to 25, y-axis 10 to −10]   **(b)** [graph: range 0 to 20, y-axis 18 to −18]   **(b)** [graph: range 0 to 15, y-axis 5 to −5]

23. (a) The motion is damped.   The bob has mass $m = 20$ kg with a damping factor of 0.7 kg/sec.
(b) 20 m downward
(c) [graph: range 0 to 25, y-axis 20 to −20]
(d) 18.33 m   **(e)** $d \to 0$

25. (a) The motion is damped.   The bob has mass $m = 40$ kg with a damping factor of 0.6 kg/sec.
(b) 30 m downward
(c) [graph: range 0 to 35, y-axis 30 to −30]
(d) 28.47 m   **(e)** $d \to 0$

27. (a) The motion is damped.   The bob has mass $m = 15$ kg with a damping factor of 0.9 kg/sec.
(b) 15 m downward
(c) [graph: range 0 to 30, y-axis 15 to −15]
(d) 12.53 m   **(e)** $d \to 0$

29.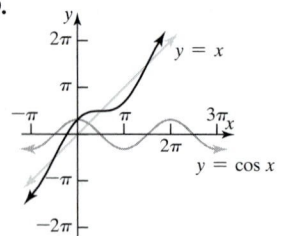

31.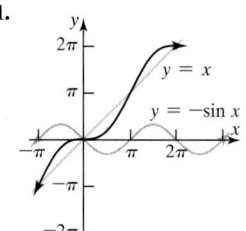

33.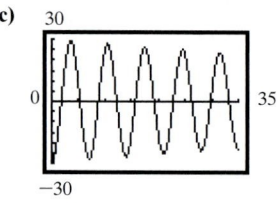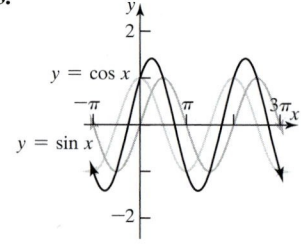

35.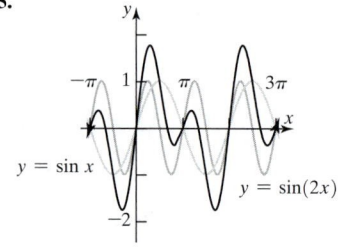

37. (a)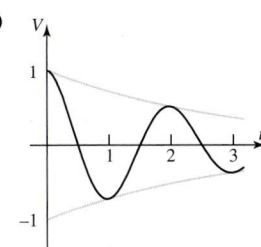

(b) At $t = 0, 2$; at $t = 1, t = 3$

(c) During the approximate intervals $0.35 < t < 0.67, 1.29 < t < 1.75,$ and $2.19 < t \leq 3$

39.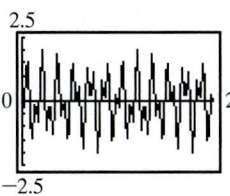

41.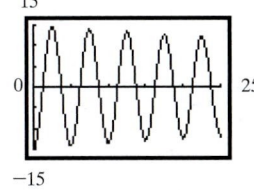

43. $y = \dfrac{1}{x} \sin x$

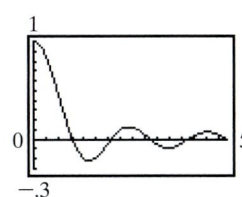

$y = \dfrac{1}{x^2} \sin x$

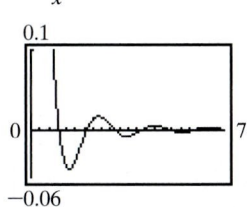

$y = \dfrac{1}{x^3} \sin x$

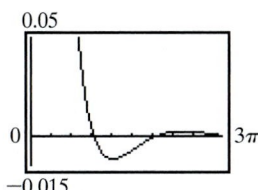

Review Exercises *(page 823)*

1. $\alpha = 70°, b \approx 3.42, a \approx 9.40$ **3.** $a \approx 4.58, \alpha = 66.4°, \beta \approx 23.6°$ **5.** $\gamma = 100°, b \approx 0.65, c \approx 1.29$ **7.** $\beta \approx 56.8°, \gamma \approx 23.2°, b \approx 4.25$
9. No triangle **11.** $b \approx 3.32, \alpha \approx 62.8°, \gamma \approx 17.2°$ **13.** No triangle **15.** $c \approx 2.32, \alpha \approx 16.1°, \beta \approx 123.9°$ **17.** $\beta = 36.2°, \gamma = 63.8°,$
$c = 4.55$ **19.** $\alpha = 39.6°, \beta = 18.6°, \gamma = 121.9°$ **21.** Two triangles: $\beta_1 \approx 13.4°, \gamma_1 \approx 156.6°, c_1 \approx 6.86$ or $\beta_2 \approx 166.6°, \gamma_2 \approx 3.4°, c_2 \approx 1.02$
23. $a = 5.23, \beta = 46.0°, \gamma = 64.0°$ **25.** 1.93 **27.** 18.79 **29.** 6 **31.** 3.80 **33.** 0.32 **35.** 839.10 ft **37.** 23.32 ft **39.** 2.15 mi
41. 204.07 mi **43. (a)** 2.59 mi **(b)** 2.92 mi **(c)** 2.53 mi **45. (a)** 131.78 mi **(b)** 23.1° **(c)** 0.21 hr **47.** 8798.67 sq ft

49. 1.92 sq in. **51.** 76.94 in. **53. (a)** Simple harmonic **(b)** 6 ft **(c)** π sec **(d)** $\dfrac{1}{\pi}$ oscillation/sec

55. (a) Simple harmonic **(b)** 2 ft **(c)** 2 sec **(d)** $\dfrac{1}{2}$ oscillation/sec **57.** $d = -4 \cos\left(\dfrac{2\pi}{3}t\right)$

59. (a) $d = -15e^{-0.75t/80} \cos\left(\sqrt{\dfrac{4\pi^2}{25} - \dfrac{0.5625}{6400}} \, t\right)$

(b)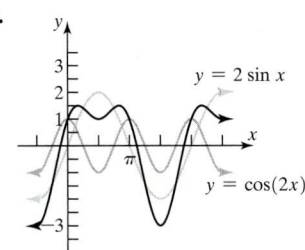

61. (a) The motion is damped. The bob has mass

$m = 20$ kg with a damping factor of 0.6 kg/sec.

(b) 15 m downward

(c)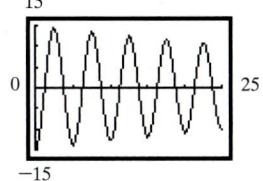

(d) 13.92 m **(e)** $d \to 0$

63.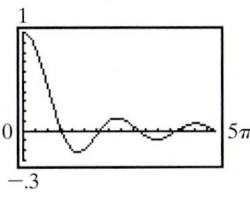

Cumulative Review Exercises *(page 828)*

1. $\left\{\dfrac{1}{3}, 1\right\}$ **2.** $(x + 5)^2 + (y - 1)^2 = 9$ **3.** $\{x \mid x \le -1 \text{ or } x \ge 4\}$ **4.**

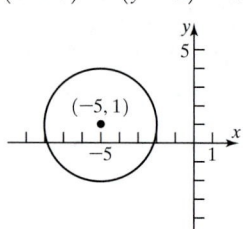   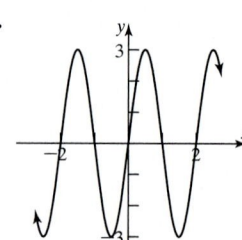   **5.**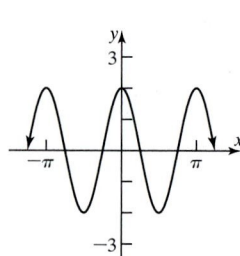

6. (a) $-\dfrac{2\sqrt{5}}{5}$ (b) $\dfrac{\sqrt{5}}{5}$ (c) $-\dfrac{4}{5}$ (d) $-\dfrac{3}{5}$ (e) $\sqrt{\dfrac{5 - \sqrt{5}}{10}}$ (f) $-\sqrt{\dfrac{5 + \sqrt{5}}{10}}$

7. (a) 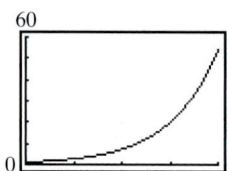   (b) 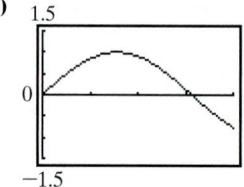   (c) 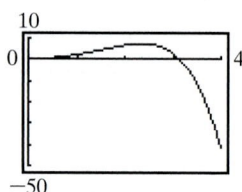   (d)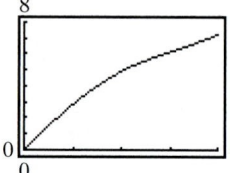

8. (a) 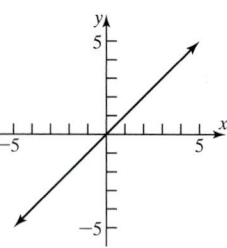   (b) 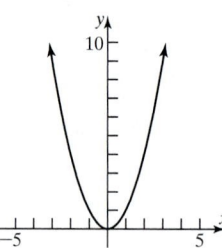   (c)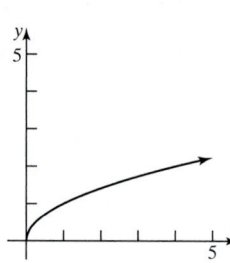

(d) 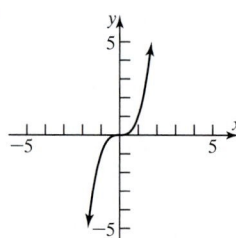   (e) 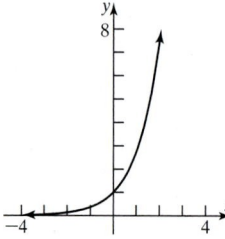   (f)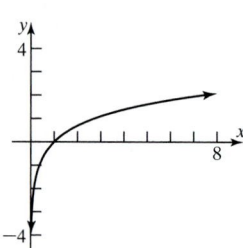

(g) 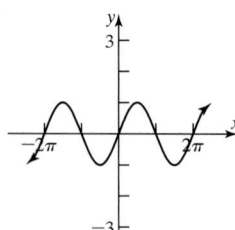   (h) 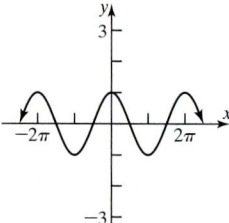   (i)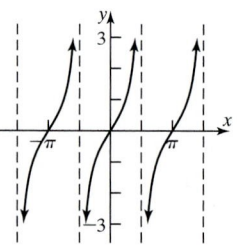

9. Two triangles: $\alpha_1 \approx 59.0°, \beta_1 = 81.0°, b_1 \approx 23.05$ or $\alpha_2 \approx 121.0°, \beta_2 \approx 19.0°, b_2 \approx 7.59$ **10.** $\left\{-2i, 2i, \dfrac{1}{3}, 1, 2\right\}$

11. $R(x) = \dfrac{(2x + 1)(x - 4)}{(x + 5)(x - 3)}$; domain: $\{x | x \neq -5, x \neq 3\}$

intercepts: $\left(-\dfrac{1}{2}, 0\right)$, $(4, 0)$, $\left(0, \dfrac{4}{15}\right)$

no symmetry
vertical asymptotes: $x = -5, x = 3$
horizontal asymptote: $y = 2$

intersects: $\left(\dfrac{26}{11}, 2\right)$

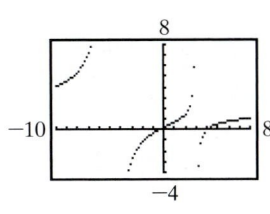
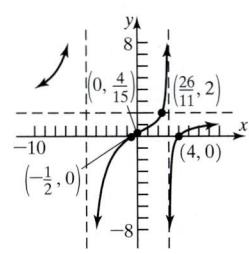

12. $\{2.26\}$ **13.** $\{1\}$

C H A P T E R   1 1   Polar Coordinates; Vectors

11.1 Concepts and Vocabulary *(page 837)*

1. pole; polar axis   **2.** -2 **3.** $(-\sqrt{3}, -1)$ **4.** F   **5.** T   **6.** T   **7.** $x = r \cos \theta; y = r \sin \theta$

11.1 Exercises *(page 838)*

1. A   **3.** C   **5.** B   **7.** A

9. 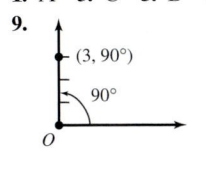   **11.**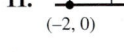

13. 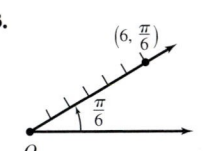   **15.** 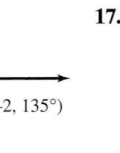   **17.**   **19.**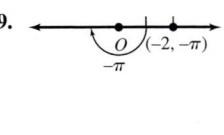

21. 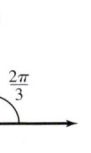   **23.** 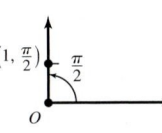   **25.** 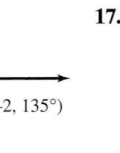   **27.**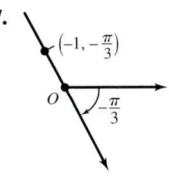

(a) $\left(5, -\dfrac{4\pi}{3}\right)$ **(a)** $(2, -2\pi)$ **(a)** $\left(1, -\dfrac{3\pi}{2}\right)$ **(a)** $\left(3, -\dfrac{5\pi}{4}\right)$

(b) $\left(-5, \dfrac{5\pi}{3}\right)$ **(b)** $(-2, \pi)$ **(b)** $\left(-1, \dfrac{3\pi}{2}\right)$ **(b)** $\left(-3, \dfrac{7\pi}{4}\right)$

(c) $\left(5, \dfrac{8\pi}{3}\right)$ **(c)** $(2, 2\pi)$ **(c)** $\left(1, \dfrac{5\pi}{2}\right)$ **(c)** $\left(3, \dfrac{11\pi}{4}\right)$

29. $(0, 3)$ **31.** $(-2, 0)$ **33.** $(-3\sqrt{3}, 3)$ **35.** $(\sqrt{2}, -\sqrt{2})$ **37.** $\left(-\dfrac{1}{2}, \dfrac{\sqrt{3}}{2}\right)$ **39.** $(2, 0)$ **41.** $(-2.57, 7.05)$ **43.** $(-4.98, -3.85)$ **45.** $(3, 0)$

47. $(1, \pi)$ **49.** $\left(\sqrt{2}, -\dfrac{\pi}{4}\right)$ **51.** $\left(2, \dfrac{\pi}{6}\right)$ **53.** $(2.47, -1.02)$ **55.** $(9.30, 0.47)$ **57.** $r^2 = \dfrac{3}{2}$ **59.** $r \cos^2 \theta - 4 \sin \theta = 0$ **61.** $r^2 \sin 2\theta = 1$

63. $r \cos \theta = 4$ **65.** $x^2 + y^2 - x = 0$ or $\left(x - \dfrac{1}{2}\right)^2 + y^2 = \dfrac{1}{4}$ **67.** $(x^2 + y^2)^{3/2} - x = 0$ **69.** $x^2 + y^2 = 4$ **71.** $y^2 = 8(x + 2)$

73. $d = \sqrt{(r_2 \cos \theta_2 - r_1 \cos \theta_1)^2 + (r_2 \sin \theta_2 - r_1 \sin \theta_1)^2}$
$= \sqrt{(r_2^2 \cos^2 \theta_2 - 2r_2 \cos \theta_2 r_1 \cos \theta_1 + r_1^2 \cos^2 \theta_1) + (r_2^2 \sin^2 \theta_2 - 2r_2 \sin \theta_2 r_1 \sin \theta_1 + r_1^2 \sin^2 \theta_1)}$
$= \sqrt{r_1^2 + r_2^2 - 2r_1 r_2 (\cos \theta_2 \cos \theta_1 + \sin \theta_2 \sin \theta_1)}$
$= \sqrt{r_1^2 + r_2^2 - 2r_1 r_2 \cos(\theta_2 - \theta_1)}$

11.2 Concepts and Vocabulary *(page 855)*

1. Polar equation   **2.** $r = 2 \cos \theta$ **3.** $-r$ **4.** F   **5.** F   **6.** F   **7.** Circle   **8.** Line   **9.** Heart shaped   **10.** Airplane propeller

11.2 Exercises *(page 855)*

1. $x^2 + y^2 = 16$;

Circle, radius 4, center at pole

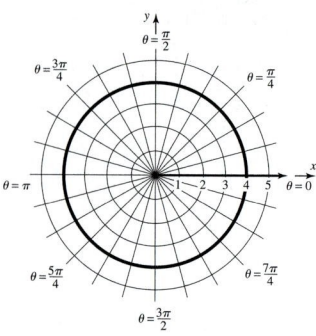

3. $y = \sqrt{3}x$; Line through pole, making an angle of $\dfrac{\pi}{3}$ with polar axis

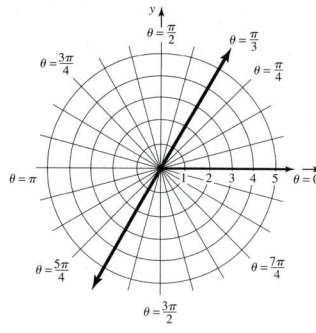

5. $y = 4$; Horizontal line 4 units above the pole

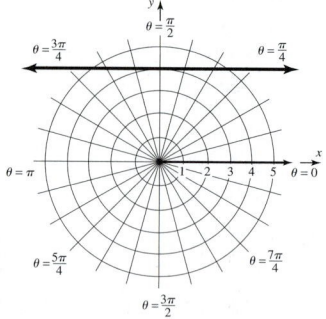

7. $x = -2$; Vertical line 2 units to the left of the pole

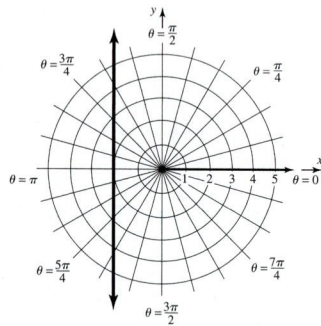

9. $(x - 1)^2 + y^2 = 1$; Circle, radius 1, center $(1, 0)$ in rectangular coordinates

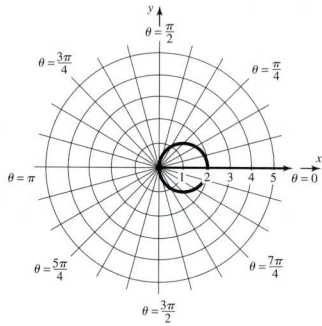

11. $x^2 + (y + 2)^2 = 4$; Circle, radius 2, center at $(0, -2)$ in rectangular coordinates

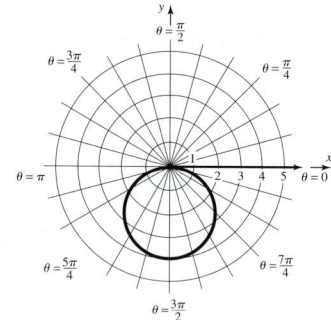

13. $(x - 2)^2 + y^2 = 4$; Circle, radius 2, center at $(2, 0)$ in rectangular coordinates

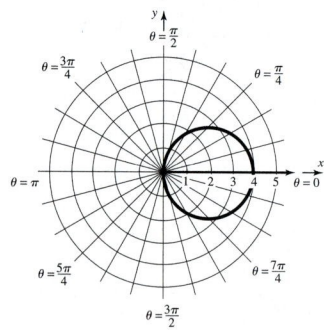

15. $x^2 + (y + 1)^2 = 1$; Circle, radius 1, center at $(0, -1)$ in rectangular coordinates

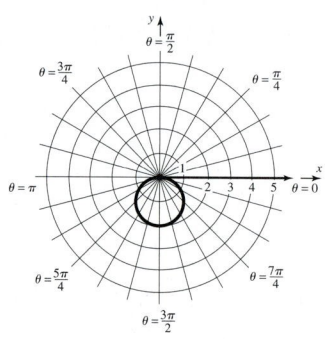

17. E   **19.** F   **21.** H   **23.** D
25. D   **27.** F   **29.** A
31. Cardioid

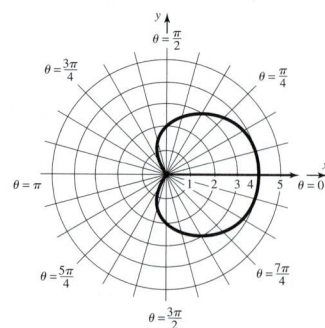

33. Cardioid

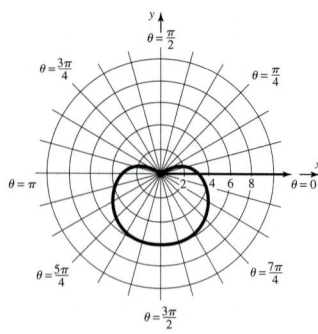

35. Limaçon without inner loop

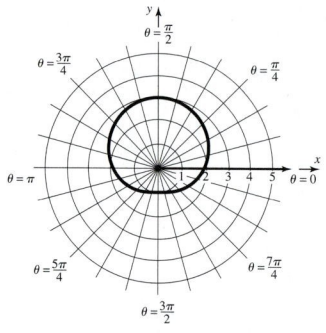

37. Limaçon without inner loop

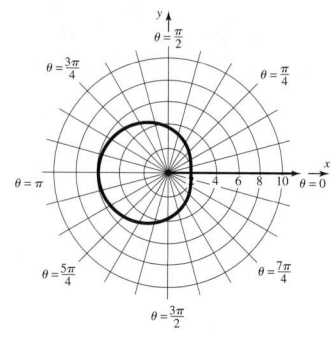

39. Limaçon with inner loop

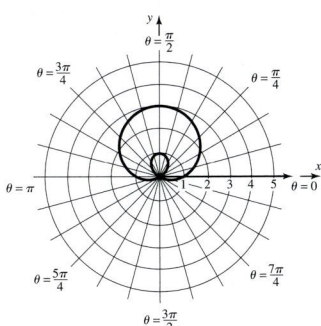

41. Limaçon with inner loop

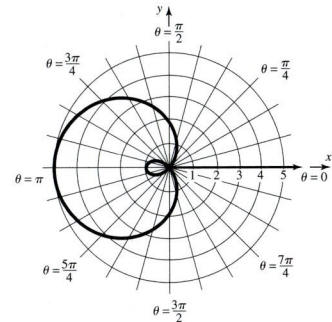

43. Rose

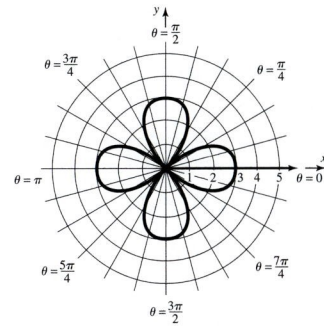

45. Rose

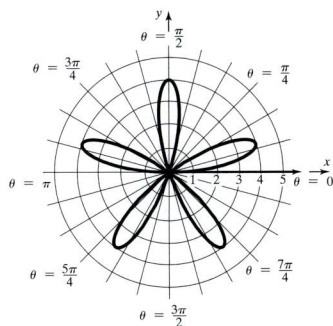

47. Lemniscate

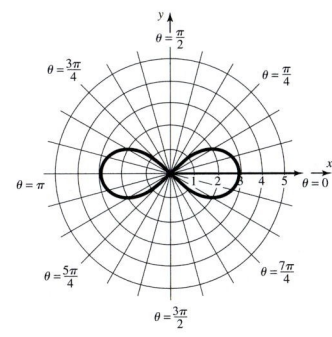

49. Spiral

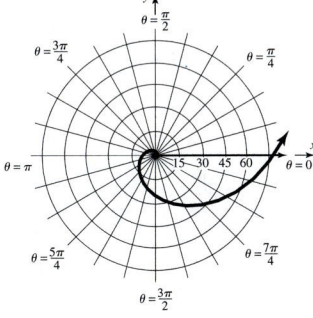

51. Cardioid

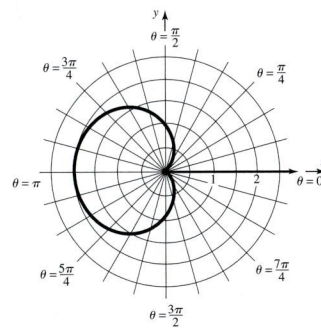

53. Limaçon with inner loop

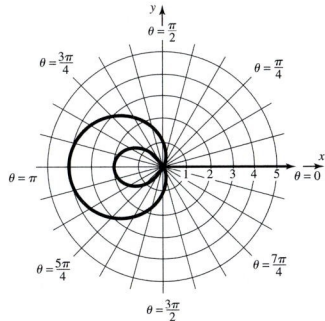

55.

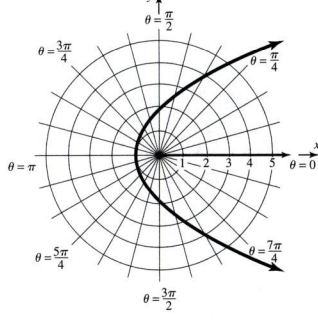

57.

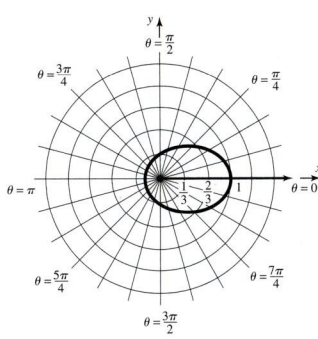

59.

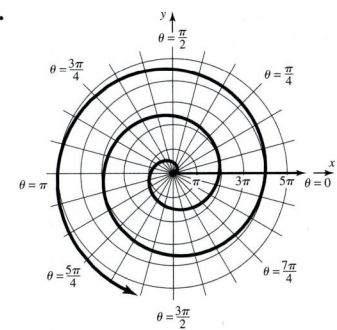

61.

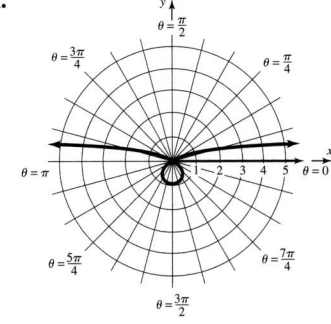

63.

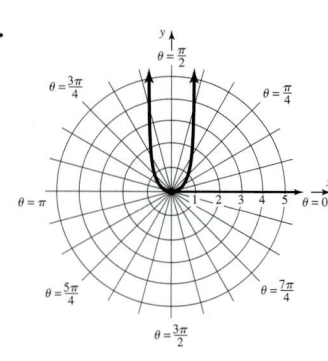

65. $r \sin \theta = a$
$y = a$

67. $r = 2a \sin \theta$
$r^2 = 2ar \sin \theta$
$x^2 + y^2 = 2ay$
$x^2 + y^2 - 2ay = 0$
$x^2 + (y - a)^2 = a^2$
Circle, radius a, center at $(0, a)$
in rectangular coordinates

69.
$r = 2a \cos \theta$
$r^2 = 2ar \cos \theta$
$x^2 + y^2 = 2ax$
$x^2 - 2ax + y^2 = 0$
$(x - a)^2 + y^2 = a^2$
Circle, radius a, center at $(a, 0)$
in rectangular coordinates

71. (a) $r^2 = \cos \theta$: $r^2 = \cos(\pi - \theta)$
$r^2 = -\cos \theta$
Not equivalent; test fails.
$(-r)^2 = \cos(-\theta)$
$r^2 = \cos \theta$
New test works.

(b) $r^2 = \sin \theta$: $r^2 = \sin(\pi - \theta)$
$r^2 = \sin \theta$
Test works.
$(-r)^2 = \sin(-\theta)$
$r^2 = -\sin \theta$
Not equivalent; new test fails.

Historical Problems *(page 864)*

1. (a) $1 + 4i, 1 + i$ **(b)** $-1, 2 + i$

11.3 Concepts and Vocabulary *(page 864)*

1. Magnitude; modulus; argument   **2.** DeMoivre's   **3.** Three   **4.** T   **5.** F   **6.** F   **8.** $z^2 = r^2[\cos(2\theta) + i \sin(2\theta)]$
9. $\sqrt{r}\left[\cos\left(\dfrac{\theta}{2} + \pi k\right) + i \sin\left(\dfrac{\theta}{2} + \pi k\right)\right]; k = 0, 1$

11.3 Exercises *(page 865)*

1.

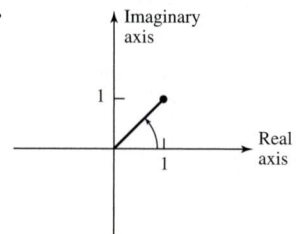

$\sqrt{2}(\cos 45° + i \sin 45°)$

3.

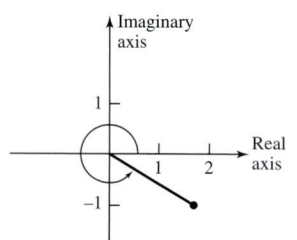

$2(\cos 330° + i \sin 330°)$

5.

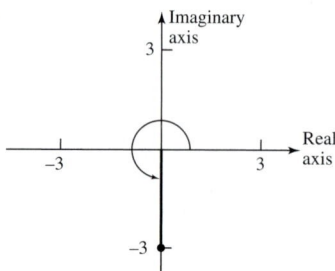

$3(\cos 270° + i \sin 270°)$

7.

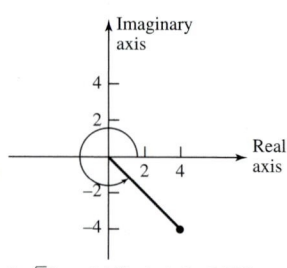

$4\sqrt{2}(\cos 315° + i \sin 315°)$

9.

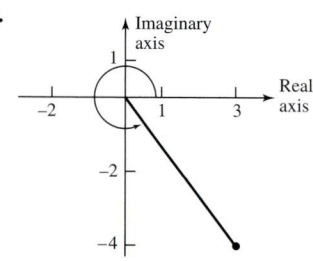

$5(\cos 306.87° + i \sin 306.87°)$

11.

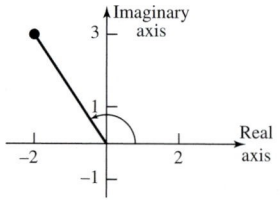

$\sqrt{13}(\cos 123.69° + i \sin 123.69°)$

13. $-1 + \sqrt{3}i$ **15.** $2\sqrt{2} - 2\sqrt{2}i$ **17.** $-3i$ **19.** $-0.03 + 0.20i$ **21.** $1.97 + 0.35i$

23. $zw = 8(\cos 60° + i \sin 60°); \dfrac{z}{w} = \dfrac{1}{2}(\cos 20° + i \sin 20°)$ **25.** $zw = 12(\cos 40° + i \sin 40°); \dfrac{z}{w} = \dfrac{3}{4}(\cos 220° + i \sin 220°)$

27. $zw = 4\left(\cos\dfrac{9\pi}{40} + i \sin\dfrac{9\pi}{40}\right); \dfrac{z}{w} = \cos\dfrac{\pi}{40} + i \sin\dfrac{\pi}{40}$ **29.** $zw = 4\sqrt{2}(\cos 15° + i \sin 15°); \dfrac{z}{w} = \sqrt{2}(\cos 75° + i \sin 75°)$

31. $-32 + 32\sqrt{3}i$ **33.** $32i$ **35.** $\dfrac{27}{2} + \dfrac{27\sqrt{3}}{2}i$ **37.** $-\dfrac{25\sqrt{2}}{2} + \dfrac{25\sqrt{2}}{2}i$ **39.** $-4 + 4i$ **41.** $-23 + 14.14i$

43. $\sqrt[6]{2}(\cos 15° + i \sin 15°), \sqrt[6]{2}(\cos 135° + i \sin 135°), \sqrt[6]{2}(\cos 255° + i \sin 255°)$

45. $\sqrt[4]{8}(\cos 75° + i \sin 75°), \sqrt[4]{8}(\cos 165° + i \sin 165°), \sqrt[4]{8}(\cos 255° + i \sin 255°), \sqrt[4]{8}(\cos 345° + i \sin 345°)$

47. $2(\cos 67.5° + i \sin 67.5°), 2(\cos 157.5° + i \sin 157.5°), 2(\cos 247.5° + i \sin 247.5°), 2(\cos 337.5° + i \sin 337.5°)$

49. $\cos 18° + i \sin 18°, \cos 90° + i \sin 90°, \cos 162° + i \sin 162°, \cos 234° + i \sin 234°, \cos 306° + i \sin 306°$

51. $1, i, -1, -i$

53. Look at formula (8); $|z_k| = \sqrt[n]{r}$ for all k.

55. Look at formula (8). The z_k are spaced apart by an angle of $\dfrac{2\pi}{n}$.

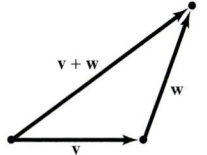

11.4 Concepts and Vocabulary *(page 876)*

1. Unit   **2.** Scalar   **3.** Horizontal; vertical   **4.** T   **5.** T   **6.** F

11.4 Exercises *(page 876)*

1.

3. $3\mathbf{v}$

5.

7.

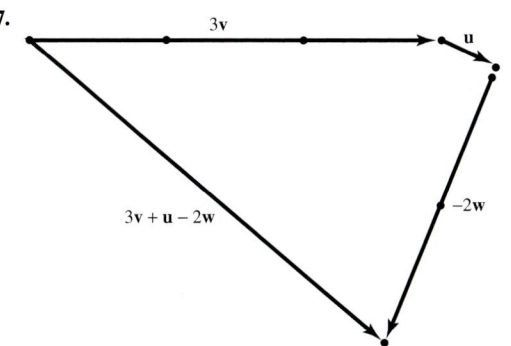

9. T   **11.** F   **13.** F   **15.** T   **17.** 12   **19.** $\mathbf{v} = 3\mathbf{i} + 4\mathbf{j}$ **21.** $\mathbf{v} = 2\mathbf{i} + 4\mathbf{j}$

23. $\mathbf{v} = 8\mathbf{i} - \mathbf{j}$ **25.** $\mathbf{v} = -\mathbf{i} + \mathbf{j}$ **27.** 5   **29.** $\sqrt{2}$ **31.** $\sqrt{13}$ **33.** $-\mathbf{j}$

35. $\sqrt{89}$ **37.** $\sqrt{34} - \sqrt{13}$ **39.** $\mathbf{i}$ **41.** $\dfrac{3}{5}\mathbf{i} - \dfrac{4}{5}\mathbf{j}$ **43.** $\dfrac{\sqrt{2}}{2}\mathbf{i} - \dfrac{\sqrt{2}}{2}\mathbf{j}$

45. $\mathbf{v} = \dfrac{8\sqrt{5}}{5}\mathbf{i} + \dfrac{4\sqrt{5}}{5}\mathbf{j}$ or $\mathbf{v} = -\dfrac{8\sqrt{5}}{5}\mathbf{i} - \dfrac{4\sqrt{5}}{5}\mathbf{j}$

47. $\{-2 + \sqrt{21}, -2 - \sqrt{21}\}$ **49.** $\mathbf{v} = \dfrac{5}{2}(\mathbf{i} + \sqrt{3}\mathbf{j})$

51. $\mathbf{v} = 7(-\mathbf{i} + \sqrt{3}\mathbf{j})$ **53.** $\mathbf{v} = \dfrac{25}{2}(\sqrt{3}\mathbf{i} - \mathbf{j})$ **55.** $\mathbf{F} = 20(\sqrt{3}\mathbf{i} + \mathbf{j})$

57. $\mathbf{F} = (20\sqrt{3} + 30\sqrt{2})\mathbf{i} + (20 - 30\sqrt{2})\mathbf{j}$; magnitude ≈ 80.26 N; direction: $-16.22°$

59. Tension in right cable: 1000 lb; Tension in left cable: 845.24 lb   **61.** Tension in right part: 1088.42 lb; Tension in left part: 1089.07 lb

63.

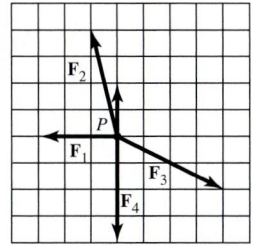

Historical Problems *(page 884)*

1. $(a\mathbf{i} + b\mathbf{j}) \cdot (c\mathbf{i} + d\mathbf{j}) = ac + bd$

real part $[(\overline{a + bi})(c + di)] =$ real part$[(a - bi)(c + di)] =$ real part$[ac + adi - bci - bdi^2] = ac + bd$

11.5 Concepts and Vocabulary *(page 884)*

1. Zero **2.** Orthogonal **3.** Parallel **4.** F **5.** T **6.** F **7.** The vectors intersect at a 90° angle.

11.5 Exercises *(page 884)*

1. 0; 90°; orthogonal **3.** 4; 36.87°; neither **5.** $\sqrt{3} - 1$; 75°; neither **7.** 24; 16.26°; neither **9.** 0; 90°; orthogonal

11. $\dfrac{2}{3}$ **13.** $\mathbf{v}_1 = \dfrac{5}{2}\mathbf{i} - \dfrac{5}{2}\mathbf{j}, \mathbf{v}_2 = -\dfrac{1}{2}\mathbf{i} - \dfrac{1}{2}\mathbf{j}$ **15.** $\mathbf{v}_1 = -\dfrac{1}{5}\mathbf{i} - \dfrac{2}{5}\mathbf{j}, \mathbf{v}_2 = \dfrac{6}{5}\mathbf{i} - \dfrac{3}{5}\mathbf{j}$ **17.** $\mathbf{v}_1 = \dfrac{14}{5}\mathbf{i} + \dfrac{7}{5}\mathbf{j}, \mathbf{v}_2 = \dfrac{1}{5}\mathbf{i} - \dfrac{2}{5}\mathbf{j}$

19. 496.66 mi/hr; 38.46° west of south **21.** 8.63° off direct heading across the current, upstream; 1.52 min

23. Force required to keep Sienna from rolling down the hill: 737.62 lb; Force perpendicular to the hill: 5248.42 lb.

25. $\mathbf{v} = (250\sqrt{2} - 30)\mathbf{i} + (250\sqrt{2} + 30\sqrt{3})\mathbf{j}$; 518.78 km/hr; N38.59°E

27. $\mathbf{v} = 3\mathbf{i} + 20\mathbf{j}$; 20.22 mi/hr; N8.53°E (Assuming boat traveling north and current traveling east.) **29.** 3 ft-lb

31. $1000\sqrt{3}$ ft-lb ≈ 1732.05 ft-lb **33.** Let $\mathbf{u} = a_1\mathbf{i} + b_1\mathbf{j}, \mathbf{v} = a_2\mathbf{i} + b_2\mathbf{j}, \mathbf{w} = a_3\mathbf{i} + b_3\mathbf{j}$. Compute $\mathbf{u} \cdot (\mathbf{v} + \mathbf{w})$ and $\mathbf{u} \cdot \mathbf{v} + \mathbf{u} \cdot \mathbf{w}$.

35. $\cos \alpha = \dfrac{\mathbf{v} \cdot \mathbf{i}}{\|\mathbf{v}\| \, \|\mathbf{i}\|} = \mathbf{v} \cdot \mathbf{i}$; if $\mathbf{v} = x\mathbf{i} + y\mathbf{j}$, then $\mathbf{v} \cdot \mathbf{i} = x = \cos \alpha$ and $\mathbf{v} \cdot \mathbf{j} = y = \cos\left(\dfrac{\pi}{2} - \alpha\right) = \sin \alpha$.

37. $\mathbf{v} = a\mathbf{i} + b\mathbf{j}$; the vector projection of $\mathbf{v}$ onto $\mathbf{i}$ is $\dfrac{\mathbf{v} \cdot \mathbf{i}}{\|\mathbf{i}\|^2}\mathbf{i} = (\mathbf{v} \cdot \mathbf{i})\mathbf{i}$; $\mathbf{v} \cdot \mathbf{i} = a, \mathbf{v} \cdot \mathbf{j} = b$, so $\mathbf{v} = (\mathbf{v} \cdot \mathbf{i})\mathbf{i} + (\mathbf{v} \cdot \mathbf{j})\mathbf{j}$.

39. $(\mathbf{v} - \alpha\mathbf{w}) \cdot \mathbf{w} = \mathbf{v} \cdot \mathbf{w} - \alpha\mathbf{w} \cdot \mathbf{w} = \alpha\|\mathbf{w}\|^2 - \alpha\|\mathbf{w}\|^2 = 0$ since the dot product of any vector with itself equals the square of its magnitude.

41. $W = \mathbf{F} \cdot \overrightarrow{AB} = 0$ when $\mathbf{F}$ is orthogonal to $\overrightarrow{AB}$.

Review Exercises *(page 887)*

1. $\left(\dfrac{3\sqrt{3}}{2}, \dfrac{3}{2}\right)$ **3.** $(1, \sqrt{3})$ **5.** $(0, 3)$ **7.** $\left(3\sqrt{2}, \dfrac{3\pi}{4}\right), \left(-3\sqrt{2}, -\dfrac{\pi}{4}\right)$

9. $\left(2, -\dfrac{\pi}{2}\right), \left(-2, \dfrac{\pi}{2}\right)$

11. $(5, 0.93), (-5, 4.07)$

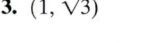

13. $x^2 + y^2 - 2y = 0$

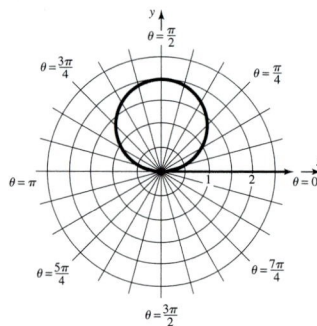

15. $x^2 + y^2 = 25$

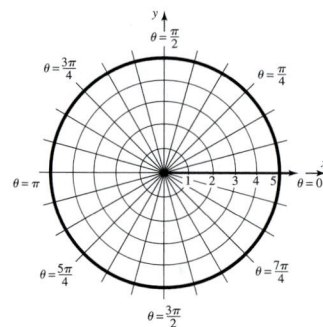

17. $x + 3y = 6$

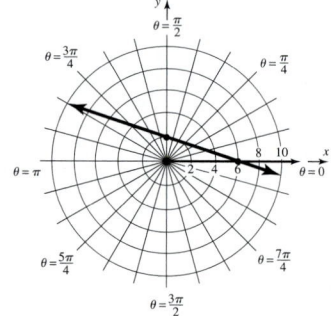

19.

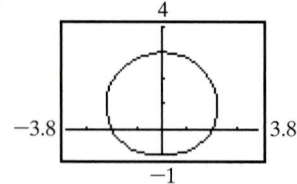

21.

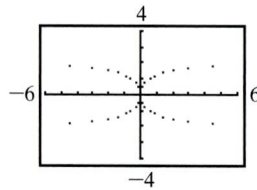

23. Circle; radius 2, center at $(2, 0)$ in rectangular coordinates; symmetric with respect to the polar axis

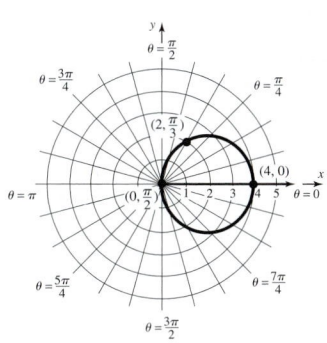

25. Cardioid; symmetric with respect to the line $\theta = \dfrac{\pi}{2}$

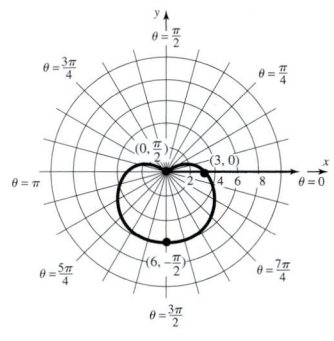

27. Limaçon without inner loop; symmetric with respect to the polar axis

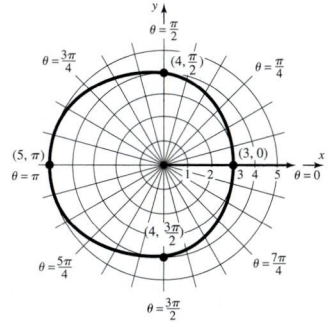

29. $\sqrt{2}(\cos 225° + i \sin 225°)$ **31.** $5(\cos 323.1° + i \sin 323.1°)$

33. $-\sqrt{3} + i$

35. $-\dfrac{3}{2} + \left(\dfrac{3\sqrt{3}}{2}\right)i$

37. $0.10 - 0.02$

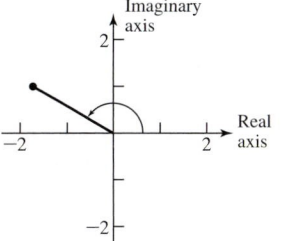

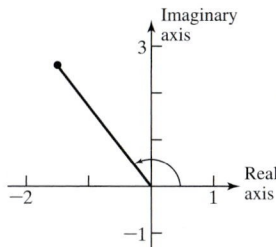

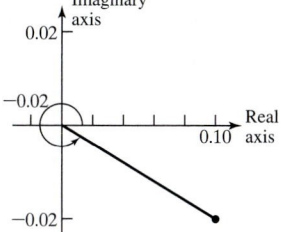

39. $zw = \cos 130° + i \sin 130°; \dfrac{z}{w} = \cos 30° + i \sin 30°$ **41.** $zw = 6(\cos 0 + i \sin 0) = 6; \dfrac{z}{w} = \dfrac{3}{2}\left(\cos \dfrac{8\pi}{5} + i \sin \dfrac{8\pi}{5}\right)$

43. $zw = 5(\cos 5° + i \sin 5°); \dfrac{z}{w} = 5(\cos 15° + i \sin 15°)$ **45.** $\dfrac{27}{2} + \dfrac{27\sqrt{3}}{2}i$ **47.** $4i$ **49.** 64 **51.** $-527 - 336i$

53. $3, 3(\cos 120° + i \sin 120°), 3(\cos 240° + i \sin 240°)$

55.

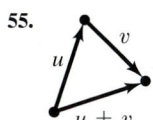

57.

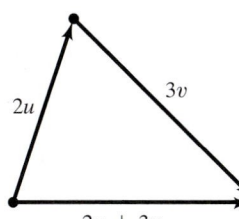

59. $\mathbf{v} = 2\mathbf{i} - 4\mathbf{j}; \|\mathbf{v}\| = 2\sqrt{5}$ **61.** $\mathbf{v} = -\mathbf{i} + 3\mathbf{j}; \|\mathbf{v}\| = \sqrt{10}$ **63.** $2\mathbf{i} - 2\mathbf{j}$
65. $-20\mathbf{i} + 13\mathbf{j}$ **67.** $\sqrt{5}$ **69.** $\sqrt{5} + 5 \approx 7.24$
71. $\dfrac{-2\sqrt{5}}{5}\mathbf{i} + \dfrac{\sqrt{5}}{5}\mathbf{j}$ **73.** $\mathbf{v} = \dfrac{3}{2} + \dfrac{3\sqrt{3}}{2}\mathbf{i}$ **75.** $\mathbf{v} \cdot \mathbf{w} = -11; \theta \approx 169.70°$
77. $\mathbf{v} \cdot \mathbf{w} = -4; \theta \approx 153.43°$ **79.** Parallel   **81.** Parallel   **83.** Perpendicular
85. $\mathbf{v}_1 = \dfrac{4}{5}\mathbf{i} - \dfrac{3}{5}\mathbf{j}; \mathbf{v}_2 = \dfrac{6}{5}\mathbf{i} + \dfrac{8}{5}\mathbf{j}$ **87.** $\dfrac{9}{10}(3\mathbf{i} + \mathbf{j})$ **89.** $\sqrt{29} \approx 5.39$ mi/hr; 0.4 mi
91. Left cable: 1843.21 lb; right cable: 1630.41 lb   **93.** 50 foot-pounds

Cumulative Review Exercises *(page 890)*

1. $\{-3, 3\}$ **2.** $y = \dfrac{\sqrt{3}}{3}x$

3. $x^2 + (y - 1)^2 = 9$

4. $\left\{x \,\middle|\, x < \dfrac{1}{2}\right\}$ or $\left(-\infty, \dfrac{1}{2}\right)$ **5.** Symmetry with respect to the y-axis

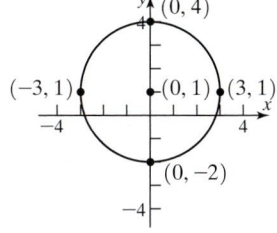

6.

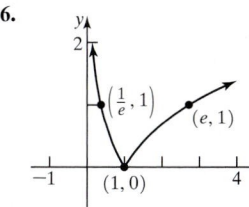

7.

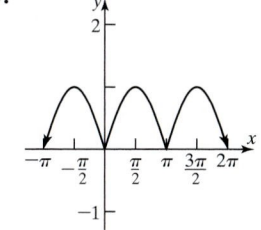

8.

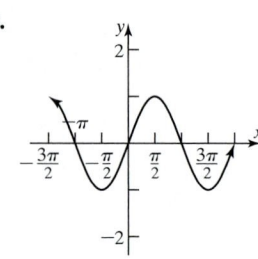

9. $-\dfrac{\pi}{6}$ **10.**

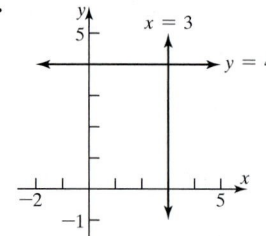

11.

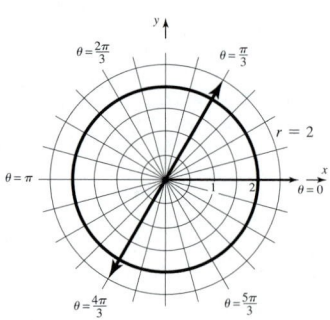

C H A P T E R   1 2   Analytic Geometry

12.2 Concepts and Vocabulary *(page 901)*

1. parabola   **2.** paraboloid   **3.** latus rectum   **4.** T   **5.** F   **6.** T

12.2 Exercises *(page 902)*

1. B   **3.** E   **5.** H   **7.** C   **9.** E   **11.** D   **13.** C

15. $y^2 = 16x$

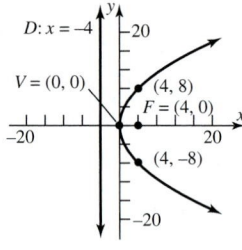

17. $x^2 = -12y$

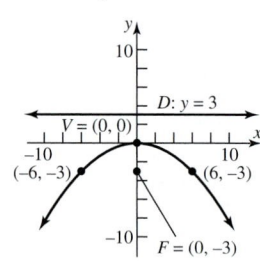

19. $y^2 = -8x$

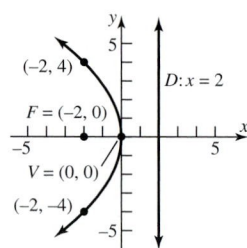

21. $x^2 = 2y$

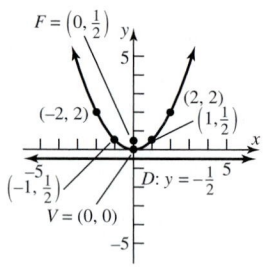

23. $(x - 2)^2 = -8(y + 3)$

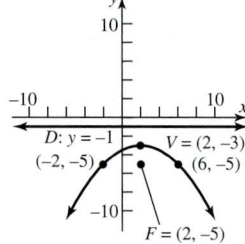

25. $x^2 = \dfrac{4}{3}y$

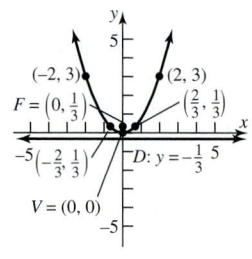

27. $(x + 3)^2 = 4(y - 3)$

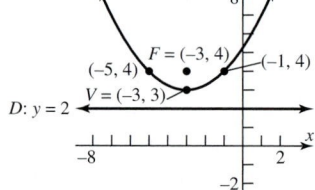

29. $(y + 2)^2 = -8(x + 1)$

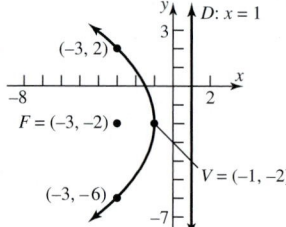

31. Vertex: $(0, 0)$; Focus: $(0, 1)$;
Directrix: $y = -1$

(a)

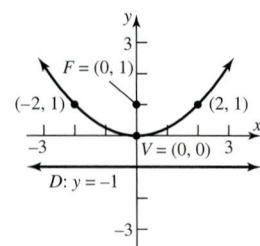

(b)

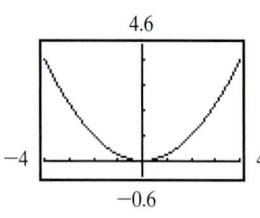

33. Vertex: $(0, 0)$; Focus: $(-4, 0)$;

Directrix: $x = 4$

(a)

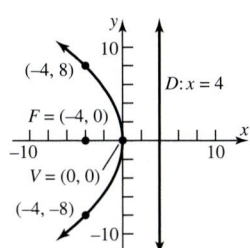

(b)

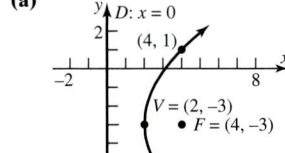

35. Vertex: $(-1, 2)$; Focus: $(1, 2)$;

Directrix: $x = -3$

(a)

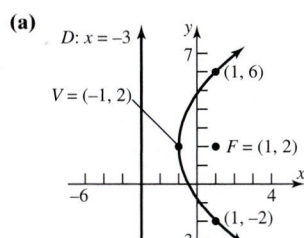

(b)

37. Vertex: $(3, -1)$; Focus: $\left(3, -\frac{5}{4}\right)$;

Directrix: $y = -\frac{3}{4}$

(a)

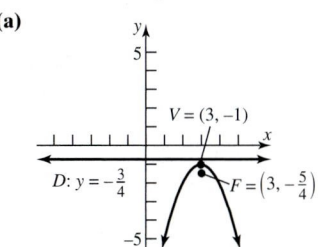

(b)

39. Vertex: $(2, -3)$; Focus: $(4, -3)$;
Directrix: $x = 0$

(a)

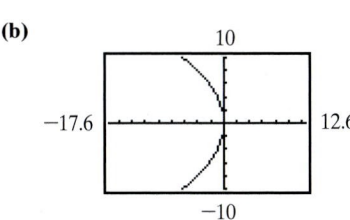

(b)

41. Vertex: $(0, 2)$; Focus: $(-1, 2)$;
Directrix: $x = 1$

(a)

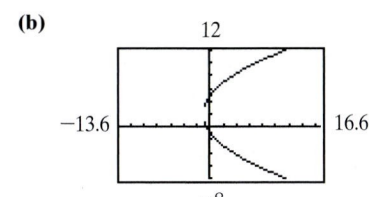

(b)

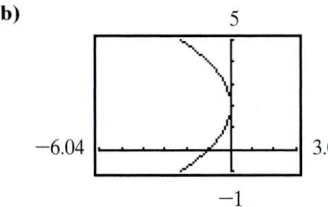

43. Vertex: $(-4, -2)$; Focus: $(-4, -1)$;
Directrix: $y = -3$

(a)

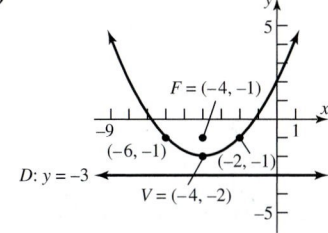

(b)

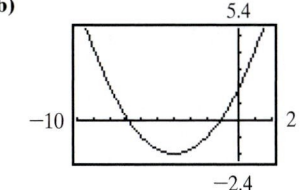

45. Vertex: $(-1, -1)$; Focus: $\left(-\frac{3}{4}, -1\right)$; Directrix: $x = -\frac{5}{4}$

(a)

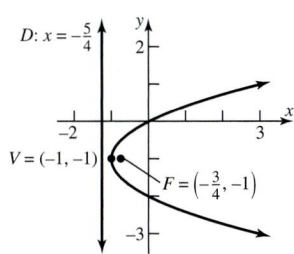

(b)

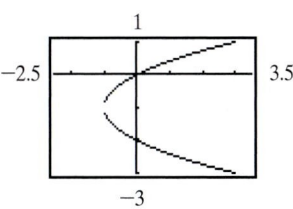

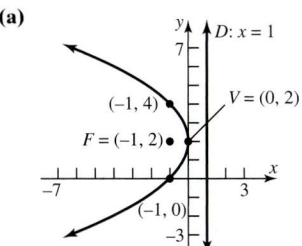

47. Vertex: $(2, -8)$; Focus: $\left(2, -\dfrac{31}{4}\right)$; Directrix: $y = -\dfrac{33}{4}$

49. $(y - 1)^2 = x$

51. $(y - 1)^2 = -(x - 2)$

(a)

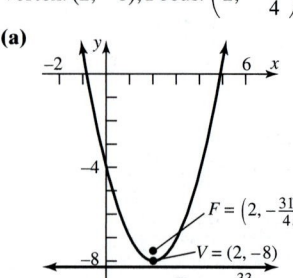

(b)

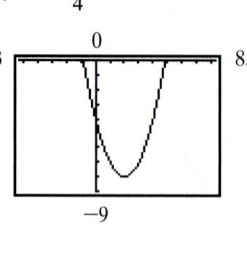

53. $x^2 = 4(y - 1)$ **55.** $y^2 = \dfrac{1}{2}(x + 2)$ **57.** 1.5625 ft from the base of the dish, along the axis of symmetry   **59.** 1 in. from the vertex

61. 20 ft   **63.** 0.78125 ft   **65.** 4.17 ft from the base along the axis of symmetry   **67.** 24.31 ft, 18.75 ft, 7.64 ft

69. $Ax^2 + Ey = 0, A \neq 0, E \neq 0$ This is the equation of a parabola with vertex at $(0, 0)$ and axis of symmetry the y-axis.

$$Ax^2 = -Ey$$

The focus is $\left(0, -\dfrac{E}{4A}\right)$; the directrix is the line $y = \dfrac{E}{4A}$. The parabola opens up if $-\dfrac{E}{A} > 0$

$$x^2 = -\dfrac{E}{A}y$$

and down if $-\dfrac{E}{A} < 0$.

71. $Ax^2 + Dx + Ey + F = 0, A \neq 0$

$$Ax^2 + Dx = -Ey - F$$

$$x^2 + \dfrac{D}{A}x = -\dfrac{E}{A}y - \dfrac{F}{A}$$

$$\left(x + \dfrac{D}{2A}\right)^2 = -\dfrac{E}{A}y - \dfrac{F}{A} + \dfrac{D^2}{4A^2}$$

$$\left(x + \dfrac{D}{2A}\right)^2 = -\dfrac{E}{A}y + \dfrac{D^2 - 4AF}{4A^2}$$

(a) If $E \neq 0$, then the equation may be written as

$$\left(x + \dfrac{D}{2A}\right)^2 = -\dfrac{E}{A}\left(y - \dfrac{D^2 - 4AF}{4AE}\right)$$

This is the equation of a parabola with vertex at

$\left(-\dfrac{D}{2A}, \dfrac{D^2 - 4AF}{4AE}\right)$ and axis of symmetry parallel to the y-axis.

(b)–(d) If $E = 0$, the graph of the equation contains no points if $D^2 - 4AF < 0$, is a single vertical line if $D^2 - 4AF = 0$, and is two vertical lines if $D^2 - 4AF > 0$.

12.3 Concepts and Vocabulary *(page 914)*

1. ellipse   **2.** major   **3.** $(0, -5); (0, 5)$ **4.** F   **5.** T   **6.** T

12.3 Exercises *(page 914)*

1. C   **3.** B   **5.** C   **7.** D

9. Vertices: $(-5, 0), (5, 0)$

Foci: $(-\sqrt{21}, 0), (\sqrt{21}, 0)$

(a)

11. Vertices: $(0, -5), (0, 5)$

Foci: $(0, -4), (0, 4)$

(a)

13. $\dfrac{x^2}{4} + \dfrac{y^2}{16} = 1$

Vertices: $(0, -4), (0, 4)$

Foci: $(0, -2\sqrt{3}), (0, 2\sqrt{3})$

(a)

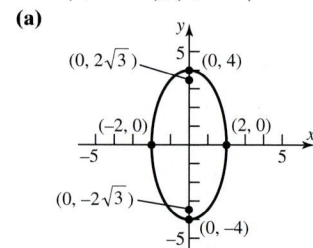

(b)

(b)

(b)

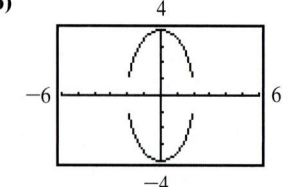

15. $\dfrac{x^2}{8} + \dfrac{y^2}{2} = 1$

Vertices: $(-2\sqrt{2}, 0)$, $(2\sqrt{2}, 0)$

Foci: $(-\sqrt{6}, 0)$, $(\sqrt{6}, 0)$

(a)

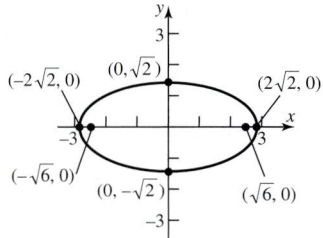

(b)

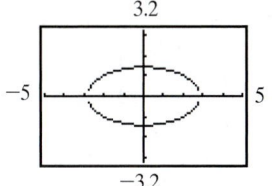

17. $\dfrac{x^2}{16} + \dfrac{y^2}{16} = 1$

Vertices: $(-4, 0)$, $(4, 0)$, $(0, -4)$, $(0, 4)$

Focus: $(0, 0)$

(a)

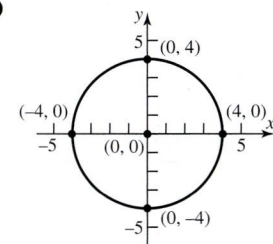

(b)

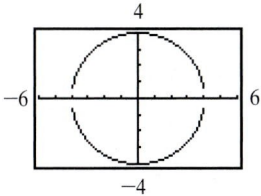

19. $\dfrac{x^2}{25} + \dfrac{y^2}{16} = 1$

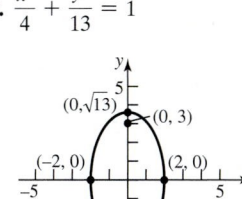

21. $\dfrac{x^2}{9} + \dfrac{y^2}{25} = 1$

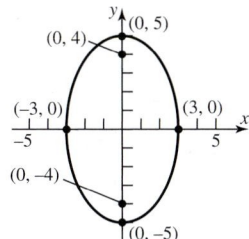

23. $\dfrac{x^2}{9} + \dfrac{y^2}{5} = 1$

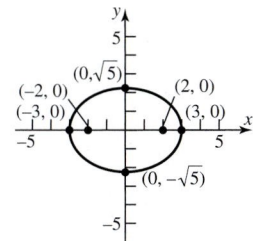

25. $\dfrac{x^2}{4} + \dfrac{y^2}{13} = 1$

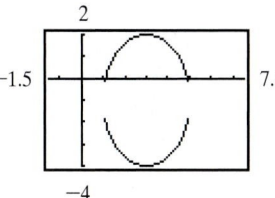

27. $x^2 + \dfrac{y^2}{16} = 1$

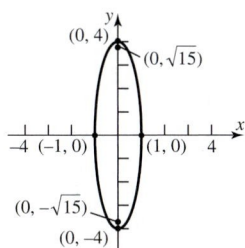

29. $\dfrac{(x+1)^2}{4} + (y-1)^2 = 1$ **31.** $(x-1)^2 + \dfrac{y^2}{4} = 1$

33. Center: $(3, -1)$; Vertices: $(3, -4)$, $(3, 2)$

Foci: $(3, -1 - \sqrt{5})$, $(3, -1 + \sqrt{5})$

(a)

(b)

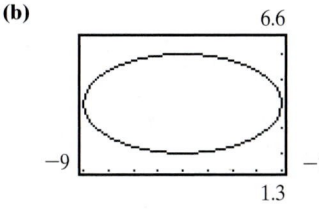

35. $\dfrac{(x+5)^2}{16} + \dfrac{(y-4)^2}{4} = 1$

Center: $(-5, 4)$; Vertices: $(-9, 4)$, $(-1, 4)$

Foci: $(-5 - 2\sqrt{3}, 4)$, $(-5 + 2\sqrt{3}, 4)$

(a)

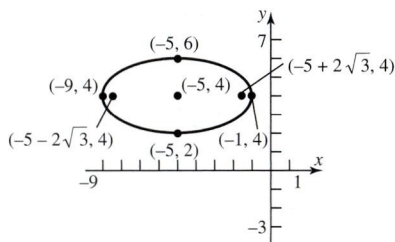

(b)

37. $\dfrac{(x+2)^2}{4} + (y-1)^2 = 1$

Center: $(-2, 1)$; Vertices: $(-4, 1)$, $(0, 1)$
Foci: $(-2 - \sqrt{3}, 1)$, $(-2 + \sqrt{3}, 1)$

(a)

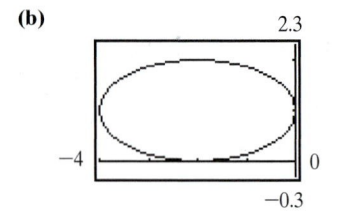

(b)

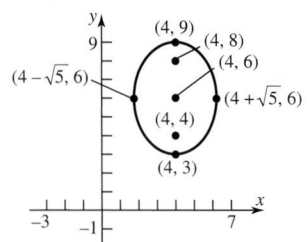

39. $\dfrac{(x-2)^2}{3} + \dfrac{(y+1)^2}{2} = 1$

Center: $(2, -1)$; Vertices: $(2 - \sqrt{3}, -1)$,
$(2 + \sqrt{3}, -1)$; Foci: $(1, -1)$, $(3, -1)$

(a)

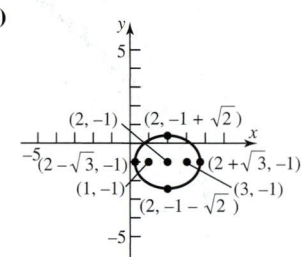

(b)

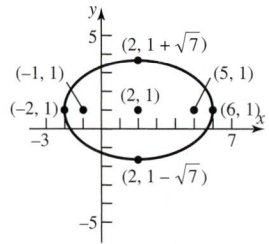

41. $\dfrac{(x-1)^2}{4} + \dfrac{(y+2)^2}{9} = 1$

Center: $(1, -2)$; Vertices: $(1, -5)$, $(1, 1)$
Foci: $(1, -2 - \sqrt{5})$, $(1, -2 + \sqrt{5})$

(a)

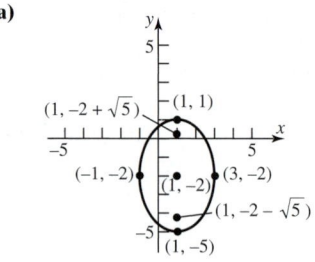

(b)

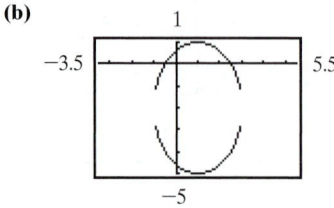

43. $x^2 + \dfrac{(y+2)^2}{4} = 1$

Center: $(0, -2)$; Vertices: $(0, -4)$, $(0, 0)$
Foci: $(0, -2 - \sqrt{3})$, $(0, -2 + \sqrt{3})$

(a)

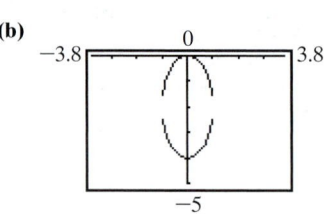

(b)

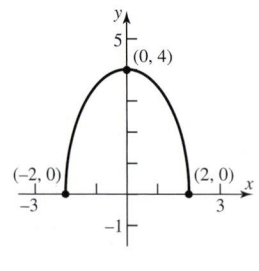

45. $\dfrac{(x-2)^2}{25} + \dfrac{(y+2)^2}{21} = 1$

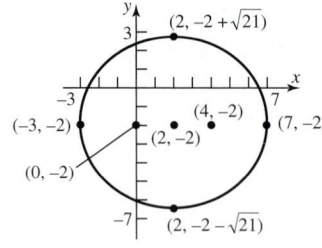

47. $\dfrac{(x-4)^2}{5} + \dfrac{(y-6)^2}{9} = 1$

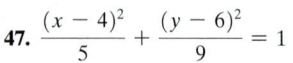

49. $\dfrac{(x-2)^2}{16} + \dfrac{(y-1)^2}{7} = 1$

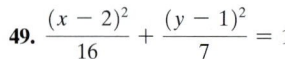

51. $\dfrac{(x-1)^2}{10} + (y-2)^2 = 1$

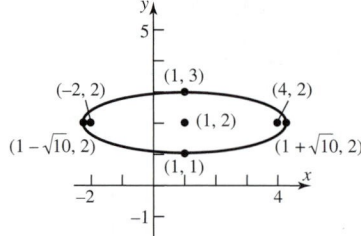

53. $\dfrac{(x-1)^2}{9} + (y-2)^2 = 1$

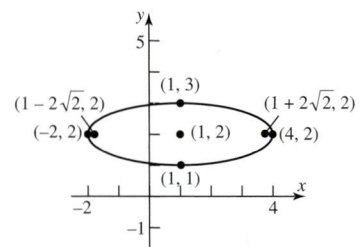

55.

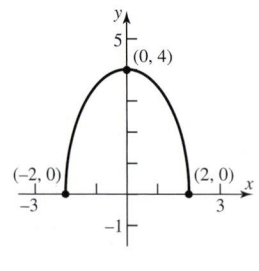

57.

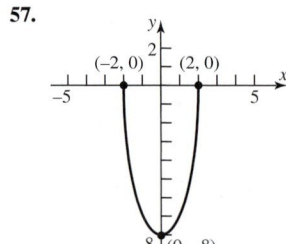

59. $\dfrac{x^2}{100} + \dfrac{y^2}{36} = 1$ **61.** 43.3 ft   **63.** 24.65 ft, 21.65 ft, 13.82 ft   **65.** 0 ft, 12.99 ft, 15 ft, 12.99 ft, 0 ft

67. 91.5 million mi; $\dfrac{x^2}{(93)^2} + \dfrac{y^2}{8646.75} = 1$ **69.** perihelion: 460.6 million mi; mean distance: 483.8 million mi; $\dfrac{x^2}{483.8^2} + \dfrac{y^2}{233,524.2} = 1$

71. 30 ft

73. (a) $Ax^2 + Cy^2 + F = 0$ If A and C are of the same sign and F is of opposite sign, then the equation takes the form

$$Ax^2 + Cy^2 = -F \qquad \dfrac{x^2}{\left(-\dfrac{F}{A}\right)} + \dfrac{y^2}{\left(-\dfrac{F}{C}\right)} = 1, \text{where } -\dfrac{F}{A} \text{ and } -\dfrac{F}{C} \text{ are positive. This is the equation of an ellipse}$$

with center at $(0, 0)$.

(b) If $A = C$, the equation may be written as $x^2 + y^2 = -\dfrac{F}{A}$.

This is the equation of a circle with center at $(0, 0)$ and radius equal to $\sqrt{-\dfrac{F}{A}}$.

12.4 Concepts and Vocabulary *(page 929)*

1. hyperbola   **2.** transverse axis   **3.** $y = \dfrac{3}{2}x$; $y = -\dfrac{3}{2}x$ **4.** F   **5.** T   **6.** F

12.4 Exercises *(page 929)*

1. B   **3.** A   **5.** B   **7.** C

9. $x^2 - \dfrac{y^2}{8} = 1$

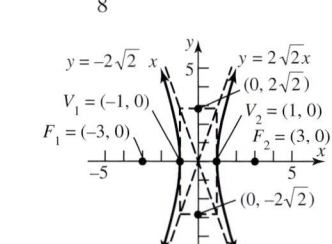

11. $\dfrac{y^2}{16} - \dfrac{x^2}{20} = 1$

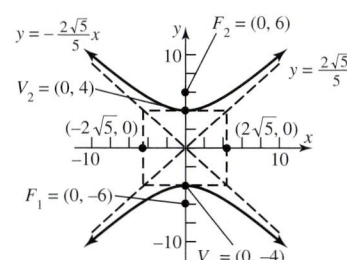

13. $\dfrac{x^2}{9} - \dfrac{y^2}{16} = 1$

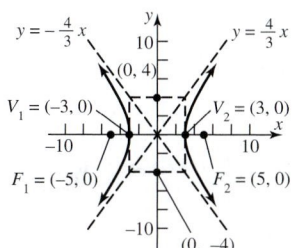

15. $\dfrac{y^2}{36} - \dfrac{x^2}{9} = 1$

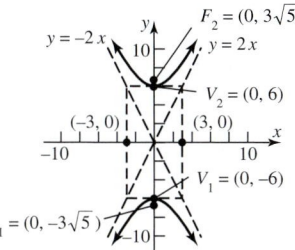

17. $\dfrac{x^2}{8} - \dfrac{y^2}{8} = 1$

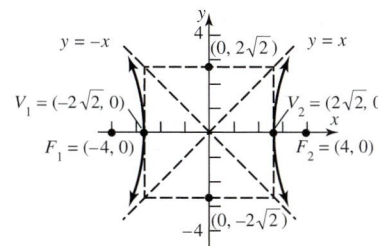

19. $\dfrac{x^2}{25} - \dfrac{y^2}{9} = 1$

Center: $(0, 0)$

Transverse axis: x-axis

Vertices: $(-5, 0)$, $(5, 0)$

Foci: $(-\sqrt{34}, 0)$, $(\sqrt{34}, 0)$

Asymptotes: $y = \pm\dfrac{3}{5}x$

(a)

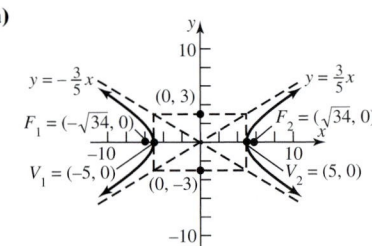

(b)

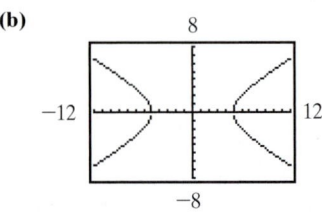

21. $\dfrac{x^2}{4} - \dfrac{y^2}{16} = 1$

Center: $(0, 0)$

Transverse axis: x-axis

Vertices: $(-2, 0)$, $(2, 0)$

Foci: $(-2\sqrt{5}, 0)$, $(2\sqrt{5}, 0)$

Asymptotes: $y = \pm2x$

(a)

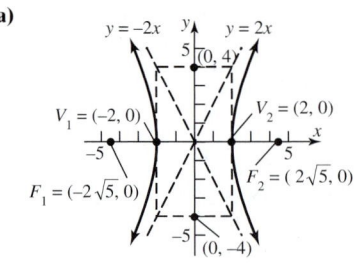

(b)

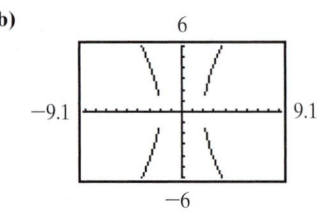

23. $\dfrac{y^2}{9} - x^2 = 1$

Center: $(0, 0)$

Transverse axis: y-axis

Vertices: $(0, -3)$, $(0, 3)$

Foci: $(0, -\sqrt{10})$, $(0, \sqrt{10})$

Asymptotes: $y = \pm3x$

(a)

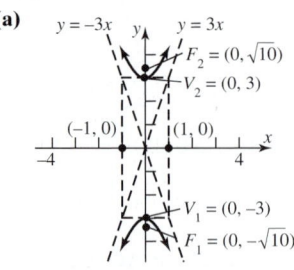

(b)

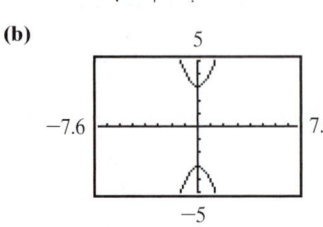

25. $\dfrac{y^2}{25} - \dfrac{x^2}{25} = 1$

Center: $(0, 0)$

Transverse axis: y-axis

Vertices: $(0, -5)$, $(0, 5)$

Foci: $(0, -5\sqrt{2})$, $(0, 5\sqrt{2})$

Asymptotes: $y = \pm x$

(a)

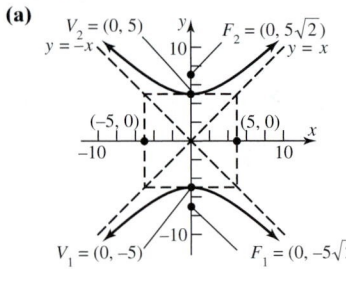

(b)

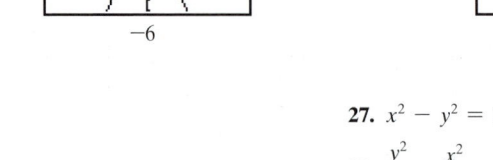

27. $x^2 - y^2 = 1$

29. $\dfrac{y^2}{36} - \dfrac{x^2}{9} = 1$

31. $\dfrac{(x-4)^2}{4} - \dfrac{(y+1)^2}{5} = 1$

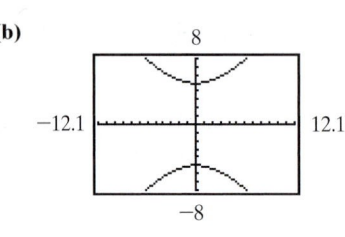

33. $\dfrac{(y+4)^2}{4} - \dfrac{(x+3)^2}{12} = 1$

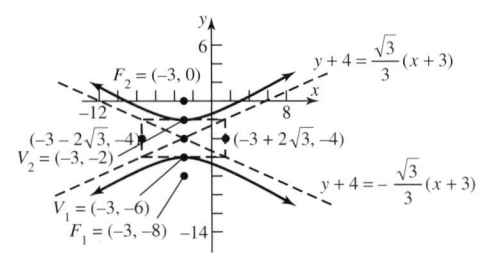

35. $(x - 5)^2 - \dfrac{(y - 7)^2}{3} = 1$

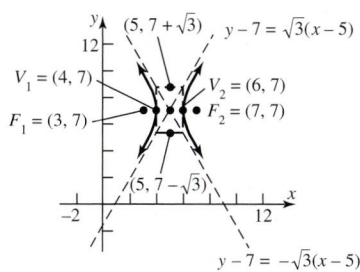

37. $\dfrac{(x - 1)^2}{4} - \dfrac{(y + 1)^2}{9} = 1$

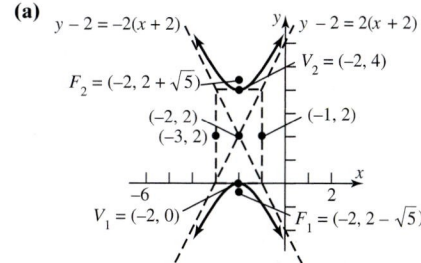

39. $\dfrac{(x - 2)^2}{4} - \dfrac{(y + 3)^2}{9} = 1$

Center: $(2, -3)$

Transverse axis: Parallel to x-axis

Vertices: $(0, -3), (4, -3)$

Foci: $(2 - \sqrt{13}, -3), (2 + \sqrt{13}, -3)$

Asymptotes: $y + 3 = \pm\dfrac{3}{2}(x - 2)$

(a)

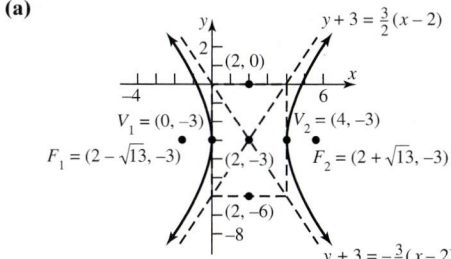

(b)

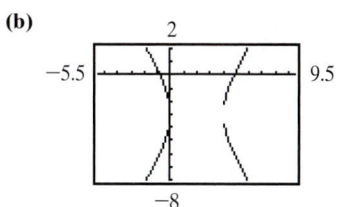

41. $\dfrac{(y - 2)^2}{4} - (x + 2)^2 = 1$

Center: $(-2, 2)$

Transverse axis: Parallel to y-axis

Vertices: $(-2, 0), (-2, 4)$

Foci: $(-2, 2 - \sqrt{5}), (-2, 2 + \sqrt{5})$

Asymptotes: $y - 2 = \pm 2(x + 2)$

(a)

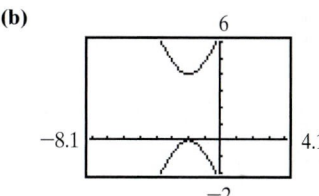

(b)

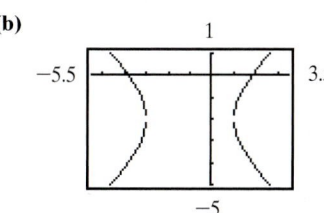

43. $\dfrac{(x + 1)^2}{4} - \dfrac{(y + 2)^2}{4} = 1$

Center: $(-1, -2)$

Transverse axis: Parallel to x-axis

Vertices: $(-3, -2), (1, -2)$

Foci: $(-1 - 2\sqrt{2}, -2), (-1 + 2\sqrt{2}, -2)$

Asymptotes: $y + 2 = \pm(x + 1)$

(a)   **(b)**

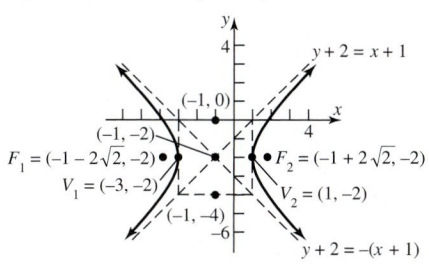

45. $(x - 1)^2 - (y + 1)^2 = 1$

Center: $(1, -1)$
Transverse axis: Parallel to x-axis
Vertices: $(0, -1)$, $(2, -1)$
Foci: $(1 - \sqrt{2}, -1)$, $(1 + \sqrt{2}, -1)$
Asymptotes: $y + 1 = \pm(x - 1)$

(a)

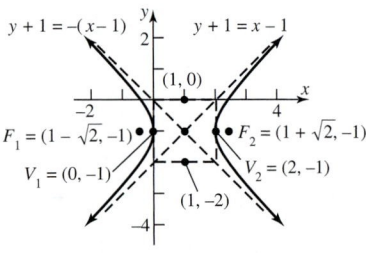

(b)

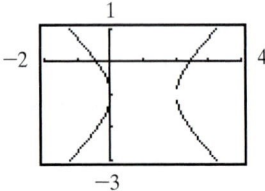

47. $\dfrac{(y - 2)^2}{4} - (x + 1)^2 = 1$

Center: $(-1, 2)$
Transverse axis: Parallel to y-axis
Vertices: $(-1, 0)$, $(-1, 4)$
Foci: $(-1, 2 - \sqrt{5})$, $(-1, 2 + \sqrt{5})$
Asymptotes: $y - 2 = \pm 2(x + 1)$

(a)

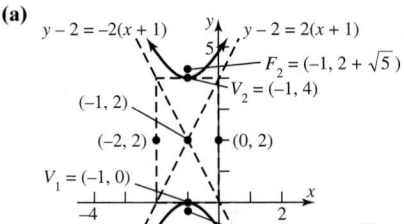

(b)

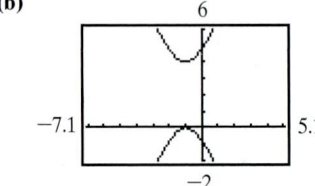

49. $\dfrac{(x - 3)^2}{4} - \dfrac{(y + 2)^2}{16} = 1$

Center: $(3, -2)$
Transverse axis: Parallel to x-axis
Vertices: $(1, -2)$, $(5, -2)$
Foci: $(3 - 2\sqrt{5}, -2)$, $(3 + 2\sqrt{5}, -2)$

Asymptotes: $y + 2 = \pm 2(x - 3)$

(a)

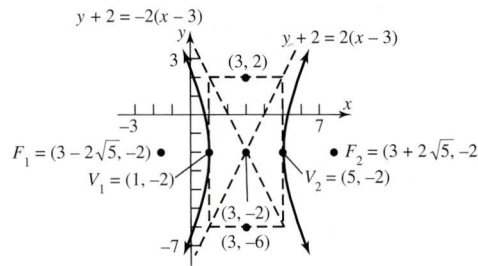

(b)

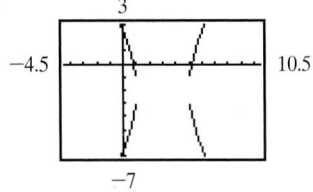

51. $\dfrac{(y - 1)^2}{4} - (x + 2)^2 = 1$

Center: $(-2, 1)$
Transverse axis: Parallel to y-axis
Vertices: $(-2, -1)$, $(-2, 3)$
Foci: $(-2, 1 - \sqrt{5})$, $(-2, 1 + \sqrt{5})$

Asymptotes: $y - 1 = \pm 2(x + 2)$

(a)

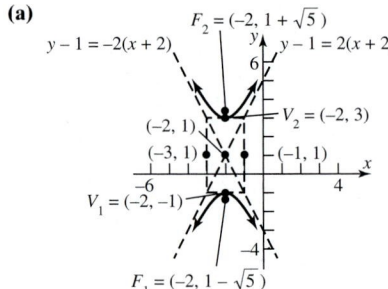

(b)

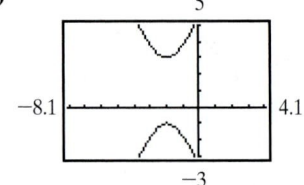

53.

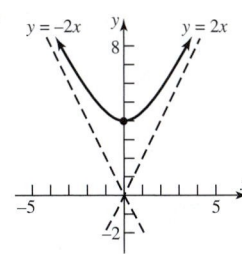

55.

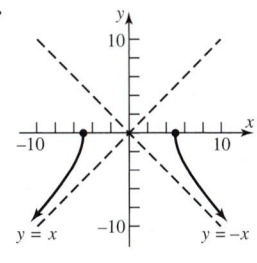

57. (a) The ship will reach shore at a point 64.66 mi from the master station.
(b) 0.00086 sec
(c) $(104, 50)$
59. (a) 450 ft
61. If e is close to 1, narrow hyperbola; if e is very large, wide hyperbola

63. $\dfrac{x^2}{4} - y^2 = 1$; asymptotes $y = \pm\dfrac{1}{2}x$, $y^2 - \dfrac{x^2}{4} = 1$; asymptotes $y = \pm\dfrac{1}{2}x$

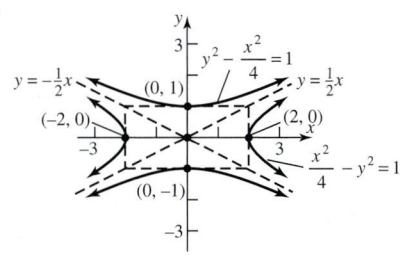

65. $Ax^2 + Cy^2 + F = 0$ If A and C are of opposite sign and $F \neq 0$, this equation may be written as $\dfrac{x^2}{\left(-\dfrac{F}{A}\right)} + \dfrac{y^2}{\left(-\dfrac{F}{C}\right)} = 1$,

$Ax^2 + Cy^2 = -F$ where $-\dfrac{F}{A}$ and $-\dfrac{F}{C}$ are opposite in sign. This is the equation of a hyperbola with center $(0, 0)$.

The transverse axis is the x-axis if $-\dfrac{F}{A} > 0$; the transverse axis is the y-axis if $-\dfrac{F}{A} < 0$.

12.5 Concepts and Vocabulary *(page 940)*

1. $\cot(2\theta) = \dfrac{A - C}{B}$ **2.** Hyperbola **3.** Ellipse **4.** T **5.** T **6.** F

12.5 Exercises *(page 940)*

1. Parabola **3.** Ellipse **5.** Hyperbola **7.** Hyperbola **9.** Circle **11.** $x = \dfrac{\sqrt{2}}{2}(x' - y'), y = \dfrac{\sqrt{2}}{2}(x' + y')$

13. $x = \dfrac{\sqrt{2}}{2}(x' - y'), y = \dfrac{\sqrt{2}}{2}(x' + y')$ **15.** $x = \dfrac{1}{2}(x' - \sqrt{3}y'), y = \dfrac{1}{2}(\sqrt{3}x' + y')$ **17.** $x = \dfrac{\sqrt{5}}{5}(x' - 2y'), y = \dfrac{\sqrt{5}}{5}(2x' + y')$

19. $x = \dfrac{\sqrt{13}}{13}(3x' - 2y'), y = \dfrac{\sqrt{13}}{13}(2x' + 3y')$

21. $\theta = 45°$ (see Problem 11)

$x'^2 - \dfrac{y'^2}{3} = 1$

Hyperbola
Center at origin
Transverse axis is the x'-axis.
Vertices at $(\pm 1, 0)$

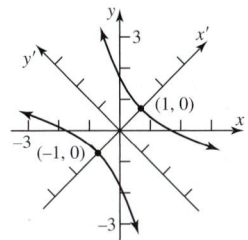

23. $\theta = 45°$ (see Problem 13)

$x'^2 + \dfrac{y'^2}{4} = 1$

Ellipse
Center at $(0, 0)$
Major axis is the y'-axis.
Vertices at $(0, \pm 2)$

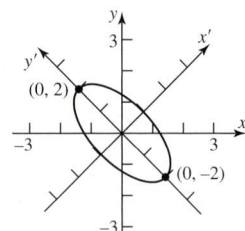

25. $\theta = 60°$ (see Problem 15)

$\dfrac{x'^2}{4} + y'^2 = 1$

Ellipse
Center at $(0, 0)$
Major axis is the x'-axis.
Vertices at $(\pm 2, 0)$

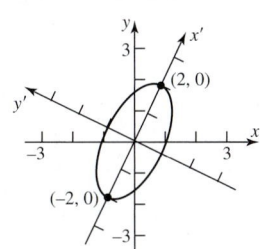

27. $\theta \approx 63°$ (see Problem 17)

$y'^2 = 8x'$

Parabola
Vertex at $(0, 0)$
Focus at $(2, 0)$

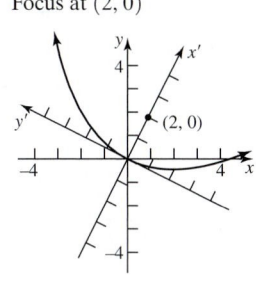

29. $\theta \approx 34°$ (see Problem 19)

$$\frac{(x'-2)^2}{4} + y'^2 = 1$$

Ellipse

Center at $(2, 0)$

Major axis is the x'-axis.

Vertices at $(4, 0)$ and $(0, 0)$

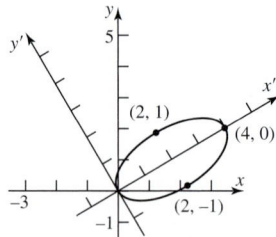

31. $\cot(2\theta) = \dfrac{7}{24}$;

$$\theta = \sin^{-1}\left(\frac{3}{5}\right) \approx 37°$$

$$(x'-1)^2 = -6\left(y' - \frac{1}{6}\right)$$

Parabola

Vertex at $\left(1, \dfrac{1}{6}\right)$

Focus at $\left(1, -\dfrac{4}{3}\right)$

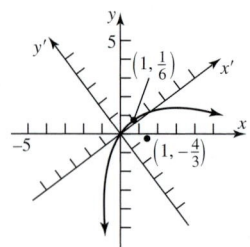

33. Hyperbola

35. Hyperbola

37. Parabola

39. Ellipse

41. Ellipse

43. Refer to equation (6):

$$A' = A\cos^2\theta + B\sin\theta\cos\theta + C\sin^2\theta$$
$$B' = B(\cos^2\theta - \sin^2\theta) + 2(C - A)(\sin\theta\cos\theta)$$
$$C' = A\sin^2\theta - B\sin\theta\cos\theta + C\cos^2\theta$$
$$D' = D\cos\theta + E\sin\theta$$
$$E' = -D\sin\theta + E\cos\theta$$
$$F' = F$$

45. Use Problem 43 to find $B'^2 - 4A'C'$. After much cancellation, $B'^2 - 4A'C' = B^2 - 4AC$.

47. The distance between P_1 and P_2 in the $x'y'$-plane equals $\sqrt{(x_2' - x_1')^2 + (y_2' - y_1')^2}$.

Assuming $x' = x\cos\theta - y\sin\theta$ and $y' = x\sin\theta + y\cos\theta$, then $(x_2' - x_1')^2 = (x_2\cos\theta - y_2\sin\theta - x_1\cos\theta + y_1\sin\theta)^2$
$= \cos^2\theta(x_2 - x_1)^2 - 2\sin\theta\cos\theta(x_2 - x_1)(y_2 - y_1) + \sin^2\theta(y_2 - y_1)^2$, and
$(y_2' - y_1')^2 = (x_2\sin\theta + y_2\cos\theta - x_1\sin\theta - y_1\cos\theta)^2 = \sin^2\theta(x_2 - x_1)^2 + 2\sin\theta\cos\theta(x_2 - x_1)(y_2 - y_1) + \cos^2\theta(y_2 - y_1)^2$.
Therefore, $(x_2' - x_1')^2 + (y_2' - y_1')^2 = \cos^2\theta(x_2 - x_1)^2 + \sin^2\theta(x_2 - x_1)^2 + \sin^2\theta(y_2 - y_1)^2 + \cos^2\theta(y_2 - y_1)^2$
$= (x_2 - x_1)^2(\cos^2\theta + \sin^2\theta) + (y_2 - y_1)^2(\sin^2\theta + \cos^2\theta) = (x_2 - x_1)^2 + (y_2 - y_1)^2$.

12.6 Concepts and Vocabulary *(page 947)*

1. $\dfrac{1}{2}$; ellipse; parallel; 4; below   **2.** 1; < 1; > 1 **3.** T   **4.** T

12.6 Exercises *(page 947)*

1. Parabola; directrix is perpendicular to the polar axis 1 unit to the right of the pole.   **3.** Hyperbola; directrix is parallel to the polar axis $\dfrac{4}{3}$ units below the pole.   **5.** Ellipse; directrix is perpendicular to the polar axis $\dfrac{3}{2}$ units to the left of the pole.

7. Parabola; directrix is perpendicular to the polar axis 1 unit to the right of the pole; vertex is at $\left(\dfrac{1}{2}, 0\right)$.

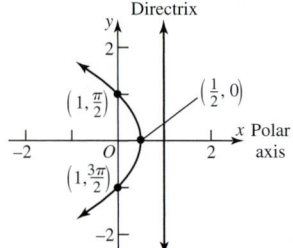

9. Ellipse; directrix is parallel to the polar axis $\dfrac{8}{3}$ units above the pole; vertices are at $\left(\dfrac{8}{7}, \dfrac{\pi}{2}\right)$ and $\left(8, \dfrac{3\pi}{2}\right)$.

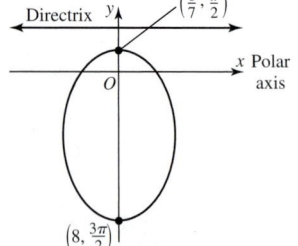

11. Hyperbola; directrix is perpendicular to the polar axis $\dfrac{3}{2}$ units to the left of the pole; vertices are at $(-3, 0)$ and $(1, \pi)$.

13. Ellipse; directrix is parallel to the polar axis 8 units below the pole; vertices are at $\left(8, \frac{\pi}{2}\right)$ and $\left(\frac{8}{3}, \frac{3\pi}{2}\right)$.

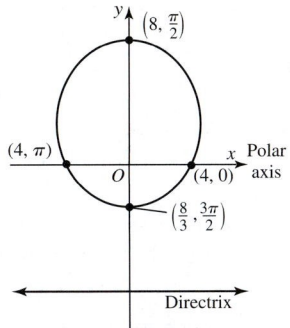

15. Ellipse; directrix is parallel to the polar axis 3 units below the pole; vertices are at $\left(6, \frac{\pi}{2}\right)$ and $\left(\frac{6}{5}, \frac{3\pi}{2}\right)$.

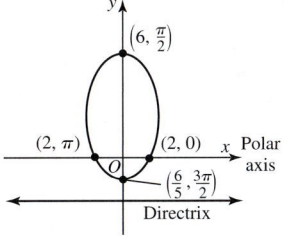

17. Ellipse; directrix is perpendicular to the polar axis 6 units to the left of the pole; vertices are at $(6, 0)$ and $(2, \pi)$.

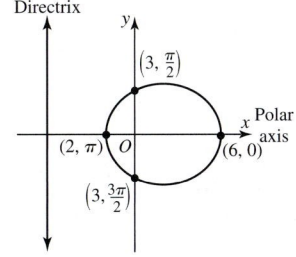

19. $y^2 + 2x - 1 = 0$ **21.** $16x^2 + 7y^2 + 48y - 64 = 0$ **23.** $3x^2 - y^2 + 12x + 9 = 0$ **25.** $4x^2 + 3y^2 - 16y - 64 = 0$

27. $9x^2 + 5y^2 - 24y - 36 = 0$ **29.** $3x^2 + 4y^2 - 12x - 36 = 0$ **31.** $r = \dfrac{1}{1 + \sin \theta}$ **33.** $r = \dfrac{12}{5 - 4 \cos \theta}$ **35.** $r = \dfrac{12}{1 - 6 \sin \theta}$

37. Use $d(D, P) = p - r \cos \theta$ in the derivation of equation (a) in Table 5.

39. Use $d(D, P) = p + r \sin \theta$ in the derivation of equation (a) in Table 5.

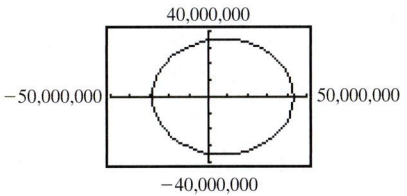

12.7 Concepts and Vocabulary *(page 960)*

1. Plane curve; parameter   **2.** Ellipse   **3.** Cycloid   **4.** F   **5.** T

12.7 Exercises *(page 960)*

1.

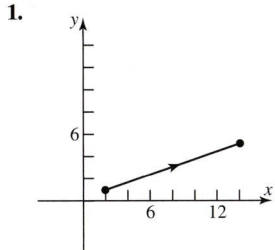

$x - 3y + 1 = 0$

3.

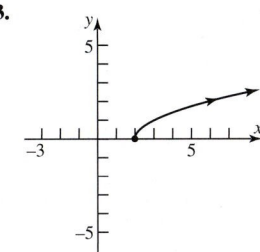

$y = \sqrt{x - 2}$

5.

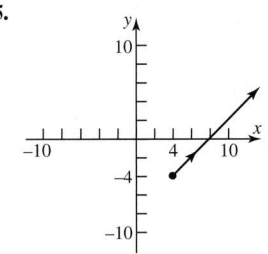

$x = y + 8$

7.

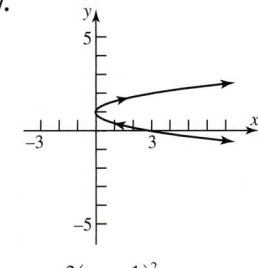

$x = 3(y - 1)^2$

9.

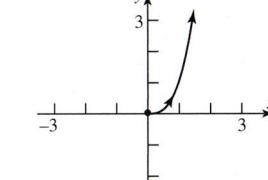

$2y = 2 + x$

11.

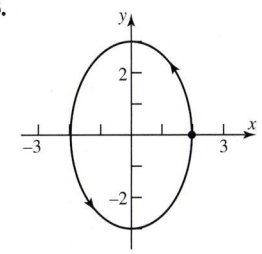

$y = x^3$

13.

$\dfrac{x^2}{4} + \dfrac{y^2}{9} = 1$

15.

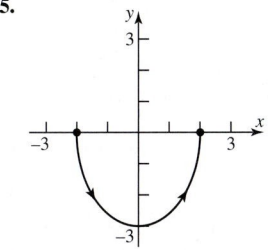

$\dfrac{x^2}{4} + \dfrac{y^2}{9} = 1$

17.

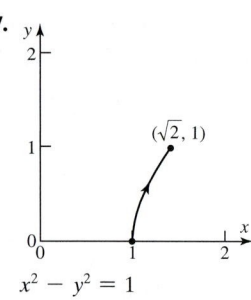

$x^2 - y^2 = 1$

19.

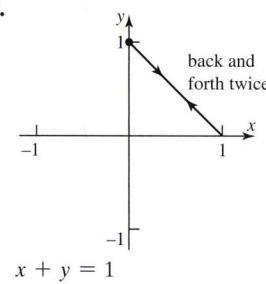

back and forth twice

$x + y = 1$

21. (a) $x = 3$
$y = -16t^2 + 50t + 6$
(b) 3.24 sec   **(c)** 1.5625 sec; 45.0625 ft
(d)

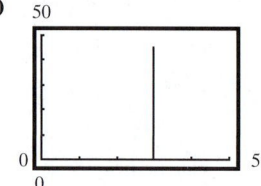

23. (a) Train: $x_1 = t^2, y_1 = 1$;
Bill: $x_2 = 5(t - 5), y_2 = 3$
(b) Bill won't catch the train.
(c)

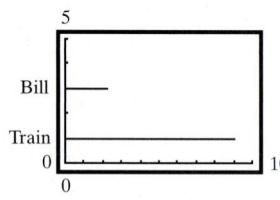

25. (a) $x = (145 \cos 20°)t$
$y = -16t^2 + (145 \sin 20°)t + 5$
(b) 3.20 sec   **(c)** 1.55 sec; 43.43 ft
(d) 435.65 ft
(e)

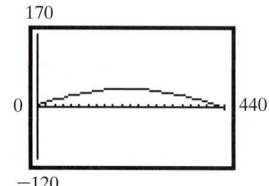

27. (a) $x = (40 \cos 45°)t$
$y = -4.9t^2 + (40 \sin 45°)t + 300$
(b) 11.23 sec   **(c)** 2.89 sec; 340.82 m
(d) 317.52 m
(e)

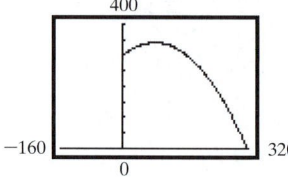

29. (a) Paseo: $x = 40t - 5, y = 0$; Bonneville: $x = 0, y = 30t - 4$ **(b)** $d = \sqrt{(40t - 5)^2 + (30t - 4)^2}$
(c)   **(d)** 0.2 mi; 7.68 min   **(e)** Turn axes off to see the graph:

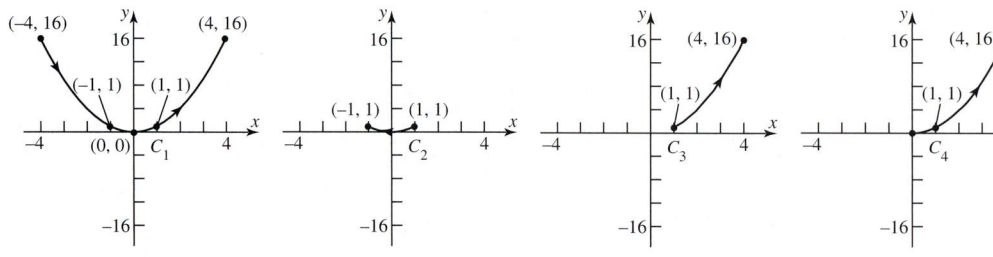

31. $x = t$ $x = \dfrac{t + 1}{4}$
$y = 4t - 1$ or $y = t$

33. $x = t$ $x = t^3$
$y = t^2 + 1$ or $y = t^6 + 1$

35. $x = t$ $x = \sqrt[3]{t}$
$y = t^3$ or $y = t$

37. $x = t^4$ $x = t^6$
$y = t^6$ or $y = t^9$

39. $x = t + 2, y = t, 0 \le t \le 5$

41. $x = 3 \cos t, y = 2 \sin t, 0 \le t \le 2\pi$

43. $x = 2 \cos(\pi t), y = -3 \sin(\pi t), 0 \le t \le 2$ **45.** $x = 2 \sin(2\pi t), y = 3 \cos(2\pi t), 0 \le t \le 1$

47.

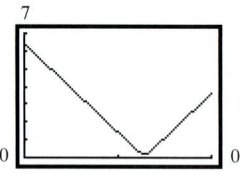

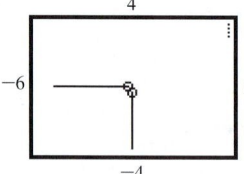

49. The orientation is from (x_1, y_1) to (x_2, y_2).

51.

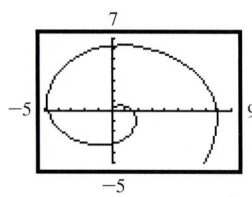

53.

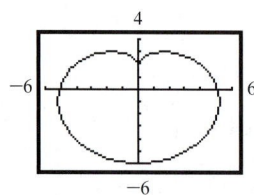

55. (a)

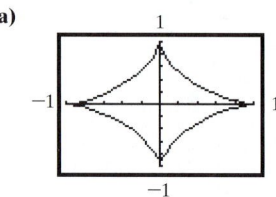

(b) $x^{2/3} + y^{2/3} = 1$

Historical Problems *(page 971)*

1. $x = 6$ units, $y = 8$ units

12.8 Exercises *(page 971)*

1.
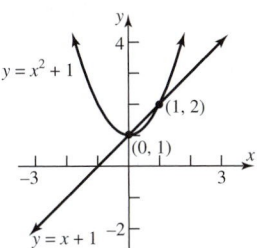

3.
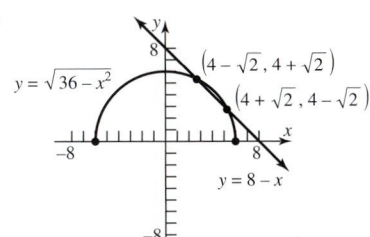

5.
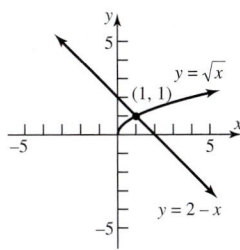

7.
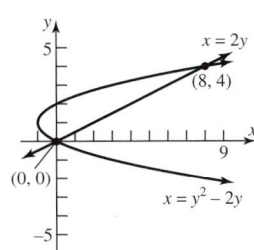

9.
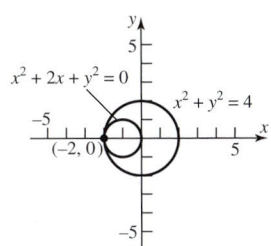

11.
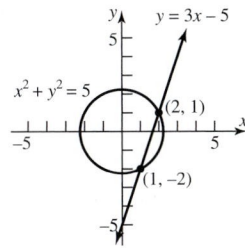

13.
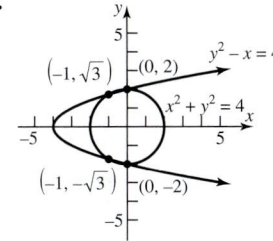

15.
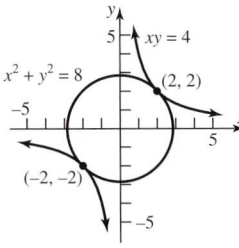

17. No points of intersection.
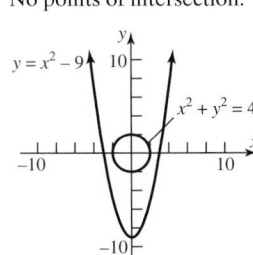

19.
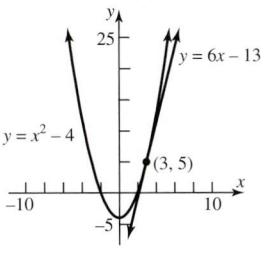

21. $x = 1, y = 4; x = -1, y = -4; x = 2\sqrt{2}, y = \sqrt{2}; x = -2\sqrt{2}, y = -\sqrt{2}$ **23.** $x = 0, y = 1; x = -\dfrac{2}{3}, y = -\dfrac{1}{3}$

25. $x = 0, y = -1; x = \dfrac{5}{2}, y = -\dfrac{7}{2}$ **27.** $x = 2, y = \dfrac{1}{3}; x = \dfrac{1}{2}, y = \dfrac{4}{3}$ **29.** $x = 3, y = 2; x = 3, y = -2; x = -3, y = 2; x = -3, y = -2$

31. $x = \dfrac{1}{2}, y = \dfrac{3}{2}; x = \dfrac{1}{2}, y = -\dfrac{3}{2}; x = -\dfrac{1}{2}, y = \dfrac{3}{2}; x = -\dfrac{1}{2}, y = -\dfrac{3}{2}$ **33.** $x = \sqrt{2}, y = 2\sqrt{2}; x = -\sqrt{2}, y = -2\sqrt{2}$

35. No real solution exists. **37.** $x = \dfrac{8}{3}, y = \dfrac{2\sqrt{10}}{3}; x = -\dfrac{8}{3}, y = \dfrac{2\sqrt{10}}{3}; x = \dfrac{8}{3}, y = -\dfrac{2\sqrt{10}}{3}; x = -\dfrac{8}{3}, y = -\dfrac{2\sqrt{10}}{3}$

39. $x = 1, y = \dfrac{1}{2}; x = -1, y = \dfrac{1}{2}; x = 1, y = -\dfrac{1}{2}; x = -1, y = -\dfrac{1}{2}$ **41.** No real solution exists.

43. $x = \sqrt{3}, y = \sqrt{3}; x = -\sqrt{3}, y = -\sqrt{3}; x = 2, y = 1; x = -2, y = -1$ **45.** $x = 0, y = -2; x = 0, y = 1; x = 2, y = -1$

47. $x = 2, y = 8$ **49.** $x = 81, y = 3$

51.
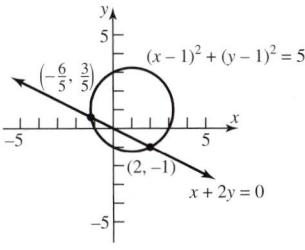

53.
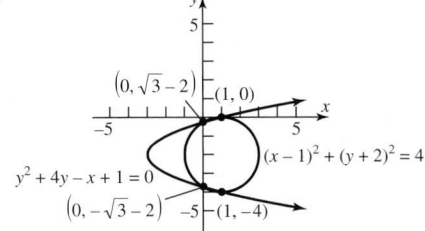

55.
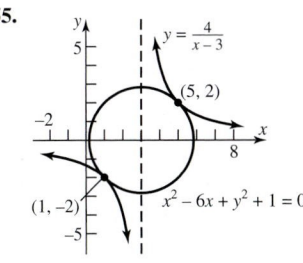

57. $x = 0.48$, $y = 0.62$ **59.** $x = -1.65$, $y = -0.89$ **61.** $x = 0.58$, $y = 1.86$; $x = 1.81$, $y = 1.05$; $x = 0.58$, $y = -1.86$; $x = 1.81$,

$y = -1.05$ **63.** $x = 2.35$, $y = 0.85$ **65.** 3 and 1; -3 and -1 **67.** 2 and 2; -2 and -2 **69.** $\frac{1}{2}$ and $\frac{1}{3}$ **71.** 5   **73.** 5 in. by 3 in.

75. 2 cm and 4 cm   **77.** Tortoise: 7 mi/hr, hare: $7\frac{1}{2}$ mi/hr   **79.** 12 cm by 18 cm   **81.** $x = 60$ ft; $y = 30$ ft

83. $l = \dfrac{P + \sqrt{P^2 - 16A}}{4}$; $w = \dfrac{P - \sqrt{P^2 - 16A}}{4}$ **85.** $y = 4x - 4$ **87.** $y = 2x + 1$ **89.** $y = -\dfrac{1}{3}x + \dfrac{7}{3}$ **91.** $y = 2x - 3$

93. $r_1 = \dfrac{-b + \sqrt{b^2 - 4ac}}{2a}$; $r_2 = \dfrac{-b - \sqrt{b^2 - 4ac}}{2a}$ **95.** **(a)** 4.27 ft by 4.27 ft or 0.093 ft by 0.093 ft

Review Exercises *(page 976)*

1. Parabola; vertex $(0, 0)$, focus $(-4, 0)$, directrix $x = 4$ **3.** Hyperbola; center $(0, 0)$, vertices $(5, 0)$ and $(-5, 0)$, foci $(\sqrt{26}, 0)$ and

$(-\sqrt{26}, 0)$, asymptotes $y = \dfrac{1}{5}x$ and $y = -\dfrac{1}{5}x$ **5.** Ellipse; center $(0, 0)$, vertices $(0, 5)$ and $(0, -5)$, foci $(0, 3)$ and $(0, -3)$

7. $x^2 = -4(y - 1)$: Parabola; vertex $(0, 1)$, focus $(0, 0)$, directrix $y = 2$ **9.** $\dfrac{x^2}{2} - \dfrac{y^2}{8} = 1$: Hyperbola; center $(0, 0)$, vertices $(\sqrt{2}, 0)$ and

$(-\sqrt{2}, 0)$, foci $(\sqrt{10}, 0)$ and $(-\sqrt{10}, 0)$, asymptotes $y = 2x$ and $y = -2x$ **11.** $(x - 2)^2 = 2(y + 2)$: Parabola; vertex $(2, -2)$,

focus $\left(2, -\dfrac{3}{2}\right)$, directrix $y = -\dfrac{5}{2}$ **13.** $\dfrac{(y - 2)^2}{4} - (x - 1)^2 = 1$: Hyperbola; center $(1, 2)$, vertices $(1, 4)$ and $(1, 0)$, foci $(1, 2 + \sqrt{5})$ and

$(1, 2 - \sqrt{5})$, asymptotes $y - 2 = \pm 2(x - 1)$ **15.** $\dfrac{(x - 2)^2}{9} + \dfrac{(y - 1)^2}{4} = 1$: Ellipse; center $(2, 1)$, vertices $(5, 1)$ and $(-1, 1)$,

foci $(2 + \sqrt{5}, 1)$ and $(2 - \sqrt{5}, 1)$ **17.** $(x - 2)^2 = -4(y + 1)$: Parabola; vertex $(2, -1)$, focus $(2, -2)$, directrix $y = 0$

19. $\dfrac{(x - 1)^2}{4} + \dfrac{(y + 1)^2}{9} = 1$: Ellipse; center $(1, -1)$, vertices $(1, 2)$ and $(1, -4)$, foci $(1, -1 + \sqrt{5})$ and $(1, -1 - \sqrt{5})$

21. $y^2 = -8x$

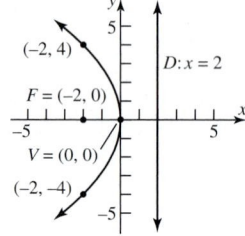

23. $\dfrac{y^2}{4} - \dfrac{x^2}{12} = 1$

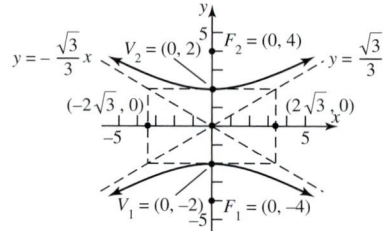

25. $\dfrac{x^2}{16} + \dfrac{y^2}{7} = 1$

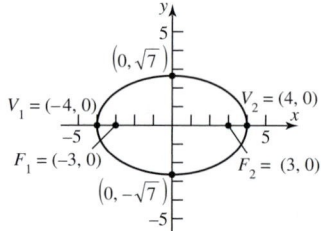

27. $(x - 2)^2 = -4(y + 3)$

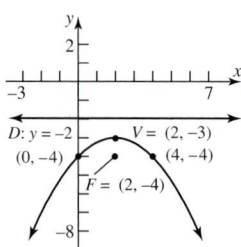

29. $(x + 2)^2 - \dfrac{(y + 3)^2}{3} = 1$

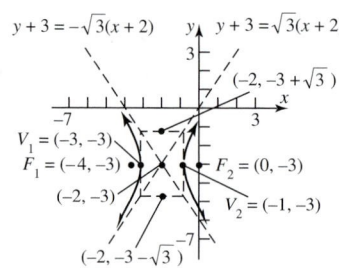

31. $\dfrac{(x + 4)^2}{16} + \dfrac{(y - 5)^2}{25} = 1$

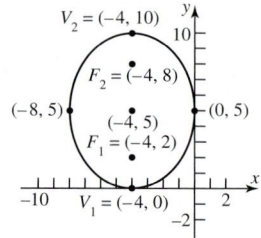

33. $\dfrac{(x+1)^2}{9} - \dfrac{(y-2)^2}{7} = 1$

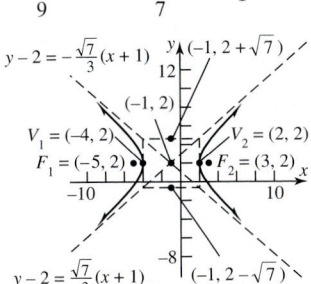

35. $\dfrac{(x-3)^2}{9} - \dfrac{(y-1)^2}{4} = 1$

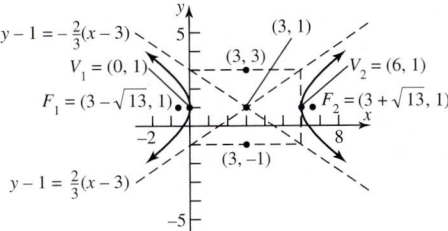

37. Parabola   **39.** Ellipse   **41.** Parabola   **43.** Hyperbola   **45.** Ellipse

47. $x'^2 - \dfrac{y'^2}{9} = 1$

Hyperbola

Center at the origin

Transverse axis the x'-axis

Vertices at $(\pm 1, 0)$

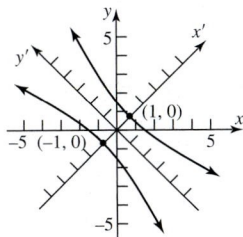

49. $\dfrac{x'^2}{2} + \dfrac{y'^2}{4} = 1$

Ellipse

Center at origin

Major axis the y'-axis

Vertices at $(0, \pm 2)$

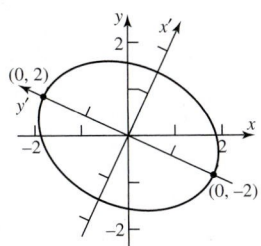

51. $y'^2 = -\dfrac{4\sqrt{13}}{13}x'$

Parabola

Vertex at the origin

Focus on the x'-axis at $\left(-\dfrac{\sqrt{13}}{13}, 0\right)$

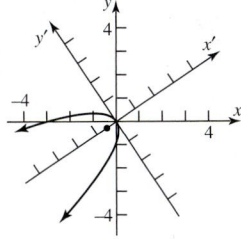

53. Parabola; directrix is perpendicular to the polar axis 4 units to the left of the pole; vertex is $(2, \pi)$.

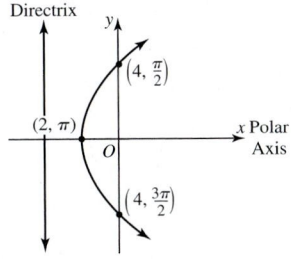

55. Ellipse; directrix is parallel to the polar axis 6 units below the pole; vertices are $\left(6, \dfrac{\pi}{2}\right)$ and $\left(2, \dfrac{3\pi}{2}\right)$.

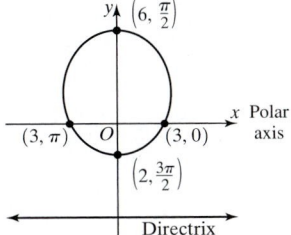

57. Hyperbola; directrix is perpendicular to the polar axis 1 unit to the right of the pole; vertices are $\left(\dfrac{2}{3}, 0\right)$ and $(-2, \pi)$.

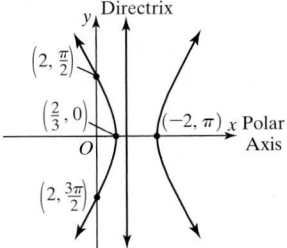

59. $y^2 - 8x - 16 = 0$ **61.** $3x^2 - y^2 - 8x + 4 = 0$

63.

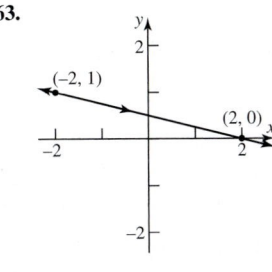

$x + 4y = 2$

65.

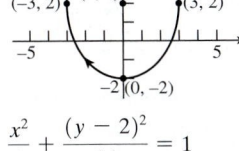

$\dfrac{x^2}{9} + \dfrac{(y-2)^2}{16} = 1$

67.

$1 + y = x$

69. $x = t, y = -2t + 4, -\infty < t < \infty; x = \dfrac{t-4}{-2}, y = t, -\infty < t < \infty$ **71.** $x = 4\cos\left(\dfrac{\pi}{2}t\right), y = 3\sin\left(\dfrac{\pi}{2}t\right), 0 \le t \le 4$

73. $x = -\dfrac{2}{5}, y = -\dfrac{11}{5}; x = -2, y = 1$ **75.** $x = 2\sqrt{2}, y = \sqrt{2}; x = -2\sqrt{2}, y = -\sqrt{2}$ **77.** $x = 0, y = 0; x = -3, y = 3; x = 3, y = 3$

79. $x = \sqrt{2}, y = -\sqrt{2}; x = -\sqrt{2}, y = \sqrt{2}; x = \dfrac{4}{3}\sqrt{2}, y = -\dfrac{2}{3}\sqrt{2}; x = -\dfrac{4}{3}\sqrt{2}, y = \dfrac{2}{3}\sqrt{2}$ **81.** $x = 1, y = -1$

83. $\dfrac{x^2}{5} - \dfrac{y^2}{4} = 1$ **85.** The ellipse $\dfrac{x^2}{16} + \dfrac{y^2}{7} = 1$ **87.** $\dfrac{1}{4}$ ft or 3 in.   **89.** 19.72 ft, 18.86 ft, 14.91 ft

91. (a) 45.24 mi from the Master Station   **(b)** 0.000645 sec   **(c)** (66, 20)

93. (a) $x = (100\cos 35°)t$ **(b)** 3.6866 sec   **(c)** 1.7924 sec; 57.4 ft   **(d)** 302 ft   **(e)**
$\quad\quad y = -16t^2 + (100\sin 35°)t + 6$

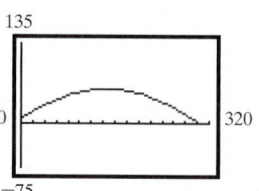

Cumulative Review Exercises *(page 979)*

1. $\theta = \dfrac{\pi}{12} \pm \pi k, k$ is any integer; $\theta = \dfrac{5\pi}{12} \pm \pi k, k$ is any integer   **2.** $\theta = \dfrac{\pi}{6}$

3. $r = 8\sin\theta$ **4.** $\left\{ x \,\middle|\, x \ne \dfrac{3\pi}{4} \pm \pi k, k \text{ is an integer} \right\}$ **5.** $x = 2, y = 1$ **6.** $x = -3, y = 9$ or $x = 2, y = 4$

7.

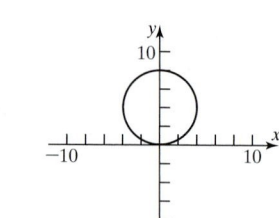

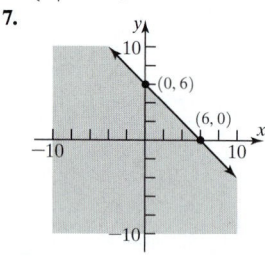

8. $-3 \le x \le 2$ or $[-3, 2]$
9. $\theta = 22.5°$
10. (a) $y = 2x - 2$ **(b)** $(x - 2)^2 + y^2 = 4$
   (c) $\dfrac{x^2}{9} + \dfrac{y^2}{4} = 1$ **(d)** $y = 2(x - 1)^2$
   (e) $\dfrac{y^2}{1} - \dfrac{x^2}{3} = 1$ **(f)** $y = 4^x$

C H A P T E R 1 3   Sequences; Induction; The Binomial Theorem

13.1 Concepts and Vocabulary *(page 994)*

1. Sequence **2.** 3; 15 **3.** 20 **4.** T **5.** T **6.** T

13.1 Exercises *(page 994)*

1. 1, 2, 3, 4, 5 **3.** $\dfrac{1}{3}, \dfrac{2}{4}, \dfrac{3}{5}, \dfrac{4}{6}, \dfrac{5}{7}$ **5.** 1, -4, 9, -16, 25 **7.** $\dfrac{1}{2}, \dfrac{2}{5}, \dfrac{2}{7}, \dfrac{8}{41}, \dfrac{8}{61}$ **9.** $-\dfrac{1}{6}, \dfrac{1}{12}, -\dfrac{1}{20}, \dfrac{1}{30}, -\dfrac{1}{42}$ **11.** $\dfrac{1}{e}, \dfrac{2}{e^2}, \dfrac{3}{e^3}, \dfrac{4}{e^4}, \dfrac{5}{e^5}$ **13.** $\dfrac{n}{n+1}$

15. $\dfrac{1}{2^{n-1}}$ **17.** $(-1)^{n+1}$ **19.** $(-1)^{n+1}n$ **21.** $a_1 = 2, a_2 = 5, a_3 = 8, a_4 = 11, a_5 = 14$ **23.** $a_1 = -2, a_2 = 0, a_3 = 3, a_4 = 7, a_5 = 12$

25. $a_1 = 5, a_2 = 10, a_3 = 20, a_4 = 40, a_5 = 80$ **27.** $a_1 = 3, a_2 = \dfrac{3}{2}, a_3 = \dfrac{1}{2}, a_4 = \dfrac{1}{8}, a_5 = \dfrac{1}{40}$ **29.** $a_1 = 1, a_2 = 2, a_3 = 2, a_4 = 4, a_5 = 8$

31. $a_1 = A, a_2 = A + d, a_3 = A + 2d, a_4 = A + 3d, a_5 = A + 4d$

33. $a_1 = \sqrt{2}, a_2 = \sqrt{2 + \sqrt{2}}, a_3 = \sqrt{2 + \sqrt{2 + \sqrt{2}}}, a_4 = \sqrt{2 + \sqrt{2 + \sqrt{2 + \sqrt{2}}}}, a_5 = \sqrt{2 + \sqrt{2 + \sqrt{2 + \sqrt{2 + \sqrt{2}}}}}$

35. $3 + 4 + 5 + 6 + 7$ **37.** $\dfrac{1}{2} + 2 + \dfrac{9}{2} + \cdots + 32$ **39.** $1 + \dfrac{1}{3} + \dfrac{1}{9} + \cdots + \dfrac{1}{3^n}$ **41.** $\dfrac{1}{3} + \dfrac{1}{9} + \cdots + \dfrac{1}{3^n}$

43. $\ln 2 - \ln 3 + \ln 4 - \cdots + (-1)^n \ln n$ **45.** $\displaystyle\sum_{k=1}^{20} k$ **47.** $\displaystyle\sum_{k=1}^{13} \dfrac{k}{k+1}$ **49.** $\displaystyle\sum_{k=0}^{6} (-1)^k\left(\dfrac{1}{3^k}\right)$ **51.** $\displaystyle\sum_{k=1}^{n} \dfrac{3^k}{k}$

53. $\displaystyle\sum_{k=0}^{n} (a + kd)$ or $\displaystyle\sum_{k=1}^{n+1} [a + (k-1)d]$ **55.** 50 **57.** 21 **59.** 90 **61.** 26 **63.** 42 **65.** 96

67. (a) \$2930 **(b)** 14 payments have been made. **(c)** 36 payments. \$3584.62 **(d)** \$584.62 **69. (a)** 2162 **(b)** After 26 months.

71. (a) $a_0 = 0, a_n = (1.02)a_{n-1} + 500$ **(b)** After 82 quarters **(c)** \$156,116.15

73. (a) $a_0 = 150{,}000, a_n = (1.005)a_{n-1} - 899.33$

(b) \$149,850.67

(c)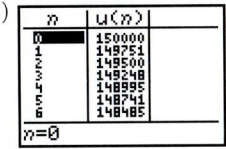

(d) After 58 payments or 4 years and 10 months later

(e) After 360 payments of \$899.33, plus last payment of \$895.10

(f) \$173,754.57

(g) (a) $a_0 = 150{,}000, a_n = (1.005)a_{n-1} - 999.33$

(b) \$149,750.67   (c)

(d) After 37 payments or 3 years and 1 month later

(e) After 279 payments of \$999.33, plus last payment of \$353.69(1.005) = \$355.46

(f) \$128,169.20

75. 21   **77.** A Fibonacci Sequence

13.2 Concepts and Vocabulary *(page 1002)*

1. Arithmetic   **2.** F   **3.** F

13.2 Exercises *(page 1002)*

1. $d = 1; 5, 6, 7, 8$ **3.** $d = 2; -3, -1, 1, 3$ **5.** $d = -2; 4, 2, 0, -2$ **7.** $d = -\dfrac{1}{3}; \dfrac{1}{6}, -\dfrac{1}{6}, -\dfrac{1}{2}, -\dfrac{5}{6}$ **9.** $d = \ln 3; \ln 3, 2\ln 3, 3\ln 3, 4\ln 3$

11. $a_n = 3n - 1; a_5 = 14$ **13.** $a_n = 8 - 3n; a_5 = -7$ **15.** $a_n = \dfrac{1}{2}(n - 1); a_5 = 2$ **17.** $a_n = \sqrt{2}n; a_5 = 5\sqrt{2}$ **19.** $a_{12} = 24$

21. $a_{10} = -26$ **23.** $a_8 = a + 7b$ **25.** $a_1 = -13; d = 3; a_n = a_{n-1} + 3$ **27.** $a_1 = -53; d = 6; a_n = a_{n-1} + 6$

29. $a_1 = 28; d = -2; a_n = a_{n-1} - 2$ **31.** $a_1 = 25; d = -2; a_n = a_{n-1} - 2$ **33.** n^2 **35.** $\dfrac{n}{2}(9 + 5n)$ **37.** 1260   **39.** 324

41. sum(seq(3.45n+4.12,n,1,20,1)) 806.9

43. sum(seq(2.4n+.4,n,1,15,1)) 294

45. sum(seq(2.58n+2.32,n,1,25,1)) 896.5

47. $-\dfrac{3}{2}$

49. 1185 seats

51. 210 of beige and 190 blue

53. 30 rows

Historical Problems *(page 1011)*

1. $1\dfrac{2}{3}$ loaves, $10\dfrac{5}{6}$ loaves, 20 loaves, $29\dfrac{1}{6}$ loaves, $38\dfrac{1}{3}$ loaves

13.3 Concepts and Vocabulary *(page 1011)*

1. Geometric   **2.** $\dfrac{a}{1 - r}$ **3.** T   **4.** F   **5.** T

13.3 Exercises *(page 1012)*

1. $r = 3; 3, 9, 27, 81$ **3.** $r = \dfrac{1}{2}; -\dfrac{3}{2}, -\dfrac{3}{4}, -\dfrac{3}{8}, -\dfrac{3}{16}$ **5.** $r = 2; \dfrac{1}{4}, \dfrac{1}{2}, 1, 2$ **7.** $r = 2^{1/3}; 2^{1/3}, 2^{2/3}, 2, 2^{4/3}$ **9.** $r = \dfrac{3}{2}; \dfrac{1}{2}, \dfrac{3}{4}, \dfrac{9}{8}, \dfrac{27}{16}$

11. Arithmetic; $d = 1$ **13.** Neither   **15.** Arithmetic; $d = -\dfrac{2}{3}$ **17.** Neither   **19.** Geometric; $r = \dfrac{2}{3}$ **21.** Geometric; $r = 2$

23. Geometric; $r = 3^{1/2}$ **25.** $a_5 = 162; a_n = 2 \cdot 3^{n-1}$ **27.** $a_5 = 5; a_n = 5 \cdot (-1)^{n-1}$ **29.** $a_5 = 0; a_n = 0$ **31.** $a_5 = 4\sqrt{2}; a_n = (\sqrt{2})^n$

33. $a_7 = \dfrac{1}{64}$ **35.** $a_9 = 1$ **37.** $a_8 = 0.00000004$ **39.** $-\dfrac{1}{4}(1 - 2^n)$ **41.** $2\left[1 - \left(\dfrac{2}{3}\right)^n\right]$ **43.** $1 - 2^n$

45. (1/4)sum(seq(2^n,n,0,14,1)) 8191.75

47. sum(seq((2/3)^n,n,1,15,1)) 1.995432683

49. -1sum(seq(2^n,n,0,14,1)) -32767

51. $\dfrac{3}{2}$ **53.** 16

55. $\dfrac{8}{5}$ **57.** $\dfrac{20}{3}$

59. $\dfrac{18}{5}$ **61.** -4 **63.** \$21,879.11 **65.** (a) 0.77 ft (b) 8th (c) 15.88 ft (d) 20 ft **67.** A: \$25,250 per year in 5th year, \$112,742 total;

B: \$24,761 per year in 5th year, \$116,801 total **69.** Option 2 results in the most: \$16,038,304; Option 1 results in the least: \$14,700,000.
71. 1.845×10^{19} **73.** 10 **75.** \$72.67 per share **77.** Yes. A constant sequence is both arithmetic and geometric. For example, $3, 3, 3, \ldots$
is an arithmetic sequence with $a_1 = 3$ and $d = 0$ and is a geometric sequence with $a_1 = 3$ and $r = 1$.

13.4 Concepts and Vocabulary *(page 1017)*

1. Mathematical Induction

13.4 Exercises *(page 1017)*

1. (I) $n = 1: 2(1) = 2$ and $1(1 + 1) = 2$
 (II) If $2 + 4 + 6 + \cdots + 2k = k(k + 1)$, then $2 + 4 + 6 + \cdots + 2k + 2(k + 1) = (2 + 4 + 6 + \cdots + 2k) + 2(k + 1)$
 $= k(k + 1) + 2(k + 1) = k^2 + 3k + 2 = (k + 1)(k + 2) = (k + 1)[(k + 1) + 1]$.

3. (I) $n = 1: 1 + 2 = 3$ and $\dfrac{1}{2}(1)(1 + 5) = \dfrac{1}{2}(6) = 3$

 (II) If $3 + 4 + 5 + \cdots + (k + 2) = \dfrac{1}{2}k(k + 5)$, then $3 + 4 + 5 + \cdots + (k + 2) + [(k + 1) + 2]$

 $= [3 + 4 + 5 + \cdots + (k + 2)] + (k + 3) = \dfrac{1}{2}k(k + 5) + k + 3 = \dfrac{1}{2}(k^2 + 7k + 6) = \dfrac{1}{2}(k + 1)(k + 6)$

 $= \dfrac{1}{2}(k + 1)[(k + 1) + 5]$.

5. (I) $n = 1: 3(1) - 1 = 2$ and $\dfrac{1}{2}(1)[3(1) + 1] = \dfrac{1}{2}(4) = 2$

 (II) If $2 + 5 + 8 + \cdots + (3k - 1) = \dfrac{1}{2}k(3k + 1)$, then $2 + 5 + 8 + \cdots + (3k - 1) + [3(k + 1) - 1]$

 $= [2 + 5 + 8 + \cdots + (3k - 1)] + (3k + 2) = \dfrac{1}{2}k(3k + 1) + (3k + 2) = \dfrac{1}{2}(3k^2 + 7k + 4) = \dfrac{1}{2}(k + 1)(3k + 4)$

 $= \dfrac{1}{2}(k + 1)[3(k + 1) + 1]$.

7. (I) $n = 1: 2^{1-1} = 1$ and $2^1 - 1 = 1$
 (II) If $1 + 2 + 2^2 + \cdots + 2^{k-1} = 2^k - 1$, then $1 + 2 + 2^2 + \cdots + 2^{k-1} + 2^{((k+1)-1)} = (1 + 2 + 2^2 + \cdots + 2^{k-1}) + 2^k$
 $= 2^k - 1 + 2^k = 2(2^k) - 1 = 2^{k+1} - 1$.

9. (I) $n = 1: 4^{1-1} = 1$ and $\dfrac{1}{3}(4^1 - 1) = \dfrac{1}{3}(3) = 1$

 (II) If $1 + 4 + 4^2 + \cdots + 4^{k-1} = \dfrac{1}{3}(4^k - 1)$, then $1 + 4 + 4^2 + \cdots + 4^{k-1} + 4^{(k+1)-1} = (1 + 4 + 4^2 + \cdots + 4^{k-1}) + 4^k$

 $= \dfrac{1}{3}(4^k - 1) + 4^k = \dfrac{1}{3}[4^k - 1 + 3(4^k)] = \dfrac{1}{3}[4(4^k) - 1] = \dfrac{1}{3}(4^{k+1} - 1)$.

11. (I) $n = 1: \dfrac{1}{1 \cdot 2} = \dfrac{1}{2}$ and $\dfrac{1}{1 + 1} = \dfrac{1}{2}$

 (II) If $\dfrac{1}{1 \cdot 2} + \dfrac{1}{2 \cdot 3} + \dfrac{1}{3 \cdot 4} + \cdots + \dfrac{1}{k(k + 1)} = \dfrac{k}{k + 1}$, then $\dfrac{1}{1 \cdot 2} + \dfrac{1}{2 \cdot 3} + \dfrac{1}{3 \cdot 4} + \cdots + \dfrac{1}{k(k + 1)} + \dfrac{1}{(k + 1)[(k + 1) + 1]}$

 $= \left[\dfrac{1}{1 \cdot 2} + \dfrac{1}{2 \cdot 3} + \dfrac{1}{3 \cdot 4} + \cdots + \dfrac{1}{k(k + 1)}\right] + \dfrac{1}{(k + 1)(k + 2)} = \dfrac{k}{k + 1} + \dfrac{1}{(k + 1)(k + 2)} = \dfrac{k(k + 2) + 1}{(k + 1)(k + 2)}$

 $= \dfrac{k^2 + 2k + 1}{(k + 1)(k + 2)} = \dfrac{(k + 1)^2}{(k + 1)(k + 2)} = \dfrac{k + 1}{k + 2} = \dfrac{k + 1}{(k + 1) + 1}$.

13. (I) $n = 1: 1^2 = 1$ and $\dfrac{1}{6} \cdot 1 \cdot 2 \cdot 3 = 1$

 (II) If $1^2 + 2^2 + 3^2 + \cdots + k^2 = \dfrac{1}{6}k(k + 1)(2k + 1)$, then $1^2 + 2^2 + 3^2 + \cdots + k^2 + (k + 1)^2$

 $= (1^2 + 2^2 + 3^2 + \cdots + k^2) + (k + 1)^2 = \dfrac{1}{6}k(k + 1)(2k + 1) + (k + 1)^2 = \dfrac{1}{6}(2k^3 + 9k^2 + 13k + 6)$

 $= \dfrac{1}{6}(k + 1)(k + 2)(2k + 3) = \dfrac{1}{6}(k + 1)[(k + 1) + 1][2(k + 1) + 1]$.

15. (I) $n = 1: 5 - 1 = 4$ and $\frac{1}{2}(1)(9 - 1) = \frac{1}{2} \cdot 8 = 4$

(II) If $4 + 3 + 2 + \cdots + (5 - k) = \frac{1}{2}k(9 - k)$, then $4 + 3 + 2 + \cdots + (5 - k) + [5 - (k + 1)]$

$= [4 + 3 + 2 + \cdots + (5 - k)] + 4 - k = \frac{1}{2}k(9 - k) + 4 - k = \frac{1}{2}(9k - k^2 + 8 - 2k) = \frac{1}{2}(-k^2 + 7k + 8)$

$= \frac{1}{2}(k + 1)(8 - k) = \frac{1}{2}(k + 1)[9 - (k + 1)]$.

17. (I) $n = 1: 1 \cdot (1 + 1) = 2$ and $\frac{1}{3} \cdot 1 \cdot 2 \cdot 3 = 2$

(II) If $1 \cdot 2 + 2 \cdot 3 + 3 \cdot 4 + \cdots + k(k + 1) = \frac{1}{3}k(k + 1)(k + 2)$, then $1 \cdot 2 + 2 \cdot 3 + 3 \cdot 4 + \cdots + k(k + 1)$

$+ (k + 1)[(k + 1) + 1] = [1 \cdot 2 + 2 \cdot 3 + 3 \cdot 4 + \cdots + k(k + 1)] + (k + 1)(k + 2)$

$= \frac{1}{3}k(k + 1)(k + 2) + (k + 1)(k + 2) = \frac{1}{3}(k + 1)(k + 2)(k + 3) = \frac{1}{3}(k + 1)[(k + 1) + 1][(k + 1) + 2]$.

19. (I) $n = 1: 1^2 + 1 = 2$ which is divisible by 2.
(II) If $k^2 + k$ is divisible by 2, then $(k + 1)^2 + (k + 1) = k^2 + 2k + 1 + k + 1 = (k^2 + k) + 2k + 2$. Since $k^2 + k$ is divisible by 2 and $2k + 2$ is divisible by 2, $(k + 1)^2 + (k + 1)$ is divisible by 2.

21. (I) $n = 1: 1^2 - 1 + 2 = 2$ which is divisible by 2.
(II) If $k^2 - k + 2$ is divisible by 2, then $(k + 1)^2 - (k + 1) + 2 = k^2 + 2k + 1 - k - 1 + 2 = (k^2 - k + 2) + 2k$. Since $k^2 - k + 2$ is divisible by 2 and $2k$ is divisible by 2, $(k + 1)^2 - (k + 1) + 2$ is divisible by 2.

23. (I) $n = 1$: If $x > 1$, then $x^1 = x > 1$.
(II) Assume, for some natural number k, that if $x > 1$, then $x^k > 1$. Multiply both sides of the inequality $x^k > 1$ by x. Then $x^{k+1} > x > 1$.

25. (I) $n = 1: a - b$ is a factor of $a^1 - b^1 = a - b$.
(II) If $a - b$ is a factor of $a^k - b^k$, show that $a - b$ is a factor of $a^{k+1} - b^{k+1}$: $a^{k+1} - b^{k+1} = a(a^k - b^k) + b^k(a - b)$. Since $a - b$ is a factor of $a^k - b^k$ and $a - b$ is a factor of $a - b$, then $a - b$ is a factor of $a^{k+1} - b^{k+1}$.

27. $n = 1: 1^2 - 1 + 41 = 41$ which is a prime number.
$n = 41: 41^2 - 41 + 41 = 1681 = 41^2$ which is not prime.

29. (I) $n = 1: ar^{1-1} = a \cdot 1 = a$ and $a \cdot \dfrac{1 - r^1}{1 - r} = a$, because $r \neq 1$.

(II) If $a + ar + ar^2 + \cdots + ar^{k-1} = a\left(\dfrac{1 - r^k}{1 - r}\right)$, then $a + ar + ar^2 + \cdots + ar^{k-1} + ar^{(k+1)-1} = (a + ar + ar^2 + \cdots + ar^{k-1}) + ar^k$

$= a\left(\dfrac{1 - r^k}{1 - r}\right) + ar^k = \dfrac{a(1 - r^k) + ar^k(1 - r)}{1 - r} = \dfrac{a - ar^k + ar^k - ar^{k+1}}{1 - r} = a\left(\dfrac{1 - r^{k+1}}{1 - r}\right)$.

31. (I) $n = 4$: The number of diagonals in a convex polygon of 4 sides is 2 and $\frac{1}{2} \cdot 4 \cdot (4 - 3) = 2$

(II) If the number of diagonals in a convex polygon of $k \geq 4$ sides is $\frac{1}{2}k(k - 3)$ then that of $(k + 1)$ sides is increased by

$(k + 1) - 2 = k - 1$. Thus, the number of diagonals in a convex polygon of $(k + 1)$ sides is $\frac{1}{2}k(k - 3) + (k - 1)$

$= \frac{1}{2}[k^2 - 3k + 2k - 2] = \frac{1}{2}[k^2 - k - 2] = \frac{1}{2}(k + 1)(k - 2) = \frac{1}{2}(k + 1)[(k + 1) - 3]$.

13.5 Concepts and Vocabulary *(page 1024)*

1. Pascal triangle   **2.** 15   **3.** F

13.5 Exercises *(page 1025)*

1. 10   **3.** 21   **5.** 50   **7.** 1   **9.** 1.866×10^{15} **11.** 1.483×10^{13} **13.** $x^5 + 5x^4 + 10x^3 + 10x^2 + 5x + 1$

15. $x^6 - 12x^5 + 60x^4 - 160x^3 + 240x^2 - 192x + 64$ **17.** $81x^4 + 108x^3 + 54x^2 + 12x + 1$

19. $x^{10} + 5x^8y^2 + 10x^6y^4 + 10x^4y^6 + 5x^2y^8 + y^{10}$ **21.** $x^3 + 6\sqrt{2}x^{5/2} + 30x^2 + 40\sqrt{2}x^{3/2} + 60x + 24\sqrt{2}x^{1/2} + 8$

23. $a^5x^5 + 5a^4bx^4y + 10a^3b^2x^3y^2 + 10a^2b^3x^2y^3 + 5ab^4xy^4 + b^5y^5$ **25.** 17,010   **27.** $-101,376$ **29.** 41,472

31. $2835x^3$ **33.** $314,928x^7$ **35.** 495   **37.** 3360   **39.** 1.00501

41. $\dbinom{n}{n - 1} = \dfrac{n!}{(n - 1)![n - (n - 1)]!} = \dfrac{n!}{(n - 1)!1!} = \dfrac{n \cdot (n - 1)!}{(n - 1)!} = n$; $\dbinom{n}{n} = \dfrac{n!}{n!(n - n)!} = \dfrac{n!}{n!0!} = \dfrac{n!}{n!} = 1$

43. $2^n = (1 + 1)^n = \dbinom{n}{0}1^n + \dbinom{n}{1}(1)^{n-1}(1) + \cdots + \dbinom{n}{n}1^n = \dbinom{n}{0} + \dbinom{n}{1} + \cdots + \dbinom{n}{n}$ **45.** 1

47. We use Mathematical Induction to prove the Binomial Theorem. First, we show that formula (2) is true for $n = 1$.

$$(x + a)^1 = x + a = \binom{1}{0}x^1 + \binom{1}{1}a^1$$

Next we suppose that formula (2) is true for some k. That is, we assume that

$$(x + a)^k = \binom{k}{0}x^k + \binom{k}{1}ax^{k-1} + \cdots + \binom{k}{j-1}a^{j-1}x^{k-j+1} + \binom{k}{j}a^j x^{k-j} + \cdots + \binom{k}{k}a^k$$

Now we calculate $(x + a)^{k+1}$.

$$(x + a)^{k+1} = (x + a)(x + a)^k = x(x + a)^k + a(x + a)^k$$

$$= x\left[\binom{k}{0}x^k + \binom{k}{1}ax^{k-1} + \cdots + \binom{k}{j-1}a^{j-1}x^{k-j+1} + \binom{k}{j}a^j x^{k-j} + \cdots + \binom{k}{k}a^k\right]$$

$$+ a\left[\binom{k}{0}x^k + \binom{k}{1}ax^{k-1} + \cdots + \binom{k}{j-1}a^{j-1}x^{k-j+1} + \binom{k}{j}a^j x^{k-j} + \cdots + \binom{k}{k-1}a^{k-1}x + \binom{k}{k}a^k\right]$$

$$= \binom{k}{0}x^{k+1} + \binom{k}{1}ax^k + \cdots + \binom{k}{j-1}a^{j-1}x^{k-j+2} + \binom{k}{j}a^j x^{k-j+1} + \cdots + \binom{k}{k}a^k x$$

$$+ \binom{k}{0}ax^k + \binom{k}{1}a^2 x^{k-1} + \cdots + \binom{k}{j-1}a^j x^{k-j+1} + \binom{k}{j}a^{j+1}x^{k-j} + \cdots + \binom{k}{k-1}a^k x + \binom{k}{k}a^{k+1}$$

$$= \binom{k}{0}x^{k+1} + \left[\binom{k}{1} + \binom{k}{0}\right]ax^k + \cdots + \left[\binom{k}{j} + \binom{k}{j-1}\right]a^j x^{k-j+1}$$

$$+ \cdots + \left[\binom{k}{k} + \binom{k}{k-1}\right]a^k x + \binom{k}{k}a^{k+1}$$

Because,

$$\binom{k}{0} = 1 = \binom{k+1}{0}, \binom{k}{1} + \binom{k}{0} = \binom{k+1}{1}, \cdots, \binom{k}{j} + \binom{k}{j-1} = \binom{k+1}{j}, \cdots, \binom{k}{k} = 1 = \binom{k+1}{k+1},$$

we have

$$(x + a)^{k+1} = \binom{k+1}{0}x^{k+1} + \binom{k+1}{1}ax^k + \cdots + \binom{k+1}{j}a^j x^{k-j+1} + \cdots + \binom{k+1}{k+1}a^{k+1}$$

Conditions I and II of the Principle of Mathematical Induction are satisfied, so formula (2) is true for all n.

Review Exercises (page 1027)

1. $-\dfrac{4}{3}, \dfrac{5}{4}, -\dfrac{6}{5}, \dfrac{7}{6}, -\dfrac{8}{7}$ **3.** $2, 1, \dfrac{8}{9}, 1, \dfrac{32}{25}$ **5.** $3, 2, \dfrac{4}{3}, \dfrac{8}{9}, \dfrac{16}{27}$ **7.** $2, 0, 2, 0, 2$ **9.** 48 **11.** $\displaystyle\sum_{k=1}^{13}(-1)^{k+1}\dfrac{1}{k}$ **13.** Arithmetic; $d = 1; S_n = \dfrac{n}{2}(n + 11)$

15. Neither   **17.** Geometric; $r = 8; S_n = \dfrac{8}{7}(8^n - 1)$ **19.** Arithmetic; $d = 4; S_n = 2n(n - 1)$ **21.** Geometric; $r = \dfrac{1}{2}; S_n = 6\left[1 - \left(\dfrac{1}{2}\right)^n\right]$

23. Neither   **25.** 115 **27.** 75 **29.** 0.49977 **31.** 35 **33.** $\dfrac{1}{10^{10}}$ **35.** $9\sqrt{2}$ **37.** $a_n = 5n - 4$ **39.** $a_n = n - 10$ **41.** $\dfrac{9}{2}$ **43.** $\dfrac{4}{3}$ **45.** 8

47. (I) $n = 1: 3 \cdot 1 = 3$ and $\dfrac{3}{2}(2) = 3$

(II) If $3 + 6 + 9 + \cdots + 3k = \dfrac{3k}{2}(k + 1)$, then $3 + 6 + 9 + \cdots + 3k + 3(k + 1) = (3 + 6 + 9 + \cdots + 3k) + (3k + 3)$

$$= \dfrac{3k}{2}(k + 1) + (3k + 3) = \dfrac{3k^2}{2} + \dfrac{9k}{2} + \dfrac{6}{2} = \dfrac{3}{2}(k^2 + 3k + 2) = \dfrac{3}{2}(k + 1)(k + 2) = \dfrac{3(k + 1)}{2}[(k + 1) + 1].$$

49. (I) $n = 1: 2 \cdot 3^{1-1} = 2$ and $3^1 - 1 = 2$

(II) If $2 + 6 + 18 + \cdots + 2 \cdot 3^{k-1} = 3^k - 1$, then $2 + 6 + 18 + \cdots + 2 \cdot 3^{k-1} + 2 \cdot 3^{(k+1)-1} = (2 + 6 + 18 + \cdots + 2 \cdot 3^{k-1}) + 2 \cdot 3^k$

$$= 3^k - 1 + 2 \cdot 3^k = 3 \cdot 3^k - 1 = 3^{k+1} - 1.$$

51. (I) $n = 1: 1^2 = 1$ and $\dfrac{1}{2}(6 - 3 - 1) = \dfrac{1}{2}(2) = 1$

(II) If $1^2 + 4^2 + 7^2 + \cdots + (3k - 2)^2 = \dfrac{1}{2}k(6k^2 - 3k - 1)$, then $1^2 + 4^2 + 7^2 + \cdots + (3k - 2)^2 + [3(k + 1) - 2]^2$

$$= [1^2 + 4^2 + 7^2 + \cdots + (3k - 2)^2] + (3k + 1)^2 = \dfrac{1}{2}k(6k^2 - 3k - 1) + (3k + 1)^2 = \dfrac{1}{2}(6k^3 - 3k^2 - k) + (9k^2 + 6k + 1)$$

$$= \dfrac{1}{2}(6k^3 + 15k^2 + 11k + 2) = \dfrac{1}{2}(k + 1)(6k^2 + 9k + 2) = \dfrac{1}{2}(k + 1)[6(k + 1)^2 - 3(k + 1) - 1].$$

53. 10 **55.** $x^5 + 10x^4 + 40x^3 + 80x^2 + 80x + 32$ **57.** $32x^5 + 240x^4 + 720x^3 + 1080x^2 + 810x + 243$ **59.** 144 **61.** 84

63. (a) 8 (b) 1100 **65.** (a) $20\left(\dfrac{3}{4}\right)^3 = \dfrac{135}{16}$ ft   (b) $20\left(\dfrac{3}{4}\right)^n$ ft   (c) 13 times   (d) 140 ft

67. (a) $4975 **(b)** 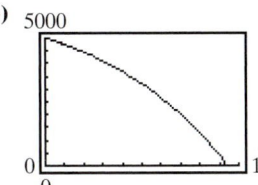 **(c)** At the beginning of the thirty-third month the balance is less than $4000. At this time, 32 payments of $100 each have been made.
(d) 94 payments; $9311.01
(e) $4311.01

Cumulative Review Exercises *(page 1030)*

1. $-3, 3, -3i, 3i$
2. (a) 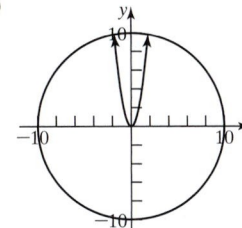 **(b)** $\left(\sqrt{\dfrac{-1 + \sqrt{3601}}{18}}, 3 \cdot \left(\dfrac{-1 + \sqrt{3601}}{18} \right) \right); \left(-\sqrt{\dfrac{-1 + \sqrt{3601}}{18}}, 3 \cdot \left(\dfrac{-1 + \sqrt{3601}}{18} \right) \right)$

(c) The circle and the parabola intersect at
$$\left(\sqrt{\dfrac{-1 + \sqrt{3601}}{18}}, 3 \cdot \left(\dfrac{-1 + \sqrt{3601}}{18} \right) \right); \left(-\sqrt{\dfrac{-1 + \sqrt{3601}}{18}}, 3 \cdot \left(\dfrac{-1 + \sqrt{3601}}{18} \right) \right).$$

3. $\left\{ \ln\left(\dfrac{5}{2} \right) \right\}$ **4.** $\left\{ \dfrac{3\pi}{2} \right\}$ **5.** $\dfrac{2\pi}{3}$ **6. (a)** $-\dfrac{\sqrt{15}}{4}$ **(b)** $-\dfrac{\sqrt{15}}{15}$ **(c)** $-\dfrac{\sqrt{15}}{8}$ **(d)** $\dfrac{7}{8}$ **(e)** $\sqrt{\dfrac{1 + \dfrac{\sqrt{15}}{4}}{2}}$

7. $\dfrac{x^2}{7} + \dfrac{y^2}{16} = 1; x = \sqrt{7} \sin t; y = 4 \cos t, 0 \le t < 2\pi$ **8.** $(x + 1)^2 = 4(y - 2)$ **9.** $r = 8 \sin \theta; x^2 + (y - 4)^2 = 16$

C H A P T E R 1 4 Counting and Probability

14.1 Concepts and Vocabulary *(page 1036)*

1. Union **2.** Intersection **3.** T **4.** T

14.1 Exercises *(page 1037)*

1. $\{1, 3, 5, 6, 7, 9\}$ **3.** $\{1, 5, 7\}$ **5.** $\{1, 6, 9\}$ **7.** $\{1, 2, 4, 5, 6, 7, 8, 9\}$ **9.** $\{1, 2, 4, 5, 6, 7, 8, 9\}$ **11.** $\{0, 2, 6, 7, 8\}$
13. $\{0, 1, 2, 3, 5, 6, 7, 8, 9\}$ **15.** $\{0, 1, 2, 3, 5, 6, 7, 8, 9\}$ **17.** $\{0, 1, 2, 3, 4, 6, 7, 8\}$ **19.** $\{0\}$
21. $\varnothing, \{a\}, \{b\}, \{c\}, \{d\}, \{a, b\}, \{a, c\}, \{a, d\}, \{b, c\}, \{b, d\}, \{c, d\}, \{a, b, c\}, \{b, c, d\}, \{a, c, d\}, \{a, b, d\}, \{a, b, c, d\}$
23. 25 **25.** 40 **27.** 25 **29.** 37 **31.** 18 **33.** 5 **35.** 175; 125 **37. (a)** 15 **(b)** 15 **(c)** 15 **(d)** 25 **(e)** 40
39. (a) 61,528 thousand **(b)** 11,085 thousand **(c)** 70,071 thousand

14.2 Concepts and Vocabulary *(page 1046)*

1. Permutation **2.** Combination **3.** T **4.** F

14.2 Exercises *(page 1046)*

1. 30 **3.** 24 **5.** 1 **7.** 1680 **9.** 28 **11.** 35 **13.** 1 **15.** 10,400,600 **17.** $\{abc, abd, abe, acb, acd, ace, adb, adc, ade, aeb, aec, aed$
$bac, bad, bae, bca, bcd, bce, bda, bdc, bde, bea, bec, bed$
$cab, cad, cae, cba, cbd, cbe, cda, cdb, cde, cea, ceb, ced$
$dab, dac, dae, dba, dbc, dbe, dca, dcb, dce, dea, deb, dec$
$eab, eac, ead, eba, ebc, ebd, eca, ecb, ecd, eda, edb, edc\}$; 60
19. $\{123, 124, 132, 134, 142, 143, 213, 214, 231, 234, 241, 243, 312, 314, 321, 324, 341, 342, 412, 413, 421, 423, 431, 432\}$; 24
21. $\{abc, abd, abe, acd, ace, ade, bcd, bce, bde, cde\}$; 10 **23.** $\{123, 124, 134, 234\}$; 4 **25.** 15 **27.** 16 **29.** 8 **31.** 24
33. 60 **35.** 18,278 **37.** 35 **39.** 1024 **41.** 9000 **43.** 120 **45.** 480 **47.** 132,860 **49.** 336 **51.** 90,720
53. (a) 63 **(b)** 35 **(c)** 1 **55.** 1.157×10^{76} **57.** 362,880 **59.** 660 **61.** 15

Historical Problems *(page 1058)*

1. (a) $\{AA, ABA, BAA, ABBA, BBAA, BABA, BBB, ABBB, BABB, BBAB\}$
(b) $P(A \text{ wins}) = \dfrac{C(4, 2) + C(4, 3) + C(4, 4)}{2^4} = \dfrac{6 + 4 + 1}{16} = \dfrac{11}{16}$

$P(B \text{ wins}) = \dfrac{C(4, 3) + C(4, 4)}{2^4} = \dfrac{4 + 1}{16} = \dfrac{5}{16}$

The outcomes listed in part (a) are not equally likely.

14.3 Concepts and Vocabulary *(page 1059)*

1. Equally likely **2.** Complement **3.** F **4.** T **5.** $\sum P(e_i) = 1; 0 \le P(e_i) \le 1$

14.3 Exercises *(page 1059)*

1. $0, 0.01, 0.35, 1$ **3.** Probability model **5.** Not a probability model

7. $S = \{HH, HT, TH, TT\}; P(HH) = \dfrac{1}{4}, P(HT) = \dfrac{1}{4}, P(TH) = \dfrac{1}{4}, P(TT) = \dfrac{1}{4}$

9. $S = \{HH1, HH2, HH3, HH4, HH5, HH6, HT1, HT2, HT3, HT4, HT5, HT6, TH1, TH2, TH3, TH4, TH5, TH6, TT1, TT2, TT3, TT4, TT5, TT6\};$ each outcome has the probability of $\dfrac{1}{24}$.

11. $S = \{HHH, HHT, HTH, HTT, THH, THT, TTH, TTT\};$ each outcome has the probability of $\dfrac{1}{8}$.

13. $S = \{$1 Yellow, 1 Red, 1 Green, 2 Yellow, 2 Red, 2 Green, 3 Yellow, 3 Red, 3 Green, 4 Yellow, 4 Red, 4 Green$\};$ each outcome has the probability of $\dfrac{1}{12}$; thus, $P(2 \text{ Red}) + P(4 \text{ Red}) = \dfrac{1}{12} + \dfrac{1}{12} = \dfrac{1}{6}$.

15. $S = \{$1 Yellow Forward, 1 Yellow Backward, 1 Red Forward, 1 Red Backward, 1 Green Forward, 1 Green Backward, 2 Yellow Forward, 2 Yellow Backward, 2 Red Forward, 2 Red Backward, 2 Green Forward, 2 Green Backward, 3 Yellow Forward, 3 Yellow Backward, 3 Red Forward, 3 Red Backward, 3 Green Forward, 3 Green Backward, 4 Yellow Forward, 4 Yellow Backward, 4 Red Forward, 4 Red Backward, 4 Green Forward, 4 Green Backward$\};$ each outcome has the probability of $\dfrac{1}{24}$; thus,

$P(1 \text{ Red Backward}) + P(1 \text{ Green Backward}) = \dfrac{1}{24} + \dfrac{1}{24} = \dfrac{1}{12}$.

17. $S = \{$11 Red, 11 Yellow, 11 Green, 12 Red, 12 Yellow, 12 Green, 13 Red, 13 Yellow, 13 Green, 14 Red, 14 Yellow, 14 Green, 21 Red, 21 Yellow, 21 Green, 22 Red, 22 Yellow, 22 Green, 23 Red, 23 Yellow, 23 Green, 24 Red, 24 Yellow, 24 Green, 31 Red, 31 Yellow, 31 Green, 32 Red, 32 Yellow, 32 Green, 33 Red, 33 Yellow, 33 Green, 34 Red, 34 Yellow, 34 Green, 41 Red, 41 Yellow, 41 Green, 42 Red, 42 Yellow, 42 Green, 43 Red, 43 Yellow, 43 Green, 44 Red, 44 Yellow, 44 Green$\};$ each outcome has the probability of $\dfrac{1}{48}$; thus, $E = \{$22 Red, 22 Green, 24 Red, 24 Green$\}; P(E) = \dfrac{n(E)}{n(S)} = \dfrac{4}{48} = \dfrac{1}{12}$.

19. A, B, C, F **21.** B **23.** $P(H) = \dfrac{4}{5}; P(T) = \dfrac{1}{5}$ **25.** $P(1) = P(3) = P(5) = \dfrac{2}{9}; P(2) = P(4) = P(6) = \dfrac{1}{9}$ **27.** $\dfrac{3}{10}$ **29.** $\dfrac{1}{2}$ **31.** $\dfrac{1}{6}$

33. $\dfrac{1}{8}$ **35.** $\dfrac{1}{4}$ **37.** $\dfrac{1}{6}$ **39.** $\dfrac{1}{18}$ **41.** 0.55 **43.** 0.70 **45.** 0.30 **47.** 0.747 **49.** 0.7 **51.** $\dfrac{17}{20}$ **53.** $\dfrac{11}{20}$ **55.** $\dfrac{1}{2}$ **57.** $\dfrac{3}{10}$ **59.** $\dfrac{2}{5}$

61. (a) 0.57 **(b)** 0.95 **(c)** 0.83 **(d)** 0.38 **(e)** 0.29 **(f)** 0.05 **(g)** 0.78 **(h)** 0.71 **63. (a)** $\dfrac{25}{33}$ **(b)** $\dfrac{25}{33}$ **65.** 0.167

67. 0.000033069 **69. (a)** $\dfrac{5}{16}$ **(b)** $\dfrac{1}{32}$ **71. (a)** 0.00463 **(b)** 0.126 **73.** $7.02 \times 10^{-6}; 0.183$ **75.** 0.1

14.4 Exercises *(page 1065)*

1. (a)

Cause of Death	Probability
Accidents and adverse effects	0.436
Homicide and legal intervention	0.180
Suicide	0.135
Malignant neoplasms	0.055
Diseases of heart	0.035
Human immunodeficiency virus infection	0.006
Congenital anomalies	0.015
Chronic obstructive pulmonary diseases	0.008
Pneumonia and influenza	0.007
Cerebrovascular diseases	0.006
All other causes	0.118

(b) 0.135 **(c)** 0.190 **(d)** 0.810

3. (a)

Response	Probability
Never	0.0262
Rarely	0.0678
Sometimes	0.1156
Most of the time	0.2632
Always	0.5272

(b) 0.0262 **(c)** 0.0940

5. (a)

Weight (in grams)	Probability
< 500	0.0015
500–999	0.0057
1000–1499	0.0073
1500–1999	0.0150
2000–2499	0.0463
2500–2999	0.1650
3000–3499	0.3702
3500–3999	0.2884
4000–4499	0.0851
4500–4999	0.0139
≥ 5000	0.0016

(b) 0.2884 **(c)** 0.3736 **(d)** 0.9984

7. 0.40

9. 0.40

11. 0.1143; 0.8857

13. 0.3733; 0.2433; 0.7567

15. 0.2133; 0.7867

17. (a)

Location	Probability
Left	0.4194
Left center	0.3387
Center	0.1935
Right center	0.0484
Right	0

(b) 0.4194
(c) 0.1935
(d) 0

Review Exercises *(page 1069)*

1. $\varnothing$, {Dave}, {Joanne}, {Erica}, {Dave, Joanne}, {Dave, Erica}, {Joanne, Erica}, {Dave, Joanne, Erica}

3. {1, 3, 5, 6, 7, 8} **5.** {3, 7} **7.** {1, 2, 4, 6, 8, 9} **9.** {1, 2, 4, 5, 6, 9} **11.** 17 **13.** 29 **15.** 7 **17.** 25 **19.** 336 **21.** 56

23. 60 **25.** 128 **27.** 3024 **29.** 70 **31.** 91 **33.** 1,600,000 **35.** 216,000 **37.** 1260 **39. (a)** 381,024 **(b)** 1260

41. (a)

Age	Probability
20–24	0.1709
25–29	0.1410
30–34	0.1344
35–39	0.1227
40–44	0.0991
45–49	0.0816
50–54	0.0602
55–59	0.0460
60–64	0.0389
65–69	0.0321
70–74	0.0293
75–79	0.0249
80–84	0.0190

(b) 0.1410 **(c)** 0.3119 **(d)** 0.8291

43. (a) $8.634628387 \times 10^{45}$ **(b)** 0.6531 **(c)** 0.3469

45. (a) 0.054 **(b)** 0.946

47. $\dfrac{4}{9}$

49. 0.2; 0.26

51. (a) 0.2461 **(b)** 9.766×10^{-4}

53. (a) 0.0065 **(b)** 0.0294 **(c)** 0.3922

Cumulative Review Exercises *(page 1072)*

1. $\left\{ \dfrac{1}{3} - \dfrac{\sqrt{2}}{3}i, \dfrac{1}{3} + \dfrac{\sqrt{2}}{3}i \right\}$

2.

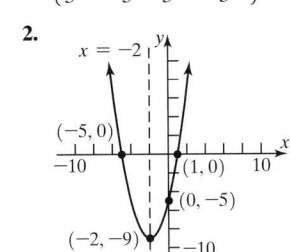

3.

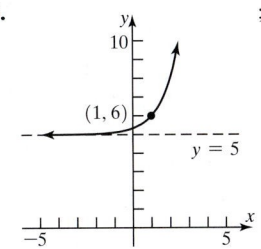

4. $\{x | 3.99 \le x \le 4.01\}$ or $[3.99, 4.01]$ **5.** $-\dfrac{1}{5}, 3, -\dfrac{1}{2} - \dfrac{\sqrt{7}}{2}i, -\dfrac{1}{2} + \dfrac{\sqrt{7}}{2}i$

6.

; Domain: all real numbers
Range: $\{y | y > 5\}$
Horizontal asymptote: $y = 5$

7. $2x$ **8.** $\left\{ \dfrac{8}{3} \right\}$

9.

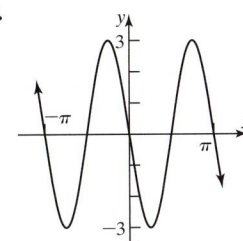

10.

; Area ≈ 14.46 square units

11. $x = 2, y = -5, z = 3$

12. 125; 700

INDEX

CONICS
Parabola

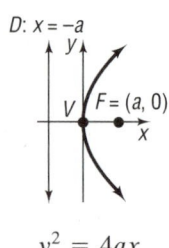

$$y^2 = 4ax \qquad y^2 = -4ax \qquad x^2 = 4ay \qquad x^2 = -4ay$$

Ellipse

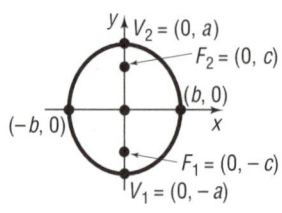

$$\frac{x^2}{a^2} + \frac{y^2}{b^2} = 1, \quad a > b, \quad c^2 = a^2 - b^2 \qquad\qquad \frac{x^2}{b^2} + \frac{y^2}{a^2} = 1, \quad a > b, \quad c^2 = a^2 - b^2$$

Hyperbola

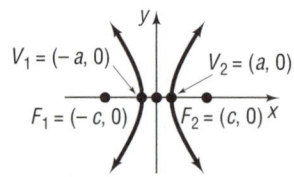

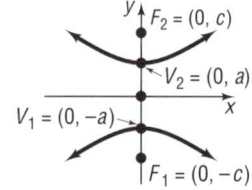

$$\frac{x^2}{a^2} - \frac{y^2}{b^2} = 1, \quad c^2 = a^2 + b^2 \qquad\qquad \frac{y^2}{a^2} - \frac{x^2}{b^2} = 1, \quad c^2 = a^2 + b^2$$

$$\text{Asymptotes: } y = \frac{b}{a}x, \quad y = -\frac{b}{a}x \qquad\qquad \text{Asymptotes: } y = \frac{a}{b}x, \quad y = -\frac{a}{b}x$$

PROPERTIES OF LOGARITHMS

$$\log_a (MN) = \log_a M + \log_a N$$

$$\log_a \left(\frac{M}{N}\right) = \log_a M - \log_a N$$

$$\log_a M^r = r \log_a M$$

$$\log_a M = \frac{\log M}{\log a} = \frac{\ln M}{\ln a}$$

PERMUTATIONS/COMBINATIONS

$$0! = 1 \qquad 1! = 1$$

$$n! = n(n - 1) \cdots (3)(2)(1)$$

$$P(n, r) = \frac{n!}{(n - r)!}$$

$$C(n, r) = \binom{n}{r} = \frac{n!}{(n - r)!r!}$$

BINOMIAL THEOREM

$$(a + b)^n = a^n + \binom{n}{1}ba^{n-1} + \binom{n}{2}b^2a^{n-2} + \cdots + \binom{n}{1}b^{n-1}a + b^n$$

ARITHMETIC SEQUENCE

$$a + (a + d) + (a + 2d) + \cdots + [a + (n - 1)d] = na + \frac{n(n - 1)}{2}d$$

GEOMETRIC SEQUENCE

$$a + ar + ar^2 + \cdots + ar^{n-1} = a\frac{1 - r^n}{1 - r}$$

GEOMETRIC SERIES

$$\text{If } |r| < 1, a + ar + ar^2 + \cdots = \sum_{k=1}^{\infty} ar^{k-1} = \frac{a}{1 - r}$$